Looking for ways to integ[rate] the Web into your curriculum?

W9-DCV-529

JRN ACADEMY
50 WALDEN
BUS, OHIO 43229

ClassZone℠

www.mcdougallittell.com

ClassZone, McDougal Littell's textbook-companion Web site, is the solution! Online teaching support for you and engaging, interactive content for your students!

ClassZone is your online guide to *McDougal Littell Algebra 2.*

- **Student Help** provides online home-work support with Extra Examples, Problem Solving help, and more.
- **Career Links** and **Application Links** extend real-life connections.
- **Data Updates** keep information current.
- **Teacher Center** provides lesson plan-ning support and teaching ideas.

Log on to ClassZone at www.mcdougallittell.com

With the purchase of *McDougal Littell Algebra 2,* you have immediate access to ClassZone.

Teacher Access Code

MCD9JDFP8MZT8

Use this code to create your own user name and password. Then, access both teacher only and student resources.

Student Access Code

MCDX5KYPV55YV

Give this code to your class. Each student creates a unique user name and password to access resources for students.

McDougal Littell

ALGEBRA 2

TEACHER'S EDITION

Ron Larson
Laurie Boswell
Timothy D. Kanold
Lee Stiff

McDougal Littell
A HOUGHTON MIFFLIN COMPANY
Evanston, Illinois • Boston • Dallas

CONTENTS

▸ ABOUT THE AUTHORS   T3

▸ REVIEWERS   T4

▸ TABLE OF CONTENTS   T6

▸ FOCUS ON CAREERS   T21

▸ STUDENT HELP OVERVIEW   T22

▸ PROGRAM OVERVIEW   T24

▸ PROGRAM RESOURCES   T30

▸ PACING THE COURSE   T36

▸ SERIES OVERVIEW   T38

▸ STUDENT EDITION   1

▸ TEACHER'S EDITION INDEX   IN1

▸ ADDITIONAL ANSWERS   AA1

About the Cover

Algebra 2 brings math to life with many real-life applications. The cover illustrates some of the applications used in this book. Examples of mathematics in architecture, bicycling, amusement parks, and animal motion are shown on pages 46, 342, 809, 811, and 818. Circling the globe are three key aspects of Algebra 2—the *equations*, *graphs*, and *applications* that you will use in this course. They will help you understand how mathematics relates to the world. As you explore the applications presented in the book, try to make your own connections between mathematics and the world around you!

Copyright © 2001 by McDougal Littell Inc. All rights reserved.

No part of this work may be reproduced or transmitted in any form or by any means, electronic or mechanical, including photocopy and recording, or by any information storage or retrieval system without the prior written permission of McDougal Littell Inc. unless such copying is expressly permitted by federal copyright law. Address inquiries to Manager, Rights and Permissions, McDougal Littell Inc., P.O. Box 1667, Evanston, IL 60204.

ISBN: 0-395-97890-4   456789–DWO–04 03 02

Internet Web Site: http://www.mcdougallittell.com

About the Authors

▶ **Ron Larson** is a professor of mathematics at Penn State University at Erie. He is the author of a broad range of mathematics textbooks for middle school, high school, and college students. He is one of the pioneers in the use of multimedia and the Internet to enhance the learning of mathematics. Dr. Larson is a member of the National Council of Teachers of Mathematics and is a frequent speaker at NCTM and other national and regional mathematics meetings.

▶ **Laurie Boswell** is a mathematics teacher at Profile Junior-Senior High School in Bethlehem, New Hampshire. She is active in NCTM and local mathematics organizations. A recipient of the 1986 Presidential Award for Excellence in Mathematics Teaching, she is also the 1992 Tandy Technology Scholar and the 1991 recipient of the Richard Balomenos Mathematics Education Service Award presented by the New Hampshire Association of Teachers of Mathematics.

▶ **Timothy D. Kanold** is Director of Mathematics and a teacher at Adlai E. Stevenson High School in Lincolnshire, IL. In 1995 he received the Award of Excellence from the Illinois State Board of Education for outstanding contributions to education. He served on NCTM's Professional Standards for Teaching Mathematics Commission. A 1986 recipient of the Presidential Award for Excellence in Mathematics Teaching, he served as president of the Council of Presidential Awardees of Mathematics.

▶ **Lee Stiff** is a professor of mathematics education in the College of Education and Psychology of North Carolina State University at Raleigh and has taught mathematics at the high school and middle school levels. He served on the NCTM Board of Directors and was elected President of NCTM for the years 2000–2002. He is the 1992 recipient of the W. W. Rankin Award for Excellence in Mathematics Education presented by the North Carolina Council of Teachers of Mathematics.

▶ REVIEWERS

Jimmy Bostock
Mathematics Department Chair
Hepzibah High School
Hepzibah, GA

Phyllis Carter
Mathematics Department Head
Parkway North High School
Creve Coeur, MO

Theresa Cepaitis
Instructional Specialist, Secondary
    Mathematics
Anne Arundel County Public Schools
Annapolis, MD

Linda M. Fulmore
Assistant Principal, Instruction
Camelback High School
Phoenix, AZ

David Sanchez
Mathematics Teacher
Orosi High School
Orosi, CA

▶ CALIFORNIA TEACHER PANEL

Courteney Dawe
Mathematics Teacher
Placerita Junior High School
Valencia, CA

Dave Dempster
Mathematics Teacher
Temecula Valley High School
Temecula, CA

Pauline Embree
Mathematics Department Chair
Rancho San Joaquin Middle School
Irvine, CA

Tom Griffith
Mathematics Teacher
Scripps Ranch High School
San Diego, CA

Diego Gutierrez
Mathematics Teacher
Crawford High School
San Diego, CA

Roger Hitchcock
Mathematics Teacher
Buchanan High School
Clovis, CA

Joseph Jacobs
Mathematics Teacher
Luther Burbank High School
Sacramento, CA

Louise McComas
Mathematics Teacher
Fremont High School
Sunnyvale, CA

Viola Okoro
Mathematics Teacher
Laguna Creek High School
Elk Grove, CA

Jon Simon
Mathematics Department Chair
Casa Grande High School
Petaluma, CA

▶ CLEVELAND TEACHER PANEL

Laura Anfang
University Supervisor, Teacher of
    Secondary Math Methods
John Carroll University
University Heights, OH

Patricia Benedict
Mathematics Department Liaison
Cleveland Heights High School
Cleveland Heights, OH

Carol Caroff
Mathematics Department
    Chair/Teacher
Solon High School
Solon, OH

Fred Dillon
Mathematics Teacher
Strongsville High School
Strongsville, OH

Bill Hunt
Past President, Ohio Council
    of Teachers of Mathematics
Strongsville, OH

Dr. Margie Raub Hunt
Executive Director, Ohio Council
    of Teachers of Mathematics
Strongsville, OH

Robert Jones
Mathematics Supervisor (K–12)
Cleveland City School District
Cleveland, OH

Andrea Kopco
Mathematics Teacher
Midpark High School
Cleveland, OH

Sandy Sikorski
Mathematics Teacher
Berea High School
Berea, OH

Gil Stevens
Mathematics Teacher
Brunswick High School
Brunswick, OH

▶ KANSAS TEACHER PANEL

Rosemary Arb
Mathematics Department Chair
North Kansas City High School
Kansas City, MO

Jerry Belshe
Mathematics Teacher
Indian Woods Middle School
Overland Park, KS

Carol Edwards
Mathematics Teacher
Belton High School
Belton, MO

Robert Franks
Mathematics Teacher
Park Hill High School
Kansas City, MO

Cynthia Hardy
Mathematics Teacher
Oxford Middle School
Overland Park, KS

Julie Knittle
Secondary Mathematics Resource Specialist
Shawnee Mission District Office
Shawnee Mission, KS

Glenda Morrison
Mathematics Teacher
Central High School
Kansas City, MO

Mitch Shea
Mathematics Teacher
Leavenworth High School
Leavenworth, KS

Ginny Taylor
Mathematics Department Chair
Santa Fe Trail Junior High School
Olathe, KS

▶ STUDENT REVIEW PANEL

Jacquelyn J. Arnold
Dunbar High School
Ohio

Michael Aznavour
William S. Hart High School
California

Kristopher Cadena
Las Cruces High School
New Mexico

Jonathan Ellenberger
Plymouth North High School
Massachusetts

Kari Frick
Standley Lake High School
Colorado

Nour Ghazi
Watertown High School
Massachusetts

Rebekah Good
Broadneck Sr. High School
Maryland

Lauren Henderson
Harriton High School
Pennsylvania

Amanda L. Hood
Robert S. Alexander High School
Georgia

Amanda Jakuszanek
Stevenson High School
Michigan

Jennifer Kelley
Lakota West High School
Ohio

David Lai
Holmdel High School
New Jersey

Melissa Machit
Polytechnic High School
California

Tracy A. Marynowski
Mother McAuley High School
Illinois

Ben McLain
Clarke Central High School
Georgia

Shoshana Muhammad
Skyline High School
California

Dayle Pivetta
Churchill High School
Michigan

Khandicia Nichelle Randolph
Whitney Young Magnet High School
Illinois

Tomekia Yoshida Mattice Reed
Murrah High School
Mississippi

Marianne Roller
Cedar Cliff High School
Pennsylvania

Brice Roncace
Meridian High School
Idaho

Joshua D. Rosenfeld
South High School
California

Nancy Ruiz
Santa Ana High School
California

Mahdi Salehi
El Camino Real High School
California

Paul Jared Schmidt
California

Tara Schoop
Glenbard East High School
Illinois

Sammy Shreibati
Westmoor High School
California

Michele D. Unruh
Truman High School
Missouri

Kristal M. Watrous
Mead High School
Washington

CHAPTER

1

Equations and Inequalities

APPLICATION HIGHLIGHTS

Railroads *1, 35*
Geography *4*
Money Exchange *6*
Real Estate *21*
Photo Framing *21*
Benefit Concert *27*
Gardening *28*
Weather Balloons *36*
Auto Maintenance *44*
Traffic Enforcement *44*

MATH & HISTORY

Problem Solving *40*

INTERNET CONNECTIONS

www.mcdougallittell.com

Application Links
1, 35, 38, 40, 46

Data Updates
6, 16

Student Help
6, 15, 18, 20, 25, 31, 38, 43, 48, 51, 54

Career Links
16, 21, 23, 30

Extra Challenge
10, 16, 24, 32, 39, 47, 55

▶ **CHAPTER STUDY GUIDE**   **2**

1.1 Real Numbers and Number Operations   **3**

1.2 Algebraic Expressions and Models   **11**
   QUIZ 1, *17*
   ▶ CALCULATOR ACTIVITY: *Evaluating Expressions, 18*

1.3 Solving Linear Equations   **19**
   ▶ GRAPHING CALCULATOR: *Using Tables to Solve Equations, 25*

1.4 Rewriting Equations and Formulas   **26**

1.5 Problem Solving Using Algebraic Models:
   EXPLORING DATA AND STATISTICS   **33**
   QUIZ 2, *40*

1.6 Solving Linear Inequalities   **41**
   ▶ GRAPHING CALCULATOR: *Solving an Inequality, 48*

1.7 Solving Absolute Value Equations and Inequalities   **50**
   ▶ CONCEPT ACTIVITY: *Absolute Value Equations and Inequalities, 49*
   QUIZ 3, *56*

ASSESSMENT

Skill Review, *2*
Quizzes, *17, 40, 56*
Test Preparation Questions, *10, 16, 24, 31, 39, 47, 55*
Chapter Summary, Review, and Test, *57*
Chapter Standardized Test, *62*

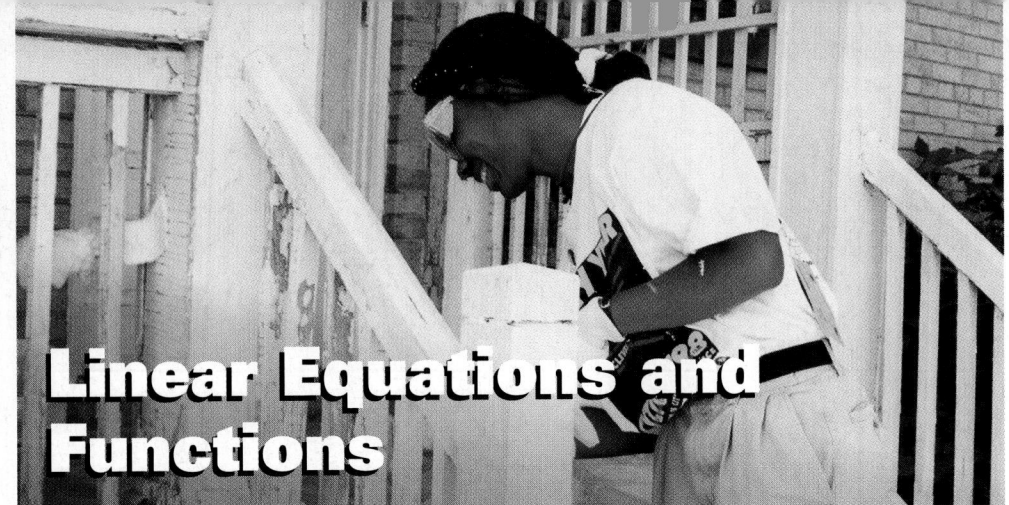

CHAPTER
2

Linear Equations and Functions

APPLICATION HIGHLIGHTS

Youth Service *65, 105*
Ballooning *70*
Ladder Safety *78*
Deserts *78*
Buying a Computer *83*
Fundraising *85*
Politics *93*
Communication *110*
Wages *116*
Billiards *124*

MATH & HISTORY

Transatlantic Voyages *89*

INTERNET CONNECTIONS

www.mcdougallittell.com

Application Links
65, 78, 89, 119, 126

Data Updates
93, 119

Student Help
70, 77, 88, 90, 92, 105, 107, 110, 118, 119, 121, 126, 127

Career Links
68, 102, 112

Extra Challenge
74, 81, 88, 98, 105, 113, 120, 127

▶ **CHAPTER STUDY GUIDE**   **66**

2.1 **Functions and Their Graphs**   **67**

2.2 **Slope and Rate of Change**   **75**

2.3 **Quick Graphs of Linear Equations**   **82**
   QUIZ 1,   *89*
   ▶ GRAPHING CALCULATOR: *Graphing Equations, 90*

2.4 **Writing Equations of Lines**   **91**

2.5 **Correlation and Best-Fitting Lines:** EXPLORING DATA AND STATISTICS   **100**
   ▶ CONCEPT ACTIVITY: *Fitting a Line to a Set of Data, 99*
   QUIZ 2,   *106*
   ▶ GRAPHING CALCULATOR: *Using Linear Regression, 107*

2.6 **Linear Inequalities in Two Variables**   **108**

2.7 **Piecewise Functions**   **114**
   ▶ GRAPHING CALCULATOR: *Graphing Piecewise Functions, 121*

2.8 **Absolute Value Functions**   **122**
   QUIZ 3,   *128*

ASSESSMENT

Skill Review, *66*
Quizzes, *89, 106, 128*
Test Preparation Questions, *74, 81, 88, 98, 105, 113, 120, 127*
Chapter Summary, Review, and Test, *129*
Chapter Standardized Test, *134*

Systems of Linear Equations and Inequalities

APPLICATION HIGHLIGHTS

Cross-Training *137, 154*
Vacation Costs *141*
Fitness *144*
Catering *151*
Heart Rate *158*
Bicycle Production *165*
Landscaping *172*
Transportation *174*
Sports *180*
Voting *183*

MATH & HISTORY

Linear Programming in
  World War II *169*

INTERNET CONNECTIONS

www.mcdougallittell.com

Application Links
  137, 169, 174, 183

Data Updates
  154, 161

Student Help
  140, 146, 153, 160, 161,
  167, 171, 172, 176, 178

Career Links
  151, 158, 172

Extra Challenge
  145, 154, 162, 168,
  175, 183

PROJECT

Chapters 1–3 Project *194*

▶ **CHAPTER STUDY GUIDE**   **138**

3.1 Solving Linear Systems by Graphing   **139**
   ▶ GRAPHING CALCULATOR: *Graphing Systems of Equations, 146*

3.2 Solving Linear Systems Algebraically   **148**
   ▶ CONCEPT ACTIVITY: *Combining Equations in a Linear System, 147*
   QUIZ 1, *155*

3.3 Graphing and Solving Systems of Linear Inequalities   **156**

3.4 Linear Programming: EXPLORING DATA AND STATISTICS   **163**
   QUIZ 2, *169*

3.5 Graphing Linear Equations in Three Variables   **170**
   ▶ GRAPHING CALCULATOR: *Graphing Linear Equations in*
     Three Variables, 176

3.6 Solving Systems of Linear Equations in Three Variables   **177**
   QUIZ 3, *184*

ASSESSMENT

Skill Review, *138*
Quizzes, *155, 169, 184*
Test Preparation Questions, *145, 154, 162, 168, 175, 183*
Chapter Summary, Review, and Test, *185*
Chapter Standardized Test, *190*
Cumulative Practice, Chapters 1–3, *192*
Project, Chapters 1–3, *194*

CHAPTER
4

Matrices and Determinants

APPLICATION HIGHLIGHTS

Music Sales *197, 204*
Health Care Plans *202*
College Costs *205*
Softball *210*
Exercise *212*
Bermuda Triangle *215*
Atomic Weights *217*
Sailing *219*
Cryptography *225*
Investing *232*

MATH & HISTORY

Systems of Equations *236*

INTERNET CONNECTIONS

www.mcdougallittell.com

Application Links
197, 210, 225, 232, 236

Data Updates
205, 212

Student Help
204, 207, 209, 216, 224, 234

Career Links
202, 217, 234

Extra Challenge
206, 213, 220, 229, 235

▶ **CHAPTER STUDY GUIDE**   **198**

4.1 Matrix Operations: EXPLORING DATA AND STATISTICS   **199**
    ▶ **GRAPHING CALCULATOR:** *Using Matrix Operations, 207*

4.2 Multiplying Matrices   **208**

4.3 Determinants and Cramer's Rule   **214**
    QUIZ 1,   *221*

4.4 Identity and Inverse Matrices   **223**
    ▶ **CONCEPT ACTIVITY:** *Investigating Identity and Inverse Matrices, 222*

4.5 Solving Systems Using Inverse Matrices   **230**
    QUIZ 2,   *236*

    EXTENSION: *Solving Systems Using Augmented Matrices*   **237**

ASSESSMENT

Skill Review, *198*
Quizzes, *221, 236*
Test Preparation Questions, *206, 213, 220, 229, 235*
Chapter Summary, Review, and Test, *239*
Chapter Standardized Test, *244*

CHAPTER
5

Quadratic Functions

APPLICATION HIGHLIGHTS

Volcanoes *247, 297*
Civil Engineering *252*
Crafts *258*
Business *259*
History *266*
Landscape Design *284*
Agriculture *285*
Entertainment *294*
Carpentry *300*
Fuel Economy *307*

MATH & HISTORY

Telescopes *270*

INTERNET CONNECTIONS

www.mcdougallittell.com

Application Links
247, 254, 268, 270, 311

Data Updates
280, 297, 305, 309

Student Help
254, 259, 265, 271,
279, 285, 290, 292,
304, 308, 310

Career Links
252, 279, 296, 304, 308

Extra Challenge
255, 263, 269, 280, 289,
297, 305, 312

▶ **CHAPTER STUDY GUIDE**   **248**

5.1 Graphing Quadratic Functions   **249**

5.2 Solving Quadratic Equations by Factoring   **256**

5.3 Solving Quadratic Equations by Finding Square Roots **264**
QUIZ 1, *270*
   ▶ GRAPHING CALCULATOR: *Solving Quadratic Equations, 271*

5.4 Complex Numbers   **272**

5.5 Completing the Square   **282**
   ▶ CONCEPT ACTIVITY: *Using Algebra Tiles to Complete the Square, 281*
   ▶ GRAPHING CALCULATOR: *Finding Maximums and Minimums, 290*

5.6 The Quadratic Formula and the Discriminant   **291**
QUIZ 2, *298*

5.7 Graphing and Solving Quadratic Inequalities   **299**

5.8 Modeling with Quadratic Functions:
EXPLORING DATA AND STATISTICS   **306**
QUIZ 3, *312*

ASSESSMENT

Skill Review, *248*
Quizzes, *270, 298, 312*
Test Preparation Questions, *255, 263, 269, 280, 289, 297, 305, 311*
Chapter Summary, Review, and Test, *313*
Chapter Standardized Test, *318*

Polynomials and Polynomial Functions

APPLICATION HIGHLIGHTS

Space Exploration *321, 327*
Photography *330*
Biology *332*
Farming *340*
Publishing *340*
Archeology *347*
Accounting *355*
Physiology *368*
Manufacturing *375*
Boating *382*

MATH & HISTORY

Solving Polynomial
   Equations *351*

INTERNET CONNECTIONS

www.mcdougallittell.com

Application Links
   321, 325, 351, 357, 377

Data Updates
   325, 327, 342, 370, 385, 395

Student Help
   *324, 332, 337, 343, 349, 354,
   363, 367, 372, 374, 381*

Career Links
   327, 330, 335, 342, 347, 355

Extra Challenge
   *328, 336, 343, 350, 358, 364,
   371, 378, 385*

PROJECT

Chapters 4–6 Project *396*

▶ **CHAPTER STUDY GUIDE**   **322**

6.1 Using Properties of Exponents   **323**

6.2 Evaluating and Graphing Polynomial Functions **329**
   ▶ GRAPHING CALCULATOR: *Setting a Good Viewing Window, 337*

6.3 Adding, Subtracting, and Multiplying Polynomials **338**
   QUIZ 1, *344*

6.4 Factoring and Solving Polynomial Equations   **345**

6.5 The Remainder and Factor Theorems   **352**

6.6 Finding Rational Zeros   **359**
   QUIZ 2, *365*

6.7 Using the Fundamental Theorem of Algebra   **366**
   ▶ GRAPHING CALCULATOR: *Solving Polynomial Equations, 372*

6.8 Analyzing Graphs of Polynomial Functions   **373**

6.9 Modeling with Polynomial Functions:
   EXPLORING DATA AND STATISTICS   **380**
   ▶ CONCEPT ACTIVITY: *Exploring Finite Differences, 379*
   QUIZ 3, *386*

ASSESSMENT

Skill Review, *322*
Quizzes, *344, 365, 386*
Test Preparation Questions, *328, 336, 343, 350, 358, 364, 371, 378, 385*
Chapter Summary, Review, and Test, *387*
Chapter Standardized Test, *392*
Cumulative Practice, Chapters 1–6, *394*
Project, Chapters 4–6, *396*

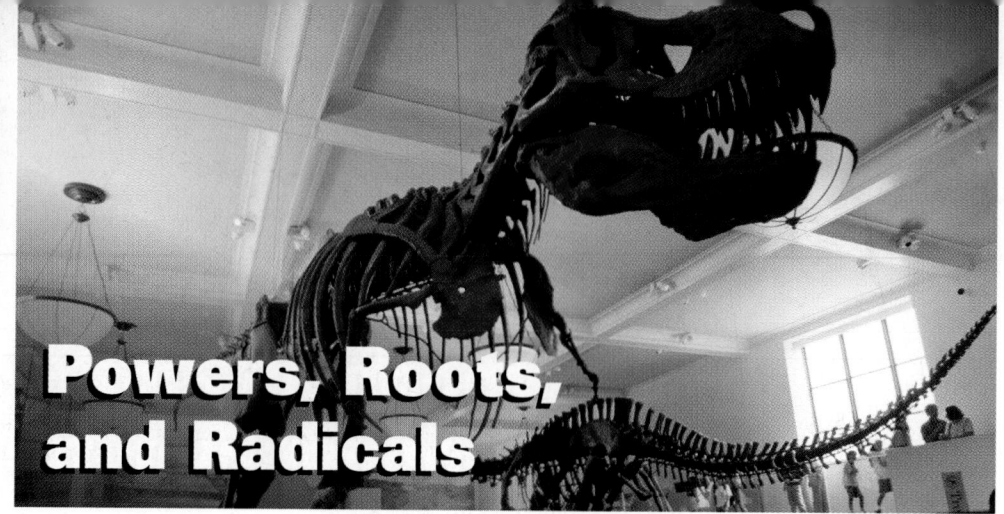

CHAPTER
7

Powers, Roots, and Radicals

APPLICATION HIGHLIGHTS

Dinosaurs *399, 442*
Space Science *403*
Nautical Science *403*
Biology *410*
Business *417*
Science *423*
Astronomy *425*
Amusement Parks *433*
Elephants *433*
Basketball *445*

MATH & HISTORY

Tsunamis *444*

INTERNET CONNECTIONS

www.mcdougallittell.com

Application Links
399, 403, 425, 440, 444

Data Updates
427, 445

Student Help
*405, 408, 409, 419, 428,
430, 435, 438, 446, 453*

Career Links
419, 427, 433, 435, 450

Extra Challenge
*406, 413, 420, 428, 436,
443, 451*

▶ **CHAPTER STUDY GUIDE**   **400**

7.1 *n*th Roots and Rational Exponents   **401**

7.2 Properties of Rational Exponents   **407**
     QUIZ 1, *414*

7.3 Power Functions and Function Operations   **415**

7.4 Inverse Functions   **422**
     ▶ CONCEPT ACTIVITY: *Exploring Inverse Functions, 421*
     QUIZ 2, *429*
     ▶ GRAPHING CALCULATOR: *Graphing Inverse Functions, 430*

7.5 Graphing Square Root and Cube Root Functions   **431**

7.6 Solving Radical Equations   **437**

7.7 Statistics and Statistical Graphs: EXPLORING DATA AND STATISTICS   **445**
     QUIZ 3, *452*
     ▶ GRAPHING CALCULATOR: *Statistics and Statistical Graphs, 453*

ASSESSMENT

Skill Review, *400*
Quizzes, *414, 429, 452*
Test Preparation Questions, *406, 413, 420, 428, 436, 443, 451*
Chapter Summary, Review, and Test, *455*
Chapter Standardized Test, *460*

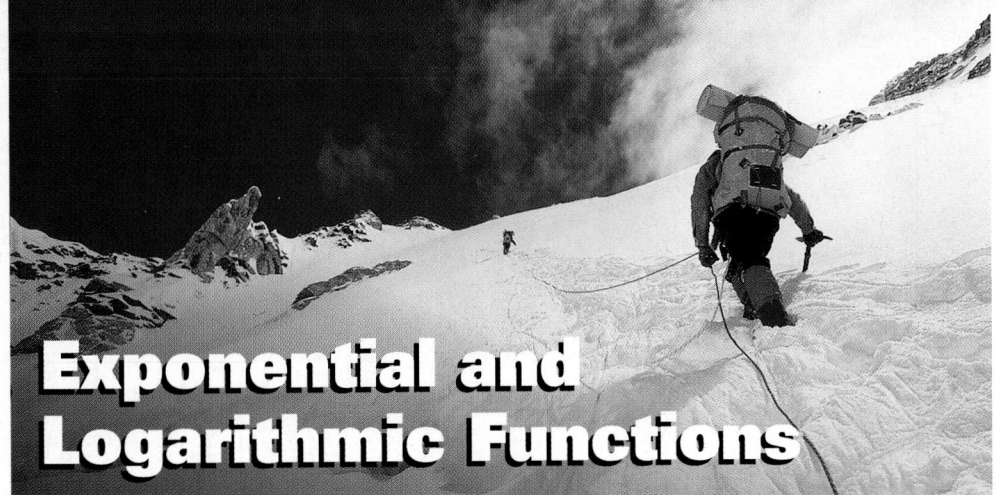

CHAPTER
8

Exponential and Logarithmic Functions

APPLICATION HIGHLIGHTS

Mountain Climbing *463, 484*
Internet Hosts *467*
Compound Interest *468*
Automobiles *476*
Endangered Species *482*
Acoustics *495*
Cooking *502*
Seismology *504*
Communications *510*
Botany *519*

MATH & HISTORY

Logarithms *499*

INTERNET CONNECTIONS

www.mcdougallittell.com
Application Links
463, 467, 491, 497, 499, 507

Data Updates
478, 522

Student Help
467, 471, 476, 484, 487, 497, 500, 502, 515, 518, 519

Career Links
468, 482, 495

Extra Challenge
472, 492, 498, 507

▶ **CHAPTER STUDY GUIDE**   **464**

8.1 Exponential Growth   **465**

8.2 Exponential Decay   **474**
  ▶ CONCEPT ACTIVITY: *Exponential Growth and Decay, 473*

8.3 The Number *e*   **480**
  QUIZ 1, *485*

8.4 Logarithmic Functions   **486**

8.5 Properties of Logarithms   **493**
  ▶ GRAPHING CALCULATOR: *Graphing Logarithmic Functions, 500*

8.6 Solving Exponential and Logarithmic Equations   **501**
  QUIZ 2, *508*

8.7 Modeling with Exponential and Power Functions:
  EXPLORING DATA AND STATISTICS   **509**

8.8 Logistic Growth Functions   **517**
  QUIZ 3, *522*

ASSESSMENT

Skill Review, *464*
Quizzes, *485, 508, 522*
Test Preparation Questions, *472, 479, 485, 492, 498, 507, 516, 522*
Chapter Summary, Review, and Test, *523*
Chapter Standardized Test, *528*

CHAPTER 9

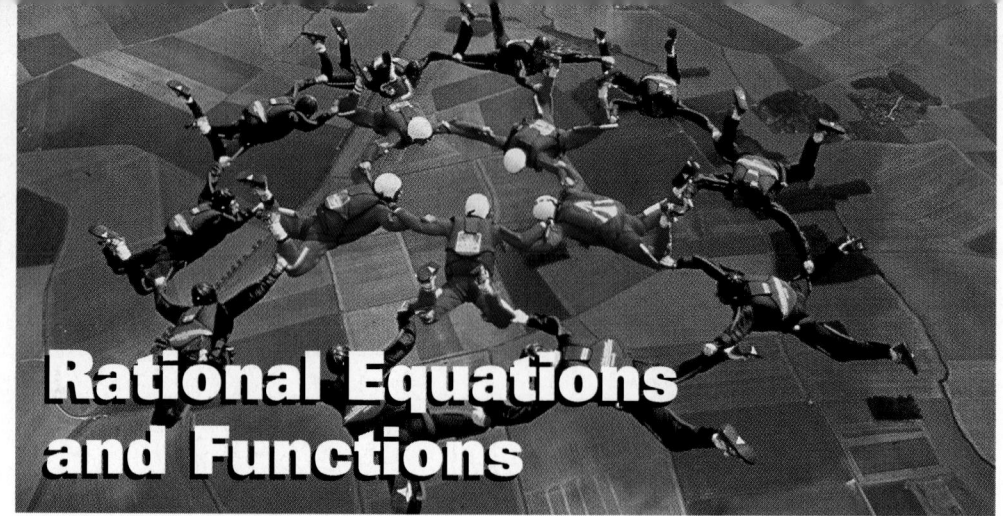

Rational Equations and Functions

APPLICATION HIGHLIGHTS

Skydiving *531, 557*
Oceanography *535*
Biology *535*
Business *542*
Manufacturing *549*
Farmland *559*
Statistics *563*
Photography *564*
Chemistry *570*
Rodeos *570*

MATH & HISTORY

Deep Water Diving *574*

INTERNET CONNECTIONS

www.mcdougallittell.com
Application Links
531, 536, 574

Data Updates
552, 572

Student Help
538, 541, 546, 551, 556, 561, 564, 568

Career Links
551, 559, 566

Extra Challenge
539, 545, 552, 560, 567, 573

PROJECT

Chapters 7–9 Project *584*

▶ **CHAPTER STUDY GUIDE**   **532**

9.1 **Inverse and Joint Variation:** EXPLORING DATA AND STATISTICS   **534**
 ▶ CONCEPT ACTIVITY: *Investigating Inverse Variation, 533*

9.2 **Graphing Simple Rational Functions**   **540**
 ▶ GRAPHING CALCULATOR: *Graphing Rational Functions, 546*

9.3 **Graphing General Rational Functions**   **547**
 QUIZ 1,   *553*

9.4 **Multiplying and Dividing Rational Expressions**   **554**
 ▶ GRAPHING CALCULATOR: *Operations with Rational Expressions, 561*

9.5 **Addition, Subtraction, and Complex Fractions**   **562**

9.6 **Solving Rational Equations**   **568**
 QUIZ 2,   *574*

ASSESSMENT

Skill Review, *532*
Quizzes, *553, 574*
Test Preparation Questions, *539, 545, 552, 560, 567, 573*
Chapter Summary, Review, and Test, *575*
Chapter Standardized Test, *580*
Cumulative Practice, Chapters 1–9, *582*
Project, Chapters 7–9, *584*

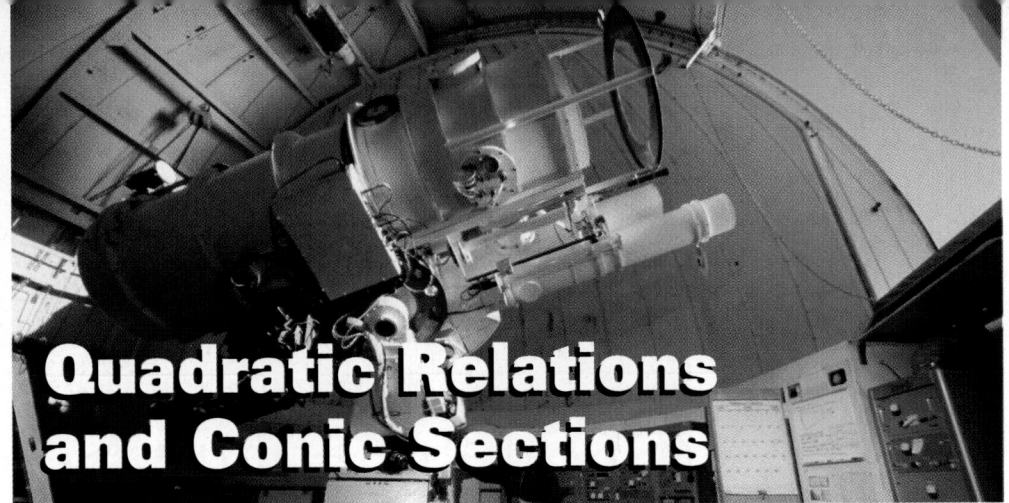

CHAPTER

10

Quadratic Relations and Conic Sections

APPLICATION HIGHLIGHTS

Telescopes *587, 627*
Archeology *591*
Solar Energy *597*
Ocean Navigation *603*
Landscaping *611*
Photography *617*
Sculpture *617*
Communications *625*
Seismology *634*
Astronomy *640*

MATH & HISTORY

History of Conic Sections
631

INTERNET CONNECTIONS

www.mcdougallittell.com

Application Links
587, 597, 617, 620, 631

Data Updates

Student Help
*593, 596, 602, 608,
610, 619, 624, 637*

Career Links
593, 606, 634, 636

Extra Challenge
*594, 600, 606, 614,
620, 630, 637*

▶ **CHAPTER STUDY GUIDE**   **588**

10.1 The Distance and Midpoint Formulas:
EXPLORING DATA AND STATISTICS   **589**

10.2 Parabolas   **595**

10.3 Circles   **601**
QUIZ 1, *607*
▶ GRAPHING CALCULATOR: *Graphing Circles, 608*

10.4 Ellipses   **609**

10.5 Hyperbolas   **615**
QUIZ 2, *621*

10.6 Graphing and Classifying Conics   **623**
▶ CONCEPT ACTIVITY: *Exploring Conic Sections, 622*

10.7 Solving Quadratic Systems   **632**
QUIZ 3, *638*

EXTENSION: *Eccentricity of Conic Sections*   **639**

ASSESSMENT

Skill Review, *588*
Quizzes, *607, 621, 638*
Test Preparation Questions, *594, 600, 606, 614, 620, 630, 637*
Chapter Summary, Review, and Test, *641*
Chapter Standardized Test, *646*

Sequences and Series

**APPLICATION
HIGHLIGHTS**

Fractals *649, 685*
Seating Capacity *662*
Honeycombs *664*
Cellular Phones *669*
Tennis *671*
Computer Science *671*
Ball Bounce *677*
Tourism *679*
Fish *683*
Tree Farm *685*

MATH & HISTORY

Fibonacci Sequence *687*

**INTERNET
CONNECTIONS**

www.mcdougallittell.com

Application Links
649, 669, 687

Data Updates
669, 672

Student Help
*656, 658, 664, 668, 676,
685, 688*

Career Links
664, 671, 679, 683, 685

Extra Challenge
657, 665, 672, 680, 686

▶ **CHAPTER STUDY GUIDE** **650**

11.1 An Introduction to Sequences and Series **651**
  ▶ GRAPHING CALCULATOR: *Working with Sequences, 658*

11.2 Arithmetic Sequences and Series **659**

11.3 Geometric Sequences and Series **666**
  QUIZ 1, *673*

11.4 Infinite Geometric Series **675**
  ▶ CONCEPT ACTIVITY: *Investigating an Infinite Geometric Series, 674*

11.5 Recursive Rules for Sequences **681**
  QUIZ 2, *687*
  ▶ SPREADSHEET SOFTWARE: *Evaluating Recursive Rules, 688*

  EXTENSION: *Mathematical Induction* **689**

ASSESSMENT

Skill Review, *650*
Quizzes, *673, 687*
Test Preparation Questions, *657, 665, 672, 680, 686*
Chapter Summary, Review, and Test, *691*
Chapter Standardized Test, *696*

CHAPTER
12

Probability and Statistics

APPLICATION HIGHLIGHTS

Concerts *699, 713*
Criminology *702*
Sports *703*
Internet *717*
Entertainment *718*
Business *725*
Home Electronics *726*
Baseball *730*
Endangered Species *732*
Health Statistics *747*

MATH & HISTORY

Probability Theory *737*

INTERNET CONNECTIONS

www.mcdougallittell.com

Application Links
699, 713, 721, 737, 741

Data Updates
719, 735, 751

Student Help
*703, 704, 706, 709, 711,
718, 723, 726, 729, 735,
743, 745, 751, 754*

Career Links
702, 728, 735, 748

Extra Challenge
*707, 714, 722, 729, 736,
744, 751*

PROJECT

Chapters 10–12 Project *764*

▶ **CHAPTER STUDY GUIDE**   **700**

12.1 The Fundamental Counting Principle and Permutations   **701**

12.2 Combinations and the Binomial Theorem   **708**
QUIZ 1, *715*

12.3 An Introduction to Probability   **716**
▶ GRAPHING CALCULATOR: *Generating Random Numbers, 723*

12.4 Probability of Compound Events   **724**

12.5 Probability of Independent and Dependent Events   **730**
QUIZ 2, *737*

12.6 Binomial Distributions   **739**
▶ CONCEPT ACTIVITY: *Investigating Binomial Distributions, 738*
▶ GRAPHING CALCULATOR: *Constructing a Binomial Distribution, 745*

12.7 Normal Distributions: EXPLORING DATA AND STATISTICS   **746**
QUIZ 3, *752*

EXTENSION: *Expected Value*   **753**

ASSESSMENT

Skill Review, *700*
Quizzes, *715, 737, 752*
Test Preparation Questions, *707, 714, 722, 729, 736, 744, 751*
Chapter Summary, Review, and Test, *755*
Chapter Standardized Test, *760*
Cumulative Practice, Chapters 1–12, *762*
Project, Chapters 10–12, *764*

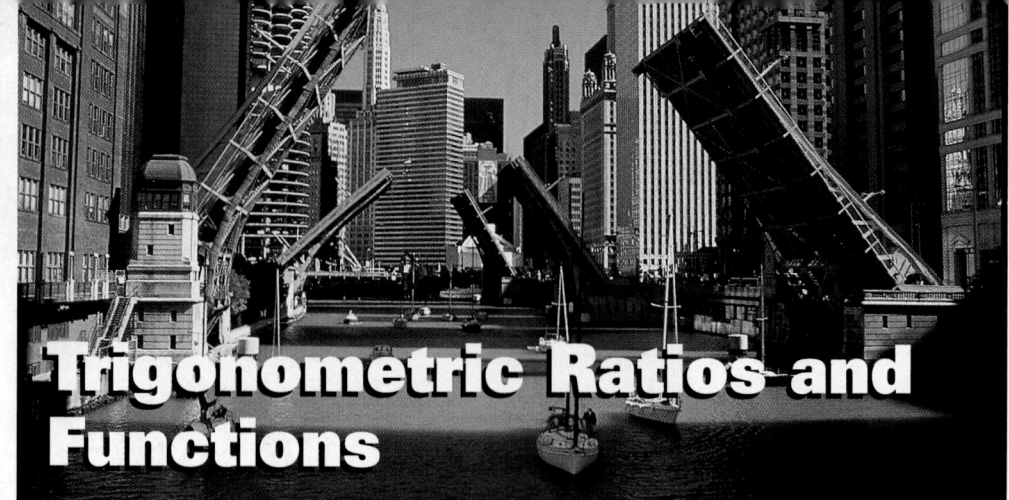

Trigonometric Ratios and Functions

APPLICATION HIGHLIGHTS

Drawbridges *767, 796*
Bicycles *778*
Space Needle Building *779*
Golf *787*
Marching Band *787*
Rock Salt *794*
Construction *794*
Astronomy *801*
Softball *808*
Aviation *814*

MATH & HISTORY

Columbus's Voyage *775*

INTERNET CONNECTIONS

www.mcdougallittell.com

Application Links
767, 775, 778, 817

Data Updates
787, 815

Student Help
770, 777, 786, 796,
804, 811, 817, 820

Career Links
789, 801, 805, 811

Extra Challenge
774, 782, 790, 797, 806,
812, 818

▶ **CHAPTER STUDY GUIDE**   **768**

13.1 Right Triangle Trigonometry   **769**

13.2 General Angles and Radian Measure   **776**
QUIZ 1, *783*

13.3 Trigonometric Functions of Any Angle   **784**

13.4 Inverse Trigonometric Functions   **792**
▶ CONCEPT ACTIVITY: *Investigating Inverse Trigonometric Functions, 791*
QUIZ 2, *798*

13.5 The Law of Sines   **799**

13.6 The Law of Cosines   **807**

13.7 Parametric Equations and Projectile Motion:
EXPLORING DATA AND STATISTICS   **813**
QUIZ 3, *819*
▶ GRAPHING CALCULATOR: *Graphing Parametric Equations, 820*

ASSESSMENT

Skill Review, *768*
Quizzes, *783, 798, 819*
Test Preparation Questions, *774, 782, 790, 797, 806, 812, 818*
Chapter Summary, Review, and Test, *821*
Chapter Standardized Test, *826*

CHAPTER

14

Trigonometric Graphs, Identities, and Equations

▶ **CHAPTER STUDY GUIDE**   **830**

14.1 Graphing Sine, Cosine, and Tangent Functions   **831**
   ▶ GRAPHING CALCULATOR: *Graphing Trigonometric Functions, 838*

14.2 Translations and Reflections of Trigonometric Graphs   **840**
   ▶ CONCEPT ACTIVITY: *Translating and Reflecting
      Trigonometric Graphs, 839*
   QUIZ 1, *847*

14.3 Verifying Trigonometric Identities   **848**

14.4 Solving Trigonometric Equations   **855**

14.5 Modeling with Trigonometric Functions:
   EXPLORING DATA AND STATISTICS   **862**
   QUIZ 2, *868*

14.6 Using Sum and Difference Formulas   **869**

14.7 Using Double- and Half-Angle Formulas   **875**
   QUIZ 3, *882*

ASSESSMENT
Skill Review, *830*
Quizzes, *847, 868, 882*
Test Preparation Questions, *837, 846, 854, 861, 867, 874, 881*
Chapter Summary, Review, and Test, *883*
Chapter Standardized Test, *888*
Cumulative Practice, Chapters 1–14, *890*
Project, Chapters 13–14, *892*

**APPLICATION
HIGHLIGHTS**

Ferris Wheel *829, 842*
Music *833*
Rappelling *843*
Physical Fitness *851*
Meteorology *863*
Biomechanics *871*
Auto Engineering *871*
Sports *878*
Inca Dwellings *881*
Optics *881*

MATH & HISTORY

Music and Math *868*

**INTERNET
CONNECTIONS**

www.mcdougallittell.com
Application Links
  829, 842, 860, 868, 878
Data Updates
Student Help
  *836, 838, 845, 851, 857,
  862, 873, 877*
Career Links
  836, 858, 871
Extra Challenge
  *837, 846, 854, 861, 867,
  874, 881*

PROJECT

Chapters 13–14 Project *892*

▶Student Resources

DERIVATIONS OF KEY FORMULAS   895–904

SKILLS REVIEW HANDBOOK   905–939

Real Numbers *905–913*
Operations with Signed Numbers • Converting Decimals, Fractions, and Percents • Calculating Percents • Least Common Denominator • Writing Ratios and Solving Proportions • Significant Digits • Scientific Notation

Geometry *914–923*
Perimeter, Area, and Volume • Triangle Relationships • Symmetry • Transformations • Similar Figures

Logical Reasoning *924–929*
Logical Argument • If-Then Statements • Counterexamples • Justify Reasoning

Problem Solving *929–932*
Translating Phrases into Algebraic Expressions • Additional Problem Solving Strategies

Graphing *933–936*
Points in the Coordinate Plane • Bar, Circle, and Line Graphs

Algebra *936–939*
Opposites • Multiplying Binomials • Factoring • Least Common Denominator

EXTRA PRACTICE FOR CHAPTERS 1–14   940–960

TABLES   961–970

Symbols *961*
Measures *962*
Formulas *963–968*
Properties *969–970*

GLOSSARY   971–980

INDEX   981–998

CREDITS   999–1000

SELECTED ANSWERS   SA1–SA64

►Who Uses Mathematics in Real Life?

Here are some careers that use the mathematics in Algebra 2.

FOCUS ON CAREERS

Physical Therapist, *p. 16*
Real Estate Broker, *p. 21*
Stockbroker, *p. 23*
Sports Statisticians, *p. 30*
Forester, *p. 68*
Pediatrician, *p. 102*
Nutritionist, *p. 112*
Caterer, *p. 151*
Personal Trainer, *p. 158*
Biotechnician, *p. 172*
Health Services Manager, *p. 202*
Chemist, *p. 217*
Dentist, *p. 234*
Civil Engineer, *p. 252*
Electrician, *p. 279*
Web Developer, *p. 296*
Set Designer, *p. 304*
Automotive Designer, *p. 308*

Ornithologist, *p. 327*
Photographer, *p. 330*
Nurse, *p. 335*
Gerontologist, *p. 342*
Archeologist, *p. 347*
Accountant, *p. 355*
Paleontologist, *p. 419*
Investment Banker, *p. 427*
Amusement Ride Designer, *p. 433*
Coast Guard, *p. 435*
Chemical Engineer, *p. 450*
Financial Planner, *p. 468*
Marine Biologist, *p. 482*
Sound Technician, *p. 495*
Hospital Administrator, *p. 551*
Farmer, *p. 559*
Pharmacist, *p. 566*
Accident Reconstructionist, *p. 593*
Air Traffic Controller, *p. 606*

Seismologist, *p. 634*
Police Officer, *p. 636*
Entomologist, *p. 664*
Computer Programmer, *p. 671*
Economist, *p. 679*
Fishery Biologist, *p. 683*
Physician, *p. 685*
Police Detective, *p. 702*
Botanist, *p. 728*
Teacher, *p. 735*
Market Researcher, *p. 748*
Cartographer, *p. 789*
Astronomer, *p. 801*
Sculptor, *p. 805*
Surveyor, *p. 811*
Musician, *p. 836*
Meteorologist, *p. 858*
Auto Mechanic, *p. 871*

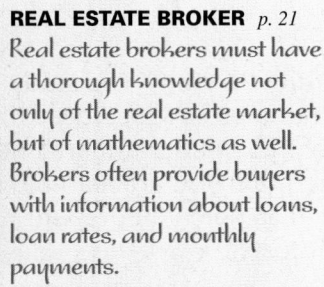

PALEONTOLOGIST *p. 419*
A paleontologist is a scientist who studies fossils of dinosaurs and other prehistoric life forms. Most paleontologists work as college professors.

WEB DEVELOPER *p. 296*
Web developers use hypertext markup language (HTML) to create electronic pages for the World Wide Web. A Web browser translates HTML into pages that can be viewed on a computer screen.

CIVIL ENGINEER *p. 252*
Civil engineers design bridges, roads, buildings, and other structures. In 1996 civil engineers held about 196,000 jobs in the United States.

PHYSICAL THERAPIST *p. 16*
Physical therapists help restore function, improve mobility, and relieve pain in patients with injuries or disease.

REAL ESTATE BROKER *p. 21*
Real estate brokers must have a thorough knowledge not only of the real estate market, but of mathematics as well. Brokers often provide buyers with information about loans, loan rates, and monthly payments.

STUDENT HELP

▶ *Your textbook contains* many special elements to help you learn. It provides several study helps that may be new to you. For example, every chapter begins with a Study Guide.

Chapter Preview The Study Guide starts with a short description of what you will be learning.

Key Vocabulary This list highlights important new terms that will be introduced in the chapter and reviews terms that you already know.

Skill Review These exercises review key skills that you'll apply in the chapter. They will help you identify any topics that you need to review.

Study Strategy The study strategies suggest ideas to help you better understand the math you are learning as well as help you prepare for tests.

CHAPTER 2

Study Guide

PREVIEW

What's the chapter about?

Chapter 2 is about **linear equations and functions**. In Chapter 2 you'll learn

- how to graph ordered pairs, relations, functions, linear equations and inequalities in two variables, piecewise functions, and absolute value functions.
- how to write equations of lines.
- how to solve real-life problems using graphs and equations.

KEY VOCABULARY

▶ **Review**
- graph, p. 3
- linear equation, p. 19
- solution, p. 19
- linear inequality in one variable, p. 41
- absolute value, p. 50

▶ **New**
- relation, p. 67
- function, p. 67
- ordered pair, p. 67
- coordinate plane, p. 67
- linear function, p. 69
- slope, p. 75

- slope-intercept form, p. 82
- standard form, p. 84
- direct variation, p. 94
- scatter plot, p. 100
- linear inequality in two variables, p. 108
- piecewise function, p. 114

PREPARE

Are you ready for the chapter?

SKILL REVIEW Do these exercises to review key skills that you'll apply in this chapter. See the given **reference page** if there is something you don't understand.

STUDENT HELP

▶ **Study Tip**
"Student Help" boxes throughout the chapter give you study tips and tell you where to look for extra help in this book and on the Internet.

Evaluate the expression for the given values of *x* and *y*. (Review Example 3, p. 12)

1. $\frac{y-7}{x-3}$; $x = 2, y = 5$
2. $\frac{5-y}{6-x}$; $x = 4, y = 1$
3. $\frac{8-y}{3-x}$; $x = -1, y = -4$

Solve the equation for *y*. (Review Example 1, p. 26)

4. $3x + y = 4$
5. $x - 2y = 10$
6. $5x + 6y = -60$

Solve the inequality. (Review Examples 1 and 2, p. 42)

7. $2x + 9 < 18$
8. $6 - 0.5y \le 19$
9. $2x + 3 > 6x - 7$

STUDY STRATEGY

Here's a study strategy!

Skills File

In a notebook, make a file of the skills you learn throughout this course. On the left side of the paper, write an important skill and the lesson that it comes from. On the right side of the paper, give an example of the skill in use. Go back now and make a skills file for Chapter 1. Then continue with Chapter 2.

66 Chapter 2

Also, in every lesson you will find a variety of Student Help notes.

STUDENT HELP

 In the Book

Study Tip The study tips will help you avoid common errors.

Skills Review Here you can find where to review skills you've studied in earlier math classes.

Look Back Here are references to material in earlier lessons that may help you understand the lesson.

Extra Practice Your book contains more exercises to practice the skills you are learning.

Homework Help Here you can find suggestions about which Examples may help you solve Exercises.

 On the Internet

Homework Help: *Extra Examples* These are places where you can find additional examples on the Web site.

Homework Help: *Problem Solving Help* Here you can find additional suggestions for solving an exercise.

Keystroke Help These provide the exact keystroke sequences for many different kinds of calculators.

STUDENT HELP

► Study Tip
For an expression like $2x - 3$, think of the expression as $2x + (-3)$, so the terms are $2x$ and -3.

STUDENT HELP

► Skills Review
For help with opposites, see p. 936.

GOAL 2 SIMPLIFYING ALGEBRAIC EXPRESSIONS

For an expression such as $2x + 3$, the parts that are added together, $2x$ and 3, are called **terms**. When a term is the product of a number and a power of a variable, such as $2x$ or $4x^3$, the number is the **coefficient** of the power.

Terms such as $3x^2$ and $-5x^2$ are **like terms** because they have the same variable part. **Constant terms** such as -4 and 2 are also like terms. The distributive property lets you *combine like terms* that have variables by adding the coefficients.

EXAMPLE 5 *Simplifying by Combining Like Terms*

a. $7x + 4x = (7 + 4)x$ Distributive property
$ = 11x$ Add coefficients.

b. $3n^2 + n - n^2 = (3n^2 - n^2) + n$ Group like terms.
$ = 2n^2 + n$ Combine like terms.

c. $2(x + 1) - 3(x - 4) = 2x + 2 - 3x + 12$ Distributive property
$ = (2x - 3x) + (2 + 12)$ Group like terms.
$ = -x + 14$ Combine like terms.

....

Two algebraic expressions are **equivalent** if they have the same value for all values of their variable(s). For instance, the expressions $7x + 4x$ and $11x$ are equivalent, as are the expressions $5x - (6x + y)$ and $-x - y$. A statement such as $7x + 4x = 11x$

GUIDED PRACTICE

Vocabulary Check ✓   **1.** Explain the difference between a simple linear inequality and a compound linear inequality.

Concept Check ✓   **2.** Tell whether this statement is *true* or *false:* Multiplying both sides of an inequality by the same number always produces an equivalent inequality. Explain.

      3. Explain the difference between solving $2x < 7$ and solving $-2x < 7$.

Skill Check ✓   Solve the inequality. Then graph your solution.

4. $x - 5 < 8$ **5.** $3x \geq 15$ **6.** $-x + 4 > 3$

7. $\frac{1}{2}x \leq 6$ **8.** $x + 8 > -2$ **9.** $-x - 3 < -5$

Graph the inequality.

10. $-2 \leq x < 5$ **11.** $x \geq 3$ or $x < -3$

12. 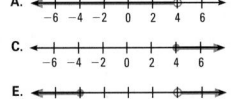 **WINTER DRIVING** You are moving to Montana and need to lower the freezing point of the cooling system in the car from Example 6 to $-50°C$. This will also raise the boiling point to $140°C$. Write a compound inequality that models this situation. Then write the inequality in degrees Fahrenheit.

PRACTICE AND APPLICATIONS

STUDENT HELP

► Extra Practice
to help you master skills is on p. 941.

MATCHING INEQUALITIES Match the inequality with its graph.

13. $x \geq 4$ **14.** $x < 4$ **15.** $-4 < x \leq 4$

16. $x \geq 4$ or $x < -4$ **17.** $-4 \leq x \leq 4$ **18.** $x > 4$ or $x \leq -4$

A.   -6 -4 -2 0 2 4 6   B.   -6 -4 -2 0 2 4 6

C.   -6 -4 -2 0 2 4 6   D.   -6 -4 -2 0 2 4 6

E.   -6 -4 -2 0 2 4 6   F.   -6 -4 -2 0 2 4 6

CHECKING SOLUTIONS Decide whether the given number is a solution of the inequality.

19. $2x + 9 < 16; 4$ **20.** $10 - x \geq 3; 7$ **21.** $7x - 12 < 8; 3$

22. $-\frac{1}{3}x - 2 \leq -4; 9$ **23.** $-3 < 2x \leq 6; 3$ **24.** $-8 < x - 11 < -6; 5$

STUDENT HELP

► HOMEWORK HELP
Examples 1, 2: Exs. 13, 14, 19–22, 25–36
Example 3: Exs. 49–51
Examples 4, 5: Exs. 15–18, 23, 24, 37–48
Example 6: Exs. 52–54
Example 7: Exs. 55, 56

SIMPLE INEQUALITIES Solve the inequality. Then graph your solution.

25. $4x + 5 > 25$ **26.** $7 - n \leq 19$ **27.** $5 - 2x \geq 27$

28. $\frac{1}{2}x - 4 > -6$ **29.** $\frac{3}{2}x - 7 < 2$ **30.** $5 + \frac{1}{3}n \leq 6$

31. $4x - 1 > 14 - x$ **32.** $-n + 6 < 7n + 4$ **33.** $4.7 - 2.1x > -7.9$

34. $2(n - 4) \leq 6$ **35.** $2(4 - x) > 8$ **36.** $5 - 5x > 4(3 - x)$

1.6 *Solving Linear Inequalities*   **45**

Plot a course to success with McDougal Littell Algebra 2!

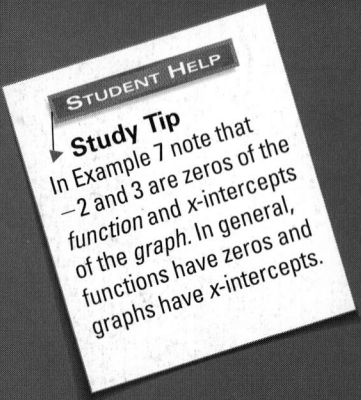

STUDENT HELP

Study Tip
In Example 7 note that −2 and 3 are zeros of the *function* and x-intercepts of the *graph*. In general, functions have zeros and graphs have x-intercepts.

HOW... can I make algebra understandable to all my students?

WHERE... can my students go for help with their homework?

WHAT... resources are available to help me succeed as a teacher?

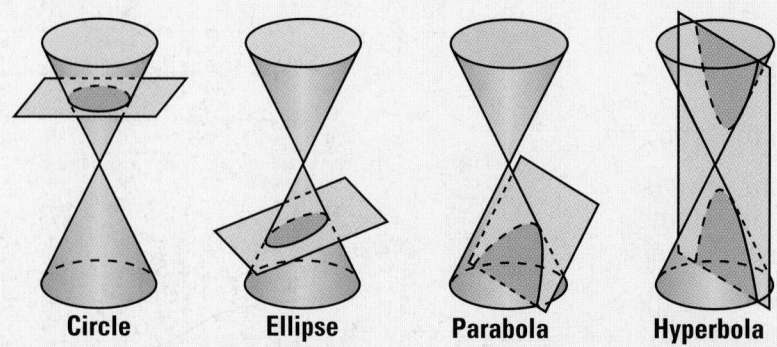

Circle   Ellipse   Parabola   Hyperbola

FIND THE ANSWERS you're looking for in McDougal Littell Algebra 2!

HOW?

Important concepts are made understandable to all students through instructional diagrams and graphics, interactive activities, and numerous examples throughout the text.

Completing the Square		
Expression	Number of 1-tiles needed to complete the square	Expression written as a square
$x^2 + 2x + \underline{?}$	?	?
$x^2 + 4x + \underline{?}$	?	?
$x^2 + 6x + \underline{?}$	9	$x^2 + 6x + 9 = (x + 3)^2$
$x^2 + 8x + \underline{?}$	?	?
$x^2 + 10x + \underline{?}$	?	?

WHERE?

Students receive support when they are learning on their own through Student Help notes throughout the book, including homework exercises correlated to the examples and Homework Help on the Internet.

You will need one x^2-tile and six x-tiles.

WHAT?

A variety of resources help you adapt the program to your teaching styles and to the needs of your students.

pylon

52 m

McDougal Littell Algebra 2 provides clear, visual explanations that every student can follow.

In the quadratic formula, the expression $b^2 - 4ac$ under the radical sign is called the **discriminant** of the associated equation $ax^2 + bx + c = 0$.

$$x = \frac{-b \pm \sqrt{b^2 - 4ac}}{2a} \quad \longleftarrow \quad \text{discriminant}$$

You can use the discriminant of a quadratic equation to determine the equation's number and type of solutions.

CONCEPT SUMMARY BOXES present ideas clearly and concisely to help students master key concepts.

NUMBER AND TYPE OF SOLUTIONS OF A QUADRATIC EQUATION

Consider the quadratic equation $ax^2 + bx + c = 0$.

- If $b^2 - 4ac > 0$, then the equation has two real solutions.
- If $b^2 - 4ac = 0$, then the equation has one real solution.
- If $b^2 - 4ac < 0$, then the equation has two imaginary solutions.

EXAMPLES are numerous, easy to follow, and correspond to the exercises.

EXAMPLE 4 *Using the Discriminant*

Find the discriminant of the quadratic equation and give the number and type of solutions of the equation.

a. $x^2 - 6x + 10 = 0$ **b.** $x^2 - 6x + 9 = 0$ **c.** $x^2 - 6x + 8 = 0$

SOLUTION

EQUATION	DISCRIMINANT	SOLUTION(S)
$ax^2 + bx + c = 0$	$b^2 - 4ac$	$x = \frac{-b \pm \sqrt{b^2 - 4ac}}{2a}$
a. $x^2 - 6x + 10 = 0$	$(-6)^2 - 4(1)(10) = -4$	Two imaginary: $3 \pm i$
b. $x^2 - 6x + 9 = 0$	$(-6)^2 - 4(1)(9) = 0$	One real: 3
c. $x^2 - 6x + 8 = 0$	$(-6)^2 - 4(1)(8) = 4$	Two real: 2, 4

· · · · · · · · · ·

In Example 4 notice that the number of real solutions of $x^2 - 6x + c = 0$ can be changed just by changing the value of c. A graph can help you see why this occurs. By changing c, you can move the graph of

$$y = x^2 - 6x + c$$

up or down in the coordinate plane. If the graph is moved too high, it won't have an x-intercept and the equation $x^2 - 6x + c = 0$ won't have a real-number solution.

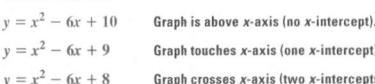

$y = x^2 - 6x + 10$ Graph is above x-axis (no x-intercept).

$y = x^2 - 6x + 9$ Graph touches x-axis (one x-intercept).

$y = x^2 - 6x + 8$ Graph crosses x-axis (two x-intercepts).

5.6 *The Quadratic Formula and the Discriminant*   293

STUDY GUIDES ▶

in every chapter will help students review key prerequisites and prepare for the chapter.

CHAPTER 5

Study Guide

PREVIEW

What's the chapter about?

Chapter 5 is about **quadratic functions, equations, and inequalities.** Many real-life situations can be modeled by quadratic functions. In Chapter 5 you'll learn

• four ways to solve quadratic equations.

• how to graph quadratic functions and inequalities.

KEY VOCABULARY

▶ Review	▶ New	
• linear equation, p. 19	• quadratic function, p. 249	• completing the square, p. 282
• linear inequality, pp. 41, 108	• parabola, p. 249	• quadratic formula, p. 291
• absolute value, p. 50	• factoring, p. 256	• discriminant, p. 293
• linear function, p. 69	• quadratic equation, p. 257	• quadratic inequality, pp. 299, 301
• x-intercept, p. 84	• zero of a function, p. 259	• best-fitting quadratic model, p. 308
• best-fitting line, p. 101	• square root, p. 264	
• vertex, p. 122	• complex number, p. 272	

PREPARE

Are you ready for the chapter?

SKILL REVIEW Do these exercises to review key skills that you'll apply in this chapter. See the given reference page if there is something you don't understand.

STUDENT HELP

↪ **Study Tip**
"Student Help" boxes throughout the chapter give you study tips and tell you where to look for extra help in this book and on the Internet.

Solve the equation. (Review Examples 1–3, pp. 19 and 20)

1. $3x - 5 = 0$ $\frac{5}{3}$ 2. $4(x + 6) = 12$ -3 3. $2x + 1 = -x + 7$ 2

Graph the inequality. (Review Example 3, p. 109) 4–6. See margin.

4. $x + y > 5$ 5. $3x - 2y \le 12$ 6. $y \ge -2x$

Graph the function and label the vertex. (Review Example 1, p. 123) 7–9. See margin.

7. $y = |x| + 2$ 8. $y = |x - 3|$ 9. $y = -2|x + 1| - 4$

STUDY STRATEGY

Here's a study strategy!

Troubleshoot

After you complete each lesson, look back and identify your trouble spots, such as concepts you didn't understand or homework problems you had difficulty solving. Review the material given in the lesson and try to solve any difficult problems again. If you're still having trouble, seek the help of another student or your teacher.

248 Chapter 5

---

3 APPLY

◯ ASSIGNMENT GUIDE

BASIC
Day 1: pp. 142–145 Exs.12–28 even, 32–40 even, 41–45 odd
Day 2: pp. 142–145 Exs. 33–39 odd, 42–50 even, 53, 54, 64, 67–79 odd

AVERAGE
Day 1: pp. 142–145 Exs.12–46 even, 53–55
Day 2: 33–39 odd, 49–52, 57–63 odd, 64, 67–79 odd

ADVANCED
Day 1: pp. 142–145 Exs.12–48 even, 49–53
Day 2: pp. 142–145 Exs. 54–64, 65, 67–79 odd

BLOCK SCHEDULE
pp. 142–145 Exs.12–50 even, 52–56, 60–64, 67–79 odd

EXERCISE LEVELS
Level A: *Easier*
11–28, 32–40, 54–55
Level B: *More Difficult*
29–31, 41–53, 56–62, 64
Level C: *Most Difficult*
63, 65

✔ HOMEWORK CHECK
To quickly check student understanding of key concepts, go over the following exercises: Exs. 12, 22, 36, 42, 54. See also the Daily Homework Quiz.
• Blackline Master (*Chapter 3 Resource Book*, p. 24)
• 📀 Transparency (p. 20)

GUIDED PRACTICE

Vocabulary Check ✔

1. Complete this statement: A(n) ? of a system is an ordered pair (x, y) that satisfies each ...

Concept Check ✔

2. How can you use the graph of a linear system ... system has? See margin.

3. Explain why a linear system in two variab ... See margin.

Skill Check ✔

Check whether the ordered pair (5, 6) is a s ...

2. *Sample answer:* If the graph consists of two non-parallel lines, then there is a single solution. If the graph consists of two parallel lines, there is no solution. If the graphs of the two lines coincide, there are infinitely many solutions.

3. *Sample answer:* If two lines share two points in common, they are the same line, and every point on that line is a solution.

4. $-2x + 4y = -14$ 5. $7x - 2y = $
   $3x + y = 21$ no $-x + 3y = $

Graph the linear system. How many solutio ... 7–9. Se ...

7. $2x - y = 4$ 8. $14x + 3y = $
   $-6x + 3y = -18$ 0 $7x - 5y = $

10. 🏫 **SCHOOL OUTING** Your school is pl ... community park. The park rents bicycles ... $6 per hour. The total budget per person i ... spend doing each activity? (2, 3)

PRACTICE AND APPLICATIONS

STUDENT HELP

↪ **Extra Practice**
to help you master skills is on p. 943.

CHECKING A SOLUTION Check whether the ordered pair is a solution of the system.

11. (6, −1) yes 12. (3, 0) no 13. (−2, −8) no
    $4x - y = 25$ $-x + 2y = 3$ $2x - y = 52$
    $-3x - 2y = -16$ $10x + y = 30$ $9x - y = -10$

14. (−3, −5) yes 15. (−4, 1) yes 16. (10, 8) yes
    $-x - y = 8$ $-4x + 3y = 19$ $-3x - y = -38$
    $2x + 5y = -31$ $5x - 7y = -27$ $-8x + 8y = -16$

17. (1, −1) no 18. (−2, −7) yes 19. (0, 2) no
    $-3x + y = -4$ $5x - y = -3$ $17x + 8y = 16$
    $7x + 2y = -5$ $x + 3y = -23$ $-x - 4y = 8$

20–31. Estimates may vary. An exact solution is given. For systems with infinitely many solutions, a sample answer is given.

See margin for graphs.

GRAPH AND CHECK Graph the linear system and estimate the solution. Then check the solution algebraically.

20. $2x + y = 13$ (3, 7) 21. $x + 2y = 9$ (7, 1) 22. $-2x + y = 5$ (−1, 3)
    $5x - 2y = 1$ $-x + 6y = -1$ $x + y = 2$

23. $3x + 4y = -10$ (2, −4) 24. $2x + y = -11$ 25. $y = 5x$ (1, 5)
    $-7x - y = -10$ $-6x - 3y = 33$ $y = x + 4$
    Sample answer: (−5, −1)

26. $-x + 3y = 3$ 27. $2x + y = -2$ 28. $3x = 12$ (4, 0)
    $2x - 6y = -6$ $x - 2y = 19$ (3, −8) $-x + 8y = -4$
    Sample answer: (6, 3)

STUDENT HELP

↪ **HOMEWORK HELP**
Example 1: Exs. 11–19
Example 2: Exs. 20–31
Example 3: Exs. 32–52
Example 4: Exs. 54–59

29. $3x - y = 8$ 30. $y = \frac{1}{6}x - 2$ (12, 0) 31. $-x + 4y = 10$ (−2, 2)
    $\frac{1}{3}x - \frac{1}{6}y = 1$ (2, −2) $y = -\frac{1}{6}x + 2$ $4x - y = -10$

142 Chapter 3 *Systems of Linear Equations and Inequalities*

142

◀ **PRACTICE SETS**

▪ include exercises of three difficulty levels—Levels A, B, and C—which are labeled in the Teacher's Edition.
▪ Also available: Practice Worksheets (Levels A, B, and C)

CHALLENGE PROBLEMS

(not pictured) are included in every practice set, with extra challenge online at www.mcdougallittell.com

WHERE... can my students go for help with their homework?

Student Help features throughout the program provide students with a variety of tips and homework support.

 EXTRA EXAMPLES

show where students can find additional support on the Internet at www.mcdougallittell.com

 KEYSTROKE HELP

guides students to the exact keystroke sequences for many different kinds of calculators (not pictured).

 PROBLEM SOLVING

provides online homework help and suggestions for solving exercises.

 **Solving a Quadratic Equation**

STUDENT HELP
► **HOMEWORK HELP**
Visit our Web site
www.mcdougallittell.com
for extra examples.

Solve $\frac{1}{3}(x + 5)^2 = 7$.

SOLUTION

$\frac{1}{3}(x + 5)^2 = 7$	Write original equation.
$(x + 5)^2 = 21$	Multiply each side by 3.
$x + 5 = \pm\sqrt{21}$	Take square roots of each side.
$x = -5 \pm \sqrt{21}$	Subtract 5 from each side.

▶ The solutions are $-5 + \sqrt{21}$ and $-5 - \sqrt{21}$.

✓**CHECK** Check the solutions either by substituting them into the original equation or by graphing $y = \frac{1}{3}(x + 5)^2 - 7$ and observing the x-intercepts.

5.3 *Solving Quadratic Equations by Finding Square Roots* **265**

---

48. Based on your graphs, can 1000 pounds of theater equipment be supported by a $\frac{1}{2}$ inch manila rope? by a $\frac{1}{2}$ inch wire rope?

49. ⬤ **HEALTH** For a person of height h (in inches), a healthy weight W (in pounds) is one that satisfies this system of inequalities:

$$W \geq \frac{19h^2}{703} \quad \text{and} \quad W \leq \frac{25h^2}{703}$$

Graph the system for $0 \leq h \leq 80$. What is the range of healthy weights for a person 67 inches tall? ▶ Source: *Parade Magazine*

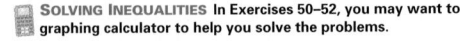 **SOLVING INEQUALITIES In Exercises 50–52, you may want to use a graphing calculator to help you solve the problems.**

STUDENT HELP
► **HOMEWORK HELP**
Visit our Web site
www.mcdougallittell.com
for help with problem
solving in Exs. 50–52.

50. ⬤ **FORESTRY** *Sawtimber* is a term for trees that are suitable for sawing into lumber, plywood, and other products. For the years 1983–1995, the unit value y (in 1994 dollars per million board feet) of one type of sawtimber harvested in California can be modeled by

$$y = 0.125x^2 - 569x + 848{,}000, \quad 400 \leq x \leq 2200$$

where x is the volume of timber harvested (in millions of board feet).
▶ Source: California Department of Forestry and Fire Protection

  a. For what harvested timber volumes is the value of the timber at least $400,000 per million board feet?

  b. **LOGICAL REASONING** What happens to the unit value of the timber as the volume harvested increases? Why would you expect this to happen?

304 **Chapter 5** *Quadratic Functions*

SKILLS REVIEW

guides students to review skills they've studied in earlier math classes to help them succeed with current concepts.

Business

EXAMPLE 5 *Using Composition of Functions*

A clothing store advertises that it is having a 25% off sale. For one day only, the store advertises an additional savings of 10%.

  a. Use composition of functions to find the total percent discount.

  b. What would be the sale price of a $40 sweater?

SOLUTION

> **STUDENT HELP**
> ↳ **Skills Review**
> For help with calculating percents, see p. 907.

  a. Let x represent the price. The sale price for a 25% discount can be represented by the function $f(x) = x - 0.25x = 0.75x$. The reduced sale price for an additional 10% discount can be represented by the function $g(x) = x - 0.10x = 0.90x$.

$$g(f(x)) = g(0.75x) = 0.90(0.75x) = 0.675x$$

  ▶ The total percent discount is $100\% - 67.5\% = 32.5\%$.

  b. Let $x = 40$. Then $g(f(x)) = g(f(40)) = 0.675(40) = 27$.

  ▶ The sale price of the sweater is $27.

LOOK BACK

contains references to material in earlier lessons that may help students understand the lesson.

> **STUDENT HELP**
> ↳ **Look Back**
> For help with solving linear systems with many or no solutions, see p. 150.

EXAMPLE 2 *Solving a System with No Solution*

Solve the system.

$x + y + z = 2$	**Equation 1**
$3x + 3y + 3z = 14$	**Equation 2**
$x - 2y + z = 4$	**Equation 3**

SOLUTION

When you multiply the first equation by -3 and add the result to the second equation, you obtain a false equation.

$$\begin{aligned} -3x - 3y - 3z &= -6 \\ \underline{3x + 3y + 3z} &= \underline{14} \\ 0 &= 8 \end{aligned}$$

  Add -3 times the first equation to the second.

  New Equation 1

▶ Because you obtained a false equation, you can conclude that the original system of equations has no solution.

EXAMPLE 3 *Solving a System with Many Solutions*

Solve the system.

$x + y + z = 2$	**Equation 1**
$x + y - z = 2$	**Equation 2**
$2x + 2y + z = 4$	**Equation 3**

EXTRA PRACTICE

is provided in the book for every section, in addition to the large number of exercises found with each lesson.

NUMEROUS EXAMPLES

in every lesson are correlated to the exercises and support students with their homework.

PRACTICE AND APPLICATIONS

> **STUDENT HELP**
> ↳ **Extra Practice**
> to help you master skills is on p. 949.

PROPERTIES OF RATIONAL EXPONENTS Simplify the expression.

22. $3^{5/3} \cdot 3^{1/3}$ **23.** $\left(5^{2/3}\right)^{1/2}$ **24.** $4^{1/4} \cdot 64^{1/4}$ **25.** $\dfrac{1}{36^{-1/2}}$

26. $\dfrac{7^{1/5}}{7^{3/5}}$ **27.** $\dfrac{70^{1/3}}{14^{1/3}}$ **28.** $\left(2^{1/4} \cdot 2^{1/3}\right)^6$ **29.** $\left(\dfrac{5^2}{8^2}\right)^{-1/2}$

30. $\dfrac{6^{2/3} \cdot 4^{2/3}}{3^{2/3}}$ **31.** $\dfrac{125^{2/9} \cdot 125^{1/9}}{5^{1/4}}$ **32.** $\dfrac{12^{10/8}}{12^{-3/8}}$ **33.** $\left(10^{3/4} \cdot 4^{3/4}\right)^{-4}$

PROPERTIES OF RADICALS Simplify the expression.

34. $\sqrt{64} \cdot \sqrt[3]{64}$ **35.** $\sqrt[3]{8} \cdot \sqrt[3]{2}$ **36.** $\sqrt[4]{5} \cdot \sqrt[4]{5}$ **37.** $\left(\sqrt[6]{6} \cdot \sqrt[6]{6}\right)^{12}$

38. $\dfrac{\sqrt[3]{7}}{\sqrt[3]{7}}$ **39.** $\dfrac{\sqrt[3]{4}}{\sqrt[3]{32}}$ **40.** $\dfrac{\sqrt[3]{8} \cdot \sqrt[3]{16}}{\sqrt[3]{2}}$ **41.** $\dfrac{\sqrt[3]{9} \cdot \sqrt[3]{6}}{\sqrt[3]{2} \cdot \sqrt[3]{2}}$

> **STUDENT HELP**
> ↳ **HOMEWORK HELP**
> **Example 1:** Exs. 22–33
> **Example 2:** Exs. 34–41
> **Example 3:** Exs. 42–49
> **Example 4:** Exs. 50–55
> **Example 5:** Exs. 56–67
> **Example 6:** Exs. 68–75
> **Example 7:** Exs. 76–81
> **Example 8:** Exs. 90–93
> **Example 9:** Exs. 94–97

SIMPLEST FORM Write the expression in simplest form.

42. $\sqrt{50}$ **43.** $\sqrt[5]{1215}$ **44.** $\sqrt[3]{18} \cdot \sqrt[3]{15}$ **45.** $3\sqrt[4]{24} \cdot 5\sqrt[4]{2}$

46. $\sqrt[3]{\dfrac{1}{7}}$ **47.** $\dfrac{2}{\sqrt[4]{81}}$ **48.** $\sqrt[4]{\dfrac{80}{9}}$ **49.** $\dfrac{\sqrt[3]{4}}{\sqrt[3]{8}}$

COMBINING ROOTS AND RADICALS Perform the indicated operation.

50. $\sqrt[5]{6} + 5\sqrt[5]{6}$ **51.** $5(5)^{1/7} - 7(5)^{1/7}$ **52.** $-\sqrt[8]{4} + 5\sqrt[8]{4}$

53. $160^{1/2} - 10^{1/2}$ **54.** $\sqrt[3]{375} + \sqrt[3]{81}$ **55.** $2\sqrt[5]{176} + 5\sqrt[5]{11}$

7.2 *Properties of Rational Exponents*   **411**

WHAT... resources are available to help me succeed as a teacher?

The McDougal Littell resources are conveniently organized and include a variety of materials to help you adapt the program to your teaching style and to the specific needs of your students!

TEACHER'S RESOURCE PACKAGE

- Chapter Resource Books (one for each chapter, organized by lesson)
- Basic Skills Workbook: Diagnosis and Remediation (TE)
- Practice Workbook with Examples (TE)
- Standardized Test Practice Workbook (TE)
- Warm-Up Transparencies and Daily Homework Quiz
- Worked-Out Solution Key

Finally! Resources organized the way you teach... one lesson at a time!

McDougal Littell's Chapter Resource Books allow you to easily carry the resources you have for a chapter in one manageable book. The materials are organized by lesson so that you can see everything you have available for a specific section.

CHAPTER RESOURCE BOOKS INCLUDE:

- Tips for New Teachers
- Parent Guide for Student Success
- Prerequisite Skills Review
- Strategies for Reading Mathematics
- Lesson Plans
- Lesson Plans for Block Scheduling
- Activity Support Masters
- Graphing Calculator Activities with Keystrokes
- Practice (Levels A, B, and C)
- Reteaching with Practice
- Quick Catch-Up for Absent Students
- Cooperative Learning Activities
- Interdisciplinary Applications
- Real-Life Applications: When Will I Ever Use This?
- Math and History Applications
- Quizzes
- Chapter Review Games and Activities
- Chapter Tests (Levels A, B, and C)
- SAT/ACT Chapter Test
- Alternative Assessment with Rubric and Math Journal
- Project with Rubric
- Cumulative Review
- Resource Book Answers

A number of transparency packages give you many easy-to-use options for reviewing homework, starting class, and teaching a lesson.

Answer Transparencies for Checking Homework

Warm-Up Transparencies and Daily Homework Quiz

- Warm-Up Exercises
- Daily Homework Quizzes

Extra Example Transparencies with Standardized Test Practice

- Extra Examples
- Checkpoint Exercises
- Standardized Test Practice Questions

Starting Points: Alternative Lesson Opener Transparency Package

There are many ways to introduce a lesson — and we provide alternative ideas for teachers on ready-to-use transparencies.

- Application Lesson Openers
- Graphing Calculator Lesson Openers
- Activity Lesson Openers
- Visual Approach Lesson Openers

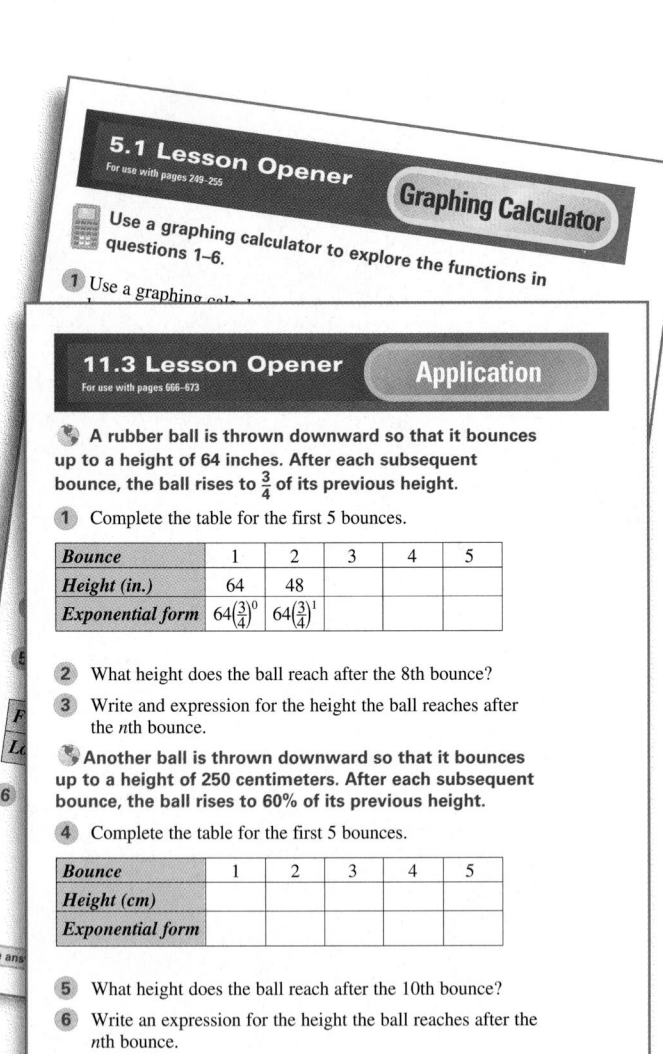

5.1 Lesson Opener
For use with pages 249–255

Graphing Calculator

Use a graphing calculator to explore the functions in questions 1–6.

1 Use a graphing cal...

11.3 Lesson Opener
For use with pages 666–673

Application

A rubber ball is thrown downward so that it bounces up to a height of 64 inches. After each subsequent bounce, the ball rises to $\frac{3}{4}$ of its previous height.

1 Complete the table for the first 5 bounces.

Bounce	1	2	3	4	5
Height (in.)	64	48			
Exponential form	$64\left(\frac{3}{4}\right)^0$	$64\left(\frac{3}{4}\right)^1$			

2 What height does the ball reach after the 8th bounce?

3 Write and expression for the height the ball reaches after the nth bounce.

Another ball is thrown downward so that it bounces up to a height of 250 centimeters. After each subsequent bounce, the ball rises to 60% of its previous height.

4 Complete the table for the first 5 bounces.

Bounce	1	2	3	4	5
Height (cm)					
Exponential form					

5 What height does the ball reach after the 10th bounce?

6 Write an expression for the height the ball reaches after the nth bounce.

See answers in Chapter Resource Book

74

WORKBOOKS

- **Basic Skills Workbook: Diagnosis and Remediation**
- **Practice Workbook with Examples**
- **Standardized Test Practice Workbook**

SPANISH RESOURCES

Resources in Spanish Include:

- **Reteaching with Practice**
- **Quizzes**
- **Chapter Tests**
- **SAT/ACT Chapter Tests**
- **Alternative Assessments with Rubrics**
- **Math Journal**
- **Glossary**

How... can I incorporate technology into my classroom?

McDougal Littell technology resources help you and your students meaningfully use technology to enhance lessons and build understanding.

Technology for
PLANNING AND TEACHING

Online Lesson Planner
- Create customized lesson plans.
- Adjust to schedule changes.
- Adapt the program to local and state objectives.

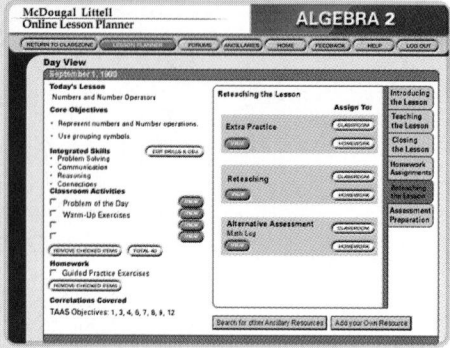

Electronic Teacher Tools
This handy tool provides all your teaching resources on one CD-ROM.

Electronic Lesson Presentations
The Electronic Lesson Presentations help you introduce concepts through colorful diagrams and animated instructional techniques.

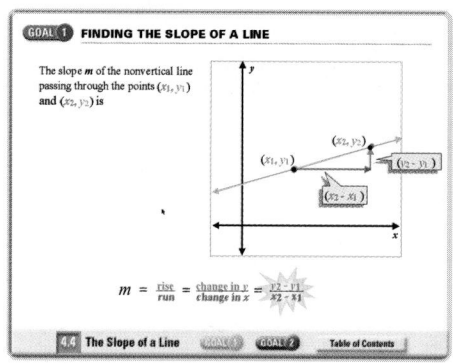

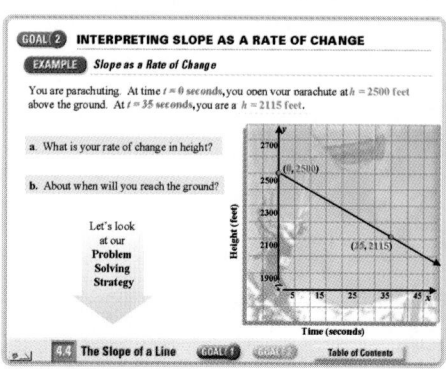

Technology for
STUDENT SUPPORT

Personal Student Tutor
This tutorial program is correlated to the McDougal Littell series. It provides:
- animated examples
- student hints
- exercises that are automatically graded
- student progress reports

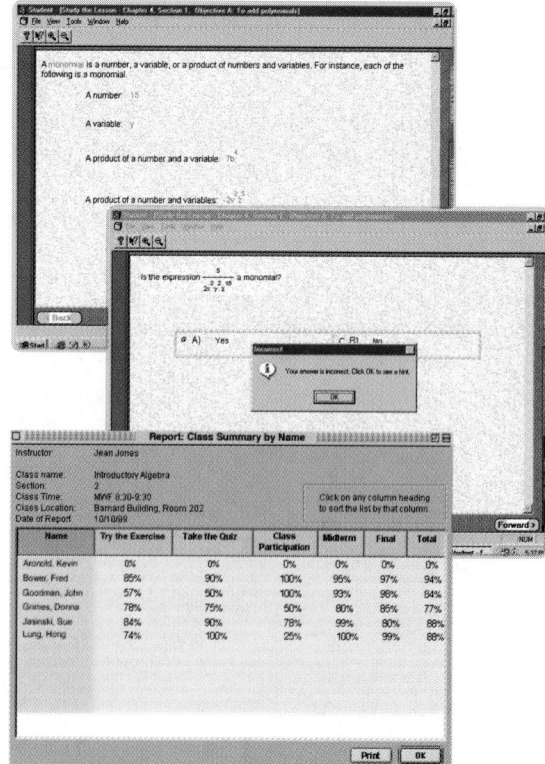

Technology for

ClassZone

ClassZone is the companion Web site to McDougal Littell Algebra 2 that includes Student Help, career links, data updates, and more. To access ClassZone, go to www.mcdougallittell.com.

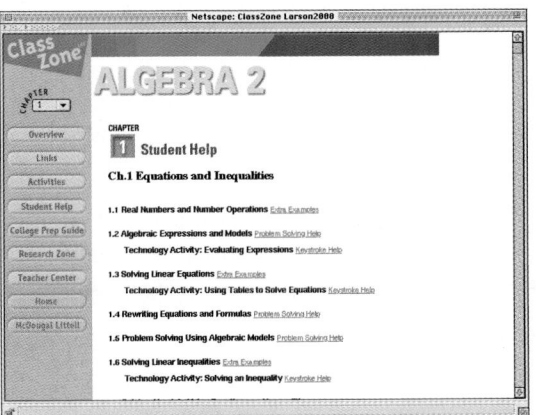

Time-Saving Test and Practice Generator

- Develop tests and practice sheets.
- Instantly create answer keys.

Instant Replay: Video Review Games

- Video Review Games provide a fun, interactive way to help students review the chapter — as well as keep previously learned skills sharp.
- Real-life Motivators provide a great start to each chapter and help students see how the concepts they are about to learn are applied to real situations.

PACING THE COURSE

COURSE PACING CHART

The Pacing Chart below shows the number of days allotted for each chapter. The Regular Schedule requires 160 days. The Block Schedule requires 80 days. These time frames include days for review and assessment: 2 days per chapter for the Regular Schedule and 1 day per chapter for the Block Schedule. Semester and trimester divisions are indicated by red and blue rules, respectively. Recommended pacing for Algebra 1 over two years is also provided.

Chapter	1	2	3	4	5	Trimester 6	7	Semester 8	9	Trimester 10	11	12	13	14
Regular Schedule	10	11	10	9	14	13	13	14	10	11	9	11	12	13
Block Schedule	5	6	5	5	7	6.5	6.5	7	5	5.5	4.5	5.5	5.5	6
Two-Year Pacing	20	22	20	18	28	26	26	28	20	22	18	22	24	26

End of year one

Assignments are provided with each lesson for a basic course, an average course, an advanced course, and a block-scheduled course. Each of the four courses covers all twelve chapters.

BASIC COURSE

The basic course is intended for students who enter with below-average mathematical and problem-solving skills. Assignments include:
- substantial work with the skills and concepts presented in the lesson
- straightforward applications of these skills and concepts
- test preparation and mixed review exercises

AVERAGE COURSE

The average course is intended for students who enter with typical mathematical and problem-solving skills. Assignments include:
- substantial work with the skills and concepts presented in the lesson
- application of these skills and concepts
- test preparation and mixed review exercises

ADVANCED COURSE

The advanced course is intended for students who enter with above-average mathematical and problem-solving skills. Assignments include:
- substantial work with the skills and concepts presented in the lesson
- more complex applications and challenge exercises
- test preparation and mixed review exercises

BLOCK-SCHEDULED COURSE

The block-scheduled course is intended for schools that use a block schedule. The exercises assigned are comparable to the exercises for the average course.

TWO-YEAR PACING

The two-year pacing allows more time for pre-course review using the prerequisite skills material. The schedule also provides more opportunities to use activities or reteaching and practice materials.

PACING EACH CHAPTER

The Pacing Chart for each chapter is located on the interleaved pages preceding the chapter. Part of the Pacing Chart for Chapter 3 is shown here. The regular schedule chart provides pacing for the basic, average, and advanced courses. The block schedule chart provides pacing for the block-scheduled course.

REGULAR SCHEDULE

Day 7

3.5

STARTING OPTIONS
- Homework Check
- Warm-Up or Daily Quiz

TEACHING OPTIONS
- Motivating the Lesson
- Les. Opener (Activity)
- Graphing Calc. Activity
- Examples 1–4
- Technology Activity
- Closure Question
- Guided Practice Exs.

APPLY/HOMEWORK
- See Assignment Guide.
- See the CRB: Practice, Reteach, Apply, Extend

ASSESSMENT OPTIONS
- Checkpoint Exercises
- Daily Quiz (3.5)
- Stand. Test Practice

Each day provides a number of options for the lesson, including starting, teaching, applying, and assessing.

The Chapter Resource Book includes a number of follow-up options for the lesson. These options include three levels of practice, reteaching examples with practice, inter-disciplinary and real-life applications, and challenge exercises.

A black dot indicates that the option can be found in the Student Edition, a green dot in the Chapter Resource Book, and a red dot in the Teacher's Edition.

BLOCK SCHEDULE

Day 4

3.5 & 3.6

DAY 5 START OPTIONS
- Homework Check
- W-Up 3.5 or D. Quiz 3.4

TEACHING 3.5 OPTIONS
- Motivating the Lesson
- Les. Opener (Activity)
- Graphing Calc. Activity
- Examples 1–4
- Technology Activity
- Closure Question
- Guided Practice Exs.

BEGINNING 3.6 OPTIONS
- Warm-Up (Les. 3.6)
- Motivating the Lesson
- Les. Opener (Visual)
- Examples 1–4
- Closure Question
- Guided Pract. Exs.

APPLY/HOMEWORK
- See Assignment Guide.
- See the CRB: Practice, Reteach, Apply, Extend

ASSESSMENT OPTIONS
- Checkpoint Exercises
- Daily Quiz (Les. 3.5, 3.6)
- Stand. Test Practice
- Quiz (3.5, 3.6)

ASSIGNMENT GUIDE

An Assignment Guide for each lesson is provided at point-of-use at the beginning of the exercise set.

Assignments are provided for basic, average, advanced, and block-scheduled courses.

ASSIGNMENT GUIDE

BASIC
Day 1: pp. 152–155 Exs. 12–20 even, 24–32 even, 36–44 even, 50
Day 2: pp. 152–155 Exs. 33–41 odd, 51–55 odd, 63, 64, 67–81 odd, Quiz 1 Exs. 1–19

AVERAGE
Day 1: pp. 152–155 Exs. 12–20 even, 24–34 even, 38–52 even
Day 2: pp. 152–155 Exs. 33–51 odd, 54–58 even, 63, 64, 67–81 odd, Quiz 1 Exs. 1–19

ADVANCED
Day 1: pp. 152–155 Exs. 12–54 even
Day 2: pp. 152–155 Exs. 47–57 odd, 59–65, 67–81 odd, Quiz 1 Exs. 1–19

BLOCK SCHEDULE
pp. 152–155 Exs. 12–54 even, 63, 64, 67–81 odd, Quiz 1 Exs. 1–19

What other books come in the McDougal Littell series?

The same solid approach and features that are found in McDougal Littell Algebra 2 to help you and your students succeed are also found in the Algebra 1 and Geometry books.

McDougal Littell ALGEBRA 1

1 Connections to Algebra

2 Properties of Real Numbers

3 Solving Linear Equations

4 Graphing Linear Equations and Functions

5 Writing Linear Functions

6 Solving and Graphing Linear Inequalities

7 Systems of Linear Equations and Inequalities

8 Exponents and Exponential Functions

9 Quadratic Equations and Functions

10 Polynomials and Factoring

11 Rational Equations and Functions

12 Radicals and Connections to Geometry

McDougal Littell GEOMETRY

1 Basics of Geometry

2 Reasoning and Proof

3 Perpendicular and Parallel Lines

4 Congruent Triangles

5 Properties of Triangles

6 Quadrilaterals

7 Transformations

8 Similarity

9 Right Triangles and Trigonometry

10 Circles

11 Area of Polygons and Circles

12 Surface Area and Volume

McDougal Littell

ALGEBRA 2

STUDENT EDITION

PLANNING THE CHAPTER

Equations and Inequalities

GOALS		NCTM	ITED	SAT9	Terra-Nova	Local
1.1 pp. 3–10	**GOAL 1** Use a number line to graph and order real numbers. **GOAL 2** Identify properties of and use operations with real numbers.	1, 10	MCWN, MIG		10, 12, 14, 48, 49, 51	
1.2 pp. 11–18	**GOAL 1** Evaluate algebraic expressions. **GOAL 2** Simplify algebraic expressions by combining like terms. *TECHNOLOGY ACTIVITY: 1.2 Evaluate expressions on a scientific calculator or graphing calculator.*	2	MCWN	40	11, 16, 49, 51, 52	
1.3 pp. 19–25	**GOAL 1** Solve linear equations. **GOAL 2** Use linear equations to solve real-life problems. *TECHNOLOGY ACTIVITY: 1.3 Solve equations using the Table feature of a graphing calculator.*	2, 6, 8, 9	RQWN	2, 4	11, 16, 18, 49, 51, 52	
1.4 pp. 26–32	**GOAL 1** Rewrite equations with more than one variable. **GOAL 2** Rewrite common formulas.	2, 9	MCWN	4	16, 49, 51	
1.5 pp. 33–40	**GOAL 1** Use a general problem solving plan to solve real-life problems. **GOAL 2** Use other problem solving strategies to help solve real-life problems.	6, 8, 9	RQWN	1, 2, 11, 19, 38, 40	16, 17, 18	
1.6 pp. 41–48	**GOAL 1** Solve simple inequalities. **GOAL 2** Solve compound inequalities. *TECHNOLOGY ACTIVITY: 1.6 Solve inequalities using the Test feature of a graphing calculator.*	2, 9	MCWN	3	11, 16, 49, 51	
1.7 pp. 49–56	*CONCEPT ACTIVITY: 1.7 Investigate absolute value equations and inequalities.* **GOAL 1** Solve absolute value equations and inequalities. **GOAL 2** Use absolute value equations and inequalities to solve real-life problems.	2, 6, 8, 9	MCWN, RQWN	2, 3	11, 16, 17, 18, 49, 51, 52	

RESOURCES

CHAPTER RESOURCE BOOKLETS

CHAPTER SUPPORT

Tips for New Teachers	p. 1	Prerequisite Skills Review	p. 5
Parent Guide for Student Success	p. 3	Strategies for Reading Mathematics	p. 7

LESSON SUPPORT

	1.1	1.2	1.3	1.4	1.5	1.6	1.7
Lesson Plans (regular and block)	p. 9	p. 22	p. 36	p. 50	p. 64	p. 78	p. 91
Warm-Up Exercises and Daily Quiz	p. 11	p. 24	p. 38	p. 52	p. 66	p. 80	p. 93
Activity Support Masters							p. 94
Lesson Openers	p. 12	p. 25	p. 39	p. 53	p. 67	p. 81	p. 95
Graphing Calculator Activities & Keystrokes		p. 26	p. 40	p. 54		p. 82	p. 96
Practice (3 levels)	p. 13	p. 27	p. 42	p. 56	p. 68	p. 83	p. 98
Reteaching with Practice	p. 16	p. 30	p. 45	p. 59	p. 71	p. 86	p. 101
Quick Catch-Up for Absent Students	p. 18	p. 32	p. 47	p. 61	p. 73	p. 88	p. 103
Cooperative Learning Activities	p. 19						
Interdisciplinary Applications		p. 33		p. 62		p. 89	
Real-Life Applications	p. 20		p. 48		p. 74		p. 104
Math & History Applications					p. 75		
Challenge: Skills and Applications	p. 21	p. 34	p. 49	p. 63	p. 76	p. 90	p. 105

REVIEW AND ASSESSMENT

Quizzes	pp. 35, 77	Alternative Assessment with Math Journal	p. 114
Chapter Review Games and Activities	p. 106	Project with Rubric	p. 116
Chapter Test (3 levels)	pp. 107–112	Cumulative Review	p. 118
SAT/ACT Chapter Test	p. 113	Resource Book Answers	p. A1

TRANSPARENCIES

	1.1	1.2	1.3	1.4	1.5	1.6	1.7
Warm-Up Exercises and Daily Quiz	p. 2	p. 3	p. 4	p. 5	p. 6	p. 7	p. 8
Alternative Lesson Opener Transparencies	p. 1	p. 2	p. 3	p. 4	p. 5	p. 6	p. 7
Examples/Standardized Test Practice	✓	✓	✓	✓	✓	✓	✓
Answer Transparencies	✓	✓	✓	✓	✓	✓	✓

TECHNOLOGY

- Electronic Teaching Tools
- Online Lesson Planner
- Internet Support
- Personal Student Tutor
- Test and Practice Generator
- Instant Replay: Video Review Games
- Electronic Lesson Presentations (Lesson 1.5)

ADDITIONAL RESOURCES

- Basic Skills Workbook: Diagnosis and Remediation
- Worked-Out Solution Key
- Resources in Spanish
- Standardized Test Practice Workbook
- Practice Workbook with Examples

PACING THE CHAPTER

REGULAR SCHEDULE

Day 1

1.1

STARTING OPTIONS
- Prereq. Skills Review
- Strategies for Reading
- Warm-Up

TEACHING OPTIONS
- Motivating the Lesson
- Les. Opener (Visual)
- Examples 1–7
- Closure Question
- Guided Practice Exs.

APPLY/HOMEWORK
- See Assignment Guide.
- See the CRB: Practice, Reteach, Apply, Extend

ASSESSMENT OPTIONS
- Checkpoint Exercises
- Daily Quiz (1.1)
- Stand. Test Practice

Day 2

1.2

STARTING OPTIONS
- Homework Check
- Warm-Up or Daily Quiz

TEACHING OPTIONS
- Motivating the Lesson
- Les. Opener (Appl.)
- Graphing Calc. Activity
- Examples 1–6
- Technology Activity
- Closure Question
- Guided Practice Exs.

APPLY/HOMEWORK
- See Assignment Guide.
- See the CRB: Practice, Reteach, Apply, Extend

ASSESSMENT OPTIONS
- Checkpoint Exercises
- Daily Quiz (1.2)
- Stand. Test Practice
- Quiz (1.1–1.2)

Day 3

1.3

STARTING OPTIONS
- Homework Check
- Warm-Up or Daily Quiz

TEACHING OPTIONS
- Motivating the Lesson
- Les. Opener (Visual)
- Graphing Calc. Activity
- Examples 1–6
- Technology Activity
- Closure Question
- Guided Practice Exs.

APPLY/HOMEWORK
- See Assignment Guide.
- See the CRB: Practice, Reteach, Apply, Extend

ASSESSMENT OPTIONS
- Checkpoint Exercises
- Daily Quiz (1.3)
- Stand. Test Practice

Day 4

1.4

STARTING OPTIONS
- Homework Check
- Warm-Up or Daily Quiz

TEACHING OPTIONS
- Motivating the Lesson
- Les. Opener (Activity)
- Graphing Calc. Activity
- Examples 1–4
- Guided Practice Exs. 1–2, 4–9

APPLY/HOMEWORK
- See Assignment Guide.
- See the CRB: Practice, Reteach, Apply, Extend

ASSESSMENT OPTIONS
- Checkpoint Exercises, p. 27

Day 5

1.4 (cont.)

STARTING OPTIONS
- Homework Check

TEACHING OPTIONS
- Examples 5–6
- Closure Question
- Guided Practice Exs. 3, 10–11

APPLY/HOMEWORK
- See Assignment Guide.
- See the CRB: Practice, Reteach, Apply, Extend

ASSESSMENT OPTIONS
- Checkpoint Exercises, p. 28
- Daily Quiz (1.4)
- Stand. Test Practice

Day 6

1.5

STARTING OPTIONS
- Homework Check
- Warm-Up or Daily Quiz

TEACHING OPTIONS
- Motivating the Lesson
- Les. Opener (Application)
- Examples 1–6
- Closure Question
- Guided Practice Exs.

APPLY/HOMEWORK
- See Assignment Guide.
- See the CRB: Practice, Reteach, Apply, Extend

ASSESSMENT OPTIONS
- Checkpoint Exercises
- Daily Quiz (1.5)
- Stand. Test Practice
- Quiz (1.3–1.5)

Day 9

Review

DAY 9 START OPTIONS
- Homework Check

REVIEWING OPTIONS
- Chapter 1 Summary
- Chapter 1 Review
- Chapter Review Games and Activities

APPLY/HOMEWORK
- Chapter 1 Test (practice)
- Ch. Standardized Test (practice)

Day 10

Assess

DAY 10 START OPTIONS
- Homework Check

ASSESSMENT OPTIONS
- Chapter 1 Test
- SAT/ACT Ch. 1 Test
- Alternative Assessment

APPLY/HOMEWORK
- Skill Review, p. 66

BLOCK SCHEDULE

Day 7

1.6

STARTING OPTIONS
- Homework Check
- Warm-Up or Daily Quiz

TEACHING OPTIONS
- Motivating the Lesson
- Les. Opener (Activity)
- Graphing Calc. Activity
- Examples 1–7
- Technology Activity
- Closure Question
- Guided Practice Exs.

APPLY/HOMEWORK
- See Assignment Guide.
- See the CRB: Practice, Reteach, Apply, Extend

ASSESSMENT OPTIONS
- Checkpoint Exercises
- Daily Quiz (1.6)
- Stand. Test Practice

Day 8

1.7

STARTING OPTIONS
- Homework Check
- Warm-Up or Daily Quiz

TEACHING OPTIONS
- Motivating the Lesson
- Concept Act. & Wksht.
- Les. Opener (Calc.)
- Graphing Calc. Activity
- Examples 1–5
- Closure Question
- Guided Practice Exs.

APPLY/HOMEWORK
- See Assignment Guide.
- See the CRB: Practice, Reteach, Apply, Extend

ASSESSMENT OPTIONS
- Checkpoint Exercises
- Daily Quiz (1.7)
- Stand. Test Practice
- Quiz (1.6–1.7)

Day 1

1.1 & 1.2

DAY 1 START OPTIONS
- Prereq. Skills Review
- Strategies for Reading
- Warm-Up (Les. 1.1)

TEACHING 1.1 OPTIONS
- Motivating the Lesson
- Les. Opener (Visual)
- Examples 1–7
- Closure Question
- Guided Practice Exs.

TEACHING 1.2 OPTIONS
- Warm-Up (Les. 1.2)
- Motivating the Lesson
- Les. Opener (Appl.)
- Graphing Calc. Activity
- Examples 1–6
- Technology Activity
- Closure Question
- Guided Practice Exs.

APPLY/HOMEWORK
- See Assignment Guide.
- See the CRB: Practice, Reteach, Apply, Extend

ASSESSMENT OPTIONS
- Checkpoint Exercises
- Daily Quiz (Les. 1.1, 1.2)
- Stand. Test Practice
- Quiz (1.1–1.2)

Day 2

1.3 & 1.4

DAY 2 START OPTIONS
- Homework Check
- W-Up 1.3 or D. Quiz 1.2

TEACHING 1.3 OPTIONS
- Motivating the Lesson
- Les. Opener (Visual)
- Graphing Calc. Activity
- Examples 1–6
- Technology Activity
- Closure Question
- Guided Practice Exs.

BEGINNING 1.4 OPTIONS
- Warm-Up (Les. 1.4)
- Motivating the Lesson
- Les. Opener (Activity)
- Graphing Calc. Activity
- Examples 1–4
- Guided Practice Exs. 1–2, 4–9

APPLY/HOMEWORK
- See Assignment Guide.
- See the CRB: Practice, Reteach, Apply, Extend

ASSESSMENT OPTIONS
- Checkpoint Exercises
- Daily Quiz (Les. 1.3)
- Stand. Test Prac. (1.3)

Day 3

1.4 & 1.5

DAY 3 START OPTIONS
- Homework Check
- Daily Quiz (Les. 1.3)

FINISHING 1.4 OPTIONS
- Examples 5–6
- Closure Question
- Guided Practice Exs. 3, 10–11

TEACHING 1.5 OPTIONS
- Warm-Up (Les 1.5)
- Motivating the Lesson
- Les. Opener (Application)
- Examples 1–6
- Closure Question
- Guided Practice Exs.

APPLY/HOMEWORK
- See Assignment Guide.
- See the CRB: Practice, Reteach, Apply, Extend

ASSESSMENT OPTIONS
- Checkpoint Exercises
- Daily Quiz (Les. 1.4, 1.5)
- Stand. Test Practice
- Quiz (1.3–1.5)

Day 4

1.6 & 1.7

DAY 4 START OPTIONS
- Homework Check
- W-Up 1.6 or D. Quiz 1.5

TEACHING 1.6 OPTIONS
- Motivating the Lesson
- Les. Opener (Activity)
- Graphing Calc. Activity
- Examples 1–7
- Technology Activity
- Closure Question
- Guided Practice Exs.

TEACHING 1.7 OPTIONS
- Warm-Up (Les. 1.7)
- Motivating the Lesson
- Concept Act. & Wksht.
- Les. Opener (Calc.)
- Graphing Calc. Activity
- Examples 1–5
- Closure Question
- Guided Practice Exs.

APPLY/HOMEWORK
- See Assignment Guide.
- See the CRB: Practice, Reteach, Apply, Extend

ASSESSMENT OPTIONS
- Checkpoint Exercises
- Daily Quiz (Les. 1.6, 1.7)
- Stand. Test Practice
- Quiz (1.6–1.7)

Day 5

Review/Assess

DAY 5 START OPTIONS
- Homework Check

REVIEWING OPTIONS
- Chapter 1 Summary
- Chapter 1 Review
- Chapter Review Games and Activities
- Chapter 1 Test (practice)
- Ch. Standardized Test (practice)

ASSESSMENT OPTIONS
- Chapter 1 Test
- SAT/ACT Ch. 1 Test
- Alternative Assessment

APPLY/HOMEWORK
- Skill Review, p. 66

MEETING INDIVIDUAL NEEDS

BEFORE THE CHAPTER

The *Chapter 1 Resource Book* has the following materials to distribute and use before the chapter:

- **Parent Guide for Student Success**
- **Prerequisite Skills Review (pictured below)**
- **Strategies for Reading Mathematics**

PREREQUISITE SKILLS *Pages 5–6*

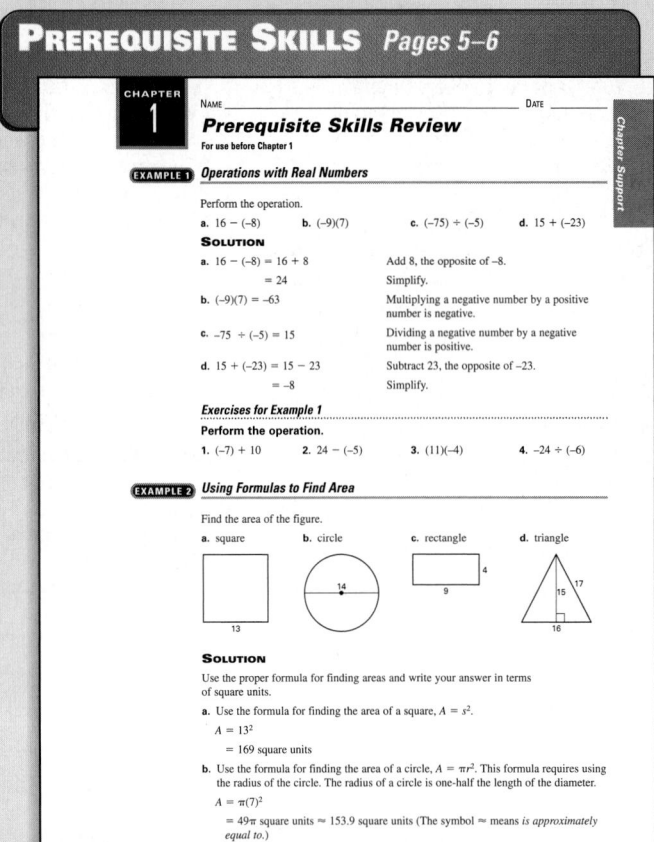

PREREQUISITE SKILLS REVIEW These two pages support the Study Guide on page 2. They help students prepare for Chapter 1 by providing worked-out examples and practice for the following skills needed in the chapter:

- **Perform operations with real numbers.**
- **Find areas using formulas.**

DURING EACH LESSON

The *Chapter 1 Resource Book* has the following alternatives for introducing the lesson:

- **Lesson Openers (pictured below)**
- **Graphing Calculator Activities with Keystrokes**

LESSON OPENER *Page 67*

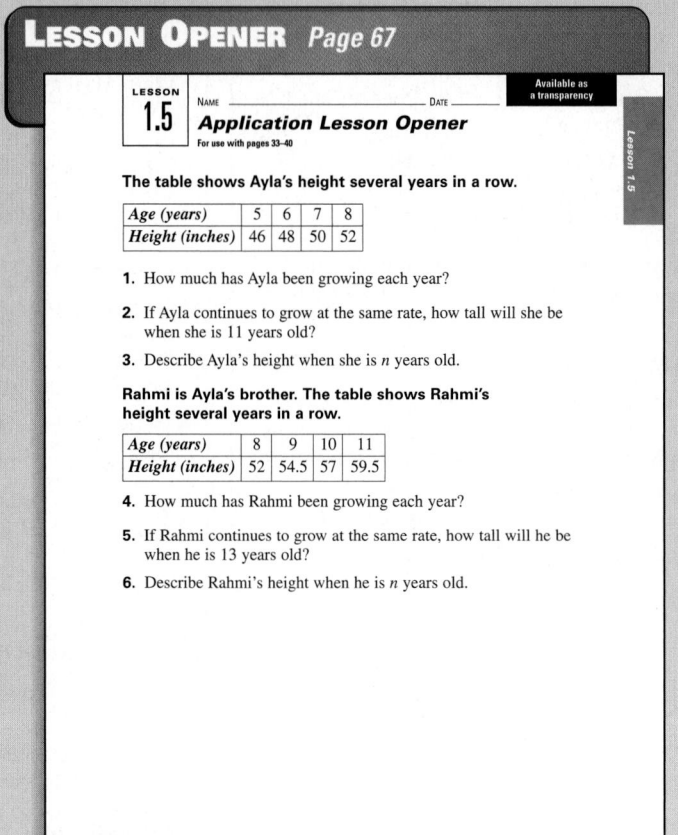

APPLICATION LESSON OPENER This Lesson Opener provides an alternative way to start Lesson 1.5 in the form of a real-life application. Students see how data tables and algebraic models relate to changes in a person's height.

The *Chapter 1 Resource Book* has a variety of materials to follow-up each lesson. They include the following:

- **Practice (3 levels)**
- **Reteaching with Practice**
- **Quick Catch-Up for Absent Students (pictured below)**
- **Challenge: Skills and Applications**
- **Interdisciplinary Applications**
- **Real-Life Applications**

The *Chapter 1 Resource Book* has the following review and assessment materials:

- **Quizzes**
- **Chapter Review Games and Activities**
- **Chapter Test (3 levels) (pictured below)**
- **SAT/ACT Chapter Test**
- **Alternative Assessment with Rubrics and Math Journal**
- **Project with Rubric**
- **Cumulative Review**

QUICK CATCH-UP *Page 47*

LESSON 1.3

NAME _____ DATE _____

Quick Catch-Up for Absent Students
For use with pages 19–25

The items checked below were covered in class on (date missed) _____

Lesson 1.3: Solving Linear Equations

____ **Goal 1:** Solve linear equations. (pp. 19–20)

Material Covered:

____ Example 1: Solving an Equation with a Variable on One Side

____ Example 2: Solving an Equation with a Variable on Both Sides

____ Example 3: Using the Distributive Property

____ Student Help: Skills Review

____ Example 4: Solving an Equation with Fractions

Vocabulary:

equation, p. 19   linear equation in one variable, p. 19

solution of an equation, p. 19   equivalent equations, p. 19

____ **Goal 2:** Use linear equations to solve real-life problems. (p. 21)

Material Covered:

____ Example 5: Writing and Using a Linear Equation

____ Example 6: Writing and Using Geometric Formula

Activity 1.3: Using Tables to Solve Equations (p. 25)

____ **Goal:** Solve linear equations using the *Table* feature of a graphing calculator.

____ Student Help: Keystroke Help

____ Student Help: Study Tip

____ Other (specify) _____
_____

Homework and Additional Learning Support

____ Textbook (specify) pp. 22–24

____ Internet: Extra Examples at www.mcdougallittell.com

____ *Reteaching with Practice* worksheet (specify exercises) _____

____ *Personal Student Tutor* for Lesson 1.3

Lesson 1.3

CHAPTER TEST *Pages 109–110*

CHAPTER 1

NAME _____ DATE _____

Chapter Test B
For use after Chapter 1

Graph the numbers on a number line. Then write the numbers in increasing order.

1. $3, \frac{1}{2}, -2, 0.7, -\sqrt{9}$
2. $-\frac{4}{3}, 2.6, -1.9, \sqrt{5}, 0, 1.2$

Identify the property shown.

3. $6 + 0 = 6$
4. $8 \cdot (2 \cdot 3) = (8 \cdot 2) \cdot 3$

Select and perform an operation to answer the question.

5. What is the difference of -36 and 3?

6. What is the quotient of 12 and $-\frac{1}{2}$?

Evaluate the expression.

7. $22 + 3 \cdot 2 - 17$
8. $7^2 - 4(8 - 3) \div 2$
9. $-4xy + 2x^2$ when $x = -1$ and $y = 3$
10. $\frac{1}{2}x + \frac{2}{7}y$ when $x = 18$ and $y = 7$

Simplify the expression.

11. $4x + 2y - 2x + 5y + x$
12. $3(x^2 - 2x) - 5(2x^2 - x)$

Solve the equation.

13. $5x - 13 = 12$
14. $\frac{1}{5}a + \frac{2}{5} = -\frac{1}{2}a - \frac{3}{4}$
15. $-1.1(x + 3) = 5.5$
16. $3(x + 2) = 8(x - 1) + 5$
17. $|x + 5| = 9$
18. $|3b + 4| = 13$

Solve the equation for y.

19. $3x + y = 4$
20. $9x - 2y = 3$
21. $3x + 2xy = 10$
22. $\frac{1}{3}x - \frac{1}{4}y = 6$

Answers

1. _____
2. _____
3. _____
4. _____
5. _____
6. _____
7. _____
8. _____
9. _____
10. _____
11. _____
12. _____
13. _____
14. _____
15. _____
16. _____
17. _____
18. _____
19. _____
20. _____
21. _____
22. _____

Review and Assess

QUICK CATCH-UP FOR ABSENT STUDENTS You can use this form to let students know what they have missed when they've been absent from class. It allows you to quickly check off which Examples and other elements of Lesson 1.3 were covered on a given day and provides space for filling in the homework assignment.

CHAPTER TEST There are three versions of this two-page Chapter Test, one each for basic (A), average (B), and advanced (C) students. Level B contains more advanced work with solving inequalities than Level A, whereas Level C contains more advanced work with solving equations than Level B.

TECHNOLOGY RESOURCE

Students who have missed class can find extra examples for Lessons 1.3, 1.6, and 1.7 on the Internet at www.mcdougallittell.com.

TECHNOLOGY RESOURCE

Teachers can use the Time-Saving Test and Practice Generator to create customized review and assessment materials for Chapter 1.

CHAPTER GOALS

Students begin the chapter by using a number line to graph and order real numbers and by identifying the properties of real numbers in operations. After evaluating and simplifying algebraic expressions, students will solve linear equations. They will also rewrite equations with more than one variable, including formulas. To set up and solve real-life applications, students will use a general five-step problem solving plan, and will implement various strategies such as drawing a model or looking for a pattern. Finally, students will use these skills to solve simple and compound inequalities as well as absolute value equations and inequalities.

APPLICATION NOTE

Some of the earliest trains used horses or mules to pull their cargo before steam-driven trains took over. Today, diesel-electric trains are widely used for long-haul of materials, while high-speed electric trains are used to transport commuters or travelers.

Japan began operating the famous "bullet" train, the Shinkansen, in 1964. Each 16 car train covers the 210 miles from Tokyo to Nagoya in 80 minutes, giving it an average speed of about 160 miles per hour.

The newest train design, the Maglev (for magnetic levitation), eliminates the need for steel wheels on a steel track. It glides silently over a raised track or guideway, held just above the surface by a magnetic field. The Maglev system is attractive because it offers clean, comfortable, economical, low maintenance rapid travel.

Additional information about high speed trains is available at **www.mcdougallittell.com.**

EQUATIONS AND INEQUALITIES

▶ *How fast can trains travel?*

APPLICATION: High-Speed Trains

Many countries now have high-speed trains. Some of these trains can travel 150 mi/h or faster. The diagram below shows the average speeds for a number of the world's fastest trains.

Fastest Trains in Each Country

Spain	209.1 km/h
U.S.A.	97.7 mi/h
Japan	261.8 km/h
Italy	164.9 km/h
France	254.3 km/h
England	111.8 mi/h

To find the fastest train, you need to convert all the speeds to the same units of measure. The formula to convert kilometers per hour to miles per hour is:

$$\text{speed in mi/h} = 0.621(\text{speed in km/h})$$

Think & Discuss 1–3. See margin.

1. Use the given formula to convert all the train speeds to miles per hour.

2. How would you convert miles per hour to kilometers per hour? Write a formula to do this.

3. Convert all the train speeds to kilometers per hour.

Learn More About It

You will calculate the average speed of another Japanese high-speed train in Example 1 on p. 33.

 APPLICATION LINK Visit www.mcdougallittell.com for more information on high-speed trains.

ADDITIONAL RESOURCES
Another way to begin the chapter is to show the video clip of a real-life motivator for Chapter 1 on the *Instant Replay: Video Review Games.*

PROJECTS
A project covering Chapters 1–3 appears on pages 194 and 195 of the Student Edition. An additional project for Chapter 1 is available in the *Chapter 1 Resource Book,* p. 116.

TECHNOLOGY

Software
• *Electronic Teaching Tools*
• *Online Lesson Planner*
• *Personal Student Tutor*
• *Test and Practice Generator*
• *Electronic Lesson Presentations* (Lesson 1.5)

Video
• *Instant Replay: Video Review Games*

 **Internet Connections**
www.mcdougallittell.com
• **Application Links**
  1, 35, 38, 40, 46
• **Data Updates**
  6
• **Student Help**
  6, 15, 18, 20, 25, 31, 38, 43, 48, 51, 54
• **Career Links**
  16, 21, 23, 30
• **Extra Challenge**
  10, 16, 24, 32, 39, 47, 55

1. Speeds are given to the nearest tenth.
   Spain: 129.9 mi/h
   Japan: 162.6 mi/h
   Italy: 102.4 mi/h
   France: 157.9 mi/h

2. $\text{speed in km/h} = \dfrac{\text{speed in mi/h}}{0.621}$

3. Speeds are given to the nearest tenth.
   United States: 157.3 km/h
   England: 180.0 km/h

Study Guide

PREPARE

DIAGNOSTIC TOOLS

The **Skill Review** exercises can help you diagnose whether students have the following skills needed in Chapter 1:

- Perform operations with positive and negative numbers.

- Find the area of a two-dimensional figure.

The following resources are available for students who need additional help with these skills:

- Prerequisite Skills Review
  (*Chapter 1 Resource Book,* p. 5;
  Warm-Up Transparencies, p. 1)

- ▢ *Personal Student Tutor*

ADDITIONAL RESOURCES

The following resources are provided to help you prepare for the upcoming chapter and customize review materials:

- *Chapter 1 Resource Book*
  Tips for New Teachers (p. 1)
  Parent Guide (p. 3)
  Lesson Plans (every lesson)
  Lesson Plans for Block Scheduling (every lesson)

- ▢ *Electronic Teaching Tools*

- ▢ *Online Lesson Planner*

- ▢ *Test and Practice Generator*

ENGLISH LEARNERS

You may want to have English learners begin keeping an algebra glossary to describe in their own words the many algebraic terms, properties, and rules they will encounter in this book. Encourage them to use the glossary as a reference guide, consulting it as necessary to recall the meanings of terms previously introduced.

PREVIEW

What's the chapter about?

Chapter 1 is about **expressions, equations, and inequalities.** In Chapter 1 you'll learn

- how to evaluate and simplify numerical and algebraic expressions.

- how to solve linear and absolute value equations and inequalities.

- how to use algebra to model and solve real-life problems.

KEY VOCABULARY

• graph of a real number, p. 3	• mathematical model, p. 12	• algebraic model, p. 33
• numerical expression, p. 11	• terms of an expression, p. 13	• linear inequality in one variable, p. 41
• order of operations, p. 11	• linear equation in one variable, p. 19	
• variable, p. 12		• compound inequality, p. 43
• algebraic expression, p. 12	• verbal model, p. 33	• absolute value, p. 50

PREPARE

Are you ready for the chapter?

SKILL REVIEW Do these exercises to review key skills that you'll apply in this chapter. See the given **reference page** if there is something you don't understand.

┌─ **STUDENT HELP**

↳ **Study Tip**
"Student Help" boxes throughout the chapter give you study tips and tell you where to look for extra help in this book and on the Internet.

Perform the operation. (Skills Review, p. 905)

1. $-3 + 14$　**11**　　**2.** $7(-10)$　-70　　**3.** $-1 - (-9)$　**8**　　**4.** $-45 \div (-5)$　**9**

5. $(-12)(-2)$　**24**　　**6.** $8 - 15$　-7　　**7.** $30 \div (-3)$　-10　　**8.** $-6 + (-2)$　-8

Find the area of the figure. (Skills Review, p. 914)

9. triangle
60 units²

10. square
121 units²

11. rectangle
165 units²

12. circle
20.25π units², or about 63.6 units²

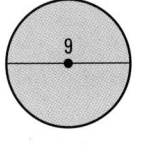

STUDY STRATEGY

Here's a study strategy!

Vocabulary File

Make a flashcard file of vocabulary words.

On the front of an index card, write an important vocabulary word or phrase. On the back of the card write the definition given in the book, along with your own definition if that helps you. Also include symbols, diagrams, examples, and your own notes.

1.1

Real Numbers and Number Operations

What you should learn

GOAL 1 Use a number line to graph and order real numbers.

GOAL 2 Identify properties of and use operations with real numbers, as applied in **Exs. 64 and 65**.

Why you should learn it

▼ To solve **real-life** problems, such as how to exchange money in **Example 7**.

GOAL 1 USING THE REAL NUMBER LINE

The numbers used most often in algebra are the *real numbers*. Some important subsets of the real numbers are listed below.

SUBSETS OF THE REAL NUMBERS

WHOLE NUMBERS   0, 1, 2, 3, . . .

INTEGERS   . . . , −3, −2, −1, 0, 1, 2, 3, . . .

RATIONAL NUMBERS   Numbers such as $\frac{3}{4}$, $\frac{1}{3}$, and $\frac{-4}{1}$ (or −4) that can be written as the ratio of two integers. When written as decimals, rational numbers terminate or repeat. For example, $\frac{3}{4} = 0.75$ and $\frac{1}{3} = 0.333. . . .$

IRRATIONAL NUMBERS   Real numbers that are not rational, such as $\sqrt{2}$ and π. When written as decimals, irrational numbers neither terminate nor repeat.

The three dots in the lists of the whole numbers and the integers above indicate that the lists continue without end.

Real numbers can be pictured as points on a line called a *real number line*. The numbers increase from left to right, and the point labeled 0 is the **origin**.

The point on a number line that corresponds to a real number is the **graph** of the number. Drawing the point is called *graphing* the number or *plotting* the point. The number that corresponds to a point on a number line is the **coordinate** of the point.

EXAMPLE 1   *Graphing Numbers on a Number Line*

Graph the real numbers $-\frac{4}{3}$, $\sqrt{2}$, and 2.7.

SOLUTION

First, recall that $-\frac{4}{3}$ is $-1\frac{1}{3}$, so $-\frac{4}{3}$ is between −2 and −1. Then, approximate $\sqrt{2}$ as a decimal to the nearest tenth: $\sqrt{2} \approx 1.4$. (The symbol $\approx$ means *is approximately equal to*.) Finally, graph the numbers.

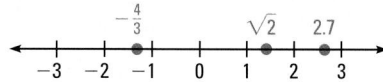

1.1 *Real Numbers and Number Operations*   3

1 PLAN

PACING
Basic: 1 day
Average: 1 day
Advanced: 1 day
Block Schedule: 0.5 block with 1.2

LESSON OPENER
VISUAL APPROACH
An alternative way to approach Lesson 1.1 is to use the Visual Approach Lesson Opener:
• Blackline Master (*Chapter 1 Resource Book*, p. 12)
•  Transparency (p. 1)

MEETING INDIVIDUAL NEEDS
• *Chapter 1 Resource Book*
  Prerequisite Skills Review (p. 5)
  Practice Level A (p. 13)
  Practice Level B (p. 14)
  Practice Level C (p. 15)
  Reteaching with Practice (p. 16)
  Absent Student Catch-Up (p. 18)
  Challenge (p. 21)
• *Resources in Spanish*
• *Personal Student Tutor*

NEW-TEACHER SUPPORT
See the Tips for New Teachers on pp. 1–2 of the *Chapter 1 Resource Book* for additional notes about Lesson 1.1.

WARM-UP EXERCISES

Transparency Available

Simplify.
1. $12 \times (-28)$ −336
2. $-23 + (-15)$ −38
3. $28 \div (-12)$ $-\frac{7}{3}$

Order the numbers from least to greatest.
4. 0.314, 0.0978, 0.309, $0.\overline{31}$
  0.0978, 0.309, $0.\overline{31}$, 0.314
5. $\frac{1}{2}, \frac{1}{3}, \frac{2}{5}, \frac{3}{16}$ $\frac{3}{16}, \frac{1}{3}, \frac{2}{5}, \frac{1}{2}$

3

People use all sorts of numbers in the real world. Fractions represent parts of wholes, negative numbers represent drops in quantities such as temperature, and 12π in. represents the circumference of a pizza with a 6 in. radius. Numbers and their properties and operations are the topics of today's lesson.

 EXTRA EXAMPLE 1
Graph the real numbers -1.8, $\sqrt{3}$, and $\frac{5}{2}$.

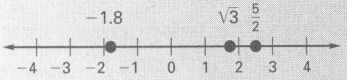

EXTRA EXAMPLE 2
Use a number line to order the real numbers.
a. -4 and 1 $-4 < 1$
b. -5 and -7 $-7 < -5$

EXTRA EXAMPLE 3
Here are the record low temperatures for five Northeastern states.
Connecticut: $-32°F$
Maine: $-48°F$
Maryland: $-40°F$
New Jersey: $-34°F$
Vermont: $-50°F$
a. Order the temperatures from lowest to highest.
$-50°F, -48°F, -40°F, -34°F, -32°F$
b. Which states have record low temperatures below $-40°F$?
Vermont and Maine

 CHECKPOINT EXERCISES

For use after Examples 1–3:
1. Graph the real numbers
$-1.75, -\sqrt{5}, -\frac{5}{3}, -\frac{12}{5}, -\sqrt{10}$.
Write the numbers from least to greatest.

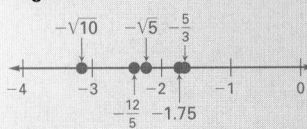

$-\sqrt{10}, -\frac{12}{5}, -\sqrt{5}, -1.75, -\frac{5}{3}$

A number line can be used to order real numbers. The *inequality symbols* $<$, $\leq$, $>$, and $\geq$ can be used to show the order of two numbers.

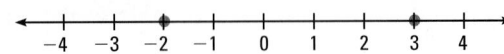 **Ordering Real Numbers**

Use a number line to order the real numbers.

a. -2 and 3 **b.** -1 and -3

SOLUTION

a. Begin by graphing both numbers.

$$\longleftarrow \overset{}{\underset{-4}{|}} \; \overset{}{\underset{-3}{|}} \; \overset{\bullet}{\underset{-2}{|}} \; \overset{}{\underset{-1}{|}} \; \overset{}{\underset{0}{|}} \; \overset{}{\underset{1}{|}} \; \overset{}{\underset{2}{|}} \; \overset{\bullet}{\underset{3}{|}} \; \overset{}{\underset{4}{|}} \longrightarrow$$

Because -2 is to the left of 3, it follows that -2 *is less than* 3, which can be written as $-2 < 3$. This relationship can also be written as $3 > -2$, which is read as "3 *is greater than* -2."

b. Begin by graphing both numbers.

$$\longleftarrow \overset{}{\underset{-4}{|}} \; \overset{\bullet}{\underset{-3}{|}} \; \overset{}{\underset{-2}{|}} \; \overset{\bullet}{\underset{-1}{|}} \; \overset{}{\underset{0}{|}} \; \overset{}{\underset{1}{|}} \; \overset{}{\underset{2}{|}} \; \overset{}{\underset{3}{|}} \; \overset{}{\underset{4}{|}} \longrightarrow$$

Because -3 is to the left of -1, it follows that -3 *is less than* -1, which can be written as $-3 < -1$. (You can also write $-1 > -3$.)

 Geography

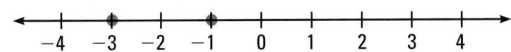 **Ordering Elevations**

Here are the elevations of five locations in Imperial Valley, California.

Alamorio: -135 feet

Curlew: -93 feet

Gieselmann Lake: -162 feet

Moss: -100 feet

Orita: -92 feet

a. Order the elevations from lowest to highest.

b. Which locations have elevations below -100 feet?

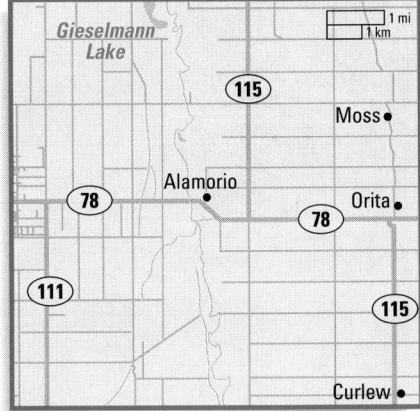

SOLUTION

a. From lowest to highest, the elevations are as follows.

Location	Gieselmann Lake	Alamorio	Moss	Curlew	Orita
Elevation (ft)	-162	-135	-100	-93	-92

b. Gieselmann Lake and Alamorio have elevations below -100 feet.

GOAL 2 USING PROPERTIES OF REAL NUMBERS

When you add or multiply real numbers, there are several properties to remember.

> **CONCEPT SUMMARY**
>
> ### PROPERTIES OF ADDITION AND MULTIPLICATION
>
> Let a, b, and c be real numbers.
>
> | Property | Addition | Multiplication |
> |---|---|---|
> | **CLOSURE** | $a + b$ is a real number. | ab is a real number. |
> | **COMMUTATIVE** | $a + b = b + a$ | $ab = ba$ |
> | **ASSOCIATIVE** | $(a + b) + c = a + (b + c)$ | $(ab)c = a(bc)$ |
> | **IDENTITY** | $a + 0 = a, 0 + a = a$ | $a \cdot 1 = a, 1 \cdot a = a$ |
> | **INVERSE** | $a + (-a) = 0$ | $a \cdot \frac{1}{a} = 1, a \neq 0$ |
>
> The following property involves both addition and multiplication.
>
> | **DISTRIBUTIVE** | $a(b + c) = ab + ac$ |
> |---|---|

EXAMPLE 4 *Identifying Properties of Real Numbers*

Identify the property shown.

a. $(3 + 9) + 8 = 3 + (9 + 8)$ **b.** $14 \cdot 1 = 14$

SOLUTION

a. Associative property of addition   **b.** Identity property of multiplication

. . . .

The **opposite**, or *additive inverse*, of any number a is $-a$. The **reciprocal**, or *multiplicative inverse*, of any nonzero number a is $\frac{1}{a}$. Subtraction is defined as *adding the opposite*, and division is defined as *multiplying by the reciprocal*.

$$a - b = a + (-b) \qquad \text{Definition of subtraction}$$
$$\frac{a}{b} = a \cdot \frac{1}{b}, b \neq 0 \qquad \text{Definition of division}$$

STUDENT HELP

↳ **Study Tip**
• If a is positive, then its opposite, $-a$, is negative.
• The opposite of 0 is 0.
• If a is negative, then its opposite, $-a$, is positive.

EXAMPLE 5 *Operations with Real Numbers*

a. The difference of 7 and -10 is:

$$7 - (-10) = 7 + 10 \qquad \text{Add 10, the opposite of } -10.$$
$$= 17 \qquad \text{Simplify.}$$

b. The quotient of -24 and $\frac{1}{3}$ is:

$$\frac{-24}{\frac{1}{3}} = -24 \cdot 3 \qquad \text{Multiply by 3, the reciprocal of } \frac{1}{3}.$$
$$= -72 \qquad \text{Simplify.}$$

EXTRA EXAMPLE 4
Identify the property shown.
a. $14 + 7 = 7 + 14$ Commutative property of addition
b. $5 \cdot \frac{1}{5} = 1$ Inverse property of multiplication

EXTRA EXAMPLE 5
a. The difference of -3 and -15 is: 12
b. The quotient of -18 and $\frac{1}{6}$ is: -108

✔ **CHECKPOINT EXERCISES**
For use after Examples 4 and 5:
1. Identify what property the statement illustrates. Then simplify both sides of the equation to show the statement is true.
$4(11 + 9) = 4 \cdot 11 + 4 \cdot 9$
Distributive property; $4(20) = 80$ and $44 + 36 = 80$

MATHEMATICAL REASONING
To understand the closure property better, it may help students to see examples. The irrational numbers are closed under addition, for example, $\sqrt{2} + \sqrt{2} = 2\sqrt{2}$, which is irrational, but not under multiplication, for example, $\sqrt{2} \cdot \sqrt{2} = 2$, which is rational. Ask students for which operations the sets of odd integers and negative integers are closed and not closed.
odd integers: closed under multiplication, not closed under addition; negative integers: closed under addition, not closed under multiplication

EXTRA EXAMPLE 6

Perform the given operation. Give the answer with the appropriate unit of measure.

a. 685 ft + 225 ft   **910 ft**

b. $(2.25 \text{ h})\left(\dfrac{60 \text{ km}}{1 \text{ h}}\right)$ **135 km**

c. $\dfrac{\$9}{4 \text{ lb}}$ **$2.25 /lb**

d. $\left(\dfrac{66 \text{ ft}}{1 \text{ sec}}\right)\left(\dfrac{3600 \text{ sec}}{1 \text{ h}}\right)\left(\dfrac{1 \text{ mi}}{5280 \text{ ft}}\right)$
   45 mi/h

EXTRA EXAMPLE 7

You are exchanging $500 for French francs. The exchange rate is 6 francs per dollar. Assume that you use other money to pay the exchange fee.

a. How much will you receive in francs?   **3000 francs**

b. When you return, you have 270 francs left. How much can you get in dollars? Assume that you use other money to pay the exchange fee.   **$45**

 CHECKPOINT EXERCISES

For use after Examples 6 and 7:

1. In Spain, the unit of currency is the peseta. How many pesetas are equivalent to $225 at an exchange rate of $1 for 150 pesetas?
   33,750 pesetas

FOCUS ON VOCABULARY

What two subsets of the real numbers combine to form all of the real numbers?   **rational numbers and irrational numbers**

CLOSURE QUESTION

When converting units with unit analysis, how do you choose whether to use a particular conversion factor or its reciprocal?
Sample answer: Choose the one that lets you "cancel" units until only the desired units remain.

DAILY PUZZLER

With what whole number(s) can you replace b so that 1,527,bb0 is divisible by −60?   **0, 6**

STUDENT HELP

HOMEWORK HELP
Visit our Web site
www.mcdougallittell.com
for extra examples.

FOCUS ON APPLICATIONS

MONEY EXCHANGE In 1997, 17,700,000 United States citizens visited Mexico and spent $7,200,000,000. That same year 8,433,000 Mexican citizens visited the United States and spent $4,289,000,000.

STUDENT HELP

DATA UPDATE
Visit our Web site
www.mcdougallittell.com

When you use the operations of addition, subtraction, multiplication, and division in real life, you should use *unit analysis* to check that your units make sense.

EXAMPLE 6 *Using Unit Analysis*

Perform the given operation. Give the answer with the appropriate unit of measure.

a. 345 miles − 187 miles = 158 miles

b. $(1.5 \text{ hours})\left(\dfrac{50 \text{ miles}}{1 \text{ hour}}\right) = 75 \text{ miles}$

c. $\dfrac{24 \text{ dollars}}{3 \text{ hours}} = 8 \text{ dollars per hour}$

d. $\left(\dfrac{88 \text{ feet}}{1 \text{ second}}\right)\left(\dfrac{3600 \text{ seconds}}{1 \text{ hour}}\right)\left(\dfrac{1 \text{ mile}}{5280 \text{ feet}}\right) = 60 \text{ miles per hour}$

EXAMPLE 7 *Operations with Real Numbers in Real Life*

MONEY EXCHANGE You are exchanging $400 for Mexican pesos. The exchange rate is 8.5 pesos per dollar, and the bank charges a 1% fee to make the exchange.

a. How much money should you take to the bank if you do not want to use part of the $400 to pay the exchange fee?

b. How much will you receive in pesos?

c. When you return from Mexico you have 425 pesos left. How much can you get in dollars? Assume that you use other money to pay the exchange fee.

SOLUTION

a. To find 1% of $400, multiply to get:

$$1\% \times \$400 = 0.01 \times \$400 \qquad \text{Rewrite 1\% as 0.01.}$$
$$= \$4 \qquad \text{Simplify.}$$

▶ You need to take $400 + $4 = $404 to the bank.

b. To find the amount you will receive in pesos, multiply $400 by the exchange rate.

$$(400 \text{ dollars})\left(\dfrac{8.5 \text{ pesos}}{1 \text{ dollar}}\right) = (400 \times 8.5) \text{ pesos}$$
$$= 3400 \text{ pesos}$$

▶ You receive 3400 pesos for $400.

c. To find the amount in dollars, divide 425 pesos by the exchange rate.

$$\dfrac{425 \text{ pesos}}{8.5 \text{ pesos per dollar}} = (425 \text{ pesos})\left(\dfrac{1 \text{ dollar}}{8.5 \text{ pesos}}\right)$$
$$= \dfrac{425}{8.5} \text{ dollars}$$
$$= \$50$$

▶ You receive $50 for 425 pesos.

GUIDED PRACTICE

Vocabulary Check ✓

Concept Check ✓

1. a number that can be written as the ratio of two integers; a real number that is not rational

Skill Check ✓

1. What is a rational number? What is an irrational number?   *See margin.*

2. Give an example of each of the following: a whole number, an integer, a rational number, and an irrational number.   *Sample answer:* $0, -2, \frac{5}{4}, \sqrt{7}$

3. Which of the following is false? Explain.   *C; negative integers are not whole numbers.*

   A. No integer is an irrational number.

   B. Every integer is a rational number.

   C. Every integer is a whole number.

Graph the numbers on a number line. Then decide which number is the greatest.   *4–7. See margin for graphs.*

4. $-3, 4, 0, -8, -10$ 4

5. $\frac{3}{2}, -1, -\frac{5}{2}, 3, -5$ 3

6. $1, -2.5, 4.5, -0.5, 6$ 6

7. $3.2, -0.7, \frac{3}{4}, -\frac{3}{2}, 0$ 3.2

Identify the property shown.

8. $5 + 2 = 2 + 5$
   commutative property of addition

9. $6 + (-6) = 0$
   inverse property of addition

10. $24 \cdot 1 = 24$
    identity property of multiplication

11. $8 \cdot 10 = 10 \cdot 8$
    commutative property of multiplication

12. $13 + 0 = 13$
    identity property of addition

13. $7\left(\frac{1}{7}\right) = 1$
    inverse property of multiplication

14. Find the product. Give the answer with the appropriate unit of measure. Explain your reasoning.

$$\left(\frac{90 \text{ miles}}{1 \text{ hour}}\right)\left(\frac{5280 \text{ feet}}{1 \text{ mile}}\right)\left(\frac{1 \text{ hour}}{60 \text{ minutes}}\right)\left(\frac{1 \text{ minute}}{60 \text{ seconds}}\right)$$

132 ft/sec; canceling like units from numerators and denominators leaves units of feet in the numerator and seconds in the denominator.

PRACTICE AND APPLICATIONS

STUDENT HELP

▶ **Extra Practice**
to help you master skills is on p. 940.

USING A NUMBER LINE Graph the numbers on a number line. Then decide which number is greater and use the symbol < or > to show the relationship.
15–26. See margin for graphs.

15. $\frac{1}{2}, -5$ $\frac{1}{2} > -5$

16. $4, \frac{3}{4}$ $4 > \frac{3}{4}$

17. $2.3, -0.6$ $2.3 > -0.6$

18. $0.3, -2.1$ $0.3 > -2.1$

19. $-\frac{5}{3}, \sqrt{3}$ $-\frac{5}{3} < \sqrt{3}$

20. $0, -\sqrt{10}$ $0 > -\sqrt{10}$

21. $-\frac{9}{4}, 3$ $-\frac{9}{4} > -3$

22. $-\frac{3}{2}, -\frac{11}{3}$ $-\frac{3}{2} > -\frac{11}{3}$

23. $\sqrt{5}, 2$ $\sqrt{5} > 2$

24. $-2, \sqrt{2}$ $-2 < \sqrt{2}$

25. $\sqrt{8}, 2.5$ $\sqrt{8} > 2.5$

26. $-4.5, -\sqrt{24}$ $-4.5 > -\sqrt{24}$

STUDENT HELP

▶ **HOMEWORK HELP**
Examples 1, 2: Exs. 15–32
Example 3: Exs. 55, 56
Example 4: Exs. 33–42
Example 5: Exs. 43–50
Example 6: Exs. 51–54
Example 7: Exs. 57–65

ORDERING NUMBERS Graph the numbers on a number line. Then write the numbers in increasing order.   *27–32. See margin for graphs.*

27. $-\frac{1}{2}, 2, \frac{13}{4}, -3, -6$ $-6, -3, -\frac{1}{2}, 2, \frac{13}{4}$

28. $\sqrt{15}, -4, -\frac{2}{9}, -1, 6$ $-4, -1, -\frac{2}{9}, \sqrt{15}, 6$

29. $-\sqrt{5}, -\frac{5}{2}, 0, 3, -\frac{1}{3}$ $-\frac{5}{2}, -\sqrt{5}, -\frac{1}{3}, 0, 3$

30. $\frac{1}{6}, 2.7, -1.5, -8, -\sqrt{7}$ $-8, -\sqrt{7}, -1.5, \frac{1}{6}, 2.7$

31. $0, -\frac{12}{5}, -\sqrt{12}, 0.3, -1.5$ $-\sqrt{12}, -\frac{12}{5}, -1.5, 0, 0.3$

32. $0.8, \sqrt{10}, -2.4, -\sqrt{6}, \frac{9}{2}$ $-\sqrt{6}, -2.4, 0.8, \sqrt{10}, \frac{9}{2}$

3 APPLY

ASSIGNMENT GUIDE

BASIC
Day 1: pp. 7–10 Exs. 15–31 odd, 34–50 even, 51–56, 66, 68–82 even

AVERAGE
Day 1: pp. 7–10 Exs. 15–31 odd, 34–50 even, 51–56, 57–63 odd, 66, 68–82 even

ADVANCED
Day 1: pp. 7–10 Exs. 15–31 odd, 34–50 even, 51–56, 57–65 odd, 66, 67, 68–82 even

BLOCK SCHEDULE WITH 1.2
pp. 7–10 Exs. 15–31 odd, 34–50 even, 51–56, 57–63 odd, 66, 68–82 even

EXERCISE LEVELS
Level A: *Easier*
15–32, 43–50, 68–75
Level B: *More Difficult*
33–42, 51–65, 76–83
Level C: *Most Difficult*
66, 67

✔ **HOMEWORK CHECK**
To quickly check student understanding of key concepts, go over the following exercises: Exs. 19, 31, 38, 46, 53, 56. See also the Daily Homework Quiz:

• Blackline Master (*Chapter 1 Resource Book*, p. 24)

• Transparency (p. 3)

4.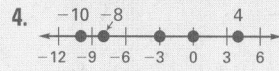

5.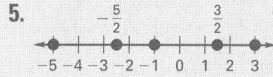

6.

7.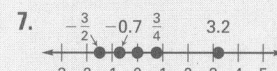

15–32. See Additional Answer beginning on page AA1.

! **COMMON ERROR**

EXERCISES 33–38 Some students may confuse the associative property with the commutative property. To help students understand and remember the difference, present the properties using the words *associate,* which indicates who you are in a group with, and *commute,* which indicates moving or changing position.

EXERCISES 40 AND 42 Some students may miss the differences between these statements and the associative properties of addition and multiplication. You may want to guide students through rewriting these statements in terms of addition and multiplication and then simplifying so that they can see that the two sides of each statement are not equivalent.

ENGLISH LEARNERS

EXERCISES 43–50 You may want to work with English learners to review mathematical symbols and their verbal equivalents before they complete Exercises 43–50. Have volunteers show what symbol(s) they can use when determining a *sum, difference, product,* or *quotient.* Then review other symbols and their verbal equivalents, including $<$, $>$, $\leq$, and $\geq$.

39. Yes; the associative property of addition is true for all real numbers a, b, and c.

40. No. *Sample answer:* For example, $(3 - 4) - 5 = -1 - 5 = -6$, but $3 - (4 - 5) = 3 - (-1) = 3 + 1 = 4$.

41. Yes; the associative property of multiplication is true for all real numbers a, b, and c.

42. No. *Sample answer:* For example, $(18 \div 6) \div 3 = 3 \div 3 = 1$, but $18 \div (6 \div 3) = 18 \div 2 = 9$.

55. Honolulu, HI; New Orleans, LA; Jackson, MS; Seattle-Tacoma, WA; Norfolk, VA; Atlanta, GA; Detroit, MI; Milwaukee, WI; Albany, NY; Helena, MT; three

56. Mark O'Meara, Jim Furyk, Paul Azinger, Tiger Woods, Jay Haas, Jeff Maggert, Lee Janzen, Jumbo Ozaki, Corey Pavin, Vijay Singh

FOCUS ON PEOPLE

VIJAY SINGH, a world-class golfer from Fiji, won the 1998 PGA Championship. His final score of -9 was 2 better than that of the second-place finisher. Singh, whose first name means "victory" in Hindi, has won several PGA Tour events.

IDENTIFYING PROPERTIES Identify the property shown.

33. $-8 + 8 = 0$
inverse property of addition

34. $(3 \cdot 5) \cdot 10 = 3 \cdot (5 \cdot 10)$
associative property of multiplication

35. $7 \cdot 9 = 9 \cdot 7$
commutative property of multiplication

36. $(9 + 2) + 4 = 9 + (2 + 4)$
associative property of addition

37. $12(1) = 12$
identity property of multiplication

38. $2(5 + 11) = 2 \cdot 5 + 2 \cdot 11$
distributive property

LOGICAL REASONING Tell whether the statement is true for all real numbers a, b, and c. Explain your answers. 39–42. See margin.

39. $(a + b) + c = a + (b + c)$

40. $(a - b) - c = a - (b - c)$

41. $(a \cdot b) \cdot c = a \cdot (b \cdot c)$

42. $(a \div b) \div c = a \div (b \div c)$

OPERATIONS Select and perform an operation to answer the question.

43. What is the sum of 32 and -7?
$32 + (-7) = 25$

44. What is the sum of -9 and -6?
$-9 + (-6) = -15$

45. What is the difference of -5 and 8?
$-5 - 8 = -13$

46. What is the difference of -1 and -10?
$-1 - (-10) = 9$

47. What is the product of 9 and -4?
$9 \cdot (-4) = -36$

48. What is the product of -7 and -3?
$-7 \cdot (-3) = 21$

49. What is the quotient of -5 and $-\frac{1}{2}$?
$-5 \div \left(-\frac{1}{2}\right) = 10$

50. What is the quotient of -14 and $\frac{7}{4}$?
$-14 \div \frac{7}{4} = -8$

UNIT ANALYSIS Give the answer with the appropriate unit of measure.

51. $8\frac{1}{6}$ feet $+ 4\frac{5}{6}$ feet   **13 ft**

52. $27\frac{1}{2}$ liters $- 18\frac{5}{8}$ liters $8\frac{7}{8}$ **L**

53. $(8.75 \text{ yards}) \left(\dfrac{\$70}{1 \text{ yard}}\right)$ **\$612.50**

54. $\left(\dfrac{50 \text{ feet}}{1 \text{ second}}\right) \left(\dfrac{1 \text{ mile}}{5280 \text{ feet}}\right) \left(\dfrac{3600 \text{ seconds}}{1 \text{ hour}}\right)$
$34\frac{1}{11}$ **mi/h, or about 34.09 mi/h**

55. **STATISTICS CONNECTION** The lowest temperatures ever recorded in various cities are shown. List the cities in decreasing order based on their lowest temperatures. How many of these cities have a record low temperature below $-25°$F? ▶ Source: National Climatic Data Center   **See margin.**

City	Low temp.	City	Low temp.
Albany, NY	$-28°$F	Jackson, MS	$2°$F
Atlanta, GA	$-8°$F	Milwaukee, WI	$-26°$F
Detroit, MI	$-21°$F	New Orleans, LA	$11°$F
Helena, MT	$-42°$F	Norfolk, VA	$-3°$F
Honolulu, HI	$53°$F	Seattle-Tacoma, WA	$0°$F

56. 🌐 **MASTERS GOLF** The table shows the final scores of 10 competitors in the 1998 Masters Golf Tournament. List the players in increasing order based on their golf scores. ▶ Source: *Sports Illustrated* **See margin.**

Player	Score	Player	Score
Paul Azinger	-6	Lee Janzen	$+6$
Tiger Woods	-3	Jeff Maggert	$+1$
Jay Haas	-2	Mark O'Meara	-9
Jim Furyk	-7	Corey Pavin	$+9$
Vijay Singh	$+12$	Jumbo Ozaki	$+8$

first digit   check digit

BAR CODES Using the operations on this bar code produces $(0 + 5 + 1 + 0 + 6 + 8)(3) + (2 + 2 + 5 + 4 + 5) = 78$. The next highest multiple of 10 is 80, and $80 - 78 = 2$, which is the check digit.

BAR CODES **In Exercises 57 and 58, use the following information.**
All packaged products sold in the United States have a Universal Product Code (UPC), or bar code, such as the one shown at the left. The following operations are performed on the first eleven digits, and the result should equal the twelfth digit, called the *check digit*.   **57, 58. See margin.**

- Add the digits in the odd-numbered positions. Multiply by 3.
- Add the digits in the even-numbered positions.
- Add the results of the first two steps.
- Subtract the result of the previous step from the next highest multiple of 10.

57. Does a UPC of 0 76737 20012 9 check? Explain.

58. Does a UPC of 0 41800 48700 3 check? Explain.

59. SOCIAL STUDIES ⟩ CONNECTION   Two of the tallest buildings in the world are the Sky Central Plaza in Guangzhou, China, which reaches a height of 1056 feet, and the Petronas Tower I in Kuala Lumpur, Malaysia, which reaches a height of 1483 feet. Find the heights of both buildings in yards, in inches, and in miles. Give your answers to four significant digits. **See margin.**

▶ Source: Council on Tall Buildings and Urban Habitat

Petronas Tower I

60. **ELEVATOR SPEED** The elevator in the Washington Monument takes 75 seconds to travel 500 feet to the top floor. What is the speed of the elevator in miles per hour? Give your answer to two significant digits.
▶ Source: National Park Service   **4.5 mi/h**

TRAVEL **In Exercises 61–63, use the following information.**
You are taking a trip to Switzerland. You are at the bank exchanging $600 for Swiss francs. The exchange rate is 1.5 francs per dollar, and the bank charges a 1.5% fee to make the exchange.

61. You brought $10 extra with you to pay the exchange fee. Do you have enough to pay the fee?   **yes**

62. How much will you receive in Swiss francs for your $600?   **900 francs**

63. After your trip, you have 321 Swiss francs left. How much is this amount in dollars? Assume that you use other money to pay the exchange fee.   **$214**

HISTORY ⟩ CONNECTION   **In Exercises 64 and 65, use the following information.**
In 1862, James Glaisher and Henry Coxwell went up too high in a hot-air balloon. At 25,000 feet, Glaisher passed out. To get the balloon to descend, Coxwell grasped a valve, but his hands were too numb to pull the cord. He was able to pull the cord with his teeth. The balloon descended, and both men made it safely back. The temperature of air drops about 3°F for each 1000 foot increase in altitude.

64. How much had the temperature dropped from the sea level temperature when Glaisher and Coxwell reached an altitude of 25,000 feet?   **75°F**

65. If the temperature at sea level was 60°F, what was the temperature at 25,000 feet?   **−15°F**

STUDENT HELP

Skills Review
For help with significant digits, see p. 911.

57. Yes; the result of performing the given operations is 9, the check digit.

58 No; the result of performing the given operations is 4, which does not equal the check digit, 3.

59 Sky Central Plaza: 352 yd, 12,672 in., 0.2 mi; Petronas Tower I: about 494.3 yd, 17,796 in., about 0.2809 mi

STUDENT HELP NOTES

Skills Review As students review significant digits on p. 911, remind them that they do not have to round the results of intermediary steps in the calculation.

ADDITIONAL PRACTICE AND RETEACHING

For Lesson 1.1:
- Practice Levels A, B, and C (*Chapter 1 Resource Book,* p. 13)
- Reteaching with Practice (*Chapter 1 Resource Book,* p. 16)
- See Lesson 1.1 of the *Personal Student Tutor*

For more Mixed Review:
- Search the *Test and Practice Generator* for key words or specific lessons.

9

DAILY HOMEWORK QUIZ

📠 *Transparency Available*

1. Graph the numbers on a number line. Then write the numbers in increasing order.
$-\sqrt{8}, -0.8, -\frac{19}{5}, 1.9, -3$

$$-\frac{19}{5}, -3, -\sqrt{8}, -0.8, 1.9$$

2. What property is illustrated by the statement $\frac{2}{3} \cdot \frac{3}{2} = 1$?
inverse property of multiplication

3. What is the quotient of $\frac{3}{2}$ and -6? $\quad -\frac{1}{4}$

4. Give the product with the appropriate unit of measure.
$\left(\dfrac{\$9}{1 \text{ hour}}\right)\left(\dfrac{8 \text{ hours}}{1 \text{ day}}\right)\left(\dfrac{5 \text{ days}}{1 \text{ week}}\right)$
\$360/wk

EXTRA CHALLENGE NOTE
↳ Challenge problems for Lesson 1.1 are available in **blackline** format in the *Chapter 1 Resource Book,* p. 21 and at **www.mcdougallittell.com.**

ADDITIONAL TEST PREPARATION
1. WRITING Describe the relationship between the whole numbers, integers, and rational numbers. *Sample answer:*
The integers consist of the whole numbers and their opposites. The rational numbers consist of the integers and all other numbers that can be written as the ratio of two integers.

Test Preparation 📝

66. MULTI-STEP PROBLEM You are taking a trip through the provinces of Alberta and British Columbia in Canada. You are at Quesnel Lake when you decide to visit some of the national parks. You visit the following places in order: Kamloops, Revelstoke, Lethbridge, and Red Deer. After you visit Red Deer, you return to Quesnel Lake. **Estimates may vary.**

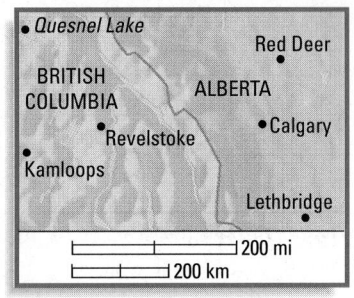

   a. Using the scale on the map, estimate the distance traveled (in kilometers) for the entire trip. Approximately where was the "halfway point" of your trip?
      about 1750 km; Lethbridge

   b. Your car gets 12 kilometers per liter of gasoline. If your gas tank holds 60 liters and the cost of gasoline is \$.29 per liter, about how much will you spend on gasoline for the entire trip? How many times will you have to stop for gasoline if you begin the trip with a full tank? **about \$42; 2**

   c. If you drive at an average speed of 88 kilometers per hour, how many hours will you spend driving on your trip? **about 19.9 h**

★ **Challenge**

67. LOGICAL REASONING Show that $a + (a + 2) = 2(a + 1)$ for all values of a by justifying the steps using the properties of addition and multiplication.

$$a + (a + 2) = (a + a) + 2$$
   a. ?   **associative property of addition**

$$= (1 \cdot a + 1 \cdot a) + 2$$
   b. ?   **identity property of multiplication**

$$= (1 + 1)a + 2$$
   c. ?   **distributive property**

$$= 2a + 2 \cdot 1$$
   d. ?   **identity property of multiplication**

$$= 2(a + 1)$$
   e. ?   **distributive property**

EXTRA CHALLENGE
↳ www.mcdougallittell.com

MIXED REVIEW

OPERATIONS WITH SIGNED NUMBERS **Perform the operation.**
(Skills Review, p. 905)

68. $4 - 12$ **−8**   **69.** $(-7)(-9)$ **63**   **70.** $-20 \div 5$ **−4**   **71.** $6(-5)$ **−30**

72. $-14 + 9$ **−5**   **73.** $6 - (-13)$ **19**   **74.** $56 \div (-7)$ **−8**   **75.** $-16 + (-18)$ **−34**

ALGEBRAIC EXPRESSIONS **Write the given phrase as an algebraic expression.**
(Skills Review, p. 929 for 1.2)

76. 7 more than a number $x + 7$ **77.** 3 less than a number $x - 3$

78. 6 times a number $6x$ **79.** $\frac{1}{4}$ of a number $\frac{1}{4}x$

GEOMETRY CONNECTION   **Find the area of the figure.** (Skills Review, p. 914)

80. Triangle with base 6 inches and height 4 inches   **12 in.²**

81. Triangle with base 7 inches and height 3 inches   **10.5 in.²**

82. Rectangle with sides 5 inches and 7 inches   **35 in.²**

83. Rectangle with sides 25 inches and 30 inches   **750 in.²**

1.2 Algebraic Expressions and Models

What you should learn

GOAL 1 Evaluate algebraic expressions.

GOAL 2 Simplify algebraic expressions by combining like terms, as applied in **Example 6**.

Why you should learn it

▼ To solve **real-life** problems, such as finding the population of Hawaii in **Ex. 57**.

GOAL 1 EVALUATING ALGEBRAIC EXPRESSIONS

A **numerical expression** consists of numbers, operations, and grouping symbols. In Lesson 1.1 you worked with addition, subtraction, multiplication, and division. In this lesson you will work with *exponentiation*, or raising to a power.

Exponents are used to represent repeated factors in multiplication. For instance, the expression 2^5 represents the number that you obtain when 2 is used as a factor 5 times.

$$2^5 = \underbrace{2 \cdot 2 \cdot 2 \cdot 2 \cdot 2}_{\text{5 factors of 2}} \qquad \text{2 to the fifth power}$$

The number 2 is the **base**, the number 5 is the **exponent**, and the expression 2^5 is a **power**. The exponent in a power represents the number of times the base is used as a factor. For a number raised to the first power, you do not usually write the exponent 1. For instance, you usually write 2^1 simply as 2.

EXAMPLE 1 Evaluating Powers

a. $(-3)^4 = (-3) \cdot (-3) \cdot (-3) \cdot (-3) = 81$

b. $-3^4 = -(3 \cdot 3 \cdot 3 \cdot 3) = -81$

· · · · · · · · · ·

In Example 1, notice how parentheses are used in part (a) to indicate that the base is -3. In the expression -3^4, however, the base is 3, not -3. An **order of operations** helps avoid confusion when evaluating expressions.

ORDER OF OPERATIONS

1. First, do operations that occur within grouping symbols.

2. Next, evaluate powers.

3. Then, do multiplications and divisions from left to right.

4. Finally, do additions and subtractions from left to right.

EXAMPLE 2 Using Order of Operations

$$
\begin{aligned}
-4 + 2(-2 + 5)^2 &= -4 + 2(3)^2 && \text{Add within parentheses.} \\
&= -4 + 2(9) && \text{Evaluate power.} \\
&= -4 + 18 && \text{Multiply.} \\
&= 14 && \text{Add.}
\end{aligned}
$$

1.2 *Algebraic Expressions and Models* **11**

LESSON OPENER
APPLICATION
An alternative way to approach Lesson 1.2 is to use the Application Lesson Opener:

• Blackline Master (*Chapter 1 Resource Book*, p. 25)
•  Transparency (p. 2)

MEETING INDIVIDUAL NEEDS
• *Chapter 1 Resource Book*
  Prerequisite Skills Review (p. 5)
  Practice Level A (p. 27)
  Practice Level B (p. 28)
  Practice Level C (p. 29)
  Reteaching with Practice (p. 30)
  Absent Student Catch-Up (p. 32)
  Challenge (p. 34)
• *Resources in Spanish*
• *Personal Student Tutor*

NEW-TEACHER SUPPORT
See the Tips for New Teachers on pp. 1–2 of the *Chapter 1 Resource Book* for additional notes about Lesson 1.2.

WARM-UP EXERCISES
Transparency Available
Simplify.
1. $-(7 \cdot 7 \cdot 7)$ -343
2. $(-3)(-3)(-3)$ -27
3. $-(3 - 4)$ 1
4. $(6 + 3 - 19)x$ $-10x$
5. $(-11 - (-4))y$ $-7y$

MOTIVATING THE LESSON
Have students name everyday events where order is important. Point out that for mathematics to be a precise language, we must all use the same order to perform operations.

EXTRA EXAMPLE 1
Evaluate the power.
a. $(-2)^6$ 64
b. -2^6 -64

EXTRA EXAMPLE 2
Evaluate $-8 + 5(1 - (-3))^3$. 312

EXTRA EXAMPLE 3
Evaluate $-4x^2 + 6x - 5$ when $x = -3$. -59

EXTRA EXAMPLE 4
You have $55 to buy digital video discs (DVDs) that cost $12 each. Write an expression for how much money you have left after buying n discs. Evaluate the expression when $n = 3$ and $n = 4$. $55 - 12n$; $19, $7

CHECKPOINT EXERCISES
For use after Examples 1 and 2:
1. Evaluate $5^2 - 6(2 + (-1))^4$. 19

For use after Example 3:
2. Evaluate $2x^3 + 3x^2 + 27$ when $x = -4$. -53

For use after Example 4:
3. Write an expression for the total monthly cost of phone service if you pay a $5 fee and 8¢ per minute. Find the cost if you talk 6 hours during the month. $5 + 0.08n$; $33.80

STUDENT HELP NOTES
Skills Review As students review operations on signed numbers on p. 905, remind them to pay special attention to whether a negative symbol is inside or outside of parentheses.

A **variable** is a letter that is used to represent one or more numbers. Any number used to replace a variable is a **value of the variable**. An expression involving variables is called an **algebraic expression**.

When the variables in an algebraic expression are replaced by numbers, you are *evaluating* the expression, and the result is called the **value of the expression**. To evaluate an algebraic expression, use the following flow chart.

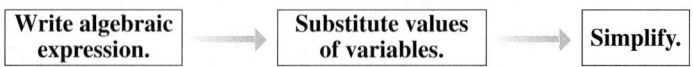

EXAMPLE 3 *Evaluating an Algebraic Expression*

Evaluate $-3x^2 - 5x + 7$ when $x = -2$.

$$-3x^2 - 5x + 7 = -3(-2)^2 - 5(-2) + 7 \qquad \text{Substitute } -2 \text{ for } x.$$
$$= -3(4) - 5(-2) + 7 \qquad \text{Evaluate power.}$$
$$= -12 + 10 + 7 \qquad \text{Multiply.}$$
$$= 5 \qquad \text{Add.}$$

STUDENT HELP

Skills Review
For help with operations with signed numbers, see p. 905.

An expression that represents a real-life situation is a **mathematical model**. When you create the expression, you are *modeling* the real-life situation.

Movies

EXAMPLE 4 *Writing and Evaluating a Real-Life Model*

You have $50 and are buying some movies on videocassettes that cost $15 each. Write an expression that shows how much money you have left after buying n movies. Evaluate the expression when $n = 2$ and $n = 3$.

SOLUTION

| **VERBAL MODEL** | $\boxed{\text{Original amount}}$ | $-$ | $\boxed{\text{Price per movie}}$ | $\cdot$ | Number of movies bought |

LABELS
Original amount = **50**   (dollars)

Price per movie = **15**   (dollars per movie)

Number of movies bought = **n**   (movies)

ALGEBRAIC MODEL $50 - 15\,n$

When you buy **2** movies, you have $50 - 15(2) = \$20$ left.

When you buy **3** movies, you have $50 - 15(3) = \$5$ left.

UNIT ANALYSIS You can use unit analysis to check your verbal model.

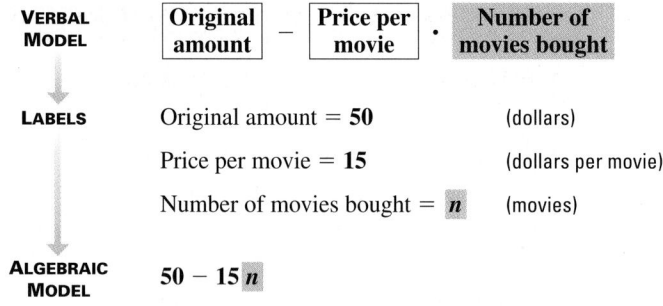

GOAL 2 SIMPLIFYING ALGEBRAIC EXPRESSIONS

For an expression such as $2x + 3$, the parts that are added together, $2x$ and 3, are called **terms**. When a term is the product of a number and a power of a variable, such as $2x$ or $4x^3$, the number is the **coefficient** of the power.

Terms such as $3x^2$ and $-5x^2$ are **like terms** because they have the same variable part. **Constant terms** such as -4 and 2 are also like terms. The distributive property lets you *combine like terms* that have variables by adding the coefficients.

STUDENT HELP

→ **Study Tip**
For an expression like $2x - 3$, think of the expression as $2x + (-3)$, so the terms are $2x$ and -3.

STUDENT HELP

→ **Skills Review**
For help with opposites, see p. 936.

EXAMPLE 5 *Simplifying by Combining Like Terms*

a. $7x + 4x = (7 + 4)x$ **Distributive property**

$= 11x$ **Add coefficients.**

b. $3n^2 + n - n^2 = (3n^2 - n^2) + n$ **Group like terms.**

$= 2n^2 + n$ **Combine like terms.**

c. $2(x + 1) - 3(x - 4) = 2x + 2 - 3x + 12$ **Distributive property**

$= (2x - 3x) + (2 + 12)$ **Group like terms.**

$= -x + 14$ **Combine like terms.**

· · · · · · · · · ·

Two algebraic expressions are **equivalent** if they have the same value for all values of their variable(s). For instance, the expressions $7x + 4x$ and $11x$ are equivalent, as are the expressions $5x - (6x + y)$ and $-x - y$. A statement such as $7x + 4x = 11x$ that equates two equivalent expressions is called an **identity**.

EXAMPLE 6 *Using a Real-Life Model*

MUSIC You want to buy either a CD or a cassette as a gift for each of 10 people. CDs cost $13 each and cassettes cost $8 each. Write an expression for the total amount you must spend. Then evaluate the expression when 4 of the people get CDs.

SOLUTION

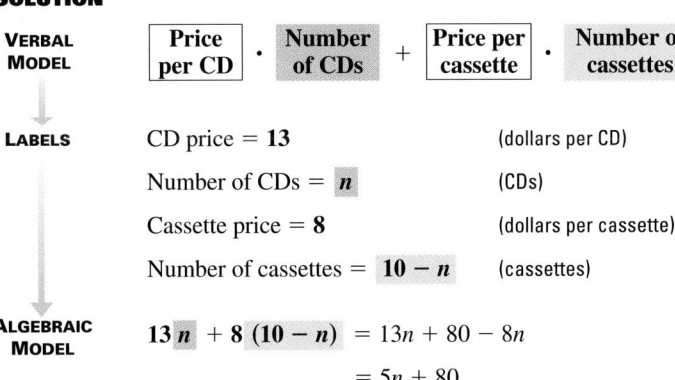

VERBAL MODEL		Price per CD	·	Number of CDs	+	Price per cassette	·	Number of cassettes

LABELS   CD price $= 13$ (dollars per CD)

Number of CDs $= n$ (CDs)

Cassette price $= 8$ (dollars per cassette)

Number of cassettes $= 10 - n$ (cassettes)

ALGEBRAIC MODEL $13n + 8(10 - n) = 13n + 80 - 8n$

$= 5n + 80$

▶ When $n = 4$, the total cost is $5(4) + 80 = 20 + 80 = \$100$.

1.2 Algebraic Expressions and Models   **13**

FOCUS ON PEOPLE

→ **HEITARO NAKAJIMA** could be called the inventor of the compact disc (CD). He was head of the research division of the company that developed the first CDs in 1982. A CD usually has a diameter of 12 centimeters, just the right size to hold Beethoven's Ninth Symphony.

Closure Question *Sample answer:*
(1) Simplify within grouping symbols.
(2) Evaluate powers. (3) Multiply and divide from left to right. (4) Add and subtract from left to right.

 EXTRA EXAMPLE 5
Simplify the expression.
a. $10y + 15y$ $25y$
b. $6m^2 - 12m - 7m^2$ $-m^2 - 12m$
c. $3(x - 2) - 5(x - 8)$ $-2x + 34$

EXTRA EXAMPLE 6
You want to buy either scented lotion or bath soap for 8 people. The lotions are $6 each and the soaps are $5 each. Write an expression for the total amount you must spend. Evaluate the expression when 5 of the people get lotion. Let ℓ represent the number of lotions; $6\ell + 5(8 - \ell)$, or $\ell + 40$; $45.

✓ CHECKPOINT EXERCISES
For use after Example 5:
1. Simplify $7(x^2 - 3) - 3(x + 4)$.
$7x^2 - 3x - 33$

For use after Example 6:
2. Write an expression for the total amount of juice in 15 cans if some hold 8 oz and some hold 12 oz. What is the total if 9 of the cans hold 8 oz? Let z represent the number of 8 oz cans; $8z + 12(15 - z)$, or $180 - 4z$; 144 oz.

FOCUS ON VOCABULARY
There are many vocabulary terms in this lesson. You may wish to have students make a list of the terms with their definitions.

CLOSURE QUESTION
State the order of operations. See sample answer below.

DAILY PUZZLER
You are buying 10 pounds of cherries and strawberries for a party. Cherries cost three times as much per pound as strawberries. You plan to have one fourth of the total weight in cherries. Later, you decide to have three times this much of the total in cherries. By what factor does this multiply the cost from your original plan? $\frac{5}{3}$

ASSIGNMENT GUIDE

BASIC
Day 1: pp. 14–17 Exs. 15–36,
37–55 odd, 62–67, 69–89
odd, Quiz 1 Exs. 1–14

AVERAGE
Day 1: pp. 14–17 Exs. 15–36,
37–55 odd, 57–61 odd,
62–67, 69–89 odd,
Quiz 1 Exs. 1–14

ADVANCED
Day 1: pp. 14–17 Exs. 15–36,
37–55 odd, 56, 57–61 odd,
62–68, 69–89 odd,
Quiz 1 Exs. 1–14

BLOCK SCHEDULE WITH 1.1
pp. 14–17 Exs. 15–36, 37–55 odd,
57–61 odd, 62–67, 69–89 odd,
Quiz 1 Exs. 1–14

EXERCISE LEVELS
Level A: *Easier*
15–44, 47–52, 75–77, 82–89
Level B: *More Difficult*
45–46, 53–66, 69–74, 78–81
Level C: *Most Difficult*
67, 68

✔ HOMEWORK CHECK
To quickly check student under-
standing of key concepts, go
over the following exercises:
Exs. 16, 22, 30, 36, 43, 49. See
also the Daily Homework Quiz:
- Blackline Master (*Chapter 1
Resource Book*, p. 38)
- Transparency (p. 4)

! COMMON ERROR
EXERCISES 21 AND 22 If students confuse these expressions,
you may want to have them rewrite
Exercise 21 as $-1 \cdot 2^5$. This way
students can see why the negative
sign is not part of the power.

5. The first line should read
$4x - (3y + 7x) = 4x - 3y - 7x$.
The other steps are then
"$= 4x - 7x - 3y$," "$= (4 - 7)x - 3y$,"
and "$= -3x - 3y$."

14

GUIDED PRACTICE

Vocabulary Check ✔
Concept Check ✔

1. base→ 8^4 ←exponent
the base, 8, is the factor
to be multiplied by itself;
the exponent, 4,
represents the number of
times the base, 8, is to be
used as a factor.

Skill Check ✔

3. First evaluate the power
8^2, or 64. Then divide by
4 to get 16. Then add and
subtract from left to right:
$3 - 16 = -13$, and
$-13 + 1 = -12$

4. The first line should read
$5 + 2(16 \div 2)^2 =$
$5 + 2(8)^2$. The other steps
are then
"$= 5 + 2(64),$"
"$= 5 + 128,$" and
"$= 133.$"

5. See margin.

1. Copy 8^4 and label the base and the exponent. What does each number represent? See margin.
2. Identify the terms of $6x^3 - 17x + 5$. $6x^3, -17x, 5$
3. Explain how the order of operations is used to evaluate $3 - 8^2 \div 4 + 1$. See margin.

ERROR ANALYSIS Find the error. Then write the correct steps. 4–5. See margin.

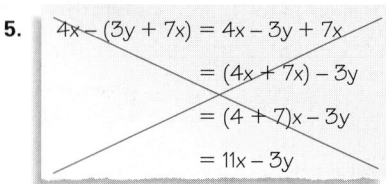

4. $5 + 2(16 \div 2)^2 = 5 + 2(16 \div 4)$
$= 5 + 2(4)$
$= 5 + 8$
$= 13$

5. $4x - (3y + 7x) = 4x - 3y + 7x$
$= (4x + 7x) - 3y$
$= (4 + 7)x - 3y$
$= 11x - 3y$

Evaluate the expression for the given value of x.

6. $x - 8$ when $x = 2$ -6
7. $3x + 14$ when $x = -3$ 5
8. $x(x + 4)$ when $x = 5$ 45
9. $x^2 - 9$ when $x = 6$ 27

Simplify the expression.

10. $9y - 14y$ $-5y$
11. $11x + 6y - 2x + 3y$ $9x + 9y$
12. $3(x + 4) - (6 + 2x)$ $x + 6$
13. $3x^2 - 5x + 5x^2 - 3x$ $8x^2 - 8x$

14. 🌐 **RETAIL BUYING** When you arrive at the music store to buy the CDs and cassettes for the 10 people mentioned in Example 6, you find that the store is having a sale. CDs now cost $11 each and cassettes now cost $7 each. Write an expression for the new total amount you will spend. Then evaluate the expression when 6 of the people get CDs. $4n + 70$; $94

PRACTICE AND APPLICATIONS

STUDENT HELP
▶ **Extra Practice**
to help you master
skills is on p. 940.

WRITING WITH EXPONENTS Write the expression using exponents.

15. eight to the third power 8^3
16. x to the fifth power x^5
17. 5 to the nth power 5^n
18. $x \cdot x \cdot x \cdot x \cdot x \cdot x \cdot x$ x^7

EVALUATING POWERS Evaluate the power.

19. 4^4 256
20. $(-4)^4$ 256
21. -2^5 -32
22. $(-2)^5$ -32
23. 5^3 125
24. 3^5 243
25. 2^8 256
26. 8^2 64

USING ORDER OF OPERATIONS Evaluate the expression.

27. $13 + 20 - 9$ 24
28. $14 \cdot 3 - 2$ 40
29. $6 \cdot 2 + 35 \div 5$ 19
30. $-6 + 3(-3 + 7)^2$ 42
31. $24 - 8 \cdot 12 \div 4$ 0
32. $16 \div (2 + 6) \cdot 10$ 20

EVALUATING EXPRESSIONS Evaluate the expression for the given value of x.

33. $x - 12$ when $x = 7$ -5
34. $6x + 9$ when $x = 4$ 33
35. $25x(x - 4)$ when $x = -1$ 125
36. $x^2 + 5 - x$ when $x = 5$ 25

STUDENT HELP
▶ **HOMEWORK HELP**
Example 1: Exs. 15–26
Example 2: Exs. 27–32
Example 3: Exs. 33–46
Example 4: Exs. 56–61
Example 5: Exs. 47–52
Example 6: Exs. 56–61

EVALUATING EXPRESSIONS Evaluate the expression for the given values of *x* and *y*.

37. $x^4 + 3y$ when $x = 2$ and $y = -8$ −8

38. $(3x)^2 - 7y^2$ when $x = 3$ and $y = 2$ 53

39. $9x + 8y$ when $x = 4$ and $y = 5$ 76

40. $5\left(\dfrac{x}{y}\right) - x$ when $x = 6$ and $y = \dfrac{2}{3}$ 39

41. $\dfrac{x^2}{2y + 1}$ when $x = -3$ and $y = 2$ $\dfrac{9}{5}$

42. $\dfrac{(x + 3)^2}{3y - 2}$ when $x = 2$ and $y = 4$ $\dfrac{5}{2}$

43. $\dfrac{x + y}{x - y}$ when $x = -4$ and $y = 9$ $-\dfrac{5}{13}$

44. $\dfrac{2x + y}{3y + x}$ when $x = 10$ and $y = 6$ $\dfrac{13}{14}$

45. $\dfrac{4(x - 2y)}{x + y}$ when $x = 4$ and $y = -2$ 16

46. $\dfrac{4y - x}{3(2x + y)}$ when $x = -3$ and $y = 3$ $-\dfrac{5}{3}$

SIMPLIFYING EXPRESSIONS Simplify the expression.

47. $7x^2 + 12x - x^2 - 40x$ $6x^2 - 28x$

48. $4x^2 + x - 3x - 6x^2$ $-2x^2 - 2x$

49. $12(n - 3) + 4(n - 13)$ $16n - 88$

50. $5(n^2 + n) - 3(n^2 - 2n)$ $2n^2 + 11n$

51. $4x - 2y + y - 9x$ $-5x - y$

52. $8(y - x) - 2(x - y)$ $-10x + 10y$

STUDENT HELP

→ **Skills Review**
For help with area, see p. 914.

→ **HOMEWORK HELP**
Visit our Web site www.mcdougallittell.com for help with problem solving in Ex. 56.

GEOMETRY CONNECTION Write an expression for the area of the figure. Then evaluate the expression for the given value(s) of the variable(s).

53. $n = 40$ $\frac{1}{2}n(n + 10)$; 1000

54. $a = 8, b = 3$ $a(a + b)$; 88

55. $x = 12, y = 5$ $(x + y)^2$; 289

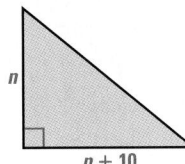

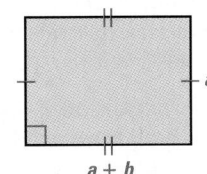

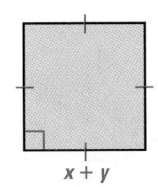

56. 🌐 **AVERAGE SALARIES** In 1980, a public high school principal's salary was approximately $30,000. From 1980 through 1996, the average salary of principals at public high schools increased by an average of $2500 per year. Use the verbal model and labels below to write an algebraic model that gives a public high school principal's average salary *t* years after 1980. Evaluate the expression when $t = 5$, 10, and 15.

▶ Source: Educational Research Service

$30 + 2.5t$; $42,500, $55,000, $67,500

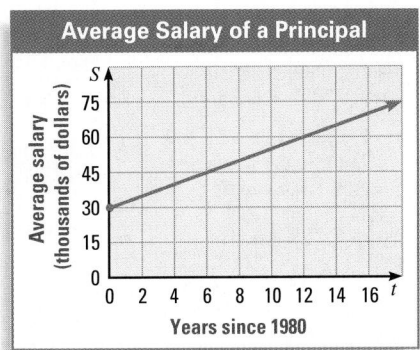

Average Salary of a Principal

| VERBAL MODEL | Salary in 1980 | + | Average increase per year | · | Years since 1980 |

↓

LABELS	Salary in 1980 = **30**	(thousands of dollars)
	Average increase per year = **2.5**	(thousands of dollars per year)
	Years since 1980 = *t*	(years)

1.2 Algebraic Expressions and Models   **15**

TEACHING TIPS
EXERCISES 27–32 To help students remember the order of operations, the mnemonic device PEMDAS ("Please Excuse My Dear Aunt Sally") can be helpful: P for parentheses, E for exponents, M and D for multiplication and division, and A and S for addition and subtraction.

MATHEMATICAL REASONING
When using a calculator, students should check the manual to see how the calculator interprets the order of operations. For example, the expression $60 \div 3 \times 4$ evaluates to 80. Depending on the calculator, however, if 4 is first stored as *x*, then $60 \div 3x$ may evaluate to 80 or to 5. The expression $3x$ is *implied* multiplication. Some calculators interpret $60 \div 3x$ as $60 \div (3 \times x)$, which accounts for the calculator returning a value of 5.

STUDENT HELP NOTES

→ **Skills Review** As students review area on p. 914, remind them that when an area expression is evaluated, the result is always in square units.

→ **Homework Help** Students can find help for Ex. 56 at **www.mcdougallittell.com**. The information can be printed out for students who don't have access to the Internet.

ENGLISH LEARNERS
EXERCISES 56–68 The ability to translate word problems into equations is an essential skill in algebra, but can be extremely difficult for students learning English. When they answer word problems, make sure that English learners can identify and understand signal words and phrases (such as *average, in thousands, increase, per,* and *either/or*) in these exercises successfully.

CAREER NOTE
EXERCISE 58 Additional information about a career as a physical therapist is available at **www.mcdougallittell.com.**

68. The three columns of the table represent the numbers of dollars needed for $15 T-shirts, for $25 sweatshirts, and the total amount needed. Following are the row entries:
8(15) = 120; 0(25) = 0; 120
7(15) = 105; 1(25) = 25; 130
6(15) = 90; 2(25) = 50; 140
5(15) = 75; 3(25) = 75; 150
4(15) = 60; 4(25) = 100; 160
3(15) = 45; 5(25) = 125; 170
2(15) = 30; 6(25) = 150; 180
1(15) = 15; 7(25) = 175; 190
0(15) = 0; 8(25) = 200; 200
Sample answer: For each sweatshirt ordered, the total amount of money needed increases by $10; 120 + 10$n$, where n is the number of sweatshirts ordered; 120 represents the cost ($) of eight T-shirts and 10n represents the additional $10 needed each time a sweatshirt is ordered instead of a T-shirt.

ADDITIONAL PRACTICE AND RETEACHING

For Lesson 1.2:
• Practice Levels A, B, and C (*Chapter 1 Resource Book,* p. 27)
• Reteaching with Practice (*Chapter 1 Resource Book,* p. 30)
• ⬚ See Lesson 1.2 of the *Personal Student Tutor*

For more Mixed Review:
• ⬚ Search the *Test and Practice Generator* for key words or specific lessons.

FOCUS ON CAREERS

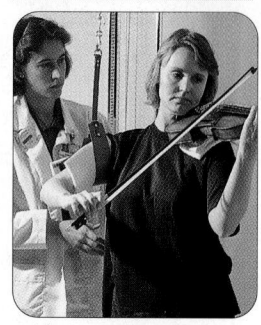

 **PHYSICAL THERAPIST**
Physical therapists help restore function, improve mobility, and relieve pain in patients with injuries or disease.

⬚ **CAREER LINK**
www.mcdougallittell.com

Test Preparation

57. about 1,200,000; about 238,000

58. 8100t + 115,000, where t is the number of years since 1996; 228,400 jobs

59. 149 + 3.85(12)n, where n is the number of movies rented each month; $426.20

60. 37,148 + 15,000t, where t is the number of years you have owned the car; 97,148 mi

61. [4n + 8(3 − n)]15, or 360 − 60n, where n is the numbers of hours spent walking; $240

★ **Challenge**

EXTRA CHALLENGE
⮡ www.mcdougallittell.com

---

57. **SOCIAL STUDIES CONNECTION** For 1980 through 1998, the population (in thousands) of Hawaii can be modeled by 13.2t + 965 where t is the number of years since 1980. What was the population of Hawaii in 1998? What was the population increase from 1980 to 1998? ▶ Source: U.S. Bureau of the Census **See margin.**

58. 🌐 **PHYSICAL THERAPY** In 1996 there were approximately 115,000 physical therapy jobs in the United States. The number of jobs is expected to increase by 8100 each year. Write an expression that gives the total number of physical therapy jobs each year since 1996. Evaluate the expression for the year 2010.
🌐 **DATA UPDATE** of U.S. Bureau of Labor Statistics data at www.mcdougallittell.com **See margin.**

59. 🌐 **MOVIE RENTALS** You buy a VCR for $149 and plan to rent movies each month. Each rental costs $3.85. Write an expression that gives the total amount you spend during the first twelve months that you own the VCR, including the price of the VCR. Evaluate the expression if you rent 6 movies each month. **See margin.**

60. 🌐 **USED CARS** You buy a used car with 37,148 miles on the odometer. Based on your regular driving habits, you plan to drive the car 15,000 miles each year that you own it. Write an expression for the number of miles that appears on the odometer at the end of each year. Evaluate the expression to find the number of miles that will appear on the odometer after you have owned the car for 4 years. **See margin.**

61. 🌐 **WALK-A-THON** You are taking part in a charity walk-a-thon where you can either walk or run. You walk at 4 kilometers per hour and run at 8 kilometers per hour. The walk-a-thon lasts 3 hours. Money is raised based on the total distance you travel in the 3 hours. Your sponsors donate $15 for each kilometer you travel. Write an expression that gives the total amount of money you raise. Evaluate the expression if you walk for 2 hours and run for 1 hour. **See margin.**

QUANTITATIVE COMPARISON In Exercises 62–67, choose the statement that is true about the given quantities.

Ⓐ The quantity in column A is greater.

Ⓑ The quantity in column B is greater.

Ⓒ The two quantities are equal.

Ⓓ The relationship cannot be determined from the given information.

	Column A	Column B	
62.	2^6	$(-2)^6$	C
63.	-4^4	$(-4)^4$	B
64.	x^4	x^5	D
65.	$3(x - 2)$ when $x = 4$	$3x - 6$ when $x = 4$	C
66.	$x + 10(x^2 - 3)$ when $x = 3$	x^6 when $x = 2$	B
67.	$2(x^2 - 1)$	$2x^2 - 1$	B

68. **MATH CLUB SHIRTS** The math club is ordering shirts for its 8 members. The club members have a choice of either a $15 T-shirt or a $25 sweatshirt. Make a table showing the total amount of money needed for each possible combination of T-shirts and sweatshirts that the math club can order. Describe any patterns you see. Write an expression that gives the total cost of the shirts. Explain what each term in the expression represents. **See margin.**

MIXED REVIEW

LEAST COMMON DENOMINATOR Find the least common denominator.
(Skills Review, p. 908)

69. $\frac{1}{2}, \frac{3}{4}, \frac{4}{5}$ 20

70. $\frac{1}{2}, \frac{3}{4}, \frac{5}{6}$ 12

71. $\frac{1}{3}, \frac{2}{5}, -\frac{14}{15}$ 15

72. $\frac{1}{4}, -\frac{3}{8}, \frac{7}{12}$ 24

73. $\frac{1}{3}, \frac{4}{5}, \frac{6}{7}$ 105

74. $\frac{1}{2}, \frac{3}{4}, -\frac{1}{16}$ 16

USING A NUMBER LINE Graph the numbers on a number line. Then decide which number is greater and use the symbol < or > to show the relationship. (Review 1.1)   75–77. See margin for graphs.

75. $-\sqrt{3}, -3$ $-\sqrt{3} > -3$ **76.** $-\frac{1}{2}, -\frac{11}{2}$ $-\frac{1}{2} > -\frac{11}{2}$ **77.** $2.75, \frac{7}{2}$ $2.75 < \frac{7}{2}$

IDENTIFYING PROPERTIES Identify the property shown. (Review 1.1)

78. $(7 \cdot 9)8 = 7(9 \cdot 8)$
associative property of multiplication

79. $-13 + 13 = 0$
inverse property of addition

80. $27 + 6 = 6 + 27$
commutative property of addition

81. $19 \cdot 1 = 19$
identity property of multiplication

FINDING RECIPROCALS Give the reciprocal of the number. (Review 1.1 for 1.3)

82. -22 $-\frac{1}{22}$

83. $\frac{7}{8}$ $\frac{8}{7}$

84. 12 $\frac{1}{12}$

85. $-\frac{5}{4}$ $-\frac{4}{5}$

86. $\frac{11}{16}$ $\frac{16}{11}$

87. $-\frac{1}{9}$ -9

88. 37 $\frac{1}{37}$

89. -14 $-\frac{1}{14}$

QUIZ 1

Self-Test for Lessons 1.1 and 1.2

Graph the numbers on a number line. Then write the numbers in increasing order. (Lesson 1.1)   1, 2. See margin for graphs.

1. $\frac{9}{2}, -2.5, 0, -\frac{3}{4}, 1$ $-2.5, -\frac{3}{4}, 0, 1, \frac{9}{2}$

2. $\frac{10}{3}, 0.8, \frac{15}{8}, -1.5, -0.25$ $-1.5, -0.25, 0.8, \frac{15}{8}, \frac{10}{3}$

Identify the property shown. (Lesson 1.1)

3. $5(3 - 7) = 5 \cdot 3 - 5 \cdot 7$
distributive property

4. $(8 + 6) + 4 = 8 + (6 + 4)$
associative property of addition

Evaluate the expression for the given value(s) of the variable(s). (Lesson 1.2)

5. $12x - 21$ when $x = 3$ 15

6. $7x - (9x + 5)$ when $x = \frac{1}{3}$ $-\frac{17}{3}$

7. $x^2 + 5x - 8$ when $x = -3$ -14

8. $x^3 + 4(x - 1)$ when $x = 4$ 76

9. $x^2 - 11x + 40y - 14$ when $x = 5$ and $y = -2$ -124

Simplify the expression. (Lesson 1.2)

10. $3x - 2y - 9y + 4 + 5x$ $8x - 11y + 4$

11. $3(x - 2) - (4 + x)$ $2x - 10$

12. $5x^2 - 3x + 8x - 6 - 7x^2$ $-2x^2 + 5x - 6$

13. $4(x + 2x) - 2(x^2 - x)$ $-2x^2 + 14x$

14. COMPUTER DISKS You are buying a total of 15 regular floppy disks and high capacity storage disks for your computer. Regular floppy disks cost $.35 each and high capacity disks cost $13.95 each. Write an expression for the total amount you spend on computer disks. (Lesson 1.2)
$0.35n + 13.95(15 - n)$, or $209.25 - 13.60n$, where n is the number of regular floppy disks bought

1.2 Algebraic Expressions and Models   **17**

Additional Test Preparation *Sample answer:*
It means they have the same value for all values
of the variable(s). *Sample answer:* $7x$, $3x - (-4x)$,
$2x^2 + x - 2(x^2 - 3x)$

4 ASSESS

DAILY HOMEWORK QUIZ

✎ *Transparency Available*

1. Write the expression "the pth power of 10" using exponents. 10^p

2. Evaluate -3^6. -729

3. Evaluate the expression $18 \div (6 - 3)^2 - 8$. -6

4. Evaluate $(3x^2 - 2) \div (5x)$ when $x = 2$.   1

5. Evaluate $\frac{y(3 - 2x)}{x - y}$ when $x = -2$ and $y = -1$.   7

6. Simplify the expression $5(-3p^2 - 2) + 2(p - p^2)$. $-17p^2 + 2p - 10$

EXTRA CHALLENGE NOTE

→ Challenge problems for Lesson 1.2 are available in **blackline** format in the *Chapter 1 Resource Book*, p. 34 and at **www.mcdougallittell.com**.

ADDITIONAL TEST PREPARATION

1. OPEN ENDED What does it mean for algebraic expressions to be equivalent? Give an example of three algebraic expressions that are equivalent.
See sample answer below.

ADDITIONAL RESOURCES

An alternative Quiz for Lessons 1.1 and 1.2 is available in the *Chapter Resource Book*, p.35.

75.

76.

77.

1–2. See Additional Answers beginning on page AA1.

1 Planning the Activity

PURPOSE
To explore the procedures for evaluating numeric expressions on scientific and graphing calculators.

MATERIALS
- scientific or graphing calculator for each student
- Keystroke blackline (*Chapter 1 Resource Book,* p. 26)

PACING
- Example — 10 min
- Exercises — 20 min

▶ LINK TO LESSON
Students can work Examples 2 and 3 of Lesson 1.2 with calculators for extra reinforcement.

2 Managing the Activity

COOPERATIVE LEARNING
Students can complete the activity in pairs, with one student using a scientific calculator and the other a graphing calculator. Then both can evaluate other expressions using each type of calculator.

3 Closing the Activity

★ KEY DISCOVERY
Entering the proper keys in the correct sequence, including the proper use of parentheses, ensures that a calculator evaluates an expression as intended.

ACTIVITY ASSESSMENT
Write the graphing calculator key sequence to evaluate the expression $3\left(\dfrac{-12}{14-6}\right)^2$. Then evaluate the expression.

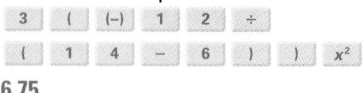

6.75

● ACTIVITY 1.2

Using Technology

Evaluating Expressions

You can use a scientific calculator or a graphing calculator to evaluate expressions. Keystrokes for evaluating several expressions are shown below. Because the keystrokes shown may not agree precisely with the keystrokes for *your* calculator, you should make sure you know how to evaluate the expressions using your own calculator.

STUDENT HELP

INTERNET KEYSTROKE HELP

See keystrokes for several models of calculators at www.mcdougallittell.com

▶ EXAMPLE

EXPRESSION	CALCULATOR	KEYSTROKES	RESULT
a. $-3^2 + 4$	Scientific	3 x^2 +/− + 4 =	−5
$-3^2 + 4$	Graphing	(−) 3 x^2 + 4 ENTER	−5
b. $(-3)^2 + 4$	Scientific	3 +/− x^2 + 4 =	13
$(-3)^2 + 4$	Graphing	((−) 3) x^2 + 4 ENTER	13
c. $(24 \div 2)^3$	Scientific	(24 ÷ 2) y^x 3 =	1728
$(24 \div 2)^3$	Graphing	(24 ÷ 2) ^ 3 ENTER	1728
d. $\dfrac{5}{4 + 3 \cdot 2}$	Scientific	5 ÷ (4 + 3 × 2) =	0.5
$\dfrac{5}{4 + 3 \cdot 2}$	Graphing	5 ÷ (4 + 3 × 2) ENTER	0.5

On a scientific calculator, notice the difference between the change sign key, +/− , and the subtraction key, − . Likewise, on a graphing calculator, the negation key, (−) , and the subtraction key, − , do not perform the same operation.

▶ EXERCISES

Write an expression that corresponds to the calculator keystrokes. Then evaluate the expression.

1. Scientific: 4 +/− x^2 − 5 = $(-4)^2 - 5$; 11

2. Scientific: 7 ÷ (3 +/− − 5) = $\dfrac{7}{-3 - 5}$; −0.875

3. Graphing: (1 + 4) ^ 6 ENTER $(1 + 4)^6$; 15,625

4. Graphing: 3 × (5 − 2) ENTER $3(5 - 2)$; 9

Use a calculator to evaluate the expression. Round the result to three decimal places.

5. $3(5.3 - 4.1)^2$ 4.32

6. $(-2.6 - 12.5)^4$ 51,988.560

7. $(0.21 + 5.23)^3$ 160.989

8. $\dfrac{4}{3}\pi(5.5)^3$ 696.910

9. $\dfrac{9.2 - 4.5}{0.6}$ 7.833

10. $\dfrac{7.3}{-6.2 - 3.6}$ −0.745

11. $1024(1 + 0.42)^5$ 5912.099

12. $\dfrac{1 + 3 \cdot 4^2}{7.25}$ 6.759

13. $\left(\dfrac{2^3 + 1}{2 \cdot 5}\right)^2$ 0.81

1.3

Solving Linear Equations

What you should learn

GOAL 1 Solve linear equations.

GOAL 2 Use linear equations to solve **real-life** problems, such as finding how much a broker must sell in **Example 5**.

Why you should learn it

▼ To solve **real-life** problems, such as finding the temperature at which dry ice changes to a gas in **Ex. 43**.

GOAL 1 SOLVING A LINEAR EQUATION

An **equation** is a statement in which two expressions are equal. A **linear equation** in one variable is an equation that can be written in the form $ax = b$ where a and b are constants and $a \neq 0$. A number is a **solution** of an equation if the statement is true when the number is substituted for the variable.

Two equations are **equivalent** if they have the same solutions. For instance, the equations $x - 4 = 1$ and $x = 5$ are equivalent because both have the number 5 as their only solution. The following *transformations*, or changes, produce equivalent equations and can be used to solve an equation.

TRANSFORMATIONS THAT PRODUCE EQUIVALENT EQUATIONS

ADDITION PROPERTY OF EQUALITY
Add the same number to both sides:
If $a = b$, then $a + c = b + c$.

SUBTRACTION PROPERTY OF EQUALITY
Subtract the same number from both sides:
If $a = b$, then $a - c = b - c$.

MULTIPLICATION PROPERTY OF EQUALITY
Multiply both sides by the same nonzero number: If $a = b$ and $c \neq 0$, then $ac = bc$.

DIVISION PROPERTY OF EQUALITY
Divide both sides by the same nonzero number: If $a = b$ and $c \neq 0$, then $a \div c = b \div c$.

EXAMPLE 1 *Solving an Equation with a Variable on One Side*

Solve $\frac{3}{7}x + 9 = 15$.

SOLUTION

Your goal is to isolate the variable on one side of the equation.

$\frac{3}{7}x + 9 = 15$ **Write original equation.**

$\frac{3}{7}x = 6$ **Subtract 9 from each side.**

$x = \frac{7}{3}(6)$ **Multiply each side by $\frac{7}{3}$, the reciprocal of $\frac{3}{7}$.**

$x = 14$ **Simplify.**

▶ The solution is 14.

✓ **CHECK**   Check $x = 14$ in the original equation.

$\frac{3}{7}(14) + 9 \stackrel{?}{=} 15$ **Substitute 14 for x.**

$15 = 15$ ✓   **Solution checks.**

1.3 Solving Linear Equations   **19**

---

1 PLAN

PACING
Basic: 1 day
Average: 1 day
Advanced: 1 day
Block Schedule: 0.5 block with 1.4

LESSON OPENER
VISUAL APPROACH
An alternative way to approach Lesson 1.3 is to use the Visual Approach Lesson Opener:
- Blackline Master (*Chapter 1 Resource Book,* p. 39)
- Transparency (p. 3)

MEETING INDIVIDUAL NEEDS
- *Chapter 1 Resource Book*
  Prerequisite Skills Review (p. 5)
  Practice Level A (p. 42)
  Practice Level B (p. 43)
  Practice Level C (p. 44)
  Reteaching with Practice (p. 45)
  Absent Student Catch-Up (p. 47)
  Challenge (p. 49)
- *Resources in Spanish*
- *Personal Student Tutor*

NEW-TEACHER SUPPORT
See the Tips for New Teachers on pp. 1–2 of the *Chapter 1 Resource Book* for additional notes about Lesson 1.3.

WARM-UP EXERCISES

 *Transparency Available*

Verify that the statement is true when $x = \frac{2}{3}$.

1. $6(1 - x) = 11x - 2(x + 2)$ $2 = 2$
2. $2x - 4(x - 1) = 2(2x - 1) + 3x$
   $\frac{8}{3} = \frac{8}{3}$

Multiply the expression by the LCD of the denominators and simplify.

3. $\frac{2}{5}x - \frac{1}{4}$ $8x - 5$
4. $\frac{11}{6} - \frac{3}{2}x + \frac{2}{7}x$ $77 - 51x$

2 TEACH

MOTIVATING THE LESSON

One phone company charges 6¢ per minute with a $10 monthly fee; another charges 7¢ per minute with a $5 monthly fee. Solving the equation $10 + 0.06n = 5 + 0.07n$ gives the number of minutes for which the costs are equal. Solving equations like this can help in making decisions in many real-life situations.

EXTRA EXAMPLE 1
Solve $\frac{2}{9}x + 8 = 16$.   36

EXTRA EXAMPLE 2
Solve $12n - 3 = 4n + 21$.   3

EXTRA EXAMPLE 3
Solve $5(x - 2) = -4(2x + 7) + x$ $-\frac{3}{2}$

EXTRA EXAMPLE 4
Solve $\frac{2}{3}x + \frac{1}{5} = 2x - \frac{3}{10}$. $\frac{3}{8}$

CHECKPOINT EXERCISES

For use after Examples 1 and 2:
1. Solve $2 + 5x = 7x - 16$ 9

For use after Example 3:
2. Solve $6(3 - x) = -5(2x + 9) + 18$
  -11.25

For use after Example 4:
3. $\frac{1}{4}x - \frac{1}{2} = \frac{3}{4}x + \frac{2}{3}$ $-\frac{7}{3}$

STUDENT HELP NOTES

→ **Homework Help** Students can find extra examples at **www.mcdougallittell.com** that parallel the examples in the student edition.

→ **Skills Review** As students review finding the LCD on page 939, remind them that the LCD is the least common multiple of the denominators.

STUDENT HELP

HOMEWORK HELP
Visit our Web site
www.mcdougallittell.com
for extra examples.

EXAMPLE 2   *Solving an Equation with a Variable on Both Sides*

Solve $5n + 11 = 7n - 9$.

SOLUTION

$5n + 11 = 7n - 9$	Write original equation.
$11 = 2n - 9$	Subtract $5n$ from each side.
$20 = 2n$	Add 9 to each side.
$10 = n$	Divide each side by 2.

▶ The solution is 10. Check this in the original equation.

EXAMPLE 3   *Using the Distributive Property*

Solve $4(3x - 5) = -2(-x + 8) - 6x$.

SOLUTION

$4(3x - 5) = -2(-x + 8) - 6x$	Write original equation.
$12x - 20 = 2x - 16 - 6x$	Distributive property
$12x - 20 = -4x - 16$	Combine like terms.
$16x - 20 = -16$	Add $4x$ to each side.
$16x = 4$	Add 20 to each side.
$x = \frac{1}{4}$	Divide each side by 16.

▶ The solution is $\frac{1}{4}$. Check this in the original equation.

EXAMPLE 4   *Solving an Equation with Fractions*

STUDENT HELP

→ **Skills Review**
For help with finding the LCD, see p. 939.

Solve $\frac{1}{3}x + \frac{1}{4} = x - \frac{1}{6}$.

SOLUTION

$\frac{1}{3}x + \frac{1}{4} = x - \frac{1}{6}$	Write original equation.
$12\left(\frac{1}{3}x + \frac{1}{4}\right) = 12\left(x - \frac{1}{6}\right)$	Multiply each side by the LCD, 12.
$4x + 3 = 12x - 2$	Distributive property
$3 = 8x - 2$	Subtract $4x$ from each side.
$5 = 8x$	Add 2 to each side.
$\frac{5}{8} = x$	Divide each side by 8.

▶ The solution is $\frac{5}{8}$. Check this in the original equation.

20   **Chapter 1** *Equations and Inequalities*

FOCUS ON CAREERS

GOAL 2   USING LINEAR EQUATIONS IN REAL LIFE

EXAMPLE 5   *Writing and Using a Linear Equation*

REAL ESTATE A real estate broker's base salary is $18,000. She earns a 4% commission on total sales. How much must she sell to earn $55,000 total?

SOLUTION

VERBAL MODEL

| Total income | = | Base salary | + | Commission rate | · | Total sales |

LABELS

Total income = **55,000**   (dollars)

Base salary = **18,000**   (dollars)

Commission rate = **0.04**   (percent in decimal form)

Total sales = **x**   (dollars)

ALGEBRAIC MODEL

$55{,}000 = 18{,}000 + 0.04\,x$ Write linear equation.

$37{,}000 = 0.04x$ Subtract 18,000 from each side.

$925{,}000 = x$ Divide each side by 0.04.

▶ The broker must sell real estate worth a total of $925,000 to earn $55,000.

REAL ESTATE BROKER

Real estate brokers must have a thorough knowledge not only of the real estate market, but of mathematics as well. Brokers often provide buyers with information about loans, loan rates, and monthly payments.

CAREER LINK
www.mcdougallittell.com

EXAMPLE 6   *Writing and Using a Geometric Formula*

Photo Framing

You have a 3 inch by 5 inch photo that you want to enlarge, mat, and frame. You want the width of the mat to be 2 inches on all sides. You want the perimeter of the framed photo to be 44 inches. By what percent should you enlarge the photo?

SOLUTION

Let x be the percent (in decimal form) of enlargement relative to the original photo. So, the dimensions of the enlarged photo (in inches) are $3x$ by $5x$. Draw a diagram.

PROBLEM SOLVING STRATEGY

VERBAL MODEL

| **Perimeter** | = | 2 · | **Width** | + | 2 · | **Length** |

LABELS

Perimeter = **44**   (inches)

Width = **$4 + 3x$**   (inches)

Length = **$4 + 5x$**   (inches)

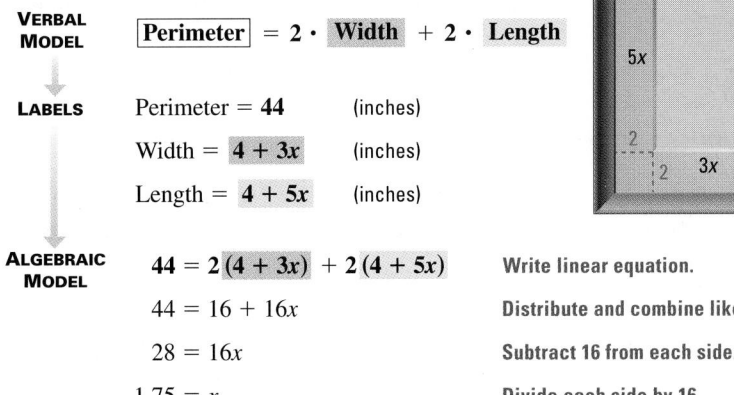

ALGEBRAIC MODEL

$44 = 2\,(4 + 3x) + 2\,(4 + 5x)$ Write linear equation.

$44 = 16 + 16x$ Distribute and combine like terms.

$28 = 16x$ Subtract 16 from each side.

$1.75 = x$ Divide each side by 16.

▶ You should enlarge the photo to 175% of its original size.

1.3 *Solving Linear Equations*   **21**

Closure Question *Sample answer:*
To solve a linear equation, you find a value for the variable that makes the equation true. To simplify a linear expression, you rewrite the expression in a form that is   equivalent for all variable values, but you don't find a specific variable value.

EXTRA EXAMPLE 5
A car salesperson's base salary is $21,000. She earns a 5% commission on sales. How much must she sell to earn $65,000 total? **$880,000**

EXTRA EXAMPLE 6
You plan to enlarge, mat, and frame a 5 in. by 7 in. photo. The mat will be 1.5 in. wide on all sides. The perimeter of the framed photo will be 42 in. By what percent should you enlarge the photo? **125%**

✔ **CHECKPOINT EXERCISES**

For use after Example 5:

1. A restaurant server earns a base salary of $6 per hour plus tips. If he averages $12 per hour in tips, how many hours must he work to earn a total of $333? **18.5 hr**

For use after Example 6:

2. You make a 1 ft wide boundary in a rectangular shape around your flower garden with border tiles. You have enough tiles so that the outside perimeter of the tiles can be 38 ft. If the original garden is 4 ft by 8 ft by what percent of its original size can you expand the garden to fit the border? **125%**

FOCUS ON VOCABULARY
What is a *transformation* of an equation? *Sample answer:* a change that you make to an equation to get a different, but equivalent, equation

CLOSURE QUESTION
How does solving a linear equation differ from simplifying a linear expression?
See sample answer below.

DAILY PUZZLER
Eva weighs 2.5 times as much as her sister, but if their 10 lb cat sits on Eva's side of a seesaw and their 70 lb dog sits on her sister's side, they exactly balance. How much does Eva weigh? **100 lb**

3 APPLY

ASSIGNMENT GUIDE

BASIC
Day 1: pp. 22–24 Exs. 18–42 even, 43, 44, 50, 57–73 odd

AVERAGE
Day 1: pp. 22–24 Exs. 18–42 even, 43, 44, 46, 47, 50, 57–77 odd

ADVANCED
Day 1: pp. 22–24 Exs. 18–42 even, 43, 44, 46, 47, 50–56, 57–77 odd

BLOCK SCHEDULE WITH 1.4
pp. 22–24 Exs. 18–42 even, 43, 44, 46, 47, 50, 57–77 odd

EXERCISE LEVELS
Level A: *Easier*
17–30, 33, 34, 41, 42, 57–68
Level B: *More Difficult*
31, 32, 35–40, 43–50, 69–78
Level C: *Most Difficult*
51–56

✔ HOMEWORK CHECK
To quickly check student understanding of key concepts, go over the following exercises: Exs. 24, 28, 30, 36, 42, 44. See also the Daily Homework Quiz:

- Blackline Master (*Chapter 1 Resource Book,* p. 52)
- 📖 Transparency (p. 5)

! COMMON ERROR
EXERCISE 31 Students often have trouble distributing the negative sign in front of parentheses. Remind students that they are actually distributing –1, which may make the computation easier for them.

5, 20–22. See Additional Answers beginning on page AA1.

GUIDED PRACTICE

Vocabulary Check ✔

1. What is an equation?   a statement of the equality of two expressions

Concept Check ✔

2. What does it mean for two equations to be equivalent? Give an example of two equivalent equations.   They have the same solution.
Sample answer: $3x - 1 = 8$ and $3x = 9$

3. How does an equation such as $2(x + 3) = 10$ differ from an identity such as $2(x + 3) = 2x + 6$?   An equation such as $2(x + 3) = 10$ is true for only one value of x, while an identity such as $2(x + 3) = 2x + 6$ is true for all values of x.

ERROR ANALYSIS   Describe the error(s). Then write the correct steps.

4. In the second line, the right side must also be multiplied by 30: $30\left(\frac{1}{5}x + \frac{1}{6}\right) = 30(-2)$.
The other steps are then "$6x + 5 = -60$," "$6x = -65$," and "$x = -\frac{65}{6}$."

5. See margin.

4.
$$\frac{1}{5}x + \frac{1}{6} = -2$$
$$30\left(\frac{1}{5}x + \frac{1}{6}\right) = -2$$
$$6x + 5 = -2$$
$$6x = -7$$
$$x = -\frac{7}{6}$$

4–5. See margin.

5.
$$2(x + 3) = -3(-x + 1)$$
$$2x + 6 = 3x - 3$$
$$5x + 6 = -3$$
$$5x = -9$$
$$x = -\frac{5}{9}$$

6. Describe the transformation(s) you would use to solve $2x - 8 = 14$.
Add 8 to each side; then divide each side by 2.

Skill Check ✔   **Solve the equation.**

7. $x + 4 = 9$ 5

8. $4x = 24$ 6

9. $2x - 3 = 7$ 5

20. Add 9 to each side; then divide each side by 2.

10. $0.2x - 8 = 0.6$ 43

11. $\frac{1}{3}x + \frac{1}{2} = \frac{11}{12}$ $\frac{5}{4}$

12. $\frac{3}{4}x - \frac{2}{3} = \frac{5}{6}$ 2

21. Subtract 2 from each side; then multiply each side by 3.

13. $1.5x + 9 = 4.5$ -3

14. $6x - 4 = 2x + 10$ $\frac{7}{2}$

15. $2(x + 2) = 3(x - 8)$ 28

22. Add 5 to each side; then divide each side by -1.

16. 🌎 **REAL ESTATE SALES** The real estate broker's base salary from Example 5 has been raised to $21,000 and the commission rate has been increased to 5%. How much real estate does the broker have to sell now to earn $70,000? $980,000

PRACTICE AND APPLICATIONS

STUDENT HELP

▶ **Extra Practice**
to help you master skills is on p. 940.

DESCRIBING TRANSFORMATIONS   Describe the transformation(s) you would use to solve the equation.   Multiply each side by 6.   Multiply each side by $-\frac{7}{4}$.
Subtract 5 from each side.
17. $x + 5 = -7$

18. $\frac{1}{6}x = 3$

19. $-\frac{4}{7}x = 6$

20–22. See margin.
20. $2x - 9 = 0$

21. $\frac{x}{3} + 2 = 89$

22. $3 = -x - 5$

SOLVING EQUATIONS   Solve the equation. Check your solution.

STUDENT HELP

▶ **HOMEWORK HELP**
Examples 1–4: Exs. 17–40
Examples 5, 6: Exs. 43–49

23. $4x + 7 = 27$ 5

24. $7s - 29 = -15$ 2

25. $3a + 13 = 9a - 8$ $\frac{7}{2}$

26. $m - 30 = 6 - 2m$ 12

27. $15n + 9 = 21$ $\frac{4}{5}$

28. $2b + 11 = 15 - 6b$ $\frac{1}{2}$

29. $2(x + 6) = -2(x - 4)$ -1

30. $4(-3x + 1) = -10(x - 4) - 14x$ 3

31. $-(x + 2) - 2x = -2(x + 1)$ 0

32. $-4(3 + x) + 5 = 4(x + 3)$ $-\frac{19}{8}$

SOLVING EQUATIONS Solve the equation. Check your solution.

33. $\frac{7}{2}x - 1 = 2x + 5$ **4**

34. $\frac{1}{2}x - \frac{5}{3} = -\frac{1}{2}x + \frac{19}{4}$ $\frac{77}{12}$

35. $\frac{3}{4}\left(\frac{4}{5}x - 2\right) = \frac{11}{4}$ $\frac{85}{12}$

36. $-\frac{2}{3}\left(\frac{6}{5}x - \frac{7}{10}\right) = \frac{17}{20}$ $-\frac{23}{48}$

37. $2.7n + 4.3 = 12.94$ **3.2**

38. $-4.2n - 6.5 = -14.06$ **1.8**

39. $3.1(x + 2) - 1.5x = 5.2(x - 4)$ **7.5**

40. $2.5(x - 3) + 1.7x = 10.8(x + 1.5)$
−3.59

GEOMETRY ▶ CONNECTION **Find the dimensions of the figure.**

41. Area = 504 **length: 36, width: 14**

42. Perimeter = 23 **side lengths: 6, 7, 10**

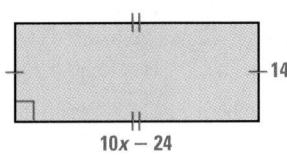
14
10x − 24

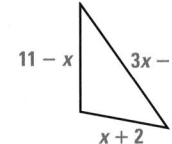
11 − x   3x − 2
x + 2

In Exercises 43 and 44, use the following formula.

$$\text{degrees Fahrenheit} = \frac{9}{5}(\text{degrees Celsius}) + 32$$

43. 🌐 **DRY ICE** Dry ice is solid carbon dioxide. Dry ice does not melt — it changes directly from a solid to a gas. Dry ice changes to a gas at $-109.3°F$. What is this temperature in degrees Celsius? **−78.5°C**

44. 🌐 **VETERINARY MEDICINE** The normal body temperature of a dog is $38.6°C$. Your dog's temperature is $101.1°F$. Does your dog have a fever? Explain. **See margin.**

45. 🌐 **CAR REPAIR** The bill for the repair of your car was $390. The cost for parts was $215. The cost for labor was $35 per hour. How many hours did the repair work take? **5 h**

46. 🌐 **SUMMER JOBS** You have two summer jobs. In the first job, you work 28 hours per week and earn $7.25 per hour. In the second job, you earn $6.50 per hour and can work as many hours as you want. If you want to earn $255 per week, how many hours must you work at your second job? **8 h**

47. 🌐 **STOCKBROKER** A stockbroker earns a base salary of $40,000 plus 5% of the total value of the stocks, mutual funds, and other investments that the stockbroker sells. Last year, the stockbroker earned $71,750. What was the total value of the investments the stockbroker sold? **$635,000**

48. 🌐 **WORD PROCESSING** You are writing a term paper. You want to include a table that has 5 columns and is 360 points wide. $\left(\text{A point is } \frac{1}{72} \text{ of an inch.}\right)$ You want the first column to be 200 points wide and the remaining columns to be equal in width. How wide should each of the remaining columns be? **40 points**

49. 🌐 **WALKWAY CONSTRUCTION** You are building a walkway of uniform width around a 100 foot by 60 foot swimming pool. After completing the walkway, you want to put a fence along the outer edge of the walkway. You have 450 feet of fencing to enclose the walkway. What is the maximum width of the walkway? **16.25 ft**

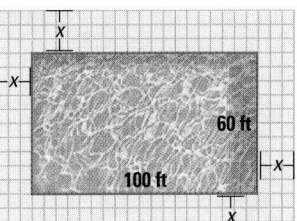
x
x
60 ft
100 ft
x

FOCUS ON
CAREERS

REAL LIFE
STOCKBROKER
Stockbrokers buy and sell stocks, bonds, and other securities for clients as discussed in Ex. 47. Stockbrokers typically study economics in college.

🌐 **CAREER LINK**
www.mcdougallittell.com

44. No; 101.1°F ≈ 38.4°C, so the dog's temperature is slightly below normal.

CAREER NOTE
EXERCISE 47 Additional information about a career as a stockbroker is available at **www.mcdougallittell.com.**

ADDITIONAL PRACTICE AND RETEACHING

For Lesson 1.3:
• Practice Levels A, B, and C (*Chapter 1 Resource Book,* p. 42)
• Reteaching with Practice (*Chapter 1 Resource Book,* p. 45)
• 🖥 See Lesson 1.3 of the *Personal Student Tutor*

For more Mixed Review:
• 🖥 Search the *Test and Practice Generator* for key words or specific lessons.

4 **ASSESS**

DAILY HOMEWORK QUIZ

🖥 *Transparency Available*

Solve the equation. Check your solution.

1. $3n + 11 = -16$ -9

2. $2x - 5 = 4 - 3x$ $\frac{9}{5}$

3. $7(1 - s) = -2(5s + 4) + 6s$ 5

4. $\frac{3}{2}\left(\frac{3}{5} - \frac{3}{4}x\right) = \frac{3}{10}$ $\frac{8}{15}$

5. A triangle with a perimeter of 50 has side lengths $x - 1$, $2x - 3$, and $3x - 6$. What are the side lengths?   9, 17, 24

6. Using the formula F = 1.8C + 32, find the temperature in degrees Celsius of an oven that is 500°F.   260°C

EXTRA CHALLENGE NOTE

↳ Challenge problems for Lesson 1.3 are available in **blackline** format in the *Chapter 1 Resource Book,* p. 49 and at **www.mcdougallittell.com.**

ADDITIONAL TEST PREPARATION

1. WRITING Describe how to solve the linear equation $3x + 7 = 13 + x$. *Sample answer:* **First subtract x from each side to get the variable on one side: $2x + 7 = 13$. Then subtract 7 from each side to get $2x = 6$. Finally, divide both sides by 2 to get $x = 3$.**

2. OPEN ENDED Write a question that can be solved using a linear equation. *Sample answer:* **Gwen can high jump 5 ft 1 in. If she improves by one half inch for each of the next several weeks, how long will it take her to jump 5 ft 5 in.? (Answer: 8 wk)**

Test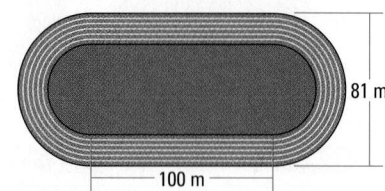
Preparation

50. MULTI-STEP PROBLEM You are in charge of constructing a fence around the running track at a high school. The fence is to be built around the track so that there is a uniform gap between the outside edge of the track and the fence.

(diagram: oval track, 81 m height, 100 m width)

a. What is the maximum width of the gap between the track and the fence if no more than 630 meters of fencing is used? (*Hint:* Use the equation for the circumference of a circle, $C = 2\pi r$, to help you.)   **about 27.9 m**

b. You are charging the school $10.50 for each meter of fencing. The school has $5250 in its budget to spend on the fence. How many meters of fencing can you use with this budget?   **500 m**

c. CRITICAL THINKING Explain whether or not it is geometrically reasonable to put up the new fence with the given budget.   **Yes; with 500 m of fencing, a fence can be built with a 7.25 m gap between the track and the fence.**

★ **Challenge**

EXTRA CHALLENGE
→ www.mcdougallittell.com

SOLVING EQUATIONS Solve the equation. If there is no solution, write *no solution*. If the equation is an identity, write *all real numbers*.

51. $5(x - 4) = 5x + 12$ no solution

52. $3(x + 5) = 3x + 15$ all real numbers

53. $7x + 14 - 3x = 4x + 14$ all real numbers

54. $11x - 3 + 2x = 6(x + 4) + 7x$ no solution

55. $-2(4 - 3x) + 7 = -2x + 6 + 8x$ no solution

56. $5(2 - x) = 3 - 2x + 7 - 3x$ all real numbers

MIXED REVIEW

GEOMETRY ▷ **CONNECTION**   Find the area of the figure. (Skills Review, p. 914)

57. Circle with radius 5 inches 25π in.^{2}, or about 78.5 in.^{2}

58. Square with side 4 inches   16 in.^{2}

59. Circle with radius 7 inches 49π in.^{2}, or about 154 in.^{2}

60. Square with side 9 inches   81 in.^{2}

EVALUATING EXPRESSIONS Evaluate the expression. (Review 1.2 for 1.4)

61. $24 - (9 + 7)$ 8

62. $-16 + 3(8 - 4)$ -4

63. $-3 + 6(1 - 3)^2$ 21

64. $2(3 - 5)^3 + 4(-4 + 7)$ -4

65. $2x + 3$ when $x = 4$ 11

66. $8(x - 2) + 3x$ when $x = 6$ 50

67. $5x - 7 + 2x$ when $x = -3$ -28

68. $6x - 3(2x + 4)$ when $x = 5$ -12

SIMPLIFYING EXPRESSIONS Simplify the expression. (Review 1.2)

69. $3(7 + x) - 8x$ $21 - 5x$

70. $2(8 + x) + 2x - x$ $16 + 3x$

71. $4x - (6 - 3x)$ $7x - 6$

72. $2x - 3(4x + 7)$ $-10x - 21$

73. $3(x + 9) + 2(4 - x)$ $x + 35$

74. $-4(x - 3) - 2(x + 7)$ $-6x - 2$

75. $2(x^2 + 2) - x + x^2 + 7$ $3x^2 - x + 11$

76. $2(x^2 - 81) - 3x^2$ $-x^2 - 162$

77. $x^2 - 5x + 3(x^2 + 7x)$ $4x^2 + 16x$

78. $4x^2 - 2(x^2 - 3x) + 6x + 8$ $2x^2 + 12x + 8$

ACTIVITY 1.3

Using Technology

Using Tables to Solve Equations

You can use the *Table* feature of a graphing calculator to solve linear equations.

▶ **EXAMPLE**

Use the *Table* feature of a graphing calculator to solve the equation $7x - 2 = 4x + 13$.

▶ **SOLUTION**

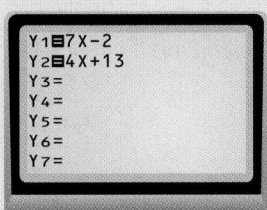

```
Y1≡7X-2
Y2≡4X+13
Y3=
Y4=
Y5=
Y6=
Y7=
```

X	Y₁	Y₂
0	-2	13
1	5	17
2	12	21
3	19	25
4	26	29
5	33	33

1 To use the *Table* feature to solve the equation, let y_1 equal the left side of the equation, and let y_2 equal the right side as shown above.

2 Then set the starting x-value of the table to 0 and the step value (the value by which the x-values increase) to 1. The table should look similar to the one shown above.

3 Scroll through the table until you find an x-value for which both sides of the equation have the same y-value or until the difference in the y-values changes sign. If both of the y-values are the same, that x-value is the solution of the equation. For the given equation, the solution is $x = 5$.

X	Y₁	Y₂
0	-2	13
1	5	17
2	12	21
3	19	25
4	26	29
5	33	33

▶ **EXERCISES**

Use the table shown to decide whether the statement is *true* or *false*. Explain your reasoning.

False; $y_1 = y_2$ when $x = -2$, not when $x = 2$.
1. The solution of $4 - 5x = 16 + x$ is 2.

True; $y_1 = y_2$ when $x = 3$.
2. The solution of $3x + 4 = x + 10$ is 3.

X	Y₁	Y₂
-3	19	13
-2	14	14
-1	9	15
0	4	16
1	-1	17
2	-6	18

X	Y₁	Y₂
-2	-2	8
-1	1	9
0	4	10
1	7	11
2	10	12
3	13	13

Use the *Table* feature of a graphing calculator to solve the equation.

3. $2x + 4 = -3x - 6$ -2   **4.** $-4x + 4 = -x - 5$ 3   **5.** $-2x - 5 = 3 - 10x$ 1

6. $-4x + 10 = 4 - 10x$ -1   **7.** $15x - 3 = 15 - 3x$ 1   **8.** $2x - 18 = -5x - 4$ 2

STUDENT HELP

KEYSTROKE HELP

See keystrokes for several models of calculators at www.mcdougallittell.com

STUDENT HELP

Study Tip
In Step 2, if the values of y_1 and y_2 become farther apart, you should reset the step value to −1. In Step 3, if the difference in y-values changes sign between x_1 and x_2 ($x_1 < x_2$), then the solution is between x_1 and x_2 and you should reset the starting x-value to x_1 and use a step value of 0.1.

---

1 Planning the Activity

PURPOSE
To use a graphing calculator's *Table* feature to solve linear equations.

MATERIALS
- graphing calculator for each student
- Keystroke blackline (*Chapter 1 Resource Book,* p. 40)

PACING
- Example — 10 min
- Exercises — 10 min

▶ **LINK TO LESSON**
Students can use the *Table* approach to solve Examples 2 and 3 of Lesson 1.3. In Example 3, they will need to start with a step value of 0.25 or 0.5.

2 Managing the Activity

CLASSROOM MANAGEMENT
To clarify when a step should be positive or negative, you can point out that because each side of the equation is linear, it represents a line. Sketch a general pair of intersecting lines to picture when you need to step forward or backward to find the common point.

3 Closing the Activity

★ **KEY DISCOVERY**
You can solve a linear equation by finding a number for which each side evaluates to the same result.

ACTIVITY ASSESSMENT
Explain how to solve the equation $-7x + 5 = 2x - 4$ using the *Table* feature of a graphing calculator. Then use this feature to find the solution.
Sample answer: Enter $-7x + 5$ as y_1 and $2x - 4$ as y_2. Then look at the table to find the x-value where y_1 and y_2 are equal; 1.

LESSON OPENER
ACTIVITY
An alternative way to approach Lesson 1.4 is to use the Activity Lesson Opener:
- Blackline Master (*Chapter 1 Resource Book*, p. 53)
- Transparency (p. 4)

MEETING INDIVIDUAL NEEDS
- *Chapter 1 Resource Book*
  Prerequisite Skills Review (p. 5)
  Practice Level A (p. 56)
  Practice Level B (p. 57)
  Practice Level C (p. 58)
  Reteaching with Practice (p. 59)
  Absent Student Catch-Up (p. 61)
  Challenge (p. 63)
- *Resources in Spanish*
- *Personal Student Tutor*

NEW-TEACHER SUPPORT
See the Tips for New Teachers on pp. 1–2 of the *Chapter 1 Resource Book* for additional notes about Lesson 1.4.

WARM-UP EXERCISES

Transparency Available

Solve the equation.

1. $10x - 14 = 6$ **2**

2. $12x + 1.3 = 9x - 8$ **−3.1**

3. $\frac{3}{2}x + \frac{2}{3} = \frac{7}{2}x + \frac{1}{6}$ **$\frac{1}{4}$**

Evaluate the expression for $m = -5$.

4. $18m - 25$ **−115**

5. $14 - m^2$ **−11**

What you should learn

GOAL 1 Rewrite equations with more than one variable.

GOAL 2 Rewrite common formulas, as applied in **Example 5**.

Why you should learn it

▼ To solve **real-life** problems, such as finding how much you should charge for tickets to a benefit concert in **Example 4**.

1.4 Rewriting Equations and Formulas

GOAL 1 **EQUATIONS WITH MORE THAN ONE VARIABLE**

In Lesson 1.3 you solved equations with one variable. Many equations involve more than one variable. You can solve such an equation for one of its variables.

EXAMPLE 1 *Rewriting an Equation with More Than One Variable*

Solve $7x - 3y = 8$ for y.

SOLUTION

$7x - 3y = 8$	Write original equation.
$-3y = -7x + 8$	Subtract $7x$ from each side.
$y = \frac{7}{3}x - \frac{8}{3}$	Divide each side by -3.

▷ ACTIVITY
Developing Concepts

Equations with More Than One Variable

Given the equation $2x + 5y = 4$, use each method below to find y when $x = -3, -1, 2,$ and 6. Tell which method is more efficient. $2, \frac{6}{5}, 0, -\frac{8}{5}$; Method 2

Method 1 Substitute $x = -3$ into $2x + 5y = 4$ and solve for y. Repeat this process for the other values of x.

Method 2 Solve $2x + 5y = 4$ for y. Then evaluate the resulting expression for y using each of the given values of x.

EXAMPLE 2 *Calculating the Value of a Variable*

Given the equation $x + xy = 1$, find the value of y when $x = -1$ and $x = 3$.

SOLUTION

Solve the equation for y.

$x + xy = 1$	Write original equation.
$xy = 1 - x$	Subtract x from each side.
$y = \frac{1 - x}{x}$	Divide each side by x.

Then calculate the value of y for each value of x.

When $x = -1$: $y = \dfrac{1 - (-1)}{-1} = -2$ **When $x = 3$:** $y = \dfrac{1 - 3}{3} = -\dfrac{2}{3}$

Benefit Concert

EXAMPLE 3 *Writing an Equation with More Than One Variable*

You are organizing a benefit concert. You plan on having only two types of tickets: adult and child. Write an equation with more than one variable that represents the revenue from the concert. How many variables are in your equation?

SOLUTION

PROBLEM SOLVING STRATEGY

VERBAL MODEL	Total revenue	=	Adult ticket price	·	Number of adults	+	Child ticket price	·	Number of children

LABELS

Total revenue = R (dollars)

Adult ticket price = p_1 (dollars per adult)

Number of adults = A (adults)

Child ticket price = p_2 (dollars per child)

Number of children = C (children)

ALGEBRAIC MODEL $R = p_1 A + p_2 C$

This equation has five variables. The variables p_1 and p_2 are read as "p sub one" and "p sub two." The small lowered numbers 1 and 2 are subscripts used to indicate the two different price variables.

EXAMPLE 4 *Using an Equation with More Than One Variable*

BENEFIT CONCERT For the concert in Example 3, your goal is to sell $25,000 in tickets. You plan to charge $25.25 per adult and expect to sell 800 adult tickets. You need to determine what to charge for child tickets. How much should you charge per child if you expect to sell 200 child tickets? 300 child tickets? 400 child tickets?

FOCUS ON APPLICATIONS

SOLUTION

First solve the equation $R = p_1 A + p_2 C$ from Example 3 for p_2.

$R = p_1 A + p_2 C$ **Write original equation.**

$R - p_1 A = p_2 C$ **Subtract $p_1 A$ from each side.**

$\dfrac{R - p_1 A}{C} = p_2$ **Divide each side by C.**

Now substitute the known values of the variables into the equation.

If $C = 200$, the child ticket price is $p_2 = \dfrac{25,000 - 25.25(800)}{200} = \$24$.

If $C = 300$, the child ticket price is $p_2 = \dfrac{25,000 - 25.25(800)}{300} = \$16$.

If $C = 400$, the child ticket price is $p_2 = \dfrac{25,000 - 25.25(800)}{400} = \$12$.

 BENEFIT CONCERT

Farm Aid, a type of benefit concert, began in 1985. Since that time Farm Aid has distributed more than $13,000,000 to family farms throughout the United States.

Extra Example 3 *Sample answer:*
Let R = total revenue, p_1 = basic hat price, B = number of basic hats sold, p_2 = fancy hat price, F = number of fancy hats sold; $R = p_1 B + p_2 F$.

---

2 TEACH

MOTIVATING THE LESSON
Ask students to recall the distance formula $d = rt$. Ask them how they would find r if they knew d and t, or t if they knew d and r. Tell them that there is a more efficient way than just substituting and then solving.

ACTIVITY NOTE
Some students may prefer substituting first. Tell them that by first solving an equation in a general form, it must be solved only once.

 EXTRA EXAMPLE 1
Solve $11x - 9y = -4$ for y.
$y = \dfrac{11}{9}x + \dfrac{4}{9}$

EXTRA EXAMPLE 2
Given the equation $xy - x = 4$, find the value of y when $x = -4$ and $x = 2$. **0, 3**

EXTRA EXAMPLE 3
You are selling two types of hats, basic and fancy. Write an equation with more than one variable that represents the total revenue. **See sample answer below.**

EXTRA EXAMPLE 4
You expect to sell 125 of the basic hats from Extra Example 3 at $8 each. To meet your goal of $1600 in sales, what would you need to charge for fancy hats if you can sell 50 fancy hats? 60 fancy hats? **$12; $10**

✓ **CHECKPOINT EXERCISES**
For use after Examples 1 and 2:
1. Solve $xy - y = 10$ for x.
Find the value of x when $y = 2$ and -5. $x = \dfrac{y + 10}{y}$; **6, −1**

For use after Examples 3 and 4:
2. Assume in Extra Examples 3 and 4 that you charge $7 for basic hats and $12 for fancy hats. If you sell 120 basic hats, how many fancy hats must you sell to reach your goal of $1600? to reach a goal of $1800? **64; 80**

EXTRA EXAMPLE 5
The formula for the area of a triangle is $A = \frac{1}{2}bh$. Solve for h.
$$h = \frac{2A}{b}$$

EXTRA EXAMPLE 6
You have 21 ft of plastic border to enclose a flower bed that is in the shape of an equilateral triangle. Express the area of the flower bed in terms of its height alone. $A = 3.5h$

 CHECKPOINT EXERCISES
For use after Example 5:
1. Solve Einstein's energy formula $E = mc^2$ for m, the mass.
$$m = \frac{E}{c^2}$$

For use after Example 6:
2. If $z = x - y$ and $w = \frac{y-z}{z}$, express w in terms of x alone when $y = 5$. $w = \frac{10-x}{x-5}$

STUDENT HELP NOTES
Skills Review As students review perimeter on p. 914, remind them that although the formulas are given solved for one variable, they can be rewritten in terms of the other variables.

FOCUS ON VOCABULARY
What is a formula?
See sample answer below.

CLOSURE QUESTION
How can rewriting formulas help you solve them? *Sample answer:* It makes it easier to find multiple solutions when more than one value will be substituted in the formula.

DAILY PUZZLER
Chen's age is the sum of the ages of his house, his dog, and his cat. His niece is six years older than the dog, three year older than the cat, and nine years younger than the house. Give Chen's age C in terms of his niece's age N. $C = 3N$

 **GOAL 2** **REWRITING COMMON FORMULAS**

Throughout this course you will be using many formulas. Several are listed below.

COMMON FORMULAS

	FORMULA	VARIABLES
Distance	$d = rt$	d = distance, r = rate, t = time
Simple Interest	$I = Prt$	I = interest, P = principal, r = rate, t = time
Temperature	$F = \frac{9}{5}C + 32$	F = degrees Fahrenheit, C = degrees Celsius
Area of Triangle	$A = \frac{1}{2}bh$	A = area, b = base, h = height
Area of Rectangle	$A = \ell w$	A = area, ℓ = length, w = width
Perimeter of Rectangle	$P = 2\ell + 2w$	P = perimeter, ℓ = length, w = width
Area of Trapezoid	$A = \frac{1}{2}(b_1 + b_2)h$	A = area, b_1 = one base, b_2 = other base, h = height
Area of Circle	$A = \pi r^2$	A = area, r = radius
Circumference of Circle	$C = 2\pi r$	C = circumference, r = radius

EXAMPLE 5 *Rewriting a Common Formula*

STUDENT HELP
→ **Skills Review**
For help with perimeter, see p. 914.

The formula for the perimeter of a rectangle is $P = 2\ell + 2w$. Solve for w.

SOLUTION

$P = 2\ell + 2w$	Write perimeter formula.
$P - 2\ell = 2w$	Subtract 2ℓ from each side.
$\frac{P - 2\ell}{2} = w$	Divide each side by 2.

Gardening

EXAMPLE 6 *Applying a Common Formula*

You have 40 feet of fencing with which to enclose a rectangular garden. Express the garden's area in terms of its length only.

SOLUTION

Use the formula for the area of a rectangle, $A = \ell w$, and the result of Example 5.

$A = \ell w$	Write area formula.
$A = \ell\left(\frac{P - 2\ell}{2}\right)$	Substitute $\frac{P - 2\ell}{2}$ for w.
$A = \ell\left(\frac{40 - 2\ell}{2}\right)$	Substitute 40 for P.
$A = \ell(20 - \ell)$	Simplify.

Focus on Vocabulary *Sample answer:*
A formula is an equation that relates two or more variables. It expresses established mathematical relationships that usually have real-life applications.

GUIDED PRACTICE

Vocabulary Check ✓

1. Complete this statement: $A = \ell w$ is an example of a(n) _?_. **formula**

Concept Check ✓

2. Which of the following are equations with more than one variable? **B and C**

    A. $2x + 5 = 9 - 5x$ **B.** $4x + 10y = 62$ **C.** $x - 8 = 3y + 7$

3. Use the equation from Example 3. Describe how you would solve for A.
    Subtract p_2C from each side; then divide each side by p_1.

Skill Check ✓ **Solve the equation for y.**

4. $4x + 8y = 17$ **5.** $5x - 3y = 9$ $y = \frac{5}{3}x - 3$ **6.** $5y - 3x = 15$ $y = \frac{3}{5}x + 3$
    $y = -\frac{1}{2}x + \frac{17}{8}$

7. $\frac{3}{4}x + 5y = 20$ **8.** $xy + 2x = 8$ $y = \frac{8 - 2x}{x}$ **9.** $\frac{2}{3}x - \frac{1}{2}y = 12$
    $y = -\frac{3}{20}x + 4$ $y = \frac{4}{3}x - 24$

In Exercises 10 and 11, use the following information.
The area A of an ellipse is given by the formula $A = \pi ab$ where a and b are half the lengths of the major and minor axes. (The longer chord is the major axis.)

10. Solve the formula for a. $a = \frac{A}{\pi b}$

11. Use the result from Exercise 10 to find the length of the major axis of an ellipse whose area is 157 square inches and whose minor axis is 10 inches long. (Use 3.14 for π.)   **20 in.**

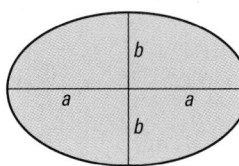

PRACTICE AND APPLICATIONS

STUDENT HELP

Extra Practice
to help you master skills is on p. 940.

EXPLORING METHODS Find the value of y for the given value of x using two methods. First, substitute the value of x into the equation and then solve for y. Second, solve for y and then substitute the value of x into the equation.

12. $4x + 9y = 30; x = 3$ **2**
13. $5x - 7y = 12; x = 1$ **−1**

14. $xy + 3x = 25; x = 5$ **2**
15. $9y - 4x = -16; x = 8$ $\frac{16}{9}$

16. $-y - 2x = -11; x = -4$ **19**
17. $-x = 3y - 55; x = 20$ $\frac{35}{3}$

18. $x = 24 + xy; x = -12$ **3**
19. $-xy + 3x = 30; x = 15$ **1**

20. $-4x + 7y + 7 = 0; x = 7$ **3**
21. $6x - 5y - 44 = 0; x = 4$ **−4**

22. $\frac{1}{2}x - \frac{4}{5}y = 19; x = 6$ **−20**
23. $\frac{3}{4}x = -\frac{9}{11}y + 12; x = 10$ $\frac{11}{2}$

REWRITING FORMULAS Solve the formula for the indicated variable.

24. Circumference of a Circle
    Solve for r: $C = 2\pi r$ $r = \frac{C}{2\pi}$

25. Volume of a Cone
    Solve for h: $V = \frac{1}{3}\pi r^2 h$ $h = \frac{3V}{\pi r^2}$

26. Area of a Triangle
    Solve for b: $A = \frac{1}{2}bh$ $b = \frac{2A}{h}$

27. Investment at Simple Interest
    Solve for P: $I = Prt$ $P = \frac{I}{rt}$

28. Celsius to Fahrenheit
    Solve for C: $F = \frac{9}{5}C + 32$
    $C = \frac{5}{9}(F - 32)$

29. Area of a Trapezoid
    Solve for b_2: $A = \frac{1}{2}(b_1 + b_2)h$
    $b_2 = \frac{2A}{h} - b_1$

STUDENT HELP

HOMEWORK HELP
Examples 1, 2: Exs. 12–23
Examples 3, 4: Exs. 33–39
Examples 5, 6: Exs. 24–32, 40–42

---

3 APPLY

ASSIGNMENT GUIDE

BASIC
Day 1: pp. 29–31 Exs. 12–20 even, 24–26, 31, 33–42
Day 2: pp. 29–32 Exs. 21–23, 32, 43, 44, 47–65 odd

AVERAGE
Day 1: pp. 29–31 Exs. 12–28 even, 30, 32–34
Day 2: pp. 29–32 Exs. 25–31 odd, 35–39, 41, 43, 44, 47–65 odd

ADVANCED
Day 1: pp. 29–31 Exs. 12–28 even, 30, 32–36
Day 2: pp. 29–32 Exs. 21–31 odd, 35–39, 41, 43, 44, 45, 47–65 odd

BLOCK SCHEDULE WITH 1.3
pp. 29–32 Exs. 12–22 even, 24–39, 41, 43, 44, 47–65 odd

EXERCISE LEVELS
Level A: *Easier*
12–21
Level B: *More Difficult*
22–40, 46–67
Level C: *Most Difficult*
41–45

✓ **HOMEWORK CHECK**
To quickly check student understanding of key concepts, go over the following exercises: Exs. 14, 20, 22, 26, 31. See also the Daily Homework Quiz:
• Blackline Master (*Chapter 1 Resource Book*, p. 66)
• Transparency (p. 6)

! **COMMON ERROR**
EXERCISES 28 AND 29
Students may multiply the terms in parentheses by the fraction as a first step in the solution. Though this will work, it is often easier to clear fractions first when a variable expression in parentheses is multiplied by a fraction.

CAREER NOTE
EXERCISES 35 AND 36
Additional information
about a career as a sports
statistician is available at
www.mcdougallittell.com.

30. $p = \dfrac{A}{2\pi w}; \dfrac{11}{2\pi}$ cm,
or about 1.75 cm

31. $h = \dfrac{S - 2\pi r^2}{2\pi r}; \dfrac{35 - 6\pi}{2\pi}$,
or about 2.57 in.

32. $r = \dfrac{P - 2x}{2\pi}; \dfrac{110}{\pi}$ yd, or
about 35.0 yd

38. 5; *R* is the total revenue
(in dollars), p_1 is sun
visor price (in dollars), *V*
is the number of sun
visors sold, p_2 is the
baseball cap price (in
dollars), and *C* is the
number of baseball caps
sold.

39. *Sample answer:* 210 sun
visors, 550 baseball
caps; 490 sun visors, 430
baseball caps; 700 sun
visors, 340 baseball caps

**FOCUS ON
CAREERS**

**SPORTS
STATISTICIANS**
are employed by many
professional sports teams,
leagues, and news organi-
zations. They collect and
analyze team and individual
data on items such as
scoring.

CAREER LINK
www.mcdougallittell.com

GEOMETRY CONNECTION In Exercises 30–32, solve the formula for the indicated variable. Then evaluate the rewritten formula for the given values. (Include units of measure in your answer.)

30. Area of a circular
ring: $A = 2\pi p w$
Solve for *p*. Find *p*
when $A = 22$ cm^2
and $w = 2$ cm.

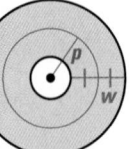

31. Surface area of a
cylinder:
$S = 2\pi rh + 2\pi r^2$
Solve for *h*. Find *h*
when $S = 105$ in.^{2}
and $r = 3$ in.

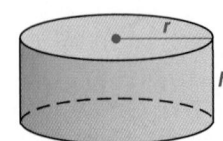

32. Perimeter of a track:
$P = 2\pi r + 2x$
Solve for *r*. Find *r* when
$P = 440$ yd and
$x = 110$ yd.

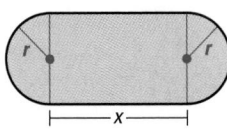

HONEYBEES In Exercises 33 and 34, use the following information.
A forager honeybee spends about three weeks becoming accustomed to the immediate surroundings of its hive and spends the rest of its life collecting pollen and nectar. The total number of miles *T* a forager honeybee flies in its lifetime *L* (in days) can be modeled by $T = m(L - 21)$ where *m* is the number of miles it flies each day.

33. Solve the equation $T = m(L - 21)$ for *L*. $L = \dfrac{T}{m} + 21$

34. A forager honeybee's flight muscles last only about 500 miles; after that the bee dies. Some forager honeybees fly about 55 miles per day. Approximately how many days do these bees live?   about 30 days

BASEBALL In Exercises 35 and 36, use the following information.
The Pythagorean Theorem of Baseball is a formula for approximating a team's ratio of wins to games played. Let *R* be the number of runs the team scores during the season, *A* be the number of runs allowed to opponents, *W* be the number of wins, and *T* be the total number of games played. Then the formula

$$\dfrac{W}{T} \approx \dfrac{R^2}{R^2 + A^2}$$

approximates the team's ratio of wins to games played. ▶ Source: *Inside Sports*

35. Solve the formula for *W*. $W \approx \dfrac{TR^2}{R^2 + A^2}$

36. The 1998 New York Yankees scored 965 runs and allowed 656. How many of its 162 games would you estimate the team won?   111 games

FUNDRAISER In Exercises 37–39, use the following information.
Your tennis team is having a fundraiser. You are going to help raise money by selling sun visors and baseball caps.

$$R = p_1 V + p_2 C$$

37. Write an equation that represents the total amount of money you raise.

38. How many variables are in the equation? What does each represent?
38, 39. See margin.

39. Your team raises a total of $4480. Give three possible combinations of sun visors and baseball caps that could have been sold if the price of a sun visor is $3.00 and the price of a baseball cap is $7.00.

40. **GEOMETRY CONNECTION** The formula for the area of a circle is $A = \pi r^2$. The formula for the circumference of a circle is $C = 2\pi r$. Write a formula for the area of a circle in terms of its circumference. $A = \dfrac{C^2}{4\pi}$

STUDENT HELP

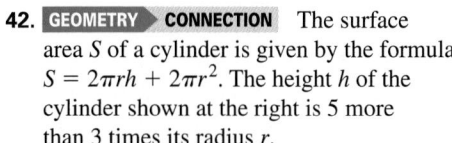

HOMEWORK HELP
Visit our Web site
www.mcdougallittell.com
for help with problem
solving in Exs. 41 and 42.

41. **GEOMETRY** **CONNECTION** The formula
for the height h of an equilateral triangle is
$h = \frac{\sqrt{3}}{2}b$ where b is the length of a side.
Write a formula for the area of an equilateral
triangle in terms of the following.

a. the length of a side only $A = \frac{\sqrt{3}}{4}b^2$

b. the height only $A = \frac{\sqrt{3}}{3}h^2$

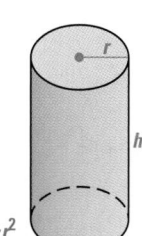

42. **GEOMETRY** **CONNECTION** The surface
area S of a cylinder is given by the formula
$S = 2\pi rh + 2\pi r^2$. The height h of the
cylinder shown at the right is 5 more
than 3 times its radius r.

a. Write a formula for the surface area of
the cylinder in terms of its radius. $S = 10\pi r + 8\pi r^2$

b. Find the surface area of the cylinder for
$r = 3, 4,$ and 6. 102π, or about 320; 168π, or about 528; 348π, or about 1090

Test
Preparation

QUANTITATIVE COMPARISON **In Exercises 43 and 44, choose the statement**
that is true about the given quantities.

Ⓐ The quantity in column A is greater.

Ⓑ The quantity in column B is greater.

Ⓒ The two quantities are equal.

Ⓓ The relationship cannot be determined from the given information.

STUDENT HELP

Skills Review
For help with the
Pythagorean theorem,
see p. 917.

Column A	Column B
43. C $V = \ell wh$	$V = \ell wh$
7 cm / 4 cm / 3 cm	5 cm / 7 cm / 3 cm
44. B $V = \pi r^2 h$	$V = \pi r^2 h$
4 in. / 6 in.	6 in. / 4 in.

1.4 *Rewriting Equations and Formulas*   **31**

STUDENT HELP NOTES

Homework Help Students can
find help for Exs. 41–42 at
www.mcdougallittell.com.
The information can be printed
out for students who don't have
access to the Internet.

ADDITIONAL PRACTICE
AND RETEACHING

For Lesson 1.4:

• Practice Levels A, B, and C
(*Chapter 1 Resource Book,* p. 56)

• Reteaching with Practice
(*Chapter 1 Resource Book,* p. 59)

• ⊡ See Lesson 1.4 of the
Personal Student Tutor

For more Mixed Review:

• ⊡ Search the *Test and Practice
Generator* for key words or
specific lessons.

DAILY HOMEWORK QUIZ

🖳 *Transparency Available*

Find y by substituting the value of x into the equation and solving for y.

1. $4x - 2xy = 8$; $x = 4$ **1**

2. $3x - 5y - 11 = 0$; $x = -3$ **-4**

Solve for y. Then find the value of y for the given value of x.

3. $-2x + 5y + 4 = 0$; $x = -3$

$y = \frac{2x - 4}{5}$; -2

4. $\frac{2}{5}x = -\frac{4}{3}y + \frac{1}{2}$; $x = 5$

$y = \frac{15 - 12x}{40}$; $-\frac{9}{8}$

5. The surface area of a rectangular prism is given by the formula $S = 2(hw + lh + lw)$. Solve the formula for w. Then evaluate the formula for $l = 4$ cm, $h = 7$ cm, and $S = 188$ cm^2.

$w = \frac{S - 2lh}{2(l + h)}$; 6 cm

EXTRA CHALLENGE NOTE

→ Challenge problems for Lesson 1.4 are available in **blackline** format in the *Chapter 1 Resource Book*, p. 63 and at **www.mcdougallittell.com**.

ADDITIONAL TEST PREPARATION

1. WRITING Describe how to solve the formula $d = v_0 t + \frac{1}{2}at^2$ for a. *Sample answer:* First subtract $v_0 t$ from both sides. Then multiply both sides by 2 to clear the fraction. Then divide both sides by t^2 to isolate a.

2. OPEN ENDED Describe a real-life situation where you would need to be able to solve the area formula of a circle for r. *Sample answer:* A load of mulch is meant to cover 300 ft^2 to a depth of 3 in. How wide a circular garden will it cover?

★ **Challenge**

45. 🌐 **FUEL EFFICIENCY** The more aerodynamic a vehicle is, the less fuel the vehicle's engine must use to overcome air resistance. To design vehicles that are as fuel efficient as possible, automotive engineers use the formula

$$R = 0.00256 \times D_C \times F_A \times s^2$$

where R is the air resistance (in pounds), D_C is the drag coefficient, F_A is the frontal area of the vehicle (in square feet), and s is the speed of the vehicle (in miles per hour). The formula assumes that there is no wind.

a. Rewrite the formula to find the drag coefficient in terms of the other variables.

EXTRA CHALLENGE
⮡ www.mcdougallittell.com

b. Find the drag coefficient of a car when the air resistance is 50 pounds, the frontal area is 25 square feet, and the speed of the car is 45 miles per hour.

a. $D_C = \dfrac{R}{0.00256 \times F_A \times s^2}$ **b.** about 0.386

MIXED REVIEW

WRITING EXPRESSIONS Write an expression to answer the question. (Skills Review, p. 929)

46. You buy x birthday cards for $1.85 each. How much do you spend?   **1.85x**

47. You have $30 and spend x dollars. How much money do you have left?   **30 − x**

48. You drive 55 miles per hour for x hours. How many miles do you drive?   **55x**

49. You have $250 in your bank account and you deposit x dollars. How much money do you now have in your account?   **250 + x**

50. You spend $42 on x music cassettes. How much does each cassette cost? $\frac{42}{x}$

51. A certain ball bearing weighs 2 ounces. A box contains x ball bearings. What is the total weight of the ball bearings?   **2x**

UNIT ANALYSIS Give the answer with the appropriate unit of measure. (Review 1.1)

52. $\left(\dfrac{7 \text{ meters}}{1 \text{ minute}}\right)(60 \text{ minutes})$ **420 m**

53. $\left(\dfrac{168 \text{ hours}}{1 \text{ week}}\right)(52 \text{ weeks})$ **8736 h**

54. $4\frac{1}{4}$ feet $+ 7\frac{3}{4}$ feet   **12 ft**

55. $13\frac{1}{4}$ liters $- 8\frac{7}{8}$ liters $4\frac{3}{8}$ **L**

56. $\left(\dfrac{3 \text{ yards}}{1 \text{ second}}\right)(12 \text{ seconds}) - 10 \text{ yards}$ **26 yd**

57. $\left(\dfrac{15 \text{ dollars}}{1 \text{ hour}}\right)(8 \text{ hours}) + 45 \text{ dollars}$ **$165**

SOLVING EQUATIONS Solve the equation. Check your solution. (Review 1.3)

58. $3d + 16 = d - 4$ **−10**

59. $5 - x = 23 + 2x$ **−6**

60. $10(y - 1) = y + 4$ $\frac{14}{9}$

61. $p - 16 + 4 = 4(2 - p)$ **4**

62. $-10x = 5x + 5$ $-\frac{1}{3}$

63. $12z = 4z - 56$ **−7**

64. $\frac{2}{3}x - 7 = 1$ **12**

65. $-\frac{3}{4}x + 19 = -11$ **40**

66. $\frac{1}{4}x + \frac{3}{8} = \frac{1}{5} - \frac{1}{5}x$ $-\frac{7}{18}$

67. $\frac{5}{4}x - \frac{3}{4} = \frac{5}{6}x + \frac{1}{2}$ **3**

1.5

Problem Solving Using Algebraic Models

 **GOAL 1** USING A PROBLEM SOLVING PLAN

What you should learn

GOAL 1 Use a general problem solving plan to solve **real-life** problems, as in **Example 2**.

GOAL 2 Use other problem solving strategies to help solve **real-life** problems, as in **Ex. 22**.

Why you should learn it

▼ To solve **real-life** problems, such as finding the average speed of the Japanese Bullet Train in **Example 1**.

One of your major goals in this course is to learn how to use algebra to solve real-life problems. You have solved simple problems in previous lessons, and this lesson will provide you with more experience in problem solving.

As you have seen, it is helpful when solving real-life problems to first write an equation in words *before* you write it in mathematical symbols. This word equation is called a **verbal model**. The verbal model is then used to write a mathematical statement, which is called an **algebraic model**. The key steps in this problem solving plan are shown below.

| Write a verbal model. | → | Assign labels. | → | Write an algebraic model. | → | Solve the algebraic model. | → | Answer the question. |

EXAMPLE 1 *Writing and Using a Formula*

The Bullet Train runs between the Japanese cities of Osaka and Fukuoka, a distance of 550 kilometers. When it makes no stops, it takes 2 hours and 15 minutes to make the trip. What is the average speed of the Bullet Train?

SOLUTION

You can use the formula $d = rt$ to write a verbal model.

VERBAL MODEL
$$\boxed{\text{Distance}} = \boxed{\text{Rate}} \cdot \boxed{\text{Time}}$$

LABELS
Distance = **550**   (kilometers)

Rate = **r**   (kilometers per hour)

Time = **2.25**   (hours)

ALGEBRAIC MODEL
$550 = \boxed{r}\,(2.25)$ Write algebraic model.

$\dfrac{550}{2.25} = r$ Divide each side by 2.25.

$244 \approx r$ Use a calculator.

▶ The Bullet Train's average speed is about 244 kilometers per hour.

UNIT ANALYSIS You can use unit analysis to check your verbal model.

$$550 \text{ kilometers} \approx \frac{244 \text{ kilometers}}{\text{hour}} \cdot 2.25 \text{ hours}$$

1.5 *Problem Solving Using Algebraic Models*   **33**

---

1 ***PLAN***

PACING
Basic: 1 day
Average: 1 day
Advanced: 1 day
Block Schedule: 0.5 block with 1.6

 LESSON OPENER
APPLICATION
An alternative way to approach Lesson 1.5 is to use the Application Lesson Opener:

• Blackline Master (*Chapter 1 Resource Book,* p. 67)
• Transparency (p. 5)

MEETING INDIVIDUAL NEEDS
• *Chapter 1 Resource Book*
  Prerequisite Skills Review (p. 5)
  Practice Level A (p. 68)
  Practice Level B (p. 69)
  Practice Level C (p. 70)
  Reteaching with Practice (p. 71)
  Absent Student Catch-Up (p.73)
  Challenge (p. 76)
• *Resources in Spanish*
• *Personal Student Tutor*

NEW-TEACHER SUPPORT
See the Tips for New Teachers on pp. 1–2 of the *Chapter 1 Resource Book* for additional notes about Lesson 1.5.

WARM-UP EXERCISES

Transparency Available

Solve each equation for t.

1. $I = prt$ $t = \dfrac{I}{pr}$

2. $d = rt$ $t = \dfrac{d}{r}$

Complete the unit analysis.

3. 25 feet per second to miles per hour
about 17 miles per hour

4. $4.50 for 2.5 pounds to dollars per ounce.
about $.11 per ounce

MOTIVATING THE LESSON

MOTIVATING THE LESSON
Ask students if any of them have baked cookies from scratch using a recipe. Point out that the recipe is a verbal model that orders the cooking process. Emphasize that verbal models are important in math also to understand a problem clearly and to order it. With a good "recipe," we can then translate the verbal model into the more precise language of algebra.

EXTRA EXAMPLE 1
On August 15, 1995 the Concorde flew 35,035 mi from New York City to New York City in 31 h 27 min. What was the average speed? **about 1114 mi/h**

EXTRA EXAMPLE 2
A shower head advertises a maximum flow rate of 2.5 gal/min. Find the flow rate if it fills a 22 gal bathtub in 9.5 min. Is this within the advertised limit?
about 2.3 gal/min; yes

EXTRA EXAMPLE 3
You drove 280 mi, using 15 gal of gasoline that cost $1.15 per gallon. If you get 24 mi/gal on the highway and 16 in the city, how much did you spend for fuel for highway driving and how much for city driving? **$5.75, $11.50**

 CHECKPOINT EXERCISES

For use after Example 1:

1. If the Concorde flies at a rate of 1114 mi/h, how long will it take it to fly 3469 mi from New York City to London?
about 3.11 h, or 3 h 7 min

For use after Examples 2 and 3:

2. On a weekend trip, you average 35 mi/h in the city and 60 mi/h on the highway. If the 150 mi trip took 3 h, how much of the time were you driving in the city? **1.2 h, or 1 h 12 min**

Water Conservation

EXAMPLE 2   *Writing and Using a Simple Model*

A water-saving faucet has a flow rate of at most 9.6 cubic inches per second. To test whether your faucet meets this standard, you time how long it takes the faucet to fill a 470 cubic inch pot, obtaining a time of 35 seconds. Find your faucet's flow rate. Does it meet the standard for water conservation?

SOLUTION

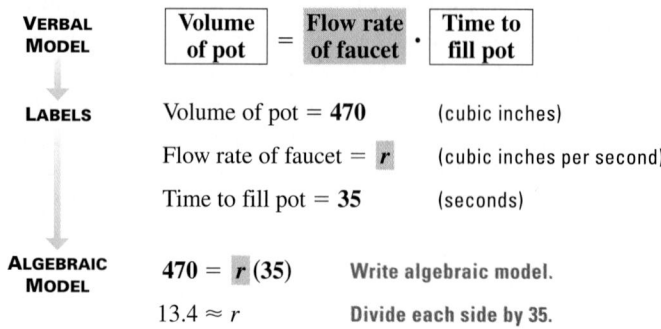

PROBLEM SOLVING STRATEGY

| VERBAL MODEL | $\boxed{\text{Volume of pot}} = \boxed{\text{Flow rate of faucet}} \cdot \boxed{\text{Time to fill pot}}$ |

LABELS   Volume of pot $= 470$ (cubic inches)

Flow rate of faucet $= r$ (cubic inches per second)

Time to fill pot $= 35$ (seconds)

ALGEBRAIC MODEL $470 = r\,(35)$ Write algebraic model.

$13.4 \approx r$ Divide each side by 35.

▶ The flow rate is about 13.4 in.^{3}/sec, which does not meet the standard.

Gasoline Cost

EXAMPLE 3   *Writing and Using a Model*

You own a lawn care business. You want to know how much money you spend on gasoline to travel to out-of-town clients. In a typical week you drive 600 miles and use 40 gallons of gasoline. Gasoline costs $1.25 per gallon, and your truck's fuel efficiency is 21 miles per gallon on the highway and 13 miles per gallon in town.

SOLUTION

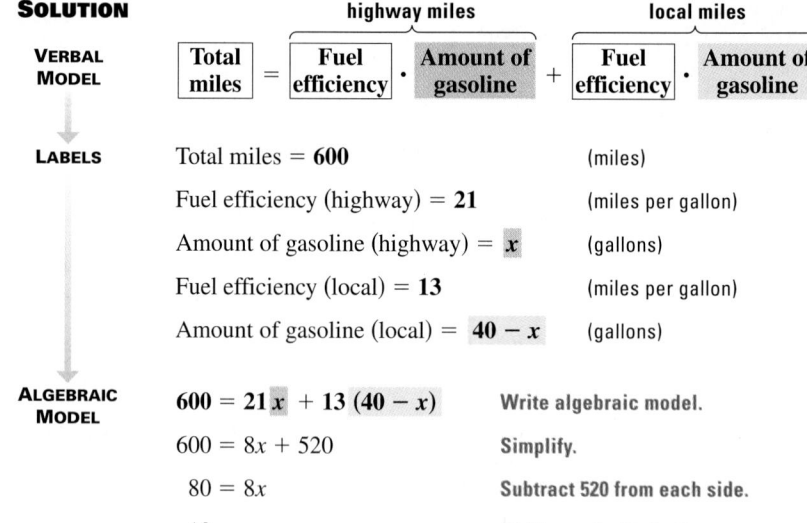

highway miles   local miles

| VERBAL MODEL | $\boxed{\text{Total miles}} = \overbrace{\boxed{\text{Fuel efficiency}} \cdot \boxed{\text{Amount of gasoline}}}^{\text{highway miles}} + \overbrace{\boxed{\text{Fuel efficiency}} \cdot \boxed{\text{Amount of gasoline}}}^{\text{local miles}}$ |

LABELS   Total miles $= 600$ (miles)

Fuel efficiency (highway) $= 21$ (miles per gallon)

Amount of gasoline (highway) $= x$ (gallons)

Fuel efficiency (local) $= 13$ (miles per gallon)

Amount of gasoline (local) $= 40 - x$ (gallons)

ALGEBRAIC MODEL $600 = 21x + 13\,(40 - x)$ Write algebraic model.

$600 = 8x + 520$ Simplify.

$80 = 8x$ Subtract 520 from each side.

$10 = x$ Divide each side by 8.

▶ In a typical week you use 10 gallons of gasoline to travel to out-of-town clients. The cost of the gasoline is (10 gallons)($1.25 per gallon) = $12.50.

┌─ **STUDENT HELP** ─┐

Study Tip
The solutions of the equations in Examples 2 and 3 are 13.4 and 10, respectively. However, these are not the answers to the questions asked. In Example 2 you must compare 13.4 to 9.6, and in Example 3 you must multiply 10 by $1.25. Be certain to answer the question asked.

RAILROADS In 1862, two companies were given the rights to build a railroad from Omaha, Nebraska to Sacramento, California. The Central Pacific Company began from Sacramento in 1863. Twenty-four months later, the Union Pacific Company began from Omaha.

APPLICATION LINK
www.mcdougallittell.com

GOAL 2 USING OTHER PROBLEM SOLVING STRATEGIES

When you are writing a verbal model to represent a real-life problem, remember that you can use other problem solving strategies, such as *draw a diagram, look for a pattern*, or *guess, check, and revise*, to help create the verbal model.

EXAMPLE 4 *Drawing a Diagram*

RAILROADS Use the information under the photo at the left. The Central Pacific Company averaged 8.75 miles of track per month. The Union Pacific Company averaged 20 miles of track per month. The photo shows the two companies meeting in Promontory, Utah, as the 1590 miles of track were completed. When was the photo taken? How many miles of track did each company build?

SOLUTION

Begin by drawing and labeling a diagram, as shown below.

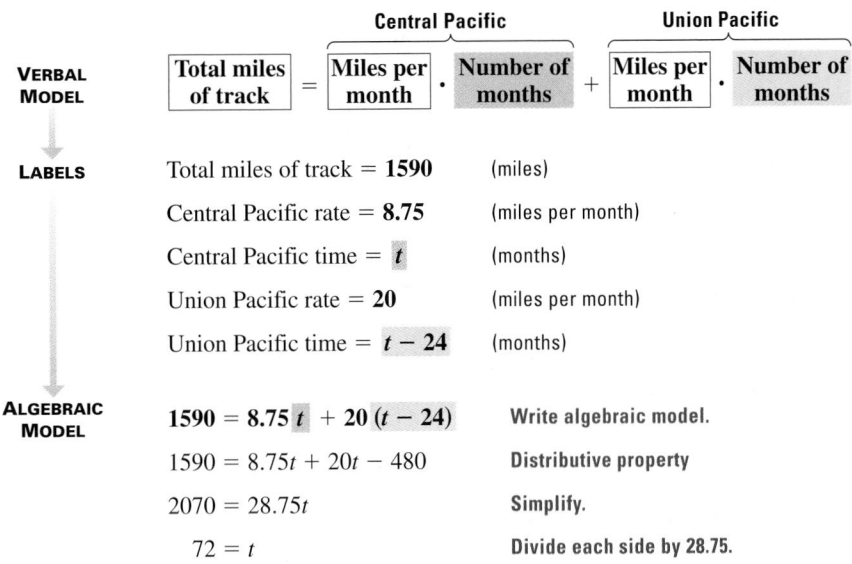

VERBAL MODEL

| Total miles of track | = | Miles per month | · | Number of months | + | Miles per month | · | Number of months |

Central Pacific ────────   Union Pacific ────────

LABELS

Total miles of track = **1590** (miles)

Central Pacific rate = **8.75** (miles per month)

Central Pacific time = **t** (months)

Union Pacific rate = **20** (miles per month)

Union Pacific time = **$t - 24$** (months)

ALGEBRAIC MODEL

$1590 = 8.75\,t + 20\,(t - 24)$ Write algebraic model.

$1590 = 8.75t + 20t - 480$ Distributive property

$2070 = 28.75t$ Simplify.

$72 = t$ Divide each side by 28.75.

▶ The construction took 72 months (6 years) from the time the Central Pacific Company began in 1863. So, the photo was taken in 1869. The number of miles of track built by each company is as follows.

Central Pacific: $\dfrac{8.75 \text{ miles}}{\text{month}} \cdot 72 \text{ months} = 630$ miles

Union Pacific: $\dfrac{20 \text{ miles}}{\text{month}} \cdot (72 - 24) \text{ months} = 960$ miles

STUDENT HELP

Skills Review For help with additional problem solving strategies, see p. 930.

1.5 Problem Solving Using Algebraic Models **35**

EXTRA EXAMPLE 4
A fire truck is called to a scene. Three minutes later, a second truck is called. The first truck averages only 30 mi/h, but the second averages 60 mi/h. The trucks travel a total of 12 mi and arrive at the same time. How long from the first call did the trucks take to arrive? How far did each travel? **10 min; first: 5 mi, second: 7 mi**

CHECKPOINT EXERCISES
For use after Example 4:
1. In Extra Example 4, how far did each truck travel if the second truck responded 5 min after the first, averaging 72 mi/h to the first truck's 36 mi/h? Did the time change? **first: 6 mi, second: 6 mi; no**

COMMON ERROR
EXAMPLE 4 Students may become overwhelmed by complex distance formula problems such as this. Students need to look first at the general picture. In this case, it is that the total distance is the sum of the distances by each company. In other cases, it may be that the total time is the sum of two different times, that two distances or times are equal, and so on. Also, students should realize that they could have labeled the times as t for Union Pacific and $t + 24$ for Central Pacific. In this case, they would find the meeting date from the Union Pacific start date.

APPLICATION NOTE
EXAMPLE 4 Additional information about railroads is available at **www.mcdougallittell.com**.

STUDENT HELP NOTES
Skills Review As students review problem solving strategies on p. 930, encourage them to experiment with different strategies.

EXTRA EXAMPLE 5
The table gives the heights from the floor to the first few steps of a flight of stairs. Determine the height of the 14th step.

Step	Landing	1	2	3
Height (inches)	4	12	20	28

116 in, or 9 ft 8 in.

EXTRA EXAMPLE 6
A store sells spherical helium-filled balloons. Each balloon contains 590 in.^{3} of helium. What is the radius of a balloon?
about 5.2 in.

 CHECKPOINT EXERCISES
For use after Example 5:
1. The table gives the heights above the stage of the seats in a concert hall balcony. How high above the stage are the seats in row RR?

Row	Height (ft)
AA	18
BB	19.25
CC	20.5
DD	21.75
EE	23

39.25 ft

For use after Example 6:
2. You are making a circular target for a parachute landing out of 64 yd^2 of material. How large can the radius of the target be? about 4.5 yd

FOCUS ON VOCABULARY
What is the relationship between a verbal model and an algebraic model? See sample answer at right.

CLOSURE QUESTION
After you have set up and solved an algebraic model for a problem description, what remains to be done? See sample answer at right.

Heights

EXAMPLE 5 *Looking for a Pattern*

The table gives the heights to the top of the first few stories of a tall building. Determine the height to the top of the 15th story.

Story	Lobby	1	2	3	4
Height to top of story (feet)	20	32	44	56	68

SOLUTION
Look at the differences in the heights given in the table. After the lobby, the height increases by 12 feet per story.

Heights: 20   32   44   56   68
          +12 +12 +12 +12

You can use the observed pattern to write a model for the height.

PROBLEM SOLVING STRATEGY

VERBAL MODEL

Height to top of a story	=	Height of lobby	+	Height per story	·	Story number

LABELS
Height to top of a story = h (feet)

Height of lobby = **20**   (feet)

Height per story = **12**   (feet per story)

Story number = n (stories)

ALGEBRAIC MODEL
$h = 20 + 12\,n$ Write algebraic model.

$= 20 + 12(15)$ Substitute 15 for n.

$= 200$ Simplify.

▶ The height to the top of the 15th story is 200 feet.

EXAMPLE 6 *Guess, Check, and Revise*

WEATHER BALLOONS A spherical weather balloon needs to hold 175 cubic feet of helium to be buoyant enough to lift an instrument package to a desired height. To the nearest tenth of an foot, what is the radius of the balloon?

SOLUTION
Use the formula for the volume of a sphere, $V = \frac{4}{3}\pi r^3$.

$175 = \frac{4}{3}\pi r^3$ Substitute 175 for V.

$42 \approx r^3$ Divide each side by $\frac{4}{3}\pi$.

You need to find a number whose cube is 42. As a first guess, try $r = 4$. This gives $4^3 = 64$. Because $64 > 42$, your guess of 4 is too high. As a second guess, try $r = 3.5$. This gives $(3.5)^3 = 42.875$, and $42.875 \approx 42$. So, the balloon's radius is about 3.5 feet.

WEATHER BALLOONS
Hundreds of weather balloons are launched daily from weather stations. The balloons typically carry about 40 pounds of instruments. Balloons usually reach an altitude of about 90,000 feet.

FOCUS ON APPLICATIONS

Focus on Vocabulary *Sample answer:*
A verbal model sets up an equation for a problem using words. These words are then given labels using numbers and variables and written in mathematical language as an algebraic model.

Closure Question *Sample answer:*
Check that the solution answers the question asked. If it doesn't, use the solution to derive the answer.

GUIDED PRACTICE

Vocabulary Check ✓
Concept Check ✓
Skill Check ✓

1. What is a verbal model? What is it used for?   See margin.

2. Describe the steps of the problem solving plan.   See margin.

3. How does this diagram help you set up the algebraic model in Example 3?
   See margin.

```
              40 gallons
  ├──────────────────────────────────────┤
  ├─ x gal ─┼──────── (40 − x) gal ────────┤
    highway               local
```

1. an equation written in words; to write an algebraic model by translating the words into mathematical symbols

2. See margin.

3. The diagram helps you see how to express the numbers of gallons used in town in terms of x, the label given to the number of gallons used on the highway.

SCIENCE CONNECTION   In Exercises 4–7, use the following information.

To study life in Arctic waters, scientists worked in an underwater building called a *Sub-Igloo* in Resolute Bay, Canada. The water pressure at the floor of the Sub-Igloo was 2184 pounds per square foot. Water pressure is zero at the water's surface and increases by 62.4 pounds per square foot for each foot of depth.

4. Write a verbal model for the water pressure.
   water pressure = (pressure per ft of depth)(depth)

5. Assign labels to the parts of the verbal model. Indicate the units of measure.
   water pressure = 2184 (lb/ft^2); pressure per ft of depth = 62.4 (lb/ft^2 per ft); depth = d (ft)

6. Use the labels to translate the verbal model into an algebraic model.
   $2184 = 62.4d$

7. Solve the algebraic model to find the depth of the Sub-Igloo's floor.   35 ft

PRACTICE AND APPLICATIONS

STUDENT HELP

→ **Extra Practice**
to help you master skills is on p. 940.

🌐 **BOAT TRIP** In Exercises 8–11, use the following information.

You are on a boat on the Seine River in France. The boat's speed is 32 kilometers per hour. The Seine has a length of 764 kilometers, but only 547 kilometers can be navigated by boats. How long will your boat ride take if you travel the entire navigable portion of the Seine? Use the following verbal model.

$$\boxed{\text{Distance}} = \boxed{\text{Rate}} \cdot \boxed{\text{Time}}$$

12. metronomic marking = 80 (beats/min), length of musical piece = t (min), number of measures in musical piece = 180 (measures), number of beats per measure = 3 (beats/measure)

8. Assign labels to the parts of the verbal model.
   distance = 547 (km), rate = 32 (km/h), time = t (h)

9. Use the labels to translate the verbal model into an algebraic model. $547 = 32t$

10. Solve the algebraic model. $\frac{547}{32}$, or about 17.1

11. Answer the question.   about 17 h

🌐 **MUSIC** In Exercises 12–14, use the following information.

A *metronome* is a device similar to a clock and is used to maintain the tempo of a musical piece. Suppose one particular piece has 180 measures with 3 beats per measure and a metronome marking of 80 beats per minute. Determine the length (in minutes) of the musical piece by using the following verbal model.

Metronome marking	·	Length of musical piece	=	Number of measures in musical piece	·	Number of beats per measure

STUDENT HELP

→ HOMEWORK HELP
Examples 1–3: Exs. 8–17
Examples 4–6: Exs. 18–27

12. Assign labels to the parts of the verbal model.   See margin.

13. Use the labels to translate the verbal model into an algebraic model. $80t = (180)(3)$

14. Answer the question. Use unit analysis to check your answer.   6 min 45 sec

○ **ASSIGNMENT GUIDE**

BASIC
Day 1: pp. 37–39 Exs. 8–18, 28, 29, 31–41 odd, Quiz 2 Exs. 1–8

AVERAGE
Day 1: pp. 37–39 Exs. 8–21, 28, 29, 31–41 odd, Quiz 2 Exs. 1–8

ADVANCED
Day 1: pp. 37–39 Exs. 8–21, 24–30, 31–41 odd, Quiz 2 Exs. 1–8

BLOCK SCHEDULE WITH 1.6
pp. 37–39 Exs. 8–21, 28, 29, 31–41 odd, Quiz 2 Exs. 1–8

EXERCISE LEVELS
Level A: *Easier*
8–18, 31–38

Level B: *More Difficult*
19–29, 39–42

Level C: *Most Difficult*
30

✓ **HOMEWORK CHECK**
To quickly check student understanding of key concepts, go over the following exercises: Exs. 8, 10, 14, 15, 18. See also the Daily Homework Quiz:

• Blackline Master (*Chapter 1 Resource Book*, p. 80)
• 🖨 Transparency (p. 7)

2. A good answer should include all of these points:
• Write a verbal model (a word equation).
• Assign labels to all of the quantities involved in the verbal model.
• Use the labels to translate the verbal model into an algebraic model (an algebraic equation).
• Solve the equation.
• Answer the question.

APPLICATION NOTE
EXERCISE 19 Additional information about the Chunnel is available at **www.mcdougallittell.com.**

STUDENT HELP NOTES

Homework Help Students can find help for Ex. 24 at **www.mcdougallittell.com.** The information can be printed out for students who don't have access to the Internet.

ADDITIONAL PRACTICE AND RETEACHING

For Lesson 1.5:
- Practice Levels A, B, and C (*Chapter 1 Resource Book,* p. 68)
- Reteaching with Practice (*Chapter 1 Resource Book,* p. 71)
- ⬛ See Lesson 1.5 of the *Personal Student Tutor*

For more Mixed Review:
- ⬛ Search the *Test and Practice Generator* for key words or specific lessons.

38

FOCUS ON APPLICATIONS

THE CHUNNEL
A tunnel under the English Channel was an engineering possibility for over a century before its completion. High speed trains traveling up to 300 kilometers per hour link London, England to Paris, France and Brussels, Belgium.

APPLICATION LINK
www.mcdougallittell.com

---

15. total calories = (calories/gram of fat)(number of grams of fat) + (calories/gram of protein)(number of grams of protein) + (calories/gram of carbohydrate)(number of grams of carbohydrate)

16. total calories = T, calories/gram of fat = 9, number of grams of fat = f, calories/gram of protein = 4, number of grams of protein = p, calories/gram of carbohydrate = 4, number of grams of carbohydrate = c; $T = 9f + 4p + 4c$

STUDENT HELP

HOMEWORK HELP
Visit our Web site www.mcdougallittell.com for help with problem solving in Ex. 24.

CALORIE INTAKE **In Exercises 15–17, use the following information.**
To determine the total number of calories of a food, you must add the number of calories provided by the grams of fat, the grams of protein, and the grams of carbohydrates. There are 9 Calories per gram of fat. A gram of protein and a gram of carbohydrates each have about 4 Calories. ▶ Source: U.S. Department of Agriculture
15, 16. See margin.

15. Write a verbal model that gives the total number of calories of a certain food.

16. Assign labels to the parts of the verbal model. Use the labels to translate the verbal model into an algebraic model.

17. One cup of raisins has 529.9 Calories and contains 0.3 gram of fat and 127.7 grams of carbohydrates. Solve the algebraic model to find the number of grams of protein in the raisins. Use unit analysis to check your answer. **4.1 g**

18. **BORROWING MONEY** You have borrowed $529 from your parents to buy a mountain bike. Your parents are not charging you interest, but they want to be repaid as soon as possible. You can afford to repay them $20 per week. How long will it take you to repay your parents? **27 weeks**

19. **THE CHUNNEL** The Chunnel connects the United Kingdom and France by a railway tunnel under the English Channel. The British started tunneling 2.5 months before the French and averaged 0.63 kilometer per month. The French averaged 0.47 kilometer per month. When the two sides met, they had tunneled 37.9 kilometers. How many kilometers of tunnel did each country build? If the French started tunneling on February 28, 1988, approximately when did the two sides meet? **Great Britain: 22.4 km, France: 15.5 km; Dec. 1, 1990**

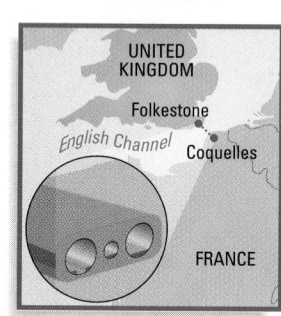

20. **FLYING LESSONS** You are taking flying lessons to get a private pilot's license. The cost of the introductory lesson is $\frac{5}{8}$ the cost of each additional lesson, which is $80. You have a total of $375 to spend on the flying lessons. How many lessons can you afford? How much money will you have left? **5 lessons, including the introductory lesson; $5**

21. **TYPING PAPERS** Some of your classmates ask you to type their history papers throughout a 7 week summer course. How much should you charge per page if you want to earn enough to pay for the flying lessons in Exercise 20 and have $75 left over for spending money? You estimate that you can type 40 pages per week. Assume that you have to take 9 flying lessons plus the introductory lesson and that you already have $375 to spend on the lessons. **$1.68 per page**

22. **WOODSHOP** You are working on a project in woodshop. You have a wooden rod that is 72 inches long. You need to cut the rod so that one piece is 6 inches longer than the other piece. How long should each piece be? **33 in., 39 in.**

23. **GARDENING** You have 480 feet of fencing to enclose a rectangular garden. You want the length of the garden to be 30 feet greater than the width. Find the length and width of the garden if you use all of the fencing. **length: 135 ft, width: 105 ft**

24. **WINDOW DISPLAYS** You are creating a window display at a toy store using wooden blocks. The display involves stacking blocks in triangular forms. You begin the display with 1 block, which is your first "triangle," and then stack 3 blocks, two on the bottom and one on the top, to get the next triangle. You create the next three triangles by stacking 6 blocks, then 10 blocks, and then 15 blocks. How many blocks will you need for the ninth triangle? **45**

SCIENCE CONNECTION In Exercises 25–27, use the following information.

As part of a science experiment, you drop a ball from various heights and measure how high it bounces on the first bounce. The results of six drops are given below.

Drop height (m)	0.5	1.5	2	2.5	4	5
First bounce height (m)	0.38	1.15	1.44	1.90	2.88	3.85

25. How high will the ball bounce if you drop it from a height of 6 meters? **4.5 m**

26. To continue the experiment, you must find the number of bounces the ball will make before it bounces less than a given number of meters. Your experiment shows the ball's bounce height is always the same percent of the height from which it fell before the bounce. Find the average percent that the ball bounces each time. **75%**

27. Find the number of times the ball bounces before it bounces less than 1 meter if it is dropped from a height of 3 meters. **4 bounces**

Test Preparation

28. **MULTIPLE CHOICE** You work at a clothing store earning $7.50 per hour. At the end of the year, you figure out that on a weekly basis you averaged 5 hours of overtime for which you were paid time and a half. How much did you make for the entire year? (Assume that a regular workweek is 40 hours.) **D**

   (A) $15,600 (B) $17,550 (C) $18,000 (D) $18,525 (E) $19,500

29. **MULTIPLE CHOICE** You are taking piano lessons. The cost of the first lesson is one and one half times the cost of each additional lesson. You spend $260 for six lessons. How much did the first lesson cost? **D**

   (A) $52 (B) $40 (C) $43.33 (D) $60 (E) $34.67

★ Challenge

30. 🌐 **OWNING A BUSINESS** You have started a business making papier-mâché sculptures. The cost to make a sculpture is $.75. Your sculptures sell for $14.50 each at a craft store. You receive 50% of the selling price. Each sculpture takes about 2 hours to complete. If you spend 14 hours per week making sculptures, about how many weeks will you work to earn a profit of $360? **8 weeks**

EXTRA CHALLENGE
↳ www.mcdougallittell.com

MIXED REVIEW

LOGICAL REASONING Tell whether the compound statement is *true* or *false*. (Skills Review, p. 924)

31. $-3 < 5$ and $-3 > -5$ **true**

32. $1 > -2$ or $1 < -2$ **true**

33. $-4 > -5$ and $1 < -2$ **false**

34. $-2.7 > -2.5$ or $156 > 165$ **false**

ORDERING NUMBERS Write the numbers in increasing order. (Review 1.1)

35. $-1, -5, 4, -10, -55$
   $-55, -10, -5, -1, 4$

36. $-\frac{2}{3}, \frac{5}{8}, \frac{1}{100}, -2, 1$ $-2, -\frac{2}{3}, \frac{1}{100}, \frac{5}{8}, 1$

37. $-1.2, 2, -2.9, 2.09, -2.1$
   $-2.9, -2.1, -1.2, 2, 2.09$

38. $-\sqrt{3}, 1, \sqrt{10}, \sqrt{2}, \frac{8}{5}$ $-\sqrt{3}, 1, \sqrt{2}, \frac{8}{5}, \sqrt{10}$

SOLVING EQUATIONS Solve the equation. Check your solution. (Review 1.3 for 1.6)

39. $6x + 5 = 17$ **2**

40. $5x - 4 = 7x + 12$ **−8**

41. $2(3x - 1) = 5 - (x + 3)$ $\frac{4}{7}$

42. $\frac{2}{3}x + \frac{1}{4} = 2x - \frac{5}{6}$ $\frac{13}{16}$

---

4 ASSESS

DAILY HOMEWORK QUIZ

📄 *Transparency Available*

Cyclists A and B each average 30 km/h for the first hour of a 100 km race. At the end of the hour B has a mishap, and loses 12 min. Cyclist A finishes the remainder of the race at an average rate of 25 km/h. If cyclist B averages 27.5 km/h after resuming, will B catch A before the finish line?

1. Write a verbal model for the distance at which A and B will again be even.
   (rate for A)(time for A) = (rate for B)(time for B)

2. Assign labels to the verbal model. rate for A = 25 km/h, time for A = t h, rate for B = 27.5 km/h, time for B = $t - 0.2$ h

3. Use the labels to translate the verbal model into an algebraic model. $25t = 27.5(t - 0.2)$

4. Solve the algebraic model. $t = 2.2$ h

5. Answer the question. Explain. Yes; 2.2 h after the mishap, the cyclists will be 25(2.2) = 27.5(2.0) = 55 km farther along. Since the race had 70 km remaining, B will catch A.

EXTRA CHALLENGE NOTE
↳ Challenge problems for Lesson 1.5 are available in **blackline** format in the *Chapter 1 Resource Book*, p. 76 and at **www.mcdougallittell.com**.

ADDITIONAL TEST PREPARATION

1. **WRITING** Describe the five-step problem solving plan. See page 33.

Solve the equation. Check your solution. (Lesson 1.3)

1. $5x - 9 = 11$ **4**

2. $6y + 8 = 3y - 16$ **−8**

3. $\frac{1}{4}z + \frac{2}{3} = \frac{1}{2}z - \frac{3}{4}$ **$\frac{17}{3}$**

4. $0.4(x - 50) = 0.2x + 12$ **160**

Solve the equation for y. Then find the value of y when x = 2. (Lesson 1.4)

5. $3x + 5y = 9$ **$y = -\frac{3}{5}x + \frac{9}{5}; \frac{3}{5}$**

6. $4x - 3y = 14$ **$y = \frac{4}{3}x - \frac{14}{3}; -2$**

7. The formula for the area of a rhombus is $A = \frac{1}{2}d_1d_2$ where d_1 and d_2 are the lengths of the diagonals. Solve the formula for d_1. (Lesson 1.4)   **$d_1 = \frac{2A}{d_2}$**

8.  **GIRL SCOUT COOKIES** Your sister is selling Girl Scout cookies that cost $2.80 per box. Your family bought 6 boxes. How many more boxes of cookies must your sister sell in order to collect $154? (Lesson 1.5)   **49 boxes**

MATH & History — Problem Solving

APPLICATION LINK
www.mcdougallittell.com

THEN

MANY CULTURES, such as the Egyptians, Greeks, Hindus, and Arabs, solved problems by using the *rule of false position*. This technique was similar to the problem solving strategy of guess, check, and revise. As an example of how to use the rule of false position, consider this problem taken from the Ahmes papyrus:

You want to divide 700 loaves of bread among four people in the ratio $\frac{2}{3}:\frac{1}{2}:\frac{1}{3}:\frac{1}{4}$. Choose a number divisible by the denominators 2, 3, and 4, such as 48. Then evaluate

$\frac{2}{3}(48) + \frac{1}{2}(48) + \frac{1}{3}(48) + \frac{1}{4}(48)$, *which has a value of 84.*

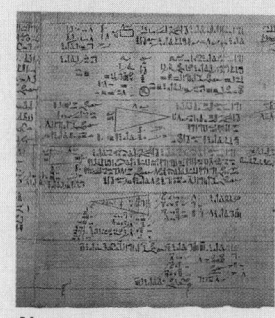
Ahmes papyrus

1. The next step is to multiply 48 by a number so that when the resulting product is substituted for 48 in the expression above, you get a new expression whose value is 700. By what number should you multiply 48? How did you use the original expression's value of 84 to get your answer?   **$8\frac{1}{3}$; Divide 84 into 700.**

2. Use your result from Exercise 1 to find the number of loaves for each person.   **267, 200, 133, 100**

NOW

TODAY, we would model this problem using $\frac{2}{3}x + \frac{1}{2}x + \frac{1}{3}x + \frac{1}{4}x = 700$. Equations like this can now be solved with symbolic manipulation software.

Brahmagupta solves linear equations in India.

Francois Viéte introduces symbolic algebra.

432 B.C.

Greeks solve quadratic equations geometrically.

A.D. 628

1591

1988

Symbolic and graphical manipulation software is introduced.

ADDITIONAL RESOURCES
An alternative Quiz for Lessons 1.3–1.5 is available in the *Chapter 1 Resource Book,* p. 77.

A **blackline** master with additional Math & History exercises is available in the *Chapter 1 Resource Book,* p. 75.

1.6 Solving Linear Inequalities

What you should learn

GOAL 1 Solve simple inequalities.

GOAL 2 Solve compound inequalities, as applied in **Example 6**.

Why you should learn it

▼ To model **real-life** situations, such as amusement park fees in **Ex. 50**.

3. *Sample answer:* Adding the same number to both sides of a true inequality produces another true inequality. Subtracting the same number from both sides of a true inequality produces another true inequality. Multiplying or dividing both sides of a true inequality by the same positive number produces another true inequality. Multiplying or dividing both sides of a true inequality by the same negative number does not produce another true inequality.

GOAL 1 SOLVING SIMPLE INEQUALITIES

Inequalities have properties that are similar to those of equations, but the properties differ in some important ways.

> ● **ACTIVITY**
> Developing Concepts
>
> ### Investigating Properties of Inequalities
>
> ❶ Write two true inequalities involving integers, one using $<$ and one using $>$.
> *Sample answer:* $0 < 4$ and $-3 > -10$
> ❷ Add, subtract, multiply, and divide each side of your inequalities by 2 and -2. In each case, decide whether the new inequality is true or false.
> Check inequalities; all are true except when you multiply or divide by -2.
> ❸ Write a general conclusion about the operations you can perform on a true inequality to produce another true inequality. **See margin.**

Inequalities such as $x \le 1$ and $2n - 3 > 9$ are examples of **linear inequalities** in one variable. A **solution** of an inequality in one variable is a value of the variable that makes the inequality true. For instance, -2, 0, 0.872, and 1 are some of the many solutions of $x \le 1$.

In the activity you may have discovered some of the following properties of inequalities. You can use these properties to solve an inequality because each transformation produces a new inequality having the same solutions as the original.

TRANSFORMATIONS THAT PRODUCE EQUIVALENT INEQUALITIES

- *Add* the same number to both sides.
- *Subtract* the same number from both sides.
- *Multiply* both sides by the same *positive* number.
- *Divide* both sides by the same *positive* number.
- *Multiply* both sides by the same *negative* number and *reverse* the inequality.
- *Divide* both sides by the same *negative* number and *reverse* the inequality.

The **graph** of an inequality in one variable consists of all points on a real number line that correspond to solutions of the inequality. To graph an inequality in one variable, use an open dot for $<$ or $>$ and a solid dot for $\le$ or $\ge$. For example, the graphs of $x < 3$ and $x \ge -2$ are shown below.

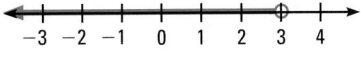
Graph of $x < 3$

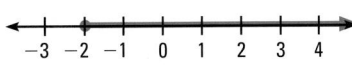
Graph of $x \ge -2$

1.6 *Solving Linear Inequalities* **41**

MEETING INDIVIDUAL NEEDS
- *Chapter 1 Resource Book*
  Prerequisite Skills Review (p. 5)
  Practice Level A (p. 83)
  Practice Level B (p. 84)
  Practice Level C (p. 85)
  Reteaching with Practice (p. 86)
  Absent Student Catch-Up (p. 88)
  Challenge (p. 90)
- *Resources in Spanish*
- *Personal Student Tutor*

WARM-UP EXERCISES

 Transparency Available
Solve the equation.
1. $7x - 4 = 5x + 2$ **3**
2. $10 - 3y = 16y - 9$ **1**
3. $\frac{3}{5}m - \frac{3}{4} = \frac{1}{2} + \frac{2}{5}m$ $\frac{25}{4}$

Fill in the blank with $<$, $>$, or $=$.
4. -16 ____ -19 **>**
5. $-4 - 8$ ____ $16 + (-4)$ **<**

A store has a rack of shirts under a "30% to 60% off" sign. Ask students what a shirt that originally cost $20 is if it is 30% off or 60% off. Point out that the shirt might also cost any price between these prices. For a problem like this, where there is a range of possibilities, students will need to use inequalities.

ACTIVITY NOTE
Students may tend to write the original inequalities just with positive numbers, but encourage them to write inequalities that include negative numbers also, so they can see that their results hold for all integers.

 EXTRA EXAMPLE 1
Solve $11y - 9 > 13$. $y > 2$

EXTRA EXAMPLE 2
Solve $7x + 9 \geq 10x - 12$. $x \leq 7$

EXTRA EXAMPLE 3
The percent of households h with cable television is modeled by $h = 2.3y + 44$, where y is the number of years since 1998. Describe the years when the percent is less than 53.2.
the years before 2002

 CHECKPOINT EXERCISES
For use after Examples 1 and 2:
1. Solve $6.4t - 8.1 < 12.7$. $t < 3.25$

For use after Example 3:
2. To receive an "A," Sachi must have at least a 90 average on her last three exams. Her average is $\frac{s + 174}{3}$, where s is her final exam score. Describe the final scores that will give her an "A." $s \geq 96$

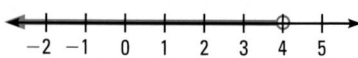 **Solving an Inequality with a Variable on One Side**

Solve $5y - 8 < 12$.

SOLUTION

$5y - 8 < 12$	Write original inequality.
$5y < 20$	Add 8 to each side.
$y < 4$	Divide each side by 5.

▶ The solutions are all real numbers less than 4, as shown in the graph at the right.

✓**CHECK** As a check, try several numbers that are less than 4 in the original inequality. Also, try checking some numbers that are greater than or equal to 4 to see that they are *not* solutions of the original inequality.

EXAMPLE 2 *Solving an Inequality with a Variable on Both Sides*

Solve $2x + 1 \leq 6x - 1$.

SOLUTION

$2x + 1 \leq 6x - 1$	Write original inequality.
$-4x + 1 \leq -1$	Subtract $6x$ from each side.
$-4x \leq -2$	Subtract 1 from each side.
$x \geq \dfrac{1}{2}$	Divide each side by -4 and reverse the inequality.

▶ The solutions are all real numbers greater than or equal to $\frac{1}{2}$. Check several numbers greater than or equal to $\frac{1}{2}$ in the original inequality.

STUDENT HELP

→ **Study Tip**
Don't forget that when you multiply or divide both sides of an inequality by a negative number, you must *reverse* the inequality to maintain a true statement. For instance, to reverse $\leq$, replace it with $\geq$.

 **Fish**

EXAMPLE 3 *Using a Simple Inequality*

The weight w (in pounds) of an Icelandic saithe is given by

$$w = 10.4t - 2.2$$

where t is the age of the fish in years. Describe the ages of a group of Icelandic saithe that weigh up to 29 pounds. ▶ Source: Marine Research Institute

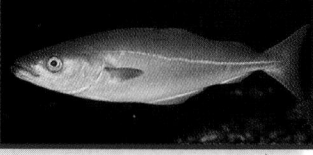
Icelandic saithe

SOLUTION

$w \leq 29$	Weights are at most 29 pounds.
$10.4t - 2.2 \leq 29$	Substitute for w.
$10.4t \leq 31.2$	Add 2.2 to each side.
$t \leq 3$	Divide each side by 10.4.

▶ The ages are less than or equal to 3 years.

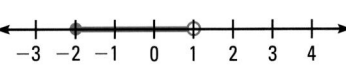

GOAL 2 SOLVING COMPOUND INEQUALITIES

STUDENT HELP

→ **Study Tip**
The inequality $a < x < b$ is read as "x is between a and b." The inequality $a \le x \le b$ is read as "x is between a and b, inclusive."

A **compound inequality** is two simple inequalities joined by "and" or "or." Here are two examples.

$$-2 \le x < 1 \qquad\qquad x < -1 \text{ or } x \ge 2$$

All real numbers that are greater than or equal to −2 *and* less than 1.

All real numbers that are less than −1 *or* greater than or equal to 2.

EXAMPLE 4 *Solving an "And" Compound Inequality*

Solve $-2 \le 3t - 8 \le 10$.

SOLUTION

To solve, you must isolate the variable between the two inequality signs.

$-2 \le 3t - 8 \le 10$	**Write original inequality.**
$6 \le 3t \le 18$	**Add 8 to each expression.**
$2 \le t \le 6$	**Divide each expression by 3.**

▶ Because t is between 2 and 6, inclusive, the solutions are all real numbers greater than or equal to 2 *and* less than or equal to 6. Check several of these numbers in the original inequality. The graph is shown below.

EXAMPLE 5 *Solving an "Or" Compound Inequality*

Solve $2x + 3 < 5$ or $4x - 7 > 9$.

STUDENT HELP

INTERNET **HOMEWORK HELP**
Visit our Web site
www.mcdougallittell.com
for extra examples.

SOLUTION

A solution of this compound inequality is a solution of *either* of its simple parts, so you should solve each part separately.

SOLUTION OF FIRST INEQUALITY		**SOLUTION OF SECOND INEQUALITY**	
$2x + 3 < 5$	**Write first inequality.**	$4x - 7 > 9$	**Write second inequality.**
$2x < 2$	**Subtract 3 from each side.**	$4x > 16$	**Add 7 to each side.**
$x < 1$	**Divide each side by 2.**	$x > 4$	**Divide each side by 4.**

▶ The solutions are all real numbers less than 1 *or* greater than 4. Check several of these numbers to see that they satisfy one of the simple parts of the original inequality. The graph is shown below.

1.6 *Solving Linear Inequalities* **43**

☑ **EXTRA EXAMPLE 4**
Solve $-9 < t + 4 < 10$. $-13 < t < 6$

EXTRA EXAMPLE 5
Solve $6x + 9 < 3$ or $3x - 8 > 13$.
$x < -1$ or $x > 7$

✔ **CHECKPOINT EXERCISES**
For use after Example 4:
1. Solve $-12 < 3x - 3 < 15$.
$-3 < x < 6$

For use after Example 5:
2. Solve $-2x + 7 < 3$ or $3x + 5 < 2$.
$x < -1$ or $x > 2$

STUDENT HELP NOTES

→ **Homework Help** Students can find extra examples at **www.mcdougallittell.com** that parallel the examples in the student edition.

EXTRA EXAMPLE 6

Milk will keep until its expiration date and will not freeze when stored at a minimum temperature of –1°C and a maximum temperature of 5°C. The temperature C satisfies the inequality $-1 < C < 5$. Write the inequality in degrees Fahrenheit. **30.2 < F < 41**

EXTRA EXAMPLE 7

The feeding instructions on your dog's food recommend $2\frac{1}{2}$ to $3\frac{1}{4}$ lb of food weekly.

a. Write the conditions that represent underfeeding or overfeeding your dog as a compound inequality.
$w < 2\frac{1}{2}$ or $w > 3\frac{1}{4}$, where w is the number of pounds

b. Rewrite the conditions in ounces of dog food (1 lb = 16 oz). $o < 40$ or $o > 52$

 CHECKPOINT EXERCISES

For use after Examples 6 and 7:

1. For your workout, you want your heart rate in beats per minute to be not less than 130, but not more than 160. Write an inequality for your target heart rate.
$130 \leq r \leq 160$, where r is the number of beats per minute

FOCUS ON VOCABULARY

How do the solutions in a compound inequality using *and* differ from those of a compound inequality using *or*? **Sample answer: A solution of an *and* compound inequality must satisfy both simple inequalities. A solution of an *or* compound inequality need satisfy only one of the simple inequalities.**

CLOSURE QUESTION

Compare solving linear inequalities with solving linear equations.
See sample answer at right.

Automotive Maintenance

EXAMPLE 6 *Using an "And" Compound Inequality*

You have added enough antifreeze to your car's cooling system to lower the freezing point to −35°C and raise the boiling point to 125°C. The coolant will remain a liquid as long as the temperature C (in degrees Celsius) satisfies the inequality $-35 < C < 125$. Write the inequality in degrees Fahrenheit.

SOLUTION

Let F represent the temperature in degrees Fahrenheit, and use the formula $C = \frac{5}{9}(F - 32)$.

$-35 < C < 125$	Write original inequality.
$-35 < \frac{5}{9}(F - 32) < 125$	Substitute $\frac{5}{9}(F - 32)$ for C.
$-63 < F - 32 < 225$	Multiply each expression by $\frac{9}{5}$, the reciprocal of $\frac{5}{9}$.
$-31 < F < 257$	Add 32 to each expression.

▶ The coolant will remain a liquid as long as the temperature stays between −31°F and 257°F.

EXAMPLE 7 *Using an "Or" Compound Inequality*

TRAFFIC ENFORCEMENT You are a state patrol officer who is assigned to work traffic enforcement on a highway. The posted minimum speed on the highway is 45 miles per hour and the posted maximum speed is 65 miles per hour. You need to detect vehicles that are traveling outside the posted speed limits.

a. Write these conditions as a compound inequality.

b. Rewrite the conditions in kilometers per hour.

SOLUTION

a. Let m represent the vehicle speeds in miles per hour. The speeds that you need to detect are given by:

$$m < 45 \text{ or } m > 65$$

b. Let k be the vehicle speeds in kilometers per hour. The relationship between miles per hour and kilometers per hour is given by the formula $m \approx 0.621k$. You can rewrite the conditions in kilometers per hour by substituting $0.621k$ for m in each inequality and then solving for k.

$m < 45$	or	$m > 65$
$0.621k < 45$	or	$0.621k > 65$
$k < 72.5$	or	$k > 105$

▶ You need to detect vehicles whose speeds are less than 72.5 kilometers per hour or greater than 105 kilometers per hour.

FOCUS ON APPLICATIONS

POLICE RADAR Police radar guns emit a continuous radio wave of known frequency. The radar gun compares the frequency of the wave reflected from a vehicle to the frequency of the transmitted wave and then displays the vehicle's speed.

Closure Question *Sample answer:*
To solve both you use transformations that produce equivalent statements. Multiplying or dividing by a negative number reverses the sign of an inequality but not of an equation. A linear equation has a single solution. **A linear inequality has a range of values representing possible solutions.**

GUIDED PRACTICE

Vocabulary Check ✓

1. Explain the difference between a simple linear inequality and a compound linear inequality. **1–3. See margin.**

Concept Check ✓

2. Tell whether this statement is *true* or *false:* Multiplying both sides of an inequality by the same number always produces an equivalent inequality. Explain.

3. Explain the difference between solving $2x < 7$ and solving $-2x < 7$.

Skill Check ✓

Solve the inequality. Then graph your solution. **4–9. See margin for graphs.**

4. $x - 5 < 8$ $x < 13$ 5. $3x \geq 15$ $x \geq 5$ 6. $-x + 4 > 3$ $x < 1$

7. $\frac{1}{2}x \leq 6$ $x \leq 12$ 8. $x + 8 > -2$ $x > -10$ 9. $-x - 3 < -5$ $x > 2$

Graph the inequality. **10, 11. See margin.**

10. $-2 \leq x < 5$ 11. $x \geq 3$ or $x < -3$

12. 🌐 **WINTER DRIVING** You are moving to Montana and need to lower the freezing point of the cooling system in the car from Example 6 to $-50°C$. This will also raise the boiling point to $140°C$. Write a compound inequality that models this situation. Then write the inequality in degrees Fahrenheit.
$-50 < C < 140$; $-58 < F < 284$

Margin answers:

1. A simple linear inequality has only one inequality symbol, used to describe the relationship between two quantities or expressions. A compound linear inequality consists of two simple linear inequalities joined by "or" or "and."

2. False; multiplying both sides of an inequality by the same negative number does not produce an equivalent inequality.

3. To solve $2x < 7$, divide each side by 2 $\left(x < \frac{7}{2} \right)$; to solve $-2x < 7$, divide each side by -2 and reverse the inequality $\left(x > -\frac{7}{2} \right)$.

PRACTICE AND APPLICATIONS

STUDENT HELP

▶ **Extra Practice** to help you master skills is on p. 941.

MATCHING INEQUALITIES Match the inequality with its graph.

13. $x \geq 4$ **C**   14. $x < 4$ **A**   15. $-4 < x \leq 4$ **D**

16. $x \geq 4$ or $x < -4$ **B**   17. $-4 \leq x \leq 4$ **F**   18. $x > 4$ or $x \leq -4$ **E**

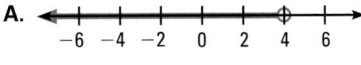

A. (number line from -6 to 6)
B. (number line from -6 to 6)
C. (number line from -6 to 6)
D. (number line from -6 to 6)
E. (number line from -6 to 6)
F. (number line from -6 to 6)

STUDENT HELP

▶ **HOMEWORK HELP**
Examples 1, 2: Exs. 13, 14, 19–22, 25–36
Example 3: Exs. 49–51
Examples 4, 5: Exs. 15–18, 23, 24, 37–48
Example 6: Exs. 52–54
Example 7: Exs. 55, 56

CHECKING SOLUTIONS Decide whether the given number is a solution of the inequality.

19. $2x + 9 < 16$; 4   no   20. $10 - x \geq 3$; 7   yes   21. $7x - 12 < 8$; 3   no

22. $-\frac{1}{3}x - 2 \leq -4$; 9   yes   23. $-3 < 2x \leq 6$; 3   yes   24. $-8 < x - 11 < -6$; 5   no

SIMPLE INEQUALITIES Solve the inequality. Then graph your solution.
25–36. See margin for graphs.

25. $4x + 5 > 25$ $x > 5$ 26. $7 - n \leq 19$ $n \geq -12$ 27. $5 - 2x \geq 27$ $x \leq -11$

28. $\frac{1}{2}x - 4 > -6$ $x > -4$ 29. $\frac{3}{2}x - 7 < 2$ $x < 6$ 30. $5 + \frac{1}{3}n \leq 6$ $n \leq 3$

31. $4x - 1 > 14 - x$ $x > 3$ 32. $-n + 6 < 7n + 4$ $n > \frac{1}{4}$ 33. $4.7 - 2.1x > -7.9$ $x < 6$

34. $2(n - 4) \leq 6$ $n \leq 7$ 35. $2(4 - x) > 8$ $x < 0$ 36. $5 - 5x > 4(3 - x)$ $x < -7$

3 APPLY

ASSIGNMENT GUIDE

BASIC
Day 1: pp. 45–47 Exs. 13–24, 26–48 even, 49, 51, 58, 62–68 even

AVERAGE
Day 1: pp. 45–47 Exs. 13–24, 26–48 even, 49, 51, 58–60, 61–69 odd

ADVANCED
Day 1: pp. 45–47 Exs. 13–24, 26–48 even, 49, 51, 55–60, 61–69

BLOCK SCHEDULE WITH 1.5
pp. 45–47 Exs. 13–24, 26–48 even, 49, 51, 58–60, 61–69 odd

EXERCISE LEVELS
Level A: *Easier*
13–51, 61–64
Level B: *More Difficult*
52–58, 65–69
Level C: *Most Difficult*
59, 60

✔ **HOMEWORK CHECK**
To quickly check student understanding of key concepts, go over the following exercises: Exs. 14, 17, 20, 24, 32, 38. See also the Daily Homework Quiz:
• Blackline Master (*Chapter 1 Resource Book,* p. 93)
• 📄 Transparency (p. 8)

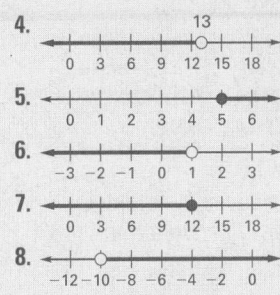

4.
5.
6.
7.
8.

9–11, 25–36. See Additional Answers beginning on page AA1.

! COMMON ERROR

EXERCISES 37–42 For a compound inequality written in this form, students may perform transformations on only two of the three algebraic expressions. Remind them that they must perform the transformations on all three expressions. If they are unclear about this, have them try simple numerical examples or write the inequality as two simple inequalities joined by *and*.

APPLICATION NOTE

EXERCISES 52–54 Additional information about Mars is available at **www.mcdougallittell.com**.

37.
0 3 **5** 6 9 12 15 18

38.
-4 -3 -2 -1 0 1 2

39.
-6 -5 -4 -3 -2 -1 0

40.
-3 -2 -1 0 1 $\frac{3}{2}$ 2 3

41.
-3 -2 -1 0 1 2 3

42.
-12 -6 0 6 12 18 **21** 24

43.
-2 0 2 **3** 4 6 8 10

44.
-5 -4 -3 -2 -1 0 1

45.
-6 -5 -4 -3 -2 **-0.52** -1 0

46–48. See Additional Answers beginning on page AA1.

ADDITIONAL PRACTICE AND RETEACHING

For Lesson 1.6:

• Practice Levels A, B, and C (*Chapter 1 Resource Book,* p. 83)

• Reteaching with Practice (*Chapter 1 Resource Book,* p. 86)

• ▢ See Lesson 1.6 of the *Personal Student Tutor*

For more Mixed Review:

• ▢ Search the *Test and Practice Generator* for key words or specific lessons.

46

COMPOUND INEQUALITIES Solve the inequality. Then graph your solution.
37–48. See margin for graphs.

37. $-2 \leq x - 7 \leq 11$
    $5 \leq x \leq 18$

38. $-16 \leq 3x - 4 \leq 2$
    $-4 \leq x \leq 2$

39. $-5 \leq -n - 6 \leq 0$
    $-6 \leq n \leq -1$

40. $-2 < -2n + 1 \leq 7$
    $-3 \leq n < \frac{3}{2}$

41. $-7 < 6x - 1 < 5$
    $-1 < x < 1$

42. $-8 < \frac{2}{3}x - 4 < 10$
    $-6 < x < 21$

43. $x + 2 \leq 5$ or $x - 4 \geq 2$
    $x \leq 3$ or $x \geq 6$

44. $3x + 2 < -10$ or $2x - 4 > -4$
    $x < -4$ or $x > 0$

45. $-5x - 4 < -1.4$ or $-2x + 1 > 11$
    $x < -5$ or $x > -0.52$

46. $x - 1 \leq 5$ or $x + 3 \geq 10$
    $x \leq 6$ or $x \geq 7$

47. $-0.1 \leq 3.4x - 1.8 < 6.7$
    $0.5 \leq x < 2.5$

48. $0.4x + 0.6 < 2.2$ or $0.6x > 3.6$
    $x < 4$ or $x > 6$

49. 🌐 **COMMISSION** Your salary is $1250 per week and you receive a 5% commission on your sales each week. What are the possible amounts (in dollars) that you can sell each week to earn at least $1500 per week?
Your sales must be greater than or equal to $5000.

50. 🌐 **PARK FEES** You have $50 and are going to an amusement park. You spend $25 for the entrance fee and $15 for food. You want to play a game that costs $.75. Write and solve an inequality to find the possible numbers of times you can play the game. If you play the game the maximum number of times, will you have spent the entire $50? Explain.

51. 🐢 **GRADES** A professor announces that course grades will be computed by taking 40% of a student's project score (0–100 points) and adding 60% of the student's final exam score (0–100 points). If a student gets an 86 on the project, what scores can she get on the final exam to get a course grade of at least 90?
Her score must be between 93 and 100, inclusive.

SCIENCE ▸ CONNECTION In Exercises 52–54, use the following information.

The international standard for scientific temperature measurement is the Kelvin scale. A Kelvin temperature can be obtained by adding 273.15 to a Celsius temperature. The daytime temperature on Mars ranges from $-89.15°C$ to $-31.15°C$. ▸ Source: NASA

52. Write the daytime temperature range on Mars as a compound inequality in degrees Celsius. $-89.15 \leq C \leq -31.15$

53. Rewrite the compound inequality in degrees Kelvin. $184 \leq K \leq 242$

54. **RESEARCH** Find the high and low temperatures in your area for any particular day. Write three compound inequalities representing the temperature range in degrees Fahrenheit, in degrees Celsius, and in degrees Kelvin.
Sample answer: high: 77°F, low: 59°F; $59 \leq F \leq 77$; $15 \leq C \leq 25$; $288.15 \leq K \leq 298.15$

🌐 **WINTER** In Exercises 55 and 56, use the following information.

The Ontario Winter Severity Index (OWSI) is a weekly calculation used to determine the severity of winter conditions. The OWSI for deer is given by

$$I = \frac{p}{30} + \frac{d}{30} + c$$

where p represents the average Snow Penetration Gauge reading (in centimeters), d represents the average snow depth (in centimeters), and c represents the *chillometer reading*, which is a measure of the cold (in kilowatt-hours) based on temperature and wind chill. An extremely mild winter occurs when $I < 5$ on average, and an extremely severe winter occurs when $I > 6.5$ on average. A deer can tolerate a maximum snow penetration of 50 centimeters. Assume the average snow depth is 60 centimeters. ▸ Source: Snow Network for Ontario Wildlife

55. What weekly chillometer readings will produce extremely severe winter readings? $c > 2.83$

56. What weekly chillometer readings will produce extremely mild winter readings? $c < 1.33$

50. You can play the game at most 13 times. No; 13 games will cost $9.75, leaving you with $.25 from the original $50 ($40 were spent on the entrance fee and food).

FOCUS ON APPLICATIONS

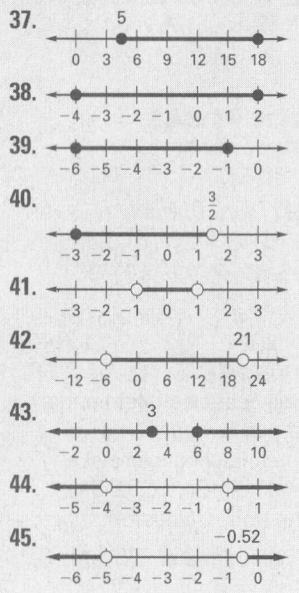

🔴 **MARS** is the fourth planet from the sun. A Martian year is 687 Earth days long, but a Martian day is only 40 minutes longer than an Earth day. Mars is also much colder than Earth, as discussed in Exs. 52–54.

📡 APPLICATION LINK
www.mcdougallittell.com

57. _Writing_ The first transformation listed in the box on page 41 can be written symbolically as follows: If a, b, and c are real numbers and $a > b$, then $a + c > b + c$. Write similar statements for the other transformations.   **See margin.**

Test Preparation

57. If a, b, and c are real numbers and $a > b$, then $a - c > b - c$.
If a, b, and c are real numbers such that $a > b$ and $c > 0$, then $ac > bc$.
If a, b, and c are real numbers such that $a > b$ and $c > 0$, then $\dfrac{a}{c} > \dfrac{b}{c}$.
If a, b, and c are real numbers such that $a > b$ and $c < 0$, then $ac < bc$.
If a, b, and c are real numbers such that $a > b$ and $c < 0$, then $\dfrac{a}{c} < \dfrac{b}{c}$.

59. _Sample answer:_
$x > x + 1$. When you subtract x from each side of $x > x + 1$, the result is $0 > 1$, a false statement; therefore, $x > x + 1$ has no solution.

★ **Challenge**

58. MULTI-STEP PROBLEM You are vacationing at Lake Tahoe, California. You decide to spend a day sightseeing in other places. You want to go from Lake Tahoe to Sacramento, from Sacramento to Sonora, and then from Sonora back to Lake Tahoe. You know that it is about 85 miles from Lake Tahoe to Sacramento and about 75 miles from Sacramento to Sonora.   **See margin.**

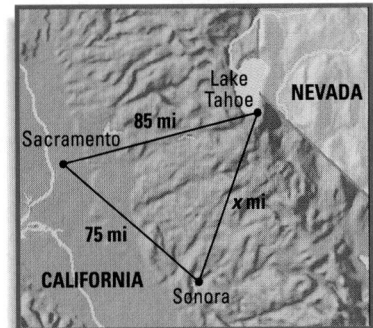

a. The triangle inequality theorem states that the sum of the lengths of any two sides of a triangle is greater than the length of the third side. Write a compound inequality that represents the distance from Sonora to Lake Tahoe.

b. CRITICAL THINKING You are reading a brochure which states that the distance between Sonora and Lake Tahoe is 170 miles. You know that the distance is a misprint. How can you be so sure? Explain.

c. You keep a journal of the distances you have traveled. Many of your distances represent triangular circuits. Your friend is reading your journal and states that you must have recorded a wrong distance for one of these circuits. To which one of the following is your friend referring? Explain.

A. 35 miles, 65 miles, 45 miles   **B.** 15 miles, 50 miles, 64 miles

C. 49 miles, 78 miles, 28 miles   **D.** 55 miles, 72 miles, 41 miles

59. Write an inequality that has no solution. Show why it has no solution.   **See margin.**

60. Write an inequality whose solutions are all real numbers. Show why the solutions are all real numbers.   _Sample answer:_ $x < x + 1$. When you subtract x from each side of $x < x + 1$, the result is $0 < 1$, a true statement; therefore, all real numbers are solutions of $x < x + 1$.

MIXED REVIEW

IDENTIFYING PROPERTIES Identify the property shown. (Review 1.1)

61. $(7 \cdot 3) \cdot 11 = 7 \cdot (3 \cdot 11)$ associative property of multiplication

62. $34 + (-34) = 0$ inverse property of addition

63. $37 + 29 = 29 + 37$ commutative property of addition

64. $3(9 + 4) = 3(9) + 3(4)$ distributive property

SOLVING EQUATIONS Solve the equation. Check your solution.
(Review 1.3 for 1.7)

65. $5x + 4 = -2(x + 3)$ $-\dfrac{10}{7}$

66. $2(3 - x) = 16(x + 1)$ $-\dfrac{5}{9}$

67. $-(x - 1) + 10 = -3(x - 3)$ -1

68. $\dfrac{1}{8}x + \dfrac{3}{2} = \dfrac{3}{4}x - 1$ 4

69. 🌎 **CONCERT TRIP** You are going to a concert in another town 48 miles away. You can average 40 miles per hour on the road you plan to take to the concert. What is the minimum number of hours before the concert starts that you should leave to get to the concert on time? (Review 1.5) $1\frac{1}{5}$ h, or 1 h 12 min

1.6 _Solving Linear Inequalities_   **47**

---

4 ASSESS

DAILY HOMEWORK QUIZ

📄 _Transparency Available_

Decide whether the given number is a solution of the inequality.

1. $14 > -3x - 4$; -6 no

2. $-4 \le -5 - 2x \le 3$; -3 yes

Solve the inequality. Then graph your solution.

3. $\dfrac{2}{3}p - 4 < -2$ $p < 3$

4. $-3 \le 2x - 5 \le 3$ $1 \le x \le 4$

5. $-3t - 5 < -8$ or $-4t + 3 > 7$
$t < -1$ or $t > 1$

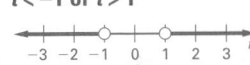

┌─ **EXTRA CHALLENGE NOTE** ─┐
→ Challenge problems for Lesson 1.6 are available in **blackline** format in the _Chapter 1 Resource Book,_ p. 90 and at **www.mcdougallittell.com.**

ADDITIONAL TEST PREPARATION

1. WRITING Compare the solutions and the graph of $-3 < 2x - 1 < 5$ with those of $-3 \le 2x - 1 \le 5$.
Sample answer: The solution of the first inequality is all real numbers between, but not including, -1 and 3. The graph is a number line shaded between -1 and 3, with open circles at -1 and 3. The solutions of the second inequality are all those of the first, plus -1 and 3. The graph is the same except that there are shaded circles at -1 and 3.

58. See Additional Answers beginning on page AA1.

1 Planning the Activity

PURPOSE
To use the *Test* feature on a graphing calculator to solve linear inequalities.

MATERIALS
• graphing calculator for each student
• Keystroke blackline (*Chapter 1 Resource Book*, p. 82)

PACING
• Example — 5 min
• Exercises — 15 min

▶ LINK TO LESSON
Students can use the procedures of this activity to check Examples 1 and 2 of Lesson 1.6.

2 Managing the Activity

CLASSROOM MANAGEMENT
Make sure students understand that a truth function value of 0 represents *false,* and a value of 1 represents *true.*

ALTERNATIVE APPROACH
Have students work an example using both the *Test* feature and the *Table* feature of a graphing calculator. Then they can see how the *x*-value in a table where the *y*-columns are equal relates to a change in the truth function graph.

3 Closing the Activity

★ KEY DISCOVERY
A truth function graph has a value of 1 for solutions of inequalities, and 0 for values that are not solutions.

ACTIVITY ASSESSMENT
Explain how to use the *Test* feature of a graphing calculator to solve a linear inequality.
See sample answer at right.

⊙ ACTIVITY 1.6

Using Technology

Solving an Inequality

Most graphing calculators are able to evaluate whether a statement is true or false. If a statement is true, the calculator returns a 1; if a statement is false, it returns a 0. You can use this *Test* feature of a graphing calculator to solve a linear inequality.

▶ EXAMPLE
Use the *Test* feature of a graphing calculator to solve the inequality $3x + 2 > -4$.

STUDENT HELP

KEYSTROKE HELP
See keystrokes for several models of calculators at www.mcdougallittell.com

▶ SOLUTION

1 To solve the inequality, you must find the values of *x* for which the inequality is true. Enter the inequality as the *truth function* $y = (3x + 2 > -4)$, as shown in the calculator screen below.

2 In the graph below you can see that the *y*-values are 1 for all *x*-values greater than -2. So, the solutions are given by $x > -2$. Check several solutions in the original inequality.

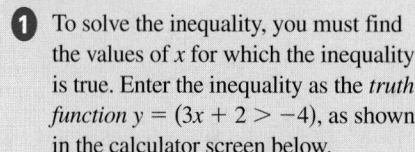

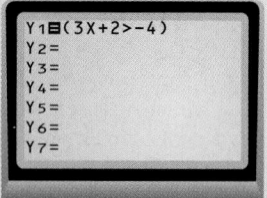 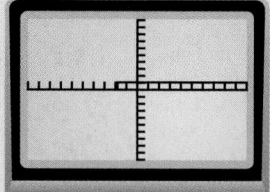

▶ EXERCISES

In Exercises 1 and 2, the *Test* feature of a graphing calculator was used to create the graph. Use the graph to solve the inequality. Check several solutions in the original inequality.

1. $y = (4x - 5 \le 11)$ $x \le 4$

2. $y = (5x + 6 \ge -14)$ $x \ge -4$

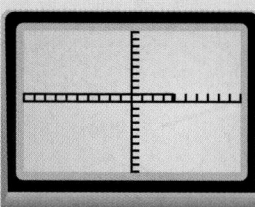 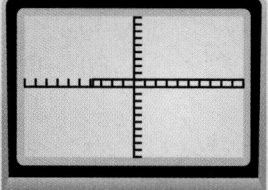

Use the *Test* feature of a graphing calculator to solve the inequality. Check several solutions in the original inequality.

3. $2x - 7 > -1$ $x > 3$ **4.** $4x + 2 < 18$ $x < 4$ **5.** $0.5x + 2 \le -1$ $x \le -6$

6. $-x + 5 \ge -3$ $x \le 8$ **7.** $-6x + 3 > -9$ $x < 2$ **8.** $-0.5x - 1.5 \le 3$ $x \ge -9$

9. $5x < 4x + 6$ $x < 6$ **10.** $4 - x \ge 2 - \frac{1}{2}x$ $x \le 4$ **11.** $3x - 4 \le 2x + 5$ $x \le 9$

12. $2x - 1 < \frac{7}{3} + \frac{4}{3}x$ $x < 5$ **13.** $5 - 5x > 12 - 4x$ $x < -7$ **14.** $8 - 4x \le 5 - x$ $x \ge 1$

Activity Assessment *Sample answer:*
Enter the inequality into the function list, choose an appropriate window, and graph the inequality. A truth function graph displays. The *x*-values where the graph has a *y*-value of 1 are solutions. The *x*-values where the graph has a *y*-value of 0 are not solutions.

● ACTIVITY 1.7

Developing Concepts

SET UP
Work with a partner.

MATERIALS
11 index cards numbered from −5 to 5

Absolute Value Equations and Inequalities

▶ **QUESTION** What does the solution of an absolute value equation or inequality look like on a number line?

The *absolute value* of a number x, written $|x|$, is the distance on a number line that the number is from 0. Because both 2 and −2 are 2 units from 0, $|2| = 2$ and $|-2| = 2$. Notice that the absolute value of a number is always positive or 0.

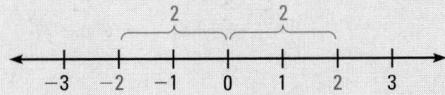

▶ **EXPLORING THE CONCEPT**

❶ You should work with a partner. Each pair of partners should have a set of index cards numbered from −5 to 5. The cards should be placed face up in numerical order to form a number line.

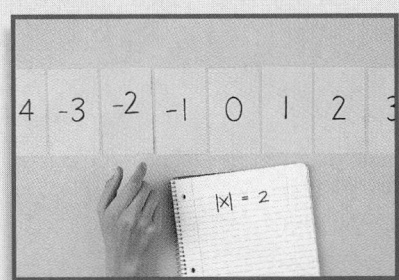 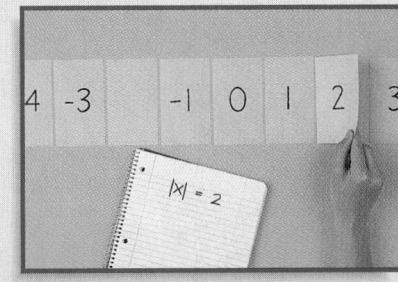

❷ For each absolute value equation or inequality below, one partner should turn over the cards whose numbers are solutions. The other partner should then agree or disagree as to whether the solutions are correct. Once in agreement, both of you should graph the solutions on a number line. You should take turns turning over the cards and checking the solutions.   **a–i. See margin for graphs.**

a. $|x| = 4$ 4, −4   b. $|x| \leq 4$ $-4 \leq x \leq 4$ c. $|x| \geq 4$ $x \geq 4$ or $x \leq -4$

d. $|3x| = 9$ 3, −3   e. $|3x| \leq 9$ $-3 \leq x \leq 3$ f. $|3x| \geq 9$ $x \leq -3$ or $x \geq 3$

g. $|x - 1| = 2$ 3, −1   h. $|x - 1| \leq 2$ $-1 \leq x \leq 3$ i. $|x - 1| \geq 2$ $x \leq -1$ or $x \geq 3$

▶ **DRAWING CONCLUSIONS**   See margin.

1. Describe the nature of the solutions of the absolute value equations in parts (a), (d), and (g). Do you think that all absolute value equations will have solutions of this nature? Will all absolute value equations have the same number of solutions?

2. Describe the nature of the solutions of the absolute value inequalities in parts (b), (e), and (h), all of which involve the $\leq$ sign. What difference, if any, would there be if the inequalities involved the $<$ sign?

3. Describe the nature of the solutions of the absolute value inequalities in parts (c), (f), and (i), all of which involve the $\geq$ sign. What difference, if any, would there be if the inequalities involved the $>$ sign?

1. *Sample answer:* One solution is positive and one solution is negative; no (for example, $|x - 5| = 1$ has two positive solutions); no (for example, $|x - 5| = 0$ has only one solution).

2. *Sample answer:* The solutions are compound "and" inequalities; the solutions would also use the $<$ sign.

3. *Sample answer:* The solutions are compound "or" inequalities; the solutions would also use the $>$ sign.

---

① Planning the Activity

PURPOSE
To model solutions to absolute value equations and inequalities.

MATERIALS
• 11 index cards numbered from −5 to 5 for each pair of students
• Activity Support Master (*Chapter 1 Resource Book,* p. 94)

PACING
• Exploring the Concept — 15 min
• Drawing Conclusions — 10 min

▶ ***LINK TO LESSON***
Examples 1–3 of Lesson 1.7 correspond to the Activity exercise of the same number.

② Managing the Activity

ALTERNATIVE APPROACH
Have students use a graphing calculator to graph $y = |x|$ and $y = 4$. The graphs are equal for $x = -4$ and $x = 4$. The graph of $y = |x|$ is below the graph of $y = 4$ between these points, and above the graph of $y = 4$ outside these points. Have students make the same comparisons for Parts d–f and g–i.

③ Closing the Activity

★ ***KEY DISCOVERY***
An absolute value equation has 0, 1, or 2 solutions. An absolute value inequality with "<" or "≤" has solutions between two values. An absolute value inequality with ">" or "≥" has solutions outside the two values.

ACTIVITY ASSESSMENT
JOURNAL Discuss how the solutions of $|x| = d$, $|x| \leq d$, and $|x| \geq d$ ($d > 0$) relate to distances on a number line.
See sample answer at left.

Step 2, a–i. See Additional Answers beginning on page AA1.

Activity Assessment *Sample answer:*
The solutions of $|x| = d$ are those points that have a distance of exactly d units from the origin. The solutions of $|x| \leq d$ are those points that have a distance of at most d units from the origin. The solutions of $|x| \geq d$ are those points that have a distance of at least d units from the origin.

LESSON OPENER
CALCULATOR ACTIVITY
An alternative way to approach
Lesson 1.7 is to use the Calculator
Activity Lesson Opener:
• Blackline Master (*Chapter 1 Resource Book,* p. 95)
• Transparency (p. 7)

MEETING INDIVIDUAL NEEDS
• *Chapter 1 Resource Book*
Prerequisite Skills Review (p. 5)
Practice Level A (p. 98)
Practice Level B (p. 99)
Practice Level C (p. 100)
Reteaching with Practice (p. 101)
Absent Student Catch-Up (p. 103)
Challenge (p. 105)
• *Resources in Spanish*
• Personal Student Tutor

NEW-TEACHER SUPPORT
See the Tips for New Teachers on
pp. 1–2 of the *Chapter 1 Resource Book* for additional notes about
Lesson 1.7.

WARM-UP EXERCISES
Transparency Available
Solve the inequality.
1. $18x + 7 > -14 + 6x$ $x > -\frac{7}{4}$
2. $\frac{3}{4}x - 3 \le 9$ $x \le 16$
3. $5(6 - x) \ge -15$ $x \le 9$
4. $x + 5 > 12$ or $x - 7 < -9$
$x > 7$ or $x < -2$
5. $-4 \le 2x - 4 \le 10$ $0 \le x \le 7$

1.7

Solving Absolute Value Equations and Inequalities

GOAL 1 SOLVING EQUATIONS AND INEQUALITIES

What you should learn

GOAL 1 Solve absolute value equations and inequalities.

GOAL 2 Use absolute value equations and inequalities to solve **real-life** problems, such as finding acceptable weights in **Example 4.**

Why you should learn it

▼ To solve **real-life** problems, such as finding recommended weight ranges for sports equipment in **Ex. 72.**

The **absolute value** of a number x, written $|x|$, is the distance the number is from 0 on a number line. Notice that the absolute value of a number is always nonnegative.

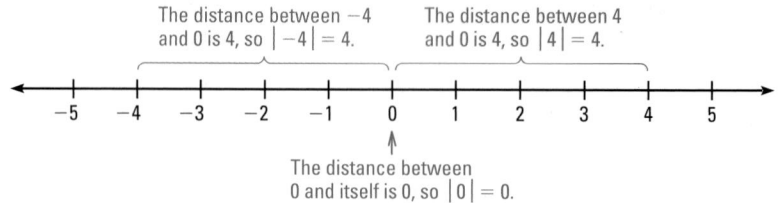

The distance between -4 and 0 is 4, so $|-4| = 4$. The distance between 4 and 0 is 4, so $|4| = 4$.

The distance between 0 and itself is 0, so $|0| = 0$.

The absolute value of x can be defined algebraically as follows.

$$|x| = \begin{cases} x, & \text{if } x \text{ is positive} \\ 0, & \text{if } x = 0 \\ -x, & \text{if } x \text{ is negative} \end{cases}$$

To solve an absolute value equation of the form $|x| = c$ where $c > 0$, use the fact that x can have two possible values: a positive value c or a negative value $-c$. For instance, if $|x| = 5$, then $x = 5$ or $x = -5$.

SOLVING AN ABSOLUTE VALUE EQUATION

The absolute value equation $|ax + b| = c$, where $c > 0$, is equivalent to the compound statement $ax + b = c$ or $ax + b = -c$.

EXAMPLE 1 *Solving an Absolute Value Equation*

Solve $|2x - 5| = 9$.

SOLUTION

Rewrite the absolute value equation as two linear equations and then solve each linear equation.

$\|2x - 5\| = 9$	Write original equation.
$2x - 5 = 9$ or $2x - 5 = -9$	Expression can be 9 or -9.
$2x = 14$ or $2x = -4$	Add 5 to each side.
$x = 7$ or $x = -2$	Divide each side by 2.

▶ The solutions are 7 and -2. Check these by substituting each solution into the original equation.

An absolute value inequality such as $|x - 2| < 4$ can be solved by rewriting it as a compound inequality, in this case as $-4 < x - 2 < 4$.

TRANSFORMATIONS OF ABSOLUTE VALUE INEQUALITIES

- The inequality $|ax + b| < c$, where $c > 0$, means that $ax + b$ is *between* $-c$ and c. This is equivalent to $-c < ax + b < c$.
- The inequality $|ax + b| > c$, where $c > 0$, means that $ax + b$ is *beyond* $-c$ and c. This is equivalent to $ax + b < -c$ or $ax + b > c$.

In the first transformation, $<$ can be replaced by $\leq$. In the second transformation, $>$ can be replaced by $\geq$.

EXAMPLE 2   *Solving an Inequality of the Form* $|ax + b| < c$

Solve $|2x + 7| < 11$.

STUDENT HELP

HOMEWORK HELP
Visit our Web site
www.mcdougallittell.com
for extra examples.

SOLUTION

$\|2x + 7\| < 11$	**Write original inequality.**
$-11 < 2x + 7 < 11$	**Write equivalent compound inequality.**
$-18 < 2x < 4$	**Subtract 7 from each expression.**
$-9 < x < 2$	**Divide each expression by 2.**

▶ The solutions are all real numbers greater than -9 and less than 2. Check several solutions in the original inequality. The graph is shown below.

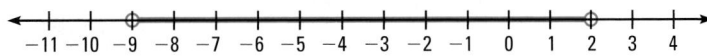

EXAMPLE 3   *Solving an Inequality of the Form* $|ax + b| \geq c$

Solve $|3x - 2| \geq 8$.

SOLUTION

This absolute value inequality is equivalent to $3x - 2 \leq -8$ or $3x - 2 \geq 8$.

SOLVE FIRST INEQUALITY		**SOLVE SECOND INEQUALITY**
$3x - 2 \leq -8$	**Write inequality.**	$3x - 2 \geq 8$
$3x \leq -6$	**Add 2 to each side.**	$3x \geq 10$
$x \leq -2$	**Divide each side by 3.**	$x \geq \dfrac{10}{3}$

▶ The solutions are all real numbers less than or equal to -2 or greater than or equal to $\frac{10}{3}$. Check several solutions in the original inequality. The graph is shown below.

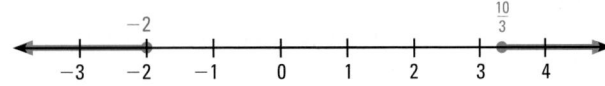

---

2 TEACH

MOTIVATING THE LESSON
Ask students if they think they always get *exactly* 12 oz in a 12 oz can of juice or *exactly* 5 lb of flour in a 5 lb bag of flour. Point out that the amounts don't have to be exact, but regulations require them to be within a certain amount, or *tolerance,* from the advertised value. *Absolute value inequalities* can be used to describe these tolerances.

EXTRA EXAMPLE 1
Solve $|6x - 3| = 15$. $-2, 3$

EXTRA EXAMPLE 2
Solve $|4x - 9| \leq 21$. $-3 \leq x \leq 7.5$

EXTRA EXAMPLE 3
Solve $|3x - 2| > 18$.
$x < -\dfrac{16}{3}$ or $x > \dfrac{20}{3}$

✔ **CHECKPOINT EXERCISES**
For use after Example 1:
1. Solve $|2 - 4x| = 10$. -2 and 3
For use after Example 2:
2. Solve $|-x + 5| < 6$. $-1 < x < 11$
For use after Example 3:
3. Solve $|-3x + 10| \geq 7$.
$x \leq 1$ or $x \geq \dfrac{17}{3}$

STUDENT HELP NOTES

↪ **Homework Help** Students can find extra examples at **www.mcdougallittell.com** that parallel the examples in the student edition.

A manufacturer has a 0.6 oz tolerance for a bottle of salad dressing advertised as 16 oz. Write and solve an absolute value inequality that describes the acceptable volumes for "16 oz" bottles.

$|x - 16| \leq 0.6$, where x represents the actual volume; $15.4 \leq x \leq 16.6$

EXTRA EXAMPLE 5

A city ordinance states that pools must be enclosed by a fence that is from 3 ft to 6 ft high. Write an absolute value inequality describing fences that don't meet this ordinance.

$|h - 4.5| > 1.5$, where h represents the height of the fence

 CHECKPOINT EXERCISES

For use after Examples 4 and 5:

1. A manufacturer has a tolerance of 0.36 lb for a bag of potting soil advertised as 9.6 lb. Write and solve an absolute value inequality that describes unacceptable weights for a "9.6 lb" bag.

$|w - 9.6| > 0.36$, where w represents the actual weight; $w > 9.96$ or $w < 9.24$

CLOSURE QUESTION

How are absolute value inequalities containing a "<" or "≤" symbol solved differently from those containing a ">" or "≥" symbol?
See sample answer below.

DAILY PUZZLER

A company's "18 oz" package of 24 cookies has a tolerance of 0.36 oz. Each cookie has a tolerance of 0.05 oz. Assume that each cookie in a bag is within tolerance. Can half the cookies be above the target weight and half below and still not meet the tolerance for the bag? Give an example.

Yes. *Sample answer:* If half the cookies average 0.045 oz below tolerance and half average 0.01 oz above, the bag will fall 0.06 oz below tolerance.

In manufacturing applications, the maximum acceptable deviation of a product from some ideal or average measurement is called the *tolerance*.

Manufacturing

EXAMPLE 4   *Writing a Model for Tolerance*

A cereal manufacturer has a tolerance of 0.75 ounce for a box of cereal that is supposed to weigh 20 ounces. Write and solve an absolute value inequality that describes the acceptable weights for "20 ounce" boxes.

SOLUTION

PROBLEM SOLVING STRATEGY

VERBAL MODEL

$$\left| \boxed{\text{Actual weight}} - \boxed{\text{Ideal weight}} \right| \leq \boxed{\text{Tolerance}}$$

LABELS

Actual weight $= x$ (ounces)

Ideal weight $= 20$ (ounces)

Tolerance $= 0.75$ (ounces)

ALGEBRAIC MODEL

$|x - 20| \leq 0.75$ Write algebraic model.

$-0.75 \leq x - 20 \leq 0.75$ Write equivalent compound inequality.

$19.25 \leq x \leq 20.75$ Add 20 to each expression.

▶ The weights can range between 19.25 ounces and 20.75 ounces, inclusive.

EXAMPLE 5   *Writing an Absolute Value Model*

QUALITY CONTROL You are a quality control inspector at a bowling pin company. A regulation pin must weigh between 50 ounces and 58 ounces, inclusive. Write an absolute value inequality describing the weights you should reject.

SOLUTION

VERBAL MODEL

$$\left| \boxed{\text{Weight of pin}} - \boxed{\text{Average of extreme weights}} \right| > \boxed{\text{Tolerance}}$$

LABELS

Weight of pin $= w$ (ounces)

Average of extreme weights $= \dfrac{50 + 58}{2} = 54$ (ounces)

Tolerance $= 58 - 54 = 4$ (ounces)

ALGEBRAIC MODEL

$|w - 54| > 4$

▶ You should reject a bowling pin if its weight w satisfies $|w - 54| > 4$.

 BOWLING Bowling pins are made from maple wood, either solid or laminated. They are given a tough plastic coating to resist cracking. The lighter the pin, the easier it is to knock down.

Closure Question *Sample answer:*

Those with a "<" or "≤" symbol are represented by a compound *and* inequality. They have solutions between two values. Those with a ">" or "≥" symbol are represented by a compound *or* inequality. They have solutions outside two values.

GUIDED PRACTICE

Vocabulary Check ✓

1. What is the absolute value of a number?
the number's distance from zero on a number line

Concept Check ✓

2. The absolute value of a number cannot be negative. How, then, can the absolute value of a be $-a$? If a is a negative number, then $-a$ is a positive number.

3. Give an example of the absolute value of a number. How many other numbers have this absolute value? State the number or numbers. $|5| = 5$; one; -5

Skill Check ✓

17. $x - 8 = 11$ or
$x - 8 = -11$

18. $5 - 2x = 13$ or
$5 - 2x = -13$

19. $6n + 1 = \frac{1}{2}$ or
$6n + 1 = -\frac{1}{2}$

20. $5n - 4 = 16$ or
$5n - 4 = -16$

21. $2x + 1 = 5$ or
$2x + 1 = -5$

Decide whether the given number is a solution of the equation.

4. $|3x + 8| = 20$; -4 no
5. $|11 - 4x| = 7$; 1 yes
6. $|2x - 9| = 11$; -1 yes

7. $|-x + 9| = 4$; -5 no
8. $|6 + 3x| = 0$; -2 yes
9. $|-5x - 3| = 8$; -1 no

Rewrite the absolute value inequality as a compound inequality.

10. $|x + 8| < 5$
$-5 < x + 8 < 5$

11. $|11 - 2x| \geq 13$
$11 - 2x \leq -13$ or $11 - 2x \geq 13$

12. $|9 - x| > 21$
$9 - x < -21$ or $9 - x > 21$

13. $|x + 5| \leq 9$
$-9 \leq x + 5 \leq 9$

14. $|10 - 3x| \geq 17$
$10 - 3x \leq -17$ or $10 - 3x \geq 17$

15. $\left|\frac{1}{4}x + 10\right| < 18$
$-18 < \frac{1}{4}x + 10 < 18$

16. 🌐 **TOLERANCE** Suppose the tolerance for the "20 ounce" cereal boxes in Example 4 is now 0.45 ounce. Write and solve an absolute value inequality that describes the new acceptable weights of the boxes. $|x - 20| \leq 0.45$; the weights can range between 19.55 oz and 20.45 oz, inclusive.

PRACTICE AND APPLICATIONS

STUDENT HELP

▶ **Extra Practice**
to help you master skills is on p. 941.

22. $2 - x = 3$ or
$2 - x = -3$

23. $15 - 2x = 8$ or
$15 - 2x = -8$

24. $\frac{1}{2}x + 4 = 6$ or
$\frac{1}{2}x + 4 = -6$

25. $\frac{2}{3}x - 9 = 18$ or
$\frac{2}{3}x - 9 = -18$

REWRITING EQUATIONS Rewrite the absolute value equation as two linear equations. 17–25. See margin.

17. $|x - 8| = 11$
18. $|5 - 2x| = 13$
19. $|6n + 1| = \frac{1}{2}$

20. $|5n - 4| = 16$
21. $|2x + 1| = 5$
22. $|2 - x| = 3$

23. $|15 - 2x| = 8$
24. $\left|\frac{1}{2}x + 4\right| = 6$
25. $\left|\frac{2}{3}x - 9\right| = 18$

CHECKING A SOLUTION Decide whether the given number is a solution of the equation.

26. $|4x + 1| = 11$; 3 no
27. $|8 - 2n| = 2$; -5 no
28. $\left|6 + \frac{1}{2}x\right| = 14$; -40 yes

29. $\left|\frac{1}{5}x - 2\right| = 4$; 10 no
30. $|4n + 7| = 1$; 2 no
31. $|-3x + 5| = 7$; 4 yes

SOLVING EQUATIONS Solve the equation.

32. $|11 + 2x| = 5$ $-3, -8$
33. $|10 - 4x| = 2$ $2, 3$
34. $|22 - 3n| = 5$ $\frac{17}{3}, 9$

35. $|2n - 5| = 7$ $6, -1$
36. $|8x + 1| = 23$ $\frac{11}{4}, -3$
37. $|30 - 7x| = 4$ $\frac{26}{7}, \frac{34}{7}$

38. $\left|\frac{1}{4}x - 5\right| = 8$ $52, -12$
39. $\left|\frac{2}{3}x + 2\right| = 10$ $12, -18$
40. $\left|\frac{1}{2}x - 3\right| = 2$ $10, 2$

STUDENT HELP

▶ **HOMEWORK HELP**
Example 1: Exs. 17–40
Examples 2, 3:
        Exs. 41–58
Examples 4, 5:
        Exs. 65–76

REWRITING INEQUALITIES Rewrite the absolute value inequality as a compound inequality.

41. $|3 + 4x| \leq 15$
$-15 \leq 3 + 4x \leq 15$

42. $|4n - 12| > 16$
$4n - 12 < -16$ or $4n - 12 > 16$

43. $|3x + 2| < 7$
$-7 < 3x + 2 < 7$

44. $|2x - 1| \geq 12$
$2x - 1 \leq -12$ or $2x - 1 \geq 12$

45. $|8 - 3n| \leq 18$
$-18 \leq 8 - 3n \leq 18$

46. $|11 + 4x| < 23$
$-23 < 11 + 4x < 23$

1.7 *Solving Absolute Value Equations and Inequalities*   53

APPLY 3

ASSIGNMENT GUIDE

BASIC
Day 1: pp. 53–56 Exs. 18–58 even, 59, 61, 65, 66, 77–79, 90–106 even, Quiz 3 Exs. 1–18

AVERAGE
Day 1: pp. 53–56 Exs. 18–58 even, 59–65 odd, 66, 67, 77–85, 90–106 even, Quiz 3 Exs. 1–18

ADVANCED
Day 1: pp. 53–56 Exs. 18–58 even, 59–65 odd, 66, 67–75 odd, 76–89, 90–106 even, Quiz 3 Exs. 1–18

BLOCK SCHEDULE
pp. 53–56 Exs. 18–58 even, 59–65 odd, 66, 67, 77–85, 90–106 even, Quiz 3 Exs. 1–18

EXERCISE LEVELS
Level A: *Easier*
17–40, 59–66, 90–100
Level B: *More Difficult*
41–58, 67–85, 100–106
Level C: *Most Difficult*
86–89

✓ **HOMEWORK CHECK**
To quickly check student understanding of key concepts, go over the following exercises: Exs. 18, 28, 34, 42, 50, 58. See also the Daily Homework Quiz:

• Blackline Master (*Chapter 2 Resource Book*, p. 11)

• 📑 Transparency (p. 10)

COMMON ERROR

EXERCISES 47–58 Some students may initially have trouble solving these inequalities. Refer students to Examples 2 and 3, and provide extra examples from the Web site to help these students.

MATHEMATICAL REASONING

The equation $|x| = 4$ means "the numbers whose distance on a number line from 0 is 4." So, these numbers are 4 and −4. Notice that this equation could be written as $|x - 0| = 4$. The equation $|x - 20| = 4$ means "the numbers whose distance on a number line from 20 is 4." So, these numbers are 16 and 24. Using this reasoning, have students write an interpretation of the tolerance requirement $|w - 10| \le 0.25$ for a 10-lb bag of potting soil. **See sample answer below.**

STUDENT HELP NOTES

→ **Keystroke Help**
Keystrokes for several models of calculators are available at **www.mcdougallittell.com**.

GRAPHING CALCULATOR NOTE

EXERCISES 59–64 For help with these exercises, review Activity 1.7 on page 49.

47. $-9 < x < 7$;

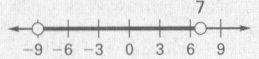

48. $-7 \le x \le 31$;

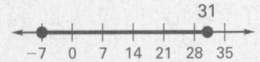

49. $x \le 6$ or $x \ge 26$;

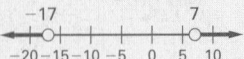

50. $x < -17$ or $x > 7$;

51–56. See the next page.
57–58, 65–66.
See Additional Answers beginning on page AA1.

54

STUDENT HELP

KEYSTROKE HELP
Visit our Web site www.mcdougallittell.com to see keystrokes for several models of calculators.

59. $-4 < x < 2$

60. $0 \le x \le 1$

61. $x < -3$ or $x > 7$

62. $-4 \le x \le 8$

63. $x < 1$ or $x > 4$

64. $x \le -6$ or $x \ge 7$

FOCUS ON APPLICATIONS

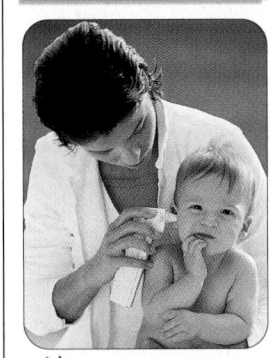

→ **BODY TEMPERATURE**
Doctors routinely use ear thermometers to measure body temperature. The first ear thermometers were used in 1990. The thermometers use an infrared sensor and microprocessors.

SOLVING AND GRAPHING Solve the inequality. Then graph your solution.
47–58. See margin.

47. $|x + 1| < 8$ **48.** $|12 - x| \le 19$ **49.** $|16 - x| \ge 10$

50. $|x + 5| > 12$ **51.** $|x - 8| \le 5$ **52.** $|x - 16| > 24$

53. $|14 - 3x| > 18$ **54.** $|4x + 10| < 20$ **55.** $|8x + 28| \ge 32$

56. $\left|20 + \frac{1}{2}x\right| > 6$ **57.** $|7x + 5| < 23$ **58.** $|11 + 6x| \le 47$

SOLVING INEQUALITIES Use the *Test* feature of a graphing calculator to solve the inequality. Most calculators use *abs* for absolute value. For example, you enter $|x + 1|$ as abs(x + 1). 59–64. See margin.

59. $|x + 1| < 3$ **60.** $\left|\frac{2}{3}x - \frac{1}{3}\right| \le \frac{1}{3}$ **61.** $|2x - 4| > 10$

62. $\left|\frac{1}{2}x - 1\right| \le 3$ **63.** $|4x - 10| > 6$ **64.** $|1 - 2x| \ge 13$

🌐 **PALM WIDTHS** In Exercises 65 and 66, use the following information. In a sampling conducted by the United States Air Force, the right-hand dimensions of 4000 Air Force men were measured. The gathering of such information is useful when designing control panels, keyboards, gloves, and so on.
65, 66. See margin for graphs.

65. Ninety-five percent of the palm widths p were within 0.26 inch of 3.49 inches. Write an absolute value inequality that describes these values of p. Graph the inequality.
$|p - 3.49| \le 0.26$

66. Ninety-nine percent of the palm widths p were within 0.37 inch of 3.49 inches. Write an absolute value inequality that describes these values of p. Graph the inequality.
$|p - 3.49| \le 0.37$

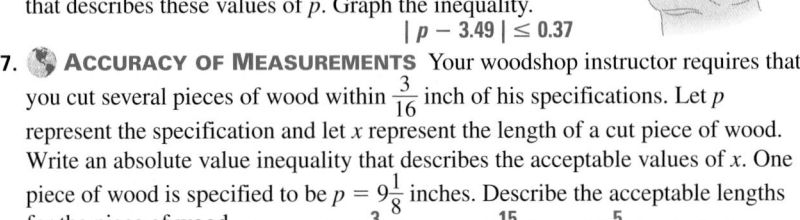

67. 🌐 **ACCURACY OF MEASUREMENTS** Your woodshop instructor requires that you cut several pieces of wood within $\frac{3}{16}$ inch of his specifications. Let p represent the specification and let x represent the length of a cut piece of wood. Write an absolute value inequality that describes the acceptable values of x. One piece of wood is specified to be $p = 9\frac{1}{8}$ inches. Describe the acceptable lengths for the piece of wood. $|x - p| \le \frac{3}{16}$; between $8\frac{15}{16}$ in. and $9\frac{5}{16}$ in., inclusive.

68. 🌐 **BASKETBALL** The length of a standard basketball court can vary from 84 feet to 94 feet, inclusive. Write an absolute value inequality that describes the possible lengths of a standard basketball court. $|l - 89| \le 5$

69. 🌐 **BODY TEMPERATURE** Physicians consider an adult's normal body temperature to be within 1°F of 98.6°F, inclusive. Write an absolute value inequality that describes the range of normal body temperatures. $|t - 98.6| \le 1$

🌐 **WEIGHING FLOUR** In Exercises 70 and 71, use the following information. A 16 ounce bag of flour probably does not weigh exactly 16 ounces. Suppose the actual weight can be between 15.6 ounces and 16.4 ounces, inclusive.

70. Write an absolute value inequality that describes the acceptable weights for a "16 ounce" bag of flour. $|w - 16| \le 0.4$

71. A case of flour contains 24 of these "16 ounce" bags. What is the greatest possible weight of the flour in a case? What is the least possible weight? Write an absolute value inequality that describes the acceptable weights of a case.
393.6 oz; 374.4 oz; $|c - 384| \le 9.6$

Mathematical Reasoning *Sample answer:*
The weights in pounds that are acceptable are those whose distance on a number line from 10 is not more than 0.25, or those from 9.75 lb (9 lb 12 oz) to 10.25 lb (10 lb 4 oz).

72. volleyball:
$|v - 270| \le 10$,
basketball:
$|b - 625| \le 25$,
water polo:
$|w - 425| \le 25$,
lacrosse:
$|l - 145.5| \le 3.5$,
football:
$|f - 14.5| \le 0.5$

73. volleyball:
$|v - 270| > 10$,
basketball:
$|b - 625| > 25$,
water polo:
$|w - 425| > 25$,
lacrosse:
$|l - 145.5| > 3.5$,
football:
$|f - 14.5| > 0.5$

Test Preparation

75. 2 L: $|c - 2000| > 9$,
1 L: $|c - 1000| > 5$,
500 mL: $|c - 500| > 2$

76. $|h - (2.26 \cdot 51.6 + 66.4)|$
≤ 3.42; between 180 cm
and 186 cm, inclusive

★ **Challenge**

EXTRA CHALLENGE
→ www.mcdougallittell.com

🌐 **SPORTS EQUIPMENT** **In Exercises 72 and 73, use the table giving the recommended weight ranges for the balls from five different sports.**

72. Write an absolute value inequality for the weight range of each ball.

73. For each ball, write an absolute value inequality describing the weights of balls that are *outside* the recommended range.

Sport	Weight range of ball used
Volleyball	260–280 grams
Basketball	600–650 grams
Water polo	400–450 grams
Lacrosse	142–149 grams
Football	14–15 ounces

74. **SCIENCE CONNECTION** Green plants can live in the ocean only at depths of 0 feet to 100 feet. Write an absolute value inequality describing the range of possible depths for green plants in an ocean. $|d - 50| \le 50$

75. 🌐 **BOTTLING** A juice bottler has a tolerance of 9 milliliters in a two liter bottle, of 5 milliliters in a one liter bottle, and of 2 milliliters in a 500 milliliter bottle. For each size of bottle, write an absolute value inequality describing the capacities that are outside the acceptable range. **See margin.**

76. **SCIENCE CONNECTION** To determine height from skeletal remains, scientists use the equation $H = 2.26f + 66.4$ where H is the person's height (in centimeters) and f is the skeleton's femur length (in centimeters). The equation has a margin of error of ± 3.42 centimeters. Suppose a skeleton's femur length is 51.6 centimeters. Write an absolute value inequality that describes the person's height. Then solve the inequality to find the range of possible heights. **See margin.**

77. **MULTIPLE CHOICE** Which of the following are solutions of $|3x - 7| = 14$? B

Ⓐ $x = \frac{7}{3}$ or $x = 7$ ⓑ $x = -\frac{7}{3}$ or $x = 7$

ⓒ $x = \frac{7}{3}$ or $x = -7$ Ⓓ $x = -\frac{7}{3}$ or $x = -7$

78. **MULTIPLE CHOICE** Which of the following is equivalent to $|2x - 9| < 3$? B

Ⓐ $-3 \le x \le 6$ ⓑ $3 < x < 6$

ⓒ $3 \le x \le 6$ Ⓓ $-3 < x < -6$

79. **MULTIPLE CHOICE** Which of the following is equivalent to $|3x + 5| \ge 19$? C

Ⓐ $x \le -\frac{14}{3}$ or $x \ge 8$ ⓑ $x < -8$ or $x > \frac{14}{3}$

ⓒ $x \le -8$ or $x \ge \frac{14}{3}$ Ⓓ $x < -\frac{14}{3}$ or $x > 8$

SOLVING INEQUALITIES Solve the inequality. If there is no solution, write *no solution*.

80. $|2x + 3| \ge -13$ all real numbers **81.** $|5x + 2| \le -2$ no solution

82. $|3x - 8| < -10$ no solution **83.** $|4x - 2| > -6$ all real numbers

84. $|6 - 2x| > -8$ all real numbers **85.** $|7 - 3x| \le -14$ no solution

SOLVING INEQUALITIES Solve for *x*. Assume *a* and *b* are positive numbers.

86. $|x + a| < b$ $-b - a < x < b - a$ **87.** $|x - a| > b$ $x < a - b$ or $x > a + b$

88. $|x + a| \ge a$ $x \le -2a$ or $x \ge 0$ **89.** $|x - a| \le a$ $0 \le x \le 2a$

1.7 *Solving Absolute Value Equations and Inequalities* **55**

! **COMMON ERROR**
EXERCISES 80–85 Students may become confused trying to solve these problems methodically. It will help them to use their observational skills and the definition of absolute value as distance. For Exs. 80, 83, and 84, they should see that a distance is always more than any negative number, so any real number will be a solution. For Exs. 81, 82, and 85, they should see that a distance is never a negative number, so it cannot be less than any negative number. So, no real number can be a solution.

51. $3 \le x \le 13$;

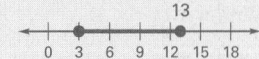

52. $x < -8$ or $x > 40$;

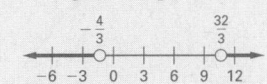

53. $x < -\frac{4}{3}$ or $x > \frac{32}{3}$;

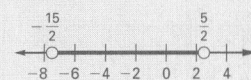

54. $-\frac{15}{2} < x < \frac{5}{2}$;

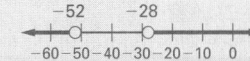

55. $x \le -\frac{15}{2}$ or $x \ge \frac{1}{2}$;

56. $x < -52$ or $x > -28$;

ADDITIONAL PRACTICE AND RETEACHING

For Lesson 1.7:
• Practice Levels A, B, and C (*Chapter 1 Resource Book,* p. 98)
• Reteaching with Practice (*Chapter 1 Resource Book,* p. 101)
• 💻 See Lesson 1.7 of the *Personal Student Tutor*

For more Mixed Review:
• 💻 Search the *Test and Practice Generator* for key words or specific lessons.

📝 **Transparency Available**

1. Rewrite $|2 - 5x| = 6$ as two linear equations. $2 - 5x = -6$ or $2 - 5x = 6$

2. Is -15 a solution of $\left|20 - \frac{5}{3}x\right| = 5$? no

3. Solve $|4 - 6x| = 2$. $\frac{1}{3}, 1$

4. Rewrite $|5 + 4x| < 7$ as a compound inequality. $-7 < 5 + 4x < 7$

Solve the inequality. Then graph your solution.

5. $|9 - 4x| < 5$ $1 < x < \frac{7}{2}$

6. $|2x + 10| \geq 4$ $x \leq -7$ or $x \geq -3$

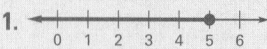

EXTRA CHALLENGE NOTE

→ Challenge problems for Lesson 1.7 are available in **blackline** format in the *Chapter 1 Resource Book,* p. 105 and at **www.mcdougallittell.com.**

ADDITIONAL TEST PREPARATION

1. WRITING The definition for an absolute value equation states that $|ax + b| = c$, where $c > 0$. Why is the condition that $c > 0$ necessary?

Sample answer: Absolute value represents distance on a number line, and the distance between any non-identical points is always positive.

1. [number line from 0 to 6]

2–4, 11–16. See the next page.

MIXED REVIEW

LOGICAL REASONING Tell whether the statement is *true* or *false*. If the statement is false, explain why. (Skills Review, p. 926)

90. A triangle is a right triangle if and only if it has a right angle. true

91. $2x = 14$ if and only if $x = -7$. False; if $x = -7$, then $2x = 2(-7) = -14$, not 14.

92. All rectangles are squares. False; a rectangle is a square only if its length and width are equal.

EVALUATING EXPRESSIONS Evaluate the expression for the given value(s) of the variable(s). (Review 1.2 for 2.1)

93. $5x - 9$ when $x = 6$ 21

94. $-2y + 4$ when $y = 14$ -24

95. $11c + 6$ when $c = -3$ -27

96. $-8a - 3$ when $a = -4$ 29

97. $a - 11b + 2$ when $a = 61$ and $b = 7$ -14

98. $15x + 8y$ when $x = \frac{1}{2}$ and $y = \frac{1}{3}$ $\frac{61}{6}$

99. $\frac{1}{5}\left(8g + \frac{1}{3}h\right)$ when $g = 6$ and $h = 6$ 10

100. $\frac{1}{5}(p + q) - 7$ when $p = 5$ and $q = 3$ $-\frac{27}{5}$

SOLVING INEQUALITIES Solve the inequality. (Review 1.6)

101. $6x + 9 > 11$ $x > \frac{1}{3}$

102. $15 - 2x \geq 45$ $x \leq -15$

103. $-3x - 5 \leq 10$ $x \geq -5$

104. $13 + 4x < 9$ $x < -1$

105. $-18 < 2x + 10 < 6$ $-14 < x < -2$

106. $x + 2 \leq -1$ or $4x \geq 8$ $x \leq -3$ or $x \geq 2$

QUIZ 3
Self-Test for Lessons 1.6 and 1.7

Solve the inequality. Then graph your solution. (Lesson 1.6) 1–4. See margin for graphs.

1. $4x - 3 \leq 17$ $x \leq 5$

2. $2y - 9 > 5y + 12$ $y < -7$

3. $-8 < 3x + 4 < 22$ $-4 < x < 6$

4. $3x - 5 \leq -11$ or $2x - 3 > 3$ $x \leq -2$ or $x > 3$

Solve the equation. (Lesson 1.7)

5. $|x + 5| = 4$ $-1, -9$

6. $|x - 3| = 2$ $5, 1$

7. $|6 - x| = 9$ $-3, 15$

8. $|4x - 7| = 13$ $5, -\frac{3}{2}$

9. $|3x + 4| = 20$ $\frac{16}{3}, -8$

10. $|15 - 3x| = 12$ $1, 9$

Solve the inequality. Then graph your solution. (Lesson 1.7) 11–16. See margin for graphs.

11. $|y + 2| \geq 3$

12. $|x + 6| < 4$

13. $|x - 3| > 7$

14. $|2y - 5| \leq 3$

15. $|2x - 3| > 1$

16. $|4x + 5| \geq 13$

11. $y \leq -5$ or $y \geq 1$

12. $-10 < x < -2$

13. $x < -4$ or $x > 10$

14. $1 \leq y \leq 4$

15. $x < 1$ or $x > 2$

16. $x \leq -\frac{9}{2}$ or $x \geq 2$

17. 🌐 **FUEL EFFICIENCY** Your car gets between 20 miles per gallon and 28 miles per gallon of gasoline and has a 16 gallon gasoline tank. Write a compound inequality that represents your fuel efficiency. How many miles can you travel on one tank of gasoline? (Lesson 1.6) $20 \leq e \leq 28$; between 320 mi and 448 mi, inclusive

18. 🌐 **MANUFACTURING TOLERANCE** The ideal diameter of a certain type of ball bearing is 30 millimeters. The manufacturer has a tolerance of 0.045 millimeter. Write an absolute value inequality that describes the acceptable diameters for these ball bearings. Then solve the inequality to find the range of acceptable diameters. (Lesson 1.7) $|d - 30| \leq 0.045$; between 29.955 mm and 30.045 mm, inclusive

Chapter Summary

WHAT did you learn?

Graph and order real numbers. **(1.1)**

Identify properties of and perform operations with real numbers. **(1.1)**

Evaluate and simplify algebraic expressions. **(1.2)**

Solve equations.
- linear equations **(1.3)**

- absolute value equations **(1.7)**

Rewrite equations and common formulas with more than one variable. **(1.4)**

Use a problem solving plan and strategies to solve real-life problems. **(1.5)**

Solve and graph inequalities in one variable.
- linear inequalities **(1.6)**

- absolute value inequalities **(1.7)**

Write and use algebraic models to solve real-life problems. **(1.2–1.7)**

WHY did you learn it?

Analyze record low temperatures. **(p. 8)**

Learn how to exchange money. **(p. 6)**

Find the population of Hawaii. **(p. 16)**

Find the temperature in degrees Celsius at which dry ice changes from a solid to a gas. **(p. 23)**

Solve problems that involve tolerance. **(p. 52)**

Find how much you should charge for tickets to a benefit concert. **(p. 27)**

Find the average speed of the Bullet Train. **(p. 33)**

Decide how to spend your money at an amusement park. **(p. 46)**

Describe recommended weight ranges for balls used in various sports. **(p. 55)**

Use femur length to find a range of possible heights for a person. **(p. 55)**

How does Chapter 1 fit into the BIGGER PICTURE of algebra?

Chapter 1 provides a review of skills and strategies you learned in Algebra 1 and a foundation for continuing your study of algebra and its applications. The primary use of algebra is to model and solve real-life problems. You will use algebra in this way throughout the course, in future courses, and perhaps in a future career.

STUDY STRATEGY

How did you make and use a vocabulary file?

Here is an example of one flashcard for your vocabulary file, following the **Study Strategy** on page 2.

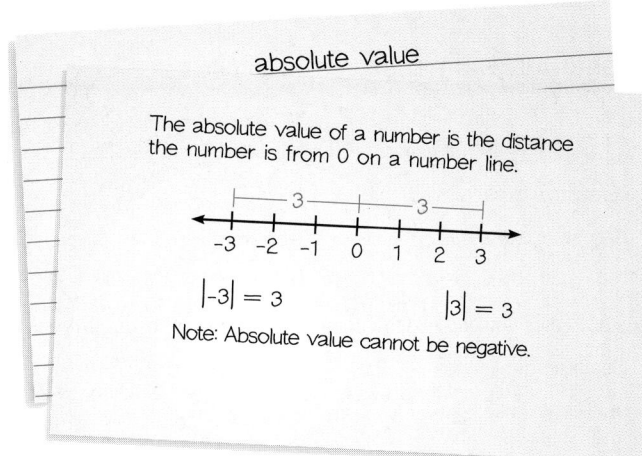

absolute value

The absolute value of a number is the distance the number is from 0 on a number line.

$|-3| = 3$ $|3| = 3$

Note: Absolute value cannot be negative.

2.

3.

4.

11.

12.

13.

14.

15.

16.

57

ADDITIONAL RESOURCES

The following resources are avail-able to help review the material in this chapter.

- Chapter Review Games and Activities (*Chapter 1 Resource Book*, p. 106)
- *Instant Replay: Video Review Games*
- Personal Student Tutor
- Cumulative Review, Ch. 1 (*Chapter 1 Resource Book*, p. 118)

1.
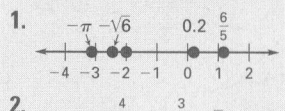

2.
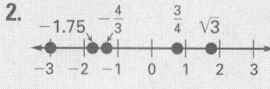

Chapter Review

VOCABULARY

- whole numbers, p. 3
- integers, p. 3
- rational numbers, p. 3
- irrational numbers, p. 3
- origin, p. 3
- graph of a real number, p. 3
- coordinate, p. 3
- opposite, p. 5
- reciprocal, p. 5
- numerical expression, p. 11

- base, p. 11
- exponent, p. 11
- power, p. 11
- order of operations, p. 11
- variable, p. 12
- value of a variable, p. 12
- algebraic expression, p. 12
- value of an expression, p. 12
- mathematical model, p. 12
- terms of an expression, p. 13

- coefficient, p. 13
- like terms, p. 13
- constant terms, p. 13
- equivalent expressions, p. 13
- identity, p. 13
- equation, p. 19
- linear equation, p. 19
- solution of an equation, p. 19
- equivalent equations, p. 19

- verbal model, p. 33
- algebraic model, p. 33
- linear inequality in one variable, p. 41
- solution of a linear inequality in one variable, p. 41
- graph of a linear inequality in one variable, p. 41
- compound inequality, p. 43
- absolute value, p. 50

1.1 REAL NUMBERS AND NUMBER OPERATIONS

Examples on pp. 3–6

> **EXAMPLE** You can use a number line to graph and order real numbers.
>
>
>
> Increasing order (left to right):
> $-4, -1, 0.3, \sqrt{7}$
>
> Properties of real numbers include the closure, commutative, associative, identity, inverse, and distributive properties.

Graph the numbers on a number line. Then write the numbers in increasing order.

1, 2. See margin for graphs.

1. $-2, 0.2, -\pi, -\sqrt{6}, \frac{6}{5}$
   $-\pi, -\sqrt{6}, -2, 0.2, \frac{6}{5}$

2. $\frac{3}{4}, \sqrt{3}, -1.75, -3, -\frac{4}{3}$
   $-3, -1.75, -\frac{4}{3}, \frac{3}{4}, \sqrt{3}$

Identify the property shown.

3. $4(5 + 1) = 4 \cdot 5 + 4 \cdot 1$ distributive property

4. $8 + (-8) = 0$ additive inverse property

1.2 ALGEBRAIC EXPRESSIONS AND MODELS

Examples on pp. 11–13

> **EXAMPLES** You can use order of operations to evaluate expressions.
>
> Numerical expression: $8(3 + 4^2) - 12 \div 2 = 8(3 + 16) - 6 = 8(19) - 6 = 152 - 6 = 146$
>
> Algebraic expression: $3x^2 - 1$ when $x = -5$
>
> $3(-5)^2 - 1 = 3(25) - 1 = 75 - 1 = 74$
>
> Sometimes you can use the distributive property to simplify an expression.
>
> Combine like terms: $2x^2 - 4x + 10x - 1 = 2x^2 + (-4 + 10)x - 1 = 2x^2 + 6x - 1$

Evaluate the expression.

5. $-3 - 6 \div 2 - 12$ **−18**

6. $-5 \div 1 + 2(7 - 10)^2$ **13**

7. $7x - 3x - 8x^3$ when $x = -1$ **4**

8. $3ab^2 + 5a^2b - 1$ when $a = 2$ and $b = -2$

Simplify the expression.

9. $7y - 2x + 5x - 3y + 2x$ **5x + 4y**

10. $4(3 - x) + 5(x - 6)$ **x − 18**

11. $6x^2 - 3x + 5x^2 + 2x$ **11x² − x**

12. $2(x^2 + x) - 3(x^2 - 4x)$ **−x² + 14x**

1.3 SOLVING LINEAR EQUATIONS

EXAMPLE You can use properties of real numbers and transformations that produce equivalent equations to solve linear equations.

Solve: $-2(x - 4) = 12$

$$-2x + 8 = 12$$

$$-2x = 4$$

$$x = -2$$

Then check: $-2(-2 - 4) \overset{?}{=} 12$

$$-2(-6) \overset{?}{=} 12$$

$$12 = 12 \checkmark$$

Solve the equation. Check your solution.

13. $-5x + 3 = 18$ **−3**

14. $\frac{2}{3}n - 5 = 1$ **9**

15. $\frac{1}{2}y = -\frac{3}{4}y - 40$ **−32**

16. $2 - 3a = 4 + a$ $-\frac{1}{2}$

17. $8(z - 6) = -16$ **4**

18. $-4x - 4 = 3(2 - x)$ **−10**

1.4 REWRITING EQUATIONS AND FORMULAS

Examples on pp. 26–28

EXAMPLES You can solve an equation that has more than one variable, such as a formula, for one of its variables.

Solve the equation for y.

$$2x - 3y = 6$$

$$-3y = -2x + 6$$

$$y = \frac{2}{3}x - 2$$

Solve the formula for the area of a trapezoid for h.

$$A = \frac{1}{2}(b_1 + b_2)h$$

$$2A = (b_1 + b_2)h$$

$$\frac{2A}{b_1 + b_2} = h$$

Solve the equation for y.

19. $5x - y = 10$ $y = 5x - 10$

20. $x + 4y = -8$ $y = -\frac{1}{4}x - 2$

21. $0.1x + 0.5y = 3.5$ $y = -0.2x + 7$

22. $2x = 3y + 9$ $y = \frac{2}{3}x - 3$

23. $5x - 6y + 12 = 0$ $y = \frac{5}{6}x + 2$

24. $x - 2xy = 1$ $y = \frac{x-1}{2x}$

Solve the formula for the indicated variable.

25. Perimeter of a Rectangle $l = \frac{P - 2w}{2}$

Solve for ℓ: $P = 2\ell + 2w$

26. Celsius to Fahrenheit $C = \frac{5}{9}(F - 32)$

Solve for C: $F = \frac{9}{5}C + 32$

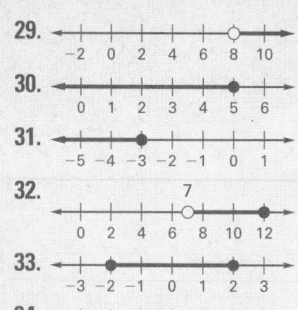

29.

30.

31.

32. 7

33.

34.

PROBLEM SOLVING USING ALGEBRAIC MODELS

Examples on
pp. 33–36

EXAMPLE You can use a problem solving plan in which you write a verbal model, assign labels, write and solve an algebraic model, and then answer the question.

How far can you drive at 55 miles per hour for 4 hours?

VERBAL MODEL
$$\boxed{\text{Distance}} = \boxed{\text{Rate}} \cdot \boxed{\text{Time}}$$

LABELS
Distance = d (miles), Rate = **55** (miles per hour), Time = **4** (hours)

ALGEBRAIC MODEL
$d = 55 \cdot 4 = 220$

▶ You can drive 220 miles.

27. How long will it take to drive 325 miles at 55 miles per hour? **about 5 h 55 min**

28. While on vacation, you take a taxi from the airport to your hotel for $21.85. The taxi costs $2.95 plus $1.35 per mile. How far is it from the airport to the hotel? **14 mi**

SOLVING LINEAR INEQUALITIES

Examples on
pp. 41–44

EXAMPLES You can use transformations to solve inequalities. Reverse the inequality when you multiply or divide both sides by a negative number.

$4x + 1 < 7x - 5$

$-3x < -6$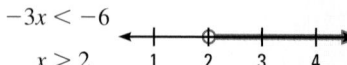

$x > 2$

$0 \leq 6 - 2n \leq 10$

$-6 \leq -2n \leq 4$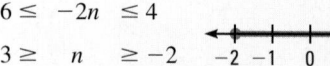

$3 \geq n \geq -2$

Solve the inequality. Then graph your solution. **29–34. See margin for graphs.**

29. $2x - 10 > 6$ $x > 8$ **30.** $12 - 5x \geq -13$ $x \leq 5$ **31.** $-3x + 4 \geq 2x + 19$ $x \leq -3$

32. $0 < x - 7 \leq 5$ $7 < x \leq 12$ **33.** $-3 \leq 2y + 1 \leq 5$ **34.** $3a + 1 < -2$ or $3a + 1 > 7$

                                                  $-2 \leq y \leq 2$ $a < -1$ or $a > 2$

SOLVING ABSOLUTE VALUE EQUATIONS AND INEQUALITIES

Examples on
pp. 50–52

EXAMPLES To solve an absolute value equation, rewrite it as two linear equations. To solve an absolute value inequality, rewrite it as a compound inequality.

$|x + 3| = 5$

$x + 3 = 5$ or $x + 3 = -5$

$x = 2$ or $x = -8$

$|x - 7| \geq 2$

$x - 7 \geq 2$ or $x - 7 \leq -2$

$x \geq 9$ or $x \leq 5$

Solve the equation or inequality.

35. $|x + 1| = 4$ $-5, 3$ **36.** $|2x - 1| = 15$ $-7, 8$ **37.** $|10 - 6x| = 26$ $-\frac{8}{3}, 6$

38. $|x + 8| > 0$ **39.** $|2x - 5| < 9$ $-2 < x < 7$ **40.** $|3x + 4| \geq 2$

$x < -8$ or $x > -8$; equivalently, $x \neq -8$ $x \leq -2$ or $x \geq -\frac{2}{3}$

Chapter 1 *Equations and Inequalities*

Chapter Test

ADDITIONAL RESOURCES
• *Chapter 1 Resource Book*
  Chapter Test (3 levels) (p. 107)
  SAT/ACT Chapter Test (p. 113)
  Alternative Assessment (p. 114)
• Test and Practice Generator

Graph the numbers on a number line. Then write the numbers in increasing order. 1–3. See margin for graphs.

1. $-0.98, -0.9, -1, -1.95$
$-1.95, -1, -0.98, -0.9$

2. $\frac{2}{3}, -\frac{3}{2}, -\frac{2}{3}, 0, \frac{3}{2}$ $-\frac{3}{2}, -\frac{2}{3}, 0, \frac{2}{3}, \frac{3}{2}$

3. $\sqrt{4}, 4, 2\frac{3}{4}, \sqrt{10}, \frac{7}{2}$ $\sqrt{4}, 2\frac{3}{4}, \sqrt{10}, \frac{7}{2}, 4$

Identify the property shown.

4. $7(11 + 9) = 7 \cdot 11 + 7 \cdot 9$
distributive property

5. $8xy = 8yx$ commutative property of multiplication

6. $50 + 0 = 50$ additive identity

Select and perform an operation to answer the question.

7. What is the product of -5 and -3? **15**

8. What is the difference of 29 and -20? **49**

Evaluate the expression.

9. $18 - 7 \cdot 15 \div 3$ **-17**

10. $36 - 5^2 \cdot 2 + 7$ **-7**

11. $12 - 3(1 - 17) \div 4$ **24**

12. $-4x^2 + 6xy$ when $x = -2$ and $y = 5$ **-76**

13. $\frac{3}{5}x - \frac{7}{2}y$ when $x = 3$ and $y = 4$ **$-12\frac{1}{5}$**

Simplify the expression.

14. $-2x + 4y - 10 + x$ $-x + 4y - 10$

15. $4y + 6x - 3(x - 2y)$ **$3x + 10y$**

16. $5(x^2 - 9x) - 2(3x + 4) + 7$ $5x^2 - 51x - 1$

Solve the equation.

17. $7x + 12 = -16$ **-4**

18. $1.2x = 2.3x - 2.2$ **2**

19. $4x + 21 = 7(x + 9)$ **-14**

20. $|x - 4| = 15$ **$-11, 19$**

21. $|5x + 11| = 9$ $-4, -\frac{2}{5}$

22. $|13 + 2x| = 5$ **$-9, -4$**

Solve the equation for y.

23. $5x + y = 7$ $y = 7 - 5x$

24. $6x - 3y = 1$ $y = 2x - \frac{1}{3}$

25. $2xy + x = 12$ $y = \frac{12 - x}{2x}$

Solve the inequality. Then graph your solution. 26–31. See margin for graphs.

26. $4x - 5 \le 15$ $x \le 5$

27. $3 < 2x + 11 < 17$ $-4 < x < 3$

28. $8x < 1$ or $x - 9 > -5$ $x < \frac{1}{8}$ or $x > 4$

29. $|3x - 1| > 7$ $x < -2$ or $x > \frac{8}{3}$

30. $|x + 3| \ge 4$ $x \le -7$ or $x \ge 1$

31. $|1 - 2x| \le 3$ $-1 \le x \le 2$

32. **GEOMETRY CONNECTION** The formula for the volume of a cylinder is $V = \pi r^2 h$. Solve the formula for h. How tall is a cylindrical can with radius 3 centimeters and volume 200 cubic centimeters? $h = \frac{V}{\pi r^2}$; **7.074 cm**

33. **PHONE CALLS** A company charges $.09 per minute for any long distance call, along with a $5 monthly fee. Your monthly bill shows that you owe $27.23. For how many minutes of long distance calls were you charged? **247 min**

34. **SAVING MONEY** You plan to save $15 per week from your allowance to buy a snowboard for $400. How many *months* will it take? **about 6 months**

35. **HOT WATER LAKE** Boiling Lake is a small lake on the island of Dominica. The water temperature of the lake is between 180°F and 197°F. Write a compound inequality for this temperature range. Graph the inequality. $180 \le T \le 197$; See margin for graph.

36. **BASKETBALL BOUNCE** If manufactured correctly, a basketball should bounce from 48 inches to 56 inches when dropped from a height of 6 feet. Determine the tolerance for the bounce height of a basketball and write an absolute value inequality for acceptable bounce heights. 4 in.; $|h - 52| \le 4$

Chapter Test **61**

1.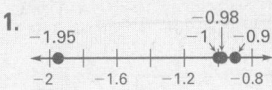

2.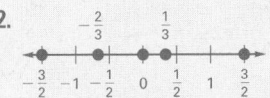

3.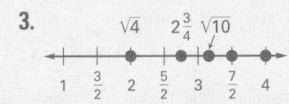

26.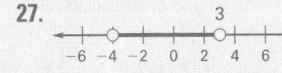

27.

28.

29.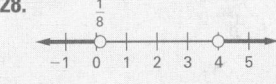

30.

31.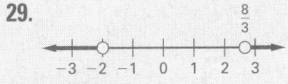

35. Water Temperature (°F)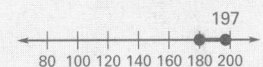

ADDITIONAL RESOURCES
• *Chapter 1 Resource Book*
  Chapter Test (3 levels) (p. 107)
  SAT/ACT Chapter Test (p. 113)
  Alternative Assessment (p. 114)
• 🖵 *Test and Practice Generator*

CHAPTER
1

Chapter Standardized Test

● **TEST-TAKING STRATEGY** Draw an arrow on your test booklet next to questions that you do not answer. This will enable you to find the questions quickly when you go back.

1. **MULTIPLE CHOICE** Which list of numbers is written in increasing order? **C**

   Ⓐ $-\sqrt{7}, -3, -\frac{5}{2}, 0, \frac{3}{4}$

   Ⓑ $-6, -4.5, -4.8, 1, 1.9$

   Ⓒ $-\sqrt{3}, -\frac{8}{5}, -\frac{1}{2}, 0, \frac{1}{8}$

   Ⓓ $-\frac{11}{2}, -\sqrt{2}, -\frac{1}{7}, -\frac{1}{4}, \frac{5}{2}$

   Ⓔ $-0.5, -\sqrt{2}, -\frac{7}{2}, -\sqrt{13}, -13$

2. **MULTIPLE CHOICE** Which property is illustrated by the statement $6(8 + 4) = 6(4 + 8)$? **D**

   Ⓐ Distributive property

   Ⓑ Associative property of addition

   Ⓒ Associative property of multiplication

   Ⓓ Commutative property of addition

   Ⓔ Commutative property of multiplication

3. **MULTIPLE CHOICE** Which expression *cannot* be simplified? **A**

   Ⓐ $8x - 8$ Ⓑ $8x - x$

   Ⓒ $8x + 5x$ Ⓓ $(8 + 5)x$

   Ⓔ $-(x + x)$

4. **MULTIPLE CHOICE** Which number does $(7 + 1)^2 - 16 \div 2 + 6 \div 3$ equal? **E**

   Ⓐ $\frac{62}{3}$ Ⓑ $\frac{23}{3}$ Ⓒ 10

   Ⓓ 2   Ⓔ 58

5. **MULTIPLE CHOICE** Which number does $4x^2 - 5x + 3$ equal when $x = -3$? **C**

   Ⓐ -48 Ⓑ 24   Ⓒ 54

   Ⓓ -54 Ⓔ -18

6. **MULTIPLE CHOICE** Which number is the solution of the equation $-4x + 8 = x - 7$? **C**

   Ⓐ -3 Ⓑ -5 Ⓒ 3

   Ⓓ 5   Ⓔ $-\frac{1}{5}$

7. **MULTIPLE CHOICE** A real estate broker earns a salary of $21,000 plus 2.5% of the value of any real estate sold. Last year the broker earned $52,000. What was the total value of all real estate sold by the broker? **D**

   Ⓐ $12,400   Ⓑ $31,000

   Ⓒ $124,000   Ⓓ $1,240,000

   Ⓔ $12,400,000

8. **MULTIPLE CHOICE** Which gives the equation $C = 2\pi r$ solved for r? **B**

   Ⓐ $r = 2\pi C$ Ⓑ $r = \frac{C}{2\pi}$

   Ⓒ $r = \frac{2\pi}{C}$ Ⓓ $r = \frac{2C}{\pi}$

   Ⓔ $r = 2\frac{\pi}{2C}$

9. **MULTIPLE CHOICE** Which inequality is the solution of $6x - 3 \geq 7 + 4x$? **E**

   Ⓐ $x \geq -5$ Ⓑ $x > 5$ Ⓒ $x \leq 5$

   Ⓓ $x \leq -5$ Ⓔ $x \geq 5$

10. **MULTIPLE CHOICE** Which number is *not* a solution of the inequality $-3 \leq -6x + 3 \leq 9$? **E**

    Ⓐ -1 Ⓑ 0   Ⓒ 0.5

    Ⓓ 1   Ⓔ 2

11. **MULTIPLE CHOICE** Which number is a solution of the equation $|5x - 2| = 8$? **B**

    Ⓐ $\frac{6}{5}$ Ⓑ $-\frac{6}{5}$ Ⓒ $\frac{1}{2}$

    Ⓓ -2 Ⓔ -1

12. **MULTIPLE CHOICE** Which graph represents $|2x - 11| > 3$? **E**

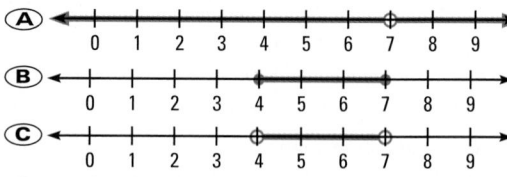

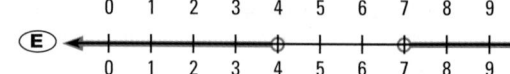

   (**A**) The quantity in column A is greater.

   (**B**) The quantity in column B is greater.

   (**C**) The two quantities are equal.

   (**D**) The relationship cannot be determined from the given information.

	Column A	Column B	
13.	$x^2 + (2x - 15) - 9x + 52 \div 2$ when $x = 1$	$x^2 + (2x - 15) - 9x + 52 \div 2$ when $x = -1$	B
14.	a if $6a + 7 = -5$	b if $b - 5 = 2b - 7$	B
15.	t if $I = Prt$, $I = \$100$, $P = \$1000$, and $r = 4\%$	t if $I = Prt$, $I = \$200$, $P = \$2000$, and $r = 4\%$	C

16. MULTI-STEP PROBLEM You buy a new car with a fuel efficiency of 31 miles per gallon on the highway and 26 miles per gallon in town. The gas tank holds 12.9 gallons. How far can you travel on the highway in your new car with a full tank of gas?

   a. Write a verbal model for this problem.   **distance traveled = (rate of fuel consumption) × (fuel consumed)**

   b. Assign labels to each part of the verbal model.
      distance traveled = d (mi); rate of fuel consumption = 31 (mi/gal); fuel used = 12.9 (gal)

   c. Use the labels to translate the verbal model into an algebraic model.
      d = (31)(12.9)

   d. Solve the algebraic model.   **399.9**

   e. Answer the question.   **You can travel almost 400 mi.**

   f. How far can you travel *in town* in your new car with a full tank of gas?   **about 335.4 mi**

17. MULTI-STEP PROBLEM The table below gives the average weight range for different types of dogs.

Dog	Average weight range (pounds)
Beagle	18–30
Bloodhound	80–100
Bulldog	40–50
Great Dane	120–150
Mastiff	165–185

▶ Source: American Kennel Club

   a. For each type of dog, write the average weight range as a compound inequality.

   b. For each type of dog, write the average weight range as an absolute value inequality.

   c. *Writing* Choose one type of dog from the table. Explain how the two inequalities you wrote for the average weight range for this type of dog are related.

17. a. Beagle: $18 \leq w \leq 30$;
Bloodhound: $80 \leq w \leq 100$;
Bulldog: $40 \leq w \leq 50$;
Great Dane: $120 \leq w \leq 150$;
Mastiff: $165 \leq w \leq 185$

b. Beagle: $|w - 24| \leq 6$;
Bloodhound: $|w - 90| \leq 10$;
Bulldog: $|w - 45| \leq 5$;
Great Dane: $|w - 135| \leq 15$;
Mastiff: $|w - 175| \leq 10$

c. *Sample answer:* Using the Beagle as an example, the linear inequality $18 \leq w \leq 30$ is the solution of the absolute value inequality $|w - 24| \leq 6$. Given the first inequality, to find the second, find the midpoint between 18 and 30. The midpoint, 24, is the quantity subtracted from w inside the absolute value bars. Half the distance between 18 and 30, which is 6, is the amount w is allowed to vary from 24.

GOALS

LESSON	GOALS	NCTM	ITED	SAT9	Terra-Nova	Local
2.1 pp. 67–74	**GOAL 1** Represent relations and functions. **GOAL 2** Graph and evaluate linear functions.	2, 3, 10	MIG	20, 38	11, 14, 16, 49, 51, 52	
2.2 pp. 75–81	**GOAL 1** Find slopes of lines and classify parallel and perpendicular lines. **GOAL 2** Use slope to solve real-life problems.	2, 3, 6, 8, 9	RQGE	1, 2, 4	11, 14, 16, 17, 49, 51, 52	
2.3 pp. 82–90	**GOAL 1** Use the slope-intercept form of a linear equation to graph linear equations. **GOAL 2** Use the standard form of a linear equation to graph linear equations. TECHNOLOGY ACTIVITY: 2.3 *Graph equations of the form y = f(x) on a graphing calculator.*	2, 3, 8	MIG, IIG	4, 38	14, 16, 18, 49, 51, 52	
2.4 pp. 91–98	**GOAL 1** Write linear equations. **GOAL 2** Write direct variation equations.	2, 8	MIRA, RQRA	4, 9, 11, 21	16, 18	
2.5 pp. 99–107	CONCEPT ACTIVITY: 2.5 *Investigate how to approximate the best-fitting line for a set of data.* **GOAL 1** Use a scatter plot to identify the correlation shown by a set of data. **GOAL 2** Approximate the best-fitting line for a set of data. TECHNOLOGY ACTIVITY: 2.5 *Find and graph an equation of the best-fitting line on a graphing calculator.*	2, 3, 5	MIG, IIG, MCS	8, 9, 10, 11, 19, 23, 38	14, 15, 16	
2.6 pp. 108–113	**GOAL 1** Graph linear inequalities in two variables. **GOAL 2** Use linear inequalities to solve real-life problems.	2, 3, 6, 8, 9	MIG, IIG, RQWN	2, 3, 38	14, 16, 17, 18, 49, 51, 52	
2.7 pp. 114–121	**GOAL 1** Represent piecewise functions. **GOAL 2** Use piecewise functions to model real-life quantities. TECHNOLOGY ACTIVITY: 2.7 *Graph a piecewise function on a graphing calculator.*	2, 3, 6, 8, 9	RQGE	21	16, 18	
2.8 pp. 122–128	**GOAL 1** Represent absolute value functions. **GOAL 2** Use absolute value functions to model real-life situations.	1, 2, 6, 8, 9	RQWN	21, 22	16, 18	

RESOURCES

CHAPTER RESOURCE BOOKLETS

CHAPTER SUPPORT

Tips for New Teachers	p. 1	Prerequisite Skills Review	p. 5
Parent Guide for Student Success	p. 3	Strategies for Reading Mathematics	p. 7

LESSON SUPPORT

	2.1	2.2	2.3	2.4	2.5	2.6	2.7	2.8
Lesson Plans (regular and block)	p. 9	p. 21	p. 36	p. 51	p. 63	p. 78	p. 90	p. 104
Warm-Up Exercises and Daily Quiz	p. 11	p. 23	p. 38	p. 53	p. 65	p. 80	p. 92	p. 106
Activity Support Masters					p. 66			
Lesson Openers	p. 12	p. 24	p. 39	p. 54	p. 67	p. 81	p. 93	p. 107
Graphing Calculator Activities & Keystrokes		p. 25	p. 40		p. 68		p. 94	p. 108
Practice (3 levels)	p. 13	p. 27	p. 41	p. 55	p. 69	p. 82	p. 96	p. 110
Reteaching with Practice	p. 16	p. 30	p. 44	p. 58	p. 72	p. 85	p. 99	p. 113
Quick Catch-Up for Absent Students	p. 18	p. 32	p. 46	p. 60	p. 74	p. 87	p. 101	p. 115
Cooperative Learning Activities		p. 33						
Interdisciplinary Applications	p. 19		p. 47		p. 75		p. 102	
Real-Life Applications		p. 34		p. 61		p. 88		p. 116
Math & History Applications			p. 48					
Challenge: Skills and Applications	p. 20	p. 35	p. 49	p. 62	p. 76	p. 89	p. 103	p. 117

REVIEW AND ASSESSMENT

Quizzes	pp. 50, 77	Alternative Assessment with Math Journal	p. 126
Chapter Review Games and Activities	p. 118	Project with Rubric	p. 128
Chapter Test (3 levels)	pp. 119–124	Cumulative Review	p. 130
SAT/ACT Chapter Test	p. 125	Resource Book Answers	p. A1

TRANSPARENCIES

	2.1	2.2	2.3	2.4	2.5	2.6	2.7	2.8
Warm-Up Exercises and Daily Quiz	p. 10	p. 11	p. 12	p. 13	p. 14	p. 15	p. 16	p. 17
Alternative Lesson Opener Transparencies	p. 8	p. 9	p. 10	p. 11	p. 12	p. 13	p. 14	p. 15
Examples/Standardized Test Practice	✓	✓	✓	✓	✓	✓	✓	✓
Answer Transparencies	✓	✓	✓	✓	✓	✓	✓	✓

TECHNOLOGY

- Electronic Teaching Tools
- Online Lesson Planner
- Internet Support
- Personal Student Tutor
- Test and Practice Generator
- Instant Replay: Video Review Games
- Electronic Lesson Presentations (Lesson 2.3)

ADDITIONAL RESOURCES

- Basic Skills Workbook: Diagnosis and Remediation
- Worked-Out Solution Key
- Resources in Spanish
- Standardized Test Practice Workbook
- Practice Workbook with Examples

PACING THE CHAPTER

REGULAR SCHEDULE

Day 1

2.1

STARTING OPTIONS
- Prereq. Skills Review
- Strategies for Reading
- Homework Check
- Warm-Up or Daily Quiz

TEACHING OPTIONS
- Motivating the Lesson
- Les. Opener (Visual)
- Examples 1–6
- Closure Question
- Guided Practice Exs.

APPLY/HOMEWORK
- See Assignment Guide.
- See the CRB: Practice, Reteach, Apply, Extend

ASSESSMENT OPTIONS
- Checkpoint Exercises
- Daily Quiz (2.1)
- Stand. Test Practice

Day 2

2.2

STARTING OPTIONS
- Homework Check
- Warm-Up or Daily Quiz

TEACHING OPTIONS
- Motivating the Lesson
- Les. Opener (Activity)
- Graphing Calc. Activity
- Examples 1–6
- Closure Question
- Guided Practice Exs.

APPLY/HOMEWORK
- See Assignment Guide.
- See the CRB: Practice, Reteach, Apply, Extend

ASSESSMENT OPTIONS
- Checkpoint Exercises
- Daily Quiz (2.2)
- Stand. Test Practice

Day 3

2.3

STARTING OPTIONS
- Homework Check
- Warm-Up or Daily Quiz

TEACHING OPTIONS
- Motivating the Lesson
- Les. Opener (Appl.)
- Graphing Calc. Activity
- Examples 1–5
- Technology Activity
- Closure Question
- Guided Practice Exs.

APPLY/HOMEWORK
- See Assignment Guide.
- See the CRB: Practice, Reteach, Apply, Extend

ASSESSMENT OPTIONS
- Checkpoint Exercises
- Daily Quiz (2.3)
- Stand. Test Practice
- Quiz (2.1–2.3)

Day 4

2.4

STARTING OPTIONS
- Homework Check
- Warm-Up or Daily Quiz

TEACHING OPTIONS
- Motivating the Lesson
- Les. Opener (Application)
- Examples 1–4
- Guided Practice Exs. 2, 4–11

APPLY/HOMEWORK
- See Assignment Guide.
- See the CRB: Practice, Reteach, Apply, Extend

ASSESSMENT OPTIONS
- Checkpoint Exercises, pp. 92–93

Day 5

2.4 (cont.)

STARTING OPTIONS
- Homework Check

TEACHING OPTIONS
- Examples 5–7
- Closure Question
- Guided Practice Exs. 1, 3, 12

APPLY/HOMEWORK
- See Assignment Guide.
- See the CRB: Practice, Reteach, Apply, Extend

ASSESSMENT OPTIONS
- Checkpoint Exercises, pp. 93–94
- Daily Quiz (2.4)
- Stand. Test Practice

Day 6

2.5

STARTING OPTIONS
- Homework Check
- Warm-Up or Daily Quiz

TEACHING OPTIONS
- Motivating the Lesson
- Concept Act. & Wksht.
- Les. Opener (Appl.)
- Graphing Calc. Activity
- Examples 1–3
- Technology Activity
- Closure Question
- Guided Practice Exs.

APPLY/HOMEWORK
- See Assignment Guide.
- See the CRB: Practice, Reteach, Apply, Extend

ASSESSMENT OPTIONS
- Checkpoint Exercises
- Daily Quiz (2.5)
- Stand. Test Practice
- Quiz (2.4–2.5)

Day 9

2.8

STARTING OPTIONS
- Homework Check
- Warm-Up or Daily Quiz

TEACHING OPTIONS
- Motivating the Lesson
- Les. Opener (Calc.)
- Graphing Calc. Activity
- Examples 1–4
- Closure Question
- Guided Practice Exs.

APPLY/HOMEWORK
- See Assignment Guide.
- See the CRB: Practice, Reteach, Apply, Extend

ASSESSMENT OPTIONS
- Checkpoint Exercises
- Daily Quiz (2.8)
- Stand. Test Practice
- Quiz (2.6–2.8)

Day 10

Review

DAY 10 START OPTIONS
- Homework Check

REVIEWING OPTIONS
- Chapter 2 Summary
- Chapter 2 Review
- Chapter Review Games and Activities

APPLY/HOMEWORK
- Chapter 2 Test (practice)
- Ch. Standardized Test (practice)

Day 11

Assess

DAY 11 START OPTIONS
- Homework Check

ASSESSMENT OPTIONS
- Chapter 2 Test
- SAT/ACT Ch. 2 Test
- Alternative Assessment

APPLY/HOMEWORK
- Skill Review, p. 138

Day 1

2.1 & 2.2

DAY 1 START OPTIONS
- Prereq. Skills Review
- Strategies for Reading
- Homework Check
- W-Up 2.1 or D. Quiz 1.7

TEACHING 2.1 OPTIONS
- Motivating the Lesson
- Les. Opener (Visual)
- Examples 1–6
- Closure Question
- Guided Practice Exs.

TEACHING 2.2 OPTIONS
- Warm-Up (Les. 2.2)
- Motivating the Lesson
- Les. Opener (Activity)
- Graphing Calc. Activity
- Examples 1–6
- Closure Question
- Guided Practice Exs.

APPLY/HOMEWORK
- See Assignment Guide.
- See the CRB: Practice, Reteach, Apply, Extend

ASSESSMENT OPTIONS
- Checkpoint Exercises
- Daily Quiz (Les. 2.1, 2.2)

Day 2

2.3

DAY 2 START OPTIONS
- Homework Check
- Warm-Up or Daily Quiz

TEACHING 2.3 OPTIONS
- Motivating the Lesson
- Les. Opener (Appl.)
- Graphing Calc. Activity
- Examples 1–5
- Technology Activity
- Closure Question
- Guided Practice Exs.

APPLY/HOMEWORK
- See Assignment Guide.
- See the CRB: Practice, Reteach, Apply, Extend

ASSESSMENT OPTIONS
- Checkpoint Exercises
- Daily Quiz (Les. 2.3)
- Stand. Test Practice
- Quiz (2.1–2.3)

Day 3

2.4

DAY 3 START OPTIONS
- Homework Check
- Warm-Up or Daily Quiz

TEACHING 2.4 OPTIONS
- Motivating the Lesson
- Les. Opener (Appl.)
- Examples 1–7
- Closure Question
- Guided Practice Exs.

APPLY/HOMEWORK
- See Assignment Guide.
- See the CRB: Practice, Reteach, Apply, Extend

ASSESSMENT OPTIONS
- Checkpoint Exercises
- Daily Quiz (Les. 2.4)
- Stand. Test Practice

Day 4

2.5 & 2.6

DAY 4 START OPTIONS
- Homework Check
- W-Up 2.5 or D. Quiz 2.4

TEACHING 2.5 OPTIONS
- Motivating the Lesson
- Concept Act. & Wksht.
- Les. Opener (Appl.)
- Technology Activities
- Examples 1–3
- Closure Question
- Guided Practice Exs.

TEACHING 2.6 OPTIONS
- Warm-Up (Les. 2.6)
- Motivating the Lesson
- Les. Opener (Activity)
- Examples 1–4
- Closure Question
- Guided Practice Exs.

APPLY/HOMEWORK
- See Assignment Guide.
- See the CRB: Practice, Reteach, Apply, Extend

ASSESSMENT OPTIONS
- Checkpoint Exercises
- Daily Quiz (Les. 2.5, 2.6)
- Quiz (2.4–2.5)

Day 5

2.7 & 2.8

DAY 5 START OPTIONS
- Homework Check
- W-Up 2.7 or D. Quiz 2.6

TEACHING 2.7 OPTIONS
- Les. Opener (Appl.)
- Technology Activities
- Examples 1–6
- Closure Question
- Guided Practice Exs.

TEACHING 2.8 OPTIONS
- Warm-Up (Les. 2.8)
- Motivating the Lesson
- Les. Opener (Calc.)
- Graphing Calc. Activity
- Examples 1–4
- Closure Question
- Guided Practice Exs.

APPLY/HOMEWORK
- See Assignment Guide.
- See the CRB: Practice, Reteach, Apply, Extend

ASSESSMENT OPTIONS
- Checkpoint Exercises
- Daily Quiz (Les. 2.7, 2.8)
- Quiz (2.6–2.8)

Day 6

Review/Assess

DAY 6 START OPTIONS
- Homework Check

REVIEWING OPTIONS
- Chapter 2 Summary
- Chapter 2 Review
- Chapter Review Games and Activities
- Chapter 2 Test (practice)
- Ch. Standardized Test (practice)

ASSESSMENT OPTIONS
- Chapter 2 Test
- SAT/ACT Ch. 2 Test
- Alternative Assessment

APPLY/HOMEWORK
- Skill Review, p. 138

Day 7

2.6

STARTING OPTIONS
- Homework Check
- Warm-Up or Daily Quiz

TEACHING OPTIONS
- Motivating the Lesson
- Les. Opener (Activity)
- Examples 1–4
- Closure Question
- Guided Practice Exs.

APPLY/HOMEWORK
- See Assignment Guide.
- See the CRB: Practice, Reteach, Apply, Extend

ASSESSMENT OPTIONS
- Checkpoint Exercises
- Daily Quiz (2.6)
- Stand. Test Practice

Day 8

2.7

STARTING OPTIONS
- Homework Check
- Warm-Up or Daily Quiz

TEACHING OPTIONS
- Les. Opener (Application)
- Graphing Calc. Activity
- Examples 1–6
- Technology Activity
- Closure Question
- Guided Practice Exs.

APPLY/HOMEWORK
- See Assignment Guide.
- See the CRB: Practice, Reteach, Apply, Extend

ASSESSMENT OPTIONS
- Checkpoint Exercises
- Daily Quiz (2.7)
- Stand. Test Practice

BEFORE THE CHAPTER

The *Chapter 2 Resource Book* has the following materials to distribute and use before the chapter:

- **Parent Guide for Student Success (pictured below)**
- **Prerequisite Skills Review**
- **Strategies for Reading Mathematics**

PARENT GUIDE *Pages 3–4*

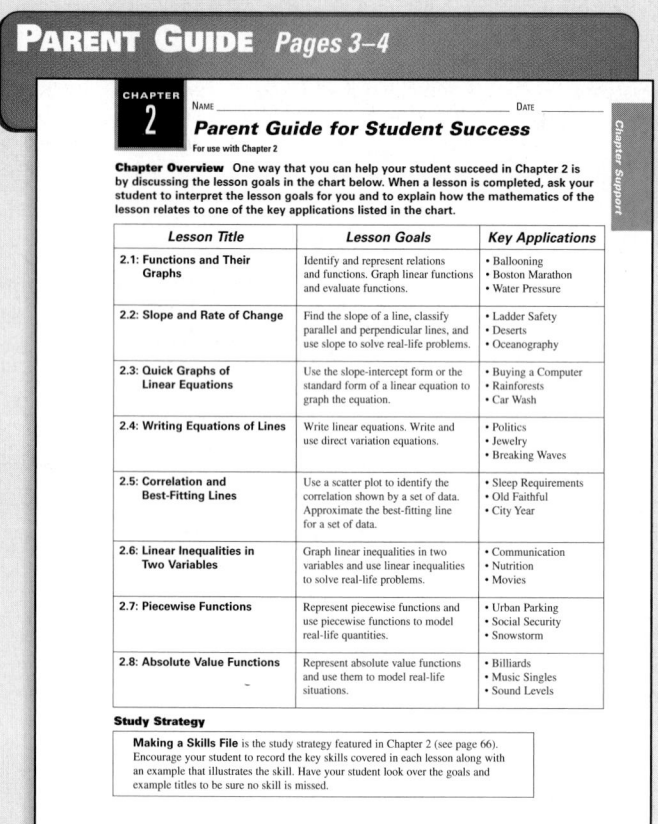

PARENT GUIDE FOR STUDENT SUCCESS The first page summarizes the content of Chapter 2. Parents are encouraged to have their students explain how the material relates to key applications in the chapter, such as water pressure. The second page (not shown) provides exercises and an activity that parents can do with their students. In the activity, parents and students represent a real-life situation with a piecewise function.

TECHNOLOGY RESOURCE

Another source of help for students is the Personal Student Tutor, which has animated examples and practice exercises correlated to the material in each chapter.

DURING EACH LESSON

The *Chapter 2 Resource Book* has the following alternatives for introducing the lesson:

- **Lesson Openers**
- **Graphing Calculator Activities with Keystrokes (below)**

CALCULATOR ACTIVITY *Pages 25–26*

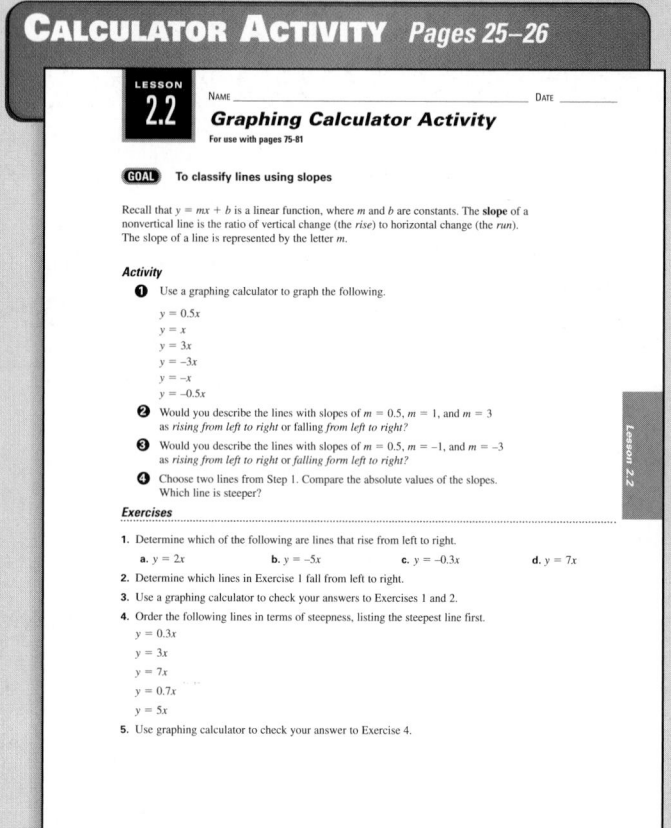

GRAPHING CALCULATOR ACTIVITY This activity uses a graphing calculator to develop an understanding of slopes, one of the concepts covered in Lesson 2.2. Keystrokes for this activity are provided on a separate sheet.

TECHNOLOGY RESOURCE

Teachers can use the Electronic Lesson Presentations CD-ROM to present Lesson 2.3 using computer animation to step through the Examples.

The *Chapter 2 Resource Book* has a variety of materials to follow-up each lesson. They include the following:

- **Practice (3 levels)**
- **Reteaching with Practice**
- **Quick Catch-Up for Absent Students**
- **Challenge: Skills and Applications**
- **Interdisciplinary Applications (pictured below)**
- **Real-Life Applications**

The *Chapter 2 Resource Book* has the following review and assessment materials:

- **Quizzes**
- **Chapter Review Games and Activities**
- **Chapter Test (3 levels)**
- **SAT/ACT Chapter Test (pictured below)**
- **Alternative Assessment with Rubrics and Math Journal**
- **Project with Rubric**
- **Cumulative Review**

APPLICATION *Page 19*

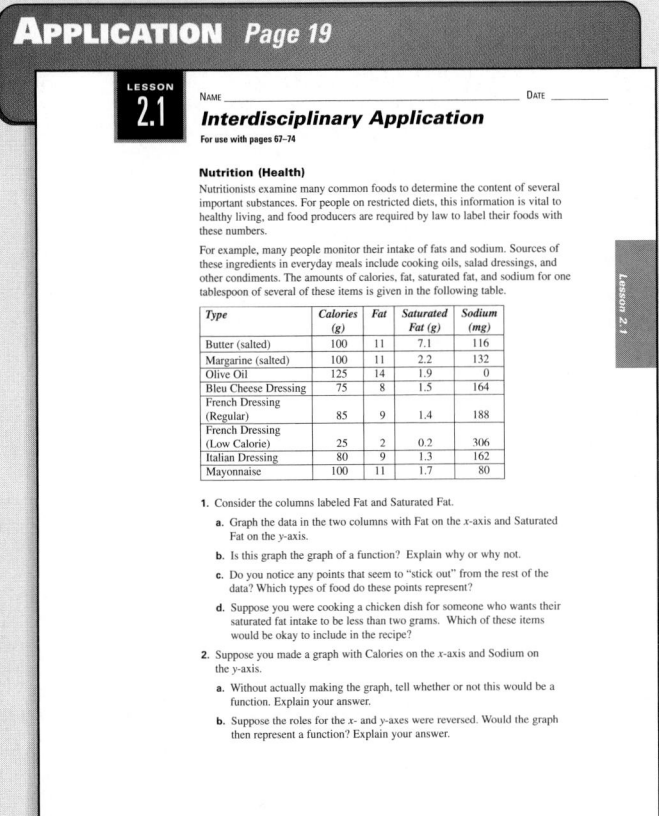

SAT/ACT CHAPTER TEST *Page 125*

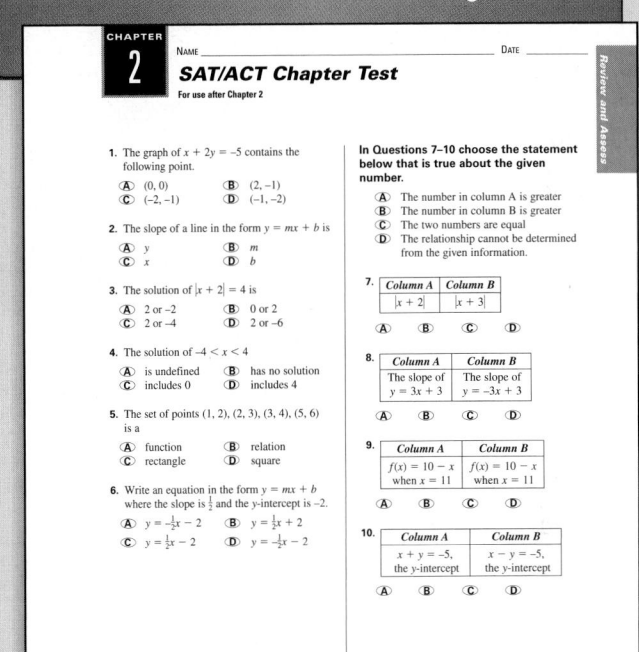

INTERDISCIPLINARY APPLICATION This application makes a connection between nutrition and graphing functions, the topic of Lesson 2.1.

SAT/ACT CHAPTER TEST This test covers the material in Chapter 2 in a standardized test format. Students taking this form of the test should be reminded to read all of the answer choices before deciding which is the correct one.

CHAPTER GOALS

Students begin the chapter by identifying and representing relations and functions, and by graphing and evaluating linear functions. They find the slope of a line, and identify parallel and perpendicular lines from their slopes. Students generalize slope as a rate of change. They graph linear equations using both slope-intercept and standard forms, and identify and graph horizontal and vertical lines. They write equations of lines using the slope and intercept, a point and the slope, or two points. Students write direct variation equations. They explore positive and negative correlation using scatter plots, and approximate best-fitting lines. Students then complete the chapter by graphing linear inequalities in two variables, piecewise functions, and absolute value functions, while using all of these to model real-life applications.

APPLICATION NOTE

City Year provides community service to many citizens. City Year corp members work with elderly or homeless persons, with community centers, and with various community-building groups. Corp members must adhere to high standards and meet rigorous expectations for their conduct and performance.

As the graph shows, membership has expanded steadily. Members feel that they have gained life-long citizenship skills from their active participation in this service program. City Year is sponsored by the federal government, corporations, foundations, and individuals.

Additional information about youth service is available at **www.mcdougallittell.com**.

LINEAR EQUATIONS AND FUNCTIONS

▶ *How can you predict membership enrollments for an organization?*

APPLICATION: Youth Service

City Year is a national youth service program that began with 57 members in Boston in 1989. The organization has expanded since then and currently has sites in cities across the country.

Think & Discuss

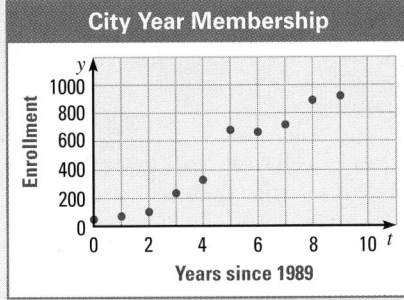

City Year Membership

(graph: Enrollment (y-axis, 0 to 1000) vs. Years since 1989 (t-axis, 0 to 10))

1. What trend do you see in the membership enrollment for City Year?   **1, 2. See margin.**

2. Based on the graph, how can you predict future membership enrollments?

Learn More About It

You will predict the number of members in City Year in 2010 in Exercises 24 and 25 on p. 105.

APPLICATION LINK Visit www.mcdougallittell.com for more information about youth service.

65

ADDITIONAL RESOURCES
Another way to begin the chapter is to show the video clip of a real-life motivator for Chapter 2 on the *Instant Replay: Video Review Games.*

PROJECTS
A project covering Chapters 1–3 appears on pages 194 and 195 of the Student Edition. An additional project for Chapter 2 is available in the *Chapter 2 Resource Book,* p. 128.

TECHNOLOGY

 **Software**
- *Electronic Teaching Tools*
- *Online Lesson Planner*
- *Personal Student Tutor*
- *Test and Practice Generator*
- *Electronic Lesson Presentations* (Lesson 2.3)

Video
- *Instant Replay: Video Review Games*

Internet Connections
www.mcdougallittell.com
- **Application Links**
  65, 78, 89, 119, 126
- **Data Updates**
  93, 119
- **Student Help**
  70, 77, 88, 90, 92, 105, 107, 110, 118, 119, 121, 126, 127
- **Career Links**
  68, 102, 112
- **Extra Challenge**
  74, 81, 88, 98, 105, 113, 120, 127

1. *Sample answer:* Enrollment tends to increase throughout the period.
2. *Sample answer:* Estimate the growth trend with a line, and use the line to predict future enrollment.

PREPARE

DIAGNOSTIC TOOLS
The **Skill Review** exercises can help you diagnose whether students have the following skills needed in Chapter 2:

- Evaluate algebraic expressions.
- Rewrite an equation with more than one variable.
- Solve an inequality.

The following resources are available for students who need additional help with these skills:

- Prerequisite Skills Review (*Chapter 2 Resource Book,* p. 5; *Warm-Up Transparencies,* p. 9)
- Reteaching with Practice (Chapter Resource Books for Lessons 1.2, 1.4, and 1.6)
- 🖥 *Personal Student Tutor*

ADDITIONAL RESOURCES
The following resources are provided to help you prepare for the upcoming chapter and customize review materials:

- *Chapter 2 Resource Book*
  Tips for New Teachers (p. 1)
  Parent Guide (p. 3)
  Lesson Plans (every lesson)
  Lesson Plans for Block Scheduling (every lesson)
- 🖥 *Electronic Teaching Tools*
- 🖥 *Online Lesson Planner*
- 🖥 *Test and Practice Generator*

PREVIEW

What's the chapter about?

Chapter 2 is about **linear equations and functions**. In Chapter 2 you'll learn

- how to graph ordered pairs, relations, functions, linear equations and inequalities in two variables, piecewise functions, and absolute value functions.
- how to write equations of lines.
- how to solve real-life problems using graphs and equations.

KEY VOCABULARY

▶ **Review**
- graph, p. 3
- linear equation, p. 19
- solution, p. 19
- linear inequality in one variable, p. 41
- absolute value, p. 50

▶ **New**
- relation, p. 67
- function, p. 67
- ordered pair, p. 67
- coordinate plane, p. 67
- linear function, p. 69
- slope, p. 75

- slope-intercept form, p. 82
- standard form, p. 84
- direct variation, p. 94
- scatter plot, p. 100
- linear inequality in two variables, p. 108
- piecewise function, p. 114

PREPARE

Are you ready for the chapter?

SKILL REVIEW Do these exercises to review key skills that you'll apply in this chapter. See the given **reference page** if there is something you don't understand.

STUDENT HELP

↳ **Study Tip**
"Student Help" boxes throughout the chapter give you study tips and tell you where to look for extra help in this book and on the Internet.

Evaluate the expression for the given values of *x* and *y*. (Review Example 3, p. 12)

1. $\frac{y-7}{x-3}$; $x = 2, y = 5$ 2

2. $\frac{5-y}{6-x}$; $x = 4, y = 1$ 2

3. $\frac{8-y}{3-x}$; $x = -1, y = -4$ 3

Solve the equation for *y*. (Review Example 1, p. 26)

4. $3x + y = 4$ $y = -3x + 4$

5. $x - 2y = 10$ $y = \frac{1}{2}x - 5$

6. $5x + 6y = -60$ $y = -\frac{5}{6}x - 10$

Solve the inequality. (Review Examples 1 and 2, p. 42)

7. $2x + 9 < 18$ $x < \frac{9}{2}$

8. $6 - 0.5y \le 19$ $y \ge -26$

9. $2x + 3 > 6x - 7$ $x < \frac{5}{2}$

STUDY STRATEGY

Here's a study strategy!

Skills File

In a notebook, make a file of the skills you learn throughout this course. On the left side of the paper, write an important skill and the lesson that it comes from. On the right side of the paper, give an example of the skill in use. Go back now and make a skills file for Chapter 1. Then continue with Chapter 2.

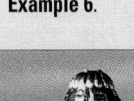

2.1

Functions and Their Graphs

What you should learn

GOAL 1 Represent relations and functions.

GOAL 2 Graph and evaluate linear functions, as applied in **Exs. 55 and 56**.

Why you should learn it

▼ To model **real-life** quantities, such as the distance a hot air balloon travels in **Example 6**.

STUDENT HELP

▶ **Study Tip**
Although the origin *O* is not usually labeled, it is understood to be the point (0, 0).

GOAL 1 REPRESENTING RELATIONS AND FUNCTIONS

A **relation** is a *mapping*, or pairing, of input values with output values. The set of input values is the **domain**, and the set of output values is the **range**. A relation is a **function** provided there is exactly one output for each input. It is not a function if at least one input has more than one output.

Relations (and functions) between two quantities can be represented in many ways, including mapping diagrams, tables, graphs, equations, and verbal descriptions.

EXAMPLE 1 *Identifying Functions*

Identify the domain and range. Then tell whether the relation is a function.

a.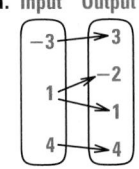

b.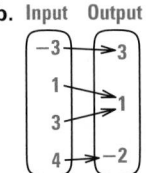

SOLUTION

a. The domain consists of -3, 1, and 4, and the range consists of -2, 1, 3, and 4. The relation is not a function because the input 1 is mapped onto both -2 and 1.

b. The domain consists of -3, 1, 3, and 4, and the range consists of -2, 1, and 3. The relation is a function because each input in the domain is mapped onto exactly one output in the range.

· · · · · · · · · ·

A relation can be represented by a set of **ordered pairs** of the form (x, y). In an ordered pair the first number is the **x-coordinate** and the second number is the **y-coordinate**. To graph a relation, plot each of its ordered pairs in a **coordinate plane**, such as the one shown. A coordinate plane is divided into four **quadrants** by the **x-axis** and the **y-axis**. The axes intersect at a point called the **origin**.

Quadrant II $x < 0, y > 0$	**Quadrant I** $x > 0, y > 0$
Quadrant III $x < 0, y < 0$	**Quadrant IV** $x > 0, y < 0$

y-axis

x-axis

origin (0, 0)

$-10\ -9\ -8\ -7\ -6\ -5\ -4\ -3\ -2$ O $1\ 2\ 3\ 4\ 5\ 6\ 7\ 8\ 9\ 10$ *x*

1 PLAN

PACING
Basic: 1 day
Average: 1 day
Advanced: 1 day
Block Schedule: 0.5 block with 2.2

LESSON OPENER
VISUAL APPROACH
An alternative way to approach Lesson 2.1 is to use the Visual Approach Lesson Opener:
• Blackline Master (*Chapter 2 Resource Book*, p. 12)
• Transparency (p. 8)

MEETING INDIVIDUAL NEEDS
• *Chapter 2 Resource Book*
Prerequisite Skills Review (p. 5)
Practice Level A (p. 13)
Practice Level B (p. 14)
Practice Level C (p. 15)
Reteaching with Practice (p. 16)
Absent Student Catch-Up (p. 18)
Challenge (p. 20)
• *Resources in Spanish*
• Personal Student Tutor

NEW-TEACHER SUPPORT
See the Tips for New Teachers on pp. 1–2 of the *Chapter 2 Resource Book* for additional notes about Lesson 2.1.

WARM-UP EXERCISES
Transparency Available
Evaluate the expression when $x = -2$.
1. $4x - 2$ -10
2. $\frac{1}{3}x + \frac{5}{3}$ 1
3. $2x^2 - x + 4$ 14
4. $-4x^2 - 6x + 12$ 8
5. $\frac{1}{2}|x - 3| - \frac{3}{2}$ 1

MOTIVATING THE LESSON
The earnings for students with jobs are likely based on an hourly wage. The earnings are a *function* of the hours worked. More specifically, they are a *linear function* of the hours worked. Tell students they will now explore exactly what a function is, and what makes a function linear.

EXTRA EXAMPLE 1
Identify the domain and range. Then tell whether the relation is a function.

a. Input Output **b.** Input Output

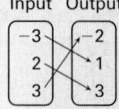 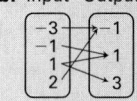

a. domain: –3, 2, 3; range: –2, 1, 3; function
b. domain: –3, –1, 1, 2; range: –1, 1, 3; not a function

EXTRA EXAMPLE 2
Graph the relations shown in Extra Example 1.

a.

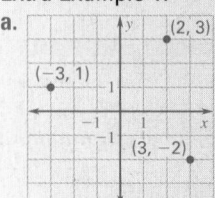

b.

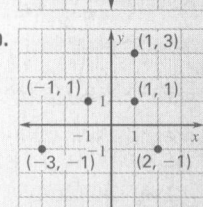

✓ CHECKPOINT EXERCISES
For use after Examples 1 and 2:

1. Identify the domain and range. Then tell whether the relation is a function, and write the relation as a set of ordered pairs.

Input Output   domain: –4, 1, 2, 7; range: –5, 0, 2; function; (–4, 2), (1, –5), (2, –5), (7, 0)

STUDENT HELP

▶ **Skills Review**
For help with plotting points in a coordinate plane, see p. 933.

FOCUS ON CAREERS

FORESTER
A forester manages, develops, and protects natural resources. To measure the diameter of trees, a forester uses a special tool called diameter tape.

CAREER LINK
www.mcdougallittell.com

EXAMPLE 2 *Graphing Relations*

Graph the relations given in Example 1.

SOLUTION

a. Write the relation as a set of ordered pairs: (−3, 3), (1, −2), (1, 1), (4, 4). Then plot the points in a coordinate plane.

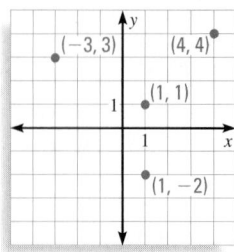

b. Write the relation as a set of ordered pairs: (−3, 3), (1, 1), (3, 1), (4, −2). Then plot the points in a coordinate plane.

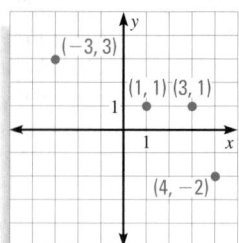

· · · · · · · · · ·

In Example 2 notice that the graph of the relation that is not a function (the graph on the left) has two points that lie on the same vertical line. You can use this property as a graphical test for functions.

> **VERTICAL LINE TEST FOR FUNCTIONS**
>
> A relation is a function if and only if no vertical line intersects the graph of the relation at more than one point.

Variables other than *x* and *y* are often used when working with relations in real-life situations, as shown in the next example.

EXAMPLE 3 *Using the Vertical Line Test in Real Life*

FORESTRY The graph shows the ages *a* and diameters *d* of several pine trees at Lundbreck Falls in Canada. Are the diameters of the trees a function of their ages? Explain.

▶ Source: National Geographical Data Center

SOLUTION

The diameters of the trees are not a function of their ages because there is at least one vertical line that intersects the graph at more than one point. For example, a vertical line intersects the graph at the points (75, 1.22) and (75, 1.58). So, at least two trees have the same age but different diameters.

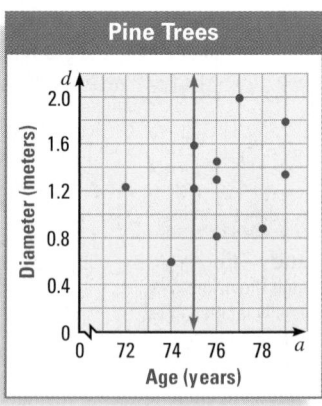

Pine Trees

GOAL 2 GRAPHING AND EVALUATING FUNCTIONS

Many functions can be represented by an **equation** in two variables, such as $y = 2x - 7$. An ordered pair (x, y) is a **solution** of such an equation if the equation is true when the values of x and y are substituted into the equation. For instance, $(2, -3)$ is a solution of $y = 2x - 7$ because $-3 = 2(2) - 7$ is a true statement.

In an equation, the input variable is called the **independent variable**. The output variable is called the **dependent variable** and depends on the value of the input variable. For the equation $y = 2x - 7$, the independent variable is x and the dependent variable is y.

The **graph** of an equation in two variables is the collection of all points (x, y) whose coordinates are solutions of the equation.

GRAPHING EQUATIONS IN TWO VARIABLES

To graph an equation in two variables, follow these steps:

STEP ❶ Construct a table of values.

STEP ❷ Graph enough solutions to recognize a pattern.

STEP ❸ Connect the points with a line or a curve.

EXAMPLE 4 *Graphing a Function*

Graph the function $y = x + 1$.

SOLUTION

❶ Begin by constructing a table of values.

Choose *x*.	−2	−1	0	1	2
Evaluate *y*.	−1	0	1	2	3

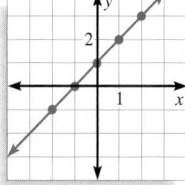

❷ Plot the points. Notice the five points lie on a line.

❸ Draw a line through the points.

· · · · · · · · · ·

The function in Example 4 is a **linear function** because it is of the form

$$y = mx + b \qquad \text{Linear function}$$

where m and b are constants. The graph of a linear function is a line. By naming a function "f" you can write the function using **function notation**.

$$f(x) = mx + b \qquad \text{Function notation}$$

The symbol $f(x)$ is read as "the value of f at x," or simply as "f of x." Note that $f(x)$ is another name for y. The domain of a function consists of the values of x for which the function is defined. The range consists of the values of $f(x)$ where x is in the domain of f. Functions do not have to be represented by the letter f. Other letters such as g or h can also be used.

STUDENT HELP

↳ **Study Tip**
When you see function notation $f(x)$, remember that it means "the value of f at x." It does not mean "f times x."

EXTRA EXAMPLE 3
The graph shows the ages a and heights h of players on a high school basketball team. Are the heights of the players a function of their ages? Explain.

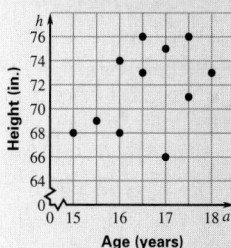

Basketball Players

No; there are several vertical lines that intersect the graph at more than one point.

EXTRA EXAMPLE 4
Graph the function $y = -x - 1$.

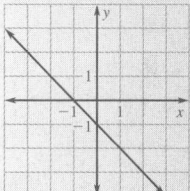

✔ **CHECKPOINT EXERCISES**

For use after Example 3:

1. The ordered pairs (0, 24), (3, 18), (4, 16), (6, 12), (9, 9), (10, 8), (12, 4), (15, 0) give numbers of grapefruits x and oranges y that can be put in a gift box. Use a graph to decide if the number of grapefruits is a function of the number of oranges. Explain.
Yes; the graph passes the vertical line test.

For use after Example 4:

2. Graph the function $y = x - 2$.

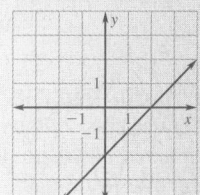

2.1 *Functions and Their Graphs* **69**

EXTRA EXAMPLE 5

Decide whether the function is linear. Then evaluate the function when $x = 3$.

a. $f(x) = x^2 + 4x - 1$ no; 20

b. $g(x) = -3x + 4$ yes; -5

EXTRA EXAMPLE 6

At 2.4 calories burned per pound of weight each hour, the calories c burned in h hours by a 110 pound person walking briskly can be modeled by $c = 110(2.4)h$.

a. For a walk of up to 4 hours, identify the domain and range.

domain: $0 \le h \le 4$;
range: $0 \le c \le 1056$

b. Graph the function. Use the graph to estimate how long it takes to burn 400 calories.

Calories Burned

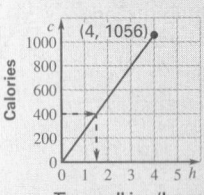

about 1.5 h

 CHECKPOINT EXERCISES

For use after Example 5:

1. Is the function $h(x) = 2 + |x|$ linear? Evaluate the function when $x = -4$.   no; 6

For use after Example 6:

2. In Oak Park, houses will be from 1450 to 2100 square feet. The cost C of building at $75 per square foot can be modeled by $C = 75f$, where f is the number of square feet. Give the domain and range of $C(f)$.

domain: $1450 \le f \le 2100$;
range: $108{,}750 \le C \le 157{,}500$

FOCUS ON VOCABULARY

Which type of variable, *independent* or *dependent,* corresponds to values in a function's *range?*

dependent

CLOSURE QUESTION

When is a relation a function?

Sample answer: when each input has exactly one output

STUDENT HELP

HOMEWORK HELP
Visit our Web site
www.mcdougallittell.com
for extra examples.

FOCUS ON
PEOPLE

PICCARD AND JONES are the first pilots to fly around the world in a balloon. Piccard is a medical doctor in Switzerland specializing in psychiatry, and Jones is a member of the Royal Air Force in the United Kingdom.

EXAMPLE 5   *Evaluating Functions*

Decide whether the function is linear. Then evaluate the function when $x = -2$.

a. $f(x) = -x^2 - 3x + 5$ **b.** $g(x) = 2x + 6$

SOLUTION

a. $f(x)$ is not a linear function because it has an x^2-term.

$f(x) = -x^2 - 3x + 5$	Write function.
$f(-2) = -(-2)^2 - 3(-2) + 5$	Substitute -2 for x.
$= 7$	Simplify.

b. $g(x)$ is a linear function because it has the form $g(x) = mx + b$.

$g(x) = 2x + 6$	Write function.
$g(-2) = 2(-2) + 6$	Substitute -2 for x.
$= 2$	Simplify.

. . . .

In Example 5 the domain of each function is all real numbers. In real-life problems the domain is restricted to the numbers that make sense in the real-life context.

EXAMPLE 6   *Using a Function in Real Life*

BALLOONING In March of 1999, Bertrand Piccard and Brian Jones attempted to become the first people to fly around the world in a balloon. Based on an average speed of 97.8 kilometers per hour, the distance d (in kilometers) that they traveled can be modeled by $d = 97.8t$ where t is the time (in hours). They traveled a total of about 478 hours. The rules governing the record state that the minimum distance covered must be at least 26,700 kilometers. ▶ Source: Breitling

a. Identify the domain and range and determine whether Piccard and Jones set the record.

b. Graph the function. Then use the graph to approximate how long it took them to travel 20,000 kilometers.

SOLUTION

a. Because their trip lasted 478 hours, the domain is $0 \le t \le 478$. The distance they traveled was $d = 97.8(478) \approx 46{,}700$ kilometers, so the range is $0 \le d \le 46{,}700$. Since $46{,}700 > 26{,}700$, they did set the record.

b. The graph of the function is shown. Note that the graph ends at (478, 46,700). To find how long it took them to travel 20,000 kilometers, start at 20,000 on the d-axis and move right until you reach the graph. Then move down to the t-axis. It took them about 200 hours to travel 20,000 kilometers.

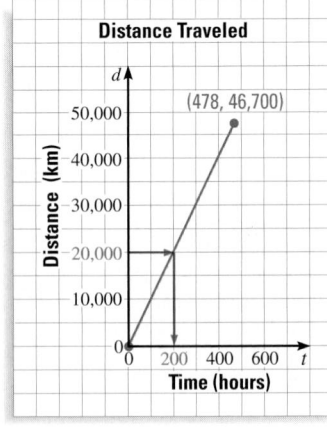

Distance Traveled

GUIDED PRACTICE

Vocabulary Check ✓

Concept Check ✓

1. What are the domain and range of a relation? **The domain is the set of input values and the range is the set of output values.**

2. Explain why a vertical line, rather than a horizontal line, is used to determine if a graph represents a function. **See margin.**

3. Explain the process for graphing an equation. **See margin.**

Skill Check ✓

2. Sample answer: A relation is not a function if any input values map to more than a single output value. If this is the case, a vertical line at that input value will contain more than one point of the graph.

3. Sample answer: First, construct a table of values for the equation. Next, plot enough of these points that a pattern can be seen. Then connect the points with a line or curve.

4. Identify the domain and range of the relation shown. Then tell whether the relation is a function. **See margin.**

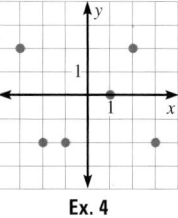
Ex. 4

Graph the function. 5–10. See margin for graphs.

5. $y = x - 1$ 6. $y = 4x$ 7. $y = 2x + 5$

8. $y = x$ 9. $y = -2x$ 10. $y = -x + 9$

Evaluate the function when $x = 3$.

11. $f(x) = x$ **3**   12. $f(x) = 6x$ **18**   13. $f(x) = x^2$ **9**

14. $g(x) = 2x + 7$ **13**   15. $h(x) = -x^2 + 10$ **1**   16. $j(x) = x^3 - 7x$ **6**

🌐 **HIGHWAY DRIVING** In Exercises 17 and 18, use the following information. A car has a 16 gallon gas tank. On a long highway trip, gas is used at a rate of about 2 gallons per hour. The gallons of gas g in the car's tank can be modeled by the equation $g = 16 - 2t$ where t is the time (in hours).

17. Identify the domain and range of the function. Then graph the function. **domain: $0 \le t \le 8$; range: $0 \le g \le 16$; See margin for graph.**

18. At the end of the trip there are 2 gallons of gas left. How long was the trip? **7 hours**

PRACTICE AND APPLICATIONS

STUDENT HELP

▸ **Extra Practice** to help you master skills is on p. 941.

4. domain: $-3, -2, -1, 1, 2, 3$; range: $-2, 0, 2$; function

DOMAIN AND RANGE Identify the domain and range.

19. 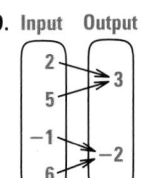 domain: $-1, 2, 5, 6$; range: $-2, 3$

20. 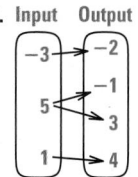 domain: $-3, 1, 5$; range: $-2, -1, 3, 4$

21. 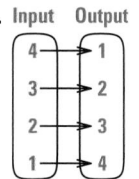 domain: $1, 2, 3, 4$; range: $1, 2, 3, 4$

GRAPHS Graph the relation. Then tell whether the relation is a function.

22. 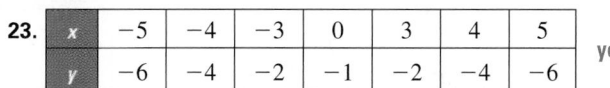 **no**   22–24. See margin for graphs.

x	0	0	2	2	4	4
y	-4	4	-3	3	-1	1

STUDENT HELP

▸ **HOMEWORK HELP**
Example 1: Exs. 19–27
Example 2: Exs. 22–27
Example 3: Exs. 30–32, 51–54
Example 4: Exs. 34–42
Example 5: Exs. 43–50
Example 6: Exs. 55–58

23. 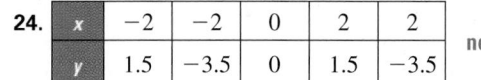 **yes**

x	-5	-4	-3	0	3	4	5
y	-6	-4	-2	-1	-2	-4	-6

24.

x	-2	-2	0	2	2
y	1.5	-3.5	0	1.5	-3.5

no

---

3 APPLY

ASSIGNMENT GUIDE

BASIC
Day 1: pp. 71–74 Exs. 20–48 even, 49, 51, 59–62, 65–81 odd

AVERAGE
Day 1: pp. 71–74 Exs. 20–48 even, 49, 51–62, 65–81 odd

ADVANCED
Day 1: pp. 71–74 Exs. 20–48 even, 49–63, 64–81

BLOCK SCHEDULE WITH 2.2
pp. 71–74 Exs. 20–48 even, 49, 51–62, 65–81 odd

EXERCISE LEVELS
Level A: *Easier*
19–42, 49–54, 59–62, 76–81
Level B: *More Difficult*
43–48, 55–58, 64–75
Level C: *Most Difficult*
63

✓ **HOMEWORK CHECK**
To quickly check student understanding of key concepts, go over the following exercises: Exs. 20, 22, 26, 30, 38, 44. See also the Daily Homework Quiz:

• Blackline Master (*Chapter 2 Resource Book*, p. 23)

• 📠 Transparency (p. 11)

5.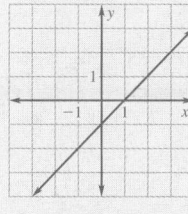

6.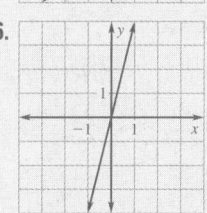

7–10, 17, 22–24.
See Additional Answers beginning on page AA1.

72

! COMMON ERROR

Exercise 45 Students may think that an absolute value function is linear. However, an absolute value function is not linear because it is not of the form $y = mx + b$. Tell students that they can think of $f(x) = |x| - 5$ as combining two linear functions: $g(x) = x - 5$ for $x \geq 0$ and $h(x) = -x - 5$ for $x < 0$, but $f(x)$ itself is not linear.

25. Input Output
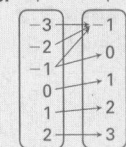

26. Input Output
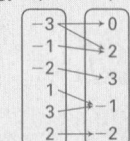

27. Input Output
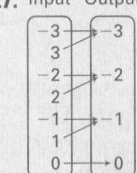

34.
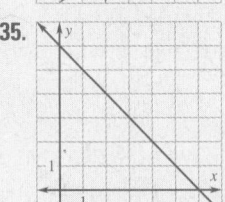

35.

36–42. See Additional Answers beginning on page AA1.

---

28. Yes; no. *Sample answer:* A function is always a relation, but a relation is not always a function. Any set of ordered pairs is a relation, but only those sets that do not map the same input value to more than one output value are functions.

STUDENT HELP

Skills Review
For help with if-then statements, see p. 926.

29. If a relation is a function, then no vertical line intersects the graph of the relation at more than one point. If no vertical line intersects the graph of a relation at more than one point, then the relation is a function.

33. $y = 3$ maps each input value to a single output value, namely, 3, while $x = 3$ matches the input value 3 to infinitely many output values.

49. 125; the volume of a cube with sides of length 5 units

50. $\frac{32\pi}{3}$, or about 33.5; the volume of a sphere with radius 2 units

51. No. *Sample answer:* Not every age corresponds to exactly one place. For example, there were 24-year-olds with finishes of first and third.

MAPPING DIAGRAMS Use a mapping diagram to represent the relation. Then tell whether the relation is a function.   25–27. See margin for diagrams.

25. no
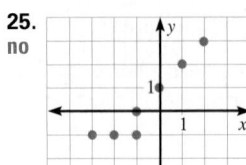

26. no
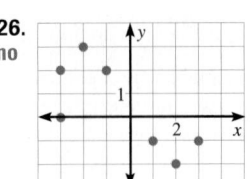

27. yes
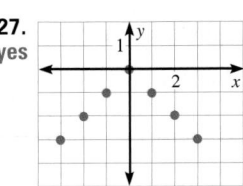

28. *Writing* Is a function always a relation? Is a relation always a function? Explain your reasoning.   See margin.

29. LOGICAL REASONING Rewrite the vertical line test as two if-then statements. See margin.

VERTICAL LINE TEST Use the vertical line test to determine whether the relation is a function.

30. no
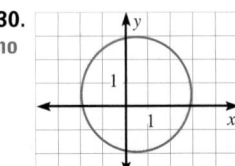

31. yes
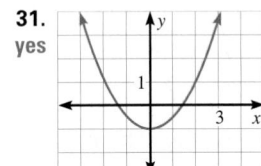

32. no
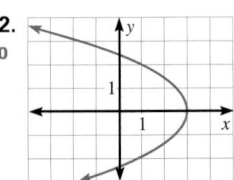

33. CRITICAL THINKING Why does $y = 3$ represent a function, but $x = 3$ does not? See margin.

GRAPHING FUNCTIONS Graph the function.   34–42. See margin for graphs.

34. $y = x - 3$

35. $y = -x + 6$

36. $y = 2x + 7$

37. $y = -5x + 1$

38. $y = 3x - 4$

39. $y = -2x - 3$

40. $y = 10x$

41. $y = 5$

42. $y = -\frac{2}{3}x + 4$

EVALUATING FUNCTIONS Decide whether the function is linear. Then evaluate the function for the given value of x.

43. $f(x) = x - 11$; $f(4)$ linear; -7

44. $f(x) = 2$; $f(-4)$ linear; 2

45. $f(x) = |x| - 5$; $f(-6)$ not linear; 1

46. $f(x) = 9x^3 - x^2 + 2$; $f(2)$ not linear; 70

47. $f(x) = -\frac{2}{3}x^2 - x + 5$; $f(6)$ not linear; -25

48. $f(x) = -3 + 4x$; $f\left(-\frac{1}{2}\right)$ linear; -5

49. GEOMETRY CONNECTION   The volume of a cube with side length s is given by the function $V(s) = s^3$. Find $V(5)$. Explain what $V(5)$ represents.

50. GEOMETRY CONNECTION   The volume of a sphere with radius r is given by the function $V(r) = \frac{4}{3}\pi r^3$. Find $V(2)$. Explain what $V(2)$ represents.

51. **BOSTON MARATHON** The graph shows the ages and finishing places of the top three competitors in each of the four categories of the 100th Boston Marathon. Is the finishing place of a competitor a function of his or her age? Explain your reasoning.
► Source: Boston Athletic Association

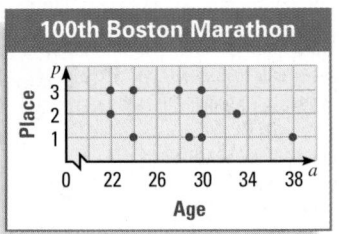

100th Boston Marathon

52. Yes; each Congress number corresponds to one number of Independents.

52. 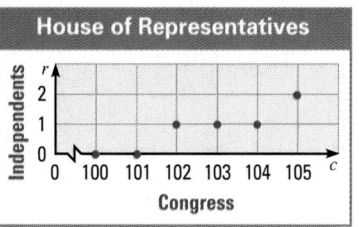 **HOUSE OF REPRESENTATIVES** The graph shows the number of Independent representatives for the 100th–105th Congresses. Is the number of Independent representatives a function of the Congress number? Explain your reasoning.

▶ Source: The Office of the Clerk, United States House of Representatives

House of Representatives

STATISTICS CONNECTION In Exercises 53 and 54, use the table which shows the number of shots attempted and the number of shots made by 9 members of the Utah Jazz basketball team in Game 1 of the 1998 NBA Finals. ▶ Source: NBA

Player	Shots attempted, x	Shots made, y
Bryon Russell	12	6
Karl Malone	25	9
Greg Foster	5	1
Jeff Hornacek	10	2
John Stockton	12	9
Howard Eisley	6	4
Chris Morris	6	3
Greg Ostertag	1	1
Shandon Anderson	5	3

55. domain: $0 \le d \le 130$;
range: $1 \le p \le 4\frac{31}{33}$;
See margin for graph.

57. domain: $20\frac{7}{8} \le c \le 25$;
range: $6\frac{5}{8} \le s \le 8$;
See margin for graph.

53. Identify the domain and range of the relation. Then graph the relation.
domain: 1, 5, 6, 10, 12, 25; range: 1, 2, 3, 4, 6, 9; See margin for graph.
54. Is the relation a function? Explain.
No. *Sample answer:* The input value 6 is mapped to two different output values, 3 and 4.

WATER PRESSURE In Exercises 55 and 56, use the information below and in the caption to the photo.
Water pressure can be measured in atmospheres, where 1 atmosphere equals 14.7 pounds per square inch. At sea level the water pressure is 1 atmosphere, and it increases by 1 atmosphere for every 33 feet in depth. Therefore, the water pressure p can be modeled as a function of the depth d by this equation:

$$p = \frac{1}{33}d + 1, \quad 0 \le d \le 130$$

55. Identify the domain and range of the function. Then graph the function. See margin.

56. What is the water pressure at a depth of 100 feet? $4\frac{1}{33}$ atmospheres $\approx$ 59.2 lb/in.^{2}

CAP SIZES In Exercises 57 and 58, use the following information.
Your cap size is based on your head circumference (in inches). For head circumferences from $20\frac{7}{8}$ inches to 25 inches, cap size s can be modeled as a function of head circumference c by this equation:

$$s = \frac{c - 1}{3}$$

57. Identify the domain and range of the function. Then graph the function. See margin.
58. If you wear a size 7 cap, what is your head circumference? 22 in.

FOCUS ON APPLICATIONS

WATER PRESSURE
Scuba divers must equalize the pressure on the inside of their bodies with the water pressure on the outside of their bodies. The maximum safe depth for recreational divers is 130 feet.

APPLICATION NOTE
EXERCISE 52 Even if a relation is a function, that does not mean there is a pattern to its output values. The number of Independent representatives in Congress does not depend on the Congress number—it depends on whom the people elect.

53.   Jazz Shooting

55.   Pressure Versus Depth

57.   Cap Size

ADDITIONAL PRACTICE AND RETEACHING

For Lesson 2.1:
• Practice Levels A, B, and C (*Chapter 2 Resource Book*, p. 13)
• Reteaching with Practice (*Chapter 2 Resource Book*, p. 16)
• See Lesson 2.1 of the *Personal Student Tutor*

For more Mixed Review:
• Search the *Test and Practice Generator* for key words or specific lessons.

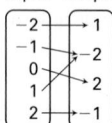

DAILY HOMEWORK QUIZ

📄 *Transparency Available*

1. Identify the domain and range.

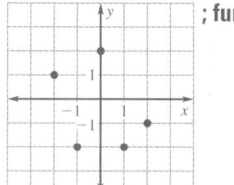

Input   Output

D: –2, –1, 0, 1, 2; R: –2, –1, 1, 2

2. Graph the relation in Exercise 1. Then use the vertical line test to tell whether the relation is a function.

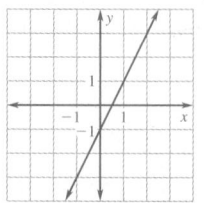

; function

3. Graph the function $y = 2x - 1$.

4. Decide whether the function $f(x) = 4x^2 - x + 1$ is linear. Find the value of $f(3)$.   not linear; 34

EXTRA CHALLENGE NOTE

→ Challenge problems for Lesson 2.1 are available in **blackline** format in the *Chapter 2 Resource Book*, p. 20 and at **www.mcdougallittell.com**.

ADDITIONAL TEST PREPARATION

1. WRITING Write the vertical line test for functions in your own words.
See sample answer at right.

2. OPEN ENDED Write your own real-life linear function. Identify its domain and range.
See sample answer at right.

63. See Additional Answers beginning on page AA1.

Test Preparation

QUANTITATIVE COMPARISON In Exercises 59–62, choose the statement that is true about the given quantities.

Ⓐ The quantity in column A is greater.

Ⓑ The quantity in column B is greater.

Ⓒ The two quantities are equal.

Ⓓ The relationship cannot be determined from the given information.

	Column A	Column B
C 59.	$f(x) = 3x + 10$ when $x = 0$	$f(x) = 2x - 4$ when $x = 7$
B 60.	$f(x) = x^2 - 4x - 11$ when $x = 6$	$f(x) = x^2 - 3x + 5$ when $x = 4$
A 61.	$f(x) = x^3 - 7x + 1$ when $x = -3$	$f(x) = -x^3 - 4$ when $x = 2$
C 62.	$f(x) = 2x + 8$ when $x = \frac{3}{2}$	$f(x) = -8x + 9$ when $x = -\frac{1}{4}$

★ **Challenge**

EXTRA CHALLENGE

↳ www.mcdougallittell.com

63. 🌐 **TELEPHONE KEYPADS** For the numbers 2 through 9 on a telephone keypad, draw two mapping diagrams: one mapping numbers onto letters, and the other mapping letters onto numbers. Are both relations functions? Explain.

See margin for diagrams; Letters to digits is a function, since each letter is mapped to a single digit. Digits to letters is not a function, since each digit corresponds to three or four different letters.

MIXED REVIEW

EVALUATING EXPRESSIONS Evaluate the expression for the given values of *x* and *y*. (Review 1.2 for 2.2)

64. $\dfrac{y - 6}{x - 9}$ when $x = -3$ and $y = -2$ $\frac{2}{3}$

65. $\dfrac{y - 11}{x - 2}$ when $x = -4$ and $y = 5$ 1

66. $\dfrac{y - (-5)}{x - 3}$ when $x = 2$ and $y = 5$ –10

67. $\dfrac{y - (-1)}{x - (-4)}$ when $x = 6$ and $y = 4$ $\frac{1}{2}$

68. $\dfrac{4 - y}{1 - x}$ when $x = 2$ and $y = 3$ –1

69. $\dfrac{10 - y}{14 - x}$ when $x = 6$ and $y = 8$ $\frac{1}{4}$

SOLVING EQUATIONS Solve the equation. Check your solution. (Review 1.3)

70. $2x + 13 = 31$ 9

71. $-2.4x + 11.8 = 29.8$ –7.5

72. $x + 17 = 10 - 3x$ $-\frac{7}{4}$

73. $\frac{5}{2} - 7x = 40 + x$ $-4\frac{11}{16}$

74. $-\frac{1}{3}(x - 15) = -48$ 159

75. $6x + 5 = 0.5(x + 6) - 4$ $-\frac{12}{11}$

CHECKING SOLUTIONS Decide whether the given number is a solution of the inequality. (Review 1.6)

76. $3x - 4 < 10; 5$ no

77. $\frac{1}{2}x - 8 \le 0; 16$ yes

78. $10 - x \ge 6; 2$ yes

79. $3 + 2x > -5; -2$ yes

80. $-5 \le x + 8 < 15; \frac{3}{2}$ yes

81. $x - 2.7 < -1$ or $3x > 6.9; 2.5$ yes

74   **Chapter 2** *Linear Equations and Functions*

Additional Test Preparation *Sample answer:*

1. If a relation is a function, then no vertical line intersects the graph of the relation at more than one point. If no vertical line intersects the graph of a relation at more than one point, then the relation is a function.

2. If gas costs $1.40 per gallon, the cost *C* to fill the gas tank of a car with a 16 gallon tank is $C = 1.4g$, where *g* is the number of gallons. The domain is $0 \le g \le 16$, and the range is $0 \le C \le 22.4$.

2.2

Slope and Rate of Change

What you should learn

GOAL 1 Find slopes of lines and classify parallel and perpendicular lines.

GOAL 2 Use slope to solve **real-life** problems, such as how to safely adjust a ladder in **Example 5**.

Why you should learn it

▼ To model **real-life** quantities, such as the average rate of change in the temperature of the Grand Canyon in **Ex. 52**.

GOAL 1 **FINDING SLOPES OF LINES**

The **slope** of a nonvertical line is the ratio of vertical change (the *rise*) to horizontal change (the *run*).

The slope of a line is represented by the letter m. Just as two points determine a line, two points are all that are needed to determine a line's slope. The slope of a line is the same regardless of which two points are used.

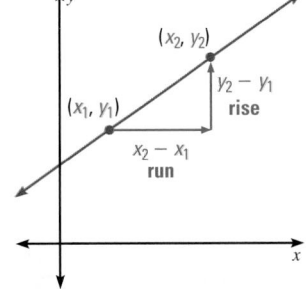

THE SLOPE OF A LINE

The slope of the nonvertical line passing through the points (x_1, y_1) and (x_2, y_2) is:

$$m = \frac{y_2 - y_1}{x_2 - x_1} = \frac{\text{rise}}{\text{run}}$$

When calculating the slope of a line, be careful to subtract the coordinates in the correct order.

EXAMPLE 1 *Finding the Slope of a Line*

Find the slope of the line passing through $(-3, 5)$ and $(2, 1)$.

SOLUTION Let $(x_1, y_1) = (-3, 5)$ and $(x_2, y_2) = (2, 1)$.

$m = \dfrac{y_2 - y_1}{x_2 - x_1}$ ◄── Rise: Difference of *y*-values
◄── Run: Difference of *x*-values

$\quad = \dfrac{1 - 5}{2 - (-3)}$ Substitute values.

$\quad = \dfrac{-4}{2 + 3}$ Simplify.

$\quad = -\dfrac{4}{5}$ Simplify.

. . . .

In Example 1 notice that the line *falls* from left to right and that the slope of the line is *negative*. This suggests one of the important uses of slope—to decide whether y decreases, increases, or is constant as x increases.

STUDENT HELP

Look Back
For help with evaluating expressions, see p. 12.

---

1 PLAN

PACING
Basic: 1 day
Average: 1 day
Advanced: 1 day
Block Schedule: 0.5 block with 2.1

LESSON OPENER
ACTIVITY
An alternative way to approach Lesson 5.1 is to use the Activity Lesson Opener:
• Blackline Master (*Chapter 2 Resource Book,* p. 24)
• Transparency (p. 9)

MEETING INDIVIDUAL NEEDS
• *Chapter 2 Resource Book*
Prerequisite Skills Review (p. 5)
Practice Level A (p. 27)
Practice Level B (p. 28)
Practice Level C (p. 29)
Reteaching with Practice (p. 30)
Absent Student Catch-Up (p. 32)
Challenge (p. 35)
• *Resources in Spanish*
• *Personal Student Tutor*

NEW-TEACHER SUPPORT
See the Tips for New Teachers on pp. 1–2 of the *Chapter 2 Resource Book* for additional notes about Lesson 2.2.

WARM-UP EXERCISES
Transparency Available
Evaluate the expression.

1. $\dfrac{-1 - 4}{4 - (-3)}$ $-\dfrac{5}{7}$

2. $\dfrac{0 - (-2)}{-1 - (-2)}$ 2

3. $\dfrac{3 - 4}{2 + 1}$ $-\dfrac{1}{3}$

4. $\dfrac{2 - (-2)}{3 - 6}$ $-\dfrac{4}{3}$

ENGLISH LEARNERS
Use a drawing as a visual aid to preteach *slope, vertical, nonvertical, horizontal, run, rise, steepness, steeper, less steep, positive slope,* and *negative slope.*

MOTIVATING THE LESSON
Students use the word slope in real life to describe the steepness of a road, a mountain, a roof, and so on. Tell them that this meaning agrees with the mathematical meaning, but that there's more to the idea of slope. Mathematically speaking, slope refers more generally to a *rate of change,* which gives it many more real-life applications.

EXTRA EXAMPLE 1
Find the slope of the line passing through $(-2, -4)$ and $(3, -1)$. $\frac{3}{5}$

EXTRA EXAMPLE 2
Without graphing, tell whether the line through the given points *rises, falls, is horizontal,* or *is vertical.*
a. $(-2, 3)$, $(1, 5)$ rises
b. $(1, -2)$, $(3, -2)$ is horizontal

EXTRA EXAMPLE 3
Tell which line is steeper.
Line 1: through $(1, -4)$ and $(5, 2)$
Line 2: through $(-2, -5)$ and $(1, -2)$ Line 1

CHECKPOINT EXERCISES

For use after Example 1:
1. Find the slope of the line passing through $(1, -5)$ and $(-2, 3)$. $-\frac{8}{3}$

For use after Example 2:
2. Without graphing, tell whether the line through $(-1, -4)$ and $(6, -5)$ *rises, falls, is horizontal,* or *is vertical.* falls

For use after Example 3:
3. Which line is steeper?
Line 1: through $(-1, -3)$ and $(-3, -2)$
Line 2: through $(3, -4)$ and $(0, -3)$ Line 1

CONCEPT QUESTION
EXAMPLE 2 What can you say about the *x*-coordinates of two points that lie on a vertical line?
They are the same.

76

---

* A line with a *positive* slope *rises* from left to right. ($m > 0$)
* A line with a *negative* slope *falls* from left to right. ($m < 0$)
* A line with a slope of *zero* is *horizontal.* ($m = 0$)
* A line with an *undefined* slope is *vertical.* (m is undefined.)

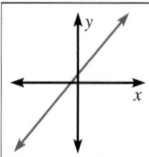

Positive slope | Negative slope | Zero slope | Undefined slope

EXAMPLE 2 *Classifying Lines Using Slope*

Without graphing tell whether the line through the given points *rises, falls, is horizontal,* or *is vertical.*

a. $(3, -4)$, $(1, -6)$ **b.** $(2, -1)$, $(2, 5)$

SOLUTION

a. $m = \dfrac{-6 - (-4)}{1 - 3} = \dfrac{-2}{-2} = 1$ Because $m > 0$, the line rises.

b. $m = \dfrac{5 - (-1)}{2 - 2} = \dfrac{6}{0}$ Because m is undefined, the line is vertical.

· · · · · · · · · ·

STUDENT HELP

Study Tip
You can think of horizontal lines as "flat" and vertical lines as "infinitely steep."

The slope of a line tells you more than whether the line rises, falls, is horizontal, or is vertical. It also tells you the steepness of the line. For two lines with **positive slopes,** the line with the greater slope is steeper. For two lines with **negative slopes,** the line with the slope of greater absolute value is steeper.

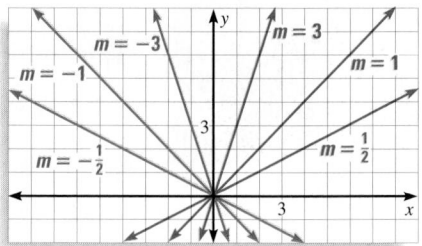

EXAMPLE 3 *Comparing Steepness of Lines*

Tell which line is steeper.

Line 1: through $(2, 3)$ and $(4, 7)$ Line 2: through $(-1, 2)$ and $(4, 5)$

SOLUTION

The slope of line 1 is $m_1 = \dfrac{7 - 3}{4 - 2} = 2$ and the slope of line 2 is $m_2 = \dfrac{5 - 2}{4 - (-1)} = \dfrac{3}{5}$.

▶ Because the lines have positive slopes and $m_1 > m_2$, line 1 is steeper than line 2.

Two lines in a plane are **parallel** if they do not intersect. Two lines in a plane are **perpendicular** if they intersect to form a right angle. Slope can be used to determine whether two different (nonvertical) lines are parallel or perpendicular.

SLOPES OF PARALLEL AND PERPENDICULAR LINES

Consider two different nonvertical lines ℓ_1 and ℓ_2 with slopes m_1 and m_2.

PARALLEL LINES The lines are parallel if and only if they have the same slope.

$$m_1 = m_2$$

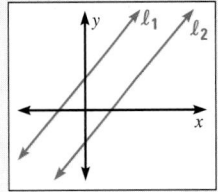

PERPENDICULAR LINES The lines are perpendicular if and only if their slopes are negative reciprocals of each other.

$$m_1 = -\frac{1}{m_2} \text{ or } m_1 m_2 = -1$$

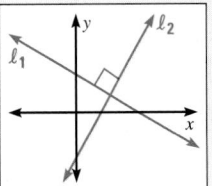

EXAMPLE 4 *Classifying Parallel and Perpendicular Lines*

STUDENT HELP

HOMEWORK HELP
Visit our Web site
www.mcdougallittell.com
for extra examples.

Tell whether the lines are *parallel*, *perpendicular*, or *neither*.

a. Line 1: through $(-3, 3)$ and $(3, -1)$
Line 2: through $(-2, -3)$ and $(2, 3)$

b. Line 1: through $(-3, 1)$ and $(3, 4)$
Line 2: through $(-4, -3)$ and $(4, 1)$

SOLUTION

a. The slopes of the two lines are:

$$m_1 = \frac{-1-3}{3-(-3)} = \frac{-4}{6} = -\frac{2}{3}$$

$$m_2 = \frac{3-(-3)}{2-(-2)} = \frac{6}{4} = \frac{3}{2}$$

Because $m_1 m_2 = -\frac{2}{3} \cdot \frac{3}{2} = -1$, m_1 and m_2 are negative reciprocals of each other. Therefore, you can conclude that the lines are perpendicular.

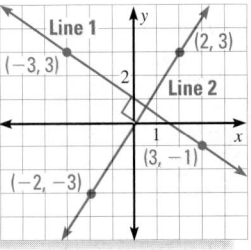

b. The slopes of the two lines are:

$$m_1 = \frac{4-1}{3-(-3)} = \frac{3}{6} = \frac{1}{2}$$

$$m_2 = \frac{1-(-3)}{4-(-4)} = \frac{4}{8} = \frac{1}{2}$$

Because $m_1 = m_2$ (and the lines are different), you can conclude that the lines are parallel.

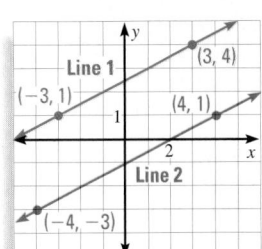

2.2 *Slope and Rate of Change* **77**

EXTRA EXAMPLE 4
Tell whether the lines are *parallel, perpendicular,* or *neither.*
a. Line 1: through $(1, -2)$ and $(3, -2)$
Line 2: through $(-5, 4)$ and $(0, 4)$ **parallel**
b. Line 1: through $(-2, -2)$ and $(4, 1)$
Line 2: through $(-3, -3)$ and $(1, 5)$ **neither**

CHECKPOINT EXERCISES
For use after Example 4:
1. Which line is perpendicular to the line through $(-3, 1)$ and $(4, -2)$?
Line 1: through $(4, -3)$ and $(1, 4)$
Line 2: through $(-3, -3)$ and $(0, 4)$ **Line 2**

MATHEMATICAL REASONING
You plan to sail toward a point 3 miles east and 5 miles north. To sail in a direction perpendicular to this, you could sail 5 miles west and 3 miles north or 5 miles east and 3 miles south. In each case, the east-west and north-south numbers are switched from the original, and one direction is reversed. This is no coincidence. In the original direction, the slope is $\frac{5}{3}$. In the new directions, the slope is $\frac{3}{-5} = -\frac{3}{5}$ or $\frac{-3}{5} = -\frac{3}{5}$, the negative reciprocal of the original slope. Suppose you plan to sail toward a point 4 miles west and 3 miles south. For your friend to sail at right angles to this, what are two possible destination points?   **3 mi W and 4 mi N or 3 mi E and 4 mi S**

STUDENT HELP NOTES
→ **Homework Help** Students can find extra examples at **www.mcdougallittell.com** that parallel the examples in the student edition.

77

EXTRA EXAMPLE 5

The slope of a road, or grade, is usually expressed as a percent. For example, if a road has a grade of 3%, it rises 3 feet for every 100 feet of horizontal distance.

a. Find the grade of a road that rises 75 feet over a horizontal distance of 2000 feet. **3.75%**

b. Find the horizontal length x of a road with a grade of 4% if the road rises 50 feet over its length. **1250 ft**

EXTRA EXAMPLE 6

The number of U.S. cell phone subscribers increased from 16 million in 1993 to 44 million in 1996. Find the average rate of change and use it to estimate the number of subscribers in 1997. $9\frac{1}{3}$ **million per year; about 53.3 million**

CHECKPOINT EXERCISES

For use after Example 5:

1. A water park slide drops 8 feet over a horizontal distance of 24 feet. Find its (positive) slope. Then find the drop over a 54 foot section with the same slope. $\frac{1}{3}$; **18 ft**

For use after Example 6:

2. The average local monthly U.S. cell phone bill decreased from \$61.48 in 1993 to \$47.70 in 1996. Find the average rate of change and use it to estimate the average monthly bill in 1997.
−\$4.59 per year; \$43.11

FOCUS ON VOCABULARY

What does it mean for two lines to be perpendicular? **They intersect in a right angle.**

CLOSURE QUESTION

How can you tell from a line's graph if it has positive, negative, or zero slope? **The slope is positive if the line rises from left to right, negative if it falls from left to right, and zero if it is horizontal.**

REAL LIFE
Ladder Safety

EXAMPLE 5 *Geometrical Use of Slope*

In a home repair manual the following ladder safety guideline is given.

Adjust the ladder until the distance from the base of the ladder to the wall is at least one quarter of the height where the top of the ladder hits the wall. For example, a ladder that hits the wall at a height of 12 feet should have its base at least 3 feet from the wall.

a. Find the maximum recommended slope for a ladder.

b. Find the minimum distance a ladder's base should be from a wall if you need the ladder to reach a height of 20 feet.

SOLUTION

a. A ladder that hits the wall at a height of 12 feet with its base about 3 feet from the wall has slope $m = \dfrac{\text{rise}}{\text{run}} = \dfrac{12}{3} = 4$. The maximum recommended slope is 4.

b. Let x represent the minimum distance that the ladder's base should be from the wall for the ladder to safely reach a height of 20 feet.

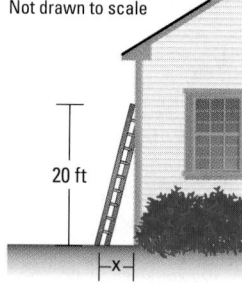
Not drawn to scale

20 ft

$\dfrac{\text{rise}}{\text{run}} = \dfrac{4}{1}$ Write a proportion.

$\dfrac{20}{x} = \dfrac{4}{1}$ The rise is 20 and the run is x.

$20 = 4x$ Cross multiply.

$5 = x$ Solve for x.

▶ The ladder's base should be at least 5 feet from the wall.

··········

In real-life problems slope is often used to describe an *average rate of change*. These rates involve units of measure, such as miles per hour or dollars per year.

STUDENT HELP

↳ **Skills Review**
For help with solving proportions, see p. 910.

FOCUS ON APPLICATIONS

REAL LIFE
↳ **DESERTS** Animals in the Mojave Desert must cope with extreme temperatures. Many reptiles burrow into the ground to escape high temperatures.

APPLICATION LINK
www.mcdougallittell.com

EXAMPLE 6 *Slope as a Rate of Change*

DESERTS In the Mojave Desert in California, temperatures can drop quickly from day to night. Suppose the temperature drops from 100°F at 2 P.M. to 68°F at 5 A.M. Find the average rate of change and use it to determine the temperature at 10 P.M.

SOLUTION

Average rate of change $= \dfrac{\text{Change in temperature}}{\text{Change in time}}$

$= \dfrac{68°F - 100°F}{5 \text{ A.M.} - 2 \text{ P.M.}} = \dfrac{-32°F}{15 \text{ hours}} \approx -2°F \text{ per hour}$

Because 10 P.M. is 8 hours after 2 P.M., the temperature changed $8(-2°F) = -16°F$. That means the temperature at 10 P.M. was about $100°F - 16°F = 84°F$.

GUIDED PRACTICE

Vocabulary Check ✓

1. Describe what is meant by the slope of a nonvertical line. Explain how your description relates to the definition of slope. **See margin.**

Concept Check ✓

2. What type of line has a slope of zero? What type of line has a slope that is undefined? **horizontal; vertical**

3. How can you decide, using slope, whether two nonvertical lines are parallel? whether two nonvertical lines are perpendicular? **They are parallel if their slopes are equal; they are perpendicular if their slopes are negative reciprocals.**

Skill Check ✓

Find the slope of the line passing through the given points. Then tell whether the line *rises, falls, is horizontal,* or *is vertical*.

4. $(4, 2)$, $(14, 3)$ $\frac{1}{10}$; rises
5. $(8, 4)$, $(8, 1)$ **undefined; vertical**
6. $(-3, 4)$, $(3, -5)$ $-\frac{3}{2}$; falls
7. $(-2, 4)$, $(-6, 8)$ -1; falls
8. $(-7, 3)$, $(4, 3)$ 0; horizontal
9. $(6, 9)$, $(-2, -7)$ 2; rises

Tell which line is steeper.

10. Line 1: through $(-5, 0)$ and $(3, 4)$
    Line 2: through $(0, 4)$ and $(1, 6)$ **line 2**
11. Line 1: through $(2, 4)$ and $(1, 7)$
    Line 2: through $(5, 2)$ and $(3, 12)$ **line 2**

Tell whether the lines are *parallel, perpendicular,* or *neither*.

12. Line 1: through $(1, 5)$ and $(-4, -2)$
    Line 2: through $(3, 0)$ and $(-2, -7)$ **parallel**
13. Line 1: through $(2, -2)$ and $(-2, 7)$
    Line 2: through $(4, -5)$ and $(5, 1)$ **neither**
14. Line 1: through $(3, 6)$ and $(2, -1)$
    Line 2: through $(-1, 2)$ and $(6, 1)$ **perpendicular**
15. Line 1: through $(9, 0)$ and $(3, 4)$
    Line 2: through $(-5, 6)$ and $(4, 0)$ **parallel**

16. 🌐 **AVERAGE SPEED** You are driving through Europe. At 9:00 A.M. you are 420 kilometers from Rome. At 3:00 P.M. you are 108 kilometers from Rome. Find your average speed. **52 km/h**

PRACTICE AND APPLICATIONS

STUDENT HELP

▶ **Extra Practice**
to help you master
skills is on p. 941.

ESTIMATING SLOPE Estimate the slope of the line.

17. 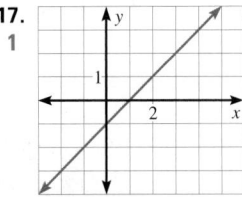 1

18. 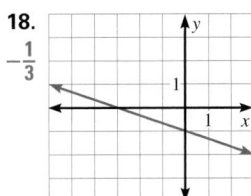 $-\frac{1}{3}$

19. **undefined**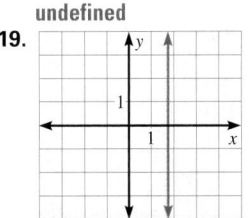

STUDENT HELP

▶ **HOMEWORK HELP**
Example 1: Exs. 17–31
Example 2: Exs. 20–31
Example 3: Exs. 32–35, 37–40
Example 4: Exs. 41–44
Example 5: Exs. 48–50
Example 6: Exs. 45–47, 51, 52

FINDING SLOPE Find the slope of the line passing through the given points. Then tell whether the line *rises, falls, is horizontal,* or *is vertical*.

20. $(3, 2)$, $(-4, 3)$ $-\frac{1}{7}$; falls
21. $(1, -4)$, $(2, 6)$ 10; rises
22. $(14, -3)$, $(4, 11)$ $-\frac{7}{5}$; falls
23. $(-10, -12)$, $(2, -6)$ $\frac{1}{2}$; rises
24. $(-7, 3)$, $(-2, 3)$ 0; horizontal
25. $(6, -6)$, $(-6, 6)$ -1; falls
26. $(4, 2)$, $(-18, 1)$ $\frac{1}{22}$; rises
27. $(-9, 8)$, $(-9, 2)$ **undefined; vertical**
28. $(3, 4)$, $\left(2, -\frac{5}{4}\right)$ $\frac{21}{4}$; rises
29. $\left(0, \frac{7}{2}\right)$, $\left(2, \frac{5}{2}\right)$ $-\frac{1}{2}$; falls
30. $\left(\frac{1}{5}, -1\right)$, $\left(\frac{3}{5}, -2\right)$ $-\frac{5}{2}$; falls
31. $\left(\frac{4}{3}, -\frac{9}{5}\right)$, $\left(\frac{4}{3}, -\frac{8}{5}\right)$ **undefined; vertical**

3 APPLY

⊙ **ASSIGNMENT GUIDE**

BASIC
Day 1: pp. 79–81 Exs. 17–36, 38–48 even, 49, 51, 54, 59–71 odd

AVERAGE
Day 1: pp. 79–81 Exs. 17–47, 48–56 even, 59–71 odd

ADVANCED
Day 1: pp. 79–81 Exs. 17–50, 52–58, 59–71 odd

BLOCK SCHEDULE WITH 2.1
pp. 79–81 Exs. 17–47, 48–56 even, 59–71 odd

EXERCISE LEVELS
Level A: *Easier*
17–28, 32–47, 59–66
Level B: *More Difficult*
29–31, 48–52, 54, 67–71
Level C: *Most Difficult*
53, 55–58

✔ **HOMEWORK CHECK**
To quickly check student understanding of key concepts, go over the following exercises: Exs. 18, 22, 31, 38, 42, 46. See also the Daily Homework Quiz:

• Blackline Master (*Chapter 2 Resource Book,* p. 38)

• 🖴 Transparency (p. 12)

1. *Sample answer:* The slope is a measure of the rate of change of *y* with respect to *x*. The slope is positive if *y* increases as *x* increases, and is negative if *y* decreases as *x* increases. This corresponds to the definition of slope = (change in *y*)/(change in *x*).

COMMON ERROR

EXERCISE 38 Remind students that the line with the steeper negative slope is not the one whose slope is the greatest, but the one whose slope has the greatest absolute value.

APPLICATION NOTE

EXERCISE 51 Point out that the rate of increase in height is an average rate. The increase in any particular year may be less than the average or much greater than the average, depending on the eruption magnitude and frequency.

36. *Sample answer:* A horizontal line has an equation like $y = k$, where k is some constant. So every point on the line has a y-coordinate of k. Pick two points on the line, say (x_1, k) and (x_2, k). The slope of the line is $\frac{\text{rise}}{\text{run}} = \frac{k - k}{x_2 - x_1} = \frac{0}{x_2 - x_1} = 0$. A vertical line has an equation like $x = c$, where c is some constant, so every point on the line has an x-coordinate of c. Pick two points on the line, say (c, y_1) and (c, y_2). The slope of the line is $\frac{\text{rise}}{\text{run}} = \frac{y_2 - y_1}{c - c} = \frac{y_2 - y_1}{0}$, which is undefined.

ADDITIONAL PRACTICE AND RETEACHING

For Lesson 2.2:

- Practice Levels A, B, and C (*Chapter 2 Resource Book,* p. 27)
- Reteaching with Practice (*Chapter 2 Resource Book,* p. 30)
- See Lesson 2.2 of the *Personal Student Tutor*

For more Mixed Review:

- Search the *Test and Practice Generator* for key words or specific lessons.

80

50. Yes; each slanted half of the rpof rises 12 feet over its 36 feet of the apartment building's width, which gives it a slope of $\frac{12}{36} = \frac{1}{3}$, the same as the $\frac{4}{12}$ required by the building code.

FOCUS ON PEOPLE

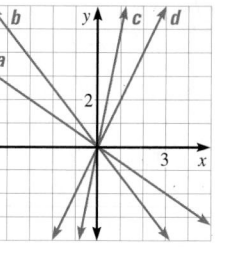

JACQUES COUSTEAU was famous for his work in oceanography, which is discussed in Ex. 51. Cousteau invented the aqua-lung, the one-man submarine, and the first underwater diving station.

MATCHING SLOPES AND LINES Match the given slopes with the given lines.

32. $-\frac{5}{4}$ *b*

33. 5 *c*

34. 2 *d*

35. $-\frac{2}{3}$ *a*

36. **LOGICAL REASONING** Use the formula for slope to verify that a horizontal line has a slope of zero and that a vertical line has an undefined slope. **See margin.**

DETERMINING STEEPNESS Tell which line is steeper.

37. Line 1: through $(-2, 6)$ and $(2, 8)$
    Line 2: through $(0, -4)$ and $(5, -3)$
    line 1

38. Line 1: through $(4, 1)$ and $(-8, 6)$
    Line 2: through $(-2, 4)$ and $(-1, -8)$
    line 2

39. Line 1: through $(3, -10)$ and $(2, -10)$
    Line 2: through $(-6, 8)$ and $(2, 12)$
    line 2

40. Line 1: through $(-5, 6)$ and $(-2, -9)$
    Line 2: through $\left(1, \frac{1}{2}\right)$ and $\left(\frac{5}{4}, 1\right)$
    line 1

TYPES OF LINES Tell whether the lines are *parallel, perpendicular,* or *neither.*

41. Line 1: through $(-1, 9)$ and $(-6, -6)$
    Line 2: through $(-7, -23)$ and $(0, -2)$
    parallel

42. Line 1: through $(4, -3)$ and $(-8, 1)$
    Line 2: through $(5, 11)$ and $(8, 20)$
    perpendicular

43. Line 1: through $(0, 3)$ and $(0, -7)$
    Line 2: through $(-6, -4)$ and $(12, -4)$
    perpendicular

44. Line 1: through $(1, 10)$ and $(5, 15)$
    Line 2: through $\left(\frac{3}{2}, \frac{3}{2}\right)$ and $(4, 2)$
    neither

AVERAGE RATE OF CHANGE Find the average rate of change in y for the given xy-pairs. State the unit of measure for the average rate of change.

45. $(4, 3)$ and $(8, 27)$ x is measured in hours and y is measured in dollars   **6; dollars/h**

46. $(0, 5)$ and $(3, 17)$ x is measured in seconds and y is measured in meters   **4; m/sec**

47. $(2, 10)$ and $(4, 16)$ x is measured in years and y is measured in inches   **3; in./year**

48. **HISTORY CONNECTION**   Aqueducts were once used to carry water from rivers using gravity. Water flowing too quickly might damage an aqueduct, but water flowing too slowly might not keep the aqueduct clear. One of the best and most common designs for an aqueduct was to raise it 3 meters for every kilometer in length. What is the slope of an aqueduct built with this design?
    ▶ Source: *Roman Aqueducts and Water Supply*   **0.003**

49. **LEANING TOWER OF PISA** The top of the Leaning Tower of Pisa is about 55.9 meters above the ground. As of 1997 its top was leaning about 5.2 meters off-center. Approximate the slope of the tower.
    ▶ Source: Endex Engineering   **10.75**

50. **PITCH OF A ROOF** Building codes require the minimum slope, or pitch, of a roof with asphalt shingles to be such that it rises at least 4 feet for every 12 feet of horizontal distance. A 72 foot wide apartment building has a 12 foot high roof. Does it meet the building code? Explain.   **See margin.**

51. **OCEANOGRAPHY** Loihi is the name of an underwater volcano that has formed twenty miles off the coast of Hawaii. The peak of the volcano is currently 3100 feet below sea level. Oceanographers estimate that it will take about 50,000 years before the peak breaks the water. If this holds true, what will be the rate of change in the volcano's height? Explain.
    ▶ Source: United States Geological Survey   **0.062 ft/year; this is the ratio of the number of vertical feet the volcano must grow to the length of time it will take to grow that high.**

53. No; no; the only
possible difference is
the ease of calculation
with some selections
over others. Check
lines and points. A
good response will
show calculations for
at least 4 pairs of
points.

**Test
Preparation**

STUDENT HELP

▶ **Skills Review**
For help with the
Pythagorean theorem,
see p. 917.

★ **Challenge**

52. 🌐 **GRAND CANYON** You are camping at the Grand Canyon. When you
pitch your tent at 1:00 P.M. the temperature is 81°F. When you wake up at
6:00 A.M. the temperature is 47°F. What is the average rate of change in the
temperature? Estimate the temperature when you went to sleep at 9:00 P.M. **2°F/h; 65°F**

53. **CRITICAL THINKING** Does it make a difference what two points on a line you
choose when finding slope? Does it make a difference which point is (x_1, y_1) and
which point is (x_2, y_2) in the formula for slope? Draw a line and calculate its
slope using several pairs of points to support your answer. **See margin.**

54. **MULTI-STEP PROBLEM** You are in charge of
building a wheelchair ramp for a doctor's office.
Federal regulations require that the ramp must
extend 12 inches for every 1 inch of rise.
The ramp needs to rise to a height of 18 inches.

▶ Source: *Uniform Federal Accessibility Standards*

18 in.

   a. How far should the end of the ramp be from the base of the building? **18 ft**

   b. Use the Pythagorean theorem to determine the length of the ramp. **about 18.1 ft**

   c. Some northern states require that outdoor ramps extend 20 inches for every
1 inch of rise because of the added problems of winter weather. Under this
regulation, what should be the length of the ramp? **about 30.0 ft**

   d. *Writing* How does changing the slope of the ramp affect the required length
of the ramp? **See margin.**

MISSING COORDINATES **Find the value of *k* so that the line through the given
points has the given slope. Check your solution.**

55. $(5, k)$ and $(k, 7)$, $m = 1$ **6**

56. $(-3, 2k)$ and $(k, 6)$, $m = 4$ **−1**

57. $(-2, k)$ and $(k, 4)$, $m = 3$ $-\dfrac{1}{2}$

58. $(9, -k)$ and $(3k, -1)$, $m = -\dfrac{1}{3}$ **2**

MIXED REVIEW

54d. *Sample answer:* The
steeper the ramp, the
shorter it will be. If
regulations require
more run for the
amount of rise, the
ramp must get
longer, as it did from
answers (b) to (c)
above.

IDENTIFYING PROPERTIES **Identify the property shown.** (Review 1.1)

59. $12 + (-12) = 0$ additive inverse property

60. $(16 + 5) + 10 = 16 + (5 + 10)$ associative property of addition

61. $8(2 + 13) = 8 \cdot 2 + 8 \cdot 13$ distributive property

62. $22 \cdot \dfrac{1}{22} = 1$ multiplicative inverse property

REWRITING EQUATIONS **Solve the equation for *y*.** (Review 1.4 for 2.3)

63. $8x + y = 15$ $15 - 8x$

64. $-2x - y = 11$ $-2x - 11$

65. $\dfrac{8}{3}x + 2y = 16$ $8 - \dfrac{4}{3}x$

66. $-6y + \dfrac{4}{5}x = 10$ $\dfrac{2}{15}x - \dfrac{5}{3}$

SOLVING EQUATIONS **Solve the equation.** (Review 1.7)

67. $|9 + 2x| = 7$ $-8, -1$

68. $|4 - 6x| = 2$ $\dfrac{1}{3}, 1$

69. $|-3x + 1| = 4$ $-1, \dfrac{5}{3}$

70. $|0.25x - 9| = 6$ **12, 60**

71. 🌐 **MIXED NUTS** A 16 ounce can of mixed nuts costs $5.82, but peanuts cost
only $.25 per ounce. The can contains 7 ounces of peanuts and 9 ounces of other
nuts. What is the cost per ounce of the other nuts? (Review 1.5 for 2.3) **about $.45/oz**

2.2 *Slope and Rate of Change* **81**

4 ASSESS

DAILY HOMEWORK QUIZ

🖥 *Transparency Available*

Find the slope of the line through
the given points. Tell whether the
line *rises, falls, is horizontal*, or
is vertical.

1. $(2, -4)$, $(1, 5)$ **−9; falls**

2. $(3, 4)$, $(-5, 4)$ **0; is horizontal**

Tell which line is steeper.

3. Line 1: through $(-3, 5)$ and
$(2, -5)$; Line 2: through $(4, -1)$
and $(0, 3)$ **Line 1**

4. Line 1: through $(2, 4)$ and $(-1, 7)$
Line 2: through $(3, -1)$ and $(4, 1)$
Line 2

5. Tell whether the lines are *parallel, perpendicular*, or *neither.*
Line 1: through $(-2, 6)$ and
$(2, -4)$; Line 2: through $(-2, -1)$
and $(3, 1)$ **perpendicular**

6. Find the average rate of
change in *y* where *x* is in seconds and *y* is in feet for the
points $(5, 6)$ and $(3, 2)$. **2 ft/sec**

EXTRA CHALLENGE NOTE

↳ Challenge problems for
Lesson 2.2 are available in
blackline format in the *Chapter 2
Resource Book*, p. 35 and at
www.mcdougallittell.com.

**ADDITIONAL TEST
PREPARATION**

1. WRITING Compare the appearance of a line with a slope of 2 to
that of a line with a slope of −2.
Sample answer: The "steepness"
is the same, but the first line rises
from left to right, and the second
falls from left to right.

2. WRITING Give two points on
each of two (nonhorizontal and
nonvertical) lines that are perpendicular. *Sample answer:*
Line 1: $(-2, 3)$ and $(2, 5)$; Line 2:
$(-1, 5)$ and $(2, -1)$

⟶ LESSON OPENER
APPLICATION

An alternative way to approach Lesson 2.3 is to use the Application Lesson Opener:

- Blackline Master (*Chapter 2 Resource Book,* p. 39)
- Transparency (p. 10)

MEETING INDIVIDUAL NEEDS

- **Chapter 2 Resource Book**
  Prerequisite Skills Review (p. 5)
  Practice Level A (p. 41)
  Practice Level B (p. 42)
  Practice Level C (p. 43)
  Reteaching with Practice (p. 44)
  Absent Student Catch-Up (p. 46)
  Challenge (p. 49)
- **Resources in Spanish**
- 🖥 **Personal Student Tutor**

NEW-TEACHER SUPPORT

See the Tips for New Teachers on pp. 1–2 of the *Chapter 2 Resource Book* for additional notes about Lesson 2.3.

WARM-UP EXERCISES

🖥 **Transparency Available**

Solve for y when $x = 0$.

1. $2x + 2y = 12$ **6**

2. $y - 4x = 8$ **8**

3. $200y + 400x = 1200$ **6**

Solve for y.

4. $2x + y = 150$ $y = -2x + 150$

5. $8x - 3y = 6$ $y = \frac{8}{3}x - 2$

ENGLISH LEARNERS

The use of *let* is common in algebra, but may be confusing for English learners. Mention that *let* $x = 0$ means "use $x = 0$."

2.3 Quick Graphs of Linear Equations

What you should learn

GOAL 1 Use the slope-intercept form of a linear equation to graph linear equations.

GOAL 2 Use the standard form of a linear equation to graph linear equations, as applied in **Example 5.**

Why you should learn it

▼ To identify relationships between **real-life** variables, such as the sales of student and adult basketball tickets in **Ex. 63.**

REAL LIFE

GOAL 1 SLOPE-INTERCEPT FORM

In Lesson 2.1 you graphed a linear equation by creating a table of values, plotting the corresponding points, and drawing a line through the points. In this lesson you will study two quicker ways to graph a linear equation.

If the graph of an equation intersects the y-axis at the point $(0, b)$, then the number b is the **y-intercept** of the graph. To find the y-intercept of a line, let $x = 0$ in an equation for the line and solve for y.

▶ ACTIVITY
Developing Concepts

Investigating Slope and *y*-intercept

Equation	Points on graph of equation	Slope	y-intercept
$y = 2x + 3$	$(0, ?), (1, ?)$	?	?
$y = -x + 2$	$(0, ?), (1, ?)$	?	?
$y = \frac{1}{2}x - 4$	$(0, ?), (1, ?)$	?	?
$y = -2x$	$(0, ?), (1, ?)$	?	?
$y = 7$	$(0, ?), (1, ?)$	?	?

❶ Copy and complete the table. See margin.

❷ What do you notice about each equation and the slope of the line?
The coefficient of x is the slope of its graph.

❸ What do you notice about each equation and the y-intercept of the line?
The constant term of the equation is the y-intercept of its graph.

The **slope-intercept form** of a linear equation is $y = mx + b$. As you saw in the activity, a line with equation $y = mx + b$ has slope m and y-intercept b.

GRAPHING EQUATIONS IN SLOPE-INTERCEPT FORM

The slope-intercept form of an equation gives you a quick way to graph the equation.

STEP ❶ Write the equation in slope-intercept form by solving for y.

STEP ❷ Find the y-intercept and use it to plot the point where the line crosses the y-axis.

STEP ❸ Find the slope and use it to plot a second point on the line.

STEP ❹ Draw a line through the two points.

1. See Additional Answers beginning on page AA1.

EXAMPLE 1   *Graphing with the Slope-Intercept Form*

Graph $y = \frac{3}{4}x - 2$.

SOLUTION

❶ The equation is already in slope-intercept form.

❷ The y-intercept is -2, so plot the point $(0, -2)$ where the line crosses the y-axis.

❸ The slope is $\frac{3}{4}$, so plot a second point on the line by moving 4 units to the right and 3 units up. This point is $(4, 1)$.

❹ Draw a line through the two points.

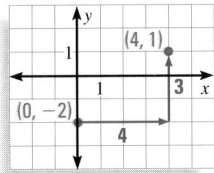 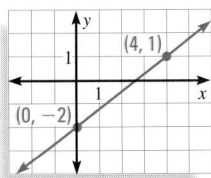

. . . .

In a real-life context the y-intercept often represents an initial amount and, as you saw in Lesson 2.2, the slope often represents a rate of change.

Buying a Computer

EXAMPLE 2   *Using the Slope-Intercept Form*

You are buying an $1100 computer on layaway. You make a $250 deposit and then make weekly payments according to the equation $a = 850 - 50t$ where a is the amount you owe and t is the number of weeks.

a. What is the original amount you owe on layaway?

b. What is your weekly payment?

c. Graph the model.

SOLUTION

a. First rewrite the equation as $a = -50t + 850$ so that it is in slope-intercept form. Then you can see that the a-intercept is 850. So, the original amount you owe on layaway (the amount when $t = 0$) is $850.

b. From the slope-intercept form you can also see that the slope is $m = -50$. This means that the amount you owe is changing at a rate of $-$50 per week. In other words, your weekly payment is $50.

c. The graph of the model is shown. Notice that the line stops when it reaches the t-axis (at $t = 17$) so the computer is completely paid for at that point.

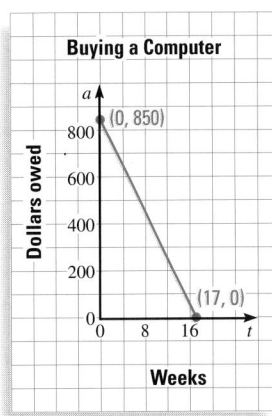

2.3 *Quick Graphs of Linear Equations*   **83**

2 TEACH

MOTIVATING THE LESSON
If you plot total income as a function of ticket sales, the resulting line has a slope equal to the price of a ticket. It also has a *y-intercept* of zero. Tell students they will learn more about the slope and y-intercept of a line in this lesson.

EXTRA EXAMPLE 1
Graph $y = \frac{1}{2}x + 1$.

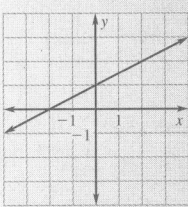

EXTRA EXAMPLE 2
To buy a $1200 stereo, you pay a $200 deposit and then make weekly payments according to the equation $a = 1000 - 40t$, where a is the amount you owe and t is the number of weeks.
a. How much do you owe originally? **$1000**
b. What is your weekly payment? **$40**
c. Graph the model.

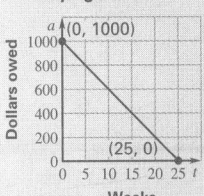

CHECKPOINT EXERCISES
For use after Examples 1 and 2:
1. The number of gallons g of water in your storage tank is given by $g = 500 - 20t$, with t in days. Using a graph, find the daily rate of water use, and determine in how many days the tank will become empty. **20 gal/day; 25 days**

EXTRA EXAMPLE 3

Graph $3x - 2y = 6$ using standard form. Then rewrite the equation in slope-intercept form.

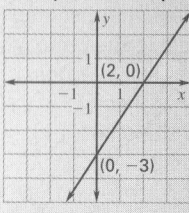

$$y = \frac{3}{2}x - 3$$

☑ **CHECKPOINT EXERCISES**

For use after Example 3:

1. Graph $2x + 5y = 10$ using standard form.

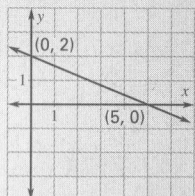

GOAL 2 **STANDARD FORM**

The **standard form** of a linear equation is $Ax + By = C$ where A and B are not both zero. A quick way to graph an equation in standard form is to plot its intercepts (when they exist). You found the y-intercept of a line in Goal 1. The **x-intercept** of a line is the x-coordinate of the point where the line intersects the x-axis.

GRAPHING EQUATIONS IN STANDARD FORM

The standard form of an equation gives you a quick way to graph the equation:

STEP ① Write the equation in standard form.

STEP ② Find the x-intercept by letting $y = 0$ and solving for x. Use the x-intercept to plot the point where the line crosses the x-axis.

STEP ③ Find the y-intercept by letting $x = 0$ and solving for y. Use the y-intercept to plot the point where the line crosses the y-axis.

STEP ④ Draw a line through the two points.

EXAMPLE 3 *Drawing Quick Graphs*

Graph $2x + 3y = 12$.

SOLUTION

Method 1 USE STANDARD FORM

① The equation is already written in standard form.

② $2x + 3(\mathbf{0}) = 12$ **Let $y = 0$.**

        $x = 6$ **Solve for x.**

The x-intercept is 6, so plot the point $(6, 0)$.

③ $2(\mathbf{0}) + 3y = 12$ **Let $x = 0$.**

        $y = 4$ **Solve for y.**

The y-intercept is 4, so plot the point $(0, 4)$.

④ Draw a line through the two points.

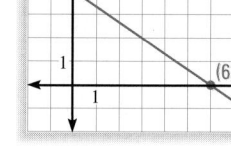

STUDENT HELP

▸ **Look Back**
For help with solving an equation for y, see p. 26.

Method 2 USE SLOPE-INTERCEPT FORM

① $2x + 3y = 12$

      $3y = -2x + 12$

      $y = -\frac{2}{3}x + 4$ **Slope-intercept form**

② The y-intercept is 4, so plot the point $(0, 4)$.

③ The slope is $-\frac{2}{3}$, so plot a second point by moving 3 units to the right and 2 units down. This point is $(3, 2)$.

④ Draw a line through the two points.

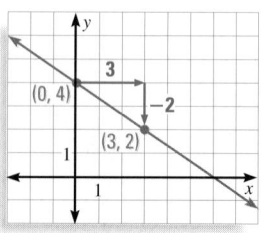

The equation of a vertical line cannot be written in slope-intercept form because the slope of a vertical line is not defined. Every linear equation, however, can be written in standard form—even the equation of a vertical line.

HORIZONTAL AND VERTICAL LINES

HORIZONTAL LINES   The graph of $y = c$ is a horizontal line through $(0, c)$.

VERTICAL LINES   The graph of $x = c$ is a vertical line through $(c, 0)$.

EXAMPLE 4   *Graphing Horizontal and Vertical Lines*

Graph **(a)** $y = 3$ and **(b)** $x = -2$.

SOLUTION

a. The graph of $y = 3$ is a horizontal line that passes through the point $(0, 3)$. Notice that every point on the line has a y-coordinate of 3.

b. The graph of $x = -2$ is a vertical line that passes through the point $(-2, 0)$. Notice that every point on the line has an x-coordinate of -2.

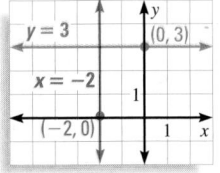

EXAMPLE 5   *Using the Standard Form*

Fundraising

The school band is selling sweatshirts and T-shirts to raise money. The goal is to raise $1200. Sweatshirts sell for a profit of $2.50 each and T-shirts for $1.50 each. Describe numbers of sweatshirts and T-shirts the band can sell to reach the goal.

SOLUTION

First write a model for the problem.

PROBLEM SOLVING STRATEGY

| **VERBAL MODEL** | Profit per sweatshirt | · | Number of sweatshirts | + | Profit per T-shirt | · | Number of T-shirts | = | Total Profit |

LABELS

Profit per sweatshirt = **$2.50**   Number of sweatshirts = s

Profit per T-shirt = **$1.50**   Number of T-shirts = t

Total profit = **$1200**

ALGEBRAIC MODEL

$2.5\,s + 1.5\,t = 1200$

STUDENT HELP

Study Tip
Finding the intercepts of a line before you draw the line can help you determine reasonable scales for the x-axis and the y-axis.

The graph of $2.5s + 1.5t = 1200$ is a line that intersects the s-axis at $(480, 0)$ and intersects the t-axis at $(0, 800)$. Points with integer coordinates on the line segment joining $(480, 0)$ and $(0, 800)$ represent ways to reach the goal. For instance, the band can sell 300 sweatshirts and 300 T-shirts.

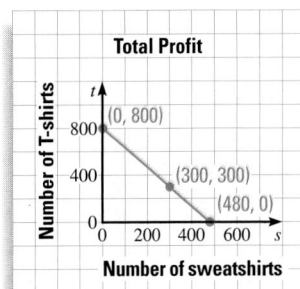

Total Profit

Number of T-shirts

Number of sweatshirts

Closure Question *Sample answer:*
The y-intercept is obvious, and a second point can easily be located using the slope.

 EXTRA EXAMPLE 4
Graph **(a)** $y = -2$ and **(b)** $x = 3$.

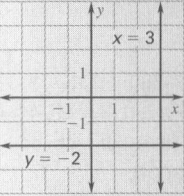

EXTRA EXAMPLE 5
Students are selling tickets to a school play. The goal is to raise $600. Tickets are $4 for adults and $3 for students. Describe the numbers of adult and student tickets sold that will reach the goal.   **points with integer coordinates on the line $4a + 3s = 600$, with a and s the numbers of adult and student tickets, such that the points are on the segment joining the a-intercept $(150, 0)$ and the s-intercept $(0, 200)$**

CHECKPOINT EXERCISES
For use after Example 4:
1. Graph **(a)** $y = 1$ and **(b)** $x = -1$.

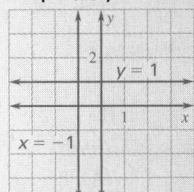

For use after Example 5:
2. If x represents the number of 4 in. high bricks and y the number of 6 in. high bricks it takes to reach the top of a 60 in. wall, what are the endpoints of the segment containing the numbers of 4 in. and 6 in. bricks that will reach the top?   **(15, 0) and (0, 10)**

FOCUS ON VOCABULARY
What is the name of the point on a line where its y-value is zero?
the x-intercept

CLOSURE QUESTION
Give an advantage of graphing a line using the slope-intercept form of its equation.
See sample answer at left.

ASSIGNMENT GUIDE

BASIC
Day 1: pp. 86–89 Exs. 16–18,
20–36 even, 37–40, 44, 46,
52–60 even, 63, 65, 66,
69–85 odd,
Quiz 1 Exs. 1–11

AVERAGE
Day 1: pp. 86–89 Exs. 16–18,
20–36 even, 37–40, 44, 46,
52–60 even, 61, 63, 65, 66,
69–85 odd, 86,
Quiz 1 Exs. 1–11

ADVANCED
Day 1: pp. 86–89 Exs. 16–18,
20–36 even, 37–40, 44, 46,
52–60 even, 61, 63–67,
69–85 odd, 86,
Quiz 1 Exs. 1–11

BLOCK SCHEDULE
pp. 86–89 Exs. 16–18, 20–36
even, 37–40, 44, 46, 52–60 even,
61, 63–67, 69–85 odd, 86,
Quiz 1 Exs. 1–11

EXERCISE LEVELS
Level A: *Easier*
16–57, 65–66, 74–85
Level B: *More Difficult*
58–61, 63–64, 68–73, 86
Level C: *Most Difficult*
62, 67

✔ **HOMEWORK CHECK**
To quickly check student under-
standing of key concepts, go
over the following exercises:
Exs. 16, 22, 28, 34, 38, 46. See
also the Daily Homework Quiz:
• Blackline Master (*Chapter 2
Resource Book,* p. 53)
• 📖 Transparency (p. 13)

10–15, 19–30.
See Additional Answers beginning on page AA1.

GUIDED PRACTICE

Vocabulary Check ✔

1. What are the slope-intercept and standard forms of a linear equation?   **See margin.**

Concept Check ✔

2. Which of the two quick-graph techniques discussed in the lesson would you use to graph $y = -2x + 4$? Explain.   **See margin.**

3. Which of the two quick-graph techniques discussed in the lesson would you use to graph $3x + 4y = 24$? Explain.   **See margin.**

Skill Check ✔

1. The slope-intercept form of the equation is $y = mx + b$, where m is the slope and b is the y-intercept of the graph of the line. The standard form of the equation of a line is $Ax + By = C$.

2. The slope-intercept technique; the equation is in slope-intercept form.

Find the slope and y-intercept of the line.

4. $y = x + 10$ 1; 10
5. $y = -2x - 7$ $-2; -7$
6. $2x - 3y = 18$ $\frac{2}{3}; -6$

Find the intercepts of the line.   **7–9. See margin.**

7. $x - y = 11$
8. $5x - 2y = 20$
9. $y = 5x - 15$

Graph the equation.   **10–15. See margin.**

10. $y = 2x + 1$
11. $y = \frac{1}{3}x - 4$
12. $y = 7$

13. $x = -5$
14. $2x - 6y = 6$
15. $5x + 3y = -15$

PRACTICE AND APPLICATIONS

STUDENT HELP

▶ **Extra Practice**
to help you master
skills is on p. 941.

3. The standard form technique; the equation is in standard form and the intercepts are easily found.

7. x-intercept: 11;
y-intercept: -11

8. x-intercept: 4;
y-intercept: -10

9. x-intercept: 3;
y-intercept: -15

MATCHING GRAPHS **Match the equation with its graph.**

16. $y = -5x + 10$ B
17. $y = -\frac{1}{2}x - 5$ A
18. $y = 4x - 12$ C

A.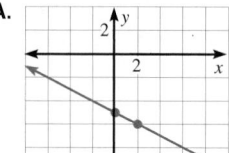
B.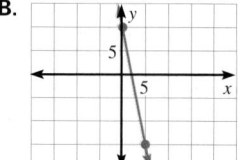
C.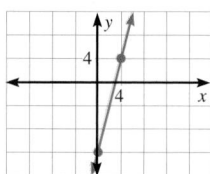

USING SLOPE AND y-INTERCEPT **Draw the line with the given slope and y-intercept.**   **19–24. See margin.**

19. $m = 3, b = -2$
20. $m = -2, b = 0$
21. $m = 1, b = 1$

22. $m = \frac{1}{2}, b = 5$
23. $m = 0, b = -7$
24. $m = -\frac{3}{7}, b = 14$

STUDENT HELP

▶ **HOMEWORK HELP**
Example 1: Exs. 16–36,
52–57
Example 2: Exs. 58–60
Example 3: Exs. 37–57
Example 4: Exs. 49–57
Example 5: Exs. 61–63

SLOPE-INTERCEPT FORM **Graph the equation.**   **25–30. See margin.**

25. $y = -x + 5$
26. $y = 4x + 1$
27. $y = \frac{4}{5}x - 1$

28. $y = 2x - 3$
29. $y = -\frac{5}{2}x - 3$
30. $y = 5x - \frac{5}{2}$

FINDING SLOPE AND y-INTERCEPT **Find the slope and y-intercept of the line.**

31. $y = 6x + 10$ 6; 10
32. $y = -9x$ $-9; 0$
33. $y = 100$ 0; 100

34. $2x + y = 14$ $-2; 14$
35. $8x - 2y = 14$ 4; -7
36. $x + 10y = 7$ $-\frac{1}{10}; \frac{7}{10}$

MATCHING GRAPHS Match the equation with its graph.

37. $x - 4y = -8$ **B**   **38.** $3x + 6y = -9$ **C**   **39.** $2x - 3y = -12$ **A**

A. 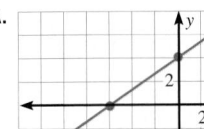   **B.** 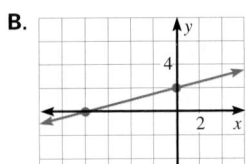   **C.**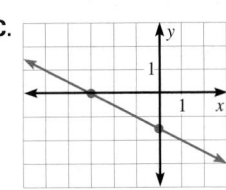

USING INTERCEPTS Draw the line with the given intercepts.   40–42. See margin.

40. x-intercept: 3
y-intercept: 5

41. x-intercept: 2
y-intercept: -6

42. x-intercept: -4
y-intercept: $-\frac{1}{2}$

STANDARD FORM Graph the equation. Label any intercepts.   43–51. See margin.

43. $2x + y = 8$ **44.** $x + 2y = 8$ **45.** $3x + 4y = -10$

46. $3x - y = 3$ **47.** $5x - 6y = -2$ **48.** $3x + 0.2y = 2$

49. $y = 6$ **50.** $x = -5$ **51.** $y = -\frac{1}{2}$

CHOOSE A METHOD Graph the equation using any method.   52–57. See margin.

52. $y = 3x + 7$ **53.** $x = -10$ **54.** $2x - 7y = 14$

55. $y = \frac{3}{4}$ **56.** $5x + 10y = 30$ **57.** $y = \frac{5}{2}x - 2$

58. 🌐 **IRS** The amount a (in billions of dollars) of annual taxes collected by the Internal Revenue Service can be modeled by $a = 57.1t + 488$ where t represents the number of years since 1980. Graph the equation.
▶ Source: *Statistical Abstract of the United States*   See margin.

59. 🌐 **PLACING AN AD** The cost C (in dollars) of placing a color advertisement in a newspaper can be modeled by $C = 7n + 20$ where n is the number of lines in the ad. Graph the equation. What do the slope and C-intercept represent?
See margin.

60. 🌐 **RAINFORESTS** The area A (in millions of hectares) of land covered by rainforests can be modeled by $A = 718.3 - 4.6t$ where t represents the number of years since 1990. Graph the equation. What are three predicted future areas of land covered by rainforests? ▶ Source: Food and Agriculture Organization   See margin.

61. 🌐 **CAR WASH** A car wash charges $8 per wash and $12 per wash-and-wax. After a busy day sales totaled $3464. Use the verbal model to write an equation that shows the different numbers of washes and wash-and-waxes that could have been done. Then graph the equation. $8w + 12x = 3464$; See margin for graph.

Price per wash	·	Number of washes	+	Price per wash-and-wax	·	Number of wash-and-waxes	=	Total sales

62. 🌐 **SAILING** The owner of a sailboat takes passengers to an island 5 miles away to go snorkeling. A sailboat averages about 9 miles per hour when using its sails and about 14 miles per hour when using its motor. Write an equation that shows the numbers of *minutes* the sailboat can use its sails and its motor to get to the island. Then graph the equation. $\frac{3}{20}s + \frac{7}{30}m = 5$; See margin for graph.

FOCUS ON
APPLICATIONS

RAINFORESTS
In Brazil the rate of rainforest destruction is 2.2 million hectares per year. Brazil recently passed a law giving its government the authority to protect forests.

! **COMMON ERROR**

EXERCISE 62 This exercise presents several opportunities for error. Remind students of the distance formula $d = rt$. In this case, the total distance d, or 5, is the sum of two distances: that by sail and that by motor, so $d = r_1t_1 + r_2t_2$. Students must also watch out for the unit change. Remind them to use dimensional analysis to ensure that they make the conversion from miles per hour to miles per minute properly.

40.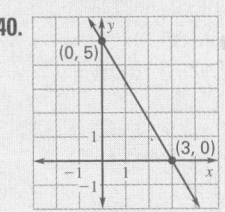

41.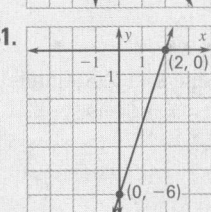

42.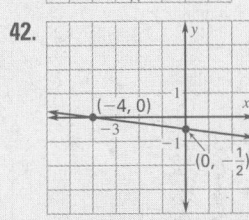

43–62. See Additional Answers beginning on page AA1.

ADDITIONAL PRACTICE AND RETEACHING

For Lesson 2.3:
• Practice Levels A, B, and C (*Chapter 2 Resource Book,* p. 41)
• Reteaching with Practice (*Chapter 2 Resource Book,* p. 44)
• 🖥 See Lesson 2.3 of the *Personal Student Tutor*

For more Mixed Review:
• 🖥 Search the *Test and Practice Generator* for key words or specific lessons.

87

DAILY HOMEWORK QUIZ

Transparency Available

1. Draw the line with slope $m = -2$ and y-intercept $b = -1$.

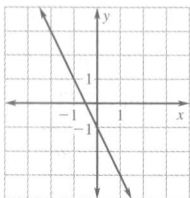

2. Graph $y = \frac{1}{2}x - 2$.

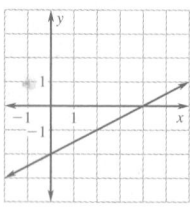

3. Find the slope and the y-intercept of the line $-3x + y = -8$.
   3; −8

4. Graph $3x - 2y = 6$. Label any intercepts.

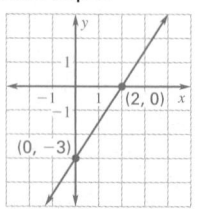

EXTRA CHALLENGE NOTE
→ Challenge problems for Lesson 2.3 are available in **blackline** format in the *Chapter 2 Resource Book*, p. 49 and at **www.mcdougallittell.com**.

ADDITIONAL TEST PREPARATION

1. **WRITING** What do the slope and y-intercept represent in a linear model for a real-life application?
   See sample answer at right.

63, 64. See next page.
68–73. See Additional Answers beginning on page AA1.

88

STUDENT HELP

HOMEWORK HELP
Visit our Web site www.mcdougallittell.com for help with problem solving in Ex. 63.

Test Preparation

63. $2.5s + 6a = 7000$;
    Sample answer: 1600 student tickets, 500 adult; 880 student, 800 adult; 400 student, 1000 adult; See margin for graph.

★ **Challenge**

64. To find the x-intercept, set y equal to 0 in the equation, and solve for x. To find the y-intercept, set x equal to 0 in the equation, and solve for y; horizontal; vertical.

86. Let T = the total number of pages read in h hours. Then $T = 120h$. It would take 8 h and 44 min.

---

63. 🌐 **TICKET PRICES** Student tickets at a high school basketball game cost $2.50 each. Adult tickets cost $6.00 each. The ticket sales at the first game of the season totaled $7000. Write a model that shows the numbers of student and adult tickets that could have been sold. Then graph the model and determine three combinations of student and adult tickets that satisfy the model. **See margin.**

64. *Writing* Explain how to find the intercepts of a line if they exist. What kind of line has no x-intercept? What kind of line has no y-intercept? **See margin.**

65. **MULTIPLE CHOICE** You have an individual retirement account (IRA). The amount a you have deposited into your account after t years can be modeled by $a = 4500 + 2000t$. How much money do you put into your IRA every year? **B**

   Ⓐ $1000   Ⓑ $2000   Ⓒ $2500   Ⓓ $4500   Ⓔ $6500

66. **MULTIPLE CHOICE** What is the slope-intercept form of $4x - 6y = 18$? **B**

   Ⓐ $x = \frac{3}{2}y + \frac{9}{2}$ Ⓑ $y = \frac{2}{3}x - 3$ Ⓒ $-y = \frac{4}{6}x + 3$

   Ⓓ $6y = -4x + 18$ Ⓔ $4x = 6y + 18$

67. **CALCULATING SLOPE** For the line $y = 7x + 6$, show that the slope is 7 regardless of the points (x_1, y_1) and (x_2, y_2) you use to calculate the slope. (*Hint:* Substitute x_1 and x_2 into the equation to obtain expressions for y_1 and y_2.)

$y_1 = 7x_1 + 6$ and $y_2 = 7x_2 + 6$, so $\dfrac{\text{rise}}{\text{run}} = \dfrac{y_2 - y_1}{x_2 - x_1} = \dfrac{(7x_2 + 6) - (7x_1 + 6)}{x_2 - x_1} = \dfrac{7x_2 - 7x_1}{x_2 - x_1} = 7.$

MIXED REVIEW

SOLVING INEQUALITIES Solve the inequality. Then graph your solution. (**Review 1.6**) 68–73. See margin for graphs.

68. $9 + x \leq 21$ $x \leq 12$ 69. $-\frac{2}{3}x + 3 < 11$ $x > -12$

70. $2x - 11 > 34 - x$ $x > 15$ 71. $64 - 3x \geq 19 - 2x$ $x \leq 45$

72. $-5 < 2x - 0.5 \leq 23$ 73. $x + 12 \leq 5$ or $3x - 21 \geq 0$
    $-2.25 < x \leq 11.75$ $x \leq -7$ or $x \geq 7$

EVALUATING FUNCTIONS Evaluate the function for the given value of x. (**Review 2.1**)

74. $f(x) = \frac{1}{2}x - 13$; $f(8)$ −9   75. $f(x) = x^2 - 3x + 2$; $f(5)$ 12

76. $f(x) = -x^3 + 8x^2 + 3$; $f(-7)$ 738   77. $f(x) = 10 - 2x$; $f(1)$ 8

78. $f(x) = |x + 17|$; $f(-5)$ 12   79. $f(x) = 12x^2 - 19$; $f\left(\frac{1}{2}\right)$ −16

FINDING SLOPE Find the slope of the line passing through the given points. (**Review 2.2 for 2.4**)

80. $(3, 2), (7, 2)$ 0   81. $(16, -3), (2, 9)$ $-\frac{6}{7}$

82. $(-12, -9), (1, -8)$ $\frac{1}{13}$ 83. $(-1, -1), (-1, -5)$ undefined

84. $(5, -2), (-3, 2)$ $-\frac{1}{2}$ 85. $(-4, 7), (2, -5)$ −2

86. 🌐 **READING SPEED** You can read a novel at a rate of 2 pages per minute. Write a model that shows the number of pages you can read in h hours. Then find how long it will take you to read a 1048 page novel. (**Review 1.5 for 2.4**)

Additional Test Preparation *Sample answer:*
The slope represents the rate of change of the dependent variable with respect to the independent variable. The y-intercept represents the starting value or the initial amount of the dependent variable.

1. domain: $-2, -1, 0, 1, 2$; range: $-2, 1$; function

2. domain: $1, 2, 3, 4$; range: $1, 2, 3, 4$; not a function

3. domain: $-3, -1, 0, 1, 2$; range: $-3, -2, 0, 1$; function

Identify the domain and range. Then tell whether the relation is a function. (Lesson 2.1)

1.
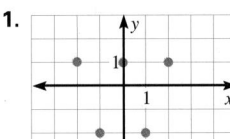

2.
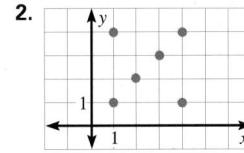

3.
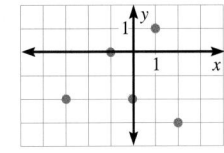

Evaluate the function for the given value of x. (Lesson 2.1)

4. $f(x) = -2x - 13; f(4)$ -21

5. $f(x) = 5x^2 - x + 9; f(-5)$ 139

Tell whether the lines are *parallel, perpendicular,* or *neither*. (Lesson 2.2)

6. Line 1: through $(2, 10)$ and $(1, 5)$
Line 2: through $(3, -7)$ and $(8, -8)$
perpendicular

7. Line 1: through $(4, 5)$ and $(9, -2)$
Line 2: through $(6, -6)$ and $(-2, -1)$
neither

Graph the equation. (Lesson 2.3)   8–10. See margin.

8. $y = 3x + 5$ **9.** $2x - 3y = 10$ **10.** $y = -11$

11. **BICYCLING** There is an annual seven day bicycle ride across Iowa that covers about 468 miles. If a participant rides each day from 8:00 A.M. to 5:00 P.M., stopping only 1 hour for lunch, what is the rider's average speed in miles per hour? (**Lesson 2.2**)   about 8.36 mi/h

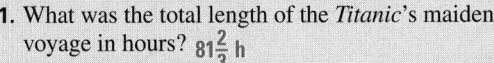
MATH & History   Transatlantic Voyages

APPLICATION LINK
www.mcdougallittell.com

THEN   **AT 2:00 P.M. ON APRIL 11, 1912,** the *Titanic* left Cobh, Ireland, on her maiden voyage to New York City. At 11:40 P.M. on April 14, the *Titanic* struck an iceberg and sank, having covered only about 2100 miles of the approximately 3400 mile trip.

1. What was the total length of the *Titanic*'s maiden voyage in hours? $81\frac{2}{3}$ h

2. What was the *Titanic*'s average speed in miles per hour?   25.7 mi/h

3. Write an equation relating the *Titanic*'s distance from New York City and the number of hours traveled. Identify the domain and range. $d = 3400 - 25.7t$, where *d* is the distance in miles and *t* is the time in hours; domain: $0 \le t \le 81.7$; $1300 \le d \le 3400$

4. Graph the equation from Exercise 3.   See margin.

NOW   **TODAY,** ocean liners still cross the Atlantic Ocean. The *Queen Elizabeth 2*, or *QE2*, is one of the fastest with a top speed of 32.5 knots (about 37 miles per hour).

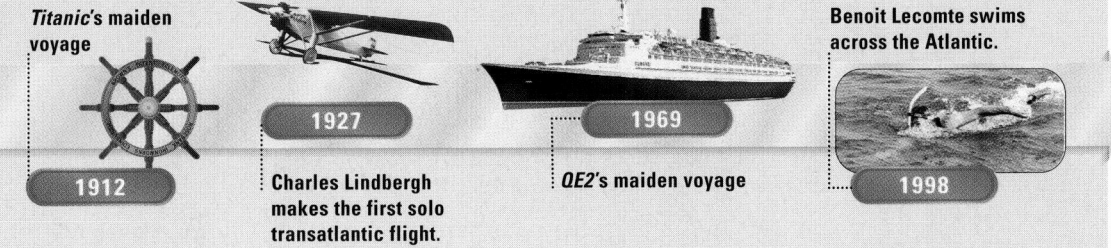

Titanic's maiden voyage

1912

1927
Charles Lindbergh makes the first solo transatlantic flight.

1969
QE2's maiden voyage

Benoit Lecomte swims across the Atlantic.
1998

ADDITIONAL RESOURCES
An alternative Quiz for Lessons 2.1–2.3 is available in the *Chapter 2 Resource Book*, p. 50.

A **blackline** master with additional Math & History Applications is available in the *Chapter 2 Resource Book*, p. 48.

63.
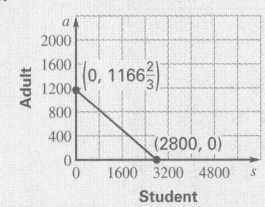
Ticket Sales
$(0, 1166\frac{2}{3})$
$(2800, 0)$

64. To find the *x*-intercept, set *y* equal to 0 in the equation, and solve for *x*. To find the *y*-intercept, set *x* equal to 0 in the equation, and solve for *y*; horizontal; vertical

8.
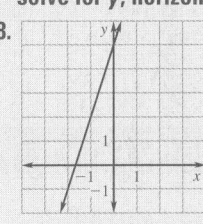

9.
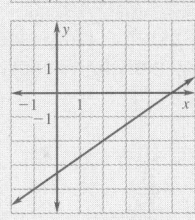

10.
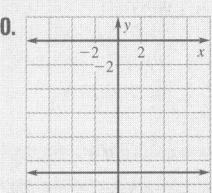

4.
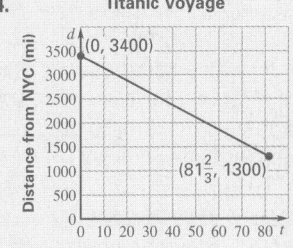
Titanic Voyage
$(0, 3400)$
$(81\frac{2}{3}, 1300)$

1 Planning the Activity

PURPOSE
To graph equations of the form $y = f(x)$ using a graphing calculator.

MATERIALS
- graphing calculator for each student
- Keystroke blackline (*Chapter 2 Resource Book,* p. 40)

PACING
- Example — 10 min
- Exercises — 15 min

▶ LINK TO LESSON
Students will not be able to use a graphing calculator to graph a line using standard form as in Method 1 of Example 3 in Lesson 2.3. Instead, they will first have to rewrite the equation in slope-intercept form as in Method 2.

2 Managing the Activity

COOPERATIVE LEARNING
Have pairs of students discuss their choice of a good viewing window for Exs. 6–8.

ALTERNATIVE APPROACH
For Exs. 6–8, students can first graph using the standard viewing window. Then by examination, tracing, or use of the "zoom" features, they can revise their window choice as desired.

3 Closing the Activity

★ KEY DISCOVERY
By first writing linear equations in slope-intercept form, you can use a graphing calculator to graph and explore lines quickly.

ACTIVITY ASSESSMENT
JOURNAL Describe how to use a graphing calculator to graph lines. See sample answer at right.

1–8. See Additional Answers beginning on page AA1.

▷ ACTIVITY 2.3

Using Technology

Graphing Equations

You can use a graphing calculator to graph equations of the form $y = f(x)$.

▶ EXAMPLE

Use a graphing calculator to graph the equation $x + 6y = 30$.

STUDENT HELP

KEYSTROKE HELP

See keystrokes for several models of calculators at www.mcdougallittell.com

▶ SOLUTION

① First solve the equation for y so that it can be entered into the calculator.

$$x + 6y = 30$$
$$6y = -x + 30$$
$$y = -\frac{1}{6}x + 5$$

② When you have fractional coefficients, you must use parentheses. So, enter the equation as $y = -(1/6)x + 5$.

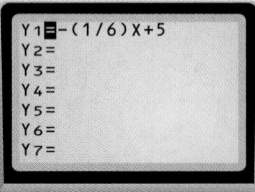

③ Finally, set a viewing window for the graph by entering the least and greatest x- and y-values and the x- and y-scales. The *standard viewing window* is $-10 \le x \le 10$ and $-10 \le y \le 10$, both with a scale of 1. The viewing window you choose should show all of the important features of the graph, such as the intercepts. The settings for the viewing window and the corresponding graph of the equation $y = -\frac{1}{6}x + 5$ are shown.

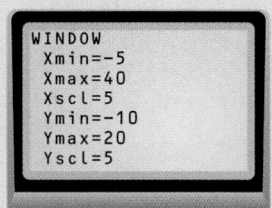
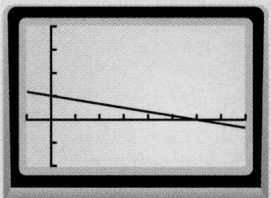

▶ EXERCISES

Use a graphing calculator to graph the equation in the standard viewing window.

1–3. See margin.

1. $y + 11 = 16 - 3x$ **2.** $2x - y = 6$ **3.** $x - 3y = -2$

Use a graphing calculator to graph the equation in the indicated viewing window.

4–5. See margin.

4. $17 - 2x = -y$ Xmin = −2, Xmax = 12, Xscl = 2, Ymin = −20, Ymax = 2, Yscl = 5

5. $y + 4 = 2x + 1$ Xmin = −2, Xmax = 5, Xscl = 1, Ymin = −4, Ymax = 1, Yscl = 1

Use a graphing calculator to graph the equation. Choose a viewing window that shows the x- and y-intercepts. 6–8. See margin.

6. $7x = 3y + 20$ **7.** $1.54x + 2.1y = 63.4$ **8.** $\frac{7}{10}x = 5y - 104$

Activity Assessment *Sample answer:*
First solve the equation for y. Then enter the equation into the function list, choose an appropriate viewing window, and graph.

2.4

Writing Equations of Lines

PACING
Basic: 2 days
Average: 2 days
Advanced: 2 days
Block Schedule: 1 block

What you should learn

GOAL 1 Write linear equations.

GOAL 2 Write direct variation equations, as applied in **Example 7**.

Why you should learn it

▼ To model **real-life** quantities, such as the number of calories you burn while dancing in **Ex. 64**.

GOAL 1 WRITING LINEAR EQUATIONS

In Lesson 2.3 you learned to find the slope and y-intercept of a line whose equation is given. In this lesson you will study the reverse process. That is, you will learn to write an equation of a line using one of the following: the slope and y-intercept of the line, the slope and a point on the line, or two points on the line.

CONCEPT SUMMARY

WRITING AN EQUATION OF A LINE

SLOPE-INTERCEPT FORM Given the slope m and the y-intercept b, use this equation:

$$y = mx + b$$

POINT-SLOPE FORM Given the slope m and a point (x_1, y_1), use this equation:

$$y - y_1 = m(x - x_1)$$

TWO POINTS Given two points (x_1, y_1) and (x_2, y_2), use the formula

$$m = \frac{y_2 - y_1}{x_2 - x_1}$$

to find the slope m. Then use the point-slope form with this slope and either of the given points to write an equation of the line.

Every nonvertical line has only one slope and one y-intercept, so the slope-intercept form is unique. The point-slope form, however, depends on the point that is used. Therefore, in this book equations of lines will be simplified to slope-intercept form so a unique solution may be given.

EXAMPLE 1   *Writing an Equation Given the Slope and the y-intercept*

Write an equation of the line shown.

SOLUTION

From the graph you can see that the slope is $m = \frac{3}{2}$. You can also see that the line intersects the y-axis at the point $(0, -1)$, so the y-intercept is $b = -1$.

Because you know the slope and the y-intercept, you should use the slope-intercept form to write an equation of the line.

$y = mx + b$ Use slope-intercept form.

$y = \frac{3}{2}x - 1$ Substitute $\frac{3}{2}$ for m and -1 for b.

▶ An equation of the line is $y = \frac{3}{2}x - 1$.

LESSON OPENER
APPLICATION

An alternative way to approach Lesson 2.4 is to use the Application Lesson Opener:

- Blackline Master (*Chapter 2 Resource Book*, p. 54)
- Transparency (p. 11)

MEETING INDIVIDUAL NEEDS

- **Chapter 3 Resource Book**
  Prerequisite Skills Review (p. 5)
  Practice Level A (p. 55)
  Practice Level B (p. 56)
  Practice Level C (p. 57)
  Reteaching with Practice (p. 58)
  Absent Student Catch-Up (p. 60)
  Challenge (p. 62)
- **Resources in Spanish**
- **Personal Student Tutor**

NEW-TEACHER SUPPORT

See the Tips for New Teachers on pp. 1–2 of the *Chapter 2 Resource Book* for additional notes about Lesson 2.4.

WARM-UP EXERCISES

Transparency Available

1. Find the slope and the y-intercept of $y = 2x - 4$. 2; –4
2. Find the slope and the y-intercept of $y = \frac{3}{2}x + 1$. $\frac{3}{2}$; 1
3. Find the slope of a line parallel to the line in Exercise 1. 2
4. Find the slope of a line perpendicular to the line in Exercise 2. $-\frac{2}{3}$

2.4 *Writing Equations of Lines*   **91**

Cycling burns 2.5 calories per hour for each pound of body weight. A 100 pound student will burn about 250 calories per hour. This is an example of *direct variation,* one type of linear equation that students will write in this lesson.

EXTRA EXAMPLE 1
Write an equation of the line shown.

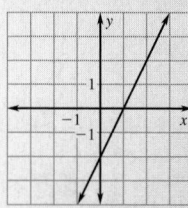

$y = 2x - 2$

EXTRA EXAMPLE 2
Write an equation of the line that passes through (–3, 4) and has a slope of $\frac{2}{3}$. $y = \frac{2}{3}x + 6$

EXTRA EXAMPLE 3
Write an equation of the line that passes through (2, –3) and is **(a)** perpendicular to and **(b)** parallel to the line $y = 2x - 3$.
(a) $y = -\frac{1}{2}x - 2$; (b) $y = 2x - 7$

✔ CHECKPOINT EXERCISES
For use after Example 1:
1. Write the equation of the line shown. $y = -\frac{3}{2}x - 2$

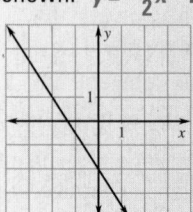

For use after Examples 2 and 3:
2. Write an equation of the line that passes through (4, –1) and is **(a)** perpendicular to and **(b)** parallel to the line $y = -x + 4$. (a) $y = x - 5$; (b) $y = -x + 3$

EXAMPLE 2 *Writing an Equation Given the Slope and a Point*

Write an equation of the line that passes through (2, 3) and has a slope of $-\frac{1}{2}$.

SOLUTION

Because you know the slope and a point on the line, you should use the point-slope form to write an equation of the line. Let $(x_1, y_1) = (2, 3)$ and $m = -\frac{1}{2}$.

$y - y_1 = m(x - x_1)$	Use point-slope form.
$y - 3 = -\frac{1}{2}(x - 2)$	Substitute for m, x_1, and y_1.

Once you have used the point-slope form to find an equation, you can simplify the result to the slope-intercept form.

$y - 3 = -\frac{1}{2}(x - 2)$	Write point-slope form.
$y - 3 = -\frac{1}{2}x + 1$	Distributive property
$y = -\frac{1}{2}x + 4$	Write in slope-intercept form.

✓**CHECK** You can check the result graphically. Draw the line that passes through the point (2, 3) with a slope of $-\frac{1}{2}$. Notice that the line has a y-intercept of 4, which agrees with the slope-intercept form found above.

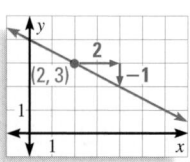

EXAMPLE 3 *Writing Equations of Perpendicular and Parallel Lines*

STUDENT HELP

HOMEWORK HELP
Visit our Web site
www.mcdougallittell.com
for extra examples.

Write an equation of the line that passes through (3, 2) and is **(a)** perpendicular and **(b)** parallel to the line $y = -3x + 2$.

SOLUTION

a. The given line has a slope of $m_1 = -3$. So, a line that is perpendicular to this line must have a slope of $m_2 = -\frac{1}{m_1} = \frac{1}{3}$. Because you know the slope and a point on the line, use the point-slope form with $(x_1, y_1) = (3, 2)$ to find an equation of the line.

$y - y_1 = m_2(x - x_1)$	Use point-slope form.
$y - 2 = \frac{1}{3}(x - 3)$	Substitute for m_2, x_1, and y_1.
$y - 2 = \frac{1}{3}x - 1$	Distributive property
$y = \frac{1}{3}x + 1$	Write in slope-intercept form.

b. For a parallel line use $m_2 = m_1 = -3$ and $(x_1, y_1) = (3, 2)$.

$y - y_1 = m_2(x - x_1)$	Use point-slope form.
$y - 2 = -3(x - 3)$	Substitute for m_2, x_1, and y_1.
$y - 2 = -3x + 9$	Distributive property
$y = -3x + 11$	Write in slope-intercept form.

FOCUS ON
PEOPLE

BARBARA
JORDAN was the
first African-American
woman elected to Congress
from a southern state. She
was a member of the House
of Representatives from 1973
to 1979.

EXAMPLE 4 *Writing an Equation Given Two Points*

Write an equation of the line that passes through $(-2, -1)$ and $(3, 4)$.

SOLUTION

The line passes through $(x_1, y_1) = (-2, -1)$ and $(x_2, y_2) = (3, 4)$, so its slope is:

$$m = \frac{y_2 - y_1}{x_2 - x_1} = \frac{4 - (-1)}{3 - (-2)} = \frac{5}{5} = 1$$

Because you know the slope and a point on the line, use the point-slope form to find an equation of the line.

$y - y_1 = m(x - x_1)$	Use point-slope form.
$y - (-1) = 1[x - (-2)]$	Substitute for m, x_1, and y_1.
$y + 1 = x + 2$	Simplify.
$y = x + 1$	Write in slope-intercept form.

EXAMPLE 5 *Writing and Using a Linear Model*

POLITICS In 1970 there were 160 African-American women in elected public office in the United States. By 1993 the number had increased to 2332. Write a linear model for the number of African-American women who held elected public office at any given time between 1970 and 1993. Then use the model to predict the number of African-American women who will hold elected public office in 2010.

DATA UPDATE of Joint Center for Political and Economic Studies data at www.mcdougallittell.com

SOLUTION

The average rate of change in officeholders is $m = \dfrac{2332 - 160}{1993 - 1970} \approx 94.4$.

You can use the average rate of change as the slope in your linear model.

PROBLEM
SOLVING
STRATEGY

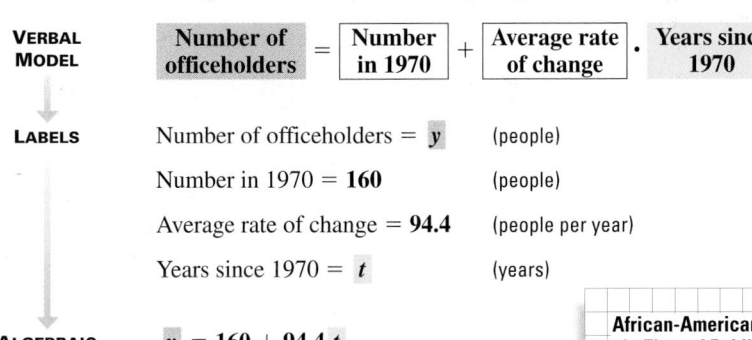

VERBAL MODEL

$$\boxed{\text{Number of officeholders}} = \boxed{\text{Number in 1970}} + \boxed{\text{Average rate of change}} \cdot \boxed{\text{Years since 1970}}$$

LABELS

Number of officeholders = y	(people)
Number in 1970 = **160**	(people)
Average rate of change = **94.4**	(people per year)
Years since 1970 = t	(years)

ALGEBRAIC MODEL

$$y = 160 + 94.4\, t$$

In 2010, which is 40 years since 1970, you can predict that there will be

$$y = 160 + 94.4(40) \approx 3936$$

African-American women in elected public office. You can graph the model to check your prediction visually.

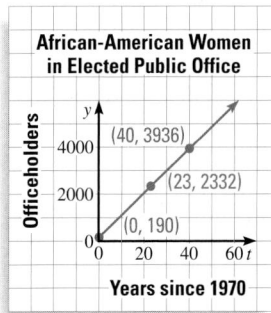

African-American Women
in Elected Public Office

(40, 3936)
(23, 2332)
(0, 190)

Officeholders

Years since 1970

EXTRA EXAMPLE 4
Write an equation of the line that passes through (1, 5) and (4, 2).
$y = -x + 6$

EXTRA EXAMPLE 5
In 1984, Americans purchased an average of 113 meals or snacks per person at restaurants. By 1996, this number was 131. Write a linear model for the number of meals or snacks purchased per person annually. Then use the model to predict the number of meals or snacks that will be purchased per person in 2006. $y = 1.5x + 113$, where y = the number of meals and x = years since 1984; about 146

CHECKPOINT EXERCISES
For use after Example 4:
1. Write an equation of the line that passes through (3, 0) and (−3, 1). $y = -\frac{1}{6}x + \frac{1}{2}$

For use after Example 5:
2. In 1991, there were 57 million cats as pets in the United States. By 1998, this number was 61 million. Write a linear model for the number of cats as pets. Then use the model to predict the number of cats as pets in 2010. $y = 0.57x + 57$, where y = millions of cats and x = years since 1991; about 68 million

DATA UPDATE
Updated data for Example 5 are available at
www.mcdougallittell.com.

EXTRA EXAMPLE 6
The variables x and y vary directly, and $y = 15$ when $x = 3$.
a. Write and graph an equation relating x and y. $y = 5x$

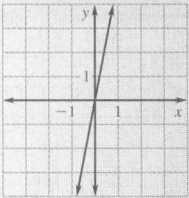

b. Find y when $x = 9$.   **45**

EXTRA EXAMPLE 7
Tell whether the data show direct variation. If so, write an equation relating x and y.
a. (3, 12), (6, 24), (9, 38), (12, 50), (15, 65), where $(x, y) =$ (Hours worked, Wages ($)) **no**
b. (8, 2.40), (16, 4.80), (24, 7.20), (32, 9.60), (40, 12.00), where $(x, y) =$ (Weight of nuts (oz), Price ($)) **yes; $y = 0.30x$**

 CHECKPOINT EXERCISES
For use after Examples 6 and 7:
1. The variables x and y vary directly, and $y = 19.22$ when $x = 6.2$.
   a. Write an equation relating x and y. $y = 3.1x$
   b. Find y when $x = 4.3$.   **13.33**

FOCUS ON VOCABULARY
How can you tell from the graph of a line that it is the graph of a direct variation? **The line will go through the origin, (0, 0), but the line will not be horizontal or vertical.**

CLOSURE QUESTION
Describe how to determine that a set of data values exhibits direct variation, and how to find the constant of variation. *Sample answer:* **For direct variation, the ratio of y to x is constant, and equals k.**

DAILY PUZZLER
If y varies directly as x and $y = kx$, does x vary directly as y? If so, what is the constant of variation?
yes; $\frac{1}{k}$

GOAL 2 WRITING DIRECT VARIATION EQUATIONS

Two variables x and y show **direct variation** provided $y = kx$ and $k \neq 0$. The nonzero constant k is called the **constant of variation,** and y is said to *vary directly* with x. The graph of $y = kx$ is a line through the origin.

EXAMPLE 6 *Writing and Using a Direct Variation Equation*

The variables x and y vary directly, and $y = 12$ when $x = 4$.

a. Write and graph an equation relating x and y.   **b.** Find y when $x = 5$.

SOLUTION

a. Use the given values of x and y to find the constant of variation.

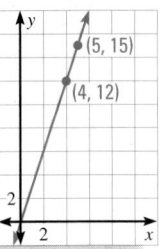

$y = kx$ Write direct variation equation.

$12 = k(4)$ Substitute 12 for y and 4 for x.

$3 = k$ Solve for k.

The direct variation equation is $y = 3x$. The graph of $y = 3x$ is shown.

b. When $x = 5$, the value of y is $y = 3(5) = 15$.

· · · · · · · · · ·

The equation for direct variation can be rewritten as $\dfrac{y}{x} = k$. This tells you that a set of data pairs (x, y) shows direct variation if the quotient of y and x is constant.

Jewelry

EXAMPLE 7 *Identifying Direct Variation*

Tell whether the data show direct variation. If so, write an equation relating x and y.

a.

14-karat Gold Chains (1 gram per inch)					
Length, x (inches)	16	18	20	24	30
Price, y (dollars)	288	324	360	432	540

b.

Loose Diamonds (round, colorless, very small flaws)					
Weight, x (carats)	0.5	0.7	1.0	1.5	2.0
Price, y (dollars)	2250	3430	6400	11,000	20,400

SOLUTION For each data set, check whether the quotient of y and x is constant.

a. For the 14-karat gold chains, $\dfrac{288}{16} = \dfrac{324}{18} = \dfrac{360}{20} = \dfrac{432}{24} = \dfrac{540}{30} = 18$. The data do show direct variation, and the direct variation equation is $y = 18x$.

b. For the loose diamonds, $\dfrac{2250}{0.5} = 4500$, but $\dfrac{3430}{0.7} = 4900$. The data do not show direct variation.

GUIDED PRACTICE

Vocabulary Check ✓

Concept Check ✓

1. Define the constant of variation for two variables x and y that vary directly.
   See margin.
2. How can you find an equation of a line given the slope and the y-intercept of the line? given the slope and a point on the line? given two points on the line?
   See margin.
3. Give a real-life example of two quantities that vary directly.
   Sample answer: the cost of a bag of apples and the weight of fruit in the bag

Skill Check ✓

Write an equation of the line that has the given properties.

4. slope: $\frac{2}{5}$, y-intercept: 2 $y = \frac{2}{5}x + 2$
5. slope: 2, passes through $(0, -4)$
   $y = 2x - 4$
6. slope: -3, passes through $(5, 2)$
   $y = -3x + 17$
7. slope: $-\frac{3}{4}$, passes through $(-7, 0)$
   See margin.
8. passes through $(4, 8)$ and $(1, 2)$
   $y = 2x$
9. passes through $(0, 2)$ and $(-5, 0)$
   See margin.
10. Write an equation of the line that passes through $(1, -6)$ and is perpendicular to the line $y = 3x + 7$. $y = -\frac{1}{3}x - \frac{17}{3}$
11. Write an equation of the line that passes through $(3, 9)$ and is parallel to the line $y = 5x - 15$. $y = 5x - 6$
12. 🌐 **LAW OF SUPPLY** The *law of supply* states that the quantity supplied of an item varies directly with the price of that item. Suppose that for \$4 per tape 5 million cassette tapes will be supplied. Write an equation that relates the number c (in millions) of cassette tapes supplied to the price p (in dollars) of the tapes. Then determine how many cassette tapes will be supplied for \$5 per tape.
    $c = 1.25p$; **6.25 million cassettes**

PRACTICE AND APPLICATIONS

STUDENT HELP

▶ **Extra Practice**
to help you master skills is on p. 942.

7. $y = -\frac{3}{4}x - \frac{21}{4}$

9. $y = \frac{2}{5}x + 2$

STUDENT HELP

▶ **HOMEWORK HELP**
Example 1: Exs. 13–18
Example 2: Exs. 19–24
Example 3: Exs. 25–28
Example 4: Exs. 29–40
Example 5: Exs. 59–62
Example 6: Exs. 43–54
Example 7: Exs. 55–58, 63–68

SLOPE-INTERCEPT FORM Write an equation of the line that has the given slope and y-intercept.

13. $m = 5$, $b = -3$
    $y = 5x - 3$
14. $m = -3$, $b = -4$
    $y = -3x - 4$
15. $m = -4$, $b = 0$
    $y = -4x$
16. $m = 0$, $b = 4$
    $y = 4$
17. $m = \frac{3}{5}$, $b = 6$
    $y = \frac{3}{5}x + 6$
18. $m = -\frac{3}{4}$, $b = \frac{7}{3}$
    $y = -\frac{3}{4}x + \frac{7}{3}$

POINT-SLOPE FORM Write an equation of the line that passes through the given point and has the given slope.

19. $(0, 4)$, $m = 2$
    $y = 2x + 4$
20. $(1, 0)$, $m = 3$
    $y = 3x - 3$
21. $(-6, 5)$, $m = 0$
    $y = 5$
22. $(9, 3)$, $m = -\frac{2}{3}$
    $y = -\frac{2}{3}x + 9$
23. $(3, -2)$, $m = -\frac{4}{3}$
    $y = -\frac{4}{3}x + 2$
24. $(7, -4)$, $m = \frac{2}{5}$
    $y = \frac{2}{5}x - \frac{34}{5}$
25. Write an equation of the line that passes through $(1, -1)$ and is perpendicular to the line $y = -\frac{1}{2}x + 6$. $y = 2x - 3$

26. Write an equation of the line that passes through $(6, -10)$ and is perpendicular to the line that passes through $(4, -6)$ and $(3, -4)$. $y = \frac{1}{2}x - 13$

27. Write an equation of the line that passes through $(2, -7)$ and is parallel to the line $x = 5$. $x = 2$

28. Write an equation of the line that passes through $(4, 6)$ and is parallel to the line that passes through $(6, -6)$ and $(10, -4)$. $y = \frac{1}{2}x + 4$

---

3 APPLY

◯ **ASSIGNMENT GUIDE**

BASIC
Day 1: pp. 95–96 Exs. 14–22 even, 29–41 odd, 42
Day 2: pp. 95–98 Exs. 25, 26–40 even, 43–59 odd, 69, 71–93 odd

AVERAGE
Day 1: pp. 95–96 Exs. 14–28 even, 25–51 odd
Day 2: pp. 95–98 Exs. 44–58 even, 59–65 odd, 69, 71–93 odd

ADVANCED
Day 1: pp. 95–97 Exs. 14–28 even, 25–57 odd, 60, 62
Day 2: pp. 95–98 Exs. 56, 58, 59–69 odd, 70, 71–93 odd

BLOCK SCHEDULE
pp. 95–98 Exs. 14–24 even, 25–61 odd, 67, 69, 71–93 odd

EXERCISE LEVELS
Level A: *Easier*
13–24, 29–40, 43–48, 77–94
Level B: *More Difficult*
25–28, 41, 42, 49–68, 71–76
Level C: *Most Difficult*
69, 70

✓ **HOMEWORK CHECK**
To quickly check student understanding of key concepts, go over the following exercises:
Exs. 14, 22, 25, 29, 35, 43. See also the Daily Homework Quiz:

• Blackline Master (*Chapter 2 Resource Book*, p. 65)

• 🖨 Transparency (p. 14)

❗ **COMMON ERROR**

EXERCISE 27 Students may be confused by the undefined slope. Remind students that because $x = 5$ is a vertical line, any other vertical line will be parallel.

MATHEMATICAL REASONING
EXERCISES 55–58 Two variables x and y vary directly if $y = kx$ for some constant k. Substituting 0 for x gives $y = k \cdot 0$, or $y = 0$. So, $(0, 0)$ is a solution of every direct variation. This shows that the graph of every direct variation passes through the origin. While every direct variation equation is linear (it is of the form $y = mx + b$, where $m = k$ and $b = 0$), only linear equations with $b = 0$ are direct variation equations.

29. $y = \frac{3}{2}x - \frac{1}{2}$

30. $y = 2x + 16$

31. $y = -\frac{1}{2}x - \frac{15}{2}$

32. $y = 4x + 30$

33. $y = -x + 8$

34. $y = -3x$

29.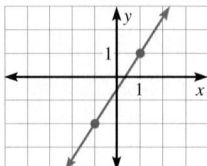

30.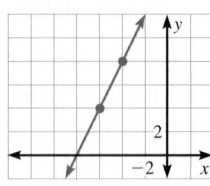

31.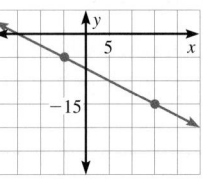

32.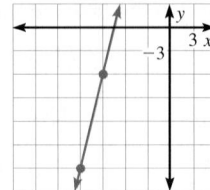

33.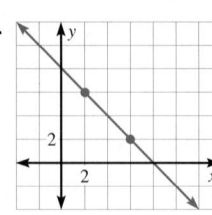

34.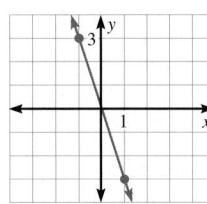

41. $3 = \left(-\frac{1}{2}\right)(2) + b$;
$3 = -1 + b$; $b = 4$.
The equation is
$y = -\frac{1}{2}x + 4$,
the same as in Example 2. The slope-intercept equation of a line is unique.

42. $m = \frac{-1 - 4}{-2 - 3} = \frac{-5}{-5} = 1$,
so $y - 4 = 1(x - 3)$;
$y = x + 1$; This is the same equation as in Example 4, since the slope-intercept equation of a line is unique.

WRITING EQUATIONS Write an equation of the line that passes through the given points.

35. $(8, 5)$, $(11, 14)$
$y = 3x - 19$

36. $(-5, 9)$, $(-4, 7)$
$y = -2x - 1$

37. $(-8, 8)$, $(0, 1)$ $y = -\frac{7}{8}x + 1$

38. $(2, 0)$, $(4, -6)$
$y = -3x + 6$

39. $(-20, -10)$, $(5, 15)$
$y = x + 10$

40. $(-2, 0)$, $(0, 6)$
$y = 3x + 6$

41. **LOGICAL REASONING** Redo Example 2 by substituting the given point and slope into $y = mx + b$. Then solve for b to write an equation of the line. Explain why using this method does not change the equation of the line.   See margin.

42. **LOGICAL REASONING** Redo Example 4 by substituting $(3, 4)$ for (x_1, y_1) into $y - y_1 = m(x - x_1)$. Then rewrite the equation in slope-intercept form. Explain why using the point $(3, 4)$ does not change the equation of the line.

RELATING VARIABLES The variables x and y vary directly. Write an equation that relates the variables. Then find y when $x = 8$.

43. $x = 2$, $y = 7$ $y = \frac{7}{2}x$; 28

44. $x = -6$, $y = 15$ $y = -\frac{5}{2}x$; -20

45. $x = -3$, $y = 9$ $y = -3x$; -24

46. $x = 24$, $y = 4$ $y = \frac{1}{6}x$; $\frac{4}{3}$

47. $x = 1$, $y = \frac{1}{2}$ $y = \frac{1}{2}x$; 4

48. $x = 0.8$, $y = 1.6$ $y = 2x$; 16

RELATING VARIABLES The variables x and y vary directly. Write an equation that relates the variables. Then find x when $y = -5$.

$y = \frac{1}{5}x$; -25

49. $x = 6$, $y = 3$ $y = \frac{1}{2}x$; -10

50. $x = 9$, $y = 15$ $y = \frac{5}{3}x$; -3

51. $x = -5$, $y = -1$

52. $x = 100$, $y = 2$ $y = \frac{1}{50}x$; -250

53. $x = \frac{5}{2}$, $y = \frac{5}{4}$ $y = \frac{1}{2}x$; -10

54. $x = -0.3$, $y = 2.2$ $y = -\frac{22}{3}x$; $\frac{15}{22}$

IDENTIFYING DIRECT VARIATION Tell whether the data show direct variation. If so, write an equation relating x and y.

55. yes; $y = \frac{1}{2}x$

56. no

57. yes; $y = -x$

58. yes; $y = -2x$

55.

x	2	4	6	8	10
y	1	2	3	4	5

56.

x	1	2	3	4	5
y	5	4	3	2	1

57.

x	3	6	9	12	15
y	-3	-6	-9	-12	-15

58.

x	-5	-4	-3	-2	-1
y	10	8	6	4	2

59. 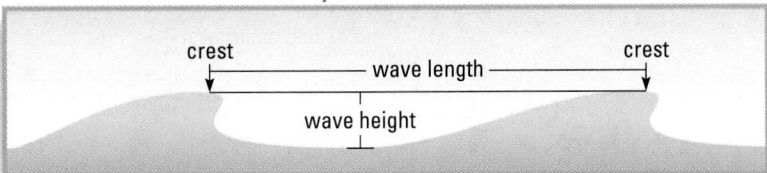 **POPULATION OF OREGON** From 1990 to 1996 the population of Oregon increased by about 60,300 people per year. In 1996 the population was about 3,204,000. Write a linear model for the population P of Oregon from 1990 to 1996. Let t represent the number of years since 1990. Then estimate the population of Oregon in 2014. ▶ Source: *Statistical Abstract of the United States*
$P = 60,300t + 2,842,200$; 4,289,400

60. 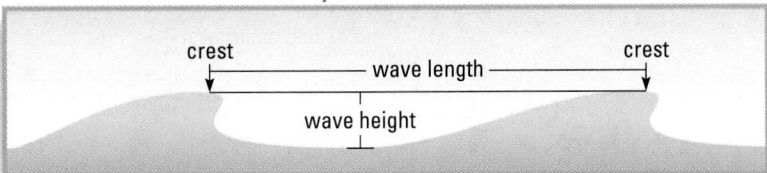 **AIRFARE** In 1998 an airline offered a special airfare of $201 to fly from Cincinnati to Washington, D.C., a distance of 386 miles. Special airfares offered for longer flights increased by about $.138 per mile. Write a linear model for the special airfares a based on the total number of miles t of the flight. Estimate the airfare offered for a flight from Boston to Sacramento, a distance of 2629 miles.
$a = 0.138(t - 386) + 201$, or $a = 0.138t + 147.732$; about $511

61. 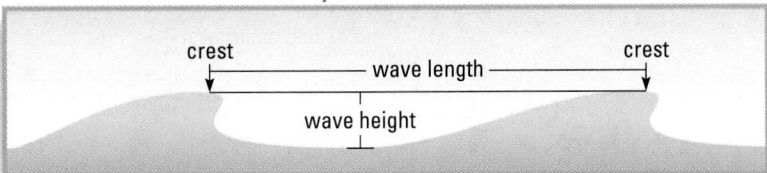 **BOOKSTORE SALES** In 1990 retail sales at bookstores were about $7.4 billion. In 1997 retail sales at bookstores were about $11.8 billion. Write a linear model for retail sales s (in billions of dollars) at bookstores from 1990 through 1997. Let t represent the number of years since 1990. Then estimate the retail sales at bookstores in 2012. ▶ Source: *American Booksellers Association*
$s = 0.629t + 7.4$; about $21.2 billion

62. **SCIENCE** ▶ **CONNECTION** The velocity of sound in dry air increases as the temperature increases. At 40°C sound travels at a rate of about 355 meters per second. At 49°C it travels at a rate of about 360 meters per second. Write a linear model for the velocity v (in meters per second) of sound based on the temperature T (in degrees Celsius). Then estimate the velocity of sound at 60°C.
▶ Source: *CRC Handbook of Chemistry and Physics* $V = \frac{5}{9}C + 333$; about 366 m/sec

63. 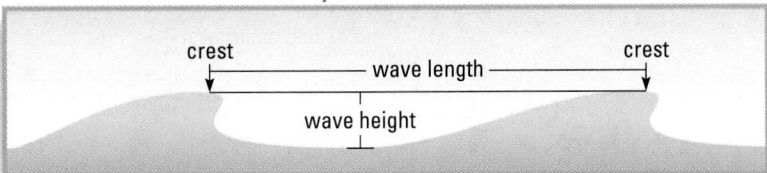 **BREAKING WAVES** The height h (in feet) at which a wave breaks varies directly with the wave length l (in feet), which is the distance from the crest of one wave to the crest of the next. A wave that breaks at a height of 4 feet has a wave length of 28 feet. Write a linear model that gives h as a function of l. Then estimate the wave length of a wave that breaks at a height of 5.5 feet.
▶ Source: Rhode Island Sea Grant $h = \frac{1}{7}l$; 38.5 ft

crest — wave length — crest

wave height

64. 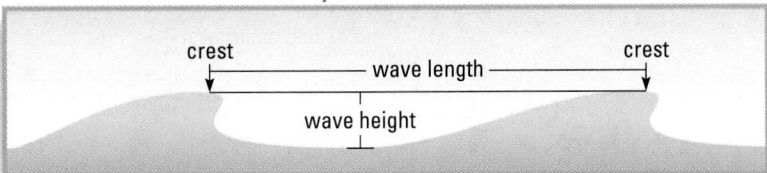 **DANCING** The number C of calories a person burns performing an activity varies directly with the time t (in minutes) the person spends performing the activity. A 160 pound person can burn 73 Calories by dancing for 20 minutes. Write a linear model that gives C as a function of t. Then estimate how long a 160 pound person should dance to burn 438 Calories. ▶ Source: *Health Journal*
$C = 3.65t$; 120 min or 2 h

65. 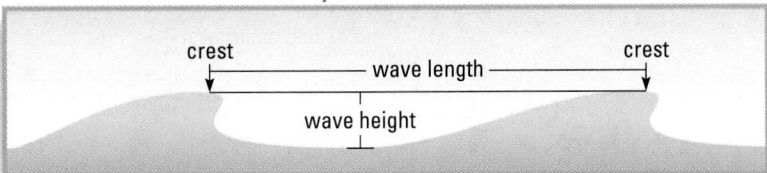 **HAILSTONES** Hailstones are formed when frozen raindrops are caught in updrafts and carried into high clouds containing water droplets. As a rule of thumb, the radius r (in inches) of a hailstone varies directly with the time t (in seconds) that the hailstone is in a high cloud. After a hailstone has been in a high cloud for 60 seconds, its radius is 0.25 inch. Write a linear model that gives r as a function of t. Then estimate how long a hailstone was in a high cloud if its radius measures 2.75 inches. ▶ Source: National Oceanic and Atmospheric Administration $r = \frac{1}{240}t$; 11 min

66. **GEOMETRY** ▶ **CONNECTION** When the length of a rectangle is fixed, the area A (in square inches) of the rectangle varies directly with its width w (in inches). When the width of a particular rectangle is 12 inches, its area is 36 square inches. Write an equation that gives A as a function of w. Then find A when w is 7.5 inches.
$A = 3w$; 22.5 in.^{2}

2.4 *Writing Equations of Lines*

FOCUS ON APPLICATIONS

HAILSTONES The largest hailstone ever recorded fell at Coffeyville, Kansas. It weighed 1.67 pounds and had a radius of about 2.75 inches.

APPLICATION NOTE
EXERCISE 59 Remind students that finding a linear model for the population gives the *average* rate of change per year in the population. The actual change will likely vary from year to year.

EXERCISE 64 Point out to students that the actual number of calories burned by a particular person of a given weight will vary depending upon the person's metabolism, dancing style, and so on.

ADDITIONAL PRACTICE AND RETEACHING

For Lesson 2.4:
• Practice Levels A, B, and C (*Chapter 2 Resource Book,* p. 55)
• Reteaching with Practice (*Chapter 2 Resource Book,* p. 58)
• See Lesson 2.4 of the *Personal Student Tutor*

For more Mixed Review:
• Search the *Test and Practice Generator* for key words or specific lessons.

DAILY HOMEWORK QUIZ

🎞 *Transparency Available*

1. Write an equation of the line with slope 4 and y-intercept -2.3. $y = 4x - 2.3$

2. Write an equation of the line with slope -1 that passes through $(2, -3)$. $y = -x - 1$

3. Write an equation of the line that passes through $(3, -5)$ and is perpendicular to the line through $(1, 4)$ and $(3, -2)$. $y = \frac{1}{3}x - 6$

4. Write an equation of the line that passes through $(-2, 5)$ and $(2, -3)$. $y = -2x + 1$

The variables x and y vary directly. Write an equation that relates the variables. Then find y when $x = 3$.

5. $x = 4$, $y = 10$ $y = 2.5x$; 7.5

6. $x = 6$, $y = \frac{1}{2}$ $y = \frac{1}{12}x$; $\frac{1}{4}$

EXTRA CHALLENGE NOTE

→ Challenge problems for Lesson 2.4 are available in **blackline** format in the *Chapter 2 Resource Book,* p. 62 and at **www.mcdougallittell.com**.

ADDITIONAL TEST PREPARATION

1. WRITING Describe the characteristics of direct variation, including the form of the equation and a description of the graph. *Sample answer:* A direct variation is of the form $y = kx$, where k is the constant of variation. As x increases by one unit, y increases k units for positive k, or decreases $|k|$ units for negative k. The graph is a line with slope k passing through the origin.

69a, 69b, 86–94.
See Additional Answers beginning on page AA1.

STATISTICS CONNECTION Tell whether the data show direct variation. If so, write an equation relating x and y.

67.

Applesauce				
Ounces, x	8	16	24	36	48
Price, y	$.89	$1.25	$1.39	$2.09	$2.49

no

68.

Fresh Apples				
Pounds, x	1	1.5	2	2.5	3
Price, y	$.89	$1.34	$1.78	$2.23	$2.49

no

Test Preparation

69. MULTI-STEP PROBLEM Besides slope-intercept and point-slope forms, another form that can be used to write equations of lines is *intercept form*: $\frac{x}{a} + \frac{y}{b} = 1$.
69a. and b. See margin for graphs.

a. Graph $\frac{x}{5} + \frac{y}{3} = 1$. **b.** Graph $\frac{x}{-2} + \frac{y}{9} = 1$.

c. *Writing* Geometrically, what do a and b represent in the intercept form of a linear equation? a is the x-intercept; b is the y-intercept.

69d. $\frac{x}{3} + \frac{y}{4} = 1$

d. Write an equation of the line shown using intercept form.

e. Write an equation of the line with x-intercept -5 and y-intercept -8 using intercept form. $\frac{x}{-5} + \frac{y}{-8} = 1$

f. Write an equation of the line that passes through $(0, -3)$ and $(2, 0)$ using intercept form. $\frac{x}{2} + \frac{y}{-3} = 1$

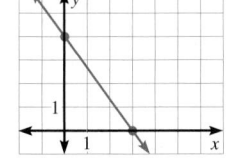

★ Challenge

70. SLOPE-INTERCEPT FORM Derive the slope-intercept form of a linear equation from the slope formula using $(0, b)$ as the coordinates of the point where the line crosses the y-axis and an arbitrary point (x, y).
$m = \frac{y - b}{x - 0}$; $m = \frac{y - b}{x}$; $mx = y - b$; $y = mx + b$

MIXED REVIEW

SOLVING EQUATIONS Solve the equation. (Review 1.7)

71. $|x - 10| = 17$ $-7, 27$ **72.** $|7 - 2x| = 5$ $1, 6$ **73.** $|-x - 9| = 1$ $-10, -8$

74. $|4x + 1| = 0.5$ $-\frac{3}{8}, -\frac{1}{8}$ **75.** $|22x + 6| = 9.2$ $-\frac{38}{55}, \frac{8}{55}$ **76.** $|5.2x + 7| = 3.8$ $-2.08, -0.615$

FINDING SLOPE Find the slope of the line passing through the given points. (Review 2.2 for 2.5)

77. $(1, -7), (2, 7)$ 14 **78.** $(-1, -1), (-5, -4)$ $\frac{3}{4}$ **79.** $(2, 4), (5, 10)$ 2

80. $(5, -2), (-3, -1)$ $-\frac{1}{8}$ **81.** $(-2, 4), (2, 4)$ 0 **82.** $(-4, -1), (5, -4)$ $-\frac{1}{3}$

83. $(0, -8), (-9, 10)$ -2 **84.** $(6, 11), (6, -5)$ undefined **85.** $(-11, 4), (-4, 11)$ 1

GRAPHING EQUATIONS Graph the equation. (Review 2.3 for 2.5) 86–94. See margin.

86. $y = \frac{3}{4}x - 5$ **87.** $y = -\frac{1}{5}x + 2$ **88.** $y = -\frac{3}{7}x + 2$

89. $3x + 7y = 42$ **90.** $2x - 8y = -15$ **91.** $-5x + 3y = 10$

92. $x = 0$ **93.** $y = -3$ **94.** $y = x$

● ACTIVITY 2.5

Developing Concepts

Fitting a Line to a Set of Data

GROUP ACTIVITY
Work in a small group.

MATERIALS
• overhead projector
• overhead transparency
• metric ruler
• meter stick
• graph paper

▶ **QUESTION** How can you approximate the *best-fitting line* for a set of data?

▶ **EXPLORING THE CONCEPT** See margin.

❶ Draw a line segment 15 centimeters long on an overhead transparency. Place the transparency on an overhead projector. First measure the distance (in centimeters) from the overhead projector to the screen, and then measure the length (in centimeters) of the line segment as it appears on the screen. Record the data in a table like the one shown.

❷ Repeat **Step 1** using nine other locations of the overhead projector.

❸ Graph the data pairs (x, y). Describe the pattern of the data.

Distance from projector to screen (cm), x	Length of line segment on screen (cm), y
200	?
210	?
220	?
230	?
240	?
250	?
260	?
270	?
280	?
290	?

❹ Use a ruler to draw a line that lies as close as possible to all of the points on the graph. The line does not have to pass through any of the points. Your line is an approximation of the best-fitting line for the data.

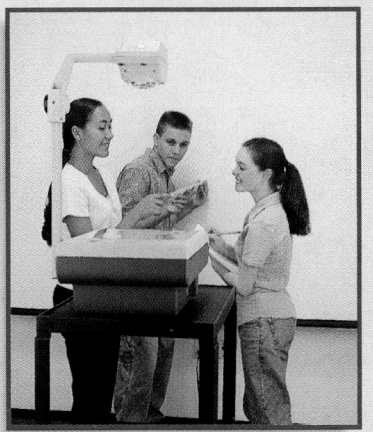

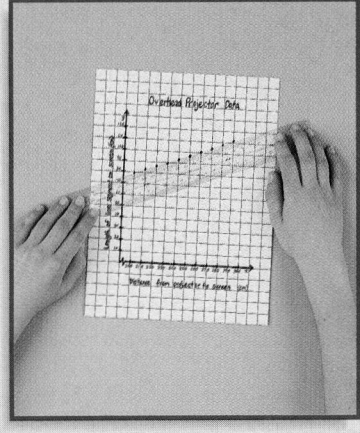

▶ **DRAWING CONCLUSIONS** See margin.

1. Find the slope of your line. What does the slope represent?

2. Find the y-intercept of your line. What does the y-intercept represent?

3. Use the slope and y-intercept to write an equation of your line.

4. Use your equation from Exercise 3 to estimate the length of the line segment as it appears on the screen if the distance from the overhead projector to the screen is 300 centimeters.

5. Test your prediction from Exercise 4. How accurate was your prediction?

6. Did every group in your class have the same line? Did every group have the same prediction?

2.5 *Concept Activity*   **99**

Steps 1–4, Exs. 1–6. See Additional Answers beginning on page AA1.

---

1 Planning the Activity

PURPOSE
To approximate the best-fitting line for data, and to use the line to make predictions.

MATERIALS
• overhead projector
• overhead transparency
• metric ruler, meter stick
• graph paper
• Activity Support Master (*Chapter 2 Resource Book,* p. 66)

PACING
• Exploring the Concept — 15 min
• Drawing Conclusions — 15 min

▶ *LINK TO LESSON*
In Examples 2 and 3 of Lesson 2.5, students sketch a line to best fit data in a scatter plot.

2 Managing the Activity

COOPERATIVE LEARNING
Have one student find the distances, a second sketch the graph and draw a line, a third find the slope of the line, and a fourth find the equation of the line. Then the group can make and test their predictions together.

CLASSROOM MANAGEMENT
Different groups may graph the data using different scales. You may want to decide on a common scale for each axis.

3 Closing the Activity

★ **KEY DISCOVERY**
The best-fitting line for data does not have to pass through any data points, but it should be as close as possible to many points.

ACTIVITY ASSESSMENT
The equation of the best-fitting line is $y = 2.5x + 304$. Predict the y-coordinate of a data point with an x-value of 82. **509**

> **LESSON OPENER**
> **APPLICATION**
> An alternative way to approach
> Lesson 2.5 is to use the Application
> Lesson Opener:
> • Blackline Master (*Chapter 2 Resource Book,* p. 67)
> •  Transparency (p. 12)

MEETING INDIVIDUAL NEEDS
• *Chapter 2 Resource Book*
   Prerequisite Skills Review (p. 5)
   Practice Level A (p. 69)
   Practice Level B (p. 70)
   Practice Level C (p. 71)
   Reteaching with Practice (p. 72)
   Absent Student Catch-Up (p. 74)
   Challenge (p. 76)
• *Resources in Spanish*
• *Personal Student Tutor*

NEW-TEACHER SUPPORT
See the Tips for New Teachers on
pp. 1–2 of the *Chapter 2 Resource Book* for additional notes about
Lesson 2.5.

WARM-UP EXERCISES

Transparency Available

1. Find the slope of the line
through (1, 4) and (–5, –2).   **1**

Find an equation of the line
through each pair of points.

2. (–2, 5) and (1, –1) $y = -2x + 1$
3. (100, 500) and (150, 1000)
   $y = 10x - 500$
4. (0.26, 8.5) and (0.36, 9.8)
   $y = 13x + 5.12$

What you should learn

GOAL 1   Use a scatter plot
to identify the correlation
shown by a set of data.

GOAL 2   Approximate
the best-fitting line for a
set of data, as applied in
Example 3.

Why you should learn it

▼ To identify **real-life** trends
in data, such as when and for
how long Old Faithful will
erupt in **Ex. 23**.

Correlation and Best-Fitting Lines

GOAL 1   **SCATTER PLOTS AND CORRELATION**

A **scatter plot** is a graph used to determine whether there is a relationship between
paired data. In many real-life situations, scatter plots follow patterns that are
approximately linear. If y tends to increase as x increases, then the paired data are
said to have a **positive correlation**. If y tends to decrease as x increases, then the
paired data are said to have a **negative correlation**. If the points show no linear
pattern, then the paired data are said to have **relatively no correlation**.

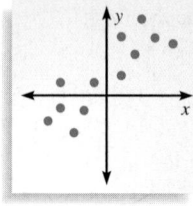

Positive correlation

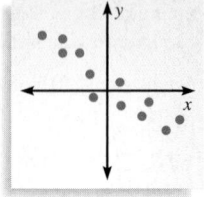

Negative correlation

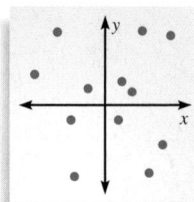

Relatively no correlation

EXAMPLE 1   *Determining Correlation*

MUSIC   Describe the correlation shown by each scatter plot.

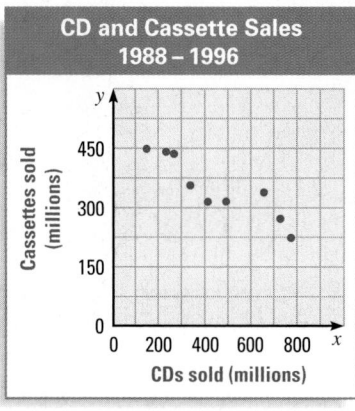
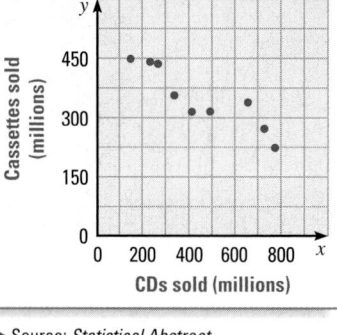

▶ Source: *Statistical Abstract
of the United States*

▶ Sources: *Electronic Market Data Book,
Recording Industry Association of America*

SOLUTION

The first scatter plot shows a negative correlation, which means that as CD sales
increased, the sales of cassettes tended to decrease.

The second scatter plot shows a positive correlation, which means that as CD sales
increased, the sales of CD players tended to increase.

GOAL 2 APPROXIMATING BEST-FITTING LINES

When data show a positive or negative correlation, you can approximate the data with a line. Finding the line that *best* fits the data is tedious to do by hand. (See page 107 for a description of how to use technology to find the best-fitting line.) You can, however, approximate the best-fitting line using the following graphical approach.

APPROXIMATING A BEST-FITTING LINE: GRAPHICAL APPROACH

STEP ① Carefully *draw a scatter plot* of the data.

STEP ② *Sketch the line* that appears to follow most closely the pattern given by the points. There should be as many points above the line as below it.

STEP ③ *Choose two points* on the line, and estimate the coordinates of each point. These two points do not have to be original data points.

STEP ④ *Find an equation of the line* that passes through the two points from Step 3. This equation models the data.

EXAMPLE 2 *Fitting a Line to Data*

Walking Speeds

Researchers have found that as you increase your walking speed (in meters per second), you also increase the length of your step (in meters). The table gives the average walking speeds and step lengths for several people. Approximate the best-fitting line for the data.

▶ Source: *Biomechanics and Energetics of Muscular Exercise*

Speed	0.8	0.85	0.9	1.3	1.4	1.6	1.75	1.9
Step	0.5	0.6	0.6	0.7	0.7	0.8	0.8	0.9
Speed	2.15	2.5	2.8	3.0	3.1	3.3	3.35	3.4
Step	0.9	1.0	1.05	1.15	1.25	1.15	1.2	1.2

SOLUTION

① Begin by drawing a scatter plot of the data.

② Next, sketch the line that appears to best fit the data.

③ Then, choose two points on the line. From the scatter plot shown, you might choose $(0.9, 0.6)$ and $(2.5, 1)$.

④ Finally, find an equation of the line. The line that passes through the two points has a slope of:

$$m = \frac{1 - 0.6}{2.5 - 0.9} = \frac{0.4}{1.6} = 0.25$$

Use the point-slope form to write the equation.

$y - y_1 = m(x - x_1)$ **Use point-slope form.**

$y - 0.6 = 0.25(x - 0.9)$ **Substitute for m, x_1, and y_1.**

$y = 0.25x + 0.375$ **Simplify.**

Walking Speeds

2.5 Correlation and Best-Fitting Lines **101**

---

2 TEACH

MOTIVATING THE LESSON
Ask students if any of them has a black-and-white television at home. As the number of color television sets sold has gone up, the number of black-and-white sets sold has gone down. Tell students that this is an example *of negative correlation*, which they will explore in this lesson.

EXTRA EXAMPLE 1
Describe the correlation shown by the scatter plot.

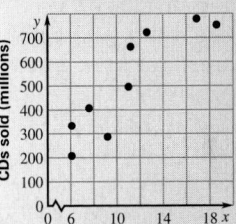

CD and Music Video Sales 1989–1997

positive correlation

EXTRA EXAMPLE 2
The data pairs give the average speed of an airplane during the first 10 minutes of a flight, with *x* in minutes and *y* in miles per hour. Approximate the best-fitting line for the data.
(1, 180), (2, 250), (3, 290), (4, 310), (5, 400), (6, 420), (7, 410), (8, 490), (9, 520), (10, 510)
Sample answer: $y = 37.5x + 172$

CHECKPOINT EXERCISES
For use after Examples 1 and 2:
1. Describe the correlation shown by the scatter plot. Then approximate the best-fitting line for the data.

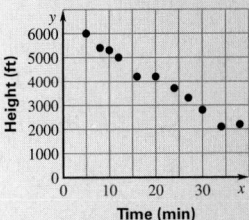

Balloon Height and Time

negative correlation; *Sample answer:* $y = -118x + 6430$

101

EXTRA EXAMPLE 3

The data pairs give the number of U.S. births from 1990 to 1997, where *x* is years since 1990 and *y* is in thousands.

(0, 4158), (1, 4111), (2, 4065), (3, 4000), (4, 3953), (5, 3900), (6, 3891), (7, 3895)

a. Approximate the best-fitting line for the data. *Sample answer:* $y = -41.5x + 4140$

b. Use the fitted line to estimate the number of births in the year 2000. **about 3,725,000**

 CHECKPOINT EXERCISES

For use after Example 3:

1. The data pairs give the U.S. production of beef from 1990 to 1997, where *x* is years since 1990 and *y* is billions of pounds.

(0, 22.7), (1, 22.9), (2, 23.1), (3, 23.0), (4, 24.4), (5, 25.2), (6, 25.5), (7, 25.5)

a. Approximate the best-fitting line for the data. *Sample answer:* $y = 0.480x + 22.4$

b. Use the fitted line to estimate the beef production in the year 2000. **about 27,200,000,000 pounds**

FOCUS ON VOCABULARY

Correlation does not imply *causation*. A correlation between data sets does not mean that one variable affects the other.

CLOSURE QUESTION

How do you use the best-fitting line to make a prediction?
Sample answer: Substitute the desired domain value for *x* and calculate the corresponding value for *y*.

DAILY PUZZLER

The points of a scatter plot all lie on a horizontal line. What kind of correlation does this show? Explain.
No correlation; no matter the value of the first variable, the value of the second is always the same.

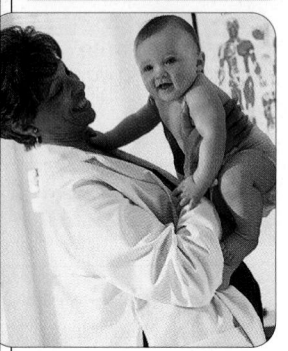

PEDIATRICIAN
A pediatrician is a medical doctor who specializes in children's health. About 7% of all medical doctors are pediatricians.

CAREER LINK
www.mcdougallittell.com

EXAMPLE 3 *Using a Fitted Line*

SLEEP REQUIREMENTS The table shows the age *t* (in years) and the number *h* of hours slept per day by 24 infants who were less than one year old.

Infant Sleep Requirements							
Age, *t*	0.03	0.05	0.05	0.08	0.11	0.19	0.21	0.26
Sleep, *h*	15.0	15.8	16.4	16.2	14.8	14.7	14.5	15.4
Age, *t*	0.34	0.35	0.35	0.44	0.52	0.69	0.70	0.75
Sleep, *h*	15.2	15.3	14.4	13.9	14.4	13.2	14.1	14.2
Age, *t*	0.80	0.82	0.86	0.91	0.94	0.97	0.98	0.98
Sleep, *h*	13.4	13.2	13.9	13.1	13.7	12.7	13.7	13.6

a. Approximate the best-fitting line for the data.

b. Use the fitted line to estimate the number of hours that a 6 month old infant sleeps per day.

SOLUTION

a. *Draw* a scatter plot of the data.

Sketch the line that appears to best fit the data.

Choose two points on the line. From the scatter plot shown, you might choose:

$$(0, 15.5) \text{ and } (0.52, 14.4)$$

Find an equation of the line. The line that passes through the two points has a slope of:

$$m = \frac{14.4 - 15.5}{0.52 - 0} = \frac{-1.1}{0.52} \approx -2.12$$

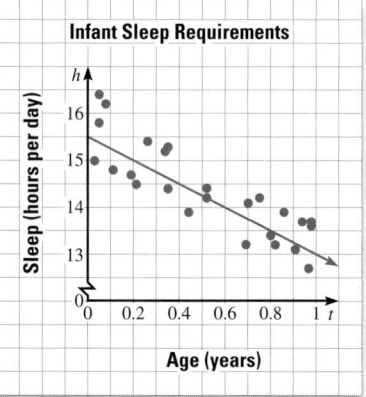

Infant Sleep Requirements

Because the *h*-intercept was chosen as one of the two points for determining the line, you can use the slope-intercept form to approximate the best-fitting line as follows:

$h = mt + b$	Use slope-intercept form.
$h = -2.12t + 15.5$	Substitute for *m* and *b*.

▶ An equation of the line is $h = -2.12t + 15.5$. Notice that a newborn infant sleeps about 15.5 hours per day and tends to sleep less as he or she gets older.

b. To estimate the number of hours that a 6 month old infant sleeps, use the model from part (a) and the fact that 6 months = 0.5 years.

$h = -2.12t + 15.5$	Write linear model.
$h = -2.12(0.5) + 15.5$	Substitute 0.5 for *t*.
$h \approx 14.4$	Simplify.

▶ A 6 month old infant sleeps about 14.4 hours per day.

GUIDED PRACTICE

Vocabulary Check ✓

Concept Check ✓

1. Explain the meaning of the terms positive correlation, negative correlation, and relatively no correlation. **See margin.**

2. Suppose you were given the shoe sizes *s* and the heights *h* of one hundred 25 year old men. Do you think that *s* and *h* would have a *positive correlation*, a *negative correlation*, or *relatively no correlation*? Explain. **See margin.**

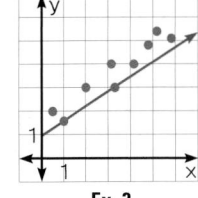

Ex. 3

3. **ERROR ANALYSIS** Explain why the line shown at the right is not a good fit for the data. **See margin.**

Skill Check ✓

2. *Sample answer:* A positive correlation; taller men tend to have larger feet.

3. *Sample answer:* Two data points lie on the line and all the rest are above the line. There should be about as many data points above the line as below the line.

4. A positive correlation; the *y*-values tend to increase as the *x*-values increase.

4. Does the scatter plot at the right show a *positive correlation*, a *negative correlation*, or *relatively no correlation*? Explain. **See margin.**

5. Look back at Example 2. Estimate the step length of a person who walks at a speed of 4 meters per second. **about 1.4 m**

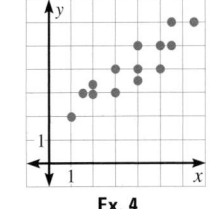

Ex. 4

🌐 **FM RADIO STATIONS** In Exercises 6 and 7, use the table below which gives the number of FM radio stations from 1989 to 1995. ▶ Source: *Statistical Abstract of the United States*

Years since 1989	0	1	2	3	4	5	6
FM radio stations	4269	4392	4570	4785	4971	5109	5730

6. Approximate the best-fitting line for the data. *Sample answer:* $y = 222t + 4166$

7. If the pattern continues, how many FM radio stations will there be in 2010? *Sample answer:* about 8830

PRACTICE AND APPLICATIONS

STUDENT HELP

▶ **Extra Practice**
to help you master skills is on p. 942.

DETERMINING CORRELATION Tell whether *x* and *y* have a *positive correlation*, a *negative correlation*, or *relatively no correlation*.

8.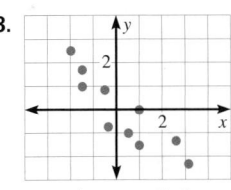

negative correlation

9.

positive correlation

10.

relatively no correlation

DRAWING SCATTER PLOTS Draw a scatter plot of the data. Then tell whether the data have a *positive correlation*, a *negative correlation*, or *relatively no correlation*. **11–12. See margin for plots.**

STUDENT HELP

▶ **HOMEWORK HELP**
Example 1: Exs. 8–14, 22, 23
Example 2: Exs. 16–21
Example 3: Exs. 24–27

11.

x	1	2	3	3	5	5	6	7	8	9
y	1	3	3	4	4	5	7	6	8	7

positive correlation

12.

x	1	1	3	4	4	5	7	7	8	8
y	8	2	5	8	3	5	3	5	1	8

relatively no correlation

---

3 APPLY

ASSIGNMENT GUIDE

BASIC
Day 1: pp. 103–106 Exs. 8–24 even, 25, 28, 31–43 odd, Quiz 2 Exs. 1–9

AVERAGE
Day 1: pp. 103–106 Exs. 8–24 even, 25–28, 31–43 odd, Quiz 2 Exs. 1–9

ADVANCED
Day 1: pp. 103–106 Exs. 8–24 even, 25–29, 31–43 odd, Quiz 2 Exs. 1–9

BLOCK SCHEDULE WITH 2.6
pp. 103–106 Exs. 8–24 even, 25–28, 31–43 odd, Quiz 2 Exs. 1–9

EXERCISE LEVELS
Level A: *Easier*
8–14, 22, 23, 34–43
Level B: *More Difficult*
15–21, 24–28, 30–33
Level C: *Most Difficult*
29

HOMEWORK CHECK ✔
To quickly check student understanding of key concepts, go over the following exercises: Exs. 8, 12, 16, 20, 24. See also the Daily Homework Quiz:

• Blackline Master (*Chapter 2 Resource Book*, p. 80)

• 💻 Transparency (p. 15)

1. *Sample answer:* If *y* tends to increase as *x* increases, the data have a positive correlation. If *y* tends to decrease as *x* increases, the data have a negative correlation. If the data points do not show a linear pattern, there is relatively no correlation.

11, 12. See Additional Answers beginning on page AA1.

APPLICATION NOTE

EXERCISE 22 As the table illustrates, temperature usually decreases as altitude increases. Sometimes in winter, however, a shallow layer of cold air may become trapped near the ground under warmer air, especially in valleys or adjacent to mountains. This is called a "temperature inversion," a condition that can lead to high levels of air pollution.

EXERCISE 23 Students may or may not be aware that "Old Faithful" received its name because of the regularity of its eruptions.

13.
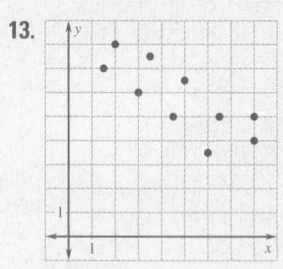

14.
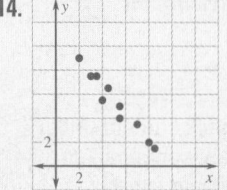

19.
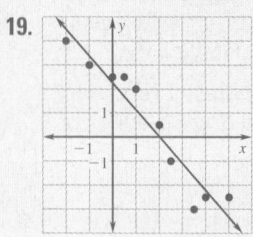

20.
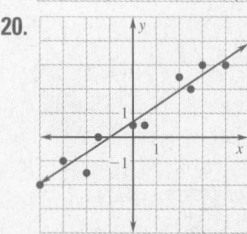

21.
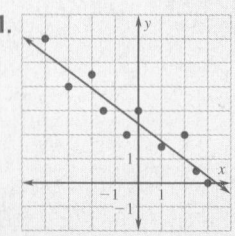

15. *Sample answer:* List the data points so that the values of *x* are in increasing order. If the *y*-values mostly increase along with the *x*-values, there is a positive correlation. If the *y*-values mostly decrease as the *x*-values increase, there is a negative correlation. Otherwise, there is relatively no correlation.

FOCUS ON
APPLICATIONS

▶ REAL LIFE **OLD FAITHFUL,** shown here between eruptions, is just one of the 200–250 geysers located in Yellowstone National Park. There are more geysers and hot springs in Yellowstone than in the rest of the world combined.

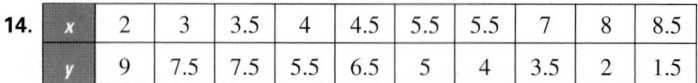
DRAWING SCATTER PLOTS Draw a scatter plot of the data. Then tell whether the data have a *positive correlation*, a *negative correlation*, or *relatively no correlation*. 13–14. See margin for plots.

13.
x	1.5	2	3	3.5	4.5	5	6	6.5	8	8
y	7	8	6	7.5	5	6.5	3.5	5	5	4

negative correlation

14.
x	2	3	3.5	4	4.5	5.5	5.5	7	8	8.5
y	9	7.5	7.5	5.5	6.5	5	4	3.5	2	1.5

negative correlation

15. **LOGICAL REASONING** Explain how you can determine the type of correlation for data by examining the data in a table as opposed to drawing a scatter plot.

APPROXIMATING BEST-FITTING LINES Copy the scatter plot. Then approximate the best-fitting line for the data. 16–18. Sample answers are given.

16.
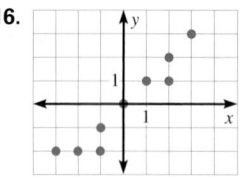
$y = 0.88x - 0.10$

17.
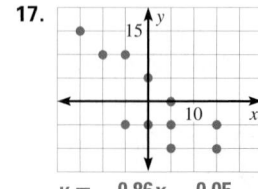
$y = -0.86x - 0.05$

18.
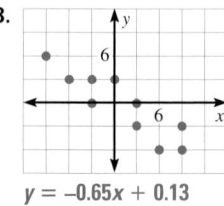
$y = -0.65x + 0.13$

FITTING A LINE TO DATA Draw a scatter plot of the data. Then approximate the best-fitting line for the data. 19–21. See margin for plots. Sample best-fitting lines are given.

19.
x	−2	−1	0	0.5	1	2	2.5	3.5	4	5
y	4	3	2.5	2.5	2	0.5	−1	−3	−2.5	−2.5

$y = -1.11x + 2.27$

20.
x	−4	−3	−2	−1.5	0	0.5	2	2.5	3	4
y	−2	−1	−1.5	0	0.5	0.5	2.5	2	3	3

$y = 0.66x + 0.60$

21.
x	−4	−3	−2	−1.5	−0.5	0	1	2	2.5	3
y	6	4	4.5	3	2	3	1.5	2	0.5	0

$y = -0.73x + 2.47$

22. 🌐 **HIGH ALTITUDE TEMPERATURES** The table shows the temperature for various elevations based on a temperature of 59°F at sea level. Draw a scatter plot of the data and describe the correlation shown. See margin for plot; negative correlation.

Elevation (ft)	1000	5000	10,000	15,000	20,000	30,000
Temperature (°F)	56	41	23	5	−15	−47

23. 🌐 **OLD FAITHFUL** Old Faithful is a geyser in Yellowstone National Park. The table shows the duration of eruptions and the time interval between eruptions for a typical day. Draw a scatter plot of the data and describe the correlation shown. See margin for plot; positive correlation.

Duration (min)	4.4	3.9	4	4	3.5	4.1	2.3	4.7	1.7	4.9	1.7	4.6	3.4
Interval (min)	78	74	68	76	80	84	50	93	55	76	58	74	75

22, 23. See Additional Answers beginning on page AA1.

STUDENT HELP

HOMEWORK HELP
Visit our Web site
www.mcdougallittell.com
for help with problem
solving in Exs. 24–27.

🌐 **CITY YEAR** In Exercises 24 and 25, use the table below which gives the enrollment for the City Year national youth service program from 1989 to 1998.

Years since 1989	0	1	2	3	4	5	6	7	8	9
Enrollment	57	76	107	234	371	688	678	716	894	918

24. Approximate the best-fitting line for the data.   *Sample answer:* $y = 110t - 22$

25. If the pattern continues, how many people will enroll in City Year in 2010?   about 2290

BIOLOGY ▸ **CONNECTION** In Exercises 26 and 27, use the table below which gives the average life expectancy (in years) of a person based on various years of birth. ▸ Source: National Center for Health Statistics

Year of birth	1900	1910	1920	1930	1940
Life expectancy	47.3	50	54.1	59.7	62.9
Year of birth	1950	1960	1970	1980	1990
Life expectancy	68.2	69.7	70.8	73.7	75.4

26. Approximate the best-fitting line for the data.   *Sample answer:* $y = 0.326x - 571$

27. Predict the life expectancy for someone born in 2010.   about 84.3 years

Test Preparation

28. **MULTI-STEP PROBLEM** The table below gives the numbers (in thousands) of black-and-white and color televisions sold in the United States for various years from 1955–1995. ▸ Source: Electronic Industries Association   28a, b. See margin for plots.

28c. Negatively correlated; as sales of color televisions increased, the sales of black-and-white televisions decreased.

29. *Sample answer:* One possibility would be the way the price of a gallon of gasoline varies over time, since the fluctuations in the price are so erratic and cannot be predicted. Another possibility would be the sales of some new piece of technology that showed up on the scene and then died out very quickly when it was replaced by something else.

Year	Black-and-white TVs sold (thousands)	Color TVs sold (thousands)
1955	7,738	20
1960	5,709	120
1965	8,753	2,694
1970	4,704	5,320
1975	4,955	6,486
1980	6,684	10,897
1985	3,684	16,995
1990	1,411	20,384
1995	480	25,600

a. Draw a scatter plot of the data pairs (*year, black-and-white TVs sold*). Then describe the correlation shown by the scatter plot.   negative correlation

b. Draw a scatter plot of the data pairs (*year, color TVs sold*). Then describe the correlation shown by the scatter plot.   positive correlation

c. **CRITICAL THINKING** Based on your answers to parts (a) and (b), are black-and-white television sales and color television sales *positively correlated*, *negatively correlated*, or *neither*? Explain.   See margin.

★ **Challenge**

29. **BEST-FITTING LINES** Describe a set of real-life data where the best-fitting line could *not* be used to make a prediction. Explain.   See margin.

STUDENT HELP NOTES
▸ **Homework Help** Students can find help for Exs. 24–27 at **www.mcdougallittell.com**. The information can be printed out for students who don't have access to the Internet.

APPLICATION NOTE
EXERCISE 27 Students may notice that life expectancy increased by an average of over 4 years per decade from 1900–1950, but at less than 2 years per decade from 1950–1990. If this trend continues, the prediction based on the best-fitting line could be somewhat high.

28a.
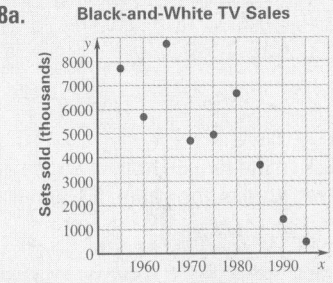
Black-and-White TV Sales

28b.
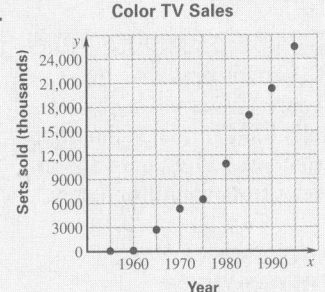
Color TV Sales

ADDITIONAL PRACTICE AND RETEACHING

For Lesson 2.5:
• Practice Levels A, B, and C (*Chapter 2 Resource Book,* p. 69)
• Reteaching with Practice (*Chapter 2 Resource Book,* p. 72)
• 🖥 See Lesson 2.5 of the *Personal Student Tutor*

For more Mixed Review:
• 🖥 Search the *Test and Practice Generator* for key words or specific lessons.

1. Draw a scatter plot of the data. Tell whether x and y have a *positive correlation,* a *negative correlation,* or *relatively no correlation.*
$(-3, 1)$, $(-3, 2)$, $(-2, 0)$, $(-1, 2)$, $(0, 0)$, $(1, -2)$, $(2, -1)$, $(2, -2)$, $(3, -1)$, $(3, -3)$
negative correlation

2. Draw a scatter plot of the data. Approximate the best-fitting line.
$(2, 10)$, $(2, 12)$, $(4, 11)$, $(6, 7)$, $(6, 9)$, $(8, 6)$, $(8, 8)$, $(10, 6)$, $(12, 3)$, $(12, 6)$

Sample answer: $y = -\frac{2}{3}x + 12\frac{2}{3}$, using $(4, 10)$ and $(10, 6)$

3. Use the results from Exercise 2 to predict the value of y when $x = 14$. **about $3\frac{1}{3}$**

EXTRA CHALLENGE NOTE

→ Challenge problems for Lesson 2.5 are available in **blackline** format in the *Chapter 2 Resource Book,* p. 76 and at **www.mcdougallittell.com.**

ADDITIONAL TEST PREPARATION

1. OPEN ENDED Describe two real-life data sets that you would expect to have a positive correlation, though not necessarily a direct linear relationship.
See sample answer below.

ADDITIONAL RESOURCES

An alternative Quiz for Lessons 2.4 and 2.5 is available in the *Chapter 2 Resource Book,* p. 77.

30–33, 38–43, 9.
   See Additional Answers beginning on page AA1.

MIXED REVIEW

SOLVING INEQUALITIES Solve the inequality. Then graph your solution. (Review 1.6 for 2.6) 30–33. See margin for graphs.

30. $2x - 9 \geq 14$ $x \geq \frac{23}{2}$

31. $3(x + 7) < -x + 10$ $x < -\frac{11}{4}$

32. $17 \leq 2x - 7 \leq 29$ $12 \leq x \leq 18$

33. $x - 4 < 0$ or $x - 6 \geq 4$ $x < 4$ or $x \geq 10$

DETERMINING STEEPNESS Tell which line is steeper. (Review 2.2)

34. Line 1: through $(-3, 4)$ and $(1, 6)$
Line 2: through $(1, -5)$ and $(6, 2)$
line 2

35. Line 1: through $(6, 1)$ and $(-4, 4)$
Line 2: through $(-2, 3)$ and $(1, -6)$
line 2

36. Line 1: through $(2, 4)$ and $(1, 7)$
Line 2: through $(-5, 8)$ and $(3, 8)$
line 1

37. Line 1: through $(4, 3)$ and $(1, -9)$
Line 2: through $(-2, -4)$ and $(3, -7)$
line 1

GRAPHING EQUATIONS Graph the equation. (Review 2.3 for 2.6) 38–43. See margin.

38. $y = \frac{1}{3}x + 5$

39. $y = -10x + 9$

40. $y = \frac{7}{3}$

41. $-x + 2y = -8$

42. $4x + 2y = 1$

43. $x = 12$

QUIZ 2

Self-Test for Lessons 2.4 and 2.5

Write an equation of the line that passes through the given point and has the given slope. (Lesson 2.4)

1. $(0, 6)$, $m = \frac{2}{3}$ $y = \frac{2}{3}x + 6$

2. $(-4, -3)$, $m = 2$ $y = 2x + 5$

3. $(2, -7)$, $m = -\frac{1}{5}$ $y = -\frac{1}{5}x - \frac{33}{5}$

4. Write an equation of the line that passes through $(1, -2)$ and is perpendicular to the line that passes through $(4, 2)$ and $(0, 4)$. (Lesson 2.4) $y = 2x - 4$

Tell whether x and y have a *positive correlation*, a *negative correlation*, or *relatively no correlation*. (Lesson 2.5)

5. relatively no correlation   **6.** negative correlation   **7.** positive correlation

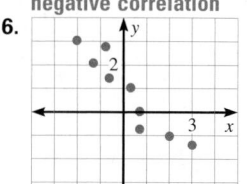

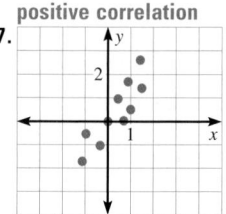

8. 🌐 **WAVES** The water depth d (in feet) at which a wave breaks varies directly with the height h (in feet) of the wave. A 6.5 foot wave breaks at a water depth of 8.45 feet. Write a linear model that gives d as a function of h. If a wave breaks at a depth of 5.2 feet, what is its height? (Lesson 2.4) $d = 1.3h$; 4 feet

9. 🌐 **HEIGHTS OF CHILDREN** The table gives the average heights of children for ages 1–10. Draw a scatter plot of the data and approximate the best-fitting line for the data. (Lesson 2.5)   *Sample answer: $h = 6.63t + 71.5$*

Age (years)	1	2	3	4	5	6	7	8	9	10
Height (cm)	73	85	93	100	107	113	120	124	130	135

Additional Test Preparation *Sample answer:*
the size of a house and its price; the height and weight of students in a class; blood cholesterol levels and risk of heart disease

● ACTIVITY 2.5

Using Technology

Using Linear Regression

Many graphing calculators have a _linear regression_ feature that can be used to find the best-fitting line for a set of data.

▶ **EXAMPLE**

The table gives the price p (in cents) of a first-class stamp over time where t is the number of years since 1970. Use the linear regression feature of a graphing calculator to find an equation of the best-fitting line for the data.

t	1	4	5	8	11	11	15	18	21	25	29
p	8	10	13	15	18	20	22	25	29	32	33

STUDENT HELP

KEYSTROKE HELP

See keystrokes for several models of calculators at www.mcdougallittell.com

▶ **SOLUTION**

1 Use the *Stat Edit* feature to enter the data into two *lists* such as L_1 and L_2. Enter the years since 1970 in L_1 and the prices in L_2.

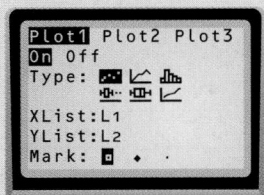

2 Find an equation of the best-fitting line by using the *Stat Calc* feature. The linear regression equation can be rounded to $p = 0.947t + 7.71$.

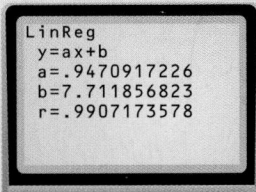

3 To see that the data have a positive correlation, graph the data pairs. To do this, make a scatter plot using the *Stat Plot* feature.

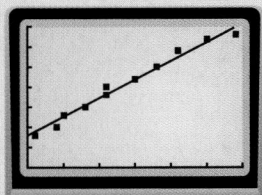

4 Graph the regression equation in an appropriate viewing window. The graph shows that the line fits the data well.

▶ **EXERCISES**

Use a graphing calculator to find and graph an equation of the best-fitting line.
1–2. See margin for graphs.

1.

x	50	75	80	100	150	175	210	250	260	320
y	0.3	0.5	0.6	0.7	0.75	0.85	1.05	0.9	1.1	1.15

$y = 0.0028x + 0.32$

2.

x	4	7	8.5	10	11	14	15	16	18	19
y	150	450	600	600	900	1100	1250	1400	1400	1650

$y = 97.8x - 248$

2.5 *Technology Activity* **107**

---

1 Planning the Activity

PURPOSE
To use a graphing calculator to find the best-fitting line for a set of data.

MATERIALS
• graphing calculator per student
• Keystroke blackline (*Chapter 2 Resource Book*, p. 68)

PACING
• Example — 10 min
• Exercises — 10 min

▶ **LINK TO LESSON**
Students can return to Examples 2 and 3 of Lesson 2.5 to check the best-fitting line using a graphing calculator.

2 Managing the Activity

COOPERATIVE LEARNING
Have pairs of students discuss their choice of a good viewing window for Step 4 of the Example. You may want to pair students who are proficient with calculators with those who are less skilled to keep the class working together.

CLASSROOM MANAGEMENT
Students may wonder what "r" is. Tell them that it is the *correlation coefficient*. The closer its value to ±1, the better the line fits the data.

3 Closing the Activity

★ **KEY DISCOVERY**
A graphing calculator's linear regression feature finds the equation of the best-fitting line.

ACTIVITY ASSESSMENT
Use a graphing calculator to find an equation of the best-fitting line for the data pairs given.
(−2, 3.1), (−1, 3.5), (0, 3.9), (1, 4.5), (2, 4.9), (3, 5.5), (4, 5.9)
$y = 0.479x + 3.99$

1, 2. See Additional Answers beginning on page AA1.

LESSON OPENER
ACTIVITY
An alternative way to approach
Lesson 2.6 is to use the Activity
Lesson Opener:

- Blackline Master (*Chapter 2
Resource Book,* p. 81)
- Transparency (p. 13)

MEETING INDIVIDUAL NEEDS
- *Chapter 2 Resource Book*
  Prerequisite Skills Review (p. 5)
  Practice Level A (p. 82)
  Practice Level B (p. 83)
  Practice Level C (p. 84)
  Reteaching with Practice (p. 85)
  Absent Student Catch-Up (p. 87)
  Challenge (p. 89)
- *Resources in Spanish*
- Personal Student Tutor

NEW-TEACHER SUPPORT
See the Tips for New Teachers on
pp. 1–2 of the *Chapter 2 Resource
Book* for additional notes about
Lesson 2.6.

WARM-UP EXERCISES

Transparency Available

Graph each line on the same
coordinate grid.

1. $y = 2$
2. $x = -1$
3. $-x + y = 1$
4. $y = 3x + 1$

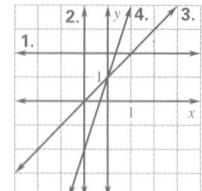

2.6 Linear Inequalities in Two Variables

What you should learn

GOAL 1 Graph linear inequalities in two variables.

GOAL 2 Use linear inequalities to solve **real-life** problems, such as finding the number of minutes you can call relatives using a calling card in **Example 4.**

Why you should learn it

▼ To model **real-life** data, such as blood pressures in your arm and ankle in **Ex. 45.**

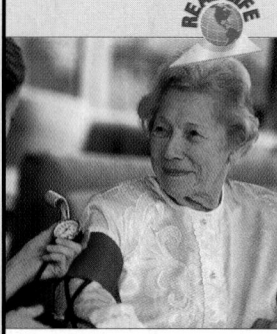

3. the blue dots lie on or above the line; the red dots are below the line.

4. *Sample answer:* Graph the related line, solid if the inequality is ≤ or ≥; dashed if the inequality is < or >. Test a point not on the line to see if it is a solution of the inequality and find out which region of the plane to shade.

GOAL 1 GRAPHING LINEAR INEQUALITIES

A **linear inequality** in two variables is an inequality that can be written in one of the following forms:

$$Ax + By < C, \quad Ax + By \leq C, \quad Ax + By > C, \quad Ax + By \geq C$$

An ordered pair (x, y) is a **solution** of a linear inequality if the inequality is true when the values of x and y are substituted into the inequality. For instance, $(-6, 2)$ is a solution of $y \geq 3x - 9$ because $2 \geq 3(-6) - 9$ is a true statement.

EXAMPLE 1 *Checking Solutions of Inequalities*

Check whether the given ordered pair is a solution of $2x + 3y \geq 5$.

a. $(0, 1)$ **b.** $(4, -1)$ **c.** $(2, 1)$

SOLUTION

ORDERED PAIR	SUBSTITUTE	CONCLUSION
a. $(0, 1)$	$2(0) + 3(1) = 3 \not\geq 5$	$(0, 1)$ is not a solution.
b. $(4, -1)$	$2(4) + 3(-1) = 5 \geq 5$	$(4, -1)$ is a solution.
c. $(2, 1)$	$2(2) + 3(1) = 7 \geq 5$	$(2, 1)$ is a solution.

▶ ACTIVITY

Developing Concepts

Investigating the Graph of an Inequality

1 Copy the scatter plot.

2 Test each circled point to see whether it is a solution of $x + y \geq 1$. If it is a solution, color it blue. If it is not a solution, color it red.

3 Graph the line $x + y = 1$. What relationship do you see between the colored points and the line? **See margin.**

4 Describe a general strategy for graphing an inequality in two variables. **See margin.**

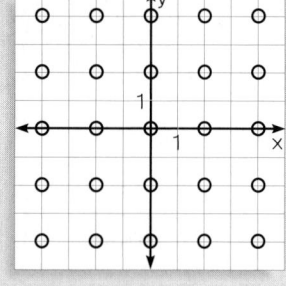

The **graph** of a linear inequality in two variables is the graph of all solutions of the inequality. The boundary line of the inequality divides the coordinate plane into two **half-planes:** a shaded region which contains the points that are solutions of the inequality, and an unshaded region which contains the points that are not.

GRAPHING A LINEAR INEQUALITY

The graph of a linear inequality in two variables is a half-plane. To graph a linear inequality, follow these steps:

STEP ❶ Graph the boundary line of the inequality. Use a dashed line for < or > and a solid line for ≤ or ≥.

STEP ❷ To decide which side of the boundary line to shade, test a point *not* on the boundary line to see whether it is a solution of the inequality. Then shade the appropriate half-plane.

STUDENT HELP

↳ **Look Back**
For help with inequalities in one variable, see p. 42.

EXAMPLE 2   *Graphing Linear Inequalities in One Variable*

Graph (**a**) $y < -2$ and (**b**) $x \le 1$ in a coordinate plane.

SOLUTION

a. Graph the boundary line $y = -2$. Use a dashed line because $y < -2$.

Test the point $(0, 0)$. Because $(0, 0)$ is *not* a solution of the inequality, shade the half-plane below the line.

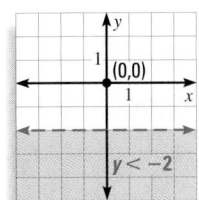

b. Graph the boundary line $x = 1$. Use a solid line because $x \le 1$.

Test the point $(0, 0)$. Because $(0, 0)$ *is* a solution of the inequality, shade the half-plane to the left of the line.

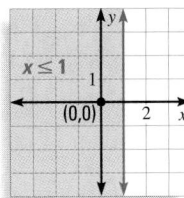

EXAMPLE 3   *Graphing Linear Inequalities in Two Variables*

Graph (**a**) $y < 2x$ and (**b**) $2x - 5y \ge 10$.

STUDENT HELP

↳ **Study Tip**
Because your test point must *not* be on the boundary line, you may not always be able to use $(0, 0)$ as a convenient test point. In such cases test a different point, such as $(1, 1)$ or $(1, 0)$.

SOLUTION

a. Graph the boundary line $y = 2x$. Use a dashed line because $y < 2x$.

Test the point $(1, 1)$. Because $(1, 1)$ *is* a solution of the inequality, shade the half-plane below the line.

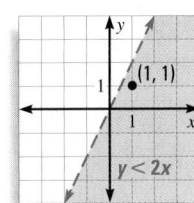

b. Graph the boundary line $2x - 5y = 10$. Use a solid line because $2x - 5y \ge 10$.

Test the point $(0, 0)$. Because $(0, 0)$ is *not* a solution of the inequality, shade the half-plane below the line.

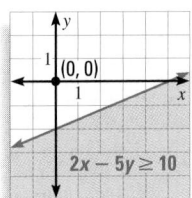

MOTIVATING THE LESSON
Ask students to imagine going to a store with up to $100 to spend on CDs and videos. How many of each can they buy? Tell students they can use a graph to find the solutions of this *inequality.*

EXTRA EXAMPLE 1
Check whether the ordered pair is a solution of $4x - 2y \ge 8$.
a. $(3, 3)$ no   **b.** $(-2, -9)$ yes

EXTRA EXAMPLE 2
Graph (**a**) $y \ge 2$ and (**b**) $x > -1$.
a.

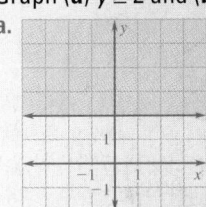

b.

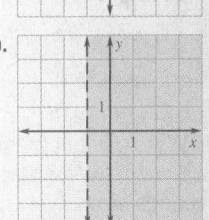

EXTRA EXAMPLE 3
Graph $4x + 2y \ge 8$.

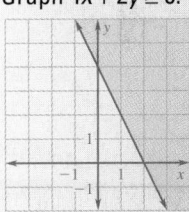

☑ CHECKPOINT EXERCISES
For use after Example 1:
1. Check whether the ordered pair is a solution of $3x + y \le 6$.
  a. $(2, 0)$ yes   **b.** $(-1, 10)$ no

For use after Examples 2 and 3:
2. Graph $3x - y < 3$.

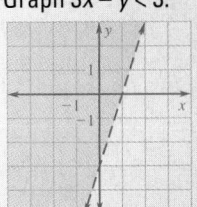

EXTRA EXAMPLE 4
You have $200 to spend on CDs and music videos. CDs cost $10 each and music videos cost $15.

a. Write a linear inequality in two variables to represent the number of CDs *x* and music videos *y* you can buy.

$10x + 15y \le 200$

b. Graph the inequality. Discuss three possible solutions for the real-life situation.

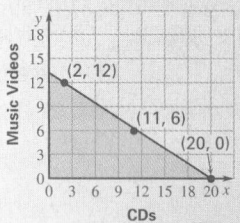

What $200 will Buy

Sample answers: You can buy 2 CDs and 12 videos for $200, 11 CDs and 6 videos for $200, or 12 CDs and 4 videos for $180.

 CHECKPOINT EXERCISES

For use after Example 4:

1. You have $15 for fresh fruit for a party. Write a linear inequality in two variables for how many pounds of strawberries *x* at $1.25 per pound and cherries *y* at $2.40 per pound you can buy. What are the intercepts of the graph?
$1.25x + 2.40y \le 15$;
(12, 0) and (0, 6.25)

FOCUS ON VOCABULARY
What is the difference between a *half-plane* for a > or < inequality and one for a ≤ or ≥ inequality?
The half-plane for ≤ or ≥ contains its boundary.

CLOSURE QUESTION
Describe the graph of a linear inequality in two variables. *Sample answer:* It consists of a boundary line and the shaded half-plane that contains the solutions. The boundary line may (solid) or may not (dashed) belong to the solution.

110

Communication

EXAMPLE 4 *Writing and Using a Linear Inequality*

You have relatives living in both the United States and Mexico. You are given a prepaid phone card worth $50. Calls within the continental United States cost $.16 per minute and calls to Mexico cost $.44 per minute.

a. Write a linear inequality in two variables to represent the number of minutes you can use for calls within the United States and for calls to Mexico.

b. Graph the inequality and discuss three possible solutions in the context of the real-life situation.

SOLUTION

PROBLEM SOLVING STRATEGY

a. VERBAL MODEL

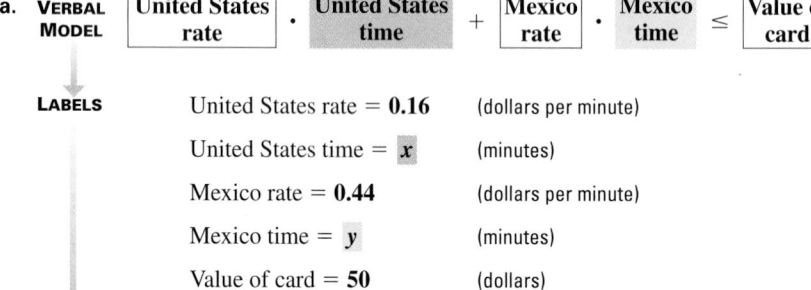

| United States rate | · | United States time | + | Mexico rate | · | Mexico time | ≤ | Value of card |

LABELS

United States rate = **0.16**   (dollars per minute)

United States time = *x*   (minutes)

Mexico rate = **0.44**   (dollars per minute)

Mexico time = *y*   (minutes)

Value of card = **50**   (dollars)

ALGEBRAIC MODEL

$0.16\,x + 0.44\,y \le 50$

STUDENT HELP

HOMEWORK HELP
Visit our Web site
www.mcdougallittell.com
for extra examples.

b. *Graph* the boundary line $0.16x + 0.44y = 50$. Use a solid line because $0.16x + 0.44y \le 50$.

Test the point (0, 0). Because (0, 0) *is* a solution of the inequality, shade the half-plane below the line. Finally, because *x* and *y* cannot be negative, restrict the graph to points in the first quadrant.

Possible solutions are points within the shaded region shown.

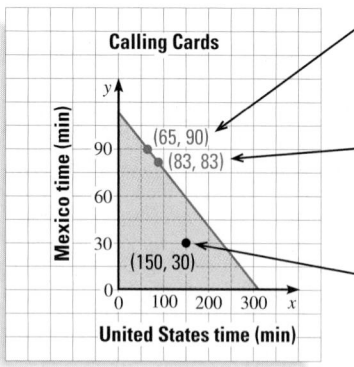

Calling Cards

One solution is to spend 65 minutes on calls within the United States and 90 minutes on calls to Mexico. The total cost will be $50.

To split the time evenly, you could spend 83 minutes on calls within the United States and 83 minutes on calls to Mexico. The total cost will be $49.80.

You could instead spend 150 minutes on calls within the United States and only 30 minutes on calls to Mexico. The total cost will be $37.20.

GUIDED PRACTICE

Vocabulary Check ✓

Concept Check ✓

1. *Sample answer:* The graph of a linear equation is a line in the plane, while the graph of a linear inequality is a half-plane with a line as its boundary.

Skill Check ✓

2. Dashed; solid; *Sample answer:* The points for which $Ax + By = C$ are solutions of the latter inequality and are included as part of the graph by using a solid line, but are not solutions of $Ax + By < C$.

1. Compare the graph of a linear inequality with the graph of a linear equation. See margin.
2. Would you use a dashed line or a solid line for the graph of $Ax + By < C$? for the graph of $Ax + By \le C$? Explain. See margin.

Tell whether the statement is *true* or *false*. Explain.

3. The point $\left(\frac{4}{3}, 0\right)$ is a solution of $3x - y > 4$. See margin.

4. The graph of $y < 3x + 5$ is the half-plane below the line $y = 3x + 5$. See margin.

GRAPHING INEQUALITIES Graph the inequality in a coordinate plane.
5–12. See margin.

5. $x > 5$ 6. $y < -4$ 7. $3x \le 1$ 8. $-y \ge \frac{4}{3}$

9. $y \ge -x + 7$ 10. $y > \frac{2}{3}x - 1$ 11. $2x - 3y < 6$ 12. $x + 5y \le -10$

13. 🌐 **CALLING CARDS** Look back at Example 4. Suppose you have relatives living in China instead of Mexico. Calls to China cost $.75 per minute. Write and graph a linear inequality showing the number of minutes you can use for calls within the United States and for calls to China. Then discuss three possible solutions in the context of the real-life situation. See margin.

PRACTICE AND APPLICATIONS

STUDENT HELP

▶ **Extra Practice**
to help you master skills is on p. 942.

3. False; $3\left(\frac{4}{3}\right) - 0 = 4$, so $\left(\frac{4}{3}, 0\right)$ is not a solution of the inequality.

STUDENT HELP

▶ **HOMEWORK HELP**
Example 1: Exs. 14–17
Example 2: Exs. 18–23, 33–44
Example 3: Exs. 24–44
Example 4: Exs. 45–51

CHECKING SOLUTIONS Check whether the given ordered pairs are solutions of the inequality.

14. $x \le -5$; $(0, 2)$, $(-5, 1)$ no; yes   15. $2y \ge 7$; $(1, -6)$, $(0, 4)$ no; yes

16. $y < -9x + 7$; $(-2, 2)$, $(3, -8)$ yes; no   17. $19x + y \ge -0.5$; $(2, 3)$, $(-1, 0)$ yes; no

INEQUALITIES IN ONE VARIABLE Graph the inequality in a coordinate plane.
18–23. See margin.

18. $x \le 6$ 19. $-x \ge 20$ 20. $10x \ge \frac{10}{3}$

21. $-3y < 21$ 22. $8y > -4$ 23. $y < 0.75$

MATCHING GRAPHS Match the inequality with its graph.

24. $2x - y \ge 4$ B   25. $-2x - y < 4$ C   26. $2x + y \le 4$ A

A. 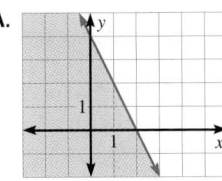   B. 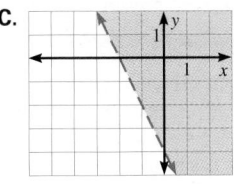   C.

INEQUALITIES IN TWO VARIABLES Graph the inequality. 27–32. See margin.

27. $y \le 3x + 11$ 28. $y > -4 - x$ 29. $y < 0.75x - 5$

30. $3x + 12y > 4$ 31. $9x - 9y > -36$ 32. $\frac{3}{2}x + \frac{2}{3}y > 1$

2.6 *Linear Inequalities in Two Variables*   **111**

---

3 APPLY

◯ **ASSIGNMENT GUIDE**

BASIC
Day 1: pp. 111–113 Exs. 14–40 even, 45, 48, 49, 52, 57–73 odd

AVERAGE
Day 1: pp. 111–113 Exs. 14–44 even, 45–49, 52, 57–73 odd

ADVANCED
Day 1: pp. 111–113 Exs. 14–44 even, 45–55, 57–73 odd

BLOCK SCHEDULE WITH 2.5
pp. 111–113 Exs. 14–44 even, 45–49, 52, 57–73 odd

EXERCISE LEVELS
Level A: *Easier*
14–19, 24–29, 33–35, 56–67
Level B: *More Difficult*
20–23, 30–32, 36–51, 68–74
Level C: *Most Difficult*
52–55

✔ **HOMEWORK CHECK**
To quickly check student understanding of key concepts, go over the following exercises: Exs. 16, 18, 24, 28, 34, 36. See also the Daily Homework Quiz:
• Blackline Master (*Chapter 2 Resource Book*, p. 92)
• 📖 Transparency (p. 16)

! COMMON ERROR
EXERCISES 36 AND 42
If a student graphs using the intercepts, he or she may see the $\le$ or $\ge$ inequality and shade the wrong side. Remind the student to check a point on one side of the boundary to make sure the correct side is shaded.

4–13, 18–23, 27–32.
See Additional Answers beginning on page AA1.

APPLICATION NOTE
EXERCISE 45 Point out that when arteries become clogged with plaque, it is commonly called *arteriosclerosis*.

CAREER NOTE
Additional information about nutritionists is available at **www.mcdougallittell.com.**

36.
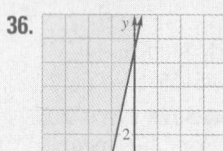

37.
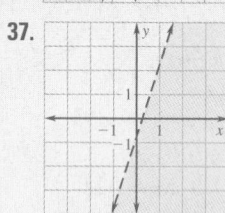

38.
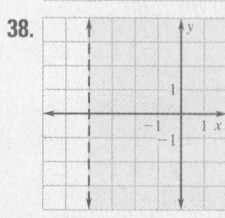

39–46, 48, 50.
See Additional Answers beginning on page AA1.

ADDITIONAL PRACTICE AND RETEACHING

For Lesson 2.6
• Practice Levels A, B, and C (*Chapter 2 Resource Book,* p. 82)
• Reteaching with Practice (*Chapter 2 Resource Book,* p. 85)
• See Lesson 2.6 of the *Personal Student Tutor*

For more Mixed Review:
• Search the *Test and Practice Generator* for key words or specific lessons.

112

49. *Sample answer:* You can attend 5 matinees and no evening showings for a total of $22.50, 2 of each for a total cost of $24, or 3 evening showings at a cost of $22.50.

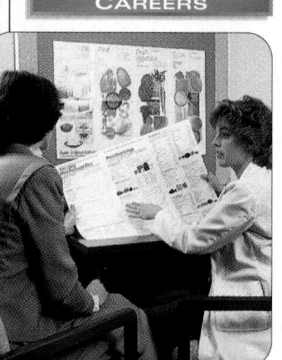

FOCUS ON CAREERS

NUTRITIONISTS A nutritionist plans nutrition programs and promotes healthy eating habits. Over one half of all nutritionists work in hospitals, nursing homes, or physician's offices.

CAREER LINK
www.mcdougallittell.com

MATCHING GRAPHS Match the inequality with its graph.

33. $x + y > 2$ **C**   **34.** $x \geq 2$ **A**   **35.** $y \leq -x + 2$ **B**

A. 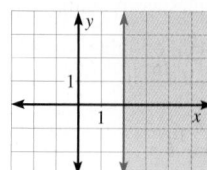   B. 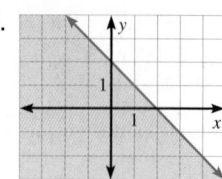   C.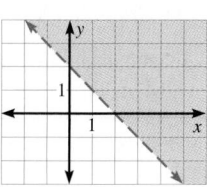

GRAPHING INEQUALITIES Graph the inequality in a coordinate plane.

36–44. See margin.

36. $9x - 2y \leq -18$ **37.** $y < 3x - \dfrac{3}{4}$ **38.** $5x > -20$

39. $y \geq \dfrac{1}{5}x + 10$ **40.** $4y \leq -6$ **41.** $2x + 3y > 4$

42. $6x \geq -\dfrac{1}{3}y$ **43.** $0.25x + 3y > 19$ **44.** $x + y < 0$

45. **HEALTH RISKS** By comparing the blood pressure in your ankle with the blood pressure in your arm, a physician can determine whether your arteries are becoming clogged with plaque. If the blood pressure in your ankle is less than 90% of the blood pressure in your arm, you may be at risk for heart disease. Write and graph an inequality that relates the unacceptable blood pressure in your ankle to the blood pressure in your arm. **$y < 0.9x$; See margin for graph.**

NUTRITION In Exercises 46 and 47, use the following information.
Teenagers should consume at least 1200 milligrams of calcium per day. Suppose you get calcium from two different sources, skim milk and cheddar cheese. One cup of skim milk supplies 296 milligrams of calcium, and one slice of cheddar cheese supplies 338 milligrams of calcium. ▶ Source: *Nutrition in Exercise and Sport*

46. Write and graph an inequality that represents the amounts of skim milk and cheddar cheese you need to consume to meet your daily requirement of calcium. **$296m + 338c \geq 1200$; See margin for graph.**

47. Determine how many cups of skim milk you should drink if you have eaten two slices of cheddar cheese. **about 1.77 cups**

MOVIES In Exercises 48 and 49, use the following information.
You receive a gift certificate for $25 to your local movie theater. Matinees are $4.50 each and evening shows are $7.50 each.

48. Write and graph an inequality that represents the numbers of matinees and evening shows you can attend. **$4.5m + 7.5e \leq 25$; See margin for graph.**

49. Give three possible combinations of the numbers of matinees and evening shows you can attend. **See margin.**

FOOTBALL In Exercises 50 and 51, use the following information.
In one of its first five games of a season, a football team scored a school record of 63 points. In all of the first five games, points came from touchdowns worth 7 points and field goals worth 3 points.

50. Write and graph an inequality that represents the numbers of touchdowns and field goals the team could have scored in any of the first five games. **$7t + 3f \leq 63$; See margin for graph.**

51. Give five possible numbers of points scored, including the number of touchdowns and the number of field goals, for the first five games.
Sample answer: 9 touchdowns and no field goals for 63 points; 5 touchdowns and 1 field goal for 38 points; 2 touchdowns and 3 field goals for 23 points; 3 touchdowns and 3 field goals for 30 points; 4 touchdowns and 6 field goals for 46 points

112   **Chapter 2** *Linear Equations and Functions*

52. MULTI-STEP PROBLEM You want to open your own truck rental company. You do some research and find that the majority of truck rental companies in your area charge a flat fee of $29.99, plus $.29 for every mile driven. You want to charge less so that you can advertise your lower rate and get more business.
52a, b, and d. See margin for graphs.

a. Write and graph an equation for the cost of renting a truck from other truck rental companies. $c = 29.99 + 0.29m$

b. Shade the region of the coordinate plane where the amount you will charge must fall.

c. To charge less than your competitors, will you offer a lower flat fee, a lower rate per mile, or both? Explain your choice.   See margin.

d. Write and graph an equation for the cost of renting a truck from your company in the same coordinate plane used in part (a).   *Sample answer:*
   $c = 27.77 + 0.29m$

e. **CRITICAL THINKING** Why can't you offer a lower rate per mile but a higher flat fee and still always charge less?   See margin.

★ **Challenge**

VISUAL THINKING In Exercises 53–55, use the graph shown.
   $4x + 9y \le 36$, or $y \le -\frac{4}{9}x + 4$

53. Write the inequality whose graph is shown.

54. Explain how you came up with the inequality.   See margin.

EXTRA CHALLENGE
www.mcdougallittell.com

55. What real-life situation could the first– quadrant portion of the graph represent?

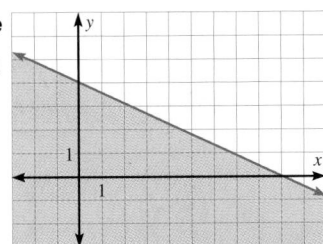

Sample answers: x is the number of g of carbohydrate and protein and y is the number of g of fat in a food that has 36 or fewer cal., or if x is the number of hours spent walking at 4 mi/h and y is the number of hours spent bike riding at 9 mi/h, then $4x + 9y \le 36$ represents those combinations of (x, y) that correspond to 36 or fewer miles total exercise.

MIXED REVIEW

SCIENTIFIC NOTATION Write the number in scientific notation. (Skills Review, p. 913)

56. 10,000,000 1.0×10^7 **57.** 1,650,000,000 1.65×10^9 **58.** 203,000 2.03×10^5

59. 0.00067 6.7×10^{-4} **60.** 0.0000009 9×10^{-7} **61.** 0.0808 8.08×10^{-2}

GRAPHING EQUATIONS Graph the equation. (Review 2.3 for 2.7)
62–67. See margin.

62. $y = \frac{5}{2}x - 5$ **63.** $y = -5x - 1$ **64.** $y = -\frac{1}{2}x + 6$

65. $x - y = 4$ **66.** $2x + y = 6$ **67.** $-4x + y = 4$

WRITING EQUATIONS Write an equation of the line that passes through the given points. (Review 2.4 for 2.7) $y = -\frac{6}{5}x + 7$ $y = -\frac{8}{9}x + \frac{46}{9}$

68. (2, 2), (5, 5) $y = x$ **69.** (0, 7), (5, 1)   **70.** (−1, 6), (8, −2)

71. (3, 2), (3, −4) $x = 3$ **72.** (1, 9), (−10, −6)   **73.** (4, −8), (−7, −8)
   $y = -8$

72. $y = \frac{15}{11}x + \frac{84}{11}$

74. 🌎 **GARDENING** The horizontal middle of the United States is at about 40°N latitude. As a rule of thumb, plants will bloom earlier south of 40°N latitude and later north of 40°N latitude. The function $w = \frac{3}{5}(l - 40)$ gives the number of weeks w (earlier or later) that plants at latitude l°N will bloom compared with those at 40°N. The equation is valid from 35°N to 45°N latitude. Identify the domain and range of the function and then graph the function. (Review 2.1)
   domain: $35 \le l \le 45$; range: $-3 \le w \le 3$

4 ASSESS

DAILY HOMEWORK QUIZ

📷 *Transparency Available*

Graph the inequality in a coordinate plane.

1. $2y \le 6$

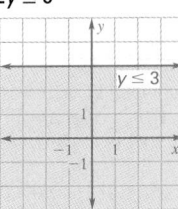

2. $x + 2y > 2$

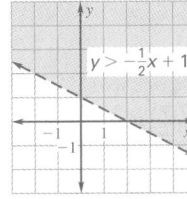

3. $1.5x - 3y \le -9$

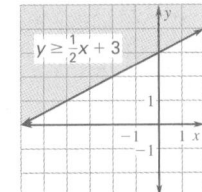

EXTRA CHALLENGE NOTE

Challenge problems for Lesson 2.6 are available in **blackline** format in the *Chapter 2 Resource Book*, p. 89 and at **www.mcdougallittell.com.**

ADDITIONAL TEST PREPARATION

1. WRITING Describe how to graph the linear inequality $2x + 3y \le 6$. *Sample answer:* **Graph the boundary line $2x + 3y = 6$, which has intercepts (3, 0) and (0, 2), with a solid line, since the inequality is "$\le$." Test the point (0, 0) in the inequality. Since (0, 0) is a solution, shade the half-plane below the boundary line.**

52a–e, 54, 62–67, 74.
   See Additional Answers beginning on page AA1.

LESSON OPENER
APPLICATION
An alternative way to approach Lesson 2.7 is to use the Application Lesson Opener:
- Blackline Master (*Chapter 2 Resource Book,* p. 93)
- Transparency (p. 14)

MEETING INDIVIDUAL NEEDS
- ***Chapter 2 Resource Book***
  Prerequisite Skills Review (p. 5)
  Practice Level A (p. 96)
  Practice Level B (p. 97)
  Practice Level C (p. 98)
  Reteaching with Practice (p. 99)
  Absent Student Catch-Up (p. 101)
  Challenge (p. 103)
- ***Resources in Spanish***
- **Personal Student Tutor**

NEW-TEACHER SUPPORT
See the Tips for New Teachers on pp. 1–2 of the *Chapter 2 Resource Book* for additional notes about Lesson 2.7.

WARM-UP EXERCISES

Transparency Available

1. Evaluate $f(x) = 3x - 2$ when $x = -2$. **−8**

2. Evaluate $h(x) = \frac{3}{2}x + \frac{5}{2}$ when $x = -5$. **−5**

3. Graph the lines $-x + y = -2$ and $y = \frac{1}{3}x - 2$ on the same coordinate grid.

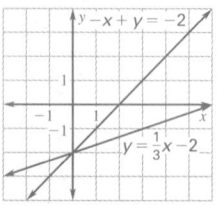

2.7 Piecewise Functions

GOAL 1 REPRESENTING PIECEWISE FUNCTIONS

What you should learn

GOAL 1 Represent piecewise functions.

GOAL 2 Use piecewise functions to model **real-life** quantities, such as the amount you earn at a summer job in **Example 6.**

Why you should learn it

▼ To solve **real-life** problems, such as determining the cost of ordering silk-screen T-shirts in **Exs. 54 and 55.**

Up to now in this chapter a function has been represented by a single equation. In many real-life problems, however, functions are represented by a combination of equations, each corresponding to a part of the domain. Such functions are called **piecewise functions.** For example, the piecewise function given by

$$f(x) = \begin{cases} 2x - 1, & \text{if } x \le 1 \\ 3x + 1, & \text{if } x > 1 \end{cases}$$

is defined by two equations. One equation gives the values of $f(x)$ when x is less than or equal to 1, and the other equation gives the values of $f(x)$ when x is greater than 1.

EXAMPLE 1 *Evaluating a Piecewise Function*

Evaluate $f(x)$ when (**a**) $x = 0$, (**b**) $x = 2$, and (**c**) $x = 4$.

$$f(x) = \begin{cases} x + 2, & \text{if } x < 2 \\ 2x + 1, & \text{if } x \ge 2 \end{cases}$$

SOLUTION

a. $f(x) = x + 2$ **Because 0 < 2, use first equation.**

   $f(0) = 0 + 2 = 2$ **Substitute 0 for x.**

b. $f(x) = 2x + 1$ **Because 2 ≥ 2, use second equation.**

   $f(2) = 2(2) + 1 = 5$ **Substitute 2 for x.**

c. $f(x) = 2x + 1$ **Because 4 ≥ 2, use second equation.**

   $f(4) = 2(4) + 1 = 9$ **Substitute 4 for x.**

EXAMPLE 2 *Graphing a Piecewise Function*

Graph this function: $f(x) = \begin{cases} \frac{1}{2}x + \frac{3}{2}, & \text{if } x < 1 \\ -x + 3, & \text{if } x \ge 1 \end{cases}$

SOLUTION

To the left of $x = 1$, the graph is given by $y = \frac{1}{2}x + \frac{3}{2}$.

To the right of and including $x = 1$, the graph is given by $y = -x + 3$.

The graph is composed of two rays with common initial point (1, 2).

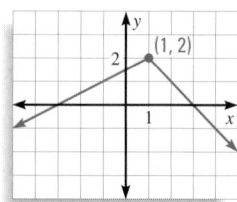

EXAMPLE 3 Graphing a Step Function

Graph this function: $f(x) = \begin{cases} 1, & \text{if } 0 \le x < 1 \\ 2, & \text{if } 1 \le x < 2 \\ 3, & \text{if } 2 \le x < 3 \\ 4, & \text{if } 3 \le x < 4 \end{cases}$

SOLUTION

The graph of the function is composed of four line segments. For instance, the first line segment is given by the equation $y = 1$ and represents the graph when x is greater than or equal to 0 and less than 1.

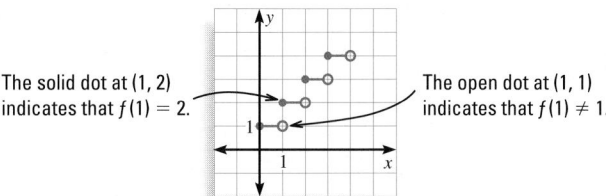

The solid dot at (1, 2) indicates that $f(1) = 2$.

The open dot at (1, 1) indicates that $f(1) \ne 1$.

· · · · · · · · · ·

The function in Example 3 is called a **step function** because its graph resembles a set of stair steps. Another example of a step function is the *greatest integer function*. This function is denoted by $g(x) = [\![x]\!]$. For every real number x, $g(x)$ is the greatest integer less than or equal to x. The graph of $g(x)$ is shown at the right. Note that in Example 3 the function f could have been written as $f(x) = [\![x]\!] + 1$, $0 \le x < 4$.

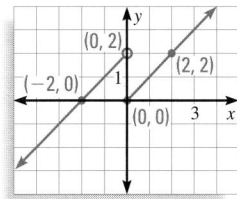

EXAMPLE 4 Writing a Piecewise Function

Write equations for the piecewise function whose graph is shown.

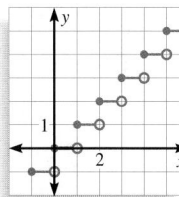

SOLUTION

To the left of $x = 0$, the graph is part of the line passing through $(-2, 0)$ and $(0, 2)$. An equation of this line is given by:

$$y = x + 2$$

To the right of and including $x = 0$, the graph is part of the line passing through $(0, 0)$ and $(2, 2)$. An equation of this line is given by:

$$y = x$$

▶ The equations for the piecewise function are:

$$f(x) = \begin{cases} x + 2, & \text{if } x < 0 \\ x, & \text{if } x \ge 0 \end{cases}$$

Note that $f(x) = x + 2$ does *not* correspond to $x = 0$ because there is an *open* dot at $(0, 2)$, but $f(x) = x$ *does* correspond to $x = 0$ because there is a *solid* dot at $(0, 0)$.

Checkpoint Exercises *Sample answer:*
a ray from an open circle at (10, 15) left, and a ray of
slope 2 up and right from a closed circle at (10, 19)

EXTRA EXAMPLE 1
Evaluate $f(x)$ when (**a**) $x = 0$, (**b**) $x = 3$, and (**c**) $x = 6$.

$$f(x) = \begin{cases} 3x + 2, & \text{if } x \le 3 \\ x - 1, & \text{if } x > 3 \end{cases}$$

a. 2; b. 11; c. 5

EXTRA EXAMPLE 2
Graph this function:

$$f(x) = \begin{cases} \dfrac{2}{3}x + \dfrac{2}{3}, & \text{if } x > 2 \\ -x + 1, & \text{if } x \le 2 \end{cases}$$

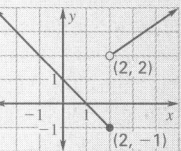

EXTRA EXAMPLE 3
Graph this function:

$$f(x) = \begin{cases} 1, & \text{if } -4 \le x < 3 \\ 2, & \text{if } -3 \le x < 2 \\ 3, & \text{if } -2 \le x < 1 \\ 4, & \text{if } -1 \le x < 0 \end{cases}$$

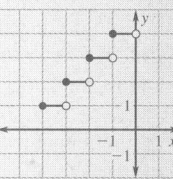

EXTRA EXAMPLE 4
Write equations for the piecewise function whose graph is shown.

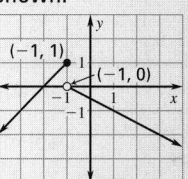

$$f(x) = \begin{cases} x + 2, & \text{if } x \le -1 \\ -\dfrac{1}{2}x - \dfrac{1}{2}, & \text{if } x > -1 \end{cases}$$

✔ CHECKPOINT EXERCISES
For use after Examples 1–4:
1. Describe the graph of $f(x)$:

$$f(x) = \begin{cases} 15, & \text{if } x < 10 \\ 2x - 1, & \text{if } x \ge 10 \end{cases}$$

See sample answer at left.

EXTRA EXAMPLE 5

Shipping costs $6 on purchases up to $50, $8 on purchases over $50 up to $100, and $10 on purchases over $100 up to $200. Write a piecewise function for these charges. Give the domain and range.

$$f(x) = \begin{cases} 6, & \text{if } 0 < x \le 50 \\ 8, & \text{if } 50 < x \le 100 \\ 10, & \text{if } 100 < x \le 200 \end{cases}$$

D: $0 < x \le 200$; R: 6, 8, 10

EXTRA EXAMPLE 6

A plane descends from 5000 ft at 250 ft/min for 6 min. Over the next 8 min, it descends at 150 ft/min. Write a piecewise function for the altitude A in terms of the time t. What is the plane's altitude after 12 min?

$$A(t) = \begin{cases} 5000 - 250t, & \text{if } 0 \le t \le 6 \\ 3500 - 150t, & \text{if } 6 < t \le 14 \end{cases}$$

2600 ft

 CHECKPOINT EXERCISES

For use after Examples 5 and 6:

1. A salesperson earns a 10% commission on the first $40,000 in sales, and 6% on sales above this amount. Write a piecewise function for the total commission C in terms of sales s. What is the commission on $75,000?

$$C(s) = \begin{cases} 0.1s, & \text{if } 0 \le s \le 40{,}000 \\ 0.06s + 1600, & \text{if } s > 40{,}000 \end{cases}$$

$6100

CLOSURE QUESTION

A phone company charges in 6 second blocks. What will a graph of the charges look like? **a step function with open circles at the left of each step and solid circles at the right**

DAILY PUZZLER

A parking garage advertises $3 for up to one hour, $2 each for the second through sixth hours, and $12 for over 6 hours. What is wrong with these rates? **It costs more to park between 5 and 6 hours than over 6 hours.**

116

Urban Parking

EXAMPLE 5 *Using a Step Function*

a. Write and graph a piecewise function for the parking charges shown on the sign.

b. What are the domain and range of the function?

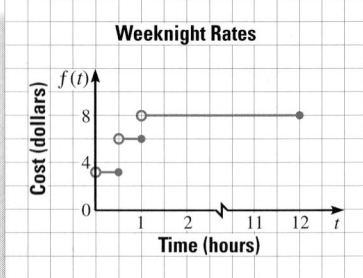

Garage Rates (Weekends)
$3 per half hour
$8 maximum for 12 hours

SOLUTION

a. For times up to one half hour, the charge is $3. For each additional half hour (or portion of a half hour), the charge is an additional $3 until you reach $8. Let t represent the number of hours you park. The piecewise function and graph are:

$$f(t) = \begin{cases} 3, & \text{if } 0 < t \le 0.5 \\ 6, & \text{if } 0.5 < t \le 1 \\ 8, & \text{if } 1 < t \le 12 \end{cases}$$

b. The domain is $0 < t \le 12$, and the range consists of 3, 6, 8

Wages

EXAMPLE 6 *Using a Piecewise Function*

You have a summer job that pays time and a half for overtime. That is, if you work more than 40 hours per week, your hourly wage for the extra hours is 1.5 times your normal hourly wage of $7.

a. Write and graph a piecewise function that gives your weekly pay P in terms of the number h of hours you work.

b. How much will you get paid if you work 45 hours?

SOLUTION

a. For up to 40 hours your pay is given by $7h$. For over 40 hours your pay is given by:

$$7(40) + 1.5(7)(h - 40) = 10.5h - 140$$

▸ The piecewise function is:

$$P(h) = \begin{cases} 7h, & \text{if } 0 \le h \le 40 \\ 10.5h - 140, & \text{if } h > 40 \end{cases}$$

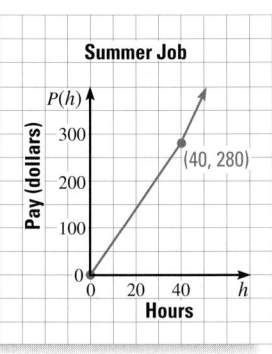

The graph of the function is shown. Note that for up to 40 hours the rate of change is $7 per hour, but for over 40 hours the rate of change is $10.50 per hour.

b. To find how much you will get paid for working 45 hours, use the equation $P(h) = 10.5h - 140$.

$$P(45) = 10.5(45) - 140 = 332.5$$

▸ You will earn $332.50.

GUIDED PRACTICE

Vocabulary Check ✓ **1.** Define piecewise function and step function. Give an example of each.
See margin.

Concept Check ✓ **2.** Look back at Example 3. What does a solid dot on the graph of a step function indicate? What does an open dot indicate? **See margin.**

Tell whether the statement is *True* or *False*. Explain.

3. In the graph of a piecewise function, the separate pieces are always connected.
See margin.

4. $f(x) = \begin{cases} 2, & \text{if } 1 \le x < 2 \\ 4, & \text{if } 2 \le x < 3 \\ 6, & \text{if } 3 \le x < 4 \end{cases}$ can be rewritten as $f(x) = 2[\![x]\!]$, $1 \le x < 4$. **See margin.**

Skill Check ✓ Evaluate $f(x) = \begin{cases} 3x - 1, & \text{if } x \le 4 \\ 2x + 7, & \text{if } x > 4 \end{cases}$ for the given value of *x*.

5. $x = 10$ 27 **6.** $x = -\frac{1}{3}$ -2 **7.** $x = 4$ 11 **8.** $x = -2$ -7

Graph the function. 9–10. See margin.

9. $f(x) = \begin{cases} 2x + 1, & \text{if } x < 1 \\ -x + 4, & \text{if } x \ge 1 \end{cases}$ **10.** $f(x) = \begin{cases} 4, & \text{if } 0 \le x < 2 \\ 5, & \text{if } 2 \le x < 4 \\ 6, & \text{if } 4 \le x < 6 \end{cases}$

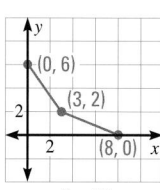
Ex. 11

12. $f(x) = \begin{cases} 3, & \text{if } 0 < x \le 0.5 \\ 6, & \text{if } 0.5 < x \le 1 \\ 9, & \text{if } 1 < x \le 1.5 \\ 12, & \text{if } 1.5 < x \le 2 \\ 15, & \text{if } 2 < x \le 2.5 \\ 18, & \text{if } 2.5 < x \le 12 \end{cases}$

11. Write equations for the piecewise function whose graph is shown.
See margin.

12. 🌐 **PARKING RATES** The weekday parking rates for a garage are shown. Write and graph a piecewise function for the weekday parking charges at that garage. **See margin for graph.**

> **Garage Rates** (Weekdays)
> **$3** per half hour
> **$18** maximum for 12 hours

PRACTICE AND APPLICATIONS

STUDENT HELP

▶ **Extra Practice**
to help you master
skills is on p. 942.

EVALUATING FUNCTIONS Evaluate the function for the given value of *x*.

$f(x) = \begin{cases} 5x - 1, & \text{if } x < -2 \\ x - 9, & \text{if } x \ge -2 \end{cases}$

$h(x) = \begin{cases} \frac{1}{2}x - 10, & \text{if } x \le 6 \\ -x - 1, & \text{if } x > 6 \end{cases}$

13. $f(-4)$ -21 **14.** $f(-2)$ -11

15. $f(0)$ -9 **16.** $f(5)$ -4

17. $h(1)$ -9.5 **18.** $h(-10)$ -15

19. $h(6)$ -7 **20.** $h(0)$ -10

GRAPHING FUNCTIONS Graph the function. 21–26. See margin.

21. $f(x) = \begin{cases} 2x, & \text{if } x \ge 1 \\ -x + 3, & \text{if } x < 1 \end{cases}$ **22.** $f(x) = \begin{cases} x + 6, & \text{if } x \le -3 \\ -\frac{2}{3}x - 3, & \text{if } x > -3 \end{cases}$

23. $f(x) = \begin{cases} 2x + 13, & \text{if } x \ge -5 \\ x + \frac{1}{2}, & \text{if } x < -5 \end{cases}$ **24.** $f(x) = \begin{cases} -x, & \text{if } x > 2 \\ x - 4, & \text{if } x \le 2 \end{cases}$

25. $f(x) = \begin{cases} 3x - 14, & \text{if } x \le 4 \\ -2x + 6, & \text{if } x > 4 \end{cases}$ **26.** $f(x) = \begin{cases} x - 8, & \text{if } x < 9 \\ \frac{1}{3}x - 2, & \text{if } x \ge 9 \end{cases}$

STUDENT HELP

▶ **HOMEWORK HELP**
Example 1: Exs. 13–20
Example 2: Exs. 21–26
Example 3: Exs. 27–32
Example 4: Exs. 35–40
Examples 5 and 6:
Exs. 50–59

---

3 APPLY

◯ **ASSIGNMENT GUIDE**

BASIC
Day 1: pp. 117–120 Exs. 14–26
even, 27–41 odd, 60, 61,
63–71 odd

AVERAGE
Day 1: pp. 117–120 Exs. 14–26
even, 27–43 odd, 50–52,
60, 61, 63–71 odd

ADVANCED
Day 1: pp. 117–120 Exs. 14–26
even, 27–47 odd, 50–55,
60–62, 63–71 odd

BLOCK SCHEDULE WITH 2.8
pp. 117–120 Exs. 14–26 even,
27–43 odd, 50–52, 60, 61,
63–71 odd

EXERCISE LEVELS
Level A: *Easier*
13–30, 35–37, 41–51
Level B: *More Difficult*
31–34, 38–40, 52–55, 60–61, 63–71
Level C: *Most Difficult*
56–59, 62

✔ **HOMEWORK CHECK**
To quickly check student understanding of key concepts, go
over the following exercises:
Exs. 14, 24, 27, 33, 35, 41. See
also the Daily Homework Quiz:

• Blackline Master (*Chapter 2
Resource Book,* p. 106)

• 🖥 Transparency (p. 17)

1. *Sample answer:* A piecewise
function is defined by different
rules over different portions of its
domain. An example is

$f(x) = \begin{cases} x + 4, & \text{if } x < -2 \\ 4, & \text{if } x = -2 \\ x + 6, & \text{if } x > -2 \end{cases}$

A step function is a piecewise
function consisting of horizontal
line segments ascending or
descending like a staircase. An
example is the cost of a first-class
postage stamp as a function of time.
2–4, 9–12, 21–26.
See Additional Answers beginning on page AA1.

➜ **Keystroke Help**
Keystrokes for several models of calculators are available at **www.mcdougallittell.com**.

⚠ **COMMON ERROR**
EXERCISES 41–49 Remind students to use "dot" mode or its equivalent for these exercises. In "connected" mode, the calculator will connect the steps.

27.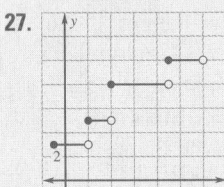

28.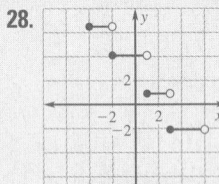

29.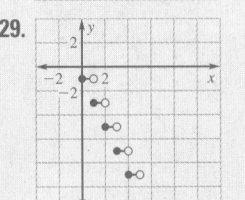

30.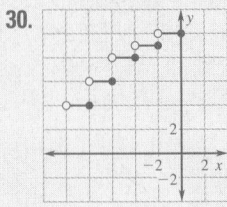

31.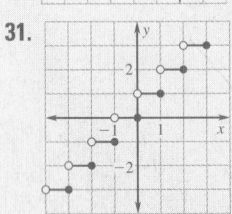

32, 35–49.
See Additional Answers beginning on page AA1.

GRAPHING STEP FUNCTIONS Graph the step function.   27–30. See margin.

27. $f(x) = \begin{cases} 3, & \text{if } -1 \le x < 2 \\ 5, & \text{if } 2 \le x < 4 \\ 8, & \text{if } 4 \le x < 9 \\ 10, & \text{if } 9 \le x < 12 \end{cases}$

28. $f(x) = \begin{cases} 6.5, & \text{if } -4 \le x < -2 \\ 4.1, & \text{if } -2 \le x < 1 \\ 0.9, & \text{if } 1 \le x < 3 \\ -2.1, & \text{if } 3 \le x < 6 \end{cases}$

29. $f(x) = \begin{cases} -1, & \text{if } 0 \le x < 1 \\ -3, & \text{if } 1 \le x < 2 \\ -5, & \text{if } 2 \le x < 3 \\ -7, & \text{if } 3 \le x < 4 \\ -9, & \text{if } 4 \le x < 5 \end{cases}$

30. $f(x) = \begin{cases} 4, & \text{if } -10 < x \le -8 \\ 6, & \text{if } -8 < x \le -6 \\ 8, & \text{if } -6 < x \le -4 \\ 9.1, & \text{if } -4 < x \le -2 \\ 10, & \text{if } -2 < x \le 0 \end{cases}$

SPECIAL STEP FUNCTIONS Graph the special step function. Then explain how you think the function got its name.   31, 32. See margin for graphs.

31. Sample answer: The function graphs each x-value to the smallest integer that is not less than it, giving a sort of upper limit to the x-values in each interval.

32. Sample answer: The function maps each x-value to the integer it rounds to.

34. Sample answer: Each open circle on the graph would be replaced by a closed circle, and each closed circle by an open circle, since a < sign does not include the endpoint and goes with an open circle, while a ≤ sign does include the endpoint and goes with a closed circle.

31. CEILING FUNCTION

$f(x) = \lceil x \rceil = \begin{cases} \ldots \\ 1, & \text{if } 0 < x \le 1 \\ 2, & \text{if } 1 < x \le 2 \\ 3, & \text{if } 2 < x \le 3 \\ \ldots \end{cases}$

32. ROUNDING FUNCTION

$f(x) = \text{ROUND}(x) = \begin{cases} \ldots \\ 1, & \text{if } 0.5 \le x < 1.5 \\ 2, & \text{if } 1.5 \le x < 2.5 \\ 3, & \text{if } 2.5 \le x < 3.5 \\ \ldots \end{cases}$

33. CRITICAL THINKING Look back at Example 2. How would the graph of the function change if < was replaced with ≤ and ≥ was replaced with >? Explain your answer.   Sample answer: The graph would not change, since the two parts of the piecewise function both give $f(1) = 2$.

34. CRITICAL THINKING Look back at Example 3. How would the graph of the function change if each ≤ was replaced with < and each < was replaced with ≤? Explain your answer.

WRITING PIECEWISE FUNCTIONS Write equations for the piecewise function whose graph is shown.   35–40. See margin.

35.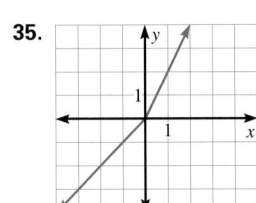

36.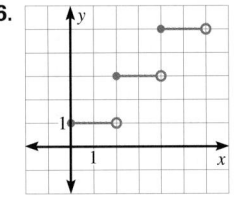

37.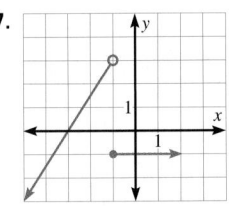

38.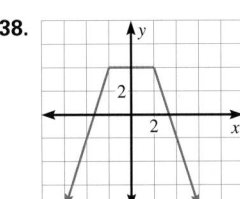

39.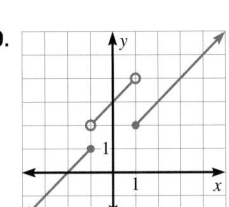

40.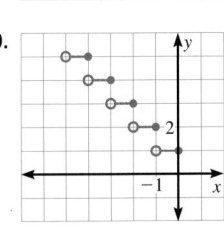

🌐 **KEYSTROKE HELP** Visit our Web site www.mcdougallittell.com to see keystrokes for several models of calculators.

📱 **GREATEST INTEGER FUNCTION** On many graphing calculators $\llbracket x \rrbracket$ is denoted by int(x). Use a graphing calculator to graph the function.   41–49. See margin.

41. $g(x) = \llbracket x \rrbracket$

42. $g(x) = \llbracket 2x \rrbracket$

43. $g(x) = \llbracket x \rrbracket - 1$

44. $g(x) = \llbracket x + 3 \rrbracket$

45. $g(x) = 6\llbracket x \rrbracket$

46. $g(x) = \llbracket 3x \rrbracket + 4$

47. $g(x) = 4\llbracket x + 7 \rrbracket$

48. $g(x) = -\llbracket x \rrbracket$

49. $g(x) = 3\llbracket x - 2 \rrbracket + 5$

POSTAL RATES In Exercises 50 and 51, use the following information.
As of January 10, 1999, the cost C (in dollars) of sending next-day mail using the United States Postal Service, depending on the weight x (in ounces) of a package up to five pounds, is given by the function below.

 DATA UPDATE of United States Postal Service data at www.mcdougallittell.com

$$C(x) = \begin{cases} 11.75, & \text{if } 0 < x \le 8 \\ 15.75, & \text{if } 8 < x \le 32 \\ 18.50, & \text{if } 32 < x \le 48 \\ 21.25, & \text{if } 48 < x \le 64 \\ 24.00, & \text{if } 64 < x \le 80 \end{cases}$$

50. Graph the function.   See margin.

51. Identify the domain and range of the function.
   domain: $0 < x \le 80$; range: 11.75, 15.75, 18.50, 21.25, 24.00

PHOTOCOPY RATES In Exercises 52 and 53, use the function given for the cost C (in dollars) of making x photocopies at a copy shop.

$$C(x) = \begin{cases} 0.15x, & \text{if } 0 < x \le 25 \\ 0.10x, & \text{if } 26 \le x \le 100 \\ 0.07x, & \text{if } 101 \le x \le 500 \\ 0.05x, & \text{if } 501 \le x \end{cases}$$

STUDENT HELP

INTERNET

HOMEWORK HELP
Visit our Web site
www.mcdougallittell.com
for help with problem
solving in Exs. 52 and 53.

52. Graph the function.   See margin.

53. VISUAL THINKING Use your graph to explain why it would not be cost-effective to make 450 photocopies.   450 photocopies cost more than 501 would.

SILK-SCREEN T-SHIRTS In Exercises 54 and 55, use the following silk-screen shop charges.

• An initial charge of $20 to create the silk screen

• $17.00 per shirt for orders of 50 or fewer shirts

• $15.80 per shirt for orders of more than 50 shirts   54, 55. See margin.

54. Write a piecewise function that gives the cost C for an order of x shirts.

55. Graph the function.

SOCIAL SECURITY In Exercises 56 and 57, use the following information.
The amount of Social Security tax you pay, part of your Federal Insurance Contributions Act (FICA) deductions, depends on your annual income. As of 1999 you pay 6.2% of your income if it is less than $72,600. If your income is at least $72,600, you pay a fixed amount of $4501.20.

DATA UPDATE of Social Security Administration data at www.mcdougallittell.com

56. Write and graph a piecewise function that gives the Social Security tax.   See margin.

57. How much Social Security tax do you pay if you make $30,000 per year? $1860

SNOWSTORM In Exercises 58 and 59, use the following information.
During a nine hour snowstorm it snows at a rate of 1 inch per hour for the first two hours, at a of rate of 2 inches per hour for the next six hours, and at a rate of 1 inch per hour for the final hour.

58. Write and graph a piecewise function that gives the depth of the snow during the snowstorm.   See margin.

59. How many inches of snow accumulated from the storm?   15 in.

FOCUS ON APPLICATIONS

SNOWSTORM
By weighing snow at the end of a snowstorm you can determine the water content of the snow. This information is one of the factors used to determine avalanche warnings.

INTERNET

APPLICATION LINK
www.mcdougallittell.com

2.7   *Piecewise Functions*   **119**

55, 56, 58. See Additional Answers
beginning on page AA1.

STUDENT HELP NOTES

→ **Homework Help** Students can find help for Exs. 52 and 53 at **www.mcdougallittell.com.** The information can be printed out for students who don't have access to the Internet.

DATA UPDATE
Updated data for Exs. 56 and 57 are available at **www.mcdougallittell.com.**

APPLICATION NOTE
EXERCISES 58 AND 59
Additional information about snowstorms is available at **www.mcdougallittell.com.**

50.
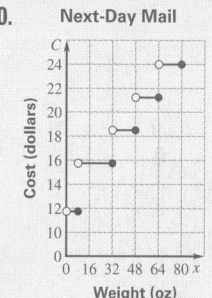
Next-Day Mail

52.
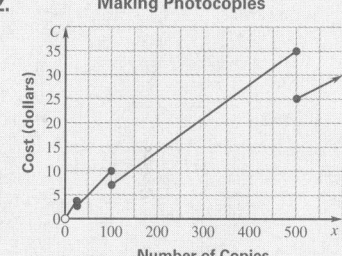
Making Photocopies

54. $C(x) = \begin{cases} 17x + 20, & \text{if } 0 < x \le 50 \\ 15.80x + 20, & \text{if } x > 50 \end{cases}$

ADDITIONAL PRACTICE AND RETEACHING

For Lesson 2.7:

• Practice Levels A, B, and C (*Chapter 2 Resource Book,* p. 96)

• Reteaching with Practice (*Chapter 2 Resource Book,* p. 99)

• See Lesson 2.7 of the *Personal Student Tutor*

For more Mixed Review:

• Search the *Test and Practice Generator* for key words or specific lessons.

DAILY HOMEWORK QUIZ

📋 *Transparency Available*

1. Evaluate

$$f(x) = \begin{cases} x, & \text{if } x > 3 \\ 2x - 1, & \text{if } x \le 3 \end{cases}$$

for $f(4)$ and $f(0)$.   **4; −1**

2. Graph

$$f(x) = \begin{cases} -x, & \text{if } x \le 1 \\ \dfrac{3}{2}x - \dfrac{3}{2}, & \text{if } x > 1 \end{cases}.$$

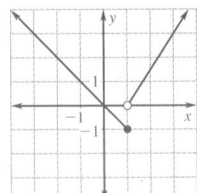

3. Write equations for the step function whose graph is shown.

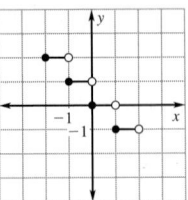

$$f(x) = \begin{cases} 2, & \text{if } -2 \le x < -1 \\ 1, & \text{if } -1 \le x < 0 \\ 0, & \text{if } 0 \le x < 1 \\ -1, & \text{if } 1 \le x < 2 \end{cases}$$

⌐ **EXTRA CHALLENGE NOTE**

↳ Challenge problems for Lesson 2.7 are available in **blackline** format in the *Chapter 2 Resource Book*, p. 103 and at **www.mcdougallittell.com**.

ADDITIONAL TEST PREPARATION

1. WRITING What are the defining characteristics of piecewise functions and step functions?
See sample answer at right.

62, 69, 70. See Additional Answers beginning on page AA1.

120

Test Preparation

QUANTITATIVE COMPARISON In Exercises 60 and 61, choose the statement that is true about the given quantities.

- Ⓐ The quantity in column A is greater.
- Ⓑ The quantity in column B is greater.
- Ⓒ The two quantities are equal.
- Ⓓ The relationship cannot be determined from the given information.

	Column A	Column B
60. A	$f(3)$ where $f(x) = \begin{cases} 2x - 7, & \text{if } x \le 1 \\ -x + 9, & \text{if } x > 1 \end{cases}$	$f(2)$ where $f(x) = \begin{cases} x + 2, & \text{if } x < 8 \\ 3x - 3, & \text{if } x \ge 8 \end{cases}$
61. B	$f(0)$ where $f(x) = \begin{cases} 5x + 1, & \text{if } x < 9 \\ 6x - 4, & \text{if } x \ge 9 \end{cases}$	$f(-4)$ where $f(x) = \begin{cases} 9, & \text{if } x \le -4 \\ 11, & \text{if } x > -4 \end{cases}$

★ **Challenge**

62. 🤿 **SCUBA DIVING** The time t (in minutes) that a person may safely scuba dive without having to decompress while surfacing is determined by the depth d (in feet) of the dive. Using the information below, write and graph a piecewise inequality that describes the time limits for scuba divers at various depths. **See margin.**

- For depths from 40 feet (the minimum depth requiring decompression) to $53\frac{1}{3}$ feet, the time must not exceed 600 minutes minus ten times the depth.

- For depths greater than $53\frac{1}{3}$ feet to less than 90 feet, the time must not exceed 120 minutes minus the depth.

- For depths from 90 feet to 130 feet (the maximum safe depth for a recreational diver), the time must not exceed 75 minutes minus one half the depth.

⌐ **EXTRA CHALLENGE**
↳ www.mcdougallittell.com

MIXED REVIEW

SOLVING EQUATIONS Solve the equation. (Review 1.7 for 2.8)

63. $|9 + 4x| = 15$ $\frac{3}{2}, -6$ **64.** $|7x + 3| = 11$ $-2, \frac{8}{7}$ **65.** $|21 - 2x| = 9$ 6, 15

66. $|2x + 8| = 1$ $-\frac{9}{2}, -\frac{7}{2}$ **67.** $\left|\frac{1}{2}x - 5\right| = 11$ $-12, 32$ **68.** $\left|1 - \frac{3}{4}x\right| = 6$ $-\frac{20}{3}, \frac{28}{3}$

SCATTER PLOTS Draw a scatter plot of the data. Then tell whether the data **69–70.** have a *positive*, a *negative*, or *relatively no correlation*. (Review 2.5)   **See margin for plots.**

69.

x	−8	−8	−7	−6	−5	−4	−4	−2	−2	−1
y	−2	−8	−5	−7	−1	−4	−8	−1	−3	−7

relatively no correlation

70.

x	1	1.5	1.5	2.5	3	3.5	5	5.5	7	8
y	9	8	6	5	6	4	2	3	1	2

negative correlation

71. $n = -\frac{1}{40}T + 2.5$; 2.5 in.

71. 🌐 **SLEEPING BAGS** To be comfortable, sleeping bags rated for −40°F have 3.5 inches of insulation, and those rated for 40°F have 1.5 inches. Write a linear model for the amount a of insulation needed to be comfortable at temperature T. How much insulation would you need to be comfortable at 0°F? (Review 2.4)

120	**Chapter 2** *Linear Equations and Functions*

Additional Test Preparation *Sample answer:*
Piecewise functions are combinations of functions, each defined over a different portion of the total domain. A step function is a piecewise function for which the graph takes a "step" up or down to a new horizontal portion of the graph for each part of the domain.

ACTIVITY 2.7

Using Technology

STUDENT HELP

KEYSTROKE HELP

See keystrokes for several models of calculators at www.mcdougallittell.com

STUDENT HELP

Look Back
For help with truth functions, see p. 48.

Graphing Piecewise Functions

You can use a graphing calculator to graph a piecewise function.

▶ **EXAMPLE**

Use a graphing calculator to graph the function f given below. Use the *Trace* feature to evaluate $f(x)$ when $x = 4$ and when $x = 10$.

$$f(x) = \begin{cases} 3x - 1, & \text{if } x < 2 \\ 7, & \text{if } 2 \le x \le 5 \\ 2x - 3, & \text{if } x > 5 \end{cases}$$

▶ **SOLUTION**

1 Put the calculator in dot mode so that the calculator does not connect separate pieces of the graph.

2 Enter the piecewise function by entering each piece of the function multiplied by a truth function specifying the values of x for which the piece applies. Add the products together. For any particular x-value, all of the products will evaluate to 0 except for the product whose truth function is satisfied.

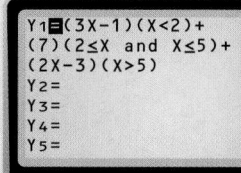

3 Choose an appropriate viewing window and graph the function. Note that the calculator does not distinguish between open and solid dots. Using the *Trace* feature, you can find that $f(4) \approx 7$ and that $f(10) \approx 17$, which you can check by substituting $x = 4$ and $x = 10$ into the function by hand.

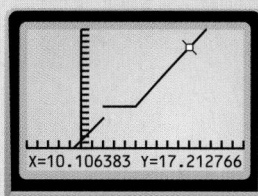

▶ **EXERCISES**

Use a graphing calculator to graph the piecewise function. Then use the *Trace* feature to evaluate the function when $x = 2$. **1–6. See margin for graphs.**

1. $f(x) = \begin{cases} 2x + 1, & \text{if } x < 0 \\ 2x + 2, & \text{if } x \ge 0 \end{cases}$ 6

2. $f(x) = \begin{cases} 2x + 3, & \text{if } x \le 3 \\ 3 - x, & \text{if } x > 3 \end{cases}$ 7

3. $f(x) = \begin{cases} \frac{1}{2}x + 1, & \text{if } x \le 2 \\ x - 2, & \text{if } x > 2 \end{cases}$ 2

4. $f(x) = \begin{cases} 3, & \text{if } x < 2 \\ 2x - 1, & \text{if } x \ge 2 \end{cases}$ 3

5. $f(x) = \begin{cases} x + 3, & \text{if } x \le 1 \\ 2(x + 1), & \text{if } 1 < x \le 3 \\ 11 - x, & \text{if } x > 3 \end{cases}$ 6

6. $f(x) = \begin{cases} 5 - x, & \text{if } x \le 4 \\ 0.25x, & \text{if } 4 < x \le 8 \\ 10 - x, & \text{if } x > 8 \end{cases}$ 3

2.7 *Technology Activity*   **121**

---

1 Planning the Activity

PURPOSE
To use a graphing calculator to graph a piecewise function.

MATERIALS
- graphing calculator for each student
- Keystroke blackline (*Chapter 2 Resource Book,* p. 95)

PACING
- Example — 10 min
- Exercises — 15 min

▶ **LINK TO LESSON**
Students can use a graphing calculator to graph the functions in Exs. 2, 3, 5, and 6 of Lesson 2.7.

2 Managing the Activity

CLASSROOM MANAGEMENT
Remind students that a truth function returns a value of 1 if a condition is true and 0 if it is false. Encourage them to substitute a value from each section of the domain into the expression shown in Step 2 to verify that this function returns the correct values. Also, remind students that a window with decimal x-values will allow tracing at the endpoints of each function.

3 Closing the Activity

⭐ **KEY DISCOVERY**
A graphing calculator can be used to graph a piecewise function by using truth functions with each separate equation.

ACTIVITY ASSESSMENT
Show that the piecewise function $Y_1 = (2x + 4)(x < 1) + (-3x + 7)(x \ge 1)$ has a value of 4 when $x = 1$.
See sample answer at left.

1–6. See Additional Answers beginning on page AA1.

Activity Assessment *Sample answer:*
Because $x = 1$, $x < 1$ is false, and has a value of 0;
$x \ge 1$ is true, and has a value of 1. So, the expression is
$(2x + 4)(0) + (-3x + 7)(1) = -3x + 7 = -3(1) + 7 = 4$.

Absolute Value Functions

PACING
Basic: 1 day
Average: 1 day
Advanced: 1 day
Block Schedule: 0.5 block with 2.7

LESSON OPENER
CALCULATOR ACTIVITY
An alternative way to approach Lesson 2.8 is to use the Calculator Activity Lesson Opener:
- Blackline Master (*Chapter 2 Resource Book*, p. 107)
- Transparency (p. 15)

MEETING INDIVIDUAL NEEDS
- *Chapter 2 Resource Book*
  Prerequisite Skills Review (p. 5)
  Practice Level A (p. 110)
  Practice Level B (p. 111)
  Practice Level C (p. 112)
  Reteaching with Practice (p. 113)
  Absent Student Catch-Up (p. 115)
  Challenge (p. 117)
- *Resources in Spanish*
- Personal Student Tutor

NEW-TEACHER SUPPORT
See the Tips for New Teachers on pp. 1–2 of the *Chapter 2 Resource Book* for additional notes about Lesson 2.8.

WARM-UP EXERCISES

Transparency Available

Evaluate the expression for $x = -6$.

1. $|x|$ 6

2. $-|x - 3|$ -9

3. $|1 - x| + 4$ 11

4. $2|x - 1.5| + 0.5$ 15.5

5. $-3|x + 4| - 1$ -7

1–3. See Additional Answers beginning on page AA1.

What you should learn

GOAL 1 Represent absolute value functions.

GOAL 2 Use absolute value functions to model **real-life** situations, such as playing pool in **Example 4**.

Why you should learn it

▼ To solve **real-life** problems, such as when an orchestra should reach a desired sound level in **Exs. 44 and 45**.

Step 1. It affects the steepness of the rays, and whether the graph is above or below the *x*-axis; (0, 0).

Step 2. A non-zero value of *h* causes a horizontal shift in the graph; (*h*, 0).

Step 3. A nonzero value of *k* causes a vertical shift in the graph; (0, *k*).

GOAL 1 **REPRESENTING ABSOLUTE VALUE FUNCTIONS**

In Lesson 1.7 you learned that the absolute value of x is defined by:

$$|x| = \begin{cases} x, & \text{if } x > 0 \\ 0, & \text{if } x = 0 \\ -x, & \text{if } x < 0 \end{cases}$$

The graph of this piecewise function consists of two rays, is V-shaped, and opens up. The corner point of the graph, called the **vertex,** occurs at the origin.

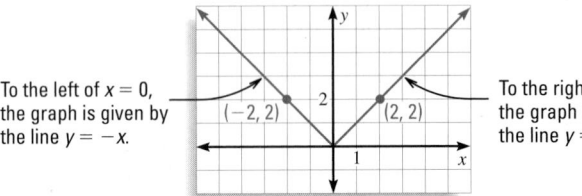

To the left of $x = 0$, the graph is given by the line $y = -x$.

$(-2, 2)$ $(2, 2)$

To the right of $x = 0$, the graph is given by the line $y = x$.

Notice that the graph of $y = |x|$ is symmetric in the *y*-axis because for every point (x, y) on the graph, the point $(-x, y)$ is also on the graph.

ACTIVITY

Developing Concepts

Graphs of Absolute Value Functions

Steps 1–3. See margin.

❶ In the same coordinate plane, graph $y = a|x|$ for $a = -2, -\frac{1}{2}, \frac{1}{2}$, and 2. What effect does a have on the graph of $y = a|x|$? What is the vertex of the graph of $y = a|x|$? Steps 1–3. See margin for graphs.

❷ In the same coordinate plane, graph $y = |x - h|$ for $h = -2, 0$, and 2. What effect does h have on the graph of $y = |x - h|$? What is the vertex of the graph of $y = |x - h|$?

❸ In the same coordinate plane, graph $y = |x| + k$ for $k = -2, 0$, and 2. What effect does k have on the graph of $y = |x| + k$? What is the vertex of the graph of $y = |x| + k$?

Although in the activity you investigated the effects of a, h, and k on the graph of $y = a|x - h| + k$ separately, these effects can be combined. For example, the graph of $y = 2|x - 4| + 3$ is shown in red along with the graph of $y = |x|$ in blue. Notice that the vertex of the red graph is (4, 3) and that the red graph is narrower than the blue graph.

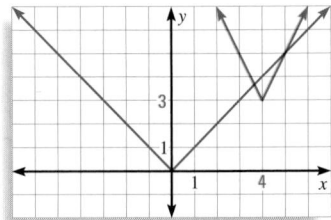

122   **Chapter 2** *Linear Equations and Functions*

GRAPHING ABSOLUTE VALUE FUNCTIONS

The graph of $y = a|x - h| + k$ has the following characteristics.

- The graph has vertex (h, k) and is symmetric in the line $x = h$.
- The graph is V-shaped. It opens up if $a > 0$ and down if $a < 0$.
- The graph is wider than the graph of $y = |x|$ if $|a| < 1$.
  The graph is narrower than the graph of $y = |x|$ if $|a| > 1$.

STUDENT HELP

► **Skills Review**
For help with
symmetry, see p. 919.

To graph an absolute value function you may find it helpful to plot the vertex and one other point. Use symmetry to plot a third point and then complete the graph.

EXAMPLE 1 *Graphing an Absolute Value Function*

Graph $y = -|x + 2| + 3$.

SOLUTION

To graph $y = -|x + 2| + 3$, plot the vertex at $(-2, 3)$. Then plot another point on the graph, such as $(-3, 2)$. Use symmetry to plot a third point, $(-1, 2)$. Connect these three points with a V-shaped graph. Note that $a = -1 < 0$ and $|a| = 1$, so the graph opens down and is the same width as the graph of $y = |x|$.

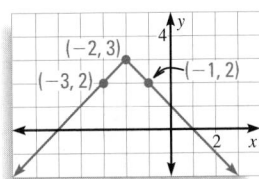

EXAMPLE 2 *Writing an Absolute Value Function*

Write an equation of the graph shown.

SOLUTION

The vertex of the graph is $(0, -3)$, so the equation has the form:

$$y = a|x - 0| + (-3) \quad \text{or} \quad y = a|x| - 3$$

To find the value of a, substitute the coordinates of the point $(2, 1)$ into the equation and solve.

$y = a\lvert x\rvert - 3$	Write equation.
$1 = a\lvert 2\rvert - 3$	Substitute 1 for *y* and 2 for *x*.
$1 = 2a - 3$	Simplify.
$4 = 2a$	Add 3 to each side.
$2 = a$	Divide each side by 2.

▶ An equation of the graph is $y = 2|x| - 3$.

✓ **CHECK** Notice the graph opens up and is narrower than the graph of $y = |x|$, so 2 is a reasonable value for *a*.

MOTIVATING THE LESSON
Ask students if they have ever tried a bank shot in miniature golf. Tell them that the path of the ball can be an example of an *absolute value function.*

EXTRA EXAMPLE 1
Graph $y = -|x - 1| + 1$.

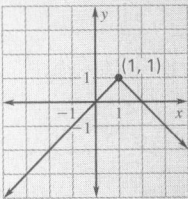

EXTRA EXAMPLE 2
Write an equation of the graph shown.

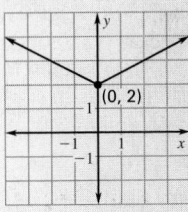

$y = \frac{1}{2}|x| + 2$

CHECKPOINT EXERCISES
For use after Example 1:
1. Graph $y = |x - 2| - 3$.

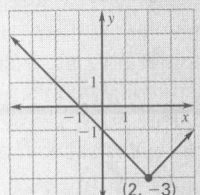

For use after Example 2:
2. Write an equation of the graph shown.

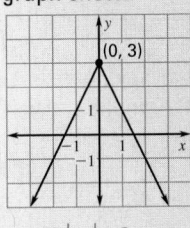

$y = -2|x| + 3$

EXTRA EXAMPLE 3

The front of a roof with its outer edges 8 feet above the ground can be modeled by $y = -\frac{2}{3}|x - 9| + 14$, with x and y in feet. Graph the function. Interpret the domain and range in this context.

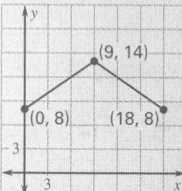

The domain is $0 \le x \le 18$, so the roof is 18 ft wide. The range is $8 \le y \le 14$, so the top of the roof is 14 ft high.

EXTRA EXAMPLE 4

You want to shoot the eight ball into the corner pocket on a pool table 10 feet long and 5 feet wide. The ball is at (2, 1); the pocket is at (10, 0). You plan to bank off the side at (6, 5).
a. Write an equation for the path of the ball. $y = -|x - 6| + 5$
b. Do you make your shot? **no**

 CHECKPOINT EXERCISES

For use after Examples 3 and 4:

1. You hit a handball off a wall 20 feet in front of you at a point 8 feet to your right. Use a sketch to write an equation for the ball's path. Your friend is 15 feet from the wall. If the ball bounces directly toward her, how far to your right is she located?
$y = -\frac{5}{2}|x - 8| + 20; 14$ ft

CLOSURE QUESTION

For the graph of $y = a|x - h| + k$, tell how to find the vertex, the direction the graph opens, and the slopes of the branches. **The vertex is (h, k). The graph opens up for $a > 0$ and down for $a < 0$. The branches have slopes a and $-a$.**

124

REAL LIFE
Camping

EXAMPLE 3 *Interpreting an Absolute Value Function*

The front of a camping tent can be modeled by the function

$$y = -1.4|x - 2.5| + 3.5$$

where x and y are measured in feet and the x-axis represents the ground.

a. Graph the function.

b. Interpret the domain and range of the function in the given context.

SOLUTION

a. The graph of the function is shown. The vertex is (2.5, 3.5) and the graph opens down. It is narrower than the graph of $y = |x|$.

b. The domain is $0 \le x \le 5$, so the tent is 5 feet wide. The range is $0 \le y \le 3.5$, so the tent is 3.5 feet tall.

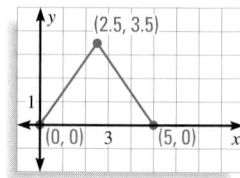

REAL LIFE
Billiards

EXAMPLE 4 *Interpreting an Absolute Value Graph*

While playing pool, you try to shoot the eight ball into the corner pocket as shown. Imagine that a coordinate plane is placed over the pool table. The eight ball is at $\left(5, \frac{5}{4}\right)$ and the pocket you are aiming for is at (10, 5). You are going to bank the ball off the side at (6, 0).

a. Write an equation for the path of the ball.

b. Do you make your shot?

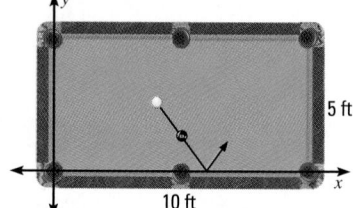
5 ft

10 ft

SOLUTION

a. The vertex of the path of the ball is (6, 0), so the equation has the form $y = a|x - 6|$. Substitute the coordinates of the point $\left(5, \frac{5}{4}\right)$ into the equation and solve for a.

$$\frac{5}{4} = a|5 - 6| \qquad \text{Substitute } \tfrac{5}{4} \text{ for } y \text{ and } 5 \text{ for } x.$$

$$\frac{5}{4} = a \qquad \text{Solve for } a.$$

▶ An equation for the path of the ball is $y = \frac{5}{4}|x - 6|$.

b. You will make your shot if the point (10, 5) lies on the path of the ball.

$$5 \stackrel{?}{=} \frac{5}{4}|10 - 6| \qquad \text{Substitute 5 for } y \text{ and 10 for } x.$$

$$5 = 5 \checkmark \qquad \text{Simplify.}$$

▶ The point (10, 5) satisfies the equation, so you do make your shot.

GUIDED PRACTICE

Vocabulary Check ✓

1. What do the coordinates (h, k) represent on the graph of $y = a|x - h| + k$? **the vertex of the graph**

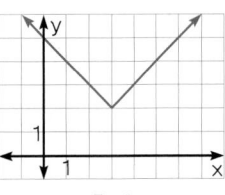
Ex. 3

Concept Check ✓

2. How do you know if the graph of $y = a|x - h| + k$ opens up or down? How do you know if it is wider, narrower, or the same width as the graph of $y = |x|$? **See margin.**

3. **ERROR ANALYSIS** Explain why the graph shown is not the graph of $y = |x + 3| + 2$. **See margin.**

Skill Check ✓

2. If a is positive, it opens up; if a is negative, the graph opens down. If $|a| < 1$, the graph is wider than that of $y = |x|$. If $|a| = 1$, the graph has the same shape as that of $y = |x|$. If $|a| > 1$, the graph is narrower.

3. The vertex should be at $(-3, 2)$, not $(3, 2)$. The general form of the equation for an absolute value graph is $y = a|x - h| + k$, and $|x + 3| = |x - (-3)|$, so $h = -3$.

Graph the function. Then identify the vertex, tell whether the graph opens up or down, and tell whether the graph is wider, narrower, or the same width as the graph of $y = |x|$. **4–9. See margin for graphs.**

4. $y = \frac{1}{2}|x|$ (0, 0); opens up; wider

5. $y = |x + 5|$ (−5, 0); opens up; same width

6. $y = |x| - 10$ (0, −10); opens up; same width

7. $y = |x| + 5$ (0, 5); opens up; same width

8. $y = 2|x + 6| - 10$ (−6, −10); opens up; narrower

9. $y = -\left|x - \frac{1}{2}\right| - 14$ See margin.

10. Write an equation for the function whose graph is shown. $y = |x - 4| + 1$

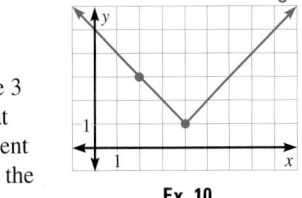
Ex. 10

11. 🌐 **CAMPING** Suppose that the tent in Example 3 is 7 feet wide and 5 feet tall. Write a function that models the front of the tent. Let the x-axis represent the ground. Then graph the function and identify the domain and range of the function.

Sample answer: $y = -\frac{10}{7}|x - 3.5| + 5$; domain: $0 \le x \le 7$; range: $0 \le y \le 5$; See margin for graph.

PRACTICE AND APPLICATIONS

STUDENT HELP

▶ **Extra Practice**
to help you master skills is on p. 942.

9. $\left(\frac{1}{2}, -14\right)$; opens down; same width

EXAMINING THE EFFECT OF a Match the function with its graph.

12. $f(x) = 3|x|$ **B**

13. $f(x) = -3|x|$ **C**

14. $f(x) = \frac{1}{3}|x|$ **A**

A.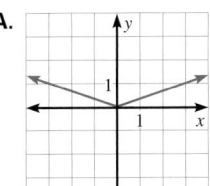

B.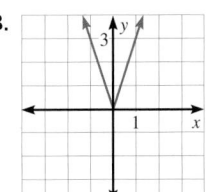

C.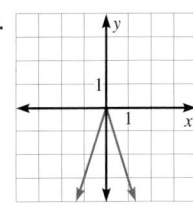

STUDENT HELP

▶ **HOMEWORK HELP**
Example 1: Exs. 12–25
Example 2: Exs. 34–39
Example 3: Exs. 40–45
Example 4: Exs. 46–48

EXAMINING THE EFFECTS OF h AND k Match the function with its graph.

15. $y = |x - 2|$ **C**

16. $y = |x| - 2$ **A**

17. $y = |x + 2|$ **B**

A.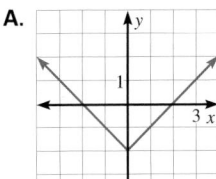

B.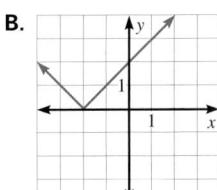

C.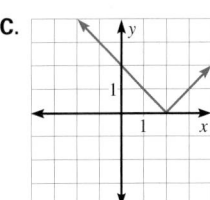

3 APPLY

⬤ **ASSIGNMENT GUIDE**

BASIC
Day 1: pp. 125–128 Exs. 12–17, 18–38 even, 40, 41, 49, 50, 57–65 odd, Quiz 3 Exs. 1–14

AVERAGE
Day 1: pp. 125–128 Exs. 12–17, 18–38 even, 40–45, 49, 50, 57–65 odd, Quiz 3 Exs. 1–14

ADVANCED
Day 1: pp. 125–128 Exs. 12–17, 18–38 even, 40–45, 49–55, 57–65 odd, Quiz 3 Exs. 1–14

BLOCK SCHEDULE WITH 2.7
pp. 125–128 Exs. 12–17, 18–38 even, 40–45, 49, 50, 57–65 odd, Quiz 3 Exs. 1–14

EXERCISE LEVELS
Level A: *Easier*
12–17, 26–33, 40–43, 56–64
Level B: *More Difficult*
18–25, 34–39, 44–47, 49–54, 65, 66
Level C: *Most Difficult*
48, 55

✔ **HOMEWORK CHECK**
To quickly check student understanding of key concepts, go over the following exercises: Exs. 12, 15, 20, 26, 34. See also the Daily Homework Quiz:

• Blackline Master (*Chapter 3 Resource Book*, p. 11)
• 📄 Transparency (p. 19)

4.
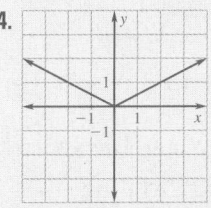

5–9, 11. See Additional Answers beginning on page AA1.

STUDENT HELP NOTES

→ **Keystroke Help**
Keystrokes for several models of calculators are available at **www.mcdougallittell.com**.

APPLICATION NOTE
EXERCISES 40 AND 41
Additional information about music singles is available at **www.mcdougallittell.com**.

18.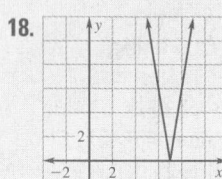

19.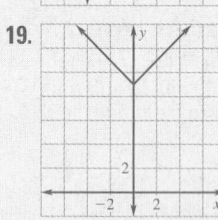

20.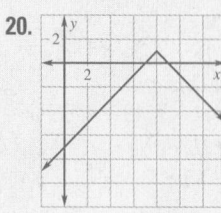

21.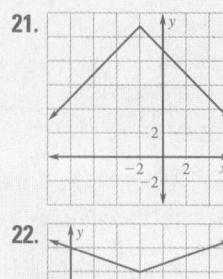

22.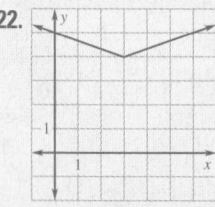

23–25, 40, 42.
See Additional Answers beginning on page AA1.

20. (8, 1); opens down; same width

STUDENT HELP

KEYSTROKE HELP
Visit our Web site www.mcdougallittell.com to see keystrokes for several models of calculators.

38. $y = -\frac{1}{3}|x - 2| + 6$

FOCUS ON APPLICATIONS

MUSIC SINGLES
A musical group's single will change position in the charts from week to week. The Beatles were at No. 1 most often with a total of 22 hit singles.

APPLICATION LINK
www.mcdougallittell.com

GRAPHING ABSOLUTE VALUE FUNCTIONS Graph the function. Then identify the vertex, tell whether the graph opens up or down, and tell whether the graph is wider, narrower, or the same width as the graph of $y = |x|$.

18–25. See margin for graphs.

18. $y = 6|x - 7|$
(7, 0); opens up; narrower

19. $y = |x| + 9$
(0, 9); opens up; same width

20. $y = -|x - 8| + 1$

21. $y = -|x + 2| + 11$
(−2, 11); opens down; same width

22. $y = \frac{1}{3}|x - 3| + 4$
(3, 4); opens up; wider

23. $y = -2|x + 9| + 3$
(−9, 3); opens down; narrower

24. $y = |x| - \frac{5}{2}$
$\left(0, -\frac{5}{2}\right)$; opens up; same width

25. $y = -\frac{1}{2}|x + 6|$ (−6, 0); opens down; wider

🖩 **ABSOLUTE VALUE** On many graphing calculators $|x|$ is denoted by ABS(x). Use a graphing calculator to graph the absolute value function. Then use the *Trace* feature to find the corresponding x-value(s) for the given y-value.

26. $y = |x| + 4; y = 10$ −6, 6

27. $y = |x + 14|; y = 9$ −23, −5

28. $y = 15|x|; y = \frac{3}{2}$ $-\frac{1}{10}, \frac{1}{10}$

29. $y = |x + \frac{4}{7}| - 5; y = 0$ $-\frac{39}{7}, \frac{31}{7}$

30. $y = -|x - 2| + 5; y = 0.5$ −2.5, 6.5

31. $y = -3.2|x| + 7; y = -2$
−2.8125, 2.8125

32. $y = -3.75|x + 1.5| - 5; y = -5$ −1.5

33. $y = 1.5|x - 3| + 6; y = 8.25$ 1.5, 4.5

WRITING EQUATIONS Write an equation of the graph shown. $y = \frac{1}{2}|x + 2|$

34.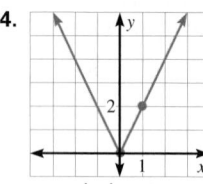
$y = 2|x|$

35.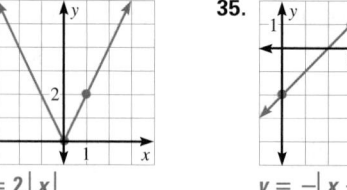
$y = -|x - 3| + 1$

36.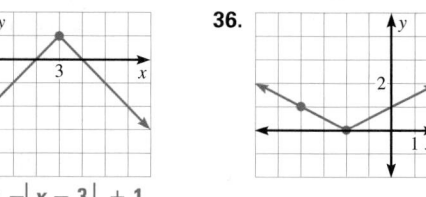

37.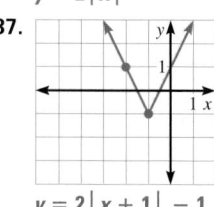
$y = 2|x + 1| - 1$

38.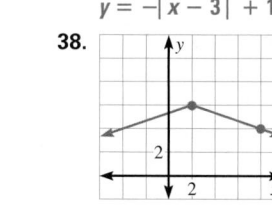

39.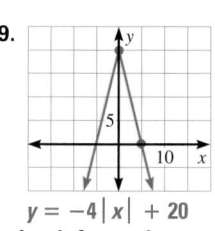
$y = -4|x| + 20$

🌐 **MUSIC SINGLES** In Exercises 40 and 41, use the following information.
A musical group's new single is released. Weekly sales s (in thousands) increase steadily for a while and then decrease as given by the function $s = -2|t - 20| + 40$ where t is the time (in weeks).

40. Graph the function. See margin.

41. What was the maximum number of singles sold in one week? 40,000

🌐 **RAINSTORMS** In Exercises 42 and 43, use the following information.
A rainstorm begins as a drizzle, builds up to a heavy rain, and then drops back to a drizzle. The rate r (in inches per hour) at which it rains is given by the function $r = -0.5|t - 1| + 0.5$ where t is the time (in hours).

42. Graph the function. See margin.

43. For how long does it rain and when does it rain the hardest?
2 h; 1 h after the rain started

🌐 **SOUND LEVELS** In Exercises 44 and 45, use the following information.

Suppose a musical piece calls for an orchestra to start at *fortissimo* (about 90 decibels), decrease in loudness to *pianissimo* (about 50 decibels) in four measures, and then increase back to *fortissimo* in another four measures. The sound level s (in decibels) of the musical piece can be modeled by the function $s = 10|m - 4| + 50$ where m is the number of measures.

44. Graph the function for $0 \le m \le 8$.   **See margin.**

45. After how many measures should the orchestra be at the loudness of *mezzo forte* (about 70 decibels)?   **after 2 measures and again after 6 measures**

46. 🌐 **MINIATURE GOLF** You are trying to make a hole-in-one on the miniature golf green shown. Imagine that a coordinate plane is placed over the golf green. The golf ball is at (2.5, 2) and the hole is at (9.5, 2). You are going to bank the ball off the side wall of the green at (6, 8). Write an equation for the path of the ball and determine if you make your shot. $y = -\dfrac{12}{7}|x - 6| + 8$; yes

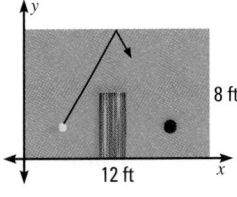

8 ft

12 ft

47. 🌐 **REFLECTING SUNLIGHT** You are sitting in a boat on a lake. You can get a sunburn from sunlight that hits you directly and from sunlight that reflects off the water. Sunlight reflects off the water at the point (2, 0) and hits you at the point (3.5, 3). Write and graph the function that shows the path of the sunlight.
$y = 2|x - 2|$; **See margin for graph.**

48. $y = -\dfrac{1706}{145}|x| + 853$,
or $y = -11.8|x| + 853$

STUDENT HELP

INTERNET

HOMEWORK HELP
Visit our Web site
www.mcdougallittell.com
for help with problem
solving in Ex. 48.

48. 🌐 **TRANSAMERICA PYRAMID** The Transamerica Pyramid, shown at the right, is an office building in San Francisco. It stands 853 feet tall and is 145 feet wide at its base. Imagine that a coordinate plane is placed over a side of the building. In the coordinate plane, each unit represents one foot, and the origin is at the center of the building's base. Write an absolute value function whose graph is the V-shaped outline of the sides of the building, ignoring the "shoulders" of the building.   **See margin.**

49. **MULTIPLE CHOICE** Which statement is true about the graph of the function $y = -|x + 2| + 3$?   **C**

(A) Its vertex is at (2, 3).   (B) Its vertex is at (−2, −3).

(C) It opens down.   (D) It is wider than the graph of $y = |x|$.

Test 🎱
Preparation

55. *Sample answer:*
$|ab| = |a| \cdot |b|$, but
$|a + b| \ne |a| + |b|$
for all values of a and b.
For example,
$|3 + 6| = 9 =$
$|3| + |6|$, but
$|-3 + 6| = 3 \ne$
$|-3| + |6| = 9$.

50. **MULTIPLE CHOICE** Which function is represented by the graph shown?   **D**

(A) $y = -|x - 10| + 2$

(B) $y = -|x + 10| - 2$

(C) $y = -|x - 2| - 10$

(D) $y = -|x + 2| + 10$

★ **Challenge**

GRAPHING Graph the functions.   **51–54. See margin.**

51. $y = |2x|$ and $y = 2|x|$

52. $y = |-5x|$ and $y = 5|x|$

53. $y = |x + 6|$ and $y = |x| + 6$

54. $y = |x + (-3)|$ and $y = |x| + 3$

EXTRA CHALLENGE
www.mcdougallittell.com

55. Based on your answers to Exercises 51–54, do you think $|ab| = |a| \cdot |b|$ and $|a + b| = |a| + |b|$ are true statements? Explain.   **See margin.**

! **COMMON ERROR**

EXERCISE 46 Noting that the vertex is (6, 8) and the initial path of the ball has a slope of $\dfrac{12}{7}$, students might write the equation $y = \dfrac{12}{7}|x - 6| + 8$. Remind them that a downward-opening absolute value graph has a negative coefficient.

STUDENT HELP NOTES

↳ **Homework Help** Students can find help for Ex. 48 at **www.mcdougallittell.com**. The information can be printed out for students who don't have access to the Internet.

44.   Orchestra Directions
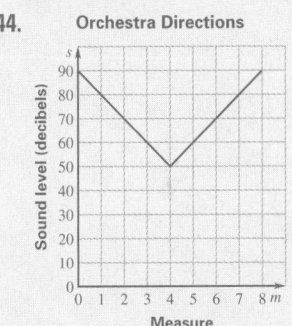

47.
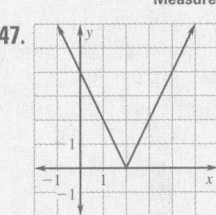

51–54. See Additional Answers beginning on page AA1.

ADDITIONAL PRACTICE AND RETEACHING

For Lesson 2.8
• Practice Levels A, B, and C (*Chapter 2 Resource Book,* p. 110)
• Reteaching with Practice (*Chapter 2 Resource Book,* p. 113)
• 🖥 See Lesson 2.8 of the *Personal Student Tutor*

For more Mixed Review:
• 🖥 Search the *Test and Practice Generator* for key words or specific lessons.

DAILY HOMEWORK QUIZ

📖 **Transparency Available**

Graph the function. Identify the vertex, and tell whether the graph is wider, narrower, or the same width as the graph of $y = |x|$.

1. Graph $y = -2|x + 1|$.

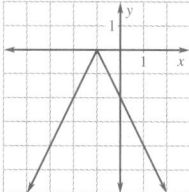

$(-1, 0)$; narrower

2. Graph $y = \frac{1}{3}|x - 2| + 1$.

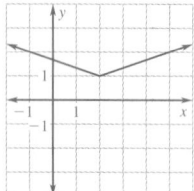

$(2, 1)$; wider

3. Use a graph to write an equation of the absolute value function that has vertex $(1, 3)$, and branches that pass through $(-2, 1)$ and $(7, -1)$.

$y = -\frac{2}{3}|x - 1| + 3$

EXTRA CHALLENGE NOTE

Challenge problems for Lesson 2.8 are available in **blackline** format in the *Chapter 2 Resource Book*, p. 117 and at **www.mcdougallittell.com**.

ADDITIONAL TEST PREPARATION

1. WRITING Compare the graphs of $y = 2|x + 1|$ and $y = |x + 1|$.

See sample answer at right.

59–65. See next page.

MIXED REVIEW

REWRITING EQUATIONS Solve the equation for *y*. (Review 1.4)

56. $3x - 5y = 8$ $y = \frac{3}{5}x - \frac{8}{5}$

57. $6x + 2y = -9$ $y = -3x - \frac{9}{2}$

58. $-\frac{1}{5}x - \frac{3}{2}y = 1$ $y = -\frac{2}{15}x - \frac{2}{3}$

GRAPHING EQUATIONS Graph the equation. (Review 2.3 for 3.1)
59–64. See margin.

59. $y = x - 5$

60. $y = 6x + 7$

61. $y = -\frac{1}{2}x + 10$

62. $x + y = 8$

63. $4x + y = 2$

64. $3x - y = -1$

FITTING A LINE TO DATA Draw a scatter plot of the data. Then approximate the best-fitting line for the data. (Review 2.5) 65–66. See margin for plots.

65.

x	−2	−1.5	−1	−0.5	0.5	1	1	1.5	2	2
y	−5	−3	−1	−2	1	−1	2	4	3	3

$y = 1.87x - 0.46$

66.

x	−2	−1	0	0.5	1	2	2.5	3.5	4	4.5
y	5	3	3.5	1.5	2	0	−2	−3.5	−2	−3.5

$y = -1.35x + 2.42$

QUIZ 3

Self-Test for Lessons 2.6–2.8

Graph the inequality in a coordinate plane. (Lesson 2.6) 1–4. See margin.

1. $y \le -8$

2. $2x \ge -5$

3. $y > 3x - 4$

4. $2x + 5y < 15$

Evaluate the function for the given value of *x*. (Lesson 2.7)

5. $f(5)$ where $f(x) = \begin{cases} 3x + 9, & \text{if } x \le 3 \\ 2x - 3, & \text{if } x > 3 \end{cases}$ 7

6. $f(0)$ where $f(x) = \begin{cases} 10, & \text{if } -1 \le x < 0 \\ 5, & \text{if } 0 \le x < 1 \\ 0, & \text{if } 1 \le x < 2 \end{cases}$ 5

Graph the function. (Lesson 2.8) 7–9. See margin.

7. $y = 3|x - 2|$

8. $y = -|x| + 6$

9. $y = -5|x + 3| - 8$

Write an equation of the graph shown. (Lesson 2.8)

10. $y = \frac{3}{2}|x - 2|$

12. $y = \frac{1}{3}|x + 1| + 2$

10.

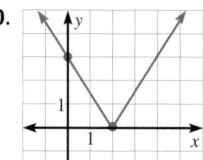

11.

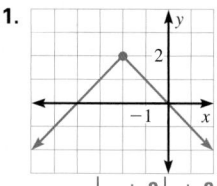

$y = -|x + 2| + 2$

12.

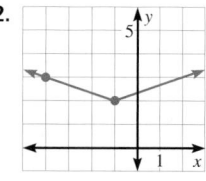

13. 🌎 **BEACH SNACKS** You and four friends have $15 to spend on snacks at the beach. A medium box of popcorn costs $2.50 and a medium soft drink costs $1.25. Write and graph an inequality that represents the numbers of medium boxes of popcorn and medium soft drinks you can buy. (Lesson 2.6)
$2.5p + 1.25d \le 15$; See margin for graph.

14. 🌎 **RENTING A CAR** A local car rental company charges a weekly rate of $200 with 1000 free miles. Each additional mile is $.20. Write and graph a piecewise function that shows the car rental charge. If you drive 1200 miles in one week, how much will the rental car cost? (Lesson 2.7)
$f(x) = 200,$ if $0 < x \le 1000$
          $0.2x,$ if $x > 1000$; See margin for graph.; $240

Additional Test Preparation *Sample answer:*
Both graphs have a vertex at $(-1, 0)$, and open upward, but the first graph has *y*-values twice as large as the second graph for each corresponding *x*-value, which means that its V-shape is narrower than that of the second graph.

Chapter Summary

WHAT did you learn?

Represent relations and functions. (2.1)

Graph and evaluate linear functions. (2.1)

Find and use the slope of a line. (2.2)

Write linear equations. (2.4)

Write direct variation equations. (2.4)

Use a scatter plot to identify the correlation shown
by a set of data. (2.5)

Approximate the best-fitting line for a set of data.
(2.5)

Graph linear equations, inequalities, and functions.
  • linear equations (2.3)

  • linear inequalities in two variables (2.6)
  • piecewise functions (2.7)
  • absolute value functions (2.8)

Use linear equations, inequalities, and functions to
solve real-life problems. (2.3–2.8)

WHY did you learn it?

Determine if the diameters of trees are a function of
their ages. (p. 68)

Model the distance a hot-air balloon travels. (p. 70)

Find the average rate of change in temperature. (p. 81)

Predict the number of African-American women who
will hold elected public office in 2010. (p. 93)

Model calories burned while dancing. (p. 97)

Identify the relationship between when and for how
long Old Faithful will erupt. (p. 104)

Predict how many people will enroll in City Year
in 2010. (p. 105)

Identify relationships between sales of student and
adult basketball tickets. (p. 88)

Model blood pressures in your arm and ankle. (p. 112)

Determine the cost of ordering T-shirts. (p. 119)

Model the sound level of an orchestra. (p. 127)

Determine how much your summer job will pay.
(p. 116)

How does Chapter 2 fit into the BIGGER PICTURE of algebra?

Your study of functions began in Chapter 2 and will continue throughout Algebra 2
and in future mathematics courses. To represent different kinds of functions with
graphs and equations is a very important part of algebra. A relationship between two
variables or two sets of data is often linear, but as you will see later in this course, it
can also be quadratic, cubic, exponential, logarithmic, or trigonometric.

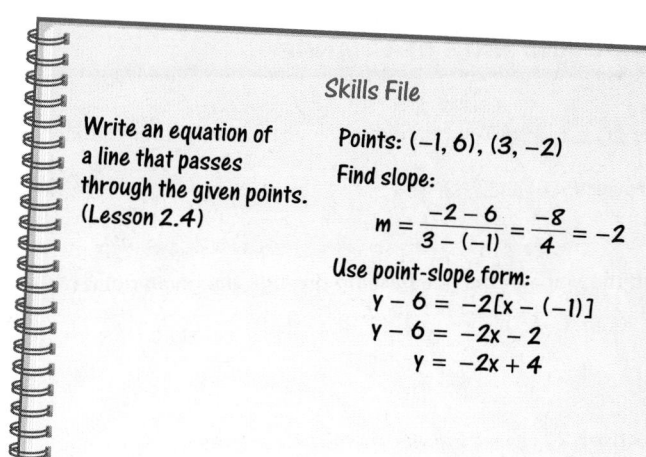

STUDY STRATEGY

How did you make and use a skills file?

Here is an example of a skill from
Lesson 2.4 for your skills file,
following the **Study Strategy** on
page 66.

Skills File

Write an equation of
a line that passes
through the given points.
(Lesson 2.4)

Points: $(-1, 6), (3, -2)$
Find slope:

$$m = \frac{-2 - 6}{3 - (-1)} = \frac{-8}{4} = -2$$

Use point-slope form:
$$y - 6 = -2[x - (-1)]$$
$$y - 6 = -2x - 2$$
$$y = -2x + 4$$

129

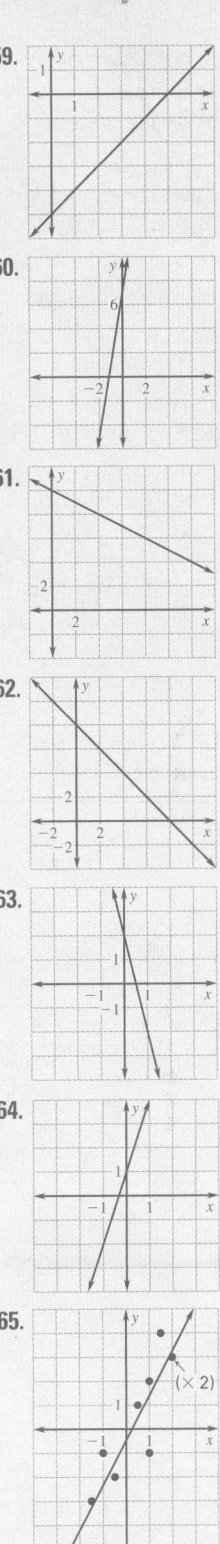

59.

60.

61.

62.

63.

64.

65.

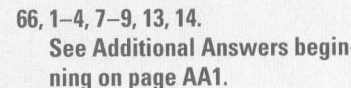

66, 1–4, 7–9, 13, 14.
See Additional Answers beginning on page AA1.

ADDITIONAL RESOURCES
The following resources are available to help review the material in this chapter.

• Chapter Review Games and Activities (*Chapter 2 Resource Book,* p. 118)

• *Instant Replay: Video Review Games*

• ▢ *Personal Student Tutor*

• Cumulative Review, Chs. 1–2 (*Chapter 2 Resource Book,* p. 130)

1.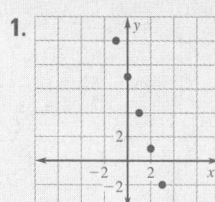

2.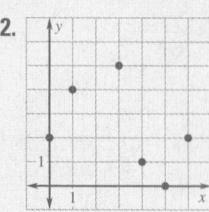

VOCABULARY

• relation, p. 67
• domain, p. 67
• range, p. 67
• function, p. 67
• ordered pair, p. 67
• coordinate plane, p. 67
• equation in two variables, p. 69
• solution of an equation in two variables, p. 69
• independent variable, p. 69

• dependent variable, p. 69
• graph of an equation in two variables, p. 69
• linear function, p. 69
• function notation, p. 69
• slope, p. 75
• parallel lines, p. 77
• perpendicular lines, p. 77
• *y*-intercept, p. 82
• slope-intercept form of a linear equation, p. 82

• standard form of a linear equation, p. 84
• *x*-intercept, p. 84
• direct variation, p. 94
• constant of variation, p. 94
• scatter plot, p. 100
• positive correlation, p. 100
• negative correlation, p. 100
• relatively no correlation, p. 100

• linear inequality in two variables, p. 108
• solution of a linear inequality in two variables, p. 108
• graph of a linear inequality in two variables, p. 108
• half-plane, p. 108
• piecewise function, p. 114
• step function, p. 115
• vertex of an absolute value graph, p. 122

2.1 FUNCTIONS AND THEIR GRAPHS

Examples on pp. 67–70

> **EXAMPLE** You can represent a relation with a table of values or a graph of ordered pairs.

x	0	1	−2	3	1
y	1	−1	0	0	2

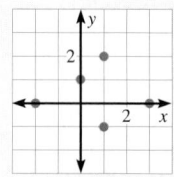

This relation is not a function because $x = 1$ is paired with both $y = -1$ and $y = 2$.

Graph the relation. Then tell whether the relation is a function.   1, 2. See margin for graphs.

1.
x	−1	0	1	2	3
y	10	7	4	1	−2
yes

2.
x	6	1	0	4	3	5
y	2	4	2	1	5	0
yes

2.2 SLOPE AND RATE OF CHANGE

Examples on pp. 75–78

> **EXAMPLE** You can find the slope of a line passing through two given points.
>
> **Points:** $(5, 0)$ and $(-3, 4)$ **Slope:** $m = \dfrac{y_2 - y_1}{x_2 - x_1} = \dfrac{4 - 0}{-3 - 5} = \dfrac{4}{-8} = -\dfrac{1}{2}$

Find the slope of the line passing through the given points.

3. $(3, 6), (-6, 0)$ $\dfrac{2}{3}$ 4. $(2, 4), (-2, 4)$ 0   5. $(-7, 2), (-1, -4)$ −1   6. $(5, 1), (5, 4)$ undefined

---

2.3 QUICK GRAPHS OF LINEAR EQUATIONS

Examples on pp. 82–85

EXAMPLES You can graph a linear equation in slope-intercept or in standard form.

$y = -3x + 1$

slope $= -3$

y-intercept $= 1$

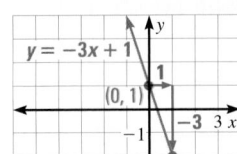

$4x - 3y = 12$

x-intercept $= 3$

y-intercept $= -4$

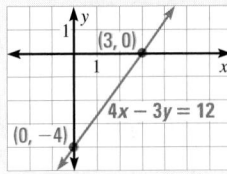

Graph the equation. 7–10. See margin.

7. $y = -x + 3$ **8.** $y = \frac{1}{2}x - 7$ **9.** $4x + y = 2$ **10.** $-4x + 8y = -16$

2.4 WRITING EQUATIONS OF LINES

Examples on pp. 91–94

EXAMPLES You can write an equation of a line using (**a**) the slope and y-intercept, (**b**) the slope and a point on the line, or (**c**) two points on the line.

a. Slope-intercept form, $m = 2$ and $b = -3$: $y = 2x - 3$

b. Point-slope form, $m = 2$ and $(x_1, y_1) = (2, 1)$: $y - 1 = 2(x - 2)$
$y = 2x - 3$

c. Points $(0, -3)$ and $(2, 1)$:   slope $= \dfrac{1 - (-3)}{2 - 0} = 2$

Use either slope-intercept form or point-slope form: $y = 2x - 3$

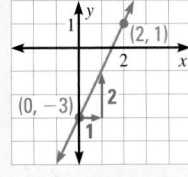

Write an equation of the line that has the given properties.

11. slope: -1, y-intercept: 2
$y = -x + 2$

12. slope: 3, point: $(-4, 1)$
$y = 3x + 13$

13. points: $(3, -8)$, $(8, 2)$
$y = 2x - 14$

2.5 CORRELATION AND BEST-FITTING LINES

Examples on pp. 100–102

EXAMPLE You can graph paired data to see what relationship, if any, exists. The table shows the price p (in dollars per pound) of bread where t is the number of years since 1990.

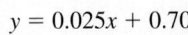

t	0	1	2	3	4	5	6
p	0.70	0.72	0.74	0.76	0.75	0.84	0.87

Approximate the best-fitting line using $(4, 0.80)$ and $(6, 0.85)$,

$m = \dfrac{0.85 - 0.80}{6 - 4} = 0.025$ $y - 0.80 = 0.025(x - 4)$

$y = 0.025x + 0.70$

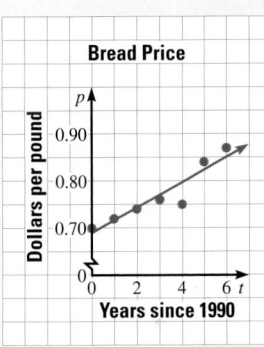

Bread Price

7.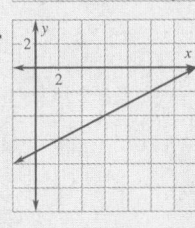

8.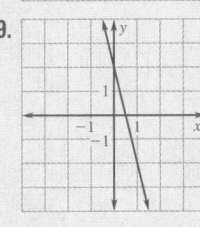

9.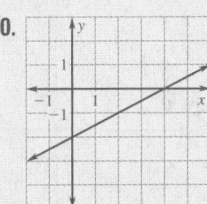

10.

Chapter Review **131**

15.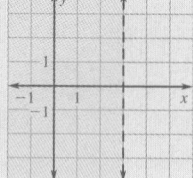

16.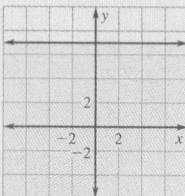

17.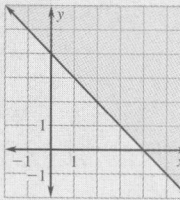

18.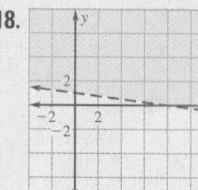

19.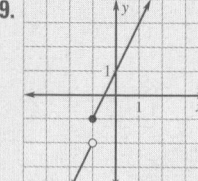

20.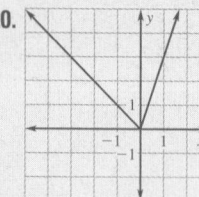

21.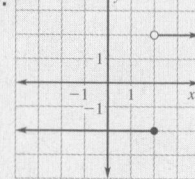

22–25. See Additional Answers beginning on page AA1.

Approximate the best-fitting line for the data.

14.

x	14	11	21	3	4	19	10	1	17	6
y	4	6	1	10	9	0	5	10	2	7

$y = -0.509x + 10.8$

2.6 LINEAR INEQUALITIES IN TWO VARIABLES

Examples on pp. 108–110

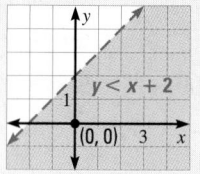

EXAMPLE You can graph a linear inequality in two variables in a coordinate plane.

To graph $y < x + 2$, first graph the boundary line $y = x + 2$. Use a dashed line since the symbol is $<$, not $\leq$. Test the point $(0, 0)$. Since $(0, 0)$ *is a* solution of the inequality, shade the half-plane that contains it.

Graph the inequality in a coordinate plane. 15–18. See margin.

15. $2x < 6$ **16.** $y \leq 7$ **17.** $y \geq -x + 4$ **18.** $x + 8y > 8$

2.7 PIECEWISE FUNCTIONS

Examples on pp. 114–116

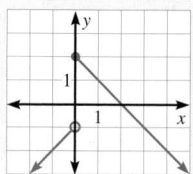

EXAMPLE You can graph a piecewise function by graphing each piece separately.

$$y = \begin{cases} x - 1, & \text{if } x < 0 \\ -x + 2, & \text{if } x \geq 0 \end{cases}$$

Graph $y = x - 1$ to the left of $x = 0$.
Graph $y = -x + 2$ to the right of and including $x = 0$.

Graph the function. 19–21. See margin.

19. $y = \begin{cases} 2x, & \text{if } x < -1 \\ 2x + 1, & \text{if } x \geq -1 \end{cases}$ **20.** $y = \begin{cases} -x, & \text{if } x \leq 0 \\ 3x, & \text{if } x > 0 \end{cases}$ **21.** $y = \begin{cases} -2, & \text{if } x \leq 2 \\ 2, & \text{if } x > 2 \end{cases}$

2.8 ABSOLUTE VALUE FUNCTIONS

Examples on pp. 122–124

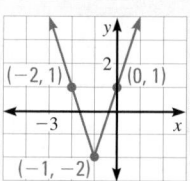

EXAMPLE You can graph an absolute value function using symmetry.

The graph of $y = 3|x + 1| - 2$ has vertex $(-1, -2)$. Plot a second point such as $(0, 1)$. Use symmetry to plot a third point, $(-2, 1)$. Note that $a = 3 > 0$ and $|a| > 1$, so the graph opens up and is narrower than the graph of $y = |x|$.

Graph the function. 22–25. See margin.

22. $y = -|x| + 1$ **23.** $y = |x - 4| + 3$ **24.** $y = 2|x| - 5$ **25.** $y = 3|x + 6| - 2$

ADDITIONAL RESOURCES
- **Chapter 2 Resource Book**
  Chapter Test (3 levels) (p. 119)
  SAT/ACT Chapter Test (p. 125)
  Alternative Assessment (p. 126)
- **Test and Practice Generator**

Graph the relation. Then tell whether the relation is a function.   1, 2. See margin for art.

yes

1.

x	−4	−2	0	2	4
y	−1	0	1	2	3

2.

x	2	−3	4	0	−3	1
y	2	−2	0	2	3	−1

no

Evaluate the function for the given value of x.

3. $f(x) = 80 - 3x$; $f(5)$ 65   **4.** $f(x) = x^2 + 4x - 7$; $f(-1)$ −10   **5.** $f(x) = 3|x - 4| + 2$; $f(2)$ 8

Graph the equation.   6–10. See margin.

6. $y = -\frac{2}{3}x + 2$ **7.** $y = -3$ **8.** $5x - 2y = 10$ **9.** $x = 4$

Write an equation of the line with the given characteristics.

10. slope: $\frac{3}{4}$, y-intercept: −5   **11.** slope: −1, point: (2, −4)   **12.** points: (−2, 5), (−6, 8) $y = -\frac{3}{4}x + \frac{7}{2}$
See margin. $y = -x - 2$
13. Write an equation of the line that passes through (−3, 2) and is parallel to the line $x - y = 7$. $x - y = -5$, or $y = x + 5$

14. Write an equation of the line that passes through (1, 4) and is perpendicular to the line $y = -3x + 1$.
$y = \frac{1}{3}x + \frac{11}{3}$

Graph the inequality in a coordinate plane.   15–18. See margin.

15. $x + 4y \leq 0$ **16.** $y > 3x - 1$ **17.** $x - y > 3$ **18.** $-x \geq 2$

Graph the function.   19–24. See margin.

19. $f(x) = \begin{cases} -2x + 3, & \text{if } x \leq 1 \\ x, & \text{if } x > 1 \end{cases}$ **20.** $f(x) = \begin{cases} 2, & \text{if } -4 < x \leq -2 \\ 5, & \text{if } -2 < x \leq 0 \\ 7, & \text{if } 0 < x \leq 2 \\ 10, & \text{if } 2 < x \leq 4 \end{cases}$ **21.** $f(x) = \begin{cases} x - 2, & \text{if } x \leq 0 \\ x + 2, & \text{if } x > 0 \end{cases}$

22. $y = -|x + 3|$ **23.** $y = 2|x| - 1$ **24.** $y = -\frac{1}{3}|x - 2| + 2$

25. ROLLER COASTERS One of the world's faster roller coasters is located in a theme park in Valencia, California. Riders go from 0 to 100 miles per hour in 7 seconds. Find the acceleration of the roller coaster during this time interval in miles per second squared.   about 0.00397 mi/sec²

26. MIRROR LENGTH To be able to see your complete reflection in a mirror that is hanging on a wall, the mirror must have a minimum length of m inches. The value of m varies directly with your height h (in inches). A person 71 inches tall requires a 35.5 inch mirror. Write a linear model that gives m as a function of h. Then find the minimum mirror length required for a person who is 66 inches tall. $m = \frac{1}{2}h$; 33 in.

27. PATENTS The table shows the number p (in thousands) of patents issued to United States residents where t is the number of years since 1985. Draw a scatter plot of the data and describe the correlation shown. Then approximate the best-fitting line for the data.   See margin for plot; the scatter plot
▶ Source: Statistical Abstract of the United States   shows a positive correlation, which means as the number of years since
1985 increased, the number of patents issued tended to increase; $p = 2.42t + 41.7$.

t	0	1	2	3	4	5	6	7	8	9	10
p	43.3	42.0	47.7	44.6	54.6	52.8	57.7	58.7	61.1	64.2	64.4

1.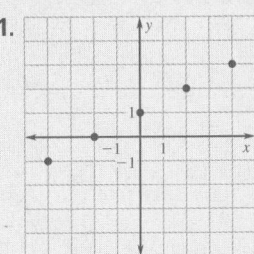

2.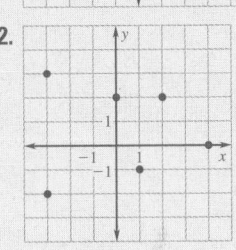

6.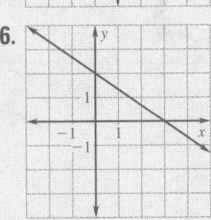

7.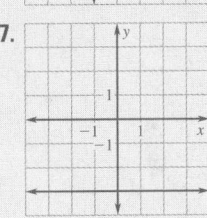

8.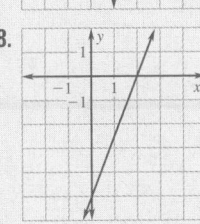

9.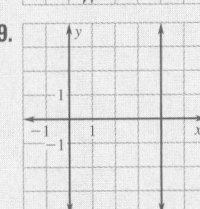

10, 15–20. See next page.
21–24, 27. See Additional Answers beginning on page AA1.

ADDITIONAL RESOURCES
- **Chapter 2 Resource Book**
  Chapter Test (3 levels) (p. 119)
  SAT/ACT Chapter Test (p. 125)
  Alternative Assessment (p. 126)
- 🖥 **Test and Practice Generator**

10. $y = \frac{3}{4}x - 5$

15.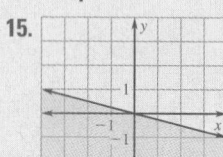

16.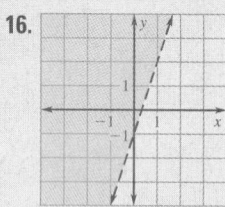

17.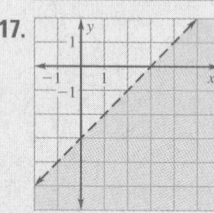

18.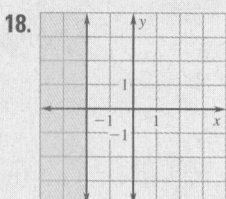

19.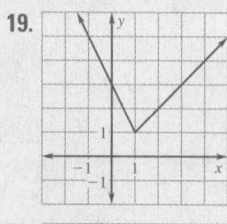

20.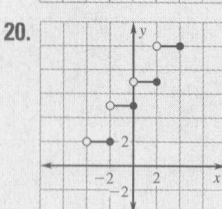

134

Chapter Standardized Test

● **TEST-TAKING STRATEGY** Read the test questions carefully. Also try to find short cuts that will help you move through the questions quicker.

1. MULTIPLE CHOICE Which of the following relations is *not* a function? **C**

Ⓐ 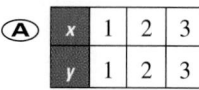 Ⓑ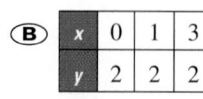

Ⓒ 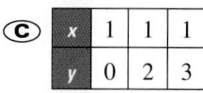 Ⓓ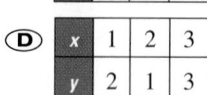

Ⓔ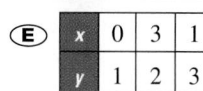

2. MULTIPLE CHOICE If $f(x) = -x^2 - 7x - 22$, what is $f(-5)$? **C**

Ⓐ -82　　Ⓑ -32　　Ⓒ -12

Ⓓ 12　　Ⓔ 38

3. MULTIPLE CHOICE What is the slope of the line that passes through $(-4, -9)$ and $(0, 5)$? **E**

Ⓐ $-\frac{7}{2}$　　Ⓑ $-\frac{2}{7}$　　Ⓒ $\frac{2}{7}$

Ⓓ 1　　Ⓔ $\frac{7}{2}$

4. MULTIPLE CHOICE Which function is represented by the graph shown? **C**

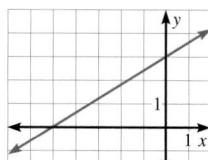

Ⓐ $3x - 5y = 15$
Ⓑ $3x - 5y = 0$
Ⓒ $3x - 5y = -15$
Ⓓ $3x + 5y = 15$
Ⓔ $3x + 5y = -15$

5. MULTIPLE CHOICE What is the y-intercept of the line $y = 4x - 3$? **D**

Ⓐ 1　　Ⓑ 3　　Ⓒ 4

Ⓓ -3　　Ⓔ -4

6. MULTIPLE CHOICE The variables x and y vary directly, and $y = 20$ when $x = 5$. Which equation relates the variables? **E**

Ⓐ $y = \frac{1}{5}x$　　Ⓑ $y = \frac{1}{4}x$　　Ⓒ $y = 5x$

Ⓓ $y = 20x$　　Ⓔ $y = 4x$

7. MULTIPLE CHOICE What is the equation of the line that passes through $(-4, -1)$ and $(0, 7)$? **C**

Ⓐ $y = 2x + 9$　　Ⓑ $y = \frac{1}{2}x + 7$

Ⓒ $y = 2x + 7$　　Ⓓ $y = -\frac{1}{2}x + 7$

Ⓔ $y = -2x + 7$

8. MULTIPLE CHOICE What is the equation of the line that contains $(3, 3)$ and is perpendicular to the line $y = -2x + 3$? **A**

Ⓐ $y = \frac{1}{2}x + \frac{3}{2}$　　Ⓑ $y = -2x + 9$

Ⓒ $y = -\frac{1}{2}x + \frac{3}{2}$　　Ⓓ $y = 2x + 9$

Ⓔ $y = \frac{1}{2}x$

9. MULTIPLE CHOICE Which inequality is represented by the graph shown? **E**

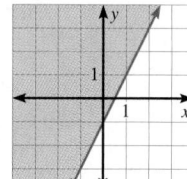

Ⓐ $y < 2x - 1$
Ⓑ $y > 2x - 1$
Ⓒ $y \le 2x - 1$
Ⓓ $y \neq 2x - 1$
Ⓔ $y \ge 2x - 1$

10. MULTIPLE CHOICE If $f(x) = \begin{cases} 2x - 3, & \text{if } x < 4 \\ -x + 6, & \text{if } x \ge 4 \end{cases}$, what is $f(4)$? **A**

Ⓐ 2　　Ⓑ 4

Ⓒ 5　　Ⓓ 10

Ⓔ 11

11. MULTIPLE CHOICE Which function is represented by the graph shown? **D**

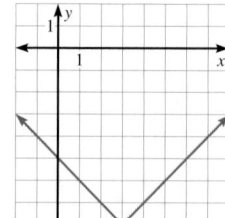

Ⓐ $y = |x - 3| + 8$
Ⓑ $y = |x + 3| - 8$
Ⓒ $y = |x + 3| + 8$
Ⓓ $y = |x - 3| - 8$
Ⓔ $y = -|x - 3| - 8$

QUANTITATIVE COMPARISON In Exercises 12 and 13, choose the statement that is true about the given quantities.

  Ⓐ The quantity in column A is greater.

  Ⓑ The quantity in column B is greater.

  Ⓒ The two quantities are equal.

  Ⓓ The relationship cannot be determined from the given information.

	Column A	Column B	
12.	slope of the line that passes through $(-6, 1)$ and $(2, 8)$	slope of the line that passes through $(0, 5)$ and $(-4, -9)$	**B**
13.	$f(-3)$ where $f(x) = x^2 - 7x - 24$	$f(-3)$ where $f(x) = \begin{cases} 2x, & \text{if } x \le 0 \\ -2x, & \text{if } x > 0 \end{cases}$	**A**

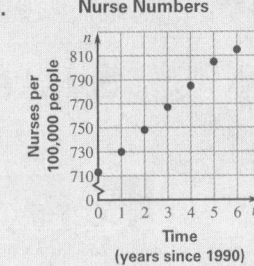

14. **MULTI-STEP PROBLEM** You are planting an herb garden. The garden has 120 inches of row space, the amount of space needed *between* rows of plants. Parsley seeds need 15 inches of row space and garlic cloves need 12 inches of row space.

  a. If you plant only parsley seeds, at most how many rows can you plant?   **9 rows**

  b. If you plant only garlic cloves, at most how many rows can you plant?   **11 rows**

  c. Write a model that shows the maximum number of rows you can plant if you plant both herbs and leave 12 inches of row space between the parsley and the garlic. $15p + 12g = 135$

  d. If you plant five rows of parsley seeds, how many rows of garlic cloves can you plant?   **5 rows**

15. **MULTI-STEP PROBLEM** The table gives the number n of nurses per 100,000 people in the United States where t is the number of years since 1990.

t	0	1	2	3	4	5	6
n	713	730	748	767	785	805	815

  a. Draw a scatter plot of the data.   **See margin.**

  b. Describe the correlation shown by the scatter plot.   **positive correlation**

  c. Approximate the best-fitting line for the data. $n = 18.1t + 715$

  d. Use your equation from part (c) to predict the number of nurses per 100,000 people in the United States in 2010.   **about 1077 nurses per 100,000 people in the United States**

16. **MULTI-STEP PROBLEM** While playing pool, you try to shoot the eight ball into the upper right corner pocket. Imagine that a coordinate plane is placed over the pool table. The eight ball is at $(4, 3)$ and the pocket you are aiming for is at $(10, 5)$. You are trying to decide at which point to bank the ball off the side.

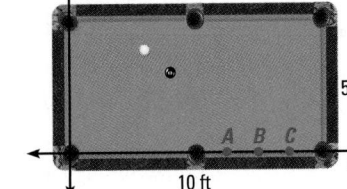

  a. Write an equation for the path of the ball if you aim for the point $(6.25, 0)$. $y = 1.33|x - 6.25|$

  b. Write an equation for the path of the ball if you aim for the point $(7.5, 0)$. $y = 0.857|x - 7.5|$

  c. Write an equation for the path of the ball if you aim for the point $(8.75, 0)$. $y = 0.632|x - 8.75|$

  d. Which point should you aim for to make your shot? $(6.25, 0)$

CHAPTER

3

PLANNING THE CHAPTER
Systems of Linear Equations and Inequalities

LESSON	GOALS	NCTM	ITED	SAT9	Terra-Nova	Local
3.1 *pp. 139–146*	**GOAL 1** Graph and solve systems of linear equations in two variables. **GOAL 2** Use linear systems to solve real-life problems. TECHNOLOGY ACTIVITY: 3.1 *Solve linear systems on a graphing calculator.*	1, 2, 3, 6, 8, 9, 10	MIG, RQWN	2, 4	11, 14, 16, 17, 18, 49, 51, 52	
3.2 *pp. 147–155*	CONCEPT ACTIVITY: 3.2 *Investigate how the sum of two linear equations relates to the graph of the system.* **GOAL 1** Use algebraic methods to solve linear systems. **GOAL 2** Use linear systems to model real-life situations.	1, 2, 3, 8, 9, 10	RQWN, MIG, IIG	2, 4, 38	11, 16, 18, 49, 51, 52	
3.3 *pp. 156–162*	**GOAL 1** Graph a system of linear inequalities to find the solutions of the system. **GOAL 2** Use systems of linear inequalities to solve real-life problems.	1, 2, 3, 6, 8, 9, 10	MIG, IIG, RQGE	2, 3	11, 14, 16, 17, 18, 49, 51, 52	
3.4 *pp. 163–169*	**GOAL 1** Solve linear programming problems. **GOAL 2** Use linear programming to solve real-life problems.	1, 2, 6, 8, 9, 10	RQGE	2, 3	11, 16, 17, 18, 49, 51, 52	
3.5 *pp. 170–176*	**GOAL 1** Graph linear equations in three variables and evaluate linear functions of two variables. **GOAL 2** Use functions of two variables to model real-life situations. TECHNOLOGY ACTIVITY: 3.5 *Graph linear equations in three variables on a graphing calculator.*	1, 2, 3, 8, 9, 10	RQGE, MIG, IIG	4	11, 14, 16, 18, 49, 51, 52	
3.6 *pp. 177–184*	**GOAL 1** Solve systems of linear equations in three variables. **GOAL 2** Use linear systems in three variables to model real-life situations.	1, 2, 8, 9, 10	RQWN	4, 38	11, 16, 18, 49, 51, 52	

RESOURCES

CHAPTER RESOURCE BOOKLETS

CHAPTER SUPPORT

Tips for New Teachers	p. 1	Prerequisite Skills Review	p. 5
Parent Guide for Student Success	p. 3	Strategies for Reading Mathematics	p. 7

LESSON SUPPORT

	3.1	3.2	3.3	3.4	3.5	3.6
Lesson Plans (regular and block)	p. 9	p. 22	p. 37	p. 51	p. 67	p. 80
Warm-Up Exercises and Daily Quiz	p. 11	p. 24	p. 39	p. 53	p. 69	p. 82
Activity Support Masters						
Lesson Openers	p. 12	p. 25	p. 40	p. 54	p. 70	p. 83
Graphing Calculator Activities & Keystrokes	p. 13	p. 26	p. 41	p. 55	p. 71	
Practice (3 levels)	p. 14	p. 28	p. 42	p. 57	p. 72	p. 84
Reteaching with Practice	p. 17	p. 31	p. 45	p. 60	p. 75	p. 87
Quick Catch-Up for Absent Students	p. 19	p. 33	p. 47	p. 62	p. 77	p. 89
Cooperative Learning Activities			p. 48			
Interdisciplinary Applications	p. 20		p. 49		p. 78	
Real-Life Applications		p. 34		p. 63		p. 90
Math & History Applications				p. 64		
Challenge: Skills and Applications	p. 21	p. 35	p. 50	p. 65	p. 79	p. 91

REVIEW AND ASSESSMENT

Quizzes	pp. 36, 66	Alternative Assessment with Math Journal	p. 100
Chapter Review Games and Activities	p. 92	Project with Rubric	p. 102
Chapter Test (3 levels)	pp. 93–98	Cumulative Review	p. 104
SAT/ACT Chapter Test	p. 99	Resource Book Answers	p. A1

TRANSPARENCIES

	3.1	3.2	3.3	3.4	3.5	3.6
Warm-Up Exercises and Daily Quiz	p. 19	p. 20	p. 21	p. 22	p. 23	p. 24
Alternative Lesson Opener Transparencies	p. 16	p. 17	p. 18	p. 19	p. 20	p. 21
Examples/Standardized Test Practice	✓	✓	✓	✓	✓	✓
Answer Transparencies	✓	✓	✓	✓	✓	✓

TECHNOLOGY

- Electronic Teaching Tools
- Online Lesson Planner
- Internet Support
- Personal Student Tutor
- Test and Practice Generator
- Instant Replay: Video Review Games
- Electronic Lesson Presentations (Lesson 3.2)

ADDITIONAL RESOURCES

- Basic Skills Workbook: Diagnosis and Remediation
- Worked-Out Solution Key
- Resources in Spanish
- Standardized Test Practice Workbook
- Practice Workbook with Examples

REGULAR SCHEDULE

Day 1

3.1

STARTING OPTIONS
- Prereq. Skills Review
- Strategies for Reading
- Homework Check
- Warm-Up or Daily Quiz

TEACHING OPTIONS
- Motivating the Lesson
- Les. Opener (Application)
- Graphing Calc. Activity
- Examples 1–3
- Guided Practice Exs. 1–9

APPLY/HOMEWORK
- See Assignment Guide.
- See the CRB: Practice, Reteach, Apply, Extend

ASSESSMENT OPTIONS
- Checkpoint Exercises, p. 140

Day 2

3.1 (cont.)

STARTING OPTIONS
- Homework Check

TEACHING OPTIONS
- Example 4
- Technology Activity
- Closure Question
- Guided Practice Ex. 10

APPLY/HOMEWORK
- See Assignment Guide.
- See the CRB: Practice, Reteach, Apply, Extend

ASSESSMENT OPTIONS
- Checkpoint Exercises, p. 141
- Daily Quiz (3.1)
- Stand. Test Practice

Day 3

3.2

STARTING OPTIONS
- Homework Check
- Warm-Up or Daily Quiz

TEACHING OPTIONS
- Motivating the Lesson
- Concept Activity
- Les. Opener (Application)
- Graphing Calc. Activity
- Examples 1–3
- Guided Practice Exs. 1–2, 4–9

APPLY/HOMEWORK
- See Assignment Guide.
- See the CRB: Practice, Reteach, Apply, Extend

ASSESSMENT OPTIONS
- Checkpoint Exercises, pp. 149–150

Day 4

3.2 (cont.)

STARTING OPTIONS
- Homework Check

TEACHING OPTIONS
- Examples 4–5
- Closure Question
- Guided Practice Exs. 3, 10

APPLY/HOMEWORK
- See Assignment Guide.
- See the CRB: Practice, Reteach, Apply, Extend

ASSESSMENT OPTIONS
- Checkpoint Exercises, pp. 150–151
- Daily Quiz (3.2)
- Stand. Test Practice
- Quiz (3.1–3.2)

Day 5

3.3

STARTING OPTIONS
- Homework Check
- Warm-Up or Daily Quiz

TEACHING OPTIONS
- Motivating the Lesson
- Les. Opener (Activity)
- Graphing Calc. Activity
- Examples 1–3
- Closure Question
- Guided Practice Exs.

APPLY/HOMEWORK
- See Assignment Guide.
- See the CRB: Practice, Reteach, Apply, Extend

ASSESSMENT OPTIONS
- Checkpoint Exercises
- Daily Quiz (3.3)
- Stand. Test Practice

Day 6

3.4

STARTING OPTIONS
- Homework Check
- Warm-Up or Daily Quiz

TEACHING OPTIONS
- Motivating the Lesson
- Les. Opener (Calculator)
- Graphing Calc. Activity
- Examples 1–3
- Closure Question
- Guided Practice Exs.

APPLY/HOMEWORK
- See Assignment Guide.
- See the CRB: Practice, Reteach, Apply, Extend

ASSESSMENT OPTIONS
- Checkpoint Exercises
- Daily Quiz (3.4)
- Stand. Test Practice
- Quiz (3.3–3.4)

Day 9

Review

DAY 9 START OPTIONS
- Homework Check

REVIEWING OPTIONS
- Chapter 3 Summary
- Chapter 3 Review
- Chapter Review Games and Activities

APPLY/HOMEWORK
- Chapter 3 Test (practice)
- Ch. Standardized Test (practice)

Day 10

Assess

DAY 10 START OPTIONS
- Homework Check

ASSESSMENT OPTIONS
- Chapter 3 Test
- SAT/ACT Ch. 3 Test
- Alternative Assessment

APPLY/HOMEWORK
- Skill Review, p. 198

Day 1
3.1
DAY 1 START OPTIONS
- Prereq. Skills Review
- Strategies for Reading
- Homework Check
- Warm-Up or Daily Quiz

TEACHING 3.1 OPTIONS
- Motivating the Lesson
- Les. Opener (Appl.)
- Graphing Calc. Activity
- Examples 1–4
- Technology Activity
- Closure Question
- Guided Practice Exs.

APPLY/HOMEWORK
- See Assignment Guide.
- See the CRB: Practice, Reteach, Apply, Extend

ASSESSMENT OPTIONS
- Checkpoint Exercises
- Daily Quiz (Les. 3.1)
- Stand. Test Practice

Day 2
3.2
DAY 2 START OPTIONS
- Homework Check
- Warm-Up or Daily Quiz

TEACHING 3.2 OPTIONS
- Motivating the Lesson
- Concept Activity
- Les. Opener (Appl.)
- Graphing Calc. Activity
- Examples 1–5
- Closure Question
- Guided Practice Exs.

APPLY/HOMEWORK
- See Assignment Guide.
- See the CRB: Practice, Reteach, Apply, Extend

ASSESSMENT OPTIONS
- Checkpoint Exercises
- Daily Quiz (Les. 3.2)
- Stand. Test Practice
- Quiz (3.1–3.2)

Day 3
3.3 & 3.4
DAY 3 START OPTIONS
- Homework Check
- W-Up 3.3 or D. Quiz 3.2

TEACHING 3.3 OPTIONS
- Motivating the Lesson
- Les. Opener (Activity)
- Graphing Calc. Activity
- Examples 1–3
- Closure Question
- Guided Practice Exs.

TEACHING 3.4 OPTIONS
- Warm-Up (Les. 3.4)
- Motivating the Lesson
- Les. Opener (Calc.)
- Graphing Calc. Activity
- Examples 1–3
- Closure Question
- Guided Practice Exs.

APPLY/HOMEWORK
- See Assignment Guide.
- See the CRB: Practice, Reteach, Apply, Extend

ASSESSMENT OPTIONS
- Checkpoint Exercises
- Daily Quiz (Les. 3.3, 3.4)
- Stand. Test Practice
- Quiz (3.3–3.4)

Day 4
3.5 & 3.6
STARTING OPTIONS
- Homework Check
- W-Up 3.5 or D. Quiz 3.4

TEACHING 3.5 OPTIONS
- Motivating the Lesson
- Les. Opener (Activity)
- Graphing Calc. Activity
- Examples 1–4
- Technology Activity
- Closure Question
- Guided Practice Exs.

TEACHING 3.6 OPTIONS
- Warm-Up (Les. 3.6)
- Motivating the Lesson
- Les. Opener (Visual)
- Examples 1–4
- Closure Question
- Guided Practice Exs.

APPLY/HOMEWORK
- See Assignment Guide.
- See the CRB: Practice, Reteach, Apply, Extend

ASSESSMENT OPTIONS
- Checkpoint Exercises
- Daily Quiz (Les. 3.5, 3.6)
- Stand. Test Practice
- Quiz (3.5–3.6)

Day 5
Review/Assess
DAY 5 START OPTIONS
- Homework Check

REVIEWING OPTIONS
- Chapter 3 Summary
- Chapter 3 Review
- Chapter Review Games and Activities
- Chapter 3 Test (practice)
- Ch. Standardized Test (practice)

ASSESSMENT OPTIONS
- Chapter 3 Test
- SAT/ACT Ch. 3 Test
- Alternative Assessment

APPLY/HOMEWORK
- Skill Review, p. 198

Day 7
3.5
STARTING OPTIONS
- Homework Check
- Warm-Up or Daily Quiz

TEACHING OPTIONS
- Motivating the Lesson
- Les. Opener (Activity)
- Graphing Calc. Activity
- Examples 1–4
- Technology Activity
- Closure Question
- Guided Practice Exs.

APPLY/HOMEWORK
- See Assignment Guide.
- See the CRB: Practice, Reteach, Apply, Extend

ASSESSMENT OPTIONS
- Checkpoint Exercises
- Daily Quiz (3.5)
- Stand. Test Practice

Day 8
3.6
STARTING OPTIONS
- Homework Check
- Warm-Up or Daily Quiz

TEACHING OPTIONS
- Motivating the Lesson
- Les. Opener (Visual)
- Examples 1–4
- Closure Question
- Guided Practice Exs.

APPLY/HOMEWORK
- See Assignment Guide.
- See the CRB: Practice, Reteach, Apply, Extend

ASSESSMENT OPTIONS
- Checkpoint Exercises
- Daily Quiz (3.6)
- Stand. Test Practice
- Quiz (3.5–3.6)

BEFORE THE CHAPTER

The *Chapter 3 Resource Book* has the following materials to distribute and use before the chapter:

- **Parent Guide for Student Success**
- **Prerequisite Skills Review**
- **Strategies for Reading Mathematics (pictured below)**

STRATEGIES FOR READING Pages 7–8

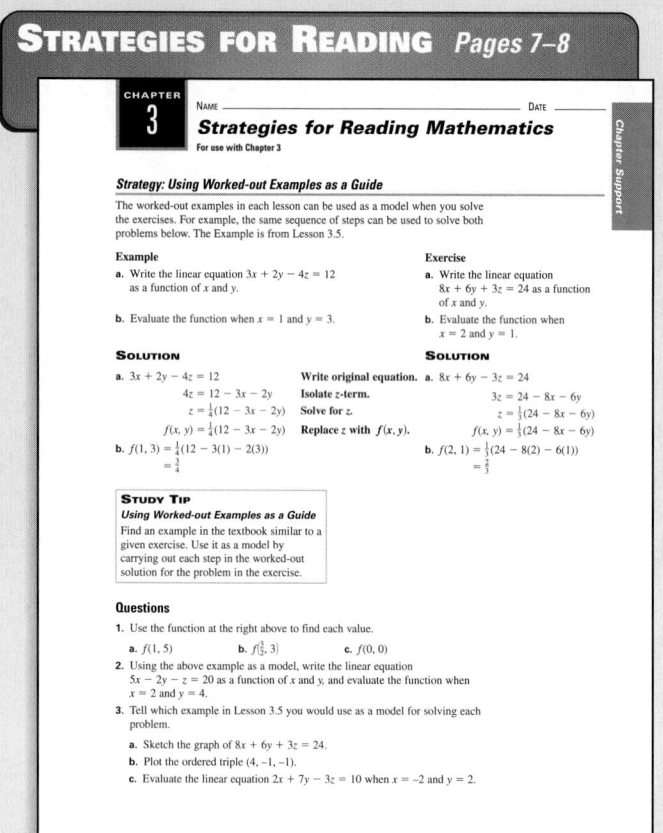

STRATEGIES FOR READING MATHEMATICS These two pages give students tips about using worked-out examples as a guide as they prepare for Chapter 3 and provide a visual glossary of key vocabulary words in the chapter, such as system of linear equations in two variables and *z*-axis.

DURING EACH LESSON

The *Chapter 3 Resource Book* has the following alternatives for introducing the lesson:

- **Lesson Openers (pictured below)**
- **Graphing Calculator Activities with Keystrokes**

LESSON OPENER Page 40

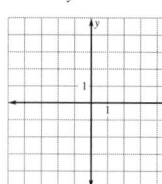

ACTIVITY LESSON OPENER This Lesson Opener provides an alternative way to start Lesson 3.3 in the form of an activity. In this activity, students use colored pencils to graph linear equations as they learn about graphing and solving systems of linear inequalities.

The *Chapter 3 Resource Book* has a variety of materials to follow-up each lesson. They include the following:

- **Practice (3 levels)**
- **Reteaching with Practice (pictured below)**
- **Quick Catch-Up for Absent Students**
- **Challenge: Skills and Applications**
- **Interdisciplinary Applications**
- **Real-Life Applications**

RETEACHING WITH PRACTICE *Pages 31–32*

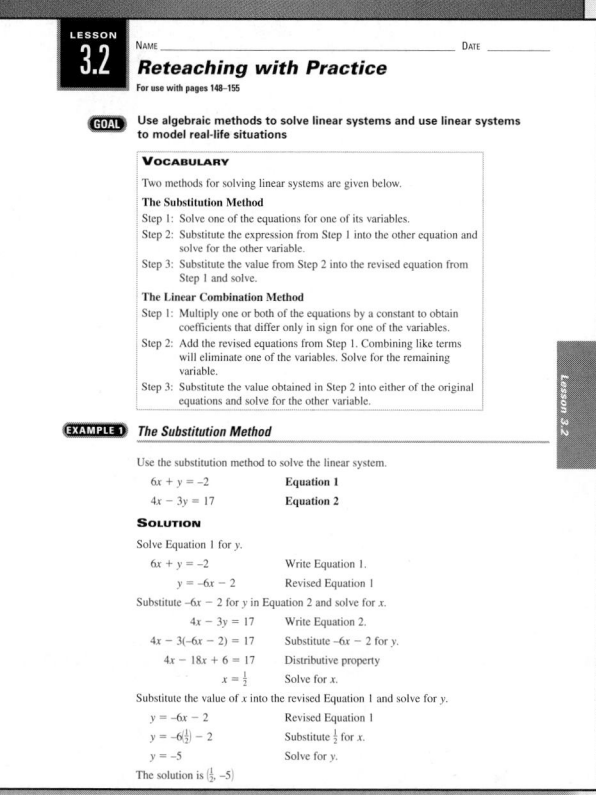

LESSON 3.2

NAME _____ DATE _____

Reteaching with Practice
For use with pages 148–155

GOAL Use algebraic methods to solve linear systems and use linear systems to model real-life situations

VOCABULARY

Two methods for solving linear systems are given below.

The Substitution Method
Step 1: Solve one of the equations for one of its variables.
Step 2: Substitute the expression from Step 1 into the other equation and solve for the other variable.
Step 3: Substitute the value from Step 2 into the revised equation from Step 1 and solve.

The Linear Combination Method
Step 1: Multiply one or both of the equations by a constant to obtain coefficients that differ only in sign for one of the variables.
Step 2: Add the revised equations from Step 1. Combining like terms will eliminate one of the variables. Solve for the remaining variable.
Step 3: Substitute the value obtained in Step 2 into either of the original equations and solve for the other variable.

EXAMPLE 1 *The Substitution Method*

Use the substitution method to solve the linear system.

$6x + y = -2$ **Equation 1**
$4x - 3y = 17$ **Equation 2**

SOLUTION

Solve Equation 1 for y.

$6x + y = -2$	Write Equation 1.
$y = -6x - 2$	Revised Equation 1

Substitute $-6x - 2$ for y in Equation 2 and solve for x.

$4x - 3y = 17$	Write Equation 2.
$4x - 3(-6x - 2) = 17$	Substitute $-6x - 2$ for y.
$4x - 18x + 6 = 17$	Distributive property
$x = \frac{1}{2}$	Solve for x.

Substitute the value of x into the revised Equation 1 and solve for y.

$y = -6x - 2$	Revised Equation 1
$y = -6(\frac{1}{2}) - 2$	Substitute $\frac{1}{2}$ for x.
$y = -5$	Solve for y.

The solution is $(\frac{1}{2}, -5)$.

Lesson 3.2

RETEACHING WITH PRACTICE On this two-page worksheet, additional examples and practice reinforce the key concept of Lesson 3.2: solving linear systems algebraically.

TECHNOLOGY RESOURCE
Students can use the Personal Student Tutor to find additional reteaching and practice for the skills in Lesson 3.2 and in the rest of Chapter 3.

The *Chapter 3 Resource Book* has the following review and assessment materials:

- **Quizzes**
- **Chapter Review Games and Activities (pictured below)**
- **Chapter Test (3 levels)**
- **SAT/ACT Chapter Test**
- **Alternative Assessment with Rubric and Math Journal**
- **Project with Rubric**
- **Cumulative Review**

CHAPTER REVIEW GAMES *Page 92*

CHAPTER 3

NAME _____ DATE _____

Chapter Review Games and Activities
For use after Chapter 3

A Random Random Random Review

1. Members of the class can be divided into two teams as they enter the classroom. Even numbered students are to sit on one side of the room and odd numbered students on the other side.

2. Then use a calculator or a random number table to generate random numbers to form groups of 4 or 5 within each team. (This may vary according to class size.)

3. Next, the teacher should generate a random number from one to six. That number will correspond to the section number in the Chapter 3 Review exercises on pages 186–188. For example, if the number 4 was generated, section 3.4 would be chosen. An additional random number should next be generated from 13–16 since these numbers are the problem numbers in the section. This will then be the problem that all teams will work on together in their groups.

4. The teacher should determine an appropriate amount of time for groups to solve the problem.

5. After sufficient time has elapsed, the teacher should generate another random number to determine which person on the team is chosen to explain their answer to the problem (for example: ask each team to determine a person whose birthday is closest to the day this activity is taking place, or whose phone number or social security number ends in a randomly chosen digit.) If that randomly-chosen person can answer and explain the problem correctly, their team receives one point. If not, the other team has the opportunity to select a random person to explain the answer. If the second team gives the correct answer, they receive the point, and it is also their turn to answer the next question since teams alternate going first in presenting the answers.

6. The next and succeeding problems are chosen according to the directions in step 3. It is suggested to end the activity approximately 10 minutes before the end of the period. The team with the most points is determined the winner.

7. The teacher should choose an appropriate "prize" for the winners. Examples of prizes may be that the homework before the exam would consist of randomly-generated problems from the Review Exercises that were not completed during this activity. The winning team would not be required to complete the entire assignment, or possibly receive a bonus point on their upcoming exam.

Alternative Plan

Students can be assigned to complete the entire Review section for homework the night before this activity, and their answers can be checked according to the directions above in number 3. In this manner, there would be time for more problems to be reviewed, but the random selection makes it fair.

Review and Assess

CHAPTER REVIEW GAMES AND ACTIVITIES On this worksheet, students form teams and answer questions using randomly generated numbers as a motivating means of reviewing the material in Chapter 3.

TECHNOLOGY RESOURCE
Teachers can use the video Instant Replay: Video Review Games as another motivating way to review the material in Chapter 3.

CHAPTER GOALS

In this chapter, students will learn to solve systems of two linear equations in two variables algebraically and by graphing. Included are those systems with one solution, no solutions, and many solutions. They will also learn to graph the solutions of systems of linear inequalities. Systems of linear equations and inequalities are used to model and solve real-life problems. The work with linear equations is extended to linear programming problems, which are used to solve real-life optimization problems. Students then graph linear equations in three variables and consider the related functions of two variables. Real-life problems are modeled using linear systems of three equations and three variables and functions of two variables. Finally, systems of linear equations in three variables are solved algebraically.

APPLICATION NOTE

Although cross-training has gained widespread popularity in recent years, athletic contests involving combinations of different events have been around for hundreds of years. Two examples are the decathlon and pentathlon. Both of these are Olympic events. The decathlon is a combination of track and field events. The pentathlon is comprised of a horse riding course, fencing, pistol shooting, swimming, and cross-country running.

Additional information about cross-training is available at **www.mcdougallittell.com**.

SYSTEMS OF LINEAR EQUATIONS AND INEQUALITIES

▶ *How can you combine swimming and inline skating to burn 300 Calories?*

CHAPTER 3

APPLICATION: Cross-Training

Cross-training involves doing a combination of two or more types of exercise. Since different exercises use different muscle groups, cross-training is a good way to get a well-rounded workout.

Think & Discuss

You burn about 12 Calories per minute swimming and about 8 Calories per minute inline skating. You want to do a combination of both activities for a total of 30 minutes and 300 Calories burned.

Minutes swimming, s	Minutes inline skating, i	$12s + 8i$
0	? 30	? 240
5	? 25	? 260
10	? 20	? 280
15	? 15	? 300
20	? 10	? 320
25	? 5	? 340
30	? 0	? 360

1. Copy the table above. Complete the second column so that $s + i = 30$. **See above.**
2. What does the expression $12s + 8i$ represent? Complete this column in your table.
   Calories burned; See above.
3. How long should you spend doing each activity?
   Swim for 15 min; skate for 15 min

Learn More About It

You will find another cross-training combination in Ex. 57 on p. 154.

 APPLICATION LINK Visit www.mcdougallittell.com for more information about cross-training.

137

ADDITIONAL RESOURCES
Another way to begin the chapter is to show the video clip of a real-life motivator for Chapter 3 on the *Instant Replay: Video Review Games.*

PROJECTS
A project covering Chapters 1–3 appears on pages 194 and 195 of the Student Edition. An additional project for Chapter 3 is available in the *Chapter 3 Resource Book*, p. 102.

TECHNOLOGY

Software
- *Electronic Teaching Tools*
- *Online Lesson Planner*
- *Personal Student Tutor*
- *Test and Practice Generator*
- *Electronic Lesson Presentations* (Lesson 3.2)

Video
- *Instant Replay: Video Review Games*

Internet Connections
www.mcdougallittell.com
- **Application Links**
  137, 169, 174, 183
- **Data Updates**
  154, 161
- **Student Help**
  140, 146, 150, 153, 160, 161, 167, 171, 172, 176, 178
- **Career Links**
  151, 158, 172
- **Extra Challenge**
  145, 154, 162, 168, 175, 183

ENGLISH LEARNERS
Keep in mind that while practical applications are intended to help students understand the relationship between algebra and the world around them, the vocabulary contained in these applications can make understanding the application difficult for English learners. In this application, make sure students understand the meanings of the words *cross-training, exercises, muscle, well-rounded, workout, calories,* and *inline skating.*

Study Guide

DIAGNOSTIC TOOLS

The **Skill Review** exercises can help you diagnose whether students have the following skills needed in Chapter 3:

- Identify solutions to linear equations and inequalities.
- Solve linear equations and inequalities.

The following resources are available for students who need additional help with these skills:

- Prerequisite Skills Review (*Chapter 3 Resource Book*, p. 5; *Warm-Up Transparencies*, p. 8)
- Reteaching with Practice (Chapter Resource Books for Lessons 3.1–3.6)
- 🖳 *Personal Student Tutor*

ADDITIONAL RESOURCES

The following resources are provided to help you prepare for the upcoming chapter and customize review materials:

- ***Chapter 3 Resource Book***
  Tips for New Teachers (p. 1)
  Parent Guide (p. 3)
  Lesson Plans (every lesson)
  Lesson Plans for Block Scheduling (every lesson)
- 🖳 *Electronic Teaching Tools*
- 🖳 *Online Lesson Planner*
- 🖳 *Test and Practice Generator*

7–12. See Additional Answers beginning on page AA1.

What's the chapter about?

Chapter 3 is about **systems of linear equations and inequalities**. In Chapter 3 you'll learn

- how to solve linear systems in two or three variables by graphing and by using algebraic methods.
- how to write and use linear systems to solve real-life problems.

KEY VOCABULARY

▶ Review

- solution of an equation in two variables, p. 69
- solution of a linear inequality in two variables, p. 108

▶ New

- system of two linear equations in two variables, p. 139

- substitution method, p. 148
- linear combination method, p. 149
- system of linear inequalities, p. 156
- linear programming, p. 163
- three-dimensional coordinate system, p. 170

- ordered triple, p. 170
- linear equation in three variables, p. 171
- function of two variables, p. 171
- system of three linear equations in three variables, p. 177

Are you ready for the chapter?

SKILL REVIEW Do these exercises to review key skills that you'll apply in this chapter. See the given **reference page** if there is something you don't understand.

▶ **Study Tip**
"Student Help" boxes throughout the chapter give you study tips and tell you where to look for extra help in this book and on the Internet.

Check whether the ordered pair is a solution of the given equation or inequality. (Review p. 69; Example 1, p. 108)

1. $y = \frac{2}{3}x - 4$, $(0, 4)$ **no**
2. $x = -3$, $(-3, 1)$ **yes**
3. $5x + y = 10$, $(1, 5)$ **yes**

4. $y \geq 0$, $(-4, 5)$ **yes**
5. $2x - 3y > 6$, $(6, 2)$ **no**
6. $x + y \leq 3$, $(-7, 9)$ **yes**

Graph in a coordinate plane. (Review Examples 1–5, pp. 83–85; Examples 2 and 3, p. 109)
7–12. See margin.

7. $y = \frac{1}{2}x + 1$
8. $2x + 5y = 20$
9. $y = 3$

10. $x \geq -2$
11. $y < -x$
12. $x - 3y \geq 9$

Here's a study strategy!

Building on Previous Skills

Many of the ideas and skills you will learn in Chapter 3 directly build upon those in Chapter 2. As you study Chapter 3, make a list of important ideas and skills. To help you understand the new material, review the related ideas and skills from Chapter 2. Write these in a second column.

3.1

Solving Linear Systems by Graphing

What you should learn

GOAL 1 Graph and solve systems of linear equations in two variables.

GOAL 2 Use linear systems to solve **real-life** problems, such as choosing the least expensive long-distance telephone service in **Ex. 64**.

Why you should learn it

▼ To solve **real-life** problems, such as how to stay within a budget on a vacation in Florida in **Example 4**.

REAL LIFE

GOAL 1 GRAPHING AND SOLVING A SYSTEM

A **system of two linear equations** in two variables x and y consists of two equations of the following form.

$$Ax + By = C \qquad \text{Equation 1}$$
$$Dx + Ey = F \qquad \text{Equation 2}$$

A **solution** of a system of linear equations in two variables is an ordered pair (x, y) that satisfies each equation.

EXAMPLE 1   Checking Solutions of a Linear System

Check whether (a) $(2, 2)$ and (b) $(0, -1)$ are solutions of the following system.

$$3x - 2y = 2 \qquad \text{Equation 1}$$
$$x + 2y = 6 \qquad \text{Equation 2}$$

SOLUTION

a. $3(2) - 2(2) = 2 ✓$ **Equation 1 checks.**

$\quad 2 + 2(2) = 6 ✓$ **Equation 2 checks.**

▶ Since $(2, 2)$ is a solution of each equation, it is a solution of the system.

b. $3(0) - 2(-1) = 2 ✓$ **Equation 1 checks.**

$\quad 0 + 2(-1) = -2 \neq 6$ **Equation 2 does not check.**

▶ Since $(0, -1)$ is not a solution of Equation 2, it is not a solution of the system.

EXAMPLE 2   Solving a System Graphically

Solve the system.

$$2x - 3y = 1 \qquad \text{Equation 1}$$
$$x + y = 3 \qquad \text{Equation 2}$$

SOLUTION

Begin by graphing both equations as shown at the right. From the graph, the lines appear to intersect at $(2, 1)$. You can check this algebraically as follows.

$\quad 2(2) - 3(1) = 1 ✓$ **Equation 1 checks.**

$\quad 2 + 1 = 3 ✓$ **Equation 2 checks.**

▶ The solution is $(2, 1)$.

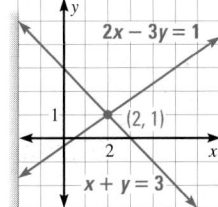

---

1 PLAN

PACING
Basic: 2 days
Average: 2 days
Advanced: 2 days
Block Schedule: 1 block

LESSON OPENER
APPLICATION
An alternative way to approach Lesson 3.1 is to use the Application Lesson Opener:

• Blackline Master (*Chapter 3 Resource Book,* p. 12)

• Transparency (p. 16)

MEETING INDIVIDUAL NEEDS
• *Chapter 3 Resource Book*
Prerequisite Skills Review (p. 5)
Practice Level A (p. 14)
Practice Level B (p. 15)
Practice Level C (p. 16)
Reteaching with Practice (p. 17)
Absent Student Catch-Up (p. 19)
Challenge (p. 21)

• *Resources in Spanish*

• *Personal Student Tutor*

NEW-TEACHER SUPPORT
See the Tips for New Teachers on pp. 1–2 of the *Chapter 3 Resource Book* for additional notes about Lesson 3.1.

WARM-UP EXERCISES
Transparency Available

Decide whether the point is a solution of the equation.

1. $2x - 3y = 6$, $(1, -3)$ no
2. $-4x + y = -5$, $(2, 3)$ yes

Match each line with its equation.

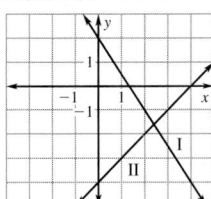

3. $3x + 2y = 4$ I
4. $x - y = 4$ II

MOTIVATING THE LESSON
Making a family budget and keeping to it is an important life skill. For example, a family may need to project vacation expenses to make wise choices and stay within budget. The mathematics in today's lesson can help in making such decisions.

 EXTRA EXAMPLE 1
Check whether **(a)** (1, 4) and **(b)** (−5, 0) are solutions of the following system.
$x - 3y = -5$
$-2x + 3y = 10$
(a) no, (b) yes

EXTRA EXAMPLE 2
Solve the system.
$2x - 2y = -8$
$2x + 2y = 4$ (−1, 3)

EXTRA EXAMPLE 3
Tell how many solutions the linear system has.
a. $2x + 4y = 12$
  $x + 2y = 6$
  infinitely many solutions
b. $x - y = 5$
  $2x - 2y = 9$ no solutions

 CHECKPOINT EXERCISES
For use after Examples 1 and 2:
1. Solve the system and check your solution.
  $4x - 3y = -15$
  $x + 2y = -1$ (−3, 1)

For use after Example 3:
2. Tell how many solutions the linear system has.
  $2x + y = 4$
  $x + \frac{1}{2}y = 1$ no solutions

The system in Example 2 has exactly one solution. It is also possible for a system of linear equations to have infinitely many solutions or no solution.

STUDENT HELP

INTERNET HOMEWORK HELP
Visit our Web site
www.mcdougallittell.com
for extra examples.

EXAMPLE 3 *Systems with Many or No Solutions*

Tell how many solutions the linear system has.

a. $3x - 2y = 6$
  $6x - 4y = 12$

b. $3x - 2y = 6$
  $3x - 2y = 2$

SOLUTION

a. The graph of the equations is the same line. So, each point on the line is a solution and the system has infinitely many solutions.

b. The graphs of the equations are two parallel lines. Because the two lines have no point of intersection, the system has no solution.

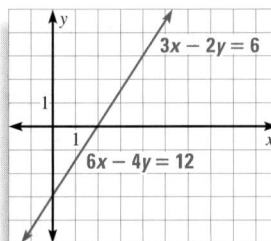

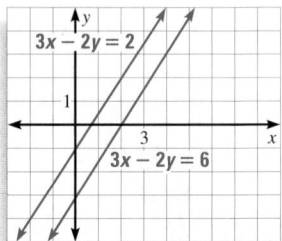

CONCEPT SUMMARY   **NUMBER OF SOLUTIONS OF A LINEAR SYSTEM**

The relationship between the graph of a linear system and the system's number of solutions is described below.

GRAPHICAL INTERPRETATION	ALGEBRAIC INTERPRETATION
The graph of the system is a pair of lines that intersect in one point.	The system has exactly one solution.
The graph of the system is a single line.	The system has infinitely many solutions.
The graph of the system is a pair of parallel lines so that there is no point of intersection.	The system has no solution.

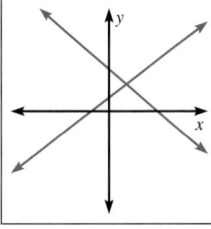

Exactly one solution   **Infinitely many solutions**   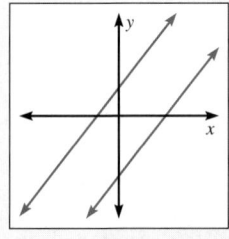 **No solution**

GOAL 2 USING LINEAR SYSTEMS IN REAL LIFE

EXAMPLE 4 *Writing and Using a Linear System*

VACATION COSTS Your family is planning a 7 day trip to Florida. You estimate that it will cost $275 per day in Tampa and $400 per day in Orlando. Your total budget for the 7 days is $2300. How many days should you spend in each location?

SOLUTION

You can use a verbal model to write a system of two linear equations in two variables.

PROBLEM
SOLVING
STRATEGY

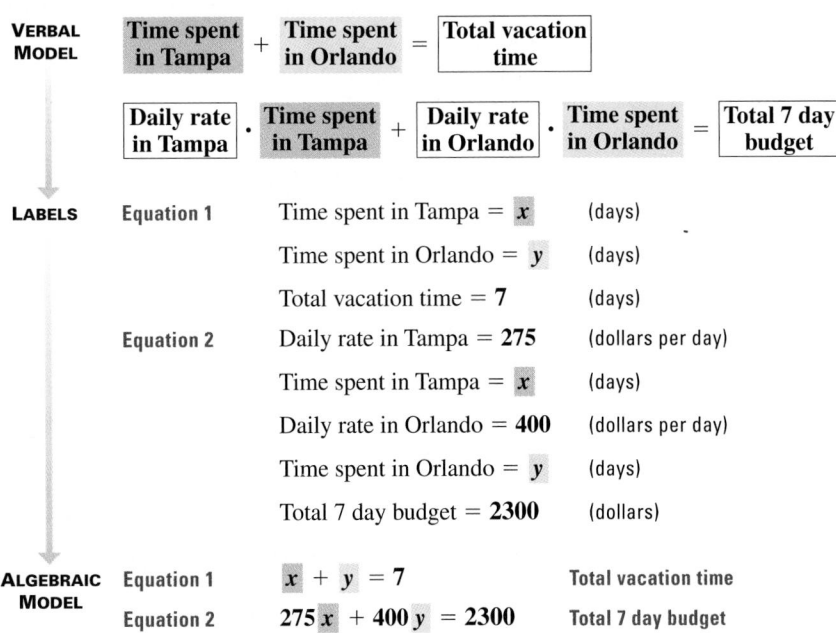

VERBAL MODEL

| Time spent in Tampa | + | Time spent in Orlando | = | Total vacation time |

| Daily rate in Tampa | · | Time spent in Tampa | + | Daily rate in Orlando | · | Time spent in Orlando | = | Total 7 day budget |

LABELS

Equation 1
- Time spent in Tampa $= x$ (days)
- Time spent in Orlando $= y$ (days)
- Total vacation time $= 7$ (days)

Equation 2
- Daily rate in Tampa $= 275$ (dollars per day)
- Time spent in Tampa $= x$ (days)
- Daily rate in Orlando $= 400$ (dollars per day)
- Time spent in Orlando $= y$ (days)
- Total 7 day budget $= 2300$ (dollars)

ALGEBRAIC MODEL

Equation 1 $x + y = 7$ Total vacation time

Equation 2 $275x + 400y = 2300$ Total 7 day budget

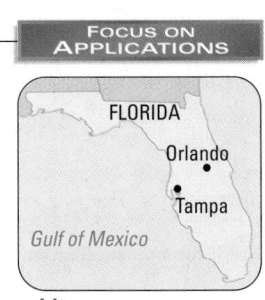

FOCUS ON
APPLICATIONS

FLORIDA
Orlando
Tampa
Gulf of Mexico

VACATION COSTS The daily costs given in Example 4 take into account money spent at tourist attractions. *Amusement Business* estimates that a family of four would spend about $188.50 at a theme park in Florida.

To solve the system, graph each equation as shown at the right.

Notice that you need to graph the equations only in the first quadrant because only positive values of x and y make sense in this situation.

The lines appear to intersect at the point (4, 3). You can check this algebraically as follows.

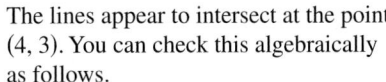

$4 + 3 = 7$ ✓   **Equation 1 checks.**

$275(4) + 400(3) = 2300$ ✓   **Equation 2 checks.**

▶ The solution is (4, 3), which means that you should plan to spend 4 days in Tampa and 3 days in Orlando.

3.1 *Solving Linear Systems by Graphing*   **141**

Closure Question *Sample answer:*
If the graph of the system is a pair of lines that intersect in one point, the coordinates of the point are the solution to the system. If the lines are parallel, the system has no solution, and if the graph is a single line, the system has infinitely many solutions.

EXTRA EXAMPLE 4
You plan to work 200 hours this summer mowing lawns and baby-sitting. You need to make a total of $1300. Baby-sitting pays $6 per hour and lawn mowing pays $8 per hour. How many hours should you work at each job?
150 hours baby-sitting and 50 hours mowing lawns

✔ CHECKPOINT EXERCISES
For use after Example 4:
1. You make small wreaths and large wreaths to sell at a craft fair. Small wreaths sell for $8.00 and large wreaths sell for $12.00. You think you can sell 40 wreaths all together and want to make $400. How many of each type of wreath should you bring to the fair?
20 large wreaths and 20 small wreaths

APPLICATION NOTE
EXAMPLE 4 Like most real-life applications of linear systems, at least one of the equations in this system is not easy to graph by hand. You can use this example to motivate the need for an algebraic method to solve linear systems.

FOCUS ON VOCABULARY
How are a solution to a linear system and a solution to an linear equation related? **A solution to a linear system is a solution to each linear equation in the system. That is, its coordinates must satisfy both equations in the system.**

CLOSURE QUESTION
Explain how to use a graph to determine how many solutions there are for a system of linear equations.
See margin for sample answer.

DAILY PUZZLER
If 2 cubes and 1 cone weigh 15 lb, and 3 cubes and 2 cones weigh 24 lb, how much do 1 cube and 3 cones weigh? **15 lb**

141

GUIDED PRACTICE

ASSIGNMENT GUIDE

BASIC
Day 1: pp. 142–145 Exs.12–28 even, 32–40 even, 41–45 odd
Day 2: pp. 142–145 Exs. 33–39 odd, 42–50 even, 53, 54, 64, 67–79 odd

AVERAGE
Day 1: pp. 142–145 Exs.12–46 even, 53–55
Day 2: 33–39 odd, 49–52, 57–63 odd, 64, 67–79 odd

ADVANCED
Day 1: pp. 142–145 Exs.12–48 even, 49–53
Day 2: pp. 142–145 Exs. 54–64, 65, 67–79 odd

BLOCK SCHEDULE
pp. 142–145 Exs.12–50 even, 52–56, 60–64, 67–79 odd

EXERCISE LEVELS
Level A: *Easier*
11–28, 32–40, 54–55
Level B: *More Difficult*
29–31, 41–53, 56–62, 64
Level C: *Most Difficult*
63, 65

✔ **HOMEWORK CHECK**
To quickly check student understanding of key concepts, go over the following exercises: Exs. 12, 22, 36, 42, 54. See also the Daily Homework Quiz:
• Blackline Master (*Chapter 3 Resource Book,* p. 24)
• 📓 Transparency (p. 20)

Vocabulary Check ✔

1. Complete this statement: A(n) _?_ of a system of linear equations in two variables is an ordered pair (x, y) that satisfies each equation.   **solution**

Concept Check ✔

2. How can you use the graph of a linear system to decide how many solutions the system has?   **See margin.**

3. Explain why a linear system in two variables cannot have exactly two solutions. **See margin.**

Skill Check ✔

2. *Sample answer:* If the graph consists of two non-parallel lines, then there is a single solution. If the graph consists of two parallel lines, there is no solution. If the graphs of the two lines coincide, there are infinitely many solutions.

3. *Sample answer:* If two lines share two points in common, they are the same line, and every point on that line is a solution.

Check whether the ordered pair (5, 6) is a solution of the system.

4. $-2x + 4y = -14$
$3x + y = 21$ **no**

5. $7x - 2y = 23$
$-x + 3y = 13$ **yes**

6. $x + y = 11$
$-x - y = -11$ **yes**

Graph the linear system. How many solutions does it have?
7–9. See margin for graphs.

7. $2x - y = 4$
$-6x + 3y = -18$ **0**

8. $14x + 3y = 16$
$7x - 5y = 34$ **1**

9. $21x - 7y = 7$
$-3x + y = -1$
infinitely many

10. 🌐 **SCHOOL OUTING** Your school is planning a 5 hour outing at the community park. The park rents bicycles for $8 per hour and inline skates for $6 per hour. The total budget per person is $34. How many hours should students spend doing each activity?   **(2, 3)**

PRACTICE AND APPLICATIONS

STUDENT HELP

▶ **Extra Practice**
to help you master skills is on p. 943.

CHECKING A SOLUTION Check whether the ordered pair is a solution of the system.

11. $(6, -1)$ **yes**
$4x - y = 25$
$-3x - 2y = -16$

12. $(3, 0)$ **no**
$-x + 2y = 3$
$10x + y = 30$

13. $(-2, -8)$ **no**
$2x - y = 52$
$9x - y = -10$

14. $(-3, -5)$ **yes**
$-x - y = 8$
$2x + 5y = -31$

15. $(-4, 1)$ **yes**
$-4x + 3y = 19$
$5x - 7y = -27$

16. $(10, 8)$ **yes**
$-3x - y = -38$
$-8x + 8y = -16$

17. $(1, -1)$ **no**
$-3x + y = -4$
$7x + 2y = -5$

18. $(-2, -7)$ **yes**
$5x - y = -3$
$x + 3y = -23$

19. $(0, 2)$ **no**
$17x + 8y = 16$
$-x - 4y = 8$

20-31. Estimates may vary. An exact solution is given. For systems with infinitely many solutions, a sample answer is given.

See margin for graphs.

GRAPH AND CHECK Graph the linear system and estimate the solution. Then check the solution algebraically.

20. $2x + y = 13$ **(3, 7)**
$5x - 2y = 1$

21. $x + 2y = 9$ **(7, 1)**
$-x + 6y = -1$

22. $-2x + y = 5$ **(-1, 3)**
$x + y = 2$

23. $3x + 4y = -10$ **(2, -4)**
$-7x - y = -10$

24. $2x + y = -11$
$-6x - 3y = 33$
Sample answer: **(-5, -1)**

25. $y = 5x$ **(1, 5)**
$y = x + 4$

STUDENT HELP

▶ **HOMEWORK HELP**
Example 1: Exs. 11–19
Example 2: Exs. 20–31
Example 3: Exs. 32–52
Example 4: Exs. 54–59

26. $-x + 3y = 3$
$2x - 6y = -6$
Sample answer: **(6, 3)**

27. $2x + y = -2$
$x - 2y = 19$ **(3, -8)**

28. $3x - y = 12$ **(4, 0)**
$-x + 8y = -4$

29. $3x - y = 8$
$\frac{1}{3}x - \frac{1}{6}y = 1$ **(2, -2)**

30. $y = \frac{1}{6}x - 2$ **(12, 0)**
$y = -\frac{1}{6}x + 2$

31. $-x + 4y = 10$ **(-2, 2)**
$4x - y = -10$

32. infinitely many solutions; the two lines coincide.

33. no solution; the two lines are parallel and have no points in common.

34. one solution; the two lines intersect in a single point.

INTERPRETING A GRAPH The graph of a system of two linear equations is shown. Tell whether the linear system has *infinitely many solutions, one solution,* or *no solution.* Explain your reasoning. 32-34. See margin.

32.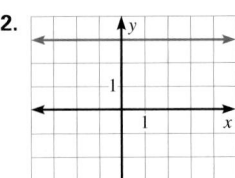

33.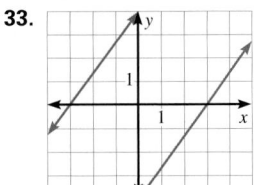

34.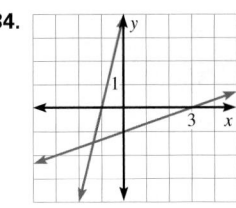

MATCHING GRAPHS Match the linear system with its graph. Tell how many solutions the system has.

35. $2x - y = -5$ E; 1
$x + 2y = 0$

36. $-2x + 3y = 12$ F; 0
$2x - 3y = 6$

37. $2x - y = 5$ B; infinitely many
$-4x + 2y = -10$

38. $x + 5y = -12$ D; 1
$x - 5y = 8$

39. $-x + 5y = 8$ A; 0
$2x - 10y = 7$

40. $4x - 7y = 27$ C; 1
$-6x - 9y = -21$

A.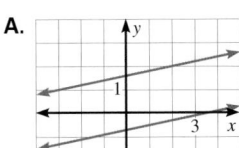

B.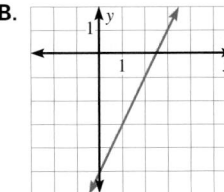

C.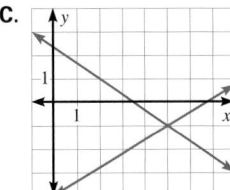

D.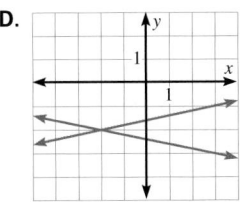

E.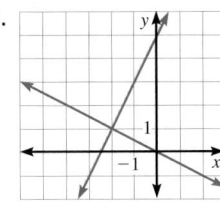

F.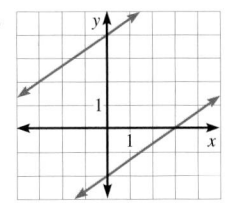

STUDENT HELP

▶ **Skills Review**
For help with graphing, see p. 933.

NUMBER OF SOLUTIONS Graph the linear system and tell how many solutions it has. If there is exactly one solution, estimate the solution and check it algebraically. 41–52. Estimates made from graphs may vary. Exact solutions are given. See margin for graphs.

41. $x = 5$
$x + y = 1$
one solution; $(5, -4)$

42. $7x + y = 10$
$3x - 2y = -3$
one solution; $(1, 3)$

43. $y = \frac{1}{2}x - 5$ no solutions
$y = \frac{1}{2}x + 3$

44. $y = -5 - x$
$x + 3y = -15$
one solution; $(0, -5)$

45. $\frac{1}{3}x + 7y = 2$
$\frac{2}{3}x + 14y = 4$
infinitely many solutions

46. $-4y = 24x + 4$
$y = -6x - 1$
infinitely many solutions

47. $2x - y = 7$
$y = 2x + 8$
no solutions

48. $y = \frac{3}{4}x + 3$
$y = 3x - 6$
one solution; $(4, 6)$

49. $6x - 2y = -2$
$-3x - 7y = 17$
one solution; $(-1, -2)$

50. $\frac{1}{2}x + 3y = 6$
$\frac{1}{3}x - 5y = -3$
one solution; $(6, 1)$

51. $-6x + 2y = 8$
$y = 3x + 4$
infinitely many solutions

52. $\frac{3}{4}x + y = 5$
$3x + 4y = 2$
no solutions

3.1 *Solving Linear Systems by Graphing* **143**

! **COMMON ERROR**
EXERCISES 20–31 Students may make mistakes graphing these equations. Stress the importance of checking their solutions in both equations to help them find graphing mistakes.

STUDENT HELP NOTES

▶ **Skills Review** As students review graphing lines, remind them that some lines may be easier to graph using *x*- and *y*-intercepts. Other lines may be easier to graph by first converting the equation to slope-intercept form and using the slope and *y*-intercept.

7.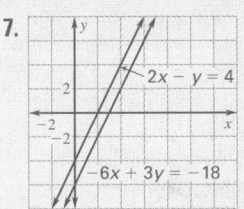

8.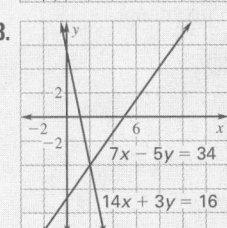

9.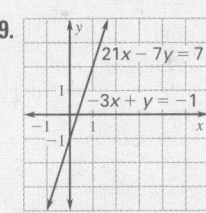

20–31, 41–52.
See Additional Answers beginning on page AA1.

APPLICATION NOTE

EXERCISE 56 Students may be confused as to whether the variables should stand for the number of miles walked and jogged, or the number of minutes walked and jogged. Recommend that they look at what they are trying to find (in this case, the time) and use that to determine the variables (in this case, the number of minutes walked and the number of minutes jogged).

ENGLISH LEARNERS

EXERCISES 54–59 Although English learners may understand the mathematical concepts presented in Chapter 3, many will have difficulty translating the challenging language of the word problems into the appropriate equations. Consider pairing students with English proficient partners for the duration of the chapter to help them understand unfamiliar vocabulary and interpret word problems.

63, 64a, 65, 76–79.
  See Additional Answers beginning on page AA1.

ADDITIONAL PRACTICE AND RETEACHING

For Lesson 3.1:

• Practice Levels A, B, and C (*Chapter 3 Resource Book,* p. 14)

• Reteaching with Practice (*Chapter 3 Resource Book,* p. 17)

• ⊞ See Lesson 3.1 of the *Personal Student Tutor*

For more Mixed Review:

• ⊞ Search the *Test and Practice Generator* for key words or specific lessons.

144

PEDOMETER
Worn at the hip, a pedometer counts the steps taken and multiplies by stride length to calculate the distance traveled. In Ex. 56 you would need to change the stride length setting when switching from walking to jogging.

53.a. *Sample answer:*
$2x + 3y = 5$
$4x - 7y = -3$

53.b. *Sample answer:*
$2x + 3y = 5$
$-4x - 6y = 2$

53.c. *Sample answer:*
$2x + 3y = 5$
$4x + 6y = 10$

54. $6.95x + 19.95y = 60.75$;
$2x + 4y = 14$; you ordered 3 paperbacks and 2 hardcovers.

55. $l + m = 125$
$0.1l + 0.5m = 32.5$; buy 75 latex balloons and 50 mylar balloons.

56. $j + w = 30$
$0.1j + 0.05w = 2.5$;
10 min.; 20 min.

57. $d + 1.25h = 6$
$720d + 1440h = 6480$; you can buy 4 high-density disks and 1 double-density disk.

58. $r + a = 5$
$4.25r + 5.5a = 25$; buy 3 packages of alkaline batteries and 2 packages of regular batteries

59. Let f = the travel time in hours of the first bus let s = the travel time in hours of the second bus; $f = s + \frac{1}{12}$,
$30f = 40s$; 10 miles from the airport

---

53. **CRITICAL THINKING** Write a system of two linear equations that has the given number of solutions.   53-59. See margin.

   a. one solution   **b.** no solution   **c.** infinitely many solutions

54. 🌐 **BOOK CLUB** You enroll in a book club in which you can earn bonus points to use toward the purchase of books. Each paperback book you order costs $6.95 and earns you 2 bonus points. Each hardcover book costs $19.95 and earns you 4 bonus points. The first order you place comes to a total of $60.75 and earns you 14 bonus points. How many of each type of book did you order? Use the verbal model to write and solve a system of linear equations.

| Price of paperback book | · | Number of paperback books | + | Price of hardcover book | · | Number of hardcover books | = | Total cost of order |

| Bonus points for paperback book | · | Number of paperback books | + | Bonus points for hardcover book | · | Number of hardcover books | = | Total number of bonus points |

55. 🌐 **DECORATION COSTS** You are on the prom decorating committee and are in charge of buying balloons. You want to use both latex and mylar balloons. The latex balloons cost $.10 each and the mylar balloons cost $.50 each. You need 125 balloons and you have $32.50 to spend. How many of each can you buy? Use a verbal model to write and solve a system of linear equations.

56. 🌐 **FITNESS** For 30 minutes you do a combination of walking and jogging. At the end of your workout your pedometer displays a total of 2.5 miles. You know that you walk 0.05 mile per minute and jog 0.1 mile per minute. For how much time were you walking? For how much time were you jogging? Use a verbal model to write and solve a system of linear equations.

57. 🌐 **FLOPPY DISK STORAGE** You want to copy some documents on your friend's computer. The documents use 6480 kilobytes(K) of disk space. You go to a store and see a sign advertising double-density disks and high-density disks. If you have $6 to spend, how many of each type of disk can you buy to get the disk space you need? Use a verbal model to write and solve a system of linear equations.

Floppy Disks

DOUBLE DENSITY 720K

HIGH DENSITY 1440K

$1.00 each $1.25 each

58. 🌐 **BATTERY POWER** Your portable stereo requires 10 size D batteries. You have $25 to spend on 5 packages of 2 batteries each and would like to maximize your battery power. Each regular package of batteries costs $4.25 and each alkaline package of batteries costs $5.50 (because alkaline batteries last longer). How many packages of each type of battery should you buy? Use a verbal model to write and solve a system of linear equations.

59. 🌐 **AIRPORT SHUTTLE** A bus station 15 miles from the airport runs a shuttle service to and from the airport. The 9:00 A.M. bus leaves for the airport traveling 30 miles per hour. The 9:05 A.M. bus leaves for the airport traveling 40 miles per hour. Write a system of linear equations to represent distance as a function of time for each bus. Graph and solve the system. How far from the airport will the 9:05 A.M. bus catch up to the 9:00 A.M. bus?

63. a triangle; $(-3, 5)$, $(-5, -5)$ and $(2, 0)$; *Sample answer:* I graphed the lines carefully and found the apparent points of their intersections from the graph. It was easy to see that two of the lines had the same x-intercept, so that was one point. The other points I checked algebraically in the equations to make sure they were solutions.
See margin for graph.

Test Preparation

★ **Challenge**

EXTRA CHALLENGE
www.mcdougallittell.com

TYPES OF SYSTEMS In Exercises 60–62, use the following definitions to tell whether the system is *consistent and independent*, *consistent and dependent*, or *inconsistent*.

A system that has at least one solution is *consistent*. A consistent system that has exactly one solution is *independent*, and a consistent system that has infinitely many solutions is *dependent*. If a system has no solution, the system is *inconsistent*.

consistent and dependent	consistent and independent	inconsistent
60. $-5x + 2y = 12$	**61.** $-3x - 3y = -6$	**62.** $2x - y = -12$
$10x - 4y = -24$	$7x + 4y = 20$	$-6x + 3y = 8$

63. **GEOMETRY** **CONNECTION** Graph the equations $x + y = 2$, $-5x + y = 20$, and $-\frac{5}{7}x + y = -\frac{10}{7}$. What geometric figure do the graphs of the equations form? What are the coordinates of the vertices of the figure? Explain the steps you used to find the coordinates. **See margin.**

64. **MULTI-STEP PROBLEM** You are choosing between two long-distance telephone companies. Company A charges \$.09 per minute plus a \$4 monthly fee. Company B charges \$.11 per minute with no monthly fee.

    a. Let x be the number of minutes you call long distance in one month, and let y be the total cost of long-distance phone service. Write and graph two equations representing the cost of each company's service for one month.
$y = 0.09x + 4$; $y = 0.11x$; See margin.

    b. Estimate the coordinates of the point where the two graphs intersect. Check your estimate algebraically. **(200, 22)**

    c. *Writing* What does the point of intersection you found in part (b) represent? How can it help you decide which long-distance company to use?
See margin.

65. **BUYING A DIGITAL CAMERA** The school yearbook staff is purchasing a digital camera. Recently the staff received two ads in the mail. The ad for Store 1 states that all digital cameras are 15% off. The ad for Store 2 gives a \$300 coupon to use when purchasing any digital camera. Assume that the lowest priced digital camera is \$700. Write and graph two equations that describe the prices at both stores. When does Store 1 have a better deal than Store 2?

    Sample answer: Let x = the regular price of a digital camera, and let y = the sale price. For Store 1, $y = 0.85x$. For Store 2, $y = x - 300$. Store 1 has a better deal if the price of the camera is over \$2000. See margin for graph.

MIXED REVIEW

64.c. Sample answer: It represents a charge of \$22 for 200 minutes of long distance service. If you make more than 200 minutes of long distance calls most months, then the first company with a 9¢ per minute charge is less expensive for you. If you make fewer than 200 minutes worth of long distance calls most months, the second company, with no monthly fee, is less expensive for you.

SOLVING EQUATIONS Solve the equation. Check your solution. (Review 1.3 for 3.2)

66. $4x + 11 = 39$ 7   **67.** $\frac{1}{2}x - 10 = 8$ 36   **68.** $6x - 8 = 3x + 16$ 8

69. $-9x - 2 = x + 1$ -0.3 **70.** $2(3x - 5) = 7(x + 2)$ -24 **71.** $10(x + 1) = \frac{1}{2}(x - 18)$ -2

CHECKING SOLUTIONS Check whether the ordered pairs are solutions of the inequality. (Review 2.6)

72. $12x + 4y \geq 3$; $(1, -3)$, $(0, 2)$
    no; yes

73. $-x - y \leq -10$; $(-3, -7)$, $(5, 4)$
    no; no

74. $15 > 2x - 2y$; $(10, 3)$, $(-5, 7)$
    yes; yes

75. $6x + \frac{1}{2}y \leq -5$; $(2, -6)$, $(-1, 7)$
    no; no

GRAPHING ABSOLUTE VALUE FUNCTIONS Graph the function. (Review 2.8)
    76–79. See margin.

76. $y = |x| - 5$ **77.** $y = |x - 9|$

78. $y = -|x - 8| + 3$ **79.** $y = |7 - x| + 4$

DAILY HOMEWORK QUIZ

🖹 *Transparency Available*

1. Check whether $(2, -5)$ is a solution. **no**
$7x + 4y = -6$
$6x + 5y = -11$

2. Graph the system and estimate the solution.
$x + y = 1$
$x - 3y = 5$
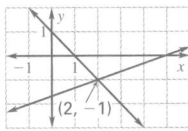

3. Tell how many solutions the system has. **infinitely many**
$7x + 5y = 10$
$y = -\frac{7}{5}x + 2$

4. A store sells packs of AA batteries for \$1.89 and packs of AAA batteries for \$1.53, tax included. Jeff bought 8 packs and spent a total of \$14.04. How many packs of each kind of battery did he buy?
5 packs of AA's, 3 packs of AAA's

EXTRA CHALLENGE NOTE
↳ Challenge problems for Lesson 3.1 are available in **blackline** format in the *Chapter 3 Resource Book*, p. 21 and at **www.mcdougallittell.com.**

ADDITIONAL TEST PREPARATION

1. **OPEN ENDED** One equation in a linear system is $3x - 2y = 5$. Write a second equation so that the system has 1 solution, no solutions, and infinitely many solutions. *Sample answer:*
$2x - 2y = 5$; $3x - 2y = 1$; $9x - 6y = 15$

2. **WRITING** Sketch some possible graphs that show systems of equations with different numbers of solutions. Explain how you can determine the number of solutions from the graph. **Check work.**

1 Planning the Activity

PURPOSE
To solve a linear system using a graphing calculator.

MATERIALS
- graphing calculator
- Activity Support Master
  (*Chapter 3 Resource Book*, p. 13)

PACING
- Activity — 20 min

▶ LINK TO LESSON
Students can use the techniques of this activity to check Examples 2 and 4 in Lesson 3.1.

2 Managing the Activity

CLASSROOM MANAGEMENT
Students can work on this activity with a partner. In Step 1, encourage students to retain the fraction forms of the slope and *y*-intercept rather than rounding them to decimal form. This will ensure the most accurate solution.

ALTERNATIVE APPROACH
This activity can be done as a class demonstration using a graphing calculator with an overhead display.

3 Closing the Activity

★ KEY DISCOVERY
A graphing calculator can be used to find an accurate estimate of the solution to a linear system.

ACTIVITY ASSESSMENT
Write a procedure you can use to find the solution to a linear system using a graphing calculator.
Sample answer: Write the equations in slope-intercept form. Graph them on the calculator. Adjust the viewing window to show the intersection. Use the *intersect* feature to find the coordinates of the intersection.

3. $\left(\frac{141}{19}, \frac{119}{19}\right)$, or about (7.42, 6.26)

4. $\left(\frac{8}{17}, \frac{31}{17}\right)$, or about (0.47, 1.82)

5. $\left(-\frac{116}{21}, \frac{47}{21}\right)$, or about (−5.52, 2.24)

◐ ACTIVITY 3.1
Using Technology

Graphing Systems of Equations

In Lesson 3.1 you learned how to *estimate* the solution of a linear system by graphing. With a graphing calculator, you can get an answer that is very close to, and sometimes *exactly* equal to, the actual solution.

▶ EXAMPLE
Solve the linear system using a graphing calculator.

$$5x + 3y = -15$$
$$4x - 2y = 45$$

STUDENT HELP

INTERNET **KEYSTROKE HELP**

See keystrokes for several models of calculators at www.mcdougallittell.com

▶ SOLUTION

1 Solve each equation for *y*.

$$5x + 3y = -15$$
$$3y = -5x - 15$$
$$y = -\frac{5}{3}x - 5$$
$$4x - 2y = 45$$
$$-2y = -4x + 45$$
$$y = 2x - \frac{45}{2}$$

2 Enter the equations. It's a good idea to use parentheses to enter fractions.

```
Y1=(-5/3)X-5
Y2=2X-(45/2)
Y3=
Y4=
Y5=
Y6=
Y7=
```

3 Using a standard viewing window, graph the equations.

4 Use the *Intersect* feature to find the point where the graphs intersect.

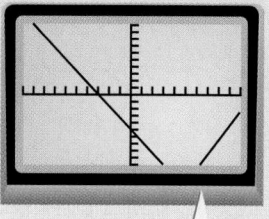

```
Intersection
X=4.7727273 Y=-12.95455
```

If the graphs do not intersect on the screen, set a different viewing window.

▶ The solution is about (4.77, −12.95).

▶ EXERCISES

Solve the linear system using a graphing calculator.

1. $y = x + 4$ (−1, 3)
   $y = 2x + 5$

2. $y = -2x + 13$ (2.25, 8.5)
   $y = 6x - 5$

3. $3x - y = 16$
   $-5x + 8y = 13$

4. $5x + 2y = 6$
   $x - 3y = -5$

5. $6x + 9y = -13$
   $-x + 2y = 10$

6. $2x + 8y = -53$
   $3x + 4y = 26$
   (26.25, −13.1875)

Group Activity for use with Lesson 3.2

Combining Equations in a Linear System

PURPOSE
To discover the relationship between the graph of a linear system and the graph of the line whose equation is the sum of the equations in the system.

GROUP ACTIVITY
Work with a partner.

MATERIALS
• graph paper
• ruler

▶ **QUESTION**

For a system of two linear equations with exactly one solution, how is the graph of the *sum* of the equations related to the graph of the system?

▶ **EXPLORING THE CONCEPT**

1 Graph the system and label the point of intersection.

$$3x - y = -5$$
$$3x + y = -1$$

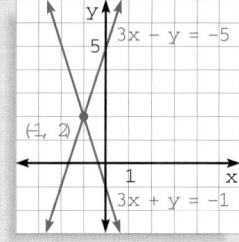

2 Add the two equations in the system. Graph the resulting equation in the same coordinate plane you used to graph the system.

$$3x - y = -5$$
$$\underline{3x + y = -1}$$
$$6x = -6$$
$$x = -1$$

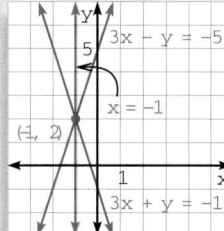

3 Note how the graph of the sum of the equations is related to the graph of the system.

▶ **DRAWING CONCLUSIONS**

3. *Sample answer:* The sum of the two equations is $(A + D)x + (B + E)y = C + F$. Since (p, q) is a solution of the system of equations, $Ap + Bq = C$ and $Dp + Eq = F$. At the point (p, q), the left-hand side of the sum equation is $(A + D)p + (B + E)q = (Ap + Bq) + (Dp + Eq) = C + F$. So, the point (p, q) is a solution of the sum equation as well.

1. Repeat the steps above for each system.   1a–f. See margin.

  a. $10x + 4y = 24$
     $-6x - 4y = 8$

  b. $x - 2y = 2$
     $-x + 4y = -20$

  c. $6x - 3y = 27$
     $2x + y = -9$

  d. $x - y = -3$
     $-2x + 5y = 6$

  e. $7x + 20y = 0$
     $-3x + 6y = 4$

  f. $2x - y = -3$
     $x + 2y = 6$

2. What seems to be true about the graph of the sum of two equations in a system if the system has exactly one solution?   The sum of the equations is a line whose graph contains the point of intersection.

3. Consider the following general system that has a single solution (p, q).   See margin.

$$Ax + By = C$$
$$Dx + Ey = F$$

Use this system to justify your conclusion from Exercise 2 algebraically.

3.2 *Concept Activity*   **147**

MATERIALS
• graph paper
• ruler
• pencil
• Activity Support Master (*Chapter 3 Resource Book,* p. 26)

PACING
• Exploring the Concept — 5 min
• Drawing Conclusions — 15 min

▶ **LINK TO LESSON**
When discussing Example 2 on page 149, show students how this activity enables them to know that the solution to the system is on the line $y = -6$ and hence has y-coordinate -6.

COOPERATIVE LEARNING
Students can work in small groups and take turns graphing the equations and checking each other's graphs. They should discuss Exercises 2 and 3 and write a single response.

ALTERNATIVE APPROACH
A graphing calculator can be used to graph the equations.

★ **KEY DISCOVERY**
The graph of the sum of the equations in a linear system with one solution is a line through the intersection point.

ACTIVITY ASSESSMENT
In a linear system with one solution, how is the solution to the system related to the sum of the equations?
See sample answer at left.

Activity Assessment *Sample answer:*
The solution to the system is a solution to the sum of the equations in the system.

1a–f. See Additional Answers beginning on page AA1.

LESSON OPENER
APPLICATION

An alternative way to approach Lesson 3.2 is to use the Application Lesson Opener:

- Blackline Master (*Chapter 3 Resource Book,* p. 25)
- Transparency (p. 17)

MEETING INDIVIDUAL NEEDS

- *Chapter 3 Resource Book*
  Prerequisite Skills Review (p. 5)
  Practice Level A (p. 28)
  Practice Level B (p. 29)
  Practice Level C (p. 30)
  Reteaching with Practice (p. 31)
  Absent Student Catch-Up (p. 33)
  Challenge (p. 35)
- *Resources in Spanish*
- *Personal Student Tutor*

NEW-TEACHER SUPPORT

See the Tips for New Teachers on pp. 1–2 of the *Chapter 3 Resource Book* for additional notes about Lesson 3.2.

WARM-UP EXERCISES

Transparency Available

Solve each equation for the indicated variable.

1. $2x - y = 5$, y $y = 2x - 5$
2. $-x + 2y = 3$, x $x = 2y - 3$
3. $3x - 4y = 12$, y $y = \frac{3}{4}x - 3$
4. $3x - 4y = 12$, x $x = \frac{4}{3}y + 4$

What you should learn

GOAL 1 Use algebraic methods to solve linear systems.

GOAL 2 Use linear systems to model **real-life** situations, such as catering an event in **Example 5**.

Why you should learn it

▼ To solve **real-life** problems, such as how to plan a 40 minute workout in **Ex. 57**.

3.2 Solving Linear Systems Algebraically

GOAL 1 USING ALGEBRAIC METHODS TO SOLVE SYSTEMS

In this lesson you will study two algebraic methods for solving linear systems. The first method is called *substitution*.

THE SUBSTITUTION METHOD

STEP 1 Solve one of the equations for one of its variables.

STEP 2 Substitute the expression from Step 1 into the other equation and solve for the other variable.

STEP 3 Substitute the value from Step 2 into the revised equation from Step 1 and solve.

EXAMPLE 1 *The Substitution Method*

Solve the linear system using the substitution method.

$$3x + 4y = -4 \quad \text{Equation 1}$$
$$x + 2y = 2 \quad \text{Equation 2}$$

SOLUTION

1 Solve Equation 2 for x.

$x + 2y = 2$	Write Equation 2.
$x = -2y + 2$	Revised Equation 2

2 Substitute the expression for x into Equation 1 and solve for y.

$3x + 4y = -4$	Write Equation 1.
$3(-2y + 2) + 4y = -4$	Substitute $-2y + 2$ for x.
$y = 5$	Solve for y.

3 Substitute the value of y into revised Equation 2 and solve for x.

$x = -2y + 2$	Write revised Equation 2.
$x = -2(5) + 2$	Substitute 5 for y.
$x = -8$	Simplify.

▶ The solution is $(-8, 5)$.

✓**CHECK** Check the solution by substituting back into the original equations.

$3x + 4y = -4$	Write original equations.	$x + 2y = 2$
$3(-8) + 4(5) \stackrel{?}{=} -4$	Substitute for x and y.	$-8 + 2(5) \stackrel{?}{=} 2$
$-4 = -4$ ✓	Solution checks.	$2 = 2$ ✓

CHOOSING A METHOD In Step 1 of Example 1, you could have solved for either x or y in either Equation 1 or Equation 2. It was easiest to solve for x in Equation 2 because the x-coefficient is 1. In general you should solve for a variable whose coefficient is 1 or -1.

$$x - 5y = 11 \longleftarrow \text{Solve for } x. \qquad 4x - 2y = -1$$
$$2x + 7y = -3 \qquad\qquad\qquad 3x - y = 8 \longleftarrow \text{Solve for } y.$$

If neither variable has a coefficient of 1 or -1, you can still use substitution. In such cases, however, the *linear combination* method may be better. The goal of this method is to add the equations to obtain an equation in one variable.

THE LINEAR COMBINATION METHOD

STEP ❶ Multiply one or both of the equations by a constant to obtain coefficients that differ only in sign for one of the variables.

STEP ❷ Add the revised equations from Step 1. Combining like terms will eliminate one of the variables. Solve for the remaining variable.

STEP ❸ Substitute the value obtained in Step 2 into either of the original equations and solve for the other variable.

EXAMPLE 2 *The Linear Combination Method: Multiplying One Equation*

Solve the linear system using the $2x - 4y = 13$ **Equation 1**
linear combination method. $4x - 5y = 8$ **Equation 2**

SOLUTION

❶ Multiply the first equation by -2 so that the x-coefficients differ only in sign.

$$2x - 4y = 13 \qquad \times \ -2 \qquad -4x + 8y = -26$$
$$4x - 5y = 8 \qquad\qquad\qquad\qquad \underline{4x - 5y = 8}$$

❷ Add the revised equations and solve for y. $3y = -18$
                                                  $y = -6$

❸ Substitute the value of y into one of the original equations. Solve for x.

$2x - 4y = 13$	**Write Equation 1.**
$2x - 4(-6) = 13$	**Substitute -6 for y.**
$2x + 24 = 13$	**Simplify.**
$x = -\dfrac{11}{2}$	**Solve for x.**

▶ The solution is $\left(-\dfrac{11}{2}, -6\right)$.

✓ **CHECK** You can check the solution algebraically using the method shown in Example 1. You can also use a graphing calculator to check the solution.

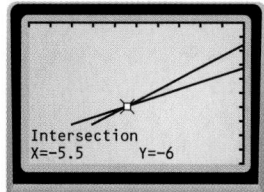

Intersection
X=-5.5   Y=-6

STUDENT HELP

↳ **Study Tip**
In Example 2, one x-coefficient is a multiple of the other. In this case, it is easier to eliminate the x-terms because you need to multiply only one equation by a constant.

---

2 TEACH

MOTIVATING THE LESSON
Tell students to think of the different ways they could get to school in the morning such as taking a bus, walking, riding a bike, or taking a subway. Why did they choose the method they did? There are often different ways to accomplish a single task and each method has its own advantages and disadvantages. Today's lesson focuses on two methods other than graphing for solving linear systems.

EXTRA EXAMPLE 1
Solve the linear system using the substitution method.
$3x - y = 13$
$2x + 2y = -10$ $(2, -7)$

EXTRA EXAMPLE 2
Solve the linear system using the linear combination method.
$2x - 6y = 19$
$-3x + 2y = 10$ $\left(-7, -\dfrac{11}{2}\right)$

✓ **CHECKPOINT EXERCISES**
For use after Example 1:
1. Solve the linear system using the substitution method.
$-x + 3y = 1$
$4x + 6y = 8$ $\left(1, \dfrac{2}{3}\right)$

For use after Example 2:
2. Solve the linear system using the linear combination method.
$2x + 3y = -1$
$-5x + 5y = 15$ $(-2, 1)$

ENGLISH LEARNERS
To help English learners understand and remember the mathematical concept of *substitution*, discuss non-mathematical meanings of the word. You may want to give the example of a substitute teacher to help them understand the general concept. Then relate the idea of substitution to the algebraic context.

EXTRA EXAMPLE 3
Solve the linear system using the linear combination method.
$9x - 5y = -7$
$-6x + 4y = 2$ $(-3, -4)$

EXTRA EXAMPLE 4
Solve the linear system.
a. $9x - 3y = 15$
$-3x + y = -5$
infinitely many solutions
b. $6x - 4y = 14$
$-3x + 2y = 7$ no solution

EXTRA EXAMPLE 5
A citrus fruit company plans to make 13.25 lb gift boxes of oranges and grapefruits. Each box is to have a retail value of $21.00. Each orange weighs 0.50 lb and has a retail value of $.75, while each grapefruit weighs 0.75 lb and has a retail value of $1.25. How many oranges and grapefruits should be included in the box?
13 oranges and 9 grapefruits

 CHECKPOINT EXERCISES

For use after Examples 3 and 4:
1. Solve the linear system.
a. $6x + 15y = 9$
$4x + 10y = 8$ no solution
b. $-9x + 14y = -2$
$11x - 6y = 8$ $\left(1, \frac{1}{2}\right)$
c. $3x - 5y = 4$
$-6x + 10y = -8$
infinitely many solutions

For use after Example 5:
2. You are planting a 160 ft² garden with shrubs and perennial plants. Each shrub costs $42 and requires 16 ft² of space. Each perennial plant costs $6 and requires 8 ft² of space. You plan to spend a total of $270. How many of each type of plant should you buy to fill the garden? **5 shrubs and 10 perennial plants**

EXAMPLE 3   *The Linear Combination Method: Multiplying Both Equations*

Solve the linear system using the linear combination method.
$7x - 12y = -22$ **Equation 1**
$-5x + 8y = 14$ **Equation 2**

SOLUTION
Multiply the first equation by 2 and the second equation by 3 so that the coefficients of y differ only in sign.

$7x - 12y = -22$ × 2 ⟹ $14x - 24y = -44$
$-5x + 8y = 14$ × 3 ⟹ $-15x + 24y = 42$

Add the revised equations $-x = -2$
and solve for x. $x = 2$

Substitute the value of x into one of the original equations. Solve for y.

$-5x + 8y = 14$ **Write Equation 2.**
$-5(2) + 8y = 14$ **Substitute 2 for x.**
$y = 3$ **Solve for y.**

▶ The solution is $(2, 3)$. Check the solution algebraically or graphically.

EXAMPLE 4   *Linear Systems with Many or No Solutions*

Solve the linear system.

a. $x - 2y = 3$ b. $6x - 10y = 12$
$2x - 4y = 7$ $-15x + 25y = -30$

SOLUTION
a. Since the coefficient of x in the first equation is 1, use substitution.

Solve the first equation for x.

$x - 2y = 3$
$x = 2y + 3$

Substitute the expression for x into the second equation.

$2x - 4y = 7$ **Write second equation.**
$2(2y + 3) - 4y = 7$ **Substitute 2y + 3 for x.**
$6 = 7$ **Simplify.**

▶ Because the statement $6 = 7$ is never true, there is *no solution.*

b. Since no coefficient is 1 or -1, use the linear combination method.

Multiply the first equation by 5 and the second equation by 2.

$6x - 10y = 12$ × 5 ⟹ $30x - 50y = 60$
$-15x + 25y = -30$ × 2 ⟹ $-30x + 50y = -60$

Add the revised equations. $0 = 0$

▶ Because the equation $0 = 0$ is always true, there are *infinitely many solutions.*

EXAMPLE 5 *Using a Linear System as a Model*

CATERING A caterer is planning a party for 64 people. The customer has $150 to spend. A $39 pan of pasta feeds 14 people and a $12 sandwich tray feeds 6 people. How many pans of pasta and how many sandwich trays should the caterer make?

SOLUTION

PROBLEM SOLVING STRATEGY

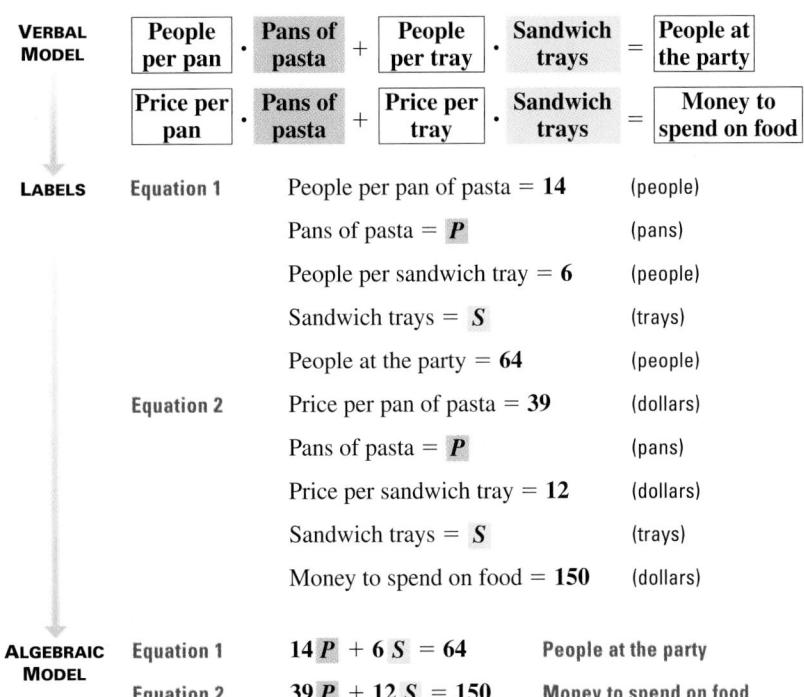

VERBAL MODEL							
People per pan	·	Pans of pasta	+	People per tray	·	Sandwich trays	= People at the party
Price per pan	·	Pans of pasta	+	Price per tray	·	Sandwich trays	= Money to spend on food

LABELS

Equation 1	People per pan of pasta = **14**	(people)
	Pans of pasta = **P**	(pans)
	People per sandwich tray = **6**	(people)
	Sandwich trays = **S**	(trays)
	People at the party = **64**	(people)
Equation 2	Price per pan of pasta = **39**	(dollars)
	Pans of pasta = **P**	(pans)
	Price per sandwich tray = **12**	(dollars)
	Sandwich trays = **S**	(trays)
	Money to spend on food = **150**	(dollars)

ALGEBRAIC MODEL

Equation 1	$14P + 6S = 64$	People at the party
Equation 2	$39P + 12S = 150$	Money to spend on food

Use the linear combination method to solve the system.

Multiply Equation 1 by -2 so that the coefficients of S differ only in sign.

$$14P + 6S = 64 \qquad \times -2 \qquad -28P - 12S = -128$$
$$39P + 12S = 150 \qquad\qquad\qquad 39P + 12S = 150$$

Add the revised equations and solve for P.
$$11P = 22$$
$$P = 2$$

Substitute the value of P into one of the original equations and solve for S.

$14P + 6S = 64$	Write Equation 1.
$14(2) + 6S = 64$	Substitute 2 for P.
$28 + 6S = 64$	Multiply.
$S = 6$	Solve for S.

▶ The caterer should make 2 pans of pasta and 6 sandwich trays for the party.

FOCUS ON CAREERS

CATERER
A caterer prepares food for special events. When planning a meal, a caterer needs to consider both the cost of the food and the number of guests.

CAREER LINK
www.mcdougallittell.com

MATHEMATICAL REASONING
Part **(a)** of Example 4 involves proof by contradiction. The assumption that the system has a solution leads to the false equation 6 = 7. Thus, the assumption is false, which means that there is no solution.

CAREER NOTE
EXAMPLE 5 Additional information about catering is available at **www.mcdougallittell.com.**

FOCUS ON VOCABULARY
Encourage students to think of the common meaning of the word *substitute:* to replace one thing for another. This should help them remember the substitution method where they are replacing one of the variables with an equivalent expression. Similarly, think of *combining* as grouping together items. In the linear combination method, the terms of the equations are grouped together or added to form a new equation.

CLOSURE QUESTION
When using the linear combination method for solving a linear system, why would you want to have the coefficients of one of the variables be opposites? If these coefficients are opposites, the sum of the equations will be an equation with only one variable.

DAILY PUZZLER
One student claims that 3 cubes and 5 cones weigh 20 oz, while 9 cubes and 15 cones weigh 40 oz. Explain why the student is mistaken. This situation leads to the system $3x + 5y = 20$, $9x + 15y = 40$, which has no solution.

ASSIGNMENT GUIDE

BASIC
Day 1: pp. 152–155 Exs. 12–20
even, 24–32 even, 36–44
even, 50
Day 2: pp. 152–155 Exs. 33–41
odd, 51–55 odd, 63, 64,
67–81 odd,
Quiz 1 Exs. 1–19

AVERAGE
Day 1: pp. 152–155 Exs. 12–20
even, 24–34 even,
38–52 even
Day 2: pp. 152–155 Exs. 33–51
odd, 54–58 even, 63, 64,
67–81 odd,
Quiz 1 Exs. 1–19

ADVANCED
Day 1: pp. 152–155 Exs. 12–54
even
Day 2: pp. 152–155 Exs. 47–57
odd, 59–65, 67–81 odd,
Quiz 1 Exs. 1–19

BLOCK SCHEDULE
pp. 152–155 Exs. 12–54 even, 63,
64, 67–81 odd, Quiz 1 Exs. 1–19

EXERCISE LEVELS
Level A: *Easier*
11–49, 54–58, 63–64
Level B: *More Difficult*
50–53, 59–62
Level C: *Most Difficult*
65

✓ HOMEWORK CHECK
To quickly check student under-
standing of key concepts, go
over the following exercises:
Exs. 14, 24, 38, 50, 54. See
also the Daily Homework Quiz:
• Blackline Master (*Chapter 3
Resource Book*, p. 39)
• Transparency (p. 21)

GUIDED PRACTICE

Vocabulary Check ✓

1. Complete this statement: To solve a linear system where one of the coefficients is 1 or -1, it is usually easiest to use the ? method.   **substitution**

Concept Check ✓

2. Read Step 3 in the box on page 148. Why do you think it recommends substituting into the revised equation from Step 1 instead of one of the original equations? *Sample answer:* **because the revised equation is already solved for the other variable, and is the most direct way to find its value.**

3. When solving a linear system algebraically, how do you know when there is no solution? How do you know when there are infinitely many solutions? **See margin.**

Skill Check ✓

3. There is no solution if you come up with an equation that is never true, like $1 = -2$. There are infinitely many solutions if you come up with an equation that is always true, like $4 = 4$.

Solve the system using the substitution method.

4. $x + 3y = -2$ **(−2, 0)**
   $-4x - 5y = 8$

5. $3x + 2y = 10$ **(4, −1)**
   $2x - y = 9$

6. $-3x + y = -7$ **(2, −1)**
   $5x - 2y = 12$

Solve the system using the linear combination method.

7. $-3x + 2y = -6$ **(6, 6)**
   $5x - 2y = 18$

8. $5x - 2y = 12$ $\left(1, -\dfrac{7}{2}\right)$
   $-9x - 8y = 19$

9. $4x - 3y = 0$ **(3, 4)**
   $-10x + 7y = -2$

10. 🌐 **BUSINESS** Selling frozen yogurt at a fair, you make $565 and use 250 cones. A single-scoop cone costs $2 and a double-scoop cone costs $2.50. How many of each type of cone did you sell?
120 single-scoop cones and 130 double-scoop cones

PRACTICE AND APPLICATIONS

> **STUDENT HELP**
> ➤ **Extra Practice**
> to help you master
> skills is on p. 943.

SUBSTITUTION METHOD Solve the system using the substitution method.

11. $2x + 3y = 5$ **(4, −1)**
    $x - 5y = 9$

12. $-2x + y = 6$ **no**
    $4x - 2y = 5$ **solution**

13. $-x + 2y = 3$ **(3, 3)**
    $4x - 5y = -3$

14. $5x + 3y = 4$ $\left(\dfrac{22}{5}, -6\right)$
    $5x + y = 16$

15. $4x + 6y = 15$ $\left(0, \dfrac{5}{2}\right)$
    $-x + 2y = 5$

16. $3x - y = 4$ $\left(\dfrac{3}{2}, \dfrac{1}{2}\right)$
    $5x + 3y = 9$

17. $\dfrac{1}{2}x + y = 9$ **(−2.4, 10.2)**
    $7x + 4y = 24$

18. $-3x + y = 2$ **(−1, −1)**
    $8x - 15y = 7$

19. $5x + 6y = -45$ **(3, −10)**
    $x - \dfrac{1}{2}y = 8$

20. $-x - 4y = -3$ $\left(\dfrac{57}{7}, -\dfrac{9}{7}\right)$
    $2x + y = 15$

21. $x + 2y = 2$ **(−2, 2)**
    $7x - 3y = -20$

22. $3x - y = 4$
    $-9x + 3y = -12$
    infinitely many solutions

31. $\left(\dfrac{18}{41}, \dfrac{605}{82}\right)$, or about (0.439, 7.378)

LINEAR COMBINATION METHOD Solve the system using the linear combination method.

> **STUDENT HELP**
> ➤ **HOMEWORK HELP**
> **Example 1:** Exs. 11–22, 35–49
> **Examples 2, 3:** Exs. 23–49
> **Example 4:** Exs. 11–49
> **Example 5:** Exs. 54–62

23. $3x + 5y = -16$ $\left(-\dfrac{11}{3}, -1\right)$
    $3x - 2y = -9$

24. $3x + 2y = 6$ **(−2, 6)**
    $-6x - 3y = -6$

25. $-6x + 5y = 4$ $\left(0, \dfrac{4}{5}\right)$
    $7x - 10y = -8$

26. $7x - 4y = -3$ **(−1, −1)**
    $2x + 5y = -7$

27. $-9x + 6y = 0$
    $-12x + 8y = 0$
    infinitely many solutions

28. $5x + 6y = -16$ $\left(-5, \dfrac{3}{2}\right)$
    $2x + 10y = 5$

29. $21x - 8y = -1$ $\left(\dfrac{1}{3}, 1\right)$
    $9x + 5y = 8$

30. $-15x - 2y = -31$
    $4x + 6y = 11$ $\left(2, \dfrac{1}{2}\right)$

31. $\dfrac{1}{4}x + 5y = 37$ **See margin.**
    $-4x + 2y = 13$

32. $7x + 2y = -3$
    $-14x - 4y = 6$
    infinitely many solutions

33. $6x - y = -2$ **no**
    $-18x + 3y = 4$ **solution**

34. $-5x + 2y = -10$ **(4, 5)**
    $3x - 6y = -18$

152 **Chapter 3** *Systems of Linear Equations and Inequalities*

35. $-5x + 7y = 11$
$-5x + 3y = 19$ $(-5, -2)$

36. $x - y = 3$
$-2x + 2y = -6$
infinitely many solutions

37. $2x - 5y = 10$ $(5, 0)$
$-3x + 4y = -15$

38. $-3x + y = 11$ $(-6, -7)$
$5x - 2y = -16$

39. $-4x - 6y = 11$ no
$6x + 9y = -3$ solution

40. $x - 4y = -2$ $\left(5, \frac{7}{4}\right)$
$-3x + 8y = -1$

41. $2x + 5y = 17$ $\left(-\frac{69}{11}, \frac{65}{11}\right)$
$-5x - 7y = -10$

42. $-3x + 7y = 6$ $\left(\frac{19}{8}, \frac{15}{8}\right)$
$5x - y = 10$

43. $-2x + 3y = 20$ $\left(-\frac{25}{4}, 2.5\right)$
$4x + 4y = -15$

44. $3x - 7y = 20$ $(-5, -5)$
$-11x + 10y = 5$

45. $x - y = 17$ $(20, 3)$
$\frac{1}{2}x - 3y = 1$

46. $4x + 9y = -10$ $(2, -2)$
$-8x - 12y = 8$

47. $12x + 3y = 16$
$-36x - 9y = 32$
no solution

48. $-x + 5y = 17$
$2x - 10y = -34$
infinitely many solutions

49. $\frac{1}{3}x + y = 9$ $(9, 6)$
$-2x + 2y = -6$

50. *Writing* Explain how you can tell whether the system has infinitely many solutions or no solution without trying to solve the system. **50a–b. See margin.**

a. $5x - 2y = 6$
$-10x + 4y = -12$

b. $-2x + y = 8$
$-6x + 3y = 12$

GEOMETRY CONNECTION Find the coordinates of the point where the diagonals of the quadrilateral intersect.

51. $(2, 3)$
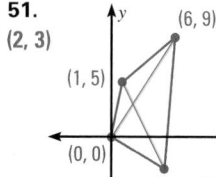

52. $(3, 3)$
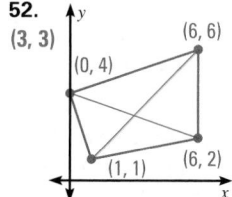

53. $(2, 2)$
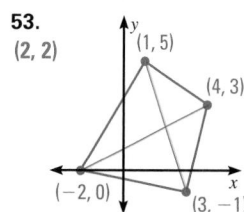

54. **BREAKING EVEN** You are starting a business selling boxes of hand-painted greeting cards. To get started, you spend $36 on paint and paintbrushes that you need. You buy boxes of plain cards for $3.50 per box, paint the cards, and then sell them for $5 per box. How many boxes must you sell for your earnings to equal your expenses? What will your earnings and expenses equal when you break even? **24; $120**

55. **HOME ELECTRONICS** To connect a VCR to a television set, you need a cable with special connectors at both ends. Suppose you buy a 6 foot cable for $15.50 and a 3 foot cable for $10.25. Assuming that the cost of a cable is the sum of the cost of the two connectors and the cost of the cable itself, what would you expect to pay for a 4 foot cable? Explain how you got your answer.

56. **SCIENCE CONNECTION** Weights of atoms and molecules are measured in *atomic mass units* (u). A molecule of C_2H_6 (ethane) is made up of 2 carbon atoms and 6 hydrogen atoms and weighs 30.07 u. A molecule of C_3H_8 (propane) is made up of 3 carbon atoms and 8 hydrogen atoms and weighs 44.097 u. Find the weights of a carbon atom and a hydrogen atom.

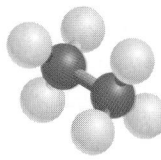

Ethane molecule

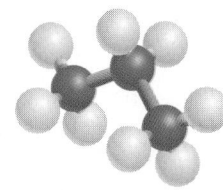

Propane molecule

A carbon atom is about 12.011 u, a hydrogen atom about 1.008 u.

3.2 *Solving Linear Systems Algebraically* **153**

50.a. *Sample answer:* the second equation is equal to -2 times the first, so there are infintely many solutions.

50.b. *Sample answer:* the left-hand side of the second equation is equal to 3 times the left-hand side of the first, but $12 \neq 3 \cdot 8$, and so there are no solutions.

STUDENT HELP

HOMEWORK HELP
Visit our Web site
www.mcdougallittell.com
for help with Exs. 51–53.

55. $12; *Sample answer:* let $x = $ the cost per foot of the cable itself and $y = $ the cost of one connector. Then $6x + 2y = 15.5$ and $3x + 2y = 10.25$. Subtracting the second equation from the first, find $x = 1.75$. Then a 4-foot cable with connectors will cost $10.25 + 1.75 = $12.

! COMMON ERROR
EXERCISES 5–50 Computational errors are very common in these types of problems. Stress to students that they should test the solution in both equations to help avoid mistakes.

STUDENT HELP NOTES
Homework Help Students can find help for Exs. 51–53 at **www.mcdougallittell.com.** The information can be printed out for students who don't have access to the Internet.

MULTIPLE REPRESENTATIONS
EXERCISES 11–49 Systems of equations can be solved graphically, the focus of Lesson 3.1 or algebraically, the focus of this lesson. To help students see the relationship between the graphical and algebraic representations of linear systems, have them solve some of these exercises graphically. This will be of particular importance in helping students see graphically what it means for a system to have no solutions (parallel lines) or an infinite number of solutions (the same line).

DATA UPDATE
Updated data for Exs. 59–62 is available at **www.mcdougallittell.com**.

GRAPHING CALCULATOR NOTE
EXERCISES 59–62 A graphing calculator is used to find the equations and intersection point of two lines of best fit. This exercise can be done by hand by estimating the lines of best fit. If done by hand, additional time will need to be allotted for these exercises.

59.

Olympic Times for Men's 100 m Freestyle

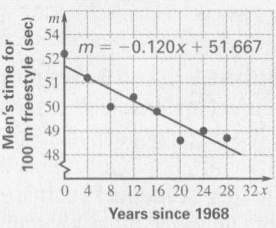

$m = -0.120x + 51.667$

Years since 1968

Olympic Times for Women's 100 m Freestyle

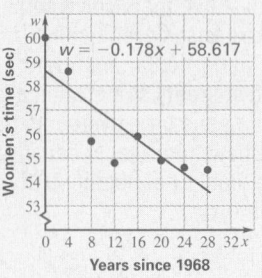

$w = -0.178x + 58.617$

Years since 1968

75–80. See Additional Answers beginning on page AA1.

ADDITIONAL PRACTICE AND RETEACHING

For Lesson 3.2:

• Practice Levels A, B, and C (*Chapter 3 Resource Book,* p. 28)

• Reteaching with Practice (*Chapter 3 Resource Book,* p. 31)

• See Lesson 3.2 of the *Personal Student Tutor*

For more Mixed Review:

• Search the *Test and Practice Generator* for key words or specific lessons.

154

SWIMMING
One way swimmers improve their racing times is by training at high altitudes. Many elite swimmers train at the Olympic Training Center in Colorado Springs, Colorado, at an altitude of 6035 feet above sea level.

61. (119.83, 37.288); 120 years after 1968, in the year 2088 summer olympics, the men's and women's times in the 100 m freestyle will both be about 37.3 sec.

Test Preparation

62. *Sample answer:* Athletic performance cannot be expected to improve at the same linear rate indefinitely. First of all, there is some limit to what the human body can accomplish, and eventually the graph of performance times would tend to become tangent to the horizontal line at this value. Further, a line would have an *x*-intercept, implying some future time at which the race would be won in 0 sec, which is impossible.

★ Challenge

57. CROSS-TRAINING You want to burn 380 Calories during 40 minutes of exercise. You burn about 8 Calories per minute inline skating and 12 Calories per minute swimming. How long should you spend doing each activity?
25 min skating and 15 min swimming

58. RENTING AN APARTMENT Two friends rent an apartment for $975 per month. Since one bedroom is 60 square feet larger than the other bedroom, each person's rent contribution is based on bedroom size. Each person agrees to pay $3.25 per square foot of bedroom area. Let *x* be the area (in square feet) of the larger bedroom, and let *y* be the area (in square feet) of the smaller bedroom. Write and solve a system of linear equations to find the area of each bedroom.
$x = y + 60$, $3.25x + 3.25y = 975$; the smaller bedroom is 120 ft^2, the larger one is 180 ft^2

SWIMMING In Exercises 59–62, use the table below of winning times in the Olympic 100 meter freestyle swimming event for the period 1968–1996.

Years since 1968, x	0	4	8	12	16	20	24	28
Men's time (sec), m	52.2	51.2	50.0	50.4	49.8	48.6	49.0	48.7
Women's time (sec), w	60.0	58.6	55.7	54.8	55.9	54.9	54.6	54.5

DATA UPDATE of USA Swimming data at www.mcdougallittell.com

59. Use a graphing calculator to make scatter plots of the data pairs (x, m) and (x, w). **See margin.**

60. For each scatter plot, find an equation of the line of best fit. Graph the equations, as shown.
$m = -0.120x + 51.667$; $w = -0.178x + 58.617$

61. Find the coordinates of the intersection point of the lines. Describe what this point represents.
61.–62. See margin.

62. CRITICAL THINKING Why might a linear model not be appropriate for projecting winning times far into the future?

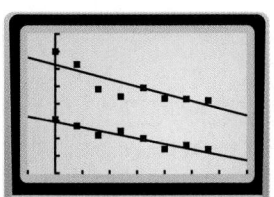

QUANTITATIVE COMPARISON In Exercises 63 and 64, choose the statement that is true about the given quantities.

Ⓐ The quantity in column A is greater.

Ⓑ The quantity in column B is greater.

Ⓒ The two quantities are equal.

Ⓓ The relationship cannot be determined from the given information.

Column A	Column B	
63. The *x*-coordinate of the solution of: $7x - y = 19$ $10x + 2y = 34$	3	C
64. −5	The *y*-coordinate of the solution of: $-2x + 6y = -26$ $x + 3y = 11$	B

65. CRITICAL THINKING Find values of *r*, *s*, and *t* that produce the solution(s).

$$-3x - 5y = 9$$
$$rx + sy = t$$

Sample answer:
$r = 6$, $s = 10$, $t = 5$
a. no solution

Sample answer: $r = 6$, $s = 10$, $t = -18$
b. infinitely many solutions

Sample answer:
$r = 2$, $s = 1$, $t = 1$
c. a solution of $(2, -3)$

MIXED REVIEW

ABSOLUTE VALUE EQUATIONS Solve the equation. **(Review 1.7)**

66. $|6x| = 12$ −2, 2   **67.** $|x + 5| = 3$ −8, −2   **68.** $|2x - 1| = 7$ −3, 4

69. $|4x + 1| = 5$ $-\frac{3}{2}$, 1   **70.** $|3x - 2| = 8$ −2, $\frac{10}{3}$ **71.** $|-x + 10| = 14$ 24, −4

WRITING EQUATIONS Write an equation of the line. **(Review 2.4)**

72. $y = -\frac{5}{3}x - \frac{1}{3}$

72. 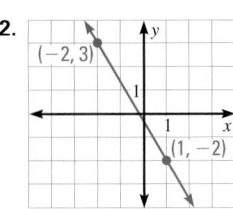   **73.**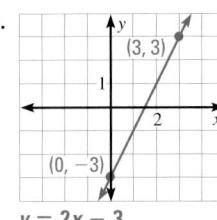
$y = 2x - 3$
74.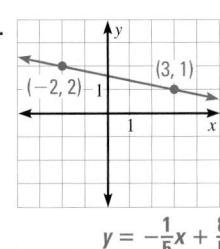
$y = -\frac{1}{5}x + \frac{8}{5}$

GRAPHING INEQUALITIES Graph the inequality in a coordinate plane.
(Review 2.6 for 3.3) 75–80. See margin.

75. $y < 4$ **76.** $x \geq -2$ **77.** $3x - y \geq 0$

78. $y < -x + 4$ **79.** $4x - y < 5$ **80.** $y \geq -2x - 1$

81.  **CONSUMER ECONOMICS** You plan to buy a pair of jeans for $25 and some T-shirts for $12 each. You have only $60 to spend. Write and solve an inequality for the number of T-shirts you can buy. **(Review 1.6 for 3.3)**
$12x + 25 \leq 60$; $x \leq \frac{35}{12}$

QUIZ 1

Self-Test for Lessons 3.1 and 3.2

Use a graph to solve the system. **(Lesson 3.1)**

1. $y = 2x + 5$ (−2, 1)
$y = -2x - 3$

2. $y = -4x + 1$ (1, −3)
$y = x - 4$

3. $-3x + 2y = 4$ no
$6x - 4y = 14$ solutions

4. $-2x - y = -2$ $\left(\frac{7}{3}, -\frac{8}{3}\right)$
$3x - 3y = 15$

5. $y = -x + 5$ (1, 4)
$3x - y = -1$

6. $4x + 5y = -9$ (−1, −1)
$x + 3y = -4$

Tell how many solutions the linear system has. **(Lessons 3.1 and 3.2)**

7. $6x + 6y = 3$ infinitely
$4x + 4y = 2$ many solutions

8. $-2x + y = 13$ 1
$x - 4y = -31$

9. $-5x + 7y = 10$
$15x - 21y = 22$
no solutions

10. $3x - 3y = 3$ 1
$-4x + y = -21$

11. $x - 6y = 6$ 1
$-3x + 2y = -2$

12. $-4x + 8y = 24$
$-x + 2y = 6$
infinitely many solutions

Solve the system using any algebraic method. **(Lesson 3.2)**

13. $-2x + 2y = -5$ $\left(-\frac{5}{4}, -\frac{15}{4}\right)$
$x + y = -5$

14. $-3x + 2y = -6$ (6, 6)
$5x - 2y = 18$

15. $-4x - y = -1$
$12x + 3y = 3$
infinitely many solutions

16. $-3x - 4y = -2$ $\left(-4, \frac{7}{2}\right)$
$x + 2y = 3$

17. $3x - 8y = 11$ no solution
$-6x + 16y = -5$

18. $3x - 8y = -7$ $\left(-\frac{33}{29}, \frac{13}{29}\right)$
$-5x - 6y = 3$

19.  **THEATER** Tickets for your school's play are $3 for students and $5 for non-students. On opening night 937 tickets are sold and $3943 is collected. How many tickets were sold to students? to non-students? **(Lesson 3.2)** 371; 566

---

4 ASSESS

DAILY HOMEWORK QUIZ

Transparency Available

1. Use substitution to solve.
$x + 5y = 33$
$4x + 3y = 13$ (−2, 7)

2. Use the linear combination method to solve. (5, −1)
$-2x + 3y = -13$
$6x + 2y = 28$

3. A tank was being filled with water at the rate of 1.5 gal/min. By 10 A.M., it contained 50 gal. Several minutes later, the water going into the tank was turned off, the plug was pulled, and the tank drained at the rate of 2.3 gal/min. The tank was empty at 11:08 A.M. How many minutes did it take to drain the tank? 40 min

EXTRA CHALLENGE NOTE

Challenge problems for Lesson 3.2 are available in **blackline** format in the *Chapter 3 Resource Book*, p. 35 and at **www.mcdougallittell.com.**

ADDITIONAL TEST PREPARATION

1. WRITING Use the linear combination method to solve the system below. Explain each step.
$3x - 5y = 12$
$7x + 10y = -37$
(−1, −3); check explanations.

2. OPEN ENDED Write a system of equations that you would solve using substitution and one that you would solve using the linear combination method. Explain why you chose each system. Check work.

ADDITIONAL RESOURCES

An alternative quiz for Lessons 3.1 and 3.2 is available in the *Chapter 3 Resource Book*, p. 36.

LESSON OPENER
ACTIVITY
An alternative way to approach Lesson 3.3 is to use the Activity Lesson Opener:
- Blackline Master (*Chapter 3 Resource Book*, p. 40)
- Transparency (p. 18)

MEETING INDIVIDUAL NEEDS
- *Chapter 3 Resource Book*
  Prerequisite Skills Review (p. 5)
  Practice Level A (p. 42)
  Practice Level B (p. 43)
  Practice Level C (p. 44)
  Reteaching with Practice (p. 45)
  Absent Student Catch-Up (p. 47)
  Challenge (p. 50)
- *Resources in Spanish*
- Personal Student Tutor

NEW-TEACHER SUPPORT
See the Tips for New Teachers on pp. 1–2 of the *Chapter 3 Resource Book* for additional notes about Lesson 3.3.

WARM-UP EXERCISES

Transparency Available

1. The admission to a carnival is $5.00. Each ride is $.40. You can spend no more than $12. Write and solve an inequality to find the number of rides you can go on.
   $5 + 0.4x \le 12$; $x \le 17.5$; you can go on up to 17 rides
2. Graph $2x - y > 2$ in the coordinate plane.

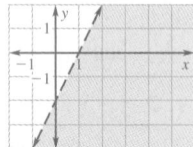

3.3 Graphing and Solving Systems of Linear Inequalities

What you should learn

GOAL 1 Graph a system of linear inequalities to find the solutions of the system.

GOAL 2 Use systems of linear inequalities to solve **real-life** problems, such as finding a person's target heart rate zone in **Example 3**.

Why you should learn it

▼ To solve **real-life** problems, such as finding out how a moose can satisfy its nutritional requirements in **Ex. 58**.

GOAL 1 GRAPHING A SYSTEM OF INEQUALITIES

The following is a **system of linear inequalities** in two variables.

$$x + y \le 6 \qquad \text{Inequality 1}$$
$$2x - y > 4 \qquad \text{Inequality 2}$$

A **solution** of a system of linear inequalities is an ordered pair that is a solution of each inequality in the system. For example, $(3, -1)$ is a solution of the system above. The **graph** of a system of linear inequalities is the graph of all solutions of the system.

● ACTIVITY
Developing Concepts

Investigating Graphs of Systems of Inequalities

The coordinate plane shows the four regions determined by the lines $3x - y = 2$ and $2x + y = 1$. Use the labeled points to help you match each region with one of the systems of inequalities.

a. $3x - y \le 2$
   $2x + y \le 1$
   Region 1

b. $3x - y \ge 2$
   $2x + y \ge 1$
   Region 3

c. $3x - y \ge 2$
   $2x + y \le 1$
   Region 4

d. $3x - y \le 2$
   $2x + y \ge 1$
   Region 2

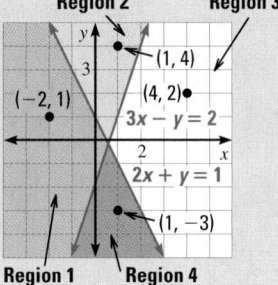

As you saw in the activity, a system of linear inequalities defines a region in a plane. Here is a method for graphing the region.

GRAPHING A SYSTEM OF LINEAR INEQUALITIES

To graph a system of linear inequalities, do the following for each inequality in the system:

- Graph the line that corresponds to the inequality. Use a dashed line for an inequality with $<$ or $>$ and a solid line for an inequality with $\le$ or $\ge$.

- Lightly shade the half-plane that is the graph of the inequality. Colored pencils may help you distinguish the different half-planes.

The graph of the system is the region common to all of the half-planes. If you used colored pencils, it is the region that has been shaded with *every* color.

EXAMPLE 1 · Graphing a System of Two Inequalities

STUDENT HELP

Look Back
For help with graphing a linear inequality, see p. 109.

Graph the system.

$$y \geq -3x - 1 \quad \text{Inequality 1}$$
$$y < x + 2 \quad \text{Inequality 2}$$

SOLUTION

Begin by graphing each linear inequality. Use a different color for each half-plane. For instance, you can use red for Inequality 1 and blue for Inequality 2. The graph of the system is the region that is shaded purple.

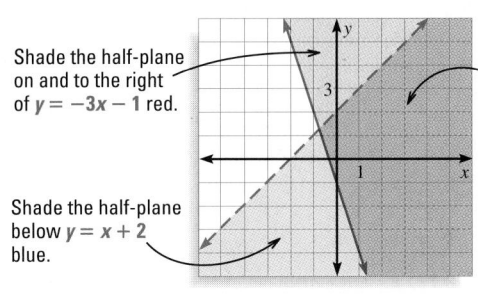

Shade the half-plane on and to the right of $y = -3x - 1$ red.

Shade the half-plane below $y = x + 2$ blue.

The graph of the system is the intersection of the red and blue regions.

· · · · · · · · · ·

You can also graph a system of three or more linear inequalities.

EXAMPLE 2 · Graphing a System of Three Inequalities

Graph the system.

$$x \geq 0 \quad \text{Inequality 1}$$
$$y \geq 0 \quad \text{Inequality 2}$$
$$4x + 3y \leq 24 \quad \text{Inequality 3}$$

SOLUTION

Inequality 1 and Inequality 2 restrict the solutions to the first quadrant. Inequality 3 is the half-plane that lies on and below the line $4x + 3y = 24$. The graph of the system of inequalities is the triangular region shown below.

STUDENT HELP

Study Tip
From this point on, only the solution region will be shaded on graphs of systems of linear inequalities.

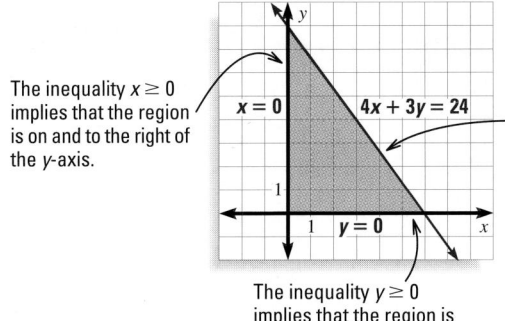

The inequality $x \geq 0$ implies that the region is on and to the right of the y-axis.

The inequality $4x + 3y \leq 24$ implies that the region is on and below the line $4x + 3y = 24$.

The inequality $y \geq 0$ implies that the region is on and above the x-axis.

3.3 *Graphing and Solving Systems of Linear Inequalities* **157**

---

2 TEACH

MOTIVATING THE LESSON
In real-life applications, variables can often have a range of acceptable values. For example, a college class may have between 20 and 140 students registered. The attendance is each class may also fluctuate. As you will see in today's lesson, situations such as this can be represented using systems of linear inequalities.

ACTIVITY NOTE
This activity will help students understand how the solution for a system of linear inequalities is represented in its graph.

EXTRA EXAMPLE 1
Graph the system.
$$x - 2y \leq 3$$
$$y > 3x - 4$$

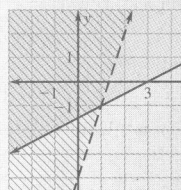

EXTRA EXAMPLE 2
Graph the system.
$$x \leq 0$$
$$y \geq 0$$
$$x - y \geq -2$$

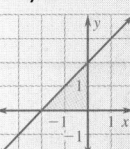

✔ CHECKPOINT EXERCISES
For use after Examples 1 and 2:
1. Graph the system.
$$x \geq 0$$
$$y > 2x - 1$$
$$y \leq 2x + 3$$

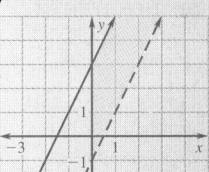

EXTRA EXAMPLE 3

At one college each class has between 20 and 140 students. From past data on attendance, it is expected that anywhere from 75% to 95% of students attend class on any one day.

a. Write and graph a system of linear inequalities that describes the information.

$x \geq 20$, $x \leq 140$, $y \geq 0.75x$, $y \leq 0.95x$

b. A class has 120 students registered in a classroom that seats 110 students. Will there be enough chairs? **yes**

CHECKPOINT EXERCISES

For use after Example 3:

1. If fewer than 50 people buy a ticket, a flight will be canceled. A maximum of 220 tickets are sold. Data have shown that from 70% to 90% of ticket holders take a flight.

a. Write and graph a system of linear inequalities that describes the information.

$x \geq 50$, $x \leq 220$, $y \leq 0.9x$, $y \geq 0.7x$

b. The plane needs 1 attendant for every 40 people. If 143 people are ticketed, will 3 attendants be enough? **no**

CLOSURE QUESTION

What is the procedure used to graph a system of linear inequalities?

Graph each inequality. The points common to all the individual graphs represent the graph of the system.

FOCUS ON CAREERS

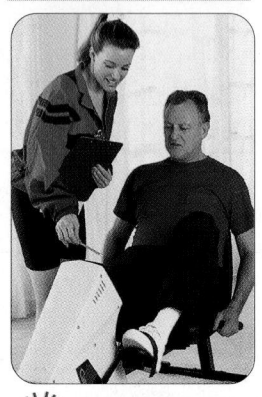

PERSONAL TRAINER

A personal trainer can help you assess your fitness level and set exercise goals. As described in Example 3, one way to do this is by monitoring your heart rate.

CAREER LINK
www.mcdougallittell.com

GOAL 2 USING SYSTEMS OF INEQUALITIES IN REAL LIFE

You can use a system of linear inequalities to describe a real-life situation, as shown in the following example.

EXAMPLE 3 *Writing and Using a System of Inequalities*

HEART RATE A person's theoretical maximum heart rate is $220 - x$ where x is the person's age in years ($20 \leq x \leq 65$). When a person exercises, it is recommended that the person strive for a heart rate that is at least 70% of the maximum and at most 85% of the maximum.

a. You are making a poster for health class. Write and graph a system of linear inequalities that describes the information given above.

b. A 40-year-old person has a heart rate of 150 (heartbeats per minute) when exercising. Is the person's heart rate in the target zone?

SOLUTION

a. Let y represent the person's heart rate. From the given information, you can write the following four inequalities.

$x \geq 20$	Person's age must be at least 20.
$x \leq 65$	Person's age can be at most 65.
$y \geq 0.7(220 - x)$	Target rate is at least 70% of maximum rate.
$y \leq 0.85(220 - x)$	Target rate is at most 85% of maximum rate.

The graph of the system is shown below.

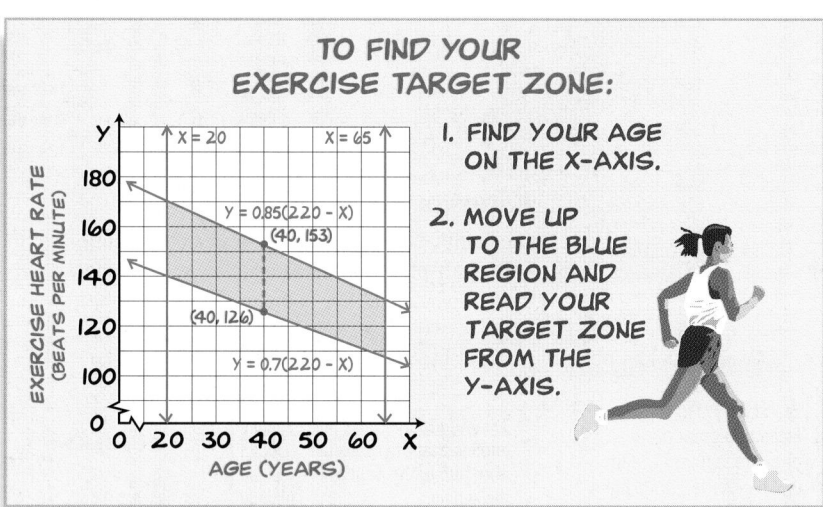

b. From the graph you can see that the target zone for a 40-year-old person is between 126 and 153, inclusive. That is,

$$126 \leq y \leq 153.$$

▶ A 40-year-old person who has a heart rate of 150 is within the target zone.

GUIDED PRACTICE

Vocabulary Check ✓

1. What must be true in order for an ordered pair to be a solution of a system of linear inequalities? **It must satisfy every inequality in the system.**

Concept Check ✓

2. Look back at Example 1 on page 157. Explain why the ordered pair $(-1, -5)$ is *not* a solution of the system. **because it does not satisfy Inequality 1: $-5 \ngeq 2$**

3. The line $y = 3$ should be solid, and the region above the line $x + y = 5$ should be shaded, not the region below.

3. **ERROR ANALYSIS** Explain what is wrong with the graph of the following system of inequalities.

$$y \leq 3$$
$$x + y \geq 5$$

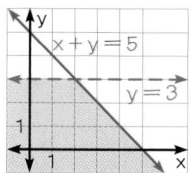

Skill Check ✓

Tell whether the ordered pair is a solution of the following system.

$$x \geq -1$$
$$y > 2x + 2$$

4. $(-1, 2)$ **yes**   **5.** $(0, 0)$ **no**   **6.** $(1, 4)$ **no**   **7.** $(2, 7)$ **yes**

Graph the system of linear inequalities. **8–10. See margin.**

8. $x \geq -1$
$y > 2x + 2$

9. $x + y \leq 3$
$y > 1$

10. $x > 0$
$y \leq x - 5$

11. 🌐 **FLIGHT ATTENDANTS** To be a flight attendant, you must be at least 18 years old and at most 55 years old, and you must be between 60 and 74 inches tall, inclusive. Let x represent a person's age (in years) and let y represent a person's height (in inches). Write and graph a system of linear inequalities showing the possible ages and heights for flight attendants. **See margin.**

PRACTICE AND APPLICATIONS

STUDENT HELP

▶ **Extra Practice**
to help you master skills is on p. 943.

CHECKING A SOLUTION Tell whether the ordered pair is a solution of the system.

12. $(25, -5)$ **yes**

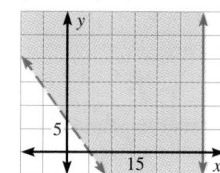

13. $(2, 3)$ **no**

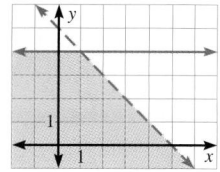

14. $(2, 6)$ **yes**

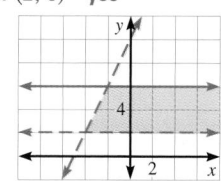

STUDENT HELP

▶ **HOMEWORK HELP**
Example 1: Exs. 12, 13, 15–17, 21, 22, 27–38
Example 2: Exs. 14, 18–20, 23–26, 39–50
Example 3: Exs. 51–58

FINDING A SOLUTION Give an ordered pair that is a solution of the system. **15–20. Sample answers are given.**

15. $x - y \geq 3$
$y < 15$ **(13, 10)**

16. $x + y < 6$
$x \geq -2$ **(0, 0)**

17. $4x > y$
$x \leq 12$ **(-2, -10)**

18. $x \geq -7$
$y < 10$
$x < y$ **(0, 1)**

19. $y > -5$
$x > 3$
$2x + y < 13$ **(4, 2)**

20. $y \geq -x$
$y \geq 0$
$x < 0$ **(-3, 5)**

○ **ASSIGNMENT GUIDE**

BASIC
Day 1: pp. 159–162 Exs. 12–14, 16–32 even, 40–46 even, 51, 60–61, 67–77 odd

AVERAGE
Day 1: pp. 159–162 Exs. 12–46 even, 51–54, 59–61, 67–77 odd

ADVANCED
Day 1: pp. 159–162 Exs. 12–50 even, 51–55, 59–65, 67–77 odd

BLOCK SCHEDULE
pp. 159–162 Exs. 12–46 even, 51–54, 59–61, 67–77 odd (with 3.4)

EXERCISE LEVELS
Level A: *Easier*
12–38, 51, 60–61
Level B: *More Difficult*
39–50, 52–59, 62
Level C: *Most Difficult*
63–65

✔ **HOMEWORK CHECK**
To quickly check student understanding of key concepts, go over the following exercises: Exs. 12, 16, 24, 30, 40. See also the Daily Homework Quiz:

• Blackline Master (*Chapter 3 Resource Book*, p. 53)
• 🖨 Transparency (p. 22)

8.

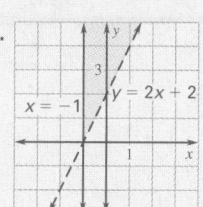

9–11. See Additional Answers beginning on page AA1.

! **COMMON ERROR**

EXERCISES 27–51, 53, 54, 56–58 Students will often shade the wrong half-plane when graphing an inequality. Have students test a point in the solution set in each of the inequalities of the system to check their answers. Also, have them double check that dashed and solid lines have been used correctly.

STUDENT HELP NOTES

→ **Homework Help** Students can find help for Exs. 21–26 at **www.mcdougallittell.com.** The information can be printed out for students who don't have access to the Internet.

27.

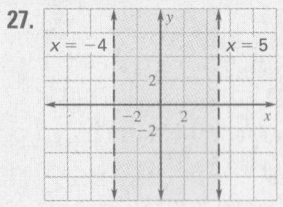

28.

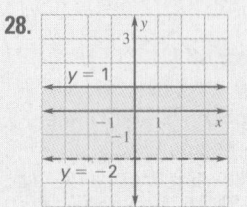

29–50. See Additional Answers beginning on page AA1.

STUDENT HELP

HOMEWORK HELP
Visit our Web site www.mcdougallittell.com for help with Exs. 21–26.

MATCHING SYSTEMS AND GRAPHS Match the system of linear inequalities with its graph.

21. $y \leq 4$ **C**
$x > -2$

22. $y > -4$ **B**
$x > -2$

23. $y > x$ **F**
$x > -3$
$y \geq 0$

24. $y > x$ **E**
$y > -3$
$x \leq 0$

25. $x \leq 3$ **A**
$y > 1$
$y \geq -x + 1$

26. $y > -1$ **D**
$x \geq -1$
$y \geq -x + 1$

A.
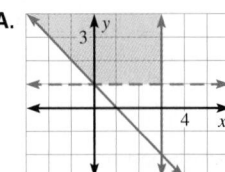

B.
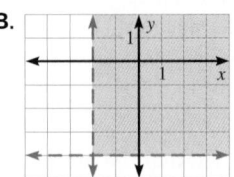

C.
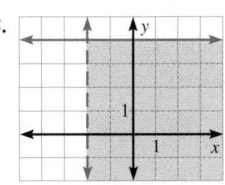

D.
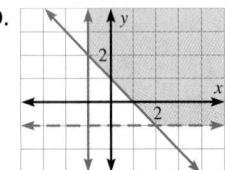

E.
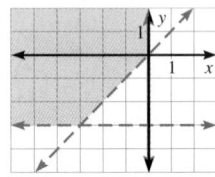

F.
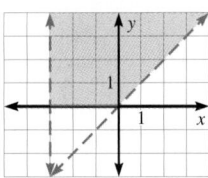

SYSTEMS OF TWO INEQUALITIES Graph the system of linear inequalities.
27–38. See margin for graphs.

27. $x < 5$
$x > -4$

28. $y > -2$
$y \leq 1$

29. $x \geq 0$
$x + y < 11$

30. $x + y \geq -2$
$-5x + y < -3$

31. $y \geq -4$
$y < -2x + 10$

32. $y > 2x - 7$
$4x + 4y < -12$

33. $y < x + 4$
$y \geq -2x + 1$

34. $x + y > -8$
$x + y \leq 6$

35. $y > -3x$
$x \leq 5y$

36. $x - y > 7$
$2x + y < 8$

37. $7x + y > 0$
$3x - 2y \leq 5$

38. $-x < y$
$x + 3y > 8$

SYSTEMS OF THREE OR MORE INEQUALITIES Graph the system of linear inequalities. 39–50. See margin for graphs.

39. $y < 4$
$x > -3$
$y > x$

40. $y \geq 1$
$x \leq 6$
$y < 2x - 5$

41. $2x - 3y > -6$
$5x - 3y < 3$
$x + 3y > -3$

42. $x - 4y > 0$
$x + y \leq 1$
$x + 3y > -1$

43. $2x + 1 \geq y$
$x < 5$
$y < x + 2$

44. $5x - 3y \leq 4$
$x + y < 8$
$y > 3$

45. $x \geq y - 2$
$x + y > 1$
$x < 10$

46. $y \geq 0$
$x - 4y < 2$
$y < x$

47. $x - y \geq 0$
$y < 2x$
$5x + 6y \geq 1$

48. $y \geq 0$
$x \leq 9$
$x + y < 15$
$y < x$

49. $x + y \leq 4$
$x + y \geq -1$
$x - y \geq -2$
$x - y \leq 2$

50. $y < 5$
$y > -6$
$2x + y \geq -1$
$y \leq x + 3$

51. 🌐 **POOL CHEMICALS** You are a lifeguard at a community pool, and you are in charge of maintaining the proper pH (amount of acidity) and chlorine levels. The water test-kit says that the pH level should be between 7.4 and 7.6 pH units and the chlorine level should be between 1.0 and 1.5 PPM (parts per million). Let p be the pH level and let c be the chlorine level (in PPM). Write and graph a system of inequalities for the pH and chlorine levels the water should have.
$7.4 \le p \le 7.6, 1.0 \le c \le 1.5$; **See margin.**

🔢 **HEALTH** **In Exercises 52–54, use the following information.**
For a healthy person who is 4 feet 10 inches tall, the recommended lower weight limit is about 91 pounds and increases by about 3.7 pounds for each additional inch of height. The recommended upper weight limit is about 119 pounds and increases by about 4.9 pounds for each additional inch of height.
▶ Source: Dietary Guidelines Advisory Committee

52. Let x be the number of inches by which a person's height exceeds 4 feet 10 inches and let y be the person's weight in pounds. Write a system of inequalities describing the possible values of x and y for a healthy person.
$y \ge 3.7x + 91; y \le 4.9x + 119$

53. Use a graphing calculator to graph the system of inequalities from Exercise 52.
See margin.

54. What is the recommended weight range for someone 6 feet tall?
between 142.8 lb and 187.6 lb

🌐 **SHOE SALE** **In Exercises 55 and 56, use the shoe store ad shown below.**

55. Let x be the regular footwear price and y be the discount price. Write a system of inequalities for the regular footwear prices and possible sale prices. $20 \le x \le 80$;
$0.75x \le y, y \le 0.9x$

56. Graph the system you wrote in Exercise 55. Use your graph to estimate the range of possible sale prices for shoes that are regularly priced at $65. **See margin for graph; $48.75 to $58.50**

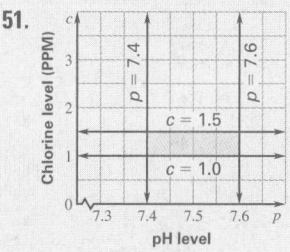
HUGE ONE-DAY SALE!
Save 10%–25% on all athletic footwear.
(Regular price: $20–$80)

57. 🌐 **WEIGHTLIFTING RECORDS** The men's world weightlifting records for the 105-kg-and-over weight category are shown in the table. The combined lift is the sum of the snatch lift and the clean and jerk lift. Let s be the weight lifted in the snatch and let j be the weight lifted in the clean and jerk. Write and graph a system of inequalities to describe the weights you could lift to break the records for both the snatch and combined lifts, but *not* the clean and jerk lift.

$s > 205.5$;
$j \le 262.5$;
$s + j > 465.0$;
See margin.

Men's +105 kg World Weightlifting Records		
Snatch	**Clean & Jerk**	**Combined**
205.5 kg	262.5 kg	465.0 kg

🔗 **DATA UPDATE** of International Weightlifting Federation data at www.mcdougallittell.com

58. **BIOLOGY** ▸ **CONNECTION** Each day, an average adult moose can process about 32 kilograms of terrestrial vegetation (twigs and leaves) and aquatic vegetation. From this food, it needs to obtain about 1.9 grams of sodium and 11,000 Calories of energy. Aquatic vegetation has about 0.15 gram of sodium per kilogram and about 193 Calories of energy per kilogram, while terrestrial vegetation has minimal sodium and about four times more energy than aquatic vegetation. Write and graph a system of inequalities describing the amounts t and a of terrestrial and aquatic vegetation, respectively, for the daily diet of an average adult moose.
▶ Source: *Biology by Numbers* $0.15a \ge 1.9$; $965t + 193a \ge 11,000$; $a + t \le 32$; **See margin for graph.**

3.3 *Graphing and Solving Systems of Linear Inequalities* **161**

STUDENT HELP

🌐 **KEYSTROKE HELP** Visit our Web site www.mcdougallittell.com to see keystrokes for several models of calculators.

FOCUS ON PEOPLE

🔗 **RONNY WELLER** of Germany set the world records for the snatch lift and combined lift in 1998. His records for these lifts are listed in the table for Ex. 57. Weller has won 3 Olympic medals: one gold, one silver, and one bronze.

GRAPHING CALCULATOR NOTE
EXERCISES 52–54 The graphing calculator is used in these exercises to help students graph equations with non-integer slopes. The use of the graphing calculator reduces the time needed to graph the system and increases the accuracy of the solution.

STUDENT HELP NOTES
→ **Keystroke Help** Keystrokes for several models of calculators are available in **blackline** master format in the *Chapter 3 Resource Book*, p. 41 and at **www.mcdougallittell.com.**

51.

Graph: Chlorine level (PPM) vs pH level, with $p = 7.4$, $p = 7.6$, $c = 1.5$, $c = 1.0$

53, 56–59. See Additional Answers beginning on page AA1.

ADDITIONAL PRACTICE AND RETEACHING

For Lesson 3.3:
• Practice Levels A, B, and C (*Chapter 3 Resource Book*, p. 42)
• Reteaching with Practice (*Chapter 3 Resource Book*, p. 45)
• 🖥 See Lesson 3.3 of the *Personal Student Tutor*

For more Mixed Review:
• 🖥 Search the *Test and Practice Generator* for key words or specific lessons.

DAILY HOMEWORK QUIZ

📖 **Transparency Available**

1. Name all the solutions of the system that have integer coordinates. **(2, 2) and (3, 2)**

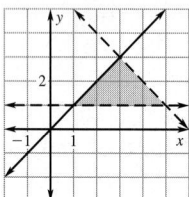

Graph each system.

2. $-x + 2y < 3$
$3x + y \leq 7$

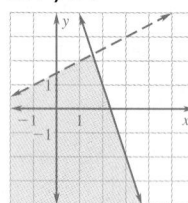

3. $y \geq -2$
$2x + 3y > 1$
$y \leq 2x + 4$

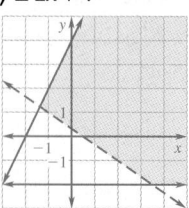

EXTRA CHALLENGE NOTE

→ Challenge problems for Lesson 3.3 are available in **blackline** format in the *Chapter 3 Resource Book*, p. 50 and at **www.mcdougallittell.com**.

ADDITIONAL TEST PREPARATION

1. OPEN ENDED Write a real-life problem that can be represented by a system of inequalities. Write the system of inequalities and graph its solution set. Check work.

162

---

59. CRITICAL THINKING Write a system of three linear inequalities that has no solution. Graph the system to show that it has no solution.
Sample answer: $x + y \leq 1$; $x + y \geq 4$; $x \geq 0$; See margin for graph.

Test  **Preparation**

60. MULTIPLE CHOICE Which system of inequalities is graphed at the right? **A**

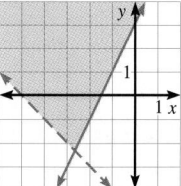

(A) $x + y > -5$
$-2x + y \geq 3$

(B) $x + y > -5$
$-2x + y < 3$

(C) $x + y > -5$
$-2x + y \leq 3$

(D) $x + y > -5$
$-2x + y > 3$

61. MULTIPLE CHOICE Which ordered pair is *not* a solution of the following system of inequalities? **C**

$$3x + 2y \geq -2$$
$$x - y < 3$$

(A) $(0, 0)$ **(B)** $(-1, 2)$ **(C)** $(4, 1)$ **(D)** $(2, 2)$

★ **Challenge**

62. $x \geq 1$; $x \leq 8$;
$y \geq -5$; $y \leq 3$

63. $y \leq x + 3$;
$y \leq 3$; $y \geq x - 6$; $y \geq -1$

64. $y \leq 4x + 27$;
$y \leq -x + 12$; $y \geq \frac{2}{3}x + 7$

EXTRA CHALLENGE
→ www.mcdougallittell.com

WRITING A SYSTEM Write a system of linear inequalities for the region.

62. 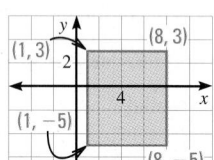   **63.** 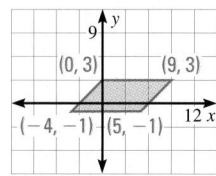   **64.**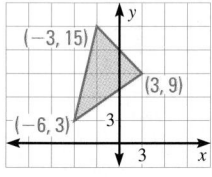

62–64. See margin.

65. VISUAL THINKING Write a system of linear inequalities whose graph is a pentagon and its interior. *Sample answer:* $x \geq 0$; $y \geq 0$; $y \leq 2$; $y \leq -x + 4$; $x \leq 3$

MIXED REVIEW

EVALUATING EXPRESSIONS Evaluate the expression for the given values of *x* and *y*. (Review 1.2 for 3.4)

66. $2x + 7y$ when $x = 5$ and $y = -3$ **−11**   **67.** $-4x - 3y$ when $x = -6$ and $y = -1$ **27**

68. $10x - 3y$ when $x = -4$ and $y = 2$ **−46**   **69.** $-y + 8x$ when $y = -3$ and $x = -2$ **−13**

DETERMINING CORRELATION Tell whether *x* and *y* have a *positive correlation*, a *negative correlation*, or *relatively no correlation*. (Review 2.5)

70. 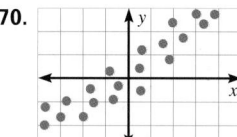   **71.**   **72.**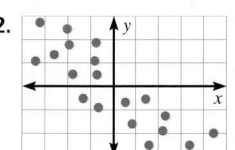

positive correlation   relatively no correlation   negative correlation

CHOOSING A METHOD Solve the system using any algebraic method. (Review 3.2)

73. $13x + 5y = 2$ $\left(\frac{58}{57}, -\frac{128}{57}\right)$ **74.** $-2x + 7y = 10$ **(9, 4)** **75.** $5x + 6y = -12$ **no**
$x - 4y = 10$ $x - 3y = -3$ $10x + 12y = 24$ **solution**

76. $-7x + 5y = 0$ $\left(\frac{5}{7}, 1\right)$ **77.** $-4x - 10y = 12$ **(−8, 2)** **78.** $6x - 8y = -18$ **infinitely**
$14x - 8y = 2$ $x + 5y = 2$ $-3x + 4y = 9$ **many solutions**

59. See Additional Answers beginning on page AA1.

Linear Programming

GOAL 1 USING LINEAR PROGRAMMING

What you should learn

GOAL 1 Solve linear programming problems.

GOAL 2 Use linear programming to solve **real-life** problems, such as purchasing file cabinets so as to maximize storage capacity in **Ex. 22.**

Why you should learn it

▼ To solve **real-life** problems, such as how a bicycle manufacturer can maximize profit in **Example 3.**

1. at O, C = 0;
   at P, C = 20;
   at R, C = 30;
   at S, C = 18;
   at T, C = 8;
   at U, C = 18;
   at V, C = 12

Many real-life problems involve a process called **optimization**, which means finding the maximum or minimum value of some quantity. In this lesson you will study one type of optimization process called *linear programming.*

Linear programming is the process of optimizing a linear **objective function** subject to a system of linear inequalities called **constraints**. The graph of the system of constraints is called the **feasible region.**

◯ ACTIVITY

Developing Concepts

Investigating Linear Programming

1 Evaluate the objective function $C = 2x + 4y$ for each labeled point in the feasible region at the right. **See margin.**

2 At which labeled point does the maximum value of C occur? At which labeled point does the minimum value of C occur? **R; O**

3 What are the maximum and minimum values of C on the entire feasible region? Try other points in the region to see if you can find values of C that are greater or lesser than those you found in **Step 2.** **30; 0; can't be done**

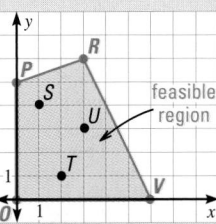

feasible region

Constraints:
$x \geq 0$
$y \geq 0$
$-x + 3y \leq 15$
$2x + y \leq 12$

In the activity you may have discovered that the optimal values of the objective function occurred at vertices of the feasible region.

OPTIMAL SOLUTION OF A LINEAR PROGRAMMING PROBLEM

If an objective function has a maximum or a minimum value, then it must occur at a vertex of the feasible region. Moreover, the objective function will have both a maximum and a minimum value if the feasible region is bounded.

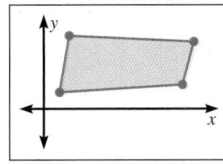

Bounded region

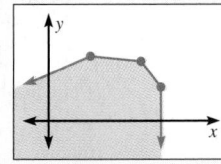

Unbounded region

3.4 *Linear Programming* 163

1 PLAN

PACING
Basic: 1 day
Average: 1 day
Advanced: 1 day
Block Schedule: 0.5 block with 3.3

LESSON OPENER
GRAPHING CALCULATOR
An alternative way to approach Lesson 3.4 is to use the Graphing Calculator Lesson Opener:
- Blackline Master (*Chapter 3 Resource Book*, p. 54)
- Transparency (p. 19)

MEETING INDIVIDUAL NEEDS
- *Chapter 3 Resource Book*
  Prerequisite Skills Review (p. 5)
  Practice Level A (p. 57)
  Practice Level B (p. 58)
  Practice Level C (p. 59)
  Reteaching with Practice (p. 60)
  Absent Student Catch-Up (p. 62)
  Challenge (p. 65)
- *Resources in Spanish*
- *Personal Student Tutor*

NEW-TEACHER SUPPORT
See the Tips for New Teachers on pp. 1–2 of the *Chapter 3 Resource Book* for additional notes about Lesson 3.4.

WARM-UP EXERCISES

Transparency Available

Solve each linear system.

1. $x = 2$
    $x + y = 5$ **(2, 3)**
2. $x - 2y = 5$
    $-x + y = -1$ **(−3, −4)**
3. $2x + 3y = 8$
    $x - 6y = -6$ $\left(2, \frac{4}{3}\right)$

MOTIVATING THE LESSON
In many situations you want to make the most of an opportunity. In today's lesson you will learn linear programming, which is a method for determining ways to do this.

ACTIVITY NOTE
Students discover that the maximum value and minimum value of an objective function occur at vertices of the feasible region.

 EXTRA EXAMPLE 1
Find the minimum value and the maximum value of $C = -x + 3y$ subject to the following constraints.
$x \geq 2$
$x \leq 5$
$y \geq 0$
$y \leq -2x + 12$ min: –5, max: 22

EXTRA EXAMPLE 2
Find the minimum value and the maximum value of $C = x + 5y$ subject to the following constraints.
$x \geq 0$
$y \leq 2x + 2$
$5 \geq x + y$ no minimum value, max: 21

 CHECKPOINT EXERCISES
For use after Examples 1 and 2:
1. Find the minimum value and the maximum value of $C = 2x - y$ subject to the following constraints.
$x \geq 0$
$y \geq x + 2$
$y \leq -x + 6$ min: –6, max: 0

MATHEMATICAL REASONING
Discuss whether the statements in the box at the bottom of page 163 are true for nonlinear objective functions. Students should be able to devise nonlinear objective functions that have neither a maximum nor a minimum on a given region.

164

EXAMPLE 1   *Solving a Linear Programming Problem*

Find the minimum value and the maximum value of

$C = 3x + 4y$ **Objective function**

subject to the following constraints.

$x \geq 0$
$y \geq 0$ **Constraints**
$x + y \leq 8$

SOLUTION

The feasible region determined by the constraints is shown. The three vertices are $(0, 0)$, $(8, 0)$, and $(0, 8)$. To find the minimum and maximum values of C, evaluate $C = 3x + 4y$ at each of the three vertices.

At $(0, 0)$: $C = 3(0) + 4(0) = 0$ ⟵ **Minimum**

At $(8, 0)$: $C = 3(8) + 4(0) = 24$

At $(0, 8)$: $C = 3(0) + 4(8) = 32$ ⟵ **Maximum**

The minimum value of C is 0. It occurs when $x = 0$ and $y = 0$. The maximum value of C is 32. It occurs when $x = 0$ and $y = 8$.

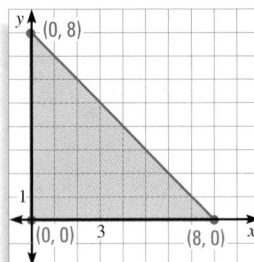

EXAMPLE 2   *A Region that is Unbounded*

Find the minimum value and the maximum value of

$C = 5x + 6y$ **Objective function**

subject to the following constraints.

$x \geq 0$
$y \geq 0$ **Constraints**
$x + y \geq 5$
$3x + 4y \geq 18$

┌─ **STUDENT HELP**
↳ **Study Tip**
You can find the coordinates of each vertex in the feasible region by solving systems of two linear equations. In Example 2 the vertex (2, 3) is the solution of this system:
$x + y = 5$
$3x + 4y = 18$

SOLUTION

The feasible region determined by the constraints is shown. The three vertices are $(0, 5)$, $(2, 3)$, and $(6, 0)$. First evaluate $C = 5x + 6y$ at each of the vertices.

At $(0, 5)$: $C = 5(0) + 6(5) = 30$

At $(2, 3)$: $C = 5(2) + 6(3) = 28$

At $(6, 0)$: $C = 5(6) + 6(0) = 30$

If you evaluate several other points in the feasible region, you will see that as the points get farther from the origin, the value of the objective function increases without bound. Therefore, the objective function has no maximum value. Since the value of the objective function is always at least 28, the minimum value is 28.

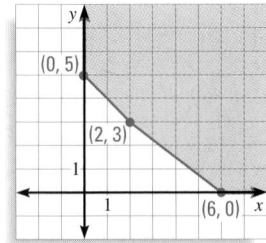

BICYCLES In China bicycles are a popular means of transportation. In 1999 China had an estimated 700–800 million bicycles.

GOAL 2 LINEAR PROGRAMMING IN REAL LIFE

EXAMPLE 3 *Using Linear Programming to Find the Maximum Profit*

BICYCLE MANUFACTURING Two manufacturing plants make the same kind of bicycle. The table gives the hours of general labor, machine time, and technical labor required to make one bicycle in each plant. For the two plants combined, the manufacturer can afford to use up to 4000 hours of general labor, up to 1500 hours of machine time, and up to 2300 hours of technical labor per week. Plant A earns a profit of $60 per bicycle and Plant B earns a profit of $50 per bicycle. How many bicycles per week should the manufacturer make in each plant to maximize profit?

Resource	Hours per bicycle in Plant A	Hours per bicycle in Plant B
General labor	10	1
Machine time	1	3
Technical labor	5	2

SOLUTION

Write an objective function. Let a and b represent the number of bicycles made in Plant A and Plant B, respectively. Because the manufacturer wants to maximize the profit P, the objective function is:

$$P = 60a + 50b$$

Write the constraints in terms of a and b. The constraints are given below and the feasible region determined by the constraints is shown at the right.

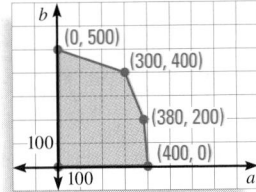

$10a + b \leq 4000$	**General labor: up to 4000 hours**
$a + 3b \leq 1500$	**Machine time: up to 1500 hours**
$5a + 2b \leq 2300$	**Technical labor: up to 2300 hours**
$a \geq 0$	**Cannot produce a negative amount**
$b \geq 0$	**Cannot produce a negative amount**

Calculate the profit at each vertex of the feasible region.

At (0, 500): $P = 60(0) + 50(500) = 25{,}000$

At (300, 400): $P = 60(300) + 50(400) = 38{,}000$ ◄——— **Maximum**

At (380, 200): $P = 60(380) + 50(200) = 32{,}800$

At (400, 0): $P = 60(400) + 50(0) = 24{,}000$

At (0, 0): $P = 60(0) + 50(0) = 0$

▶ The maximum profit is obtained by making 300 bicycles in Plant A and 400 bicycles in Plant B.

Focus on Vocabulary *Sample answer:*
The objective function is the function that you want to minimize or maximize. The constraints are the linear inequalities that determine the feasible region.

Closure Question *Sample answer:*
The maximum value of the objective function will occur at a vertex of the feasible region and so will the minimum value.

📋 **EXTRA EXAMPLE 3**
A furniture manufacturer makes chairs and sofas from prepackaged parts. The table gives the number of packages of wood parts, stuffing, and material required for each chair or sofa. The packages are delivered weekly and the manufacturer has room to store 1300 packages of wood parts, 2000 packages of stuffing, and 800 packages of fabric. The manufacturer earns $200 per chair and $350 per sofa. How many chairs and sofas should they make each week to maximize profit?

Material	Chair	Sofa
wood	2 boxes	3 boxes
stuffing	4 boxes	3 boxes
fabric	1 box	2 boxes

200 chairs and 300 sofas

✔ **CHECKPOINT EXERCISES**
For use after Example 3:
1. Refer to Extra Example 3. The factory leases another warehouse to hold an additional 400 packages of stuffing. Now how many chairs and sofas should they make each week to maximize profit? They should still make 200 chairs and 300 sofas.

FOCUS ON VOCABULARY
Define the terms *objective function* and *constraints* for linear programming problems. See below for sample answer.

CLOSURE QUESTION
Why do you need to find the vertices of the feasible region when using linear programming?
See below for sample answer.

DAILY PUZZLER
In a linear programming problem, the feasible region is a square in the first quadrant with sides of length a and two sides along the axes. What is the maximum value of the objective function $C = 2x + 3y$ in terms of a? $5a$

ASSIGNMENT GUIDE

BASIC
Day 1: pp. 166–169 Exs. 9–17 odd,
25–26, 29–47 odd,
Quiz 2 Exs. 1–7

AVERAGE
Day 1: pp. 166–169 Exs. 9–23 odd,
25–26, 29–47 odd,
Quiz 2 Exs. 1–7

ADVANCED
Day 1: pp. 166–169 Exs. 9–23 odd,
25–26, 29–47 odd,
Quiz 2 Exs. 1–7

BLOCK SCHEDULE
pp. 166–169 Exs. 9–23 odd,
25–26, 29–47 odd,
Quiz 2 Exs. 1–7 (with 3.3)

EXERCISE LEVELS

Level A: *Easier*
9–14

Level B: *More Difficult*
15–26

Level C: *Most Difficult*
27

✔ **HOMEWORK CHECK**
To quickly check student under-
standing of key concepts, go
over the following exercises:
Exs. 9, 13, 21. See also the Daily
Homework Quiz:
- Blackline Master (*Chapter 3
Resource Book*, p. 69)
- Transparency (p. 23)

GUIDED PRACTICE

Vocabulary Check ✔

Concept Check ✔

1. Define linear programming.   **1–3. See margin.**

2. How is the objective function used in a linear programming problem? How is the
system of constraints used?

3. In a linear programming problem, which ordered pairs should be tested to find a
minimum or maximum value?

Skill Check ✔

1. Linear programming is the
process of optimizing a linear
objective function subject to a
set of linear inequalities
known as constraints.

2. The value of the objective
function is tested at various
points of the feasible region
to determine where it is a
maximum and/or minimum;
the system of constraints
defines the feasible region.

3. If the feasible region is
bounded, the maximum and
minimum values of the
objective function must occur
at a vertex, so the value of the
objective function is checked
at each vertex.

In Exercises 4 and 5, use the feasible region at the right.

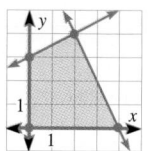

4. What are the vertices of the feasible region?
(0, 0); (0, 3); (2, 4); (4, 0)

5. What are the minimum and maximum values of the
objective function $C = 5x + 7y$?   minimum is 0;
maximum is 38

**Find the minimum and maximum values of the objective function subject to
the given constraints.**

6. Objective function: $C = x + y$; **Constraints:** $y \leq 5, y \geq 0, y - 2x \geq 0$
max of 7.5 at (2.5, 5); no minimum—feasible region is unbounded
7. Objective function: $C = 2x - y$; **Constraints:** $x \geq 0, x + y \leq 20, y \geq 3$
max of 31 at (17, 3); min of −20 at (0, 20)
8. 🌎 **PLANNING A FUNDRAISER** Your club plans to raise money by selling two
sizes of fruit baskets. The plan is to buy small baskets for $10 and sell them for
$16 and to buy large baskets for $15 and sell them for $25. The club president
estimates that you will not sell more than 100 baskets. Your club can afford to
spend up to $1200 to buy the baskets. Find the number of small and large fruit
baskets you should buy in order to maximize profit.   **Buy 80 large baskets and sell
them for a profit of $800.**

PRACTICE AND APPLICATIONS

STUDENT HELP

▶ **Extra Practice**
to help you master
skills is on p. 943.

9. min of −40 at (0, 40);
max of 40 at (40, 0)

10. min of −34 at (−2, −6);
max of 27 at (1, 5)

11. min of 10 at (2, 1); no
max—feasible region is
unbounded

STUDENT HELP

▶ **HOMEWORK HELP**
Examples 1, 2: Exs. 9–20
Example 3: Exs. 21–24

CHECKING VERTICES **Find the minimum and maximum values of
the objective function for the given feasible region.**   **9–11. See margin.**

9. $C = x - y$

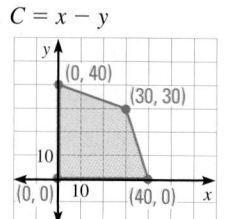

10. $C = 2x + 5y$

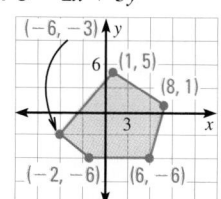

11. $C = 4x + 2y$

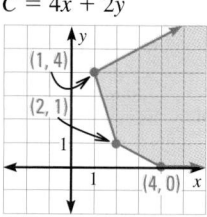

FINDING VALUES **In Exercises 12–20, find the minimum and maximum values
of the objective function subject to the given constraints.**

12. Objective function:
$C = 2x + 3y$

Constraints: min of 0 at
$x \geq 0$ (0, 0);
$y \geq 0$ max of 27
$x + y \leq 9$ at (0, 9)

13. Objective function:
$C = x + 4y$

Constraints: min of 6 at
$x \geq 2$ (2, 1); max
$x \leq 5$ of 29 at
$y \geq 1$ (5, 6)
$y \leq 6$

14. Objective function:
$C = 2x + y$

Constraints: min of
$x \geq -5$ −12 at
$x \leq 0$ (−5, −2);
$y \geq -2$ max of 2
$y \leq 2$ at (0, 2)

TOM FIRST AND TOM SCOTT started a business in 1990, mixing juice in a blender and selling it from their boat. Like the problem described in Ex. 21, they had to figure out the best way of using available resources so as to maximize profit.

21. Make 37.5 gallons of Orangeade and 31.25 gallons of Berry−fruity for a profit of $31.25.

22. Purchase 2 of type A and 9 of type B for a total storage capacity of 186 ft³.

23. Make 14 jars of tomato sauce and 4 jars of salsa for a profit of $34.

24. Eat about 1.887 servings of pinto beans and none of rice for a total cost of $1.08.

STUDENT HELP

HOMEWORK HELP
Visit our Web site
www.mcdougallittell.com
for help with problem
solving in Ex. 24.

15. Objective function:
$$C = 10x + 7y$$

Constraints:
$0 \le x \le 60$
$0 \le y \le 45$
$5x + 6y \le 420$

min of 0 at (0, 0); max of 740 at (60, 20)

16. Objective function:
$$C = -2x + y$$

Constraints:
$x \ge 0$
$y \ge 0$
$x + y \ge 7$
$5x + 2y \ge 20$

min of −14 at (7, 0); no max as feasible region is unbounded

17. Objective function:
$$C = 4x + 6y$$

Constraints: See margin.
$-x + y \le 11$
$x + y \le 27$
$2x + 5y \le 90$

18. Objective function:
$$C = 5x + 4y$$

Constraints:
$x \ge 0$
$y \ge 0$
$y \le 8$
$x + y \le 14$
$5x + y \le 50$

min of 0 at (0, 0); max of 65 at (9, 5)

19. Objective function:
$$C = 4x + 3y$$

Constraints:
$x \ge 0$
$2x + 3y \ge 6$
$3x - 2y \le 9$
$x + 5y \le 20$

min of 6 at (0, 2); max of 29 at (5, 3)

20. Objective function:
$$C = 10x + 3y$$

Constraints: See margin.
$x \ge 0$
$y \ge 0$
$-x + y \ge 0$
$2x + y \ge 4$
$2x + y \le 13$

21. 🌐 **JUICE BLENDS** A juice company makes two kinds of juice: Orangeade and Berry-fruity. One gallon of Orangeade is made by mixing 2.5 quarts of orange juice and 1.5 quarts of raspberry juice, while one gallon of Berry-fruity is made by mixing 3 quarts of raspberry juice and 1 quart of orange juice. A profit of $.50 is made on every gallon of Orangeade sold, and a profit of $.40 is made on every gallon of Berry-fruity sold. If the company has 150 gallons of raspberry juice and 125 gallons of orange juice on hand, how many gallons of each type of juice should be made to maximize profit?

22. 🌐 **FILE CABINETS** An office manager is purchasing file cabinets and wants to maximize storage space. The office has 60 square feet of floor space for the cabinets and $600 in the budget to purchase them. Cabinet A requires 3 square feet of floor space, has a storage capacity of 12 cubic feet, and costs $75. Cabinet B requires 6 square feet of floor space, has a storage capacity of 18 cubic feet, and costs $50. How many of each cabinet should the office manager buy?

23. 🌐 **HOME CANNING** You have 180 tomatoes and 15 onions left over from your garden. You want to use these to make jars of tomato sauce and jars of salsa to sell at a farm stand. A jar of tomato sauce requires 10 tomatoes and 1 onion, and a jar of salsa requires 5 tomatoes and $\frac{1}{4}$ onion. You'll make a profit of $2 on every jar of tomato sauce sold and a profit of $1.50 on every jar of salsa sold. The farm stand wants at least three times as many jars of tomato sauce as jars of salsa. How many jars of each should you make to maximize profit?

24. 🌐 **NUTRITION** You are planning a dinner of pinto beans and brown rice. You want to consume at least 2100 Calories and 44 grams of protein per day, but no more than 2400 milligrams of sodium and 73 grams of fat. So far today, you have consumed 1600 Calories, 24 grams of protein, 2370 milligrams of sodium, and 65 grams of fat. Pinto beans cost $.57 per cup and brown rice costs $.78 per cup. How many cups of pinto beans and brown rice should you make to minimize cost while satisfying your nutritional requirements?

Contents	1 cup pinto beans	1 cup brown rice (with salt)
Calories	265	230
Protein (g)	15	5
Sodium (mg)	3	10
Fat (g)	1	1

! **COMMON ERROR**

EXERCISES 5 AND 9 Students may forget to consider vertices that lie on the axes. Caution students that the maximum value or minimum value of the objective function may occur at one of these points.

! **COMMON ERROR**

EXERCISES 15–24 Due to the length and complexity of these problems, errors are very common. To avoid needless errors, encourage students to keep their work very organized and to write down each step. They should use graph paper to graph each feasible region and use the graph to check the coordinates of the vertices found by solving systems of linear equations.

STUDENT HELP NOTES

→ **Homework Help** Students can find help for Ex. 24 at **www.mcdougallittell.com**. The information can be printed out for students who don't have access to the Internet.

17. no min, since feasible region is unbounded; max of 132 at (15, 12)

20. min of 12 at (0, 4); max of $\frac{169}{3}$ at $\left(\frac{13}{3}, \frac{13}{3}\right)$

27, 28–33, 42–47. See Additional Answers beginning on page AA1.

ADDITIONAL PRACTICE AND RETEACHING

For Lesson 3.4:
• Practice Levels A, B, and C (*Chapter 3 Resource Book*, p. 57)
• Reteaching with Practice (*Chapter 3 Resource Book*, p. 60)
• 🖥 See Lesson 3.4 of the *Personal Student Tutor*

For more Mixed Review:
• 🖥 Search the *Test and Practice Generator* for key words or specific lessons.

DAILY HOMEWORK QUIZ

Transparency Available

1. Find the minimum and maximum values of $C = -2x + 3y$ for the given feasible region.

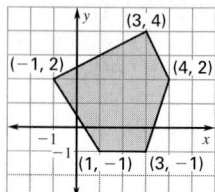

min of −9 at (3, −1);
max of 8 at (−1, 2)

2. Find the minimum and maximum values of $C = 7x + 5y$ subject to the given constraints.
$x \geq 1$ min of 17
$y \geq 2$ at (1, 2);
$x + 3y \leq 13$ max of 45
$x + y \leq 7$ at (5, 2)

3. A cook plans to make burgers that contain ground meat and mushrooms. He will buy from 4 to 5 lb of meat at $4.50/lb and from 0.5 to 0.8 lb of mushrooms at $2.70/lb. He wants to use at most 0.15 times as much chopped mushrooms as ground meat. What are the greatest and least costs possible for the burgers?
greatest: $24.53; least: $19.35

EXTRA CHALLENGE NOTE

Challenge problems for Lesson 3.4 are available in **blackline** format in the *Chapter 3 Resource Book,* p. 65 and at **www.mcdougallittell.com.**

ADDITIONAL TEST PREPARATION

1. WRITING Find the maximum value of the function $C = 3x - y$ for the given constraints.
$x \geq 0$
$y \geq \frac{3}{2}x$
$x + 2y \leq 4$
Explain each step as you proceed.
1.5; check explanations.

Test Preparation

25. MULTIPLE CHOICE Given the feasible region shown, what is the maximum value of the objective function $C = 2x + 6y$? **D**

(A) 0   (B) 60   (C) 200
(D) 276   (E) 326

26. MULTIPLE CHOICE Given the constraints $y \geq 0$, $y \leq x + 8$, and $y \geq 2x + 8$, what is the minimum value of the objective function $C = -2x - y$? **A**

(A) −8   (B) 16   (C) −16   (D) 8

★ Challenge

27. CONSECUTIVE VERTICES Find the value of the objective function at each vertex of the feasible region and at two points on each line segment connecting two vertices. What can you conclude? Sample points are chosen. See margin.

a. Objective function:
$C = 2x + 2y$
Constraints:
$y \leq 4$
$x \leq 5$
$x + y \leq 6$

b. Objective function:
$C = 5x - y$
Constraints:
$y \geq -1$
$x \leq 3$
$-5x + y \leq 4$

EXTRA CHALLENGE
www.mcdougallittell.com

MIXED REVIEW

GRAPHING EQUATIONS Graph the equation. Label any intercepts. **(Review 2.3 for 3.5)** 28–33. See margin.

28. $x - y = 10$
29. $3x + 4y = -12$
30. $y = -3x + 2$
31. $5x - 15y = 15$
32. $y = -\frac{3}{4}x + 2$
33. $y = -\frac{1}{2}x + 7$

EVALUATING FUNCTIONS Evaluate the function for the given value of *x*. **(Review 2.7)**

$$f(x) = \begin{cases} 3x - 1, & \text{if } x < -2 \\ x - 5, & \text{if } x \geq -2 \end{cases} \qquad g(x) = \begin{cases} -7x, & \text{if } x \leq 0 \\ 2x + 1, & \text{if } x > 0 \end{cases}$$

34. $f(0)$ −5
35. $f(-2)$ −7
36. $f(-10)$ −31
37. $f(-1)$ −6
38. $g(1)$ 3
39. $g(-5)$ 35
40. $g(-1)$ 7
41. $g(7)$ 15

GRAPHING SYSTEMS OF INEQUALITIES Graph the system of linear inequalities. **(Review 3.3)** 42–47. See margin.

42. $x > 2$
$y < 6$

43. $x + y \leq 5$
$y > 0$

44. $x < -1$
$x - y \geq 4$

45. $y < 5$
$x \geq -1$
$y \geq 1$

46. $-x + y > 2$
$y > 0$
$2x + y \leq 3$

47. $x + y \leq 6$
$-\frac{1}{2}x + y \leq 3$
$y \leq 3$

48. AMUSEMENT CENTER You have 30 tokens for playing video games and pinball. It costs 3 tokens to play a video game and 2 tokens to play pinball. You want to play an equal number of video games and pinball games. Use an algebraic model to find how many games of each you can play. **(Review 1.5)**
Play 6 games of each.

Graph the system of linear inequalities. (Lesson 3.3)   1–3. See margin.

1. $y > -2$
$x \geq -4$
$y \leq -x + 1$

2. $y > -5$
$x \leq 2$
$y \leq x + 2$

3. $x \leq 3$
$y < 2$
$y > -x + 1$

Find the minimum and maximum values of the objective function $C = 5x + 2y$ subject to the given constraints. (Lesson 3.4)

4. Constraints:
$x \leq -2$
$x \geq -4$
$y \geq 1$
$y \leq 6$
min of -18 at $(-4, 1)$; max of 2 at $(-2, 6)$

5. Constraints:
$x \geq 0$
$y \geq 2$
$2x + y \leq 10$
$x - 3y \geq -3$
min of 19 at $(3, 2)$; max of 24 at $(4, 2)$

6. Constraints:
$x \geq 0$
$y \geq 0$
$y \leq 8$
$x + y \leq 14$
min of 0 at $(0, 0)$; max of 70 at $(14, 0)$

7.  **MAXIMUM INCOME** You are stenciling wooden boxes to sell at a fair. It takes you 2 hours to stencil a small box and 3 hours to stencil a large box. You make a profit of $10 for a small box and $20 for a large box. If you have no more than 30 hours available to stencil and want at least 12 boxes to sell, how many of each size box should you stencil to maximize your profit? (Lesson 3.4)
6 small boxes and 6 large boxes

MATH & History

Linear Programming in World War II

INTERNET APPLICATION LINK
www.mcdougallittell.com

THEN

DURING WORLD WAR II, the need for efficient transportation of supplies inspired mathematician George Dantzig to develop linear programming.

The LST was a ship used during World War II that carried 3 ton trucks and 25 ton tanks. The upper deck could carry 27 trucks, but no tanks. The tank deck could carry 500 tons, but no more than 33 trucks.

1. What is the maximum number of tanks that an LST could hold?
20
2. What is the maximum number of trucks that an LST could hold?
60
3. Suppose an LST was to be loaded with as many tanks and trucks as possible, and at least three times as many trucks as tanks. What is the maximum number of tanks and trucks that could be loaded?
16 tanks and 60 trucks

U.S. Marines loading military supplies on an LST (Landing Ship, Tank)

NOW

IN 1984 mathematician Narendra Karmarkar developed a new time-saving linear programming method. Today his method is used by industries that deal with allocation of resources, such as telephone companies, airlines, and manufacturers.

1947
George Dantzig develops simplex method.

L.V. Kantorovich and T.C. Koopmans receive Nobel Prize for their linear programming work.
1975

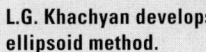
A Soviet Discovery Rocks World of Mathematics
1979
L.G. Khachyan develops ellipsoid method.

1984
N. Karmarkar devises a polynomial-time algorithm.

3.4 *Linear Programming*   **169**

ADDITIONAL RESOURCES
An alternative quiz for Lessons 3.3 and 3.4 is available in the *Chapter 3 Resource Book*, p. 66.

MATH & HISTORY NOTE
See the enrichment master Math & History Applications on p. 64 of the *Chapter 3 Resource Book* for additional exercises.
27–33, 42–47. See Additional Answers beginning on page AA1.

1.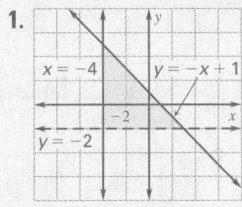

2.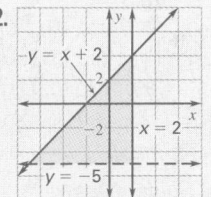

3.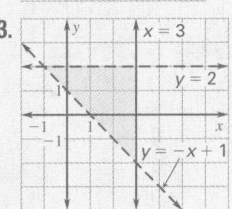

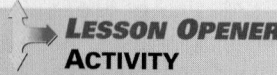

LESSON OPENER
ACTIVITY
An alternative way to approach Lesson 3.5 is to use the Activity Lesson Opener:
- Blackline Master (*Chapter 3 Resource Book*, p. 70)
- Transparency (p. 20)

MEETING INDIVIDUAL NEEDS
- *Chapter 3 Resource Book*
  Prerequisite Skills Review (p. 5)
  Practice Level A (p. 72)
  Practice Level B (p. 73)
  Practice Level C (p. 74)
  Reteaching with Practice (p. 75)
  Absent Student Catch-Up (p. 77)
  Challenge (p. 79)
- *Resources in Spanish*
- *Personal Student Tutor*

NEW-TEACHER SUPPORT
See the Tips for New Teachers on pp. 1–2 of the *Chapter 3 Resource Book* for additional notes about Lesson 3.5.

WARM-UP EXERCISES

Transparency Available

Find the *x*- and *y*-intercepts of the graph of each equation.

1. $2x + 4y = 20$ **10, 5**
2. $x - 3y = -15$ **-15, 5**

Use the linear equation to write *y* as a function of *x*.

3. $2x - 6y = 24$ $f(x) = \frac{1}{3}x - 4$
4. $3x + 5y = 12$ $f(x) = \frac{1}{5}(12 - 3x)$

What you should learn

GOAL 1 Graph linear equations in three variables and evaluate linear functions of two variables.

GOAL 2 Use functions of two variables to model **real-life** situations, such as finding the cost of planting a lawn in **Example 4**.

Why you should learn it

▼ To solve **real-life** problems, such as finding how many times to air a radio commercial in **Ex. 53**.

3.5 Graphing Linear Equations in Three Variables

GOAL 1 GRAPHING IN THREE DIMENSIONS

Solutions of equations in three variables can be pictured with a **three-dimensional coordinate system**. To construct such a system, begin with the *xy*-coordinate plane in a horizontal position. Then draw the **z-axis** as a vertical line through the origin.

In much the same way that points in a two-dimensional coordinate system are represented by ordered pairs, each point in space can be represented by an **ordered triple** (x, y, z).

Drawing the point represented by an ordered triple is called *plotting* the point.

The three axes, taken two at a time, determine three coordinate planes that divide space into eight **octants**. The first octant is the one for which all three coordinates are positive.

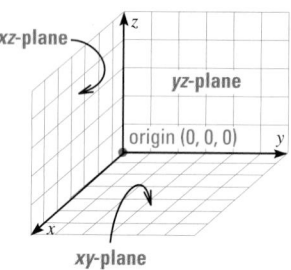

EXAMPLE 1 *Plotting Points in Three Dimensions*

Plot the ordered triple in a three-dimensional coordinate system.

a. $(-5, 3, 4)$ **b.** $(3, -4, -2)$

SOLUTION

a. To plot $(-5, 3, 4)$, it helps to first find the point $(-5, 3)$ in the *xy*-plane. The point $(-5, 3, 4)$ lies four units above.

b. To plot $(3, -4, -2)$, find the point $(3, -4)$ in the *xy*-plane. The point $(3, -4, -2)$ lies two units below.

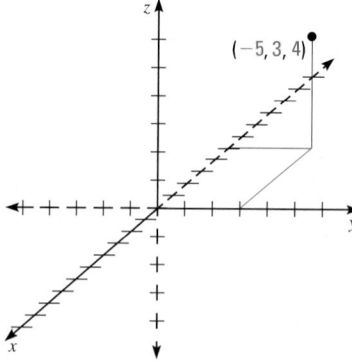

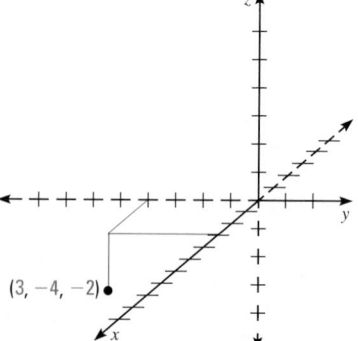

A **linear equation in three variables** x, y, and z is an equation of the form

$$ax + by + cz = d$$

where a, b, and c are not all zero. An ordered triple (x, y, z) is a *solution* of this equation if the equation is true when the values of x, y, and z are substituted into the equation. The *graph* of an equation in three variables is the graph of all its solutions. The graph of a linear equation in three variables is a plane.

EXAMPLE 2 *Graphing a Linear Equation in Three Variables*

STUDENT HELP

HOMEWORK HELP
Visit our Web site
www.mcdougallittell.com
for extra examples.

Sketch the graph of $3x + 2y + 4z = 12$.

SOLUTION

Begin by finding the points at which the graph intersects the axes. Let $x = 0$ and $y = 0$, and solve for z to get $z = 3$. This tells you that the z-intercept is 3, so plot the point $(0, 0, 3)$. In a similar way, you can find that the x-intercept is 4 and the y-intercept is 6. After plotting $(0, 0, 3)$, $(4, 0, 0)$, and $(0, 6, 0)$, you can connect these points with lines to form the triangular region of the plane that lies in the first octant.

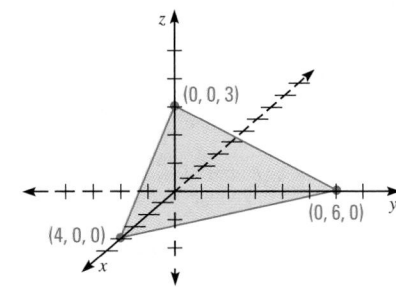

· · · · · · · · ·

A linear equation in x, y, and z can be written as a **function of two variables**. To do this, solve the equation for z. Then replace z with $f(x, y)$.

EXAMPLE 3 *Evaluating a Function of Two Variables*

a. Write the linear equation $3x + 2y + 4z = 12$ as a function of x and y.

b. Evaluate the function when $x = 1$ and $y = 3$. Interpret the result geometrically.

SOLUTION

a. $3x + 2y + 4z = 12$ **Write original equation.**

$\quad\quad 4z = 12 - 3x - 2y$ **Isolate z-term.**

$\quad\quad z = \frac{1}{4}(12 - 3x - 2y)$ **Solve for z.**

$\quad f(x, y) = \frac{1}{4}(12 - 3x - 2y)$ **Replace z with $f(x, y)$.**

STUDENT HELP

Study Tip
Remember that just as the notation $f(x)$ means the value of f at x, $f(x, y)$ means the value of f at the point (x, y).

b. $f(1, 3) = \frac{1}{4}(12 - 3(1) - 2(3)) = \frac{3}{4}$. This tells you that the graph of f contains the point $\left(1, 3, \frac{3}{4}\right)$.

3.5 Graphing Linear Equations in Three Variables   **171**

---

2 TEACH

EXTRA EXAMPLE 1
Plot $(3, -1, -5)$.

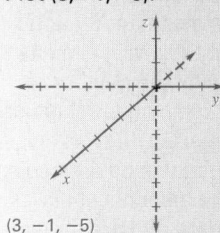

$(3, -1, -5)$

EXTRA EXAMPLE 2
Graph $3x - 12y + 5z = 30$.

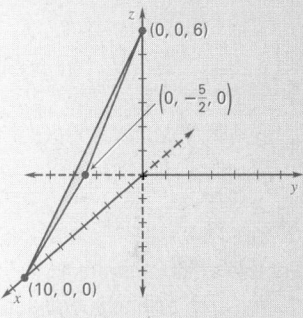

EXTRA EXAMPLE 3
a. Write $3x - 12y + 5z = 30$ as a function of x and y.
$f(x, y) = \frac{1}{5}(30 - 3x + 12y)$
b. Evaluate the function when $x = -2$ and $y = 2$. $f(-2, 2) = 12$

✓ **CHECKPOINT EXERCISES**
For use after Examples 1–3:
1. a. Graph $x - 2y + 2z = 6$.

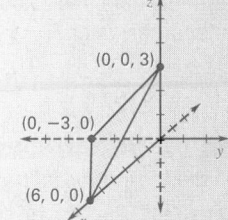

b. Write the equation in part **(a)** as a function of x and y. Then evaluate $f(0, -2)$.
$f(x, y) = \frac{1}{2}(6 - x + 2y);$
$f(0, -2) = 1$

EXTRA EXAMPLE 4

A family is planning a vacation at a resort. The airfare will be $1200, lodging is $120 per night and family-style meals are $40 each.

a. Write a model for the total amount they will spend as a function of the number of nights and number of meals.
$C = 1200 + 120x + 40y$

b. Evaluate the model for several different amounts of nights and meals, and organize your results in a table.

	Meals		
	2	**5**	**10**
1	$1400	$1520	$1720
2	$1520	$1640	$1840
3	$1640	$1760	$1960
5	$1880	$2000	$2200

(Nights labeled along left side)

✔ CHECKPOINT EXERCISES

For use after Example 4:

1. You are packing a food supply crate for a canoe trip. The crate weighs 12 lb and it will be filled with boxes of granola bars, each weighing 1.5 lb, and boxes of macaroni each weighing 0.75 lb. Write a model for the total weight of the crate as a function of the number of boxes of granola bars and macaroni. How much will a crate with 15 boxes of granola bars and 25 boxes of macaroni weigh?
$W = 12 + 1.5x + 0.75y$; 53.25 lb

FOCUS ON VOCABULARY

Explain the term *ordered triple*.
See margin for sample answer.

CLOSURE QUESTION

Name a point on the graph of f if $f(-2, 3) = 7$. $(-2, 3, 7)$

DAILY PUZZLER

You want to draw a cube whose vertices have 0, 2, and −2 as their only coordinates. How many such cubes are there? 9 cubes

172

FOCUS ON CAREERS

▶ **BIOTECHNICIAN**
There are about 40 grass species used as turf, two of which are discussed in Example 4. Turfgrass biotechnicians manipulate genetic traits of existing grass seed to breed improved varieties of grass.

🌐 **CAREER LINK**
www.mcdougallittell.com

STUDENT HELP

🌐 **HOMEWORK HELP**
You can also use a spreadsheet to evaluate a function of two variables. For help with how to do this visit www.mcdougallittell.com

EXAMPLE 4 *Modeling a Real-Life Situation*

LANDSCAPING You are planting a lawn and decide to use a mixture of two types of grass seed: bluegrass and rye. The bluegrass costs $2 per pound and the rye costs $1.50 per pound. To spread the seed you buy a spreader that costs $35.

a. Write a model for the total amount you will spend as a function of the number of pounds of bluegrass and rye.

b. Evaluate the model for several different amounts of bluegrass and rye, and organize your results in a table.

SOLUTION

a. Your total cost involves two variable costs (for the two types of seed) and one fixed cost (for the spreader).

VERBAL MODEL

Total cost	=	Blue-grass cost	·	Blue-grass amount	+	Rye cost	·	Rye amount	+	Spreader cost

LABELS

Total cost = C	(dollars)
Bluegrass cost = **2**	(dollars per pound)
Bluegrass amount = x	(pounds)
Rye cost = **1.5**	(dollars per pound)
Rye amount = y	(pounds)
Spreader cost = **35**	(dollars)

ALGEBRAIC MODEL

$$C = 2x + 1.5y + 35$$

b. To evaluate the function of two variables, substitute values of x and y into the function. For instance, when $x = 10$ and $y = 20$, the total cost is:

$C = 2x + 1.5y + 35$	**Write original function.**
$= 2(10) + 1.5(20) + 35$	**Substitute for *x* and *y*.**
$= 85$	**Simplify.**

The table shows the total cost for several different values of x and y.

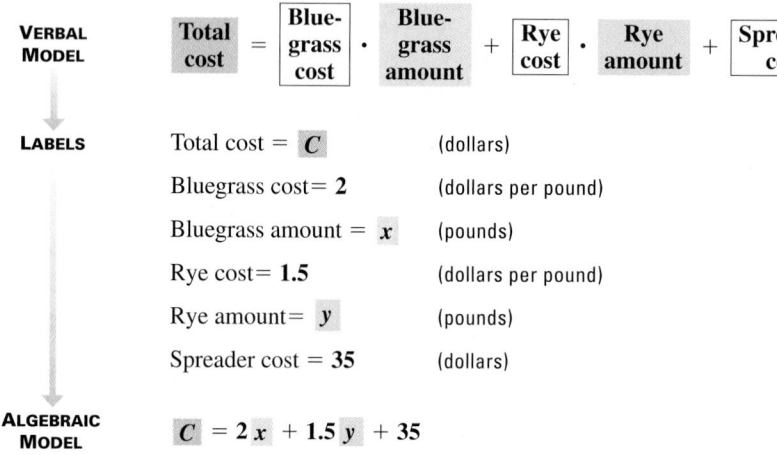

		Rye (lb)			
	0	**10**	**20**	**30**	**40**
10		$70	$85	$100	$115
20		$90	$105	$120	$135
30		$110	$125	$140	$155
40		$130	$145	$160	$175

(Bluegrass (lb) labeled along left side)

Focus on Vocabulary *Sample answer:*
An ordered triple is a group of three numbers (x, y, z) that represents a point in space.

GUIDED PRACTICE

Vocabulary Check ✓

1. Write the general form of a linear equation in three variables. How is the solution of such an equation represented? **1, 3–4. See margin.**

Concept Check ✓

2. **LOGICAL REASONING** Tell whether this statement is *true* or *false*: The graph of a linear equation in three variables consists of three different lines. **false**

3. How are octants and quadrants similar?

4. Describe how you would graph a linear equation in three variables.

Skill Check ✓

1. $ax + by + cz = d$; the solution of such an equation is a plane in three-dimensional space; to graph it, find the three intercepts and shade the triangular region lying in one octant.

3. *Sample answer:* octants and quadrants are both distinct regions bounded by the axes, and defined by the signs of the x-, y- and z- coordinates.

4. To graph a linear equation in three variables, find the three intercepts and shade the triangular region lying in one octant.

15. $f(x, y) = -2x - \frac{1}{2}y - 4; -17$

5. Draw a three-dimensional coordinate system and plot the ordered triple $(2, -4, -6)$.
See margin.

6. Write the coordinates of the vertices A, B, C, and D of the rectangular prism shown, given that one vertex is the point $(2, 3, 4)$.
$A(2, 0, 4)$; $B(2, 3, 0)$; $C(0, 3, 4)$; $D(0, 0, 4)$

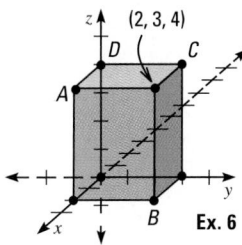
Ex. 6

Sketch the graph of the equation. Label the points where the graph crosses the x-, y-, and z-axes. **7–12. See margin.**

7. $8x + 4y + 2z = 16$ 8. $2x + 4y + 5z = 20$ 9. $3x + 3y + 7z = 21$

10. $10x + 2y + 5z = 10$ 11. $9x + 3y + 3z = 27$ 12. $4x + y + 2z = 8$

Write the linear equation as a function of x and y. Then evaluate the function for the given values.

13. $6x + 6y + 3z = 9$, $f(1, 2)$ $f(x, y) = -2x - 2y + 3; -3$

14. $-2x - y + z = 7$, $f(-3, 2)$ $f(x, y) = 2x + y + 7; 3$

15. $8x + 2y + 4z = -16$, $f(5, 6)$ **See margin.**

16. $5x - 10y - 5z = 15$, $f(2, 2)$ $f(x, y) = x - 2y - 3; -5$

17. 🌐 **TRAIL MIX** You are making bags of a trail mix called GORP (Good Old Raisins and Peanuts). The raisins cost $2.25 per pound and the peanuts cost $2.95 per pound. The package of bags for the trail mix costs $2.65. Write a model for the total cost as a function of the number of pounds of raisins and peanuts you buy. Evaluate the model for 5 lb of raisins and 8 lb of peanuts.
$C = 2.25r + 2.95p + 2.65; \$37.50$

PRACTICE AND APPLICATIONS

STUDENT HELP

▶ **Extra Practice**
to help you master skills is on p. 943.

PLOTTING POINTS Plot the ordered triple in a three-dimensional coordinate system. **18–25. See margin.**

18. $(2, 4, 0)$ 19. $(4, -1, -6)$ 20. $(5, -2, -2)$ 21. $(0, 6, -3)$

22. $(3, 4, -2)$ 23. $(-2, 1, 1)$ 24. $(5, -1, 5)$ 25. $(-3, 2, -7)$

STUDENT HELP

▶ **HOMEWORK HELP**
Example 1: Exs. 18–25
Example 2: Exs. 26–37

continued on p. 174

SKETCHING GRAPHS Sketch the graph of the equation. Label the points where the graph crosses the x-, y-, and z-axes. **26–37. See margin.**

26. $x + y + z = 7$ 27. $5x + 4y + 2z = 20$ 28. $x + 6y + 4z = 12$

29. $12x + 3y + 8z = 24$ 30. $2x + 18y + 3z = 36$ 31. $7x + 9y + 21z = 63$

32. $7x + 7y + 2z = 14$ 33. $6x + 4y + 3z = 10$ 34. $3x + 5y + 3z = 15$

35. $\frac{1}{2}x + 4y - 3z = 8$ 36. $5x + y + 2z = -4$ 37. $-2x + 9y + 3z = 18$

3.5 *Graphing Linear Equations in Three Variables*   **173**

3 APPLY

○ **ASSIGNMENT GUIDE**

BASIC
Day 1: pp. 173–175 Exs.18–36 even, 39–45 odd, 49, 50, 53, 57–67 odd

AVERAGE
Day 1: pp. 173–175 Exs.18–36 even, 39–53 odd, 57–67 odd

ADVANCED
Day 1: pp. 173–175 Exs.18–36 even, 39–51 odd, 52–56, 57–67 odd

BLOCK SCHEDULE
pp. 173–175 Exs.18–36 even, 39–53 odd, 57–67 odd, (with 3.6)

EXERCISE LEVELS
Level A: *Easier*
18–45, 48–52
Level B: *More Difficult*
46–47, 53
Level C: *Most Difficult*
54–56

✔ **HOMEWORK CHECK**
To quickly check student understanding of key concepts, go over the following exercises:
Exs. 18, 26, 39. See also the Daily Homework Quiz:
• Blackline Master (*Chapter 3 Resource Book*, p. 82)
• 📖 Transparency (p. 24)

5.
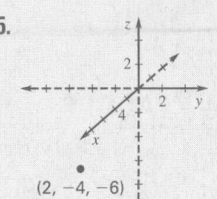
$(2, -4, -6)$

7–12, 18–25, 26–37.
See Additional Answers beginning on page AA1.

STUDENT HELP NOTES

→ **Toolbox** As students look back to p. 914, remind them that volume is measured in cubic units.

APPLICATION NOTE

EXERCISE 5 In this application, students will be able decide when it is better to buy a monthly pass for public transportation and when it is better to pay for rides individually.

48–52, 53b. See Additional Answers beginning on page AA1.

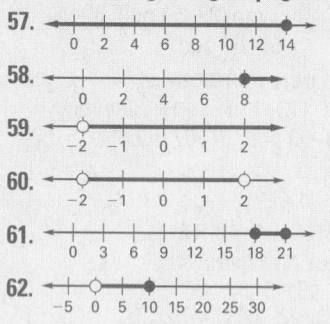

57. (number line: 0 2 4 6 8 10 12 14, point at 14)

58. (number line: 0 2 4 6 8, point at 7)

59. (number line: −2 −1 0 1 2, open circle at −1)

60. (number line: −2 −1 0 1 2, open circles at −1 and 1)

61. (number line: 0 3 6 9 12 15 18 21, points at 15 and 18)

62. (number line: −5 0 5 10 15 20 25 30, open circle at 0, point at 5)

ADDITIONAL PRACTICE AND RETEACHING

For Lesson 3.5:

- Practice Levels A, B, and C
  (*Chapter 3 Resource Book,* p. 72)

- Reteaching with Practice
  (*Chapter 3 Resource Book,* p. 75)

- See Lesson 3.5 of the
  Personal Student Tutor

For more Mixed Review:

- Search the *Test and Practice Generator* for key words or specific lessons.

174

STUDENT HELP

→ **HOMEWORK HELP**
continued from p. 173

Example 3: Exs. 38–45
Example 4: Exs. 48–52

41. $f(x, y) = -\frac{6}{5}x + \frac{3}{10}y + \frac{18}{5}$; 12

STUDENT HELP

→ **Skills Review**
For help with volume, see p. 914.

42. $f(x, y) = -\frac{1}{7}x - \frac{2}{7}y - 2$; $\frac{11}{7}$

43. $f(x, y) = -\frac{1}{6}x - \frac{1}{4}y + \frac{1}{5}$; $\frac{1}{2}$

45. $f(x, y) = -\frac{1}{9}x + \frac{2}{3}y - \frac{4}{3}$; $\frac{121}{18}$

FOCUS ON APPLICATIONS

▶ **TRANSPORTATION**
The Massachusetts Bay Transit Authority (MBTA) is the nation's oldest subway system. On an average weekday, the MBTA serves about 1.2 million passengers on its bus, ferry, and train lines.

🌐 **APPLICATION LINK**
www.mcdougallittell.com

EVALUATING FUNCTIONS Write the linear equation as a function of *x* and *y*. Then evaluate the function for the given values. $f(x, y) = \frac{2}{5}x + y + 3$; $\frac{8}{5}$

38. $6x + 2y + 3z = 18$, $f(2, 1)$
$f(x, y) = -2x - \frac{2}{3}y + 6$; $\frac{4}{3}$

39. $-2x - 5y + 5z = 15$, $f\left(\frac{3}{2}, -2\right)$

40. $x + 6y + z = 10$, $f(-4, -1)$
$f(x, y) = -x - 6y + 10$; 20

41. $3x - \frac{3}{4}y + \frac{5}{2}z = 9$, $f(-3, 16)$
41–43, 45. See margin.

42. $-x - 2y - 7z = 14$, $f(-5, -10)$

43. $10x + 15y + 60z = 12$, $f\left(-3, \frac{4}{5}\right)$

44. $x - 5y - z = 14$, $f(3, 6)$
$f(x, y) = x - 5y - 14$; −41

45. $-x + 6y - 9z = 12$, $f\left(-\frac{1}{2}, 12\right)$

46. **GEOMETRY CONNECTION** Use the given point (4, 7, 2) to find the volume of the rectangular prism. **56**

47. **GEOMETRY CONNECTION** Use the given point (5, 6, −2) to find the volume of the rectangular prism. **60**

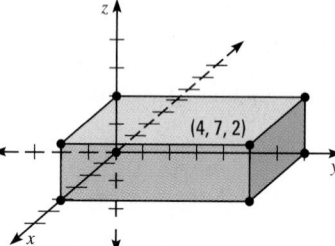
(4, 7, 2)

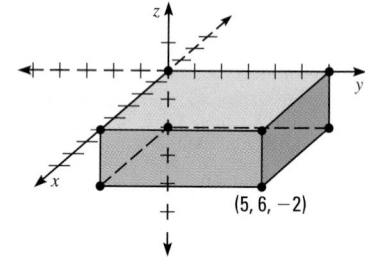
(5, 6, −2)

48. 🐠 **HOME AQUARIUM** You want to buy an aquarium and stock it with goldfish and angelfish. The pet store sells goldfish for $.40 each and angelfish for $4 each. The aquarium starter kit costs $65. Write a model for the amount you will spend as a function of the number of goldfish and angelfish you buy. Make a table that shows the total cost for several different numbers of goldfish and angelfish.
$C = 0.4g + 4a + 65$; See margin.

49. 🏺 **POTTERY** A craft store has paint-your-own pottery sessions available. You pick out a piece of pottery that ranges in price from $8 to $50 and pick out paint colors for $1.50 per color. The craft store charges a base fee of $16 for sitting time, brushes, glaze, and kiln time. Write a model for the total cost of making a piece of pottery as a function of the price of the pottery and the number of paint colors you use. Make a table that shows the total cost for several different pieces of pottery and numbers of paint colors. $C = 1.5n + p + 16$; See margin.

50. 💐 **FLOWER ARRANGEMENT** You are buying tulips, carnations, and a glass vase to make a flower arrangement. The flower shop sells tulips for $.70 each and carnations for $.30 each. The glass vase costs $12. Write a model for the total cost of the flower arrangement as a function of the number of tulips and carnations you use. Make a table that shows the total cost for several different numbers of tulips and carnations. $C = 0.7t + 0.3c + 12$; See margin.

51. 🚌 **TRANSPORTATION** Every month you buy a local bus pass for $20 that is worth $.60 toward the fare for the local bus, the express bus, or the subway. The local bus costs $.60, the express bus costs $1.50, and the subway costs $.85. Write a model for the total cost of transportation in a month as a function of the number of times you take the express bus and the number of times you take the subway. Evaluate the model for 8 express bus rides and 10 subway rides. Make a table that shows the total cost for several different numbers of rides.
$C = 0.9e + 0.25s + 20$; $29.70; See margin.

52.  **AFTER-SCHOOL JOBS** Several days after school you are a lifeguard at a community pool. On weekends you baby-sit to earn extra money. Lifeguarding pays $8 per hour and baby-sitting pays $6 per hour. You also get a weekly allowance of $10 for doing chores around the house. Write an equation for your total weekly earnings as a function of the number of hours you lifeguard and baby-sit. Make a table that shows several different amounts of weekly earnings.
$S = 8l + 6b + 10$; See margin.

53. **MULTI-STEP PROBLEM** You are deciding how many times to air a 60 second commercial on a radio station. The station charges $100 for a 60 second spot during off-peak listening hours and $350 for a 60 second spot during peak listening hours. The company you have hired to make your commercial charges $500.

  a. Write a model for the total amount that will be spent making and airing the commercial as a function of the number of times it is aired during off-peak and peak listening hours. $C = 100x + 350y + 500$, where x = number of off-peak commercial spots and y = number of peak commercial spots

  b. Evaluate the model for several different numbers of off-peak airings and peak airings. Organize your results in a table. **b–c. See margin.**

  c. *Writing* Suppose your advertising budget is $4000. Using the table you made in part (b), can you air the commercial 8 times during off-peak hours and 8 times during peak hours? What combination of off-peak and peak airings would you recommend? Explain.

Test Preparation

53.c. *Sample answer:* 8 of each kind of spot costs $4100, so you don't have enough money. I would spend $3950 on 10 off-peak spots and 7 peak ones. This way, your commercial gets 17 airings rather than 16, and you are $50 under budget.

★ **Challenge**

EXTRA CHALLENGE
www.mcdougallittell.com

WRITING EQUATIONS **Write an equation of the plane having the given x-, y-, and z-intercepts. Explain the method you used.**

54. x-intercept: 4
y-intercept: -2
z-intercept: 4
$x - 2y + z = 4$

55. x-intercept: $\frac{3}{2}$
y-intercept: 12
z-intercept: 6
$8x + y + 2z = 12$

56. x-intercept: 4
y-intercept: -6
z-intercept: -9
$27x - 18y - 12z = 108$

Sample explanation: I found the least common multiple of the given intercepts and divided each one into the LCM to find the coefficients of the variables.

MIXED REVIEW

SOLVING INEQUALITIES **Solve the inequality. Then graph the solution.**
(Review 1.6)

$x \le 14$; See margin.
57. $3 + x \le 17$

$x \ge 8$; See margin.
58. $2x + 5 \ge 21$

$x > -2$; See margin.
59. $-x + 3 < 3x + 11$

60. $-13 < 6x - 1 < 11$
$-2 < x < 2$; See margin.

61. $24 \le 2x - 12 \le 30$
$18 \le x \le 21$; See margin.

62. $-3 < 2x - 3 \le 17$
$0 < x \le 10$; See margin.

TYPES OF LINES **Tell whether the lines are *parallel*, *perpendicular*, or *neither*.**
(Review 2.2)

63. Line 1: through $(1, 7)$ and $(-3, -5)$
Line 2: through $(-6, 20)$ and $(0, 2)$
neither

64. Line 1: through $(4, -4)$ and $(-16, 1)$
Line 2: through $(1, 5)$ and $(5, 21)$
perpendicular

65. Line 1: through $(-2, 1)$ and $(0, 3)$
Line 2: through $(2, 1)$ and $(0, -1)$
parallel

66. Line 1: through $(0, 6)$ and $(5, -2)$
Line 2: through $(-1, -1)$ and $(7, 4)$
perpendicular

67.  **HOME CARPENTRY** You have budgeted $48.50 to purchase red oak and poplar boards to make a bookcase. Each red oak board costs $3.95 and each poplar board costs $3.10. You need a total of 14 boards for the bookcase. Write and solve a system of equations to find the number of red oak boards and the number of poplar boards you should buy. **(Review 3.1, 3.2 for 3.6)**
$3.95r + 3.1p = 48.5$; $r + p = 14$; buy 6 red oak boards and 8 poplar boards.

DAILY HOMEWORK QUIZ

🖳 *Transparency Available*

1. Plot $(3, 2, -4)$ in a three-dimensional coordinate system.

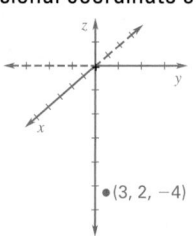

2. Sketch the graph of $-15x + 10y - 6z = -30$. Label the points where the graph crosses the x-, y-, and z-axes.

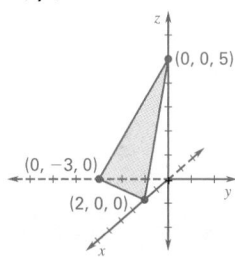

3. Write $8x - 5y + 2z = 9$ as a function of x and y. Then evaluate $f(-1, 6)$.
$f(x, y) = -4x + \frac{5}{2}y + \frac{9}{2}$; $\frac{47}{2}$

EXTRA CHALLENGE NOTE

→ Challenge problems for Lesson 3.5 are available in **blackline** format in the *Chapter 3 Resource Book,* p. 79 and at **www.mcdougallittell.com.**

ADDITIONAL TEST PREPARATION

1. OPEN ENDED Write a real-life problem that can be modeled by a function in two variables. Write the function and make a table to show the value of the function for several different values of the variables.
Check work.

175

1 Planning the Activity

PURPOSE
To graph linear equations in three variables on a graphing calculator.

MATERIALS
- a TI-92 graphing calculator or a computer with 3-D graphing software
- Activity Support Master (*Chapter 3 Resource Book*, p. 71)

PACING
- Activity — 30 min

▶ LINK TO LESSON
You can have students check some of the graphs they drew for homework in Lesson 3.5 such as Exs. 26–37 on p. 173. Seeing the graphs on the calculator or computer may help some students visualize them better.

2 Managing the Activity

COOPERATIVE LEARNING
Students can work in pairs at a single calculator or computer. Have students take turns writing the equations as functions and entering them.

ALTERNATIVE APPROACH
This can be done as a class using a single TI-92 or computer and an overhead display.

3 Closing the Activity

★ KEY DISCOVERY
Graphing calculators or computers can be used to draw 3-dimensional graphs.

ACTIVITY ASSESSMENT
Describe the procedure used to graph a linear equation in three variables on a graphing calculator or computer. **Write the equation as a function of two variables. Enter the function into the computer or calculator and adjust the viewing window.**

◗ ACTIVITY 3.5
Using Technology

MATERIALS
TI-92 graphing calculator or computer with 3-D graphing software

STUDENT HELP

KEYSTROKE HELP

See keystrokes for several models of calculators at www.mcdougallittell.com

---

Graphing Linear Equations in Three Variables

Some graphing calculators can be used to graph a linear equation in three variables. The instructions for graphing on a TI-92 are given below.

▶ EXAMPLE
Use a graphing calculator (or a computer) to graph the equation $3x + 5y + 6z = 30$.

▶ SOLUTION

1 Solve the equation for z.

$3x + 5y + 6z = 30$	Write equation.
$6z = 30 - 3x - 5y$	Isolate the z-term.
$z = 5 - \frac{1}{2}x - \frac{5}{6}y$	Solve for z.

2 Enter the equation in the [Z=] editor.

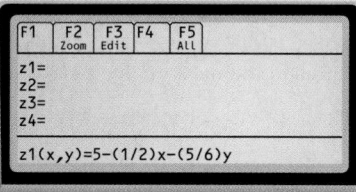

3 Display the axes in box format and turn the labels on.

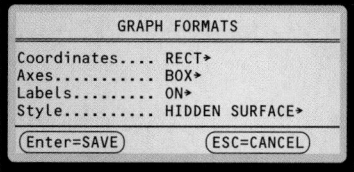

4 Set the window values as shown.

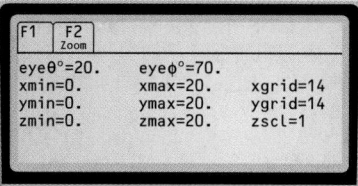

5 Graph the equation. You can use the *Evaluate* feature to evaluate z for values of x and y.

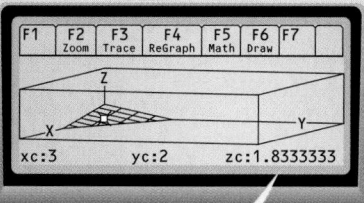

When $x = 3$ and $y = 2$, $z \approx 1.83$.

▶ EXERCISES

Use a graphing calculator (or a computer) to graph the equation. Then evaluate z for the given values of x and y.

1. $4x + 18y + 3z = 54$; $x = 6, y = 4$ -14 **2.** $3x + y + z = 24$; $x = 1.5, y = 19$ 0.5

3. $x + 3y + 10z = 45$; $x = 20, y = 7$ 0.4 **4.** $7x + 6y + 2z = 61$; $x = 4, y = 4$ 4.5

5. $4x + 13y - 5z = 26$; $x = 14, y = 6$ 21.6 **6.** $3x - 25y + 20z = 35$; $x = 5, y = 0$ 1

3.6

Solving Systems of Linear Equations in Three Variables

What you should learn

GOAL 1 Solve systems of linear equations in three variables.

GOAL 2 Use linear systems in three variables to model **real-life** situations, such as a high school swimming meet in **Example 4**.

Why you should learn it

▼ To solve **real-life** problems, such as finding the number of athletes who placed first, second, and third in a track meet in **Ex. 35**.

GOAL 1 SOLVING A SYSTEM IN THREE VARIABLES

In Lessons 3.1 and 3.2 you learned how to solve a system of two linear equations in two variables. In this lesson you will learn how to solve a **system of three linear equations** in three variables. Here is an example.

$$x + 2y - 3z = -3 \qquad \textbf{Equation 1}$$

$$2x - 5y + 4z = 13 \qquad \textbf{Equation 2}$$

$$5x + 4y - z = 5 \qquad \textbf{Equation 3}$$

A **solution** of such a system is an ordered triple (x, y, z) that is a solution of all three equations. For instance, $(2, -1, 1)$ is a solution of the system above.

$$2 + 2(-1) - 3(1) = 2 - 2 - 3 = -3 \checkmark$$

$$2(2) - 5(-1) + 4(1) = 4 + 5 + 4 = 13 \checkmark$$

$$5(2) + 4(-1) - 1 = 10 - 4 - 1 = 5 \checkmark$$

From Lesson 3.5 you know that the graph of a linear equation in three variables is a plane. Three planes in space can intersect in different ways.

If the planes intersect in a single point, as shown below, the system has exactly one solution.

If the planes intersect in a line, as shown below, the system has infinitely many solutions.

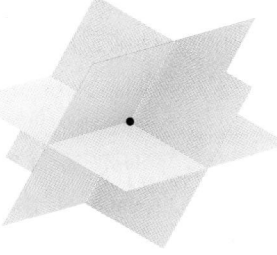

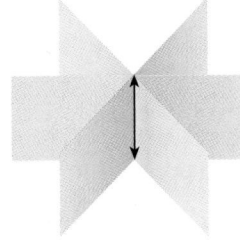

If the planes have no point of intersection, the system has no solution. In the example on the left, the planes intersect pairwise, but all three have no points in common. In the example on the right, the planes are parallel.

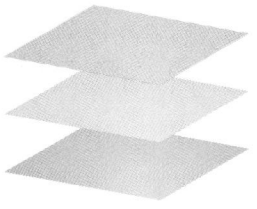

MEETING INDIVIDUAL NEEDS
- *Chapter 3 Resource Book*
  Prerequisite Skills Review (p. 5)
  Practice Level A (p. 84)
  Practice Level B (p. 85)
  Practice Level C (p. 86)
  Reteaching with Practice (p. 87)
  Absent Student Catch-Up (p. 89)
  Challenge (p. 91)
- *Resources in Spanish*
- *Personal Student Tutor*

WARM-UP EXERCISES

Transparency Available

Solve the linear system.
1. $x - 5y = 10$
   $3x - 15y = 15$ no solution
2. $y = 2x + 3$
   $-4x + 2y = 6$
   infinitely many solutions
3. $x - y = -5$
   $x + 3y = 11$ $(-1, 4)$

In the coordinate plane, the solution to a linear system in two variables is the intersection of two lines. In today's lesson you will see that the solution to a linear system in three variables is the intersection of three planes.

EXTRA EXAMPLE 1
Solve the system.
$x + 3y - z = -11$
$2x + y + z = 1$
$5x - 2y + 3z = 21$
$(2, -4, 1)$

CHECKPOINT EXERCISES
For use after Example 1:
1. Solve the system.
$2x + 3y + 7z = -3$
$x - 6y + z = -4$
$-x - 3y + 8z = 1$
$\left(-2, \frac{1}{3}, 0\right)$

TEACHING TIPS
Point out to students that in Example 1 each one of the three original equations must be used at least once when eliminating the variable in Step 1. Also, the same variable must be eliminated with both pairs of equations.

STUDENT HELP NOTES
→ **Homework Help** Students can find extra examples at **www.mcdougallittell.com** that parallel the examples in the student edition.

The linear combination method you learned in Lesson 3.2 can be extended to solve a system of linear equations in three variables.

THE LINEAR COMBINATION METHOD (3-VARIABLE SYSTEMS)

STEP ❶ Use the linear combination method to rewrite the linear system in three variables as a linear system in *two* variables.

STEP ❷ Solve the new linear system for both of its variables.

STEP ❸ Substitute the values found in Step 2 into one of the original equations and solve for the remaining variable.

Note: If you obtain a false equation, such as $0 = 1$, in any of the steps, then the system has no solution. If you do not obtain a false solution, but obtain an identity, such as $0 = 0$, then the system has infinitely many solutions.

EXAMPLE 1 *Using the Linear Combination Method*

STUDENT HELP

HOMEWORK HELP
Visit our Web site
www.mcdougallittell.com
for extra examples.

Solve the system.

$3x + 2y + 4z = 11$	**Equation 1**
$2x - y + 3z = 4$	**Equation 2**
$5x - 3y + 5z = -1$	**Equation 3**

SOLUTION

❶ Eliminate one of the variables in two of the original equations.

$$\begin{array}{r} 3x + 2y + 4z = 11 \\ 4x - 2y + 6z = 8 \\ \hline 7x + 10z = 19 \end{array}$$

Add 2 times the second equation to the first.

New Equation 1

$$\begin{array}{r} 5x - 3y + 5z = -1 \\ -6x + 3y - 9z = -12 \\ \hline -x - 4z = -13 \end{array}$$

Add -3 times the second equation to the third.

New Equation 2

❷ Solve the new system of linear equations in two variables.

$$\begin{array}{r} 7x + 10z = 19 \\ -7x - 28z = -91 \\ \hline -18z = -72 \end{array}$$

New Equation 1
Add 7 times new Equation 2.

$z = 4$ Solve for z.

$x = -3$ Substitute into new Equation 1 or 2 to find x.

❸ Substitute $x = -3$ and $z = 4$ into an original equation and solve for y.

$2x - y + 3z = 4$	**Equation 2**
$2(-3) - y + 3(4) = 4$	Substitute -3 for x and 4 for z.
$y = 2$	Solve for y.

▶ The solution is $x = -3$, $y = 2$, and $z = 4$, or the ordered triple $(-3, 2, 4)$. Check this solution in each of the original equations.

STUDENT HELP

→ **Look Back**
For help with solving linear systems with many or no solutions, see p. 150.

EXAMPLE 2 *Solving a System with No Solution*

Solve the system.

$$x + y + z = 2 \qquad \text{Equation 1}$$
$$3x + 3y + 3z = 14 \qquad \text{Equation 2}$$
$$x - 2y + z = 4 \qquad \text{Equation 3}$$

SOLUTION

When you multiply the first equation by -3 and add the result to the second equation, you obtain a false equation.

$$
\begin{array}{ll}
-3x - 3y - 3z = -6 & \text{Add } -3 \text{ times the first} \\
\underline{3x + 3y + 3z = 14} & \text{equation to the second.} \\
\qquad\qquad\quad 0 = 8 & \text{New Equation 1}
\end{array}
$$

▶ Because you obtained a false equation, you can conclude that the original system of equations has no solution.

EXAMPLE 3 *Solving a System with Many Solutions*

Solve the system.

$$x + y + z = 2 \qquad \text{Equation 1}$$
$$x + y - z = 2 \qquad \text{Equation 2}$$
$$2x + 2y + z = 4 \qquad \text{Equation 3}$$

SOLUTION

Rewrite the linear system in three variables as a linear system in two variables.

$$
\begin{array}{ll}
x + y + z = 2 & \text{Add the first equation} \\
\underline{x + y - z = 2} & \text{to the second.} \\
\quad 2x + 2y = 4 & \text{New Equation 1}
\end{array}
$$

$$
\begin{array}{ll}
x + y - z = 2 & \text{Add the second equation} \\
\underline{2x + 2y + z = 4} & \text{to the third.} \\
\quad 3x + 3y = 6 & \text{New Equation 2}
\end{array}
$$

The result is a system of linear equations in two variables.

$$2x + 2y = 4 \qquad \text{New Equation 1}$$
$$3x + 3y = 6 \qquad \text{New Equation 2}$$

Solve the new system by adding -3 times the first equation to 2 times the second equation. This produces the identity $0 = 0$. So, the system has infinitely many solutions.

Describe the solution. One way to do this is to divide new Equation 1 by 2 to get $x + y = 2$, or $y = -x + 2$. Substituting this into original Equation 1 produces $z = 0$. So, any ordered triple of the form

$$(x, -x + 2, 0)$$

is a solution of the system. For instance, $(0, 2, 0)$, $(1, 1, 0)$, and $(2, 0, 0)$ are all solutions.

EXTRA EXAMPLE 2
Solve the system.
$$-x + 2y + z = 3$$
$$2x + 2y + z = 5$$
$$4x + 4y + 2z = 6$$
no solution

EXTRA EXAMPLE 3
Solve the system.
$$-2x + 4y + z = 1$$
$$3x - 3y - z = 2$$
$$5x - y - z = 8$$
$$(x, 3 - x, 6x - 11)$$

✓ CHECKPOINT EXERCISES

For use after Example 2:
1. Solve the system.
$$2x - 3y + 4z = 5$$
$$2x - 2y + 6z = 4$$
$$3x - 3y + 9z = 8$$
no solution

For use after Example 3:
2. Solve the system.
$$x + y + 2z = 10$$
$$-x + 2y + z = 5$$
$$-x + 4y + 3z = 15$$
$$(y, y, -y + 5)$$

STUDENT HELP NOTES

→ **Look Back** As students look back to p. 150, remind them that if a false statement occurs when solving a linear system, the system has no solution. If the procedure produces an identity, the system has infinitely many solutions.

EXTRA EXAMPLE 4

A theater group sold a total of 440 tickets for $3940. Each regular ticket costs $5, each premium ticket costs $15, and each elite ticket costs $25. The number of regular tickets was three times the number of premium and elite tickets combined. How many of each type of ticket were sold?

330 regular tickets, 46 premium tickets, and 64 elite tickets

CHECKPOINT EXERCISES

For use after Example 4:

1. A quilt maker plans to make 14 quilts this year from 113 yd of fabric. A small quilt requires 4 yd of fabric. A medium quilt requires 7 yd of fabric and a large quilt requires 11 yd of fabric. She plans to make twice as many large quilts as small quilts. How many of each type of quilt should she make? **3 small quilts, 5 medium quilts, and 6 large quilts**

FOCUS ON VOCABULARY

What is a *solution* of a system of equations in three variables?
It is an ordered triple that satisfies each equation in the system.

CLOSURE QUESTION

How do you decide if a linear system in three variables has one, infinitely many, or no solutions using the equations? using the graphs?
See margin for sample answer.

DAILY PUZZLER

I reached into a bag of coins and grabbed a handful. I had 22 coins with a total value of $1.90. There were as many coins of greatest value as coins of least value. What kinds of coins did I get?
pennies, nickels, and quarters

> In yesterday's swim meet, **Roosevelt High** dominated in the individual events, with 24 individual-event placers scoring a total of 56 points. A first-place finish scores 5 points, a second-place finish scores 3 points, and a third-place finish scores 1 point. Having as many third-place finishers as first- and second-place finishers combined really shows the team's depth.

EXAMPLE 4 *Writing and Solving a Linear System*

SPORTS Use a system of equations to model the information in the newspaper article. Then solve the system to find how many swimmers finished in each place.

SOLUTION

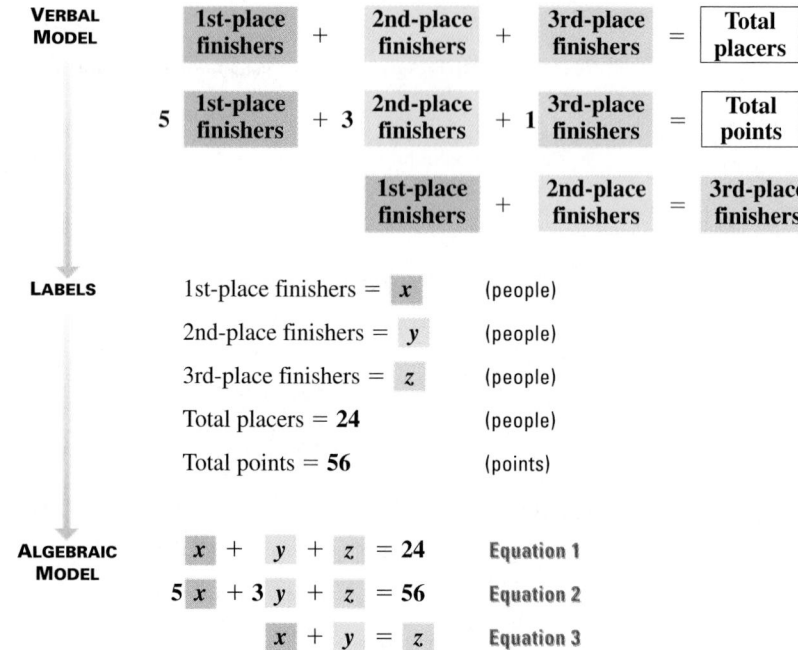

VERBAL MODEL

| 1st-place finishers | + | 2nd-place finishers | + | 3rd-place finishers | = | Total placers |

| 5 | 1st-place finishers | + 3 | 2nd-place finishers | + 1 | 3rd-place finishers | = | Total points |

| | 1st-place finishers | + | 2nd-place finishers | = | 3rd-place finishers |

LABELS

1st-place finishers = x (people)

2nd-place finishers = y (people)

3rd-place finishers = z (people)

Total placers = **24**   (people)

Total points = **56**   (points)

ALGEBRAIC MODEL

$x + y + z = 24$ Equation 1

$5x + 3y + z = 56$ Equation 2

$x + y = z$ Equation 3

Substitute the expression for z from Equation 3 into Equation 1.

$x + y + z = 24$ **Write Equation 1.**

$x + y + (x + y) = 24$ **Substitute $x + y$ for z.**

$2x + 2y = 24$ **New Equation 1**

Substitute the expression for z from Equation 3 into Equation 2.

$5x + 3y + z = 56$ **Write Equation 2.**

$5x + 3y + (x + y) = 56$ **Substitute $x + y$ for z.**

$6x + 4y = 56$ **New Equation 2**

You now have a system of two equations in two variables.

$2x + 2y = 24$ **New Equation 1**
$6x + 4y = 56$ **New Equation 2**

▶ When you solve this system you get $x = 4$ and $y = 8$. Substituting these values into original Equation 3 gives you $z = 12$. There were 4 first-place finishers, 8 second-place finishers, and 12 third-place finishers.

Closure Question *Sample answer:*
You solve the system of equations. You may get a single solution. If you get a false statement, there is no solution, and if you get an identity, there are infinitely many solutions. If the planes intersect in one point, there is one solution; if they intersect in a line, there are infinitely many solutions; and if there is no point common to all three planes, there is no solution.

GUIDED PRACTICE

Vocabulary Check ✓

Concept Check ✓

1. Sample answer:
$x + 2y - z = 7$
$2x - y - z = 12$
$3x + 11y - 4z = 20$

2. When you come up with an impossibility such as $0 = 3$, that tells you that the original system of equations is inconsistent and there are no solutions. If there are many solutions, you would obtain an identity, such as $0 = 0$.

4. Solve one of the equations for one variable in terms of the other two, and then substitute this expression into each of the other two equations, obtaining a system of two equations in two variables.

1. Give an example of a system of three linear equations in three variables.
   See margin.
2. **ERROR ANALYSIS** A student correctly solves a system of equations in three variables and obtains the equation $0 = 3$. The student concludes that the system has infinitely many solutions. Explain the error in the student's reasoning.
   See margin.
3. Look back at the intersecting planes on page 177. How else can three planes intersect so that the system has infinitely many solutions?
   Two or more of the planes could coincide.
4. Explain how to use the substitution method to solve a system of three linear equations in three variables. See margin.

Skill Check ✓ **Decide whether the given ordered triple is a solution of the system.**

5. $(1, 4, 2)$ no
$-2x - y + 5z = 12$
$3x + 2y - z = -7$
$-5x + 4y + 2z = -17$

6. $(7, -1, 0)$ yes
$-4x + 6y - z = -34$
$-2x - 5y + 8z = -9$
$5x + 2y - 4z = 33$

7. $(-2, 3, 3)$ no
$5x - 2y + z = -13$
$x + 4y + 3z = 19$
$-3x + y + 6z = 15$

Use the indicated method to solve the system.

8. linear combination
$x + 5y - z = 16$ (6, 2, 0)
$3x - 3y + 2z = 12$
$2x + 4y + z = 20$

9. substitution
$-2x + y + 3z = -8$
$3x + 4y - 2z = 9$
$x + 2y + z = 4$
$(5, -1, 1)$

10. any method
$9x + 5y - z = -11$
$6x + 4y + 2z = 2$
$2x - 2y + 4z = 4$
$(-2, 2, 3)$

11. 🌐 **INVESTMENTS** Your aunt receives an inheritance of $20,000. She wants to put some of the money into a savings account that earns 2% interest annually and invest the rest in certificates of deposit (CDs) and bonds. A broker tells her that CDs pay 5% interest annually and bonds pay 6% interest annually. She wants to earn $1000 interest per year, and she wants to put twice as much money in CDs as in bonds. How much should she put in each type of investment?
She should invest $2000 in savings, $12,000 in CDs, and $6000 in bonds.

PRACTICE AND APPLICATIONS

STUDENT HELP

▶ **Extra Practice**
to help you master skills is on p. 944.

14. $\left(-\dfrac{22}{13}, \dfrac{29}{13}, \dfrac{6}{13}\right)$

16. All points of the form $(5z + 2, -3z + 3, z)$

STUDENT HELP

▶ **HOMEWORK HELP**
Example 1: Exs. 12–17, 24–33
Examples 2, 3: Exs. 12–33
Example 4: Exs. 18–23, 34–39

LINEAR COMBINATION METHOD Solve the system using the linear combination method.

12. $3x + 2y - z = 8$
$-3x + 4y + 5z = -14$
$x - 3y + 4z = -14$
$(1, 1, -3)$

13. $x + 2y + 5z = -1$
$2x - y + z = 2$
$3x + 4y - 4z = 14$
$(2, 1, -1)$

14. $3x + 2y - 3z = -2$
$7x - 2y + 5z = -14$
$2x + 4y + z = 6$
See margin.

15. $5x - 4y + 4z = 18$
$-x + 3y - 2z = 0$
$4x - 2y + 7z = 3$
$(6, 0, -3)$

16. $x + y - 2z = 5$
$x + 2y + z = 8$
$2x + 3y - z = 13$
See margin.

17. $-5x + 3y + z = -15$
$10x + 2y + 8z = 18$
$15x + 5y + 7z = 9$
$(1, -4, 2)$

SUBSTITUTION METHOD Solve the system using the substitution method.

18. $-2x + y + 6z = 1$
$3x + 2y + 5z = 16$
$7x + 3y - 4z = 11$
$(4, -3, 2)$

19. $x - 6y - 2z = -8$
$-x + 5y + 3z = 2$
$3x - 2y - 4z = 18$
$(4, 3, -3)$

20. $x + y + z = 4$
$5x + 5y + 5z = 12$
$x - 4y + z = 9$
no solutions

21. $x - 3y + 6z = 21$
$3x + 2y - 5z = -30$
$2x - 5y + 2z = -6$
$(-3, 2, 5)$

22. $x + y - 2z = 5$
$x + 2y + z = 8$
$2x + 3y - z = 1$
no solutions

23. $2x - 3y + z = 10$
$y + 2z = 13$
$z = 5$
$(7, 3, 5)$

3.6 *Solving Systems of Linear Equations in Three Variables* **181**

⚪ **ASSIGNMENT GUIDE**

BASIC
Day 1: pp. 181–184 Exs.12–27 multiples of 3, 34, 42–69 multiples of 3,
Quiz 3 Exs. 1–14

AVERAGE
Day 1: pp. 181–184 Exs.12–30 multiples of 3, 35–39, 42, 45–71 odd,
Quiz 3 Exs. 1–14

ADVANCED
Day 1: pp. 181–184 Exs.12–33 multiples of 3, 36–44, 45–71 odd,
Quiz 3 Exs. 1–14

BLOCK SCHEDULE
pp. 181–184 Exs.12–30 multiples of 3, 35–39, 42, 45–71 odd,
Quiz 3 Exs. 1–14 (with 3.5)

EXERCISE LEVELS
Level A: *Easier*
12–23

Level B: *More Difficult*
24–39, 42

Level C: *Most Difficult*
40–41, 43–44

✔ **HOMEWORK CHECK**
To quickly check student understanding of key concepts, go over the following exercises:
Exs. 12, 18, 24, 34. See also the Daily Homework Quiz:

• Blackline Master (*Chapter 4 Resource Book*, p. 11)

• 📄 Transparency (p. 26)

CHOOSING A METHOD Solve the system using any algebraic method.

24. $2x - 2y + z = 3$ $\quad (-5, -6, 1)$
$5y - z = -31$
$x + 3y + 2z = -21$

25. $17x - y + 2z = -9$ $\quad \left(-\frac{2}{7}, 0, -\frac{29}{14}\right)$
$x + y - 4z = 8$
$3x - 2y - 12z = 24$

26. $-2x + y + z = -2$ $\quad (7, 9, 3)$
$5x + 3y + 3z = 71$
$4x - 2y - 3z = 1$

27. $x - 9y + 4z = 1$ $\quad (2, 1, 2)$
$-4x + 18y - 8z = -6$
$2x + y - 4z = -3$

28. $2x + y + 2z = 7$ $\quad$ all points of
$2x - y + 2z = 1$ $\quad$ the form
$5x + y + 5z = 13$ $\quad (2 - z, 3, z)$

29. $7x - 3y + 4z = -14$ $\quad (-1, 1, -1)$
$8x + 2y - 24z = 18$
$6x - 10y + 8z = -24$

30. $12x + 6y + 7z = -35$ $\quad (10, -20, -5)$
$7x - 5y - 6z = 200$
$x + y = -10$

31. $7x - 10y + 8z = -50$ $\quad (6, 6, -4)$
$-2x - 5y + 12z = -90$
$3x + 4y + 4z = 26$

32. $-2x - 3y - 6z = -26$ $\quad (10, -10, 6)$
$5x + 5y + 4z = 24$
$3x + 4y - 5z = -40$

33. $3x + 3y + z = 30$ $\quad \left(\frac{128}{13}, -\frac{113}{26}, 13.5\right)$
$10x - 3y - 7z = 17$
$-6x + 7y + 3z = -49$

34. 🌐 **FIELD TRIP** You and two friends buy snacks for a field trip. Using the information given in the table, determine the price per pound for mixed nuts, granola, and dried fruit. **A lb of mixed nuts costs about $3.15, a lb of granola about $2.75 and a lb of dried fruit about $2.89.**

Shopper	Mixed nuts	Granola	Dried fruit	Total price
You	1 lb	$\frac{1}{2}$ lb	$\frac{1}{2}$ lb	$5.97
Kenny	$1\frac{1}{3}$ lb	$\frac{1}{4}$ lb	$\frac{3}{2}$ lb	$9.22
Vanessa	$\frac{1}{3}$ lb	$1\frac{1}{2}$ lb	2 lb	$10.96

35. $f + s + t = 20$
$5f + 3s + t = 68$
$s = f + t$,
There were 7 first place finishers, 10 second place finishers, and 3 third place finishers.

35. 🌐 **TRACK MEET** Use a system of linear equations to model the data in the following newspaper article. Solve the system to find how many athletes finished in each place.

Lawrence High prevailed in Saturday's track meet with the help of 20 individual-event placers earning a combined 68 points. A first-place finish earns 5 points, a second-place finish earns 3 points, and a third-place finish earns 1 point. Lawrence had a strong second-place showing, with as many second-place finishers as first- and third-place finishers combined.

36. 🌐 **CHINESE RESTAURANT** Jeanette, Raj, and Henry go to a Chinese restaurant for lunch and order three different luncheon combination platters. Jeanette orders 2 portions of fried rice and 1 portion of chicken chow mein. Raj orders 1 portion of fried rice, 1 portion of chicken chow mein, and 1 portion of sautéed broccoli. Henry orders 1 portion of sautéed broccoli and 2 portions of chicken chow mein. Jeanette's platter costs $5, Raj's costs $5.25, and Henry's costs $5.75. How much does 1 portion of chicken chow mein cost? **Chicken chow mein is $2 per order.**

FOCUS ON APPLICATIONS

VOTER REGISTRATION

In November of 1996 there were 10.8 million people aged 18–20 years old in the United States. Of these, 5 million people were registered voters and 3.4 million actually voted.

APPLICATION LINK
www.mcdougallittell.com

FURNITURE SALE In Exercises 37 and 38, use the furniture store ad shown at the right.

$1300 Sofa and love seat
$1400 Sofa and two chairs
$1600 Sofa, love seat, and one chair

Sam's Furniture Store

37. Write a system of equations for the three combinations of furniture.
$s + l = 1300$; $s + 2c = 1400$; $s + l + c = 1600$

38. What is the price of each piece of furniture? A chair costs $300, a sofa $800, and a love seat $500.

39. SOCIAL STUDIES CONNECTION For several political parties, the table shows the approximate percent of votes for the party's presidential candidate that were cast in 1996 by voters in two regions of the United States. Write and solve a system of equations to find the *total* number of votes for each party (Democrat, Republican, and Other). Use the fact that a total of about 100 million people voted in 1996. ▶ Source: *Statistical Abstract of the United States*
Democrat: 50 million, Republican: 40 million, Other parties: 10 million

Region	Democrat (%)	Republican (%)	Other parties (%)	Total voters (millions)
Northeast	20	15	20	18
South	30	35	25	31.5

40. **GOING IN REVERSE** Which values should be given to a, b, and c so that the linear system shown has $(-1, 2, -3)$ as its only solution? $a = 12$, $b = -4$, $c = 10$

$$x + 2y - 3z = a$$
$$-x - y + z = b$$
$$2x + 3y - 2z = c$$

41. **CRITICAL THINKING** Write a system of three linear equations in three variables that has the given number of solutions. See margin. Sample answers are given.

a. one solution **b.** no solution **c.** infinitely many solutions

42. **MULTI-STEP PROBLEM** You have $25 to spend on picking 21 pounds of three different types of apples in an orchard. The Empire apples cost $1.40 per pound, the Red Delicious apples cost $1.10 per pound, and the Golden Delicious apples cost $1.30 per pound. You want twice as many Red Delicious apples as the other two kinds combined.
$e + r + g = 21$; $1.4e + 1.1r + 1.3g = 25$; $r = 2(e + g)$

a. Write a system of equations to represent the given information.

b. How many pounds of each type of apple should you buy? See margin.

c. *Writing* Create your own situation in which you are buying three different types of fruit. State the total amount of fruit you need, the price of each type of fruit, the amount of money you have to spend, and the desired ratio of one type of fruit to the other two types. Write a system of equations representing your situation. Then solve your system to find the number of pounds of each type of fruit you should buy. See margin.

★ Challenge

SYSTEMS OF FOUR EQUATIONS Solve the system of equations. Describe what you are doing at each step in your solution process.

43. $w + x + y + z = 6$ See margin.
$3w - x + y - z = -3$
$2w + 2x - 2y + z = 4$
$2w - x - y + z = -4$

44. $2w - x + 5y + z = -3$
$3w + 2x + 2y - 6z = -32$
$w + 3x + 3y - z = -47$
$5w - 2x - 3y + 3z = 49$
$w = 2$, $x = -12$, $y = -4$, $z = 1$

EXTRA CHALLENGE
www.mcdougallittell.com

Test Preparation

42b. You should pick 5 lb of empire apples, 2 lb of golden delicious, and 14 lb of red delicious.

43. $w = -\frac{2}{19}$, $x = \frac{123}{38}$, $y = \frac{65}{38}$, $z = \frac{22}{19}$

41 a. $x + y + z = 3$
$2x - 2y + 5z = 23$
$4x + 3z = 1$

41 b. $x + y + z = 3$
$2x - 2y + 5z = 23$
$4x - 4y + 10z = 11$

41c. $x + y + z = 3$
$2x - 2y + 5z = 23$
$3x - y + 6z = 26$

42c. *Sample answer:* You need 4 pounds of berries to make berry tarts for a party. Strawberries cost $1.50 per pound, raspberries cost $4.00 per pound and blueberries cost $2.00 per pound. You have $8 to spend, and plan to use as many pounds of strawberries as of blueberries and raspberries combined.
$s + r + b = 4$
$s = r + b$
$1.5s + 4r + 2b = 8$
Buy 2 lb of strawberries, $\frac{1}{2}$ lb of raspberries and 1.5 lb of blueberries.

54–71. See Additional Answers beginning on page AA1.
1–6. See Additional Answers beginning on page AA1.

ADDITIONAL PRACTICE AND RETEACHING

For Lesson 3.6:
• Practice Levels A, B, and C (*Chapter 3 Resource Book*, p. 84)
• Reteaching with Practice (*Chapter 3 Resource Book*, p. 87)
• ☐ See Lesson 3.6 of the *Personal Student Tutor*

For more Mixed Review:
• ☐ Search the *Test and Practice Generator* for key words or specific lessons.

DAILY HOMEWORK QUIZ

📖 **Transparency Available**

Solve each system.

1. $-x + 5y + 8z = 45$
$3x + 2y + 2z = 4$
$5x + 2y - z = -18$
$(-2, -1, 6)$

2. $2x + 3y + 4z = 5$
$2x - 3y + 2z = 29$
$-7x + 5y + z = -50$
$(4, -5, 3)$

3. Jennifer has 15 coins worth a total of $1.45. All the coins are nickels, dimes, or quarters. The number of quarters is $\frac{1}{4}$ the number of nickels and dimes combined. Solve a linear system to find how many of each kind of coin she has.
10 nickels, 2 dimes, 3 quarters

EXTRA CHALLENGE NOTE
↳ Challenge problems for Lesson 3.6 are available in **blackline** format in the *Chapter 3 Resource Book*, p. 91 and at **www.mcdougallittell.com.**

ADDITIONAL TEST PREPARATION

1. OPEN ENDED Write a real-life problem that can be solved using a linear system in three variables. Solve the system and interpret your answer.
Check work.

2. WRITING Describe the different types of linear systems in three variables. *Sample answer:* **A system with one solution has a graph whose planes intersect in one point. A system with an infinite number of solutions has a graph whose planes intersect in a line or an entire plane, and a system with no solution has a graph where the planes do not have a common intersection.**

MIXED REVIEW

54. $-9 < x < 31$;

55. $x \le -14.5$ or $x \ge 11.5$;

56. $x \le -56$ or $x \ge -16$;

57. $x < -\frac{3}{2}$ or $x > -\frac{1}{4}$;

58. $-5 < x < 15$;

59. $-\frac{29}{3} \le x \le \frac{31}{3}$;

60. $x \le -2$ or $x \ge 6$;

61. $-\frac{22}{3} < x < 6$;

62. $x < 4$ or $x > 6$;

See margin for graphs.

PERFORMING AN OPERATION Perform the indicated operation. (Review 1.1 for 4.1)

45. $-10 + 21$ **11**

46. $15 - (-1)$ **16**

47. $12 \cdot 7$ **84**

48. $-2 - (-20)$ **18**

49. $-9 + (-7)$ **-16**

50. $-8(-6)$ **48**

51. $-\frac{1}{2} + \frac{4}{5}$ **$\frac{3}{10}$**

52. $-\frac{1}{3}\left(-\frac{2}{7}\right)$ **$\frac{2}{21}$**

53. $\frac{3}{4} - 3$ **$-\frac{9}{4}$**

SOLVING AND GRAPHING Solve the inequality. Then graph your solution. (Review 1.7)   54–62. See margin.

54. $|11 - x| < 20$

55. $|2x + 3| \ge 26$

56. $\left|18 + \frac{1}{2}x\right| \ge 10$

57. $|7 + 8x| > 5$

58. $|5 - x| < 10$

59. $|3x - 1| \le 30$

60. $|-3x + 6| \ge 12$

61. $|6x + 4| < 40$

62. $|15 - 3x| > 3$

PLOTTING POINTS Plot the ordered triple in a three-dimensional coordinate system. (Review 3.5)   63–71. See margin.

63. $(3, 6, 0)$

64. $(-3, -6, -4)$

65. $(-5, 9, 2)$

66. $(-9, 4, -7)$

67. $(6, -2, -6)$

68. $(-8, 5, -6)$

69. $(0, -3, -3)$

70. $(2, 2, -2)$

71. $(-4, -7, -3)$

QUIZ 3   *Self-Test for Lessons 3.5 and 3.6*

Sketch the graph of the equation. Label the points where the graph crosses the x-, y-, and z-axes. (Lesson 3.5)   1– 6. See margin.

1. $2x + 5y + 3z = 15$

2. $x + 4y + 16z = 8$

3. $3x + y + z = 10$

4. $3x + 12y + 6z = 9$

5. $5x - 2y + z = 15$

6. $-x + 9y - 3z = 18$

7. $f(x, y) =$
$\frac{1}{3}x - \frac{1}{6}y + 6$; $\frac{20}{3}$

8. $f(x, y) = \frac{1}{2}x + y + 2$; 4

9. $f(x, y) = 20x - 3y - 15$; 66

10. $f(x, y) =$
$\frac{1}{3}x - \frac{1}{6}y + 4$; $\frac{41}{6}$

Write the linear equation as a function of x and y. Then evaluate the function for the given values. (Lesson 3.5)   7– 10. See margin.

7. $-x + \frac{1}{2}y + 3z = 18$, $f(2, 0)$

8. $4x + 8y - 8z = -16$, $f(-4, 4)$

9. $20x - 3y - z = 15$, $f(3, -7)$

10. $-2x + y + 6z = 24$, $f(12, 7)$

Solve the system using any algebraic method. (Lesson 3.6)

11. $2x + 4y + 3z = 10$
$3x - y + 6z = 15$
$5x + 2y - z = 25$
(5, 0, 0)

12. $3x - 2y + 3z = 11$
$5x + 2y - 2z = 4$
$-x + y + z = -7$
(2, −4, −1)

13. $x - 2y + 3z = -9$
$2x + 5y + z = 10$
$3x - 6y + 9z = 12$
no solutions

14. 🌐 **STATE ORCHESTRA** Fifteen band members from your school were selected to play in the state orchestra. Twice as many students who play a wind instrument were selected as students who play a string or percussion instrument. Of the students selected, one fifth play a string instrument. How many students playing each type of instrument were selected to play in the state orchestra? (Lesson 3.6) **3 string players, 10 woodwinds, and 2 percussionists were selected.**

1–6. See Additional Answers beginning on page AA1.

Chapter Summary

WHAT did you learn?

Solve systems of linear equations in two variables.
- by graphing **(3.1)**
- using algebraic methods **(3.2)**

Graph and solve systems of linear inequalities. **(3.3)**

Solve linear programming problems. **(3.4)**

Graph linear equations in three variables. **(3.5)**

Model real-life problems with functions of two variables. **(3.5)**

Solve systems of linear equations in three variables. **(3.6)**

Identify the number of solutions of a linear system. **(3.1, 3.2, 3.6)**

Solve real-life problems.
- using a system of linear equations **(3.1, 3.2, 3.6)**
- using a system of linear inequalities **(3.3, 3.4)**

WHY did you learn it?

Plan a vacation within a budget. **(p. 141)**
Find the weights of atoms in a molecule. **(p. 153)**

Describe conditions that will satisfy nutritional requirements of wildlife. **(p. 161)**

Plan a meal that minimizes cost while satisfying nutritional requirements. **(p. 167)**

Find the volume of a geometric figure graphed in a three-dimensional coordinate system. **(p. 174)**

Evaluate advertising costs of a commercial. **(p. 175)**

Use regional data to find the number of voters for different political parties in the United States. **(p. 183)**

See if a bus catches up to another one before arriving at a common destination. **(p. 144)**

Find the break-even point of a business. **(p. 153)**
Display possible sale prices for shoes. **(p. 161)**

How does Chapter 3 fit into the BIGGER PICTURE of algebra?

Linear algebra is an important branch of mathematics that begins with solving linear systems. It has widespread applications to other areas of mathematics and to real-life problems, especially in business and the sciences. You will continue your study of linear algebra in the next chapter with matrices.

STUDY STRATEGY

Did you recognize when new skills related to previously learned skills?

The two-column list you made, following the **Study Strategy** on page 138, may resemble this one.

Building on Previous Skills

Chapter 3

Graph a system of linear equations or inequalities.

Check a solution of a system.

Tell the number of solutions a system has.

Plot an ordered triple.

Graph $ax + by + cx = d$.

Function notation: $f(x, y)$

Chapter 2

Graph a linear equation or inequality.

Check a solution of an equation or inequality.

Decide if lines are parallel.

Plot an ordered pair.

Graph $Ax + By = C$.

Function notation: $f(x)$

185

ADDITIONAL RESOURCES

The following resources are available to help review the material in this chapter.

- Chapter Review Games and Activities (*Chapter 3 Resource Book*, p. 92)
- *Instant Replay: Video Review Games*
- Personal Student Tutor
- Cumulative Review, Chs. 1–3 (*Chapter 3 Resource Book*, p. 104)

1.

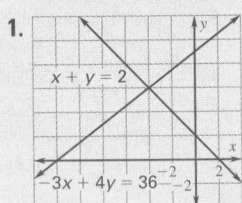

2.

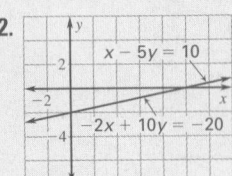

3.

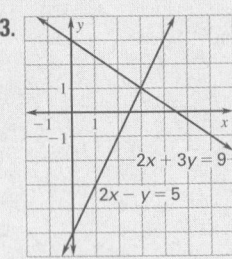

4.

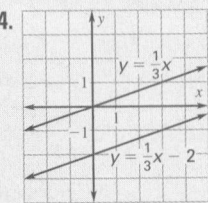

VOCABULARY

- system of two linear equations in two variables, p. 139
- solution of a system of linear equations, p. 139
- substitution method, p. 148
- linear combination method, p. 149

- System of linear inequalities in two variables, p. 156
- solution of a system of linear inequalities, p. 156
- graph of a system of linear inequalities, p. 156
- optimization, p. 163
- linear programming, p. 163

- objective function, p. 163
- constraints, p. 163
- feasible region, p. 163
- three-dimensional coordinate system, p. 170
- z-axis, p. 170
- ordered triple, p. 170
- octants, p. 170

- linear equation in three variables, p. 171
- function of two variables, p. 171
- system of three linear equations in three variables, p. 177
- solution of a system of three linear equations, p. 177

3.1 SOLVING LINEAR SYSTEMS BY GRAPHING

Examples on pp. 139–141

> **EXAMPLE** You can solve a system of two linear equations in two variables by graphing.
>
> $x + 2y = -4$ **Equation 1**
> $3x + 2y = 0$ **Equation 2**
>
> From the graph, the lines appear to intersect at $(2, -3)$. You can check this algebraically as follows.
>
> $2 + 2(-3) = -4$ ✓ **Equation 1 checks.**
>
> $3(2) + 2(-3) = 0$ ✓ **Equation 2 checks.**

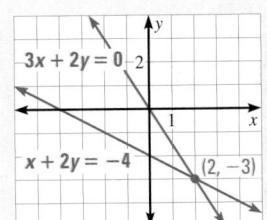

Graph the linear system and tell how many solutions it has. If there is exactly one solution, estimate the solution and check it algebraically. 1–4. See margin for graphs.

1. $x + y = 2$
$-3x + 4y = 36$
one solution; $(-4, 6)$

2. $x - 5y = 10$
$-2x + 10y = -20$
infinitely many solutions

3. $2x - y = 5$
$2x + 3y = 9$
one solution; $(3, 1)$

4. $y = \frac{1}{3}x$
$y = \frac{1}{3}x - 2$
no solution

3.2 SOLVING LINEAR SYSTEMS ALGEBRAICALLY

Examples on pp. 148–151

> **EXAMPLE 1** You can use the substitution method to solve a system algebraically.
>
> ① Solve the first equation for x.
> ② Substitute the value of x into the second equation and solve for y.
>
> $x - 4y = -25$ 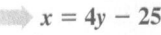 $x = 4y - 25$ ⟹ $2(4y - 25) + 12y = 10$
> $2x + 12y = 10$ $y = 3$
>
> When you substitute $y = 3$ into one of the original equations, you get $x = -13$.

EXAMPLE 2 You can also use the linear combination method to solve a system of equations algebraically.

❶ Multiply the first equation by 3 and add to the second equation. Solve for x.

$$x - 4y = -25 \quad\Rightarrow\quad 3x - 12y = -75$$
$$2x + 12y = 10 \quad\Rightarrow\quad \underline{2x + 12y = 10}$$
$$5x = -65$$
$$x = -13$$

❷ Substitute $x = -13$ into the original first equation and solve for y.

$$-13 - 4y = -25$$
$$-4y = -12$$
$$y = 3$$

Solve the system using any algebraic method.

5. $9x - 5y = -30$
 $x + 2y = 12$
 (0, 6)

6. $x + 3y = -2$
 $x + y = 2$
 (4, −2)

7. $2x + 3y = -7$
 $-4x - 5y = 13$
 (−2,−1)

8. $3x + 3y = 0$
 $-2x + 6y = -24$
 (3, −3)

3.3

GRAPHING AND SOLVING SYSTEMS OF LINEAR INEQUALITIES

Examples on pp. 156–158

EXAMPLE You can use a graph to show all the solutions of a system of linear inequalities.

$$x \geq 0$$
$$y \geq 0$$
$$x + 2y < 10$$

Graph each inequality. The graph of the system is the region common to *all* of the shaded half-planes and includes any solid boundary line.

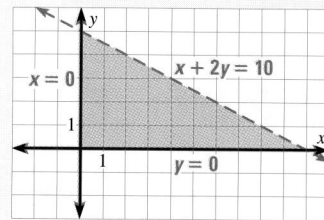

Graph the system of linear inequalities. 9–12. See margin.

9. $y < -3x + 3$
 $y > x - 1$

10. $x \geq 0$
 $y \geq 0$
 $-x + 2y < 8$

11. $x \geq -2$
 $x \leq 5$
 $y \geq -1$
 $y \leq 3$

12. $x + y \leq 8$
 $2x - y > 0$
 $y \leq 4$

3.4

LINEAR PROGRAMMING

Examples on pp. 163–165

EXAMPLE You can find the minimum and maximum values of the objective function $C = 6x + 5y$ subject to the constraints graphed below. They must occur at vertices of the feasible region.

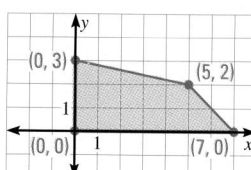

At (0, 0): $C = 6(0) + 5(0) = 0$ ◄——— Minimum

At (0, 3): $C = 6(0) + 5(3) = 15$

At (5, 2): $C = 6(5) + 5(2) = 40$

At (7, 0): $C = 6(7) + 5(0) = 42$ ◄——— Maximum

9.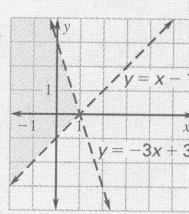

10.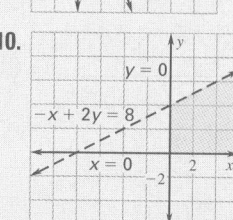

11.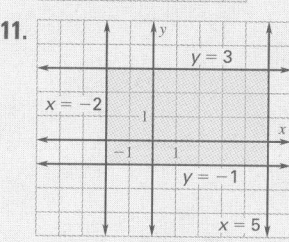

12.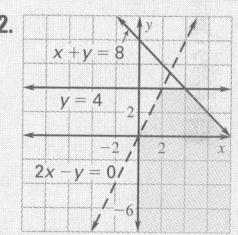

13. max of 50 at (10, 0); min of 0 at (0, 0)
14. max of 25 at (5, 0); min of 0 at (0, 0)
15. max of 38 at (4, 9); min of 5 at (1, 0)
16. max of 35 at (5, 5); min of 0 at (0, 0)

17.
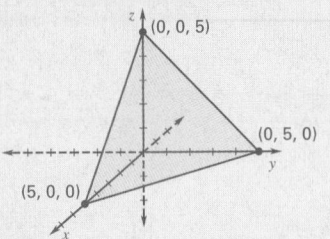

18.
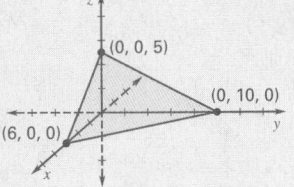

19.
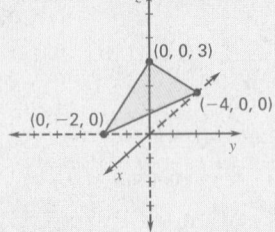

3.4 continued

Find the minimum and maximum values of the objective function
$C = 5x + 2y$ **subject to the given constraints.** 13–16. See margin.

13. $x \geq 0$
$y \geq 0$
$x + y \leq 10$

14. $x \geq 0$
$y \geq 0$
$4x + 5y \leq 20$

15. $x \geq 1; x \leq 4$
$y \geq 0; y \leq 9$

16. $y \leq 6; x + y \leq 10$
$x \geq 0; x - y \leq 0$

3.5 | **GRAPHING LINEAR EQUATIONS IN THREE VARIABLES** | Examples on pp. 170–172

EXAMPLE You can sketch the graph of an equation in
three variables in a three-dimensional coordinate system.

To graph $3x + 4y - 3z = 12$, find x-, y-, and z-intercepts.

If $y = 0$ and $z = 0$, then $x = 4$. Plot (4, 0, 0).

If $x = 0$ and $z = 0$, then $y = 3$. Plot (0, 3, 0).

If $x = 0$ and $y = 0$, then $z = -4$. Plot (0, 0, -4).

Draw the plane that contains (4, 0, 0), (0, 3, 0), and (0, 0, -4).

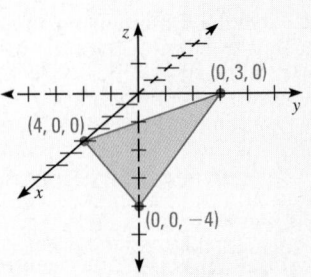

Sketch the graph of the equation. Label the points where the graph crosses the
x-, y-, and z-axes. 17–19. See margin.

17. $x + y + z = 5$

18. $5x + 3y + 6z = 30$

19. $3x + 6y - 4z = -12$

3.6 | **SOLVING SYSTEMS OF LINEAR EQUATIONS IN THREE VARIABLES** | Examples on pp. 177–180

EXAMPLE You can use algebraic methods to solve a system of linear equations in
three variables. First rewrite it as a system in two variables.

❶ Add the first and second equations.

$x - 3y + z = 22$ ⟶ $x - 3y + z = 22$
$2x - 2y - z = -9$ ⟶ $\underline{2x - 2y - z = -9}$
$x + y + 3z = 24$ $\qquad 3x - 5y = 13$

❷ Multiply the second equation by 3
and add to the third equation.
$6x - 6y - 3z = -27$
$\underline{x + y + 3z = 24}$
$7x - 5y = -3$

❸ Solve the new system.
$3x - 5y = 13$
$\underline{-7x + 5y = 3}$
$-4x = 16$
$x = -4$ and $y = -5$

When you substitute $x = -4$ and $y = -5$ into one of the original
equations, you get the value of the last variable: $z = 11$.

Solve the system using any algebraic method.

20. $x + 2y - z = 3$ (2, 0, -1)
$-x + y + 3z = -5$
$3x + y + 2z = 4$

21. $2x - 4y + 3z = 1$ $\left(-\frac{1}{2}, 1, 2\right)$
$6x + 2y + 10z = 19$
$-2x + 5y - 2z = 2$

22. $x + y + z = 3$
$x + y - z = 3$
$2x + 2y + z = 6$
all points of the form $(x, 3 - x, 0)$

Chapter Test

ADDITIONAL RESOURCES
• **Chapter 3 Resource Book**
Chapter Test (3 levels) (p. 93)
SAT/ACT Chapter Test (p. 99)
Alternative Assessment (p. 100)
• ⊞ **Test and Practice Generator**

Graph the linear system and tell how many solutions it has. If there is exactly one solution, estimate the solution and check it algebraically. 1–4. See margin for graphs.

1. $x + y = 1$
$2x - 3y = 12$
one solution; $(3, -2)$

2. $y = -\frac{1}{3}x + 4$
$y = 6$
one solution; $(-6, 6)$

3. $y = 2x + 2$
$y = 2x - 3$
no solution

4. $\frac{1}{2}x + 5y = 2$
$-x - 10y = -4$
infinitely many solutions

Solve the system using any algebraic method.

5. $3x + 6y = -9$
$x + 2y = -3$
infinitely many solutions

6. $x - y = -5$ $(3, 8)$
$x + y = 11$

7. $7x + y = -17$ $(-2, -3)$
$3x - 10y = 24$

8. $8x + 3y = -2$
$-5x + y = -3$
$\left(\frac{7}{23}, -\frac{34}{23}\right)$

Graph the system of linear inequalities. 9–12. See margin.

9. $2x + y \geq 1$
$x \leq 3$

10. $x \geq 0$
$y < x$
$y > -x$

11. $x + 2y \geq -6$
$x + 2y \leq 2$
$y \geq -1$

12. $x + y < 7$
$2x - y \geq 5$
$x \geq -2$

Find the minimum and maximum values of the objective function subject to the given constraints.

13. Objective function: $C = 7x + 4y$ max of 42 at $(6, 0)$; min of 0 at $(0, 0)$

Constraints: $x \geq 0$
$y \geq 0$
$4x + 3y \leq 24$

14. Objective function: $C = 3x + 4y$ max of 36 at $(4, 6)$; no min—feasible region is unbounded

Constraints: $x + y \leq 10$
$-x + y \leq 5$
$2x + 4y \leq 32$

Plot the ordered triple in a three-dimensional coordinate system. 15–18. See margin.

15. $(-1, 3, 2)$

16. $(0, 4, -2)$

17. $(-5, -1, 2)$

18. $(6, -2, 1)$

Sketch the graph of the equation. Label the points where the graph crosses the x-, y-, and z-axes. 19–21. See margin.

19. $2x + 3y + 5z = 30$

20. $4x + y + 2z = 8$

21. $3x + 12y - 6z = 24$

22. Write the linear equation $2x - 5y + z = 9$ as a function of x and y. Then evaluate the function when $x = 10$ and $y = 3$. $f(x, y) = -2x + 5y + 9$; 4

Solve the system using any algebraic method.

23. $x + 2y - 6z = 23$ $(1, 2, -3)$
$x + 3y + z = 4$
$2x + 5y - 4z = 24$

24. $x + y + 2z = 1$ no solutions
$x - y + z = 0$
$3x + 3y + 6z = 4$

25. $x + 3y - z = 1$ $(-5, 2, 0)$
$-4x - 2y + 5z = 16$
$7x + 10y + 6z = -15$

26. 🔵 **CRAFT SUPPLIES** You are buying beads and string to make a necklace. The string costs \$1.50, a package of 10 decorative beads costs \$.50, and a package of 25 plain beads costs \$.75. You can spend only \$7.00 and you need 150 beads. How many packages of each type of bead should you buy?

Buy 5 packages of decorative beads, 4 packages of plain beads and the string for a total cost of \$7.00.

27. 🔵 **BUSINESS** An appliance store manager is ordering chest and upright freezers. One chest freezer costs \$250 and delivers a \$40 profit. One upright freezer costs \$400 and delivers a \$60 profit. Based on previous sales, the manager expects to sell at least 100 freezers. Total profit must be at least \$4800. Find the least number of each type of freezer the manager should order to minimize costs.

Order 120 chest freezers. This will give a profit of \$4800 at a cost of \$30,000.

1.
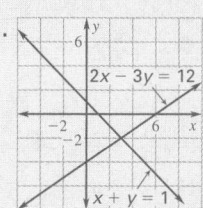
$2x - 3y = 12$
$x + y = 1$

2.
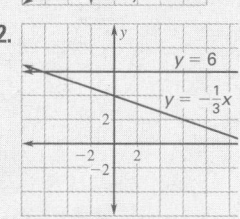
$y = 6$
$y = -\frac{1}{3}x$

3.
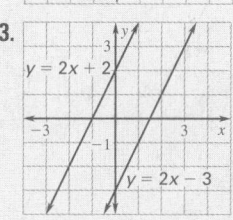
$y = 2x + 2$
$y = 2x - 3$

4.
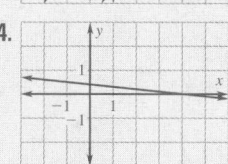

9–12, 15–18, 19–21.
See Additional answers beginning on page AA1.

Chapter Standardized Test

> **TEST-TAKING STRATEGY** If you find yourself spending too much time on one test question and getting frustrated, move on to the next question. You can revisit a difficult problem later with a fresh perspective.

ADDITIONAL RESOURCES
- *Chapter 3 Resource Book*
  Chapter Test (3 levels) (p. 93)
  SAT/ACT Chapter Test (p. 99)
  Alternative Assessment (p. 100)
- 🖥 *Test and Practice Generator*

1. MULTIPLE CHOICE Which ordered pair is a solution of the following system of linear equations? **C**

$$2x - 5y = -12$$
$$-x + 4y = 9$$

Ⓐ $(-6, 0)$ Ⓑ $(3, 3)$ Ⓒ $(-1, 2)$

Ⓓ $(-9, 0)$ Ⓔ $(2, 2)$

2. MULTIPLE CHOICE How many solutions does the following system have? **E**

$$8x - 4y = 20$$
$$2x - y = 5$$

Ⓐ 0   Ⓑ 1   Ⓒ 2

Ⓓ 4   Ⓔ infinitely many

3. MULTIPLE CHOICE A total of $6500 is invested in two funds. One fund pays 4% interest annually and the other fund pays 6% interest annually. The combined annual interest earned is $350. How much of the $6500 is invested in one of the funds? **A**

Ⓐ $2000   Ⓑ $2500   Ⓒ $3250

Ⓓ $4000   Ⓔ $5500

4. MULTIPLE CHOICE Which ordered pair is *not* a solution of the following system of linear inequalities? **C**

$$x \geq -2$$
$$y \geq -3$$
$$y < 3x + 3$$

Ⓐ $(4, -3)$ Ⓑ $(0, 0)$ Ⓒ $(1, 6)$

Ⓓ $(5, 17)$ Ⓔ $(-1, -1)$

5. MULTIPLE CHOICE What is the minimum value of the objective function $C = 4x + 3y$ subject to the following constraints? **C**

$$x \geq 0$$
$$y \geq 0$$
$$2x + 3y \leq 18$$
$$3x + y \geq 6$$

Ⓐ 0   Ⓑ 2   Ⓒ 8

Ⓓ 18   Ⓔ 36

6. MULTIPLE CHOICE Which linear equation is graphed below? **D**

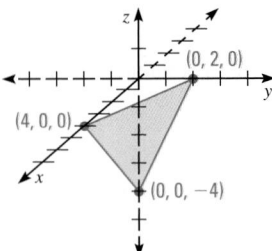

Ⓐ $x - 2y - z = 4$ Ⓑ $x - 2y + z = -4$

Ⓒ $x + 2y - z = -4$ Ⓓ $x + 2y - z = 4$

Ⓔ $-x + 2y + z = 4$

7. MULTIPLE CHOICE At which point does the graph of $15x - 6y - 3z = 30$ cross the y-axis? **E**

Ⓐ $(0, -6, 0)$ Ⓑ $(2, 0, 0)$

Ⓒ $(0, -3, 0)$ Ⓓ $(0, 0, -10)$

Ⓔ $(0, -5, 0)$

8. MULTIPLE CHOICE Which ordered triple is a solution of the following linear system? **A**

$$2x + 5y + 3z = 10$$
$$3x - y + 4z = 8$$
$$5x - 2y + 7z = 12$$

Ⓐ $(7, 1, -3)$ Ⓑ $(7, -1, -3)$

Ⓒ $(7, 1, 3)$ Ⓓ $(7, -1, 3)$

Ⓔ $(-7, 1, -3)$

9. MULTIPLE CHOICE A cashier at a restaurant made the chart below for popular lunch combinations. What is the individual price of soup? **B**

Lunch Combinations
Soup + Salad = $4.25
Soup + Sandwich = $4.75
Salad + Sandwich = $5.50

Ⓐ $1.50   Ⓑ $1.75   Ⓒ $2.25

Ⓓ $2.50   Ⓔ $3.00

QUANTITATIVE COMPARISON In Exercises 10 and 11, choose the statement that is true about the given quantities.

(A) The quantity in column A is greater.

(B) The quantity in column B is greater.

(C) The two quantities are equal.

(D) The relationship cannot be determined from the given information.

Column A	Column B	
10. $f(x, y) = \frac{1}{5}(20 - 2x + y)$, $f(4, 8)$	$f(x, y) = \frac{1}{5}(20 - 2x + y)$, $f(-1, 3)$	**B**
11. $f(x, y) = \frac{1}{2}(10 + 4x - 3y)$, $f(-2, -1)$	$f(x, y) = \frac{1}{2}(10 + 4x - 3y)$, $f(2, 1)$	**B**

12. MULTI-STEP PROBLEM Use the following system of linear equations.

$$x - 2y = 2 \qquad \text{Equation 1}$$
$$5x - 4y = -8 \qquad \text{Equation 2}$$

a. Solve the system by graphing. $(-4, -3)$

b. Solve the system using the substitution method. Show your work. **12b–d. See margin.**

c. Solve the system using the linear combination method. Show your work.

d. *Writing* Which method do you prefer for solving this system? Explain.

13. MULTI-STEP PROBLEM Write an equation which, when paired with $-2x + 3y = 12$ to form a system, has the given number of solutions.

a. exactly one solution $x + 4y = 5$

b. no solution $4x - 6y = 5$

c. infinitely many solutions $4x - 6y = -24$

d. *Writing* Explain how you wrote each of the equations in parts (a)–(c). **See margin.**

14. MULTI-STEP PROBLEM The cholesterol in your blood is necessary, but too much cholesterol can lead to health problems. A blood cholesterol test gives three readings: LDL "bad" cholesterol, HDL "good" cholesterol, and total cholesterol (LDL + HDL). It is recommended that your LDL cholesterol be less than 130 milligrams per deciliter, HDL cholesterol be at least 35 milligrams per deciliter, and total cholesterol be no more than 200 milligrams per deciliter.

a. Write a system of three linear inequalities for the recommended cholesterol readings. Let x represent HDL cholesterol and y represent LDL cholesterol. $0 \le y < 130$; $x \ge 35$; $x + y \le 200$

b. Graph the system. Label any vertices of the solution region. **See margin for graph.**

c. Are the cholesterol readings at the right within recommendations? **not within recommendations**

d. Give an example of blood cholesterol test results in which the LDL cholesterol is too high, but HDL and total cholesterol readings are fine. Write a system of linear inequalities to describe all the examples of this type. **LDL: 135 mg/dL; HDL: 40 mg/dL; Total: 175 mg/dL; $y \ge 130$; $x \ge 35$; $x + y \le 200$**

e. Another recommendation is that the ratio of total cholesterol to HDL cholesterol be less than 4. Find a point in your solution region from part (b) that meets this recommendation and show that it does. **LDL $= 120 \frac{mg}{dL}$, HDL $= 40$. Then $\frac{total}{HDL} = 3$, and $3 < 4$.**

LDL: 120 mg/dL
HDL: 90 mg/dL
Total: 210 mg/dL

12b. *Sample answer:* $x = 2y + 2$, so
$5(2y + 2) - 4y = 6y + 10 = -8$.
$6y = -18$, or $y = -3$.
$x = 2(-3) + 2 = -4$.
The solution is $(-4, -3)$.

12c. *Sample answer:* Multiply the first equation by -2, and add it to the second equation:

$$-2x + 4y = -4$$
$$\underline{5x - 4y = -8}$$
$$3x = -12$$

So $x = -4$. Then $2y = x - 2 = -4 - 2 = -6$, so $y = -3$.

12d. *Sample answer:* I prefer the linear combination method, since it works with any system and usually yields a solution easily, especially when only one equation needs to be multiplied, as here.

13d. For part (a), I started with a different linear combination of x and y which would give a different slope, then seeing that $(-3, 2)$ is a solution of the given equation, I calculated the value of the right-hand side to make it a solution of my equation as well. For part (b), I made the left-hand side -2 times the left-hand side of the original equation, then picked a completely different number for the right-hand side. For part (c), I made the whole equation -2 times the original equation.

14b.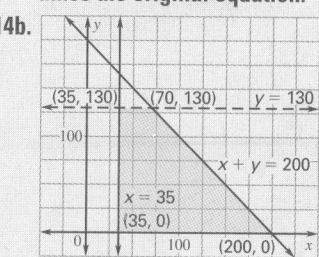

Cumulative Practice

ADDITIONAL RESOURCES

A Cumulative Review covering Chapters 1–3 is available in the *Chapter 3 Resource Book,* p. 104.

1.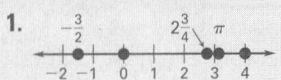

2.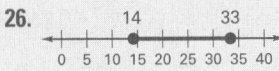

3.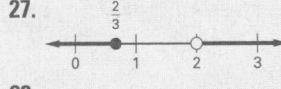

25.

26.

27.

28.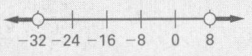

29. $x \le -\frac{3}{7}$ or $x \ge 3$;

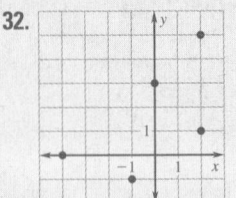

30. $x < -32$ or $x > 8$;

31.

32.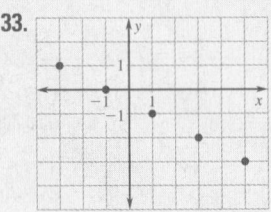

33.

Plot the numbers on a number line. Write the numbers in increasing order. (1.1) 1–3. See margin for graphs.

1. $0, \pi, 2\frac{3}{4}, -\frac{3}{2}, 4$ $-\frac{3}{2}, 0, 2\frac{3}{4}, \pi, 4$ **2.** $\frac{5}{2}, -\frac{1}{10}, -2, \sqrt{5}, 1.9$ **3.** $-4.25, -\frac{16}{3}, -\sqrt{9}, -0.4, -1$

$-2, -\frac{1}{10}, 1.9, \sqrt{5}, \frac{5}{2}$ $-\frac{16}{3}, -4.25, -\sqrt{9}, -1, -0.4$

Identify the property shown. (1.1)

4. $8 \cdot \frac{1}{8} = 1$ multiplicative inverse property   **5.** $-1(9 + 7) = (-1)9 + (-1)7$ distributive property   **6.** $-6 \cdot (-3 \cdot 4) = (-6 \cdot (-3)) \cdot 4$ associative property of multiplication

Evaluate the expression. (1.2)

7. $12 \div 2 - 4 \cdot 7$ -22 **8.** $-8 + 3(1 - 5)^2$ 40 **9.** $17 - 2^4 \div 8 + 1$ 16 **10.** $-2(16 + 7) \div -10$ 4.6

Simplify the expression. (1.2)

11. $18a + 7a - 9a + 11$ **12.** $10x - (4y - x) + y$ **13.** $6(n^2 - n) - 5n^2 + 8n$ $n^2 + 2n$
$16a + 11$ $11x - 3y$

Solve the equation. (1.3, 1.7)

14. $\frac{5}{8}x - 9 = 21$ 48 **15.** $-75 = 9x - 3$ -8 **16.** $4(2x - 1) = -20$ -2 **17.** $3 - x = 5x + 27$ -4

18. $|x| = 9$ $-9, 9$ **19.** $|4x + 1| = 39$ $-10, 9.5$ **20.** $|7 - 2x| = 15$ $-4, 11$ **21.** $|x - 10| = 0$ 10

Solve the formula for the indicated variable. (1.4)

22. Distance $r = \frac{d}{t}$ **23. Volume of a Cylinder** $h = \frac{V}{\pi r^2}$ **24. Area of a Trapezoid** $h = \frac{2A}{b_1 + b_2}$
Solve for r: $d = rt$ Solve for h: $V = \pi r^2 h$ Solve for h: $A = \frac{1}{2}(b_1 + b_2)h$

Solve the inequality. Then graph the solution. (1.6, 1.7) 25–31. See margin for graphs.

25. $14 - 5x > -6$ $x < 4$ **26.** $1 \le x - 13 \le 20$ $14 \le x \le 33$ **27.** $3x - 2 \le 0$ or $x + 6 > 8$ $x \le \frac{2}{3}$ or $x > 2$

28. $|x - 7| \le 1$ $6 \le x \le 8$ **29.** $|7x - 9| \ge 12$ See margin.   **30.** $\left|\frac{1}{4}x + 3\right| > 5$ See margin.   **31.** $|-5x| < 10$ $-2 < x < 2$

Graph the relation. Then tell whether the relation is a function. (2.1) 32, 33. See margin for graphs.

32.

x	2	−4	2	−1	0
y	1	0	5	−1	3

No

33.

x	−3	−1	1	3	5
y	1	0	−1	−2	−3

Yes

Graph in a coordinate plane. (2.1, 2.3, 2.6–2.8) 34–48. See margin.

34. $y = -2x + 5$ **35.** $x - 3y = 6$ **36.** $y = 2$ **37.** $x = -4$

38. $y > \frac{2}{5}x - 2$ **39.** $y \le -1$ **40.** $4x + 3y \le 24$ **41.** $y > -x$

42. $f(x) = 4|x|$ **43.** $f(x) = |x| - 3$ **44.** $f(x) = 2|x + 2|$ **45.** $f(x) = -|x - 5| + 1$

46. $f(x) = \begin{cases} 2x, & \text{if } x \le 0 \\ -2x, & \text{if } x > 0 \end{cases}$ **47.** $f(x) = \begin{cases} \frac{1}{2}x + 1, & \text{if } x \le -2 \\ x + 1, & \text{if } x > -2 \end{cases}$ **48.** $f(x) = \begin{cases} 4, & \text{if } -5 \le x < 0 \\ -4, & \text{if } 0 \le x \le 5 \end{cases}$

Graph the system. Describe the solution(s). (3.1, 3.3) 49–52. See margin.

49. $4x - 2y = 8$ **50.** $y = x$ **51.** $2x - y > 1$ **52.** $x \ge 0$
  $4x + y = 2$ $y = x - 3$ $x < 3$ $y \ge 0$
          $y = x + 5$ $x + y \le 8$

Tell whether the lines are *perpendicular, parallel,* or *neither.* (2.2)

53. Line 1: through (0, 7) and (3, 6)
Line 2: through (−2, −9) and (0, −3)
perpendicular

54. Line 1: through (−6, −3) and (0, 1)   **neither**
Line 2: through (0, −5) and (4, −2)

Write an equation of the line with the given characteristics. (2.4)

55. slope: −3, y-intercept: 7
$y = -3x + 7$

56. vertical line through (2, 5)
$x = 2$

57. x-intercept: −2, y-intercept: 1 $y = \frac{1}{2}x + 1$

Evaluate the function for the given value(s). (2.1, 2.7, 2.8, 3.5)

58. $f(x) = 5x - 17$, $f(-3)$ **−32**   **59.** $f(x) = x^2 - 2x + 11$, $f(2)$ **11**   **60.** $f(x) = \begin{cases} x - 4, & \text{if } x \le 0 \\ x + 2, & \text{if } x > 0 \end{cases}$, $f(-2)$ **−6**

61. $f(x) = -\left|12 - 8x\right|$, $f(1)$
−4

62. $f(x, y) = 8x - 5y$, $f(3, -2)$
34

63. $f(x, y) = 2(-x + y)$, $f(-1, 0)$ **2**

Solve the system using any algebraic method. (3.2, 3.6)

64. $-x + 5y = 8$
$-3x + 15y = 24$
infinitely many solutions

65. $x - 3y = 7$ **(4, −1)**
$2x + y = 7$

66. $x + y - z = 7$
$-x + 2y + 2z = 3$
$3x - y - z = 1$
(1, 4, −2)

67. $2x + y + z = 4$ **(0, −1, 5)**
$x - y - 2z = -9$
$2x - y + z = 6$

Graph in a three-dimensional coordinate system. (3.5)   68–71. See margin.

68. (1, −4, 2)   **69.** (−2, 3, −5)   **70.** $x + 2y + 3z = 6$ **71.** $10x + 4y + 5z = 20$

72. **SWEATER SALE** You pay $38.50 for a sweater that is marked 30% off the regular price. What is the regular price of the sweater? How much did you save by buying it on sale? (1.5) **$55; $16.50**

73. **BODY TEMPERATURE** Although the average body temperature of a healthy baby is 98.6°F, the temperature can vary from 97°F to 100°F. Write an inequality to describe the range of healthy temperatures. On a number line, graph the inequality and mark the average body temperature of a healthy baby. (1.6) $97 \le T \le 100$; **See margin.**

74. **HIGHWAY TRAVEL** If you drive at a constant speed then the distance you travel d varies directly with the time t. Suppose you use cruise control and drive 180 miles in 3 hours. Write an equation to show the relationship between d and t. What is the constant of variation and what does it represent? (2.4) $d = 60t$; **60; speed in mph**

75. **SOLID WASTE** The table gives the amount of material recovered from solid waste (in millions of tons) in the United States from 1988 to 1996. Make a scatter plot of the data and approximate a best-fitting line. Predict the amount of material recovered in the United States in 2002. (2.5)

Sample answer:
$m = 4.20t + 24.5$; **83.3 million tons; See margin.**

Years since 1988, *t*	0	1	2	3	4	5	6	7	8
Material, *m*	23.5	29.9	33.6	37.0	40.6	43.8	50.9	55.1	57.3

▶ Source: *Statistical Abstract of the United States*

76. **AUTO RENTAL** An automobile rental agency charges $60 per day with unlimited mileage. A second agency charges $45 per day plus $.25 per mile after the first 100 miles. For a one-day rental, after how many miles will the first agency be less expensive? (3.1, 3.2)   **160 mi**

77. **STIR-FRY RECIPE** A restaurant serves a stir-fry dish containing vegetables and beef. The recipe calls for no more than twice as many pounds of vegetables as beef. The owner buys vegetables at $1.39 per pound and beef at $1.79 per pound and will order a total of 150 pounds. To minimize the cost yet satisfy the recipe, how much of each food should the owner order? What will be the total cost? (3.4)

order 100 lb of vegetables and 50 lb of beef at a total cost of $228.50

34.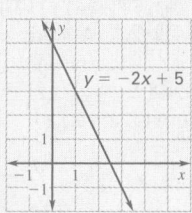

35.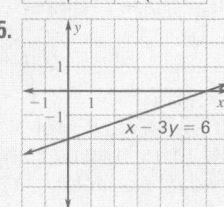

36.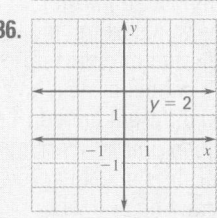

37.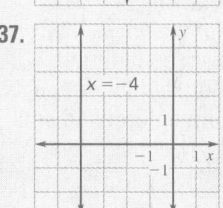

38.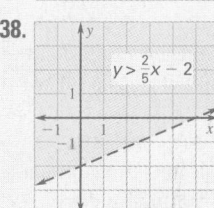

39.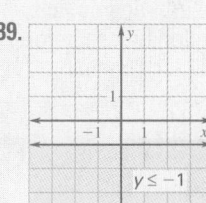

40.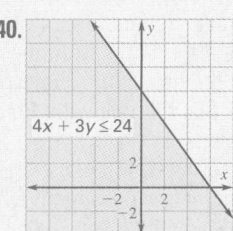

41–52, 68–71, 73, 75.
See Additional Answers beginning on page AA1.

- Graph a system of linear equations.
- Write a system of linear equations to model a real-life situation.

MANAGING THE PROJECT
CLASSROOM MANAGEMENT
The Chapter 3 Project may be completed by individual students or by students working with a partner. If students work with a partner, they should work together to choose the object and each should draw it and experiment with different horizon lines and vanishing points. Working together, the partners should agree upon the final drawing and write the equations for the lines. The partners can work on different parts of the report, but they should discuss each part beforehand and edit each others' work before a final draft is made.

GUIDING STUDENTS' WORK
It will be important for students to pick a fairly simple object to draw in Step 1 of the Investigation. Objects that are cubes or rectangles or a combination of cubes and rectangles work best. Some students may need to practice drawing cubic or rectangular wooden blocks before progressing to a more complex structure.

CONCLUDING THE PROJECT
Have students display their final drawings and instructions on a bulletin board. Have students follow another student's set of instructions and compare the drawings. Then use questions like the following for class discussion.
- How can you tell which point is a vanishing point in a perspective drawing?
- How is domain used to define the drawing?
- Why is perspective drawing useful?
- In which careers might perspective drawing be useful?

194

PROJECT
Applying Chapters 1–3

Drawing with Linear Perspective

OBJECTIVE Use linear equations to represent a drawing made with linear perspective.

Materials: graph paper, ruler

During the Renaissance, artists turned to mathematics to develop *perspective*, a method for realistically depicting a three-dimensional object on a two-dimensional surface. A drawing with linear perspective has all slanted lines converging toward a point or points on the horizon. These points are called *vanishing points*. The painting below and on the left has all slanted lines converging toward a single vanishing point at the far end of the road. The painting on the right has all slanted lines converging toward one of two vanishing points, one on either side of the building.

The Avenue at Middelharnis, painted in 1689 by Meindert Hobbema

Corner of George and Hunter Streets, Sydney, painted in 1849 by A. Torning

HOW TO DRAW AN OBJECT IN TWO-POINT PERSPECTIVE

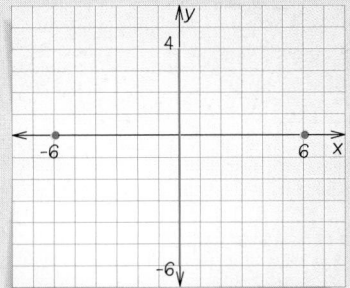

1 Use the *x*-axis as the horizon. Select two points equidistant from the origin and on the *x*-axis as the vanishing points. Draw a vertical segment to represent the front edge of the object—in this case, the front edge of a building.

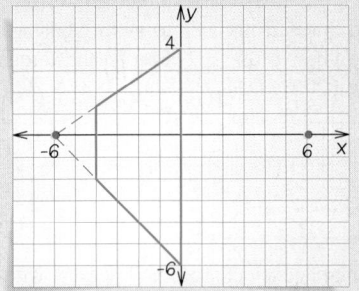

2 To draw the left wall of the building, draw segments from the endpoints of the front edge toward the vanishing point on the left. Connect the segments with a vertical line to represent the end of the wall.

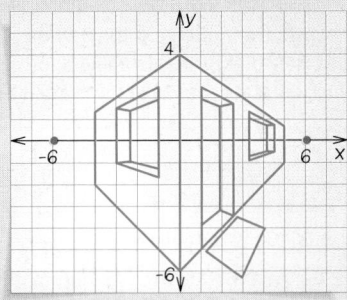

3 Continue drawing slanted lines that are to the left of the front edge toward the vanishing point on the left, and lines that are to the right of the front edge toward the vanishing point on the right.

INVESTIGATION

1. Choose an object that has many parallel edges, such as a building, courtyard, or computer. Use the method given on the previous page to draw the object in two-point perspective.

2. The higher the horizon line, the higher the vantage point from which you view the drawing. The farther apart the vanishing points are, the more gradual the change from near to far in the drawing.

2. Experiment with using a lower or higher horizon line, as well as vanishing points that are farther apart or closer together, until your drawing has the look you want. How does the placement of the horizon line and the vanishing points affect the way your drawing looks?

3. Write an equation for each line in your drawing. Include the domain to indicate the length of the line. For example, the upper left edge of the building on the previous page is defined by $y = \frac{2}{3}x + 4$ for $-4 \le x \le 0$.

PRESENT YOUR RESULTS

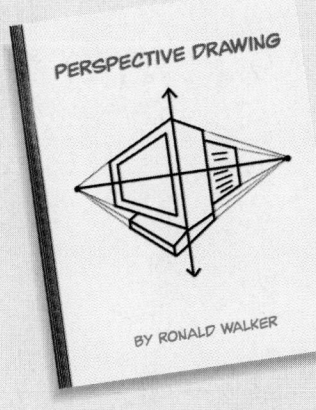

Write a report to present your results.

- Include your drawing and any preliminary sketches you did.

- Include your answers to Exercises 1–3 above.

- Write a set of instructions for how to draw the object just as you have drawn it. Include the equations you wrote.

- Tell how this project has helped you mathematically.

Test your results.

- Trade drawing instructions with a partner (do not trade actual drawings). Follow the instructions to create your partner's drawing.

- Compare your drawing with the original.

EXTENSION

Another way to suggest a three-dimensional object on a two-dimensional surface is to add shadowing. Select a point for a light source and decide where the shadows cast by your object would fall. Write a system of linear inequalities to indicate each shaded region. Add these to your report.

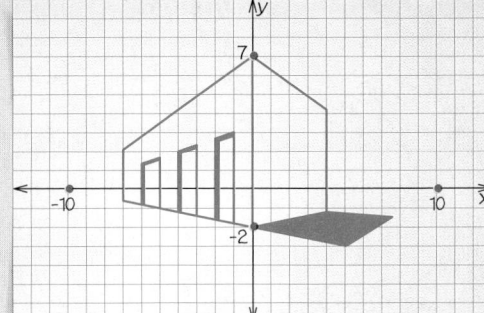

System of inequalities for the building's shadow:

$$y \le \frac{1}{5}x - 2$$

$$y \ge -\frac{1}{5}x - 2$$

$$y \ge \frac{3}{5}x - 6$$

$$y \le -\frac{3}{35}x - \frac{6}{7}$$

GRADING THE PROJECT
RUBRIC FOR CHAPTER PROJECT

4 Students' two-point perspective drawing is complete and correct. Sketches including various vanishing points and horizon lines are provided. A clear set of instructions for drawing the object is provided that include linear equations with appropriate domains. The written report demonstrates the understanding of the effects of different vanishing points and horizon lines on perspective drawings and explains how the project helped them mathematically.

3 Students' two-point perspective drawing is included but may be lacking in some details. Some sketches including various vanishing points and horizon lines are provided but the effects of various vanishing points and horizon lines may not be fully explained in the report. A set of instructions for drawing the object is provided including linear equations that may contain small errors or may be missing the domain restrictions. The written report explains how the project helped them mathematically.

2 The two-point perspective drawing contains serious errors or is incomplete. Not all of the questions are answered in the report. The instructions and equations are incomplete and or inaccurate.

1 The drawing is inaccurate and shows that the student does not have an understanding of the method presented. The drawing cannot be reconstructed from the instructions if they are provided. The report is incomplete or not understandable. The project should be returned with a new deadline for completion. The student should speak with the teacher as soon as possible so that he or she understands the purpose and format of the project.

PLANNING THE CHAPTER

Matrices and Determinants

GOALS

LESSON			NCTM	ITED	SAT9	Terra-Nova	Local
4.1 pp. 199–207	GOAL 1	Add and subtract matrices, multiply a matrix by a scalar, and solve matrix equations.	1, 2, 8, 9, 10	MCWN, RQWN, MIG, IIG		11, 16, 18, 49, 51, 52	
	GOAL 2	Use matrices in real-life situations.					
		TECHNOLOGY ACTIVITY: 4.1 *Perform matrix operations on a graphing calculator.*					
4.2 pp. 208–213	GOAL 1	Multiply two matrices.	1, 2, 8, 9, 10	MCWN, RQWN		11, 16, 18, 49, 51, 52	
	GOAL 2	Use matrix multiplication in real-life situations.					
4.3 pp. 214–221	GOAL 1	Evaluate determinants of 2×2 and 3×3 matrices.	1, 2	MIG, IIG	4, 30	11, 16, 49, 51, 52	
	GOAL 2	Use Cramer's rule to solve systems of linear equations.					
4.4 pp. 222–229		**CONCEPT ACTIVITY: 4.4** *Investigate properties of identity and inverse matrices.*	1, 2, 8, 9, 10	MCWN, RQWN		16, 18	
	GOAL 1	Find and use inverse matrices.					
	GOAL 2	Use inverse matrices in real-life situations.					
4.5 pp. 230–236	GOAL 1	Solve systems of linear equations using inverse matrices.	1, 2, 6, 9, 10	RQWN	2, 4	11, 16, 17	
	GOAL 2	Use systems of linear equations to solve real-life problems.					
Extension pp. 237–238	GOAL 1	Solve systems of linear equations using elementary row operations on augmented matrices.	1, 2	MCWN	4	11, 16, 49, 51, 52	

RESOURCES

CHAPTER RESOURCE BOOKLETS

CHAPTER SUPPORT

Tips for New Teachers	p. 1	Prerequisite Skills Review	p. 5
Parent Guide for Student Success	p. 3	Strategies for Reading Mathematics	p. 7

LESSON SUPPORT

	4.1	4.2	4.3	4.4	4.5
Lesson Plans (regular and block)	p. 9	p. 22	p. 36	p. 52	p. 65
Warm-Up Exercises and Daily Quiz	p. 11	p. 24	p. 38	p. 54	p. 67
Activity Support Masters					
Lesson Openers	p. 12	p. 25	p. 39	p. 55	p. 68
Graphing Calculator Activities & Keystrokes	p. 13	p. 26	p. 40	p. 56	p. 69
Practice (3 levels)	p. 14	p. 28	p. 42	p. 57	p. 70
Reteaching with Practice	p. 17	p. 31	p. 45	p. 60	p. 73
Quick Catch-Up for Absent Students	p. 19	p. 33	p. 47	p. 62	p. 75
Cooperative Learning Activities			p. 48		
Interdisciplinary Applications	p. 20		p. 49		p. 76
Real-Life Applications		p. 34		p. 63	
Math & History Applications					p. 77
Challenge: Skills and Applications	p. 21	p. 35	p. 50	p. 64	p. 78

REVIEW AND ASSESSMENT

Quizzes	pp. 51, 79	Alternative Assessment with Math Journal	p. 88
Chapter Review Games and Activities	p. 80	Project with Rubric	p. 90
Chapter Test (3 levels)	pp. 81–86	Cumulative Review	p. 92
SAT/ACT Chapter Test	p. 87	Resource Book Answers	p. A1

TRANSPARENCIES

	4.1	4.2	4.3	4.4	4.5
Warm-Up Exercises and Daily Quiz	p. 26	p. 27	p. 28	p. 29	p. 30
Alternative Lesson Opener Transparencies	p. 22	p. 23	p. 24	p. 25	p. 26
Examples/Standardized Test Practice	✓	✓	✓	✓	✓
Answer Transparencies	✓	✓	✓	✓	✓

TECHNOLOGY

- Electronic Teaching Tools
- Online Lesson Planner
- Internet Support
- Personal Student Tutor
- Test and Practice Generator
- Instant Replay: Video Review Games
- Electronic Lesson Presentations (Lesson 4.2)

ADDITIONAL RESOURCES

- Basic Skills Workbook: Diagnosis and Remediation
- Worked-Out Solution Key
- Resources in Spanish
- Standardized Test Practice Workbook
- Practice Workbook with Examples

REGULAR SCHEDULE

Day 1

4.1

STARTING OPTIONS
- Prereq. Skills Review
- Strategies for Reading
- Homework Check
- Warm-Up or Daily Quiz

TEACHING OPTIONS
- Les. Opener (Application)
- Graphing Calc. Activity
- Examples 1–4
- Guided Practice Exs. 1–4, 6–9

APPLY/HOMEWORK
- See Assignment Guide.
- See the CRB: Practice, Reteach, Apply, Extend

ASSESSMENT OPTIONS
- Checkpoint Exercises, pp. 200–201

Day 2

4.1 (cont.)

STARTING OPTIONS
- Homework Check

TEACHING OPTIONS
- Examples 5–6
- Technology Activity
- Closure Question
- Guided Practice Exs. 5, 10

APPLY/HOMEWORK
- See Assignment Guide.
- See the CRB: Practice, Reteach, Apply, Extend

ASSESSMENT OPTIONS
- Checkpoint Exercises, p. 202
- Daily Quiz (4.1)
- Stand. Test Practice

Day 3

4.2

STARTING OPTIONS
- Homework Check
- Warm-Up or Daily Quiz

TEACHING OPTIONS
- Les. Opener (Visual)
- Graphing Calc. Activity
- Examples 1–5
- Closure Question
- Guided Practice Exs.

APPLY/HOMEWORK
- See Assignment Guide.
- See the CRB: Practice, Reteach, Apply, Extend

ASSESSMENT OPTIONS
- Checkpoint Exercises
- Daily Quiz (4.2)
- Stand. Test Practice

Day 4

4.3

STARTING OPTIONS
- Homework Check
- Warm-Up or Daily Quiz

TEACHING OPTIONS
- Les. Opener (Activity)
- Graphing Calc. Activity
- Examples 1–2, 4
- Guided Practice Exs. 1–10

APPLY/HOMEWORK
- See Assignment Guide.
- See the CRB: Practice, Reteach, Apply, Extend

ASSESSMENT OPTIONS
- Checkpoint Exercises, pp. 215–216

Day 5

4.3 (cont.)

STARTING OPTIONS
- Homework Check

TEACHING OPTIONS
- Examples 3, 5
- Closure Question
- Guided Practice Ex. 11

APPLY/HOMEWORK
- See Assignment Guide.
- See the CRB: Practice, Reteach, Apply, Extend

ASSESSMENT OPTIONS
- Checkpoint Exercises, pp. 216–217
- Daily Quiz (4.3)
- Stand. Test Practice
- Quiz (4.1–4.3)

Day 6

4.4

STARTING OPTIONS
- Homework Check
- Warm-Up or Daily Quiz

TEACHING OPTIONS
- Motivating the Lesson
- Concept Activity
- Les. Opener (Calculator)
- Graphing Calc. Activity
- Examples 1–6
- Closure Question
- Guided Practice Exs.

APPLY/HOMEWORK
- See Assignment Guide.
- See the CRB: Practice, Reteach, Apply, Extend

ASSESSMENT OPTIONS
- Checkpoint Exercises
- Daily Quiz (4.4)
- Stand. Test Practice

Day 9

Assess

DAY 9 START OPTIONS
- Homework Check

ASSESSMENT OPTIONS
- Chapter 4 Test
- SAT/ACT Ch. 4 Test
- Alternative Assessment

APPLY/HOMEWORK
- Skill Review, p. 248

Day 1

4.1

DAY 1 START OPTIONS
- Prereq. Skills Review
- Strategies for Reading
- Homework Check
- Warm-Up or Daily Quiz

TEACHING 4.1 OPTIONS
- Les. Opener (Appl.)
- Graphing Calc. Activity
- Examples 1–6
- Technology Activity
- Closure Question
- Guided Practice Exs.

APPLY/HOMEWORK
- See Assignment Guide.
- See the CRB: Practice, Reteach, Apply, Extend

ASSESSMENT OPTIONS
- Checkpoint Exercises
- Daily Quiz (Les. 4.1)
- Stand. Test Practice

Day 2

4.2

DAY 2 START OPTIONS
- Homework Check
- Warm-Up or Daily Quiz

TEACHING 4.2 OPTIONS
- Les. Opener (Visual)
- Graphing Calc. Activity
- Examples 1–5
- Closure Question
- Guided Practice Exs.

APPLY/HOMEWORK
- See Assignment Guide.
- See the CRB: Practice, Reteach, Apply, Extend

ASSESSMENT OPTIONS
- Checkpoint Exercises
- Daily Quiz (Les. 4.2)
- Stand. Test Practice

Day 3

4.3

DAY 3 START OPTIONS
- Homework Check
- Warm-Up or Daily Quiz

TEACHING 4.3 OPTIONS
- Les. Opener (Activity)
- Graphing Calc. Activity
- Examples 1–5
- Closure Question
- Guided Practice Exs.

APPLY/HOMEWORK
- See Assignment Guide.
- See the CRB: Practice, Reteach, Apply, Extend

ASSESSMENT OPTIONS
- Checkpoint Exercises
- Daily Quiz (Les. 4.3)
- Stand. Test Practice
- Quiz (4.1–4.3)

Day 4

4.4 & 4.5

DAY 4 START OPTIONS
- Homework Check
- W-Up 4.4 or D. Quiz 4.3

TEACHING 4.4 OPTIONS
- Motivating the Lesson
- Concept Activity
- Les. Opener (Calc.)
- Graphing Calc. Activity
- Examples 1–6
- Closure Question
- Guided Practice Exs.

TEACHING 4.5 OPTIONS
- Warm-Up (Les. 4.5)
- Les. Opener (Appl.)
- Graphing Calc. Activity
- Examples 1–4
- Closure Question
- Guided Practice Exs.

APPLY/HOMEWORK
- See Assignment Guide.
- See the CRB: Practice, Reteach, Apply, Extend

ASSESSMENT OPTIONS
- Checkpoint Exercises
- Daily Quiz (Les. 4.4, 4.5)
- Stand. Test Practice
- Quiz (4.4–4.5)

Day 5

Review/Assess

DAY 5 START OPTIONS
- Homework Check

REVIEWING OPTIONS
- Chapter 4 Summary
- Chapter 4 Review
- Chapter Review Games and Activities
- Chapter 4 Test (practice)
- Ch. Standardized Test (practice)

ASSESSMENT OPTIONS
- Chapter 4 Test
- SAT/ACT Ch. 4 Test
- Alternative Assessment

APPLY/HOMEWORK
- Skill Review, p. 248

Day 7

4.5

STARTING OPTIONS
- Homework Check
- Warm-Up or Daily Quiz

TEACHING OPTIONS
- Les. Opener (Application)
- Graphing Calc. Activity
- Examples 1–4
- Closure Question
- Guided Practice Exs.

APPLY/HOMEWORK
- See Assignment Guide.
- See the CRB: Practice, Reteach, Apply, Extend

ASSESSMENT OPTIONS
- Checkpoint Exercises
- Daily Quiz (4.5)
- Stand. Test Practice
- Quiz (4.4–4.5)

Day 8

Review

DAY 8 START OPTIONS
- Homework Check

REVIEWING OPTIONS
- Chapter 4 Summary
- Chapter 4 Review
- Chapter Review Games and Activities

APPLY/HOMEWORK
- Chapter 4 Test (practice)
- Ch. Standardized Test (practice)

BEFORE THE CHAPTER

The *Chapter 4 Resource Book* has the following materials to distribute and use before the chapter:

- **Parent Guide for Student Success**
- **Prerequisite Skills Review (pictured below)**
- **Strategies for Reading Mathematics**

PREREQUISITE SKILLS *Pages 5–6*

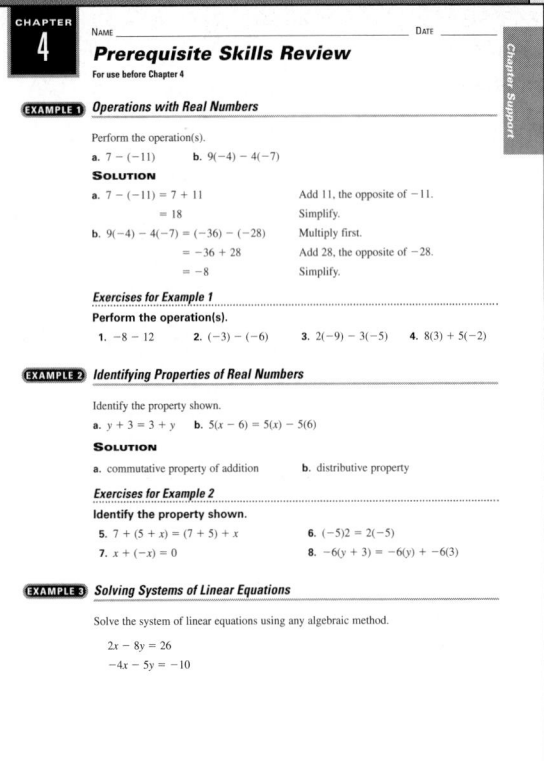

CHAPTER 4

NAME _____ DATE _____

Prerequisite Skills Review
For use before Chapter 4

EXAMPLE 1 *Operations with Real Numbers*

Perform the operation(s).

a. $7 - (-11)$ b. $9(-4) - 4(-7)$

SOLUTION

a. $7 - (-11) = 7 + 11$ Add 11, the opposite of -11.

$= 18$ Simplify.

b. $9(-4) - 4(-7) = (-36) - (-28)$ Multiply first.

$= -36 + 28$ Add 28, the opposite of -28.

$= -8$ Simplify.

Exercises for Example 1

Perform the operation(s).

1. $-8 - 12$ 2. $(-3) - (-6)$ 3. $2(-9) - 3(-5)$ 4. $8(3) + 5(-2)$

EXAMPLE 2 *Identifying Properties of Real Numbers*

Identify the property shown.

a. $y + 3 = 3 + y$ b. $5(x - 6) = 5(x) - 5(6)$

SOLUTION

a. commutative property of addition   b. distributive property

Exercises for Example 2

Identify the property shown.

5. $7 + (5 + x) = (7 + 5) + x$ 6. $(-5)2 = 2(-5)$

7. $x + (-x) = 0$ 8. $-6(y + 3) = -6(y) + -6(3)$

EXAMPLE 3 *Solving Systems of Linear Equations*

Solve the system of linear equations using any algebraic method.

$2x - 8y = 26$

$-4x - 5y = -10$

PREREQUISITE SKILLS REVIEW These two pages support the Study Guide on page 198. They help students prepare for Chapter 4 by providing worked-out examples and practice for the following skills needed in the chapter:

- **Perform operations with real numbers.**
- **Identify properties of real numbers.**
- **Solve systems of linear equations.**

TECHNOLOGY RESOURCE

Teachers can use the Real-Life Motivator clip for Chapter 4 from the video Instant Replay: Video Review Games as an alternative way to introduce the chapter.

DURING EACH LESSON

The *Chapter 4 Resource Book* has the following alternatives for introducing the lesson:

- **Lesson Openers (pictured below)**
- **Graphing Calculator Activities with Keystrokes**

LESSON OPENER *Page 25*

LESSON 4.2

NAME _____ DATE _____   Available as a transparency

Visual Approach Lesson Opener
For use with pages 208–213

SET UP: You will need: • red, green, blue, and yellow colored pencils

$$A = \begin{bmatrix} 4 & -2 & 3 \\ 0 & 1 & -1 \end{bmatrix} \quad B = \begin{bmatrix} 1 & 5 \\ -3 & 2 \\ 0 & -4 \end{bmatrix} \quad AB = \begin{bmatrix} & \\ & \end{bmatrix}$$

1. Use a red pencil to mark the first row of A and the first column of B. Are they the same length?

2. You can multiply the first row of A by the first column of B as follows: $4(1) + (-2)(-3) + 3(0)$. Do the computations. The answer is the entry in the first row and the first column of the product matrix. Write it there with a red pencil.

3. Mark the first row of A and the second column of B with a blue pencil. Find the product. Write it with a blue pencil in the first row and second column of the product matrix.

4. Mark the second row of A and the first column of B with a green pencil. Find the product and write it with a green pencil in the product matrix.

5. Mark the second row of A and the second column of B with a yellow pencil. Find the product and write it with a yellow pencil in the product matrix.

6. Use a red pencil to mark the first row of C and the first column of D. Are they the same length? Do you think the product CD is defined?

$$C = \begin{bmatrix} 8 & -5 \\ -1 & 2 \end{bmatrix} \quad D = \begin{bmatrix} 4 & -3 \\ 0 & 1 \\ 2 & -6 \end{bmatrix}$$

VISUAL APPROACH LESSON OPENER This Lesson Opener uses visuals as an alternative way to start Lesson 4.2. Students use colored pencils to identify rows and columns of matrices as an introduction to multiplying matrices.

TECHNOLOGY RESOURCE

Teachers can use the Electronic Lesson Presentations CD-ROM to present Lesson 4.2 using computer animation to step through the Examples.

FOLLOWING EACH LESSON

The *Chapter 4 Resource Book* has a variety of materials to follow-up each lesson. They include the following:

- **Practice (3 levels)**
- **Reteaching with Practice**
- **Quick Catch-Up for Absent Students**
- **Challenge: Skills and Applications (pictured below)**
- **Interdisciplinary Applications**
- **Real-Life Applications**

CHALLENGE *Page 21*

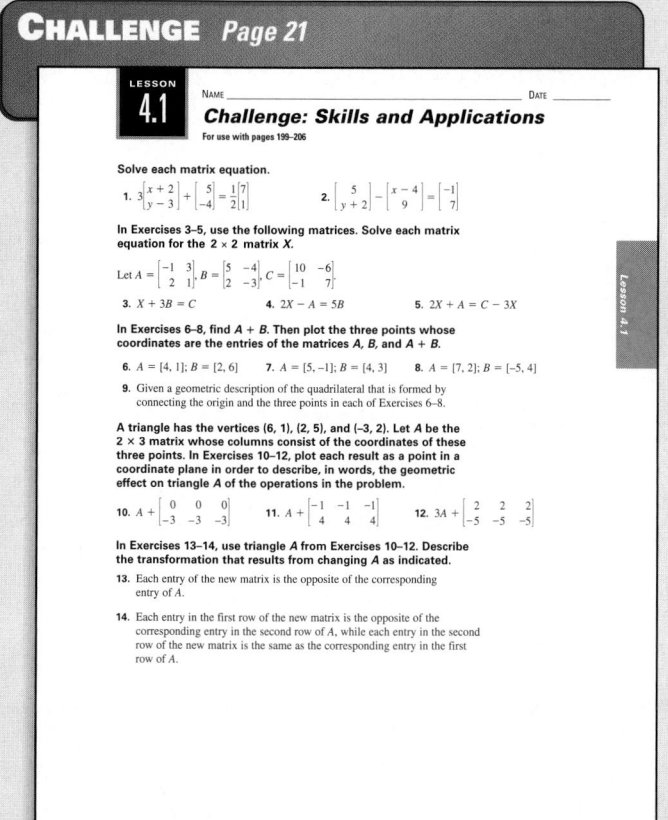

LESSON 4.1

NAME _____ DATE _____

Challenge: Skills and Applications
For use with pages 199–206

Solve each matrix equation.

1. $3\begin{bmatrix} x+2 \\ y-3 \end{bmatrix} + \begin{bmatrix} 5 \\ -4 \end{bmatrix} = \frac{1}{2}\begin{bmatrix} 7 \\ 1 \end{bmatrix}$

2. $\begin{bmatrix} 5 \\ y+2 \end{bmatrix} - \begin{bmatrix} x-4 \\ 9 \end{bmatrix} = \begin{bmatrix} -1 \\ 7 \end{bmatrix}$

In Exercises 3–5, use the following matrices. Solve each matrix equation for the 2 × 2 matrix *X*.

Let $A = \begin{bmatrix} -1 & 3 \\ 2 & 1 \end{bmatrix}$, $B = \begin{bmatrix} 5 & -4 \\ 2 & -3 \end{bmatrix}$, $C = \begin{bmatrix} 10 & -6 \\ -1 & 7 \end{bmatrix}$

3. $X + 3B = C$ 4. $2X - A = 5B$ 5. $2X + A = C - 3X$

In Exercises 6–8, find *A* + *B*. Then plot the three points whose coordinates are the entries of the matrices *A*, *B*, and *A* + *B*.

6. $A = [4, 1]$; $B = [2, 6]$ 7. $A = [5, -1]$; $B = [4, 3]$ 8. $A = [7, 2]$; $B = [-5, 4]$

9. Given a geometric description of the quadrilateral that is formed by connecting the origin and the three points in each of Exercises 6–8.

A triangle has the vertices (6, 1), (2, 5), and (–3, 2). Let *A* be the 2 × 3 matrix whose columns consist of the coordinates of these three points. In Exercises 10–12, plot each result as a point in a coordinate plane in order to describe, in words, the geometric effect on triangle *A* of the operations in the problem.

10. $A + \begin{bmatrix} 0 & 0 & 0 \\ -3 & -3 & -3 \end{bmatrix}$ 11. $A + \begin{bmatrix} -1 & -1 & -1 \\ 4 & 4 & 4 \end{bmatrix}$ 12. $3A + \begin{bmatrix} 2 & 2 & 2 \\ -5 & -5 & -5 \end{bmatrix}$

In Exercises 13–14, use triangle *A* from Exercises 10–12. Describe the transformation that results from changing *A* as indicated.

13. Each entry of the new matrix is the opposite of the corresponding entry of *A*.

14. Each entry in the first row of the new matrix is the opposite of the corresponding entry in the second row of *A*, while each entry in the second row of the new matrix is the same as the corresponding entry in the first row of *A*.

Lesson 4.1

CHALLENGE: SKILLS AND APPLICATIONS Each lesson has a page of challenge exercises that extend the material in the lesson. The challenge exercises for Lesson 4.1 involve solving more complicated matrices and appling matrices to transformational geometry.

ASSESSING THE CHAPTER

The *Chapter 4 Resource Book* has the following review and assessment materials:

- **Quizzes**
- **Chapter Review Games and Activities**
- **Chapter Test (3 levels)**
- **SAT/ACT Chapter Test**
- **Alternative Assessment with Rubric and Math Journal**
- **Project with Rubric (pictured below)**
- **Cumulative Review**

PROJECT WITH RUBRIC *Pages 90–91*

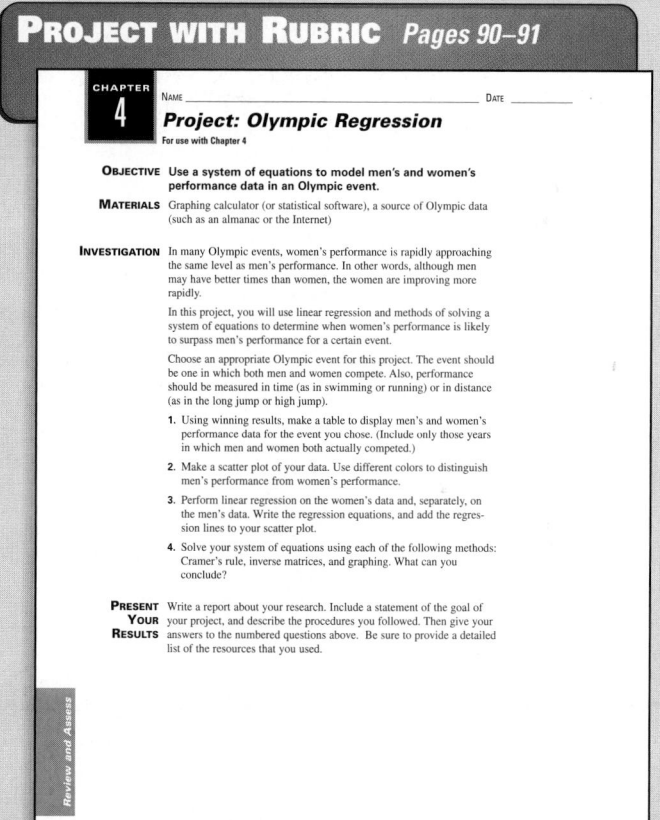

CHAPTER 4

NAME _____ DATE _____

Project: Olympic Regression
For use with Chapter 4

OBJECTIVE Use a system of equations to model men's and women's performance data in an Olympic event.

MATERIALS Graphing calculator (or statistical software), a source of Olympic data (such as an almanac or the Internet)

INVESTIGATION In many Olympic events, women's performance is rapidly approaching the same level as men's performance. In other words, although men may have better times than women, the women are improving more rapidly.

In this project, you will use linear regression and methods of solving a system of equations to determine when women's performance is likely to surpass men's performance for a certain event.

Choose an appropriate Olympic event for this project. The event should be one in which both men and women compete. Also, performance should be measured in time (as in swimming or running) or in distance (as in the long jump or high jump).

1. Using winning results, make a table to display men's and women's performance data for the event you chose. (Include only those years in which men and women both actually competed.)

2. Make a scatter plot of your data. Use different colors to distinguish men's performance from women's performance.

3. Perform linear regression on the women's data and, separately, on the men's data. Write the regression equations, and add the regression lines to your scatter plot.

4. Solve your system of equations using each of the following methods: Cramer's rule, inverse matrices, and graphing. What can you conclude?

PRESENT YOUR RESULTS Write a report about your research. Include a statement of the goal of your project, and describe the procedures you followed. Then give your answers to the numbered questions above. Be sure to provide a detailed list of the resources that you used.

Review and Assess

PROJECT WITH RUBRIC The Project for Chapter 4 provides students with the opportunity to apply the concepts they have learned in the chapter in a new way. In this project, students use what they have learned about matrix operations and properties to compare men's and women's performances in the Olympics. Teacher's notes and a scoring rubric are provided on a separate sheet.

CHAPTER GOALS

Chapter 4 involves manipulating matrices and finding determinants. Students will learn to add, subtract, and multiply matrices by a scalar and another matrix. They will then use these operations to solve real-world problems. Students will find determinants of 2×2 and 3×3 matrices and then use Cramer's rule to solve systems. Students will find inverse matrices and use them to solve systems of equations and real-world problems. The chapter concludes by having students solve systems using augmented matrices.

APPLICATION NOTE

Most of the revenue for music sales is from the sale of compact discs. Compact discs replaced the long-playing phonograph record as the principal medium for music storage. It offers several advantages: no background noise, a more uniform and accurate sound, a wider range of levels of sound, and since nothing mechanical touches the surface of the disc when it is played it lasts longer. .

The discs themselves are 12 cm in diameter and are covered by a thin, reflective metallic layer and protected by a clear plastic coating. To play music stored on a CD, a low-powered laser beam reads the data through the reflective rear surface of the disk.

Additional information about music sales is available at **www.mcdougallittell.com.**

MATRICES AND DETERMINANTS

▶ *How can you organize data about music sales?*

APPLICATION: Music Sales

When you shop for music, your purchases matter. Music sales by people age 15 to 19 years accounted for about 16% of the $13.7 billion total music purchases in 1998. The tables below give the dollar value of the music sold by age category in 1997 and 1998.

1997 Music Sales	
Age (years)	Sales (billions)
10–14	108.58
15–19	204.96
20–24	168.36
25–29	142.74
30–34	134.2
35–39	141.52
40–44	107.36
45+	201.3

1998 Music Sales	
Age (years)	Sales (billions)
10–14	124.67
15–19	216.46
20–24	167.14
25–29	156.18
30–34	156.18
35–39	172.52
40–44	113.71
45+	247.97

Think & Discuss

1. From 1997 to 1998, how much did sales in the 15–19 age category increase? How much did sales in the 20–24 age category decrease?
   $11.5 billion; $1.22 billion
2. Make a new table that gives the change in the sales in each age category from 1997 to 1998. (Use negative numbers to indicate decreases.) **See margin.**

Learn More About It

You will find the total Hispanic music sales for 1996 and 1997 in Exercise 40 on p. 204.

 APPLICATION LINK Visit www.mcdougallittell.com for more information about music sales.

ADDITIONAL RESOURCES
Another way to begin the chapter is to show the video clip of a real-life motivator for Chapter 4 on the *Instant Replay: Video Review Games.*

PROJECTS
A project covering Chapters 4–6 appears on page 396 of the Student Edition. An additional project for Chapter 4 is available in the *Chapter 4 Resource Book,* p. 90.

TECHNOLOGY

Software
- *Electronic Teaching Tools*
- *Online Lesson Planner*
- *Personal Student Tutor*
- *Test and Practice Generator*
- *Electronic Lesson Presentations* (Lesson 4.2)

Video
- *Instant Replay: Video Review Games*

Internet Connections
www.mcdougallittell.com
- **Application Links**
  197, 210, 225, 232, 236
- **Student Help**
  204, 207, 209, 216, 224, 234
- **Career Links**
  202, 217, 234
- **Extra Challenge**
  213, 220, 229, 235

2.

Change in Music Sales	
Age (yrs)	Change (billions)
10–14	16.09
15–19	11.5
20–24	−1.22
25–29	13.44
30–34	21.98
35–39	31
40–44	6.35
45+	46.67

PREPARE

DIAGNOSTIC TOOLS

The **Skill Review** exercises can help you diagnose whether students have the following skills needed in Chapter 4:

- Perform operations with positive and negative numbers.
- Identify algebraic properties.
- Solve a system of linear equations using an algebraic method.

The following resources are available for students who need additional help with these skills:

- Prerequisite Skills Review (*Chapter 4 Resource Book*, p. 5; *Warm-Up Transparencies*, p. 25)
- Reteaching with Practice (Chapter Resource Books for Lessons 1.1 and 3.2)
- ▢ *Personal Student Tutor*

ADDITIONAL RESOURCES

The following resources are provided to help you prepare for the upcoming chapter and customize review materials:

- *Chapter 4 Resource Book*
  Tips for New Teachers (p. 1)
  Parent Guide (p. 3)
  Lesson Plans (every lesson)
  Lesson Plans for Block Scheduling (every lesson)
- ▢ *Electronic Teaching Tools*
- ▢ *Online Lesson Planner*
- ▢ *Test and Practice Generator*

PREVIEW

What's the chapter about?

Chapter 4 is about **matrices and determinants**. You can use matrices to organize numerical data. In Chapter 4 you'll learn

- how to add, subtract, and multiply matrices, and how to evaluate determinants.
- how to solve linear systems using Cramer's rule and inverse matrices.

KEY VOCABULARY

▶ **Review**
- multiplicative inverse, p. 5
- coefficient, p. 13
- constant, p. 13
- solution of a system of linear equations, p. 139

▶ **New**
- matrix, p. 199
- equal matrices, p. 199
- scalar, p. 200
- scalar multiplication, p. 200

- determinant, p. 214
- Cramer's rule, p. 216
- identity matrix, p. 223
- inverse matrix, p. 223

PREPARE

Are you ready for the chapter?

SKILL REVIEW Do these exercises to review key skills that you'll apply in this chapter. See the given **reference page** if there is something you don't understand.

STUDENT HELP

↳ **Study Tip**
"Student Help" boxes throughout the chapter give you study tips and tell you where to look for extra help in this book and on the Internet.

Perform the operation(s). (Skills Review, p. 905)

1. $4 + (-5)$ -1 **2.** $-10 - 3$ -13 **3.** $(-1)6 - 4(2)$ -14 **4.** $12(5) + 2(-10)$ 40

Identify the property shown. (Review Example 4, p. 5)
commutative property of multiplication   distributive property of subtraction
5. $3(-9) = -9(3)$ **6.** $7 + y = y + 7$ **7.** $4(x - 2) = 4x - 4(2)$
                              commutative property of addition

Solve the system of linear equations using any algebraic method. (Review p. 149)

8. $2x = 30$
$x + 4y = 27$
$(15, 3)$

9. $x - y = 7$
$3x - 7y = 61$
$(-3, -10)$

10. $x - 2y = 24$
$x + 3y = 20$
$\left(\dfrac{112}{5}, -\dfrac{4}{5}\right)$

11. $3x + y = -8$
$8x - 3y = -10$
$(-2, -2)$

STUDY STRATEGY

Here's a study strategy!

Writing Out the Steps

In this chapter you will perform calculations with numbers in matrices. When you work with matrices, point your fingers at the numbers you will add, subtract, or multiply. Before calculating, write the numbers and the operation symbols in the correct location of the solution matrix. By doing this step, you are less likely to make a mistake. If you do make a mistake, you can easily trace backwards to find it.

What you should learn

GOAL 1 Add and subtract matrices, multiply a matrix by a scalar, and solve matrix equations.

GOAL 2 Use matrices in **real-life** situations, such as organizing data about health care plans in **Example 5**.

Why you should learn it

▼ To organize **real-life** data, such as the data for Hispanic CD, cassette, and music video sales in **Exs. 39–41**.

Matrix Operations

GOAL 1 USING MATRIX OPERATIONS

A **matrix** is a rectangular arrangement of numbers in rows and columns. For instance, matrix A below has two rows and three columns. The **dimensions** of this matrix are 2×3 (read "2 by 3"). The numbers in a matrix are its **entries**. In matrix A, the entry in the second row and third column is **5**.

$$A = \begin{bmatrix} 6 & 2 & -1 \\ -2 & 0 & 5 \end{bmatrix} \Big\} \text{ 2 rows}$$

3 columns

Some *matrices* (the plural of *matrix*) have special names because of their dimensions or entries.

NAME	DESCRIPTION	EXAMPLE
Row matrix	A matrix with only 1 row	$\begin{bmatrix} 3 & -2 & 0 & 4 \end{bmatrix}$
Column matrix	A matrix with only 1 column	$\begin{bmatrix} 1 \\ 3 \end{bmatrix}$
Square matrix	A matrix with the same number of rows and columns	$\begin{bmatrix} 4 & -1 & 5 \\ 2 & 0 & 1 \\ 1 & -3 & 6 \end{bmatrix}$
Zero matrix	A matrix whose entries are all zeros	$\begin{bmatrix} 0 & 0 \\ 0 & 0 \\ 0 & 0 \end{bmatrix}$

Two matrices are **equal** if their dimensions are the same and the entries in corresponding positions are equal.

EXAMPLE 1 *Comparing Matrices*

a. The following matrices are equal because corresponding entries are equal.

$$\begin{bmatrix} 5 & 0 \\ -\frac{4}{4} & \frac{3}{4} \end{bmatrix} = \begin{bmatrix} 5 & 0 \\ -1 & 0.75 \end{bmatrix}$$

b. The following matrices are not equal because corresponding entries in the second row are not equal.

$$\begin{bmatrix} -2 & 6 \\ 0 & -3 \end{bmatrix} \neq \begin{bmatrix} -2 & 6 \\ 3 & 0 \end{bmatrix}$$

· · · · · · · · · ·

To add or subtract matrices, you simply add or subtract corresponding entries. You can add or subtract matrices only if they have the same dimensions.

4.1 *Matrix Operations*   **199**

1 PLAN

PACING
Basic: 2 days
Average: 2 days
Advanced: 2 days
Block Schedule: 1 block

LESSON OPENER
APPLICATION
An alternative way to approach Lesson 4.1 is to use the Application Lesson Opener:
• Blackline Master (*Chapter 4 Resource Book*, p. 12)
• Transparency (p. 22)

MEETING INDIVIDUAL NEEDS
• *Chapter 4 Resource Book*
Prerequisite Skills Review (p. 5)
Practice Level A (p. 14)
Practice Level B (p. 15)
Practice Level C (p. 16)
Reteaching with Practice (p. 17)
Absent Student Catch-Up (p. 19)
Challenge (p. 21)
• *Resources in Spanish*
• **Personal Student Tutor**

NEW-TEACHER SUPPORT
See the Tips for New Teachers on pp. 1–2 of the *Chapter 4 Resource Book* for additional notes about Lesson 4.1.

WARM-UP EXERCISES
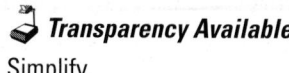
Transparency Available
Simplify.
1. $12 \times 5 + (-4)$ 56
2. $-3 - 9 - 28$ −40
3. $(-3) + (-9) + (-28)$ −40
4. $7(-8)$ −56
5. $2 + 4 + 7 - 12$ 1

ENGLISH LEARNERS
While students fluent in English will understand mathematical phrases, many English learners will not. If necessary, write two matrices on the board and point out their *corresponding entries*.

EXTRA EXAMPLE 1
Tell whether the following matrices are equal or not equal.

a. $\begin{bmatrix} 2.5 & -1 \\ -0.2 & 3 \end{bmatrix}$ and $\begin{bmatrix} \frac{5}{2} & -1 \\ -\frac{1}{5} & 3 \end{bmatrix}$ **equal**

b. $\begin{bmatrix} 1 & 2 \\ 3 & 5 \end{bmatrix}$ and $\begin{bmatrix} 0 & -1 \\ 8 & 3 \end{bmatrix}$ **not equal**

EXTRA EXAMPLE 2
Perform the indicated operation, if possible.

a. $\begin{bmatrix} 0 \\ 4 \\ 3 \end{bmatrix} + \begin{bmatrix} 1 \\ -9 \\ 8 \end{bmatrix}$ $\begin{bmatrix} 1 \\ -5 \\ 11 \end{bmatrix}$

b. $\begin{bmatrix} 3 & 3 \\ 9 & -5 \end{bmatrix} - \begin{bmatrix} 2 & 0 \\ -3 & 5 \end{bmatrix}$ $\begin{bmatrix} 1 & 3 \\ 12 & -10 \end{bmatrix}$

c. $[9 \ -7] + \begin{bmatrix} 1 & 3 \\ 4 & -8 \end{bmatrix}$ **not possible**

EXTRA EXAMPLE 3
Perform the indicated operation(s).

a. $6\begin{bmatrix} 2 & -3 \\ 8 & 4 \end{bmatrix}$ $\begin{bmatrix} 12 & -18 \\ 48 & 24 \end{bmatrix}$

b. $-1\begin{bmatrix} 4 & -7 \\ 3 & 3 \\ 2 & -9 \end{bmatrix} + \begin{bmatrix} 3 & 6 \\ 9 & -8 \\ 1 & -4 \end{bmatrix}$ $\begin{bmatrix} -1 & 13 \\ 6 & -11 \\ -1 & 5 \end{bmatrix}$

CHECKPOINT EXERCISES

For use after Example 1:

1. Tell whether the matrices are equal or not equal.

$\begin{bmatrix} 7 & 3 \\ -3 & 8 \end{bmatrix}$ and $\begin{bmatrix} 3 & 7 \\ 8 & -3 \end{bmatrix}$ **not equal**

For use after Example 2:

2. Perform the indicated operation.

$\begin{bmatrix} 7 & 8 \\ -9 & 0 \end{bmatrix} + \begin{bmatrix} 1 & 0 \\ 1 & 3 \end{bmatrix}$ $\begin{bmatrix} 8 & 8 \\ -8 & 3 \end{bmatrix}$

For use after Example 3:

3. Perform the indicated operation(s).

$2\begin{bmatrix} 6 & -2 \\ 6 & 3 \end{bmatrix} - \begin{bmatrix} 1 & -5 \\ 7 & 5 \end{bmatrix}$ $\begin{bmatrix} 11 & 1 \\ 5 & 1 \end{bmatrix}$

EXAMPLE 2 *Adding and Subtracting Matrices*

Perform the indicated operation, if possible.

a. $\begin{bmatrix} 3 \\ -4 \\ 7 \end{bmatrix} + \begin{bmatrix} 1 \\ 0 \\ 3 \end{bmatrix}$ **b.** $\begin{bmatrix} 8 & 3 \\ 4 & 0 \end{bmatrix} - \begin{bmatrix} 2 & -7 \\ 6 & -1 \end{bmatrix}$ **c.** $\begin{bmatrix} 2 & 0 \\ 3 & 4 \end{bmatrix} + \begin{bmatrix} 1 \\ 5 \end{bmatrix}$

SOLUTION

a. Since the matrices have the same dimensions, you can add them.

$$\begin{bmatrix} 3 \\ -4 \\ 7 \end{bmatrix} + \begin{bmatrix} 1 \\ 0 \\ 3 \end{bmatrix} = \begin{bmatrix} 3+1 \\ -4+0 \\ 7+3 \end{bmatrix} = \begin{bmatrix} 4 \\ -4 \\ 10 \end{bmatrix}$$

b. Since the matrices have the same dimensions, you can subtract them.

$$\begin{bmatrix} 8 & 3 \\ 4 & 0 \end{bmatrix} - \begin{bmatrix} 2 & -7 \\ 6 & -1 \end{bmatrix} = \begin{bmatrix} 8-2 & 3-(-7) \\ 4-6 & 0-(-1) \end{bmatrix} = \begin{bmatrix} 6 & 10 \\ -2 & 1 \end{bmatrix}$$

c. Since the dimensions of $\begin{bmatrix} 2 & 0 \\ 3 & 4 \end{bmatrix}$ are 2×2 and the dimensions of $\begin{bmatrix} 1 \\ 5 \end{bmatrix}$ are 2×1, you cannot add the matrices.

· · · · · · · · · ·

In matrix algebra, a real number is often called a **scalar**. To multiply a matrix by a scalar, you multiply each entry in the matrix by the scalar. This process is called **scalar multiplication**.

EXAMPLE 3 *Multiplying a Matrix by a Scalar*

Perform the indicated operation(s), if possible.

a. $3\begin{bmatrix} -2 & 0 \\ 4 & -7 \end{bmatrix}$ **b.** $-2\begin{bmatrix} 1 & -2 \\ 0 & 3 \\ -4 & 5 \end{bmatrix} + \begin{bmatrix} -4 & 5 \\ 6 & -8 \\ -2 & 6 \end{bmatrix}$

> **STUDENT HELP**
>
> ↳ **Study Tip**
> The order of operations for matrix expressions is similar to that for real numbers. In particular, you perform scalar multiplication before matrix addition and subtraction, as shown in part (b) of Example 3.

SOLUTION

a. $3\begin{bmatrix} -2 & 0 \\ 4 & -7 \end{bmatrix} = \begin{bmatrix} 3(-2) & 3(0) \\ 3(4) & 3(-7) \end{bmatrix} = \begin{bmatrix} -6 & 0 \\ 12 & -21 \end{bmatrix}$

b. $-2\begin{bmatrix} 1 & -2 \\ 0 & 3 \\ -4 & 5 \end{bmatrix} + \begin{bmatrix} -4 & 5 \\ 6 & -8 \\ -2 & 6 \end{bmatrix} = \begin{bmatrix} -2(1) & -2(-2) \\ -2(0) & -2(3) \\ -2(-4) & -2(5) \end{bmatrix} + \begin{bmatrix} -4 & 5 \\ 6 & -8 \\ -2 & 6 \end{bmatrix}$

$$= \begin{bmatrix} -2 & 4 \\ 0 & -6 \\ 8 & -10 \end{bmatrix} + \begin{bmatrix} -4 & 5 \\ 6 & -8 \\ -2 & 6 \end{bmatrix}$$

$$= \begin{bmatrix} -6 & 9 \\ 6 & -14 \\ 6 & -4 \end{bmatrix}$$

You can use what you know about matrix operations and matrix equality to solve a matrix equation.

EXAMPLE 4 **Solving a Matrix Equation**

Solve the matrix equation for x and y: $2\left(\begin{bmatrix} 3x & -1 \\ 8 & 5 \end{bmatrix} + \begin{bmatrix} 4 & 1 \\ -2 & -y \end{bmatrix}\right) = \begin{bmatrix} 26 & 0 \\ 12 & 8 \end{bmatrix}$

SOLUTION

Simplify the left side of the equation.

$$2\left(\begin{bmatrix} 3x & -1 \\ 8 & 5 \end{bmatrix} + \begin{bmatrix} 4 & 1 \\ -2 & -y \end{bmatrix}\right) = \begin{bmatrix} 26 & 0 \\ 12 & 8 \end{bmatrix}$$

$$2\begin{bmatrix} 3x + 4 & 0 \\ 6 & 5 - y \end{bmatrix} = \begin{bmatrix} 26 & 0 \\ 12 & 8 \end{bmatrix}$$

$$\begin{bmatrix} 6x + 8 & 0 \\ 12 & 10 - 2y \end{bmatrix} = \begin{bmatrix} 26 & 0 \\ 12 & 8 \end{bmatrix}$$

Equate corresponding entries and solve the two resulting equations.

$$6x + 8 = 26 \qquad\qquad 10 - 2y = 8$$
$$x = 3 \qquad\qquad\qquad y = 1$$

· · · · · · · · · ·

STUDENT HELP

▶ **Look Back**
For help with properties of real numbers, see p. 5.

In Example 4, you could have distributed the scalar 2 to each matrix inside the parentheses before adding the matrices.

$$2\left(\begin{bmatrix} 3x & -1 \\ 8 & 5 \end{bmatrix} + \begin{bmatrix} 4 & 1 \\ -2 & -y \end{bmatrix}\right) = 2\begin{bmatrix} 3x & -1 \\ 8 & 5 \end{bmatrix} + 2\begin{bmatrix} 4 & 1 \\ -2 & -y \end{bmatrix}$$

$$= \begin{bmatrix} 6x & -2 \\ 16 & 10 \end{bmatrix} + \begin{bmatrix} 8 & 2 \\ -4 & -2y \end{bmatrix}$$

$$= \begin{bmatrix} 6x + 8 & 0 \\ 12 & 10 - 2y \end{bmatrix}$$

This illustrates one of several properties of matrix operations stated below.

CONCEPT SUMMARY
PROPERTIES OF MATRIX OPERATIONS

Let A, B, and C be matrices with the same dimensions and let c be a scalar.

When adding matrices, you can regroup them and change their order without affecting the result.

ASSOCIATIVE PROPERTY OF ADDITION	$(A + B) + C = A + (B + C)$
COMMUTATIVE PROPERTY OF ADDITION	$A + B = B + A$

Multiplication of a sum or difference of matrices by a scalar obeys the distributive property.

DISTRIBUTIVE PROPERTY OF ADDITION	$c(A + B) = cA + cB$
DISTRIBUTIVE PROPERTY OF SUBTRACTION	$c(A - B) = cA - cB$

4.1 *Matrix Operations* **201**

EXTRA EXAMPLE 4
Solve the matrix equation for x and y:

$$4\left(\begin{bmatrix} 8 & 0 \\ -1 & 2y \end{bmatrix} + \begin{bmatrix} 4 & -2x \\ 1 & 6 \end{bmatrix}\right) = \begin{bmatrix} 48 & -48 \\ 0 & 8 \end{bmatrix} \quad x = 6, \, y = -2$$

☑ **CHECKPOINT EXERCISES**
For use after Example 4:
1. Solve the matrix equation for x and y.

$$3\left(\begin{bmatrix} 10 & 2 \\ 5 & 4y \end{bmatrix} - \begin{bmatrix} x & 5 \\ -1 & 1 \end{bmatrix}\right) = \begin{bmatrix} 0 & -9 \\ 18 & 21 \end{bmatrix} \quad x = 10, \, y = 2$$

EXTRA EXAMPLE 5
Use matrices to organize the following information about car insurance rates.
This year For 1 car, Comprehensive, collision, and basic insurance cost $612.15, $518.29, and $486.91. For 2 cars, comprehensive, collision, and basic insurance cost $1150.32, $984.16, and $892.51.
Next year For 1 car, comprehensive, collision, and basic insurance cost $616.28, $520.39, and $490.05. For 2 cars, comprehensive, collision, and basic insurance cost $1155.84, $987.72, and $895.13.

THIS YEAR (A)
1 car   2 cars

$$\begin{bmatrix} \$612.15 & \$1150.32 \\ \$518.29 & \$984.16 \\ \$486.91 & \$892.51 \end{bmatrix}$$

NEXT YEAR (B)
1 car   2 cars

$$\begin{bmatrix} \$616.28 & \$1155.84 \\ \$520.39 & \$987.72 \\ \$490.05 & \$895.13 \end{bmatrix}$$

Checkpoint Exercises for Example 5 on next page.

201

EXTRA EXAMPLE 6

The rates in Extra Example 5 on page 201 are for a specific policy-holder, depending on what types of cars they own and their previous driving record. Use the matrices to write a matrix that shows the monthly changes in car insurance payments from this year to next year.

$$\frac{1}{12}(B - A) = \begin{bmatrix} \$.34 & \$.46 \\ \$.18 & \$.30 \\ \$.26 & \$.22 \end{bmatrix}$$

CHECKPOINT EXERCISES

For use after Examples 5 and 6:

1. Condominium owners must pay yearly fees to cover the cost of maintenance, land-scaping, and remodeling. The fees this year are $96, $18, and $66 for a 1-bedroom unit, and $128, $24, and $88 for a 2-bedroom unit. The fees next year are $105, $20, and $73 for a 1-bedroom unit, and $141, $26, and $97 for a 2-bedroom unit. Use matrices to organize the information. Then use the matrices to find the monthly changes in fees from this year to next year.

THIS YEAR (*A*)
1 BR 2 BR

$$\begin{bmatrix} \$96 & \$128 \\ \$18 & \$24 \\ \$66 & \$88 \end{bmatrix}$$

NEXT YEAR (*B*)
1 BR 2 BR

$$\begin{bmatrix} \$105 & \$141 \\ \$20 & \$26 \\ \$73 & \$97 \end{bmatrix} ; \begin{bmatrix} \$.75 & \$1.08 \\ \$.17 & \$.17 \\ \$.58 & \$.75 \end{bmatrix}$$

CLOSURE QUESTION

What does it mean for a matrix to be a 4 × 3 matrix? **The matrix has 4 rows and 3 columns.**

REAL LIFE

Health Care

EXAMPLE 5   *Using Matrices to Organize Data*

Use matrices to organize the following information about health care plans.

This Year *For individuals, Comprehensive, HMO Standard, and HMO Plus cost $694.32, $451.80, and $489.48, respectively. For families, the Comprehensive, HMO Standard, and HMO Plus plans cost $1725.36, $1187.76, and $1248.12.*

Next Year *For individuals, Comprehensive, HMO Standard, and HMO Plus will cost $683.91, $463.10, and $499.27, respectively. For families, the Comprehensive, HMO Standard, and HMO Plus plans will cost $1699.48, $1217.45, and $1273.08.*

SOLUTION

One way to organize the data is to use 3 × 2 matrices, as shown.

	THIS YEAR (*A*)		NEXT YEAR (*B*)	
	Individual	Family	Individual	Family
Comprehensive	$694.32	$1725.36	$683.91	$1699.48
HMO Standard	$451.80	$1187.76	$463.10	$1217.45
HMO Plus	$489.48	$1248.12	$499.27	$1273.08

You can also organize the data using 2 × 3 matrices where the row labels are levels of coverage (individual and family) and the column labels are the types of plans (Comprehensive, HMO Standard, and HMO Plus).

EXAMPLE 6   *Using Matrix Operations*

HEALTH CARE A company offers the health care plans in Example 5 to its employees. The employees receive monthly paychecks from which health care payments are deducted. Use the matrices in Example 5 to write a matrix that shows the monthly changes in health care payments from this year to next year.

SOLUTION

Begin by subtracting matrix *A* from matrix *B* to determine the yearly changes in health care payments. Then multiply the result by $\frac{1}{12}$ and round answers to the nearest cent to find the monthly changes.

$$\frac{1}{12}(B - A) = \frac{1}{12}\left(\begin{bmatrix} 683.91 & 1699.48 \\ 463.10 & 1217.45 \\ 499.27 & 1273.08 \end{bmatrix} - \begin{bmatrix} 694.32 & 1725.36 \\ 451.80 & 1187.76 \\ 489.48 & 1248.12 \end{bmatrix} \right)$$

$$= \frac{1}{12} \begin{bmatrix} -10.41 & -25.88 \\ 11.30 & 29.69 \\ 9.79 & 24.96 \end{bmatrix}$$

$$\approx \begin{bmatrix} -\$.87 & -\$2.16 \\ \$.94 & \$2.47 \\ \$.82 & \$2.08 \end{bmatrix}$$

▶ The monthly deductions for the Comprehensive plan will decrease, but the monthly deductions for the other two plans will increase.

FOCUS ON CAREERS

 HEALTH SERVICES MANAGER

Health services managers in health maintenance organizations (HMOs) plan and organize the delivery of health care.

CAREER LINK
www.mcdougallittell.com

GUIDED PRACTICE

Vocabulary Check ✓

Concept Check ✓

2. Yes; the dimensions of both matrices are 3 × 2 and all the corresponding entries are equal.

3. The matrices must have the same dimensions.

Skill Check ✓

4. $\begin{bmatrix} 6 & -6 \\ -12 & 10 \\ 12 & -22 \end{bmatrix}$; no;

part (b) of Example 3 is $-2A + B$, which is not the same as $-2(A + B)$.

1–5. See margin.

1. What is a matrix? Describe and give an example of a row matrix, a column matrix, and a square matrix.

2. Are the two matrices equal? Explain. $\begin{bmatrix} -6 & \frac{1}{2} \\ 4 & -5 \\ 3 & 5 \end{bmatrix} \stackrel{?}{=} \begin{bmatrix} -6 & 0.5 \\ 4 & -5 \\ 3 & 5 \end{bmatrix}$

3. To add or subtract two matrices, what must be true?

4. Use the matrices at the right to find $-2(A + B)$. Is your answer the same as that for part (b) of Example 3? Explain. $\quad A = \begin{bmatrix} 1 & -2 \\ 0 & 3 \\ -4 & 5 \end{bmatrix} \quad B = \begin{bmatrix} -4 & 5 \\ 6 & -8 \\ -2 & 6 \end{bmatrix}$

5. Rework Example 5 by organizing the data using 2 × 3 matrices.

Perform the indicated operation(s), if possible.

6. $\begin{bmatrix} 20 \\ -22 \\ 9 \end{bmatrix} - \begin{bmatrix} -11 \\ -10 \\ -6 \end{bmatrix} \begin{bmatrix} 31 \\ -12 \\ 15 \end{bmatrix}$

7. See margin. $\begin{bmatrix} -6 & -7 & 4 \\ -4 & 0 & -1 \end{bmatrix} + \begin{bmatrix} -1 & -5 & 8 \\ 9 & 12 & -9 \end{bmatrix}$

8. $-4\begin{bmatrix} 2 & 0 \\ -4 & -5 \end{bmatrix} \begin{bmatrix} -8 & 0 \\ 16 & 20 \end{bmatrix}$

9. See margin. $6\begin{bmatrix} -5 & -1 \\ 2 & 0 \end{bmatrix} - 5\begin{bmatrix} -1 & 0 \\ 4 & -3 \end{bmatrix}$

10. 🌐 **HEALTH CARE** In Example 5, suppose the annual health care costs given in matrix B increase by 4% the following year. Write a matrix that shows the new *monthly* payment. $\begin{bmatrix} \$59.27 & \$147.29 \\ \$40.14 & \$105.51 \\ \$43.27 & \$110.33 \end{bmatrix}$

PRACTICE AND APPLICATIONS

STUDENT HELP

➤ **Extra Practice**
to help you master skills is on p. 944.

21. Not possible; the two matrices do not have the same dimensions.

22. $\begin{bmatrix} -\frac{3}{2} & -\frac{1}{2} \\ \frac{5}{2} & 3 \end{bmatrix}$

STUDENT HELP

➤ **HOMEWORK HELP**
Example 1: Exs. 11–14
Example 2: Exs. 15–22
Example 3: Exs. 23–32
Example 4: Exs. 33–36
Examples 5, 6: Exs. 37–47

COMPARING MATRICES Tell whether the matrices are *equal* or *not equal*.

11. $\begin{bmatrix} 5 & -1 & 7 \end{bmatrix}, \begin{bmatrix} 5 \\ -1 \\ 7 \end{bmatrix}$ not equal

12. $\begin{bmatrix} 1 & 0 & -8 \\ 8 & 0 & 1 \end{bmatrix}, \begin{bmatrix} 1 & 0 & -8 \\ 8 & 0 & 1 \end{bmatrix}$ equal

13. $\begin{bmatrix} 4 & 0 \\ 2 & -4 \end{bmatrix}, \begin{bmatrix} 4 & 0 \\ -2 & -4 \end{bmatrix}$ not equal

14. $\begin{bmatrix} 2 & 1.5 & 4.25 \\ 0.5 & -0.5 & 0 \end{bmatrix}, \begin{bmatrix} 2 & \frac{3}{2} & \frac{17}{4} \\ \frac{1}{2} & -\frac{1}{2} & 0 \end{bmatrix}$ equal

ADDING AND SUBTRACTING MATRICES Perform the indicated operation, if possible. If not possible, state the reason.

15. $\begin{bmatrix} 1 & -4 \\ -7 & 2 \end{bmatrix} + \begin{bmatrix} 3 & 5 \\ -5 & 2 \end{bmatrix} \begin{bmatrix} 4 & 1 \\ -12 & 4 \end{bmatrix}$

16. $\begin{bmatrix} 4 & -2 \\ 0 & -6 \end{bmatrix} + \begin{bmatrix} 4 \\ -1 \end{bmatrix}$ Not possible; the two matrices do not have the same dimensions.

17. $\begin{bmatrix} -8 & -2 \\ 6 & -6 \end{bmatrix} - \begin{bmatrix} -4 & 5 \\ 1 & -1 \end{bmatrix} \begin{bmatrix} -4 & -7 \\ 5 & -5 \end{bmatrix}$

18. $\begin{bmatrix} -3 & 5 \\ 0 & -1 \end{bmatrix} + \begin{bmatrix} 2 & -7 \\ -4 & 9 \end{bmatrix} \begin{bmatrix} -1 & -2 \\ -4 & 8 \end{bmatrix}$

19. $\begin{bmatrix} 1.2 & 3.5 \\ 0.2 & 5.1 \end{bmatrix} + \begin{bmatrix} 4.1 & 8.7 \\ 2.6 & 5.3 \end{bmatrix} \begin{bmatrix} 5.3 & 12.2 \\ 2.8 & 10.4 \end{bmatrix}$

20. $\begin{bmatrix} 7 & -1 & 4 \\ 11 & -9 & 2 \end{bmatrix} + \begin{bmatrix} -3 & 6 & 3 \\ 10 & 1 & -5 \end{bmatrix}$

21. $\begin{bmatrix} 1 & 6 \\ -1 & -6 \\ 2 & 8 \end{bmatrix} - \begin{bmatrix} 7 & -3 & 9 \\ -2 & -7 & 9 \\ 11 & -1 & 2 \end{bmatrix}$

22. $\begin{bmatrix} \frac{1}{2} & \frac{1}{4} \\ 3 & 8 \end{bmatrix} - \begin{bmatrix} 2 & \frac{3}{4} \\ \frac{1}{2} & 5 \end{bmatrix} \begin{bmatrix} 4 & 5 & 7 \\ 21 & -8 & -3 \end{bmatrix}$

21, 22. See margin.

---

3 APPLY

⬤ **ASSIGNMENT GUIDE**

BASIC
Day 1: pp. 203–205 Exs. 12–34 even, 37, 38
Day 2: pp. 203–206 Exs. 19, 21, 27, 35, 36, 39–42, 48, 51–65 odd

AVERAGE
Day 1: pp. 203–205 Exs. 12–36 even, 37–41
Day 2: pp. 203–206 Exs. 29–35 odd, 39–44, 48, 51–65 odd

ADVANCED
Day 1: pp. 203–205 Exs. 12–24 even. 37–41
Day 2: pp. 203–206 Exs. 42–49, 50–66 even

BLOCK SCHEDULE
pp. 203–206 Exs. 12–36 even, 29–41 odd, 45–48, 51–65 odd

EXERCISE LEVELS
Level A: *Easier*
11–32, 37, 38
Level B: *More Difficult*
33–36, 39–42, 48
Level C: *Most Difficult*
43–47, 49

✔ **HOMEWORK CHECK**
To quickly check student understanding of key concepts, go over the following exercises: Exs. 14, 18, 26, 32, 34. See also the Daily Homework Quiz:
• Blackline Master (*Chapter 4 Resource Book*, p. 24)
• 📖 Transparency (p. 27)

1, 5. See next page.

7. $\begin{bmatrix} -7 & -12 & 12 \\ 5 & 12 & -10 \end{bmatrix}$

9. $\begin{bmatrix} -25 & -6 \\ -8 & -15 \end{bmatrix}$

STUDENT HELP NOTES

→ **Homework Help** Students can find help for Exs. 33–36 at **www.mcdougallittell.com**. The information can be printed out for students who don't have access to the Internet.

! COMMON ERROR

EXERCISES 37–41 Students may confuse the rows of a matrix with the columns of a matrix. To help them understand the difference, remind them that columns in architecture are vertical pillars, or columns in newspapers run up and down. So the columns of a matrix are the vertical parts of the matrix.

1. *Sample answer:* a rectangular arrangement of numbers in rows and columns; row matrix: a matrix with only one row, such as $[5\ 2\ -1]$, column matrix: a matrix with only one column, such as $\begin{bmatrix} 6 \\ -2 \end{bmatrix}$, square matrix: a matrix with the same number of rows and columns, such as $\begin{bmatrix} 1 & -1 & 0 \\ 0 & 5 & 3 \\ 1 & 8 & -6 \end{bmatrix}$.

5.

	This Year (A)		
	Compre-hensive	HMO Standard	HMO Plus
Ind.	$694.32	$451.80	$489.48
Fam.	$1725.36	$1187.76	$1248.12

	Next Year (B)		
Ind.	$683.91	$463.10	$499.27
Fam.	$1699.48	$1217.45	$1273.08

37.

	Before	
	Wins	Losses
Braves	59	29
Mariners	37	51
Cubs	48	39

	After	
Braves	47	27
Mariners	39	34
Cubs	42	34

39. See next page.

---

23. $\begin{bmatrix} 4 & 12 & -28 \\ 16 & 0 & -24 \end{bmatrix}$

24. $\begin{bmatrix} -10 & -30 \\ 15 & 5 \end{bmatrix}$

25. $\begin{bmatrix} 4 & 12 & 36 \\ -20 & 20 & 60 \\ -12 & -20 & -44 \end{bmatrix}$

26. $\begin{bmatrix} 0 & 0 \\ 18 & 18 \\ -3 & -4 \end{bmatrix}$

27. $\begin{bmatrix} -1 & -1 & 2 \\ \frac{1}{8} & \frac{3}{11} & -5 \end{bmatrix}$

STUDENT HELP

INTERNET HOMEWORK HELP
Visit our Web site **www.mcdougallittell.com** for help with Exs. 33–36.

28. $\begin{bmatrix} -21.5 & 8.5 \\ 3 & -12.75 \\ -12 & 11 \\ 25 & -20 \end{bmatrix}$

29. $\begin{bmatrix} 8 & -8 \\ 12 & -3 \\ -16 & 23 \end{bmatrix}$

30. $\begin{bmatrix} -11 & -25 & -9 \\ 11 & -8 & -27 \end{bmatrix}$

31. $\begin{bmatrix} 22 & -30 \\ -22 & -18 \end{bmatrix}$

32. $\begin{bmatrix} -29 & 5 & 14 \\ 30 & -8 & -16 \end{bmatrix}$

37–41. Matrices can also be written with the rows and columns switched.

38. $\begin{bmatrix} 106 & 56 \\ 76 & 85 \\ 90 & 73 \end{bmatrix}$

40. $\begin{bmatrix} 47,056 & \$613,138 \\ 33,098 & \$266,974 \\ 115 & \$2,176 \end{bmatrix}$

41. $\begin{bmatrix} 5,498 & \$76,256 \\ 2,500 & \$22,316 \\ 25 & \$344 \end{bmatrix}$

MULTIPLYING BY A SCALAR Perform the indicated operation.
23–28. See margin.

23. $-4\begin{bmatrix} -1 & -3 & 7 \\ -4 & 0 & 6 \end{bmatrix}$

24. $5\begin{bmatrix} -2 & -6 \\ 3 & 1 \end{bmatrix}$

25. $4\begin{bmatrix} 1 & 3 & 9 \\ -5 & 5 & 15 \\ -3 & -5 & -11 \end{bmatrix}$

26. $-9\begin{bmatrix} 0 & 0 \\ -2 & -2 \\ \frac{1}{3} & \frac{4}{9} \end{bmatrix}$

27. $\frac{1}{2}\begin{bmatrix} -2 & -2 & 4 \\ \frac{1}{4} & \frac{6}{11} & -10 \end{bmatrix}$

28. $2.5\begin{bmatrix} -8.6 & 3.4 \\ 1.2 & -5.1 \\ -4.8 & 4.4 \\ 10 & -8 \end{bmatrix}$

COMBINING MATRIX OPERATIONS Perform the indicated operations.
29–32. See margin.

29. $\begin{bmatrix} 12 & -8 \\ 0 & 5 \\ 0 & 3 \end{bmatrix} + 4\begin{bmatrix} -1 & 0 \\ 3 & -2 \\ -4 & 5 \end{bmatrix}$

30. $2\begin{bmatrix} -6 & -10 & 2 \\ 4 & -7 & -4 \end{bmatrix} - \begin{bmatrix} -1 & 5 & 13 \\ -3 & -6 & 19 \end{bmatrix}$

31. $2\begin{bmatrix} 7 & -7 \\ -1 & 3 \end{bmatrix} + 4\begin{bmatrix} 2 & -4 \\ -5 & -6 \end{bmatrix}$

32. $3\begin{bmatrix} -7 & 1 & 0 \\ 8 & -6 & -2 \end{bmatrix} - 2\begin{bmatrix} 4 & -1 & -7 \\ -3 & -5 & 5 \end{bmatrix}$

SOLVING MATRIX EQUATIONS Solve the matrix equation for *x* and *y*.

33. $\begin{bmatrix} -2x & -8 \\ -10 & -9 \end{bmatrix} = \begin{bmatrix} 6 & y \\ -10 & -9 \end{bmatrix}$ $x = -3, y = -8$

34. $\begin{bmatrix} 3x & -2 \\ -1 & 8 \end{bmatrix} + \begin{bmatrix} -4 & 0 \\ -7 & -8 \end{bmatrix} = \begin{bmatrix} -16 & -2 \\ y & 0 \end{bmatrix}$ $x = -4, y = -8$

35. $2x\begin{bmatrix} -3 & 4 \\ -11 & 5 \end{bmatrix} = \begin{bmatrix} 12 & -16 \\ y & -20 \end{bmatrix}$ $x = -2, y = 44$

36. $\begin{bmatrix} 4 & -3 \\ 8 & -7 \\ 1 & 2 \end{bmatrix} + \begin{bmatrix} -5 & x \\ -7 & 7 \\ 4 & -9 \end{bmatrix} = \begin{bmatrix} -1 & -8 \\ y & 0 \\ 5 & -7 \end{bmatrix}$ $x = -5, y = 1$

🌎 **BASEBALL STATISTICS** In Exercises 37 and 38, use the following information about three Major League Baseball teams' wins and losses in 1998 before and after the All-Star Game. ▸ Source: CNN/SI 37, 38. See margin.

Before *The Atlanta Braves had 59 wins and 29 losses, the Seattle Mariners had 37 wins and 51 losses, and the Chicago Cubs had 48 wins and 39 losses.*

After *The Atlanta Braves had 47 wins and 27 losses, the Seattle Mariners had 39 wins and 34 losses, and the Chicago Cubs had 42 wins and 34 losses.*

37. Use matrices to organize the information.

38. Using your matrices from Exercise 37, write a matrix that shows the total numbers of wins and losses for the three teams in 1998.

🌎 **HISPANIC MUSIC** In Exercises 39–41, use the following information.
The figures below give the number (in millions) of Hispanic CD, cassette, and music video units shipped to all market channels and the dollar value (in millions) of those shipments (at suggested list prices). ▸ Source: Recording Industry Association of America
39–41. See margins.

1996 *Number of units—CDs: 20,779; cassettes: 15,299; and music videos: 45. Dollar value—CDs: 268,441; cassettes: 122,329; and music videos: 916.*

1997 *Number of units—CDs: 26,277; cassettes: 17,799; and music videos: 70. Dollar value—CDs: 344,697; cassettes: 144,645; and music videos: 1,260.*

39. Use matrices to organize the information.

40. Write a matrix that gives the total numbers of units shipped and total values for both years.

41. Write a matrix that gives the change in units shipped and dollar value from 1996 to 1997.

Cost of a 4-Year College

COLLEGE COSTS Since 1980, college costs have risen substantially. After adjusting for inflation, costs rose by 48% at public 4-year colleges and 71% at private 4-year colleges between 1980 and 1995.

43. $2V + M$;
$\begin{bmatrix} 146.8 & 148.4 \\ 146.1 & 147.8 \\ 146.8 & 148.4 \\ 146.2 & 148.1 \end{bmatrix}$

46. $\begin{bmatrix} -0.6 & -1.2 & -0.3 \\ -1.0 & -0.7 & -0.1 \\ -0.4 & 1.4 & 0.7 \\ 0.1 & 0.8 & 0.3 \\ 0.4 & 0.6 & 0.2 \end{bmatrix}$

47. South: 18–65, over 65, Mountain: 0–17, 18–65, over 65, Pacific: 0–17, 18–65, over 65

42. **COLLEGE COSTS** The matrices below show the average yearly cost (in dollars) of tuition and room and board at colleges in the United States from 1995 through 1997. Use matrix addition to write a matrix showing the totals of these costs. ▶ Source: U.S. Department of Education **See margin.**

	TUITION			ROOM AND BOARD		
	1995	**1996**	**1997**	**1995**	**1996**	**1997**
Public 2-year college	1,192	1,239	1,283	2,944	2,978	3,128
Public 4-year college	2,681	2,848	2,986	3,990	4,166	4,345
Private 2-year college	6,914	7,094	7,190	4,256	4,469	4,699
Private 4-year college	11,481	12,243	12,920	5,121	5,368	5,555

PSAT SCORES In Exercises 43 and 44, use the following information.
Eligibility for a National Merit Scholarship is based on a student's PSAT score. Through 1996, this total score was found by doubling a student's verbal score and adding this value to the student's mathematics score. Let V represent the average verbal scores and let M represent the average mathematics scores earned by sophomores and juniors at Central High for tests taken in 1993 through 1996.

	VERBAL SCORES (V)		MATHEMATICS SCORES (M)	
	Sophomores	**Juniors**	**Sophomores**	**Juniors**
1993	48.9	49.0	49.0	50.4
1994	48.9	48.9	48.3	50.0
1995	48.7	48.8	49.4	50.8
1996	48.2	48.6	49.8	50.9

43. Write an expression in terms of V and M that you could use to determine the average total PSAT scores for sophomores and juniors at Central High from 1993 through 1996. Then evaluate the expression.

44. Use the matrix from Exercise 43 to determine the average total PSAT score for juniors at Central High in 1996.   **148.1**

U.S. POPULATION In Exercises 45–47, use the following information.
The matrices show the number of people (in thousands) who lived in each region of the United States in 1991 and the number of people (in thousands) projected to live in each region in 2010. The regional populations are separated into three age categories.

DATA UPDATE of U.S. Bureau of the Census data at www.mcdougallittell.com

	1991			2010		
	0–17	**18–65**	**Over 65**	**0–17**	**18–65**	**Over 65**
Northeast	12,142	31,791	7,043	12,493	33,822	7,377
Midwest	15,814	36,554	7,857	15,840	41,095	8,980
South	22,504	53,471	10,942	25,428	67,337	14,832
Mountain	3,993	8,461	1,580	5,094	12,420	2,707
Pacific	10,693	25,001	4,331	13,655	31,125	5,511

45. The total population in 1991 was 252,177,000 and the projected total population in 2010 is 297,716,000. Rewrite the matrices to give the information as percents of the total population. (*Hint:* Multiply each matrix by the reciprocal of the total population (in thousands), and then multiply by 100.) **See margin.**

46. Write a matrix that gives the projected change in the percent of the population in each region and age group from 1991 to 2010.

47. Based on the result of Exercise 46, which region(s) and age group(s) are projected to show relative growth from 1991 to 2010?

4.1 *Matrix Operations* **205**

39.
	1996	
	No. shipped (in millions)	**Dollar value (in millions)**
CDs	20,779	268,441
Cassettes	15,299	122,329
Music videos	45	916

	1997	
	No. shipped (in millions)	**Dollar value (in millions)**
CDs	26,277	344,697
Cassettes	17,799	144,645
Music videos	70	1,260

42. $\begin{bmatrix} 4,136 & 4,217 & 4,411 \\ 6,671 & 7,014 & 7,331 \\ 11,170 & 11,563 & 11,889 \\ 16,602 & 17,611 & 18,475 \end{bmatrix}$

45.
	1991		
	0–17	**18–65**	**Over 65**
Northeast	4.8	12.6	2.8
Midwest	6.3	14.5	3.1
South	8.9	21.2	4.3
Mountain	1.6	3.4	0.6
Pacific	4.2	9.9	1.7

	2010		
	0–17	**18–65**	**Over 65**
Northeast	4.2	11.4	2.5
Midwest	5.3	13.8	3.0
South	8.5	22.6	5.0
Mountain	1.7	4.2	0.9
Pacific	4.6	10.5	1.9

ADDITIONAL PRACTICE AND RETEACHING

For Lesson 4.1:
- Practice Levels A, B, and C (*Chapter 4 Resource Book*, p. 14)
- Reteaching with Practice (*Chapter 4 Resource Book*, p. 17)
- See Lesson 4.1 of the *Personal Student Tutor*

For more Mixed Review:
- Search the *Test and Practice Generator* for key words or specific lessons.

DAILY HOMEWORK QUIZ

🖳 *Transparency Available*

Perform the indicated operation, if possible. If not possible, state the reason.

1. $\begin{bmatrix} 1 & -2 \\ 3 & 6 \end{bmatrix} + \begin{bmatrix} 3 & 5 \\ -1 & 4 \end{bmatrix}$ $\begin{bmatrix} 4 & 3 \\ 2 & 10 \end{bmatrix}$

2. $[5\ 6\ 1] - \begin{bmatrix} 0 & 0 \\ 3 & -3 \end{bmatrix}$ not possible;

the two matrices do not have the same dimensions,

3. $2\begin{bmatrix} 4 & -1 \\ 2 & 1 \end{bmatrix}$ $\begin{bmatrix} 8 & -2 \\ 4 & 2 \end{bmatrix}$

4. $3\begin{bmatrix} -2 & 3 & 8 \\ 1 & 0 & 4 \end{bmatrix} - \begin{bmatrix} -1 & 3 & 1 \\ -2 & 6 & 0 \end{bmatrix}$

$\begin{bmatrix} -5 & 6 & 23 \\ 5 & -6 & 12 \end{bmatrix}$

5. Solve the matrix equation for x and y.

$x\begin{bmatrix} 2 & -3 \\ 3 & 6 \end{bmatrix} = \begin{bmatrix} 4 & -6 \\ y & 12 \end{bmatrix}$ 2, 6

┌─ **EXTRA CHALLENGE NOTE** ─┐

→ Challenge problems for Lesson 4.1 are available in **blackline** format in the *Chapter 4 Resource Book,* p. 21 and at **www.mcdougallittell.com.**

ADDITIONAL TEST PREPARATION

1. WRITING If matrix A is equal to matrix B and

$A = \begin{bmatrix} a & 0.5 \\ c & -3 \end{bmatrix}$ and $B = \begin{bmatrix} -5 & \frac{1}{2} \\ a & -3 \end{bmatrix}$,

what is true about a and c?
They are equal, and $a = c = -5$.

48c, 50–52. See Additional Answers beginning on page AA1.

48. MULTI-STEP PROBLEM The matrices show the number of hardcover volumes sold and the average price per volume (in dollars) for different subject areas.
▶ Source: *The Bowker Annual*

48a. $\begin{bmatrix} -46,000 & 12.17 \\ 111,000 & 15.42 \\ 2,000 & -4.06 \\ -20,000 & -4.38 \end{bmatrix}$;

111,000 more; $4.06 less

	1995 (A)		1996 (B)	
	Volumes sold	Average price per volume	Volumes sold	Average price per volume
Art	1,116,000	41.23	1,070,000	53.40
Law	716,000	73.09	827,000	88.51
Music	251,000	43.27	253,000	39.21
Travel	199,000	38.30	179,000	33.92

a. Calculate $B - A$. How many more (or fewer) law volumes were sold in 1996 than in 1995? How much more (or less) did the average music book cost in 1996 than in 1995?

b. Calculate $B + A$. Does the "volumes sold" column in $B + A$ give you meaningful information? Does the "average price per volume" column in $B + A$ give you meaningful information? Explain. **See margin.**

c. *Writing* What conclusions can you make about the number of volumes sold and the average price per volume of these books from 1995 to 1996? **See margin.**

★ **Challenge**

49. GEOMETRY CONNECTION A triangle has vertices (2, 2), (8, 2), and (5, 6). Assign a letter to each vertex and organize the triangle's vertices in a matrix. When you multiply the matrix by 4, what does the "new" triangle look like? How are the two triangles related? Use a graph to help you.

$\begin{bmatrix} 2 & 2 \\ 8 & 2 \\ 5 & 6 \end{bmatrix}$; the "new" triangle has the vertices $A' = (8, 8)$, $B' = (32, 8)$, and $C' = (20, 24)$; two triangles are similar, with each side of $\triangle A'B'C'$ being parallel to and 4 times as long as the corresponding side of $\triangle ABC$.

MIXED REVIEW

48b. $\begin{bmatrix} 2,186,000 & 94.63 \\ 1,543,000 & 161.60 \\ 504,000 & 82.48 \\ 378,000 & 72.22 \end{bmatrix}$;

yes; no; for each subject area, adding the numbers of volumes sold in 1995 and 1996 gives the total number of volumes sold over the two-year period, but adding the average prices for 1995 and 1996 does not result in an average price for all the volumes sold over the two-year period.

TRANSFORMING FIGURES Draw the figure produced by each transformation of the figure shown. (Skills Review, p. 921)
50–52. See margin.

50. a 270° counterclockwise rotation about the origin

51. a translation by 2 units right and 4 units down

52. a reflection over the y-axis

MULTIPLYING REAL NUMBERS Find the product. (Skills Review, p. 905)

53. $-4(-5)$ 20
54. $8(-2)$ −16
55. $-7(-1)$ 7
56. $\frac{1}{2}(-7)$ $-\frac{7}{2}$
57. $\frac{5}{6} \cdot \frac{3}{7}$ $\frac{5}{14}$
58. $3.2(2.4 + 8.1)$ 33.6

CHECKING SOLUTIONS Check whether the ordered pairs are solutions of the inequality. (Review 2.6)

59. $x + 2y \le -3$; (0, 3), (−5, 1) no, yes
60. $5x - y > 2$; (−5, 0), (5, 23) no, no
61. $-8x - 3y < 5$; (−1, 1), (3, −9) no, yes
62. $21x - 10y > 4$; (2, 3), (−1, 0) yes, no

FINDING A SOLUTION Give an ordered pair that is a solution of the system. (Lesson 3.3) 63−65. Sample answers are given.

63. $x + y < 10$ (1, 2)
   $y > 1$
64. $x - y \ge 3$ (0, −3)
   $y < 12$
65. $3x > y$ (5, 5)
   $x \le 15$

● ACTIVITY 4.1

Using Technology

Using Matrix Operations

You can use a graphing calculator to perform matrix operations.

▶ **EXAMPLE**

The matrices show music sales (in thousands) for a chain of stores during a two-month period. The music formats are CDs and cassette tapes. The categories of music are Rock (R), Country (C), Jazz (J), and Easy Listening (E).

	MAY				**JUNE**			
	R	C	J	E	R	C	J	E
CDs	32	16	3	8	24	15	3	7
Tapes	28	12	5	15	25	10	4	15

a. Find the total sales for each format and category for May and June.

b. Estimate the sales for July if they are expected to decrease 3% from the June sales.

▶ **SOLUTION**

a. To find the total sales in each format and category for May and June, let matrix *A* represent May sales and let matrix *B* represent June sales. Use a graphing calculator to enter each matrix. Then find the sum of matrix *A* and matrix *B*.

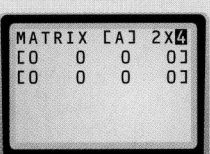

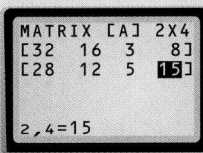

For each matrix, enter the dimensions of the matrix.

Then enter the data row by row until all 8 entries have been filled for each matrix.

To find the sum, select each matrix from the matrix menu and add.

b. To estimate the sales for July, multiply matrix *B* by the scalar 0.97. The calculator shows, for instance, that July sales for rock CDs are expected to be 23,280.

```
.97[B]
[[23.28 14.5...
 [24.25 9.7 ...
```

▶ **EXERCISES**

Use a graphing calculator to perform the indicated operation(s). 1–5. See margin.

1. $\begin{bmatrix} -4.9 & 3.1 \\ 9.3 & -2.5 \end{bmatrix} + \begin{bmatrix} 11.5 & -9.2 \\ 6.03 & 4.22 \end{bmatrix}$

2. $\begin{bmatrix} 418 & 418 \\ 452 & 452 \\ 146 & 146 \end{bmatrix} - \begin{bmatrix} 512 & 359 \\ 428 & 184 \\ 735 & 299 \end{bmatrix}$

3. $1.043 \begin{bmatrix} 6.2 & 1.6 \\ 22.1 & 7 \end{bmatrix}$

4. $5.08 \begin{bmatrix} 0.06 & 1.25 \\ 3.32 & 0.01 \end{bmatrix} - 0.07 \begin{bmatrix} 1.009 & 0.052 \\ 11.2 & 34.15 \end{bmatrix}$

5. ● **MUSIC** Write a matrix that gives the change in sales from May to June for the chain of stores discussed in the example. For which categories of CDs and cassettes did sales increase from May to June? For which did sales decrease?

STUDENT HELP

INTERNET **KEYSTROKE HELP**

See keystrokes for several models of calculators at www.mcdougallittell.com

1. $\begin{bmatrix} 6.6 & -6.1 \\ 15.33 & 1.72 \end{bmatrix}$

2. $\begin{bmatrix} -94 & 59 \\ 24 & 268 \\ -589 & -153 \end{bmatrix}$

3. $\begin{bmatrix} 6.4666 & 1.6688 \\ 23.0503 & 7.301 \end{bmatrix}$

4. $\begin{bmatrix} 0.23417 & 6.34636 \\ 16.0816 & -2.3397 \end{bmatrix}$

5. $\begin{bmatrix} -8 & -1 & 0 & -1 \\ -3 & -2 & -1 & 0 \end{bmatrix}$;

none; Rock CDs, Country CDs, Easy Listening CDs, Rock tapes, Country tapes, Jazz tapes

---

1 Planning the Activity

PURPOSE
To use matrices to model sales and to perform matrix operations using a graphing calculator.

MATERIALS
• graphing calculator
• Keystroke blackline (*Chapter 7 Resource Book*, p. 13)

PACING
• Activity — 20 min

▶ **LINK TO LESSON**
Students may wish to use graphing calculators to perform the calculations in Example 6 in Lesson 4.1.

2 Managing the Activity

COOPERATIVE LEARNING
Students can work with a partner to complete the activity. This will enable students to check each other's work as they do the activity.

ALTERNATIVE APPROACH
A graphing calculator for the overhead projector can be used to demonstrate the solution to the Example. Students can then do the exercises themselves.

3 Closing the Activity

★ **KEY DISCOVERY**
Graphing calculators can be used to perform matrix operations.

ACTIVITY ASSESSMENT
Sales are projected to have a 10% increase from May of this year to May of next year. How could you use matrices to model this situation? Multiply the matrix for May in the Example by the scalar 1.1 to get a

matrix of $\begin{bmatrix} 35.2 & 17.6 & 3.3 & 8.8 \\ 30.8 & 13.2 & 5.5 & 16.5 \end{bmatrix}$.

LESSON OPENER
VISUAL APPROACH
An alternative way to approach
Lesson 4.2 is to use the Visual
Approach Lesson Opener:
• Blackline Master (*Chapter 4 Resource Book*, p. 25)
• Transparency (p. 23)

MEETING INDIVIDUAL NEEDS
• *Chapter 4 Resource Book*
  Prerequisite Skills Review (p. 5)
  Practice Level A (p. 28)
  Practice Level B (p. 29)
  Practice Level C (p. 30)
  Reteaching with Practice (p. 31)
  Absent Student Catch-Up (p. 33)
  Challenge (p. 35)
• *Resources in Spanish*
• Personal Student Tutor

NEW-TEACHER SUPPORT
See the Tips for New Teachers on
pp. 1–2 of the *Chapter 4 Resource
Book* for additional notes about
Lesson 4.2.

WARM-UP EXERCISES

Transparency Available

Simplify.
1. $2 \times 5 + 6 \times (-8)$ −38
2. $(6 + 4)(10 - 3)$ 70
3. $-4(3x - 8)$ $-12x + 32$
4. $\frac{1}{2}\left(\frac{3}{4}x + \frac{2}{5}\right)$ $\frac{3}{8}x + \frac{1}{5}$
5. $3.2(6.1x - 4.7)$ $19.52x - 15.04$

ENGLISH LEARNERS
The complexity of the language in
Example 2 may make learning the
concepts difficult for English learners. You may want to describe the
steps of the solution.

208

4.2

What you should learn

GOAL 1 Multiply two matrices.

GOAL 2 Use matrix multiplication in **real-life** situations, such as finding the number of calories burned in **Ex. 40.**

Why you should learn it

▼ To solve **real-life** problems, such as calculating the cost of softball equipment in **Example 5.**

Multiplying Matrices

GOAL 1 **MULTIPLYING TWO MATRICES**

The product of two matrices A and B is defined provided the number of columns in A is equal to the number of rows in B.

If A is an $m \times n$ matrix and B is an $n \times p$ matrix, then the product AB is an $m \times p$ matrix.

$$\begin{array}{ccccc} A & \cdot & B & = & AB \\ m \times n & & n \times p & & m \times p \end{array}$$
$$\uparrow \qquad \uparrow$$
$$\text{equal}$$
$$\text{dimensions of } AB$$

EXAMPLE 1 *Describing Matrix Products*

State whether the product AB is defined. If so, give the dimensions of AB.

a. A: 2×3, B: 3×4

b. A: 3×2, B: 3×4

SOLUTION

a. Because A is a 2×3 matrix and B is a 3×4 matrix, the product AB is defined and is a 2×4 matrix.

b. Because the number of columns in A (two) does not equal the number of rows in B (three), the product AB is not defined.

EXAMPLE 2 *Finding the Product of Two Matrices*

Find AB if $A = \begin{bmatrix} -2 & 3 \\ 1 & -4 \\ 6 & 0 \end{bmatrix}$ and $B = \begin{bmatrix} -1 & 3 \\ -2 & 4 \end{bmatrix}$.

SOLUTION

Because A is a 3×2 matrix and B is a 2×2 matrix, the product AB is defined and is a 3×2 matrix. To write the entry in the first row and first column of AB, multiply corresponding entries in the first row of A and the first column of B. Then add. Use a similar procedure to write the other entries of the product.

$$AB = \begin{bmatrix} -2 & 3 \\ 1 & -4 \\ 6 & 0 \end{bmatrix} \begin{bmatrix} -1 & 3 \\ -2 & 4 \end{bmatrix}$$

$$= \begin{bmatrix} (-2)(-1) + (3)(-2) & (-2)(3) + (3)(4) \\ (1)(-1) + (-4)(-2) & (1)(3) + (-4)(4) \\ (6)(-1) + (0)(-2) & (6)(3) + (0)(4) \end{bmatrix}$$

$$= \begin{bmatrix} -4 & 6 \\ 7 & -13 \\ -6 & 18 \end{bmatrix}$$

STUDENT HELP

HOMEWORK HELP
Visit our Web site
www.mcdougallittell.com
for extra examples.

EXAMPLE 3 — *Finding the Product of Two Matrices*

If $A = \begin{bmatrix} 3 & 2 \\ -1 & 0 \end{bmatrix}$ and $B = \begin{bmatrix} 1 & -4 \\ 2 & 1 \end{bmatrix}$, find each product.

a. AB **b.** BA

SOLUTION

a. $AB = \begin{bmatrix} 3 & 2 \\ -1 & 0 \end{bmatrix}\begin{bmatrix} 1 & -4 \\ 2 & 1 \end{bmatrix} = \begin{bmatrix} 7 & -10 \\ -1 & 4 \end{bmatrix}$

b. $BA = \begin{bmatrix} 1 & -4 \\ 2 & 1 \end{bmatrix}\begin{bmatrix} 3 & 2 \\ -1 & 0 \end{bmatrix} = \begin{bmatrix} 7 & 2 \\ 5 & 4 \end{bmatrix}$

· · · · · · · · ·

Notice in Example 3 that $AB \neq BA$. Matrix multiplication is not, in general, commutative.

EXAMPLE 4 — *Using Matrix Operations*

If $A = \begin{bmatrix} 2 & 1 \\ -1 & 3 \end{bmatrix}$, $B = \begin{bmatrix} -2 & 0 \\ 4 & 2 \end{bmatrix}$, and $C = \begin{bmatrix} 1 & 1 \\ 3 & 2 \end{bmatrix}$, simplify each expression.

a. $A(B + C)$ **b.** $AB + AC$

SOLUTION

a. $A(B + C) = \begin{bmatrix} 2 & 1 \\ -1 & 3 \end{bmatrix}\left(\begin{bmatrix} -2 & 0 \\ 4 & 2 \end{bmatrix} + \begin{bmatrix} 1 & 1 \\ 3 & 2 \end{bmatrix}\right)$

$= \begin{bmatrix} 2 & 1 \\ -1 & 3 \end{bmatrix}\begin{bmatrix} -1 & 1 \\ 7 & 4 \end{bmatrix} = \begin{bmatrix} 5 & 6 \\ 22 & 11 \end{bmatrix}$

b. $AB + AC = \begin{bmatrix} 2 & 1 \\ -1 & 3 \end{bmatrix}\begin{bmatrix} -2 & 0 \\ 4 & 2 \end{bmatrix} + \begin{bmatrix} 2 & 1 \\ -1 & 3 \end{bmatrix}\begin{bmatrix} 1 & 1 \\ 3 & 2 \end{bmatrix}$

$= \begin{bmatrix} 0 & 2 \\ 14 & 6 \end{bmatrix} + \begin{bmatrix} 5 & 4 \\ 8 & 5 \end{bmatrix} = \begin{bmatrix} 5 & 6 \\ 22 & 11 \end{bmatrix}$

· · · · · · · · ·

Notice in Example 4 that $A(B + C) = AB + AC$, which is true in general. This and other properties of matrix multiplication are summarized below.

CONCEPT SUMMARY — **PROPERTIES OF MATRIX MULTIPLICATION**

Let A, B, and C be matrices and let c be a scalar.

ASSOCIATIVE PROPERTY OF MATRIX MULTIPLICATION	$A(BC) = (AB)C$
LEFT DISTRIBUTIVE PROPERTY	$A(B + C) = AB + AC$
RIGHT DISTRIBUTIVE PROPERTY	$(A + B)C = AC + BC$
ASSOCIATIVE PROPERTY OF SCALAR MULTIPLICATION	$c(AB) = (cA)B = A(cB)$

4.2 *Multiplying Matrices*   **209**

Extra Example 4

a. $\begin{bmatrix} 3 & 3 \\ -12 & -9 \end{bmatrix}$ b. $\begin{bmatrix} 3 & 3 \\ -12 & -9 \end{bmatrix}$

---

2 TEACH

EXTRA EXAMPLE 1
State whether AB is defined. If so, give the dimensions.
a. $A: 2 \times 4$, $B: 4 \times 3$ Yes; 2×3
b. $A: 1 \times 4$, $B: 1 \times 4$ No

EXTRA EXAMPLE 2
Find AB if $A = \begin{bmatrix} -1 & 5 \\ 5 & 2 \\ 0 & -4 \end{bmatrix}$ and

$B = \begin{bmatrix} 4 & -3 \\ 6 & 8 \end{bmatrix}$. $\begin{bmatrix} 26 & 43 \\ 32 & 1 \\ -24 & -32 \end{bmatrix}$

EXTRA EXAMPLE 3
$A = \begin{bmatrix} 4 & 1 \\ 0 & -2 \end{bmatrix}$ and $B = \begin{bmatrix} -4 & -3 \\ 1 & 2 \end{bmatrix}$

a. Find AB. $\begin{bmatrix} -15 & -10 \\ -2 & -4 \end{bmatrix}$

b. Find BA. $\begin{bmatrix} -16 & 2 \\ 4 & -3 \end{bmatrix}$

EXTRA EXAMPLE 4
$A = \begin{bmatrix} 2 & -2 \\ 1 & 4 \end{bmatrix}$, $B = \begin{bmatrix} 0 & 1 \\ -3 & -2 \end{bmatrix}$,

and $C = \begin{bmatrix} 0 & 3 \\ 2 & -1 \end{bmatrix}$

a. Find $B(A + C)$.   See margin.
b. Find $BA + BC$.   See margin.

CHECKPOINT EXERCISES
For use after Example 1:
1. Is AB defined? If so, give the dimensions.
$A: 2 \times 2$, $B: 4 \times 2$ No

For use after Example 2:
2. Find AB if $A = [8 \ \ 0 \ \ -4]$ and

$B = \begin{bmatrix} 4 \\ 6 \\ -3 \end{bmatrix}$.   [44]

For use after Examples 3 and 4:
3. If $A = \begin{bmatrix} 1 & 1 \\ -2 & -2 \end{bmatrix}$, $B = \begin{bmatrix} 3 & 3 \\ -2 & 0 \end{bmatrix}$,

and $C = \begin{bmatrix} 0 & 4 \\ -4 & 1 \end{bmatrix}$, find $A(BC)$.

$\begin{bmatrix} -12 & 7 \\ 24 & -14 \end{bmatrix}$

209

EXTRA EXAMPLE 5

Two lacrosse teams submit equipment lists to their sponsors.
Women's team: 5 sticks, 15 balls, and 16 uniforms.
Men's team: 8 sticks, 22 balls, and 17 uniforms.
Each stick costs $55, each ball cost $6, and each uniform costs $35. Use matrix multiplication to find the total cost of the equipment for each team.

$$\begin{bmatrix} 5 & 15 & 16 \\ 8 & 22 & 17 \end{bmatrix} \begin{bmatrix} 55 \\ 6 \\ 35 \end{bmatrix} = \begin{bmatrix} 925 \\ 1167 \end{bmatrix}$$

Women: $925; Men: $1167

CHECKPOINT EXERCISES

For use after Example 5:

1. The soccer teams submitted equipment lists to their sponsors.
Women's team: 6 balls, 16 uniforms
Men's team: 7 balls, 15 uniforms
Each ball costs $45 and each uniform costs $38. Use matrix multiplication to find the total cost of the equipment for each team.

$$\begin{bmatrix} 6 & 16 \\ 7 & 15 \end{bmatrix} \begin{bmatrix} 45 \\ 38 \end{bmatrix} = \begin{bmatrix} 878 \\ 885 \end{bmatrix}$$

Women: $878; Men: $885

CLOSURE QUESTION

If A is a 3×4 matrix and B is a 2×3 matrix, which product, AB or BA, is defined? Explain.

BA; The number of columns of matrix *B* is equal to the number of rows of matrix *A,* so *BA* is defined. The resulting matrix will be a 2×4 matrix.

DAILY PUZZLER

Consider the product
$$\frac{a_n}{a_{n+1}} \cdot \frac{a_{n+1}}{a_{n+2}} \cdot \frac{a_{n+2}}{a_{n+3}} \cdot \frac{a_{n+3}}{a_{n+4}} \cdot \frac{a_{n+4}}{a_{n+5}} \cdots$$
What is the product if the multiplication stops when $\dfrac{a_{n+99}}{a_{n+100}}$ is the last factor? $\dfrac{a_n}{a_{n+100}}$

210

DOT RICHARDSON helped lead the United States to the first women's softball gold medal in the 1996 Olympics by playing shortstop.

APPLICATION LINK
www.mcdougallittell.com

Matrix multiplication is useful in business applications because an *inventory* matrix, when multiplied by a *cost per item* matrix, results in a *total cost* matrix.

$$\underset{m \times n}{\begin{bmatrix} \text{Inventory} \\ \text{matrix} \end{bmatrix}} \cdot \underset{n \times p}{\begin{bmatrix} \text{Cost per item} \\ \text{matrix} \end{bmatrix}} = \underset{m \times p}{\begin{bmatrix} \text{Total cost} \\ \text{matrix} \end{bmatrix}}$$

For the total cost matrix to be meaningful, the column labels for the inventory matrix must match the row labels for the cost per item matrix.

EXAMPLE 5   *Using Matrices to Calculate the Total Cost*

SPORTS Two softball teams submit equipment lists for the season.

Women's team	Men's team
12 bats	15 bats
45 balls	38 balls
15 uniforms	17 uniforms

Each bat costs $21, each ball costs $4, and each uniform costs $30. Use matrix multiplication to find the total cost of equipment for each team.

SOLUTION

To begin, write the equipment lists and the costs per item in matrix form. Because you want to use matrix multiplication to find the total cost, set up the matrices so that the columns of the equipment matrix match the rows of the cost matrix.

EQUIPMENT

	Bats	Balls	Uniforms
Women's team	12	45	15
Men's team	15	38	17

COST

	Dollars
Bats	21
Balls	4
Uniforms	30

The total cost of equipment for each team can now be obtained by multiplying the equipment matrix by the cost per item matrix. The equipment matrix is 2×3 and the cost per item matrix is 3×1, so their product is a 2×1 matrix.

$$\begin{bmatrix} 12 & 45 & 15 \\ 15 & 38 & 17 \end{bmatrix} \begin{bmatrix} 21 \\ 4 \\ 30 \end{bmatrix} = \begin{bmatrix} 12(21) + 45(4) + 15(30) \\ 15(21) + 38(4) + 17(30) \end{bmatrix} = \begin{bmatrix} 882 \\ 977 \end{bmatrix}$$

The labels for the product matrix are as follows.

TOTAL COST

	Dollars
Women's team	882
Men's team	977

▶ The total cost of equipment for the women's team is $882, and the total cost of equipment for the men's team is $977.

GUIDED PRACTICE

Vocabulary Check ✓

1. Complete this statement: The product of matrices A and B is defined provided the number of __?__ in A is equal to the number of __?__ in B. **columns, rows**

Concept Check ✓

2. Matrix A is 6×1. Matrix B is 1×2. Which of the products is defined, AB or BA? Explain. **See margin.**

3. Tell whether the matrix equation is *true* or *false*. Explain. **See margin.**

$$\begin{bmatrix} 5 & 3 \\ -3 & 5 \end{bmatrix}\begin{bmatrix} 2 & 0 \\ 0 & 1 \end{bmatrix} = \begin{bmatrix} 2 & 0 \\ 0 & 1 \end{bmatrix}\begin{bmatrix} 5 & 3 \\ -3 & 5 \end{bmatrix}$$

Skill Check ✓

State whether the product *AB* is defined. If so, give the dimensions of *AB*.

4. A: 3×2, B: 2×3 **defined; 3 × 3**

5. A: 3×3, B: 3×3 **defined; 3 × 3**

6. A: 3×2, B: 3×2 **not defined**

Find the product.

7. $\begin{bmatrix} 1 & 0 \\ -2 & -1 \end{bmatrix}\begin{bmatrix} 2 & 0 \\ 1 & 3 \end{bmatrix}$

8. $\begin{bmatrix} 4 & 4 \end{bmatrix}\begin{bmatrix} -2 \\ -3 \end{bmatrix}$ $\begin{bmatrix} -20 \end{bmatrix}$

9. $\begin{bmatrix} -3 & 3 \\ 3 & -2 \\ 0 & -1 \end{bmatrix}\begin{bmatrix} 1 & 0 \\ -2 & -1 \end{bmatrix}$

7 and 9. See margin.

10. **SOFTBALL EQUIPMENT** Use matrix multiplication to find the total cost of equipment in Example 5 if the women's team needs 16 bats, 42 balls, and 16 uniforms and the men's team needs 14 bats, 43 balls, and 15 uniforms. **women's team: \$984, men's team: \$916**

2. *AB* is defined because the number of columns in *A* (1) is equal to the number of rows in *B* (1). *BA* is not defined because the number of columns in *B* (2) is not equal to the number of rows in *A* (6).

PRACTICE AND APPLICATIONS

STUDENT HELP

Extra Practice to help you master skills is on p. 944.

MATRIX PRODUCTS **State whether the product *AB* is defined. If so, give the dimensions of *AB*.**

11. A: 1×3, B: 3×2 **defined; 1 × 2**

12. A: 2×4, B: 4×3 **defined; 2 × 3**

13. A: 4×2, B: 3×5 **not defined**

14. A: 5×5, B: 5×4 **defined; 5 × 4**

15. A: 3×4, B: 4×1 **defined; 3 × 1**

16. A: 3×3, B: 2×4 **not defined**

FINDING MATRIX PRODUCTS **Find the product. If it is not defined, state the reason.**

17. $\begin{bmatrix} -\frac{1}{6} & \frac{1}{2} & -\frac{1}{3} \end{bmatrix}\begin{bmatrix} 12 \\ 0 \\ -12 \end{bmatrix}$ $[2]$

18. $\begin{bmatrix} 7.3 & 1.5 \\ 1.8 & 0 \\ 2.9 & 3.2 \end{bmatrix}\begin{bmatrix} -4.2 & 2.6 & -8.7 \end{bmatrix}$ **See margin.**

19. $\begin{bmatrix} 1 & -4 \\ 3 & -2 \end{bmatrix}\begin{bmatrix} 4 & -1 \\ 0 & -3 \end{bmatrix}$ $\begin{bmatrix} 4 & 11 \\ 12 & 3 \end{bmatrix}$

20. $\begin{bmatrix} -6 & -2 \\ 0 & 3 \end{bmatrix}\begin{bmatrix} -1 & 4 \\ -5 & 3 \end{bmatrix}$ $\begin{bmatrix} 16 & -30 \\ -15 & 9 \end{bmatrix}$

21. $\begin{bmatrix} 2 & -8 & 1 \\ 0 & -5 & 2 \end{bmatrix}\begin{bmatrix} 0 & 1 & -2 \\ 8 & -2 & -5 \end{bmatrix}$ **See margin.**

22. $\begin{bmatrix} 6.0 & 0 \\ -0.2 & 0.2 \\ 2.9 & 0.3 \end{bmatrix}\begin{bmatrix} 1 & 0 \\ 1.5 & -0.5 \end{bmatrix}$ $\begin{bmatrix} 6.0 & 0 \\ 0.1 & -0.1 \\ 3.35 & -0.15 \end{bmatrix}$

23. $\begin{bmatrix} -1 & -0.5 & 1.25 \\ 1 & -1.5 & -0.25 \end{bmatrix}\begin{bmatrix} 1.2 \\ 0.2 \\ 0 \end{bmatrix}$ $\begin{bmatrix} -1.3 \\ 0.9 \end{bmatrix}$

24. $\begin{bmatrix} -6 & 1 & 1 \\ -2 & 3 & 8 \\ 0.1 & 7 & 1 \end{bmatrix}\begin{bmatrix} 0 & -1 & 3 \\ -7 & -2 & 4 \\ -1 & 3 & 4 \end{bmatrix}$ **24–26. See margin.**

25. $\begin{bmatrix} 6 & -2 \\ 1 & 4 \\ 0 & 5 \end{bmatrix}\begin{bmatrix} -4 & -2 & 5 \\ 4 & -6 & -1 \end{bmatrix}$

26. $\begin{bmatrix} 0 & 1 & 0 \\ 6 & -3 & -1 \\ -2 & 5 & 3 \end{bmatrix}\begin{bmatrix} 5 & -7 & 4 \\ 3 & 12 & 6 \\ -4 & -5 & -12 \end{bmatrix}$

18. Not defined; the number of columns in the left matrix (2) does not equal the number of rows in the right matrix (1).

21. Not defined; the number of columns in the left matrix (3) does not equal the number of rows in the right matrix (2).

STUDENT HELP

HOMEWORK HELP
Example 1: Exs. 11–16
Examples 2, 3: Exs. 17–26
Example 4: Exs. 27–32
Example 5: Exs. 35–40

3 APPLY

ASSIGNMENT GUIDE

BASIC
Day 1: pp. 211–213 Exs. 12–28 even, 31, 33, 35–37, 42, 45–61 odd

AVERAGE
Day 1: pp. 211–213 Exs. 12–34 even, 35–39, 42, 45–61 odd

ADVANCED
Day 1: pp. 211–213 Exs. 12–34 even, 35–39, 42, 45–61 odd

BLOCK SCHEDULE
pp. 211–213 Exs. 12–34 even, 35–44, 45–61 odd

EXERCISE LEVELS

Level A: ***Easier***
11–22, 35–37

Level B: ***More Difficult***
23–34, 38–42

Level C: ***Most Difficult***
43

✓ HOMEWORK CHECK

To quickly check student understanding of key concepts, go over the following exercises: Exs. 14, 16, 18, 20, 28. See also the Daily Homework Quiz:

- Blackline Master (*Chapter 4 Resource Book,* p. 38)
- Transparency (p. 28)

3. False. Matrix multiplication is not commutative; the left side of the equation is equal to $\begin{bmatrix} 10 & 3 \\ -6 & 5 \end{bmatrix}$ and the right side of the equation is equal to $\begin{bmatrix} 10 & 6 \\ -3 & 5 \end{bmatrix}$.

7. $\begin{bmatrix} 2 & 0 \\ -5 & -3 \end{bmatrix}$

9. $\begin{bmatrix} -9 & -3 \\ 7 & 2 \\ 2 & 1 \end{bmatrix}$

24–26. See Additional Answers beginning on page AA1.

DATA UPDATE
Updated data for Exs. 35 and 36 is available at www.mcdougallittell.com.

35.

Grain Production

	Wheat	Rice	Maize
China	0.201	0.348	0.180
India	0.220	0.215	0.017
C.I.S.	0.073	0.001	0.005
U.S.	0.113	0.014	0.405

36. Organize the total world production data in a 3 × 1 matrix. Then multiply the matrix from Ex. 35 by this matrix:

$$\begin{bmatrix} 0.201 & 0.348 & 0.180 \\ 0.220 & 0.215 & 0.017 \\ 0.073 & 0.001 & 0.005 \\ 0.113 & 0.014 & 0.405 \end{bmatrix} \begin{bmatrix} 608,846 \\ 570,906 \\ 586,923 \end{bmatrix} =$$

$$\begin{bmatrix} 426,699.474 \\ 266,668.601 \\ 47,951.279 \\ 314,496.097 \end{bmatrix}$$ China produced

about 426,700,000 metric tons of grain, India produced about 266,669,000 metric tons of grain, the Commonwealth of Independent States produced about 47,951,000 metric tons of grain, and the United States produced about 314,496,000 metric tons of grain.

38. $\begin{bmatrix} 59 \\ 60 \\ 62 \end{bmatrix}$

ADDITIONAL PRACTICE AND RETEACHING

For Lesson 4.2:
• Practice Levels A, B, and C (*Chapter 4 Resource Book*, p. 28)
• Reteaching with Practice (*Chapter 4 Resource Book*, p. 31)
• ⊞ See Lesson 4.2 of the *Personal Student Tutor*

For more Mixed Review:
• ⊞ Search the *Test and Practice Generator* for key words or specific lessons.

212

---

27. $\begin{bmatrix} 16 & -16 \\ 16 & -8 \end{bmatrix}$

28. $\begin{bmatrix} 8 & 2 \\ 4 & 13 \end{bmatrix}$

29. $\begin{bmatrix} 8 & -5 & 8 \\ -1 & 1 & 1 \\ 7 & -30 & -35 \end{bmatrix}$

30. $\begin{bmatrix} 16 & 24 & -16 \\ 16 & 20 & -24 \\ -3 & -4 & 6 \end{bmatrix}$

STUDENT HELP

DATA UPDATE Visit our Web site www.mcdougallittell.com

31. $\begin{bmatrix} 0 & -30 \\ 12 & -51 \end{bmatrix}$

32. $\begin{bmatrix} 4 & 16 \\ -4 & 32 \end{bmatrix}$

37. Matrix B

$\begin{bmatrix} 6 \\ 5 \\ 4 \end{bmatrix}$

FOCUS ON APPLICATIONS

EXERCISE A 120 pound person walking at a moderate pace of 3 mi/h would burn about 64 Cal in 20 min. At a brisk pace of 4.5 mi/h, the person would burn about 82 Cal in 20 min.

SIMPLIFYING EXPRESSIONS Using the given matrices, simplify the expression. 27–32. See margin.

$$A = \begin{bmatrix} 4 & -2 \\ 6 & -1 \end{bmatrix}, B = \begin{bmatrix} 1 & 0 \\ -2 & 4 \end{bmatrix}, C = \begin{bmatrix} -1 & 3 \\ -2 & 1 \end{bmatrix}, D = \begin{bmatrix} 3 & -2 & 1 \\ -1 & 2 & 4 \\ -2 & -3 & 3 \end{bmatrix}, E = \begin{bmatrix} -2 & 5 & 6 \\ -1 & 4 & 2 \\ 3 & 1 & -4 \end{bmatrix}$$

27. $2AB$

28. $AB + AC$

29. $D(D + E)$

30. $(E + D)E$

31. $-3(AC)$

32. $0.5(AB) + 2AC$

SOLVING MATRIX EQUATIONS Solve for *x* and *y*.

33. $\begin{bmatrix} -2 & 1 & 2 \\ 3 & 2 & 4 \\ 0 & -2 & 4 \end{bmatrix} \begin{bmatrix} 1 \\ x \\ 3 \end{bmatrix} = \begin{bmatrix} 6 \\ 19 \\ y \end{bmatrix}$ $x = 2, y = 8$

34. $\begin{bmatrix} 4 & 1 & 3 \\ -2 & x & 1 \end{bmatrix} \begin{bmatrix} 9 & -2 \\ 2 & 1 \\ -1 & 4 \end{bmatrix} = \begin{bmatrix} y & 5 \\ -13 & 11 \end{bmatrix}$ $x = 3, y = 35$

🌎 **AGRICULTURE** In Exercises 35 and 36, use the following information.

The percents of the total 1997 world production of wheat, rice, and maize are shown in the matrix for the four countries that grow the most grain: China, India, the Commonwealth of Independent States (formerly the Soviet Union), and the United States. The total 1997 world production (in thousands of metric tons) of wheat, rice, and maize is 608,846, 570,906, and 586,923, respectively.

GRAIN PRODUCTION

	Wheat	Rice	Maize
China	20.1%	34.8%	18%
India	22%	21.5%	1.7%
C.I.S.	7.3%	0.1%	0.5%
U.S.	11.3%	1.4%	40.5%

▶ Source: Food and Agriculture Organization of the United Nations 35, 36. See margin.

35. Rewrite the matrix to give the percents as decimals.

36. Show how matrix multiplication can be used to determine how many metric tons of all three grains were produced in each of the four countries.

🌎 **CLASS DEBATE** In Exercises 37–39, use the following information.

Three teams participated in a debating competition. The final score for each team is based on how many students ranked first, second, and third in a debate. The results of 12 debates are shown in matrix *A*. 37, 38. See margin.

MATRIX A

	1st	2nd	3rd
Team 1	3	5	4
Team 2	5	2	5
Team 3	4	6	2

37. Teams earn 6 points for each first place, 5 points for each second place, and 4 points for each third place. Organize this information into a matrix *B*.

38. Find the product *AB*.

39. LOGICAL REASONING Which team won the competition? How many points did the winning team score? Team 3; 62 points

40. 🏃 **EXERCISE** The numbers of calories burned by people of different weights doing different activities for 20 minutes are shown in the matrix. Show how matrix multiplication can be used to write the total number of calories burned by a 120 pound person and a 150 pound person who each bicycled for 40 minutes, jogged for 10 minutes, and then walked for 60 minutes.

CALORIES BURNED

	120 lb person	150 lb person
Bicycling	109	136
Jogging	127	159
Walking	64	79

$\begin{bmatrix} 2 & 0.5 & 3 \end{bmatrix} \begin{bmatrix} 109 & 136 \\ 127 & 159 \\ 64 & 79 \end{bmatrix} = \begin{bmatrix} 473.5 & 588.5 \end{bmatrix}$

▶ Source: *Medicine and Science in Sports and Exercise*

The 120 lb person burned about 474 Cal and the 150 lb person burned about 589 Cal.

41. *Writing* Describe the process you use when multiplying any two matrices. **See margin.**

42. MULTIPLE CHOICE What is the product of $\begin{bmatrix} 0 & -1 \\ -4 & -2 \end{bmatrix}$ and $\begin{bmatrix} 7 & -2 \\ -1 & 0 \end{bmatrix}$? **C**

Ⓐ $\begin{bmatrix} 2 & 0 \\ -26 & -24 \end{bmatrix}$ Ⓑ $\begin{bmatrix} 8 & 0 \\ -30 & 6 \end{bmatrix}$ Ⓒ $\begin{bmatrix} 1 & 0 \\ -26 & 8 \end{bmatrix}$ Ⓓ $\begin{bmatrix} 1 & 0 \\ -30 & 8 \end{bmatrix}$

43. MULTIPLE CHOICE If A is a 2×3 matrix and B is a 3×2 matrix, what are the dimensions of BA? **B**

Ⓐ 2×2 Ⓑ 3×3 Ⓒ 3×2 Ⓓ 2×3 Ⓔ BA not defined

44. ROTATIONAL MATRIX Matrix A is a $90°$ rotational matrix. Matrix B contains the coordinates of the triangle's vertices shown in the graph. **See margin.**

$$A = \begin{bmatrix} 0 & -1 \\ 1 & 0 \end{bmatrix} \qquad B = \begin{bmatrix} -7 & -4 & -4 \\ 4 & 8 & 2 \end{bmatrix}$$

a. Calculate AB. Graph the coordinates of the vertices given by AB. What rotation does AB represent in the graph?

b. Find the $180°$ and $270°$ rotations of the original triangle by using repeated multiplication of the $90°$ rotational matrix. What are the coordinates of the vertices of the rotated triangles?

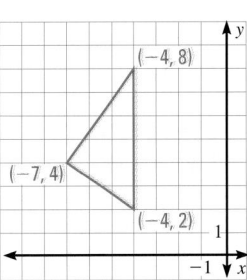

MIXED REVIEW

44a. $\begin{bmatrix} -4 & -8 & -2 \\ -7 & -4 & -4 \end{bmatrix}$; a 90° rotation of the original triangle; See margin for graph.

44b. a 180° rotation: $\begin{bmatrix} 7 & 4 & 4 \\ -4 & -8 & -2 \end{bmatrix}$; a 270° rotation: $\begin{bmatrix} 4 & 8 & 2 \\ 7 & 4 & 4 \end{bmatrix}$; 180° rotation: $(7, -4)$, $(4, -8)$, $(4, -2)$; 270° rotation: $(4, 7)$, $(8, 4)$, $(2, 4)$

61. $\left(-\dfrac{49}{37}, -\dfrac{52}{37} \right)$

62. $\left(-\dfrac{7}{5}, -\dfrac{36}{5} \right)$

CALCULATING AREA Find the area of the figure. (Skills Review, p. 914)

45. 180 m² 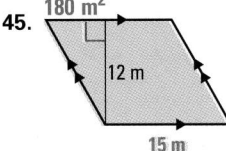 12 m, 15 m

46. 12 m² 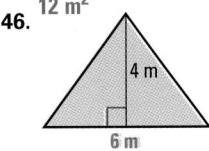 4 m, 6 m

47. 9π ft², or about 28.26 ft² 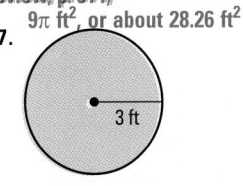 3 ft

WRITING EQUATIONS Write an equation of the line with the given properties. (Review 2.4)

48. slope: $\dfrac{1}{2}$, passes through $(1, 8)$ $y = \dfrac{1}{2}x + \dfrac{15}{2}$

49. slope: $-\dfrac{1}{4}$, passes through $(0, 4)$ $y = -\dfrac{1}{4}x + 4$

50. passes through $(3, -6)$ and $(2, 10)$ $y = -16x + 42$

51. passes through $(1, 5)$ and $(4, 14)$ $y = 3x + 2$

52. x-intercept: -7, y-intercept: -5 $y = -\dfrac{5}{7}x - 5$

53. x-intercept: 4, y-intercept: -6 $y = \dfrac{3}{2}x - 6$

SOLVING SYSTEMS Solve the system of linear equations using any algebraic method. (Review 3.2 for 4.3)

54. $4x + y = 6$ $(4, -10)$
    $-3x - 2y = 8$

55. $2x + y = -9$ $(-7, 5)$
    $3x + 5y = 4$

56. $-9x + 5y = 1$ $(-4, -7)$
    $3x - 2y = 2$

57. $2x - 2y = 8$ no solution
    $x - y = 1$

58. $-3x + 4y = -1$ $\left(1, \dfrac{1}{2}\right)$
    $6x + 2y = 7$

59. $5x + 2y = -10$ $(0, -5)$
    $-3x - 8y = 40$

60. $7x + 3y = 11$ $(-1, 6)$
    $-2x + 5y = 32$

61. $5x - 4y = -1$
    $2x - 9y = 10$

62. $-x + 7y = -49$
    $12x + y = -24$

4.2 *Multiplying Matrices* **213**

4 ASSESS

DAILY HOMEWORK QUIZ

🖾 *Transparency Available*

Find the product. If it is not defined, state the reason.

1. $\begin{bmatrix} 5 & 1 & -2 \\ 2 & -3 & 0 \end{bmatrix} \begin{bmatrix} 1 & 0 \\ -1 & 2 \\ 4 & -3 \end{bmatrix} \begin{bmatrix} -4 & 8 \\ 5 & -6 \end{bmatrix}$

2. $\begin{bmatrix} 2 & 7 & 1 \end{bmatrix} \begin{bmatrix} 2 & 1 \\ 4 & -7 \end{bmatrix}$

Not defined; the number of columns in the left matrix (3) does not equal the number of rows in the right matrix (2).

3. Solve for x and y.

$$\begin{bmatrix} 1 & 3 & -2 \\ -1 & 3 & 2 \\ 2 & -4 & 0 \end{bmatrix} \begin{bmatrix} 1 \\ -1 \\ x \end{bmatrix} = \begin{bmatrix} -8 \\ 2 \\ y \end{bmatrix}$$

$x = 3$, $y = 6$

EXTRA CHALLENGE NOTE
↳ Challenge problems for Lesson 4.2 are available in **blackline** format in the *Chapter 4 Resource Book,* p. 35 and at **www.mcdougallittell.com.**

ADDITIONAL TEST PREPARATION

1. WRITING Explain in words how to find the product of two matrices, $\begin{bmatrix} a & b \\ c & d \end{bmatrix} \begin{bmatrix} e & f \\ g & h \end{bmatrix}$.

Sample answer: Multiply the first row of the first matrix with the first column of the second matrix to get $ae + bg$ for the entry in the first row, first column of the product matrix. Following this pattern, $af + bh$ will be the entry in the first row, second column; $ce + dg$ will be the entry in the second row, first column; and $cf + dh$ will be the entry in the second row, second column of the product matrix.

41, 44a. See Additional Answers beginning on page AA1.

LESSON OPENER
ACTIVITY
An alternative way to approach Lesson 4.3 is to use the Activity Lesson Opener:

- Blackline Master (*Chapter 4 Resource Book*, p. 39)
- Transparency (p. 24)

MEETING INDIVIDUAL NEEDS

- *Chapter 4 Resource Book*
  Prerequisite Skills Review (p. 5)
  Practice Level A (p. 42)
  Practice Level B (p. 43)
  Practice Level C (p. 44)
  Reteaching with Practice (p. 45)
  Absent Student Catch-Up (p. 47)
  Challenge (p. 50)
- *Resources in Spanish*
- Personal Student Tutor

NEW-TEACHER SUPPORT
See the Tips for New Teachers on pp. 1–2 of the *Chapter 4 Resource Book* for additional notes about Lesson 4.3.

WARM-UP EXERCISES

Transparency Available

Solve the system of equations.

1. $2x + 3y = 7$ (2, 1)
  $x + y = 3$
2. $x = -2y + 4$ (6, −1)
  $x = -y + 5$
3. $y = 2x + 1$ (2, 5)
  $y = 4x - 3$

Simplify.

4. $2(8) + (-3)(6)$ −2
5. $5(0) + 2(4) + (-7)(-1)$ 15

What you should learn

GOAL 1 Evaluate determinants of 2 × 2 and 3 × 3 matrices.

GOAL 2 Use Cramer's rule to solve systems of linear equations, as applied in **Example 5**.

Why you should learn it

▼ To solve **real-life** problems, such as finding the area of the Golden Triangle of India in **Ex. 58**.

Determinants and Cramer's Rule

GOAL 1 EVALUATING DETERMINANTS

Associated with each square matrix is a real number called its **determinant**. The determinant of a matrix A is denoted by det A or by $|A|$.

THE DETERMINANT OF A MATRIX

DETERMINANT OF A 2 × 2 MATRIX

$$\det \begin{bmatrix} a & b \\ c & d \end{bmatrix} = \begin{vmatrix} a & b \\ c & d \end{vmatrix} = ad - cb$$

The determinant of a 2 × 2 matrix is the difference of the products of the entries on the diagonals.

DETERMINANT OF A 3 × 3 MATRIX

1 Repeat the first two columns to the right of the determinant.

2 Subtract the sum of the products in red from the sum of the products in blue.

$$\det \begin{bmatrix} a & b & c \\ d & e & f \\ g & h & i \end{bmatrix} = \begin{vmatrix} a & b & c \\ d & e & f \\ g & h & i \end{vmatrix} \begin{matrix} a & b \\ d & e \\ g & h \end{matrix} = (aei + bfg + cdh) - (gec + hfa + idb)$$

EXAMPLE 1 *Evaluating Determinants*

Evaluate the determinant of the matrix.

a. $\begin{bmatrix} 1 & 3 \\ 2 & 5 \end{bmatrix}$

b. $\begin{bmatrix} 2 & -1 & 3 \\ -2 & 0 & 1 \\ 1 & 2 & 4 \end{bmatrix}$

SOLUTION

a. $\begin{vmatrix} 1 & 3 \\ 2 & 5 \end{vmatrix} = 1(5) - 2(3) = 5 - 6 = -1$

b. $\begin{vmatrix} 2 & -1 & 3 \\ -2 & 0 & 1 \\ 1 & 2 & 4 \end{vmatrix} \begin{matrix} 2 & -1 \\ -2 & 0 \\ 1 & 2 \end{matrix} = [0 + (-1) + (-12)] - (0 + 4 + 8) = -13 - 12$
$$= -25$$

· · · · · · · · ·

You can use a determinant to find the area of a triangle whose vertices are points in a coordinate plane.

AREA OF A TRIANGLE

The area of a triangle with vertices (x_1, y_1), (x_2, y_2), and (x_3, y_3) is given by

$$\text{Area} = \pm\frac{1}{2}\begin{vmatrix} x_1 & y_1 & 1 \\ x_2 & y_2 & 1 \\ x_3 & y_3 & 1 \end{vmatrix}$$

where the symbol ± indicates that the appropriate sign should be chosen to yield a positive value.

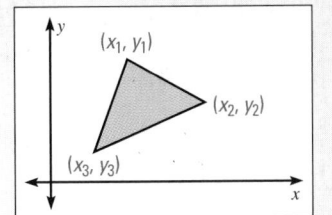

EXAMPLE 2 *The Area of a Triangle*

The area of the triangle shown is:

$$\text{Area} = \pm\frac{1}{2}\begin{vmatrix} 1 & 2 & 1 \\ 4 & 0 & 1 \\ 6 & 2 & 1 \end{vmatrix}$$

$$= \pm\frac{1}{2}[(0 + 12 + 8) - (0 + 2 + 8)] = 5$$

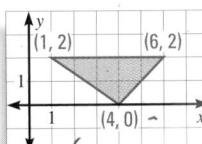

EXAMPLE 3 *The Area of a Triangular Region*

BERMUDA TRIANGLE The Bermuda Triangle is a large triangular region in the Atlantic Ocean. Many ships and airplanes have been lost in this region. The triangle is formed by imaginary lines connecting Bermuda, Puerto Rico, and Miami, Florida. Use a determinant to estimate the area of the Bermuda Triangle.

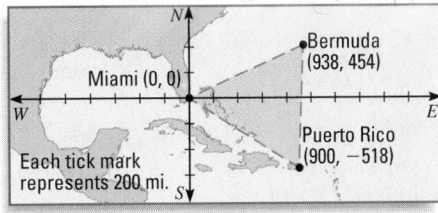

SOLUTION

The approximate coordinates of the Bermuda Triangle's three vertices are $(938, 454)$, $(900, -518)$, and $(0, 0)$. So, the area of the region is as follows:

$$\text{Area} = \pm\frac{1}{2}\begin{vmatrix} 938 & 454 & 1 \\ 900 & -518 & 1 \\ 0 & 0 & 1 \end{vmatrix}$$

$$= \pm\frac{1}{2}[(-485{,}884 + 0 + 0) - (0 + 0 + 408{,}600)]$$

$$= 447{,}242$$

▶ The area of the Bermuda Triangle is about 447,000 square miles.

FOCUS ON APPLICATIONS

BERMUDA TRIANGLE
The U.S.S. *Cyclops*, shown above, disappeared in the Bermuda Triangle in March, 1918.

2 TEACH

EXTRA EXAMPLE 1
Evaluate the determinant.

a. $\begin{bmatrix} 7 & 2 \\ 2 & 3 \end{bmatrix}$ b. $\begin{bmatrix} 4 & 3 & 1 \\ 5 & -7 & 0 \\ 1 & -2 & 2 \end{bmatrix}$

17   −89

EXTRA EXAMPLE 2
Find the area of the triangle.

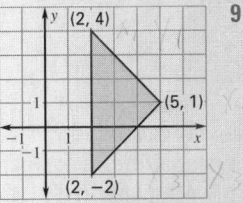

EXTRA EXAMPLE 3
The cities of San Francisco $(-30, 32)$, Oakland $(-16, 40)$, and San Jose $(0, 0)$ form a triangular-shaped area. Use a determinant to estimate the area formed by the three cities.
about 344 square miles

✓ **CHECKPOINT EXERCISES**

For use after Example 1:
1. Evaluate the determinant.

$$\begin{bmatrix} 4 & 2 & -1 \\ 1 & 1 & 0 \\ 2 & -3 & 5 \end{bmatrix} \quad 15$$

For use after Example 2:
2. Find the area of a triangle with vertices $(5, -2)$, $(3, 3)$, and $(-4, 1)$. **19.5 square units**

For use after Example 3:
3. Use the three cities from Extra Example 3. In kilometers, San Francisco is at $(-48.3, 51.5)$, Oakland is at $(-25.7, 64.4)$, and San Jose is at $(0, 0)$. Estimate the area formed by the cities in square kilometers.
about 893 km^2

EXTRA EXAMPLE 4

Use Cramer's rule to solve this system:

$2x + y = 1$
$3x - 2y = -23$ $(-3, 7)$

 CHECKPOINT EXERCISES

For use after Example 4:

1. Use Cramer's rule to solve this system:

$4x - 6y = 4$
$x + 5y = 14$ $(4, 2)$

STUDENT HELP NOTES

↳ **Homework Help** Students can find extra examples at **www.mcdougallittell.com** that parallel the examples in the student edition.

MATHEMATICAL REASONING

A determinant may be expanded by minors of any row or column. Ask your students what the sign of each partial product will be. **If the sum of the row and the column number of the element is odd, multiply the product by −1.**

EXTRA EXAMPLE 5

The total number of protons in three compounds are shown. Use a linear system and Cramer's rule to find the number of protons for carbon (C), hydrogen (H), and oxygen (O).

Compound	Formula	Protons
Methane	CH_4	10
Glycerol	$C_3H_8O_3$	50
Water	H_2O	10

$C + 4H = 10$; $3C + 8H + 30 = 50$;
$2H + O = 10$; $C = 6$, $H = 1$, $O = 8$

Checkpoint Exercises for Example 5 on next page.

216

GOAL 2  **USING CRAMER'S RULE**

You can use determinants to solve a system of linear equations. The method, called **Cramer's rule** and named after the Swiss mathematician Gabriel Cramer (1704−1752), uses the **coefficient matrix** of the linear system.

LINEAR SYSTEM	COEFFICIENT MATRIX
$ax + by = e$ $cx + dy = f$	$\begin{bmatrix} a & b \\ c & d \end{bmatrix}$

CRAMER'S RULE FOR A 2 X 2 SYSTEM

Let A be the coefficient matrix of this linear system:

$$ax + by = e$$
$$cx + dy = f$$

If $\det A \neq 0$, then the system has exactly one solution. The solution is:

$$x = \frac{\begin{vmatrix} e & b \\ f & d \end{vmatrix}}{\det A} \quad \text{and} \quad y = \frac{\begin{vmatrix} a & e \\ c & f \end{vmatrix}}{\det A}$$

In Cramer's rule, notice that the denominator for x and y is the determinant of the coefficient matrix of the system. The numerators for x and y are the determinants of the matrices formed by using the column of constants as replacements for the coefficients of x and y, respectively.

EXAMPLE 4 *Using Cramer's Rule for a 2 × 2 System*

Use Cramer's rule to solve this system: $\begin{aligned} 8x + 5y &= 2 \\ 2x - 4y &= -10 \end{aligned}$

STUDENT HELP

🔻 **HOMEWORK HELP**
Visit our Web site **www.mcdougallittell.com** for extra examples.

SOLUTION

Evaluate the determinant of the coefficient matrix.

$$\begin{vmatrix} 8 & 5 \\ 2 & -4 \end{vmatrix} = -32 - 10 = -42$$

Apply Cramer's rule since the determinant is not 0.

$$x = \frac{\begin{vmatrix} 2 & 5 \\ -10 & -4 \end{vmatrix}}{-42} = \frac{-8 - (-50)}{-42} = \frac{42}{-42} = -1$$

$$y = \frac{\begin{vmatrix} 8 & 2 \\ 2 & -10 \end{vmatrix}}{-42} = \frac{-80 - 4}{-42} = \frac{-84}{-42} = 2$$

▶ The solution is $(-1, 2)$.

✔ **CHECK** Check this solution in the original equations.

$8(-1) + 5(2) \stackrel{?}{=} 2 \qquad 2(-1) - 4(2) \stackrel{?}{=} -10$
$\qquad\qquad 2 = 2 ✓ \qquad\qquad\qquad -10 = -10 ✓$

216 **Chapter 4** *Matrices and Determinants*

CRAMER'S RULE FOR A 3 × 3 SYSTEM

Let A be the coefficient matrix of this linear system:

$$ax + by + cz = j$$
$$dx + ey + fz = k$$
$$gx + hy + iz = l$$

If $\det A \neq 0$, then the system has exactly one solution. The solution is:

$$x = \dfrac{\begin{vmatrix} j & b & c \\ k & e & f \\ l & h & i \end{vmatrix}}{\det A}, \quad y = \dfrac{\begin{vmatrix} a & j & c \\ d & k & f \\ g & l & i \end{vmatrix}}{\det A}, \quad \text{and} \quad z = \dfrac{\begin{vmatrix} a & b & j \\ d & e & k \\ g & h & l \end{vmatrix}}{\det A}$$

EXAMPLE 5 **Using Cramer's Rule for a 3 × 3 System**

SCIENCE CONNECTION The atomic weights of three compounds are shown. Use a linear system and Cramer's rule to find the atomic weights of carbon (C), hydrogen (H), and oxygen (O).

Compound	Formula	Atomic weight
Methane	CH_4	16
Glycerol	$C_3H_8O_3$	92
Water	H_2O	18

SOLUTION

Write a linear system using the formula for each compound. Let C, H, and O represent the atomic weights of carbon, hydrogen, and oxygen.

$$C + 4H \qquad = 16$$
$$3C + 8H + 3O = 92$$
$$\qquad 2H + \ O = 18$$

Evaluate the determinant of the coefficient matrix.

$$\begin{vmatrix} 1 & 4 & 0 \\ 3 & 8 & 3 \\ 0 & 2 & 1 \end{vmatrix} = (8 + 0 + 0) - (0 + 6 + 12) = -10$$

Apply Cramer's rule since the determinant is not 0.

$$C = \dfrac{\begin{vmatrix} 16 & 4 & 0 \\ 92 & 8 & 3 \\ 18 & 2 & 1 \end{vmatrix}}{-10} = \dfrac{-120}{-10} = 12 \qquad \textit{Atomic weight of carbon}$$

$$H = \dfrac{\begin{vmatrix} 1 & 16 & 0 \\ 3 & 92 & 3 \\ 0 & 18 & 1 \end{vmatrix}}{-10} = \dfrac{-10}{-10} = 1 \qquad \textit{Atomic weight of hydrogen}$$

$$O = \dfrac{\begin{vmatrix} 1 & 4 & 16 \\ 3 & 8 & 92 \\ 0 & 2 & 18 \end{vmatrix}}{-10} = \dfrac{-160}{-10} = 16 \qquad \textit{Atomic weight of oxygen}$$

▶ The weights of carbon, hydrogen, and oxygen are 12, 1, and 16, respectively.

FOCUS ON CAREERS

CHEMIST
Chemists research and put to practical use knowledge about chemicals. Research on the chemistry of living things sparks advances in medicine, agriculture, and other fields.

CAREER LINK
www.mcdougallittell.com

CHECKPOINT EXERCISES

For use after Example 5:

1. Use the chart below to check your answer to Extra Example 5.

Compound	Formula	Protons
Carbon Dioxide	CO_2	22
Glycogen	$C_6H_{12}O_6$	96
Water	H_2O	10

$C + 2O = 22$; $6C + 12H + 6O = 96$;
$2H + O = 10$; $C = 6$, $H = 1$, $O = 8$

FOCUS ON VOCABULARY
What is the difference between $\begin{bmatrix} a & b \\ c & d \end{bmatrix}$ and $\begin{vmatrix} a & b \\ c & d \end{vmatrix}$? $\begin{vmatrix} a & b \\ c & d \end{vmatrix}$ is the determinant of matrix $\begin{bmatrix} a & b \\ c & d \end{bmatrix}$.

CLOSURE QUESTION
How do you find the determinant of a 2 × 2 matrix?
See the box on page 214.

DAILY PUZZLER
In the decimal expansion of $\frac{1}{13}$, what digit is in the 2000th place? 7

CAREER NOTE
EXAMPLE 5 Additional information about a career as a chemist is available at **www.mcdougallittell.com.**

ASSIGNMENT GUIDE

BASIC
Day 1: pp. 218–219 Exs. 12–24 even, 31–41 odd, 54
Day 2: pp. 218–221 Exs. 26–38 even, 43–49 odd, 55–57, 62, 63, 65–79 odd, Quiz 1 Exs 1–17

AVERAGE
Day 1: pp. 218–219 Exs. 12–36 even, 31–49 odd, 54–57
Day 2: pp. 218–221 Exs. 30–44 even, 50–53, 59, 61–63, 65–79 odd, Quiz 1 Exs 1–17

ADVANCED
Day 1: pp. 218–219 Exs. 12–28 even. 31–41 odd, 45–49 odd, 54–57
Day 2: pp. 218–221 Exs. 29, 34–38 even, 51–53 odd, 58–64, 65–79 odd, Quiz 1 Exs 1–17

BLOCK SCHEDULE
pp. 203–206 Exs. 12–28 even, 31–49 odd, 54–58, 62, 63, 65–79 odd, Quiz 1 Exs 1–17

EXERCISE LEVELS
Level A: *Easier*
12–35, 54–57
Level B: *More Difficult*
36–53, 58–60, 62, 63
Level C: *Most Difficult*
61, 64

✔ **HOMEWORK CHECK**
To quickly check student understanding of key concepts, go over the following exercises: Exs. 14, 22, 31, 37, 45. See also the Daily Homework Quiz:

• Blackline Master (*Chapter 4 Resource Book*, p. 54)
• Transparency (p. 29)

GUIDED PRACTICE

Vocabulary Check ✔
Concept Check ✔

2. Yes; *Sample answer:*

$$\begin{vmatrix} 5 & 1 \\ 0 & 2 \end{vmatrix} = \begin{vmatrix} 6 & 1 \\ 2 & 2 \end{vmatrix} = 10$$

3a. The calculation should be $6 - (-5) = 11$.

3b. The calculation should be $12 - (-2) = 14$.

Skill Check ✔

1. Explain Cramer's rule and how it is used.   **See margin.**

2. Can two different matrices have the same determinant? If so, give an example.

3. **ERROR ANALYSIS** Find the error in each calculation.

a.
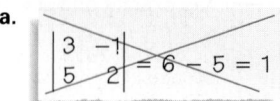

b.
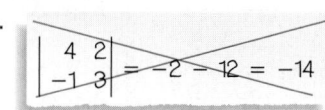

4. To use Cramer's rule to solve a linear system, what must be true of the determinant of the coefficient matrix?   **The determinant must not be 0.**

Evaluate the determinant of the matrix.

5. $\begin{bmatrix} 0 & 1 \\ 6 & 2 \end{bmatrix}$ —6

6. $\begin{bmatrix} -1 & 4 \\ 5 & -1 \end{bmatrix}$ —19

7. $\begin{bmatrix} 8 & -2 \\ -2 & 4 \end{bmatrix}$ 28

Use Cramer's rule to solve the linear system.

8. $6x - 8y = 4$ $(-26, -20)$
$4x - 5y = -4$

9. $2x + 7y = -3$ $(-5, 1)$
$3x - 8y = -23$

10. $12x - 2y = 2$ $\left(\frac{31}{26}, \frac{80}{13}\right)$
$-14x + 11y = 51$

11. **SCHOOL SPIRIT** You are making a large pennant for your school football team. A diagram of the pennant is shown at the right. The coordinates given are measured in inches. How many square inches of material will you need to make the pennant?   **1750 in.²**

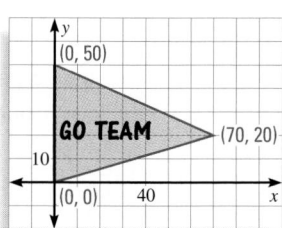

PRACTICE AND APPLICATIONS

STUDENT HELP

→ **Extra Practice**
to help you master skills is on p. 945.

2 × 2 DETERMINANTS Evaluate the determinant of the matrix.

12. $\begin{bmatrix} -4 & 2 \\ 5 & -2 \end{bmatrix}$ —2

13. $\begin{bmatrix} 8 & 0 \\ -1 & 3 \end{bmatrix}$ 24

14. $\begin{bmatrix} 9 & 3 \\ -2 & 1 \end{bmatrix}$ 15

15. $\begin{bmatrix} -7 & 11 \\ -7 & 2 \end{bmatrix}$ 63

16. $\begin{bmatrix} 4 & 0 \\ -3 & 4 \end{bmatrix}$ 16

17. $\begin{bmatrix} 1 & 8 \\ 5 & 9 \end{bmatrix}$ —31

18. $\begin{bmatrix} -6 & 5 \\ -3 & 9 \end{bmatrix}$ —39

19. $\begin{bmatrix} 0 & -3 \\ 8 & 10 \end{bmatrix}$ 24

20. $\begin{bmatrix} 12 & 2 \\ -5 & 8 \end{bmatrix}$ 106

3 × 3 DETERMINANTS Evaluate the determinant of the matrix.

STUDENT HELP

→ **HOMEWORK HELP**
Example 1: Exs. 12–29
Example 2: Exs. 30–35
Example 3: Exs. 54–58
Example 4: Exs. 36–44, 59
Example 5: Exs. 45–53, 60

21. $\begin{bmatrix} 12 & 4 & -1 \\ -2 & 3 & 2 \\ 5 & 8 & 1 \end{bmatrix}$ —77

22. $\begin{bmatrix} 5 & -9 & 4 \\ 4 & 2 & 1 \\ 0 & 1 & 1 \end{bmatrix}$ 57

23. $\begin{bmatrix} 0 & 5 & 2 \\ 10 & 13 & -4 \\ -5 & 4 & -1 \end{bmatrix}$ 360

24. $\begin{bmatrix} 1 & 16 & -2 \\ 20 & 4 & 2 \\ 7 & 1 & -4 \end{bmatrix}$ 1502

25. $\begin{bmatrix} -4 & 0 & -1 \\ 0 & 8 & 9 \\ 0 & 5 & 2 \end{bmatrix}$ 116

26. $\begin{bmatrix} 8 & 2 & 9 \\ 12 & 3 & 9 \\ 3 & 13 & 4 \end{bmatrix}$ 441

27. $\begin{bmatrix} 3 & 12 & -1 \\ 10 & 9 & 0 \\ -5 & 6 & -2 \end{bmatrix}$ 81

28. $\begin{bmatrix} -3 & 2 & 20 \\ -10 & 9 & 18 \\ 11 & 15 & 12 \end{bmatrix}$ —3858

29. $\begin{bmatrix} 15 & 4 & -10 \\ -10 & 0 & 6 \\ -8 & 2 & -14 \end{bmatrix}$ —732

AREA OF A TRIANGLE Find the area of the triangle with the given vertices.

30. $A(0, 1), B(2, 7), C(5, 5)$ **11**

31. $A(3, 6), B(3, 0), C(1, 3)$ **6**

32. $A(6, -1), B(2, 2), C(4, 8)$ **15**

33. $A(-4, 2), B(3, -1), C(-2, -2)$ **11**

34. $A(2, -6), B(-1, -4), C(0, 2)$ **10**

35. $A(1, 3), B(-2, 6), C(-1, 1)$ **6**

USING CRAMER'S RULE Use Cramer's rule to solve the linear system.

36. $2x + y = 3$ **(2, −1)**
$5x + 6y = 4$

37. $7x - 5y = 11$ **(−2, −5)**
$3x + 10y = -56$

38. $9x + 2y = 7$ **(3, −10)**
$4x - 3y = 42$

39. $x + 7y = -3$ **(4, −1)**
$3x - 5y = 17$

40. $-x - 12y = 44$ **(−8, −3)**
$12x - 15y = -51$

41. $4x - 3y = 18$ **(6, 2)**
$8x - 7y = 34$

42. $4x - 5y = 13$
$2x - 7y = 24$ $\left(-\dfrac{29}{18}, -\dfrac{35}{9}\right)$

43. $8x - 9y = 32$
$-5x + 7y = 40$ $\left(\dfrac{584}{11}, \dfrac{480}{11}\right)$

44. $3x + 10y = 50$
$12x + 15y = 64$ $\left(-\dfrac{22}{15}, \dfrac{136}{25}\right)$

SOLVING SYSTEMS Use Cramer's rule to solve the linear system.

47. $\left(-\dfrac{2}{3}, -34, -12\right)$

53. $\left(-\dfrac{1}{44}, -\dfrac{69}{22}, -\dfrac{481}{88}\right)$

45. **(0, 5, 4)**
$x + 2y - 3z = -2$
$x - y + z = -1$
$3x + 4y - 4z = 4$

46. **(2, 0, 1)**
$x + 3y - z = 1$
$-2x - 6y + z = -3$
$3x + 5y - 2z = 4$

47. **See margin.**
$3x + 2y - 5z = -10$
$6x - z = 8$
$-y + 3z = -2$

48. **(−1, 6, −2)**
$x + 2y + z = 9$
$x + y + z = 3$
$5x - 2z = -1$

49. **(4, 3, −2)**
$4x + y + 6z = 7$
$3x + 3y + 2z = 17$
$-x - y + z = -9$

50. **(2, −1, 5)**
$x + 4y - z = -7$
$2x - y + 2z = 15$
$-3x + y - 3z = -22$

51. $2x + y + z = 5$
$x + 4y - 2z = 9$
$6x + 5y = 16$ $\left(\dfrac{1}{11}, \dfrac{34}{11}, \dfrac{19}{11}\right)$

52. $-x + 2y + 7z = 13$
$2x - y - 2z = -2$
$3x + 5y + 2z = -14$ **(0, −4, 3)**

53. $-3x + y + 2z = -14$
$9x - y + 2z = -8$
$8x + 5y - 4z = 6$ **See margin.**

54. 🌐 **BIRDS** Black-necked stilts are birds that live throughout Florida and surrounding areas but breed mostly in the triangular region shown on the map. Estimate the area of this region. The coordinates given are measured in miles. **11,340 mi²**

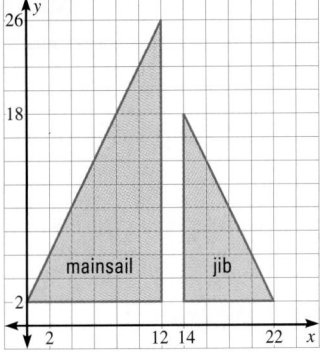

Each tick mark represents 50 miles.

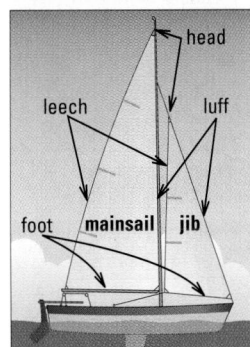

FOCUS ON APPLICATIONS

head
leech
luff
foot **mainsail** **jib**

SAILING The edges of a sail are called the luff, leech, and foot. The luff length of the jib is usually 80% to 90% of the distance from the deck to the head of the jib.

🌐 **SAILING** In Exercises 55–57, use the following information.
On a Marconi-rigged sloop, there are two triangular sails, a mainsail and a jib. These sails are shown in a coordinate plane at the right. The coordinates in the plane are measured in feet.

55. Find the area of the mainsail shown. **144 ft²**

56. Find the area of the jib shown. **64 ft²**

57. Suppose you are making a scale model of the sailboat with the sails shown using a scale of 1 in. = 6 ft. What is the area of the model's mainsail? **4 in.²**

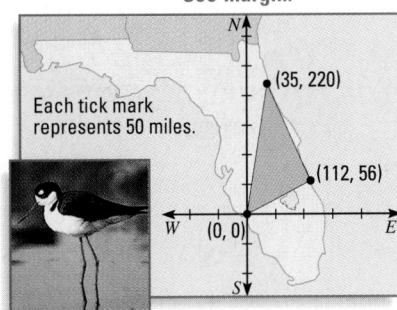

mainsail jib

!**COMMON ERROR**
EXERCISES 21–29 When finding the determinant of a 3×3 matrix, students must find the difference of the sum of the products of the diagonals in one direction and the sum of the products of the diagonals in the other direction. Students may become confused with the steps involved, so it may help to have them write all of the steps.

APPLICATION NOTE
EXERCISES 55–59 A marconi-rigged sloop is obviously a small sailing vessel. Other larger types of sailing vessels include barks, brigantines, and square-rigged ships.

1. *Sample answer:* Cramer's rule is a method of solving a system of linear equations by using the determinant of the coefficient matrix associated with the system. First find the determinant of the coefficient matrix and check that it does not equal 0. Each coordinate of the solution of the system is found as a fraction. The denominator of the fraction is the determinant of the coefficient matrix. The numerator of the fraction is the determinant of the matrix formed by using the coefficient matrix and replacing the column of coefficients of the corresponding variable with the column of constants from the system.

ADDITIONAL PRACTICE AND RETEACHING

For Lesson 4.3:
• Practice Levels A, B, and C (*Chapter 4 Resource Book,* p. 42)
• Reteaching with Practice (*Chapter 4 Resource Book,* p. 45)
• See Lesson 4.3 of the *Personal Student Tutor*

For more Mixed Review:
• Search the *Test and Practice Generator* for key words or specific lessons.

219

DAILY HOMEWORK QUIZ

📄 **Transparency Available**

Evaluate the determinant of the matrix.

1. $\begin{bmatrix} 3 & 3 \\ 4 & -2 \end{bmatrix}$ −18

2. $\begin{bmatrix} 8 & 3 & -2 \\ -1 & 3 & 2 \\ 5 & -1 & 1 \end{bmatrix}$ 101

3. Find the area of the triangle with vertices at $A(0, 2)$, $B(2, 8)$, and $C(5, 6)$. 11

4. Use Cramer's rule to solve the linear system.
$7x - 3y = 10$
$3x + 8y = 60$ (4, 6)

EXTRA CHALLENGE NOTE

↳ Challenge problems for Lesson 4.3 are available in **blackline** format in the *Chapter 4 Resource Book,* p. 50 and at **www.mcdougallittell.com.**

ADDITIONAL TEST PREPARATION

1. WRITING The determinant of $\begin{bmatrix} a & b \\ c & d \end{bmatrix}$ is $ad - bc$. When will the determinant be negative?
Sample answer: When the product *bc* is larger than the product *ad.*

58. SOCIAL STUDIES CONNECTION The Golden Triangle refers to a large triangular region in India. The Taj Mahal is one of the many wonders that lie within the boundaries of this triangle. The triangle is formed by imaginary lines that connect the cities of New Delhi, Jaipur, and Agra. Use the coordinates on the map and a determinant to estimate the area of the Golden Triangle. The coordinates given are measured in miles.
7400 mi²

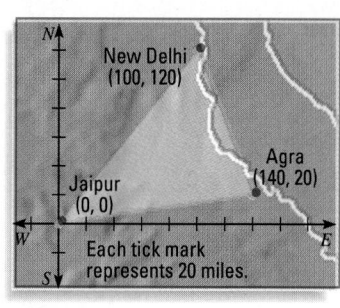

New Delhi (100, 120)
Agra (140, 20)
Jaipur (0, 0)
Each tick mark represents 20 miles.

59. 🌐 **BUYING GASOLINE** You fill up your car with 10 gallons of premium gasoline and fill a small gas can with 2 gallons of regular gasoline for your lawn mower. You pay the cashier $13.56. The price of premium gasoline is 12 cents more per gallon than the price of regular gasoline. Use a linear system and Cramer's rule to find the price per gallon for regular and premium gasoline.
regular: $1.03 per gal, premium: $1.15 per gal

60. SCIENCE CONNECTION The atomic weights of three compounds are shown.
S = 32, N = 14, F = 19

Compound	Formula	Atomic weight
Tetrasulphur tetranitride	S_4N_4	184
Sulphur hexaflouride	SF_6	146
Dinitrogen tetraflouride	N_2F_4	104

Use a linear system and Cramer's rule to find the atomic weights of sulphur (S), nitrogen (N), and flourine (F).

61. The determinant is multiplied by −1.
Proof for 2 × 2 matrices:

$-1 \begin{vmatrix} a & b \\ c & d \end{vmatrix} = -1(ad - bc) =$

$bc - ad = \begin{vmatrix} b & a \\ d & c \end{vmatrix}$

Test Preparation

61. LOGICAL REASONING Explain what happens to the determinant of a matrix when you switch two rows or two columns.

QUANTITATIVE COMPARISON In Exercises 62 and 63, choose the statement that is true about the given quantities.

Ⓐ The quantity in column A is greater.

Ⓑ The quantity in column B is greater.

Ⓒ The two quantities are equal.

Ⓓ The relationship cannot be determined from the given information.

	Column A	Column B	
62.	The area of a triangle with vertices $(-3, 4)$, $(4, 2)$, and $(1, -2)$	The area of a triangle with vertices $(4, 2)$, $(1, -2)$, and $(3, -4)$	A
63.	$\det \begin{bmatrix} -5 & 6 \\ -2 & 10 \end{bmatrix}$	$\det \begin{bmatrix} -7 & 1 \\ 3 & 5 \end{bmatrix}$	C

★ **Challenge**

EXTRA CHALLENGE
↳ www.mcdougallittell.com

64. DETERMINANT RELATIONSHIPS Let $A = \begin{bmatrix} 2 & -1 \\ 3 & 2 \end{bmatrix}$ and $B = \begin{bmatrix} 3 & 5 \\ -2 & -4 \end{bmatrix}$.

a. How is det *AB* related to det *A* and det *B*? det AB = (det A)(det B)

b. How is det *kA* related to det *A* if *k* is a constant? Check your answer using matrix *B* and several other 2 × 2 matrices. det $kA = k^2$det A

MIXED REVIEW

(Review 2.1)

EVALUATING FUNCTIONS Find the indicated value of $f(x)$. (Review 2.1)

65. $f(x) = x - 10, f(7)$ -3

66. $f(x) = 3x + 7, f(-2)$ 1

67. $f(x) = -x^2 + 5, f(-1)$ 4

68. $f(x) = x^2 - 2x - 4, f(7)$ 31

69. $f(x) = x^2 + 4x - 1, f\left(\frac{1}{2}\right)$ $\frac{5}{4}$

70. $f(x) = x^5 - 2x - 10, f(3)$ 227

GRAPHING SYSTEMS Graph the system of linear inequalities. (Review 3.3)

71–76. See margin.

71. $x < 3$
$x > -2$

72. $y \geq 2x - 3$
$y > -5x - 8$

73. $y > -x - 5$
$y > 3x + 1$

74. $x + y > 3$
$4x + y < 4$

75. $2x - y \geq 2$
$5x - y \geq 2$

76. $4x - 3y > 1$
$-x + y \geq 4$

MULTIPLYING MATRICES Find the product. (Review 4.2 for 4.4)

77. $\begin{bmatrix} -24 & 14 \\ 33 & -8 \end{bmatrix}$

78. $\begin{bmatrix} -4 & -29 \\ -40 & -92 \end{bmatrix}$

79. $\begin{bmatrix} -104 & 35 \\ 32 & -4 \end{bmatrix}$

77. $\begin{bmatrix} -2 & -4 \\ 5 & 1 \end{bmatrix}\begin{bmatrix} 6 & -1 \\ 3 & -3 \end{bmatrix}$

78. $\begin{bmatrix} 7 & -1 \\ 4 & -10 \end{bmatrix}\begin{bmatrix} 0 & -3 \\ 4 & 8 \end{bmatrix}$

79. $\begin{bmatrix} 11 & -2 \\ 0 & 4 \end{bmatrix}\begin{bmatrix} -8 & 3 \\ 8 & -1 \end{bmatrix}$

80. $\begin{bmatrix} 3 & -5 \\ -7 & 2 \end{bmatrix}\begin{bmatrix} 10 & 9 \\ 12 & 16 \end{bmatrix}$

81. $\begin{bmatrix} 0.5 & 3 \\ 0.2 & 1 \end{bmatrix}\begin{bmatrix} 0 & 0.6 \\ 4 & 0.8 \end{bmatrix}$

82. $\begin{bmatrix} -2 & 1.3 \\ 1.5 & -3 \end{bmatrix}\begin{bmatrix} 1.6 & 6 \\ -4 & 1.9 \end{bmatrix}$

$\begin{bmatrix} -30 & -53 \\ -46 & -31 \end{bmatrix}$

$\begin{bmatrix} 12 & 2.7 \\ 4 & 0.92 \end{bmatrix}$

$\begin{bmatrix} -8.4 & -9.53 \\ 14.4 & 3.3 \end{bmatrix}$

QUIZ 1

Self-Test for Lessons 4.1–4.3

Perform the indicated operation(s). (Lessons 4.1, 4.2)

1. $\begin{bmatrix} -5 & 4 & 15 \\ 2 & -14 & 1 \end{bmatrix}$

2. $\begin{bmatrix} -5 & -7 \\ 0 & -1 \end{bmatrix}$

3. $\begin{bmatrix} -2 & -2 \\ -18 & -12 \end{bmatrix}$

4. $\begin{bmatrix} -4 & -2 & 22 \\ 3 & -18 & 20 \\ 17 & -4 & 1 \end{bmatrix}$

15. $\left(\frac{7}{3}, 10, -\frac{4}{3}\right)$

1. $\begin{bmatrix} -2 & 5 & 10 \\ 4 & -6 & 8 \end{bmatrix} + \begin{bmatrix} -3 & -1 & 5 \\ -2 & -8 & -7 \end{bmatrix}$

2. $\begin{bmatrix} -8 & 0 \\ 5 & -2 \end{bmatrix} - \begin{bmatrix} -3 & 7 \\ 5 & -1 \end{bmatrix}$

3. $-2\begin{bmatrix} 7 & -2 \\ 4 & 9 \end{bmatrix} + 2\begin{bmatrix} 6 & -3 \\ -5 & 3 \end{bmatrix}$

4. $\begin{bmatrix} 4 & -6 & 10 \\ 3 & 6 & 0 \\ 9 & -4 & 5 \end{bmatrix} - 4\begin{bmatrix} 2 & -1 & -3 \\ 0 & 6 & -5 \\ -2 & 0 & 1 \end{bmatrix}$

5. $\begin{bmatrix} 8 & -1 \\ 6 & -2 \end{bmatrix}\begin{bmatrix} 3 & 7 \\ -2 & 0 \end{bmatrix}\begin{bmatrix} 26 & 56 \\ 22 & 42 \end{bmatrix}$

6. $\begin{bmatrix} 2 & -1 & 3 \\ 2 & 4 & 0 \end{bmatrix}\begin{bmatrix} 1 & 0 \\ 9 & -3 \\ 4 & -6 \end{bmatrix}\begin{bmatrix} 5 & -15 \\ 38 & -12 \end{bmatrix}$

Evaluate the determinant of the matrix. (Lesson 4.3)

7. $\begin{bmatrix} -4 & 3 \\ -6 & 2 \end{bmatrix}$ 10

8. $\begin{bmatrix} 9 & -3 \\ 6 & -2 \end{bmatrix}$ 0

9. $\begin{bmatrix} -1 & 2 & 3 \\ 5 & 0 & -2 \\ 6 & 8 & 1 \end{bmatrix}$ 70

10. $\begin{bmatrix} 12 & 5 & -6 \\ 2 & 2 & 3 \\ 1 & 0 & -3 \end{bmatrix}$ -15

Use Cramer's rule to solve the linear system. (Lesson 4.3)

11. $-8x + y = -6$ $(1, 2)$
$-5x + 4y = 3$

12. $3x - 2y = 10$ $\left(\frac{4}{9}, -\frac{13}{3}\right)$
$-6x + y = -7$

13. $5x + 4y = 12$ $\left(2, \frac{1}{2}\right)$
$3x - 6y = 3$

14. $4x + y + 6z = 2$
$2x + 2y + 4z = 1$
$-x - y + z = -5$ $\left(\frac{5}{2}, 1, -\frac{3}{2}\right)$

15. $x + y + 4z = 7$
$2x - 3y - z = -24$
$-4x + 2y + 2z = 8$

16. $3x + 3y - 2z = -18$
$-5x - 2y - 3z = -1$
$7x + y + 6z = 14$
$(0, -4, 3)$

17. 🌎 **GARDENING** You are planning to turn a triangular region of your yard into a garden. The vertices of the triangle are $(0, 0)$, $(5, 2)$, and $(3, 6)$ where the coordinates are measured in feet. Find the area of the triangular region. (Lesson 4.3)
12 ft^2

4.3 *Determinants and Cramer's Rule* **221**

ADDITIONAL RESOURCES
An alternative Quiz for Lessons 4.1–4.3 is available in the *Chapter 4 Resource Book,* p. 51)

71.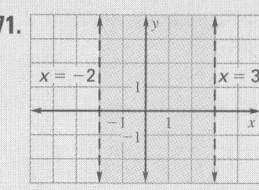

72.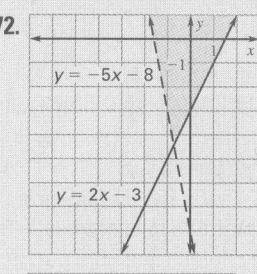

73.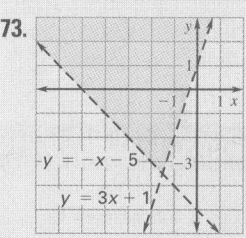

74.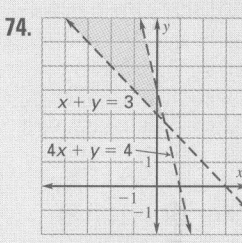

75.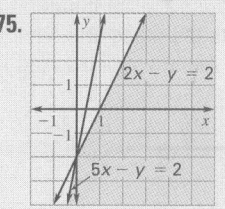

76.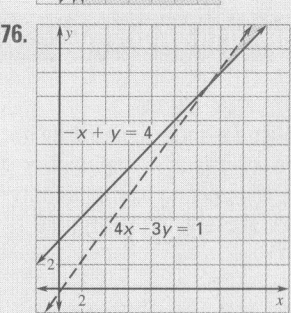

1 Planning the Activity

PURPOSE
To investigate the properties of and relationships between 2 × 2 matrices, their inverses, and the 2 × 2 identity matrix.

MATERIALS
- pencil
- paper

PACING
- Exploring the Concept — 15 min
- Drawing Conclusions — 15 min

▶ LINK TO LESSON
Lesson 4.4 uses the identity and inverse matrices in Examples 1–3. Students should recall the conclusions reached in this activity as they study those Examples.

2 Managing the Activity

COOPERATIVE LEARNING
Partners should share the calculations when multiplying the matrices. When working on the Drawing Conclusions section, have each pair write their answers, then compare them with other pairs of students.

ALTERNATIVE APPROACH
You can demonstrate the concepts in Steps 1–4 to the class and have them answer the questions asked based on what they observe. Students can then work in pairs to solve Exercises 1–6.

3 Closing the Activity

★ KEY DISCOVERY
Multiplying a square matrix by its inverse will result in the identity matrix. Multiplying by the identity matrix will result in the original matrix.

ACTIVITY ASSESSMENT
What entries in the matrix will make the statement true?

$$\begin{bmatrix} 1 & -1 \\ 3 & 8 \end{bmatrix} \begin{bmatrix} ? & ? \\ ? & ? \end{bmatrix} = \begin{bmatrix} 1 & -1 \\ 3 & 8 \end{bmatrix} \begin{bmatrix} 1 & 0 \\ 0 & 1 \end{bmatrix}$$

222

● ACTIVITY 4.4
Developing Concepts

GROUP ACTIVITY
Work with a partner.

MATERIALS
- pencil
- paper

Step 1. $\begin{bmatrix} 1 & 3 \\ 2 & 5 \end{bmatrix}, \begin{bmatrix} -4 & 0 \\ -7 & 6 \end{bmatrix},$

$\begin{bmatrix} 0.1 & 0.8 \\ 0.6 & 0.3 \end{bmatrix}$; multiplying each

matrix by $I = \begin{bmatrix} 1 & 0 \\ 0 & 1 \end{bmatrix}$ results

in the original matrix.

Step 2. $\begin{bmatrix} 1 & 3 \\ 2 & 5 \end{bmatrix}, \begin{bmatrix} -4 & 0 \\ -7 & 6 \end{bmatrix},$

$\begin{bmatrix} 0.1 & 0.8 \\ 0.6 & 0.3 \end{bmatrix}$; yes

Step 3. $\begin{bmatrix} 1 & 0 \\ 0 & 1 \end{bmatrix}, \begin{bmatrix} 1 & 0 \\ 0 & 1 \end{bmatrix}$; each

product is the identity matrix

$I = \begin{bmatrix} 1 & 0 \\ 0 & 1 \end{bmatrix}$.

Step 4. $\begin{bmatrix} -5 & 3 \\ 2 & -1 \end{bmatrix}$

2. Multiplying a 2 × 2 matrix

by $I = \begin{bmatrix} 1 & 0 \\ 0 & 1 \end{bmatrix}$ results in the

original matrix, just as multiplying a real number by 1 results in the original number.

3. $\begin{bmatrix} 1 & 0 & 0 \\ 0 & 1 & 0 \\ 0 & 0 & 1 \end{bmatrix}$

4. The product of a matrix and its inverse is the identity matrix; the product of a nonzero real number and its reciprocal is 1.

Investigating Identity and Inverse Matrices

▶ **QUESTION** What are some properties of identity and inverse matrices?

▶ **EXPLORING THE CONCEPT**

❶ Let $A = \begin{bmatrix} 1 & 3 \\ 2 & 5 \end{bmatrix}$, $B = \begin{bmatrix} -4 & 0 \\ -7 & 6 \end{bmatrix}$, and $C = \begin{bmatrix} 0.1 & 0.8 \\ 0.6 & 0.3 \end{bmatrix}$. Consider the 2 × 2 *identity matrix* $I = \begin{bmatrix} 1 & 0 \\ 0 & 1 \end{bmatrix}$. Find AI, BI, and CI. What do you notice?

❷ Find IA, IB, and IC using the matrices from **Step 1**. Is multiplication by the identity matrix commutative?

❸ Let $D = \begin{bmatrix} 7 & 5 \\ 4 & 3 \end{bmatrix}$. The *inverse* of D is $E = \begin{bmatrix} 3 & -5 \\ -4 & 7 \end{bmatrix}$. Find DE and ED. What do you notice?

❹ Use matrix multiplication to decide which of the following is the inverse of the matrix A in **Step 1**: $\begin{bmatrix} 5 & -3 \\ -2 & 1 \end{bmatrix}$, $\begin{bmatrix} -5 & 3 \\ 2 & -1 \end{bmatrix}$, or $\begin{bmatrix} -1 & 2 \\ 3 & -5 \end{bmatrix}$.

▶ **DRAWING CONCLUSIONS**

1. For any 2 × 2 matrix A, what is true of the products AI and IA where I is the 2 × 2 identity matrix? Justify your answer mathematically. **See margin.**

   (*Hint:* Let $A = \begin{bmatrix} a & b \\ c & d \end{bmatrix}$, and compute AI and IA.)

2. How is the relationship between $I = \begin{bmatrix} 1 & 0 \\ 0 & 1 \end{bmatrix}$ and other 2 × 2 matrices similar to the relationship between 1 and other real numbers?

3. What do you think is the identity matrix for the set of 3 × 3 matrices? Check your answer by multiplying your proposed identity matrix by several 3 × 3 matrices.

4. What is the relationship between a matrix, its inverse, and the identity matrix? How is this relationship like the one that exists between a nonzero real number, its reciprocal, and 1?

5. Does every nonzero matrix have an inverse? Explain. (*Hint:* Consider a 2 × 2 matrix whose first row contains all nonzero entries and whose second row contains all zero entries.) **See margin.**

6. Find the inverse of $F = \begin{bmatrix} 2 & 7 \\ 1 & 4 \end{bmatrix}$ by finding values of a, b, c, and d such that

$$\begin{bmatrix} 2 & 7 \\ 1 & 4 \end{bmatrix} \begin{bmatrix} a & b \\ c & d \end{bmatrix} = \begin{bmatrix} 1 & 0 \\ 0 & 1 \end{bmatrix}. \quad \begin{bmatrix} 4 & -7 \\ -1 & 2 \end{bmatrix}$$

222 **Chapter 4** *Matrices and Determinants*

1, 5. See Additional Answers beginning on page AA1.

Identity and Inverse Matrices

What you should learn

GOAL **1** Find and use inverse matrices.

GOAL **2** Use inverse matrices in **real-life** situations, such as encoding a message in **Example 5**.

Why you should learn it

▼ To solve **real-life** problems, such as decoding names of landmarks in **Exs. 44–48**.

The artist Jim Sanborn uses cryptograms in his work, such as *Kryptos* above.

GOAL **1** USING INVERSE MATRICES

The number 1 is the multiplicative identity for real numbers because $1 \cdot a = a$ and $a \cdot 1 = a$. For matrices, the $n \times n$ **identity matrix** is the matrix that has 1's on the main diagonal and 0's elsewhere.

2 × 2 IDENTITY MATRIX

$$I = \begin{bmatrix} 1 & 0 \\ 0 & 1 \end{bmatrix}$$

3 × 3 IDENTITY MATRIX

$$I = \begin{bmatrix} 1 & 0 & 0 \\ 0 & 1 & 0 \\ 0 & 0 & 1 \end{bmatrix}$$

If A is any $n \times n$ matrix and I is the $n \times n$ identity matrix, then $IA = A$ and $AI = A$.

Two $n \times n$ matrices are **inverses** of each other if their product (in *both* orders) is the $n \times n$ identity matrix. For example, matrices A and B below are inverses of each other.

$$AB = \begin{bmatrix} 3 & -1 \\ -5 & 2 \end{bmatrix} \begin{bmatrix} 2 & 1 \\ 5 & 3 \end{bmatrix} = \begin{bmatrix} 1 & 0 \\ 0 & 1 \end{bmatrix} = I \quad BA = \begin{bmatrix} 2 & 1 \\ 5 & 3 \end{bmatrix} \begin{bmatrix} 3 & -1 \\ -5 & 2 \end{bmatrix} = \begin{bmatrix} 1 & 0 \\ 0 & 1 \end{bmatrix} = I$$

The symbol used for the inverse of A is A^{-1}.

THE INVERSE OF A 2 X 2 MATRIX

The inverse of the matrix $A = \begin{bmatrix} a & b \\ c & d \end{bmatrix}$ is

$$A^{-1} = \frac{1}{|A|} \begin{bmatrix} d & -b \\ -c & a \end{bmatrix} = \frac{1}{ad-cb} \begin{bmatrix} d & -b \\ -c & a \end{bmatrix}$$ provided $ad - cb \neq 0$.

EXAMPLE 1 *Finding the Inverse of a 2 × 2 Matrix*

Find the inverse of $A = \begin{bmatrix} 3 & 1 \\ 4 & 2 \end{bmatrix}$.

SOLUTION

$$A^{-1} = \frac{1}{6-4} \begin{bmatrix} 2 & -1 \\ -4 & 3 \end{bmatrix} = \frac{1}{2} \begin{bmatrix} 2 & -1 \\ -4 & 3 \end{bmatrix} = \begin{bmatrix} 1 & -\frac{1}{2} \\ -2 & \frac{3}{2} \end{bmatrix}$$

✔ **CHECK** You can check the inverse by showing that $AA^{-1} = I = A^{-1}A$.

$$\begin{bmatrix} 3 & 1 \\ 4 & 2 \end{bmatrix} \begin{bmatrix} 1 & -\frac{1}{2} \\ -2 & \frac{3}{2} \end{bmatrix} = \begin{bmatrix} 1 & 0 \\ 0 & 1 \end{bmatrix} \text{ and } \begin{bmatrix} 1 & -\frac{1}{2} \\ -2 & \frac{3}{2} \end{bmatrix} \begin{bmatrix} 3 & 1 \\ 4 & 2 \end{bmatrix} = \begin{bmatrix} 1 & 0 \\ 0 & 1 \end{bmatrix}$$

STUDENT HELP

► **Look Back**
For help with multiplicative inverses of real numbers, see p. 5.

4.4 Identity and Inverse Matrices **223**

1 PLAN

PACING
Basic: 1 day
Average: 1 day
Advanced: 1 day
Block Schedule: 0.5 block with 4.5

LESSON OPENER
GRAPHING CALCULATOR
An alternative way to approach Lesson 4.4 is to use the Graphing Calculator Lesson Opener:
• Blackline Master (*Chapter 4 Resource Book*, p. 55)
• Transparency (p. 25)

MEETING INDIVIDUAL NEEDS
• *Chapter 4 Resource Book*
  Prerequisite Skills Review (p. 5)
  Practice Level A (p. 57)
  Practice Level B (p. 58)
  Practice Level C (p. 59)
  Reteaching with Practice (p. 60)
  Absent Student Catch-Up (p. 62)
  Challenge (p. 64)
• *Resources in Spanish*
• *Personal Student Tutor*

NEW-TEACHER SUPPORT
See the Tips for New Teachers on pp. 1–2 of the *Chapter 4 Resource Book* for additional notes about Lesson 4.4.

WARM-UP EXERCISES

Transparency Available
Find the product.

1. $\begin{bmatrix} 9 & 0 \\ -6 & 3 \end{bmatrix} \begin{bmatrix} 0 & 7 \\ 2 & -5 \end{bmatrix}$ $\begin{bmatrix} 0 & 63 \\ 6 & -57 \end{bmatrix}$

2. $\begin{bmatrix} 1 & 4 \\ 6 & -6 \end{bmatrix} \begin{bmatrix} 8 & -7 \\ -8 & 2 \end{bmatrix}$ $\begin{bmatrix} -24 & 1 \\ 96 & -54 \end{bmatrix}$

3. $[-5 \ 1 \ 4] \begin{bmatrix} 0 \\ -1 \\ 9 \end{bmatrix}$ $[35]$

4. $\begin{bmatrix} 0 \\ -1 \\ 9 \end{bmatrix} [-5 \ 1 \ 4]$ $\begin{bmatrix} 0 & 0 & 0 \\ 5 & -1 & -4 \\ -45 & 9 & 36 \end{bmatrix}$

MOTIVATING THE LESSON
Ask students to recall the multiplication identity and multiplicative inverse for real numbers. Have students state what these did when used with multiplication. Tell students that identities and inverses occur for multiplication of square matrices as well.

EXTRA EXAMPLE 1
Find the inverse of $A = \begin{bmatrix} 3 & 2 \\ 2 & 4 \end{bmatrix}$.

$$\begin{bmatrix} \frac{1}{2} & -\frac{1}{4} \\ -\frac{1}{4} & \frac{3}{8} \end{bmatrix}$$

EXTRA EXAMPLE 2
Solve the matrix equation $AX = B$ for the 2×2 matrix X.
$$\begin{bmatrix} -3 & 4 \\ 5 & -7 \end{bmatrix} X = \begin{bmatrix} 3 & 8 \\ 2 & -2 \end{bmatrix} \quad \begin{bmatrix} -29 & -48 \\ -21 & -34 \end{bmatrix}$$

EXTRA EXAMPLE 3
Use a graphing calculator to find the inverse of
$$A = \begin{bmatrix} 0 & 1 & -1 \\ 2 & 0 & -1 \\ -1 & 3 & -3 \end{bmatrix}. \quad A^{-1} = \begin{bmatrix} 3 & 0 & -1 \\ 7 & -1 & -2 \\ 6 & -1 & -2 \end{bmatrix}$$

☑ CHECKPOINT EXERCISES
For use after Example 1:
1. Find the inverse of
$$A = \begin{bmatrix} 6 & 1 \\ -8 & -2 \end{bmatrix}. \quad A^{-1} = \begin{bmatrix} \frac{1}{2} & \frac{1}{4} \\ -2 & -\frac{3}{2} \end{bmatrix}$$

For use after Example 2:
2. Solve the matrix equation $AX = B$ for the 2×2 matrix X.
$$\begin{bmatrix} 1 & -2 \\ 4 & -7 \end{bmatrix} X = \begin{bmatrix} 7 & 2 \\ -1 & 9 \end{bmatrix} \quad \begin{bmatrix} -51 & 4 \\ -29 & 1 \end{bmatrix}$$

For use after Example 3:
3. Use a graphing calculator to find the inverse of
$$A = \begin{bmatrix} 1 & 2 & 3 \\ 3 & -1 & -2 \\ 3 & 1 & 1 \end{bmatrix}. \quad \begin{bmatrix} 1 & 1 & -1 \\ -9 & -8 & 11 \\ 6 & 5 & -7 \end{bmatrix}$$

STUDENT HELP

HOMEWORK HELP
Visit our Web site
www.mcdougallittell.com
for extra examples.

EXAMPLE 2 *Solving a Matrix Equation*

Solve the matrix equation $AX = B$ for the 2×2 matrix X.

$$\overset{A}{\begin{bmatrix} 4 & -1 \\ -3 & 1 \end{bmatrix}} X = \overset{B}{\begin{bmatrix} 8 & -5 \\ -6 & 3 \end{bmatrix}}$$

SOLUTION
Begin by finding the inverse of A.

$$A^{-1} = \frac{1}{4-3} \begin{bmatrix} 1 & 1 \\ 3 & 4 \end{bmatrix} = \begin{bmatrix} 1 & 1 \\ 3 & 4 \end{bmatrix}$$

To solve the equation for X, multiply both sides of the equation by A^{-1} *on the left.*

$$\begin{bmatrix} 1 & 1 \\ 3 & 4 \end{bmatrix} \begin{bmatrix} 4 & -1 \\ -3 & 1 \end{bmatrix} X = \begin{bmatrix} 1 & 1 \\ 3 & 4 \end{bmatrix} \begin{bmatrix} 8 & -5 \\ -6 & 3 \end{bmatrix} \qquad A^{-1}AX = A^{-1}B$$

$$\begin{bmatrix} 1 & 0 \\ 0 & 1 \end{bmatrix} X = \begin{bmatrix} 2 & -2 \\ 0 & -3 \end{bmatrix} \qquad IX = A^{-1}B$$

$$X = \begin{bmatrix} 2 & -2 \\ 0 & -3 \end{bmatrix} \qquad X = A^{-1}B$$

✓**CHECK** You can check the solution by multiplying A and X to see if you get B.

· · · · · · · · · ·

Some matrices do not have an inverse. You can tell whether a matrix has an inverse by evaluating its determinant. If $\det A = 0$, then A does not have an inverse. If $\det A \neq 0$, then A has an inverse.

The inverse of a 3×3 matrix is difficult to compute by hand. A calculator that will compute inverse matrices is useful in this case.

EXAMPLE 3 *Finding the Inverse of a 3 × 3 Matrix*

Use a graphing calculator to find the inverse of A. Then use the calculator to verify your result.

$$A = \begin{bmatrix} 1 & -1 & 0 \\ 1 & 0 & -1 \\ 6 & -2 & -3 \end{bmatrix}$$

SOLUTION
Enter the matrix A into the graphing calculator and calculate A^{-1}. Then compute AA^{-1} and $A^{-1}A$ to verify that you obtain the 3×3 identity matrix.

STUDENT HELP

KEYSTROKE HELP
Visit our Web site
www.mcdougallittell.com
to see keystrokes for
several models of
calculators.

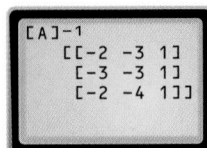

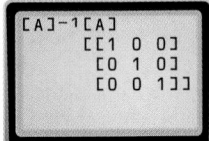

GOAL 2 USING INVERSE MATRICES IN REAL LIFE

A *cryptogram* is a message written according to a secret code. (The Greek word *kruptos* means *hidden* and the Greek word *gramma* means *letter*.) The following technique uses matrices to encode and decode messages.

First assign a number to each letter in the alphabet with 0 assigned to a blank space.

__ = 0	E = 5	J = 10	O = 15	T = 20	Y = 25
A = 1	F = 6	K = 11	P = 16	U = 21	Z = 26
B = 2	G = 7	L = 12	Q = 17	V = 22	
C = 3	H = 8	M = 13	R = 18	W = 23	
D = 4	I = 9	N = 14	S = 19	X = 24	

Then convert the message to numbers partitioned into 1×2 *uncoded row matrices*.

To *encode* a message, choose a 2×2 matrix A that has an inverse and multiply the uncoded row matrices by A *on the right* to obtain *coded row matrices*.

Cryptography

EXAMPLE 4 *Converting a Message*

Use the list above to convert the message GET HELP to row matrices.

SOLUTION

$$\begin{matrix} \text{G} & \text{E} & \text{T} & \_\_ & \text{H} & \text{E} & \text{L} & \text{P} \end{matrix}$$
$$[7\ \ 5][20\ \ 0][8\ \ 5][12\ \ 16]$$

FOCUS ON APPLICATIONS

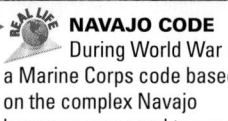

NAVAJO CODE
During World War II, a Marine Corps code based on the complex Navajo language was used to send messages.

APPLICATION LINK
www.mcdougallittell.com

EXAMPLE 5 *Encoding a Message*

CRYPTOGRAPHY Use $A = \begin{bmatrix} 2 & 3 \\ -1 & -2 \end{bmatrix}$ to encode the message GET HELP.

SOLUTION

The coded row matrices are obtained by multiplying each of the uncoded row matrices from Example 4 by the matrix A *on the right*.

UNCODED ROW MATRIX	ENCODING MATRIX A	CODED ROW MATRIX
$[7\ \ 5]$	$\begin{bmatrix} 2 & 3 \\ -1 & -2 \end{bmatrix}$	$= [9\ \ 11]$
$[20\ \ 0]$	$\begin{bmatrix} 2 & 3 \\ -1 & -2 \end{bmatrix}$	$= [40\ \ 60]$
$[8\ \ 5]$	$\begin{bmatrix} 2 & 3 \\ -1 & -2 \end{bmatrix}$	$= [11\ \ 14]$
$[12\ \ 16]$	$\begin{bmatrix} 2 & 3 \\ -1 & -2 \end{bmatrix}$	$= [8\ \ 4]$

▶ The coded message is 9, 11, 40, 60, 11, 14, 8, 4.

4.4 *Identity and Inverse Matrices* **225**

GRAPHING CALCULATOR NOTE
EXAMPLE 3 Students need to use the x^{-1} key on their graphing calculator to find the inverse of a matrix. Also, they cannot use the A key or any other letter to represent the matrix. They must press MATRX to type the letter of a matrix under the NAMES menu.

EXTRA EXAMPLE 4
Use the table on page 225 to convert the message SURPRISE to row matrices.
[19 21] [18 16] [18 9] [19 5]

EXTRA EXAMPLE 5
Use $A = \begin{bmatrix} 2 & 1 \\ -2 & 3 \end{bmatrix}$ to encode the message SURPRISE.
[−4 82] [4 66] [18 45] [28 34]

CHECKPOINT EXERCISES
For use after Example 4:
1. Use the table on page 225 to convert the message THANK YOU to row matrices. [20 8] [1 14] [11 0] [25 15] [21 0]

For use after Example 5:
2. Use $A = \begin{bmatrix} 1 & 4 \\ 2 & 6 \end{bmatrix}$ to encode the message matrix THANK YOU.
[36 128] [29 88] [11 44] [55 190] [21 84]

EXTRA EXAMPLE 6

Use the inverse of $A = \begin{bmatrix} -1 & 1 \\ 2 & -3 \end{bmatrix}$

and the table on page 225 to decode this message: −4, −1, −12, 7, −18, 18, 11, −20, −12, 7, 42, −63, −16, 16   NEVER GIVE UP

 CHECKPOINT EXERCISES

For use after Example 6:

1. Use the inverse of $A = \begin{bmatrix} 2 & -3 \\ 1 & -1 \end{bmatrix}$

and the table on page 225 to decode this message: 59, −81, 45, −65, 9, −9, 28, −42, 48, −68, 10, −15, 22, −27, 13, −18, 49, −69, 44, −59
VOTE IN THE ELECTION

FOCUS ON VOCABULARY
The *identity matrix* and the term identity are closely related and are used together in the lesson to help students understand the meaning of an identity matrix. The meaning of an inverse and an *inverse matrix* can be looked at in a similar way. An inverse matrix is used to get the identity matrix, much like working with inverse operations to get the multiplicative identity 1.

CLOSURE QUESTION
How are a square matrix, its identity matrix, and the inverse matrix related?
Sample answer: When a matrix is multiplied by its inverse matrix, the result is the identity matrix.

DAILY PUZZLER
Find all pairs of prime numbers whose sum equals 1999.   2 and 1997

4. *Sample answer:* In order for *AX* to be defined, the number of rows in *X* must equal the number of columns in *A*, so *X* has 2 rows. The number of columns in the product, *B*, must equal the number of columns in *X*, so *X* has 2 columns.

FOCUS ON PEOPLE

 **ALAN TURING,** an English mathematician, helped break codes used by the German military during World War II.

DECODING USING MATRICES Decoding the cryptogram created in Example 5 would be difficult for people who do not know the matrix *A*. When larger coding matrices are used, decoding is even more difficult. But for an authorized receiver who knows the matrix *A*, decoding is simple. The receiver only needs to multiply the coded row matrices by A^{-1} *on the right* to retrieve the uncoded row matrices.

EXAMPLE 6   *Decoding a Message*

CRYPTOGRAPHY Use the inverse of $A = \begin{bmatrix} 3 & -1 \\ -2 & 1 \end{bmatrix}$ to decode this message:

−4, 3, −23, 12, −26, 13, 15, −5, 31, −5, −38, 19, −21, 12, 20, 0, 75, −25

SOLUTION
First find A^{-1}: $A^{-1} = \dfrac{1}{3-2} \begin{bmatrix} 1 & 1 \\ 2 & 3 \end{bmatrix} = \begin{bmatrix} 1 & 1 \\ 2 & 3 \end{bmatrix}$

To decode the message, partition it into groups of two numbers to form coded row matrices. Then multiply each coded row matrix by A^{-1} *on the right* to obtain the uncoded row matrices.

CODED ROW MATRIX	DECODING MATRIX A^{-1}		UNCODED ROW MATRIX
$\begin{bmatrix} -4 & 3 \end{bmatrix}$	$\begin{bmatrix} 1 & 1 \\ 2 & 3 \end{bmatrix}$	=	$\begin{bmatrix} 2 & 5 \end{bmatrix}$
$\begin{bmatrix} -23 & 12 \end{bmatrix}$	$\begin{bmatrix} 1 & 1 \\ 2 & 3 \end{bmatrix}$	=	$\begin{bmatrix} 1 & 13 \end{bmatrix}$
$\begin{bmatrix} -26 & 13 \end{bmatrix}$	$\begin{bmatrix} 1 & 1 \\ 2 & 3 \end{bmatrix}$	=	$\begin{bmatrix} 0 & 13 \end{bmatrix}$
$\begin{bmatrix} 15 & -5 \end{bmatrix}$	$\begin{bmatrix} 1 & 1 \\ 2 & 3 \end{bmatrix}$	=	$\begin{bmatrix} 5 & 0 \end{bmatrix}$
$\begin{bmatrix} 31 & -5 \end{bmatrix}$	$\begin{bmatrix} 1 & 1 \\ 2 & 3 \end{bmatrix}$	=	$\begin{bmatrix} 21 & 16 \end{bmatrix}$
$\begin{bmatrix} -38 & 19 \end{bmatrix}$	$\begin{bmatrix} 1 & 1 \\ 2 & 3 \end{bmatrix}$	=	$\begin{bmatrix} 0 & 19 \end{bmatrix}$
$\begin{bmatrix} -21 & 12 \end{bmatrix}$	$\begin{bmatrix} 1 & 1 \\ 2 & 3 \end{bmatrix}$	=	$\begin{bmatrix} 3 & 15 \end{bmatrix}$
$\begin{bmatrix} 20 & 0 \end{bmatrix}$	$\begin{bmatrix} 1 & 1 \\ 2 & 3 \end{bmatrix}$	=	$\begin{bmatrix} 20 & 20 \end{bmatrix}$
$\begin{bmatrix} 75 & -25 \end{bmatrix}$	$\begin{bmatrix} 1 & 1 \\ 2 & 3 \end{bmatrix}$	=	$\begin{bmatrix} 25 & 0 \end{bmatrix}$

From the uncoded row matrices you can read the message as follows.

$\begin{bmatrix} 2 & 5 \end{bmatrix}\begin{bmatrix} 1 & 13 \end{bmatrix}\begin{bmatrix} 0 & 13 \end{bmatrix}\begin{bmatrix} 5 & 0 \end{bmatrix}\begin{bmatrix} 21 & 16 \end{bmatrix}\begin{bmatrix} 0 & 19 \end{bmatrix}\begin{bmatrix} 3 & 15 \end{bmatrix}\begin{bmatrix} 20 & 20 \end{bmatrix}\begin{bmatrix} 25 & 0 \end{bmatrix}$
  B E A M _ M E _ U P _ S C O T T Y _

GUIDED PRACTICE

Vocabulary Check ✓

Concept Check ✓

1. $\begin{bmatrix} 1 & 0 \\ 0 & 1 \end{bmatrix}$; $\begin{bmatrix} 1 & 0 & 0 \\ 0 & 1 & 0 \\ 0 & 0 & 1 \end{bmatrix}$

2. $AB = BA = \begin{bmatrix} 1 & 0 \\ 0 & 1 \end{bmatrix}$

Skill Check ✓

3. *Sample answer:* To find the inverse of a 2 × 2 matrix *A*, first calculate det *A* and check that it is not 0. Then switch the upper left and lower right entries of *A*. Negate the lower left and upper right entries of *A*. Multiply this new matrix by the reciprocal of the det *A*.

1. What is the identity matrix for 2×2 matrices? for 3×3 matrices?

2. For two 2×2 matrices A and B to be inverses of each other, what must be true of AB and BA?

3. Explain how to find the inverse of a 2×2 matrix.

4. How do you know that the matrix X in Example 2 must be 2×2? **See margin.**

5. If $B = \begin{bmatrix} 8 & -4 \\ -2 & 1 \end{bmatrix}$, does B have an inverse? Explain. **No; det $B = 0$**

Find the inverse of the matrix.

6. $\begin{bmatrix} -4 & 3 \\ -3 & 2 \end{bmatrix}$ $\begin{bmatrix} 2 & -3 \\ 3 & -4 \end{bmatrix}$

7. $\begin{bmatrix} -3 & 2 \\ 0 & -1 \end{bmatrix}$ $\begin{bmatrix} -\frac{1}{3} & -\frac{2}{3} \\ 0 & -1 \end{bmatrix}$

8. $\begin{bmatrix} -1 & 0 \\ 6 & 4 \end{bmatrix}$ $\begin{bmatrix} -1 & 0 \\ \frac{3}{2} & \frac{1}{4} \end{bmatrix}$

9. $\begin{bmatrix} \frac{1}{2} & 4 \\ -2 & \frac{1}{4} \end{bmatrix}$ **9–11. See margin.**

10. $\begin{bmatrix} 0.5 & 3 \\ 2.5 & 4 \end{bmatrix}$

11. $\begin{bmatrix} 1.6 & 2 \\ 3.2 & 0.2 \end{bmatrix}$

12. 🌐 **DECODING A MESSAGE** Use the coding information on pages 225 and 226 and the inverse of the matrix D to decode the following message. **GREETINGS**

$D = \begin{bmatrix} -5 & 3 \\ -2 & 1 \end{bmatrix}$ $-71, 39, -35, 20, -118, 69, -84, 49, -95, 57$

PRACTICE AND APPLICATIONS

STUDENT HELP

▶ **Extra Practice**
to help you master skills is on p. 945.

18. $\begin{bmatrix} -\frac{1}{15} & -\frac{2}{15} \\ -\frac{4}{15} & \frac{7}{15} \end{bmatrix}$

21. $\begin{bmatrix} \frac{1}{2} & \frac{1}{2} \\ \frac{3}{2} & \frac{11}{6} \end{bmatrix}$

FINDING INVERSES Find the inverse of the matrix.

13. $\begin{bmatrix} 4 & -5 \\ -3 & 4 \end{bmatrix}$ $\begin{bmatrix} 4 & 5 \\ 3 & 4 \end{bmatrix}$

14. $\begin{bmatrix} 6 & 2 \\ 8 & 3 \end{bmatrix}$ $\begin{bmatrix} \frac{3}{2} & -1 \\ -4 & 3 \end{bmatrix}$

15. $\begin{bmatrix} 1 & 8 \\ 1 & 7 \end{bmatrix}$ $\begin{bmatrix} -7 & 8 \\ 1 & -1 \end{bmatrix}$

16. $\begin{bmatrix} -6 & 17 \\ 1 & -3 \end{bmatrix}$ $\begin{bmatrix} -3 & -17 \\ -1 & -6 \end{bmatrix}$

17. $\begin{bmatrix} 7 & 2 \\ 3 & 1 \end{bmatrix}$ $\begin{bmatrix} 1 & -2 \\ -3 & 7 \end{bmatrix}$

18. $\begin{bmatrix} -7 & -2 \\ -4 & 1 \end{bmatrix}$

19. $\begin{bmatrix} -6 & -7 \\ 2 & 2 \end{bmatrix}$ $\begin{bmatrix} 1 & \frac{7}{2} \\ -1 & -3 \end{bmatrix}$

20. $\begin{bmatrix} 5 & -4 \\ -4 & 4 \end{bmatrix}$ $\begin{bmatrix} 1 & 1 \\ 1 & \frac{5}{4} \end{bmatrix}$

21. $\begin{bmatrix} 11 & -3 \\ -9 & 3 \end{bmatrix}$

22. $\begin{bmatrix} \frac{3}{2} & 1 \\ -2 & 1 \end{bmatrix}$ $\begin{bmatrix} \frac{2}{5} & -\frac{1}{5} \\ \frac{4}{5} & \frac{3}{5} \end{bmatrix}$

23. $\begin{bmatrix} 2.2 & 2.5 \\ 8 & 10 \end{bmatrix}$ $\begin{bmatrix} 5 & -1.25 \\ -4 & 1.1 \end{bmatrix}$

24. $\begin{bmatrix} \frac{4}{5} & \frac{3}{4} \\ -1 & \frac{5}{2} \end{bmatrix}$ $\begin{bmatrix} \frac{10}{11} & -\frac{3}{11} \\ \frac{4}{11} & \frac{16}{55} \end{bmatrix}$

SOLVING EQUATIONS Solve the matrix equation. **25–32. See margin.**

25. $\begin{bmatrix} -5 & -13 \\ 0 & 5 \end{bmatrix} X = \begin{bmatrix} 3 & 1 \\ -4 & 0 \end{bmatrix}$

26. $\begin{bmatrix} 5 & -1 \\ 8 & 2 \end{bmatrix} X = \begin{bmatrix} 17 & 20 \\ 26 & 20 \end{bmatrix}$

27. $\begin{bmatrix} 2 & 4 \\ 0 & 1 \end{bmatrix} X = \begin{bmatrix} 4 & 0 & 6 \\ 3 & -1 & 5 \end{bmatrix}$

28. $\begin{bmatrix} -5 & -3 \\ 4 & 1 \end{bmatrix} X = \begin{bmatrix} -12 & -5 & 18 \\ 4 & -3 & -13 \end{bmatrix}$

29. $\begin{bmatrix} 3 & 7 \\ 1 & 4 \end{bmatrix} X + \begin{bmatrix} 8 & 5 \\ 1 & 15 \end{bmatrix} = \begin{bmatrix} 7 & -3 \\ -2 & -9 \end{bmatrix}$

30. $\begin{bmatrix} -7 & -9 \\ 4 & 5 \end{bmatrix} X + \begin{bmatrix} 3 & 4 \\ 4 & -3 \end{bmatrix} = \begin{bmatrix} 1 & 9 \\ 6 & -6 \end{bmatrix}$

31. $\begin{bmatrix} -1 & 2 \\ -4 & 6 \end{bmatrix} X - \begin{bmatrix} 2 & 1 \\ 3 & 0 \end{bmatrix} = \begin{bmatrix} 3 & -2 \\ 1 & -1 \end{bmatrix}$

32. $\begin{bmatrix} 4 & -3 \\ 6 & -2 \end{bmatrix} X - \begin{bmatrix} -1 & 1 \\ 5 & 7 \end{bmatrix} = \begin{bmatrix} 4 & 6 \\ 8 & 2 \end{bmatrix}$

STUDENT HELP

▶ HOMEWORK HELP
Example 1: Exs. 13–24, 33
Example 2: Exs. 25–32
Example 3: Exs. 34–39
Examples 4, 5: Exs. 40–43
Example 6: Exs. 44–48

4.4 *Identity and Inverse Matrices* **227**

---

3 APPLY

ASSIGNMENT GUIDE

BASIC
Day 1: pp. 227–229 Exs. 14–20 even, 26–28, 33, 34, 37, 41, 51, 52, 55–63 odd

AVERAGE
Day 1: pp. 227–229 Exs. 14–22 even, 26–29, 33, 37, 40–44, 51, 52, 55–65 odd

ADVANCED
Day 1: pp. 227–229 Exs. 14–30 even, 33–45 odd, 49–53, 55–65 odd

BLOCK SCHEDULE
pp. 227–229 Exs. 14–22 even, 26–29, 33, 37, 40–44, 51, 52, 55–65 odd (with 4.5)

EXERCISE LEVELS
Level A: *Easier*
14–22, 25–28, 33, 37
Level B: *More Difficult*
23–24, 29–32, 35–36, 38–47
Level C: *Most Difficult*
48–50, 53

✔ **HOMEWORK CHECK**
To quickly check student understanding of key concepts, go over the following exercises:
Exs. 14, 20, 26, 33, 37. See also the Daily Homework Quiz:
• Blackline Master (*Chapter 4 Resource Book*, p. 67)
• 📻 Transparency (p. 30)

9–11, 25–32. See next page.

GRAPHING CALCULATOR NOTE

EXERCISES 37–39 Remind students that the 3×3 identity matrix has 1s along the main diagonal and 0s everywhere else. Students can also use the fraction option under the MATH key to put the entries in a matrix into fraction form.

9. $\begin{bmatrix} \frac{2}{65} & -\frac{32}{65} \\ \frac{16}{65} & \frac{4}{65} \end{bmatrix}$ 10. $\begin{bmatrix} -0.7\overline{2} & 0.5\overline{4} \\ 0.\overline{45} & -0.\overline{09} \end{bmatrix}$

11. $\begin{bmatrix} -0.0329 & 0.3289 \\ 0.5263 & -0.2632 \end{bmatrix}$

25. $\begin{bmatrix} \frac{37}{25} & -\frac{1}{5} \\ -\frac{4}{5} & 0 \end{bmatrix}$ 26. $\begin{bmatrix} \frac{10}{3} & \frac{10}{3} \\ -\frac{1}{3} & -\frac{10}{3} \end{bmatrix}$

27. $\begin{bmatrix} -4 & 2 & -7 \\ 3 & -1 & 5 \end{bmatrix}$ 28. $\begin{bmatrix} 0 & -2 & -3 \\ 4 & 5 & -1 \end{bmatrix}$

29. $\begin{bmatrix} \frac{17}{5} & \frac{136}{5} \\ -\frac{8}{5} & \frac{64}{5} \end{bmatrix}$ 30. $\begin{bmatrix} 8 & -2 \\ -6 & 1 \end{bmatrix}$

31. $\begin{bmatrix} 11 & -2 \\ 8 & -1.5 \end{bmatrix}$ 32. $\begin{bmatrix} \frac{33}{10} & \frac{13}{10} \\ \frac{17}{5} & \frac{3}{5} \end{bmatrix}$

37. $\begin{bmatrix} -0.0654 & -0.0131 & 0.1634 \\ 0.0131 & 0.2026 & -0.0327 \\ 0.1503 & -0.1699 & 0.1242 \end{bmatrix}$

ADDITIONAL PRACTICE AND RETEACHING

For Lesson 4.4:

• Practice Levels A, B, and C (*Chapter 4 Resource Book,* p. 57)

• Reteaching with Practice (*Chapter 4 Resource Book,* p. 60)

• ⊞ See Lesson 4.4 of the *Personal Student Tutor*

For more Mixed Review:

• ⊞ Search the *Test and Practice Generator* for key words or specific lessons.

228

---

38. $\begin{bmatrix} 0 & -3 & -1 \\ 0.5 & -21.5 & -7.5 \\ -0.167 & 3.5 & 1.167 \end{bmatrix}$

39. $\begin{bmatrix} 12 & -7 & 3 \\ -20 & 12 & -5 \\ 1.5 & -1 & 0.5 \end{bmatrix}$

40. $-20, 55, 2, -4, 13, -21, -12, 36, -8, 20, -13, 40, 5, -10$

41. $39, 98, 26, 77, 20, 60, 13, 31, 23, 51$

42. $-33, 12, 37, -8, -60, 20, 60, -15, -11, 5, 14, 0, -34, 13$

43. $36, -14, 16, 0, 125, -50, -26, 14, 10, 4, 24, -8, -95, 48$

49a. $\begin{bmatrix} -1 & -4 & -2 \\ 1 & 2 & 3 \end{bmatrix}$;

$\begin{bmatrix} -1 & -2 & -3 \\ -1 & -4 & -2 \end{bmatrix}$;

90° rotation; See margin for graph.

STUDENT HELP

▶ **Skills Review**
For help with transformations, see p. 921.

49b. *Sample answer:* Find A^{-1} and then multiply AAT by A^{-1} on the left: $A^{-1}AAT = IAT = AT$. Now multiply AT by A^{-1} on the left: $A^{-1}AT = IT = T$

IDENTIFYING INVERSES Tell whether the matrices are inverses of each other.

33. $\begin{bmatrix} 10 & -3 \\ 3 & -1 \end{bmatrix}$ and $\begin{bmatrix} 1 & 3 \\ 3 & -10 \end{bmatrix}$ **no**

34. $\begin{bmatrix} 0 & 2 & -1 \\ 5 & 2 & 3 \\ 7 & 3 & 4 \end{bmatrix}$ and $\begin{bmatrix} -2 & -10 & 8 \\ 11 & 7 & -5 \\ 1 & 12 & -10 \end{bmatrix}$ **no**

35. $\begin{bmatrix} 11 & 2 & -8 \\ 4 & 1 & -3 \\ -8 & -1 & 6 \end{bmatrix}$ and $\begin{bmatrix} 3 & -4 & 2 \\ 0 & 2 & 1 \\ 4 & -5 & 3 \end{bmatrix}$ **yes**

36. $\begin{bmatrix} 10 & 2 & -25 \\ 4 & 1 & -10 \\ -9 & -2 & 23 \end{bmatrix}$ and $\begin{bmatrix} 3 & 4 & 5 \\ -2 & 5 & 0 \\ 1 & 2 & 2 \end{bmatrix}$ **yes**

▦ **FINDING INVERSES** Use a graphing calculator to find the inverse of the matrix *A*. Check the result by showing that $AA^{-1} = I$ and $A^{-1}A = I$.

37. $A = \begin{bmatrix} -3 & 4 & 5 \\ 1 & 5 & 0 \\ 5 & 2 & 2 \end{bmatrix}$ 38. $A = \begin{bmatrix} -7 & 0 & -6 \\ -4 & 1 & 3 \\ 11 & -3 & -9 \end{bmatrix}$ 39. $A = \begin{bmatrix} 2 & 1 & -2 \\ 5 & 3 & 0 \\ 4 & 3 & 8 \end{bmatrix}$

See margin.

ENCODING Use the code on page 225 and the matrix to encode the message.

40. JOB WELL DONE

$A = \begin{bmatrix} 1 & -2 \\ -2 & 5 \end{bmatrix}$

41. STAY THERE

$A = \begin{bmatrix} 1 & 2 \\ 1 & 3 \end{bmatrix}$

42. COME TO DINNER

$A = \begin{bmatrix} 4 & -1 \\ -3 & 1 \end{bmatrix}$

43. HAPPY BIRTHDAY

$A = \begin{bmatrix} 5 & -2 \\ -4 & 2 \end{bmatrix}$

🌐 **TRAVEL** In Exercises 44–48, use the following information.

Your friend is traveling abroad and is sending you postcards with encoded messages. You must decipher what landmarks your friend has visited. Use the inverse of matrix *D* to decode each message. Each message represents a landmark in the country where your friend is traveling. Use the coding information on pages 225 and 226 to help you.

$$D = \begin{bmatrix} 2 & -3 \\ -1 & 2 \end{bmatrix}$$

44. $-1, 4, 30, -41, 39, -58, 22, -33, 31, -46, 23, -34, 1, 1$ **BESHTAK PALACE**

45. $21, -31, 22, -26, -9, 19, -20, 40, -3, 11, 20, -24, 10, -15$ **KARNAK TEMPLE**

46. $39, -58, -2, 12, 0, 9, -19, 38, 13, -9, -16, 33, 10, -15$ **TAHRIR SQUARE**

47. $32, -44, 10, -15, -4, 15, 9, -13, 40, -60, 22, -25, 7, -6, 4, 6$ **THE GREAT SPHINX**

48. Using the decoded messages, tell what country your friend is visiting. **Egypt**

49. GEOMETRY ▶ CONNECTION Use the matrices shown. The columns of matrix *T* give the coordinates of the vertices of a triangle. Matrix *A* is a transformation matrix.

$$A = \begin{bmatrix} 0 & -1 \\ 1 & 0 \end{bmatrix} \qquad T = \begin{bmatrix} 1 & 2 & 3 \\ 1 & 4 & 2 \end{bmatrix}$$

a. Find *AT* and *AAT*. Then draw the original triangle and the two transformed triangles. What transformation does *A* represent? **See margin.**

b. Suppose you start with the triangle determined by *AAT* and want to reverse the transformation process to produce the triangle determined by *AT* and then the triangle determined by *T*. Describe how you can do this.

50. *Writing* Describe the process used to solve a matrix equation.   **See margin.**

50. *Sample answer:* Write the equation in the form $AX = B$. Find A^{-1} and multiply both sides of the equation by A^{-1} on the left.

51. **MULTIPLE CHOICE** What is the inverse of $\begin{bmatrix} -2 & -2 \\ 7 & 6 \end{bmatrix}$?   **D**

Ⓐ $\begin{bmatrix} \frac{3}{13} & -\frac{1}{13} \\ \frac{7}{26} & -\frac{1}{13} \end{bmatrix}$ Ⓑ $\begin{bmatrix} -3 & 1 \\ -\frac{7}{2} & 1 \end{bmatrix}$ Ⓒ $\begin{bmatrix} \frac{3}{13} & -\frac{1}{13} \\ -\frac{7}{26} & \frac{1}{13} \end{bmatrix}$ Ⓓ $\begin{bmatrix} 3 & 1 \\ -\frac{7}{2} & -1 \end{bmatrix}$

52. **MULTIPLE CHOICE** What is the solution of $\begin{bmatrix} 5 & 2 \\ 3 & 1 \end{bmatrix} X = \begin{bmatrix} 4 & 43 \\ 2 & 25 \end{bmatrix}$?   **A**

Ⓐ $\begin{bmatrix} 0 & 7 \\ 2 & 4 \end{bmatrix}$ Ⓑ $\begin{bmatrix} 0 & 7 \\ -2 & 4 \end{bmatrix}$ Ⓒ $\begin{bmatrix} 0 & 2 \\ 7 & 4 \end{bmatrix}$ Ⓓ $\begin{bmatrix} 0 & 4 \\ 2 & 7 \end{bmatrix}$ Ⓔ $\begin{bmatrix} 0 & 7 \\ 4 & 2 \end{bmatrix}$

★ **Challenge**

53. 🌐 **CODE BREAKER** You are a code breaker and intercept the encoded message 45, −35, 38, −30, 18, −18, 35, −30, 81, −60, 42, −28, 75, −55, 2, −2, 22, −21, 15, −10 that you know is being sent to someone named John. You can conclude that $\begin{bmatrix} 45 & -35 \end{bmatrix} A^{-1} = \begin{bmatrix} 10 & 15 \end{bmatrix}$ and $\begin{bmatrix} 38 & -30 \end{bmatrix} A^{-1} = \begin{bmatrix} 8 & 14 \end{bmatrix}$ where A^{-1} is the inverse of the encoding matrix A, 10 represents J, 15 represents O, 8 represents H, and 14 represents N.

Let $A^{-1} = \begin{bmatrix} w & x \\ y & z \end{bmatrix}$.

$w = 1, x = -2, y = 1, z = -3$

a. Write and solve two systems of equations to find w, x, y, and z.

b. Find A^{-1}, and decode the rest of the message. $\begin{bmatrix} 1 & -2 \\ 1 & -3 \end{bmatrix}$; **RETURN TO BASE**

 EXTRA CHALLENGE
➜ www.mcdougallittell.com

MIXED REVIEW

SOLVING SYSTEMS Solve the system of linear equations using any algebraic method. (Review 3.2, 3.6 for 4.5)

54. $3x + 5y = 12$ $(-1, 3)$
    $x + 4y = 11$

55. $4x - 12y = 2$ **all real numbers**
    $-2x + 6y = -1$

56. $-5x + 7y = 33$ $(-1, 4)$
    $4x - 9y = -40$

57. $7x + y + 3z = 22$
    $2x - 2y + 9z = -10$
    $-3x - 5y - 10z = 8$
    $(4, 0, -2)$

58. $x + 3z = 6$ $(3, -2, 1)$
    $-2x + 3y + z = -11$
    $3x - y + 2z = 13$

59. $2x + y - 4z = 4$ $\left(\frac{1}{2}, 4, \frac{1}{4}\right)$
    $4x - 3y + 8z = -8$
    $-2x + 7y - 12z = 24$

MATRIX OPERATIONS Perform the indicated operation, if possible. If not possible, state the reason. (Review 4.1)

61. Not possible; the matrices have different dimensions.

63. $\begin{bmatrix} 17 & -3 & -1 \\ 0 & 25 & 31 \end{bmatrix}$

64. $\begin{bmatrix} 9 & 2 & -2 \\ -5 & 13 & -5 \end{bmatrix}$

60. $\begin{bmatrix} -4 & 2 \\ -5 & -1 \end{bmatrix} + \begin{bmatrix} 4 & -3 \\ -2 & 0 \end{bmatrix} \begin{bmatrix} 0 & -1 \\ -7 & -1 \end{bmatrix}$

61. $\begin{bmatrix} 8 & -6 \\ -2 & 5 \end{bmatrix} - \begin{bmatrix} -2 & -3 & -5 \\ 2 & -1 & 3 \end{bmatrix}$

62. $-8\begin{bmatrix} -1 & 3 & 4 \\ -6 & 8 & 0 \end{bmatrix} \begin{bmatrix} 8 & -24 & -32 \\ 48 & -64 & 0 \end{bmatrix}$

63. $\begin{bmatrix} 7 & -5 & 8 \\ -9 & 13 & 16 \end{bmatrix} - \begin{bmatrix} -10 & -2 & 9 \\ -9 & -12 & -15 \end{bmatrix}$

64. $\begin{bmatrix} 6 & -2 & -1 \\ -3 & 5 & 4 \end{bmatrix} + \begin{bmatrix} 3 & 4 & -1 \\ -2 & 8 & -9 \end{bmatrix}$

65. $\frac{1}{2}\begin{bmatrix} 4 & 10 & 2 \\ 6 & 8 & 16 \end{bmatrix} \begin{bmatrix} 2 & 5 & 1 \\ 3 & 4 & 8 \end{bmatrix}$

66. 🌐 **CATERING** You are in charge of catering for a school function. To limit the cost, you will serve only two entrees. One is a vegetarian dish that costs $6 and the other is a chicken dish that costs $8. If there will be 150 people at the function and your budget for the food is $1000, how many of each type of entree will be served? (Review 3.1, 3.2)   **100 vegetarian, 50 chicken**

DAILY HOMEWORK QUIZ

🖑 *Transparency Available*

1. Find the inverse of

$\begin{bmatrix} 8 & -2 \\ -6 & 1 \end{bmatrix}$. $\begin{bmatrix} -\frac{1}{4} & -\frac{1}{2} \\ -\frac{3}{2} & -2 \end{bmatrix}$

2. Solve the matrix equation.

$\begin{bmatrix} 4 & 2 \\ 5 & 2 \end{bmatrix} X = \begin{bmatrix} 2 & -2 \\ 1 & 1 \end{bmatrix} \begin{bmatrix} -1 & 3 \\ 3 & -7 \end{bmatrix}$

3. Use the coding information on page 225 and the inverse of matrix D to decode the following message.

$D = \begin{bmatrix} -4 & 2 \\ 3 & -2 \end{bmatrix}$

−37, 12, 0, −6, −20, 10   **SMILE**

EXTRA CHALLENGE NOTE
➜ Challenge problems for Lesson 4.4 are available in **blackline** format in the *Chapter 4 Resource Book*, p. 64 and at **www.mcdougallittell.com.**

ADDITIONAL TEST PREPARATION

1. **OPEN ENDED** If the inverse of matrix $\begin{bmatrix} a & b \\ c & d \end{bmatrix}$ is $\frac{1}{ad - bc}\begin{bmatrix} d & -b \\ -c & a \end{bmatrix}$, when will the matrix not have an inverse? **when $ad - bc = 0$**

49a.
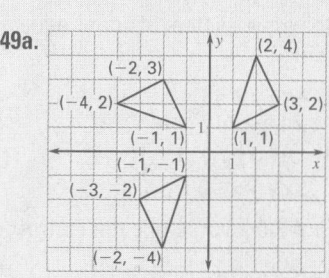

LESSON OPENER
APPLICATION
An alternative way to approach Lesson 4.5 is to use the Application Lesson Opener:

• Blackline Master (*Chapter 4 Resource Book,* p. 68)
• Transparency (p. 26)

MEETING INDIVIDUAL NEEDS
• *Chapter 4 Resource Book*
  Prerequisite Skills Review (p. 5)
  Practice Level A (p. 70)
  Practice Level B (p. 71)
  Practice Level C (p. 72)
  Reteaching with Practice (p. 73)
  Absent Student Catch-Up (p. 75)
  Challenge (p. 78)
• *Resources in Spanish*
• 🖥 *Personal Student Tutor*

NEW-TEACHER SUPPORT
See the Tips for New Teachers on pp. 1–2 of the *Chapter 4 Resource Book* for additional notes about Lesson 4.5.

WARM-UP EXERCISES

🖥 **Transparency Available**

Solve the system of linear equations.

1. $4x - 2y = 8$ (4, 4)
    $x + 2y = 12$

2. $-3x - 2y = 12$ (−2, −3)
    $3x + y = -9$

Find the inverse of each matrix.

3. $\begin{bmatrix} 4 & -5 \\ 0 & 8 \end{bmatrix}$ $\begin{bmatrix} \frac{1}{4} & \frac{5}{32} \\ 0 & \frac{1}{8} \end{bmatrix}$

4. $\begin{bmatrix} 2 & 1 \\ 9 & 5 \end{bmatrix}$ $\begin{bmatrix} 5 & -1 \\ -9 & 2 \end{bmatrix}$

What you should learn

GOAL 1 Solve systems of linear equations using inverse matrices.

GOAL 2 Use systems of linear equations to solve **real-life** problems, such as determining how much money to invest in **Example 4.**

Why you should learn it

▼ To solve **real-life** problems, such as planning a stained glass project in **Ex. 42.**

Step 1. a linear system:
$$5x - 4y = 8$$
$$x + 2y = 6$$

Step 2.
$$\begin{bmatrix} 2 & -1 \\ -4 & 9 \end{bmatrix}\begin{bmatrix} x \\ y \end{bmatrix} = \begin{bmatrix} -4 \\ 1 \end{bmatrix}$$

4.5 Solving Systems Using Inverse Matrices

GOAL 1 SOLVING SYSTEMS USING MATRICES

In Lesson 4.3 you learned how to solve a system of linear equations using Cramer's rule. Here you will learn to solve a system using inverse matrices.

▶ **ACTIVITY**

Developing Concepts

Investigating Matrix Equations

Steps 1 and 2. See margin.

1 Write the left side of the matrix equation as a single matrix. Then equate corresponding entries of the matrices. What do you obtain?

$$\begin{bmatrix} 5 & -4 \\ 1 & 2 \end{bmatrix}\begin{bmatrix} x \\ y \end{bmatrix} = \begin{bmatrix} 8 \\ 6 \end{bmatrix}$$ **Matrix equation**

2 Use what you learned in **Step 1** to write the following linear system as a matrix equation.

$$2x - y = -4 \qquad \text{Equation 1}$$
$$-4x + 9y = 1 \qquad \text{Equation 2}$$

In the activity you learned that a linear system can be written as a matrix equation $AX = B$. The matrix A is the coefficient matrix of the system, X is the **matrix of variables,** and B is the **matrix of constants**.

EXAMPLE 1 *Writing a Matrix Equation*

Write the system of linear equations as a matrix equation.

$$-3x + 4y = 5 \qquad \text{Equation 1}$$
$$2x - y = -10 \qquad \text{Equation 2}$$

SOLUTION

$$\overset{A}{\begin{bmatrix} -3 & 4 \\ 2 & -1 \end{bmatrix}} \overset{X}{\begin{bmatrix} x \\ y \end{bmatrix}} = \overset{B}{\begin{bmatrix} 5 \\ -10 \end{bmatrix}}$$

· · · · · · · · · ·

Once you have written a linear system as $AX = B$, you can solve for X by multiplying each side of the matrix by A^{-1} *on the left.*

$AX = B$	Write original matrix equation.
$A^{-1}AX = A^{-1}B$	Multiply each side by A^{-1}.
$IX = A^{-1}B$	$A^{-1}A = I$
$X = A^{-1}B$	$IX = X$

SOLUTION OF A LINEAR SYSTEM Let $AX = B$ represent a system of linear equations. If the determinant of A is nonzero, then the linear system has exactly one solution, which is $X = A^{-1}B$.

STUDENT HELP

Look Back
For help with systems of linear equations, see p. 150.

EXAMPLE 2 *Solving a Linear System*

Use matrices to solve the linear system in Example 1.

$$-3x + 4y = 5 \qquad \text{Equation 1}$$
$$2x - y = -10 \qquad \text{Equation 2}$$

SOLUTION

Begin by writing the linear system in matrix form, as in Example 1. Then find the inverse of matrix A.

$$A^{-1} = \frac{1}{3-8}\begin{bmatrix} -1 & -4 \\ -2 & -3 \end{bmatrix} = \begin{bmatrix} \frac{1}{5} & \frac{4}{5} \\ \frac{2}{5} & \frac{3}{5} \end{bmatrix}$$

Finally, multiply the matrix of constants by A^{-1}.

$$X = A^{-1}B = \begin{bmatrix} \frac{1}{5} & \frac{4}{5} \\ \frac{2}{5} & \frac{3}{5} \end{bmatrix}\begin{bmatrix} 5 \\ -10 \end{bmatrix} = \begin{bmatrix} -7 \\ -4 \end{bmatrix} = \begin{bmatrix} x \\ y \end{bmatrix}$$

▶ The solution of the system is $(-7, -4)$. Check this solution in the original equations.

EXAMPLE 3 *Using a Graphing Calculator*

 Use a matrix equation and a graphing calculator to solve the linear system.

$$2x + 3y + z = -1 \qquad \text{Equation 1}$$
$$3x + 3y + z = 1 \qquad \text{Equation 2}$$
$$2x + 4y + z = -2 \qquad \text{Equation 3}$$

STUDENT HELP

Study Tip
Remember that you can use the method shown in Examples 2 and 3 provided A has an inverse. If A does not have an inverse, then the system has either no solution or infinitely many solutions, and you should use a different technique.

SOLUTION

The matrix equation that represents the system is $\begin{bmatrix} 2 & 3 & 1 \\ 3 & 3 & 1 \\ 2 & 4 & 1 \end{bmatrix}\begin{bmatrix} x \\ y \\ z \end{bmatrix} = \begin{bmatrix} -1 \\ 1 \\ -2 \end{bmatrix}$.

Using a graphing calculator, you can solve the system as shown.

```
MATRIX [A] 3X3
[2    3    1    ]
[3    3    1    ]
[2    4    1    ]
```
Enter matrix A.

```
MATRIX [B] 3X1
[-1         ]
[1          ]
[-2         ]
```
Enter matrix B.

```
[A]-1[B]
         [[2 ]
          [-1]
          [-2]]
```
Multiply B by A⁻¹.

▶ The solution is $(2, -1, -2)$. Check this solution in the original equations.

EXTRA EXAMPLE 1
Write the system of linear equations as a matrix equation.
$-2x - 5y = -19$
$3x + 2y = 1$
$\begin{bmatrix} -2 & -5 \\ 3 & 2 \end{bmatrix}\begin{bmatrix} x \\ y \end{bmatrix} = \begin{bmatrix} -19 \\ 1 \end{bmatrix}$

EXTRA EXAMPLE 2
Use matrices to solve the linear system in Extra Example 1.
$(-3, 5)$

EXTRA EXAMPLE 3
Use a matrix equation and a graphing calculator to solve the linear system.
$x + y + 2z = 3$
$2x - y + 3z = -4$
$4x - 3y - z = -18$ $(-2, 3, 1)$

CHECKPOINT EXERCISES
For use after Examples 1 and 2:
1. Write and solve the system of linear equations using matrices.
$2x + y = -13$
$x - 3y = 11$
$\begin{bmatrix} 2 & 1 \\ 1 & -3 \end{bmatrix}\begin{bmatrix} x \\ y \end{bmatrix} = \begin{bmatrix} -13 \\ 11 \end{bmatrix}$; $(-4, -5)$

For use after Example 3:
2. Use a matrix equation and a graphing calculator to solve the linear system.
$2x + 3y - z = 14$
$4x + 5y + 2z = 34$
$-x + 3y - 4z = 20$ $(-6, 10, 4)$

GRAPHING CALCULATOR NOTE
EXAMPLE 3 After a matrix has been entered into a graphing calculator, the key sequence of `2nd` `QUIT` must be entered to get out of the edit mode for matrices. Then students can find $A^{-1}B$.

EXTRA EXAMPLE 4
You have $8,000. You want to invest the money in a stock mutual fund, a bond mutual fund, and a money market fund. The expected annual returns for these funds are given in the table in Example 4. You want your investments to obtain an overall annual return of $8\frac{1}{4}$%.

A financial planner recommends that you invest 4 times as much in stocks as you do in a money market fund, and the same amount in stocks as in the money market fund and bonds combined. How much should you invest in each fund?
$4000 in stocks; $3000 in bonds; and $1000 in a money market

 CHECKPOINT EXERCISES
For use after Example 4:
1. Use the conditions of Extra Example 4 with $20,000 to invest. $10,000 in stocks; $7500 in bonds; and $2500 in a money market

APPLICATION NOTE
EXAMPLE 4 Additional information about investing is available at **www.mcdougallittell.com.**

FOCUS ON VOCABULARY
Students should understand the difference between the three matrices discussed in this lesson: the *coefficient matrix*, the *matrix of variables*, and the *matrix of constants*. After students have completed the activity, ask them to identify the three types of matrices in Examples 1–4.

CLOSURE QUESTION
How can you use inverse matrices to solve a system of equations?
See margin.

DAILY PUZZLER
What is the sum of the numerator and the denominator of the simplified fraction form of 0.06818181... 47

INVESTING
Each year students across the country in grades 4 through 12 invest a hypothetical $100,000 in stocks to compete in the Stock Market Game. Students can enter their transactions using the internet.

APPLICATION LINK
www.mcdougallittell.com

GOAL 2 USING LINEAR SYSTEMS IN REAL LIFE

EXAMPLE 4 *Writing and Using a Linear System*

INVESTING You have $10,000 to invest. You want to invest the money in a stock mutual fund, a bond mutual fund, and a money market fund. The expected annual returns for these funds are given in the table.

You want your investment to obtain an overall annual return of 8%. A financial planner recommends that you invest the same amount in stocks as in bonds and the money market combined. How much should you invest in each fund?

Investment	Expected return
Stock mutual fund	10%
Bond mutual fund	7%
Money market (MM) fund	5%

SOLUTION

VERBAL MODEL

$$\boxed{\text{Stock amount}} + \boxed{\text{Bond amount}} + \boxed{\text{MM amount}} = \boxed{\text{Total invested}}$$

$$0.10\boxed{\text{Stock amount}} + 0.07\boxed{\text{Bond amount}} + 0.05\boxed{\text{MM amount}} = 0.08\boxed{\text{Total invested}}$$

$$\boxed{\text{Stock amount}} = \boxed{\text{Bond amount}} + \boxed{\text{MM amount}}$$

LABELS

Stock amount = s Money market amount = m

Bond amount = b Total invested = **10,000**

ALGEBRAIC MODEL

$s + b + m = 10{,}000$ Equation 1

$0.10\,s + 0.07\,b + 0.05\,m = 0.08\,(10{,}000)$ Equation 2

$s = b + m$ Equation 3

First rewrite the equations above in standard form and then in matrix form.

$$\begin{aligned} s + b + m &= 10{,}000 \\ 0.10s + 0.07b + 0.05m &= 800 \\ s - b - m &= 0 \end{aligned} \qquad \begin{bmatrix} 1 & 1 & 1 \\ 0.1 & 0.07 & 0.05 \\ 1 & -1 & -1 \end{bmatrix} \begin{bmatrix} s \\ b \\ m \end{bmatrix} = \begin{bmatrix} 10{,}000 \\ 800 \\ 0 \end{bmatrix}$$

Enter the coefficient matrix A and the matrix of constants B into a graphing calculator. Then find the solution $X = A^{-1}B$.

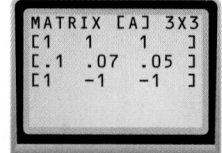
```
MATRIX [A] 3X3
[1     1    1    ]
[.1   .07  .05  ]
[1    -1   -1   ]
```

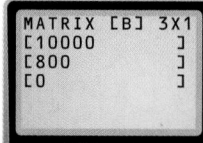
```
MATRIX [B] 3X1
[10000        ]
[800          ]
[0            ]
```

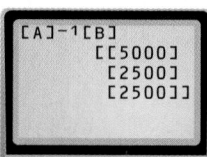
```
[A]⁻¹[B]
       [[5000]
        [2500]
        [2500]]
```

▶ You should invest $5000 in the stock mutual fund, $2500 in the bond mutual fund, and $2500 in the money market fund.

Closure Question *Sample answer:*
Write the system as a single matrix equation. Find the inverse of the coefficient matrix and multiply it with the matrix of constants to find the solution.

GUIDED PRACTICE

Vocabulary Check ✓

Concept Check ✓

Skill Check ✓

1. *Sample answer:* A matrix of variables is a column matrix containing only the variables of the equations in a linear system. A constant matrix is a column matrix containing only the constant terms of the equations in a linear system. To solve a linear system that has been written as a matrix equation, solve for the matrix of variables by multiplying (on the left) the matrix of constants by the inverse of the coefficient matrix.

1. What are a matrix of variables and a matrix of constants, and how are they used to solve a system of linear equations? **See margin.**

2. If $|A| \neq 0$, what is the solution of $AX = B$ in terms of A and B? $X = A^{-1}B$

3. Explain why the solution of $AX = B$ is *not* $X = BA^{-1}$. **See margin.**

Write the linear system as a matrix equation.
4–6. See margin.

4. $x + y = 8$
   $2x - y = 6$

5. $x + 3y = 9$
   $4x - 2y = 7$

6. $x + y + z = 10$
   $5x - y = 1$
   $3x + 4y + z = 8$

Use an inverse matrix to solve the linear system.

7. $x + y = 2$ $(-5, 7)$
   $7x + 8y = 21$

8. $-x - 2y = 3$ $\left(-\frac{13}{2}, \frac{7}{4}\right)$
   $2x + 8y = 1$

9. $4x + 3y = 6$ $\left(\frac{21}{13}, -\frac{2}{13}\right)$
   $6x - 2y = 10$

10. 🌐 **INVESTING** Look back at Example 4 on page 232. Suppose you have $60,000 to invest and you want an overall annual return of 9%. Use the expected annual returns shown to determine how much you should invest in each fund. Assume you are investing as much in stocks as in bonds and the money market combined.

Investment	Expected return
Stock mutual fund	12%
Bond mutual fund	8%
Money market fund	5%

stock mutual fund: $30,000, bond mutual fund: $10,000, money market fund: $20,000

PRACTICE AND APPLICATIONS

STUDENT HELP

▶ **Extra Practice**
to help you master skills is on p. 945.

3. *Sample answer:* Because matrix multiplication is not commutative, AXA^{-1} cannot be simplified to $AA^{-1}X = X$; therefore, the matrix equation cannot be solved by multiplying both sides by A^{-1} on the right.

WRITING MATRIX EQUATIONS Write the linear system as a matrix equation.
11–22. See margin.

11. $x + y = 5$
    $3x - 4y = 8$

12. $x + 2y = 6$
    $4x - y = 5$

13. $5x - 3y = 9$
    $-4x + 2y = 10$

14. $2x - 5y = -11$
    $-3x + 7y = 15$

15. $x + 8y = 4$
    $4x - 5y = -11$

16. $2x - 5y = 4$
    $x - 3y = 1$

17. $x - 4y + 5z = -4$
    $2x + y - 7z = -23$
    $-4x + 5y + 2z = 38$

18. $3x - y + 4z = 16$
    $2x + 4y - z = 10$
    $x - y + 3z = 31$

19. $0.5x + 3.1y - 0.2z = 5.9$
    $1.2x - 2.5y + 0.7z = 2.2$
    $0.3x + 4.8y - 4.3z = 4.8$

20. $x + z = 9$
    $-x - y + 2z = 6$
    $2x + 7y - z = -4$

21. $8y - 10z = -23$
    $6y - 12z = 14$
    $-9x + 5z = 0$

22. $x + y - z = 0$
    $2x - z = 1$
    $y + z = 2$

SOLVING SYSTEMS Use an inverse matrix to solve the linear system.

STUDENT HELP

▶ HOMEWORK HELP
Example 1: Exs. 11–22
Example 2: Exs. 23–31
Example 3: Exs. 32–39
Example 4: Exs. 40–44

23. $3x + y = 8$ $(5, -7)$
    $5x + 2y = 11$

24. $x + y = -1$ $(-20, 19)$
    $11x + 12y = 8$

25. $2x + 7y = -53$ $(5, -9)$
    $x + 3y = -22$

26. $7x + 5y = 8$ $(4, -4)$
    $4x + 3y = 4$

27. $5x - 7y = 54$ $(1, -7)$
    $2x - 4y = 30$

28. $-5x - 7y = -9$ $(6, -3)$
    $2x + 3y = 3$

29. $x + 2y = -9$ $(-1, -4)$
    $-2x - 3y = 14$

30. $2x + 4y = -26$
    $2x + 5y = -31$
    $(-3, -5)$

31. $9x - 5y = 43$ $(-3, -14)$
    $-2x + 2y = -22$

---

③ **APPLY**

⭕ **ASSIGNMENT GUIDE**

BASIC
Day 1: pp. 233–236 Exs. 12–18 even, 24–28 even, 32, 34, 40, 41, 45, 47–65 odd, Quiz 2 Exs. 1–8

AVERAGE
Day 1: pp. 233–236 Exs. 12–34 even, 40–42, 45, 47–65 odd, Quiz 2 Exs. 1–8

ADVANCED
Day 1: pp. 233–236 Exs. 12–42 even, 44–46, 47–65 odd, Quiz 2 Exs. 1–8

BLOCK SCHEDULE
pp. 233–236 Exs. 12–34 even, 40–42, 45, 47–65 odd, Quiz 2 Exs. 1–8 (with 4.4)

EXERCISE LEVELS
Level A: *Easier*
11–29, 34–39
Level B: *More Difficult*
30–33, 40–45
Level C: *Most Difficult*
46

✓ **HOMEWORK CHECK**
To quickly check student understanding of key concepts, go over the following exercises:
Exs. 12, 18, 24, 32, 34. See also the Daily Homework Quiz:
• Blackline Master (*Chapter 5 Resource Book*, p. 11)
• 🖥 Transparency (p. 32)

4. $\begin{bmatrix} 1 & 1 \\ 2 & -1 \end{bmatrix}\begin{bmatrix} x \\ y \end{bmatrix} = \begin{bmatrix} 8 \\ 6 \end{bmatrix}$

5. $\begin{bmatrix} 1 & 3 \\ 4 & -2 \end{bmatrix}\begin{bmatrix} x \\ y \end{bmatrix} = \begin{bmatrix} 9 \\ 7 \end{bmatrix}$

6. $\begin{bmatrix} 1 & 1 & 1 \\ 5 & -1 & 0 \\ 3 & 4 & 1 \end{bmatrix}\begin{bmatrix} x \\ y \\ z \end{bmatrix} = \begin{bmatrix} 10 \\ 1 \\ 8 \end{bmatrix}$

11–22. See Additional Answers beginning on page AA1.

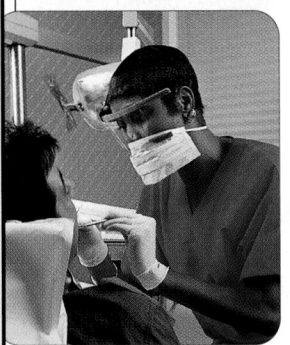

STUDENT HELP NOTES

Homework Help Students can find help for Exs. 32 and 33 at **www.mcdougallittell.com.** The information can be printed out for students who don't have access to the Internet.

GRAPHING CALCULATOR NOTE

EXERCISES 34–39 If students enter the coefficient matrix as *B*, the matrix of constants as *A*, and then try to find AB^{-1}, they will get an error on their graphing calculator. Make sure that students take the inverse of the coefficient matrix and multiply by the matrix of constants, which in this case is $B^{-1}A$.

CAREER NOTE

EXERCISE 41 Additional information about a career as a dentist is available at **www.mcdougallittell.com.**

ENGLISH LEARNERS

EXERCISE 41 Many students may require definitions of *amalgams, fillings, powdered, alloys, tin,* and *copper* to correctly conceptualize and respond to Exercise 41.

55–60. See Additional Answers beginning on page AA1.

ADDITIONAL PRACTICE AND RETEACHING

For Lesson 4.5:

• Practice Levels A, B, and C (*Chapter 4 Resource Book*, p. 70)

• Reteaching with Practice (*Chapter 4 Resource Book*, p. 73)

• See Lesson 4.5 of the *Personal Student Tutor*

For more Mixed Review:

• Search the *Test and Practice Generator* for key words or specific lessons.

234

STUDENT HELP

HOMEWORK HELP
Visit our Web site
www.mcdougallittell.com
for help with Exs. 32 and 33.

FOCUS ON CAREERS

DENTIST
Dentists diagnose, prevent, and treat problems of the teeth and mouth. Dental amalgams have been used for more than 150 years to restore the teeth of over 100 million Americans.

CAREER LINK
www.mcdougallittell.com

SOLVING SYSTEMS Use the given inverse of the coefficient matrix to solve the linear system.

32. $2y - z = -2$ (−82, 51, 104)
$5x + 2y + 3z = 4$
$7x + 3y + 4z = -5$

$$A^{-1} = \begin{bmatrix} -1 & -11 & 8 \\ 1 & 7 & -5 \\ 1 & 14 & -10 \end{bmatrix}$$

33. $x - y - 3z = 9$ (−61, 179, −83)
$5x + 2y + z = -30$
$-3x - y = 4$

$$A^{-1} = \begin{bmatrix} 1 & 3 & 5 \\ -3 & -9 & -16 \\ 1 & 4 & 7 \end{bmatrix}$$

SOLVING SYSTEMS Use an inverse matrix and a graphing calculator to solve the linear system.

34. $3x + 2y = 13$ (5, −1, 0)
$3x + 2y + z = 13$
$2x + y + 3z = 9$

35. (4, 3, 1)
$-x + y - 3z = -4$
$3x - 2y + 8z = 14$
$2x - 2y + 5z = 7$

36. (2, −2, −5)
$3x + 5y - 5z = 21$
$-4x + 8y - 5z = 1$
$2x - 5y + 6z = -16$

37. $2x + z = 2$ (2, 3, −2)
$5x - y + z = 5$
$-x + 2y + 2z = 0$

38. $4x + 3y + z = 14$
$6x + y = 9$
$3x + 5y + 3z = 21$
(1, 3, 1)

39. $x + y - 3z = -17$
$2x + z = 12$
$-7x - 2y + z = -11$
(3, −2, 6)

40. **SKATING PARTY** You are planning a birthday party for your younger brother at a skating rink. The cost of admission is $3.50 per adult and $2.25 per child, and there is a limit of 20 people. You have $50 to spend. Use an inverse matrix to determine how many adults and how many children you can invite.
4 adults, 16 children

41. **DENTAL FILLINGS** Dentists use various amalgams for silver fillings. The matrix shows the percents (expressed as decimals) of powdered alloys used in preparing three different amalgams. Suppose a dentist has 5483 grams of silver, 2009 grams of tin, and 129 grams of copper. How much of each amalgam can be made? 2239.8 g of A, 1313.6 g of B, 4067.6 g of C

PERCENT ALLOY BY WEIGHT

	Amalgam		
	A	B	C
Silver	0.70	0.72	0.73
Tin	0.26	0.25	0.27
Copper	0.04	0.03	0.00

42. **STAINED GLASS** You are making mosaic tiles from three types of stained glass. You need 6 square feet of glass for the project and you want there to be as much iridescent glass as red and blue glass combined. The cost of a sheet of glass having an area of 0.75 square foot is $6.50 for iridescent, $4.50 for red, and $5.50 for blue. How many sheets of each type should you purchase if you plan to spend $45 on the project? 4 sheets of iridescent, 3 sheets of red, 1 sheet of blue

43. **WALKWAY LIGHTING** A walkway lighting package includes a transformer, a certain length of wire, and a certain number of lights on the wire. The price of each lighting package depends on the length of wire and the number of lights on the wire. transformer: $10.00, wire: $.20 per ft, light: $1.00

• A package that contains a transformer, 25 feet of wire, and 5 lights costs $20.
• A package that contains a transformer, 50 feet of wire, and 15 lights costs $35.
• A package that contains a transformer, 100 feet of wire, and 20 lights costs $50.

Write and solve a system of equations to find the cost of a transformer, the cost per foot of wire, and the cost of a light. Assume the cost of each item is the same in each lighting package.

44. salaries: $151,515, equipment maintenance: $30,303, general expenses: $18,182

44.  **CONSTRUCTION BUSINESS** You are an accountant for a construction business and are planning next year's budget. You have $200,000 to spend on salaries, equipment maintenance, and other general expenses. Based on previous financial records of the business, you expect to spend five times as much on salaries as on equipment maintenance, and you expect general expenses to be 10% of the amount spent on the other two categories combined. Write and solve a system of equations to find the amount you should budget for each category.

Test
Preparation

45. **MULTI-STEP PROBLEM** A company sells different sizes of gift baskets with a varying assortment of meat and cheese. A basic basket with 2 cheeses and 3 meats costs $15, a big basket with 3 cheeses and 5 meats costs $24, and a super basket with 7 cheeses and 10 meats costs $50.

45a. The average unit price for each choice of meat and cheese is $3.

   a. Write and solve a system of equations using the information about the basic and big baskets.

45b. The average unit price for each choice of cheese is $2 and for each choice of meat is $3.60.

   b. Write and solve a system of equations using the information about the big and super baskets.

   c. *Writing* Compare the results from parts (a) and (b) and make a conjecture about why there is a discrepancy.   See margin.

★ **Challenge**

46. **SOLVING SYSTEMS OF FOUR EQUATIONS** Solve the linear system using the given inverse of the coefficient matrix. $(20, -1, -12, -8)$

$$\begin{aligned} w + 6x + 3y - 3z &= 2 \\ 2w + 7x + y + 2z &= 5 \\ w + 5x + 3y - 3z &= 3 \\ -6x - 2y + 3z &= 6 \end{aligned} \qquad A^{-1} = \begin{bmatrix} 40 & -3 & -33 & 9 \\ 1 & 0 & -1 & 0 \\ -39 & 3 & 33 & -8 \\ -24 & 2 & 20 & -5 \end{bmatrix}$$

EXTRA CHALLENGE
www.mcdougallittell.com

MIXED REVIEW

45c. *Sample answer:* The average unit price for cheese and for meat is different in parts (a) and (b); perhaps the super basket has a more expensive assortment of meat and a less expensive assortment of cheese than the other baskets.

EVALUATING FUNCTIONS Evaluate $f(x)$ or $g(x)$ for the given value of x. (Review 2.7)

$$f(x) = \begin{cases} \frac{3}{4}x - 8, & \text{if } x \le 8 \\ -x + 6, & \text{if } x > 8 \end{cases} \qquad g(x) = \begin{cases} \frac{1}{8}x + 8, & \text{if } x < -1 \\ 2x - 1, & \text{if } x \ge -1 \end{cases}$$

47. $f(8)$ -2 **48.** $f(11)$ -5 **49.** $f(-2)$ $-\frac{19}{2}$ **50.** $f(0)$ -8

51. $g(3)$ 5 **52.** $g(0)$ -1 **53.** $g(-1)$ -3 **54.** $g(-3)$ $\frac{61}{8}$

GRAPHING FUNCTIONS Graph the function and label the vertex. (Review 2.8 for 5.1) 55–60. See margin.

55. $y = |x - 5|$ **56.** $y = |x| + 8$ **57.** $y = -|x - 8| - 9$

58. $y = |x - 5| + 4$ **59.** $y = -|x + 3| + 4$ **60.** $y = -|x + 6| - 2$

FINDING INVERSES Find the inverse of the matrix. (Review 4.4)

61. $\begin{bmatrix} 7 & -4 \\ -5 & 3 \end{bmatrix}$ $\begin{bmatrix} 3 & 4 \\ 5 & 7 \end{bmatrix}$ **62.** $\begin{bmatrix} 5 & 2 \\ 2 & 1 \end{bmatrix}$ $\begin{bmatrix} 1 & -2 \\ -2 & 5 \end{bmatrix}$ **63.** $\begin{bmatrix} 8 & 17 \\ -1 & -2 \end{bmatrix}$ $\begin{bmatrix} -2 & -17 \\ 1 & 8 \end{bmatrix}$

64. $\begin{bmatrix} 11 & -5 \\ 3 & -1 \end{bmatrix}$ $\begin{bmatrix} -\frac{1}{4} & \frac{5}{4} \\ -\frac{3}{4} & \frac{11}{4} \end{bmatrix}$ **65.** $\begin{bmatrix} 7 & 4 \\ 3 & 2 \end{bmatrix}$ $\begin{bmatrix} 1 & -2 \\ -\frac{3}{2} & \frac{7}{2} \end{bmatrix}$ **66.** $\begin{bmatrix} 6 & -2 \\ 7 & -2 \end{bmatrix}$ $\begin{bmatrix} -1 & 1 \\ -\frac{7}{2} & 3 \end{bmatrix}$

DAILY HOMEWORK QUIZ

 **Transparency Available**

1. Write the linear system as a matrix equation.
$x + y = 3$
$2x - 5y = 7$
$\begin{bmatrix} 1 & 1 \\ 2 & -5 \end{bmatrix} \begin{bmatrix} x \\ y \end{bmatrix} = \begin{bmatrix} 3 \\ 7 \end{bmatrix}$

2. Use an inverse matrix to solve the linear system in Exercise 1.
$\left(\frac{22}{7}, -\frac{1}{7} \right)$

3. Use an inverse matrix and a graphing calculator to solve the linear system.
$2x + 8y + z = -5$
$2x + y + z = 2$
$5x - 3y + 2z = 12$ $(3, -1, -3)$

EXTRA CHALLENGE NOTE

→ Challenge problems for Lesson 4.5 are available in **blackline** format in the *Chapter 4 Resource Book,* p. 78 and at **www.mcdougallittell.com**.

ADDITIONAL TEST PREPARATION

1. WRITING Explain why the matrix equation
$\begin{bmatrix} 2 & 3 \\ 4 & -1 \end{bmatrix} \begin{bmatrix} x \\ y \end{bmatrix} = \begin{bmatrix} 12 \\ 8 \end{bmatrix}$ represents the system of equations with $2x + 3y = 12$ and $4x - y = 8$.

Since $\begin{bmatrix} 2 & 3 \\ 4 & -1 \end{bmatrix} \begin{bmatrix} x \\ y \end{bmatrix} = \begin{bmatrix} 2x + 3y \\ 4x - y \end{bmatrix}$

the equation is $\begin{bmatrix} 2x + 3y \\ 4x - y \end{bmatrix} = \begin{bmatrix} 12 \\ 8 \end{bmatrix}$.

This means that $2x + 3y = 12$ and $4x - y = 8$.

ADDITIONAL RESOURCES

An alternative quiz for Lessons 4.4 and 4.5 is available in the *Chapter 4 Resource Book,* p. 79.

A **blackline** master with additional Math & History exercises is available in the *Chapter 4 Resource Book,* p. 77.

1. *Sample answer:* The information can be organized into a matrix equation, where x = the number of dou per bundle of top-grade ears of rice, y = the number of dou per bundle of medium-grade ears of rice, and z = the number of dou per bundle of low-grade ears of rice:

Grades of Rice

$$\begin{matrix}\text{Top}\\\text{Medium}\\\text{Low}\end{matrix}\begin{bmatrix}3 & 2 & 1\\2 & 3 & 1\\1 & 2 & 3\end{bmatrix}\begin{bmatrix}x\\y\\z\end{bmatrix}=\begin{bmatrix}39\\34\\26\end{bmatrix}$$

top-grade ears: 9.25 dou per bundle, medium-grade ears: 4.25 dou per bundle, low-grade ears: 2.75 dou per bundle

2. *Sample answer:* The first three rows of counting rods correspond to the coefficient matrix, but the order of the entries is different: each row of counting rods, read from left to right, is the same as the corresponding column of the coefficient matrix, read from top to bottom. The last row of counting rods, read from left to right, corresponds to the constant matrix.

Find the inverse of the matrix. (Lesson 4.4)

$$2.\begin{bmatrix}-3 & -5\\-4 & -7\end{bmatrix}$$

1. $\begin{bmatrix}4 & 1\\7 & 2\end{bmatrix}\begin{bmatrix}2 & -1\\-7 & 4\end{bmatrix}$ 2. $\begin{bmatrix}-7 & 5\\4 & -3\end{bmatrix}$ 3. $\begin{bmatrix}-6 & 1\\9 & -3\end{bmatrix}$ 4. $\begin{bmatrix}6 & 5\\8 & 7\end{bmatrix}$

Use an inverse matrix to solve the linear system. (Lesson 4.5)

$$3.\begin{bmatrix}-\frac{1}{3} & -\frac{1}{9}\\-1 & -\frac{2}{3}\end{bmatrix}$$

5. $4x + 7y = 24$ $(-1, 4)$ 6. $-9x + 13y = 3$ $(4, 3)$ 7. $8x + 7y = 3$ $(3, -3)$
   $x + 2y = 7$ $2x - 3y = -1$ $-2x - 2y = 0$

$$4.\begin{bmatrix}\frac{7}{2} & -\frac{5}{2}\\-4 & 3\end{bmatrix}$$

8. 🌐 **BUYING FLATWARE** The price of flatware varies depending on the number of place settings you buy as well as other items included in the set. Suppose a set with 4 place settings costs \$142 and a set with 8 place settings and a serving set costs \$351. Find the cost of a place setting and a serving set. Assume that the cost of each item is the same for each flatware set. (Lesson 4.5)
place setting: \$35.50, serving set: \$67.00

MATH & History — **Systems of Equations**

APPLICATION LINK
www.mcdougallittell.com

THEN

OVER 2000 YEARS AGO, a method for solving systems of equations using rectangular arrays of counting rods was presented in *Nine Chapters on the Mathematical Art,* an early Chinese mathematics text. A problem from this book is given below.

> *Three bundles of top-grade ears of rice, two bundles of medium-grade ears of rice, and one bundle of low-grade ears of rice make 39 dou (of rice by volume); two bundles of top-grade ears of rice, three bundles of medium-grade ears of rice, and one bundle of low-grade ears of rice make 34 dou; one bundle of top-grade ears of rice, two bundles of medium-grade ears of rice, and three bundles of low-grade ears of rice make 26 dou. How many dou are there in a bundle of each grade of rice?*

Top-grade ears of rice
Medium-grade ears of rice
Low-grade ears of rice

Total produce

1, 2. See margin.
1. Use matrices to organize the given information and solve the problem.

2. How is the arrangement of counting rods similar to your matrices? How is it different?

NOW

TODAY, computers use matrices representing systems with many variables to solve complicated problems like predicting weather and designing aircraft.

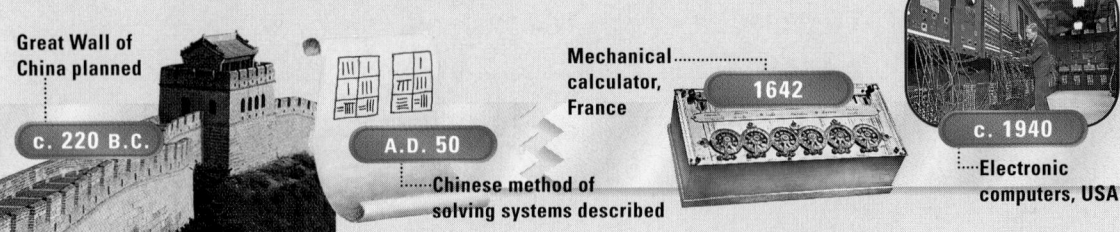

Great Wall of China planned
c. 220 B.C.

A.D. 50
Chinese method of solving systems described

Mechanical calculator, France
1642

c. 1940
Electronic computers, USA

GOAL Solve systems of linear equations using elementary row operations on augmented matrices.

Why you should learn it

The use of augmented matrices allows you to solve a linear system by suppressing the variables and working only with the coefficients and constants.

Solving Systems Using Augmented Matrices

GOAL **USING ELEMENTARY ROW OPERATIONS**

A matrix containing the coefficient matrix and the matrix of constants for a system of linear equations is called the **augmented matrix** of the system.

$$
\begin{array}{cc}
\text{LINEAR SYSTEM} & \text{AUGMENTED MATRIX} \\
\begin{aligned}
x - 2y &= 7 \\
-3x + 5y &= -4
\end{aligned}
&
\left[\begin{array}{cc:c}
1 & -2 & 7 \\
-3 & 5 & -4
\end{array}\right]
\end{array}
$$

Recall from Chapter 3 that equations in a system can be multiplied by a constant, or a multiple of one equation can be added to another equation. Similar operations can be performed on the rows of an augmented matrix to solve the corresponding system.

ELEMENTARY ROW OPERATIONS

Two augmented matrices are *row-equivalent* if their corresponding systems have the same solution(s). Any of these row operations performed on an augmented matrix will produce a matrix that is row-equivalent to the original:

- Interchange two rows.
- Multiply a row by a nonzero constant.
- Add a multiple of one row to another row.

To solve a system, use elementary row operations to transform the original augmented matrix into a matrix having 1's along the main diagonal and 0's below the main diagonal. A matrix of this form is said to be in *triangular form*.

EXAMPLE 1 *Using Row Operations to Solve a Two-Variable System*

$$
\begin{array}{cc}
\text{LINEAR SYSTEM} & \text{AUGMENTED MATRIX} \\
\begin{aligned}
x - 2y &= 7 \\
-3x + 5y &= -4
\end{aligned}
&
\left[\begin{array}{cc:c}
1 & -2 & 7 \\
-3 & 5 & -4
\end{array}\right]
\end{array}
$$

Add 3 times the first equation to the second equation. You get this system:

$$
\begin{aligned}
x - 2y &= 7 \\
-y &= 17
\end{aligned}
$$

Add 3 times the first row, R_1, to the second row, R_2.

$$
3R_1 + R_2 \rightarrow
\left[\begin{array}{cc:c}
1 & -2 & 7 \\
0 & -1 & 17
\end{array}\right]
$$

Multiply the second equation by -1.

$$
\begin{aligned}
x - 2y &= 7 \\
y &= -17
\end{aligned}
$$

Multiply the second row by -1.

$$
(-1)R_2 \rightarrow
\left[\begin{array}{cc:c}
1 & -2 & 7 \\
0 & 1 & -17
\end{array}\right]
$$

The second row of the matrix tells you that $y = -17$. Substitute -17 for y in the equation for the first row: $x - 2(-17) = 7$, or $x = -27$. The solution is $(-27, -17)$.

Chapter 4 *Extension* **237**

EXTRA EXAMPLE 1
Use an augmented matrix to solve the linear system.
$x + y = 4$
$3x + y = 8$ (2, 2)

EXTRA EXAMPLE 2
Use an augmented matrix to solve the linear system.
$2x + 3y + z = 5$
$4x + y + 2z = 5$
$-x + 3y - z = 10$ (9, 1, −16)

CHECKPOINT EXERCISES
For use after Example 1:
1. Use an augmented matrix to solve the linear system.
$5x + 2y = 2$
$3x - y = -12$ (−2, 6)

For use after Example 2:
2. Use an augmented matrix to solve the linear system.
$7x + 3y + 2z = -3$
$-2x + y + z = 9$
$-x + 5y - z = 3$ (−2, 1, 4)

MATHEMATICAL REASONING
Matrices can be used to solve a system of *n* equations in *n* unknowns. The matrix of the system must be transformed into triangular form. Ask the students how many zeros will form the triangle in a matrix for a system of 4 equations and 4 unknowns. **6**

EXAMPLE 2 *Using Row Operations to Solve a Three-Variable System*

LINEAR SYSTEM	AUGMENTED MATRIX

$$2x + 4y + 5z = 5$$
$$x + 3y + 3z = 2$$
$$2x + 4y + 6z = 2$$

$$\begin{bmatrix} 2 & 4 & 5 & : & 5 \\ 1 & 3 & 3 & : & 2 \\ 2 & 4 & 6 & : & 2 \end{bmatrix}$$

Add -1 times the first equation to the third equation. You get this system:

Add -1 times the first row to the third row.

$$2x + 4y + 5z = 5$$
$$x + 3y + 3z = 2$$
$$z = -3$$

$$(-1)R_1 + R_3 \rightarrow \begin{bmatrix} 2 & 4 & 5 & : & 5 \\ 1 & 3 & 3 & : & 2 \\ 0 & 0 & 1 & : & -3 \end{bmatrix}$$

Add -0.5 times the first equation to the second equation.

Add -0.5 times the first row to the second row.

$$2x + 4y + 5z = 5$$
$$y + 0.5z = -0.5$$
$$z = -3$$

$$-0.5R_1 + R_2 \rightarrow \begin{bmatrix} 2 & 4 & 5 & : & 5 \\ 0 & 1 & 0.5 & : & -0.5 \\ 0 & 0 & 1 & : & -3 \end{bmatrix}$$

Multiply the first equation by 0.5.

Multiply the first row by 0.5.

$$x + 2y + 2.5z = 2.5$$
$$y + 0.5z = -0.5$$
$$z = -3$$

$$0.5R_1 \rightarrow \begin{bmatrix} 1 & 2 & 2.5 & : & 2.5 \\ 0 & 1 & 0.5 & : & -0.5 \\ 0 & 0 & 1 & : & -3 \end{bmatrix}$$

The third row of the matrix tells you that $z = -3$. Substitute -3 for z in the equation for the second row, $y + 0.5z = -0.5$, to obtain $y + 0.5(-3) = -0.5$, or $y = 1$. Then substitute -3 for z and 1 for y in the equation for the first row, $x + 2y + 2.5z = 2.5$, to obtain $x + 2(1) + 2.5(-3) = 2.5$, or $x = 8$. The solution is $(8, 1, -3)$.

EXERCISES

SOLVING SYSTEMS **Use an augmented matrix to solve the linear system.**

1. $6x + 4y = 8$ $(-2, 5)$
$3x + 3y = 9$

2. $x + y = 2$ $(-5, 7)$
$7x + 8y = 21$

3. $x + 2y = -9$ $(-1, -4)$
$-2x - 3y = 14$

4. $x - 3y = 5$ $(-4, -3)$
$-2x - 4y = 20$

5. $3x + 2y = 2$ $(4, -5)$
$5x - 6y = 50$

6. $x + y = -1$ $(5, -6)$
$7x + 9y = -19$

7. $-2x - y = -5$ $(2, 1)$
$6x + 5y = 17$

8. $9x - 4y = 2$ $\left(\frac{1}{3}, \frac{1}{4}\right)$
$-6x - 16y = -6$

9. $-12x + 15y = 3$ $\left(0, \frac{1}{5}\right)$
$-7x - 20y = -4$

10. $2x + 6y + 3z = 2$ $(4, -1, 0)$
$x + 3y + z = 1$
$x + 5y + 2z = -1$

11. $2x + 6y + 3z = 8$ $(16, -5, 2)$
$x + 5y + 5z = 1$
$x + 3y + z = 3$

12. $2x + 10y = 28$ $(-6, 4, 4)$
$x + 3y + 4z = 22$
$x + 5y - z = 10$

13. $x + 4y - 2z = 3$ $(-5, 2, 0)$
$x + 3y + 7z = 1$
$2x + 9y + z = 8$

14. $x - y + 3z = 6$ $(-11, -8, 3)$
$x - 2y = 5$
$2x - 2y + 5z = 9$

15. $x + 2z = 4$ $(-16, 12, 10)$
$x + y + z = 6$
$3x + 3y + 4z = 28$

16. **CRITICAL THINKING** Try using an augmented matrix to solve the given system. What happens? What can you say about the system's solution(s)?

All entries in R_2 and R_3 become 0; the system has an infinite number of solutions, the solutions of the equation $x - 2y + 7z = 6$.

$$x - 2y + 7z = 6$$
$$5x - 10y + 35z = 30$$
$$3x - 6y + 21z = 18$$

Chapter Summary

WHAT did you learn?

Use matrices and determinants in real-life
situations. **(4.1–4.5)**

Perform matrix operations.
- add and subtract matrices **(4.1)**

- multiply a matrix by a scalar **(4.1)**

- multiply two matrices **(4.2)**

Evaluate the determinant of a matrix. **(4.3)**

Find the inverse of a matrix. **(4.4)**

Solve matrix equations. **(4.1, 4.4, 4.5)**

Solve systems of linear equations.
- using Cramer's rule **(4.3)**
- using inverse matrices **(4.5)**

Use systems of linear equations to solve real-life
problems. **(4.3, 4.5)**

WHY did you learn it?

Organize data, such as the number and dollar value of
Hispanic music products shipped. **(p. 204)**

Find the total cost of college tuition plus room and
board. **(p. 205)**

Write U.S. population data as percents of total
population. **(p. 205)**

Calculate the cost of softball equipment. **(p. 210)**

Find the area of the Golden Triangle. **(p. 220)**

Encode or decode a cryptogram. **(pp. 225, 226)**

Extend the process of equation solving to equations
whose solutions are matrices. **(pp. 201, 224)**

Find the cost of gasoline. **(p. 220)**
Calculate a budget. **(p. 235)**

Decide how much money to invest in each of three
types of mutual funds. **(p. 232)**

How does Chapter 4 fit into the BIGGER PICTURE of algebra?

Your study of linear algebra has continued with Chapter 4. Matrices are used
throughout linear algebra, especially for solving linear systems. This introduction
to matrices, their uses, and properties of matrix operations also connects to your
past. For example, instead of multiplying real numbers, or variables that represent
real numbers, you multiplied matrices. You also saw properties, such as the
commutative property of multiplication, that apply to real numbers but not
to matrices.

STUDY STRATEGY

Did you write out the steps?

Here is an example of several
steps you can write when
multiplying two matrices,
following the **Study Strategy**
on page 198.

Writing Out the Steps

$$\begin{bmatrix} 1 & -6 \\ -3 & 2 \end{bmatrix}\begin{bmatrix} -9 & 0 \\ -2 & 7 \end{bmatrix}$$

$$= \begin{bmatrix} 1(-9) + (-6)(-2) & 1(0) + (-6)7 \\ -3(-9) + 2(-2) & -3(0) + 2(7) \end{bmatrix}$$

$$= \begin{bmatrix} -9 + 12 & 0 - 42 \\ 27 - 4 & 0 + 14 \end{bmatrix}$$

$$= \begin{bmatrix} 3 & -42 \\ 23 & 14 \end{bmatrix}$$

ADDITIONAL RESOURCES

The following resources are available to help review the material in this chapter.

- Chapter Review Games and Activities (*Chapter 4 Resource Book*, p. 80)
- *Instant Replay: Video Review Games*
- ☐ *Personal Student Tutor*
- Cumulative Review, Chs. 1–4 (*Chapter 4 Resource Book*, p. 92)

2. Not possible; the two matrices do not have the same dimensions.

4. Not possible; the two matrices do not have the same dimensions.

5. $\begin{bmatrix} 8 & 12 & -2 \\ 20 & -10 & 4 \\ 0 & 22 & 2 \end{bmatrix}$

Chapter Review

VOCABULARY

- matrix, p. 199
- dimensions of a matrix, p. 199
- entries of a matrix, p. 199
- row matrix, p. 199

- column matrix, p. 199
- square matrix, p. 199
- zero matrix, p. 199
- equal matrices, p. 199

- scalar, p. 200
- determinant, p. 214
- Cramer's rule, p. 216
- coefficient matrix, p. 216

- identity matrix, p. 223
- inverse matrix, p. 223
- matrix of variables, p. 230
- matrix of constants, p. 230

4.1 MATRIX OPERATIONS

Examples on pp. 199–202

EXAMPLES You can add or subtract matrices that have the same dimensions by adding or subtracting corresponding entries.

$$\begin{bmatrix} 5 & -2 \\ 0 & 6 \end{bmatrix} + \begin{bmatrix} 9 & 1 \\ -4 & 4 \end{bmatrix} = \begin{bmatrix} 5+9 & -2+1 \\ 0+(-4) & 6+4 \end{bmatrix} = \begin{bmatrix} 14 & -1 \\ -4 & 10 \end{bmatrix}$$

You cannot subtract these matrices because they have different dimensions.

$$\begin{bmatrix} -2 & 1 \\ 0 & -3 \end{bmatrix} - \begin{bmatrix} 1 & -5 & -4 \\ 2 & 7 & 1 \end{bmatrix}$$

To do scalar multiplication, multiply each entry in the matrix by the scalar.

$$-3\begin{bmatrix} -12 & -6 \\ 3 & 1 \\ 2 & 8 \end{bmatrix} = \begin{bmatrix} (-3)(-12) & (-3)(-6) \\ (-3)(3) & (-3)(1) \\ (-3)(2) & (-3)(8) \end{bmatrix} = \begin{bmatrix} 36 & 18 \\ -9 & -3 \\ -6 & -24 \end{bmatrix}$$

To solve this matrix equation, equate corresponding entries and solve for x and y.

$$\begin{bmatrix} x+2 & 2 \\ -1 & 9 \end{bmatrix} = \begin{bmatrix} -6 & 2 \\ -1 & 3y \end{bmatrix} \qquad \begin{aligned} x+2 &= -6 & 3y &= 9 \\ x &= -8 & y &= 3 \end{aligned}$$

Perform the indicated operation if possible. If not possible, state the reason.

1. $\begin{bmatrix} 15 & 4 \\ 3 & 12 \end{bmatrix} - \begin{bmatrix} 0 & 9 \\ 2 & 7 \end{bmatrix}$ $\begin{bmatrix} 15 & -5 \\ 1 & 5 \end{bmatrix}$

2. $\begin{bmatrix} 3 & -2 \\ -4 & 1 \end{bmatrix} - \begin{bmatrix} 5 \\ -3 \end{bmatrix}$ See margin.

3. $\begin{bmatrix} 6 & 10 \\ 9 & 6 \\ 4 & -1 \end{bmatrix} + \begin{bmatrix} 2 & 1 \\ 0 & 7 \\ 4 & 7 \end{bmatrix}$ $\begin{bmatrix} 8 & 11 \\ 9 & 13 \\ 8 & 6 \end{bmatrix}$

4. $\begin{bmatrix} 0 & 1 & 5 \\ -2 & 3 & 1 \\ 1 & 2 & -4 \end{bmatrix} + \begin{bmatrix} 1 & -2 \\ 4 & 1 \\ 2 & -3 \end{bmatrix}$ **4 and 5.** See margin.

5. $2\begin{bmatrix} 4 & 6 & -1 \\ 10 & -5 & 2 \\ 0 & 11 & 1 \end{bmatrix}$

6. $\frac{1}{2}\begin{bmatrix} -2 & 0 \\ 4 & 8 \\ -6 & -2 \end{bmatrix}\begin{bmatrix} -1 & 0 \\ 2 & 4 \\ -3 & -1 \end{bmatrix}$

Solve the matrix equation for x and y.

7. $\begin{bmatrix} 1 & 14 \\ -5x & 10 \end{bmatrix} = \begin{bmatrix} y-9 & 14 \\ 5 & 10 \end{bmatrix}$ $x = -1, y = 10$

8. $\begin{bmatrix} 3 & 4y \\ -1 & 13 \end{bmatrix} + \begin{bmatrix} -6 & 5 \\ 8 & 0 \end{bmatrix} = \begin{bmatrix} -3 & -7 \\ x & 13 \end{bmatrix}$ $x = 7, y = -3$

9. $\begin{bmatrix} 2 & 3y \\ 4 & -1 \end{bmatrix} + \begin{bmatrix} 0 & -4 \\ x & -2 \end{bmatrix} = \begin{bmatrix} 2 & 11 \\ 3 & -3 \end{bmatrix}$ $x = -1, y = 5$

10. $\begin{bmatrix} 7y & -2 \\ -3 & 5 \end{bmatrix} - \begin{bmatrix} 1 & 5 \\ x & -3 \end{bmatrix} = \begin{bmatrix} 6 & -7 \\ -2 & 8 \end{bmatrix}$ $x = -1, y = 1$

MULTIPLYING MATRICES

Examples on pp. 208–210

EXAMPLE You can multiply a matrix with n columns by a matrix with n rows.

$$\begin{bmatrix} -6 & 1 \\ 5 & -2 \end{bmatrix}\begin{bmatrix} 6 & 3 \\ 0 & 1 \end{bmatrix} = \begin{bmatrix} (-6)(6)+(1)(0) & (-6)(3)+(1)(1) \\ (5)(6)+(-2)(0) & (5)(3)+(-2)(1) \end{bmatrix} = \begin{bmatrix} -36 & -17 \\ 30 & 13 \end{bmatrix}$$

Write the product. If it is not defined, state the reason.

11. $\begin{bmatrix} 12 \\ -4 \end{bmatrix}\begin{bmatrix} -10 & -7 \end{bmatrix}$
$\begin{bmatrix} -120 & -84 \\ 40 & 28 \end{bmatrix}$

12. $\begin{bmatrix} 2 & 15 \\ -3 & 10 \end{bmatrix}\begin{bmatrix} -5 & 12 \\ 1 & 0 \end{bmatrix}$
$\begin{bmatrix} 5 & 24 \\ 25 & -36 \end{bmatrix}$

13. $\begin{bmatrix} 1 & 7 \\ 0 & 9 \end{bmatrix}\begin{bmatrix} 3 & -1 & 8 \\ 2 & -4 & 8 \end{bmatrix}$
$\begin{bmatrix} 17 & -29 & 64 \\ 18 & -36 & 72 \end{bmatrix}$

DETERMINANTS AND CRAMER'S RULE

Examples on pp. 214–217

EXAMPLES You can evaluate the determinant of a 2 × 2 or a 3 × 3 matrix. Find products of the entries on the diagonals and subtract.

$$\det\begin{bmatrix} -2 & -6 \\ 1 & 4 \end{bmatrix} = \begin{vmatrix} -2 & -6 \\ 1 & 4 \end{vmatrix} = -2(4) - 1(-6) = -8 + 6 = -2$$

$$\det\begin{bmatrix} 2 & 1 & 5 \\ -1 & 6 & 3 \\ 2 & -4 & 2 \end{bmatrix} = \begin{vmatrix} 2 & 1 & 5 \\ -1 & 6 & 3 \\ 2 & -4 & 2 \end{vmatrix}\begin{matrix} 2 & 1 \\ -1 & 6 \\ 2 & -4 \end{matrix} = (24 + 6 + 20) - [60 + (-24) + (-2)] = 16$$

You can find the area of a triangle with vertices (x_1, y_1), (x_2, y_2), and (x_3, y_3) using

$$\text{Area} = \pm\frac{1}{2}\begin{vmatrix} x_1 & y_1 & 1 \\ x_2 & y_2 & 1 \\ x_3 & y_3 & 1 \end{vmatrix}$$

where $\pm$ indicates you should choose the sign that yields a positive value.

You can use Cramer's rule to solve a system of linear equations. First find the determinant of the coefficient matrix and then use Cramer's rule to solve for x and y.

$$\begin{aligned} 3x - 4y &= 12 \\ x + 2y &= 14 \end{aligned} \qquad \det\begin{bmatrix} 3 & -4 \\ 1 & 2 \end{bmatrix} = \begin{vmatrix} 3 & -4 \\ 1 & 2 \end{vmatrix} = 3(2) - 1(-4) = 6 + 4 = 10$$

$$x = \frac{\begin{vmatrix} 12 & -4 \\ 14 & 2 \end{vmatrix}}{10} = \frac{12(2) - 14(-4)}{10} = \frac{80}{10} = 8 \qquad y = \frac{\begin{vmatrix} 3 & 12 \\ 1 & 14 \end{vmatrix}}{10} = \frac{3(14) - 1(12)}{10} = \frac{30}{10} = 3$$

Evaluate the determinant of the matrix.

14. $\begin{bmatrix} -9 & 1 \\ 3 & 2 \end{bmatrix}$ −21

15. $\begin{bmatrix} 6 & -3 \\ 2 & 1 \end{bmatrix}$ 12

16. $\begin{bmatrix} 3 & 1 & 0 \\ 2 & 1 & 1 \\ 0 & 3 & 4 \end{bmatrix}$ −5

17. $\begin{bmatrix} 2 & -3 & 4 \\ 0 & 1 & -2 \\ 1 & 2 & -3 \end{bmatrix}$ 4

18. Find the area of a triangle with vertices $A(0, 1)$, $B(2, 4)$, and $C(1, 8)$. $\frac{11}{2}$ square units

Use Cramer's rule to solve the linear system.

19. $7x - 4y = -3$
$2x + 5y = -7$ $(-1, -1)$

20. $2x + y = -2$
$x - 2y = 19$ $(3, -8)$

21. $5x - 4y + 4z = 18$
$-x + 3y - 2z = 0$ $(6, 0, -3)$
$4x - 2y + 7z = 3$

IDENTITY AND INVERSE MATRICES

Examples on
pp. 222–226

EXAMPLES You can find the inverse of an $n \times n$ matrix provided its determinant does not equal zero.

The inverse of $A = \begin{bmatrix} a & b \\ c & d \end{bmatrix}$ is $A^{-1} = \dfrac{1}{|A|} \begin{bmatrix} d & -b \\ -c & a \end{bmatrix}$.

If $A = \begin{bmatrix} 7 & 3 \\ 5 & 2 \end{bmatrix}$, then $A^{-1} = \dfrac{1}{7(2) - 5(3)} \begin{bmatrix} 2 & -3 \\ -5 & 7 \end{bmatrix} = -1 \begin{bmatrix} 2 & -3 \\ -5 & 7 \end{bmatrix} = \begin{bmatrix} -2 & 3 \\ 5 & -7 \end{bmatrix}$.

You can use the inverse of a matrix A to solve a matrix equation $AX = B$: $X = A^{-1}B$.

$$\begin{bmatrix} 1 & 3 \\ 2 & 7 \end{bmatrix} X = \begin{bmatrix} 3 & 0 \\ 5 & 2 \end{bmatrix}$$

$$A^{-1} = \frac{1}{7 - 6} \begin{bmatrix} 7 & -3 \\ -2 & 1 \end{bmatrix} = \begin{bmatrix} 7 & -3 \\ -2 & 1 \end{bmatrix}$$

$$X = \begin{bmatrix} 7 & -3 \\ -2 & 1 \end{bmatrix} \begin{bmatrix} 3 & 0 \\ 5 & 2 \end{bmatrix} = \begin{bmatrix} 6 & -6 \\ -1 & 2 \end{bmatrix}$$

Find the inverse of the matrix.

22. $\begin{bmatrix} 2 & 3 \\ 7 & 11 \end{bmatrix}$ $\begin{bmatrix} 11 & -3 \\ -7 & 2 \end{bmatrix}$ **23.** $\begin{bmatrix} 2 & 2 \\ 1 & 3 \end{bmatrix}$ $\begin{bmatrix} \frac{3}{4} & -\frac{1}{2} \\ -\frac{1}{4} & \frac{1}{2} \end{bmatrix}$ **24.** $\begin{bmatrix} -3 & 6 \\ 2 & -4 \end{bmatrix}$ no inverse **25.** $\begin{bmatrix} 6 & -1 \\ -5 & 1 \end{bmatrix}$ $\begin{bmatrix} 1 & 1 \\ 5 & 6 \end{bmatrix}$

Solve the matrix equation.

26. $\begin{bmatrix} 5 & 3 \\ 3 & 2 \end{bmatrix} X = \begin{bmatrix} 0 & 9 \\ -1 & 4 \end{bmatrix}$ $\begin{bmatrix} 3 & 6 \\ -5 & -7 \end{bmatrix}$ **27.** $\begin{bmatrix} -7 & -5 \\ 4 & 3 \end{bmatrix} X + \begin{bmatrix} 8 & -2 \\ 6 & 1 \end{bmatrix} = \begin{bmatrix} 9 & -3 \\ 6 & 2 \end{bmatrix}$ $\begin{bmatrix} -3 & -2 \\ 4 & 3 \end{bmatrix}$

SOLVING SYSTEMS USING INVERSE MATRICES

Examples on
pp. 230–232

EXAMPLE You can use inverse matrices to solve a system of linear equations.

$\begin{aligned} x + 3y &= 10 \\ 2x + 5y &= -2 \end{aligned}$ **Write in matrix form.** → $\underset{A}{\begin{bmatrix} 1 & 3 \\ 2 & 5 \end{bmatrix}} \underset{X}{\begin{bmatrix} x \\ y \end{bmatrix}} = \underset{B}{\begin{bmatrix} 10 \\ -2 \end{bmatrix}}$

Then $X = A^{-1}B = \dfrac{1}{1(5) - 2(3)} \begin{bmatrix} 5 & -3 \\ -2 & 1 \end{bmatrix} \begin{bmatrix} 10 \\ -2 \end{bmatrix} = -1 \begin{bmatrix} 56 \\ -22 \end{bmatrix} = \begin{bmatrix} -56 \\ 22 \end{bmatrix}$.

The solution is $(-56, 22)$.

Use an inverse matrix to solve the linear system.

28. $\begin{aligned} 9x + 8y &= -6 \\ -x - y &= 1 \end{aligned}$ $(2, -3)$ **29.** $\begin{aligned} x - 3y &= -2 \\ 5x + 3y &= 17 \end{aligned}$ $\left(\frac{5}{2}, \frac{3}{2}\right)$ **30.** $\begin{aligned} 4x - 14y &= -15 \\ 18x - 12y &= 9 \end{aligned}$ $\left(\frac{3}{2}, \frac{3}{2}\right)$

Use an inverse matrix and a graphing calculator to solve the linear system.

31. $\begin{aligned} x - y - 4z &= 3 \\ -x + 3y - z &= -1 \\ x - y + 5z &= 3 \end{aligned}$ $(4, 1, 0)$ **32.** $\begin{aligned} 4x + 10y - z &= -3 \\ 11x + 28y - 4z &= 1 \\ -6x - 15y + 2z &= -1 \end{aligned}$ $(19, -9, -11)$ **33.** $\begin{aligned} 5x - 3y + 5z &= -1 \\ 3x + 2y + 4z &= 11 \\ 2x - y + 3z &= 4 \end{aligned}$ $(-3, 2, 4)$

Chapter 4 *Matrices and Determinants*

Chapter Test

ADDITIONAL RESOURCES
• *Chapter 4 Resource Book*
  Chapter Test (3 levels) (p. 81)
  SAT/ACT Chapter Test (p. 87)
  Alternative Assessment (p. 88)
• 🖥 *Test and Practice Generator*

Perform the indicated operation(s).
See margin.

1. $\begin{bmatrix} 2 & 5 & -4 \\ 3 & 0 & -2 \end{bmatrix} + \begin{bmatrix} 3 & 2 & 7 \\ -2 & -5 & 7 \end{bmatrix}$

2. $0.25 \begin{bmatrix} 8 & 20 & -12 \\ -8 & -4 & 36 \end{bmatrix}$ $\begin{bmatrix} 2 & 5 & -3 \\ -2 & -1 & 9 \end{bmatrix}$

3. $-4\left(\begin{bmatrix} 1 & 10 \\ -4 & -6 \end{bmatrix} - \begin{bmatrix} 4 & 8 \\ -3 & -8 \end{bmatrix} \right)$ $\begin{bmatrix} 12 & -8 \\ 4 & -8 \end{bmatrix}$

4. $\begin{bmatrix} 4 & 1 & 4 \\ -1 & 8 & -3 \\ 4 & 3 & 0 \end{bmatrix} \begin{bmatrix} -2 & 18 \\ 2 & 0 \\ 6 & -2 \end{bmatrix}$

5. $\begin{bmatrix} -6 & 1 \\ 9 & 2 \end{bmatrix} \begin{bmatrix} 3 & 0 \\ -5 & 4 \end{bmatrix} \begin{bmatrix} -23 & 4 \\ 17 & 8 \end{bmatrix}$

6. $\begin{bmatrix} 0 & 1 & 0 \\ 2 & -1 & 1 \\ 0 & 2 & -1 \end{bmatrix} \begin{bmatrix} -1 & 2 & 0 \\ 4 & 6 & 0 \\ 1 & 0 & 1 \end{bmatrix}$ See margin.

Solve the matrix equation for *x* and *y*.

7. $\begin{bmatrix} -1 & y+6 \\ x-4 & 3 \end{bmatrix} = \begin{bmatrix} -1 & 8 \\ -9 & 3 \end{bmatrix}$ $x = -5, y = 2$

8. $\begin{bmatrix} -22 & 9 \\ 1 & -y \end{bmatrix} = \begin{bmatrix} 2x & 9 \\ 1 & 4 \end{bmatrix}$ $x = -11, y = -4$

9. $3\begin{bmatrix} x & 1 \\ 8 & -4 \end{bmatrix} = \begin{bmatrix} -15 & 3 \\ y & -12 \end{bmatrix}$ $x = -5, y = 24$

Evaluate the determinant of the matrix.

10. $\begin{bmatrix} 7 & -9 \\ -3 & 4 \end{bmatrix} 1$

11. $\begin{bmatrix} -2 & -1 \\ 1 & -1 \end{bmatrix} 3$

12. $\begin{bmatrix} 4 & 0 & 1 \\ 1 & 5 & 3 \\ 2 & 2 & 0 \end{bmatrix} -32$

13. $\begin{bmatrix} -1 & 3 & 4 \\ 6 & 0 & -2 \\ 0 & -5 & 1 \end{bmatrix} -128$

Find the area of the triangle with the given vertices.

14. $A(2, 1), B(5, 3), C(7, 1)$ 5

15. $A(-1, 0), B(-3, 3), C(0, 4) \frac{11}{2}$

16. $A(-3, 2), B(-1, 4), C(-4, 3)$ 2

Use Cramer's rule to solve the linear system.

17. $2x + y = 12$ (9, −6)
    $5x + 3y = 27$

18. $-4x + 5y = -10$ (5, 2)
    $5x - 6y = 13$

19. $x + y = 2$ $\left(\frac{3}{2}, \frac{1}{2}, 1 \right)$
    $2y - z = 0$
    $-x - y + z = -1$

20. $5x - 2y + 7z = 12$ (7, 1, −3)
    $2x + 5y + 3z = 10$
    $3x - y + 4z = 8$

Find the inverse of the matrix.

21. $\begin{bmatrix} 4 & 5 \\ 3 & 9 \end{bmatrix}$ See margin.

22. $\begin{bmatrix} -1 & -2 \\ 1 & 1 \end{bmatrix} \begin{bmatrix} 1 & 2 \\ -1 & -1 \end{bmatrix}$

23. $\begin{bmatrix} -6 & 4 \\ 6 & -5 \end{bmatrix} \begin{bmatrix} -\frac{5}{6} & -\frac{2}{3} \\ -1 & -1 \end{bmatrix}$

24. $\begin{bmatrix} 1 & 0 \\ 0 & -5 \end{bmatrix} \begin{bmatrix} 1 & 0 \\ 0 & -\frac{1}{5} \end{bmatrix}$

Solve the matrix equation.

25. $\begin{bmatrix} 8 & 7 \\ 1 & 1 \end{bmatrix} X = \begin{bmatrix} 3 & -6 \\ -2 & 9 \end{bmatrix}$ See margin.

26. $\begin{bmatrix} 2 & 5 \\ 2 & 6 \end{bmatrix} X = \begin{bmatrix} 1 & 0 \\ 0 & 1 \end{bmatrix} \begin{bmatrix} 3 & -\frac{5}{2} \\ -1 & 1 \end{bmatrix}$

27. $\begin{bmatrix} 1 & 0 \\ -6 & 2 \end{bmatrix} X = \begin{bmatrix} 10 & 6 & 8 \\ 4 & 12 & 2 \end{bmatrix}$ $\begin{bmatrix} 10 & 6 & 8 \\ 32 & 24 & 25 \end{bmatrix}$

Use an inverse matrix to solve the linear system.

28. $x - y = 5$ (6, 1)
    $-2x + 3y = -9$

29. $3x + 2y = -8$ (−4, 2)
    $-2x + 5y = 18$

30. $2x - 7y = 6$ (−4, −2)
    $-3x + 11y = -10$

31. 🔷 **STAINED GLASS** You are making a stained glass panel using different colors as shown. The coordinates given are measured in inches. Find the area of the red triangle. $\frac{13}{2}$

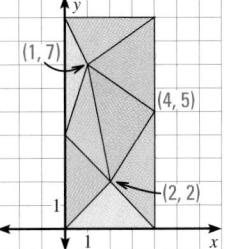

32. 🔷 **DECODING** Use the inverse of $A = \begin{bmatrix} 2 & -1 \\ 3 & -1 \end{bmatrix}$ and the coding information on pages 225 and 226 to decode the message below. **AN APPLE A DAY**

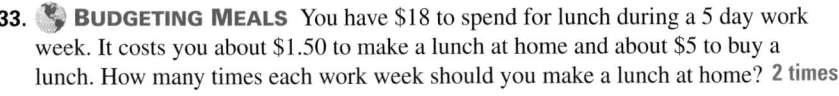
44, −15, 3, −1, 80, −32, 39, −17, 3, −1, 12, −4, 77, −26

33. 🔷 **BUDGETING MEALS** You have $18 to spend for lunch during a 5 day work week. It costs you about $1.50 to make a lunch at home and about $5 to buy a lunch. How many times each work week should you make a lunch at home? **2 times**

Test and Practice Generator sidebar answers:

1. $\begin{bmatrix} 5 & 7 & 3 \\ 1 & -5 & 5 \end{bmatrix}$

6. $\begin{bmatrix} 4 & 6 & 0 \\ -5 & -2 & 1 \\ 7 & 12 & -1 \end{bmatrix}$

21. $\begin{bmatrix} \frac{3}{7} & -\frac{5}{21} \\ -\frac{1}{7} & \frac{4}{21} \end{bmatrix}$

25. $\begin{bmatrix} 17 & -69 \\ -19 & 78 \end{bmatrix}$

ADDITIONAL RESOURCES
• *Chapter 4 Resource Book*
  Chapter Test (3 levels) (p. 81)
  SAT/ACT Chapter Test (p. 87)
  Alternative Assessment (p. 88)

• 🖥 *Test and Practice Generator*

CHAPTER
4

Chapter Standardized Test

▶ **TEST-TAKING STRATEGY** During a test it is important to stay mentally focused, but also physically relaxed. If you start to get tense, put your pencil down and take some deep breaths. This may help you regain control.

1. MULTIPLE CHOICE Which matrix equals

$$2\left(\begin{bmatrix} 2 & -7 \\ -4 & 3 \end{bmatrix} + \begin{bmatrix} 0 & -5 \\ -3 & 6 \end{bmatrix}\right)? \ \textbf{B}$$

(A) $\begin{bmatrix} 4 & -4 \\ -2 & 18 \end{bmatrix}$ **(B)** $\begin{bmatrix} 4 & -24 \\ -14 & 18 \end{bmatrix}$

(C) $\begin{bmatrix} 6 & -22 \\ -12 & 20 \end{bmatrix}$ **(D)** $\begin{bmatrix} -24 & 60 \\ -6 & 30 \end{bmatrix}$

(E) $\begin{bmatrix} 42 & -104 \\ -18 & 76 \end{bmatrix}$

2. MULTIPLE CHOICE What are the values of x and y in the matrix equation? **D**

$$2x\begin{bmatrix} -2 & -1 \\ -10 & 5 \end{bmatrix} = \begin{bmatrix} -16 & -8 \\ y & 40 \end{bmatrix}$$

(A) $x = 8, y = -80$ **(B)** $x = -4, y = -80$

(C) $x = 4, y = 80$ **(D)** $x = 4, y = -80$

(E) $x = 2, y = -40$

3. MULTIPLE CHOICE What is the product of

$$\begin{bmatrix} -1 & 0 & 4 \\ -2 & 1 & 3 \\ 3 & 2 & -1 \end{bmatrix} \text{ and } \begin{bmatrix} 1 & -2 \\ 0 & 1 \\ 5 & -1 \end{bmatrix}? \ \textbf{A}$$

(A) $\begin{bmatrix} 19 & -2 \\ 13 & 2 \\ -2 & -3 \end{bmatrix}$ **(B)** $\begin{bmatrix} 19 & -2 \\ 13 & -6 \\ -2 & -3 \end{bmatrix}$

(C) $\begin{bmatrix} 19 & -2 \\ 13 & 2 \\ -2 & -7 \end{bmatrix}$ **(D)** $\begin{bmatrix} 19 & -2 \\ 13 & 2 \\ 2 & -3 \end{bmatrix}$

(E) $\begin{bmatrix} 21 & -2 \\ 13 & 8 \\ -2 & -3 \end{bmatrix}$

4. MULTIPLE CHOICE What is the determinant of

$$\begin{bmatrix} 2 & 1 & 5 \\ -3 & -1 & 2 \\ 0 & 4 & -2 \end{bmatrix}? \ \textbf{A}$$

(A) -78 **(B)** -34 **(C)** -16

(D) 34 **(E)** 78

5. MULTIPLE CHOICE What is the inverse of

$$\begin{bmatrix} 9 & -5 \\ 7 & -4 \end{bmatrix}? \ \textbf{D}$$

(A) $\begin{bmatrix} 4 & -5 \\ -7 & -9 \end{bmatrix}$ **(B)** $\begin{bmatrix} -4 & 5 \\ 7 & 9 \end{bmatrix}$

(C) $\begin{bmatrix} -4 & -5 \\ -7 & -9 \end{bmatrix}$ **(D)** $\begin{bmatrix} 4 & -5 \\ 7 & -9 \end{bmatrix}$

(E) $\begin{bmatrix} -4 & 5 \\ -7 & 9 \end{bmatrix}$

6. MULTIPLE CHOICE Which matrix has no inverse? **B**

(A) $\begin{bmatrix} 6 & 0 \\ 0 & 5 \end{bmatrix}$ **(B)** $\begin{bmatrix} 4 & 6 \\ -6 & -9 \end{bmatrix}$

(C) $\begin{bmatrix} -2 & 4 \\ 3 & 6 \end{bmatrix}$ **(D)** $\begin{bmatrix} 1 & 1 \\ 0 & 1 \end{bmatrix}$

(E) $\begin{bmatrix} -4 & 4 \\ 2 & 1 \end{bmatrix}$

7. MULTIPLE CHOICE What is the area of the triangle in square units? **D**

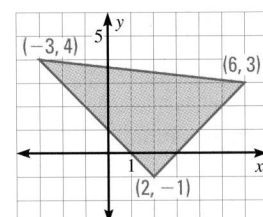

(A) 40 **(B)** 30 **(C)** 26

(D) 20 **(E)** 13

8. MULTIPLE CHOICE What is the solution of the linear system? **C**

$$7x + 5y = 6$$
$$4x + 3y = 3$$

(A) $x = 3, y = 3$ **(B)** $x = -3, y = -3$

(C) $x = 3, y = -3$ **(D)** $x = -3, y = 3$

(E) $x = 8, y = -10$

9. QUANTITATIVE COMPARISON Choose the statement that is true about the given quantities.   **A**

(A) The quantity in column A is greater.

(B) The quantity in column B is greater.

(C) The two quantities are equal.

(D) The relationship cannot be determined from the given information.

Column A	Column B
$\det \begin{bmatrix} 0 & 1 \\ -8 & 2 \end{bmatrix}$	$\det \begin{bmatrix} -2 & -1 \\ 3 & 5 \end{bmatrix}$

10. MULTI-STEP PROBLEM The School Spirit club ordered shirts to sell at basketball games. The number of each size and type of shirt they ordered is shown in the matrix at the right.

a. At the first basketball game the club sold 17 T-shirts (4 small, 5 medium, 6 large, and 2 extra large) and 12 sweatshirts (3 medium, 5 large, and 4 extra large). Write a matrix that gives the number of each size and type of shirt sold at the game. Then write a matrix that shows the number of shirts left.

b. The wholesale price of a T-shirt is $8, and the club sells them for $10 each. The wholesale price of a sweatshirt is $20, and the club sells them for $25 each. Write a row matrix for the number of each type of shirt (T-shirt or sweatshirt) sold at the game from part (a). Write a column matrix for the profit on each type of shirt. Multiply the row matrix by the column matrix and interpret the result.

10. a.
	T-shirt	Sweatshirt
S	4	0
M	5	3
L	6	5
XL	2	4

	T-shirt	Sweatshirt
S	21	20
M	20	17
L	94	35
XL	48	16

FUNDRAISING

	T-shirt	Sweatshirt
S	25	20
M	25	20
L	100	40
XL	50	20

$[17, \ 12]; \begin{bmatrix} 2 \\ 5 \end{bmatrix}; [94]$, **a profit of $94 was made at the first game.**

11. MULTI-STEP PROBLEM You and a friend are planning a secret meeting.

You agree that $A = \begin{bmatrix} 1 & 1 \\ 1 & 2 \end{bmatrix}$ will be your coding matrix. Use the coding information on pages 225 and 226.

a. Use matrix A to encode the message SATURDAY.   **20, 21, 41, 62, 22, 26, 26, 51**

b. Use the inverse of A to decode your answer to part (a).   **SATURDAY**

c. Your friend replied with the message below. Use the inverse of A to decode it.   **SOCCER FIELD**

> 34, 49, 6, 9, 23, 41, 6, 12, 14, 19, 16, 20

d. Use matrix A to encode your answer to part (c).   **34, 49, 6, 9, 23, 41, 6, 12, 14, 19, 16, 20**

12. MULTI-STEP PROBLEM Use the following linear system.

$$x - 2y = 1$$
$$3x - 5y = 4$$

a. Use Cramer's rule to solve the system.   **(3, 1)**

b. Use inverse matrices to solve the system.   **(3, 1)**

c. Use the substitution or linear combination method to solve the system.   **(3, 1)**

d. Solve the system by graphing. Label the solution on your graph.   **See margin.**

e. *Writing* Which method do you prefer for solving this linear system?   **See margin.**

12d.
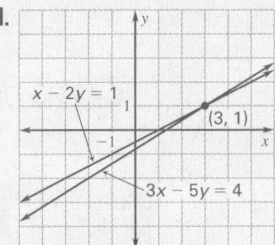

12e. **Here are some reasons different students may prefer each method:**
- **Some students may prefer Cramer's rule because it involves only simple calculations of determinants; there are no algebraic manipulations. Because the determinant of the coefficient matrix is 1, the denominators for x and y are 1, making Cramer's rule especially easy to use.**
- **Some students may prefer using inverse matrices because this method finds both the x and the y values at the same time. There is no need to remember multiple steps, as the whole answer is contained within the calculations of $A^{-1}B$. Because the determinant of the coefficient matrix is 1, the inverse is especially easy to find, making all the computations fairly straightforward.**
- **Students who are very comfortable with algebraic manipulations may prefer using the substitution or linear combination method. Both methods are easy to apply to this system.**
- **Students who are visually oriented may prefer the graphing method; however, these equations are not especially easy to graph because they have fractional y-intercepts. In addition, because the slopes are close in value, it is difficult to identify the intersection point of the lines from the graph.**

PLANNING THE CHAPTER
Quadratic Functions

LESSON	GOALS	NCTM	ITED	SAT9	Terra-Nova	Local
5.1 *pp. 249–255*	**GOAL 1** Graph quadratic functions. **GOAL 2** Use quadratic functions to solve real-life problems.	1, 2, 3, 6, 8, 9, 10	MIG, IIG, SAPE, RQGE	2, 22, 43	14, 16, 17, 18	
5.2 *pp. 256–263*	**GOAL 1** Factor quadratic expressions and solve quadratic equations by factoring. **GOAL 2** Find zeros of quadratic functions.	1, 2	SAPE	43	11, 16, 49, 51, 52	
5.3 *pp. 264–271*	**GOAL 1** Solve quadratic equations by finding square roots. **GOAL 2** Use quadratic equations to solve real-life problems. **TECHNOLOGY ACTIVITY: 5.3** *Solve quadratic equations having real-number solutions on a graphing calculator.*	1, 2, 3, 6, 8, 9, 10	SAPE, RQWN	7	11, 16, 17, 18, 49, 51, 52	
5.4 *pp. 272–280*	**GOAL 1** Solve quadratic equations with complex solutions and perform operations with complex numbers. **GOAL 2** Apply complex numbers to fractal geometry.	1, 2, 3, 9	MCWN, RQGE, SAPE		11, 14, 16, 49, 51, 52	
5.5 *pp. 281–290*	**CONCEPT ACTIVITY: 5.5** *Investigate completing the square using algebra tiles.* **GOAL 1** Solve quadratic equations by completing the square. **GOAL 2** Use completing the square to write quadratic functions in vertex form. **TECHNOLOGY ACTIVITY: 5.5** *Find maximum and minimum values of quadratic functions on a graphing calculator.*	1, 2, 9, 10	SAPE, MIG, IIG	7, 43	11, 14, 16, 49, 51, 52	
5.6 *pp. 291–298*	**GOAL 1** Solve quadratic equations using the quadratic formula. **GOAL 2** Use the quadratic formula in real-life situations.	1, 2, 8, 9, 10	SAPE, RQWN	2, 7	11, 16, 18, 49, 51, 52	
5.7 *pp. 299–305*	**GOAL 1** Graph quadratic inequalities in two variables. **GOAL 2** Solve quadratic inequalities in one variable.	1, 2, 3	MIG, IIG	38	11, 14, 16, 49, 51, 52	
5.8 *pp. 306–312*	**GOAL 1** Write quadratic functions given characteristics of their graphs. **GOAL 2** Use technology to find quadratic models for data.	1, 2, 3, 5, 9, 10	SAPE	11, 21, 23	16	

RESOURCES

CHAPTER RESOURCE BOOKLETS

CHAPTER SUPPORT

Tips for New Teachers	p. 1	Prerequisite Skills Review	p. 5
Parent Guide for Student Success	p. 3	Strategies for Reading Mathematics	p. 7

LESSON SUPPORT

	5.1	5.2	5.3	5.4	5.5	5.6	5.7	5.8
Lesson Plans (regular and block)	p. 9	p. 24	p. 36	p. 51	p. 63	p. 77	p. 92	p. 105
Warm-Up Exercises and Daily Quiz	p.11	p. 26	p. 38	p. 53	p. 65	p. 79	p. 94	p. 107
Activity Support Masters					p. 66			
Lesson Openers	p. 12	p. 27	p. 39	p. 54	p. 67	p. 80	p. 95	p. 108
Graphing Calculator Activities & Keystrokes	p. 13		p. 40		p. 68	p. 81	p. 96	p. 109
Practice (3 levels)	p. 15	p. 28	p. 41	p. 55	p. 69	p. 83	p. 97	p. 110
Reteaching with Practice	p. 18	p. 31	p. 44	p. 58	p. 72	p. 86	p. 100	p. 113
Quick Catch-Up for Absent Students	p. 20	p. 33	p. 46	p. 60	p. 74	p. 88	p. 102	p. 115
Cooperative Learning Activities	p. 21							
Interdisciplinary Applications		p. 34		p. 61		p. 89		p. 116
Real-Life Applications	p. 22		p. 47		p. 75		p. 103	
Math & History Applications			p. 48					
Challenge: Skills and Applications	p. 23	p. 35	p. 49	p. 62	p. 76	p. 90	p. 104	p. 117

REVIEW AND ASSESSMENT

Quizzes	pp. 50, 91	Alternative Assessment with Math Journal	p. 126
Chapter Review Games and Activities	p. 118	Project with Rubric	p. 128
Chapter Test (3 levels)	pp. 119–124	Cumulative Review	p. 130
SAT/ACT Chapter Test	p. 125	Resource Book Answers	p. A1

TRANSPARENCIES

	5.1	5.2	5.3	5.4	5.5	5.6	5.7	5.8
Warm-Up Exercises and Daily Quiz	p. 32	p. 33	p. 34	p. 35	p. 36	p. 37	p. 38	p. 39
Alternative Lesson Opener Transparencies	p. 27	p. 28	p. 29	p. 30	p. 31	p. 32	p. 33	p. 34
Examples/Standardized Test Practice	✓	✓	✓	✓	✓	✓	✓	✓
Answer Transparencies	✓	✓	✓	✓	✓	✓	✓	✓

TECHNOLOGY

- Electronic Teaching Tools
- Online Lesson Planner
- Internet Support
- Personal Student Tutor
- Test and Practice Generator
- Instant Replay: Video Review Games
- Electronic Lesson Presentations (Lesson 5.1)

ADDITIONAL RESOURCES

- Basic Skills Workbook: Diagnosis and Remediation
- Worked-Out Solution Key
- Resources in Spanish
- Standardized Test Practice Workbook
- Practice Workbook with Examples

PACING THE CHAPTER

REGULAR SCHEDULE

Day 1

5.1

STARTING OPTIONS
- Prereq. Skills Review
- Strategies for Reading
- Homework Check
- Warm-Up or Daily Quiz

TEACHING OPTIONS
- Motivating the Lesson
- Les. Opener (Calc.)
- Graphing Calc. Activity
- Examples 1–3
- Guided Practice
  Exs. 1–2, 4–9

APPLY/HOMEWORK
- See Assignment Guide.
- See the CRB: Practice,
  Reteach, Apply, Extend

ASSESSMENT OPTIONS
- Checkpoint Exercises,
  pp. 250–251

Day 2

5.1 (cont.)

STARTING OPTIONS
- Homework Check

TEACHING OPTIONS
- Examples 4–6
- Closure Question
- Guided Practice
  Exs. 3, 10–16

APPLY/HOMEWORK
- See Assignment Guide.
- See the CRB: Practice,
  Reteach, Apply, Extend

ASSESSMENT OPTIONS
- Checkpoint Exercises,
  pp. 251–252
- Daily Quiz (5.1)
- Stand. Test Practice

Day 3

5.2

STARTING OPTIONS
- Homework Check
- Warm-Up or Daily Quiz

TEACHING OPTIONS
- Motivating the Lesson
- Les. Opener (Visual)
- Examples 1–4
- Guided Practice
  Exs. 2, 4–9

APPLY/HOMEWORK
- See Assignment Guide.
- See the CRB: Practice,
  Reteach, Apply, Extend

ASSESSMENT OPTIONS
- Checkpoint Exercises,
  p. 257

Day 4

5.2 (cont.)

STARTING OPTIONS
- Homework Check

TEACHING OPTIONS
- Examples 5–8
- Closure Question
- Guided Practice
  Exs. 1, 3, 10–22

APPLY/HOMEWORK
- See Assignment Guide.
- See the CRB: Practice,
  Reteach, Apply, Extend

ASSESSMENT OPTIONS
- Checkpoint Exercises,
  pp. 258–259
- Daily Quiz (5.2)
- Stand. Test Practice

Day 5

5.3

STARTING OPTIONS
- Homework Check
- Warm-Up or Daily Quiz

TEACHING OPTIONS
- Motivating the Lesson
- Les. Opener (Appl.)
- Graphing Calc. Activity
- Examples 1–4
- Technology Activity
- Closure Question
- Guided Practice Exs.

APPLY/HOMEWORK
- See Assignment Guide.
- See the CRB: Practice,
  Reteach, Apply, Extend

ASSESSMENT OPTIONS
- Checkpoint Exercises
- Daily Quiz (5.3)
- Stand. Test Practice
- Quiz (5.1–5.3)

Day 6

5.4

STARTING OPTIONS
- Homework Check
- Warm-Up or Daily Quiz

TEACHING OPTIONS
- Motivating the Lesson
- Les. Opener (Activity)
- Examples 1–4
- Guided Practice
  Exs. 1, 3–9

APPLY/HOMEWORK
- See Assignment Guide.
- See the CRB: Practice,
  Reteach, Apply, Extend

ASSESSMENT OPTIONS
- Checkpoint Exercises,
  pp. 273–274

Day 9

5.5 (cont.)

STARTING OPTIONS
- Homework Check

TEACHING OPTIONS
- Examples 4–7
- Technology Activity
- Closure Question
- Guided Practice
  Exs. 3, 16–22

APPLY/HOMEWORK
- See Assignment Guide.
- See the CRB: Practice,
  Reteach, Apply, Extend

ASSESSMENT OPTIONS
- Checkpoint Exercises,
  pp. 284–285
- Daily Quiz (5.5)
- Stand. Test Practice

Day 10

5.6

STARTING OPTIONS
- Homework Check
- Warm-Up or Daily Quiz

TEACHING OPTIONS
- Motivating the Lesson
- Les. Opener (Visual)
- Graphing Calc. Activity
- Examples 1–5
- Closure Question
- Guided Practice Exs.

APPLY/HOMEWORK
- See Assignment Guide.
- See the CRB: Practice,
  Reteach, Apply, Extend

ASSESSMENT OPTIONS
- Checkpoint Exercises
- Daily Quiz (5.6)
- Stand. Test Practice
- Quiz (5.4–5.6)

Day 11

5.7

STARTING OPTIONS
- Homework Check
- Warm-Up or Daily Quiz

TEACHING OPTIONS
- Motivating the Lesson
- Les. Opener (Activity)
- Graphing Calc. Activity
- Examples 1–7
- Closure Question
- Guided Practice Exs.

APPLY/HOMEWORK
- See Assignment Guide.
- See the CRB: Practice,
  Reteach, Apply, Extend

ASSESSMENT OPTIONS
- Checkpoint Exercises
- Daily Quiz (5.7)
- Stand. Test Practice

Day 12

5.8

STARTING OPTIONS
- Homework Check
- Warm-Up or Daily Quiz

TEACHING OPTIONS
- Les. Opener
  (Application)
- Graphing Calc. Activity
- Examples 1–4
- Closure Question
- Guided Practice Exs.

APPLY/HOMEWORK
- See Assignment Guide.
- See the CRB: Practice,
  Reteach, Apply, Extend

ASSESSMENT OPTIONS
- Checkpoint Exercises
- Daily Quiz (5.8)
- Stand. Test Practice
- Quiz (5.7–5.8)

Day 13

Review

DAY 13 START OPTIONS
- Homework Check

REVIEWING OPTIONS
- Chapter 5 Summary
- Chapter 5 Review
- Chapter Review
  Games and Activities

APPLY/HOMEWORK
- Chapter 5 Test
  (practice)
- Ch. Standardized Test
  (practice)

Day 14

Assess

DAY 14 START OPTIONS
- Homework Check

ASSESSMENT OPTIONS
- Chapter 5 Test
- SAT/ACT Ch. 5 Test
- Alternative Assessment

APPLY/HOMEWORK
- Skill Review, p. 322

BLOCK SCHEDULE

Day 7

5.4 (cont.)

STARTING OPTIONS
- Homework Check

TEACHING OPTIONS
- Examples 5–7
- Closure Question
- Guided Practice
  Exs. 2, 10–16

APPLY/HOMEWORK
- See Assignment Guide.
- See the CRB: Practice, Reteach, Apply, Extend

ASSESSMENT OPTIONS
- Checkpoint Exercises, pp. 274–276
- Daily Quiz (5.4)
- Stand. Test Practice

Day 8

5.5

STARTING OPTIONS
- Homework Check
- Warm-Up or Daily Quiz

TEACHING OPTIONS
- Motivating the Lesson
- Concept Act. & Wksht.
- Les. Opener (Application)
- Graphing Calc. Activity
- Examples 1–3
- Guided Practice
  Exs. 1–2, 4–15

APPLY/HOMEWORK
- See Assignment Guide.
- See the CRB: Practice, Reteach, Apply, Extend

ASSESSMENT OPTIONS
- Checkpoint Exercises, p. 283

Day 1

5.1

DAY 1 START OPTIONS
- Prereq. Skills Review
- Strategies for Reading
- Homework Check
- Warm-Up or Daily Quiz

TEACHING 5.1 OPTIONS
- Motivating the Lesson
- Les. Opener (Calc.)
- Graphing Calc. Activity
- Examples 1–6
- Closure Question
- Guided Practice Exs.

APPLY/HOMEWORK
- See Assignment Guide.
- See the CRB: Practice, Reteach, Apply, Extend

ASSESSMENT OPTIONS
- Checkpoint Exercises
- Daily Quiz (Les. 5.1)
- Stand. Test Practice

Day 2

5.2

DAY 2 START OPTIONS
- Homework Check
- Warm-Up or Daily Quiz

TEACHING 5.2 OPTIONS
- Motivating the Lesson
- Les. Opener (Visual)
- Examples 1–8
- Closure Question
- Guided Practice Exs.

APPLY/HOMEWORK
- See Assignment Guide.
- See the CRB: Practice, Reteach, Apply, Extend

ASSESSMENT OPTIONS
- Checkpoint Exercises
- Daily Quiz (Les. 5.2)
- Stand. Test Practice

Day 3

5.3 & 5.4

DAY 3 START OPTIONS
- Homework Check
- W-Up 5.3 or D. Quiz 5.2

TEACHING 5.3 OPTIONS
- Motivating the Lesson
- Les. Opener (Appl.)
- Graphing Calc. Activity
- Examples 1–4
- Technology Activity
- Closure Question
- Guided Practice Exs.

BEGINNING 5.4 OPTIONS
- Warm-Up (Les. 5.4)
- Motivating the Lesson
- Les. Opener (Activity)
- Examples 1–4
- Guided Practice
  Exs. 1, 3–9

APPLY/HOMEWORK
- See Assignment Guide.
- See the CRB: Practice, Reteach, Apply, Extend

ASSESSMENT OPTIONS
- Checkpoint Exercises
- Daily Quiz (Les. 5.3)
- Stand. Test Prac. (5.3)
- Quiz (5.1–5.3)

Day 4

5.4 & 5.5

DAY 4 START OPTIONS
- Homework Check
- Daily Quiz (Les. 5.3)

FINISHING 5.4 OPTIONS
- Examples 5–7
- Closure Question
- Guided Practice
  Exs. 2, 10–16

BEGINNING 5.5 OPTIONS
- Warm-Up (Les. 5.5)
- Motivating the Lesson
- Concept Act. & Wksht.
- Les. Opener (Appl.)
- Graphing Calc. Activity
- Examples 1–3
- Guided Practice
  Exs. 1–2, 4–15

APPLY/HOMEWORK
- See Assignment Guide.
- See the CRB: Practice, Reteach, Apply, Extend

ASSESSMENT OPTIONS
- Checkpoint Exercises
- Daily Quiz (Les. 5.4)
- Stand. Test Prac. (5.4)

Day 5

5.5 & 5.6

DAY 5 START OPTIONS
- Homework Check
- Daily Quiz (Les. 5.4)

FINISHING 5.5 OPTIONS
- Examples 4–7
- Technology Activity
- Closure Question
- Guided Practice
  Exs. 3, 16–22

TEACHING 5.6 OPTIONS
- Warm-Up (Les. 5.6)
- Motivating the Lesson
- Les. Opener (Visual)
- Graphing Calc. Activity
- Examples 1–5
- Closure Question
- Guided Practice Exs.

APPLY/HOMEWORK
- See Assignment Guide.
- See the CRB: Practice, Reteach, Apply, Extend

ASSESSMENT OPTIONS
- Checkpoint Exercises
- Daily Quiz (Les. 5.5, 5.6)
- Stand. Test Practice
- Quiz (5.4–5.6)

Day 6

5.7 & 5.8

DAY 6 START OPTIONS
- Homework Check
- W-Up 5.7 or D. Quiz 5.6

TEACHING 5.7 OPTIONS
- Motivating the Lesson
- Les. Opener (Activity)
- Graphing Calc. Activity
- Examples 1–7
- Closure Question
- Guided Practice Exs.

TEACHING 5.8 OPTIONS
- Warm-Up (Les. 5.8)
- Les. Opener (Appl.)
- Graphing Calc. Activity
- Examples 1–4
- Closure Question
- Guided Practice Exs.

APPLY/HOMEWORK
- See Assignment Guide.
- See the CRB: Practice, Reteach, Apply, Extend

ASSESSMENT OPTIONS
- Checkpoint Exercises
- Daily Quiz (Les. 5.7, 5.8)
- Stand. Test Practice
- Quiz (5.7–5.8)

Day 7

Review/Assess

DAY 7 START OPTIONS
- Homework Check

REVIEWING OPTIONS
- Chapter 5 Summary
- Chapter 5 Review
- Chapter Review
  Games and Activities
- Chapter 5 Test
  (practice)
- Ch. Standardized Test
  (practice)

ASSESSMENT OPTIONS
- Chapter 5 Test
- SAT/ACT Ch. 5 Test
- Alternative Assessment

APPLY/HOMEWORK
- Skill Review, p. 322

BEFORE THE CHAPTER

The *Chapter 5 Resource Book* has the following materials to distribute and use before the chapter:

- **Parent Guide for Student Success (pictured below)**
- **Prerequisite Skills Review**
- **Strategies for Reading Mathematics**

PARENT GUIDE *Pages 3–4*

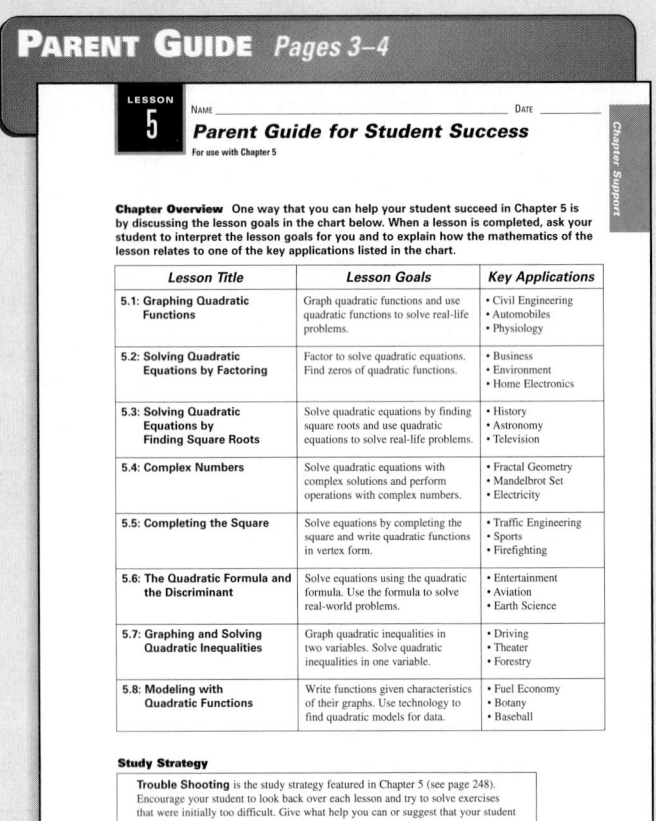

PARENT GUIDE FOR STUDENT SUCCESS The first page summarizes the content of Chapter 5. Parents are encouraged to have their students explain how the material relates to key applications in the chapter, such as aviation. The second page (not shown) provides exercises and an activity that parents can do with their students. In the activity, parents and students find the dimensions of a rug by using a quadratic equation.

DURING EACH LESSON

The *Chapter 5 Resource Book* has the following alternatives for introducing the lesson:

- **Lesson Openers (pictured below)**
- **Graphing Calculator Activities with Keystrokes**

LESSON OPENER *Page 12*

Available as a transparency

LESSON 5.1 Graphing Calculator Lesson Opener
For use with pages 249–255

1. Use a graphing calculator to graph each function. Estimate the lowest point on each graph.

Function	$y = x^2 - 3$	$y = x^2 + 0$	$y = x^2 + 2$
Low point			

2. Look for a pattern. Without graphing, guess the lowest point on the graph of $y = x^2 + 7$.

3. Use a graphing calculator to graph each function. Estimate the lowest point on each graph.

Function	$y = (x - 3)^2$	$y = (x - 1)^2$	$y = (x + 2)^2$
Low point			

4. Look for a pattern. Without graphing, guess the lowest point on the graph of $y = (x - 6)^2$.

5. Use a graphing calculator to graph each function. Estimate the lowest point on each graph.

Function	$y = (x - 2)^2 + 1$	$y = (x - 2)^2 - 3$
Low point		

6. Describe the effects of h and k on the graph of $y = (x - h)^2 + k$.

GRAPHING CALCULATOR LESSON OPENER This Lesson Opener provides an alternative way to start Lesson 5.1 through the use of a graphing calculator. Students graph quadratic functions to develop an understanding of their behaviors and properties.

 TECHNOLOGY RESOURCE

Teachers can use the Electronic Lesson Presentations CD-ROM to present Lesson 5.1 using computer animation to step through the Examples.

The *Chapter 5 Resource Book* has a variety of materials to follow-up each lesson. They include the following:

- **Practice (3 levels) (pictured below)**
- **Reteaching with Practice**
- **Quick Catch-Up for Absent Students**
- **Challenge: Skills and Applications**
- **Interdisciplinary Applications**
- **Real-Life Applications**

PRACTICE Page 16

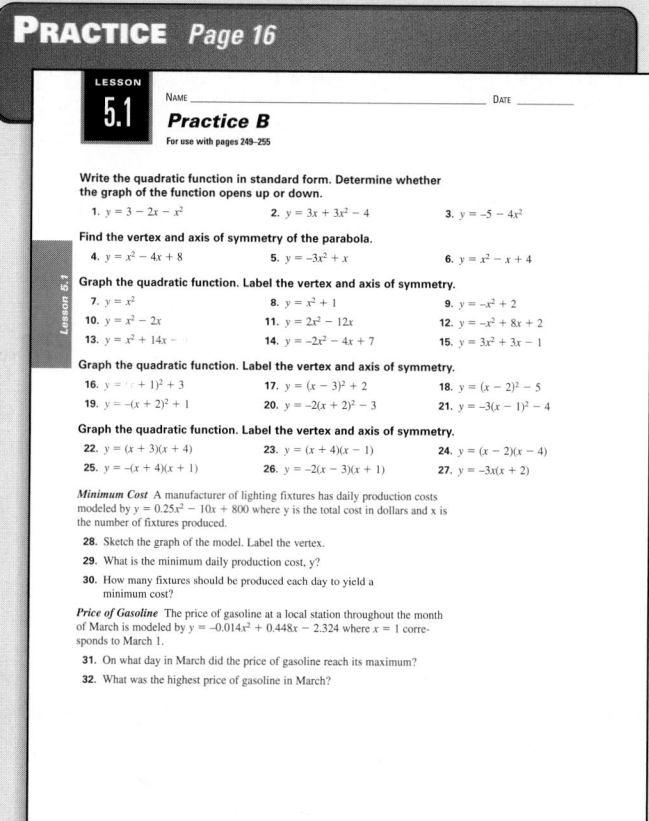

LESSON 5.1

NAME _____ DATE _____

Practice B
For use with pages 249–255

Write the quadratic function in standard form. Determine whether the graph of the function opens up or down.

1. $y = 3 - 2x - x^2$
2. $y = 3x + 3x^2 - 4$
3. $y = -5 - 4x^2$

Find the vertex and axis of symmetry of the parabola.

4. $y = x^2 - 4x + 8$
5. $y = -3x^2 + x$
6. $y = x^2 - x + 4$

Graph the quadratic function. Label the vertex and axis of symmetry.

7. $y = x^2$
8. $y = x^2 + 1$
9. $y = -x^2 + 2$
10. $y = x^2 - 2x$
11. $y = 2x^2 - 12x$
12. $y = -x^2 + 8x + 2$
13. $y = x^2 + 14x -$
14. $y = -2x^2 - 4x + 7$
15. $y = 3x^2 + 3x - 1$

Graph the quadratic function. Label the vertex and axis of symmetry.

16. $y = (x + 1)^2 + 3$
17. $y = (x - 3)^2 + 2$
18. $y = (x - 2)^2 - 5$
19. $y = -(x + 2)^2 + 1$
20. $y = -2(x + 2)^2 - 3$
21. $y = -3(x - 1)^2 - 4$

Graph the quadratic function. Label the vertex and axis of symmetry.

22. $y = (x + 3)(x + 4)$
23. $y = (x + 4)(x - 1)$
24. $y = (x - 2)(x - 4)$
25. $y = -(x + 4)(x + 1)$
26. $y = -2(x - 3)(x + 1)$
27. $y = -3x(x + 2)$

Minimum Cost A manufacturer of lighting fixtures has daily production costs modeled by $y = 0.25x^2 - 10x + 800$ where y is the total cost in dollars and x is the number of fixtures produced.

28. Sketch the graph of the model. Label the vertex.
29. What is the minimum daily production cost, y?
30. How many fixtures should be produced each day to yield a minimum cost?

Price of Gasoline The price of gasoline at a local station throughout the month of March is modeled by $y = -0.014x^2 + 0.448x - 2.324$ where $x = 1$ corresponds to March 1.

31. On what day in March did the price of gasoline reach its maximum?
32. What was the highest price of gasoline in March?

PRACTICE Each lesson has three levels of practice: basic (A), average (B), and advanced (C). This Practice B involves more advanced work with graphing quadratic functions than Practice A, but does not have as much work with application problems as Practice C.

TECHNOLOGY RESOURCE

Teachers can use the Time-Saving Test and Practice Generator to create additional practice worksheets for Lesson 5.1 and for the other lessons in Chapter 5.

The *Chapter 5 Resource Book* has the following review and assessment materials:

- **Quizzes**
- **Chapter Review Games and Activities**
- **Chapter Test (3 levels)**
- **SAT/ACT Chapter Test**
- **Alternative Assessment with Math Journal (below)**
- **Project with Rubric**
- **Cumulative Review**

ALTERNATIVE ASSESSMENT Pages 126–127

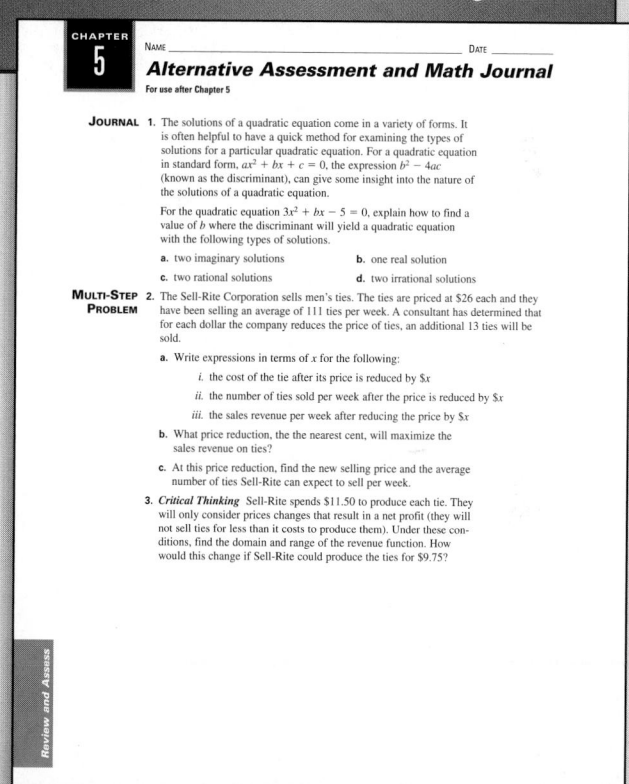

CHAPTER 5

NAME _____ DATE _____

Alternative Assessment and Math Journal
For use after Chapter 5

JOURNAL

1. The solutions of a quadratic equation come in a variety of forms. It is often helpful to have a quick method for examining the types of solutions for a particular quadratic equation. For a quadratic equation in standard form, $ax^2 + bx + c = 0$, the expression $b^2 - 4ac$ (known as the discriminant), can give some insight into the nature of the solutions of a quadratic equation.

For the quadratic equation $3x^2 + bx - 5 = 0$, explain how to find a value of b where the discriminant will yield a quadratic equation with the following types of solutions.

   a. two imaginary solutions
   b. one real solution
   c. two rational solutions
   d. two irrational solutions

MULTI-STEP PROBLEM

2. The Sell-Rite Corporation sells men's ties. The ties are priced at $26 each and they have been selling an average of 111 ties per week. A consultant has determined that for each dollar the company reduces the price of ties, an additional 13 ties will be sold.

   a. Write expressions in terms of x for the following:
      i. the cost of the tie after its price is reduced by $x
      ii. the number of ties sold per week after the price is reduced by $x
      iii. the sales revenue per week after reducing the price by $x
   b. What price reduction, the the nearest cent, will maximize the sales revenue on ties?
   c. At this price reduction, find the new selling price and the average number of ties Sell-Rite can expect to sell per week.

3. *Critical Thinking* Sell-Rite spends $11.50 to produce each tie. They will only consider prices changes that result in a net profit (they will not sell ties for less than it costs to produce them). Under these conditions, find the domain and range of the revenue function. How would this change if Sell-Rite could produce the ties for $9.75?

ALTERNATIVE ASSESSMENT WITH RUBRIC AND MATH JOURNAL
The journal exercise asks students to demonstrate their understanding of quadratic equations in writing. The Multi-Step Problem has students pull together a variety of concepts from Chapter 5 to solve a problem about the relationships between price, revenue, and products sold after a price reduction. Answers and a scoring rubric are provided on a separate sheet.

CHAPTER GOALS

Chapter 5 leads students through all the major topics involving quadratic functions. First quadratic equations are graphed and then the standard, vertex, and intercept forms are introduced. These forms are used throughout the chapter and students will learn to convert between them. Quadratic expressions are factored. Then quadratic equations are solved by factoring, finding square roots, completing the square, or using the quadratic formula. Factoring is also used to find the zeros of quadratic functions. Quadratic models are written using a graph or quadratic regression and are used to solve real-life problems. Students will solve quadratic equations with complex solutions and perform operations with complex numbers. The discriminant is used to determine the number and nature of the solutions to a quadratic equation. Complex numbers are applied to fractal geometry. Finally, quadratic inequalities in two variables, including systems of inequalities are graphed, and the graphs of quadratic functions are used to solve quadratic inequalities in one variable.

APPLICATION NOTE

The movement of an object dropped or thrown near Earth's surface is governed by gravity. The height is dependent only on the vertical velocity with which the object is thrown and its initial height. Point out to students that the height of a falling object is not dependent on its weight. Note however, that air resistance can affect the height of the object. A leaf will take longer to fall to Earth than a pebble because the shape of the leaf will make it more prone to air resistance.

Additional information about volcanoes is available at **www.mcdougallittell.com.**

QUADRATIC FUNCTIONS

▶ *How can you model the height of lava from an erupting volcano?*

246

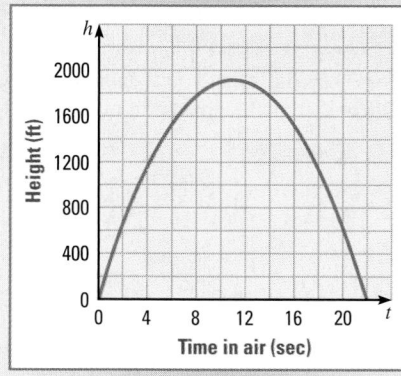

CHAPTER 5

APPLICATION: Volcanoes

Volcanic eruptions can eject lava hundreds of feet into the air, creating spectacular but dangerous "lava fountains." As the lava cools and hardens, it may accumulate to form the cone shape of a volcano.

Think & Discuss

The highest recorded lava fountain occurred during a 1959 eruption at Kilauea Iki Crater in Hawaii. The graph models the height of a typical lava fragment in the fountain while the fragment was in the air.

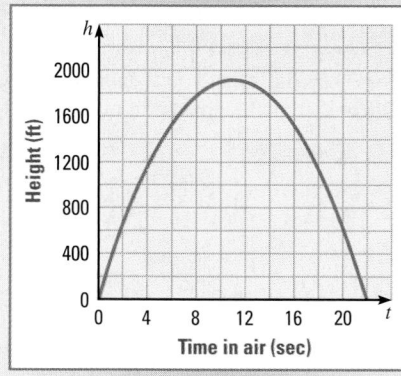

1. Estimate the lava fragment's maximum height above the ground.   about 1900 ft

2. For how long was the lava fragment in the air? How did you use the graph to get your answer?
   about 22 sec; See margin for explanation.

Learn More About It

You will work with an equation modeling the height of lava in Exercise 80 on p. 297.

 APPLICATION LINK Visit www.mcdougallittell.com for more information about volcanoes.

ADDITIONAL RESOURCES
Another way to begin the chapter is to show the video clip of a real-life motivator for Chapter 5 on the *Instant Replay: Video Review Games.*

PROJECTS
A project covering Chapters 4–6 appears on page 396 of the Student Edition. An additional project for Chapter 5 is available in the *Chapter 5 Resource Book,* p. 128.

TECHNOLOGY
 **Software**
- *Electronic Teaching Tools*
- *Online Lesson Planner*
- *Personal Student Tutor*
- *Test and Practice Generator*
- *Electronic Lesson Presentations* (Lesson 5.1)

Video
- *Instant Replay: Video Review Games*

Internet Connections
www.mcdougallittell.com
- **Application Links**
  247, 254, 268, 270, 311
- **Data Updates**
  280, 297, 305, 309
- **Student Help**
  254, 259, 265, 271, 279, 285, 290, 292, 304, 308, 310
- **Career Links**
  252, 279, 296, 304, 308
- **Extra Challenge**
  255, 263, 269, 280, 289, 297, 305, 312

2. Get the answer by measuring the distance between the points where the graph hits the *x*-axis.

247

247

DIAGNOSTIC TOOLS

The **Skill Review** exercises can help you diagnose whether students have the following skills needed in Chapter 5:

- Solve linear equations.
- Graph linear inequalities.
- Graph absolute value functions.

The following resources are available for students who need additional help with these skills:

- Prerequisite Skills Review (*Chapter 5 Resource Book,* p. 5; *Warm-Up Transparencies,* p. 31)
- Reteaching with Practice (Chapter Resource Books for Lessons 1.3, 2.6, and 2.8)
- 🖿 *Personal Student Tutor*

ADDITIONAL RESOURCES

The following resources are provided to help you prepare for the upcoming chapter and customize review materials:

- ***Chapter 5 Resource Book***
  Tips for New Teachers (p. 1)
  Parent Guide (p. 3)
  Lesson Plans (every lesson)
  Lesson Plans for Block Scheduling (every lesson)
- 🖿 *Electronic Teaching Tools*
- 🖿 *Online Lesson Planner*
- 🖿 *Test and Practice Generator*

4.
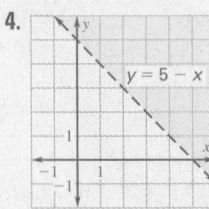

5.
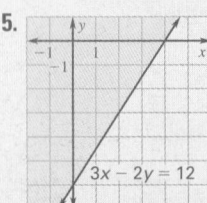

6–9. See Additional Answers beginning on page AA1.

PREVIEW

What's the chapter about?

Chapter 5 is about **quadratic functions, equations, and inequalities.** Many real-life situations can be modeled by quadratic functions. In Chapter 5 you'll learn

- four ways to solve quadratic equations.
- how to graph quadratic functions and inequalities.

KEY VOCABULARY

▶ **Review**
- linear equation, p. 19
- linear inequality, pp. 41, 108
- absolute value, p. 50
- linear function, p. 69
- x-intercept, p. 84
- best-fitting line, p. 101
- vertex, p. 122

▶ **New**
- quadratic function, p. 249
- parabola, p. 249
- factoring, p. 256
- quadratic equation, p. 257
- zero of a function, p. 259
- square root, p. 264
- complex number, p. 272

- completing the square, p. 282
- quadratic formula, p. 291
- discriminant, p. 293
- quadratic inequality, pp. 299, 301
- best-fitting quadratic model, p. 308

PREPARE

Are you ready for the chapter?

SKILL REVIEW Do these exercises to review key skills that you'll apply in this chapter. See the given **reference page** if there is something you don't understand.

┌─── **STUDENT HELP** ───
│
│ ↳ **Study Tip**
│ "Student Help" boxes throughout the chapter give you study tips and tell you where to look for extra help in this book and on the Internet.

Solve the equation. (Review Examples 1–3, pp. 19 and 20)

1. $3x - 5 = 0$ $\frac{5}{3}$
2. $4(x + 6) = 12$ -3
3. $2x + 1 = -x + 7$ 2

Graph the inequality. (Review Example 3, p. 109) 4–6. See margin.

4. $x + y > 5$
5. $3x - 2y \leq 12$
6. $y \geq -2x$

Graph the function and label the vertex. (Review Example 1, p. 123) 7–9. See margin.

7. $y = |x| + 2$
8. $y = |x - 3|$
9. $y = -2|x + 1| - 4$

STUDY STRATEGY

Here's a study strategy!

Troubleshoot

After you complete each lesson, look back and identify your trouble spots, such as concepts you didn't understand or homework problems you had difficulty solving. Review the material given in the lesson and try to solve any difficult problems again. If you're still having trouble, seek the help of another student or your teacher.

5.1

Graphing Quadratic Functions

What you should learn

GOAL 1 Graph quadratic functions.

GOAL 2 Use quadratic functions to solve **real-life** problems, such as finding comfortable temperatures in **Example 5**.

Why you should learn it

▼ To model **real-life** objects, such as the cables of the Golden Gate Bridge in **Example 6**.

GOAL 1 **GRAPHING A QUADRATIC FUNCTION**

A **quadratic function** has the form $y = ax^2 + bx + c$ where $a \neq 0$. The graph of a quadratic function is U-shaped and is called a **parabola**.

For instance, the graphs of $y = x^2$ and $y = -x^2$ are shown at the right. The origin is the lowest point on the graph of $y = x^2$ and the highest point on the graph of $y = -x^2$. The lowest or highest point on the graph of a quadratic function is called the **vertex**.

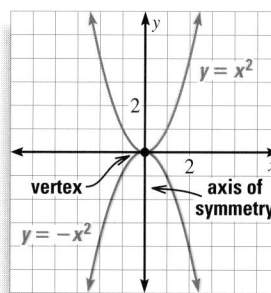

The graphs of $y = x^2$ and $y = -x^2$ are symmetric about the y-axis, called the *axis of symmetry*. In general, the **axis of symmetry** for the graph of a quadratic function is the vertical line through the vertex.

ACTIVITY

Developing Concepts

Investigating Parabolas

Steps 1, 2. See margin.

❶ Use a graphing calculator to graph each of these functions in the same viewing window: $y = \frac{1}{2}x^2$, $y = x^2$, $y = 2x^2$, and $y = 3x^2$.

❷ Repeat **Step 1** for these functions: $y = -\frac{1}{2}x^2$, $y = -x^2$, $y = -2x^2$, and $y = -3x^2$.

❸ What are the vertex and axis of symmetry of the graph of $y = ax^2$? (0, 0); $x = 0$

❹ Describe the effect of a on the graph of $y = ax^2$.
The graph opens up if $a > 0$, the graph opens down if $a < 0$.

In the activity you examined the graph of the simple quadratic function $y = ax^2$. The graph of the more general function $y = ax^2 + bx + c$ is described below.

CONCEPT SUMMARY

THE GRAPH OF A QUADRATIC FUNCTION

The graph of $y = ax^2 + bx + c$ is a parabola with these characteristics:

• The parabola opens up if $a > 0$ and opens down if $a < 0$. The parabola is wider than the graph of $y = x^2$ if $|a| < 1$ and narrower than the graph of $y = x^2$ if $|a| > 1$.

• The x-coordinate of the vertex is $-\dfrac{b}{2a}$.

• The axis of symmetry is the vertical line $x = -\dfrac{b}{2a}$.

5.1 Graphing Quadratic Functions **249**

Step 1–2. See Additional Answers beginning on page AA1.

---

1 PLAN

PACING
Basic: 2 days
Average: 2 days
Advanced: 2 days
Block Schedule: 1 block

LESSON OPENER
GRAPHING CALCULATOR
An alternative way to approach Lesson 5.1 is to use the Graphing Calculator Lesson Opener:

• Blackline Master (*Chapter 5 Resource Book*, p. 12)
• Transparency (p. 27)

MEETING INDIVIDUAL NEEDS
• *Chapter 5 Resource Book*
  Prerequisite Skills Review (p. 5)
  Practice Level A (p. 15)
  Practice Level B (p. 16)
  Practice Level C (p. 17)
  Reteaching with Practice (p. 18)
  Absent Student Catch-Up (p. 20)
  Challenge (p. 23)
• *Resources in Spanish*
• *Personal Student Tutor*

NEW-TEACHER SUPPORT
See the Tips for New Teachers on pp. 1–2 of the *Chapter 5 Resource Book* for additional notes about Lesson 5.1.

WARM-UP EXERCISES

 Transparency Available

Evaluate when $x = -1$, 0, and 2.
1. $y = 2x^2 - 3x + 5$ 10, 5, 7
2. $y = 3(x - 7)^2 - 6$ 186, 141, 69
3. $y = -(x + 3)(2x - 7)$ 18, 21, 15

ENGLISH LEARNERS
Make sure that English learners understand parabolas. Draw a parabola that *opens up* and another that *opens down* to point out the distinction between the two. Then draw a narrow parabola and a wider one, pointing out which is *narrower* and which is *wider*.

249

Tell students to think of the shape of a satellite dish. The cross section of a satellite dish and many other real-life objects can be modeled by a quadratic function. Quadratic functions and their graphs are the focus of today's lesson.

EXTRA EXAMPLE 1
Graph $y = -x^2 + x + 12$.

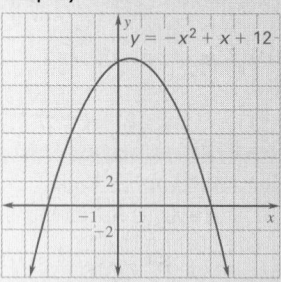

EXTRA EXAMPLE 2
Graph $y = 2(x-1)^2 + 3$.

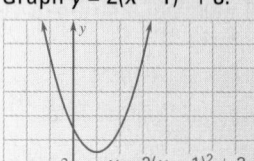

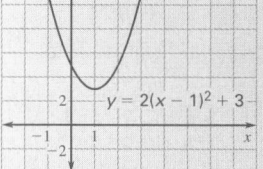

✔ CHECKPOINT EXERCISES
For use after Example 1:

1. Graph $y = \frac{1}{2}x^2 - x - 6$.

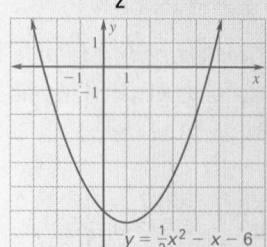

For use after Example 2:

2. Graph $y = -(x+5)^2 + 2$.

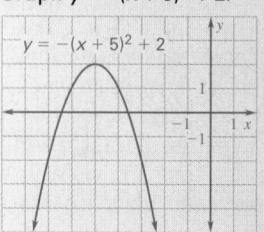

250

EXAMPLE 1 *Graphing a Quadratic Function*

Graph $y = 2x^2 - 8x + 6$.

SOLUTION

Note that the coefficients for this function are $a = 2$, $b = -8$, and $c = 6$. Since $a > 0$, the parabola opens up.

Find and plot the vertex. The *x*-coordinate is:

$$x = -\frac{b}{2a} = -\frac{-8}{2(2)} = 2$$

The *y*-coordinate is:

$$y = 2(2)^2 - 8(2) + 6 = -2$$

So, the vertex is $(2, -2)$.

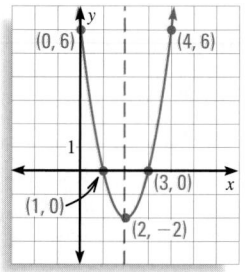

Draw the axis of symmetry $x = 2$.

Plot two points on one side of the axis of symmetry, such as $(1, 0)$ and $(0, 6)$. Use symmetry to plot two more points, such as $(3, 0)$ and $(4, 6)$.

Draw a parabola through the plotted points.

. . . .

The quadratic function $y = ax^2 + bx + c$ is written in **standard form**. Two other useful forms for quadratic functions are given below.

STUDENT HELP

▸ **Skills Review**
For help with symmetry, see p. 919.

VERTEX AND INTERCEPT FORMS OF A QUADRATIC FUNCTION	
FORM OF QUADRATIC FUNCTION	**CHARACTERISTICS OF GRAPH**
Vertex form: $y = a(x - h)^2 + k$	The vertex is (h, k).
	The axis of symmetry is $x = h$.
Intercept form: $y = a(x - p)(x - q)$	The *x*-intercepts are *p* and *q*.
	The axis of symmetry is halfway between $(p, 0)$ and $(q, 0)$.
For both forms, the graph opens up if $a > 0$ and opens down if $a < 0$.	

EXAMPLE 2 *Graphing a Quadratic Function in Vertex Form*

Graph $y = -\frac{1}{2}(x + 3)^2 + 4$.

SOLUTION

The function is in vertex form $y = a(x - h)^2 + k$ where $a = -\frac{1}{2}$, $h = -3$, and $k = 4$. Since $a < 0$, the parabola opens down. To graph the function, first plot the vertex $(h, k) = (-3, 4)$. Draw the axis of symmetry $x = -3$ and plot two points on one side of it, such as $(-1, 2)$ and $(1, -4)$. Use symmetry to complete the graph.

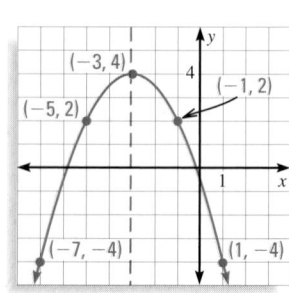

STUDENT HELP

▸ **Look Back**
For help with graphing functions, see p. 123.

EXAMPLE 3 *Graphing a Quadratic Function in Intercept Form*

Graph $y = -(x + 2)(x - 4)$.

SOLUTION

The quadratic function is in intercept form $y = a(x - p)(x - q)$ where $a = -1$, $p = -2$, and $q = 4$. The *x*-intercepts occur at $(-2, 0)$ and $(4, 0)$. The axis of symmetry lies halfway between these points, at $x = 1$. So, the *x*-coordinate of the vertex is $x = 1$ and the *y*-coordinate of the vertex is:

$$y = -(1 + 2)(1 - 4) = 9$$

The graph of the function is shown.

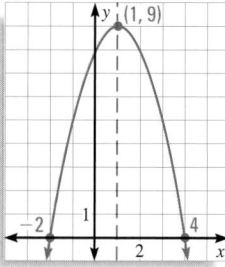

· · · · · · · · · ·

STUDENT HELP

▶ **Skills Review**
For help with multiplying algebraic expressions, see p. 937.

You can change quadratic functions from intercept form or vertex form to standard form by multiplying algebraic expressions. One method for multiplying expressions containing two terms is *FOIL*. Using this method, you add the products of the *F*irst terms, the *O*uter terms, the *I*nner terms, and the *L*ast terms. Here is an example:

$$\overset{\text{F \quad O \quad I \quad L}}{(x + 3)(x + 5) = x^2 + 5x + 3x + 15 = x^2 + 8x + 15}$$

Methods for changing from standard form to intercept form or vertex form will be discussed in Lessons 5.2 and 5.5.

EXAMPLE 4 *Writing Quadratic Functions in Standard Form*

Write the quadratic function in standard form.

a. $y = -(x + 4)(x - 9)$　　　　**b.** $y = 3(x - 1)^2 + 8$

SOLUTION

a. $y = -(x + 4)(x - 9)$	Write original function.
$= -(x^2 - 9x + 4x - 36)$	Multiply using FOIL.
$= -(x^2 - 5x - 36)$	Combine like terms.
$= -x^2 + 5x + 36$	Use distributive property.
b. $y = 3(x - 1)^2 + 8$	Write original function.
$= 3(x - 1)(x - 1) + 8$	Rewrite $(x - 1)^2$.
$= 3(x^2 - x - x + 1) + 8$	Multiply using FOIL.
$= 3(x^2 - 2x + 1) + 8$	Combine like terms.
$= 3x^2 - 6x + 3 + 8$	Use distributive property.
$= 3x^2 - 6x + 11$	Combine like terms.

5.1 *Graphing Quadratic Functions*　　**251**

EXTRA EXAMPLE 3
Graph $y = 2(x - 3)(x + 1)$.

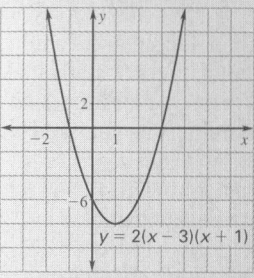

EXTRA EXAMPLE 4
Write the quadratic function in standard form.

a. $y = \frac{1}{2}(x - 6)(x - 4)$

  $y = \frac{1}{2}x^2 - 5x + 12$

b. $y = -4(x - 7)^2 + 2$
  $y = -4x^2 + 56x - 194$

✓ **CHECKPOINT EXERCISES**
For use after Example 3:
1. Graph $y = 2x(x - 4)$.

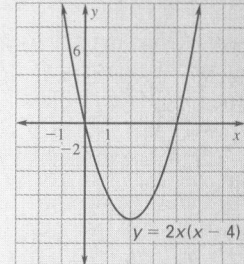

For use after Example 4:
2. Write the quadratic function in standard form.

a. $y = \frac{1}{2}(x + 2)^2 - 3$

  $y = \frac{1}{2}x^2 + 2x - 1$

b. $y = -3(x + 1)(x - 5)$
  $y = -3x^2 + 12x + 15$

MATHEMATICAL REASONING
Students need to understand that only parabolas that open up or down are functions. Parabolas that open to the left or right do not represent functions because, except for the vertex, two values in the range are paired with one value in the domain. Ask students to give an example of an equation for a parabola that opens to the left or right.
Sample answer: $x = 3y^2 + 2y + 1$

EXTRA EXAMPLE 5

Suppose that a group of high school students conducted an experiment to determine the number of hours of study that leads to the highest score on a comprehensive year-end exam. The exam score y for each student who studied for x hours can be modeled by $y = -0.853x^2 + 17.48x + 6.923$. Which amount of studying produced the highest score on the exam? What is the highest score the model predicts?
about 10 h; about 96

EXTRA EXAMPLE 6

The path of a ball thrown by a baseball player forms a parabola with equation:
$y = \dfrac{-3}{2401}(x-49)^2 + 8.5$, where x is the horizontal distance in feet of the ball from the player and y is the height in feet of the ball.
a. How far does the ball travel before it again reaches the same height from which it was thrown? **98 ft**
b. How high was the ball at its highest point? **8.5 ft**

 CHECKPOINT EXERCISES

For use after Examples 5 and 6:
1. The archway that forms the ceiling of a tunnel can be modeled by the equation $y = -0.0355x^2 + 0.923x + 10$ where x is the horizontal distance in feet from one wall of the tunnel to the other and y is the height in feet of the ceiling from the floor of the tunnel. How many feet from the walls of the tunnel does the ceiling reach its maximum height? What is the maximum height of the tunnel?
13 ft; about 16 ft

CLOSURE QUESTION

Give an example of a quadratic equation in vertex form. What is the vertex of the graph of this equation?
Check students' work.

252

Temperature

EXAMPLE 5 *Using a Quadratic Model in Standard Form*

Researchers conducted an experiment to determine temperatures at which people feel comfortable. The percent y of test subjects who felt comfortable at temperature x (in degrees Fahrenheit) can be modeled by:

$$y = -3.678x^2 + 527.3x - 18,807$$

What temperature made the greatest percent of test subjects comfortable? At that temperature, what percent felt comfortable? ▶ Source: *Design with Climate*

SOLUTION

Since $a = -3.678$ is negative, the graph of the quadratic function opens down and the function has a maximum value. The maximum value occurs at:

$$x = -\frac{b}{2a} = -\frac{527.3}{2(-3.678)} \approx 72$$

The corresponding value of y is:

$$y = -3.678(72)^2 + 527.3(72) - 18,807 \approx 92$$

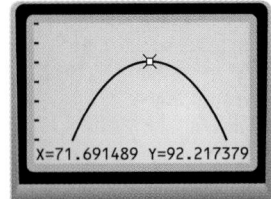

X=71.691489 Y=92.217379

▶ The temperature that made the greatest percent of test subjects comfortable was about 72°F. At that temperature about 92% of the subjects felt comfortable.

EXAMPLE 6 *Using a Quadratic Model in Vertex Form*

CIVIL ENGINEERING The Golden Gate Bridge in San Francisco has two towers that rise 500 feet above the road and are connected by suspension cables as shown. Each cable forms a parabola with equation

$$y = \frac{1}{8960}(x - 2100)^2 + 8$$

where x and y are measured in feet.
▶ Source: Golden Gate Bridge, Highway and Transportation District

a. What is the distance d between the two towers?

b. What is the height ℓ above the road of a cable at its lowest point?

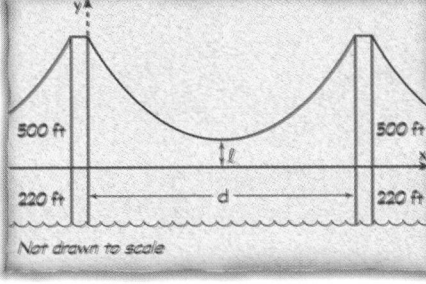

500 ft   500 ft

220 ft   ◄── d ──►   220 ft

Not drawn to scale

FOCUS ON CAREERS

CIVIL ENGINEER
Civil engineers design bridges, roads, buildings, and other structures. In 1996 civil engineers held about 196,000 jobs in the United States.

 CAREER LINK
www.mcdougallittell.com

SOLUTION

a. The vertex of the parabola is (2100, 8), so a cable's lowest point is 2100 feet from the left tower shown above. Since the heights of the two towers are the same, the symmetry of the parabola implies that the vertex is also 2100 feet from the right tower. Therefore, the towers are $d = 2(2100) = 4200$ feet apart.

b. The height ℓ above the road of a cable at its lowest point is the y-coordinate of the vertex. Since the vertex is (2100, 8), this height is $\ell = 8$ feet.

GUIDED PRACTICE

Vocabulary Check ✔
Concept Check ✔

Skill Check ✔

1. Complete this statement: The graph of a quadratic function is called a(n) _?_ .
   parabola
2. Does the graph of $y = 3x^2 - x - 2$ open up or down? Explain.
   up; since $a = 3$ and is greater than 0, the parabola opens up.
3. Is $y = -2(x - 5)(x - 8)$ in standard form, vertex form, or intercept form?
   intercept form

Graph the quadratic function. Label the vertex and axis of symmetry.

4. $y = x^2 - 4x + 7$ **5.** $y = 2(x + 1)^2 - 4$ **6.** $y = -(x + 2)(x - 1)$
   4–9. See margin.
7. $y = -\frac{1}{3}x^2 - 2x - 3$ **8.** $y = -\frac{3}{5}(x - 4)^2 + 6$ **9.** $y = \frac{5}{2}x(x - 3)$

Write the quadratic function in standard form.

10. $y = (x + 1)(x + 2)$ **11.** $y = -2(x + 4)(x - 3)$ **12.** $y = 4(x - 1)^2 + 5$
   $y = x^2 + 3x + 2$ $y = -2x^2 - 2x + 24$ $y = 4x^2 - 8x + 9$
13. $y = -(x + 2)^2 - 7$ **14.** $y = -\frac{1}{2}(x - 6)(x - 8)$ **15.** $y = \frac{2}{3}(x - 9)^2 - 4$
   $y = -x^2 - 4x - 11$ $y = -\frac{1}{2}x^2 + 7x - 24$ $y = \frac{2}{3}x^2 - 12x + 50$

16. **SCIENCE** **CONNECTION**   The equation given in Example 5 is based on temperature preferences of both male and female test subjects. Researchers also analyzed data for males and females separately and obtained the equations below.

> **Males:** $y = -4.290x^2 + 612.6x - 21{,}773$
>
> **Females:** $y = -6.224x^2 + 908.9x - 33{,}092$

What was the most comfortable temperature for the males? for the females?
71.4°F; 73°F

PRACTICE AND APPLICATIONS

STUDENT HELP

▶ **Extra Practice**
to help you master
skills is on p. 945.

MATCHING GRAPHS Match the quadratic function with its graph.

17. $y = (x + 2)(x - 3)$ **C**   **18.** $y = -(x - 3)^2 + 2$ **A**   **19.** $y = x^2 - 6x + 11$ **B**

A. 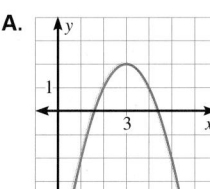   **B.** 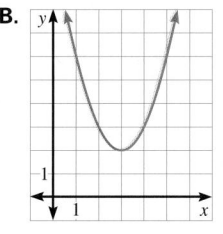   **C.**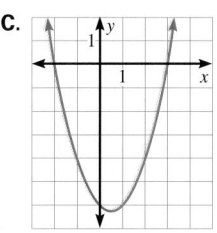

GRAPHING WITH STANDARD FORM Graph the quadratic function. Label the vertex and axis of symmetry. 20–25. See margin.

20. $y = x^2 - 2x - 1$ **21.** $y = 2x^2 - 12x + 19$ **22.** $y = -x^2 + 4x - 2$

23. $y = -3x^2 + 5$ **24.** $y = \frac{1}{2}x^2 + 4x + 5$ **25.** $y = -\frac{1}{6}x^2 - x - 3$

GRAPHING WITH VERTEX FORM Graph the quadratic function. Label the vertex and axis of symmetry. 26–31. See margin.

26. $y = (x - 1)^2 + 2$ **27.** $y = -(x - 2)^2 - 1$ **28.** $y = -2(x + 3)^2 - 4$

29. $y = 3(x + 4)^2 + 5$ **30.** $y = -\frac{1}{3}(x + 1)^2 + 3$ **31.** $y = \frac{5}{4}(x - 3)^2$

STUDENT HELP

▶ HOMEWORK HELP
Example 1: Exs. 17–25
Example 2: Exs. 17–19,
   26–31
Example 3: Exs. 17–19,
   32–37
Example 4: Exs. 38–49
Examples 5, 6: Exs. 51–54

5.1 Graphing Quadratic Functions   **253**

---

3 APPLY

○ **ASSIGNMENT GUIDE**

BASIC
Day 1: pp. 253–254 Exs. 17–29
   odd, 20–40 even, 50
Day 2: pp. 253–255 Exs. 25, 31, 37,
   41–51 odd, 55, 57–77 odd

AVERAGE
Day 1: pp. 253–254 Exs. 17–19,
   20–44 even, 50
Day 2: pp. 253–255 Exs. 25, 31, 37,
   41–55 odd, 57–77 odd, 78

ADVANCED
Day 1: pp. 253–254 Exs. 17–19,
   20–42 even
Day 2: pp. 253–255 Exs. 25, 31, 37,
   43–55 odd, 56, 57–77 odd,
   78

BLOCK SCHEDULE
pp. 253–255 Exs. 17–19, 20–44
even, 50, 41–55 odd, 57–77 odd,
78

EXERCISE LEVELS
Level A: *Easier*
   17–23, 26–29, 32–50
Level B: *More Difficult*
   24–25, 30–31, 51–55
Level C: *Most Difficult*
   56

✔ **HOMEWORK CHECK**
To quickly check student understanding of key concepts, go
over the following exercises:
Exs. 18, 20, 26, 32, 48, 50. See
also the Daily Homework Quiz:

• Blackline Master (*Chapter 5
  Resource Book*, p. 26)
• Transparency (p. 33)

4–9, 20–31. See Additional Answers
   beginning on page AA1.

32. $y = (x - 2)(x - 6)$ **33.** $y = 4(x + 1)(x - 1)$ **34.** $y = -(x + 3)(x + 5)$

35. $y = \frac{1}{3}(x + 4)(x + 1)$ **36.** $y = -\frac{1}{2}(x - 3)(x + 2)$ **37.** $y = -3x(x - 2)$

WRITING IN STANDARD FORM Write the quadratic function in standard form.
38–49. See margin.

38. $y = (x + 5)(x + 2)$ **39.** $y = -(x + 3)(x - 4)$ **40.** $y = 2(x - 1)(x - 6)$

41. $y = -3(x - 7)(x + 4)$ **42.** $y = (5x + 8)(4x + 1)$ **43.** $y = (x + 3)^2 + 2$

44. $y = -(x - 5)^2 + 11$ **45.** $y = -6(x - 2)^2 - 9$ **46.** $y = 8(x + 7)^2 - 20$

47. $y = -(9x + 2)^2 + 4x$ **48.** $y = -\frac{7}{3}(x + 6)(x + 3)$ **49.** $y = \frac{1}{2}(8x - 1)^2 - \frac{3}{2}$

50. VISUAL THINKING In parts (a) and (b), use a graphing calculator to examine how *b* and *c* affect the graph of $y = ax^2 + bx + c$.

    a. Graph $y = x^2 + c$ for $c = -2, -1, 0, 1$, and 2. Use the same viewing window for all the graphs. How do the graphs change as *c* increases?
The graphs move upward as *c* increases.

    b. Graph $y = x^2 + bx$ for $b = -2, -1, 0, 1$, and 2. Use the same viewing window for all the graphs. How do the graphs change as *b* increases?
The graphs move to the left as *b* increases.

51. AUTOMOBILES The engine torque *y* (in foot-pounds) of one model of car is given by   about 3,090 revolutions per min; about 74.7 foot-pounds

$$y = -3.75x^2 + 23.2x + 38.8$$

where *x* is the speed of the engine (in thousands of revolutions per minute). Find the engine speed that maximizes torque. What is the maximum torque?

52. SPORTS Although a football field appears to be flat, its surface is actually shaped like a parabola so that rain runs off to either side. The cross section of a field with synthetic turf can be modeled by

$$y = -0.000234(x - 80)^2 + 1.5$$

where *x* and *y* are measured in feet. What is the field's width? What is the maximum height of the field's surface? ▶ Source: Boston College
160 ft; 1.5 ft

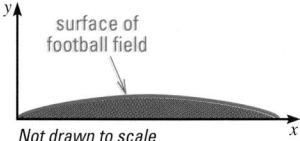
surface of football field
Not drawn to scale

53. PHYSIOLOGY Scientists determined that the rate *y* (in calories per minute) at which you use energy while walking can be modeled by

$$y = 0.00849(x - 90.2)^2 + 51.3, \quad 50 \le x \le 150$$

where *x* is your walking speed (in meters per minute). Graph the function on the given domain. Describe how energy use changes as walking speed increases. What speed minimizes energy use? ▶ Source: *Bioenergetics and Growth*

54. BIOLOGY CONNECTION The woodland jumping mouse can hop surprisingly long distances given its small size. A relatively long hop can be modeled by

$$y = -\frac{2}{9}x(x - 6)$$

where *x* and *y* are measured in feet. How far can a woodland jumping mouse hop? How high can it hop?
▶ Source: University of Michigan Museum of Zoology 6 ft; 2 ft

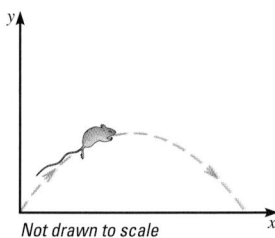
Not drawn to scale

APPLICATION NOTE

EXERCISE 51 Additional information about torque is available at **www.mcdougallittell.com**.

STUDENT HELP NOTES

→ **Homework Help** Students can find help for Ex. 54 at **www.mcdougallittell.com**. The information can be printed out for students who don't have access to the Internet.

32.
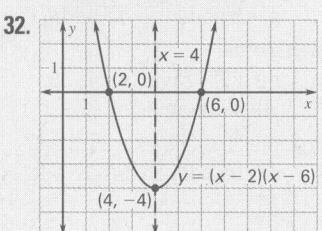

33.
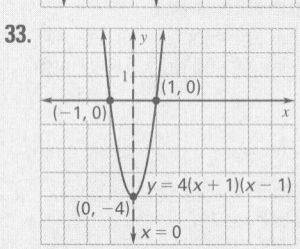

34.
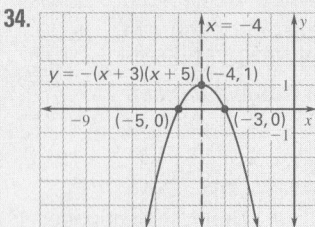

35–37. See Additional Answers beginning on page AA1.

ADDITIONAL PRACTICE AND RETEACHING

For Lesson 5.1:

• Practice Levels A, B, and C (*Chapter 5 Resource Book*, p. 15)

• Reteaching with Practice (*Chapter 5 Resource Book*, p. 18)

• ▣ See Lesson 5.1 of the *Personal Student Tutor*

For more Mixed Review:

• ▣ Search the *Test and Practice Generator* for key words or specific lessons.

254

REAL LIFE **TORQUE**, the focus of Ex. 51, is the "twisting force" produced by the crankshaft in a car's engine. As torque increases, a car is able to accelerate more quickly.

INTERNET **APPLICATION LINK**
www.mcdougallittell.com

38. $y = x^2 + 7x + 10$

39. $y = -x^2 + x + 12$

40. $y = 2x^2 - 14x + 12$

41. $y = -3x^2 + 9x + 84$

42. $y = 20x^2 + 37x + 8$

43. $y = x^2 + 6x + 11$

44. $y = -x^2 + 10x - 14$

45. $y = -6x^2 + 24x - 33$

46. $y = 8x^2 + 112x + 372$

47. $y = -81x^2 - 32x - 4$

48. $y = -\frac{7}{3}x^2 - 21x - 42$

49. $y = 32x^2 - 8x - 1$

53. *Sample answer:* The energy use decreases until about 90 meters per minute and then increases.

STUDENT HELP

INTERNET **HOMEWORK HELP**
Visit our Web site www.mcdougallittell.com for help with problem solving in Ex. 54.

55. MULTI-STEP PROBLEM A kernel of popcorn contains water that expands when the kernel is heated, causing it to pop. The equations below give the "popping volume" y (in cubic centimeters per gram) of popcorn with moisture content x (as a percent of the popcorn's weight). ▶ Source: *Cereal Chemistry*

Hot-air popping: $y = -0.761x^2 + 21.4x - 94.8$

Hot-oil popping: $y = -0.652x^2 + 17.7x - 76.0$

a. For hot-air popping, what moisture content maximizes popping volume? What is the maximum volume? **about 14%; about 56 cm³ per gram**

b. For hot-oil popping, what moisture content maximizes popping volume? What is the maximum volume? **about 13.6%; about 44 cm³ per gram**

55d. Sample answer: Hot-air popping produces a greater volume than hot-oil popping.

c. The moisture content of popcorn typically ranges from 8% to 18%. Graph the equations for hot-air and hot-oil popping on the interval $8 \le x \le 18$.
See margin.

d. *Writing* Based on the graphs from part (c), what general statement can you make about the volume of popcorn produced from hot-air popping versus hot-oil popping for any moisture content in the interval $8 \le x \le 18$?

 **★ Challenge**

EXTRA CHALLENGE
→ www.mcdougallittell.com

56. LOGICAL REASONING Write $y = a(x - h)^2 + k$ and $y = a(x - p)(x - q)$ in standard form. Knowing that the vertex of the graph of $y = ax^2 + bx + c$ occurs at $x = -\dfrac{b}{2a}$, show that the vertex for $y = a(x - h)^2 + k$ occurs at $x = h$ and that the vertex for $y = a(x - p)(x - q)$ occurs at $x = \dfrac{p + q}{2}$. See margin.

MIXED REVIEW

SOLVING LINEAR EQUATIONS Solve the equation. (Review 1.3 for 5.2)

57. $x - 2 = 0$ **2**

58. $2x + 5 = 0$ **−2.5**

59. $-4x - 7 = 21$ **−7**

60. $3x + 9 = -x + 1$ **−2**

61. $6(x + 8) = 18$ **−5**

62. $5(4x - 1) = 2(x + 3)$ $\frac{11}{18}$

63. $0.6x = 0.2x + 2.8$ **7**

64. $\dfrac{7x}{8} - \dfrac{3x}{5} = \dfrac{11}{2}$ **20**

65. $\dfrac{5x}{12} + \dfrac{1}{4} = \dfrac{x}{6} - \dfrac{1}{2}$ **−3**

GRAPHING IN THREE DIMENSIONS Sketch the graph of the equation. Label the points where the graph crosses the *x*-, *y*-, and *z*-axes. (Review 3.5)
66–71. See margin.

66. $x + y + z = 4$

67. $x + y + 2z = 6$

68. $3x + 4y + z = 12$

69. $5x + 5y + 2z = 10$

70. $2x + 7y + 3z = 42$

71. $x + 3y - 3z = 9$

USING CRAMER'S RULE Use Cramer's rule to solve the linear system. (Review 4.3)

72. $x + y = 1$ **(−3, 4)**
      $-5x + y = 19$

73. $2x + y = 5$ **(2, 1)**
      $3x - 4y = 2$

74. $7x - 10y = -15$
      $x + 2y = -9$
      (−5, −2)

75. $5x + 2y + 2z = 4$
      $3x + y - 6z = -4$
      $-x - y - z = 1$
      (2, −4, 1)

76. $x + 3y + z = 5$ **(−1, 0, 6)**
      $-x + y + z = 7$
      $2x - 7y + 5z = 28$

77. $2x - 3y - 9z = 11$
      $6x + y - z = 45$
      $9x - 2y + 4z = 56$
      (7, 2.5, −0.5)

78. 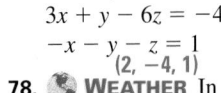 **WEATHER** In January, 1996, rain and melting snow caused the depth of the Susquehanna River in Pennsylvania to rise from 7 feet to 22 feet in 14 hours. $1\frac{1}{14}$ ft Find the average rate of change in the depth during that time. (Review 2.2) per hour

Additional Test Preparation Sample answer:
Students' answers should include the following concepts:
- A quadratic function in the form $y = a(x - p)(x - q)$ has its *x*-intercepts at p and q.
- The axis of symmetry is halfway between $(p, 0)$ and $(q, 0)$.
- The graph opens up if $a > 0$ and opens down if $a < 0$.

---

4 ASSESS

DAILY HOMEWORK QUIZ

📖 *Transparency Available*

1. Graph $y = 2x^2 - 4x + 2$.

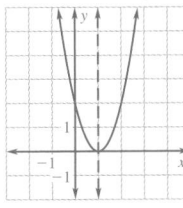

2. Graph $y = -\dfrac{1}{2}(x - 3)^2 + 5$.

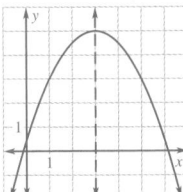

3. Graph $y = -(x - 3)(x + 1)$.

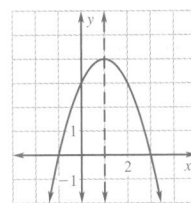

Write the quadratic function in standard form.

4 $y = -2(x - 4)^2 + 3$
    $y = -2x^2 + 16x - 32$

5. $y = -(x + 7)(x - 2)$
    $y = -x^2 - 5x + 14$

EXTRA CHALLENGE NOTE
→ Challenge problems for Lesson 5.1 are available in **blackline** format in the *Chapter 5 Resource Book*, p. 23 and at **www.mcdougallittell.com**.

ADDITIONAL TEST PREPARATION

1. WRITING You are given a quadratic equation in intercept form. Explain how you can use what you know about this type of equation to draw its graph.
See margin.

55c, 56, 66–71. See Additional Answers beginning on page AA1.

255

> ▶ **LESSON OPENER**
> **VISUAL APPROACH**
> An alternative way to approach
> Lesson 5.2 is to use the Visual
> Approach Lesson Opener:
> • Blackline Master (*Chapter 5 Resource Book*, p. 27)
> • 🎞 Transparency (p. 28)

MEETING INDIVIDUAL NEEDS
• *Chapter 5 Resource Book*
Prerequisite Skills Review (p. 5)
Practice Level A (p. 28)
Practice Level B (p. 29)
Practice Level C (p. 30)
Reteaching with Practice (p. 31)
Absent Student Catch-Up (p. 33)
Challenge (p. 35)
• *Resources in Spanish*
• 🖥 *Personal Student Tutor*

NEW-TEACHER SUPPORT
See the Tips for New Teachers on
pp. 1–2 of the *Chapter 5 Resource Book* for additional notes about
Lesson 5.2.

WARM-UP EXERCISES

🎞 **Transparency Available**

Solve the equation.

1. $3x - 4 = 0$ $\frac{4}{3}$

2. $2x - 11 = -15$ -2

3. $-3x - 5 = 2x$ -1

4. $2(x - 3) = 6$ 6

ENGLISH LEARNERS
Help English learners distinguish
among *monomial, binomial,* and
trinomial. Explain that *mono-* means
"one," *bi-* means "two," and *tri-*
means "three." Provide examples
such as *monorail, bicycle,* and
tricycle.

5.2

Solving Quadratic Equations by Factoring

𝘞𝘩𝘢𝘵 *you should learn*

GOAL 1 Factor quadratic expressions and solve quadratic equations by factoring.

GOAL 2 Find zeros of quadratic functions, as applied in **Example 8**.

𝘞𝘩𝘺 *you should learn it*

▼ To solve **real-life** problems, such as finding appropriate dimensions for a mural in **Ex. 97**.

GOAL 1 **FACTORING QUADRATIC EXPRESSIONS**

You know how to write $(x + 3)(x + 5)$ as $x^2 + 8x + 15$. The expressions $x + 3$ and $x + 5$ are **binomials** because they have two terms. The expression $x^2 + 8x + 15$ is a **trinomial** because it has three terms. You can use **factoring** to write a trinomial as a product of binomials. To factor $x^2 + bx + c$, find integers m and n such that:

$$x^2 + bx + c = (x + m)(x + n)$$
$$= x^2 + (m + n)x + mn$$

So, the *sum* of m and n must equal b and the *product* of m and n must equal c.

EXAMPLE 1 *Factoring a Trinomial of the Form $x^2 + bx + c$*

Factor $x^2 - 12x - 28$.

SOLUTION

You want $x^2 - 12x - 28 = (x + m)(x + n)$ where $mn = -28$ and $m + n = -12$.

Factors of -28 (m, n)	$-1, 28$	$1, -28$	$-2, 14$	$2, -14$	$-4, 7$	$4, -7$
Sum of factors $(m + n)$	27	-27	12	-12	3	-3

▶ The table shows that $m = 2$ and $n = -14$. So, $x^2 - 12x - 28 = (x + 2)(x - 14)$.

· · · · · · · · · ·

To factor $ax^2 + bx + c$ when $a \neq 1$, find integers $k, l, m,$ and n such that:

$$ax^2 + bx + c = (kx + m)(lx + n)$$
$$= klx^2 + (kn + lm)x + mn$$

Therefore, k and l must be factors of a, and m and n must be factors of c.

EXAMPLE 2 *Factoring a Trinomial of the Form $ax^2 + bx + c$*

Factor $3x^2 - 17x + 10$.

SOLUTION

You want $3x^2 - 17x + 10 = (kx + m)(lx + n)$ where k and l are factors of 3 and m and n are (negative) factors of 10. Check possible factorizations by multiplying.

$$(3x - 10)(x - 1) = 3x^2 - 13x + 10 \qquad (3x - 1)(x - 10) = 3x^2 - 31x + 10$$
$$(3x - 5)(x - 2) = 3x^2 - 11x + 10 \qquad (3x - 2)(x - 5) = 3x^2 - 17x + 10 \checkmark$$

▶ The correct factorization is $3x^2 - 17x + 10 = (3x - 2)(x - 5)$.

┌─ **STUDENT HELP**
↳ **Skills Review**
   For help with factoring,
   see p. 938.

As in Example 2, factoring quadratic expressions often involves trial and error. However, some expressions are easy to factor because they follow special patterns.

SPECIAL FACTORING PATTERNS

PATTERN NAME	PATTERN	EXAMPLE
Difference of Two Squares	$a^2 - b^2 = (a + b)(a - b)$	$x^2 - 9 = (x + 3)(x - 3)$
Perfect Square Trinomial	$a^2 + 2ab + b^2 = (a + b)^2$	$x^2 + 12x + 36 = (x + 6)^2$
	$a^2 - 2ab + b^2 = (a - b)^2$	$x^2 - 8x + 16 = (x - 4)^2$

EXAMPLE 3 *Factoring with Special Patterns*

Factor the quadratic expression.

a. $4x^2 - 25 = (2x)^2 - 5^2$ **Difference of two squares**

            $= (2x + 5)(2x - 5)$

b. $9y^2 + 24y + 16 = (3y)^2 + 2(3y)(4) + 4^2$ **Perfect square trinomial**

                   $= (3y + 4)^2$

c. $49r^2 - 14r + 1 = (7r)^2 - 2(7r)(1) + 1^2$ **Perfect square trinomial**

                 $= (7r - 1)^2$

. . . .

A **monomial** is an expression that has only one term. As a first step to factoring, you should check to see whether the terms have a common monomial factor.

EXAMPLE 4 *Factoring Monomials First*

Factor the quadratic expression.

a. $5x^2 - 20 = 5(x^2 - 4)$

         $= 5(x + 2)(x - 2)$

c. $2u^2 + 8u = 2u(u + 4)$

b. $6p^2 + 15p + 9 = 3(2p^2 + 5p + 3)$

               $= 3(2p + 3)(p + 1)$

d. $4x^2 + 4x - 4 = 4(x^2 + x - 1)$

. . . .

You can use factoring to solve certain *quadratic equations*. A **quadratic equation** in one variable can be written in the form $ax^2 + bx + c = 0$ where $a \neq 0$. This is called the **standard form** of the equation. If the left side of $ax^2 + bx + c = 0$ can be factored, then the equation can be solved using the *zero product property*.

ZERO PRODUCT PROPERTY

Let A and B be real numbers or algebraic expressions. If $AB = 0$, then $A = 0$ or $B = 0$.

5.2 *Solving Quadratic Equations by Factoring*   **257**

STUDENT HELP

▶ **Study Tip**
It is not always possible to factor a trinomial into a product of two binomials with integer coefficients. For instance, the trinomial $x^2 + x - 1$ in part (d) of Example 4 cannot be factored. Such trinomials are called *irreducible*.

---

2 TEACH

MOTIVATING THE LESSON
Ask students what it means to *factor* a number. Tell them that in today's lesson, they will learn to factor expressions.

EXTRA EXAMPLE 1
Factor $x^2 - 2x - 48$. $(x - 8)(x + 6)$

EXTRA EXAMPLE 2
Factor $4y^2 - 4y - 3$.
$(2y + 1)(2y - 3)$

EXTRA EXAMPLE 3
Factor the quadratic expression.
a. $16y^2 - 225$ $(4y + 15)(4y - 15)$
b. $4z^2 - 12z + 9$ $(2z - 3)^2$
c. $36w^2 + 60w + 25$ $(6w + 5)^2$

EXTRA EXAMPLE 4
Factor the quadratic expression.
a. $14x^2 + 2x - 12$ $2(x + 1)(7x - 6)$
b. $3v^2 - 18v$ $3v(v - 6)$
c. $12x^2 + 3x + 3$ $3(4x^2 + x + 1)$
d. $4u^2 - 36$ $4(u + 3)(u - 3)$

CHECKPOINT EXERCISES
For use after Examples 1 and 2:
1. Factor $5x^2 + 17x + 14$.
$(5x + 7)(x + 2)$

For use after Example 3:
2. Factor $64x^2 - 9$.
$(8x + 3)(8x - 3)$
3. Factor $16x^2 + 8x + 1$. $(4x + 1)^2$

For use after Example 4:
4. Factor $30u^2 - 57u + 21$.
$3(2u - 1)(5u - 7)$

257

EXTRA EXAMPLE 5
Solve.
a. $9t^2 - 12t + 4 = 0$ $\frac{2}{3}$
b. $3x - 6 = x^2 - 10$ $-1, 4$

EXTRA EXAMPLE 6
A painter is making a rectangular canvas for her next painting. She wants the length of the canvas to be 4 ft more than twice the width of the canvas. The area of the canvas must be 30 ft². What should the dimensions of the canvas be?
3 ft wide by 10 ft long

 CHECKPOINT EXERCISES

For use after Example 5:
1. Solve $2w^2 - 10w = 23w - w^2$.
   0, 11

For use after Example 6:
2. A yearbook editor is designing a page layout. The outside dimensions of the page are 9 in. wide by 12 in. long. The white border around the rectangular printed matter on the page is twice as wide on the sides as it is at the top and bottom of the page. The area of the printed matter is 50 in.². What are the dimensions of the printed matter? **5 in. wide by 10 in. long**

EXAMPLE 5 *Solving Quadratic Equations*

STUDENT HELP

▶ **Look Back**
For help with solving equations, see p. 19.

Solve (**a**) $x^2 + 3x - 18 = 0$ and (**b**) $2t^2 - 17t + 45 = 3t - 5$.

SOLUTION

a. $x^2 + 3x - 18 = 0$ Write original equation.

$(x + 6)(x - 3) = 0$ Factor.

$x + 6 = 0$ or $x - 3 = 0$ Use zero product property.

$x = -6$ or $x = 3$ Solve for x.

▶ The solutions are -6 and 3. Check the solutions in the original equation.

b. $2t^2 - 17t + 45 = 3t - 5$ Write original equation.

$2t^2 - 20t + 50 = 0$ Write in standard form.

$t^2 - 10t + 25 = 0$ Divide each side by 2.

$(t - 5)^2 = 0$ Factor.

$t - 5 = 0$ Use zero product property.

$t = 5$ Solve for t.

▶ The solution is 5. Check the solution in the original equation.

 *Crafts*

EXAMPLE 6 *Using a Quadratic Equation as a Model*

You have made a rectangular stained glass window that is 2 feet by 4 feet. You have 7 square feet of clear glass to create a border of uniform width around the window. What should the width of the border be?

SOLUTION

 **PROBLEM SOLVING STRATEGY**

VERBAL MODEL

$$\boxed{\text{Area of border}} = \boxed{\begin{array}{c}\text{Area of border}\\\text{and window}\end{array}} - \boxed{\begin{array}{c}\text{Area of}\\\text{window}\end{array}}$$

LABELS
Width of border $= x$ (feet)
Area of border $= 7$ (square feet)
Area of border and window $= (2 + 2x)(4 + 2x)$ (square feet)
Area of window $= 2 \cdot 4 = 8$ (square feet)

ALGEBRAIC MODEL

$7 = (2 + 2x)(4 + 2x) - 8$ Write algebraic model.

$0 = 4x^2 + 12x - 7$ Write in standard form.

$0 = (2x + 7)(2x - 1)$ Factor.

$2x + 7 = 0$ or $2x - 1 = 0$ Use zero product property.

$x = -3.5$ or $x = 0.5$ Solve for x.

▶ Reject the negative value, -3.5. The border's width should be 0.5 ft, or 6 in.

GOAL 2 FINDING ZEROS OF QUADRATIC FUNCTIONS

In Lesson 5.1 you learned that the x-intercepts of the graph of $y = a(x - p)(x - q)$ are p and q. The numbers p and q are also called **zeros** of the function because the function's value is zero when $x = p$ and when $x = q$. If a quadratic function is given in standard form $y = ax^2 + bx + c$, you may be able to find its zeros by using factoring to rewrite the function in intercept form.

EXAMPLE 7 *Finding the Zeros of a Quadratic Function*

Find the zeros of $y = x^2 - x - 6$.

SOLUTION

Use factoring to write the function in intercept form.

$$y = x^2 - x - 6$$
$$= (x + 2)(x - 3)$$

▸ The zeros of the function are -2 and 3.

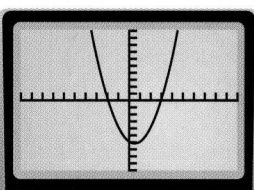

✓ **CHECK** Graph $y = x^2 - x - 6$. The graph passes through $(-2, 0)$ and $(3, 0)$, so the zeros are -2 and 3.

. . . .

From Lesson 5.1 you know that the vertex of the graph of $y = a(x - p)(x - q)$ lies on the vertical line halfway between $(p, 0)$ and $(q, 0)$. In terms of zeros, the function has its maximum or minimum value when x equals the *average* of the zeros.

EXAMPLE 8 *Using the Zeros of a Quadratic Model*

BUSINESS You maintain a music-oriented Web site that allows subscribing customers to download audio and video clips of their favorite bands. When the subscription price is $16 per year, you get 30,000 subscribers. For each $1 increase in price, you expect to lose 1000 subscribers. How much should you charge to maximize your annual revenue? What is your maximum revenue?

SOLUTION

> **Revenue** = **Number of subscribers** · **Subscription price**

Let R be your annual revenue and let x be the number of $1 price increases.

$$R = (30{,}000 - 1000x)(16 + x)$$
$$= (-1000x + 30{,}000)(x + 16)$$
$$= -1000(x - 30)(x + 16)$$

The zeros of the revenue function are 30 and -16. The value of x that maximizes R is the average of the zeros, or $x = \dfrac{30 + (-16)}{2} = 7$.

▸ To maximize revenue, charge $16 + $7 = $23 per year for a subscription. Your maximum revenue is $R = -1000(7 - 30)(7 + 16) = $529{,}000$.

STUDENT HELP

↱ **Study Tip**
In Example 7 note that -2 and 3 are zeros of the *function* and x-intercepts of the *graph*. In general, functions have zeros and graphs have x-intercepts.

STUDENT HELP

HOMEWORK HELP
Visit our Web site
www.mcdougallittell.com
for extra examples.

 EXTRA EXAMPLE 7
Find the zeros of $y = 3x^2 + 14x - 5$.
$-5, \frac{1}{3}$

EXTRA EXAMPLE 8
You own an amusement park that averages 75,000 visitors per year who each pay a $12 admission charge. You plan to lower the admission price to attract new customers. It has been shown that each $1 decrease in price results in 15,000 new visitors. What admission should you charge to maximize your annual revenue? What is the maximum revenue? $8.50, $1,083,750; For whole dollar changes either $8.00 or $9.00 will bring in $1,080,000

☑ **CHECKPOINT EXERCISES**
For use after Example 7:
1. Find the zeros of $y = x^2 + 8x + 15$. $-5, -3$

For use after Example 8:
2. Refer to Extra Example 8. Another study showed that for every $1 increase in price, there would be 5000 fewer visitors. What admission should you charge to maximize your annual revenue? What is the maximum revenue? $13.50; $911,250; For whole dollar changes either $13 or $14 will bring in $910,000.

FOCUS ON VOCABULARY
Fill in the blank. The _____ of a function are the same as the x-intercepts of its graph. **zeros**

CLOSURE QUESTION
What must be true about a quadratic equation before you can solve it using the zero product property? The equation must have a zero on one side and the other side must be factored.

DAILY PUZZLER
A father's age is 3 more than four times the age of his son. If the product of their ages is 351, what is the father's age? **39**

ASSIGNMENT GUIDE

BASIC
Day 1: pp. 260–263 Exs. 23–31 odd, 35–43 odd, 47–51 odd, 57–69 odd, 74, 80
Day 2: pp. 260–263 Exs. 32–34, 44–46, 53–55, 71–95 odd, 102, 107–135 odd

AVERAGE
Day 1: pp. 260–263 Exs. 23–31 odd, 35–43 odd, 47–51 odd, 57–69 odd, 74, 80, 90
Day 2: pp. 260–263 Exs. 32–34, 44–46, 53–55, 71–95 odd, 102, 107–135 odd, 136

ADVANCED
Day 1: pp. 260–263 Exs. 23–31 odd, 35–43 odd, 47–51 odd, 57–69 odd, 74, 80, 90
Day 2: pp. 260–263 Exs. 32–34, 44–46, 53–55, 71–95 odd, 102, 105, 107–135 odd, 136

BLOCK SCHEDULE
pp. 260–263 Exs. 23–31 odd, 32–34, 35–43 odd, 44–46, 47–51 odd, 57–69 odd, 74, 80, 90–96 even, 107–135 odd, 136

EXERCISE LEVELS
Level A: *Easier*
23–34, 47–66, 74–77, 80–85
Level B: *More Difficult*
35–46, 67–73, 78–79, 86–98
Level C: *Most Difficult*
99–105

✔ HOMEWORK CHECK
To quickly check student understanding of key concepts, go over the following exercises: Exs. 23, 37, 51, 63, 69, 74, 80. See also the Daily Homework Quiz:

- Blackline Master (*Chapter 5 Resource Book*, p. 38)
- 📄 Transparency (p. 34)

260

GUIDED PRACTICE

Vocabulary Check ✔
Sample answer: numbers where the value of the function is zero
1. What is a zero of a function $y = f(x)$?

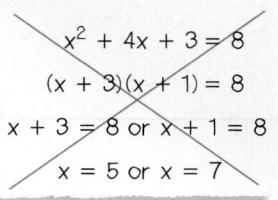

$$x^2 + 4x + 3 = 8$$
$$(x + 3)(x + 1) = 8$$
$$x + 3 = 8 \text{ or } x + 1 = 8$$
$$x = 5 \text{ or } x = 7$$

Concept Check ✔
2. In Example 2, how do you know that m and n must be *negative* factors of 10?

2. The x-term is negative and its absolute value is greater than the absolute value of the constant term.

3. **ERROR ANALYSIS** A student solved $x^2 + 4x + 3 = 8$ as shown. Explain the student's mistake. Then solve the equation correctly.

Skill Check ✔

3. The student did not set the factors equal to zero; −5, 1

Factor the expression.
4. $x^2 - x - 2$ $(x-2)(x+1)$
5. $2x^2 + x - 3$ $(2x+3)(x-1)$
6. $x^2 - 16$ $(x-4)(x+4)$
7. $y^2 + 2y + 1$ $(y+1)^2$
8. $p^2 - 4p + 4$ $(p-2)^2$
9. $q^2 + q$ $q(q+1)$

Solve the equation.
10. $(x+3)(x-1) = 0$ $-3, 1$
11. $x^2 - 2x - 8 = 0$ $-2, 4$
12. $3x^2 + 10x + 3 = 0$ $-\frac{1}{3}, -3$
13. $4u^2 - 1 = 0$ $-\frac{1}{2}, \frac{1}{2}$
14. $v^2 - 14v = -49$ 7
15. $5w^2 = 30w$ $0, 6$

Write the quadratic function in intercept form and give the function's zeros.
16–21. See margin.
16. $y = x^2 - 6x + 5$
17. $y = x^2 + 6x + 8$
18. $y = x^2 - 1$
19. $y = x^2 + 10x + 25$
20. $y = 2x^2 - 2x - 24$
21. $y = 3x^2 - 8x + 4$

16. $y = (x - 5)(x - 1)$; 1, 5
17. $y = (x + 4)(x + 2)$; −4, −2
18. $y = (x - 1)(x + 1)$; −1, 1
19. $y = (x + 5)^2$; −5
20. $y = 2(x - 4)(x + 3)$; −3, 4
21. $y = (3x - 2)(x - 2)$; $\frac{2}{3}$, 2

22. 🌐 **URBAN PLANNING** You have just planted a rectangular flower bed of red roses in a park near your home. You want to plant a border of yellow roses around the flower bed as shown. Since you bought the same number of red and yellow roses, the areas of the border and inner flower bed will be equal. What should the width x of the border be? 2 ft

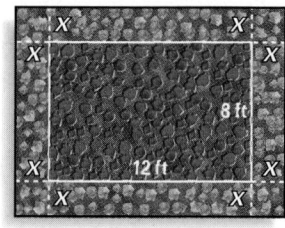

PRACTICE AND APPLICATIONS

> **STUDENT HELP**
>
> ▶ **Extra Practice**
> to help you master skills is on p. 945.

FACTORING $x^2 + bx + c$ Factor the trinomial. If the trinomial cannot be factored, say so.

23. $x^2 + 5x + 4$
$(x + 4)(x + 1)$
24. $x^2 + 9x + 14$
$(x + 7)(x + 2)$
25. $x^2 + 13x + 40$
$(x + 5)(x + 8)$
26. $x^2 - 4x + 3$
$(x - 3)(x - 1)$
27. $x^2 - 8x + 12$
$(x - 6)(x - 2)$
28. $x^2 - 16x + 51$
cannot be factored
29. $a^2 + 3a - 10$
$(a + 5)(a - 2)$
30. $b^2 + 6b - 27$
$(b + 9)(b - 3)$
31. $c^2 + 2c - 80$
$(c + 10)(c - 8)$
32. $p^2 - 5p - 6$
$(p - 6)(p + 1)$
33. $q^2 - 7q - 10$
cannot be factored
34. $r^2 - 14r - 72$
$(r - 18)(r + 4)$

FACTORING $ax^2 + bx + c$ Factor the trinomial. If the trinomial cannot be factored, say so.

35. $2x^2 + 7x + 3$
$(2x + 1)(x + 3)$
36. $3x^2 + 17x + 10$
$(3x + 2)(x + 5)$
37. $8x^2 + 18x + 9$
$(4x + 3)(2x + 3)$
38. $5x^2 - 7x + 2$
$(5x - 2)(x - 1)$
39. $6x^2 - 9x + 5$
cannot be factored
40. $10x^2 - 19x + 6$
$(5x - 2)(2x - 3)$
41. $3k^2 + 32k - 11$
$(3k - 1)(k + 11)$
42. $11m^2 + 14m - 16$
$(11m - 8)(m + 2)$
43. $18n^2 + 9n - 14$
$(3n - 2)(6n + 7)$
44. $7u^2 - 4u - 3$
$(7u + 3)(u - 1)$
45. $12v^2 - 25v - 7$
$(3v - 7)(4v + 1)$
46. $4w^2 - 13w - 27$
cannot be factored

Chapter 5 *Quadratic Functions*

STUDENT HELP

▶ HOMEWORK HELP
Example 1: Exs. 23–34
Example 2: Exs. 35–46
Example 3: Exs. 47–55
Example 4: Exs. 56–64
Example 5: Exs. 65–79
Example 6: Exs. 90, 91,
　　　　　97, 98
Example 7: Exs. 80–88
Example 8: Exs. 99–101

FACTORING WITH SPECIAL PATTERNS Factor the expression.

47. $x^2 - 25$　$(x - 5)(x + 5)$　**48.** $x^2 + 4x + 4$　$(x + 2)^2$　**49.** $x^2 - 6x + 9$　$(x - 3)^2$

50. $4r^2 - 4r + 1$　$(2r - 1)^2$　**51.** $9s^2 + 12s + 4$　$(3s + 2)^2$　**52.** $16t^2 - 9$　$(4t - 3)(4t + 3)$

53. $49 - 100a^2$　　**54.** $25b^2 - 60b + 36$　　**55.** $81c^2 + 198c + 121$
　　$(7 - 10a)(7 + 10a)$　　$(5b - 6)^2$　　$(9c + 11)^2$

FACTORING MONOMIALS FIRST Factor the expression.

56. $5x^2 + 5x - 10$　　**57.** $18x^2 - 2$　　$3(x + 9)^2$
　　$5(x + 2)(x - 1)$　　$2(3x - 1)(3x + 1)$　**58.** $3x^2 + 54x + 243$

59. $8y^2 - 28y - 60$　　**60.** $112a^2 - 168a + 63$　**61.** $u^2 + 7u$　$u(u + 7)$
　　$4(2y + 3)(y - 5)$　　$7(4a - 3)^2$

62. $6t^2 - 36t$　$6t(t - 6)$　**63.** $-v^2 + 2v - 1$　$-(v - 1)^2$　**64.** $2d^2 + 12d - 16$
　　　　　　　　　　　　　　　　　　　　$2(d^2 + 6d - 8)$

EQUATIONS IN STANDARD FORM Solve the equation.

65. $x^2 - 3x - 4 = 0$　$-1, 4$　**66.** $x^2 + 19x + 88 = 0$　**67.** $5x^2 - 13x + 6 = 0$　$\frac{3}{5}, 2$
　　　　　　　　　　　　　　　　　$-11, -8$

68. $8x^2 - 6x - 5 = 0$　$-\frac{1}{2}, \frac{5}{4}$　**69.** $k^2 + 24k + 144 = 0$　-12　**70.** $9m^2 - 30m + 25 = 0$　$\frac{5}{3}$

71. $81n^2 - 16 = 0$　$-\frac{4}{9}, \frac{4}{9}$　**72.** $40a^2 + 4a = 0$　$-\frac{1}{10}, 0$　**73.** $-3b^2 + 3b + 90 = 0$
　　　　　　　　　　　　　　　　　　　　　　　　　　　　$-5, 6$

EQUATIONS NOT IN STANDARD FORM Solve the equation.

74. $x^2 + 9x = -20$　$-5, -4$　　　**75.** $16x^2 = 8x - 1$　$\frac{1}{4}$

76. $5p^2 - 25 = 4p^2 + 24$　$-7, 7$　**77.** $2y^2 - 4y - 8 = -y^2 + y$　$-1, \frac{8}{3}$

80. $y = (x - 2)(x - 1)$; 1, 2　**78.** $2q^2 + 4q - 1 = 7q^2 - 7q + 1$　$\frac{1}{5}, 2$　**79.** $(w + 6)^2 = 3(w + 12) - w^2$　$-\frac{9}{2}, 0$

81. $y = (x + 4)(x + 3)$;
　　$-4, -3$
FINDING ZEROS Write the quadratic function in intercept form and give the
function's zeros.　80–88. See margin.

82. $y = (x + 7)(x - 5)$; $-7, 5$　**80.** $y = x^2 - 3x + 2$　　**81.** $y = x^2 + 7x + 12$　　**82.** $y = x^2 + 2x - 35$

83. $y = (x - 2)(x + 2)$; $-2, 2$　**83.** $y = x^2 - 4$　　　**84.** $y = x^2 + 20x + 100$　**85.** $y = x^2 - 3x$

84. $y = (x + 10)^2$; -10
　　　　　　　　　　　86. $y = 3x^2 - 12x - 15$　**87.** $y = -x^2 + 16x - 64$　**88.** $y = 2x^2 - 9x + 4$

85. $y = x(x - 3)$; 0, 3

86. $y = 3(x - 5)(x + 1)$; $-1, 5$　**89. LOGICAL REASONING** Is there a formula for factoring the *sum* of two squares?
　　　　　　　　　　　You will investigate this question in parts (a) and (b).　See margin.
87. $y = -(x - 8)^2$; 8

88. $y = (2x - 1)(x - 4)$; $\frac{1}{2}$, 4　　**a.** Consider the sum of squares $x^2 + 9$. If this sum can be factored, then there are
　　　　　　　　　　　　integers m and n such that $x^2 + 9 = (x + m)(x + n)$. Write two equations
　　　　　　　　　　　　relating the sum and the product of m and n to the coefficients in $x^2 + 9$.

89. a. $m + n = 0$, $mn = 9$　　**b.** Show that there are no integers m and n that satisfy both equations you wrote
89. b. If $m + n = 0$, then　　　　in part (a). What can you conclude?
　　$m = -n$. Substituting in
　　$mn = 9$, $(-n)(n) = 9$,　**90. QUILTING** You have made a quilt
　　$-n^2 = 9$, and $n^2 = -9$.　　　that is 4 feet by 5 feet. You want to use
　　There is no number　　　　the remaining 10 square feet of fabric
　　such that $n^2 = -9$.　　　　to add a decorative border of uniform
　　Therefore, $x^2 + 9$ is not　　width. What should the width of the
　　factorable.　　　　　　border be?　0.5 ft

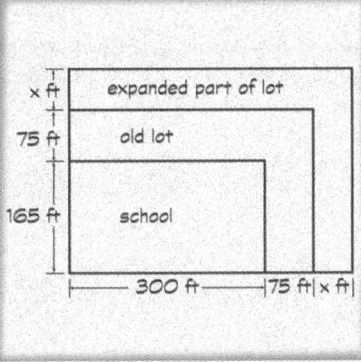

91. CONSTRUCTION A high school
　　wants to double the size of its parking
　　lot by expanding the existing lot as
　　shown. By what distance x should the
　　lot be expanded?　60 ft

! COMMON ERROR
EXERCISES 23–64 Factoring
trinomials is a skill that comes with
practice. Have students first factor
out any common factors and look
for special patterns. Then look for
factors of the constant term c
whose sum or difference is b. To
avoid errors students should multiply the factors to make sure the
product is the original expression.

TEACHING TIPS
The reasoning used in Ex. 89 is
called proof by contradiction. In
proof by contradiction, you assume
the opposite of what you want to
prove is true. Then you show that
this assumption leads to a contradiction and hence must be false.
This means that what you wanted
to prove must be true. In this case,
you assume that the difference of
two squares can be factored and
show that this leads to a system of
two equations with no solutions.
This means that the assumption
that the sum of two squares is factorable is false so that the sum of
two squares is not factorable.

96b.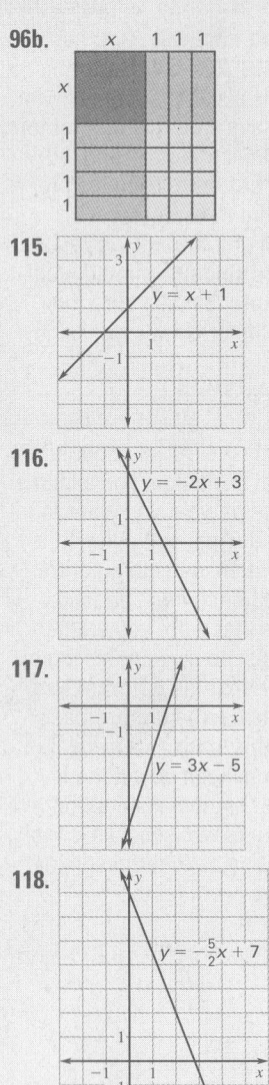

115.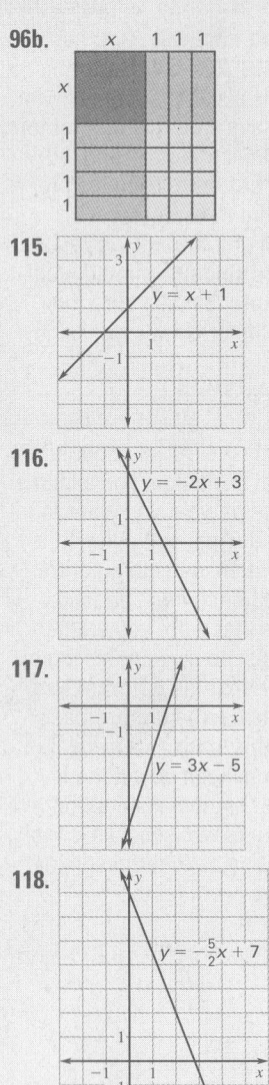

116.

117.

118.

119–126. See Additional Answers beginning on page AA1.

ADDITIONAL PRACTICE AND RETEACHING

For Lesson 5.2:
• Practice Levels A, B, and C (*Chapter 5 Resource Book,* p. 28)
• Reteaching with Practice (*Chapter 5 Resource Book,* p. 31)
• 🖥 See Lesson 5.2 of the *Personal Student Tutor*

For more Mixed Review:
• 🖥 Search the *Test and Practice Generator* for key words or specific lessons.

262

Skills Review
For help with areas of geometric figures, see p. 914.

96a. *Sample answer:* The area of the rectangle equals the sum of the areas of its parts. The area of the rectangle also equals the product of the lengths of the sides so $x^2 + 5x + 6 = (x + 2)(x + 3)$.

FOCUS ON APPLICATIONS

▶ **ENVIRONMENT**
Ecology gardens are often used to conduct research with different plant species under a variety of growing conditions.

GEOMETRY CONNECTION Find the value of *x*.

92. Area of rectangle = 40   **5**

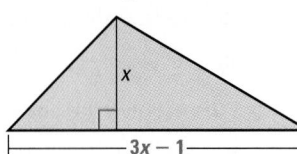
x
$x + 3$

93. Area of rectangle = 105   **7**

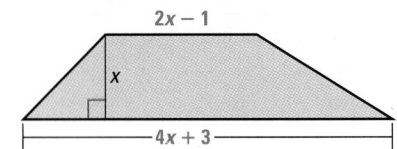
x
$2x + 1$

94. Area of triangle = 22   **4**

x
$3x - 1$

95. Area of trapezoid = 114   **6**

$2x - 1$
x
$4x + 3$

96. VISUAL THINKING Use the diagram shown at the right.
   a. Explain how the diagram models the factorization $x^2 + 5x + 6 = (x + 2)(x + 3)$.
   b. Draw a diagram that models the factorization $x^2 + 7x + 12 = (x + 3)(x + 4)$.   **See margin.**

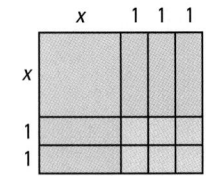
x 1 1 1
x
1
1

97. ART CONNECTION As part of Black History Month in February, an artist is creating a mural on the side of a building. A painting of Dr. Martin Luther King, Jr., will occupy the center of the mural and will be surrounded by a border of uniform width showing other prominent African-Americans. The side of the building is 50 feet wide by 30 feet high, and the artist wants to devote 25% of the available space to the border. What should the width of the border be?   **2.5 ft**

98. ENVIRONMENT A student environmental group wants to build an ecology garden as shown. The area of the garden should be 800 square feet to accommodate all the species of plants the group wants to grow. A construction company has donated 120 feet of iron fencing to enclose the garden. What should the dimensions of the garden be?   **20 ft by 40 ft**

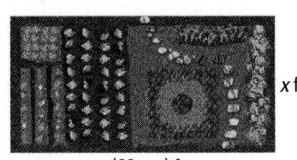
x ft
$(60 - x)$ ft

99. ATHLETIC WEAR A shoe store sells about 200 pairs of a new basketball shoe each month when it charges $60 per pair. For each $1 increase in price, about 2 fewer pairs per month are sold. How much per pair should the store charge to maximize monthly revenue? What is the maximum revenue?   **$80; $12,800**

100. HOME ELECTRONICS The manager of a home electronics store is considering repricing a new model of digital camera. At the current price of $680, the store sells about 70 cameras each month. Sales data from other stores indicate that for each $20 decrease in price, about 5 more cameras per month would be sold. How much should the manager charge for a camera to maximize monthly revenue? What is the maximum revenue?   **$480; $57,600**

101. HISTORY CONNECTION Big Bertha, a cannon used in World War I, could fire shells incredibly long distances. The path of a shell could be modeled by $y = -0.0196x^2 + 1.37x$ where x was the horizontal distance traveled (in miles) and y was the height (in miles). How far could Big Bertha fire a shell? What was the shell's maximum height? ▶ Source: World War I: Trenches on the Web   **about 70 mi; about 24 mi**

102. MULTIPLE CHOICE Suppose $x^2 + 4x + c = (x + m)(x + n)$ where c, m, and n are integers. Which of the following are *not* possible values of m and n? **C**

Ⓐ $m = 2, n = 2$ 　　　Ⓑ $m = -1, n = 5$

Ⓒ $m = -2, n = -2$ 　　Ⓓ $m = 1, n = 3$

103. MULTIPLE CHOICE What are all solutions of $2x^2 - 11x + 16 = x^2 - 3x$? **D**

Ⓐ $2, 6$ 　　Ⓑ -4 　　Ⓒ $-4, 4$ 　　Ⓓ 4

104. MULTIPLE CHOICE Given that 4 is a zero of $y = 3x^2 + bx - 8$, what is the value of b? **B**

Ⓐ -40 　　Ⓑ -10 　　Ⓒ -8 　　Ⓓ 2

★ **Challenge**

105. MULTICULTURAL MATHEMATICS The following problem is from the *Chiu chang suan shu*, an ancient Chinese mathematics text. Solve the problem. (*Hint:* Use the Pythagorean theorem.) **8 ch'ih by 6 ch'ih**

> *A rod of unknown length is used to measure the dimensions of a rectangular door. The rod is 4 ch'ih longer than the width of the door, 2 ch'ih longer than the height of the door, and the same length as the door's diagonal. What are the dimensions of the door? (Note: 1 ch'ih is slightly greater than 1 foot.)*

EXTRA CHALLENGE
➜ www.mcdougallittell.com

MIXED REVIEW

ABSOLUTE VALUE Solve the equation or inequality. (Review 1.7)

106. $|x| = 3$ $-3, 3$ 　　**107.** $|x - 2| = 6$ $-4, 8$ 　　**108.** $|4x - 9| = 2$ $1.75, 2.75$

109. $|-5x + 4| = 14$ $-2, 3.6$ 　　**110.** $|7 - 3x| = -8$ no solution 　　**111.** $|x + 1| < 3$ $-4 < x < 2$

112. $|2x - 5| \le 1$ $2 \le x \le 3$ 　　**113.** $|x - 4| > 7$ $x < -3$ or $x > 11$ 　　**114.** $\left|\frac{1}{3}x + 1\right| \ge 2$ $x \le -9$ or $x \ge 3$

GRAPHING LINEAR EQUATIONS Graph the equation. (Review 2.3)
115–126. See margin.

115. $y = x + 1$ 　　**116.** $y = -2x + 3$ 　　**117.** $y = 3x - 5$

118. $y = -\frac{5}{2}x + 7$ 　　**119.** $x + y = 4$ 　　**120.** $2x - y = 6$

121. $3x + 4y = -12$ 　　**122.** $-5x + 3y = 15$ 　　**123.** $y = 2$

124. $y = -3$ 　　**125.** $x = -1$ 　　**126.** $x = 4$

GRAPHING QUADRATIC FUNCTIONS Graph the function. (Review 5.1 for 5.3)
127–135. See margin.

127. $y = x^2 - 2$ 　　**128.** $y = 2x^2 - 5$ 　　**129.** $y = -x^2 + 3$

130. $y = (x + 1)^2 - 4$ 　　**131.** $y = -(x - 2)^2 + 1$ 　　**132.** $y = -3(x + 3)^2 + 7$

133. $y = \frac{1}{4}x^2 - 1$ 　　**134.** $y = \frac{1}{2}(x - 4)^2 - 6$ 　　**135.** $y = -\frac{2}{3}(x + 1)(x - 3)$

136. COMMUTING You can take either the subway or the bus to your after-school job. A round trip from your home to where you work costs $2 on the subway and $3 on the bus. You prefer to take the bus as often as possible but can afford to spend only $50 per month on transportation. If you work 22 days each month, how many of these days can you take the bus? (Review 1.5) **6 days**

Additional Test Preparation *Sample answer:*

1. Check to see that answers includes at least two of the following special factoring patterns: difference of two squares, perfect square trinomial with addition, perfect square trinomial with subtraction.

2. Factor the function to find the zeros. Average the zeros to find the x-coordinate of the maximum or minimum point. Evaluate the function at this value of x to find its maximum or minimum value.

---

4 ASSESS

DAILY HOMEWORK QUIZ

Transparency Available

Factor the quadratic expression.

1. $x^2 - 14x - 15$ $(x - 15)(x + 1)$

2. $5x^2 + 4x - 12$ $(5x - 6)(x + 2)$

3. $36x^2 - 49$ $(6x - 7)(6x + 7)$

4. $25x^2 - 10x + 1$ $(5x - 1)^2$

Solve.

5. $16x^2 + 24x + 9 = 0$ $-\frac{3}{4}$

6. $14x^2 + 11x + 3 = 2x^2 - 3x + 3$ $0, -\frac{7}{6}$

7. Find the zeros of $y = 8x^2 - 18x$. $0, \frac{9}{4}$

EXTRA CHALLENGE NOTE
➜ Challenge problems for Lesson 5.2 are available in **blackline** format in the *Chapter 5 Resource Book,* p. 35 and at **www.mcdougallittell.com**.

ADDITIONAL TEST PREPARATION

1. OPEN ENDED Name and give an example of two of the three special factoring patterns. Then describe how to factor them. See margin.

2. WRITING How can you find the maximum or minimum value of a quadratic function? See margin.

127–135. See Additional Answers beginning on page AA1.

PACING
Basic: 1 day
Average: 1 day
Advanced: 1 day
Block Schedule: 0.5 block with 5.4

LESSON OPENER
APPLICATION
An alternative way to approach Lesson 5.3 is to use the Application Lesson Opener:

• Blackline Master (*Chapter 5 Resource Book,* p. 39)
•  Transparency (p. 29)

MEETING INDIVIDUAL NEEDS
• *Chapter 5 Resource Book*
Prerequisite Skills Review (p. 5)
Practice Level A (p. 41)
Practice Level B (p. 42)
Practice Level C (p. 43)
Reteaching with Practice (p. 44)
Absent Student Catch-Up (p. 46)
Challenge (p. 49)
• *Resources in Spanish*
• Personal Student Tutor

NEW-TEACHER SUPPORT
See the Tips for New Teachers on pp. 1–2 of the *Chapter 5 Resource Book* for additional notes about Lesson 5.3.

WARM-UP EXERCISES

Transparency Available

Solve the equation.

1. $5x - 3 = 17$ 4
2. $0 = -12 + 3t$ 4

Find the value of y when $x = 0, 1,$ and 2.

3. $y = -16x^2 + 24$ 24, 8, −40
4. $y = -18x^2 + 321$ 321, 303, 249

5.3 Solving Quadratic Equations by Finding Square Roots

What you should learn

GOAL 1 Solve quadratic equations by finding square roots.

GOAL 2 Use quadratic equations to solve **real-life** problems, such as finding how long a falling stunt man is in the air in **Example 4**.

Why you should learn it

▼ To model **real-life** quantities, such as the height of a rock dropped off the Leaning Tower of Pisa in **Ex. 69**.

GOAL 1 SOLVING QUADRATIC EQUATIONS

A number r is a **square root** of a number s if $r^2 = s$. A positive number s has two square roots denoted by $\sqrt{s}$ and $-\sqrt{s}$. The symbol $\sqrt{\ }$ is a **radical sign**, the number s beneath the radical sign is the **radicand**, and the expression $\sqrt{s}$ is a **radical**.

For example, since $3^2 = 9$ and $(-3)^2 = 9$, the two square roots of 9 are $\sqrt{9} = 3$ and $-\sqrt{9} = -3$. You can use a calculator to approximate $\sqrt{s}$ when s is not a perfect square. For instance, $\sqrt{2} \approx 1.414$.

ACTIVITY
Developing Concepts

Investigating Properties of Square Roots

1 Evaluate the two expressions. What do you notice about the square root of a product of two numbers? **See margin.**

  a. $\sqrt{36}, \sqrt{4} \cdot \sqrt{9}$ **b.** $\sqrt{8}, \sqrt{4} \cdot \sqrt{2}$ **c.** $\sqrt{30}, \sqrt{3} \cdot \sqrt{10}$

2 Evaluate the two expressions. What do you notice about the square root of a quotient of two numbers? **See margin.**

  a. $\sqrt{\dfrac{4}{9}}, \dfrac{\sqrt{4}}{\sqrt{9}}$ **b.** $\sqrt{\dfrac{25}{2}}, \dfrac{\sqrt{25}}{\sqrt{2}}$ **c.** $\sqrt{\dfrac{19}{7}}, \dfrac{\sqrt{19}}{\sqrt{7}}$

In the activity you may have discovered the following properties of square roots. You can use these properties to simplify expressions containing square roots.

PROPERTIES OF SQUARE ROOTS ($a > 0$, $b > 0$)

Product Property: $\sqrt{ab} = \sqrt{a} \cdot \sqrt{b}$ **Quotient Property:** $\sqrt{\dfrac{a}{b}} = \dfrac{\sqrt{a}}{\sqrt{b}}$

Step 1a–c. The answers show that $\sqrt{ab} = \sqrt{a} \cdot \sqrt{b}$.

Step 1a. 6, 6
Step 1b. 2.8, 2.8
Step 1c. 5.5, 5.5

Step 2a–c. The answers show that $\sqrt{\dfrac{a}{b}} = \dfrac{\sqrt{a}}{\sqrt{b}}$.

Step 2a. $\dfrac{2}{3}, \dfrac{2}{3}$
Step 2b. 3.5, 3.5
Step 2c. 1.6, 1.6

A square-root expression is considered simplified if (1) no radicand has a perfect-square factor other than 1, and (2) there is no radical in a denominator.

EXAMPLE 1 *Using Properties of Square Roots*

Simplify the expression.

a. $\sqrt{24} = \sqrt{4} \cdot \sqrt{6} = 2\sqrt{6}$ **b.** $\sqrt{6} \cdot \sqrt{15} = \sqrt{90} = \sqrt{9} \cdot \sqrt{10} = 3\sqrt{10}$

c. $\sqrt{\dfrac{7}{16}} = \dfrac{\sqrt{7}}{\sqrt{16}} = \dfrac{\sqrt{7}}{4}$ **d.** $\sqrt{\dfrac{7}{2}} = \dfrac{\sqrt{7}}{\sqrt{2}} \cdot \dfrac{\sqrt{2}}{\sqrt{2}} = \dfrac{\sqrt{14}}{2}$

In part (d) of Example 1, the square root in the denominator of $\dfrac{\sqrt{7}}{\sqrt{2}}$ was eliminated by multiplying both the numerator and the denominator by $\sqrt{2}$. This process is called **rationalizing the denominator**.

You can use square roots to solve some types of quadratic equations. For instance, if $s > 0$, then the quadratic equation $x^2 = s$ has two real-number solutions: $x = \sqrt{s}$ and $x = -\sqrt{s}$. These solutions are often written in condensed form as $x = \pm\sqrt{s}$. The symbol $\pm\sqrt{s}$ is read as "plus or minus the square root of s."

EXAMPLE 2 *Solving a Quadratic Equation*

Solve $2x^2 + 1 = 17$.

SOLUTION

Begin by writing the equation in the form $x^2 = s$.

$2x^2 + 1 = 17$	Write original equation.
$2x^2 = 16$	Subtract 1 from each side.
$x^2 = 8$	Divide each side by 2.
$x = \pm\sqrt{8}$	Take square roots of each side.
$x = \pm2\sqrt{2}$	Simplify.

▶ The solutions are $2\sqrt{2}$ and $-2\sqrt{2}$.

✓**CHECK** You can check the solutions algebraically by substituting them into the original equation. Since this equation is equivalent to $2x^2 - 16 = 0$, you can also check the solutions by graphing $y = 2x^2 - 16$ and observing that the graph's x-intercepts appear to be about $2.8 \approx 2\sqrt{2}$ and $-2.8 \approx -2\sqrt{2}$.

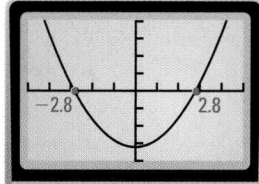

EXAMPLE 3 *Solving a Quadratic Equation*

STUDENT HELP

HOMEWORK HELP
Visit our Web site
www.mcdougallittell.com
for extra examples.

Solve $\dfrac{1}{3}(x + 5)^2 = 7$.

SOLUTION

$\dfrac{1}{3}(x + 5)^2 = 7$	Write original equation.
$(x + 5)^2 = 21$	Multiply each side by 3.
$x + 5 = \pm\sqrt{21}$	Take square roots of each side.
$x = -5 \pm \sqrt{21}$	Subtract 5 from each side.

▶ The solutions are $-5 + \sqrt{21}$ and $-5 - \sqrt{21}$.

✓**CHECK** Check the solutions either by substituting them into the original equation or by graphing $y = \dfrac{1}{3}(x + 5)^2 - 7$ and observing the x-intercepts.

5.3 *Solving Quadratic Equations by Finding Square Roots* **265**

---

2 TEACH

MOTIVATING THE LESSON
Ask students which takes longer to fall to the ground, a cotton ball or a rubber ball the same size and shape. Remind students that as long as air resistance is ignored, all objects close to the Earth's surface fall at the same rate. Today's lesson will focus on using square roots to solve quadratic equations such as those that model the time it takes for an object to fall on Earth.

ACTIVITY NOTE
In doing this activity, students should discover the properties of square roots given in the box. Students will need to use calculators to evaluate the expressions. After the activity, ask students to identify the values of a and b from the properties for each problem.

EXTRA EXAMPLE 1
Simplify the expression.
a. $\sqrt{500}$ $10\sqrt{5}$
b. $3\sqrt{12} \cdot \sqrt{6}$ $18\sqrt{2}$
c. $\sqrt{\dfrac{25}{3}}$ $\dfrac{5\sqrt{3}}{3}$
d. $\sqrt{\dfrac{2}{11}}$ $\dfrac{\sqrt{22}}{11}$

EXTRA EXAMPLE 2
Solve $3 - 5x^2 = -9$.
$-\dfrac{2\sqrt{15}}{5}$ and $\dfrac{2\sqrt{15}}{5}$

EXTRA EXAMPLE 3
Solve $3(x - 2)^2 = 21$.
$2 + \sqrt{7}$ and $2 - \sqrt{7}$

✓ **CHECKPOINT EXERCISES**
For use after Example 1:
Simplify the expression.
1. $\sqrt{6} \cdot \sqrt{8}$ $4\sqrt{3}$
2. $\sqrt{\dfrac{49}{5}}$ $\dfrac{7\sqrt{5}}{5}$

For use after Examples 2 and 3:
3. Solve $4x^2 - 6 = 42$.
$-2\sqrt{3}$ and $2\sqrt{3}$
4. Solve $\dfrac{1}{5}(x - 4)^2 = 6$.
$4 + \sqrt{30}$ and $4 - \sqrt{30}$

EXTRA EXAMPLE 4
The tallest building in the United States is in Chicago, Illinois. It is 1450 ft tall.
a. How long would it take a penny to drop from the top of this building? **about 9.5 sec**
b. How fast would the penny be traveling when it hits the ground if the speed is given by $s = 32t$ where t is the number of seconds since the penny was dropped? **about 304 ft/sec**

 CHECKPOINT EXERCISES

For use after Example 4:
1. How long will it take an object dropped from a 550-foot tall tower to land on the roof of a 233-foot tall building?
about 4.5 sec

FOCUS ON VOCABULARY
Since $\sqrt{s}$ is read "the square root of s" students may think that $\sqrt{s}$ is the only square root of s. Stress to students that a *square root* of s is any number r so that $r^2 = s$. All positive numbers s have two square roots, $\sqrt{s}$ and $-\sqrt{s}$.

CLOSURE QUESTION
For what purpose would you use the product or quotient properties of square roots when solving quadratic equations using square roots? **to simplify the resulting radical expression**

DAILY PUZZLER
An object dropped from the top of a building passed you on the fifth floor 3 sec later. If the fifth floor is 98 ft from the ground, how tall is the building? **242 ft**

 **GOAL 2** **USING QUADRATIC MODELS IN REAL LIFE**

When an object is dropped, its speed continually increases, and therefore its height above the ground decreases at a faster and faster rate. The height h (in feet) of the object t seconds after it is dropped can be modeled by the function

$$h = -16t^2 + h_0$$

where h_0 is the object's initial height. This model assumes that the force of air resistance on the object is negligible. Also, the model works only on Earth. For planets with stronger or weaker gravity, different models are used (see Exercise 71).

Movies

EXAMPLE 4 *Modeling a Falling Object's Height with a Quadratic Function*

A stunt man working on the set of a movie is to fall out of a window 100 feet above the ground. For the stunt man's safety, an air cushion 26 feet wide by 30 feet long by 9 feet high is positioned on the ground below the window.

a. For how many seconds will the stunt man fall before he reaches the cushion?

b. A movie camera operating at a speed of 24 frames per second records the stunt man's fall. How many frames of film show the stunt man falling?

SOLUTION

a. The stunt man's initial height is $h_0 = 100$ feet, so his height as a function of time is given by $h = -16t^2 + 100$. Since the top of the cushion is 9 feet above the ground, you can determine how long it takes the stunt man to reach the cushion by finding the value of t for which $h = 9$. Here are two methods:

Method 1: Make a table of values.

t	0	1	2	3
h	100	84	36	-44

▶ From the table you can see that $h = 9$ at a value of t between $t = 2$ and $t = 3$. It takes between 2 sec and 3 sec for the stunt man to reach the cushion.

Method 2: Solve a quadratic equation.

$h = -16t^2 + 100$	Write height function.
$9 = -16t^2 + 100$	Substitute 9 for h.
$-91 = -16t^2$	Subtract 100 from each side.
$\dfrac{91}{16} = t^2$	Divide each side by –16.
$\sqrt{\dfrac{91}{16}} = t$	Take positive square root.
$2.4 \approx t$	Use a calculator.

▶ It takes about 2.4 seconds for the stunt man to reach the cushion.

b. The number of frames of film that show the stunt man falling is given by the product (2.4 sec)(24 frames/sec), or about 57 frames.

GUIDED PRACTICE

Vocabulary Check ✔

1. Explain what it means to "rationalize the denominator" of a quotient containing square roots. **1, 2. See margin.**

Concept Check ✔

2. State the product and quotient properties of square roots in words.

3. How many real-number solutions does the equation $x^2 = s$ have when $s > 0$? when $s = 0$? when $s < 0$? **2; 1; 0**

Skill Check ✔

1. Sample answer: to eliminate a radical from the denominator of a fraction

2. Sample answer: The product property says that the square root of a product equals the product of the square roots. The quotient property says that the square root of a quotient equals the quotient of the square roots.

Simplify the expression.

4. $\sqrt{49}$ **7**

5. $\sqrt{12}$ **$2\sqrt{3}$**

6. $\sqrt{45}$ **$3\sqrt{5}$**

7. $\sqrt{3} \cdot \sqrt{27}$ **9**

8. $\sqrt{\dfrac{16}{25}}$ **$\dfrac{4}{5}$**

9. $\sqrt{\dfrac{7}{9}}$ **$\dfrac{\sqrt{7}}{3}$**

10. $\dfrac{1}{\sqrt{3}}$ **$\dfrac{\sqrt{3}}{3}$**

11. $\sqrt{\dfrac{5}{2}}$ **$\dfrac{\sqrt{10}}{2}$**

Solve the equation.

12. $x^2 = 64$ **$-8, 8$**

13. $x^2 - 9 = 16$ **$-5, 5$**

14. $4x^2 + 7 = 23$ **$-2, 2$**

15. $\dfrac{x^2}{6} - 2 = 0$ **$-2\sqrt{3}, 2\sqrt{3}$**

16. $5(x - 1)^2 = 50$ **$-\sqrt{10} + 1, \sqrt{10} + 1$**

17. $\dfrac{1}{2}(x + 8)^2 = 14$ **$-2\sqrt{7} - 8, 2\sqrt{7} - 8$**

18. 🌐 **ENGINEERING** At an engineering school, students are challenged to design a container that prevents an egg from breaking when dropped from a height of 50 feet. Write an equation giving a container's height h (in feet) above the ground after t seconds. How long does the container take to hit the ground? **$h = -16t^2 + 50$; about 1.8 sec**

PRACTICE AND APPLICATIONS

STUDENT HELP

▶ **Extra Practice**
to help you master skills is on p. 946.

USING THE PRODUCT PROPERTY Simplify the expression.

19. $\sqrt{18}$ **$3\sqrt{2}$**

20. $\sqrt{48}$ **$4\sqrt{3}$**

21. $\sqrt{27}$ **$3\sqrt{3}$**

22. $\sqrt{52}$ **$2\sqrt{13}$**

23. $\sqrt{72}$ **$6\sqrt{2}$**

24. $\sqrt{175}$ **$5\sqrt{7}$**

25. $\sqrt{98}$ **$7\sqrt{2}$**

26. $\sqrt{605}$ **$11\sqrt{5}$**

27. $2\sqrt{7} \cdot \sqrt{7}$ **14**

28. $\sqrt{8} \cdot \sqrt{2}$ **4**

29. $\sqrt{3} \cdot \sqrt{12}$ **6**

30. $3\sqrt{20} \cdot 6\sqrt{5}$ **180**

31. $\sqrt{12} \cdot \sqrt{2}$ **$2\sqrt{6}$**

32. $\sqrt{6} \cdot \sqrt{10}$ **$2\sqrt{15}$**

33. $4\sqrt{3} \cdot \sqrt{21}$ **$12\sqrt{7}$**

34. $\sqrt{8} \cdot \sqrt{6} \cdot \sqrt{3}$ **12**

USING THE QUOTIENT PROPERTY Simplify the expression.

35. $\sqrt{\dfrac{1}{9}}$ **$\dfrac{1}{3}$**

36. $\sqrt{\dfrac{4}{49}}$ **$\dfrac{2}{7}$**

37. $\sqrt{\dfrac{36}{25}}$ **$\dfrac{6}{5}$**

38. $\sqrt{\dfrac{100}{81}}$ **$\dfrac{10}{9}$**

39. $\sqrt{\dfrac{3}{16}}$ **$\dfrac{\sqrt{3}}{4}$**

40. $\sqrt{\dfrac{11}{64}}$ **$\dfrac{\sqrt{11}}{8}$**

41. $\sqrt{\dfrac{75}{36}}$ **$\dfrac{5\sqrt{3}}{6}$**

42. $\sqrt{\dfrac{40}{169}}$ **$\dfrac{2\sqrt{10}}{13}$**

43. $\dfrac{2}{\sqrt{3}}$ **$\dfrac{2\sqrt{3}}{3}$**

44. $\dfrac{5}{\sqrt{17}}$ **$\dfrac{5\sqrt{17}}{17}$**

45. $\sqrt{\dfrac{6}{5}}$ **$\dfrac{\sqrt{30}}{5}$**

46. $\sqrt{\dfrac{144}{11}}$ **$\dfrac{12\sqrt{11}}{11}$**

47. $\sqrt{\dfrac{7}{8}}$ **$\dfrac{\sqrt{14}}{4}$**

48. $\sqrt{\dfrac{18}{13}}$ **$\dfrac{3\sqrt{26}}{13}$**

49. $\sqrt{\dfrac{45}{32}}$ **$\dfrac{3\sqrt{10}}{8}$**

50. $\sqrt{\dfrac{15}{7}} \cdot \sqrt{\dfrac{4}{3}}$ **$\dfrac{2\sqrt{35}}{7}$**

STUDENT HELP

▶ **HOMEWORK HELP**
Example 1: Exs. 19–50
Example 2: Exs. 51–59
Example 3: Exs. 60–68
Example 4: Exs. 69–73

SOLVING QUADRATIC EQUATIONS Solve the equation.

51. $x^2 = 121$ **$-11, 11$**

52. $x^2 = 90$ **$-3\sqrt{10}, 3\sqrt{10}$**

53. $3x^2 = 108$ **$-6, 6$**

54. $2x^2 + 5 = 41$ **$-3\sqrt{2}, 3\sqrt{2}$**

55. $-x^2 - 12 = -87$ **$-5\sqrt{3}, 5\sqrt{3}$**

56. $7 - 10u^2 = 1$ **$-\dfrac{\sqrt{15}}{5}, \dfrac{\sqrt{15}}{5}$**

57. $\dfrac{v^2}{25} - 1 = 11$ **$-10\sqrt{3}, 10\sqrt{3}$**

58. $6 - \dfrac{p^2}{8} = -4$ **$-4\sqrt{5}, 4\sqrt{5}$**

59. $\dfrac{5q^2}{6} - \dfrac{q^2}{3} = 72$ **$-12, 12$**

---

3 APPLY

ASSIGNMENT GUIDE

BASIC
Day 1: 267–270 Exs. 20–64 even, 67, 69, 74, 77–91 odd, Quiz 1 Exs. 1–11

AVERAGE
Day 1: 267–270 Exs. 20–68 even, 69–73 odd, 74, 77–91 odd, Quiz 1 Exs. 1–11

ADVANCED
Day 1: 267–270 Exs. 19–73 odd, 74, 75, 77–91 odd, Quiz 1 Exs. 1–11

BLOCK SCHEDULE
pp. 267–270 Exs. 20–68 even, 69–73 odd, 74, 77–91 odd, Quiz 1 Exs. 1–11 (with 5.4)

EXERCISE LEVELS
Level A: *Easier*
19–56, 60–63
Level B: *More Difficult*
57–59, 64–74
Level C: *Most Difficult*
75

✔ **HOMEWORK CHECK**
To quickly check student understanding of key concepts, go over the following exercises: Exs. 20, 30, 40, 48, 54, 60. See also the Daily Homework Quiz:
• Blackline Master (*Chapter 5 Resource Book,* p. 53)
• 📄 Transparency (p. 35)

EXERCISE 70 This exercise shows students how the *Table* feature of the calculator can be used to get a quick estimate of the value of a quadratic function. Making this table by hand would be quite time consuming.

APPLICATION NOTE
EXERCISE 71 Additional information about astronomy is available at **www.mcdougallittell.com**.

STUDENT HELP NOTES

→ **Skills Review** As students review the Pythagorean theorem on p. 917 remind them that the Pythagorean theorem can be used to find the length of the diagonal of a rectangle.

ADDITIONAL PRACTICE AND RETEACHING

For Lesson 5.3:

• Practice Levels A, B, and C (*Chapter 5 Resource Book,* p. 41)

• Reteaching with Practice (*Chapter 5 Resource Book,* p. 44)

• ⊞ See Lesson 5.3 of the *Personal Student Tutor*

For more Mixed Review:

• ⊞ Search the *Test and Practice Generator* for key words or specific lessons.

FOCUS ON
APPLICATIONS

ASTRONOMY The acceleration due to gravity on the moon is about 5.3 ft/sec². This means that the moon's gravity is only about one sixth as strong as Earth's.

APPLICATION LINK
www.mcdougallittell.com

STUDENT HELP

→ **Skills Review**
For help with the Pythagorean theorem, see p. 917.

SOLVING QUADRATIC EQUATIONS Solve the equation.

60. $2(x - 3)^2 = 8$ **1, 5**

61. $4(x + 1)^2 = 100$ **−6, 4**

62. $-3(x + 2)^2 = -18$ $-2 + \sqrt{6}, -2 - \sqrt{6}$

63. $5(x - 7)^2 = 135$ $-3\sqrt{3} + 7, 3\sqrt{3} + 7$

64. $8(x + 4)^2 = 9$ $\frac{-3\sqrt{2}}{4} - 4, \frac{3\sqrt{2}}{4} - 4$

65. $2(a - 6)^2 - 45 = 53$ **−1, 13**

66. $\frac{1}{4}(b - 8)^2 = 7$ $-2\sqrt{7} + 8, 2\sqrt{7} + 8$

67. $(2r - 5)^2 = 81$ **−2, 7**

68. $\frac{(s + 1)^2}{10} - \frac{12}{5} = \frac{15}{2}$ $-3\sqrt{11} - 1, 3\sqrt{11} - 1$

69. **HISTORY CONNECTION** According to legend, in 1589 the Italian scientist Galileo Galilei dropped two rocks of different weights from the top of the Leaning Tower of Pisa. He wanted to show that the rocks would hit the ground at the same time. Given that the tower's height is about 177 feet, how long would it have taken for the rocks to hit the ground?   **about 3.3 sec**

70. ▢ **ORNITHOLOGY** Many birds drop shellfish onto rocks to break the shell and get to the food inside. Crows along the west coast of Canada use this technique to eat whelks (a type of sea snail). Suppose a crow drops a whelk from a height of 20 feet, as shown.

▶ Source: *Cambridge Encyclopedia of Ornithology*

a. Write an equation giving the whelk's height h (in feet) after t seconds. $h = -16t^2 + 20$

b. Use the *Table* feature of a graphing calculator to find h when $t = 0, 0.1, 0.2, 0.3, \ldots, 1.4, 1.5$. (You'll need to scroll down the table to see all the values.) To the nearest tenth of a second, how long does it take for the whelk to hit the ground? Check your answer by solving a quadratic equation.   **1.1 sec**

20 ft

71. 🌐 **ASTRONOMY** On *any* planet, the height h (in feet) of a falling object t seconds after it is dropped can be modeled by

$$h = -\frac{g}{2}t^2 + h_0$$

Earth: 3.5 sec; Mars: 5.8 sec; Jupiter: 2.2 sec; Neptune: 3.3 sec; Pluto: 13.8 sec

where h_0 is the object's initial height and g is the acceleration (in feet per second squared) due to the planet's gravity. For each planet in the table, find the time it takes for a rock dropped from a height of 200 feet to hit the ground.

Planet	Earth	Mars	Jupiter	Neptune	Pluto
g (ft/sec²)	32	12	81	36	2.1

▶ Source: STARLab, Stanford University

72. 🌐 **OCEANOGRAPHY** The equation $h = 0.019s^2$ gives the height h (in feet) of the largest ocean waves when the wind speed is s knots. How fast is the wind blowing if the largest waves are 15 feet high? ▶ Source: *Encyclopaedia Britannica*   **about 28.1 knots**

73. 🌐 **TELEVISION** The *aspect ratio* of a TV screen is the ratio of the screen's width to its height. For most TVs, the aspect ratio is 4:3. What are the width and height of the screen for a 27 inch TV? (*Hint:* Use the Pythagorean theorem and the fact that TV sizes such as 27 inches refer to the length of the screen's diagonal.)   **16.2 in. by 21.6 in.**

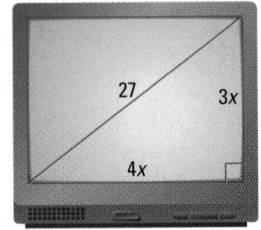

27   3x

4x

Test Preparation

74a. about 88.4 mi/h

74b. No; *Sample answer:* When $P = 40$ lb/ft^2, speed is 125 mi/h which is not $2 \cdot 88.4$.

74c. *Sample answer:* The wind speed value is squared in the formula and squaring increases the pressure value quickly.

★ **Challenge**

75a. about 60.6 sec

75b. 146 sec

75c. *Sample answer:* The water drains more slowly as the time increases.

EXTRA CHALLENGE

www.mcdougallittell.com

84. $\begin{bmatrix} 12 & -20 & 4 \\ -16 & 16 & -32 \end{bmatrix}$

85. $\begin{bmatrix} 81 & 57 \\ -40 & -31 \end{bmatrix}$

86. $y = x^2 + 3x - 10$

87. $y = x^2 - 9x + 8$

88. $y = 2x^2 + 15x + 28$

89. $y = 16x^2 - 81$

90. $y = x^2 - 6x + 10$

91. $y = 5x^2 + 60x + 168$

---

74. **MULTI-STEP PROBLEM** Building codes often require that buildings be able to withstand a certain amount of wind pressure. The pressure P (in pounds per square foot) from wind blowing at s miles per hour is given by $P = 0.00256s^2$.

▶ Source: *The Complete How to Figure It*

a. You are designing a two-story library. Buildings this tall are often required to withstand wind pressure of 20 lb/ft^2. Under this requirement, how fast can the wind be blowing before it produces excessive stress on a building?

b. To be safe, you design your library so that it can withstand wind pressure of 40 lb/ft^2. Does this mean that the library can survive wind blowing at twice the speed you found in part (a)? Justify your answer mathematically.

c. *Writing* Use the pressure formula to explain why even a relatively small increase in wind speed could have potentially serious effects on a building.

75. **SCIENCE CONNECTION** For a bathtub with a rectangular base, *Torricelli's law* implies that the height h of water in the tub t seconds after it begins draining is given by

$$h = \left(\sqrt{h_0} - \frac{2\pi d^2 \sqrt{3}}{lw} t \right)^2$$

where l and w are the tub's length and width, d is the diameter of the drain, and h_0 is the water's initial height. (All measurements are in inches.) Suppose you completely fill a tub with water. The tub is 60 inches long by 30 inches wide by 25 inches high and has a drain with a 2 inch diameter.

a. Find the time it takes for the tub to go from being full to half-full.

b. Find the time it takes for the tub to go from being half-full to empty.

c. **CRITICAL THINKING** Based on your results, what general statement can you make about the speed at which water drains?

MIXED REVIEW

SOLVING SYSTEMS Solve the linear system by graphing. (Review 3.1)

76. $x + y = 5$ (2, 3)
    $-x + 2y = 4$

77. $x - y = -1$ (1, 2)
    $3x + y = 5$

78. $-3x + y = 7$ (−1, 4)
    $2x + y = 2$

79. $2x - 3y = 9$ (−3, −5)
    $4x - 3y = 3$

80. $x + 4y = 4$ (4, 0)
    $3x - 2y = 12$

81. $2x + 3y = 6$ (6, −2)
    $x - 6y = 18$

MATRIX OPERATIONS Perform the indicated operation(s). (Review 4.1)

82. $\begin{bmatrix} 6 & -1 \\ 8 & 2 \end{bmatrix} + \begin{bmatrix} -5 & -4 \\ 10 & -2 \end{bmatrix}\begin{bmatrix} 1 & -5 \\ 18 & 0 \end{bmatrix}$

83. $\begin{bmatrix} 7 & 3 \\ -2 & 0 \end{bmatrix} - \begin{bmatrix} -6 & 4 \\ 9 & -1 \end{bmatrix}\begin{bmatrix} 13 & -1 \\ -11 & 1 \end{bmatrix}$

84. $-4\begin{bmatrix} -3 & 5 & -1 \\ 4 & -4 & 8 \end{bmatrix}$

85. $-2\begin{bmatrix} 12 & 10 \\ 20 & -9 \end{bmatrix} + 7\begin{bmatrix} 15 & 11 \\ 0 & -7 \end{bmatrix}$

WRITING IN STANDARD FORM Write the quadratic function in standard form. (Review 5.1 for 5.4)

86. $y = (x + 5)(x - 2)$

87. $y = (x - 1)(x - 8)$

88. $y = (2x + 7)(x + 4)$

89. $y = (4x + 9)(4x - 9)$

90. $y = (x - 3)^2 + 1$

91. $y = 5(x + 6)^2 - 12$

5.3 *Solving Quadratic Equations by Finding Square Roots* **269**

---

4 ASSESS

DAILY HOMEWORK QUIZ

📖 **Transparency Available**

Simplify the expression.

1. $\sqrt{63}$ $3\sqrt{7}$

2. $\sqrt{\dfrac{3}{25}}$ $\dfrac{\sqrt{3}}{5}$

3. $\sqrt{8} \cdot \sqrt{10}$ $4\sqrt{5}$

4. $\sqrt{\dfrac{5}{3}}$ $\dfrac{\sqrt{15}}{3}$

5. Solve $3x^2 + 2 = 62$. $2\sqrt{5}, -2\sqrt{5}$

6. Solve $\dfrac{1}{2}(x + 3)^2 = 5$. $-3 \pm \sqrt{10}$

EXTRA CHALLENGE NOTE
→ Challenge problems for Lesson 5.3 are available in **blackline** format in the *Chapter 5 Resource Book*, p. 49 and at **www.mcdougallittell.com**.

ADDITIONAL TEST PREPARATION

1. **OPEN ENDED** Write a quadratic equation that you would solve by finding square roots and show how to solve it. Why would you choose to solve this equation using square roots rather than factoring?
   Sample answer:
   $2x^2 + 5 = 21$; $2x^2 = 16$, $x^2 = 8$, $x = \pm\sqrt{8}$ or $\pm 2\sqrt{2}$; to solve $2x^2 + 5 = 21$ by factoring, rewrite it as $2(x^2 - 8) = 0$, which is not easily factored because 8 is not a perfect square.

2. **WRITING** How do you know if a radical expression is simplified? There should be no radicand with a perfect square factor other than 1, and there should be no radical in the denominator.

269

QUIZ 1

ADDITIONAL RESOURCES
An alternative Quiz for Lessons 5.1–5.3 is available in the *Chapter 5 Resource Book,* p. 50.

A **blackline** master with additional Math & History exercises is available in the *Chapter 5 Resource Book,* p. 48.

1.
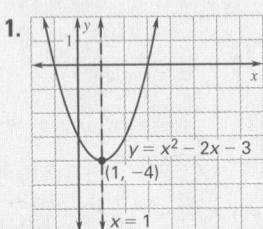

2.
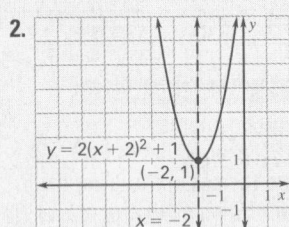

3.
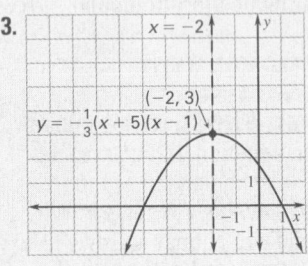

2.
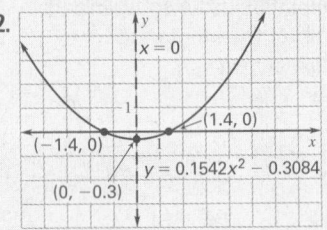

Graph the function. (Lesson 5.1)   1–3. See margin.

1. $y = x^2 - 2x - 3$ **2.** $y = 2(x + 2)^2 + 1$ **3.** $y = -\frac{1}{3}(x + 5)(x - 1)$

Solve the equation. (Lesson 5.2)

4. $x^2 - 6x - 27 = 0$ $-3, 9$ **5.** $4x^2 + 21x + 20 = 0$ **6.** $7t^2 - 4t = 3t^2 - 1$ $\frac{1}{2}$
$\quad\quad\quad\quad\quad\quad\quad\quad\quad\quad\quad\quad -4, -\frac{5}{4}$

Simplify the expression. (Lesson 5.3)

7. $\sqrt{54}$ $3\sqrt{6}$ **8.** $7\sqrt{2} \cdot \sqrt{10}$ $14\sqrt{5}$ **9.** $\sqrt{\frac{36}{5}}$ $\frac{6\sqrt{5}}{5}$ **10.** $\frac{4}{\sqrt{12}}$ $\frac{2\sqrt{3}}{3}$

11. 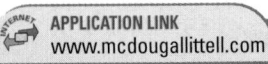 **SWIMMING** The drag force F (in pounds) of water on a swimmer can be modeled by $F = 1.35s^2$ where s is the swimmer's speed (in miles per hour). How fast must you swim to generate a drag force of 10 pounds? (Lesson 5.3)
about 2.7 mi/h

MATH & History

Telescopes

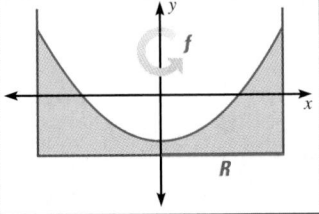

APPLICATION LINK
www.mcdougallittell.com

THEN

THE FIRST TELESCOPE is thought to have been made in 1608 by Hans Lippershey, a Dutch optician. Lippershey's telescope, called a *refracting telescope*, used lenses to magnify objects. Another type of telescope is a *reflecting telescope*. Reflecting telescopes magnify objects with parabolic mirrors, traditionally made from glass.

NOW

RECENTLY "liquid mirrors" for telescopes have been made by spinning reflective liquids, such as mercury. A cross section of the surface of a spinning liquid is a parabola with equation

$$y = \frac{\pi^2 f^2}{16}x^2 - \frac{\pi^2 f^2 R^2}{32}$$

where f is the spinning frequency (in revolutions per second) and R is the radius (in feet) of the container.

1. Write an equation for the surface of a liquid before it is spun. What does the equation tell you about the location of the x-axis relative to the liquid? $y = 0$; it lines up with the liquid's surface.

3. $-\frac{R\sqrt{2}}{2}, \frac{R\sqrt{2}}{2}$;

no, the x-intercepts are in terms of the radius only.

2. Suppose mercury is spun with a frequency of 0.5 revolution/sec in a container with radius 2 feet. Write and graph an equation for the mercury's surface. $y = 0.1542x^2 - 0.3084$; See margin for graph.

3. Find the x-intercepts of the graph of $y = \frac{\pi^2 f^2}{16}x^2 - \frac{\pi^2 f^2 R^2}{32}$. Does changing the spinning frequency affect the x-intercepts? Explain. See margin.

Galileo first uses a refracting telescope for astronomical purposes.

1609

1668

Isaac Newton builds first reflecting telescope.

Maria Mitchell is first to use a telescope to discover a comet.

1847

1987

Liquid mirrors are first used to do astronomical research.

● ACTIVITY 5.3

Using Technology

Solving Quadratic Equations

You can use a graphing calculator to solve quadratic equations having real-number solutions.

STUDENT HELP

INTERNET KEYSTROKE HELP

See keystrokes for several models of calculators at www.mcdougallittell.com

▶ **EXAMPLE**

Solve $2(x - 3)^2 = 5$.

▶ **SOLUTION**

1 Write the equation in the form $f(x) = 0$.

$$2(x - 3)^2 = 5$$

$$2(x - 3)^2 - 5 = 0$$

Therefore, the solutions of the original equation are the zeros of the function $y = 2(x - 3)^2 - 5$, or equivalently, the x-intercepts of this function's graph.

2 Enter $y = 2(x - 3)^2 - 5$ into your graphing calculator.

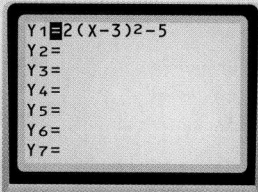

```
Y1=2(X-3)2-5
Y2=
Y3=
Y4=
Y5=
Y6=
Y7=
```

3 Graph the function you entered in **Step 2**. Use your calculator's *Zero* or *Root* feature to find the x-intercepts of the graph. (*Root* is another word for a solution of an equation, in this case $2(x - 3)^2 - 5 = 0$.)

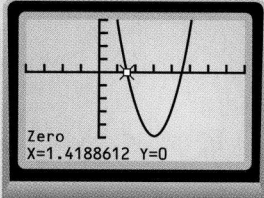

```
Zero
X=1.4188612  Y=0
```

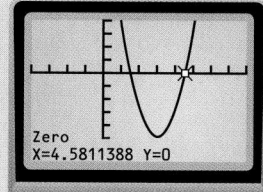

```
Zero
X=4.5811388  Y=0
```

▶ The solutions are about 1.42 and about 4.58.

▶ **EXERCISES**

Use a graphing calculator to solve the equation.

1. $3x^2 - 7 = 0$ $-1.53, 1.53$

2. $-2x^2 + 9 = 3$ $-1.73, 1.73$

3. $5x^2 + 2 = 6x^2 - 4$ $-2.45, 2.45$

4. $1.2x^2 - 5.6 = 0.8x^2 - 2.3$ $-2.87, 2.87$

5. $(x + 1)^2 - 3 = 0$ $-2.73, 0.73$

6. $-\frac{1}{3}(x - 4)^2 = -8$ $-0.90, 8.90$

7. $x^2 + 2x - 6 = 0$ $-3.65, 1.65$

8. $2x^2 + 8x + 3 = 4x^2 + 5x - 1$ $-0.85, 2.35$

9. 🌐 **MANUFACTURING** A company sells ground coffee in cans having a radius of 2 inches and a height of 6 inches. The company wants to manufacture a larger can that has the same height but holds twice as much coffee. Write an equation you can use to find the larger can's radius. (*Hint:* Use the formula $V = \pi r^2 h$ for the volume of a cylinder.) Solve the equation with a graphing calculator. $48\pi = 6\pi r^2$; $r \approx 2.8$ in.

5.3 *Technology Activity* **271**

---

1 Planning the Activity

PURPOSE
To solve quadratic equations using a graphing calculator.

MATERIALS
• graphing calculator
• Keystroke blackline (*Chapter 5 Resource Book,* p. 40)

PACING
• Activity — 20 min

▶ **LINK TO LESSON**
Have students go back and check the solution to Example 3 on page 265 using a graphing calculator.

2 Managing the Activity

CLASSROOM MANAGEMENT
You may want to have students solve the Example algebraically as well and round the radical answer to the nearest thousandth. In this way, students will be able to see the connection between the two methods.

ALTERNATIVE APPROACH
The class can complete this activity using computers with graphing software if graphing calculators are not available.

3 Closing the Activity

⭐ **KEY DISCOVERY**
The solutions to a quadratic equation are the same as the x-intercepts of the graph of its corresponding quadratic function.

ACTIVITY ASSESSMENT
What are some of the advantages and disadvantages of using a graphing calculator to find the solutions to a quadratic equation?
Sample answer: The graphing calculator method has the advantage in that it is fast. The major disadvantage is that the solutions may be approximate.

PACING
Basic: 2 days
Average: 2 days
Advanced: 2 days
Block Schedule: 0.5 block with 5.3,
0.5 block with 5.5

LESSON OPENER
ACTIVITY
An alternative way to approach
Lesson 5.4 is to use the Activity
Lesson Opener:

• Blackline Master (*Chapter 5
Resource Book,* p. 54)
• Transparency (p. 30)

MEETING INDIVIDUAL NEEDS
• *Chapter 5 Resource Book*
Prerequisite Skills Review (p. 5)
Practice Level A (p. 55)
Practice Level B (p. 56)
Practice Level C (p. 57)
Reteaching with Practice (p. 58)
Absent Student Catch-Up (p. 60)
Challenge (p. 62)
• *Resources in Spanish*
• 🖥 *Personal Student Tutor*

NEW-TEACHER SUPPORT
See the Tips for New Teachers on
pp. 1–2 of the *Chapter 5 Resource
Book* for additional notes about
Lesson 5.4.

WARM-UP EXERCISES

🖥 **Transparency Available**

Simplify.

1. $\sqrt{200}$ $10\sqrt{2}$
2. $\sqrt{75}$ $5\sqrt{3}$
3. $\sqrt{20}$ $2\sqrt{5}$
4. $\sqrt{98}$ $7\sqrt{2}$

What you should learn

GOAL 1 Solve quadratic
equations with complex
solutions and perform
operations with complex
numbers.

GOAL 2 Apply complex
numbers to fractal geometry.

Why you should learn it

▼ To solve problems, such
as determining whether a
complex number belongs to
the Mandelbrot set
in **Example 7**.

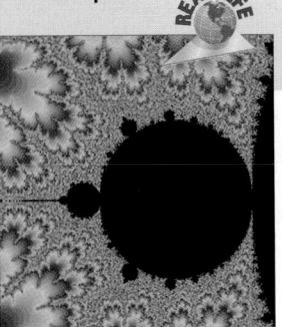

REAL LIFE

5.4

Complex Numbers

GOAL 1 **OPERATIONS WITH COMPLEX NUMBERS**

Not all quadratic equations have real-number solutions. For instance, $x^2 = -1$ has
no real-number solutions because the square of any real number x is never negative.
To overcome this problem, mathematicians created an expanded system of numbers
using the **imaginary unit** i, defined as $i = \sqrt{-1}$. Note that $i^2 = -1$. The
imaginary unit i can be used to write the square root of *any* negative number.

THE SQUARE ROOT OF A NEGATIVE NUMBER

PROPERTY	EXAMPLE
1. If r is a positive real number, then $\sqrt{-r} = i\sqrt{r}$.	$\sqrt{-5} = i\sqrt{5}$
2. By Property (1), it follows that $(i\sqrt{r})^2 = -r$.	$(i\sqrt{5})^2 = i^2 \cdot 5 = -5$

EXAMPLE 1   *Solving a Quadratic Equation*

Solve $3x^2 + 10 = -26$.

SOLUTION

$3x^2 + 10 = -26$	Write original equation.
$3x^2 = -36$	Subtract 10 from each side.
$x^2 = -12$	Divide each side by 3.
$x = \pm\sqrt{-12}$	Take square roots of each side.
$x = \pm i\sqrt{12}$	Write in terms of i.
$x = \pm 2i\sqrt{3}$	Simplify the radical.

▶ The solutions are $2i\sqrt{3}$ and $-2i\sqrt{3}$.

· · · · · · · · · ·

A **complex number** written in **standard form** is
a number $a + bi$ where a and b are real numbers.
The number a is the *real part* of the complex number,
and the number bi is the *imaginary part*. If $b \neq 0$,
then $a + bi$ is an **imaginary number**. If $a = 0$ and
$b \neq 0$, then $a + bi$ is a **pure imaginary number**.
The diagram shows how different types of complex
numbers are related.

Complex Numbers $(a + bi)$

Real Numbers $(a + 0i)$	Imaginary Numbers $(a + bi, b \neq 0)$
$-1 \qquad \frac{5}{2}$	$2 + 3i \quad 5 - 5i$
3	**Pure Imaginary Numbers** $(0 + bi, b \neq 0)$
$\pi \quad \sqrt{2}$	$-4i \qquad 6i$

Just as every real number corresponds to a point on the real number line, every complex number corresponds to a point in the **complex plane**. As shown in the next example, the complex plane has a horizontal axis called the *real axis* and a vertical axis called the *imaginary axis*.

EXAMPLE 2 *Plotting Complex Numbers*

Plot the complex numbers in the complex plane.

a. $2 - 3i$ **b.** $-3 + 2i$ **c.** $4i$

SOLUTION

a. To plot $2 - 3i$, start at the origin, move 2 units to the right, and then move 3 units down.

b. To plot $-3 + 2i$, start at the origin, move 3 units to the left, and then move 2 units up.

c. To plot $4i$, start at the origin and move 4 units up.

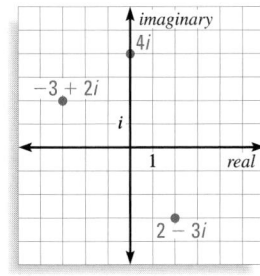

· · · · · · · · · ·

Two complex numbers $a + bi$ and $c + di$ are equal if and only if $a = c$ and $b = d$. For instance, if $x + yi = 8 - i$, then $x = 8$ and $y = -1$.

To add (or subtract) two complex numbers, add (or subtract) their real parts and their imaginary parts separately.

Sum of complex numbers: $(a + bi) + (c + di) = (a + c) + (b + d)i$

Difference of complex numbers: $(a + bi) - (c + di) = (a - c) + (b - d)i$

EXAMPLE 3 *Adding and Subtracting Complex Numbers*

Write the expression as a complex number in standard form.

a. $(4 - i) + (3 + 2i)$ **b.** $(7 - 5i) - (1 - 5i)$ **c.** $6 - (-2 + 9i) + (-8 + 4i)$

SOLUTION

a. $(4 - i) + (3 + 2i) = (4 + 3) + (-1 + 2)i$ Definition of complex addition

$= 7 + i$ Standard form

b. $(7 - 5i) - (1 - 5i) = (7 - 1) + (-5 + 5)i$ Definition of complex subtraction

$= 6 + 0i$ Simplify.

$= 6$ Standard form

c. $6 - (-2 + 9i) + (-8 + 4i) = [(6 + 2) - 9i] + (-8 + 4i)$ Subtract.

$= (8 - 9i) + (-8 + 4i)$ Simplify.

$= (8 - 8) + (-9 + 4)i$ Add.

$= 0 - 5i$ Simplify.

$= -5i$ Standard form

5.4 *Complex Numbers* **273**

2 TEACH

MOTIVATING THE LESSON
When students first learned about numbers, they knew only counting numbers. Then they learned about fractions and then negative numbers. Along with the fractions and the negative numbers came new rules for adding, subtracting, dividing, and multiplying. In today's lesson, they will learn about another set of numbers, complex numbers, and their rules of operation.

EXTRA EXAMPLE 1
Solve $2x^2 + 26 = -10$.
$-3i\sqrt{2}$ and $3i\sqrt{2}$

EXTRA EXAMPLE 2
Plot the complex numbers in the complex plane.
a. $-4 - i$ **b.** 5
c. $1 + 3i$

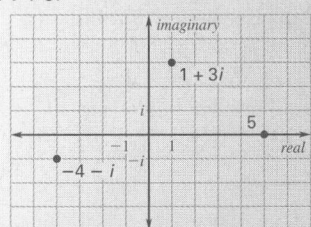

EXTRA EXAMPLE 3
Write the expression as a complex number in standard form.
a. $(-1 + 2i) + (3 + 3i)$ $2 + 5i$
b. $(2 - 3i) - (3 - 7i)$ $-1 + 4i$
c. $2i - (3 + i) + (2 - 3i)$ $-1 - 2i$

✔ CHECKPOINT EXERCISES
For use after Example 1:
1. Solve $-\frac{1}{2}(x + 1)^2 = 5$.
$-1 + i\sqrt{10}$ and $-1 - i\sqrt{10}$

For use after Example 2:
2. In which quadrant of the complex plane is $-3 + 5i$?
second quadrant

For use after Example 3:
3. Write $(3 - 5i) - (9 + 2i)$ as a complex number in standard form. $-6 - 7i$

EXTRA EXAMPLE 4
Write the expression as a complex number in standard form.
a. $-i(3 + i)$ $1 - 3i$
b. $(2 + 3i)(-6 - 2i)$ $-6 - 22i$
c. $(1 + 2i)(1 - 2i)$ 5

EXTRA EXAMPLE 5
Write the quotient $\frac{2 - 7i}{1 + i}$ in standard form. $-\frac{5}{2} - \frac{9}{2}i$

 CHECKPOINT EXERCISES
For use after Example 4:
Write the expression as a complex number in standard form.
1. $3i(9 - i)$ $3 + 27i$
2. $(-1 + 4i)(3 - 6i)$ $21 + 18i$
For use after Example 5:
3. Write the quotient $\frac{3 + 11i}{-1 - 2i}$ in standard form. $-5 - i$

CONCEPT QUESTION
Why do you multiply both the numerator and denominator of the quotient by the complex conjugate of the denominator? **This changes the denominator to a real number that can then be distributed to each of the terms of the numerator. Also, you need to multiply both the numerator and denominator so as not to change the value of the quotient.**

To multiply two complex numbers, use the distributive property or the FOIL method just as you do when multiplying real numbers or algebraic expressions. Other properties of real numbers that also apply to complex numbers include the associative and commutative properties of addition and multiplication.

EXAMPLE 4 *Multiplying Complex Numbers*

Write the expression as a complex number in standard form.

a. $5i(-2 + i)$ **b.** $(7 - 4i)(-1 + 2i)$ **c.** $(6 + 3i)(6 - 3i)$

SOLUTION

a. $5i(-2 + i) = -10i + 5i^2$ **Distributive property**

$\qquad\qquad\quad = -10i + 5(-1)$ **Use $i^2 = -1$.**

$\qquad\qquad\quad = -5 - 10i$ **Standard form**

b. $(7 - 4i)(-1 + 2i) = -7 + 14i + 4i - 8i^2$ **Use FOIL.**

$\qquad\qquad\qquad\qquad = -7 + 18i - 8(-1)$ **Simplify and use $i^2 = -1$.**

$\qquad\qquad\qquad\qquad = 1 + 18i$ **Standard form**

c. $(6 + 3i)(6 - 3i) = 36 - 18i + 18i - 9i^2$ **Use FOIL.**

$\qquad\qquad\qquad\quad = 36 - 9(-1)$ **Simplify and use $i^2 = -1$.**

$\qquad\qquad\qquad\quad = 45$ **Standard form**

· · · · · · · · · ·

In part (c) of Example 4, notice that the two factors $6 + 3i$ and $6 - 3i$ have the form $a + bi$ and $a - bi$. Such numbers are called **complex conjugates**. The product of complex conjugates is always a real number. You can use complex conjugates to write the quotient of two complex numbers in standard form.

EXAMPLE 5 *Dividing Complex Numbers*

Write the quotient $\frac{5 + 3i}{1 - 2i}$ in standard form.

SOLUTION

The key step here is to multiply the numerator and the denominator by the complex conjugate of the denominator.

$$\frac{5 + 3i}{1 - 2i} = \frac{5 + 3i}{1 - 2i} \cdot \frac{1 + 2i}{1 + 2i} \qquad \text{\textbf{Multiply by } } 1 + 2i, \text{ \textbf{the conjugate of} } 1 - 2i.$$

$$= \frac{5 + 10i + 3i + 6i^2}{1 + 2i - 2i - 4i^2} \qquad \text{\textbf{Use FOIL.}}$$

$$= \frac{-1 + 13i}{5} \qquad \text{\textbf{Simplify.}}$$

$$= -\frac{1}{5} + \frac{13}{5}i \qquad \text{\textbf{Standard form}}$$

BENOIT MANDELBROT
was born in Poland in 1924, came to the United States in 1958, and is now a professor at Yale University. He pioneered the study of fractal geometry in the 1970s.

GOAL 2 USING COMPLEX NUMBERS IN FRACTAL GEOMETRY

In the hands of a person who understands *fractal geometry*, the complex plane can become an easel on which stunning pictures called *fractals* are drawn. One very famous fractal is the *Mandelbrot set*, named after mathematician Benoit Mandelbrot. The Mandelbrot set is the black region in the complex plane below. (The points in the colored regions are *not* part of the Mandelbrot set.)

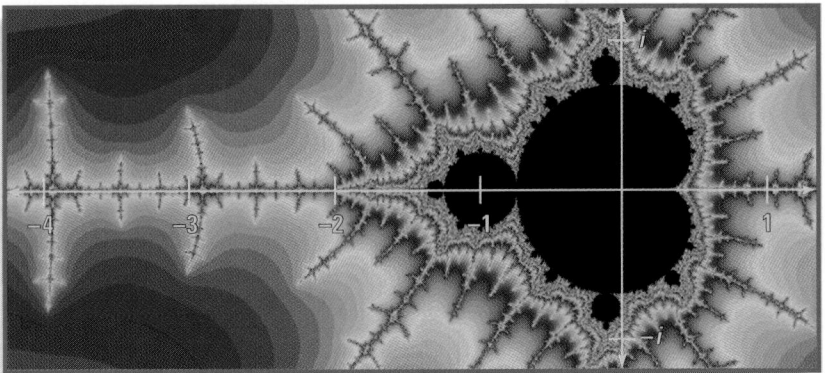

To understand how the Mandelbrot set is constructed, you need to know how the *absolute value* of a complex number is defined.

ABSOLUTE VALUE OF A COMPLEX NUMBER

The **absolute value** of a complex number $z = a + bi$, denoted $|z|$, is a nonnegative *real* number defined as follows:

$$|z| = \sqrt{a^2 + b^2}$$

Geometrically, the absolute value of a complex number is the number's distance from the origin in the complex plane.

EXAMPLE 6 *Finding Absolute Values of Complex Numbers*

Find the absolute value of each complex number. Which number is farthest from the origin in the complex plane?

a. $3 + 4i$ **b.** $-2i$ **c.** $-1 + 5i$

SOLUTION

a. $|3 + 4i| = \sqrt{3^2 + 4^2} = \sqrt{25} = 5$

b. $|-2i| = |0 + (-2i)| = \sqrt{0^2 + (-2)^2} = 2$

c. $|-1 + 5i| = \sqrt{(-1)^2 + 5^2} = \sqrt{26} \approx 5.10$

Since $-1 + 5i$ has the greatest absolute value, it is farthest from the origin in the complex plane.

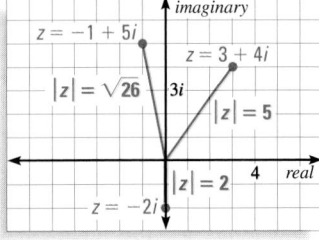

EXTRA EXAMPLE 6
Find the absolute value of each complex number. Which number is closest to the origin in the complex plane?
a. $-2 + 5i$ $\sqrt{29}$ about 5.39
b. $-6i$ 6
c. $5 - 3i$ $\sqrt{34}$ about 5.83
     $-2 + 5i$ is closest to the origin.

CHECKPOINT EXERCISES
For use after Example 6:
1. Find the absolute value of $-3 - 7i$. $\sqrt{58}$ about 7.62

EXTRA EXAMPLE 7
Tell whether the complex number c belongs to the Mandelbrot set.
a. $c = -0.5i$ yes
b. $c = -3$ no
c. $c = -2 + i$ no

 CHECKPOINT EXERCISES
For use after Example 7:
1. Tell whether $c = 2 - i$ belongs to the Mandelbrot set.   no

FOCUS ON VOCABULARY
Describe the *complex plane.*
The complex plane is a plane where each complex number has a corresponding point. The horizontal axis is the real axis and the vertical axis is the imaginary axis.

CLOSURE QUESTION
Describe the procedure for each of the four basic operations on complex numbers.
For adding or subtracting, add or subtract the corresponding real and imaginary parts. For multiplying, use the FOIL method and simplify. For dividing, multiply the numerator and denominator by the complex conjugate of the denominator and simplify.

DAILY PUZZLER
Two complex conjugates have a sum of 4 and a product of 13. What are the numbers? $2 + 3i$ and $2 - 3i$

15.

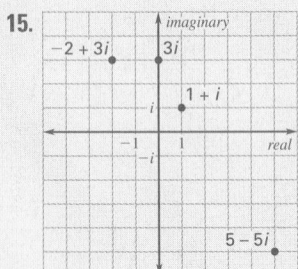

29–36.

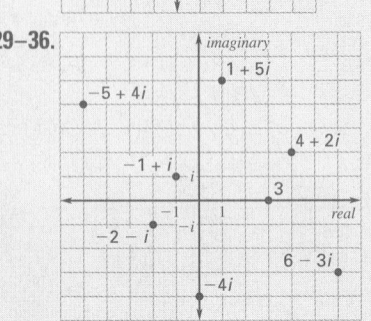

The following result shows how absolute value can be used to tell whether a given complex number belongs to the Mandelbrot set.

COMPLEX NUMBERS IN THE MANDELBROT SET

To determine whether a complex number c belongs to the Mandelbrot set, consider the function $f(z) = z^2 + c$ and this infinite list of complex numbers:

$$z_0 = 0, \; z_1 = f(z_0), \; z_2 = f(z_1), \; z_3 = f(z_2), \ldots$$

- If the absolute values $|z_0|, |z_1|, |z_2|, |z_3|, \ldots$ are all less than some fixed number N, then c belongs to the Mandelbrot set.

- If the absolute values $|z_0|, |z_1|, |z_2|, |z_3|, \ldots$ become infinitely large, then c does not belong to the Mandelbrot set.

EXAMPLE 7 *Determining if a Complex Number Is in the Mandelbrot Set*

Tell whether the complex number c belongs to the Mandelbrot set.

a. $c = i$ **b.** $c = 1 + i$ **c.** $c = -2$

SOLUTION

a. Let $f(z) = z^2 + i$.

$z_0 = 0$ $|z_0| = 0$

$z_1 = f(0) = 0^2 + i = i$ $|z_1| = 1$

$z_2 = f(i) = i^2 + i = -1 + i$ $|z_2| = \sqrt{2} \approx 1.41$

$z_3 = f(-1 + i) = (-1 + i)^2 + i = -i$ $|z_3| = 1$

$z_4 = f(-i) = (-i)^2 + i = -1 + i$ $|z_4| = \sqrt{2} \approx 1.41$

At this point the absolute values alternate between 1 and $\sqrt{2}$, and so all the absolute values are less than $N = 2$. Therefore, $c = i$ belongs to the Mandelbrot set.

b. Let $f(z) = z^2 + (1 + i)$.

$z_0 = 0$ $|z_0| = 0$

$z_1 = f(0) = 0^2 + (1 + i) = 1 + i$ $|z_1| \approx 1.41$

$z_2 = f(1 + i) = (1 + i)^2 + (1 + i) = 1 + 3i$ $|z_2| \approx 3.16$

$z_3 = f(1 + 3i) = (1 + 3i)^2 + (1 + i) = -7 + 7i$ $|z_3| \approx 9.90$

$z_4 = f(-7 + 7i) = (-7 + 7i)^2 + (1 + i) = 1 - 97i$ $|z_4| \approx 97.0$

The next few absolute values in the list are (approximately) 9409, 8.85×10^7, and 7.84×10^{15}. Since the absolute values are becoming infinitely large, $c = 1 + i$ does not belong to the Mandelbrot set.

c. Let $f(z) = z^2 + (-2)$, or $f(z) = z^2 - 2$. You can show that $z_0 = 0$, $z_1 = -2$, and $z_n = 2$ for $n > 1$. Therefore, the absolute values of $z_0, z_1, z_2, z_3, \ldots$ are all less than $N = 3$, and so $c = -2$ belongs to the Mandelbrot set.

GUIDED PRACTICE

Vocabulary Check ✓

1. Complete this statement: For the complex number $3 - 7i$, the real part is _?_ and the imaginary part is _?_. **3, −7i**

Concept Check ✓

2. ERROR ANALYSIS A student thinks that the complex conjugate of $-5 + 2i$ is $5 - 2i$. Explain the student's mistake, and give the correct complex conjugate of $-5 + 2i$. **See margin.**

3. Geometrically, what does the absolute value of a complex number represent? *Sample answer:* distance from the origin

Skill Check ✓

Solve the equation.

2. *Sample answer:* The real part should be the same and the imaginary part should be the opposite of the given imaginary part; $-5 - 2i$

4. $x^2 = -9$ **−3i, 3i**

5. $2x^2 + 3 = -13$ **−2i√2, 2i√2**

6. $(x - 1)^2 = -7$ **1 − i√7, 1 + i√7**

Write the expression as a complex number in standard form.

7. $(1 + 5i) + (6 - 2i)$ **7 + 3i**

8. $(4 + 3i) - (-2 + 4i)$ **6 − i**

9. $(1 - i)(7 + 2i)$ **9 − 5i**

10. $\dfrac{3 - 4i}{1 + i}$ **$\dfrac{-1 - 7i}{2}$**

Find the absolute value of the complex number.

11. $1 + i$ **√2**

12. $3i$ **3**

13. $-2 + 3i$ **√13**

14. $5 - 5i$ **5√2**

15. Plot the numbers in Exercises 11–14 in the same complex plane. **See margin.**

16. FRACTAL GEOMETRY Tell whether $c = 1 - i$ belongs to the Mandelbrot set. Use absolute value to justify your answer.
Sample answer: It does not because the absolute values become infinitely larger.

PRACTICE AND APPLICATIONS

STUDENT HELP

▸ **Extra Practice**
to help you master
skills is on p. 946.

SOLVING QUADRATIC EQUATIONS Solve the equation.

17. $x^2 = -4$ **−2i, 2i**

18. $x^2 = -11$ **−i√11, i√11**

19. $3x^2 = -81$ **−3i√3, 3i√3**

20. $2x^2 + 9 = -41$ **−5i, 5i**

21. $5x^2 + 18 = 3$ **−i√3, i√3**

22. $-x^2 - 4 = 14$ **−3i√2, 3i√2**

23. $8r^2 + 7 = 5r^2 + 4$ **−i, i**

24. $3s^2 - 1 = 7s^2$ **$-\frac{1}{2}i, \frac{1}{2}i$**

25. $(t - 2)^2 = -16$ **2 + 4i, 2 − 4i**

26. $-6(u + 5)^2 = 120$ **$-5 - 2i\sqrt{5}, -5 + 2i\sqrt{5}$**

27. $-\frac{1}{8}(v + 3)^2 = 7$ **$-3 - 2i\sqrt{14}, -3 + 2i\sqrt{14}$**

28. $9(w - 4)^2 + 1 = 0$ **$4 - \frac{1}{3}i, 4 + \frac{1}{3}i$**

PLOTTING COMPLEX NUMBERS Plot the numbers in the same complex plane.
29–36. See margin.

29. $4 + 2i$

30. $-1 + i$

31. $-4i$

32. 3

33. $-2 - i$

34. $1 + 5i$

35. $6 - 3i$

36. $-5 + 4i$

STUDENT HELP

▸ **HOMEWORK HELP**
Example 1: Exs. 17–28
Example 2: Exs. 29–36
Example 3: Exs. 37–46
Example 4: Exs. 47–55
Example 5: Exs. 56–63
Example 6: Exs. 64–71
Example 7: Exs. 72–79

ADDING AND SUBTRACTING Write the expression as a complex number in standard form.

37. $(2 + 3i) + (7 + i)$ **9 + 4i**

38. $(6 + 2i) + (5 - i)$ **11 + i**

39. $(-4 + 7i) + (-4 - 7i)$ **−8**

40. $(-1 - i) + (9 - 3i)$ **8 − 4i**

41. $(8 + 5i) - (1 + 2i)$ **7 + 3i**

42. $(2 - 6i) - (-10 + 4i)$ **12 − 10i**

43. $(-0.4 + 0.9i) - (-0.6 + i)$ **0.2 − 0.1i**

44. $(25 + 15i) - (25 - 6i)$ **21i**

45. $-i + (8 - 2i) - (5 - 9i)$ **3 + 6i**

46. $(30 - i) - (18 + 6i) + 30i$ **12 + 23i**

ASSIGNMENT GUIDE

BASIC
Day 1: pp. 277–280 Exs. 18–24 even, 30–42 even, 48–72 even, 80–85
Day 2: pp. 277–280 Exs. 26–28 even, 43–46, 86, 92, 97–99, 101–113 odd

AVERAGE
Day 1: pp. 277–280 Exs. 18–72 even, 80–86
Day 2: pp. 277–280 Exs. 74, 76, 87–94, 97–99, 102–114 even

ADVANCED
Day 1: pp. 277–280 Exs. 18–72 even, 80–86
Day 2: pp. 277–280 Exs. 74, 76, 87–99, 100–114 even

BLOCK SCHEDULE
pp. 277–280 Exs. 18–72 even, 80–86 (with 5.3)
pp. 277–280 Exs. 74, 76, 87–94, 97–99, 102–114 even (with 5.5)

EXERCISE LEVELS
Level A: *Easier*
17–46
Level B: *More Difficult*
47–94, 97–99
Level C: *Most Difficult*
95–96, 100

✓ HOMEWORK CHECK
To quickly check student understanding of key concepts, go over the following exercises:
Exs. 20, 30, 40, 48, 56, 64. See also the Daily Homework Quiz:

• Blackline Master (*Chapter 5 Resource Book,* p. 65)

• Transparency (p. 36)

STUDENT HELP NOTES

→ **Skills Review** As students review disproving statements by counterexample on page 927, remind them that one counterexample is sufficient to disprove a statement.

TEACHING TIPS
You may want to extend Exs. 84 and 85 by either showing students a short proof of one of these true statements or asking students to write a proof. This may help students remember that although examples can help them decide if a statement is true or not, the only way to know that a mathematical statement is always true is to have a proof.

86a.

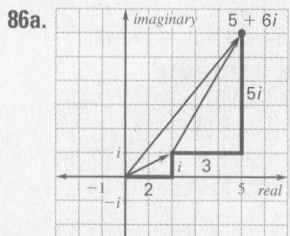

86b.

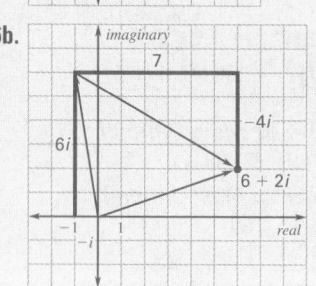

61. $-\frac{87}{97} + \frac{26}{97}i$

62. $\frac{10}{11} + \frac{\sqrt{10}}{11}i$

63. $\frac{17}{19} - \frac{6\sqrt{2}}{19}i$

72. *Sample answer:* It does not because the absolute values become infinitely large.

73. *Sample answer:* It does because the absolute values are equal to or less than $N = 1$.

74. *Sample answer:* It does because the absolute values are less than $N = 2$.

75. *Sample answer:* It does not because the absolute values become infinitely large.

STUDENT HELP

→ **Skills Review**
For help with disproving statements by counterexample, see p. 927.

76. *Sample answer:* It does not because the absolute values become infinitely large.

77. *Sample answer:* It does not because the absolute values become infinitely large.

78. *Sample answer:* It does because the absolute values are less than $N = 1$.

79. *Sample answer:* It does because the absolute values are less than $N = 1$.

MULTIPLYING Write the expression as a complex number in standard form.

47. $i(3 + i)$ $-1 + 3i$

48. $4i(6 - i)$ $4 + 24i$

49. $-10i(4 + 7i)$ $70 - 40i$

50. $(5 + i)(8 + i)$ $39 + 13i$

51. $(-1 + 2i)(11 - i)$ $-9 + 23i$

52. $(2 - 9i)(9 - 6i)$ $-36 - 93i$

53. $(7 + 5i)(7 - 5i)$ 74

54. $(3 + 10i)^2$ $-91 + 60i$

55. $(15 - 8i)^2$ $161 - 240i$

DIVIDING Write the expression as a complex number in standard form.

56. $\frac{8}{1 + i}$ $4 - 4i$

57. $\frac{2i}{1 - i}$ $-1 + i$

58. $\frac{-5 - 3i}{4i}$ $-\frac{3}{4} + \frac{5}{4}i$

59. $\frac{3 + i}{3 - i}$ $\frac{4}{5} + \frac{3}{5}i$

60. $\frac{2 + 5i}{5 + 2i}$ $\frac{20}{29} + \frac{21}{29}i$

61. $\frac{-7 + 6i}{9 - 4i}$

62. $\frac{\sqrt{10}}{\sqrt{10} - i}$

63. $\frac{6 - i\sqrt{2}}{6 + i\sqrt{2}}$

61–63. See margin.

ABSOLUTE VALUE Find the absolute value of the complex number.

64. $3 - 4i$ 5

65. $5 + 12i$ 13

66. $-2 - i$ $\sqrt{5}$

67. $-7 + i$ $5\sqrt{2}$

68. $2 + 5i$ $\sqrt{29}$

69. $4 - 8i$ $4\sqrt{5}$

70. $-9 + 6i$ $3\sqrt{13}$

71. $\sqrt{11} + i\sqrt{5}$ 4

MANDELBROT SET Tell whether the complex number c belongs to the Mandelbrot set. Use absolute value to justify your answer.

72–79. See margin.

72. $c = 1$

73. $c = -1$

74. $c = -i$

75. $c = -1 - i$

76. $c = 2$

77. $c = -1 + i$

78. $c = -0.5$

79. $c = 0.5i$

LOGICAL REASONING In Exercises 80–85, tell whether the statement is *true* or *false*. If the statement is false, give a counterexample.

80. Every complex number is an imaginary number. false; *Sample answer:* 1 is complex but not imaginary.

81. Every irrational number is a complex number. true

82. All real numbers lie on a single line in the complex plane. true

83. The sum of two imaginary numbers is always an imaginary number. false; *Sample answer:* $(6 + 3i) + (-5 - 3i) = 1$, which is not imaginary

84. Every real number equals its complex conjugate. true

85. The absolute values of a complex number and its complex conjugate are always equal. true

86. **VISUAL THINKING** The graph shows how you can geometrically add two complex numbers (in this case, $3 + 2i$ and $1 + 4i$) to find their sum (in this case, $4 + 6i$). Find each of the following sums by drawing a graph.

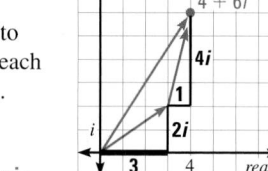

a. $(2 + i) + (3 + 5i)$ $5 + 6i$; See margin for graph.

b. $(-1 + 6i) + (7 - 4i)$ $6 + 2i$; See margin for graph.

COMPARING REAL AND COMPLEX NUMBERS Tell whether the property is true for (a) the set of real numbers and (b) the set of complex numbers.

87. If r, s, and t are numbers in the set, then $(r + s) + t = r + (s + t)$. true; true

88. If r is a number in the set and $|r| = k$, then $r = k$ or $r = -k$. true; false

89. If r and s are numbers in the set, then $r - s = s - r$. false; false

90. If r, s, and t are numbers in the set, then $r(s + t) = rs + rt$. true; true

91. If r and s are numbers in the set, then $|r + s| = |r| + |s|$. false; false

FOCUS ON
CAREERS

92. CRITICAL THINKING Evaluate $\sqrt{-4} \cdot \sqrt{-9}$ and $\sqrt{36}$. Does the rule $\sqrt{a} \cdot \sqrt{b} = \sqrt{ab}$ on page 264 hold when a and b are negative numbers? **no**

93. *Writing* Give both an algebraic argument and a geometric argument explaining why the definitions of absolute value on pages 50 and 275 are consistent when applied to real numbers. **See margin.**

94. EXTENSION: ADDITIVE AND MULTIPLICATIVE INVERSES The *additive inverse* of a complex number z is a complex number z_a such that $z + z_a = 0$. The *multiplicative inverse* of z is a complex number z_m such that $z \cdot z_m = 1$. Find the additive and multiplicative inverses of each complex number.
See margin.

a. $z = 1 + i$ **b.** $z = 3 - i$ **c.** $z = -2 + 8i$

🌐 **ELECTRICITY** In Exercises 95 and 96, use the following information.

Electrical circuits may contain several types of components such as resistors, inductors, and capacitors. The resistance of each component to the flow of electrical current is the component's *impedance*, denoted by Z. The value of Z is a real number R for a resistor of R ohms (Ω), a pure imaginary number Li for an inductor of L ohms, and a pure imaginary number $-Ci$ for a capacitor of C ohms. Examples are given in the table.

Component	Symbol	Z
Resistor	⌇⌇⌇ 3Ω	3
Inductor	⎛⎛⎛ 5Ω	$5i$
Capacitor	⊣⊢ 6Ω	$-6i$

95. 🌐 **SERIES CIRCUITS** A *series circuit* is a type of circuit found in switches, fuses, and circuit breakers. In a series circuit, there is only one pathway through which current can flow. To find the total impedance of a series circuit, add the impedances of the components in the circuit. What is the impedance of each series circuit shown below? (*Note:* The symbol ⊘ denotes an alternating current source and does not affect the calculation of impedance.)

a.

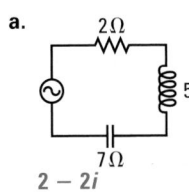

$2 - 2i$

b.

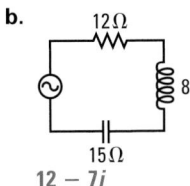

$12 - 7i$

c.

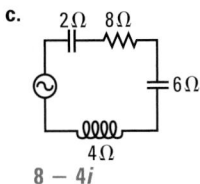

$8 - 4i$

96. 🌐 **PARALLEL CIRCUITS** *Parallel circuits* are used in household lighting and appliances. In a parallel circuit, there is more than one pathway through which current can flow. To find the impedance Z of a parallel circuit with two pathways, first calculate the impedances Z_1 and Z_2 of the pathways separately by treating each pathway as a series circuit. Then apply this formula:

$$Z = \frac{Z_1 Z_2}{Z_1 + Z_2}$$

What is the impedance of each parallel circuit shown below?

a.

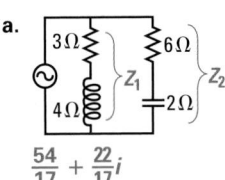

$\frac{54}{17} + \frac{22}{17}i$

b.

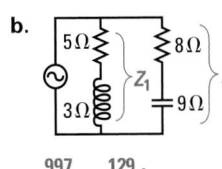

$\frac{997}{205} + \frac{129}{205}i$

c.

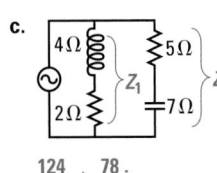

$\frac{124}{29} + \frac{78}{29}i$

5.4 *Complex Numbers*   **279**

Margin (left):

🔺 **ELECTRICIAN**
An electrician installs, maintains, and repairs electrical systems. This often involves working with the types of circuits described in Exs. 95 and 96.

💻 **CAREER LINK**
www.mcdougallittell.com

93. *Sample answer:* algebraic: a real number can be written as $a + 0i$. Then $|z| = \sqrt{a^2 + 0^2} = \sqrt{a^2} = |a|$; geometric: in both definitions, the absolute value is the distance from the point to the origin.

STUDENT HELP

🌐 **HOMEWORK HELP**
Visit our Web site www.mcdougallittell.com for help with problem solving in Exs. 95 and 96.

94a. $-1 - i; \frac{1}{2} - \frac{1}{2}i$

94b. $-3 + i; \frac{3}{10} + \frac{1}{10}i$

94c. $2 - 8i; -\frac{1}{34} - \frac{2}{17}i$

Margin (right):

CAREER NOTE
EXERCISES 95 AND 96
Additional information about a career as an electrician is available at **www.mcdougallittell.com**.

STUDENT HELP NOTES
→ **Homework Help** Students can find help for Exs. 95 and 96 at **www.mcdougallittell.com**. The information can be printed out for students who do not have access to the Internet.

ADDITIONAL PRACTICE AND RETEACHING

For Lesson 5.4:
• Practice Levels A, B, and C (*Chapter 5 Resource Book,* p. 55)
• Reteaching with Practice (*Chapter 5 Resource Book,* p. 58)
• 🖥 See Lesson 5.4 of the *Personal Student Tutor*

For more Mixed Review:
• 🖥 Search the *Test and Practice Generator* for key words or specific lessons.

DAILY HOMEWORK QUIZ

Transparency Available

1. Solve $4x^2 + 9 = -17$. $\pm\dfrac{i\sqrt{26}}{2}$

2. Plot $3 - 2i$, $-1 + i$, and $3i$ in the same complex plane.

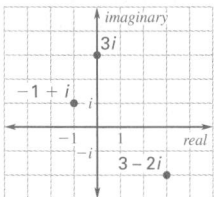

3. Write $(8 - i) - (3 + 2i) + (-4 + 6i)$ in standard form. $1 + 3i$

4. Write $(2 - 3i)(-1 + 4i)$ in standard form. $10 + 11i$

5. Write $\dfrac{3 + 2i}{2 + i}$ in standard form. $\dfrac{8}{5} + \dfrac{1}{5}i$

6. Find the absolute value of $4 + 2i$, $-8i$, and $-3 + 7i$. Which is farthest from the origin? about 4.5, 8, about 7.6; $-8i$

EXTRA CHALLENGE NOTE

Challenge problems for Lesson 5.4 are available in **blackline** format in the *Chapter 5 Resource Book*, p. 62 and at **www.mcdougallittell.com**.

ADDITIONAL TEST PREPARATION

1. OPEN ENDED Pick two complex numbers and show how to find their sum, difference, product, and quotient. Then find the absolute value of each number. **Check students' work.**

2. WRITING In one or two paragraphs explain complex numbers, imaginary numbers and pure imaginary numbers, and their relationships to each other and to real numbers. **See margin.**

114. See Additional Answers beginning on page AA1.

280

Test Preparation

QUANTITATIVE COMPARISON In Exercises 97–99, choose the statement that is true about the given quantities.

(A) The quantity in column A is greater.

(B) The quantity in column B is greater.

(C) The two quantities are equal.

(D) The relationship cannot be determined from the given information.

	Column A	Column B	
97.	$|5 + 4i|$	$|3 - 6i|$	B
98.	$|-6 + 8i|$	$|-10i|$	C
99.	$|2 + bi|$ where $b < -1$	$|\sqrt{3} + ci|$ where $0 < c < 1$	A

★ **Challenge**

100. POWERS OF *i* In this exercise you will investigate a pattern that appears when the imaginary unit i is raised to successively higher powers.

a. Copy and complete the table. $i^4 = 1$; $i^5 = i$; $i^6 = -1$; $i^7 = -i$; $i^8 = 1$

Power of i	i^1	i^2	i^3	i^4	i^5	i^6	i^7	i^8
Simplified form	i	-1	$-i$	?	?	?	?	?

100. b. *Sample answer:* the pattern is i, -1, $-i$, 1; $i^9 = i$, $i^{10} = -1$, $i^{11} = -i$, $i^{12} = 1$

b. *Writing* Describe the pattern you observe in the table. Verify that the pattern continues by evaluating the next four powers of i.

c. Use the pattern you described in part (b) to evaluate i^{26} and i^{83}. -1; $-i$

EXTRA CHALLENGE
www.mcdougallittell.com

MIXED REVIEW

EVALUATING FUNCTIONS Evaluate $f(x)$ for the given value of x. **(Review 2.1)**

101. $f(x) = 4x - 1$ when $x = 3$ **11**   **102.** $f(x) = x^2 - 5x + 8$ when $x = -4$ **44**

103. $f(x) = |-x + 6|$ when $x = 9$ **3**   **104.** $f(x) = 2$ when $x = -30$ **2**

SOLVING SYSTEMS Use an inverse matrix to solve the system. **(Review 4.5)**

105. $3x + y = 5$ **(1, 2)** **106.** $x + y = 2$ **(−5, 7)** **107.** $x - 2y = 10$ **(4, −3)**
    $5x + 2y = 9$ $7x + 8y = 21$ $3x + 4y = 0$

SOLVING QUADRATIC EQUATIONS Solve the equation. **(Review 5.3 for 5.5)**

108. $(x + 4)^2 = 1$ **−5, −3** **109.** $(x + 2)^2 = 36$ **−8, 4** **110.** $(x - 11)^2 = 25$ **6, 16**

Write in standard form.

111. $5 - \sqrt{10}$, $5 + \sqrt{10}$

112. $-7 - 2\sqrt{3}$, $-7 + 2\sqrt{3}$

113. $6 - \sqrt{7}$, $6 + \sqrt{7}$

111. $-(x - 5)^2 = -10$ **112.** $2(x + 7)^2 = 24$ **113.** $3(x - 6)^2 - 8 = 13$

114. STATISTICS CONNECTION The table shows the cumulative number N (in thousands) of DVD players sold in the United States from the end of February, 1997, to time t (in months). Make a scatter plot of the data. Approximate the equation of the best-fitting line. **(Review 2.5)** $y = 35.33x - 14.58$; See margin for graph.

t	1	2	3	4	5	6	7	8	9	10	11	12
N	34	69	96	125	144	178	213	269	307	347	383	416

DATA UPDATE of *DVD Insider* data at www.mcdougallittell.com

280 Chapter 5 *Quadratic Functions*

Additional Test Preparation *Sample answer:*

2. Complex numbers are numbers of the form $a + bi$, where a and b are real numbers. The real part is a. The imaginary part is bi, where $i = \sqrt{-1}$. Real numbers are complex numbers whose imaginary parts are zero while imaginary numbers are complex numbers whose imaginary parts are not zero. Pure imaginary numbers are imaginary numbers whose real part is zero.

ACTIVITY 5.5

Developing Concepts

GROUP ACTIVITY
Work with a partner.

MATERIALS
algebra tiles

Using Algebra Tiles to Complete the Square

▶ **QUESTION** Given b, what is the value of c that makes $x^2 + bx + c$ a perfect square trinomial?

▶ **EXPLORING THE CONCEPT**

① Use algebra tiles to model the expression $x^2 + 6x$.

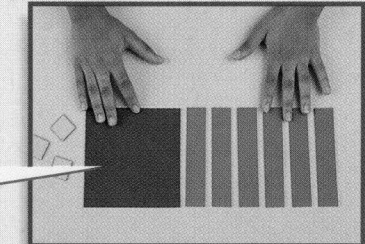

> You will need one x^2-tile and six x-tiles.

② Arrange the tiles in a square. Your arrangement will be incomplete in one corner.

> You want the length and width of your "square" to be equal.

1. $x^2 + 2x + 1$, 1, $(x + 1)^2$;
   $x^2 + 4x + 4$, 4, $(x + 2)^2$;
   $x^2 + 6x + 9$, 9, $(x + 3)^2$;
   $x^2 + 8x + 16$, 16, $(x + 4)^2$;
   $x^2 + 10x + 25$, 25, $(x + 5)^2$

2. a. $d = \frac{1}{2}b$

2. b. $c = d^2$

2. c. Find the square of half of the coefficient.

③ Determine the number of 1-tiles needed to complete the square.

> By adding nine 1-tiles, you can see that $x^2 + 6x + 9 = (x + 3)^2$.

Completing the Square

Expression	Number of 1-tiles needed to complete the square	Expression written as a square
$x^2 + 2x +$ _?_	?	?
$x^2 + 4x +$ _?_	?	?
$x^2 + 6x +$ _?_	9	$x^2 + 6x + 9 = (x + 3)^2$
$x^2 + 8x +$ _?_	?	?
$x^2 + 10x +$ _?_	?	?

▶ **DRAWING CONCLUSIONS**
1, 2. See margin.
1. Copy and complete the table at the left by following the steps above.

2. Look for patterns in the last column of your table. Consider the general statement $x^2 + bx + c = (x + d)^2$.

   a. How is d related to b in each case?

   b. How is c related to d in each case?

   c. How can you obtain the numbers in the second column of the table directly from the coefficients of x in the expressions from the first column?

5.5 *Concept Activity* **281**

---

1 Planning the Activity

PURPOSE
To complete the square using algebra tiles.

MATERIALS
• Algebra tiles
• Activity Support Master (*Chapter 5 Resource Book,* p. 66)

PACING
• Exploring the Concept — 10 min
• Drawing Conclusions — 10 min

▶ **LINK TO LESSON**
When discussing Example 1 on page 282, remind students of the physical process they used with the model. This process is reflected in the algebraic process of dividing b by 2 and then squaring it to get the constant term.

2 Managing the Activity

COOPERATIVE LEARNING
Students can work with a partner. Each student should have a turn manipulating the tiles while the other records the work in the chart.

ALTERNATIVE APPROACH
This activity can be done as a class using the algebra tiles on an overhead projector.

3 Closing the Activity

★ **KEY DISCOVERY**
By adding a constant, you can turn any quadratic expression into a perfect square trinomial.

ACTIVITY ASSESSMENT
JOURNAL Pick one of the expressions for which you completed the square. For this expression sketch the algebra tile model for Steps 1–3 of this activity and explain each step of the process for completing the square.

PACING
Basic: 2 days
Average: 2 days
Advanced: 2 days
Block Schedule: 0.5 block with 5.4,
0.5 block with 5.6

LESSON OPENER
APPLICATION
An alternative way to approach
Lesson 5.5 is to use the Application
Lesson Opener:

• Blackline Master (*Chapter 5
Resource Book*, p. 67)
•  Transparency (p. 31)

MEETING INDIVIDUAL NEEDS
• *Chapter 5 Resource Book*
Prerequisite Skills Review (p. 5)
Practice Level A (p. 69)
Practice Level B (p. 70)
Practice Level C (p. 71)
Reteaching with Practice (p. 72)
Absent Student Catch-Up (p. 74)
Challenge (p. 76)
• *Resources in Spanish*
• Personal Student Tutor

NEW-TEACHER SUPPORT
See the Tips for New Teachers on
pp. 1–2 of the *Chapter 5 Resource
Book* for additional notes about
Lesson 5.5.

WARM-UP EXERCISES
Transparency Available

Solve the equation.
1. $(x-2)^2 = 16$ $-2, 6$
2. $3(x+5)^2 = 24$ $-5 \pm 2\sqrt{2}$
3. $11(x-7)^2 - 3 = 19$ $7 \pm \sqrt{2}$

5.5

What you should learn

GOAL 1 Solve quadratic
equations by completing the
square.

GOAL 2 Use completing the
square to write quadratic
functions in vertex form, as
applied in **Example 7**.

Why you should learn it

▼ To solve **real-life**
problems, such as finding
where to position a fire
hose in **Ex. 91**.

Completing the Square

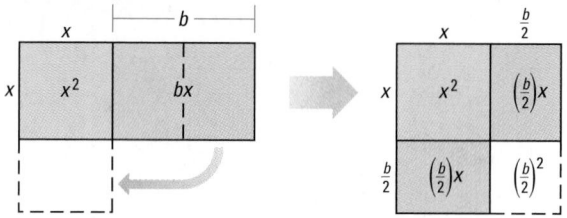

GOAL 1 **SOLVING QUADRATIC EQUATIONS BY
COMPLETING THE SQUARE**

Completing the square is a process that allows you to write an expression of the
form $x^2 + bx$ as the square of a binomial. This process can be illustrated using an
area model, as shown below.

You can see that to complete the square for $x^2 + bx$, you need to add $\left(\dfrac{b}{2}\right)^2$, the area
of the incomplete corner of the square in the second diagram. This diagram models
the following rule:

$$x^2 + bx + \left(\frac{b}{2}\right)^2 = \left(x + \frac{b}{2}\right)^2$$

EXAMPLE 1 *Completing the Square*

Find the value of c that makes $x^2 - 7x + c$ a perfect square trinomial. Then write the
expression as the square of a binomial.

SOLUTION
In the expression $x^2 - 7x + c$, note that $b = -7$. Therefore:

$$c = \left(\frac{b}{2}\right)^2 = \left(\frac{-7}{2}\right)^2 = \frac{49}{4}$$

Use this value of c to write $x^2 - 7x + c$ as a perfect square trinomial, and then as the
square of a binomial.

$$x^2 - 7x + c = x^2 - 7x + \frac{49}{4} \qquad \text{Perfect square trinomial}$$

$$= \left(x - \frac{7}{2}\right)^2 \qquad \text{Square of a binomial: } \left(x + \frac{b}{2}\right)^2$$

· · · · · · · · · ·

In Lesson 5.2 you learned how to solve quadratic equations by factoring. However,
many quadratic equations, such as $x^2 + 10x - 3 = 0$, contain expressions that cannot
be factored. Completing the square is a method that lets you solve *any* quadratic
equation, as the next example illustrates.

EXAMPLE 2 *Solving a Quadratic Equation if the Coefficient of x^2 Is 1*

Solve $x^2 + 10x - 3 = 0$ by completing the square.

SOLUTION

$$x^2 + 10x - 3 = 0 \qquad \text{Write original equation.}$$

$$x^2 + 10x = 3 \qquad \text{Write the left side in the form } x^2 + bx.$$

$$x^2 + 10x + 5^2 = 3 + 25 \qquad \text{Add } \left(\frac{10}{2}\right)^2 = 5^2 = 25 \text{ to each side.}$$

$$(x + 5)^2 = 28 \qquad \text{Write the left side as a binomial squared.}$$

$$x + 5 = \pm\sqrt{28} \qquad \text{Take square roots of each side.}$$

$$x = -5 \pm \sqrt{28} \qquad \text{Solve for } x.$$

$$x = -5 \pm 2\sqrt{7} \qquad \text{Simplify.}$$

▶ The solutions are $-5 + 2\sqrt{7}$ and $-5 - 2\sqrt{7}$.

✓ **CHECK** You can check the solutions by substituting them back into the original equation. Alternatively, you can graph $y = x^2 + 10x - 3$ and observe that the x-intercepts are about $0.29 \approx -5 + 2\sqrt{7}$ and $-10.29 \approx -5 - 2\sqrt{7}$.

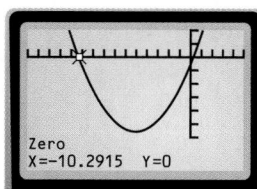

Zero
X=-10.2915   Y=0

· · · · · · · · · ·

If the coefficient of x^2 in a quadratic equation is not 1, you should divide each side of the equation by this coefficient before completing the square.

EXAMPLE 3 *Solving a Quadratic Equation if the Coefficient of x^2 Is Not 1*

Solve $3x^2 - 6x + 12 = 0$ by completing the square.

SOLUTION

$$3x^2 - 6x + 12 = 0 \qquad \text{Write original equation.}$$

$$x^2 - 2x + 4 = 0 \qquad \text{Divide each side by the coefficient of } x^2.$$

$$x^2 - 2x = -4 \qquad \text{Write the left side in the form } x^2 + bx.$$

$$x^2 - 2x + (-1)^2 = -4 + 1 \qquad \text{Add } \left(\frac{-2}{2}\right)^2 = (-1)^2 = 1 \text{ to each side.}$$

$$(x - 1)^2 = -3 \qquad \text{Write the left side as a binomial squared.}$$

$$x - 1 = \pm\sqrt{-3} \qquad \text{Take square roots of each side.}$$

$$x = 1 \pm \sqrt{-3} \qquad \text{Solve for } x.$$

$$x = 1 \pm i\sqrt{3} \qquad \text{Write in terms of the imaginary unit } i.$$

▶ The solutions are $1 + i\sqrt{3}$ and $1 - i\sqrt{3}$.

✓ **CHECK** Because the solutions are imaginary, you cannot check them graphically. However, you can check the solutions algebraically by substituting them back into the original equation.

5.5 *Completing the Square* **283**

STUDENT HELP

↳ **Study Tip**
In Example 2 note that you must add 25 to *both sides* of the equation $x^2 + 10x = 3$ when completing the square.

---

2 TEACH

MOTIVATING THE LESSON
When reconstructing an accident, investigators can use the length of skid marks to determine the speed of the car. The length of the marks is the distance it took for the car to stop. In today's lesson, you will learn a method for solving quadratic equations called completing the square. This method will allow you to solve problems such as finding the speed of a car when you know its stopping distance.

EXTRA EXAMPLE 1
Find the value of c that makes $x^2 - 3x + c$ a perfect square trinomial. Then write the expression as the square of a binomial. $\frac{9}{4}; \left(x - \frac{3}{2}\right)^2$

EXTRA EXAMPLE 2
Solve $x^2 + 6x - 8 = 0$ by completing the square. $-3 \pm \sqrt{17}$

EXTRA EXAMPLE 3
Solve $5x^2 - 10x + 30 = 0$ by completing the square. $1 \pm i\sqrt{5}$

✓ **CHECKPOINT EXERCISES**

For use after Example 1:
1. Find the value of c that makes $x^2 - 11x + c$ a perfect square trinomial. Then write the expression as the square of a binomial. $\frac{121}{4}; \left(x - \frac{11}{2}\right)^2$

For use after Example 2:
2. Solve $x^2 + 4x - 1 = 0$ by completing the square. $-2 \pm \sqrt{5}$

For use after Example 3:
3. Solve $3x^2 - 12x + 16 = 0$ by completing the square. $2 \pm \frac{2i\sqrt{3}}{3}$

EXTRA EXAMPLE 4
Under certain road conditions, the formula for a car's stopping distance is given by $d = 0.1s^2 + 1.1s$. If a driver leaves 5 car lengths, approximately 75 ft, between him and the driver in front of him, what is the maximum speed he can drive and still stop safely? **about 22 mi/h**

EXTRA EXAMPLE 5
You have 30 ft of chain link fence to make a rectangular enclosure for your dog. A pet store owner recommended that an enclosure for one dog be at least 48 ft^2 in area. What should the dimensions of the enclosure be to make the area 48 ft^2? **about 10.4 ft by 4.6 ft**

  CHECKPOINT EXERCISES
For use after Example 4:

1. Under certain road conditions, the formula for a car's stopping distance is $d = 0.12s^2 + 1.1s$. What speed limit should be posted at a toll both where drivers have 100 ft to come to a complete stop? **25 mi/h**

For use after Example 5:

2. You are making a fence to enclose a rose garden on the side of your garden shed. You have 10 ft of fence and want to make the garden have an area of 9 ft^2. What are the possible dimensions of the garden?

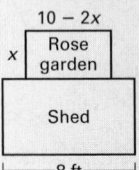

about 3.8 ft by 2.4 ft or 1.2 ft by 7.6 ft

Traffic Engineering

EXAMPLE 4   *Using a Quadratic Equation to Model Distance*

On dry asphalt the distance d (in feet) needed for a car to stop is given by

$$d = 0.05s^2 + 1.1s$$

where s is the car's speed (in miles per hour). What speed limit should be posted on a road where drivers round a corner and have 80 feet to come to a stop?

SOLUTION

$d = 0.05s^2 + 1.1s$	Write original equation.
$80 = 0.05s^2 + 1.1s$	Substitute 80 for d.
$1600 = s^2 + 22s$	Divide each side by the coefficient of s^2.
$1600 + 121 = s^2 + 22s + 11^2$	Add $\left(\frac{22}{2}\right)^2 = 11^2 = 121$ to each side.
$1721 = (s + 11)^2$	Write the right side as a binomial squared.
$\pm\sqrt{1721} = s + 11$	Take square roots of each side.
$-11 \pm \sqrt{1721} = s$	Solve for s.
$s \approx 30$ or $s \approx -52$	Use a calculator.

▶ Reject the solution -52 because a car's speed cannot be negative. The posted speed limit should be at most 30 miles per hour.

Landscape Design

EXAMPLE 5   *Using a Quadratic Equation to Model Area*

You want to plant a rectangular garden along part of a 40 foot side of your house. To keep out animals, you will enclose the garden with wire mesh along its three open sides. You will also cover the garden with mulch. If you have 50 feet of mesh and enough mulch to cover 100 square feet, what should the garden's dimensions be?

SOLUTION

Draw a diagram. Let x be the length of the sides of the garden perpendicular to the house. Then $50 - 2x$ is the length of the third fenced side of the garden.

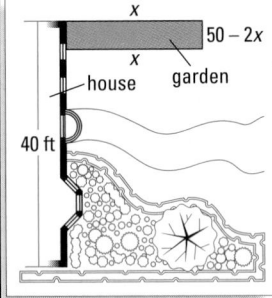

$x(50 - 2x) = 100$	Length $\times$ Width = Area
$50x - 2x^2 = 100$	Distributive property
$-2x^2 + 50x = 100$	Write the x^2-term first.
$x^2 - 25x = -50$	Divide each side by -2.
$x^2 - 25x + (-12.5)^2 = -50 + 156.25$	Complete the square.
$(x - 12.5)^2 = 106.25$	Write as a binomial squared.
$x - 12.5 = \pm\sqrt{106.25}$	Take square roots of each side.
$x = 12.5 \pm \sqrt{106.25}$	Solve for x.
$x \approx 22.8$ or $x \approx 2.2$	Use a calculator.

▶ Reject $x = 2.2$ since $50 - 2x = 45.6$ is greater than the house's length. If $x = 22.8$, then $50 - 2x = 4.4$. The garden should be about 22.8 feet by 4.4 feet.

GOAL 2 WRITING QUADRATIC FUNCTIONS IN VERTEX FORM

Given a quadratic function in standard form, $y = ax^2 + bx + c$, you can use completing the square to write the function in vertex form, $y = a(x - h)^2 + k$.

STUDENT HELP

HOMEWORK HELP
Visit our Web site
www.mcdougallittell.com
for extra examples.

EXAMPLE 6 *Writing a Quadratic Function in Vertex Form*

Write the quadratic function $y = x^2 - 8x + 11$ in vertex form. What is the vertex of the function's graph?

SOLUTION

$$y = x^2 - 8x + 11 \qquad \text{Write original function.}$$

$$y + \underline{?} = (x^2 - 8x + \underline{?}) + 11 \qquad \text{Prepare to complete the square for } x^2 - 8x.$$

$$y + 16 = (x^2 - 8x + 16) + 11 \qquad \text{Add } \left(\frac{-8}{2}\right)^2 = (-4)^2 = 16 \text{ to each side.}$$

$$y + 16 = (x - 4)^2 + 11 \qquad \text{Write } x^2 - 8x + 16 \text{ as a binomial squared.}$$

$$y = (x - 4)^2 - 5 \qquad \text{Solve for } y.$$

▶ The vertex form of the function is $y = (x - 4)^2 - 5$. The vertex is $(4, -5)$.

Agriculture

EXAMPLE 7 *Finding the Maximum Value of a Quadratic Function*

The amount s (in pounds per acre) of sugar produced from sugarbeets can be modeled by the function

$$s = -0.0655n^2 + 7.855n + 5562$$

where n is the amount (in pounds per acre) of nitrogen fertilizer used. How much fertilizer should you use to maximize sugar production? What is the maximum amount of sugar you can produce?

▶ Source: Sugarbeet Research and Education Board of Minnesota and North Dakota

SOLUTION

The optimal amount of fertilizer and the maximum amount of sugar are the coordinates of the vertex of the function's graph. One way to find the vertex is to write the function in vertex form.

$$s = -0.0655n^2 + 7.855n + 5562$$

$$s = -0.0655(n^2 - 120n) + 5562$$

$$s - 0.0655(\underline{?}) = -0.0655(n^2 - 120n + \underline{?}) + 5562$$

$$s - 0.0655(3600) = -0.0655(n^2 - 120n + 3600) + 5562$$

$$s - 236 = -0.0655(n - 60)^2 + 5562$$

$$s = -0.0655(n - 60)^2 + 5798$$

▶ The vertex is approximately $(60, 5798)$. To maximize sugar production, you should use about 60 pounds per acre of nitrogen fertilizer. The maximum amount of sugar you can produce is about 5800 pounds per acre.

EXTRA EXAMPLE 6
Write the quadratic function $y = x^2 + 6x + 16$ in vertex form. What is the vertex of the function's graph? $y = (x + 3)^2 + 7$; $(-3, 7)$

EXTRA EXAMPLE 7
An agricultural researcher finds that the height h (in inches) of one type of pepper plant can be modeled by the function $h = -0.88r^2 + 8.8r + 20$ where r is the amount of rainfall (in inches) that fell during the growing season. How much rain would maximize the height of the pepper plants? What is the maximum height? 5 in.; 42 in.

✓ CHECKPOINT EXERCISES

For use after Example 6:

1. Write the quadratic function $y = x^2 + 3x + 3$ in vertex form. What is the vertex of the function's graph?

$$y = \left(x + \frac{3}{2}\right)^2 + \frac{3}{4}; \left(-\frac{3}{2}, \frac{3}{4}\right)$$

For use after Example 7:

2. The number of people n who attend a traveling circus can be modeled by the function $n = -1.540x^2 + 249.5x + 2371$, where x is the number of times the circus was advertised on the local radio station during the month prior to the show. How many times should the circus advertise to maximize the number of people attending the show? What is the maximum number of people? 81 times; 12,477 people

CLOSURE QUESTION
Why was completing the square used to find the maximum value of a function? See margin.

DAILY PUZZLER
The length of a rectangle is one more than three times its width. What are the dimensions of the rectangle if its area is 30? 3 ft by 10 ft

285

Closure Question *Sample answer:*
Completing the square was used to put the function into vertex form so that the vertex could be found. This was necessary because the maximum or minimum value of a quadratic function always occurs at the vertex.

ASSIGNMENT GUIDE

BASIC
Day 1: pp. 286–288 Exs. 24–28 even, 32–40 even, 48–52 even, 56–60 even, 64–74 even
Day 2: pp. 286–289 Exs. 29, 31, 43, 45, 53, 61, 82–92 even, 96–98, 101–117 odd

AVERAGE
Day 1: pp. 286–288 Exs. 24–28 even, 32–40 even, 48–52 even, 56–60 even, 64–74 even
Day 2: pp. 286–289 Exs. 29, 31, 43, 45, 53, 61, 82–96 even, 101–117 odd

ADVANCED
Day 1: pp. 286–288 Exs. 24–72
Day 2: pp. 286–289 Exs. 74–84 even, 85–88, 89–95 odd, 96–100, 101–117 odd

BLOCK SCHEDULE
pp. 286–288 Exs. 24–28 even, 32–40 even, 48–52 even, 56–60 even, 64–74 even (with 5.4)
pp. 286–289 Exs. 29, 31, 43, 45, 53, 61, 82–96 even, 101–117 odd (with 5.6)

EXERCISE LEVELS
Level A: *Easier*
23–28, 32–43, 47–50, 63–68, 73–80
Level B: *More Difficult*
29–31, 44–46, 51–62, 69–72, 81–98
Level C: *Most Difficult*
99–100

✓ HOMEWORK CHECK
To quickly check student understanding of key concepts, go over the following exercises:
Exs. 26, 38, 48, 56, 64, 74. See also the Daily Homework Quiz:
- Blackline Master (*Chapter 5 Resource Book,* p. 79)
- Transparency (p. 37)

GUIDED PRACTICE

Vocabulary Check ✓

1. Describe what it means to "complete the square" for an expression of the form $x^2 + bx$. *Sample answer: writing the expression as the square of a binomial*

Concept Check ✓

2. Which method for solving quadratic equations—factoring or completing the square—is more general? Explain. **2, 3. See margin.**

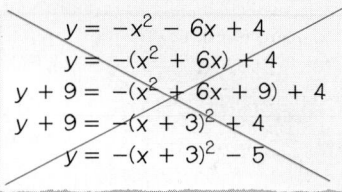

$y = -x^2 - 6x + 4$
$y = -(x^2 + 6x) + 4$
$y + 9 = -(x^2 + 6x + 9) + 4$
$y + 9 = -(x + 3)^2 + 4$
$y = -(x + 3)^2 - 5$

3. **ERROR ANALYSIS** A student tried to write $y = -x^2 - 6x + 4$ in vertex form as shown. Explain the student's mistake. Then write the correct vertex form of the function.

Skill Check ✓

2. *Sample answer:* completing the square since not every quadratic equation can be solved by factoring

3. *Sample answer:* The number −9 should have been added to the left side since −1(9) = −9; $y = -(x + 3)^2 + 13$

Find the value of *c* that makes the expression a perfect square trinomial. Then write the expression as the square of a binomial.

4. $x^2 + 2x + c$ 1; $(x + 1)^2$
5. $x^2 + 14x + c$ 49; $(x + 7)^2$
6. $x^2 - 6x + c$ 9; $(x - 3)^2$
7. $x^2 - 10x + c$ 25; $(x - 5)^2$
8. $x^2 + 5x + c$ $\frac{25}{4}$; $\left(x + \frac{5}{2}\right)^2$
9. $x^2 - 13x + c$ $\frac{169}{4}$; $\left(x - \frac{13}{2}\right)^2$

Solve the equation by completing the square.

10. $x^2 + 4x = -1$ $-2 - \sqrt{3}, -2 + \sqrt{3}$
11. $x^2 - 2x = 4$ $1 - \sqrt{5}, 1 + \sqrt{5}$
12. $x^2 - 16x + 76 = 0$ $8 - 2i\sqrt{3}, 8 + 2i\sqrt{3}$
13. $x^2 + 8x + 9 = 0$ $-4 - \sqrt{7}, -4 + \sqrt{7}$
14. $2x^2 + 12x = 4$ $-3 - \sqrt{11}, -3 + \sqrt{11}$
15. $3x^2 - 12x + 93 = 0$ $2 - 3i\sqrt{3}, 2 + 3i\sqrt{3}$

Write the quadratic function in vertex form and identify the vertex.
16, 19–20. See margin.

16. $y = x^2 + 12x$
17. $y = x^2 - 4x + 7$ $y = (x - 2)^2 + 3$; (2, 3)
18. $y = x^2 - 8x + 31$ $y = (x - 4)^2 + 15$; (4, 15)
19. $y = x^2 + 10x + 17$
20. $y = -x^2 + 14x - 45$
21. $y = 2x^2 + 4x - 4$ $y = 2(x + 1)^2 - 6$; (−1, −6)

16. $y = (x + 6)^2 - 36$; (−6, −36)
19. $y = (x + 5)^2 - 8$; (−5, −8)
20. $y = -(x - 7)^2 + 4$; (7, 4)

22. **LANDSCAPE DESIGN** Suppose the homeowner in Example 5 has 60 feet of wire mesh to put around the garden and enough mulch to cover an area of 140 square feet. What should the dimensions of the garden be? **about 27.4 ft by 5.1 ft**

PRACTICE AND APPLICATIONS

STUDENT HELP

→ **Extra Practice** to help you master skills is on p. 946.

REWRITING EXPRESSIONS Write the expression as the square of a binomial.

23. $x^2 + 16x + 64$ $(x + 8)^2$
24. $x^2 + 20x + 100$ $(x + 10)^2$
25. $x^2 - 24x + 144$ $(x - 12)^2$
26. $x^2 - 38x + 361$ $(x - 19)^2$
27. $x^2 + x + 0.25$ $(x + 0.5)^2$
28. $x^2 - 1.4x + 0.49$ $(x - 0.7)^2$
29. $x^2 - 3x + \frac{9}{4}$ $\left(x - \frac{3}{2}\right)^2$
30. $x^2 + \frac{1}{6}x + \frac{1}{144}$ $\left(x + \frac{1}{12}\right)^2$
31. $x^2 - \frac{4}{9}x + \frac{4}{81}$ $\left(x - \frac{2}{9}\right)^2$

37. $\frac{121}{4}$; $\left(x - \frac{11}{2}\right)^2$
38. $\frac{529}{4}$; $\left(x - \frac{23}{2}\right)^2$
39. $\frac{225}{4}$; $\left(x + \frac{15}{2}\right)^2$
44. $\frac{1}{49}$; $\left(x - \frac{1}{7}\right)^2$
45. $\frac{25}{9}$; $\left(x + \frac{5}{3}\right)^2$
46. $\frac{289}{256}$; $\left(x + \frac{17}{16}\right)^2$

COMPLETING THE SQUARE Find the value of *c* that makes the expression a perfect square trinomial. Then write the expression as the square of a binomial.

32. $x^2 - 12x + c$ 36; $(x - 6)^2$
33. $x^2 + 18x + c$ 81; $(x + 9)^2$
34. $x^2 + 26x + c$ 169; $(x + 13)^2$
35. $x^2 - 44x + c$ 484; $(x - 22)^2$
36. $x^2 + 9x + c$ $\frac{81}{4}$; $\left(x + \frac{9}{2}\right)^2$
37. $x^2 - 11x + c$
38. $x^2 - 23x + c$
39. $x^2 + 15x + c$
40. $x^2 - 0.2x + c$ 0.01; $(x - 0.1)^2$
41. $x^2 - 5.8x + c$ 8.41; $(x - 2.9)^2$
42. $x^2 + 1.6x + c$ 0.64; $(x + 0.8)^2$
43. $x^2 + 9.4x + c$ 22.09; $(x + 4.7)^2$
44. $x^2 - \frac{2}{7}x + c$
45. $x^2 + \frac{10}{3}x + c$
46. $x^2 + \frac{17}{8}x + c$

STUDENT HELP

HOMEWORK HELP
Example 1: Exs. 23–46
Example 2: Exs. 47–54, 63–64
Example 3: Exs. 55–72
Example 4: Exs. 89–91
Example 5: Exs. 92, 93
Example 6: Exs. 73–84
Example 7: Exs. 94, 95

COEFFICIENT OF x^2 IS 1 **Solve the equation by completing the square.**

47. $x^2 + 2x = 9$ $-1 + \sqrt{10}, -1 - \sqrt{10}$

48. $x^2 - 12x = -28$ $6 + 2\sqrt{2}, 6 - 2\sqrt{2}$

49. $x^2 + 20x + 104 = 0$ $-10 + 2i, -10 - 2i$

50. $x^2 + 3x - 1 = 0$ $\frac{-3 + \sqrt{13}}{2}, \frac{-3 - \sqrt{13}}{2}$

51. $u^2 - 4u = 2u + 35$ $3 - 2\sqrt{11}, 3 + 2\sqrt{11}$

52. $v^2 - 17v + 200 = 13v - 43$ $15 - 3i\sqrt{2}, 15 + 3i\sqrt{2}$

53. $m^2 + 1.8m - 1.5 = 0$ $-0.9 - \sqrt{2.31}, -0.9 + \sqrt{2.31}$

54. $n^2 - \frac{4}{3}n - \frac{14}{9} = 0$ $\frac{2}{3} - \sqrt{2}, \frac{2}{3} + \sqrt{2}$

COEFFICIENT OF x^2 IS NOT 1 **Solve the equation by completing the square.**

55. $2x^2 - 12x = -14$ $3 + \sqrt{2}, 3 - \sqrt{2}$

56. $-3x^2 + 24x = 27$ $4 - \sqrt{7}, 4 + \sqrt{7}$

57. $6x^2 + 84x + 300 = 0$ $-7 - i, -7 + i$

58. $4x^2 + 40x + 280 = 0$ $-5 - 3i\sqrt{5}, -5 + 3i\sqrt{5}$

59. $-4r^2 + 21r = r + 13$

60. $3s^2 - 26s + 2 = 5s^2 + 1$

61. $0.4t^2 + 0.7t = 0.3t - 0.2$

62. $\frac{w^2}{24} - \frac{w}{2} + \frac{13}{6} = 0$ $6 - 4i, 6 + 4i$

SOLVING BY ANY METHOD **Solve the equation by factoring, by finding square roots, or by completing the square.**

63. $x^2 + 4x - 12 = 0$ $-6, 2$

64. $x^2 - 6x - 15 = 0$ $3 - 2\sqrt{6}, 3 + 2\sqrt{6}$

65. $9x^2 - 23 = 0$ $-\frac{\sqrt{23}}{3}, \frac{\sqrt{23}}{3}$

66. $2x^2 + 9x + 7 = 0$ $-\frac{7}{2}, -1$

67. $3x^2 + x = 2x - 6$ $\frac{1 - i\sqrt{71}}{6}, \frac{1 + i\sqrt{71}}{6}$

68. $4(x + 8)^2 = 144$ $-14, -2$

69. $7k^2 + 10k - 100 = 2k^2 + 55$ $-1 - 4\sqrt{2}, -1 + 4\sqrt{2}$

70. $14b^2 - 19b + 4 = -11b^2 + 11b - 5$ $\frac{3}{5}$

71. $0.01p^2 - 0.22p + 2.9 = 0$ $11 - 13i, 11 + 13i$

72. $\frac{q^2}{4} - \frac{9q^2}{20} = 18$ $-3i\sqrt{10}, 3i\sqrt{10}$

WRITING IN VERTEX FORM **Write the quadratic function in vertex form and identify the vertex.** 73–84. See margin.

73. $y = x^2 - 6x + 11$

74. $y = x^2 - 2x - 9$

75. $y = x^2 + 16x + 14$

76. $y = x^2 + 26x + 68$

77. $y = x^2 - 3x - 2$

78. $y = x^2 + 7x - 1$

79. $y = -x^2 + 20x - 80$

80. $y = -x^2 - 14x - 47$

81. $y = 3x^2 - 12x + 1$

82. $y = -2x^2 - 2x - 7$

83. $y = 1.4x^2 + 5.6x + 3$

84. $y = \frac{2}{3}x^2 - \frac{4}{5}x$

GEOMETRY CONNECTION **Find the value of x.**

85. Area of rectangle = 100

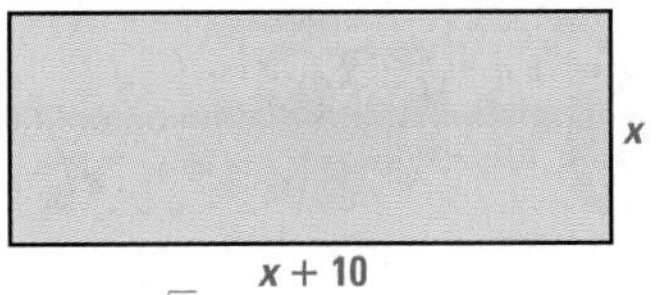

$-5 + 5\sqrt{5}$, or ≈ 6.18

86. Area of triangle = 40

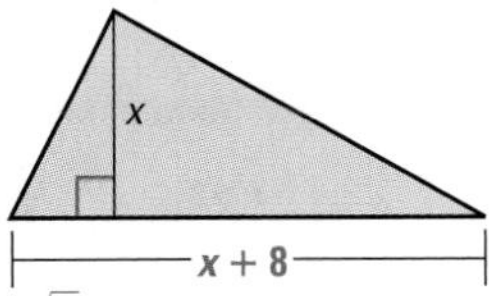

$4\sqrt{6} - 4$, or ≈ 5.80

87. Area of trapezoid = 70

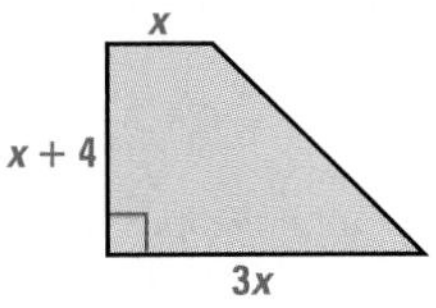

$\sqrt{39} - 2$, or ≈ 4.24

88. Area of parallelogram = 54

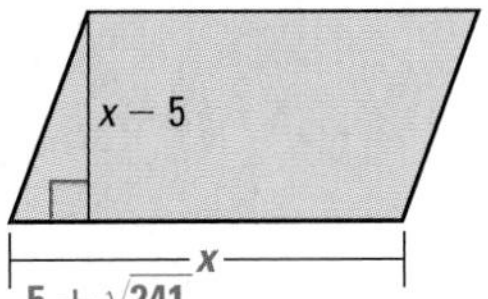

$\frac{5 + \sqrt{241}}{2}$, or ≈ 10.26

59. $\frac{5 - 2\sqrt{3}}{2}, \frac{5 + 2\sqrt{3}}{2}$

60. $\frac{-13 - 3\sqrt{19}}{2}, \frac{-13 + 3\sqrt{19}}{2}$

61. $\frac{-1 - i}{2}, \frac{-1 + i}{2}$

73. $y = (x - 3)^2 + 2$; $(3, 2)$

74. $y = (x - 1)^2 - 10$; $(1, -10)$

75. $y = (x + 8)^2 - 50$; $(-8, -50)$

76. $y = (x + 13)^2 - 101$; $(-13, -101)$

77. $y = \left(x - \frac{3}{2}\right)^2 - \frac{17}{4}$; $\left(\frac{3}{2}, -\frac{17}{4}\right)$

78. $y = \left(x + \frac{7}{2}\right)^2 - \frac{53}{4}$; $\left(-\frac{7}{2}, -\frac{53}{4}\right)$

79. $y = -(x - 10)^2 + 20$; $(10, 20)$

80. $y = -(x + 7)^2 + 2$; $(-7, 2)$

STUDENT HELP

Skills Review
For help with areas of geometric figures, see p. 914.

81. $y = 3(x - 2)^2 - 11$; $(2, -11)$

82. $y = -2\left(x + \frac{1}{2}\right)^2 - \frac{13}{2}$; $\left(-\frac{1}{2}, -\frac{13}{2}\right)$

83. $y = 1.4(x + 2)^2 - 2.6$; $(-2, -2.6)$

84. $y = \frac{2}{3}\left(x - \frac{3}{5}\right)^2 - \frac{6}{25}$; $\left(\frac{3}{5}, -\frac{6}{25}\right)$

! COMMON ERROR

EXERCISES 79–84 Students frequently ignore the coefficient of the x^2 term. Remind students that if they added a number inside parentheses when completing the square, they must multiply that number by the number outside of the parentheses before adding it to the other side of the equation.

STUDENT HELP NOTES

Skills Review As students review the areas of triangles, trapezoids, and parallelograms on page 914, remind them that the heights of these figures must be perpendicular to the base.

99a.

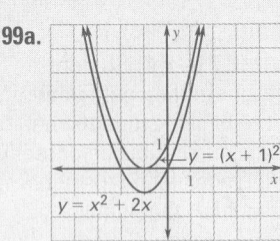

$y = (x + 1)^2$
$y = x^2 + 2x$

99b.

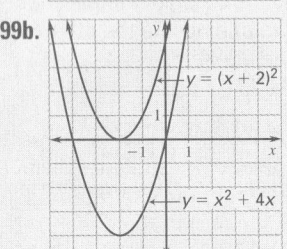

$y = (x + 2)^2$
$y = x^2 + 4x$

99c.

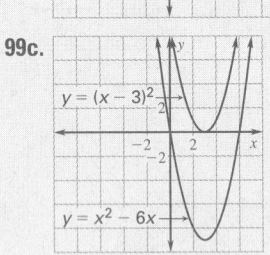

$y = (x - 3)^2$
$y = x^2 - 6x$

113.

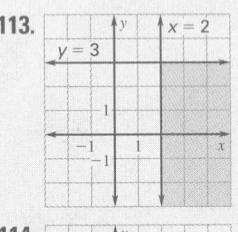

$y = 3$ $x = 2$

114.

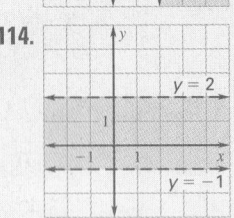

$y = 2$
$y = -1$

115–118. See Additional Answers beginning on page AA1.

ADDITIONAL PRACTICE AND RETEACHING

For Lesson 5.5:
- Practice Levels A, B, and C (*Chapter 5 Resource Book,* p. 69)
- Reteaching with Practice (*Chapter 5 Resource Book,* p. 72)
- ⊞ See Lesson 5.5 of the *Personal Student Tutor*

For more Mixed Review:
- ⊞ Search the *Test and Practice Generator* for key words or specific lessons.

288

FOCUS ON PEOPLE

JACKIE JOYNER-KERSEE became one of the greatest female athletes in history despite having severe asthma as a child and as an adult. She has won six Olympic medals: three gold, one silver, and two bronze.

89. 🖟 **TRAFFIC ENGINEERING** For a road covered with dry, packed snow, the formula for a car's stopping distance given in Example 4 becomes:

$$d = 0.08s^2 + 1.1s$$

Show that, in snowy conditions, a driver cannot safely round the corner in Example 4 when traveling at the calculated speed limit of 30 miles per hour. What is a safe speed limit if the road is covered with snow?
$d = 0.08(30)^2 + 1.1(30) = 105$ ft; about 25.5 mi/h

90. 🖟 **SPORTS** Jackie Joyner-Kersee won the women's heptathlon during the 1992 Olympics in Barcelona, Spain. Her throw in the shot put, one of the seven events in the heptathlon, can be modeled by

$$y = -0.0241x^2 + x + 5.5$$

where x is the shot put's horizontal distance traveled (in feet) and y is its corresponding height (in feet). How long was Joyner-Kersee's throw?   about 46.4 ft

91. 🖟 **FIREFIGHTING** In firefighting, a good water stream can be modeled by

$$y = -0.003x^2 + 0.62x + 3$$

where x is the water's horizontal distance traveled (in feet) and y is its corresponding height (in feet). If a firefighter is aiming a good water stream at a building's window 25 feet above the ground, at what two distances can the firefighter stand from the building?   45.50 ft; 161.16 ft

92. 🖟 **CORRALS** You have 240 feet of wooden fencing to form two adjacent rectangular corrals as shown. You want each corral to have an area of 1000 square feet.

a. Show that $w = 80 - \frac{4}{3}\ell$.   *Not drawn to scale*
$4\ell + 3w = 240$; $3w = 240 - 4\ell$; $w = 80 - \frac{4}{3}\ell$
b. Use your answer from part (a) to find the possible dimensions of each corral.
42.25 ft by 23.67 ft or 17.75 ft by 56.33 ft

93. 🖟 **POTTERY** You are taking a pottery class. As an assignment, you are given a lump of clay whose volume is 200 cubic centimeters and asked to make a cylindrical pencil holder. The pencil holder should be 9 centimeters high and have an inner radius of 3 centimeters. What thickness x should your pencil holder have if you want to use all the clay? (*Hint:* The volume of clay equals the difference of the volumes of two cylinders.)   about 1 cm

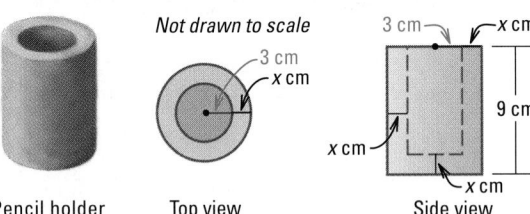

Pencil holder   Top view   Side view

94. **BIOLOGY** **CONNECTION** When a gray kangaroo jumps, its path through the air can be modeled by

$$y = -0.0267x^2 + 0.8x$$

where x is the kangaroo's horizontal distance traveled (in feet) and y is its corresponding height (in feet). How high can a gray kangaroo jump? How far can it jump?   about 6 ft; about 30 ft

288   **Chapter 5** *Quadratic Functions*

95. SCIENCE CONNECTION In a fireplace, the heat loss q (in Btu/ft^3) resulting from hot gases escaping through the chimney can be modeled by

$$q = -0.00002T^2 + 0.0203T - 1.24$$

where T is the temperature (in degrees Fahrenheit) of the gases. (This model assumes an indoor temperature of 65°F.) For what gas temperature is heat loss maximized? What is the maximum heat loss? ▶ Source: *Workshop Math*
507.5°F; 3.91 Btu/ft^3

Test Preparation

96. MULTIPLE CHOICE If $x^2 - 28x + c$ is a perfect square trinomial, what is the value of c? **C**

Ⓐ -14 Ⓑ 28 Ⓒ 196 Ⓓ 784

97. MULTIPLE CHOICE What are the solutions of $x^2 + 12x + 61 = 0$? **B**

Ⓐ $-1, -11$ Ⓑ $-6 \pm 5i$ Ⓒ $-6 \pm \sqrt{97}$ Ⓓ $-6 \pm i\sqrt{61}$

98. MULTIPLE CHOICE What is the vertex form of $y = 2x^2 - 8x + 3$? **A**

Ⓐ $y = 2(x - 2)^2 - 5$ Ⓑ $y = 2(x - 2)^2 + 3$

Ⓒ $y = 2(x - 4)^2 - 29$ Ⓓ $y = 2(x - 4)^2 + 3$

★ **Challenge**

CRITICAL THINKING Exercises 99 and 100 should be done together.

99. Graph the two functions in the same coordinate plane. a–c. See margin.

a. $y = x^2 + 2x$ **b.** $y = x^2 + 4x$ **c.** $y = x^2 - 6x$
    $y = (x + 1)^2$ $y = (x + 2)^2$ $y = (x - 3)^2$

EXTRA CHALLENGE
www.mcdougallittell.com

100. Compare the graphs of $y = x^2 + bx$ and $y = \left(x + \dfrac{b}{2}\right)^2$. What happens to the graph of $y = x^2 + bx$ when you complete the square for $x^2 + bx$?
Sample answer: The vertex moves up from the position of the other vertex so that the new vertex lies on the *x*-axis.

MIXED REVIEW

EVALUATING EXPRESSIONS Evaluate $b^2 - 4ac$ for the given values of a, b, and c. (Review 1.2 for 5.6)

101. $a = 1, b = 5, c = 2$ 17   **102.** $a = 3, b = -8, c = 7$ -20

103. $a = -5, b = 0, c = 2.6$ 52   **104.** $a = 11, b = 4, c = -1$ 60

105. $a = 16, b = -24, c = 9$ 0   **106.** $a = -1.4, b = 2, c = -0.5$ 1.2

EQUATIONS OF LINES Write an equation in slope-intercept form of the line through the given point and having the given slope. (Review 2.4)

107. $(3, 1), m = 2$ **108.** $(2, -4), m = 1$ **109.** $(-7, 10), m = -5$
    $y = 2x - 5$ $y = x - 6$ $y = -5x - 25$

110. $(-8, -8), m = -3$ **111.** $(6, 9), m = \dfrac{1}{3}$ **112.** $(11, -2), m = -\dfrac{5}{4}$
    $y = -3x - 32$ $y = \dfrac{1}{3}x + 7$ $y = -\dfrac{5}{4}x + \dfrac{47}{4}$

SYSTEMS OF LINEAR INEQUALITIES Graph the system of inequalities. (Review 3.3)   113–118. See margin.

113. $x \geq 2$ **114.** $y > -1$ **115.** $x \geq 0$
    $y \leq 3$ $y < 2$ $x + y < 4$

116. $y < x - 2$ **117.** $3x - 2y < 8$ **118.** $y \leq 2x + 3$
    $x - 3y \leq 6$ $2x + y > 0$ $y \geq 2x - 3$

5.5 *Completing the Square*   **289**

DAILY HOMEWORK QUIZ

📠 **Transparency Available**

1. Find the value of c that makes $x^2 - 8x + c$ a perfect square trinomial. Then write the expression as the square of a binomial.   16; $(x - 4)^2$

2. Solve $x^2 - 12x + 4 = 0$ by completing the square. $6 \pm 4\sqrt{2}$

3. Solve $2x^2 + 8x - 1 = 0$.
$-2 \pm \dfrac{3\sqrt{2}}{2}$

4. Write $y = x^2 - 4x + 3$ in vertex form. What is the vertex of the function's graph?
$y = (x - 2)^2 - 1$; $(2, -1)$

EXTRA CHALLENGE NOTE

→ Challenge problems for Lesson 5.5 are available in **blackline** format in the *Chapter 5 Resource Book,* p. 76 and at **www.mcdougallittell.com.**

ADDITIONAL TEST PREPARATION

1. OPEN ENDED Give an example of a quadratic equation that can be solved using completing the square with the coefficient of x^2 being 1 and an example whose coefficient of x^2 is not 1. Show how the method is used in each problem. *Sample answers:*
$x^2 + 8x - 9 = 0$, $x^2 + 8x = 9$,
$x^2 + 8x + 4^2 = 9 + 16$,
$(x + 4)^2 = 25$, $x + 4 = \pm 5$,
$x = 1$ or $x = -9$;
$2x^2 - 8x + 6 = 0$, $x^2 - 4x + 3 = 0$,
$x^2 - 4x = -3$,
$x^2 - 4x + (-2)^2 = -3 + 4$
$(x - 2)^2 = 1$, $x - 2 = \pm 1$, $x = 3$, or $x = 1$

1 Planning the Activity

PURPOSE
To use a graphing calculator to find the maximum or minimum of a quadratic function.

MATERIALS
- graphing calculator
- Keystroke blackline (*Chapter 5 Resource Book,* p. 68)

PACING
- Activity — 20 min

▶ LINK TO LESSON
Have students use a graphing calculator to check the result of Example 7 on page 285.

2 Managing the Activity

CLASSROOM MANAGEMENT
Students will need to use some trial and error to find an appropriate viewing window for Ex. 10. Have them make a guess for a good upper bound for *C* and *F* based on the situation. A good first guess might be 500 cars for each value. They will graph this and see that some adjustment needs to be made. A good viewing window is $0 \le x \le 500$, $0 \le y \le 3000$.

3 Closing the Activity

★ KEY DISCOVERY
Graphing calculators can be used to find the maximum or minimum values of quadratic functions.

ACTIVITY ASSESSMENT
Write two quadratic functions, one with a maximum value and one with a minimum value. Find the maximum and minimum values.
Check students' work.

● ACTIVITY 5.5

Using Technology

STUDENT HELP

INTERNET **KEYSTROKE HELP**
See keystrokes for several models of calculators at www.mcdougallittell.com

1. minimum; −4.25; 2.5
2. maximum; 5; 4
3. minimum; 4; −3
4. minimum; −5; −4
5. maximum; 8.125; −0.75
6. maximum; −2.125; −3.75
7. minimum; 2.375; 3.7
8. minimum; −4; −1
9. maximum; 8.65; 2.3
10. 80 cars/mi; 1997 cars/h

STUDENT HELP

↳ **Study Tip**
To find the minimum value of a function, select the *Minimum* feature instead of the *Maximum* feature from the menu in **Step 1**.

Finding Maximums and Minimums

You can use a graphing calculator to find maximum or minimum values of quadratic functions.

▶ EXAMPLE
Find the maximum value of $y = -x^2 - 7x - 6$ and the value of x where it occurs.

▶ SOLUTION

❶ Graph the given function and select the *Maximum* feature.

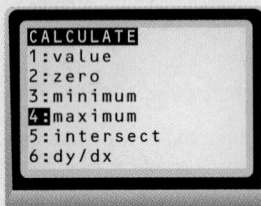

❷ Move the cursor to the left of the maximum point. Press ENTER.

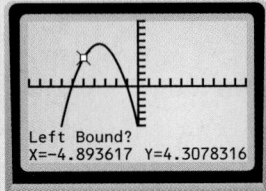

❸ Move the cursor to the right of the maximum point. Press ENTER.

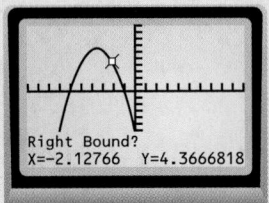

❹ Put the cursor approximately on the maximum point. Press ENTER.

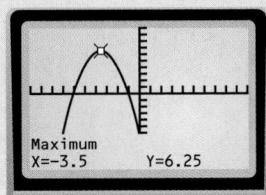

▶ The maximum value of the function is $y = 6.25$ and occurs at $x = -3.5$.

▶ EXERCISES

Tell whether the function has a maximum value or a minimum value. Then find the maximum or minimum value and the value of x where it occurs.

1–10. Estimates may vary. See margin.

1. $y = x^2 - 5x + 2$
2. $y = -x^2 + 8x - 11$
3. $y = x^2 + 6x + 13$

4. $y = 3x^2 + 24x + 43$
5. $y = -2x^2 - 3x + 7$
6. $y = -1.2x^2 - 9x - 19$

7. $y = 0.4x^2 - 3x + 8$
8. $y = \frac{1}{2}x^2 + x - \frac{7}{2}$
9. $y = -\frac{8}{5}x^2 + \frac{22}{3}x + \frac{1}{4}$

10. 🚗 **TRAFFIC FLOW** On a typical single-lane highway, the traffic flow F (in cars per hour) can be modeled by $F = -0.313C^2 + 50C$ where C is the traffic concentration (in cars per mile). For what traffic concentration is traffic flow maximized? What is the maximum traffic flow?

▶ Source: *Towing Icebergs, Falling Dominoes, and Other Adventures in Applied Mathematics*

290　**Chapter 5**　*Quadratic Functions*

5.6

The Quadratic Formula and the Discriminant

GOAL 1 **SOLVING EQUATIONS WITH THE QUADRATIC FORMULA**

What you should learn

GOAL 1 Solve quadratic equations using the quadratic formula.

GOAL 2 Use the quadratic formula in **real-life** situations, such as baton twirling in **Example 5**.

Why you should learn it

▼ To solve **real-life** problems, such as finding the speed and duration of a thrill ride in **Ex. 84**.

In Lesson 5.5 you solved quadratic equations by completing the square for *each equation separately*. By completing the square *once* for the general equation $ax^2 + bx + c = 0$, you can develop a formula that gives the solutions of *any* quadratic equation. The formula for the solutions is called the **quadratic formula**. A derivation of the quadratic formula appears on page 895.

THE QUADRATIC FORMULA

Let a, b, and c be real numbers such that $a \neq 0$. The solutions of the quadratic equation $ax^2 + bx + c = 0$ are:

$$x = \frac{-b \pm \sqrt{b^2 - 4ac}}{2a}$$

Remember that *before* you apply the quadratic formula to a quadratic equation, you must write the equation in standard form, $ax^2 + bx + c = 0$.

EXAMPLE 1 *Solving a Quadratic Equation with Two Real Solutions*

Solve $2x^2 + x = 5$.

SOLUTION

$2x^2 + x = 5$	Write original equation.
$2x^2 + x - 5 = 0$	Write in standard form.
$x = \dfrac{-b \pm \sqrt{b^2 - 4ac}}{2a}$	Quadratic formula
$x = \dfrac{-1 \pm \sqrt{1^2 - 4(2)(-5)}}{2(2)}$	$a = 2, b = 1, c = -5$
$x = \dfrac{-1 \pm \sqrt{41}}{4}$	Simplify.

▶ The solutions are

$$x = \frac{-1 + \sqrt{41}}{4} \approx 1.35$$

and

$$x = \frac{-1 - \sqrt{41}}{4} \approx -1.85.$$

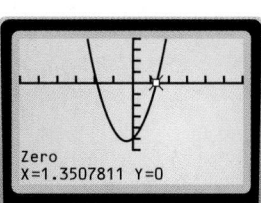

Zero
X=1.3507811 Y=0

✓**CHECK** Graph $y = 2x^2 + x - 5$ and note that the x-intercepts are about 1.35 and about -1.85.

5.6 *The Quadratic Formula and the Discriminant*   291

1 **PLAN**

PACING
Basic: 1 day
Average: 1 day
Advanced: 1 day
Block Schedule: 0.5 block with 5.5

LESSON OPENER
VISUAL APPROACH
An alternative way to approach Lesson 5.6 is to use the Visual Approach Lesson Opener:
• Blackline Master (*Chapter 5 Resource Book*, p. 80)
• Transparency (p. 32)

MEETING INDIVIDUAL NEEDS
• *Chapter 5 Resource Book*
  Prerequisite Skills Review (p. 5)
  Practice Level A (p. 83)
  Practice Level B (p. 84)
  Practice Level C (p. 85)
  Reteaching with Practice (p. 86)
  Absent Student Catch-Up (p. 88)
  Challenge (p. 90)
• *Resources in Spanish*
• *Personal Student Tutor*

NEW-TEACHER SUPPORT
See the Tips for New Teachers on pp. 1–2 of the *Chapter 5 Resource Book* for additional notes about Lesson 5.6.

WARM-UP EXERCISES

Transparency Available

Evaluate the expression $b^2 - 4ac$ for the given values of a, b, and c.
1. $a = 1, b = 3, c = -1$ **13**
2. $a = 2, b = -2, c = 0$ **4**
3. $a = -1, b = 0, c = 5$ **20**
4. $a = -2, b = 2, c = -3$ **−20**

MOTIVATING THE LESSON
Remind students that they can solve quadratic equations using factoring or by completing the square. Factoring can *not* be used for every quadratic equation while completing the square can be a complicated procedure. Today's lesson introduces a method for solving quadratic equations that works for every quadratic equation and is usually less complicated than completing the square.

 EXTRA EXAMPLE 1
Solve $3x^2 + 8x = 35$. $-5, \frac{7}{3}$

EXTRA EXAMPLE 2
Solve $12x - 5 = 2x^2 + 13$. 3

EXTRA EXAMPLE 3
Solve $-2x^2 = -2x + 3$. $\frac{1}{2} \pm \frac{i\sqrt{5}}{2}$

 CHECKPOINT EXERCISES

For use after Example 1:
1. Solve $2x^2 + x = x^2 - 2x + 4$.
   $-4, 1$

For use after Example 2:
2. Solve $x^2 + 64 = 16x$. 8

For use after Example 3:
3. Solve $x^2 = 2x - 5$. $1 \pm 2i$

STUDENT HELP NOTES
Homework Help Students can find extra examples at **www.mcdougallittell.com** that parallel the examples in the student edition.

STUDENT HELP

▶ **HOMEWORK HELP**
Visit our Web site
www.mcdougallittell.com
for extra examples.

EXAMPLE 2   *Solving a Quadratic Equation with One Real Solution*

Solve $x^2 - x = 5x - 9$.

SOLUTION

$$x^2 - x = 5x - 9$$ Write original equation.

$$x^2 - 6x + 9 = 0$$ $a = 1, b = -6, c = 9$

$$x = \frac{6 \pm \sqrt{(-6)^2 - 4(1)(9)}}{2(1)}$$ Quadratic formula

$$x = \frac{6 \pm \sqrt{0}}{2}$$ Simplify.

$$x = 3$$ Simplify.

▶ The solution is 3.

✓ **CHECK** Graph $y = x^2 - 6x + 9$ and note that the only x-intercept is 3. Alternatively, substitute 3 for x in the original equation.

$$3^2 - 3 \stackrel{?}{=} 5(3) - 9$$

$$9 - 3 \stackrel{?}{=} 15 - 9$$

$$6 = 6 ✓$$

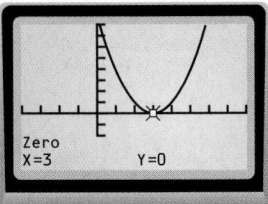

EXAMPLE 3   *Solving a Quadratic Equation with Two Imaginary Solutions*

Solve $-x^2 + 2x = 2$.

SOLUTION

$$-x^2 + 2x = 2$$ Write original equation.

$$-x^2 + 2x - 2 = 0$$ $a = -1, b = 2, c = -2$

$$x = \frac{-2 \pm \sqrt{2^2 - 4(-1)(-2)}}{2(-1)}$$ Quadratic formula

$$x = \frac{-2 \pm \sqrt{-4}}{-2}$$ Simplify.

$$x = \frac{-2 \pm 2i}{-2}$$ Write using the imaginary unit i.

$$x = 1 \pm i$$ Simplify.

▶ The solutions are $1 + i$ and $1 - i$.

✓ **CHECK** Graph $y = -x^2 + 2x - 2$ and note that there are no x-intercepts. So, the original equation has no real solutions. To check the imaginary solutions $1 + i$ and $1 - i$, substitute them into the original equation. The check for $1 + i$ is shown.

$$-(1 + i)^2 + 2(1 + i) \stackrel{?}{=} 2$$

$$-2i + 2 + 2i \stackrel{?}{=} 2$$

$$2 = 2 ✓$$

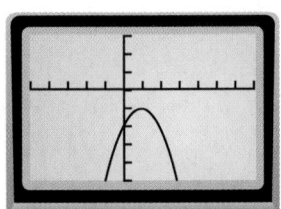

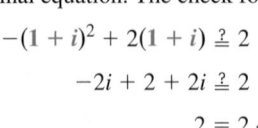

In the quadratic formula, the expression $b^2 - 4ac$ under the radical sign is called the **discriminant** of the associated equation $ax^2 + bx + c = 0$.

$$x = \frac{-b \pm \sqrt{b^2 - 4ac}}{2a} \longleftarrow \text{discriminant}$$

You can use the discriminant of a quadratic equation to determine the equation's number and type of solutions.

> **NUMBER AND TYPE OF SOLUTIONS OF A QUADRATIC EQUATION**
>
> Consider the quadratic equation $ax^2 + bx + c = 0$.
>
> - If $b^2 - 4ac > 0$, then the equation has two real solutions.
> - If $b^2 - 4ac = 0$, then the equation has one real solution.
> - If $b^2 - 4ac < 0$, then the equation has two imaginary solutions.

EXAMPLE 4 *Using the Discriminant*

Find the discriminant of the quadratic equation and give the number and type of solutions of the equation.

a. $x^2 - 6x + 10 = 0$ **b.** $x^2 - 6x + 9 = 0$ **c.** $x^2 - 6x + 8 = 0$

SOLUTION

EQUATION	DISCRIMINANT	SOLUTION(S)
$ax^2 + bx + c = 0$	$b^2 - 4ac$	$x = \dfrac{-b \pm \sqrt{b^2 - 4ac}}{2a}$
a. $x^2 - 6x + 10 = 0$	$(-6)^2 - 4(1)(10) = -4$	Two imaginary: $3 \pm i$
b. $x^2 - 6x + 9 = 0$	$(-6)^2 - 4(1)(9) = 0$	One real: 3
c. $x^2 - 6x + 8 = 0$	$(-6)^2 - 4(1)(8) = 4$	Two real: 2, 4

· · · · · · · · · ·

In Example 4 notice that the number of real solutions of $x^2 - 6x + c = 0$ can be changed just by changing the value of c. A graph can help you see why this occurs. By changing c, you can move the graph of

$$y = x^2 - 6x + c$$

up or down in the coordinate plane. If the graph is moved too high, it won't have an x-intercept and the equation $x^2 - 6x + c = 0$ won't have a real-number solution.

$y = x^2 - 6x + 10$	Graph is above x-axis (no x-intercept).
$y = x^2 - 6x + 9$	Graph touches x-axis (one x-intercept).
$y = x^2 - 6x + 8$	Graph crosses x-axis (two x-intercepts).

EXTRA EXAMPLE 4
Find the discriminant of the quadratic equation and give the number and type of solutions of the equation.
a. $9x^2 + 6x + 1 = 0$
  0; one real solution
b. $9x^2 + 6x - 4 = 0$
  180; two real solutions
c. $9x^2 + 6x + 5 = 0$
  -144; two imaginary solutions

CHECKPOINT EXERCISES
For use after Example 4:
1. Find the discriminant of $5x^2 + 3x + 1 = 0$ and give the number and type of solutions of the equation.
  -11; two imaginary solutions

CONCEPT QUESTION
How are the discriminant and the graph of a quadratic equation related? If the discriminant is negative, the graph does not intersect the x-axis. If the discriminant is 0, the graph intersects the x-axis in one point. If the discriminant is positive, the graph intersects the x-axis in two points.

MATHEMATICAL REASONING
Tell students that the solutions to a quadratic equation can be used to find the quadratic equation. Ask students to determine how the sums and products of the solutions are related to the general quadratic equation.
The solutions of $ax^2 + bx + c = 0$ are $\dfrac{-b \pm \sqrt{b^2 - 4ac}}{2a}$, $a \neq 0$. The sum of the solutions equals $-\dfrac{b}{a}$ and the product of the solutions equals $\dfrac{c}{a}$.

EXTRA EXAMPLE 5

The water in a large fountain leaves the spout with a vertical velocity of 30 ft per second. After going up in the air it lands in a basin 6 ft below the spout. If the spout is 10 ft above the ground, how long does it take a single drop of water to travel from the spout to the basin? Use the model $h = -16t^2 + v_0 t + h_0$.

about 2.1 sec

 CHECKPOINT EXERCISES

For use after Example 5:

1. A man tosses a penny up into the air above a 100-feet deep well with a velocity of 5 ft/sec. The penny leaves the man's hand at a height of 4 ft. How long will it take the penny to reach the bottom of the well? Use the model $h = -16t^2 + v_0 t + h_0$.

about 2.7 sec

FOCUS ON VOCABULARY

What is the *quadratic formula* and what is it used for?

$x = \dfrac{-b \pm \sqrt{b^2 - 4ac}}{2a}$; the quadratic

formula is used to find the solutions to a quadratic equation in the form $ax^2 + bx + c = 0$.

CLOSURE QUESTION

Describe how to use a discriminant to determine the number of solutions of a quadratic equation.

If the discriminant is negative, the equation has two imaginary solutions. If the discriminant is zero there is one real solution. If the discriminant is positive, there are two real solutions.

DAILY PUZZLER

A friend says he wrote a function whose zeros total 6 but whose graph does not intersect the *x*-axis. Can this be true? Explain.

Yes; *sample answer:* the zeros of the function could be two imaginary numbers with opposite imaginary parts such as 3 + *i*, and 3 – *i*.

 **2** **USING THE QUADRATIC FORMULA IN REAL LIFE**

In Lesson 5.3 you studied the model $h = -16t^2 + h_0$ for the height of an object that is *dropped*. For an object that is *launched or thrown*, an extra term $v_0 t$ must be added to the model to account for the object's initial vertical velocity v_0.

Models $h = -16t^2 + h_0$ **Object is dropped.**

                 $h = -16t^2 + v_0 t + h_0$ **Object is launched or thrown.**

Labels h = height   (feet)

                 t = time in motion   (seconds)

                 h_0 = initial height   (feet)

                 v_0 = initial vertical velocity   (feet per second)

The initial vertical velocity of a launched object can be positive, negative, or zero. If the object is launched upward, its initial vertical velocity is positive ($v_0 > 0$). If the object is launched downward, its initial vertical velocity is negative ($v_0 < 0$). If the object is launched parallel to the ground, its initial vertical velocity is zero ($v_0 = 0$).

Entertainment

EXAMPLE 5   *Solving a Vertical Motion Problem*

A baton twirler tosses a baton into the air. The baton leaves the twirler's hand 6 feet above the ground and has an initial vertical velocity of 45 feet per second. The twirler catches the baton when it falls back to a height of 5 feet. For how long is the baton in the air?

SOLUTION

Since the baton is thrown (not dropped), use the model $h = -16t^2 + v_0 t + h_0$ with $v_0 = 45$ and $h_0 = 6$. To determine how long the baton is in the air, find the value of t for which $h = 5$.

$h = -16t^2 + v_0 t + h_0$ **Write height model.**

$5 = -16t^2 + 45t + 6$ **h = 5, v_0 = 45, h_0 = 6**

$0 = -16t^2 + 45t + 1$ **a = −16, b = 45, c = 1**

$t = \dfrac{-45 \pm \sqrt{2089}}{-32}$ **Quadratic formula**

$t \approx -0.022$ or $t \approx 2.8$ **Use a calculator.**

▶ Reject the solution -0.022 since the baton's time in the air cannot be negative. The baton is in the air for about 2.8 seconds.

GUIDED PRACTICE

Vocabulary Check ✓

1. In the quadratic formula, what is the expression $b^2 - 4ac$ called?   **the discriminant**

Concept Check ✓

2. How many solutions does a quadratic equation have if its discriminant is positive? if its discriminant is zero? if its discriminant is negative?
2 real; 1 real; 2 imaginary

3. Describe a real-life situation in which you can use the model $h = -16t^2 + v_0t + h_0$ but not the model $h = -16t^2 + h_0$.
Sample answer: **when an object is thrown upward**

Skill Check ✓

Use the quadratic formula to solve the equation.

18. $\dfrac{-3 + \sqrt{17}}{2}, \dfrac{-3 - \sqrt{17}}{2}$

4. $x^2 - 4x + 3 = 0$ **3, 1**

5. $x^2 + x - 1 = 0$ $\dfrac{-1 + \sqrt{5}}{2}, \dfrac{-1 - \sqrt{5}}{2}$

19. $1 + \sqrt{5}, 1 - \sqrt{5}$

6. $2x^2 + 3x + 5 = 0$ $\dfrac{-3 + i\sqrt{31}}{4}, \dfrac{-3 - i\sqrt{31}}{4}$ 7. $9x^2 + 6x - 1 = 0$ $\dfrac{-1 + \sqrt{2}}{3}, \dfrac{-1 - \sqrt{2}}{3}$

20. $-5 + \sqrt{3}, -5 - \sqrt{3}$

8. $-x^2 + 8x = 1$ $4 + \sqrt{15}, 4 - \sqrt{15}$

9. $5x^2 - 2x + 37 = x^2 + 2x$ $\dfrac{1}{2} + 3i, \dfrac{1}{2} - 3i$

21. $-3 + 7i, -3 - 7i$

22. $\dfrac{7 + 3i\sqrt{3}}{2}, \dfrac{7 - 3i\sqrt{3}}{2}$

Find the discriminant of the quadratic equation and give the number and type of solutions of the equation.

10. $x^2 + 5x + 2 = 0$
17; 2 real

11. $x^2 + 2x + 5 = 0$
−16; 2 imaginary

12. $4x^2 - 4x + 1 = 0$
0; 1 real

23. $\dfrac{-3 + \sqrt{29}}{10}, \dfrac{-3 - \sqrt{29}}{10}$

13. $-2x^2 + 3x - 7 = 0$
−47; 2 imaginary

14. $9x^2 + 12x + 4 = 0$
0; 1 real

15. $5x^2 - x - 13 = 0$
261; 2 real

25. $\dfrac{-1 + i\sqrt{7}}{4}, \dfrac{-1 - i\sqrt{7}}{4}$

16. 🌐 **BASKETBALL** A basketball player passes the ball to a teammate who catches it 11 ft above the court, just above the rim of the basket, and slam-dunks it through the hoop. (This play is called an "alley-oop.") The first player releases the ball 5 ft above the court with an initial vertical velocity of 21 ft/sec. How long is the ball in the air before being caught, assuming it is caught as it rises?
0.42 sec

PRACTICE AND APPLICATIONS

STUDENT HELP

▸ **Extra Practice**
to help you master
skills is on p. 946.

26. $\dfrac{2}{3} + \dfrac{i\sqrt{2}}{6}, \dfrac{2}{3} - \dfrac{i\sqrt{2}}{6}$

27. $-1, \dfrac{9}{7}$

28. $\dfrac{-1 + 3i}{4}, \dfrac{-1 - 3i}{4}$

29. $\dfrac{-9 + \sqrt{33}}{8}, \dfrac{-9 - \sqrt{33}}{8}$

30. $\dfrac{2}{3} + 3i, \dfrac{2}{3} - 3i$

31. $-\dfrac{2}{5} + \dfrac{\sqrt{26}}{10}, -\dfrac{2}{5} - \dfrac{\sqrt{26}}{10}$

STUDENT HELP

▸ **HOMEWORK HELP**
Examples 1–3: Exs. 17–55
Example 4: Exs. 56–64
Example 5: Exs. 74–80

EQUATIONS IN STANDARD FORM Use the quadratic formula to solve the equation.   18–23, 25–31. See margin.

17. $x^2 - 5x - 14 = 0$ **−2, 7**

18. $x^2 + 3x - 2 = 0$

19. $x^2 - 2x - 4 = 0$

20. $x^2 + 10x + 22 = 0$

21. $x^2 + 6x + 58 = 0$

22. $-x^2 + 7x - 19 = 0$

23. $5x^2 + 3x - 1 = 0$

24. $3x^2 - 11x - 4 = 0$ $-\dfrac{1}{3}$, 4

25. $2x^2 + x + 1 = 0$

26. $6p^2 - 8p + 3 = 0$

27. $-7q^2 + 2q + 9 = 0$

28. $8r^2 + 4r + 5 = 0$

29. $-4t^2 - 9t - 3 = 0$

30. $9u^2 - 12u + 85 = 0$

31. $10v^2 + 8v - 1 = 0$

EQUATIONS NOT IN STANDARD FORM Use the quadratic formula to solve the equation.   36–45. See margin.

32. $x^2 + 4x = -20$ **−2 + 4i, −2 − 4i**

33. $x^2 - 2x = 99$ **−9, 11**

34. $x^2 + 14 = 10x$ $5 + \sqrt{11}, 5 - \sqrt{11}$

35. $x^2 = 8x - 35$ $4 + i\sqrt{19}, 4 - i\sqrt{19}$

36. $-x^2 - 3x = -7$

37. $-x^2 = 16x + 46$

38. $3x^2 + 6x = -2$

39. $8x^2 - 8x = 1$

40. $5x^2 + 9x = -x^2 + 5x + 1$

41. $40x - 7x^2 = 101 - 3x^2$

42. $-16k^2 = 20k^2 + 24k + 5$

43. $13n^2 + 11n - 9 = 4n^2 - n - 4$

44. $3(d - 1)^2 = 4d + 2$

45. $3.5y^2 + 2.6y - 8.2 = -0.4y^2 - 6.9y$

5.6 *The Quadratic Formula and the Discriminant*   **295**

---

3 APPLY

ASSIGNMENT GUIDE

BASIC
Day 1: 295–298 Exs. 18–40 even, 46–60 even, 65–68, 75, 77, 81–83, 85–101 odd, Quiz 2 Exs. 1–22

AVERAGE
Day 1: pp. 295–298 Exs. 18–40 even, 46–64 even, 65–69, 75–79 odd, 81–83, 85–101 odd, Quiz 2 Exs. 1–22

ADVANCED
Day 1: pp. 295–298 Exs. 18–40 even, 46–64 even, 65–69, 74, 75–79 odd, 81–84, 85–101 odd, Quiz 2 Exs. 1–22

BLOCK SCHEDULE
pp. 295–298 Exs. 18–40 even, 46–64 even, 65–69, 75–79 odd, 81–83, 85–101 odd, Quiz 2 Exs. 1–22 (with 5.5)

EXERCISE LEVELS
Level A: *Easier*
17–39, 46–51, 56–67
Level B: *More Difficult*
40–45, 52–55, 68–83
Level C: *Most Difficult*
84

✔ **HOMEWORK CHECK**
To quickly check student understanding of key concepts, go over the following exercises: Exs. 18, 32, 46, 56, 66. See also the Daily Homework Quiz:

• Blackline Master (*Chapter 5 Resource Book*, p. 94)

• 🖨 Transparency (p. 38)

36–45. See next page.

TEACHING TIPS

EXERCISES 46–55 You may wish to have students list the method they used and explain why they used it rather than another method.

CAREER NOTE

EXERCISE 76 Additional information about a career as a Web developer is available at **www.mcdougallittell.com.**

DATA UPDATE

Updated data for Ex. 79 is available at **www.mcdougallittell.com.**

36. $\dfrac{-3 + \sqrt{37}}{2}, \dfrac{-3 - \sqrt{37}}{2}$

37. $-8 + 3\sqrt{2}, -8 - 3\sqrt{2}$

38. $-1 + \dfrac{\sqrt{3}}{3}, -1 - \dfrac{\sqrt{3}}{3}$

39. $\dfrac{1}{2} + \dfrac{\sqrt{6}}{4}, \dfrac{1}{2} - \dfrac{\sqrt{6}}{4}$

40. $-\dfrac{1}{3} + \dfrac{\sqrt{10}}{6}, -\dfrac{1}{3} - \dfrac{\sqrt{10}}{6}$

41. $5 + \dfrac{i}{2}, 5 - \dfrac{i}{2}$

42. $-\dfrac{1}{3} + \dfrac{i}{6}, -\dfrac{1}{3} - \dfrac{i}{6}$

43. $\dfrac{1}{3}, -\dfrac{5}{3}$

44. $\dfrac{5 + \sqrt{22}}{3}, \dfrac{5 - \sqrt{22}}{3}$

45. $\dfrac{-9.5 + \sqrt{218.17}}{7.8},$

$\dfrac{-9.5 - \sqrt{218.17}}{7.8}$

52. $1 + \sqrt{2}, 1 - \sqrt{2}$

54. $2 + \dfrac{\sqrt{5}}{3}, 2 - \dfrac{\sqrt{5}}{3}$

68. $c < 1; c = 1; c > 1$

69. $c < 4; c = 4; c > 4$

70. $c < 25; c = 25; c > 25$

71. $c < 16; c = 16; c > 16$

72. $c < 9; c = 9; c > 9$

73. $c < 36; c = 36; c > 36$

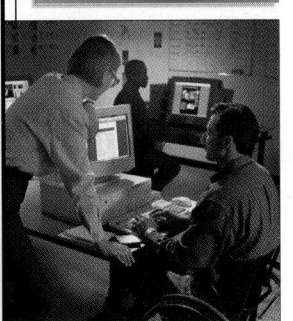

FOCUS ON CAREERS

WEB DEVELOPER Web developers use *hypertext markup language* (HTML) to create electronic pages for the World Wide Web. A *Web browser* translates HTML into pages that can be viewed on a computer screen.

CAREER LINK
www.mcdougallittell.com

ADDITIONAL PRACTICE AND RETEACHING

For Lesson 5.6:

• Practice Levels A, B, and C (*Chapter 5 Resource Book,* p. 83)

• Reteaching with Practice (*Chapter 5 Resource Book,* p. 86)

• See Lesson 5.6 of the *Personal Student Tutor*

For more Mixed Review:

• Search the *Test and Practice Generator* for key words or specific lessons.

SOLVING BY ANY METHOD Solve the equation by factoring, by finding square roots, or by using the quadratic formula.

46. $6x^2 - 12 = 0$ $-\sqrt{2}, \sqrt{2}$

47. $x^2 - 3x - 15 = 0$ $\dfrac{3 + \sqrt{69}}{2}, \dfrac{3 - \sqrt{69}}{2}$

48. $x^2 + 4x + 29 = 0$ $-2 + 5i, -2 - 5i$

49. $x^2 - 18x + 32 = 0$ $2, 16$

50. $4x^2 + 28x = -49$ $-\dfrac{7}{2}$

51. $3(x + 4)^2 = -27$ $-4 + 3i, -4 - 3i$

52. $-2u^2 + 5 = 3u^2 - 10u$

53. $11m^2 - 1 = 7m^2 + 2$ $\dfrac{\sqrt{3}}{2}, -\dfrac{\sqrt{3}}{2}$

54. $-9v^2 + 35v - 30 = 1 - v$

55. $20p^2 + 6p = 6p^2 - 13p + 3$ $-\dfrac{3}{2}, \dfrac{1}{7}$

USING THE DISCRIMINANT Find the discriminant of the quadratic equation and give the number and type of solutions of the equation.

56. $x^2 - 4x + 10 = 0$
   -24; 2 imaginary

57. $x^2 + 3x - 6 = 0$
   33; 2 real

58. $x^2 + 14x + 49 = 0$
   0; 1 real

59. $3x^2 - 10x - 5 = 0$
   160; 2 real

60. $64x^2 - 16x + 1 = 0$
   0; 1 real

61. $-2x^2 - 5x - 4 = 0$
   -7; 2 imaginary

62. $7r^2 - 3 = 0$
   84; 2 real

63. $s^2\sqrt{5} + s + \sqrt{5} = 0$
   -19; 2 imaginary

64. $-4t^2 + 20t - 25 = 0$
   0; 1 real

VISUAL THINKING In Exercises 65–67, the graph of a quadratic function $y = ax^2 + bx + c$ is shown. Tell whether the discriminant of $ax^2 + bx + c = 0$ is *positive, negative,* or *zero.*

65.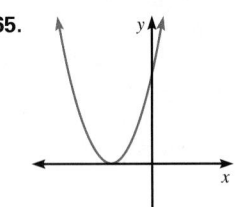

zero

66.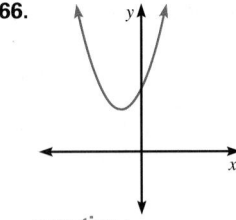

negative

67.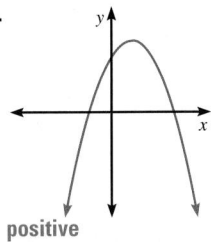

positive

THE CONSTANT TERM Find all values of *c* for which the equation has (a) two real solutions, (b) one real solution, and (c) two imaginary solutions.
68–73. See margin.

68. $x^2 - 2x + c = 0$

69. $x^2 + 4x + c = 0$

70. $x^2 + 10x + c = 0$

71. $x^2 - 8x + c = 0$

72. $x^2 + 6x + c = 0$

73. $x^2 - 12x + c = 0$

74. **CRITICAL THINKING** Explain why the height model $h = -16t^2 + v_0t + h_0$ applies not only to launched or thrown objects, but to dropped objects as well. (*Hint:* What is the initial vertical velocity of a dropped object?)
   Sample answer: The initial velocity substituted into the formula can be zero.

75. **DIVING** In July of 1997, the first Cliff Diving World Championships were held in Brontallo, Switzerland. Participants performed acrobatic dives from heights of up to 92 feet. Suppose a cliff diver jumps from this height with an initial upward velocity of 5 feet per second. How much time does the diver have to perform acrobatic maneuvers before hitting the water?
   ▶ Source: World High Diving Federation   **about 2.56 sec**

76. **WORLD WIDE WEB** A Web developer is creating a Web site devoted to mountain climbing. Each page on the Web site will have frames along its top and left sides showing the name of the site and links to different parts of the site. These frames will take up one third of the computer screen. What will the width *x* of the frames be on the screen shown? **about 1.56 in.**

77. 🌐 **VOLLEYBALL** In a volleyball game, a player on one team spikes the ball over the net when the ball is 10 feet above the court. The spike drives the ball downward with an initial vertical velocity of -55 feet per second. Players on the opposing team must hit the ball back over the net before the ball touches the court. How much time do the opposing players have to hit the spiked ball?
about 0.17 sec

78. 🌐 **AVIATION** The length l (in feet) of runway needed for a small airplane to land is given by $l = 0.1s^2 - 3s + 22$ where s is the airplane's speed (in feet per second). If a pilot is landing a small airplane on a runway 2000 feet long, what is the maximum speed at which the pilot can land? **about 156.4 ft/sec**

79. 🌐 **TELECOMMUNICATIONS** For the years 1989–1996, the amount A (in billions of dollars) spent on long distance telephone calls in the United States can be modeled by $A = 0.560t^2 + 0.488t + 51$ where t is the number of years since 1989. In what year did the amount spent reach \$60 billion? **1993**

🌐 DATA UPDATE of *Statistical Abstract of the United States* data at www.mcdougallittell.com

80. 🌐 **EARTH SCIENCE** The volcanic cinder cone Puu Puai in Hawaii was formed in 1959 when a massive "lava fountain" erupted at Kilauea Iki Crater, shooting lava hundreds of feet into the air. When the eruption was most intense, the height h (in feet) of the lava t seconds after being ejected from the ground could be modeled by $h = -16t^2 + 350t$. ▶ Source: Volcano World

a. What was the initial vertical velocity of the lava? What was the lava's maximum height above the ground? **350 ft/sec; about 1914 ft**

b. CHOOSING A METHOD For how long was the lava in the air? Solve the problem either by factoring or by using the quadratic formula. **21.875 sec**

Test 🎲 Preparation

84. a. maximum height occurs when $t = \frac{v_0}{32}$; $v_0 = 32\sqrt{10}$ ft/sec

84. b. $t = \sqrt{10}$ sec ≈ 3.16 sec; *Sample answer:* if $t = 2$ sec, then v_0 would need to increase by about 10.7% to 112 ft/sec.

★ **Challenge**

QUANTITATIVE COMPARISON In Exercises 81–83, choose the statement that is true about the given quantities.

(A) The quantity in column A is greater.

(B) The quantity in column B is greater.

(C) The two quantities are equal.

(D) The relationship cannot be determined from the given information.

	Column A	Column B	
81.	Discriminant of $x^2 - 6x - 1 = 0$	Discriminant of $x^2 + 5x - 4 = 0$	B
82.	Discriminant of $x^2 + 2kx + 1 = 0$	Discriminant of $kx^2 + 3x - k = 0$	B
83.	Least zero of $f(x) = x^2 - 10x + 23$	Greatest zero of $f(x) = x^2 - 2x - 2$	A

84. 🌐 **THRILL RIDES** The Stratosphere Tower in Las Vegas is 921 feet tall and has a "needle" at its top that extends even higher into the air. A thrill ride called the Big Shot catapults riders 160 feet up the needle and then lets them fall back to the launching pad. ▶ Source: Stratosphere Tower **84. a, b. See margin.**

a. The height h (in feet) of a rider on the Big Shot can be modeled by $h = -16t^2 + v_0t + 921$ where t is the elapsed time (in seconds) after launch and v_0 is the initial vertical velocity (in feet per second). Find v_0 using the fact that the maximum value of h is $921 + 160 = 1081$ feet.

b. A brochure for the Big Shot states that the ride up the needle takes 2 seconds. Compare this time with the time given by the model $h = -16t^2 + v_0t + 921$ where v_0 is the value you found in part (a). Discuss the model's accuracy.

EXTRA CHALLENGE
▶ www.mcdougallittell.com

DAILY HOMEWORK QUIZ

📝 *Transparency Available*

1. Solve $3x^2 + x - 8 = 0$.
about 1.47, about -1.81

2. Solve $8x^2 + 5x = -3x - 2$. -0.5

3. Solve $x^2 + 4x = -9$ $-2 \pm i\sqrt{5}$

4. Find the discriminant of $x^2 - 4x + 7 = 0$ and give the number and type of solutions of the equation.
-12; two imaginary

EXTRA CHALLENGE NOTE
↳ Challenge problems for Lesson 5.6 are available in **blackline** format in the *Chapter 5 Resource Book*, p. 90 and at **www.mcdougallittell.com.**

ADDITIONAL TEST PREPARATION

1. WRITING Explain the relationship between the discriminant of a quadratic equation, the number and type of solutions of the quadratic equation, and the x-intercepts of its corresponding graph. **If the discriminant of the equation is negative, the equation will have two imaginary solutions and the graph will have no x-intercepts. If the discriminant is zero, the equation will have one real solution and the graph will have one x-intercept. If the discriminant is positive, the equation will have two real solutions and the graph will have two x-intercepts.**

ADDITIONAL RESOURCES
An alternative Quiz for Lessons
5.4–5.6 is available in the *Chapter 5
Resource Book*, p. 91.

91.

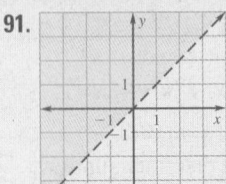

92.

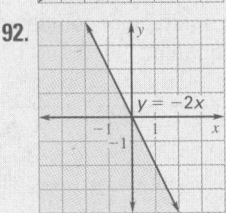

93.

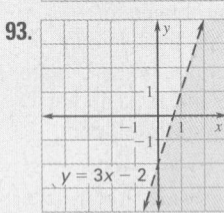

94.

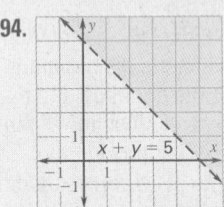

95.

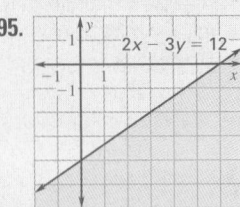

96.

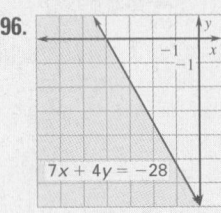

97.

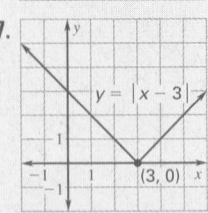

98–102, 5–10.
See Additional Answers beginning on page AA1.

SOLVING LINEAR INEQUALITIES Solve the inequality. Then graph your
solution. (Review 1.6 for 5.7)

85. $3x + 6 > 12$ $x > 2$

86. $16 - 7x \geq -5$ $x \leq 3$

87. $-2(x + 9) \leq 8$ $x \geq -13$

88. $10x + 3 < 6x - 1$ $x < -1$

89. $4 \leq 5x - 11 \leq 29$ $3 \leq x \leq 8$

90. $\frac{3}{2}x + 20 \leq 14$ or $1 > 8 - x$
$x \leq -4$ or $x > 7$

GRAPHING LINEAR INEQUALITIES Graph the inequality. (Review 2.6 for 5.7)
91–96. See margin.

91. $y > x$

92. $y \leq -2x$

93. $y < 3x - 2$

94. $x + y > 5$

95. $2x - 3y \geq 12$

96. $7x + 4y \leq -28$

ABSOLUTE VALUE FUNCTIONS Graph the function. (Review 2.8)
97–102. See margin.

97. $y = |x - 3|$

98. $y = |x| + 2$

99. $y = -2|x| - 1$

100. $y = 3|x + 4|$

101. $y = |x + 2| + 3$

102. $y = \frac{1}{2}|x - 5| - 4$

QUIZ 2

Self-Test for Lessons 5.4–5.6

Write the expression as a complex number in standard form. (Lesson 5.4)

1. $(7 + 5i) + (-2 + 11i)$ $5 + 16i$

2. $(-1 + 8i) - (3 - 2i)$ $-4 + 10i$

3. $(4 - i)(6 + 7i)$ $31 + 22i$

4. $\frac{1 - 3i}{5 + i}$ $\frac{1}{13} - \frac{8}{13}i$

Plot the numbers in the same complex plane and find their absolute values.
(Lesson 5.4)   5–10. See margin for graph.

5. $2 + 4i$ $2\sqrt{5}$

6. $-5i$ 5

7. $-3 + i$ $\sqrt{10}$

8. $4 + 3i$ 5

9. -4 4

10. $-\frac{3}{2} - \frac{7}{2}i$ $\frac{\sqrt{58}}{2}$

Solve the quadratic equation by completing the square. (Lesson 5.5)

11. $x^2 + 8x = -14$ $-4 + \sqrt{2}, -4 - \sqrt{2}$

12. $x^2 - 2x + 17 = 0$ $1 + 4i, 1 - 4i$

13. $4p^2 - 40p - 8 = 0$
$5 + 3\sqrt{3}, 5 - 3\sqrt{3}$

14. $3q^2 + 20q = -2q^2 - 19$
$-2 + \frac{\sqrt{5}}{5}, -2 - \frac{\sqrt{5}}{5}$

Write the quadratic function in vertex form. (Lesson 5.5)

15. $y = x^2 + 6x + 1$
$y = (x + 3)^2 - 8$

16. $y = x^2 - 18x + 50$
$y = (x - 9)^2 - 31$

17. $y = -2x^2 + 8x - 7$
$y = -2(x - 2)^2 + 1$

Use the quadratic formula to solve the equation. (Lesson 5.6)

18. $x^2 + 2x - 10 = 0$
$-1 + \sqrt{11}, -1 - \sqrt{11}$

19. $x^2 - 16x + 73 = 0$ $8 + 3i, 8 - 3i$

20. $\frac{3 + i\sqrt{7}}{2}, \frac{3 - i\sqrt{7}}{2}$
20. $3w^2 + 3w = 4w^2 + 4$

21. $\frac{-4 + 2\sqrt{6}}{5}, \frac{-4 - 2\sqrt{6}}{5}$
21. $14 + 2y - 25y^2 = 42y + 6$

22. 🌐 **ENTERTAINMENT** A juggler throws a ball into the air, releasing it 5 feet
above the ground with an initial vertical velocity of 15 ft/sec. She catches the
ball with her other hand when the ball is 4 feet above the ground. Using the
model $h = -16t^2 + v_0t + h_0$, find how long the ball is in the air. (Lesson 5.6)
about 1 sec

5.7 Graphing and Solving Quadratic Inequalities

What you should learn

GOAL 1 Graph quadratic inequalities in two variables.

GOAL 2 Solve quadratic inequalities in one variable, as applied in **Example 7**.

Why you should learn it

▼ To solve **real-life** problems, such as finding the weight of theater equipment that a rope can support in **Exs. 47 and 48**.

GOAL 1 QUADRATIC INEQUALITIES IN TWO VARIABLES

In this lesson you will study four types of **quadratic inequalities in two variables**.

$$y < ax^2 + bx + c \qquad y \le ax^2 + bx + c$$
$$y > ax^2 + bx + c \qquad y \ge ax^2 + bx + c$$

The graph of any such inequality consists of all solutions (x, y) of the inequality. The steps used to graph a quadratic inequality are very much like those used to graph a linear inequality. (See Lesson 2.6.)

GRAPHING A QUADRATIC INEQUALITY IN TWO VARIABLES

To graph one of the four types of quadratic inequalities shown above, follow these steps:

STEP 1 Draw the parabola with equation $y = ax^2 + bx + c$. Make the parabola *dashed* for inequalities with $<$ or $>$ and *solid* for inequalities with $\le$ or $\ge$.

STEP 2 Choose a point (x, y) inside the parabola and check whether the point is a solution of the inequality.

STEP 3 If the point from Step 2 is a solution, shade the region inside the parabola. If it is not a solution, shade the region outside the parabola.

EXAMPLE 1 Graphing a Quadratic Inequality

Graph $y > x^2 - 2x - 3$.

SOLUTION

Follow Steps 1–3 listed above.

1 Graph $y = x^2 - 2x - 3$. Since the inequality symbol is $>$, make the parabola dashed.

2 Test a point inside the parabola, such as $(1, 0)$.

$$y > x^2 - 2x - 3$$
$$0 \overset{?}{>} 1^2 - 2(1) - 3$$
$$0 > -4 \checkmark$$

So, $(1, 0)$ is a solution of the inequality.

3 Shade the region inside the parabola.

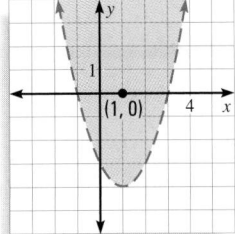

---

1 PLAN

PACING
Basic: 1 day
Average: 1 day
Advanced: 1 day
Block Schedule: 0.5 block with 5.8

LESSON OPENER
ACTIVITY
An alternative way to approach Lesson 5.7 is to use the Activity Lesson Opener:
- Blackline Master (*Chapter 5 Resource Book*, p. 95)
- Transparency (p. 33)

MEETING INDIVIDUAL NEEDS
- *Chapter 5 Resource Book*
  Prerequisite Skills Review (p. 5)
  Practice Level A (p. 97)
  Practice Level B (p. 98)
  Practice Level C (p. 99)
  Reteaching with Practice (p. 100)
  Absent Student Catch-Up (p. 102)
  Challenge (p. 104)
- *Resources in Spanish*
- Personal Student Tutor

NEW-TEACHER SUPPORT
See the Tips for New Teachers on pp. 1–2 of the *Chapter 5 Resource Book* for additional notes about Lesson 5.7.

WARM-UP EXERCISES

Transparency Available
Solve and graph.

1. $2x - 7 \le 11$ $x \le 9$
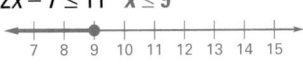

2. $3 - 6(x - 1) > 9$ $x < 0$
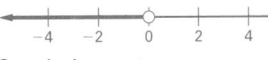

3. Graph the system:
$y < 3x + 1$
$y \ge -x$
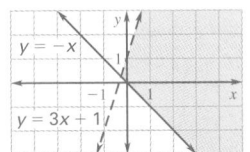

MOTIVATING THE LESSON
In real-life problems, rather than having values that are exact, you might have values that are in a range. For example, you may want to know how much you can charge for puppet show tickets if you want to make a profit of at least $300. These types of problems involve inequalities. In today's lesson you will learn how to solve problems involving quadratic inequalities.

EXTRA EXAMPLE 1
Graph $y \le 2x^2 - 5x - 3$.

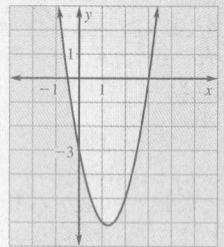

EXTRA EXAMPLE 2
You are making a photo album. Each album page needs to be able to hold 6 square pictures. If the length of one side of each picture is x, then $A \ge 6x^2$ is the area of one album page.
a. Graph this function.

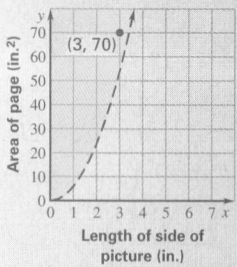

b. If you have an album page that has an area of 70 square inches, will it be able to accommodate 6 pictures with 3-inch sides? **Yes**

Checkpoint Exercises for Examples 1 and 2 on next page.

300

Carpentry

EXAMPLE 2 *Using a Quadratic Inequality as a Model*

You are building a wooden bookcase. You want to choose a thickness d (in inches) for the shelves so that each is strong enough to support 60 pounds of books without breaking. A shelf can safely support a weight of W (in pounds) provided that:

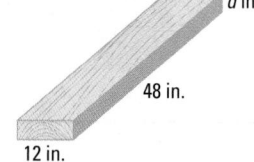

$$W \le 300d^2$$

a. Graph the given inequality.

b. If you make each shelf 0.75 inch thick, can it support a weight of 60 pounds?

SOLUTION

STUDENT HELP
Look Back
For help with graphing inequalities in two variables, see p. 108.

a. Graph $W = 300d^2$ for nonnegative values of d. Since the inequality symbol is $\le$, make the parabola solid. Test a point inside the parabola, such as $(0.5, 240)$.

$$W \le 300d^2$$
$$240 \overset{?}{\le} 300(0.5)^2$$
$$240 \nleq 75$$

Since the chosen point is not a solution, shade the region outside (below) the parabola.

b. The point $(0.75, 60)$ lies in the shaded region of the graph from part (a), so $(0.75, 60)$ is a solution of the given inequality. Therefore, a shelf that is 0.75 inch thick *can* support a weight of 60 pounds.

· · · · · · · · · ·

Graphing a *system* of quadratic inequalities is similar to graphing a system of linear inequalities. First graph each inequality in the system. Then identify the region in the coordinate plane common to all the graphs. This region is called the *graph of the system*.

EXAMPLE 3 *Graphing a System of Quadratic Inequalities*

Graph the system of quadratic inequalities.

$$y \ge x^2 - 4 \qquad \text{Inequality 1}$$
$$y < -x^2 - x + 2 \qquad \text{Inequality 2}$$

SOLUTION

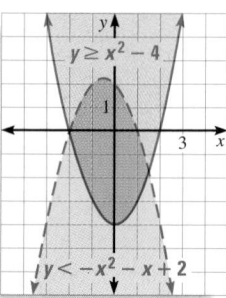

Graph the inequality $y \ge x^2 - 4$. The graph is the red region inside and including the parabola $y = x^2 - 4$.

Graph the inequality $y < -x^2 - x + 2$. The graph is the blue region inside (but not including) the parabola $y = -x^2 - x + 2$.

Identify the **purple region** where the two graphs overlap. This region is the graph of the system.

300 **Chapter 5** *Quadratic Functions*

GOAL 2 QUADRATIC INEQUALITIES IN ONE VARIABLE

One way to solve a **quadratic inequality in one variable** is to use a graph.

- To solve $ax^2 + bx + c < 0$ (or $ax^2 + bx + c \le 0$), graph $y = ax^2 + bx + c$ and identify the x-values for which the graph lies *below* (or *on and below*) the x-axis.

- To solve $ax^2 + bx + c > 0$ (or $ax^2 + bx + c \ge 0$), graph $y = ax^2 + bx + c$ and identify the x-values for which the graph lies *above* (or *on and above*) the x-axis.

EXAMPLE 4 *Solving a Quadratic Inequality by Graphing*

STUDENT HELP

↳ **Look Back**
For help with solving inequalities in one variable, see p. 41.

Solve $x^2 - 6x + 5 < 0$.

SOLUTION

The solution consists of the x-values for which the graph of $y = x^2 - 6x + 5$ lies below the x-axis. Find the graph's x-intercepts by letting $y = 0$ and using factoring to solve for x.

$$0 = x^2 - 6x + 5$$

$$0 = (x - 1)(x - 5)$$

$$x = 1 \text{ or } x = 5$$

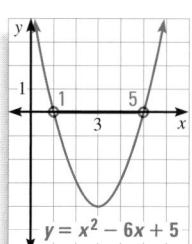

Sketch a parabola that opens up and has 1 and 5 as x-intercepts. The graph lies below the x-axis between $x = 1$ and $x = 5$.

▶ The solution of the given inequality is $1 < x < 5$.

EXAMPLE 5 *Solving a Quadratic Inequality by Graphing*

Solve $2x^2 + 3x - 3 \ge 0$.

SOLUTION

The solution consists of the x-values for which the graph of $y = 2x^2 + 3x - 3$ lies on and above the x-axis. Find the graph's x-intercepts by letting $y = 0$ and using the quadratic formula to solve for x.

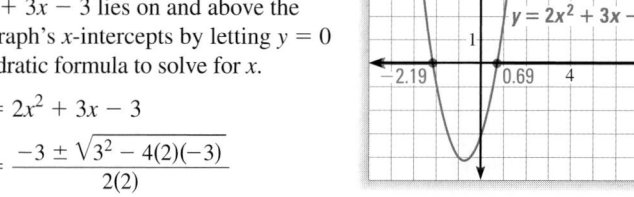

$$0 = 2x^2 + 3x - 3$$

$$x = \frac{-3 \pm \sqrt{3^2 - 4(2)(-3)}}{2(2)}$$

$$x = \frac{-3 \pm \sqrt{33}}{4}$$

$$x \approx 0.69 \text{ or } x \approx -2.19$$

Sketch a parabola that opens up and has 0.69 and -2.19 as x-intercepts. The graph lies on and above the x-axis to the left of (and including) $x = -2.19$ and to the right of (and including) $x = 0.69$.

▶ The solution of the given inequality is approximately $x \le -2.19$ or $x \ge 0.69$.

5.7 *Graphing and Solving Quadratic Inequalities* **301**

📋 **EXTRA EXAMPLE 3**
Graph the system of inequalities.
$y \le -x^2 + 9$
$y \ge x^2 + 5x - 6$

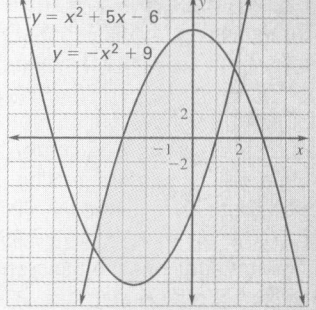

EXTRA EXAMPLE 4
Solve $x^2 - 5x + 6 \ge 0$.
$x \le 2$ or $x \ge 3$

EXTRA EXAMPLE 5
Solve $x^2 - 11x + 5 \le 0$.
$0.48 \le x \le 10.52$

✓ **CHECKPOINT EXERCISES**
For use after Examples 1–3:
1. Is the point $(-1, 4)$ a solution to the system
$y > x^2 + 4x$
$y \le 3x^2$? no

For use after Example 4:
2. Solve $-x^2 - 9x + 36 > 0$.
$-12 < x < 3$

For use after Example 5:
3. Solve $-3x^2 + x + 7 < 0$.
$x < -1.37$ or $x > 1.70$

EXTRA EXAMPLE 6
Solve $2x^2 - x > 3$.
$x < -1$ or $x > \frac{3}{2}$

EXTRA EXAMPLE 7
Suppose a study was conducted to test the average reading comprehension of a person x years of age. The study found that the number of points $P(x)$ scored on a reading comprehension test could be modeled by: $P(x) = -0.017x^2 + 1.9x + 31$, $5 \leq x \leq 95$. At what ages does the average person score greater than 60 points on the test? **between the ages of 18 years and 94 years**

 CHECKPOINT EXERCISES
For use after Examples 6 and 7:
1. Solve $3x^2 + 11x \leq 4$ algebraically. Check your answer using a graph.
$-4 \leq x \leq \frac{1}{3}$

APPLICATION NOTE
EXAMPLE 7 Point out the restrictions on the domain of this function, $16 \leq x \leq 70$. The domain is restricted in this application because the data they used to make the model was gotten from the subjects tested who were between the ages of 16 and 70.

FOCUS ON VOCABULARY
What is a *quadratic inequality in one variable*? See margin for sample answer.

CLOSURE QUESTION
What is the procedure used to solve a quadratic inequality in two variables? See margin.

DAILY PUZZLER
A mother's age today is 3 times that of her daughter's. In 4 years, the product of the mother's and daughter's ages will be at least 500. What are the possible ages of the daughter? The daughter must be at least 11.

302

You can also use an algebraic approach to solve a quadratic inequality in one variable, as demonstrated in Example 6.

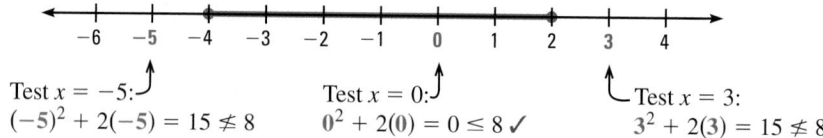 **Solving a Quadratic Inequality Algebraically**

Solve $x^2 + 2x \leq 8$.

SOLUTION
First write and solve the equation obtained by replacing the inequality symbol with an equals sign.

$x^2 + 2x \leq 8$	Write original inequality.
$x^2 + 2x = 8$	Write corresponding equation.
$x^2 + 2x - 8 = 0$	Write in standard form.
$(x + 4)(x - 2) = 0$	Factor.
$x = -4$ or $x = 2$	Zero product property

The numbers -4 and 2 are called the *critical x-values* of the inequality $x^2 + 2x \leq 8$. Plot -4 and 2 on a number line, using solid dots because the values satisfy the inequality. The critical x-values partition the number line into three intervals. Test an x-value in each interval to see if it satisfies the inequality.

Test $x = -5$:
$(-5)^2 + 2(-5) = 15 \nleq 8$

Test $x = 0$:
$0^2 + 2(0) = 0 \leq 8$ ✓

Test $x = 3$:
$3^2 + 2(3) = 15 \nleq 8$

▶ The solution is $-4 \leq x \leq 2$.

EXAMPLE 7 **Using a Quadratic Inequality as a Model**

DRIVING For a driver aged x years, a study found that the driver's reaction time $V(x)$ (in milliseconds) to a visual stimulus such as a traffic light can be modeled by:

$$V(x) = 0.005x^2 - 0.23x + 22, \quad 16 \leq x \leq 70$$

At what ages does a driver's reaction time tend to be greater than 25 milliseconds?
▶ Source: *Science Probe!*

SOLUTION
You want to find the values of x for which:

$$V(x) > 25$$
$$0.005x^2 - 0.23x + 22 > 25$$
$$0.005x^2 - 0.23x - 3 > 0$$

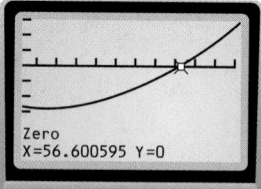

Zero
X=56.600595 Y=0

Graph $y = 0.005x^2 - 0.23x - 3$ on the domain $16 \leq x \leq 70$. The graph's x-intercept is about 57, and the graph lies above the x-axis when $57 < x \leq 70$.

▶ Drivers over 57 years old tend to have reaction times greater than 25 milliseconds.

FOCUS ON APPLICATIONS

 DRIVING Driving simulators help drivers safely improve their reaction times to hazardous situations they may encounter on the road.

302 **Chapter 5** *Quadratic Functions*

Focus on Vocabulary *Sample answer:*
It is an inequality that can be arranged to have a quadratic expression in one variable on one side of an inequality sign and a zero on the other side.

Closure Question *Sample answer:*
Sketch a graph of the parabola that forms the boundary of the solution using a solid or dashed line as appropriate. Test a point to see if you should shade above or below the parabola.

GUIDED PRACTICE

Vocabulary Check ✓

1. Give one example each of a quadratic inequality in one variable and a quadratic inequality in two variables. **1–3. See margin.**

Concept Check ✓

2. How does the graph of $y > x^2$ differ from the graph of $y \geq x^2$?

3. Explain how to solve $x^2 - 3x - 4 > 0$ graphically and algebraically.

Skill Check ✓

Graph the inequality. **4–6. See margin for graphs.**

4. $y \geq x^2 + 2$
5. $y \leq -2x^2$
6. $y < x^2 - 5x + 4$

Graph the system of inequalities. **7–9. See margin for graphs.**

7. $y \leq -x^2 + 3$
   $y \geq x^2 + 2x - 4$

8. $y \geq -x^2 + 3$
   $y \geq x^2 + 2x - 4$

9. $y \geq -x^2 + 3$
   $y \leq x^2 + 2x - 4$

Solve the inequality.

10. $x^2 - 4 < 0$ $\;-2 < x < 2$
11. $x^2 - 4 \geq 0$
    $x \leq -2$ or $x \geq 2$
12. $x^2 - 4 > 3x$
    $x < -1$ or $x > 4$

13. 🌐 **ARCHITECTURE** The arch of the Sydney Harbor Bridge in Sydney, Australia, can be modeled by $y = -0.00211x^2 + 1.06x$ where x is the distance (in meters) from the left pylons and y is the height (in meters) of the arch above the water. For what distances x is the arch above the road? **about 55.1 m and 447.3 m**

pylon
52 m

PRACTICE AND APPLICATIONS

STUDENT HELP

→ **Extra Practice**
to help you master skills is on p. 947.

MATCHING GRAPHS **Match the inequality with its graph.**

14. $y \geq x^2 - 4x + 1$ **B**
15. $y < x^2 - 4x + 1$ **C**
16. $y \leq -x^2 - 4x + 1$ **A**

A.
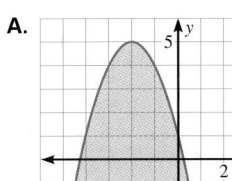

B.
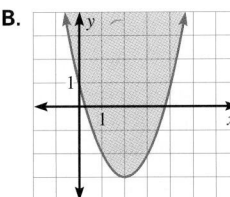

C.
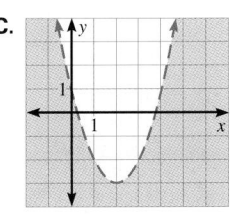

STUDENT HELP

→ **HOMEWORK HELP**
Example 1: Exs. 14–28
Example 2: Exs. 47–49
Example 3: Exs. 29–34, 49
Examples 4, 5: Exs. 35–40
Example 6: Exs. 41–46
Example 7: Exs. 50, 51

GRAPHING QUADRATIC INEQUALITIES **Graph the inequality.** **17–28. See margin.**

17. $y \geq 3x^2$
18. $y \leq -x^2$
19. $y > -x^2 + 5$
20. $y < x^2 - 3x$
21. $y \leq x^2 + 8x + 16$
22. $y \leq -x^2 + x + 6$
23. $y \geq 2x^2 - 2x - 5$
24. $y \geq -2x^2 - x + 3$
25. $y > -3x^2 + 5x - 4$
26. $y < -\frac{1}{2}x^2 - 2x + 4$
27. $y > \frac{4}{3}x^2 - 12x + 29$
28. $y < 0.6x^2 + 3x + 2.4$

3 APPLY

○ **ASSIGNMENT GUIDE**

BASIC
Day 1: pp. 303–305 Exs. 14–16, 18–24 even, 30–46 even, 47–48, 52, 55–65 odd

AVERAGE
Day 1: pp. 303–305 Exs. 14–16, 18–46 even, 47–49, 52, 55–69 odd

ADVANCED
Day 1: pp. 303–305 Exs. 14–16, 18–46 even, 47–49, 52, 53–69 odd

BLOCK SCHEDULE
pp. 303–305 Exs. 14–16, 18–46 even, 47–49, 51–53, 55–69 odd (with 5.8)

EXERCISE LEVELS
Level A: *Easier*
14–26, 35–46
Level B: *More Difficult*
27–34, 47–52
Level C: *Most Difficult*
53

✓ **HOMEWORK CHECK**
To quickly check student understanding of key concepts, go over the following exercises: Exs. 14, 20, 30, 38, 42. See also the Daily Homework Quiz:
• Blackline Master (*Chapter 5 Resource Book,* p. 107)
• Transparency (p. 39)

4–9, 17–28. See Additional Answers beginning on page AA1.

Side margin answers:

1. *Sample answer:*
   one variable:
   $-x^2 - 5x + 7 > 0$;
   two variables:
   $-y \geq x^2 - 5x + 7$

2. *Sample answer:* $y \geq x^2$ includes points on the graph of $y = x^2$ while $y > x^2$ does not.

3. *Sample answer:* graphical: graph $y = x^2 - 3x - 4$ using a dotted line; find the x-intercepts and determine where the graph lies above the x-axis; algebraic: factor $x^2 - 3x - 4$ and graph the critical x-values on a number line; determine where the solutions lie on the number line.

! COMMON ERROR

EXERCISES 29–34 Shading incorrectly is a common error in these problems. Suggest students test one point in the solution in both of the inequalities to help avoid this error.

CAREER NOTE
EXERCISES 47 AND 48
Additional information about a career as a set designer is available at **www.mcdougallittell.com**.

STUDENT HELP NOTES

→ **Homework Help** Students can find help for Exs. 50–52 at **www.mcdougallittell.com**. The information can be printed out for students who don't have access to the Internet.

GRAPHING CALCULATOR NOTE
EXERCISES 50–52 The graphing calculator is recommended for these problems so that the student can solve them graphically. If a graphing calculator is not available, the problems can be solved algebraically using the quadratic formula.

29–34, 47, 49, 52a.
  See Additional Answers beginning on page AA1.

ADDITIONAL PRACTICE AND RETEACHING

For Lesson 5.7:

• Practice Levels A, B, and C (*Chapter 5 Resource Book*, p. 97)

• Reteaching with Practice (*Chapter 5 Resource Book*, p. 100)

• See Lesson 5.7 of the *Personal Student Tutor*

For more Mixed Review:

• Search the *Test and Practice Generator* for key words or specific lessons.

304

FOCUS ON CAREERS

SET DESIGNER
A set designer creates the scenery, or *sets*, used in a theater production. The designer may make scale models of the sets before they are actually built.

CAREER LINK
www.mcdougallittell.com

GRAPHING SYSTEMS Graph the system of inequalities. 29–34. See margin.

29. $y \geq x^2$
$y \leq x^2 + 3$

30. $y < -3x^2$
$y \geq -\frac{1}{2}x^2 - 5$

31. $y > x^2 - 6x + 9$
$y < -x^2 + 6x - 3$

32. $y \geq x^2 + 2x + 1$
$y \geq x^2 - 4x + 4$

33. $y < 3x^2 + 2x - 5$
$y \geq -2x^2 + 1$

34. $y \leq 2x^2 - 9x + 8$
$y > -x^2 - 6x - 4$

SOLVING BY GRAPHING Solve the inequality by graphing.

35. $x^2 + x - 2 < 0$
$-2 < x < 1$

36. $2x^2 - 7x + 3 \geq 0$
$x \leq \frac{1}{2}$ or $x \geq 3$

37. $-x^2 - 2x + 8 \leq 0$
$x \leq -4$ or $x \geq 2$

38. $-x^2 + x + 5 > 0$
$-1.8 < x < 2.8$

39. $3x^2 + 24x \geq -41$
$x \leq -5.5$ or $x \geq -2.5$

40. $-\frac{3}{4}x^2 + 4x - 8 < 0$
no real solutions

SOLVING ALGEBRAICALLY Solve the inequality algebraically.

41. $x^2 + 3x - 18 \geq 0$
$x \leq -6$ or $x \geq 3$

42. $3x^2 - 16x + 5 \leq 0$
$\frac{1}{3} \leq x \leq 5$

43. $4x^2 < 25$ $-\frac{5}{2} < x < \frac{5}{2}$

44. $-x^2 - 12x < 32$
$x < -8$ or $x > -4$

45. $2x^2 - 4x - 5 > 0$
$x < -0.9$ or $x > 2.9$

46. $\frac{1}{2}x^2 + 3x \leq -6$
no real solutions

🌎 **THEATER** In Exercises 47 and 48, use the following information.
You are a member of a theater production crew. You use manila rope and wire rope to support lighting, scaffolding, and other equipment. The weight W (in pounds) that can be safely supported by a rope with diameter d (in inches) is given below for both types of rope. ▶ Source: *Workshop Math*

  Manila rope: $W \leq 1480d^2$ **Wire rope:** $W \leq 8000d^2$

47. Graph the inequalities in separate coordinate planes for $0 \leq d \leq 1\frac{1}{2}$. See margin.

48. Based on your graphs, can 1000 pounds of theater equipment be supported by a $\frac{1}{2}$ inch manila rope? by a $\frac{1}{2}$ inch wire rope? no; yes

49. 🩺 **HEALTH** For a person of height h (in inches), a healthy weight W (in pounds) is one that satisfies this system of inequalities: See margin for graph;
121 ≤ W ≤ 160

$$W \geq \frac{19h^2}{703} \quad \text{and} \quad W \leq \frac{25h^2}{703}$$

Graph the system for $0 \leq h \leq 80$. What is the range of healthy weights for a person 67 inches tall? ▶ Source: *Parade Magazine*

📟 **SOLVING INEQUALITIES** In Exercises 50–52, you may want to use a graphing calculator to help you solve the problems.

50. 🌲 **FORESTRY** *Sawtimber* is a term for trees that are suitable for sawing into lumber, plywood, and other products. For the years 1983–1995, the unit value y (in 1994 dollars per million board feet) of one type of sawtimber harvested in California can be modeled by

$$y = 0.125x^2 - 569x + 848{,}000, \quad 400 \leq x \leq 2200$$

where x is the volume of timber harvested (in millions of board feet).
▶ Source: California Department of Forestry and Fire Protection

a. For what harvested timber volumes is the value of the timber at least $400,000 per million board feet? $400 \leq x \leq 1012.6$

b. **LOGICAL REASONING** What happens to the unit value of the timber as the volume harvested increases? Why would you expect this to happen?
decreases; there is an oversupply of timber

STUDENT HELP

HOMEWORK HELP
Visit our Web site
www.mcdougallittell.com
for help with problem solving in Exs. 50–52.

51. 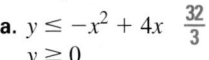 **MEDICINE** In 1992 the average income I (in dollars) for a doctor aged x years could be modeled by: about 39 to 61 years old

$$I = -425x^2 + 42,500x - 761,000$$

For what ages did the average income for a doctor exceed \$250,000?

> **DATA UPDATE** of *American Almanac of Jobs and Salaries* data at www.mcdougallittell.com

Test Preparation

52. **MULTI-STEP PROBLEM** A study of driver reaction times to audio stimuli found that the reaction time $A(x)$ (in milliseconds) of a driver can be modeled by

$$A(x) = 0.0051x^2 - 0.319x + 15, \quad 16 \le x \le 70$$

where x is the driver's age (in years). ▶ Source: *Science Probe!*

52b. Sample answer: A(x) is always less than V(x).

52c. Sample answer: siren; since audio stimuli reaction time is less than visual stimuli reaction time

a. Graph $y = A(x)$ on the given domain. Also graph $y = V(x)$, the reaction-time model for visual stimuli from Example 7, in the same coordinate plane.
See margin for graph.
b. For what values of x in the interval $16 \le x \le 70$ is $A(x) < V(x)$?

c. *Writing* Based on your results from part (b), do you think a driver would react more quickly to a traffic light changing from green to yellow or to the siren of an approaching ambulance? Explain.

★ **Challenge**

53. **GEOMETRY** ▶ **CONNECTION**   The area A of the region bounded by a parabola and a horizontal line is given by $A = \frac{2}{3}bh$ where b and h are as defined in the diagram. Find the area of the region determined by each pair of inequalities.

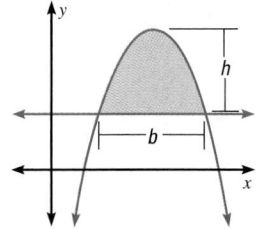

EXTRA CHALLENGE
→ www.mcdougallittell.com

a. $y \le -x^2 + 4x$ $\frac{32}{3}$ **b.** $y \ge x^2 - 4x - 5$ about 56
　　$y \ge 0$ $y \le 3$

MIXED REVIEW

SOLVING FOR A VARIABLE Solve the equation for y. **(Review 1.4)**

54. $3x + y = 1$ $y = 1 - 3x$ **55.** $8x - 2y = 10$ **56.** $-2x + 5y = 9$
　　　　　　　　　　　　　　　　　　$y = 4x - 5$　　　　　　$y = \frac{9}{5} + \frac{2}{5}x$

57. $\frac{1}{6}x + \frac{1}{3}y = -\frac{11}{12}$ **58.** $xy - x = 2$ $y = \frac{2}{x} + 1$ **59.** $\frac{x - 3y}{4} = 7x$ $y = -9x$
　　$y = \frac{11}{4} - \frac{1}{2}x$

SOLVING SYSTEMS Solve the system of linear equations. **(Review 3.6 for 5.8)**

60. $5x - 3y - 2z = -17$ $(-1, 2, 3)$ **61.** $x - 4y + z = -14$ $(2, 3, -4)$
　　$-x + 7y - 3z = 6$　　　　　　　　　　$2x + 3y + 7z = -15$
　　$3x + 2y + 4z = 13$　　　　　　　　　　$-3x + 5y - 5z = 29$

COMPLEX NUMBERS Write the expression as a complex number in standard form. **(Review 5.4)**

62. $(3 + 4i) + (10 - i)$ $13 + 3i$ **63.** $(-11 - 2i) + (5 + 2i)$ -6

64. $(9 + i) - (4 - i)$ $5 + 2i$ **65.** $(5 - 3i) - (-1 + 2i)$ $6 - 5i$

66. $6i(8 + i)$ $-6 + 48i$ **67.** $(7 + 3i)(2 - 5i)$ $29 - 29i$

68. $\frac{1}{3 - i}$ $\frac{3}{10} + \frac{1}{10}i$ **69.** $\frac{4 - 3i}{9 + 2i}$ $\frac{6}{17} - \frac{7}{17}i$

---

4 ASSESS

DAILY HOMEWORK QUIZ

🖥 *Transparency Available*

1. Graph $y \ge x^2 - 4x + 3$.

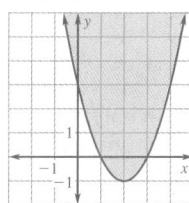

2. Graph the system of quadratic inequalities.
$$y \ge x^2 - 9$$
$$y < -x^2 - x + 3$$

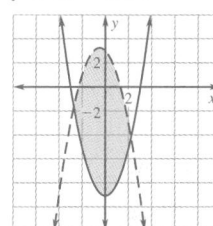

3. Solve $x^2 - 10x + 24 > 0$.
$x < 4$ or $x > 6$

EXTRA CHALLENGE NOTE

→ Challenge problems for Lesson 5.7 are available in **blackline** format in the *Chapter 5 Resource Book,* p. 104 and at **www.mcdougallittell.com.**

ADDITIONAL TEST PREPARATION

1. OPEN ENDED Give an example of a system of quadratic inequalities with no solution.
Sample answer:
$$y > x^2$$
$$y < -x^2$$

2. WRITING Describe how to solve $-x^2 - 6x + 7 \ge 0$ algebraically.   First change $\ge$ to =. Then solve the equation to get $x = -7$ and $x = 1$. Test points to the left of -7, between -7 and 1, and to the right of 1. The solution is $-7 \le x \le 1$.

305

LESSON OPENER
APPLICATION
An alternative way to approach Lesson 5.8 is to use the Application Lesson Opener:

• Blackline Master (*Chapter 5 Resource Book*, p. 108)
• Transparency (p. 34)

MEETING INDIVIDUAL NEEDS
• *Chapter 5 Resource Book*
  Prerequisite Skills Review (p. 5)
  Practice Level A (p. 110)
  Practice Level B (p. 111)
  Practice Level C (p. 112)
  Reteaching with Practice (p. 113)
  Absent Student Catch-Up (p. 115)
  Challenge (p. 117)
• *Resources in Spanish*
• *Personal Student Tutor*

NEW-TEACHER SUPPORT
See the Tips for New Teachers on pp. 1–2 of the *Chapter 5 Resource Book* for additional notes about Lesson 5.8.

WARM-UP EXERCISES

 Transparency Available

Solve the system of linear equations.

1. $2x - y + z = 2$
   $x + y + z = 3$
   $-3x - 2y + z = -4$
   $x = 1, y = 1, z = 1$

2. $-x + y = -2$
   $x + 3y - z = -5$
   $2x - y + z = 6$
   $x = 1, y = -1, z = 3$

EXPLORING DATA AND STATISTICS

5.8

What you should learn

GOAL 1 Write quadratic functions given characteristics of their graphs.

GOAL 2 Use technology to find quadratic models for data, such as the fuel economy data in **Examples 3 and 4**.

Why you should learn it

▼ To solve **real-life** problems, such as determining the effect of wind on a runner's performance in **Ex. 36**.

Modeling with Quadratic Functions

GOAL 1 **WRITING QUADRATIC FUNCTIONS**

In Lesson 5.1 you learned how to graph a given quadratic function. In this lesson you will write quadratic functions when given information about their graphs.

EXAMPLE 1 *Writing a Quadratic Function in Vertex Form*

Write a quadratic function for the parabola shown.

SOLUTION

Because you are given the vertex $(h, k) = (2, -3)$, use the vertex form of the quadratic function.

$$y = a(x - h)^2 + k$$
$$y = a(x - 2)^2 - 3$$

Use the other given point, $(4, 1)$, to find a.

$1 = a(4 - 2)^2 - 3$	Substitute 4 for *x* and 1 for *y*.
$1 = 4a - 3$	Simplify coefficient of *a*.
$4 = 4a$	Add 3 to each side.
$1 = a$	Divide each side by 4.

▶ A quadratic function for the parabola is $y = (x - 2)^2 - 3$.

EXAMPLE 2 *Writing a Quadratic Function in Intercept Form*

Write a quadratic function for the parabola shown.

SOLUTION

Because you are given the *x*-intercepts $p = -2$ and $q = 3$, use the intercept form of the quadratic function.

$$y = a(x - p)(x - q)$$
$$y = a(x + 2)(x - 3)$$

Use the other given point, $(-1, 2)$, to find a.

$2 = a(-1 + 2)(-1 - 3)$	Substitute −1 for *x* and 2 for *y*.
$2 = -4a$	Simplify coefficient of *a*.
$-\frac{1}{2} = a$	Divide each side by −4.

▶ A quadratic function for the parabola is $y = -\frac{1}{2}(x + 2)(x - 3)$.

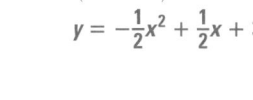

ACTIVITY

Developing
Concepts

Writing a Quadratic in Standard Form

In this activity you will write a quadratic function in standard form, $y = ax^2 + bx + c$, for the parabola in Example 2.

① The parabola passes through $(-2, 0)$, $(-1, 2)$, and $(3, 0)$. Substitute the coordinates of each point into $y = ax^2 + bx + c$ to obtain three equations in a, b, and c. For instance, the equation for $(-2, 0)$ is: $a - b + c = 2;$

$9a + 3b + c = 0$

$0 = a(-2)^2 + b(-2) + c$, or

$0 = 4a - 2b + c$

Step 2. $\left(-\frac{1}{2}, \frac{1}{2}, 3\right);$

$y = -\frac{1}{2}x^2 + \frac{1}{2}x + 3$

② Solve the system from **Step 1** to find a, b, and c. Use these values to write a quadratic function in standard form for the parabola.

③ As a check of your work, use multiplication to write the function

$y = -\frac{1}{2}(x + 2)(x - 3)$ from Example 2 in standard form. Your answer

should match the function you wrote in **Step 2**. $\quad y = -\frac{1}{2}x^2 + \frac{1}{2}x + 3$

REAL LIFE

Fuel Economy

EXAMPLE 3 *Finding a Quadratic Model for a Data Set*

A study compared the speed x (in miles per hour) and the average fuel economy y (in miles per gallon) for cars. The results are shown in the table. Find a quadratic model in standard form for the data. ▶ Source: *Transportation Energy Data Book*

Speed, x	15	20	25	30	35	40
Fuel economy, y	22.3	25.5	27.5	29.0	28.8	30.0
Speed, x	45	50	55	60	65	70
Fuel economy, y	29.9	30.2	30.4	28.8	27.4	25.3

SOLUTION

Plot the data pairs (x, y) in a coordinate plane.

Draw the parabola you think best fits the data.

Estimate the coordinates of three points on the parabola, such as $(20, 25)$, $(40, 30)$, and $(60, 28)$.

Substitute the coordinates of the points into the model $y = ax^2 + bx + c$ to obtain a system of three linear equations.

$$400a + 20b + c = 25$$
$$1600a + 40b + c = 30$$
$$3600a + 60b + c = 28$$

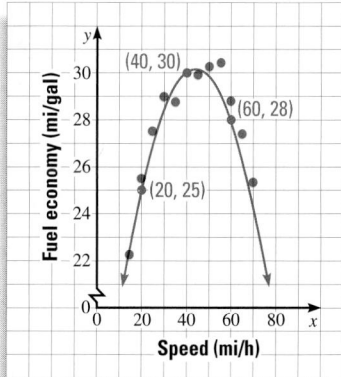

STUDENT HELP

▶ **Look Back**
For help with solving systems of three linear equations, see pp. 177, 217, and 231.

Solve the linear system. The solution is $a = -0.00875$, $b = 0.775$, and $c = 13$.

▶ A quadratic model for the data is $y = -0.00875x^2 + 0.775x + 13$.

Write a quadratic function for the parabola shown.

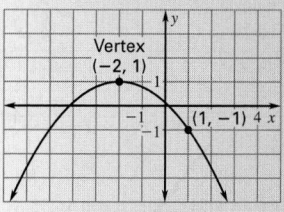

$y = -\frac{2}{9}(x + 2)^2 + 1$

EXTRA EXAMPLE 2
Write a quadratic function for the parabola shown.

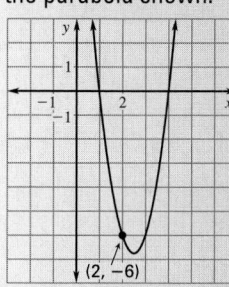

$y = 3(x - 1)(x - 4)$

EXTRA EXAMPLE 3
A group of students dropped a rubber ball and measured the height in inches of the ball for each of its successive bounces. The results are shown in the table. Find a quadratic model in standard form for the data using the first 3 points.

Bounce	1	2	3	4	5
Height	47	39	26	21	6

$y = -2.5x^2 - 0.5x + 50$

 **CHECKPOINT EXERCISES**

For use after Examples 1 and 2:

1. Write an equation for the parabola with vertex $(-1, 4)$ that goes through the point $(2, 7)$. $\quad y = \frac{1}{3}(x + 1)^2 + 4$

For use after Example 3:

2. Refer to Extra Example 3. Find a quadratic model for the data using the first, third, and fifth points. $y = 0.125x^2 - 11x + 57.875$

EXTRA EXAMPLE 4

A bank adjusts its interest rates for new certificates of deposits daily. The table shows the interest rates on the first of the month for January through May.

Month	0	1	2	3	4
Rate	3.9	4.4	4.6	3.8	3.1

a. Find the best-fitting quadratic model for the data.
$y = -0.243x^2 + 0.751x + 3.91$

b. According to the model, during which month did the certificates of deposit have the highest interest rate. What was that rate? **during February; 4.5%**

 CHECKPOINT EXERCISES

For use after Example 4:

Refer to Extra Example 3.

1. Find the best-fitting quadratic model for the data in the table.

Bounce	1	2	3	4	5
Height	47	39	26	21	6

$y = -0.429x^2 - 7.43x + 54.8$

CAREER NOTE
EXAMPLE 4 Additional information about a career in automotive design is available at **www.mcdougallittell.com.**

STUDENT HELP NOTES
→ **Keystroke Help** Keystrokes for several models of calculators are available in **blackline** format in the *Chapter 5 Resource Book*, p. 109 and at **www.mcdougallittell.com.**

FOCUS ON VOCABULARY
What is a *best-fitting quadratic model*? It is the quadratic model found using quadratic regression.

CLOSURE QUESTION
Give four ways to find a quadratic model for a set of data points.
See margin.

FOCUS ON CAREERS

 AUTOMOTIVE DESIGNER
Automotive designers help conceive of and develop new cars. They have to consider such factors as a car's appearance, performance, and fuel economy (the focus of Example 4).

CAREER LINK
www.mcdougallittell.com

STUDENT HELP

KEYSTROKE HELP
Visit our Web site www.mcdougallittell.com to see keystrokes for several models of calculators.

GOAL 2 **USING TECHNOLOGY TO FIND QUADRATIC MODELS**

In Chapter 2 you used a graphing calculator to perform linear regression on a data set in order to find a linear model for the data. A graphing calculator can also be used to perform *quadratic regression*. Quadratic regression produces a more accurate quadratic model than the procedure in Example 3 because it uses *all* the data points. The model given by quadratic regression is called the **best-fitting quadratic model**.

EXAMPLE 4 *Using Quadratic Regression to Find a Model*

FUEL ECONOMY Use the fuel economy data given in Example 3 to complete parts (a) and (b).

a. Use a graphing calculator to find the best-fitting quadratic model for the data.

b. Find the speed that maximizes a car's fuel economy.

SOLUTION

a. Enter the data into two lists of a graphing calculator.

Make a scatter plot of the data. Note that the points show a parabolic trend.

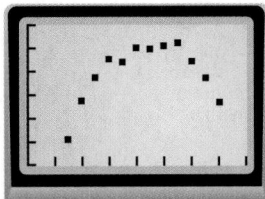

Use the quadratic regression feature to find the best-fitting quadratic model for the data.

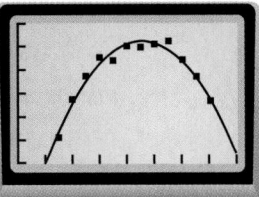

Check how well the model fits the data by graphing the model and the data in the same viewing window.

▶ The best-fitting quadratic model is $y = -0.00820x^2 + 0.746x + 13.5$.

b. You can find the speed that maximizes fuel economy by using the *Maximum* feature of a graphing calculator, as shown at the right.

You can also find the speed algebraically using the formula for the *x*-coordinate of a parabola's vertex from Lesson 5.1:

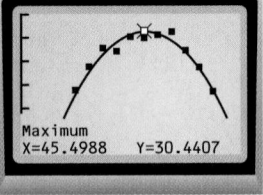

$$x = -\frac{b}{2a} = -\frac{0.746}{2(-0.00820)} \approx 45$$

▶ The speed that maximizes a car's fuel economy is about 45 miles per hour.

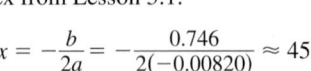
Closure Question *Sample answer:*
Use the vertex and a point in the vertex form of the equation; use the intercepts and a point in the intercept form; use three points and solve the system of equations to find the coefficients; or use quadratic regression on a calculator.

GUIDED PRACTICE

Vocabulary Check ✓

1. Complete this statement: When you perform quadratic regression on a set of data, the quadratic model you obtain is called the _?_.   **best-fitting quadratic model**

Concept Check ✓

2. How many points are needed to determine a parabola if one of the points is the vertex? if none of the points is the vertex?   **2; 3**

Skill Check ✓

Write a quadratic function in the specified form for the parabola shown.

3. vertex form

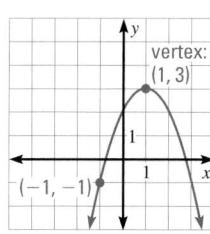

$y = -1(x - 1)^2 + 3$

4. intercept form

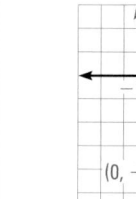

$y = 2(x + 1)(x - 2)$

5. standard form

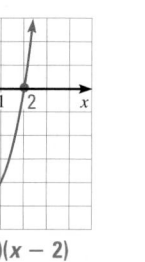

$y = x^2 + 3x - 2$

6. 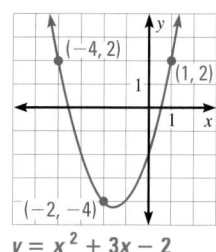 **REAL ESTATE** The table shows the average sale price p of a house in Suffolk County, Massachusetts, for various years t since 1988. Use a system of equations to write a quadratic model for the data. Check your model by performing quadratic regression on a graphing calculator.
$p = 1.83t^2 - 19.6t + 173$

Years since 1988, t	0	2	4	6	8	10
Average sale price (thousands of dollars), p	165	154.5	124.5	115	128	165

🌐 DATA UPDATE of *Boston Globe* data at www.mcdougallittell.com

PRACTICE AND APPLICATIONS

STUDENT HELP

▶ **Extra Practice**
to help you master skills is on p. 947.

11. $y = \frac{1}{3}(x + 4)^2 + 6$

15. $y = -\frac{3}{2}(x + 6)^2 - 7$

STUDENT HELP

▶ **HOMEWORK HELP**
Example 1: Exs. 7–15, 34
Example 2: Exs. 16–24, 35
Example 3: Exs. 25–33, 36–38
Example 4: Exs. 37, 38

WRITING THE VERTEX FORM Write a quadratic function in vertex form for the parabola shown.

$y = (x - 2)^2 - 2$

7.

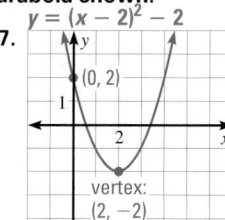

$y = -2(x + 1)^2 + 4$

8.

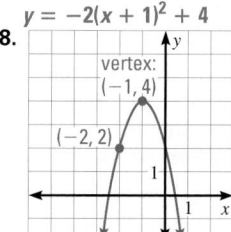

$y = -\frac{3}{4}(x - 1)^2$

9.

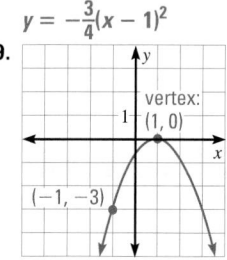

WRITING THE VERTEX FORM Write a quadratic function in vertex form whose graph has the given vertex and passes through the given point.

$y = (x - 2)^2 - 1$ **See margin.** $y = -\frac{1}{2}(x - 4)^2 + 5$

10. vertex: $(2, -1)$
    point: $(4, 3)$

11. vertex: $(-4, 6)$
    point: $(-1, 9)$

12. vertex: $(4, 5)$
    point: $(8, -3)$

13. vertex: $(0, 0)$ $y = -3x^2$
    point: $(-2, -12)$

14. vertex: $(1, -10)$
    point: $(-3, 54)$
    $y = 4(x - 1)^2 - 10$

15. vertex: $(-6, -7)$
    point: $(0, -61)$
    See margin.

5.8 *Modeling with Quadratic Functions*   **309**

---

3 APPLY

ASSIGNMENT GUIDE

BASIC
Day 1: pp. 309–312 Exs. 7–12, 16–21, 25–28, 34, 36, 39, 41–47 odd, Quiz 3 Exs. 1–10

AVERAGE
Day 1: pp. 309–312 Exs. 8–30 even, 34–36, 39, 42–48 even, Quiz 3 Exs. 1–10

ADVANCED
Day 1: pp. 309–312 Exs. 8–38 even, 39, 40, 42–48 even, Quiz 3 Exs. 1–10

BLOCK SCHEDULE
pp. 309–312 Exs. 8–30 even, 34–36, 39, 42–48 even, Quiz 3 Exs. 1–10 (with 5.7)

EXERCISE LEVELS
Level A: *Easier*
7–24, 34–36
Level B: *More Difficult*
25–33, 37–39
Level C: *Most Difficult*
40

✓ **HOMEWORK CHECK**
To quickly check student understanding of key concepts, go over the following exercises:
Exs. 8, 16, 20, 26, 28. See also the Daily Homework Quiz:

• Blackline Master (*Chapter 6 Resource Book,* p. 11)
• 📖 Transparency (p. 41)

DATA UPDATE
Updated data for Ex. 6 is available at **www.mcdougallittell.com**.

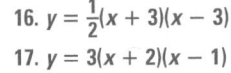

Homework Help Students can find help for Exs. 34 and 35 at **www.mcdougallittell.com**. The information can be printed out for students who don't have access to the Internet.

APPLICATION NOTE
EXERCISE 38 Additional information about Mark McGwire is available at **www.mcdougallittell.com**.

16. $y = \frac{1}{2}(x + 3)(x - 3)$

17. $y = 3(x + 2)(x - 1)$

18. $y = -1(x - 0)(x - 4)$

19. $y = -1(x - 1)(x - 4)$

20. $y = \frac{2}{3}(x + 2)(x - 2)$

21. $y = 2(x + 1)(x - 6)$

22. $y = -5(x + 10)(x + 8)$

23. $y = \frac{7}{5}(x - 3)(x - 9)$

24. $y = -3(x + 5)(x - 0)$

25. $y = -x^2 + x + 4$

26. $y = 2x^2 - 15x + 29$

27. $y = -\frac{3}{4}x^2 - \frac{11}{4}x + 1$

28. $y = x^2 - x + 3$

29. $y = -x^2 + 5x - 2$

30. $y = 3x^2 + 7x + 1$

31. $y = -2x^2 - 4x + 9$

32. $y = -\frac{1}{4}x^2 + x + \frac{5}{4}$

33. $y = \frac{5}{2}x^2 + 6x - 8$

WRITING THE INTERCEPT FORM Write a quadratic function in intercept form for the parabola shown.

16.
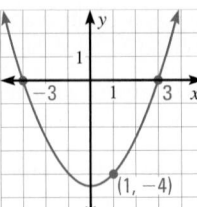

17.
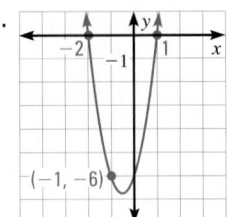

18.
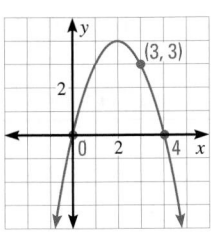

WRITING THE INTERCEPT FORM Write a quadratic function in intercept form whose graph has the given x-intercepts and passes through the given point.
19–24. See margin.

19. x-intercepts: 1, 4
point: $(3, 2)$

20. x-intercepts: -2, 2
point: $(-4, 8)$

21. x-intercepts: -1, 6
point: $(1, -20)$

22. x-intercepts: -10, -8
point: $(-7, -15)$

23. x-intercepts: 3, 9
point: $(14, 77)$

24. x-intercepts: -5, 0
point: $(-3, 18)$

WRITING THE STANDARD FORM Write a quadratic function in standard form for the parabola shown.   25–27. See margin.

25.
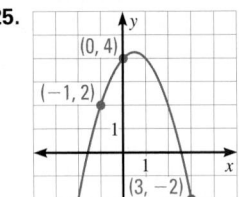

26.
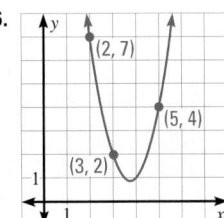

27.
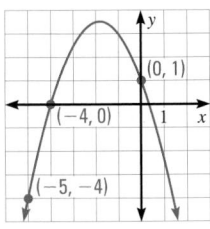

WRITING THE STANDARD FORM Write a quadratic function in standard form whose graph passes through the given points.   28–33. See margin.

28. $(-1, 5), (0, 3), (3, 9)$

29. $(1, 2), (3, 4), (6, -8)$

30. $(-2, -1), (1, 11), (2, 27)$

31. $(-4, -7), (-3, 3), (3, -21)$

32. $(-3, -4), (-1, 0), (9, -10)$

33. $(-6, 46), (2, 14), (4, 56)$

STUDENT HELP

HOMEWORK HELP
Visit our Web site www.mcdougallittell.com for help with problem solving in Exs. 34 and 35.

34. 🌍 **BOTANY** *Amaranth* is a type of vegetable commonly grown in Asia, West Africa, and the Caribbean. When amaranth plants are grown in rows, the height that the plants attain is a quadratic function of the spacing between plants within a row. According to one study, the minimum height of the plants, about 16 cm, occurred when the plants were spaced about 27 cm apart. The study also found that the plants grew to about 20 cm when spaced about 40 cm apart. Write a quadratic model giving the plant height h as a function of the spacing s.
▶ Source: Center for New Crops and Plant Products, Perdue University $h = \frac{4}{169}(s - 27)^2 + 16$

35. 🌍 **TRANSPORTATION** The surfaces of some roads are shaped like parabolas to allow rain to run off to either side. (This is also true of football fields; see Exercise 52 on page 254.) Write a quadratic model for the surface of the road shown.
▶ Source: Massachusetts Highway Department
$y = -0.00168(x - 0)(x - 24)$

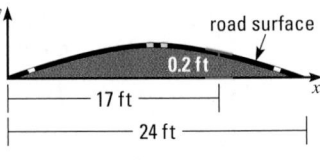
Not drawn to scale

ADDITIONAL PRACTICE AND RETEACHING

For Lesson 5.8:
• Practice Levels A, B, and C (*Chapter 5 Resource Book,* p. 110)
• Reteaching with Practice (*Chapter 5 Resource Book,* p. 113)
• ⊞ See Lesson 5.8 of the *Personal Student Tutor*

For more Mixed Review:
• ⊞ Search the *Test and Practice Generator* for key words or specific lessons.

310

FOCUS ON
PEOPLE

MARK MCGWIRE
hit 70 home runs during the 1998 Major League Baseball season, breaking Roger Maris's record of 61. McGwire's longest home run traveled 545 ft (166 m).

APPLICATION LINK
www.mcdougallittell.com

37. $s = -0.0807p^2 + 55.2p + 330$;
$k = -0.0000609p^2 + 0.626p + 125$

38. $d = -0.0771A^2 + 6.5803A + 2.4614$;
$d = -0.0738A^2 + 6.4304A + 0.6928$;
$d = -0.0700A^2 + 6.2284A - 0.2623$

Test Preparation

39a. 1.35, 1.68, 2.03, 2.37, 2.725, 3.07, 3.4; no, the ratios keep increasing as the diameter increases.

39b. 0.0675, 0.0672, 0.0678, 0.0678, 0.0681, 0.0681, 0.068; the ratios are approximately equal.

36. **RUNNING** The table shows how wind affects a runner's performance in the 200 meter dash. Positive wind speeds correspond to tailwinds, and negative wind speeds correspond to headwinds. Positive changes in finishing time mean worsened performance, and negative changes mean improved performance. Use a system of equations to write a quadratic model for the change t in finishing time as a function of the wind speed s. ▸ Source: *The Physics of Sports*

$t = 0.0119s^2 - 0.309s - 0.0005$

Wind speed (m/sec), s	−6	−4	−2	0	2	4	6
Change in finishing time (sec), t	2.28	1.42	0.67	0	−0.57	−1.05	−1.42

37. **AGRICULTURE** Researchers compared protein intake to average shoulder and kidney weight for a group of pigs. The results are shown in the table. Use systems of equations to write quadratic models for the shoulder weight s and kidney weight k as a function of the protein intake p. Check your models using the quadratic regression feature of a graphing calculator.
▸ Source: *Livestock Research for Rural Development*   **See margin.**

Protein intake (g/day), p	195	238	297	341	401	427
Shoulder weight (g), s	8130	8740	9680	9690	9810	8990
Kidney weight (g), k	239	287	288	334	379	373

38. **BASEBALL** The table shows the distance (in meters) traveled by a baseball hit at various angles and with different types of spin. (In each case the initial speed of the ball off the bat is assumed to be 40 m/sec.) Use systems of equations to write three quadratic models—one for each type of spin—that give the distance d as a function of the angle A. Check your models using the quadratic regression feature of a graphing calculator. ▸ Source: *The Physics of Sports*

Angle	10°	15°	30°	36°	42°	45°	48°	54°	60°
Distance (backspin)	61.2	83.0	130.4	139.4	143.2	142.7	140.7	132.8	119.7
Distance (no spin)	58.3	79.7	126.9	136.6	140.6	140.9	139.3	132.5	120.5
Distance (topspin)	56.1	76.3	122.8	133.2	138.3	139.0	137.8	132.1	120.9

39. **MULTI-STEP PROBLEM** The table shows the time t it takes to boil a potato whose smallest diameter (that is, whose shortest distance through the center) is d. ▸ Source: Dr. Peter Barham, University of Bristol

Diameter (mm), d	20	25	30	35	40	45	50
Boiling time (min), t	27	42	61	83	109	138	170

a. Find the ratios $\frac{t}{d}$. Does boiling time vary directly with diameter? Explain.

b. Find the ratios $\frac{t}{d^2}$. What do you notice?

c. Use the result of part (b) to write a quadratic model for t as a function of d. Find the time needed to boil a potato whose smallest diameter is 55 mm.
$t \approx 0.0680d^2$; about 206 min

5.8 *Modeling with Quadratic Functions*   311

DAILY HOMEWORK QUIZ

Transparency Available

1. Write a quadratic function in vertex form for the parabola shown.

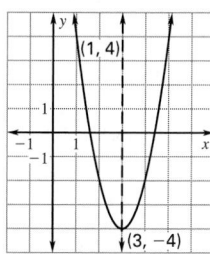

(1, 4)
(3, −4)

$y = 2(x - 3)^2 - 4$

2. Write a quadratic function in intercept form for the parabola shown.

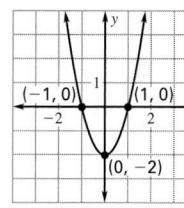

(−1, 0)   (1, 0)
(0, −2)

$y = 2(x - 1)(x + 1)$

3. Write a quadratic function in standard form for the parabola whose graph passes through (2, −2), (3, 4), (0, −2).
$y = 2x^2 - 4x - 2$

EXTRA CHALLENGE NOTE

Challenge problems for Lesson 5.8 are available in **blackline** format in the *Chapter 5 Resource Book,* p. 117 and at **www.mcdougallittell.com.**

ADDITIONAL TEST PREPARATION

1. **WRITING** Describe the steps for finding a quadratic model for a data set without using a graphing calculator. Show how to use these steps to find a quadratic model for the three points (5, 33), (10, 53), (35, 3). **See steps listed in Example 3.** $y = -0.2x^2 + 7x + 3$

1.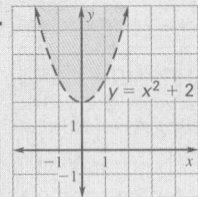

$y = x^2 + 2$

2.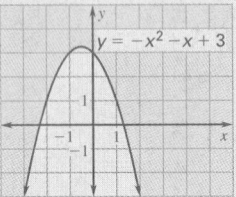

$y = -x^2 - x + 3$

3.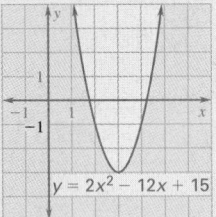

$y = 2x^2 - 12x + 15$

4.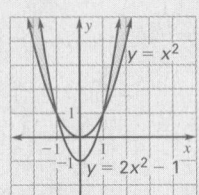

$y = x^2$

$y = 2x^2 - 1$

5.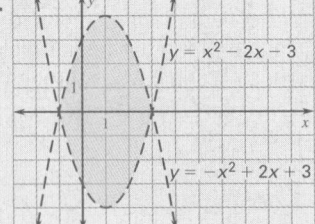

$y = x^2 - 2x - 3$

$y = -x^2 + 2x + 3$

6.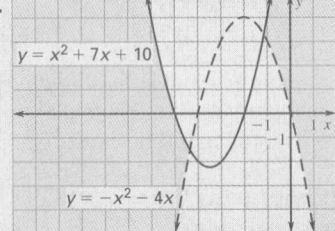

$y = x^2 + 7x + 10$

$y = -x^2 - 4x$

★ **Challenge**

40. GEOMETRY CONNECTION Let R be the maximum number of regions into which a circle can be divided using n chords. For example, the diagram shows that $R = 4$ when $n = 2$. Copy and complete the table. Then write a quadratic model giving R as a function of n.

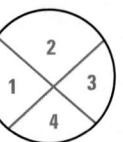

EXTRA CHALLENGE

www.mcdougallittell.com

n	0	1	2	3	4	5	6
R	?	?	4	?	?	?	?

$R = 1, 2, 4, 7, 11, 16, 22;\ R = 0.5n^2 + 0.5n + 1$

MIXED REVIEW

EVALUATING EXPRESSIONS Evaluate the expression for the given value of the variable. **(Review 1.2 for 6.1)**

41. $x^2 - 4$ when $x = 3$ 5

42. x^5 when $x = 2$ 32

43. $3u^3 + 10$ when $u = -4$ −182

44. $-v^4 + 2v + 7$ when $v = -1$ 4

SOLVING SYSTEMS Solve the system using either the substitution method or the linear combination method. **(Review 3.2)**

45. $x - y = 4$ (3, −1)
$x + y = 2$

46. $2x - y = 0$ (1, 2)
$5x + 3y = 11$

47. $3x + 2y = -2$ (−4, 5)
$4x + 7y = 19$

48. 🌐 **HEALTH** You belong to a health maintenance organization (HMO). Each year, you pay the HMO an insurance premium of $1800. In addition, you pay $15 for each visit to your doctor's office and $10 for each prescription. Write an equation for the annual cost C of your health plan as a function of your number v of office visits and number p of prescriptions. **(Review 3.5)**
$C = 1800 + 15v + 10p$

QUIZ 3

Self-Test for Lessons 5.7 and 5.8

Graph the inequality. (Lesson 5.7)   1–3. See margin.

1. $y > x^2 + 2$

2. $y \geq -x^2 - x + 3$

3. $y \leq 2x^2 - 12x + 15$

Graph the system of inequalities. (Lesson 5.7)   4–6. See margin.

4. $y \geq x^2$
$y \leq 2x^2 - 1$

5. $y > x^2 - 2x - 3$
$y < -x^2 + 2x + 3$

6. $y > -x^2 - 4x$
$y \leq x^2 + 7x + 10$

Write a quadratic function in the specified form whose graph has the given characteristics. (Lesson 5.8)

$y = 2(x - 5)^2 - 2$

$y = -1(x + 3)(x - 1)$

$y = \frac{3}{4}x^2 + x$

7. vertex form
vertex: $(5, -2)$
point on graph: $(4, 0)$

8. intercept form
x-intercepts: $-3, 1$
point on graph: $(2, -5)$

9. standard form
points on graph:
$(-4, 8), (-2, 1), (2, 5)$

10. 🌐 **COMPUTERS** Using an algorithm called *insertion sort*, a common minicomputer can sort N numbers from least to greatest in t milliseconds where $t = 0.00339N^2 + 0.00143N - 5.95$. How many numbers can the minicomputer sort in less than 1 second (1000 milliseconds)? Write your answer as an inequality. **(Lesson 5.7)** $0.00339N^2 + 0.00143N - 5.95 < 1000;\ 0 < N < 544$

Chapter Summary

WHAT did you learn?

Graph quadratic functions. **(5.1)**

Write quadratic functions in standard, intercept, and vertex forms. **(5.1, 5.2, 5.5)**

Find zeros of quadratic functions. **(5.2)**

Solve quadratic equations.
- by factoring **(5.2)**
- by finding square roots **(5.3)**
- by completing the square **(5.5)**
- by using the quadratic formula **(5.6)**

Perform operations with complex numbers. **(5.4)**

Find the discriminant of a quadratic equation. **(5.6)**

Graph quadratic inequalities in two variables. **(5.7)**

Solve quadratic inequalities in one variable. **(5.7)**

Find quadratic models for data. **(5.8)**

WHY did you learn it?

Model the suspension cables on the Golden Gate Bridge. **(p. 252)**

Find the amount of fertilizer that maximizes the sugar yield from sugarbeets. **(p. 285)**

Determine what subscription price to charge for a Web site in order to maximize revenue. **(p. 259)**

Calculate dimensions for a mural. **(p. 262)**
Find a falling rock's time in the air. **(p. 268)**
Tell how a firefighter should position a hose. **(p. 288)**
Find the speed and duration of a thrill ride. **(p. 297)**

Determine whether a complex number belongs to the Mandelbrot set. **(p. 276)**

Identify the number and type of solutions of a quadratic equation. **(p. 293)**

Calculate the weight that a rope can support. **(p. 304)**

Relate a driver's age and reaction time. **(p. 302)**

Determine the effect of wind on a runner's performance. **(p. 311)**

How does Chapter 5 fit into the BIGGER PICTURE of algebra?

In Chapter 5 you saw the relationship between the *solutions* of the quadratic equation $ax^2 + bx + c = 0$, the *zeros* of the quadratic function $y = ax^2 + bx + c$, and the *x-intercepts* of this function's graph. You'll continue to see this relationship with other types of functions. Also, the graph of a quadratic function—a parabola—is one of the four conic sections. You'll study all the conic sections in Chapter 10.

STUDY STRATEGY

How did you troubleshoot?

Here is an example of a trouble spot identified and eliminated, following the **Study Strategy** on page 248.

Troubleshoot

Trouble spot: Changing a quadratic function from standard form to vertex form by completing the square.

How to eliminate: Remember to add the same constant to *both* sides of the equation for the function.

Example:
$$y = x^2 + 10x - 3$$
$$y + 25 = (x^2 + 10x + 25) - 3$$
$$y + 25 = (x + 5)^2 - 3$$
$$y = (x + 5)^2 - 28$$

313

ADDITIONAL RESOURCES

The following resources are available to help review the material in this chapter.

- Chapter Review Games and Activities (*Chapter 5 Resource Book*, p. 118)
- *Instant Replay: Video Review Games*
- Personal Student Tutor
- Cumulative Review, Chs. 1–5 (*Chapter 5 Resource Book*, p. 130)

1.

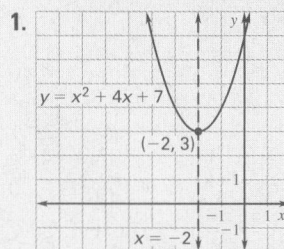

2.

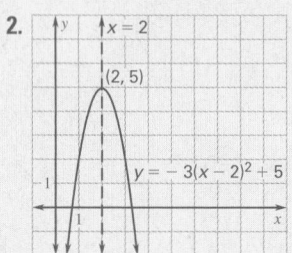

3.

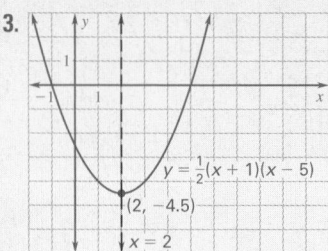

Chapter Review

VOCABULARY

- quadratic function, p. 249
- parabola, p. 249
- vertex of a parabola, p. 249
- axis of symmetry, p. 249
- standard form of a quadratic function, p. 250
- vertex form of a quadratic function, p. 250
- intercept form of a quadratic function, p. 250
- binomial, p. 256

- trinomial, p. 256
- factoring, p. 256
- monomial, p. 257
- quadratic equation, p. 257
- standard form of a quadratic equation, p. 257
- zero product property, p. 257
- zero of a function, p. 259
- square root, p. 264
- radical sign, p. 264

- radicand, p. 264
- radical, p. 264
- rationalizing the denominator, p. 265
- imaginary unit i, p. 272
- complex number, p. 272
- standard form of a complex number, p. 272
- imaginary number, p. 272
- pure imaginary number, p. 272

- complex plane, p. 273
- complex conjugates, p. 274
- absolute value of a complex number, p. 275
- completing the square, p. 282
- quadratic formula, p. 291
- discriminant, p. 293
- quadratic inequality, pp. 299, 301
- best-fitting quadratic model, p. 308

5.1   GRAPHING QUADRATIC FUNCTIONS

Examples on pp. 249–252

> **EXAMPLE** You can graph a quadratic function given in standard form, vertex form, or intercept form. For instance, the same function is given below in each of these forms, and its graph is shown.
>
> **Standard form:** $y = x^2 + 2x - 3$;
>
> $\quad$ axis of symmetry: $x = -\dfrac{b}{2a} = -\dfrac{2}{2(1)} = -1$
>
> **Vertex form:** $y = (x + 1)^2 - 4$; vertex: $(-1, -4)$
>
> **Intercept form:** $y = (x + 3)(x - 1)$; x-intercepts: $-3, 1$

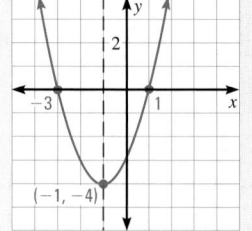

Graph the quadratic function. **1–3. See margin.**

1. $y = x^2 + 4x + 7$ $\qquad$ **2.** $y = -3(x - 2)^2 + 5$ $\qquad$ **3.** $y = \frac{1}{2}(x + 1)(x - 5)$

5.2–5.3   SOLVING BY FACTORING AND BY FINDING SQUARE ROOTS

Examples on pp. 256–259, 264–266

> **EXAMPLES** You can use factoring or square roots to solve quadratic equations.
>
> **Solving by factoring:**
>
> $x^2 - 4x - 21 = 0$
>
> $(x + 3)(x - 7) = 0$
>
> $x + 3 = 0 \quad$ or $\quad x - 7 = 0$
>
> $\qquad x = -3 \quad$ or $\qquad x = 7$
>
> **Solving by finding square roots:**
>
> $4x^2 - 7 = 65$
>
> $4x^2 = 72$
>
> $x^2 = 18$
>
> $x = \pm\sqrt{18} = \pm 3\sqrt{2}$

Solve the quadratic equation.

4. $x^2 + 11x + 24 = 0$ $-8, -3$ **5.** $x^2 - 8x + 16 = 0$ 4   **6.** $2x^2 + 3x + 1 = 0$ $-\frac{1}{2}, -1$

7. $3u^2 = -4u + 15$ $-3, \frac{5}{3}$ **8.** $25v^2 - 30v = -9$ $\frac{3}{5}$ **9.** $2x^2 = 200$ $-10, 10$

10. $5x^2 - 2 = 13$ $-\sqrt{3}, \sqrt{3}$ **11.** $4(t + 6)^2 = 160$ **12.** $-(k - 1)^2 + 7 = -43$
                                     $-6 - 2\sqrt{10}, -6 + 2\sqrt{10}$ $1 - 5\sqrt{2}, 1 + 5\sqrt{2}$

5.4   COMPLEX NUMBERS

Examples on pp. 272–276

> **EXAMPLES** You can add, subtract, multiply, and divide complex numbers. You can also find the absolute value of a complex number.
>
> **Addition:** $(1 + 8i) + (2 - 3i) = (1 + 2) + (8 - 3)i = 3 + 5i$
>
> **Subtraction:** $(1 + 8i) - (2 - 3i) = (1 - 2) + (8 + 3)i = -1 + 11i$
>
> **Multiplication:** $(1 + 8i)(2 - 3i) = 2 - 3i + 16i - 24i^2 = 2 + 13i - 24(-1) = 26 + 13i$
>
> **Division:** $\dfrac{1 + 8i}{2 - 3i} = \dfrac{1 + 8i}{2 - 3i} \cdot \dfrac{2 + 3i}{2 + 3i} = \dfrac{-22 + 19i}{13} = -\dfrac{22}{13} + \dfrac{19}{13}i$
>
> **Absolute value:** $\left| 1 + 8i \right| = \sqrt{1^2 + 8^2} = \sqrt{65}$

In Exercises 13–16, write the expression as a complex number in standard form.

13. $(7 - 4i) + (-2 + 5i)$ $5 + i$ **14.** $(2 + 11i) - (6 - i)$ $-4 + 12i$

15. $(3 + 10i)(4 - 9i)$ $102 + 13i$ **16.** $\dfrac{8 + i}{1 - 2i}$ $\frac{6}{5} + \frac{17}{5}i$

17. Find the absolute value of $6 + 9i$. $3\sqrt{13}$

5.5   COMPLETING THE SQUARE

Examples on pp. 282–285

> **EXAMPLES** You can use completing the square to solve quadratic equations and change quadratic functions from standard form to vertex form.
>
> **Solving an equation:**
>
> $x^2 + 6x + 13 = 0$
>
> $x^2 + 6x = -13$
>
> $x^2 + 6x + 9 = -13 + 9$
>
> $(x + 3)^2 = -4$
>
> $x + 3 = \pm\sqrt{-4}$
>
> $x = -3 \pm 2i$
>
> **Writing a function in vertex form:**
>
> $y = x^2 + 6x + 13$
>
> $y + \underline{\ ?\ } = (x^2 + 6x + \underline{\ ?\ }) + 13$
>
> $y + 9 = (x^2 + 6x + 9) + 13$
>
> $y + 9 = (x + 3)^2 + 13$
>
> $y = (x + 3)^2 + 4$
>
> Note that the vertex is $(-3, 4)$.

Solve the quadratic equation by completing the square.

18. $x^2 + 4x = 3$ **19.** $x^2 - 10x + 26 = 0$ **20.** $2w^2 + w - 7 = 0$
     $-2 - \sqrt{7}, -2 + \sqrt{7}$ $5 + i, 5 - i$ $-\frac{1}{4} + \frac{\sqrt{57}}{4}, -\frac{1}{4} - \frac{\sqrt{57}}{4}$

Write the quadratic function in vertex form and identify the vertex.

21. $y = x^2 - 8x + 17$ **22.** $y = -x^2 - 2x - 6$ **23.** $y = 4x^2 + 16x + 23$
     $y = (x - 4)^2 + 1; (4, 1)$ $y = -(x + 1)^2 - 5; (-1, -5)$ $y = 4(x + 2)^2 + 7; (-2, 7)$

Chapter Review   **315**

27.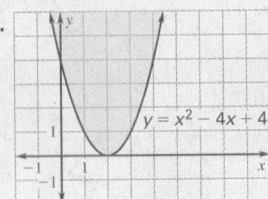

$y = x^2 - 4x + 4$

28.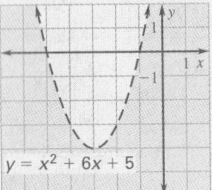

$y = x^2 + 6x + 5$

29.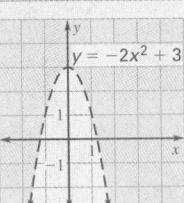

$y = -2x^2 + 3$

1.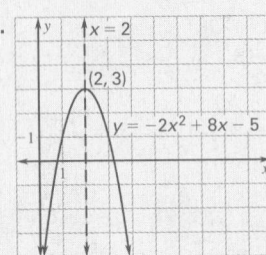

$x = 2$

$(2, 3)$

$y = -2x^2 + 8x - 5$

2.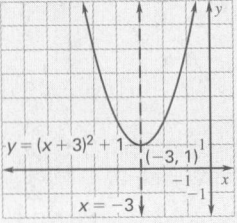

$y = (x + 3)^2 + 1$ $(-3, 1)$

$x = -3$

3.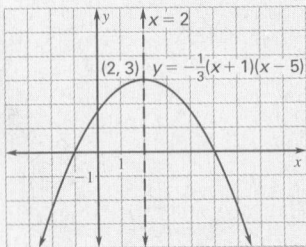

$x = 2$

$(2, 3)$ $y = -\frac{1}{3}(x + 1)(x - 5)$

Examples on pp. 291–294

5.6 **THE QUADRATIC FORMULA AND THE DISCRIMINANT**

EXAMPLE You can use the quadratic formula to solve any quadratic equation.

$$3x^2 - 5x = -1$$

$$3x^2 - 5x + 1 = 0$$

$$x = \frac{-b \pm \sqrt{b^2 - 4ac}}{2a} = \frac{5 \pm \sqrt{(-5)^2 - 4(3)(1)}}{2(3)} = \frac{5 \pm \sqrt{13}}{6}$$

Use the quadratic formula to solve the equation.

24. $x^2 - 8x + 5 = 0$
$4 - \sqrt{11}, 4 + \sqrt{11}$

25. $9x^2 = 1 - 7x$
$-\frac{7}{18} + \frac{\sqrt{85}}{18}, -\frac{7}{18} - \frac{\sqrt{85}}{18}$

26. $5v^2 + 6v + 7 = v^2 - 4v$
$-\frac{5}{4} + \frac{\sqrt{3}}{4}i, -\frac{5}{4} - \frac{\sqrt{3}}{4}i$

Examples on pp. 299–302

5.7 **GRAPHING AND SOLVING QUADRATIC INEQUALITIES**

EXAMPLES You can graph a quadratic inequality in two variables and solve a quadratic inequality in one variable.

Graphing an inequality in two variables: To graph $y < -x^2 + 4$, draw the dashed parabola $y = -x^2 + 4$. Test a point inside the parabola, such as $(0, 0)$. Since $(0, 0)$ is a solution of the inequality, shade the region inside the parabola.

Solving an inequality in one variable: To solve $-x^2 + 4 < 0$, graph $y = -x^2 + 4$ and identify the x-values where the graph lies below the x-axis. Or, solve $-x^2 + 4 = 0$ to find the critical x-values -2 and 2, then test an x-value in each interval determined by -2 and 2 to find the solution. The solution is $x < -2$ or $x > 2$.

$y < -x^2 + 4$

$(0, 0)$

$-x^2 + 4 < 0$

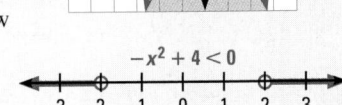

Graph the quadratic inequality. 27–29. See margin.

27. $y \geq x^2 - 4x + 4$

28. $y < x^2 + 6x + 5$

29. $y > -2x^2 + 3$

Solve the quadratic inequality.
$x \leq \frac{-7 - \sqrt{33}}{4}$ or $x \geq \frac{-7 + \sqrt{33}}{4}$

30. $x^2 - 3x - 4 \leq 0$ $-1 \leq x \leq 4$ **31.** $2x^2 + 7x + 2 \geq 0$

32. $9x^2 > 49$ $x < -\frac{7}{3}$ or $x > \frac{7}{3}$

Examples on pp. 306–308

5.8 **MODELING WITH QUADRATIC FUNCTIONS**

EXAMPLE You can write a quadratic function given characteristics of its graph.

To find a function for the parabola with vertex $(1, -3)$ and passing through $(0, -1)$, use the vertex form $y = a(x - h)^2 + k$ with $(h, k) = (1, -3)$ to write $y = a(x - 1)^2 - 3$. Use the point $(0, -1)$ to find a: $-1 = a(0 - 1)^2 - 3$, so $-1 = a - 3$, and therefore $a = 2$. The function is $y = 2(x - 1)^2 - 3$.

Write a quadratic function whose graph has the given characteristics.

$y = (x - 6)^2 + 1$
33. vertex: $(6, 1)$
point on graph: $(4, 5)$

$y = -2(x + 4)(x - 3)$
34. x-intercepts: $-4, 3$
point on graph: $(1, 20)$

$y = 0.5x^2 + 1.5x - 4$
35. points on graph:
$(-5, 1), (-4, -2), (3, 5)$

316 Chapter 5 *Quadratic Functions*

Chapter Test

ADDITIONAL RESOURCES
- *Chapter 5 Resource Book*
  Chapter Test (3 levels) (p. 119)
  SAT/ACT Chapter Test (p. 125)
  Alternative Assessment (p. 126)
- *Test and Practice Generator*

Graph the quadratic function. 1–3. See margin.

1. $y = -2x^2 + 8x - 5$ **2.** $y = (x + 3)^2 + 1$ **3.** $y = -\frac{1}{3}(x + 1)(x - 5)$

4. Write $y = 4(x - 3)^2 - 7$ in standard form. $y = 4x^2 - 24x + 29$

Factor the expression.

5. $x^2 - x - 20$ $(x - 5)(x + 4)$ **6.** $9x^2 + 6x + 1$ $(3x + 1)^2$ **7.** $3u^2 - 108$ $3(u + 6)(u - 6)$

8. Write $y = x^2 - 10x + 16$ in intercept form and give the function's zeros. $y = (x - 8)(x - 2)$; 2, 8

9. Simplify the radical expressions $\sqrt{500}$ and $\sqrt{\frac{8}{3}}$. $10\sqrt{5}$; $\frac{2\sqrt{6}}{3}$

10. Plot these numbers in the same complex plane: $4 + 2i$, $-5 + i$, and $-3i$.   See margin.

Write the expression as a complex number in standard form.

11. $(3 + i) + (1 - 5i)$ $4 - 4i$ **12.** $(-4 + 2i) - (7 - 3i)$ $-11 + 5i$ **13.** $(8 + i)(6 + 2i)$ $46 + 22i$ **14.** $\frac{9 + 2i}{1 - 4i}$ $\frac{1}{17} + \frac{38}{17}i$

15. Is $c = -0.5i$ in the Mandelbrot set? Use absolute value to justify your answer.   yes; all absolute values are less than $N = 1$.

Find the value of c that makes the expression a perfect square trinomial. Then write the expression as the square of a binomial.

16. $x^2 - 4x + c$ 4; $(x - 2)^2$ **17.** $x^2 + 11x + c$ $\frac{121}{4}$; $\left(x + \frac{11}{2}\right)^2$ **18.** $x^2 - 0.6x + c$ 0.09; $(x - 0.3)^2$

19. Write $y = x^2 + 18x - 4$ in vertex form and identify the vertex. $y = (x + 9)^2 - 85$; $(-9, -85)$

Solve the quadratic equation using any appropriate method.

20. $7x^2 - 3 = 11$ $-\sqrt{2}, \sqrt{2}$ **21.** $5x^2 - 60x + 180 = 0$ 6   **22.** $4x^2 + 28x - 15 = 0$ $-\frac{15}{2}, \frac{1}{2}$

23. $m^2 + 8m = -3$ $-4 - \sqrt{13}, -4 + \sqrt{13}$ **24.** $3(p - 9)^2 = 81$ $9 - 3\sqrt{3}, 9 + 3\sqrt{3}$ **25.** $6t^2 - 2t + 2 = 4t^2 + t$ $\frac{3}{4} + \frac{\sqrt{7}}{4}i, \frac{3}{4} - \frac{\sqrt{7}}{4}i$

26. Find the discriminant of $7x^2 - x + 10 = 0$. What does the discriminant tell you about the number and type of solutions of the equation?   two imaginary

Graph the quadratic inequality.   27–29. See margin.

27. $y \geq x^2 + 1$ **28.** $y \leq -x^2 + 4x + 2$ **29.** $y < 2x^2 + 12x + 15$

Solve the quadratic inequality.

30. $-x^2 + x + 6 \geq 0$ $-2 \leq x \leq 3$ **31.** $2x^2 - 9 > 23$ $x < -4$ or $x > 4$ **32.** $x^2 - 7x < -4$ $\frac{7 - \sqrt{33}}{2} < x < \frac{7 + \sqrt{33}}{2}$

Write a quadratic function whose graph has the given characteristics.

33. vertex: $(-3, 2)$ $y = -5(x + 3)^2 + 2$ **34.** x-intercepts: 1, 8 $y = \frac{1}{3}(x - 1)(x - 8)$ **35.** points on graph: $y = x^2 - 8x + 14$
point on graph: $(-1, -18)$ point on graph: $(2, -2)$ $(1, 7), (4, -2), (5, -1)$

36. WATERFALLS Niagara Falls in New York is 167 feet high. How long does it take for water to fall from the top to the bottom of Niagara Falls?   about 3.23 seconds

37. INSURANCE An insurance company charges a 35-year-old nonsmoker an annual premium of \$118 for a \$100,000 term life insurance policy. The premiums for 45-year-old and 55-year-old nonsmokers are \$218 and \$563, respectively. Write a quadratic model for the premium p as a function of age a. $p = 1.225a^2 - 88a + 1697.375$

10.
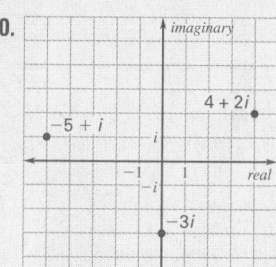

27.
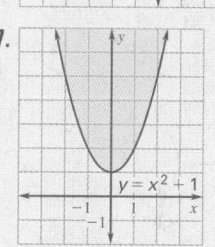

28.
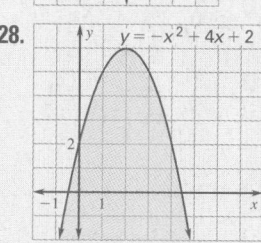

29.
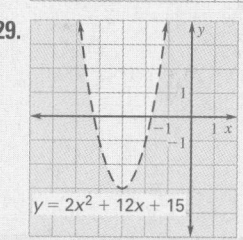

Chapter Standardized Test

▶ **TEST-TAKING STRATEGY** When checking your answer to a question, try using a method different from the one you used to get the answer. If you use the same method to find *and* check an answer, you may make the same mistake twice.

ADDITIONAL RESOURCES
- *Chapter 5 Resource Book*
  Chapter Test (3 levels) (p. 119)
  SAT/ACT Chapter Test (p. 125)
  Alternative Assessment (p. 126)
- 🖳 *Test and Practice Generator*

1. **MULTIPLE CHOICE** What is the vertex of the graph of $y = 2(x - 3)^2 - 7$? **B**

   Ⓐ $(3, 7)$　　Ⓑ $(3, -7)$　　Ⓒ $(-3, -7)$

   Ⓓ $(-3, 7)$　　Ⓔ $(2, 3)$

2. **MULTIPLE CHOICE** What is a correct factorization of $4x^2 + 4x - 35$? **E**

   Ⓐ $(4x + 5)(x - 7)$　　Ⓑ $(4x - 5)(x + 7)$

   Ⓒ $(2x + 5)(2x - 7)$　　Ⓓ $(2x + 35)(2x - 1)$

   Ⓔ $(2x - 5)(2x + 7)$

3. **MULTIPLE CHOICE** What are the zeros of $y = x^2 - 13x + 40$? **D**

   Ⓐ $-5, -8$　　Ⓑ $5, -8$　　Ⓒ $4, 10$

   Ⓓ $5, 8$　　Ⓔ $-4, -10$

4. **MULTIPLE CHOICE** What are all solutions of $4(x - 1)^2 - 3 = 25$? **C**

   Ⓐ 3　　Ⓑ 8　　Ⓒ $1 \pm \sqrt{7}$

   Ⓓ $0.5, 3$　　Ⓔ $1 \pm 2\sqrt{7}$

5. **MULTIPLE CHOICE** What does the product $(-12 + 8i)(10 - i)$ equal? **D**

   Ⓐ $-128 + 68i$　　Ⓑ $-128 + 92i$

   Ⓒ $-112 + 68i$　　Ⓓ $-112 + 92i$

   Ⓔ $-120 - 8i^2$

6. **MULTIPLE CHOICE** If $x^2 + 8x + c$ is a perfect square trinomial, what is the value of c? **C**

   Ⓐ 4　　Ⓑ 8　　Ⓒ 16

   Ⓓ 32　　Ⓔ 64

7. **MULTIPLE CHOICE** How many real and imaginary solutions does the equation $3x^2 + 2x - 7 = 0$ have? **A**

   Ⓐ 2 real solutions, no imaginary solutions

   Ⓑ 1 real solution, no imaginary solutions

   Ⓒ 1 real solution, 1 imaginary solution

   Ⓓ no real solutions, 2 imaginary solutions

   Ⓔ no real solutions, 1 imaginary solution

8. **MULTIPLE CHOICE** What is the solution of $x^2 + 7x - 8 > 0$? **B**

   Ⓐ $x = -8$ or $x = 1$　　Ⓑ $x < -8$ or $x > 1$

   Ⓒ $-8 < x < 1$　　Ⓓ $x < -1$ or $x > 8$

   Ⓔ $-1 < x < 8$

9. **MULTIPLE CHOICE** Which quadratic inequality is graphed? **D**

   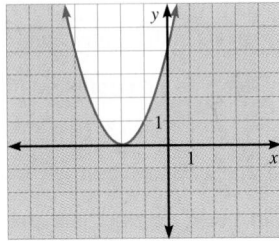

   Ⓐ $y \geq x^2 + 2$　　Ⓑ $y \leq x^2 + 2$

   Ⓒ $y \geq x^2 - 2$　　Ⓓ $y \leq (x + 2)^2$

   Ⓔ $y \leq (x - 2)^2$

10. **MULTIPLE CHOICE** Which quadratic function *cannot* be represented by the graph shown? **E**

    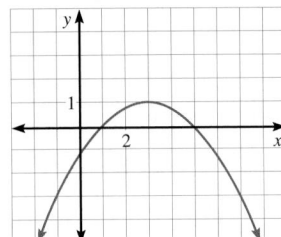

    Ⓐ $y = -\frac{1}{4}(x - 1)(x - 5)$

    Ⓑ $y = -\frac{1}{4}(x - 3)^2 + 1$

    Ⓒ $y = -\frac{1}{4}x^2 + \frac{3}{2}x - \frac{5}{4}$

    Ⓓ $y = -\frac{1}{4}(x^2 - 6x + 5)$

    Ⓔ $y = -\frac{1}{4}(x + 1)(x + 5)$

QUANTITATIVE COMPARISON In Exercises 11 and 12, choose the statement that is true about the given quantities.

Ⓐ The quantity in column A is greater.

Ⓑ The quantity in column B is greater.

Ⓒ The two quantities are equal.

Ⓓ The relationship cannot be determined from the given information.

	Column A	Column B	
11.	$\left|-3 + 2i\right|$	$\left|1 - 4i\right|$	**B**
12.	Discriminant of $x^2 - 7x - 24 = 0$	Discriminant of $5x^2 - 14x + 10 = 0$	**A**

13. **MULTI-STEP PROBLEM** An engineer designing a curved road must make the curve's radius large enough so that car passengers are not pulled to one side as they round the curve at the posted speed limit. The minimum radius r (in feet) that should be used is given by $r = 0.334s^2$ where s is the expected speed of traffic (in miles per hour).

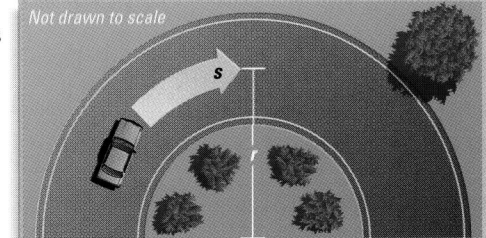

Not drawn to scale

a. If the expected speed of traffic around a curve is 30 miles per hour, what should the minimum radius of the curve be? **300.6 ft**

b. How fast can a car comfortably round a curve with a radius of 400 feet? **about 34.6 miles per hour**

c. Consider a semicircular curve, as shown in the diagram. The radius of the curve is measured from the center of the semicircles formed by the road's inner and outer edges to a point on the road halfway between these edges. Write an equation giving the area A of the pavement needed for the road as a function of the radius r. Assume the road is 24 feet wide. **$A = 24\pi r$**

d. Use your answer from part (c) to write an equation giving the area A of pavement needed as a function of the expected speed s of traffic. **$A = 8.016\pi s^2$**

e. **CRITICAL THINKING** What type of function is the function from part (c)? What type of function is the function from part (d)? **linear; quadratic**

14. **MULTI-STEP PROBLEM** You and your friend are playing tennis. Your friend lobs the ball high into the air, hitting it 3 feet above the court with an initial vertical velocity of 40 feet per second. You back up and prepare to hit an overhead smash to win the point.

a. Use the model $h = -16t^2 + v_0t + h_0$ to write an equation giving the height of the lobbed tennis ball as a function of time. **$h = -16t^2 + 40t + 3$**

b. At what time t does the ball reach its maximum height above the court? What is the maximum height? **1.25 seconds; 28 ft**

c. If you plan to hit the smash when the ball falls to a height of 8 feet above the court, how long do you have to prepare for the shot? **about 2.37 seconds**

d. If you plan to hit the smash when the ball is between 6 feet and 9 feet above the court (inclusive), what are your possible preparation times? **from about 2.34 to 2.42 seconds**

e. Suppose you hit the smash when the ball is 8 feet above the court. It takes 0.1 second for the ball you smashed to hit the court on your friend's side of the net. What was the ball's initial vertical velocity coming off your racket? **78.4 feet per second downward**

PLANNING THE CHAPTER

Polynomials and Polynomial Functions

GOALS		NCTM	ITED	SAT9	Terra-Nova	Local
LESSON						
6.1 pp. 323–328	GOAL 1 Use properties of exponents to evaluate and simplify expressions involving powers.	1, 2, 6, 8, 9, 10	SAPE, RQWN		10, 11, 16, 17, 18, 49, 51, 52	
	GOAL 2 Use exponents and scientific notation to solve real-life problems.					
6.2 pp. 329–337	GOAL 1 Evaluate a polynomial function.	1, 2, 3	MCWN, MIG, IIG, SAPE	6, 11	11, 14, 16, 49, 51, 52	
	GOAL 2 Graph a polynomial function.					
	TECHNOLOGY ACTIVITY: 6.2 *Set a good viewing window to graph polynomial functions on a graphing calculator.*					
6.3 pp. 338–344	GOAL 1 Add, subtract, and multiply polynomials.	1, 2, 6, 8, 9, 10	MCWN, RQWN, SAPE	6	11, 16, 17, 18, 49, 51, 52	
	GOAL 2 Use polynomial operations in real-life problems.					
6.4 pp. 345–351	GOAL 1 Factor polynomial expressions.	1, 2, 8, 9, 10	MCWN, SAPE		11, 16, 18, 49, 51, 52	
	GOAL 2 Use factoring to solve polynomial equations.					
6.5 pp. 352–358	GOAL 1 Divide polynomials and relate the result to the remainder theorem and the factor theorem.	1, 2, 6, 8, 9, 10	MCWN, SAPE, RQWN	2	11, 16, 17, 18, 49, 51, 52	
	GOAL 2 Use polynomial division in real-life problems.					
6.6 pp. 359–365	GOAL 1 Find the rational zeros of a polynomial function.	1, 2, 6, 8, 9, 10	MCWN, SAPE, RQWN	2, 5	11, 16, 17, 18, 49, 51, 52	
	GOAL 2 Use polynomial equations to solve real-life problems.					
6.7 pp. 366–372	GOAL 1 Use the fundamental theorem of algebra to determine the number of zeros of a polynomial function.	1, 2, 8	SAPE, MIG, IIG		11, 16, 18, 49, 51, 52	
	GOAL 2 Use technology to approximate the real zeros of a polynomial function.					
	TECHNOLOGY ACTIVITY: 6.7 *Find the real solutions of polynomial equations on a graphing calculator.*					
6.8 pp. 373–378	GOAL 1 Analyze the graph of a polynomial function.	2, 3, 7, 9, 10	MIG, IIG, RQGE	2, 6, 43	14, 15, 16	
	GOAL 2 Use the graph of a polynomial function to answer questions about real-life situations.					
6.9 pp. 379–386	CONCEPT ACTIVITY: 6.9 *Investigate finite differences for polynomial functions.*	1, 2, 5, 8, 9, 10	MCS	6, 9, 20, 21, 23	15, 16, 18	
	GOAL 1 Use finite differences to determine the degree of a polynomial function that will fit a set of data.					
	GOAL 2 Use technology to find polynomial models for real-life data.					

RESOURCES

CHAPTER RESOURCE BOOKLETS

CHAPTER SUPPORT

Tips for New Teachers	p. 1	Prerequisite Skills Review	p. 5
Parent Guide for Student Success	p. 3	Strategies for Reading Mathematics	p. 7

LESSON SUPPORT

	6.1	6.2	6.3	6.4	6.5	6.6	6.7	6.8	6.9
Lesson Plans (regular and block)	p. 9	p. 21	p. 34	p. 49	p. 63	p. 75	p. 88	p. 102	p. 116
Warm-Up Exercises and Daily Quiz	p. 11	p. 23	p. 36	p. 51	p. 65	p. 77	p. 90	p. 104	p. 118
Activity Support Masters									p. 119
Lesson Openers	p. 12	p. 24	p. 37	p. 52	p. 66	p. 78	p. 91	p. 105	p. 120
Graphing Calculator Activities & Keystrokes		p. 25	p. 38				p. 92	p. 106	
Practice (3 levels)	p. 13	p. 26	p. 40	p. 53	p. 67	p. 79	p. 94	p. 108	p. 121
Reteaching with Practice	p. 16	p. 29	p. 43	p. 56	p. 70	p. 82	p. 97	p. 111	p. 124
Quick Catch-Up for Absent Students	p. 18	p. 31	p. 45	p. 58	p. 72	p. 84	p. 99	p. 113	p. 126
Cooperative Learning Activities				p. 59					
Interdisciplinary Applications		p. 32		p. 60		p. 85		p. 114	
Real-Life Applications	p. 19		p. 46		p. 73		p. 100		p. 127
Math & History Applications				p. 61					
Challenge: Skills and Applications	p. 20	p. 33	p. 47	p. 62	p. 74	p. 86	p. 101	p. 115	p. 128

REVIEW AND ASSESSMENT

Quizzes	pp. 48, 87	Alternative Assessment with Math Journal	p. 137
Chapter Review Games and Activities	p. 129	Project with Rubric	p. 139
Chapter Test (3 levels)	pp. 130–135	Cumulative Review	p. 141
SAT/ACT Chapter Test	p. 136	Resource Book Answers	p. A1

TRANSPARENCIES

	6.1	6.2	6.3	6.4	6.5	6.6	6.7	6.8	6.9
Warm-Up Exercises and Daily Quiz	p. 41	p. 42	p. 43	p. 44	p. 45	p. 46	p. 47	p. 48	p. 49
Alternative Lesson Opener Transparencies	p. 35	p. 36	p. 37	p. 38	p. 39	p. 40	p. 41	p. 42	p. 43
Examples/Standardized Test Practice	✓	✓	✓	✓	✓	✓	✓	✓	✓
Answer Transparencies	✓	✓	✓	✓	✓	✓	✓	✓	✓

TECHNOLOGY

- Electronic Teaching Tools
- Online Lesson Planner
- Internet Support
- Personal Student Tutor
- Test and Practice Generator
- Instant Replay: Video Review Games
- Electronic Lesson Presentations (Lesson 6.2)

ADDITIONAL RESOURCES

- Basic Skills Workbook: Diagnosis and Remediation
- Worked-Out Solution Key
- Resources in Spanish
- Standardized Test Practice Workbook
- Practice Workbook with Examples

Day 2

6.2

STARTING OPTIONS
- ...iew
- ...Reading
- ...k Check
- ...Up or Daily Quiz

TEACHING OPTIONS
- Motivating the Lesson
- Les. Opener (Appl.)
- Examples 1–4
- Closure Question
- Guided Practice Exs.

APPLY/HOMEWORK
- See Assignment Guide.
- See the CRB: Practice, Reteach, Apply, Extend

ASSESSMENT OPTIONS
- Checkpoint Exercises
- Daily Quiz (6.1)
- Stand. Test Practice

Day 3

6.3

STARTING OPTIONS
- Homework Check
- Warm-Up or Daily Quiz

TEACHING OPTIONS
- Motivating the Lesson
- Les. Opener (Visual)
- Graphing Calc. Activity
- Examples 1–5
- Technology Activity
- Closure Question
- Guided Practice Exs.

APPLY/HOMEWORK
- See Assignment Guide.
- See the CRB: Practice, Reteach, Apply, Extend

ASSESSMENT OPTIONS
- Checkpoint Exercises
- Daily Quiz (6.2)
- Stand. Test Practice

Day 3

6.3

STARTING OPTIONS
- Homework Check
- Warm-Up or Daily Quiz

TEACHING OPTIONS
- Les. Opener (Activity)
- Graphing Calc. Activity
- Examples 1–6
- Guided Practice Exs. 1–11

APPLY/HOMEWORK
- See Assignment Guide.
- See the CRB: Practice, Reteach, Apply, Extend

ASSESSMENT OPTIONS
- Checkpoint Exercises, p. 339

Day 4

6.3 (cont.)

STARTING OPTIONS
- Homework Check

TEACHING OPTIONS
- Examples 7–8
- Closure Question
- Guided Practice Ex. 12

APPLY/HOMEWORK
- See Assignment Guide.
- See the CRB: Practice, Reteach, Apply, Extend

ASSESSMENT OPTIONS
- Checkpoint Exercises, p. 340
- Daily Quiz (6.3)
- Stand. Test Practice
- Quiz (6.1–6.3)

Day 5

6.4

STARTING OPTIONS
- Homework Check
- Warm-Up or Daily Quiz

TEACHING OPTIONS
- Motivating the Lesson
- Les. Opener (Application)
- Examples 1–3
- Guided Practice Exs. 1–2, 4–10

APPLY/HOMEWORK
- See Assignment Guide.
- See the CRB: Practice, Reteach, Apply, Extend

ASSESSMENT OPTIONS
- Checkpoint Exercises, p. 346

Day 6

6.4 (cont.)

STARTING OPTIONS
- Homework Check

TEACHING OPTIONS
- Examples 4–5
- Closure Question
- Guided Practice Exs. 3, 11–17

APPLY/HOMEWORK
- See Assignment Guide.
- See the CRB: Practice, Reteach, Apply, Extend

ASSESSMENT OPTIONS
- Checkpoint Exercises, p. 347
- Daily Quiz (6.4)
- Stand. Test Practice

Day 9

6.7

STARTING OPTIONS
- Homework Check
- Warm-Up or Daily Quiz

TEACHING OPTIONS
- Motivating the Lesson
- Les. Opener (Calc.)
- Graphing Calc. Activity
- Examples 1–5
- Technology Activity
- Closure Question
- Guided Practice Exs.

APPLY/HOMEWORK
- See Assignment Guide.
- See the CRB: Practice, Reteach, Apply, Extend

ASSESSMENT OPTIONS
- Checkpoint Exercises
- Daily Quiz (6.7)
- Stand. Test Practice

Day 10

6.8

STARTING OPTIONS
- Homework Check
- Warm-Up or Daily Quiz

TEACHING OPTIONS
- Motivating the Lesson
- Les. Opener (Visual)
- Graphing Calc. Activity
- Examples 1–3
- Closure Question
- Guided Practice Exs.

APPLY/HOMEWORK
- See Assignment Guide.
- See the CRB: Practice, Reteach, Apply, Extend

ASSESSMENT OPTIONS
- Checkpoint Exercises
- Daily Quiz (6.8)
- Stand. Test Practice

Day 11

6.9

STARTING OPTIONS
- Homework Check
- Warm-Up or Daily Quiz

TEACHING OPTIONS
- Concept Act. & Wksht.
- Les. Opener (Application)
- Examples 1–4
- Closure Question
- Guided Practice Exs.

APPLY/HOMEWORK
- See Assignment Guide.
- See the CRB: Practice, Reteach, Apply, Extend

ASSESSMENT OPTIONS
- Checkpoint Exercises
- Daily Quiz (6.9)
- Stand. Test Practice
- Quiz (6.7–6.9)

Day 12

Review

DAY 12 START OPTIONS
- Homework Check

REVIEWING OPTIONS
- Chapter 6 Summary
- Chapter 6 Review
- Chapter Review Games and Activities

APPLY/HOMEWORK
- Chapter 6 Test (practice)
- Ch. Standardized Test (practice)

Day 13

Assess

DAY 13 START OPTIONS
- Homework Check

ASSESSMENT OPTIONS
- Chapter 6 Test
- SAT/ACT Ch. 6 Test
- Alternative Assessment

APPLY/HOMEWORK
- Skill Review, p. 400

BLOCK SCHEDULE

Day 7

6.5

STARTING OPTIONS
- Homework Check
- Warm-Up or Daily Quiz

TEACHING OPTIONS
- Motivating the Lesson
- Les. Opener (Activity)
- Examples 1–5
- Closure Question
- Guided Practice Exs.

APPLY/HOMEWORK
- See Assignment Guide.
- See the CRB: Practice, Reteach, Apply, Extend

ASSESSMENT OPTIONS
- Checkpoint Exercises
- Daily Quiz (6.5)
- Stand. Test Practice

Day 8

6.6

STARTING OPTIONS
- Homework Check
- Warm-Up or Daily Quiz

TEACHING OPTIONS
- Motivating the Lesson
- Les. Opener (Visual)
- Examples 1–3
- Closure Question
- Guided Practice Exs.

APPLY/HOMEWORK
- See Assignment Guide.
- See the CRB: Practice, Reteach, Apply, Extend

ASSESSMENT OPTIONS
- Checkpoint Exercises
- Daily Quiz (6.6)
- Stand. Test Practice
- Quiz (6.4–6.6)

Day 1

6.1 & 6.2

DAY 1 START OPTIONS
- Prereq. Skills Review
- Strategies for Reading
- Homework Check
- W-Up 6.1 or D. Quiz 5.8

TEACHING 6.1 OPTIONS
- Motivating the Lesson
- Les. Opener (Appl.)
- Examples 1–4
- Closure Question
- Guided Practice Exs.

TEACHING 6.2 OPTIONS
- Warm-Up (Les. 6.2)
- Motivating the Lesson
- Les. Opener (Visual)
- Technology Activities
- Examples 1–5
- Closure Question
- Guided Practice Exs.

APPLY/HOMEWORK
- See Assignment Guide.
- See the CRB: Practice, Reteach, Apply, Extend

ASSESSMENT OPTIONS
- Checkpoint Exercises
- Daily Quiz (Les. 6.1, 6.2)
- Stand. Test Practice

Day 2

6.3

DAY 2 START OPTIONS
- Homework Check
- Warm-Up or Daily Quiz

TEACHING 6.3 OPTIONS
- Les. Opener (Activity)
- Graphing Calc. Activity
- Examples 1–8
- Closure Question
- Guided Practice Exs.

APPLY/HOMEWORK
- See Assignment Guide.
- See the CRB: Practice, Reteach, Apply, Extend

ASSESSMENT OPTIONS
- Checkpoint Exercises
- Daily Quiz (Les. 6.3)
- Stand. Test Practice
- Quiz (6.1–6.3)

Day 3

6.4

DAY 3 START OPTIONS
- Homework Check
- Warm-Up or Daily Quiz

TEACHING 6.4 OPTIONS
- Motivating the Lesson
- Les. Opener (Appl.)
- Examples 1–5
- Closure Question
- Guided Practice Exs.

APPLY/HOMEWORK
- See Assignment Guide.
- See the CRB: Practice, Reteach, Apply, Extend

ASSESSMENT OPTIONS
- Checkpoint Exercises
- Daily Quiz (6.4)
- Stand. Test Practice

Day 4

6.5 & 6.6

DAY 4 START OPTIONS
- Homework Check
- W-Up 6.5 or D. Quiz 6.4

TEACHING 6.5 OPTIONS
- Motivating the Lesson
- Les. Opener (Activity)
- Examples 1–5
- Closure Question
- Guided Practice Exs.

TEACHING 6.6 OPTIONS
- Warm-Up (Les. 6.6)
- Motivating the Lesson
- Les. Opener (Visual)
- Examples 1–3
- Closure Question
- Guided Practice Exs.

APPLY/HOMEWORK
- See Assignment Guide.
- See the CRB: Practice, Reteach, Apply, Extend

ASSESSMENT OPTIONS
- Checkpoint Exercises
- Daily Quiz (Les. 6.5, 6.6)
- Stand. Test Practice
- Quiz (6.4–6.6)

Day 5

6.7 & 6.8

DAY 5 START OPTIONS
- Homework Check
- W-Up 6.7 or D. Quiz 6.6

TEACHING 6.7 OPTIONS
- Motivating the Lesson
- Les. Opener (Calc.)
- Technology Activities
- Examples 1–5
- Closure Question
- Guided Practice Exs.

TEACHING 6.8 OPTIONS
- Warm-Up (Les. 6.8)
- Motivating the Lesson
- Les. Opener (Visual)
- Graphing Calc. Activity
- Examples 1–3
- Closure Question
- Guided Practice Exs.

APPLY/HOMEWORK
- See Assignment Guide.
- See the CRB: Practice, Reteach, Apply, Extend

ASSESSMENT OPTIONS
- Checkpoint Exercises
- Daily Quiz (Les. 6.7, 6.8)
- Stand. Test Practice

Day 6

6.9 & Review

DAY 6 START OPTIONS
- Homework Check
- Warm-Up or Daily Quiz

TEACHING 6.9 OPTIONS
- Concept Act. & Wksht.
- Les. Opener (Appl.)
- Examples 1–4
- Closure Question
- Guided Practice Exs.

REVIEWING OPTIONS
- Chapter 6 Summary
- Chapter 6 Review
- Chapter Review Games and Activities

APPLY/HOMEWORK
- See Assignment Guide.
- See the CRB: Practice, Reteach, Apply, Extend
- Chapter 6 Test (prac.)
- Ch. Standardized Test (prac.)

ASSESSMENT OPTIONS
- Checkpoint Exercises
- Daily Quiz (6.9)
- Stand. Test Practice
- Quiz (6.7–6.9)

Day 7

Assess & 7.1

(Day 7 = Ch. 7 Day 1)

ASSESSMENT OPTIONS
- Chapter 6 Test
- SAT/ACT Ch. 6 Test
- Alternative Assessment

CH. 7 START OPTIONS
- Skill Review, p. 400
- Prereq. Skills Review
- Strategies for Reading

TEACHING 7.1 OPTIONS
- Warm-Up (Les. 7.1)
- Motivating the Lesson
- Les. Opener (Appl.)
- Examples 1–6
- Closure Question
- Guided Practice Exs.

APPLY/HOMEWORK
- See Assignment Guide.
- See the CRB: Practice, Reteach, Apply, Extend

ASSESSMENT OPTIONS
- Checkpoint Exercises
- Daily Quiz (Les. 7.1)
- Stand. Test Practice

BEFORE THE CHAPTER

The *Chapter 6 Resource Book* has the following materials to distribute and use before the chapter:

- **Parent Guide for Student Success**
- **Prerequisite Skills Review (pictured below)**
- **Strategies for Reading Mathematics**

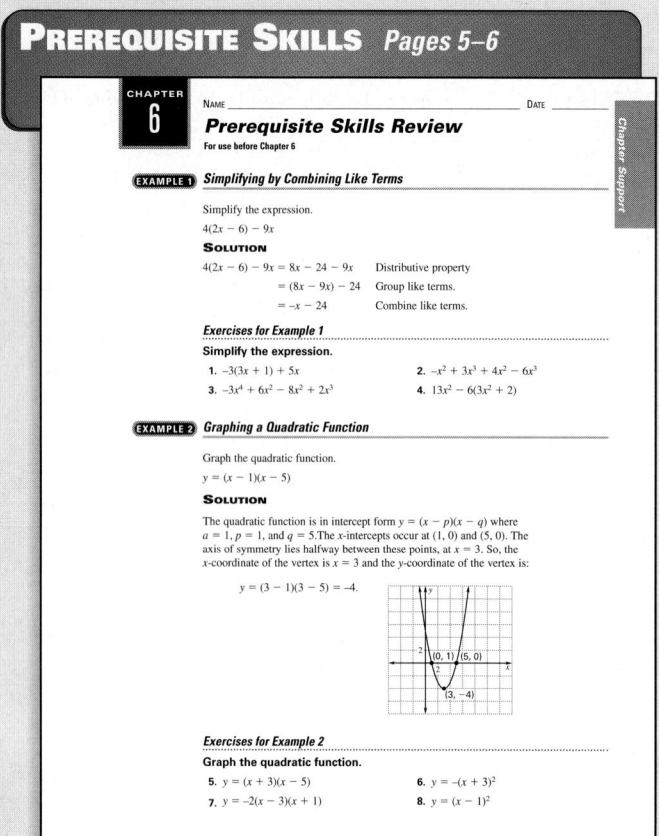

PREREQUISITE SKILLS Pages 5–6

CHAPTER 6

NAME _____ DATE _____

Prerequisite Skills Review
For use before Chapter 6

Chapter Support

EXAMPLE 1 *Simplifying by Combining Like Terms*

Simplify the expression.

$4(2x - 6) - 9x$

SOLUTION

$4(2x - 6) - 9x = 8x - 24 - 9x$ Distributive property
$= (8x - 9x) - 24$ Group like terms.
$= -x - 24$ Combine like terms.

Exercises for Example 1

Simplify the expression.

1. $-3(3x + 1) + 5x$
2. $-x^2 + 3x^3 + 4x^2 - 6x^3$
3. $-3x^4 + 6x^2 - 8x^2 + 2x^3$
4. $13x^2 - 6(3x^2 + 2)$

EXAMPLE 2 *Graphing a Quadratic Function*

Graph the quadratic function.

$y = (x - 1)(x - 5)$

SOLUTION

The quadratic function is in intercept form $y = (x - p)(x - q)$ where
$a = 1$, $p = 1$, and $q = 5$. The x-intercepts occur at $(1, 0)$ and $(5, 0)$. The
axis of symmetry lies halfway between these points, at $x = 3$. So, the
x-coordinate of the vertex is $x = 3$ and the y-coordinate of the vertex is:

$y = (3 - 1)(3 - 5) = -4.$

Exercises for Example 2

Graph the quadratic function.

5. $y = (x + 3)(x - 5)$
6. $y = -(x + 3)^2$
7. $y = -2(x - 3)(x + 1)$
8. $y = (x - 1)^2$

PREREQUISITE SKILLS REVIEW These two pages support the
Study Guide on page 322. They help students prepare for
Chapter 6 by providing worked-out examples and practice for
the following skills needed in the chapter:

- **Simplify expressions by combining like terms.**
- **Graph quadratic functions.**
- **Solve polynomial equations.**

TECHNOLOGY RESOURCE

Students can use the Personal Student Tutor to find
additional reteaching and practice for skills from earlier
chapters that are used in Chapter 6.

DURING EACH LESSON

The *Chapter 6 Resource Book* has the following alternatives
for introducing the lesson:

- **Lesson Openers (pictured below)**
- **Graphing Calculator Activities with Keystrokes**

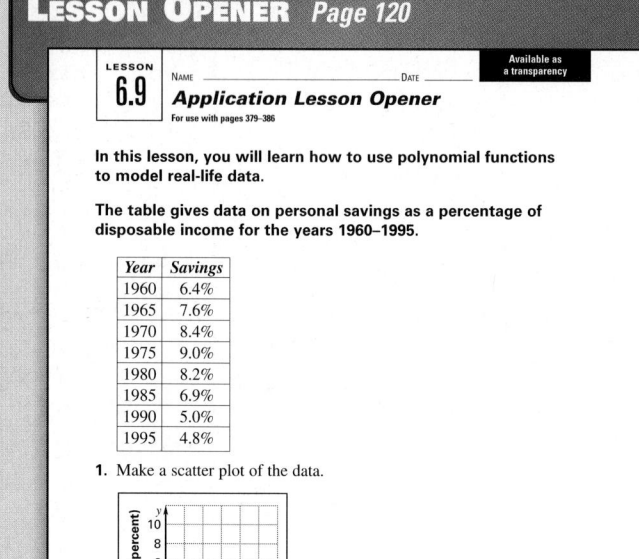

LESSON OPENER Page 120

LESSON 6.9

NAME _____ DATE _____

Available as
a transparency

Application Lesson Opener
For use with pages 379–386

In this lesson, you will learn how to use polynomial functions
to model real-life data.

The table gives data on personal savings as a percentage of
disposable income for the years 1960–1995.

Year	Savings
1960	6.4%
1965	7.6%
1970	8.4%
1975	9.0%
1980	8.2%
1985	6.9%
1990	5.0%
1995	4.8%

1. Make a scatter plot of the data.

2. Suppose you want to model the data using a polynomial function.
   Do you think it would be best to use a linear function, a quadratic
   function, or a cubic function? Explain.

3. Sketch the polynomial function you think best fits the data.

APPLICATION LESSON OPENER This Lesson Opener provides
an alternative way to start Lesson 6.9 in the form of a real-life
application. Students see how polynomial functions relate to
real-life data, such as savings.

FOLLOWING EACH LESSON

The *Chapter 6 Resource Book* has a variety of materials to follow-up each lesson. They include the following:

- **Practice (3 levels)**
- **Reteaching with Practice**
- **Quick Catch-Up for Absent Students**
- **Challenge: Skills and Applications**
- **Interdisciplinary Applications**
- **Real-Life Applications (pictured below)**

APPLICATION *Page 19*

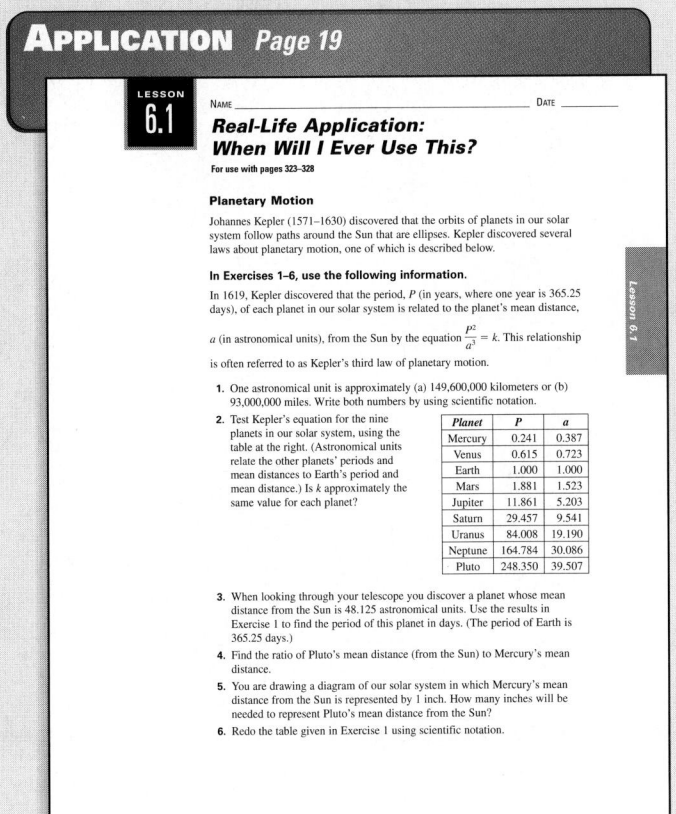

REAL-LIFE APPLICATION When students ask "When will I ever use properties of exponents?," you can have them work through this Real-Life Application in which they apply the content of Lesson 6.1 to paths of planets in our solar system.

ASSESSING THE CHAPTER

The *Chapter 6 Resource Book* has the following review and assessment materials:

- **Quizzes (pictured below)**
- **Chapter Review Games and Activities**
- **Chapter Test (3 levels)**
- **SAT/ACT Chapter Test**
- **Alternative Assessment with Rubrics and Math Journal**
- **Project with Rubric**
- **Cumulative Review**

QUIZ *Page 48*

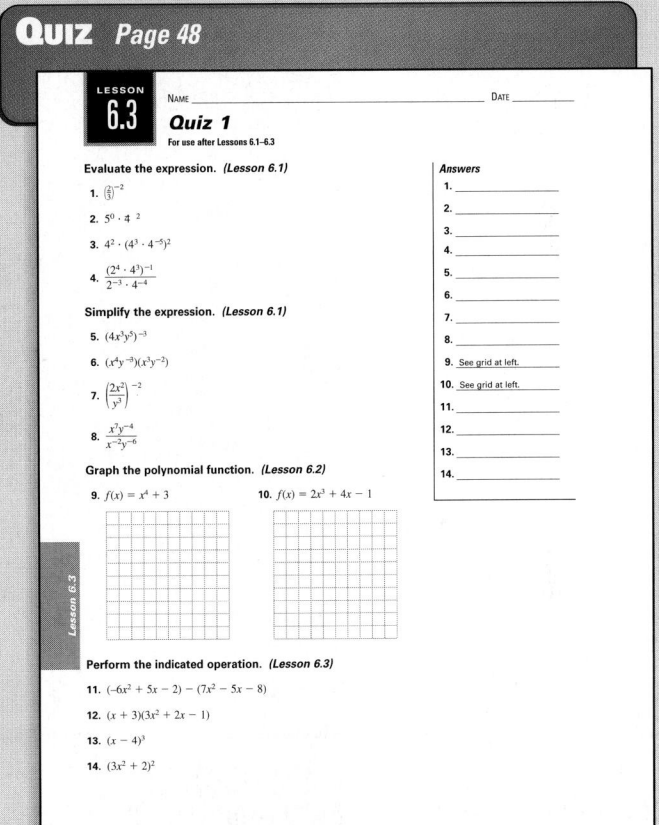

QUIZ Lessons 6.1–6.3 are covered on this quiz, which is an alternate form of the Quiz on page 344 of the textbook. You might want to assign the quiz in the textbook for homework and then use this version for assessment.

TECHNOLOGY RESOURCE

Teachers can use the Time-Saving Test and Practice Generator to create a customized quiz for Lessons 6.1 through 6.3, or for any other combination of lessons in Chapter 6.

CHAPTER GOALS

Students will use properties of exponents and scientific notation to simplify algebraic expressions and to model real-life problems. They will use synthetic substitution to evaluate polynomial expressions. They will also graph polynomial functions and investigate their end behavior. They will add, subtract, multiply and divide polynomial functions and will use factoring, synthetic division, and the rational zero theorem to find zeros of polynomial functions. Students will see how the fundamental theorem of algebra can be used to determine the number of solutions of a polynomial equation and will use graphing calculators to approximate the real zeros of a polynomial function. They will use zeros to write polynomial functions and they will use x-intercepts and turning points to graph polynomial functions. Finally, students will use finite differences to determine the degree of a polynomial that will fit a set of data. Throughout the chapter, students will apply the skills they have learned to solve real-life problems.

APPLICATION NOTE

The three main rocket engines that thrust the space shuttle to its orbiting altitude are propelled by liquid hydrogen and liquid oxygen and are reusable. They are very large and each engine uses about 450 kg of fuel each second. When they stop firing (after about 8 minutes), another propulsion system is activated to keep the shuttle in its orbit. The three main engines can then be reused for up to 55 flights.

Additional information about space exploration is available at **www.mcdougallittell.com**.

POLYNOMIALS AND POLYNOMIAL FUNCTIONS

▶ *What type of function models the speed of a space shuttle?*

CHAPTER 6

APPLICATION: Space Exploration

The space shuttle's engines produce immense power, equivalent to the output of 23 Hoover Dams. All this power is needed to accelerate the shuttle to more than 17,000 miles per hour in about eight minutes after launch. This speed allows the shuttle to achieve and maintain an orbit 240 miles above Earth's surface.

Think & Discuss

The table below gives the time (in seconds) after launch and the corresponding average speed (in feet per second) of the shuttle.

Time (sec)	Speed (ft/sec)
20	463.4
40	979.3
60	1421.3
80	2283.5

1. Make a scatter plot of the data. Estimate how long it takes the shuttle to reach a speed of 1000 feet per second. **about 41 sec; See margin for plot.**

2. Would either a linear function or a quadratic function be a good model for the data? Explain.
 A quadratic function would be a good model because the data lies on a curve.

Learn More About It

You will model the speed of the space shuttle in Exercise 49 on p. 385.

 APPLICATION LINK Visit www.mcdougallittell.com for more information on space exploration.

321

ADDITIONAL RESOURCES
Another way to begin the chapter is to show the video clip of a real-life motivator for Chapter 6 on the *Instant Replay: Video Review Games.*

PROJECTS
A project covering Chapters 4–6 appears on pages 396–397 of the Student Edition. An additional project for Chapter 6 is available in the *Chapter 6 Resource Book,* p. 139.

TECHNOLOGY

 **Software**
- *Electronic Teaching Tools*
- *Online Lesson Planner*
- *Personal Student Tutor*
- *Test and Practice Generator*
- *Electronic Lesson Presentations* (Lesson 6.2)

Video
- *Instant Replay: Video Review Games*

 **Internet Connections**
www.mcdougallittell.com
- **Application Links**
 321, 325, 351, 357, 377
- **Data Updates**
 325, 327, 342, 370, 385, 395
- **Student Help**
 324, 332, 337, 343, 349, 354, 363, 367, 372, 374, 381
- **Career Links**
 327, 330, 335, 342, 347, 355
- **Extra Challenge**
 328, 336, 343, 350, 358, 364, 371, 378, 385

1. **Shuttle Speed After Launch**

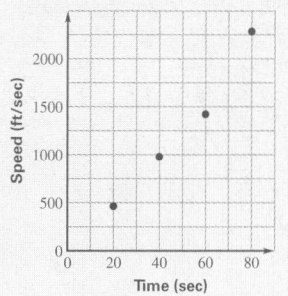

DIAGNOSTIC TOOLS
The **Skill Review** exercises can help you diagnose whether students have the following skills needed in Chapter 6:

- Simplifying polynomial expressions.
- Writing quadratic functions in standard form.
- Graphing quadratic functions and solving quadratic equations.

The following resources are available for students who need additional help with these skills:

- Prerequisite Skills Review (*Chapter 6 Resource Book*, p. 5; *Warm-Up Transparencies*, p. 40)
- Reteaching with Practice (Chapter Resource Books for Lessons 6.1–6.9)
- ▣ *Personal Student Tutor*

ADDITIONAL RESOURCES
The following resources are provided to help you prepare for the upcoming chapter and customize review materials:

- ***Chapter 6 Resource Book***
  Tips for New Teachers (p. 1)
  Parent Guide (p. 3)
  Lesson Plans (every lesson)
  Lesson Plans for Block Scheduling (every lesson)
- ▣ *Electronic Teaching Tools*
- ▣ *Online Lesson Planner*
- ▣ *Test and Practice Generator*

4–6. See Additional Answers beginning on page AA1.

PREVIEW

What's the chapter about?

Chapter 6 is about **polynomials, polynomial equations, and polynomial functions**. In Chapter 6 you'll learn

- how to perform operations on polynomials and solve polynomial equations.
- how to evaluate, graph, and find zeros of polynomial functions.

KEY VOCABULARY

▶ **Review**
- power, p. 11
- *x*-intercept, p. 84
- zeros of a function, p. 259

▶ **New**
- polynomial function, p. 329

- end behavior, p. 331
- polynomial long division, p. 352
- synthetic division, p. 353
- rational zero theorem, p. 359

- fundamental theorem of algebra, p. 366
- local maximum, p. 374
- local minimum, p. 374
- finite differences, p. 380

PREPARE

Are you ready for the chapter?

SKILL REVIEW Do these exercises to review key skills that you'll apply in this chapter. See the given **reference page** if there is something you don't understand.

STUDENT HELP

↳ **Study Tip**
"Student Help" boxes throughout the chapter give you study tips and tell you where to look for extra help in this book and on the Internet.

Simplify the expression. (Review Example 5, p. 13)

1. $4x^2 - 2x + x - x^2$
$3x^2 - x$

2. $2(8x + 5) - 19x$
$-3x + 10$

3. $-x^3 - 5x^4 - 3x^3 + 7x^2$
$-5x^4 - 4x^3 + 7x^2$

Graph the quadratic function. (Review Examples 1–3, pp. 250 and 251) 4–6. See margin.

4. $y = -3(x - 2)^2$

5. $y = (x + 1)(x - 5)$

6. $y = 2(x + 6)(x + 4)$

Write the quadratic function in standard form. (Review Example 4, p. 251)

7. $y = (x - 1)^2 - 7$
$y = x^2 - 2x - 6$

8. $y = 2(x + 4)^2$
$y = 2x^2 + 16x + 32$

9. $y = -(x - 2)(x + 8)$
$y = -x^2 - 6x + 16$

Solve the equation. (Review Example 5, p. 258)

10. $x^2 + 6x - 27 = 0$
$-9, 3$

11. $x^2 + 20x + 100 = 0$
-10

12. $2x^2 + 5x - 12 = 0$
$-4, \dfrac{3}{2}$

STUDY STRATEGY

Here's a study strategy!

Making a Flow Chart

A flow chart is a diagram that shows the possible paths and steps you can follow to solve a problem.

After you complete the chapter, make a flow chart that shows how to find all the zeros of a polynomial function. Include techniques and theorems you learned in Chapter 6 which you can use with various types of polynomial functions.

6.1

Using Properties of Exponents

What you should learn

GOAL 1 Use properties of exponents to evaluate and simplify expressions involving powers.

GOAL 2 Use exponents and scientific notation to solve **real-life** problems, such as finding the per capita GDP of Denmark in **Example 4**.

Why you should learn it

▼ To simplify **real-life** expressions, such as the ratio of a state's park space to total area in **Ex. 57**.

Lake Clark National Park, Alaska

GOAL 1 PROPERTIES OF EXPONENTS

Recall that the expression a^n, where n is a positive integer, represents the product that you obtain when a is used as a factor n times. In the activity you will investigate two properties of exponents.

▶ ACTIVITY

Developing Concepts

Products and Quotients of Powers

1. How many factors of 2 are there in the product $2^3 \cdot 2^4$? Use your answer to write the product as a single power of 2. $7; 2^7$

2. Write each product as a single power of 2 by counting the factors of 2. Use a calculator to check your answers.

   a. $2^2 \cdot 2^5$ 2^7 **b.** $2^1 \cdot 2^6$ 2^7 **c.** $2^3 \cdot 2^6$ 2^9 **d.** $2^4 \cdot 2^4$ 2^8

3. Complete this equation: $2^m \cdot 2^n = 2^?$ 2^{m+n}

4. Write each quotient as a single power of 2 by first writing the numerator and denominator in "expanded form" (for example, $2^3 = 2 \cdot 2 \cdot 2$) and then canceling common factors. Use a calculator to check your answers.

   a. $\dfrac{2^3}{2^1}$ 2^2 **b.** $\dfrac{2^5}{2^2}$ 2^3 **c.** $\dfrac{2^7}{2^3}$ 2^4 **d.** $\dfrac{2^6}{2^2}$ 2^4

5. Complete this equation: $\dfrac{2^m}{2^n} = 2^?$ 2^{m-n}

In the activity you may have discovered two of the following properties of exponents.

CONCEPT SUMMARY

PROPERTIES OF EXPONENTS

Let a and b be real numbers and let m and n be integers.

PRODUCT OF POWERS PROPERTY	$a^m \cdot a^n = a^{m+n}$
POWER OF A POWER PROPERTY	$(a^m)^n = a^{mn}$
POWER OF A PRODUCT PROPERTY	$(ab)^m = a^m b^m$
NEGATIVE EXPONENT PROPERTY	$a^{-m} = \dfrac{1}{a^m}, a \neq 0$
ZERO EXPONENT PROPERTY	$a^0 = 1, a \neq 0$
QUOTIENT OF POWERS PROPERTY	$\dfrac{a^m}{a^n} = a^{m-n}, a \neq 0$
POWER OF A QUOTIENT PROPERTY	$\left(\dfrac{a}{b}\right)^m = \dfrac{a^m}{b^m}, b \neq 0$

6.1 *Using Properties of Exponents*   323

---

1 PLAN

PACING
Basic: 1 day
Average: 1 day
Advanced: 1 day
Block Schedule: 0.5 block with 6.2

➤ LESSON OPENER
APPLICATION

An alternative way to approach Lesson 6.1 is to use the Application Lesson Opener:

- Blackline Master (*Chapter 6 Resource Book,* p. 12)
- Transparency (p. 35)

MEETING INDIVIDUAL NEEDS

- *Chapter 6 Resource Book*
  Prerequisite Skills Review (p. 5)
  Practice Level A (p. 13)
  Practice Level B (p. 14)
  Practice Level C (p. 15)
  Reteaching with Practice (p. 16)
  Absent Student Catch-Up (p. 18)
  Challenge (p. 20)
- *Resources in Spanish*
- *Personal Student Tutor*

NEW-TEACHER SUPPORT

See the Tips for New Teachers on pp. 1–2 of the *Chapter 6 Resource Book* for additional notes about Lesson 6.1.

WARM-UP EXERCISES

Transparency Available

Evaluate each expression.

1. 2^4 16   **2.** $(-2)^4$ 16
3. -2^4 −16   **4.** 2^3 8
5. $\left(\dfrac{1}{2}\right)^5$ $\dfrac{1}{32}$

ENGLISH LEARNERS

The names for the properties of exponents are very similar; English learners may have difficulty differentiating them. Mention that in English, word order is often important in determining meaning. Then talk about how the words in each name relate to the property.

MOTIVATING THE LESSON
Scientists typically use scientific notation to express very large and very small quantities. Tell students that the properties of exponents they will study in this lesson make it easy to compute with and compare quantities expressed in this way.

ACTIVITY NOTE
Calculator If students are using graphing calculators for Steps 1 and 2, they may find it helpful to use parentheses when they enter and evaluate the given expressions.

 EXTRA EXAMPLE 1
Evaluate each expression.
a. $(3^4)^2$ 6561
b. $\left(\dfrac{5}{8}\right)^3$ $\dfrac{125}{512}$
c. $(-2)^{-3}(-2)^9$ 64

EXTRA EXAMPLE 2
Evaluate each expression.
a. $\left(\dfrac{a^2}{b^{-3}}\right)^3$ $a^6 b^9$
b. $(-y^2)^5 y^2 y^{-12}$ -1
c. $\dfrac{rs^2}{(rs^{-1})^3}$ $\dfrac{s^5}{r^2}$

 CHECKPOINT EXERCISES
For use after Example 1:
1. Evaluate $(10^2)^4(10)^{-2}$.
  1,000,000

For use after Example 2:
2. Simplify $\left(\dfrac{a^2 b}{a^{-1} b^5}\right)^3 \cdot \dfrac{a^9}{b^{12}}$

! COMMON ERROR
EXAMPLE 1C Many students have difficulty understanding that a number with a negative exponent does not necessarily represent a negative number. It may help to evaluate some expressions that use positive bases and negative exponents to show that the results are always positive.

The properties of exponents can be used to evaluate numerical expressions and to simplify algebraic expressions. In this book we assume that any base with a zero or negative exponent is nonzero. A simplified algebraic expression contains only positive exponents.

EXAMPLE 1 *Evaluating Numerical Expressions*

a. $(2^3)^4 = 2^{3 \cdot 4}$ Power of a power property

$= 2^{12}$ Simplify exponent.

$= 4096$ Evaluate power.

b. $\left(\dfrac{3}{4}\right)^2 = \dfrac{3^2}{4^2}$ Power of a quotient property

$= \dfrac{9}{16}$ Evaluate powers.

c. $(-5)^{-6}(-5)^4 = (-5)^{-6+4}$ Product of powers property

$= (-5)^{-2}$ Simplify exponent.

$= \dfrac{1}{(-5)^2}$ Negative exponent property

$= \dfrac{1}{25}$ Evaluate power.

STUDENT HELP
▸ **Study Tip**
When you multiply powers, do not multiply the bases. For example, $2^3 \cdot 2^5 \neq 4^8$.

EXAMPLE 2 *Simplifying Algebraic Expressions*

a. $\left(\dfrac{r}{s^{-5}}\right)^2 = \dfrac{r^2}{(s^{-5})^2}$ Power of a quotient property

$= \dfrac{r^2}{s^{-10}}$ Power of a power property

$= r^2 s^{10}$ Negative exponent property

b. $(7b^{-3})^2 b^5 b = 7^2 (b^{-3})^2 b^5 b$ Power of a product property

$= 49 b^{-6} b^5 b$ Power of a power property

$= 49 b^{-6+5+1}$ Product of powers property

$= 49 b^0$ Simplify exponent.

$= 49$ Zero exponent property

c. $\dfrac{(xy^2)^2}{x^3 y^{-1}} = \dfrac{x^2 (y^2)^2}{x^3 y^{-1}}$ Power of a product property

$= \dfrac{x^2 y^4}{x^3 y^{-1}}$ Power of a power property

$= x^{2-3} y^{4-(-1)}$ Quotient of powers property

$= x^{-1} y^5$ Simplify exponents.

$= \dfrac{y^5}{x}$ Negative exponent property

STUDENT HELP

HOMEWORK HELP
Visit our Web site
www.mcdougallittell.com
for extra examples.

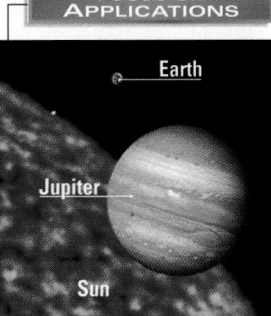

FOCUS ON
APPLICATIONS

Earth

Jupiter

Sun

 **ASTRONOMY**
Jupiter is the largest planet in the solar system. It has a radius of 71,400 km—over 11 times as great as Earth's, but only about one tenth as great as the sun's.

 APPLICATION LINK
www.mcdougallittell.com

STUDENT HELP

▸ **Skills Review**
For help with scientific notation, see p. 913.

GOAL 2 USING PROPERTIES OF EXPONENTS IN REAL LIFE

EXAMPLE 3 *Comparing Real-Life Volumes*

ASTRONOMY The radius of the sun is about 109 times as great as Earth's radius. How many times as great as Earth's volume is the sun's volume?

SOLUTION

Let r represent Earth's radius.

$$\frac{\text{Sun's volume}}{\text{Earth's volume}} = \frac{\frac{4}{3}\pi(109r)^3}{\frac{4}{3}\pi r^3}$$ The volume of a sphere is $\frac{4}{3}\pi r^3$.

$$= \frac{\frac{4}{3}\pi 109^3 r^3}{\frac{4}{3}\pi r^3}$$ Power of a product property

$$= 109^3 r^0$$ Quotient of powers property

$$= 109^3$$ Zero exponent property

$$= 1,295,029$$ Evaluate power.

▸ The sun's volume is about 1.3 million times as great as Earth's volume.

....

A number is expressed in **scientific notation** if it is in the form $c \times 10^n$ where $1 \leq c < 10$ and n is an integer. For instance, the width of a molecule of water is about 2.5×10^{-8} meter, or 0.000000025 meter. When working with numbers in scientific notation, the properties of exponents listed on page 323 can help make calculations easier.

 REAL LIFE
Economics

EXAMPLE 4 *Using Scientific Notation in Real Life*

In 1997 Denmark had a population of 5,284,000 and a gross domestic product (GDP) of $131,400,000,000. Estimate the per capita GDP of Denmark.

DATA UPDATE of UN/ECE Statistical Division data at www.mcdougallittell.com

SOLUTION

"Per capita" means per person, so divide the GDP by the population.

$$\frac{\text{GDP}}{\text{Population}} = \frac{131,400,000,000}{5,284,000}$$ Divide GDP by population.

$$= \frac{1.314 \times 10^{11}}{5.284 \times 10^6}$$ Write in scientific notation.

$$= \frac{1.314}{5.284} \times 10^5$$ Quotient of powers property

$$\approx 0.249 \times 10^5$$ Use a calculator.

$$= 24,900$$ Write in standard notation.

▸ The per capita GDP of Denmark in 1997 was about $25,000 per person.

6.1 Using Properties of Exponents **325**

 EXTRA EXAMPLE 3
A circular component used in the manufacture of a microprocessor has a diameter of 200 mm and a thickness of 0.01 mm. What is its volume? 100π mm^3

EXTRA EXAMPLE 4
The red blood cells, white blood cells, and platelets found in human blood are all generated from the same stem cells. In laboratory experiments, scientists have found that as few as 10 stem cells can grow into 1,200,000,000,000 platelets in just four weeks. The number of white blood cells generated was $\frac{1}{40}$ the number of platelets. How many white blood cells were generated? **30 billion**

✓ **CHECKPOINT EXERCISES**

For use after Example 3:

1. The diameter of the moon is about one-fourth that of Earth's diameter. What fractional part of Earth's volume is the moon's volume? $\frac{1}{64}$

For use after Example 4:

2. An average adult has about 528,000,000 ft of blood vessels in his or her body. How many times greater is this than the circumference of Earth, which is about 25,000 mi? **about 4 times greater**

CLOSURE QUESTION
Which properties of exponents require you to check that two or more bases are the same before applying the property? **product of powers property and quotient of powers property**

DAILY PUZZLER
A certain whole number is greater than 10^6 but less than 10^9. Can the sum of its digits be equal to 84? Explain. **no; the number can have at most 9 digits, which means that the sum of its digits is at most 81.**

325

ASSIGNMENT GUIDE

BASIC
Day 1: pp. 326–328 Exs. 16–46 even, 49–53 odd, 57, 61–81 odd

AVERAGE
Day 1: pp. 326–328 Exs. 16–46 even, 48–53, 57, 61–81 odd

ADVANCED
Day 1: pp. 326–328 Exs. 16–46 even, 48–59, 61–81 odd

BLOCK SCHEDULE
pp. 326–328 Exs. 16–46 even, 48–53, 57, 61–81 odd (with 6.2)

EXERCISE LEVELS

Level A: *Easier*
16–27, 32–47

Level B: *More Difficult*
28–31, 48–57

Level C: *Most Difficult*
58, 59

✔ HOMEWORK CHECK

To quickly check student understanding of key concepts, go over the following exercises: Exs. 18, 30, 38, 44, 48. See also the Daily Homework Quiz:

• Blackline Master (*Chapter 6 Resource Book,* p. 23)
• Transparency (p. 42)

GUIDED PRACTICE

Vocabulary Check ✔

1. State the name of the property illustrated.

   a. $a^m \cdot a^n = a^{m+n}$
   product of powers property

   b. $(a^m)^n = a^{mn}$
   power of a power property

   c. $(ab)^m = a^m b^m$
   power of a product property

Concept Check ✔

2. **ERROR ANALYSIS** Describe the mistake made in simplifying the expression.

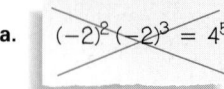

a. $(-2)^2(-2)^3 = 4^5$

b. $\dfrac{x^8}{x^2} = x^4$

c. $x^4 \cdot x^3 = x^{12}$

Skill Check ✔

2a. The bases were multiplied; should be $(-2)^5$.

2b. The exponents were divided when they should have been subtracted; should be x^6.

2c. The exponents were multiplied rather than added; should be x^7.

Evaluate the expression. Tell which properties of exponents you used. 3–8. Check properties.

3. $6 \cdot 6^2$ 216

4. $(9^6)(9^2)^{-3}$ 1

5. $(2^3)^2$ 64

6. $\left(\dfrac{3}{2^{-2}}\right)\left(\dfrac{1}{2}\right)^2$ 3

7. $\left(\dfrac{3}{5}\right)^{-2}$ $\dfrac{25}{9}$

8. $\dfrac{7^{-5}}{7^{-3}}$ $\dfrac{1}{49}$

9–14. Check properties.

Simplify the expression. Tell which properties of exponents you used.

9. $z^{-2} \cdot z^{-4} \cdot z^6$ 1

10. $yz^{-2}(x^2y)^3z$ $\dfrac{x^6y^4}{z}$

11. $(4x^3)^{-2}$ $\dfrac{1}{16x^6}$

12. $\left(\dfrac{2}{x^{-3}}\right)^6$ $64x^{18}$

13. $\dfrac{3y^6}{y^3}$ $3y^3$

14. $\dfrac{(xy)^4}{xy^{-1}}$ x^3y^5

15. 🌐 **ASTRONOMY** Earth has a radius of about 6.38×10^3 kilometers. The sun has a radius of about 6.96×10^5 kilometers. Use the formula for the volume of a sphere given on page 325 to calculate the volume of the sun and the volume of Earth. Divide the volumes. Do you get the same result as in Example 3?
sun's volume: 1.41×10^{18} km^3; Earth's volume: 1.09×10^{12} km^3; ratio is about 1,298,000; the results match.

PRACTICE AND APPLICATIONS

STUDENT HELP

→ **Extra Practice**
to help you master skills is on p. 947.

EVALUATING NUMERICAL EXPRESSIONS **Evaluate the expression. Tell which properties of exponents you used.** 16–31. Check properties.

16. $4^2 \cdot 4^4$ 4096

17. $(5^{-2})^3$ $\dfrac{1}{15,625}$

18. $(-9)(-9)^3$ 6561

19. $(8^2)^3$ 262,144

20. $\dfrac{5^2}{5^5}$ $\dfrac{1}{125}$

21. $\left(\dfrac{3}{7}\right)^3$ $\dfrac{27}{343}$

22. $\left(\dfrac{5}{9}\right)^{-3}$ $\dfrac{729}{125}$

23. $11^{-2} \cdot 11^0$ $\dfrac{1}{121}$

24. $\dfrac{4^{-2}}{4^{-3}}$ 4

25. $\left(\dfrac{1}{8}\right)^{-4}$ 4096

26. $(2^{-4})^{-2}$ 256

27. $\dfrac{2^2}{2^{-9}}$ 2048

28. $\dfrac{6^2}{(6^{-2} \cdot 5^1)^{-2}}$ $\dfrac{25}{36}$

29. $6^0 \cdot 6^3 \cdot 6^{-4}$ $\dfrac{1}{6}$

30. $\left(\dfrac{1}{10}\right)^3\left(\dfrac{1}{10}\right)^{-3}$ 1

31. $\left(\left(\dfrac{2}{5}\right)^{-3}\right)^2$ $\dfrac{15,625}{64}$

SIMPLIFYING ALGEBRAIC EXPRESSIONS **Simplify the expression. Tell which properties of exponents you used.** 32–47. Check properties.

32. $x^8 \cdot \dfrac{1}{x^3}$ x^5

33. $(2^3x^2)^5$ $32,768x^{10}$

34. $(x^2y^2)^{-1}$ $\dfrac{1}{x^2y^2}$

35. $\dfrac{x^5}{x^{-2}}$ x^7

STUDENT HELP

→ **HOMEWORK HELP**
Example 1: Exs. 16–31
Example 2: Exs. 32–51
Examples 3, 4: Exs. 52–56

36. $\dfrac{x^5y^2}{x^4y^0}$ xy^2

37. $(x^4y^7)^{-3}$ $\dfrac{1}{x^{12}y^{21}}$

38. $\dfrac{x^{11}y^{10}}{x^{-3}y^{-1}}$ $x^{14}y^{11}$

39. $-3x^{-4}y^0$ $-\dfrac{3}{x^4}$

40. $(10x^3y^5)^{-3}$ $\dfrac{1}{1000x^9y^{15}}$

41. $\dfrac{x^{-1}y}{xy^{-2}}$ $\dfrac{y^3}{x^2}$

42. $(4x^2y^5)^{-2}$ $\dfrac{1}{16x^4y^{10}}$

43. $\dfrac{2x^2y}{6xy^{-1}}$ $\dfrac{1}{3}xy^2$

44. $\dfrac{5x^3y^9}{20x^2y^{-2}}$ $\dfrac{1}{4}xy^{11}$

45. $\dfrac{xy^9}{3y^{-2}} \cdot \dfrac{-7y}{21x^5}$ $-\dfrac{y^{12}}{9x^4}$

46. $\dfrac{y^{10}}{2x^3} \cdot \dfrac{20x^{14}}{xy^6}$ $10x^{10}y^4$

47. $\dfrac{12xy}{7x^4} \cdot \dfrac{7x^5y^2}{4y}$ $3x^2y^2$

326   **Chapter 6** *Polynomials and Polynomial Functions*

GEOMETRY ▶ **CONNECTION** Write an expression for the area or volume of the figure in terms of *x*.

48. $A = \frac{\sqrt{3}}{4}s^2$ $A = \frac{\sqrt{3}}{16}x^2$

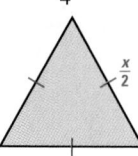

49. $A = \pi r^2$ $A = 16\pi x^2$

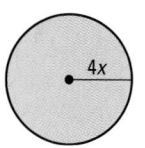

50. $V = \pi r^2 h$ $V = 4\pi x^3$

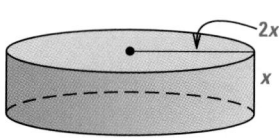

51. $V = \frac{4}{3}\pi r^3$ $V = \frac{4}{81}\pi x^3$

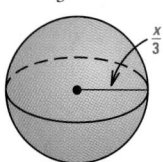

SCIENTIFIC NOTATION In Exercises 52–56, use scientific notation.

52. 🌐 **NATIONAL DEBT** On June 8, 1999, the national debt of the United States was about $5,608,000,000,000. The population of the United States at that time was about 273,000,000. Suppose the national debt was divided evenly among everyone in the United States. How much would each person owe? **about $2.05 × 10⁴**

    📡 DATA UPDATE of Bureau of the Public Debt and U.S. Census Bureau data at www.mcdougallittell.com

53. France $2.13 × 10⁴
    Germany $2.24 × 10⁴
    Ireland $1.95 × 10⁴
    Luxembourg $3.24 × 10⁴
    Netherlands $2.14 × 10⁴
    Sweden $2.00 × 10⁴

53. **SOCIAL STUDIES** ▶ **CONNECTION** The table shows the population and gross domestic product (GDP) in 1997 for each of six different countries. Calculate the per capita GDP for each country.

    📡 DATA UPDATE of UN/ECE Statistical Division data at www.mcdougallittell.com

Country	Population	GDP (U.S. dollars)
France	58,607,000	1,249,600,000,000
Germany	82,061,000	1,839,300,000,000
Ireland	3,661,000	71,300,000,000
Luxembourg	420,000	13,600,000,000
The Netherlands	15,600,000	333,400,000,000
Sweden	8,849,000	177,300,000,000

FOCUS ON CAREERS

ORNITHOLOGIST
An ornithologist is a scientist who studies the history, classification, biology, and behavior of birds.

📡 CAREER LINK
www.mcdougallittell.com

54. **BIOLOGY** ▶ **CONNECTION** A red blood cell has a diameter of approximately 0.00075 centimeter. Suppose one of the arteries in your body has a diameter of 0.0456 centimeter. How many red blood cells could fit across the artery?
    about 6.1 × 10¹ red blood cells

55. 🌐 **SPACE EXPLORATION** On February 17, 1998, *Voyager 1* became the most distant manmade object in space, at a distance of 10,400,000,000 kilometers from Earth. How long did it take *Voyager 1* to travel this distance given that it traveled an average of 1,390,000 kilometers per day? ▶ Source: NASA
    about 7.48 × 10³ days

56. 🌐 **ORNITHOLOGY** Some scientists estimate that there are about 8600 species of birds in the world. The mean number of birds per species is approximately 12,000,000. About how many birds are there in the world?
    about 1.03 × 10¹¹ birds

6.1 *Using Properties of Exponents*   **327**

APPLICATION NOTE
EXERCISE 52 The national deficit is the difference in any fiscal year between the amount of money the government takes in and the amount it spends. The National or Public Debt is based on the accumulated deficits plus any budget surpluses. An update of the Public Debt Outstanding is produced every morning.

EXERCISE 53 A country's GDP measures the total value of its goods and services produced for a given time. The GDP does not include the contributions of any economies that are outside the mainstream economy and do not report their activity to the government.

DATA UPDATE
Updated data for Exercise 53 is available at **www.mcdougallittell.com.**

CAREER NOTE
EXERCISE 56 Additional information about ornithologists is available at **www.mcdougallittell.com.**

ADDITIONAL PRACTICE AND RETEACHING

For Lesson 6.1:
• Practice Levels A, B, and C
  (*Chapter 6 Resource Book*, p. 13)
• Reteaching with Practice
  (*Chapter 6 Resource Book*, p. 16)
• 🖥 See Lesson 6.1 of the *Personal Student Tutor*

For more Mixed Review:
• 🖥 Search the *Test and Practice Generator* for key words or specific lessons.

DAILY HOMEWORK QUIZ

📋 **Transparency Available**

Evaluate the expression.

1. $(-2)^{-3}(-2)^7$ **16**

2. $\dfrac{7}{7^{-2}} \cdot 3^5 \cdot 3^{-5}$ **343**

Simplify the expression.

3. $\dfrac{4x^{-2}y^3}{32x^5y^{-6}}$ $\dfrac{y^9}{8x^7}$

4. $\dfrac{28x^4y^2}{7xy^5}$ $\dfrac{4x^3}{y^3}$

5. Write an expression for the volume of the cone in terms of x. $\dfrac{3}{2}\pi x^3$

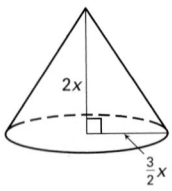

EXTRA CHALLENGE NOTE

→ Challenge problems for Lesson 6.1 are available in **blackline** format in the *Chapter 6 Resource Book,* p. 20 and at **www.mcdougallittell.com**.

ADDITIONAL TEST PREPARATION

1. WRITING How does the power of a power property of exponents differ from the product of powers property? **The power of a power property is used to raise a power to a power. The product of powers property is used when two expressions with the same base are being multiplied.**

57a–b, 58–68.
   See Additional Answers beginning on page AA1.

Test Preparation

57. MULTI-STEP PROBLEM Suppose you live in a state that has a total area of 5.38×10^7 acres and 4.19×10^5 acres of park space. You think that the state should set aside more land for parks. The table shows the total area and the amount of park space for several states.   **57a, b. See margin.**

State	Total area (acres)	Amount of park space (acres)
Alaska	393,747,200	3,250,000
California	101,676,000	1,345,000
Connecticut	3,548,000	176,000
Kansas	52,660,000	29,000
Ohio	28,690,000	204,000
Pennsylvania	29,477,000	283,000

▶ Source: *Statistical Abstract of the United States*

57c. A good answer should include the fact that the writer's state has 0.78% park space, and should compare that percentage to the percentages for other states.

a. Write the total area and the amount of park space for each state in scientific notation.

b. For each state, divide the amount of park space by the total area.

c. *Writing* You want to ask the state legislature to increase the amount of park space in your state. Use your results from parts (a) and (b) to write a letter that explains why your state needs more park space.

★ **Challenge**

LOGICAL REASONING In Exercises 58 and 59, refer to the properties of exponents on page 323.

58. Show how the negative exponent property can be derived from the quotient of powers property and the zero exponent property. **See margin.**

59. Show how the quotient of powers property can be derived from the product of powers property and the negative exponent property. **See margin.**

EXTRA CHALLENGE
→ www.mcdougallittell.com

MIXED REVIEW

GRAPHING Graph the equation. (Review 2.3, 5.1 for 6.2) **60–68. See margin.**

60. $y = -4$ **61.** $y = -x - 3$ **62.** $y = 3x + 1$

63. $y = -2x + 5$ **64.** $y = 3x^2 + 2$ **65.** $y = -2x(x + 6)$

66. $y = x^2 - 2x - 6$ **67.** $y = 2x^2 - 4x + 10$ **68.** $y = -2(x - 3)^2 + 8$

SOLVING QUADRATIC EQUATIONS Solve the equation. (Review 5.3)

69. $2x^2 = 32$ ± 4 **70.** $-3x^2 = -24$ $\pm 2\sqrt{2}$ **71.** $25x^2 = 16$ $\pm\dfrac{4}{5}$

72. $3x^2 - 8 = 100$ ± 6 **73.** $13 - 5x^2 = 8$ ± 1 **74.** $4x^2 - 5 = 9$ $\pm\dfrac{\sqrt{14}}{2}$

75. $-x^2 + 9 = 2x^2 - 6 \pm\sqrt{5}$ **76.** $12 + 2x^2 = 5x^2 - 8$ $\pm\dfrac{2\sqrt{15}}{3}$ **77.** $-2x^2 + 7 = x^2 - 2$ $\pm\sqrt{3}$

OPERATIONS WITH COMPLEX NUMBERS Write the expression as a complex number in standard form. (Review 5.4)

78. $(9 + 4i) + (9 - i)$ **$18 + 3i$**   **79.** $(-5 + 3i) - (-2 - i)$ **$-3 + 4i$**   **80.** $(10 - i) - (4 + 7i)$ **$6 - 8i$**

81. $-i(7 + 2i)$ **$2 - 7i$**   **82.** $-11i(5 + i)$ **$11 - 55i$**   **83.** $(3 + i)(9 + i)$ **$26 + 12i$**

6.2 Evaluating and Graphing Polynomial Functions

What you should learn

GOAL 1 Evaluate a polynomial function.

GOAL 2 Graph a polynomial function, as applied in **Example 5**.

Why you should learn it

▼ To find values of **real-life** functions, such as the amount of prize money awarded at the U.S. Open Tennis Tournament in **Ex. 86**.

GOAL 1 EVALUATING POLYNOMIAL FUNCTIONS

A **polynomial function** is a function of the form

$$f(x) = a_n x^n + a_{n-1} x^{n-1} + \cdots + a_1 x + a_0$$

where $a_n \neq 0$, the exponents are all whole numbers, and the coefficients are all real numbers. For this polynomial function, a_n is the **leading coefficient**, a_0 is the **constant term**, and n is the **degree**. A polynomial function is in **standard form** if its terms are written in descending order of exponents from left to right.

You are already familiar with some types of polynomial functions. For instance, the linear function $f(x) = 3x + 2$ is a polynomial function of degree 1. The quadratic function $f(x) = x^2 + 3x + 2$ is a polynomial function of degree 2. Here is a summary of common types of polynomial functions.

Degree	Type	Standard form
0	Constant	$f(x) = a_0$
1	Linear	$f(x) = a_1 x + a_0$
2	Quadratic	$f(x) = a_2 x^2 + a_1 x + a_0$
3	Cubic	$f(x) = a_3 x^3 + a_2 x^2 + a_1 x + a_0$
4	Quartic	$f(x) = a_4 x^4 + a_3 x^3 + a_2 x^2 + a_1 x + a_0$

EXAMPLE 1 *Identifying Polynomial Functions*

Decide whether the function is a polynomial function. If it is, write the function in standard form and state its degree, type, and leading coefficient.

a. $f(x) = \frac{1}{2}x^2 - 3x^4 - 7$

b. $f(x) = x^3 + 3^x$

c. $f(x) = 6x^2 + 2x^{-1} + x$

d. $f(x) = -0.5x + \pi x^2 - \sqrt{2}$

SOLUTION

a. The function is a polynomial function. Its standard form is $f(x) = -3x^4 + \frac{1}{2}x^2 - 7$. It has degree 4, so it is a quartic function. The leading coefficient is -3.

b. The function is not a polynomial function because the term 3^x does not have a variable base and an exponent that is a whole number.

c. The function is not a polynomial function because the term $2x^{-1}$ has an exponent that is not a whole number.

d. The function is a polynomial function. Its standard form is $f(x) = \pi x^2 - 0.5x - \sqrt{2}$. It has degree 2, so it is a quadratic function. The leading coefficient is π.

PACING
Basic: 1 day
Average: 1 day
Advanced: 1 day
Block Schedule: 0.5 block with 6.1

LESSON OPENER
VISUAL APPROACH
An alternative way to approach Lesson 6.2 is to use the Visual Approach Lesson Opener:
- Blackline Master (*Chapter 6 Resource Book*, p. 24)
- Transparency (p. 36)

MEETING INDIVIDUAL NEEDS
- *Chapter 6 Resource Book*
  Prerequisite Skills Review (p. 5)
  Practice Level A (p. 26)
  Practice Level B (p. 27)
  Practice Level C (p. 28)
  Reteaching with Practice (p. 29)
  Absent Student Catch-Up (p. 31)
  Challenge (p. 33)
- *Resources in Spanish*
- Personal Student Tutor

NEW-TEACHER SUPPORT
See the Tips for New Teachers on pp. 1–2 of the *Chapter 6 Resource Book* for additional notes about Lesson 6.2.

WARM-UP EXERCISES

Transparency Available

Identify each function as linear or quadratic.

1. $f(x) = 2x^2 + x - 6$ quadratic
2. $f(x) = 5x + 3$ linear

Find $f(x)$ when $x = -2$.

3. $f(x) = 2x - 9$ -13
4. $f(x) = x^2 - 5x + 7$ 21
5. $f(x) = 3x^3 + 10$ -14

ACTIVITY NOTE
Graphing Calculator Students can readily see the end behavior of the graphs by using the Trace feature of the calculator to examine the function values for x-values to the left and right of the origin.

EXTRA EXAMPLE 1
Decide whether the function is a polynomial function. If it is, write the function in standard form and state its degree, type, and leading coefficient.
a. $f(x) = 2x^2 - x^{-2}$ no
b. $f(x) = -0.8x^3 + x^4 - 5$ yes;
$y = x^4 - 0.8x^3 - 5$; 4; quartic; 1

EXTRA EXAMPLE 2
Use synthetic division to evaluate $f(x) = 3x^5 - x^4 - 5x + 10$ when $x = -2$. -92

 CHECKPOINT EXERCISES
For use after Example 1:
1. State the degree, type, and leading coefficient of $f(x) = 8x^3 - 9$.   3; cubic; 8

For use after Example 2:
2. Use synthetic division to evaluate $f(x) = 5x^3 + x^2 - 4x + 1$ when $x = 4$.   321

One way to evaluate a polynomial function is to use direct substitution. For instance, $f(x) = 2x^4 - 8x^2 + 5x - 7$ can be evaluated when $x = 3$ as follows.

$$f(3) = 2(3)^4 - 8(3)^2 + 5(3) - 7$$
$$= 162 - 72 + 15 - 7$$
$$= 98$$

Another way to evaluate a polynomial function is to use **synthetic substitution**.

> **STUDENT HELP**
> ⮕ **Study Tip**
> In Example 2, note that the row of coefficients for $f(x)$ must include a coefficient of 0 for the "missing" x^3-term.

EXAMPLE 2   *Using Synthetic Substitution*

Use synthetic substitution to evaluate $f(x) = 2x^4 - 8x^2 + 5x - 7$ when $x = 3$.

SOLUTION

Write the value of x and the coefficients of $f(x)$ as shown. Bring down the leading coefficient. **Multiply by 3** and write the result in the next column. **Add** the numbers in that column and write the sum below the line. Continue to multiply and add, as shown.

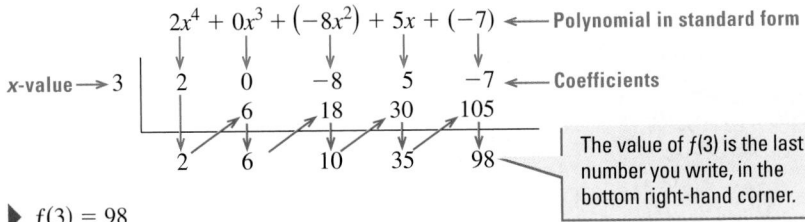

$2x^4 + 0x^3 + (-8x^2) + 5x + (-7)$ ⟵ Polynomial in standard form

x-value ⟶ 3

Coefficients

The value of $f(3)$ is the last number you write, in the bottom right-hand corner.

▶ $f(3) = 98$

....

Using synthetic substitution is equivalent to evaluating the polynomial in *nested form*.

$f(x) = 2x^4 + 0x^3 - 8x^2 + 5x - 7$ Write original function.

$= (2x^3 + 0x^2 - 8x + 5)x - 7$ Factor x out of first 4 terms.

$= ((2x^2 + 0x - 8)x + 5)x - 7$ Factor x out of first 3 terms.

$= (((2x + 0)x - 8)x + 5)x - 7$ Factor x out of first 2 terms.

EXAMPLE 3   *Evaluating a Polynomial Function in Real Life*

PHOTOGRAPHY The time t (in seconds) it takes a camera battery to recharge after flashing n times can be modeled by $t = 0.000015n^3 - 0.0034n^2 + 0.25n + 5.3$. Find the recharge time after 100 flashes. ▶ Source: *Popular Photography*

SOLUTION

100	0.000015	−0.0034	0.25	5.3
		0.0015	−0.19	6
	0.000015	−0.0019	0.06	11.3

▶ The recharge time is about 11 seconds.

> **FOCUS ON CAREERS**
>
>
>
> ⮕ **PHOTOGRAPHER**
> Some photographers work in advertising, some work for newspapers, and some are self-employed. Others specialize in aerial, police, medical, or scientific photography.
> **CAREER LINK**
> www.mcdougallittell.com

GOAL 2 GRAPHING POLYNOMIAL FUNCTIONS

STUDENT HELP

► **Look Back**
For help with graphing functions, see pp. 69 and 250.

The **end behavior** of a polynomial function's graph is the behavior of the graph as x approaches positive infinity $(+\infty)$ or negative infinity $(-\infty)$. The expression $x \to +\infty$ is read as "x approaches positive infinity."

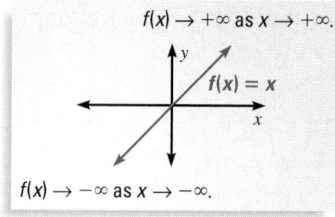

$f(x) \to +\infty$ as $x \to +\infty$.

$f(x) = x$

$f(x) \to -\infty$ as $x \to -\infty$.

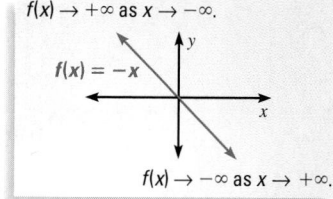

$f(x) \to +\infty$ as $x \to -\infty$.

$f(x) = -x$

$f(x) \to -\infty$ as $x \to +\infty$.

1a. $f(x) \to -\infty$ as $x \to -\infty$ and $f(x) \to +\infty$ as $x \to +\infty$

1b. $f(x) \to +\infty$ as $x \to -\infty$ and $f(x) \to +\infty$ as $x \to +\infty$

1c. $f(x) \to -\infty$ as $x \to -\infty$ and $f(x) \to +\infty$ as $x \to +\infty$

1d. $f(x) \to +\infty$ as $x \to -\infty$ and $f(x) \to +\infty$ as $x \to +\infty$

1e. $f(x) \to +\infty$ as $x \to -\infty$ and $f(x) \to -\infty$ as $x \to +\infty$

1f. $f(x) \to -\infty$ as $x \to -\infty$ and $f(x) \to -\infty$ as $x \to +\infty$

1g. $f(x) \to +\infty$ as $x \to -\infty$ and $f(x) \to -\infty$ as $x \to +\infty$

1h. $f(x) \to -\infty$ as $x \to -\infty$ and $f(x) \to -\infty$ as $x \to +\infty$

2. If the leading coefficient is positive, the values of the function approach $+\infty$; if the coefficient is negative, the values of the function approach $-\infty$.

3. When the function's degree is odd, the ends will go in opposite directions. When the function's degree is even, the ends go in the same direction.

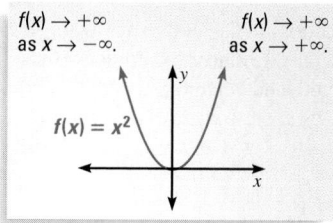

$f(x) \to +\infty$ as $x \to -\infty$. $\qquad f(x) \to +\infty$ as $x \to +\infty$.

$f(x) = x^2$

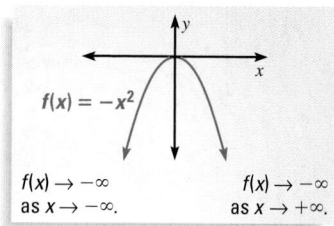

$f(x) = -x^2$

$f(x) \to -\infty$ as $x \to -\infty$. $\qquad f(x) \to -\infty$ as $x \to +\infty$.

▶ ACTIVITY

Developing Concepts

📟 Investigating End Behavior

① Use a graphing calculator to graph each function. Then complete these statements: $f(x) \to \underline{\ ?\ }$ as $x \to -\infty$ and $f(x) \to \underline{\ ?\ }$ as $x \to +\infty$. **1–3. See margin.**

a. $f(x) = x^3$ **b.** $f(x) = x^4$ **c.** $f(x) = x^5$ **d.** $f(x) = x^6$

e. $f(x) = -x^3$ **f.** $f(x) = -x^4$ **g.** $f(x) = -x^5$ **h.** $f(x) = -x^6$

② How does the sign of the leading coefficient affect the behavior of a polynomial function's graph as $x \to +\infty$?

③ How is the behavior of a polynomial function's graph as $x \to +\infty$ related to its behavior as $x \to -\infty$ when the function's degree is odd? when it is even?

In the activity you may have discovered that the end behavior of a polynomial function's graph is determined by the function's degree and leading coefficient.

CONCEPT SUMMARY   END BEHAVIOR FOR POLYNOMIAL FUNCTIONS

The graph of $f(x) = a_nx^n + a_{n-1}x^{n-1} + \cdots + a_1x + a_0$ has this end behavior:

- For $a_n > 0$ and n even, $f(x) \to +\infty$ as $x \to -\infty$ and $f(x) \to +\infty$ as $x \to +\infty$.
- For $a_n > 0$ and n odd, $f(x) \to -\infty$ as $x \to -\infty$ and $f(x) \to +\infty$ as $x \to +\infty$.
- For $a_n < 0$ and n even, $f(x) \to -\infty$ as $x \to -\infty$ and $f(x) \to -\infty$ as $x \to +\infty$.
- For $a_n < 0$ and n odd, $f(x) \to +\infty$ as $x \to -\infty$ and $f(x) \to -\infty$ as $x \to +\infty$.

6.2 *Evaluating and Graphing Polynomial Functions*   **331**

EXTRA EXAMPLE 3
To disarm a business security panel, you must push 4 buttons, no two of which have the same number or letter. The total number of ways to set the system is modeled by $f(b) = b^4 - 6b^3 + 11b^2 - 6b$, where b is the number of buttons on the panel. Find how many ways the system can be set if there are 12 buttons on the panel. **11,880 ways**

☑ **CHECKPOINT EXERCISES**
For use after Example 3:
1. You are hiking in a wet area where a 10-foot board has been extended on logs to create a temporary bridge. How much the bridge sags as you walk on it can be modeled by $y = -4x^4 + 30x^3 - 200x^2$, where x is the distance in feet from the beginning of the board and y is the distance (in thousandths of an inch) below the level position of the board. How far below its level position is the board when you are 4 feet from its beginning? **about 2.3 inches**

CAREER NOTE
Additional information about photographers is available at **www.mcdougallittell.com.**

EXTRA EXAMPLE 4
Graph $f(x) = x^3 + 2x^2 - x + 3$.

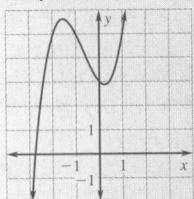

EXTRA EXAMPLE 5
The number of new words students in a language course were asked to learn each week is modeled by $y = 0.0003x^3 + 50$, where x is the number of weeks since the course began. Graph the model. Use the graph to estimate the number of words the students must learn in week 32.

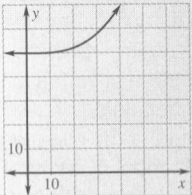

about 60 words

✓ **CHECKPOINT EXERCISES**

For use after Examples 4 and 5:

1. A tourist agency found that the number of tickets it typically sells for trips to Mexico during the first 40 weeks of the year is modeled by $y = 0.003x^3 - 0.15x^2 + 2x + 8$, where x is the week number. Graph the model. Use the graph to estimate the number of tickets to Mexico sold in week 35.

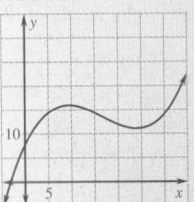

about 23 tickets

CLOSURE QUESTION
Which term of a polynomial function is most important in determining the end behavior of the function?
The term of highest degree

332

STUDENT HELP

HOMEWORK HELP
Visit our Web site
www.mcdougallittell.com
for extra examples.

EXAMPLE 4 *Graphing Polynomial Functions*

Graph (**a**) $f(x) = x^3 + x^2 - 4x - 1$ and (**b**) $f(x) = -x^4 - 2x^3 + 2x^2 + 4x$.

SOLUTION

a. To graph the function, make a table of values and plot the corresponding points. Connect the points with a smooth curve and check the end behavior.

x	−3	−2	−1	0	1	2	3
f(x)	−7	3	3	−1	−3	3	23

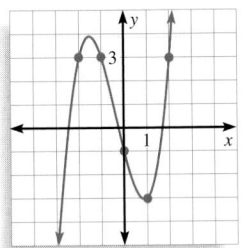

The degree is odd and the leading coefficient is positive, so $f(x) \to -\infty$ as $x \to -\infty$ and $f(x) \to +\infty$ as $x \to +\infty$.

b. To graph the function, make a table of values and plot the corresponding points. Connect the points with a smooth curve and check the end behavior.

x	−3	−2	−1	0	1	2	3
f(x)	−21	0	−1	0	3	−16	−105

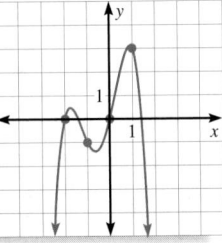

The degree is even and the leading coefficient is negative, so $f(x) \to -\infty$ as $x \to -\infty$ and $f(x) \to -\infty$ as $x \to +\infty$.

Biology

EXAMPLE 5 *Graphing a Polynomial Model*

A rainbow trout can grow up to 40 inches in length. The weight y (in pounds) of a rainbow trout is related to its length x (in inches) according to the model $y = 0.0005x^3$. Graph the model. Use your graph to estimate the length of a 10 pound rainbow trout.

SOLUTION

Make a table of values. The model makes sense only for positive values of x.

x	0	5	10	15	20	25	30	35	40
y	0	0.0625	0.5	1.69	4	7.81	13.5	21.4	32

Plot the points and connect them with a smooth curve, as shown at the right. Notice that the leading coefficient of the model is positive and the degree is odd, so the graph rises to the right.

Read the graph backwards to see that $x \approx 27$ when $y = 10$.

▶ A 10 pound trout is approximately 27 inches long.

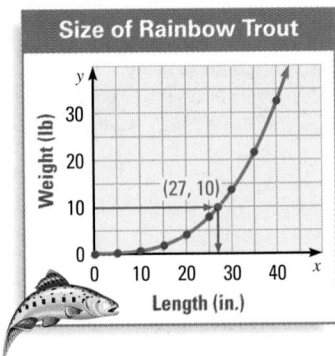

Size of Rainbow Trout

(27, 10)

Weight (lb) / Length (in.)

GUIDED PRACTICE

Vocabulary Check ✓

1. Identify the degree, type, leading coefficient, and constant term of the polynomial function $f(x) = 5x - 2x^3$. **3, cubic, −2, 0**

Concept Check ✓

2. Complete the synthetic substitution shown at the right. Describe each step of the process.
See margin.

$$
\begin{array}{r|rrrr}
-2 & 3 & 1 & -9 & 2 \\
   &   & ? & ? & ? \\
\hline
   & 3 & ? & ? & 0
\end{array}
$$

3. Describe the graph of a constant function.
a horizontal line

Skill Check ✓

Decide whether each function is a polynomial function. If it is, use synthetic substitution to evaluate the function when $x = -1$.

4. $f(x) = x^4\sqrt{5} - x$ yes; $\sqrt{5} + 1$

5. $f(x) = x^3 + x^2 - x^{-3} + 3$ no

6. $f(x) = 6^{2x} - 12x$ no

7. $f(x) = 14 - 21x^2 + 5x^4$ yes; −2

Describe the end behavior of the graph of the polynomial function by completing the statements $f(x) \to \underline{\ ?\ }$ as $x \to -\infty$ and $f(x) \to \underline{\ ?\ }$ as $x \to +\infty$.
8–13. See margin.

8. $f(x) = x^3 - 5x$

9. $f(x) = -x^5 - 3x^3 + 2$

10. $f(x) = x^4 - 4x^2 + x$

11. $f(x) = x + 12$

12. $f(x) = -x^2 + 3x + 1$

13. $f(x) = -x^8 + 9x^5 - 2x^4$

14. 🌐 **VIDEO RENTALS** The total revenue (actual and projected) from home video rentals in the United States from 1985 to 2005 can be modeled by

$$R = 1.8t^3 - 76t^2 + 1099t + 2600$$

where R is the revenue (in millions of dollars) and t is the number of years since 1985. Graph the function. ▶ Source: *The Wall Street Journal Almanac*
See margin.

PRACTICE AND APPLICATIONS

STUDENT HELP

▶ **Extra Practice**
to help you master skills is on p. 947.

CLASSIFYING POLYNOMIALS Decide whether the function is a polynomial function. If it is, write the function in standard form and state the degree, type, and leading coefficient.   **15–26. See margin.**

15. $f(x) = 12 - 5x$

16. $f(x) = 2x + \frac{3}{5}x^4 + 9$

17. $f(x) = x + \pi$

18. $f(x) = x^2\sqrt{2} + x - 5$

19. $f(x) = x - 3x^{-2} - 2x^3$

20. $f(x) = -2$

21. $f(x) = x^2 - x + 1$

22. $f(x) = 22 - 19x + 2^x$

23. $f(x) = 36x^2 - x^3 + x^4$

24. $f(x) = 3x^2 - 2x^{-x}$

25. $f(x) = 3x^3$

26. $f(x) = -6x^2 + x - \frac{3}{x}$

STUDENT HELP

▶ **HOMEWORK HELP**
Example 1: Exs. 15–26
Example 2: Exs. 37–46
Example 3: Exs. 81, 82
Example 4: Exs. 47–79
Example 5: Exs. 83–86

DIRECT SUBSTITUTION Use direct substitution to evaluate the polynomial function for the given value of x.

27. $f(x) = 2x^3 + 5x^2 + 4x + 8, x = -2$ **4**

28. $f(x) = 2x^3 - x^4 + 5x^2 - x, x = 3$ **15**

29. $f(x) = x + \frac{1}{2}x^3, x = 4$ **36**

30. $f(x) = x^2 - x^5 + 1, x = -1$ **3**

31. $f(x) = 5x^4 - 8x^3 + 7x^2, x = 1$ **4**

32. $f(x) = x^3 + 3x^2 - 2x + 5, x = -3$ **11**

33. $f(x) = 11x^3 - 6x^2 + 2, x = 0$ **2**

34. $f(x) = x^4 - 2x + 7, x = 2$ **19**

35. $f(x) = 7x^3 + 9x^2 + 3x, x = 10$ **7930**

36. $f(x) = -x^5 - 4x^3 + 6x^2 - x, x = -2$ **90**

6.2 *Evaluating and Graphing Polynomial Functions*   333

3 APPLY

🔵 **ASSIGNMENT GUIDE**

BASIC
Day 1: pp. 333–336 Exs. 16–32 even, 38–44 even, 50–60 even, 65–71 odd, 81, 83, 87, 91–107 odd

AVERAGE
Day 1: pp. 333–336 Exs. 16–58 even, 65–73 odd, 80, 82–84, 87, 91–107 odd

ADVANCED
Day 1: pp. 333–336 Exs. 16–60 even, 65–75 odd, 80–89, 91–107 odd

BLOCK SCHEDULE
pp. 333–336 Exs. 16–58 even, 65–73 odd, 80, 82–84, 87, 91–107 odd (with 6.1)

EXERCISE LEVELS
Level A: *Easier*
15–46, 65–73

Level B: *More Difficult*
47–64, 74–87

Level C: *Most Difficult*
88

✓ **HOMEWORK CHECK**
To quickly check student understanding of key concepts, go over the following exercises:
Exs. 16, 24, 28, 38, 50. See also the Daily Homework Quiz:

• Blackline Master (*Chapter 6 Resource Book*, p. 36)

• 📷 Transparency (p. 43)

2. See Additional Answers beginning on page AA1.

8. $f(x) \to -\infty$ as $x \to -\infty$ and $f(x) \to +\infty$ as $x \to +\infty$

9. $f(x) \to +\infty$ as $x \to -\infty$ and $f(x) \to -\infty$ as $x \to +\infty$

10. $f(x) \to +\infty$ as $x \to -\infty$ and $f(x) \to +\infty$ as $x \to +\infty$

11. $f(x) \to -\infty$ as $x \to -\infty$ and $f(x) \to +\infty$ as $x \to +\infty$

12. $f(x) \to -\infty$ as $x \to -\infty$ and $f(x) \to -\infty$ as $x \to +\infty$

13. $f(x) \to -\infty$ as $x \to -\infty$ and $f(x) \to -\infty$ as $x \to +\infty$

14–26. See next page.

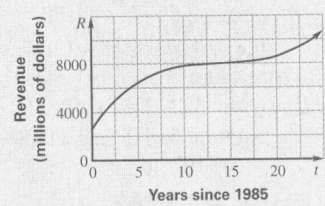

! **COMMON ERROR**

EXERCISES 37–46 Students often forget to include zeros for the coefficients of missing terms. Remind students to write each polynomial in standard form, including all powers of *x* from *n* to 0, where *n* is the degree of the polynomial.

14.

Total Revenue from Home Video Rentals

(graph: Revenue (millions of dollars) R vs Years since 1985 t)

15. polynomial; $f(x) = -5x + 12$; 1; linear; –5

16. polynomial; $f(x) = \frac{3}{5}x^4 + 2x + 9$; 4; quartic; $\frac{3}{5}$

17. polynomial; $f(x) = x + \pi$; 1; linear; 1

18. polynomial; $f(x) = x^2\sqrt{2} + x - 5$; 2; quadratic; $\sqrt{2}$

19. not a polynomial

20. polynomial; $f(x) = -2$; 0; constant; –2

21. polynomial; $f(x) = x^2 - x + 1$; 2; quadratic; 1

22. not a polynomial

23. polynomial; $f(x) = x^4 - x^3 + 36x^2$; 4; quartic; 1

24. not a polynomial

25. polynomial; $f(x) = 3x^3$; 3; cubic; 3

26. not a polynomial

47–48. See Additional Answers beginning on page AA1.

53. $f(x) \to -\infty$ as $x \to -\infty$ and $f(x) \to -\infty$ as $x \to +\infty$

54. $f(x) \to -\infty$ as $x \to -\infty$ and $f(x) \to -\infty$ as $x \to +\infty$

55. $f(x) \to -\infty$ as $x \to -\infty$ and $f(x) \to +\infty$ as $x \to +\infty$

56. $f(x) \to +\infty$ as $x \to -\infty$ and $f(x) \to -\infty$ as $x \to +\infty$

57. $f(x) \to -\infty$ as $x \to -\infty$ and $f(x) \to -\infty$ as $x \to +\infty$

58. $f(x) \to -\infty$ as $x \to -\infty$ and $f(x) \to +\infty$ as $x \to +\infty$

59. $f(x) \to +\infty$ as $x \to -\infty$ and $f(x) \to -\infty$ as $x \to +\infty$

60. $f(x) \to -\infty$ as $x \to -\infty$ and $f(x) \to +\infty$ as $x \to +\infty$

61. $f(x) \to +\infty$ as $x \to -\infty$ and $f(x) \to +\infty$ as $x \to +\infty$

62. $f(x) \to +\infty$ as $x \to -\infty$ and $f(x) \to +\infty$ as $x \to +\infty$

63. $f(x) \to +\infty$ as $x \to -\infty$ and $f(x) \to -\infty$ as $x \to +\infty$

64. $f(x) \to +\infty$ as $x \to -\infty$ and $f(x) \to +\infty$ as $x \to +\infty$

SYNTHETIC SUBSTITUTION Use synthetic substitution to evaluate the polynomial function for the given value of *x*.

37. $f(x) = 5x^3 + 4x^2 + 8x + 1, x = 2$ 73

38. $f(x) = -3x^3 + 7x^2 - 4x + 8, x = 3$ –22

39. $f(x) = x^3 + 3x^2 + 6x - 11, x = -5$ –91

40. $f(x) = x^3 - x^2 + 12x + 15, x = -1$ 1

41. $f(x) = -4x^3 + 3x - 5, x = 2$ –31

42. $f(x) = -x^4 + x^3 - x + 1, x = -3$ –104

43. $f(x) = 2x^4 + x^3 - 3x^2 + 5x, x = -1$ –7

44. $f(x) = 3x^5 - 2x^2 + x, x = 2$ 90

45. $f(x) = 2x^3 - x^2 + 6x, x = 5$ 255

46. $f(x) = -x^4 + 8x^3 + 13x - 4, x = -2$ –110

END BEHAVIOR PATTERNS Graph each polynomial function in the table. Then copy and complete the table to describe the end behavior of the graph of each function.   47, 48. See margin.

47.

Function	As $x \to -\infty$	As $x \to +\infty$
$f(x) = -5x^3$	?	?
$f(x) = -x^3 + 1$	?	?
$f(x) = 2x - 3x^3$	?	?
$f(x) = 2x^2 - x^3$	?	?

48.

Function	As $x \to -\infty$	As $x \to +\infty$
$f(x) = x^4 + 3x^3$	?	?
$f(x) = x^4 + 2$	?	?
$f(x) = x^4 - 2x - 1$	?	?
$f(x) = 3x^4 - 5x^2$	?	?

MATCHING Use what you know about end behavior to match the polynomial function with its graph.

49. $f(x) = 4x^6 - 3x^2 + 5x - 2$ C

50. $f(x) = -2x^3 + 5x^2$ D

51. $f(x) = -x^4 + 1$ B

52. $f(x) = 6x^3 + 1$ A

A.
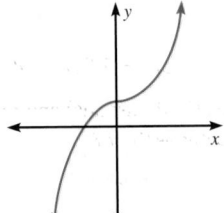

B.
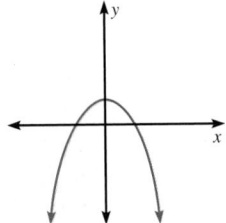

C.
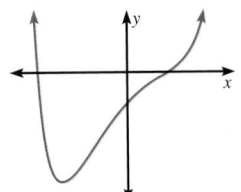

D.
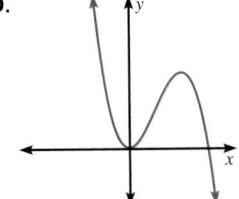

DESCRIBING END BEHAVIOR Describe the end behavior of the graph of the polynomial function by completing these statements: $f(x) \to \underline{?}$ as $x \to -\infty$ and $f(x) \to \underline{?}$ as $x \to +\infty$.   53–64. See margin.

53. $f(x) = -5x^4$

54. $f(x) = -x^2 + 1$

55. $f(x) = 2x$

56. $f(x) = -10x^3$

57. $f(x) = -x^6 + 2x^3 - x$

58. $f(x) = x^5 + 2x^2$

59. $f(x) = -3x^5 - 4x^2 + 3$

60. $f(x) = x^7 - 3x^3 + 2x$

61. $f(x) = 3x^6 - x - 4$

62. $f(x) = 3x^8 - 4x^3$

63. $f(x) = -6x^3 + 10x$

64. $f(x) = x^4 - 5x^3 + x - 1$

GRAPHING POLYNOMIALS **Graph the polynomial function.** 65–79. See margin.

65. $f(x) = -x^3$

66. $f(x) = -x^4$

67. $f(x) = x^5 + 2$

68. $f(x) = x^4 - 4$

69. $f(x) = x^4 + 6x^2 - 5$

70. $f(x) = 2 - x^3$

71. $f(x) = x^5 - 2$

72. $f(x) = -x^4 + 3$

73. $f(x) = -x^3 + 3x$

74. $f(x) = -x^3 + 2x^2 - 4$

75. $f(x) = -x^5 + x^2 + 1$

76. $f(x) = x^3 - 3x - 1$

77. $f(x) = x^5 + 3x^3 - x$

78. $f(x) = x^4 - 2x - 3$

79. $f(x) = -x^4 + 2x - 1$

80. Answers may vary. Any polynomial function of odd degree that has a positive leading coefficient will work; *sample answer:* $f(x) = 4x^3$.

80. **CRITICAL THINKING** Give an example of a polynomial function f such that $f(x) \rightarrow -\infty$ as $x \rightarrow -\infty$ and $f(x) \rightarrow +\infty$ as $x \rightarrow +\infty$.

83. $f(x) \rightarrow -\infty$ as $x \rightarrow -\infty$ and $f(x) \rightarrow -\infty$ as $x \rightarrow +\infty$; less; the graph will tend to go down over time.

81. **SHOPPING** The retail space in shopping centers in the United States from 1972 to 1996 can be modeled by   about 4272.9 million ft^2

$$S = -0.0068t^3 - 0.27t^2 + 150t + 1700$$

where S is the amount of retail space (in millions of square feet) and t is the number of years since 1972. How much retail space was there in 1990?

85. $f(x) \rightarrow +\infty$ as $x \rightarrow -\infty$ and $f(x) \rightarrow +\infty$ as $x \rightarrow +\infty$; more; the graph will tend to go up over time.

82. **CABLE TELEVISION** The average monthly cable TV rate from 1980 to 1997 can be modeled by   about $8.55

$$R = -0.0036t^3 + 0.13t^2 - 0.073t + 7.7$$

where R is the monthly rate (in dollars) and t is the number of years since 1980. What was the monthly rate in 1983?

NURSING In Exercises 83 and 84, use the following information.
From 1985 to 1995, the number of graduates from nursing schools in the United States can be modeled by

$$y = -0.036t^4 + 0.605t^3 - 1.87t^2 - 4.67t + 82.5$$

where y is the number of graduates (in thousands) and t is the number of years since 1985. ▶ Source: *Statistical Abstract of the United States*

FOCUS ON CAREERS

83. Describe the end behavior of the graph of the function. From the end behavior, would you expect the number of nursing graduates in the year 2010 to be more than or less than the number of nursing graduates in 1995? Explain.   See margin.

84. Graph the function for $0 \le t \le 10$. Use the graph to find the first year in which there were over 82,500 nursing graduates.   1992; See margin.

TENNIS In Exercises 85 and 86, use the following information.
The amount of prize money for the women's U.S. Open Tennis Tournament from 1970 to 1997 can be modeled by

$$P = 1.141t^2 - 5.837t + 14.31$$

NURSE
Although the majority of nurses work in hospitals, nurses also work in doctors' offices, in private homes, at nursing homes, and in other community settings.
CAREER LINK
www.mcdougallittell.com

where P is the prize money (in thousands of dollars) and t is the number of years since 1970. ▶ Source: U.S. Open

85. Describe the end behavior of the graph of the function. From the end behavior, would you expect the amount of prize money in the year 2005 to be more than or less than the amount in 1995? Explain.   See margin.

86. Graph the function for $0 \le t \le 40$. Use the graph to estimate the amount of prize money in the year 2005.   about $1,208,000; See margin.

TEACHING TIPS
EXERCISES 65–79 Encourage students to use a graphing calculator for these exercises. Even when sketching graphs on graph paper, students will find a graphing calculator helpful in determining what values of x and y are needed to give a good idea of the behavior of the function.

CAREER NOTE
EXERCISES 83 AND 84
Additional information about nurses is available at **www.mcdougallittell.com.**

APPLICATION NOTE
EXERCISES 83 AND 84
Depending on the program from which they graduated, nursing graduates can earn a diploma, an associate degree, a baccalaureate degree, a master's degree, or a doctoral degree. From 1992 to 1996, the number of male registered nurses increased from 4.3% to 5.4% of the total number of registered nurses in the United States.

65–79, 84, 86.
See Additional Answers beginning on page AA1.

ADDITIONAL PRACTICE AND RETEACHING

For Lesson 6.2:
• Practice Levels A, B, and C (*Chapter 6 Resource Book,* p. 26)
• Reteaching with Practice (*Chapter 6 Resource Book,* p. 29)
• See Lesson 6.2 of the *Personal Student Tutor*

For more Mixed Review:
• Search the *Test and Practice Generator* for key words or specific lessons.

DAILY HOMEWORK QUIZ

Transparency Available

1. Tell whether $f(x) = \frac{3x}{5} - x^3 + 4x^2 + 7$ is a polynomial function. If it is, write it in standard form and state the degree, type, and leading coefficient.

   yes; $f(x) = -x^3 + 4x^2 + \frac{3}{5}x + 7$;

   3; cubic; –1

2. Use synthetic division to evaluate $f(x) = 2x^4 + x^3 - 3x^2 + 6$ for $x = -2$. 18

3. Use what you know about end behavior to match the graph with one of the polynomial functions. b

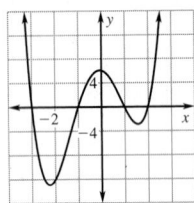

   (a) $f(x) = -2x^3 + 4x^2 + 6$
   (b) $f(x) = x^4 + x^3 - 7x^2 - x + 6$
   (c) $f(x) = -x^6 + 3x^5 - x^2 + 6$
   (d) $f(x) = x^5 - x^3 + 5x^2 + 6$

EXTRA CHALLENGE NOTE

Challenge problems for Lesson 6.2 are available in **blackline** format in the *Chapter 6 Resource Book*, p. 33 and at **www.mcdougallittell.com**.

ADDITIONAL TEST PREPARATION

1. **OPEN ENDED** Give an example of a function that is a polynomial function and an example of a function that is not a polynomial function. **See margin.**

2. **WRITING** For a particular function, $f(x) \to +\infty$ as $x \to -\infty$. Explain the meaning of this notation in terms of the size of the values of x and $f(x)$. **See margin.**

87c–d, 88. See Additional Answers beginning on page AA1.

 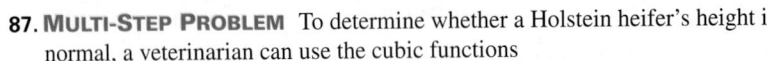
87a. 50.7 in. to 52.3 in.

87b. for both, $f(x) \to -\infty$ as $x \to -\infty$ and $f(x) \to +\infty$ as $x \to +\infty$

87c. See margin.

87d. See margin.

★ **Challenge**

EXTRA CHALLENGE

www.mcdougallittell.com

---

87. **MULTI-STEP PROBLEM** To determine whether a Holstein heifer's height is normal, a veterinarian can use the cubic functions

$$L = 0.0007t^3 - 0.061t^2 + 2.02t + 30$$

$$H = 0.001t^3 - 0.08t^2 + 2.3t + 31$$

where L is the minimum normal height (in inches), H is the maximum normal height (in inches), and t is the age (in months).

▶ Source: *Journal of Dairy Science*

a. What is the normal height range for an 18-month-old Holstein heifer?

b. Describe the end behavior of each function's graph.

c. Graph the two height functions.

d. *Writing* Suppose a veterinarian examines a Holstein heifer that is 43 inches tall. About how old do you think the cow is? How did you get your answer?

A heifer is a young cow that has not yet had calves.

EXAMINING END BEHAVIOR Use a spreadsheet or a graphing calculator to evaluate the polynomial functions $f(x) = x^3$ and $g(x) = x^3 - 2x^2 + 4x + 5$ for the given values of x.

88. Copy and complete the table. See margin.

89. Use the results of Exercise 88 to complete this statement:

As $x \to +\infty$, $\frac{f(x)}{g(x)} \to \underline{?}$.

Explain how this statement shows that the functions f and g have the same end behavior as $x \to +\infty$.

1; Eventually the combined values of the terms after the leading term will be negligible compared to the value of the leading term.

x	$f(x)$	$g(x)$	$\frac{f(x)}{g(x)}$
50	?	?	?
100	?	?	?
500	?	?	?
1000	?	?	?
5000	?	?	?

MIXED REVIEW

90. $-2x + 5$

91. $7x$

92. $-4x^2 - 1$

93. $x^2 + 4x - 11$

94. $-3x^2 + 4x - 1$

95. $-x^2 - x + 2$

96. $y = -4x^2 + 16x - 11$

97. $y = -2x^2 - 2x + 60$

98. $y = 2x^2 - 6x - 56$

99. $y = 4x^2 - 24x + 12$

100. $y = -x^2 - 10x - 13$

101. $y = -3x^2 + 30x - 72$

SIMPLIFYING EXPRESSIONS Simplify the expression. (Review 1.2 for 6.3)

90. $x + 3 - 2x - x + 2$ 91. $-2x^2 + 3x + 4x + 2x^2$ 92. $-3x^2 + 1 - (x^2 + 2)$

93. $x^2 + x + 1 + 3(x - 4)$ 94. $4x - 2x^2 + 3 - x^2 - 4$ 95. $x^2 - 1 - (2x^2 + x - 3)$

STANDARD FORM Write the quadratic function in standard form. (Review 5.1 for 6.3)

96. $y = -4(x - 2)^2 + 5$ 97. $y = -2(x + 6)(x - 5)$ 98. $y = 2(x - 7)(x + 4)$

99. $y = 4(x - 3)^2 - 24$ 100. $y = -(x + 5)^2 + 12$ 101. $y = -3(x - 5)^2 + 3$

SOLVING QUADRATIC EQUATIONS Solve the equation. (Review 5.4)

102. $x^2 = -9$ $\pm 3i$ 103. $x^2 = -5$ $\pm\sqrt{5}\,i$ 104. $-3x^2 + 1 = 7$ $\pm\sqrt{2}\,i$

105. $4x^2 + 15 = 3$ $\pm\sqrt{3}\,i$ 106. $6x^2 + 5 = 2x^2 + 1 \pm i$ 107. $x^2 = 7x^2 + 1$ $\pm\frac{\sqrt{6}}{6}i$

108. $x^2 - 4 = -3x^2 - 24$ 109. $3x^2 + 5 = 5x^2 + 10$ $\pm\frac{\sqrt{10}}{2}i$ 110. $5x^2 + 2 = -2x^2 + 1$ $\pm\frac{\sqrt{7}}{7}i$
   $\pm\sqrt{5}\,i$

Additional Test Preparation *Sample answer:*

1. $f(x) = 4x^4 - 5x^3 + x + 15$ is a polynominal function,

   but $f(x) = \frac{3}{x^2}$ is not.

2. As the negative values of x take on greater and greater absolute values, the values of $f(x)$ become larger and larger positive numbers.

Setting a Good Viewing Window

Graphing Calculator Activity for use with Lesson 6.2

When you graph a polynomial function with a graphing calculator, you must choose a viewing window that displays the important characteristics of the graph. Use what you know about end behavior to find such a viewing window.

▶ **EXAMPLE**

Graph $f(x) = 0.2x^3 - 5x^2 + 38x - 97$.

▶ **SOLUTION**

STUDENT HELP

INTERNET KEYSTROKE HELP

See keystrokes for several models of calculators at www.mcdougallittell.com

① Graph the function using the standard viewing window.

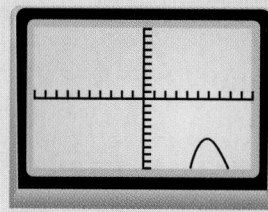

$-10 \le x \le 10, -10 \le y \le 10$

② Adjust the horizontal scale and the vertical scale until you see the graph's end behavior and any points where it turns. A good viewing window for this graph is $-10 \le x \le 20$ and $-20 \le y \le 10$.

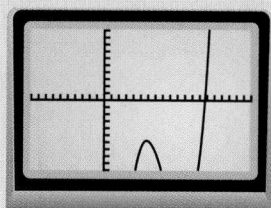

$-10 \le x \le 20, -10 \le y \le 10$

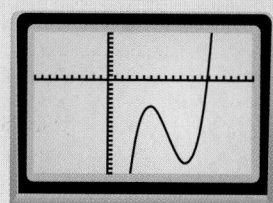

$-10 \le x \le 20, -20 \le y \le 10$

▶ **EXERCISES**

Find intervals for *x* and *y* that describe a good viewing window for the graph of the polynomial function. 1–7. Accept reasonable ranges; See margin.

1. $-10 \le x \le 10, -10 \le y \le 100$

2. $-10 \le x \le 30, 0 \le y \le 3000$

3. $-5 \le x \le 5, -5 \le y \le 10$

4. $-5 \le x \le 5, -5 \le y \le 30$

5. $-5 \le x \le 5, 0 \le y \le 20$

6. $0 \le x \le 5, -5 \le y \le 5$

7. $0 \le x \le 15, 0 \le y \le 300,000$

1. $f(x) = x^3 + 6x^2 - 11x + 3$

2. $f(x) = -x^3 + 25x^2 + 4$

3. $f(x) = x^4 - 5x^2 + 6$

4. $f(x) = -x^4 - 3x^3 + x^2 - x + 5$

5. $f(x) = -x^5 + 5x^3 - 4x + 10$

6. $f(x) = x^5 - 10x^4 + 35x^3 - 50x^2 + 24x$

7. 🌐 **EDUCATION** For 1983 to 1996, the amount *P* (in millions of dollars) spent by public elementary and secondary schools and the amount *R* (in millions of dollars) spent by private elementary and secondary schools can be modeled by

$$P = 11.7x^4 - 340x^3 + 2931x^2 + 1560x + 182,000$$

$$R = 0.422x^4 - 9.84x^3 + 44.9x^2 + 779x + 15,900$$

where *x* is the number of years since 1983. Find intervals for the horizontal and vertical axes that describe a good viewing window for the graphs of both functions. ▶ Source: U.S. National Center for Education Statistics

6.2 *Technology Activity* **337**

---

1 Planning the Activity

PURPOSE
To choose an appropriate window when using a graphing calculator to graph a polynomial function.

MATERIALS
• graphing calculators
• Keystroke blackline (*Chapter 6 Resource Book,* p. 25)

PACING
• Activity — 25 min

▶ **LINK TO LESSON**
Students can apply the rules they learned for determining end behavior to find an appropriate viewing window. In the example, since $f(x)$ eventually increases as *x* increases and $f(x)$ eventually decreases as *x* decreases, intervals should be chosen that show this clearly.

2 Managing the Activity

COOPERATIVE LEARNING
Have each student determine viewing windows for the exercises, then compare his or her window with another student's. Each pair of students should decide which window is better.

CLASSROOM MANAGEMENT
Students can work in groups of 3 or 4 to be sure they agree on the end behavior of each function before they graph the function.

3 Closing the Activity

⭐ **KEY DISCOVERY**
The end behavior and all turning points on the graph should be visible in the viewing window.

ACTIVITY ASSESSMENT
How would you adjust the viewing window if the graph looks as though turning points have been chopped off at the bottom?
See sample answer at left.

Activity Assessment *Sample answer:*
Press the WINDOW key and use a smaller value for Ymin.

LESSON OPENER
ACTIVITY
An alternative way to approach Lesson 6.3 is to use the Activity Lesson Opener:

• Blackline Master (*Chapter 6 Resource Book*, p. 37)
• Transparency (p. 37)

MEETING INDIVIDUAL NEEDS
• *Chapter 6 Resource Book*
  Prerequisite Skills Review (p. 5)
  Practice Level A (p. 40)
  Practice Level B (p. 41)
  Practice Level C (p. 42)
  Reteaching with Practice (p. 43)
  Absent Student Catch-Up (p. 45)
  Challenge (p. 47)
• *Resources in Spanish*
• Personal Student Tutor

NEW-TEACHER SUPPORT
See the Tips for New Teachers on pp. 1–2 of the *Chapter 6 Resource Book* for additional notes about Lesson 6.3.

WARM-UP EXERCISES

Transparency Available
Simplify.
1. $2(x-5)$ $2x-10$
2. $-4(x^2-5x+1)$ $-4x^2+20x-4$
3. $(3x)(2x)$ $6x^2$
4. $a \cdot a^3 \cdot a^3$ a^7
5. $7m^2-12m^2$ $-5m^2$

6.3 Adding, Subtracting, and Multiplying Polynomials

What you should learn

GOAL 1 Add, subtract, and multiply polynomials.

GOAL 2 Use polynomial operations in **real-life** problems, such as finding net farm income in **Example 7**.

Why you should learn it

▼ To combine **real-life** polynomial models into a new model, such as the model for the power needed to keep a bicycle moving at a certain speed in **Ex. 66**.

REAL LIFE

GOAL 1 ADDING, SUBTRACTING, AND MULTIPLYING

To add or subtract polynomials, add or subtract the coefficients of like terms. You can use a vertical or horizontal format.

EXAMPLE 1 *Adding Polynomials Vertically and Horizontally*

Add the polynomials.

a.
$$\begin{array}{r} 3x^3 + 2x^2 - x - 7 \\ + \ x^3 - 10x^2 \quad\ + 8 \\ \hline 4x^3 - 8x^2 - x + 1 \end{array}$$

b. $(9x^3 - 2x + 1) + (5x^2 + 12x - 4) = 9x^3 + 5x^2 - 2x + 12x + 1 - 4$
$$= 9x^3 + 5x^2 + 10x - 3$$

EXAMPLE 2 *Subtracting Polynomials Vertically and Horizontally*

Subtract the polynomials.

a.
$$\begin{array}{r} 8x^3 - 3x^2 - 2x + 9 \\ - (2x^3 + 6x^2 - \ x + 1) \end{array}$$
→
$$\begin{array}{r} 8x^3 - 3x^2 - 2x + 9 \\ -2x^3 - 6x^2 + \ x - 1 \\ \hline 6x^3 - 9x^2 - \ x + 8 \end{array}$$
Add the opposite.

b. $(2x^2 + 3x) - (3x^2 + x - 4) = 2x^2 + 3x - 3x^2 - x + 4$ Add the opposite.
$$= -x^2 + 2x + 4$$

· · · · · · · · · ·

To multiply two polynomials, each term of the first polynomial must be multiplied by each term of the second polynomial.

EXAMPLE 3 *Multiplying Polynomials Vertically*

Multiply the polynomials.

$$\begin{array}{r} -x^2 + 2x + \ 4 \\ \times \qquad\quad x - \ 3 \\ \hline 3x^2 - 6x - 12 \\ -x^3 + 2x^2 + 4x \\ \hline -x^3 + 5x^2 - 2x - 12 \end{array}$$

Multiply $-x^2 + 2x + 4$ by -3.

Multiply $-x^2 + 2x + 4$ by x.

Combine like terms.

STUDENT HELP

Look Back
For help with simplifying expressions, see p. 251.

EXAMPLE 4   *Multiplying Polynomials Horizontally*

Multiply the polynomials.

$$(x - 3)(3x^2 - 2x - 4) = (x - 3)3x^2 - (x - 3)2x - (x - 3)4$$
$$= 3x^3 - 9x^2 - 2x^2 + 6x - 4x + 12$$
$$= 3x^3 - 11x^2 + 2x + 12$$

EXAMPLE 5   *Multiplying Three Binomials*

STUDENT HELP

▶ **Look Back**
For help with multiplying binomials, see p. 251.

Multiply the polynomials.

$$(x - 1)(x + 4)(x + 3) = (x^2 + 3x - 4)(x + 3)$$
$$= (x^2 + 3x - 4)x + (x^2 + 3x - 4)3$$
$$= x^3 + 3x^2 - 4x + 3x^2 + 9x - 12$$
$$= x^3 + 6x^2 + 5x - 12$$

· · · · · · · · · ·

Some binomial products occur so frequently that it is worth memorizing their *special product patterns.* You can verify these products by multiplying.

SPECIAL PRODUCT PATTERNS

SUM AND DIFFERENCE   Example

$(a + b)(a - b) = a^2 - b^2$ $(x + 3)(x - 3) = x^2 - 9$

SQUARE OF A BINOMIAL

$(a + b)^2 = a^2 + 2ab + b^2$ $(y + 4)^2 = y^2 + 8y + 16$

$(a - b)^2 = a^2 - 2ab + b^2$ $(3t^2 - 2)^2 = 9t^4 - 12t^2 + 4$

CUBE OF A BINOMIAL

$(a + b)^3 = a^3 + 3a^2b + 3ab^2 + b^3$ $(x + 1)^3 = x^3 + 3x^2 + 3x + 1$

$(a - b)^3 = a^3 - 3a^2b + 3ab^2 - b^3$ $(p - 2)^3 = p^3 - 6p^2 + 12p - 8$

EXAMPLE 6   *Using Special Product Patterns*

Multiply the polynomials.

a. $(4n - 5)(4n + 5) = (4n)^2 - 5^2$ **Sum and difference**
$$= 16n^2 - 25$$

b. $(9y - x^2)^2 = (9y)^2 - 2(9y)(x^2) + (x^2)^2$ **Square of a binomial**
$$= 81y^2 - 18x^2y + x^4$$

c. $(ab + 2)^3 = (ab)^3 + 3(ab)^2(2) + 3(ab)(2)^2 + 2^3$ **Cube of a binomial**
$$= a^3b^3 + 6a^2b^2 + 12ab + 8$$

---

2 TEACH

EXTRA EXAMPLE 1
Add the polynomials.

a. $5x^2 + \ x - 7$
   $\underline{-3x^2 - 6x - 1}$
      $2x^2 - 5x - 8$

b. $(x^4 + 2x^3 + 8) + (2x^4 - 9)$
    $3x^4 + 2x^3 - 1$

EXTRA EXAMPLE 2
Subtract the polynomials.

a. $3x^3 + 8x^2 - x - \ 5$
   $\underline{-(5x^3 - \ x^2 \quad\ + 17)}$
    $-2x^3 + 9x^2 - x - 22$

b. $(9x^4 - 12x^3 + x^2 - 8) -$
    $(3x^4 - 12x^3 - x)$
    $6x^4 + x^2 + x - 8$

EXTRA EXAMPLE 3
Multiply the polynomials.
      $4x^2 + \ x - 5$
$\underline{\times \qquad\qquad 2x + 1}$
$8x^3 + 6x^2 - 9x - 5$

EXTRA EXAMPLE 4
Multiply $(x + 2)(5x^2 + 3x - 1)$.
$5x^3 + 13x^2 + 5x - 2$

EXTRA EXAMPLE 5
Multiply $(x - 2)(x - 1)(x + 3)$.
$x^3 - 7x + 6$

EXTRA EXAMPLE 6
Multiply the polynomials.
a. $(3x - 2)(3x + 2)$. $9x^2 - 4$
b. $(5a + 2)^2$ $25a^2 + 20a + 4$
c. $(2m - 3)^3$
    $8m^3 - 36m^2 + 54m - 27$

✔ **CHECKPOINT EXERCISES**

For use after Example 1:
1. Add.
     $10x^3 - 5x^2 - 2x + 9$
$\underline{+ \ \ 5x^3 - \ x^2 - 2x}$
    $15x^3 - 6x^2 - 4x + 9$

For use after Example 2:
2. Subtract. $(2x^4 + 2x^2 + x - 3) -$
    $(3x^4 - 2x^3 + x + 7)$
    $-x^4 + 2x^3 + 2x^2 - 10$

For use after Examples 3–6:
3. Multiply $(mn - 4)^3$.
    $m^3n^3 - 12m^2n^2 + 48mn - 64$

EXTRA EXAMPLE 7
From 1985 through 1996, the number of flu shots given in one city can be modeled by $A = -11.33x^4 - 8.325x^3 + 2194x^2 - 4190x + 7592$ for adults and by $C = -6.87x^4 + 106x^3 - 251x^2 + 135x + 540$ for children, where x is the number of years since 1985. Write a model for the total number T of flu shots given in these years. $T = -18.2x^4 + 97.675x^3 + 1943x^2 - 4055x + 8132$

EXTRA EXAMPLE 8
From 1990 through 1996, the number of students S enrolled during the fall semester and the average number of credits C carried by each student can be modeled by $S = 134.56x^2 - 1417x + 26,628$ and $C = 0.25x + 15$, where x represents the number of years since 1990. Write a model for the total number of credits T carried by students x years after 1990. $T = 33.64x^3 + 1664.15x^2 - 14,598x + 399,420$

CHECKPOINT EXERCISES
For use after Examples 7 and 8:
1. From 1990 through 1999, the amount spent per week for food by a typical employee of a company is $T = 0.0668x^3 - 0.561x^2 + 2.08x + 50$, where x is the number of years since 1990. The amount per employee spent for food prepared at home is $H = 0.185x^2 + 1.11x + 25$. The number of employees is $E = 2.5x + 47$. Write a model for the total amount N employees spent per week for food not prepared at home. $N = 0.167x^4 + 1.2746x^3 - 32.637x^2 + 108.09x + 1175$

CLOSURE QUESTION
How do you add or subtract two polynomials? Add or subtract the coefficients of like terms.

FOCUS ON APPLICATIONS

▶ **FARMING** The number of farms in the United States has been decreasing steadily since the 1930s. However, the average size of farms has been increasing.

Publishing

GOAL 2 USING POLYNOMIAL OPERATIONS IN REAL LIFE

EXAMPLE 7 *Subtracting Polynomial Models*

FARMING From 1985 through 1995, the gross farm income G and farm expenses E (in billions of dollars) in the United States can be modeled by

$$G = -0.246t^2 + 7.88t + 159 \quad \text{and} \quad E = 0.174t^2 + 2.54t + 131$$

where t is the number of years since 1985. Write a model for the *net* farm income N for these years. ▶ Source: U.S. Department of Agriculture

SOLUTION

To find a model for the net farm income, subtract the expenses model from the gross income model.

$$\begin{array}{r} -0.246t^2 + 7.88t + 159 \\ -\ (0.174t^2 + 2.54t + 131) \\ \hline -0.420t^2 + 5.34t + 28 \end{array}$$

▶ The net farm income can be modeled by $N = -0.42t^2 + 5.34t + 28$.

The graphs of the models are shown. Although G and E both increase, the net income N eventually decreases because E increases faster than G.

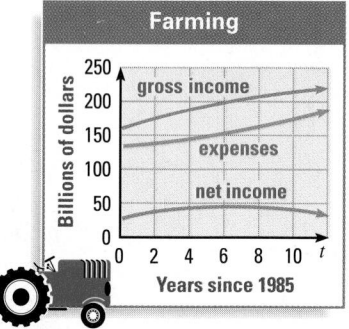

Farming

EXAMPLE 8 *Multiplying Polynomial Models*

From 1982 through 1995, the number of softbound books N (in millions) sold in the United States and the average price per book P (in dollars) can be modeled by

$$N = 1.36t^2 + 2.53t + 1076 \quad \text{and} \quad P = 0.314t + 3.42$$

where t is the number of years since 1982. Write a model for the total revenue R received from the sales of softbound books. What was the total revenue from softbound books in 1990? ▶ Source: Book Industry Study Group, Inc.

SOLUTION

To find a model for R, multiply the models for N and P.

$$\begin{array}{r} 1.36t^2 + 2.53t + 1076 \\ \times 0.314t + 3.42 \\ \hline 4.6512t^2 + 8.6526t + 3679.92 \\ 0.42704t^3 + 0.79442t^2 + 337.864t \\ \hline 0.42704t^3 + 5.44562t^2 + 346.5166t + 3679.92 \end{array}$$

▶ The total revenue can be modeled by $R = 0.427t^3 + 5.45t^2 + 347t + 3680$. The graph of the revenue model is shown at the right. By substituting $t = 8$ into the model for R, you can calculate that the revenue was about $7020 million, or $7.02 billion, in 1990.

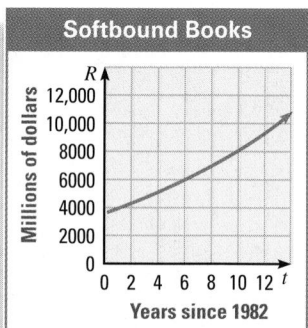

Softbound Books

GUIDED PRACTICE

Vocabulary Check ✔

Concept Check ✔

2. The negative sign was not distributed over all of the second polynomial.

Skill Check ✔

8. $2x^3 + 4x^2 + 3x + 6$

9. $4x^4 + 10x^3 + 27x^2 - 41x - 70$

10. $2x^3 + 9x^2 - 6x - 5$

21. $-7x^3 - x^2 + 2x - 11$
22. $-13x^2 - 15x - 4$

1. When you add or subtract polynomials, you add or subtract the coefficients of $\underline{\ ?\ }$.
  like terms

2. **ERROR ANALYSIS** Describe the error in the subtraction shown below.

$$(x^2 - 3x + 4) - (x^2 + 7x - 2) = x^2 - 3x + 4 - x^2 + 7x - 2$$
$$= 4x + 2$$

3. When you multiply a polynomial of degree 2 by a polynomial of degree 4, what is the degree of the product?   6

Perform the indicated operation.

4. $(4x^2 + 3) + (3x^2 + 8)$ $7x^2 + 11$

5. $(2x^3 - 4x^2 + 5) + (-x^2 - 3x + 1)$
    $2x^3 - 5x^2 - 3x + 6$

6. $(x^2 + 7x - 5) - (3x^2 + 1)$
    $-2x^2 + 7x - 6$

7. $(x^2 + 1) - (3x^2 - 4x + 3)$
    $-2x^2 + 4x - 2$

8. $(x + 2)(2x^2 + 3)$

9. $(x^2 + 3x + 10)(4x^2 - 2x - 7)$

10. $(x - 1)(2x + 1)(x + 5)$

11. $(-3x + 1)^3$ $-27x^3 + 27x^2 - 9x + 1$

12. **GEOMETRY** **CONNECTION** Write a polynomial model in standard form for the volume of the rectangular prism shown at the right. $x^3 - 2x^2 - 9x + 18$

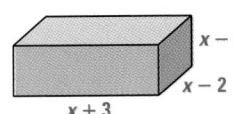
$x - 3$
$x - 2$
$x + 3$

PRACTICE AND APPLICATIONS

Extra Practice
to help you master skills is on p. 948.

23. $9x^3 - 3x^2 + 3x - 1$
24. $8x^3 + 7x^2 + 51x + 1$
25. $x^3 + 7x^2 + 8x + 14$
26. $12x^4 + 2x^3 - 10x + 1$

33. $x^3 - x^2 - 3x + 27$
34. $x^3 + x^2 - 59x - 24$
35. $6x^4 + 13x^3 - 3x^2 + 5x$
36. $12x^3 - 32x^2 - 6x + 2$
37. $x^3 + 6x^2 - 46x + 99$
38. $4x^4 - 24x^3 + 35x^2 + 6x - 9$

STUDENT HELP

HOMEWORK HELP
Examples 1, 2: Exs. 13–26
Examples 3, 4: Exs. 27–44
Example 5: Exs. 45–52
Example 6: Exs. 53–61
Example 7: Exs. 64, 65, 69
Example 8: Exs. 66–68

ADDING AND SUBTRACTING POLYNOMIALS **Find the sum or difference.**

13. $(8x^2 + 1) + (3x^2 - 2)$ $11x^2 - 1$

14. $(3x^3 + 10x + 5) - (x^3 - 4x + 6)$
    $2x^3 + 14x - 1$

15. $(x^2 - 6x + 5) - (x^2 + x - 2)$ $-7x + 7$

16. $(16 - 13x) + (10x - 11)$ $-3x + 5$

17. $(7x^3 - 1) - (15x^3 + 4x^2 - x + 3)$
    $-8x^3 - 4x^2 + x - 4$

18. $8x + (14x + 3 - 41x^2 + x^3)$
    $x^3 - 41x^2 + 22x + 3$

19. $(4x^2 - 11x + 10) + (5x - 31)$
    $4x^2 - 6x - 21$

20. $(9x^3 - 4 + x^2 + 8x) - (7x^3 - 3x + 7)$
    $2x^3 + x^2 + 11x - 11$

21. $(-3x^3 + x - 11) - (4x^3 + x^2 - x)$

22. $(6x^2 - 19x + 5) - (19x^2 - 4x + 9)$

23. $(10x^3 - 4x^2 + 3x) - (x^3 - x^2 + 1)$

24. $(50x - 3) + (8x^3 + 7x^2 + x + 4)$

25. $(10x - 3 + 7x^2) + (x^3 - 2x + 17)$

26. $(3x^3 - 5x^4 - 10x + 1) + (17x^4 - x^3)$

MULTIPLYING POLYNOMIALS **Find the product of the polynomials.**

27. $x(x^2 + 6x - 7)$
    $x^3 + 6x^2 - 7x$

28. $10x^2(x - 5)$ $10x^3 - 50x^2$

29. $-4x(x^2 - 8x + 3)$
    $-4x^3 + 32x^2 - 12x$

30. $5x(3x^2 - x + 3)$
    $15x^3 - 5x^2 + 15x$

31. $(x - 4)(x - 7)$
    $x^2 - 11x + 28$

32. $(x + 9)(x - 2)$
    $x^2 + 7x - 18$

33. $(x + 3)(x^2 - 4x + 9)$

34. $(x + 8)(x^2 - 7x - 3)$

35. $(2x + 5)(3x^2 - x^2 + x)$

36. $(6x + 2)(2x^2 - 6x + 1)$

37. $(x + 11)(x^2 - 5x + 9)$

38. $(4x^2 - 1)(x^2 - 6x + 9)$

39. $(x - 1)(x^3 + 2x^2 + 2)$
    $x^4 + x^3 - 2x^2 + 2x - 2$

40. $(x + 1)(5x^3 - x^2 + x - 4)$
    $5x^4 + 4x^3 - 3x - 4$

41. $(3x^2 - 2)(x^2 + 4x + 3)$
    $3x^4 + 12x^3 + 7x^2 - 8x - 6$

42. $(-x^3 - 2)(x^2 + 3x - 3)$
    $-x^5 - 3x^4 + 3x^3 - 2x^2 - 6x + 6$

43. $(x^2 + x + 4)(2x^2 - x + 1)$
    $2x^4 + x^3 + 8x^2 - 3x + 4$

44. $(x^2 - x - 3)(x^2 + 4x + 2)$
    $x^4 + 3x^3 - 5x^2 - 14x - 6$

6.3 *Adding, Subtracting, and Multiplying Polynomials*   **341**

ASSIGNMENT GUIDE

BASIC
Day 1: pp. 341–344 Exs. 14–40 even, 46–50 even 54–58 even, 62, 64
Day 2: pp. 341–344 Exs. 23–63 odd, 65–66, 70–71, 73–87 odd, Quiz 1 Exs. 1–27

AVERAGE
Day 1: pp. 341–344 Exs. 14–40 even, 46–50 even, 54–64 even
Day 2: pp. 341–344 Exs. 21–63 odd, 65–68, 70–71, 73–87 odd, Quiz 1 Exs. 1–27

ADVANCED
Day 1: pp. 341–344 Exs. 14–64 even
Day 2: pp. 341–344 Exs. 21–63 odd, 65–72, 73–87 odd, Quiz 1 Exs. 1–27

BLOCK SCHEDULE
pp. 341–344 Exs. 14–64 even, 65–68, 70–71, 73–87 odd, Quiz 1 Exs. 1–27

EXERCISE LEVELS
Level A: *Easier*
13–32, 53–58, 64–65, 70
Level B: *More Difficult*
33–52, 59–63, 66–69, 71
Level C: *Most Difficult*
72

HOMEWORK CHECK ✔
To quickly check student understanding of key concepts, go over the following exercises: Exs. 14, 26, 32, 40, 48, 58. See also the Daily Homework Quiz:

• Blackline Master (*Chapter 6 Resource Book*, p. 51)
• Transparency (p. 44)

APPLICATION NOTE

EXERCISE 64 By graphing both functions, students can see the trend in purchases of different types of vehicles since 1980. A big influence on the number of trucks sold in the United States has been the increased popularity of Sport Utility Vehicles, which are classified as trucks.

EXERCISE 65 The U.S. Bureau of the Census refers to the population 85 years and older as the "oldest old." An examination of the two graphs will confirm that this group is increasing at a much greater rate than the total population in the United States. In fact, from 1960 to 1994 the oldest old population increased 274% while the total population increased 45%.

CAREER NOTE

EXERCISE 65 Additional information about gerontology is available at **www.mcdougallittell.com**.

DATA UPDATE
Updated data for Exercise 65 is available at **www.mcdougallittell.com**.

┌─────────────────────────────┐
STUDENT HELP NOTES

→ **Homework Help** Students can find help with problem solving for Ex. 69 at **www.mcdougallittell.com**. The information can be printed out for students who don't have access to the Internet.
└─────────────────────────────┘

ADDITIONAL PRACTICE AND RETEACHING

For Lesson 6.3:
- Practice Levels A, B, and C (*Chapter 6 Resource Book*, p. 40)
- Reteaching with Practice (*Chapter 6 Resource Book*, p. 43)
- See Lesson 6.3 of the *Personal Student Tutor*

For more Mixed Review:
- Search the *Test and Practice Generator* for key words or specific lessons.

342

MULTIPLYING THREE BINOMIALS Find the product of the binomials.

45. $(x + 9)(x - 2)(x - 7)$ $x^3 - 67x + 126$ **46.** $(x + 3)(x - 4)(x - 5)$
                                                                        $x^3 - 6x^2 - 7x + 60$

47. $(x + 5)(x + 7)(-x + 1)$ **48.** $(2x - 3)(x + 7)(x + 6)$
    $-x^3 - 11x^2 - 23x + 35$ $2x^3 + 23x^2 + 45x - 126$

49. $(x - 9)(x - 2)(3x + 2)$ **50.** $(x - 1)(-2x - 5)(x - 8)$
    $3x^3 - 31x^2 + 32x + 36$ $-2x^3 + 13x^2 + 29x - 40$

51. $(2x + 1)(3x + 1)(x + 4)$ **52.** $(4x - 1)(2x - 1)(3x - 2)$
    $6x^3 + 29x^2 + 21x + 4$ $24x^3 - 34x^2 + 15x - 2$

SPECIAL PRODUCTS Find the product.

55. $64x^3 - 144x^2 + 108x - 27$ **53.** $(x + 7)(x - 7)$ $x^2 - 49$ **54.** $(x + 4)^2$ $x^2 + 8x + 16$ **55.** $(4x - 3)^3$

58. $4y^2 + 20xy + 25x^2$ **56.** $(10x + 3)(10x - 3)$ **57.** $\left(6 - x^2\right)^2$ **58.** $(2y + 5x)^2$
                                            $100x^2 - 9$ $x^4 - 12x^2 + 36$

59. $27x^3 + 189x^2 + 441x + 343$ **59.** $(3x + 7)^3$ **60.** $(7y - x)^2$ **61.** $(2x + 3y)^3$

60. $49y^2 - 14xy + x^2$

61. $8x^3 + 36x^2y + 54xy^2 + 27y^3$

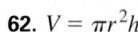 **GEOMETRY CONNECTION**   Write the volume of the figure as a polynomial in standard form.

62. $V = \pi r^2 h$ **63.** $V = lwh$

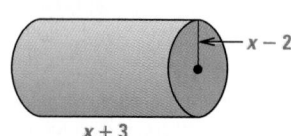   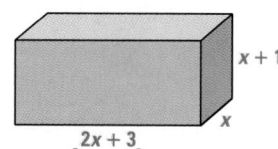

$V = \pi x^3 - \pi x^2 - 8\pi x + 12\pi$ $V = 2x^3 + 5x^2 + 3x$

64. 🌎 **MOTOR VEHICLE SALES** For 1983 through 1996, the number of cars C (in thousands) and the number of trucks and buses T (in thousands) sold that were manufactured in the United States can be modeled by

$$C = -1.63t^4 + 49.5t^3 - 476t^2 + 1370t + 6705$$

$$T = -1.052t^4 + 31.6t^3 - 296t^2 + 1097t + 2290$$

where t is the number of years since 1983. Find a model that represents the total number of vehicles sold that were manufactured in the United States. How many vehicles were sold in 1990? $V = -2.682t^4 + 81.1t^3 - 772t^2 + 2467t + 8995$; about 9,814,000 vehicles

65. **SOCIAL STUDIES CONNECTION**   For 1980 through 1996, the population P (in thousands) of the United States and the number of people S (in thousands) age 85 and over can be modeled by

$$P = -0.804t^4 + 26.9t^3 - 262t^2 + 3010t + 227{,}000$$

$$S = 0.0206t^4 - 0.670t^3 + 6.42t^2 + 213t + 7740$$

where t is the number of years since 1980. Find a model that represents the number of people in the United States under the age of 85. How many people were under the age of 85 in 1995?

📡 **DATA UPDATE** of U.S. Bureau of the Census data at www.mcdougallittell.com
    $Y = -0.8246t^4 + 27.57t^3 - 268.42t^2 + 2797t + 219{,}260$; about 252 million people

66. 🌎 **BICYCLING** The equation $P = 0.00267sF$ gives the power P (in horsepower) needed to keep a certain bicycle moving at speed s (in miles per hour), where F is the force of road and air resistance (in pounds). On level ground this force is given by $F = 0.0116s^2 + 0.789$. Write a polynomial function (in terms of s only) for the power needed to keep the bicycle moving at speed s on level ground. How much power does a cyclist need to exert to keep the bicycle moving at 10 miles per hour?
$P = 0.000030972s^3 + 0.00210663s$; about 0.05 horsepower

FOCUS ON CAREERS

GERONTOLOGIST
A gerontologist studies the biological, psychological, and sociological phenomena associated with old age. As people's life expectancies have increased, demand for gerontologists has grown.

📡 **CAREER LINK**
www.mcdougallittell.com

67. $W = -0.0004128t^5 -$
$0.03414t^4 + 1.3539t^3$
$-12.8387t^2 + 51.9t +$
833; about 1,086,000
degrees

68. $R = -0.1809t^4 + 0.075t^3$
$+ 41.953t^2 + 509.76t +$
6110; about \$10,461
million

69. $4000(1 + r)^3 +$
$5000(1 + r)^2 +$
$7000(1 + r)$; $10,000r^3 +$
$43,000r^2 + 72,000r +$
39,000

STUDENT HELP

HOMEWORK HELP
Visit our Web site
www.mcdougallittell.com
for help with problem
solving in Ex. 69.

Test Preparation

★ **Challenge**

72b. $x^n - 1 = (x - 1)(x^{n-1} + x^{n-2} + \cdots + x + 1)$
Multiply: $x(x^{n-1}) + (-1)(x^{n-1}) + x(x^{n-2}) + \cdots - 1$
(Pairs of middle terms will cancel.)

EXTRA CHALLENGE
↪ www.mcdougallittell.com

67. 🌐 **EDUCATION** For 1980 through 1995, the number of degrees D (in thousands) earned by people in the United States and the percent of degrees P earned by women can be modeled by

$$D = -0.096t^4 + 3t^3 - 27t^2 + 91t + 1700$$
$$P = 0.43t + 49$$

where t is the number of years since 1980. Find a model that represents the number of degrees W (in thousands) earned by women from 1980 to 1995. How many degrees were earned by women in 1991? ▶ Source: U.S. Bureau of the Census

68. 🌐 **PUBLISHING** From 1985 through 1993, the number of hardback books N (in millions) sold in the United States and the average price per book P (in dollars) can be modeled by

$$N = -0.27t^3 + 3.9t^2 + 7.9t + 650$$
$$P = 0.67t + 9.4$$

where t is the number of years since 1985. Write a model that represents the total revenue R (in millions of dollars) received from the sales of hardback books. What was the revenue in 1991?

69. 🌐 **PERSONAL FINANCE** Suppose two brothers each make three deposits in accounts earning the same annual interest rate r (expressed as a decimal).

🦅 EagleBank		
Porter, Mark J.		#05-8922-4310
Date	**Transaction**	**Amount**
1/1/97	Deposit	\$6000.00
1/1/98	Deposit	\$8000.00
1/1/99	Deposit	\$9000.00

🌐 WorldBank		
Porter, Tom R.		#12-4600-2541
Date	**Transaction**	**Amount**
1/1/97	Deposit	\$4000.00
1/1/98	Deposit	\$5000.00
1/1/99	Deposit	\$7000.00

Mark's account is worth $6000(1 + r)^3 + 8000(1 + r)^2 + 9000(1 + r)$ on January 1, 2000. Find the value of Tom's account on January 1, 2000. Then find the total value of the two accounts on January 1, 2000. Write the total value as a polynomial in standard form. **See margin.**

70. MULTIPLE CHOICE What is the sum of $2x^4 + 5x^3 - 8x^2 - x + 10$ and $8x^4 - 4x^3 + x^2 - x + 2$? **C**

  Ⓐ $10x^4 + x^3 - 9x^2 + 12$ Ⓑ $10x^4 + x^3 - 9x^2 - 2x + 12$

  Ⓒ $10x^4 + x^3 - 7x^2 - 2x + 12$ Ⓓ $10x^4 + 9x^3 - 7x^2 - 2x + 12$

71. MULTIPLE CHOICE $(3x - 8)^3 = $ __?__ **A**

  Ⓐ $27x^3 - 216x^2 + 576x - 512$ Ⓑ $27x^3 - 216x^2 + 576x + 512$

  Ⓒ $27x^3 - 72x^2 + 576x - 512$ Ⓓ $27x^3 - 216x^2 + 72x - 512$

72. FINDING A PATTERN Look at the following polynomials and their factorizations.

$$x^2 - 1 = (x - 1)(x + 1)$$
$$x^3 - 1 = (x - 1)(x^2 + x + 1)$$
$$x^4 - 1 = (x - 1)(x^3 + x^2 + x + 1)$$

  a. Factor $x^5 - 1$ and $x^6 - 1$. Check your answers by multiplying.
$x^5 - 1 = (x - 1)(x^4 + x^3 + x^2 + x + 1)$; $x^6 - 1 = (x - 1)(x^5 + x^4 + x^3 + x^2 + x + 1)$
  b. In general, how can $x^n - 1$ be factored? Show that this factorization works by multiplying the factors.

6.3 *Adding, Subtracting, and Multiplying Polynomials*   **343**

4 ASSESS

DAILY HOMEWORK QUIZ

🗐 *Transparency Available*

Find the sum or difference.

1. $(9x^2 + 7x^3 - x^4 - 9) +$
$(5x^4 - 11x^3 + 6x^2)$
$4x^4 - 4x^3 + 15x^2 - 9$

2. $(6 - 4x^2 + 5x^3) -$
$(7x^2 - x^4 - 12x^3)$
$x^4 + 17x^3 - 11x^2 + 6$

Find the product.

3. $(2x - 3)(2x - 5)$ $4x^2 - 16x + 15$

4. $(x - 1)(-4x^3 + x^2 - 10)$
$-4x^4 + 5x^3 - x^2 - 10x + 10$

5. $(5x + 2y)^3$
$125x^3 + 150x^2y + 60xy^2 + 8y^3$

6. $(x - 6)(2x + 1)(5x - 3)$
$10x^3 - 61x^2 + 3x + 18$

EXTRA CHALLENGE NOTE
→ Challenge problems for
Lesson 6.3 are available in
blackline format in the *Chapter 6 Resource Book*, p. 47 and at
www.mcdougallittell.com.

ADDITIONAL TEST PREPARATION

1. WRITING Explain how to cube a binomial. **Cube the first term. Add three times the product of the square of the first term and the second term. Add three times the product of the first term and the square of the second term. Add the cube of the second term.**

ADDITIONAL RESOURCES
An alternative quiz for Lessons
6.1–6.3 is available in the *Chapter 6
Resource Book*, p. 48.

13.

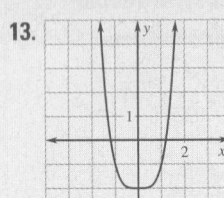

14.

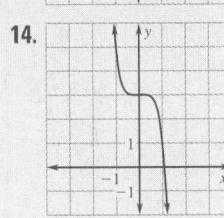

15.

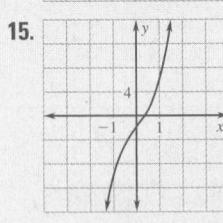

16.

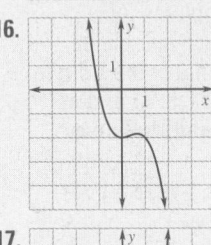

17.

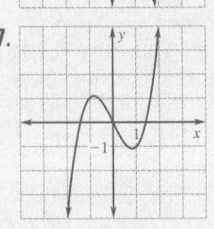

18. See Additional Answers beginning on page AA1.

344

MIXED REVIEW

SOLVING QUADRATIC EQUATIONS Solve the equation. (Review 5.2 for 6.4)

73. $4x^2 - 36 = 0$ ± 3 **74.** $x^2 + 3x - 40 = 0$ $-8, 5$ **75.** $x^2 + 16x + 64 = 0$ -8

76. $x^2 - x - 56 = 0$ $-7, 8$ **77.** $2x^2 - 7x - 15 = 0$ $-\frac{3}{2}, 5$ **78.** $6x^2 + 10x - 4 = 0$ $-2, \frac{1}{3}$

WRITING QUADRATIC FUNCTIONS Write a quadratic function in standard
form whose graph passes through the given points. (Review 5.8)

79. $(-4, 0), (2, 0), (1, 6)$ $y = -\frac{6}{5}x^2 - \frac{12}{5}x + \frac{48}{5}$ **80.** $(10, 0), (1, 0), (4, 3)$ $y = -\frac{1}{6}x^2 + \frac{11}{6}x - \frac{5}{3}$

81. $(-6, 0), (6, 0), (-3, -9)$ $y = \frac{1}{3}x^2 - 12$ **82.** $(-3, 0), (5, 0), (-2, 7)$ $y = -x^2 + 2x + 15$

SIMPLIFYING ALGEBRAIC EXPRESSIONS Simplify the expression. Tell which
properties of exponents you used. (Review 6.1)

83. $x^5 \cdot \frac{1}{x^2} \cdot x^3$ **84.** $\frac{x^4 y^5}{xy^3}$ $x^3 y^2$ **85.** $-5^{-2} y^0$ $-\frac{1}{25}$

86. $(4x^{-3})^4 \cdot \left(\frac{x^6}{2}\right)^2$ 64 **87.** $\frac{3x^5 y^8}{6xy^{-3}}$ $\frac{1}{2}x^4 y^{11}$ **88.** $\frac{6x^4 y^2}{30x^2 y^{-1}}$ $\frac{1}{5}x^2 y^3$

QUIZ 1

Self-Test for Lessons 6.1–6.3

Evaluate the expression. (Lesson 6.1)

1. $7^0 \cdot 5^{-3}$ $\frac{1}{125}$ **2.** $\left(\frac{4}{9}\right)^{-2}$ $\frac{81}{16}$ **3.** $\left(\frac{5}{3^2}\right)^2$ $\frac{25}{81}$

4. $3^2 \cdot (3^2 \cdot 2^4)^{-1}$ $\frac{1}{16}$ **5.** $(8^2 \cdot 8^{-3})^2 \cdot 8^2$ 1 **6.** $\frac{(2^5 \cdot 3^2)^{-1}}{2^{-2} \cdot 3^2}$ $\frac{1}{648}$

Simplify the expression. (Lesson 6.1)

7. $(-5)^{-2} y^0$ $\frac{1}{25}$ **8.** $(3x^3 y^6)^{-2}$ $\frac{1}{9x^6 y^{12}}$ **9.** $(x^3 y^{-5})(x^2 y)^2$ $\frac{x^7}{y^3}$

10. $(x^2 y^{-3})(xy^2)$ $\frac{x^3}{y}$ **11.** $\left(\frac{2x}{y^2}\right)^{-3}$ $\frac{y^6}{8x^3}$ **12.** $\frac{x^6 y^{-2}}{x^{-1} y^5}$ $\frac{x^7}{y^7}$

Graph the polynomial function. (Lesson 6.2)   13–18. See margin.

13. $f(x) = x^4 - 2$ **14.** $f(x) = -2x^5 + 3$ **15.** $f(x) = 3x^3 + 5x - 2$

16. $f(x) = -x^3 + x^2 - 2$ **17.** $f(x) = x^3 - 2x$ **18.** $f(x) = -x^4 - 3x + 6$

Perform the indicated operation. (Lesson 6.3)

19. $(7x^3 + 8x - 11) + (3x^2 - x + 8)$ $7x^3 + 3x^2 + 7x - 3$ **20.** $(-2x^2 + 4x) + (5x^2 - x - 11)$ $3x^2 + 3x - 11$

21. $(-5x^2 + 12x - 9) - (-7x^2 - 6x - 7)$ $2x^2 + 18x - 2$ **22.** $(3x^2 + 4x - 1) - (-x^3 + 2x + 5)$ $x^3 + 3x^2 + 2x - 6$

23. $(x + 5)(4x^2 - x - 1)$ $4x^3 + 19x^2 - 6x - 5$ **24.** $(x - 3)(x + 2)(2x + 5)$ $2x^3 + 3x^2 - 17x - 30$

25. $(x - 6)^3$ $x^3 - 18x^2 + 108x - 216$ **26.** $(2x^2 + 3)^2$ $4x^4 + 12x^2 + 9$

27. ASTRONOMY Suppose NASA launches a spacecraft that can travel at a
speed of 25,000 miles per hour in space. How long would it take the spacecraft
to reach Jupiter if Jupiter is about 495,000,000 miles away? Use scientific
notation to get your answer. (Lesson 6.1)   about 1.98×10^4 hours (about 825 days)

Factoring and Solving Polynomial Equations

What you should learn

GOAL 1 Factor polynomial expressions.

GOAL 2 Use factoring to solve polynomial equations, as applied in **Ex. 87**.

Why you should learn it

▼ To solve **real-life** problems, such as finding the dimensions of a block discovered at an underwater archeological site in **Example 5**.

1. If the cube had no missing pieces, its volume would be a^3. The volume of the missing piece is b^3. Thus, the volume of the figure is $a^3 - b^3$ which is equal to the sum of the volumes of solids I, II, and III.

2. solid I; $a^2(a - b)$;
   solid II; $ab(a - b)$;
   solid III; $b^2(a - b)$

GOAL 1 FACTORING POLYNOMIAL EXPRESSIONS

In Chapter 5 you learned how to factor the following types of quadratic expressions.

TYPE	EXAMPLE
General trinomial	$2x^2 - 5x - 12 = (2x + 3)(x - 4)$
Perfect square trinomial	$x^2 + 10x + 25 = (x + 5)^2$
Difference of two squares	$4x^2 - 9 = (2x + 3)(2x - 3)$
Common monomial factor	$6x^2 + 15x = 3x(2x + 5)$

In this lesson you will learn how to factor other types of polynomials.

● ACTIVITY

Developing Concepts

The Difference of Two Cubes

Use the diagram to answer the questions.

❶ Explain why $a^3 - b^3 = \boxed{\text{Volume of solid I}} + \boxed{\text{Volume of solid II}} + \boxed{\text{Volume of solid III}}$.

❷ For each of solid I, solid II, and solid III, write an algebraic expression for the solid's volume. Leave your expressions in factored form.

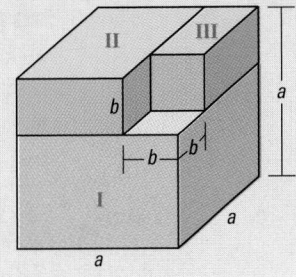

❸ Substitute your expressions from **Step 2** into the equation from **Step 1**. Use the resulting equation to factor $a^3 - b^3$ completely.

3. $a^3 - b^3 = a^2(a - b) + ab(a - b) + b^2(a - b)$
   $= (a - b)(a^2 + ab + b^2)$

In the activity you may have discovered how to factor the difference of two cubes. This factorization and the factorization of the sum of two cubes are given below.

SPECIAL FACTORING PATTERNS

SUM OF TWO CUBES	Example
$a^3 + b^3 = (a + b)(a^2 - ab + b^2)$	$x^3 + 8 = (x + 2)(x^2 - 2x + 4)$
DIFFERENCE OF TWO CUBES	
$a^3 - b^3 = (a - b)(a^2 + ab + b^2)$	$8x^3 - 1 = (2x - 1)(4x^2 + 2x + 1)$

1 PLAN

PACING
Basic: 2 days
Average: 2 days
Advanced: 2 days
Block Schedule: 1 block

LESSON OPENER
APPLICATION
An alternative way to approach Lesson 6.4 is to use the Application Lesson Opener:

- Blackline Master (*Chapter 6 Resource Book,* p. 52)
-  Transparency (p. 38)

MEETING INDIVIDUAL NEEDS
- ***Chapter 6 Resource Book***
  Prerequisite Skills Review (p. 5)
  Practice Level A (p. 53)
  Practice Level B (p. 54)
  Practice Level C (p. 55)
  Reteaching with Practice (p. 56)
  Absent Student Catch-Up (p. 58)
  Challenge (p. 62)
- ***Resources in Spanish***
- ***Personal Student Tutor***

NEW-TEACHER SUPPORT
See the Tips for New Teachers on pp. 1–2 of the *Chapter 6 Resource Book* for additional notes about Lesson 6.4.

WARM-UP EXERCISES

Transparency Available

Factor.
1. $4x^2 - 24x$ $4x(x - 6)$
2. $2x^2 + 11x - 21$ $(2x - 3)(x + 7)$
3. $4x^2 - 36x + 81$ $(2x - 9)^2$

Solve.
4. $x^2 + 10x + 25 = 0$ -5
5. $6x^2 + x = 15$ $\frac{3}{2}, -\frac{5}{3}$

MOTIVATING THE LESSON
A manufacturer of shipping cartons who needs to make cartons for a specific use often has to use special relationships between the length, width, height, and volume to find the exact dimensions of the carton. The dimensions can usually be found by writing and solving a polynomial equation. This lesson looks at how factoring can be used to solve such equations.

ACTIVITY NOTE
Partner Activity Students can work with a partner. When they arrive at a factored form of the expression, they can choose values for a and b. One student can evaluate $a^3 - b^3$ and the other the factored form. The results should be equal.

EXTRA EXAMPLE 1
Factor each polynomial.
a. $125 + x^3$ $(25 - 5x + x^2)(5 + x)$
b. $64a^4 - 27a$
   $a(4a - 3)(16a^2 + 12a + 9)$

EXTRA EXAMPLE 2
Factor the polynomial
$x^2 y^2 - 3x^2 - 4y^2 + 12$.
$(y^2 - 3)(x + 2)(x - 2)$

EXTRA EXAMPLE 3
Factor each polynomial.
a. $25x^4 - 36$ $(5x^2 - 6)(5x^2 + 6)$
b. $a^2 b^2 - 8ab^3 + 16b^4$
   $b^2(a - 4b)^2$

 CHECKPOINT EXERCISES
For use after Example 1:
1. Factor $x^3 + 343$.
   $(x + 7)(x^2 - 7x + 49)$

For use after Example 2:
2. Factor $bx^2 + 2a + 2b + ax^2$.
   $(2 + x^2)(a + b)$

For use after Example 3:
3. Factor $x^4 + 4x^2 - 21$.
   $(x^2 + 7)(x^2 - 3)$

EXAMPLE 1 *Factoring the Sum or Difference of Cubes*

Factor each polynomial.

a. $x^3 + 27$

b. $16u^5 - 250u^2$

SOLUTION

a. $x^3 + 27 = x^3 + 3^3$ Sum of two cubes

           $= (x + 3)(x^2 - 3x + 9)$

b. $16u^5 - 250u^2 = 2u^2(8u^3 - 125)$ Factor common monomial.

           $= 2u^2[(2u)^3 - 5^3]$ Difference of two cubes

           $= 2u^2(2u - 5)(4u^2 + 10u + 25)$

· · · · · · · · · ·

For some polynomials, you can **factor by grouping** pairs of terms that have a common monomial factor. The pattern for this is as follows.

$$ra + rb + sa + sb = r(a + b) + s(a + b)$$
$$= (r + s)(a + b)$$

EXAMPLE 2 *Factoring by Grouping*

Factor the polynomial $x^3 - 2x^2 - 9x + 18$.

SOLUTION

$x^3 - 2x^2 - 9x + 18 = x^2(x - 2) - 9(x - 2)$ Factor by grouping.

           $= (x^2 - 9)(x - 2)$

           $= (x + 3)(x - 3)(x - 2)$ Difference of squares

· · · · · · · · · ·

An expression of the form $au^2 + bu + c$ where u is any expression in x is said to be in **quadratic form**. The factoring techniques you studied in Chapter 5 can sometimes be used to factor such expressions.

EXAMPLE 3 *Factoring Polynomials in Quadratic Form*

Factor each polynomial.

a. $81x^4 - 16$

b. $4x^6 - 20x^4 + 24x^2$

SOLUTION

a. $81x^4 - 16 = (9x^2)^2 - 4^2$ **b.** $4x^6 - 20x^4 + 24x^2 = 4x^2(x^4 - 5x^2 + 6)$

           $= (9x^2 + 4)(9x^2 - 4)$ $= 4x^2(x^2 - 2)(x^2 - 3)$

           $= (9x^2 + 4)(3x + 2)(3x - 2)$

GOAL 2 SOLVING POLYNOMIAL EQUATIONS BY FACTORING

In Chapter 5 you learned how to use the zero product property to solve factorable quadratic equations. You can extend this technique to solve some higher-degree polynomial equations.

EXAMPLE 4 *Solving a Polynomial Equation*

Solve $2x^5 + 24x = 14x^3$.

SOLUTION

$2x^5 + 24x = 14x^3$	Write original equation.
$2x^5 - 14x^3 + 24x = 0$	Rewrite in standard form.
$2x(x^4 - 7x^2 + 12) = 0$	Factor common monomial.
$2x(x^2 - 3)(x^2 - 4) = 0$	Factor trinomial.
$2x(x^2 - 3)(x + 2)(x - 2) = 0$	Factor difference of squares.
$x = 0, x = \sqrt{3}, x = -\sqrt{3}, x = -2,$ or $x = 2$	Zero product property

▶ The solutions are $0, \sqrt{3}, -\sqrt{3}, -2$, and 2. Check these in the original equation.

STUDENT HELP

▸ Study Tip
In the solution of Example 4, do not divide both sides of the equation by a variable or a variable expression. Doing so will result in the loss of solutions.

EXAMPLE 5 *Solving a Polynomial Equation in Real Life*

ARCHEOLOGY In 1980 archeologists at the ruins of Caesara discovered a huge hydraulic concrete block with a volume of 330 cubic yards. The block's dimensions are x yards high by $13x - 11$ yards long by $13x - 15$ yards wide. What is the height?

SOLUTION

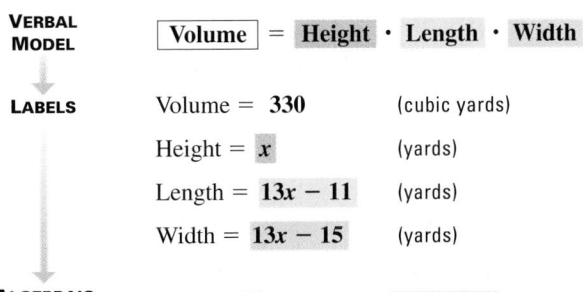

VERBAL MODEL		
	Volume = **Height** · **Length** · **Width**	

LABELS		
	Volume = **330**	(cubic yards)
	Height = x	(yards)
	Length = $13x - 11$	(yards)
	Width = $13x - 15$	(yards)

ALGEBRAIC MODEL

$$330 = x\,(13x - 11)\,(13x - 15)$$

$0 = 169x^3 - 338x^2 + 165x - 330$	Write in standard form.
$0 = 169x^2(x - 2) + 165(x - 2)$	Factor by grouping.
$0 = (169x^2 + 165)(x - 2)$	

▶ The only real solution is $x = 2$, so $13x - 11 = 15$ and $13x - 15 = 11$. The block is 2 yards high. The dimensions are 2 yards by 15 yards by 11 yards.

FOCUS ON CAREERS

ARCHEOLOGIST
Archeologists excavate, classify, and date items used by ancient people. They may specialize in a particular geographical region and/or time period.

CAREER LINK
www.mcdougallittell.com

6.4 *Factoring and Solving Polynomial Equations* **347**

EXTRA EXAMPLE 4
Solve $2y^5 - 18y = 0$. $0, \sqrt{3}, -\sqrt{3}$

EXTRA EXAMPLE 5
An optical company is going to make a glass prism that has a volume of 15 cm³. The height will be h cm, and the base will be a right triangle with legs of length $(h - 2)$ cm and $(h - 3)$ cm. What will be the height? 5 cm

✓ CHECKPOINT EXERCISES
For use after Examples 4 and 5:
1. You are building a bin to hold cedar mulch for your garden. The bin will hold 162 ft³ of mulch. The dimensions of the bin are x ft by $5x - 6$ ft by $5x - 9$ ft. How tall will the bin be? 3 ft

CAREER NOTE
EXAMPLE 5 Additional information about archeologists is available at **www.mcdougallittell.com**.

FOCUS ON VOCABULARY
When would you factor a polynomial by grouping? Polynomials that have pairs of terms with a common monomial factor can be factored by grouping the pairs having the common factor.

CLOSURE QUESTION
How can you use the zero product property to solve polynomial equations of degree 3 or more? Write the equation in standard form. Then use the rules for factoring to rewrite the polynomial in factored form. Set each factor equal to zero and solve.

DAILY PUZZLER
For what integer d will $x(x - 3) \times (x - 7) - d$ give a polynomial that can be factored by grouping to give a factor of $x - 10$? 210

ASSIGNMENT GUIDE

BASIC
Day 1: pp. 348–351 Exs. 18–22
even, 24–27, 28–40 even,
46–68 even
Day 2: pp. 348–351 Exs. 29–59
odd, 63–83 odd, 88–90,
95–99 odd

AVERAGE
Day 1: pp. 348–351 Exs. 18–27,
28–68 even
Day 2: pp. 348–351 Exs. 29–61
odd, 75–85 odd, 88–90,
95–99 odd

ADVANCED
Day 1: pp. 348–351 Exs. 18–27,
28–74 even, 81–83
Day 2: pp. 348–351 Exs. 29–79
odd, 84–93, 95–99 odd

BLOCK SCHEDULE
pp. 348–351 Exs. 18–27, 28–68
even, 67–85 odd, 88–90,
95–99 odd

EXERCISE LEVELS
Level A: *Easier*
18–35, 45–53, 63–67
Level B: *More Difficult*
36–44, 54–62, 68–85, 88–90
Level C: *Most Difficult*
86, 87, 91–93

✔ **HOMEWORK CHECK**
To quickly check student under-
standing of key concepts, go
over the following exercises:
Exs. 20, 24, 28, 40, 46, 64. See
also the Daily Homework Quiz:
• Blackline Master (*Chapter 6
  Resource Book*, p. 65)
• Transparency (p. 45)

4b. *Sample answer:* The graph of
$y = x^2 - x + 1$ does not intersect
the x-axis, so $x^2 - x + 1$ is not
factorable.

GUIDED PRACTICE

Vocabulary Check ✔
Concept Check ✔

1. Give an example of a polynomial in quadratic form that contains an x^3-term.
   Sample answer: $x^6 + 2x^3 + 1$
2. State which factoring method you would use to factor each of the following.

   a. $6x^3 - 2x^2 + 9x - 3$ **b.** $8x^3 - 125$ **c.** $16x^4 - 9$
       factoring by grouping   difference of two cubes   polynomial in quadratic form

3. You can't divide by $2x^2$, which contains a variable. 0 is also a solution.

3. **ERROR ANALYSIS** What is wrong with the solution at the right?

4. **a.** Factor the polynomial $x^3 + 1$ into the product of a linear binomial and a quadratic trinomial.
       $(x + 1)(x^2 - x + 1)$
   b. Show that you can't factor the quadratic trinomial from part (a). **See margin.**

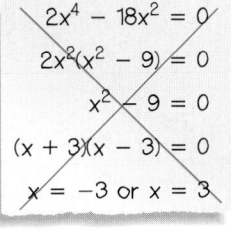

$2x^4 - 18x^2 = 0$
$2x^2(x^2 - 9) = 0$
$x^2 - 9 = 0$
$(x + 3)(x - 3) = 0$
$x = -3$ or $x = 3$

Ex. 3

Skill Check ✔

Factor the polynomial using any method.

12. $-\frac{7}{3}, \pm 2$

15. $\pm\frac{\sqrt{6}}{3}$

5. $x^6 + 125$
   $(x^2 + 5)(x^4 - 5x^2 + 25)$

6. $4x^3 + 16x^2 + x + 4$
   $(4x^2 + 1)(x + 4)$

7. $x^4 - 1$
   $(x + 1)(x - 1)(x^2 + 1)$

8. $2x^3 - 3x^2 - 10x + 15$
   $(x^2 - 5)(2x - 3)$

9. $5x^3 - 320$
   $5(x - 4)(x^2 + 4x + 16)$

10. $x^4 + 7x^2 + 10$
    $(x^2 + 2)(x^2 + 5)$

Find the real-number solutions of the equation.

11. $x^3 - 27 = 0$ 3

12. $3x^3 + 7x^2 - 12x = 28$

13. $x^3 + 2x^2 - 9x = 18$ $-2, \pm 3$

14. $54x^3 = -2$ $-\frac{1}{3}$

15. $9x^4 - 12x^2 + 4 = 0$

16. $16x^8 = 81$ $\pm\frac{\sqrt{6}}{2}$

17. 🌐 **BUSINESS** The revenue R (in thousands of dollars) for a small business can be modeled by

$$R = t^3 - 8t^2 + t + 82$$

where t is the number of years since 1990. In what year did the revenue reach $90,000? **1998**

PRACTICE AND APPLICATIONS

STUDENT HELP
↳ **Extra Practice**
to help you master
skills is on p. 948.

MONOMIAL FACTORS Find the greatest common factor of the terms in the polynomial.

18. $14x^2 + 8x + 72$ 2

19. $3x^4 - 12x^3$ $3x^3$

20. $7x + 28x^2 - 35x^3$ $7x$

21. $24x^4 - 6x$ $6x$

22. $39x^5 + 13x^3 - 78x^2$ $13x^2$

23. $145x^9 - 17$ 1

24. $6x^6 - 3x^4 - 9x^2$ $3x^2$

25. $72x^9 + 15x^6 + 9x^3$ $3x^3$

26. $6x^4 - 18x^3 + 15x^2$ $3x^2$

MATCHING Match the polynomial with its factorization.

27. $3x^2 + 11x + 6$ C

28. $x^3 - 4x^2 + 4x - 16$ D

29. $125x^3 - 216$ F

30. $2x^7 - 32x^3$ A

31. $2x^5 + 4x^4 - 4x^3 - 8x^2$ E

32. $2x^3 - 32x$ B

A. $2x^3(x + 2)(x - 2)(x^2 + 4)$

B. $2x(x + 4)(x - 4)$

C. $(3x + 2)(x + 3)$

D. $(x^2 + 4)(x - 4)$

E. $2x^2(x^2 - 2)(x + 2)$

F. $(5x - 6)(25x^2 + 30x + 36)$

SUM OR DIFFERENCE OF CUBES Factor the polynomial. 33–40. See margin.

33. $x^3 - 8$ **34.** $x^3 + 64$ **35.** $216x^3 + 1$ **36.** $125x^3 - 8$

37. $1000x^3 + 27$ **38.** $27x^3 + 216$ **39.** $32x^3 - 4$ **40.** $2x^3 + 54$

GROUPING Factor the polynomial by grouping. 41–49. See margin.

41. $x^3 + x^2 + x + 1$ **42.** $10x^3 + 20x^2 + x + 2$ **43.** $x^3 + 3x^2 + 10x + 30$

44. $x^3 - 2x^2 + 4x - 8$ **45.** $2x^3 - 5x^2 + 18x - 45$ **46.** $-2x^3 - 4x^2 - 3x - 6$

47. $3x^3 - 6x^2 + x - 2$ **48.** $2x^3 - x^2 + 2x - 1$ **49.** $3x^3 - 2x^2 - 9x + 6$

QUADRATIC FORM Factor the polynomial. 50–58. See margin.

50. $16x^4 - 1$ **51.** $x^4 + 3x^2 + 2$ **52.** $x^4 - 81$

53. $81x^4 - 256$ **54.** $4x^4 - 5x^2 - 9$ **55.** $x^4 + 10x^2 + 16$

56. $81 - 16x^4$ **57.** $32x^6 - 2x^2$ **58.** $6x^5 - 51x^3 - 27x$

CHOOSING A METHOD Factor using any method. 59–67. See margin.

59. $18x^3 - 2x^2 + 27x - 3$ **60.** $6x^3 + 21x^2 + 15x$ **61.** $4x^4 + 39x^2 - 10$

62. $8x^3 - 12x^2 - 2x + 3$ **63.** $8x^3 - 64$ **64.** $3x^4 - 300x^2$

65. $3x^4 - 24x$ **66.** $5x^4 + 31x^2 + 6$ **67.** $3x^4 + 9x^3 + x^2 + 3x$

SOLVING EQUATIONS Find the real-number solutions of the equation.

68. $x^3 - 3x^2 = 0$ 0, 3   **69.** $2x^3 - 6x^2 = 0$ 0, 3   **70.** $3x^4 + 15x^2 - 72 = 0$ $\pm\sqrt{3}$

71. $x^3 + 27 = 0$ -3 **72.** $x^3 + 2x^2 - x = 2$ $\pm1, -2$ **73.** $x^4 + 7x^3 - 8x - 56 = 0$ $-7, 2$

74. $2x^4 - 26x^2 + 72 = 0$ $\pm2, \pm3$ **75.** $3x^7 - 243x^3 = 0$ 0, ±3 **76.** $x^3 + 3x^2 - 2x - 6 = 0$ $-3, \pm\sqrt{2}$

77. $8x^3 - 1 = 0$ $\frac{1}{2}$ **78.** $x^3 + 8x^2 = -16x$ $-4, 0$ **79.** $x^3 - 5x^2 + 5x - 25 = 0$ 5

80. $3x^4 + 3x^3 = 6x^2 + 6x$ $-1, 0, \pm\sqrt{2}$ **81.** $x^4 + x^3 - x = 1$ ±1 **82.** $4x^4 + 20x^2 = -25$ none

83. $-2x^6 = 16$ none   **84.** $3x^7 = 81x^4$ 0, 3   **85.** $2x^5 - 12x^3 = -16x$ 0, $\pm2, \pm\sqrt{2}$

86. *Writing* You have now factored several different types of polynomials. Explain which factoring techniques or patterns are useful for factoring binomials, trinomials, and polynomials with more than three terms.

87. 🌐 **PACKAGING** A candy factory needs a box that has a volume of 30 cubic inches. The width should be 2 inches less than the height and the length should be 5 inches greater than the height. What should the dimensions of the box be? about 3.16 in. by 1.16 in. by 8.16 in.

88. 🌐 **MANUFACTURING** A manufacturer wants to build a rectangular stainless steel tank with a holding capacity of 500 gallons, or about 66.85 cubic feet. If steel that is one half inch thick is used for the walls of the tank, then about 5.15 cubic feet of steel is needed. The manufacturer wants the outside dimensions of the tank to be related as follows:

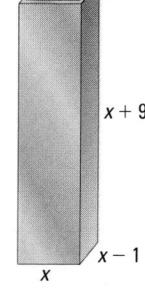

$x + 9$

$x - 1$

x

• The width should be one foot less than the length.

• The height should be nine feet more than the length.

What should the outside dimensions of the tank be?

Left margin:

STUDENT HELP

↳ **HOMEWORK HELP**
Example 1: Exs. 18–40, 59–67
Example 2: Exs. 18–32, 41–49, 59–67
Example 3: Exs. 18–32, 50–67
Example 4: Exs. 68–85
Example 5: Exs. 87–92

86. Essay should include the following points:

• For two terms, finding a common factor and using sum/difference of cubes

• For 3 terms, looking for a quadratic pattern

• For 4 or more terms, grouping and looking for a common factor

STUDENT HELP

🌐 **HOMEWORK HELP**
Visit our Web site www.mcdougallittell.com for help with problem solving in Ex. 88.

88. width: 2 ft, length: 3 ft, height: 12 ft

Right margin:

❗ **COMMON ERROR**
EXERCISES 39 AND 40 Some students may think that these binomials cannot be factored using the sum or difference of two cubes. Ask these students what common numerical factor can be factored out as a first step.

TEACHING TIPS
EXERCISES 41–49 Encourage students to be on the lookout for polynomials that can be factored by grouping in more than one way. For example, the polynomial in Ex. 41 can be factored by using the grouping $(x^3 + x^2) + (x + 1)$ or by using $(x^3 + 1) + (x^2 + x)$.

STUDENT HELP NOTES
↳ **Homework Help** Students can find help with problem solving for Ex. 88 at **www.mcdougallittell.com**. The information can be printed out for students who don't have access to the Internet.

33. $(x - 2)(x^2 + 2x + 4)$
34. $(x + 4)(x^2 - 4x + 16)$
35. $(6x + 1)(36x^2 - 6x + 1)$
36. $(5x - 2)(25x^2 + 10x + 4)$
37. $(10x + 3)(100x^2 - 30x + 9)$
38. $27(x + 2)(x^2 - 2x + 4)$
39. $4(2x - 1)(4x^2 + 2x + 1)$
40. $2(x + 3)(x^2 - 3x + 9)$
41–67. See Additional Answers beginning on page AA1.

ADDITIONAL PRACTICE AND RETEACHING

For Lesson 6.4:
• Practice Levels A, B, and C (*Chapter 6 Resource Book,* p. 53)
• Reteaching with Practice (*Chapter 6 Resource Book,* p. 56)
• 🖥 See Lesson 6.4 of the *Personal Student Tutor*

For more Mixed Review:
• 🖥 Search the *Test and Practice Generator* for key words or specific lessons.

DAILY HOMEWORK QUIZ

📋 *Transparency Available*

Find the greatest common factor of the terms in the polynomial.

1. $14x^2 + 35x^3 - 21x$ **7x**

2. $60x^4 + 15x^3 - 30x^2$ **15x²**

Factor the polynomial.

3. $8x^3 + 125$
$(2x + 5)(4x^2 - 10x + 25)$

4. $5x^3 + 10x^2 - x - 2$
$(x + 2)(5x^2 - 1)$

5. $200x^6 - 2x^4$
$2x^4(10x + 1)(10x - 1)$

EXTRA CHALLENGE NOTE

↪ Challenge problems for Lesson 6.4 are available in **blackline** format in the *Chapter 6 Resource Book,* p. 62 and at **www.mcdougallittell.com.**

ADDITIONAL TEST PREPARATION

1. OPEN ENDED Give an example of a binomial that can be factored either as the difference of two squares or as the difference of two cubes. Show the complete factorization of your binomial. Answers may vary.
Sample answer: $x^6 - 1$;
$(x + 1)(x - 1)(x^2 + x + 1)(x^2 - x + 1)$

89. 🌐 **CITY PARK** For the city park commission, you are designing a marble planter in which to plant flowers. You want the length of the planter to be six times the height and the width to be three times the height. The sides should be one foot thick. Since the planter will be on the sidewalk, it does not need a bottom. What should the outer dimensions of the planter be if it is to hold 4 cubic feet of dirt?   **6 ft by 3 ft by 1 ft**

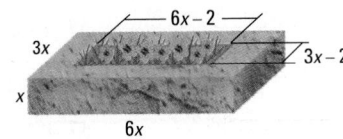

🌐 **SCULPTURE** In Exercises 90 and 91, refer to the sculpture shown in the picture.

90. The "cube" portion of the sculpture is actually a rectangular prism with dimensions x feet by $5x - 10$ feet by $2x - 1$ feet. The volume of the prism is 25 cubic feet. What are the dimensions of the prism? $2\frac{1}{2}$ **ft by** $2\frac{1}{2}$ **ft by 4 ft**

"Charred Sphere, Cube, and Pyramid" by David Nash

91. Suppose a pyramid like the one in the sculpture is $3x$ feet high and has a square base measuring $x - 5$ feet on each side. If the volume is 250 cubic feet, what are the dimensions of the pyramid? (Use the formula $V = \frac{1}{3}Bh$.)
base: 5 ft by 5 ft; height: 30 ft

92. 🪨 **CRAFTS** Suppose you have 250 cubic inches of clay with which to make a rectangular prism for a sculpture. If you want the height and width each to be 5 inches less than the length, what should the dimensions of the prism be?
5 in. by 5 in. by 10 in.

Test Preparation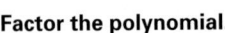

93. MULTIPLE CHOICE The expression $(3x - 4)(9x^2 + 12x + 16)$ is the factorization of which of the following?   **C**

   Ⓐ $27x^3 - 8$ Ⓑ $27x^3 + 36x^2$ Ⓒ $27x^3 - 64$ Ⓓ $27x^3 + 64$

94. MULTIPLE CHOICE Which of the following is the factorization of $x^3 - 8$?   **D**

   Ⓐ $(x - 2)(x^2 + 4x + 4)$ Ⓑ $(x + 2)(x^2 - 2x + 4)$

   Ⓒ $(x + 2)(x^2 - 4x + 4)$ Ⓓ $(x - 2)(x^2 + 2x + 4)$

95. MULTIPLE CHOICE What are the real solutions of the equation $x^5 = 81x$?   **D**

   Ⓐ $x = \pm3, \pm3i$ Ⓑ $x = 0, \pm9$

   Ⓒ $x = 0, \pm3, \pm3i$ Ⓓ $x = 0, \pm3$

★ Challenge

96. GEOMETRY CONNECTION Explain how the figure shown at the right can be used as a geometric factoring model for the sum of two cubes.   **See margin.**

$$a^3 + b^3 = (a + b)(a^2 - ab + b^2)$$

Factor the polynomial.

97. $30x^2y + 36x^2 - 20xy - 24x$ **2x(3x − 2)(5y + 6)**

98. $2x^7 - 127x$ **x(2x⁶ − 127)**

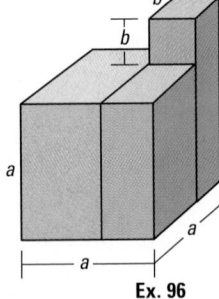

Ex. 96

EXTRA CHALLENGE

↪ www.mcdougallittell.com

MIXED REVIEW

SIMPLIFYING EXPRESSIONS Simplify the expression. (Review 6.1 for 6.5)

99. $\dfrac{6x^3y^9}{36x^3y^{-2}} \cdot \dfrac{y^{11}}{6}$

100. $\dfrac{5^{-2}x^2y^{-1}}{5^2xy^3} \cdot \dfrac{x}{625y^4}$

101. $\dfrac{7^2x^{-3}y^2}{49x^{-3}y^{-2}} \cdot y^4$

SYNTHETIC SUBSTITUTION Use synthetic substitution to evaluate the polynomial function for the given value of *x*. (Review 6.2 for 6.5)

102. $f(x) = 3x^4 + 2x^3 - x^2 - 12x + 1, x = 3$ **253**

103. $f(x) = 2x^5 - x^3 + 7x + 1, x = 3$ **481**

104. 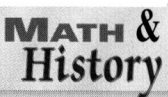 **SEWING** At the fabric store you are buying solid fabric at $4 per yard, print fabric at $6 per yard, and a pattern for $8. Write an equation for the amount you spend as a function of the amount of solid and print fabric you buy. (Review 3.5)
$T = 4s + 6p + 8$

MATH & History

Solving Polynomial Equations

APPLICATION LINK
www.mcdougallittell.com

THEN

IN 2000 B.C. the Babylonians solved polynomial equations by referring to tables of values. One such table gave the values of $y^3 + y^2$. To be able to use this table, the Babylonians sometimes had to manipulate the equation, as shown below.

$$ax^3 + bx^2 = c \qquad \text{Write original equation.}$$

$$\dfrac{a^3x^3}{b^3} + \dfrac{a^2x^2}{b^2} = \dfrac{a^2c}{b^3} \qquad \text{Multiply by } \dfrac{a^2}{b^3}.$$

$$\left(\dfrac{ax}{b}\right)^3 + \left(\dfrac{ax}{b}\right)^2 = \dfrac{a^2c}{b^3} \qquad \text{Re-express cubes and squares.}$$

Then they would find $\dfrac{a^2c}{b^3}$ in the $y^3 + y^2$ column of the table.

Because they knew that the corresponding *y*-value was equal to $\dfrac{ax}{b}$, they could conclude that $x = \dfrac{by}{a}$.

1. Calculate $y^3 + y^2$ for $y = 1, 2, 3, \ldots, 10$. Record the values in a table. See margin.

Use your table and the method discussed above to solve the equation.

2. $x^3 + x^2 = 252$ **6**
3. $x^3 + 2x^2 = 288$ **6**
4. $3x^3 + x^2 = 90$ **3**
5. $2x^3 + 5x^2 = 2500$ **10**
6. $7x^3 + 6x^2 = 1728$ **6**
7. $10x^3 + 3x^2 = 297$ **3**

NOW

TODAY computers use polynomial equations to accomplish many things, such as making robots move.

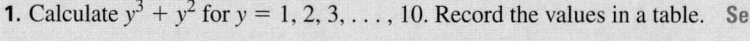

2000 B.C.
Babylonians use tables.

A.D. 1100
····Chinese solve cubic equations.

1545
Cardano solves cubic equations.

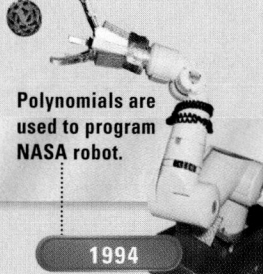

Polynomials are used to program NASA robot.

1994

6.4 *Factoring and Solving Polynomial Equations* **351**

MATH & HISTORY NOTE
See the enrichment master Math & History Applications on p. 61 of the *Chapter 6 Resource Book* for additional exercises.

96. The volume of the solid figure is the sum of the volumes of three rectangular prisms:
(Vol. right–hand/back) +
(Vol right–hand/front) +
(Vol. left–hand). This sum can be represented and simplified algebraically as follows.
$b^2(a + b) + ab(a - b) + a^2(a - b) =$
$b^2(a + b) + a(a - b)(b + a) =$
$b^2(a + b) + a(a - b)(a + b) =$
$(a + b)[b^2 + a(a - b)] =$
$(a + b)(a^2 - ab + b^2)$
The solid can also be thought of as a cube that measures *a* units along each side together with a cube that measures *b* units along each side. Thought of in this way, the volume is $a^3 + b^3$. Therefore, $a^3 + b^3 = (a + b)(a^2 - ab + b^2)$.

1.

y	$y^3 + y^2$
1	2
2	12
3	36
4	80
5	150
6	252
7	392
8	576
9	810
10	1100

LESSON OPENER
ACTIVITY
An alternative way to approach Lesson 6.5 is to use the Activity Lesson Opener:
- Blackline Master (*Chapter 6 Resource Book,* p. 66)
- Transparency (p. 39)

MEETING INDIVIDUAL NEEDS
- *Chapter 6 Resource Book*
  Prerequisite Skills Review (p. 5)
  Practice Level A (p. 67)
  Practice Level B (p. 68)
  Practice Level C (p. 69)
  Reteaching with Practice (p. 70)
  Absent Student Catch-Up (p. 72)
  Challenge (p. 74)
- *Resources in Spanish*
- *Personal Student Tutor*

NEW-TEACHER SUPPORT
See the Tips for New Teachers on pp. 1–2 of the *Chapter 6 Resource Book* for additional notes about Lesson 6.5.

WARM-UP EXERCISES

Transparency Available
Simplify.
1. $\dfrac{3x^3}{x^2}$ $3x$ **2.** $\dfrac{5x^2}{x^2}$ 5
Find the missing factor.
3. $x^2 - 2x - 63 = (x - 9)(?)$ $x + 7$
4. $2x^2 + 13x + 15 = (x + 5)(?)$
  $2x + 3$

6.5 The Remainder and Factor Theorems

What you should learn
GOAL 1 Divide polynomials and relate the result to the remainder theorem and the factor theorem.

GOAL 2 Use polynomial division in **real-life** problems, such as finding a production level that yields a certain profit in **Example 5**.

Why you should learn it
▼ To combine two **real-life** models into one new model, such as a model for money spent at the movies each year in **Ex. 62**.

GOAL 1 DIVIDING POLYNOMIALS

When you divide a polynomial $f(x)$ by a divisor $d(x)$, you get a quotient polynomial $q(x)$ and a remainder polynomial $r(x)$. We write this as $\dfrac{f(x)}{d(x)} = q(x) + \dfrac{r(x)}{d(x)}$. The degree of the remainder must be less than the degree of the divisor.

Example 1 shows how to divide polynomials using a method called **polynomial long division**.

EXAMPLE 1   *Using Polynomial Long Division*

Divide $2x^4 + 3x^3 + 5x - 1$ by $x^2 - 2x + 2$.

SOLUTION

Write division in the same format you would use when dividing numbers. Include a "0" as the coefficient of x^2.

> At each stage, divide the term with the highest power in what's left of the dividend by the first term of the divisor. This gives the next term of the quotient.

$$\frac{2x^4}{x^2} \qquad \frac{7x^3}{x^2} \qquad \frac{10x^2}{x^2}$$

$$2x^2 + 7x + 10$$

$$x^2 - 2x + 2 \overline{)\,2x^4 + 3x^3 + 0x^2 + 5x - 1}$$

$$\underline{2x^4 - 4x^3 + 4x^2} \quad \longleftarrow \text{Subtract } 2x^2(x^2 - 2x + 2).$$

$$7x^3 - 4x^2 + 5x$$

$$\underline{7x^3 - 14x^2 + 14x} \quad \longleftarrow \text{Subtract } 7x(x^2 - 2x + 2).$$

$$10x^2 - 9x - 1$$

$$\underline{10x^2 - 20x + 20} \quad \longleftarrow \text{Subtract } 10(x^2 - 2x + 2).$$

$$11x - 21 \quad \longleftarrow \text{remainder}$$

Write the result as follows.

$$\blacktriangleright \quad \frac{2x^4 + 3x^3 + 5x - 1}{x^2 - 2x + 2} = 2x^2 + 7x + 10 + \frac{11x - 21}{x^2 - 2x + 2}$$

✓**CHECK** You can check the result of a division problem by multiplying the divisor by the quotient and adding the remainder. The result should be the dividend.

$$(2x^2 + 7x + 10)(x^2 - 2x + 2) + 11x - 21$$

$$= 2x^2(x^2 - 2x + 2) + 7x(x^2 - 2x + 2) + 10(x^2 - 2x + 2) + 11x - 21$$

$$= 2x^4 - 4x^3 + 4x^2 + 7x^3 - 14x^2 + 14x + 10x^2 - 20x + 20 + 11x - 21$$

$$= 2x^4 + 3x^3 + 5x - 1 \checkmark$$

> **◉ ACTIVITY**
> **Developing Concepts**
> ## Investigating Polynomial Division
>
> Let $f(x) = 3x^3 - 2x^2 + 2x - 5$.
>
> **❶** Use long division to divide $f(x)$ by $x - 2$. What is the quotient? What is the remainder? $3x^2 + 4x + 10 + \dfrac{15}{x-2}$; $3x^2 + 4x + 10$; 15
>
> **❷** Use synthetic substitution to evaluate $f(2)$. How is $f(2)$ related to the remainder? What do you notice about the other constants in the last row of the synthetic substitution?
> 15; they are equal; they match the coefficients of the quotient.

In the activity you may have discovered that $f(2)$ gives you the remainder when $f(x)$ is divided by $x - 2$. This result is generalized in the *remainder theorem*.

REMAINDER THEOREM

If a polynomial $f(x)$ is divided by $x - k$, then the remainder is $r = f(k)$.

┌─────────────────────┐
│ **STUDENT HELP** │

▶ Study Tip
Notice that synthetic division could *not* have been used to divide the polynomials in Example 1 because the divisor, $x^2 - 2x + 2$, is not of the form $x - k$.
└─────────────────────┘

You may also have discovered in the activity that synthetic substitution gives the coefficients of the quotient. For this reason, synthetic substitution is sometimes called **synthetic division**. It can be used to divide a polynomial by an expression of the form $x - k$.

EXAMPLE 2   *Using Synthetic Division*

Divide $x^3 + 2x^2 - 6x - 9$ by **(a)** $x - 2$ and **(b)** $x + 3$.

SOLUTION

a. Use synthetic division for $k = 2$.

$$
\begin{array}{r|rrrr}
2 & 1 & 2 & -6 & -9 \\
  &   & 2 & 8 & 4 \\
\hline
  & 1 & 4 & 2 & -5
\end{array}
$$

▶ $\dfrac{x^3 + 2x^2 - 6x - 9}{x - 2} = x^2 + 4x + 2 + \dfrac{-5}{x - 2}$

b. To find the value of k, rewrite the divisor in the form $x - k$. Because $x + 3 = x - (-3)$, $k = -3$.

$$
\begin{array}{r|rrrr}
-3 & 1 & 2 & -6 & -9 \\
   &   & -3 & 3 & 9 \\
\hline
   & 1 & -1 & -3 & 0
\end{array}
$$

▶ $\dfrac{x^3 + 2x^2 - 6x - 9}{x + 3} = x^2 - x - 3$

---

2 TEACH

MOTIVATING THE LESSON
In some business situations, producing and selling fewer units may yield the same profit as producing and selling more units. Division of polynomials can often be helpful in analyzing such situations.

 EXTRA EXAMPLE 1
Divide $y^4 + 2y^2 - y + 5$ by $y^2 - y + 1$. $y^2 + y + 2 + \dfrac{3}{y^2 - y + 1}$

EXTRA EXAMPLE 2
Divide $x^3 - x^2 - 2x + 8$ by each binomial.
a. $x - 1$ $x^2 - 2 + \dfrac{6}{x - 1}$
b. $x + 2$ $x^2 - 3x + 4$

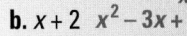 **CHECKPOINT EXERCISES**
For use after Example 1:
1. Use long division to divide $2x^2 + 13x - 7$ by $x + 6$.
$2x + 1 + \dfrac{-13}{x + 6}$

For use after Example 2 :
2. Use synthetic division to divide $x^3 - 3x^2 - 7x + 6$ by each binomial.
a. $x + 2$ $x^2 - 5x + 3$
b. $x - 4$ $x^2 + x - 3 + \dfrac{-6}{x - 4}$

ENGLISH LEARNERS
The description of how to perform polynomial long division may be hard to understand for some English learners. Carefully walk students through the solution to Example 1, explaining each step as you solve the problem step-by-step on the board. Verify students' understanding before continuing.

MATHEMATICAL REASONING
In discussing the remainder theorem, point out that if $f(x) = q(x)(x - k) + r(x)$, where $q(x)$ and $r(x)$ are the quotient and remainder polynomials, then $r(x)$ will be a constant, since its degree will be less than that of $x - k$. Substituting k for x shows that the value of this constant is $f(k)$.

353

EXTRA EXAMPLE 3
Factor $f(x) = 3x^3 + 13x^2 + 2x - 8$ given that $f(-4) = 0$.
$(x + 4)(x + 1)(3x - 2)$

EXTRA EXAMPLE 4
One zero of $f(x) = x^3 + 6x^2 + 3x - 10$ is $x = -5$. Find the other zeros of the function. -2 and 1

 CHECKPOINT EXERCISES
For use after Example 3:
1. Factor $f(x) = 3x^3 + 14x^2 - 28x - 24$ given that $f(-6) = 0$.
$(x + 6)(x - 2)(3x + 2)$

For use after Example 4 :
2. One zero of $f(x) = 2x^3 - 9x^2 - 32x - 21$ is $x = 7$. Find the other zeros of the function.
-1 and $-\frac{3}{2}$

STUDENT HELP NOTES
→ **Homework Help** Students can find extra examples at **www.mcdougallittell.com** that parallel the examples in the student edition.

In part (b) of Example 2, the remainder is 0. Therefore, you can rewrite the result as:

$$x^3 + 2x^2 - 6x - 9 = (x^2 - x - 3)(x + 3)$$

This shows that $x + 3$ is a factor of the original dividend.

FACTOR THEOREM

A polynomial $f(x)$ has a factor $x - k$ if and only if $f(k) = 0$.

Recall from Chapter 5 that the number k is called a *zero* of the function f because $f(k) = 0$.

STUDENT HELP

HOMEWORK HELP
Visit our Web site www.mcdougallittell.com for extra examples.

EXAMPLE 3 *Factoring a Polynomial*

Factor $f(x) = 2x^3 + 11x^2 + 18x + 9$ given that $f(-3) = 0$.

SOLUTION

Because $f(-3) = 0$, you know that $x - (-3)$ or $x + 3$ is a factor of $f(x)$. Use synthetic division to find the other factors.

$$
\begin{array}{r|rrrr}
-3 & 2 & 11 & 18 & 9 \\
   &   & -6 & -15 & -9 \\
\hline
   & 2 & 5 & 3 & 0
\end{array}
$$

The result gives the coefficients of the quotient.

$$2x^3 + 11x^2 + 18x + 9 = (x + 3)(2x^2 + 5x + 3)$$
$$= (x + 3)(2x + 3)(x + 1)$$

EXAMPLE 4 *Finding Zeros of a Polynomial Function*

One zero of $f(x) = x^3 - 2x^2 - 9x + 18$ is $x = 2$. Find the other zeros of the function.

SOLUTION

To find the zeros of the function, factor $f(x)$ completely. Because $f(2) = 0$, you know that $x - 2$ is a factor of $f(x)$. Use synthetic division to find the other factors.

$$
\begin{array}{r|rrrr}
2 & 1 & -2 & -9 & 18 \\
  &   & 2 & 0 & -18 \\
\hline
  & 1 & 0 & -9 & 0
\end{array}
$$

The result gives the coefficients of the quotient.

$$f(x) = (x - 2)(x^2 - 9) \qquad \text{Write } f(x) \text{ as a product of two factors.}$$
$$= (x - 2)(x + 3)(x - 3) \qquad \text{Factor difference of squares.}$$

▶ By the factor theorem, the zeros of f are 2, -3, and 3.

GOAL 2 USING POLYNOMIAL DIVISION IN REAL LIFE

In business and economics, a function that gives the price per unit p of an item in terms of the number x of units sold is called a *demand function*.

EXAMPLE 5 **Using Polynomial Models**

ACCOUNTING You are an accountant for a manufacturer of radios. The demand function for the radios is $p = 40 - 4x^2$ where x is the number of radios produced in millions. It costs the company $15 to make a radio.

a. Write an equation giving profit as a function of the number of radios produced.

b. The company currently produces 1.5 million radios and makes a profit of $24,000,000, but you would like to scale back production. What lesser number of radios could the company produce to yield the same profit?

SOLUTION

> **PROBLEM SOLVING STRATEGY**

a. VERBAL MODEL

Profit = Revenue − Cost

$$\text{Profit} = \boxed{\frac{\text{Price}}{\text{per unit}}} \cdot \boxed{\frac{\text{Number}}{\text{of units}}} - \boxed{\frac{\text{Cost}}{\text{per unit}}} \cdot \boxed{\frac{\text{Number}}{\text{of units}}}$$

LABELS

Profit = P (millions of dollars)

Price per unit = $40 - 4x^2$ (dollars per unit)

Number of units = x (millions of units)

Cost per unit = 15 (dollars per unit)

ALGEBRAIC MODEL

$$P = (40 - 4x^2)\, x - 15x$$
$$P = -4x^3 + 25x$$

b. Substitute 24 for P in the function you wrote in part (a).

$$24 = -4x^3 + 25x$$
$$0 = -4x^3 + 25x - 24$$

You know that $x = 1.5$ is one solution of the equation. This implies that $x - 1.5$ is a factor. So divide to obtain the following:

$$-2(x - 1.5)(2x^2 + 3x - 8) = 0$$

Use the quadratic formula to find that $x \approx 1.39$ is the other positive solution.

▶ The company can make the same profit by selling 1,390,000 units.

✓ **CHECK** Graph the profit function to confirm that there are two production levels that produce a profit of $24,000,000.

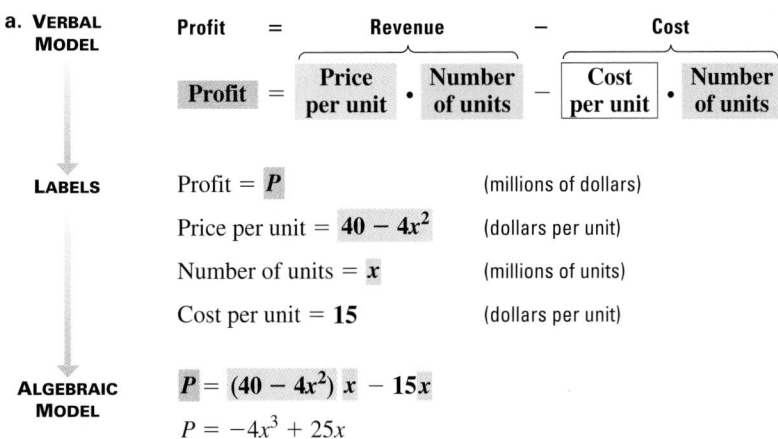

Radio Production

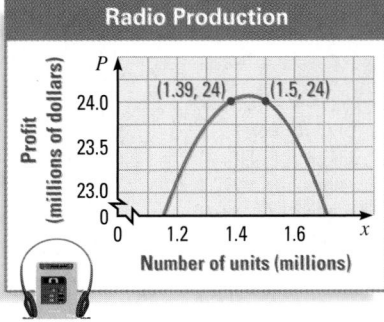

FOCUS ON CAREERS

ACCOUNTANT
Most people think of accountants as working for many clients. However, it is common for an accountant to work for a single client, such as a company or the government.

CAREER LINK
www.mcdougallittell.com

6.5 *The Remainder and Factor Theorems*   **355**

EXTRA EXAMPLE 5

A company that manufactures CD–ROM drives would like to increase its production. The demand function for the drives is $p = 75 - 3x^2$, where p is the price the company charges per unit when the company produces x million units. It costs the company $25 to produce each drive.
a. Write an equation giving the company's profit as a function of the number of CD–ROM drives it manufactures.
$P = -3x^3 + 50x$
b. The company currently manufactures 2 million CD–ROM drives and makes a profit of $76,000,000. At what other level of production would the company also make $76,000,000?   about 2.7 million CD–ROM drives

✓ CHECKPOINT EXERCISES

For use after Example 5:
1. Suppose it costs the company in Example 5 $17 to make each radio.
   a. Which equation models the company's profit as a function of the number of radios it produces?
   $P = -4x^3 + 23x$
   b. The company makes a profit of $19,000,000 when it produces one million radios. At what other production level would the company make the same profit?   about 1.736 million radios

CLOSURE QUESTION

If $f(x)$ is a polynomial that has $x - a$ as a factor, what do you know about the value of $f(a)$?
It is equal to 0.

DAILY PUZZLER

What is the least possible degree for a polynomial that has zeros at −5, 6, and 10? 3

3 APPLY

ASSIGNMENT GUIDE

BASIC
Day 1: pp. 356–358 Exs. 16–20 even, 28–34 even, 40–56 even, 59–63 odd, 64, 67–81 odd

AVERAGE
Day 1: pp. 356–358 Exs. 16–20 even, 28–58 even, 59–65 odd, 67–81 odd

ADVANCED
Day 1: pp. 356–358 Exs. 16–58 even, 59–65, 67–81 odd

BLOCK SCHEDULE
pp. 356–358 Exs. 16–20 even, 28–58 even, 59–65 odd, 67–81 odd (with 6.6)

EXERCISE LEVELS

Level A: *Easier*
15–22, 27–31

Level B: *More Difficult*
23–26, 32–56, 59–65

Level C: *Most Difficult*
57–58

✔ HOMEWORK CHECK

To quickly check student understanding of key concepts, go over the following exercises: Exs. 18, 30, 40, 48. See also the Daily Homework Quiz:

- Blackline Master (*Chapter 6 Resource Book*, p. 77)
- 📖 Transparency (p. 46)

27. $x^2 + 2x - 3 - \dfrac{12}{x-2}$

28. $x^2 - 4x + 2$

29. $4x + 1 - \dfrac{5}{x+1}$

30. $x - 2 - \dfrac{1}{x-2}$

31. $2x + 11 + \dfrac{30}{x-2}$

32. $3x + 8 + \dfrac{48}{x-6}$

GUIDED PRACTICE

Vocabulary Check ✔

Concept Check ✔

1. For any number k, the remainder obtained when a polynomial $f(x)$ is divided by $x - k$ is the value of $f(x)$ when $x = k$.

Skill Check ✔

15. $x + 9 + \dfrac{13}{x-2}$

16. $3x + 20 + \dfrac{61}{x-3}$

17. $2x - 5 + \dfrac{19}{x+4}$

18. $x - 7 + \dfrac{11}{x+1}$

19. $x + 15 + \dfrac{147}{x-10}$

20. $x^2 - 2x - 1 - \dfrac{9}{x-1}$

21. $2x^2 + 2 + \dfrac{9}{x^2-1}$

22. $x + 8 - \dfrac{8x+24}{x^2+5}$

23. $3x - 4 + \dfrac{5}{2x+3}$

24. $10x + 7 + \dfrac{5}{x^2+2x}$

25. $5x^2 - x + 3$

26. $2x - \dfrac{9}{x^3+x^2-5}$

1. State the remainder theorem.   **See margin.**

2. Write a polynomial division problem that you would use long division to solve. Then write a polynomial division problem that you would use synthetic division to solve.   *Sample answer:* $\dfrac{3x^3 + x^2 + 5x + 8}{3x-4}$; $\dfrac{3x^3 + x^2 + 5x + 8}{x-2}$

3. Write the polynomial divisor, dividend, and quotient represented by the synthetic division shown at the right.
$x + 3, x^3 - 2x^2 - 9x + 18, x^2 - 5x + 6$

$$-3 \;\big|\; \begin{array}{rrrr} 1 & -2 & -9 & 18 \\ & -3 & 15 & -18 \end{array}$$
$$\begin{array}{rrrr} 1 & -5 & 6 & 0 \end{array}$$

Divide using polynomial long division.

4. $(2x^3 - 7x^2 - 17x - 3) \div (2x + 3)$ $x^2 - 5x - 1$

5. $(x^3 + 5x^2 - 2) \div (x + 4)$ $x^2 + x - 4 + \dfrac{14}{x+4}$

6. $(-3x^3 + 4x - 1) \div (x - 1)$ $-3x^2 - 3x + 1$

7. $(-x^3 + 2x^2 - 2x + 3) \div (x^2 - 1)$ $-x + 2 + \dfrac{-3x+5}{x^2-1}$

Divide using synthetic division.

8. $(x^3 - 8x + 3) \div (x + 3)$ $x^2 - 3x + 1$

9. $(x^4 - 16x^2 + x + 4) \div (x + 4)$ $x^3 - 4x^2 + 1$

10. $(x^2 + 2x + 15) \div (x - 3)$ $x + 5 + \dfrac{30}{x-3}$

11. $(x^2 + 7x - 2) \div (x - 2)$ $x + 9 + \dfrac{16}{x-2}$

Given one zero of the polynomial function, find the other zeros.

12. $f(x) = x^3 - 8x^2 + 4x + 48; 4$ $-2, 6$

13. $f(x) = 2x^3 - 14x^2 - 56x - 40; 10$ $-2, -1$

14. 🌐 **BUSINESS** Look back at Example 5. If the company produces 1 million radios, it will make a profit of $21,000,000. Find another number of radios that the company could produce to make the same profit.   **about 1.85 million radios**

PRACTICE AND APPLICATIONS

STUDENT HELP

▶ **Extra Practice**
to help you master skills is on p. 948.

USING LONG DIVISION **Divide using polynomial long division.**   15–26. See margin.

15. $(x^2 + 7x - 5) \div (x - 2)$

16. $(3x^2 + 11x + 1) \div (x - 3)$

17. $(2x^2 + 3x - 1) \div (x + 4)$

18. $(x^2 - 6x + 4) \div (x + 1)$

19. $(x^2 + 5x - 3) \div (x - 10)$

20. $(x^3 - 3x^2 + x - 8) \div (x - 1)$

21. $(2x^4 + 7) \div (x^2 - 1)$

22. $(x^3 + 8x^2 - 3x + 16) \div (x^2 + 5)$

23. $(6x^2 + x - 7) \div (2x + 3)$

24. $(10x^3 + 27x^2 + 14x + 5) \div (x^2 + 2x)$

25. $(5x^4 + 14x^3 + 9x) \div (x^2 + 3x)$

26. $(2x^4 + 2x^3 - 10x - 9) \div (x^3 + x^2 - 5)$

STUDENT HELP

▶ **HOMEWORK HELP**
Example 1: Exs. 15–26
Example 2: Exs. 27–38
Example 3: Exs. 39–46
Example 4: Exs. 47–54
Example 5: Exs. 60–62

USING SYNTHETIC DIVISION **Divide using synthetic division.**   27–38. See margin.

27. $(x^3 - 7x - 6) \div (x - 2)$

28. $(x^3 - 14x + 8) \div (x + 4)$

29. $(4x^2 + 5x - 4) \div (x + 1)$

30. $(x^2 - 4x + 3) \div (x - 2)$

31. $(2x^2 + 7x + 8) \div (x - 2)$

32. $(3x^2 - 10x) \div (x - 6)$

33. $(x^2 + 10) \div (x + 4)$

34. $(x^2 + 3) \div (x + 3)$

35. $(10x^4 + 5x^3 + 4x^2 - 9) \div (x + 1)$

36. $(x^4 - 6x^3 - 40x + 33) \div (x - 7)$

37. $(2x^4 - 6x^3 + x^2 - 3x - 3) \div (x - 3)$

38. $(4x^4 + 5x^3 + 2x^2 - 1) \div (x + 1)$

FACTORING Factor the polynomial given that $f(k) = 0$.   39–46. See margin.

39. $f(x) = x^3 - 5x^2 - 2x + 24; k = -2$ **40.** $f(x) = x^3 - 3x^2 - 16x - 12; k = 6$

41. $f(x) = x^3 - 12x^2 + 12x + 80; k = 10$ **42.** $f(x) = x^3 - 18x^2 + 95x - 126; k = 9$

43. $f(x) = x^3 - x^2 - 21x + 45; k = -5$ **44.** $f(x) = x^3 - 11x^2 + 14x + 80; k = 8$

45. $f(x) = 4x^3 - 4x^2 - 9x + 9; k = 1$ **46.** $f(x) = 2x^3 + 7x^2 - 33x - 18; k = -6$

FINDING ZEROS Given one zero of the polynomial function, find the other zeros.
49–52. See margin.

47. $f(x) = 9x^3 + 10x^2 - 17x - 2; -2\frac{1}{9}, 1$ **48.** $f(x) = x^3 + 11x^2 - 150x - 1512; -14$ $-9, 12$

49. $f(x) = 2x^3 + 3x^2 - 39x - 20; 4$ **50.** $f(x) = 15x^3 - 119x^2 - 10x + 16; 8$

51. $f(x) = x^3 - 14x^2 + 47x - 18; 9$ **52.** $f(x) = 4x^3 + 9x^2 - 52x + 15; -5$

53. $f(x) = x^3 + x^2 + 2x + 24; -3$ $\frac{1 \pm i\sqrt{7}}{}$ **54.** $f(x) = 5x^3 - 27x^2 - 17x - 6; 6$ $\frac{-3 \pm i\sqrt{11}}{10}$

GEOMETRY **CONNECTION** You are given an expression for the volume of the rectangular prism. Find an expression for the missing dimension.

55. $V = 3x^3 + 8x^2 - 45x - 50$ $3x - 10$ **56.** $V = 2x^3 + 17x^2 + 40x + 25$ $2x + 5$

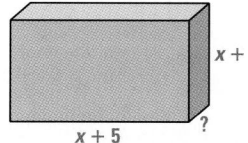
$x + 1$
$x + 5$?

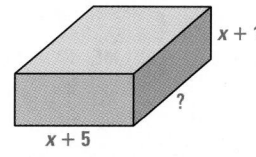
$x + 1$
$x + 5$?

POINTS OF INTERSECTION Find all points of intersection of the two graphs given that one intersection occurs at $x = 1$.

57.

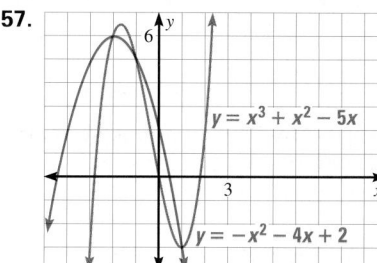
$y = x^3 + x^2 - 5x$
$y = -x^2 - 4x + 2$
$(-2, 6), (-1, 5), (1, -3)$

58.

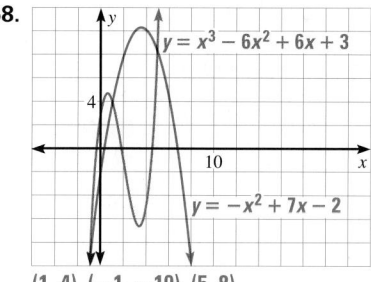
$y = x^3 - 6x^2 + 6x + 3$
$y = -x^2 + 7x - 2$
$(1, 4), (-1, -10), (5, 8)$

59. LOGICAL REASONING You divide two polynomials and obtain the result $5x^2 - 13x + 47 - \frac{102}{x + 2}$. What is the dividend? How did you find it?   **See margin.**

60. 🌐 **COMPANY PROFIT** The demand function for a type of camera is given by the model $p = 100 - 8x^2$ where p is measured in dollars per camera and x is measured in millions of cameras. The production cost is $25 per camera. The production of 2.5 million cameras yielded a profit of $62.5 million. What other number of cameras could the company sell to make the same profit?
about 0.92 million cameras

61. 🌐 **FUEL CONSUMPTION** From 1980 to 1991, the total fuel consumption T (in billions of gallons) by cars in the United States and the average fuel consumption A (in gallons per car) can be modeled by   **See margin.**

$$T = -0.026x^3 + 0.47x^2 - 2.2x + 72 \qquad \text{and} \qquad A = -8.4x + 580$$

where x is the number of years since 1980. Find a function for the number of cars from 1980 to 1991. About how many cars were there in 1990?

6.5 *The Remainder and Factor Theorems*

---

49. $-5, -\frac{1}{2}$

50. $-\frac{2}{5}, \frac{1}{3}$

51. $\frac{5 \pm \sqrt{17}}{2}$

52. $\frac{11 \pm \sqrt{73}}{8}$

59. $5x^3 - 3x^2 + 21x - 8$; I multiplied the quotient by the denominator $x + 2$ and added -102.

61. Answers may vary depending on rounding. $C = 0.0031x^2 + 0.1578x + 11.155 + \frac{6398}{8.4x - 580}$; about 144 million cars

FOCUS ON APPLICATIONS

ALTERNATIVE FUEL
Joshua and Kaia Tickell built the Green Grease Machine, which converts used restaurant vegetable oil into biodiesel fuel. The Tickells use the fuel in their motor home, the Veggie Van, as an alternative to the fuel referred to in **Ex. 61**.

🌐 **APPLICATION LINK**
www.mcdougallittell.com

❗ **COMMON ERROR**
EXERCISES 8–11, 27–38
Some students may use $-k$ in the synthetic division when the binomial divisor is $x - k$ and k is positive. Remind students that if $x - k$ is a factor of a polynomial function, then $x - k = 0$ and $x = k$ is a zero of the function. Therefore, they should use k in the synthetic division.

❗ **COMMON ERROR**
EXERCISES 53 AND 54
Students who are using a graphing calculator may assume that since the graphs of these functions cross the x-axis in only one point, there are no other zeros. Remind them to ask whether there might be complex zeros.

APPLICATION NOTE
EXERCISE 61 Additional information about alternative fuels is available at
www.mcdougallittell.com.

33. $x - 4 + \frac{26}{x + 4}$

34. $x - 3 + \frac{12}{x + 3}$

35. $10x^3 - 5x^2 + 9x - 9$

36. $x^3 + x^2 + 7x + 9 + \frac{96}{x - 7}$

37. $2x^3 + x - \frac{3}{x - 3}$

38–46. See Additional Answers beginning on page AA1.

ADDITIONAL PRACTICE AND RETEACHING

For Lesson 6.5:
• Practice Levels A, B, and C (*Chapter 6 Resource Book,* p. 67)
• Reteaching with Practice (*Chapter 6 Resource Book,* p. 70)
• 🖥 See Lesson 6.5 of the *Personal Student Tutor*

For more Mixed Review:
• 🖥 Search the *Test and Practice Generator* for key words or specific lessons.

DAILY HOMEWORK QUIZ

📄 *Transparency Available*

1. Divide using polynomial long division. $(x^2 - 7x - 12) \div (x + 5)$
$x - 12 + \dfrac{48}{x + 5}$

2. Divide using synthetic division. $(2x^3 + 6x^2 - 9) \div (x + 2)$
$2x^2 + 2x - 4 - \dfrac{1}{x + 2}$

3. Factor the polynomial $f(x) = x^3 - 2x^2 - 19x + 20$ given that $f(5) = 0$. $(x - 5)(x + 4)(x - 1)$

4. Given that -3 is a zero of $f(x) = x^3 - 6x^2 - 13x + 42$, find the other zeros.   2, 7

EXTRA CHALLENGE NOTE

↳ Challenge problems for Lesson 6.5 are available in **blackline** format in the *Chapter 6 Resource Book*, p. 74 and at **www.mcdougallittell.com**.

ADDITIONAL TEST PREPARATION

1. OPEN ENDED Give an example of a polynomial division that can be performed using synthetic division. Give an example that could not be performed using synthetic division.
Sample answer: $(4x^3 - 5x + 7) \div (x - 9)$; $(4x^3 - 5 + 7) \div (x^2 + 5x - 9)$

65–69. See Additional Answers beginning on page AA1.

62. Answers may vary depending on rounding.
$A = -1.1686x^2 + 137.4865x - 13,097.384 + \dfrac{3,240,124}{2.61x + 247}$; about $21

62.  **MOVIES** The amount M (in millions of dollars) spent at movie theaters from 1989 to 1996 can be modeled by

$$M = -3.05x^3 + 70.2x^2 - 225x + 5070$$

where x is the number of years since 1989. The United States population P (in millions) from 1989 to 1996 can be modeled by the following function:

$$P = 2.61x + 247$$

Find a function for the average annual amount spent per person at movie theaters from 1989 to 1996. On average, about how much did each person spend at movie theaters in 1989? ▸ *Source: Statistical Abstract of the United States*

Test Preparation

63. MULTIPLE CHOICE What is the result of dividing $x^3 - 9x + 5$ by $x - 3$?   **C**

Ⓐ $x^2 + 3x + 5$ Ⓑ $x^2 + 3x$ Ⓒ $x^2 + 3x + \dfrac{5}{x - 3}$

Ⓓ $x^2 + 3x - \dfrac{5}{x - 3}$ Ⓔ $x^2 + 3x - 18 + \dfrac{59}{x - 3}$

64. MULTIPLE CHOICE Which of the following is a factor of the polynomial $2x^3 - 19x^2 - 20x + 100$?   **E**

Ⓐ $x + 10$ Ⓑ $x + 2$ Ⓒ $2x - 5$ Ⓓ $x - 5$ Ⓔ $2x + 5$

★ **Challenge**

65. COMPARING METHODS Divide the polynomial $12x^3 - 8x^2 + 5x + 2$ by $2x + 1$, $3x + 1$, and $4x + 1$ using long division. Then divide the same polynomial by $x + \frac{1}{2}$, $x + \frac{1}{3}$, and $x + \frac{1}{4}$ using synthetic division. What do you notice about the remainders and the coefficients of the quotients from the two types of division?   See margin.

EXTRA CHALLENGE
↳ www.mcdougallittell.com

MIXED REVIEW

CHECKING SOLUTIONS Check whether the given ordered pairs are solutions of the inequality. **(Review 2.6)** 66–69. See margin.

66. $x + 7y \le -8$; $(6, -2)$, $(-2, -3)$ **67.** $2x + 5y \ge 1$; $(-2, 4)$, $(8, -3)$

68. $9x - 4y > 7$; $(-1, -4)$, $(2, 2)$ **69.** $-3x - 2y < -6$; $(2, 0)$, $(1, 4)$

QUADRATIC FORMULA Use the quadratic formula to solve the equation. **(Review 5.6 for 6.6)** 73–78. See margin.

70. $x^2 - 5x + 3 = 0$ $\dfrac{5 \pm \sqrt{13}}{2}$ **71.** $x^2 - 8x + 3 = 0$ $4 \pm \sqrt{13}$ **72.** $x^2 - 10x + 15 = 0$ $5 \pm \sqrt{10}$

73. $4x^2 - 7x + 1 = 0$ **74.** $-6x^2 - 9x + 2 = 0$ **75.** $5x^2 + x - 2 = 0$

76. $2x^2 + 3x + 5 = 0$ **77.** $-5x^2 - x - 8 = 0$ **78.** $3x^2 + 3x + 1 = 0$

73. $\dfrac{7 \pm \sqrt{33}}{8}$

74. $\dfrac{-9 \pm \sqrt{129}}{12}$

75. $\dfrac{-1 \pm \sqrt{41}}{10}$

76. $\dfrac{-3 \pm i\sqrt{31}}{4}$

77. $\dfrac{-1 \pm i\sqrt{159}}{10}$

78. $\dfrac{-3 \pm i\sqrt{3}}{6}$

POLYNOMIAL OPERATIONS Perform the indicated operation. **(Review 6.3)**

79. $(x^2 - 3x + 8) - (x^2 + x - 1)$ $-4x + 9$

80. $(14x^2 - 15x + 3) + (11x - 7)$ $14x^2 - 4x - 4$

81. $(8x^3 - 1) - (22x^3 + 2x^2 - x - 5)$ $-14x^3 - 2x^2 + x + 4$

82. $(x + 5)(x^2 - x + 5)$ $x^3 + 4x^2 + 25$

83. 🍽 **CATERING** You are helping your sister plan her wedding reception. The guests have chosen whether they would like the chicken dish or the vegetarian dish. The caterer charges $24 per chicken dish and $21 per vegetarian dish. After ordering the dinners for the 120 guests, the caterer's bill comes to $2766. How many guests requested chicken? **(Lesson 3.2)**   82 guests

6.6

Finding Rational Zeros

What you should learn

GOAL 1 Find the rational zeros of a polynomial function.

GOAL 2 Use polynomial equations to solve **real-life** problems, such as finding the dimensions of a monument in **Ex. 60**.

Why you should learn it

▼ To model **real-life** quantities, such as the volume of a representation of the Louvre pyramid in **Example 3**.

GOAL 1 USING THE RATIONAL ZERO THEOREM

The polynomial function

$$f(x) = 64x^3 + 120x^2 - 34x - 105$$

has $-\dfrac{3}{2}$, $-\dfrac{5}{4}$, and $\dfrac{7}{8}$ as its zeros. Notice that the numerators of these zeros (-3, -5, and 7) are factors of the constant term, -105. Also notice that the denominators (2, 4, and 8) are factors of the leading coefficient, 64. These observations are generalized by the *rational zero theorem*.

THE RATIONAL ZERO THEOREM

If $f(x) = a_n x^n + \cdots + a_1 x + a_0$ has *integer* coefficients, then every rational zero of f has the following form:

$$\frac{p}{q} = \frac{\text{factor of constant term } a_0}{\text{factor of leading coefficient } a_n}$$

EXAMPLE 1   *Using the Rational Zero Theorem*

Find the rational zeros of $f(x) = x^3 + 2x^2 - 11x - 12$.

SOLUTION

List the possible rational zeros. The leading coefficient is 1 and the constant term is -12. So, the possible rational zeros are:

$$x = \pm\frac{1}{1},\ \pm\frac{2}{1},\ \pm\frac{3}{1},\ \pm\frac{4}{1},\ \pm\frac{6}{1},\ \pm\frac{12}{1}$$

Test these zeros using synthetic division.

Test $x = 1$:

```
1 |  1    2   -11   -12
  |       1    3    -8
  ----------------------
     1    3   -8   -20
```

Test $x = -1$:

```
-1 |  1    2   -11   -12
   |      -1   -1    12
   ----------------------
      1    1   -12    0
```

Since -1 is a zero of f, you can write the following:

$$f(x) = (x + 1)(x^2 + x - 12)$$

Factor the trinomial and use the factor theorem.

$$f(x) = (x + 1)(x^2 + x - 12) = (x + 1)(x - 3)(x + 4)$$

▶ The zeros of f are -1, 3, and -4.

6.6 *Finding Rational Zeros*   **359**

1 PLAN

PACING
Basic: 1 day
Average: 1 day
Advanced: 1 day
Block Schedule: 0.5 block with 6.5

LESSON OPENER
VISUAL APPROACH
An alternative way to approach Lesson 6.6 is to use the Visual Approach Lesson Opener:
• Blackline Master (*Chapter 6 Resource Book,* p. 78)
•  Transparency (p. 40)

MEETING INDIVIDUAL NEEDS
• *Chapter 6 Resource Book*
  Prerequisite Skills Review (p. 5)
  Practice Level A (p. 79)
  Practice Level B (p. 80)
  Practice Level C (p. 81)
  Reteaching with Practice (p. 82)
  Absent Student Catch-Up (p. 84)
  Challenge (p. 86)
• *Resources in Spanish*
• *Personal Student Tutor*

NEW-TEACHER SUPPORT
See the Tips for New Teachers on pp. 1–2 of the *Chapter 6 Resource Book* for additional notes about Lesson 6.6.

WARM-UP EXERCISES

Transparency Available

Factor each polynomial.
1. $6x^2 + 13x - 28$ $(3x - 4)(2x + 7)$
2. $5x^3 - 22x^2 + 12x - 16$ if $f(4) = 0$
   $(5x^2 - 2x + 4)(x - 4)$
Find the zeros of each function.
3. $f(x) = 2x^2 + 3x - 2$ -2 and $\dfrac{1}{2}$
4. $f(x) = x^4 - 9x^2 + 20$ $-2, 2,$
   $-\sqrt{5}, \sqrt{5}$
5. Use synthetic division to find
   $f\!\left(\dfrac{5}{4}\right)$ when $f(x) = 8x^3 - 22x^2 +$
   $47x - 40$. **0**

EXTRA EXAMPLE 1
Find the rational zeros of $f(x) = x^3 - 4x^2 - 11x + 30$. $-3, 2, 5$

EXTRA EXAMPLE 2
Find all real zeros of $f(x) = 15x^4 - 68x^3 - 7x^2 + 24x - 4$.
$-\dfrac{2}{3}, \dfrac{1}{5}, \dfrac{5 - \sqrt{17}}{2}, \dfrac{5 + \sqrt{17}}{2}$

CHECKPOINT EXERCISES

For use after Example 1:

1. Find the rational zeros of $f(x) = x^3 - x^2 - 9x + 9$. $-3, 1, 3$

For use after Example 2:

2. Find all real zeros of $f(x) = x^3 - 7x^2 + 10x + 6$.
$3, 2 + \sqrt{6}, 2 - \sqrt{6}$

CONCEPT QUESTION
EXAMPLE 2 Why is any zero of g also a zero of f?
$f(x) = (2x + 3) \cdot g(x)$ and hence $f(k) = (2k + 3) \cdot g(k)$. If k is a zero of g, then $g(k) = 0$, which means that $(2k + 3) \cdot g(k) = 0$.

MATHEMATICAL REASONING
Students may find it helpful to examine a product such as $(3x + 5)(2x + 7)(11x + 6)$ to get a better grasp of the rational zero theorem. Note which integers are multiplied to get the leading coefficient (3, 2, and 11) and which to get the constant term (5, 7, and 6). Relate these to the numerators and denominators of the rational zeros $\left(-\dfrac{5}{3}, -\dfrac{7}{2}, \text{ and } -\dfrac{6}{11}\right)$.

In Example 1, the leading coefficient is 1. When the leading coefficient is not 1, the list of possible rational zeros can increase dramatically. In such cases the search can be shortened by sketching the function's graph—either by hand or by using a graphing calculator.

EXAMPLE 2 *Using the Rational Zero Theorem*

Find all real zeros of $f(x) = 10x^4 - 3x^3 - 29x^2 + 5x + 12$.

SOLUTION

List the possible rational zeros of f: $\pm\dfrac{1}{1}, \pm\dfrac{2}{1}, \pm\dfrac{3}{1}, \pm\dfrac{4}{1}, \pm\dfrac{6}{1},$
$\pm\dfrac{12}{1}, \pm\dfrac{3}{2}, \pm\dfrac{1}{5}, \pm\dfrac{2}{5}, \pm\dfrac{3}{5}, \pm\dfrac{6}{5}, \pm\dfrac{12}{5}, \pm\dfrac{1}{10}, \pm\dfrac{3}{10}, \pm\dfrac{12}{10}.$

Choose values to check.

With so many possibilities, it is worth your time to sketch the graph of the function. From the graph, it appears that some reasonable choices are $x = -\dfrac{3}{2}, x = -\dfrac{3}{5}, x = \dfrac{4}{5},$ and $x = \dfrac{3}{2}$.

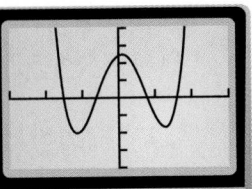

Check the chosen values using synthetic division.

$$
-\tfrac{3}{2} \,\bigg|\; \begin{array}{rrrrr} 10 & -3 & -29 & 5 & 12 \\ & -15 & 27 & 3 & -12 \\ \hline 10 & -18 & -2 & 8 & 0 \end{array}
$$

$\longleftarrow -\dfrac{3}{2}$ is a zero.

Factor out a binomial using the result of the synthetic division.

$$f(x) = \left(x + \dfrac{3}{2}\right)(10x^3 - 18x^2 - 2x + 8) \qquad \text{Rewrite as a product of two factors.}$$

$$= \left(x + \dfrac{3}{2}\right)(2)(5x^3 - 9x^2 - x + 4) \qquad \text{Factor 2 out of the second factor.}$$

$$= (2x + 3)(5x^3 - 9x^2 - x + 4) \qquad \text{Multiply the first factor by 2.}$$

Repeat the steps above for $g(x) = 5x^3 - 9x^2 - x + 4$.

Any zero of g will also be a zero of f. The possible *rational* zeros of g are $x = \pm 1, \pm 2, \pm 4, \pm\dfrac{1}{5}, \pm\dfrac{2}{5},$ and $\pm\dfrac{4}{5}$. The graph of f shows that $\dfrac{4}{5}$ may be a zero.

$$
\tfrac{4}{5} \,\bigg|\; \begin{array}{rrrr} 5 & -9 & -1 & 4 \\ & 4 & -4 & -4 \\ \hline 5 & -5 & -5 & 0 \end{array}
$$

$\longleftarrow \dfrac{4}{5}$ is a zero.

So $f(x) = (2x + 3)\left(x - \dfrac{4}{5}\right)(5x^2 - 5x - 5) = (2x + 3)(5x - 4)(x^2 - x - 1)$

Find the remaining zeros of f by using the quadratic formula to solve $x^2 - x - 1 = 0$.

▶ The real zeros of f are $-\dfrac{3}{2}, \dfrac{4}{5}, \dfrac{1 + \sqrt{5}}{2},$ and $\dfrac{1 - \sqrt{5}}{2}$.

GOAL 2 SOLVING POLYNOMIAL EQUATIONS IN REAL LIFE

EXAMPLE 3 *Writing and Using a Polynomial Model*

Crafts

You are designing a candle-making kit. Each kit will contain 25 cubic inches of candle wax and a mold for making a model of the pyramid-shaped building at the Louvre Museum in Paris, France. You want the height of the candle to be 2 inches less than the length of each side of the candle's square base. What should the dimensions of your candle mold be?

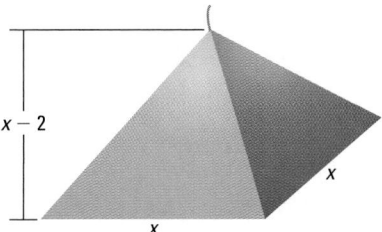

SOLUTION

The volume is $V = \frac{1}{3}Bh$ where B is the area of the base and h is the height.

PROBLEM SOLVING STRATEGY

VERBAL MODEL	$\boxed{\text{Volume}} = \frac{1}{3} \cdot \boxed{\text{Area of base}} \cdot \boxed{\text{Height}}$

LABELS	Volume $= \mathbf{25}$	(cubic inches)
	Side of square base $= \boldsymbol{x}$	(inches)
	Area of base $= \boldsymbol{x^2}$	(square inches)
	Height $= \boldsymbol{x - 2}$	(inches)

ALGEBRAIC MODEL	$25 = \frac{1}{3}x^2(x-2)$	Write algebraic model.
	$75 = x^3 - 2x^2$	Multiply each side by 3 and simplify.
	$0 = x^3 - 2x^2 - 75$	Subtract 75 from each side.

The possible rational solutions are $x = \pm\frac{1}{1}, \pm\frac{3}{1}, \pm\frac{5}{1}, \pm\frac{15}{1}, \pm\frac{25}{1}, \pm\frac{75}{1}$.

Use the possible solutions. Note that in this case, it makes sense to test only positive x-values.

$$
\begin{array}{r|rrrr}
1 & 1 & -2 & 0 & -75 \\
  &   & 1 & -1 & -1 \\
\hline
  & 1 & -1 & -1 & -76
\end{array}
\qquad
\begin{array}{r|rrrr}
3 & 1 & -2 & 0 & -75 \\
  &   & 3 & 3 & 9 \\
\hline
  & 1 & 1 & 3 & -66
\end{array}
$$

$$
\begin{array}{r|rrrr}
5 & 1 & -2 & 0 & -75 \\
  &   & 5 & 15 & 75 \\
\hline
  & 1 & 3 & 15 & 0
\end{array}
\quad \longleftarrow \text{5 is a solution.}
$$

So $x = 5$ is a solution. The other two solutions, which satisfy $x^2 + 3x + 15 = 0$, are

$x = \dfrac{-3 \pm i\sqrt{51}}{2}$ and can be discarded because they are imaginary.

▶ The base of the candle mold should be 5 inches by 5 inches. The height of the mold should be $5 - 2 = 3$ inches.

FOCUS ON PEOPLE

I.M. PEI designed the pyramid at the Louvre. His geometric architecture can be seen in Boston, New York, Dallas, Los Angeles, Taiwan, Beijing, and Singapore.

6.6 *Finding Rational Zeros* **361**

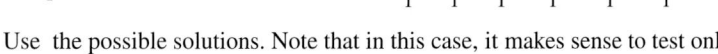

EXTRA EXAMPLE 3
A rectangular column of cement is to have a volume of 20.25 ft³. The base is to be square, with sides 3 ft less than half the height of the column. What should the dimensions of the column be? **1.5 ft by 1.5 ft by 9 ft**

✓ **CHECKPOINT EXERCISES**
For use after Example 3:
1. A company that makes salsa wants to change the size of its cylindrical salsa cans. The radius of the new can will be 5 cm less than the height. The container will hold 144π cm³ of salsa. What are the dimensions of the new container? **radius: 4 cm; height: 9 cm**

APPLICATION NOTE
EXAMPLE 3 The pyramid build by I. M. Pei is at the entrance to the Louvre and caused considerable controversy. Many thought its modern glass facade would create too great a visual contrast to the surrounding classical Renaissance architecture. Because many also feared that the new entrance would obstruct the view of the main structure, the pyramid's actual height is less than originally planned. The pyramid is 60 feet tall and stands on a 90-foot square base.

FOCUS ON VOCABULARY
What is a rational zero of a polynomial function? **a rational number for which the value of the function is zero**

CLOSURE QUESTION
How can you use the graph of a polynomial function to help determine its real roots? **See margin.**

DAILY PUZZLER
The sum of the coefficients of a polynomial function is 0. Does the polynomial function have a rational zero? Explain. **yes; 1 is a zero of the function.**

Closure Question *Sample answer:*
Examine the graph to get an estimate of where it crosses the *x*-axis. Test the possible rational zeros that are close to your estimates to identify any rational zeros. Divide the polynomial by the product of all binomials of the form *x* – *r*, where *r* is a rational zero. See if it is possible to identify the zeros of the quotient polynomial.

 361

ASSIGNMENT GUIDE

BASIC
Day 1: pp. 362–365 Exs. 16–30
even, 34–40 even, 46–48,
59, 65–66, 71–83 odd,
Quiz 2 Exs. 1–19

AVERAGE
Day 1: pp. 362–365 Exs. 16–30
even, 34–50 even, 59–61,
65–66, 71–83 odd,
Quiz 2 Exs. 1–19

ADVANCED
Day 1: pp. 362–365 Exs. 16–54
even, 60–64 even, 65–70,
71–83 odd, Quiz 2 Exs. 1–19

BLOCK SCHEDULE
pp. 362–365 Exs. 16–30 even,
34–50 even, 59–61, 65–66, 71–83
odd, Quiz 2 Exs. 1–19 (with 6.5)

EXERCISE LEVELS

Level A: *Easier*
15–30, 33–46

Level B: *More Difficult*
31–32, 47–62, 65–66

Level C: *Most Difficult*
63–64, 67–70

✔ HOMEWORK CHECK

To quickly check student understanding of key concepts, go
over the following exercises:
Exs. 20, 24, 34, 48. See also the
Daily Homework Quiz:

• Blackline Master (*Chapter 6
  Resource Book*, p. 90)
• 📄 Transparency (p. 47)

15. $\pm 1, \pm 2, \pm 3, \pm 4, \pm 6, \pm 8, \pm 12, \pm 24$
16. $\pm 1, \pm \frac{1}{2}$
17. $\pm \frac{1}{2}, \pm 1, \pm 2, \pm 4, \pm 8, \pm 16$
18. $\pm 1, \pm 2, \pm 3, \pm 4, \pm 5, \pm 6, \pm 10, \pm 12,$
$\pm 15, \pm 20, \pm 30, \pm 60, \pm \frac{1}{2}, \pm \frac{3}{2}, \pm \frac{5}{2}, \pm \frac{15}{2}$

GUIDED PRACTICE

Vocabulary Check ✔

Concept Check ✔

2a. yes; coefficients are
all integers.

2b. no; coefficients are
not all integers.

Skill Check ✔

2c. no; coefficients are
not all integers.

5. $\pm 1, \pm 2, \pm 3, \pm 4, \pm 6,$
$\pm 8, \pm 9, \pm 12, \pm 18,$
$\pm 24, \pm 36, \pm 72$

6. $\pm 1, \pm 2, \pm 3, \pm 5, \pm 6,$
$\pm 10, \pm 15, \pm 30,$
$\pm \frac{1}{2}, \pm \frac{3}{2}, \pm \frac{5}{2}, \pm \frac{15}{2}$

1. Complete this statement of the rational zero theorem: If a polynomial function has integer coefficients, then every rational zero of the function has the form $\frac{p}{q}$, where p is a factor of the ? and q is a factor of the ? .
constant term, leading coefficient

2. For each polynomial function, decide whether you can use the rational zero theorem to find its zeros. Explain why or why not.

a. $f(x) = 6x^2 - 8x + 4$ **b.** $f(x) = 0.3x^2 + 2x + 4.5$ **c.** $f(x) = \frac{1}{4}x^2 - x + \frac{7}{8}$

3. Describe a method you can use to shorten the list of possible rational zeros when using the rational zero theorem.   **Make a graph.**

List the possible rational zeros of f using the rational zero theorem.
5, 6. See margin.
4. $f(x) = x^3 + 14x^2 + 41x - 56$
$\pm 1, \pm 2, \pm 4, \pm 7, \pm 8, \pm 14, \pm 28, \pm 56$
5. $f(x) = x^3 - 17x^2 + 54x + 72$
6. $f(x) = 2x^3 + 7x^2 - 7x + 30$
7. $f(x) = 5x^4 + 12x^3 - 16x^2 + 10$
$\pm 1, \pm 2, \pm 5, \pm 10, \pm \frac{1}{5}, \pm \frac{2}{5}$

Find all the real zeros of the function.

8. $f(x) = x^3 - 3x^2 - 6x + 8$ $-2, 1, 4$
9. $f(x) = x^3 + 4x^2 - x - 4$ $-4, -1, 1$
10. $f(x) = 2x^3 - 5x^2 - 2x + 5$ $-1, 1, \frac{5}{2}$
11. $f(x) = 2x^3 - x^2 - 15x + 18$ $-3, \frac{3}{2}, 2$
12. $f(x) = x^3 + 4x^2 + x - 6$ $-3, -2, 1$
13. $f(x) = x^3 + 5x^2 - x - 5$ $-5, -1, 1$

14. 🌐 **CRAFTS** Suppose you have 18 cubic inches of wax and you want to make a candle in the shape of a pyramid with a square base. If you want the height of the candle to be 3 inches greater than the length of each side of the base, what should the dimensions of the candle be?   **base: 3 in. by 3 in.; height: 6 in.**

PRACTICE AND APPLICATIONS

STUDENT HELP

↪ **Extra Practice**
to help you master
skills is on p. 948.

LISTING RATIONAL ZEROS **List the possible rational zeros of f using the rational zero theorem.** **15–22. See margin.**

15. $f(x) = x^4 + 2x^2 - 24$
16. $f(x) = 2x^3 + 5x^2 - 6x - 1$
17. $f(x) = 2x^5 + x^2 + 16$
18. $f(x) = 2x^3 + 9x^2 - 53x - 60$
19. $f(x) = 6x^4 - 3x^3 + x + 10$
20. $f(x) = 4x^3 + 5x^2 - 3$
21. $f(x) = 8x^2 - 12x - 3$
22. $f(x) = 3x^4 + 2x^3 - x + 15$

USING SYNTHETIC DIVISION **Use synthetic division to decide which of the following are zeros of the function: 1, −1, 2, −2.**

STUDENT HELP

↪ **HOMEWORK HELP**
Example 1: Exs. 15–32
Example 2: Exs. 33–58
Example 3: Exs. 59–64

23. $f(x) = x^3 + 7x^2 - 4x - 28$ $-2, 2$
24. $f(x) = x^3 + 5x^2 + 2x - 8$ $-2, 1$
25. $f(x) = x^4 + 3x^3 - 7x^2 - 27x - 18$
$-2, -1$
26. $f(x) = 2x^4 - 9x^3 + 8x^2 + 9x - 10$
$-1, 1, 2$
27. $f(x) = x^4 + 3x^3 + 3x^2 - 3x - 4$
$-1, 1$
28. $f(x) = 3x^4 + 3x^3 + 2x^2 + 5x - 10$ none
29. $f(x) = x^3 - 3x^2 + 4x - 12$ none
30. $f(x) = x^3 + x^2 - 11x + 10$ 2
31. $f(x) = x^6 - 2x^4 - 11x^2 + 12$
$-2, -1, 1, 2$
32. $f(x) = x^5 - x^4 - 2x^3 - x^2 + x + 2$
$-1, 1, 2$

FINDING REAL ZEROS Find all the real zeros of the function.

33. $f(x) = x^3 - 8x^2 - 23x + 30$ $-3, 1, 10$ **34.** $f(x) = x^3 + 2x^2 - 11x - 12$ $-4, -1, 3$

35. $f(x) = x^3 - 7x^2 + 2x + 40$ $-2, 4, 5$ **36.** $f(x) = x^3 + x^2 - 2x - 2$ $-\sqrt{2}, -1, \sqrt{2}$

37. $f(x) = x^3 + 72 - 5x^2 - 18x$ $-4, 3, 6$ **38.** $f(x) = x^3 + 9x^2 - 4x - 36$ $-9, -2, 2$

39. $f(x) = x^4 - 5x^3 + 7x^2 + 3x - 10$ $-1, 2$ **40.** $f(x) = x^4 + x^3 + x^2 - 9x - 10$ $-1, 2$

41. $f(x) = x^4 + x^3 - 11x^2 - 9x + 18$ $-3, -2, 1, 3$ **42.** $f(x) = x^4 - 3x^3 + 6x^2 - 2x - 12$ $-1, 2$

43. $f(x) = x^5 + x^4 - 9x^3 - 5x^2 - 36$ $-3, -2, 3$ **44.** $f(x) = x^5 - x^4 - 7x^3 + 11x^2 - 8x + 12$ $-3, 2$

ELIMINATING POSSIBLE ZEROS Use the graph to shorten the list of possible rational zeros. Then find all the real zeros of the function.

45. $f(x) = 4x^3 - 12x^2 - x + 15$ $-1, \frac{3}{2}, \frac{5}{2}$ **46.** $f(x) = -3x^3 + 20x^2 - 36x + 16$ $\frac{2}{3}, 2, 4$

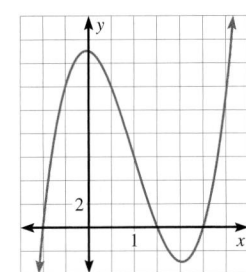
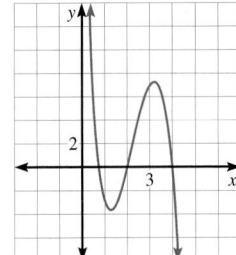

FINDING REAL ZEROS Find all the real zeros of the function.

47. $f(x) = 2x^3 + 4x^2 - 2x - 4$ $-2, -1, 1$ **48.** $f(x) = 2x^3 - 5x^2 - 14x + 8$ $-2, \frac{1}{2}, 4$

49. $f(x) = 2x^3 - 5x^2 - x + 6$ $-1, \frac{3}{2}, 2$ **50.** $f(x) = 2x^3 + x^2 - 50x - 25$ $-5, -\frac{1}{2}, 5$

51. $f(x) = 2x^3 - x^2 - 32x + 16$ $-4, \frac{1}{2}, 4$ **52.** $f(x) = 3x^3 + 12x^2 + 3x - 18$ $-3, -2, 1$

53. $f(x) = 2x^4 + 3x^3 - 3x^2 + 3x - 5$ $-\frac{5}{2}, 1$ **54.** $f(x) = 3x^4 - 8x^3 - 5x^2 + 16x - 5$

55. $f(x) = 2x^4 + x^3 - x^2 - x - 1$ $-1, 1$ **56.** $f(x) = 3x^4 + 11x^3 + 11x^2 + x - 2$ $-2, -1, \frac{1}{3}$

57. $f(x) = 2x^5 + x^4 - 32x - 16$ $-2, -\frac{1}{2}, 2$ **58.** $f(x) = 3x^5 + x^4 - 243x - 81$ $-3, -\frac{1}{3}, 3$

59. 🌐 **HEALTH PRODUCT SALES** From 1990 to 1994, the mail order sales of health products in the United States can be modeled by **1993**

$$S = 10t^3 + 115t^2 + 25t + 2505$$

where S is the sales (in millions of dollars) and t is the number of years since 1990. In what year were about $3885 million of health products sold? (*Hint:* First substitute 3885 for S, then divide both sides by 5.)

60. 🌐 **MONUMENTS** You are designing a monument and a base as shown at the right. You will use 90 cubic feet of concrete for both pieces. Find the value of x. **3 ft**

61. 🌐 **MOLTEN GLASS** At a factory, molten glass is poured into molds to make paperweights. Each mold is a rectangular prism whose height is 3 inches greater than the length of each side of the square base. A machine pours 20 cubic inches of liquid glass into each mold. What are the dimensions of the mold?
2 in. by 2 in. by 5 in.

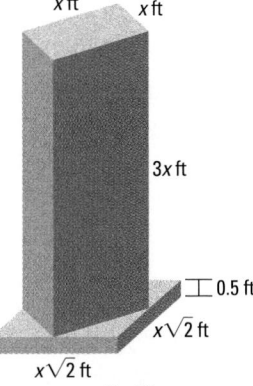

x ft x ft

$3x$ ft

0.5 ft

$x\sqrt{2}$ ft

$x\sqrt{2}$ ft

Ex. 60

EXERCISES 4–7, 15–22
It is easy for students to confuse the values of p and q. Give this example. If $2x - 3$ is a factor of a polynomial, then $x = \frac{3}{2}$ is a root of the function. Ask students which coefficient of the polynomial will have 3 as a factor and which coefficient will have 2 as a factor.

TEACHING TIPS
EXERCISE 54 The polynomial for this function can be factored as $(x^2 - 3x + 1)(3x^2 + x - 5)$. The factorization can be accomplished by trial and error.

STUDENT HELP NOTES
→ **Homework Help** Students can find help with problem solving for Ex. 60 at **www.mcdougallittell.com**. The information can be printed out for students who don't have access to the Internet.

19. $\pm 1, \pm 2, \pm 5, \pm 10, \pm \frac{1}{2}, \pm \frac{5}{2}, \pm \frac{1}{3}, \pm \frac{2}{3}, \pm \frac{5}{3}, \pm \frac{10}{3}, \pm \frac{1}{6}, \pm \frac{5}{6}$

20. $\pm 1, \pm 3, \pm \frac{1}{2}, \pm \frac{3}{2}, \pm \frac{1}{4}, \pm \frac{3}{4}$

21–22. See Additional Answers beginning on page AA1.

ADDITIONAL PRACTICE AND RETEACHING

For Lesson 6.6:
• Practice Levels A, B, and C (*Chapter 6 Resource Book*, p. 79)
• Reteaching with Practice (*Chapter 6 Resource Book*, p. 82)
• See Lesson 6.6 of the *Personal Student Tutor*

For more Mixed Review:
• Search the *Test and Practice Generator* for key words or specific lessons.

STUDENT HELP

INTERNET
HOMEWORK HELP
Visit our Web site www.mcdougallittell.com for help with problem solving in Ex. 60.

54. $\frac{-1 \pm \sqrt{61}}{6}, \frac{3 \pm \sqrt{5}}{2}$

FOCUS ON APPLICATIONS

MOLTEN GLASS
In order for glass to melt so that it can be poured into a mold, it must be heated to temperatures between 1000°C and 2000°C.

4 ASSESS

DAILY HOMEWORK QUIZ

✏️ *Transparency Available*

1. List the possible rational zeros of $f(x) = 6x^3 + 7x^2 + 3x - 4$.
$\pm 1, \pm 2, \pm 4, \pm\frac{1}{2}, \pm\frac{1}{3}, \pm\frac{2}{3}, \pm\frac{1}{6}, \pm\frac{4}{3}$

2. Use synthetic division to decide which of the numbers $2, -2, 3,$ and -3 are zeros of $f(x) = 2x^3 + 4x^2 - 18x - 36$.
$-3, -2, 3$

Find all the real zeros of the function.

3. $f(x) = 2x^3 + 3x^2 - 10x - 15$
$-\sqrt{5}, -\frac{3}{2}, \sqrt{5}$

4. $f(x) = 6x^3 - 37x^2 + 37x - 10$
$\frac{1}{2}, \frac{2}{3}, 5$

EXTRA CHALLENGE NOTE

➤ Challenge problems for Lesson 6.6 are available in **blackline** format in the *Chapter 6 Resource Book,* p. 86 and at **www.mcdougallittell.com**.

ADDITIONAL TEST PREPARATION

1. WRITING A student is working with a polynomial function that has integer coefficients. The leading coefficient is 1. The student claims that if there are any rational zeros, they must be integers. Do you agree? Explain.

yes; if $\frac{p}{q}$ is a rational number that is a zero of the polynomial, and if p and q have no common factor greater than 1, then q is a factor of the leading coefficient 1 and must therefore be equal to 1 or −1. This implies that $\frac{p}{q}$ is an integer.

FOCUS ON APPLICATIONS

SAND SCULPTURE The tallest sand sculptures built were over 20 feet tall and each consisted of hundreds of tons of sand.

Test Preparation

★ **Challenge**

67. $-2, -1, 1$; B
68. $-2, 1$; A
69. -1; C

EXTRA CHALLENGE
➤ www.mcdougallittell.com

---

62. 🌐 **SAND CASTLES** You are designing a kit for making sand castles. You want one of the molds to be a cone that will hold 48π cubic inches of sand. What should the dimensions of the cone be if you want the height to be 5 inches more than the radius of the base? **4 in. radius, 9 in. height**

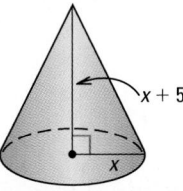

63. 🌐 **SWIMMING POOLS** You are designing an in-ground lap swimming pool with a volume of 2000 cubic feet. The width of the pool should be 5 feet more than the depth, and the length should be 35 feet more than the depth. What should the dimensions of the pool be?
5 ft deep, 10 ft wide, 40 ft long

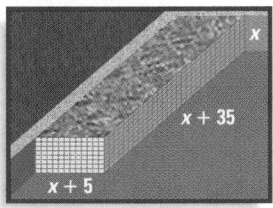

64. 🌐 **WHEELCHAIR RAMPS** You are building a solid concrete wheelchair ramp. The width of the ramp is three times the height, and the length is 5 feet more than 10 times the height. If 150 cubic feet of concrete is used, what are the dimensions of the ramp? **2 ft high, 6 ft wide, 25 ft long**

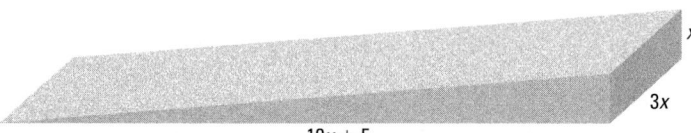

QUANTITATIVE COMPARISON In Exercises 65 and 66, choose the statement that is true about the given quantities.

Ⓐ The quantity in column A is greater.
Ⓑ The quantity in column B is greater.
Ⓒ The two quantities are equal.
Ⓓ The relationship cannot be determined from the given information.

	Column A	Column B
65. C	The number of possible rational zeros of $f(x) = x^4 - 3x^2 + 5x + 12$	The number of possible rational zeros of $f(x) = x^2 - 13x + 20$
66. A	The greatest real zero of $f(x) = x^3 + 2x^2 - 5x - 6$	The greatest real zero of $f(x) = x^4 + 3x^3 - 2x^2 - 6x + 4$

Find the real zeros of the function. Then match each function with its graph.

67. $f(x) = x^3 + 2x^2 - x - 2$ **68.** $g(x) = x^3 - 3x + 2$ **69.** $h(x) = x^3 - x^2 + 2$

A. 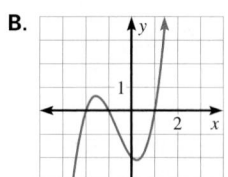   B.   C.

70. CRITICAL THINKING Is it possible for a cubic function to have more than three real zeros? Is it possible for a cubic function to have no real zeros? Explain.
See margin.

MIXED REVIEW

SOLVING QUADRATIC EQUATIONS Solve the equation. (Review 5.2 for 6.7)

71. $x^2 - 6x + 9 = 0$ **3**

72. $x^2 + 12 = 10x - 13$ **5**

73. $x - 1 = x^2 - x$ **1**

74. $x^2 + 18 = 12x - x^2$ **3**

75. $2x^2 - 20x = x^2 - 100$ **10**

76. $x^2 - 12x + 49 = 6x - 32$ **9**

WRITING QUADRATIC FUNCTIONS Write a quadratic function in intercept form whose graph has the given *x*-intercepts and passes through the given point. (Review 5.8 for 6.7)

77. $y = -\frac{5}{9}(x + 3)(x - 3)$

78. $y = \frac{2}{3}(x + 5)(x - 1)$

79. $y = -2(x + 1)(x - 5)$

80. $y = \frac{7}{414}(x - 12)(x - 7)$

81. $y = -\frac{1}{63}(x + 12)(x + 6)$

82. $y = \frac{4}{5}(x - 2)(x - 8)$

83. $y = -\frac{1}{3}(x - 4)(x - 10)$

84. $y = x(x + 6)$

85. $y = (x + 1)(x + 9)$

77. *x*-intercepts: $-3, 3$
point: $(0, 5)$

78. *x*-intercepts: $-5, 1$
point: $(-2, -6)$

79. *x*-intercepts: $-1, 5$
point: $(0, 10)$

80. *x*-intercepts: $12, 7$
point: $(-11, 7)$

81. *x*-intercepts: $-12, -6$
point: $(9, -5)$

82. *x*-intercepts: $2, 8$
point: $(3, -4)$

83. *x*-intercepts: $4, 10$
point: $(7, 3)$

84. *x*-intercepts: $-6, 0$
point: $(2, 16)$

85. *x*-intercepts: $-9, -1$
point: $(1, 20)$

86. 🖼 **PICTURE FRAMES** You have a picture that you want to frame, but first you have to put a mat around it. The picture is 12 inches by 16 inches. The area of the mat is 204 square inches. If the mat extends beyond the picture the same amount in each direction, what will the final dimensions of the picture and mat be? (Review 5.2) width of mat: 3 in.; overall dimensions: 18 in. by 22 in.

QUIZ 2

Self-Test for Lessons 6.4–6.6

Factor the polynomial. (Lesson 6.4)

1. $5x^3 + 135$ $5(x + 3)(x^2 - 3x + 9)$

2. $6x^3 + 12x^2 + 12x + 24$ $6(x + 2)(x^2 + 2)$

3. $4x^5 - 16x$ $4x(x^2 + 2)(x^2 - 2)$

4. $3x^3 - x^2 - 15x + 5$ $(x^2 - 5)(3x - 1)$

Find the real-number solutions of the equation. (Lesson 6.4)

5. $7x^4 = 252x^2$ $0, \pm 6$

6. $16x^6 = 54x^3$ $0, \frac{3}{2}$

7. $6x^5 - 18x^4 + 12x^3 = 36x^2$ $0, 3$

8. $2x^3 + 5x^2 = 8x + 20$ $-\frac{5}{2}, -2, 2$

Divide. Use synthetic division when possible. (Lesson 6.5)

10. $x - \frac{10}{3} + \frac{80}{3(3x + 2)}$

11. $4x - 7 + \frac{11x - 11}{x^2 - 3}$

12. $12x^3 - 7x^2 +$
$10x - 10 + \frac{5}{x + 1}$

13. $x + \frac{2x^2 + 6x + 6}{x^3 - 3}$

14. $5x^3 - 23x^2 +$
$115x - 576 + \frac{2875}{x + 5}$

9. $(x^2 + 7x - 44) \div (x - 4)$ $x + 11$

10. $(3x^2 - 8x + 20) \div (3x + 2)$

11. $(4x^3 - 7x^2 - x + 10) \div (x^2 - 3)$

12. $(12x^4 + 5x^3 + 3x^2 - 5) \div (x + 1)$

13. $(x^4 + 2x^2 + 3x + 6) \div (x^3 - 3)$

14. $(5x^4 + 2x^3 - x - 5) \div (x + 5)$

Find all the real zeros of the function. (Lesson 6.6)

15. $f(x) = x^3 - 4x^2 - 7x + 28$ $\pm\sqrt{7}, 4$

16. $f(x) = x^3 - 6x^2 + 21x - 26$ **2**

17. $f(x) = 2x^3 + 15x^2 + 22x - 15$ $-5, -3, \frac{1}{2}$

18. $f(x) = 2x^3 + 7x^2 - 28x + 12$ $-6, \frac{1}{2}, 2$

19. 🌐 **DESIGNING A PATIO** You are a landscape artist designing a patio. The square patio floor is to be made from 128 cubic feet of concrete. The thickness of the floor is 15.5 feet less than each side length of the patio. What are the dimensions of the patio floor? (Lesson 6.6) 16 ft by 16 ft by 0.5 ft

6.6 Finding Rational Zeros **365**

ADDITIONAL RESOURCES
An alternative quiz for Lessons 6.4–6.6 is available in the *Chapter 6 Resource Book*, p. 87.

70. no, no; If a cubic polynomial had 4 or more distinct real zeros, then there would be 4 or more binomials of the form $x - a$ that divide the polynomial to give a zero remainder. This would imply that the polynomial has degree 4 or greater. However, this is impossible since the polynomial is a cubic polynomial. So a cubic polynomial has at most 3 real zeros. As $x \to -\infty$ and $x \to +\infty$, the values of a cubic polynomial approach $-\infty$ and $+\infty$, respectively, or else $+\infty$ and $-\infty$. At some value of *x*, therefore, the graph is below the *x*-axis, and at some other value of *x*, the graph is above the *x*-axis. This means that the graph crosses the *x*-axis somewhere between these two values, and the *x*-coordinate of the point where the graph crosses the *x*-axis is a zero.

➡️ **LESSON OPENER**
GRAPHING CALCULATOR
An alternative way to approach Lesson 6.7 is to use the Graphing Calculator Lesson Opener:

- Blackline Master (*Chapter 6 Resource Book,* p. 91)
- Transparency (p. 41)

MEETING INDIVIDUAL NEEDS
- ***Chapter 6 Resource Book***
  Prerequisite Skills Review (p. 5)
  Practice Level A (p. 94)
  Practice Level B (p. 95)
  Practice Level C (p. 96)
  Reteaching with Practice (p. 97)
  Absent Student Catch-Up (p. 99)
  Challenge (p. 101)
- ***Resources in Spanish***
- ⬛ ***Personal Student Tutor***

NEW-TEACHER SUPPORT
See the Tips for New Teachers on pp. 1–2 of the *Chapter 6 Resource Book* for additional notes about Lesson 6.7.

WARM-UP EXERCISES
🔲 **Transparency Available**

Name the rational zeros of each function.

1. $f(x) = x^2 - 6x + 9$ 3
2. $f(x) = 2x^2 - x - 15$ $-\frac{5}{2}, 3$
3. $f(x) = x^3 + x^2 + 7x - 9$ 1
4. $f(x) = 2x^3 + x^2 + x - 1$ $\frac{1}{2}$
5. $f(x) = x^4 - x^2 + 2x - 4$ none

What you should learn

GOAL 1 Use the fundamental theorem of algebra to determine the number of zeros of a polynomial function.

GOAL 2 Use technology to approximate the real zeros of a polynomial function, as applied in **Example 5**.

Why you should learn it

▼ To solve **real-life** problems, such as finding the American Indian, Aleut, and Eskimo population in **Ex. 59.**

1c. 1; $\dfrac{-1 \pm i\sqrt{3}}{2}$; 3; 1 is rational,

$\dfrac{-1 \pm i\sqrt{3}}{2}$ are imaginary.

6.7 Using the Fundamental Theorem of Algebra

GOAL 1 **THE FUNDAMENTAL THEOREM OF ALGEBRA**

The following important theorem, called the fundamental theorem of algebra, was first proved by the famous German mathematician Carl Friedrich Gauss (1777–1855).

> **THE FUNDAMENTAL THEOREM OF ALGEBRA**
>
> If $f(x)$ is a polynomial of degree n where $n > 0$, then the equation $f(x) = 0$ has at least one root in the set of complex numbers.

In the following activity you will investigate how the number of solutions of $f(x) = 0$ is related to the degree of the polynomial $f(x)$.

▶ ACTIVITY
Developing Concepts

Investigating the Number of Solutions

1 Solve each polynomial equation. State how many solutions the equation has, and classify each as rational, irrational, or imaginary.

a. $2x - 1 = 0$
0.5; 1; rational

b. $x^2 - 2 = 0$
$\pm\sqrt{2}$; 2; both irrational

c. $x^3 - 1 = 0$
See margin.

Make a conjecture about the relationship between the degree of a polynomial $f(x)$ and the number of solutions of $f(x) = 0$.
If $f(x)$ has degree $n > 1$, then $f(x) = 0$ has n solutions.

2 Solve the equation $x^3 + x^2 - x - 1 = 0$. How many different solutions are there? How can you reconcile this number with your conjecture?
$-1, 1; 2; -1$ is a solution twice.

The equation $x^3 - 6x^2 - 15x + 100 = 0$, which can be written as $(x + 4)(x - 5)^2 = 0$, has only two distinct solutions: -4 and 5. Because the factor $x - 5$ appears twice, however, you can count the solution 5 twice. So, with 5 counted as a **repeated solution**, this *third*-degree equation can be said to have *three* solutions: -4, 5, and 5.

In general, when all real and imaginary solutions are counted (with all repeated solutions counted individually), an nth-degree polynomial equation has *exactly n* solutions. Similarly, any nth-degree polynomial function has exactly n zeros.

EXAMPLE 1 *Finding the Number of Solutions or Zeros*

a. The equation $x^3 + 3x^2 + 16x + 48 = 0$ has three solutions: -3, $4i$, and $-4i$.

b. The function $f(x) = x^4 + 6x^3 + 12x^2 + 8x$ has four zeros: -2, -2, -2, and 0.

EXAMPLE 2 *Finding the Zeros of a Polynomial Function*

Find all the zeros of $f(x) = x^5 - 2x^4 + 8x^2 - 13x + 6$.

SOLUTION

The possible rational zeros are ± 1, ± 2, ± 3, and ± 6. Using synthetic division, you can determine that 1 is a repeated zero and that -2 is also a zero. You can write the function in factored form as follows:

$$f(x) = (x - 1)(x - 1)(x + 2)(x^2 - 2x + 3)$$

Complete the factorization, using the quadratic formula to factor the trinomial.

$$f(x) = (x - 1)(x - 1)(x + 2)[x - (1 + i\sqrt{2})][x - (1 - i\sqrt{2})]$$

▶ This factorization gives the following five zeros:

$$1, 1, -2, 1 + i\sqrt{2}, \text{ and } 1 - i\sqrt{2}$$

The graph of f is shown at the right. Note that only the *real* zeros appear as x-intercepts. Also note that the graph only *touches* the x-axis at the repeated zero $x = 1$, but *crosses* the x-axis at the zero $x = -2$.

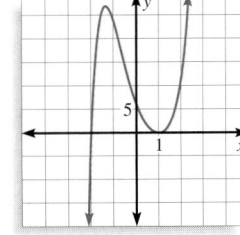

· · · · · · · · ·

The graph in Example 2 illustrates the behavior of the graph of a polynomial function near its zeros. When a factor $x - k$ is raised to an odd power, the graph crosses the x-axis at $x = k$. When a factor $x - k$ is raised to an even power, the graph is tangent to the x-axis at $x = k$.

In Example 2 the zeros $1 + i\sqrt{2}$ and $1 - i\sqrt{2}$ are complex conjugates. The complex zeros of a polynomial function with *real* coefficients always occur in complex conjugate pairs. That is, if $a + bi$ is a zero, then $a - bi$ must also be a zero.

EXAMPLE 3 *Using Zeros to Write Polynomial Functions*

STUDENT HELP

HOMEWORK HELP
Visit our Web site
www.mcdougallittell.com
for extra examples.

Write a polynomial function f of least degree that has real coefficients, a leading coefficient of 1, and 2 and $1 + i$ as zeros.

SOLUTION

Because the coefficients are real and $1 + i$ is a zero, $1 - i$ must also be a zero. Use the three zeros and the factor theorem to write $f(x)$ as a product of three factors.

$f(x) = (x - 2)[x - (1 + i)][x - (1 - i)]$	**Write $f(x)$ in factored form.**
$= (x - 2)[(x - 1) - i][(x - 1) + i]$	**Regroup terms.**
$= (x - 2)[(x - 1)^2 - i^2]$	**Multiply.**
$= (x - 2)[x^2 - 2x + 1 - (-1)]$	**Expand power and use $i^2 = -1$.**
$= (x - 2)(x^2 - 2x + 2)$	**Simplify.**
$= x^3 - 2x^2 + 2x - 2x^2 + 4x - 4$	**Multiply.**
$= x^3 - 4x^2 + 6x - 4$	**Combine like terms.**

✔**CHECK** You can check this result by evaluating $f(x)$ at each of its three zeros.

6.7 *Using the Fundamental Theorem of Algebra* **367**

---

2 TEACH

MOTIVATING THE LESSON
When you are solving a polynomial equation, how do you know you have found all the solutions? Tell students that in this lesson they will learn about a theorem that will help them decide when they have located all the solutions.

ACTIVITY NOTE
Graphing Calculator Students can use graphing calculators to confirm the number of real roots.

EXTRA EXAMPLE 1
State the number of solutions and tell what they are.
a. $x^2 - 14x + 49 = 0$ 1 solution; 7
b. $x^4 + 3x^3 - 8x^2 - 22x - 24 = 0$
  4 solutions; -4, 3, $-1 + i$, and $-1 - i$

EXTRA EXAMPLE 2
Find all the zeros of $f(x) = x^3 + x^2 - x + 15$. $-3, 1 + 2i, 1 - 2i$

EXTRA EXAMPLE 3
Write a polynomial function of least degree that has real coefficients, a leading coefficient of 1, and 1, $-2 + i$, and $-2 - i$ as zeros.
$f(x) = x^3 + 3x^2 + x - 5$

✔ **CHECKPOINT EXERCISES**
For use after Example 1:
1. State the number of zeros of $f(x) = x^3 - 3x + 52$ and tell what they are. 3 zeros; -4, $2 - 3i$, and $2 + 3i$.

For use after Example 2:
2. Find all the zeros of $f(x) = x^4 + 5x^2 - 6$. $1, -1, i\sqrt{6}, -i\sqrt{6}$

For use after Example 3:
3. Write a polynomial function of least degree that has real coefficients, a leading coefficient of 1, and 5, $2i$, and $-2i$ as zeros. $f(x) = x^3 - 5x^2 + 4x - 20$

367

EXTRA EXAMPLE 4
Approximate the real zeros of
$f(x) = x^3 - 4x^2 - 5x + 14$.
–2 and about 1.59 and 4.41

EXTRA EXAMPLE 5
A rectangular piece of sheet
metal is 10 in. long and
10 in. wide. Squares of side
length x are cut from the corners
and the remaining piece is folded
to make a box. The volume of the
box is modeled by $V(x) = 4x^3 -
40x^2 + 100x$. What size square can
be cut from the corners to give
a box with a volume of 25 in.^{3}?
a square with side length about
0.28 in. or about 3.70 in.

 CHECKPOINT EXERCISES
For use after Example 4:
1. Approximate the real zeros
of $f(x) = x^3 - 3x^2 - 2x + 6$.
3 and about –1.41 and 1.41

For use after Example 5:
2. In Extra Example 5, what size
square will give a volume of
30 in.^{3}? a square with side
length about 0.35 in. or about
3.55 in.

FOCUS ON VOCABULARY
Explain what the conjugate of a
complex number is. the complex
number obtained by replacing the
number by which i is multiplied with
its opposite

CLOSURE QUESTION
What information does the degree
of a polynomial function give you
about its zeros? It tells how many
zeros there are, including repeated
zeros.

DAILY PUZZLER
On a scale, 2 green balls and 1 red
ball balance 2 yellow balls. Three
red balls balance 4 yellow balls.
How many green balls will balance
1 red ball? four green balls

GOAL 2 USING TECHNOLOGY TO APPROXIMATE ZEROS

The rational zero theorem gives you a way to find the rational zeros of a polynomial function with integer coefficients. To find the *real* zeros of *any* polynomial function, you may need to use technology.

EXAMPLE 4 *Approximating Real Zeros*

Approximate the real zeros of $f(x) = x^4 - 2x^3 - x^2 - 2x - 2$.

SOLUTION

There are several ways to use a graphing calculator to approximate the real zeros of a function. One way is to use the *Zero* (or *Root*) feature as shown below.

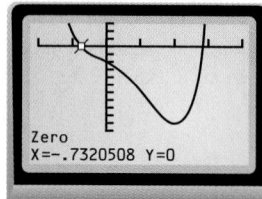 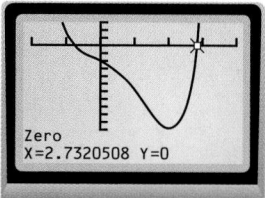

▶ From these screens, you can see that the real zeros are about -0.73 and 2.73.

Because the polynomial function has degree 4, you know that there must be two other zeros. These may be repeats of the real zeros, or they may be imaginary zeros. In this particular case, the two other zeros are imaginary: $x = \pm i$.

EXAMPLE 5 *Approximating Real Zeros of a Real-Life Function*

PHYSIOLOGY For one group of people it was found that a person's score S on the Harvard Step Test was related to his or her amount of hemoglobin x (in grams per 100 milliliters of blood) by the following model:

$$S = -0.015x^3 + 0.6x^2 - 2.4x + 19$$

The normal range of hemoglobin is 12–18 grams per 100 milliliters of blood. Approximate the amount of hemoglobin for a person who scored 75.

SOLUTION

You can solve the equation

$$75 = -0.015x^3 + 0.6x^2 - 2.4x + 19$$

by rewriting it as $0 = -0.015x^3 + 0.6x^2 - 2.4x - 56$ and then using a graphing calculator to approximate the real zeros of $f(x) = -0.015x^3 + 0.6x^2 - 2.4x - 56$. From the graph you can see that there are three real zeros: $x \approx -7.3$, $x \approx 16.4$, and $x \approx 30.9$.

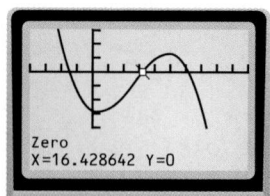

▶ The person's hemoglobin is probably about 16.4 grams per 100 milliliters of blood, since this is the only zero within the normal range.

FOCUS ON APPLICATIONS

HARVARD STEP TEST When taking the Harvard Step Test, a person steps up and down a 20 inch platform for 5 minutes. The person's score is determined by his or her heart rate in the first few minutes after stopping.

GUIDED PRACTICE

Vocabulary Check ✓
Concept Check ✓

1. If $f(x)$ is a polynomial of positive degree, then $f(x) = 0$ has at least one root in the set of complex numbers.

Skill Check ✓

3. 2 real zeroes; no imaginary zeroes; the existence of an imaginary zero would imply the existence of two distinct imaginary zeros, which would not be consistent with the fact that $f(x)$ has degree 3. The real number 2 is a repeated zero.

1. State the fundamental theorem of algebra.

2. Two zeros of $f(x) = x^3 - 6x^2 - 16x + 96$ are 4 and -4. Explain why the third zero must also be a real number. **See margin.**

3. The graph of $f(x) = x^3 - x^2 - 8x + 12$ is shown at the right. How many real zeros does the function have? How many imaginary zeros does the function have? Explain your reasoning. **See margin.**

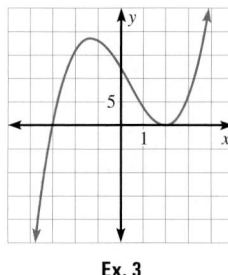

Ex. 3

Find all the zeros of the polynomial function.

4. $f(x) = x^3 - x^2 - 2x$ $-1, 0, 2$
5. $f(x) = x^4 + x^2 - 12$ $\pm\sqrt{3}, \pm 2i$
6. $f(x) = x^3 + 5x^2 - 9x - 45$ $-5, -3, 3$
7. $f(x) = x^4 - x^3 + 2x^2 - 4x - 8$ $-1, 2, \pm 2i$

Write a polynomial function of least degree that has real coefficients, the given zeros, and a leading coefficient of 1. **8–13. See margin.**

8. $3, 0, -2$
9. $1, 1, i, -i$
10. $5, 2 + 3i$
11. $1, -1, 2, -2, 3$
12. $3, -2, -1 + i$
13. $4i, 4i$

14. 🌐 **GROCERY STORE REVENUE** For the 25 years that a grocery store has been open, its annual revenue R (in millions of dollars) can be modeled by

$$R = \frac{1}{10,000}(-t^4 + 12t^3 - 77t^2 + 600t + 13,650)$$

where t is the number of years the store has been open. In what year(s) was the revenue $1.5 million? **3 and 9 years after opening**

PRACTICE AND APPLICATIONS

STUDENT HELP

▶ **Extra Practice**
to help you master skills is on p. 948.

33. $-3, -1, 3, 4.5$

34. $-4, -2, 2.5, 4$

STUDENT HELP

▶ **HOMEWORK HELP**
Example 1: Exs. 21–54
Example 2: Exs. 21–34
Example 3: Exs. 35–46
Example 4: Exs. 47–54
Example 5: Exs. 55–59

CHECKING ZEROS Decide whether the given x-value is a zero of the function.

15. $f(x) = x^3 - x^2 + 4x - 4, x = 1$ **yes**
16. $f(x) = x^3 + 3x^2 - 5x + 8, x = 4$ **no**
17. $f(x) = x^4 - x^2 - 3x + 3, x = 0$ **no**
18. $f(x) = x^3 + 5x^2 + x + 5, x = -5$ **yes**
19. $f(x) = x^3 - 4x^2 + 16x - 64, x = 4i$ **yes**
20. $f(x) = x^3 - 3x^2 + x - 3, x = -i$ **yes**

FINDING ZEROS Find all the zeros of the polynomial function.

21. $f(x) = x^4 + 5x^3 + 5x^2 - 5x - 6$ $-3, -2, -1, 1$
22. $f(x) = x^4 + 4x^3 - 6x^2 - 36x - 27$ $-3, -3, -1, 3$
23. $f(x) = x^3 - 4x^2 + 3x$ $0, 1, 3$
24. $f(x) = x^3 + 5x^2 - 4x - 20$ $-5, -2, 2$
25. $f(x) = x^4 + 7x^3 - x^2 - 67x - 60$ $-5, -4, -1, 3$
26. $f(x) = x^4 - 5x^2 - 36$ $\pm 3, \pm 2i$
27. $f(x) = x^3 - x^2 + 49x - 49$ $1, \pm 7i$
28. $f(x) = x^3 - x^2 + 25x - 25$ $1, \pm 5i$
29. $f(x) = x^4 + 6x^3 + 14x^2 + 54x + 45$ $-5, -1, \pm 3i$
30. $f(x) = x^3 + 3x^2 + 25x + 75$ $-3, \pm 5i$
31. $f(x) = x^4 - x^3 - 5x^2 - x - 6$ $-2, 3, \pm i$
32. $f(x) = x^4 + x^3 + 2x^2 + 4x - 8$ $-2, 1, \pm 2i$
33. $f(x) = 2x^4 - 7x^3 - 27x^2 + 63x + 81$ **See margin.**
34. $f(x) = 2x^4 - x^3 - 42x^2 + 16x + 160$ **See margin.**

6.7 *Using the Fundamental Theorem of Algebra* **369**

3 APPLY

🔵 **ASSIGNMENT GUIDE**

BASIC
Day 1: pp. 369–371 Exs. 16–28 even, 35–38, 47–50, 55–56, 60, 65–71 odd

AVERAGE
Day 1: pp. 369–371 Exs. 16–30 even, 36–60 even, 65–71 odd

ADVANCED
Day 1: pp. 369–371 Exs. 16–60 even, 61–63, 65–71 odd

BLOCK SCHEDULE
pp. 369–371 Exs. 16–30 even, 36–60 even, 65–71 odd (with 6.8)

EXERCISE LEVELS
Level A: *Easier*
15–18, 21–28, 35–40, 47–59
Level B: *More Difficult*
19–20, 29–34, 41–46
Level C: *Most Difficult*
60–63

✔ **HOMEWORK CHECK**
To quickly check student understanding of key concepts, go over the following exercises:
Exs. 16, 22, 36, 48. See also the Daily Homework Quiz:
• Blackline Master (*Chapter 6 Resource Book*, p. 104)
• 📖 Transparency (p. 48)

MULTIPLE REPRESENTATIONS
EXERCISES 21–34 Remind students that the graph of a polynomial function touches or crosses the x-axis at the real solutions of the corresponding polynomial equation. Students can use this relationship to narrow down their search for rational zeros by examining the graphs of polynomial functions. As the graphs of polynomial functions tend to be difficult to sketch by hand, students can use a graphing calculator to graph the functions.

2, 8–13. See Additional Answers beginning on page AA1.

! COMMON ERROR

EXERCISES 47–54 Students will get a boundary error message on their calculators if they guess the location of a zero to be outside the interval they picked when they chose the left and right bounds. Urge students to be careful when choosing the bounds and making their guesses.

DATA UPDATE
Updated data for Exercise 59 is available at **www.mcdougallittell.com.**

55. 1988
56. 1992
57. Yes; there were 2 such years, 1988 and 1993, because the graph intersects the line $S = 2000$ when t is about 1.6 and when t is about 6.3.

ADDITIONAL PRACTICE AND RETEACHING

For Lesson 6.7:
• Practice Levels A, B, and C (*Chapter 6 Resource Book*, p. 94)
• Reteaching with Practice (*Chapter 6 Resource Book*, p. 97)
• ⬚ See Lesson 6.7 of the *Personal Student Tutor*

For more Mixed Review:
• ⬚ Search the *Test and Practice Generator* for key words or specific lessons.

370

FOCUS ON
APPLICATIONS

UNITED STATES EXPORTS The United States exports more than any other country in the world. It also imports more than any other country.

35. $f(x) = x^3 - 7x^2 + 14x - 8$

36. $f(x) = x^3 - 2x^2 - 19x + 20$

37. $f(x) = x^3 - 2x^2 - 33x + 90$

38. $f(x) = x^3 + 5x^2 - 4x - 20$

39. $f(x) = x^3 + 13x^2 + 50x + 56$

40. $f(x) = x^3 - 8x^2 + x - 8$

41. $f(x) = x^3 - 5x^2 + 9x - 45$

42. $f(x) = x^4 + 32x^2 - 144$

43. $f(x) = x^4 + 10x^2 + 9$

44. $f(x) = x^4 - 6x^3 + 35x^2 - 150x + 250$

45. $f(x) = x^4 - 12x^3 + 53x^2 - 104x + 80$

46. $f(x) = x^5 + x^4 + 8x^3 + 4x^2 - 128x - 192$

WRITING POLYNOMIAL FUNCTIONS Write a polynomial function of least degree that has real coefficients, the given zeros, and a leading coefficient of 1.
35–46. See margin.

35. $2, 1, 4$

36. $1, -4, 5$

37. $-6, 3, 5$

38. $-5, 2, -2$

39. $-2, -4, -7$

40. $8, -i, i$

41. $3i, -3i, 5$

42. $2, -2, -6i$

43. $i, -3i, 3i$

44. $3 - i, 5i$

45. $4, 4, 2 + i$

46. $-2, -2, 3, -4i$

FINDING ZEROS Use a graphing calculator to graph the polynomial function. Then use the *Zero* (or *Root*) feature of the calculator to find the real zeros of the function.

47. $f(x) = x^3 - x^2 - 5x + 3$
   $-2.09, 0.57, 2.51$

48. $f(x) = 2x^3 - x^2 - 3x - 1$
   $-0.62, -0.5, 1.62$

49. $f(x) = x^3 - 2x^2 + x + 1$ -0.47

50. $f(x) = x^4 - 2x - 1$ $-0.47, 1.40$

51. $f(x) = x^4 - x^3 - 4x^2 - 3x - 2$
   $-1.27, 2.86$

52. $f(x) = x^4 - x^3 - 3x^2 - x + 1$
   $0.42, 2.37$

53. $f(x) = x^4 + 3x^2 - 2$ $-0.75, 0.75$

54. $f(x) = x^4 - x^3 - 20x^2 + 10x + 27$
   $-4.09, -0.98, 1.47, 4.60$

GRAPHING MODELS In Exercises 55–59, you may find it helpful to graph the model on a graphing calculator. 55–57. See margin.

55. **UNITED STATES EXPORTS** For 1980 through 1996, the total exports E (in billions of dollars) of the United States can be modeled by
$$E = -0.131t^3 + 5.033t^2 - 23.2t + 233$$
where t is the number of years since 1980. In what year were the total exports about $312.76 billion? ▶ Source: U.S. Bureau of the Census

56. **EDUCATION DONATIONS** For 1983 through 1995, the amount of private donations D (in millions of dollars) allocated to education can be modeled by
$$D = 1.78t^3 - 6.02t^2 + 752t + 6701$$
where t is the number of years since 1983. In what year was $14.3 billion of private donations allocated to education? ▶ Source: AAFRC Trust for Philanthropy

57. **SPORTS EQUIPMENT** For 1987 through 1996, the sales S (in millions of dollars) of gym shoes and sneakers can be modeled by
$$S = -0.982t^5 + 24.6t^4 - 211t^3 + 661t^2 - 318t + 1520$$
where t is the number of years since 1987. Were there any years in which sales were about $2 billion? Explain. ▶ Source: National Sporting Goods Association

58. **TELEVISION** For 1990 through 2000, the actual and projected amount spent on television per person per year in the United States can be modeled by
$$S = -0.213t^3 + 3.96t^2 + 10.2t + 366$$
where S is the amount spent (in dollars) and t is the number of years since 1990. During which year was $455 spent per person on television?
▶ Source: Veronis, Suhler & Associates, Inc. late 1993

59. **POPULATION** For 1890 through 1990, the American Indian, Eskimo, and Aleut population P (in thousands) can be modeled by the function
$$P = 0.00496t^3 - 0.432t^2 + 11.3t + 212$$
where t is the number of years since 1890. In what year did the population reach 722,000? DATA UPDATE of *Statistical Abstract of the United States* data at www.mcdougallittell.com
1965

60.c. 1.05, 5%, *Sample answer:* I graphed $y = x^3 + x^2 + x + 1$ and $y = 4.3$ and found the x-coordinate of the intersection point.

61.a.

zeros	sum	prod.
2, 3	5	6
−3, 1, 2	0	−6
−3, 1, ±2i	−2	−12
−3, 2, 0, 2±$\sqrt{3}$	3	0

61.b. If $f(x)$ is a polynomial with leading coefficient 1 and degree n, where $n > 0$, then the sum of the roots is the opposite of the coefficient of the x^{n-1} term.

★ **Challenge**

61.c. If $f(x)$ is a polynomial with leading coefficient 1 and degree n, where $n > 0$, then the product of the zeros is the constant term if n is even and the opposite of the constant term if n is odd.

EXTRA CHALLENGE

→ www.mcdougallittell.com

60. MULTI-STEP PROBLEM Mary plans to save $1000 each summer to buy a used car at the end of the fourth summer. At the end of each summer, she will deposit the $1000 she earned from her summer job into her bank account. The table shows the value of her deposits over the four year period. In the table, g is the growth factor $1 + r$ where r is the annual interest rate expressed as a decimal.

	End of 1st summer	End of 2nd summer	End of 3rd summer	End of 4th summer
Value of 1st deposit	1000	1000g	1000g^2	1000g^3
Value of 2nd deposit	—	1000	?	?
Value of 3rd deposit	—	—	1000	?
Value of 4th deposit	—	—	—	1000

a. Copy and complete the table.   1000g, 1000g^2; 1000g

b. Write a polynomial function of g that represents the value of Mary's account at the end of the fourth summer. $1000g^3 + 1000g^2 + 1000g + 1000$

c. *Writing* Suppose Mary wants to buy a car that costs about $4300. What growth factor does she need to obtain this amount? What annual interest rate does she need? Explain how you found your answers.   **See margin.**

61. a. Copy and complete the table.   **61.a–c. See margin.**

Function	Zeros	Sum of zeros	Product of zeros
$f(x) = x^2 - 5x + 6$	?	?	?
$f(x) = x^3 - 7x + 6$	?	?	?
$f(x) = x^4 + 2x^3 + x^2 + 8x - 12$	?	?	?
$f(x) = x^5 - 3x^4 - 9x^3 + 25x^2 - 6x$	?	?	?

b. Use your completed table to make a conjecture relating the sum of the zeros of a polynomial function with the coefficients of the polynomial function.

c. Use your completed table to make a conjecture relating the product of the zeros of a polynomial function with the coefficients of the polynomial function.

62. Show that the sum of a pair of complex conjugates is a real number.

63. Show that the product of a pair of complex conjugates is a real number.
$(a + bi)(a - bi) = a^2 - b^2$; since a and b are real, $a^2 - b^2$ is real.

MIXED REVIEW

62. $(a + bi) + (a - bi) = (a + a) + (bi - bi) = 2a$; since a is real, $2a$ must be real.

GRAPHING WITH INTERCEPT FORM Graph the quadratic function. Label the vertex, axis of symmetry, and x-intercepts. (Review 5.1 for 6.8)   **64–71. See margin.**

64. $y = -3(x - 2)(x + 2)$

65. $y = 2(x - 1)(x - 5)$

66. $y = 2(x + 4)(x - 3)$

67. $y = -(x + 1)(x - 5)$

GRAPHING POLYNOMIALS Graph the polynomial function. (Review 6.2 for 6.8)

68. $f(x) = -2x^4$

69. $f(x) = -x^3 - 4$

70. $f(x) = x^3 + 4x - 3$

71. $f(x) = x^4 - 3x^3 + x + 2$

6.7 *Using the Fundamental Theorem of Algebra*   **371**

---

4 ASSESS

DAILY HOMEWORK QUIZ

📑 *Transparency Available*

1. Decide whether $x = -3$ is a zero of $f(x) = 2x^3 - 3x^2 - 23x + 12$.
yes

2. Find all the zeros of $f(x) = x^4 + 2x^3 - 7x^2 + 2x - 8$. $-4, 2, \pm i$

3. Write a polynomial function of least degree that has real coefficients, a leading coefficient of 1, and whose zeros are $-\sqrt{2}, -1, \sqrt{2}$.
$f(x) = x^3 + x^2 - 2x - 2$

4. Use a graphing calculator to find the real zeros of $f(x) = x^4 - 6x^3 + 8x^2 + 4x - 4$. Round to the nearest hundredth.
$-0.73, 0.59, 2.73, 3.41$

EXTRA CHALLENGE NOTE

→ Challenge problems for Lesson 6.7 are available in **blackline** format in the *Chapter 6 Resource Book*, p. 101 and at **www.mcdougallittell.com**.

ADDITIONAL TEST PREPARATION

1. WRITING How can you tell from the factored form of a polynomial function whether the function has a repeated zero?
At least one of the factors will occur more than once.

64–71. See Additional Answers beginning on page AA1.

1 Planning the Activity

PURPOSE
To use a graphing calculator to solve polynomial equations that cannot be solved by factoring.

MATERIALS
- graphing calculators
- Keystroke blackline (*Chapter 6 Resource Book,* p. 92)

PACING
- Activity — 15 min

▶ LINK TO LESSON
Students can use the method of this activity to solve Example 5 on page 368 directly without rewriting the equation.

2 Managing the Activity

ALTERNATIVE APPROACH
To save time, use an overhead projector to demonstrate Steps 1–4.

3 Closing the Activity

★ KEY DISCOVERY
A polynomial equation that is not in standard form can be solved by graphing both sides of the equation. The solutions are the *x*-coordinates of the points where the graphs intersect.

ACTIVITY ASSESSMENT
JOURNAL Suppose you are given the function $f(x) = 2x^3 - x^2 - 4x + 1$. You graph the equation to find its real zeros. Your friend graphs the two equations $y = 2x^3$ and $y = x^2 + 4x - 1$ and finds the intersections of the two graphs. Explain how your results are related. **The real zeros of your function and the x-values of your friend's intersections are equal and are the solutions of $2x^3 - x^2 - 4x + 1 = 0$.**

◒ ACTIVITY 6.7
Using Technology

Solving Polynomial Equations

In Lesson 6.4 you learned to solve polynomial equations by factoring. When factoring is not possible, you can use a graphing calculator instead.

▶ EXAMPLE
Use a graphing calculator to find the real solutions of $x^3 + 4x^2 - 2x + 5 = 19$.

▶ SOLUTION

1 To solve the equation graphically, graph each side of the equation as follows.

2 When the equations are graphed in the standard viewing window, you see most of the graph of y_1, but none of the graph of y_2.

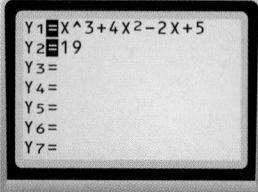

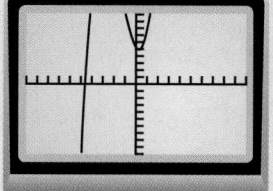

3 You know that $y_2 = 19$ is a horizontal line. Change the viewing window so that $-10 \le x \le 10$ and $-2 \le y \le 22$.

4 The graphs of y_1 and y_2 intersect at three points. Use the *Intersect* feature to find the *x*-coordinates of these points.

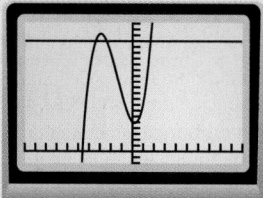

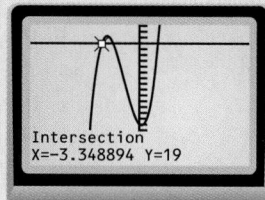

▶ The solutions are $x \approx -3.35$, $x \approx -2.40$, and $x \approx 1.74$.

▶ EXERCISES

Use a graphing calculator to find the real solutions of the equation.

1. $\frac{1}{2}x^3 - 3x^2 + x + 6 = 4$

2. $2x^3 - 8x^2 + 5x + 14 = 7$ -0.640

3. $x^3 - 5x^2 + x + 3 = 8$ 5

4. $0.3x^3 - 5x^2 + 8x + 15 = 13$

5. $x^4 - 6x^2 + 5 = 2$

6. $0.2x^4 - 3x^3 - 12x^2 + 8x + 22 = 13$

7. $17x^5 - 24x^3 + x^2 + 2x = 4$
$-1.088, -0.668, 1.191$

8. $-1.25x^5 + 3.75x^2 + 0.4x - 6 = -4$
$-0.735, 0.722, 1.326$

9. Look back at Example 5 on page 368. Use the method described above to solve the problem. Does your answer agree with the answer given in Example 5? $-7.349, 16.429, 30.921;$ yes

1. $-0.640, 1.135, 5.505$

4. $-0.219, 2.047, 14.839$

5. $-2.334, -0.742, 0.742, 2.334$

6. $-3.629, -0.629, 1.085, 18.173$

STUDENT HELP

KEYSTROKE HELP

See keystrokes for several models of calculators at www.mcdougallittell.com

6.8

Analyzing Graphs of Polynomial Functions

What you should learn

GOAL 1 Analyze the graph of a polynomial function.

GOAL 2 Use the graph of a polynomial function to answer questions about **real-life** situations, such as maximizing the volume of a box in **Example 3**.

Why you should learn it

▼ To find the maximum and minimum values of **real-life** functions, such as the function modeling orange consumption in the United States in **Ex. 36**.

GOAL 1 ANALYZING POLYNOMIAL GRAPHS

In this chapter you have learned that zeros, factors, solutions, and x-intercepts are closely related concepts. The relationships are summarized below.

CONCEPT SUMMARY   **ZEROS, FACTORS, SOLUTIONS, AND INTERCEPTS**

Let $f(x) = a_nx^n + a_{n-1}x^{n-1} + \cdots + a_1x + a_0$ be a polynomial function. The following statements are equivalent.

ZERO: k is a zero of the polynomial function f.

FACTOR: $x - k$ is a factor of the polynomial $f(x)$.

SOLUTION: k is a solution of the polynomial equation $f(x) = 0$.

If k is a real number, then the following is also equivalent.

X-INTERCEPT: k is an x-intercept of the graph of the polynomial function f.

EXAMPLE 1   *Using x-Intercepts to Graph a Polynomial Function*

Graph the function $f(x) = \frac{1}{4}(x + 2)(x - 1)^2$.

SOLUTION

Plot x-intercepts. Since $x + 2$ and $x - 1$ are factors of $f(x)$, -2 and 1 are the x-intercepts of the graph of f. Plot the points $(-2, 0)$ and $(1, 0)$.

Plot points between and beyond the x-intercepts.

x	-4	-3	-1	0	2	3
y	$-12\frac{1}{2}$	-4	1	$\frac{1}{2}$	1	5

Determine the end behavior of the graph. Because $f(x)$ has three linear factors of the form $x - k$ and a constant factor of $\frac{1}{4}$, it is a cubic function with a positive leading coefficient. Therefore, $f(x) \rightarrow -\infty$ as $x \rightarrow -\infty$ and $f(x) \rightarrow +\infty$ as $x \rightarrow +\infty$.

Draw the graph so that it passes through the points you plotted and has the appropriate end behavior.

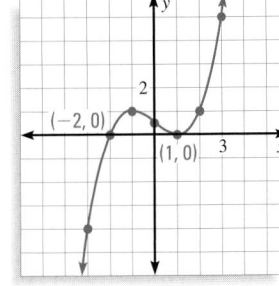

---

1 PLAN

PACING
Basic: 1 day
Average: 1 day
Advanced: 1 day
Block Schedule: 0.5 block with 6.7

LESSON OPENER
VISUAL APPROACH
An alternative way to approach Lesson 6.8 is to use the Visual Approach Lesson Opener:
• Blackline Master (*Chapter 6 Resource Book,* p. 105)
• Transparency (p. 42)

MEETING INDIVIDUAL NEEDS
• *Chapter 6 Resource Book*
  Prerequisite Skills Review (p. 5)
  Practice Level A (p. 108)
  Practice Level B (p. 109)
  Practice Level C (p. 110)
  Reteaching with Practice (p. 111)
  Absent Student Catch-Up (p. 113)
  Challenge (p. 115)
• *Resources in Spanish*
• *Personal Student Tutor*

NEW-TEACHER SUPPORT
See the Tips for New Teachers on pp. 1–2 of the *Chapter 6 Resource Book* for additional notes about Lesson 6.8.

WARM-UP EXERCISES
Transparency Available

1. If 2 is a zero of a polynomial function $f(x)$, name a factor of $f(x)$. $x-2$

2. If -2, 3, and 7 are x-intercepts of a polynomial function, what is the least possible degree of the function?   **3**

3. Does the graph of $y = 2x^2 - 5x + 4$ have a minimum or a maximum point?   **minimum**

4. Describe the end behavior of $f(x) = x(x - 6)(x + 2)$.
$f(x) \rightarrow -\infty$ as $x \rightarrow -\infty$ and $f(x) \rightarrow +\infty$ as $x \rightarrow +\infty$.

MOTIVATING THE LESSON
Companies struggle to maximize profits and minimize costs. Graphs of polynomial functions that model profit and cost data can help companies discover conditions that will result in maximum profit. This lesson focuses on using graphs to discover maximum points and minimum points on graphs.

EXTRA EXAMPLE 1
Graph $f(x) = -2(x^2 - 9)(x + 4)$.

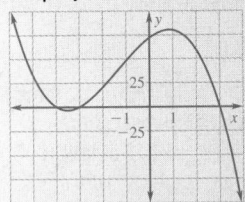

EXTRA EXAMPLE 2
Graph each function. Identify the x-intercepts, local maximums, and local minimums.
a. $f(x) = x^3 + 2x^2 - 5x + 1$
  x-intercepts: about -3.51, 0.22, 1.29; local max: $(-2.12, 11.06)$; local min: $(0.79, -1.21)$
b. $f(x) = 2x^4 - 5x^3 - 4x^2 - 6$
  x-intercepts: about -1.16 and 3.21; local min: $(2.31, -32.03)$ and $(-0.43, -6.27)$; local max: $(0, -6)$

CHECKPOINT EXERCISES
For use after Examples 1 and 2:
1. Graph $f(x) = x(2x - 5)(x + 5)$.

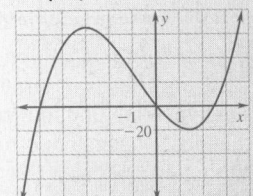

2. Use a graphing calculator to graph $f(x) = x^3 - 5x^2 + 4x + 3$. Identify the x-intercepts, local maximums, and local minimums. x-intercepts ≈ -0.46, 1.76, 3.70; local max: $(0.46, 3.88)$; local min: $(2.87, -3.06)$

TURNING POINTS Another important characteristic of graphs of polynomial functions is that they have *turning points* corresponding to local maximum and minimum values. The y-coordinate of a turning point is a **local maximum** of the function if the point is higher than all nearby points. The y-coordinate of a turning point is a **local minimum** if the point is lower than all nearby points.

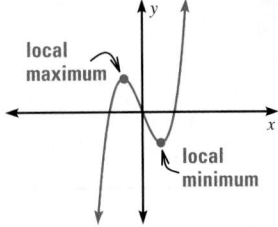

TURNING POINTS OF POLYNOMIAL FUNCTIONS

The graph of every polynomial function of degree n has *at most $n - 1$* turning points. Moreover, if a polynomial function has n distinct real zeros, then its graph has *exactly $n - 1$* turning points.

Recall that in Chapter 5 you used technology to find the maximums and minimums of quadratic functions. In Example 2 you will use technology to find turning points of higher-degree polynomial functions. If you take calculus, you will learn symbolic techniques for finding maximums and minimums.

 EXAMPLE 2   *Finding Turning Points*

Graph each function. Identify the x-intercepts and the points where the local maximums and local minimums occur.

a. $f(x) = x^3 - 3x^2 + 2$ **b.** $f(x) = x^4 - 4x^3 - x^2 + 12x - 2$

SOLUTION

a. Use a graphing calculator to graph the function.

Notice that the graph has three x-intercepts and two turning points. You can use the graphing calculator's *Zero*, *Maximum*, and *Minimum* features to approximate the coordinates of the points.

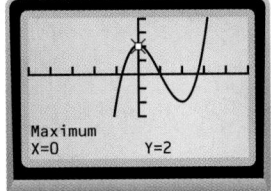

▶ The x-intercepts of the graph are $x \approx -0.73$, $x = 1$, and $x \approx 2.73$. The function has a local maximum at $(0, 2)$ and a local minimum at $(2, -2)$.

b. Use a graphing calculator to graph the function.

Notice that the graph has four x-intercepts and three turning points. You can use the graphing calculator's *Zero*, *Maximum*, and *Minimum* features to approximate the coordinates of the points.

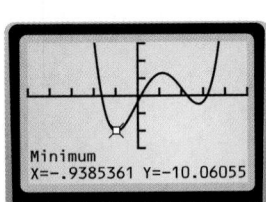

▶ The x-intercepts of the graph are $x \approx -1.63$, $x \approx 0.17$, $x \approx 2.25$, and $x \approx 3.20$. The function has local minimums at $(-0.94, -10.06)$ and $(2.79, -2.58)$, and it has a local maximum at $(1.14, 6.14)$.

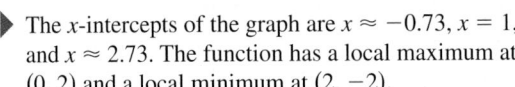
STUDENT HELP
HOMEWORK HELP
Visit our Web site
www.mcdougallittell.com
for extra examples.

 **GOAL 2** **USING POLYNOMIAL FUNCTIONS IN REAL LIFE**

In the following example, technology is used to maximize a polynomial function that models a real-life situation.

Manufacturing

EXAMPLE 3 *Maximizing a Polynomial Model*

You are designing an open box to be made of a piece of cardboard that is 10 inches by 15 inches. The box will be formed by making the cuts shown in the diagram and folding up the sides so that the flaps are square. You want the box to have the greatest volume possible. How long should you make the cuts? What is the maximum volume? What will the dimensions of the finished box be?

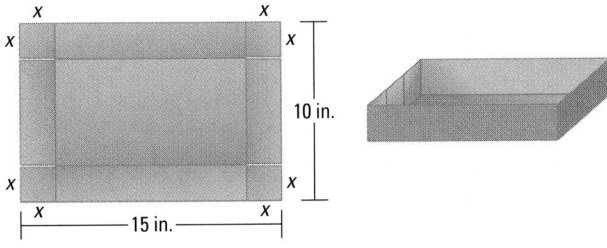

SOLUTION

PROBLEM SOLVING STRATEGY

VERBAL MODEL

| Volume | = | Width | · | Length | · | Height |

LABELS
Volume = V (cubic inches)

Width = $10 - 2x$ (inches)

Length = $15 - 2x$ (inches)

Height = x (inches)

ALGEBRAIC MODEL

$V = (10 - 2x)(15 - 2x)x$

$= (4x^2 - 50x + 150)x$

$= 4x^3 - 50x^2 + 150x$

To find the maximum volume, graph the volume function on a graphing calculator as shown at the right. When you use the *Maximum* feature, you consider only the interval $0 < x < 5$ because this describes the physical restrictions on the size of the flaps. From the graph, you can see that the maximum volume is about 132 and occurs when $x \approx 1.96$.

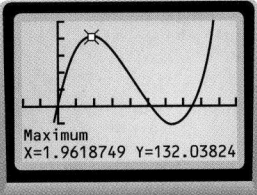
Maximum
X=1.9618749 Y=132.03824

▶ You should make the cuts approximately 2 inches long. The maximum volume is about 132 cubic inches. The dimensions of the box with this volume will be $x = 2$ inches by $10 - 2x = 6$ inches by $15 - 2x = 11$ inches.

6.8 *Analyzing Graphs of Polynomial Functions* **375**

 EXTRA EXAMPLE 3
You want to make a rectangular box that is x cm high, $(x + 5)$ cm long, and $(10 - x)$ cm wide. What is the greatest volume possible? What will the dimensions of the box be? about 264 cm³; about 11.1 cm by 3.9 cm by 6.1 cm

CHECKPOINT EXERCISES
For use after Example 3:
1. A rectangular box is to be x in. by $(12 - x)$ in. by $(15 - x)$ in. What is the greatest volume possible? What will the dimensions of the box be? about 354 in.³; about 4.4 in. by 7.6 in. by 10.6 in.

APPLICATION NOTE
EXAMPLE 3 The two positive zeros represent the conditions in which the cuts would reduce the length and width to zero.

FOCUS ON VOCABULARY
In Example 2, why is the point (0, 2) called a local maximum and not simply a maximum? There are points on the graph higher than (0, 2), but (0, 2) is higher than the nearby points of the graph.

CLOSURE QUESTION
What is the greatest number of local maximums and minimums that a cubic function can have? 2 (1 of each)

DAILY PUZZLER
Suppose $f(x)$ has degree n, where $n \geq 2$, and that it has n real zeros, none equal to 0. If you multiply all the x-coordinates of the turning points together with the x-intercepts, how does the product compare with the corresponding product for $g(x) = 2f(x)$? It is $\frac{1}{2^{n-1}}$ times as great as the product for $g(x)$.

375

ASSIGNMENT GUIDE

BASIC
Day 1: pp. 376–378 Exs. 14–26
even, 30–32, 35–36,
42–43, 45–59 odd

AVERAGE
Day 1: pp. 376–378 Exs. 14–36
even, 37–39, 42–43,
45–59 odd

ADVANCED
Day 1: pp. 376–378 Exs. 14–36
even, 37–44, 45–59 odd

BLOCK SCHEDULE
pp. 376–378 Exs. 14–36 even,
37–39, 42–43, 45–59 odd (with 6.7)

EXERCISE LEVELS
Level A: *Easier*
13–28
Level B: *More Difficult*
29–43
Level C: *Most Difficult*
44

✔ **HOMEWORK CHECK**
To quickly check student under-
standing of key concepts, go
over the following exercises:
Exs. 16, 24, 26, 30. See also the
Daily Homework Quiz:

• Blackline Master (*Chapter 6
Resource Book,* p. 118)
• Transparency (p. 49)

4.
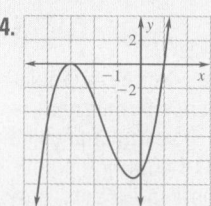

5–7, 13–22. See Additional Answers
beginning on page AA1.

GUIDED PRACTICE

Vocabulary Check ✔
Concept Check ✔

1. the *y*-coordinate of a point
of the graph that is higher
than all nearby points.

1. Explain what a local maximum of a function is.

2. Let *f* be a fourth-degree polynomial function with these zeros: 6, −2, 2*i*, and −2*i*.

   a. How many distinct linear factors does $f(x)$ have?   **4**

   b. How many distinct solutions does $f(x) = 0$ have?   **4**

   c. What are the *x*-intercepts of the graph of *f*?   **−2, 6**

3. Let *f* be a fifth-degree polynomial function with five distinct real zeros. How many turning points does the graph of *f* have?   **4**

Skill Check ✔

Graph the function.   **4–7. See margin.**

4. $f(x) = (x − 1)(x + 3)^2$

5. $f(x) = (x − 1)(x + 1)(x − 3)$

6. $f(x) = \frac{1}{8}(x + 1)(x − 1)(x − 3)$

7. $f(x) = \frac{1}{5}(x − 3)^2(x + 1)^2$

Use a graphing calculator to graph the function. Identify the *x*-intercepts and the points where the local maximums and local minimums occur.

8. $f(x) = 3x^4 − 5x^2 + 2x + 1$

9. $f(x) = x^3 − 3x^2 + x + 1$

10. $f(x) = −2x^3 + x^2 + 4x$

11. $f(x) = x^5 + x^4 − 4x^3 − 3x^2 + 5x$

8. *x*-intercepts: −1.40,
−0.29; local maximum:
(0.21, 1.21); local
minimums: (−1, −3),
(0.79, 0.63)

9. *x*-intercepts: −0.41, 1,
2.41; local maximum:
(0.18, 1.09); local
minimum: (1.82, −1.09)

10. *x*-intercepts: −1.19, 0,
1.69; local maximum:
(1, 3); local minimum:
(−0.67, −1.63)

11. *x*-intercepts: 0, 1, 1.51;
local maximums:
(−1.59, −3.23),
(0.49, 1.35); local
minimums: (−1, −4),
(1.30, −0.79)

12. 🌎 **MANUFACTURING** In Example 3, suppose you used a piece of cardboard that is 18 inches by 18 inches. Then the volume of the box would be given by this function:

$$V = 4x^3 − 72x^2 + 324x$$

Using a graphing calculator, you would obtain the graph shown at the right.

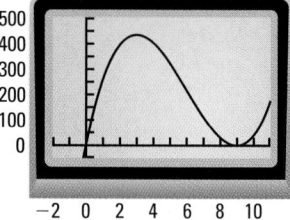

   a. What is the domain of the volume function? Explain.   **See margin.**

   b. Use the graph to estimate the length of the cut that will maximize the volume of the box.   **3 in.**

   c. Estimate the maximum volume the box can have.   *Sample answer*: 430 in.³

PRACTICE AND APPLICATIONS

STUDENT HELP

▶**Extra Practice**
to help you master
skills is on p. 948.

12a. 0 < *x* < 9, because
the flaps can't be
more than 9 in.

GRAPHING POLYNOMIAL FUNCTIONS **Graph the function.**   **13–22. See margin.**

13. $f(x) = (x − 1)^3(x + 1)$

14. $f(x) = \frac{1}{10}(x + 3)(x − 1)(x − 4)$

15. $f(x) = \frac{1}{8}(x + 4)(x + 2)(x − 3)$

16. $f(x) = 2(x + 2)^2(x + 4)^2$

17. $f(x) = 5(x − 1)(x − 2)(x − 3)$

18. $f(x) = \frac{1}{12}(x + 4)(x − 3)(x + 1)^2$

19. $f(x) = (x + 1)(x^2 − 3x + 3)$

20. $f(x) = (x + 2)(2x^2 − 2x + 1)$

21. $f(x) = (x − 2)(x^2 + x + 1)$

22. $f(x) = (x − 3)(x^2 − x + 1)$

STUDENT HELP

▶ HOMEWORK HELP
Example 1: Exs. 13–22
Example 2: Exs. 23–34
Example 3: Exs. 35–40

ANALYZING GRAPHS Estimate the coordinates of each turning point and state whether each corresponds to a local maximum or a local minimum. Then list all the real zeros and determine the least degree that the function can have.

23–34. Decimal values are approximate. See margin.

23.
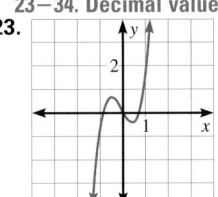

24.
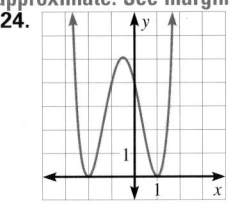

25.
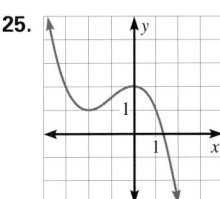

26.
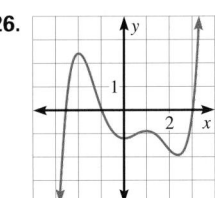

27.
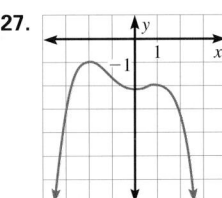

28.
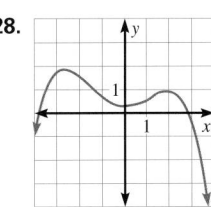

🖩 **USING GRAPHS** Use a graphing calculator to graph the polynomial function. Identify the *x*-intercepts and the points where the local maximums and local minimums occur. 29–34. See margin.

29. $f(x) = 3x^3 - 9x + 1$

30. $f(x) = -\frac{1}{3}x^3 + x - \frac{2}{3}$

31. $f(x) = -\frac{1}{4}x^4 + 2x^2$

32. $f(x) = x^5 - 6x^3 + 9x$

33. $f(x) = x^5 - 5x^3 + 4x$

34. $f(x) = x^4 - 2x^3 - 3x^2 + 5x + 2$

35. 🌐 **SWIMMING** The polynomial function

$$S = -241t^7 + 1062t^6 - 1871t^5 + 1647t^4 - 737t^3 + 144t^2 - 2.432t$$

models the speed S (in meters per second) of a swimmer doing the breast stroke during one complete stroke, where t is the number of seconds since the start of the stroke. Graph the function. At what time is the swimmer going the fastest?
at about $t = 0.8$ sec into the stroke; See margin.

36. 🌐 **FOOD** The average amount of oranges (in pounds) eaten per person each year in the United States from 1991 to 1996 can be modeled by

$$f(x) = 0.298x^3 - 2.73x^2 + 7.05x + 8.45$$

where x is the number of years since 1991. Graph the function and identify any turning points on the interval $0 \le x \le 5$. What real-life meaning do these points have? See margin.

🌐 **QUONSET HUTS** In Exercises 37–39, use the following information.
A quonset hut is a dwelling shaped like half a cylinder. Suppose you have 600 square feet of material with which to build a quonset hut. $\quad l = \dfrac{600 - \pi r^2}{\pi r}$

37. The formula for surface area is $S = \pi r^2 + \pi r l$ where r is the radius of the semicircle and l is the length of the hut. Substitute 600 for S and solve for l.

38. The formula for the volume of the hut is $V = \frac{1}{2}\pi r^2 l$. Write an equation for the volume V of the quonset hut as a polynomial function of r by substituting the expression for l from Exercise 37 into the volume formula. $\quad V = 300r - 0.5\pi r^3$

39. Use the function you wrote in Exercise 38 to find the maximum volume of a quonset hut with a surface area of 600 square feet. What are the hut's dimensions?
1600 ft³; $r \approx 7.98$ ft, $l \approx 15.97$ ft, or about 16 ft long, 16 ft wide, and 8 ft high

FOCUS ON APPLICATIONS

QUONSET HUTS were invented during World War II. They were temporary structures that could be assembled quickly and easily. After the war they were sold as homes for about $1000 each.

🌐 **APPLICATION LINK** www.mcdougallittell.com

APPLICATION NOTE
EXERCISES 37–39
Additional information about quonset huts is available at **www.mcdougallittell.com.**

EXERCISE 40 Annual production of butter in the United States has dropped from about 2 billion pounds before World War II to about 1.1 billion pounds in the early 1990's. Competition from lower-priced margarine and the emphasis on low-fat diets are both responsible for the decline. In a free market economy, the price of an item is affected by the demand for it. With a declining market for butter, production has decreased and prices have increased.

23. (–0.5, 0.5) max, (0.5, –0.3) min; –0.9, 0, 0.6; 3
24. (–2, 0) min, (–0.5, 5) max, (1, 0) min; –2, 1; 4
25. (–2, 1) min, (0, 2) max; 1.4; 3
26. (–2, 2.4) max, (0, –1.2) min, (1, –1) max, (2.4, –2) min; –2.5, –1, 3; 5
27. (–2, –1) max, (0, –2.2) min, (1, –2) max; none; 4
28. (–2.8, 1.9) max; (0, 0.25) min, (2, 1) max; –3.8; 2.8; 4
29–36. See Additional Answers beginning on page AA1.

ADDITIONAL PRACTICE AND RETEACHING

For Lesson 6.8:
• Practice Levels A, B, and C (*Chapter 6 Resource Book,* p. 108)
• Reteaching with Practice (*Chapter 6 Resource Book,* p. 111)
• 🖥 See Lesson 6.8 of the *Personal Student Tutor*

For more Mixed Review:
• 🖥 Search the *Test and Practice Generator* for key words or specific lessons.

DAILY HOMEWORK QUIZ

Transparency Available

1. Graph the function $f(x) = \frac{1}{2}(x-1)(x+2)(x-3)^2$.

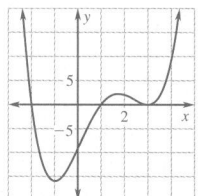

2. Graph the function $f(x) = 2x^3 - 5x^2 + 1$. Identify the x-intercepts and the points where the local maximums and minimums occur.

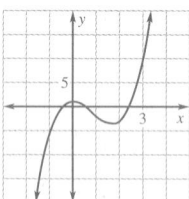

x-intercepts: about −0.41, 0.5, about 2.41; local max. at (0, 1), local min. at about (1.67, −3.63)

EXTRA CHALLENGE NOTE

Challenge problems for Lesson 6.8 are available in **blackline** format in the *Chapter 6 Resource Book,* p. 115 and at **www.mcdougallittell.com**.

ADDITIONAL TEST PREPARATION

1. WRITING If a polynomial function has *n* distinct turning points, must it have *n* + 1 distinct real zeros? Explain.

no; Explanations may vary. *Sample answer:* The function $f(x) = (x-1)(x-2)(x-3) + 5$ has two distinct turning points, but it has only one real zero.

44. See Additional Answers beginning on page AA1.

40. reaches a local minimum at (2.71, 50.03); the producer price index declined from 1991 to a low of about 50.03 around September 1993, after which it began to increase.

Test Preparation

41. Check students' graphs. A polynomial with 3 turning points must be of degree 4 or higher. (It must also be of even degree.)

★ **Challenge**

EXTRA CHALLENGE
www.mcdougallittell.com

55. $y = -(x-1)^2 + 4$

56. $y = -(x+2)^2 + 6$

57. $y = \frac{5}{24}(x+5)(x-5)$

58. $y = \frac{4}{9}(x+2)(x-4)$

40. 🌐 **CONSUMER ECONOMICS** The producer price index of butter from 1991 to 1997 can be modeled by $P = -0.233x^4 + 2.64x^3 - 6.59x^2 - 3.93x + 69.1$ where *x* is the number of years since 1991. Graph the function and identify any turning points on the interval $0 \le x \le 6$. What real-life meaning do these points have?

41. CRITICAL THINKING Sketch the graph of a polynomial function that has three turning points. Label each turning point as a local maximum or local minimum. What must be true about the degree of the polynomial function that has such a graph? Explain your reasoning. **See margin.**

In Exercises 42 and 43, use the graph of the polynomial function *f* shown at the right.

42. MULTIPLE CHOICE What is the local maximum of *f* on the interval $-2 \le x \le -1$?   **A**

(A) $f(x) \approx 3.7$ (B) $f(x) \approx 1.4$

(C) $f(x) \approx -1.4$ (D) $f(x) \approx -3.7$

43. MULTIPLE CHOICE What is the local maximum of *f* on the interval $-1 \le x \le 1$?   **B**

(A) $f(x) \approx 3.7$ (B) $f(x) \approx 1.4$ (C) $f(x) \approx -1.4$ (D) $f(x) \approx -3.7$

44. GRAPHING OPPOSITES Sketch the graph of $y = f(x)$ for this function:

$$f(x) = x^3 + 4x^2$$

Then sketch the graph of $y = -f(x)$. Explain how the graphs, the x-intercepts, the local maximums, and the local minimums are related. Finally, sketch the graph of $y = f(-x)$. Compare it with the others. **See margin.**

MIXED REVIEW

RELATING VARIABLES The variables *x* and *y* vary directly. Write an equation that relates the variables. (Review 2.4)

45. $x = 1, y = 7$ $y = 7x$ **46.** $x = -4, y = 6$ $y = -\frac{3}{2}x$ **47.** $x = 12, y = 3$ $y = \frac{1}{4}x$

48. $x = 2, y = -5$ $y = -\frac{5}{2}x$ **49.** $x = -5, y = 3$ $y = -\frac{3}{5}x$ **50.** $x = -6, y = -15$ $y = \frac{5}{2}x$

MATRIX PRODUCTS Let *A* and *B* be matrices with the given dimensions. State whether the product *AB* is defined. If so, give the dimensions of *AB*. (Review 4.2)

51. A: 4×3, B: 3×1 yes; 4×1 **52.** A: 2×4, B: 4×5 yes; 2×5

53. A: 4×3, B: 2×4 no   **54.** A: 6×6, B: 6×5 yes; 6×5

WRITING QUADRATIC FUNCTIONS Write a quadratic function whose graph passes through the given points. (Review 5.8 for 6.9)

55. vertex: (1, 4); point: (4, −5)   **56.** vertex: (−2, 6); point: (0, 2)

57. points: (−5, 0), (5, 0), (7, 5)   **58.** points: (−2, 0), (4, 0), (1, −4)

59. 🌐 **PLANT GROWTH** You have a kudzu vine in your back yard. On Monday, the vine is 30 inches long. The following Thursday, the vine is 60 inches long. What is the average rate of change in the length of the vine? (Lesson 2.2)
10 in./day

● ACTIVITY 6.9

Developing Concepts

Exploring Finite Differences

GROUP ACTIVITY
Work with a partner.

MATERIALS
• Paper
• Pencil

▶ **QUESTION** How are the finite differences for a polynomial function related to the function's degree?

▶ **EXPLORING THE CONCEPT**

The number of paths that lead, through a sequence of upward and rightward movements only, from the bottom left corner of a grid to the top right corner depends on the grid's dimensions.

For an $n \times 1$ grid, the number of paths is given by $f(n) = n + 1$.

For an $n \times 2$ grid, the number of paths is given by $g(n) = \frac{1}{2}(n + 1)(n + 2)$.

For an $n \times 3$ grid, the number of paths is given by $h(n) = \frac{1}{6}(n + 1)(n + 2)(n + 3)$.

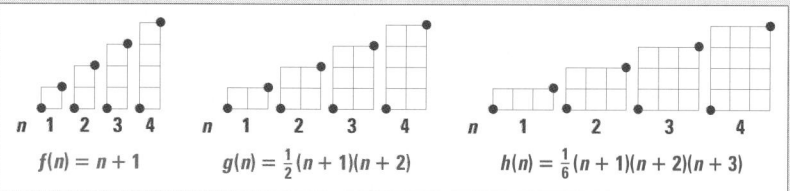

For each function given above, follow **Steps 1–4**. The steps for the first function f have been done for you. Steps 1–4. See margin.

❶ Evaluate the function when $n = 1, 2, 3, 4, 5,$ and 6.

n	1	2	3	4	5	6
$f(n)$	2	3	4	5	6	7

1 1 1 1 1

First-order differences are constant.

❷ For each pair of consecutive function values, find the difference of the values. These numbers are called *first-order differences*.

❸ Are the first-order differences constant? If so, stop here. Otherwise, find the differences in each pair of consecutive first-order differences. These numbers are called *second-order differences*.

❹ Are the second-order differences constant? If so, stop here. Otherwise, find the differences in each pair of consecutive second-order differences. These numbers are called *third-order differences*.

▶ **DRAWING CONCLUSIONS**

1. Repeat **Steps 1–4** for each function. See margin.

a. $f(n) = 3n + 1$ **b.** $f(n) = n^2 + 2n - 3$ **c.** $f(n) = 2n^3 - n + 4$

2. For each function giving the number of paths through a grid and for each function in Exercise 1, state the degree of the function and the number of times differences were calculated before a row of constant, nonzero differences was obtained. What do you notice? See margin.

3. Which order differences do you think will be constant for the function $f(n) = n^4 + n$? Explain. Then find the differences to see if you are correct.
fourth; see margin for explanation.

2. degrees: 1, 2, 3; 1, 2, 3; number of times differences were calculated before arriving at a row of constant, nonzero differences: 1, 2, 3; 1, 2, 3; the degree equals the number of times differences were calculated.

Exploring the Concept, Drawing Conclusions 1, 3.
See Additional Answers beginning on page AA1.

---

1 Planning the Activity

PURPOSE
To explore nth-order differences for polynomial functions at equally-spaced values of the independent variable in order to predict which order of differences will be constant for a function.

MATERIALS
• paper and pencil
• Activity Support Master (*Chapter 6 Resource Book*, p. 119)

PACING
• Exploring the Concept — 15 min
• Drawing Conclusions — 15 min

▶ **LINK TO LESSON**
As students study Example 2 in Lesson 6.9, ask if they can see the relationship between the formula for the number of paths on an $n \times 2$ grid and the triangular numbers. The formula for the number of paths of an $n \times 2$ grid gives the value of the $(n + 1)$st triangular number.

2 Managing the Activity

ALTERNATIVE APPROACH
To save time and ensure that all students understand the procedure, demonstrate the algorithm for the $n \times 1$ and $n \times 2$ grids.

3 Closing the Activity

★ **KEY DISCOVERY**
A polynomial of degree n has constant nth order differences for equally-spaced values of the independent variable.

ACTIVITY ASSESSMENT
Given a polynomial of degree n and equally-spaced values of the independent variable, how many times would you have to calculate differences before you got a row of constant nonzero differences? n times

PACING
Basic: 1 day
Average: 1 day
Advanced: 1 day
Block Schedule: 0.5 block with
Ch. Rev.

LESSON OPENER
APPLICATION
An alternative way to approach
Lesson 6.9 is to use the Application
Lesson Opener:

• Blackline Master (*Chapter 6
  Resource Book*, p. 120)
• Transparency (p. 43)

MEETING INDIVIDUAL NEEDS

• *Chapter 6 Resource Book*
  Prerequisite Skills Review (p. 5)
  Practice Level A (p. 121)
  Practice Level B (p. 122)
  Practice Level C (p. 123)
  Reteaching with Practice (p. 124)
  Absent Student Catch-Up (p. 126)
  Challenge (p. 128)
• *Resources in Spanish*
• *Personal Student Tutor*

NEW-TEACHER SUPPORT
See the Tips for New Teachers on
pp. 1–2 of the *Chapter 6 Resource
Book* for additional notes about
Lesson 6.9.

WARM-UP EXERCISES

 Transparency Available

1. Write a cubic function that
  has real coefficients, zeros
  at −2, 1, 3, and a leading
  coefficient of 1.
  $f(x) = (x + 2)(x - 1)(x - 3)$
2. Find $f(2)$ for $f(x) = 3x^3 - 4x^2 + x + 7$. **17**
3. Solve the linear system.
  $a + b + c = 9$
  $4a + 2b + c = 5$
  $9a + 3b + c = 7$
  $a = 3, b = -13, c = 19$

EXPLORING DATA AND STATISTICS

6.9

What you should learn

GOAL 1 Use finite
differences to determine the
degree of a polynomial
function that will fit a set
of data.

GOAL 2 Use technology to
find polynomial models for
real-life data, as applied in
Example 4.

Why you should learn it

▼ To model **real-life**
quantities, such as the speed
of the space shuttle
in **Ex. 49**.

Modeling with Polynomial Functions

GOAL 1 **USING FINITE DIFFERENCES**

You know that two points determine a line and that three points determine a parabola.
In Example 1 you will see that four points determine the graph of a cubic function.

EXAMPLE 1 *Writing a Cubic Function*

Write the cubic function whose graph is shown at the right.

SOLUTION

Use the three given x-intercepts to write the following:

$$f(x) = a(x + 3)(x - 2)(x - 5)$$

To find a, substitute the coordinates of the fourth point.

$$-15 = a(0 + 3)(0 - 2)(0 - 5), \text{ so } a = -\frac{1}{2}$$

▶ $f(x) = -\frac{1}{2}(x + 3)(x - 2)(x - 5)$

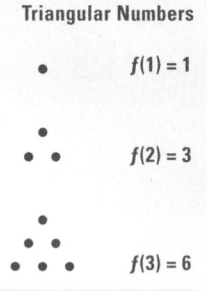

✓ **CHECK** Check the graph's end behavior. The degree of f
is odd and $a < 0$, so $f(x) \to +\infty$ as $x \to -\infty$ and $f(x) \to -\infty$ as $x \to +\infty$.

· · · · · · · · · ·

To decide whether y-values for equally-spaced x-values can be modeled by a
polynomial function, you can use **finite differences**.

EXAMPLE 2 *Finding Finite Differences*

The first three triangular numbers are shown at the right.
A formula for the nth triangular number is $f(n) = \frac{1}{2}(n^2 + n)$.
Show that this function has constant second-order differences.

Triangular Numbers

• $f(1) = 1$

•
• • $f(2) = 3$

•
• •
• • • $f(3) = 6$

SOLUTION

Write the first several triangular numbers. Find the first-order
differences by subtracting consecutive triangular numbers.
Then find the second-order differences by subtracting
consecutive first-order differences.

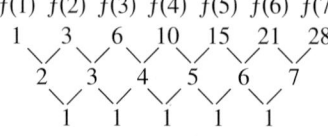

$f(1)$	$f(2)$	$f(3)$	$f(4)$	$f(5)$	$f(6)$	$f(7)$
1	3	6	10	15	21	28

Function values for
equally-spaced n-values

First-order differences: 2   3   4   5   6   7

Second-order differences: 1   1   1   1   1

In Example 2 notice that the function has degree *two* and that the *second*-order differences are constant. This illustrates the first property of finite differences.

> ## PROPERTIES OF FINITE DIFFERENCES
>
> 1. If a polynomial function $f(x)$ has degree n, then the nth-order differences of function values for equally spaced x-values are nonzero and constant.
> 2. Conversely, if the nth-order differences of equally-spaced data are nonzero and constant, then the data can be represented by a polynomial function of degree n.

The following example illustrates the second property of finite differences.

EXAMPLE 3 **Modeling with Finite Differences**

STUDENT HELP

HOMEWORK HELP
Visit our Web site
www.mcdougallittell.com
for extra examples.

The first six triangular pyramidal numbers are shown below. Find a polynomial function that gives the nth triangular pyramidal number.

      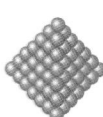

$f(1) = 1$ $f(2) = 4$ $f(3) = 10$ $f(4) = 20$ $f(5) = 35$ $f(6) = 56$ $f(7) = 84$

SOLUTION

Begin by finding the finite differences.

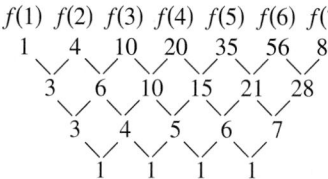

	Function values for equally-spaced n-values

$f(1)\ f(2)\ f(3)\ f(4)\ f(5)\ f(6)\ f(7)$
 1   4   10   20   35   56   84   Function values for equally-spaced n-values
   3   6   10   15   21   28   First-order differences
     3   4   5   6   7   Second-order differences
       1   1   1   1   Third-order differences

Because the third-order differences are constant, you know that the numbers can be represented by a cubic function which has the form $f(n) = an^3 + bn^2 + cn + d$.

By substituting the first four triangular pyramidal numbers into the function, you can obtain a system of four linear equations in four variables.

$a(1)^3 + b(1)^2 + c(1) + d = 1 \implies a + b + c + d = 1$

$a(2)^3 + b(2)^2 + c(2) + d = 4 \implies 8a + 4b + 2c + d = 4$

$a(3)^3 + b(3)^2 + c(3) + d = 10 \implies 27a + 9b + 3c + d = 10$

$a(4)^3 + b(4)^2 + c(4) + d = 20 \implies 64a + 16b + 4c + d = 20$

Using a calculator to solve the system gives $a = \frac{1}{6}$, $b = \frac{1}{2}$, $c = \frac{1}{3}$, and $d = 0$.

▶ The nth triangular pyramidal number is given by $f(n) = \frac{1}{6}n^3 + \frac{1}{2}n^2 + \frac{1}{3}n$.

6.9 *Modeling with Polynomial Functions*   **381**

---

 2 TEACH

EXTRA EXAMPLE 1
Write the cubic function whose graph is shown below.

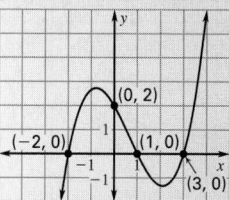

$f(x) = \frac{1}{3}(x+2)(x-1)(x-3)$

EXTRA EXAMPLE 2
An equation for a polynomial function is $f(n) = 2n^3 + n^2 + 2n + 1$. Show that this function has constant third-order differences.

$f(1)\ f(2)\ f(3)\ f(4)\ f(5)\ f(6)$
 6   25   70 153 286 481
   19   45   83 133 195
     26   38   50   62
       12   12   12

EXTRA EXAMPLE 3
The values of a polynomial function for six consecutive whole numbers are given below. Write a polynomial function for $f(n)$.
$f(1) = -2$, $f(2) = 2$, $f(3) = 12$, $f(4) = 28$, $f(5) = 50$, and $f(6) = 78$
$f(n) = 3n^2 - 5n$

✓ CHECKPOINT EXERCISES
For use after Example 1:
1. Write a cubic equation whose zeros are $-5, -1$, and 4, and whose y-intercept is $6\frac{2}{3}$.
$y = -\frac{1}{3}(x+5)(x+1)(x-4)$

For use after Examples 2 and 3:
2. The values of a polynomial function for six consecutive whole numbers are given below. Write a polynomial function for $f(n)$.
$f(1) = 6$, $f(2) = -2$, $f(3) = -32$, $f(4) = -96$, $f(5) = -206$, and $f(6) = -374$
$f(n) = -2n^3 + n^2 + 3n + 4$

EXTRA EXAMPLE 4

The table shows data on the increase in girth of a tree in a redwood forest over a 10-week period.

Week	Increase (mm)
0	0.34
2	0.35
4	0.37
6	0.31
8	0.18
10	0.18

a. Find a quartic model for the data.
$y = 0.000378x^4 - 0.00682x^3 + 0.0326x^2 - 0.0369x + 0.34$

b. Use the model you found in part (**a**) to estimate the increase in girth in week 5.
about 0.354 mm

c. In which week was the increase in girth about 0.25 mm? about week 7

 ## CHECKPOINT EXERCISES

For use after Example 4:

1. Find a cubic model for the data in Extra Example 4.
$y = 0.000729x^3 - 0.0138x^2 + 0.0494x + 0.33$

FOCUS ON VOCABULARY

How do you find second-order differences? Subtract consecutive first-order differences.

CLOSURE QUESTION

If the third-order differences of equally-spaced data are nonzero and constant, what degree polynomial function can represent the data? degree 3

DAILY PUZZLER

How many polynomial functions of the form $f(x) = x^n$ have graphs that pass through $(0, 0)$, $(-1, -1)$, and $(1, 1)$? infinitely many

FOCUS ON APPLICATIONS

MOTORBOATS often have tachometers instead of speedometers. The tachometer measures the engine speed in revolutions per minute, which can then be used to determine the speed of the boat.

GOAL 2 POLYNOMIAL MODELING WITH TECHNOLOGY

In Examples 1 and 3 you found a cubic model that *exactly* fits a set of data points. In many real-life situations, you cannot find a simple model to fit data points exactly. Instead you can use the regression feature on a graphing calculator to find an nth-degree polynomial model that best fits the data.

EXAMPLE 4 Modeling with Cubic Regression

BOATING The data in the table give the average speed y (in knots) of the *Trident* motor yacht for several different engine speeds x (in hundreds of revolutions per minute, or RPMs).

a. Find a polynomial model for the data.

b. Estimate the average speed of the *Trident* for an engine speed of 2400 RPMs.

c. What engine speed produces a boat speed of 14 knots?

Engine speed, x	9	11	13	15	17	19	21.5
Boat speed, y	6.43	7.61	8.82	9.86	10.88	12.36	15.24

SOLUTION

a. ***Enter*** the data in a graphing calculator and make a scatter plot. From the scatter plot, it appears that a cubic function will fit the data better than a linear or quadratic function.

Use cubic regression to obtain a model.

▸ $y = 0.00475x^3 - 0.194x^2 + 3.13x - 9.53$

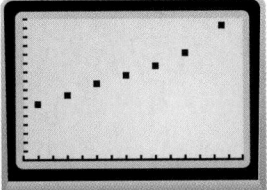

✓ **CHECK** By graphing the model in the same viewing window as the scatter plot, you can see that it is a good fit.

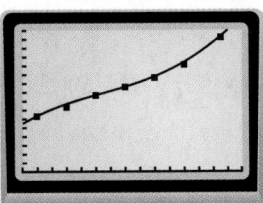

b. Substitute $x = 24$ into the model from part (a).

$$y = 0.00475(24)^3 - 0.194(24)^2 + 3.13(24) - 9.53$$

$$= 19.51$$

▸ The *Trident*'s speed for an engine speed of 2400 RPMs is about 19.5 knots.

c. Graph the model and the equation $y = 14$ on the same screen. Use the *Intersect* feature to find the point where the graphs intersect.

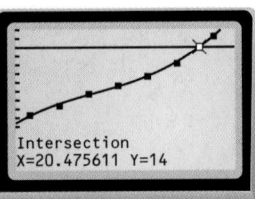

Intersection
X=20.475611 Y=14

▸ An engine speed of about 2050 RPMs produces a boat speed of 14 knots.

GUIDED PRACTICE

Vocabulary Check ✓

Concept Check ✓

1. Describe what first-order differences and second-order differences are.
The differences between $f(n)$ and $f(n + 1)$; the differences of adjacent first-order differences

2. How many points do you need to determine a quartic function? **5**

3. Why can't you use finite differences to find a model for the data in Example 4?
Because the points will not lie exactly on the curve generated by the model.

Skill Check ✓

4. Write the cubic function whose graph passes through $(3, 0)$, $(-1, 0)$, $(-2, 0)$, and $(1, 2)$. $f(x) = -\frac{1}{6}(x - 3)(x + 1)(x + 2)$

Show that the nth-order finite differences for the given function of degree n are nonzero and constant.   5–8. See margin.

5. $f(x) = 5x^2 - 2x + 1$

6. $f(x) = x^3 + x^2 - 1$

7. $f(x) = x^4 + 2x$

8. $f(x) = 2x^3 - 12x^2 - 5x + 3$

Use finite differences to determine the degree of the polynomial function that will fit the data.

9.
3

x	1	2	3	4	5	6
f(x)	−1	3	3	5	15	39

10.
2

x	1	2	3	4	5	6
f(x)	0	8	12	12	8	0

Find a polynomial function that fits the data.

$f(x) = x^3 - 4x^2 + 2x$

$f(x) = -x^3 + 5x^2 + x + 1$ **11.**

x	1	2	3	4	5	6
f(x)	6	15	22	21	6	−29

12.

x	1	2	3	4	5	6
f(x)	−1	−4	−3	8	35	84

13. **GEOMETRY CONNECTION** Find a polynomial function that gives the number of diagonals of a polygon with n sides. $d = \frac{1}{2}n^2 - \frac{3}{2}n$

Number of sides, n	3	4	5	6	7	8
Number of diagonals, d	0	2	5	9	14	20

PRACTICE AND APPLICATIONS

STUDENT HELP

▶ **Extra Practice**
to help you master skills is on p. 949.

15. $f(x) = -\frac{1}{2}(x + 1)(x - 2)(x - 3)$

16. $f(x) = \frac{1}{5}(x + 3)(x + 1)(x - 3)$

WRITING CUBIC FUNCTIONS Write the cubic function whose graph is shown.
15, 16. See margin.

14.

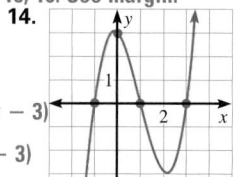

15.

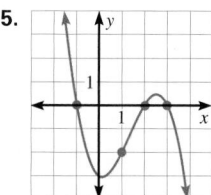

16.

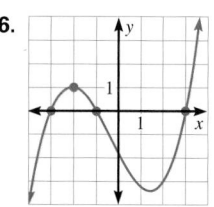

$f(x) = (x + 1)(x - 1)(x - 3)$

FINDING A CUBIC MODEL Write a cubic function whose graph passes through the given points.   17–22. See margin.

17. $(-1, 0), (-2, 0), (0, 0), (1, -3)$

18. $(3, 0), (2, 0), (-3, 0), (1, -1)$

19. $(1, 0), (3, 0), (-2, 0), (2, 1)$

20. $(-1, 0), (-4, 0), (4, 0), (0, 3)$

21. $(3, 0), (2, 0), (-1, 0), (1, 4)$

22. $(0, 0), (-3, 0), (5, 0), (-2, 3)$

6.9 Modeling with Polynomial Functions   **383**

ASSIGNMENT GUIDE

BASIC
Day 1: pp. 383–386 Exs. 14–16, 18, 24, 32, 43, 47, 49–65 odd, Quiz 3 Exs. 1–21

AVERAGE
Day 1: pp. 383–386 Exs. 14–26 even, 32–34, 44–47, 49–65 odd, Quiz 3 Exs. 1–21

ADVANCED
Day 1: pp. 383–386 Exs. 14–26 even, 32–35, 44–48, 49–67 odd, Quiz 3 Exs. 1–21

BLOCK SCHEDULE
pp. 383–386 Exs. 14–26 even, 32–34, 44–47, 49–65 odd, Quiz 3 Exs. 1–21 (with Ch. 6 Review)

EXERCISE LEVELS
Level A: *Easier*
14–31

Level B: *More Difficult*
32–47

Level C: *Most Difficult*
48–51

✓ HOMEWORK CHECK
To quickly check student understanding of key concepts, go over the following exercises: Exs. 16, 18, 24, 32. See also the Daily Homework Quiz:

• Blackline Master (*Chapter 7 Resource Book*, p. 11)

• Transparency (p. 51)

5. $f(1)\ f(2)\ f(3)\ f(4)\ f(5)\ f(6)$
    4   17   40   73   116 169
      13   23   33   43   53
        10   10   10   10

6–8, 17–22. See Additional Answers beginning on page AA1.

! **COMMON ERROR**

EXERCISES 32–43 Students who have difficulty finding a model may be calculating the finite differences incorrectly. particularly where negative numbers are involved. Remind students that each first-order difference is found by subtracting $f(n+1) - f(n)$. Differences of higher order are calculated according to a similar pattern.

GRAPHING CALCULATOR NOTE

EXERCISES 47–49 A graphing calculator will allow students to try more than one model and then decide which model best fits the data.

23–31. See Additional Answers beginning on page AA1.

44. $f(1)$ $f(2)$ $f(3)$ $f(4)$ $f(5)$ $f(6)$
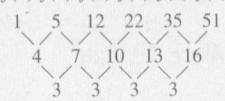

45. $f(1)$ $f(2)$ $f(3)$ $f(4)$ $f(5)$ $f(6)$
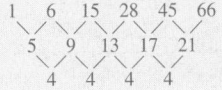

ADDITIONAL PRACTICE AND RETEACHING

For Lesson 6.9:

• Practice Levels A, B, and C (*Chapter 6 Resource Book*, p. 121)

• Reteaching with Practice (*Chapter 6 Resource Book*, p. 124)

• ⊞ See Lesson 6.9 of the *Personal Student Tutor*

For more Mixed Review:

• ⊞ Search the *Test and Practice Generator* for key words or specific lessons.

384

┌─ STUDENT HELP
↳ HOMEWORK HELP

Example 1: Exs. 14–22
Example 2: Exs. 23–31, 44, 45
Example 3: Exs. 32–43, 46
Example 4: Exs. 47–49

FINDING FINITE DIFFERENCES Show that the nth-order differences for the given function of degree n are nonzero and constant. **23–31. See margin.**

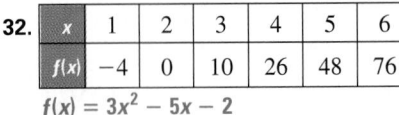
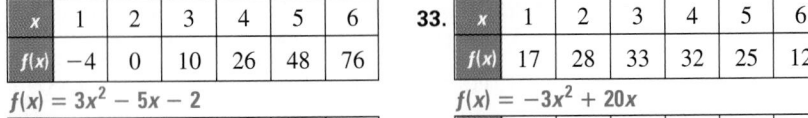

23. $f(x) = x^2 - 3x + 7$ **24.** $f(x) = 2x^3 - 5x^2 - x$ **25.** $f(x) = -x^3 + 3x^2 - 2x - 3$

26. $f(x) = x^4 - 3x^3$ **27.** $f(x) = 2x^4 - 20x$ **28.** $f(x) = -4x^2 + x + 6$

29. $f(x) = -x^4 + 5x^2$ **30.** $f(x) = 3x^3 - 5x^2 - 2$ **31.** $f(x) = -3x^2 + 4x + 2$

📟 **FINDING A MODEL** Use finite differences and a system of equations to find a polynomial function that fits the data. You may want to use a calculator.

32.

x	1	2	3	4	5	6
f(x)	−4	0	10	26	48	76

$f(x) = 3x^2 - 5x - 2$

33.

x	1	2	3	4	5	6
f(x)	17	28	33	32	25	12

$f(x) = -3x^2 + 20x$

34.

x	1	2	3	4	5	6
f(x)	−4	−6	−2	14	48	106

$f(x) = x^3 - 3x^2 - 2$

35.

x	1	2	3	4	5	6
f(x)	−2	−6	−6	4	30	78

$f(x) = x^3 - 4x^2 + x$

36.

x	1	2	3	4	5	6
f(x)	−3	−8	−15	−21	−23	−18

$f(x) = 0.5x^3 - 4x^2 + 3.5x - 3$

37.

x	1	2	3	4	5	6
f(x)	2	20	58	122	218	352

$f(x) = x^3 + 4x^2 - x - 2$

38.

x	1	2	3	4	5	6
f(x)	−5	0	9	16	15	0

$f(x) = -x^3 + 8x^2 - 12x$

39.

x	1	2	3	4	5	6
f(x)	−2	1	−4	−5	10	53

$f(x) = 2x^3 - 16x^2 + 37x - 25$

40.

x	1	2	3	4	5	6
f(x)	20	−2	−4	2	4	−10

$f(x) = -2x^3 + 22x^2 - 74x + 74$

41.

x	1	2	3	4	5	6
f(x)	2	−5	−4	−1	−2	−13

$f(x) = -x^3 + 10x^2 - 30x + 23$

42.

x	1	2	3	4	5	6
f(x)	26	−4	−2	2	2	16

$f(x) = x^4 - 15x^3 + 81x^2 - 183x + 142$

43.

x	1	2	3	4	5	6
f(x)	0	6	2	6	12	−10

$f(x) = -x^4 + 13x^3 - 58x^2 + 104x - 58$

44. PENTAGONAL NUMBERS The dot patterns show pentagonal numbers. A formula for the nth pentagonal number is $f(n) = \frac{1}{2}n(3n - 1)$. Show that this function has constant second-order differences. **See margin.**

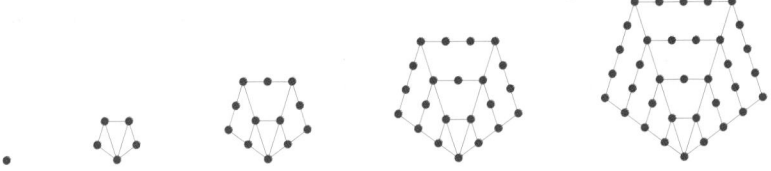

45. HEXAGONAL NUMBERS A formula for the nth hexagonal number is $f(n) = n(2n - 1)$. Show that this function has constant second-order differences.
See margin.

46. SQUARE PYRAMIDAL NUMBERS The first six square pyramidal numbers are shown. Find a polynomial function that gives the nth square pyramidal number. $f(n) = \frac{n}{6}(2n + 1)(n + 1)$

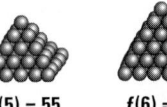
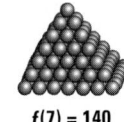

$f(1) = 1$ $f(2) = 5$ $f(3) = 14$ $f(4) = 30$ $f(5) = 55$ $f(6) = 91$ $f(7) = 140$

FOCUS ON
PEOPLE

EILEEN COLLINS was selected by NASA for the astronaut program in 1990. Since then she has become the first woman to pilot a spacecraft and the first woman to command a space shuttle.

47. $f(t) = 0.641t^3 - 4.93t^2 + 25.8t + 232$ where t is the number of years since 1989; 772,000 Girl Scouts

Test Preparation

48. $y = 0.242t^3 - 3.00t^2 + 13.5t + 140$ where t is the number of years since 1987; about $340,000

49. $y = 0.007t^3 - 0.740t^2 + 49.0t - 236$; about 101 sec

★ **Challenge**

EXTRA CHALLENGE
www.mcdougallittell.com

FINDING MODELS In Exercises 47–49, use a graphing calculator to find a polynomial model for the data.

47. **GIRL SCOUTS** The table shows the number of Girl Scouts (in thousands) from 1989 to 1996. Find a polynomial model for the data. Then estimate the number of Girl Scouts in 2000. **See margin.**

t	1989	1990	1991	1992	1993	1994	1995	1996
y	231	253.3	273.8	284.1	294.1	303.6	368.6	383.7

See margin.

48. **REAL ESTATE** The table shows the average price (in thousands of dollars) of a house in the Northeastern United States for 1987 to 1995. Find a polynomial model for the data. Then predict the average price of a house in the Northeast in 2000. **DATA UPDATE** of *Statistical Abstract of the United States* data at www.mcdougallittell.com

x	1987	1988	1989	1990	1991	1992	1993	1994	1995
f(x)	140	149	159.6	159	155.9	169	162.9	169	180

49. **SPACE EXPLORATION** The table shows the average speed y (in feet per second) of a space shuttle for different times t (in seconds) after launch. Find a polynomial model for the data. When the space shuttle reaches a speed of approximately 4400 feet per second, its booster rockets fall off. Use the model to determine how long after launch this happens. **See margin.**

t	10	20	30	40	50	60	70	80
y	202.4	463.4	748.2	979.3	1186.3	1421.3	1795.4	2283.5

50. **MULTI-STEP PROBLEM** Your friend has a dog-walking service and your cousin has a lawn-care service. You want to start a small business of your own. You are trying to decide which of the two services you should choose. The profits for the first 6 months of the year are shown in the table. **50.a, b. See margin.**

Dog-walking service	Month, t	1	2	3	4	5	6
	Profit, P	3	5	22	54	101	163
Lawn-care service	Month, t	1	2	3	4	5	6
	Profit, P	3	21	41	68	107	163

a. Use finite differences to find a polynomial model for each business.

b. *Writing* You want to choose the business that will make the greater profit in December (when $t = 12$). Explain which business you should choose and why.

51. a. Substitute the expressions $x, x + 1, x + 2, \ldots, x + 5$ for x in the function $f(x) = ax^3 + bx^2 + cx + d$ and show that third-order differences are constant. **51.a, b. See margin.**

b. The data below can be modeled by a cubic function. Set the variable expressions you found in part (a) equal to the first-, second-, and third-order differences for these values. Solve the equations to find the coefficients of the function that models the data. Check your work by substituting the original data values into the function.

x	1	2	3	4	5	6
f(x)	−1	1	−3	−7	−5	9

DAILY HOMEWORK QUIZ

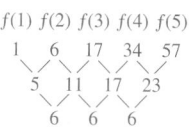 *Transparency Available*

1. Write a cubic function whose graph passes through (−3, 0), (−1, 0), (1, −8), and (4, 0).
$f(x) = \frac{1}{3}(x + 3)(x + 1)(x - 4)$

2. Show that the second-order differences for $f(x) = 3x^2 - 4x + 2$ are nonzero and constant.

$f(1)\ f(2)\ f(3)\ f(4)\ f(5)$
  1   6   17   34   57
     5   11   17   23
        6   6   6

3. Use finite differences and a system of equations to find a polynomial function that fits the data.

x	1	2	3	4	5
f(x)	2	0	4	20	54

$f(x) = x^3 - 3x^2 + 4$

EXTRA CHALLENGE NOTE

→ Challenge problems for Lesson 6.9 are available in **blackline** format in the *Chapter 6 Resource Book,* p. 128 and at **www.mcdougallittell.com**.

ADDITIONAL TEST PREPARATION

1. **WRITING** Explain when you can use finite differences to model data with a polynomial function. **If the *n*th-order differences of equally-spaced data are nonzero and constant, then the data can be represented by a polynomial function of degree *n*.**

50–51. See next page.

50a. Dog-walking:
$P = 7.5t^2 - 20.5t + 16$
Lawn-care:
$P = 0.833t^3 - 4t^2 + 24.17t - 18$

b. *Sample answer:* The lawn-care business would have more profit in December according to these models, $1135 vs. $850.

51a. $f(x) = ax^3 + bx^2 + cx + d$;
$f(x+1) = ax^3 + (3a+b)x^2 + (3a+2b+c)x + (a+b+c+d)$;
$f(x+2) = ax^3 + (6a+b)x^2 + (12a+4b+c)x + (8a+4b+2c+d)$;
$f(x+3) = ax^3 + (9a+b)x^2 + (27a+6b+c)x + (27a+9b+3c+d)$;
$f(x+4) = ax^3 + (12a+b)x^2 + (48a+8b+c)x + (64a+16b+4c+d)$;
$f(x+5) = ax^3 + (15a+b)x^2 + (75a+10b+c)x + (125a+25b+5c+d)$;
the third order differences are all equal to $6a$.

b. $f(x) = x^3 - 9x^2 + 22x - 15$

58. $-9, -3$

59. $-3 \pm \sqrt{33}$

60. $-3, 6$

61. $-2 \pm \dfrac{i\sqrt{6}}{2}$

62. $-1, 15$

63. $3 \pm \dfrac{i\sqrt{15}}{3}$

64. $(2x - 1)(4x^2 + 2x + 1)$

65. $(3x + 2)(9x^2 - 6x + 4)$

66. $8(3x + 2)(9x^2 - 6x + 4)$

67. $(2x - 5)(4x^2 + 10x + 25)$

68. $3(x - 2)(x^2 + 2x + 4)$

69. $8(x + 3)(x^2 - 3x + 9)$

70. $(3x + 10)(9x^2 - 30x + 100)$

71. $3(x + 3)(x^2 - 3x + 9)$

MIXED REVIEW

SOLVING QUADRATIC EQUATIONS Solve the equation. (Review 5.3 for 7.1)

52. $3x^2 = 6$ $\pm\sqrt{2}$

53. $16x^2 = 4$ $\pm\dfrac{1}{2}$

54. $4x^2 - 5 = 9$ $\pm\dfrac{\sqrt{14}}{2}$

55. $6x^2 + 3 = 16$ $\pm\dfrac{\sqrt{78}}{6}$

56. $-x^2 + 9 = 2x^2 - 6$ $\pm\sqrt{5}$

57. $-x^2 + 2 = x^2 + 1$ $\pm\dfrac{\sqrt{2}}{2}$

SOLVING EQUATIONS Solve the equation by completing the square. (Review 5.5)

58. $x^2 + 12x + 27 = 0$

59. $x^2 + 6x - 24 = 0$

60. $x^2 - 3x - 18 = 0$

61. $2x^2 + 8x + 11 = 0$

62. $-x^2 + 14x + 15 = 0$

63. $3x^2 - 18x + 32 = 0$

SUM OR DIFFERENCE OF CUBES Factor the polynomial. (Review 6.4)

64. $8x^3 - 1$

65. $27x^3 + 8$

66. $216x^3 + 64$

67. $8x^3 - 125$

68. $3x^3 - 24$

69. $8x^3 + 216$

70. $27x^3 + 1000$

71. $3x^3 + 81$

QUIZ 3
Self-Test for Lessons 6.7–6.9

Find all the zeros of the polynomial function. (Lesson 6.7)

1. $f(x) = 2x^3 - x^2 - 22x - 15$
   $-2.61, -0.74, 3.86$

2. $f(x) = x^3 + 3x^2 + 3x + 2$ $-2, \dfrac{-1 \pm i\sqrt{3}}{2}$

3. $f(x) = x^4 - 3x^3 - 2x^2 - 6x - 8$
   $-1, 4, \pm\sqrt{2}\,i$

4. $f(x) = 2x^4 - x^3 - 8x^2 + x + 6$
   $-\dfrac{3}{2}, -1, 1, 2$

Write a polynomial of least degree that has real coefficients, the given zeros, and a leading coefficient of 1. (Lesson 6.7)

5. $-2, -2, 2$
   $y = x^3 + 2x^2 - 4x - 8$

6. $0, 1, -3$
   $y = x^3 + 2x^2 - 3x$

7. $4, 2 + i, 2 - i$
   $y = x^3 - 8x^2 + 21x - 20$

8. $2, 5, -i$
   $y = x^4 - 7x^3 + 11x^2 - 7x + 10$

9. $4, 2 - 3i$
   $y = x^3 - 8x^2 + 29x - 52$

10. $1 - i, 2 + 2i$
    $y = x^4 - 6x^3 + 18x^2 - 24x + 16$

Graph the function. Estimate the local maximums and minimums. (Lesson 6.8)

11. $f(x) = -(x - 2)(x + 3)(x + 1)$

12. $f(x) = x(x - 1)(x + 1)(x + 2)$

13. $f(x) = 2(x - 2)(x - 3)(x - 4)$

14. $f(x) = (x + 1)(x + 3)^2$

Write a cubic function whose graph passes through the points. (Lesson 6.9)
15–18. See Margin.

15. $(-2, 0), (2, 0), (-4, 0), (-1, 3)$

16. $(-1, 0), (4, 0), (2, 0), (-3, 1)$

17. $(3, 0), (0, 0), (5, 0), (2, 6)$

18. $(1, 0), (-3, 0), (-5, 0), (-4, 10)$

11. local max $(0.79, 8.21)$, local min $(-2.12, -4.06)$

12. local max $(-0.50, 0.56)$, local min $(-1.62, -1)$, $(0.62, -1)$

13. local max $(2.42, 0.77)$, local min $(3.58, -0.77)$

14. local max $(-3, 0)$, local min $(-1.67, -1.19)$

15. $f(x) = -\dfrac{1}{3}(x + 2)(x + 4) \times (x - 2)$

16. $f(x) = -\dfrac{1}{70}(x + 1)(x - 4) \times (x - 2)$

17. $f(x) = x(x - 3)(x - 5)$

18. $f(x) = 2(x - 1)(x + 3)(x + 5)$

Find a polynomial function that models the data. (Lesson 6.9)

19.

x	1	2	3	4	5	6
f(x)	−5	−6	−1	16	51	110

$f(x) = x^3 - 3x^2 + x - 4$

20.

x	1	2	3	4	5	6
f(x)	−1	−4	−3	8	35	84

$f(x) = x^3 - 4x^2 + 2x$

21. **SOCIAL SECURITY** The table gives the number of children (in thousands) receiving Social Security for each year from 1988 to 1995. Use a graphing calculator to find a polynomial model for the data. (Lesson 6.9)

$N = -3.75x^3 + 50.9x^2 - 97.3x + 3210$ where x is the number of years since 1988

Year	1988	1989	1990	1991	1992	1993	1994	1995
Number of children	3204	3165	3187	3268	3391	3527	3654	3734

Chapter Summary

WHAT did you learn?

Use properties of exponents to evaluate and simplify expressions. **(6.1)**

Evaluate polynomial functions using direct or synthetic substitution. **(6.2)**

Sketch and analyze graphs of polynomial functions. **(6.2, 6.8)**

Add, subtract, and multiply polynomials. **(6.3)**

Factor polynomial expressions. **(6.4)**

Solve polynomial equations. **(6.4)**

Divide polynomials using long division or synthetic division. **(6.5)**

Find zeros of polynomial functions. **(6.6, 6.7)**

Use finite differences and cubic regression to find polynomial models for data. **(6.9)**

Use polynomials to solve real-life problems. **(6.1–6.9)**

WHY did you learn it?

Use scientific notation to find the ratio of a state's park space to its total area. **(p. 328)**

Estimate the amount of prize money awarded at a tennis tournament. **(p. 335)**

Find maximum or minimum values of a function such as oranges consumed in the U.S. **(p. 377)**

Write a polynomial model for the power needed to move a bicycle at a certain speed. **(p. 342)**

Find the dimensions of a block discovered by archeologists. **(p. 347)**

Find the dimensions of a sculpture. **(p. 350)**

Write a function for the average annual amount of money spent per person at the movies. **(p. 358)**

Find dimensions for a candle-wax model of the Louvre pyramid. **(p. 361)**

Write and use a polynomial model for the speed of a space shuttle. **(p. 385)**

Find the maximum volume and dimensions of a box made from a piece of cardboard. **(p. 375)**

How does Chapter 6 fit into the BIGGER PICTURE of algebra?

Chapter 6 contains the fundamental theorem of algebra. Finding the solutions of a polynomial equation is the most classic problem in all of algebra. It is equivalent to finding the zeros of a polynomial function. Real-life situations have been modeled by polynomial functions for hundreds of years.

STUDY STRATEGY

How did you make and use a flow chart?

Here is a flow chart for finding all the zeros of a polynomial function, following the **Study Strategy** on page 322.

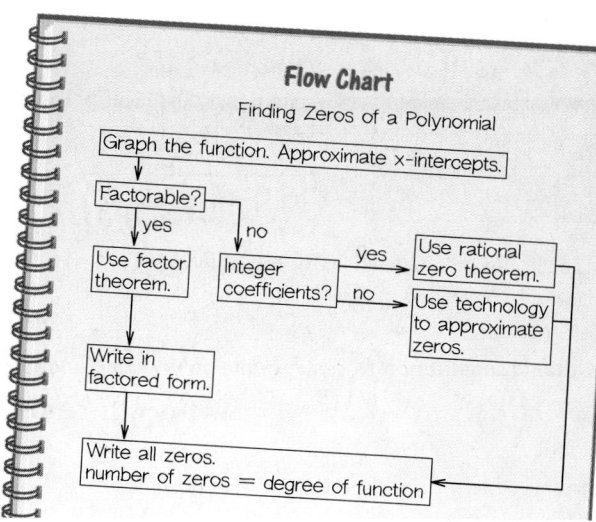

Flow Chart
Finding Zeros of a Polynomial

Graph the function. Approximate x-intercepts.

Factorable?
yes → Use factor theorem.
no → Integer coefficients?
  yes → Use rational zero theorem.
  no → Use technology to approximate zeros.

Write in factored form.

Write all zeros. number of zeros = degree of function

387

Chapter Review

ADDITIONAL RESOURCES

The following resources are available to help review the material in this chapter.

- Chapter Review Games and Activities (*Chapter 6 Resource Book,* p. 129)
- *Instant Replay: Video Review Games*
- ⊞ *Personal Student Tutor*
- Cumulative Review, Chs. 1–6 (*Chapter 5 Resource Book,* p. 141)

1. $\dfrac{96x^3}{y^3}$; negative exponent, power of a quotient, power of a product, and power of a power properties

2. 1; negative exponent, product of powers, power of a power, and zero exponent properties

3. $-\dfrac{7}{2}x^3y^6$; quotient of powers property

4. $\dfrac{1}{5}y$; negative exponent, quotient of powers, and zero exponent properties

---

VOCABULARY

- scientific notation, p. 325
- polynomial function, p. 329
- leading coefficient, p. 329
- constant term, p. 329
- degree of a polynomial function, p. 329
- standard form of a polynomial function, p. 329

- synthetic substitution, p. 330
- end behavior, p. 331
- factor by grouping, p. 346
- quadratic form, p. 346
- polynomial long division, p. 352
- remainder theorem, p. 353
- synthetic division, p. 353

- factor theorem, p. 354
- rational zero theorem, p. 359
- fundamental theorem of algebra, p. 366
- repeated solution, p. 366
- local maximum, p. 374
- local minimum, p. 374
- finite differences, p. 380

6.1 USING PROPERTIES OF EXPONENTS

Examples on pp. 323–325

> **EXAMPLE** You can use properties of exponents to evaluate numerical expressions and to simplify algebraic expressions.
>
> $$\frac{(3x^2y)^5}{9x^{10}y^6} = \frac{3^5x^{2\cdot5}y^5}{9x^{10}y^6} = \frac{243}{9}x^{10-10}y^{5-6} = 27x^0y^{-1} = \frac{27}{y} \qquad \text{all positive exponents}$$

Simplify the expression. Tell which properties of exponents you used. 1–4. See margin.

1. $\left(\dfrac{2}{3}\right)^2 \cdot \left(6xy^{-1}\right)^3$ 2. $x^4\left(x^{-5}x^3\right)^2$ 3. $\dfrac{-63xy^9}{18x^{-2}y^3}$ 4. $\dfrac{5x^2}{y^{-2}} \cdot \dfrac{1}{25x^2y}$

6.2 EVALUATING AND GRAPHING POLYNOMIAL FUNCTIONS

Examples on pp. 329–332

> **EXAMPLES** Use direct or synthetic substitution to evaluate a polynomial function.
>
> Evaluate $f(x) = x^3 - 2x - 1$ when $x = 3$ (synthetic substitution):
>
> ```
> 3 |  1   0   -2   -1
>   |      3    9   21
>   ----------------------
>      1   3    7   20  ← f(3) = 20
> ```
>
> To graph, make a table of values, plot points, and identify end behavior.

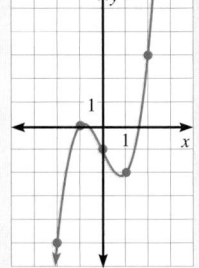

x	−3	−2	−1	0	1	2	3
f(x)	−22	−5	0	−1	−2	3	20

> The leading coefficient is positive and the degree is odd, so
> $f(x) \to -\infty$ as $x \to -\infty$ and $f(x) \to +\infty$ as $x \to +\infty$.

Use synthetic substitution to evaluate the polynomial function for the given value of x.

5. $f(x) = x^3 + 3x^2 - 12x + 7, x = 3$ 25   6. $f(x) = x^4 - 5x^3 - 3x^2 + x - 5, x = -1$ −3

Graph the polynomial function. 7–9. See margin.

7. $f(x) = -x^3 + 2$ **8.** $f(x) = x^4 - 3$ **9.** $f(x) = x^3 - 4x + 1$

7.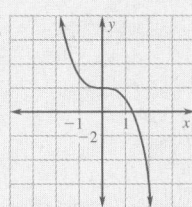

8.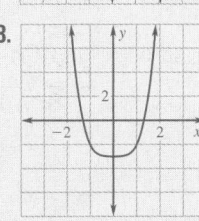

9.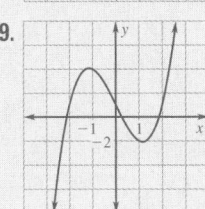

6.3 ADDING, SUBTRACTING, AND MULTIPLYING POLYNOMIALS

Examples on pp. 338–340

EXAMPLES You can add, subtract, or multiply polynomials.

$$\begin{array}{r} 4x^3 + 2x^2 + 1 \\ - (x^2 + x - 5) \\ \hline 4x^3 + x^2 - x + 6 \end{array}$$

$$\begin{aligned} (x-3)(x^2 + 5x - 1) &= (x-3)(x^2) + (x-3)(5x) + (x-3)(-1) \\ &= x^3 - 3x^2 + 5x^2 - 15x - x + 3 \\ &= x^3 + 2x^2 - 16x + 3 \end{aligned}$$

Perform the indicated operation.

10. $(3x^3 + x^2 + 1) - (x^3 + 3)$
$2x^3 + x^2 - 2$

11. $(x-3)(x^2 + x - 7)$
$x^3 - 2x^2 - 10x + 21$

12. $(x+3)(x-5)(2x+1)$
$2x^3 - 3x^2 - 32x - 15$

6.4 FACTORING AND SOLVING POLYNOMIAL EQUATIONS

Examples on pp. 345–347

EXAMPLES You can solve some polynomial equations by factoring.

Factor $8x^3 - 125$.

$$\begin{aligned} 8x^3 - 125 &= (2x)^3 - 5^3 \\ &= (2x - 5)\big((2x)^2 + (2x \cdot 5) + 5^2\big) \\ &= (2x - 5)(4x^2 + 10x + 25) \end{aligned}$$

Solve $x^3 - 3x^2 - 5x + 15 = 0$.

$$\begin{aligned} x^2(x-3) - 5(x-3) &= 0 \\ (x-3)(x^2 - 5) &= 0 \\ x = 3 \text{ or } x &= \pm\sqrt{5} \end{aligned}$$

Find the real-number solutions of the equation.

13. $x^3 + 64 = 0$ -4

14. $x^4 - 6x^2 = 27$ $-3, 3$

15. $x^3 + 3x^2 - x - 3 = 0$
$-3, -1, 1$

6.5 THE REMAINDER AND FACTOR THEOREMS

Examples on pp. 352–355

EXAMPLES You can use polynomial long division, and in some cases synthetic division, to divide polynomials.

$$\begin{array}{r} x^2 - 7x + 6 \\ x + 9 \overline{)x^3 + 2x^2 - 57x + 54} \\ \underline{x^3 + 9x^2} \\ -7x^2 - 57x \\ \underline{-7x^2 - 63x} \\ 6x + 54 \\ \underline{6x + 54} \\ 0 \end{array}$$

$$\frac{x^3 + 2x^2 - 57x + 54}{x + 9} = x^2 - 7x + 6$$

Divide $3x^3 + 2x^2 - x + 4$ by $x + 5$.

$$\begin{array}{r|rrrr} -5 & 3 & 2 & -1 & 4 \\ & & -15 & 65 & -320 \\ \hline & 3 & -13 & 64 & -316 \end{array}$$

$$\frac{3x^3 + 2x^2 - x + 4}{x + 5} = 3x^2 - 13x + 64 + \frac{-316}{x + 5}$$

Divide. Use synthetic division if possible.

16. $(x^4 + 5x^3 - x^2 - 3x - 1) \div (x - 1)$
$x^3 + 6x^2 + 5x + 2 + \dfrac{1}{x - 1}$

17. $(2x^3 - 5x^2 + 5x + 4) \div (2x - 5)$
$x^2 + \dfrac{5}{2} + \dfrac{33}{2(2x - 5)}$

Chapter Review

20.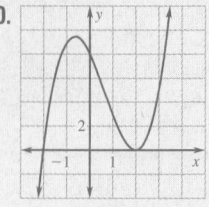

;

x-intercepts: −2, 2;
local max: (−0.67, 9.48);
local min: (2, 0)

21.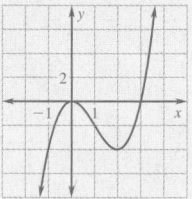

;

x-intercepts: 0, 3;
local max: (0, 0);
local min: (2, −4)

22.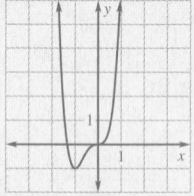

;

x-intercepts: $-\frac{4}{3}$, 0;

local max: none;
local min: (−1, −1)

23.
$f(1)\ f(2)\ f(3)\ f(4)\ f(5)\ f(6)$

2　　9　　28　　65　　126　　217

7　　19　　37　　61　　91

12　　18　　24　　30

6　　6　　6

6.6–6.7

Examples on pp. 359–361
and pp. 366–368

FINDING ZEROS OF POLYNOMIAL FUNCTIONS

EXAMPLE You can use the rational zero theorem and the fundamental theorem of algebra to find all the zeros of a polynomial function.

$f(x) = x^4 + 3x^3 - 5x^2 - 21x + 22$　　Possible rational zeros: $\dfrac{\pm 1,\ \pm 2,\ \pm 11,\ \pm 22}{1}$

Using synthetic division, you can find that the rational zeros are 1 and 2. The degree of f is 4, so f has 4 zeros. To find the other two zeros, write in factored form: $f(x) = (x - 1)(x - 2)(x^2 + 6x + 11)$. Solve $x^2 + 6x + 11 = 0$: $x = -3 \pm \sqrt{2}\,i$. So the zeros of $f(x) = x^4 + 3x^3 - 5x^2 - 21x + 22$ are $1, 2, -3 + \sqrt{2}\,i, -3 - \sqrt{2}\,i$.

Find all the real zeros of the function.

18. $f(x) = x^3 + 12x^2 + 21x + 10$　$-10, -1$

19. $f(x) = x^4 + x^3 - x^2 + x - 2$　$-2, 1$

6.8

Examples on
pp. 373–375

ANALYZING GRAPHS OF POLYNOMIAL FUNCTIONS

EXAMPLE You can identify *x*-intercepts and turning points when you analyze the graph of a polynomial function.

The graph of $f(x) = 3x^3 - 9x + 6$ has

- two *x*-intercepts, −2 and 1.
- a local maximum at $(-1, 12)$.
- a local minimum at $(1, 0)$.

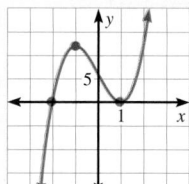

Graph the polynomial function. Identify the *x*-intercepts and the points where the local maximums and local minimums occur. 20–22. See margin.

20. $f(x) = (x - 2)^2(x + 2)$　　**21.** $f(x) = x^3 - 3x^2$　　**22.** $f(x) = 3x^4 + 4x^3$

6.9

Examples on
pp. 380–382

MODELING WITH POLYNOMIALS

EXAMPLE Sometimes you can use finite differences or cubic regression to find a polynomial model for a set of data.

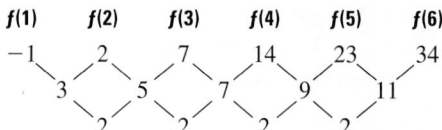

$f(1)$	$f(2)$	$f(3)$	$f(4)$	$f(5)$	$f(6)$	
−1	2	7	14	23	34	function values
	3	5	7	9	11	first-order differences
		2	2	2	2	second-order differences

Since second-order differences are nonzero and constant, the data set can be modeled by a polynomial function of degree 2. The function is $f(x) = x^2 - 2$.

23. Show that the third-order differences for the function $f(n) = n^3 + 1$ are nonzero and constant. **See margin.**

24. Write a cubic function whose graph passes through points $(1, 0), (-1, 0), (4, 0)$, and $(2, -12)$. Use cubic regression on a graphing calculator to verify your answer.
$f(x) = 2(x - 4)(x - 1)(x + 1) = 2x^3 - 8x^2 - 2x + 8$

Chapter Test

ADDITIONAL RESOURCES
- **Chapter 6 Resource Book**
  Chapter Test (3 levels) (p. 130)
  SAT/ACT Chapter Test (p. 136)
  Alternative Assessment (p. 137)
- 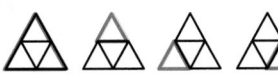 **Test and Practice Generator**

Simplify the expression. Tell which properties of exponents you used. 1–5. See margin.

1. $x^7 \cdot \dfrac{1}{x^2}$ **2.** $(3^2 x^6)^3$ **3.** $\dfrac{x^9}{x^{-2}}$ **4.** $(8x^3 y^2)^{-3}$ **5.** $\dfrac{15x^2 y}{6x^4 y^5} \cdot \dfrac{6x^3 y^2}{5xy}$

Describe the end behavior of the graph of the polynomial function. Then evaluate the function for $x = -4, -3, -2, \ldots, 4$. Then graph the function. 6–8. See margin.

6. $y = x^4 - 2x^2 - x - 1$ **7.** $y = -3x^3 - 6x^2$ **8.** $y = (x - 3)(x + 1)(x + 2)$

Perform the indicated operation.

9. $(3x^2 - 5x + 7) - (2x^2 + 9x - 1)$ **10.** $(2x - 3)(5x^2 - x + 6)$ **11.** $(x - 4)(x + 1)(x + 3)$ $x^3 - 13x - 12$
$x^2 - 14x + 8$ $10x^3 - 17x^2 + 15x - 18$

Factor the polynomial.

12. $64x^3 + 343$ **13.** $400x^2 - 25$ **14.** $x^4 + 8x^2 - 9$ **15.** $2x^3 - 3x^2 + 4x - 6$
$(4x + 7)(16x^2 - 28x + 49)$ $25(4x + 1)(4x - 1)$ $(x^2 + 9)(x + 1)(x - 1)$ $(2x - 3)(x^2 + 2)$

Solve the equation.

16. $3x^4 - 11x^2 - 20 = 0$ $\pm\dfrac{2\sqrt{3}}{3}i, \pm\sqrt{5}$ **17.** $81x^4 = 16$ $\pm\dfrac{2}{3}, \pm\dfrac{2}{3}i$ **18.** $4x^3 - 8x^2 - x + 2 = 0$ $-\dfrac{1}{2}, \dfrac{1}{2}, 2$

Divide. Use synthetic division if possible. **19.** $8x^3 - 3x^2 + 7x - 8 + \dfrac{15}{x + 1}$

19. $(8x^4 + 5x^3 + 4x^2 - x + 7) \div (x + 1)$ **20.** $(12x^3 + 31x^2 - 17x - 6) \div (x + 3)$ $12x^2 - 5x - 2$

List all the possible rational zeros of f using the rational zero theorem. Then find all the zeros of the function. $\pm1, \pm2, \pm3, \pm4, \pm6, \pm9, \pm12, \pm18, \pm36; -4, \pm3i$

21. $f(x) = x^3 - 5x^2 - 14x$ **22.** $f(x) = x^3 + 4x^2 + 9x + 36$ **23.** $f(x) = x^4 + x^3 - 2x^2 + 4x - 24$
$0, \pm1, \pm2, \pm7, \pm14; -2, 0, 7$ $\pm1, \pm2, \pm3, \pm4, \pm6, \pm8, \pm12, \pm24; -3, 2, \pm2i$

Write a polynomial function of least degree that has real coefficients, the given zeros, and a leading coefficient of 1. $f(x) = x^4 - 3x^3 + 4x$ $f(x) = x^3 - 5x^2 + 4x - 20$

24. $1, -3, 4$ **25.** $2, 2, -1, 0$ **26.** $5, 2i, -2i$ **27.** $3, -3, 2 - i$
$f(x) = x^3 - 2x^2 - 11x + 12$ $f(x) = x^4 - 4x^3 - 4x^2 + 36x - 45$

28. Use technology to approximate the real zeros of $f(x) = 0.25x^3 - 7x^2 + 15$.
about -1.428, about 1.505, about 27.923

29. Identify the x-intercepts, local maximum, and local minimum of the graph of
$f(x) = \dfrac{1}{9}(x - 3)^2(x + 3)^2$. Then describe the end behavior of the graph. See margin.

30. Show that $f(x) = x^4 - 2x + 8$ has nonzero constant fourth-order differences. See margin.

31. The table gives the number of triangles that point upward that you can find in a large triangle that is n units on a side and divided into triangles that are each one unit on a side. Find a polynomial model for $f(n)$.

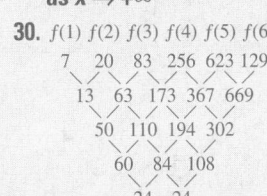

$f(2) = 4$

n	1	2	3	4	5	6	7
$f(n)$	1	4	10	20	35	56	84

$f(n) = \dfrac{1}{6}n^3 + \dfrac{1}{2}n^2 + \dfrac{1}{3}n$

32. **CELLS** An adult human body contains about 75,000,000,000,000 cells. Each is about 0.001 inch wide. If the cells were laid end to end to form a chain, about how long would the chain be in miles? Give your answer in scientific notation. about 1.1837×10^6 mi

1. x^5; quotient of powers property

2. $729x^{18}$; power of a product and power of a power properties

3. x^{11}; quotient of powers property

4. $\dfrac{1}{512x^9 y^6}$; power of a power, power of a product, and negative exponent properties

5. $\dfrac{3}{y^3}$; product of a power, quotient of a power, zero exponent, and negative exponent properties

6–8. See Additional Answers beginning on page AA1.

29. x-intercepts: $-3, 3$; local max at $(0, 9)$; local min at $(-3, 0)$ and $(3, 0)$; $f(x) \to +\infty$ as $x \to -\infty$, $f(x) \to +\infty$ as $x \to +\infty$

30. $f(1)\ f(2)\ f(3)\ f(4)\ f(5)\ f(6)$
$\quad 7\quad 20\quad 83\quad 256\quad 623\quad 1292$
$\qquad 13\quad 63\quad 173\quad 367\quad 669$
$\qquad\quad 50\quad 110\quad 194\quad 302$
$\qquad\qquad 60\quad 84\quad 108$
$\qquad\qquad\quad 24\quad 24$

ADDITIONAL RESOURCES

• *Chapter 6 Resource Book*
  Chapter Test (3 levels) (p. 130)
  SAT/ACT Chapter Test (p. 136)
  Alternative Assessment (p. 137)

• 🖳 *Test and Practice Generator*

CHAPTER
6

Chapter Standardized Test

▶ **TEST-TAKING STRATEGY** The mathematical portion of the SAT is based on concepts and skills taught in high school mathematics courses. The best way to prepare for the SAT is to keep up with your day-to-day studies.

1. MULTIPLE CHOICE What is the value of -4^0? **D**

Ⓐ 4   Ⓑ 1   Ⓒ 0

Ⓓ -1 Ⓔ -4

2. MULTIPLE CHOICE What is the value of $f(x) = 7x^4 - 3x^3 + 8x^2 + x - 9$ when $x = -1$? **A**

Ⓐ 8   Ⓑ 4   Ⓒ 2

Ⓓ -8 Ⓔ -14

3. MULTIPLE CHOICE Which statement about the end behavior of the graph of $f(x) = x^4 + 1$ is true? **A**

Ⓐ $f(x) \to +\infty$ as $x \to -\infty$.

Ⓑ $f(x) \to +\infty$ as $x \to 0$.

Ⓒ $f(x) \to -\infty$ as $x \to -\infty$.

Ⓓ $f(x) \to -\infty$ as $x \to 0$.

Ⓔ $f(x) \to -\infty$ as $x \to +\infty$.

4. MULTIPLE CHOICE For 1992 through 1995, the number of grocery stores in the United States can be modeled by $G = 0.03t^2 - 1.5t + 171$, where G is the number of stores in thousands and t is the number of years since 1990. The average sales per grocery store can be modeled by $S = 4.7t^2 + 49.1t + 2009$, where S is sales in thousands of dollars. What were the approximate total sales in millions of dollars for grocery stores in the United States in 1994? **C**

Ⓐ 3.8×10^{-1} Ⓑ 3.8×10^1

Ⓒ 3.8×10^5 Ⓓ 3.8×10^8

Ⓔ 3.8×10^{11}

5. MULTIPLE CHOICE Which polynomial has the factorization $(2x + 1)(4x^2 - 2x + 1)$? **E**

Ⓐ $2x^3 - 1$ Ⓑ $8x^3 - 1$

Ⓒ $2x^3 + 1$ Ⓓ $4x^3 + 1$

Ⓔ $8x^3 + 1$

6. MULTIPLE CHOICE What are all the *real* solutions of the equation $x^5 = 256x$? **A**

Ⓐ $0, \pm 4$ Ⓑ $4, -4$ Ⓒ $\pm 4, \pm 4i$

Ⓓ $0, \pm 4i$ Ⓔ $0, \pm 4, \pm 4i$

7. MULTIPLE CHOICE What is the quotient of $(4x^3 - 11x^2 - 9x - 5) \div (x - 4)$? **B**

Ⓐ $4x^3 + 5x^2 + 11x + 39$

Ⓑ $4x^2 + 5x + 11 + \dfrac{39}{x - 4}$

Ⓒ $4x^2 + 5x + 11 + \dfrac{39}{4x^3 - 11x^2 - 9x - 5}$

Ⓓ $4x^2 - 27x + 99 - \dfrac{401}{x - 4}$

Ⓔ $4x^2 - 27x + 99 - \dfrac{401}{x + 4}$

8. MULTIPLE CHOICE What are all the rational zeros of $f(x) = x^3 - 8x^2 + x + 42$? **E**

Ⓐ $-2, -3, -7$ Ⓑ $2, 3, 7$

Ⓒ $2, -3, -7$ Ⓓ $0, 6, 7$

Ⓔ $-2, 3, 7$

9. MULTIPLE CHOICE How many zeros does the function $f(x) = -3x^4 + x + 2$ have? **E**

Ⓐ 0   Ⓑ 1   Ⓒ 2

Ⓓ 3   Ⓔ 4

10. MULTIPLE CHOICE Which function is graphed? **D**

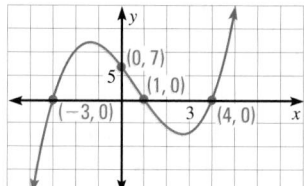

Ⓐ $f(x) = (x + 3)(x - 1)(x - 4)$

Ⓑ $f(x) = 7(x + 3)(x - 1)(x - 4)$

Ⓒ $f(x) = \dfrac{7}{12}(x - 3)(x + 1)(x + 4)$

Ⓓ $f(x) = \dfrac{7}{12}(x + 3)(x - 1)(x - 4)$

Ⓔ $f(x) = -\dfrac{7}{12}(x + 3)(x - 1)(x - 4)$

QUANTITATIVE COMPARISON In Exercises 11 and 12, choose the statement that is true about the given quantities.

   Ⓐ The quantity in column A is greater.

   Ⓑ The quantity in column B is greater.

   Ⓒ The two quantities are equal.

   Ⓓ The relationship cannot be determined from the given information.

	Column A	Column B	
11.	x^{-2}	x^2	D
12.	Degree of $f(x) = x^4 - 7x + 13$	Degree of $f(x) = 4x^3 + 2x^2 - x + 1$	A

13. MULTI-STEP PROBLEM You are designing a monument for the city park. The monument is to be a rectangular prism with dimensions $x + 1$ feet, $x - 5$ feet, and $x - 6$ feet.

  a. Write a function $f(x)$ for the volume of the monument. $f(x) = x^3 - 10x^2 + 19x + 30$

  b. Use a graphing calculator to graph $f(x)$ for $-10 \leq x \leq 20$.   See margin.

  c. *Writing* Look back at your graph from part (b). Identify the local maximums and local minimums. Do these values represent maximum and minimum possible volumes of the monument? Explain.

      13c. no; The local maximum occurs at about (1.15, 40.15) and the local minimum occurs at about (5.52, −1.63), but x must be greater than 6 for the side of length $x - 6$ to have a positive measure.

  d. If the volume of the monument is to be 220 cubic feet, what will the dimensions be?   11 ft by 5 ft by 4 ft

14. MULTI-STEP PROBLEM The numbers in the table give the volumes of the first six prisms in a sequence.

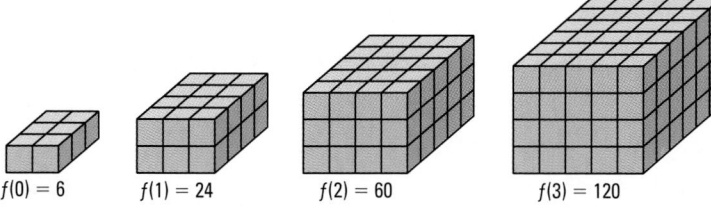

$f(0) = 6$ $f(1) = 24$ $f(2) = 60$ $f(3) = 120$

Prism (n)	0	1	2	3	4	5
Volume, $f(n)$	6	24	60	120	210	336

  a. Use finite differences to determine the degree of f.   3

  b. Use a system of equations to find a polynomial model for $f(n)$ in standard form. $f(n) = n^3 + 6n^2 + 11n + 6$

  c. *Writing* Factor the polynomial. Explain how the factors are related to the dimensions of the prism. $f(n) = (n + 1)(n + 2)(n + 3)$; for prism n, the dimensions are $(n + 1)$ by $(n + 2)$ by $(n + 3)$.

  d. Use your model to find the volume of the 50th prism in the sequence.   132,600

  e. Sketch a graph of your model and label the points that represent the first six prisms. What is the domain of the function?

    See margin for graph. The domain is all whole numbers.

13b.

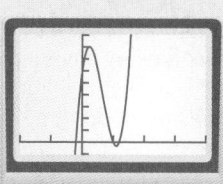

14e.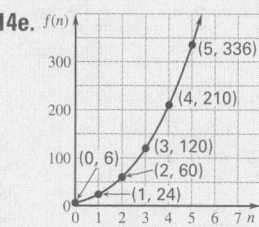

Chapter Standardized Test   **393**

394

ADDITIONAL RESOURCES
A Cumulative Review covering Chapter 1–6 is available in the *Chapter 6 Resource Book*, p. 141

5.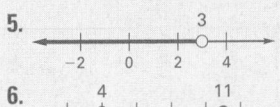

6.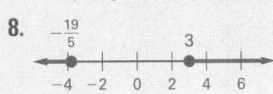

7.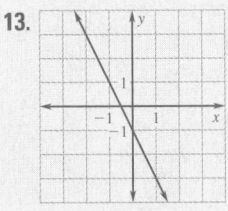

8.

13.

14.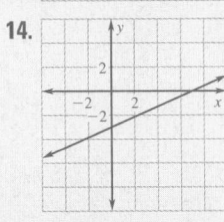

15.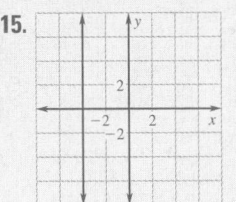

Solve the equation. (1.3, 1.7)

1. $5x + 4 = -21$ −5

2. $3(2x + 5) = 69$ 9

3. $|x - 2| = 6$ −4, 8

4. $|7 - 3x| = 23$ $-\frac{16}{3}$, 10

Solve the inequality. Then graph your solution. (1.6, 1.7) 5–8. See margin for graphs.

5. $10 - 4x > -2$ $x < 3$

6. $0 \le 2x - 8 \le 14$ $4 \le x \le 11$

7. $|x - 3| < 5$ $-2 < x < 8$

8. $|5x + 2| \ge 17$ $x \le -\frac{19}{5}$ or $x \ge 3$

Find the slope of the line passing through the given points. (2.2)

9. $(4, 1), (-2, 1)$ 0

10. $(-3, 0), (0, 2)$ $\frac{2}{3}$

11. $(-1, -5), (2, 7)$ 4

12. $(-4, 4), (1, -3)$ $-\frac{7}{5}$

Graph the equation or inequality. (2.3, 2.6–2.8) 13–21. See margin.

13. $y = -2x - 1$

14. $3x - 7y = 21$

15. $x = -4$

16. $y > \frac{3}{2}x + 2$

17. $2x + 6y \le 12$

18. $y = |x| + 3$

19. $y = -2|x + 4| - 1$

20. $f(x) = \begin{cases} -2, & \text{if } x \le 0 \\ 3, & \text{if } x > 0 \end{cases}$

21. $f(x) = \begin{cases} -x, & \text{if } x < 1 \\ x - 2, & \text{if } x \ge 1 \end{cases}$

Write an equation of the line with the given characteristics. (2.4)

22. slope: 3, y-intercept: −2 $y = 3x - 2$

23. points on line: $(-1, 9), (1, 1)$ $y = -4x + 5$

24. vertical line through $(-8, 6)$ $x = -8$

Solve the system of linear equations using any method. (3.1, 3.2, 3.6, 4.3, 4.5)

25. $x + y = 8$ (3, 5)
$2x - y = 1$

26. $3x - 4y = 5$ $\left(1, -\frac{1}{2}\right)$
$2x + 2y = 1$

27. $x + y + z = 4$ (1, 0, 3)
$x - 4y + 3z = 10$
$-4x + y + z = -1$

28. $x - y + z = 1$ (−2, −2, 1)
$-x + y + 2z = 2$
$x + y + z = -3$

Graph the ordered triple or equation in a three-dimensional coordinate system. (3.5) 29–32. See margin for graphs.

29. $(-1, -3, 0)$

30. $(2, 4, -2)$

31. $4x + 2y + z = 4$

32. $5x + 5y + 2z = 10$

Perform the indicated operation. (4.1, 4.2)

33. $\begin{bmatrix} -3 & 7 \\ 4 & -2 \end{bmatrix} + \begin{bmatrix} -8 & 1 \\ -3 & 0 \end{bmatrix}$ See margin.

34. $-6 \begin{bmatrix} 2 & 3 \\ -4 & -2 \\ -5 & -1 \end{bmatrix}$ $\begin{bmatrix} -12 & -18 \\ 24 & 12 \\ 30 & 6 \end{bmatrix}$

35. $\begin{bmatrix} 1 & -5 \\ 6 & 3 \end{bmatrix} \begin{bmatrix} 2 & -2 & 8 \\ -3 & 1 & 7 \end{bmatrix}$ See margin.

Evaluate the determinant of the matrix. (4.3)

36. $\begin{bmatrix} -5 & 2 \\ 4 & -2 \end{bmatrix}$ 2

37. $\begin{bmatrix} 0 & 1 \\ -3 & 6 \end{bmatrix}$ 3

38. $\begin{bmatrix} 3 & 9 & 1 \\ -5 & 1 & 2 \\ -2 & 4 & 8 \end{bmatrix}$ 306

39. $\begin{bmatrix} -1 & 0 & 1 \\ 3 & 7 & -2 \\ 8 & 1 & 0 \end{bmatrix}$ −55

Find the inverse of the matrix. (4.4)

40. $\begin{bmatrix} -5 & 2 \\ -7 & 3 \end{bmatrix}$ $\begin{bmatrix} -3 & 2 \\ -7 & 5 \end{bmatrix}$

41. $\begin{bmatrix} -1 & -2 \\ 4 & 7 \end{bmatrix}$ $\begin{bmatrix} 7 & 2 \\ -4 & -1 \end{bmatrix}$

42. $\begin{bmatrix} 4 & 9 \\ 2 & 4 \end{bmatrix}$ $\begin{bmatrix} -2 & \frac{9}{2} \\ 1 & -2 \end{bmatrix}$

43. $\begin{bmatrix} 4 & -2 \\ -2 & 1 \end{bmatrix}$ no inverse

Graph the equation or inequality. (5.1, 5.7, 6.2, 6.8) 44–52. See margin for graphs.

44. $y = x^2 + 8x + 16$

45. $y = -(x - 1)^2 + 3$

46. $y = 2(x + 1)(x - 3)$

47. $y \le \frac{1}{4}x^2 - 3$

48. $y < -2x^2 + 4x + 5$

49. $y = x^3 - 4x^2 + x + 7$

50. $y = -3x^4 + 9x^2 - 2$

51. $y = -(x + 2)(x - 1)(x - 2)$

52. $y = 2x^2(x - 3)^2$

Solve the equation or inequality. (5.2–5.7, 6.4)

53. $3x^2 - 7 = 2(x^2 + 3)$ $\pm\sqrt{13}$ **54.** $4x^2 + 12x + 9 = 0$ $-\frac{3}{2}$ **55.** $x^2 + 64 = 0$ $\pm 8i$

56. $x^2 + 4x = 4$ $-2 \pm 2\sqrt{2}$ **57.** $100 - x^2 \geq 0$ $-10 \leq x \leq 10$ **58.** $x^2 - 6 > -5x$ $x < -6$ or $x > 1$

59. $x^4 - 5x^2 + 4 = 0$ $\pm 2, \pm 1$ **60.** $3x^4 - 15x^3 = 0$ $0, 5$ **61.** $2x^3 + 4x^2 - 3x - 6 = 0$ $-2, \pm\frac{\sqrt{6}}{2}$

Write the expression as a complex number in standard form. (5.4)

62. $\dfrac{7 + 3i}{4 - i}$ $\dfrac{25}{17} + \dfrac{19}{17}i$ **63.** $4i(5 - 8i)$ $32 + 20i$ **64.** $(9 + 5i)(9 - 5i)$ 106 **65.** $(6 - 2i) - (-3 - 4i)$
                                                                                                               $9 + 2i$

Write a quadratic function in the specified form whose graph has the given characteristics. (5.8)

66. vertex form
vertex: (5, 3)
point on graph: (7, 11)
$y = 2(x - 5)^2 + 3$

67. intercept form
x-intercepts: $-3, -2$
point on graph: (0, -6)
$y = -(x + 3)(x + 2)$

68. standard form
points on graph:
(1, 4), (3, -4), (6, -61)
$y = -3x^2 + 8x - 1$

Simplify the expression. (6.1)

69. $(6xy^3)^2$ $36x^2y^6$ **70.** $7x^{-10}y^4$ $\dfrac{7y^4}{x^{10}}$ **71.** $\left(\dfrac{5}{4}\right)^{-2}$ $\dfrac{16}{25}$ **72.** $\dfrac{3x^2y^{-1}}{2x} \cdot \dfrac{10x^2y}{3y^{-3}}$ $5x^3y^3$

Perform the indicated operation. (6.3, 6.5)

73. $(x - 3)(x^3 - 2x^2 + 5x - 12)$
$x^4 - 5x^3 + 11x^2 - 27x + 36$

74. $(7x^3 - 9x + 2) + (5x^3 + 9x)$
$12x^3 + 2$

75. $(x^4 - 3x^3 + 8x^2 - 2) \div (x + 2)$
$x^3 - 5x^2 + 18x - 36 + \dfrac{70}{x + 2}$

Find all the zeros of the function. (6.6, 6.7)

76. $f(x) = 2x^3 - 5x^2 - 4x + 3$ $-1, \frac{1}{2}, 3$ **77.** $f(x) = x^4 - 25$ $\pm\sqrt{5}, \pm\sqrt{5}\,i$ **78.** $f(x) = x^3 + 11x^2 + x + 11$ $-11, \pm i$

Write a cubic function whose graph passes through the given points. (6.9)

79. $(-4, 0), (-1, 0), (1, 0), (-2, 6)$
$f(x) = (x + 4)(x + 1)(x - 1)$

80. $(-6, 0), (0, 0), (3, 0), (6, -144)$
$f(x) = -\frac{2}{3}x(x + 6)(x - 3)$

81. 🌐 **SIMPLE INTEREST** The formula for simple interest is $I = Prt$. Solve the
formula for r. Then find the annual interest rate if a \$1000 deposit earns \$165 of
simple interest in 3 years. (1.4) $r = \dfrac{I}{Pt}$, 5.5%

82. 🌐 **COST OF BREAD** The table gives the number of one-pound loaves of bread
you could buy for \$1.00 in the United States for various years since 1900. Make
a scatter plot of the data and describe the correlation shown. (2.5) negative correlation; See margin.

Years since 1900, t	13	30	50	70	90	97
Loaves of bread, b	17.8	11.6	6.9	4.1	1.4	1.1

🌐 **DATA UPDATE** of Bureau of Labor Statistics data at www.mcdougallittell.com

83. 🌐 **PHONE RATES** A long distance carrier charges a flat rate of \$.09 per
minute for telephone calls. A second carrier charges \$.30 for the first minute and
\$.06 for each additional minute. After how many minutes will the second carrier
be less expensive than the first carrier? (3.2) 8 min

84. 🌐 **CRYPTOGRAMS** Use the matrix $A = \begin{bmatrix} -2 & 5 \\ 1 & 8 \end{bmatrix}$ and the coding
information on page 225 to encode the message EXIT NOW. (4.4) 14, 217, 2, 205, 14, 112, -7, 259

85. SCIENCE CONNECTION   Pluto is about 3,660,000,000 mi from the sun. Light
travels through space at a speed of about 671,000,000 mi/h. Use scientific
notation to find how long it takes light from the sun to reach Pluto. (6.1) about 5.45 h

16.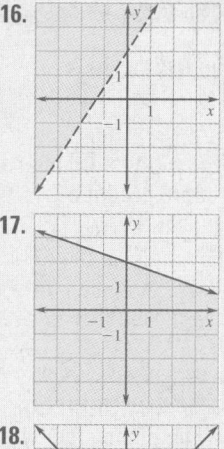

17.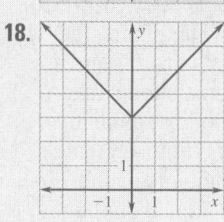

18.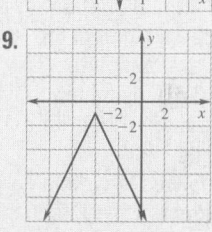

19.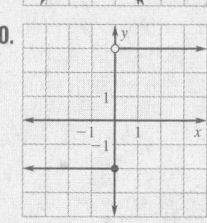

20.

21.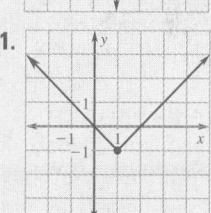

29–33, 35, 44–52, 82.
See Additional Answers beginning on page AA1.

Magic Squares

MATHEMATICAL GOALS

- Make a 3 × 3 and a 4 × 4 magic square.
- Determine which transformations of a magic square produce another magic square.
- Use quadratic, cubic, and quartic functions to find the sum of the entries in a magic square.

MANAGING THE PROJECT
CLASSROOM MANAGEMENT

This project may be completed by individuals or in pairs. If pairs are used, the students should take turns suggesting row, column, or diagonal entries for the magic square. Both students should check the results and should continue guessing and checking until the magic squares are complete. Students should discuss their answers to Questions 1–9 before recording results.

If the pair chooses to write a report, both students can write a draft and collaborate to create the finished report. Each student should include a copy of the report in his or her portfolio.

If the pair chooses to make a poster, they should discuss the elements of the poster and decide which elements of the poster each member will create.

ALTERNATIVE APPROACH

Ask for three volunteers to draw the squares for Step 1 at the chalkboard. Ask the members of the class to suggest a row entry for one square, a column entry for the second square, and a diagonal entry for the third square. The volunteers at the board should record the entries. Repeat this exercise two more times until the magic square is made, adjusting values as necessary. Lead a discussion of the techniques students used to create the magic squares. Have students complete the investigation exercises alone or in groups of two.

OBJECTIVE Explore the mathematics behind magic squares.

Materials: paper, pencil

A *magic square* is a square array of consecutive integers, usually (but not always) beginning with 1, for which the sum of the entries in each row, column, and diagonal is the same. This common sum is called the *magic constant*.

For example, a 4 × 4 magic square with a magic constant of 34 is shown. This square appears in the engraving *Melancholia*, which was created in 1514 by the German artist and mathematician Albrecht Dürer.

Magic squares were discovered in China around 2200 B.C. and later spread to India, Japan, and eventually to Europe. The challenge of creating magic squares has fascinated mathematicians and puzzle lovers for many centuries.

HOW TO MAKE A 3 × 3 MAGIC SQUARE

4	9	2
3	5	7
8	1	6

❶ Draw a 3 × 3 square. You want to use the integers 1 through 9 to fill in the square. Start by writing the middle value, 5, in the center.

❷ Continue filling in numbers until you have an arrangement where the entries in each row, column, and diagonal add up to 15.

INVESTIGATION

1. yes; The resulting matrix is a magic square with a magic constant of 21.

2. yes; The result is always a magic square. The magic constant in terms of a is $3(a + 5)$.

3. Answers will vary.

4. Yes, the result is a magic square with a magic constant of 30.

1. Think of a magic square as a matrix. Suppose a 3 × 3 matrix containing all 2's is added to the magic square in **Step 2**. Is the resulting matrix also a magic square? If so, what is the magic constant?

2. Generalize your work from Exercise 1 by adding a 3 × 3 matrix containing all a's, where a represents *any* integer, to the magic square in **Step 2**. Is the result always a magic square? If so, what is the magic constant in terms of a?

3. Use the integers 1 through 9 to make another 3 × 3 magic square. Add your square to the one in **Step 2**. Is the result a magic square? Explain. (Remember that the square's rows, columns, and diagonals must have the same sum *and* the numbers in the square must be consecutive integers.)

4. Use scalar multiplication to multiply the magic square in **Step 2** by the scalar 2. Is the result a magic square? Explain.

<table>
<tr><td>7</td><td></td><td></td><td>14</td></tr>
<tr><td></td><td>13</td><td>8</td><td></td></tr>
<tr><td></td><td>3</td><td>10</td><td></td></tr>
<tr><td>9</td><td></td><td></td><td>4</td></tr>
</table>

5. The transpose is also a magic square with a magic constant of 15.

6. The diagonal gives us a magic constant of 34, so the magic square can be completed using trial and error until a match is found.

7 2 11 14
12 13 8 1
6 3 10 15
9 16 5 4

7. The sum of the entries is 45 for the 3 × 3 magic square and 136 for the 4 × 4 magic square.

8. $S = \frac{1}{2}n^4 + \frac{1}{2}n^2$ is a quartic function.

9. $M = \frac{1}{2}n^3 + \frac{1}{2}n$ is a cubic function.

$S = \frac{1}{2}n^4 + \left(a - \frac{1}{2}\right)n^2$

$M = \frac{1}{2}n^3 + \left(a - \frac{1}{2}\right)n$

For the magic square shown, $S = 81$ and $M = 27$.

INVESTIGATION (*continued*)

5. The *transpose* of a matrix A is a matrix A^T obtained by interchanging the rows and columns of A—the first row of A becomes the first column of A^T, the second row of A becomes the second column of A^T, and so on. Find the transpose of the magic square in **Step 2**. Is the transpose also a magic square?

6. Copy and complete the 4 × 4 magic square shown. What reasoning did you use to place the remaining numbers?

7. The sum S of the first k positive integers is given by the quadratic function $S = \frac{1}{2}k^2 + \frac{1}{2}k$. Use this function to find the sum of the entries in the 3 × 3 and 4 × 4 magic squares from **Step 2** and Exercise 6. Check your answers by computing the sums directly.

8. Consider an $n \times n$ magic square that contains the integers 1 through n^2. Use the function from Exercise 7 to write a formula for the sum S of the entries in the square in terms of n. What type of function is this formula?

9. For an $n \times n$ magic square that contains the integers 1 through n^2, write a formula for the square's magic constant M in terms of n. (*Hint:* Note that the magic constant is the sum of all the entries in the square divided by the number of rows or columns.) What type of function is this formula?

PRESENT YOUR RESULTS

Write a report to present your results.

- Include the 3 × 3 and 4 × 4 magic squares you made.

- Tell whether a magic square is produced by performing each of the following operations on an $n \times n$ magic square A: adding the same integer to each entry of A, multiplying each entry of A by the same integer, adding another $n \times n$ magic square to A, and taking the transpose of A.

- Include the formulas you found for the sum of the entries and for the magic constant of an $n \times n$ magic square containing the integers 1 through n^2.

- Describe how you used your knowledge of matrices, quadratic functions, and higher-degree polynomial functions in this project.

EXTENSION

Consider an $n \times n$ magic square containing the integers a through $a + n^2 - 1$. Such a magic square is shown at the right for $n = 3$ and $a = 5$. For this type of magic square, write formulas for the sum S of the entries and for the magic constant M in terms of n and a. Verify that your formulas work for the magic square shown.

<table>
<tr><td>8</td><td>13</td><td>6</td></tr>
<tr><td>7</td><td>9</td><td>11</td></tr>
<tr><td>12</td><td>5</td><td>10</td></tr>
</table>

CONCLUDING THE PROJECT
Have one group member present the report or project to the class. The other member can present and explain the magic squares.

GRADING THE PROJECT
4 Students complete the magic squares and record the results. They complete the exercises and coherently explain their results and defend their conclusions. They extend the project by writing a formula for the sum of the entries and for the magic constant. They use the 3 × 3 square to verify their formulas. Their report or poster contains a sketch with explanatory text for each exercise. Both posters and reports are clearly presented.

3 Students complete the magic squares and answer all the questions, but one of the squares may be incorrect or they may have answered some questions incorrectly. They may not relate all the mathematical ideas to their results. Students include all the information for the report or poster and show they extended the project by verifying their formulas. The report and the poster are clear and show considerable effort.

2 Students complete the exercises but may not accurately draw the magic squares or identify which matrix transformations produce magic squares. They do not complete any of the extension activities. A report or poster is created, but either may be missing a sketch or text. Either the report or the poster does not include all the ideas necessary to convey meaning.

1 Students do not show that they understand the mathematical ideas and their work is incomplete. Their magic squares are incorrect, as are their explanations. They are unable to use formulas to find the sum of the entries and the report or poster does not convey an understanding of the material.

PLANNING THE CHAPTER

Powers, Roots, and Radicals

GOALS

LESSON		NCTM	ITED	SAT9	Terra-Nova	Local
7.1 *pp. 401–406*	**GOAL 1** Evaluate *n*th roots of real numbers using both radical notation and rational exponent notation. **GOAL 2** Use *n*th roots to solve real-life problems.	1, 2, 6, 8, 9, 10	SAPE	7	11, 16, 17, 18, 49, 51, 52	
7.2 *pp. 407–414*	**GOAL 1** Use properties of rational exponents to evaluate and simplify expressions. **GOAL 2** Use properties of rational exponents to solve real-life problems.	1, 2, 6, 8, 9, 10	SAPE, RQWN		11, 16, 17, 18, 49, 51, 52	
7.3 *pp. 415–420*	**GOAL 1** Perform operations with functions including power functions. **GOAL 2** Use power functions and function operations to solve real-life problems.	1, 2, 6, 8, 9, 10	SAPE, RQWN	2	16, 17, 18, 49, 51, 52	
7.4 *pp. 421–430*	**CONCEPT ACTIVITY: 7.4** *Investigate how a function and its inverse are related.* **GOAL 1** Find inverses of linear functions. **GOAL 2** Find inverses of nonlinear functions. **TECHNOLOGY ACTIVITY: 7.4** *Graph inverse functions on a graphing calculator.*	1, 2, 3	MIG, IIG		14, 16	
7.5 *pp. 431–436*	**GOAL 1** Graph square root and cube root functions. **GOAL 2** Use square root and cube root functions to find real-life quantities.	1, 2, 3, 8, 9, 10	SAPE	22, 34, 38	14, 16, 18, 49, 51, 52	
7.6 *pp. 437–444*	**GOAL 1** Solve equations that contain radicals or rational exponents. **GOAL 2** Use radical equations to solve real-life problems.	1, 2, 6, 8, 9, 10	SAPE, RQWN	7	11, 16, 17, 18, 49, 51, 52	
7.7 *pp. 445–454*	**GOAL 1** Use measures of central tendency and measures of dispersion to describe data sets. **GOAL 2** Use box-and-whisker plots and histograms to represent data graphically. **TECHNOLOGY ACTIVITY: 7.7** *Find statistics and draw statistical graphs on a graphing calculator.*	1, 5, 8, 9, 10	MCS, MIG, IIG	11, 13	11, 15, 18	

RESOURCES

CHAPTER RESOURCE BOOKLETS

CHAPTER SUPPORT

Tips for New Teachers	p. 1	Prerequisite Skills Review	p. 5
Parent Guide for Student Success	p. 3	Strategies for Reading Mathematics	p. 7

LESSON SUPPORT

	7.1	7.2	7.3	7.4	7.5	7.6	7.7
Lesson Plans (regular and block)	p. 9	p. 21	p. 34	p. 48	p. 63	p. 76	p. 92
Warm-Up Exercises and Daily Quiz	p. 11	p. 23	p. 36	p. 50	p. 65	p. 78	p. 94
Activity Support Masters				p. 51			
Lesson Openers	p. 12	p. 24	p. 37	p. 52	p. 66	p. 79	p. 95
Graphing Calculator Activities & Keystrokes			p. 38	p. 53	p. 67	p. 80	p. 96
Practice (3 levels)	p. 13	p. 25	p. 40	p. 54	p. 68	p. 82	p. 97
Reteaching with Practice	p. 16	p. 28	p. 43	p. 57	p. 71	p. 85	p. 100
Quick Catch-Up for Absent Students	p. 18	p. 30	p. 45	p. 59	p. 73	p. 87	p. 102
Cooperative Learning Activities						p. 88	
Interdisciplinary Applications	p. 19		p. 46		p. 74		p. 103
Real-Life Applications		p. 31		p. 60		p. 89	
Math & History Applications						p. 90	
Challenge: Skills and Applications	p. 20	p. 32	p. 47	p. 61	p. 75	p. 91	p. 104

REVIEW AND ASSESSMENT

Quizzes	pp. 33, 62	Alternative Assessment with Math Journal	p. 113
Chapter Review Games and Activities	p. 105	Project with Rubric	p. 115
Chapter Test (3 levels)	pp. 106–111	Cumulative Review	p. 117
SAT/ACT Chapter Test	p. 112	Resource Book Answers	p. A1

TRANSPARENCIES

	7.1	7.2	7.3	7.4	7.5	7.6	7.7
Warm-Up Exercises and Daily Quiz	p. 51	p. 52	p. 53	p. 54	p. 55	p. 56	p. 57
Alternative Lesson Opener Transparencies	p. 44	p. 45	p. 46	p. 47	p. 48	p. 49	p. 50
Examples/Standardized Test Practice	✓	✓	✓	✓	✓	✓	✓
Answer Transparencies	✓	✓	✓	✓	✓	✓	✓

TECHNOLOGY

- Electronic Teaching Tools
- Online Lesson Planner
- Internet Support
- Personal Student Tutor
- Test and Practice Generator
- Instant Replay: Video Review Games
- Electronic Lesson Presentations (Lesson 7.4)

ADDITIONAL RESOURCES

- Basic Skills Workbook: Diagnosis and Remediation
- Worked-Out Solution Key
- Resources in Spanish
- Standardized Test Practice Workbook
- Practice Workbook with Examples

CHAPTER

7

PACING THE CHAPTER

Resource Key
● STUDENT EDITION
● CHAPTER 7 RESOURCE BOOK
● TEACHER'S EDITION

REGULAR SCHEDULE

Day 1

7.1

STARTING OPTIONS
- Prereq. Skills Review
- Strategies for Reading
- Homework Check
- Warm-Up or Daily Quiz

TEACHING OPTIONS
- Motivating the Lesson
- Les. Opener (Appl.)
- Examples 1–6
- Closure Question
- Guided Practice Exs.

APPLY/HOMEWORK
- See Assignment Guide.
- See the CRB: Practice, Reteach, Apply, Extend

ASSESSMENT OPTIONS
- Checkpoint Exercises
- Daily Quiz (7.1)
- Stand. Test Practice

Day 2

7.2

STARTING OPTIONS
- Homework Check
- Warm-Up or Daily Quiz

TEACHING OPTIONS
- Les. Opener (Activity)
- Examples 1–5
- Guided Practice Exs. 1–12

APPLY/HOMEWORK
- See Assignment Guide.
- See the CRB: Practice, Reteach, Apply, Extend

ASSESSMENT OPTIONS
- Checkpoint Exercises, pp. 408–409

Day 3

7.2 (cont.)

STARTING OPTIONS
- Homework Check

TEACHING OPTIONS
- Examples 6–9
- Closure Question
- Guided Practice Exs. 13–21

APPLY/HOMEWORK
- See Assignment Guide.
- See the CRB: Practice, Reteach, Apply, Extend

ASSESSMENT OPTIONS
- Checkpoint Exercises, pp. 409–410
- Daily Quiz (7.2)
- Stand. Test Practice
- Quiz (7.1–7.2)

Day 4

7.3

STARTING OPTIONS
- Homework Check
- Warm-Up or Daily Quiz

TEACHING OPTIONS
- Les. Opener (Application)
- Graphing Calc. Activity
- Examples 1–5
- Closure Question
- Guided Practice

APPLY/HOMEWORK
- See Assignment Guide.
- See the CRB: Practice, Reteach, Apply, Extend

ASSESSMENT OPTIONS
- Checkpoint Exercises
- Daily Quiz (7.3)
- Stand. Test Practice

Day 5

7.4

STARTING OPTIONS
- Homework Check
- Warm-Up or Daily Quiz

TEACHING OPTIONS
- Concept Act. & Wksht.
- Les. Opener (Activity)
- Graphing Calc. Activity
- Examples 1–4
- Guided Practice Exs. 1–2, 4–10

APPLY/HOMEWORK
- See Assignment Guide.
- See the CRB: Practice, Reteach, Apply, Extend

ASSESSMENT OPTIONS
- Checkpoint Exercises, pp. 423–424

Day 6

7.4 (cont.)

STARTING OPTIONS
- Homework Check

TEACHING OPTIONS
- Examples 5–6
- Technology Activity
- Closure Question
- Guided Practice Exs. 3, 11–13

APPLY/HOMEWORK
- See Assignment Guide.
- See the CRB: Practice, Reteach, Apply, Extend

ASSESSMENT OPTIONS
- Checkpoint Exercises, p. 425
- Daily Quiz (7.4)
- Stand. Test Practice
- Quiz (7.3–7.4)

Day 9

7.6 (cont.)

STARTING OPTIONS
- Homework Check

TEACHING OPTIONS
- Examples 5–6
- Closure Question
- Guided Practice Exs. 1, 10–16

APPLY/HOMEWORK
- See Assignment Guide.
- See the CRB: Practice, Reteach, Apply, Extend

ASSESSMENT OPTIONS
- Checkpoint Exercises, pp. 439–440
- Daily Quiz (7.6)
- Stand. Test Practice

Day 10

7.7

STARTING OPTIONS
- Homework Check
- Warm-Up or Daily Quiz

TEACHING OPTIONS
- Motivating the Lesson
- Les. Opener (Application)
- Graphing Calc. Activity
- Examples 1–2
- Guided Practice Exs. 1–2, 4–5

APPLY/HOMEWORK
- See Assignment Guide.
- See the CRB: Practice, Reteach, Apply, Extend

ASSESSMENT OPTIONS
- Checkpoint Exercises, p. 446

Day 11

7.7 (cont.)

STARTING OPTIONS
- Homework Check

TEACHING OPTIONS
- Examples 3–6
- Technology Activity
- Closure Question
- Guided Practice Exs. 3, 6–9

APPLY/HOMEWORK
- See Assignment Guide.
- See the CRB: Practice, Reteach, Apply, Extend

ASSESSMENT OPTIONS
- Checkpoint Exercises, pp. 446–448
- Daily Quiz (7.7)
- Stand. Test Practice
- Quiz (7.5–7.7)

Day 12

Review

DAY 12 START OPTIONS
- Homework Check

REVIEWING OPTIONS
- Chapter 7 Summary
- Chapter 7 Review
- Chapter Review Games and Activities

APPLY/HOMEWORK
- Chapter 7 Test (practice)
- Ch. Standardized Test (practice)

Day 13

Assess

DAY 13 START OPTIONS
- Homework Check

ASSESSMENT OPTIONS
- Chapter 7 Test
- SAT/ACT Ch. 7 Test
- Alternative Assessment

APPLY/HOMEWORK
- Skill Review, p. 464

BLOCK SCHEDULE

Day 7

7.5

STARTING OPTIONS
- Homework Check
- Warm-Up or Daily Quiz

TEACHING OPTIONS
- Les. Opener (Visual)
- Graphing Calc. Activity
- Examples 1–6
- Closure Question
- Guided Practice Exs.

APPLY/HOMEWORK
- See Assignment Guide.
- See the CRB: Practice, Reteach, Apply, Extend

ASSESSMENT OPTIONS
- Checkpoint Exercises
- Daily Quiz (7.5)
- Stand. Test Practice

Day 8

7.6

STARTING OPTIONS
- Homework Check
- Warm-Up or Daily Quiz

TEACHING OPTIONS
- Motivating the Lesson
- Les. Opener (Calculator)
- Graphing Calc. Activity
- Examples 1–4
- Guided Practice Exs. 2–9

APPLY/HOMEWORK
- See Assignment Guide.
- See the CRB: Practice, Reteach, Apply, Extend

ASSESSMENT OPTIONS
- Checkpoint Exercises, p. 438

Day 1

Assess & 7.1
(Day 1 = Ch. 6 Day 7)

ASSESSMENT OPTIONS
- Chapter 6 Test
- SAT/ACT Ch. 6 Test
- Alternative Assessment

CH. 7 START OPTIONS
- Skill Review, p. 400
- Prereq. Skills Review
- Strategies for Reading

TEACHING 7.1 OPTIONS
- Warm-Up (Les. 7.1)
- Motivating the Lesson
- Les. Opener (Appl.)
- Examples 1–6
- Closure Question
- Guided Practice Exs.

APPLY/HOMEWORK
- See Assignment Guide.
- See the CRB: Practice, Reteach, Apply, Extend

ASSESSMENT OPTIONS
- Checkpoint Exercises
- Daily Quiz (Les. 7.1)
- Stand. Test Practice

Day 2

7.2

DAY 2 START OPTIONS
- Homework Check
- Warm-Up or Daily Quiz

TEACHING 7.2 OPTIONS
- Les. Opener (Activity)
- Examples 1–9
- Closure Question
- Guided Practice Exs.

APPLY/HOMEWORK
- See Assignment Guide.
- See the CRB: Practice, Reteach, Apply, Extend

ASSESSMENT OPTIONS
- Checkpoint Exercises
- Daily Quiz (Les. 7.2)
- Stand. Test Practice
- Quiz (7.1–7.2)

Day 3

7.3 & 7.4

DAY 3 START OPTIONS
- Homework Check
- W-Up 7.3 or D. Quiz 7.2

TEACHING 7.3 OPTIONS
- Les. Opener (Appl.)
- Graphing Calc. Activity
- Examples 1–5
- Closure Question
- Guided Practice Exs.

BEGINNING 7.4 OPTIONS
- Warm-Up (Les. 7.4)
- Concept Act. & Wksht.
- Les. Opener (Activity)
- Graphing Calc. Activity
- Examples 1–4
- Guided Practice Exs. 1–2, 4–10

APPLY/HOMEWORK
- See Assignment Guide.
- See the CRB: Practice, Reteach, Apply, Extend

ASSESSMENT OPTIONS
- Checkpoint Exercises
- Daily Quiz (Les. 7.3)
- Stand. Test Prac. (7.3)

Day 4

7.4 & 7.5

DAY 4 START OPTIONS
- Homework Check
- Daily Quiz (Les. 7.3)

FINISHING 7.4 OPTIONS
- Examples 5–6
- Technology Activity
- Closure Question
- Guided Practice Exs. 3, 11–13

TEACHING 7.5 OPTIONS
- Warm-Up (Les. 7.5)
- Les. Opener (Visual)
- Graphing Calc. Activity
- Examples 1–6
- Closure Question
- Guided Practice Exs.

APPLY/HOMEWORK
- See Assignment Guide.
- See the CRB: Practice, Reteach, Apply, Extend

ASSESSMENT OPTIONS
- Checkpoint Exercises
- Daily Quiz (7.4, 7.5)
- Stand. Test Practice
- Quiz (7.3–7.4)

Day 5

7.6

DAY 5 START OPTIONS
- Homework Check
- Warm-Up or Daily Quiz

TEACHING OPTIONS
- Motivating the Lesson
- Les. Opener (Calc.)
- Graphing Calc. Activity
- Examples 1–6
- Closure Question
- Guided Practice Exs.

APPLY/HOMEWORK
- See Assignment Guide.
- See the CRB: Practice, Reteach, Apply, Extend

ASSESSMENT OPTIONS
- Checkpoint Exercises
- Daily Quiz (Les. 7.6)
- Stand. Test Practice

Day 6

7.7

DAY 6 START OPTIONS
- Homework Check
- Warm-Up or Daily Quiz

TEACHING OPTIONS
- Motivating the Lesson
- Les. Opener (Appl.)
- Graphing Calc. Activity
- Examples 1–6
- Technology Activity
- Closure Question
- Guided Practice Exs.

APPLY/HOMEWORK
- See Assignment Guide.
- See the CRB: Practice, Reteach, Apply, Extend

ASSESSMENT OPTIONS
- Checkpoint Exercises
- Daily Quiz (Les. 7.7)
- Stand. Test Practice
- Quiz (7.5–7.7)

Day 7

Review/Assess

DAY 7 START OPTIONS
- Homework Check

REVIEWING OPTIONS
- Chapter 7 Summary
- Chapter 7 Review
- Chapter Review Games and Activities
- Chapter 7 Test (practice)
- Ch. Standardized Test (practice)

ASSESSMENT OPTIONS
- Chapter 7 Test
- SAT/ACT Ch. 7 Test
- Alternative Assessment

APPLY/HOMEWORK
- Skill Review, p. 464

MEETING INDIVIDUAL NEEDS

BEFORE THE CHAPTER

The *Chapter 7 Resource Book* has the following materials to distribute and use before the chapter:

- **Parent Guide for Student Success (pictured below)**
- **Prerequisite Skills Review**
- **Strategies for Reading Mathematics**

PARENT GUIDE *Pages 3–4*

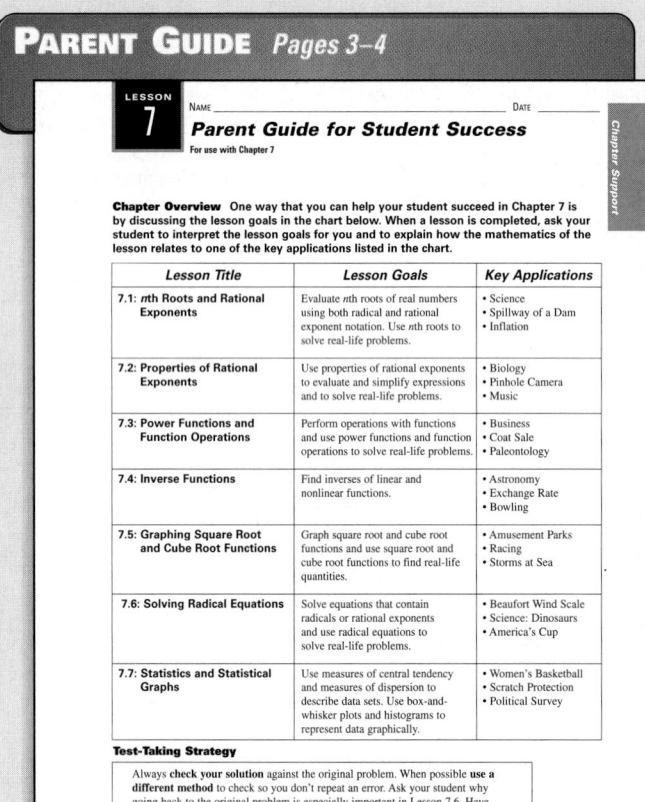

LESSON 7

NAME _____ DATE _____

Parent Guide for Student Success
For use with Chapter 7

Chapter Support

Chapter Overview One way that you can help your student succeed in Chapter 7 is by discussing the lesson goals in the chart below. When a lesson is completed, ask your student to interpret the lesson goals for you and to explain how the mathematics of the lesson relates to one of the key applications listed in the chart.

Lesson Title	Lesson Goals	Key Applications
7.1: nth Roots and Rational Exponents	Evaluate nth roots of real numbers using both radical and rational exponent notation. Use nth roots to solve real-life problems.	• Science • Spillway of a Dam • Inflation
7.2: Properties of Rational Exponents	Use properties of rational exponents to evaluate and simplify expressions and to solve real-life problems.	• Biology • Pinhole Camera • Music
7.3: Power Functions and Function Operations	Perform operations with functions and use power functions and function operations to solve real-life problems.	• Business • Coat Sale • Paleontology
7.4: Inverse Functions	Find inverses of linear and nonlinear functions.	• Astronomy • Exchange Rate • Bowling
7.5: Graphing Square Root and Cube Root Functions	Graph square root and cube root functions and use square root and cube root functions to find real-life quantities.	• Amusement Parks • Racing • Storms at Sea
7.6: Solving Radical Equations	Solve equations that contain radicals or rational exponents and use radical equations to solve real-life problems.	• Beaufort Wind Scale • Science: Dinosaurs • America's Cup
7.7: Statistics and Statistical Graphs	Use measures of central tendency and measures of dispersion to describe data sets. Use box-and-whisker plots and histograms to represent data graphically.	• Women's Basketball • Scratch Protection • Political Survey

Test-Taking Strategy

Always **check your solution** against the original problem. When possible **use a different method** to check so you don't repeat an error. Ask your student why going back to the original problem is especially important in Lesson 7.6. Have your student show you an example from the chapter of how to check using different steps than those used to solve the problem.

PARENT GUIDE FOR STUDENT SUCCESS The first page summarizes the content of Chapter 7. Parents are encouraged to have their students explain how the material relates to key applications in the chapter, such as astronomy. The second page (not shown) provides exercises and an activity that parents can do with their students. In the activity, parents and students find the mean, median, mode, range, and standard deviation of hours spent watching television. They use the results to analyze viewing habits.

DURING EACH LESSON

The *Chapter 7 Resource Book* has the following alternatives for introducing the lesson:

- **Lesson Openers (pictured below)**
- **Graphing Calculator Activities with Keystrokes**

LESSON OPENER *Page 52*

LESSON 7.4

NAME _____ DATE _____

Activity Lesson Opener
For use with pages 422–429

Available as a transparency

SET UP: Work individually.

Match each composition of functions on the left with a simplified expression on the right to reveal something you might say while riding a roller coaster. (A choice may be used more than once or not at all.)

1. $f(x) = 2x + 3$ _____ $f(g(x))$
   $g(x) = 5x - 2$ _____ $g(f(x))$
2. $f(x) = x + 5$ _____ $f(g(x))$
   $g(x) = x - 5$ _____ $g(f(x))$
3. $f(x) = 3x$ _____ $f(g(x))$
   $g(x) = \dfrac{x}{3}$ _____ $g(f(x))$
4. $f(x) = 5x + 20$ _____ $f(g(x))$
   $g(x) = \dfrac{1}{5}x - 4$ _____ $g(f(x))$
5. $f(x) = x^3 + 5$ _____ $f(g(x))$
   $g(x) = \sqrt[3]{x - 5}$ _____ $g(f(x))$
6. $f(x) = \dfrac{x^5 - 8}{10}$ _____ $f(g(x))$
   $g(x) = \sqrt[5]{10x + 8}$ _____ $g(f(x))$

A	$9x$
B	$x - 10$
C	$x^5 - 8$
E	x
G	$9x$
H	$10x + 13$
I	$x - 5$
L	x^5
O	$x + 10$
P	$x + 5$
R	$\dfrac{x}{3}$
W	$10x - 1$
U	$3x$
Y	$\sqrt[3]{x}$

7. Comment on any patterns you observe.

Lesson 7.4

ACTIVITY LESSON OPENER This Lesson Opener provides an alternative way to start Lesson 7.4 in the form of an activity. In this activity, students simplify function compositions as they learn about inverse functions.

TECHNOLOGY RESOURCE

Teachers can use the Electronic Lesson Presentations CD-ROM to present Lesson 7.4 using computer animation to step through the Examples.

The *Chapter 7 Resource Book* has a variety of materials to follow-up each lesson. They include the following:

- **Practice (3 levels)**
- **Reteaching with Practice**
- **Quick Catch-Up for Absent Students**
- **Challenge: Skills and Applications**
- **Interdisciplinary Applications (pictured below)**
- **Real-Life Applications**

APPLICATION *Page 46*

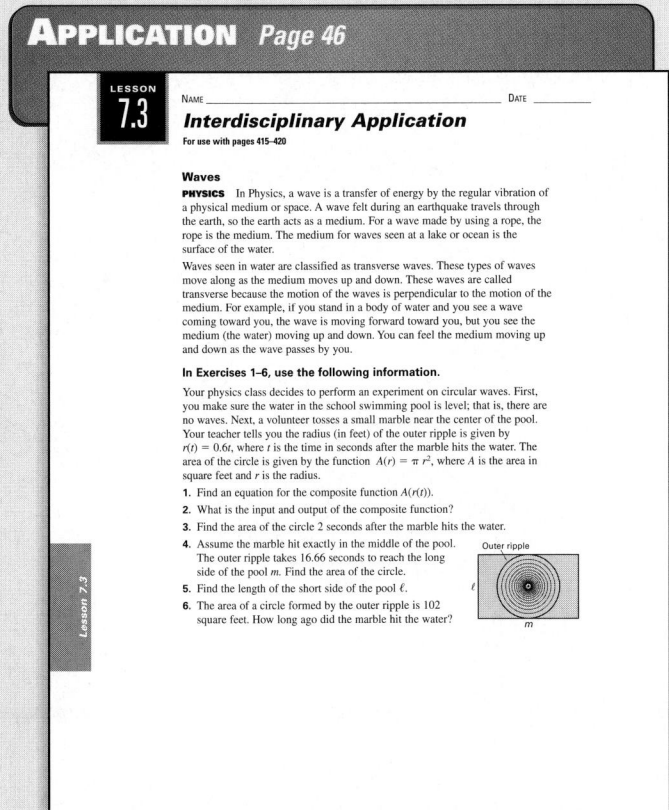

INTERDISCIPLINARY APPLICATION In this application students use function composition, a topic of Lesson 7.3, to explore the behavior of ripples that form when a marble is tossed into a pool.

The *Chapter 7 Resource Book* has the following review and assessment materials:

- **Quizzes**
- **Chapter Review Games and Activities**
- **Chapter Test (3 levels)**
- **SAT/ACT Chapter Test**
- **Alternative Assessment with Rubrics and Math Journal**
- **Project with Rubric**
- **Cumulative Review (pictured below)**

CUMULATIVE REVIEW *Pages 117–118*

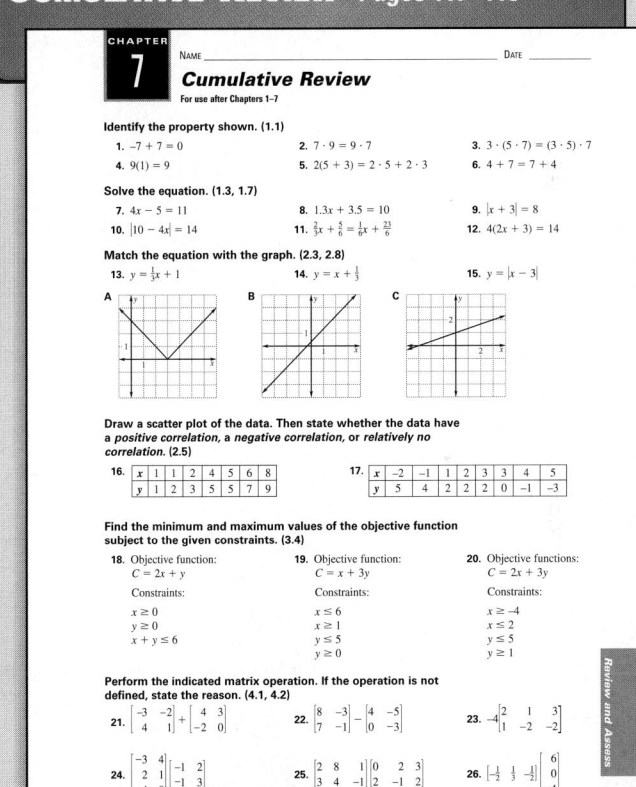

CUMULATIVE REVIEW The content of Chapters 1–7 is reviewed in this two-page cumulative review. Lesson references with each cluster of exercises guide students to the places in the textbook where they can go to review each concept.

 TECHNOLOGY RESOURCE

Students can use the Personal Student Tutor to find additional examples that review the material covered in the Cumulative Review.

CHAPTER GOALS

In this chapter, students will learn how to evaluate *n*th roots of real numbers using both radical and exponential notation. They will use properties of rational exponents to evaluate and simplify expressions, and they will evaluate power functions and perform arithmetic operations with functions as well as composition of functions. They will find inverses of functions, and they will graph square root and cube root functions. Students will solve equations that have radicals or rational exponents. They will use roots, rational exponents, power functions, function operations, and radical equations to solve real-life problems. Students will also learn to use measures of central tendency and dispersion to describe a data set and will use box-and-whisker plots and histograms to represent data.

APPLICATION NOTE

The weight of dinosaurs (or other animals) can be estimated if the animals being compared are geometrically similar. Using femur measurements to predict weight assumes the four dinosaurs in the table (Ex. 1) are similar in shape.

The hypacrosaurus (Ex. 2) is one of the duck-billed dinosaurs. This type of dinosaur had a hollow hood in its head that is assumed to be a resonating chamber for trumpeting as well as a chamber to cool the brain. Hypacrosaurus dinosaurs were nine meters long and probably ate woody flowering plants.

The heaviest dinosaur may have been Seismosaurus, which may have weighed 90,900 kilograms, or 200,000 pounds.

Additional information about dinosaurs is available at **www.mcdougallittell.com**.

POWERS, ROOTS, AND RADICALS

▶ *How can you estimate the weight of a dinosaur?*

398

APPLICATION: Dinosaurs

Scientists have determined relationships between the bone measurements and the heights and weights of living animals. By applying the relationships to dinosaur bones, scientists can estimate heights and weights of these prehistoric animals.

Think & Discuss

The table below gives the femur circumference (in millimeters) and estimated weight (in kilograms) for four different dinosaurs that walk on two feet.

Femur circumference	Estimated weight
103	50
201	310
348	1400
504	3800

The points do not appear to lie on a line.
1. Graph the ordered pairs (*femur circumference, estimated weight*) from the table. Why can't you model the data with a linear function?
   See margin for graph.
2. The femur circumference of a *Hypacrosaurus altispinus* is about 400 millimeters. Estimate a reasonable weight for this dinosaur. How did you derive your estimate?
   about 2200 kg; See margin for explanation.

Learn More About It

You will apply the relationship between femur circumference and weight to a *Tyrannosaurus rex* in Exercise 64 on p. 442.

APPLICATION LINK Visit www.mcdougallittell.com for more information about dinosaurs.

ADDITIONAL RESOURCES
Another way to begin the chapter is to show the video clip of a real-life motivator for Chapter 7 on the *Instant Replay: Video Review Games.*

PROJECTS
A project covering Chapters 7–9 appears on pages 584–585 of the Student Edition. An additional project for Chapter 7 is available in the *Chapter 7 Resource Book*, p. 115.

TECHNOLOGY
 Software
 • *Electronic Teaching Tools*
 • *Online Lesson Planner*
 • *Personal Student Tutor*
 • *Test and Practice Generator*
 • *Electronic Lesson Presentations* (Lesson 7.4)

Video
 • *Instant Replay: Video Review Games*

 Internet Connections
www.mcdougallittell.com
 • **Application Links**
  399, 403, 425, 440, 444
 • **Data Updates**
  427, 445
 • **Student Help**
  405, 408, 409, 419, 428, 430, 435, 438, 446, 453
 • **Career Links**
  419, 427, 433, 435, 450
 • **Extra Challenge**
  406, 413, 420, 428, 436, 443, 451

1.
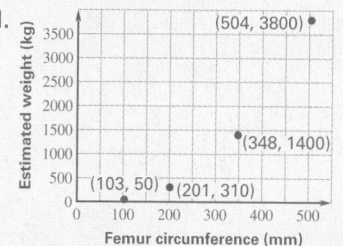

2. See Additional Answers beginning on page AA1.

Study Guide

PREPARE

DIAGNOSTIC TOOLS

The **Skill Review** exercises can help you diagnose whether students have the following skills needed in Chapter 7:

- Solve an equation for *y*.
- Factor trinomials.
- Simplify expressions involving exponents.
- Perform operations with polynomials.

The following resources are available for students who need additional help with these skills:

- Prerequisite Skills Review (*Chapter 7 Resource Book,* p. 5; *Warm-Up Transparencies,* p. 50)
- ▢ *Personal Student Tutor*

ADDITIONAL RESOURCES

The following resources are provided to help you prepare for the upcoming chapter and customize review materials:

- ***Chapter 7 Resource Book***
  Tips for New Teachers (p. 1)
  Parent Guide (p. 3)
  Lesson Plans (every lesson)
  Lesson Plans for Block Scheduling (every lesson)
- ▢ *Electronic Teaching Tools*
- ▢ *Online Lesson Planner*
- ▢ *Test and Practice Generator*

PREVIEW

What's the chapter about?

Chapter 7 is about **powers, roots, and radicals**. In Chapter 7 you'll learn

- how to use rational exponents and *n*th roots of numbers.
- how to perform operations with and find inverses of functions.
- how to graph radical functions and solve radical equations.

KEY VOCABULARY

▶ Review	▶ New	
• exponent, p. 11	• *n*th root of *a,* p. 401	• measure of central tendency, p. 445
• relation, p. 67	• power function, p. 415	• measure of dispersion, p. 446
• function, p. 67	• composition, p. 416	• box-and-whisker plot, p. 447
• square root, p. 264	• inverse function, p. 422	• histogram, p. 448
	• radical function, p. 431	• frequency distribution, p. 448

PREPARE

Are you ready for the chapter?

SKILL REVIEW Do these exercises to review key skills that you'll apply in this chapter. See the given **reference page** if there is something you don't understand.

STUDENT HELP

↳ **Study Tip**
"Student Help" boxes throughout the chapter give you study tips and tell you where to look for extra help in this book and on the Internet.

Solve the equation for *y*. (Review Example 1, p. 26)

1. $3x - 2y = 12$ $y = \frac{3x - 12}{2}$ **2.** $x + \frac{1}{2}y = 5$ $y = 10 - 2x$ **3.** $x = 4y - 1$ $y = \frac{x + 1}{4}$

Factor the trinomial. (Review Examples 1 and 2, p. 256)

4. $x^2 + 10x + 21$
$(x + 7)(x + 3)$

5. $x^2 + 5x - 36$
$(x + 9)(x - 4)$

6. $2x^2 - 16x + 30$
$2(x - 3)(x - 5)$

Simplify the expression. (Review Example 2, p. 324)

7. $(abc^2)^4$ $a^4b^4c^8$ **8.** $x^5 \cdot x^{-3}$ x^2 **9.** $\left(\frac{x^2}{y}\right)^2$ $\frac{x^4}{y^2}$ **10.** $\frac{3x}{y} \cdot \frac{3x^2y^{-2}}{12y^3}$ $\frac{3x^3}{4y^6}$

Perform the indicated operation. (Review Examples 1–6, pp. 338 and 339)

11. $5x^2(x - 8)$ $5x^3 - 40x^2$ **12.** $(3y - 2)^2$
$9y^2 - 12y + 4$
13. $(7x^2 + x) - (6x - 4)$
$7x^2 - 5x + 4$

STUDY STRATEGY

Here's a study strategy!

Quiz Yourself

After you complete a homework assignment, copy a few representative problems from the assignment on a separate piece of paper. Record the lesson number for the problems and leave space for the answers. You can use these problems to quiz yourself later, such as before a class quiz is given.

*n*th Roots and Rational Exponents

What you should learn

GOAL 1 Evaluate *n*th roots of real numbers using both radical notation and rational exponent notation.

GOAL 2 Use *n*th roots to solve **real-life** problems, such as finding the total mass of a spacecraft that can be sent to Mars in **Example 5**.

Why you should learn it

▼ To solve **real-life** problems, such as finding the number of reptile and amphibian species that Puerto Rico can support in **Ex. 67**.

GOAL 1 **EVALUATING *N*TH ROOTS**

You can extend the concept of a square root to other types of roots. For instance, 2 is a cube root of 8 because $2^3 = 8$, and 3 is a fourth root of 81 because $3^4 = 81$. In general, for an integer *n* greater than 1, if $b^n = a$, then *b* is an **nth root of *a***. An *n*th root of *a* is written as $\sqrt[n]{a}$, where *n* is the **index** of the radical.

You can also write an *n*th root of *a* as a power of *a*. For the particular case of a square root, suppose that $\sqrt{a} = a^k$. Then you can determine a value for *k* as follows:

$\sqrt{a} \cdot \sqrt{a} = a$	**Definition of square root**
$a^k \cdot a^k = a$	**Substitute a^k for $\sqrt{a}$.**
$a^{2k} = a^1$	**Product of powers property**
$2k = 1$	**Set exponents equal when bases are equal.**
$k = \dfrac{1}{2}$	**Solve for *k*.**

Therefore, you can see that $\sqrt{a} = a^{1/2}$. In a similar way you can show that $\sqrt[3]{a} = a^{1/3}$ and $\sqrt[4]{a} = a^{1/4}$. In general, $\sqrt[n]{a} = a^{1/n}$ for any integer *n* greater than 1.

REAL *N*TH ROOTS

Let *n* be an integer greater than 1 and let *a* be a real number.

- If *n* is odd, then *a* has one real *n*th root: $\sqrt[n]{a} = a^{1/n}$
- If *n* is even and $a > 0$, then *a* has two real *n*th roots: $\pm\sqrt[n]{a} = \pm a^{1/n}$
- If *n* is even and $a = 0$, then *a* has one *n*th root: $\sqrt[n]{0} = 0^{1/n} = 0$
- If *n* is even and $a < 0$, then *a* has no real *n*th roots.

EXAMPLE 1 *Finding nth Roots*

Find the indicated real *n*th root(s) of *a*.

a. $n = 3, a = -125$ **b.** $n = 4, a = 16$

SOLUTION

a. Because $n = 3$ is odd, $a = -125$ has one real cube root. Because $(-5)^3 = -125$, you can write:

$$\sqrt[3]{-125} = -5 \quad \text{or} \quad (-125)^{1/3} = -5$$

b. Because $n = 4$ is even and $a = 16 > 0$, 16 has two real fourth roots. Because $2^4 = 16$ and $(-2)^4 = 16$, you can write:

$$\pm\sqrt[4]{16} = \pm2 \quad \text{or} \quad \pm16^{1/4} = \pm2$$

---

1 PLAN

PACING
Basic: 1 day
Average: 1 day
Advanced: 1 day
Block Schedule: 0.5 block with Ch. 6 Assess.

↑ **LESSON OPENER**
APPLICATION
An alternative way to approach Lesson 7.1 is to use the Application Lesson Opener:

- Blackline Master (*Chapter 7 Resource Book*, p. 12)
-  Transparency (p. 44)

MEETING INDIVIDUAL NEEDS
- *Chapter 7 Resource Book*
  Prerequisite Skills Review (p. 5)
  Practice Level A (p. 13)
  Practice Level B (p. 14)
  Practice Level C (p. 15)
  Reteaching with Practice (p. 16)
  Absent Student Catch-Up (p. 18)
  Challenge (p. 20)
- *Resources in Spanish*
- 💻 *Personal Student Tutor*

NEW-TEACHER SUPPORT
See the Tips for New Teachers on pp. 1–2 of the *Chapter 7 Resource Book* for additional notes about Lesson 7.1.

WARM-UP EXERCISES

📄 *Transparency Available*
Evaluate the expression.
1. $\sqrt{9}$ 3   **2.** $-\sqrt{121}$ –11
3. $\left(\sqrt{25}\right)^2$ 25
Solve each equation.
4. $x^2 = 49$ 7 or –7
5. $(x-1)^2 = 64$ 9 or –7

ENGLISH LEARNERS
Help English learners decode *n*th by explaining that the suffix *-th* can be added to most numbers to convert them into adjectives.

Ask students the side length of a cube with volume 1000 cubic centimeters (10 centimeters). Finding the side length is an example of finding a third root. Finding roots is the focus of today's lesson.

 EXTRA EXAMPLE 1
Find the indicated nth root(s) of a.
a. $n = 5$, $a = -32$ -2
b. $n = 3$, $a = 64$ 4

EXTRA EXAMPLE 2
a. $16^{5/2} = \left(\sqrt{16}\right)^5 = 4^5 = 1024$

$16^{5/2} = (16^{1/2})^5 = 4^5 = 1024$

b. $64^{-2/3}$

$= \dfrac{1}{64^{2/3}} = \dfrac{1}{\left(\sqrt[3]{64}\right)^2} = \dfrac{1}{4^2} = \dfrac{1}{16}$

$64^{-2/3}$

$= \dfrac{1}{64^{2/3}} = \dfrac{1}{(64^{1/3})^2} = \dfrac{1}{4^2} = \dfrac{1}{16}$

EXTRA EXAMPLE 3
Use a graphing calculator to approximate $\sqrt[3]{3^4}$. about 4.33

EXTRA EXAMPLE 4
Solve each equation.
a. $6x^4 = 3750$ ± 5
b. $(x + 1)^3 = 18$ $\sqrt[3]{18} - 1 \approx 1.62$

 CHECKPOINT EXERCISES

For use after Example 1:
1. Find the indicated nth root(s) of a.
a. $n = 4$, $a = 625$ ± 5
b. $n = 3$, $a = -27$ -3

For use after Example 2:
2. Evaluate the expression.
a. $49^{3/2}$ 343
b. $16^{-3/4}$ $\dfrac{1}{8}$

For use after Example 3:
3. Use a graphing calculator to approximate $\sqrt[4]{3^3}$. about 2.28

For use after Example 4:
4. Solve each equation.
a. $5y^4 = 80$ ± 2
b. $(y - 1)^3 = 32$ $\sqrt[3]{32} + 1 \approx 4.17$

A rational exponent does not have to be of the form $\dfrac{1}{n}$ where n is an integer greater than 1. Other rational numbers such as $\dfrac{3}{2}$ and $-\dfrac{1}{2}$ can also be used as exponents.

RATIONAL EXPONENTS

Let $a^{1/n}$ be an nth root of a, and let m be a positive integer.

• $a^{m/n} = (a^{1/n})^m = \left(\sqrt[n]{a}\right)^m$

• $a^{-m/n} = \dfrac{1}{a^{m/n}} = \dfrac{1}{(a^{1/n})^m} = \dfrac{1}{\left(\sqrt[n]{a}\right)^m}$, $a \neq 0$

EXAMPLE 2 *Evaluating Expressions with Rational Exponents*

a. $9^{3/2} = \left(\sqrt{9}\right)^3 = 3^3 = 27$ **Using radical notation**

$9^{3/2} = (9^{1/2})^3 = 3^3 = 27$ **Using rational exponent notation**

b. $32^{-2/5} = \dfrac{1}{32^{2/5}} = \dfrac{1}{\left(\sqrt[5]{32}\right)^2} = \dfrac{1}{2^2} = \dfrac{1}{4}$ **Using radical notation**

$32^{-2/5} = \dfrac{1}{32^{2/5}} = \dfrac{1}{(32^{1/5})^2} = \dfrac{1}{2^2} = \dfrac{1}{4}$ **Using rational exponent notation**

· · · · · · · · · ·

When using a graphing calculator to approximate an nth root, you may have to rewrite the nth root using a rational exponent. Then use the calculator's power key.

EXAMPLE 3 *Approximating a Root with a Calculator*

Use a graphing calculator to approximate $\left(\sqrt[4]{5}\right)^3$.

SOLUTION First rewrite $\left(\sqrt[4]{5}\right)^3$ as $5^{3/4}$. Then enter the following:

Keystrokes: 5 [^] [(] 3 [÷] 4 [)] [ENTER]   Display: 3.343701525

▶ $\left(\sqrt[4]{5}\right)^3 \approx 3.34$

· · · · · · · · · ·

↳ **Study Tip**
To use a scientific calculator in Example 3, replace [^] with [y^x] and replace [ENTER] with [=].

To solve simple equations involving x^n, isolate the power and then take the nth root of *each* side.

EXAMPLE 4 *Solving Equations Using nth Roots*

a. $2x^4 = 162$
    $x^4 = 81$
    $x = \pm\sqrt[4]{81}$
    $x = \pm 3$

b. $(x - 2)^3 = 10$
    $x - 2 = \sqrt[3]{10}$
    $x = \sqrt[3]{10} + 2$
    $x \approx 4.15$

GOAL 2 USING *N*TH ROOTS IN REAL LIFE

Space Science

EXAMPLE 5   *Evaluating a Model with nth Roots*

The total mass *M* (in kilograms) of a spacecraft that can be propelled by a magnetic sail is, in theory, given by

$$M = \frac{0.015m^2}{fd^{4/3}}$$

where *m* is the mass (in kilograms) of the magnetic sail, *f* is the drag force (in newtons) of the spacecraft, and *d* is the distance (in astronomical units) to the sun. Find the total mass of a spacecraft that can be sent to Mars using *m* = 5000 kg, *f* = 4.52 N, and *d* = 1.52 AU. ▶ Source: *Journal of Spacecraft and Rockets*

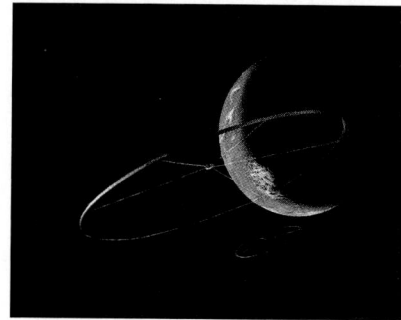

Artist's rendition of a magnetic sail

SOLUTION

$$M = \frac{0.015m^2}{fd^{4/3}} \qquad \text{Write model for total mass.}$$

$$= \frac{0.015(5000)^2}{4.52(1.52)^{4/3}} \qquad \text{Substitute for } m, f, \text{ and } d.$$

$$\approx 47{,}500 \qquad \text{Use a calculator.}$$

▶ The spacecraft can have a total mass of about 47,500 kilograms. (For comparison, the liftoff weight for a space shuttle is usually about 2,040,000 kilograms.)

EXAMPLE 6   *Solving an Equation Using an nth Root*

NAUTICAL SCIENCE The *Olympias* is a reconstruction of a trireme, a type of Greek galley ship used over 2000 years ago. The power *P* (in kilowatts) needed to propel the *Olympias* at a desired speed *s* (in knots) can be modeled by this equation:

$$P = 0.0289s^3$$

A volunteer crew of the *Olympias* was able to generate a maximum power of about 10.5 kilowatts. What was their greatest speed? ▶ Source: *Scientific American*

SOLUTION

$$P = 0.0289s^3 \qquad \text{Write model for power.}$$

$$10.5 = 0.0289s^3 \qquad \text{Substitute 10.5 for } P.$$

$$363 \approx s^3 \qquad \text{Divide each side by 0.0289.}$$

$$\sqrt[3]{363} \approx s \qquad \text{Take cube root of each side.}$$

$$7 \approx s \qquad \text{Use a calculator.}$$

▶ The greatest speed attained by the *Olympias* was approximately 7 knots (about 8 miles per hour).

FOCUS ON APPLICATIONS

NAUTICAL SCIENCE The *Olympias* was completed and first launched in 1987. A crew of 170 rowers is needed to run the ship.

APPLICATION LINK
www.mcdougallittell.com

Closure Question *Sample answer:*

Divide 7.4 by 2.175, then take the cube root of each side

to get an approximation of 1.50.

EXTRA EXAMPLE 5
The rate *r* at which an initial deposit *P* will grow to a balance *A* in *t* years with interest compounded *n* times a year is given by the formula $r = n\left[\left(\dfrac{A}{P}\right)^{1/nt} - 1\right]$.
Find *r* if *P* = $1000, *A* = $2000, *t* = 11 years, and *n* = 12.
about 6.3%

EXTRA EXAMPLE 6
A basketball has a volume of about 455.6 cubic inches. The formula for the volume of a basketball is $V = 4.18879r^3$. Find the radius of the basketball.
about 4.77 in.

CHECKPOINT EXERCISES
For use after Example 5:
1. Solve $T = \dfrac{0.0374y^5}{x^{1/3}z^{3/2}}$ for *T* when $y = 2.1$, $x = 5.5$, and $z = 2.8$.
about 0.185

For use after Example 6:
2. Solve $R = 2.178d^3$ for *d* when $R = 22.5$.   about 2.18

MATHEMATICAL REASONING
If *n* is even, the *n*th root of a negative number is not a real number. For example, the 4th root of −16 cannot be 2 because $2^4 = 16$, and it cannot be −2 because $(-2)^4 = 16$. Ask students to give another example.
Answers will vary.

APPLICATION NOTE
Additional information about nautical science is available at
www.mcdougallittell.com.

FOCUS ON VOCABULARY
b is the *n*th root of *a*. Write this using a radical sign. $b = \sqrt[n]{a}$

CLOSURE QUESTION
Describe how to solve $m = 2.175x^3$ for *x* when $m = 7.4$.   See margin.

DAILY PUZZLER
Find the square root of the fourth root of 256.   2

403

ASSIGNMENT GUIDE

BASIC
Day 1: pp. 404–406 Exs. 13–22,
24–62 even, 68, 73–87 odd

AVERAGE
Day 1: pp. 404–406 Exs. 13–28,
30–64 even, 68, 73–87 odd

ADVANCED
Day 1: pp. 404–406 Exs. 13–28,
30–68 even, 69–71,
73–89 odd

BLOCK SCHEDULE
pp. 404–406 Exs. 13–28,
30–64 even, 68, 73–87 odd,
(with Ch. 6 Assess.)

EXERCISE LEVELS
Level A: *Easier*
13–37, 41–57, 62–63
Level B: *More Difficult*
38–40, 58–61, 64–67
Level C: *Most Difficult*
68–71

✓ **HOMEWORK CHECK**
To quickly check student understanding of key concepts, go
over the following exercises:
Exs. 16, 22, 26, 32, 42, 54. See
also the Daily Homework Quiz:

• Blackline Master (*Chapter 7
  Resource Book*, p. 23)
• Transparency (p. 52)

GUIDED PRACTICE

Vocabulary Check ✓

Concept Check ✓

2a. Always true; take the
nth root of each side
of the first equation to
get the second.

Skill Check ✓

2b. Sometimes true; if $a = 1$,
then $1^{1/n} = \frac{1}{1^n}$.

1. What is the index of a radical? *n* is the index of the radical $\sqrt[n]{a}$ (the *n*th root of *a*).

2. **LOGICAL REASONING** Let *n* be an integer greater than 1. Tell whether the
given statement is *always true*, *sometimes true*, or *never true*. Explain. **See margin.**

   a. If $x^n = a$, then $x = \sqrt[n]{a}$.   **b.** $a^{1/n} = \frac{1}{a^n}$

3. Try to evaluate the expressions $-\sqrt[4]{625}$ and $\sqrt[4]{-625}$. Explain the difference in
   your results. -5; no real 4th roots; When *n* is even, there are only real *n*th roots for
   nonnegative numbers.

Evaluate the expression.

4. $\sqrt[4]{81}$ 3   5. $-\left(49^{1/2}\right)$ -7 6. $\left(\sqrt[3]{-8}\right)^5$ -32 7. $3125^{2/5}$ 25

Solve the equation.

8. $x^3 = 125$ 5   9. $3x^5 = -3$ -1 10. $(x + 4)^2 = 0$ -4 11. $x^4 - 7 = 9993$ ±10

12. 🌐 **SHOT PUT** The shot (a metal sphere) used in men's shot put has a volume
    of about 905 cubic centimeters. Find the radius of the shot. (*Hint:* Use the
    formula $V = \frac{4}{3}\pi r^3$ for the volume of a sphere.) **about 6 cm**

PRACTICE AND APPLICATIONS

STUDENT HELP

▶ **Extra Practice**
to help you master
skills is on p. 949.

USING RATIONAL EXPONENT NOTATION Rewrite the expression using
rational exponent notation.

13. $\sqrt[4]{14}$ $14^{1/4}$ 14. $\sqrt[3]{11}$ $11^{1/3}$ 15. $\left(\sqrt[7]{5}\right)^2$ $5^{2/7}$ 16. $\left(\sqrt[9]{16}\right)^5$ $16^{5/9}$ 17. $\left(\sqrt[8]{2}\right)^{11}$ $2^{11/8}$

USING RADICAL NOTATION Rewrite the expression using radical notation.

18. $6^{1/3}$ $\sqrt[3]{6}$ 19. $7^{1/4}$ $\sqrt[4]{7}$ 20. $10^{3/7}$ $\left(\sqrt[7]{10}\right)^3$ 21. $5^{2/5}$ $\left(\sqrt[5]{5}\right)^2$ 22. $8^{7/4}$ $\left(\sqrt[4]{8}\right)^7$

FINDING *N*TH ROOTS Find the indicated real *n*th root(s) of *a*.

23. $n = 2, a = 100$ ±10 24. $n = 4, a = 0$ 0   25. $n = 3, a = -8$ -2

26. $n = 7, a = 128$ 2   27. $n = 6, a = -1$ none   28. $n = 5, a = 0$ 0

EVALUATING EXPRESSIONS Evaluate the expression without using a
calculator.

29. $\sqrt[3]{64}$ 4   30. $\sqrt[3]{-1000}$ -10 31. $-\sqrt[6]{64}$ -2

32. $4^{-1/2}$ $\frac{1}{2}$ 33. $1^{1/3}$ 1   34. $-\left(256^{1/4}\right)$ -4

35. $\left(\sqrt[4]{16}\right)^2$ 4   36. $\left(\sqrt[3]{-27}\right)^{-4}$ $\frac{1}{81}$ 37. $\left(\sqrt[6]{0}\right)^3$ 0

38. $-\left(25^{-3/2}\right)$ $-\frac{1}{125}$ 39. $32^{4/5}$ 16   40. $(-125)^{-2/3}$ $\frac{1}{25}$

STUDENT HELP

▶ **HOMEWORK HELP**
Example 1: Exs. 13–28
Example 2: Exs. 29–40
Example 3: Exs. 41–52
Example 4: Exs. 53–61
Example 5: Exs. 62–64
Example 6: Exs. 65–67

🖩 **APPROXIMATING ROOTS** Evaluate the expression using a calculator.
Round the result to two decimal places when appropriate.

41. $\sqrt[5]{-16,807}$ -7 42. $\sqrt[9]{1124}$ 2.18   43. $\sqrt[8]{65,536}$ 4

44. $4^{1/10}$ 1.15   45. $10^{-1/4}$ 0.56   46. $-\left(1331^{1/3}\right)$ -11

47. $\left(\sqrt[3]{112}\right)^{-4}$ 0.0019   48. $\left(\sqrt[7]{-280}\right)^3$ -11.19 49. $\left(\sqrt[6]{6}\right)^2$ 1.82

50. $(-190)^{-4/5}$ 0.015   51. $26^{-3/4}$ 0.087   52. $522^{2/7}$ 5.98

H.W. (handwritten annotation)

SOLVING EQUATIONS Solve the equation. Round your answer to two decimal places when appropriate.

53. $x^5 = 243$ **3**
54. $6x^3 = -1296$ **−6**
55. $x^6 + 10 = 10$ **0**

56. $(x - 4)^4 = 81$ **7, 1**
57. $-x^7 = 40$ **−1.69**
58. $-12x^4 = -48$ **±1.41**

✳ **59.** $(x + 12)^3 = 21$ **−9.24**
estimate
60. $x^3 - 14 = 22$ **3.30**
61. $x^8 - 25 = -10$ **±1.40**

62. **BIOLOGY** ▷ **CONNECTION**   For mammals, the lung volume V (in milliliters) can be modeled by $V = 170m^{4/5}$ where m is the body mass (in kilograms). Find the lung volume of each mammal in the table shown.

▶ Source: *Respiration Physiology*

188.79 mL; 13,131.59 mL; 24,917.53 mL; 32,101.65 mL

Mammal	Body mass (kg)
Banded mongoose	1.14
Camel	229
Horse	510
Swiss cow	700

63. 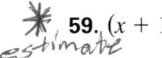 **SPILLWAY OF A DAM**   A dam's spillway capacity is an indication of how the dam will perform under certain flood conditions. The spillway capacity q (in cubic feet per second) of a dam can be calculated using the formula $q = c\ell h^{3/2}$ where c is the discharge coefficient, ℓ is the length (in feet) of the spillway, and h is the height (in feet) of the water on the spillway. A dam with a spillway 40 feet long, 5 feet deep, and 5 feet wide has a discharge coefficient of 2.79. What is the dam's maximum spillway capacity?   **1247.73 ft³/sec**

▶ Source: *Standard Handbook for Civil Engineers*

STUDENT HELP

HOMEWORK HELP
Visit our Web site
www.mcdougallittell.com
for help with problem
solving in Ex. 64.

64. 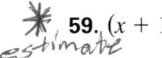 **INFLATION**   If the price of an item increases from p_1 to p_2 over a period of n years, the annual rate of inflation i (expressed as a decimal) can be modeled by $i = \left(\dfrac{p_2}{p_1}\right)^{1/n} - 1$. In 1940 the average value of a home was \$2900. In 1990 the average value was \$79,100. What was the rate of inflation for a home?

▶ Source: *Bureau of the Census*   **0.068**

65. **GEOMETRY** ▷ **CONNECTION**   The formula for the volume V of a regular dodecahedron (a solid with 12 regular pentagons as faces) is $V \approx 7.66a^3$ where a is the length of an edge of the dodecahedron. Find the length of an edge of a regular dodecahedron that has a volume of 30 cubic feet. Round your answer to two decimal places.   **1.58 ft**

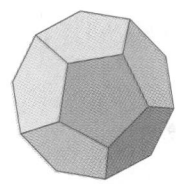

66. **GEOMETRY** ▷ **CONNECTION**   The formula for the volume V of a regular icosahedron (a solid with 20 congruent equilateral triangles as faces) is $V \approx 2.18a^3$ where a is the length of an edge of the icosahedron. Find the length of an edge of a regular icosahedron that has a volume of 21 cubic centimeters. Round your answer to two decimal places.   **2.13 cm**

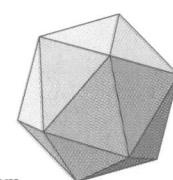

67. 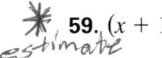 **ISLAND SPECIES**   Philip Darlington discovered a rule of thumb that relates an island's land area A (in square miles) to the number s of reptile and amphibian species the island can support by the model $A = 0.0779s^3$. The area of Puerto Rico is roughly 4000 square miles. About how many reptile and amphibian species can it support?

▶ Source: *The Song of the Dodo: Island Biogeography in an Age of Extinctions*   **about 37 species**

7.1 *nth Roots and Rational Exponents*   **405**

❗ **COMMON ERROR**
EXERCISES 18–22 Students are often confused about the index in radical notation. Remind students that the denominator of the rational exponent is the index.

❗ **COMMON ERROR**
EXERCISE 36 Some students may forget that a negative power represents a reciprocal. Remind students to find the fourth power of the reciprocal of $\sqrt[3]{-27}$, or $\left(\dfrac{1}{\sqrt[3]{-27}}\right)^4$.

STUDENT HELP NOTES
↳ **Homework Help** Students can find help for Ex. 64 at **www.mcdougallittell.com**. The information can be printed out for students who don't have access to the Internet.

APPLICATION NOTE
EXERCISE 64 This formula is based on the simple interest formula $p_2 = p_1(1 + i)^n$ where p_1 is the initial price, p_2 is the final price, i is the interest rate, and n is the number of years. You may want to show the steps to derive the formula.

$$i = \left(\frac{p_2}{p_1}\right)^{1/n} - 1$$

$$1 + i = \left(\frac{p_2}{p_1}\right)^{1/n}$$

$$(1 + i)^n = \frac{p_2}{p_1}$$

$$p_1(1 + i)^n = p_2$$

ADDITIONAL PRACTICE AND RETEACHING

For Lesson 7.1:
- Practice Levels A, B, and C (*Chapter 7 Resource Book*, p. 13)
- Reteaching with Practice (*Chapter 7 Resource Book*, p. 16)
- 🖥 See Lesson 7.1 of the *Personal Student Tutor.*

For more Mixed Review:
- 🖥 Search the *Test and Practice Generator* for key words or specific lessons.

DAILY HOMEWORK QUIZ

🔲 **Transparency Available**

Evaluate the expression.

1. $\left(\sqrt[4]{24}\right)^2$ $2\sqrt{6}$

2. $9^{5/4}$ $9 \cdot 3^{1/2}$

3. $16^{-1/4}$ $\frac{1}{2}$

Evaluate the expression using a calculator. Round the result to two decimal places.

4. $\sqrt[3]{225}$ 6.08   **5.** $12^{1/4}$ 1.86

Solve the equation. Round the result to two decimal places.

6. $2x^4 = 35$ ± 2.05

7. $5x^3 + 10 = 961$ 5.75

┌─ **EXTRA CHALLENGE NOTE** ─┐
→ Challenge problems for Lesson 7.1 are available in **blackline** format in the *Chapter 7 Resource Book*, p. 20 and at **www.mcdougallittell.com**.

ADDITIONAL TEST PREPARATION

1. WRITING Explain why $x^4 = 16$ has two real solutions while $x^3 = 8$ has only one real solution. An even power of a positive or negative number is positive, so there are 2 real solutions (± 2) for $x^4 = 16$. An odd power of a negative number is negative, while an odd power of a positive number is positive. Therefore there is only one real solution (2) for $x^3 = 8$.

69.

	$a < 0$	$a = 0$	$a > 0$
n is even	0	1	2
n is odd	1	1	1

71.

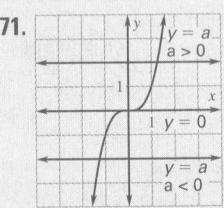

Test Preparation

70. Real roots exist whenever the line $y = a$ crosses the graph. This happens twice for $a > 0$, once for $a = 0$ and not at all for $a < 0$.

74. $x = -\frac{11}{2}, y = -\frac{9}{2}$

76. $x = \frac{18}{79}, y = -\frac{101}{79}$

78. x^2; product of powers property

79. $\frac{1}{x^{15}}$; power of a power and negative exponent properties

★ **Challenge**

80. $\frac{1}{4x^2y^6}$; power of a power, power of a product, and negative exponent properties

┌─ **EXTRA CHALLENGE** ─┐
→ www.mcdougallittell.com

---

68. MULTI-STEP PROBLEM A board foot is a unit for measuring wood. One board foot has a volume of 144 cubic inches. The Doyle log rule, given by $b = l\left(\frac{r-2}{2}\right)^2$, is a formula for approximating the number b of board feet in a log with length l (in feet) and radius r (in inches). The total volume V (in cubic inches) of wood in the main trunk of a Douglas fir can be modeled by $V = 250r^3$ where r is the radius of the trunk at the base of the tree. Suppose you need 5000 board feet from a 20 foot Douglas fir log.

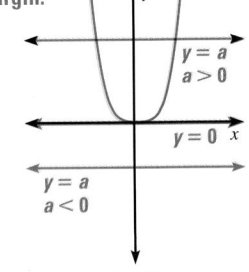

Log-sawing patterns for maximum board feet

a. What volume of wood do you need?   720,000 in.^{3}

b. What is the radius of a log that will meet your needs?   33.62 in.

c. What is the total volume of wood in the main trunk of a Douglas fir tree that will meet your needs?   9,500,000 in.^{3}

d. If you find a suitable tree, what fraction of the tree would you actually use?   7.6%

e. *Writing* How does your answer to part (d) change if you instead need only 2500 board feet?   the fraction increases to almost 10%

69. VISUAL THINKING Copy the table. Give the   See margin. number of nth roots of a for each category.

	$a < 0$	$a = 0$	$a > 0$
n is even	?	?	?
n is odd	?	?	?

70. The graph of $y = x^n$ where n is even is shown in red. Explain how the graph justifies the table for n even.

71. Draw a similar graph to justify the table for n odd.

70, 71. See margin.

Ex. 70

MIXED REVIEW

81. $\frac{5}{x^2}$; negative exponent and zero exponent properties

82. x^7; quotient of powers property

83. $\frac{1}{x^4y^2}$; negative exponent and power of a quotient properties

84. $\frac{x^2y^{10}}{2}$; quotient of powers property

85. $4x^2y$; product of powers and quotient of powers properties

SOLVING SYSTEMS Use Cramer's rule to solve the linear system. *(Review 4.3)*
See margin.

72. $x + 4y = 12$ $x = 4, y = 2$
$2x + 5y = 18$

73. $x - 2y = 11$ $x = 3, y = -4$
$2x + 5y = -14$

74. $2x - 4y = 7$
$-x + y = 1$ $x = \frac{13}{11},$

75. $-3x + 2y = -9$
$x - 4y = 2$ $x = \frac{16}{5}, y = \frac{3}{10}$

76. $-x - 8y = 10$
$10x + y = 1$ See margin.

77. $-x - y = 0$
$5x - 6y = 13$ $y = -\frac{13}{11}$

SIMPLIFYING EXPRESSIONS Simplify the expression. Tell which properties of exponents you used. *(Review 6.1 for 7.2)*   78–85. See margin.

78. $x^4 \cdot x^{-2}$ **79.** $\left(x^{-3}\right)^5$ **80.** $\left(2xy^3\right)^{-2}$ **81.** $5x^{-2}y^0$

82. $\frac{x^3}{x^{-4}}$ **83.** $\left(\frac{x^{-2}}{y}\right)^2$ **84.** $\frac{7x^3y^8}{14xy^{-2}}$ **85.** $\frac{16xy}{9x^5} \cdot \frac{9x^6y}{4y}$

FINDING ZEROS Find all the zeros of the polynomial function. *(Review 6.7)*

86. $f(x) = x^4 + 9x^3 - 5x^2 - 153x - 140$ $-1, 4, -5, -7$

87. $f(x) = x^4 + x^3 - 19x^2 + 11x + 30$ $-1, 2, 3, -5$

88. $f(x) = x^3 - 5x^2 + 16x - 80$ $5, \pm 4i$

89. $f(x) = x^3 - x^2 + 9x - 9$ $1, \pm 3i$

Properties of Rational Exponents

What you should learn

GOAL 1 Use properties of rational exponents to evaluate and simplify expressions.

GOAL 2 Use properties of rational exponents to solve **real-life** problems, such as finding the surface area of a mammal in **Example 8**.

Why you should learn it

▼ To model **real-life** quantities, such as the frequencies in the musical range of a trumpet for **Ex. 94**.

GOAL 1 PROPERTIES OF RATIONAL EXPONENTS AND RADICALS

The properties of integer exponents presented in Lesson 6.1 can also be applied to rational exponents.

CONCEPT SUMMARY — PROPERTIES OF RATIONAL EXPONENTS

Let a and b be real numbers and let m and n be rational numbers. The following properties have the same names as those listed on page 323, but now apply to rational exponents as illustrated.

	PROPERTY	EXAMPLE
1.	$a^m \cdot a^n = a^{m+n}$	$3^{1/2} \cdot 3^{3/2} = 3^{(1/2 + 3/2)} = 3^2 = 9$
2.	$(a^m)^n = a^{mn}$	$(4^{3/2})^2 = 4^{(3/2 \cdot 2)} = 4^3 = 64$
3.	$(ab)^m = a^m b^m$	$(9 \cdot 4)^{1/2} = 9^{1/2} \cdot 4^{1/2} = 3 \cdot 2 = 6$
4.	$a^{-m} = \dfrac{1}{a^m}, a \neq 0$	$25^{-1/2} = \dfrac{1}{25^{1/2}} = \dfrac{1}{5}$
5.	$\dfrac{a^m}{a^n} = a^{m-n}, a \neq 0$	$\dfrac{6^{5/2}}{6^{1/2}} = 6^{(5/2 - 1/2)} = 6^2 = 36$
6.	$\left(\dfrac{a}{b}\right)^m = \dfrac{a^m}{b^m}, b \neq 0$	$\left(\dfrac{8}{27}\right)^{1/3} = \dfrac{8^{1/3}}{27^{1/3}} = \dfrac{2}{3}$

If $m = \dfrac{1}{n}$ for some integer n greater than 1, the third and sixth properties can be written using radical notation as follows:

$$\sqrt[n]{a \cdot b} = \sqrt[n]{a} \cdot \sqrt[n]{b} \qquad \text{Product property}$$

$$\sqrt[n]{\frac{a}{b}} = \frac{\sqrt[n]{a}}{\sqrt[n]{b}} \qquad \text{Quotient property}$$

EXAMPLE 1 Using Properties of Rational Exponents

Use the properties of rational exponents to simplify the expression.

a. $5^{1/2} \cdot 5^{1/4} = 5^{(1/2 + 1/4)} = 5^{3/4}$

b. $(8^{1/2} \cdot 5^{1/3})^2 = (8^{1/2})^2 \cdot (5^{1/3})^2 = 8^{(1/2 \cdot 2)} \cdot 5^{(1/3 \cdot 2)} = 8^1 \cdot 5^{2/3} = 8 \cdot 5^{2/3}$

c. $(2^4 \cdot 3^4)^{-1/4} = [(2 \cdot 3)^4]^{-1/4} = (6^4)^{-1/4} = 6^{[4 \cdot (-1/4)]} = 6^{-1} = \dfrac{1}{6}$

d. $\dfrac{7}{7^{1/3}} = \dfrac{7^1}{7^{1/3}} = 7^{(1 - 1/3)} = 7^{2/3}$

e. $\left(\dfrac{12^{1/3}}{4^{1/3}}\right)^2 = \left[\left(\dfrac{12}{4}\right)^{1/3}\right]^2 = (3^{1/3})^2 = 3^{(1/3 \cdot 2)} = 3^{2/3}$

STUDENT HELP

→ **Look Back**
For help with properties of exponents, see p. 324.

1 PLAN

PACING
Basic: 2 days
Average: 2 days
Advanced: 2 days
Block Schedule: 1 block

→ **LESSON OPENER**
ACTIVITY
An alternative way to approach Lesson 7.2 is to use the Activity Lesson Opener:
- Blackline Master (*Chapter 7 Resource Book,* p. 24)
- Transparency (p. 45)

MEETING INDIVIDUAL NEEDS
- *Chapter 7 Resource Book*
  Prerequisite Skills Review (p. 5)
  Practice Level A (p. 25)
  Practice Level B (p. 26)
  Practice Level C (p. 27)
  Reteaching with Practice (p. 28)
  Absent Student Catch-Up (p. 30)
  Challenge (p. 32)
- *Resources in Spanish*
- *Personal Student Tutor*

NEW-TEACHER SUPPORT
See the Tips for New Teachers on pp. 1–2 of the *Chapter 7 Resource Book* for additional notes about Lesson 7.2.

WARM-UP EXERCISES

Transparency Available
Evaluate the expression.
1. $4^2 \cdot 4^3$ 1024
2. $(2^2)^3$ 64
3. $\dfrac{3^4}{3^2}$ 9
4. 2^{-3} $\dfrac{1}{8}$
5. $(2 \cdot 3)^4$ 1296

EXTRA EXAMPLE 1
Simplify.
a. $6^{1/2} \cdot 6^{1/3}$ $6^{5/6}$
b. $(27^{1/3} \cdot 6^{1/4})^2$ $9 \cdot 6^{1/2}$
c. $(4^3 \cdot 2^3)^{-1/3}$ $\dfrac{1}{8}$
d. $\dfrac{6}{6^{3/4}}$ $6^{1/4}$ **e.** $\left(\dfrac{18^{1/4}}{9^{1/4}}\right)^3$ $2^{3/4}$

EXTRA EXAMPLE 2
Simplify.
a. $\sqrt[3]{25} \cdot \sqrt[3]{5}$ 5   **b.** $\dfrac{\sqrt[3]{32}}{\sqrt[3]{4}}$ 2

EXTRA EXAMPLE 3
Write in simplest form.
a. $\sqrt[4]{64}$ $2\sqrt[4]{4}$ **b.** $\sqrt[4]{\dfrac{7}{8}}$ $\dfrac{\sqrt[4]{14}}{2}$

EXTRA EXAMPLE 4
Perform the indicated operation.
a. $5(4^{3/4}) - 3(4^{3/4})$ $2(4^{3/4})$
b. $\sqrt[3]{81} - \sqrt[3]{3}$ $2\sqrt[3]{3}$

✓ CHECKPOINT EXERCISES

For use after Example 1:
1. Simplify.
   a. $3^{1/4} \cdot 3^{3/4}$ 3
   b. $(64^{1/3} \cdot 8^{1/3})^2$ 64
   c. $(3^3 \cdot 6^3)^{-1/3}$ $\dfrac{1}{18}$
   d. $\dfrac{4}{4^{1/2}}$ 2
   e. $\left(\dfrac{54^{1/4}}{27^{1/4}}\right)^2$ $2^{1/2}$

For use after Example 2:
2. Simplify.
   a. $\sqrt[4]{27} \cdot \sqrt[4]{3}$ 3
   b. $\dfrac{\sqrt[3]{625}}{\sqrt[3]{5}}$ 5

For use after Example 3:
3. Write in simplest form.
   a. $\sqrt[3]{10{,}000}$ $10\sqrt[3]{10}$
   b. $\sqrt[4]{\dfrac{2}{3}}$ $\dfrac{\sqrt[4]{54}}{3}$

For use after Example 4:
4. Perform the indicated operation.
   a. $6(3^{2/3}) + 4(3^{2/3})$ $10(3^{2/3})$
   b. $\sqrt[3]{625} - \sqrt[3]{5}$ $4\sqrt[3]{5}$

EXAMPLE 2   *Using Properties of Radicals*

Use the properties of radicals to simplify the expression.

a. $\sqrt[3]{4} \cdot \sqrt[3]{16} = \sqrt[3]{4 \cdot 16} = \sqrt[3]{64} = 4$ **Use the product property.**

b. $\dfrac{\sqrt[4]{162}}{\sqrt[4]{2}} = \sqrt[4]{\dfrac{162}{2}} = \sqrt[4]{81} = 3$ **Use the quotient property.**

· · · · · · · · · ·

For a radical to be in **simplest form,** you must not only apply the properties of radicals, but also remove any perfect *n*th powers (other than 1) and rationalize any denominators.

EXAMPLE 3   *Writing Radicals in Simplest Form*

STUDENT HELP

HOMEWORK HELP
Visit our Web site
www.mcdougallittell.com
for extra examples.

Write the expression in simplest form.

a. $\sqrt[3]{54} = \sqrt[3]{27 \cdot 2}$ **Factor out perfect cube.**
   $= \sqrt[3]{27} \cdot \sqrt[3]{2}$ **Product property**
   $= 3\sqrt[3]{2}$ **Simplify.**

b. $\sqrt[5]{\dfrac{3}{4}} = \sqrt[5]{\dfrac{3 \cdot 8}{4 \cdot 8}}$ **Make the denominator a perfect fifth power.**
   $= \sqrt[5]{\dfrac{24}{32}}$ **Simplify.**
   $= \dfrac{\sqrt[5]{24}}{\sqrt[5]{32}}$ **Quotient property**
   $= \dfrac{\sqrt[5]{24}}{2}$ **Simplify.**

· · · · · · · · · ·

Two radical expressions are **like radicals** if they have the same index and the same radicand. For instance, $\sqrt[3]{2}$ and $4\sqrt[3]{2}$ are like radicals. To add or subtract like radicals, use the distributive property.

EXAMPLE 4   *Adding and Subtracting Roots and Radicals*

Perform the indicated operation.

a. $7\left(6^{1/5}\right) + 2\left(6^{1/5}\right) = (7 + 2)\left(6^{1/5}\right) = 9\left(6^{1/5}\right)$

b. $\sqrt[3]{16} - \sqrt[3]{2} = \sqrt[3]{8 \cdot 2} - \sqrt[3]{2}$
   $= \sqrt[3]{8} \cdot \sqrt[3]{2} - \sqrt[3]{2}$
   $= 2\sqrt[3]{2} - \sqrt[3]{2}$
   $= (2 - 1)\sqrt[3]{2}$
   $= \sqrt[3]{2}$

The properties of rational exponents and radicals can also be applied to expressions involving variables. Because a variable can be positive, negative or zero, sometimes absolute value is needed when simplifying a variable expression.

$$\sqrt[n]{x^n} = x \text{ when } n \text{ is odd} \qquad\qquad \sqrt[7]{2^7} = 2 \text{ and } \sqrt[7]{(-2)^7} = -2$$

$$\sqrt[n]{x^n} = |x| \text{ when } n \text{ is even} \qquad\qquad \sqrt[4]{5^4} = 5 \text{ and } \sqrt[4]{(-5)^4} = 5$$

Absolute value is not needed when all variables are assumed to be positive.

EXAMPLE 5 *Simplifying Expressions Involving Variables*

Simplify the expression. Assume all variables are positive.

a. $\sqrt[3]{125y^6} = \sqrt[3]{5^3(y^2)^3} = 5y^2$

b. $\left(9u^2v^{10}\right)^{1/2} = 9^{1/2}\left(u^2\right)^{1/2}\left(v^{10}\right)^{1/2} = 3u^{(2\cdot 1/2)}v^{(10\cdot 1/2)} = 3uv^5$

c. $\sqrt[4]{\dfrac{x^4}{y^8}} = \dfrac{\sqrt[4]{x^4}}{\sqrt[4]{y^8}} = \dfrac{\sqrt[4]{x^4}}{\sqrt[4]{(y^2)^4}} = \dfrac{x}{y^2}$

d. $\dfrac{6xy^{1/2}}{2x^{1/3}z^{-5}} = 3x^{(1-1/3)}y^{1/2}z^{-(-5)} = 3x^{2/3}y^{1/2}z^5$

EXAMPLE 6 *Writing Variable Expressions in Simplest Form*

STUDENT HELP

HOMEWORK HELP
Visit our Web site
www.mcdougallittell.com
for extra examples.

Write the expression in simplest form. Assume all variables are positive.

a. $\sqrt[5]{5a^5b^9c^{13}} = \sqrt[5]{5a^5b^5b^4c^{10}c^3}$ **Factor out perfect fifth powers.**

$= \sqrt[5]{a^5b^5c^{10}} \cdot \sqrt[5]{5b^4c^3}$ **Product property**

$= abc^2\sqrt[5]{5b^4c^3}$ **Simplify.**

b. $\sqrt[3]{\dfrac{x}{y^7}} = \sqrt[3]{\dfrac{xy^2}{y^7y^2}}$ **Make the denominator a perfect cube.**

$= \sqrt[3]{\dfrac{xy^2}{y^9}}$ **Simplify.**

$= \dfrac{\sqrt[3]{xy^2}}{\sqrt[3]{y^9}}$ **Quotient property**

$= \dfrac{\sqrt[3]{xy^2}}{y^3}$ **Simplify.**

EXAMPLE 7 *Adding and Subtracting Expressions Involving Variables*

Perform the indicated operation. Assume all variables are positive.

a. $5\sqrt{y} + 6\sqrt{y} = (5+6)\sqrt{y} = 11\sqrt{y}$

b. $2xy^{1/3} - 7xy^{1/3} = (2-7)xy^{1/3} = -5xy^{1/3}$

c. $3\sqrt[3]{5x^5} - x\sqrt[3]{40x^2} = 3x\sqrt[3]{5x^2} - 2x\sqrt[3]{5x^2} = (3x-2x)\sqrt[3]{5x^2} = x\sqrt[3]{5x^2}$

7.2 *Properties of Rational Exponents*   **409**

CONCEPT QUESTION
EXAMPLE 2 Can you find the quotient of two radicals with different indices? **Yes. Change both to rational exponent form and use the quotient property.**

EXTRA EXAMPLE 5
Simplify the expression. Assume all variables are positive.
a. $\sqrt[3]{27z^9}$ $3z^3$
b. $(16g^4h^2)^{1/2}$ $4g^2h$
c. $\sqrt[5]{\dfrac{x^5}{y^{10}}}$ $\dfrac{x}{y^2}$
d. $\dfrac{18rs^{2/3}}{6r^{1/4}t^{-3}}$ $3r^{3/4}s^{2/3}t^3$

EXTRA EXAMPLE 6
Write the expression in simplest form. Assume all variables are positive.
a. $\sqrt[4]{12d^4e^9f^{14}}$ $de^2f^3\sqrt[4]{12ef^2}$
b. $\sqrt[5]{\dfrac{g^2}{h^7}}$ $\dfrac{\sqrt[5]{g^2h^3}}{h^2}$

EXTRA EXAMPLE 7
Perform the indicated operations. Assume all variables are positive.
a. $8\sqrt{x} - 3\sqrt{x}$ $5\sqrt{x}$
b. $3gh^{1/4} - 6gh^{1/4}$ $-3gh^{1/4}$
c. $2\sqrt[4]{6x^5} + x\sqrt[4]{6x}$ $3x\sqrt[4]{6x}$

CHECKPOINT EXERCISES
For use after Examples 5 and 6:
1. Simplify the expression. Assume all variables are positive.
  a. $\sqrt[3]{8r^3s^5t^{10}}$ $2rst^3\sqrt[3]{s^2t}$
  b. $(625j^8k^4)^{1/4}$ $5j^2k$
  c. $\sqrt[4]{\dfrac{x^8y^{11}}{z^6}}$ $\dfrac{x^2y^2\sqrt[4]{y^3z^2}}{z^2}$
  d. $\dfrac{15d^2e^{2/3}f}{5df^{-4}}$ $3de^{2/3}f^5$

For use after Example 7:
2. Simplify the expression. Assume all variables are positive.
  a. $6\sqrt{s} - 2\sqrt{s} + \sqrt{s}$ $5\sqrt{s}$
  b. $4m^2n^{1/3} - 6m^2n^{1/3}$ $-2m^2n^{1/3}$
  c. $3\sqrt[3]{6y^7} + 2y^2\sqrt[3]{6y}$ $5y^2\sqrt[3]{6y}$

EXTRA EXAMPLE 8
The weight W in tons of a whale as a function of length L in feet can be approximated by the model $W = 0.1077L^{3/2}$. Approximate the weight of a humpback whale with length 49.17 feet.   **37.1 tons**

EXTRA EXAMPLE 9
A pilot whale is about $\frac{1}{5}$ the length of a blue whale. Is its weight also $\frac{1}{5}$ the weight of the blue whale?   **No, its weight is about $\left(\frac{1}{5}\right)^{3/2}$, or 0.0894, times the weight of the blue whale.**

 CHECKPOINT EXERCISES
For use after Examples 8 and 9:
1. The mass M in grams of a bat as a function of length L in centimeters can be approximated by the model $M = 0.402L^{3/2}$. Approximate the weight of the California mitosis bat with length 4.3 centimeters.   **3.6 g**

MATHEMATICAL REASONING
The properties of exponents also apply to rational exponents. Ask students to give the property that you use to simplify $(6^{3/2})^2 = 216$.
Power of a power property: $(a^m)^n = a^{m \cdot n}$

FOCUS ON VOCABULARY
Give a like radical for $\sqrt[4]{7}$.
Sample answer: $2\sqrt[4]{7}$

CLOSURE QUESTION
When is $\sqrt[n]{x^n}$ equal to x? When is $\sqrt[n]{x^n}$ equal to $|x|$?   **when the index n is odd; when the index n is even**

DAILY PUZZLER
When is $\sqrt[n]{x} > x$?   **if n is even, when $0 < x < 1$; if n is odd, when $0 < x < 1$ or $x < -1$**

410

Biology

EXAMPLE 8 *Evaluating a Model Using Properties of Rational Exponents*

Biologists study characteristics of various living things. One way of comparing different animals is to compare their sizes. For example, a mammal's surface area S (in square centimeters) can be approximated by the model $S = km^{2/3}$ where m is the mass (in grams) of the mammal and k is a constant. The values of k for several mammals are given in the table.

Mammal	Mouse	Cat	Large dog	Cow	Rabbit	Human
k	9.0	10.0	11.2	9.0	9.75	11.0

Approximate the surface area of a cat that has a mass of 5 kilograms (5×10^3 grams).   ▶ Source: *Scaling: Why Is Animal Size So Important?*

SOLUTION

$$
\begin{aligned}
S &= km^{2/3} & \text{Write model.}\\
&= 10.0\left(5 \times 10^3\right)^{2/3} & \text{Substitute for } k \text{ and } m.\\
&= 10.0(5)^{2/3}\left(10^3\right)^{2/3} & \text{Power of a product property}\\
&\approx 10.0(2.92)\left(10^2\right) & \text{Power of a power property}\\
&= 2920 & \text{Simplify.}
\end{aligned}
$$

▶ The cat's surface area is approximately 3000 square centimeters.

EXAMPLE 9 *Using Properties of Rational Exponents with Variables*

BIOLOGY CONNECTION You are studying a Canadian lynx whose mass is twice the mass of an average house cat. Is its surface area also twice that of an average house cat?

FOCUS ON APPLICATIONS

▶ **BIOLOGY** The average mass of a Canadian lynx is about 9.1 kilograms. The average mass of a house cat is about 4.8 kilograms.

SOLUTION

Let m be the mass of an average house cat. Then the mass of the Canadian lynx is $2m$. The surface areas of the house cat and the Canadian lynx can be approximated by:

$$S_{\text{cat}} = 10.0m^{2/3} \qquad S_{\text{lynx}} = 10.0(2m)^{2/3}$$

To compare the surface areas look at their ratio.

$$
\begin{aligned}
\frac{S_{\text{lynx}}}{S_{\text{cat}}} &= \frac{10.0(2m)^{2/3}}{10.0m^{2/3}} & \text{Write ratio of surface areas.}\\[4pt]
&= \frac{10.0\left(2^{2/3}\right)\left(m^{2/3}\right)}{10.0m^{2/3}} & \text{Power of a product property}\\[4pt]
&= 2^{2/3} & \text{Simplify.}\\[4pt]
&\approx 1.59 & \text{Evaluate.}
\end{aligned}
$$

▶ The surface area of the Canadian lynx is about one and a half times that of an average house cat, not twice as much.

GUIDED PRACTICE

Vocabulary Check ✓

Concept Check ✓

3. $5\sqrt[4]{5}$; To add or subtract like radicals, use the distributive property.

Skill Check ✓

1. List three pairs of like radicals. *Sample answers:* $5\sqrt{10}, 2\sqrt{10}; 7\sqrt[3]{4}, \sqrt[3]{4}; 9\sqrt[6]{37}, 8\sqrt[6]{37}$

2. If you know that $46,656,000 = 2^9 \cdot 3^6 \cdot 5^3$, what is the cube root of $46,656,000$? Explain your reasoning. $360; 2^3 \times 3^2 \times 5 = 360$

ERROR ANALYSIS Explain the error made in simplifying the expression.

3. $3\sqrt[4]{5} + 2\sqrt[4]{5} = 5\sqrt[4]{10}$

4. $\left(\dfrac{x}{y^8}\right)^{1/3} = \dfrac{x^{1/3}}{(y^8)^{1/3}} = \dfrac{x^{1/3}}{y^2}$ $\dfrac{x^{1/3}}{y^{8/3}}$; Use the power of a power property.

Simplify the expression.

5. $3^{1/4} \cdot 3^{3/4}$ 3

6. $\left(5^{1/3}\right)^6$ 25

7. $\sqrt[3]{16} \cdot \sqrt[3]{4}$ 4

8. $4^{-1/2}$ $\dfrac{1}{2}$

9. $\sqrt[4]{\dfrac{16}{81}}$ $\dfrac{2}{3}$

10. $\sqrt[3]{\dfrac{1}{4}}$ $\dfrac{2^{1/3}}{2}$

11. $8^{1/7} + 2\left(8^{1/7}\right)$ $3\sqrt[7]{8}$

12. $\sqrt{200} - 3\sqrt{2}$ $7\sqrt{2}$

Simplify the expression. Assume all variables are positive.

13. $x^{2/3} \cdot x^{4/3}$ x^2

14. $\left(y^{1/6}\right)^3$ $y^{1/2}$

15. $\sqrt{4a^6}$ $2a^3$

16. $b^{-1/3}$ $\dfrac{b^{2/3}}{b}$

17. $\sqrt[5]{\dfrac{x^{10}}{y^5}}$ $\dfrac{x^2}{y}$

18. $\sqrt[3]{\dfrac{x^2}{z}}$ $\dfrac{(xz)^{2/3}}{z}$

19. $2a^{1/5} - 6a^{1/5}$ $-4a^{1/5}$

20. $x\sqrt[3]{y^6} + y^2\sqrt[3]{x^3}$ $2xy^2$

21. **BIOLOGY** ▸ **CONNECTION** The average mass of a rabbit is 1.6 kilograms. Use the information given in Example 8 to approximate the surface area of a rabbit. 1333.78 cm^2

PRACTICE AND APPLICATIONS

STUDENT HELP

▸ **Extra Practice**
to help you master skills is on p. 949.

PROPERTIES OF RATIONAL EXPONENTS Simplify the expression.

22. $3^{5/3} \cdot 3^{1/3}$ 9

23. $\left(5^{2/3}\right)^{1/2}$ $5^{1/3}$

24. $4^{1/4} \cdot 64^{1/4}$ 4

25. $\dfrac{1}{36^{-1/2}}$ 6

26. $\dfrac{7^{1/5}}{7^{3/5}}$ $\dfrac{7^{3/5}}{7}$

27. $\dfrac{70^{1/3}}{14^{1/3}}$ $5^{1/3}$

28. $\left(2^{1/4} \cdot 2^{1/3}\right)^6$ $2^{7/2}$

29. $\left(\dfrac{5^2}{8^2}\right)^{-1/2}$ $\dfrac{8}{5}$

30. $\dfrac{6^{2/3} \cdot 4^{2/3}}{3^{2/3}}$ 4

31. $\dfrac{125^{2/9} \cdot 125^{1/9}}{5^{3/4}}$ $5^{1/4}$

32. $\dfrac{12^{10/8}}{12^{-3/8}}$ $12^{13/8}$

33. $\left(10^{3/4} \cdot 4^{3/4}\right)^{-4}$ $\dfrac{1}{64,000}$

PROPERTIES OF RADICALS Simplify the expression.

34. $\sqrt{64} \cdot \sqrt[3]{64}$ 32

35. $\sqrt[4]{8} \cdot \sqrt[4]{2}$ 2

36. $\sqrt[4]{5} \cdot \sqrt[4]{5}$ $5^{1/2} \approx 2.24$

37. $\left(\sqrt[3]{6} \cdot \sqrt[4]{6}\right)^{12}$ $6^7 = 279,936$

38. $\dfrac{\sqrt{7}}{\sqrt[5]{7}}$ $7^{3/10} \approx 1.79$

39. $\dfrac{\sqrt[3]{4}}{\sqrt[3]{32}}$ $\dfrac{1}{2}$

40. $\dfrac{\sqrt[6]{8} \cdot \sqrt[6]{16}}{\sqrt[6]{2}}$ 2

41. $\dfrac{\sqrt[3]{9} \cdot \sqrt[3]{6}}{\sqrt[6]{2} \cdot \sqrt[6]{2}}$ 3

SIMPLEST FORM Write the expression in simplest form.

42. $\sqrt{50}$ $5\sqrt{2}$

43. $\sqrt[5]{1215}$ $3\sqrt[5]{5}$

44. $\sqrt[3]{18} \cdot \sqrt[3]{15}$ $3\sqrt[3]{10}$

45. $3\sqrt[4]{24} \cdot 5\sqrt[4]{2}$ $30\sqrt[4]{3}$

46. $\sqrt[3]{\dfrac{1}{7}}$ $\dfrac{\sqrt[3]{49}}{7}$

47. $\dfrac{2}{\sqrt[6]{81}}$ $\dfrac{2\sqrt[3]{3}}{3}$

48. $\sqrt[4]{\dfrac{80}{9}}$ $\dfrac{2\sqrt[4]{45}}{3}$

49. $\dfrac{\sqrt[3]{4}}{\sqrt[5]{8}}$ $\sqrt[15]{2}$

COMBINING ROOTS AND RADICALS Perform the indicated operation.

50. $\sqrt[5]{6} + 5\sqrt[5]{6}$ $6\sqrt[5]{6}$

51. $5(5)^{1/7} - 7(5)^{1/7}$ $-2\sqrt[7]{5}$

52. $-\sqrt[8]{4} + 5\sqrt[8]{4}$ $4\sqrt[8]{4} = 4\sqrt[4]{2}$

53. $160^{1/2} - 10^{1/2}$ $3\sqrt{10}$

54. $\sqrt[3]{375} + \sqrt[3]{81}$ $8\sqrt[3]{3}$

55. $2\sqrt[4]{176} + 5\sqrt[4]{11}$ $9\sqrt[4]{11}$

STUDENT HELP

▸ **HOMEWORK HELP**
Example 1: Exs. 22–33
Example 2: Exs. 34–41
Example 3: Exs. 42–49
Example 4: Exs. 50–55
Example 5: Exs. 56–67
Example 6: Exs. 68–75
Example 7: Exs. 76–81
Example 8: Exs. 90–93
Example 9: Exs. 94–97

---

3 APPLY

ASSIGNMENT GUIDE

BASIC
Day 1: pp. 411–413 Exs. 22–48 even, 50–52, 55–63, 69, 71, 77
Day 2: pp. 411–413 Exs. 33, 41, 49, 53–55, 64–67, 68–90 even, 98, 101–111 odd, Quiz 1 Exs. 1–22

AVERAGE
Day 1: pp. 411–413 Exs. 22–50 even, 56–84 even, 90–91
Day 2: pp. 411–413 Exs. 33, 41, 49, 55, 67, 81, 89, 92–94, 97, 101–111 odd, Quiz 1 Exs. 1–22

ADVANCED
Day 1: pp. 411–413 Exs. 22–80 even, 81–88, 90–92
Day 2: pp. 411–413 Exs. 33, 41, 49, 55, 67, 81, 89, 93–99, 101–111 odd, Quiz 1 Exs. 1–22

BLOCK SCHEDULE
pp. 411–413 Exs. 22–84 even, 98, 101–111 odd, Quiz 1 Exs. 1–22

EXERCISE LEVELS
Level A: *Easier*
22–49, 90–93
Level B: *More Difficult*
50–89, 94–98
Level C: *Most Difficult*
99

✓ HOMEWORK CHECK
To quickly check student understanding of key concepts, go over the following exercises: Exs. 22, 34, 44, 50, 56, 68. See also the Daily Homework Quiz:
• Blackline Master (*Chapter 7 Resource Book*, p. 36)
• 🖨 Transparency (p. 53)

TEACHING TIPS
EXERCISES 82–89 Some students may have difficulty working with irrational exponents. Encourage students to apply the properties carefully to each of these exercises.

STUDENT HELP NOTES

→ **Skill Review** As students review geometric formulas on p. 914, remind them that any real numbers, including radicals, may be substituted for variables in a formula.

VARIABLE EXPRESSIONS Simplify the expression. Assume all variables are positive.

56. $x^{1/3} \cdot x^{1/5}$ $x^{8/15}$ **57.** $(y^3)^{1/6}$ $y^{1/2}$ **58.** $\sqrt[5]{32x^5}$ $2x$ **59.** $\dfrac{1}{x^{-5/4}}$ $x^{5/4}$

60. $\dfrac{x^{3/7}}{x^{1/3}}$ $x^{2/21}$ **61.** $\sqrt[4]{\dfrac{x^{12}}{y^4}}$ $\dfrac{x^3}{y}$ **62.** $\dfrac{x^{5/3}y}{xy^{-1/2}}$ $x^{2/3}y^{3/2}$ **63.** $(y \cdot y^{1/4})^{4/3}$ $y^{5/3}$

64. $\left(\sqrt[4]{x^3} \cdot \sqrt[4]{x^5}\right)^{-2}$ $\dfrac{1}{x^4}$ **65.** $\dfrac{x^{3/4}yz^{-1/3}}{x^{1/4}z^{2/3}}$ $\dfrac{x^{1/2}y}{z}$ **66.** $\dfrac{2\sqrt{x} \cdot \sqrt{x^3}}{\sqrt{9x^{10}}}$ $\dfrac{2}{3x^3}$ **67.** $\dfrac{\sqrt[3]{y^6}}{\sqrt[3]{27y} \cdot \sqrt[3]{y^{11}}}$ $\dfrac{1}{3y^2}$

SIMPLEST FORM Write the expression in simplest form. Assume all variables are positive.

68. $\sqrt{36x^3}$ $6x\sqrt{x}$ **69.** $\sqrt[4]{10x^5y^8z^{10}}$ $xy^2z^2\sqrt[4]{10xz^2}$ **70.** $\sqrt[5]{8xy^7} \cdot \sqrt[5]{6x^6}$ $xy\sqrt[5]{48x^2y^2}$ **71.** $\sqrt{xyz} \cdot \sqrt{2y^3z^4}$ $y^2z^2\sqrt{2xz}$

72. $\dfrac{4}{\sqrt[3]{x}}$ $\dfrac{4\sqrt[3]{x^2}}{x}$ **73.** $\sqrt[3]{\dfrac{x^3}{y^2}}$ $\dfrac{x\sqrt[3]{y}}{y}$ **74.** $\sqrt{\dfrac{9x^2y}{32z^3}}$ $\dfrac{3x\sqrt{2yz}}{8z^2}$ **75.** $\dfrac{\sqrt[5]{x^3}}{\sqrt[7]{x^4}}$ $x^{1/35}$

COMBINING VARIABLE EXPRESSIONS Perform the indicated operation. Assume all variables are positive.

76. $2\sqrt[5]{y} + 7\sqrt[5]{y}$ $9\sqrt[5]{y}$ **77.** $9x^{1/5} - 2x^{1/5}$ $7x^{1/5}$ **78.** $-\sqrt[4]{x} + 2\sqrt[4]{x}$ $\sqrt[4]{x}$

79. $(x^9y)^{1/3} + (xy^{1/9})^3$ $2x^3y^{1/3}$ **80.** $\sqrt{4x^5} - x\sqrt{x^3}$ $x^2\sqrt{x}$ **81.** $y\sqrt[3]{24x^5} + \sqrt[3]{-3x^2y^3}$ $(2x-1)y\sqrt[3]{3x^2}$

EXTENSION: IRRATIONAL EXPONENTS The properties you studied in this lesson can also be applied to irrational exponents. Simplify the expression. Assume all variables are positive.

82. $x^2 \cdot x^{\sqrt{3}}$ $x^{2+\sqrt{3}}$ **83.** $\left(y^{\sqrt{2}}\right)^{\sqrt{2}}$ y^2 **84.** $(xy)^\pi$ $x^\pi y^\pi$ **85.** $4^{-\sqrt{7}}$ $\dfrac{1}{4^{\sqrt{7}}}$

86. $\dfrac{x^{2\sqrt{5}}}{x^{\sqrt{5}}}$ $x^{\sqrt{5}}$ **87.** $\left(\dfrac{x^{1/\pi}}{y^{2/\pi}}\right)^\pi$ $\dfrac{x}{y^2}$ **88.** $3x^{\sqrt{2}} + x^{\sqrt{2}}$ $4x^{\sqrt{2}}$ **89.** $xy^{\sqrt{11}} - 3xy^{\sqrt{11}}$ $-2xy^{\sqrt{11}}$

STUDENT HELP

→ **Skills Review**
For help with perimeter and area, see p. 914.

90. GEOMETRY CONNECTION   Find a radical expression for the perimeter of the triangle. Simplify the expression. $5 + 3\sqrt{5}$

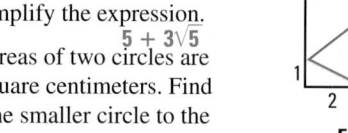
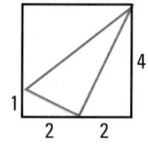
Ex. 90

91. GEOMETRY CONNECTION   The areas of two circles are 15 square centimeters and 20 square centimeters. Find the exact ratio of the radius of the smaller circle to the radius of the larger circle. $\dfrac{\sqrt{3}}{2}$

92. BIOLOGY CONNECTION   Look back at Example 8. Approximate the surface area of a human that has a mass of 68 kilograms.   about 18,300 cm²

93. 🌎 **PINHOLE CAMERA** A pinhole camera is made out of a light-tight box with a piece of film attached to one side and a pinhole on the opposite side. The optimum diameter d (in millimeters) of the pinhole can be modeled by $d = 1.9\left[(5.5 \times 10^{-4})\ell\right]^{1/2}$, where ℓ is the length of the camera box (in millimeters). What is the optimum diameter for a pinhole camera if the camera box has a length of 10 *centimeters*?   0.45 mm

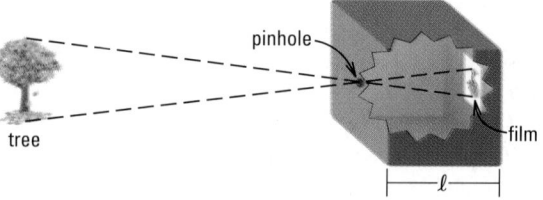
tree   pinhole   film
ℓ

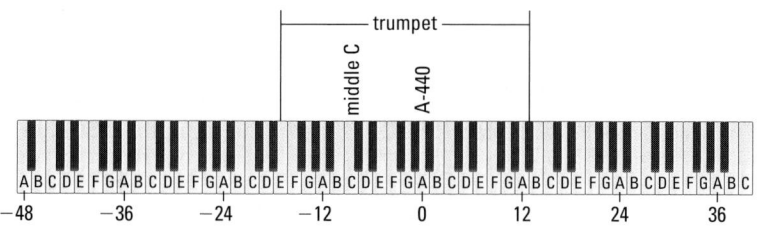

The musical note A-440 (the A above middle C) has a frequency of 440 vibrations per second. The frequency f of any note can be found using $f = 440 \cdot 2^{n/12}$ where n represents the number of black and white keys the given note is above or below A-440. For notes above A-440, $n > 0$, and for notes below A-440, $n < 0$.

94. highest: $f = 440 \cdot 2^{13/12}$,
lowest: $f = 440 \cdot 2^{-17/12}$,
ratio: $2^{5/2}$

98a. $S \approx 0.90 \text{ cm}^2$;
$V \approx 0.057 \text{ cm}^3$;
Yes, the surface area is large enough.

b. $S \approx 900,000 \text{ cm}^2$;
$V \approx 57,000,000 \text{ cm}^3$; No, the surface area is not large enough.

c. $S = 2\pi(1000)^2 rh + 2\pi(1000)^2 r^2$; Surface area is increased by a factor of 1,000,000;
$V = \pi r^2 h (1000)^3$; Volume is increased by a factor of 1,000,000,000; Giant ants do not exist because their volume increases 1000 times as fast as their surface area, so they could not meet their oxygen needs.

Test Preparation

94. Find the highest and lowest frequencies in the musical range of a trumpet. What is the exact ratio of these two frequencies? See margin.

95. **LOGICAL REASONING** Describe the pattern of the frequencies of successive notes with the same letter. Higher notes have frequencies twice as high as lower notes of the same letter.

96. **DISTANCE OF AN OBJECT** The maximum horizontal distance d that an object can travel when launched at an optimum angle of projection from an initial height h_0 can be modeled by $d = \dfrac{v_0\sqrt{(v_0)^2 + 2gh_0}}{g}$ where v_0 is the initial speed and g is the acceleration due to gravity. Simplify the model when $h_0 = 0$. When $h_0 = 0$, $d = \dfrac{(v_0)^2}{g}$.

97. **BALLOONS** You have filled two round balloons with air. One balloon has twice as much air as the other balloon. The formula for the surface area S of a sphere in terms of its volume V is $S = (4\pi)^{1/3}(3V)^{2/3}$. By what factor is the surface area of the larger balloon greater than that of the smaller balloon? $2^{2/3}$

98. **MULTI-STEP PROBLEM** A common ant absorbs oxygen at a rate of about 6.2 milliliters per second per square centimeter of exoskeleton. It needs about 24 milliliters of oxygen per second per cubic centimeter of its body. An ant is basically cylindrical in shape, so its surface area S and volume V can be approximated by the formulas for the surface area and volume of a cylinder: a–c. See margin.

$$S = 2\pi rh + 2\pi r^2 \qquad V = \pi r^2 h$$

a. Approximate the surface area and volume of an ant that is 8 millimeters long and has a radius of 1.5 millimeters. Would this ant have a surface area large enough to meet its oxygen needs?

b. Consider a "giant" ant that is 8 meters long and has a radius of 1.5 meters. Would this ant have a surface area large enough to meet its oxygen needs?

c. *Writing* Substitute 1000r for r and 1000h for h into the formulas for surface area and volume. How does increasing the radius and height by a factor of 1000 affect surface area? How does it affect volume? Use the results to explain why "giant" ants do not exist.

99. When m and n are both even, you must use an absolute value symbol around any odd power of x in the answer. For example, $\sqrt{x^6} = |x^3|$.

★ Challenge

99. **CRITICAL THINKING** Substitute different combinations of odd and even positive integers for m and n in the expression $\sqrt[n]{x^m}$. Do *not* assume that x is always positive. When is absolute value needed in simplifying the expression? See margin.

MATHEMATICAL REASONING
EXERCISE 97 This formula is derived from the formula for surface area of a sphere, $S = 4\pi r^2$, and volume of a sphere, $V = \frac{4}{3}\pi r^3$. You may want to show the derivation: Solve the volume formula for r to get $r = \left(\dfrac{3V}{4\pi}\right)^{1/3}$. Substitute this expression for r in the surface area formula to get

$$S = 4\pi\left(\left(\dfrac{3V}{4\pi}\right)^{1/3}\right)^2$$
$$= \dfrac{4\pi(3V)^{2/3}}{(4\pi)^{2/3}}$$
$$= (4\pi)^{1/3}(3V)^{2/3}$$

ADDITIONAL PRACTICE AND RETEACHING

For Lesson 7.2:

• Practice Levels A, B, and C (*Chapter 7 Resource Book*, p. 25)

• Reteaching with Practice (*Chapter 7 Resource Book*, p. 28)

• See Lesson 7.2 of the *Personal Student Tutor*

For more Mixed Review:

• Search the *Test and Practice Generator* for key words or specific lessons.

413

DAILY HOMEWORK QUIZ

📝 **Transparency Available**

Simplify the expression. Assume all variables are positive.

1. $6^{3/8} \cdot 6^{1/4}$ $6^{5/8}$

2. $\sqrt[6]{17} \cdot \sqrt[3]{17}$ $\sqrt{17}$

3. $\sqrt[4]{5} \cdot \sqrt[4]{125}$ 5

4. $y^{3/2} \cdot y^{1/3}$ $y^{11/6}$

5. $\sqrt{121z^5}$ $11z^2\sqrt{z}$

Perform the indicated operation.

6. $2\sqrt[3]{4} + 3\sqrt[3]{4}$ $5\sqrt[3]{4}$

EXTRA CHALLENGE NOTE

→ Challenge problems for Lesson 7.2 are available in **blackline** format in the *Chapter 7 Resource Book,* p. 32 and at **www.mcdougallittell.com.**

ADDITIONAL TEST PREPARATION

1. OPEN ENDED Find the ratio of the volume of a ball packed tightly inside a cube to the volume of the cube. $\frac{\pi}{6}$

ADDITIONAL RESOURCES

An alternative Quiz for Lessons 7.1–7.2 is available in the *Chapter 7 Resource Book,* p. 33.

22. No; the surface area of the Labrador retriever is about 2.08 times as great as the surface area of the Scottish terrier.

MIXED REVIEW

COMPLETING THE SQUARE Find the value of c that makes the expression a perfect square trinomial. Then write the expression as the square of a binomial. (Review 5.5)

100. $x^2 + 14x + c$ $49, (x + 7)^2$

101. $x^2 - 21x + c$ $\frac{441}{4}, \left(x - \frac{21}{2}\right)^2$

102. $x^2 - 7.6x + c$ $14.44, (x - 3.8)^2$

103. $x^2 + 9.9x + c$ $24.5, (x + 4.95)^2$

104. $x^2 + \frac{2}{3}x + c$ $\frac{1}{9}, \left(x + \frac{1}{3}\right)^2$

105. $x^2 - \frac{1}{4}x + c$ $\frac{1}{64}, \left(x - \frac{1}{8}\right)^2$

POLYNOMIAL OPERATIONS Perform the indicated operation. (Review 6.3 for 7.3)

106. $(-3x^3 + 6x) - (8x^3 + x^2 - 4x)$ $-11x^3 - x^2 + 10x$

107. $(50x - 3) + (8x^3 + 9x^2 + 2x + 4)$ $8x^3 + 9x^2 + 52x + 1$

108. $20x^2(x - 9)$ $20x^3 - 180x^2$

109. $(2x + 7)^2$ $4x^2 + 28x + 49$

LONG DIVISION Divide using long division. (Review 6.5 for 7.3)

110. $(x^3 - 28x - 48) \div (x + 4)$ $x^2 - 4x - 12$

111. $(4x^2 + 3x - 3) \div (x + 1)$ $(4x - 1) - \frac{2}{x + 1}$

112. $(4x^2 - 6x) \div (x - 2)$ $4x + 2 + \frac{4}{x - 2}$

113. $(x^4 - 2x^3 - 70x + 20) \div (x - 5)$ $x^3 + 3x^2 + 15x + 5 + \frac{45}{x - 5}$

QUIZ 1

Self-Test for Lessons 7.1 and 7.2

Evaluate the expression without using a calculator. (Lesson 7.1)

1. $8^{2/3}$ 4

2. $32^{-3/5}$ $\frac{1}{8}$

3. $-(81^{1/4})$ -3

4. $(-64)^{2/3}$ 16

Solve the equation. Round your answer to two decimal places. (Lesson 7.1)

5. $x^5 = 10$ 1.58

6. $-9x^6 = -18$ ± 1.12

7. $x^4 - 4 = 9$ ± 1.90

8. $(x + 2)^3 = -15$ -4.47

Write the expression in simplest form. (Lesson 7.2)

9. $\frac{1}{4^{-1/4}}$ $4^{1/4}$ or $2^{1/2}$

10. $\sqrt[4]{\frac{16}{3}}$ $\frac{2\sqrt[4]{27}}{3}$

11. $\frac{512^{1/3}}{8^{1/3}}$ 4

12. $\sqrt{45}$ $3\sqrt{5}$

13. $\sqrt[3]{7} \cdot \sqrt[3]{49}$ 7

14. $8^{1/5} + 2(8^{1/5})$ $3\sqrt[5]{8}$

Write the expression in simplest form. Assume all variables are positive. (Lesson 7.2)

15. $\sqrt[3]{x^2} \cdot \sqrt[4]{x}$ $x^{11/12}$

16. $(x^{1/5})^{5/2}$ $x^{1/2}$

17. $\frac{xy^{1/2}}{x^{3/4}y^{-2}}$ $x^{1/4}y^{5/2}$

18. $\sqrt[3]{5x^3y^5}$ $xy\sqrt[3]{5y^2}$

19. $\sqrt{\frac{36x}{y^3}}$ $\frac{6\sqrt{xy}}{y^2}$

20. $x(9y)^{1/2} - (x^2y)^{1/2}$ $2xy^{1/2}$

21. 🌐 **GENERATING POWER** As a rule of thumb, the power P (in horsepower) that a ship needs can be modeled by $P = \frac{d^{2/3} \cdot s^3}{c}$ where d is the ship's displacement (in tons), s is the normal speed (in knots), and c is the Admiralty coefficient. If a ship displaces 30,090 tons, has a normal speed of 22.5 knots, and has an Admiralty coefficient of 370, how much power does it need? (Lesson 7.1) about 30,000 horsepower

22. **BIOLOGY CONNECTION** The surface area S (in square centimeters) of a large dog can be approximated by the model $S = 11.2m^{2/3}$ where m is the mass (in grams) of the dog. A Labrador retriever's mass is about three times the mass of a Scottish terrier. Is its surface area also three times that of a Scottish terrier? (Lesson 7.2)

7.3 Power Functions and Function Operations

What you should learn

GOAL ① Perform operations with functions including power functions.

GOAL ② Use power functions and function operations to solve **real-life** problems, such as finding the proportion of water loss in a bird's egg in **Example 4**.

Why you should learn it

▼ To solve **real-life** problems, such as finding the height of a dinosaur in **Ex. 56**.

GOAL ① PERFORMING FUNCTION OPERATIONS

In Chapter 6 you learned how to add, subtract, multiply, and divide polynomial functions. These operations can be defined for any functions.

CONCEPT SUMMARY

OPERATIONS ON FUNCTIONS

Let f and g be any two functions. A new function h can be defined by performing any of the four basic operations (addition, subtraction, multiplication, and division) on f and g.

Operation	Definition	Example: $f(x) = 2x$, $g(x) = x + 1$
ADDITION	$h(x) = f(x) + g(x)$	$h(x) = 2x + (x + 1) = 3x + 1$
SUBTRACTION	$h(x) = f(x) - g(x)$	$h(x) = 2x - (x + 1) = x - 1$
MULTIPLICATION	$h(x) = f(x) \cdot g(x)$	$h(x) = (2x)(x + 1) = 2x^2 + 2x$
DIVISION	$h(x) = \dfrac{f(x)}{g(x)}$	$h(x) = \dfrac{2x}{x + 1}$

The domain of h consists of the x-values that are in the domains of both f and g. Additionally, the domain of a quotient does not include x-values for which $g(x) = 0$.

So far you have studied various types of functions, including linear functions, quadratic functions, and polynomial functions of higher degree. Another common type of function is a **power function**, which has the form $y = ax^b$ where a is a real number and b is a rational number.

Note that when b is a positive integer, a power function is simply a type of polynomial function. For example, $y = ax^b$ is a linear function when $b = 1$, a quadratic function when $b = 2$, and a cubic function when $b = 3$.

EXAMPLE 1   Adding and Subtracting Functions

Let $f(x) = 2x^{1/2}$ and $g(x) = -6x^{1/2}$. Find **(a)** the sum of the functions, **(b)** the difference of the functions, and **(c)** the domains of the sum and difference.

SOLUTION

a. $f(x) + g(x) = 2x^{1/2} + (-6x^{1/2}) = (2 - 6)x^{1/2} = -4x^{1/2}$

b. $f(x) - g(x) = 2x^{1/2} - (-6x^{1/2}) = [2 - (-6)]x^{1/2} = 8x^{1/2}$

c. The functions f and g each have the same domain—all nonnegative real numbers. So, the domains of $f + g$ and $f - g$ also consist of all nonnegative real numbers.

7.3 Power Functions and Function Operations   **415**

① PLAN

PACING
Basic: 1 day
Average: 1 day
Advanced: 1 day
Block Schedule: 0.5 block with 7.4

 LESSON OPENER
APPLICATION
An alternative way to approach Lesson 7.3 is to use the Application Lesson Opener:

• Blackline Master (*Chapter 7 Resource Book*, p. 37)
•  Transparency (p. 46)

MEETING INDIVIDUAL NEEDS
• *Chapter 7 Resource Book*
  Prerequisite Skills Review (p. 5)
  Practice Level A (p. 40)
  Practice Level B (p. 41)
  Practice Level C (p. 42)
  Reteaching with Practice (p. 43)
  Absent Student Catch-Up (p. 45)
  Challenge (p. 47)
• *Resources in Spanish*
• *Personal Student Tutor*

NEW-TEACHER SUPPORT
See the Tips for New Teachers on pp. 1–2 of the *Chapter 7 Resource Book* for additional notes about Lesson 7.3.

WARM-UP EXERCISES

Transparency Available
Simplify.
1. $4(x^2 + 1)$ $4x^2 + 4$
2. $(x - 2) + (x^2 + 1)$ $x^2 + x - 1$
3. $(x - 2) - (x^2 + 1)$ $-x^2 + x - 3$
4. $(x - 2)(x + 1)$ $x^2 - x - 2$
5. $(x - 2)^2$ $x^2 - 4x + 4$

2 TEACH

EXTRA EXAMPLE 1

Let $f(x) = 3x^{1/3}$ and $g(x) = 2x^{1/3}$. Find **(a)** the sum, **(b)** the difference, and **(c)** the domains.

a. $5x^{1/3}$ b. $x^{1/3}$

c. The domains are all real numbers.

EXTRA EXAMPLE 2

Let $f(x) = 4x^{1/3}$ and $g(x) = x^{1/2}$. Find **(a)** the product, **(b)** the quotient, and **(c)** the domains.

a. $4x^{5/6}$ b. $4x^{-1/6}$

c. The domain of f is real numbers; the domain of g is all nonnegative real numbers; the domain of $f \cdot g$ is all nonnegative real numbers; the domain of $\frac{f}{g}$ is all positive real numbers.

EXTRA EXAMPLE 3

Let $f(x) = 2x^{-1}$ and $g(x) = x^2 - 1$. Find **(a)** $f(g(x))$, **(b)** $g(f(x))$, **(c)** $f(f(x))$, and **(d)** the domains. See margin.

✓ **CHECKPOINT EXERCISES**

For use after Examples 1 and 2:

1. Let $f(x) = -2x^{1/2}$ and $g(x) = x^{1/2}$. Find **(a)** the sum, **(b)** the difference, **(c)** the product, **(d)** the quotient, and **(e)** the domains.

a. $-x^{1/2}$ b. $-3x^{1/2}$

c. $-2x$ d. -2

e. The domain of $f + g$, $f - g$, and $f \cdot g$ is nonnegative real numbers; the domain of $\frac{f}{g}$ is all positive real numbers.

For use after Example 3:

2. Let $f(x) = x^{-1}$ and $g(x) = x + 1$. Find **(a)** $f(g(x))$, **(b)** $g(f(x))$, **(c)** $f(f(x))$, and **(d)** the domains.

a. $\frac{1}{x+1}$ b. $\frac{1}{x}+1$ c. x

d. The domain of $f(g(x))$ is all real numbers except -1; the domain of $g(f(x))$ and $f(f(x))$ is nonzero real numbers.

416

STUDENT HELP

Look Back
For help with function operations, see p. 338.

STUDENT HELP

Study Tip
When you are writing the composition of f with g, you may want to first rewrite $f(x) = 3x^{-1}$ as

$$f(\square) = 3(\square)^{-1}$$

and then substitute $g(x) = 2x - 1$ everywhere there is a box.

EXAMPLE 2   *Multiplying and Dividing Functions*

Let $f(x) = 3x$ and $g(x) = x^{1/4}$. Find **(a)** the product of the functions, **(b)** the quotient of the functions, and **(c)** the domains of the product and quotient.

SOLUTION

a. $f(x) \cdot g(x) = (3x)(x^{1/4}) = 3x^{(1 + 1/4)} = 3x^{5/4}$

b. $\dfrac{f(x)}{g(x)} = \dfrac{3x}{x^{1/4}} = 3x^{(1 - 1/4)} = 3x^{3/4}$

c. The domain of f consists of all real numbers and the domain of g consists of all nonnegative real numbers. So, the domain of $f \cdot g$ consists of all nonnegative real numbers. Because $g(0) = 0$, the domain of $\dfrac{f}{g}$ is restricted to all *positive* real numbers.

· · · · · · · · · ·

A fifth operation that can be performed with two functions is *composition*.

COMPOSITION OF TWO FUNCTIONS

The **composition** of the function f with the function g is:

$$h(x) = f(g(x))$$

The domain of h is the set of all x-values such that x is in the domain of g and $g(x)$ is in the domain of f.

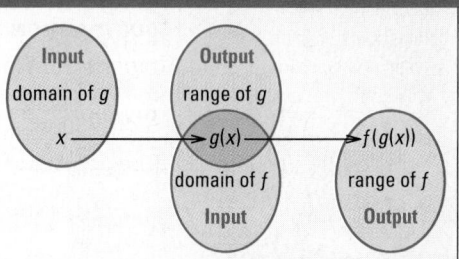

As with subtraction and division of functions, you need to pay attention to the order of functions when they are composed. In general, $f(g(x))$ is not equal to $g(f(x))$.

EXAMPLE 3   *Finding the Composition of Functions*

Let $f(x) = 3x^{-1}$ and $g(x) = 2x - 1$. Find the following.

a. $f(g(x))$ **b.** $g(f(x))$ **c.** $f(f(x))$ **d.** the domain of each composition

SOLUTION

a. $f(g(x)) = f(2x - 1) = 3(2x - 1)^{-1} = \dfrac{3}{2x - 1}$

b. $g(f(x)) = g(3x^{-1}) = 2(3x^{-1}) - 1 = 6x^{-1} - 1 = \dfrac{6}{x} - 1$

c. $f(f(x)) = f(3x^{-1}) = 3(3x^{-1})^{-1} = 3(3^{-1}x) = 3^0 x = x$

d. The domain of $f(g(x))$ consists of all real numbers except $x = \dfrac{1}{2}$ because $g\left(\dfrac{1}{2}\right) = 0$ is not in the domain of f. The domains of $g(f(x))$ and $f(f(x))$ consist of all real numbers except $x = 0$, because 0 is not in the domain of f. Note that $f(f(x))$ simplifies to x, but that result is not what determines the domain.

416   **Chapter 7**   *Powers, Roots, and Radicals*

Extra Example 3 *Sample answer:*

a. $\dfrac{2}{x^2 - 1}$ b. $\dfrac{4}{x^2} - 1$ c. x d. The domain of $f(g(x))$ is all real numbers except 1 or -1; the domain of $g(f(x))$ is all real numbers except 0; the domain of $f(f(x))$ is all real numbers except 0.

GOAL 2 USING FUNCTION OPERATIONS IN REAL LIFE

EXAMPLE 4 *Using Function Operations*

BIOLOGY CONNECTION You are doing a science project and have found research indicating that the incubation time I (in days) of a bird's egg can be modeled by $I(m) = 12m^{0.217}$ where m is the egg's mass (in grams). You have also found that during incubation the egg's rate of water loss R (in grams per day) can be modeled by $R(m) = 0.015m^{0.742}$.

You conjecture that the proportion of water loss during incubation is about the same for any size egg. Show how you can use the two power function models to verify your conjecture. ▶ Source: *Biology by Numbers*

BIOLOGY The largest bird egg is an ostrich's egg, which has a mass of about 1.65 kilograms. For comparison, an average chicken egg has a mass of only 56.7 grams.

SOLUTION

VERBAL MODEL

$$\text{Proportion of water loss} = \frac{\text{Total water loss}}{\text{Egg's mass}} = \frac{\text{Daily water loss} \cdot \text{Number of days}}{\text{Egg's mass}}$$

LABELS Daily water loss $= R(m)$ Number of days $= I(m)$ Egg's mass $= m$

ALGEBRAIC MODEL

$$\frac{R(m) \cdot I(m)}{m} = \frac{(0.015m^{0.742})(12m^{0.217})}{m}$$

$$= \frac{0.18m^{0.959}}{m}$$

$$= 0.18m^{-0.041}$$

Because $m^{-0.041}$ is approximately m^0, the proportion of water loss can be treated as $0.18m^0 = (0.18)(1) = 0.18$. So, the proportion of water loss is about 18% for any size bird's egg, and your conjecture is correct.

EXAMPLE 5 *Using Composition of Functions*

Business

A clothing store advertises that it is having a 25% off sale. For one day only, the store advertises an additional savings of 10%.

a. Use composition of functions to find the total percent discount.

b. What would be the sale price of a $40 sweater?

STUDENT HELP

▶ **Skills Review**
For help with calculating percents, see p. 907.

SOLUTION

a. Let x represent the price. The sale price for a 25% discount can be represented by the function $f(x) = x - 0.25x = 0.75x$. The reduced sale price for an additional 10% discount can be represented by the function $g(x) = x - 0.10x = 0.90x$.

$$g(f(x)) = g(0.75x) = 0.90(0.75x) = 0.675x$$

▶ The total percent discount is $100\% - 67.5\% = 32.5\%$.

b. Let $x = 40$. Then $g(f(x)) = g(f(40)) = 0.675(40) = 27$.

▶ The sale price of the sweater is $27.

Focus on Vocabulary *Sample answer:*
The composition of functions is a function of a function. The output of one function becomes the input of the other function. The product of functions is the product of the output of each function when you multiply the two functions.

Closure Question *Sample answer:*
The domain of the composition is the set of all x-values such that x is in the domain of g and $g(x)$ is in the domain of f.

EXTRA EXAMPLE 4
You do an experiment on bacteria and find that the growth rate G of the bacteria can be modeled by $G(t) = 82t^{0.25}$, and that the death rate D is $D(t) = 10.8t^{0.25}$, where t is time in hours. Find an expression for the number N of bacteria living at time t.
$N(t) = G(t) - D(t) = 71.2t^{0.25}$

EXTRA EXAMPLE 5
A computer catalog offers computers at a savings of 15% off the retail price. At the end of the month, it offers an additional 10% off its own prices.
a. Use composition of functions to find the total percent discount. **23.5%**
b. What would be the sale price of a $899 computer? **$687.74**

☑ **CHECKPOINT EXERCISES**
For use after Example 4:
1. Let $f(x) = 2.5x^{0.179}$ and $g(x) = 0.25x^{0.275}$.
Find $f(x) \div g(x)$. $10x^{-0.096}$

For use after Example 5:
2. A music store offers albums at a savings of 10% off the retail price. For a special promotion, it offers an additional 20% off its own prices.
a. Use composition of functions to find the total percent discount. **28%**
b. Find the sale price of a $20 album. **$14.40**

FOCUS ON VOCABULARY
How is the composition of functions different from the product of functions? See margin.

CLOSURE QUESTION
Describe the domain of the composition of two functions, $f(g(x))$.
See margin.

DAILY PUZZLER
Give an example of power functions $f(x)$ and $g(x)$ for which $f(g(x)) = g(f(x))$. *Sample answer:*
$f(x) = x^2$ and $g(x) = x^{1/2}$ for all positive values of x.

417

ASSIGNMENT GUIDE

BASIC
Day 1: pp. 418–420 Exs. 12–36
even, 40–48 even, 52,
57–61, 69–83 odd

AVERAGE
Day 1: pp. 418–420 Exs. 12–50
even, 52, 53, 57–61,
69–83 odd

ADVANCED
Day 1: pp. 418–420 Exs. 12–50
even, 52–54, 56–67,
69–83 odd, 84

BLOCK SCHEDULE
pp. 418–420 Exs. 12–50 even,
52–54, 56–67, 69–83 odd, 84
(with 7.4)

EXERCISE LEVELS
Level A: *Easier*
12–19, 32–33, 40–47, 52
Level B: *More Difficult*
20–31, 34–39, 48–51, 53–61
Level C: *Most Difficult*
62–67

✔ **HOMEWORK CHECK**
To quickly check student under-
standing of key concepts, go
over the following exercises:
Exs. 12, 20, 28, 32, 46. See
also the Daily Homework Quiz:
• Blackline Master (*Chapter 7
Resource Book,* p. 50)
• 📖 Transparency (p. 54)

12–17, 20–39.
See Additional Answers beginning on page AA1.

GUIDED PRACTICE

Vocabulary Check ✔

1. Complete this statement: The function $y = ax^b$ is a(n) ? function where a is a(n) ? number and b is a(n) ? number. **power function; real; rational**

Concept Check ✔

2. **LOGICAL REASONING** Tell whether the sum of two power functions is *sometimes*, *always*, or *never* a power function.
Sometimes; the sum is a power function if the exponents (*b*) are the same.
ERROR ANALYSIS Let $f(x) = x^2 + 2$ and $g(x) = 3x$. **What is wrong with the composition shown? Explain.**
The equation is actually *g*(*f*(*x*)).

4. $f(3x) = (3x)^2 + 2 =$
$9x^2 + 2$; The entire
quantity $3x$ must be
squared.

3.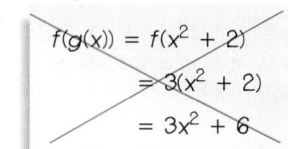

4.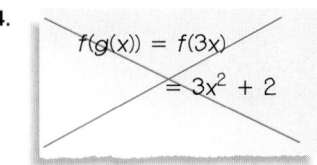

Skill Check ✔

Let $f(x) = 4x$ and $g(x) = x - 1$. **Perform the indicated operation and state the domain.**
$5x - 1$; all real numbers $3x + 1$; all real numbers $4x^2 - 4x$; all real numbers
5. $f(x) + g(x)$ **6.** $f(x) - g(x)$ **7.** $f(x) \cdot g(x)$

8. $\dfrac{4x}{x-1}$; all real numbers
except $x = 1$

8. $\dfrac{f(x)}{g(x)}$ **9.** $f(g(x))$ **10.** $g(f(x))$
                   $4x - 4$; all real numbers $4x - 1$; all real numbers

11. 🌎 **SALES BONUS** You are a sales representative for a clothing manufacturer. You are paid an annual salary plus a bonus of 2% of your sales over $200,000. Consider two functions: $f(x) = x - 200,000$ and $g(x) = 0.02x$. If $x > \$200,000$, which composition, $f(g(x))$ or $g(f(x))$, represents your bonus? Explain.
$g(f(x))$; The bonus is 0.02 times the amount over \$200,000 $(x - 200,000)$, so calculate amount first and then take 2%.

PRACTICE AND APPLICATIONS

STUDENT HELP
▶ **Extra Practice**
to help you master
skills is on p. 949.

ADDING AND SUBTRACTING FUNCTIONS Let $f(x) = x^2 - 5x + 8$ and $g(x) = x^2 - 4$. **Perform the indicated operation and state the domain.**
12–17. See margin.
12. $f(x) + g(x)$ **13.** $g(x) + f(x)$ **14.** $f(x) + f(x)$ **15.** $g(x) + g(x)$

16. $f(x) - g(x)$ **17.** $g(x) - f(x)$ **18.** $f(x) - f(x)$ **19.** $g(x) - g(x)$
                                     0; all real numbers   0; all real numbers

MULTIPLYING AND DIVIDING FUNCTIONS Let $f(x) = 2x^{2/3}$ and $g(x) = 3x^{1/2}$. **Perform the indicated operation and state the domain.**
20–27. See margin.
20. $f(x) \cdot g(x)$ **21.** $g(x) \cdot f(x)$ **22.** $f(x) \cdot f(x)$ **23.** $g(x) \cdot g(x)$

24. $\dfrac{f(x)}{g(x)}$ **25.** $\dfrac{g(x)}{f(x)}$ **26.** $\dfrac{f(x)}{f(x)}$ **27.** $\dfrac{g(x)}{g(x)}$

STUDENT HELP
▶ **HOMEWORK HELP**
Example 1: Exs. 12–19,
32–51
Example 2: Exs. 20–27,
32–51
Example 3: Exs. 28–51
Example 4: Exs. 52, 53
Example 5: Exs. 54–56

COMPOSITION OF FUNCTIONS Let $f(x) = 4x^{-5}$ and $g(x) = x^{3/4}$. **Perform the indicated operation and state the domain.** 28–31. See margin.

28. $f(g(x))$ **29.** $g(f(x))$ **30.** $f(f(x))$ **31.** $g(g(x))$

FUNCTION OPERATIONS Let $f(x) = 10x$ and $g(x) = x + 4$. **Perform the indicated operation and state the domain.** 32–39. See margin.

32. $f(x) + g(x)$ **33.** $f(x) - g(x)$ **34.** $f(x) \cdot g(x)$ **35.** $\dfrac{f(x)}{g(x)}$

36. $f(g(x))$ **37.** $g(f(x))$ **38.** $f(f(x))$ **39.** $g(g(x))$

42. $-2x^{2/3}$; all real numbers

43. $x^2 - x - 8$; all real numbers

48. $\dfrac{6}{5x - 2}$; all real numbers except $\dfrac{2}{5}$

49. $x^4 - 6x^2 + 10$; all real numbers

52. $1.446 \times 10^9 m^{-0.05}$; multiplying beats per minute by number of minutes per lifetime yields the number of beats over the entire lifetime.

53. $r(w) = 220w^{-0.266}$; about 134 breaths per minute; about 18 breaths per minute; about 11 breaths per minute

STUDENT HELP

HOMEWORK HELP
Visit our Web site www.mcdougallittell.com for help with problem solving in Ex. 53.

FOCUS ON CAREERS

PALEONTOLOGIST
A paleontologist is a scientist who studies fossils of dinosaurs and other prehistoric life forms. Most paleontologists work as college professors.

CAREER LINK
www.mcdougallittell.com

FUNCTION OPERATIONS Perform the indicated operation and state the domain.

$6x + 3$; all real numbers
40. $f + g$; $f(x) = x + 3$, $g(x) = 5x$
42, 43. See margin.

41. $x^{1/2}$; nonnegative real numbers
$f + g$; $f(x) = 3x^{1/2}$, $g(x) = -2x^{1/2}$

42. $f - g$; $f(x) = -x^{2/3}$, $g(x) = x^{2/3}$

43. $f - g$; $f(x) = x^2 - 3$, $g(x) = x + 5$

44. $f \cdot g$; $f(x) = 7x^{2/5}$, $g(x) = -2x^3$
$-14x^{17/5}$; all real numbers

45. $f \cdot g$; $f(x) = x - 4$, $g(x) = 4x^2$
$4x^3 - 16x^2$; all real numbers

46. $\dfrac{f}{g}$; $f(x) = 9x^{-1}$, $g(x) = x^{1/4}$
$9x^{-5/4}$; positive real numbers

47. $\dfrac{f}{g}$; $f(x) = x^2 - 5x$, $g(x) = x$
$x - 5$; all real numbers except 0

48. $f(g(x))$; $f(x) = 6x^{-1}$, $g(x) = 5x - 2$
48, 49. See margin.

49. $g(f(x))$; $f(x) = x^2 - 3$, $g(x) = x^2 + 1$

50. $f(f(x))$; $f(x) = 2x^{1/5}$
$2^{6/5}x^{1/25}$; all real numbers

51. $g(g(x))$; $g(x) = 9x - 2$
$81x - 20$; all real numbers

52. **HEART RATE** For a mammal, the heart rate r (in beats per minute) and the life span s (in minutes) are related to body mass m (in kilograms) by these formulas:

$$r(m) = 241m^{-0.25} \qquad s(m) = (6 \times 10^6)m^{0.2}$$

Find the relationship between body mass and average number of heartbeats in a lifetime by calculating $r(m) \cdot s(m)$. Explain the results.
▶ Source: *Physiology by Numbers* See margin.

53. **BREATHING RATE** For a mammal, the volume b (in milliliters) of air breathed in and the volume d (in milliliters) of the dead space (the portion of the lungs not filled with air) are related to body weight w (in grams) by these formulas:

$$b(w) = 0.007w \qquad d(w) = 0.002w$$

The relationship between breathing rate r (in breaths per minute) and body weight is:

$$r(w) = \frac{1.1w^{0.734}}{b(w) - d(w)}$$

Simplify $r(w)$ and calculate the breathing rate for body weights of 6.5 grams, 12,300 grams, and 70,000 grams. ▶ Source: *Respiration* See margin.

COAT SALE In Exercises 54 and 55, use the following information.
A clothing store is having a sale in which you can take $50 off the cost of any coat in the store. The store also offers 10% off your entire purchase if you open a charge account. You decide to open a charge account and buy a coat.

54. Use composition of functions to find the sale price of a $175 coat when $50 is subtracted before the 10% discount is applied. **$112.50**

55. **CRITICAL THINKING** Why doesn't the store apply the 10% discount before subtracting $50? **10% off $175 would be $17.50 rather than $12.50. There is a smaller discount after the subtraction.**

56. **PALEONTOLOGY** The height at the hip h (in centimeters) of an ornithomimid, a type of dinosaur, can be modeled by

$$h(l) = 3.49l^{1.02}$$

where l is the length (in centimeters) of the dinosaur's instep. The length of the instep can be modeled by

$$l(f) = 1.5f$$

where f is the footprint length (in centimeters). Use composition of functions to find the relationship between height and footprint length. Then find the height of an ornithomimid with a footprint length of 30 centimeters. ▶ Source: *Dinosaur Tracks*
$h = 3.49(1.5f)^{1.02}$; 169.47 cm

7.3 *Power Functions and Function Operations* **419**

! COMMON ERROR

EXERCISE 50 Some students may think the composition of f with itself is always x. Emphasize that students should substitute the entire function f carefully for x to get $f(f(x)) = 2(2x^{1/5})^{1/5} = 2^{6/5}x^{1/25}$.

STUDENT HELP NOTES

Homework Help Students can find help for Ex. 53 at **www.mcdougallittell.com.** The information can be printed out for students who don't have access to the Internet.

APPLICATION NOTE

EXERCISE 56 This exercise is an example of how composition of functions can be used to find a relationship between variables. In this case, students find a relationship between height and foot length.

CAREER NOTE

EXERCISE 56
Additional information about paleontologists is available at **www.mcdougallittell.com.**

ADDITIONAL PRACTICE AND RETEACHING

For Lesson 7.3:
• Practice Levels A, B, and C (*Chapter 7 Resource Book,* p. 40)
• Reteaching with Practice (*Chapter 7 Resource Book,* p. 43)
• See Lesson 7.3 of the *Personal Student Tutor.*

For more Mixed Review:
• Search the *Test and Practice Generator* for key words or specific lessons.

DAILY HOMEWORK QUIZ

✏️ *Transparency Available*

Let $f(x) = x^2 - 2$ and $g(x) = x^3 + 4x$. Perform the indicated operation and state the domain.

1. $f(x) + g(x)$ $x^3 + x^2 + 4x - 2$; all real numbers

2. $f(x) \cdot g(x)$ $x^5 + 2x^3 - 8x$; all real numbers

3. $f(g(x))$ $x^6 + 8x^4 + 16x^2 - 2$; all real numbers

4. $\dfrac{g(x)}{f(x)}$ $\dfrac{x^3 + 4x}{x^2 - 2}$; all real numbers except $\pm\sqrt{2}$

EXTRA CHALLENGE NOTE

↳ Challenge problems for Lesson 7.3 are available in **blackline** format in the *Chapter 7 Resource Book,* p. 47 and at **www.mcdougallittell.com.**

ADDITIONAL TEST PREPARATION

1. WRITING Explain how to find $f(g(x))$ if $f(x) = x^2$ and $g(x) = x + 2$ and give the domain of each. You replace $x + 2$ for x in $f(x)$ to get $(x + 2)^2$, or $x^2 + 4x + 4$. The domain is all real numbers for each function.

74–79. See Additional Answers beginning on page AA1.

57. *Writing* Explain how to perform the function operations $f(x) + g(x)$, $f(x) - g(x)$, $f(x) \cdot g(x)$, $\dfrac{f(x)}{g(x)}$, and $f(g(x))$ for any two functions f and g. See margin.

Test Preparation

QUANTITATIVE COMPARISON In Exercise 58–61, choose the statement that is true about the given quantities.

(A) The quantity in column A is greater.

(B) The quantity in column B is greater.

(C) The two quantities are equal.

(D) The relationship cannot be determined from the given information.

	Column A	Column B	
58.	$f(g(4))$; $f(x) = 6x$, $g(x) = 3x^2$	$f(g(2))$; $f(x) = x^{2/3}$, $g(x) = -2x$	A
59.	$g(f(-1))$; $f(x) = 5x^{-2}$, $g(x) = x$	$g(f(0))$; $f(x) = 2x + 5$, $g(x) = x^2$	B
60.	$f(f(3))$; $f(x) = 3x - 7$	$f(f(-2))$; $f(x) = 10x^3$	A
61.	$g(g(5))$; $g(x) = 16x^{-1/4}$	$g(g(7))$; $g(x) = x^2 + 8$	B

★ **Challenge**

EXTRA CHALLENGE

↳ www.mcdougallittell.com

FUNCTION COMPOSITION Find functions f and g such that $f(g(x)) = h(x)$.

62. $h(x) = (6x - 5)^3$ *Sample answer:* $f(x) = x^3$, $g(x) = 6x - 5$

63. $h(x) = \sqrt[3]{x + 2}$ *Sample answer:* $f(x) = x^{1/3}$, $g(x) = x + 2$

64. $h(x) = \dfrac{\sqrt[4]{x}}{2}$ *Sample answer:* $f(x) = x^{1/4}$, $g(x) = \dfrac{x}{16}$

65. $h(x) = 3x^2 + 7$ *Sample answer:* $f(x) = x + 7$, $g(x) = 3x^2$

66. $h(x) = |2x + 9|$ *Sample answer:* $f(x) = |x|$, $g(x) = 2x + 9$

67. $h(x) = 21x$ *Sample answer:* $f(x) = 3x$, $g(x) = 7x$

MIXED REVIEW

57. For addition and subtraction, add or subtract the expressions for f and g, and combine like terms.

For multiplication and division, multiply or divide the equations for f and g, and simplify the result.

For composition of functions $f(g(x))$, substitute the expression for $g(x)$ for the x in the expression for $f(x)$ and simplify.

REWRITING EQUATIONS Solve the equation for y. (Review 1.4 for 7.4)

68. $y - 3x = 10$ $y = 3x + 10$

69. $2x + 3y = -8$ $y = \dfrac{-2x - 8}{3}$

70. $x = -2y + 6$ $y = \dfrac{6 - x}{2}$

71. $xy + 2 = 7$ $y = \dfrac{5}{x}$

72. $\dfrac{1}{2}x - \dfrac{2}{3}y = 1$ $y = \dfrac{3x}{4} - \dfrac{3}{2}$

73. $ax + by = c$ $y = \dfrac{c - ax}{b}$

GRAPHING FUNCTIONS Graph the function. (Review 2.1 for 7.4)
74–79. See margin.

74. $y = x - 2$

75. $y = 4x - 3$

76. $y = 5x - \dfrac{2}{3}$

77. $y = -2x - 4$

78. $y = -\dfrac{1}{2}x + 7$

79. $y = -8$

SOLVING EQUATIONS Find the real-number solutions of the equation. (Review 6.4)

80. $3x^3 - 2x^2 = 0$ $0, \dfrac{2}{3}$

81. $2x^3 - 6x^2 + x = 3$ 3

82. $5x^4 + 19x^2 - 4 = 0$ $\pm\sqrt{0.2} \approx \pm 0.447$

83. $x^4 + 6x^3 + 8x + 48 = 0$ $-6, -2$

84. 🌐 **CRYPTOGRAPHY** Use the inverse of $A = \begin{bmatrix} 5 & 2 \\ 2 & 1 \end{bmatrix}$ and the code on page 225 to decode the message. (Review 4.4) congratulations

45, 21, 84, 35, 92, 37, 142, 61
62, 25, 118, 49, 103, 44, 95, 38

▶ ACTIVITY 7.4

Developing Concepts

Exploring Inverse Functions

SET UP
Work in a group of three.

MATERIALS
• graph paper
• straightedge

Step 6. $f(g(x)) = g(f(x)) = x$;

$$f(g(x)) = \frac{(2x + 3) - 3}{2} = x;$$

$$g(f(x)) = 2\left(\frac{x - 3}{2}\right) + 3 = x$$

For $f(x) = 2x + 5$:

1. See margin for graph;
$g(x) = \frac{x - 5}{2}$.

2. Graph the reflection.

3. g is the function that subtracts 5 from x and then divides by 2. Both compositions equal x. If both compositions equal x, then the functions are inverses.

For $f(x) = \frac{x - 2}{4}$:

1. See margin for graph; $g(x) = 4x + 2$.

2. Graph the reflection.

3. g is the function that multiplies x by 4 and then adds 2. Both compositions equal x. If both compositions equal x, then the functions are inverses.

For $f(x) = 5 - \frac{5}{2}x$:

1. See margin for graph; $g(x) = \frac{2}{5}(5 - x)$.

2. Graph the reflection.

3. g is the function that subtracts x from 5 and then multiplies by $\frac{2}{5}$. Both compositions equal x. If both compositions equal x, then the functions are inverses.

▶ **QUESTION** How are a function and its *inverse* related?

▶ **EXPLORING THE CONCEPT**

Use the following steps to find the inverse of $f(x) = \frac{x - 3}{2}$.

1 Choose values of x and find the corresponding values of $y = f(x)$. Plot the points and draw the line that passes through them. See margin for graph.

2 Interchange the x- and y-coordinates of the ordered pairs found in **Step 1**. Plot the new points and draw the line that passes through them. See margin for graph.

3 Write an equation of the line from **Step 2**. Call this function g. $g(x) = 2x + 3$

4 Fold your graph paper so that the graphs of f and g coincide. How are the graphs geometrically related? They are reflections of one another.

5 In words, f is the function that subtracts 3 from x and then divides the result by 2. Describe the function g in words. g is the function that multiplies x by 2 and then adds 3.

6 Predict what the compositions $f(g(x))$ and $g(f(x))$ will be. Confirm your predictions by finding $f(g(x))$ and $g(f(x))$. See margin.

The functions f and g are inverses of each other.

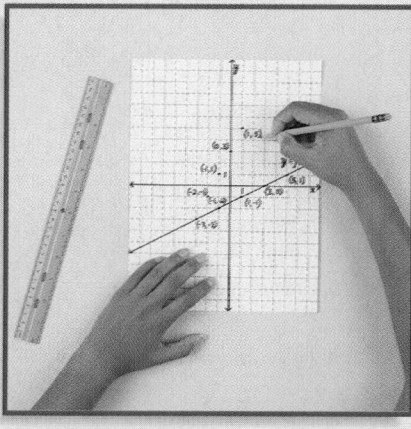 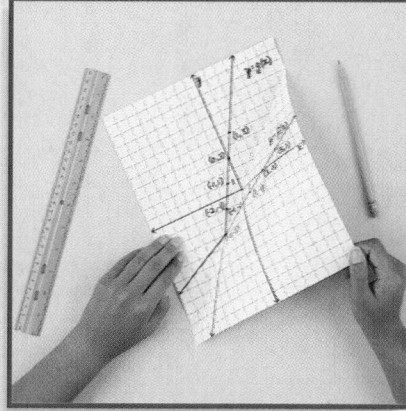

▶ **DRAWING CONCLUSIONS**

Each member in your group should choose a different function from the list below.

$$f(x) = 2x + 5 \qquad f(x) = \frac{x - 2}{4} \qquad f(x) = 5 - \frac{5}{2}x$$

1. Complete **Steps 1–3** above to find the inverse of your function. 1–3. See margin.

2. Complete **Step 4**. How can you graph the inverse of a function without first finding ordered pairs (x, y)?

3. Complete **Steps 5 and 6**. How can you test to see if the function you found in Exercise 1 is indeed the inverse of the original function?

7.4 *Concept Activity* **421**

▶ 1 Planning the Activity

PURPOSE
To find how functions and their inverses are related.

MATERIALS
• graph paper
• straightedge
• Activity Support Master (*Chapter 7 Resource Book,* p. 51)

PACING
• Exploring the Concept — 15 min
• Drawing Conclusions — 15 min

▶ **LINK TO LESSON**
The steps provide a geometrical connection for finding the inverse of a function. This connection can also be made to check answers in Examples 1, 2, and 4 of Lesson 7.4.

▶ 2 Managing the Activity

CLASSROOM MANAGEMENT
Students may have difficulty describing the inverse function. Ask students to discuss and correct their description of the function g in Step 5 with their group.

ALTERNATIVE APPROACH
You may want each member of the group to do all three exercises. Then have them compare their answers with each other.

▶ 3 Closing the Activity

★ **KEY DISCOVERY**
The graph of an inverse of a function is a reflection of the graph of the original function in the line $y = x$.

ACTIVITY ASSESSMENT
JOURNAL Two functions are graphed on the same axes. Explain how you can tell if the functions are inverses of each other.
See sample answer at left.

Activity Assessment *Sample answer:*
The graphs are reflections of each other in the line
$y = x$.

Steps 1, 2, Ex. 1. See Additional Answers beginning on page AA1.

PACING
Basic: 2 days
Average: 2 days
Advanced: 2 days
Block Schedule: 0.5 block with 7.3,
0.5 block with 7.5

LESSON OPENER
ACTIVITY
An alternative way to approach Lesson 7.4 is to use the Activity Lesson Opener:

• Blackline Master (*Chapter 7 Resource Book,* p. 52)
• Transparency (p. 47)

MEETING INDIVIDUAL NEEDS
• *Chapter 7 Resource Book*
  Prerequisite Skills Review (p. 5)
  Practice Level A (p. 54)
  Practice Level B (p. 55)
  Practice Level C (p. 56)
  Reteaching with Practice (p. 57)
  Absent Student Catch-Up (p. 59)
  Challenge (p. 61)
• *Resources in Spanish*
• *Personal Student Tutor*

NEW-TEACHER SUPPORT
See the Tips for New Teachers on pp. 1–2 of the *Chapter 7 Resource Book* for additional notes about Lesson 7.4.

WARM-UP EXERCISES

Transparency Available

Solve for *y* in each of the following.
1. $3y = 6x$ $y = 2x$
2. $2y = 4x + 2$ $y = 2x + 1$
3. $3x + y = 6$ $y = -3x + 6$
4. $2x + 6y = 6$ $y = -\frac{1}{3}x + 1$
5. $3x + 8 = 4y$ $y = \frac{3}{4}x + 2$

7.4

What you should learn

GOAL 1 Find inverses of linear functions.

GOAL 2 Find inverses of nonlinear functions, as applied in **Example 6**.

Why you should learn it

▼ To solve **real-life** problems, such as finding your bowling average in **Ex. 59**.

┌ STUDENT HELP
└→ **Look Back**
For help with solving equations for *y*, see p. 26.

Inverse Functions

GOAL 1 **FINDING INVERSES OF LINEAR FUNCTIONS**

In Lesson 2.1 you learned that a *relation* is a mapping of input values onto output values. An **inverse relation** maps the output values back to their original input values. This means that the domain of the inverse relation is the range of the original relation and that the range of the inverse relation is the domain of the original relation.

Original relation

x	-2	-1	0	1	2
y	4	2	0	-2	-4

Inverse relation

x	4	2	0	-2	-4
y	-2	-1	0	1	2

The graph of an inverse relation is the *reflection* of the graph of the original relation. The line of reflection is $y = x$.

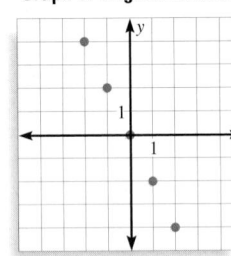

Graph of original relation   **Reflection in $y = x$**   **Graph of inverse relation**

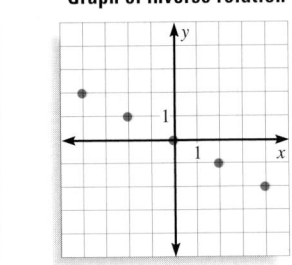

To find the inverse of a relation that is given by an equation in x and y, switch the roles of x and y and solve for y (if possible).

EXAMPLE 1   *Finding an Inverse Relation*

Find an equation for the inverse of the relation $y = 2x - 4$.

SOLUTION

$$y = 2x - 4 \qquad \text{Write original relation.}$$
$$x = 2y - 4 \qquad \text{Switch } x \text{ and } y.$$
$$x + 4 = 2y \qquad \text{Add 4 to each side.}$$
$$\frac{1}{2}x + 2 = y \qquad \text{Divide each side by 2.}$$

▶ The inverse relation is $y = \frac{1}{2}x + 2$.

· · · · · · · · · ·

In Example 1 both the original relation and the inverse relation happen to be functions. In such cases the two functions are called **inverse functions**.

STUDENT HELP

↳ **Study Tip**
The notation for an inverse function, f^{-1}, looks like a negative exponent, but it should not be interpreted that way. In other words,
$$f^{-1}(x) \neq (f(x))^{-1} = \frac{1}{f(x)}.$$

INVERSE FUNCTIONS

Functions f and g are inverses of each other provided:
$$f(g(x)) = x \quad \text{and} \quad g(f(x)) = x$$
The function g is denoted by f^{-1}, read as "f inverse."

Given any function, you can always find its inverse relation by switching x and y. For a linear function $f(x) = mx + b$ where $m \neq 0$, the inverse is itself a linear function.

EXAMPLE 2 *Verifying Inverse Functions*

Verify that $f(x) = 2x - 4$ and $f^{-1}(x) = \frac{1}{2}x + 2$ are inverses.

SOLUTION Show that $f(f^{-1}(x)) = x$ and $f^{-1}(f(x)) = x$.

$$f(f^{-1}(x)) = f\left(\frac{1}{2}x + 2\right) \qquad\qquad f^{-1}(f(x)) = f^{-1}(2x - 4)$$

$$= 2\left(\frac{1}{2}x + 2\right) - 4 \qquad\qquad = \frac{1}{2}(2x - 4) + 2$$

$$= x + 4 - 4 \qquad\qquad\qquad\quad = x - 2 + 2$$

$$= x \checkmark \qquad\qquad\qquad\qquad\quad = x \checkmark$$

Science

EXAMPLE 3 *Writing an Inverse Model*

When calibrating a spring scale, you need to know how far the spring stretches based on given weights. Hooke's law states that the length a spring stretches is proportional to the weight attached to the spring. A model for one scale is $\ell = 0.5w + 3$ where ℓ is the total length (in inches) of the spring and w is the weight (in pounds) of the object.

a. Find the inverse model for the scale.

b. If you place a melon on the scale and the spring stretches to a total length of 5.5 inches, how much does the melon weigh?

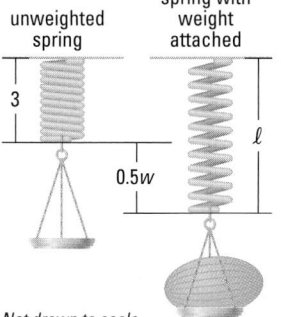
unweighted spring

spring with weight attached

Not drawn to scale

STUDENT HELP

↳ **Study Tip**
Notice that you do not switch the variables when you are finding inverses for models. This would be confusing because the letters are chosen to remind you of the real-life quantities they represent.

SOLUTION

a. $\ell = 0.5w + 3$ Write original model.

$\ell - 3 = 0.5w$ Subtract 3 from each side.

$\dfrac{\ell - 3}{0.5} = w$ Divide each side by 0.5.

$2\ell - 6 = w$ Simplify.

b. To find the weight of the melon, substitute 5.5 for ℓ.

$$w = 2\ell - 6 = 2(5.5) - 6 = 11 - 6 = 5$$

▶ The melon weighs 5 pounds.

Checkpoint Exercises *Sample answer:*

2. a. $m = \dfrac{T - 29.95}{0.05} = 20T - 599$ for $T \geq 29.95$

EXTRA EXAMPLE 1
Find an inverse of $y = -3x + 6$.
$$y = -\frac{1}{3}x + 2$$

EXTRA EXAMPLE 2
Verify that $f(x) = -3x + 6$ and $f^{-1}(x) = -\frac{1}{3}x + 2$ are inverses.
$$f(f^{-1}(x)) = f\left(-\frac{1}{3}x + 2\right)$$
$$= -3\left(-\frac{1}{3}x + 2\right) + 6$$
$$= x - 6 + 6 = x$$
$$f^{-1}(f(x)) = f^{-1}(-3x + 6)$$
$$= -\frac{1}{3}(-3x + 6) + 2$$
$$= x - 2 + 2 = x$$

EXTRA EXAMPLE 3
A model for a salary is $S = 10.50h + 50$, where S is the total salary (in dollars) for one week and h is the number of hours worked.
a. Find the inverse function for the model. $h = \dfrac{S - 50}{10.50}$
b. If a person's salary is $533, how many hours does the person work? **46 h**

✓ **CHECKPOINT EXERCISES**
For use after Examples 1 and 2:
1. a. Find the inverse of $y = \frac{1}{3}x - 1$. $y = 3x + 3$
b. Verify that this function and your answer are inverses.
$$f(f^{-1}(x)) = f(3x + 3)$$
$$= \frac{1}{3}(3x + 3) - 1$$
$$= x + 1 - 1 = x$$
$$f^{-1}f(x)) = f^{-1}\left(\frac{1}{3}x - 1\right)$$
$$= 3\left(\frac{1}{3}x - 1\right) + 3$$
$$= x - 3 + 3 = x$$

For use after Example 3:
2. A model for a telephone bill is $T = 0.05m + 29.95$, where T is the total bill, and m is the number of minutes used.
a. Find the inverse model.
See margin.
b. If the total bill is $54.15, how many minutes were used? **484**

EXTRA EXAMPLE 4
Find the inverse of the function
$f(x) = x^5$. $y = \sqrt[5]{x}$

 CHECKPOINT EXERCISES

For use after Example 4:

1. Find the inverse of the function $f(x) = x^4$, $x \geq 0$. $y = \pm\sqrt[4]{x}$

GOAL 2 FINDING INVERSES OF NONLINEAR FUNCTIONS

The graphs of the power functions $f(x) = x^2$ and $g(x) = x^3$ are shown below along with their reflections in the line $y = x$. Notice that the inverse of $g(x) = x^3$ is a function, but that the inverse of $f(x) = x^2$ is *not* a function.

STUDENT HELP

▸ **Look Back**
For help with recognizing when a relationship is a function, see p. 70.

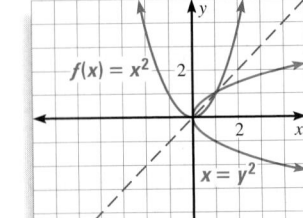
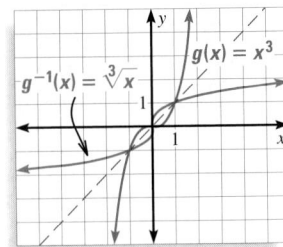

If the domain of $f(x) = x^2$ is *restricted*, say to only nonnegative real numbers, then the inverse of f *is* a function.

EXAMPLE 4 *Finding an Inverse Power Function*

Find the inverse of the function $f(x) = x^2$, $x \geq 0$.

SOLUTION

$\begin{aligned} f(x) &= x^2 && \text{Write original function.} \\ y &= x^2 && \text{Replace } f(x) \text{ with } y. \\ x &= y^2 && \text{Switch } x \text{ and } y. \\ \pm\sqrt{x} &= y && \text{Take square roots of each side.} \end{aligned}$

Because the domain of f is restricted to nonnegative values, the inverse function is $f^{-1}(x) = \sqrt{x}$. (You would choose $f^{-1}(x) = -\sqrt{x}$ if the domain had been restricted to $x \leq 0$.)

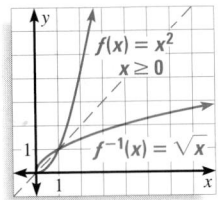

✓ **CHECK** To check your work, graph f and f^{-1} as shown. Note that the graph of $f^{-1}(x) = \sqrt{x}$ is the reflection of the graph of $f(x) = x^2$, $x \geq 0$ in the line $y = x$.

· · · · · · · · · ·

In the graphs at the top of the page, notice that the graph of $f(x) = x^2$ can be intersected twice with a horizontal line and that its inverse is *not* a function. On the other hand, the graph of $g(x) = x^3$ cannot be intersected twice with a horizontal line and its inverse *is* a function. This observation suggests the *horizontal line test*.

HORIZONTAL LINE TEST

If no horizontal line intersects the graph of a function f more than once, then the inverse of f is itself a function.

EXAMPLE 5 Finding an Inverse Function

Consider the function $f(x) = \frac{1}{2}x^3 - 2$. Determine whether the inverse of f is a function. Then find the inverse.

SOLUTION

Begin by graphing the function and noticing that no horizontal line intersects the graph more than once. This tells you that the inverse of f is itself a function. To find an equation for f^{-1}, complete the following steps.

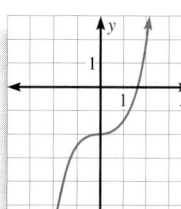

$$f(x) = \frac{1}{2}x^3 - 2 \qquad \text{Write original function.}$$

$$y = \frac{1}{2}x^3 - 2 \qquad \text{Replace } f(x) \text{ with } y.$$

$$x = \frac{1}{2}y^3 - 2 \qquad \text{Switch } x \text{ and } y.$$

$$x + 2 = \frac{1}{2}y^3 \qquad \text{Add 2 to each side.}$$

$$2x + 4 = y^3 \qquad \text{Multiply each side by 2.}$$

$$\sqrt[3]{2x + 4} = y \qquad \text{Take cube root of each side.}$$

▶ The inverse function is $f^{-1}(x) = \sqrt[3]{2x + 4}$.

EXAMPLE 6 Writing an Inverse Model

ASTRONOMY Near the end of a star's life the star will eject gas, forming a planetary nebula. The Ring Nebula is an example of a planetary nebula. The volume V (in cubic kilometers) of this nebula can be modeled by $V = (9.01 \times 10^{26})t^3$ where t is the age (in years) of the nebula. Write the inverse model that gives the age of the nebula as a function of its volume. Then determine the approximate age of the Ring Nebula given that its volume is about 1.5×10^{38} cubic kilometers.

SOLUTION

$$V = (9.01 \times 10^{26})t^3 \qquad \text{Write original model.}$$

$$\frac{V}{9.01 \times 10^{26}} = t^3 \qquad \text{Isolate power.}$$

$$\sqrt[3]{\frac{V}{9.01 \times 10^{26}}} = t \qquad \text{Take cube root of each side.}$$

$$(1.04 \times 10^{-9})\sqrt[3]{V} = t \qquad \text{Simplify.}$$

To find the age of the nebula, substitute 1.5×10^{38} for V.

$$t = (1.04 \times 10^{-9})\sqrt[3]{V} \qquad \text{Write inverse model.}$$

$$= (1.04 \times 10^{-9})\sqrt[3]{1.5 \times 10^{38}} \qquad \text{Substitute for } V.$$

$$\approx 5500 \qquad \text{Use a calculator.}$$

▶ The Ring Nebula is about 5500 years old.

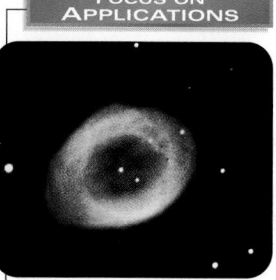

FOCUS ON APPLICATIONS

REAL LIFE ASTRONOMY
The Ring Nebula is part of the constellation Lyra. The radius of the nebula is expanding at an average rate of about 5.99×10^8 kilometers per year.

APPLICATION LINK
www.mcdougallittell.com

7.4 *Inverse Functions*   **425**

Daily Puzzler
No; *Sample answer:* a constant function, such as $f(x) = 6$, does not have an inverse since division by zero is not possible. Its reflection over the line $y = x$ is a vertical line.

EXTRA EXAMPLE 5
Consider the function $f(x) = 2x^2 - 4$. Determine whether the inverse of f is a function and then find the inverse.

The inverse is $y = \pm\sqrt{\dfrac{x + 4}{2}}$, and it is not a function.

EXTRA EXAMPLE 6
The volume of a sphere is given by $V = \dfrac{4}{3}\pi r^3$, where V is the volume and r is the radius. Write the inverse function that gives the radius as a function of the volume. Then determine the radius of a volleyball given that its volume is about 293 cubic inches. $r = \sqrt[3]{\dfrac{3V}{4\pi}}$; about 4.12 in.

✔ CHECKPOINT EXERCISES
For use after Examples 5 and 6:

1. The volume of a cylinder with height 10 feet is given by $V = 10\pi r^2$, where V is the volume, and r is the radius. Write the inverse model that gives the radius as a function of the volume. Then determine the radius of a cylinder given that its volume is 1050 cubic feet.
$r = \sqrt{\dfrac{V}{10\pi}}$; about 5.78 ft

APPLICATION NOTE
Additional information about astronomy is available at **www.mcdougallittell.com**.

FOCUS ON VOCABULARY
When are two functions f and g inverses of one another?
when $f(g(x)) = x$ and $g(f(x)) = x$

CLOSURE QUESTION
Describe the steps for finding the inverse of a relation. Write the original equation; switch x and y; solve for y.

DAILY PUZZLER
Does every function have an inverse? Explain. See margin.

425

ASSIGNMENT GUIDE

BASIC
Day 1: pp. 426–429 Exs. 14–15,
16–30 even, 33–35, 36–38,
42
Day 2: pp. 426–429 Exs. 39–41,
44–56 even, 57, 59,
63–64, 69–75 odd,
Quiz 2 Exs. 1–17

AVERAGE
Day 1: pp. 426–429 Exs. 14–56
even, 57, 58, 62
Day 2: pp. 426–429 Exs. 33–55
odd, 60, 63–64, 69–75 odd,
Quiz 2 Exs. 1–17

ADVANCED
Day 1: pp. 426–429 Exs. 14–60
even
Day 2: pp. 426–429 Exs. 15–61
odd, 62–68, 69–83 odd,
Quiz 2 Exs. 1–17

BLOCK SCHEDULE
pp. 426–429 Exs. 14–56 even, 57,
58, 62 (with 7.3)
pp. 426–429 Exs. 33–55 odd, 60,
63–64, 69–75 odd, Quiz 2 Exs.
1–17 (with 7.5)

EXERCISE LEVELS
Level A: *Easier*
14–28, 33–44, 48–56
Level B: *More Difficult*
29–32, 45–47, 57–64
Level C: *Most Difficult*
65–68

✔ HOMEWORK CHECK

To quickly check student understanding of key concepts, go
over the following exercises:
Exs. 14, 16, 26, 34, 36, 42. See
also the Daily Homework Quiz:

• Blackline Master (*Chapter 7
Resource Book*, p. 65)
• 🖐 Transparency (p. 55)

4, 5, 14–15. See Additional Answers
beginning on page AA1.

GUIDED PRACTICE

Vocabulary Check ✓

Concept Check ✓

Skill Check ✓

1. Explain how to use the horizontal line test to determine if an inverse relation is an inverse function. **If no horizontal line crosses the graph of the function more than once, then the inverse relation is an inverse function.**
2. Describe how the graph of a relation and the graph of its inverse are related. **See margin.**
3. Explain the steps in finding an equation for an inverse function. **Switch x and y in the original equation and solve for y.**

2. The graphs of a relation and its inverse are reflections of one another in the line $y = x$.

Find the inverse relation.
4–5. See margin.

4.

x	1	2	3	4	5
y	−1	−2	−3	−4	−5

5.

x	−4	−2	0	2	4
y	2	1	0	1	2

Find an equation for the inverse relation.

6. $y = 5x$ $y = \frac{x}{5}$
7. $y = 2x - 1$ $y = \frac{x+1}{2}$
8. $y = -\frac{2}{3}x + 6$ $y = -\frac{3}{2}x + 9$

Verify that f and g are inverse functions.

9. $f(x) = 8x^3, g(x) = \frac{x^{1/3}}{2}$ **Both compositions equal x.**
10. $f(x) = 6x + 3, g(x) = \frac{1}{6}x - \frac{1}{2}$ **Both compositions equal x.**

Find the inverse function.

12. $\frac{\sqrt[3]{4x-4}}{2}$

11. $f(x) = 3x^4, x \geq 0$ $\sqrt[4]{\frac{27x}{3}}$
12. $f(x) = 2x^3 + 1$
13. The graph of $f(x) = -|x| + 1$ is shown. Is the inverse of f a function? Explain. **No. Horizontal lines, such as $y = 0$, cross the graph more than once.**

Ex. 13

PRACTICE AND APPLICATIONS

STUDENT HELP
▸ **Extra Practice**
to help you master
skills is on p. 949.

INVERSE RELATIONS **Find the inverse relation.**
14, 15. See margin.

14.

x	1	4	1	0	1
y	3	−1	6	−3	9

15.

x	1	−2	4	2	−2
y	0	3	−2	2	−1

FINDING INVERSES **Find an equation for the inverse relation.**

16. $y = -2x + 5$ $y = \frac{5-x}{2}$
17. $y = 3x - 3$ $y = \frac{x+3}{3}$
18. $y = \frac{1}{2}x + 6$ $y = 2(x - 6)$
19. $y = -\frac{4}{5}x + 11$ $y = -\frac{5}{4}(x - 11)$
20. $y = 11x - 5$ $y = \frac{x+5}{11}$
21. $y = -12x + 7$ $y = \frac{-x+7}{12}$
22. $y = 3x - \frac{1}{4}$ $y = \frac{x+\frac{1}{4}}{3}$
23. $y = 8x - 13$ $y = \frac{x+13}{8}$
24. $y = -\frac{3}{7}x + \frac{5}{7}$ $y = -\frac{7}{3}x + \frac{5}{3}$

25–32. Check students' work. In each case, $f(g(x)) = x$ and $g(f(x)) = x$.

STUDENT HELP
▸ **HOMEWORK HELP**
Example 1: Exs. 14–24
Example 2: Exs. 25–32
Example 3: Exs. 57–59
Example 4: Exs. 33–41
Example 5: Exs. 42–56
Example 6: Exs. 60–62

VERIFYING INVERSES **Verify that f and g are inverse functions.**
25–32. See margin.

25. $f(x) = x + 7, g(x) = x - 7$
26. $f(x) = 3x - 1, g(x) = \frac{1}{3}x + \frac{1}{3}$
27. $f(x) = \frac{1}{2}x + 1, g(x) = 2x - 2$
28. $f(x) = -2x + 4, g(x) = -\frac{1}{2}x + 2$
29. $f(x) = 3x^3 + 1, g(x) = \left(\frac{x-1}{3}\right)^{1/3}$
30. $f(x) = \frac{1}{3}x^2, x \geq 0; g(x) = (3x)^{1/2}$
31. $f(x) = \frac{x^5 + 2}{7}, g(x) = \sqrt[5]{7x - 2}$
32. $f(x) = 256x^4, x \geq 0; g(x) = \frac{\sqrt[4]{x}}{4}$

43. $\sqrt[5]{-\frac{1}{2}x + \frac{1}{6}}$

44. $-\sqrt{\frac{-x + 2}{2}}$

46. $\sqrt[4]{x + \frac{1}{2}}$

48. Yes, inverse is a function; See margin.

49. Yes, inverse is a function; See margin.

50. No, inverse is not a function; See margin.

51. No, inverse is not a function; See margin.

52. Yes, inverse is a function; See margin.

53. Yes, inverse is a function; See margin.

54. No, inverse is not a function; See margin.

55. No, inverse is not a function; See margin.

56. No, inverse is not a function; See margin.

VISUAL THINKING Match the graph with the graph of its inverse.

33. A
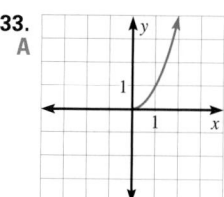

34. C
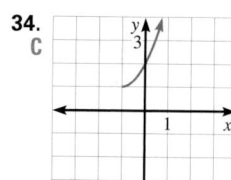

35. B
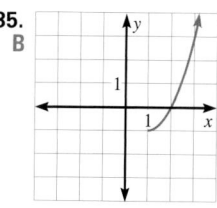

A.
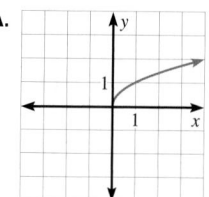

B.
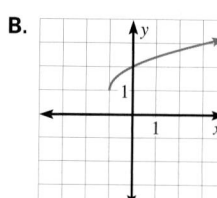

C.
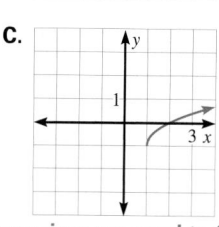

36–47. The given expressions are equal to $f^{-1}(x)$.

INVERSES OF POWER FUNCTIONS Find the inverse power function.

36. $f(x) = x^7$ $\sqrt[7]{x}$

37. $f(x) = -x^6, x \geq 0$ $\sqrt[6]{-x}$

38. $f(x) = 3x^4, x \leq 0$ $-\sqrt[4]{\frac{x}{3}}$

39. $f(x) = \frac{1}{32}x^5$ $2\sqrt[5]{x}$

40. $f(x) = 10x^3$ $\frac{\sqrt[3]{100x}}{10}$

41. $f(x) = -\frac{9}{4}x^2, x \leq 0$ $-\frac{2}{3}\sqrt{-x}$

INVERSES OF NONLINEAR FUNCTIONS Find the inverse function.

43–44, 46. See margin.

42. $f(x) = x^3 + 2$ $\sqrt[3]{x - 2}$

43. $f(x) = -2x^5 + \frac{1}{3}$

44. $f(x) = 2 - 2x^2, x \leq 0$

45. $f(x) = \frac{3}{5}x^3 - 9$ $\sqrt[3]{\frac{5}{3}x + 15}$

46. $f(x) = x^4 - \frac{1}{2}, x \geq 0$

47. $f(x) = \frac{1}{6}x^5 + \frac{2}{3}$ $\sqrt[5]{6x - 4}$

HORIZONTAL LINE TEST Graph the function f. Then use the graph to determine whether the inverse of f is a function.

48–56. See margin.

48. $f(x) = -2x + 3$

49. $f(x) = x + 3$

50. $f(x) = x^2 + 1$

51. $f(x) = -3x^2$

52. $f(x) = x^3 + 3$

53. $f(x) = 2x^3$

54. $f(x) = |x| + 2$

55. $f(x) = (x + 1)(x - 3)$

56. $f(x) = 6x^4 - 9x + 1$

57. 🌐 **EXCHANGE RATE** The Federal Reserve Bank of New York reports international exchange rates at 12:00 noon each day. On January 20, 1999, the exchange rate for Canada was 1.5226. Therefore, the formula that gives Canadian dollars in terms of United States dollars on that day is $D_{US} = 0.65677 D_C$

$$D_C = 1.5226 D_{US}$$

where D_C represents Canadian dollars and D_{US} represents United States dollars. Find the inverse of the function to determine the value of a United States dollar in terms of Canadian dollars on January 20, 1999.

🔗 DATA UPDATE of Federal Reserve Bank of New York data at www.mcdougallittell.com

58. 🌐 **TEMPERATURE CONVERSION** The formula to convert temperatures from degrees Fahrenheit to degrees Celsius is: $F = \frac{9}{5}C + 32$; 84.2°F, 50°F, 32°F

$$C = \frac{5}{9}(F - 32)$$

Write the inverse of the function, which converts temperatures from degrees Celsius to degrees Fahrenheit. Then find the Fahrenheit temperatures that are equal to 29°C, 10°C, and 0°C.

FOCUS ON CAREERS

REAL LIFE **INVESTMENT BANKER**
Investment bankers have a wide variety of job descriptions. Some buy and sell international currencies at reported exchange rates, discussed in Ex. 57.

🔗 CAREER LINK
www.mcdougallittell.com

! **COMMON ERROR**
EXERCISE 38 A common error is to think that the inverse of any power function with an even exponent is not a function. Point out that if the domain of a function is restricted, its inverse may be a function.

MATHEMATICAL REASONING
EXERCISES 48–53 You may want students to predict whether the inverse is a function before they graph the original equation. The inverses of power functions with odd powers are functions. The inverses of power functions with even powers are not functions unless the domain is restricted.

CAREER NOTE
EXERCISE 57 Additional information about investment bankers is available at **www.mcdougallittell.com**.

APPLICATION NOTE
EXERCISES 58 Students will recognize the inverse as the formula to convert Celsius to Fahrenheit temperature. This is an example of a real-life application where you do not switch the variables to find the inverse. You solve for the other variable, F.

48–56. See Additional Answers beginning on page AA1.

APPLICATION NOTE

EXERCISE 59 For some sports, the handicap is given as a score rather than as a percent. For example, your handicap in golf may be to subtract 10 strokes from your score.

EXERCISE 62 This exercise reinforces a student's understanding of the use of an inverse. Ask students why it is important to know the inverse of this function. **It will tell you how to position the supports for the shelf so the shelf can hold a particular set of books.**

┌─ **STUDENT HELP NOTES**

→ **Homework Help** Students can find help for Ex. 62 at **www.mcdougallittell.com.** This information can be printed out for students who don't have access to the Internet.

66. *Sample answer:*

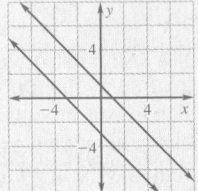

ADDITIONAL PRACTICE AND RETEACHING

For Lesson 7.4:

• Practice Levels A, B, and C (*Chapter 7 Resource Book*, p. 54)

• Reteaching with Practice (*Chapter 7 Resource Book*, p. 57)

• ⊞ See Lesson 7.4 of the *Personal Student Tutor.*

For more Mixed Review:

• ⊞ Search the *Test and Practice Generator* for key words or specific lessons.

428

---

59. 🌐 **BOWLING** In bowling a *handicap* is a change in score to adjust for differences in players' abilities. You belong to a bowling league in which each bowler's handicap h is determined by his or her average a using this formula: $a = 200 - 1.11h; 170$

$$h = 0.9(200 - a)$$

(If the bowler's average is over 200, the handicap is 0.) Find the inverse of the function. Then find your average if your handicap is 27.

60. 🎲 **GAMES** You and a friend are playing a number-guessing game. You ask your friend to think of a positive number, square the number, multiply the result by 2, and then add 3. If your friend's final answer is 53, what was the original number chosen? Use an inverse function in your solution. **5**

61. 🐟 **FISH** The weight w (in kilograms) of a hake, a type of fish, is related to its length l (in centimeters) by this function: $l = \sqrt[3]{106723.59w}$; **41.69 cm**

$$w = (9.37 \times 10^{-6})l^3$$

Find the inverse of the function. Then determine the approximate length of a hake that weighs 0.679 kilogram. ▶ Source: *Fishbyte*

Hake

62. 📚 **SHELVES** The weight w (in pounds) that can be supported by a shelf made from half-inch Douglas fir plywood can be modeled by

$$w = \left(\frac{82.9}{d}\right)^3 \qquad d = \frac{82.9}{w^{1/3}}; 20.51 \text{ in.}$$

where d is the distance (in inches) between the supports for the shelf. Find the inverse of the function. Then find the distance between the supports of a shelf that can hold a set of encyclopedias weighing 66 pounds.

Test 🎯 Preparation

STUDENT HELP

→ 🌐 **HOMEWORK HELP** Visit our Web site www.mcdougallittell.com for help with problem solving in Ex. 62.

QUANTITATIVE COMPARISON **In Exercises 63 and 64, choose the statement that is true about the given quantities.**

ⓐ The quantity in column A is greater.

ⓑ The quantity in column B is greater.

ⓒ The two quantities are equal.

ⓓ The relationship cannot be determined from the given information.

	Column A	Column B	
63.	$f^{-1}(3)$ where $f(x) = 6x + 1$	$f^{-1}(-4)$ where $f(x) = -2x + 9$	B
64.	$f^{-1}(2)$ where $f(x) = -5x^3$	$f^{-1}(0)$ where $f(x) = x^3 + 14$	A

★ Challenge

65. Check graphs; $f(x) = x$ and $g(x) = -x$ are their own inverses because the graph of each is its own reflection in the line $y = x$.

┌─ **EXTRA CHALLENGE** ─┐
→ www.mcdougallittell.com

INVERSE FUNCTIONS **Complete Exercises 65–68 to explore functions that are their own inverses.**

65. VISUAL THINKING The functions $f(x) = x$ and $g(x) = -x$ are their own inverses. Graph each function and explain why this is true.

66. Graph other linear functions that are their own inverses. **See margin.**

67. Write equations of the lines you graphed in Exercise 66. *Sample answers:* $y = -x + 1; y = -x - 3$

68. Use your equations from Exercise 67 to find a general formula for a family of linear equations that are their own inverses. $y = -x + a$, where a is a real number

ABSOLUTE VALUE FUNCTIONS Graph the absolute value function.
(Review 2.8 for 7.5) 69–76. See margin.

69. $f(x) = |x| - 1$

70. $f(x) = 2|x| + 7$

71. $f(x) = |x - 4| + 5$

72. $f(x) = -3|x + 2| - 7$

QUADRATIC FUNCTIONS Graph the quadratic function. (Review 5.1 for 7.5)

73. $f(x) = x^2 + 2$

74. $f(x) = (x + 3)^2 - 7$

75. $f(x) = 2(x + 2)^2 - 5$

76. $f(x) = -3(x - 4)^2 + 1$

SIMPLIFYING EXPRESSIONS Simplify the expression. Assume all variables are positive. (Review 7.2)

77. $\sqrt[4]{20} \cdot \sqrt[4]{\frac{4}{5}}$ 2

78. $\left(\frac{1}{9}\right)^{1/6}\left(\frac{1}{9}\right)^{1/3}$ $\frac{1}{3}$

79. $\frac{(5y)^{1/5}}{(5y)^{6/5}}$ $\frac{1}{5y}$

80. $\sqrt[6]{2x^6}$ $x\sqrt[6]{2}$

81. $3\sqrt[7]{5} + 2\sqrt[7]{5}$ $5\sqrt[7]{5}$

82. $\sqrt[3]{270} + 2\sqrt[3]{10}$ $5\sqrt[3]{10}$

83. 🌐 **SNACK FOODS** Delia, Ruth, and Amy go to the store to buy snacks. Delia buys 3 bagels and 3 apples. Ruth buys 1 pretzel, 2 bagels, and 3 apples. Amy buys 2 pretzels and 4 bagels. Delia's bill comes to $3.72, Ruth's to $5.06, and Amy's to $6.58. How much does one bagel cost? (Review 3.6) $.65

QUIZ 2

Self-Test for Lessons 7.3 and 7.4

Let $f(x) = 6x^2 - x^{1/2}$ and $g(x) = 2x^{1/2}$. Perform the indicated operation and state the domain. (Lesson 7.3)

1–4. See margin.

1. $f(x) + g(x)$ **2.** $f(x) - g(x)$ **3.** $f(x) \cdot g(x)$ **4.** $\dfrac{f(x)}{g(x)}$

Let $f(x) = 3x^{-1}$ and $g(x) = x - 8$. Perform the indicated operation and state the domain. (Lesson 7.3)   5–8. See margin.

5. $f(g(x))$ **6.** $g(f(x))$ **7.** $f(f(x))$ **8.** $g(g(x))$

Verify that f and g are inverse functions. (Lesson 7.4)

9. $f(x) = 2x - 3$, $g(x) = \frac{1}{2}x + \frac{3}{2}$ **10.** $f(x) = (x + 1)^{1/3}$, $g(x) = x^3 - 1$
Both compositions equal x.   Both compositions equal x.

Find the inverse function. (Lesson 7.4)

11. $f(x) = x + 8$ $x - 8$ **12.** $f(x) = 2x^4, x \le 0$ **13.** $f(x) = -x^5 + 6$ $\sqrt[5]{6 - x}$
See margin.

Graph the function f. Then use the graph to determine whether the inverse of f is a function. (Lesson 7.4)   14–16. See margin.

14. $f(x) = 3x^6 + 2$ **15.** $f(x) = -2x^5 + 3x - 1$ **16.** $f(x) = 6\sqrt[3]{x + 4}$

17. 🌐 **RIPPLES IN A POND** You drop a pebble into a calm pond causing ripples of concentric circles. The radius r (in feet) of the outer ripple is given by $r(t) = 0.6t$ where t is the time (in seconds) after the pebble hits the water. The area A (in square feet) of the outer ripple is given by $A(r) = \pi r^2$. Use composition of functions to find the relationship between area and time. Then find the area of the outer ripple after 2 seconds. (Lesson 7.3) $A(t) = 0.36\pi t^2$; about 4.52 ft^2

7.4 *Inverse Functions*   **429**

Margin answers (left column):

1–8, 11–13. An expression that defines the function is given.

1. $6x^2 + x^{1/2}$; nonnegative real numbers

2. $6x^2 - 3x^{1/2}$; nonnegative real numbers

3. $2x(6x^{3/2} - 1)$; nonnegative real numbers

4. $3x^{3/2} - \frac{1}{2}$; positive real numbers

5. $\frac{3}{x - 8}$; real numbers except 8

6. $\frac{3}{x} - 8$; real numbers except 0

7. x; real numbers except 0

8. $x - 16$; all real numbers

12. $\dfrac{-\sqrt[4]{8x}}{2}$

14. No, inverse is not a function; See margin.

15. No, inverse is not a function; See margin.

16. Yes, inverse is a function; See margin.

Margin answers (right column):

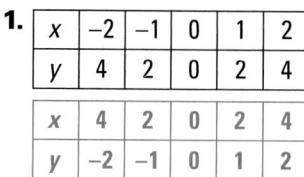

4 ASSESS

DAILY HOMEWORK QUIZ

📕 *Transparency Available*

Find the inverse relation.

1.

x	−2	−1	0	1	2
y	4	2	0	2	4

x	4	2	0	2	4
y	−2	−1	0	1	2

2. $y = 3x - 4$ $y = \frac{1}{3}(x + 4)$

Find the inverse function.

3. $f(x) = 2x^5$ $f^{-1}(x) = \sqrt[5]{\frac{x}{2}}$

4. $f(x) = 4x^3 - 1$ $f^{-1}(x) = \sqrt[3]{\frac{x + 1}{4}}$

EXTRA CHALLENGE NOTE

↳ Challenge problems for Lesson 7.4 are available in **blackline** format in the *Chapter 7 Resource Book,* p. 61 and at **www.mcdougallittell.com.**

ADDITIONAL TEST PREPARATION

1. WRITING Explain how to find the inverse of a function and how to determine if it is also a function. *Sample answer:* Write the original function. Switch x and y. Solve for y. If the graph of the original function passes the horizontal line test, its inverse will be a function.

ADDITIONAL RESOURCES

An alternative Quiz for Lessons 7.3–7.4 is available in the *Chapter 7 Resource Book,* p. 62.

69–76, Quiz 2 14–16.
See Additional Answers beginning on page AA1.

1 Planning the Activity

PURPOSE
To graph inverse functions with a graphing calculator.

MATERIALS
- graphing calculator
- Keystroke blackline (*Chapter 7 Resource Book*, p. 53)

PACING
- Example — 15 min
- Exercises — 15 min

▶ LINK TO LESSON
Students can use the Draw Inverse feature to view the graphs of functions and their inverses in Exercises 36–47 in Lesson 7.4.

2 Managing the Activity

CLASSROOM MANAGEMENT
The DrawInv feature does not distinguish between inverses that are functions and inverses that are not functions. You may want to make this distinction with your students.

ALTERNATIVE APPROACH
Ask students to find the inverse of some exercises algebraically and then use a graphing calculator to verify that they are inverse functions.

3 Closing the Activity

★ KEY DISCOVERY
The Draw Inverse feature can be used to draw the inverse of a function. The vertical line test allows you to determine whether this inverse is also a function.

ACTIVITY ASSESSMENT
Graph $y = 0.5x^2 - 3$ and its inverse with a graphing calculator. Is the inverse a function? Explain.
No; the original function does not pass the horizontal line test.

▷ ACTIVITY 7.4
Using Technology

STUDENT HELP

KEYSTROKE HELP
See keystrokes for several models of calculators at www.mcdougallittell.com

1. Yes; the inverse passes the vertical line test.
2. Yes; the inverse passes the vertical line test.
3. Yes; the inverse passes the vertical line test.
4. No; the inverse does not pass the vertical line test.
5. No; the inverse does not pass the vertical line test.
6. No; the inverse does not pass the vertical line test.
7. Yes; the inverse passes the vertical line test.
8. Yes; the inverse passes the vertical line test.
9. Yes; the inverse passes the vertical line test.
10. No; the inverse does not pass the vertical line test.
11. No; the inverse does not pass the vertical line test.
12. No; the inverse does not pass the vertical line test.

Graphing Inverse Functions

You can use a graphing calculator to graph inverse functions.

▶ EXAMPLE
Use a graphing calculator to graph the inverse of $y = 2x - 5$.

▶ SOLUTION

❶ Graph the original function. Use a viewing window such as $-15 \le x \le 15$ and $-10 \le y \le 10$.

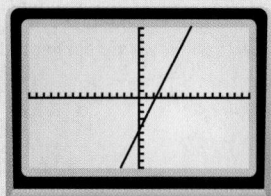

❷ Use the *Draw Inverse* feature to graph the inverse.

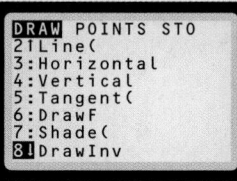 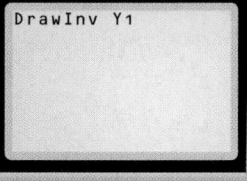

❸ Display the graphs of the original function and its inverse.

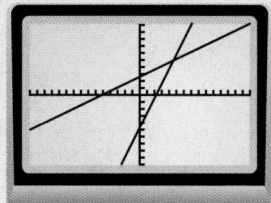

The graph of the function $y = 2x - 5$ passes the horizontal line test, so you know that the inverse of $y = 2x - 5$ is also a function. You can further verify this fact by observing that the inverse passes the vertical line test you learned in Lesson 2.1.

▶ EXERCISES

Graph the function on a graphing calculator. Then use the *Draw Inverse* feature to graph the function's inverse in the same viewing window. Is the inverse a function? Explain. 1–12. See margin.

1. $y = 6x + 4$
2. $y = 0.6x - 2$
3. $y = 0.4x + 5$
4. $y = 0.2x^2 + 1$
5. $y = x^2 - 4x + 3$
6. $y = x^2 - 3x$
7. $y = x^3 - 4$
8. $y = x^3 + x$
9. $y = 2.1x^3 - 0.4x^2 + 1$
10. $y = |x + 4|$
11. $y = -|x| + 5.7$
12. $y = 2|x + 1| - 8$

Graphing Square Root and Cube Root Functions

What you should learn

GOAL 1 Graph square root and cube root functions.

GOAL 2 Use square root and cube root functions to find **real-life** quantities, such as the power of a race car in **Ex. 48.**

Why you should learn it

▼ To solve **real-life** problems, such as finding the age of an African elephant in **Example 6.**

Steps 1, 2.

See Additional Answers beginning on page AA1 for graphs.

The absolute value of *a* determines how much the graph of $y = a\sqrt{x}$ is stretched or compressed compared with the graph of $y = \sqrt{x}$. The sign of *a* determines whether there is a reflection in the *x*-axis. The variable *a* affects the graph of $y = a\sqrt[3]{x}$ in a similar fashion as compared with the graph of $y = \sqrt[3]{x}$.

GOAL 1 GRAPHING RADICAL FUNCTIONS

In Lesson 7.4 you saw the graphs of $y = \sqrt{x}$ and $y = \sqrt[3]{x}$. These are examples of **radical functions**.

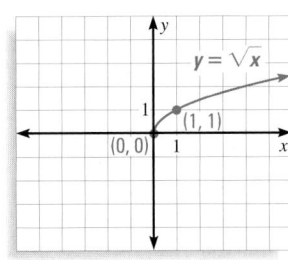

Domain: $x \ge 0$, Range: $y \ge 0$

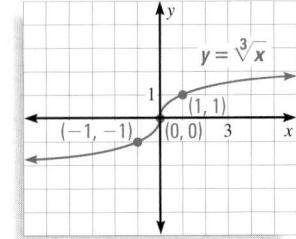

Domain and range: all real numbers

In this lesson you will learn to graph functions of the form $y = a\sqrt{x - h} + k$ and $y = a\sqrt[3]{x - h} + k$.

◐ ACTIVITY

Developing Concepts

Investigating Graphs of Radical Functions

❶ Graph $y = a\sqrt{x}$ for $a = 2, \frac{1}{2}, -3$, and -1. Use the graph of $y = \sqrt{x}$ shown above and the labeled points on the graph as a reference. Describe how *a* affects the graph. **See margin.**

❷ Graph $y = a\sqrt[3]{x}$ for $a = 2, \frac{1}{2}, -3$, and -1. Use the graph of $y = \sqrt[3]{x}$ shown above and the labeled points on the graph as a reference. Describe how *a* affects the graph. **See margin.**

In the activity you may have discovered that the graph of $y = a\sqrt{x}$ starts at the origin and passes through the point $(1, a)$. Similarly, the graph of $y = a\sqrt[3]{x}$ passes through the origin and the points $(-1, -a)$ and $(1, a)$. The following describes how to graph more general radical functions.

GRAPHS OF RADICAL FUNCTIONS

To graph $y = a\sqrt{x - h} + k$ or $y = a\sqrt[3]{x - h} + k$, follow these steps.

STEP ❶ Sketch the graph of $y = a\sqrt{x}$ or $y = a\sqrt[3]{x}$.

STEP ❷ Shift the graph *h* units horizontally and *k* units vertically.

7.5 *Graphing Square Root and Cube Root Functions* **431**

PACING
Basic: 1 day
Average: 1 day
Advanced: 1 day
Block Schedule: 0.5 block with 7.4

→ LESSON OPENER
VISUAL APPROACH
An alternative way to approach Lesson 7.5 is to use the Visual Approach Lesson Opener:
- Blackline Master (*Chapter 7 Resource Book*, p. 66)
- Transparency (p. 48)

MEETING INDIVIDUAL NEEDS
- *Chapter 7 Resource Book*
  Prerequisite Skills Review (p. 5)
  Practice Level A (p. 68)
  Practice Level B (p. 69)
  Practice Level C (p. 70)
  Reteaching with Practice (p. 71)
  Absent Student Catch-Up (p. 73)
  Challenge (p. 75)
- *Resources in Spanish*
- *Personal Student Tutor*

NEW-TEACHER SUPPORT
See the Tips for New Teachers on pp. 1–2 of the *Chapter 7 Resource Book* for additional notes about Lesson 7.5.

WARM-UP EXERCISES
Transparency Available

1. Graph $y = -2x^2 + 3$.
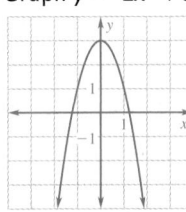

2. Graph $y = x^3 - 2$.
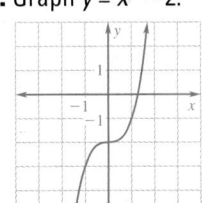

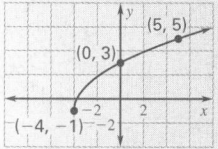

EXTRA EXAMPLE 1
Describe how to obtain the graph of $y = \sqrt[3]{x-2} + 1$ from the graph of $y = \sqrt[3]{x}$. **Shift the graph of $y = \sqrt[3]{x}$ right 2 units and up 1 unit.**

EXTRA EXAMPLE 2
Graph $y = 2\sqrt{x+4} - 1$.

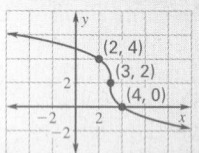

EXTRA EXAMPLE 3
Graph $y = -2\sqrt[3]{x-3} + 2$.

EXTRA EXAMPLE 4
State the domain and range of the function in (**a**) Extra Example 2 and (**b**) Extra Example 3.
See margin.

 CHECKPOINT EXERCISES

For use after Examples 1 and 2:
1. Describe how to obtain the graph of $y = \sqrt{x-3} + 2$ from the graph of $y = \sqrt{x}$. Then graph. **Shift the graph of $y = \sqrt{x}$ right 3 units and up 2 units.**

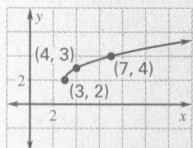

For use after Example 3.
2. Graph $y = -2\sqrt[3]{x-1} + 1$.

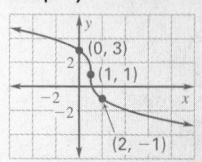

Checkpoint Exercises for Example 4 on next page.

432

STUDENT HELP
Skills Review
For help with transformations, see p. 921.

EXAMPLE 1 *Comparing Two Graphs*

Describe how to obtain the graph of $y = \sqrt{x+1} - 3$ from the graph of $y = \sqrt{x}$.

SOLUTION
Note that $y = \sqrt{x+1} - 3 = \sqrt{x-(-1)} + (-3)$, so $h = -1$ and $k = -3$. To obtain the graph of $y = \sqrt{x+1} - 3$, shift the graph of $y = \sqrt{x}$ left 1 unit and down 3 units.

EXAMPLE 2 *Graphing a Square Root Function*

Graph $y = -3\sqrt{x-2} + 1$.

SOLUTION
① Sketch the graph of $y = -3\sqrt{x}$ (shown dashed). Notice that it begins at the origin and passes through the point $(1, -3)$.

② Note that for $y = -3\sqrt{x-2} + 1$, $h = 2$ and $k = 1$. So, shift the graph right 2 units and up 1 unit. The result is a graph that starts at $(2, 1)$ and passes through the point $(3, -2)$.

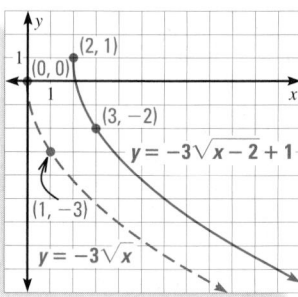

EXAMPLE 3 *Graphing a Cube Root Function*

Graph $y = 3\sqrt[3]{x+2} - 1$.

SOLUTION
① Sketch the graph of $y = 3\sqrt[3]{x}$ (shown dashed). Notice that it passes through the origin and the points $(-1, -3)$ and $(1, 3)$.

② Note that for $y = 3\sqrt[3]{x+2} - 1$, $h = -2$ and $k = -1$. So, shift the graph left 2 units and down 1 unit. The result is a graph that passes through the points $(-3, -4)$, $(-2, -1)$, and $(-1, 2)$.

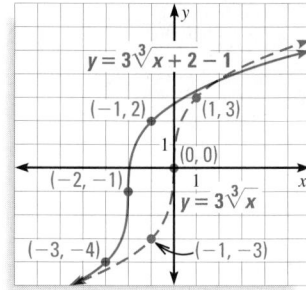

EXAMPLE 4 *Finding Domain and Range*

State the domain and range of the function in (**a**) Example 2 and (**b**) Example 3.

SOLUTION
a. From the graph of $y = -3\sqrt{x-2} + 1$ in Example 2, you can see that the domain of the function is $x \geq 2$ and the range of the function is $y \leq 1$.

b. From the graph of $y = 3\sqrt[3]{x+2} - 1$ in Example 3, you can see that the domain and range of the function are both all real numbers.

Extra Example 4
a. domain: $x \geq -4$; range: $y \geq -1$
b. domain and range: all real numbers

FOCUS ON CAREERS

AMUSEMENT RIDE DESIGNER

An amusement ride designer uses math and science to ensure the safety of the rides. Most amusement ride designers have a degree in mechanical engineering.

CAREER LINK
www.mcdougallittell.com

GOAL 2 **USING RADICAL FUNCTIONS IN REAL LIFE**

When you use radical functions in real life, the domain is understood to be restricted to the values that make sense in the real-life situation.

EXAMPLE 5 *Modeling with a Square Root Function*

 AMUSEMENT PARKS At an amusement park a ride called the *rotor* is a cylindrical room that spins around. The riders stand against the circular wall. When the rotor reaches the necessary speed, the floor drops out and the centrifugal force keeps the riders pinned to the wall.

The model that gives the speed s (in meters per second) necessary to keep a person pinned to the wall is

$$s = 4.95\sqrt{r}$$

where r is the radius (in meters) of the rotor. Use a graphing calculator to graph the model. Then use the graph to estimate the radius of a rotor that spins at a speed of 8 meters per second.

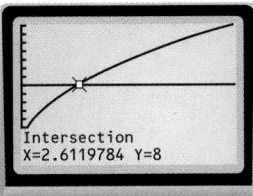

SOLUTION

Graph $y = 4.95\sqrt{x}$ and $y = 8$. Choose a viewing window that shows the point where the graphs intersect. Then use the *Intersect* feature to find the x-coordinate of that point. You get $x \approx 2.61$.

Intersection
X=2.6119784 Y=8

▶ The radius is about 2.61 meters.

EXAMPLE 6 *Modeling with a Cube Root Function*

 Biologists have discovered that the shoulder height h (in centimeters) of a male African elephant can be modeled by

$$h = 62.5\sqrt[3]{t} + 75.8$$

where t is the age (in years) of the elephant. Use a graphing calculator to graph the model. Then use the graph to estimate the age of an elephant whose shoulder height is 200 centimeters. ▶ *Source: Elephants*

SOLUTION

Graph $y = 62.5\sqrt[3]{x} + 75.8$ and $y = 200$. Choose a viewing window that shows the point where the graphs intersect. Then use the *Intersect* feature to find the x-coordinate of that point. You get $x \approx 7.85$.

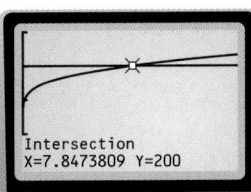

Intersection
X=7.8473809 Y=200

▶ The elephant is about 8 years old.

7.5 *Graphing Square Root and Cube Root Functions*   433

For use after Example 4:

3. State the domain and range of the function in (**a**) Checkpoint Exercise 1 and (**b**) Checkpoint Exercise 2.
   a. domain: $x \geq 3$; range: $y \geq 2$
   b. domain and range: all real numbers

EXTRA EXAMPLE 5

A model for the period of a simple pendulum as measured in time units is given by $T = 2\pi\sqrt{\dfrac{L}{g}}$, where T is the time in seconds, L is the length of the pendulum in feet, and g is 32 ft/sec^2. Use a graphing calculator to graph the model. Then use the graph to estimate the period of a pendulum that is 3 feet long. **1.92 sec**

EXTRA EXAMPLE 6

The length of a whale can be modeled by $L = 21.04\sqrt[3]{W}$, where L is the length in feet, and W is the weight in tons. Graph the model, then use the graph to find the weight of a whale that is 60 ft long. **about 23.2 tons**

CHECKPOINT EXERCISES

For use after Example 5:

1. Graph the equation $y = 3.2\sqrt{x}$. Then use the graph to estimate the value of x when y is 8.4. **6.89**

For use after Example 6:

2. Graph the equation $y = 2.7\sqrt[3]{x} + 4.1$. Then use the graph to estimate the value of x when y is 7.7. **2.37**

FOCUS ON VOCABULARY

Give an example of a radical function. *Sample answer:* $y = 3\sqrt[3]{x} + 2$

CLOSURE QUESTION

Explain how to graph $y = 2\sqrt[3]{x+3} - 1$ from the graph of $y = 2\sqrt[3]{x}$. **Sketch the graph of $y = 2\sqrt[3]{x}$. Shift the graph 3 units to the left and one unit down.**

ASSIGNMENT GUIDE

BASIC
Day 1: pp. 434–436 Exs. 15–21,
22–40 even, 46, 50,
55–69 odd

AVERAGE
Day 1: pp. 434–436 Exs. 15–21,
22–50 even, 55–69 odd

ADVANCED
Day 1: pp. 434–436 Exs. 15–21,
22–50 even, 51–53,
55–69 odd, 70

BLOCK SCHEDULE
pp. 434–436 Exs. 15–21, 22–50
even, 55–69 odd, (with 7.4)

EXERCISE LEVELS
Level A: *Easier*
15–27, 31–36, 46

Level B: *More Difficult*
28–30, 37–45, 47–50

Level C: *Most Difficult*
51–53

✔ **HOMEWORK CHECK**
To quickly check student understanding of key concepts, go
over the following exercises:
Exs. 16, 20, 22, 32, 42. See also
the Daily Homework Quiz:

• Blackline Master (*Chapter 7
Resource Book,* p. 78)

• Transparency (p. 56)

2–13, 15–18, 22–30.
See Additional Answers beginning on page AA1.

GUIDED PRACTICE

Vocabulary Check ✔ **1.** Complete this statement: Square root functions and cube root functions are examples of _?_ functions. **radical**

Concept Check ✔ **2. ERROR ANALYSIS** Explain why the graph shown at the near right is not the graph of $y = \sqrt{x - 1} + 2$. **See margin.**

3. ERROR ANALYSIS Explain why the graph shown at the far right is not the graph of $y = \sqrt[3]{x + 2} - 3$. **See margin.**

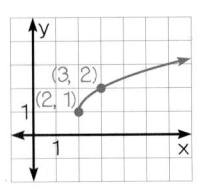
Ex. 2

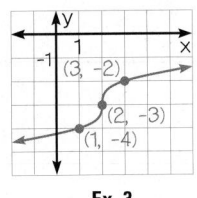
Ex. 3

Skill Check ✔ **Describe how to obtain the graph of *g* from the graph of *f*.** **4, 5. See margin.**

4. $g(x) = \sqrt{x + 5}$, $f(x) = \sqrt{x}$ **5.** $g(x) = \sqrt[3]{x} - 10$, $f(x) = \sqrt[3]{x}$

Graph the function. Then state the domain and range. **6–13. See margin.**

6. $y = -\sqrt{x}$ **7.** $y = \sqrt{x + 1}$ **8.** $y = \sqrt{x - 2}$ **9.** $y = 2\sqrt{x + 3} - 1$

10. $y = \frac{2}{3}\sqrt[3]{x}$ **11.** $y = \sqrt[3]{x} - 6$ **12.** $y = \sqrt[3]{x + 5}$ **13.** $y = -3\sqrt[3]{x - 7} - 4$

14. 📱 **ELEPHANTS** Look back at Example 6. Use a graphing calculator to graph the model. Then use the graph to estimate the age of an elephant whose shoulder height is 250 centimeters. **21.65 years old**

PRACTICE AND APPLICATIONS

STUDENT HELP

→ **Extra Practice**
to help you master
skills is on p. 950.

COMPARING GRAPHS **Describe how to obtain the graph of *g* from the graph of *f*.** **15–18. See margin.**

15. $g(x) = \sqrt{x + 14}$, $f(x) = \sqrt{x}$ **16.** $g(x) = 5\sqrt{x - 10} - 3$, $f(x) = 5\sqrt{x}$

17. $g(x) = -\sqrt[3]{x} - 10$, $f(x) = -\sqrt[3]{x}$ **18.** $g(x) = \sqrt[3]{x + 6} - 5$, $f(x) = \sqrt[3]{x}$

MATCHING GRAPHS **Match the function with its graph.**

19. $y = \sqrt{x} - 1$ **B**   **20.** $y = \sqrt{x + 1}$ **A**   **21.** $y = \sqrt{x + 1} - 1$ **C**

A. 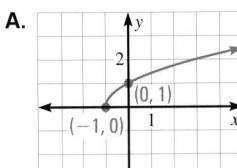   **B.** 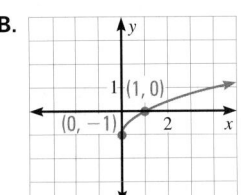   **C.**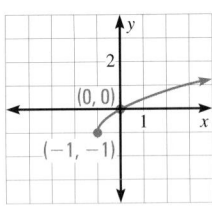

STUDENT HELP

→ **HOMEWORK HELP**
Example 1: Exs. 15–18
Example 2: Exs. 19–30
Example 3: Exs. 31–39
Example 4: Exs. 22–45
Example 5: Exs. 46, 47
Example 6: Exs. 48, 49

SQUARE ROOT FUNCTIONS **Graph the function. Then state the domain and range.** **22–30. See margin.**

22. $y = -5\sqrt{x}$ **23.** $y = \frac{1}{3}\sqrt{x}$ **24.** $y = x^{1/2} + \frac{1}{4}$

25. $y = x^{1/2} - 2$ **26.** $y = \sqrt{x + 6}$ **27.** $y = (x - 7)^{1/2}$

28. $y = (x - 1)^{1/2} + 7$ **29.** $y = 2\sqrt{x + 5} - 1$ **30.** $y = -\frac{2}{5}\sqrt{x - 3} - 2$

CUBE ROOT FUNCTIONS Graph the function. Then state the domain and range. **31–39. See margin for graphs; x, y are all real numbers.**

31. $y = \frac{1}{2}\sqrt[3]{x}$

32. $y = -2x^{1/3}$

33. $y = \sqrt[3]{x} - 7$

34. $y = \sqrt[3]{x} + \frac{3}{4}$

35. $y = \sqrt[3]{x - 5}$

36. $y = \left(x + \frac{2}{3}\right)^{1/3}$

37. $y = \frac{1}{5}x^{1/3} - 2$

38. $y = -3\sqrt[3]{x + 4}$

39. $y = 2\sqrt[3]{x - 4} + 3$

CRITICAL THINKING Find the domain and range of the function without graphing. Explain how you found your solution. **40–45. See margin.**

40. $y = \sqrt{x - 13}$

41. $y = 2\sqrt{x} - 2$

42. $y = -\sqrt{x - 3} - 7$

43. $y = \sqrt[3]{x + 8}$

44. $y = -\frac{2}{3}\sqrt[3]{x} - 5$

45. $y = 4\sqrt[3]{x + 4} + 7$

40–45. Explanations may vary.

40. $x \geq 13$, $y \geq 0$

41. $x \geq 0$, $y \geq -2$

42. $x \geq 3$, $y \leq -7$

43. x, y are all real numbers.

44. x, y are all real numbers.

45. x, y are all real numbers.

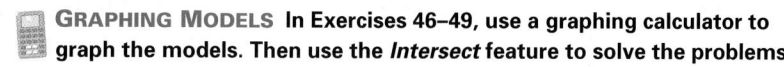 **GRAPHING MODELS** In Exercises 46–49, use a graphing calculator to graph the models. Then use the *Intersect* feature to solve the problems.

46.  **OCEAN DISTANCES** When you look at the ocean, the distance d (in miles) you can see to the horizon can be modeled by $d = 1.22\sqrt{a}$ where a is your altitude (in feet above sea level). Graph the model. Then determine at what altitude you can see 10 miles. ▶ Source: *Mathematics in Everyday Things*
67.19 ft above sea level

47. 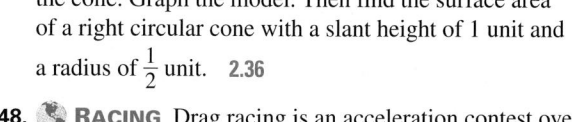 **GEOMETRY CONNECTION** In a right circular cone with a slant height of 1 unit, the radius r of the cone is given by $r = \frac{1}{\sqrt{\pi}} \sqrt{S + \frac{\pi}{4}} - \frac{1}{2}$ where S is the surface area of the cone. Graph the model. Then find the surface area of a right circular cone with a slant height of 1 unit and a radius of $\frac{1}{2}$ unit. **2.36**

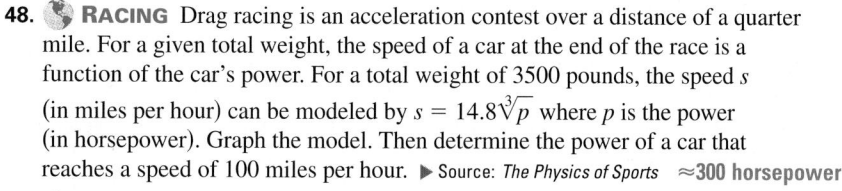

48.  **RACING** Drag racing is an acceleration contest over a distance of a quarter mile. For a given total weight, the speed of a car at the end of the race is a function of the car's power. For a total weight of 3500 pounds, the speed s (in miles per hour) can be modeled by $s = 14.8\sqrt[3]{p}$ where p is the power (in horsepower). Graph the model. Then determine the power of a car that reaches a speed of 100 miles per hour. ▶ Source: *The Physics of Sports* **≈300 horsepower**

49.  **STORMS AT SEA** The fetch f (in nautical miles) of the wind at sea is the distance over which the wind is blowing. The minimum fetch required to create a fully developed storm can be modeled by $s = 3.1\sqrt[3]{f + 10} + 11.1$ where s is the speed (in knots) of the wind. Graph the model. Then determine the minimum fetch required to create a fully developed storm if the wind speed is 25 knots.
▶ Source: *Oceanography* **80.15 nautical miles**

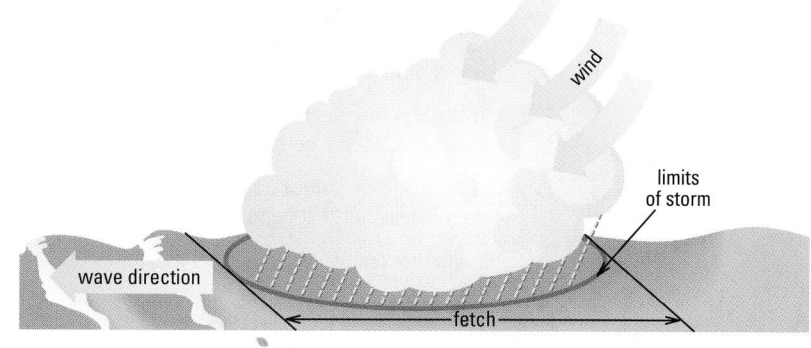

7.5 *Graphing Square Root and Cube Root Functions* **435**

STUDENT HELP

HOMEWORK HELP
Visit our Web site
www.mcdougallittell.com
for help with problem
solving in Exs. 40–45.

FOCUS ON CAREERS

COAST GUARD
Members of the Coast Guard have a variety of responsibilities. Some participate in search and rescue missions that involve rescuing people caught in storms at sea, discussed in Ex. 49.

CAREER LINK
www.mcdougallittell.com

! COMMON ERROR
EXERCISES 22–39 Students may confuse the horizontal and vertical shifts or shift horizontally in the wrong direction. Refer these students to the box on page 431 and to Examples 2 and 3 on page 432.

TEACHING TIPS
EXERCISES 40–45 Ask students to visualize the graph of the corresponding radical function of the form $y = a\sqrt[n]{x}$ in order to find the domain and range. Then visualize the domain and range adjustments as the graph moves right or left, or up or down.

STUDENT HELP NOTES
→ **Homework Help** Students can find help for Exs. 40–45 at **www.mcdougallittell.com**. The information can be printed out for students who don't have access to the Internet.

CAREER NOTE
EXERCISE 49 Additional information about the Coast Guard is available at **www.mcdougallittell.com**.

31–39. See Additional Answers beginning on page AA1.

ADDITIONAL PRACTICE AND RETEACHING

For Lesson 7.5:
• Practice Levels A, B, and C (*Chapter 7 Resource Book*, p. 68)
• Reteaching with Practice (*Chapter 7 Resource Book*, p. 71)
• See Lesson 7.5 of the *Personal Student Tutor*

For more Mixed Review:
• Search the *Test and Practice Generator* for key words or specific lessons.

DAILY HOMEWORK QUIZ

📝 *Transparency Available*

Describe how to obtain the graph of g from the graph of f.

1. $g(x) = 3\sqrt{x+2} - 1$
$f(x) = 3\sqrt{x}$ Shift the graph of f 2 units left and 1 unit down.

Graph the function. Then state the domain and range.

2. $y = 3x^{1/2}$

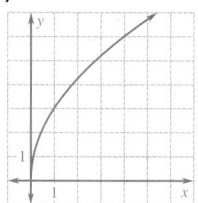

domain: $x \geq 0$, range: $y \geq 0$

3. $y = \sqrt[3]{x} + 4$

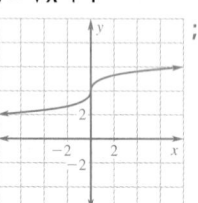

domain, range: all real numbers

EXTRA CHALLENGE NOTE

↳ Challenge problems for Lesson 7.5 are available in **blackline** format in the *Chapter 7 Resource Book,* p. 75 and at **www.mcdougallittell.com.**

ADDITIONAL TEST PREPARATION

1. WRITING Describe the steps in graphing a function of the form $y = a\sqrt{x-h} + k$. Graph $y = a\sqrt{x}$. Then shift the graph h units horizontally and k units vertically.

50. See Additional Answers beginning on page AA1.

Test Preparation

50a–c. See margin for graphs.

50. a. The graphs are reflections across the y-axis.

b. The graphs are reflections across the y-axis.

★ **Challenge**

EXTRA CHALLENGE
↳ www.mcdougallittell.com

50. d. To graph radical functions of the form
$f(x) = a\sqrt{-(x-h)} + k$
or
$g(x) = a\sqrt[3]{-(x-h)} + k,$
follow these steps:

1. Sketch the graph of $y = a\sqrt{x}$ or $y = a\sqrt[3]{x}$.

2. Reflect the graph across the y-axis.

3. Shift the graph h units horizontally and k units vertically.

67. $f(g(x)) = 2x - 5$;
$g(f(x)) = 2x - 2$

68. $f(g(x)) = x^2 + 4x + 3$;
$g(f(x)) = x^2 + 1$

69. $f(g(x)) = 9x^2 - 18x + 16$;
$g(f(x)) = 3x^2 + 18$

---

50. MULTI-STEP PROBLEM Follow the steps below to graph radical functions of the form $y = f(-x)$.   50a–d. See margin.

a. Graph $f_1(x) = \sqrt{x}$ and $f_2(x) = \sqrt{-x}$. How are the graphs related?

b. Graph $g_1(x) = \sqrt[3]{x}$ and $g_2(x) = \sqrt[3]{-x}$. How are the graphs related?

c. Graph $f_3(x) = \sqrt{-(x-2)} - 4$ and $g_3(x) = 2\sqrt[3]{-(x+1)} + 5$ using what you learned from parts (a) and (b) and what you know about the effects of a, h, and k on the graphs of $y = a\sqrt{x-h} + k$ and $y = a\sqrt[3]{x-h} + k$.

d. *Writing* Describe the steps for graphing a function of the form $f(x) = a\sqrt{-(x-h)} + k$ or $g(x) = a\sqrt[3]{-(x-h)} + k$.

ANALYZING GRAPHS Write an equation for the function whose graph is shown.

51.

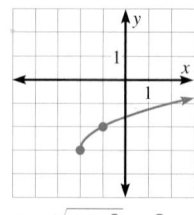

$y = \sqrt{x+2} - 3$

52.

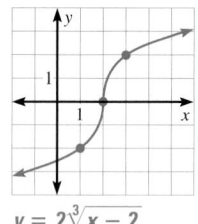

$y = 2\sqrt[3]{x} - 2$

53.

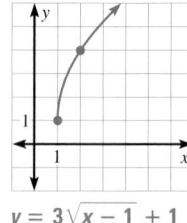

$y = 3\sqrt{x-1} + 1$

MIXED REVIEW

SOLVING EQUATIONS Solve the equation. (Review 5.3 for 7.6)

54. $2x^2 = 32$ ± 4

55. $(x+7)^2 = 10$ $\pm\sqrt{10} - 7$

56. $9x^2 + 3 = 5$ $\frac{\pm\sqrt{2}}{3}$

57. $\frac{1}{2}x^2 - 5 = 13$ ± 6

58. $\frac{1}{4}(x+6)^2 = 22$ $\pm 2\sqrt{22} - 6$

59. $2(x - 0.25)^2 = 16.5$ $\frac{\pm\sqrt{33}}{2} + \frac{1}{4}$

SPECIAL PRODUCTS Find the product. (Review 6.3 for 7.6)

60. $(x+4)^2$ $x^2 + 8x + 16$

61. $(x - 9y)^2$ $x^2 - 18xy + 81y^2$

62. $(2x^3 + 7)^2$ $4x^6 + 28x^3 + 49$

63. $(-3x + 4y^4)^2$ $9x^2 - 24xy^4 + 16y^8$

64. $(6 - 5x)^2$ $36 - 60x + 25x^2$

65. $(-1 - 2x^2)^2$ $1 + 4x^2 + 4x^4$

COMPOSITION OF FUNCTIONS Find $f(g(x))$ and $g(f(x))$. (Review 7.3)
$f(g(x)) = 2x + 7$; $g(f(x)) = 2x + 14$ 67–69. See margin.

66. $f(x) = x + 7, g(x) = 2x$

67. $f(x) = 2x + 1, g(x) = x - 3$

68. $f(x) = x^2 - 1, g(x) = x + 2$

69. $f(x) = x^2 + 7, g(x) = 3x - 3$

70. 🌐 **MANDELBROT SET** To determine whether a complex number c belongs to the Mandelbrot set, consider the function $f(z) = z^2 + c$ and the infinite list of complex numbers $z_0 = 0, z_1 = f(z_0), z_2 = f(z_1), z_3 = f(z_2), \ldots$.

• If the absolute values $|z_0|, |z_1|, |z_2|, |z_3|, \ldots$ are all less than some fixed number N, then c belongs to the Mandelbrot set.

• If the absolute values $|z_0|, |z_1|, |z_2|, |z_3|, \ldots$ become infinitely large, then c does not belong to the Mandelbrot set.

Tell whether the complex number c belongs to the Mandelbrot set. (Review 5.4)

a. $c = 3i$ no

b. $c = 2 + 2i$ no

c. $c = 6$ no

Solving Radical Equations

What you should learn

GOAL 1 Solve equations that contain radicals or rational exponents.

GOAL 2 Use radical equations to solve **real-life** problems, such as determining wind speeds that correspond to the Beaufort wind scale in **Example 6**.

Why you should learn it

▼ To solve **real-life** problems, such as determining which boats satisfy the rule for competing in the America's Cup sailboat race in **Ex. 68**.

GOAL 1 SOLVING A RADICAL EQUATION

To solve a *radical equation*—an equation that contains radicals or rational exponents—you need to eliminate the radicals or rational exponents and obtain a polynomial equation. The key step is to *raise each side of the equation to the same power*.

$$\text{If } a = b, \text{ then } a^n = b^n. \qquad \textbf{Powers property of equality}$$

Then solve the new equation using standard procedures. Before raising each side of an equation to the same power, you should isolate the radical expression on one side of the equation.

EXAMPLE 1 *Solving a Simple Radical Equation*

Solve $\sqrt[3]{x} - 4 = 0$.

SOLUTION

$\sqrt[3]{x} - 4 = 0$	Write original equation.
$\sqrt[3]{x} = 4$	Isolate radical.
$(\sqrt[3]{x})^3 = 4^3$	Cube each side.
$x = 64$	Simplify.

▶ The solution is 64. Check this in the original equation.

EXAMPLE 2 *Solving an Equation with Rational Exponents*

Solve $2x^{3/2} = 250$.

SOLUTION

Because x is raised to the $\frac{3}{2}$ power, you should isolate the power and then raise each side of the equation to the $\frac{2}{3}$ power $\left(\frac{2}{3} \text{ is the reciprocal of } \frac{3}{2}\right)$.

$2x^{3/2} = 250$	Write original equation.
$x^{3/2} = 125$	Isolate power.
$(x^{3/2})^{2/3} = 125^{2/3}$	Raise each side to $\frac{2}{3}$ power.
$x = (125^{1/3})^2$	Apply properties of roots.
$x = 5^2 = 25$	Simplify.

▶ The solution is 25. Check this in the original equation.

STUDENT HELP

Study Tip
To solve an equation of the form $x^{m/n} = k$ where k is a constant, raise both sides of the equation to the $\frac{n}{m}$ power, because
$$(x^{m/n})^{n/m} = x^1 = x.$$

1 PLAN

PACING
Basic: 2 days
Average: 2 days
Advanced: 2 days
Block Schedule: 1 block

LESSON OPENER
GRAPHING CALCULATOR
An alternative way to approach Lesson 7.6 is to use the Graphing Calculator Lesson Opener:
• Blackline Master (*Chapter 7 Resource Book,* p. 79)
• Transparency (p. 49)

MEETING INDIVIDUAL NEEDS
• *Chapter 7 Resource Book*
  Prerequisite Skills Review (p. 5)
  Practice Level A (p. 82)
  Practice Level B (p. 83)
  Practice Level C (p. 84)
  Reteaching with Practice (p. 85)
  Absent Student Catch-Up (p. 87)
  Challenge (p. 91)
• *Resources in Spanish*
• *Personal Student Tutor*

NEW-TEACHER SUPPORT
See the Tips for New Teachers on pp. 1–2 of the *Chapter 7 Resource Book* for additional notes about Lesson 7.6.

WARM-UP EXERCISES
Transparency Available
1. Multiply $(x-3)^2$. $x^2 - 6x + 9$
2. Factor $x^2 + 8x + 16$. $(x+4)^2$
Evaluate when $x = 4$.
3. $\sqrt{10x - 4}$ 6   **4.** $\sqrt[3]{3x - 4}$ 2
5. $\sqrt[3]{6x + 192} - \sqrt[3]{2x}$ 4

ENGLISH LEARNERS
Draw English learners' attention to the use of italics. Point out that in English, italic letters often signal that the text is important. Mention that in algebra, variables are also written in italics.

2 TEACH

MOTIVATING THE LESSON

The equation $t = \sqrt{\dfrac{d}{16}}$ gives the time t in seconds it takes for an object to fall a distance of d feet. If $t = 1.4$ seconds, finding the number of feet the object falls (31.36 feet) involves solving a radical equation, the focus of today's lesson.

EXTRA EXAMPLE 1
Solve $5 - \sqrt[4]{x} = 0$. 625

EXTRA EXAMPLE 2
Solve $3x^{4/3} = 243$. 27

EXTRA EXAMPLE 3
Solve $\sqrt{2x+8} - 4 = 6$. 46

EXTRA EXAMPLE 4
Solve $\sqrt{4x+28} - 3\sqrt{2x} = 0$. 2

CHECKPOINT EXERCISES

For use after Example 1:

1. Solve $\sqrt[3]{x} + 6 = 12$. 216

For use after Example 2:

2. Solve $2x^{1/4} = 8$. 256

For use after Examples 3 and 4:

3. Solve $\sqrt{12 - 2x} - 2\sqrt{x} = 0$. 2

STUDENT HELP NOTES

→ **Homework Help** Students can find extra examples at **www.mcdougallittell.com** that parallel the examples in the student edition.

STUDENT HELP

HOMEWORK HELP
Visit our Web site
www.mcdougallittell.com
for extra examples.

EXAMPLE 3 *Solving an Equation with One Radical*

Solve $\sqrt{4x - 7} + 2 = 5$.

SOLUTION

$\sqrt{4x - 7} + 2 = 5$	Write original equation.
$\sqrt{4x - 7} = 3$	Isolate radical.
$(\sqrt{4x - 7})^2 = 3^2$	Square each side.
$4x - 7 = 9$	Simplify.
$4x = 16$	Add 7 to each side.
$x = 4$	Divide each side by 4.

✓**CHECK** Check $x = 4$ in the original equation.

$\sqrt{4x - 7} + 2 = 5$	Write original equation.
$\sqrt{4(4) - 7} \stackrel{?}{=} 3$	Substitute 4 for x.
$\sqrt{9} \stackrel{?}{=} 3$	Simplify.
$3 = 3$ ✓	Solution checks.

▶ The solution is 4.

· · · · · · · · · ·

Some equations have two radical expressions. Before raising both sides to the same power, you should rewrite the equation so that each side of the equation has only one radical expression.

EXAMPLE 4 *Solving an Equation with Two Radicals*

Solve $\sqrt{3x + 2} - 2\sqrt{x} = 0$.

SOLUTION

$\sqrt{3x + 2} - 2\sqrt{x} = 0$	Write original equation.
$\sqrt{3x + 2} = 2\sqrt{x}$	Add $2\sqrt{x}$ to each side.
$(\sqrt{3x + 2})^2 = (2\sqrt{x})^2$	Square each side.
$3x + 2 = 4x$	Simplify.
$2 = x$	Solve for x.

✓**CHECK** Check $x = 2$ in the original equation.

$\sqrt{3x + 2} - 2\sqrt{x} = 0$	Write original equation.
$\sqrt{3(2) + 2} - 2\sqrt{2} \stackrel{?}{=} 0$	Substitute 2 for x.
$2\sqrt{2} - 2\sqrt{2} \stackrel{?}{=} 0$	Simplify.
$0 = 0$ ✓	Solution checks.

▶ The solution is 2.

If you try to solve $\sqrt{x} = -1$ by squaring both sides, you get $x = 1$. But $x = 1$ is not a valid solution of the original equation. This is an example of an **extraneous** (or false) **solution**. Raising both sides of an equation to the same power may introduce extraneous solutions. So, when you use this procedure it is critical that you check each solution in the *original* equation.

EXTRA EXAMPLE 5

Solve $x + 2 = \sqrt{2x + 28}$. 4

CHECKPOINT EXERCISES

For use after Example 5:

1. Solve $x - 3 = \sqrt{4x}$. 9

EXAMPLE 5 *An Equation with an Extraneous Solution*

Solve $x - 4 = \sqrt{2x}$.

STUDENT HELP

Look Back
For help with factoring, see p. 256.

SOLUTION

$x - 4 = \sqrt{2x}$	Write original equation.
$(x - 4)^2 = (\sqrt{2x})^2$	Square each side.
$x^2 - 8x + 16 = 2x$	Expand left side; simplify right side.
$x^2 - 10x + 16 = 0$	Write in standard form.
$(x - 2)(x - 8) = 0$	Factor.
$x - 2 = 0$ or $x - 8 = 0$	Zero product property
$x = 2$ or $x = 8$	Simplify.

✓**CHECK** Check $x = 2$ in the original equation.

$x - 4 = \sqrt{2x}$	Write original equation.
$2 - 4 \stackrel{?}{=} \sqrt{2(2)}$	Substitute 2 for x.
$-2 \stackrel{?}{=} \sqrt{4}$	Simplify.
$-2 \neq 2$	Solution does not check.

✓**CHECK** Check $x = 8$ in the original equation.

$x - 4 = \sqrt{2x}$	Write original equation.
$8 - 4 \stackrel{?}{=} \sqrt{2(8)}$	Substitute 8 for x.
$4 \stackrel{?}{=} \sqrt{16}$	Simplify.
$4 = 4$ ✓	Solution checks.

▶ The only solution is 8.

· · · · · · · · · ·

If you graph each side of the equation in Example 5, as shown, you can see that the graphs of $y = x - 4$ and $y = \sqrt{2x}$ intersect only at $x = 8$. This confirms that $x = 8$ is a solution of the equation, but that $x = 2$ is not.

In general, all, some, or none of the apparent solutions of a radical equation can be extraneous. When all of the apparent solutions of a radical equation are extraneous, the equation has *no solution*.

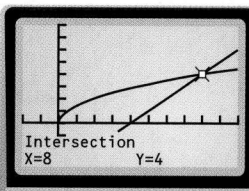

Intersection
X=8   Y=4

7.6 Solving Radical Equations   439

EXTRA EXAMPLE 6

The strings of guitars and pianos are under tension. The speed v of a wave on the string depends on the force (tension) F on the string and the mass M per unit length L according to the formula $v = \sqrt{\dfrac{F \cdot L}{M}}$. A wave travels through a string with a mass of 0.2 kilograms at a speed of 9 meters per second. It is stretched by a force of 19.6 Newtons. Find the length of the string. **about 0.83 m**

CHECKPOINT EXERCISE

For use after Example 6:

1. Solve $R = 2.4\sqrt{x + 3.9} + 7.2$ for x if $R = 19.8$. **about 23.66**

APPLICATION NOTE
Additional information about the Beaufort wind scale is available at **www.mcdougallittell.com**.

FOCUS ON VOCABULARY
What is an extraneous solution and how can you tell if a solution is extraneous? **A solution that is not a solution to the original equation; check the solution in the original equation.**

CLOSURE QUESTION
Describe how to solve $x + 2 = \sqrt{5x + 10}$. **Square both sides to get $x^2 + 4x + 4 = 5x + 10$. Add $-5x - 10$ to both sides to get $x^2 - x - 6 = 0$. Factor and solve to get solutions 3 and −2. Check for extraneous solutions. The solutions are 3 and −2.**

DAILY PUZZLER
Without solving, explain why $\sqrt{2x + 4} = -8$ has no solution.
Sample answer: The positive square root cannot be a negative number.

BEAUFORT WIND SCALE
The Beaufort wind scale was developed by Rear-Admiral Sir Francis Beaufort in 1805 so that sailors could detect approaching storms. Today the scale is used mainly by meteorologists.

APPLICATION LINK www.mcdougallittell.com

GOAL 2 SOLVING RADICAL EQUATIONS IN REAL LIFE

EXAMPLE 6 *Using a Radical Model*

BEAUFORT WIND SCALE The Beaufort wind scale was devised to measure wind speed. The Beaufort numbers B, which range from 0 to 12, can be modeled by $B = 1.69\sqrt{s + 4.45} - 3.49$ where s is the speed (in miles per hour) of the wind. Find the wind speed that corresponds to the Beaufort number $B = 11$.

Beaufort Wind Scale		
Beaufort number	Force of wind	Effects of wind
0	Calm	Smoke rises vertically.
1	Light air	Direction shown by smoke.
2	Light breeze	Leaves rustle; wind felt on face.
3	Gentle breeze	Leaves move; flags extend.
4	Moderate breeze	Small branches sway; paper blown about.
5	Fresh breeze	Small trees sway.
6	Strong breeze	Large branches sway; umbrellas difficult to use.
7	Moderate gale	Large trees sway; walking difficult.
8	Fresh gale	Twigs break; walking hindered.
9	Strong gale	Branches scattered about; slight damage to buildings.
10	Whole gale	Trees uprooted; severe damage to buildings.
11	Storm	Widespread damage.
12	Hurricane	Devastation.

SOLUTION

$$B = 1.69\sqrt{s + 4.45} - 3.49 \qquad \text{Write model.}$$

$$11 = 1.69\sqrt{s + 4.45} - 3.49 \qquad \text{Substitute 11 for } B.$$

$$14.49 = 1.69\sqrt{s + 4.45} \qquad \text{Add 3.49 to each side.}$$

$$8.57 \approx \sqrt{s + 4.45} \qquad \text{Divide each side by 1.69.}$$

$$73.4 \approx s + 4.45 \qquad \text{Square each side.}$$

$$69.0 \approx s \qquad \text{Subtract 4.45 from each side.}$$

▶ The wind speed is about 69 miles per hour.

✓ **ALGEBRAIC CHECK** Substitute 69 for s into the model and evaluate.

$$1.69\sqrt{69 + 4.45} - 3.49 \approx 1.69(8.57) - 3.49$$
$$\approx 11 \checkmark$$

✓ **GRAPHIC CHECK** You can use a graphing calculator to graph the model, and then use the *Intersect* feature to check that $x \approx 69$ when $y = 11$.

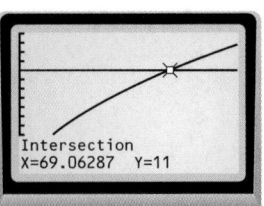

Intersection
X=69.06287 Y=11

GUIDED PRACTICE

Vocabulary Check ✓

Concept Check ✓

1. What is an extraneous solution? An extraneous solution is a solution to an equation raised to a power that is not a solution to the original equation.

2. Marcy began solving $x^{2/3} = 5$ by cubing each side. What will she have to do next? What could she have done to solve the equation in just one step?
   2–3. See margin.

3. Zach was asked to solve $\sqrt{5x-2} - \sqrt{7x-4} = 0$. His first step was to square each side. While trying to isolate x, he gave up in frustration. What could Zach have done to avoid this situation?

Skill Check ✓

2. Next, she will have to take the square root of both sides. To solve the equation in one step, she could have raised both sides to the $\frac{3}{2}$ power.

Solve the rational exponent equation. Check for extraneous solutions.

4. $3x^{1/4} = 4$ $\frac{256}{81}$ 5. $(2x+7)^{3/2} = 27$ 1   6. $x^{4/3} + 9 = 25$ 8

7. $4x^{2/3} - 6 = 10$ 8   8. $5(x-8)^{3/4} = 40$ 24   9. $(x+9)^{5/2} - 1 = 31$ -5

Solve the radical equation. Check for extraneous solutions.

10. $\sqrt[4]{x} = 3$ 81   11. $\sqrt[3]{3x} + 6 = 10$ $\frac{64}{3}$ 12. $\sqrt[5]{2x+1} + 5 = 9$ $\frac{1023}{2}$

13. $\sqrt{x-2} = x - 2$ 2, 3   14. $\sqrt[3]{x+4} = \sqrt[3]{2x-5}$ 9   15. $6\sqrt{x} - \sqrt{x-1} = 0$ no solution

16. 🌐 **BEAUFORT WIND SCALE** Use the information in Example 6 to determine the wind speed that corresponds to the Beaufort number $B = 2$.   6.10 mi/h

PRACTICE AND APPLICATIONS

STUDENT HELP

➤ **Extra Practice**
to help you master skills is on p. 950.

3. First, he should have rewritten the equation with only one radical expression on each side: $\sqrt{5x-2} = \sqrt{7x-4}$.

CHECKING SOLUTIONS Check whether the given x-value is a solution of the equation.

17. $\sqrt{x} - 3 = 6$; $x = 81$ yes   18. $4(x-5)^{1/2} = 28$; $x = 12$ no

19. $(x+7)^{3/2} - 20 = 7$; $x = 2$ yes   20. $\sqrt[3]{4x} + 11 = 5$; $x = -54$ yes

21. $2\sqrt{5x+4} + 10 = 10$; $x = 0$ no   22. $\sqrt{4x-3} - \sqrt{3x} = 0$; $x = 3$ yes

SOLVING RATIONAL EXPONENT EQUATIONS Solve the equation. Check for extraneous solutions.

23. $x^{5/2} = 32$ 4   24. $x^{1/3} - \frac{2}{5} = 0$ $\frac{8}{125}$ 25. $x^{2/3} + 15 = 24$ 27

26. $-\frac{1}{2}x^{1/5} = 10$ $-3{,}200{,}000$ 27. $4x^{3/4} = 108$ 81   28. $(x-4)^{3/2} = -6$ no solution

29. $(2x+5)^{1/2} = 4$ $\frac{11}{2}$ 30. $3(x+1)^{4/3} = 48$ 7   31. $-(x-5)^{1/4} + \frac{7}{3} = 2$ $\frac{406}{81}$

STUDENT HELP

➤ **HOMEWORK HELP**
Example 1: Exs. 17–22, 32–46
Example 2: Exs. 17–22, 23–31
Example 3: Exs. 17–22, 32–46
Example 4: Exs. 17–22, 47–54
Example 5: Exs. 23–54
Example 6: Exs. 63–69

SOLVING RADICAL EQUATIONS Solve the equation. Check for extraneous solutions.

32. $\sqrt{x} = \frac{1}{9}$ $\frac{1}{81}$ 33. $\sqrt[3]{x} + 10 = 16$ 216   34. $\sqrt[4]{2x} - 13 = -9$ 128

35. $\sqrt{x+56} = 16$ 200   36. $\sqrt[3]{x+40} = -5$ -165 37. $\sqrt{6x-5} + 10 = 3$ no solution

38. $\frac{2}{5}\sqrt{10x+6} = 12$ 89.4   39. $2\sqrt{7x+4} - 1 = 7$ $\frac{12}{7}$ 40. $-2\sqrt[5]{2x-1} + 4 = 0$ $\frac{33}{2}$

41. $x - 12 = \sqrt{16x}$ 36   42. $\sqrt[4]{x^4+1} = 3x$ $\frac{\sqrt[4]{125}}{10}$ 43. $\sqrt{x^2+5} = x + 3$ $-\frac{2}{3}$

44. $\sqrt[3]{x} = x - 6$ 8   45. $\sqrt{8x+1} = x + 2$ 1, 3   46. $\sqrt{2x+\frac{1}{6}} = x + \frac{5}{6}$ no solution

7.6 *Solving Radical Equations*   **441**

---

3 APPLY

ASSIGNMENT GUIDE

BASIC
Day 1: pp. 441–444 Exs. 17–22, 24–42 even, 63–64
Day 2: pp. 441–444 Exs. 48–52 even, 56–60 even, 70, 71–73, 81–89 odd

AVERAGE
Day 1: pp. 441–444 Exs. 17–22, 24–46 even, 56–60 even, 63–65
Day 2: pp. 441–444 Exs. 48–54 even, 61, 62, 69–73, 81–89

ADVANCED
Day 1: pp. 441–444 Exs. 17–22, 24–56 even, 63–65
Day 2: pp. 441–444 Exs. 51–61 odd, 66–79, 81–89

BLOCK SCHEDULE
pp. 441–444 Exs. 17–22, 24–60 even, 63–65, 69–73, 81–89

EXERCISE LEVELS
Level A: *Easier*
17–28, 32–41
Level B: *More Difficult*
29–31, 42–73
Level C: *Most Difficult*
74–79

✔ **HOMEWORK CHECK**
To quickly check student understanding of key concepts, go over the following exercises: Exs. 18, 24, 34, 48. See also the Daily Homework Quiz:

• Blackline Master (*Chapter 7 Resource Book*, p. 94)
• 📖 Transparency (p. 57)

GRAPHING CALCULATOR NOTE

EXERCISE 57 The intersect feature of a graphing calculator gives the coordinates of the intersection points of the two graphs, (0.10344828, 1.0923017). The *x*-coordinate of the point is the solution.

! COMMON ERROR

EXERCISE 61 Students may have difficulty graphing the left side because of the index of the radical. Remind students that an index of 4 is equivalent to raising an expression to the power $\frac{1}{4}$. Some calculators have a feature that will graph fourth roots without changing to rational exponent notation.

APPLICATION NOTE

EXERCISE 66 A plumb bob is a weight suspended at the end of a line used to help determine true vertical. Sample uses are applying wall paper or finding the depth of water.

ADDITIONAL PRACTICE AND RETEACHING

For Lesson 7.6:

• Practice Levels A, B, and C (*Chapter 7 Resource Book,* p. 82)

• Reteaching with Practice (*Chapter 7 Resource Book,* p. 85)

• See Lesson 7.6 of the *Personal Student Tutor*

For more Mixed Review:

• Search the *Test and Practice Generator* for key words or specific lessons.

442

SOLVING EQUATIONS WITH TWO RADICALS Solve the equation. Check for extraneous solutions.

47. $\sqrt{2x - 1} = \sqrt{x + 4}$ 5

48. $\sqrt[4]{6x - 5} = \sqrt[4]{x + 10}$ 3

49. $-\sqrt{8x + \frac{4}{3}} = \sqrt{2x + \frac{1}{3}}$ $-\frac{1}{6}$

50. $2\sqrt[3]{10 - 3x} = \sqrt[3]{2 - x}$ $\frac{78}{23}$

51. $\sqrt[4]{2x} + \sqrt[4]{x + 3} = 0$ no solution

52. $\sqrt{x - 6} - \sqrt{\frac{1}{3}x} = 0$ 9

53. $\sqrt{2x + 10} - 2\sqrt{x} = 0$ 5

54. $\sqrt[3]{2x + 15} - \frac{3}{2}\sqrt[3]{x} = 0$ $\frac{120}{11}$

SOLVING EQUATIONS Use the *Intersect* feature on a graphing calculator to solve the equation.

55. $\frac{3}{4}x^{1/3} = -2$ −18.96296

56. $2(x + 19)^{2/5} - 1 = 17$ 224

57. $(3.5x + 1)^{2/7} = (6.4x + 0.7)^{2/7}$ 0.10345

58. $\left(\frac{1}{5}x\right)^{3/4} = x - \frac{3}{8}$ 0.57160

59. $\sqrt{6.7x + 14} = 9.4$ 11.099

60. $\sqrt[3]{70 - 2x} - 10 = -6$ 3

61. $\sqrt[4]{x - \frac{1}{6}} = 2\sqrt[4]{3x}$ no solution

62. $\sqrt{1.1x + 2.4} = 19x - 4.2$ 0.30816

63. 🌐 **NAILS** The length *l* (in inches) of a standard nail can be modeled by

$$l = 54d^{3/2}$$

where *d* is the diameter (in inches) of the nail. What is the diameter of a standard nail that is 3 inches long?   0.146 in.

64. SCIENCE CONNECTION   Scientists have found that the body mass *m* (in kilograms) of a dinosaur that walked on two feet can be modeled by

$$m = (1.6 \times 10^{-4})C^{273/100}$$

where *C* is the circumference (in millimeters) of the dinosaur's femur. Scientists have estimated that the mass of a *Tyrannosaurus rex* might have been 4500 kilograms. What size femur would have led them to this conclusion?
▶ Source: The Zoological Society of London   535.31 mm

65. 🌐 **WOMEN IN MEDICINE** For 1970 through 1995, the percent *p* of Doctor of Medicine (MD) degrees earned each year by women can be modeled by

$$p = (0.867t^2 + 39.2t + 57.1)^{1/2}$$

where *t* is the number of years since 1970. In what year were about 36% of the degrees earned by women? ▶ Source: *Statistical Abstract of the United States*   1991

66. 🌐 **PLUMB BOBS** You work for a company that manufactures plumb bobs. The same mold is used to cast plumb bobs of different sizes. The equation

$$h = 1.5\sqrt[3]{t}, 0 \le h \le 3$$

models the relationship between the height *h* (in inches) of the plumb bob and the time *t* (in seconds) that metal alloy is poured into the mold. How long should you pour the alloy into the mold to cast a plumb bob with a height of 2 inches?
2.37 sec

FOCUS ON PEOPLE

▷ REAL LIFE **DR. ALEXA CANADY** was the first African-American woman to become a neurosurgeon in the United States. She received her MD degree, discussed in Ex. 65, in 1975.

67. 🌐 **BEAUFORT WIND SCALE** Recall from Example 6 that the Beaufort number B from the Beaufort wind scale can be modeled by

$$B = 1.69\sqrt{s + 4.45} - 3.49$$

where s is the speed (in miles per hour) of the wind. Find the wind speed that corresponds to the Beaufort number $B = 7$.   **34.078 mi/h**

68. 🌐 **AMERICA'S CUP** In order to compete in the America's Cup sailboat race, a boat must satisfy the rule

$$\frac{l + 1.25\sqrt{s} - 9.8\sqrt[3]{d}}{0.679} \le 24$$

where l is the length (in meters) of the boat, s is the area (in square meters) of the sails, and d is the volume (in cubic meters) of water displaced by the boat. If a boat has a length of 20 meters and a sail area of 300 square meters, what is the maximum allowable value for d? ▶ Source: America's Cup   **17.32 m³**

69. `GEOMETRY` `CONNECTION` You are trying to determine the height of a truncated pyramid that cannot be measured directly. The height h and slant height l of a truncated pyramid are related by the formula

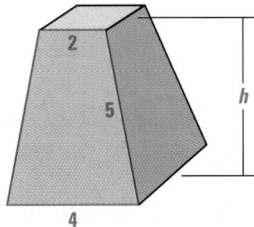

$$l = \sqrt{h^2 + \frac{1}{4}(b_2 - b_1)^2}$$

where b_1 and b_2 are the lengths of the upper and lower bases of the pyramid, respectively. If $l = 5$, $b_1 = 2$, and $b_2 = 4$, what is the height of the pyramid?   **4.90**

70. **CRITICAL THINKING** Look back at Example 5. Solve $x - 4 = -\sqrt{2x}$ instead of $x - 4 = \sqrt{2x}$. How does changing $\sqrt{2x}$ to $-\sqrt{2x}$ change the solution(s) of the equation?   In this case, $x = 2$ is a solution to the equation, but $x = 8$ is an extraneous solution.

Test Preparation

71. **MULTIPLE CHOICE** What is the solution of $\sqrt{6x - 4} = 3$?   **E**

Ⓐ $-\dfrac{1}{6}$ Ⓑ $\dfrac{5}{6}$ Ⓒ $\dfrac{7}{6}$ Ⓓ $\dfrac{5}{3}$ Ⓔ $\dfrac{13}{6}$

72. **MULTIPLE CHOICE** What is (are) the solution(s) of $\sqrt{2x - 3} = \dfrac{1}{2}x$?   **B**

Ⓐ 2   Ⓑ 2, 6   Ⓒ $\dfrac{18}{7}$ Ⓓ $\dfrac{21}{4}$ Ⓔ none

73. **MULTIPLE CHOICE** What is the solution of $\sqrt[3]{x - 7} = \sqrt[3]{\dfrac{3}{4}x + 1}$?   **E**

Ⓐ -6 Ⓑ $-\dfrac{24}{7}$ Ⓒ -4 Ⓓ 2   Ⓔ 32

★ **Challenge**

SOLVING EQUATIONS WITH TWO RADICALS Solve the equation. Check for extraneous solutions. (*Hint:* To solve these equations you will need to square each side of the equation two separate times.)

74. $\sqrt{x + 5} = 5 - \sqrt{x}$ **4**

75. $\sqrt{2x + 3} = 3 - \sqrt{2x}$ $\dfrac{1}{2}$

76. $\sqrt{x + 3} - \sqrt{x - 1} = 1$ $\dfrac{13}{4}$

77. $\sqrt{2x + 4} + \sqrt{3x - 5} = 4$ ≈2.1

EXTRA CHALLENGE
➤ www.mcdougallittell.com

78. $\sqrt{3x - 2} = 1 + \sqrt{2x - 3}$ **2, 6**

79. $\dfrac{1}{2}\sqrt{2x - 5} - \dfrac{1}{2}\sqrt{3x + 4} = 1$ **no solution**

7.6 *Solving Radical Equations*   **443**

---

4 ASSESS

DAILY HOMEWORK QUIZ

📄 *Transparency Available*

Solve the equation. Check for extraneous solutions.

1. $x^{2/3} + 4 = 20$ **64**
2. $(3x)^{1/2} - 7 = 2$ **27**
3. $\sqrt{x - 4} = \sqrt[4]{12x + 16}$ **20**

EXTRA CHALLENGE NOTE
↪ Challenge problems for Lesson 7.6 are available in **blackline** format in the *Chapter 7 Resource Book*, p. 91 and at **www.mcdougallittell.com.**

ADDITIONAL TEST PREPARATION

1. OPEN ENDED Give the steps for solving $\sqrt[3]{3x - 4} = \sqrt[3]{2x}$ along with the solution.
Sample answer: Cube both sides to get $3x - 4 = 2x$. Add $-2x + 4$ to both sides to get $x = 4$. Check for extraneous roots. The solution is 4.

ADDITIONAL RESOURCES
A **blackline** master with additional Math & History exercises is available in the *Chapter 7 Resource Book,* p. 90.

90.
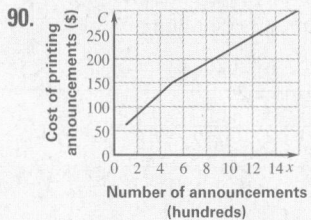
Cost of printing announcements ($) vs. Number of announcements (hundreds)

MIXED REVIEW

USING ORDER OF OPERATIONS Evaluate the expression. (Review 1.2 for 7.7)

80. $6 + 24 \div 3$ **14**   **81.** $3 \cdot 5 + 10 \div 2$ **20**   **82.** $27 - 4 \cdot 16 \div 8$ **19**

83. $2 - (10 \cdot 2)^2 \div 5$ **−78**   **84.** $8 + (3 \cdot 10) \div 6 - 1$ **12**   **85.** $11 - 8 \div 2 + 48 \div 4$ **19**

USING GRAPHS Graph the polynomial function. Identify the *x*-intercepts, local maximums, and local minimums. (Review 6.8)

86. about −0.791, 1, about 3.79; (0, 3); about (2.67, −6.48)

88. ±1; no local maximum; $\left(0, -\dfrac{1}{2}\right)$

86. $f(x) = x^3 - 4x^2 + 3$

87. $f(x) = 3x^3 - 2.5x^2 + 1.25x + 6$
about −0.95; no local maximums or minimums

88. $f(x) = \dfrac{1}{2}x^4 - \dfrac{1}{2}$

89. $f(x) = x^5 + x^3 - 6x$
0, about ±1.41; (−0.914, 4.08); (0.914, −4.08)

90. 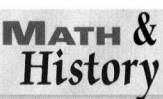 **PRINTING RATES** The cost C (in dollars) of printing x announcements (in hundreds) is given by the function shown. Graph the function. (Review 2.7)
See margin.

$$C = \begin{cases} 62 + 22(x - 1), & \text{if } 1 \le x \le 5 \\ 150 + 14(x - 5), & \text{if } x > 5 \end{cases}$$

MATH & History ▸ **Tsunamis**

APPLICATION LINK
www.mcdougallittell.com

THEN ▸ **IN AUGUST OF 1883,** a volcano erupted on the island of Krakatau, Indonesia. The eruption caused a *tsunami* (a type of wave) to form and travel into the Indian Ocean and into the Java Sea. The speed s (in kilometers per hour) that a tsunami travels can be modeled by $s = 356\sqrt{d}$ where d is the depth (in kilometers) of the water.

1. A tsunami from Krakatau hit Jakarta traveling about 60 kilometers per hour. What is the average depth of the water between Krakatau and Jakarta? **about 0.0284 km**

2. After 15 hours and 12 minutes a tsunami from Krakatau hit Port Elizabeth, South Africa, 7546 kilometers away. Find the average speed of the tsunami. **about 496 km/h**

3. Based on your answer to Exercise 2, what is the average depth of the Indian Ocean between Krakatau and Port Elizabeth? **about 1.94 km**

(Map: ASIA, Pacific Ocean, AFRICA, Krakatau, Jakarta, Indian Ocean, AUSTRALIA, Port Elizabeth)

NOW ▸ **AFTER A TRAGIC TSUNAMI** hit the Aleutian Islands in 1946, scientists began work on a tsunami warning system. Today that system is operated 24 hours a day at the Honolulu Observatory and effectively warns people when a tsunami might arrive.

Famous tsunami art created by Hokusai.
c. 1800

1883
Krakatau erupts.

Alaskan earthquake causes Pacific-wide tsunami.
1957

1995
Prototype of tsunami real-time reporting system developed.

Statistics and Statistical Graphs

GOAL ❶ MEASURES OF CENTRAL TENDENCY AND DISPERSION

What you should learn

GOAL ❶ Use measures of central tendency and measures of dispersion to describe data sets.

GOAL ❷ Use box-and-whisker plots and histograms to represent data graphically, as applied in **Exs. 40–42.**

Why you should learn it

▼ To use statistics and statistical graphs to analyze **real-life** data sets, such as the free-throw percentages for the players in the WNBA in **Examples 1–6.**

In this lesson you will use the following two data sets. They show the free-throw percentages for the players in the Women's National Basketball Association (WNBA) for 1998. ⏩ **DATA UPDATE** of WNBA data at www.mcdougallittell.com

Women's National Basketball Association Free-Throw Percentages	
Eastern Conference	**Western Conference**
46, 47, 48, 50, 50, 50, 52, 53, 55, 57, 57, 58, 60, 61, 61, 62, 63, 63, 63, 63, 63, 63, 63, 64, 64, 67, 67, 67, 69, 71, 72, 72, 72, 72, 73, 75, 75, 75, 75, 75, 76, 77, 78, 79, 79, 80, 81, 81, 82, 82, 83, 83, 85, 89, 91, 92, 100	36, 50, 50, 56, 56, 57, 57, 58, 61, 61, 62, 62, 63, 63, 64, 64, 65, 66, 66, 66, 66, 67, 69, 69, 70, 70, 70, 70, 71, 71, 71, 71, 72, 72, 73, 73, 74, 74, 74, 74, 75, 76, 76, 76, 77, 77, 78, 80, 81, 83, 83, 83, 85, 85, 87, 100, 100, 100

Statistics are numerical values used to summarize and compare sets of data. The following **measures of central tendency** are three commonly used statistics.

1. The **mean**, or *average*, of n numbers is the sum of the numbers divided by n. The mean is denoted by $\bar{x}$, which is read as "x-bar." For the data $x_1, x_2, \ldots, x_n$, the mean is $\bar{x} = \dfrac{x_1 + x_2 + \cdots + x_n}{n}$.

2. The **median** of n numbers is the middle number when the numbers are written in order. (If n is even, the median is the mean of the two middle numbers.)

3. The **mode** of n numbers is the number or numbers that occur most frequently. There may be one mode, no mode, or more than one mode.

EXAMPLE 1 *Finding Measures of Central Tendency*

Find the mean, median, and mode of the two data sets listed above.

SOLUTION

EASTERN CONFERENCE: Mean: $\bar{x} = \dfrac{46 + 47 + \cdots + 100}{57} = \dfrac{3931}{57} \approx 69.0$

Median: 69   Mode: 63

WESTERN CONFERENCE: Mean: $\bar{x} = \dfrac{36 + 50 + \cdots + 100}{58} = \dfrac{4106}{58} \approx 70.8$

Median: 71   Modes: 66, 70, 71, 74

All three measures of central tendency for the Western Conference are greater than those for the Eastern Conference. So, the Western Conference has better free-throw percentages overall.

❶ PLAN

PACING
Basic: 2 days
Average: 2 days
Advanced: 2 days
Block Schedule: 1 block

➡ LESSON OPENER
APPLICATION
An alternative way to approach Lesson 7.7 is to use the Application Lesson Opener:
• Blackline Master (*Chapter 7 Resource Book,* p. 95)
• Transparency (p. 50)

MEETING INDIVIDUAL NEEDS
• *Chapter 7 Resource Book*
  Prerequisite Skills Review (p. 5)
  Practice Level A (p. 97)
  Practice Level B (p. 98)
  Practice Level C (p. 99)
  Reteaching with Practice (p. 100)
  Absent Student Catch-Up (p. 102)
  Challenge (p. 104)
• *Resources in Spanish*
• ▦ *Personal Student Tutor*

NEW-TEACHER SUPPORT
See the Tips for New Teachers on pp. 1–2 of the *Chapter 7 Resource Book* for additional notes about Lesson 7.7.

WARM-UP EXERCISES

⬙ *Transparency Available*

Evaluate the expression. Express your answer to the nearest hundredth.

1. $\dfrac{6 + 9 + 12 + 15}{4}$ 10.5

2. $(15 - 10.5)^2$ 20.25

3. $\sqrt{72.25}$ 8.5

4. $\sqrt{\dfrac{328}{12}}$ 5.23

5. $\sqrt{\dfrac{(12 - 10.5)^2 + (15 - 10.5)^2}{2}}$ 3.35

EXTRA EXAMPLE 1
The number of games won by teams in the Eastern Conference for the 1997–98 regular season of the National Hockey League is shown in the chart below.

Eastern Conference
36, 39, 40, 34, 48, 33, 25, 30, 37, 17, 42, 40, 24

Find the mean, median, and mode for the data set.
mean: 34.2; median: 36; mode: 40

EXTRA EXAMPLE 2
Find the range of the number of wins in the data set in Extra Example 1.   **31 games**

EXTRA EXAMPLE 3
Find the standard deviation for the number of wins in the data set in Extra Example 1.   **8.12**

CHECKPOINT EXERCISES
For use after Examples 1–3:
1. Find the mean, median, mode, range, and standard deviation for the data set below.

Test Scores
92, 94, 87, 76, 69, 82, 62, 90, 76, 82, 85, 87, 64, 61, 95, 87

mean: 80.6; median: 83.5; mode: 87; range: 34; standard deviation: 10.99

Measures of central tendency tell you what the *center* of the data is. Other commonly used statistics are called **measures of dispersion**. They tell you how *spread out* the data are. One simple measure of dispersion is the **range**, which is the difference between the greatest and least data values.

EXAMPLE 2   *Finding Ranges of Data Sets*

The ranges of the free-throw percentages in the two data sets on the previous page are:

EASTERN CONFERENCE: Range $= 100 - 46 = 54$

WESTERN CONFERENCE: Range $= 100 - 36 = 64$

Because the Western Conference's range of free-throw percentages is greater, its free-throw percentages are more spread out.

. . . .

Another measure of dispersion is **standard deviation**, which describes the typical difference (or *deviation*) between the mean and a data value.

STANDARD DEVIATION OF A SET OF DATA

The standard deviation σ (read as "sigma") of $x_1, x_2, \ldots, x_n$ is:

$$\sigma = \sqrt{\frac{(x_1 - \overline{x})^2 + (x_2 - \overline{x})^2 + \cdots + (x_n - \overline{x})^2}{n}}$$

EXAMPLE 3   *Finding Standard Deviations of Data Sets*

STUDENT HELP

> **HOMEWORK HELP**
> Visit our Web site
> www.mcdougallittell.com
> for extra examples.

The standard deviations of the free-throw percentages in the two data sets on the previous page are:

EASTERN CONFERENCE: $\sigma \approx \sqrt{\dfrac{(46 - 69.0)^2 + (47 - 69.0)^2 + \cdots + (100 - 69.0)^2}{57}}$

$\approx \sqrt{\dfrac{8660}{57}}$

$\approx \sqrt{152}$

≈ 12.3

WESTERN CONFERENCE: $\sigma \approx \sqrt{\dfrac{(36 - 70.8)^2 + (50 - 70.8)^2 + \cdots + (100 - 70.8)^2}{58}}$

$\approx \sqrt{\dfrac{7910}{58}}$

$\approx \sqrt{136}$

≈ 11.7

Because the Eastern Conference's standard deviation is greater, its free-throw percentages are more spread out *about the mean*.

GOAL 2 USING STATISTICAL GRAPHS

Although statistics are useful in describing a data set, sometimes a graph of the data can be more informative. One type of statistical graph is a **box-and-whisker plot**. The "box" encloses the middle half of the data set and the "whiskers" extend to the minimum and maximum data values.

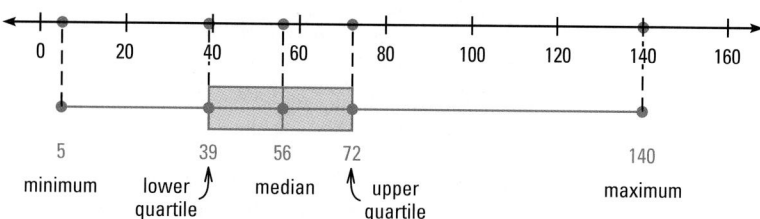

The median divides the data set into two halves. The **lower quartile** is the median of the lower half, and the **upper quartile** is the median of the upper half. You can use the following steps to draw a box-and-whisker plot.

1. Order the data from least to greatest.

2. Find the minimum and maximum values.

3. Find the median.

4. Find the lower and upper quartiles.

5. Plot these five numbers below a number line.

6. Draw the box, the whiskers, and a line segment through the median.

EXAMPLE 4 — *Drawing Box-and-Whisker Plots*

Draw a box-and-whisker plot of each data set on page 445.

SOLUTION

EASTERN CONFERENCE

The minimum is 46 and the maximum is 100. The median is 69. The lower quartile is 61 and the upper quartile is 78.5.

WESTERN CONFERENCE

The minimum is 36 and the maximum is 100. The median is 71. The lower quartile is 64 and the upper quartile is 76.

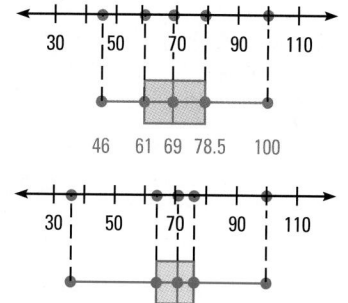

Like the computations in Examples 2 and 3, the box-and-whisker plots show you that the Western Conference's free-throw percentages are more spread out overall (comparing the entire plots) and that the Eastern Conference's free-throw percentages are more spread out about the mean (comparing the boxes).

FOCUS ON PEOPLE

SANDY BRONDELLO was ranked first in the WNBA in 1998 for free-throw percentage (among players who attempted at least 10 free throws). As a player for the Detroit Shock, she made 96 out of 104 free throws for a free-throw percentage of 92.

7.7 *Statistics and Statistical Graphs* **447**

EXTRA EXAMPLE 4
Draw a box-and-whisker plot for the set of data in Extra Example 1.

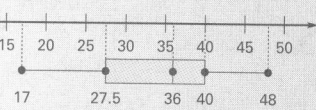

CHECKPOINT EXERCISES
For use after Example 4:
1. Draw a box-and-whisker plot for the set of data in Checkpoint Exercise 1, page 446.

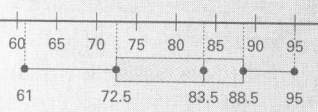

EXTRA EXAMPLE 5
Make a frequency distribution for the data set in Extra Example 1. Use four intervals beginning with the interval 11–20.

Eastern Conference		
Interval	Tally	Frequency
11–20	I	1
21–30	III	3
31–40	JHT II	7
41–50	II	2

CHECKPOINT EXERCISES
For use after Example 5:
1. Make a frequency distribution for the data set in Checkpoint Exercise 1, page 446. Use four intervals beginning with the interval 61–70.

Test Score		
Interval	Tally	Frequency
61–70	IIII	4
71–80	II	2
81–90	JHT II	7
91–100	III	3

EXTRA EXAMPLE 6
Draw a histogram for the data set in Extra Example 1.

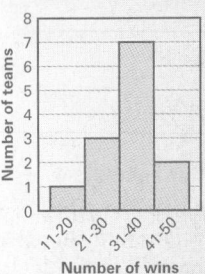

 CHECKPOINT EXERCISES

For use after Example 6:
1. Draw a histogram for the data set in Checkpoint Exercise 1, page 446.

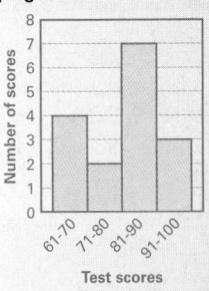

FOCUS ON VOCABULARY
Name the measures of central tendency and dispersion used in this lesson. mean, median, mode, range, standard deviation

CLOSURE QUESTION
When is finding the mean or median more useful than finding the standard deviation? when you need to know where the center of the data is

DAILY PUZZLER
Give a data set with the same mean and median. Sample answer: 5, 10, 10, 15, 15, 20; mean = median = 12.5

Another way to display numerical data is with a special type of bar graph called a **histogram**. In a histogram data are grouped into intervals of *equal* width. The number of data values in each interval is the **frequency** of the interval. To draw a histogram, begin by making a **frequency distribution**, which shows the frequency of each interval.

EXAMPLE 5   *Making Frequency Distributions*

Make a frequency distribution of each data set on page 445. Use seven intervals beginning with the interval 31–40.

SOLUTION   Begin by writing the seven intervals. Then tally the data values by interval. Finally, count the tally marks to get the frequencies.

Eastern Conference			Western Conference		
Interval	Tally	Frequency	Interval	Tally	Frequency
31–40		0	31–40	I	1
41–50	Ⅲ I	6	41–50	II	2
51–60	Ⅲ II	7	51–60	Ⅲ	5
61–70	Ⅲ Ⅲ Ⅲ I	16	61–70	Ⅲ Ⅲ Ⅲ Ⅲ	20
71–80	Ⅲ Ⅲ Ⅲ II	17	71–80	Ⅲ Ⅲ Ⅲ Ⅲ	20
81–90	Ⅲ III	8	81–90	Ⅲ II	7
91–100	III	3	91–100	III	3

EXAMPLE 6   *Drawing Histograms*

Draw a histogram of each data set on page 445.

STUDENT HELP

▶ Skills Review
For help with statistical graphs, see p. 934.

SOLUTION   Use the frequency distributions in Example 5. Draw a horizontal axis, divide it into seven equal sections, and label the sections with the intervals. Then draw a vertical axis for measuring the frequencies. Finally, draw bars of appropriate heights to represent the frequencies of the intervals.

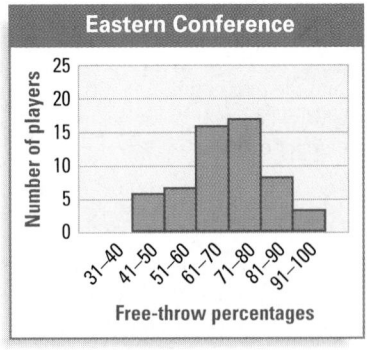

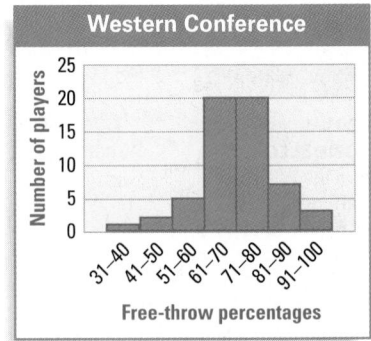

GUIDED PRACTICE

Vocabulary Check ✓

Concept Check ✓

1. The mean is the average. It is calculated by dividing the sum of the numbers by n, the number of numbers.

1. Define mean, median, mode, and range of a set of n numbers.   **See margin.**

2. Give an example of two sets of four numbers, each with a mean of 5. Choose the numbers so that one set has a range of 3 and the other set has a range of 7. Which has the greater standard deviation?   *Sample answers: 3, 5, 6, 6; 2, 4, 5, 9; second set*

3. The following box-and-whisker plots represent two sets of data. Which data set has the greater range? Explain.   **Set A has the greater range because $10 - 3 = 7 > 11 - 5 = 6$.**

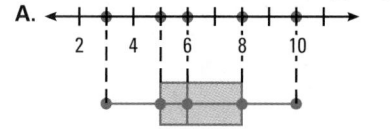 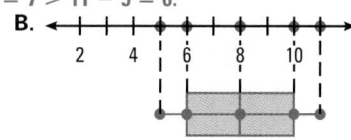

Skill Check ✓

The median is the middle number when the numbers are written in ascending (or descending) order. If n is even, the median is the mean of the two middle numbers.

The mode is the number or numbers that occur most frequently.

The range is the difference between the greatest and least data values.

🌐 **HISTORY SCORES** In Exercises 4–9, use the following data set of scores received by students on a history exam.

$$68, 72, 76, 81, 84, 86, 86, 86, 89, 91, 95, 99$$

4. Find the mean, median, and mode of the data.   **mean: about 84.4; median: 86; mode: 86**

5. Find the range of the data.   **31**

6. Find the standard deviation of the data.   **8.67**

7. Draw a box-and-whisker plot of the data.   **See margin.**

8. Make a frequency distribution of the data. Use five intervals beginning with 61–68.

9. Draw a histogram of the data.   **See margin.**   **See margin.**

PRACTICE AND APPLICATIONS

> **STUDENT HELP**
>
> ▶ **Extra Practice**
> to help you master skills is on p. 950.

> **STUDENT HELP**
>
> ▶ **HOMEWORK HELP**
> **Example 1:** Exs. 10–15, 33, 36, 39
> **Examples 2, 3:** Exs. 16–21, 34, 37
> **Example 4:** Exs. 22–27, 40, 43
> **Examples 5, 6:** Exs. 28–32, 41, 44

MEASURES OF CENTRAL TENDENCY Find the mean, median, and mode of the data set.

10. 6, 7, 9, 9, 9, 10   **8.33, 9, 9**

11. 43, 46, 47, 47, 51, 54, 59   **49.57, 47, 47**

12. 88, 83, 91, 82, 78, 81, 91, 91   **85.625, 85.5, 91**

13. 220, 250, 210, 290, 310, 230, 230   **about 249, 230, 230**

14. 2.9, 2.1, 2.6, 2.9, 3.0, 2.5, 3.4   **2.77, 2.9, 2.9**

15. 0, 0, 0.1, 0.2, 0.3, 0.5, 0.5, 0.6, 1.0   **0.356; 0.3; 0, 0.5 (two modes)**

MEASURES OF DISPERSION Find the range and standard deviation of the data set.

16. 10, 20, 30, 40, 50, 60   **50, 17.1**

17. 1, 1, 3, 4, 5, 9   **8, 2.73**

18. 10, 12, 7, 11, 20, 7, 6, 8, 9   **14, 4**

19. 1202, 1229, 1012, 1014, 1120, 1429   **417, 143**

20. 3.1, 2.7, 6.0, 5.6, 2.3, 2.0, 1.3   **4.7, 1.68**

21. 19.4, 16.3, 12.7, 24.8, 19.2, 15.4   **12.1, 3.82**

BOX-AND-WHISKER PLOTS Draw a box-and-whisker plot of the data set.
22–27. See margin.

22. 1, 2, 2, 4, 5, 7

23. 19, 19, 19, 89, 93, 95

24. 47, 88, 89, 61, 70, 71, 79

25. 40, 100, 20, 40, 100, 70, 90

26. 1.7, 8.5, 1.2, 3.8, 8.5, 5.2, 6.9

27. 61.2, 23.0, 72.7, 74.3, 19.1, 6.6, 28.4

3 APPLY

○ **ASSIGNMENT GUIDE**

BASIC
Day 1: pp. 449–451 Exs. 10–30 even, 38
Day 2: pp. 449–452 Exs. 32–34, 39–41, 46–48, 51–65 odd, Quiz 3 Exs. 1–12

AVERAGE
Day 1: pp. 449–451 Exs. 10–30 even, 31, 38
Day 2: pp. 449–452 Exs. 32–37, 39–44, 46–48, 51–65 odd, Quiz 3 Exs. 1–12

ADVANCED
Day 1: pp. 449–451 Exs. 10–30 even, 31–34, 38
Day 2: pp. 449–452 Exs. 35–37, 39–41, 43–49, 51–65 odd, Quiz 3 Exs. 1–12

BLOCK SCHEDULE
pp. 449–451 Exs. 10–30 even, 31, 32–37, 39–44, 46–48, 51–65 odd, Quiz 3 Exs. 1–12

EXERCISE LEVELS
Level A: *Easier*
10–29, 32–34, 39, 42
Level B: *More Difficult*
30–31, 35–38, 40–41, 43–49
Level C: *Most Difficult*
50

✓ **HOMEWORK CHECK**
To quickly check student understanding of key concepts, go over the following exercises: Exs. 10, 16, 22, 28, 38. See also the Daily Homework Quiz:

• Blackline Master (*Chapter 8 Resource Book*, p. 11)
• 🖨 Transparency (p. 59)

! **COMMON ERROR**
EXERCISE 15 A common error is to not count 0 when finding the number of data values. Make sure students count each 0.

7–9, 22–27. See Additional Answers beginning on page AA1.

CAREER NOTE
EXERCISES 33–35
Additional information about chemical engineers is available at www.mcdougallittell.com.

MATHEMATICAL REASONING
EXERCISE 38 Values that skew the mean are called outliers. An outlier is a value that is more than 1.5 times the difference between the first and third quartile or $1.5 \times (199,900 - 119,900) = \$120,000$ in this case. $750,000 is an outlier. Outliers are sometimes excluded from a data set so that a more accurate interpretation of the data can be made. Without this data value, the mean home price drops to $146,917.

28.
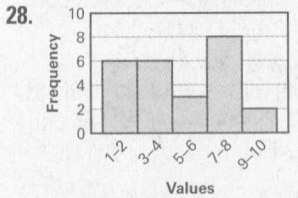

29.

10–19	5
20–29	5
30–39	7
40–49	2
50–59	1

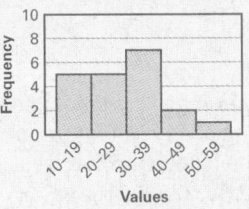

30–32. See Additional Answers beginning on page AA1.

FOCUS ON CAREERS

CHEMICAL ENGINEER
Chemical engineers work in a variety of fields such as chemical manufacturing, electronics, and photographic equipment. Some test methods of production, as discussed in Exs. 33–35.

CAREER LINK
www.mcdougallittell.com

28.

1–2	6
3–4	6
5–6	3
7–8	8
9–10	2

See margin for graph.

33. machine 1: 2.59, 2.59, none; machine 2: 2.59, 2.59, none

34. machine 1: 0.25, 0.09; machine 2: 3.01, 1.11

38. The highest selling price makes the mean price higher than six out of seven of the home prices. Reporting the median rather than the mean eliminates the huge effect of one outlying value.

39. The mode is the most appropriate measure because it would indicate that most people have a positive opinion on the issue. Because the categories are not part of an ordered scale, means and medians are not meaningful.

HISTOGRAMS Use the given intervals to make a frequency distribution of the data set. Then draw a histogram of the data set. 28–32. See margin.

28. Use five intervals beginning with 1–2.
1, 1, 1, 2, 2, 2, 3, 3, 4, 4, 4, 4, 5, 6, 6, 7, 7, 7, 7, 8, 8, 8, 8, 9, 9

29. Use five intervals beginning with 10–19.
10, 12, 14, 15, 15, 23, 26, 26, 27, 28, 37, 37, 37, 37, 38, 39, 39, 40, 48, 58

30. Use ten intervals beginning with 0–9.
66, 9, 43, 28, 5, 7, 90, 9, 78, 6, 69, 43, 28, 55, 13, 64, 45, 10, 54, 96

31. Use fifteen intervals beginning with 0.0–0.4.
2.2, 5.6, 2.2, 4.4, 2.2, 6.6, 2.4, 5.8, 2.4, 4.8, 2.4, 6.1, 5.4, 5.4

32. Use eight intervals beginning with 15.5–20.4.
22.2, 22.2, 22.6, 24.3, 24.3, 24.8, 54.5, 54.4, 54.1, 52.3, 52.3, 52.9

SCRATCH PROTECTION In Exercises 33–35, use the following information. Computer chips have a silicon dioxide coating that gives them scratch protection. Phosphorus is contained in this coating and is a key to its effectiveness. A company that produces silicon dioxide has two machines that vary from run to run on the phosphorus content. The phosphorus contents for four runs on each machine are shown below.

Machine 1		Machine 2	
Run number	Percent of coating that is phosphorous (by weight)	Run number	Percent of coating that is phosphorous (by weight)
1	2.63	1	1.09
2	2.72	2	3.06
3	2.47	3	4.10
4	2.55	4	2.12

33. Find the mean, median, and mode of each data set.

34. Find the range and standard deviation of each data set.

35. **CRITICAL THINKING** Based on the data, which machine is more consistent? Explain your reasoning. Machine 1 is more consistent because it has a smaller range and standard deviation.

REAL ESTATE In Exercises 36–38, suppose a real estate agent has sold seven homes priced at $104,900, $119,900, $134,900, $142,000, $179,900, $199,900, and $750,000.

36. Find the mean, median, and mode of the selling prices. $233,071.43; $142,000; none

37. Find the range and standard deviation of the selling prices. $645,100; $213,243.66

38. **CRITICAL THINKING** How does the selling price $750,000 affect the mean? How does this explain why the most commonly used measure of central tendency for housing prices is the median rather than the mean or the mode?

39. **POLITICAL SURVEY** A random survey of 1000 people asked their opinions on a controversial issue. The opinions are categorized as 0 = a negative opinion, 1 = a positive opinion, and 2 = a neutral opinion. The results of the survey are 212 category 0 responses, 627 category 1 responses, and 161 category 2 responses. Which measure of central tendency is most appropriate to represent the data? Explain.

41.

Age	Pres	VP
30–39	0	1
40–49	8	12
50–59	24	21
60–69	10	12
70–79	0	1

42. *Sample answers:* The range of ages of Vice Presidents is greater than that of Presidents.

Most Presidents and Vice Presidents are in their forties, fifties or sixties when they first take office.

45. *Sample answer:* You cannot conclude that one conference consistently has larger (or smaller) margins of victory than the other.

HISTORY CONNECTION In Exercises 40–42, use the tables below which give the ages of the Presidents and Vice Presidents of the United States.

Ages of the first 42 Presidents of the United States when they first took office
42, 43, 46, 46, 47, 48, 49, 49, 50, 51, 51, 51, 51, 51, 52, 52, 54, 54, 54, 54, 55, 55, 55, 55, 56, 56, 56, 57, 57, 57, 57, 58, 60, 61, 61, 61, 62, 64, 64, 65, 68, 69

Ages of the first 47 Vice Presidents of the United States when they first took office
36, 40, 41, 42, 42, 42, 44, 45, 45, 46, 48, 49, 49, 50, 50, 51, 51, 51, 52, 52, 52, 52, 52, 53, 53, 53, 53, 56, 56, 56, 57, 57, 58, 59, 60, 60, 61, 64, 64, 65, 65, 66, 66, 68, 69, 69, 71

▶ Source: *Facts About the Presidents*

40. Draw a box-and-whisker plot of each data set. **40–42. See margin.**

41. Make a frequency distribution of each data set using five intervals beginning with 30–39. Then draw a histogram of each data set.

42. VISUAL THINKING What is one conclusion you can draw about the ages of the presidents and vice presidents based on your graphs?

STATISTICS CONNECTION In Exercises 43–45, use the tables below which give the margins of victory for each championship game in the AFC and in the NFC for the 1966–1998 seasons. **43–45. See margin.**

AFC Championship margins of victory
24, 33, 4, 10, 10, 21, 4, 17, 11, 6, 17, 3, 29, 14, 7, 20, 14, 16, 17, 17, 3, 5, 11, 16, 48, 3, 19, 17, 4, 4, 14, 3, 13

NFC Championship margins of victory
7, 4, 34, 20, 7, 11, 23, 17, 4, 30, 11, 17, 28, 9, 13, 1, 14, 3, 23, 24, 17, 7, 25, 27, 2, 31, 10, 17, 10, 11, 17, 13, 3

43. Draw a box-and-whisker plot of each data set.

44. Make a frequency distribution of each data set using five intervals beginning with 1–10. Then draw a histogram of each data set.

45. VISUAL THINKING What is one conclusion you *cannot* draw about the margins of victory in the AFC and NFC championship games based on your graphs?

46. RESEARCH Research two sets of data that you can compare. Find the measures of central tendency and measures of dispersion of each data set. Then draw a box-and-whisker plot and a histogram of each data set. What do you observe? **Answers will vary. Check students' work.**

Test Preparation

47. MULTIPLE CHOICE What is the mean of 2, 2, 6, 7, 9, 10? **B**

Ⓐ 2   Ⓑ 6   Ⓒ 6.5   Ⓓ 7   Ⓔ 7.2

48. MULTIPLE CHOICE What is the median of 0.5, 0.6, 0.7, 1.2, 1.5, 1.5? **B**

Ⓐ 0.7   Ⓑ 0.95   Ⓒ 1   Ⓓ 1.2   Ⓔ 1.5

49. MULTIPLE CHOICE What is (are) the mode(s) of 12, 13, 13, 15, 16, 16? **C**

Ⓐ 13   Ⓑ 14   Ⓒ 13, 16   Ⓓ 17   Ⓔ none

★ **Challenge**

50. ALTERNATE FORMULA The formula for standard deviation can also be written as follows: **See margin.**

$$\sigma = \sqrt{\frac{x_1{}^2 + x_2{}^2 + \cdots + x_n{}^2}{n} - \bar{x}^2}$$

For n = 3, show that this formula is equivalent to the formula given on page 446. (*Hint:* You will need to show that $x_1 + x_2 + x_3 = 3\bar{x}$.)

EXTRA CHALLENGE
www.mcdougallittell.com

7.7 Statistics and Statistical Graphs **451**

APPLICATION NOTE
EXERCISE 41
A good choice of intervals for this data starts with an age of 30 since the minimum age for the President and Vice President is 35.

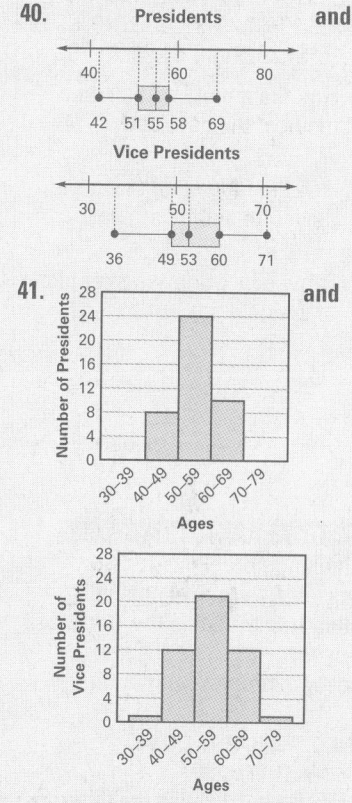

43–44, 50. See Additional Answers beginning on page AA1.

ADDITIONAL PRACTICE AND RETEACHING

For Lesson 7.7:

• Practice Levels A, B, and C (*Chapter 7 Resource Book,* p. 97)

• Reteaching with Practice (*Chapter 7 Resource Book,* p. 100)

• See Lesson 7.7 of the *Personal Student Tutor*

For more Mixed Review:

• Search the *Test and Practice Generator* for key words or specific lessons.

DAILY HOMEWORK QUIZ

📋 *Transparency Available*

Use the following data set of prices for bread: $.59, $.75, $.79, $.99, $.99, $1.09, $1.19, $1.19, $1.19, $1.25, $1.39, $1.49, $1.55, $1.79

1. Find the mean, median, and mode of the data set.
$1.16, $1.19, $1.19

2. Find the range and standard deviation of the data set.
$1.20, $.32

3. Draw a box-and-whisker plot of the data set.

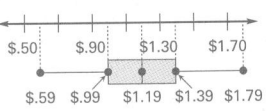

EXTRA CHALLENGE NOTE

→ Challenge problems for Lesson 7.7 are available in **blackline** format in the *Chapter 7 Resource Book*, p. 104 and at **www.mcdougallittell.com**.

ADDITIONAL TEST PREPARATION

1. WRITING Explain the difference between the mean, median, and mode. **The mean is the average of the data, while the median is the middle data value. The median may be near the mean, but it doesn't have to be. The mode is the data value that appears the most in the data list.**

61–66, Quiz 3: 1–3, 9, 11.
See Additional Answers beginning on page AA1.

12.

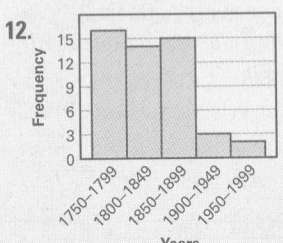

MIXED REVIEW

EVALUATING EXPRESSIONS Evaluate the expression for the given value of *x*. (Review 1.2 for 8.1)

51. $x^5 - 8$ when $x = 2$ 24

52. $3x^3 + 7$ when $x = \frac{3}{7}$ $\frac{2482}{343} \approx 7.24$

53. $(7x)^3 + 17$ when $x = -1$ -326

54. $\frac{4x}{x^4 - 1}$ when $x = 0.5$ $-\frac{32}{15} \approx -2.13$

55. product of powers

56. power of a power and negative exponent

57. product of powers and negative exponent

58. power of a power

59. zero exponent and negative exponent

60. zero exponent and product of powers

PROPERTIES OF EXPONENTS Evaluate the expression. Tell which properties of exponents you used. (Review 6.1 for 8.1)   See margin for properties.

55. $3^3 \cdot 3^4$ $3^7 = 2187$

56. $\left(4^{-3}\right)^2$ $\frac{1}{4096}$

57. $(-2)(-2)^{-3}$ $\frac{1}{4}$

58. $\left(5^{-2}\right)^{-2}$ 625

59. $10^{-2} \cdot 10^0$ $\frac{1}{100}$

60. $7^0 \cdot 7^2 \cdot 7^{-2}$ 1

GRAPHING POLYNOMIALS Graph the polynomial function. (Review 6.2)
61–66. See margin.

61. $f(x) = 3x^5$

62. $f(x) = x^3 - 4$

63. $f(x) = -x^4 + 3x^2 - 5$

64. $f(x) = -x^3 + 2x$

65. $f(x) = x^4 - 6x^2 - 9$

66. $f(x) = x^5 - 2x + 4$

QUIZ 3

Self Test for Lessons 7.5–7.7

1–3. See margin for graphs.
Graph the function. Then state the domain and range. (Lesson 7.5)

1. $y = \sqrt{x + 8}$
$x \geq -8, y \geq 0$

2. $y = (x + 7)^{1/2} - 2$
$x \geq -7, y \geq -2$

3. $y = 3\sqrt[3]{x - 6}$
x and y are all real numbers.

Solve the equation. Check for extraneous solutions. (Lesson 7.6)

4. $\sqrt[4]{2x} = 5$ 312.5

5. $\sqrt{3x + 7} = x - 1$
6 (-1 is an extraneous solution.)

6. $\sqrt[3]{2x} - 2\sqrt[3]{x} = 0$ 0

Find the mean, median, mode, range, and standard deviation of the data set. (Lesson 7.7)
4.4, 5.5, 6, 9, 2.8   23.9, 21, none, 31, 9.99

7. 0, 1, 2, 2, 5, 6, 6, 6, 7, 9

8. 15, 32, 18, 21, 26, 12, 43

9. GEOMETRY CONNECTION   The radius *r* of a sphere is related to the volume *V* of the sphere by the formula $r = 0.620\sqrt[3]{V}$. Graph the formula. Then estimate the volume of a sphere with a radius of 10 units. (Lesson 7.5)   4196 cubic units;
See margin for graph.

10. 🌎 ASTRONOMY   Kepler's third law of planetary motion states that a planet's orbital period *P* (in days) is related to its orbit's semi-major axis *a* (in millions of kilometers) by the formula $P = 0.199a^{3/2}$. The orbital period of Mars is about 1.88 *years*. What is its semi-major axis? (Lesson 7.6)   228.24 million km

HISTORY CONNECTION   In Exercises 11 and 12, use the following data set of years when each state was admitted to statehood. (Lesson 7.7)

1819, 1959, 1912, 1836, 1850, 1876, 1788, 1787, 1845, 1788,
1959, 1890, 1818, 1816, 1846, 1861, 1792, 1812, 1820, 1788,
1788, 1837, 1858, 1817, 1821, 1889, 1867, 1864, 1788, 1787,
1912, 1788, 1789, 1889, 1803, 1907, 1859, 1787, 1790, 1788,
1889, 1796, 1845, 1896, 1791, 1788, 1889, 1863, 1848, 1890

11. Draw a box-and-whisker plot of the data set.   See margin for graph.

12. Make a frequency distribution of the data set using five intervals beginning with 1750–1799. Then draw a histogram of the data set.   See margin.

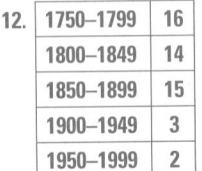

12.	
1750–1799	16
1800–1849	14
1850–1899	15
1900–1949	3
1950–1999	2

▶ ACTIVITY 7.7
Using Technology

Statistics and Statistical Graphs

You can use a graphing calculator to find statistics and draw statistical graphs.

▶ **EXAMPLE**

The fat content and number of calories in several different sandwiches available at a restaurant are shown in the tables below. Use a graphing calculator to (**a**) find the mean, median, range, and standard deviation of the fat content in the sandwiches, (**b**) draw a box-and-whisker plot of the fat content in the sandwiches, and (**c**) draw a histogram of the number of calories in the sandwiches.

Sandwich	Fat (g)	Calories
Hamburger	9	260
Cheeseburger	13	320
Quarter-pound hamburger	21	420
Quarter-pound cheeseburger	30	530
Double cheeseburger	31	560

Sandwich	Fat (g)	Calories
Bacon cheeseburger	34	590
Fried chicken	25	500
Grilled chicken	20	440
Breaded fish on deluxe roll	28	560
Breaded fish on plain bun	25	450

STUDENT HELP

INTERNET **KEYSTROKE HELP**

See keystrokes for several models of calculators at www.mcdougallittell.com

▶ **SOLUTION**

a. Use the *Stat Edit* feature to enter the data in a list. Then use the *Stat Calc* menu to choose 1-variable statistics.

The mean is $\bar{x}$ and the standard deviation is σx. By scrolling you can find the median (Med). The range is the difference of maxX and minX.

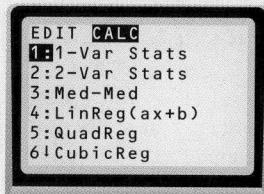
```
EDIT CALC
1:1-Var Stats
2:2-Var Stats
3:Med-Med
4:LinReg(ax+b)
5:QuadReg
6↓CubicReg
```

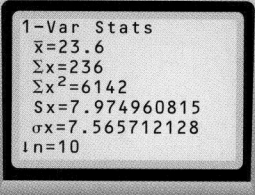
```
1-Var Stats
x̄=23.6
Σx=236
Σx²=6142
Sx=7.974960815
σx=7.565712128
↓n=10
```

STUDENT HELP

↳ **Study Tip**
The lower quartile is also known as the first quartile (Q1 on a graphing calculator), and the upper quartile is also known as the third quartile (Q3 on a graphing calculator).

b. Use the *Stat Plot* menu to choose the type of plot (box-and-whisker), the list of data, and the frequency for the data. Then set an appropriate viewing window.

Draw the box-and-whisker plot. Use the *Trace* feature to view the minimum (9), the lower quartile (20), the median (25), the upper quartile (30), and the maximum (34).

```
Plot1 Plot2 Plot3
On Off
Type: ⌐ ⌐ ⌐
      ⌐ ⌐ ⌐
XList:L1
Freq:1
```

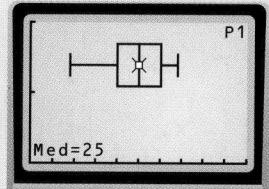

---

1 **Planning the Activity**

PURPOSE
To find and represent statistics with a graphing calculator.

MATERIALS
• graphing calculator
• Keystroke blackline (*Chapter 7 Resource Book,* p. 96)

PACING
• Activity — 35 min

▶ **LINK TO LESSON**
Students can check their answers to some of the exercises in Lesson 7.7 using the graphing calculator.

2 **Managing the Activity**

CLASSROOM MANAGEMENT
Urge students to enter the data slowly and carefully. The graphing calculator cannot tell them if they enter data incorrectly.
    Point out that the data for the Exercises is different from the data in the Example. Students must enter the new data before they begin the Exercises.

COOPERATIVE LEARNING
To save time, one student can enter the data for fat content while a partner enters the calorie data on a second calculator. At the end of the activity, students can take turns showing each other how they found their results using their calculators.

3 **Closing the Activity**

★ **KEY DISCOVERY**
You can use statistics and a graphing calculator to help you draw conclusions about sets of data.

ACTIVITY ASSESSMENT
What are some advantages and disadvantages of finding statistics with a graphing calculator?
See sample answer at left.

Activity Assessment *Sample answer:*
The calculator saves time and does not make mistakes calculating, but if you enter data incorrectly and don't check your work, all your conclusions could be incorrect.

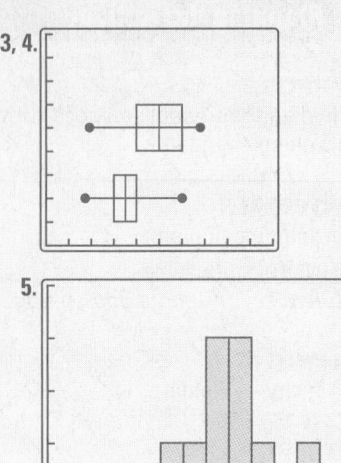

3, 4.

5.

STUDENT HELP

↳ **Study Tip**
The interval width for a histogram is determined by the *x*-scale when setting an appropriate viewing window.

c. Use the *Stat Edit* feature to enter the data, and use the *Stat Plot* menu to choose the type of plot (histogram), the list of data, and the frequency for the data. Then set an appropriate viewing window.

Draw the histogram. Use the *Trace* feature to view the minimum of each interval, the maximum of each interval, and the frequency of each interval.

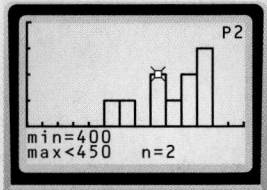

▶ **EXERCISES**

In Exercises 1–5, use the tables below which give the fat content and the number of calories in several different sandwiches available at a competing restaurant.

Sandwich	Fat (g)	Calories
Plain hamburger	16	360
Hamburger with everything	20	420
Bacon cheeseburger	30	580
Small hamburger	10	270
Small cheeseburger	17	360

Sandwich	Fat (g)	Calories
Small bacon cheeseburger	19	380
Grilled chicken	8	310
Breaded chicken	18	440
Chicken club	20	470
Spicy chicken	15	410

2. The restaurant in the exercises (second restaurant) has a lower mean and median fat content and a lower standard deviation than the restaurant in the example (first restaurant). Overall, the second restaurant has sandwiches with less fat than the first restaurant.

4. The second restaurant has sandwiches with a lower fat content. The box-and-whisker plots make it easy to compare the medians and ranges, so it is clear that the second restaurant's sandwiches have less fat.

5. The histograms show that half of the sandwiches in the first restaurant contain over 500 calories, while only 1 out of 10 sandwiches at the second restaurant contain over 500 calories.

1. Use a graphing calculator to find the mean, median, range, and standard deviation of the fat content in the sandwiches. **17.3, 17.5, 22, 5.71**

2. How does the mean fat content in the sandwiches from the restaurant above compare with the mean fat content in the sandwiches from the restaurant in the example? Which restaurant's sandwiches have a higher median fat content? Which restaurant's sandwiches have a lower standard deviation of fat content? Based on these statistics, write a sentence that compares the fat content in the sandwiches of the two restaurants. **See margin.**

3. Use a graphing calculator to draw a box-and-whisker plot of the fat content in the sandwiches. **3–5. See margin for graph.**

4. Use a graphing calculator to draw a box-and-whisker plot of the fat content in the sandwiches from the restaurant in the example in the same viewing window as the plot from Exercise 3. Which restaurant's sandwiches have a lower fat content? Write a sentence explaining how the box-and-whisker plots help you analyze the data.

5. Use a graphing calculator to draw a histogram of the number of calories in the sandwiches. Use an interval width of 50 for each. Write a sentence comparing the number of calories in the sandwiches from the restaurant above with the number of calories in the sandwiches from the restaurant in the example on page 453.

Chapter Summary

WHAT did you learn?

Evaluate *n*th roots of real numbers. **(7.1)**

Use properties of rational exponents to evaluate and simplify expressions. **(7.2)**

Perform function operations. **(7.3)**

Find inverses of linear and nonlinear functions. **(7.4)**

Graph square root and cube root functions. **(7.5)**

Solve equations that contain radicals or rational exponents. **(7.6)**

Use roots and rational exponents in real-life problems. **(7.1–7.6)**

Use power functions, inverse functions, and radical functions to solve real-life problems. **(7.3–7.6)**

Use measures of central tendency and measures of dispersion to describe data sets. **(7.7)**

Represent data graphically with box-and-whisker plots and histograms. **(7.7)**

WHY did you learn it?

Find the number of reptile and amphibian species that Puerto Rico can support. **(p. 405)**

Model frequencies in the musical range of a trumpet. **(p. 413)**

Find the height of a dinosaur. **(p. 419)**

Find your bowling average. **(p. 428)**

Find the age of an African elephant. **(p. 433)**

Determine which boats satisfy the rule for competing in the America's Cup. **(p. 443)**

Find surface areas of mammals. **(p. 410)**

Find wind speeds that correspond to Beaufort wind scale numbers. **(p. 440)**

Analyze data sets such as the free-throw percentages for the players in the WNBA. **(pp. 445 and 446)**

Graph data sets such as the ages of the Presidents and Vice Presidents of the United States. **(p. 451)**

How does Chapter 7 fit into the BIGGER PICTURE of algebra?

In Chapter 7 you saw the familiar ideas of squares and square roots extended. This was a significant step in your study of powers and roots as you used exponents that were *not* whole numbers in expressions, functions, and many real-life problems. You will continue to build on these ideas as long as you study mathematics.

STUDY STRATEGY

How did you quiz yourself?

Here is an example of a quiz that was written for Lesson 7.3 and used before a class quiz was given, following the **Study Strategy** on page 400.

Quiz Yourself

Let $f(x) = -2x$ and $g(x) = x - 4$. Perform the indicated operation and state the domain. (Lesson 7.3)

1. $f(x) + g(x)$

2. $f(x) - g(x)$

3. $f(x) \cdot g(x)$

4. $\dfrac{f(x)}{g(x)}$

5. $g(f(x))$

6. $f(f(x))$

455

Chapter Review

ADDITIONAL RESOURCES
The following resources are available to help review the material in this chapter.

- Chapter Review Games and Activities (*Chapter 7 Resource Book,* p. 105)
- *Instant Replay: Video Review Games*
- 🖾 *Personal Student Tutor*
- Cumulative Review, Chs. 1–7 (*Chapter 7 Resource Book,* p. 117)

VOCABULARY

- *n*th root of *a*, p. 401
- index, p. 401
- simplest form, p. 408
- like radicals, p. 408
- power function, p. 415
- composition, p. 416
- inverse relation, p. 422

- inverse function, p. 422
- radical function, p. 431
- extraneous solution, p. 439
- statistics, p. 445
- measure of central tendency, p. 445
- mean, p. 445

- median, p. 445
- mode, p. 445
- measure of dispersion, p. 446
- range, p. 446
- standard deviation, p. 446
- box-and-whisker plot, p. 447

- lower quartile, p. 447
- upper quartile, p. 447
- histogram, p. 448
- frequency, p. 448
- frequency distribution, p. 448

7.1 *N*TH ROOTS AND RATIONAL EXPONENTS

Examples on pp. 401–403

EXAMPLES You can evaluate *n*th roots using radicals or rational exponents.

Radical notation: $27^{-2/3} = \dfrac{1}{27^{2/3}} = \dfrac{1}{(\sqrt[3]{27})^2} = \dfrac{1}{3^2} = \dfrac{1}{9}$

Rational exponent notation: $27^{-2/3} = \dfrac{1}{27^{2/3}} = \dfrac{1}{(27^{1/3})^2} = \dfrac{1}{3^2} = \dfrac{1}{9}$

Evaluate the expression without using a calculator.

1. $\sqrt[4]{16}$ **2**
2. $(\sqrt[3]{64})^2$ **16**
3. $9^{-5/2}$ $\dfrac{1}{243}$
4. $216^{1/3}$ **6**
5. $\sqrt[5]{-32}$ **−2**

6. Find the real *n*th root(s) of *a* if $n = 4$ and $a = 81$.   **±3**

7. Find the real *n*th root(s) of *a* if $n = 5$ and $a = -1$.   **−1**

8. Find the real *n*th root(s) of *a* if $n = 7$ and $a = 0$.   **0**

7.2 PROPERTIES OF RATIONAL EXPONENTS

Examples on pp. 407–410

EXAMPLES You can use properties of rational exponents to simplify expressions.

$\sqrt[3]{12} \cdot \sqrt[3]{4} = \sqrt[3]{12 \cdot 4} = \sqrt[3]{48} = \sqrt[3]{8 \cdot 6} = \sqrt[3]{8} \cdot \sqrt[3]{6} = 2\sqrt[3]{6}$

$\dfrac{(x^{1/2}y)^2}{x^{1/2}y^{3/4}} = \dfrac{x^{(1/2 \cdot 2)}y^2}{x^{1/2}y^{3/4}} = \dfrac{xy^2}{x^{1/2}y^{3/4}} = x^{(1-1/2)}y^{(2-3/4)} = x^{1/2}y^{5/4}$

Simplify the expression. Assume all variables are positive.

9. $5^{1/4} \cdot 5^{-9/4}$ $\dfrac{1}{25}$
10. $(100^{1/3})^{3/4}$ $10^{1/2}$
11. $\sqrt[3]{\dfrac{16}{1000}}$ $\dfrac{\sqrt[3]{2}}{5}$
12. $5\sqrt[3]{17} - 4\sqrt[3]{17}$ $\sqrt[3]{17}$

13. $(81x)^{1/4}$ $3x^{1/4}$
14. $\dfrac{(4x)^2}{(4x)^{1/2}}$ $8x^{3/2}$
15. $\sqrt[6]{6x^6y^7z^{10}}$ $xyz\sqrt[6]{6yz^4}$
16. $\sqrt[3]{4a^6} + a\sqrt[3]{108a^3}$ $4a^2\sqrt[3]{4}$

Chapter 7 *Powers, Roots, and Radicals*

7.3 POWER FUNCTIONS AND FUNCTION OPERATIONS

Examples on pp. 415–417

EXAMPLES You can add, subtract, multiply, or divide any two functions f and g. You can also find the composition of any two functions.

Let $f(x) = 2x^{1/2}$ and $g(x) = x^4$

Addition: $f(x) + g(x) = 2x^{1/2} + x^4$

Multiplication: $f(x) \cdot g(x) = 2x^{1/2} \cdot x^4 = 2x^{9/2}$

Composition: $f(g(x)) = f(x^4) = 2(x^4)^{1/2} = 2x^2$

Let $f(x) = 2x - 4$ and $g(x) = x - 2$. Perform the indicated operation.

17. $f(x) + g(x)$
$3x - 6$

18. $f(x) - g(x)$
$x - 2$

19. $f(x) \cdot g(x)$
$2x^2 - 8x + 8$

20. $\dfrac{f(x)}{g(x)}$ 2

21. $f(g(x))$
$2x - 8$

7.4 INVERSE FUNCTIONS

Examples on pp. 422–425

EXAMPLES You can find the inverse relation of any function. To verify that two functions are inverses of each other, show that $f(f^{-1}(x)) = f^{-1}(f(x)) = x$.

$f(x) = y = 2x - 5$

$x = 2y - 5$

$x + 5 = 2y$

$\dfrac{1}{2}x + \dfrac{5}{2} = y = f^{-1}(x)$

$f(f^{-1}(x)) = 2\left(\dfrac{1}{2}x + \dfrac{5}{2}\right) - 5 = x + 5 - 5 = x$

$f^{-1}(f(x)) = \dfrac{1}{2}(2x - 5) + \dfrac{5}{2} = x - \dfrac{5}{2} + \dfrac{5}{2} = x$

Find the inverse function.

22. $f(x) = -2x + 1$
$f^{-1}(x) = -\dfrac{1}{2}x + \dfrac{1}{2}$

23. $f(x) = -x^4, x \geq 0$
$f^{-1}(x) = (-x)^{1/4}, x \leq 0$

24. $f(x) = 5x^3 + 7$
$f^{-1}(x) = \sqrt[3]{\dfrac{x-7}{5}}$

25. Verify that $f(x) = -2x^5$ and $g(x) = \sqrt[5]{-\dfrac{x}{2}}$ are inverse functions. **Both compositions equal x.**

7.5 GRAPHING SQUARE ROOT AND CUBE ROOT FUNCTIONS

Examples on pp. 431–433

EXAMPLE You can graph a square root function by starting with the graph of $y = \sqrt{x}$. You can graph a cube root function by starting with the graph of $y = \sqrt[3]{x}$.

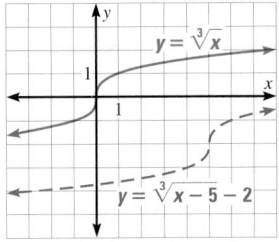

To graph $y = \sqrt[3]{x - 5} - 2$, first sketch $y = \sqrt[3]{x}$ (shown in red). Then shift the graph right 5 units and down 2 units. From the graph of $y = \sqrt[3]{x - 5} - 2$, you can see that the domain and range of the function are both all real numbers.

Graph the function. Then state the domain and range. 26–29. See margin for graphs.

26. $y = (x - 7)^{1/3}$
x and y are all real numbers

27. $y = \sqrt{x} + 6$
$x \geq 0; y \geq 6$

28. $y = -2(x - 3)^{1/2}$
$x \geq 3; y \leq 0$

29. $y = 3\sqrt[3]{x + 4} - 9$
x and y are all real numbers

Chapter Review **457**

26.
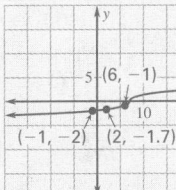

27.
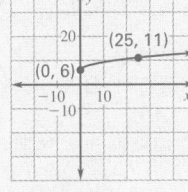

28.
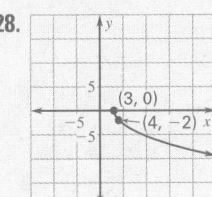

29.
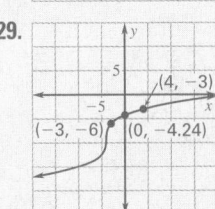

34.

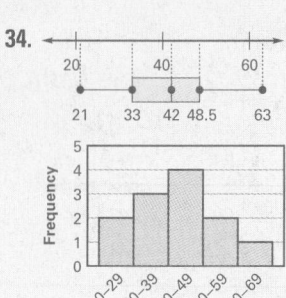

7.6

SOLVING RADICAL EQUATIONS

Examples on pp. 437–440

EXAMPLES You can solve equations that contain radicals or rational exponents by raising each side of the equation to the same power.

$$\sqrt{x-4} = 6 \qquad\qquad\qquad 4x^{2/3} = 100$$

$$\left(\sqrt{x-4}\right)^2 = 6^2 \quad \text{Square each side.} \qquad x^{2/3} = 25$$

$$x - 4 = 36 \qquad\qquad \left(x^{2/3}\right)^{3/2} = 25^{3/2} \quad \text{Raise each side to } \tfrac{3}{2} \text{ power.}$$

$$x = 40 \qquad\qquad\qquad\qquad\qquad x = 125$$

Solve the equation. Check for extraneous solutions.

30. $3(x+1)^{1/5} + 5 = 11$ **31**   **31.** $\sqrt[3]{5x+3} - \sqrt[3]{4x} = 0$ **−3**   **32.** $\sqrt{4x} = x - 8$ **16**

7.7

STATISTICS AND STATISTICAL GRAPHS

Examples on pp. 445–448

EXAMPLES The table shows the normal daily high temperatures (in degrees Fahrenheit) for Phoenix, Arizona, from 1961 to 1990.

Jan.	Feb.	Mar.	Apr.	May	June	July	Aug.	Sept.	Oct.	Nov.	Dec.
65.9	70.7	75.5	84.5	93.6	103.5	105.9	103.7	98.3	88.1	74.9	66.2

MEAN Find the average of the numbers: $\dfrac{65.9 + 70.7 + \cdots + 66.2}{12} = \dfrac{1030.8}{12} = 85.9$

MEDIAN Write the numbers in increasing order and locate the middle number(s): 65.9, 66.2, 70.7, 74.9, 75.5, **84.5**, **88.1**, 93.6, 98.3, 103.5, 103.7, 105.9

There are two middle numbers, so find their mean: $\dfrac{84.5 + 88.1}{2} = 86.3$

MODE Find the number(s) that occur most frequently: none

RANGE Find the difference between the greatest and least numbers: $105.9 - 65.9 = 40$

STANDARD DEVIATION Use the formula: $\sqrt{\dfrac{(65.9 - 85.9)^2 + (70.7 - 85.9)^2 + \cdots + (66.2 - 85.9)^2}{12}} \approx 14.4$

BOX-AND-WHISKER PLOT Find the quartiles: $\dfrac{70.7 + 74.9}{2} = 72.8$ and $\dfrac{98.3 + 103.5}{2} = 100.9$

Plot the minimum, maximum, median, and quartiles. Then draw the box and the whiskers (not shown).

HISTOGRAM Using five intervals beginning with 60–69, tally the data values for each interval. Then draw a histogram of the data set (not shown).

In Exercises 33 and 34, use the following data set of employees' ages at a small company: 21, 25, 30, 36, 39, 40, 44, 45, 46, 51, 51, 63.

33. Find the mean, median, mode, range, and standard deviation of the data set.   **40.9, 42, 51, 42, about 11.3**

34. Draw a box-and-whisker plot and a histogram of the data set. For the histogram, use five intervals beginning with 20–29. **See margin for graphs.**

Chapter Test

ADDITIONAL RESOURCES
- **Chapter 7 Resource Book**
  Chapter Test (3 levels) (p. 106)
  SAT/ACT Chapter Test (p. 112)
  Alternative Assessment (p. 113)
- ☐ **Test and Practice Generator**

Evaluate the expression without using a calculator.

1. $\sqrt[3]{-1000}$ **−10**   2. $4^{5/2}$ **32**   3. $(-64)^{2/3}$ **16**   4. $243^{-1/5}$ $\frac{1}{3}$ 5. $\sqrt[4]{16}$ **2**

Simplify the expression. Assume all variables are positive.

6. $(2^{1/3} \cdot 5^{1/2})^4$ $50\sqrt[3]{2}$ 7. $\sqrt[3]{27x^3y^6z^9}$ $3xy^2z^3$ 8. $\dfrac{3xy^{-1}}{12x^{1/2}y}$ $\dfrac{x^{1/2}}{4y^2}$

9. $\left(\dfrac{81x^2}{y}\right)^{3/4}$ $\dfrac{27x^{3/2}}{y^{3/4}}$ 10. $\sqrt{18} + \sqrt{200}$ $13\sqrt{2}$

Perform the indicated operation and state the domain.

11. $f + g$; $f(x) = x - 8$, $g(x) = 3x$
    $4x - 8$, x is all real numbers

12. $f - g$; $f(x) = 2x^{1/4}$, $g(x) = 5x^{1/4}$ $-3x^{1/4}$, $x \geq 0$

13. $f \cdot g$; $f(x) = 5x + 7$, $g(x) = x - 9$
    $5x^2 - 38x - 63$, x is all real numbers

14. $\dfrac{f}{g}$; $f(x) = x^{-1/5}$, $g(x) = x^{3/5}$ $\dfrac{1}{x^{4/5}}$, x is all real numbers except 0

15. $f(g(x))$; $f(x) = 4x^2 - 5$, $g(x) = -x$
    $4x^2 - 5$, x is all real numbers

16. $g(f(x))$; $f(x) = x^2 + 3x$, $g(x) = 2x + 1$
    $2x^2 + 6x + 1$, x is all real numbers

Find the inverse function.

17. $f(x) = \frac{1}{3}x - 4$
    $f^{-1}(x) = 3x + 12$

18. $f(x) = -5x + 5$
    $f^{-1}(x) = -\frac{1}{5}x + 1$

19. $f(x) = \frac{3}{4}x^2$, $x \geq 0$
    $f^{-1}(x) = \frac{2}{3}\sqrt{3x}$

20. $f(x) = x^5 - 2$
    $f^{-1}(x) = (x + 2)^{1/5}$

Graph the function. Then state the domain and range.
21–24. See margin.

21. $f(x) = \sqrt{x - 6}$ 22. $f(x) = \sqrt[3]{x} + 3$ 23. $f(x) = 3(x + 4)^{1/3} - 2$ 24. $f(x) = -2x^{1/2} + 4$

Solve the equation. Check for extraneous solutions.

25. $x^{5/2} - 10 = 22$ **4**   26. $(x + 8)^{1/4} + 1 = 0$ **no solution**   27. $\sqrt[3]{7x - 9} + 11 = 14$ $\frac{36}{7}$ 28. $\sqrt{4x + 15} - 3\sqrt{x} = 0$ **3**

29. **BIOLOGY** **CONNECTION**   Some biologists study the structure of animals. By studying a series of antelopes, biologists have found that the length l (in millimeters) of an antelope's bone can be modeled by

$$l = 24.1d^{2/3}$$

where d is the midshaft diameter of the bone (in millimeters). If the bone of an antelope has a midshaft diameter of 20 millimeters, what is the length of the bone? ▶ Source: *On Size and Life*   **177.57 mm**

🌐 **ACADEMY AWARDS** In Exercises 30–33, use the tables below which give the ages of the Academy Award winners for best actress and for best actor from 1980 to 1998.

Best actress
21, 25, 26, 29, 31, 33, 33, 34, 34, 38, 39, 41, 42, 45, 49, 49, 61, 72, 80

Best actor
30, 32, 35, 37, 37, 38, 39, 42, 43, 45, 45, 46, 51, 52, 52, 54, 60, 61, 76

30. Find the mean, median, mode, range, and standard deviation of each data set.   30–33. See margin.

31. Draw a box-and-whisker plot of each data set.

32. Make a frequency distribution of each data set using six intervals beginning with 21–30. Then draw a histogram of each data set.

33. *Writing* Compare the ages of the best actresses with the ages of the best actors. Use statistics and statistical graphs to support your statements.

Chapter Test   **459**

21. 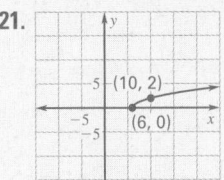   , $x \geq 6$, $y \geq 0$

22. 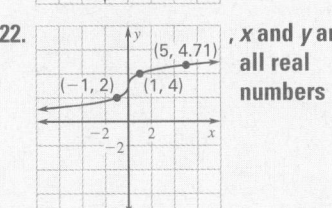   , x and y are all real numbers

23. 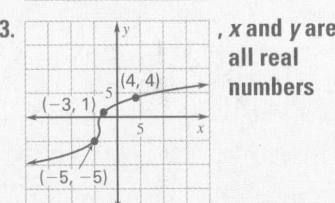   , x and y are all real numbers

24. 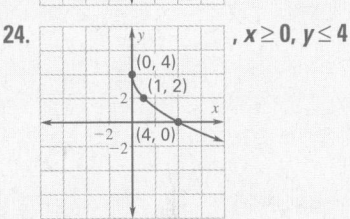   , $x \geq 0$, $y \leq 4$

30. Actresses: 41.16; 38; 33, 34, 49 (3 modes); 59; 15.2
    Actors: 46.05; 45; 37, 45, 52 (3 modes); 46; 11.14

31. 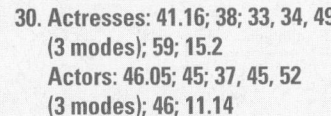   and
    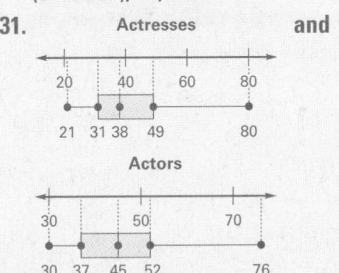

32–33. See Additional Answers beginning on page AA1.

ADDITIONAL RESOURCES

• *Chapter 7 Resource Book*
  Chapter Test (3 levels) (p. 106)
  SAT/ACT Chapter Test (p. 112)
  Alternative Assessment (p. 113)

• ▣ *Test and Practice Generator*

▶ **TEST-TAKING STRATEGY** Some college entrance exams allow the optional use of calculators. If you do use a calculator, make sure it is one you are familiar with and have used before.

1. MULTIPLE CHOICE If $x^4 = 625$, what does x equal? **C**

Ⓐ 5   Ⓑ −5   Ⓒ ±5

Ⓓ 25   Ⓔ ±25

2. MULTIPLE CHOICE What is the simplified form of the expression $\sqrt{18} + \sqrt{200} + \sqrt{2} - \sqrt{8}$? **A**

Ⓐ $12\sqrt{2}$ Ⓑ $14\sqrt{2}$

Ⓒ $18\sqrt{2}$ Ⓓ $14\sqrt{2} - \sqrt{8}$

Ⓔ $4\sqrt{2} - 4\sqrt{8}$

3. MULTIPLE CHOICE What is the simplified form of the expression $\sqrt[3]{54x^3y^6z^{10}}$? (Assume all variables are positive.) **D**

Ⓐ $xy^2\sqrt[3]{54z^{10}}$ Ⓑ $xy^2z^3\sqrt[3]{54z}$

Ⓒ $3y^3z^7\sqrt[3]{5}$ Ⓓ $3xy^2z^3\sqrt[3]{2z}$

Ⓔ $18xy^3z^7$

4. MULTIPLE CHOICE Which of the following is true if $f(x) = 3x^{-1/2}$, $g(x) = 6x^{3/4}$, and $h(x) = 18x^{1/4}$? **C**

Ⓐ $h(x) = f(x) + g(x)$ Ⓑ $h(x) = f(x) - g(x)$

Ⓒ $h(x) = f(x) \cdot g(x)$ Ⓓ $h(x) = \dfrac{f(x)}{g(x)}$

Ⓔ $h(x) = f(g(x))$

5. MULTIPLE CHOICE If $f(x) = x^2 - 3x + 7$ and $g(x) = x^2 + 2$, what is $f(g(x))$? **B**

Ⓐ $x^4 + x^2 + 17$ Ⓑ $x^4 + x^2 + 5$

Ⓒ $x^4 + x^2 - 9$ Ⓓ $x^4 + x^2 - 3$

Ⓔ $x^4 + 7x^2 + 5$

6. MULTIPLE CHOICE Which function is the inverse of $f(x) = \dfrac{1}{2}x - 5$? **C**

Ⓐ $f^{-1}(x) = 2x - 10$

Ⓑ $f^{-1}(x) = 2x + 5$

Ⓒ $f^{-1}(x) = 2x + 10$

Ⓓ $f^{-1}(x) = \dfrac{1}{2}x + 5$

Ⓔ $f^{-1}(x) = \dfrac{1}{2}x + \dfrac{5}{2}$

7. MULTIPLE CHOICE Which function is graphed? **C**

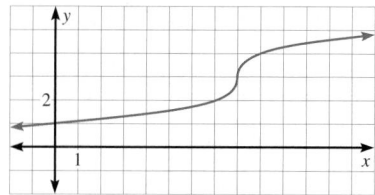

Ⓐ $y = \sqrt[3]{x - 3} + 8$ Ⓑ $y = \sqrt[3]{x + 3} + 8$

Ⓒ $y = \sqrt[3]{x - 8} + 3$ Ⓓ $y = \sqrt[3]{x + 8} - 3$

Ⓔ $y = \sqrt[3]{x + 8} + 3$

8. MULTIPLE CHOICE What is the solution of the equation $(3x + 5)^{1/2} - 3 = 4$? **E**

Ⓐ $-\dfrac{4}{3}$ Ⓑ $\dfrac{8}{3}$ Ⓒ $\dfrac{11}{3}$

Ⓓ $\dfrac{14}{3}$ Ⓔ $\dfrac{44}{3}$

9. MULTIPLE CHOICE What is the solution of the equation $4\sqrt[3]{x - 5} = 20$? **B**

Ⓐ 120   Ⓑ 130   Ⓒ 220

Ⓓ 2005   Ⓔ 4101

10. MULTIPLE CHOICE What is the median of 6, 4, 4, 10, 5, 12, 1? **B**

Ⓐ 4   Ⓑ 5   Ⓒ 6

Ⓓ 7   Ⓔ 10

11. MULTIPLE CHOICE Which data set matches the box-and-whisker plot shown? **D**

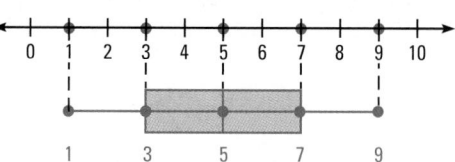

Ⓐ 1, 1, 3, 5, 6, 7, 9   Ⓑ 1, 2, 3, 5, 7, 8, 9

Ⓒ 1, 2, 4, 5, 6, 8, 9   Ⓓ 1, 3, 4, 5, 6, 7, 9

Ⓔ 1, 3, 5, 5, 7, 8, 9

QUANTITATIVE COMPARISON In Exercises 12 and 13, choose the statement that is true about the given quantities.

(A) The quantity in column A is greater.

(B) The quantity in column B is greater.

(C) The two quantities are equal.

(D) The relationship cannot be determined from the given information.

	Column A	Column B	
12.	$f(8)$ where $f(x) = x^{-2/3}$	$f(2)$ where $f(x) = x^{-2}$	C
13.	$f(f(0))$ where $f(x) = 5x - 2$	$f(f(0))$ where $f(x) = x^3 + 1$	B

14. **MULTI-STEP PROBLEM** The metabolic rate r (in kilocalories per day) of a mammal can be modeled by $r = km^{3/4}$ where k is a constant and m is the mass (in kilograms) of the mammal. The specific metabolic rate s (the rate per unit mass) can be modeled by $s = \dfrac{km^{3/4}}{m}$. ▶ Source: *Scaling: Why is Animal Size so Important?*

a. A 922 kilogram cow has a metabolic rate of about 11,700 kilocalories per day. What is the value of k in the model for metabolic rate?   **about 69.9**

b. Using the k-value from part (a), simplify the model given for specific metabolic rate. $s = \dfrac{69.9}{m^{1/4}}$

c. What is the specific metabolic rate of a 922 kilogram cow?   **about 12.7 kilocalories per day per kilogram**

d. What is the specific metabolic rate of a 16 *gram* mouse?   **about 197 kilocalories per day per kilogram**

e. *Writing* How does the specific metabolic rate change with decreasing body mass?   **See margin.**

15. **MULTI-STEP PROBLEM** Follow the steps below to find the relationship between the number of pedal revolutions of a bicycle and the distance traveled.

a. The rear wheel of a bicycle has a diameter of 70 centimeters. Write the function that describes the distance d traveled by the bicycle in terms of the number w of rear-wheel revolutions. (*Hint:* When $w = 1$, the distance traveled by the bicycle is equal to the circumference of the rear wheel.) $d = f(w) = 220w$

b. The gear ratio of a bicycle is calculated by dividing the number of teeth in the chainwheel by the number of teeth in the freewheel. The number w of rear-wheel revolutions is equal to the product of the gear ratio and the number p of pedal revolutions. A bicycle in first gear has 24 teeth in the chainwheel and 32 teeth in the freewheel. Write the function that describes w in terms of p.   gear ratio $(g) = \dfrac{\text{\# of chainwheel teeth}}{\text{\# of freewheel teeth}}$ so $w(p) = gp = 0.75p$

c. Use composition of functions to find the relationship between d and p. $d = f(w(p)) = 220gp$

d. Shifting gears on a bicycle changes the gear ratio. Use the table below to find how the distance traveled per pedal revolution changes as you shift gears.   **1st gear: about 165 cm per pedal revolution**

Gear	Number of teeth in chainwheel	Number of teeth in freewheel	
5th	24	19	about 278 cm per pedal revolution
10th	40	22	about 400 cm per pedal revolution
15th	50	19	about 579 cm per pedal revolution

Distance traveled per pedal revolution increases as gear number increases; it becomes harder to pedal.

14e. Specific metabolic rate increases as body mass decreases because the rate is proportional to $m^{-1/4}$. In other words, as mass decreases, the denominator gets smaller so the rate increases.

PLANNING THE CHAPTER
Exponential and Logarithmic Functions

LESSON	GOALS	NCTM	ITED	SAT9	Terra-Nova	Local
8.1 *pp. 465–472*	**GOAL 1** Graph exponential growth functions. **GOAL 2** Use exponential growth functions to model real-life situations.	1, 2, 3, 8, 9, 10	SAPE, MIG, IIG	2, 22	14, 16, 18	
8.2 *pp. 473–479*	**CONCEPT ACTIVITY: 8.2** *Investigate exponential growth and exponential decay.* **GOAL 1** Graph exponential decay functions. **GOAL 2** Use exponential decay functions to model real-life situations.	1, 2, 3, 8, 9, 10	SAPE, MIG, IIG	2	14, 16, 18	
8.3 *pp. 480–485*	**GOAL 1** Use the number *e* as the base of exponential functions. **GOAL 2** Use the natural base *e* in real-life situations.	1, 2, 8, 9, 10	SAPE	2	11, 16, 18	
8.4 *pp. 486–492*	**GOAL 1** Evaluate logarithmic functions. **GOAL 2** Graph logarithmic functions.	1, 2, 3	MIG, IIG		11, 14, 16, 49, 51, 52	
8.5 *pp. 493–500*	**GOAL 1** Use properties of logarithms. **GOAL 2** Use properties of logarithms to solve real-life problems. **TECHNOLOGY ACTIVITY: 8.5** *Graph logarithmic functions on a graphing calculator.*	1, 2, 3, 8, 9, 10	MIG, IIG	2	14, 16, 17, 18	
8.6 *pp. 501–508*	**GOAL 1** Solve exponential equations. **GOAL 2** Solve logarithmic equations.	1, 2	SAPE		11, 16, 49, 51, 52	
8.7 *pp. 509–516*	**GOAL 1** Model data with exponential functions. **GOAL 2** Model data with power functions.	1, 2, 5, 10	SAPE	19, 23	15, 16	
8.8 *pp. 517–522*	**GOAL 1** Evaluate and graph logistic growth functions. **GOAL 2** Use logistic growth functions to model real-life quantities.	1, 2, 8, 9, 10	MIG, IIG	22, 23	11, 14, 16, 18, 49, 51, 52	

RESOURCES

CHAPTER RESOURCE BOOKLETS

CHAPTER SUPPORT

Tips for New Teachers	p. 1	Prerequisite Skills Review	p. 5
Parent Guide for Student Success	p. 3	Strategies for Reading Mathematics	p. 7

LESSON SUPPORT

	8.1	8.2	8.3	8.4	8.5	8.6	8.7	8.8
Lesson Plans (regular and block)	p. 9	p. 22	p. 37	p. 50	p. 65	p. 79	p. 92	p. 105
Warm-Up Exercises and Daily Quiz	p. 11	p. 24	p. 39	p. 52	p. 67	p. 81	p. 94	p. 107
Activity Support Masters		p. 25						
Lesson Openers	p. 12	p. 26	p. 40	p. 53	p. 68	p. 82	p. 95	p. 108
Graphing Calculator Activities & Keystrokes	p. 13	p. 27		p. 54	p. 69		p. 96	p. 109
Practice (3 levels)	p. 14	p. 29	p. 41	p. 56	p. 70	p. 83	p. 97	p. 110
Reteaching with Practice	p. 17	p. 32	p. 44	p. 59	p. 73	p. 86	p. 100	p. 113
Quick Catch-Up for Absent Students	p. 19	p. 34	p. 46	p. 61	p. 75	p. 88	p. 102	p. 115
Cooperative Learning Activities				p. 62				
Interdisciplinary Applications		p. 35		p. 63		p. 89		p. 116
Real-Life Applications	p. 20		p. 47		p. 76		p. 103	
Math & History Applications					p. 77			
Challenge: Skills and Applications	p. 21	p. 36	p. 48	p. 64	p. 78	p. 90	p. 104	p. 117

REVIEW AND ASSESSMENT

Quizzes	pp. 49, 91	Alternative Assessment with Math Journal	p. 126
Chapter Review Games and Activities	p. 118	Project with Rubric	p. 128
Chapter Test (3 levels)	pp. 119–124	Cumulative Review	p. 130
SAT/ACT Chapter Test	p. 125	Resource Book Answers	p. A1

TRANSPARENCIES

	8.1	8.2	8.3	8.4	8.5	8.6	8.7	8.8
Warm-Up Exercises and Daily Quiz	p. 59	p. 60	p. 61	p. 62	p. 63	p. 64	p. 65	p. 66
Alternative Lesson Opener Transparencies	p. 51	p. 52	p. 53	p. 54	p. 55	p. 56	p. 57	p. 58
Examples/Standardized Test Practice	✓	✓	✓	✓	✓	✓	✓	✓
Answer Transparencies	✓	✓	✓	✓	✓	✓	✓	✓

TECHNOLOGY

- Electronic Teaching Tools
- Online Lesson Planner
- Internet Support
- Personal Student Tutor
- Test and Practice Generator
- Instant Replay: Video Review Games
- Electronic Lesson Presentations (Lesson 8.6)

ADDITIONAL RESOURCES

- Basic Skills Workbook: Diagnosis and Remediation
- Worked-Out Solution Key
- Resources in Spanish
- Standardized Test Practice Workbook
- Practice Workbook with Examples

Resource Key
- ● STUDENT EDITION
- ● CHAPTER 8 RESOURCE BOOK
- ● TEACHER'S EDITION

REGULAR SCHEDULE

Day 1

8.1

STARTING OPTIONS
- Prereq. Skills Review
- Strategies for Reading
- Homework Check
- Warm-Up or Daily Quiz

TEACHING OPTIONS
- Les. Opener (Appl.)
- Graphing Calc. Activity
- Examples 1–4
- Closure Question
- Guided Practice Exs.

APPLY/HOMEWORK
- See Assignment Guide.
- See the CRB: Practice, Reteach, Apply, Extend

ASSESSMENT OPTIONS
- Checkpoint Exercises
- Daily Quiz (8.1)
- Stand. Test Practice

Day 2

8.2

STARTING OPTIONS
- Homework Check
- Warm-Up or Daily Quiz

TEACHING OPTIONS
- Concept Act. & Wksht.
- Les. Opener (Visual)
- Graphing Calc. Activity
- Examples 1–4
- Closure Question
- Guided Practice Exs.

APPLY/HOMEWORK
- See Assignment Guide.
- See the CRB: Practice, Reteach, Apply, Extend

ASSESSMENT OPTIONS
- Checkpoint Exercises
- Daily Quiz (8.2)
- Stand. Test Practice

Day 3

8.3

STARTING OPTIONS
- Homework Check
- Warm-Up or Daily Quiz

TEACHING OPTIONS
- Les. Opener (Application)
- Examples 1–3
- Guided Practice Exs. 1–15

APPLY/HOMEWORK
- See Assignment Guide.
- See the CRB: Practice, Reteach, Apply, Extend

ASSESSMENT OPTIONS
- Checkpoint Exercises, p. 481

Day 4

8.3 *(cont.)*

STARTING OPTIONS
- Homework Check

TEACHING OPTIONS
- Examples 4–5
- Closure Question
- Guided Practice Ex. 16

APPLY/HOMEWORK
- See Assignment Guide.
- See the CRB: Practice, Reteach, Apply, Extend

ASSESSMENT OPTIONS
- Checkpoint Exercises, p. 482
- Daily Quiz (8.3)
- Stand. Test Practice
- Quiz (8.1–8.3)

Day 5

8.4

STARTING OPTIONS
- Homework Check
- Warm-Up or Daily Quiz

TEACHING OPTIONS
- Motivating the Lesson
- Les. Opener (Activity)
- Graphing Calc. Activity
- Examples 1–4
- Guided Practice Exs. 1–12

APPLY/HOMEWORK
- See Assignment Guide.
- See the CRB: Practice, Reteach, Apply, Extend

ASSESSMENT OPTIONS
- Checkpoint Exercises, p. 487

Day 6

8.4 *(cont.)*

STARTING OPTIONS
- Homework Check

TEACHING OPTIONS
- Examples 5–8
- Closure Question
- Guided Practice Exs. 13–15

APPLY/HOMEWORK
- See Assignment Guide.
- See the CRB: Practice, Reteach, Apply, Extend

ASSESSMENT OPTIONS
- Checkpoint Exercises, pp. 488–489
- Daily Quiz (8.4)
- Stand. Test Practice

Day 9

8.6

STARTING OPTIONS
- Homework Check
- Warm-Up or Daily Quiz

TEACHING OPTIONS
- Motivating the Lesson
- Les. Opener (Calculator)
- Examples 1–4
- Guided Practice Exs. 4–9, 16

APPLY/HOMEWORK
- See Assignment Guide.
- See the CRB: Practice, Reteach, Apply, Extend

ASSESSMENT OPTIONS
- Checkpoint Exercises, p. 502

Day 10

8.6 *(cont.)*

STARTING OPTIONS
- Homework Check

TEACHING OPTIONS
- Examples 5–8
- Closure Question
- Guided Practice Exs. 1–3, 10–15, 17–18

APPLY/HOMEWORK
- See Assignment Guide.
- See the CRB: Practice, Reteach, Apply, Extend

ASSESSMENT OPTIONS
- Checkpoint Exercises, pp. 503–504
- Daily Quiz (8.6)
- Stand. Test Practice
- Quiz (8.4–8.6)

Day 11

8.7

STARTING OPTIONS
- Homework Check
- Warm-Up or Daily Quiz

TEACHING OPTIONS
- Les. Opener (Visual)
- Graphing Calc. Activity
- Examples 1–6
- Closure Question
- Guided Practice Exs.

APPLY/HOMEWORK
- See Assignment Guide.
- See the CRB: Practice, Reteach, Apply, Extend

ASSESSMENT OPTIONS
- Checkpoint Exercises
- Daily Quiz (8.7)
- Stand. Test Practice

Day 12

8.8

STARTING OPTIONS
- Homework Check
- Warm-Up or Daily Quiz

TEACHING OPTIONS
- Motivating the Lesson
- Les. Opener (Calc.)
- Graphing Calc. Activity
- Examples 1–5
- Closure Question
- Guided Practice Exs.

APPLY/HOMEWORK
- See Assignment Guide.
- See the CRB: Practice, Reteach, Apply, Extend

ASSESSMENT OPTIONS
- Checkpoint Exercises
- Daily Quiz (8.8)
- Stand. Test Practice
- Quiz (8.7–8.8)

Day 13

Review

DAY 13 START OPTIONS
- Homework Check

REVIEWING OPTIONS
- Chapter 8 Summary
- Chapter 8 Review
- Chapter Review Games and Activities

APPLY/HOMEWORK
- Chapter 8 Test (practice)
- Ch. Standardized Test (practice)

Day 14

Assess

DAY 14 START OPTIONS
- Homework Check

ASSESSMENT OPTIONS
- Chapter 8 Test
- SAT/ACT Ch. 8 Test
- Alternative Assessment

APPLY/HOMEWORK
- Skill Review, p. 532

Day 1

8.1 & 8.2

STARTING OPTIONS
- Prereq. Skills Review
- Strategies for Reading
- Homework Check
- W-Up 8.1 or D. Quiz 7.7

TEACHING 8.1 OPTIONS
- Les. Opener (Appl.)
- Graphing Calc. Activity
- Examples 1–4
- Closure Question
- Guided Practice Exs.

TEACHING 8.2 OPTIONS
- Warm-Up (Les. 8.2)
- Concept Act. & Wksht.
- Les. Opener (Visual)
- Graphing Calc. Activity
- Examples 1–4
- Closure Question
- Guided Practice Exs.

APPLY/HOMEWORK
- See Assignment Guide.
- See the CRB: Practice, Reteach, Apply, Extend

ASSESSMENT OPTIONS
- Checkpoint Exercises
- Daily Quiz (Les. 8.1, 8.2)

Day 2

8.3

DAY 3 START OPTIONS
- Homework Check
- Warm-Up or Daily Quiz

TEACHING 8.3 OPTIONS
- Les. Opener (Appl.)
- Examples 1–5
- Closure Question
- Guided Practice Exs.

APPLY/HOMEWORK
- See Assignment Guide.
- See the CRB: Practice, Reteach, Apply, Extend

ASSESSMENT OPTIONS
- Checkpoint Exercises
- Daily Quiz (Les. 8.3)
- Stand. Test Practice
- Quiz (8.1–8.3)

Day 3

8.4

STARTING OPTIONS
- Homework Check
- Warm-Up or Daily Quiz

TEACHING 8.4 OPTIONS
- Motivating the Lesson
- Les. Opener (Activity)
- Graphing Calc. Activity
- Examples 1–8
- Closure Question
- Guided Practice Exs.

APPLY/HOMEWORK
- See Assignment Guide.
- See the CRB: Practice, Reteach, Apply, Extend

ASSESSMENT OPTIONS
- Checkpoint Exercises
- Daily Quiz (Les. 8.4)
- Stand. Test Practice

Day 4

8.5

DAY 4 START OPTIONS
- Homework Check
- Warm-Up or Daily Quiz

TEACHING 8.5 OPTIONS
- Motivating the Lesson
- Les. Opener (Activity)
- Graphing Calc. Activity
- Examples 1–5
- Technology Activity
- Closure Question
- Guided Practice Exs.

APPLY/HOMEWORK
- See Assignment Guide.
- See the CRB: Practice, Reteach, Apply, Extend

ASSESSMENT OPTIONS
- Checkpoint Exercises
- Daily Quiz (Les. 8.5)
- Stand. Test Practice

Day 5

8.6

DAY 5 START OPTIONS
- Homework Check
- Warm-Up or Daily Quiz

TEACHING 8.6 OPTIONS
- Motivating the Lesson
- Les. Opener (Calc.)
- Examples 1–8
- Closure Question
- Guided Practice Exs.

APPLY/HOMEWORK
- See Assignment Guide.
- See the CRB: Practice, Reteach, Apply, Extend

ASSESSMENT OPTIONS
- Checkpoint Exercises
- Daily Quiz (Les. 8.6)
- Stand. Test Practice
- Quiz (8.4–8.6)

Day 6

8.7 & 8.8

DAY 6 START OPTIONS
- Homework Check
- W-Up 8.7 or D. Quiz 8.6

TEACHING 8.7 OPTIONS
- Les. Opener (Visual)
- Graphing Calc. Activity
- Examples 1–6
- Closure Question
- Guided Practice Exs.

TEACHING 8.8 OPTIONS
- Warm-Up (Les. 8.8)
- Motivating the Lesson
- Les. Opener (Calc.)
- Graphing Calc. Activity
- Examples 1–5
- Closure Question
- Guided Practice Exs.

APPLY/HOMEWORK
- See Assignment Guide.
- See the CRB: Practice, Reteach, Apply, Extend

ASSESSMENT OPTIONS
- Checkpoint Exercises
- Daily Quiz (Les. 8.7, 8.8)
- Quiz (8.7–8.8)

Day 7

Review/Assess

DAY 7 START OPTIONS
- Homework Check

REVIEWING OPTIONS
- Chapter 8 Summary
- Chapter 8 Review
- Chapter Review Games and Activities
- Chapter 8 Test (practice)
- Ch. Standardized Test (practice)

ASSESSMENT OPTIONS
- Chapter 8 Test
- SAT/ACT Ch. 8 Test
- Alternative Assessment

APPLY/HOMEWORK
- Skill Review, p. 532

Day 7

8.5

STARTING OPTIONS
- Homework Check
- Warm-Up or Daily Quiz

TEACHING OPTIONS
- Motivating the Lesson
- Les. Opener (Activity)
- Graphing Calc. Activity
- Examples 1–3
- Guided Practice Exs. 1–12

APPLY/HOMEWORK
- See Assignment Guide.
- See the CRB: Practice, Reteach, Apply, Extend

ASSESSMENT OPTIONS
- Checkpoint Exercises, p. 494

Day 8

8.5 (cont.)

STARTING OPTIONS
- Homework Check

TEACHING OPTIONS
- Examples 4–5
- Technology Activity
- Closure Question
- Guided Practice Ex. 13

APPLY/HOMEWORK
- See Assignment Guide.
- See the CRB: Practice, Reteach, Apply, Extend

ASSESSMENT OPTIONS
- Checkpoint Exercises, pp. 494–495
- Daily Quiz (8.5)
- Stand. Test Practice

BEFORE THE CHAPTER

The *Chapter 8 Resource Book* has the following materials to distribute and use before the chapter:

- **Parent Guide for Student Success**
- **Prerequisite Skills Review (pictured below)**
- **Strategies for Reading Mathematics**

PREREQUISITE SKILLS *Pages 5–6*

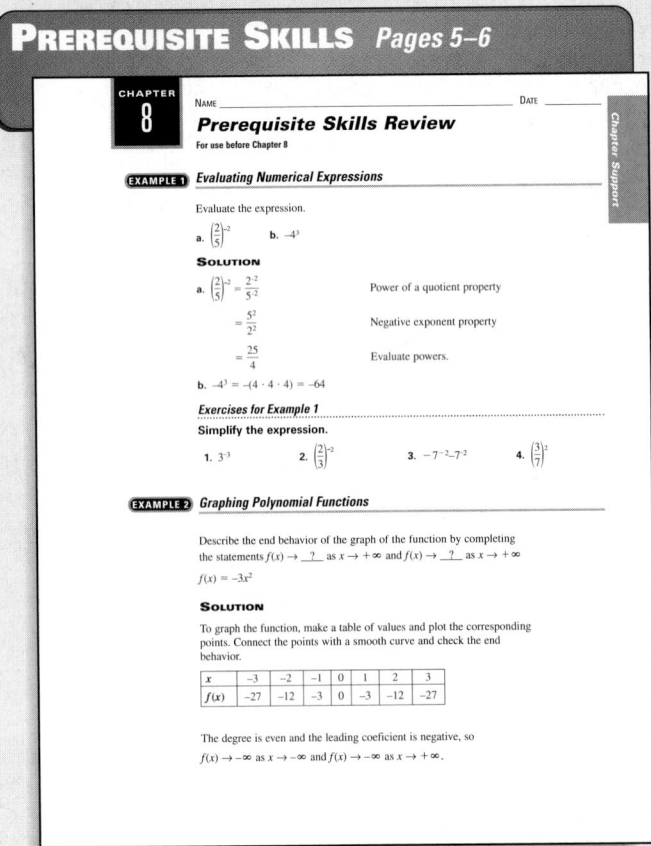

PREREQUISITE SKILLS REVIEW These two pages support the Study Guide on page 464. They help students prepare for Chapter 8 by providing worked-out examples and practice for the following skills needed in the chapter:

- **Evaluate numerical expressions.**
- **Graph polynomial functions.**
- **Fit lines to data.**

DURING EACH LESSON

The *Chapter 8 Resource Book* has the following alternatives for introducing the lesson:

- **Lesson Openers**
- **Graphing Calculator Activities with Keystrokes (below)**

CALCULATOR ACTIVITY *Pages 27–28*

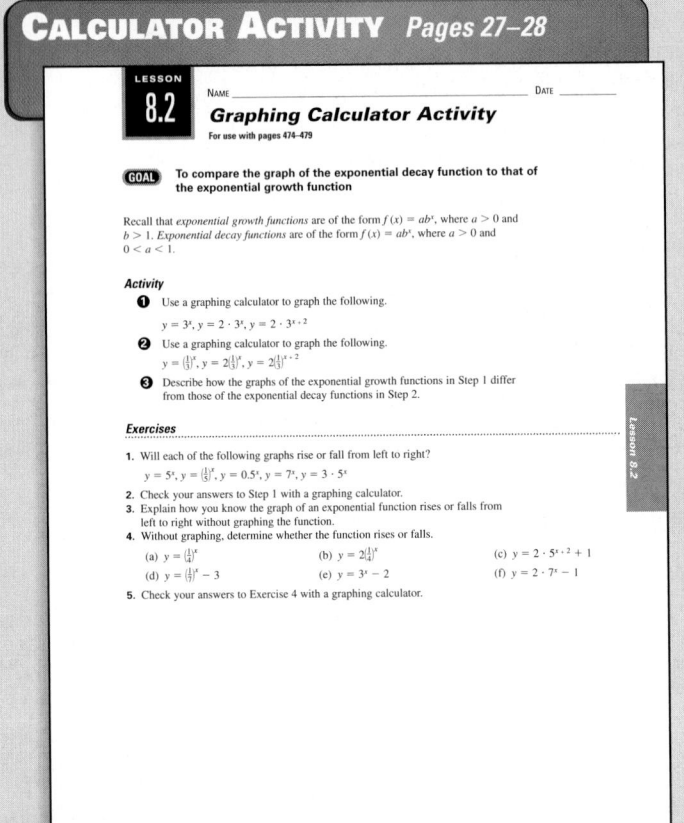

GRAPHING CALCULATOR ACTIVITY This activity uses a graphing calculator to develop an understanding of exponential decay functions, one of the concepts covered in Lesson 8.2. Keystrokes for this activity are provided on a separate sheet.

The *Chapter 8 Resource Book* has a variety of materials to follow-up each lesson. They include the following:

- **Practice (3 levels) (pictured below)**
- **Reteaching with Practice**
- **Quick Catch-Up for Absent Students**
- **Challenge: Skills and Applications**
- **Interdisciplinary Applications**
- **Real-Life Applications**

PRACTICE *Page 84*

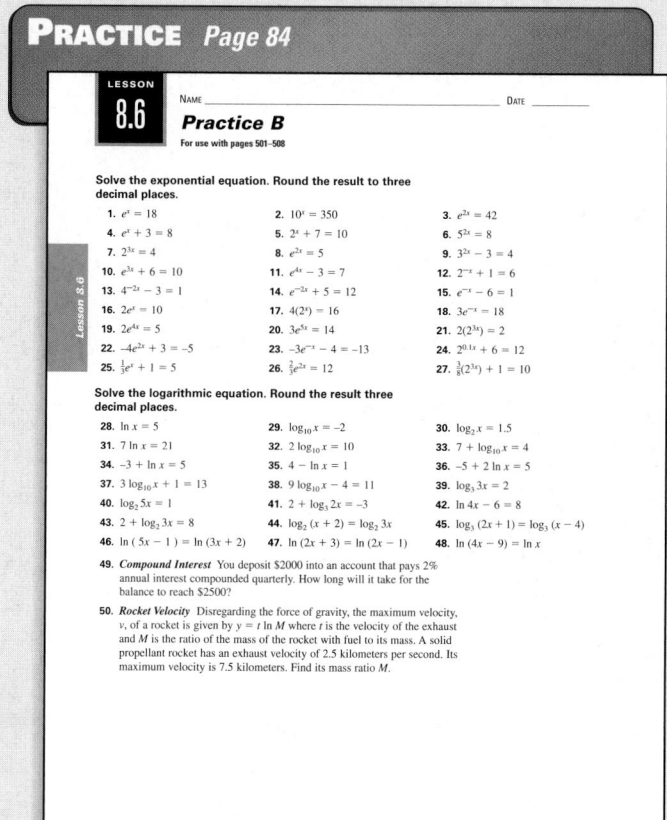

LESSON 8.6

NAME _____ DATE _____

Practice B
For use with pages 501–508

Solve the exponential equation. Round the result to three decimal places.

1. $e^x = 18$
2. $10^x = 350$
3. $e^{2x} = 42$
4. $e^x + 3 = 8$
5. $2^x + 7 = 10$
6. $5^{2x} = 8$
7. $2^{3x} = 4$
8. $e^{2x} = 5$
9. $3^{2x} - 3 = 4$
10. $e^{3x} + 6 = 10$
11. $e^{4x} - 3 = 7$
12. $2^{-x} + 1 = 6$
13. $4^{-2x} - 3 = 1$
14. $e^{-2x} + 5 = 12$
15. $e^{-x} - 6 = 1$
16. $2e^x = 10$
17. $4(2^x) = 16$
18. $3e^{-x} = 18$
19. $2e^{4x} = 5$
20. $3e^{5x} = 14$
21. $2(2^{3x}) = 2$
22. $-4e^{2x} + 3 = -5$
23. $-3e^{-x} - 4 = -13$
24. $2^{0.1x} + 6 = 12$
25. $\frac{1}{5}e^x + 1 = 5$
26. $\frac{2}{3}e^{2x} = 12$
27. $\frac{3}{8}(2^{3x}) + 1 = 10$

Solve the logarithmic equation. Round the result three decimal places.

28. $\ln x = 5$
29. $\log_{10} x = -2$
30. $\log_2 x = 1.5$
31. $7 \ln x = 21$
32. $2 \log_{10} x = 10$
33. $7 + \log_{10} x = 4$
34. $-3 + \ln x = 5$
35. $4 - \ln x = 1$
36. $-5 + 2 \ln x = 5$
37. $3 \log_{10} x + 1 = 13$
38. $9 \log_{10} x - 4 = 11$
39. $\log_3 3x = 2$
40. $\log_3 5x = 1$
41. $2 + \log_3 2x = -3$
42. $\ln 4x - 6 = 8$
43. $2 + \log_2 3x = 8$
44. $\log_3 (x + 2) = \log_3 3x$
45. $\log_2 (2x + 1) = \log_2 (x - 4)$
46. $\ln (5x - 1) = \ln (3x + 2)$
47. $\ln (2x + 3) = \ln (2x - 1)$
48. $\ln (4x - 9) = \ln x$

49. **Compound Interest** You deposit $2000 into an account that pays 2% annual interest compounded quarterly. How long will it take for the balance to reach $2500?

50. **Rocket Velocity** Disregarding the force of gravity, the maximum velocity, v, of a rocket is given by $y = t \ln M$ where t is the velocity of the exhaust and M is the ratio of the mass of the rocket with fuel to its mass. A solid propellant rocket has an exhaust velocity of 2.5 kilometers per second. Its maximum velocity is 7.5 kilometers. Find its mass ratio M.

PRACTICE Each lesson has three levels of practice: basic (A), average (B), and advanced (C). This Practice B involves more advanced work with exponential and logarithmic equations than Practice A, but does not have as much work with application problems as Practice C.

TECHNOLOGY RESOURCE

Teachers can use the Time-Saving Test and Practice Generator to create additional practice worksheets for Lesson 8.6 and for the other lessons in Chapter 8.

The *Chapter 8 Resource Book* has the following review and assessment materials:

- **Quizzes**
- **Chapter Review Games and Activities (pictured below)**
- **Chapter Test (3 levels)**
- **SAT/ACT Chapter Test**
- **Alternative Assessment with Rubrics and Math Journal**
- **Project with Rubric**
- **Cumulative Review**

CHAPTER REVIEW GAMES *Page 118*

CHAPTER 8

NAME _____ DATE _____

Chapter Review Games and Activities
For use after Chapter 8

Match the graph with the type of function by placing the appropriate lower case letter in the blank in front of the numbered name of the function, and match the equations with a capital letter in the blank behind the name of the function. You must use all graphs, therefore, one of the functions will have two answers.

_____ 1. Exponential growth _____
_____ 2. Exponential decay _____
_____ 3. Logistic growth _____
_____ 4. Logarithmic _____

a $y = \left(\frac{1}{2}\right)3^x$

b $y = -5\left(\frac{2}{3}\right)^x$

c $y = 4\left(\frac{2}{3}\right)^x$

d $y = \ln (x - 2)$

e $y = \dfrac{5}{(1 + 5e^{-x})}$

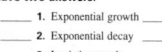

A

B

C

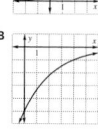
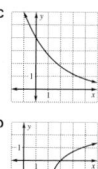

D

E

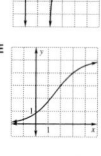

CHAPTER REVIEW GAMES AND ACTIVITIES On this worksheet, students match graphs with exponential and logarithmic functions to review the material in Chapter 8.

TECHNOLOGY RESOURCE

Teachers can use the video Instant Replay: Video Review Games as another motivating way to review the material in Chapter 8.

CHAPTER GOALS

Students will graph general exponential functions. They will identify and graph exponential growth and exponential decay functions, and use them to model real-life situations such as compounding interest and depreciating the value of goods. Students will be introduced to the natural base e and will simplify and evaluate expressions involving e. They will graph natural base functions, including those that model real-life situations. Students will learn the definition of logarithm to the base b, which they will use to evaluate logarithmic expressions and functions, including those involving common logarithms and natural logarithms. They will examine the inverse relationship between logarithmic and exponential functions and will graph logarithmic functions, which they will use to solve problems. Students will use the change-of-base formula and properties of logarithms to expand and condense logarithmic functions. They will solve exponential and logarithmic equations. Students will write exponential and power functions, using exponential and power regression to model real-life problems. Finally, students will evaluate and graph logistic functions, solve logistic equations, and use logistic growth functions to model real-life problems.

APPLICATION NOTE

Rapid ascents at elevations above 2400 m (8000 ft) can cause dizziness, shortness of breath, headache, nausea, confusion, and can contribute to death when untreated. The air is so thin at the 29,028 foot summit of Mount Everest that climbers carry bottled oxygen. In 1978, Reinhold Messner and Peter Habeler made the first ascent of Everest without bottled oxygen.

Additional information about atmospheric pressure is available at **www.mcdougallittell.com**.

EXPONENTIAL AND LOGARITHMIC FUNCTIONS

▶ *How does altitude affect the air?*

APPLICATION: Mountain Climbing

A surprising fact that you may not know is that air has weight! The weight of the air above you produces what scientists call *atmospheric pressure*.

Mountain climbers need to be aware of changes in atmospheric pressure because as the pressure decreases, so does the amount of oxygen they have to breathe.

Think & Discuss

The graph below shows the relationship between atmospheric pressure and altitude.

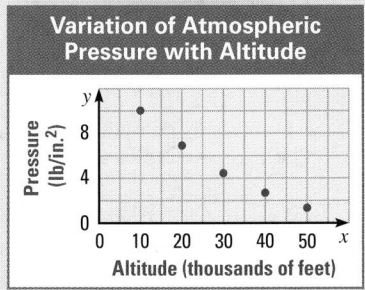

Variation of Atmospheric Pressure with Altitude

Pressure (lb/in.²) vs. Altitude (thousands of feet)

1. Describe what happens to the atmospheric pressure as the altitude increases.
   Atmospheric pressure decreases as altitude increases.
2. Mount McKinley in Alaska is 20,320 feet high. Estimate the atmospheric pressure at its peak.
   about 7 lb/in.²

Learn More About It

You will find the atmospheric pressure at the peak of Mount Everest in Exercise 79 on p. 484.

APPLICATION LINK Visit www.mcdougallittell.com for more information about atmospheric pressure.

ADDITIONAL RESOURCES
Another way to begin the chapter is to show the video clip of a real-life motivator for Chapter 8 on the *Instant Replay: Video Review Games.*

PROJECTS
A project covering Chapters 7–9 appears on pages 584–585 of the Student Edition. An additional project for Chapter 8 is available in the *Chapter 8 Resource Book*, p. 128.

TECHNOLOGY

 **Software**
• *Electronic Teaching Tools*
• *Online Lesson Planner*
• *Personal Student Tutor*
• *Test and Practice Generator*
• *Electronic Lesson Presentations* (Lesson 8.6)

Video
• *Instant Replay: Video Review Games*

Internet Connections
www.mcdougallittell.com
• **Application Links**
  463, 467, 491, 497, 499, 507
• **Data Updates**
  478, 522
• **Student Help**
  467, 471, 476, 484, 487, 497, 500, 502, 515, 518, 519
• **Career Links**
  468, 482, 495
• **Extra Challenge**
  472, 479, 485, 492, 498, 507, 516, 522

PREPARE

DIAGNOSTIC TOOLS
The **Skill Review** exercises can help you diagnose whether students have the following skills needed in Chapter 8:

• Evaluate expressions with exponents.

• Describe the end behavior of graphs.

• Draw scatter plots.

The following resources are available for students who need additional help with these skills:

• Prerequisite Skills Review (*Chapter 8 Resource Book,* p. 5; *Warm-Up Transparencies,* p. 58)

• Reteaching with Practice (Chapter Resource Books for Lessons 2.5, 6.1, and 6.2)

• ▣ *Personal Student Tutor*

ADDITIONAL RESOURCES
The following resources are provided to help you prepare for the upcoming chapter and customize review materials:

• *Chapter 8 Resource Book*
  Tips for New Teachers (p. 1)
  Parent Guide (p. 3)
  Lesson Plans (every lesson)
  Lesson Plans for Block Scheduling (every lesson)

• ▣ *Electronic Teaching Tools*

• ▣ *Online Lesson Planner*

• ▣ *Test and Practice Generator*

ENGLISH LEARNERS
Remembering English-language mathematical terms can be even more challenging for English learners than for other students. Before beginning Lesson 8.1, briefly review the terms *power, base, exponent, domain, range, function, logarithm,* and *end behavior.*

PREVIEW

What's the chapter about?

Chapter 8 is about **exponential and logarithmic functions**. These functions are inverses of each other. In Chapter 8 you'll learn

• how to graph and use exponential, logarithmic, and logistic growth functions.

• how to use the number *e* and the definition and properties of logarithms.

• how to solve exponential and logarithmic equations.

KEY VOCABULARY

▶ **Review**
• base, p. 11
• inverse function, p. 422

▶ **New**
• exponential function, p. 465

• asymptote, p. 465
• exponential growth function, p. 466
• exponential decay function, p. 474
• natural base *e*, p. 480

• logarithm of *y* with base *b*, p. 486
• common logarithm, p. 487
• natural logarithm, p. 487
• logistic growth function, p. 517

PREPARE

Are you ready for the chapter?

SKILL REVIEW Do these exercises to review key skills that you'll apply in this chapter. See the given **reference page** if there is something you don't understand.

STUDENT HELP

↳ **Study Tip**
"Student Help" boxes throughout the chapter give you study tips and tell you where to look for extra help in this book and on the Internet.

Evaluate the expression. (Review Example 1, p. 11; Example 1, p. 324)

1. 4^{-3} $\frac{1}{64}$ **2.** $\left(\frac{1}{3}\right)^2$ $\frac{1}{9}$ **3.** $\left(\frac{3}{4}\right)^0$ 1 **4.** -5^2 -25 **5.** $\left(\frac{5}{2}\right)^{-1}$ $\frac{2}{5}$

Describe the end behavior of the graph of the function by completing the statements $f(x) \to \underline{?}$ **as** $x \to -\infty$ **and** $f(x) \to \underline{?}$ **as** $x \to +\infty$. (Review Example 4, p. 332)

6. $f(x) = 2x^3$ **7.** $f(x) = -x^2$ **8.** $f(x) = 4x^4$ **9.** $f(x) = -5x^3$

6–9. See margin.

Draw a scatter plot of the data. Then approximate an equation of the best-fitting line. (Review Example 2, p. 101)

10.

x	1	2	3	4	5	6	7	8	9	10
y	2.2	2.9	3.0	4.1	4.2	4.3	4.8	5.0	5.9	5.9

Sample answer: $y = 0.403x + 2.013$

STUDY STRATEGY

Here's a study strategy!

Study Group

Form a study group. Have each group member take lessons from the chapter and summarize the important concepts and skills in those lessons. Then have each member lead a discussion on how to solve the types of problems in his or her lessons.

6. $f(x) \to -\infty$ as $x \to -\infty$; $f(x) \to +\infty$ as $x \to +\infty$
7. $f(x) \to -\infty$ as $x \to -\infty$; $f(x) \to -\infty$ as $x \to +\infty$
8. $f(x) \to +\infty$ as $x \to -\infty$; $f(x) \to +\infty$ as $x \to +\infty$
9. $f(x) \to +\infty$ as $x \to -\infty$; $f(x) \to -\infty$ as $x \to +\infty$

8.1

Exponential Growth

1 PLAN

PACING
Basic: 1 day
Average: 1 day
Advanced: 1 day
Block Schedule: 0.5 block with 8.2

What you should learn

GOAL 1 Graph exponential growth functions.

GOAL 2 Use exponential growth functions to model **real-life** situations, such as Internet growth in **Example 3**.

Why you should learn it

▼ To solve **real-life** problems, such as finding the amount of energy generated from wind turbines in **Exs. 49–51**.

GOAL 1 GRAPHING EXPONENTIAL GROWTH FUNCTIONS

An **exponential function** involves the expression b^x where the **base** b is a positive number other than 1. In this lesson you will study exponential functions for which $b > 1$. To see the basic shape of the graph of an exponential function such as $f(x) = 2^x$, you can make a table of values and plot points, as shown below.

x	$f(x) = 2^x$
-3	$2^{-3} = \frac{1}{8}$
-2	$2^{-2} = \frac{1}{4}$
-1	$2^{-1} = \frac{1}{2}$
0	$2^0 = 1$
1	$2^1 = 2$
2	$2^2 = 4$
3	$2^3 = 8$

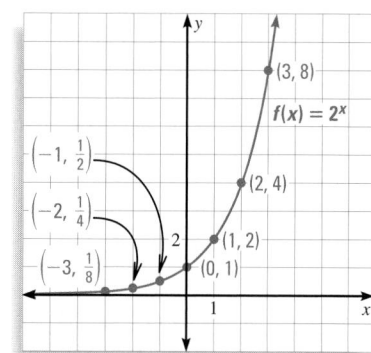

Notice the end behavior of the graph. As $x \to +\infty$, $f(x) \to +\infty$, which means that the graph moves up to the right. As $x \to -\infty$, $f(x) \to 0$, which means that the graph has the line $y = 0$ as an *asymptote*. An **asymptote** is a line that a graph approaches as you move away from the origin.

▶ ACTIVITY

Developing Concepts

Investigating Graphs of Exponential Functions

1–3. See margin.

① Graph $y = \frac{1}{3} \cdot 2^x$ and $y = 3 \cdot 2^x$. Compare the graphs with the graph of $y = 2^x$.

② Graph $y = -\frac{1}{5} \cdot 2^x$ and $y = -5 \cdot 2^x$. Compare the graphs with the graph of $y = 2^x$.

③ Describe the effect of a on the graph of $y = a \cdot 2^x$ when a is positive and when a is negative.

In the activity you may have observed the following about the graph of $y = a \cdot 2^x$:

- The graph passes through the point $(0, a)$. That is, the y-intercept is a.
- The x-axis is an asymptote of the graph.
- The domain is all real numbers.
- The range is $y > 0$ if $a > 0$ and $y < 0$ if $a < 0$.

8.1 *Exponential Growth* **465**

1–3. See Additional Answers beginning on page AA1.

LESSON OPENER
APPLICATION
An alternative way to approach Lesson 8.1 is to use the Application Lesson Opener:
- Blackline Master (*Chapter 8 Resource Book,* p. 12)
- Transparency (p. 51)

MEETING INDIVIDUAL NEEDS
- ***Chapter 8 Resource Book***
  Prerequisite Skills Review (p. 5)
  Practice Level A (p. 14)
  Practice Level B (p. 15)
  Practice Level C (p. 16)
  Reteaching with Practice (p. 17)
  Absent Student Catch-Up (p. 19)
  Challenge (p. 21)
- ***Resources in Spanish***
- ***Personal Student Tutor***

NEW-TEACHER SUPPORT
See the Tips for New Teachers on pp. 1–2 of the *Chapter 8 Resource Book* for additional notes about Lesson 8.1.

WARM-UP EXERCISES

Transparency Available

Describe the end behavior of each graph as $x \to \infty$.

1. $f(x) = 3x^3$ $f(x) \to \infty$
2. $f(x) = -3x^3$ $f(x) \to -\infty$
3. $f(x) = \frac{1}{3}x^2$ $f(x) \to \infty$

State the domain and range of each function.

4. $y = x^2$ domain: all real numbers; range: $y \geq 0$
5. $y = \sqrt{x}$ domain: $x \geq 0$; range: $y \geq 0$

EXTRA EXAMPLE 1
Graph the function.
a. $y = \frac{2}{3} \cdot 2^x$.

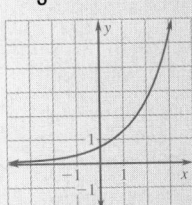

b. $y = -2 \cdot 2^x$

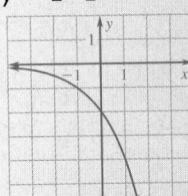

EXTRA EXAMPLE 2
Graph $y = 2 \cdot 3^{x-2} + 1$. State the domain and range.

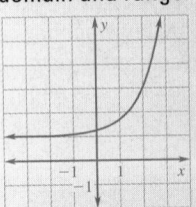

domain: all real numbers;
range: $y > 1$

 CHECKPOINT EXERCISES
For use after Examples 1 and 2:
1. Graph $y = -3^x - 2$. State the domain and the range.

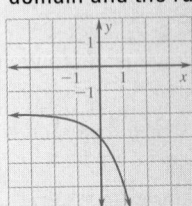

domain: all real numbers;
range: $y < -2$

The characteristics of the graph of $y = a \cdot 2^x$ listed on the previous page are true of the graph of $y = ab^x$. If $a > 0$ and $b > 1$, the function $y = ab^x$ is an **exponential growth function**.

EXAMPLE 1 *Graphing Exponential Functions of the Form $y = ab^x$*

STUDENT HELP
▸ **Look Back**
For help with end behavior of graphs, see p. 331.

Graph the function.

a. $y = \frac{1}{2} \cdot 3^x$

b. $y = -\left(\frac{3}{2}\right)^x$

SOLUTION

a. Plot $\left(0, \frac{1}{2}\right)$ and $\left(1, \frac{3}{2}\right)$. Then, from left to right, draw a curve that begins just above the x-axis, passes through the two points, and moves up to the right.

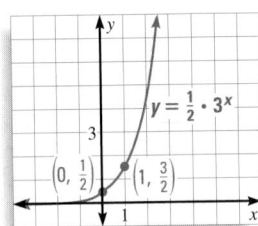

b. Plot $(0, -1)$ and $\left(1, -\frac{3}{2}\right)$. Then, from left to right, draw a curve that begins just below the x-axis, passes through the two points, and moves down to the right.

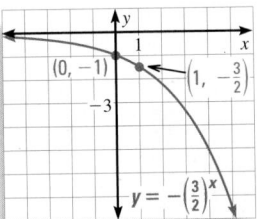

. . . .

To graph a general exponential function,

$$y = ab^{x-h} + k,$$

begin by sketching the graph of $y = ab^x$. Then translate the graph horizontally by h units and vertically by k units.

EXAMPLE 2 *Graphing a General Exponential Function*

Graph $y = 3 \cdot 2^{x-1} - 4$. State the domain and range.

SOLUTION

Begin by lightly sketching the graph of $y = 3 \cdot 2^x$, which passes through $(0, 3)$ and $(1, 6)$. Then translate the graph 1 unit to the right and 4 units down. Notice that the graph passes through $(1, -1)$ and $(2, 2)$. The graph's asymptote is the line $y = -4$. The domain is all real numbers, and the range is $y > -4$.

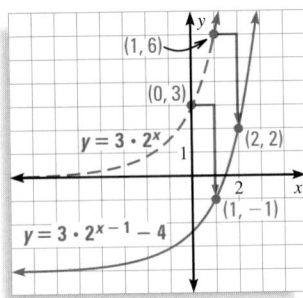

GOAL 2 USING EXPONENTIAL GROWTH MODELS

When a real-life quantity increases by a fixed percent each year (or other time period), the amount y of the quantity after t years can be modeled by this equation:

$$y = a(1 + r)^t$$

In this model, a is the initial amount and r is the percent increase expressed as a decimal. The quantity $1 + r$ is called the **growth factor**.

EXAMPLE 3 *Modeling Exponential Growth*

STUDENT HELP

HOMEWORK HELP
Visit our Web site
www.mcdougallittell.com
for extra examples.

INTERNET HOSTS In January, 1993, there were about 1,313,000 Internet hosts. During the next five years, the number of hosts increased by about 100% per year.
▶ Source: Network Wizards

a. Write a model giving the number h (in millions) of hosts t years after 1993. About how many hosts were there in 1996?

b. Graph the model.

c. Use the graph to estimate the year when there were 30 million hosts.

SOLUTION

a. The initial amount is $a = 1.313$ and the percent increase is $r = 1$. So, the exponential growth model is:

$h = a(1 + r)^t$ \qquad Write exponential growth model.

$= 1.313(1 + 1)^t$ \qquad Substitute for a and r.

$= 1.313 \cdot 2^t$ \qquad Simplify.

Using this model, you can estimate the number of hosts in 1996 ($t = 3$) to be $h = 1.313 \cdot 2^3 \approx 10.5$ million.

b. The graph passes through the points $(0, 1.313)$ and $(1, 2.626)$. It has the t-axis as an asymptote. To make an accurate graph, plot a few other points. Then draw a smooth curve through the points.

c. Using the graph, you can estimate that the number of hosts was 30 million sometime during 1997 ($t \approx 4.5$).

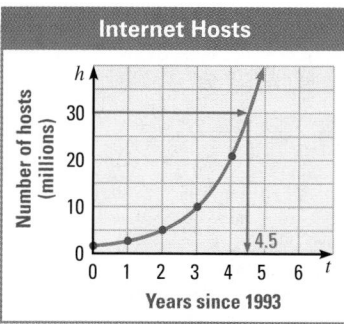

Internet Hosts

. . . .

In Example 3 notice that the annual percent increase was 100%. This translated into a growth factor of 2, which means that the number of Internet hosts doubled each year.

People often confuse percent increase and growth factor, especially when a percent increase is 100% or more. For example, a percent increase of 200% means that a quantity *tripled*, because the growth factor is $1 + 2 = 3$. When you hear or read reports of how a quantity has changed, be sure to pay attention to whether a percent increase or a growth factor is being discussed.

FOCUS ON APPLICATIONS

INTERNET HOSTS
A *host* is a computer that stores information you can access through the Internet. For example, Web sites are stored on host computers.

APPLICATION LINK
www.mcdougallittell.com

8.1 *Exponential Growth* **467**

EXTRA EXAMPLE 3
In 1980 about 2,180,000 U.S. workers worked at home. During the next ten years, the number of workers working at home increased 5% per year.
a. Write a model giving the number w (in millions) of workers working at home t years after 1980. $w = 2.18 \cdot 1.05^t$
b. Graph the model.

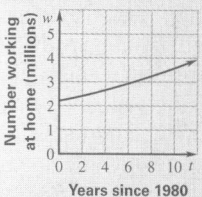

c. Use the graph to estimate the year when there were about 3.22 million workers who worked at home. **1988**

✓ **CHECKPOINT EXERCISES**
For use after Example 3:
1. In 1990 the cost of tuition at a state university was $4300. During the next 8 years, the tuition rose 4% each year.
a. Write a model that gives the tuition y (in dollars) t years after 1990.
$y = 4300 \cdot 1.04^t$
b. Graph the model.

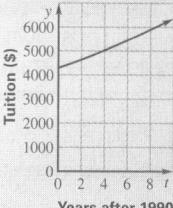

MATHEMATICAL REASONING
A graph that passes the vertical line test is the graph of a function for which each value of x has a unique value of y. The exponential functions in this lesson also have a unique value of x for each value of y. For this reason they are called one-to-one functions. What kind of graphical test will tell you whether or not a function is one-to-one?
If the function passes both a vertical line test and a horizontal line test, then it is a one-to-one function.

467

EXTRA EXAMPLE 4
You deposit $1500 in an account that pays 6% annual interest. Find the balance after 1 year if the interest is compounded
a. annually $1590
b. semiannually $1591.35
c. quarterly $1592.05

✔ **CHECKPOINT EXERCISES**
For use after Example 4:
1. You deposit $2000 to an account that pays 8% annual interest. How much more does the account earn in one year if the interest is compounded monthly rather than annually? about $6.00

CAREER NOTE
EXAMPLE 4 Additional information about financial planners can be found at **www.mcdougallittell.com**.

FOCUS ON VOCABULARY
Why are the *y*-values of an exponential growth function either always greater than or less than the asymptote of the function? An asymptote is a line that the graph of an exponential function approaches but never reaches.

CLOSURE QUESTION
If the population of a town increased by 30% per year over a period of 10 years, by how many times did the population increase in the ten-year period? 13.8 times

DAILY PUZZLER
Rhonda hears a rumor at 8:00 A.M. She immediately tells her two best friends the rumor. One hour later Rhonda's friends have each told two of their friends. This pattern continues each hour, with each friend reporting the rumor to two friends who have not already heard it. By 8:00 P.M. that evening, how many people have heard the rumor?
8191

COMPOUND INTEREST Exponential growth functions are used in real-life situations involving *compound interest*. Compound interest is interest paid on the initial investment, called the *principal*, and on previously earned interest. (Interest paid only on the principal is called *simple interest*.)

Although interest earned is expressed as an *annual* percent, the interest is usually compounded more frequently than once per year. Therefore, the formula $y = a(1 + r)^t$ must be modified for compound interest problems.

COMPOUND INTEREST

Consider an initial principal P deposited in an account that pays interest at an annual rate r (expressed as a decimal), compounded n times per year. The amount A in the account after t years can be modeled by this equation:

$$A = P\left(1 + \frac{r}{n}\right)^{nt}$$

EXAMPLE 4 *Finding the Balance in an Account*

FINANCE You deposit $1000 in an account that pays 8% annual interest. Find the balance after 1 year if the interest is compounded with the given frequency.

a. annually   b. quarterly   c. daily

SOLUTION

a. With interest compounded annually, the balance at the end of 1 year is:

$$A = 1000\left(1 + \frac{0.08}{1}\right)^{1 \cdot 1} \qquad P = 1000, r = 0.08, n = 1, t = 1$$

$$= 1000(1.08)^1 \qquad \text{Simplify.}$$

$$= 1080 \qquad \text{Use a calculator.}$$

▶ The balance at the end of 1 year is $1080.

b. With interest compounded quarterly, the balance at the end of 1 year is:

$$A = 1000\left(1 + \frac{0.08}{4}\right)^{4 \cdot 1} \qquad P = 1000, r = 0.08, n = 4, t = 1$$

$$= 1000(1.02)^4 \qquad \text{Simplify.}$$

$$\approx 1082.43 \qquad \text{Use a calculator.}$$

▶ The balance at the end of 1 year is $1082.43.

c. With interest compounded daily, the balance at the end of 1 year is:

$$A = 1000\left(1 + \frac{0.08}{365}\right)^{365 \cdot 1} \qquad P = 1000, r = 0.08, n = 365, t = 1$$

$$\approx 1000(1.000219)^{365} \qquad \text{Simplify.}$$

$$\approx 1083.28 \qquad \text{Use a calculator.}$$

▶ The balance at the end of 1 year is $1083.28.

FOCUS ON CAREERS

FINANCIAL PLANNER
Financial planners interview clients to determine their assets, liabilities, and financial objectives. They analyze this information and develop an individual financial plan.

CAREER LINK
www.mcdougallittell.com

GUIDED PRACTICE

Vocabulary Check ✓

Concept Check ✓

1. What is an asymptote? *An asymptote is a line that a graph approaches more and more closely.*

2. Given the general exponential function $f(x) = ab^{x-h} + k$, describe the effects of a, h, and k on the graph of the function. *See margin.*

3. For what values of b does $y = b^x$ represent exponential growth? *b > 1*

Skill Check ✓

Graph the function. State the domain and range. *4–9. See margin.*

4. $y = 4^x$

5. $y = 3^{x-1}$

6. $y = 2^{x+2}$

7. $y = 5^x - 3$

8. $y = 5^{x+1} + 2$

9. $y = 2^{x-3} + 1$

2. If $a < 0$, the graph lies below the line $y = k$, and approaches it asymptotically from below. If $a > 0$, the graph lies above the line $y = k$, and approaches it asymptotically from above. The graph of $y = ab^{x-h} + k$ is the same as that of $y = ab^x$ translated horizontally h units and vertically k units.

10. What is the asymptote of the graph of $y = 3 \cdot 4^{x-1} + 2$? What is the value of y when $x = 2$? *y = 2; 14*

11. 🌐 **POPULATION** The population of Winnemucca, Nevada, can be modeled by $P = 6191(1.04)^t$ where t is the number of years since 1990. What was the population in 1990? By what percent did the population increase each year? *6191; 4%*

12. 🌐 **ACCOUNT BALANCE** You deposit $500 in an account that pays 3% annual interest. Find the balance after 2 years if the interest is compounded with the given frequency.

a. annually *$530.45*   **b.** quarterly *$530.80*   **c.** daily *$530.92*

PRACTICE AND APPLICATIONS

STUDENT HELP

▶ **Extra Practice**
to help you master skills is on p. 950.

INVESTIGATING GRAPHS Identify the y-intercept and the asymptote of the graph of the function.

13. $y = 5^x$ *1; the x-axis*

14. $y = -2 \cdot 4^x$ *−2; the x-axis*

15. $y = 4 \cdot 2^x$ *4; the x-axis*

16. $y = 2^x - 1$ *0; the line y = −1*

17. $y = 3 \cdot 2^x$ *3/2; the x-axis*

18. $y = 2 \cdot 3^{x-4}$ *2/81; the x-axis*

MATCHING GRAPHS Match the function with its graph.

19. $y = 2 \cdot 5^x$ *C*

20. $y = 3 \cdot 4^x$ *E*

21. $y = -2 \cdot 5^x$ *B*

22. $y = \frac{1}{3} \cdot 4^x$ *A*

23. $y = 3^{x-2}$ *F*

24. $y = 3^x - 2$ *D*

STUDENT HELP

▶ **HOMEWORK HELP**
Example 1: Exs. 13–15, 19–22, 25–33
Example 2: Exs. 16–18, 23, 24, 34–42
Example 3: Exs. 43–54, 56, 58, 66
Example 4: Exs. 55, 57, 59–65, 67

A.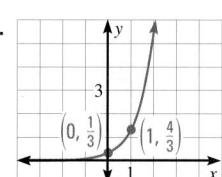

B.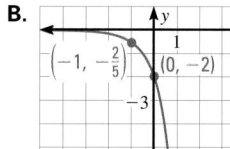

C.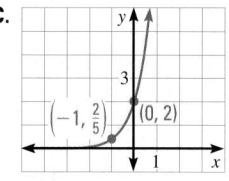

D.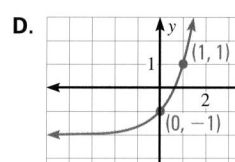

E.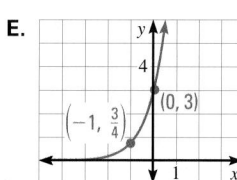

F.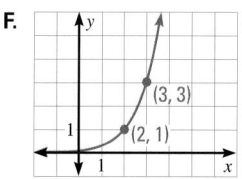

---

3 APPLY

○ **ASSIGNMENT GUIDE**

BASIC
Day 1: pp. 469–472 Exs. 14–30 even, 34–40 even, 43–45, 55, 68–69, 71–85 odd, 100

AVERAGE
Day 1: pp. 469–472 Exs. 14–40 even, 43–48, 55, 66–70, 71–89 odd, 100

ADVANCED
Day 1: pp. 469–472 Exs. 14–40 even, 43–48, 55, 62–70, 71–99 odd, 100

BLOCK SCHEDULE WITH 8.2
pp. 469–472 Exs. 14–40 even, 43–48, 55, 66–70, 71–89 odd, 100

EXERCISE LEVELS
Level A: *Easier*
13–16, 19–30, 34–39
Level B: *More Difficult*
17–18, 31–33, 40–69
Level C: *Most Difficult*
70

✔ **HOMEWORK CHECK**
To quickly check student understanding of key concepts, go over the following exercises: Exs. 16, 22, 28, 36, 43. See also the Daily Homework Quiz:
• Blackline Master (*Chapter 8 Resource Book*, p. 24)
• 📄 Transparency (p. 60)

! **COMMON ERROR**
EXERCISES 23 AND 24 It is easy for students to confuse these graphs. Point out to students that $y = 3^{x-2}$ will have the same range as $y = 3^x$, but $y = 3^x - 2$ will not.

4–9. See Additional Answers beginning on page AA1.

470

! COMMON ERROR
EXERCISE 26 Students may need reminding that $y = -2^x$ is not equivalent to $y = (-2)^x$.

MATHEMATICAL REASONING
EXERCISES 43–45 Do you think that gas will continue to be consumed at this rate? Explain.
Sample answer: Natural gas cannot continue to be consumed at this rate due to a finite supply.

APPLICATION NOTE
EXERCISES 49–51 A projection model predicted that by the year 2000, wind turbines would produce an amount of electricity equivalent to what could be generated by three to four power plants. Ask students to research whether this projection model was accurate.

25.

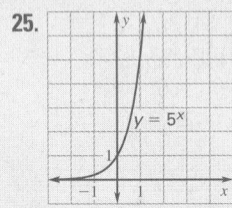

26.

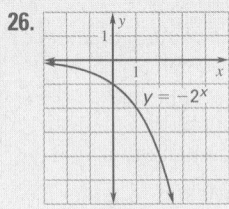

27.

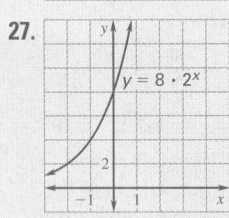

28.

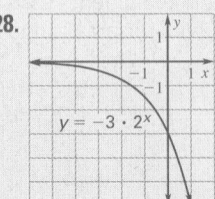

29–32. See next page.
33–42, 44, 47, 50, 53.
   See Additional Answers beginning on page AA1.

34. domain: all real numbers; range: $y < 0$
35. domain: all real numbers; range: $y > 0$
36. domain: all real numbers; range: $y > 0$
37. domain: all real numbers; range: $y > 0$
38. domain: all real numbers; range: $y > 1$
39. domain: all real numbers; range: $y > 3$
40. domain: all real numbers; range: $y < -2$
41. domain: all real numbers; range: $y > 1$
42. domain: all real numbers; range: $y > -3$

51. $t \approx 5.98$; very near the end of 1985 (reading from graphs, students are likely to say 1986)

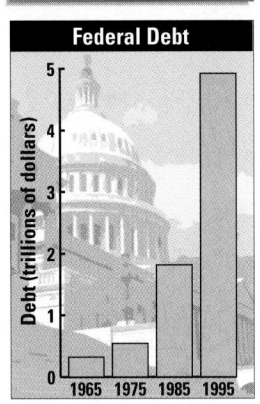

FOCUS ON APPLICATIONS

Federal Debt

FEDERAL DEBT When the government has an annual deficit it must borrow money. The accumulation of this borrowing is called the Federal debt.

GRAPHING FUNCTIONS Graph the function. 25–33. See margin.

25. $y = 5^x$
26. $y = -2^x$
27. $y = 8 \cdot 2^x$
28. $y = -3 \cdot 2^x$
29. $y = -2 \cdot 5^x$
30. $y = -(2.5)^x$
31. $y = 6\left(\dfrac{5}{4}\right)^x$
32. $y = -\dfrac{2}{3} \cdot 3^x$
33. $y = -\dfrac{1}{5}(1.5)^x$

34–42. See margin for graphs.
GRAPHING FUNCTIONS Graph the function. State the domain and range.

34. $y = -2 \cdot 3^{x+2}$
35. $y = 4 \cdot 5^{x-1}$
36. $y = 7 \cdot 3^{x-2}$
37. $y = 3 \cdot 4^{x-1}$
38. $y = 3^{x+1} + 1$
39. $y = 2^{x-3} + 3$
40. $y = -3 \cdot 6^{x+2} - 2$
41. $y = 4 \cdot 2^{x-3} + 1$
42. $y = 8 \cdot 2^{x-3} - 3$

🌐 **NATURAL GAS** In Exercises 43–45, use the following information.
The amount g (in trillions of cubic feet) of natural gas consumed in the United States from 1940 to 1970 can be modeled by

$$g = 2.91(1.07)^t$$

where t is the number of years since 1940. ▶ Source: *Wind Energy Comes of Age*

43. Identify the initial amount, the growth factor, and the annual percent increase.
   2.91 trillion ft³; 1.07; 7%

44. Graph the function. **See margin.**

45. Estimate the natural gas consumption in 1955. **8.03 trillion ft³**

🌐 **COMPUTER CHIPS** In Exercises 46–48, use the following information.
From 1971 to 1995, the average number n of transistors on a computer chip can be modeled by

$$n = 2300(1.59)^t$$

where t is the number of years since 1971.

46. Identify the initial amount, the growth factor, and the annual percent increase.
   2300; 1.59; 59%

47. Graph the function. **See margin.**

48. Estimate the number of transistors on a computer chip in 1998. **about 630 million transistors**

🌐 **WIND ENERGY** In Exercises 49–51, use the following information.
In 1980 wind turbines in Europe generated about 5 gigawatt-hours of energy. Over the next 15 years, the amount of energy increased by about 59% per year.

49. Write a model giving the amount E (in gigawatt-hours) of energy t years after 1980. About how much wind energy was generated in 1984? **$E = 5(1.59)^t$; about 32 gigawatt-hours**

50. Graph the model. **See margin.**

51. Estimate the year when 80 gigawatt-hours of energy were generated. **See margin.**

🌐 **FEDERAL DEBT** In Exercises 52–54, use the following information.
In 1965 the federal debt of the United States was $322.3 billion. During the next 30 years, the debt increased by about 10.2% each year. ▶ Source: U.S. Bureau of the Census

52. Write a model giving the amount D (in billions of dollars) of debt t years after 1965. About how much was the federal debt in 1980? **$D = 322.3(1.102)^t$; about $1.384 trillion**

53. Graph the model. **See margin.**

54. Estimate the year when the federal debt was $2,120 billion. **during 1984**

STUDENT HELP

KEYSTROKE HELP
Visit our Web site www.mcdougallittell.com to see keystrokes for several models of calculators.

55. **EARNING INTEREST** You deposit $2500 in a bank that pays 4% interest compounded annually. Use the process below and a graphing calculator to determine the balance of your account each year.

   a. Enter the initial deposit, 2500, into the calculator. Then enter the formula ANS + ANS × 0.04 to find the balance after one year.   **$2600**

   b. What is the balance after five years? (*Hint:* The balance after each year will be displayed each time you press the ⬛ ENTER key.)   **$3041.63**

   c. How would you enter the formula in part (a) if the interest is compounded quarterly? What do you have to do to find the balance after one year?
                ANS + ANS × 0.01; push "ENTER" four times.
   d. Find the balance after 5 years if the interest is compounded quarterly. Compare this result with your answer to part (b).   **$3050.48; this is $8.85 more.**

WRITING MODELS In Exercises 56–58, write an exponential growth model that describes the situation.

56. **COIN COLLECTING** You buy a commemorative coin for $110. Each year t, the value V of the coin increases by 4%. $V = 110(1.04)^t$

57. **SAVINGS ACCOUNT** You deposit $400 in an account that pays 2% annual interest compounded quarterly. $A = 400(1.005)^{4t}$ where t is the number of years

58. **ANTIQUES** You purchase an antique table for $525. Each year t, the value V of the table increases by 5%. $V = 525(1.05)^t$

ACCOUNT BALANCE In Exercises 59–61, use the following information.
You deposit $1600 in a bank account. Find the balance after 3 years for each of the following situations.

59. The account pays 2.5% annual interest compounded monthly.   **$1724.48**

60. The account pays 1.75% annual interest compounded quarterly.   **$1686.05**

61. The account pays 4% annual interest compounded yearly.   **$1799.78**

DEPOSITING FUNDS In Exercises 62–64, use the following information.
You want to have $2500 after 2 years. Find the amount you should deposit for each of the situations described below.

62. The account pays 2.25% annual interest compounded monthly.   **$2390.09**

63. The account pays 2% annual interest compounded quarterly.   **$2402.21**

64. The account pays 5% annual interest compounded yearly.   **$2267.57**

65. No; Michelle will have more. All $800 of her money will start to earn interest right away, while only part of Juan's $800 will earn interest each year.

65. **CRITICAL THINKING** Juan and Michelle each have $800. Juan plans to invest $200 for each of the next four years, while Michelle plans to invest all $800 now. Both accounts pay 3% annual interest compounded monthly. Will they have the same amount of money after four years? If not, explain why.

66. **LAND VALUE** You have inherited land that was purchased for $30,000 in 1960. The value V of the land increased by approximately 5% per year.

   a. Write a model for the value of the land t years after 1960. $V = 30,000(1.05)^t$

   b. What is the approximate value of the land in the year 2010?   **about $344,000**

67. No; $8000(1.06)^t \neq 4000(1.05)^t + 4000(1.07)^t$. In fact, the split accounts scheme will outperform the single account after the first year.

67. **LOGICAL REASONING** Is investing $4000 at 5% annual interest and $4000 at 7% annual interest equivalent to investing $8000 (the total of the two principals) at 6% annual interest (the average of the two interest rates)? Explain.

STUDENT HELP NOTES

➤ **Keystroke Help** Keystrokes for several models of calculators are available in **blackline** format in the *Chapter 8 Resource Book,* p. 13 and at **www.mcdougallittell.com.**

29.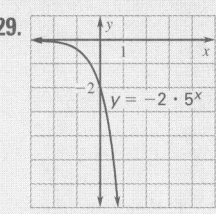
   $y = -2 \cdot 5^x$

30.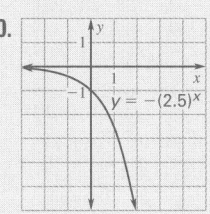
   $y = -(2.5)^x$

31.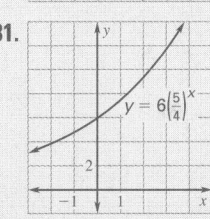
   $y = 6\left(\frac{5}{4}\right)^x$

32.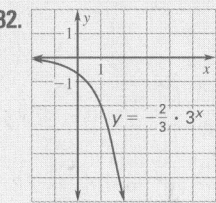
   $y = -\frac{2}{3} \cdot 3^x$

ADDITIONAL PRACTICE AND RETEACHING

For Lesson 8.1:
- Practice Levels A, B, and C (*Chapter 8 Resource Book,* p. 14)
- Reteaching with Practice (*Chapter 8 Resource Book,* p. 17)
- ⬛ See Lesson 8.1 of the *Personal Student Tutor*

For more Mixed Review:
- ⬛ Search the *Test and Practice Generator* for key words or specific lessons.

DAILY HOMEWORK QUIZ

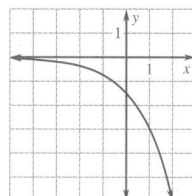 *Transparency Available*

1. Identify the y-intercept and the asymptote of the graph of $y = 0.5 \cdot 3^{x+2}$. **4.5; the x-axis**

2. Graph $y = -1.5 \cdot 2^x$.

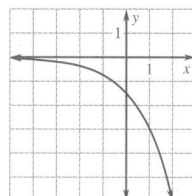

3. Graph $y = 2^{x+1} - 3$. State the domain and range.

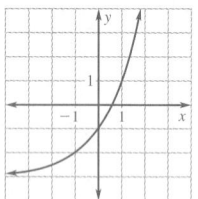

domain: all real numbers;
range: $y > -3$

EXTRA CHALLENGE NOTE

Challenge problems for Lesson 8.1 are available in **blackline** format in the *Chapter 8 Resource Book*, p. 21 and at **www.mcdougallittell.com.**

ADDITIONAL TEST PREPARATION

1. WRITING Describe the relationship between the graph of $y = 5^x$ and the graph of $y = 5^{x+2}$.
When each point on the graph of $y = 5^x$ is shifted 2 units left, the resulting graph is the graph of $y = 5^{x+2}$.

2. OPEN ENDED Describe a function whose graph has $y = 0$ as an asymptote and for which $f(x) \rightarrow -\infty$ as $x \rightarrow \infty$.
Exponential functions of the form $y = a \cdot b^x$ where $a < 0$, satisfy the conditions.

Test  **Preparation**

68. MULTIPLE CHOICE The student enrollment E of a high school was 1240 in 1990 and increased by 15% per year until 1996. Which exponential growth model shows the school's student enrollment in terms of t, the number of years since 1990? **B**

    Ⓐ $E = 15(1240)^t$ **Ⓑ** $E = 1240(1.15)^t$ **Ⓒ** $E = 1240(15)^t$

    Ⓓ $E = 0.15(1240)^t$ **Ⓔ** $E = 1.15(1240)^t$

69. MULTIPLE CHOICE Which function is graphed below? **D**

    Ⓐ $f(x) = -3^{x+1} + 6$

    Ⓑ $f(x) = -2 \cdot 3^{x+1} + 6$

    Ⓒ $f(x) = 2 \cdot 3^{x+1} + 6$

    Ⓓ $f(x) = 2 \cdot 3^{x-1} + 6$

    Ⓔ $f(x) = 3^{x+1} + 6$

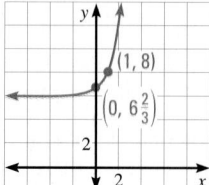

★ **Challenge**

70. IRRATIONAL EXPONENTS Use a calculator to evaluate the following powers. Round the results to five decimal places.

$$3^{14/10}, \ 3^{141/100}, \ 3^{1,414/1,000}, \ 3^{14,142/10,000}, \ 3^{141,421/100,000}, \ 3^{1,414,213/1,000,000}$$

EXTRA CHALLENGE
www.mcdougallittell.com

Each of these powers has a rational exponent. Explain how you can use these powers to define $3^{\sqrt{2}}$, which has an irrational exponent.
4.65554; 4.70697; 4.72770; 4.72873; 4.72879; 4.72880; These successive powers approach a limit, which can be defined to be $3^{\sqrt{2}}$.

MIXED REVIEW

91. $4x^2 + 6x - 11$; all real numbers

92. $-4x^2 + 6x - 11$; all real numbers

93. $24x^3 - 44x^2$; all real numbers

94. $4x^2 - 6x + 11$; all real numbers

95. $24x^2 - 11$; all real numbers

96. $144x^2 - 528x + 484$; all real numbers

97. $\dfrac{6x - 11}{4x^2}$; all non-zero real numbers

98. $\dfrac{4x^2}{6x - 11}$; all real numbers except $\dfrac{11}{6}$

99. $36x - 77$; all real numbers

EVALUATING POWERS Evaluate the expression. **(Review 1.2 for 8.2)**

71. $\left(\dfrac{1}{2}\right)^3$ $\dfrac{1}{8}$ **72.** $\left(\dfrac{3}{7}\right)^3$ $\dfrac{27}{343}$ **73.** $\left(\dfrac{1}{2}\right)^5$ $\dfrac{1}{32}$ **74.** $\left(\dfrac{5}{8}\right)^4$ $\dfrac{625}{4096}$

75. $\left(\dfrac{7}{12}\right)^3$ $\dfrac{343}{1728}$ **76.** $\left(\dfrac{2}{3}\right)^4$ $\dfrac{16}{81}$ **77.** $\left(\dfrac{4}{5}\right)^2$ $\dfrac{16}{25}$ **78.** $\left(\dfrac{3}{10}\right)^5$ $\dfrac{243}{100,000}$

EVALUATING EXPRESSIONS Evaluate the expression using a calculator. Round the result to two decimal places when appropriate. **(Review 7.1)**

79. $8^{3/8}$ 2.18   **80.** $15,625^{1/6}$ 5   **81.** $-243^{1/5}$ -3 **82.** $1024^{1/5}$ 4

83. $10^{1/2}$ 3.16   **84.** $106^{1/3}$ 4.73   **85.** $\sqrt[4]{81}$ 3   **86.** $\sqrt[7]{100}$ 1.93

87. $\sqrt[3]{28}$ 3.04   **88.** $\sqrt[4]{120}$ 3.31   **89.** $\sqrt[4]{9}$ 1.73   **90.** $\sqrt[6]{180}$ 2.38

OPERATIONS WITH FUNCTIONS Let $f(x) = 6x - 11$ and $g(x) = 4x^2$. **Perform the indicated operation and state the domain. (Review 7.3)** 91–99. See margin.

91. $f(x) + g(x)$ **92.** $f(x) - g(x)$ **93.** $f(x) \cdot g(x)$

94. $g(x) - f(x)$ **95.** $f(g(x))$ **96.** $g(f(x))$

97. $\dfrac{f(x)}{g(x)}$ **98.** $\dfrac{g(x)}{f(x)}$ **99.** $f(f(x))$

100. FENCING You want to build a rectangular pen for your dog using 40 feet of fencing. The area of the pen should be 90 square feet. What should the dimensions of the pen be? **(Review 5.2)**
$10 - \sqrt{10}$ by $10 + \sqrt{10}$, or about 6.84 ft by 13.16 ft

▶ ACTIVITY 8.2

Developing Concepts

Exponential Growth and Decay

SET UP
Work with a partner.

MATERIALS
• paper
• pencil
• graph paper

▶ **QUESTION** What relationships exist between exponential growth and exponential decay when a piece of paper is folded repeatedly?

▶ **EXPLORING THE CONCEPT**

① Fold a rectangular piece of paper in half. The fold divides the paper into two regions, each of which has half the area of the paper.

② Fold the paper in half again. Into how many regions has the original piece of paper been folded? What fraction of the paper's area does each region have? $4; \frac{1}{4}$

③ Continue to fold the paper until it is no longer possible to make another fold. After each fold, record in a table like the one shown the fold number, the number of regions into which the paper has been folded, and the fraction of the paper's area that each region has.

Fold number	0	1	2	3	4	5	
Number of regions	1	2	?	?	?	?	4; 8; 16; 32;
Fractional area of each region	1	$\frac{1}{2}$	?	?	?	?	$\frac{1}{4}, \frac{1}{8}, \frac{1}{16}, \frac{1}{32}$

④ Make two scatter plots of the data in the table. The first scatter plot will have ordered pairs of the form (*fold number, number of regions*) and the second will have ordered pairs of the form (*fold number, fractional area of each region*).   **See margin.**

▶ **DRAWING CONCLUSIONS**

1. The first scatter plot is an example of exponential growth. Write an equation for the graph. $y = 2^x$

2. Use the equation from Exercise 1 to determine the number of regions there would be after 8 folds.   256

3. The second scatter plot is an example of exponential decay. Write an equation for the graph. $y = \left(\frac{1}{2}\right)^x$

4. Use the equation from Exercise 3 to determine the fractional area of each region after 8 folds. $\frac{1}{256}$

5. Multiply the exponential expressions from Exercise 2 and Exercise 4. Explain why the product should be 1.

5. $(2^x)\left(\frac{1}{2}\right)^x = \left(\frac{2}{2}\right)^x = 1$;

The original area is one, and at each stage the number of regions times the area of each region must continue to equal one whole.

8.2 *Concept Activity*   **473**

---

1 Planning the Activity

PURPOSE
To model the relationship between exponential growth and exponential decay.

MATERIALS
• Activity Support Master (*Chapter 8 Resource Book,* p. 25)

PACING
• Exploring the Concept — 10 min
• Drawing Conclusions — 15 min

▶ **LINK TO LESSON**
Students will see a complete graph of the exponential decay function they created in this Activity on p. 474.

2 Managing the Activity

COOPERATIVE LEARNING
To save time, have each partner graph one of the scatter plots in Step 4 and write the equation for the graph. Students can then check each other's work.

CLASSROOM MANAGEMENT
Suggest that students graph the equations they wrote in Exercises 1 and 3 to confirm that these model the data points.

3 Closing the Activity

★ **KEY DISCOVERY**
In an exponential decay function modeled by $y = ab^x$ the value of b is between 0 and 1.

ACTIVITY ASSESSMENT
In the exponential decay function that you graphed in Step 4, as x increases, what number do the values of $f(x)$ approach?   zero

Step 4. See Additional Answers beginning on page AA1.

473

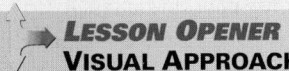

LESSON OPENER
VISUAL APPROACH
An alternative way to approach
Lesson 8.2 is to use the Visual
Approach Lesson Opener:

• Blackline Master (*Chapter 8
  Resource Book,* p. 26)
• Transparency (p. 52)

MEETING INDIVIDUAL NEEDS
• *Chapter 8 Resource Book*
  Prerequisite Skills Review (p. 5)
  Practice Level A (p. 29)
  Practice Level B (p. 30)
  Practice Level C (p. 31)
  Reteaching with Practice (p. 32)
  Absent Student Catch-Up (p. 34)
  Challenge (p. 36)
• *Resources in Spanish*
• Personal Student Tutor

NEW-TEACHER SUPPORT
See the Tips for New Teachers on
pp. 1–2 of the *Chapter 8 Resource
Book* for additional notes about
Lesson 8.2.

WARM-UP EXERCISES
Transparency Available
Identify the value of *b* in each
exponential function, $f(x) = ab^x$.

1. $f(x) = 3 \cdot \left(\frac{1}{2}\right)^x$ $\frac{1}{2}$

2. $f(x) = 3.5^x$ 3.5

3. $f(x) = 5 \cdot (-2)^{x+1}$ -2

State the domain and range of
each function.

4. $y = 5 \cdot 3^x$ domain: all real
numbers; range: $y > 0$

5. $y = -\frac{1}{4} \cdot 2^x$ domain: all real
numbers; range: $y < 0$

8.2

What you should learn

GOAL 1 Graph exponential
decay functions.

GOAL 2 Use exponential
decay functions to model
real-life situations, such as
the decline of record sales in
Exs. 47–49.

Why you should learn it

▼ To solve **real-life**
problems, such as finding the
depreciated value of a car in
Example 4.

Exponential Decay

GOAL 1 GRAPHING EXPONENTIAL DECAY FUNCTIONS

In Lesson 8.1 you studied exponential growth functions. In this lesson you will study
exponential decay functions, which have the form $f(x) = ab^x$ where $a > 0$ and
$0 < b < 1$.

EXAMPLE 1 *Recognizing Exponential Growth and Decay*

State whether $f(x)$ is an exponential growth or exponential decay function.

a. $f(x) = 5\left(\frac{2}{3}\right)^x$ **b.** $f(x) = 8\left(\frac{3}{2}\right)^x$ **c.** $f(x) = 10(3)^{-x}$

SOLUTION

a. Because $0 < b < 1$, f is an exponential decay function.

b. Because $b > 1$, f is an exponential growth function.

c. Rewrite the function as $f(x) = 10\left(\frac{1}{3}\right)^x$. Because $0 < b < 1$, f is an exponential
decay function.

· · · · · · · · · ·

To see the basic shape of the graph of an exponential decay function, you can make a
table of values and plot points, as shown below.

x	$f(x) = \left(\frac{1}{2}\right)^x$
-3	$\left(\frac{1}{2}\right)^{-3} = 8$
-2	$\left(\frac{1}{2}\right)^{-2} = 4$
-1	$\left(\frac{1}{2}\right)^{-1} = 2$
0	$\left(\frac{1}{2}\right)^{0} = 1$
1	$\left(\frac{1}{2}\right)^{1} = \frac{1}{2}$
2	$\left(\frac{1}{2}\right)^{2} = \frac{1}{4}$
3	$\left(\frac{1}{2}\right)^{3} = \frac{1}{8}$

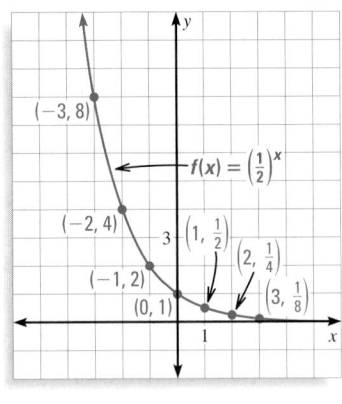

Notice the end behavior of the graph. As $x \to -\infty$, $f(x) \to +\infty$, which means that the
graph moves up to the left. As $x \to +\infty$, $f(x) \to 0$, which means that the graph has
the line $y = 0$ as an asymptote.

Recall that in general the graph of an exponential function $y = ab^x$ passes through the point $(0, a)$ and has the x-axis as an asymptote. The domain is all real numbers, and the range is $y > 0$ if $a > 0$ and $y < 0$ if $a < 0$.

EXAMPLE 2 *Graphing Exponential Functions of the Form* $y = ab^x$

Graph the function.

a. $y = 3\left(\frac{1}{4}\right)^x$

b. $y = -5\left(\frac{2}{3}\right)^x$

SOLUTION

a. Plot $(0, 3)$ and $\left(1, \frac{3}{4}\right)$.

Then, from *right* to *left*, draw a curve that begins just above the x-axis, passes through the two points, and moves up to the left.

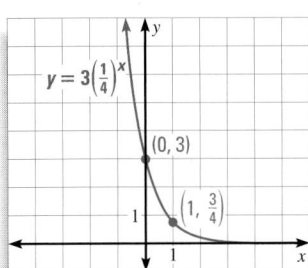

b. Plot $(0, -5)$ and $\left(1, -\frac{10}{3}\right)$.

Then, from *right* to *left*, draw a curve that begins just below the x-axis, passes through the two points, and moves down to the left.

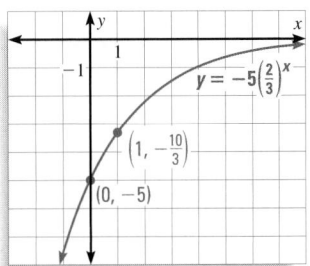

· · · · · · · · · ·

Remember that to graph a general exponential function, $y = ab^{x-h} + k$, begin by sketching the graph of $y = ab^x$. Then translate the graph horizontally by h units and vertically by k units.

EXAMPLE 3 *Graphing a General Exponential Function*

Graph $y = -3\left(\frac{1}{2}\right)^{x+2} + 1$. State the domain and range.

SOLUTION

Begin by lightly sketching the graph of $y = -3\left(\frac{1}{2}\right)^x$, which passes through $(0, -3)$ and $\left(1, -\frac{3}{2}\right)$. Then translate the graph 2 units to the left and 1 unit up. Notice that the graph passes through $(-2, -2)$ and $\left(-1, -\frac{1}{2}\right)$. The graph's asymptote is the line $y = 1$. The domain is all real numbers, and the range is $y < 1$.

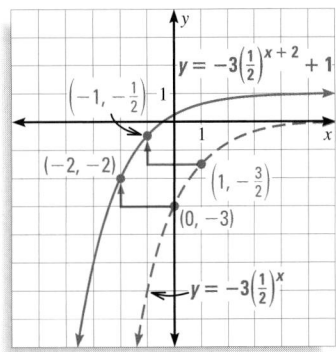

8.2 *Exponential Decay* **475**

2 TEACH

EXTRA EXAMPLE 1
State whether $f(x)$ is an exponential growth or exponential decay function.
a. $f(x) = \frac{1}{3}(2)^{-x}$ exp. decay
b. $f(x) = 4\left(\frac{5}{8}\right)^x$ exp. decay

EXTRA EXAMPLE 2
Graph the function.
a. $y = -2\left(\frac{1}{3}\right)^x$
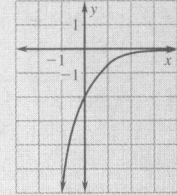
b. $y = 4\left(\frac{2}{5}\right)^x$
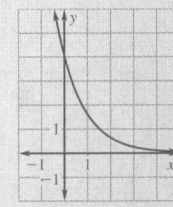

EXTRA EXAMPLE 3
Graph $y = 5\left(\frac{1}{8}\right)^{x+1} - 2$. State the domain and range. domain: all real numbers; range: $y > -2$
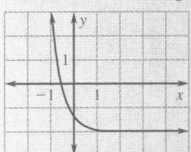

CHECKPOINT EXERCISES
For use after Examples 1–3:
1. Graph $y = 2\left(\frac{4}{5}\right)^{x-2} + 3$. State whether the function is an exponential growth or decay function. State the domain and the range of the function. decay; domain: all real numbers; range: $y > 3$
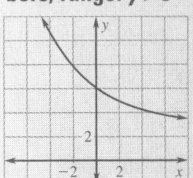

EXTRA EXAMPLE 4

There are 40,000 homes in your city. Each year 10% of the homes are expected to disconnect from septic systems and connect to the sewer system.

a. Write an exponential decay model for the number of homes that still use septic systems. Use the model to estimate the number of homes using septic systems after 5 years.
$y = 40,000(0.9)^t$, where t is the number of years since homes have been able to connect to the sewer system; about 23,620 homes

b. Graph the model and estimate when about 17,200 homes will still not be connected to the sewer system. after about 8 yr

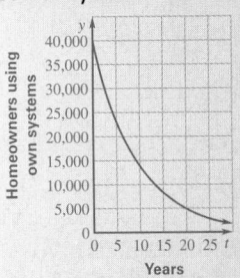

☑ **CHECKPOINT EXERCISES**

For use after Example 4:

1. A new car costs $23,000. The value decreases by 15% each year. Write an exponential decay model for the car's value. Use the model to estimate the value after 3 years.
$y = 23,000(0.85)^t$, where t is the number of years after buying the car; about $14,125

2. Graph the model. Predict when the value will be half of the original value.
after about 4.3 yr

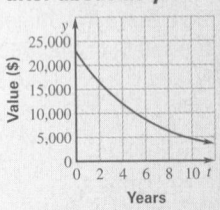

CLOSURE QUESTION
Describe the behavior of y as $x \to \infty$ for the graph of $y = 3(0.25)^{x+1} + 2$.

GOAL 2 **USING EXPONENTIAL DECAY MODELS**

When a real-life quantity decreases by a fixed percent each year (or other time period), the amount y of the quantity after t years can be modeled by the equation

$$y = a(1 - r)^t$$

where a is the initial amount and r is the percent decrease expressed as a decimal. The quantity $1 - r$ is called the **decay factor**.

Automobiles

EXAMPLE 4 *Modeling Exponential Decay*

You buy a new car for $24,000. The value y of the car decreases by 16% each year.

a. Write an exponential decay model for the value of the car. Use the model to estimate the value after 2 years.

b. Graph the model.

c. Use the graph to estimate when the car will have a value of $12,000.

STUDENT HELP

HOMEWORK HELP
Visit our Web site
www.mcdougallittell.com
for extra examples.

SOLUTION

a. Let t be the number of years since you bought the car. The exponential decay model is:

$y = a(1 - r)^t$	Write exponential decay model.
$= 24,000(1 - 0.16)^t$	Substitute for a and r.
$= 24,000(0.84)^t$	Simplify.

When $t = 2$, the value is
$y = 24,000(0.84)^2 \approx \$16,934$.

b. The graph of the model is shown at the right. Notice that it passes through the points $(0, 24,000)$ and $(1, 20,160)$. The asymptote of the graph is the line $y = 0$.

c. Using the graph, you can see that the value of the car will drop to $12,000 after about 4 years.

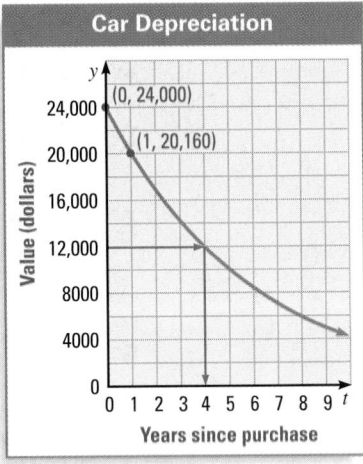

In Example 4 notice that the percent decrease, 16%, tells you how much value the car *loses* from one year to the next. The decay factor, 0.84, tells you what fraction of the car's value *remains* from one year to the next. The closer the percent decrease for some quantity is to 0%, the more the quantity is conserved or retained over time. The closer the percent decrease is to 100%, the more the quantity is used or lost over time.

Closure Question *Sample answer:*
The value of y decreases, approaching an asymptote of $y = 2$.

GUIDED PRACTICE

Vocabulary Check ✓

1. In the exponential decay model $y = 1500(0.65)^t$, identify the initial amount, the decay factor, and the percent decrease. **1500; 0.65; 35%**

Concept Check ✓

2. What is the asymptote of the graph of the function $y = 2\left(\frac{1}{5}\right)^{x-2} + 3$? **$y = 3$**

4. domain: all real numbers; range: $y < 0$

3. For what values of b does $y = b^x$ represent exponential decay? **$0 < b < 1$**

Skill Check ✓

Graph the function. State the domain and range. 4–9. See margin for graphs.

5. domain: all real numbers; range: $y > 0$

4. $y = -(0.5)^x$ **5.** $y = 2\left(\frac{1}{3}\right)^x$ **6.** $y = 4\left(\frac{2}{3}\right)^x$

6. domain: all real numbers; range: $y > 0$

7. $y = -5\left(\frac{2}{3}\right)^{x-2}$ **8.** $y = -4(0.25)^{x+1}$ **9.** $y = 5\left(\frac{1}{2}\right)^x + 2$

7. domain: all real numbers; range: $y < 0$

10. 🌐 **RADIOACTIVE DECAY** The amount y (in grams) of a sample of iodine-131 after t days is given by $y = 50(0.92)^t$.

8. domain: all real numbers; range: $y < 0$

a. Identify the initial amount of the substance. **50 g**

9. domain: all real numbers; range: $y > 2$

b. What percent of the substance decays each day? **8%**

PRACTICE AND APPLICATIONS

STUDENT HELP

▸ **Extra Practice**
to help you master skills is on p. 950.

11. exponential decay

12. exponential growth

13. exponential decay

14. exponential growth

15. exponential growth

16. exponential growth

17. exponential decay

18. exponential growth

IDENTIFYING FUNCTIONS Tell whether the function represents *exponential growth* or *exponential decay*. 11–18. See margin.

11. $f(x) = 4\left(\frac{3}{8}\right)^x$ **12.** $f(x) = 10 \cdot 3^x$ **13.** $f(x) = 8 \cdot 7^{-x}$ **14.** $f(x) = 8 \cdot 7^x$

15. $f(x) = 5\left(\frac{1}{8}\right)^{-x}$ **16.** $f(x) = 3\left(\frac{4}{3}\right)^x$ **17.** $f(x) = 8\left(\frac{2}{3}\right)^x$ **18.** $f(x) = 5(0.25)^{-x}$

MATCHING GRAPHS Match the function with its graph.

19. $y = (0.25)^x$ **F**   **20.** $y = -3^{x-1} + 3$ **E**   **21.** $y = -\left(\frac{1}{3}\right)^{x-1} + 3$ **D**

22. $y = \left(\frac{1}{2}\right)^{x-1}$ **B**   **23.** $y = -(0.25)^x$ **C**   **24.** $y = (0.5)^x - 1$ **A**

A. 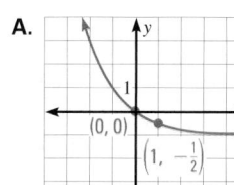   **B.** 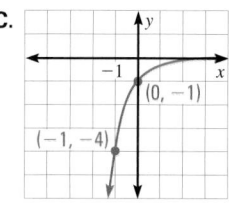   **C.**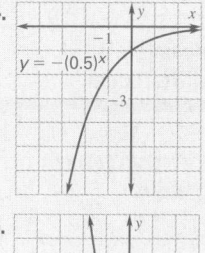

STUDENT HELP

▸ **HOMEWORK HELP**
Example 1: Exs. 11–18
Example 2: Exs. 19, 23, 25–33
Example 3: Exs. 20–22, 24, 34–42
Example 4: Exs. 43–56

D. 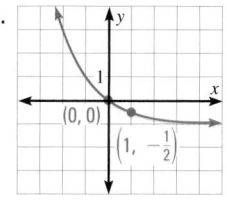   **E.** 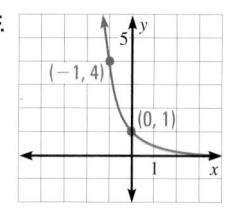   **F.**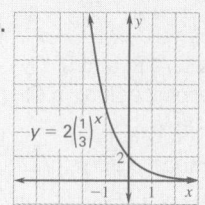

3 APPLY

ASSIGNMENT GUIDE

BASIC
Day 1: pp. 477–479 Exs. 12–40 even, 44, 50–52, 57–67 odd

AVERAGE
Day 1: pp. 477–479 Exs. 12–40 even, 44, 47–52, 57–67 odd

ADVANCED
Day 1: pp. 477–479 Exs. 12–40 even, 44, 47–58, 59–67 odd

BLOCK SCHEDULE WITH 8.1
pp. 477–479 Exs. 12–40 even, 44, 47–52, 57, 59–67 odd

EXERCISE LEVELS
Level A: *Easier*
11–14, 19–33, 43–45
Level B: *More Difficult*
15–18, 34–42, 46–57
Level C: *Most Difficult*
58

✓ **HOMEWORK CHECK**
To quickly check student understanding of key concepts, go over the following exercises:
Exs. 12, 16, 20, 28, 34, 44. See also the Daily Homework Quiz:

• Blackline Master (*Chapter 8 Resource Book*, p. 39)
• 📄 Transparency (p. 61)

4.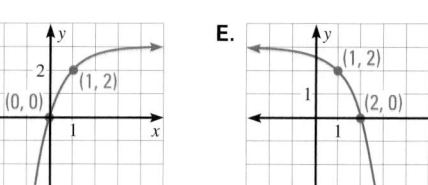

5.

6–9. See Additional Answers beginning on page AA1.

EXERCISES 25–42 In an exponential decay function, the ratio of y-values of successive points is constant. The slopes between these points reflect this decay. For $f(x) = 0.5^x$, $f(-3) = 8$, $f(-2) = 4$, and $f(-1) = 2$. The slope between $(-3, 8)$ and $(-2, 4)$ is -4. The slope between $(-2, 4)$ and $(-1, 2)$ is $-4(0.5) = -2$. Suppose $A(-4, a)$, $B(-3, b)$, $C(-2, c)$, and $D(-1, d)$ lie on the graph of an exponential decay function. If the slope between A and B is x, what is the slope between B and C?

between C and D? $\dfrac{bx}{a}$; $\dfrac{b^2x}{a^2}$

EXERCISES 25–42 If $|y|$ gets smaller as x increases for a function $f(x)$, must the graph of $f(x)$ fall from left to right? Explain.

No; *Sample answer:* For example, $f(x) = -0.5^x$ rises from left to right.

APPLICATION NOTE

EXERCISE 45 The exponential decay model is based on an average adult and is used to determine frequency of dosage.

EXERCISE 46 The exponential decay function models the behavior of the isotope plutonium-239, which has a half-life of 24,360 years.

DATA UPDATE

Updated data for Exs. 47–49 are available at **www.mcdougallittell.com**.

ADDITIONAL PRACTICE AND RETEACHING

For Lesson 8.2:

- Practice Levels A, B, and C (*Chapter 8 Resource Book,* p. 29)
- Reteaching with Practice (*Chapter 8 Resource Book,* p. 32)
- See Lesson 8.2 of the *Personal Student Tutor*

For more Mixed Review:

- Search the *Test and Practice Generator* for key words or specific lessons.

478

GRAPHING FUNCTIONS Graph the function. 25–33. See margin.

25. $y = 3\left(\dfrac{1}{2}\right)^x$ **26.** $y = 2\left(\dfrac{1}{5}\right)^x$ **27.** $y = -2\left(\dfrac{1}{4}\right)^x$

28. $y = -5\left(\dfrac{1}{2}\right)^x$ **29.** $y = 4\left(\dfrac{1}{3}\right)^x$ **30.** $y = 5\left(\dfrac{1}{4}\right)^x$

31. $y = -3\left(\dfrac{2}{3}\right)^x$ **32.** $y = -5(0.75)^x$ **33.** $y = 3\left(\dfrac{3}{8}\right)^x$

34–42. See margin for graphs.
GRAPHING FUNCTIONS Graph the function. State the domain and range.

34. domain: all real numbers; range: $y < 1$

35. domain: all real numbers; range: $y > 0$

36. domain: all real numbers; range: $y > 0$

37. domain: all real numbers; range: $y > 0$

38. domain: all real numbers; range: $y > 0$

39. domain: all real numbers; range: $y > 3$

40. domain: all real numbers; range: $y < 0$

41. domain: all real numbers; range: $y > -2$

42. domain: all real numbers; range: $y > -1$

34. $y = -\left(\dfrac{1}{2}\right)^x + 1$ **35.** $y = \left(\dfrac{2}{3}\right)^{x-1}$ **36.** $y = 4\left(\dfrac{1}{2}\right)^{x+1}$

37. $y = \left(\dfrac{1}{3}\right)^{x-2}$ **38.** $y = 2\left(\dfrac{1}{3}\right)^{x-1}$ **39.** $y = (0.25)^x + 3$

40. $y = -3\left(\dfrac{1}{3}\right)^{x-1}$ **41.** $y = \left(\dfrac{1}{3}\right)^x - 2$ **42.** $y = \left(\dfrac{2}{3}\right)^x - 1$

WRITING MODELS In Exercises 43–45, write an exponential decay model that describes the situation.

43. **STEREO SYSTEM** You buy a stereo system for $780. Each year t, the value V of the stereo system decreases by 5%. $V = 780(0.95)^t$

44. **BEVERAGES** You drink a beverage with 120 milligrams of caffeine. Each hour h, the amount c of caffeine in your system decreases by about 12%.
$c = 120(0.88)^h$

45. **MEDICINE** An adult takes 400 milligrams of ibuprofen. Each hour h, the amount i of ibuprofen in the person's system decreases by about 29%. $i = 400(0.71)^h$

46. **RADIOACTIVE DECAY** One hundred grams of plutonium is stored in a container. The amount P (in grams) of plutonium present after t years can be modeled by this equation:

$$P = 100(0.99997)^t$$

How much plutonium is present after 20,000 years? about 54.88 g

RECORD ALBUMS In Exercises 47–49, use the following information.
The number A (in millions) of record albums sold each year in the United States from 1982 to 1993 can be modeled by

$$A = 265(0.39)^t$$

where t represents the number of years since 1982.

DATA UPDATE of Recording Industry Association of America data at www.mcdougallittell.com

47. Identify the initial amount, the decay factor, and the annual percent decrease.
265; 0.39; 61%

48. Graph the model. See margin.

49. Estimate when the number of records sold was 1 million. about 1988

DEPRECIATION In Exercises 50–52, use the following information.
You buy a new car for $22,000. The value of the car decreases by 12.5% each year.

50. Write an exponential decay model for the value of the car. Use the model to estimate the value after 3 years. $V = 22,000(0.875)^t$; $14,738

51. Graph the model. See margin.

52. Estimate when the car will have a value of $8000. after about 7.6 years

25–42, 48, 51. See Additional Answers beginning on page AA1.

COMPUTERS In Exercises 53–55, use the following information.
You buy a new computer for $2100. The value of the computer decreases by about 50% annually.

53. Write an exponential decay model for the value of the computer. Use the model to estimate the value after 2 years. $V = 2100(0.5)^t$; $525

54. Graph the model.   **See margin.**

55. Estimate when the computer will have a value of $600.   **after about 22 months**

56. SCIENCE ▶ CONNECTION   During normal breathing, about 12% of the air in the lungs is replaced after one breath. Write an exponential decay model for the amount of the original air left in the lungs if the initial amount of air in the lungs is 500 milliliters. How much of the original air is present after 240 breaths?
   $y = 500(0.88)^n$; about 2.4×10^{-11} mL

57. **MULTI-STEP PROBLEM** A new automobile worth $18,354 depreciates by about 17% each year. The payoff amount on a loan after making n monthly payments is given by the model

$$A(n) = \left(A_0 - \frac{P}{r}\right)(1 + r)^n + \frac{P}{r}$$

where A_0 is the original amount of the loan, P is the monthly payment, and r is the monthly interest rate expressed as a decimal.

57b. $A(n) =$

$$\left(18{,}354 - \frac{280}{\frac{0.085}{12}}\right) \cdot$$

$$\left(1 + \frac{0.085}{12}\right)^n + \frac{280}{\frac{0.085}{12}}$$

Test Preparation

57c. *Sample answer:* It would make the most sense to sell the car after the 5th year, when the value is more than the amount owed, so that you could sell the car for enough money to pay off the rest of the loan.

   a. Write an exponential decay model for the value V of the automobile t years after it is purchased. $V = 18{,}354(0.83)^t$

   b. Write a model for the payoff amount on a loan of $18,354 with a monthly payment of $280 and an *annual* interest rate of 8.5%. (*Hint:* The model for the payoff amount uses the monthly interest rate, not the annual interest rate.)
   See margin.

   c. *Writing* Make a table showing the value of the car and the payoff amount on the loan for 5 years. When would it make sense to sell the car? Explain.
   See margin for table.

 **Challenge**

58. **CRITICAL THINKING** Is the product of two exponential decay functions always another exponential decay function? Is the quotient of two exponential decay functions always another exponential decay function? Justify your answers.
See margin.

MIXED REVIEW

GRAPHING FUNCTIONS Graph the function. (Review 7.5)   59–64. See margin.

59. $y = (x + 1)^{1/3}$ **60.** $y = \sqrt[3]{x} + 1$ **61.** $y = -3x^{1/3}$

62. $y = \sqrt{x} + 4$ **63.** $y = -\sqrt{x + 5}$ **64.** $y = \sqrt[3]{x} + \frac{1}{4}$

USING A DATA SET Find the mean, the median, the mode, and the range for the set of data. (Review 7.7)

65. 11, 18, 13, 15, 17, 15, 23, 20, 12   **66.** 25, 30, 32, 42, 31, 33, 36, 22
      16; 15; 15; 12   31.375; 31.5; none; 20

67. FINANCE You deposit $2000 in a bank account. Find the balance after 4 years for each of the following situations. (Review 8.1 for 8.3)

   a. The account pays 7% annual interest compounded quarterly. $2639.86

   b. The account pays 5% annual interest compounded monthly. $2441.79

8.2 *Exponential Decay*   **479**

DAILY HOMEWORK QUIZ

📄 *Transparency Available*

Tell whether the function represents *exponential growth* or *exponential decay.*

1. $f(x) = 0.2 \cdot 1.5^x$
   exponential growth

2. $f(x) = 10 \cdot 3^{-x}$
   exponential decay

3. Graph $y = 3 \cdot 0.5^x$.

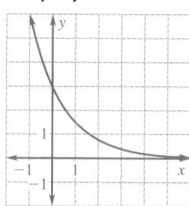

4. Graph $y = -2\left(\frac{1}{4}\right)^{x+2} + 2$.
   State the domain and range.

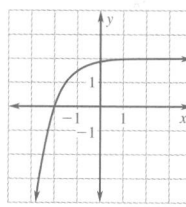

   domain: all real numbers; range: $y < 2$

EXTRA CHALLENGE NOTE
Challenge problems for Lesson 8.2 are available in **blackline** format in the *Chapter 8 Resource Book,* p. 36 and at **www.mcdougallittell.com.**

ADDITIONAL TEST PREPARATION

1. **WRITING** The graphs of an exponential growth function and an exponential decay function both lie in the first and second quadrants. Describe the difference in the graphs.   The graph of the growth function rises from left to right. The graph of the decay function falls from left to right.

54, 57c, 58–64.
   See Additional Answers beginning on page AA1.

LESSON OPENER
APPLICATION
An alternative way to approach Lesson 8.3 is to use the Application Lesson Opener:
- Blackline Master (*Chapter 8 Resource Book,* p. 40)
-  Transparency (p. 53)

MEETING INDIVIDUAL NEEDS
- *Chapter 8 Resource Book*
  Prerequisite Skills Review (p. 5)
  Practice Level A (p. 41)
  Practice Level B (p. 42)
  Practice Level C (p. 43)
  Reteaching with Practice (p. 44)
  Absent Student Catch-Up (p. 46)
  Challenge (p. 48)
- *Resources in Spanish*
- Personal Student Tutor

NEW-TEACHER SUPPORT
See the Tips for New Teachers on pp. 1–2 of the *Chapter 8 Resource Book* for additional notes about Lesson 8.3.

WARM-UP EXERCISES

Transparency Available

Simplify. Round to the nearest hundredth.

1. $\left(1+\frac{1}{2}\right)^2$ 2.25 **2.** $\left(1+\frac{1}{3}\right)^3$ 2.37

3. $\left(1+\frac{1}{4}\right)^4$ 2.44 **4.** $\left(1+\frac{1}{5}\right)^5$ 2.49

State the domain and range of the function.

5. $y = 3\left(\frac{1}{2}\right)^x$ domain: all real numbers; range: $y > 0$

6. $y = -3(2)^x$ domain: all real numbers; range: $y < 0$

8.3

The Number *e*

What you should learn

GOAL 1 Use the number *e* as the base of exponential functions.

GOAL 2 Use the natural base *e* in **real-life** situations, such as finding the air pressure on Mount Everest in **Ex. 79.**

Why you should learn it

▼ To solve **real-life** problems, such as finding the number of listed endangered species in **Example 5.**

The grizzly bear was first listed as threatened in 1975 and remains an endangered species today.

GOAL 1 **USING THE NATURAL BASE *e***

The history of mathematics is marked by the discovery of special numbers such as counting numbers, zero, negative numbers, π, and imaginary numbers. In this lesson you will study one of the most famous numbers of modern times. Like π and i, the number e is denoted by a letter. The number is called the **natural base *e*,** or the **Euler number,** after its discoverer, Leonhard Euler (1707–1783).

ACTIVITY
Developing Concepts
Investigating the Natural Base *e*

1 Copy the table and use a calculator to complete the table.

n	10^1	10^2	10^3	10^4	10^5	10^6
$\left(1+\frac{1}{n}\right)^n$	2.594	?	?	?	?	?

2.705 2.717 2.718 2.718 2.718

2 Do the values in the table appear to be approaching a fixed decimal number? If so, what is the number rounded to three decimal places? **yes; 2.718**

In the activity you may have discovered that as n gets larger and larger, the expression $\left(1+\frac{1}{n}\right)^n$ gets closer and closer to $2.71828\ldots$, which is the value of e.

THE NATURAL BASE *e*

The natural base e is irrational. It is defined as follows:

As n approaches $+\infty$, $\left(1+\frac{1}{n}\right)^n$ approaches $e \approx 2.718281828459$.

EXAMPLE 1 *Simplifying Natural Base Expressions*

Simplify the expression.

a. $e^3 \cdot e^4$

b. $\dfrac{10e^3}{5e^2}$

c. $\left(3e^{-4x}\right)^2$

SOLUTION

a. $e^3 \cdot e^4 = e^{3+4}$
$= e^7$

b. $\dfrac{10e^3}{5e^2} = 2e^{3-2}$
$= 2e$

c. $\left(3e^{-4x}\right)^2 = 3^2 e^{(-4x)(2)}$
$= 9e^{-8x} = \dfrac{9}{e^{8x}}$

continued his
mathematical research
despite losing sight in one
eye in 1735. He published
more than 500 books and
papers during his lifetime.
Euler's use of *e* appeared
in his book *Mechanica*,
published in 1736.

EXAMPLE 2 *Evaluating Natural Base Expressions*

Use a calculator to evaluate the expression: **a.** e^2 **b.** $e^{-0.06}$

SOLUTION

EXPRESSION	KEYSTROKES	DISPLAY
a. e^2	2nd [e^x] 2 ENTER	7.389056
b. $e^{-0.06}$	2nd [e^x] (−) .06 ENTER	0.941765

· · · · · · · · · ·

A function of the form $f(x) = ae^{rx}$ is called a *natural base exponential function*. If $a > 0$ and $r > 0$, the function is an exponential growth function, and if $a > 0$ and $r < 0$, the function is an exponential decay function. The graphs of the basic functions $y = e^x$ and $y = e^{-x}$ are shown below.

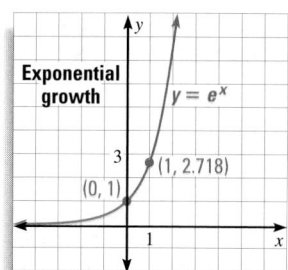

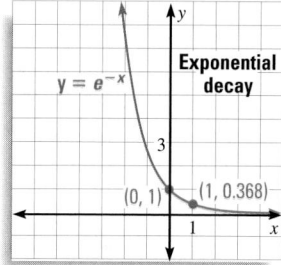

EXAMPLE 3 *Graphing Natural Base Functions*

Graph the function. State the domain and range.

a. $y = 2e^{0.75x}$

b. $y = e^{-0.5(x-2)} + 1$

SOLUTION

a. Because $a = 2$ is positive and $r = 0.75$ is positive, the function is an exponential growth function. Plot the points $(0, 2)$ and $(1, 4.23)$ and draw the curve.

b. Because $a = 1$ is positive and $r = -0.5$ is negative, the function is an exponential decay function. Translate the graph of $y = e^{-0.5x}$ to the right 2 units and up 1 unit.

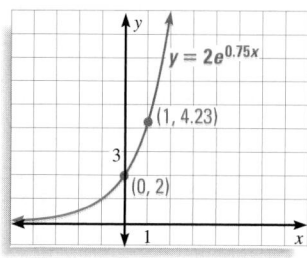

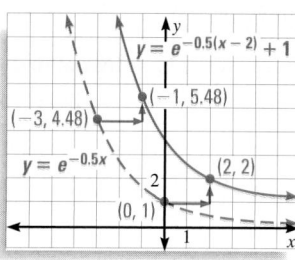

The domain is all real numbers, and the range is all positive real numbers.

The domain is all real numbers, and the range is $y > 1$.

8.3 *The Number e* **481**

EXTRA EXAMPLE 1
Simplify the expression.
a. $\dfrac{24e^8}{8e^5}$ $3e^3$ **b.** $(2e^{-5x})^{-2} \cdot \dfrac{e^{10x}}{4}$

EXTRA EXAMPLE 2
Use a graphing calculator to evaluate the expression.
a. e^3 20.085537
b. $e^{-0.12}$ 0.88692044

EXTRA EXAMPLE 3
Graph the function. State the domain and range.
a. $y = -3e^{0.5x}$ domain: all real numbers; range: $y < 0$

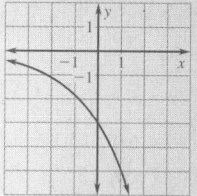

b. $y = e^{0.4(x+1)} - 2$ domain: all real numbers; range: $y > -2$

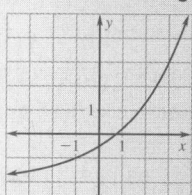

CHECKPOINT EXERCISES
For use after Example 1:
1. Simplify $\dfrac{3e^2 \cdot 4e^4}{6e^8} \cdot \dfrac{2}{e^2}$

For use after Example 2:
2. Use a graphing calculator to evaluate $e^{0.85}$. 2.339647

For use after Example 3:
3. Graph $y = e^{0.61x} + 3$. State the domain and the range.
domain: all real numbers; range: $y > 3$

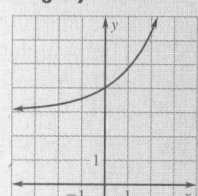

EXTRA EXAMPLE 4
You deposit $1500 in an account that pays 7.5% annual interest compounded continuously. What is the balance after 1 year? **about $1616.83**

EXTRA EXAMPLE 5
The atmospheric pressure P (in pounds per square inch) of an object d miles above sea level can be modeled by $P = 14.7e^{-0.21d}$.
a. How much pressure per square inch would you experience at the summit of Mount Washington, 6288 feet above sea level? **about 11.45 lb/in.²**
b. Graph the model. Estimate your height above sea level if you experience 13.23 lb/in.² of pressure. **about 0.5 mi**

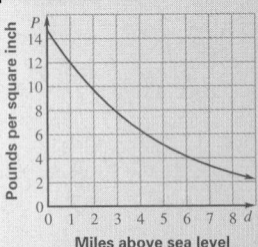

CHECKPOINT EXERCISES
For use after Examples 4 and 5:
1. The radioactive decay of radon-222 can be modeled by the function $A = Ce^{-0.1813t}$, where A is the amount remaining, C is the original amount, and t is the time in days. If there are 15 mg of radon-222 sealed in a glass tube, how much will remain in the tube after 8 days? If 10 mg of radon-222 remain after 5 days, how much was originally there? **about 3.5 mg; about 24.8 mg**

CLOSURE QUESTION
Compare 8% interest compounded quarterly to 7.75% compounded continuously. **8% compounded quarterly is 8.24% annual interest; 7.75% compounded continuously is 8.06% annual interest.**

FOCUS ON CAREERS

 MARINE BIOLOGIST
A marine biologist studies salt-water plants and animals. Those who work for the U.S. Fish and Wildlife Service help maintain populations of manatees, walruses, and other endangered species.

CAREER LINK
www.mcdougallittell.com

 **GOAL 2** **USING e IN REAL LIFE**

In Lesson 8.1 you learned that the amount A in an account earning interest compounded n times per year for t years is given by

$$A = P\left(1 + \frac{r}{n}\right)^{nt}$$

where P is the principal and r is the annual interest rate expressed as a decimal. As n approaches positive infinity, the compound interest formula approximates the following formula for *continuously compounded interest:*

$$A = Pe^{rt}$$

Finance

EXAMPLE 4 *Finding the Balance in an Account*

You deposit $1000 in an account that pays 8% annual interest compounded continuously. What is the balance after 1 year?

SOLUTION

Note that $P = 1000$, $r = 0.08$, and $t = 1$. So, the balance at the end of 1 year is:

$$A = Pe^{rt} = 1000e^{0.08(1)} \approx \$1083.29$$

In Example 4 of Lesson 8.1, you found that the balance from daily compounding is $1083.28. So, continuous compounding earned only an additional $.01.

EXAMPLE 5 *Using an Exponential Model*

ENDANGERED SPECIES Since 1972 the U.S. Fish and Wildlife Service has kept a list of endangered species in the United States. For the years 1972–1998, the number s of species on the list can be modeled by

$$s = 119.6e^{0.0917t}$$

where t is the number of years since 1972.

a. What was the number of endangered species in 1972?

b. Graph the model.

c. Use the graph to estimate when the number of endangered species reached 1000.

SOLUTION

a. In 1972, when $t = 0$, the model gives:

$$s = 119.6e^0 = 119.6$$

So, there were about 120 endangered species on the list in 1972.

b. The graph of the model is shown.

c. Use the *Intersect* feature to determine that s reaches 1000 when $t \approx 23$, which is about 1995.

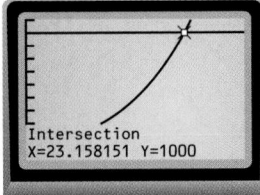

Intersection
X=23.158151 Y=1000

GUIDED PRACTICE

Vocabulary Check ✓

1. What is the Euler number? Give an approximation of the Euler number rounded to three decimal places. **The Euler number, e, is the limit of $\left(1+\frac{1}{n}\right)^n$ as $n \to +\infty$; 2.718.**

Concept Check ✓

2. Tell whether the function $f(x) = \frac{1}{4}e^{2x}$ is an example of *exponential growth* or *exponential decay*. Explain. **exponential growth, since $\frac{1}{4} > 0$ and $e^2 > 1$.**

3. Is it possible to express e as a ratio of two integers? Explain. **no, since e is irrational**

Skill Check ✓

Simplify the expression.

4. $e^2 \cdot e^6$ e^8 5. $e^{-2} \cdot 3e^7$ $3e^5$ 6. $(2e^{5x})^2$ $4e^{10x}$ 7. $(4e^{-2})^3$ $\dfrac{64}{e^6}$

8. $\left(\frac{1}{2}e^{-2}\right)^4$ $\dfrac{1}{16e^8}$ 9. $\sqrt{36e^{4x}}$ $6e^{2x}$ 10. $\dfrac{e^x}{e^{2x}}$ $\dfrac{1}{e^x}$ 11. $\dfrac{12e^4}{36e^{-2}}$ $\dfrac{e^6}{3}$

12. What is the horizontal asymptote of the graph of $f(x) = 2e^x - 2$? $y = -2$

Graph the function.   13–15. See margin.

13. $y = e^{-2x}$ 14. $y = \frac{1}{2}e^x$ 15. $y = \frac{1}{8}e^{2x}$

16. 🌐 **ENDANGERED SPECIES** Use the model in Example 5 to estimate the number of endangered species in 1998.   **about 1298**

PRACTICE AND APPLICATIONS

STUDENT HELP

▶ **Extra Practice**
to help you master
skills is on p. 950.

49. exponential decay
50. exponential growth
51. exponential decay
52. exponential growth
53. exponential growth
54. exponential decay
55. exponential growth
56. exponential decay
57. exponential decay
58. exponential growth
59. exponential decay
60. exponential growth

STUDENT HELP

▶ **HOMEWORK HELP**
Example 1: Exs. 17–32
Example 2: Exs. 33–48
Example 3: Exs. 49–75
Example 4: Exs. 76–78
Example 5: Exs. 79, 80

SIMPLIFYING EXPRESSIONS   Simplify the expression.

17. $e^2 \cdot e^4$ e^6 18. $e^{-3} \cdot e^5$ e^2 19. $(3e^{-3x})^{-1}$ $\dfrac{e^{3x}}{3}$ 20. $(3e^{4x})^2$ $9e^{8x}$

21. $3e^{-2} \cdot e^6$ $3e^4$ 22. $\left(\frac{1}{4}e^{-2}\right)^3$ $\dfrac{1}{64e^6}$ 23. $e^x \cdot e^{-3x} \cdot e^5$ e^{-2x+5} 24. $\sqrt{4e^{2x}}$ $2e^x$

25. $(100e^{0.5x})^{-2}$ $\dfrac{1}{10,000e^x}$ 26. $e^x \cdot 4e^{2x+1}$ $4e^{3x+1}$ 27. $\dfrac{e^x}{2e}$ $\dfrac{e^{x-1}}{2}$ 28. $\dfrac{5e^x}{e^{5x}}$ $\dfrac{5}{e^{4x}}$

29. $\sqrt[3]{27e^{6x}}$ $3e^{2x}$ 30. $(32e^{-4x})^3$ $\dfrac{32,768}{e^{12x}}$ 31. $\dfrac{6e^{3x}}{4e}$ $\dfrac{3}{2}e^{3x-1}$ 32. $\sqrt[3]{64e^{9x}}$ $4e^{3x}$

🖩 **EVALUATING EXPRESSIONS**   Use a calculator to evaluate the expression. Round the result to three decimal places.

33. e^3 20.086   34. $e^{-2/3}$ 0.513   35. $e^{1.7}$ 5.474   36. $e^{1/2}$ 1.649

37. $e^{-1/4}$ 0.779   38. $e^{3.2}$ 24.533   39. e^8 2980.958   40. e^{-3} 0.050

41. e^{-4} 0.018   42. $2e^{1/2}$ 3.297   43. $-4e^{-3}$ -0.199 44. $0.5e^{3.2}$ 12.266

45. $-1.2e^5$ -178.096 46. $0.02e^{-0.3}$ 0.015   47. $225e^{-50}$ 4.340×10^{-20} 48. $-8.95e^{1/5}$ -10.932

GROWTH OR DECAY?   Tell whether the function is an example of *exponential growth* or *exponential decay*.   49–60. See margin.

49. $f(x) = 5e^{-3x}$ 50. $f(x) = \frac{1}{8}e^{5x}$ 51. $f(x) = e^{-4x}$ 52. $f(x) = \frac{1}{6}e^{2x}$

53. $f(x) = \frac{1}{4}e^{2x}$ 54. $f(x) = e^{-8x}$ 55. $f(x) = e^{3x}$ 56. $f(x) = \frac{1}{4}e^{-x}$

57. $f(x) = e^{-6x}$ 58. $f(x) = \frac{3}{8}e^{7x}$ 59. $f(x) = e^{-9x}$ 60. $f(x) = e^{8x}$

---

3 APPLY

⬤ **ASSIGNMENT GUIDE**

BASIC
Day 1: pp. 483–484 Exs. 18–44 even, 50–60 even, 61–66
Day 2: pp. 483–485 Exs. 45–48, 68–78 even, 81, 82, 84–94 even, Quiz 1 Exs. 1–16

AVERAGE
Day 1: pp. 483–484 Exs. 18–60 even, 61–66, 68–74 even
Day 2: pp. 483–485 Exs. 75–78, 81–83, 84–94 even, Quiz 1 Exs. 1–16

ADVANCED
Day 1: pp. 483–484 Exs. 18–60 even, 61–72
Day 2: pp. 483–485 Exs. 73–79, 81–83, 84–94 even, Quiz 1 Exs. 1–16

BLOCK SCHEDULE
pp. 483–485 Exs. 18–60 even, 61–66, 68–74 even, 75–78, 81–83, 84–94 even, Quiz 1 Exs. 1–16

EXERCISE LEVELS
Level A: *Easier*
17–28, 33–44, 49–60, 76
Level B: *More Difficult*
29–32, 45–48, 61–75, 77–82
Level C: *Most Difficult*
83

✔ **HOMEWORK CHECK**
To quickly check student understanding of key concepts, go over the following exercises: Exs. 18, 28, 36, 50, 62, 68. See also the Daily Homework Quiz:

• Blackline Master (*Chapter 8 Resource Book,* p. 52)
• Transparency (p. 62)

❗ **COMMON ERROR**
EXERCISES 4–11 Students may not be aware that all the laws of exponents apply to e.

13–15. See Additional Answers beginning on page AA1.

483

STUDENT HELP NOTES

→ **Homework Help** Students can find help for Exs. 67–75 at **www.mcdougallittell.com.** The information can be printed out for students who don't have access to the Internet.

APPLICATION NOTE
EXERCISE 79 Since the function models an exponential decay function, the rate of decay is decreasing. Students can use the graph of the function to find for what change in altitude the air pressure is approximately halved. **Air pressure is approximately halved for each additional increase of 3.3 miles in altitude.**

EXERCISE 80 In general, the more damage that occurs at the site of a wound, the longer it will take to heal. The time it takes for a wound to heal also depends on the location of the wound, the age of the skin, and whether or not infection is introduced to the site.

67.
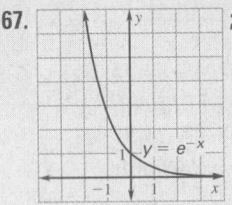
;

domain: all real numbers;
range: $y > 0$
68–75, 78. See Additional Answers beginning on page AA1.

ADDITIONAL PRACTICE AND RETEACHING

For Lesson 8.3:
- Practice Levels A, B, and C (*Chapter 8 Resource Book,* p. 41)
- Reteaching with Practice (*Chapter 8 Resource Book,* p. 44)
- ⊞ See Lesson 8.3 of the *Personal Student Tutor*

For more Mixed Review:
- ⊞ Search the *Test and Practice Generator* for key words or specific lessons.

77. $2650; $2652.25; $2653.41; $2654.19; $2654.59; *Sample answer:* The extra amount of interest earned with more and more compoundings decreases drastically, with the difference between compounding monthly and continuously being only 40¢, 0.016% of the amount initially invested.

STUDENT HELP

🌐 **HOMEWORK HELP** Visit our Web site www.mcdougallittell.com for help with Exs. 67–75.

FOCUS ON PEOPLE

▸ **MOUNT EVEREST** In 1953 Sir Edmund Hillary of New Zealand and Tenzing Norgay, a Nepalese Sherpa tribesman, became the first people to reach the top of Mount Everest.

MATCHING GRAPHS Match the function with its graph.

61. $y = 3e^{0.5x}$ **C**   **62.** $y = \frac{1}{3}e^{0.5x}$ **E**   **63.** $y = \frac{1}{2}e^{-(x-1)}$ **F**

64. $y = e^{-x} + 1$ **B**   **65.** $y = 3e^{-x} - 2$ **D**   **66.** $y = 3e^x - 2$ **A**

A.
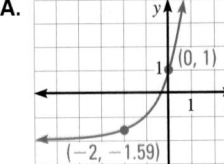

B.
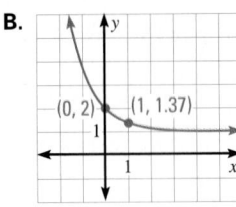

C.
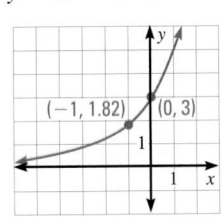

D.
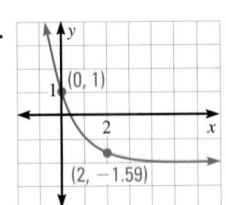

E.
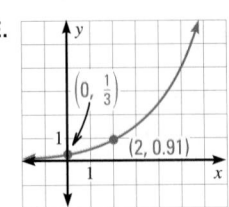

F.
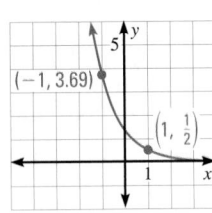

GRAPHING FUNCTIONS Graph the function. State the domain and range.
67–75. See margin.

67. $y = e^{-x}$ **68.** $y = 4e^x$ **69.** $y = \frac{1}{3}e^x$

70. $y = 3e^{2x} + 2$ **71.** $y = 1.5e^{-0.5x}$ **72.** $y = 0.1e^{2x} - 4$

73. $y = \frac{1}{3}e^{x-2} - 1$ **74.** $y = \frac{4}{3}e^{x-3} + 1$ **75.** $y = 0.5e^{-2(x-1)} - 2$

76. 🌐 **CONTINUOUS COMPOUNDING** You deposit $975 in an account that pays 5.5% annual interest compounded continuously. What is the balance after 6 years? **$1356.19**

77. 🌐 **COMPARING FORMULAS** You deposit $2500 in an account that pays 6% annual interest. Use the formulas at the top of page 482 to calculate the account balance after one year when the interest is compounded annually, semiannually, quarterly, monthly, and continuously. What do you notice? Explain. **See margin.**

78. *Writing* Compare the effects of compounding interest continuously and compounding interest daily using the formulas $A = Pe^{rt}$ and **See margin.**

$$A = P\left(1 + \frac{r}{365}\right)^{365t}.$$

79. 🌐 **MOUNT EVEREST** The air pressure P at sea level is about 14.7 pounds per square inch. As the altitude h (in feet above sea level) increases, the air pressure decreases. The relationship between air pressure and altitude can be modeled by:

$$P = 14.7e^{-0.00004h}$$

Mount Everest in Tibet and Nepal rises to a height of 29,028 feet above sea level. What is the air pressure at the peak of Mount Everest? **about 4.603 lb/in.²**

80. 🌐 **RATE OF HEALING** The area of a wound decreases exponentially with time. The area A of a wound after t days can be modeled by

$$A = A_0e^{-0.05t}$$

where A_0 is the initial wound area. If the initial wound area is 4 square centimeters, how much of the wound area is present after 14 days? **about 2 cm²**

81. MULTIPLE CHOICE What is the simplified form of $\sqrt[3]{\dfrac{8(81e^{11}x)}{3e^5x^{-2}}}$? E

Ⓐ $6e^2\sqrt[3]{x}$ Ⓑ $6x\sqrt[3]{e^6}$ Ⓒ $6\sqrt[3]{e^{16}x}$ Ⓓ $\dfrac{6e^2}{x}$ Ⓔ $6e^2x$

82. MULTIPLE CHOICE Which function is graphed at the right? B

Ⓐ $f(x) = 3e^{x-2}$ Ⓑ $f(x) = 3e^x - 2$

Ⓒ $f(x) = 3e^{-x} - 2$ Ⓓ $f(x) = 3e^{-(x+2)}$

Ⓔ $f(x) = 3e^{x+2}$

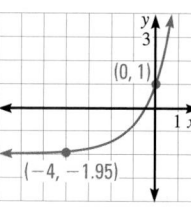
(0, 1)
(−4, −1.95)

★ **Challenge**

83. CRITICAL THINKING Find a value of n for which $\left(1 + \dfrac{1}{n}\right)^n$ gives the value of e correct to 9 decimal places. Explain the process you used to find your answer. **See margin.**

MIXED REVIEW

84. $f^{-1}(x) = \dfrac{-x}{3}$

85. $f^{-1}(x) = \dfrac{x-7}{6}$

86. $f^{-1}(x) = -\left(\dfrac{x+24}{5}\right)$

87. $f^{-1}(x) = 2x + 20$

88. $f^{-1}(x) = -\left(\dfrac{x-7}{14}\right)$

89. $f^{-1}(x) = -5x - 65$

FINDING INVERSE FUNCTIONS Find an equation for the inverse of the function. **(Review 7.4 for 8.4)**

84. $f(x) = -3x$ **85.** $f(x) = 6x + 7$ **86.** $f(x) = -5x - 24$

87. $f(x) = \dfrac{1}{2}x - 10$ **88.** $f(x) = -14x + 7$ **89.** $f(x) = -\dfrac{1}{5}x - 13$

SOLVING EQUATIONS Solve the equation. **(Review 7.6)**

90. $\sqrt{x} = 20$ **400**   **91.** $\sqrt[3]{5x-4} + 7 = 10$ **6.2**   **92.** $2(x+4)^{2/3} = 8$ **−12, 4**

93. $\sqrt{x^2-4} = x - 2$ **2**   **94.** $\sqrt{x+3} = \sqrt{2x-1}$ **4**   **95.** $\sqrt{3x-5} - 3\sqrt{x} = 0$
                                                                                          no solution

QUIZ 1

Self-Test for Lessons 8.1–8.3

1. domain: all real numbers; range: $y > -1$

2. domain: all real numbers; range: $y > 2$

3. domain: all real numbers; range: $y > 0$

4. domain: all real numbers; range: $y < 0$

5. domain: all real numbers; range: $y > 2$

6. domain: all real numbers; range: $y < 3$

1–6. See margin for graphs.
Graph the function. State the domain and range. (Lessons 8.1, 8.2)

1. $y = 4^x - 1$ **2.** $y = 3^{x+1} + 2$ **3.** $y = \dfrac{1}{2} \cdot 5^{x-1}$

4. $y = -2\left(\dfrac{1}{6}\right)^x$ **5.** $y = \left(\dfrac{5}{8}\right)^x + 2$ **6.** $y = -2 \cdot 6^{x-3} + 3$

Simplify the expression. (Lesson 8.3)

7. $2e^3 \cdot e^4$ $2e^7$ **8.** $4e^{-5} \cdot e^7$ $4e^2$ **9.** $(-3e^{2x})^2$ $9e^{4x}$ **10.** $(5e^{-3})^{-4x}$ $\dfrac{e^{12x}}{5^{4x}}$

11. $\dfrac{3e^x}{4e}$ $\dfrac{3}{4}e^{x-1}$ **12.** $\dfrac{6e^x}{e^{5x}}$ $\dfrac{6}{e^{4x}}$ **13.** $\sqrt{16e^2x}$ $4e\sqrt{x}$ **14.** $\sqrt[3]{125e^{6x}}$ $5e^{2x}$

15. Graph the function $f(x) = -4e^{2x}$. **(Lesson 8.3) See margin.**

16. 🌐 **RADIOACTIVE DECAY** One hundred grams of radium is stored in a container. The amount R (in grams) of radium present after t years can be modeled by $R = 100e^{-0.00043t}$. Graph the model. How much of the radium is present after 10,000 years? **(Lesson 8.3)**   about 1.357 g; See margin for graph.

Additional Test Preparation *Sample answer:*
Both graphs have $y = 0$ as an asymptote. The graph of $y = e^x$ approaches its asymptote as $x \to -\infty$. The graph of $y = e^{-x}$ approaches its asymptote as $x \to \infty$. Both graphs have y-intercept 1. Both graphs have all real numbers as their domain and all positive real numbers as their range. Each graph is a reflection of the other in the y-axis.

DAILY HOMEWORK QUIZ

🖊 *Transparency Available*

Simplify the expression.

1. $2e^{2x} \cdot e^x$ $2e^{3x}$

2. $\dfrac{6e^{-x}}{12e^{3x}}$ $\dfrac{1}{2e^{4x}}$

Tell whether the function represents *exponential growth* or *exponential decay*.

3. $f(x) = 0.1 \cdot e^{0.3x}$
exponential growth

4. $f(x) = 5 \cdot e^{-4x}$
exponential decay

5. Graph $y = 0.2e^{2x+1} - 2$. State the domain and range.

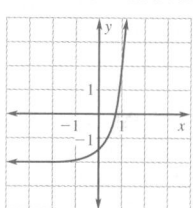

domain: all real numbers;
range: $y > -2$

EXTRA CHALLENGE NOTE
→ Challenge problems for Lesson 8.3 are available in **blackline** format in the *Chapter 8 Resource Book*, p. 48 and at **www.mcdougallittell.com.**

ADDITIONAL TEST PREPARATION

1. WRITING Compare the graphs of $y = e^x$ and $y = e^{-x}$. See sample answer at left.

ADDITIONAL RESOURCES
An alternative Quiz for Lessons 8.1–8.3 is available in the *Chapter 8 Resource Book*, p. 49.

83, 1–6, 15, 16.
See Additional Answers beginning on page AA1.

➡ **LESSON OPENER**
ACTIVITY
An alternative way to approach Lesson 8.4 is to use the Activity Lesson Opener:

- Blackline Master (*Chapter 8 Resource Book,* p. 53)
- Transparency (p. 54)

MEETING INDIVIDUAL NEEDS
- *Chapter 8 Resource Book*
  Prerequisite Skills Review (p. 5)
  Practice Level A (p. 56)
  Practice Level B (p. 57)
  Practice Level C (p. 58)
  Reteaching with Practice (p. 59)
  Absent Student Catch-Up (p. 61)
  Challenge (p. 64)
- *Resources in Spanish*
- 🖥 *Personal Student Tutor*

NEW-TEACHER SUPPORT
See the Tips for New Teachers on pp. 1–2 of the *Chapter 8 Resource Book* for additional notes about Lesson 8.4.

WARM-UP EXERCISES

🖵 *Transparency Available*
Find the value of *x*.

1. $3^x = 9$ 2·
2. $x^3 = -8$ -2
3. $10^0 = x$ 1
4. $10^x = 0.001$ -3
5. $\left(\dfrac{3}{2}\right)^{-1} = x$ $\dfrac{2}{3}$

8.4

Logarithmic Functions

GOAL 1 **EVALUATING LOGARITHMIC FUNCTIONS**

What you should learn

GOAL 1 Evaluate logarithmic functions.

GOAL 2 Graph logarithmic functions, as applied in **Example 8**.

Why you should learn it

▼ To model **real-life** situations, such as the slope of a beach in **Example 4**.

You know that $2^2 = 4$ and $2^3 = 8$. However, for what value of *x* does $2^x = 6$? Because $2^2 < 6 < 2^3$, you would expect *x* to be between 2 and 3. To find the exact *x*-value, mathematicians defined *logarithms*. In terms of a logarithm, $x = \log_2 6 \approx 2.585$. (In the next lesson you will see how this *x*-value is obtained.)

DEFINITION OF LOGARITHM WITH BASE *b*

Let *b* and *y* be positive numbers, $b \neq 1$. The **logarithm of *y* with base *b*** is denoted by $\log_b y$ and is defined as follows:

$$\log_b y = x \text{ if and only if } b^x = y$$

The expression $\log_b y$ is read as "log base *b* of *y*."

This definition tells you that the equations $\log_b y = x$ and $b^x = y$ are equivalent. The first is in *logarithmic form* and the second is in *exponential form*. Given an equation in one of these forms, you can always rewrite it in the other form.

EXAMPLE 1 *Rewriting Logarithmic Equations*

LOGARITHMIC FORM	EXPONENTIAL FORM
a. $\log_2 32 = 5$	$2^5 = 32$
b. $\log_5 1 = 0$	$5^0 = 1$
c. $\log_{10} 10 = 1$	$10^1 = 10$
d. $\log_{10} 0.1 = -1$	$10^{-1} = 0.1$
e. $\log_{1/2} 2 = -1$	$\left(\dfrac{1}{2}\right)^{-1} = 2$

. . . .

Parts (b) and (c) of Example 1 illustrate two special logarithm values that you should learn to recognize.

SPECIAL LOGARITHM VALUES

Let *b* be a positive real number such that $b \neq 1$.

LOGARITHM OF 1	$\log_b 1 = 0$ because $b^0 = 1$.
LOGARITHM OF BASE *b*	$\log_b b = 1$ because $b^1 = b$.

EXAMPLE 2 *Evaluating Logarithmic Expressions*

STUDENT HELP

HOMEWORK HELP
Visit our Web site
www.mcdougallittell.com
for extra examples.

Evaluate the expression.

a. $\log_3 81$ **b.** $\log_5 0.04$ **c.** $\log_{1/2} 8$ **d.** $\log_9 3$

SOLUTION

To help you find the value of $\log_b y$, ask yourself what power of b gives you y.

a. 3 to what power gives 81?   **b.** 5 to what power gives 0.04?

     $3^4 = 81$, so $\log_3 81 = \mathbf{4}$. $5^{-2} = 0.04$, so $\log_5 0.04 = \mathbf{-2}$.

c. $\frac{1}{2}$ to what power gives 8?   **d.** 9 to what power gives 3?

     $\left(\frac{1}{2}\right)^{-3} = 8$, so $\log_{1/2} 8 = \mathbf{-3}$. $9^{1/2} = 3$, so $\log_9 3 = \frac{1}{2}$.

· · · · · · · · ·

The logarithm with base 10 is called the **common logarithm**. It is denoted by $\log_{10}$ or simply by log. The logarithm with base e is called the **natural logarithm**. It can be denoted by $\log_e$, but it is more often denoted by ln.

COMMON LOGARITHM	NATURAL LOGARITHM
$\log_{10} x = \log x$	$\log_e x = \ln x$

Most calculators have keys for evaluating common and natural logarithms.

EXAMPLE 3 *Evaluating Common and Natural Logarithms*

EXPRESSION	KEYSTROKES	DISPLAY
a. $\log 5$	LOG 5 ENTER	0.698970
b. $\ln 0.1$	LN .1 ENTER	-2.302585

FOCUS ON APPLICATIONS

Sand particle	Diameter (mm)
Pebble	4
Granule	2
Very coarse sand	1
Coarse sand	0.5
Medium sand	0.25
Fine sand	0.125
Very fine sand	0.0625

SAND The table gives the diameters of different types of sand. Notice that the diameter of a pebble is about 64 times larger than the diameter of very fine sand.

EXAMPLE 4 *Evaluating a Logarithmic Function*

SCIENCE CONNECTION The slope s of a beach is related to the average diameter d (in millimeters) of the sand particles on the beach by this equation:

$$s = 0.159 + 0.118 \log d$$

Find the slope of a beach if the average diameter of the sand particles is 0.25 millimeter.

SOLUTION

If $d = 0.25$, then the slope of the beach is:

$s = 0.159 + 0.118 \log \mathbf{0.25}$ **Substitute 0.25 for d.**

$\approx 0.159 + 0.118(-0.602)$ **Use a calculator.**

≈ 0.09 **Simplify.**

▶ The slope of the beach is about 0.09. This is a gentle slope that indicates a rise of only 9 meters for a run of 100 meters.

---

2 TEACH

MOTIVATING THE LESSON
Seismologists use several different scales to measure earthquakes. The Richter scale measures the magnitude of seismic surface waves. The scale is logarithmic because each measure on the scale represents a magnitude ten times greater than the previous magnitude.

EXTRA EXAMPLE 1
Write the logarithmic equation in exponential form.
a. $\log_3 9 = 2$ $3^2 = 9$
b. $\log_8 1 = 0$ $8^0 = 1$
c. $\log_5 \left(\frac{1}{25}\right) = -2$ $5^{-2} = \frac{1}{25}$

EXTRA EXAMPLE 2
Evaluate the expression.
a. $\log_4 64$ 3
b. $\log_2 0.125$ -3
c. $\log_{1/4} 256$ -4
d. $\log_{32} 2$ $\frac{1}{5}$

EXTRA EXAMPLE 3
Evaluate.
a. $\log 7$ 0.84509804
b. $\ln 0.25$ -1.386294361

EXTRA EXAMPLE 4
Refer to the formula in Example 4. Find the slope of a beach if the average diameter of the sand particles is 0.5 millimeters.
about 0.12

CHECKPOINT EXERCISES

For use after Example 1:
1. Write $\log_5 25 = 2$ in exponential form. $5^2 = 25$

For use after Example 2:
2. Evaluate $\log_2 128$.   7

For use after Example 3:
3. Evaluate $\ln 1.5$.   0.4054651

For use after Example 4:
4. Use the formula given in Example 4 to find the slope of a beach if the average diameter of the sand particles is 0.18 mm.   about 0.07

EXTRA EXAMPLE 5
Simplify the expression.
a. $10^{\log x}$ *x*   **b.** $\log_5 125^x$ *3x*

EXTRA EXAMPLE 6
Find the inverse of the function.
a. $y = \log_8 x$ $y = 8^x$
b. $y = \ln(x - 3)$ $y = e^x + 3$

✓ CHECKPOINT EXERCISES
For use after Example 5:
1. Simplify $10^{\log 5x}$. *5x*
2. Simplify $\log 10{,}000^x$. *4x*

For use after Example 6:
3. Find the inverse of $y = \log_{2/5} x$.
$$y = \left(\frac{2}{5}\right)^x$$
4. Find the inverse of
$y = \ln(x - 10)$. $y = e^x + 10$

TEACHING TIPS
As students are introduced to the
inverse properties in Example 5,
remind them that a logarithm is
simply an exponent. Thus in Part (**a**),
log 2 represents the exponent *x*
that satisfies $10^x = 2$. Therefore,
when 10 is raised to that power,
the result is 2.

CONCEPT QUESTION
Why does the graph of the
exponential function $f(x) = b^x$
intersect the graph of its inverse
$f^{-1}(x) = \log_b x$ when $0 < b < 1$?
The graph of $f(x) = b^x$ intersects the
line $y = x$ when $0 < b < 1$.

MATHEMATICAL REASONING
Exponential functions are often
used to model growth. Suppose that
your rapidly growing community is
making plans to build a new high
school. They need to make sure
that the building will be big enough
to accommodate the student population as the community grows.
Why might an exponential function
not be the best model to show how
the population increases over time?
See sample answer at right.

GOAL 2 GRAPHING LOGARITHMIC FUNCTIONS

By the definition of a logarithm, it follows that the logarithmic function $g(x) = \log_b x$ is the inverse of the exponential function $f(x) = b^x$. This means that:

$$g(f(x)) = \log_b b^x = x \qquad \text{and} \qquad f(g(x)) = b^{\log_b x} = x$$

In other words, exponential functions and logarithmic functions "undo" each other.

EXAMPLE 5 *Using Inverse Properties*

Simplify the expression.

STUDENT HELP

► Look Back
For help with inverses,
see p. 422.

a. $10^{\log 2}$ **b.** $\log_3 9^x$

SOLUTION

a. $10^{\log 2} = 2$ **b.** $\log_3 9^x = \log_3 \left(3^2\right)^x = \log_3 3^{2x} = 2x$

EXAMPLE 6 *Finding Inverses*

Find the inverse of the function.

a. $y = \log_3 x$ **b.** $y = \ln(x + 1)$

SOLUTION

a. From the definition of logarithm, the inverse of $y = \log_3 x$ is $y = 3^x$.

b. $y = \ln(x + 1)$ Write original function.

$x = \ln(y + 1)$ Switch *x* and *y*.

$e^x = y + 1$ Write in exponential form.

$e^x - 1 = y$ Solve for *y*.

▶ The inverse of $y = \ln(x + 1)$ is $y = e^x - 1$.

· · · · · · · · · ·

The inverse relationship between exponential and logarithmic functions is also useful for graphing logarithmic functions. Recall from Lesson 7.4 that the graph of f^{-1} is the reflection of the graph of f in the line $y = x$.

Graphs of *f* and f^{-1} for $b > 1$

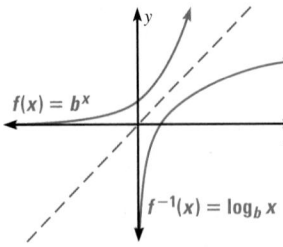

Graphs of *f* and f^{-1} for $0 < b < 1$

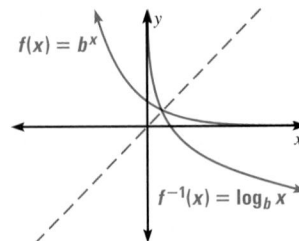

Mathematical Reasoning *Sample answer:*
Though the population is increasing rapidly now, it
will likely level off due to limiting factors such as the
difficulty of building new community infrastructure.

CONCEPT
SUMMARY

GRAPHS OF LOGARITHMIC FUNCTIONS

The graph of $y = \log_b (x - h) + k$ has the following characteristics:

- The line $x = h$ is a vertical asymptote.
- The domain is $x > h$, and the range is all real numbers.
- If $b > 1$, the graph moves up to the right. If $0 < b < 1$, the graph moves down to the right.

EXAMPLE 7 *Graphing Logarithmic Functions*

Graph the function. State the domain and range.

a. $y = \log_{1/3} x - 1$

b. $y = \log_5 (x + 2)$

SOLUTION

a. Plot several convenient points, such as $\left(\frac{1}{3}, 0\right)$ and $(3, -2)$. The vertical line $x = 0$ is an asymptote. From left to right, draw a curve that starts just to the right of the y-axis and moves down.

b. Plot several convenient points, such as $(-1, 0)$ and $(3, 1)$. The vertical line $x = -2$ is an asymptote. From left to right, draw a curve that starts just to the right of the line $x = -2$ and moves up.

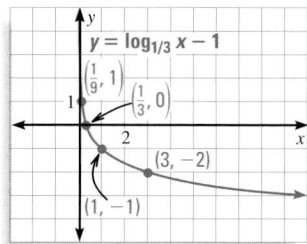

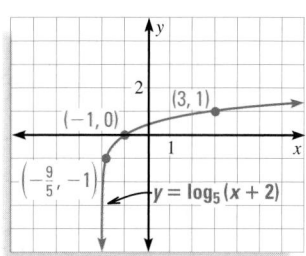

The domain is $x > 0$, and the range is all real numbers.

The domain is $x > -2$, and the range is all real numbers.

EXAMPLE 8 *Using the Graph of a Logarithmic Function*

SCIENCE CONNECTION Graph the model from Example 4, $s = 0.159 + 0.118 \log d$. Then use the graph to estimate the average diameter of the sand particles for a beach whose slope is 0.2.

SOLUTION

You can use a graphing calculator to graph the model. Then, using the *Intersect* feature, you can determine that $s = 0.2$ when $d \approx 2.23$, as shown at the right. So, the average diameter of the sand particles is about 2.23 millimeters.

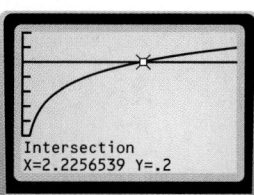

FOCUS ON
APPLICATIONS

SAND A beach with very fine sand makes an angle of about 1° with the horizontal while a beach with pebbles makes an angle of about 17°.

EXTRA EXAMPLE 7
Graph the function. State the domain and range.
a. $y = \log_{1/2} x + 4$ domain: $x > 0$; range: all real numbers

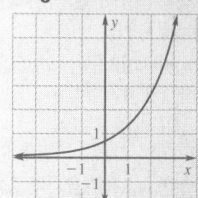

b. $y = \log_3 (x - 2)$ domain: $x > 2$; range: all real numbers

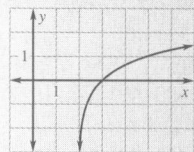

EXTRA EXAMPLE 8
Use the model from Example 4, $s = 0.159 + 0.118 \log d$, to estimate the average diameter of the sand particles for a beach whose slope is 0.15. about 0.84 mm

CHECKPOINT EXERCISES
For use after Examples 7 and 8:
1. Graph $y = \log_3 (x + 1)$. State the domain and range. domain: $x > -1$; range: all real numbers

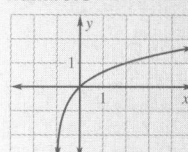

FOCUS ON VOCABULARY
Logarithms were invented to simplify arithmetic operations such as multiplication and division. Logarithm means *ratio number,* which reflects the fact that ratios are constant for powers.

CLOSURE QUESTION
How can you use the technique of switching x and y to find the inverse of $y = \log_5 x$? To find the inverse of $y = \log_5 x$, switch x and y. This gives $x = \log_5 y$. Solving for y produces the inverse function $y = 5^x$.

ASSIGNMENT GUIDE

BASIC
Day 1: pp. 490–491 Exs. 16–46 even, 77, 78
Day 2: pp. 490–492 Exs. 48–76 even, 80, 82–87, 93–107 odd

AVERAGE
Day 1: pp. 490–491 Exs. 16–34 even, 36–47, 77–79
Day 2: pp. 490–492 Exs. 48–76 even, 80, 82–89, 93–111 odd

ADVANCED
Day 1: pp. 490–491 Exs. 16–34 even, 36–47, 77–79
Day 2: pp. 490–492 Exs. 48–74, 80–92, 93–111 odd

BLOCK SCHEDULE
pp. 490–492 Exs. 16–34 even, 36–47, 48–76 even, 77–80, 82–89, 93–111 odd

EXERCISE LEVELS
Level A: *Easier*
16–31, 36–52, 56–58
Level B: *More Difficult*
32–35, 53–55, 59–87
Level C: *Most Difficult*
88–92

✓ **HOMEWORK CHECK**
To quickly check student understanding of key concepts, go over the following exercises: Exs. 16, 24, 36, 48, 58, 68. See also the Daily Homework Quiz:
• Blackline Master (*Chapter 8 Resource Book*, p. 67)
• Transparency (p. 63)

! **COMMON ERROR**
EXERCISES 16–23 It is easy for students to confuse bases, exponents, and powers. Requiring them to read each expression as *log base b of y* rather than *log b of y* will help them correctly identify the base.

2–4, 13, 14. See Additional Answers beginning on page AA1.

GUIDED PRACTICE

Vocabulary Check ✓
1. Complete this statement: The logarithm with base 10 is called the ___?___.
common logarithm

Concept Check ✓
2. Explain why the expressions $\log_3(-1)$ and $\log_1 1$ are not defined. **See margin.**

3. Explain the meaning of $\log_b y$. **See margin.**

4. ERROR ANALYSIS To simplify $\log_2 25$, a student reasoned as shown. Describe the error that the student made. **See margin.**

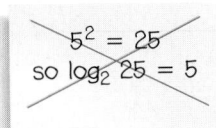

$5^2 = 25$
so $\log_2 25 = 5$

Skill Check ✓ **Rewrite the equation in exponential form.**

5. $\log_3 9 = 2$ $3^2 = 9$ **6.** $\log_5 5 = 1$ $5^1 = 5$ **7.** $\log_{1/2} 4 = -2$ $\frac{1}{2}^{-2} = 4$ **8.** $\log_{19} 1 = 0$ $19^0 = 1$

Evaluate the expression.

9. $\log_2 64$ 6   **10.** $\log_{25} 5$ $\frac{1}{2}$ **11.** $\log_6 1$ 0   **12.** $10^{\log 4}$ 4

Graph the function. State the domain and range. **13–14. See margin for graphs.**

13. $y = \log_2(x+1) - 3$ **14.** $y = \log_{1/2}(x-2) + 1$

13. domain: $x > -1$; range: all real numbers

14. domain: $x > 2$; range: all real numbers

15. 🖩 **SLOPE OF A BEACH** Using the model from Example 4 and a graphing calculator, find the average diameter of the sand particles for a beach whose slope is 0.1. **about 0.316 mm**

PRACTICE AND APPLICATIONS

STUDENT HELP
→ **Extra Practice** to help you master skills is on p. 951.

REWRITING IN EXPONENTIAL FORM Rewrite the equation in exponential form.

16. $\log_4 1024 = 5$ $4^5 = 1024$ **17.** $\log_5 \frac{1}{5} = -1$ $5^{-1} = \frac{1}{5}$ **18.** $\log_{36} \frac{1}{6} = -\frac{1}{2}$ $36^{-1/2} = \frac{1}{6}$ **19.** $\log_8 512 = 3$ $8^3 = 512$

20. $\log_{12} 144 = 2$ $12^2 = 144$ **21.** $\log_{14} 196 = 2$ $14^2 = 196$ **22.** $\log_8 4096 = 4$ $8^4 = 4096$ **23.** $\log_{105} 11{,}025 = 2$ $105^2 = 11{,}025$

EVALUATING EXPRESSIONS Evaluate the expression without using a calculator.

24. $\log_5 125$ 3   **25.** $\log_7 343$ 3   **26.** $\log_8 1$ 0   **27.** $\log_{12} 12$ 1

28. $\log_6 36$ 2   **29.** $\log_4 16$ 2   **30.** $\log_9 729$ 3   **31.** $\log_7 2401$ 4

32. $\log_{1/4} \frac{1}{4}$ 1   **33.** $\log_4 4^{-0.38}$ -0.38 **34.** $\log_4 \frac{1}{2}$ $-\frac{1}{2}$ **35.** $\log_{1/5} 25$ -2

STUDENT HELP
→ **HOMEWORK HELP**
Example 1: Exs. 16–23
Example 2: Exs. 24–35
Example 3: Exs. 36–47
Example 4: Exs. 77–79
Example 5: Exs. 48–55
Example 6: Exs. 56–64
Example 7: Exs. 65–76
Example 8: Exs. 80, 81

🖩 **EVALUATING LOGARITHMS** Use a calculator to evaluate the expression. Round the result to three decimal places.

36. $\log 8$ 0.903   **37.** $\ln 10$ 2.303   **38.** $\log \sqrt{2}$ 0.151   **39.** $\log 3.724$ 0.571

40. $\log 2.54$ 0.405   **41.** $\log 0.3$ -0.523 **42.** $\log 4.05$ 0.607   **43.** $\log 3.5$ 0.544

44. $\ln 4.6$ 1.526   **45.** $\ln 150$ 5.011   **46.** $\ln 6.9$ 1.932   **47.** $\ln 22.5$ 3.114

USING INVERSES Simplify the expression.

48. $5^{\log_5 x}$ x   **49.** $\log_2 2^x$ x   **50.** $9^{\log_9 x}$ x   **51.** $35^{\log_{35} x}$ x

52. $\log_4 16^x$ 2x   **53.** $7^{\log_7 x}$ x   **54.** $\log 100^x$ 2x   **55.** $\log_{20} 8000^x$ 3x

65. domain: $x > 0$; range: all real numbers

66. domain: $x > 0$; range: all real numbers

67. domain: $x > 0$; range: all real numbers

68. domain: $x > 0$; range: all real numbers

69. domain: $x > 0$; range: all real numbers

70. domain: $x > -1$; range: all real numbers

71. domain: $x > 2$; range: all real numbers

72. domain: $x > 2$; range: all real numbers

73. domain: $x > -4$; range: all real numbers

74. domain: $x > 0$; range: all real numbers

75. domain: $x > 0$; range: all real numbers

76. domain: $x > 0$; range: all real numbers

FOCUS ON
APPLICATIONS

TORNADOES form along a boundary between warm humid air from the Gulf of Mexico and cool dry air from the north. When thunderstorm clouds develop along this boundary, the result is violent weather which can produce a tornado.

APPLICATION LINK
www.mcdougallittell.com

FINDING INVERSES Find the inverse of the function.

56. $y = \log_9 x$ $y = 9^x$ **57.** $y = \log_{1/4} x$ $y = \left(\frac{1}{4}\right)^x$ **58.** $y = \log_5 x$ $y = 5^x$

59. $y = \log_{1/2} x$ $y = \left(\frac{1}{2}\right)^x$ **60.** $y = \log_7 49^x$ $y = \frac{x}{2}$ **61.** $y = \ln 6x$ $y = \frac{e^x}{6}$

62. $y = \ln(x - 1)$ **63.** $y = \ln(x + 2)$ **64.** $y = \ln(x - 2)$
$y = 1 + e^x$ $y = -2 + e^x$ $y = 2 + e^x$

GRAPHING FUNCTIONS Graph the function. State the domain and range.
65–76. See margin for graphs.

65. $y = \log_5 x$ **66.** $y = \ln x + 3$ **67.** $y = \log_2 x + 1$

68. $y = \ln x - 1$ **69.** $y = \log_8 x - 2$ **70.** $y = \ln(x + 1)$

71. $y = \log(x - 2)$ **72.** $y = \ln(x - 2)$ **73.** $y = \log_5(x + 4)$

74. $y = \log_{1/2} x - 1$ **75.** $y = \log_{1/4} x - 3$ **76.** $y = \ln x + 5$

77. **SCIENCE CONNECTION**   The pH of a solution is given by the formula

$$pH = -\log[H^+]$$

where $[H^+]$ is the solution's hydrogen ion concentration (in moles per liter). Find the pH of the solution.

a. lemon juice: $[H^+] = 1 \times 10^{-2.4}$ moles per liter   **2.4**

b. vinegar: $[H^+] = 1 \times 10^{-3}$ moles per liter   **3**

c. orange juice: $[H^+] = 1 \times 10^{-3.5}$ moles per liter   **3.5**

78. **GEOMETRY CONNECTION**   Part of the three-dimensional mathematical figure called the horn of Gabriel is shown. The area of the cross section (in the coordinate plane) of the horn is given by:

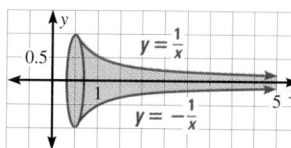

$$A = \frac{2}{\log e}$$

Approximate this area to three decimal places.   **4.605**

SEISMOLOGY In Exercises 79 and 80, use the following information.
The Richter scale is used for measuring the magnitude of an earthquake. The Richter magnitude R is given by the model

$$R = 0.67 \log(0.37E) + 1.46$$

where E is the energy (in kilowatt-hours) released by the earthquake.

79. Suppose an earthquake releases 15,500,000,000 kilowatt-hours of energy. What is the earthquake's magnitude? (Use a calculator.)   **about 8 (7.9982...)**

80. How many kilowatt-hours of energy would the earthquake in Exercise 79 have to release in order to increase its magnitude by one-half of a unit on the Richter scale? Use a graph to solve the problem.   **about 86,400,000,000 kWh**

81. **TORNADOES** Most tornadoes last less than an hour and travel less than 20 miles. The wind speed s (in miles per hour) near the center of a tornado is related to the distance d (in miles) the tornado travels by this model:

$$s = 93 \log d + 65$$

On March 18, 1925, a tornado whose wind speed was about 280 miles per hour struck the Midwest. Use a graph to estimate how far the tornado traveled.

about 205 mi

8.4 *Logarithmic Functions*   **491**

APPLICATION NOTE
EXERCISE 77 The pH is a measure of a solution's acidity. A pH of 7 is considered neutral. A pH below 7 represents an increasingly acidic solution, meaning that the lower the pH the more acidic the solution. A pH above 7 represents an increasingly alkaline solution. By comparison, the pH of toothpaste is about 9.9 and the pH of the gastric juices in the human stomach is 1.0.

EXERCISE 81 Additional information about tornadoes can be found at **www.mcdougallittell.com**.

MATHEMATICAL REASONING
The graph of $f^{-1}(x) = \log_b x$ is the reflection of the graph of $f(x) = b^x$ in the line $y = x$. If you were to place a mirror along the line $y = x$ facing either of these graphs, you would create the other graph with your mirror. This means that the two graphs have mirror symmetry or bilateral symmetry. If you held a mirror along the line $y = x$ facing the graph of $x = \ln(y - 2)$, what graph would you create with the mirror?
$y = e^x + 2$

65.

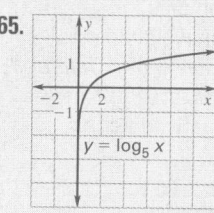

$y = \log_5 x$

66–76. See Additional Answers beginning on page AA1.

ADDITIONAL PRACTICE AND RETEACHING

For Lesson 8.4:
• Practice Levels A, B, and C (*Chapter 8 Resource Book*, p. 56)
• Reteaching with Practice (*Chapter 8 Resource Book*, p. 59)
• See Lesson 8.4 of the *Personal Student Tutor*

For more Mixed Review:
• Search the *Test and Practice Generator* for key words or specific lessons.

4 ASSESS

DAILY HOMEWORK QUIZ

📠 *Transparency Available*

1. Rewrite $\log_7 2401 = 4$ in exponential form. $7^4 = 2401$

2. Evaluate $\log_3 729$ without using a calculator. **6**

3. Simplify $\log_5 625^x$. **4x**

4. Using a calculator, evaluate ln 7.4. Round to three decimal places. **2.001**

5. Find the inverse of $y = \ln(x - 0.5)$. $y = e^x + 0.5$

6. Graph $y = \log_3 x + 3$. State the domain and range.

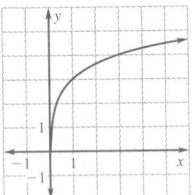

domain: $x > 0$; range: all real numbers

EXTRA CHALLENGE NOTE

↳ Challenge problems for Lesson 8.4 are available in **blackline** format in the *Chapter 8 Resource Book,* p. 64 and at **www.mcdougallittell.com.**

ADDITIONAL TEST PREPARATION

1. OPEN ENDED Give an example of an exponential function whose inverse has a graph that lies in Quadrants I and IV.
any graph of the form $y = ab^x$, where $a > 0$ and $b > 1$ or $0 < b < 1$

Test Preparation

QUANTITATIVE COMPARISON In Exercises 82–87 choose the statement that is true about the given quantities.

Ⓐ The quantity in column A is greater.

Ⓑ The quantity in column B is greater.

Ⓒ The two quantities are equal.

Ⓓ The relationship cannot be determined from the given information.

	Column A	Column B	
82.	$\log_9 9^{2/3}$	$\log 100$	B
83.	$\log_{16} 1$	0	C
84.	$\log_4 16$	$\log_8 64$	C
85.	$f(8)$ if $f(x) = \log_2 x$	4	B
86.	$f(-1)$ if $f(x) = \log_5 5^x$	-1	C
87.	$f\left(\frac{1}{2}\right)$ if $f(x) = \log_3 9^x$	$\log_3 81$	B

★ **Challenge**

EVALUATING EXPRESSIONS Evaluate the expression. (*Hint:* Each expression has the form $\log_b x$. Rewrite the base b and the x-value as powers of the same number.)

88. $\log_{16} 8$ $\frac{3}{4}$ **89.** $\log_{16} 64$ $\frac{3}{2}$ **90.** $\log_9 27$ $\frac{3}{2}$ **91.** $\log_4 512$ $\frac{9}{2}$

EXTRA CHALLENGE
↳ www.mcdougallittell.com

92. CRITICAL THINKING What pattern do you recognize in your answers to Exercises 88–91? If $b = c^n$ and $x = c^m$, then $\log_b x = \frac{m}{n}$.

MIXED REVIEW

EVALUATING NUMERICAL EXPRESSIONS Evaluate the numerical expression. (Review 6.1 for 8.5)

93. $5^2 \cdot 5^3$ 3125   **94.** $(3^{-4})^2$ $\frac{1}{6561}$ **95.** $7^0 \cdot 7^3 \cdot 7^{-2}$ 7   **96.** $\left(\frac{3}{7}\right)^{-2}$ $\frac{49}{9}$

97. $\frac{6^3}{6^4}$ $\frac{1}{6}$ **98.** $\left(\frac{3}{8}\right)^{-3}$ $\frac{512}{27}$ **99.** $(-2^3)^2$ 64   **100.** $\left(\frac{4}{5}\right)^3$ $\frac{64}{125}$

101. $\left(\frac{1}{2}\right)^{-4}$ 16   **102.** $(-3^2)^{-1}$ $-\frac{1}{9}$ **103.** $\frac{2^5}{2^9}$ $\frac{1}{16}$ **104.** $\left(\frac{7}{9}\right)^{-2}$ $\frac{81}{49}$

USING LONG DIVISION Divide using long division. (Review 6.5)

105. $(2x^2 + x - 1) \div (x + 4)$ $2x - 7 + \frac{27}{x + 4}$

106. $(x^2 - 5x + 4) \div (x - 1)$ $x - 4$

107. $(4x^3 + 3x^2 + 2x - 3) \div (x^2 + 2)$ $4x + 3 - \frac{6x + 9}{x^2 + 2}$

108. $(6x^3 - 8x^2 + 7) \div (x - 3)$ $6x^2 + 10x + 30 + \frac{97}{x - 3}$

FINDING A CUBIC MODEL Write a cubic function whose graph passes through the given points. (Review 6.9)

109. $(2, 0), (-3, 0), (0, 0), (3, -3)$ $y = -\frac{1}{6}x(x - 2)(x + 3)$

110. $(3, 0), (2, 0), (-3, 0), (0, -1)$ $y = -\frac{1}{18}(x - 3)(x - 2)(x + 3)$

111. $(4, 0), (6, 0), (-4, 0), (1, 1)$ $y = \frac{1}{75}(x - 4)(x - 6)(x + 4)$

112. $(-2, 0), (-3, 0), (3, 0), (0, 2)$ $y = -\frac{1}{9}(x + 2)(x + 3)(x - 3)$

Properties of Logarithms

What you should learn

GOAL 1 Use properties of logarithms.

GOAL 2 Use properties of logarithms to solve **real-life** problems, such as finding the energy needed for molecular transport in **Exs. 77–79.**

Why you should learn it

▼ To model **real-life** quantities, such as the loudness of different sounds in **Example 5.**

Airport workers wear hearing protection because of the loudness of jet engines.

GOAL 1 USING PROPERTIES OF LOGARITHMS

Because of the relationship between logarithms and exponents, you might expect logarithms to have properties similar to the properties of exponents you studied in Lesson 6.1.

> **▶ ACTIVITY**
> **Developing Concepts**
>
> ### Investigating a Property of Logarithms
>
> **1** Copy and complete the table one row at a time.
>
> | $\log_b u$ | $\log_b v$ | $\log_b uv$ | |
> |---|---|---|---|
> | $\log 10 = ?$ | $\log 100 = ?$ | $\log 1000 = ?$ | 1; 2; 3 |
> | $\log 0.1 = ?$ | $\log 0.01 = ?$ | $\log 0.001 = ?$ | $-1; -2; -3$ |
> | $\log_2 4 = ?$ | $\log_2 8 = ?$ | $\log_2 32 = ?$ | 2; 3; 5 |
>
> **2** Use the completed table to write a conjecture about the relationship among $\log_b u$, $\log_b v$, and $\log_b uv$. $\log_b uv = \log_b u + \log_b v$

In the activity you may have discovered one of the properties of logarithms listed below.

PROPERTIES OF LOGARITHMS

Let b, u, and v be positive numbers such that $b \neq 1$.

PRODUCT PROPERTY $\log_b uv = \log_b u + \log_b v$

QUOTIENT PROPERTY $\log_b \frac{u}{v} = \log_b u - \log_b v$

POWER PROPERTY $\log_b u^n = n \log_b u$

EXAMPLE 1   *Using Properties of Logarithms*

Use $\log_5 3 \approx 0.683$ and $\log_5 7 \approx 1.209$ to approximate the following.

a. $\log_5 \frac{3}{7}$ **b.** $\log_5 21$ **c.** $\log_5 49$

SOLUTION

a. $\log_5 \frac{3}{7} = \log_5 3 - \log_5 7 \approx 0.683 - 1.209 = -0.526$

b. $\log_5 21 = \log_5 (3 \cdot 7) = \log_5 3 + \log_5 7 \approx 0.683 + 1.209 = 1.892$

c. $\log_5 49 = \log_5 7^2 = 2 \log_5 7 \approx 2(1.209) = 2.418$

8.5 *Properties of Logarithms*   **493**

1 PLAN

PACING
Basic: 2 days
Average: 2 days
Advanced: 2 days
Block Schedule: 1 block

LESSON OPENER
ACTIVITY
An alternative way to approach Lesson 8.5 is to use the Activity Lesson Opener:
• Blackline Master (*Chapter 8 Resource Book*, p. 68)
•  Transparency (p. 55)

MEETING INDIVIDUAL NEEDS
• *Chapter 8 Resource Book*
  Prerequisite Skills Review (p. 5)
  Practice Level A (p. 70)
  Practice Level B (p. 71)
  Practice Level C (p. 72)
  Reteaching with Practice (p. 73)
  Absent Student Catch-Up (p. 75)
  Challenge (p. 78)
• *Resources in Spanish*
• *Personal Student Tutor*

NEW-TEACHER SUPPORT
See the Tips for New Teachers on pp. 1–2 of the *Chapter 8 Resource Book* for additional notes about Lesson 8.5.

WARM-UP EXERCISES
Transparency Available
Simplify.
1. $\log 100 + \log 10{,}000$ 6
2. $\log_5 25 + \log_5 125$ 5
3. $\log_5 125 - \log_5 25$ 1
4. $\log_5 25^2$ 4
5. $2 \cdot \log_5 25$ 4

CONCEPT QUESTION
What property of exponents is related to the product property of logarithms? $x^m \cdot x^n = x^{m+n}$

EXTRA EXAMPLE 1
Use $\log_9 5 \approx 0.732$ and $\log_9 11 \approx 1.091$ to approximate the following.
a. $\log_9 \frac{5}{11}$ -0.359
b. $\log_9 55$ 1.823
c. $\log_9 25$ 1.464

EXTRA EXAMPLE 2
Expand $\log_5 2x^6$. Assume x is positive. $\log_5 2 + 6 \log_5 x$

EXTRA EXAMPLE 3
Condense $2 \log_3 7 - 5 \log_3 x$.
$\log_3 \frac{49}{x^5}$

EXTRA EXAMPLE 4
Evaluate $\log_4 8$ using common and natural logarithms. 1.5

 CHECKPOINT EXERCISES
For use after Examples 1–2:
1. Expand $\log_7 \frac{y}{3x^2}$.
$\log_7 y - \log_7 3 - 2 \log_7 x$
For use after Examples 3–4:
2. Condense
$2 \log_8 x - \log_8 5 - 3 \log_8 y$.
$\log_8 \frac{x^2}{5y^3}$
3. Evaluate $\log_6 15$ using natural logarithms. 1.511

You can use the properties of logarithms to expand and condense logarithmic expressions.

STUDENT HELP

Study Tip
When you are expanding or condensing an expression involving logarithms, you may assume the variables are positive.

EXAMPLE 2   *Expanding a Logarithmic Expression*

Expand $\log_2 \frac{7x^3}{y}$. Assume x and y are positive.

SOLUTION

$$\log_2 \frac{7x^3}{y} = \log_2 7x^3 - \log_2 y \qquad \text{Quotient property}$$

$$= \log_2 7 + \log_2 x^3 - \log_2 y \qquad \text{Product property}$$

$$= \log_2 7 + 3 \log_2 x - \log_2 y \qquad \text{Power property}$$

EXAMPLE 3   *Condensing a Logarithmic Expression*

Condense $\log 6 + 2 \log 2 - \log 3$.

SOLUTION

$$\log 6 + 2 \log 2 - \log 3 = \log 6 + \log 2^2 - \log 3 \qquad \text{Power property}$$

$$= \log \left(6 \cdot 2^2\right) - \log 3 \qquad \text{Product property}$$

$$= \log \frac{6 \cdot 2^2}{3} \qquad \text{Quotient property}$$

$$= \log 8 \qquad \text{Simplify.}$$

Logarithms with any base other than 10 or e can be written in terms of common or natural logarithms using the *change-of-base formula*.

CHANGE-OF-BASE FORMULA

Let u, b, and c be positive numbers with $b \neq 1$ and $c \neq 1$. Then:

$$\log_c u = \frac{\log_b u}{\log_b c}$$

In particular, $\log_c u = \dfrac{\log u}{\log c}$ and $\log_c u = \dfrac{\ln u}{\ln c}$.

EXAMPLE 4   *Using the Change-of-Base Formula*

Evaluate the expression $\log_3 7$ using common and natural logarithms.

SOLUTION

Using common logarithms: $\log_3 7 = \dfrac{\log 7}{\log 3} \approx \dfrac{0.8451}{0.4771} \approx 1.771$

Using natural logarithms: $\log_3 7 = \dfrac{\ln 7}{\ln 3} \approx \dfrac{1.946}{1.099} \approx 1.771$

GOAL 2 USING LOGARITHMIC PROPERTIES IN REAL LIFE

Acoustics

EXAMPLE 5 *Using Properties of Logarithms*

The loudness L of a sound (in decibels) is related to the intensity I of the sound (in watts per square meter) by the equation

$$L = 10 \log \frac{I}{I_0}$$

where I_0 is an intensity of 10^{-12} watt per square meter, corresponding roughly to the faintest sound that can be heard by humans.

a. Two roommates each play their stereos at an intensity of 10^{-5} watt per square meter. How much louder is the music when both stereos are playing, compared with when just one stereo is playing?

b. Generalize the result from part (a) by using I for the intensity of each stereo.

Decibel level	Example
130	Jet airplane takeoff
120	Riveting machine
110	Rock concert
100	Boiler shop
90	Subway train
80	Average factory
70	City traffic
60	Conversational speech
50	Average home
40	Quiet library
30	Soft whisper
20	Quiet room
10	Rustling leaf
0	Threshold of hearing

SOLUTION

Let L_1 be the loudness when one stereo is playing and let L_2 be the loudness when both stereos are playing.

a. Increase in loudness $= L_2 - L_1$

$= 10 \log \dfrac{2 \cdot 10^{-5}}{10^{-12}} - 10 \log \dfrac{10^{-5}}{10^{-12}}$ **Substitute for L_2 and L_1.**

$= 10 \log \left(2 \cdot 10^7\right) - 10 \log 10^7$ **Simplify.**

$= 10 \left(\log 2 + \log 10^7 - \log 10^7\right)$ **Product property**

$= 10 \log 2$ **Simplify.**

≈ 3 **Use a calculator.**

▶ The sound is about 3 decibels louder.

b. Increase in loudness $= L_2 - L_1$

$= 10 \log \dfrac{2I}{10^{-12}} - 10 \log \dfrac{I}{10^{-12}}$

$= 10 \left(\log \dfrac{2I}{10^{-12}} - \log \dfrac{I}{10^{-12}} \right)$

$= 10 \left(\log 2 + \log \dfrac{I}{10^{-12}} - \log \dfrac{I}{10^{-12}} \right)$

$= 10 \log 2$

≈ 3

▶ Again, the sound is about 3 decibels louder. This result tells you that the loudness increases by 3 decibels when both stereos are played regardless of the intensity of each stereo individually.

FOCUS ON CAREERS

SOUND TECHNICIAN

Sound technicians operate technical equipment to amplify, enhance, record, mix, or reproduce sound. They may work in radio or television recording studios or at live performances.

CAREER LINK
www.mcdougallittell.com

EXTRA EXAMPLE 5
The Richter magnitude M of an earthquake is based on the intensity I of the earthquake and the intensity I_0 of an earthquake that can be barely felt. One formula used is $M = \log \dfrac{I}{I_0}$. If the intensity of the Los Angeles earthquake in 1994 was $10^{6.8}$ times I_0, what was the magnitude of the earthquake? What magnitude on the Richter scale does an earthquake have if its intensity is 100 times the intensity of a barely felt earthquake? **6.8; 2**

☑ **CHECKPOINT EXERCISES**
For use after Example 5:
1. Use the decibel formula from Example 5. How much louder is the sound of four subway trains passing by a point than just one subway train passing the point?
about 6 decibels louder

FOCUS ON VOCABULARY
What does the change-of-base formula allow you to do?
The formula allows you to express logarithms that are not in base 10 or base e in terms of base 10 or base e.

CLOSURE QUESTION
Use the change-of-base formula to write two equations that are equivalent to $y = \log_2 x$.
$y = \dfrac{\log x}{\log 2}$ and $y = \dfrac{\ln x}{\ln 2}$

DAILY PUZZLER
You have $31.36 in coins. You have twice as many quarters as half-dollars, twice as many dimes as quarters, twice as many nickels as dimes, and twice as many pennies as nickels. How many of each coin do you have?
16 half-dollars, 32 quarters, 64 dimes, 128 nickels, and 256 pennies

ASSIGNMENT GUIDE

BASIC
Day 1: pp. 496–497 Exs. 14–44 even, 46–51
Day 2: pp. 497–499 Exs. 52–57, 58–72 even, 74–76, 88–90, 92–102 even

AVERAGE
Day 1: pp. 496–497 Exs. 14–21, 22–44 even, 46–55
Day 2: pp. 497–499 Exs. 56–84 even, 86–90, 92–102 even

ADVANCED
Day 1: pp. 496–497 Exs. 14–21, 22–44 even, 46–57
Day 2: pp. 497–499 Exs. 58–84 even, 86–91, 92–102 even

BLOCK SCHEDULE
pp. 496–499 Exs. 14–21, 22–44 even, 46–55, 56–84 even, 86–90, 92–102 even

EXERCISE LEVELS
Level A: *Easier*
14–36, 46–51, 88–90
Level B: *More Difficult*
37–45, 52–87
Level C: *Most Difficult*
91

✔ **HOMEWORK CHECK**
To quickly check student understanding of key concepts, go over the following exercises: Exs. 16, 24, 36, 48, 62. See also the Daily Homework Quiz:
• Blackline Master (*Chapter 8 Resource Book*, p. 81)
• Transparency (p. 64)

! COMMON ERROR
EXERCISE 3 Students often assume a distributive property of logarithms. Point out that logarithms are exponents, and there is no distributive property for exponents. Remind students that they learned a property of logarithms for $\log_b u + \log_b v$ in this chapter, but not a property for $\log_b (u + v)$.

1. See next page.

496

GUIDED PRACTICE

Vocabulary Check ✔

1. Give an example of the property of logarithms. **See margin.**
  a. product property   **b.** quotient property   **c.** power property

Concept Check ✔

2. A; using first the power property and then the quotient property gives
$\log \left(\frac{7}{9}\right)^2 = 2 \log \frac{7}{9} = 2(\log 7 - \log 9).$

2. Which is equivalent to $\log \left(\frac{7}{9}\right)^2$? Explain.
  A. $2(\log 7 - \log 9)$ **B.** $\frac{2 \log 7}{\log 9}$ **C.** Neither A nor B

3. Which is equivalent to $\log_8 (5x^2 + 3)$? Explain. **See margin.**
  A. $\log_8 5x^2 + \log_8 3$ **B.** $\log_8 5x^2 \cdot \log_8 3$ **C.** Neither A nor B

3. C; none of the properties of logarithms applies to a sum of terms.

4. Describe two ways to find the value of $\log_6 11$ using a calculator. **See margin.**

4. $\log_6 11 = \frac{\ln 11}{\ln 6}$ and $\log_6 11 = \frac{\log 11}{\log 6}$

Skill Check ✔ **Use a property of logarithms to evaluate the expression.**

5. $\log_3 (3 \cdot 9)$ 3   **6.** $\log_2 4^5$ 10   **7.** $\log_3 \frac{1}{3}$ −1   **8.** $\log_5 \left(\frac{1}{5}\right)^3$ −3

Use $\log_2 7 \approx 2.81$ and $\log_2 21 \approx 4.39$ to approximate the value of the expression.

9. $\log_2 3$ 1.58   **10.** $\log_2 49$ 5.62   **11.** $\log_2 147$ 7.2   **12.** $\log_2 441$ 8.78

13. 🌐 **SOUND INTENSITY** Use the loudness of sound equation in Example 5 to find the difference in the loudness of an average office with an intensity of 1.26×10^{-7} watt per square meter and a broadcast studio with an intensity of 3.16×10^{-10} watt per square meter. **about 26 decibels**

PRACTICE AND APPLICATIONS

STUDENT HELP
▸ **Extra Practice**
to help you master skills is on p. 951.

EVALUATING EXPRESSIONS Use a property of logarithms to evaluate the expression.

14. $\log_2 (4 \cdot 16)$ 6   **15.** $\ln e^{-2}$ −2   **16.** $\log_2 4^3$ 6   **17.** $\log_5 125$ 3

18. $\log_3 9^4$ 8   **19.** $\log \frac{1}{10}$ −1   **20.** $\ln \frac{1}{e^3}$ −3   **21.** $\log (0.01)^3$ −6

38. $\log 6 + 3 \log x + \log y + \log z$

40. $\frac{1}{2} \ln x + 3 \ln y$

41. $\frac{5}{6} \log_3 12 + 9 \log_3 x$

43. $\ln 3 + 4 \ln y - 3 \ln x$

APPROXIMATING EXPRESSIONS Use log 5 ≈ 0.699 and log 15 ≈ 1.176 to approximate the value of the expression.

22. $\log 3$ 0.477   **23.** $\log 25$ 1.398   **24.** $\log 75$ 1.875   **25.** $\log 125$ 2.097

26. $\log \frac{1}{5}$ −0.699   **27.** $\log 225$ 2.352   **28.** $\log \frac{1}{15}$ −1.176   **29.** $\log \frac{1}{3}$ −0.477

STUDENT HELP
▸ **HOMEWORK HELP**
Example 1: Exs. 14–29
Example 2: Exs. 30–45
Example 3: Exs. 46–57
Example 4: Exs. 58–73
Example 5: Exs. 74–85

EXPANDING EXPRESSIONS Expand the expression.

30. $\log_2 9x$ $\log_2 9 + \log_2 x$
31. $\ln 22x$ $\ln 22 + \ln x$
32. $\log 4x^5$ $\log 4 + 5 \log x$
33. $\log_6 x^6$ $6 \log_6 x$

34. $\log_4 \frac{4}{3}$ $1 - \log_4 3$
35. $\log_3 25$ $2 \log_3 5$
36. $\log_6 \frac{10}{3}$ $\log_6 10 - \log_6 3$
37. $\ln 3xy^3$ $\ln 3 + \ln x + 3 \ln y$

38, 40, 41, 43. See margin.

38. $\log 6x^3yz$
39. $\log_8 64x^2$ $2 + 2 \log_8 x$
40. $\ln x^{1/2}y^3$
41. $\log_3 12^{5/6}x^9$

42. $\log \sqrt{x}$ $\frac{1}{2} \log x$
43. $\ln \frac{3y^4}{x^3}$
44. $\log \sqrt[4]{x^3}$ $\frac{3}{4} \log x$
45. $\log_2 \sqrt{4x}$ $1 + \frac{1}{2} \log_2 x$

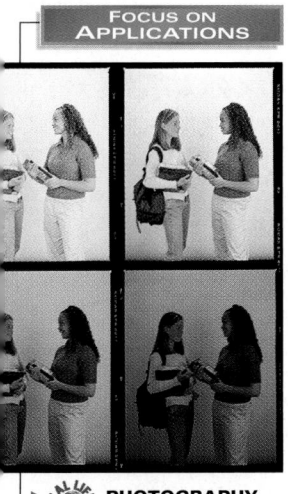

PHOTOGRAPHY
Photographers use f-stops to achieve the desired amount of light in a photo. The smaller the f-stop number, the more light the lens transmits.

APPLICATION LINK
www.mcdougallittell.com

CONDENSING EXPRESSIONS Condense the expression.

46. $\log_5 8 - \log_5 12$ $\log_5 \frac{2}{3}$

47. $\ln 16 - \ln 4$ $\ln 4$

48. $2 \log x + \log 5$ $\log 5x^2$

49. $4 \log_{16} 12 - 4 \log_{16} 2$ $\log_{16} 1296$

50. $3 \ln x + 5 \ln y$ $\ln x^3 y^5$

51. $7 \log_4 2 + 5 \log_4 x + 3 \log_4 y$ $\log_4 128x^5 y^3$

52. $\ln 20 + 2 \ln \frac{1}{2} + \ln x$ $\ln 5x$

53. $\log_3 2 + \frac{1}{2} \log_3 y$ $\log_3 2\sqrt{y}$

54. $10 \log x + 2 \log 10$ $\log 100x^{10}$

55. $3(\ln 3 - \ln x) + (\ln x - \ln 9)$ $\ln \frac{3}{x^2}$

56. $2(\log_6 15 - \log_6 5) + \frac{1}{2} \log_6 \frac{1}{25}$ $\log_6 \frac{9}{5}$ **57.** $\frac{1}{4} \log_5 81 - \left(2 \log_5 6 - \frac{1}{2} \log_5 4\right)$ $\log_5 \frac{1}{6}$

CHANGE-OF-BASE FORMULA Use the change-of-base formula to evaluate the expression.

58. $\log_5 7$ 1.209

59. $\log_7 12$ 1.277

60. $\log_3 16$ 2.524

61. $\log_9 25$ 1.465

62. $\log_2 5$ 2.322

63. $\log_6 9$ 1.226

64. $\log_3 17$ 2.579

65. $\log_5 32$ 2.153

66. $\log_2 125$ 6.966

67. $\log_6 24$ 1.774

68. $\log_4 19$ 2.124

69. $\log_{16} 81$ 1.585

70. $\log_8 \frac{22}{7}$ 0.551

71. $\log_9 \frac{5}{16}$ −0.529

72. $\log_2 \frac{4}{15}$ −1.907

73. $\log_5 \frac{32}{3}$ 1.471

PHOTOGRAPHY In Exercises 74–76, use the following information.
The f-stops on a 35 millimeter camera control the amount of light that enters the camera. Let s be a measure of the amount of light that strikes the film and let f be the f-stop. Then s and f are related by this equation:

$$s = \log_2 f^2$$

74. Expand the expression for s. $s = 2 \log_2 f$

75. The table shows the first eight f-stops on a 35 millimeter camera. Copy and complete the table. Then describe the pattern.   **See margin.**

f	1.414	2.000	2.828	4.000	5.657	8.000	11.314	16.000
s	?	?	?	?	?	?	?	?

76. Many 35 millimeter cameras have nine f-stops. What do you think the ninth f-stop is? Explain your reasoning.

SCIENCE CONNECTION In Exercises 77–79, use the following information.
The energy E (in kilocalories per gram-molecule) required to transport a substance from the outside to the inside of a living cell is given by

$$E = 1.4(\log C_2 - \log C_1)$$

where C_2 is the concentration of the substance inside the cell and C_1 is the concentration outside the cell.

77. Condense the expression for E. $E = 1.4 \log \frac{C_2}{C_1}$

78. The concentration of a particular substance inside a cell is twice the concentration outside the cell. How much energy is required to transport the substance from outside to inside the cell?   **about 0.421 kcal/g-molecule**

79. The concentration of a particular substance inside a cell is six times the concentration outside the cell. How much energy is required to transport the substance from outside to inside the cell?   **about 1.089 kcal/g-molecule**

STUDENT HELP

HOMEWORK HELP
Visit our Web site
www.mcdougallittell.com
for help with Exs. 77–79.

MATHEMATICAL REASONING
EXERCISES 58–73
Since $\log x = \frac{\ln x}{\ln 10}$, $\ln x = (\ln 10)(\log x)$, so $\ln x \approx 2.3026 \log x$. For $x > 1$, where does the graph of $y = \ln x$ lie with respect to the graph of $y = \log x$?   **above it**

APPLICATION NOTE
EXERCISES 74–76
Additional information about photography can be found at
www.mcdougallittell.com.

STUDENT HELP NOTES
Homework Help Students can find help for Exs. 77–79 at
www.mcdougallittell.com.
The information can be printed out for students who don't have access to the Internet.

1a. *Sample answer:* $\log \left(100 \cdot \frac{1}{10}\right) =$
$\log 100 + \log \frac{1}{10} = 2 + (-1) = 1$

b. *Sample answer:* $\log_2 40 - \log_2 5 =$
$\log_2 \frac{40}{5} = \log_2 8 = 3$

c. *Sample answer:* $\log (1000)^2 =$
$2 \log 1000 = 2 \cdot 3 = 6$

75. row entries for s: 1.000, 2.000, 3.000, 4.000, 5.000, 6.000, 7.000, 8.000; The first row of the table shows successive powers of $\sqrt{2}$, and the second row shows the integers, beginning with 1.

8.5 *Properties of Logarithms*   **497**

APPLICATION NOTE
EXERCISES 80–85 Humans can hear sounds as low as 0 decibels. Sounds that are more than 140 decibels cause damaging ear pain. The smallest sound, however, is about one trillion times less intense than a sound that causes damaging pain.

87. $\log 1 = \log \frac{2}{2} = \log 2 - \log 2 = 0$

log 2 and log 3 are given.

$\log 4 = \log 2^2 = 2 \log 2 \approx 0.6020$

$\log 5 = \log 10 - \log 2 \approx 0.6990$

$\log 6 = \log (2 \cdot 3) =$
  $\log 2 + \log 3 \approx 0.7781$

$\log 8 = \log 2^3 = 3 \log 2 \approx 0.9030$

$\log 9 = \log 3^2 = 2 \log 3 \approx 0.9542$

$\log 10 = 1$

$\log 12 = \log (3 \cdot 4) =$
  $\log 3 + 2 \log 2 \approx 1.0791$

$\log 15 = \log (3 \cdot 5) =$
  $\log 3 + \log 5 \approx 1.1761$

$\log 16 = \log 2^4 = 4 \log 2 \approx 1.204$

$\log 18 = \log (2 \cdot 9) =$
  $\log 2 + 2 \log 3 \approx 1.2552$

$\log 20 = \log (2 \cdot 10) =$
  $\log 2 + 1 \approx 1.3010$

log 7, log 11, log 13, log 14, log 17, and log 19 cannot be found. Those numbers with a prime factorization involving only 2, 3, and 5 can be written in terms of these logs. Other numbers cannot.

91. See Additional Answers beginning on page AA1.

ADDITIONAL PRACTICE AND RETEACHING

For Lesson 8.5:

• Practice Levels A, B, and C (*Chapter 8 Resource Book,* p. 70)

• Reteaching with Practice (*Chapter 8 Resource Book,* p. 73)

• See Lesson 8.5 of the *Personal Student Tutor*

For more Mixed Review:

• Search the *Test and Practice Generator* for key words or specific lessons.

FOCUS ON APPLICATIONS

RALPH E. ALLISON developed the first single zero-point audiometer in 1937, making the equipment usable for doctors who had previously used tuning forks to test hearing.

Test Preparation

★ **Challenge**

EXTRA CHALLENGE
www.mcdougallittell.com

🌎 **ACOUSTICS** In Exercises 80–85, use the table and the loudness of sound equation from Example 5.

80. The intensity of the sound made by a propeller aircraft is 0.316 watts per square meter. Find the decibel level of a propeller aircraft. To what sound in the table from Example 5 is a propeller aircraft's sound most similar?
about 115 decibels; between rock concert and riveting machine

81. The intensity of the sound made by Niagara Falls is 0.003 watts per square meter. Find the decibel level of Niagara Falls. To what sound in the table from Example 5 is the sound of Niagara Falls most similar? about 95 decibels; between subway train and boiler shop

82. Three groups of people are in a room, and each group is having a conversation at an intensity of 1.4×10^{-7} watt per square meter. What is the decibel level of the combined conversations in the room? about 56.2 decibels

83. Five cars are in a parking garage, and the sound made by each running car is at an intensity of 3.16×10^{-4} watt per square meter. What is the decibel level of the sound produced by all five cars in the parking garage? about 92 decibels

84. A certain sound has an intensity of I watts per square meter. By how many decibels does the sound increase when the intensity is tripled?
10 log 3, or about 4.8 decibels

85. A certain sound has an intensity of I watts per square meter. By how many decibels does the sound decrease when the intensity is halved?
10 log 0.5, or about 3 decibels less

86. CRITICAL THINKING Tell whether this statement is *true* or *false*:
$\log (u + v) = \log u + \log v$. If true, prove it. If false, give a counterexample.
false; *sample answer:* $\log (10 + 10) = \log 20 \approx 1.301$, but $\log 10 + \log 10 = 2$.

87. *Writing* Let n be an integer from 1 to 20. Use only the fact that $\log 2 \approx 0.3010$ and $\log 3 \approx 0.4771$ to find as many values of $\log n$ as you possibly can. Show how you obtained each value. What can you conclude about the values of n for which you *cannot* find $\log n$? See margin.

88. MULTIPLE CHOICE Which of the following is *not* correct? C

  (A) $\log_2 24 = \log_2 6 + \log_2 4$ (B) $\log_2 24 = \log_2 72 - \log_2 3$

  (C) $\log_2 24 = \log_2 8 + \log_2 16$ (D) $\log_2 24 = 2 \log_2 2 + \log_2 6$

89. MULTIPLE CHOICE Which of the following is equivalent to $\log_5 8$? E

  (A) $\dfrac{\log 5}{\log 8}$ (B) $\dfrac{\log 8}{\log 5}$ (C) $\dfrac{\ln 8}{\ln 5}$ (D) $\dfrac{\ln 13}{\ln 5}$ (E) Both B and C

90. MULTIPLE CHOICE Which of the following is equivalent to $4 \log_3 5$? B

  (A) $\log_3 20$ (B) $\log_3 625$ (C) $\log_3 60$ (D) $\log_3 243$ (E) Both B and C

91. LOGICAL REASONING Use the given hint and properties of exponents to prove each property of logarithms. See margin.

  a. Product property (*Hint:* Let $x = \log_b u$ and let $y = \log_b v$. Then $u = b^x$ and $v = b^y$ so that $\log_b uv = \log_b (b^x \cdot b^y)$.)

  b. Quotient property (*Hint:* Let $x = \log_b u$ and let $y = \log_b v$. Then $u = b^x$ and $v = b^y$ so that $\log_b \dfrac{u}{v} = \log_b \dfrac{b^x}{b^y}$.)

  c. Power property (*Hint:* Let $x = \log_b u$. Then $u = b^x$ and $u^n = b^{nx}$ so that $\log_b u^n = \log_b (b^{nx})$.)

  d. Change-of-base formula (*Hint:* Let $x = \log_b u$, $y = \log_b c$, and $z = \log_c u$. Then $u = b^x$, $c = b^y$, and $u = c^z$ so that $b^x = c^z$.)

MIXED REVIEW

SIMPLIFYING EXPRESSIONS Simplify the expression. (Review 6.1)

92. $3y^2 \cdot y^2$ $3y^4$ **93.** $(y^4)^3$ y^{12} **94.** $(x^3y)^4$ $x^{12}y^4$ **95.** $(-3x^2)^2$ $9x^4$

96. $4x^{-1}y$ $\dfrac{4y}{x}$ **97.** $xy^{-2}x$ $\dfrac{x^2}{y^2}$ **98.** $\dfrac{x^3}{x^{-1}}$ x^4 **99.** $\dfrac{4x^2y^7}{8xy^{-1}}$ $\dfrac{xy^8}{2}$

SOLVING RADICAL EQUATIONS Solve the equation. Check for extraneous solutions. (Review 7.6 for 8.6)

100. $\sqrt[4]{x+2} + 9 = 14$ 623   **101.** $\sqrt[3]{3x-4} = \sqrt[3]{x+10}$ 7

102. $\sqrt{3x+7} = x + 3$ $-2, -1$ **103.** $(5x)^{1/2} - 18 = 32$ 500

 **EVALUATING EXPRESSIONS** Use a calculator to evaluate the expression. Round the result to three decimal places. (Review 8.3, 8.4 for 8.6)

104. e^9 8103.084   **105.** e^{-12} 6.14×10^{-6} **106.** $e^{1.7}$ 5.474   **107.** $e^{-5.632}$ 3.581×10^{-3}

108. $\log 15$ 1.176   **109.** $\log 1.729$ 0.238 **110.** $\ln 16$ 2.773   **111.** $\ln 5.89$ 1.773

MATH & History

Logarithms

APPLICATION LINK
www.mcdougallittell.com

THEN

IN 1614, John Napier published his discovery of logarithms. This discovery allowed calculations with exponents to be performed more easily. In 1632 William Oughtred set two logarithmic scales side by side to form the first slide rule. Because the slide rule could be used to multiply, divide, raise to powers, and take roots, it eliminated the need for many tedious paper-and-pencil calculations.

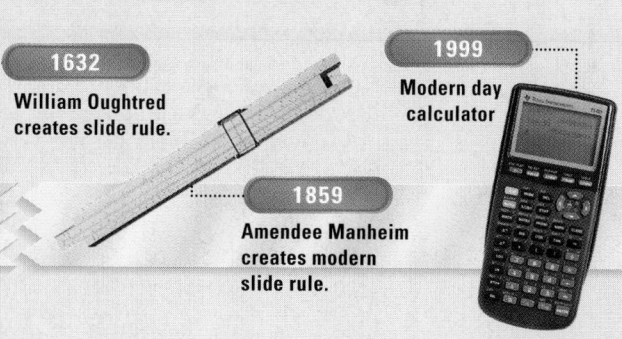

1. To approximate the logarithm of a number, look at the number on the D row and the corresponding value on the L row of the slide rule shown above. For example, log 4 ≈ 0.6. Approximate log 3 and log 5.   log 3 ≈ 0.476 and log 5 ≈ 0.7.

2. Use the product property of logarithms to find log 15.   log 15 = log 3 + log 5 ≈ 1.176

NOW

TODAY, calculators have replaced the use of slide rules but not the use of logarithms. Logarithms are still used for scaling purposes, such as the decibel scale and the Richter scale, because the numbers involved span many orders of magnitude.

1617
John Napier invents Napier's Bones.

1632
William Oughtred creates slide rule.

1859
Amendee Manheim creates modern slide rule.

1999
Modern day calculator

---

4 ASSESS

DAILY HOMEWORK QUIZ

Transparency Available

1. Use a property of logarithms to evaluate $\log_2 8^4$. 12

2. Use $\log 5 \approx 0.699$ and $\log 6 \approx 0.778$ to approximate the value of $\log \dfrac{1}{150}$. -2.176

3. Expand $\ln 7^{1/4}2x^{2/3}$.
   $\dfrac{1}{4}\ln 7 + \ln 2 + \dfrac{2}{3}\ln x$

4. Condense $4\log_5 10 - 2\log_5 50$.
   $\log_5 4$

5. Use the change-of-base formula to evaluate $\log_8 20$. 1.441

EXTRA CHALLENGE NOTE

→ Challenge problems for Lesson 8.5 are available in **blackline** format in the *Chapter 8 Resource Book*, p. 78 and at **www.mcdougallittell.com**.

ADDITIONAL TEST PREPARATION

1. **WRITING** Explain why the properties of logarithms that you learned in this lesson are defined only for $u > 0$ and $v > 0$. Logarithms are not defined for zero or negative numbers.

ADDITIONAL RESOURCES

A **blackline** master with additional Math & History exercises is available in the *Chapter 8 Resource Book*, p. 77.

---

1 Planning the Activity

PURPOSE
To graph logarithmic functions on a graphing calculator using the change-of-base formula.

MATERIALS
- graphing calculator for each student
- Keystroke blackline (*Chapter 8 Resource Book,* p. 69)

PACING
- Example — 5 min
- Exercises — 15 min

▶ LINK TO LESSON
Students should recognize that the change-of-base formula will allow them to rewrite the functions in the example using natural logarithms as well as common logarithms.

2 Managing the Activity

ALTERNATIVE APPROACH
If you are limited by time, you may want to demonstrate the example on an overhead projector. Then have students work in groups of four to complete the exercises.

CLASSROOM MANAGEMENT
Suggest that students use the TABLE function on their calculators to verify the domain and range of the functions.

3 Closing the Activity

★ KEY DISCOVERY
The change of base formula can be used to graph logarithmic functions on a graphing calculator.

ACTIVITY ASSESSMENT
JOURNAL Explain how to graph a function of the form $y = \log_b x$ on a graphing calculator.
See sample answer at right.

1–12. See Additional Answers beginning on page AA1.

▶ ACTIVITY 8.5
Using Technology

Graphing Logarithmic Functions

You can use a graphing calculator to graph logarithmic functions simply by using the LOG or LN key. To graph a logarithmic function having a base other than 10 or *e*, you need to use the change-of-base formula to rewrite the function in terms of common or natural logarithms.

▶ EXAMPLE
Use a graphing calculator to graph $y = \log_2 x$ and $y = \log_2 (x - 3) + 1$.

▶ SOLUTION

1 Rewrite each function in terms of common logarithms.

$$y = \log_2 x \qquad\qquad y = \log_2 (x - 3) + 1$$
$$= \frac{\log x}{\log 2} \qquad\qquad = \frac{\log (x - 3)}{\log 2} + 1$$

2 Enter each function into a graphing calculator.

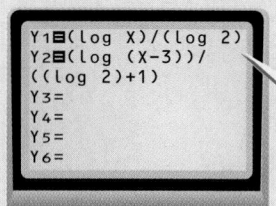

Although the calculator will correctly evaluate the function without parentheses, you can include them for clarity.

3 Graph the functions.

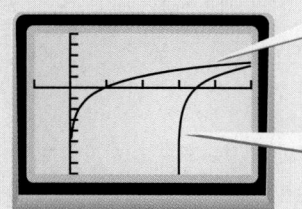

The graph of $y = \log_2 x$ passes through (1, 0), and the line $x = 0$ is a vertical asymptote.

The graph of $y = \log_2 (x - 3) + 1$ passes through (4, 1), and the line $x = 3$ is a vertical asymptote.

▶ EXERCISES

Use a graphing calculator to graph the function. Give the coordinates of a point through which the graph passes, and state the vertical asymptote of the graph.
1–12. Points may vary. Points given are sample responses. See margin for graphs.

1. $y = \log_3 x$ **2.** $y = \log_9 x$ **3.** $y = \log_4 x$

4. $y = \log_7 x$ **5.** $y = \log_5 x$ **6.** $y = \log_{11} x$

7. $y = \log_5 (x - 2)$ **8.** $y = \log_4 (x + 1)$ **9.** $y = \log_2 (x - 5) - 3$

10. $y = \log_4 (x - 7) + 9$ **11.** $y = \log_5 (x + 2) + 6$ **12.** $y = \log_7 (x - 4) + 4$

13. Compare the domains of the graphs of $y = \log x$ and $y = \log |x|$.

1. (1, 0); $x = 0$

2. (1, 0); $x = 0$

3. (1, 0); $x = 0$

4. (1, 0); $x = 0$

5. (1, 0); $x = 0$

6. (1, 0); $x = 0$

7. (3, 0); $x = 2$

8. (0, 0); $x = -1$

9. (6, −3); $x = 5$

10. (8, 9); $x = 7$

11. (−1, 6); $x = -2$

12. (5, 4); $x = 4$

13. *Sample answer:* The domain of $y = \log x$ is all real numbers greater than 0, while that of $y = \log |x|$ is all real numbers except 0. The graph of $y = \log |x|$ is the graph of $y = \log x$ and its reflection in the *y*-axis.

Activity Assessment *Sample answer:*

Rewrite the function as $y = \dfrac{\log x}{\log b}$ or $y = \dfrac{\ln x}{\ln b}$ and enter

it in the function editor. Choose an appropriate viewing window. Then graph the function.

8.6 Solving Exponential and Logarithmic Equations

GOAL 1 SOLVING EXPONENTIAL EQUATIONS

One way to solve exponential equations is to use the property that if two powers with the *same base* are equal, then their exponents must be equal.

For $b > 0$ and $b \neq 1$, if $b^x = b^y$, then $x = y$.

EXAMPLE 1 Solving by Equating Exponents

Solve $4^{3x} = 8^{x+1}$.

SOLUTION

$4^{3x} = 8^{x+1}$	Write original equation.
$(2^2)^{3x} = (2^3)^{x+1}$	Rewrite each power with base 2.
$2^{6x} = 2^{3x+3}$	Power of a power property
$6x = 3x + 3$	Equate exponents.
$x = 1$	Solve for x.

▶ The solution is 1.

✓**CHECK** Check the solution by substituting it into the original equation.

$4^{3 \cdot 1} \stackrel{?}{=} 8^{1+1}$	Substitute 1 for x.
$64 = 64$ ✓	Solution checks.

· · · · · · · · · ·

When it is not convenient to write each side of an exponential equation using the same base, you can solve the equation by taking a logarithm of each side.

EXAMPLE 2 Taking a Logarithm of Each Side

Solve $2^x = 7$.

SOLUTION

$2^x = 7$	Write original equation.
$\log_2 2^x = \log_2 7$	Take $\log_2$ of each side.
$x = \log_2 7$	$\log_b b^x = x$
$x = \dfrac{\log 7}{\log 2} \approx 2.807$	Use change-of-base formula and a calculator.

▶ The solution is about 2.807. Check this in the original equation.

What you should learn

GOAL 1 Solve exponential equations.

GOAL 2 Solve logarithmic equations, as applied in **Example 8**.

Why you should learn it

▼ To solve **real-life** problems, such as finding the diameter of a telescope's objective lens or mirror in **Ex. 69**.

---

1 PLAN

PACING
Basic: 2 days
Average: 2 days
Advanced: 2 days
Block Schedule: 1 block

LESSON OPENER
CALCULATOR ACTIVITY
An alternative way to approach Lesson 8.6 is to use the Calculator Activity Lesson Opener:
- Blackline Master (*Chapter 8 Resource Book,* p. 82)
- Transparency (p. 56)

MEETING INDIVIDUAL NEEDS
- *Chapter 8 Resource Book*
  Prerequisite Skills Review (p. 5)
  Practice Level A (p. 83)
  Practice Level B (p. 84)
  Practice Level C (p. 85)
  Reteaching with Practice (p. 86)
  Absent Student Catch-Up (p. 88)
  Challenge (p. 90)
- *Resources in Spanish*
- *Personal Student Tutor*

NEW-TEACHER SUPPORT
See the Tips for New Teachers on pp. 1–2 of the *Chapter 8 Resource Book* for additional notes about Lesson 8.6.

WARM-UP EXERCISES
Transparency Available

1. Write 27^{4x} as a power of 3. 3^{12x}
2. Evaluate $\log_5 4$. **0.861**
3. Simplify $(4^2)^{2x-3}$. 4^{4x-6}

Complete each statement.

4. If $3^x = 5$, then $\log_3 3^x = $ _____.
   $\log_3 5$
5. If $10^{\log(2x)} = 10^3$, then $\log(2x) = $ _____. **3**

How does your memory retain information? Some studies have shown that people tend to forget 20% of the information they learned the day before. You will learn about solving exponential equations in this lesson, a skill that could be useful in determining how many days you should prepare in advance for a test in order to score at least 85%.

EXTRA EXAMPLE 1
Solve $2^{4x} = 32^{x-1}$. 5

EXTRA EXAMPLE 2
Solve $4^x = 15$. about 1.95

EXTRA EXAMPLE 3
Solve $5^{x+2} + 3 = 25$. about −0.079

EXTRA EXAMPLE 4
Use the information given in Example 4. How long will it take to cool the stew to a temperature of 90°F? about 43 minutes

 CHECKPOINT EXERCISES

For use after Example 1:
1. $9^{2x} = 81^{3x-2}$. 1

For use after Example 2:
2. Solve $5^x = 18$. about 1.80

For use after Example 3:
3. Solve $8 + 10^{5x+4} = 35$.
about −0.514

For use after Example 4:
4. Solve $40e^{0.6x} = 240$.
about 2.99

CONCEPT QUESTION
EXAMPLE 1 Where is the *x*-coordinate of the intersection of the graphs of $y = 4^{3x}$ and $y = 8^{x+1}$?
$x = 1$

MATHEMATICAL REASONING
EXAMPLE 2 Which property of logarithms allows you to take the log of both sides of an equation?
If $x = y$ ($x > 0$ and $y > 0$), then $\log_b x = \log_b y$.

EXAMPLE 3 *Taking a Logarithm of Each Side*

Solve $10^{2x-3} + 4 = 21$.

SOLUTION

$10^{2x-3} + 4 = 21$	Write original equation.
$10^{2x-3} = 17$	Subtract 4 from each side.
$\log 10^{2x-3} = \log 17$	Take common log of each side.
$2x - 3 = \log 17$	$\log 10^x = x$
$2x = 3 + \log 17$	Add 3 to each side.
$x = \frac{1}{2}(3 + \log 17)$	Multiply each side by $\frac{1}{2}$.
$x \approx 2.115$	Use a calculator.

▶ The solution is about 2.115.

✔ **CHECK** Check the solution algebraically by substituting into the original equation. Or, check it graphically by graphing both sides of the equation and observing that the two graphs intersect at $x \approx 2.115$.

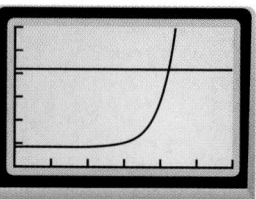

· · · · · · · · ·

Newton's law of cooling states that the temperature T of a cooling substance at time t (in minutes) can be modeled by the equation

$$T = (T_0 - T_R)e^{-rt} + T_R$$

where T_0 is the initial temperature of the substance, T_R is the room temperature, and r is a constant that represents the cooling rate of the substance.

Cooking

EXAMPLE 4 *Using an Exponential Model*

You are cooking *aleecha*, an Ethiopian stew. When you take it off the stove, its temperature is 212°F. The room temperature is 70°F and the cooling rate of the stew is $r = 0.046$. How long will it take to cool the stew to a serving temperature of 100°F?

SOLUTION

You can use Newton's law of cooling with $T = 100$, $T_0 = 212$, $T_R = 70$, and $r = 0.046$.

STUDENT HELP
▶ **HOMEWORK HELP**
Visit our Web site
www.mcdougallittell.com
for extra examples.

$T = (T_0 - T_R)e^{-rt} + T_R$	Newton's law of cooling
$100 = (212 - 70)e^{-0.046t} + 70$	Substitute for T, T_0, T_R, and r.
$30 = 142e^{-0.046t}$	Subtract 70 from each side.
$0.211 \approx e^{-0.046t}$	Divide each side by 142.
$\ln 0.211 \approx \ln e^{-0.046t}$	Take natural log of each side.
$-1.556 \approx -0.046t$	$\ln e^x = \log_e e^x = x$
$33.8 \approx t$	Divide each side by −0.046.

▶ You should wait about 34 minutes before serving the stew.

GOAL 2 SOLVING LOGARITHMIC EQUATIONS

To solve a logarithmic equation, use this property for logarithms with the *same base:*

For positive numbers b, x, and y where $b \neq 1$, $\log_b x = \log_b y$ if and only if $x = y$.

EXAMPLE 5 *Solving a Logarithmic Equation*

Solve $\log_3 (5x - 1) = \log_3 (x + 7)$.

SOLUTION

$\log_3 (5x - 1) = \log_3 (x + 7)$	**Write original equation.**
$5x - 1 = x + 7$	**Use property stated above.**
$5x = x + 8$	**Add 1 to each side.**
$x = 2$	**Solve for x.**

▶ The solution is 2.

✓**CHECK** Check the solution by substituting it into the original equation.

$\log_3 (5x - 1) = \log_3 (x + 7)$	**Write original equation.**
$\log_3 (5 \cdot 2 - 1) \stackrel{?}{=} \log_3 (2 + 7)$	**Substitute 2 for x.**
$\log_3 9 = \log_3 9$ ✓	**Solution checks.**

· · · · · · · · · ·

When it is not convenient to write both sides of an equation as logarithmic expressions with the same base, you can *exponentiate* each side of the equation.

For $b > 0$ and $b \neq 1$, if $x = y$, then $b^x = b^y$.

EXAMPLE 6 *Exponentiating Each Side*

Solve $\log_5 (3x + 1) = 2$.

SOLUTION

$\log_5 (3x + 1) = 2$	**Write original equation.**
$5^{\log_5 (3x + 1)} = 5^2$	**Exponentiate each side using base 5.**
$3x + 1 = 25$	$b^{\log_b x} = x$
$x = 8$	**Solve for x.**

▶ The solution is 8.

✓**CHECK** Check the solution by substituting it into the original equation.

$\log_5 (3x + 1) = 2$	**Write original equation.**
$\log_5 (3 \cdot 8 + 1) \stackrel{?}{=} 2$	**Substitute 8 for x.**
$\log_5 25 \stackrel{?}{=} 2$	**Simplify.**
$2 = 2$ ✓	**Solution checks.**

8.6 *Solving Exponential and Logarithmic Equations*   **503**

! COMMON ERROR
EXAMPLE 2 Students may need reminding that they must take the same log of each side.

EXTRA EXAMPLE 5
Solve $\log_4 (x + 3) = \log_4 (8x + 17)$.
−2

EXTRA EXAMPLE 6
Solve $\log_4 (x + 3) = 2$. 13

✓ CHECKPOINT EXERCISES
For use after Example 5:
1. Solve $\log_2 (2x - 1) = \log_2 (x + 5)$.
6

For use after Example 6:
2. Solve $\log_5 (x - 4) = 2$. 29

MATHEMATICAL REASONING
The logarithmic property used in Step 2 of Example 5 consists of two if-then statements. Which statement justifies Step 2? For positive numbers b, x, and y where $b \neq 1$, if $\log_b x = \log_b y$, then $x = y$.

Why can Extra Example 6 also be solved using the equation $4^2 = x + 3$? The equation is an equivalent exponential equation for the given logarithmic equation.

! COMMON ERROR
EXAMPLE 5 Students may believe that the procedure used to solve the equation can be used to solve a logarithmic equation of the form $\log_b x = \log_d y$, where $b \neq d$. Remind students that the property cited applies only when the bases are the same.

ENGLISH LEARNERS
EXAMPLE 5 You may want to explain that the term *exponentiate* is a verb formed from the noun *exponent*: when a person *exponentiates* each side of an equation, she or he *converts the terms on each side of the equation to exponents.*

EXTRA EXAMPLE 7
Solve $\log_2 x + \log_2 (x - 7) = 3$. **8**

EXTRA EXAMPLE 8
The moment magnitude M of an earthquake that releases energy E (in ergs) can be modeled by the equation $M = 0.291 \ln E + 1.17$. If the earthquake in Prince William Sound in 1964 had a moment magnitude of 8.6, how much energy did it release?
about 123 billion ergs of energy

☑ **CHECKPOINT EXERCISES**
For use after Example 7:
1. Solve $\log_6 (x + 5) + \log_6 x = 2$.
4

For use after Example 8:
2. Solve $14 = 8e^{0.02t}$. **about 28**

MULTIPLE REPRESENTATIONS
EXAMPLE 7 demonstrates a graphical check for extraneous solutions to logarithmic equations. Students may find this graphical technique simpler to use than an algebraic check in which they must calculate the logarithms of numbers.

FOCUS ON VOCABULARY
What does it mean to *exponentiate* both sides of an equation?
Write each side of the equation as the exponent of the same base.

CLOSURE QUESTION
If $\log (x + 1) + \log (x - 1) = 5$, then what expression is equal to 10^5?
$x^2 - 1$

DAILY PUZZLER
Jerome won $50 in a charity raffle. His sister Jill wants him to share his prize with her. He tells her that she can have the fractional part of his prize that is equal to the solution to the equation $(2x + 9)^{5x - 2} = 1$ if she can solve the equation. If Jill solves the equation, how much will she receive? **$20**

STUDENT HELP

Look Back
For help with the zero product property, see p. 257.

Because the domain of a logarithmic function generally does not include all real numbers, you should be sure to check for extraneous solutions of logarithmic equations. You can do this algebraically or graphically.

EXAMPLE 7   *Checking for Extraneous Solutions*

Solve $\log 5x + \log (x - 1) = 2$. Check for extraneous solutions.

SOLUTION

$\log 5x + \log (x - 1) = 2$	Write original equation.
$\log [5x(x - 1)] = 2$	Product property of logarithms
$10^{\log (5x^2 - 5x)} = 10^2$	Exponentiate each side using base 10.
$5x^2 - 5x = 100$	$10^{\log x} = x$
$x^2 - x - 20 = 0$	Write in standard form.
$(x - 5)(x + 4) = 0$	Factor.
$x = 5$ or $x = -4$	Zero product property

The solutions appear to be 5 and -4. However, when you check these in the original equation or use a graphic check as shown at the right, you can see that $x = 5$ is the only solution.

▶ The solution is 5.

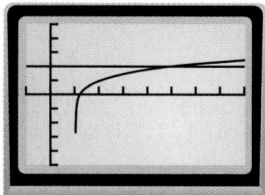

EXAMPLE 8   *Using a Logarithmic Model*

SEISMOLOGY The moment magnitude M of an earthquake that releases energy E (in ergs) can be modeled by this equation:

$$M = 0.291 \ln E + 1.17$$

On May 22, 1960, a powerful earthquake took place in Chile. It had a moment magnitude of 9.5. How much energy did this earthquake release?
▶ Source: U.S. Geological Survey National Earthquake Information Center

SOLUTION

$M = 0.291 \ln E + 1.17$	Write model for moment magnitude.
$9.5 = 0.291 \ln E + 1.17$	Substitute 9.5 for M.
$8.33 = 0.291 \ln E$	Subtract 1.17 from each side.
$28.625 \approx \ln E$	Divide each side by 0.291.
$e^{28.625} \approx e^{\ln E}$	Exponentiate each side using base e.
$2.702 \times 10^{12} \approx E$	$e^{\ln x} = e^{\log_e x} = x$

▶ The earthquake released about 2.7 trillion ergs of energy.

FOCUS ON PEOPLE

CHARLES RICHTER
developed the Richter scale in 1935 as a mathematical means of comparing the sizes of earthquakes. For large earthquakes, seismologists use a different measure called moment magnitude.

GUIDED PRACTICE

Vocabulary Check ✓

Sample answer: $2^{4x} = 8^{x-3}$; $\log(x+1) + \log(x-1) = 2.32$
1. Give an example of an exponential equation and a logarithmic equation.

Concept Check ✓

2. How is solving a logarithmic equation similar to solving an exponential equation? How is it different? **See margin.**

3. Why do logarithmic equations sometimes have extraneous solutions?
because the domain of a logarithmic function does not generally include all real numbers

Skill Check ✓

Solve the equation.

2. *Sample answer:* Both types of equations can be solved by equating exponents or the expressions whose logarithms you are trying to find; otherwise by using the inverse of the function. In the case of an exponential equation, this means taking the logarithm of each side; in the case of a logarithmic equation, raising each side to the same power.

4. $3^x = 14$ 2.402
5. $5^x = 8$ 1.292
6. $9^{2x} = 3^{x-6}$ -2

7. $10^{3x-4} = 0.1$ 1
8. $2^{3x} = 4^{x-1}$ -2
9. $10^{3x-1} + 4 = 32$ $\frac{\log 28 + 1}{3} \approx 0.816$

Solve the equation.

10. $\log x = 2.4$ 251.189
11. $\log x = 3$ 1000
12. $\log_3(2x-1) = 3$ 14

13. $12 \ln x = 44$ 39.121
14. $\log_2(x+2) = \log_2 x^2$ $-1, 2$
15. $\log 3x + \log(x+2) = 1$ $-1 + \frac{\sqrt{39}}{3} \approx 1.082$

ERROR ANALYSIS In Exercises 16 and 17, describe the error.

32. $\frac{\ln 3}{5} \approx 0.220$

34. $\ln \frac{5}{3} \approx 0.511$

35. $\frac{\log 5}{2} \approx 0.3495$

38. $\frac{\ln 5.5}{4} \approx 0.426$

39. $-\frac{1}{12} \log 94 \approx -0.164$

16.
$$4^{x+1} = 8^x$$
$$\log_4 4^{x+1} = \log_4 8^x$$
$$x + 1 = x \log_4 8$$
$$x + 1 = 2x$$
$$1 = x$$

$\log_4 8 \neq 2$ 17.
$$\log_2 5x = 8$$
$$e^{\log_2 5x} = e^8$$
$$5x = e^8$$
$$x = \frac{1}{5}e^8$$

$e^{\log_2 5x} \neq 5x$, since e^x and $\log_2 x$ are not inverse functions.

18. 🌎 **EARTHQUAKES** An earthquake that took place in Alaska on March 28, 1964, had a moment magnitude of 9.2. Use the equation given in Example 8 to determine how much energy this earthquake released. **about 960 billion ergs**

PRACTICE AND APPLICATIONS

STUDENT HELP

▸ **Extra Practice**
to help you master skills is on p. 951.

CHECKING SOLUTIONS Tell whether the *x*-value is a solution of the equation.

19. $\ln x = 27, x = e^{27}$ yes
20. $5 - \log_4 2x = 3, x = 8$ yes

21. $\ln 5x = 4, x = \frac{1}{4}e^5$ no
22. $\log_5 \frac{1}{2}x = 17, x = 2e^{17}$ no

23. $5e^x = 15, x = \ln 3$ yes
24. $e^x + 2 = 18, x = \log_2 16$ no

STUDENT HELP

▸ **HOMEWORK HELP**
Examples 1–3:
  Exs. 23–42
Example 4: Exs. 62–68
Examples 5–7:
  Exs. 19–22, 43–60
Example 8: Exs. 69, 70

SOLVING EXPONENTIAL EQUATIONS Solve the equation.
32, 34, 35, 38, 39.
See margin.

25. $10^{x-3} = 100^{4x-5}$ 1
26. $25^{x-1} = 125^{4x}$ $-\frac{1}{5}$
27. $3^{x-7} = 27^{2x}$ $-\frac{7}{5}$

28. $36^{x-9} = 6^{2x}$ no solution
29. $8^{5x} = 16^{3x+4}$ $\frac{16}{3}$
30. $e^{-x} = 6$ $\ln \frac{1}{6} \approx -1.792$

31. $2^x = 15$ 3.907
32. $1.2e^{-5x} + 2.6 = 3$
33. $4^x - 5 = 3$ $\frac{3}{2}$

34. $-5e^{-x} + 9 = 6$
35. $10^{2x} + 3 = 8$
36. $0.25^x - 0.5 = 2$ -0.661

37. $\frac{1}{4}(4)^{2x} + 1 = 5$ 1
38. $\frac{2}{3}e^{4x} + \frac{1}{3} = 4$
39. $10^{-12x} + 6 = 100$

40. $4 - 2e^x = -23$
  $\ln 13.5 \approx 2.603$
41. $3^{0.1x} - 4 = 5$ 20
42. $-16 + 0.2(10)^x = 35$
  $\log 255 \approx 2.407$

8.6 Solving Exponential and Logarithmic Equations   **505**

ASSIGNMENT GUIDE

BASIC
Day 1: pp. 505–507 Exs. 23–33, 34–42 even, 62, 63, 65
Day 2: pp. 505–508 Exs. 19–22, 44–60 even, 61, 64, 69, 71, 77–87 odd, Quiz 2 Exs. 1–17

AVERAGE
Day 1: pp. 505–508 Exs. 23–40, 62–68 even
Day 2: pp. 505–508 Exs. 19–22, 44–60 even, 61, 67–71 odd, 77–87 odd, Quiz 2 Exs. 1–17

ADVANCED
Day 1: pp. 505–508 Exs. 23–42, 62–68
Day 2: pp. 505–508 Exs. 19–22, 44–60 even, 61, 67–71 odd, 72–76, 77–87 odd, Quiz 2 Exs. 1–17

BLOCK SCHEDULE
pp. 505–508 Exs. 19–24, 26–60 even, 61–71 odd, 77–87 odd, Quiz 2 Exs. 1–17

EXERCISE LEVELS
Level A: *Easier*
25–33, 62–65
Level B: *More Difficult*
19–24, 34–61, 66–70
Level C: *Most Difficult*
71–76

✔ **HOMEWORK CHECK**
To quickly check student understanding of key concepts, go over the following exercises: Exs. 21, 24, 28, 44. See also the Daily Homework Quiz:

• Blackline Master (*Chapter 8 Resource Book*, p. 94)

• 📝 Transparency (p. 65)

MATHEMATICAL REASONING
EXERCISES 62–65 Ask students to describe a set of data points that could be modeled by a logarithmic function. Data that increase at a slower and slower rate but do not appear to be leveling off to a horizontal asymptote can be modeled by a logarithmic function.

APPLICATION NOTE
EXERCISE 67 Currents on the surface of an ocean are formed by winds and the rotation of the earth. Currents in deeper parts of the ocean are the result of the differences in density between adjacent water masses. Both evaporation and cooling increase the density of sea water.

52. $\dfrac{-3 + \sqrt{9 + 4e}}{2} \approx 0.729$

61. $x = \dfrac{\log 8}{3 \log 4 - \log 8} = 1$;

Sample answer: This method is needlessly complicated. I definitely prefer the other way.

67. Subantarctic: 8°;
Antarctic intermediate: 4°; North Atlantic deep: 2°; Antarctic bottom: 0°

SOLVING LOGARITHMIC EQUATIONS Solve the equation. Check for extraneous solutions.

43. $\ln (4x + 1) = \ln (2x + 5)$ 2

44. $\log_2 x = -1$ $\frac{1}{2}$

45. $4 \log_3 x = 28$ 2187

46. $16 \ln x = 30$ $e^{15/8}$

47. $\frac{1}{2} \log_6 16x = 3$ 2916

48. $1 - 2 \ln x = -4$ $e^{5/2}$

49. $2 \ln (-x) + 7 = 14$ $-e^{7/2}$

50. $\log_5 (2x + 15) = \log_5 3x$ 15

51. $\ln x + \ln (x - 2) = 1$ $\frac{1 + \sqrt{1 + e}}{1} \approx 2.928$

52. $\ln x + \ln (x + 3) = 1$

53. $\log_8 (11 - 6x) = \log_8 (1 - x)$ no solution

54. $15 + 2 \log_2 x = 31$ 256

55. $-5 + 2 \ln 3x = 5$ $\frac{1}{3}e^5$

56. $\log (5 - 3x) = \log (4x - 9)$ no solution

57. $6.5 \log_5 3x = 20$ 47.158

58. $\ln (x + 5) = \ln (x - 1) - \ln (x + 1)$ no solution

59. $\ln (5.6 - x) = \ln (18.4 - 2.6x)$ no solution

60. $10 \ln 100x - 3 = 117$ $0.01e^{12}$

61. *Writing* Solve the equation $4^{3x} = 8^{x + 1}$ in Example 1 by taking the common logarithm of each side of the equation. Do you prefer this method to the method shown in Example 1? Why or why not?

62. **COOKING** You are cooking chili. When you take it off the stove, it has a temperature of 205°F. The room temperature is 68°F and the cooling rate of the chili is $r = 0.03$. How long will it take to cool to a serving temperature of 95°F?   **about 54 min**

63. **FINANCE** You deposit $2000 in an account that pays 2% annual interest compounded quarterly. How long will it take for the balance to reach $2400?   **a little over 9 years**

64. **RADIOACTIVE DECAY** You have 20 grams of phosphorus-32 that decays 5% per day. How long will it take for half of the original amount to decay?   **about 13.5 days**

65. **DOUBLING TIME** You deposit $500 in an account that pays 2.5% annual interest compounded continuously. How long will it take for the balance to double?   **about 27.7 years**

66. **HISTORY CONNECTION** The first permanent English colony in America was established in Jamestown, Virginia, in 1607. From 1620 through 1780, the population P of colonial America can be modeled by the equation

$$P = 8863(1.04)^t$$

where t is the number of years since 1620. When was the population of colonial America about 345,000?   **after about 93.4 years, or in 1713**

67. **OCEANOGRAPHY** Oceanographers use the density d (in grams per cubic centimeter) of seawater to obtain information about the circulation of water masses and the rates at which waters of different densities mix. For water with a salinity of 30%, the density is related to the water temperature T (in degrees Celsius) by this equation:

$$d = 1.0245 - e^{0.1266T - 7.828}$$

Use the equation to find the temperature of each layer of water whose density is given in the diagram.

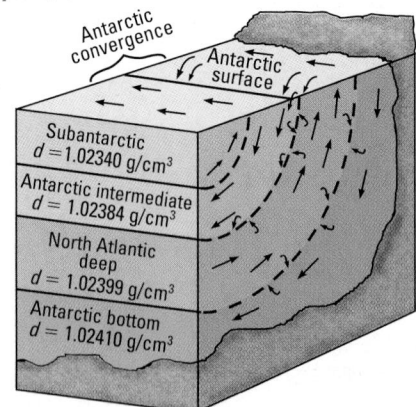

APPARENT
MAGNITUDE of a
star is a number indicating
the brightness of the star as
seen from Earth. The greater
the apparent magnitude, the
fainter the star.

 APPLICATION LINK
www.mcdougallittell.com

68. 🌐 **MUON DECAY** A muon is an elementary particle that is similar to an electron, but much heavier. Muons are unstable—they very quickly decay to form electrons and other particles. In an experiment conducted in 1943, the number m of muon decays (of an original 5000 muons) was related to the time t (in microseconds) by this model:

$$m = e^{6.331 - 0.403t}$$

After how many microseconds were 204 decays recorded? **after about 2.51μs**

69. 🌐 **ASTRONOMY** The relationship between a telescope's limiting magnitude (the apparent magnitude of the dimmest star that can be seen with the telescope) and the diameter of the telescope's objective lens or mirror can be modeled by

$$M = 5 \log D + 2$$

where M is the limiting magnitude and D is the diameter (in millimeters) of the lens or mirror. If a telescope can reveal stars with a magnitude of 12, what is the diameter of its objective lens or mirror? ▶ *Source: Practical Astronomy* **100 mm**

70. 🌐 **ALTIMETER** An altimeter is an instrument that finds the height above sea level by measuring the air pressure. The height and the air pressure are related by the model

$$h = -8005 \ln \frac{P}{101,300}$$

where h is the height (in meters) above sea level and P is the air pressure (in pascals). What is the air pressure when the height is 4000 meters above sea level?

61,461 pascals

Test
Preparation

71. **MULTI-STEP PROBLEM** A simple technique that biologists use to estimate the age of an African elephant is to measure the length of the elephant's footprint and then calculate its age using the equation

$$l = 45 - 25.7e^{-0.09a}$$

where l is the length of the footprint (in centimeters) and a is the age (in years).
▶ Source: *Journal of Wildlife Management*

a. Use the equation to find the ages of the elephants whose footprints are shown.

b. Solve the equation for a, and use this equation to find the ages of the elephants whose footprints are shown.

c. *Writing* Compare the methods you used in parts (a) and (b). Which method do you prefer? Explain.

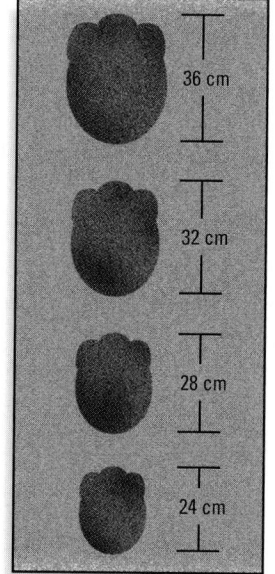

36 cm

32 cm

28 cm

24 cm

71. a. about 12, 8, 5, and 2 years

b. about 12, 8, 5, and 2 years

c. *Sample answer:* It is difficult to solve for a in part (b), so I prefer the method in part (a).

★ **Challenge**

SOLVING EQUATIONS Solve the equation.

72. $2^{x + 3} = 5^{3x - 1}$ $\dfrac{3 + \log_2 5}{3 \log_2 5 - 1} \approx 0.892$ **73.** $10^{5x + 2} = 5^{4 - x}$ $\dfrac{4 \log 5 - 2}{5 + \log 5} \approx 0.140$

74. $\log_3 (x - 6) = \log_9 2x$ $\dfrac{7 + \sqrt{13}}{2} \approx 10.606$

75. $\log_4 x = \log_8 4x$ **16**

EXTRA CHALLENGE
www.mcdougallittell.com

76. *Writing* In Exercises 72–75 you solved exponential and logarithmic equations with different bases. Describe general methods for solving such equations.

See margin.

8.6 *Solving Exponential and Logarithmic Equations*   **507**

APPLICATION NOTE
EXERCISE 68 The average life of a muon is only 2.20×10^{-6} seconds.

EXERCISE 69 Additional information about apparent magnitude of a star is available at **www.mcdougallittell.com**.

76. *Sample answer:* For an exponential equation, take the logarithm of both sides to one of the bases, use the change of base formula for the other side of the equation, or take the common logarithm or natural logarithm of both sides of the equation. For a logarithmic equation, use the change of base formula to rewrite both sides in terms of logarithms with a single base, then solve normally.

ADDITIONAL PRACTICE AND RETEACHING

For Lesson 8.6:
• Practice Levels A, B, and C (*Chapter 8 Resource Book*, p. 83)
• Reteaching with Practice (*Chapter 8 Resource Book*, p. 86)
• 🖥 See Lesson 8.6 of the *Personal Student Tutor*

For more Mixed Review:
• 🖥 Search the *Test and Practice Generator* for key words or specific lessons.

DAILY HOMEWORK QUIZ

📖 *Transparency Available*

Tell whether the *x*-value is a solution of the equation.

1. $\log_6 4x = 3$; $x = 0.25 \cdot 6^3$ **yes**

2. $2e^x - 3 = 9$; $x = \ln 12$ **no**

Solve the equation.

3. $3^{2x} = 27^{x+2}$ **−6**

4. $5e^{3x} + 2 = 17$ $\frac{\ln 3}{3} \approx 0.366$

Solve the equation. Check for extraneous solutions.

5. $\log_4 (5x - 11) = \log_4 (3 - 2x)$
   no solution

6. $-3 \ln \frac{x}{2} = 4$ $2e^{-4/3} \approx 0.527$

EXTRA CHALLENGE NOTE
→ Challenge problems for Lesson 8.6 are available in **blackline** format in the *Chapter 8 Resource Book*, p. 90 and at **www.mcdougallittell.com**.

ADDITIONAL TEST PREPARATION

1. WRITING Explain why it is sometimes necessary to discard negative solutions of logarithmic equations. **Negative numbers do not have logarithms.**

2. WRITING You solve the equation $2e^x = 8$. Your solution is ln 4. The *x*-coordinate of the intersection of the graphs of $y = 2e^x$ and $y = 8$ is approximately 1.4. How does this verify your solution of $2e^x = 8$? **Since ln 4 ≈ 1.4, the intersection verifies that the solution is ln 4.**

ADDITIONAL RESOURCES

An alternative Quiz for Lessons 8.4–8.6 is available in the *Chapter 8 Resource Book*, p. 91.

77–78, 5–7. See Additional Answers beginning on page AA1.

MIXED REVIEW

MAKING SCATTER PLOTS Draw a scatter plot of the data. Then approximate an equation of the best-fitting line. (Review 2.5 for 8.7)
77, 78. Lines may vary. See margin for graphs.

$y = 0.305x + 1.780$

77.

x	−2	−1	−0.5	0	0.5	1	2	3	3.5	4
y	1.25	1.5	1.5	2	1.75	2	2.5	2.5	2.75	3.25

$y = 0.305x + 2.640$

78.

x	−4	−3	−2.5	−2	−1.5	−1	0	1	1.5	2
y	1.5	1.75	1.75	2.25	2	2.25	2.75	2.75	3	3.5

THE SUBSTITUTION METHOD Solve the linear system using the substitution method. (Review 3.2 for 8.7)

79. $2x - y = 3$ **(4, 5)**
   $3x - 2y = 2$

80. $2x + y = 4$ **(1, 2)**
   $x + y = 3$

81. $x + 4y = -24$ **(0, −6)**
   $x - 4y = 24$

82. $x - 3y = -3$ **(3, 2)**
   $2x + y = 8$

83. $2x + y = -1$ **no solution**
   $-4x - 2y = -5$

84. $-x + 6y = -32$ **(2, −5)**
   $7x - 2y = 24$

FACTORING Factor the polynomial by grouping. (Review 6.4)

85. $3x^3 - 6x^2 + 4x - 8$ $(3x^2 + 4)(x - 2)$ **86.** $2x^3 - 5x^2 + 16x - 40$ $(x^2 + 8)(2x - 5)$

87. $7x^3 + 4x^2 + 35x + 20$ $(x^2 + 5)(7x + 4)$ **88.** $4x^3 - 3x^2 + 8x - 6$ $(x^2 + 2)(4x - 3)$

QUIZ 2
Self-Test for Lessons 8.4–8.6

Evaluate the expression without using a calculator. (Lesson 8.4)

1. $\log_2 8$ **3**

2. $\log_5 625$ **4**

3. $\log_8 512$ **3**

4. Find the inverse of the function $y = \ln (x + 3)$. (Lesson 8.4) $y = e^x - 3$

5. domain: $x > 0$; range: all real numbers

Graph the function. State the domain and range. (Lesson 8.4)   5–7. See margin for graphs.

5. $y = 1 + \log_4 x$

6. domain: $x > -3$; range: all real numbers

6. $y = \log_4 (x + 3)$

7. $y = 2 + \log_6 (x - 2)$

7. domain: $x > 2$; range: all real numbers

Use a property of logarithms to evaluate the expression. (Lesson 8.5)

8. $\log_3 (3 \cdot 27)$ **4**

9. $\log_2 \frac{1}{2}$ **−1**

10. $\ln e^2$ **2**

11. Expand the expression $\log_4 x^{1/2} y^4$. (Lesson 8.5) $\frac{1}{2}\log_4 x + 4 \log_4 y$

12. Condense the expression $2 \log_6 14 + 3 \log_6 x - \log_6 7$. (Lesson 8.5) $\log_6 28x^3$

13. Use the change-of-base formula to evaluate the expression $\log_4 22$. (Lesson 8.5)   **2.230**

Solve the equation. (Lesson 8.6)

14. $3e^x - 1 = 14$ **ln 5**   **15.** $3 \log_2 x = 28$ $2^{28/3}$ **16.** $\ln (2x + 7) = \ln (x - 4)$ **no solution**

17. 🌐 **EARTHQUAKES** An earthquake that took place in Indonesia on February 1, 1938, had a moment magnitude of 8.5. Use the model $M = 0.291 \ln E + 1.17$, where *M* is the moment magnitude and *E* is the energy (in ergs) of an earthquake, to determine how much energy the Indonesian earthquake released. (Lesson 8.6)   **about 87 billion ergs**

Modeling with Exponential and Power Functions

GOAL 1 MODELING WITH EXPONENTIAL FUNCTIONS

Just as two points determine a line, two points also determine an exponential curve.

EXAMPLE 1 *Writing an Exponential Function*

Write an exponential function $y = ab^x$ whose graph passes through $(1, 6)$ and $(3, 24)$.

SOLUTION

Substitute the coordinates of the two given points into $y = ab^x$ to obtain two equations in a and b.

$6 = ab^1$ **Substitute 6 for *y* and 1 for *x*.**

$24 = ab^3$ **Substitute 24 for *y* and 3 for *x*.**

To solve the system, solve for a in the first equation to get $a = \frac{6}{b}$, then substitute into the second equation.

$24 = \left(\frac{6}{b}\right)b^3$ **Substitute $\frac{6}{b}$ for *a*.**

$24 = 6b^2$ **Simplify.**

$4 = b^2$ **Divide each side by 6.**

$2 = b$ **Take the positive square root.**

Using $b = 2$, you then have $a = \frac{6}{b} = \frac{6}{2} = 3$. So, $y = 3 \cdot 2^x$.

. . . .

When you are given more than two points, you can decide whether an exponential model fits the points by plotting the natural logarithms of the y-values against the x-values. If the new points $(x, \ln y)$ fit a linear pattern, then the original points (x, y) fit an exponential pattern.

Graph of points (x, y)

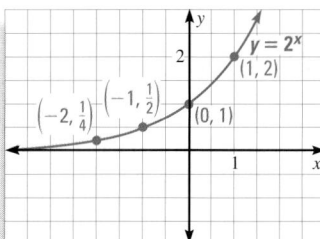

The graph is an exponential curve.

Graph of points $(x, \ln y)$

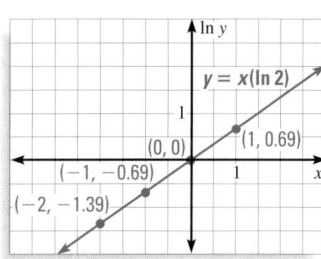

The graph is a line.

What you should learn

GOAL 1 Model data with exponential functions.

GOAL 2 Model data with power functions, as applied in **Example 5**.

Why you should learn it

▼ To solve **real-life** problems, such as finding the number of U.S. stamps issued in **Ex. 56**.

PACING
Basic: 1 day
Average: 1 day
Advanced: 1 day
Block Schedule: 0.5 block with 8.8

LESSON OPENER
VISUAL APPROACH
An alternative way to approach Lesson 8.7 is to use the Visual Approach Lesson Opener:
• Blackline Master (*Chapter 8 Resource Book*, p. 95)
• Transparency (p. 57)

MEETING INDIVIDUAL NEEDS
• *Chapter 8 Resource Book*
Prerequisite Skills Review (p. 5)
Practice Level A (p. 97)
Practice Level B (p. 98)
Practice Level C (p. 99)
Reteaching with Practice (p. 100)
Absent Student Catch-Up (p. 102)
Challenge (p. 104)
• *Resources in Spanish*
• *Personal Student Tutor*

NEW-TEACHER SUPPORT
See the Tips for New Teachers on pp. 1–2 of the *Chapter 8 Resource Book* for additional notes about Lesson 8.7.

WARM-UP EXERCISES

Transparency Available

1. The graph of $y = 2 \cdot 5^x$ passes through $(x_1, 250)$ and $(2, y_2)$. Find the values of x_1 and y_2.
$x_1 = 3$ and $y_2 = 50$

2. What is the general form of an exponential equation? $y = ab^x$

3. Write $2.7 = 4^b$ in logarithmic form. $\log_4 2.7 = b$

4. Use the properties of exponents to simplify $4^{0.5x + 3}$.
$64 \cdot 2^x$

5. Evaluate $\frac{\log 2.5}{\log 0.7}$. about -2.569

EXTRA EXAMPLE 1
Write an exponential function whose graph passes through (−1, 0.0625) and (2, 32).
$y = 0.5 \cdot 8^x$

EXTRA EXAMPLE 2
The ordered pairs represent the number of years t since buying a car and its trade-in value V in dollars. (0, 12,995), (1, 9050), (2, 6300), (3, 4495), (4, 3150), (5, 2205), (6, 1550), (7, 1085), (8, 750)
a. Draw a scatter plot of ln V versus t. Is an exponential model a good fit for the data? **yes**

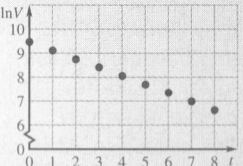

b. Find an exponential model for the original data.
Sample answer using (1, 9050) and (6, 1550): $V = 12,880(0.703)^t$

EXTRA EXAMPLE 3
Use a graphing calculator to find an exponential model for the data in Extra Example 2. Use the model to estimate the trade-in value of the car after 12 years.
$V = 13,000(0.701)^t$; about $183

 CHECKPOINT EXERCISES
For use after Example 1:
1. Write an exponential function whose graph passes through (2, 9) and (4, 20.25). $y = 4 \cdot 1.5^x$
For use after Examples 2 and 3:
2. Find an exponential model to fit the data. Use the model to estimate y when x is 15. (0, 14.7), (1, 13.5), (2, 12.9), (3, 12.4), (4, 11.9), (5, 11.4), (6, 10.9), (7, 10.4), (8, 10.0), (9, 9.6) $y = 14.3(0.956)^x$; 7.3

MATHEMATICAL REASONING
The exponential model has $y = 0$ as a horizontal asymptote. Why does this make it a good model for the cell-phone data in Example 2?

510

Communications

STUDENT HELP

Look Back
For help with scatter plots and best-fitting lines, see pp. 100–101.

EXAMPLE 2 *Finding an Exponential Model*

The table gives the number y (in millions) of cell-phone subscribers from 1988 to 1997 where t is the number of years since 1987.

t	1	2	3	4	5	6	7	8	9	10
y	1.6	2.7	4.4	6.4	8.9	13.1	19.3	28.2	38.2	48.7

▶ Source: Cellular Telecommunications Industry Association

a. Draw a scatter plot of ln y versus x. Is an exponential model a good fit for the original data?

b. Find an exponential model for the original data.

SOLUTION

a. Use a calculator to create a new table of values.

t	1	2	3	4	5	6	7	8	9	10
ln y	0.47	0.99	1.48	1.86	2.19	2.57	2.96	3.34	3.64	3.89

Then plot the new points as shown. The points lie close to a line, so an exponential model should be a good fit for the original data.

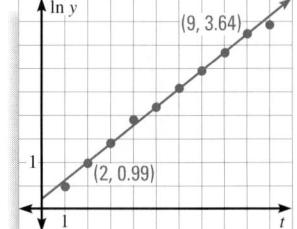

b. To find an exponential model $y = ab^t$, choose two points on the line, such as (2, 0.99) and (9, 3.64). Use these points to find an equation of the line. Then solve for y.

$\ln y = 0.379t + 0.233$ **Equation of line**

$y = e^{0.379t + 0.233}$ **Exponentiate each side using base e.**

$y = e^{0.233}(e^{0.379})^t$ **Use properties of exponents.**

$y = 1.30(1.46)^t$ **Exponential model**

· · · · · · · · · ·

A graphing calculator that performs exponential regression does essentially what is done in Example 2, but uses all of the original data.

Communications

EXAMPLE 3 *Using Exponential Regression*

Use a graphing calculator to find an exponential model for the data in Example 2. Use the model to estimate the number of cell-phone subscribers in 1998.

SOLUTION

Enter the original data into a graphing calculator and perform an exponential regression. The model is:

$$y = 1.30(1.46)^t$$

Substituting $t = 11$ (for 1998) into the model gives $y = 1.30(1.46)^{11} \approx 84$ million cell-phone subscribers.

```
ExpReg
y=a*b^x
a=1.30076406
b=1.458520596
r²=.9934944894
r=.9967419372
```

Mathematical Reasoning *Sample answer:*
Cell-phone use was expensive and not widely available when it was first used by business. Only in the late 1990s did it become affordable and attractive for personal use. These factors drove the sales of cell phones.

GOAL 2 MODELING WITH POWER FUNCTIONS

Recall from Lesson 7.3 that a power function has the form $y = ax^b$. Because there are only two constants (a and b), only two points are needed to determine a power curve through the points.

EXAMPLE 4 *Writing a Power Function*

Write a power function $y = ax^b$ whose graph passes through (2, 5) and (6, 9).

SOLUTION

Substitute the coordinates of the two given points into $y = ax^b$ to obtain two equations in a and b.

$5 = a \cdot 2^b$ **Substitute 5 for *y* and 2 for *x*.**

$9 = a \cdot 6^b$ **Substitute 9 for *y* and 6 for *x*.**

To solve the system, solve for a in the first equation to get $a = \dfrac{5}{2^b}$, then substitute into the second equation.

$9 = \left(\dfrac{5}{2^b}\right)6^b$ **Substitute $\frac{5}{2^b}$ for *a*.**

$9 = 5 \cdot 3^b$ **Simplify.**

$1.8 = 3^b$ **Divide each side by 5.**

$\log_3 1.8 = b$ **Take $\log_3$ of each side.**

$\dfrac{\log 1.8}{\log 3} = b$ **Use the change-of-base formula.**

$0.535 \approx b$ **Use a calculator.**

Using $b = 0.535$, you then have $a = \dfrac{5}{2^b} = \dfrac{5}{2^{0.535}} \approx 3.45$. So, $y = 3.45x^{0.535}$.

· · · · · · · · · ·

When you are given more than two points, you can decide whether a power model fits the points by plotting the natural logarithms of the y-values against the natural logarithms of the x-values. If the new points ($\ln x$, $\ln y$) fit a linear pattern, then the original points (x, y) fit a power pattern.

Graph of points (x, y)

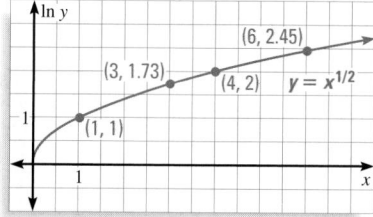

Graph of points (ln x, ln y)

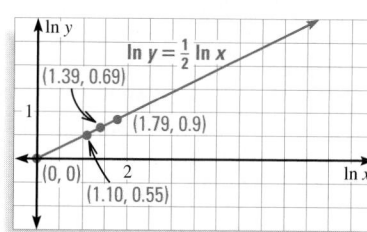

The graph is a power curve.   The graph is a line.

TEACHING TIPS
You can suggest that students use a graphing calculator to find a linear regression model for the scatter plot data in Example 2.

APPLICATION NOTE
EXAMPLE 3 With the availability of technological tools, such as graphing and software utilities that can fit linear and nonlinear functions to data, the focus on creating mathematical models has switched from performing computations to making decisions about appropriate models.

 EXTRA EXAMPLE 4
Write a power function whose graph passes through (3, 8) and (9, 12). $y = 5.33x^{0.369}$

✔ **CHECKPOINT EXERCISES**
For use after Example 4:
1. Write a power function whose graph passes through (5, 2) and (10, 6). $y = 0.156x^{1.585}$

EXTRA EXAMPLE 5
The ordered pairs (t, r) describe the circular area r (square feet) that oil from a leaking oil tanker covers t minutes after it begins leaking. (1, 28.26), (5, 706.5), (10, 2826), (15, 6358.5), (20, 11,304), (25, 17,663), (35, 34,618.5), (60, 101,736).
a. Draw a scatter plot of ln t versus ln r. Is the power model a good fit for the original data?
yes

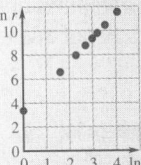

b. Find a power model for the original data. $r = 28.3t^2$

EXTRA EXAMPLE 6
Use a graphing calculator to find a power model for the data in Extra Example 5. Use the model to estimate the area that will be covered by the leaking oil after an hour and a half. $r = 28.3t^2$; **about 229,230 ft²**

 CHECKPOINT EXERCISES
For use after Examples 5 and 6:
1. Find a power model to fit the data. Use the model to estimate y when x is 20. (1, 1), (2, 4.9), (3, 12.5), (4, 24.3), (5, 40.5), (6, 61.6), (7, 87.85), (8, 119.4), (9, 156.6), (10, 199.5)
$y = x^{2.3}$; **about 983**

FOCUS ON VOCABULARY
In an exponential function, the base is constant and the exponent varies. In a power function, the base varies and the exponent is constant.

CLOSURE QUESTION
How can you use a line to determine an exponential model for a set of points (x, y)? to determine a power model for the set of points? **Plot ln y versus x; plot ln y versus ln x.**

512

Astronomy

EXAMPLE 5 *Finding a Power Model*

The table gives the mean distance x from the sun (in astronomical units) and the period y (in Earth years) of the six planets closest to the sun.

Planet	Mercury	Venus	Earth	Mars	Jupiter	Saturn
x	0.387	0.723	1.000	1.524	5.203	9.539
y	0.241	0.615	1.000	1.881	11.862	29.458

a. Draw a scatter plot of ln y versus ln x. Is a power model a good fit for the original data?

b. Find a power model for the original data.

SOLUTION

a. Use a calculator to create a new table of values.

ln x	−0.949	−0.324	0.000	0.421	1.649	2.255
ln y	−1.423	−0.486	0.000	0.632	2.473	3.383

Then plot the new points, as shown at the right. The points lie close to a line, so a power model should be a good fit for the original data.

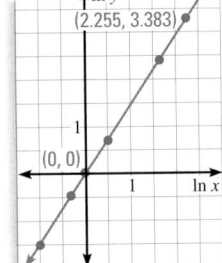

b. To find a power model $y = ax^b$, choose two points on the line, such as $(0, 0)$ and $(2.255, 3.383)$. Use these points to find an equation of the line. Then solve for y.

$\ln y = 1.5 \ln x$ **Equation of line**

$\ln y = \ln x^{1.5}$ **Power property of logarithms**

$y = x^{1.5}$ $\log_b x = \log_b y$ **if and only if** $x = y$.

. . . .

A graphing calculator that performs power regression does essentially what is done in Example 5, but uses all of the original data.

FOCUS ON PEOPLE

JOHANNES KEPLER, a German astronomer and mathematician, was the first person to observe that a planet's distance from the sun and its period were related by the power function in Examples 5 and 6.

EXAMPLE 6 *Using Power Regression*

ASTRONOMY Use a graphing calculator to find a power model for the data in Example 5. Use the model to estimate the period of Neptune, which has a mean distance from the sun of 30.043 astronomical units.

SOLUTION

Enter the original data into a graphing calculator and perform a power regression. The model is:

$$y = x^{1.5}$$

Substituting 30.043 for x in the model gives $y = (30.043)^{1.5} \approx 165$ years for the period of Neptune.

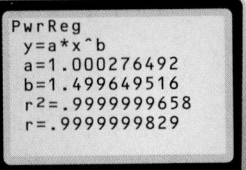

```
PwrReg
 y=a*x^b
 a=1.000276492
 b=1.499649516
 r²=.9999999658
 r=.9999999829
```

512   **Chapter 8**   *Exponential and Logarithmic Functions*

GUIDED PRACTICE

Vocabulary Check ✓

1. Complete this statement: When you are given more than two points, you can decide whether you can fit a(n) _?_ model to the points by plotting the natural logarithms of the y-values against the x-values. **exponential**

Concept Check ✓

2. How many points determine an exponential function $y = ab^x$? How many points determine a power function $y = ax^b$? **2; 2**

3. No, since 0 is not in the domain of $f(x) = \ln x$.

3. Can you use the procedure in Example 5 to find a power model for a data set where one of the points has an x-coordinate of 0? Explain why or why not.

Skill Check ✓

Write an exponential function of the form $y = ab^x$ whose graph passes through the given points.

$y = \frac{2}{9}(3)^x$ $y = 2 \cdot 2^x$

4. $(1, 3), (2, 36)$ $y = \frac{1}{4} \cdot 12^x$ 5. $(2, 2), (4, 18)$ 6. $(1, 4), (3, 16)$

7. $(2, 3.5), (1, 5.2)$ 8. $(5, 8), (3, 32)$ 9. $\left(1, \frac{1}{2}\right), \left(3, \frac{3}{8}\right)$

     $y = \frac{2704}{350} \cdot \left(\frac{35}{52}\right)^x$ $y = 256(0.5)^x$ $y = \frac{1}{\sqrt{3}} \cdot \left(\frac{\sqrt{3}}{2}\right)^x$

Write a power function of the form $y = ax^b$ whose graph passes through the given points.

10. $(3, 27), (9, 243)$ $y = 3x^2$ 11. $(1, 2), (4, 32)$ $y = 2x^2$ 12. $(4, 48), (2, 6)$ $y = \frac{3}{4}x^3$

13. $(1, 4), (3, 8)$ $y = 4x^{0.631}$ 14. $(4.5, 9.2), (1, 6.4)$ 15. $\left(2, \frac{1}{2}\right), \left(4, \frac{3}{5}\right)$

     $y = 6.4x^{0.241}$ $y = 0.417x^{0.263}$

16. 🌐 **CELL-PHONE USERS** Use the model in Example 3 to estimate the number of cell-phone users in 2005. What does your answer tell you about the model?
about 1.18 billion; *sample answer:* that the model probably isn't applicable over such a long period of time.

PRACTICE AND APPLICATIONS

STUDENT HELP

▸ **Extra Practice**
to help you master skills is on p. 951.

WRITING EXPONENTIAL FUNCTIONS Write an exponential function of the form $y = ab^x$ whose graph passes through the given points. $y = \frac{1}{512} \cdot 4^x$

17. $(1, 4), (2, 12)$ $y = \frac{4}{3} \cdot 3^x$ 18. $(2, 18), (3, 108)$ $y = \frac{1}{2} \cdot 6^x$ 19. $(6, 8), (7, 32)$

                                                        $y = 81(3)^x$

20. $(1, 7), (3, 63)$ $y = \frac{7}{3} \cdot 3^x$ 21. $(3, 8), (6, 64)$ $y = 2^x$ 22. $(-3, 3), (4, 6561)$

23. $\left(4, \frac{112}{81}\right), \left(-1, \frac{21}{2}\right)$ 24. $(3, 13.5), (5, 30.375)$ 25. $\left(2, \frac{25}{4}\right), \left(4, \frac{625}{4}\right)$

            $y = 7 \cdot \left(\frac{2}{3}\right)^x$ $y = 4(1.5)^x$ $y = \frac{1}{4} \cdot 5^x$

FINDING EXPONENTIAL MODELS Use the table of values to draw a scatter plot of $\ln y$ versus x. Then find an exponential model for the data.

26–28. See margin for graphs.

STUDENT HELP

▸ **HOMEWORK HELP**
Example 1: Exs. 17–25
Example 2: Exs. 26–28
Example 3: Exs. 54–56
Example 4: Exs. 29–37
Example 5: Exs. 38–40
Example 6: Exs. 57, 58

26.

x	1	2	3	4	5	6	7	8
y	14	28	56	112	224	448	896	1792

$y = 7(2)^x$

27.

x	1	2	3	4	5	6	7	8
y	10.2	30.5	43.4	61.2	89.7	120.6	210.4	302.5

Sample answer: $y = 9.715(1.550)^x$

28.

x	2	4	6	8	10	12	14	16
y	12.8	20.48	32.77	52.43	83.89	134.22	214.75	343.6

Sample answer: $y = 8(1.265)^x$

8.7 *Modeling with Exponential and Power Functions*   **513**

28. See Additional Answers beginning on page AA1.

---

3 APPLY

○ **ASSIGNMENT GUIDE**

BASIC
Day 1: pp. 513–516 Exs. 18–26 even, 30–38 even, 42–50 even, 55–59 odd, 61–71 odd

AVERAGE
Day 1: pp. 513–516 Exs. 18–38 even, 42–54 even, 55–59 odd, 61–79 odd

ADVANCED
Day 1: pp. 513–516 Exs. 18–38 even, 42–54 even, 55–60, 61–79 odd

BLOCK SCHEDULE WITH 8.8
pp. 513–516 Exs. 18–38 even, 42–54 even, 55–59 odd, 61–79 odd

EXERCISE LEVELS
Level A: *Easier*
17–25, 29–37

Level B: *More Difficult*
26–28, 38–59

Level C: *Most Difficult*
60

✔ **HOMEWORK CHECK**
To quickly check student understanding of key concepts, go over the following exercises: Exs. 18, 26, 30, 38, 42. See also the Daily Homework Quiz:

• Blackline Master (*Chapter 8 Resource Book,* p. 107)

• 📄 Transparency (p. 66)

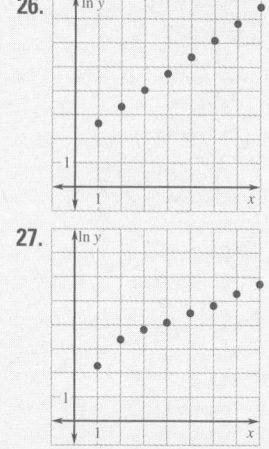

26.

27.

! COMMON ERROR
EXERCISES 38–40 Some students may find a power function for ln x and ln y, especially if they are using a graphing calculator to find regression models. Insist that students graph the points (ln x, ln y) so they have a visual model of this linear relationship. Point out that since these data fit a linear model so well, they could not also be modeled well by a power function.

MATHEMATICAL REASONING
EXERCISE 53 When a function is used to make predictions, we say that we extrapolate from the model. Two points determine a linear function, an exponential function, or a power function, but are not enough to determine a good model. The function may not be appropriate for further extrapolating data. For what domain of values would each of the three functions give nearly the same predicted value? $1 \le x \le 2$

GRAPHING CALCULATOR NOTE
EXERCISES 54–58 Using a graphing calculator will allow students to focus on the appropriateness of a model rather than on computation.

29. $y = 0.362x^{1.465}$

30. $y = 0.164x^{2.170}$

31. $y = 0.358x^{2.181}$

32. $y = 0.369x^{2.170}$

33. $y = 6.325x^{0.661}$

34. $y = 0.97x^{1.355}$

38.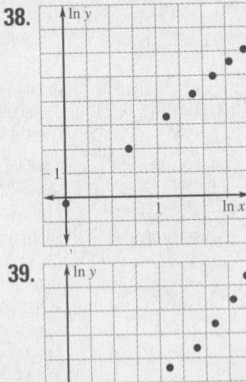

39.

53. $y = 9x - 6$; $y = \frac{3}{4} \cdot 4^x$; $y = 3x^2$; See margin for graph; *Sample answer:* The linear function grows the slowest, the quadratic is in the middle, and the exponential function grows at the fastest rate.

54. a. $y = 12.104(1.798)^x$

b. about 13,817

c. about 5.57 billion; this number is unrealistically large—almost 1 hit for every person in the world.

40, 53. See Additional Answers beginning on page AA1.

WRITING POWER FUNCTIONS Write a power function of the form $y = ax^b$ whose graph passes through the given points.

29. $(2, 1), (6, 5)$

30. $(6, 8), (12, 36)$

31. $(5, 12), (7, 25)$

32. $(3, 4), (6, 18)$

33. $(2, 10), (8, 25)$

34. $(6, 11), (24, 72)$

35. $(2.2, 10.4), (8.8, 20.3)$ $y = 7.109x^{0.482}$

36. $(2.9, 9.4), (7.3, 12.8)$ $y = 6.584x^{0.334}$

37. $(2.71, 6.42), (13.55, 29.79)$ $y = 2.481x^{0.954}$

FINDING POWER MODELS Use the table of values to draw a scatter plot of ln y versus ln x. Then find a power model for the data.
38–40. See margin for graphs.

38.

x	1	2	3	4	5	6	7
y	0.78	7.37	27.41	69.63	143.47	259.00	426.79

$y = 0.78x^{3.240}$

39.

x	1	2	3	4	5	6	7
y	1.2	5.4	9.8	14.3	25.6	41.2	65.8

$y = 1.193x^{1.962}$

40.

x	2	4	6	8	10	12	14
y	1.89	1.44	1.22	1.09	1.00	0.93	0.87

$y = 2.493x^{-0.398}$

WRITING EQUATIONS Write y as a function of x.

41. $\log y = 0.24x + 4.5$ $y = 31{,}623(1.738)^x$

42. $\log y = 0.2 \log x + 0.8$ $y = (6.310)x^{0.2}$

43. $\ln y = x + 4$ $y = 54.598e^x$

44. $\log y = -0.12 + 0.88x$ $y = (0.759)7.586^x$

45. $\log y = -0.48 \log x - 0.548$
$y = 0.283x^{-0.48}$

46. $\ln y = 2.3 \ln x + 4.7$ $y = 109.947x^{2.3}$

47. $\ln y = -2.38x + 0.98$
$y = 2.664(0.0926)^x$

48. $\log y = -1.48 + 3.751 \log x$
$y = 0.0331x^{3.751}$

49. $\ln y = -1.5x + 2.5$ $y = 12.182(0.223)^x$

50. $1.2 \log y = 3.4 \log x$ $y = x^{17/6}$

51. $\frac{1}{2} \log y = \frac{5}{6} \log x$ $y = x^{5/3}$

52. $2\frac{1}{8} \ln y = 4\frac{1}{4} \ln x + \frac{3}{8}$
$y = e^{0.1765}x^2 \approx 1.193x^2$

53. **VISUAL THINKING** Find equations of the line, the exponential curve, and the power curve that each pass through the points $(1, 3)$ and $(2, 12)$. Graph the equations in the same coordinate plane and then describe what happens when the equations are used as models to predict y-values for x-values greater than 2.

MODELING DATA In Exercises 54–58, you may wish to use a graphing calculator to perform exponential regression or power regression.

54. **NEW WEB SITE** You have just created your own Web site. You are keeping track of the number of hits (the number of visits to the site). The table shows the number y of hits in each of the first 10 months where x is the month number.

x	1	2	3	4	5	6	7	8	9	10
y	22	39	70	126	227	408	735	1322	2380	4285

a. Find an exponential model for the data.

b. According to your model, how many hits do you expect in the twelfth month?

c. According to your model, how many hits would there be in the thirty-fourth month? What is wrong with this number?

FOCUS ON APPLICATIONS

CRANES
The red-crowned crane (*Grus japonensis*) is the second-rarest crane species, with a total population in the wild of about 1700–2000 birds.

STUDENT HELP

HOMEWORK HELP
Visit our Web site
www.mcdougallittell.com
for help with Ex. 57.

55. 🌐 **CRANES** The table shows the number *C* of cranes in Izumi, Japan, from 1950 to 1990 where *t* represents the number of years since 1950.

▶ Source: Yamashina Institute of Ornithology

t	0	5	10	15	20	25	30	35	40
C	293	299	438	1573	2336	3649	5602	7610	9959

a. Draw a scatter plot of ln *C* versus *t*. Is an exponential model a good fit for the original data? **yes; See margin for graph.**

b. Find an exponential model for the original data. Estimate the number of cranes in Izumi, Japan, in the year 2000. $C = 250.31(1.104)^t$; **about 35,232**

56. 🌐 **UNITED STATES STAMPS** The table shows the cumulative number *s* of different stamps in the United States from 1889 to 1989 where *t* represents the number of years since 1889.

t	0	10	20	30	40	50	60	70	80	90	100
s	218	293	374	541	681	858	986	1138	1138	1794	2438

a. Draw a scatter plot of ln *s* versus *t*. Is an exponential model a good fit for the original data? **yes; See margin for graph.**

$s = 325.057(1.019)^t$; **about 2676**

b. Find an exponential model for the original data. Estimate the cumulative number of stamps in the United States in the year 2000.

57. 🌐 **CITIES OF ARGENTINA**
The table shows the population *y* (in millions) and the population rank *x* for nine cities in Argentina in 1991.

a. Draw a scatter plot of ln *y* versus ln *x*. Is a power model a good fit for the original data?
yes; See margin for graph.

b. Find a power model for the original data. Estimate the population of the city Vicente López, which has a population rank of 20. $y = 2.022x^{-0.582}$; **about 354,000**

City	Rank, x	Population (millions), y
Cordoba	2	1.21
Rosario	3	1.12
La Matanza	4	1.11
Mendoza	5	0.77
La Plata	6	0.64
Moron	7	0.64
San Miguel de Tucuman	8	0.62
Lomas de Zamoras	9	0.57
Mar de Plata	10	0.51

58. **SCIENCE CONNECTION** The table shows the atomic number *x* and the melting point *y* (in degrees Celsius) for the alkali metals.

Alkali metal	Lithium	Sodium	Potassium	Rubidium	Cesium
Atomic number, x	3	11	19	37	55
Melting point, y	180.5	97.8	63.7	38.9	28.5

a. Draw a scatter plot of ln *y* versus ln *x*. Is a power model a good fit for the original data? **yes; See margin for graph.**

b. Find a power model for the original data. $y = 397.610x^{-0.639}$

c. One of the alkali metals, francium, is not shown in the table. It has an atomic number of 87. Using your model, predict the melting point of francium.
about 22.9°C

8.7 *Modeling with Exponential and Power Functions* **515**

STUDENT HELP NOTES

→ **Homework Help** Students can find help for Ex. 57 at **www.mcdougallittell.com**. The information can be printed out for students who don't have access to the Internet.

55a.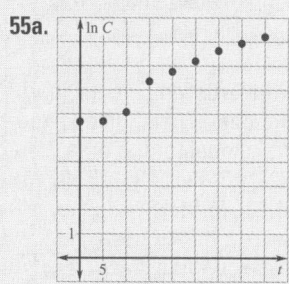

56a.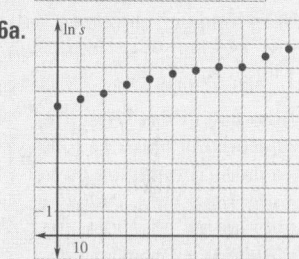

57a.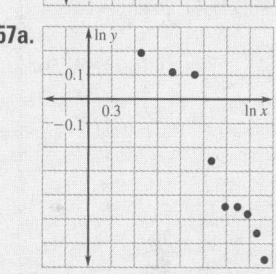

58a, 59a. See Additional Answers beginning on page AA1.

ADDITIONAL PRACTICE AND RETEACHING

For Lesson 8.7:
• Practice Levels A, B, and C (*Chapter 8 Resource Book*, p. 97)

• Reteaching with Practice (*Chapter 8 Resource Book*, p. 100)

• 🖥 See Lesson 8.7 of the *Personal Student Tutor*

For more Mixed Review:
• 🖥 Search the *Test and Practice Generator* for key words or specific lessons.

DAILY HOMEWORK QUIZ

📖 *Transparency Available*

1. Write an exponential function $y = ab^x$ whose graph passes through $(-2, 81)$ and $(2, 16)$.

$y = 36\left(\dfrac{2}{3}\right)^x$

2. Find the ordered pairs $(x, \ln y)$ for the data. Then find an exponential model for the data. $(1, 7.20)$, $(2, 12.96)$, $(3, 23.33)$, $(4, 41.99)$, $(5, 75.58)$

$(1, 1.97)$, $(2, 2.56)$, $(3, 3.15)$, $(4, 3.74)$, $(5, 4.33)$; $y = 4(1.8)^x$

3. Write a power function $y = ax^b$ whose graph passes through $(1, 2)$ and $(4, 16)$. $y = 2x^{3/2}$

4. Draw a scatter plot of the ordered pairs $(\ln x, \ln y)$ for the data. Then find a power model for the data. $(2, 12.25)$, $(3, 31.64)$, $(4, 62.03)$, $(5, 104.57)$, $(6, 160.21)$, $(7, 229.80)$

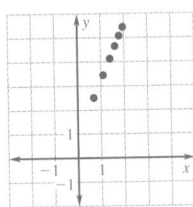

$y = 2.42x^{2.34}$

EXTRA CHALLENGE NOTE

→ Challenge problems for Lesson 8.7 are available in **blackline** format in the *Chapter 8 Resource Book*, p. 104 and at **www.mcdougallittell.com.**

ADDITIONAL TEST PREPARATION

1. WRITING Suppose that $(25, 1000)$ and $(50, 1500)$ belong to a data set that can be modeled by a power function. Explain why $(75, 2500)$ cannot be a point on the scatter plot.
See sample answer at right.

59a, 67–75. See Additional Answers beginning on page AA1.

Test Preparation

59. MULTI-STEP PROBLEM The femur is a large bone found in the leg or hind limb of an animal. Scientists use the circumference of an animal's femur to estimate the animal's weight. The table at the right shows the femur circumference C (in millimeters) and the weight W (in kilograms) of several animals.

Animal	C (mm)	W (kg)
Meadow mouse	5.5	0.047
Guinea pig	15	0.385
Otter	28	9.68
Cheetah	68.7	38
Warthog	72	90.5
Nyala	97	134.5
Grizzly bear	106.5	256
Kudu	135	301
Giraffe	173	710

▶ Source: Zoological Society of London

59. a. See margin.

b. *Sample answer:* a power model, since the scatter plot of ln W vs. C looks curved, like the graph of a logarithmic function, while the scatter plot of ln W vs. ln C looks more like a straight line.

c. $y = (0.000349)x^{2.834}$

d. raccoon: 4.4 kg; cougar: 38.7 kg; bison: 700.9 kg; hippo: 1295 kg

a. Draw two scatter plots, one of ln W versus C and another of ln W versus ln C.

b. *Writing* Looking at your scatter plots, tell which type of model you think is a better fit for the original data. Explain your reasoning.

c. Using your answer from part (b), find a model for the original data.

d. The table at the right shows the femur circumference C (in millimeters) of four animals. Use the model you found in part (c) to estimate the weight of each animal.

Animal	C (mm)
Raccoon	28
Cougar	60.25
Bison	167.5
Hippopotamus	208

★ **Challenge**

60. DERIVING FORMULAS Using $y = ab^x$ and $y = ax^b$, take the natural logarithm of both sides of each equation. What is the slope and y-intercept of the line relating x and ln y for $y = ab^x$? of the line relating ln x and ln y for $y = ax^b$?
If $y = ab^x$, then ln $y = (\ln b)x + \ln a$, so the slope is ln b and the y-intercept is ln a.
If $y = ax^b$, then ln $y = b \ln x + \ln a$, and the slope is b with a y-intercept of ln a.

MIXED REVIEW

DESCRIBING END BEHAVIOR Describe the end behavior of the graph of the polynomial function by completing the statements $f(x) \to \underline{?}$ as $x \to -\infty$ and $f(x) \to \underline{?}$ as $x \to +\infty$. (Review 6.2 for 8.8)

61. $f(x) \to +\infty$ as $x \to -\infty$; $f(x) \to -\infty$ as $x \to +\infty$

62. $f(x) \to +\infty$ as $x \to -\infty$; $f(x) \to +\infty$ as $x \to +\infty$

63. $f(x) \to -\infty$ as $x \to -\infty$; $f(x) \to -\infty$ as $x \to +\infty$

64. $f(x) \to -\infty$ as $x \to -\infty$; $f(x) \to +\infty$ as $x \to +\infty$

65. $f(x) \to +\infty$ as $x \to -\infty$; $f(x) \to +\infty$ as $x \to +\infty$

66. $f(x) \to +\infty$ as $x \to -\infty$; $f(x) \to -\infty$ as $x \to +\infty$

61. $f(x) = -x^3 + x^2 - x + 4$

62. $f(x) = x^4 - 7x^2 + 2$

63. $f(x) = -x^4 + 3x - 3$

64. $f(x) = 3x^5 - x^4 - x^2 + 1$

65. $f(x) = x^6 - 2x - 1$

66. $f(x) = -2x^5 + 3x^4 - 2x^3 + x^2 + 5$

GRAPHING FUNCTIONS Graph the function. (Review 8.3 for 8.8) 67–75. See margin.

67. $y = 4e^{-0.75x}$

68. $y = 10e^{-0.4x}$

69. $y = 2e^{x-3}$

70. $y = e^{0.5x} + 2$

71. $y = e^{-0.25x} - 4$

72. $y = 3e^{-1.5x} - 1$

73. $y = 2e^{0.25x} + 1$

74. $y = e^{x+1} - 5$

75. $y = 2.5e^{-0.6x} + 2$

CONDENSING EXPRESSIONS Condense the expression. (Review 8.5)

76. $5 \log 2 - \log 8$ log 4

77. $2 \log 9 - \log 3$ log 27

78. $\ln x + 5 \ln 3$ ln $243x$

79. $2 \ln x - \ln 4$ $\ln \dfrac{x^2}{4}$

80. $\log_2 8 + 3 \log_2 3 - \log_2 6$ $\log_2 36$

81. $\log_7 12 + 3 \log_7 4 + \log_7 5$ $\log_7 3840$

Additional Test Preparation *Sample answer:*
The point (ln 75, ln 2500) does not lie on the line determined by (ln 25, ln 1000) and (ln 50, ln 2500).

8.8

Logistic Growth Functions

What you should learn

GOAL 1 Evaluate and graph logistic growth functions.

GOAL 2 Use logistic growth functions to model **real-life** quantities, such as a yeast population in **Exs. 50 and 51**.

Why you should learn it

▼ To solve **real-life** problems, such as modeling the height of a sunflower in **Example 5**.

1. as $x \to -\infty$, $y \to 0$; as $x \to +\infty$, $y \to 100$

2. *Sample answer:* The graph has two horizontal asymptotes, the x-axis and the line $y = c$. The graph is continuously increasing, symmetric about the point where it crosses the line $y = \frac{c}{2}$. The graph grows steeply in the center, and is flat at each end.

GOAL 1 USING LOGISTIC GROWTH FUNCTIONS

In this lesson you will study a family of functions of the form

$$y = \frac{c}{1 + ae^{-rx}}$$

where a, c, and r are all positive constants. Functions of this form are called **logistic growth functions**.

EXAMPLE 1 *Evaluating a Logistic Growth Function*

Evaluate $f(x) = \frac{100}{1 + 9e^{-2x}}$ for each value of x.

a. $f(-3)$ **b.** $f(0)$ **c.** $f(2)$ **d.** $f(4)$

SOLUTION

a. $f(-3) = \frac{100}{1 + 9e^{-2(-3)}} \approx 0.0275$ **b.** $f(0) = \frac{100}{1 + 9e^{-2(0)}} = \frac{100}{10} = 10$

c. $f(2) = \frac{100}{1 + 9e^{-2(2)}} \approx 85.8$ **d.** $f(4) = \frac{100}{1 + 9e^{-2(4)}} \approx 99.7$

○ ACTIVITY

Developing Concepts

📊 Graphs of Logistic Growth Functions

1, 2. See margin for graphs.

❶ Use a graphing calculator to graph the logistic growth function from Example 1. Trace along the graph to determine the function's end behavior.

❷ Use a graphing calculator to graph each of the following. Then describe the basic shape of the graph of a logistic growth function.

a. $y = \frac{1}{1 + e^{-x}}$ **b.** $y = \frac{10}{1 + 5e^{-2x}}$ **c.** $y = \frac{5}{1 + 10e^{-2x}}$

In this chapter you learned that an exponential growth function $f(x)$ increases without bound as x increases. On the other hand, the logistic growth function $y = \frac{c}{1 + ae^{-rx}}$ has $y = c$ as an upper bound.

Logistic growth functions are used to model real-life quantities whose growth levels off because the rate of growth changes—from an increasing growth rate to a decreasing growth rate.

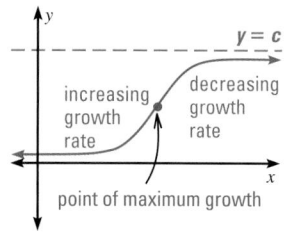

increasing growth rate

decreasing growth rate

point of maximum growth

$y = c$

---

1 PLAN

PACING
Basic: 1 day
Average: 1 day
Advanced: 1 day
Block Schedule: 0.5 block with 8.7

⟶ LESSON OPENER
CALCULATOR ACTIVITY
An alternative way to approach Lesson 8.8 is to use the Calculator Activity Lesson Opener:
● Blackline Master (*Chapter 8 Resource Book,* p. 108)
● 🖨 Transparency (p. 58)

MEETING INDIVIDUAL NEEDS
● *Chapter 8 Resource Book*
 Prerequisite Skills Review (p. 5)
 Practice Level A (p. 110)
 Practice Level B (p. 111)
 Practice Level C (p. 112)
 Reteaching with Practice (p. 113)
 Absent Student Catch-Up (p. 115)
 Challenge (p. 117)
● *Resources in Spanish*
● 🖥 *Personal Student Tutor*

NEW-TEACHER SUPPORT
See the Tips for New Teachers on pp. 1–2 of the *Chapter 8 Resource Book* for additional notes about Lesson 8.8.

WARM-UP EXERCISES

🖨 *Transparency Available*

Simplify.

1. $\frac{\ln 3}{0.36}$ about 3.05

2. $\frac{\ln 10}{10}$ about 0.23

3. Evaluate $f(4)$ for $f(x) = 1 + 2e^{-3x}$. about 1

4. What is the horizontal asymptote of the graph of $y = 1 + 2e^{-2x}$? $y = 1$

5. Solve $7 = 35e^{-4x}$. about 0.402

1, 2. See Additional Answers beginning on page AA1.

When popcorn pops, the number of popped kernels increases exponentially at first, then levels off. Logistic models, the subject of this lesson, can be used for situations like this.

ACTIVITY NOTE
Ask students to compare the function's end behavior to their results from Example 1.

EXTRA EXAMPLE 1
Evaluate $f(x) = \dfrac{10}{1 + 2e^{-0.8x}}$ for each value of x.
- **a.** $f(-2)$ about 0.917
- **b.** $f(0.9)$ about 5.07
- **c.** $f(6)$ about 9.84
- **d.** $f(20)$ about 10

EXTRA EXAMPLE 2
Graph $y = \dfrac{3}{1 + 5e^{-2x}}$.

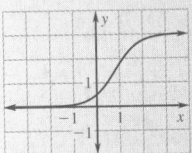

EXTRA EXAMPLE 3
Solve $\dfrac{30}{1 + 5e^{-2x}} = 10$. about 0.458

CHECKPOINT EXERCISES
For use after Example 1:
1. Evaluate $f(-10)$, $f(0)$, and $f(10)$ for $f(x) = \dfrac{8}{1 + 3e^{-x}}$.
about 0.0001; 2; about 7.999

For use after Example 2:
2. Graph $\dfrac{1}{1 + 4e^{-3.5x}}$.

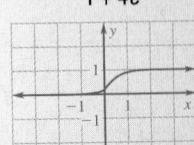

For use after Example 3:
3. Solve $\dfrac{45}{1 + 8e^{-1.5x}} = 25$.
about 1.54

518

GRAPHS OF LOGISTIC GROWTH FUNCTIONS

The graph of $y = \dfrac{c}{1 + ae^{-rx}}$ has the following characteristics:

- The horizontal lines $y = 0$ and $y = c$ are asymptotes.

- The y-intercept is $\dfrac{c}{1 + a}$.

- The domain is all real numbers, and the range is $0 < y < c$.

- The graph is increasing from left to right. To the left of its point of maximum growth, $\left(\dfrac{\ln a}{r}, \dfrac{c}{2}\right)$, the rate of increase is increasing. To the right of its point of maximum growth, the rate of increase is decreasing.

EXAMPLE 2 *Graphing a Logistic Growth Function*

Graph $y = \dfrac{6}{1 + 2e^{-0.5x}}$.

SOLUTION

Begin by sketching the upper horizontal asymptote, $y = 6$. Then plot the y-intercept at $(0, 2)$ and the point of maximum growth $\left(\dfrac{\ln 2}{0.5}, \dfrac{6}{2}\right) \approx (1.4, 3)$. Finally, from left to right, draw a curve that starts just above the x-axis, curves up to the point of maximum growth, and then levels off as it approaches the upper horizontal asymptote.

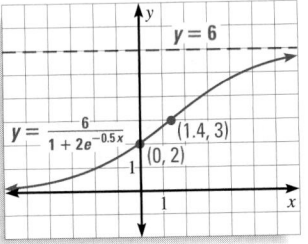

EXAMPLE 3 *Solving a Logistic Growth Equation*

Solve $\dfrac{50}{1 + 10e^{-3x}} = 40$.

STUDENT HELP

HOMEWORK HELP
Visit our Web site
www.mcdougallittell.com
for extra examples.

SOLUTION

$\dfrac{50}{1 + 10e^{-3x}} = 40$	Write original equation.
$50 = \left(1 + 10e^{-3x}\right)(40)$	Multiply each side by $1 + 10e^{-3x}$.
$50 = 40 + 400e^{-3x}$	Use distributive property.
$10 = 400e^{-3x}$	Subtract 40 from each side.
$0.025 = e^{-3x}$	Divide each side by 400.
$\ln 0.025 = -3x$	Take natural log of each side.
$-\dfrac{1}{3} \ln 0.025 = x$	Divide each side by -3.
$1.23 \approx x$	Use a calculator.

▶ The solution is about 1.23. Check this in the original equation.

GOAL 2 USING LOGISTIC GROWTH MODELS IN REAL LIFE

Logistic growth functions are often more useful as models than exponential growth functions because they account for constraints placed on the growth. An example is a bacteria culture allowed to grow under initially ideal conditions, followed by less favorable conditions that inhibit growth.

Biology

EXAMPLE 4 *Using a Logistic Growth Model*

A colony of the bacteria *B. dendroides* is growing in a petri dish. The colony's area A (in square centimeters) can be modeled by

$$A = \frac{49.9}{1 + 134e^{-1.96t}}$$

where t is the elapsed time in days. Graph the function and describe what it tells you about the growth of the bacteria colony.

SOLUTION

The graph of the model is shown. The initial area is

$$A = \frac{49.9}{1 + 134e^{-1.96(0)}} \approx 0.37 \text{ cm}^2.$$

The colony grows more and more rapidly until

$$t = \frac{\ln 134}{1.96} \approx 2.5 \text{ days}.$$

Then the rate of growth decreases. The colony's area is limited to $A = 49.9$ cm², which might possibly be the area of the petri dish.

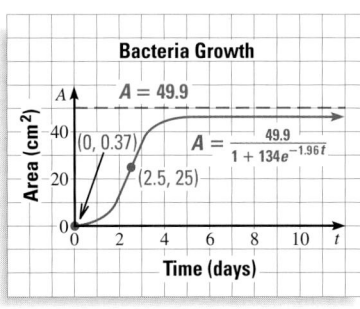

Bacteria Growth

$A = 49.9$

$A = \frac{49.9}{1 + 134e^{-1.96t}}$

(0, 0.37)

(2.5, 25)

Time (days)

Botany

EXAMPLE 5 *Writing a Logistic Growth Model*

You planted a sunflower seedling and kept track of its height h (in centimeters) over time t (in weeks). Find a model that gives h as a function of t.

t	0	1	2	3	4	5	6	7	8	9	10
h	18	33	56	90	130	170	203	225	239	247	251

SOLUTION

A scatter plot shows that the data can be modeled by a logistic growth function.

The logistic regression feature of a graphing calculator returns the values shown at the right.

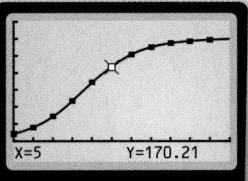

X=5   Y=170.21

▶ The model is:

$$h = \frac{256}{1 + 13e^{-0.65t}}$$

8.8 *Logistic Growth Functions* 519

STUDENT HELP

KEYSTROKE HELP
Visit our Web site
www.mcdougallittell.com
to see keystrokes for
several models of
calculators.

EXTRA EXAMPLE 4
The average monthly price P of a company's stock over the last 12 months can be modeled by $P = \frac{68.5}{1 + 19.19e^{-0.61t}}$, where t is the number of months. Graph the function and describe what it tells you about the price of the stock. **The stock increases more rapidly in price for about the first 5 months. Then the rate of growth decreases. The price begins to level off to about $68.50 in about the 9th month.**

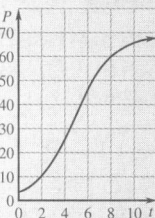

EXTRA EXAMPLE 5
The amount of contrast visible in a black and white photograph is determined by plotting the density d of the film against its exposure x. The ordered pairs (x, d) represent this relationship for a particular film. (0, 0.1), (0.25, 0.13), (0.75, 0.21), (1, 0.5), (1.25, 0.9), (1.75, 2.15), (2, 2.5), (2.25, 2.8), (2.75, 2.85), (3, 2.9) Find a model that gives d as a function of x.
$$d = \frac{2.92}{1 + 169e^{-3.5x}}$$

CHECKPOINT EXERCISES
For use after Examples 4 and 5:
1. Find a model that gives y as a function of x for the ordered pairs (x, y). (0, 15), (1, 25), (2, 60), (3, 150), (4, 270), (5, 340), (6, 380), (7, 385), (8, 388), (9, 392), (10, 393)
$$y = \frac{392}{1 + 58.3e^{-1.2x}}$$

CLOSURE QUESTION
Is the rate of change constant for a logistic function? Explain. **No; initially, the rate of growth increases. Then it increases more slowly until it levels off near the asymptote.**

3 APPLY

ASSIGNMENT GUIDE

BASIC
Day 1: pp. 520–522 Exs. 16–26, 27–41 odd, 52, 55–63 odd, Quiz 3 Exs. 1–7

AVERAGE
Day 1: pp. 520–522 Exs. 16–26, 27–43 odd, 50–52, 55–63 odd, Quiz 3 Exs. 1–7

ADVANCED
Day 1: pp. 520–522 Exs. 16–26, 27–43 odd, 45–47, 50–53, 55–63 odd, Quiz 3 Exs. 1–7

BLOCK SCHEDULE WITH 8.7
pp. 520–522 Exs. 16–26, 27–43 odd, 50–52, 55–63 odd, Quiz 3 Exs. 1–7

EXERCISE LEVELS

Level A: *Easier*
16–26

Level B: *More Difficult*
27–44, 50–52

Level C: *Most Difficult*
45–49, 53

✔ **HOMEWORK CHECK**
To quickly check student understanding of key concepts, go over the following exercises: Exs. 18, 24, 27, 37. See also the Daily Homework Quiz:

- Blackline Master (*Chapter 9 Resource Book,* p. 11)
- Transparency (p. 68)

❗ **COMMON ERROR**
EXERCISES 9–11 When finding the point of maximum growth, students commonly substitute the value of −*r* into the formula for the *x*-coordinate. Point out that if the coefficient of *x* is negative, then *r* is positive.

9–11. See Additional Answers beginning on page AA1.

GUIDED PRACTICE

Vocabulary Check ✔

1. What is the name of a function having the form $y = \dfrac{c}{1 + ae^{-rx}}$ where *c*, *a*, and *r* are positive constants?   **a logistic growth function**

Concept Check ✔

2. What is a significant difference between using exponential growth functions and using logistic growth functions as models for real-life quantities? **See margin.**

3. What is the significance of the point (ln 3, 4) on the graph of $f(x) = \dfrac{8}{1 + 3e^{-x}}$? **See margin.**

Skill Check ✔

Evaluate the function $f(x) = \dfrac{12}{1 + 5e^{-2x}}$ **for the given value of *x*.**

2. An exponential function increases without bound, while a logistic growth function approaches a finite limiting value.

4. $f(0)$ **2**   **5.** $f(-2)$ **0.0438** **6.** $f(5)$ **12.00** **7.** $f\left(-\dfrac{1}{2}\right)$ **0.822** **8.** $f(10)$ **almost 12**

3. This is the point of maximum growth for the function.

Graph the function. Identify the asymptotes, *y*-intercept, and point of maximum growth. **9–11. See margin for graphs.**

9. $f(x) = \dfrac{5}{1 + 4e^{-2.5x}}$

10. $f(x) = \dfrac{8}{1 + 3e^{-0.4x}}$

11. $f(x) = \dfrac{2}{1 + 4e^{-0.25x}}$

9. *x*-axis and $y = 5$; 1; (0.555, 2.5)

10. *x*-axis and $y = 8$; 2; (2.747, 4)

11. *x*-axis and $y = 2$; $\dfrac{2}{5}$; (5.545, 1)

Solve the equation.

12. $\dfrac{18}{1 + 2e^{-2x}} = 10$ **0.458** **13.** $\dfrac{30}{1 + 4e^{-x}} = 10$ **0.693** **14.** $\dfrac{12.5}{1 + 7e^{-0.2x}} = 9$ **14.452**

15. 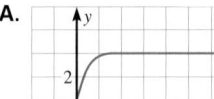 **PLANTING SEEDS** You planted a seedling and kept track of its height *h* (in centimeters) over time *t* (in weeks). Use the data in the table to find a model that gives *h* as a function of *t*. $h = \dfrac{117}{1 + 18e^{-0.73t}}$

t	0	1	2	3	4	5	6	7	8
h	5	12	26	39	51	88	94	103	112

PRACTICE AND APPLICATIONS

STUDENT HELP

▶ **Extra Practice**
to help you master skills is on p. 952.

EVALUATING FUNCTIONS Evaluate the function $f(x) = \dfrac{7}{1 + 3e^{-x}}$ **for the given value of *x*.**

16. $f(1)$ **3.328**   **17.** $f(3)$ **6.090**   **18.** $f(-1)$ **0.765**   **19.** $f(-6)$ **0.00578**

20. $f(0)$ $\dfrac{7}{4}$ **21.** $f\left(\dfrac{3}{4}\right)$ **2.896**   **22.** $f(2.2)$ **5.254**   **23.** $f(-0.9)$ **0.835**

MATCHING GRAPHS Match the function with its graph.

STUDENT HELP

▶ **HOMEWORK HELP**
Example 1: Exs. 16–23
Example 2: Exs. 24–35
Example 3: Exs. 36–44
Example 4: Exs. 45–49
Example 5: Exs. 50, 51

24. $f(x) = \dfrac{4}{1 + 2e^{-3x}}$ **C**   **25.** $f(x) = \dfrac{3}{1 + 2e^{-4x}}$ **A**   **26.** $f(x) = \dfrac{2}{1 + 3e^{-4x}}$ **B**

A. 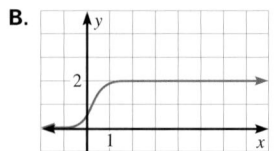   **B.** 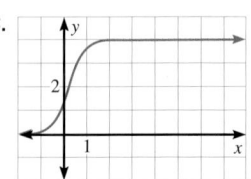   **C.**

GRAPHING FUNCTIONS Graph the function. Identify the asymptotes, *y*-intercept, and point of maximum growth. **27–35. See margin.**

27. $y = \dfrac{1}{1 + 6e^{-x}}$

28. $y = \dfrac{2}{1 + 0.4e^{-0.3x}}$

29. $y = \dfrac{5}{1 + e^{-10x}}$

30. $y = \dfrac{4}{1 + 0.08e^{-2.1x}}$

31. $y = \dfrac{4}{1 + 3e^{-3x}}$

32. $y = \dfrac{3}{1 + 3e^{-8x}}$

33. $y = \dfrac{8}{1 + e^{-1.02x}}$

34. $y = \dfrac{10}{1 + 6e^{-0.5x}}$

35. $y = \dfrac{6}{1 + 0.8e^{-2x}}$

SOLVING EQUATIONS Solve the equation. $\dfrac{\ln 18}{4} \approx 0.723$

36. $\dfrac{8}{1 + 3e^{-x}} = 5$ $\ln 5$

37. $\dfrac{10}{1 + 2e^{-4x}} = 9$

38. $\dfrac{3}{1 + 18e^{-x}} = 1$ $\ln 9$

39. $\dfrac{28}{1 + 13e^{-2x}} = 20$ 1.741

40. $\dfrac{82}{1 + 50e^{-x}} = 68$ 5.492

41. $\dfrac{36}{1 + 7e^{-10x}} = 30$ 0.356

42. $\dfrac{41}{1 + 14.9e^{-6x}} = 7$ 0.187

43. $\dfrac{9}{1 + 5e^{-0.2x}} = \dfrac{3}{4}$ −3.942

44. $\dfrac{40}{1 + 2.5e^{-0.4x}} = 6.4$ −1.855

 OWNING A VCR In Exercises 45–47, use the following information.
The number of households in the United States that own VCRs has shown logistic growth from 1980 through 1999. The number *H* (in millions) of households can be modeled by the equation

$$H = \dfrac{91.86}{1 + 22.96e^{-0.4t}}$$

where *t* is the number of years since 1980. ▶ Source: Veronis, Suhler & Associates

45. In what year were there approximately 86 million households with VCRs?
during 1994

46. Graph the model. In what year did the growth rate for the number of households stop increasing and start decreasing?   **1987; See margin for graph.**

47. What is the long-term trend in VCR ownership?   to approach 91.86 million households

 ECONOMICS In Exercises 48 and 49, use the following information.
The gross domestic product (GDP) of the United States has shown logistic growth from 1970 through 1992. The gross domestic product *G* (in billions of dollars) can be modeled by the equation

$$G = \dfrac{9200}{1 + 8.03e^{-0.121t}}$$

where *t* is the number of years since 1970. ▶ Source: U.S. Bureau of the Census

48. In what year was the GDP approximately $5000 billion?   1988

49. Graph the model. When did the GDP reach its point of maximum growth?
1987; See margin for graph.

YEAST POPULATION In Exercises 50 and 51, use the following information.
In biology class, you observed the biomass of a yeast population over a period of time. The table gives the yeast mass *y* (in grams) after *t* hours.

t	0	1	2	3	4	5	6	7	8	9
y	9.6	18.3	29.0	47.2	71.1	119.1	174.6	257.3	350.7	441.0

50. Draw a scatter plot of the data.   See margin.

51. Find a model that gives *y* as a function of *t* using the logistic regression feature of a graphing calculator. $y = \dfrac{721}{1 + 72e^{-0.526t}}$

8.8 *Logistic Growth Functions*   **521**

29–35, 46, 49, 50. See Additional Answers
beginning on page AA1.

**MATHEMATICAL REASONING
EXERCISES 27–35** The numerator of a logistic function is the horizontal asymptote of the function. Why can the logistic function

$y = \dfrac{c}{1 + ae^{-rx}}$ never be equal to *c*?

$1 + ae^{-rx}$ can never equal 1.

**GRAPHING CALCULATOR NOTE
EXERCISES 50–51** The TI-83 uses an iterative least-squares fit, to which the students have not been introduced, to draw a logistic regression. Students will notice that it takes several seconds before this kind of regression function displays on their calculators.

27.
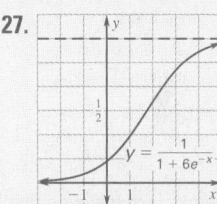

asymptotes: *x*-axis and *y* = 1;
y-intercept: $\dfrac{1}{7}$; pt. of max. growth: (1.792, 0.5)

28.
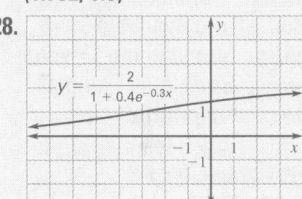

asymptotes: *x*-axis and *y* = 2;
y-intercept: $\dfrac{10}{7}$; pt. of max. growth: (−3.054, 1)

ADDITIONAL PRACTICE AND RETEACHING

For Lesson 8.8:
• Practice Levels A, B, and C (*Chapter 8 Resource Book,* p. 110)
• Reteaching with Practice (*Chapter 8 Resource Book,* p. 113)
• See Lesson 8.8 of the *Personal Student Tutor*

For more Mixed Review:
• Search the *Test and Practice Generator* for key words or specific lessons.

FOCUS ON APPLICATIONS

ECONOMICS
Gross domestic product, the focus of Exs. 48 and 49, is the value of all goods and services produced within a country during a given period.

DAILY HOMEWORK QUIZ

📖 *Transparency Available*

1. Evaluate $f(x) = \dfrac{5}{1 - 4e^{-2x}}$ for

$f(2.5)$. **5.138**

2. Graph $y = \dfrac{5}{1 + e^{-0.6x}}$. Identify

the asymptotes, y-intercept, and point of maximum growth.

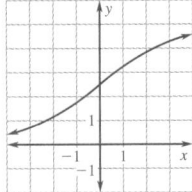

asymptotes: $y = 0$, $y = 5$;
y-intercept: 2.5; point of
maximum growth: (0, 2.5)

3. Solve the equation

$\dfrac{15}{1 + 2e^{-3x}} = 2$. $-\dfrac{\ln 3.25}{3} \approx -0.39$

EXTRA CHALLENGE NOTE
→ Challenge problems for
Lesson 8.8 are available in
blackline format in the *Chapter 8
Resource Book,* p. 117 and at
www.mcdougallittell.com.

ADDITIONAL TEST PREPARATION

1. WRITING How are the graphs
of logistic functions like the
graphs of exponential functions?
How are they different?
Sample answer: In an exponential
growth function, y increases as
x increases. This is also true of a
logistic function. In a logistic
function, however, the rate of
growth is not constant. At some
point the rate slows and growth
levels off.

2. OPEN ENDED Describe a
real-world situation that can be
modeled by a logistic function.
See sample answer at right.

52. See Additional Answers beginning on page AA1.

Test Preparation

52a. exponential model:
$P = 5.36(1.03)^t$

STUDENT HELP

🌐 **DATA UPDATE**
Visit our Web site
www.mcdougallittell.com

logistic model:

$P = \dfrac{186.45}{1 + 35.37e^{-0.03t}}$

★ Challenge

b. Exponential model gives 1896.
Logistic model gives 1918
which is closer. The data
more closely follows
a logistic pattern.

c. Exponential model gives a
value of 2.7 billion, which
is far too large. Logistic
model gives a closer
value of 175.1 million.

7. when $t \approx 10.65$, or after
about $10\frac{1}{2}$ days; See
margin for graph.

---

52. 📟 **MULTI-STEP PROBLEM** The table shows the population P (in millions) of the United States from 1800 to 1870 where t represents the number of years since 1800. ▶ Source: U.S. Bureau of the Census **52a–c. See margin.**

a. Use a graphing calculator to find an exponential growth model and a logistic growth model for the data. Then graph both models. **See margin for graphs.**

b. Use the models from part (a) to find the year when the population was about 92 million. Which of the models gives a year that is closer to 1910, the correct answer? Explain why you think that model is more accurate.

c. Use each model to predict the population in 2010. Which model gives a population closer to 297.7 million, the predicted population from the U.S. Bureau of the Census?

t	P
0	5.3
10	7.2
20	9.6
30	12.9
40	17.0
50	23.2
60	31.4
70	39.8

53. ANALYZING MODELS The graph of a logistic growth function

$y = \dfrac{c}{1 + ae^{-rx}}$ reaches its point of maximum growth where $y = \dfrac{c}{2}$.

Show that the x-coordinate of this point is $x = \dfrac{\ln a}{r}$.

Sample answer: If $y = \dfrac{c}{1 + ae^{-rx}} = \dfrac{c}{2}$, then $1 + ae^{-rx} = 2$, so $ae^{-rx} = 1$, $e^{rx} = a$,

$rx = \ln a$, $x = \dfrac{\ln a}{r}$.

MIXED REVIEW

WRITING EQUATIONS The variables x and y vary directly. Write an equation that relates the variables. (Review 2.4 for 9.1)

54. $x = 4$, $y = 36$ $y = 9x$ **55.** $x = -5$, $y = 10$ $y = -2x$ **56.** $x = 2$, $y = 13$ $y = \frac{13}{2}x$

57. $x = 40$, $y = 5$ $y = \frac{1}{8}x$ **58.** $x = 0.1$, $y = 0.9$ $y = 9x$ **59.** $x = 1$, $y = 0.2$ $y = 0.2x$

WRITING EQUATIONS Write y as a function of x. (Review 8.7)

60. $\log y = 0.9 \log x + 2.11$ $y = 128.8x^{0.9}$ **61.** $\ln y = 0.94 - 2.44x$ $y = 2.560(0.0872)^x$

62. $\log y = -1.82 + 0.4x$ **63.** $\log y = -0.75 \log x - 1.76$
$y = 0.0151(2.512)^x$ $y = 0.0174x^{-0.75}$

QUIZ 3 *Self-Test for Lessons 8.7 and 8.8*

Write an exponential function of the form $y = ab^x$ whose graph passes through the given points. (Lesson 8.7)

1. (2, 3), (5, 12)   **2.** (1, 16), (3, 45)   **3.** (5, 9), (8, 35)
$y = 1.191(1.587)^x$ $y = 9.541(1.677)^x$ $y = 0.936(1.573)^x$

Write a power function of the form $y = ax^b$ whose graph passes through the given points. (Lesson 8.7)

4. (2, 28), (8, 192)   **5.** (1, 0.5), (6, 48) $y = \frac{1}{2}x^{2.547}$ **6.** (5, 40), (2, 6)
$y = 10.693x^{1.389}$ $y = 1.429x^{2.070}$

7. 🦠 **FLU VIRUS** The spread of a virus through a student population can be

modeled by $S = \dfrac{5000}{1 + 4999e^{-0.8t}}$ where S is the total number of students infected

after t days. Graph the model and tell when the point of maximum growth in infections is reached. (Lesson 8.8)

Additional Test Preparation *Sample answer:*
2. You turn on the cold water faucet after it has been off for several hours. You run the water to get it as cold as possible to drink. The temperature of the water t seconds after turning on the faucet can be modeled by a logistic function.

Chapter Summary

7.

What did you learn?

Graph exponential functions.
- exponential growth functions (8.1)
- exponential decay functions (8.2)
- natural base functions (8.3)

Evaluate and simplify expressions.
- exponential expressions with base e (8.3)
- logarithmic expressions (8.4)

Graph logarithmic functions. (8.4)

Use properties of logarithms. (8.5)

Solve exponential and logarithmic equations. (8.6)

Model data with exponential and power functions. (8.7)

Evaluate and graph logistic growth functions. (8.8)

Use exponential, logarithmic, and logistic growth functions to model real-life situations. (8.1–8.8)

Why did you learn it?

Estimate wind energy generated by turbines. (p. 470)
Find the depreciated value of a car. (p. 476)
Find the number of endangered species. (p. 482)

Find air pressure on Mount Everest. (p. 484)
Approximate distance traveled by a tornado. (p. 491)

Estimate the average diameter of sand particles for a beach with given slope. (p. 489)

Compare loudness of sounds. (p. 495)

Use Newton's law of cooling. (p. 502)

Model the number of U.S. stamps issued. (p. 515)

Model the height of a sunflower. (p. 519)

Model a telescope's limiting magnitude. (p. 507)

How does Chapter 8 fit into the BIGGER PICTURE of algebra?

In Chapter 2 you began your study of functions and learned that quantities that increase by the same *amount* over equal periods of time are modeled by linear functions. In Chapter 8 you saw that quantities that increase by the same *percent* over equal periods of time are modeled by exponential functions.

Exponential functions and logarithmic functions are two important "families" of functions. They model many real-life situations, and they are used in advanced mathematics topics such as calculus and probability.

STUDY STRATEGY

How did you study with a group?

Here is an example of a summary prepared for Lesson 8.4 and presented to the group, following the **Study Strategy** on page 464.

> **Study Group**
>
> Lesson 8.4 Summary
>
> Definition of logarithm: $\log_b y = x$ if and only if $b^x = y$
>
> Common logarithm (base 10): $\log_{10} x = \log x$
>
> Natural logarithm (base e): $\log_e x = \ln x$
>
> Inverse functions: $f(x) = b^x$ (exponential) and
> $g(x) = \log_b x$ (logarithmic)
>
> Graph of logarithmic function $f(x) = \log_b (x - h) + k$:
> asymptote $x = h$; domain $x > h$; range all real
> numbers; up $b > 0$; down $0 < b < 1$

523

ADDITIONAL RESOURCES

The following resources are available to help review the material in this chapter.

- Chapter Review Games and Activities (*Chapter 8 Resource Book*, p. 118)
- *Instant Replay: Video Review Games*
- 📺 *Personal Student Tutor*
- Cumulative Review, Chs. 1–8 (*Chapter 8 Resource Book*, p. 130)

1.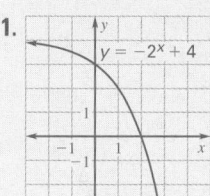

domain: all real numbers;
range: $y < 4$

2.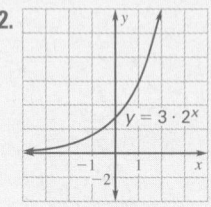

domain: all real numbers;
range: $y > 0$

3.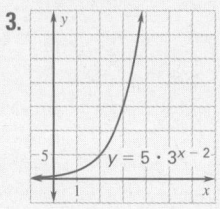

domain: all real numbers;
range: $y > 0$

4.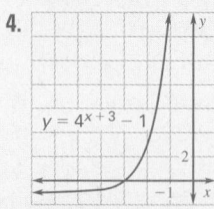

domain: all real numbers;
range: $y > -1$

9–12. See Additional Answers beginning on page AA1.

CHAPTER 8
Chapter Review

VOCABULARY

- exponential function, p. 465
- base of an exponential function, p. 465
- asymptote, p. 465
- exponential growth function, p. 466
- growth factor, p. 467

- exponential decay function, p. 474
- decay factor, p. 476
- natural base e, or Euler number, p. 480
- logarithm of y with base b, p. 486

- common logarithm, p. 487
- natural logarithm, p. 487
- change-of-base formula, p. 494
- logistic growth function, p. 517

8.1 EXPONENTIAL GROWTH

Examples on pp. 465–468

EXAMPLE An exponential growth function has the form $y = ab^x$ with $a > 0$ and $b > 1$.

To graph $y = 2 \cdot 5^{x+2} - 4$, first lightly sketch the graph of $y = 2 \cdot 5^x$, which passes through $(0, 2)$ and $(1, 10)$. Then translate the graph 2 units to the left and 4 units down. The graph passes through $(-2, -2)$ and $(-1, 6)$. The asymptote is the line $y = -4$. The domain is all real numbers, and the range is $y > -4$.

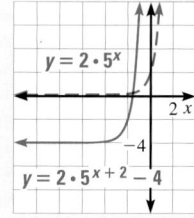

Graph the function. State the domain and range. 1–4. See margin.

1. $y = -2^x + 4$ **2.** $y = 3 \cdot 2^x$ **3.** $y = 5 \cdot 3^{x-2}$ **4.** $y = 4^{x+3} - 1$

8.2 EXPONENTIAL DECAY

Examples on pp. 474–476

EXAMPLE An exponential decay function has the form $y = ab^x$ with $a > 0$ and $0 < b < 1$.

To graph $y = 4\left(\frac{1}{3}\right)^x$, plot $(0, 4)$ and $\left(1, \frac{4}{3}\right)$. From *right* to *left* draw a curve that begins just above the x-axis, passes through the two points, and moves up. The asymptote is the line $y = 0$. The domain is all real numbers, and the range is $y > 0$.

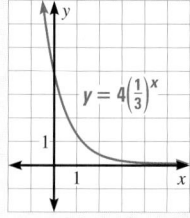

Tell whether the function represents *exponential growth* or *exponential decay*.

5. $f(x) = 5\left(\frac{3}{4}\right)^x$ **6.** $f(x) = 2\left(\frac{5}{4}\right)^x$ **7.** $f(x) = 3(6)^{-x}$ **8.** $f(x) = 4(3)^x$

exponential decay   exponential growth   exponential decay   exponential growth

Graph the function. State the domain and range. 9–12. See margin.

9. $y = \left(\frac{1}{4}\right)^x$ **10.** $y = 2\left(\frac{3}{5}\right)^{x-1}$ **11.** $y = \left(\frac{1}{2}\right)^x - 5$ **12.** $y = -3\left(\frac{3}{4}\right)^x + 2$

THE NUMBER e

Examples on pp. 480–482

EXAMPLES You can use e as the base of an exponential function. To graph such a function, use $e \approx 2.718$ and plot some points.

$f(x) = 3e^{2x}$ is an exponential growth function, since $2 > 0$.
$g(x) = 3e^{-2x}$ is an exponential decay function, since $-2 < 0$.

For both functions, the y-intercept is 3, the asymptote is $y = 0$, the domain is all real numbers, and the range is $y > 0$.

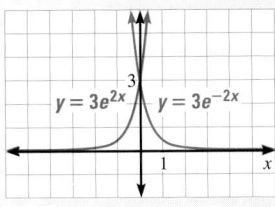

Graph the function. State the domain and range. 13–16. See margin.

13. $y = e^{x} + 5$ **14.** $y = 0.4e^{x} - 3$ **15.** $y = 4e^{-2x}$ **16.** $y = -e^{x} + 3$

LOGARITHMIC FUNCTIONS

Examples on pp. 486–489

EXAMPLES You can use the definition of logarithm to evaluate expressions:
$\log_b y = x$ if and only if $b^x = y$. The common logarithm has base 10 ($\log_{10} x = \log x$). The natural logarithm has base e ($\log_e x = \ln x$).

To evaluate $\log_8 4096$, write $\log_8 4096 = \log_8 8^4 = 4$.

To graph the logarithmic function $f(x) = 2 \log x + 1$, plot points such as $(1, 1)$ and $(10, 3)$. The vertical line $x = 0$ is an asymptote. The domain is $x > 0$, and the range is all real numbers.

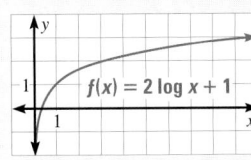

Evaluate the expression without using a calculator.

17. $\log_4 64$ 3 **18.** $\log_2 \frac{1}{8}$ -3 **19.** $\log_3 \frac{1}{9}$ -2 **20.** $\log_6 1$ 0

Graph the function. State the domain and range. 21–24. See margin.

21. $y = 3 \log_5 x$ **22.** $y = \log 4x$ **23.** $y = \ln x + 4$ **24.** $y = \log(x - 2)$

PROPERTIES OF LOGARITHMS

Examples on pp. 493–495

EXAMPLES You can use product, quotient, and power properties of logarithms.

Expand: $\log_2 \frac{3x}{y} = \log_2 3x - \log_2 y = \log_2 3 + \log_2 x - \log_2 y$

Condense: $3 \log_6 4 + \log_6 2 = \log_6 4^3 + \log_6 2 = \log_6 (64 \cdot 2) = \log_6 128$

Expand the expression.

25. $\log_3 6xy$
$\log_3 6 + \log_3 x + \log_3 y$

26. $\ln \frac{7x}{3}$
$\ln 7 + \ln x - \ln 3$

27. $\log 5x^3$
$\log 5 + 3 \log x$

28. $\log \frac{x^5 y^{-2}}{2y}$
$5 \log x - \log 2 - 3 \log y$

Condense the expression.

29. $2 \ln 3 - \ln 5$ $\ln \frac{9}{5}$ **30.** $\log_4 3 + 3 \log_4 2$ $\log_4 24$ **31.** $0.5 \log 4 + 2(\log 6 - \log 2)$
$\log 18$

13.

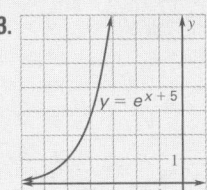

domain: all real numbers;
range: $y > 0$

14.

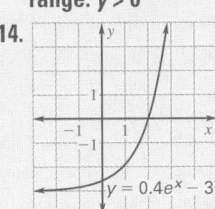

domain: all real numbers;
range: $y > -3$

15.

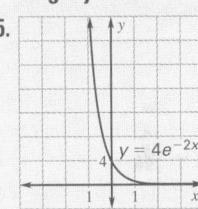

domain: all real numbers;
range: $y > 0$

16.

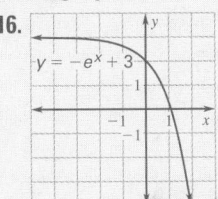

domain: all real numbers;
range: $y < 3$

21.

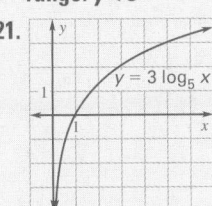

domain: $x > 0$;
range: all real numbers

22.

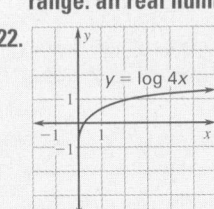

domain: $x > 0$;
range: all real numbers
23, 24. See next page.

23.

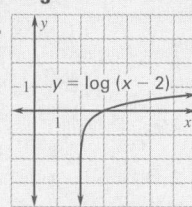

domain: $x > 0$;
range: all real numbers

24.

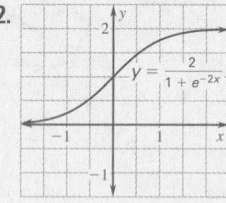

domain: $x > 2$;
range: all real numbers

42.

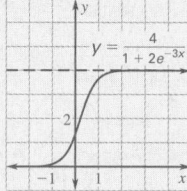

asymptotes: x-axis and $y = 2$;
y-intercept: 1; pt. of max. growth:
$(0, 1)$

43.

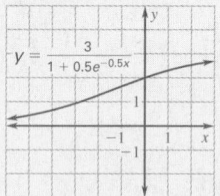

asymptotes: x-axis and $y = 4$;
y-intercept: $\frac{4}{3}$; pt. of max. growth:
$(0.231, 2)$

44.

asymptotes: x-axis and $y = 3$;
y-intercept: 2; pt. of max. growth:
$(-1.386, 1.5)$

8.6

SOLVING EXPONENTIAL AND LOGARITHMIC EQUATIONS

Examples on pp. 501–504

EXAMPLES You can solve exponential equations by equating exponents or by taking the logarithm of each side. You can solve logarithmic equations by exponentiating each side of the equation.

$$10^x = 4.3 \qquad\qquad \log_4 x = 3$$

$$\log 10^x = \log 4.3 \;\longleftarrow\; \text{Take log of each side.} \qquad 4^{\log_4 x} = 4^3 \;\longleftarrow\; \text{Exponentiate each side.}$$

$$x = \log 4.3 \approx 0.633 \qquad\qquad x = 4^3 = 64$$

Solve the equation. Check for extraneous solutions.

32. $2(3)^{2x} = 5$ 0.417 **33.** $3e^{-x} - 4 = 9$ **34.** $3 + \ln x = 8$ 148.41 **35.** $5 \log (x - 2) = 11$ 160.49
-1.466

8.7

MODELING WITH EXPONENTIAL AND POWER FUNCTIONS

Examples on pp. 509–512

EXAMPLE You can write an exponential function of the form $y = ab^x$ or a power function of the form $y = ax^b$ that passes through two given points.

To find a power function given $(3, 2)$ and $(9, 12)$, substitute the coordinates into $y = ax^b$ to get the equations $2 = a \cdot 3^b$ and $12 = a \cdot 9^b$. Solve the system of equations by substitution: $a \approx 0.333$ and $b \approx 1.631$. So, the function is $y = 0.333x^{1.631}$.

Find an exponential function of the form $y = ab^x$ whose graph passes through the given points.

$$y = 3.9605(1.499)^x$$

36. $(2, 6), (3, 8)$ $y = \left(\frac{27}{8}\right)\left(\frac{4}{3}\right)^x$ **37.** $(2, 8.9), (4, 20)$ **38.** $(2, 4.2), (4, 3.6)$ $y = 4.9(0.926)^x$

Find a power function of the form $y = ax^b$ whose graph passes through the given points.

39. $(2, 3.4), (6, 7.3)$ **40.** $(2, 12.5), (4, 33.2)$ **41.** $(0.5, 1), (10, 150)$ $y = 3.188x^{1.673}$
$y = 2.099x^{0.696}$ $y = 4.706x^{1.409}$

8.8

LOGISTIC GROWTH FUNCTIONS

Examples on pp. 517–519

EXAMPLE You can graph logistic growth functions by plotting points and identifying important characteristics of the graph.

The graph of $y = \dfrac{6}{1 + 3e^{-2x}}$ is shown. It has asymptotes $y = 0$ and $y = 6$. The y-intercept is 1.5. The point of maximum growth is $\left(\dfrac{\ln 3}{2}, \dfrac{6}{2}\right) \approx (0.55, 3)$.

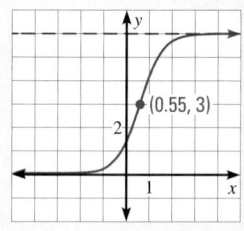

Graph the function. Identify the asymptotes, y-intercept, and point of maximum growth. 42–44. See margin.

42. $y = \dfrac{2}{1 + e^{-2x}}$ **43.** $y = \dfrac{4}{1 + 2e^{-3x}}$ **44.** $y = \dfrac{3}{1 + 0.5e^{-0.5x}}$

ADDITIONAL RESOURCES
- **Chapter 8 Resource Book**
  Chapter Test (3 levels) (p. 119)
  SAT/ACT Chapter Test (p. 125)
  Alternative Assessment (p. 126)
- ⊞ **Test and Practice Generator**

Graph the function. State the domain and range.   1–8. See margin.

1. $y = 2\left(\frac{1}{6}\right)^x$

2. $y = 4^{x-2} - 1$

3. $y = \frac{1}{2}e^x + 1$

4. $y = e^{-0.4x}$

5. $y = \log_{1/2} x$

6. $y = \ln x - 4$

7. $y = \log(x + 6)$

8. $y = \dfrac{2}{1 + 2e^{-x}}$

Simplify the expression.

9. $(2e^{-1})(3e^2)$ 6e

10. $\dfrac{-4e^x}{2e^{5x}}$ $\dfrac{-2}{e^{4x}}$

11. $\dfrac{e^6 \cdot e^x \cdot e^{-3x}}{e^{-2x+6}}$ 1

12. $\log 1000^2$ 6

13. $8^{\log_8 x}$ x

Evaluate the expression without using a calculator.

14. $\log_4 0.25$ −1

15. $\log_{1/3} 27$ −3

16. $\log 1$ 0

17. $\ln e^{-2}$ −2

18. $\log_3 243^2$
10

Solve the equation. Check for extraneous solutions.

19. $12 = 10^{x+5} - 7$ $\frac{\log 19 - 5}{} \approx -3.721$

20. $5 - \ln x = 7$ $e^{-2} \approx 0.135$

21. $\log_2 4x = \log_2 (x + 15)$ 5

22. $\dfrac{4}{1 + 2.5e^{-4x}} = 3.3$ 0.617

23. Tell whether the function $f(x) = 10(0.87)^x$ represents *exponential growth* or *exponential decay*.   **exponential decay**

24. Find the inverse of the function $y = \log_6 x$. $y = 6^x$

25. Use $\log_2 5 \approx 2.322$ to approximate $\log_2 50$ and $\log_2 0.4$.   5.644; −1.322

26. Condense the expression $3 \log_4 14 - 3 \log_4 42$. $\log_4 \frac{1}{27}$

27. Expand the expression $\ln 2y^2x$. $\ln 2 + 2 \ln y + \ln x$

28. Use the change-of-base formula to evaluate the expression $\log_7 15$.   1.392

29. Find an exponential function of the form $y = ab^x$ whose graph passes through the points $(4, 6)$ and $(7, 10)$. $y = 3.036(1.186)^x$

30. Find a power function of the form $y = ax^b$ whose graph passes through the points $(2, 3)$ and $(10, 21)$. $y = 1.298x^{1.209}$

31. 🌎 **CAR DEPRECIATION** The value of a new car purchased for $24,900 decreases by 10% per year. Write an exponential decay model for the value of the car. After about how many years will the car be worth half its purchase price? $V = 24,900(0.90)^t$; after about 6.58 years

32. 🌎 **EARNING INTEREST** You deposit $4000 in an account that pays 7% annual interest compounded continuously. Find the balance at the end of 5 years. $5676.27

33. 🌎 **COD WEIGHT** The table gives the mean weight w (in kilograms) and age x (in years) of Atlantic cod from the Gulf of Maine.

x	1	2	3	4	5	6	7	8
w	0.751	1.079	1.702	2.198	3.438	4.347	7.071	11.518

  a. Draw a scatter plot of $\ln w$ versus x. Is an exponential model a good fit for the original data?   **yes; See margin for graph.**

  b. Find an exponential model for the original data. Estimate the weight of a cod that is 9 years old. $w = 0.509(1.460)^x$; 15.34 kg

1.
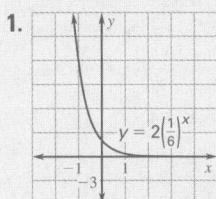
domain: all real numbers;
range: $y > 0$

2.
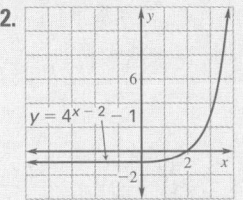
domain: all real numbers;
range: $y > -1$

3.
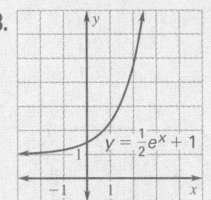
domain: all real numbers;
range: $y > 1$

4.
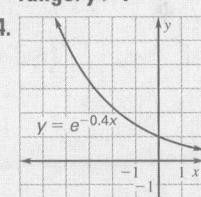
domain: all real numbers;
range: $y > 0$

5.
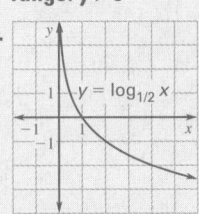
domain: $x > 0$;
range: all real numbers

6–8, 33a. See Additional Answers
beginning on page AA1.

ADDITIONAL RESOURCES
• **Chapter 8 Resource Book**
  Chapter Test (3 levels) (p. 119)
  SAT/ACT Chapter Test (p. 125)
  Alternative Assessment (p. 126)
• 🖳 **Test and Practice Generator**

CHAPTER
8

Chapter Standardized Test

▶ **TEST-TAKING STRATEGY** If you get stuck on a question, look at the answer choices for clues. Or select an answer choice and check to see if it is a reasonable answer to the question.

1. MULTIPLE CHOICE Which function is graphed? **E**

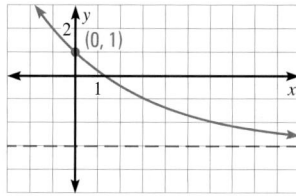

    Ⓐ $f(x) = 3(0.8)^x$

    Ⓑ $f(x) = 2(0.8)^x - 3$

    Ⓒ $f(x) = 2(0.8)^{x-3}$

    Ⓓ $f(x) = 4(0.8)^{x-3}$

    Ⓔ $f(x) = 4(0.8)^x - 3$

2. MULTIPLE CHOICE Suppose you deposit money in an investment account that pays 7% annual interest compounded continuously. About how many years will it take for your initial deposit to double? **D**

    Ⓐ 5   **Ⓑ** 7   **Ⓒ** 9

    Ⓓ 10   **Ⓔ** 14

3. MULTIPLE CHOICE Which function is the inverse of $y = \ln(x - 2)$? **A**

    Ⓐ $y = e^x + 2$ **Ⓑ** $y = e^{x+2}$

    Ⓒ $y = e^x - 2$ **Ⓓ** $y = e^{x-2}$

    Ⓔ $y = e^{-2x}$

4. MULTIPLE CHOICE Which function is graphed? **D**

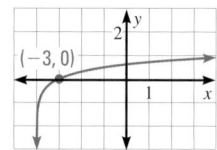

    Ⓐ $f(x) = \log 4x$

    Ⓑ $f(x) = 4 + \log x$

    Ⓒ $f(x) = -4 + \log x$

    Ⓓ $f(x) = \log(x + 4)$

    Ⓔ $f(x) = \log(x - 4)$

5. MULTIPLE CHOICE Which of the following is equivalent to $\log_2 7$? **D**

    Ⓐ 7^2 **Ⓑ** 2^7 **Ⓒ** $7 \log 2$

    Ⓓ $\dfrac{\log 7}{\log 2}$ **Ⓔ** $\dfrac{\log 2}{\log 7}$

6. MULTIPLE CHOICE Which of the following is equivalent to $\log \dfrac{xy^2}{z}$? **B**

    Ⓐ $\log x + 2 \log y + \log z$

    Ⓑ $\log x + 2 \log y - \log z$

    Ⓒ $\log z - \log x - 2 \log y$

    Ⓓ $2 \log xy - \log z$

    Ⓔ $\log z - 2 \log xy$

7. MULTIPLE CHOICE What is the solution of the equation $2^{x+14} = 16^{2x}$? **A**

    Ⓐ 2   **Ⓑ** 4   **Ⓒ** 8

    Ⓓ 16   **Ⓔ** No solution

8. MULTIPLE CHOICE What is the solution of the equation $0.5 \log_3 x = 2$? **C**

    Ⓐ 4   **Ⓑ** 64   **Ⓒ** 81

    Ⓓ $\dfrac{1}{64}$ **Ⓔ** $\dfrac{1}{81}$

9. MULTIPLE CHOICE Which function does *not* have a graph with asymptote $y = 0$? **C**

    Ⓐ $f(x) = 3^x$ **Ⓑ** $f(x) = (0.25)^{x+2}$

    Ⓒ $f(x) = \log 6x$ **Ⓓ** $f(x) = e^{-5x}$

    Ⓔ $f(x) = \dfrac{1}{1 + e^{-2x}}$

10. MULTIPLE CHOICE What type of function is $f(x) = 4e^{0.5x}$? **B**

    Ⓐ Exponential decay function

    Ⓑ Exponential growth function

    Ⓒ Logarithmic function

    Ⓓ Logistic growth function

    Ⓔ Power function

QUANTITATIVE COMPARISON In Exercises 11 and 12, choose the statement that is true about the given quantities.

(A) The quantity in column A is greater.

(B) The quantity in column B is greater.

(C) The two quantities are equal.

(D) The relationship cannot be determined from the given information.

	Column A	Column B	
11.	log 10,000	$\ln e^4$	C
12.	$\log_2 4$	$\log_4 2$	A

13. **MULTI-STEP PROBLEM** You are considering two job offers. The first offer is a salary of $32,000 with a $550 annual raise. The other offer is a salary of $29,500 with a 4% annual raise.

   a. Write a linear model for the total salary with the first offer as a function of the number t of years. $y = 550t + 32{,}000$

   b. Write an exponential model for the total salary with the other offer as a function of the number t of years. $y = 29{,}500(1.04)^t$

   c. Graph the functions in the same coordinate plane with domain $0 \le t \le 8$. Find the point of intersection of the two graphs and tell what it represents.   The intersection point (3.6, 33,984) represents when the salaries are the same.

   d. *Writing* Explain the difference in the salaries over time.
   Sample answer: After about 4 years, the salary from the second offer becomes greater than the first.

14. 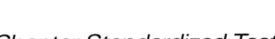 **MULTI-STEP PROBLEM** The table gives the weight w (in pounds) of an average girl for the first five years of life where t is her age in months. ▶ Source: *Your Baby & Child*

t	2	4	6	8	10	12	24	36	48	60
w	10.5	13.5	16.0	18.5	20.0	21.5	27.5	31.5	35.5	39.0

   a. Draw a scatter plot of $\ln w$ versus t.   See margin.

   b. Draw a scatter plot of $\ln w$ versus $\ln t$.   See margin.

   c. Analyze your scatter plots and decide whether an exponential model or a power model is a better fit for the original data. Explain your choice.

   d. Using your answer from part (c), find a model for the data. Check your model by using the regression feature of a graphing calculator. $y = 8.118x^{0.383}$

   e. Use your model to estimate a girl's weight at $1\frac{1}{2}$, $2\frac{1}{2}$, $3\frac{1}{2}$, and $4\frac{1}{2}$ years old.
   24.6 lb; 29.9 lb; 34.0 lb; 37.4 lb

15. **MULTI-STEP PROBLEM** Use the function $f(x) = \dfrac{5}{1 + 9e^{-x}}$.

   a. Find $f(-1)$, $f(0)$, and $f(2)$.   0.196; 0.5; 2.254

   b. Sketch a graph of the function.   See margin.

   c. Identify the asymptotes, y-intercept, and point of maximum growth.

   d. Write and solve an equation to find the value of x when $f(x)$ equals 4. Label this point on your graph. $\ln 36 \approx 3.584$

   e. *Writing* Describe how the growth represented by this function changes over time. $y = 5$ asymptotically.

13c.

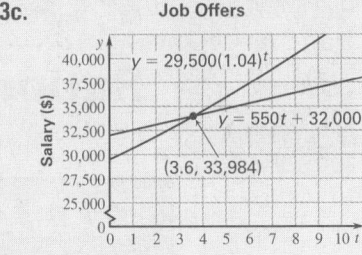

Job Offers

14a.

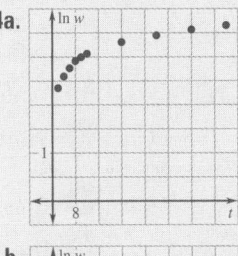

b.

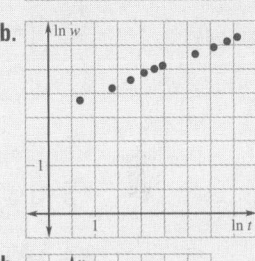

15b.

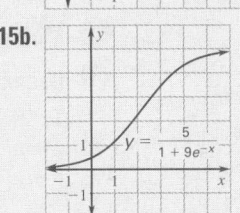

14c. A power model is a better fit for the original data. The points in the graph of $(t, \ln w)$ lie on a curve, while those in the graph of $(\ln t, \ln w)$ lie close to a straight line.

15c. asymptotes: x-axis and $y = 5$; y-intercept: $\frac{1}{2}$; pt. of max. growth: (2.197, 2.5)

   e. *Sample answer:* At first the function grows very slowly, then it grows faster and faster until the growth rate reaches its peak at the point (ln 9, 2.5). After this, the growth rate drops off until the curve again becomes almost flat, approaching

PLANNING THE CHAPTER

Rational Equations and Functions

LESSON	GOALS	NCTM	ITED	SAT9	Terra-Nova	Local
9.1 pp. 533–539	**CONCEPT ACTIVITY: 9.1** *Investigate inverse variation.* **GOAL 1** Write and use inverse variation models. **GOAL 2** Write and use joint variation models.	1, 2	RQRA, MIRA	2, 4, 11	11, 16, 17, 48	
9.2 pp. 540–546	**GOAL 1** Graph simple rational functions. **GOAL 2** Use the graph of a rational function to solve real-life problems. **TECHNOLOGY ACTIVITY: 9.2** *Graph rational functions on a graphing calculator.*	1, 2, 3, 6, 8, 9, 10	MIG, IIG, RQRA	2, 22	14, 16, 17, 18, 48	
9.3 pp. 547–553	**GOAL 1** Graph general rational functions. **GOAL 2** Use the graph of a rational function to solve real-life problems.	1, 2, 3, 6, 8, 9, 10	MIG, IIG, RQRA	2	14, 16, 17, 18, 48, 49, 51, 52	
9.4 pp. 554–561	**GOAL 1** Multiply and divide rational expressions. **GOAL 2** Use rational expressions to model real-life quantities. **TECHNOLOGY ACTIVITY: 9.4** *Simplify a rational expression and verify the result using a graphing calculator.*	1, 2, 8, 9, 10	MCWN, RQRA		11, 16, 18, 48	
9.5 pp. 562–567	**GOAL 1** Add and subtract rational expressions. **GOAL 2** Simplify complex fractions.	1, 2	MCWN, RQRA		11, 16, 49, 51, 52	
9.6 pp. 568–574	**GOAL 1** Solve rational equations. **GOAL 2** Use rational equations to solve real-life problems.	1, 2, 6, 8, 9, 10	MIRA, RQRA	2	11, 16, 17, 18, 48, 49, 51, 52	

RESOURCES

CHAPTER RESOURCE BOOKLETS

CHAPTER SUPPORT

Tips for New Teachers	p. 1	Prerequisite Skills Review	p. 5
Parent Guide for Student Success	p. 3	Strategies for Reading Mathematics	p. 7

LESSON SUPPORT

	9.1	9.2	9.3	9.4	9.5	9.6
Lesson Plans (regular and block)	p. 9	p. 24	p. 37	p. 50	p. 63	p. 78
Warm-Up Exercises and Daily Quiz	p. 11	p. 26	p. 39	p. 52	p. 65	p. 80
Activity Support Masters	p. 12					
Lesson Openers	p. 13	p. 27	p. 40	p. 53	p. 66	p. 81
Graphing Calculator Activities & Keystrokes	p. 14	p. 28		p. 54	p. 67	
Practice (3 levels)	p. 16	p. 29	p. 41	p. 55	p. 69	p. 82
Reteaching with Practice	p. 19	p. 32	p. 44	p. 58	p. 72	p. 85
Quick Catch-Up for Absent Students	p. 21	p. 34	p. 46	p. 60	p. 74	p. 87
Cooperative Learning Activities					p. 75	
Interdisciplinary Applications		p. 35		p. 61		p. 88
Real-Life Applications	p. 22		p. 47		p. 76	
Math & History Applications						p. 89
Challenge: Skills and Applications	p. 23	p. 36	p. 48	p. 62	p. 77	p. 90

REVIEW AND ASSESSMENT

Quizzes	pp. 49, 91	Alternative Assessment with Math Journal	p. 100
Chapter Review Games and Activities	p. 92	Project with Rubric	p. 102
Chapter Test (3 levels)	pp. 93–98	Cumulative Review	p. 104
SAT/ACT Chapter Test	p. 99	Resource Book Answers	p. A1

TRANSPARENCIES

	9.1	9.2	9.3	9.4	9.5	9.6
Warm-Up Exercises and Daily Quiz	p. 68	p. 69	p. 70	p. 71	p. 72	p. 73
Alternative Lesson Opener Transparencies	p. 59	p. 60	p. 61	p. 62	p. 63	p. 64
Examples/Standardized Test Practice	✓	✓	✓	✓	✓	✓
Answer Transparencies	✓	✓	✓	✓	✓	✓

TECHNOLOGY

- Electronic Teaching Tools
- Online Lesson Planner
- Internet Support
- Personal Student Tutor
- Test and Practice Generator
- Instant Replay: Video Review Games
- Electronic Lesson Presentations (Lesson 9.3)

ADDITIONAL RESOURCES

- Basic Skills Workbook: Diagnosis and Remediation
- Worked-Out Solution Key
- Resources in Spanish
- Standardized Test Practice Workbook
- Practice Workbook with Examples

PACING THE CHAPTER

REGULAR SCHEDULE

Day 1

9.1

STARTING OPTIONS
- Prereq. Skills Review
- Strategies for Reading
- Homework Check
- Warm-Up or Daily Quiz

TEACHING OPTIONS
- Concept Act. & Wksht.
- Les. Opener (Appl.)
- Graphing Calc. Activity
- Examples 1–6
- Closure Question
- Guided Practice Exs.

APPLY/HOMEWORK
- See Assignment Guide.
- See the CRB: Practice, Reteach, Apply, Extend

ASSESSMENT OPTIONS
- Checkpoint Exercises
- Daily Quiz (9.1)
- Stand. Test Practice

Day 2

9.2

STARTING OPTIONS
- Homework Check
- Warm-Up or Daily Quiz

TEACHING OPTIONS
- Les. Opener (Visual)
- Graphing Calc. Activity
- Examples 1–3
- Technology Activity
- Closure Question
- Guided Practice Exs.

APPLY/HOMEWORK
- See Assignment Guide.
- See the CRB: Practice, Reteach, Apply, Extend

ASSESSMENT OPTIONS
- Checkpoint Exercises
- Daily Quiz (9.2)
- Stand. Test Practice

Day 3

9.3

STARTING OPTIONS
- Homework Check
- Warm-Up or Daily Quiz

TEACHING OPTIONS
- Les. Opener (Visual)
- Examples 1–3
- Guided Practice Exs. 1–9

APPLY/HOMEWORK
- See Assignment Guide.
- See the CRB: Practice, Reteach, Apply, Extend

ASSESSMENT OPTIONS
- Checkpoint Exercises, p. 548

Day 4

9.3 *(cont.)*

STARTING OPTIONS
- Homework Check

TEACHING OPTIONS
- Example 4
- Closure Question
- Guided Practice Ex. 10

APPLY/HOMEWORK
- See Assignment Guide.
- See the CRB: Practice, Reteach, Apply, Extend

ASSESSMENT OPTIONS
- Checkpoint Exercises, p. 549
- Daily Quiz (9.3)
- Stand. Test Practice
- Quiz (9.1–9.3)

Day 5

9.4

STARTING OPTIONS
- Homework Check
- Warm-Up or Daily Quiz

TEACHING OPTIONS
- Motivating the Lesson
- Les. Opener (Activity)
- Graphing Calc. Activity
- Examples 1–5
- Guided Practice Exs. 1–14

APPLY/HOMEWORK
- See Assignment Guide.
- See the CRB: Practice, Reteach, Apply, Extend

ASSESSMENT OPTIONS
- Checkpoint Exercises, pp. 555–556

Day 6

9.4 *(cont.)*

STARTING OPTIONS
- Homework Check

TEACHING OPTIONS
- Examples 6–8
- Technology Activity
- Closure Question
- Guided Practice Ex. 15

APPLY/HOMEWORK
- See Assignment Guide.
- See the CRB: Practice, Reteach, Apply, Extend

ASSESSMENT OPTIONS
- Checkpoint Exercises, pp. 556–557
- Daily Quiz (9.4)
- Stand. Test Practice

Day 9

Review

DAY 9 START OPTIONS
- Homework Check

REVIEWING OPTIONS
- Chapter 9 Summary
- Chapter 9 Review
- Chapter Review Games and Activities

APPLY/HOMEWORK
- Chapter 9 Test (practice)
- Ch. Standardized Test (practice)

Day 10

Assess

DAY 10 START OPTIONS
- Homework Check

ASSESSMENT OPTIONS
- Chapter 9 Test
- SAT/ACT Ch. 9 Test
- Alternative Assessment

APPLY/HOMEWORK
- Skill Review, p. 588

BLOCK SCHEDULE

Day 7

9.5

STARTING OPTIONS
- Homework Check
- Warm-Up or Daily Quiz

TEACHING OPTIONS
- Les. Opener (Visual)
- Graphing Calc. Activity
- Examples 1–6
- Closure Question
- Guided Practice Exs.

APPLY/HOMEWORK
- See Assignment Guide.
- See the CRB: Practice, Reteach, Apply, Extend

ASSESSMENT OPTIONS
- Checkpoint Exercises
- Daily Quiz (9.5)
- Stand. Test Practice

Day 8

9.6

STARTING OPTIONS
- Homework Check
- Warm-Up or Daily Quiz

TEACHING OPTIONS
- Motivating the Lesson
- Les. Opener (Calculator)
- Examples 1–6
- Closure Question
- Guided Practice Exs.

APPLY/HOMEWORK
- See Assignment Guide.
- See the CRB: Practice, Reteach, Apply, Extend

ASSESSMENT OPTIONS
- Checkpoint Exercises
- Daily Quiz (9.6)
- Stand. Test Practice
- Quiz (9.4–9.6)

Day 1

9.1 & 9.2

DAY 1 START OPTIONS
- Prereq. Skills Review
- Strategies for Reading
- Homework Check
- W-Up 9.1 or D. Quiz 8.8

TEACHING 9.1 OPTIONS
- Concept Act. & Wksht.
- Les. Opener (Appl.)
- Graphing Calc. Activity
- Examples 1–6
- Closure Question
- Guided Practice Exs.

TEACHING 9.2 OPTIONS
- Warm-Up (Les. 9.2)
- Les. Opener (Visual)
- Graphing Calc. Activity
- Examples 1–3
- Technology Activity
- Closure Question
- Guided Practice Exs.

APPLY/HOMEWORK
- See Assignment Guide.
- See the CRB: Practice, Reteach, Apply, Extend

ASSESSMENT OPTIONS
- Checkpoint Exercises
- Daily Quiz (Les. 9.1, 9.2)
- Stand. Test Practice

Day 2

9.3

DAY 2 START OPTIONS
- Homework Check
- Warm-Up or Daily Quiz

TEACHING 9.3 OPTIONS
- Les. Opener (Visual)
- Examples 1–4
- Closure Question
- Guided Practice Exs.

APPLY/HOMEWORK
- See Assignment Guide.
- See the CRB: Practice, Reteach, Apply, Extend

ASSESSMENT OPTIONS
- Checkpoint Exercises
- Daily Quiz (Les. 9.3)
- Stand. Test Practice
- Quiz (9.1–9.3)

Day 3

9.4

DAY 3 START OPTIONS
- Homework Check
- Warm-Up or Daily Quiz

TEACHING 9.4 OPTIONS
- Motivating the Lesson
- Les. Opener (Activity)
- Graphing Calc. Activity
- Examples 1–8
- Technology Activity
- Closure Question
- Guided Practice Exs.

APPLY/HOMEWORK
- See Assignment Guide.
- See the CRB: Practice, Reteach, Apply, Extend

ASSESSMENT OPTIONS
- Checkpoint Exercises
- Daily Quiz (Les. 9.4)
- Stand. Test Practice

Day 4

9.5 & 9.6

DAY 4 START OPTIONS
- Homework Check
- W-Up 9.5 or D. Quiz 9.4

TEACHING 9.5 OPTIONS
- Les. Opener (Visual)
- Graphing Calc. Activity
- Examples 1–6
- Closure Question
- Guided Practice Exs.

TEACHING 9.6 OPTIONS
- Warm-Up (Les. 9.6)
- Motivating the Lesson
- Les. Opener (Calc.)
- Examples 1–6
- Closure Question
- Guided Practice Exs.

APPLY/HOMEWORK
- See Assignment Guide.
- See the CRB: Practice, Reteach, Apply, Extend

ASSESSMENT OPTIONS
- Checkpoint Exercises
- Daily Quiz (Les. 9.5, 9.6)
- Stand. Test Practice
- Quiz (9.4–9.6)

Day 5

Review/Assess

DAY 5 START OPTIONS
- Homework Check

REVIEWING OPTIONS
- Chapter 9 Summary
- Chapter 9 Review
- Chapter Review Games and Activities
- Chapter 9 Test (practice)
- Ch. Standardized Test (practice)

ASSESSMENT OPTIONS
- Chapter 9 Test
- SAT/ACT Ch. 9 Test
- Alternative Assessment

APPLY/HOMEWORK
- Skill Review, p. 588

MEETING INDIVIDUAL NEEDS

BEFORE THE CHAPTER

The *Chapter 9 Resource Book* has the following materials to distribute and use before the chapter:

• **Parent Guide for Student Success** (pictured below)
• **Prerequisite Skills Review**
• **Strategies for Reading Mathematics**

PARENT GUIDE *Pages 3–4*

LESSON
9

NAME _____ DATE _____

Parent Guide for Student Success
For use with Chapter 9

Chapter Overview One way that you can help your student succeed in Chapter 9 is by discussing the lesson goals in the chart below. When a lesson is completed, ask your student to interpret the lesson goals for you and to explain how the mathematics of the lesson relates to one of the key applications listed in the chart.

Lesson Title	Lesson Goals	Key Applications
9.1: Inverse and Joint Variation	Write and use inverse and joint variation models.	• Oceanography • Home Repair • Astronomy
9.2: Graphing Simple Rational Functions	Graph simple rational functions and use the graph of a rational function to solve real-life problems.	• Business • Lightning • Economics
9.3: Graphing General Rational Functions	Graph general rational functions and use the graphs to solve real-life problems.	• Manufacturing • Energy Expenditure • Hospital Costs
9.4: Multiplying and Dividing Rational Expressions	Multiply and divide rational expressions and use rational expressions to model real-life situations.	• Skydiving • Heat Generation • Farmland
9.5: Addition, Subtraction, and Complex Fractions	Add and subtract rational expressions. Simplify complex fractions.	• Statistics • Photography • Electronics
9.6: Solving Rational Equations	Solve rational equations and use rational equations to solve real-life problems.	• Chemistry • Football Statistics • Fuel Efficiency

Study Strategy

Making a Function Dictionary is the study strategy featured in Chapter 9 (see page 532). You may want to suggest that your student look back over the review material in previous chapters to identify the different types of functions to include in his or her function dictionary. Be sure your student includes the general form, a specific example, and a graph. If possible, provide graph paper to make it easier for your student to draw the graphs. Your student can add to this dictionary while working through Chapters 9–14.

PARENT GUIDE FOR STUDENT SUCCESS The first page summarizes the content of Chapter 9. Parents are encouraged to have their students explain how the material relates to key applications in the chapter, such as photography. The second page (not shown) provides exercises and an activity that parents can do with their students. In the activity, parents and students create a variation model between the number of "baskets" made into the trash can and the distance from the trash can.

DURING EACH LESSON

The *Chapter 9 Resource Book* has the following alternatives for introducing the lesson:

• **Lesson Openers** (pictured below)
• **Graphing Calculator Activities with Keystrokes**

LESSON OPENER *Page 40*

LESSON
9.3

NAME _____ DATE _____

Available as a transparency

Visual Approach Lesson Opener
For use with pages 547–553

You have learned how to graph simple rational functions. Many rational functions have been more complicated graphs than the ones you have seen so far, but these graphs often have asymptotes similar to the ones you have seen.

For each graph below, sketch any vertical or horizontal asymptotes, and give their equations.

1. $y = \dfrac{x^2 + 1}{x^2 - 1}$
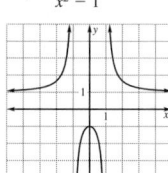

2. $y = \dfrac{2x^2}{x^2 + 1}$
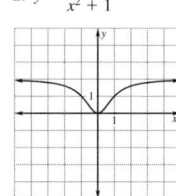

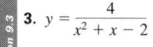

3. $y = \dfrac{4}{x^2 + x - 2}$
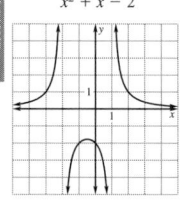

4. $y = \dfrac{x^2 + 4x + 6}{x + 2}$
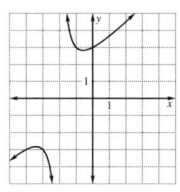

VISUAL APPROACH LESSON OPENER This Lesson Opener uses visuals as an alternative way to start Lesson 9.3. Students find vertical and horizontal asymptotes of graphs as an introduction to graphing general rational functions.

TECHNOLOGY RESOURCE

Teachers can use the Electronic Lesson Presentations CD-ROM to present Lesson 9.3 using computer animation to step through the Examples.

The *Chapter 9 Resource Book* has a variety of materials to follow-up each lesson. They include the following:

- **Practice (3 levels)**
- **Reteaching with Practice (pictured below)**
- **Quick Catch-Up for Absent Students**
- **Challenge: Skills and Applications**
- **Interdisciplinary Applications**
- **Real-Life Applications**

RETEACHING WITH PRACTICE Pages 58–59

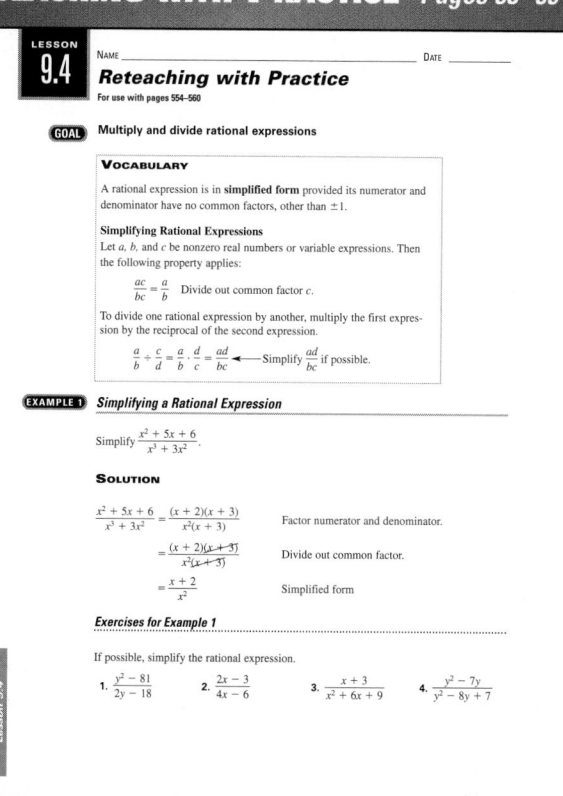

RETEACHING WITH PRACTICE On this two-page worksheet, additional examples and practice reinforce the key concepts of Lesson 9.4: multiplying and dividing rational expressions.

TECHNOLOGY RESOURCE

Students can use the Personal Student Tutor to find additional reteaching and practice for the skills in Lesson 9.4 and in the rest of Chapter 9.

The *Chapter 9 Resource Book* has the following review and assessment materials:

- **Quizzes**
- **Chapter Review Games and Activities**
- **Chapter Test (3 levels)**
- **SAT/ACT Chapter Test**
- **Alternative Assessment and Math Journal (below)**
- **Project with Rubric**
- **Cumulative Review**

ALTERNATIVE ASSESSMENT Pages 100–101

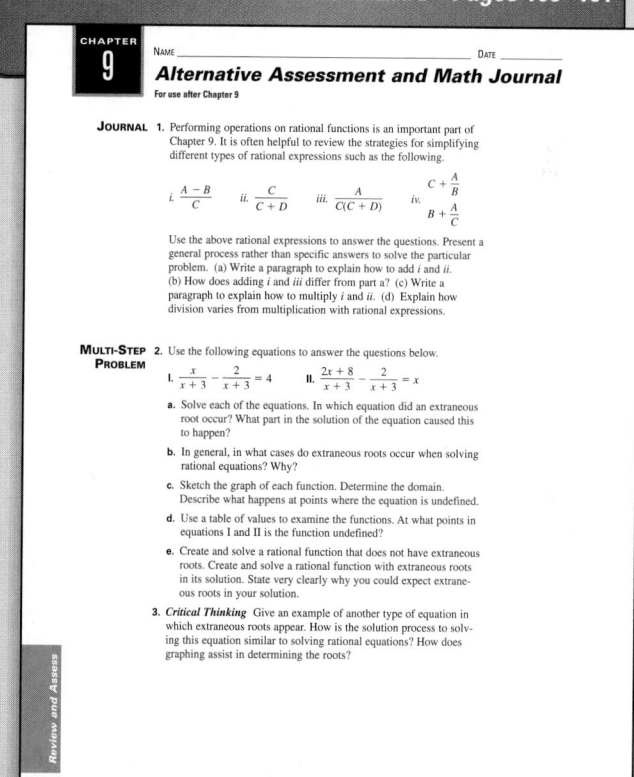

ALTERNATIVE ASSESSMENT WITH RUBRIC AND MATH JOURNAL The journal exercise asks students to demonstrate their understanding of rational functions in writing. The Multi-Step Problem has students pull together a variety of concepts from Chapter 9 to solve a problem about a rational function. Answers and a scoring rubric are provided on a separate sheet.

CHAPTER OVERVIEW

CHAPTER GOALS

The goals of this chapter include writing and using inverse variation and joint variation models in real-life problems. Students will graph rational functions, simplify complex fractions and rational expressions, and solve equations that contain rational expressions. They will use rational expressions or equations in real-life problems. In the project, students will graph data from an experiment, fit a model to the data, and compare the model to other models they have learned so far.

APPLICATION NOTE

Gravity is a downward force, and air resistance is an upward force exerted on a skydiver as he or she falls. Increasing the cross-sectional area of the skydiver increases the air resistance, which in turn decreases the falling speed. Students can see in the opening photograph that the "face-to-Earth" position has the most cross-sectional area for a skydiver with a closed parachute. The speed of the fall with a closed parachute continues to increase every second until the forces are in balance and the skydiver reaches a constant speed called terminal velocity. Opening the parachute increases the air resistance even further so that the speed of the dive can be decreased to levels that allow the skydiver to land safely.

The opening photo shows formation skydiving. On July 26, 1998, 246 divers set a new world record in formation skydiving. The skydivers held their position for 7.3 seconds.

Additional information about skydiving is available at **www.mcdougallittell.com.**

RATIONAL EQUATIONS AND FUNCTIONS

▶ *How does volume and surface area affect a skydiver's falling speed?*

530

CHAPTER 9

APPLICATION: Skydiving

How fast a skydiver falls depends on both the volume and the cross-sectional surface area of the skydiver. Different falling positions result in different ratios of volume to cross-sectional surface area: the larger the ratio, the greater the skydiver's falling speed.

Think & Discuss

The table below gives the volume and cross-sectional surface area for a 70 inch tall skydiver in each of three different positions.

Position	Volume (in.^{3})	Cross-sectional surface area (in.^{2})
Face-to-Earth	13,000	1300
Sitting	13,000	800
Headfirst	13,000	350

1. Find the ratio of volume to cross-sectional surface area for the skydiver in each of the three positions. **10, 16.25, 37.1**
2. In which position will the skydiver have the greatest falling speed? Explain. **headfirst; This position has the highest ratio of volume to cross-sectional surface area.**

Learn More About It

You will use a geometric model to write a rational expression for the ratio of a skydiver's volume to his or her cross-sectional surface area in Example 8 on p. 557 and in Ex. 15 on p. 558.

 APPLICATION LINK Visit www.mcdougallittell.com for more information on skydiving.

ADDITIONAL RESOURCES
Another way to begin the chapter is to show the video clip of a real-life motivator for Chapter 9 on the *Instant Replay: Video Review Games.*

PROJECTS
A project covering Chapters 7–9 appears on pages 584–585 of the Student Edition. An additional project for Chapter 9 is available in the *Chapter 9 Resource Book,* p. 102.

TECHNOLOGY
 **Software**
- *Electronic Teaching Tools*
- *Online Lesson Planner*
- *Personal Student Tutor*
- *Test and Practice Generator*
- *Electronic Lesson Presentations* (Lesson 9.3)

Video
- *Instant Replay: Video Review Games*

 **Internet Connections** www.mcdougallittell.com
- **Application Links** 531, 536, 574
- **Data Updates** 552, 572
- **Student Help** 538, 541, 546, 551, 556, 561, 564, 568
- **Career Links** 559, 566
- **Extra Challenge** 552, 560, 567, 573

531

Study Guide

The **Skill Review** exercises can help you diagnose whether students have the following skills needed in Chapter 9:

- Write direct variation equations.
- Multiply polynomials.
- Factor polynomials.
- Find the zeros of a function.

The following resources are available for students who need additional help with these skills:

- Prerequisite Skills Review (*Chapter 9 Resource Book,* p. 5; *Warm-Up Transparencies,* p. 67)
- ⊞ *Personal Student Tutor*

ADDITIONAL RESOURCES

The following resources are provided to help you prepare for the upcoming chapter and customize review materials:

- ***Chapter 9 Resource Book***
  Tips for New Teachers (p. 1)
  Parent Guide (p. 3)
  Lesson Plans (every lesson)
  Lesson Plans for Block Scheduling (every lesson)
- ⊞ *Electronic Teaching Tools*
- ⊞ *Online Lesson Planner*
- ⊞ *Test and Practice Generator*

ENGLISH LEARNERS

Remembering English language mathematical terms can be more challenging for English learners than for other students, particularly when a chapter contains a large number of such terms. Before beginning Chapter 9, give students time to review all the key vocabulary terms. You may also want to preteach some of the new terms students will encounter.

PREVIEW

What's the chapter about?

Chapter 9 is about **rational expressions**, **functions**, and **equations**. In Chapter 9 you'll learn

- how to simplify and perform operations with rational expressions.
- how to graph rational functions and solve rational equations.
- how to use variation models and rational models in real-life situations.

KEY VOCABULARY

▶ **Review**
- rational numbers, p. 3
- x-intercept, p. 84
- direct variation, p. 94
- zero of a function, p. 259

- degree of a polynomial function, p. 329
- asymptote, p. 465

▶ **New**
- inverse variation, p. 534
- joint variation, p. 536

- rational function, p. 540
- hyperbola, p. 540
- simplified form of a rational expression, p. 554
- complex fraction, p. 564

PREPARE

Are you ready for the chapter?

SKILL REVIEW Do these exercises to review key skills that you'll apply in this chapter. See the given **reference page** if there is something you don't understand.

STUDENT HELP

↳ **Study Tip**
"Student Help" boxes throughout the chapter give you study tips and tell you where to look for extra help in this book and on the Internet.

The variables x and y vary directly. Write an equation that relates the variables. (Review Example 6, p. 94)

1. $x = 2, y = 5$ $y = \frac{5}{2}x$ **2.** $x = 1, y = 0.1$ $y = \frac{1}{10}x$ **3.** $x = 8, y = -2$ $y = -\frac{1}{4}x$ **4.** $x = -3, y = 12$ $y = -4x$

Multiply the polynomials. (Review Example 5, p. 13; Example 4, p. 339)

5. $5(3x - 1)$ $15x - 5$

6. $(x - 1)(x + 4)^2$ $x^3 + 7x^2 + 8x - 16$

7. $-x(x^2 - 5)$ $-x^3 + 5x$

8. $x(x - 1)(x + 8)$ $x^3 + 7x^2 - 8x$

Factor the polynomial. (Review Examples 1–4, pp. 256 and 257; Examples 1–3, p. 346)

9. $x^2 - 6x + 9$ $(x - 3)^2$

10. $4x^3 - 4$ $4(x - 1)(x^2 + x + 1)$

11. $8x^3 - 162x$ $2x(2x - 9)(2x + 9)$

12. $6x^2 + 7x - 5$ $(2x - 1)(3x + 5)$

Find all the real zeros of the function. (Review Example 7, p. 259; Example 4, p. 354; Example 1, p. 359)

13. $y = x^2 + 2x$ $0, -2$

14. $y = x^2 + 2x - 15$ $-5, 3$

15. $y = x^3 - 2x^2 - 7x - 4$ $-1, 4$

STUDY STRATEGY

Here's a study strategy!

Dictionary of Functions

Make a dictionary of all the types of functions you have learned in this course. For each entry, include the general form of the function and an example of the function and its graph. Continue to add entries as you work through this chapter. Use your dictionary as a study and reference tool.

ACTIVITY 9.1

Developing Concepts

Investigating Inverse Variation

GROUP ACTIVITY
Work with a partner.

MATERIALS
• tape measure or meter stick
• centimeter ruler
• masking tape

▶ **QUESTION** What is the relationship between the distance you are standing from your partner and the apparent height of your partner?

▶ **EXPLORING THE CONCEPT**

1 Have your partner stand with his or her back against a wall. Place the end of a tape measure against the wall and between your partner's feet. Use masking tape to mark off distances of 3 meters, 4 meters, . . . , 9 meters from the wall.

2 Stand facing your partner, with your toes just touching the 3 meter mark. Hold a centimeter ruler at arm's length and line up the "0" end of the ruler with the top of your partner's head. Measure (to the nearest centimeter) the apparent height of your partner at this distance.

3 Repeat **Step 2** for each of the marked distances and record your results in a table like the one shown.

Distance (m)	3	4	5	6	7	8	9
Apparent height (cm)	?	?	?	?	?	?	?

▶ **DRAWING CONCLUSIONS**

1. Does apparent height vary directly with distance? Justify your answer mathematically.

2. Multiply the paired values of distance and apparent height together. What do you notice?

3. Based on your results from Exercise 2, write an equation relating distance and apparent height.

4. Use your equation from Exercise 3 to predict your partner's apparent height at a distance not listed in your table. Test your prediction by standing that distance from your partner and measuring his or her apparent height. How close was your prediction?

1. No; as the distance increases, the apparent height decreases.

2. The product of distance and apparent height is approximately constant.

3. Equations may vary but should have the form $dh = c$, where d is the distance between the partners, h is the height of the person standing against the wall, and c is the constant from Ex. 2.

4. Answers may vary but should be consistent with the equation from Ex. 3.

Activity Assessment *Sample answer:*
Answers should be consistent with the equation from Exercise 3.

9.1 *Concept Activity* **533**

1 Planning the Activity

PURPOSE
To find the relationship between distance and apparent height.

MATERIALS
• Tape measure or meter stick
• Centimeter ruler
• Masking tape
• Activity Support Master (*Chapter 9 Resource Book,* p. 12)

PACING
• Exploring the Concept — 10 min
• Drawing Conclusions — 10 min

▶ **LINK TO LESSON**
The product of distance and apparent height is approximately constant in this activity. When students read Example 4 of Lesson 9.1, they will notice that the product of r and ℓ is also approximately constant. Both sets of data points show inverse variation.

2 Managing the Activity

CLASSROOM MANAGEMENT
Some students may have difficulty writing the equation for Exercise 3. Suggest that they let d = distance, h = height, and c = the constant from Exercise 2. Then $d \cdot h = c$.

ALTERNATIVE APPROACH
The power regression feature of a graphing calculator can be used to find a model for the data. The model is $y = a \cdot x^{\wedge}b$, where a = constant from Exercise 2 and $b = -1$.

3 Closing the Activity

★ **KEY DISCOVERY**
The apparent height of an object varies inversely with the distance from the object.

ACTIVITY ASSESSMENT
Suppose you stand at a distance of 10 meters from your partner. What is the apparent height of your partner?
See sample answer at left.

533

1 PLAN

PACING
Basic: 1 day
Average: 1 day
Advanced: 1 day
Block Schedule: 0.5 block with 9.2

LESSON OPENER
APPLICATION
An alternative way to approach
Lesson 9.1 is to use the Application
Lesson Opener:
• Blackline Master (*Chapter 9 Resource Book,* p. 13)
• 📖 Transparency (p. 59)

MEETING INDIVIDUAL NEEDS
• *Chapter 9 Resource Book*
 Prerequisite Skills Review (p. 5)
 Practice Level A (p. 16)
 Practice Level B (p. 17)
 Practice Level C (p. 18)
 Reteaching with Practice (p. 19)
 Absent Student Catch-Up (p. 21)
 Challenge (p. 23)
• *Resources in Spanish*
• 🖥 *Personal Student Tutor*

NEW-TEACHER SUPPORT
See the Tips for New Teachers on
pp. 1–2 of the *Chapter 9 Resource Book* for additional notes about
Lesson 9.1.

WARM-UP EXERCISES

 Transparency Available

Solve for *y* in each equation.
1. $x + y = 2$ $y = 2 - x$
2. $xy = 8$ $y = \dfrac{8}{x}$
3. $2y = x^2$ $y = \dfrac{x^2}{2}$
4. $0.1 = xy$ $y = \dfrac{0.1}{x}$
5. $x = 8y$ $y = \dfrac{x}{8}$

EXPLORING DATA AND STATISTICS

9.1

What you should learn

GOAL 1 Write and use inverse variation models, as applied in **Example 4**.

GOAL 2 Write and use joint variation models, as applied in **Example 6**.

Why you should learn it

▼ To solve **real-life** problems, such as finding the speed of a whirlpool's current in **Example 3**.

Inverse and Joint Variation

GOAL 1 USING INVERSE VARIATION

In Lesson 2.4 you learned that two variables x and y show direct variation if $y = kx$ for some nonzero constant k. Another type of variation is called *inverse variation.* Two variables x and y show **inverse variation** if they are related as follows:

$$y = \frac{k}{x}, k \neq 0$$

The nonzero constant k is called the **constant of variation,** and y is said to *vary inversely* with x.

EXAMPLE 1 *Classifying Direct and Inverse Variation*

Tell whether x and y show *direct variation*, *inverse variation*, or *neither*.

GIVEN EQUATION	REWRITTEN EQUATION	TYPE OF VARIATION
a. $\dfrac{y}{5} = x$	$y = 5x$	Direct
b. $y = x + 2$		Neither
c. $xy = 4$	$y = \dfrac{4}{x}$	Inverse

EXAMPLE 2 *Writing an Inverse Variation Equation*

The variables x and y vary inversely, and $y = 8$ when $x = 3$.

a. Write an equation that relates x and y.

b. Find y when $x = -4$.

SOLUTION

a. Use the given values of x and y to find the constant of variation.

$y = \dfrac{k}{x}$ Write general equation for inverse variation.

$8 = \dfrac{k}{3}$ Substitute 8 for *y* and 3 for *x*.

$24 = k$ Solve for *k*.

▶ The inverse variation equation is $y = \dfrac{24}{x}$.

b. When $x = -4$, the value of y is:

$y = \dfrac{24}{-4}$

$= -6$

Oceanography

EXAMPLE 3 *Writing an Inverse Variation Model*

The speed of the current in a whirlpool varies inversely with the distance from the whirlpool's center. The Lofoten Maelstrom is a whirlpool located off the coast of Norway. At a distance of 3 kilometers (3000 meters) from the center, the speed of the current is about 0.1 meter per second. Describe the change in the speed of the current as you move closer to the whirlpool's center.

SOLUTION

First write an inverse variation model relating distance from center d and speed s.

$$s = \frac{k}{d} \qquad \text{Model for inverse variation}$$

$$0.1 = \frac{k}{3000} \qquad \text{Substitute 0.1 for } s \text{ and 3000 for } d.$$

$$300 = k \qquad \text{Solve for } k.$$

The model is $s = \frac{300}{d}$. The table shows some speeds for different values of d.

Distance from center (meters), d	2000	1500	500	250	50
Speed (meters per second), s	0.15	0.2	0.6	1.2	6

▶ From the table you can see that the speed of the current increases as you move closer to the whirlpool's center.

· · · · · · · · · ·

The equation for inverse variation can be rewritten as $xy = k$. This tells you that a set of data pairs (x, y) shows inverse variation if the products xy are constant or approximately constant.

EXAMPLE 4 *Checking Data for Inverse Variation*

FOCUS ON APPLICATIONS

COMMON SCOTER The common scoter migrates from the Quebec/Labrador border in Canada to coastal cities such as Portland, Maine, and Galveston, Texas. To reach its winter destination, the scoter will travel up to 2150 miles.

BIOLOGY CONNECTION The table compares the wing flapping rate r (in beats per second) to the wing length l (in centimeters) for several birds. Do these data show inverse variation? If so, find a model for the relationship between r and l.

Bird	r (beats per second)	l (cm)
Carrion crow	3.6	32.5
Common scoter	5.0	23.5
Great crested grebe	6.3	18.7
Curlew	4.0	29.2
Lesser black-backed gull	2.8	42.2

▶ Source: *Smithsonian Miscellaneous Collections*

SOLUTION

Each product rl is approximately equal to 117. For instance, $(3.6)(32.5) = 117$ and $(5.0)(23.5) = 117.5$. So, the data do show inverse variation. A model for the relationship between wing flapping rate and wing length is $r = \frac{117}{l}$.

---

2 TEACH

EXTRA EXAMPLE 1
Do x and y show *direct variation, inverse variation,* or *neither?*
a. $xy = 4.8$ inverse
b. $x = \frac{y}{1.5}$ direct

EXTRA EXAMPLE 2
x and y vary inversely, and $y = 6$ when $x = 1.5$.
a. Write an equation that relates x and y. $y = \frac{9}{x}$
b. Find y when $x = \frac{4}{3}$. 6.75

EXTRA EXAMPLE 3
The volume of gas in a container varies inversely with the amount of pressure. A gas has volume 75 in.^{3} at a pressure of 25 lb/in.^{2}. Write a model relating volume and pressure. $V = \frac{1875}{P}$

EXTRA EXAMPLE 4
Do these data show inverse variation? If so, find a model.
yes; $H = \frac{18}{W}$

W	2	4	6	8	10
H	9	4.5	3	2.25	1.8

CHECKPOINT EXERCISES
For use after Example 1:
1. Do x and y show *direct variation, inverse variation,* or *neither?*
  a. $y = 3.5x + 5$ neither
  b. $3y = 4x$ direct

For use after Examples 2 and 3:
2. x and y vary inversely, and $y = 7.5$ when $x = 2$.
  a. Find an equation that relates x and y. $y = \frac{15}{x}$
  b. Find y when $x = 1.2$. 12.5

For use after Example 4:
3. Do these data show inverse variation? If so, find a model.
  yes; $T = \frac{300}{R}$

R	10	20	30	40	50
T	30	15	10	7.5	6

EXTRA EXAMPLE 5

Write an equation.

a. y varies directly with x and inversely with z^2. $y = \dfrac{kx}{z^2}$

b. y varies inversely with x^3.

$y = \dfrac{k}{x^3}$

c. y varies directly with x^2 and inversely with z. $y = \dfrac{kx^2}{z}$

d. z varies jointly with x^2 and y.

$z = kx^2y$

e. y varies inversely with x and z.

$y = \dfrac{k}{xz}$

EXTRA EXAMPLE 6

The ideal gas law states that the volume V (in liters) varies directly with the number of molecules n (in moles) and temperature T (in Kelvin) and varies inversely with the pressure P (in kilopascals). The constant of variation is denoted by R and is called the universal gas constant.

a. Write an equation for the ideal gas law. $V = \dfrac{nRT}{P}$

b. Estimate the universal gas constant if $V = 251.6$ liters; $n = 1$ mole; $T = 288$ K; $P = 9.5$ kilopascals   **about 8.3**

✓ CHECKPOINT EXERCISES

For use after Examples 5 and 6:

1. The volume of a geometric figure varies jointly with the square of the radius of the base and the height.

   a. Write an equation for the volume. $V = kr^2h$

   b. Estimate the constant of variation if $V = 63.33$ in.^{3}; $r = 2.4$ in.; $h = 10.5$ in.
   about 1.047

CLOSURE QUESTION

If z varies directly with x and inversely with w^3, and $z = 1.6$ when $x = 8$ and $w = 4$, describe how to find the constant of variation.

Write the model $z = \dfrac{kx}{w^3}$; substitute the known values into the equation. Solve for k to get $k = 12.8$.

STUDENT HELP

Look Back
For help with direct variation, see p. 94.

Joint variation occurs when a quantity varies directly as the product of *two or more* other quantities. For instance, if $z = kxy$ where $k \neq 0$, then z varies jointly with x and y. Other types of variation are also possible, as illustrated in the following example.

EXAMPLE 5 *Comparing Different Types of Variation*

Write an equation for the given relationship.

RELATIONSHIP	EQUATION
a. y varies directly with x.	$y = kx$
b. y varies inversely with x.	$y = \dfrac{k}{x}$
c. z varies jointly with x and y.	$z = kxy$
d. y varies inversely with the square of x.	$y = \dfrac{k}{x^2}$
e. z varies directly with y and inversely with x.	$z = \dfrac{ky}{x}$

EXAMPLE 6 *Writing a Variation Model*

SCIENCE CONNECTION The *law of universal gravitation* states that the gravitational force F (in newtons) between two objects varies jointly with their masses m_1 and m_2 (in kilograms) and inversely with the square of the distance d (in meters) between the two objects. The constant of variation is denoted by G and is called the *universal gravitational constant*.

a. Write an equation for the law of universal gravitation.

b. Estimate the universal gravitational constant. Use the Earth and sun facts given at the right.

> Mass of Earth:
> $m_1 = 5.98 \times 10^{24}$ kg
>
> Mass of sun:
> $m_2 = 1.99 \times 10^{30}$ kg
>
> Mean distance between Earth and sun:
> $d = 1.50 \times 10^{11}$ m
>
> Force between Earth and sun:
> $F = 3.53 \times 10^{22}$ N

SOLUTION

a. $F = \dfrac{Gm_1m_2}{d^2}$

b. Substitute the given values and solve for G.

$$F = \frac{Gm_1m_2}{d^2}$$

$$3.53 \times 10^{22} = \frac{G(5.98 \times 10^{24})(1.99 \times 10^{30})}{(1.50 \times 10^{11})^2}$$

$$3.53 \times 10^{22} \approx G(5.29 \times 10^{32})$$

$$6.67 \times 10^{-11} \approx G$$

▶ The universal gravitational constant is about $6.67 \times 10^{-11} \dfrac{\text{N} \cdot \text{m}^2}{\text{kg}^2}$.

FOCUS ON APPLICATIONS

Earth's orbital path

p   *a*

Earth   *sun*

Not drawn to scale

EARTH AND SUN
Earth's orbit around the sun is elliptical, so its distance from the sun varies. The shortest distance p is 1.47×10^{11} meters and the longest distance a is 1.52×10^{11} meters.

 APPLICATION LINK
www.mcdougallittell.com

GUIDED PRACTICE

Vocabulary Check ✓

1. Complete this statement: If w varies directly as the product of x, y, and z, then w varies ? with x, y, and z.   **jointly**

Concept Check ✓

2. How can you tell whether a set of data pairs (x, y) shows inverse variation?
Each product xy will be approximately equal to the same number.

3. Suppose z varies jointly with x and y. What can you say about $\frac{z}{xy}$?
It will be a constant.

Skill Check ✓

Tell whether x and y show *direct variation*, *inverse variation*, or *neither*.

4. $xy = \frac{1}{4}$
inverse variation

5. $\frac{x}{y} = 5$
direct variation

6. $y = x - 3$ **neither**

7. $x = \frac{7}{y}$
inverse variation

8. $\frac{y}{x} = 12$
direct variation

9. $\frac{1}{2}xy = 9$
inverse variation

10. $y = \frac{1}{x}$
inverse variation

11. $2x + y = 4$
neither

Tell whether x varies jointly with y and z.

12. $x = 15yz$ **yes**

13. $\frac{x}{z} = 0.5y$ **yes**

14. $xy = 4z$ **no**

15. $x = \frac{yz}{2}$ **yes**

16. $x = \frac{3z}{y}$ **no**

17. $2yz = 7x$ **yes**

18. $\frac{x}{y} = 17z$ **yes**

19. $5x = 4yz$ **yes**

20. 🌎 **TOOLS** The force F needed to loosen a bolt with a wrench varies inversely with the length l of the handle. Write an equation relating F and l, given that 250 pounds of force must be exerted to loosen a bolt when using a wrench with a handle 6 inches long. How much force must be exerted when using a wrench with a handle 24 inches long? $F = \frac{1500}{l}$; **62.5 lb**

PRACTICE AND APPLICATIONS

STUDENT HELP

→ **Extra Practice**
to help you master
skills is on p. 952.

DETERMINING VARIATION Tell whether x and y show *direct variation, inverse variation*, or *neither*.

21. $xy = 10$
inverse variation

22. $xy = \frac{1}{10}$
inverse variation

23. $y = x - 1$
neither

24. $\frac{y}{9} = x$
direct variation

25. $x = \frac{5}{y}$ **inverse variation**

26. $3x = y$
direct variation

27. $x = 5y$
direct variation

28. $x + y = 2.5$
neither

INVERSE VARIATION MODELS The variables x and y vary inversely. Use the given values to write an equation relating x and y. Then find y when $x = 2$.

29. $x = 5$, $y = -2$ $y = -\frac{10}{x}$; -5

30. $x = 4$, $y = 8$ $y = \frac{32}{x}$; 16

31. $x = 7$, $y = 1$ $y = \frac{7}{x}$; 3.5

32. $x = \frac{1}{2}$, $y = 10$ $y = \frac{5}{x}$; 2.5

33. $x = -\frac{2}{3}$, $y = 6$ $y = -\frac{4}{x}$; -2

34. $x = \frac{3}{4}$, $y = \frac{3}{8}$ $y = \frac{9}{32x}$; $\frac{9}{64}$

STUDENT HELP

→ **HOMEWORK HELP**
Example 1: Exs. 21–28
Example 2: Exs. 29–34
Example 3: Exs. 51–54
Example 4: Exs. 35–38,
48, 49
Example 5: Exs. 45–47
Example 6: Exs. 55–58

INTERPRETING DATA Determine whether x and y show *direct variation, inverse variation*, or *neither*.

35.

x	y
1.5	20
2.5	12
4	7.5
5	6

inverse variation

36.

x	y
31	217
20	140
17	119
12	84

direct variation

37.

x	y
3	36
7	105
5	50
16	48

neither

38.

x	y
4	16
5	12.8
1.6	40
20	3.2

inverse variation

---

3 APPLY

○ **ASSIGNMENT GUIDE**

BASIC
Day 1: pp. 537–539 Exs. 21–28,
30–42 even, 45–50, 59,
61–67 odd

AVERAGE
Day 1: pp. 537–539 Exs. 21–28,
30–44 even, 45–53, 59,
61–67 odd

ADVANCED
Day 1: pp. 537–539 Exs. 21–28,
30–44 even, 45–47, 54–60,
61–67 odd

BLOCK SCHEDULE
pp. 537–539 Exs. 21–28, 30–44
even, 45–53, 59, 61–67 odd,
(with 9.2)

EXERCISE LEVELS
Level A: *Easier*
21–34, 45–47
Level B: *More Difficult*
35–44, 48–59
Level C: *Most Difficult*
60

✓ **HOMEWORK CHECK**
To quickly check student understanding of key concepts, go
over the following exercises:
Exs. 22, 30, 36, 40, 46. See also
the Daily Homework Quiz:

• Blackline Master (*Chapter 9
Resource Book,* p. 26)
• 🖥 Transparency (p. 69)

STUDENT HELP NOTES

→**Homework Help** Students can find help for Exs. 45–47 at **www.mcdougallittell.com.** The information can be printed out for students who don't have access to the Internet.

APPLICATION NOTE

EXERCISE 54 Another unit of measure for the intensity of sound is the decibel, or dB. An intensity of 10 watts per square meter is equivalent to 130 dB, and is considered a threshold of pain.

ENGLISH LEARNERS

EXERCISES 48–54 Some students may require definitions of *caulking, luminosity,* and *intensity* to visualize the situations described in these problems. You may want to mention to students that whenever they have a question about the language in a problem, they should refer to a dictionary or ask another student for clarification before they try to respond.

ADDITIONAL PRACTICE AND RETEACHING

For Lesson 9.1:
• Practice Levels A, B, and C (*Chapter 9 Resource Book,* p. 16)
• Reteaching with Practice (*Chapter 9 Resource Book,* p. 19)
• ⊞ See Lesson 9.1 of the *Personal Student Tutor*

For more Mixed Review:
• ⊞ Search the *Test and Practice Generator* for key words or specific lessons.

538

JOINT VARIATION MODELS The variable z varies jointly with x and y. Use the given values to write an equation relating x, y, and z. Then find z when $x = -4$ and $y = 7$.

39. $x = 3$, $y = 8$, $z = 6$ $z = \frac{1}{4}xy$; -7 **40.** $x = -12$, $y = 4$, $z = 2$ $z = -\frac{1}{24}xy$; $\frac{7}{6}$

41. $x = 1$, $y = \frac{1}{3}$, $z = 5$ $z = 15xy$; -420 **42.** $x = -6$, $y = 3$, $z = \frac{2}{5}$ $z = -\frac{1}{45}xy$; $\frac{28}{45}$

43. $x = \frac{5}{6}$, $y = \frac{3}{10}$, $z = 8$ $z = 32xy$; -896 **44.** $x = \frac{3}{8}$, $y = \frac{16}{17}$, $z = \frac{3}{2}$ $z = \frac{17}{4}xy$; -119

STUDENT HELP

▸ **HOMEWORK HELP** Visit our Web site www.mcdougallittell.com for help with Exs. 45–47.

WRITING EQUATIONS Write an equation for the given relationship.

45. x varies inversely with y and directly with z. $x = \frac{kz}{y}$

46. y varies jointly with z and the square root of x. $y = kz\sqrt{x}$

47. w varies inversely with x and jointly with y and z. $w = \frac{kyz}{x}$

🔧 **HOME REPAIR** In Exercises 48–50, use the following information.
On some tubes of caulking, the diameter of the circular nozzle opening can be adjusted to produce lines of varying thickness. The table shows the length l of caulking obtained from a tube when the nozzle opening has diameter d and cross-sectional area A.

d (in.)	A (in.²)	l (in.)
$\frac{1}{8}$	$\frac{\pi}{256}$	1440
$\frac{1}{4}$	$\frac{\pi}{64}$	360
$\frac{3}{8}$	$\frac{9\pi}{256}$	160
$\frac{1}{2}$	$\frac{\pi}{16}$	90

48. Determine whether l varies inversely with d. If so, write an equation relating l and d.   **no**

49. Determine whether l varies inversely with A. If so, write an equation relating l and A.   **yes;** $l = \frac{45\pi}{8A}$

50. Find the length of caulking you get from a tube whose nozzle opening has a diameter of $\frac{3}{4}$ inch.   **40 in.**

🌐 **ASTRONOMY** In Exercises 51–53, use the following information.
A star's diameter D (as a multiple of the sun's diameter) varies directly with the square root of the star's luminosity L (as a multiple of the sun's luminosity) and inversely with the square of the star's temperature T (in kelvins).

51. Write an equation relating D, L, T, and a constant k. $D = \frac{k\sqrt{L}}{T^2}$

52. The luminosity of Polaris is 10,000 times the luminosity of the sun. The surface temperature of Polaris is about 5800 kelvins. Using $k = 33,640,000$, find how the diameter of Polaris compares with the diameter of the sun.
The diameter of Polaris is 100 times the diameter of the sun.

53. The sun's diameter is 1,390,000 kilometers. What is the diameter of Polaris?
139,000,000 km

54. 🔊 **INTENSITY OF SOUND** The intensity I of a sound (in watts per square meter) varies inversely with the square of the distance d (in meters) from the sound's source. At a distance of 1 meter from the stage, the intensity of the sound at a rock concert is about 10 watts per square meter. Write an equation relating I and d. If you are sitting 15 meters back from the stage, what is the intensity of the sound you hear? $I = \frac{10}{d^2}$; **0.044 watts/m²**

55. SCIENCE▸ CONNECTION   The work W (in joules) done when lifting an object varies jointly with the mass m (in kilograms) of the object and the height h (in meters) that the object is lifted. The work done when a 120 kilogram object is lifted 1.8 meters is 2116.8 joules. Write an equation that relates W, m, and h. How much work is done when lifting a 100 kilogram object 1.5 meters? $W = \frac{49}{5}mh$, **1470 joules**

FOCUS ON PEOPLE

▸ ⚫ **STEPHEN HAWKING,** a theoretical physicist, has spent years studying *black holes.* A black hole is believed to be formed when a star's core collapses. The gravitational pull becomes so strong that even the star's light, as discussed in Exs. 51–53, cannot escape.

HEAT LOSS In Exercises 56 and 57, use the following information.
The heat loss h (in watts) through a single-pane glass window varies jointly with the window's area A (in square meters) and the difference between the inside and outside temperatures d (in kelvins).

56. Write an equation relating h, A, d, and a constant k. $h = kAd$

57. A single-pane window with an area of 1 square meter and a temperature difference of 1 kelvin has a heat loss of 5.7 watts. What is the heat loss through a single-pane window with an area of 2.5 square meters and a temperature difference of 20 kelvins?   **285 watts**

58. GEOMETRY CONNECTION   The area of a trapezoid varies jointly with the height and the sum of the lengths of the bases. When the sum of the lengths of the bases is 18 inches and the height is 4 inches, the area is 36 square inches. Find a formula for the area of a trapezoid. $A = \frac{1}{2}h(b_1 + b_2)$

Test Preparation

59. **MULTI-STEP PROBLEM** The load P (in pounds) that can be safely supported by a horizontal beam varies jointly with the width W (in feet) of the beam and the square of its depth D (in feet), and inversely with its length L (in feet).

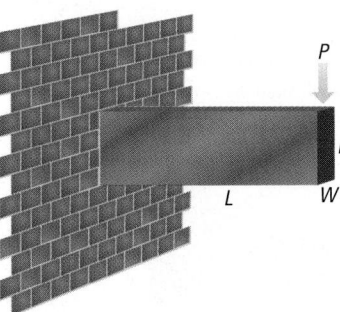

a. How does P change when the width and length of the beam are doubled?
It is unchanged.
b. How does P change when the width and depth of the beam are doubled?
It is multiplied by a factor of 8.
c. How does P change when all three dimensions are doubled?
It is multiplied by a factor of 4.
d. *Writing* Describe several ways a beam can be modified if the safe load it is required to support is increased by a factor of 4.

59d. Sample answer: Double only the depth or make the beam 4 times as wide.

★ **Challenge**

60. **LOGICAL REASONING** Suppose x varies inversely with y and y varies inversely with z. How does x vary with z? Justify your answer algebraically.
x varies directly with z; $x = \frac{k_1}{y}$ and $y = \frac{k_2}{z}$, so $x = \frac{k_1}{k_2}z$ or $x = kz$

MIXED REVIEW

SQUARE ROOT FUNCTIONS Graph the function. Then state the domain and range. (Review 7.5 for 9.2)   61–63. See margin.

61. domain: all real numbers x such that $x \geq -2$; range: all real numbers y such that $y \geq 0$

61. $y = \sqrt{x + 2}$ **62.** $y = \sqrt{x} - 4$ **63.** $y = \sqrt{x + 1} - 3$

SOLVING RADICAL EQUATIONS Solve the equation. Check for extraneous solutions. (Review 7.6)

62. domain: all real numbers x such that $x \geq 0$; range: all real numbers y such that $y \geq -4$

64. $\sqrt{x} = 22$ **484**   **65.** $\sqrt[4]{2x} + 2 = 6$ **128**   **66.** $x^{1/3} - 7 = 0$ **343**

67. $\sqrt[3]{x + 12} = 5$ **113**   **68.** $(x - 2)^{3/2} = -8$ **69.** $\sqrt{3x + 1} = \sqrt{x + 15}$ **7**
                                    no solution

63. domain: all real numbers x such that $x \geq -1$; range: all real numbers y such that $y \geq -3$

70. **COLLEGE ADMISSION** The number of admission applications received by a college was 1152 in 1990 and increased 5% per year until 1998. (Review 8.1 for 9.2)

a. Write a model giving the number A of applications t years after 1990.
$A = 1152(1.05)^t$
b. Graph the model. Use the graph to estimate the year in which there were 1400 applications.   **1994; See margin.**

63, 70. See Additional Answers beginning on page AA1.

DAILY HOMEWORK QUIZ

Transparency Available

Tell whether x and y show *direct variation, inverse variation,* or *neither.*

1. $xy = \frac{2}{3}$ **inverse**

2. $y = 7x$ **direct**

Write an equation for the given relation.

3. x varies inversely with y and $x = 2$ when $y = 12$. $y = \frac{24}{x}$

4. z varies jointly with x and y and $z = 8$ when $x = 4$ and $y = 10$. $z = \frac{xy}{5}$

5. z varies inversely with the square of x and directly with y. $z = \frac{ky}{x^2}$

EXTRA CHALLENGE NOTE

→ Challenge problems for Lesson 9.1 are available in **blackline** format in the *Chapter 9 Resource Book,* p. 23 and at **www.mcdougallittell.com.**

ADDITIONAL TEST PREPARATION

1. **OPEN ENDED** Give an example of a geometric figure in which the area varies jointly with two variables.
Sample answer: the area of a triangle $A = 0.5bh$, where b is the base and h is the height.

61.
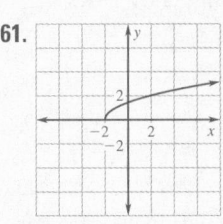

62.
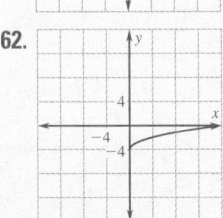

LESSON OPENER
VISUAL APPROACH
An alternative way to approach Lesson 9.2 is to use the Visual Approach Lesson Opener:
• Blackline Master (*Chapter 9 Resource Book*, p. 27)
• Transparency (p. 60)

MEETING INDIVIDUAL NEEDS
• *Chapter 9 Resource Book*
  Prerequisite Skills Review (p. 5)
  Practice Level A (p. 29)
  Practice Level B (p. 30)
  Practice Level C (p. 31)
  Reteaching with Practice (p. 32)
  Absent Student Catch-Up (p. 34)
  Challenge (p. 36)
• *Resources in Spanish*
• Personal Student Tutor

NEW-TEACHER SUPPORT
See the Tips for New Teachers on pp. 1–2 of the *Chapter 9 Resource Book* for additional notes about Lesson 9.2.

WARM-UP EXERCISES

 Transparency Available

Tell how each equation is related to the graph of $y = \sqrt{x}$.

1. $y = \sqrt{x - 1}$
  shifted 1 unit to the right

2. $y = \sqrt{x} + 1$
  shifted one unit vertically

3. $y = -\sqrt{x}$ reflected in the *x*-axis

4. $y = \sqrt{x + 2} + 1$ shifted 2 units to the left and one unit up

5. $y = 2\sqrt{x}$ *y*-values are doubled for each corresponding *x*-value

Step 1. See Additional Answers beginning on page AA1.

540

9.2 Graphing Simple Rational Functions

What you should learn

GOAL 1 Graph simple rational functions.

GOAL 2 Use the graph of a rational function to solve **real-life** problems, such as finding the average cost per calendar in **Example 3**.

Why you should learn it

▼ To solve **real-life** problems, such as finding the frequency of an approaching ambulance siren in **Exs. 47 and 48**.

Step 2. If $a > 0$, the branches of the hyperbola are in the first and third quadrants. If $a < 0$, the branches of the hyperbola are in the second and fourth quadrants.

Step 3. As $|a|$ gets bigger, the branches move farther away from the origin.

GOAL 1 **GRAPHING A SIMPLE RATIONAL FUNCTION**

A **rational function** is a function of the form

$$f(x) = \frac{p(x)}{q(x)}$$

where $p(x)$ and $q(x)$ are polynomials and $q(x) \neq 0$. In this lesson you will learn to graph rational functions for which $p(x)$ and $q(x)$ are linear. For instance, consider the following rational function:

$$y = \frac{1}{x}$$

The graph of this function is called a **hyperbola** and is shown below. Notice the following properties.

• The *x*-axis is a horizontal asymptote.

• The *y*-axis is a vertical asymptote.

• The domain and range are all nonzero real numbers.

• The graph has two symmetrical parts called **branches**. For each point (x, y) on one branch, there is a corresponding point $(-x, -y)$ on the other branch.

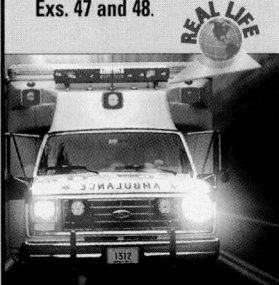

x	y
-4	$-\frac{1}{4}$
-3	$-\frac{1}{3}$
-2	$-\frac{1}{2}$
-1	-1
$-\frac{1}{2}$	-2

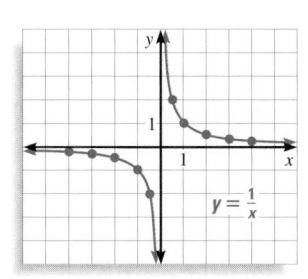

x	y
4	$\frac{1}{4}$
3	$\frac{1}{3}$
2	$\frac{1}{2}$
1	1
$\frac{1}{2}$	2

ACTIVITY

Developing Concepts

Investigating Graphs of Rational Functions

❶ Graph each function.   a–d. See margin.

  a. $y = \frac{2}{x}$ **b.** $y = \frac{3}{x}$ **c.** $y = \frac{-1}{x}$ **d.** $y = \frac{-2}{x}$

❷ Use the graphs to describe how the sign of a affects the graph of $y = \frac{a}{x}$.

❸ Use the graphs to describe how $|a|$ affects the graph of $y = \frac{a}{x}$.

All rational functions of the form $y = \dfrac{a}{x - h} + k$ have graphs that are hyperbolas with asymptotes at $x = h$ and $y = k$. To draw the graph, plot a couple of points on each side of the vertical asymptote. Then draw the two branches of the hyperbola that approach the asymptotes and pass through the plotted points.

STUDENT HELP

→ **Look Back**
For help with asymptotes, see p. 465.

EXAMPLE 1 *Graphing a Rational Function*

Graph $y = \dfrac{-2}{x + 3} - 1$. State the domain and range.

SOLUTION

Draw the asymptotes $x = -3$ and $y = -1$.

Plot two points to the left of the vertical asymptote, such as $(-4, 1)$ and $(-5, 0)$, and two points to the right, such as $(-1, -2)$ and $\left(0, -\dfrac{5}{3}\right)$.

Use the asymptotes and plotted points to draw the branches of the hyperbola.

The domain is all real numbers except -3, and the range is all real numbers except -1.

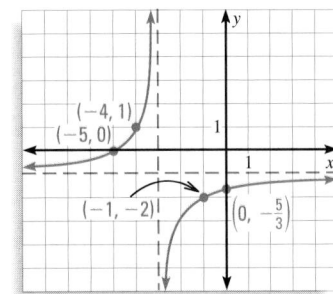

· · · · · · · · · ·

All rational functions of the form $y = \dfrac{ax + b}{cx + d}$ also have graphs that are hyperbolas. The vertical asymptote occurs at the x-value that makes the denominator zero. The horizontal asymptote is the line $y = \dfrac{a}{c}$.

EXAMPLE 2 *Graphing a Rational Function*

STUDENT HELP

HOMEWORK HELP
Visit our Web site
www.mcdougallittell.com
for extra examples.

Graph $y = \dfrac{x + 1}{2x - 4}$. State the domain and range.

SOLUTION

Draw the asymptotes. Solve $2x - 4 = 0$ for x to find the vertical asymptote $x = 2$. The horizontal asymptote is the line $y = \dfrac{a}{c} = \dfrac{1}{2}$.

Plot two points to the left of the vertical asymptote, such as $\left(0, -\dfrac{1}{4}\right)$ and $(1, -1)$, and two points to the right, such as $(3, 2)$ and $\left(4, \dfrac{5}{4}\right)$.

Use the asymptotes and plotted points to draw the branches of the hyperbola.

The domain is all real numbers except 2, and the range is all real numbers except $\dfrac{1}{2}$.

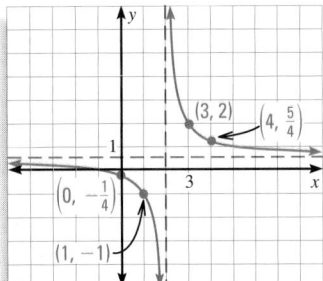

9.2 Graphing Simple Rational Functions **541**

---

2 TEACH

EXTRA EXAMPLE 1
Graph $y = \dfrac{3}{x - 1} + 2$. State the domain and range.

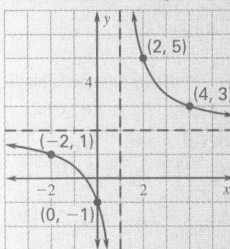

domain: all real numbers except 1;
range: all real numbers except 2

EXTRA EXAMPLE 2
Graph $y = \dfrac{x - 2}{3x + 3}$. State the domain and range.

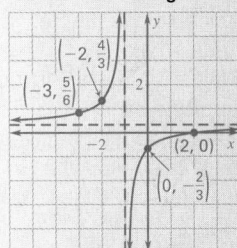

domain: all real numbers except -1; range: all real numbers except $\dfrac{1}{3}$

CHECKPOINT EXERCISES
For use after Examples 1 and 2:

1. Graph $y = \dfrac{-1}{x - 4} + 3$. State the domain and range.

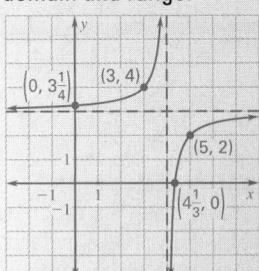

domain: all real numbers except 4; range: all real numbers except 3

 EXTRA EXAMPLE 3
The senior class is sponsoring a dinner. The cost of catering the dinner is $9.95 per person plus an $18 delivery charge.

a. Write a model that gives the average cost per person.

$$A = \frac{18.00 + 9.95x}{x}$$

b. Graph the model and use it to estimate the number of people needed to lower the cost to $11 per person.
at least 17 people

c. Describe what happens to the average cost per person as the number increases.
the average cost gets closer and closer to $9.95

☑ **CHECKPOINT EXERCISES**

For use after Example 3:

1. The speed of sound can be modeled by $1.09F + 1087.8$ ft/sec, where F is the temperature in °F.

a. Write a model that gives the time it takes you to hear thunder a mile away.

$$t = \frac{5280}{1.09F + 1087.8}$$

b. Graph the model and use it to estimate how long it takes you to hear the thunder a mile away if it is 75°.
about 4.5 seconds

c. Describe what happens to the length of time it takes for the thunder to reach your ears as the temperature decreases. **The length of time increases. For example, at 34°, it takes 4.7 seconds.**

CLOSURE QUESTION
Give an example of a rational function whose graph is a hyperbola that has a vertical asymptote at $x = 2$ and a horizontal asymptote at $y = 1$. *Sample answer:* $y = \frac{1}{x-2} + 1$

 **Business**

EXAMPLE 3 *Writing a Rational Model*

For a fundraising project, your math club is publishing a fractal art calendar. The cost of the digital images and the permission to use them is $850. In addition to these "one-time" charges, the *unit cost* of printing each calendar is $3.25.

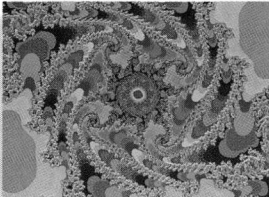

a. Write a model that gives the average cost per calendar as a function of the number of calendars printed.

b. Graph the model and use the graph to estimate the number of calendars you need to print before the average cost drops to $5 per calendar.

c. Describe what happens to the average cost as the number of calendars printed increases.

SOLUTION

a. The average cost is the total cost of making the calendars divided by the number of calendars printed.

 **PROBLEM SOLVING STRATEGY**

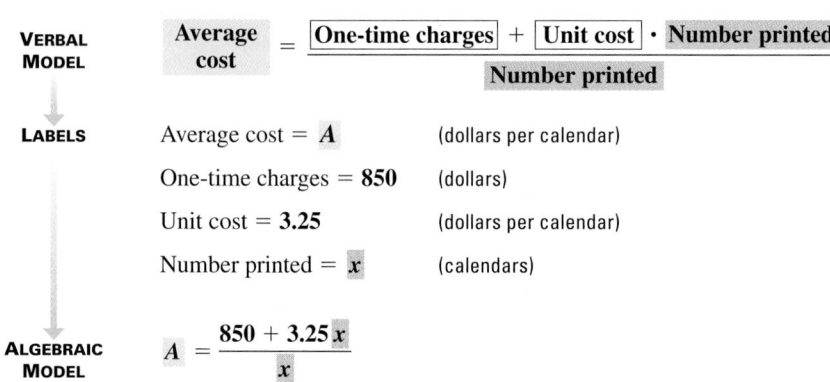

b. The graph of the model is shown at the right. The A-axis is the vertical asymptote and the line $A = 3.25$ is the horizontal asymptote. The domain is $x > 0$ and the range is $A > 3.25$. When $A = 5$ the value of x is about 500. So, you need to print about 500 calendars before the average cost drops to $5 per calendar.

c. As the number of calendars printed increases, the average cost per calendar gets closer and closer to $3.25. For instance, when $x = 5000$ the average cost is $3.42, and when $x = 10,000$ the average cost is $3.34.

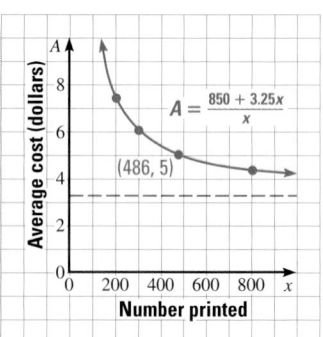

GUIDED PRACTICE

Vocabulary Check ✓

1. Complete this statement: The graph of a function of the form $y = \frac{a}{x-h} + k$ is called a(n) _?_. **hyperbola**

Concept Check ✓

2. ERROR ANALYSIS Explain why the graph shown is not the graph of $y = \frac{6}{x+3} + 7$.
The vertical asymptote should be $x = -3$.

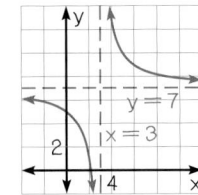

11. $y = 2$; $x = 0$; domain: all real numbers except 0; range: all real numbers except 2

12. $y = 2$; $x = 3$; domain: all real numbers except 3; range: all real numbers except 2

3. If the graph of a rational function is a hyperbola with the x-axis and the y-axis as asymptotes, what is the domain of the function? What is the range?
all nonzero real numbers; all nonzero real numbers

Identify the horizontal and vertical asymptotes of the graph of the function.

$y = 4$; $x = 3$
4. $y = \frac{2}{x-3} + 4$

$y = 2$; $x = -4$
5. $y = \frac{2x+3}{x+4}$

6. $y = \frac{x-3}{x+3}$ $y = 1$; $x = -3$

Skill Check ✓

13. $y = -2$; $x = -3$; domain: all real numbers except -3; range: all real numbers except -2

14. $y = 1$; $x = 3$; domain: all real numbers except 3; range: all real numbers except 1

15. $y = \frac{2}{3}$; $x = -\frac{1}{3}$; domain: all real numbers except $-\frac{1}{3}$; range: all real numbers except $\frac{2}{3}$

7. $y = \frac{x+5}{2x-4}$ $y = \frac{1}{2}$; $x = 2$

8. $y = \frac{3}{x+8} - 10$
$y = -10$; $x = -8$

9. $y = \frac{-4}{x-6} - 5$ $y = -5$; $x = 6$

10. 🌐 **CALENDAR FUNDRAISER** Look back at Example 3 on page 542. Suppose you decide to generate your own fractals on a computer to save money. The cost for the software (a "one-time" cost) is $125. Write a model that gives the average cost per calendar as a function of the number of calendars printed. Graph the model and use the graph to estimate the number of calendars you need to print before the average cost drops to $5 per calendar.   **See margin for graph.**

$A = \frac{125 + 3.25n}{n}$; about 70 calendars

PRACTICE AND APPLICATIONS

STUDENT HELP

▶ **Extra Practice**
to help you master skills is on p. 952.

16. $y = \frac{3}{4}$; $x = -\frac{5}{4}$; domain: all real numbers except $-\frac{5}{4}$; range: all real numbers except $\frac{3}{4}$

IDENTIFYING ASYMPTOTES Identify the horizontal and vertical asymptotes of the graph of the function. Then state the domain and range.   **11–19. See margin.**

11. $y = \frac{3}{x} + 2$

12. $y = \frac{4}{x-3} + 2$

13. $y = \frac{-2}{x+3} - 2$

14. $y = \frac{x+2}{x-3}$

15. $y = \frac{2x+2}{3x+1}$

16. $y = \frac{-3x+2}{-4x-5}$

17. $y = \frac{-22}{x+43} - 17$

18. $y = \frac{34x-2}{16x+4}$

19. $y = \frac{4}{x-6} + 19$

MATCHING GRAPHS Match the function with its graph.

20. $y = \frac{3}{x-2} + 3$ **B**

21. $y = \frac{-3}{x-2} + 3$ **C**

22. $y = \frac{x+2}{x+3}$ **A**

STUDENT HELP

▶ **HOMEWORK HELP**
Example 1: Exs. 11–31
Example 2: Exs. 11–22, 32–40
Example 3: Exs. 42–48

A.
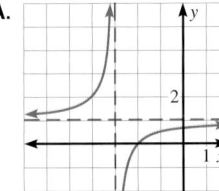

B.
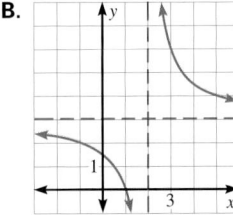

C.
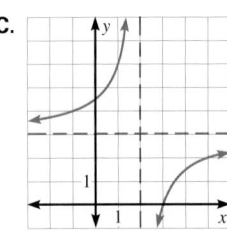

ASSIGNMENT GUIDE

BASIC
Day 1: pp. 543–545 Exs. 12–28 even, 32–38 even, 41–43, 49–51, 53–67 odd

AVERAGE
Day 1: pp. 543–545 Exs. 12–28 even, 32–38 even, 41–46, 49–51, 53–69 odd

ADVANCED
Day 1: pp. 543–545 Exs. 12–40 even, 41–46, 49–52, 53–69 odd

BLOCK SCHEDULE
pp. 543–545 Exs. 12–28 even, 32–38 even, 41–46, 49–52, 53–69 odd (with 9.1)

EXERCISE LEVELS

Level A: *Easier*
11–22, 41, 49

Level B: *More Difficult*
23–40, 42–48, 50–51

Level C: *Most Difficult*
52

✔ HOMEWORK CHECK

To quickly check student understanding of key concepts, go over the following exercises: Exs. 12, 14, 22, 26, 34. See also the Daily Homework Quiz:

- Blackline Master (*Chapter 9 Resource Book*, p. 39)
- 📄 Transparency (p. 70)

10, 17–19. See Additional Answers beginning on page AA1.

APPLICATION NOTE
EXERCISES 45 AND 46
According to Laffer's theory, when the government tax rate is zero percent, people keep all of what they earn. With zero revenue the government ceases to exist. When the tax rate is 100 percent, people have no incentive to work and thus all production stops. The government then gets 100 percent of zero money earned, which results in zero government revenue. At some tax rate *t* between zero and 100 percent, both production and government revenues are maximized.

23.
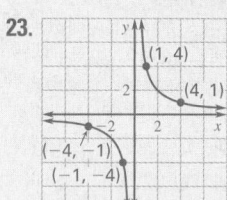

24.
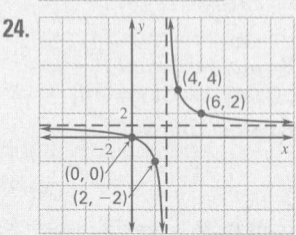

25.
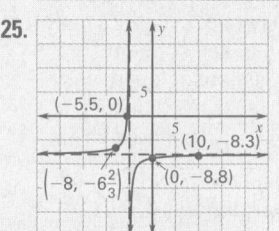

26–42, 45. See Additional Answers beginning on page AA1.

ADDITIONAL PRACTICE AND RETEACHING

For Lesson 9.2:
• Practice Levels A, B, and C (*Chapter 9 Resource Book,* p. 29)
• Reteaching with Practice (*Chapter 9 Resource Book,* p. 32)
• See Lesson 9.2 of the *Personal Student Tutor*

For more Mixed Review:
• Search the *Test and Practice Generator* for key words or specific lessons.

544

23. domain: all real numbers except 0; range: all real numbers except 0

24. domain: all real numbers except 3; range: all real numbers except 1

25. domain: all real numbers except −5; range: all real numbers except −8

26. domain: all real numbers except 7; range: all real numbers except 3

27. domain: all real numbers except −2; range: all real numbers except −6

28. domain: all real numbers except 0; range: all real numbers except 4

29. domain: all real numbers except −3; range: all real numbers except −2

FOCUS ON APPLICATIONS

LIGHTNING
The equation given in Ex. 44 uses the fact that at 0°C, sound travels at 331 meters per second. Since light travels at about 300,000 kilometers per second, a flash of lightning is seen before it is heard.

GRAPHING FUNCTIONS Graph the function. State the domain and range.
23–31. See margin.

23. $y = \dfrac{4}{x}$

24. $y = \dfrac{3}{x - 3} + 1$

25. $y = \dfrac{-4}{x + 5} - 8$

26. $y = \dfrac{1}{x - 7} + 3$

27. $y = \dfrac{6}{x + 2} - 6$

28. $y = \dfrac{5}{x} + 4$

29. $y = \dfrac{1}{4x + 12} - 2$

30. $y = \dfrac{3}{2x}$

31. $y = \dfrac{4}{3x - 6} + 5$

GRAPHING FUNCTIONS Graph the function. State the domain and range.
32–40. See margin.

32. $y = \dfrac{x + 2}{x + 3}$

33. $y = \dfrac{x}{4x + 3}$

34. $y = \dfrac{x - 7}{3x - 8}$

35. $y = \dfrac{9x + 1}{3x - 2}$

36. $y = \dfrac{-3x + 10}{4x - 12}$

37. $y = \dfrac{5x + 2}{4x}$

38. $y = \dfrac{3x}{2x - 4}$

39. $y = \dfrac{7x}{-x - 15}$

40. $y = \dfrac{-14x - 4}{2x - 1}$

41. **CRITICAL THINKING** Write a rational function that has the vertical asymptote $x = -4$ and the horizontal asymptote $y = 3$. **See margin.**

RACQUETBALL In Exercises 42 and 43, use the following information.
You've paid $120 for a membership to a racquetball club. Court time is $5 per hour.

42. Write a model that represents your average cost per hour of court time as a function of the number of hours played. Graph the model. What is an equation of the horizontal asymptote and what does the asymptote represent? **See margin.**

43. Suppose that you can play racquetball at the YMCA for $9 per hour without being a member. How many hours would you have to play at the racquetball club before your average cost per hour of court time is less than $9? **30**

44. **LIGHTNING** Air temperature affects how long it takes sound to travel a given distance. The time it takes for sound to travel one kilometer can be modeled by

$$t = \dfrac{1000}{0.6T + 331}$$

where *t* is the time (in seconds) and *T* is the temperature (in degrees Celsius). You are 1 kilometer from a lightning strike and it takes you exactly 3 seconds to hear the sound of thunder. Use a graph to find the approximate air temperature. (*Hint:* Use tick marks that are 0.1 unit apart on the *t*-axis.) **about 3.89° Celsius**

ECONOMICS In Exercises 45 and 46, use the following information.
Economist Arthur Laffer argues that beyond a certain percent p_m, increased taxes will produce less government revenue. His theory is illustrated in the graph below.

45. Using Laffer's theory, an economist models the revenue generated by one kind of tax by

$$R = \dfrac{80p - 8000}{p - 110}$$

where *R* is the government revenue (in tens of millions of dollars) and *p* is the percent tax rate ($55 \le p \le 100$). Graph the model. **See margin.**

46. Use your graph from Exercise 45 to find the tax rate that yields $600 million of revenue. **70%**

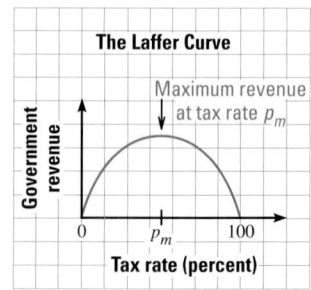

The Laffer Curve

Maximum revenue at tax rate p_m

Government revenue

Tax rate (percent)

DOPPLER EFFECT In Exercises 47 and 48, use the following information.
When the source of a sound is moving relative to a stationary listener, the frequency f_l (in hertz) heard by the listener is different from the frequency f_s (in hertz) of the sound at its source. An equation for the frequency heard by the listener is

$$f_l = \frac{740 f_s}{740 - r}$$

where r is the speed (in miles per hour) of the sound source relative to the listener.

47. The sound of an ambulance siren has a frequency of about 2000 hertz. You are standing on the sidewalk as an ambulance approaches with its siren on. Write the frequency that you hear as a function of the ambulance's speed. $f = \dfrac{1{,}480{,}000}{740 - r}$

48. Graph the function from Exercise 47 for $0 \le r \le 60$. What happens to the frequency you hear as the value of r increases? **See margin for graph.**
The frequency increases.

49. *Writing* In what line(s) is the graph of $y = \dfrac{1}{x}$ symmetric? What does this symmetry tell you about the inverse of the function $f(x) = \dfrac{1}{x}$?

49. It is symmetric in the line $y = x$ and in the line $y = -x$. Because it is symmetric in the line $y = x$, the function and its inverse are the same.

Test Preparation

50. MULTIPLE CHOICE What are the asymptotes of the graph of $y = \dfrac{-73}{x - 141} + 27$? **A**

(A) $x = 141, y = 27$ (B) $x = -141, y = 27$ (C) $x = -73, y = 27$

(D) $x = -73, y = 141$ (E) None of these

51. MULTIPLE CHOICE Which of the following is a function whose domain and range are all *nonzero* real numbers? **E**

(A) $f(x) = \dfrac{x}{2x + 1}$ (B) $f(x) = \dfrac{2x - 1}{3x - 2}$ (C) $f(x) = \dfrac{1}{x} + 1$

(D) $f(x) = \dfrac{x - 2}{x}$ (E) None of these

★ Challenge

52. EQUIVALENT FORMS Show algebraically that the function $f(x) = \dfrac{3}{x - 5} + 10$ and the function $g(x) = \dfrac{10x - 47}{x - 5}$ are equivalent.

$$f(x) = \frac{3}{x - 5} + 10 = \frac{3}{x - 5} + \frac{10x - 50}{x - 5} = \frac{10x - 47}{x - 5}$$

MIXED REVIEW

GRAPHING POLYNOMIALS Graph the polynomial function. (Review 6.2 for 9.3)
53–58. See margin.

53. $f(x) = 3x^5$ **54.** $f(x) = 4 - 2x^3$ **55.** $f(x) = x^6 - 1$

56. $f(x) = 4x^4 + 1$ **57.** $f(x) = 6x^7$ **58.** $f(x) = x^3 - 5$

FACTORING Factor the polynomial. (Review 6.4 for 9.3)

59. $8x^3 - 125$ **60.** $3x^3 + 81$ **61.** $x^3 + 3x^2 + 3x + 9$

62. $5x^3 + 10x^2 + x + 2$ **63.** $81x^4 - 1$ **64.** $4x^4 - 4x^2 - 120$

59. $(2x - 5)(4x^2 + 10x + 25)$
60. $3(x + 3)(x^2 - 3x + 9)$
61. $(x + 3)(x^2 + 3)$
62. $(x + 2)(5x^2 + 1)$
63. $(3x - 1)(3x + 1)(9x^2 + 1)$
64. $4(x^2 - 6)(x^2 + 5)$

SIMPLIFYING EXPRESSIONS Simplify the expression. (Review 8.3)

65. $\dfrac{e^x}{5e}$ $\dfrac{1}{5} e^{x-1}$ **66.** $7e^{-5}e^8$ $7e^3$ **67.** $e^x e^{4x+1}$ e^{5x+1}

68. $\dfrac{6e^x}{e^{6x}}$ $\dfrac{6}{e^{5x}}$ **69.** $e^4 e^{2x} e^{-3x}$ e^{4-x} **70.** $e^3 e^{-5}$ $\dfrac{1}{e^2}$

9.2 *Graphing Simple Rational Functions* **545**

Additional Test Preparation *Sample answer:*
The graph is the same as the graph of a hyperbola $y = \dfrac{a}{x}$. It has been shifted h units horizontally and k units vertically. There is a vertical asymptote at $x = h$ and a horizontal asymptote at $y = k$.

4 ASSESS

DAILY HOMEWORK QUIZ

Transparency Available

Identify the horizontal and vertical asymptotes of the graph of the function. State the domain and range. Graph the function.

1. $y = \dfrac{-3}{x + 1} + 2$

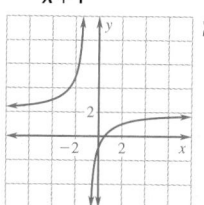

$x = -1, y = 2$; domain: all real numbers except -1, range: all real numbers except 2

2. $y = \dfrac{x + 1}{2x + 3}$

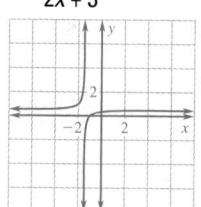

$x = \dfrac{-3}{2}, y = \dfrac{1}{2}$; domain: all real numbers except $\dfrac{-3}{2}$, range: all real numbers except $\dfrac{1}{2}$

EXTRA CHALLENGE NOTE
Challenge problems for Lesson 9.2 are available in **blackline** format in the *Chapter 9 Resource Book*, p. 36 and at **www.mcdougallittell.com.**

ADDITIONAL TEST PREPARATION

1. WRITING Describe the graph of a rational function of the form $y = \dfrac{a}{x - h} + k$. **See margin.**

48, 53–58. See Additional Answers beginning on page AA1.

545

1 Planning the Activity

PURPOSE
To graph rational functions.

MATERIALS
- graphing calculator
- Keystroke blackline (*Chapter 9 Resource Book,* p. 28)

PACING
- Activity — 20 min

▶ LINK TO LESSON
The important characteristics of the graph of a rational function include asymptotes and intercepts. When students graph the rational function in Example 2 of Lesson 9.3, ask them to verify the graph with a graphing calculator. They can also graph $x = -2$ and $x = 2$ to verify that these are asymptotes.

2 Managing the Activity

COOPERATIVE LEARNING
Ask students to graph the exercises with partners. Have them discuss their choice of a good viewing window.

ALTERNATIVE APPROACH
To save time, do this activity as a demonstration. Emphasize that a good viewing window includes the intercepts.

3 Closing the Activity

★ KEY DISCOVERY
Students can use graphing calculators to graph rational functions.

ACTIVITY ASSESSMENT
JOURNAL Why is it important to know how a graph should look before you graph it using a calculator?
See sample answer at right.

7. See Additional Answers beginning on page AA1.

546

▶ ACTIVITY 9.2
Using Technology

Graphing Rational Functions

You can use a graphing calculator to graph rational functions.

▶ EXAMPLE
Use a graphing calculator to graph $y = \dfrac{x+2}{x-2}$.

▶ SOLUTION

Begin by entering the function, using parentheses as shown.

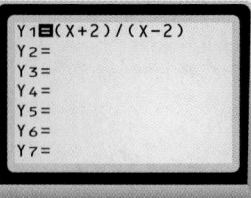

Most graphing calculators have two graphing modes: *Connected* mode and *Dot* mode. The graphs below show the function graphed in each mode.

Notice that the graph on the left has a vertical line at approximately $x = 2$. This line is *not* part of the graph—it is simply the graphing calculator's attempt at connecting the two branches of the graph.

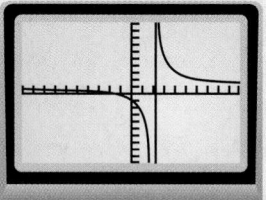

Connected mode

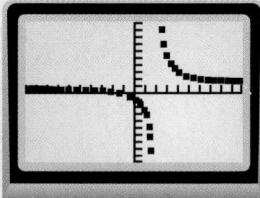

Dot mode

STUDENT HELP

INTERNET KEYSTROKE HELP

See keystrokes for several models of calculators at www.mcdougallittell.com

Be sure to choose a viewing window that shows all of the important characteristics of the graph. For instance, in the graph shown at the right, the viewing window is inadequate because it does not show the two branches of the graph of the rational function.

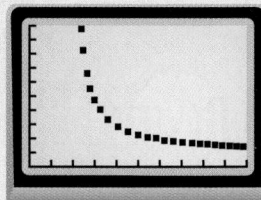

▶ EXERCISES

Use a graphing calculator to graph the rational function. Choose a viewing window that displays the important characteristics of the graph.

1–6. Check students graphs.

1. $y = \dfrac{7}{x} + 3$

2. $y = 9 - \dfrac{5}{x}$

3. $y = 5 + \dfrac{3}{x-6}$

4. $y = \dfrac{4}{x} + 6$

5. $y = \dfrac{x-5}{x+1}$

6. $y = \dfrac{8-5x}{x-5}$

7. $A = \dfrac{2+8n}{n}$; The average cost approaches \$8.

7. 🌐 **DELIVERY CHARGES** You and your friends order pizza and have it delivered to your house. The restaurant charges \$8 per pizza plus a \$2 delivery fee. Write a model that gives the average cost per pizza as a function of the number of pizzas ordered. Graph the model. Describe what happens to the average cost as the number of pizzas ordered increases. **See margin for graph.**

Activity Assessment *Sample answer:*
If you don't know what the graph looks like, you might choose an inadequate viewing window and draw the wrong conclusions about the graph.

9.3 Graphing General Rational Functions

What you should learn

GOAL 1 Graph general rational functions.

GOAL 2 Use the graph of a rational function to solve **real-life** problems, such as determining the efficiency of packaging in **Example 4**.

Why you should learn it

▼ To solve **real-life** problems, such as finding the energy expenditure of a parakeet in **Ex. 39**.

GOAL 1 GRAPHING RATIONAL FUNCTIONS

In Lesson 9.2 you learned how to graph rational functions of the form

$$f(x) = \frac{p(x)}{q(x)}$$

for which $p(x)$ and $q(x)$ are linear polynomials and $q(x) \neq 0$. In this lesson you will learn how to graph rational functions for which $p(x)$ and $q(x)$ may be higher-degree polynomials.

CONCEPT SUMMARY

GRAPHS OF RATIONAL FUNCTIONS

Let $p(x)$ and $q(x)$ be polynomials with no common factors other than 1. The graph of the rational function

$$f(x) = \frac{p(x)}{q(x)} = \frac{a_m x^m + a_{m-1}x^{m-1} + \cdots + a_1 x + a_0}{b_n x^n + b_{n-1}x^{n-1} + \cdots + b_1 x + b_0}$$

has the following characteristics.

1. The x-intercepts of the graph of f are the real zeros of $p(x)$.
2. The graph of f has a vertical asymptote at each real zero of $q(x)$.
3. The graph of f has at most one horizontal asymptote.

   • If $m < n$, the line $y = 0$ is a horizontal asymptote.

   • If $m = n$, the line $y = \frac{a_m}{b_n}$ is a horizontal asymptote.

   • If $m > n$, the graph has no horizontal asymptote. The graph's end behavior is the same as the graph of $y = \frac{a_m}{b_n}x^{m-n}$.

EXAMPLE 1 *Graphing a Rational Function (m < n)*

Graph $y = \dfrac{4}{x^2 + 1}$. State the domain and range.

SOLUTION

The numerator has no zeros, so there is no x-intercept. The denominator has no real zeros, so there is no vertical asymptote. The degree of the numerator (0) is less than the degree of the denominator (2), so the line $y = 0$ (the x-axis) is a horizontal asymptote. The *bell-shaped* graph passes through the points $(-3, 0.4)$, $(-1, 2)$, $(0, 4)$, $(1, 2)$, and $(3, 0.4)$. The domain is all real numbers, and the range is $0 < y \leq 4$.

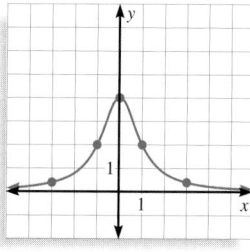

PACING
Basic: 2 days
Average: 2 days
Advanced: 2 days
Block Schedule: 1 block

LESSON OPENER
VISUAL APPROACH
An alternative way to approach Lesson 9.3 is to use the Visual Approach Lesson Opener:
• Blackline Master (*Chapter 9 Resource Book,* p. 40)
• Transparency (p. 61)

MEETING INDIVIDUAL NEEDS
• *Chapter 9 Resource Book*
Prerequisite Skills Review (p. 5)
Practice Level A (p. 41)
Practice Level B (p. 42)
Practice Level C (p. 43)
Reteaching with Practice (p. 44)
Absent Student Catch-Up (p. 46)
Challenge (p. 48)
• *Resources in Spanish*
• *Personal Student Tutor*

NEW-TEACHER SUPPORT
See the Tips for New Teachers on pp. 1–2 of the *Chapter 9 Resource Book* for additional notes about Lesson 9.3.

WARM-UP EXERCISES

Transparency Available

Find the solution(s) to each equation.
1. $(x - 3)(x + 3) = 0$ $3, -3$
2. $(x - 4)(x + 1) = 0$ $4, -1$
3. $x(x^2 - 1) = 0$ $0, 1, -1$
4. $x^2 - 4x - 5 = 0$ $5, -1$
5. $x^2 + 1 = 0$ no real solutions

EXTRA EXAMPLE 1

Graph $y = \dfrac{x}{x^2 + 1}$. State the domain and range.

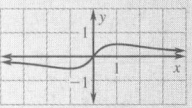

domain: all real numbers;
range: $-\dfrac{1}{2} \le y \le \dfrac{1}{2}$

EXTRA EXAMPLE 2

Graph $y = \dfrac{2x^3}{x^3 - 1}$.

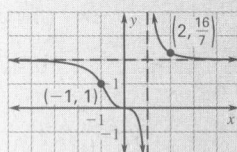

✓ CHECKPOINT EXERCISES

For use after Example 1:

1. Graph $y = -\dfrac{1}{x^3 + 1}$. State the domain and range.

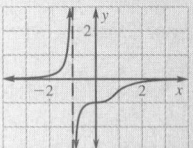

domain: all real numbers
except -1; range: all real numbers except 0

For use after Example 2:

2. Graph $y = \dfrac{x^2}{x^2 - 1}$.

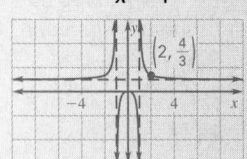

STUDENT HELP

⤷ **Look Back**
For help with finding zeros, see p. 259.

EXAMPLE 2   *Graphing a Rational Function (m = n)*

Graph $y = \dfrac{3x^2}{x^2 - 4}$.

SOLUTION

The numerator has 0 as its only zero, so the graph has one x-intercept at $(0, 0)$. The denominator can be factored as $(x + 2)(x - 2)$, so the denominator has zeros -2 and 2. This implies that the lines $x = -2$ and $x = 2$ are vertical asymptotes of the graph. The degree of the numerator (2) is equal to the degree of the denominator (2), so the horizontal asymptote is $y = \dfrac{a_m}{b_n} = 3$. To draw the graph, plot points between and beyond the vertical asymptotes.

	x	y
To the left of $x = -2$	-4	4
	-3	5.4
Between $x = -2$ and $x = 2$	-1	-1
	0	0
	1	-1
To the right of $x = 2$	3	5.4
	4	4

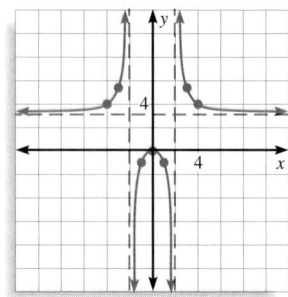

EXAMPLE 3   *Graphing a Rational Function (m > n)*

Graph $y = \dfrac{x^2 - 2x - 3}{x + 4}$.

SOLUTION

The numerator can be factored as $(x - 3)(x + 1)$, so the x-intercepts of the graph are 3 and -1. The only zero of the denominator is -4, so the only vertical asymptote is $x = -4$. The degree of the numerator (2) is greater than the degree of the denominator (1), so there is no horizontal asymptote and the end behavior of the graph of f is the same as the end behavior of the graph of $y = x^{2 - 1} = x$. To draw the graph, plot points to the left and right of the vertical asymptote.

	x	y
To the left of $x = -4$	-12	-20.6
	-9	-19.2
	-6	-22.5
To the right of $x = -4$	-2	2.5
	0	-0.75
	2	-0.5
	4	0.63
	6	2.1

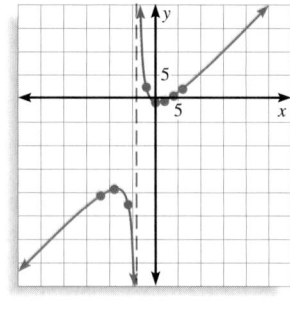

GOAL 2 USING RATIONAL FUNCTIONS IN REAL LIFE

Manufacturers often want to package their products in a way that uses the least amount of packaging material. Finding the most efficient packaging sometimes involves finding a local minimum of a rational function.

Manufacturing

EXAMPLE 4 *Finding a Local Minimum*

 A standard beverage can has a volume of 355 cubic centimeters.

a. Find the dimensions of the can that has this volume and uses the least amount of material possible.

b. Compare your result with the dimensions of an actual beverage can, which has a radius of 3.1 centimeters and a height of 11.8 centimeters.

SOLUTION

STUDENT HELP

► **Look Back**
For help with rewriting an equation with more than one variable, see p. 26.

a. The volume must be 355 cubic centimeters, so you can write the height h of each possible can in terms of its radius r.

$$V = \pi r^2 h \qquad \text{Formula for volume of cylinder}$$

$$355 = \pi r^2 h \qquad \text{Substitute 355 for } V.$$

$$\frac{355}{\pi r^2} = h \qquad \text{Solve for } h.$$

Using the least amount of material is equivalent to having a minimum surface area S. You can find the minimum surface area by writing its formula in terms of a single variable and graphing the result.

$$S = 2\pi r^2 + 2\pi r h \qquad \text{Formula for surface area of cylinder}$$

$$= 2\pi r^2 + 2\pi r \left(\frac{355}{\pi r^2} \right) \qquad \text{Substitute for } h.$$

$$= 2\pi r^2 + \frac{710}{r} \qquad \text{Simplify.}$$

Graph the function for the surface area S using a graphing calculator. Then use the *Minimum* feature to find the minimum value of S. When you do this, you get a minimum value of about 278, which occurs when $r \approx 3.84$ centimeters and

$$h \approx \frac{355}{\pi(3.84)^2} \approx 7.66 \text{ centimeters.}$$

Minimum
X=3.8372   Y=277.54

b. An actual beverage can is taller and narrower than the can with minimal surface area—probably to make it easier to hold the can in one hand.

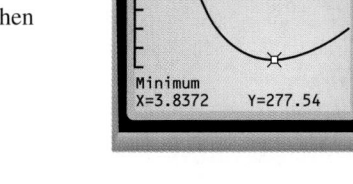

3.8 cm
7.7 cm
Can with minimal surface area

3.1 cm
11.8 cm
Actual beverage can

9.3 *Graphing General Rational Functions* **549**

Closure Question *Sample answer:*
For each h that is a zero of $q(x)$ that is not also a zero of $p(x)$, a vertical asymptote is $x = h$.

EXTRA EXAMPLE 3
Graph $y = \dfrac{x^2 - 3x - 4}{x - 2}$.

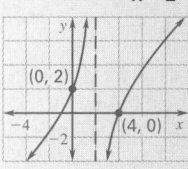
(0, 2)
−4
−2
(4, 0)   x

EXTRA EXAMPLE 4
A frozen yogurt cone has a volume of 10 cubic inches. The surface area S of a cone excluding the base is $S = \pi r \sqrt{r^2 + h^2}$, where r is the radius of the base and h is the height.
a. Find the dimensions of the cone that has this volume and the smallest surface area possible.
radius 1.9 in.; height 2.67 in.
b. Compare your results with the dimensions of an actual cone, which has a radius of 1.26 in. and a height of 6 in.   The actual cone is taller because a thinner cone is easier to hold.

✓ CHECKPOINT EXERCISES
For use after Example 3:
1. Graph $y = \dfrac{x^2 - 2x + 1}{x - 2}$.

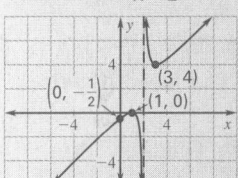
$\left(0, -\frac{1}{2}\right)$
−4
(3, 4)
(1, 0)
4   x
−4

For use after Example 4:
2. A silo is to be built in the shape of a cylinder with a volume of 100,000 cubic feet. Find the dimensions of the silo that use the least amount of material. Include the top and bottom surfaces of the silo.
$r = 25.15$ ft, $h = 50.32$ ft

CLOSURE QUESTION
Suppose you are given a rational function of the form $y = \dfrac{p(x)}{q(x)}$ with p and q polynomials with degree m and n, respectively, and $m < n$. Describe the vertical asymptotes.
See margin.

549

ASSIGNMENT GUIDE

BASIC
Day 1: pp. 550–553 Exs. 11–14, 20–25, 26–34 even
Day 2: pp. 550–553 Exs. 38–40, 47–48, 51–59 odd, Quiz 1 Exs. 1–13

AVERAGE
Day 1: pp. 550–553 Exs. 11–16, 20–25, 26–34 even
Day 2: pp. 550–553 Exs. 38–41, 46–48, 51–59 odd, Quiz 1 Exs. 1–13

ADVANCED
Day 1: pp. 550–553 Exs. 11–19, 20–25, 26–36 even
Day 2: pp. 550–553 Exs. 38–41, 43–49, 51–59 odd, Quiz 1 Exs. 1–13

BLOCK SCHEDULE
pp. 550–553 Exs. 11–16, 20–25, 26–34 even, 38–41, 46–48, 51–59 odd, Quiz 1 Exs. 1–13

EXERCISE LEVELS
Level A: *Easier*
20–25
Level B: *More Difficult*
11–19, 26–37, 39–48
Level C: *Most Difficult*
38, 49

✔ **HOMEWORK CHECK**
To quickly check student understanding of key concepts, go over the following exercises: Exs. 12, 20, 24, 26, 30. See also the Daily Homework Quiz:

• Blackline Master (*Chapter 9 Resource Book,* p. 52)
• 📄 Transparency (p. 71)

4–9. See Additional Answers beginning on page AA1.

GUIDED PRACTICE

Vocabulary Check ✔

11. *x*-intercept: 0; vertical asymptotes: $x = -3$, $x = 3$

Concept Check ✔

12. no *x*-intercept; vertical asymptote: $x = 1$

13. *x*-intercepts: $-\frac{1}{2}$, 5; vertical asymptotes: $x = -4$, $x = 4$

Skill Check ✔

14. *x*-intercept: $-\frac{3}{2}$; vertical asymptote: $x = 0$

15. *x*-intercepts: -5, 1; vertical asymptote: $x = 6$

16. *x*-intercepts: 1, $\frac{10}{3}$; no vertical asymptotes

17. *x*-intercept: -4; vertical asymptotes: $x = -\sqrt{3}$, $x = \sqrt{3}$

18. *x*-intercepts: 0, $-\frac{1}{2}$; no vertical asymptotes.

19. *x*-intercept: 3; vertical asymptote: $x = 0$

1. Let $f(x) = \dfrac{p(x)}{q(x)}$ where $p(x)$ and $q(x)$ are polynomials with no common factors other than 1. Complete this statement: The line $y = 0$ is a horizontal asymptote of the graph of f when the degree of $q(x)$ is __?__ the degree of $p(x)$. **greater than**

2. Let $f(x) = \dfrac{p(x)}{q(x)}$ where $p(x)$ and $q(x)$ are polynomials with no common factors other than 1. Describe how to find the *x*-intercepts and the vertical asymptotes of the graph of f. **The *x*-intercepts are the real zeros of $p(x)$. The graph has a vertical asymptote at each real zero of $q(x)$.**

3. Let $f(x) = \dfrac{p(x)}{q(x)}$ where $p(x)$ and $q(x)$ are both cubic polynomials with no common factors other than 1. The leading coefficient of $p(x)$ is 8 and the leading coefficient of $q(x)$ is 2. Describe the end behavior of the graph of f.
The graph approaches the horizontal asymtote of $y = 4$.

Graph the function.   4–9. See margin.

4. $y = \dfrac{6}{x^2 + 3}$

5. $y = \dfrac{x^2 - 4}{x + 1}$

6. $y = \dfrac{x^2 - 7}{x^2 + 2}$

7. $y = \dfrac{x^3}{x^2 + 7}$

8. $y = \dfrac{2x^2}{x^2 - 1}$

9. $y = \dfrac{x}{x^2 - 16}$

10. 🌐 **SOUP CANS** The can for a popular brand of soup has a volume of about 342 cubic centimeters. Find the dimensions of the can with this volume that uses the least metal possible. Compare these dimensions with the dimensions of the actual can, which has a radius of 3.3 centimeters and a height of 10 centimeters.
r is about 3.8 cm and *h* is about 7.6 cm; the actual can is taller and narrower than the can with minimal surface area.

PRACTICE AND APPLICATIONS

STUDENT HELP
▶ **Extra Practice**
to help you master skills is on p. 952.

ANALYZING GRAPHS Identify the *x*-intercepts and vertical asymptotes of the graph of the function.   11–19. See margin.

11. $y = \dfrac{x}{x^2 - 9}$

12. $y = \dfrac{2x^2 + 3}{x - 1}$

13. $y = \dfrac{2x^2 - 9x - 5}{x^2 - 16}$

14. $y = \dfrac{2x + 3}{x^3}$

15. $y = \dfrac{x^2 + 4x - 5}{x - 6}$

16. $y = \dfrac{3x^2 - 13x + 10}{x^2 + 8}$

17. $y = \dfrac{2x + 8}{3x^2 - 9}$

18. $y = \dfrac{2x^2 + x}{x^2 + 1}$

19. $y = \dfrac{x^3 - 27}{2x}$

STUDENT HELP
▶ **HOMEWORK HELP**
Examples 1–3: Exs. 11–37
Example 4: Exs. 38–45

MATCHING GRAPHS Match the function with its graph.

20. $y = \dfrac{x^2 - 7}{x^2 + 2}$ **A**

21. $y = \dfrac{-8}{x^2 - 4}$ **C**

22. $y = \dfrac{x^3}{x^2 - 4}$ **B**

A.

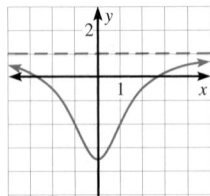

B.

C.

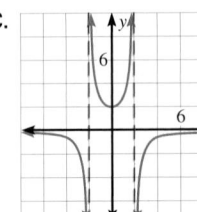

MATCHING GRAPHS Match the function with its graph.

23. $y = \dfrac{3}{x^3 - 27}$ **B**

24. $y = \dfrac{-x^3}{x^2 + 9}$ **A**

25. $y = \dfrac{x^2 + 4x}{2x - 1}$ **C**

A.

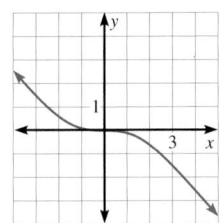

B.

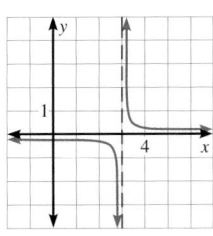

C.

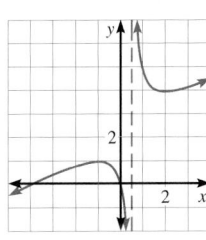

GRAPHING FUNCTIONS Graph the function.   26–37. See margin.

26. $y = \dfrac{2x^2 - 3}{x + 2}$

27. $y = \dfrac{-24}{x^2 + 8}$

28. $y = \dfrac{x^2 - 4}{x^2 + 3}$

29. $y = \dfrac{4x + 1}{x^2 - 1}$

30. $y = \dfrac{2x^2 + 3x + 1}{x^2 - 5x + 4}$

31. $y = \dfrac{-2x^2}{3x + 6}$

32. $y = \dfrac{3x^3 + 1}{4x^3 - 32}$

33. $y = \dfrac{x^2 - 11x - 12}{x^3 + 27}$

34. $y = \dfrac{4 - x}{5x^2 - 4x - 1}$

35. $y = \dfrac{-4x^2}{x^2 - 16}$

36. $y = \dfrac{x^2 - 9x + 20}{2x}$

37. $y = \dfrac{x^3 + 5x^2 - 1}{x^2 - 4x}$

38. 🌐 **GARDEN FENCING** Suppose you want to make a rectangular garden with an area of 200 square feet. You want to use the side of your house for one side of the garden and use fencing for the other three sides. Find the dimensions of the garden that minimize the length of fencing needed.
10 ft by 20 ft

🖩 **GRAPHING MODELS** In Exercises 39–45, you may find it helpful to use a graphing calculator to graph the models.

39. 🌐 **ENERGY EXPENDITURE** The total energy expenditure E (in joules per gram mass per kilometer) of a typical budgerigar parakeet can be modeled by

$$E = \dfrac{0.31v^2 - 21.7v + 471.75}{v}$$

where v is the speed of the bird (in kilometers per hour). Graph the model. What speed minimizes a budgerigar's energy expenditure? **about 39 km/h;**
▶ Source: *Introduction to Mathematics for Life Scientists* **See margin for graph.**

40. 🌐 **OCEANOGRAPHY** The mean temperature T (in degrees Celsius) of the Atlantic Ocean between latitudes 40°N and 40°S can be modeled by

$$T = \dfrac{17,800d + 20,000}{3d^2 + 740d + 1000}$$

where d is the depth (in meters). Graph the model. Use your graph to estimate the depth at which the mean temperature is 4°C. **about 1240 m; See margin for graph.**
▶ Source: *Practical Handbook of Marine Science*

41. 🌐 **HOSPITAL COSTS** For 1985 to 1995, the average daily cost per patient C (in dollars) at community hospitals in the United States can be modeled by

$$C = \dfrac{-22,407x + 462,048}{5x^2 - 122x + 1000}$$

where x is the number of years since 1985. Graph the model. Would you use this model to predict patient costs in 2005? Explain. ▶ Source: *Hospital Statistics*
See margin.

STUDENT HELP

🌐 **HOMEWORK HELP**
Visit our Web site
www.mcdougallittell.com
for help with Exs. 26–37.

41. No; This model predicts an average daily cost close to zero after 2005, and this is clearly not realistic.

FOCUS ON CAREERS

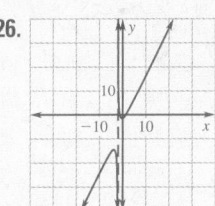

🌐 **HOSPITAL ADMINISTRATOR**
A hospital administrator oversees the quality of care and finances at a hospital. In 1996 there were approximately 329,000 hospital administrators in the United States.

🌐 **CAREER LINK**
www.mcdougallittell.com

! **COMMON ERROR**
EXERCISE 17 Students may think the zeros of the denominator are +3 and −3. Remind them to rewrite the expression as $3(x^2 - 3)$. The zeros are $\pm\sqrt{3}$.

STUDENT HELP NOTES
↪ **Homework Help** Students can find help for Exs. 23–37 at **www.mcdougallittell.com.** The information can be printed out for students who don't have access to the Internet.

GRAPHING CALCULATOR NOTE
EXERCISES 39–45 The real-world models described in these exercises cannot be simplified and will be very time-consuming to graph by hand.

TEACHING TIPS
EXERCISE 40 If students do this exercise with a calculator, have them find the intersection of the graph of $y = \dfrac{17,800x + 20,000}{3x^2 + 740x + 1000}$ and the graph of $y = 4$. Using the table feature to see y-values for each graph may be helpful.

26.

27.

28.

29–37, 39–41.
See Additional Answers beginning on page AA1.

DATA UPDATE
Updated data for
Exercise 42 is available at
www.mcdougallittell.com.

! **COMMON ERROR**
EXERCISE 49 When a rational
function has a common factor in
the numerator and denominator,
its graph contains a hole. Caution
students not to reduce the common
factor, or the hole—a characteristic
of the graph—will be lost. Refer to
Activity 9.4, p. 561.

42.
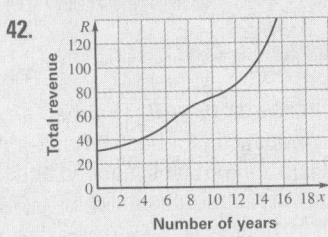

43.
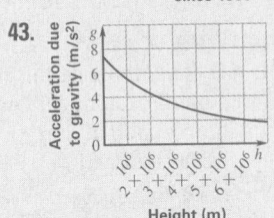

49. See Additional Answers beginning on page AA1.

**ADDITIONAL PRACTICE
AND RETEACHING**

For Lesson 9.3:

• Practice Levels A, B, and C
(*Chapter 9 Resource Book*, p. 41)

• Reteaching with Practice
(*Chapter 9 Resource Book*, p. 44)

• ⊞ See Lesson 9.3 of the
Personal Student Tutor

For more Mixed Review:

• ⊞ Search the *Test and Practice
Generator* for key words or
specific lessons.

552

---

42. 🌐 **AUTOMOTIVE INDUSTRY** For 1980 to 1995, the total revenue R
(in billions of dollars) from parking and automotive service and repair in the
United States can be modeled by

$$R = \frac{427x^2 - 6416x + 30{,}432}{-0.7x^3 + 25x^2 - 268x + 1000}$$

where x is the number of years since 1980. Graph the model. In what year was
the total revenue approximately \$75 billion?   **1990; See margin for graph.**

🌐 DATA UPDATE of U.S. Bureau of the Census data at www.mcdougallittell.com

SCIENCE ▶ CONNECTION   **In Exercises 43–45, use the following information.**
The acceleration due to gravity g' (in meters per second squared) of a falling object
at the moment it is dropped is given by the function

$$g' = \frac{3.99 \times 10^{14}}{h^2 + (1.28 \times 10^7)h + 4.07 \times 10^{13}}$$

where h is the object's altitude (in meters) above sea level.

43. Graph the function.   **See margin.**

44. What is the acceleration due to gravity for an object dropped at an altitude of
5000 kilometers?   **3.08 m/sec²**

45. Describe what happens to g' as h increases.   **g' decreases as h increases**

46. **CRITICAL THINKING** Give an example of a rational function whose graph has
two vertical asymptotes: $x = 2$ and $x = 7$.   *Sample answer:* $y = \dfrac{x}{x^2 - 9x + 14}$

Test Preparation

47. **MULTIPLE CHOICE** What is the horizontal asymptote of the graph of the
following function?   **B**

$$y = \frac{10x^2 - 1}{x^3 + 8}$$

Ⓐ $y = -10$ Ⓑ $y = 0$ Ⓒ $y = 2$

Ⓓ $y = 10$ Ⓔ No horizontal asymptote

48. **MULTIPLE CHOICE** Which of the following functions
is graphed?   **D**

Ⓐ $y = \dfrac{-5x^2}{x^2 + 9}$ Ⓑ $y = \dfrac{5x^2}{x^2 - 9}$

Ⓒ $y = \dfrac{5x^2}{x^2 + 9}$ Ⓓ $y = \dfrac{-5x^2}{x^2 - 9}$

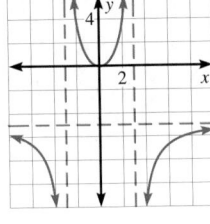

★ Challenge

49. Consider the following two functions:

$$f(x) = \frac{(x + 1)(x + 2)}{(x - 3)(x - 5)} \quad \text{and} \quad g(x) = \frac{(x + 2)(x - 3)}{(x - 3)(x - 5)}$$

Notice that the numerator and denominator of g have a common factor of $x - 3$.

a. Make a table of values for each function from $x = 2.95$ to $x = 3.05$ in
increments of 0.01.   **See margin.**

49c. $f(x)$ has a vertical asymptote
at $x = 3$. The values of $g(x)$
get close to -2.5.

b. Use your table of values to graph each function for $2.95 \le x \le 3.05$.   **See margin.**

c. As x approaches 3, what happens to the graph of $f(x)$? to the graph of $g(x)$?

EXTRA CHALLENGE
▶ www.mcdougallittell.com

d. What do you think is true about the graph of a function $g(x) = \dfrac{p(x)}{q(x)}$ where
$p(x)$ and $q(x)$ have a common factor $x - k$?
The graph has a hole at $x = k$.

MIXED REVIEW

SIMPLIFYING ALGEBRAIC EXPRESSIONS Simplify the expression. Tell which properties of exponents you used. (Review 6.1 for 9.4) **Check properties.**

50. $\dfrac{x^{-3}y}{xy^4}$ $\dfrac{1}{x^4y^3}$

51. $\dfrac{x^6y^5}{xy}$ x^5y^4

52. $\dfrac{3x^3y^3}{6x^{-1}y}$ $\dfrac{x^4y^2}{2}$

53. $\dfrac{12x^5y^{-2}}{3x^{-2}y^5}$ $\dfrac{4x^7}{y^7}$

54. $\left(\dfrac{x^2y^2}{x^3y}\right)^2\dfrac{y^2}{x^2}$

55. $\left(\dfrac{5x^3}{25xy^2}\right)^3\dfrac{x^6}{125y^6}$

60. $f(g(x)) = f(2x+6) =$
$\frac{1}{2}(2x+6) - 3 = x +$
$3 - 3 = x;\ g(f(x)) =$
$g\left(\frac{1}{2}x - 3\right) =$
$2\left(\frac{1}{2}x - 3\right) + 6 =$
$x - 6 + 6 = x$

61. $f(g(x)) = f\left(-\frac{1}{3}x + \frac{2}{3}\right) =$
$-3\left(-\frac{1}{3}x + \frac{2}{3}\right) + 2 =$
$x - 2 + 2 = x;$
$g(f(x)) = g(-3x + 2) =$
$-\frac{1}{3}(-3x + 2) + \frac{2}{3} =$
$x - \frac{2}{3} + \frac{2}{3} = x$

JOINT VARIATION MODELS The variable z varies jointly with x and y. Use the given values to write an equation relating x, y, and z. Then find z when $x = -3$ and $y = 2$. (Review 9.1)

56. $x = 3, y = -6, z = 2$ $z = -\dfrac{xy}{9}; \dfrac{2}{3}$

57. $x = -5, y = 2, z = \dfrac{3}{4}$ $z = -\dfrac{3xy}{40}; \dfrac{9}{20}$

58. $x = -8, y = 4, z = \dfrac{8}{3}$ $z = -\dfrac{xy}{12}; \dfrac{1}{2}$

59. $x = 1, y = \dfrac{1}{2}, z = 4$ $z = 8xy; -48$

VERIFYING INVERSES Verify that f and g are inverse functions. (Review 7.4)

60. $f(x) = \dfrac{1}{2}x - 3, g(x) = 2x + 6$

61. $f(x) = -3x + 2, g(x) = -\dfrac{1}{3}x + \dfrac{2}{3}$

62. $f(x) = 5x^3 + 2, g(x) = \left(\dfrac{x-2}{5}\right)^{1/3}$

63. $f(x) = 16x^4, x \geq 0; g(x) = \dfrac{\sqrt[4]{x}}{2}$

QUIZ 1

Self-Test for Lessons 9.1–9.3

62. $f(g(x)) = f\left(\left(\dfrac{x-2}{5}\right)^{1/3}\right) =$
$5\left(\left(\dfrac{x-2}{5}\right)^{1/3}\right)^3 + 2 =$
$5\left(\dfrac{x-2}{5}\right) + 2 =$
$x - 2 + 2 = x;$
$g(f(x)) = g(5x^3 + 2) =$
$\left(\dfrac{5x^3 + 2 - 2}{5}\right)^{1/3} =$
$(x^3)^{1/3} = x$

63. $f(g(x)) = f\left(\dfrac{\sqrt[4]{x}}{2}\right) =$
$16\left(\dfrac{\sqrt[4]{x}}{2}\right)^4 = 16\left(\dfrac{x}{16}\right) = x;$
$g(f(x)) = g(16x^4) =$
$\dfrac{\sqrt[4]{16x^4}}{2} = \dfrac{2x}{2} = x$

The variables x and y vary inversely. Use the given values to write an equation relating x and y. Then find y when $x = -3$. (Lesson 9.1)

1. $x = 6, y = -2$ $y = -\dfrac{12}{x}; 4$

2. $x = 11, y = 6$ $y = \dfrac{66}{x}; -22$

3. $x = \dfrac{1}{5}, y = 30$ $y = \dfrac{6}{x}; -2$

The variable x varies jointly with y and z. Use the given values to write an equation relating x, y, and z. Then find y when $x = 4$ and $z = 1$. (Lesson 9.1)

4. $x = 5, y = -5, z = 6$ $x = -\dfrac{yz}{6}; -24$

5. $x = 12, y = 6, z = \dfrac{1}{2}$ $x = 4yz; 1$

6. $x = -10, y = 2, z = 4$ $x = -\dfrac{5yz}{4}; -\dfrac{16}{5}$

Graph the function. (Lessons 9.2 and 9.3)   7–12. See margin.

7. $y = \dfrac{10}{x}$

8. $y = \dfrac{2}{x+9} - 7$

9. $y = \dfrac{3x+5}{2x-11}$

10. $y = \dfrac{6x}{x^2 - 36}$

11. $y = \dfrac{3x^2}{x^2 - 25}$

12. $y = \dfrac{x^2 - 4x - 5}{x + 2}$

13. 🌐 **HOTEL REVENUE** For 1980 to 1995, the total revenue R (in billions of dollars) from hotels and motels in the United States can be modeled by

$$R = \dfrac{2.76x + 26.88}{-0.01x + 1}$$

where x is the number of years since 1980. Graph the model. Use your graph to find the year in which the total revenue from hotels and motels was approximately $68 billion. (Lesson 9.2)   1992; See margin.

9.3 *Graphing General Rational Functions* 553

Additional Test Preparation *Sample answer:*

$f(x) = \dfrac{3x^3}{x^2 - 4}$

4 ASSESS

DAILY HOMEWORK QUIZ

Transparency Available

Graph the function. Identify any asymptotes.

1. $y = \dfrac{x^2 + 2}{x - 1}$

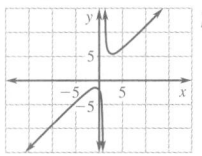

no horizontal asymptote, $x = 1$

2. $y = \dfrac{3x^2}{x^2 + 4}$

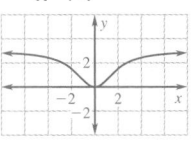

$y = 3$, no vertical asymptote

3. $y = \dfrac{2x + 7}{x^2 - 9}$

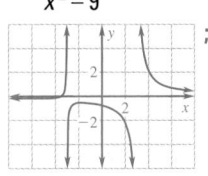

$y = 0, x = \pm 3$

EXTRA CHALLENGE NOTE

→ Challenge problems for Lesson 9.3 are available in **blackline** format in the *Chapter 9 Resource Book*, p. 48 and at **www.mcdougallittell.com.**

ADDITIONAL TEST PREPARATION

1. OPEN ENDED Write a rational function $f(x)$ so that f has vertical asymptotes $x = 2$ and $x = -2$, no horizontal asymptote, and has end behavior that can be modeled by $y = 3x$. See margin.

ADDITIONAL RESOURCES

An alternative Quiz for Lessons 9.1–9.3 is available in the *Chapter 9 Resource Book*, p. 49.

7–13. See Additional Answers beginning on page AA1.

1 PLAN

PACING
Basic: 2 days
Average: 2 days
Advanced: 2 days
Block Schedule: 1 block

> **→ LESSON OPENER**
> **ACTIVITY**
> An alternative way to approach
> Lesson 9.4 is to use the Activity
> Lesson Opener:
> • Blackline Master (*Chapter 9 Resource Book*, p. 53)
> • Transparency (p. 62)

MEETING INDIVIDUAL NEEDS
• *Chapter 9 Resource Book*
  Prerequisite Skills Review (p. 5)
  Practice Level A (p. 55)
  Practice Level B (p. 56)
  Practice Level C (p. 57)
  Reteaching with Practice (p. 58)
  Absent Student Catch-Up (p. 60)
  Challenge (p. 62)
• *Resources in Spanish*
• 🖥 *Personal Student Tutor*

NEW-TEACHER SUPPORT
See the Tips for New Teachers on
pp. 1–2 of the *Chapter 9 Resource Book* for additional notes about Lesson 9.4.

WARM-UP EXERCISES

🗎 **Transparency Available**

Factor each expression.

1. $x^2 + 3x - 4$ $(x-1)(x+4)$
2. $x^2 + 5x + 6$ $(x+2)(x+3)$
3. $4x^2 - 9$ $(2x-3)(2x+3)$
4. $6x^2 + x$ $x(6x+1)$
5. $8x^3 + 1$ $(2x+1)(4x^2 - 2x + 1)$

9.4 Multiplying and Dividing Rational Expressions

GOAL 1 WORKING WITH RATIONAL EXPRESSIONS

What you should learn

GOAL 1 Multiply and divide rational expressions.

GOAL 2 Use rational expressions to model **real-life** quantities, such as the heat generated by a runner in **Exs. 50 and 51**.

Why you should learn it

▼ To solve **real-life** problems, such as finding the average number of acres per farm in **Exs. 52 and 53**.

A rational expression is in **simplified form** provided its numerator and denominator have no common factors (other than ± 1). To simplify a rational expression, apply the following property.

SIMPLIFYING RATIONAL EXPRESSIONS

Let a, b, and c be nonzero real numbers or variable expressions. Then the following property applies:

$$\frac{a\cancel{c}}{b\cancel{c}} = \frac{a}{b} \qquad \text{Divide out common factor } c.$$

Simplifying a rational expression usually requires two steps. First, factor the numerator and denominator. Then, divide out any factors that are common to both the numerator and denominator. Here is an example:

$$\frac{x^2 + 5x}{x^2} = \frac{x(x+5)}{x \cdot x} = \frac{x+5}{x}$$

Notice that you can divide out common factors in the second expression above, but you cannot divide out like terms in the third expression.

EXAMPLE 1 *Simplifying a Rational Expression*

Simplify: $\dfrac{x^2 - 4x - 12}{x^2 - 4}$

SOLUTION

$$\frac{x^2 - 4x - 12}{x^2 - 4} = \frac{(x+2)(x-6)}{(x+2)(x-2)} \qquad \text{Factor numerator and denominator.}$$

$$= \frac{\cancel{(x+2)}(x-6)}{\cancel{(x+2)}(x-2)} \qquad \text{Divide out common factor.}$$

$$= \frac{x-6}{x-2} \qquad \text{Simplified form}$$

· · · · · · · · · ·

The rule for multiplying rational expressions is the same as the rule for multiplying numerical fractions: multiply numerators, multiply denominators, and write the new fraction in simplified form.

$$\frac{a}{b} \cdot \frac{c}{d} = \frac{ac}{bd} \quad \longleftarrow \text{Simplify } \tfrac{ac}{bd} \text{ if possible.}$$

EXAMPLE 2 *Multiplying Rational Expressions Involving Monomials*

Multiply: $\dfrac{5x^2y}{2xy^3} \cdot \dfrac{6x^3y^2}{10y}$

SOLUTION

$$\dfrac{5x^2y}{2xy^3} \cdot \dfrac{6x^3y^2}{10y} = \dfrac{30x^5y^3}{20xy^4}$$ **Multiply numerators and denominators.**

$$= \dfrac{3 \cdot 10 \cdot x \cdot x^4 \cdot y^3}{2 \cdot 10 \cdot x \cdot y \cdot y^3}$$ **Factor and divide out common factors.**

$$= \dfrac{3x^4}{2y}$$ **Simplified form**

EXAMPLE 3 *Multiplying Rational Expressions Involving Polynomials*

Multiply: $\dfrac{4x - 4x^2}{x^2 + 2x - 3} \cdot \dfrac{x^2 + x - 6}{4x}$

SOLUTION

$$\dfrac{4x - 4x^2}{x^2 + 2x - 3} \cdot \dfrac{x^2 + x - 6}{4x}$$

$$= \dfrac{4x(1 - x)}{(x - 1)(x + 3)} \cdot \dfrac{(x + 3)(x - 2)}{4x}$$ **Factor numerators and denominators.**

$$= \dfrac{4x(1 - x)(x + 3)(x - 2)}{(x - 1)(x + 3)(4x)}$$ **Multiply numerators and denominators.**

$$= \dfrac{4x(-1)(x - 1)(x + 3)(x - 2)}{(x - 1)(x + 3)(4x)}$$ **Rewrite $(1 - x)$ as $(-1)(x - 1)$.**

$$= \dfrac{4x(-1)(x - 1)(x + 3)(x - 2)}{(x - 1)(x + 3)(4x)}$$ **Divide out common factors.**

$$= -x + 2$$ **Simplified form**

EXAMPLE 4 *Multiplying by a Polynomial*

Multiply: $\dfrac{x + 3}{8x^3 - 1} \cdot (4x^2 + 2x + 1)$

STUDENT HELP

Look Back
For help with factoring a
difference of two cubes,
see p. 345.

SOLUTION

$$\dfrac{x + 3}{8x^3 - 1} \cdot (4x^2 + 2x + 1)$$

$$= \dfrac{x + 3}{8x^3 - 1} \cdot \dfrac{4x^2 + 2x + 1}{1}$$ **Write polynomial as rational expression.**

$$= \dfrac{(x + 3)(4x^2 + 2x + 1)}{(2x - 1)(4x^2 + 2x + 1)}$$ **Factor and multiply numerators and denominators.**

$$= \dfrac{(x + 3)(4x^2 + 2x + 1)}{(2x - 1)(4x^2 + 2x + 1)}$$ **Divide out common factors.**

$$= \dfrac{x + 3}{2x - 1}$$ **Simplified form**

9.4 *Multiplying and Dividing Rational Expressions*   **555**

---

2 TEACH

MOTIVATING THE LESSON
Refer students to the photograph of
the skydivers on pp. 530–531. Ask
students if they think the size of a
skydiver affects the fall. Rational
expressions in this lesson can be
used to model how a skydiver's size
affects the falling speed.

EXTRA EXAMPLE 1
Simplify: $\dfrac{x^2 - 5x - 6}{x^2 - 1} \cdot \dfrac{x - 6}{x - 1}$

EXTRA EXAMPLE 2
Multiply: $\dfrac{6x^2y^3}{2x^2y^2} \cdot \dfrac{10x^3y^4}{18y^2}$ $\dfrac{5x^3y^3}{3}$

EXTRA EXAMPLE 3
Multiply:
$\dfrac{3x - 27x^3}{3x^2 - 2x - 1} \cdot \dfrac{3x^2 - 4x + 1}{3x}$
$-(3x - 1)^2$

EXTRA EXAMPLE 4
Multiply: $\dfrac{x + 2}{27x^3 + 8} \cdot (9x^2 - 6x + 4)$
$\dfrac{x + 2}{3x + 2}$

CHECKPOINT EXERCISES
For use after Example 1:
1. Simplify: $\dfrac{x^2 + x - 2}{x^2 - 1} \cdot \dfrac{x + 2}{x + 1}$

For use after Example 2:
2. Multiply: $\dfrac{9x^3y}{3x^2y^3} \cdot \dfrac{12x^4y^5}{27y^2}$
$\dfrac{4x^5y}{3}$

For use after Example 3:
3. Multiply:
$\dfrac{4x - 2x^2}{x^2 - 5x + 6} \cdot \dfrac{x^2 - 4x + 3}{2x}$ $1 - x$

For use after Example 4:
4. Multiply:
$\dfrac{x - 3}{64x^3 - 1} \cdot (16x^2 + 4x + 1)$
$\dfrac{x - 3}{4x - 1}$

EXTRA EXAMPLE 5

Divide: $\dfrac{3}{4x-8} \div \dfrac{x^2+3x}{x^2+x-6}$ $\dfrac{3}{4x}$

EXTRA EXAMPLE 6

Divide: $\dfrac{8x^2+10x-3}{4x^2} \div (4x^2-x)$

$\dfrac{2x+3}{4x^3}$

EXTRA EXAMPLE 7

Simplify: $\dfrac{x}{x-2} \cdot (2x+3) \div \dfrac{4x^2-9}{x-2}$

$\dfrac{x}{2x-3}$

 CHECKPOINT EXERCISES

For use after Examples 5 and 6:

1. Divide: $\dfrac{6x^2+7x-3}{x-1} \div (3x-1)$

$\dfrac{2x+3}{x-1}$

For use after Example 7:

2. $\dfrac{x}{x+3} \cdot (4x+1) \div \dfrac{16x^2-1}{x+3}$

$\dfrac{x}{4x-1}$

STUDENT HELP NOTES

→ **Homework Help** Students can find extra examples at **www.mcdougallittell.com** that parallel the examples in the student edition.

STUDENT HELP

HOMEWORK HELP
Visit our Web site
www.mcdougallittell.com
for extra examples.

To divide one rational expression by another, multiply the first expression by the reciprocal of the second expression.

$$\frac{a}{b} \div \frac{c}{d} = \frac{a}{b} \cdot \frac{d}{c} = \frac{ad}{bc} \longleftarrow \text{Simplify } \frac{ad}{bc} \text{ if possible.}$$

EXAMPLE 5 *Dividing Rational Expressions*

Divide: $\dfrac{5x}{3x-12} \div \dfrac{x^2-2x}{x^2-6x+8}$

SOLUTION

$$\frac{5x}{3x-12} \div \frac{x^2-2x}{x^2-6x+8} = \frac{5x}{3x-12} \cdot \frac{x^2-6x+8}{x^2-2x} \qquad \text{Multiply by reciprocal.}$$

$$= \frac{5x}{3(x-4)} \cdot \frac{(x-2)(x-4)}{x(x-2)} \qquad \text{Factor.}$$

$$= \frac{5x(x-2)(x-4)}{3(x-4)(x)(x-2)} \qquad \text{Divide out common factors.}$$

$$= \frac{5}{3} \qquad \text{Simplified form}$$

EXAMPLE 6 *Dividing by a Polynomial*

Divide: $\dfrac{6x^2+7x-3}{6x^2} \div (2x^2+3x)$

SOLUTION

$$\frac{6x^2+7x-3}{6x^2} \div (2x^2+3x) = \frac{6x^2+7x-3}{6x^2} \cdot \frac{1}{2x^2+3x}$$

$$= \frac{(3x-1)(2x+3)}{6x^2} \cdot \frac{1}{x(2x+3)}$$

$$= \frac{(3x-1)(2x+3)}{(6x^2)(x)(2x+3)}$$

$$= \frac{3x-1}{6x^3}$$

EXAMPLE 7 *Multiplying and Dividing*

Simplify: $\dfrac{x}{x+5} \cdot (3x-5) \div \dfrac{9x^2-25}{x+5}$

SOLUTION

$$\frac{x}{x+5} \cdot (3x-5) \div \frac{9x^2-25}{x+5} = \frac{x}{x+5} \cdot \frac{3x-5}{1} \cdot \frac{x+5}{9x^2-25}$$

$$= \frac{x(3x-5)(x+5)}{(x+5)(3x-5)(3x+5)}$$

$$= \frac{x}{3x+5}$$

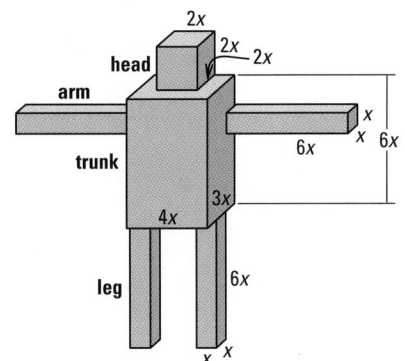

GREGORY
ROBERTSON
made a daring rescue while skydiving in 1987. To reach a novice skydiver in trouble, he increased his velocity by diving headfirst towards the other diver. Just 10 seconds before he would have hit the ground, he was able to deploy both of their chutes.

GOAL 2 **USING RATIONAL EXPRESSIONS IN REAL LIFE**

EXAMPLE 8 *Writing and Simplifying a Rational Model*

SKYDIVING A falling skydiver accelerates until reaching a constant falling speed, called the *terminal velocity*. Because of air resistance, the ratio of a skydiver's volume to his or her cross-sectional surface area affects the terminal velocity: the larger the ratio, the greater the terminal velocity.

a. The diagram shows a simplified geometric model of a skydiver with maximum cross-sectional surface area. Use the diagram to write a model for the ratio of volume to cross-sectional surface area for a skydiver.

b. Use the result of part (a) to compare the terminal velocities of two skydivers: one who is 60 inches tall and one who is 72 inches tall.

SOLUTION

a. The volume and cross-sectional surface area of each part of the skydiver are given in the table below. (Assume that the front side of the skydiver's body is parallel with the ground when falling.)

Body part	Volume	Cross-sectional surface area
Arm or leg	$V = 6x^3$	$S = 6x(x) = 6x^2$
Head	$V = 8x^3$	$S = 2x(2x) = 4x^2$
Trunk	$V = 72x^3$	$S = 6x(4x) = 24x^2$

Using these volumes and cross-sectional surface areas, you can write the ratio as:

$$\frac{\text{Volume}}{\text{Surface area}} = \frac{4(6x^3) + 8x^3 + 72x^3}{4(6x^2) + 4x^2 + 24x^2}$$

$$= \frac{104x^3}{52x^2}$$

$$= 2x$$

b. The overall height of the geometric model is $14x$. For the skydiver whose height is 60 inches, $14x = 60$, so $x \approx 4.3$. For the skydiver whose height is 72 inches, $14x = 72$, so $x \approx 5.1$. The ratio of volume to cross-sectional surface area for each skydiver is:

60 inch skydiver: $\dfrac{\text{Volume}}{\text{Surface area}} = 2x \approx 2(4.3) = 8.6$

72 inch skydiver: $\dfrac{\text{Volume}}{\text{Surface area}} = 2x \approx 2(5.1) = 10.2$

▶ The taller skydiver has the greater terminal velocity.

Closure Question *Sample answer:*
Factor all numerators and denominators and multiply them to get one expression. Divide out common factors. Multiply the factors that are left.

Daily Puzzler *Sample answer:*
No. $\dfrac{x^2 - 3x + 2}{x - 1}$ is not defined for $x = 1$ although $(x - 2)$ is.

EXTRA EXAMPLE 8
The diagram below shows a simplified version of a robotic dog.

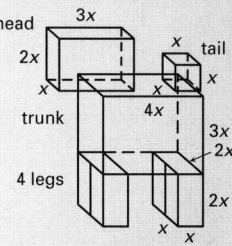

a. Use the diagram to write a model for the ratio of the volume to the cross-sectional area of the dog's feet.
$$\frac{39x^3}{4x^2} = 9.75x$$

b. Using the ratio in part **(a)**, find the value of the ratio for two robots, one with a tail 10 centimeters long and a second with a tail 15 centimeters long.
97.5; 146.25

✔ **CHECKPOINT EXERCISES**
For use after Example 8:

1. Refer to Extra Example 8. Suppose the "aerodynamics of the design" is defined as the ratio of the surface area of the front of the dog to the volume.
  a. Find the ratio of the front surface area to the volume.
  $\dfrac{12}{39x}$
  b. Find the aerodynamics of a robotic dog that is 70 centimeters high. **0.03**

FOCUS ON VOCABULARY
When is a rational expression simplified? **when the numerator and denominator have no common factors except 1 or –1**

CLOSURE QUESTION
What is the procedure for multiplying rational expressions involving polynomials. **See margin.**

DAILY PUZZLER
Suppose you simplify $\dfrac{x^2 - 3x + 2}{x - 1}$ to get $(x - 2)$. Are these two expressions equal for all values of x?
See margin.

ASSIGNMENT GUIDE

BASIC
Day 1: pp. 558–559 Exs. 16–24, 28–33, 38–41
Day 2: pp. 559–560 Exs. 44–49, 54–57, 59, 63–73 odd

AVERAGE
Day 1: pp. 558–559 Exs. 16–26, 28–35, 38–43
Day 2: pp. 559–560 Exs. 44–49, 54–59, 63–73 odd, 74–75

ADVANCED
Day 1: pp. 558–559 Exs. 16–26, 28–35, 38–43
Day 2: pp. 559–560 Exs. 44–51, 54–59, 61–71 odd, 72–73

BLOCK SCHEDULE
pp. 558–560 Exs. 16–26, 28–35, 38–43, 46–51, 54–59, 61–71 odd, 72–73

EXERCISE LEVELS
Level A: *Easier*
16–21, 28–34, 38–41
Level B: *More Difficult*
22–27, 35–37, 42–55, 57
Level C: *Most Difficult*
56, 58–59

✔ HOMEWORK CHECK
To quickly check student understanding of key concepts, go over the following exercises: Exs. 18, 24, 32, 38, 40, 46. See also the Daily Homework Quiz:

• Blackline Master (*Chapter 9 Resource Book*, p. 65)
• Transparency (p. 72)

GUIDED PRACTICE

Vocabulary Check ✔

1. Explain how you know when a rational expression is in simplified form.

Concept Check ✔

2. ERROR ANALYSIS Explain what is wrong with the simplification of the rational expression shown.

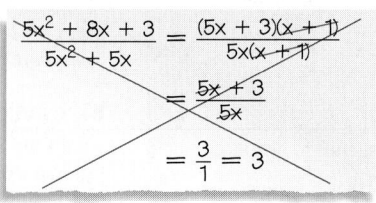

$$\frac{5x^2 + 8x + 3}{5x^2 + 5x} = \frac{(5x+3)(x+1)}{5x(x+1)}$$
$$= \frac{5x+3}{5x}$$
$$= \frac{3}{1} = 3$$

Skill Check ✔

1. A rational expression is in simplified form if its numerator and denominator have no common factors (other than ±1).

2. The $5x$'s cannot be canceled in the second line.

If possible, simplify the rational expression.

3. $\dfrac{4x^2}{4x^3 + 12x}$ $\dfrac{x}{x^2 + 3}$

4. $\dfrac{x^2 + 4x - 5}{x^2 - 1}$ $\dfrac{x + 5}{x + 1}$

5. $\dfrac{x^2 + 10x - 4}{x^2 + 10x}$ not possible

6. $\dfrac{6x^2 - 4x - 3}{3x^2 + x}$ not possible

7. $\dfrac{x^2 - 9}{2x + 1}$ not possible

8. $\dfrac{2x^3 - 32x}{x^2 + 8x + 16}$ $\dfrac{2x(x - 4)}{x + 4}$

Perform the indicated operation. Simplify the result.

9. $\dfrac{16x^3}{5y^9} \cdot \dfrac{x^5 y^8}{80x^3 y}$ $\dfrac{x^5}{25y^2}$

10. $\dfrac{7x^4 y^3}{5xy} \cdot \dfrac{2x^7}{21y^5}$ $\dfrac{2x^{10}}{15y^3}$

11. $\dfrac{x^2 + x - 6}{2x^2} \cdot \dfrac{2x + 8}{x^2 + 7x + 12}$ $\dfrac{x - 2}{x^2}$

12. $\dfrac{144}{4xy} \div \dfrac{54y^3}{3x^3 y}$ $\dfrac{2x^2}{y^3}$

13. $\dfrac{16xy}{3x^5 y^5} \div \dfrac{8x^2}{9xy^7}$ $\dfrac{6y^3}{x^5}$

14. $\dfrac{5x^2 + 10x}{x^2 - x - 6} \div \dfrac{15x^3 + 45x^2}{x^2 - 9}$ $\dfrac{1}{3x}$

15. 🌐 **SKYDIVING** Look back at Example 8 on page 557. Some skydivers wear "wings" to increase their surface area. Suppose a skydiver who is 65 inches tall is wearing wings that add $18x^2$ of surface area and an insignificant amount of volume. Calculate the skydiver's volume to surface area ratio with and without the wings. with: 6.9; without: 9.3

PRACTICE AND APPLICATIONS

STUDENT HELP

▶ **Extra Practice** to help you master skills is on p. 953.

SIMPLIFYING If possible, simplify the rational expression.

16. $\dfrac{3x^3}{12x^2 + 9x}$ $\dfrac{x^2}{4x + 3}$

17. $\dfrac{x^2 - x - 6}{x^2 + 8x + 16}$ not possible

18. $\dfrac{x^2 - 3x + 2}{x^2 + 5x - 6}$ $\dfrac{x - 2}{x + 6}$

19. $\dfrac{x^2 + 2x - 4}{x^2 + x - 6}$ not possible

20. $\dfrac{x^2 - 2x - 3}{x^2 - 7x + 12}$ $\dfrac{x + 1}{x - 4}$

21. $\dfrac{3x^2 - 3x - 6}{x^2 - 4}$ $\dfrac{3(x + 1)}{x + 2}$

22. $\dfrac{x - 2}{x^3 - 8}$ $\dfrac{1}{x^2 + 2x + 4}$

23. $\dfrac{x^3 - 27}{x^3 + 3x^2 + 9x}$ $\dfrac{x - 3}{x}$

24. $\dfrac{x^2 + 6x + 9}{x^2 - 9}$ $\dfrac{x + 3}{x - 3}$

25. $\dfrac{15x^2 - 8x - 18}{-20x^2 + 14x + 12}$ not possible

26. $\dfrac{x^3 - 2x^2 + x - 2}{3x^2 - 3x - 8}$ not possible

27. $\dfrac{x^3 + 3x^2 - 2x - 6}{x^3 + 27}$ $\dfrac{x^2 - 2}{x^2 - 3x + 9}$

MULTIPLYING Multiply the rational expressions. Simplify the result.

STUDENT HELP

▶ **HOMEWORK HELP**
Example 1: Exs. 16–27
Examples 2–4: Exs. 28–35, 44–49
Examples 5–7: Exs. 36–49
Example 8: Exs. 50–55

28. $\dfrac{4xy^3}{x^2 y} \cdot \dfrac{y}{8x}$ $\dfrac{y^3}{2x^2}$

29. $\dfrac{80x^4}{y^3} \cdot \dfrac{xy}{5x^2}$ $\dfrac{16x^3}{y^2}$

30. $\dfrac{2x^2 - 10}{x + 1} \cdot \dfrac{x + 2}{3x^2 - 15}$ $\dfrac{2(x + 2)}{3(x + 1)}$

31. $\dfrac{x - 3}{2x - 8} \cdot \dfrac{6x^2 - 96}{x^2 - 9}$ $\dfrac{3(x + 4)}{x + 3}$

32. $\dfrac{x^2 - x - 6}{4x^3} \cdot \dfrac{x + 1}{x^2 + 5x + 6}$ $\dfrac{(x + 1)(x - 3)}{4x^3(x + 3)}$

33. $\dfrac{2x^2 - 2}{x^2 - 6x - 7} \cdot (x^2 - 10x + 21)$ $2(x - 1)(x - 3)$

34. $\dfrac{x^3 + 5x^2 - x - 5}{x^2 - 25} \cdot (x + 1)$ $\dfrac{(x - 1)(x + 1)^2}{x - 5}$

35. $\dfrac{x - 3}{-x^3 + 3x^2} \cdot (x^2 + 2x + 1)$ $\dfrac{-(x + 1)^2}{x^2}$

DIVIDING Divide the rational expressions. Simplify the result.

36. $\dfrac{32x^3y}{y^9} \div \dfrac{8x^4}{y^6}$ $\dfrac{4}{xy^2}$

37. $\dfrac{2xyz}{x^2z^2} \div \dfrac{6y^3}{3xz}$ $\dfrac{1}{y^2}$

38. $\dfrac{3x^2 + x - 2}{x^2 + 3x + 2} \div \dfrac{2x}{x + 2}$ $\dfrac{3x - 2}{2x}$

39. $\dfrac{x^2 - 14x + 48}{x^2 - 6x} \div (3x - 24)$ $\dfrac{1}{3x}$

40. $\dfrac{2x^2 - 12x}{x^2 - 7x + 6} \div \dfrac{2x}{3x - 3}$ 3

41. $\dfrac{x^2 + 8x + 16}{x + 2} \div \dfrac{x^2 + 6x + 8}{x^2 - 4}$ $\dfrac{(x + 4)(x - 2)}{x + 2}$

42. $\dfrac{x^2 + 6x - 7}{3x^2} \div \dfrac{x + 7}{6x}$ $\dfrac{2(x - 1)}{x}$

43. $(x^2 + 6x - 27) \div \dfrac{3x^2 + 27x}{x + 5}$ $\dfrac{(x - 3)(x + 5)}{3x}$

COMBINED OPERATIONS Perform the indicated operations. Simplify the result.

44. $(x - 5) \div \dfrac{x^2 - 11x + 30}{x^2 + 7x + 12} \cdot (x - 6)$

45. $\dfrac{x^2 - x - 12}{8x^2} \div \dfrac{x^3 + 3x^2}{8x^3 - 2x^2} \div \dfrac{4x - 1}{x + 2}$

46. $\dfrac{x^2 + 11x}{x - 2} \div (3x^2 + 6x) \cdot \dfrac{x^2 - 4}{x + 11}$ $\dfrac{1}{3}$

47. $\dfrac{2x^2 + x - 15}{2x^2 - 11x - 21} \cdot (6x + 9) \div \dfrac{2x - 5}{3x - 21}$ $9(x + 3)$

48. $(x^3 + 8) \cdot \dfrac{x - 2}{x^2 - 2x + 4} \div \dfrac{x^2 - 4}{x - 6}$ $x - 6$

49. $\dfrac{x^2 + 12x + 20}{4x^2 - 9} \cdot \dfrac{6x^3 - 9x^2}{x^3 + 10x^2} \cdot (2x + 3)$ $\dfrac{3(x + 2)}{}$

44. $(x + 3)(x + 4)$

45. $\dfrac{(x - 4)(x + 2)}{4x^2}$

51. $H = \dfrac{k_2}{k_1V^2}$ or HV^2 is a constant. A shorter runner can run faster than a taller runner and still have the heat being generated equal the heat being released, so a shorter runner has an advantage.

52. $A = \dfrac{(43.3t + 999)(0.05t^2 + 1)}{(0.0428t + 1)(0.101t^2 + 2.20)}$

HEAT GENERATION In Exercises 50 and 51, use the following information.
Almost all of the energy generated by a long-distance runner is released in the form of heat. The rate of heat generation h_g and the rate of heat released h_r for a runner of height H can be modeled by

$$h_g = k_1 H^3 V^2 \qquad \text{and} \qquad h_r = k_2 H^2$$

where k_1 and k_2 are constants and V is the runner's speed.

50. Write the ratio of heat generated to heat released. $\dfrac{k_1HV^2}{k_2}$

51. When the ratio of heat generated to heat released equals 1, how is height related to velocity? Does this mean that a taller or a shorter runner has an advantage?
See margin.

FARMLAND In Exercises 52 and 53, use the following information.
From 1987 to 1996, the total acres of farmland L (in millions) and the total number of farms F (in hundreds of thousands) in the United States can be modeled by

$$L = \dfrac{43.3t + 999}{0.0482t + 1} \qquad \text{and} \qquad F = \dfrac{0.101t^2 + 2.20}{0.0500t^2 + 1}$$

where t represents the number of years since 1987. ▶ Source: U.S. Bureau of the Census

52. Write a model for the average number of acres A per farm as a function of the year.
See margin.

53. What was the average number of acres per farm in 1993?   468.5 acres

WEIGHT IN GOLD In Exercises 54 and 55, use the following information.
From 1990 to 1996, the price P of gold (in dollars per ounce) and the weight W of gold mined (in millions of ounces) in the United States can be modeled by

$$P = \dfrac{53.4t^2 - 243t + 385}{0.00146t^3 + 0.122t^2 - 0.586t + 1}$$

$$W = -0.0112t^5 + 0.193t^4 - 1.17t^3 + 2.82t^2 - 1.76t + 10.4$$

where t represents the number of years since 1990. ▶ Source: U.S. Bureau of the Census

54. Write a model for the total value V of gold mined as a function of the year.
See margin.

55. What was the total value of gold mined in the United States in 1994?
about $4.4 billion

54. $V = \dfrac{(53.4t^2 - 243t + 385)(-0.0112t^5 + 0.193t^4 - 1.17t^3 + 2.82t^2 - 1.76t + 10.4)}{0.00146t^3 + 0.122t^2 - 0.586t + 1}$

FOCUS ON
CAREERS

FARMER
In 1996 there were 1.3 million farmers and farm managers in the United States. In addition to knowing about crops and animals, farmers must keep up with changing technology and possess strong business skills.

CAREER LINK
www.mcdougallittell.com

! COMMON ERROR
EXERCISE 27 A common error is to forget that the sum of two cubes can be factored. Remind students that $a^3 + b^3 = (a + b)(a^2 - ab + b^2)$.

APPLICATION NOTE
EXERCISE 54 Point out that real-world applications of rational functions often cannot be factored. The model for this exercise cannot be simplified.

CAREER NOTE
Additional information about farmers is available at **www.mcdougallittell.com**.

ADDITIONAL PRACTICE AND RETEACHING

For Lesson 9.4:
- Practice Levels A, B, and C (*Chapter 9 Resource Book*, p. 55)
- Reteaching with Practice (*Chapter 9 Resource Book*, p. 58)
- See Lesson 9.4 of the *Personal Student Tutor*

For more Mixed Review:
- Search the *Test and Practice Generator* for key words or specific lessons.

559

DAILY HOMEWORK QUIZ

📠 **Transparency Available**

1. If possible, simplify the rational expression.

a. $\dfrac{x^2 - x - 6}{x^2 + 6x + 8}$ $\dfrac{x-3}{x+4}$

b. $\dfrac{x^2 - 16}{x^2 + 8x + 16}$ $\dfrac{x-4}{x+4}$

2. Perform the indicated operation. Simplify the result.

a. $\dfrac{x^2 - 1}{12x^2 + 24x} \cdot \dfrac{4}{x^2 + x}$

$\dfrac{x-1}{3x^2(x+2)}$

b. $\dfrac{x^2 + 5x + 6}{x^3 - x^2} \div \dfrac{x+3}{x^2}$ $\dfrac{x+2}{x-1}$

c. $\dfrac{6x^2 + 7x + 1}{7x + 49} \div \dfrac{2x+2}{2x+14}$ $\dfrac{6x+1}{7}$

d. $\dfrac{x+1}{6x-3} \cdot \dfrac{3x^2}{x^2 + x} \div \dfrac{x^2}{2x^2 - x}$ 1

┌─── EXTRA CHALLENGE NOTE ───┐

→ Challenge problems for Lesson 9.5 are available in **blackline** format in the *Chapter 9 Resource Book*, p. 62 and at **www.mcdougallittell.com**.

ADDITIONAL TEST PREPARATION

1. OPEN ENDED Give an example of a rational expression that can be simplified to $(x+3)$.

Sample answer: $\dfrac{x^2 + x - 6}{x - 2}$

73.

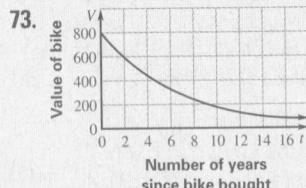

Number of years since bike bought

---

56. GEOMETRY CONNECTION Use the diagram at the right. Find the ratio of the volume of the rectangular prism to the volume of the inscribed cylinder. Write your answer in simplified form. $\dfrac{4}{\pi}$

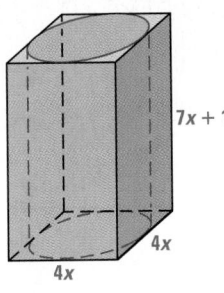

$7x + 1$

$4x$ $4x$

57. MULTI-STEP PROBLEM The surface area S and the volume V of a tin can are given by $S = 2\pi r^2 + 2\pi rh$ and $V = \pi r^2 h$ where r is the radius of the can and h is the height of the can. One measure of the *efficiency* of a tin can is the ratio of its surface area to its volume.

Test Preparation

a. Find a general formula (in simplified form) for the ratio $\dfrac{S}{V}$. $\dfrac{2(r+h)}{rh}$

b. Find the efficiency of a can when $h = 2r$. $\dfrac{3}{r}$

c. Calculate the efficiency of each can. **1.278; 0.698; 0.609**

57d. 3-pound coffee can, 2-pound coffee can, soup can; Reducing the surface area to volume ratio makes the can more efficient. Less material is used to contain the same volume.

- A soup can with $r = 2\frac{5}{8}$ inches and $h = 3\frac{7}{8}$ inches.

- A 2 pound coffee can with $r = 5\frac{1}{8}$ inches and $h = 6\frac{1}{2}$ inches.

- A 3 pound coffee can with $r = 6\frac{3}{16}$ inches and $h = 7$ inches.

d. *Writing* Rank the three cans in part (c) by efficiency (most efficient to least efficient). Explain your rankings.

★ **Challenge**

58. Find two rational functions $f(x)$ and $g(x)$ such that $f(x) \cdot g(x) = x^2$ and $\dfrac{f(x)}{g(x)} = \dfrac{(x-1)^2}{(x+2)^2}$. *Sample answer:* $f(x) = \dfrac{x(x-1)}{x+2}$; $g(x) = \dfrac{x(x+2)}{x-1}$

59. Find two rational functions $f(x)$ and $g(x)$ such that $f(x) \cdot g(x) = x - 1$ and $\dfrac{f(x)}{g(x)} = \dfrac{(x+1)^2(x-1)}{x^4}$. *Sample answer:* $f(x) = \dfrac{x^2 - 1}{x^2}$; $g(x) = \dfrac{x^2}{x+1}$

┌─── EXTRA CHALLENGE ───┐

→ www.mcdougallittell.com

MIXED REVIEW

GCFs AND LCMs Find the greatest common factor and least common multiple of each pair of numbers. (Skills Review, p. 908)

60. 96, 160 **32; 480**

61. 120, 165 **15; 1320**

62. 48, 108 **12; 432**

63. 72, 84 **12; 504**

64. 238, 51 **17; 714**

65. 480, 600 **120; 2400**

MULTIPLYING POLYNOMIALS Find the product. (Review 6.3 for 9.5)

66. $x(x^2 + 7x - 1)$ $x^3 + 7x^2 - x$

67. $(x + 7)(x - 1)$ $x^2 + 6x - 7$

68. $(x + 10)(x - 3)$ $x^2 + 7x - 30$

69. $(x + 3)(x^2 + 3x + 2)$ $x^3 + 6x^2 + 11x + 6$

70. $(2x - 2)(x^3 - 4x^2)$ $2x^4 - 10x^3 + 8x^2$

71. $x(x^2 - 4)(5 - 6x^3)$ $-6x^6 + 24x^4 + 5x^3 - 20x$

🌐 **BICYCLE DEPRECIATION** In Exercises 72 and 73, use the following information. You bought a new mountain bike for $800. The value of the bike decreases by about 14% each year. (Review 8.2)

72. Write an exponential decay model for the value of the bike. Use the model to estimate the value after 4 years. $V = 800(0.86)^t$; about $438

See margin. In 6.5 years.

73. Graph the model. Use the graph to estimate when the bike will be worth $300.

● ACTIVITY 9.4

Using Technology

Operations with Rational Expressions

You have learned how to simplify rational expressions and how to multiply and divide rational expressions. You can use a graphing calculator to verify the results of these operations numerically and graphically.

▶ **EXAMPLE**

Simplify $\dfrac{x^2 + 3x - 10}{x^2 - 5x + 6}$. Use a graphing calculator to verify the results numerically and graphically.

▶ **SOLUTION**

You can simplify the rational expression as follows.

$$\frac{x^2 + 3x - 10}{x^2 - 5x + 6} = \frac{(x-2)(x + 5)}{(x-2)(x - 3)} = \frac{x + 5}{x - 3}$$

1 Enter the original expression as y_1 and the simplified result as y_2. Use the *Path* style for y_2.

```
Y1■(X²+3X-10)/
(X²-5X+6)
-0Y2■(X+5)/(X-3)
Y3=
Y4=
Y5=
Y6=
```

Remember to use parentheses correctly.

2 Use the *Table* feature to examine corresponding values of the two expressions.

```
X    | Y1     | Y2
0    | -1.667 | -1.667
1    | -3     | -3
2    | ERROR  | -7
3    | ERROR  | ERROR
4    | 9      | 9
X=0
```

Why is only y_1 undefined at $x = 2$?

3 Put your calculator in *Connected* mode. Display your graphs in the standard viewing window.

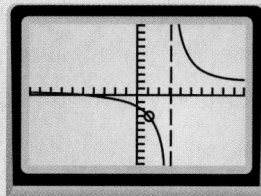

The *Path* style shows the graph of y_2 being drawn even though it coincides with the graph of y_1.

The graphing calculator allows you to check your solution in two ways. If the two expressions are equivalent, the values of y_1 and y_2 will be the same in the table *except where a common factor has been divided out*. Also, the graphs of the two expressions will coincide.

▶ **EXERCISES**

Simplify the expression. Use a graphing calculator to verify the result numerically and graphically.

1. $\dfrac{x^2 - 3x}{x^2 + x - 12}$ $\dfrac{x}{x + 4}$

2. $\dfrac{2x^2 - 10x}{x^2 - 4x - 5}$ $\dfrac{2x}{x + 1}$

3. $\dfrac{x^2 + x - 6}{x^2 + 4x + 3}$ $\dfrac{x - 2}{x + 1}$

Perform the indicated operation and simplify. Use a graphing calculator to verify the result numerically and graphically.

4. $\dfrac{x - 1}{2x^2} \cdot \dfrac{x + 2}{x - 1}$ $\dfrac{x + 2}{2x^2}$

5. $\dfrac{2x^2 - 10x}{3x + 3} \div \dfrac{x - 5}{x + 1}$ $\dfrac{2x}{3}$

6. $\dfrac{x^2 - x - 12}{x^2 + 6x + 6} \cdot \dfrac{x^2 + 3x + 2}{x^2 + 5x + 6}$ $\dfrac{(x - 4)(x + 1)}{x^2 + 6x + 6}$

STUDENT HELP

INTERNET   **KEYSTROKE HELP**

See keystrokes for several models of calculators at www.mcdougallittell.com

9.4 *Technology Activity* **561**

---

1 **Planning the Activity**

PURPOSE
To verify the results of operations with rational expressions.

MATERIALS
• graphing calculator
• Keystroke blackline (*Chapter 9 Resource Book,* p. 54)

PACING
• Activity — 25 min

▶ **LINK TO LESSON**
The graphs of the original expression and the simplified results coincide except where a common factor has been divided out. Verify this in Example 1 of Lesson 9.4.

2 **Managing the Activity**

COOPERATIVE LEARNING
Ask pairs of students to take turns simplifying and graphing each expression. Have them discuss their numerical and graphical results.

CLASSROOM MANAGEMENT
Remind students who are having difficulty graphing that both the numerator and the denominator need to be enclosed in parentheses when entered in the calculator.

3 **Closing the Activity**

★ **KEY DISCOVERY**
The graph of a rational expression and its simplified form are equivalent except where a common factor has been divided out.

ACTIVITY ASSESSMENT
JOURNAL Describe how to check the simplification of $\dfrac{x^2 - 3x + 2}{x - 1}$ with a graphing calculator.

Graph $y = \dfrac{x^2 - 3x + 2}{x - 1}$ and $y = x - 2$.
Compare corresponding values. They are identical except for where $x = 1$.

561

> **LESSON OPENER**
> **VISUAL APPROACH**
> An alternative way to approach
> Lesson 9.5 is to use the Visual
> Approach Lesson Opener:
> • Blackline Master (*Chapter 9 Resource Book,* p. 66)
> • Transparency (p. 63)

MEETING INDIVIDUAL NEEDS
• *Chapter 9 Resource Book*
 Prerequisite Skills Review (p. 5)
 Practice Level A (p. 69)
 Practice Level B (p. 70)
 Practice Level C (p. 71)
 Reteaching with Practice (p. 72)
 Absent Student Catch-Up (p. 74)
 Challenge (p. 77)
• *Resources in Spanish*
• Personal Student Tutor

NEW-TEACHER SUPPORT
See the Tips for New Teachers on
pp. 1–2 of the *Chapter 9 Resource
Book* for additional notes about
Lesson 9.5.

WARM-UP EXERCISES

Transparency Available

Simplify.

1. $\dfrac{3}{5} + \dfrac{1}{5}$ $\dfrac{4}{5}$

2. $\dfrac{3}{4} + \dfrac{1}{2}$ $\dfrac{5}{4}$

3. $\dfrac{2}{3} - \dfrac{1}{2}$ $\dfrac{1}{6}$

4. $\dfrac{\frac{1}{2}}{\frac{3}{4}}$ $\dfrac{2}{3}$

5. $\dfrac{1}{\frac{1}{2} + \frac{1}{3}}$ $\dfrac{6}{5}$

9.5 Addition, Subtraction, and Complex Fractions

GOAL 1 WORKING WITH RATIONAL EXPRESSIONS

What you should learn

GOAL 1 Add and subtract rational expressions, as applied in **Example 4.**

GOAL 2 Simplify complex fractions, as applied in **Example 6.**

Why you should learn it

▼ To solve **real-life** problems, such as modeling the total number of male college graduates in **Ex. 47.**

As with numerical fractions, the procedure used to add (or subtract) two rational expressions depends upon whether the expressions have *like* or *unlike* denominators.

To add (or subtract) two rational expressions with *like* denominators, simply add (or subtract) their numerators and place the result over the common denominator.

EXAMPLE 1 *Adding and Subtracting with Like Denominators*

Perform the indicated operation.

a. $\dfrac{4}{3x} + \dfrac{5}{3x}$

b. $\dfrac{2x}{x+3} - \dfrac{4}{x+3}$

SOLUTION

a. $\dfrac{4}{3x} + \dfrac{5}{3x} = \dfrac{4+5}{3x} = \dfrac{9}{3x} = \dfrac{3}{x}$ Add numerators and simplify expression.

b. $\dfrac{2x}{x+3} - \dfrac{4}{x+3} = \dfrac{2x-4}{x+3}$ Subtract numerators.

· · · · · · · · · ·

To add (or subtract) rational expressions with *unlike* denominators, first find the least common denominator (LCD) of the rational expressions. Then, rewrite each expression as an equivalent rational expression using the LCD and proceed as with rational expressions with like denominators.

EXAMPLE 2 *Adding with Unlike Denominators*

Add: $\dfrac{5}{6x^2} + \dfrac{x}{4x^2 - 12x}$

SOLUTION

First find the least common denominator of $\dfrac{5}{6x^2}$ and $\dfrac{x}{4x^2 - 12x}$.

It helps to factor each denominator: $6x^2 = 6 \cdot x \cdot x$ and $4x^2 - 12x = 4 \cdot x \cdot (x - 3)$.

The LCD is $12x^2(x - 3)$. Use this to rewrite each expression.

$$\dfrac{5}{6x^2} + \dfrac{x}{4x^2 - 12x} = \dfrac{5}{6x^2} + \dfrac{x}{4x(x-3)} = \dfrac{5[2(x-3)]}{6x^2[2(x-3)]} + \dfrac{x(3x)}{4x(x-3)(3x)}$$

$$= \dfrac{10x - 30}{12x^2(x-3)} + \dfrac{3x^2}{12x^2(x-3)}$$

$$= \dfrac{3x^2 + 10x - 30}{12x^2(x-3)}$$

> **STUDENT HELP**
>
> ↳ **Skills Review**
> For help with LCDs, see p. 908.

EXAMPLE 3 *Subtracting With Unlike Denominators*

Subtract: $\dfrac{x+1}{x^2+4x+4} - \dfrac{2}{x^2-4}$

SOLUTION

$$\dfrac{x+1}{x^2+4x+4} - \dfrac{2}{x^2-4} = \dfrac{x+1}{(x+2)^2} - \dfrac{2}{(x-2)(x+2)}$$

$$= \dfrac{(x+1)(x-2)}{(x+2)^2(x-2)} - \dfrac{2(x+2)}{(x-2)(x+2)(x+2)}$$

$$= \dfrac{x^2-x-2-(2x+4)}{(x+2)^2(x-2)}$$

$$= \dfrac{x^2-3x-6}{(x+2)^2(x-2)}$$

STUDENT HELP

Look Back
For help with multiplying polynomials, see p. 338.

Statistics

EXAMPLE 4 *Adding Rational Models*

The distribution of heights for American men and women aged 20–29 can be modeled by

$$y_1 = \dfrac{0.143}{1+0.008(x-70)^4} \qquad \text{American men's heights}$$

$$y_2 = \dfrac{0.143}{1+0.008(x-64)^4} \qquad \text{American women's heights}$$

where x is the height (in inches) and y is the percent (in decimal form) of adults aged 20–29 whose height is $x \pm 0.5$ inches. ▶ Source: *Statistical Abstract of the United States*

a. Graph each model. What is the most common height for men aged 20–29? What is the most common height for women aged 20–29?

b. Write a model that shows the distribution of the heights of *all* adults aged 20–29. Graph the model and find the most common height.

SOLUTION

a. From the graphing calculator screen shown at the top right, you can see that the most common height for men is 70 inches (14.3%). The second most common heights are 69 inches and 71 inches (14.2% each). For women, the curve has the same shape, but is shifted to the left so that the most common height is 64 inches. The second most common heights are 63 inches and 65 inches.

b. To find a model for the distribution of all adults aged 20–29, add the two models and divide by 2.

$$y = \dfrac{1}{2}\left[\dfrac{0.143}{1+0.008(x-70)^4} + \dfrac{0.143}{1+0.008(x-64)^4}\right]$$

From the graph shown at the bottom right, you can see that the most common height is 67 inches.

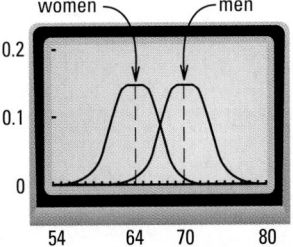
women ⟍   ⟋ men

54   64   70   80

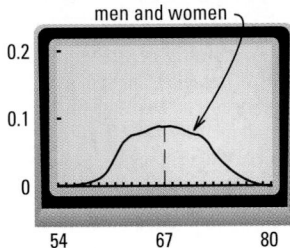
men and women ⟍

54   67   80

9.5 *Addition, Subtraction, and Complex Fractions* **563**

---

2 TEACH

EXTRA EXAMPLE 1
Perform the indicated operation.
a. $\dfrac{3}{2x} - \dfrac{7}{2x}$ $-\dfrac{2}{x}$

b. $\dfrac{3x}{x-4} + \dfrac{6}{x-4}$ $\dfrac{3x+6}{x-4}$

EXTRA EXAMPLE 2
Add: $\dfrac{4}{3x^3} + \dfrac{x}{6x^3+3x^2}$ $\dfrac{x^2+8x+4}{3x^3(2x+1)}$

EXTRA EXAMPLE 3
Subtract: $\dfrac{x+1}{x^2+6x+9} - \dfrac{1}{x^2-9}$
$\dfrac{x^2-3x-6}{(x+3)^2(x-3)}$

EXTRA EXAMPLE 4
Josh drove 42 miles and then took the train. The entire trip was 172 miles. The average speed of the train was 35 mi/h faster than the average speed of the car. Let x equal the average speed of the car and y equal the total time traveled. Then $y_1 = \dfrac{42}{x}$ is the time the car traveled and $y_2 = \dfrac{130}{x+35}$ is the time the train traveled.
a. Graph each model. What is the time Josh traveled in the car if his rate is 35 mi/h? What is his time on the train if his car rate is 35 mi/h? 1.2 h; 1.86 h
b. Write a model that shows the total time it takes for the trip. Graph the model. If the car's average speed is 50 mi/h, how long does the trip take?
$\dfrac{42}{x} + \dfrac{130}{x+35} = \dfrac{172x+1470}{x(x+35)}$; 2.37 h

☑ CHECKPOINT EXERCISES
For use after Examples 1–3:
1. Perform the indicated operation.
a. $\dfrac{2x-1}{x+1} - \dfrac{x}{x+1}$ $\dfrac{x-1}{x+1}$
b. $\dfrac{x-1}{x^2+4x+3} + \dfrac{x}{x^2-9}$
$\dfrac{2x^2-3x+3}{(x+3)(x-3)(x+1)}$

Checkpoint Exercises for Example 4 on next page.

CHECKPOINT EXERCISES

For use after Example 4:

2. Given the following rational functions:

$$y_1 = \frac{0.25}{1 + (x - 2.8)^4}$$

$$y_2 = \frac{0.25}{1 + (x - 3.5)^4}$$

a. Graph each function. For what values of x is y_1 a maximum? y_2? **2.8; 3.5**

b. Graph a model that shows the sum of the functions. For what value of x is this function a maximum? **3.1**

 EXTRA EXAMPLE 5

Simplify: $\dfrac{\frac{3}{x-4}}{\frac{1}{x-4} + \frac{3}{x+1}}$ $\dfrac{3x+3}{4x-11}$

EXTRA EXAMPLE 6

The focal length f (in centimeters) of a curved mirror is

$f = \dfrac{1}{\frac{1}{d_i} + \frac{1}{d_o}}$, where d_o is the object's distance from the mirror and d_i is the image's distance from the mirror. Simplify the complex fraction.

$f = \dfrac{d_i \cdot d_o}{d_i + d_o}$

 CHECKPOINT EXERCISES

For use after Examples 5 and 6:

1. Simplify: $\dfrac{\frac{2}{x-1}}{\frac{4}{x-1} + \frac{1}{x}}$ $\dfrac{2x}{5x-1}$

CLOSURE QUESTION

Describe how to simplify $\dfrac{2x}{x+2} + \dfrac{1}{x-3}$. Find the LCD of $(x+2)$ and $(x-3)$; rewrite each expression as an equivalent expression with the LCD $(x+2)(x-3)$ as its denominator, or $\dfrac{2x(x-3)}{(x+2)(x-3)} + \dfrac{1(x+2)}{(x+2)(x-3)}$; add the two fractions and simplify the result to get $\dfrac{2x^2 - 5x + 2}{(x+2)(x-3)}$.

564

A **complex fraction** is a fraction that contains a fraction in its numerator or denominator. To simplify a complex fraction, write its numerator and its denominator as single fractions. Then divide by multiplying by the reciprocal of the denominator.

 **STUDENT HELP**

HOMEWORK HELP
Visit our Web site
www.mcdougallittell.com
for extra examples.

EXAMPLE 5 *Simplifying a Complex Fraction*

Simplify: $\dfrac{\frac{2}{x+2}}{\frac{1}{x+2} + \frac{2}{x}}$

SOLUTION

$\dfrac{\frac{2}{x+2}}{\frac{1}{x+2} + \frac{2}{x}} = \dfrac{\frac{2}{x+2}}{\frac{3x+4}{x(x+2)}}$ **Add fractions in denominator.**

$= \dfrac{2}{x+2} \cdot \dfrac{x(x+2)}{3x+4}$ **Multiply by reciprocal.**

$= \dfrac{2x(x+2)}{(x+2)(3x+4)}$ **Divide out common factor.**

$= \dfrac{2x}{3x+4}$ **Write in simplified form.**

· · · · · · · · · ·

Another way to simplify a complex fraction is to multiply the numerator and denominator by the least common denominator of *every* fraction in the numerator and denominator.

EXAMPLE 6 *Simplifying a Complex Fraction*

PHOTOGRAPHY The focal length f of a thin camera lens is given by

$$f = \dfrac{1}{\frac{1}{p} + \frac{1}{q}}$$

where p is the distance between an object being photographed and the lens and q is the distance between the lens and the film. Simplify the complex fraction.

SOLUTION

$f = \dfrac{1}{\frac{1}{p} + \frac{1}{q}}$ **Write equation.**

$= \dfrac{pq}{pq} \cdot \dfrac{1}{\frac{1}{p} + \frac{1}{q}}$ **Multiply numerator and denominator by pq.**

$= \dfrac{pq}{q + p}$ **Simplify.**

FOCUS ON APPLICATIONS

object   lens   image

 PHOTOGRAPHY
The focal length of a camera lens is the distance between the lens and the point where light rays converge after passing through the lens.

GUIDED PRACTICE

Vocabulary Check ✓

Concept Check ✓

1.
$$\frac{4 + \dfrac{1}{x}}{2 - \dfrac{3}{2x+5}}; \quad \frac{\dfrac{5}{x+1} + 3}{x^2 + 5x + 6}$$

Skill Check ✓

3. (1) Write the numerator and denominator as single fractions. Then divide by multiplying the numerator by the reciprocal of the denominator. (2) Multiply the numerator and denomi-nator by the least common denominator of every fraction in the numerator and denominator.

24. sometimes; The LCD will be the product of the denominators if the denominators have no common factors.

1. Give two examples of a complex fraction.   *Sample answers:* See margin.

2. How is adding (or subtracting) rational expressions similar to adding (or subtracting) numerical fractions?   **You need common denominators to add (or subtract) rational expressions or numerical fractions.**

3. Describe two ways to simplify a complex fraction.

4. Why isn't $(x+1)^3$ the LCD of $\dfrac{1}{x+1}$ and $\dfrac{1}{(x+1)^2}$? What is the LCD?
   $(x+1)^3$ **is a common denominator, but not the lowest; The LCD is** $(x+1)^2$**.**

Perform the indicated operation and simplify.

5. $\dfrac{2x}{x+5} + \dfrac{7}{x+5}$ $\dfrac{2x+7}{x+5}$ 6. $\dfrac{7}{5x} + \dfrac{8}{3x}$ $\dfrac{61}{15x}$ 7. $\dfrac{x}{x-4} - \dfrac{6}{x+3}$
   $\dfrac{x^2 - 3x + 24}{(x-4)(x+3)}$

Simplify the complex fraction.

8. $\dfrac{\dfrac{x}{5} + 4}{8 + \dfrac{1}{x}}$ $\dfrac{x(x+20)}{5(8x+1)}$ 9. $\dfrac{\dfrac{x+2}{5} - 5}{8 + \dfrac{4}{x}}$ $\dfrac{x(x-23)}{20(2x+1)}$ 10. $\dfrac{\dfrac{15}{2x+2}}{\dfrac{6}{x} - \dfrac{1}{2}}$ $\dfrac{-15x}{(x-12)(x+1)}$

11. 🌐 **FINANCE** For a loan paid back over t years, the monthly payment is given by $M = \dfrac{Pi}{1 - \left(\dfrac{1}{1+i}\right)^{12t}}$ where P is the principal and i is the annual interest rate.

Show that this formula is equivalent to $M = \dfrac{Pi(1+i)^{12t}}{(1+i)^{12t} - 1}$.

$$\frac{Pi}{1 - \left(\frac{1}{1+i}\right)^{12t}} = \frac{Pi(1+i)^{12t}}{\left(1 - \left(\frac{1}{(1+i)^{12t}}\right)\right)(1+i)^{12t}} = \frac{Pi(1+i)^{12t}}{(1+i)^{12t} - 1}$$

PRACTICE AND APPLICATIONS

STUDENT HELP

▶ **Extra Practice**
to help you master skills is on p. 953.

25. always; Each denominator must be a factor of the LCD, so the LCD must have degree greater than or equal to each of the separate denominators.

STUDENT HELP

▶ **HOMEWORK HELP**
Example 1: Exs. 12–17
Examples 2, 3: Exs. 18–23, 26–37
Example 4: Exs. 47–51
Example 5: Exs. 38–46
Example 6: Exs. 52, 53

OPERATIONS WITH LIKE DENOMINATORS Perform the indicated operation and simplify.

12. $\dfrac{7}{6x} + \dfrac{11}{6x}$ $\dfrac{3}{x}$ 13. $\dfrac{23}{10x^2} - \dfrac{x}{10x^2}$ $\dfrac{23-x}{10x^2}$ 14. $\dfrac{4x}{x+1} - \dfrac{3}{x+1}$ $\dfrac{4x-3}{x+1}$

15. $\dfrac{5x^2}{x+8} + \dfrac{5x}{x+8}$ $\dfrac{5x(x+1)}{x+8}$ 16. $\dfrac{6x^2}{x-2} - \dfrac{12x}{x-2}$ $6x$ 17. $\dfrac{x}{x^2-5x} - \dfrac{5}{x^2-5x}$ $\dfrac{1}{x}$

FINDING LCDS Find the least common denominator.

18. $\dfrac{14}{4(x+1)}, \dfrac{7}{4x}$ $4x(x+1)$ 19. $\dfrac{4}{21x^2}, \dfrac{x}{3x^2-15x}$ $21x^2(x-5)$

20. $\dfrac{5x+2}{4x^2-1}, \dfrac{3}{x}, \dfrac{9x}{2x+1}$ $x(2x-1)(2x+1)$ 21. $\dfrac{1}{x(x-6)}, \dfrac{12}{x^2-3x-18}$ $x(x+3)(x-6)$

22. $\dfrac{3x+1}{x(x-7)}, \dfrac{3}{x^2-6x-7}$ $x(x+1)(x-7)$ 23. $\dfrac{1}{x^2-3x-28}, \dfrac{x}{x^2+6x+8}$
    $(x-7)(x+2)(x+4)$

LOGICAL REASONING Tell whether the statement is *always true, sometimes true,* or *never true.* Explain your reasoning.

24. The LCD of two rational expressions is the product of the denominators.   **See margin.**

25. The LCD of two rational expressions will have a degree greater than or equal to that of the denominator with the higher degree.   **See margin.**

3 APPLY

ASSIGNMENT GUIDE

BASIC
Day 1: pp. 565–567 Exs. 12–14, 18–21, 26–40 even, 52–54, 57–71 odd

AVERAGE
Day 1: pp. 565–567 Exs. 12–16, 18–25, 26–40 even, 47, 52–54, 57–71 odd

ADVANCED
Day 1: pp. 565–567 Exs. 12–16, 18–25, 26–46 even, 48–54, 57–71 odd

BLOCK SCHEDULE
pp. 565–567 Exs. 12–16, 18–25, 26–40 even, 47, 52–54, 57–71 odd (with 9.6)

EXERCISE LEVELS
Level A: *Easier*
12–21, 26–29
Level B: *More Difficult*
22–25, 30–41, 48–54
Level C: *Most Difficult*
42–47, 55–56

✓ **HOMEWORK CHECK**
To quickly check student under-standing of key concepts, go over the following exercises:
Exs. 14, 18, 20, 26, 30, 38. See also the Daily Homework Quiz:

• Blackline Master (*Chapter 9 Resource Book*, p. 80)
• 📖 Transparency (p. 73)

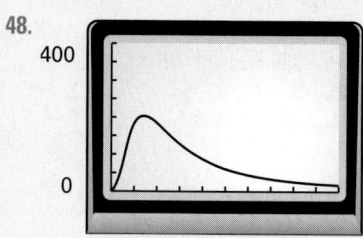

STUDENT HELP NOTES

→ **Look Back** As students look back to p. 323, remind them that x^{-1} is the reciprocal of x, or $\frac{1}{x}$, and x^{-2} is the reciprocal of x^2, or $\frac{1}{x^2}$.

CAREER NOTE
Additional information about pharmacists is available at **www.mcdougallittell.com**.

47. $M = \dfrac{357t^3 + 5500t^2 - 37100t + 485000}{(0.00418t^2 + 1)(-0.0580t + 1)}$

48.

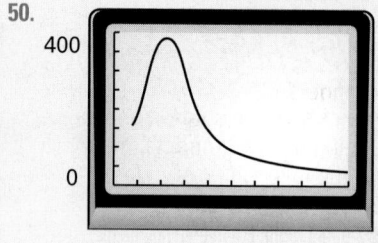

49. $A = \dfrac{391(t-1)^2 + 0.112}{0.218(t-1)^4 + 0.991(t-1)^2 + 1}$

50.

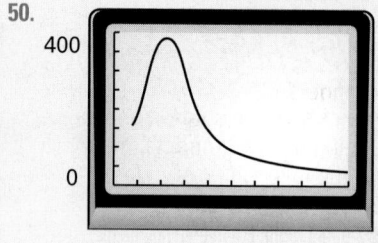

ADDITIONAL PRACTICE AND RETEACHING

For Lesson 9.5:
• Practice Levels A, B, and C (*Chapter 9 Resource Book,* p. 69)
• Reteaching with Practice (*Chapter 9 Resource Book,* p. 72)
• ▣ See Lesson 9.5 of the *Personal Student Tutor*

For more Mixed Review:
• ▣ Search the *Test and Practice Generator* for key words or specific lessons.

566

STUDENT HELP

▶ **Look Back**
For help with the negative exponents in Exs. 41 and 42, see p. 323.

34. $\dfrac{(4x+1)(x-3)}{(x-8)(x-4)}$

40. $\dfrac{x-3}{6(x-2)(x-1)}$

45. $\dfrac{3x(x-4)}{(13x+8)(x^2-4x+16)}$

46. $\dfrac{3}{2(x^2+3x+9)}$

FOCUS ON CAREERS

▶ **PHARMACIST**
In addition to mixing and dispensing prescription drugs, pharmacists advise patients and physicians on the use of medications. This includes warning of possible side effects and recommending drug dosages, as discussed in Exs. 48–51.

🖱 **CAREER LINK**
www.mcdougallittell.com

OPERATIONS WITH UNLIKE DENOMINATORS Perform the indicated operation(s) and simplify.

26. $\dfrac{6}{4x^2} + \dfrac{2}{5x}$ $\dfrac{4x+15}{10x^2}$

27. $-\dfrac{4}{7x} - \dfrac{5}{3x}$ $\dfrac{-47}{21x}$

28. $\dfrac{7}{6(x-2)} - \dfrac{x+3}{6x}$ $\dfrac{-(x^2-6x-6)}{6x(x-2)}$

29. $\dfrac{6x+1}{x^2-9} + \dfrac{4}{x-3}$ $\dfrac{10x+13}{(x-3)(x+3)}$

30. $\dfrac{10}{x^2-5x-14} + \dfrac{2}{x-7}$ $\dfrac{2(x+7)}{(x-7)(x+2)}$

31. $\dfrac{5x-1}{x^2+2x-8} - \dfrac{6}{x+4}$ $\dfrac{11-x}{(x-2)(x+4)}$

32. $\dfrac{4x^2}{3x+5} - \dfrac{10}{x+8}$ $\dfrac{2(2x^3+16x^2-15x-25)}{(x+8)(3x+5)}$

33. $\dfrac{2-5x}{x-10} + \dfrac{1}{3x+2}$ $\dfrac{-3(5x^2+x+2)}{(x-10)(3x+2)}$

34. $\dfrac{x^2+x-3}{x^2-12x+32} + \dfrac{3x}{x-8}$

35. $\dfrac{2x+1}{x^2+8x+16} - \dfrac{3}{x^2-16}$ $\dfrac{2(x^2-5x-8)}{(x-4)(x+4)^2}$

36. $\dfrac{4x}{x+1} + \dfrac{5}{2x-3} - \dfrac{4}{x}$ $\dfrac{8x^3-15x^2+9x+12}{x(x+1)(2x-3)}$

37. $\dfrac{10x}{3x^2-3} + \dfrac{4}{x-1} + \dfrac{5}{6x}$ $\dfrac{49x^2+24x-5}{6x(x-1)(x+1)}$

SIMPLIFYING COMPLEX FRACTIONS Simplify the complex fraction.

38. $\dfrac{\frac{x}{2}-5}{6+\frac{3}{x}}$ $\dfrac{x(x-10)}{6(2x+1)}$

39. $\dfrac{\frac{20}{x+1}}{\frac{1}{4}-\frac{7}{x+1}}$ $\dfrac{80}{x-27}$

40. $\dfrac{\frac{1}{2x^2-2}}{\frac{2}{x+1}+\frac{x}{x^2-2x-3}}$

41. $\dfrac{\frac{1}{x}-\frac{x}{x^{-1}+1}}{\frac{3}{x}}$ $\dfrac{-(x^3-x-1)}{3(x+1)}$

42. $\dfrac{\frac{1-x}{x^4}}{x^{-2}-\frac{2}{x^3+x^2}}$ $\dfrac{-x-1}{x^2}$

43. $\dfrac{\frac{1}{4x+3}-\frac{5}{3(4x+3)}}{\frac{x}{4x+3}}$ $\dfrac{-2}{3x}$

44. $\dfrac{\frac{4}{x^2-9}+\frac{2}{x-3}}{\frac{1}{x+3}+\frac{1}{x-3}}$ $\dfrac{x+5}{x}$

45. $\dfrac{\frac{1}{x^3+64}}{\frac{5}{x^2-16}-\frac{2}{3x^2+12x}}$

46. $\dfrac{\frac{3}{2x^2+6x+18}+\frac{x}{x^3-27}}{\frac{5x}{3x-9}-\frac{3}{x-3}}$

47. 🌐 **COLLEGE GRADUATES** From the 1984–85 school year through the 1993–94 school year, the number of female college graduates F and the total number of college graduates G in the United States can be modeled by

$$F = \dfrac{-19{,}600t + 493{,}000}{-0.0580t + 1} \quad \text{and} \quad G = \dfrac{7560t^2 + 978{,}000}{0.00418t^2 + 1}$$

where t is the number of school years since the 1984–85 school year. Write a model for the number of male college graduates. ▶ Source: U.S. Department of Education See margin.

📱 **DRUG ABSORPTION** In Exercises 48–51, use the following information.
The amount A (in milligrams) of an oral drug, such as aspirin, in a person's bloodstream can be modeled by

$$A = \dfrac{391t^2 + 0.112}{0.218t^4 + 0.991t^2 + 1}$$

where t is the time (in hours) after one dose is taken. ▶ Source: *Drug Disposition in Humans*

48. Graph the equation using a graphing calculator. See margin.

49. A second dose of the drug is taken 1 hour after the first dose. Write an equation to model the amount of the second dose in the bloodstream. See margin.

50. Write and graph a model for the total amount of the drug in the bloodstream after the second dose is taken. See margin.

51. About how long after the second dose has been taken is the greatest amount of the drug in the bloodstream? about 1.2 hours after the second dose

50. $A = \dfrac{391t^2 + 0.112}{0.218t^4 + 0.991t^2 + 1} + \dfrac{391(t-1)^2 + 0.112}{0.218(t-1)^4 + 0.991(t-1)^2 + 1}$; See margin for graph.

ELECTRONICS In Exercises 52 and 53, use the following information.

If three resistors in a parallel circuit have resistances R_1, R_2, and R_3 (all in ohms), then the total resistance R_t (in ohms) is given by this formula:

$$R_t = \frac{1}{\dfrac{1}{R_1} + \dfrac{1}{R_2} + \dfrac{1}{R_3}}$$

52. Simplify the complex fraction. $\quad R_t = \dfrac{R_1 R_2 R_3}{R_1 R_2 + R_1 R_3 + R_2 R_3}$

53. You have three resistors in a parallel circuit with resistances 6 ohms, 12 ohms, and 24 ohms. What is the total resistance of the circuit? $\frac{24}{7}$ ohms

Test Preparation

54. **MULTI-STEP PROBLEM** From 1988 through 1997, the total dollar value V (in millions of dollars) of the United States sound-recording industry can be modeled by

$$V = \frac{5783 + 1134t}{1 + 0.025t}$$

where t represents the number of years since 1988.

▶ Source: Recording Industry Association of America

a. Calculate the percent change in dollar value from 1988 to 1989. **16.7%**

b. Develop a general formula for the percent change in dollar value from year t to year $t + 1$. $\quad C = \dfrac{98942.5}{(0.025t + 1.025)(5783 + 1134t)}$

c. Enter the formula into a graphing calculator or spreadsheet. Observe the changes from year to year for 1988 through 1997. Describe what you observe from the data. **The percent change is getting smaller.**

★ **Challenge**

CRITICAL THINKING In Exercises 55 and 56, use the following expressions.

$$2 + \frac{1}{1 + \dfrac{1}{2}}, \quad 2 + \frac{1}{1 + \dfrac{1}{2 + \dfrac{2}{3}}}, \quad 2 + \frac{1}{1 + \dfrac{1}{2 + \dfrac{2}{3 + \dfrac{3}{4}}}}$$

55. The expressions form a pattern. Continue the pattern two more times. Then simplify all five expressions. $\dfrac{8}{3}, \dfrac{30}{11}, \dfrac{144}{53}, \dfrac{280}{103}, \dfrac{5760}{2119}$

EXTRA CHALLENGE
www.mcdougallittell.com

56. The expressions are getting closer and closer to some value. What is it? $e \approx 2.718$

MIXED REVIEW

SOLVING LINEAR EQUATIONS Solve the equation. (Review 1.3 for 9.6)

57. $\frac{1}{2}x - 7 = 5$ **24**
58. $6 - \frac{1}{10}x = -1$ **70**
59. $\frac{3}{4}x + \frac{1}{2} = x - \frac{5}{6}$ $\frac{16}{3}$

60. $\frac{3}{8}x + 4 = -8$ **−32**
61. $-\frac{1}{12}x - 3 = \frac{5}{2}$ **−66**
62. $2 = -\frac{4}{3}x + 10$ **6**

63. $-5x - \frac{3}{4}x = \frac{51}{2}$ $-\frac{102}{23}$
64. $2x + \frac{7}{8}x = -23$ **−8**
65. $x = 12 + \frac{5}{6}x$ **72**

SOLVING QUADRATIC EQUATIONS Solve the equation. (Review 5.2, 5.3 for 9.6)

66. $x^2 - 5x - 24 = 0$ **−3, 8**
67. $5x^2 - 8 = 4(x^2 + 3)$
68. $6x^2 + 13x - 5 = 0$ $-\frac{5}{2}, \frac{1}{3}$

69. $3(x - 5)^2 = 27$ **2, 8**
70. $2(x + 7)^2 - 1 = 49$ **−12, −2**
71. $2x(x + 6) = 7 - x$ $-7, \frac{1}{2}$

9.5 Addition, Subtraction, and Complex Fractions **567**

DAILY HOMEWORK QUIZ

📽 *Transparency Available*

Find the least common denominator.

1. $\dfrac{7}{x-3}, \dfrac{x}{2(x+3)}$ $2(x^2 - 9)$

2. $\dfrac{-5}{6x^2}, \dfrac{x^2}{2(x+3)}$ $6x^2(x+3)$

Perform the indicated operation and simplify.

3. $\dfrac{x}{x-4} + \dfrac{5}{x-4}$ $\dfrac{x+5}{x-4}$

4. $\dfrac{5}{2x^2} - \dfrac{3}{4x}$ $\dfrac{10-3x}{4x^2}$

5. $\dfrac{x+1}{x^2-x-6} + \dfrac{5}{x+2}$ $\dfrac{2(3x-7)}{(x+2)(x-3)}$

Simplify the complex fraction.

6. $\dfrac{\dfrac{2x}{3} - 2}{1 + \dfrac{12}{x}}$ $\dfrac{2x(x-3)}{3(x+12)}$

EXTRA CHALLENGE NOTE

↳ Challenge problems for Lesson 9.5 are available in **blackline** format in the *Chapter 9 Resource Book*, p. 77 and at **www.mcdougallittell.com**.

ADDITIONAL TEST PREPARATION

1. WRITING Describe how adding rational expressions is like adding numerical fractions. The procedure is the same. Find the LCD of the fractions. Express the original fractions as equivalent fractions with the LCD as the denominator. Add the numerators and simplify.

LESSON OPENER
GRAPHING CALCULATOR
An alternative way to approach
Lesson 9.6 is to use the Graphing
Calculator Lesson Opener:
- Blackline Master (*Chapter 9
  Resource Book*, p. 81)
- Transparency (p. 64)

MEETING INDIVIDUAL NEEDS
- *Chapter 9 Resource Book*
  Prerequisite Skills Review (p. 5)
  Practice Level A (p. 82)
  Practice Level B (p. 83)
  Practice Level C (p. 84)
  Reteaching with Practice (p. 85)
  Absent Student Catch-Up (p. 87)
  Challenge (p. 90)
- *Resources in Spanish*
- Personal Student Tutor

NEW-TEACHER SUPPORT
See the Tips for New Teachers on
pp. 1–2 of the *Chapter 9 Resource
Book* for additional notes about
Lesson 9.6.

WARM-UP EXERCISES

Transparency Available

Solve the equation.

1. $(x-1)(x+2) = 0$ $-2, 1$
2. $6x = -18$ -3
3. $4 - 2x = 12$ -4
4. $x^2 - 16 = 9$ $-5, 5$
5. $x^3 - 25x = 0$ $0, -5, 5$

9.6 Solving Rational Equations

What you should learn

GOAL 1 Solve rational equations.

GOAL 2 Use rational equations to solve **real-life** problems, such as finding how to dilute an acid solution in **Example 5**.

Why you should learn it

▼ To solve **real-life** problems, such as finding the year in which a certain amount of rodeo prize money was earned in **Example 6**.

STUDENT HELP

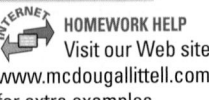
HOMEWORK HELP
Visit our Web site
www.mcdougallittell.com
for extra examples.

GOAL 1 SOLVING A RATIONAL EQUATION

To solve a rational equation, multiply each term on both sides of the equation by the LCD of the terms. Simplify and solve the resulting polynomial equation.

EXAMPLE 1 *An Equation with One Solution*

Solve: $\dfrac{4}{x} + \dfrac{5}{2} = -\dfrac{11}{x}$

SOLUTION

The least common denominator is $2x$.

$\dfrac{4}{x} + \dfrac{5}{2} = -\dfrac{11}{x}$	Write original equation.
$2x\left(\dfrac{4}{x} + \dfrac{5}{2}\right) = 2x\left(-\dfrac{11}{x}\right)$	Multiply each side by $2x$.
$8 + 5x = -22$	Simplify.
$5x = -30$	Subtract 8 from each side.
$x = -6$	Divide each side by 5.

▶ The solution is -6. Check this in the original equation.

EXAMPLE 2 *An Equation with an Extraneous Solution*

Solve: $\dfrac{5x}{x-2} = 7 + \dfrac{10}{x-2}$

SOLUTION

The least common denominator is $x - 2$.

$$\dfrac{5x}{x-2} = 7 + \dfrac{10}{x-2}$$

$$(x-2) \cdot \dfrac{5x}{x-2} = (x-2) \cdot 7 + (x-2) \cdot \dfrac{10}{x-2}$$

$$5x = 7(x-2) + 10$$

$$5x = 7x - 4$$

$$x = 2$$

▶ The solution appears to be 2. After checking it in the original equation, however, you can conclude that 2 is an extraneous solution because it leads to division by zero. So, the original equation has no solution.

EXAMPLE 3 *An Equation with Two Solutions*

Solve: $\dfrac{4x+1}{x+1} = \dfrac{12}{x^2-1} + 3$

SOLUTION

Write each denominator in factored form. The LCD is $(x+1)(x-1)$.

$$\frac{4x+1}{x+1} = \frac{12}{(x+1)(x-1)} + 3$$

$$(x+1)(x-1)\cdot\frac{4x+1}{x+1} = (x+1)(x-1)\cdot\frac{12}{(x+1)(x-1)} + (x+1)(x-1)\cdot 3$$

$$(x-1)(4x+1) = 12 + 3(x+1)(x-1)$$

$$4x^2 - 3x - 1 = 12 + 3x^2 - 3$$

$$x^2 - 3x - 10 = 0$$

$$(x+2)(x-5) = 0$$

$$x+2 = 0 \quad \text{or} \quad x-5 = 0$$

$$x = -2 \quad \text{or} \quad x = 5$$

▶ The solutions are -2 and 5. Check these in the original equation.

· · · · · · · · · ·

You can use **cross multiplying** to solve a simple rational equation for which each side of the equation is a single rational expression.

EXAMPLE 4 *Solving an Equation by Cross Multiplying*

Solve: $\dfrac{2}{x^2-x} = \dfrac{1}{x-1}$

SOLUTION

$\dfrac{2}{x^2-x} = \dfrac{1}{x-1}$	**Write original equation.**
$2(x-1) = 1(x^2-x)$	**Cross multiply.**
$2x - 2 = x^2 - x$	**Simplify.**
$0 = x^2 - 3x + 2$	**Write in standard form.**
$0 = (x-2)(x-1)$	**Factor.**
$x = 2 \text{ or } x = 1$	**Zero product property**

▶ The solutions appear to be 2 and 1. After checking them in the original equation, however, you see that the only solution is 2. The apparent solution $x = 1$ is extraneous. A graphic check shows that the graphs of the left and right sides of the equation, $y = \dfrac{2}{x^2-x}$ and $y = \dfrac{1}{x-1}$, intersect only at $x = 2$. At $x = 1$, the graphs have a common vertical asymptote.

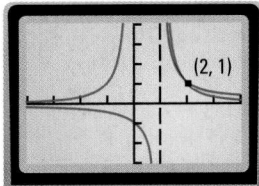
(2, 1)

MOTIVATING THE LESSON
Tell students that if baseball players want to increase their averages, solving rational equations would tell them how many hits they needed. Solving rational equations is the focus of the lesson.

MATHEMATICAL REASONING
EXAMPLE 2 An extraneous solution may be introduced if both sides of an equation are multiplied by factors that use restricted domain values of the original equation. Ask students to verify that 2 is extraneous because x cannot equal 2 in the original equation.

EXTRA EXAMPLE 1
Solve: $\dfrac{3}{x} - \dfrac{1}{2} = \dfrac{12}{x}$ -18

EXTRA EXAMPLE 2
Solve: $\dfrac{5x}{x+1} = 4 - \dfrac{5}{x+1}$
no solution

EXTRA EXAMPLE 3
Solve: $\dfrac{3x-2}{x-2} = \dfrac{6}{x^2-4} + 1$ $-3, 1$

EXTRA EXAMPLE 4
Solve: $\dfrac{3}{x^2+4x} = \dfrac{1}{x+4}$ 3

 CHECKPOINT EXERCISES
For use after Examples 1 and 2:
1. Solve: $\dfrac{2x}{x+3} = 1 - \dfrac{6}{x+3}$
no solution

For use after Example 3:
2. Solve: $\dfrac{2x+1}{x-4} = \dfrac{16}{x^2-16} + 3$
12, -3

For use after Example 4:
3. Solve: $\dfrac{6}{2x^2+2x} = \dfrac{x-2}{x+1}$ 3

EXTRA EXAMPLE 5
You have 1.4 liters of an acid solution whose acid concentration is 2.1 moles per liter. You want to dilute the solution with water so that its acid concentration is 1.5 moles per liter. How much water should you add to the solution? **0.56 L**

EXTRA EXAMPLE 6
In economics, an increasing supply curve means that as prices increase, sellers usually increase production. A decreasing demand curve means that as prices increase, consumers buy less. Suppose that a market situation is modeled by the following equations:
Supply: $y = 8 + 0.2x^2$
Demand: $y = \dfrac{40}{1 + 0.2x}$
Use the *Intersection* feature of a graphing calculator to determine the equilibrium price—the price at which the supply equals the demand. **about $6.73**

 CHECKPOINT EXERCISES
For use after Example 5:
1. You have 3.2 liters of a 54% acid solution. You want to strengthen the solution with pure acid so that its concentration is 75%. How much acid should you add to the solution? **2.7 L**

For use after Example 6:
2. Use a graph of the rational model $y = \dfrac{50x - 20}{x^2 + 18x - 1}$ to find the value of x when $y = 1.2$. **0.67**

FOCUS ON VOCABULARY
When is cross multiplying appropriate to solve a rational equation? Give an example. *Sample answer:* when the equation is of the form: fraction = fraction; $\dfrac{3x - 2}{x + 1} = \dfrac{4}{x - 1}$

CLOSURE QUESTION
Describe how to solve the rational equation $\dfrac{x - 4}{x - 1} + 2 = \dfrac{x + 3}{x + 1}$. **See margin.**

 CHEMICAL ENGRAVING
Acid mixtures are used to engrave, or *etch*, electronic circuits on silicon wafers. A wafer like the one shown above can be cut into as many as 1000 computer chips.

GOAL 2 USING RATIONAL EQUATIONS IN REAL LIFE

EXAMPLE 5 *Writing and Using a Rational Model*

CHEMISTRY You have 0.2 liter of an acid solution whose acid concentration is 16 moles per liter. You want to dilute the solution with water so that its acid concentration is only 12 moles per liter. How much water should you add to the solution?

SOLUTION

VERBAL MODEL

$$\boxed{\text{Concentration of new solution}} = \dfrac{\boxed{\text{Moles of acid in original solution}}}{\boxed{\text{Volume of original solution}} + \boxed{\text{Volume of water added}}}$$

LABELS

Concentration of new solution $= 12$ (moles per liter)

Moles of acid in original solution $= 16(0.2)$ (moles)

Volume of original solution $= 0.2$ (liters)

Volume of water added $= x$ (liters)

ALGEBRAIC MODEL

$$12 = \dfrac{16(0.2)}{0.2 + x} \qquad \text{Write equation.}$$

$$12(0.2 + x) = 16(0.2) \qquad \text{Multiply each side by } 0.2 + x.$$

$$2.4 + 12x = 3.2 \qquad \text{Simplify.}$$

$$12x = 0.8 \qquad \text{Subtract 2.4 from each side.}$$

$$x \approx 0.067 \qquad \text{Divide each side by 12.}$$

▶ You should add about 0.067 liter, or 67 milliliters, of water.

EXAMPLE 6 *Using a Rational Model*

RODEOS From 1980 through 1997, the total prize money P (in millions of dollars) at Professional Rodeo Cowboys Association events can be modeled by

$$P = \dfrac{380t + 5}{-t^2 + 31t + 1}$$

where t represents the number of years since 1980. During which year was the total prize money about $20 million? ▶ Source: Professional Rodeo Cowboys Association

STUDENT HELP

→ **Study Tip**
Example 6 can also be solved by setting the expression for P equal to 20 and solving the resulting equation algebraically.

SOLUTION

Use a graphing calculator to graph the equation $y = \dfrac{380x + 5}{-x^2 + 31x + 1}$. Then graph the line $y = 20$.

Use the *Intersect* feature to find the value of x that gives a y-value of 20. As shown at the right, this value is $x \approx 12$. So, the total prize money was about $20 million 12 years after 1980, in 1992.

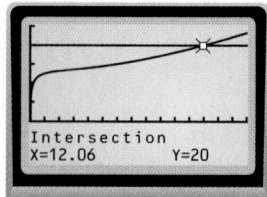

Intersection
X=12.06   Y=20

Closure Question *Sample answer:*
Find the LCD $(x - 1)(x + 1)$ of all fractions. Multiply each term by the LCD. Simplify and solve the resulting rational equation to get $-\dfrac{1}{2}$, 3.

GUIDED PRACTICE

Vocabulary Check ✔

1. Give an example of a rational equation that can be solved using cross multiplication. *Sample answer:* $\dfrac{3}{x+1} = \dfrac{5x}{x+8}$

Concept Check ✔

2. A student solved the equation $\dfrac{2}{x-3} = \dfrac{x}{x-3}$ and got the solutions 2 and 3. Which, if either, of these is extraneous? Explain how you know.
   3 is extraneous; It makes both fractions undefined.

3. Describe two methods that can be used to solve a rational equation. Which method can always be used? Why? **See margin.**

4. $x = 2$; The graphs do not intersect at the extraneous solutions.

4. Solve the equation $\dfrac{1}{x} = \dfrac{2}{x^2}$. Check the apparent solutions graphically. Explain how a graph can help you identify actual and extraneous solutions.

Skill Check ✔ Solve the equation using any method. Check each solution.

5. $\dfrac{7}{x} + \dfrac{3}{4} = \dfrac{5}{x}$ $-\dfrac{8}{3}$

6. $\dfrac{x-2}{6} = \dfrac{x-2}{x-1}$ 2, 7

7. $3x + \dfrac{x}{3} = 5$ $\dfrac{3}{2}$

8. $\dfrac{x}{x-3} = 2 - \dfrac{2}{x-3}$ 8

9. $\dfrac{5}{x-3} = \dfrac{2x}{x^2-9}$ -5

10. $\dfrac{5x}{x-1} + 5 = \dfrac{15}{x-1}$ 2

11. $\dfrac{2x}{x+3} = \dfrac{3x}{x-3}$ $-15, 0$

12. $\dfrac{2x}{x-4} = \dfrac{8}{x-4} + 3$ no solution

13. $\dfrac{2x}{2x+4} = \dfrac{3x}{x+2}$ 0

14. 🌐 **BASKETBALL STATISTICS** So far in the basketball season you have made 12 free throws out of the 20 free throws you have attempted, for a free-throw shooting percentage of 60%. How many consecutive free-throw shots would you have to make to raise your free-throw shooting percentage to 80%?   20

PRACTICE AND APPLICATIONS

STUDENT HELP

▶ **Extra Practice**
to help you master skills is on p. 953.

CHECKING SOLUTIONS Determine whether the given x-value is a solution of the equation.

15. $\dfrac{2x-3}{x+3} = \dfrac{3x}{x+4}$; $x = -1$ no

16. $\dfrac{x}{2x+1} = \dfrac{5}{4-x}$; $x = -1$ yes

17. $\dfrac{4x-3}{x-4} + 1 = \dfrac{x}{x-3}$; $x = 2$ no

18. $\dfrac{3x}{x-6} = 5 + \dfrac{18}{x-6}$; $x = 6$ no

19. $\dfrac{x}{x-3} = \dfrac{6}{x-3}$; $x = 6$ yes

20. $\dfrac{2}{x(x+2)} + \dfrac{3}{x} = \dfrac{4}{x-2}$; $x = 2$ no

STUDENT HELP

▶ **HOMEWORK HELP**
Examples 1–3: Exs. 15–32, 42–50
Example 4: Exs. 15, 16, 19, 33–50
Examples 5, 6: Exs. 54–59

LEAST COMMON DENOMINATOR Solve the equation by using the LCD. Check each solution.

21. $\dfrac{3}{2} + \dfrac{1}{x} = 2$ 2

22. $\dfrac{3}{x} + x = 4$ 1, 3

23. $\dfrac{3}{2x} - \dfrac{9}{2} = 6x$ $-1, \dfrac{1}{4}$

24. $\dfrac{8}{x+2} + \dfrac{8}{2} = 5$ 6

25. $\dfrac{3x}{x+1} + \dfrac{6}{2x} = \dfrac{7}{x}$ $-\dfrac{2}{3}, 2$

26. $\dfrac{2}{3x} + \dfrac{2}{3} = \dfrac{8}{x+6}$ 2, 3

27. $\dfrac{6x}{x+4} + 4 = \dfrac{2x+2}{x-1}$ $-\dfrac{3}{2}, 2$

28. $\dfrac{x-3}{x-4} + 4 = \dfrac{3x}{x}$ $\dfrac{7}{2}$

29. $\dfrac{7x+1}{2x+5} + 1 = \dfrac{10x-3}{3x}$ $\dfrac{5}{7}, 3$

30. $\dfrac{10}{x^2-2x} + \dfrac{4}{x} = \dfrac{5}{x-2}$ no solution

31. $\dfrac{4(x-1)}{x-1} = \dfrac{2x-2}{x+1}$ -3

32. $\dfrac{2(x+7)}{x+4} - 2 = \dfrac{2x+20}{2x+8}$ no solution

---

3 APPLY

⚪ **ASSIGNMENT GUIDE**

BASIC
Day 1: pp. 571–574 Exs. 15–18, 21–27, 33–38, 42–46 even, 58–61, 63–75 odd, Quiz 2 Exs. 1–9

AVERAGE
Day 1: pp. 571–574 Exs. 15–20, 22–32 even, 33–38, 42–46 even, 51–53, 58–61, 63–75 odd, Quiz 2 Exs. 1–9

ADVANCED
Day 1: pp. 571–574 Exs. 15–20, 22–50 even, 51–53, 57–62, 63–75 odd, Quiz 2 Exs. 1–9

BLOCK SCHEDULE
pp. 571–574 Exs. 15–20, 22–32 odd, 33–38, 42–46 even, 51–53, 58–61, 63–75 odd, Quiz 2 Exs. 1–9 (with 9.5)

EXERCISE LEVELS
Level A: *Easier*
15–22, 33–43
Level B: *More Difficult*
23–32, 44–56, 58–61
Level C: *Most Difficult*
57, 62

✔ **HOMEWORK CHECK**
To quickly check student understanding of key concepts, go over the following exercises: Exs. 16, 18, 22, 34, 44. See also the Daily Homework Quiz:

• Blackline Master (*Chapter 10 Resource Book*, p. 11)
• 🖥 Transparency (p. 75)

3. See Additional Answers beginning on page AA1.

TEACHING TIPS
EXERCISE 54 Have students graph the model $y = \frac{326 + x}{575 + x}$ and find the intersection with the graph of $y = \frac{4763}{7989}$. Since this number is less than 1, a good viewing window has $0 \le y \le 1$.

DATA UPDATE
Updated data for Exercise 54 is available at **www.mcdougallittell.com**.

! COMMON ERROR
EXERCISE 55 Students may need help modeling mathematical, real-world situations. Have them read the problem to find what they are being asked to find. Have them define x as this value. Now direct them to model the information they are given using x.

APPLICATION NOTE
EXERCISE 56 Direct students to work from the model $d = (r + c)t$ and $d = (r - c)\left(\frac{53}{60} - t\right)$, where d is the distance (5 miles), r is the rate in still water, (12 miles per hour), c is the rate of the current, and t is the time in hours. Students substitute the known values and solve for c in each equation. Then they graph to find the intersection point.

ADDITIONAL PRACTICE AND RETEACHING

For Lesson 9.6:
- Practice Levels A, B, and C (*Chapter 9 Resource Book*, p. 82)
- Reteaching with Practice (*Chapter 9 Resource Book*, p. 85)
- ⊞ See Lesson 9.6 of the *Personal Student Tutor*

For more Mixed Review:
- ⊞ Search the *Test and Practice Generator* for key words or specific lessons.

572

CROSS MULTIPLYING Solve the equation by cross multiplying. Check each solution.

33. $\frac{3}{4x} = \frac{5}{x + 2}$ $\frac{6}{17}$

34. $\frac{-3}{x + 1} = \frac{4}{x - 1}$ $-\frac{1}{7}$

35. $\frac{x}{x^2 - 8} = \frac{2}{x}$ $-4, 4$

36. $\frac{x}{2x + 7} = \frac{x - 5}{x - 1}$ $-5, 7$

37. $\frac{-2}{x - 1} = \frac{x - 8}{x + 1}$ $2, 5$

38. $\frac{2(x - 2)}{x^2 - 10x + 16} = \frac{2}{x + 2}$ no solution

39. $\frac{8(x - 1)}{x^2 - 4} = \frac{4}{x - 2}$ 4

40. $\frac{x^2 - 3}{x + 2} = \frac{x - 3}{2}$ $-1, 0$

41. $\frac{-1}{x - 3} = \frac{x - 4}{x^2 - 27}$ $-\frac{3}{2}, 5$

CHOOSING A METHOD Solve the equation using any method. Check each solution.

42. $\frac{x - 2}{x + 2} = \frac{3}{x}$ $-1, 6$

43. $\frac{3}{x + 2} = \frac{6}{x - 1}$ -5

44. $\frac{3x}{x + 1} = \frac{12}{x^2 - 1} + 2$ $-2, 5$

45. $\frac{3x + 6}{x^2 - 4} = \frac{x + 1}{x - 2}$ no solution

46. $\frac{x - 4}{x} = \frac{6}{x^2 - 3x}$ $1, 6$

47. $\frac{2x}{4 - x} = \frac{x^2}{x - 4}$ $-2, 0$

48. $\frac{2x}{x - 3} = \frac{3x}{x^2 - 9} + 2$ -6

49. $\frac{x}{2x - 6} = \frac{2}{x - 4}$ $2, 6$

50. $\frac{2}{x + 1} + \frac{x}{x - 1} = \frac{2}{x^2 - 1}$ -4

LOGICAL REASONING In Exercises 51–53, a is a nonzero real number. Tell whether the algebraic statement is *always true*, *sometimes true*, or *never true*. Explain your reasoning.

51. always; When you solve by cross multiplying, you get $x = 1$ or $x = a$ and $x = a$ makes both fractions undefined.

52. sometimes; When $a = x$, the equation has no solutions.

53. always; When you multiply each side of the equation by $x^2 - a^2$ you get $x = a$, making the fractions undefined.

51. For the equation $\frac{1}{x - a} = \frac{x}{x - a}$, $x = a$ is an extraneous solution.

52. The equation $\frac{3}{x - a} = \frac{x}{x - a}$ has exactly one solution.

53. The equation $\frac{1}{x - a} = \frac{2}{x + a} + \frac{2a}{x^2 - a^2}$ has no solution.

54. 🌐 **FOOTBALL STATISTICS** At the end of the 1998 season, the National Football League's all-time leading passer during regular season play was Dan Marino with 4763 completed passes out of 7989 attempts. In his debut 1998 season, Peyton Manning made 326 completed passes out of 575 attempts. How many consecutive completed passes would Peyton Manning have to make to equal Dan Marino's pass completion percentage?   **42**

   DATA UPDATE of National Football League data at www.mcdougallittell.com

55. 🌐 **PHONE CARDS** A telephone company offers you an opportunity to sell prepaid, 30 minute long-distance phone cards. You will have to pay the company a one-time setup fee of $200. Each phone card will cost you $5.70. How many cards would you have to sell before your average total cost per card falls to $8?   **87**

56. 🌐 **RIVER CURRENT** It takes a paddle boat 53 minutes to travel 5 miles up a river and 5 miles back, going at a steady speed of 12 miles per hour (with respect to the water). Find the speed of the current.   **2.85 mi/hr**

57. **BIOLOGY ▶ CONNECTION** The number f of flies eaten by a praying mantis in 8 hours can be modeled by

$$f = \frac{26.6d}{d + 0.0017}$$

where d is the density of flies available (in flies per cubic centimeter). Approximate the density of flies (in flies per cubic *meter*) when a praying mantis eats 15 flies in 8 hours. (*Hint:* There are 1,000,000 cm³ in 1 m³.) ▶ Source: *Biology by Numbers*   **about 2,198 flies/m³**

The praying mantis blends in with its environment.

 **FUEL EFFICIENCY** In Exercises 58 and 59, use the following information.
The cost of fueling your car for one year can be calculated using this equation:

$$\text{Fuel cost for one year} = \frac{\text{Miles driven} \times \text{Price per gallon of fuel}}{\text{Fuel efficiency rate}}$$

58. Last year you drove 9000 miles, paid $1.10 per gallon of gasoline, and spent a total of $412.50 on gasoline. What is the fuel efficiency rate of your car? **24 mi/gallon**

59. How much would you have saved if your car's fuel efficiency rate were 25 miles per gallon? **$16.50**

Test Preparation

QUANTITATIVE COMPARISON In Exercises 60 and 61, choose the statement below that is true about the given quantities.

Ⓐ The quantity in column A is greater.

Ⓑ The quantity in column B is greater.

Ⓒ The two quantities are equal.

Ⓓ The relationship cannot be determined from the given information.

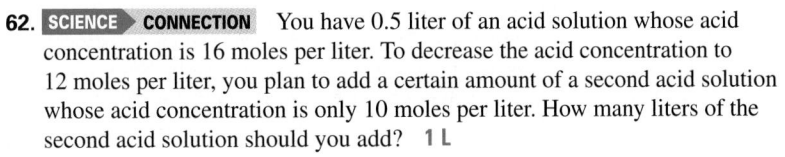

	Column A	Column B	
60.	The solution of $\frac{x^3 + 1}{x} = 2x^2$	The solution of $\frac{-2}{x + 3} = \frac{4}{x - 2}$	A
61.	The solution of $\frac{1}{x} + 3 = \frac{9}{2x}$	The solution of $\frac{1}{2} + \frac{3}{x} = \frac{43}{14}$	C

★ **Challenge**

EXTRA CHALLENGE
www.mcdougallittell.com

62. SCIENCE CONNECTION You have 0.5 liter of an acid solution whose acid concentration is 16 moles per liter. To decrease the acid concentration to 12 moles per liter, you plan to add a certain amount of a second acid solution whose acid concentration is only 10 moles per liter. How many liters of the second acid solution should you add? **1 L**

MIXED REVIEW

SLOPES OF LINES Find the slope of a line parallel to the given line and the slope of a line perpendicular to the given line. (Review 2.4 for 10.1)

63. $y = x + 3$ 1; −1 **64.** $y = 3x - 4$ $3; -\frac{1}{3}$ **65.** $y = -\frac{2}{3}x + 15$ $-\frac{2}{3}, \frac{3}{2}$

66. $y + 3 = 3x + 2$ $3; -\frac{1}{3}$ **67.** $2y - x = 7$ $\frac{1}{2}; -2$ **68.** $4x - 3y = 17$ $\frac{4}{3}; -\frac{3}{4}$

PROPERTIES OF SQUARE ROOTS Simplify the expression. (Review 5.3 for 10.1)

69. $\sqrt{48}$ $4\sqrt{3}$ **70.** $\sqrt{18}$ $3\sqrt{2}$ **71.** $\sqrt{108}$ $6\sqrt{3}$ **72.** $\sqrt{432}$ $12\sqrt{3}$

73. $\sqrt{6} \cdot \sqrt{45}$ $3\sqrt{30}$ **74.** $\sqrt{\frac{16}{72}}$ $\frac{\sqrt{2}}{3}$ **75.** $\sqrt{75} \cdot \sqrt{3}$ 15 **76.** $\sqrt{\frac{8}{49}}$ $\frac{2\sqrt{2}}{7}$

77. 🌎 **GEOLOGY** You can find the pH of a soil by using the formula

$$pH = -\log [H^+]$$

where $[H^+]$ is the soil's hydrogen ion concentration (in moles per liter). Find the pH of a layer of soil that has a hydrogen ion concentration of 1.6×10^{-7} moles per liter. (Review 8.4) **6.796**

9.6 Solving Rational Equations **573**

DAILY HOMEWORK QUIZ

🗐 *Transparency Available*

Solve the equation.

1. $\frac{x}{6} + \frac{x + 4}{2} = \frac{2}{x + 5}$ −2, −6

2. $\frac{4}{x} + 5x = \frac{21}{2}$ $\frac{1}{2}, \frac{8}{5}$

3. $\frac{15}{x} - \frac{5}{2} = \frac{2.5}{x}$ 5

4. $\frac{1}{x + 1} = \frac{2}{3x - 1}$ 3

EXTRA CHALLENGE NOTE

⤷ Challenge problems for Lesson 9.6 are available in **blackline** format in the *Chapter 9 Resource Book,* p. 90 and at **www.mcdougallittell.com.**

ADDITIONAL TEST PREPARATION

1. OPEN ENDED Give an example of a rational equation that has exactly one extraneous solution. *Sample answer:*

$$-\frac{1}{x + 2} = \frac{2}{x(x + 2)}$$

Perform the indicated operation and simplify. (Lessons 9.4 and 9.5)

1. $\dfrac{3x^3y}{2xy^2} \cdot \dfrac{10x^4y^2}{9x}$ $\dfrac{5x^5y}{3}$

2. $\dfrac{x^2 - 3x - 40}{5x} \div (x + 5)$ $\dfrac{x - 8}{5x}$

3. $\dfrac{18x}{x^2 - 5x - 36} + \dfrac{2x}{x + 4}$ $\dfrac{2x^2}{(x - 9)(x + 4)}$

4. $\dfrac{8x^2}{25x^2 - 36} - \dfrac{1}{10x + 12}$ $\dfrac{16x^2 - 5x + 6}{2(5x - 6)(5x + 6)}$

Simplify the complex fraction. (Lesson 9.5)

7. $\dfrac{-3(x - 3)(2x - 1)(2x + 1)}{(x - 1)(x + 1)}$

8. $\dfrac{2x}{(x - 5)(x + 5)}$

5. $\dfrac{\frac{8}{x} + 11}{\frac{1}{6x} - 1}$ $\dfrac{-6(11x + 8)}{6x - 1}$

6. $\dfrac{36 - \frac{1}{x^2}}{\frac{1}{6x^2} - 6}$ -6

7. $\dfrac{\frac{2}{x^2 - 1} - \frac{1}{x + 1}}{\frac{1}{12x^2 - 3}}$

8. $\dfrac{\frac{1}{x - 5} - \frac{x}{x^2 - 25}}{\frac{5}{2x}}$

9. 🌐 **AVERAGE COST** You bought a potholder weaving frame for $10. A bag of potholder material costs $4 and contains enough material to make a dozen potholders. How many dozens of potholders must you make before your average total cost per dozen falls to $4.50? (Lesson 9.6)   **20 dozens**

MATH & History

Deep Water Diving

APPLICATION LINK
www.mcdougallittell.com

THEN

IN 1530 the invention of the diving bell provided the first effective means of breathing underwater. Like many other diving devices, a diving bell uses air that is compressed by the pressure of the water. Because oxygen under high pressure (at great depths) can have toxic effects on the body, the percent of oxygen in the air must be adjusted. The recommended percent p of oxygen (by volume) in the air that a diver breathes is

$$p = \frac{660}{d + 33}$$

where d is the depth (in feet) at which the diver is working.

1. Graph the equation.   **See margin.**

2. At what depth is the recommended percent of oxygen 5%?   **99 ft.**

3. What value does the recommended percent of oxygen approach as a diver's depth increases?   **0%**

NOW

TODAY diving technology makes it easier for scientists like Dr. Sylvia Earle to study ocean life. Using one-person submarines, Earle has undertaken a five-year study of marine sanctuaries.

Sylvia Earle, marine biologist.

1530
The diving bell is invented. It is open to the water at the bottom and traps air at the top.

Augustus Siebe invents the closed hard-hat diving suit.
1837

William Beebe descends 1426 feet in a bathysphere.
1930

Sylvia Earle walks untethered on the ocean floor at a record depth of 1250 feet.
1979

ADDITIONAL RESOURCES
An alternative Quiz for Lessons 9.4–9.6 is available in the *Chapter 9 Resource Book,* p. 91.

A **blackline** master with additional Math & History exercises is available in the *Chapter 9 Resource Book,* 89.

1.

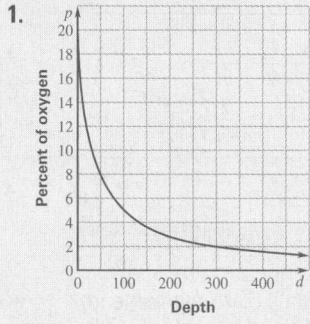

Percent of oxygen / Depth

Chapter Summary

WHAT did you learn?

Write and use variation models.
- inverse variation **(9.1)**
- joint variation **(9.1)**

Graph rational functions.
- simple rational functions **(9.2)**

- general rational functions **(9.3)**

Perform operations with rational expressions.
- multiply and divide **(9.4)**
- add and subtract **(9.5)**

Simplify complex fractions. **(9.5)**

Solve rational equations. **(9.6)**

Use rational models to solve real-life problems.
(9.1–9.6)

WHY did you learn it?

Find the speed of a whirlpool's current. **(p. 535)**
Find the heat loss through a window. **(p. 539)**

Describe the frequency of an approaching ambulance siren. **(p. 545)**
Find the energy expenditure of a parakeet. **(p. 551)**

Compare the velocities of two skydivers. **(p. 557)**
Write a model for the number of male college graduates in the United States. **(p. 566)**

Write a simplified model for the focal length of a camera lens. **(p. 564)**

Find the amount of water to add when diluting an acid solution. **(p. 570)**

Find the year in which a certain amount of rodeo prize money was earned. **(p. 570)**

How does Chapter 9 fit into the BIGGER PICTURE of algebra?

In Chapter 9 you studied rational functions. A rational function is the ratio of two polynomial functions, which you studied in Chapter 2 (linear functions), Chapter 5 (quadratic functions), and Chapter 6 (polynomial functions).

A hyperbola is the graph of one important type of rational function. In the next chapter you will learn more about hyperbolas, parabolas, circles, and ellipses, which together are called the conic sections.

STUDY STRATEGY

How did you make and use a dictionary of functions?

Here is an example of one entry in a dictionary of functions, following the **Study Strategy** on page 532.

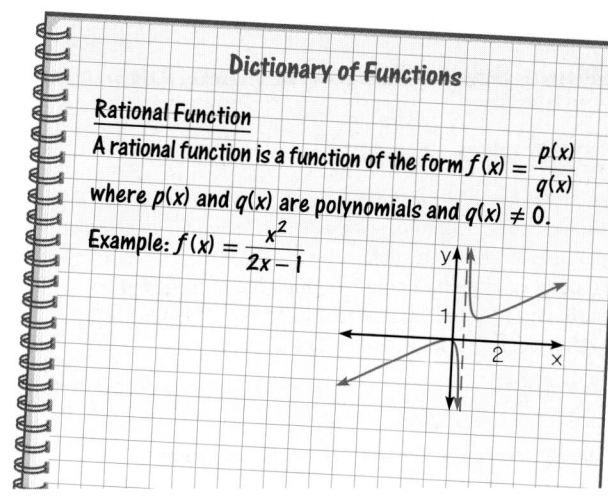

Dictionary of Functions

Rational Function

A rational function is a function of the form $f(x) = \dfrac{p(x)}{q(x)}$ where $p(x)$ and $q(x)$ are polynomials and $q(x) \neq 0$.

Example: $f(x) = \dfrac{x^2}{2x-1}$

575

The following resources are available to help review the material in this chapter.

- Chapter Review Games and Activities (*Chapter 9 Resource Book*, p. 92)
- *Instant Replay: Video Review Games*
- Personal Student Tutor

8.

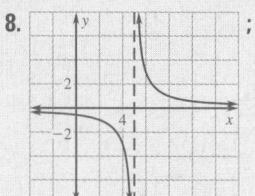

domain: all real numbers except 5; range: all real numbers except 0

9.

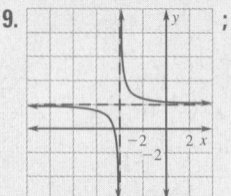

domain: all real numbers except −4; range: all real numbers except 2

10.

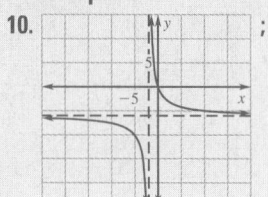

domain: all real numbers except −2; range: all real numbers except −6

11.

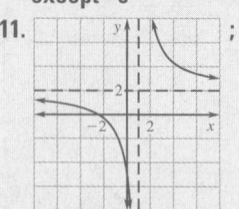

domain: all real numbers except 1; range: all real numbers except 2

CHAPTER 9

Chapter Review

VOCABULARY

- inverse variation, p. 534
- constant of variation, p. 534
- joint variation, p. 536
- rational function, p. 540
- hyperbola, p. 540
- branches of a hyperbola, p. 540
- simplified form of a rational expression, p. 554
- complex fraction, p. 564
- cross multiplying, p. 569

9.1 INVERSE AND JOINT VARIATION

Examples on pp. 534–536

EXAMPLES You can write an inverse or joint variation equation using a general equation for the variation and given values of the variables.

Inverse variation: $x = 5$, $y = 4$

$y = \dfrac{k}{x}$ *y varies inversely with x.*

$4 = \dfrac{k}{5}$ *Substitute for x and y.*

$20 = k$ *Solve for k.*

The inverse variation equation is $y = \dfrac{20}{x}$.

Joint variation: $x = 3$, $y = 8$, $z = 30$

$z = kxy$ *z varies jointly with x and y.*

$30 = k(3)(8)$ *Substitute for x, y, and z.*

$30 = 24k$ *Multiply.*

$k = \dfrac{30}{24} = \dfrac{5}{4}$ *Solve for k.*

The joint variation equation is $z = \dfrac{5}{4}xy$.

The variables *x* and *y* vary inversely. Use the given values to write an equation relating *x* and *y*. Then find *y* when *x* = 2.

1. $x = 1$, $y = 5$ $y = \dfrac{5}{x}$; 2.5

2. $x = 15$, $y = \dfrac{2}{3}$ $y = \dfrac{10}{x}$; 5

3. $x = \dfrac{1}{4}$, $y = 8$ $y = \dfrac{2}{x}$; 1

4. $x = -2$, $y = 2$ $y = -\dfrac{4}{x}$; −2

The variable *z* varies jointly with *x* and *y*. Use the given values to write an equation relating *x*, *y*, and *z*. Then find *z* when *x* = 5 and *y* = −6.

5. $x = 1$, $y = 12$, $z = 4$ $z = \dfrac{1}{3}xy$; −10

6. $x = 6$, $y = 8$, $z = -6$ $z = -\dfrac{1}{8}xy$; $\dfrac{15}{4}$

7. $x = \dfrac{3}{4}$, $y = 4$, $z = 9$ $z = 3xy$; −90

9.2 GRAPHING SIMPLE RATIONAL FUNCTIONS

Examples on pp. 540–542

EXAMPLE 1 To graph $y = \dfrac{1}{x+2} + 3$, note that the asymptotes are $x = -2$ and $y = 3$. Plot two points to the left of the vertical asymptote, such as $(-3, 2)$ and $(-4, 2.5)$, and two points to the right, such as $(-1, 4)$ and $(0, 3.5)$. Use the asymptotes and plotted points to draw the branches of the hyperbola. The domain is all real numbers except −2, and the range is all real numbers except 3.

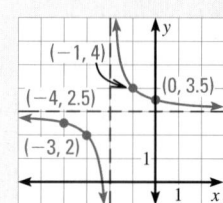

EXAMPLE 2 To graph $y = \dfrac{x+1}{x-3}$, note that when the denominator equals zero, $x = 3$. So the vertical asymptote is $x = 3$. The
horizontal asymptote, which occurs at the ratio of the x-coefficients,
is $y = 1$. Plot some points to the left and right of the vertical
asymptote. Use the asymptotes and plotted points to draw the
branches of the hyperbola. The domain is all real numbers except 3,
and the range is all real numbers except 1.

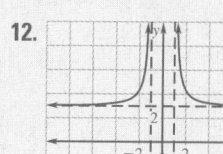

Graph the function. State the domain and range.   8–11. See margin.

8. $y = \dfrac{3}{x-5}$ **9.** $y = \dfrac{1}{x+4} + 2$ **10.** $y = \dfrac{-6x}{x+2}$ **11.** $y = \dfrac{2x+5}{x-1}$

9.3   GRAPHING GENERAL RATIONAL FUNCTIONS

Examples on pp. 547–549

EXAMPLE To graph $y = \dfrac{3x^2}{x+2}$, note that the numerator has 0 as its only real zero,
so the graph has one x-intercept at $(0, 0)$. The only zero of the denominator is -2, so the
only vertical asymptote is $x = -2$. The degree of the numerator is greater than the
degree of the denominator, so there is no horizontal asymptote.

	x	y
Plot some points to the left of $x = -2$.	-8	-32
	-4	-24
	-3	-27
Plot some points to the right of $x = -2$.	-1	3
	2	3
	4	8

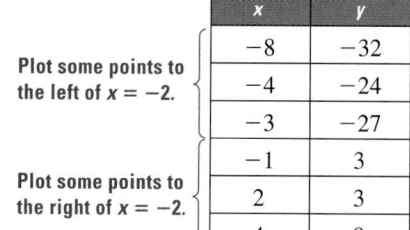
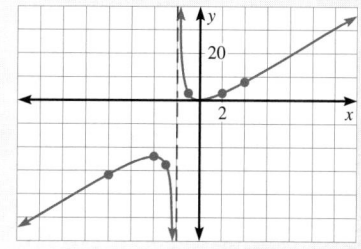

Graph the function.   12–15. See margin.

12. $y = \dfrac{3x^2 + 1}{x^2 - 1}$ **13.** $y = \dfrac{x^3}{10}$ **14.** $y = \dfrac{x}{x^2 - 4}$ **15.** $y = \dfrac{3x^2 - 4x + 1}{x^2 - 2x - 3}$

9.4   MULTIPLYING AND DIVIDING RATIONAL EXPRESSIONS

Examples on pp. 554–557

EXAMPLE Dividing rational expressions is like dividing numerical fractions.

$$\frac{x^2 - 9}{5(x+2)} \div \frac{x-3}{5(x^2-4)} = \frac{x^2-9}{5(x+2)} \cdot \frac{5(x^2-4)}{x-3}$$ Multiply by reciprocal.

$$= \frac{(x+3)(x-3)(5)(x+2)(x-2)}{5(x+2)(x-3)}$$ Factor and divide out common factors.

$$= (x+3)(x-2)$$ Simplified form

Perform the indicated operation(s). Simplify the result.

16. $\dfrac{x^2 - 3x}{4x^2 - 8x} \cdot (4x^2 - 16)$
$(x-3)(x+2)$

17. $5x \div \dfrac{1}{x-6} \cdot \dfrac{x^2-9}{x}$
$5(x-6)(x+3)(x-3)$

18. $\dfrac{x^2 - 2x - 3}{x+1} \div \dfrac{x^2 + x - 12}{x^2} \cdot \dfrac{1}{\dfrac{x^2 - x - 4}{x+4}}$

Chapter Review **577**

12.
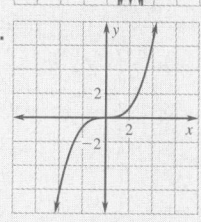

13.
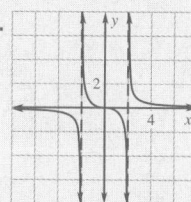

14.
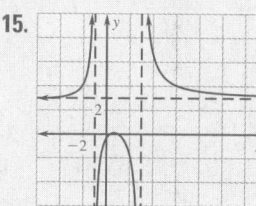

15.
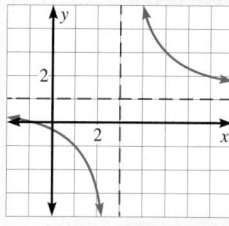

8.

9.

ADDITION, SUBTRACTION, AND COMPLEX FRACTIONS

Examples on pp. 562–564

EXAMPLES You can use the LCD to add or subtract rational expressions.

$$\frac{3}{x-3} - \frac{5}{x+2} = \frac{3(x+2)}{(x-3)(x+2)} - \frac{5(x-3)}{(x-3)(x+2)}$$ Rewrite each expression using the LCD.

$$= \frac{3(x+2) - 5(x-3)}{(x-3)(x+2)}$$ Subtract.

$$= \frac{3x+6 - 5x + 15}{(x-3)(x+2)}$$ Multiply.

$$= \frac{-2x+21}{(x-3)(x+2)}$$ Simplified form

To simplify a complex fraction, divide the numerator by the denominator.

$$\frac{\frac{2}{x}+4}{\frac{2x+1}{5x^2}} = \frac{\frac{2+4x}{x}}{\frac{2x+1}{5x^2}} = \frac{2+4x}{x} \cdot \frac{5x^2}{2x+1} = \frac{2(1+2x)(5x^2)}{x(2x+1)} = 10x$$

Perform the indicated operation(s) and simplify.

19. $\dfrac{5}{x^2(x-2)} + \dfrac{x}{x-2}$ $\dfrac{x^3+5}{x^2(x-2)}$ **20.** $\dfrac{x+5}{x-5} - \dfrac{3}{x+5}$

$\dfrac{x^2+7x+40}{(x+5)(x-5)}$

21. $\dfrac{x-2}{5x(x-1)} + \dfrac{1}{x-1} - \dfrac{3x+2}{x^2+4x-5}$

$\dfrac{-9x^2+18x-10}{5x(x-1)(x+5)}$

Simplify the complex fraction.

22. $\dfrac{\frac{x+3}{6}}{1+\frac{x}{3}}$ $\dfrac{1}{2}$ **23.** $\dfrac{\frac{x}{2}-4}{9+\frac{2}{x}}$ $\dfrac{x(x-8)}{2(9x+2)}$ **24.** $\dfrac{\frac{1}{x+1}+\frac{1}{x-1}}{\frac{x}{x+1}}$ $\dfrac{2}{x-1}$ **25.** $\dfrac{\frac{4}{5-x}}{\frac{2}{5-x}+\frac{1}{3x-15}}$ $\dfrac{12}{5}$

SOLVING RATIONAL EQUATIONS

Examples on pp. 568–570

EXAMPLES You can solve rational equations by multiplying each side of the equation by the LCD of the terms. If each side of the equation is a single rational expression, you can use cross multiplying. Check for extraneous solutions.

$$\frac{4}{x} + \frac{3}{2x} = 11$$

$$(2x)\frac{4}{x} + (2x)\frac{3}{2x} = (2x)11$$ Multiply each side by 2x.

$$8 + 3 = 22x$$

$$x = \frac{1}{2}$$

$$\frac{2}{3x+6} = \frac{x+2}{x^2-10}$$

$$2(x^2-10) = (x+2)(3x+6)$$ Cross multiply.

$$2x^2 - 20 = 3x^2 + 12x + 12$$

$$0 = x^2 + 12x + 32$$

$$0 = (x+8)(x+4)$$

$$x = -8 \text{ or } x = -4$$

Solve the equation using any method. Check each solution.

26. $\dfrac{x}{x-1} = \dfrac{2x+10}{x+11}$ $-2, 5$ **27.** $\dfrac{x+3}{x} - 1 = \dfrac{1}{x-1}$ $\dfrac{3}{2}$ **28.** $\dfrac{2}{x-2} - \dfrac{2x}{3} = \dfrac{x-3}{3}$ $0, 3$

29. $\dfrac{3x+2}{x+1} = 2 - \dfrac{2x+3}{x+1}$ no solution **30.** $\dfrac{2}{x-6} = \dfrac{-5}{x+1}$ 4 **31.** $1 + \dfrac{3}{x-3} = \dfrac{4}{x^2-9}$ $-4, 1$

Chapter Test

ADDITIONAL RESOURCES
- *Chapter 9 Resource Book*
  Chapter Test (3 levels) (p. 93)
  SAT/ACT Chapter Test (p. 99)
  Alternative Assessment (p. 100)
- 🖥 *Test and Practice Generator*

The variables x and y vary inversely. Use the given values to write an equation relating x and y. Then find y when $x = 3$.

1. $x = -4, y = 9$ $y = -\dfrac{36}{x}; -12$ **2.** $x = \dfrac{1}{2}, y = 5$ $y = \dfrac{5}{2x}; \dfrac{5}{6}$ **3.** $x = 12, y = \dfrac{2}{3}$ $y = \dfrac{8}{x}; \dfrac{8}{3}$ **4.** $x = 6, y = -1$ $y = -\dfrac{6}{x}; -2$

The variable z varies jointly with x and y. Use the given values to write an equation relating x, y, and z. Then find z when $x = -2$ and $y = 4$.

5. $x = 5, y = 4, z = 2$ $z = \dfrac{1}{10}xy; -\dfrac{4}{5}$ **6.** $x = -3, y = 2, z = 18$ $z = -3xy; 24$ **7.** $x = \dfrac{1}{3}, y = \dfrac{3}{4}, z = \dfrac{5}{2}$ $z = 10xy; -80$

Graph the function.   8–15. See margin.

8. $y = \dfrac{-1}{x + 1} - 2$ **9.** $y = \dfrac{4}{x - 2}$ **10.** $y = \dfrac{x}{2x + 5}$ **11.** $y = \dfrac{4x - 3}{x - 4}$

12. $y = \dfrac{6}{x^2 + 4}$ **13.** $y = \dfrac{-3x^2}{2x - 1}$ **14.** $y = \dfrac{x^2 - 2}{x^2 - 9}$ **15.** $y = \dfrac{x^2 - 2x + 15}{x + 1}$

Perform the indicated operation. Simplify the result.

16. $\dfrac{x^2 - 4}{x + 3} \cdot \dfrac{x^2 + 4x + 3}{2x - 4}$ $\dfrac{(x + 2)(x + 1)}{2}$ **17.** $\dfrac{4x - 8}{x^2 - 3x + 2} \div \dfrac{3x - 6}{x - 1}$ $\dfrac{4}{3(x - 2)}$ **18.** $\dfrac{x + 4}{x^2 - 25} \cdot (x^2 + 3x - 10)$ $\dfrac{(x + 4)(x - 2)}{x - 5}$

19. $\dfrac{5}{6x} + \dfrac{7}{18x}$ $\dfrac{11}{9x}$ **20.** $\dfrac{x - 1}{x - 2} - \dfrac{x - 4}{x + 1}$ $\dfrac{6x - 9}{(x - 2)(x + 1)}$ **21.** $\dfrac{3x}{x^2 - 10x + 21} + \dfrac{5}{x - 3}$ $\dfrac{8x - 35}{(x - 3)(x - 7)}$

Simplify the complex fraction.

22. $\dfrac{1 + \dfrac{3}{x}}{2 - \dfrac{5}{x^2}}$ $\dfrac{x(x + 3)}{2x^2 - 5}$ **23.** $\dfrac{\dfrac{4 + x}{10}}{\dfrac{x^2 - 16}{8}}$ $\dfrac{4}{5(x - 4)}$ **24.** $\dfrac{\dfrac{2}{x - 1} + 5}{\dfrac{x}{3}}$ $\dfrac{3(5x - 3)}{x(x - 1)}$ **25.** $\dfrac{36}{\dfrac{1}{x} + \dfrac{7}{2x}}$ $8x$

Solve the equation using any method. Check each solution.

26. $\dfrac{9}{x} + \dfrac{11}{5} = \dfrac{31}{x}$ 10   **27.** $\dfrac{-15}{x} = \dfrac{x + 16}{4}$ $-10, -6$ **28.** $\dfrac{8}{x + 3} = \dfrac{5}{x - 3}$ 13   **29.** $\dfrac{4x}{x + 3} = \dfrac{37}{x^2 - 9} - 3$ $\dfrac{16}{7}, 4$

30. **SCIENCE CONNECTION**   A lever pivots on a support called a *fulcrum*. For a balanced lever, the distance d (in feet) an object is from the fulcrum varies inversely with the object's weight w (in pounds). An object weighing 140 pounds is placed 6 feet from a fulcrum. How far from the fulcrum must a 112 pound object be placed to balance the lever?   **7.5 ft**

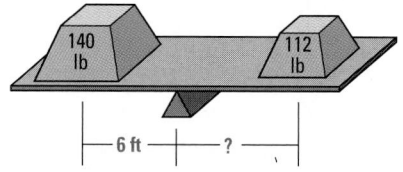

31. **GEOMETRY CONNECTION**   A sphere with radius r is inscribed in a cube as shown. Find the ratio of the volume of the cube to the volume of the sphere. Write your answer in simplified form. $\dfrac{6}{\pi}$

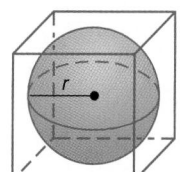

32. 🐝 **STARTING A BUSINESS**   You start a small bee-keeping business, spending $500 for equipment and bees. You figure it will cost $1.25 per pound to collect, clean, bottle, and label the honey. How many pounds of honey must you produce before your average cost per pound is $1.79?   **about 920 lb**

10.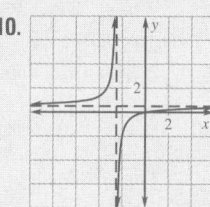

11.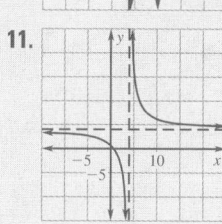

12.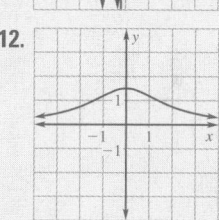

13.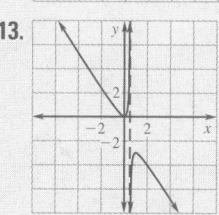

14.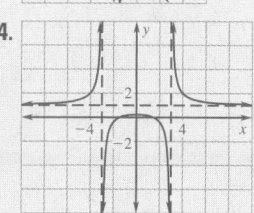

15.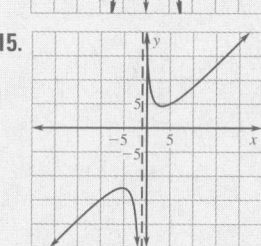

Chapter Test   **579**

▶ **TEST-TAKING STRATEGY** During a test, draw graphs and figures in your test booklet to help you solve problems. Even though you must keep your answer sheet neat, you can make any kind of mark you want in your test booklet.

ADDITIONAL RESOURCES
- *Chapter 9 Resource Book*
  Chapter Test (3 levels) (p. 93)
  SAT/ACT Chapter Test (p. 99)
  Alternative Assessment (p. 100)
- 🖳 *Test and Practice Generator*

1. MULTIPLE CHOICE The variable x varies inversely with y. When $x = 6$, $y = 6.5$. Which equation relates x and y? **A**

Ⓐ $xy = 39$ Ⓑ $xy = 11.5$ Ⓒ $xy = \frac{1}{2}$

Ⓓ $y = \frac{1}{2}x$ Ⓔ $y = 39x$

2. MULTIPLE CHOICE The variable z varies jointly with x and y. When $x = 6$ and $y = \frac{1}{3}$, $z = 30$. Which equation relates x, y, and z? **C**

Ⓐ $z = 30xy$ Ⓑ $30 = xyz$ Ⓒ $z = 15xy$

Ⓓ $z = \frac{1}{30}xy$ Ⓔ $z = \frac{1}{15}xy$

3. MULTIPLE CHOICE Which function is graphed? **B**

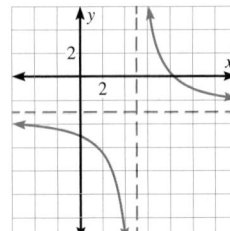

Ⓐ $y = \frac{10}{x+5} - 3$ Ⓑ $y = \frac{10}{x-5} - 3$

Ⓒ $y = \frac{10}{x+5} + 3$ Ⓓ $y = \frac{10}{x-5} + 3$

Ⓔ $y = \frac{10}{x-5}$

4. MULTIPLE CHOICE What is the quotient $(x + 2) \div \dfrac{x^2 - 9x - 22}{x^2 - 121}$? **A**

Ⓐ $x + 11$ Ⓑ $\frac{x+11}{x+2}$ Ⓒ $\frac{x+2}{x+11}$

Ⓓ $\frac{x+2}{x-11}$ Ⓔ $x + 2$

5. MULTIPLE CHOICE What are all the solutions of the equation $\frac{-10}{x-9} = \frac{x}{2}$? **E**

Ⓐ $-4, -5$ Ⓑ $4, -5$ Ⓒ 4

Ⓓ 5 Ⓔ $4, 5$

6. MULTIPLE CHOICE Which function is graphed? **A**

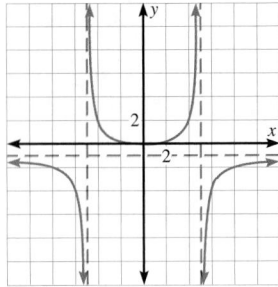

Ⓐ $y = \frac{-x^2}{x^2 - 25}$ Ⓑ $y = \frac{-3x^2}{x^2 - 16}$

Ⓒ $y = \frac{-3x^2}{x^2 - 25}$ Ⓓ $y = \frac{3x^2}{x^2 - 25}$

Ⓔ $y = \frac{-3x^2}{x^2 + 25}$

7. MULTIPLE CHOICE What is the difference $\dfrac{8x - 3}{x^2 + 2x - 35} - \dfrac{7}{x^2 - 25}$? **D**

Ⓐ $\dfrac{2(4x^2 + 15x - 17)}{(x^2 + 2x - 35)(x + 5)}$

Ⓑ $\dfrac{2(4x^2 + 15x + 32)}{(x^2 + 2x - 35)(x + 5)}$

Ⓒ $\dfrac{2(4x^2 + 15x + 17)}{(x^2 + 2x - 35)(x + 5)}$

Ⓓ $\dfrac{2(4x^2 + 15x - 32)}{(x^2 + 2x - 35)(x + 5)}$

Ⓔ $\dfrac{2(4x^2 + 15x - 32)}{(x^2 + 2x - 35)(x^2 - 25)}$

8. MULTIPLE CHOICE What is the simplified form of the following complex fraction? **B**

$$\dfrac{\dfrac{10}{x+1}}{\dfrac{1}{2} + \dfrac{3}{x+1}}$$

Ⓐ $\frac{20x}{x+7}$ Ⓑ $\frac{20}{x+7}$ Ⓒ $\frac{10}{x+7}$

Ⓓ $\frac{10(x+7)}{x+1}$ Ⓔ 20

9. QUANTITATIVE COMPARISON Choose the statement that is true about the given quantities. **B**

Ⓐ The quantity in column A is greater.

Ⓑ The quantity in column B is greater.

Ⓒ The two quantities are equal.

Ⓓ The relationship cannot be determined from the given information.

Column A	Column B
The solution of $\dfrac{x-4}{x+1} = \dfrac{7}{2}$	The solution of $\dfrac{5}{x} - \dfrac{8}{3} = \dfrac{1}{12x}$

10. MULTI-STEP PROBLEM For parts (a)–(d), graph the function and identify the point at which the horizontal and vertical asymptotes intersect. **10a–d. See margin.**

a. $y = \dfrac{2}{x}$ **(0, 0)**   **b.** $y = \dfrac{2}{x-1} + 3$ **(1, 3)**   **c.** $y = \dfrac{2}{x-1} - 3$ **(1, −3)** **d.** $y = \dfrac{2}{x+1} + 3$ **(−1, 3)**

e. Use your answers to parts (a)–(d) to predict the point of intersection of the asymptotes of the graph of $y = \dfrac{2}{x+1} - 3$. Check your prediction by graphing. **(−1, −3)**

f. CRITICAL THINKING Generalize your results for any function of the form $y = \dfrac{a}{x-h} + k$. **The asymptotes will intersect at (h, k).**

11. MULTI-STEP PROBLEM Three tennis balls fit tightly in a can as shown. Recall that the formula for the volume of a cylinder is $V = \pi r^2 h$ and the formula for the volume of a sphere is $V = \dfrac{4}{3}\pi r^3$.

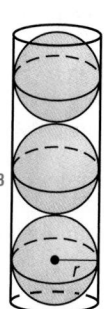

a. Write an expression for the height of the can, h, in terms of r. Rewrite the formula for the volume of a cylinder with r as the only variable. $h = 6r$; $V = 6\pi r^3$

b. Find the ratio of the volume of the three tennis balls to the volume of the can. $\dfrac{2}{3}$

c. *Writing* Do you think using a cylindrical can is an efficient way of packaging tennis balls? Explain your reasoning. **See margin.**

12. MULTI-STEP PROBLEM The length l and width w of a *golden rectangle* satisfy the equation $\dfrac{l}{w} = \dfrac{l+w}{l}$. The ratio $\dfrac{l}{w}$ is called the *golden ratio*. For centuries, golden rectangles have been known to be very pleasing to the human eye.

a. Rewrite the right side of the equation as a complex fraction by dividing each term of the numerator and denominator by w. $\dfrac{\frac{l}{w}+1}{\frac{l}{w}}$

b. Let g represent the golden ratio, so $g = \dfrac{l}{w}$. Substitute g for each occurrence of $\dfrac{l}{w}$ in the equation from part (a) and simplify the equation. $g^2 - g - 1 = 0$

c. Solve the equation from part (b) for g. (*Hint:* Use the quadratic formula.) Write an exact value and an approximate value for the golden ratio. $\dfrac{1+\sqrt{5}}{2} \approx 1.618$

d. GEOMETRY   CONNECTION   Use a ruler or graph paper to draw an accurate golden rectangle of any size. Label the dimensions of your rectangle. **See margin.**

11c. *Sample answer:* A cylinder is a fairly efficient way to package tennis balls. Only $\dfrac{1}{3}$ of the space inside the cylinder is wasted.

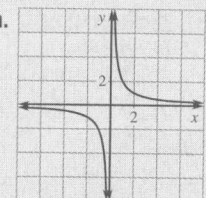

b.

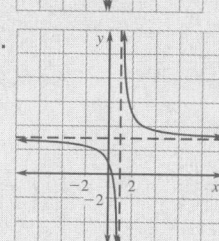

c.

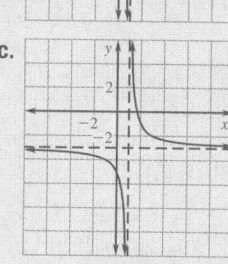

d.

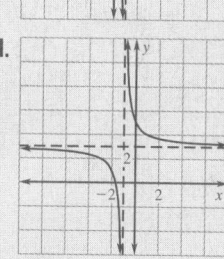

12d. $\sqrt{5}+1$

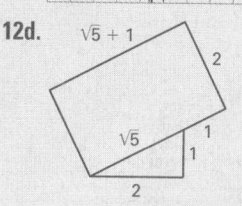

Chapter Standardized Test   **581**

CHAPTER 9

Cumulative Practice

for Chapters 1–9

7.

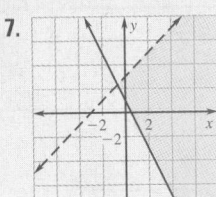

8.

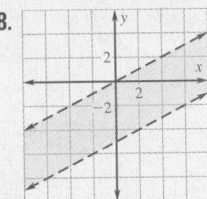

9.

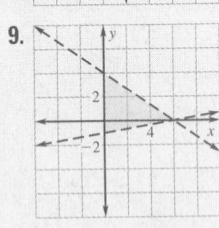

10.

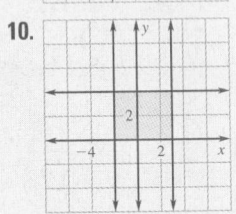

14–19.

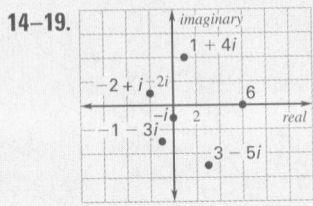

46.

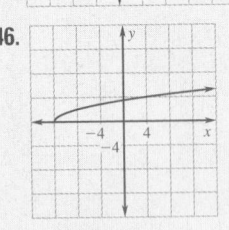

47.

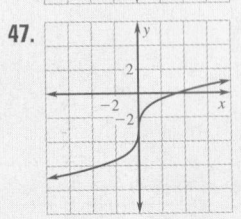

Solve the equation for y. (1.4)

1. $6x - 2y = 7$ $\quad y = \frac{6x-7}{2}$

2. $-\frac{3}{4}x - y = 9$ $\quad y = -\frac{3}{4}x - 9$

3. $\frac{1}{3}x + \frac{2}{5}y = 10$ $\quad y = \frac{150 - 5x}{6}$

4. $xy + 5x = -4$ $\quad y = \frac{-5x - 4}{x}$

Tell whether the lines are *parallel*, *perpendicular*, or *neither*. (2.2)

5. Line 1: through $(2, 1)$ and $(-6, 1)$
Line 2: through $(0, -3)$ and $(-2, -3)$ **parallel**

6. Line 1: through $(-1, 1)$ and $(5, -1)$
Line 2: through $(2, 5)$ and $(3, 2)$ **neither**

Graph the system of linear inequalities. (3.3) **7–10. See margin.**

7. $y < x + 3$
$y \geq -2x + 1$

8. $y < \frac{1}{2}x$
$y + 5 > \frac{1}{2}x$

9. $x \geq 0$
$2x + 3y < 12$
$x - 6y < 6$

10. $x \geq -2$
$x \leq 3$
$y \geq 0$
$y \leq 4$

Perform the indicated operation, if possible. If not possible, state the reason. (4.1, 4.2)

11. $\begin{bmatrix} 10 & 3 \\ -6 & -1 \end{bmatrix} - \begin{bmatrix} -2 & 5 \\ 6 & -3 \end{bmatrix}$ $\begin{bmatrix} 12 & -2 \\ -12 & 2 \end{bmatrix}$

12. $3\begin{bmatrix} 1 & 0 & 6 \\ -3 & 5 & -2 \\ 2 & 8 & -1 \end{bmatrix}$ $\begin{bmatrix} 3 & 0 & 18 \\ -9 & 15 & -6 \\ 6 & 24 & -3 \end{bmatrix}$

13. $\begin{bmatrix} 1 & -1 & -2 \\ 4 & 3 & -5 \end{bmatrix}\begin{bmatrix} 0 & 4 \\ 4 & 8 \\ -1 & 2 \end{bmatrix}$ $\begin{bmatrix} -2 & -8 \\ 17 & 30 \end{bmatrix}$

Plot the numbers in the same complex plane and find their absolute values. (5.4)
14–19. See margin for graphs.

14. $1 + 4i$ $\quad \sqrt{17}$

15. $-2 + i$ $\quad \sqrt{5}$

16. $-i$ $\quad 1$

17. 6 $\quad 6$

18. $-1 - 3i$ $\quad \sqrt{10}$

19. $3 - 5i$ $\quad \sqrt{34}$

Find all the zeros of the polynomial function. (6.6, 6.7)

20. $f(x) = x^3 + 2x^2 - 11x - 12$
$-4, -1, 3$

21. $f(x) = x^3 - 5x^2 + 5x - 25$ $\quad 5$

22. $f(x) = x^4 - 81$ $\quad -3, 3$

Simplify the expression. Assume all variables are positive. (6.1, 7.2, 8.3)

23. $\frac{3x^5}{5y} \cdot \frac{xy}{2x^2}$ $\quad \frac{3x^4}{10}$

24. $(-6x^{-2}y)^{-2}$ $\quad \frac{x^4}{36y^2}$

25. $\sqrt[4]{16a^4b^5c}$ $\quad 2ab\sqrt[4]{bc}$

26. $(9x^6)^{3/2}$ $\quad 27x^9$

27. $\left(\frac{1}{2}e^{-2}\right)^3$ $\quad \frac{1}{8e^6}$

28. $\frac{100e^{6x}}{24e^{4x}}$ $\quad \frac{25e^{2x}}{6}$

Evaluate the expression without using a calculator. (7.1, 8.4)

29. $\sqrt[6]{64}$ $\quad 2$

30. $-(100^{3/2})$ $\quad -1000$

31. $125^{-1/3}$ $\quad \frac{1}{5}$

32. $\log_2 \frac{1}{32}$ $\quad -5$

33. $\log_7 \sqrt{7}$ $\quad \frac{1}{2}$

34. $\log 0.1$ $\quad -1$

Perform the indicated operation and state the domain. (7.3)

35. $f - g$; $f(x) = 2x - 7$, $g(x) = x^2 - 20$
$-x^2 + 2x + 13$; all real numbers

36. $f \cdot g$; $f(x) = 3x^{1/4}$, $g(x) = -x^{5/4}$
$-3x^{3/2}$; all nonnegative real numbers

37. $f(g(x))$; $f(x) = x - 10$, $g(x) = -2x^2 - 5$
$-2x^2 - 15$; all real numbers

38. $g(f(x))$; $f(x) = x + 6$, $g(x) = x^2 - 7x + 3$
$x^2 + 5x - 3$; all real numbers

Find the inverse of the function. (7.4, 8.4)

39. $f(x) = \frac{1}{2}x - 6$
$f^{-1}(x) = 2(x + 6)$

40. $f(x) = x^2 + 1, x \geq 0$
$f^{-1}(x) = \sqrt{x - 1}$

41. $f(x) = \log_5 x$
$f^{-1}(x) = 5^x$

42. $f(x) = \ln 3x$ $\quad f^{-1}(x) = \frac{e^x}{3}$

Condense the expression. (8.5)

43. $\log 3 + 2 \log x + 3 \log y$ $\quad \log(3x^2y^3)$

44. $\log_7 4 + \log_7 y - 2 \log_7 3$ $\quad \log_7\left(\frac{4y}{9}\right)$

45. $2(\ln x + \ln y)$ $\quad \ln(x^2y^2)$

Graph the function. (7.5, 8.1–8.4, 8.8, 9.2, 9.3) 46–57. See margin.

46. $y = \sqrt{x + 12}$

47. $y = 2x^{1/3} - 3$

48. $y = 3\left(\dfrac{4}{3}\right)^x$

49. $y = 3\left(\dfrac{1}{2}\right)^{x + 2}$

50. $y = e^x - 5$

51. $y = \log(x - 1)$

52. $y = \ln x - 2$

53. $y = \dfrac{2}{1 + e^{-3x}}$

54. $y = \dfrac{5}{x - 2} - 1$

55. $y = \dfrac{x - 4}{2x + 1}$

56. $y = \dfrac{13}{x^2 - 4}$

57. $y = \dfrac{3x^2 + 5x - 2}{x - 1}$

Solve the equation. Check each solution. (7.6, 8.6, 8.8, 9.6)

58. $2\sqrt{x + 5} = 18$ 76

59. $\dfrac{1}{8}(x - 6)^{3/2} = 1$ 10

60. $2^{5x} = 8^{x + 6}$ 9

61. $4 \log_3 (-2x) = 10$ $\dfrac{9\sqrt{3}}{2}$

62. $2 \ln x + 5 = 7$ e

63. $\dfrac{5}{1 + 2e^{-x}} = 4 \ln 8$

64. $\dfrac{5}{2x - 3} = \dfrac{2x}{x + 4}$ $-\dfrac{5}{4}, 4$

65. $\dfrac{1}{x - 2} - \dfrac{4}{x^2 - 4} = 5$ $-\dfrac{9}{5}$

Write an exponential function of the form $y = ab^x$ whose graph passes through the given points. (8.7)

66. $(1, 2), (3, 10)$
$y = 0.894(2.236)^x$

67. $(5, 5), (6, 10)$ $y = \dfrac{5}{32}(2)^x$

68. $(2, 4), (4, 8)$ $2 = 2(\sqrt{2})^x$

69. $(0.5, 1), (5, 12)$
$y = 0.759(1.737)^x$

Write a power function of the form $y = ax^b$ whose graph passes through the given points. (8.7)

70. $(1, 1), (3, 9)$ $y = x^2$

71. $(2, 3), (10, 12)$
$y = 1.651x^{0.861}$

72. $(4, 1), (8, 7)$
$y = 0.0204x^{2.807}$

73. $(0.1, 1), (2, 2)$
$y = 1.704x^{0.231}$

Perform the indicated operations. Simplify the result. (9.4, 9.5)

74. $\dfrac{3x^2y}{x - 2} \cdot \dfrac{x^2 + x - 6}{3x - 6} \div (x^2 - 4)$ $\dfrac{x^2y(x + 3)}{(x + 2)(x - 2)^2}$

75. $\dfrac{6x}{3x + 1} + \dfrac{9}{2x} - \dfrac{x + 1}{x - 1}$ $\dfrac{6x^3 + 7x^2 - 20x - 9}{2x(x - 1)(3x + 1)}$

76. ✦ **FOOTBALL** The circumference of a standard football can vary from $20\frac{3}{4}$ inches to $21\frac{1}{4}$ inches around the middle and from $27\frac{3}{4}$ inches to $28\frac{1}{2}$ inches around the length of the football. Write two absolute value inequalities that describe the possible circumferences C_m and C_ℓ of a football. (1.7) $|C_m - 21| \le \frac{1}{4}; |C_\ell - 28\frac{1}{8}| \le \frac{3}{8}$

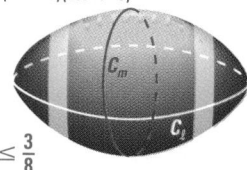

77. ✦ **WISHING WELL** A stone is dropped into a deep wishing well. The water level of the well is 200 feet below the top of the well. After how many seconds will the stone hit the water? (5.3) about 3.5 sec

🌐 **HOME RUNS** In Exercises 78–80, use the following data set of number of home runs hit by each member of the Chicago Cubs in the 1998 baseball season. (7.7)

8, 1, 17, 66, 1, 1, 8, 14, 8, 2, 0, 9, 23, 31, 0, 5, 7, 4, 0, 0, 2, 1, 0

78. Find the mean, median, mode, range, and standard deviation of the data set.
mean: 9.04; median: 4; mode: 0; range: 66; standard deviation: 14.48

79. Draw a box-and-whisker plot of the data set. See margin.

80. Make a frequency distribution of the data set. Use four intervals beginning with $1 - 10$. Then draw a histogram of the data set. See margin.

SCIENCE **CONNECTION** In Exercises 81 and 82, use the following information.
The electrical force f (in newtons) between two charged particles varies jointly with the electric charges q_1 and q_2 (in coulombs) of the particles and inversely with the square of the distance r (in meters) between the particles. (9.1)

81. Write an equation relating f, q_1, q_2, r, and a constant k. $f = \dfrac{kq_1q_2}{r^2}$

82. Using $k = 8,987,760,000$, find the electrical force between a 2 coulomb charged particle and a 3 coulomb charged particle if the two particles are 2 meters apart.   13,481,640,000 newtons

56–57, 79–80. **See Additional Answers beginning on page AA1.**

48.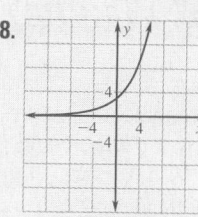

49.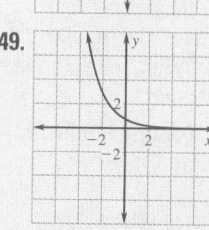

50.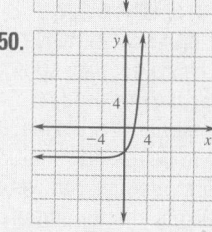

51.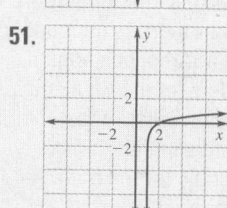

52.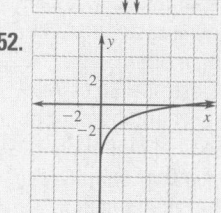

53.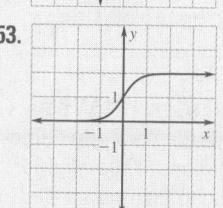

54.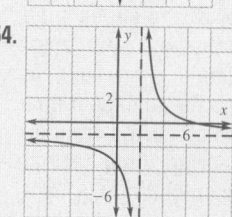

55.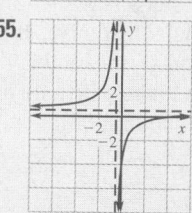

PROJECT GOALS

- Collect and organize data about learning time using a table and scatter plot.
- Model data with a linear, quadratic, polynomial, power, radical, exponential, logarithmic, or rational function.
- Select the best-fitting model for the data collected.

MANAGING THE PROJECT
CLASSROOM MANAGEMENT

The Chapters 7–9 project may be completed by partners or groups. If groups of four students are used, ask one student to be the test subject and another student to complete the table and make a scatter plot of the data. Ask a third group member to test linear, quadratic, polynomial, and power models for the data, and the fourth group member to test the radical, exponential, logarithmic, and rational models for the data. All group members should discuss and refine the models and decide which is the best-fitting model.

Group members should collaborate on the report and be able to explain how the final model was chosen. Each member should include his or her findings in a group report to be presented to the class.

GUIDING STUDENTS' WORK

Students may have difficulty deciding which is the best-fitting model. Ask students who do a paper and pencil scatter plot to check if the difference in the time is constant for a linear model, if it goes up by the same ratio for an exponential model, and so on for the various models. Encourage students who use graphing calculators or computers to find the value of r, the correlation coefficient. The closer this value is to 1 or -1, the better the model fit. Ask them to include a discussion of r in their report.

Mathematical Models of Learning

OBJECTIVE Graph data about learning time and fit a model to the data.

Materials: simple puzzle, timer

When you were first learning to read, a page in a children's book might have taken you several minutes to work through. Now when reading a book to a small child, you can read an entire page easily. How did the time needed to read a page change as you got older and better at reading?

Many scientists study learning. Some study which environments seem to promote learning and which hinder it. Some study which parts of the brain are active when people learn a new skill and which are active when they practice old ones. Others explore how learning changes over time. In this project you will measure the time it takes to do a task as it becomes more familiar to you. You will then fit a model to your data.

INVESTIGATION

1. Find or create a simple task for someone to learn. The task should not take too long to complete and should gradually get easier with practice. Note, a brainteaser puzzle that is very difficult until you realize the trick and then is very easy will not show a gradual increase in performance. A task that is too easy will show only a slight variation in performance.

Possible tasks include: assembling a small puzzle, such as a 10–30 piece jigsaw puzzle; putting 20 index cards with words on them in alphabetical order; solving 15 arithmetic problems involving order of operations where the same problems are in a different order each time; arranging shapes to match a given pattern.

2. Choose a partner as a test subject. Explain to your partner how to do the task you chose. Administer the task to your partner and then measure the time your partner takes to complete the task. Record the data in a table like the one on the right. Repeat the process nine times.

3. Make a scatter plot of your data.

4. So far in this book you have studied linear, quadratic, polynomial, power, radical, exponential, logarithmic, and rational functions. Find a mathematical model that is one of these types of functions to fit your data.

Task number	Time
1	?
2	?
3	?
4	?
5	?
6	?
7	?
8	?
9	?
10	?

PRESENT YOUR RESULTS

Write a report to present your results.

- Begin with a description of the learning task you used and explain how you chose it.

- Include your table of data, your graph, and your mathematical model.

- Explain how you chose the type of function to model your data.

- Explain how you obtained your model.

- Write a description of your data and what they show.

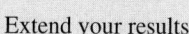

MATHEMATICAL
MODELS OF
LEARNING

Project Applying
Chapters 7-9

Corey Williams
Algebra 2
4th period

Extend your results.

- Recruit a second partner and repeat the experiment.

- Find a mathematical model to describe the learning time for the new data.

- Do you get the same or different results this time? Explain why you might expect similar or different results. (For example, you might obtain similar results because both subjects are juniors in high school who like word puzzles. You might obtain different results because one subject is several years younger than the other.)

EXTENSION

You have measured the times it takes to perform an increasingly familiar task. Now you will measure the times it takes to perform an increasingly complicated task.

Possible tasks include: assembling a puzzle cut out of paper with 2 pieces, then one with 4 pieces, then one with 8 pieces, and so on; arranging index cards with words on them in alphabetical order, first doing 5 cards, then 10 cards, then 15 cards, and so on; finding the ace of spades in a deck with 4 cards, then 8 cards, and so on.

Choose a task and recruit a volunteer to help you. Measure the times it takes your volunteer to complete the task 10 times at increasing levels of difficulty. Record the data in a table, make a scatter plot of the data, and then find a mathematical model to fit your data.

Display and discuss students' tables, graphs, and model choices in class. You may want students to perform the learning time experiment again with a person outside of class and compare the results.

GRADING THE PROJECT
RUBRIC FOR CHAPTERS 7–9 PROJECT

4 Students complete the table, do a scatter plot correctly, and find a correct linear, quadratic, polynomial, power, radical, exponential, logarithmic, or rational model for the data. They explain why the model they chose fits the data better than the other models. Students write a thorough report that includes a description of the data and what is shown. The report relates the project goals with the mathematical goals of fitting a model to data.

3 Students complete the table, do a scatter plot correctly, and find a correct linear, quadratic, polynomial, power, radical, exponential, logarithmic, or rational model for the data. They explain clearly why the model they chose fits the data better than the other models. Students also write a report that includes a description of the data, but it shows a minimal amount of work in relating the project goals with the mathematical goals of fitting a model to data.

2 Students complete the table and do the scatter plot correctly, but they do not find the correct mathematical model for the data. They also include a report that includes a description of the data, but their report shows a minimal amount of work in relating the project goals with the mathematical goals of fitting a model to data.

1 Students complete the table and do the scatter plot correctly. They do not find all the models for the data or they turn in a report that is unclear or incomplete, and the report does not relate the project goals to the mathematical goals.

PLANNING THE CHAPTER

Quadratic Relations and Conic Sections

GOALS		NCTM	ITED	SAT9	Terra-Nova	Local
LESSON						
10.1 pp. 589–594	**GOAL 1** Find the distance between two points and find the midpoint of the line segment joining two points.	1, 2, 4, 8, 9, 10	RQM	32, 33	11, 14, 18, 49	
	GOAL 2 Use the distance and midpoint formulas in real-life situations.					
10.2 pp. 595–600	**GOAL 1** Graph and write equations of parabolas.	1, 2, 3, 6, 8, 9, 10	SAPE, MIG, RQGE	22	14, 16, 17, 18	
	GOAL 2 Use parabolas to solve real-life problems.					
10.3 pp. 601–608	**GOAL 1** Graph and write equations of circles.	1, 2, 3, 6, 8, 9, 10	MIG, IIG, RQGE, SAPE	2, 3	14, 16, 17, 18	
	GOAL 2 Use circles to solve real-life problems.					
	TECHNOLOGY ACTIVITY: 10.3 *Graph the equation of a circle on a graphing calculator.*					
10.4 pp. 609–614	**GOAL 1** Graph and write equations of ellipses.	1, 2, 3, 8, 9, 10	MIG, IIG, SAPE	2	14, 16, 18	
	GOAL 2 Use ellipses in real-life situations.					
10.5 pp. 615–621	**GOAL 1** Graph and write equations of hyperbolas.	1, 2, 3, 6, 8, 9, 10	MIG, IIG, RQGE, SAPE	2, 22	14, 16, 17, 18	
	GOAL 2 Use hyperbolas to solve real-life problems.					
10.6 pp. 622–631	CONCEPT ACTIVITY: 10.6 *Investigate conic sections.*	1, 2, 3, 7	MIG, IIG, SAPE		14, 16	
	GOAL 1 Write and graph an equation of a parabola with its vertex at (h, k) and an equation of a circle, ellipse, or hyperbola with its center at (h, k).					
	GOAL 2 Classify a conic using its equation.					
10.7 pp. 632–638	**GOAL 1** Solve systems of quadratic equations.	1, 2, 6, 8, 9, 10	SAPE, RQGE	2	11, 16, 17, 18, 49, 51, 52	
	GOAL 2 Use quadratic systems to solve real-life problems.					
Extension pp. 639–640	**GOAL** Find the eccentricity of a conic section.	1, 2, 3	RQGE		11, 14, 16, 49, 51, 52	

RESOURCES

CHAPTER RESOURCE BOOKLETS

CHAPTER SUPPORT

Tips for New Teachers	p. 1	Prerequisite Skills Review	p. 5
Parent Guide for Student Success	p. 3	Strategies for Reading Mathematics	p. 7

LESSON SUPPORT

	10.1	10.2	10.3	10.4	10.5	10.6	10.7
Lesson Plans (regular and block)	p. 9	p. 21	p. 36	p. 50	p. 62	p. 76	p. 90
Warm-Up Exercises and Daily Quiz	p. 11	p. 23	p. 38	p. 52	p. 64	p. 78	p. 92
Activity Support Masters						p. 79	
Lesson Openers	p. 12	p. 24	p. 39	p. 53	p. 65	p. 80	p. 93
Graphing Calculator Activities & Keystrokes		p. 25	p. 40		p. 66		p. 94
Practice (3 levels)	p. 13	p. 27	p. 41	p. 54	p. 67	p. 81	p. 96
Reteaching with Practice	p. 16	p. 30	p. 44	p. 57	p. 70	p. 84	p. 99
Quick Catch-Up for Absent Students	p. 18	p. 32	p. 46	p. 59	p. 72	p. 86	p. 101
Cooperative Learning Activities		p. 33					
Interdisciplinary Applications		p. 34		p. 60		p. 87	
Real-Life Applications	p. 19		p. 47		p. 73		p. 102
Math & History Applications						p. 88	
Challenge: Skills and Applications	p. 20	p. 35	p. 48	p. 61	p. 74	p. 89	p. 103

REVIEW AND ASSESSMENT

Quizzes	pp. 49, 75	Alternative Assessment with Math Journal	p. 112
Chapter Review Games and Activities	p. 104	Project with Rubric	p. 114
Chapter Test (3 levels)	pp. 105–110	Cumulative Review	p. 116
SAT/ACT Chapter Test	p. 111	Resource Book Answers	p. A1

TRANSPARENCIES

	10.1	10.2	10.3	10.4	10.5	10.6	10.7
Warm-Up Exercises and Daily Quiz	p. 75	p. 76	p. 77	p. 78	p. 79	p. 80	p. 81
Alternative Lesson Opener Transparencies	p. 65	p. 66	p. 67	p. 68	p. 69	p. 70	p. 71
Examples/Standardized Test Practice	✓	✓	✓	✓	✓	✓	✓
Answer Transparencies	✓	✓	✓	✓	✓	✓	✓

TECHNOLOGY

- Electronic Teaching Tools
- Online Lesson Planner
- Internet Support
- Personal Student Tutor
- Test and Practice Generator
- Instant Replay: Video Review Games
- Electronic Lesson Presentations (Lesson 10.6)

ADDITIONAL RESOURCES

- Basic Skills Workbook: Diagnosis and Remediation
- Worked-Out Solution Key
- Resources in Spanish
- Standardized Test Practice Workbook
- Practice Workbook with Examples

PACING THE CHAPTER

REGULAR SCHEDULE

Day 1

10.1

STARTING OPTIONS
- Prereq. Skills Review
- Strategies for Reading
- Homework Check
- Warm-Up or Daily Quiz

TEACHING OPTIONS
- Les. Opener (Application)
- Examples 1–5
- Closure Question
- Guided Practice Exs.

APPLY/HOMEWORK
- See Assignment Guide.
- See the CRB: Practice, Reteach, Apply, Extend

ASSESSMENT OPTIONS
- Checkpoint Exercises
- Daily Quiz (10.1)
- Stand. Test Practice

Day 2

10.2

STARTING OPTIONS
- Homework Check
- Warm-Up or Daily Quiz

TEACHING OPTIONS
- Les. Opener (Activity)
- Graphing Calc. Activity
- Examples 1–3
- Closure Question
- Guided Practice Exs.

APPLY/HOMEWORK
- See Assignment Guide.
- See the CRB: Practice, Reteach, Apply, Extend

ASSESSMENT OPTIONS
- Checkpoint Exercises
- Daily Quiz (10.2)
- Stand. Test Practice

Day 3

10.3

STARTING OPTIONS
- Homework Check
- Warm-Up or Daily Quiz

TEACHING OPTIONS
- Motivating the Lesson
- Les. Opener (Appl.)
- Graphing Calc. Activity
- Examples 1–5
- Technology Activity
- Closure Question
- Guided Practice Exs.

APPLY/HOMEWORK
- See Assignment Guide.
- See the CRB: Practice, Reteach, Apply, Extend

ASSESSMENT OPTIONS
- Checkpoint Exercises
- Daily Quiz (10.3)
- Stand. Test Practice
- Quiz (10.1–10.3)

Day 4

10.4

STARTING OPTIONS
- Homework Check
- Warm-Up or Daily Quiz

TEACHING OPTIONS
- Motivating the Lesson
- Les. Opener (Activity)
- Examples 1–4
- Closure Question
- Guided Practice Exs.

APPLY/HOMEWORK
- See Assignment Guide.
- See the CRB: Practice, Reteach, Apply, Extend

ASSESSMENT OPTIONS
- Checkpoint Exercises
- Daily Quiz (10.4)
- Stand. Test Practice

Day 5

10.5

STARTING OPTIONS
- Homework Check
- Warm-Up or Daily Quiz

TEACHING OPTIONS
- Motivating the Lesson
- Les. Opener (Visual)
- Graphing Calc. Activity
- Examples 1–2
- Guided Practice Exs. 1–13

APPLY/HOMEWORK
- See Assignment Guide.
- See the CRB: Practice, Reteach, Apply, Extend

ASSESSMENT OPTIONS
- Checkpoint Exercises, p. 616

Day 6

10.5 (cont.)

STARTING OPTIONS
- Homework Check

TEACHING OPTIONS
- Examples 3–4
- Closure Question
- Guided Practice Ex. 14

APPLY/HOMEWORK
- See Assignment Guide.
- See the CRB: Practice, Reteach, Apply, Extend

ASSESSMENT OPTIONS
- Checkpoint Exercises, p. 617
- Daily Quiz (10.5)
- Stand. Test Practice
- Quiz (10.4–10.5)

Day 9

10.7

STARTING OPTIONS
- Homework Check
- Warm-Up or Daily Quiz

TEACHING OPTIONS
- Motivating the Lesson
- Les. Opener (Visual)
- Graphing Calc. Activity
- Examples 1–4
- Closure Question
- Guided Practice Exs.

APPLY/HOMEWORK
- See Assignment Guide.
- See the CRB: Practice, Reteach, Apply, Extend

ASSESSMENT OPTIONS
- Checkpoint Exercises
- Daily Quiz (10.7)
- Stand. Test Practice
- Quiz (10.6–10.7)

Day 10

Review

DAY 10 START OPTIONS
- Homework Check

REVIEWING OPTIONS
- Chapter 10 Summary
- Chapter 10 Review
- Chapter Review Games and Activities

APPLY/HOMEWORK
- Chapter 10 Test (practice)
- Ch. Standardized Test (practice)

Day 11

Assess

DAY 11 START OPTIONS
- Homework Check

ASSESSMENT OPTIONS
- Chapter 10 Test
- SAT/ACT Ch. 10 Test
- Alternative Assessment

APPLY/HOMEWORK
- Skill Review, p. 650

BLOCK SCHEDULE

Day 1

10.1 & 10.2

DAY 1 START OPTIONS
- Prereq. Skills Review
- Strategies for Reading
- W-Up 10.1 or D. Quiz 9.6

TEACHING 10.1 OPTIONS
- Les. Opener (Appl.)
- Examples 1–5
- Closure Question
- Guided Practice Exs.

TEACHING 10.2 OPTIONS
- Warm-Up (Les. 10.2)
- Les. Opener (Activity)
- Graphing Calc. Activity
- Examples 1–3
- Closure Question
- Guided Practice Exs.

APPLY/HOMEWORK
- See Assignment Guide.
- See the CRB: Practice, Reteach, Apply, Extend

ASSESSMENT OPTIONS
- Checkpoint Exercises
- Daily Quiz (10.1, 10.2)
- Stand. Test Practice

Day 2

10.3 & 10.4

DAY 2 START OPTIONS
- Homework Check
- W-Up 10.3 or D. Quiz 10.2

TEACHING 10.3 OPTIONS
- Motivating the Lesson
- Les. Opener (Appl.)
- Technology Activities
- Examples 1–5
- Closure Question
- Guided Practice Exs.

TEACHING 10.4 OPTIONS
- Warm-Up (Les. 10.4)
- Motivating the Lesson
- Les. Opener (Activity)
- Examples 1–4
- Closure Question
- Guided Practice Exs.

APPLY/HOMEWORK
- See Assignment Guide.
- See the CRB: Practice, Reteach, Apply, Extend

ASSESSMENT OPTIONS
- Checkpoint Exercises
- Daily Quiz (10.3, 10.4)
- Quiz (10.1–10.3)

Day 3

10.5

DAY 3 START OPTIONS
- Homework Check
- Warm-Up or Daily Quiz

TEACHING 10. 5 OPTIONS
- Motivating the Lesson
- Les. Opener (Visual)
- Graphing Calc. Activity
- Examples 1–4
- Closure Question
- Guided Practice Exs.

APPLY/HOMEWORK
- See Assignment Guide.
- See the CRB: Practice, Reteach, Apply, Extend

ASSESSMENT OPTIONS
- Checkpoint Exercises
- Daily Quiz (Les. 10.5)
- Stand. Test Practice
- Quiz (10.4–10.5)

Day 4

10.6

DAY 4 START OPTIONS
- Homework Check
- Warm-Up or Daily Quiz

TEACHING OPTIONS
- Concept Act. & Wksht.
- Les. Opener (Calc.)
- Examples 1–8
- Closure Question
- Guided Practice Exs.

APPLY/HOMEWORK
- See Assignment Guide.
- See the CRB: Practice, Reteach, Apply, Extend

ASSESSMENT OPTIONS
- Checkpoint Exercises
- Daily Quiz (Les. 10.6)
- Stand. Test Practice

Day 5

10.7 & Review

DAY 5 START OPTIONS
- Homework Check
- Warm-Up or Daily Quiz

TEACHING 10.7 OPTIONS
- Motivating the Lesson
- Les. Opener (Visual)
- Graphing Calc. Activity
- Examples 1–4
- Closure Question
- Guided Practice Exs.

REVIEWING OPTIONS
- Chapter 10 Summary
- Chapter 10 Review
- Chapter Review Games and Activities

APPLY/HOMEWORK
- See Assignment Guide.
- See the CRB: Practice, Reteach, Apply, Extend
- Chapter 10 Test (prac.)
- Ch. Stand. Test (prac.)

ASSESSMENT OPTIONS
- Checkpoint Exercises
- Daily Quiz (Les. 10.7)
- Stand. Test Practice
- Quiz (10.6–10.7)

Day 6

Assess & 11.1

(Day 6 = Ch. 11 Day 1)

ASSESSMENT OPTIONS
- Chapter 10 Test
- SAT/ACT Ch. 10 Test
- Alternative Assessment

CH. 11 START OPTIONS
- Skill Review, p. 650
- Prereq. Skills Review
- Strategies for Reading

TEACHING 11.1 OPTIONS
- Warm-Up (Les. 11.1)
- Motivating the Lesson
- Les. Opener (Appl.)
- Graphing Calc. Activity
- Examples 1–6
- Technology Activity
- Closure Question
- Guided Practice Exs.

APPLY/HOMEWORK
- See Assignment Guide.
- See the CRB: Practice, Reteach, Apply, Extend

ASSESSMENT OPTIONS
- Checkpoint Exercises
- Daily Quiz (Les. 11.1)
- Stand. Test Practice

Day 7

10.6

STARTING OPTIONS
- Homework Check
- Warm-Up or Daily Quiz

TEACHING OPTIONS
- Concept Act. & Wksht.
- Les. Opener (Calculator)
- Examples 1–5
- Guided Practice Exs. 1–2, 4–7

APPLY/HOMEWORK
- See Assignment Guide.
- See the CRB: Practice, Reteach, Apply, Extend

ASSESSMENT OPTIONS
- Checkpoint Exercises, pp. 624–625

Day 8

10.6 (cont.)

STARTING OPTIONS
- Homework Check

TEACHING OPTIONS
- Examples 6–8
- Closure Question
- Guided Practice Exs. 3, 8–12

APPLY/HOMEWORK
- See Assignment Guide.
- See the CRB: Practice, Reteach, Apply, Extend

ASSESSMENT OPTIONS
- Checkpoint Exercises, pp. 626–627
- Daily Quiz (10.6)
- Stand. Test Practice

CHAPTER 10 — MEETING INDIVIDUAL NEEDS

BEFORE THE CHAPTER

The *Chapter 10 Resource Book* has the following materials to distribute and use before the chapter:

- **Parent Guide for Student Success**
- **Prerequisite Skills Review**
- **Strategies for Reading Mathematics (pictured below)**

STRATEGIES FOR READING Pages 7–8

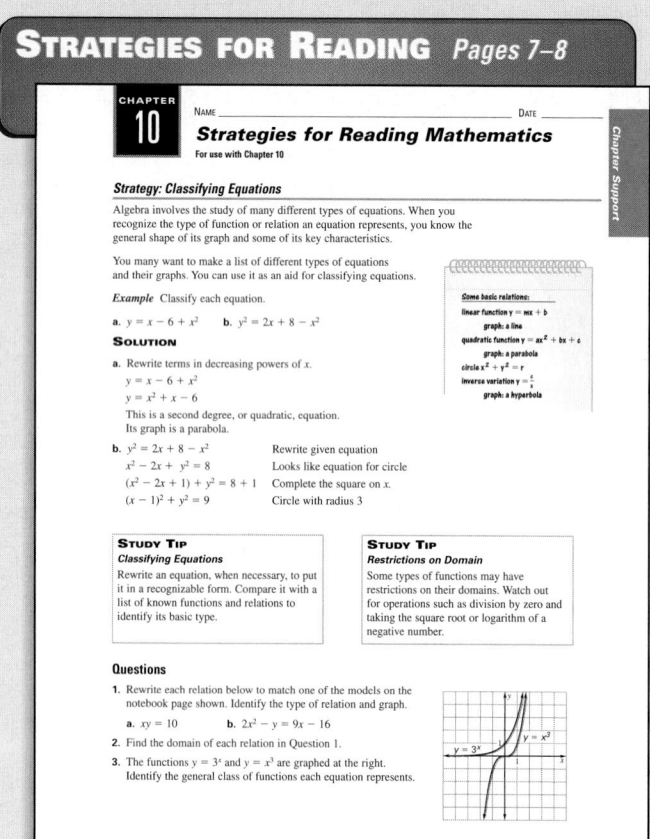

CHAPTER 10   NAME _____ DATE _____

Chapter Support

Strategies for Reading Mathematics
For use with Chapter 10

Strategy: Classifying Equations

Algebra involves the study of many different types of equations. When you recognize the type of function or relation an equation represents, you know the general shape of its graph and some of its key characteristics.

You many want to make a list of different types of equations and their graphs. You can use it as an aid for classifying equations.

Example Classify each equation.

a. $y = x - 6 + x^2$ **b.** $y^2 = 2x + 8 - x^2$

SOLUTION

a. Rewrite terms in decreasing powers of x.

$y = x - 6 + x^2$
$y = x^2 + x - 6$

This is a second degree, or quadratic, equation. Its graph is a parabola.

b. $y^2 = 2x + 8 - x^2$ Rewrite given equation
$x^2 - 2x + y^2 = 8$ Looks like equation for circle
$(x^2 - 2x + 1) + y^2 = 8 + 1$ Complete the square on x.
$(x - 1)^2 + y^2 = 9$ Circle with radius 3

Some basic relations:
linear function $y = mx + b$
  graph: a line
quadratic function $y = ax^2 + bx + c$
  graph: a parabola
circle $x^2 + y^2 = r$
inverse variation $y = \frac{c}{x}$
  graph: a hyperbola

STUDY TIP
Classifying Equations
Rewrite an equation, when necessary, to put it in a recognizable form. Compare it with a list of known functions and relations to identify its basic type.

STUDY TIP
Restrictions on Domain
Some types of functions may have restrictions on their domains. Watch out for operations such as division by zero and taking the square root or logarithm of a negative number.

Questions

1. Rewrite each relation below to match one of the models on the notebook page shown. Identify the type of relation and graph.
   a. $xy = 10$ **b.** $2x^2 - y = 9x - 16$
2. Find the domain of each relation in Question 1.
3. The functions $y = 3^x$ and $y = x^3$ are graphed at the right. Identify the general class of functions each equation represents.

STRATEGIES FOR READING MATHEMATICS These two pages give students tips about classifying equations as they prepare for Chapter 10 and provide a visual glossary of key vocabulary words in the chapter, such as distance formula and hyperbola.

DURING EACH LESSON

The *Chapter 10 Resource Book* has the following alternatives for introducing the lesson:

- **Lesson Openers (pictured below)**
- **Graphing Calculator Activities with Keystrokes**

LESSON OPENER Page 53

LESSON 10.4   NAME _____ DATE _____

Available as a transparency

Activity Lesson Opener
For use with pages 609–614

SET UP: Work individually or in a small group.
  You will need: • Cardboard (or paperboard)
          • String
          • Pencil

An *ellipse* is the set of all points P such that the sum of the distances between P and two distinct points, called the *foci* (plural of *focus*), is a constant. In this activity, you will construct ellipses.

1. Using a piece of cardboard, choose a location for the two points called the foci. These points should not be too close to the edge of your cardboard.

2. Use your pencil, a hole punch, or a sharp object to make a small hole at each focus.

3. Insert an end of the string through each focus. Tie the string ends together behind the cardboard to form a loop that goes through the foci. (Don't pull the string too tight! The string loop should be long enough that the string can be pulled an inch or two away from the cardboard.)

4. Now use the string as a guide to sketch your ellipse. As you sketch, the tip of your pencil should be pulling the string taut.

5. Repeat these steps several times, varying both the distance between the foci and the length of your string loop in order to construct several different ellipses.

6. Use the definition of an ellipse to explain why figures drawn in this manner are ellipses.

Lesson 10.4

ACTIVITY LESSON OPENER This Lesson Opener provides an alternative way to start Lesson 10.4 in the form of an activity. In this activity, students work in a group to construct ellipses from cardboard and string.

 TECHNOLOGY RESOURCE

Teachers can use the Electronic Lesson Presentations CD-ROM to present Lesson 10.6 using computer animation to step through the Examples.

The *Chapter 10 Resource Book* has a variety of materials to follow-up each lesson. They include the following:

- **Practice (3 levels)**
- **Reteaching with Practice**
- **Quick Catch-Up for Absent Students**
- **Challenge: Skills and Applications**
- **Interdisciplinary Applications**
- **Real-Life Applications (pictured below)**

APPLICATION Page 19

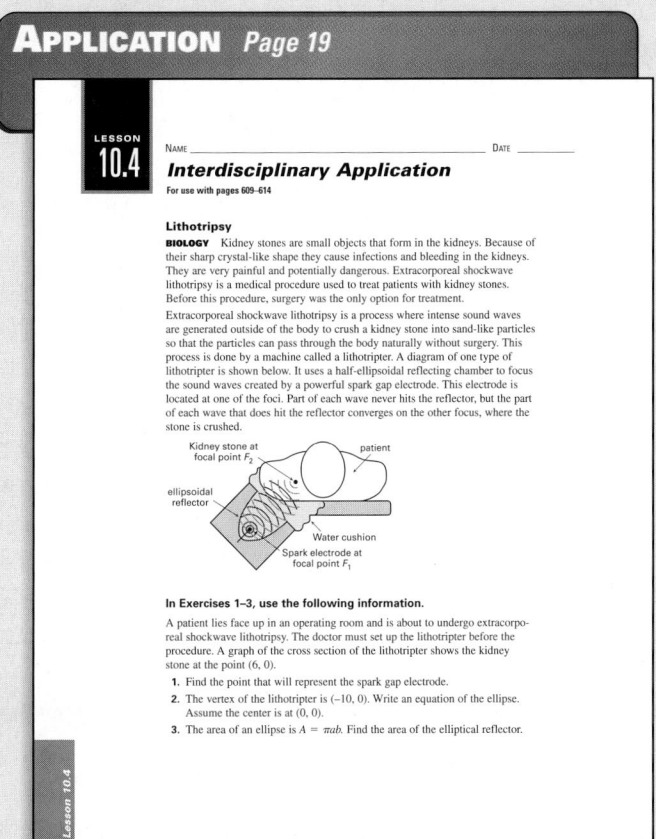

REAL-LIFE APPLICATION When students ask "When will I ever use ellipses?," you can have them work through this Real-Life Application that discusses how an ellipse is used in the design of medical equipment.

The *Chapter 10 Resource Book* has the following review and assessment materials:

- **Quizzes**
- **Chapter Review Games and Activities**
- **Chapter Test (3 levels)**
- **SAT/ACT Chapter Test (pictured below)**
- **Alternative Assessment with Rubrics and Math Journal**
- **Project with Rubric**
- **Cumulative Review**

SAT/ACT CHAPTER TEST Page 111

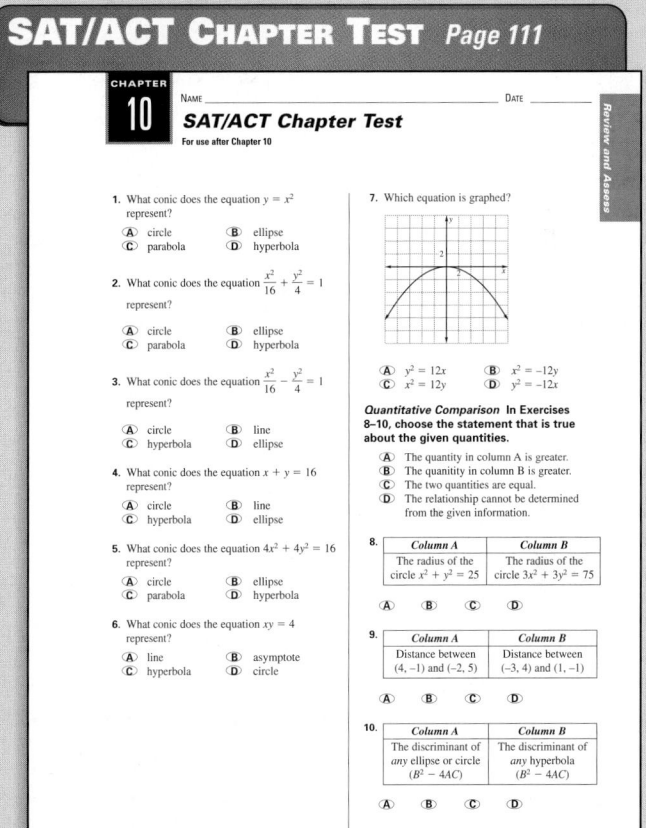

SAT/ACT CHAPTER TEST This test covers the material in Chapter 10 in a standardized test format. Students taking this form of the test should be reminded to read all of the answer choices before deciding which is the correct one.

CHAPTER OVERVIEW

CHAPTER GOALS

In Chapter 10 students first make use of the distance and midpoint formulas for line segments. Then they learn to write equations and draw graphs for the four conic sections: parabolas, circles, ellipses, and hyperbolas. They explore these quadratic relations and their graphs, first with centers at the origin and then with centers translated to other points in the coordinate plane. Finally, students learn to solve quadratic systems by using the algebraic techniques they used for systems of linear equations.

APPLICATION NOTE

The Hubble Space Telescope (HST) was launched on April 15, 1990, from the space shuttle *Discovery*. It is solar powered and contains a wide variety of cameras and other equipment used to take pictures in space. A team of about 300 workers commands the telescope, and its activities are coordinated by the Space Telescope Science Institute in Baltimore, Maryland.

In 1993, NASA launched a mission to repair the main mirror of the telescope so that it would focus properly. A corrective camera was installed, which improved the optics of the telescope. In 1996, observations from the HST indicated that there are about 50 billion more galaxies than had previously been supposed.

Additional information about telescopes is available at **www.mcdougallittell.com.**

QUADRATIC RELATIONS AND CONIC SECTIONS

▶ *What shape does a telescope mirror have?*

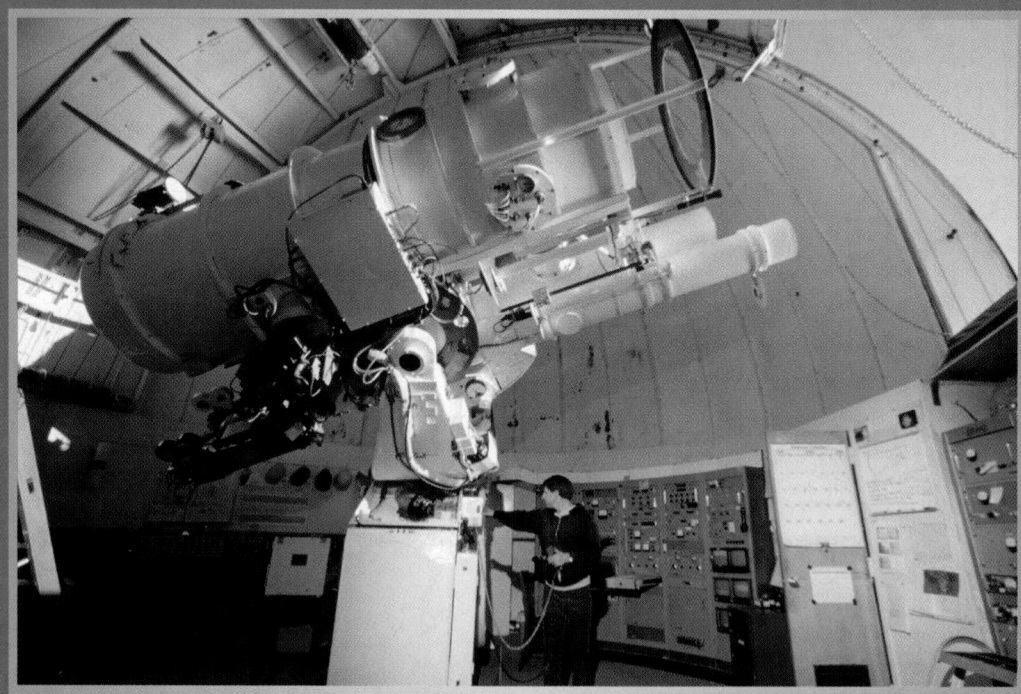

APPLICATION: Telescopes

Using powerful telescopes like the Hubble telescope, astronomers have discovered planets that exist outside our solar system. By examining the wobble in a star's motion, astronomers can detect the presence of one or more planets orbiting that star. The challenge is to see the planets, but that would require a telescope with a mirror 100 meters in diameter, 10 times larger than any existing telescope.

Think & Discuss

The diagram shows a cross section of a mirror from a telescope at the Big Bear Solar Observatory in California. An equation for the surface of the mirror, based on the coordinate system shown, is $y = \dfrac{x^2}{1040}$ where x and y are measured in centimeters.

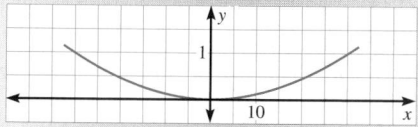

1. What shape is the cross section of the mirror?
   a parabola
2. The mirror has a diameter of 65 cm. What is the depth of the mirror? How did you get your answer?
   about 1.02 cm; substitute the radius (32.5 cm) for x in the equation and solve for y.

Learn More About It

In Exercise 67 on p. 630 you will use your knowledge of conic sections to determine what shape a telescope's mirrors have.

APPLICATION LINK Visit www.mcdougallittell.com for more information about telescopes.

587

ADDITIONAL RESOURCES
Another way to begin the chapter is to show the video clip of a real-life motivator for Chapter 10 on the *Instant Replay: Video Review Games.*

PROJECTS
A project covering Chapters 10–12 appears on pages 767–768 of the Student Edition. An additional project for Chapter 10 is available in the *Chapter 10 Resource Book,* p. 114.

TECHNOLOGY
 **Software**
- *Electronic Teaching Tools*
- *Online Lesson Planner*
- *Personal Student Tutor*
- *Test and Practice Generator*
- *Electronic Lesson Presentations* (Lesson 10.6)

Video
- *Instant Replay: Video Review Games*

Internet Connections
www.mcdougallittell.com
- **Application Links**
  587, 597, 617, 620, 631
- **Student Help**
  593, 596, 602, 608, 610, 619, 624, 637
- **Career Links**
  593, 606, 634, 636
- **Extra Challenge**
  600, 620, 630

Study Guide

DIAGNOSTIC TOOLS

The **Skill Review** exercises can help you diagnose whether students have the following skills needed in Chapter 10:

- Write linear equations using a given point and slope.
- Solve a system of linear equations algebraically.
- Graph and label quadratic functions.
- Solve quadratic equations by completing the square.

The following resources are available for students who need additional help with these skills:

- Prerequisite Skills Review (*Chapter 10 Resource Book*, p. 5; *Warm-Up Transparencies*, p. 74)
- Reteaching with Practice (Chapter Resource Books for Lessons 2.4, 3.2, 5.1, and 5.5)
- ⬛ *Personal Student Tutor*

ADDITIONAL RESOURCES

The following resources are provided to help you prepare for the upcoming chapter and customize review materials:

- *Chapter 10 Resource Book*
  Tips for New Teachers (p. 1)
  Parent Guide (p. 3)
  Lesson Plans (every lesson)
  Lesson Plans for Block Scheduling (every lesson)
- ⬛ *Electronic Teaching Tools*
- ⬛ *Online Lesson Planner*
- ⬛ *Test and Practice Generator*

ENGLISH LEARNERS

Not knowing how newly introduced mathematical terms are pronounced can make it more difficult for English learners to follow and participate in class discussions. As part of the process of introducing chapter concepts and vocabulary, model the pronunciation of *conic, parabola, ellipse, hyperbola,* and *discriminant.*

What's the chapter about?

Chapter 10 is about **conic sections**. The four conic sections are parabolas, circles, ellipses, and hyperbolas. In Chapter 10 you'll learn

- how to use the distance and midpoint formulas.
- how to graph and write equations of conics, and how to classify conics.
- how to solve systems of quadratic equations.

KEY VOCABULARY

▶ **Review**
- parabola, p. 249
- hyperbola, p. 540

▶ **New**
- distance formula, p. 589

- midpoint formula, p. 590
- circle, p. 601
- ellipse, p. 609
- hyperbola, p. 615
- conic sections, p. 623

- general second-degree equation, p. 626
- discriminant, p. 626

Are you ready for the chapter?

SKILL REVIEW Do these exercises to review key skills that you'll apply in this chapter. See the given **reference page** if there is something you don't understand.

▶ **Study Tip**
"Student Help" boxes throughout the chapter give you study tips and tell you where to look for extra help in this book and on the Internet.

---

11. $-\frac{3}{2} \pm \frac{i\sqrt{11}}{2}$

Write an equation of the line that passes through the given point and has the given slope. (Review Example 2, p. 92)

1. $(0, 4)$, $m = 2$ $y = 2x + 4$
2. $(2, -2)$, $m = \frac{1}{3}$ $y = \frac{1}{3}x - \frac{8}{3}$
3. $(-4, 1)$, $m = -\frac{3}{4}$ $y = -\frac{3}{4}x - 2$

Solve the system using any algebraic method. (Review Examples 1–3, pp. 148–150)

4. $x + 2y = 8$ $(2, 3)$
   $3x - y = 3$
5. $2x + y = 3$ $(-1, 5)$
   $3x + y = 2$
6. $4x - y = 7$ $(4, 9)$
   $5x - 2y = 2$

Graph the function. Label the vertex and axis of symmetry.
(Review Examples 1–3, pp. 250 and 251)   7–9. See margin.

7. $y = x^2 + 4$
8. $y = -3x^2$
9. $y = 2(x - 3)^2 - 1$

Solve the equation by completing the square. (Review Examples 2 and 3, p. 283)

10. $x^2 + 8x + 14 = 0$
    $-4 \pm \sqrt{2}$
11. $5x^2 + 15x = -25$
12. $x^2 - 2x = -8x + 14$
    $-3 \pm \sqrt{23}$

Here's a study strategy!

Dictionary of Graphs

Make a dictionary of graphs to use as a reference tool. Draw and label an example of each conic. Note the important characteristics of the conic, and write the conic's equation. Expand your dictionary to include all the types of graphs you have learned and continue to learn in this course.

7–9. See Additional Answers beginning on page AA1.

What you should learn

GOAL ❶ Find the distance between two points and find the midpoint of the line segment joining two points.

GOAL ❷ Use the distance and midpoint formulas in **real-life** situations, such as finding the diameter of a broken dish in **Example 5**.

Why you should learn it

▼ To solve **real-life** problems, such as finding the distance a medical helicopter must travel to a hospital in **Exs. 53–56**.

The Distance and Midpoint Formulas

GOAL ❶ USING THE DISTANCE AND MIDPOINT FORMULAS

To find the distance d between $A(x_1, y_1)$ and $B(x_2, y_2)$, you can apply the Pythagorean theorem to right triangle ABC.

$$(AB)^2 = (AC)^2 + (BC)^2$$
$$d^2 = (x_2 - x_1)^2 + (y_2 - y_1)^2$$
$$d = \sqrt{(x_2 - x_1)^2 + (y_2 - y_1)^2}$$

The third equation is called the **distance formula**.

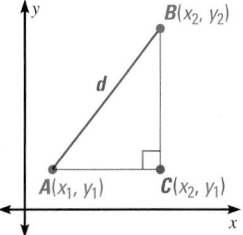

THE DISTANCE FORMULA

The distance d between the points (x_1, y_1) and (x_2, y_2) is as follows:

$$d = \sqrt{(x_2 - x_1)^2 + (y_2 - y_1)^2}$$

EXAMPLE 1 *Finding the Distance Between Two Points*

Find the distance between $(-2, 5)$ and $(3, -1)$.

SOLUTION

Let $(x_1, y_1) = (-2, 5)$ and $(x_2, y_2) = (3, -1)$.

$$d = \sqrt{(x_2 - x_1)^2 + (y_2 - y_1)^2} \qquad \text{Use distance formula.}$$
$$= \sqrt{(3 - (-2))^2 + (-1 - 5)^2} \qquad \text{Substitute.}$$
$$= \sqrt{25 + 36} \qquad \text{Simplify.}$$
$$= \sqrt{61} \approx 7.81 \qquad \text{Use a calculator.}$$

EXAMPLE 2 *Classifying a Triangle Using the Distance Formula*

Classify $\triangle ABC$ as *scalene*, *isosceles*, or *equilateral*.

SOLUTION

$$AB = \sqrt{(6 - 4)^2 + (1 - 6)^2} = \sqrt{29}$$
$$BC = \sqrt{(1 - 6)^2 + (3 - 1)^2} = \sqrt{29}$$
$$AC = \sqrt{(1 - 4)^2 + (3 - 6)^2} = 3\sqrt{2}$$

▶ Because $AB = BC$, $\triangle ABC$ is isosceles.

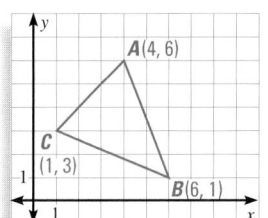

10.1 *The Distance and Midpoint Formulas* **589**

---

① PLAN

PACING
Basic: 1 day
Average: 1 day
Advanced: 1 day
Block Schedule: 0.5 block with 10.2

LESSON OPENER
APPLICATION
An alternative way to approach Lesson 10.1 is to use the Application Lesson Opener:
• Blackline Master (*Chapter 10 Resource Book*, p. 12)
• Transparency (p. 65)

MEETING INDIVIDUAL NEEDS
• *Chapter 10 Resource Book*
 Prerequisite Skills Review (p. 5)
 Practice Level A (p. 13)
 Practice Level B (p. 14)
 Practice Level C (p. 15)
 Reteaching with Practice (p. 16)
 Absent Student Catch-Up (p. 18)
 Challenge (p. 20)
• *Resources in Spanish*
• *Personal Student Tutor*

NEW-TEACHER SUPPORT
See the Tips for New Teachers on pp. 1–2 of the *Chapter 10 Resource Book* for additional notes about Lesson 10.1.

WARM-UP EXERCISES

Transparency Available

Use the Pythagorean theorem to find the length of the missing side.
1. $a = 12$, $b = 9$ $c = 15$
2. $a = 5$, $c = 13$ $b = 12$
3. $b = 15$, $c = 17$ $a = 8$

Find the mean of the two numbers.
4. 18 and 34   26
5. 18 and −34   −8

EXTRA EXAMPLE 1
Find the distance between
$(2, -4)$ and $(-5, -1)$. $\sqrt{58} \approx 7.62$

EXTRA EXAMPLE 2
Classify $\triangle DEF$ as scalene,
isosceles, or equilateral.

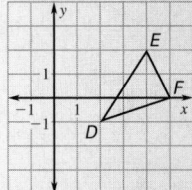

scalene, since $DE \neq DF \neq EF$

EXTRA EXAMPLE 3
Find the midpoint of the line segment joining $(6, -2)$ and $(2, -9)$.
$\left(4, -\dfrac{11}{2}\right)$

EXTRA EXAMPLE 4
Write an equation for the perpendicular bisector of the line
segment joining $C(-2, 1)$ and
$D(1, 4)$. $y = -x + 2$

☑ **CHECKPOINT EXERCISES**
For use after Examples 1 and 2:
1. Classify the polygon as a
   quadrilateral, parallelogram,
   or rhombus. **parallelogram**

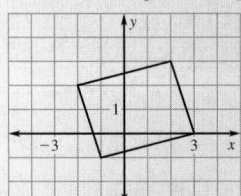

For use after Examples 3 and 4:
2. Write an equation for the perpendicular bisector of the line
   segment joining $E(6, 3)$ and
   $F(-4, -2)$. $y = -2x + \dfrac{5}{2}$

**MULTIPLE
REPRESENTATIONS**
Students should be familiar with the
physical representation of a perpendicular bisector from geometry, but
some review may be needed.

Another formula involving two points in a coordinate plane is the **midpoint formula**. Recall that the midpoint of a segment is the point on the segment that is equidistant from the two endpoints.

THE MIDPOINT FORMULA

The midpoint of the line segment joining $A(x_1, y_1)$ and $B(x_2, y_2)$ is as follows:

$$M\left(\frac{x_1 + x_2}{2}, \frac{y_1 + y_2}{2}\right)$$

Each coordinate of M is the mean of the corresponding coordinates of A and B.

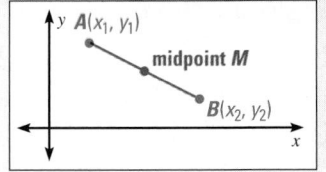

EXAMPLE 3 *Finding the Midpoint of a Segment*

Find the midpoint of the line segment joining $(-7, 1)$ and $(-2, 5)$.

SOLUTION

Let $(x_1, y_1) = (-7, 1)$ and $(x_2, y_2) = (-2, 5)$.

$$\left(\frac{x_1 + x_2}{2}, \frac{y_1 + y_2}{2}\right) = \left(\frac{-7 + (-2)}{2}, \frac{1 + 5}{2}\right)$$

$$= \left(-\frac{9}{2}, 3\right)$$

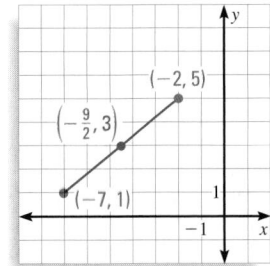

EXAMPLE 4 *Finding a Perpendicular Bisector*

Write an equation for the perpendicular bisector of the line segment joining $A(-1, 4)$ and $B(5, 2)$.

SOLUTION
First find the midpoint of the line segment:

$$\left(\frac{x_1 + x_2}{2}, \frac{y_1 + y_2}{2}\right) = \left(\frac{-1 + 5}{2}, \frac{4 + 2}{2}\right) = (2, 3)$$

Then find the slope of $\overline{AB}$:

$$m = \frac{y_2 - y_1}{x_2 - x_1} = \frac{2 - 4}{5 - (-1)} = \frac{-2}{6} = -\frac{1}{3}$$

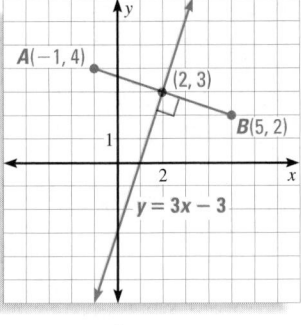

The slope of the perpendicular bisector is the negative reciprocal of $-\dfrac{1}{3}$, or $m\bot = 3$.

STUDENT HELP
↳ **Look Back**
For help with
perpendicular lines,
see p. 92.

Since you know the slope of the perpendicular bisector and a point that the bisector passes through, you can use the point-slope form to write its equation.

$$y - 3 = 3(x - 2)$$

$$y = 3x - 3$$

▶ An equation for the perpendicular bisector of $\overline{AB}$ is $y = 3x - 3$.

ARCHEOLOGISTS
use grids to system-
atically explore a site. By
labeling the grid squares,
they can record where each
artifact is found.

GOAL 2 DISTANCE AND MIDPOINT FORMULAS IN REAL LIFE

Recall from geometry that the perpendicular bisector of a chord of a circle passes
through the center of the circle. Using this theorem, you can find the center of a
circle given three points on the circle.

EXAMPLE 5 *Using the Distance and Midpoint Formulas in Real Life*

ARCHEOLOGY While on an archeological dig, you
discover a piece of a broken dish. To estimate the
original diameter of the dish, you lay the piece on a
coordinate plane and mark three points on the
circular edge, as shown. Use these points to find the
diameter of the dish. (Each unit in the coordinate
plane represents 1 inch.)

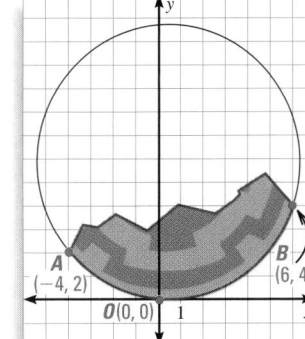

SOLUTION

Use the method illustrated in Example 4 to find the
perpendicular bisectors of $\overline{AO}$ and $\overline{OB}$.

$y = 2x + 5$ **Perpendicular bisector of $\overline{AO}$**

$y = -\dfrac{3}{2}x + \dfrac{13}{2}$ **Perpendicular bisector of $\overline{OB}$**

Both bisectors pass through the circle's center.
Therefore, the center of the circle is the solution
of the system formed by these two equations.

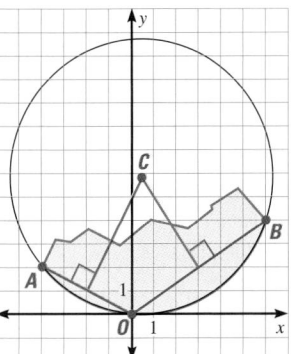

$y = 2x + 5$ **Write first equation.**

$-\dfrac{3}{2}x + \dfrac{13}{2} = 2x + 5$ **Substitute for y.**

$-3x + 13 = 4x + 10$ **Multiply each side by 2.**

$-7x = -3$ **Simplify.**

$x = \dfrac{3}{7}$ **Divide each side by -7.**

$y = 2\left(\dfrac{3}{7}\right) + 5$ **Substitute the x-value into the first equation.**

$y = \dfrac{41}{7}$ **Simplify.**

STUDENT HELP

Look Back
For help with solving
systems, see p. 148.

The center of the circle is $C\left(\dfrac{3}{7}, \dfrac{41}{7}\right)$. The radius of the circle is the distance
between C and any of the three given points.

$$CO = \sqrt{\left(0 - \dfrac{3}{7}\right)^2 + \left(0 - \dfrac{41}{7}\right)^2}$$

$$= \sqrt{\dfrac{1690}{49}}$$

$$\approx 5.87$$

▶ The dish had a diameter of about $2(5.87) = 11.74$ inches.

10.1 *The Distance and Midpoint Formulas*   **591**

EXTRA EXAMPLE 5
A circular fossil is found in a
rock. Only part of the object is
fossilized. To estimate the original diameter, you lay a centimeter grid over the fossil and mark
three points on the circular
edge, as shown. Use these
points to find the diameter.

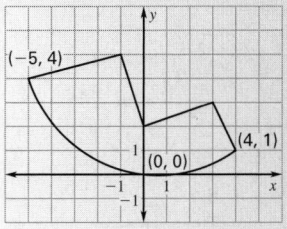

center: $\left(\dfrac{9}{14}, \dfrac{83}{14}\right)$;

diameter $\approx 2(5.963) \approx 11.93$ cm

CHECKPOINT EXERCISES
For use after Example 5:
1. Another archaeologist found
the diameter of the fossil from
Extra Example 5 in inches,
using the grid below. What is
the diameter in inches?

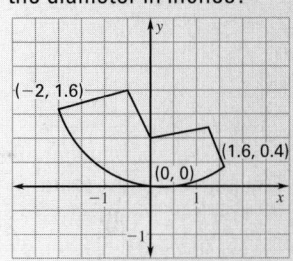

center: $\left(\dfrac{9}{35}, \dfrac{83}{35}\right)$;

diameter: $\approx 2(2.385) \approx 4.77$ in.

CLOSURE QUESTION
Describe in words how to find
the length of a segment using the
distance formula. *Sample answer:*
**Subtract the *x*-values of the two
endpoints and subtract the *y*-values
of the two endpoints. Then square
both values, find their sum, and take
the square root of the sum.**

ASSIGNMENT GUIDE

BASIC
Day 1: pp. 592–594 Exs. 18–28
even, 32–34, 41–43,
47, 48, 51, 52, 57, 59–62,
65–79 odd

AVERAGE
Day 1: pp. 592–594 Exs. 18–30
even, 32–35, 41–43, 47–52,
57, 59–62, 65–83 odd

ADVANCED
Day 1: pp. 592–594 Exs. 18–40
even, 41–52, 57–63,
65–83 odd

BLOCK SCHEDULE
pp. 592–594 Exs. 18–30 even,
32–35, 41–43, 47–52, 57, 59–62,
65–83 odd (with 10.2)

EXERCISE LEVELS

Level A: *Easier*
17–25, 41–44, 51, 52
Level B: *More Difficult*
26–40, 45–50, 53–62
Level C: *Most Difficult*
63

✔ HOMEWORK CHECK
To quickly check student under-
standing of key concepts, go
over the following exercises:
Exs. 20, 26, 32, 42, 48. See also
the Daily Homework Quiz:
• Blackline Master (*Chapter 10
Resource Book,* p. 23)
• Transparency (p. 76)

GUIDED PRACTICE

Vocabulary Check ✔

Concept Check ✔

1. distance formula: $d =$
$\sqrt{(x_2 - x_1)^2 + (y_2 - y_1)^2}$;
midpoint formula:
$M\left(\dfrac{x_1 + x_2}{2}, \dfrac{y_1 + y_2}{2}\right)$

Skill Check ✔

2. $d = \sqrt{61}$ for each. The
differences are opposite
the first pair of differences;
since they are squared, the
answer is the same.

15. $\left(-\dfrac{9}{2}, -\dfrac{1}{2}\right)$

20. $8\sqrt{2} \approx 11.31$; (2, 4)

21. $5\sqrt{5} \approx 11.18$; $\left(2, \dfrac{3}{2}\right)$

23. $2\sqrt{58} \approx 15.23$; (−2, −1)

24. $12\sqrt{2} \approx 16.97$; (4, 4)

25. $2\sqrt{13} \approx 7.21$; (5, 1)

26. $11\sqrt{13} \approx 39.66$; $\left(1, \dfrac{3}{2}\right)$

27. $\sqrt{115.25} \approx 10.74$; (1.25, −1.3)

28. $\sqrt{25.25} \approx 5.02$; (3.8, −8.75)

30. $\sqrt{\dfrac{3589}{144}} \approx 4.99$; $\left(-\dfrac{17}{12}, -\dfrac{33}{8}\right)$

1. State the distance and midpoint formulas. **See margin.**

2. Look back at Example 1. Find the distance between $(-2, 5)$ and $(3, -1)$, but this time letting $(x_1, y_1) = (3, -1)$ and $(x_2, y_2) = (-2, 5)$. How are the calculations different? Do you get the same answer? **See margin.**

3. **a.** Write a formula for the distance between a point (x, y) and the origin. $d = \sqrt{x^2 + y^2}$
   b. Write a formula for the midpoint of the segment joining a point (x, y) and the origin. $M\left(\dfrac{x}{2}, \dfrac{y}{2}\right)$

Find the distance between the two points.

4. $(2, -1), (2, 3)$ 4

5. $(-5, -2), (0, -2)$ 5

6. $(0, 6), (4, 9)$ 5

7. $(10, -2), (7, 4)$ $3\sqrt{5} \approx 6.71$

8. $(-3, 8), (5, 6)$ $2\sqrt{17} \approx 8.25$

9. $(6, -1), (-9, 8)$ $3\sqrt{34} \approx 17.49$

Find the midpoint of the line segment joining the two points.

10. $(0, 0), (-8, 14)$ (−4, 7)

11. $(0, 3), (4, 9)$ (2, 6)

12. $(1, -2), (1, 6)$ (1, 2)

13. $(1, 3), (3, 11)$ (2, 7)

14. $(-5, 4), (2, -4)$ $\left(-\dfrac{3}{2}, 0\right)$

15. $(-1, 5), (-8, -6)$

16. 🌐 **HIKING** You are going on a two-day hike. The map at the right shows the trails you plan to follow. (Each unit represents 1 mile.)

   a. You hike from the lodge to point A and decide that you will hike to the midpoint of $\overline{AB}$ before you camp for the night. At what point in the plane will you be camping? $\left(-\dfrac{1}{2}, 4\right)$

   b. How far will you hike each day?
   day 1: 6.81 mi; day 2: 9.52 mi

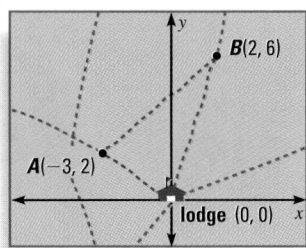

PRACTICE AND APPLICATIONS

STUDENT HELP
▸ **Extra Practice**
to help you master
skills is on p. 953.

31. $\sqrt{\dfrac{377}{8}} \approx 6.86$; $\left(\dfrac{17}{8}, \dfrac{1}{8}\right)$

STUDENT HELP
▸ **HOMEWORK HELP**
Example 1: Exs. 17–31,
47–50
Example 2: Exs. 32–40
Example 3: Exs. 17–31
Example 4: Exs. 41–46
Example 5: Exs. 51–58

20, 21; 23–28; 30–31. See margin.

USING THE FORMULAS Find the distance between the two points. Then find the midpoint of the line segment joining the two points.

17. $(0, 0), (3, 4)$ 5; $\left(\dfrac{3}{2}, 2\right)$

18. $(0, 0), (4, 12)$ $4\sqrt{10} \approx 12.65$; (2, 6)

19. $(0, 4), (8, -3)$ $\sqrt{113} \approx 10.63$; $\left(4, \dfrac{1}{2}\right)$

20. $(-2, 8), (6, 0)$

21. $(-3, -1), (7, 4)$

22. $(9, -2), (3, 6)$ 10; (6, 2)

23. $(-5, -8), (1, 6)$

24. $(-2, 10), (10, -2)$

25. $(8, 3), (2, -1)$

26. $(-10, -15), (12, 18)$

27. $(-3.5, 1.2), (6, -3.8)$

28. $(6.3, -9), (1.3, -8.5)$

29. $(-7, 2), \left(-\dfrac{11}{2}, 4\right)$ 2.5; (−6.25, 3)

30. $\left(\dfrac{2}{3}, -\dfrac{11}{4}\right), \left(-\dfrac{7}{2}, -\dfrac{11}{2}\right)$

31. $\left(-\dfrac{3}{4}, 2\right), \left(5, -\dfrac{7}{4}\right)$

GEOMETRY CONNECTION The vertices of a triangle are given. Classify the triangle as *scalene, isosceles,* or *equilateral.*

32. $(2, 0), (0, 8), (-2, 0)$ isosceles

33. $(4, 1), (1, -2), (6, -4)$ isosceles

34. $(1, 9), (-4, 2), (4, 2)$ scalene

35. $(2, 5), (8, 2), (4, -1)$ scalene

36. $(5, -1), (-4, 0), (3, 5)$ scalene

37. $(4, 4), (8, 1), (6, -5)$ scalene

38. $(0, -3), (3, 5), (-5, 2)$ isosceles

39. $(1, 1), (-4, 0), (-2, 5)$

40. $(2, 4), (3, -2), (-1, 1)$

45. $y = \frac{2}{15}x - 2.22$

46. $y = \frac{9}{14}x - \frac{113}{56}$

FINDING EQUATIONS Write an equation for the perpendicular bisector of the line segment joining the two points. $y = -\frac{4}{5}x - \frac{41}{5}$ $\quad y = \frac{4}{15}x + \frac{61}{30}$

41. $(2, 2), (6, 14)$ $\quad y = -\frac{1}{3}x + \frac{28}{3}$ 42. $(0, 0), (-8, -10)$ $\quad$ 43. $(0, -6), (-4, 9)$

44. $(3, -7), (-3, 1)$ $\quad y = \frac{3}{4}x - 3$ 45. $(-3, -7.2), (-4.2, 1.8)$ $\quad$ 46. $\left(\frac{3}{2}, -6\right), (-3, 1)$

FINDING A COORDINATE Use the given distance d between the two points to solve for x.

47. $(0, 1), (x, 4); d = \sqrt{34}$ $\quad -5; 5$ $\qquad$ 48. $(1, 3), (-6, x); d = \sqrt{74}$ $\quad -2; 8$

49. $(x, -10), (-8, 4); d = 7\sqrt{5}$ $\quad -15; -1$ $\qquad$ 50. $(0.5, x), (7, 2); d = 8.5$ $\quad 2 \pm \sqrt{30}$

STUDENT HELP

HOMEWORK HELP
Visit our Web site
www.mcdougallittell.com
for help with problem
solving in Exs. 51 and 52.

51. $\left(\frac{25}{2}, \frac{35}{2}\right); \left(\frac{75}{2}, \frac{35}{2}\right)$

🌎 **URBAN PLANNING** In Exercises 51 and 52, use the following information.
You are designing a city park like the one shown at the right. You want the park to have two fountains so that each fountain is equidistant from four of the six park entrances. The labeled points shown in the coordinate plane represent the park entrances.

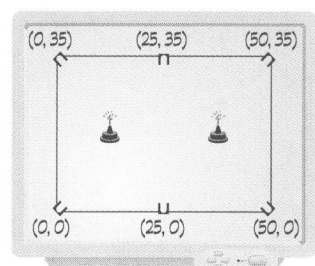

51. Where should the fountains be placed?

52. How far apart should the fountains be placed? **25**

🌎 **HELICOPTER RESCUE** In Exercises 53–56, use the following information to find the distance a medical helicopter would have to travel to St. John's Hospital from each highway intersection.
The Highway Department of Sangamon County in Illinois uses a map with a coordinate plane whose origin represents downtown Springfield. Each unit represents one mile and the letters N, S, E, and W are used to indicate the direction. For example, 3E 5S corresponds to $(3, -5)$, a point 3 miles east and 5 miles south of downtown Springfield. St. John's Hospital is located at 1E 0, or $(1, 0)$.

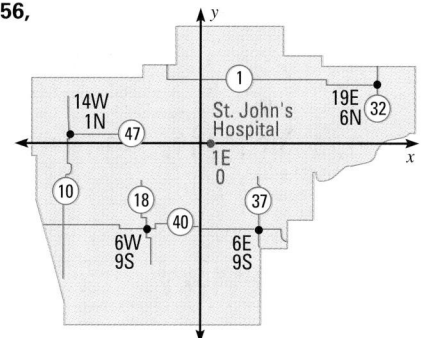

FOCUS ON CAREERS

ACCIDENT RECONSTRUCTIONIST An accident reconstructionist uses physical evidence, such as skid marks, to determine how accidents occurred.

CAREER LINK
www.mcdougallittell.com

53. Rt. 1–Rt. 32 intersection at 19E 6N
about 18.97 mi
54. Rt. 37–Rt. 40 intersection at 6E 9S
about 10.30 mi
55. Rt. 18–Rt. 40 intersection at 6W 9S
about 11.40 mi
56. Rt. 10–Rt. 47 intersection at 14W 1N
about 15.03 mi

57. 🌎 **ACCIDENT RECONSTRUCTION** When a car makes a fast, sharp turn, an accident reconstructionist can use the car's skid mark to determine its speed. The equation $v = \sqrt{ar}$ gives the car's speed v (in meters per second) as a function of the radius r of the circle (in meters) along which the car was traveling. The constant a (measured in meters per second squared) varies depending on road conditions. Find the radius of the skid mark shown below. Then use the given equation and 6.86 for a to find how fast the car was going.
▶ Source: *Mathematical Modeling* **See margin.**

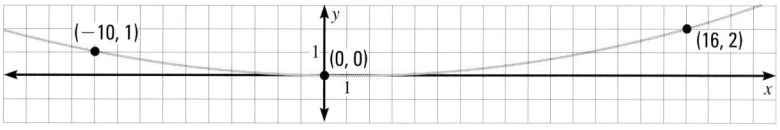

STUDENT HELP NOTES

→ **Homework Help** Students can find help for Exs. 51 and 52 at **www.mcdougallittell.com**. The information can be printed out for students who don't have access to the Internet.

CAREER NOTE
EXERCISE 57 Additional information about a career as an Accident Reconstructionist is available at **www.mcdougallittell.com**.

❗ **COMMON ERROR**
EXERCISES 17–31 Students may incorrectly write the distance formula when trying to find the length of a line segment by writing addition signs instead of subtraction signs or vice versa. To help avoid this problem, use the idea of the Pythagorean theorem to remind them that you subtract to find the lengths of the two sides of the right triangle, and you add the squares of their lengths.

57. r is about 58.56 m, v is about 20 m/sec.

ADDITIONAL PRACTICE AND RETEACHING

For Lesson 10.1:
• Practice Levels A, B, and C (*Chapter 10 Resource Book*, p. 13)
• Reteaching with Practice (*Chapter 10 Resource Book*, p. 16)
• See Lesson 10.1 of the *Personal Student Tutor*

For more Mixed Review:
• Search the *Test and Practice Generator* for key words or specific lessons.

10.1 *The Distance and Midpoint Formulas* **593**

DAILY HOMEWORK QUIZ

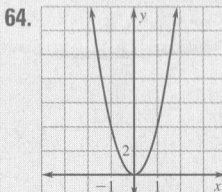 Transparency Available

Find the distance between the points and the midpoint of the segment joining them.

1. $(6, 3)$, $(-2, 1)$ $2\sqrt{17} \approx 8.25$; $(2, 2)$

2. $(7, -1)$, $(-2, -3)$ $\sqrt{85} \approx 9.22$;
$\left(\dfrac{5}{2}, -2\right)$

3. A triangle has vertices $X(-2, 0)$, $Y(2, 0)$, and $Z(0, 2\sqrt{3})$. Is $\triangle XYZ$ scalene, isosceles, or equilateral?   **equilateral**

4. Write an equation for the perpendicular bisector of the segment joining $(4, -6)$ and $(-2, 4)$.
$y = \dfrac{3}{5}x - \dfrac{8}{5}$

EXTRA CHALLENGE NOTE

→ Challenge problems for Lesson 10.1 are available in **blackline** format in the *Chapter 10 Resource Book,* p. 20 and at **www.mcdougallittell.com**.

ADDITIONAL TEST PREPARATION

1. WRITING Use the points $(-1, 5)$ and $(3, 9)$.
 a. Find the average of the x-values and call it $\bar{x}$.   **1**
 b. Find the average of the y-values and call it $\bar{y}$.   **7**
 c. Use your answers from parts (a) and (b) to write a new point $(\bar{x}, \bar{y})$.   **(1, 7)**
 d. What is the mathematical name for $(\bar{x}, \bar{y})$?   **the midpoint**

64.

65–71. See Additional Answers beginning on page AA1.

---

58. STATISTICS CONNECTION   A physician uses many tests to evaluate a patient's condition. Some of these tests yield numerical results. In these cases, the physician can treat two test results as an ordered pair and use the distance formula to determine how close to average the patient is. In the table below the serum creatinine (C) and systolic blood pressure (P) for several patients are given. Tell how far from normal each patient is, where normal is represented by the ordered pair $(C, P) = (1, 127)$. ≈ 7.07; 4; 13; ≈ 13.42; ≈ 15.13; ≈ 3.61; 3

C	2	5	1	7	3	4	1
P	120	127	140	115	112	125	130

Test  Preparation

QUANTITATIVE COMPARISON In Exercises 59–62, choose the statement that is true about the given quantities.

Ⓐ The quantity in column A is greater.

Ⓑ The quantity in column B is greater.

Ⓒ The two quantities are equal.

Ⓓ The relationship cannot be determined from the given information.

	Column A	Column B
B **59.**	Distance between $(0, 7)$ and $(1, -1)$	Distance between $(9, 2)$ and $(3, 8)$
A **60.**	Distance between $(-5, -2)$ and $(5, 2)$	Distance between $(-5, 5)$ and $(2, -2)$
A **61.**	Distance between $(-3, 0)$ and $(2, -4)$	Distance between $(7, 6)$ and $(1, 5)$
A **62.**	Distance between $(2, -5)$ and $(1, 6)$	Distance between $(0, 8)$ and $(6, 0)$

★ **Challenge**

63. FINDING A FORMULA Find formulas for the distance between a point (x, y) and each of the following: **(a)** a horizontal line $y = k$ and **(b)** a vertical line $x = h$.
a. $|y - k|$; b. $|x - h|$

MIXED REVIEW

GRAPHING FUNCTIONS Graph the quadratic function. (Review 5.1 for 10.2)
64–71. See margin.
64. $y = 4x^2$ **65.** $y = 3x^2$ **66.** $y = -3x^2$ **67.** $y = -2x^2$

68. $y = \dfrac{1}{3}x^2$ **69.** $y = -\dfrac{2}{3}x^2$ **70.** $y = -\dfrac{3}{4}x^2$ **71.** $y = \dfrac{5}{6}x^2$

SOLVING EQUATIONS Solve the equation. Check for extraneous solutions. (Review 7.6)

72. $x^{2/3} + 13 = 17$ **8**   **73.** $\sqrt{x + 100} = 25$ **525**   **74.** $\sqrt{2x} = x - 4$ **8**

75. $\sqrt{x + 2} = \sqrt{3x}$ **1**   **76.** $2\sqrt[3]{3x} = 6$ **9**   **77.** $-2x^{3/2} = -8$ $4^{2/3} \approx 2.52$

OPERATIONS WITH RATIONAL EXPRESSIONS Perform the indicated operation and simplify. (Review 9.5)

78. $\dfrac{2}{x + 1} - \dfrac{x}{x^2 - 1}$ $\dfrac{x - 2}{x^2 - 1}$ **79.** $\dfrac{4}{2x^2} + \dfrac{1}{3x}$ $\dfrac{x + 6}{3x^2}$ **80.** $\dfrac{11}{4(x - 5)} - \dfrac{x + 1}{4x}$ $\dfrac{-x^2 + 15x + 5}{4x^2 - 20x}$

81. $\dfrac{3x}{x^2} - \dfrac{x - 1}{x + 3}$ $\dfrac{-x^2 + 4x + 9}{x^2 + 3x}$ **82.** $\dfrac{2}{3x + 2} + \dfrac{5x^2}{x - 4}$ **83.** $\dfrac{1 - 3x}{x - 6} + \dfrac{2}{2x + 1}$

82. $\dfrac{15x^3 + 10x^2 + 2x - 8}{(3x + 2)(x - 4)}$

83. $\dfrac{-6x^2 + x - 11}{(x - 6)(2x + 1)}$

10.2

Parabolas

What you should learn

GOAL 1 Graph and write equations of parabolas.

GOAL 2 Use parabolas to solve **real-life** problems, such as finding the depth of a solar energy collector in **Example 3**.

Why you should learn it

▼ To model **real-life** parabolas, such as the reflector of a car's headlight in **Ex. 79**.

GOAL 1 **GRAPHING AND WRITING EQUATIONS OF PARABOLAS**

You already know that the graph of $y = ax^2$ is a parabola whose vertex $(0, 0)$ lies on its axis of symmetry $x = 0$. Every parabola has the property that any point on it is equidistant from a point called the **focus** and a line called the **directrix**.

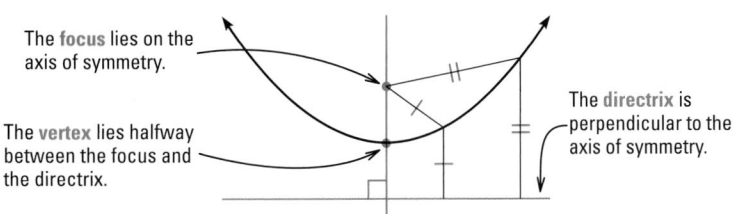

The **focus** lies on the axis of symmetry.

The **vertex** lies halfway between the focus and the directrix.

The **directrix** is perpendicular to the axis of symmetry.

In Chapter 5 you saw parabolas that have a vertical axis of symmetry and open up or down. In this lesson you will also work with parabolas that have a horizontal axis of symmetry and open left or right. In the four cases shown below, the focus and the directrix each lie $|p|$ units from the vertex.

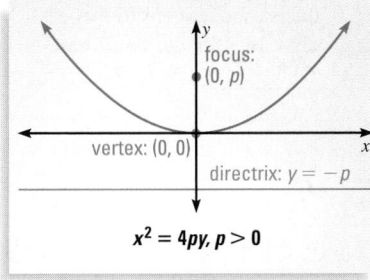

$x^2 = 4py, p > 0$

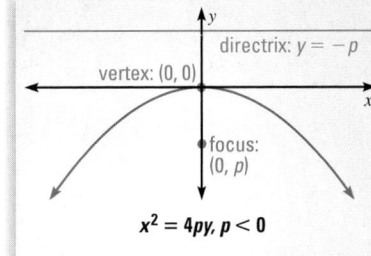

$x^2 = 4py, p < 0$

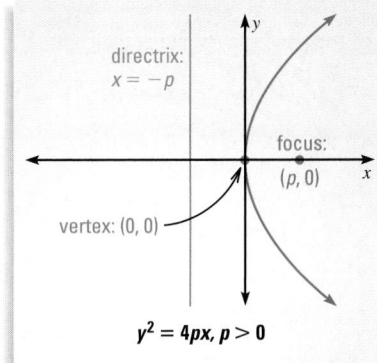

$y^2 = 4px, p > 0$

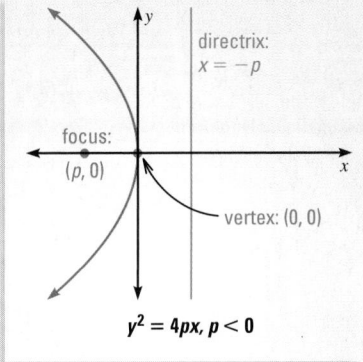

$y^2 = 4px, p < 0$

Characteristics of the parabolas shown above are given on the next page.

1 PLAN

PACING
Basic: 1 day
Average: 1 day
Advanced: 1 day
Block Schedule: 0.5 block with 10.1

▶ LESSON OPENER
ACTIVITY
An alternative way to approach Lesson 10.2 is to use the Activity Lesson Opener:
- Blackline Master (*Chapter 10 Resource Book*, p. 24)
- Transparency (p. 66)

MEETING INDIVIDUAL NEEDS
- *Chapter 10 Resource Book*
  Prerequisite Skills Review (p. 5)
  Practice Level A (p. 27)
  Practice Level B (p. 28)
  Practice Level C (p. 29)
  Reteaching with Practice (p. 30)
  Absent Student Catch-Up (p. 32)
  Challenge (p. 35)
- *Resources in Spanish*
- *Personal Student Tutor*

NEW-TEACHER SUPPORT
See the Tips for New Teachers on pp. 1–2 of the *Chapter 10 Resource Book* for additional notes about Lesson 10.2.

WARM-UP EXERCISES

Transparency Available

State whether the graph of the equation is a parabola.
1. $y^2 = 2x^2 + 3$ **No**
2. $y + x^2 = 7$ **Yes**
3. $y = 3x + 9$ **No**
4. $y = 6x^2 - x + 3$ **Yes**
5. $y^2 = 5$ **No**

EXTRA EXAMPLE 1

Identify the focus and the directrix of the parabola given by $x = \frac{3}{4}y^2$. Draw the parabola.

focus: $\left(\frac{1}{3}, 0\right)$; directrix: $x = -\frac{1}{3}$;

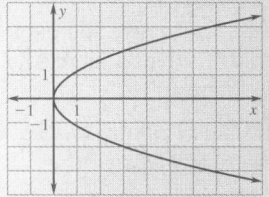

EXTRA EXAMPLE 2

Write an equation of the parabola shown below.

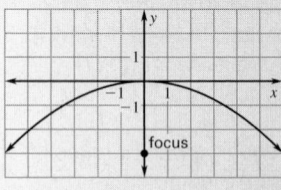

$x^2 = -12y$

CHECKPOINT EXERCISES

For use after Example 1:

1. Identify the focus and the directrix of the parabola given by $y = -\frac{1}{2}x^2$. Draw the parabola.

   focus: $\left(0, -\frac{1}{2}\right)$; directrix: $x = \frac{1}{2}$;

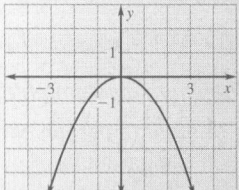

For use after Example 2:

2. Write an equation of the parabola shown below.

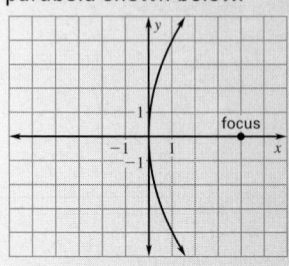

$y^2 = 16x$

STUDENT HELP

Look Back
For help with drawing parabolas, see p. 249.

STUDENT HELP

HOMEWORK HELP
Visit our Web site
www.mcdougallittell.com
for extra examples.

STANDARD EQUATION OF A PARABOLA (VERTEX AT ORIGIN)

The standard form of the equation of a parabola with vertex at (0, 0) is as follows.

EQUATION	FOCUS	DIRECTRIX	AXIS OF SYMMETRY
$x^2 = 4py$	$(0, p)$	$y = -p$	Vertical $(x = 0)$
$y^2 = 4px$	$(p, 0)$	$x = -p$	Horizontal $(y = 0)$

EXAMPLE 1 *Graphing an Equation of a Parabola*

Identify the focus and directrix of the parabola given by $x = -\frac{1}{6}y^2$. Draw the parabola.

SOLUTION

Because the variable y is squared, the axis of symmetry is horizontal. To find the focus and directrix, rewrite the equation as follows.

$$x = -\frac{1}{6}y^2 \qquad \text{Write original equation.}$$

$$-6x = y^2 \qquad \text{Multiply each side by } -6.$$

Since $4p = -6$, you know $p = -\frac{3}{2}$. The focus is $(p, 0) = \left(-\frac{3}{2}, 0\right)$ and the directrix is $x = -p = \frac{3}{2}$.

To draw the parabola, make a table of values and plot points. Because $p < 0$, the parabola opens to the left. Therefore, only negative x-values should be chosen.

x	-1	-2	-3	-4	-5
y	± 2.45	± 3.46	± 4.24	± 4.90	± 5.48

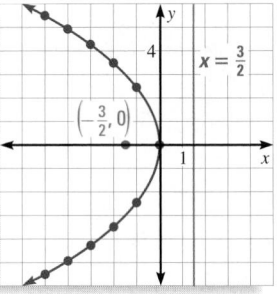

EXAMPLE 2 *Writing an Equation of a Parabola*

Write an equation of the parabola shown at the right.

SOLUTION

The graph shows that the vertex is $(0, 0)$ and the directrix is $y = -p = -2$. Substitute 2 for p in the standard equation for a parabola with a vertical axis of symmetry.

$$x^2 = 4py \qquad \text{Standard form, vertical axis of symmetry}$$

$$x^2 = 4(2)y \qquad \text{Substitute 2 for } p.$$

$$x^2 = 8y \qquad \text{Simplify.}$$

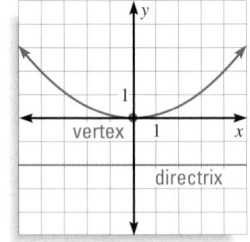

✓**CHECK** You can check this result by solving the equation for y to get $y = \frac{1}{8}x^2$ and graphing the equation using a graphing calculator.

GOAL 2 USING PARABOLAS IN REAL LIFE

Parabolic reflectors have cross sections that are parabolas. A special property of any parabolic reflector is that all incoming rays parallel to the axis of symmetry that hit the reflector are directed to the focus (Figure 1). Similarly, rays emitted from the focus that hit the reflector are directed in rays parallel to the axis of symmetry (Figure 2). These properties are the reason satellite dishes and flashlights are parabolic.

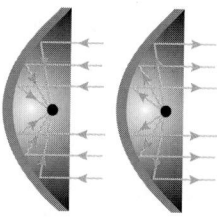

Figure 1   Figure 2

EXAMPLE 3   *Modeling a Parabolic Reflector*

SOLAR ENERGY Sunfire is a glass parabola used to collect solar energy. The sun's rays are reflected from the mirrors toward two boilers located at the focus of the parabola. When heated, the boilers produce steam that powers an alternator to produce electricity.

a. Write an equation for Sunfire's cross section.

b. How deep is the dish?

Sunfire

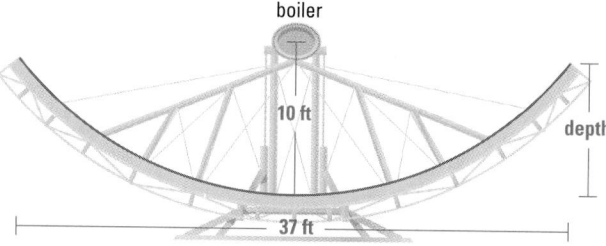

FOCUS ON PEOPLE

HOWARD FRANK BROYLES is an engineer and inventor. More than 250 high school students volunteered their time to help him build Sunfire. The project took over ten years.

APPLICATION LINK
www.mcdougallittell.com

SOLUTION

a. The boilers are 10 feet above the vertex of the dish. Because the boilers are at the focus and the focus is p units from the vertex, you can conclude that $p = 10$.

Assuming the vertex is at the origin, an equation for the parabolic cross section is as follows:

$x^2 = 4py$	**Standard form, vertical axis of symmetry**
$x^2 = 4(10)y$	**Substitute 10 for p.**
$x^2 = 40y$	**Simplify.**

b. The dish extends $\frac{37}{2} = 18.5$ feet on either side of the origin. To find the depth of the dish, substitute **18.5** for x in the equation from part (a).

$x^2 = 40y$	**Equation for the cross section**
$(18.5)^2 = 40y$	**Substitute 18.5 for x.**
$8.6 \approx y$	**Solve for y.**

▶ The dish is about 8.6 feet deep.

EXTRA EXAMPLE 3

A microphone has a parabolic reflector around it to capture sound. The microphone is placed at the focus of the parabola to reflect as much sound as possible to the microphone. A cross section of the reflector is shown below.

a. Write an equation for the cross section of the reflector.
$y = \frac{1}{20}x^2$

b. How high is the microphone above the vertex? **5 inches**

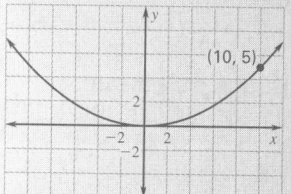

✓ CHECKPOINT EXERCISES

For use after Example 3:

1. A store uses a parabolic mirror to see all of the aisles in the store. A cross section of the mirror is shown below.

a. Write an equation for the cross section of the mirror.
$x^2 = 32y$

b. What is the focus of the cross section? **(0, 8)**

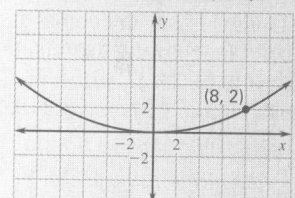

FOCUS ON VOCABULARY

The terms *focus* and *directrix* may be new to students. Emphasize that the focus is a point and the directrix is a line.

CLOSURE QUESTION

How can you tell from the standard equation of a parabola if it opens up or down, or if it opens left or right?
Sample answer: If the equation is $y^2 = 4px$, the parabola opens left or right. If the equation is $x^2 = 4py$, the parabola opens up or down.

ASSIGNMENT GUIDE

BASIC
Day 1: pp. 598–600 Exs. 16–21,
22–25, 30–33, 38–43,
54–60, 66–69, 78, 83,
85–97 odd

AVERAGE
Day 1: pp. 598–600 Exs. 16–21,
22–70 even, 78, 82, 83,
85–97 odd

ADVANCED
Day 1: pp. 598–600 Exs. 16–21,
22–78 even, 81–84,
85–103 odd

BLOCK SCHEDULE
pp. 598–600 Exs. 16–21, 22–70
even, 78, 82, 83, 85–97 odd,
103 (with 10.1)

EXERCISE LEVELS
Level A: *Easier*
16–29, 54–65
Level B: *More Difficult*
30–53, 66–81
Level C: *Most Difficult*
82–84

✔ HOMEWORK CHECK
To quickly check student under-
standing of key concepts, go
over the following exercises:
Exs. 16, 18, 24, 32, 40, 54, 66. See
also the Daily Homework Quiz:
- Blackline Master (*Chapter 10
Resource Book*, p. 38)
- Transparency (p. 77)

4.
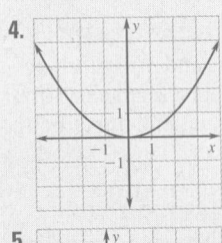

5.
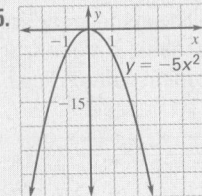

GUIDED PRACTICE

Vocabulary Check ✔

1. Complete this statement: A parabola is the set of points equidistant from a point called the __?__ and a line called the __?__. **focus; directrix**

Concept Check ✔

2. How does the graph of $x = ay^2$ differ from the graph of $y = ax^2$? **See margin.**

3. Knowing the value of a in $y = ax^2$, how can you find the focus and directrix?
See margin.

Skill Check ✔

Graph the equation. Identify the focus and directrix of the parabola. **4–9. See margin for graphs.**

4. $x^2 = 4y$ (0, 1); $y = -1$ 5. $y = -5x^2 \left(0, -\frac{1}{20}\right); y = \frac{1}{20}$ 6. $-12x = y^2$ (−3, 0); $x = 3$

7. $8y^2 = x$ $\left(\frac{1}{32}, 0\right); x = -\frac{1}{32}$ 8. $-6x = y^2 \left(-\frac{3}{2}, 0\right); x = \frac{3}{2}$ 9. $x^2 = 2y$ $\left(0, \frac{1}{2}\right); y = -\frac{1}{2}$

Write the standard form of the equation of the parabola with the given focus or directrix and vertex at (0, 0).

10. focus: (0, 3) $x^2 = 12y$ 11. focus: (5, 0) $y^2 = 20x$ 12. focus: (−6, 0) $y^2 = -24x$

13. directrix: $x = 4$ 14. directrix: $x = -1$ 15. directrix: $y = 8$
$y^2 = -16x$ $y^2 = 4x$ $x^2 = -32y$

PRACTICE AND APPLICATIONS

STUDENT HELP
↳ **Extra Practice**
to help you master
skills is on p. 953.

MATCHING Match the equation with its graph.

16. $y^2 = 4x$ D   17. $x^2 = -4y$ B   18. $x^2 = 4y$ A

19. $y^2 = -4x$ E   20. $y^2 = \frac{1}{4}x$ F   21. $x^2 = \frac{1}{4}y$ C

A.
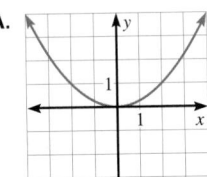

B.
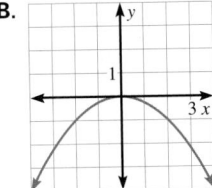

C.
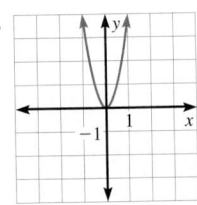

D.
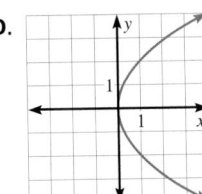

E.
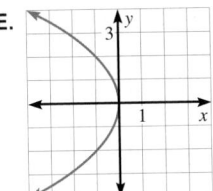

F.
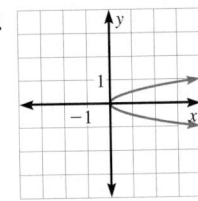

30. $\left(0, -\frac{1}{12}\right); y = \frac{1}{12}$

31. $\left(\frac{1}{8}, 0\right); x = -\frac{1}{8}$

33. $\left(-\frac{5}{2}, 0\right); x = \frac{5}{2}$

DIRECTION Tell whether the parabola opens *up, down, left,* or *right*.

22. $y = -3x^2$ down 23. $-9x^2 = 2y$ down 24. $2y^2 = -6x$ left 25. $x = 7y^2$ right

26. $x^2 = 16y$ up 27. $-3y^2 = 8x$ left 28. $-5x = -y^2$ right 29. $x^2 = \frac{4}{3}y$ up

STUDENT HELP
↳ **HOMEWORK HELP**
Example 1: Exs. 16–53
Example 2: Exs. 54–77
Example 3: Exs. 78–81

FOCUS AND DIRECTRIX Identify the focus and directrix of the parabola.
30, 31, 33. See margin.

30. $3x^2 = -y$ 31. $2y^2 = x$ 32. $x^2 = 8y$ (0, 2); $y = -2$ 33. $y^2 = -10x$

34. $y^2 = -16x$ 35. $x^2 = -36y$ 36. $-4x + 9y^2 = 0$ 37. $-28y + x^2 = 0$
(−4, 0); $x = 4$ (0, −9); $y = 9$ $\left(\frac{1}{9}, 0\right); x = -\frac{1}{9}$ (0, 7); $y = -7$

6–9. See Additional Answers
beginning on page AA1.

2. The graph of $y = ax^2$
rotated 90° clockwise is
the graph of $x = ay^2$.

3. focus: $\left(0, \frac{1}{4a}\right)$; directrix:
$y = -\frac{1}{4a}$

38. (3, 0); $x = -3$

39. $\left(0, -\dfrac{3}{2}\right)$; $y = \dfrac{3}{2}$

40. $\left(-\dfrac{1}{2}, 0\right)$; $x = \dfrac{1}{2}$

41. (6, 0); $x = -6$

42. (0, 2); $y = -2$

43. $\left(-\dfrac{7}{2}, 0\right)$; $x = \dfrac{7}{2}$

44. (0, −5); $y = 5$

45. $\left(0, \dfrac{9}{2}\right)$; $y = -\dfrac{9}{2}$

46. (0, −1); $y = 1$

47. (0, 4); $y = -4$

48. $\left(\dfrac{9}{4}, 0\right)$; $x = -\dfrac{9}{4}$

49. $\left(-\dfrac{3}{4}, 0\right)$; $x = \dfrac{3}{4}$

50. (0, 10); $y = -10$

51. (−5, 0); $x = 5$

52. $\left(0, \dfrac{1}{3}\right)$; $y = -\dfrac{1}{3}$

53. (2, 0); $x = -2$

GRAPHING Graph the equation. Identify the focus and directrix of the parabola.
38–53. See margin for graphs.

38. $y^2 = 12x$ **39.** $x^2 = -6y$ **40.** $y^2 = -2x$ **41.** $y^2 = 24x$

42. $x^2 = 8y$ **43.** $y^2 = -14x$ **44.** $x^2 = -20y$ **45.** $x^2 = 18y$

46. $x^2 = -4y$ **47.** $x^2 = 16y$ **48.** $y^2 = 9x$ **49.** $y^2 = -3x$

50. $x^2 - 40y = 0$ **51.** $x + \dfrac{1}{20}y^2 = 0$ **52.** $3x^2 = 4y$ **53.** $x - \dfrac{1}{8}y^2 = 0$

WRITING EQUATIONS Write the standard form of the equation of the parabola with the given focus and vertex at (0, 0).

54. (4, 0) $y^2 = 16x$ **55.** (−2, 0) $y^2 = -8x$ **56.** (−3, 0) $y^2 = -12x$ **57.** (0, 1) $x^2 = 4y$

58. (0, 4) $x^2 = 16y$ **59.** $\left(0, -3\right)$ $x^2 = -12y$ **60.** $\left(0, -4\right)$ $x^2 = -16y$ **61.** (−5, 0) $y^2 = -20x$

62. $\left(-\dfrac{1}{4}, 0\right)$ $y^2 = -x$ **63.** $\left(0, -\dfrac{3}{8}\right)$ $x^2 = -\dfrac{3}{2}y$ **64.** $\left(0, \dfrac{1}{2}\right)$ $x^2 = 2y$ **65.** $\left(\dfrac{5}{12}, 0\right)$ $y^2 = \dfrac{5}{3}x$

WRITING EQUATIONS Write the standard form of the equation of the parabola with the given directrix and vertex at (0, 0).

66. $y = 2$ $x^2 = -8y$ **67.** $y = -3$ $x^2 = 12y$ **68.** $x = -4$ $y^2 = 16x$ **69.** $x = 6$ $y^2 = -24x$

70. $x = -5$ $y^2 = 20x$ **71.** $y = -1$ $x^2 = 4y$ **72.** $x = 2$ $y^2 = -8x$ **73.** $y = 4$ $x^2 = -16y$

74. $x = -\dfrac{1}{2}$ $y^2 = 2x$ **75.** $x = \dfrac{3}{4}$ $y^2 = -3x$ **76.** $y = \dfrac{5}{8}$ $x^2 = -\dfrac{5}{2}y$ **77.** $y = -\dfrac{1}{12}$ $x^2 = \dfrac{1}{3}y$

78. 🌐 **COMMUNICATIONS** The cross section of a television antenna dish is a parabola. For the dish at the right, the receiver is located at the focus, 4 feet above the vertex. Find an equation for the cross section of the dish. (Assume the vertex is at the origin.) If the dish is 8 feet wide, how deep is it? $x^2 = 16y$; 1 ft

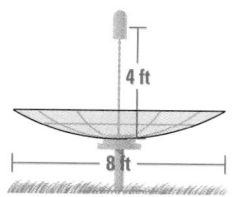
4 ft
8 ft

79. 🌐 **AUTOMOTIVE ENGINEERING** The filament of a lightbulb is a thin wire that glows when electricity passes through it. The filament of a car headlight is at the focus of a parabolic reflector, which sends light out in a straight beam. Given that the filament is 1.5 inches from the vertex, find an equation for the cross section of the reflector. If the reflector is 7 inches wide, how deep is it? $y^2 = 6x$; 2.04 in.

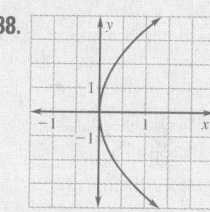
7 in.
1.5 in.

80. **HISTORY** **CONNECTION** In the drawing shown at the left, the rays of the sun are lighting a candle. If the candle flame is 12 inches from the back of the parabolic reflector and the reflector is 6 inches deep, then what is the diameter of the reflector?   about 34 in.

81. 🌐 **CAMPING** You can make a solar hot dog cooker using foil-lined cardboard shaped as a parabolic trough. The drawing at the right shows how to suspend a hot dog with a wire through the focus of each end piece. If the trough is 12 inches wide and 4 inches deep, how far from the bottom should the wire be placed?

▶ Source: *Boys' Life*   2.25 in.

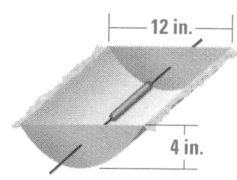
12 in.
4 in.

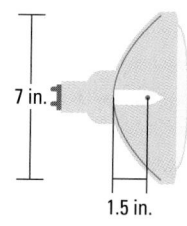

FOCUS ON APPLICATIONS

MATH IN HISTORY
Man has been aware of the reflective properties of parabolas for two thousand years. The above illustration of Archimedes' burning mirror was taken from a book printed in the 17th century.

10.2 *Parabolas*   **599**

! COMMON ERROR
EXERCISES 30–53 It may be confusing for students to understand how x, y, and p are related in the standard equations for parabolas. To help them understand, have them explain in their own words how p is used to find the focus and directrix of the parabola.

TEACHING TIPS
EXERCISES 38–53 In previous courses, students have have used factoring, the quadratic formula, and completing the square to write an equation in vertex form to graph a parabola. Relate these ideas to what is being discussed in this lesson to help them make connections.

ENGLISH LEARNERS
EXERCISE 79 Note that English learners often have difficulty translating words and word problems into correct equations. For Exercise 79, you may want to read aloud the given information one sentence at a time, pointing out on the diagram the locations of items and distances mentioned.

38.

39–53. See Additional Answers beginning on page AA1.

ADDITIONAL PRACTICE AND RETEACHING

For Lesson 10.2:
• Practice Levels A, B, and C (*Chapter 10 Resource Book*, p. 27)
• Reteaching with Practice (*Chapter 10 Resource Book*, p. 30)
• 🖥 See Lesson 10.2 of the *Personal Student Tutor*

For more Mixed Review:
• 🖥 Search the *Test and Practice Generator* for key words or specific lessons.

DAILY HOMEWORK QUIZ

📄 **Transparency Available**

Graph the parabola and identify its focus and directrix.

1. $x^2 = -10y$ (0, −2.5); $y = 2.5$

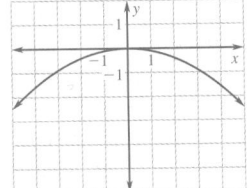

2. $y^2 = -4x$ (−1, 0); $x = 1$

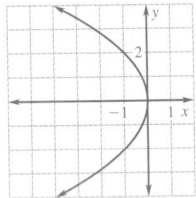

3. What is the equation of the parabola with vertex at (0, 0) and directrix $y = -\frac{1}{2}$? $x^2 = 2y$

4. What is the equation of the parabola with vertex at (0, 0) and focus at (−6, 0)? $y^2 = -24x$

EXTRA CHALLENGE NOTE

Challenge problems for Lesson 10.2 are available in **blackline** format in the *Chapter 10 Resource Book*, p. 35 and at **www.mcdougallittell.com**.

ADDITIONAL TEST PREPARATION

1. WRITING Use the equation $x = \frac{y^2}{4p}$.

a. What is the shape of the graph of the equation? **parabola that opens right or left**

b. What does the value of p tell you about the graph? **p represents the focus of the graph. If p is positive, the graph opens to the right.**

---

82. *Writing* For an equation of the form $y = ax^2$, discuss what effect increasing $|a|$ has on the focus and directrix. **As $|a|$ increases, focus and directrix move closer to the origin.**

Test Preparation

83. MULTI-STEP PROBLEM A flashlight has a parabolic reflector. An equation for the cross section of the reflector is $y^2 = \frac{32}{7}x$. The depth of the reflector is $\frac{3}{2}$ inches.

a. *Writing* Explain why the value of p must be less than the depth of the reflector of a flashlight. **If not, the bulb would extend outside of the flashlight.**

b. How wide is the beam of light projected by the flashlight? **About 5.2 in.**

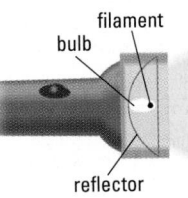

filament
bulb
reflector

c. Write an equation for the cross section of a reflector having the same depth but a wider beam than the flashlight shown. How wide is the beam of the new reflector? **$y^2 = 6x$; 6 in.**

d. Write an equation for the cross section of a reflector having the same depth but a narrower beam than the flashlight shown. How wide is the beam of the new reflector? **$y^2 = 2x$; 3.46 in.**

★ **Challenge**

EXTRA CHALLENGE
▸ www.mcdougallittell.com

84. GEOMETRY CONNECTION The *latus rectum* of a parabola is the line segment that is parallel to the directrix, passes through the focus, and has endpoints that lie on the parabola. Find the length of the latus rectum of a parabola given by $x^2 = 4py$. **$4p$**

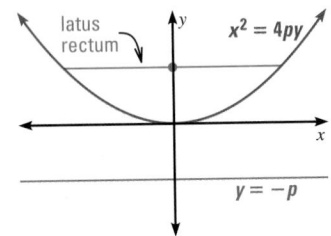

latus rectum
$x^2 = 4py$
$y = -p$

MIXED REVIEW

LOGARITHMIC AND EXPONENTIAL EQUATIONS Solve the equation. Check for extraneous solutions. (Review 8.6)

85. $8^{5x} = 16^{2x+1}$ $\frac{4}{7}$

86. $3^x = 15$ ≈ 2.465

87. $5^x = 7$ ≈ 1.209

88. $10^{3x+1} + 4 = 33$ ≈ 0.154

89. $\log_7 (3x - 5) = \log_7 8x$ no solution

90. $\log_3 (4x - 3) = 3$ $\frac{15}{2}$

OPERATIONS WITH RATIONAL EXPRESSIONS Perform the indicated operation and simplify. (Review 9.4 and 9.5)

91. $\frac{3xy^3}{x^3y} \cdot \frac{y}{6x}$ $\frac{y^3}{2x^3}$

92. $\frac{3xy^3}{2x} \div \frac{2xy^3}{3x}$ $\frac{9}{4}$

93. $\frac{x^2 - 9}{x^2 - x - 6} \cdot (x + 2)$ $x + 3$

94. $\frac{-3x}{x + 2} + \frac{4x}{x - 1}$ $\frac{x^2 + 11x}{(x + 2)(x - 1)}$

95. $\frac{x + 1}{6x^2} - \frac{x + 1}{6x^2 + 6x}$ $\frac{1}{6x^2}$

96. $\frac{x^2 - 3x + 2}{x - 1} - \frac{x^2 - 4}{x - 2}$ -4

FINDING A DISTANCE Find the distance between the two points. (Review 10.1 for 10.3)

97. (3, 4), (6, 7) $3\sqrt{2} \approx 4.243$

98. (−3, 7), (−7, 3) $4\sqrt{2} \approx 5.657$

99. (18, −4), (−2, 9) $\sqrt{569} \approx 23.854$

100. (3.7, 5.1), (2, 5) $\sqrt{2.9} \approx 1.703$

101. (−9, −31), (8, 7) $\sqrt{1733} \approx 41.629$

102. (8.8, 3.3), (1.2, 6) $\sqrt{65.05} \approx 8.065$

103. 🌐 **CONSUMER ECONOMICS** The amount A (in dollars) you pay for grapes varies directly with the amount P (in pounds) that you buy. Suppose you buy $1\frac{1}{2}$ pounds for $2.25. Write a linear model that gives A as a function of P. (Review 2.4) **$A = 1.5p$**

10.3

Circles

What you should learn

GOAL 1 Graph and write equations of circles.

GOAL 2 Use circles to solve **real-life** problems, such as determining whether you are affected by an earthquake in **Ex. 81**.

Why you should learn it

▼ To model **real-life** situations with circular models, such as the region lit by a lighthouse beam in **Example 4**.

GOAL 1 **GRAPHING AND WRITING EQUATIONS OF CIRCLES**

A **circle** is the set of all points (x, y) that are equidistant from a fixed point, called the **center** of the circle. The distance r between the center and any point (x, y) on the circle is the **radius**.

The distance formula can be used to obtain an equation of the circle whose center is the origin and whose radius is r. Because the distance between any point (x, y) on the circle and the center $(0, 0)$ is r, you can write the following.

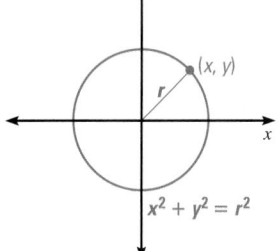

$$\sqrt{(x - 0)^2 + (y - 0)^2} = r \qquad \text{Distance formula}$$

$$(x - 0)^2 + (y - 0)^2 = r^2 \qquad \text{Square both sides.}$$

$$x^2 + y^2 = r^2 \qquad \text{Simplify.}$$

STANDARD EQUATION OF A CIRCLE (CENTER AT ORIGIN)

The **standard form of the equation of a circle** with center at $(0, 0)$ and radius r is as follows:

$$x^2 + y^2 = r^2$$

EXAMPLE A circle with center at $(0, 0)$ and radius 3 has equation $x^2 + y^2 = 9$.

EXAMPLE 1 *Graphing an Equation of a Circle*

Draw the circle given by $y^2 = 25 - x^2$.

SOLUTION

Write the equation in standard form.

$$y^2 = 25 - x^2 \qquad \text{Original equation}$$

$$x^2 + y^2 = 25 \qquad \text{Add } x^2 \text{ to each side.}$$

In this form you can see that the graph is a circle whose center is the origin and whose radius is $r = \sqrt{25} = 5$.

Plot several points that are 5 units from the origin. The points $(0, 5)$, $(5, 0)$, $(0, -5)$, and $(-5, 0)$ are most convenient.

Draw a circle that passes through the four points.

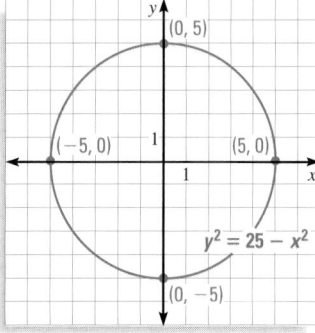

1 PLAN

PACING
Basic: 1 day
Average: 1 day
Advanced: 1 day
Block Schedule: 0.5 block with 10.4

LESSON OPENER
APPLICATION
An alternative way to approach Lesson 10.3 is to use the Application Lesson Opener:
- Blackline Master (*Chapter 10 Resource Book*, p. 39)
- Transparency (p. 67)

MEETING INDIVIDUAL NEEDS
- *Chapter 10 Resource Book*
  Prerequisite Skills Review (p. 5)
  Practice Level A (p. 41)
  Practice Level B (p. 42)
  Practice Level C (p. 43)
  Reteaching with Practice (p. 44)
  Absent Student Catch-Up (p. 46)
  Challenge (p. 48)
- *Resources in Spanish*
- *Personal Student Tutor*

NEW-TEACHER SUPPORT
See the Tips for New Teachers on pp. 1–2 of the *Chapter 10 Resource Book* for additional notes about Lesson 10.3.

WARM-UP EXERCISES

Transparency Available

Solve for x when $y = 2$.
1. $x^2 + y^2 = 13$ $-3, 3$
2. $y^2 = 12 - x^2$ $-2\sqrt{2}, 2\sqrt{2}$
3. $x^2 = 25 - y^2$ $-\sqrt{21}, \sqrt{21}$

Find the slope of the line.
4. $y = 4 - 2x$ -2
5. $3x + 4y = 5$ $-\frac{3}{4}$

2 TEACH

MOTIVATING THE LESSON
Draw a circle on the board and ask students what the figure is. When they say a circle, ask them why its a circle. Encourage them to use terms from geometry. Tell students that today's lesson deals with the algebraic aspects of circles.

EXTRA EXAMPLE 1
Draw the circle $y^2 = 4 - x^2$.

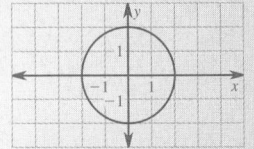

EXTRA EXAMPLE 2
$(-2, 5)$ is on a circle centered at the origin. Write the equation of the circle. $x^2 + y^2 = 29$

EXTRA EXAMPLE 3
Write an equation of the line that is tangent to the circle $x^2 + y^2 = 17$ at $(1, 4)$. $y = -\frac{1}{4}x + \frac{17}{4}$

CHECKPOINT EXERCISES
For use after Example 1:

1. Draw the circle given by $x^2 = 16 - y^2$.

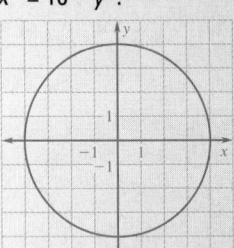

For use after Example 2:

2. $(5, 1)$ is on a circle centered at the origin. Write the equation of the circle. $x^2 + y^2 = 26$

For use after Example 3:

3. Write an equation of the line that is tangent to the circle $x^2 + y^2 = 20$ at $(-2, 4)$. $y = \frac{1}{2}x + 5$

> **STUDENT HELP**
>
> **HOMEWORK HELP**
> Visit our Web site
> www.mcdougallittell.com
> for extra examples.

> **STUDENT HELP**
>
> → **Study Tip**
> In mathematics the term radius is used in two ways. As defined on the previous page, it is the distance from the center of a circle to a point on the circle. It can also refer to the line segment that connects the center to a point on the circle.

EXAMPLE 2 *Writing an Equation of a Circle*

The point $(1, 4)$ is on a circle whose center is the origin. Write the standard form of the equation of the circle.

SOLUTION

Because the point $(1, 4)$ is on the circle, the radius of the circle must be the distance between the center and the point $(1, 4)$.

$$r = \sqrt{(1 - 0)^2 + (4 - 0)^2} \qquad \text{Use the distance formula.}$$
$$= \sqrt{1 + 16} \qquad \text{Simplify.}$$
$$= \sqrt{17}$$

Knowing that the radius is $\sqrt{17}$, you can use the standard form to find an equation of the circle.

$$x^2 + y^2 = r^2 \qquad \text{Standard form}$$
$$x^2 + y^2 = \left(\sqrt{17}\right)^2 \qquad \text{Substitute } \sqrt{17} \text{ for } r.$$
$$x^2 + y^2 = 17 \qquad \text{Simplify.}$$

· · · · · · · · ·

A theorem in geometry states that a line tangent to a circle is perpendicular to the circle's radius at the point of tangency. In the diagram, $\overleftrightarrow{AB}$ is tangent to the circle with center C at the point of tangency B, so $\overleftrightarrow{AB} \perp \overline{BC}$. This property of circles is used in the next example.

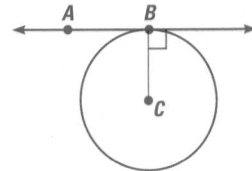

EXAMPLE 3 *Finding a Tangent Line*

Write an equation of the line that is tangent to the circle $x^2 + y^2 = 13$ at $(2, 3)$.

SOLUTION

The slope of the radius through the point $(2, 3)$ is:

$$m = \frac{3 - 0}{2 - 0} = \frac{3}{2}$$

Because the tangent line at $(2, 3)$ is perpendicular to this radius, its slope must be the negative reciprocal of $\frac{3}{2}$, or $-\frac{2}{3}$. So, an equation of the tangent line is as follows.

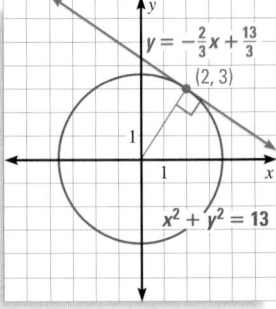

$$y - 3 = -\frac{2}{3}(x - 2) \qquad \text{Point-slope form}$$
$$y - 3 = -\frac{2}{3}x + \frac{4}{3} \qquad \text{Distributive property}$$
$$y = -\frac{2}{3}x + \frac{13}{3} \qquad \text{Add 3 to each side.}$$

▶ An equation of the tangent line is $y = -\frac{2}{3}x + \frac{13}{3}$.

602 **Chapter 10** *Quadratic Relations and Conic Sections*

FOCUS ON
APPLICATIONS

THE PHAROS OF ALEXANDRIA

was a lighthouse built in Egypt in about 280 B.C. One of the Seven Wonders of the World, it was said to be over 440 feet tall. It stood for nearly 1400 years.

GOAL 2 **USING CIRCLES IN REAL LIFE**

The regions inside and outside the circle $x^2 + y^2 = r^2$ can be described by inequalities.

> **Region inside circle:** $x^2 + y^2 < r^2$
>
> **Region outside circle:** $x^2 + y^2 > r^2$

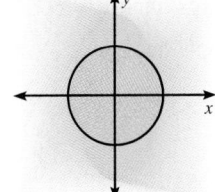

EXAMPLE 4 *Using a Circular Model*

OCEAN NAVIGATION The beam of a lighthouse can be seen for up to 20 miles. You are on a ship that is 10 miles east and 16 miles north of the lighthouse.

a. Write an inequality to describe the region lit by the lighthouse beam.

b. Can you see the lighthouse beam?

SOLUTION

a. As shown at the right the lighthouse beam can be seen from all points that satisfy this inequality:

$$x^2 + y^2 < 20^2$$

b. Substitute the coordinates of the ship into the inequality you wrote in part (a).

$x^2 + y^2 < 20^2$	Inequality from part (a)
$10^2 + 16^2 \overset{?}{<} 20^2$	Substitute for x and y.
$100 + 256 \overset{?}{<} 400$	Simplify.
$356 < 400$ ✓	The inequality is true.

▶ You can see the beam from the ship.

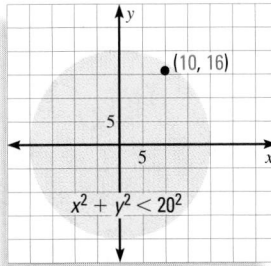

In the diagram above, the origin represents the lighthouse and the positive y-axis represents north.

EXAMPLE 5 *Using a Circular Model*

OCEAN NAVIGATION Your ship in Example 4 is traveling due south. For how many more miles will you be able to see the beam?

SOLUTION

When the ship exits the region lit by the beam, it will be at a point on the circle $x^2 + y^2 = 20^2$. Furthermore, its x-coordinate will be 10 and its y-coordinate will be negative. Find the point $(10, y)$ where $y < 0$ on the circle $x^2 + y^2 = 20^2$.

$x^2 + y^2 = 20^2$	Equation for the boundary
$10^2 + y^2 = 20^2$	Substitute 10 for x.
$y = \pm\sqrt{300} \approx \pm 17.3$	Solve for y.

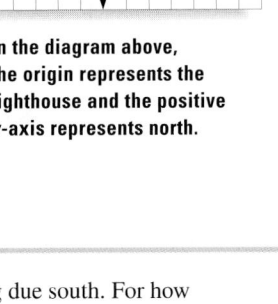

▶ Since $y < 0$, $y \approx -17.3$. The beam will be in view as the ship travels from $(10, 16)$ to $(10, -17.3)$, a distance of $|16 - (-17.3)| = 33.3$ miles.

10.3 *Circles* **603**

📖 **EXTRA EXAMPLE 4**
A street light can be seen on the ground within 30 yd of its center. You are driving and are 10 yd east and 25 yd south of the light.
a. Write an inequality to describe the region on the ground that is lit by the light. $x^2 + y^2 < 30^2$
b. Is the street light visible? $10^2 + (-25)^2 = 725 < 900$; yes

EXTRA EXAMPLE 5
In Extra Example 4 you are driving due north. For how many more yards will you be able to see the beam? about 53.28 yd

✓ **CHECKPOINT EXERCISES**
For use after Examples 4 and 5:
1. You are sky diving and are trying to land on a target with a radius of 15 m. You land 10 m west and 10 m north of the center of the target.
 a. Write an inequality to describe the region inside the target area. $x^2 + y^2 < 15^2$
 b. Do you land within the target area? yes
 c. If another skydiver lands 4 m east and 14 m north of the center of the target, who is closer to the center? You are; you land 14.14 m from the center and the other skydiver lands 14.56 m from the center.

CLOSURE QUESTION
What is the standard equation for a circle whose center is at the origin and whose radius is length r? $x^2 + y^2 = r^2$

DAILY PUZZLER
Six people try to guess the number of jelly beans in a jar. The six guesses are 104, 118, 124, 130, 98, and 84. One guess is 24 away from the correct number, and the other guesses are 2, 8, 12, 18, and 22 away from it. How many jelly beans are in the jar? 106

ASSIGNMENT GUIDE

BASIC
Day 1: pp. 604–607 Exs. 20–25,
26–44 even, 48–64 even,
71–74, 79–81, 88, 89,
91–105 odd,
Quiz 1 Exs. 1–23

AVERAGE
Day 1: pp. 604–607 Exs. 20–25,
26–44 even, 48–64 even,
71–74, 79–81, 88, 89,
91–105 odd,
Quiz 1 Exs. 1–23

ADVANCED
Day 1: pp. 604–607 Exs. 20–25,
26–44 even, 48–64 even,
71–74, 79–90, 91–105 odd,
Quiz 1 Exs. 1–23

BLOCK SCHEDULE
pp. 604–607 Exs. 20–25, 26–44
even, 48–64 even, 71–74, 79–81,
88, 89, 91–105 odd,
Quiz 1 Exs. 1–23 (with 10.4)

EXERCISE LEVELS
Level A: *Easier*
20–32, 38, 39, 47–60
Level B: *More Difficult*
33–37, 40–46, 61–82, 86–89
Level C: *Most Difficult*
83–85, 91

✔ HOMEWORK CHECK
To quickly check student understanding of key concepts, go
over the following exercises:
Exs. 20, 30, 38, 44, 52, 60, 72. See
also the Daily Homework Quiz:

• Blackline Master (*Chapter 10
Resource Book*, p. 52)
• Transparency (p. 78)

13.
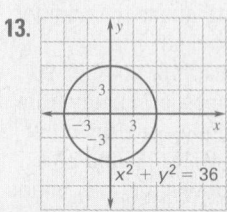
$x^2 + y^2 = 36$

14–18. See Additional Answers
beginning on page AA1.

GUIDED PRACTICE

Vocabulary Check ✔   **1.** State the definition of a circle. **The set of all points** (x, y) **equidistant from a fixed point.**

Concept Check ✔   **2. LOGICAL REASONING** Tell whether the following statement is *always true*, *sometimes true*, or *never true*: For a given circle and a given x-coordinate, there are two points on the circle with that x-coordinate.   **sometimes true**

3. How is the slope of a line tangent to a circle related to the slope of the radius at the point of tangency?   **They are negative reciprocals of each other (except if one line is vertical).**

4. ERROR ANALYSIS A student was asked to write an equation of a circle with its center at the origin and a radius of 4. The student wrote the following equation:

$$x^2 + y^2 = 4$$

What did the student do wrong? Write the correct equation.
The student failed to square the radius; $x^2 + y^2 = 16.$

Skill Check ✔   **Write the standard form of the equation of the circle that passes through the given point and whose center is the origin.**

5. $(4, 0)$ $x^2 + y^2 = 16$ **6.** $(0, -2)$ $x^2 + y^2 = 4$ **7.** $(-8, 6)$ $x^2 + y^2 = 100$ **8.** $(-5, -12)$ $x^2 + y^2 = 169$

9. $(6, -9)$ $x^2 + y^2 = 117$ **10.** $(3, 1)$ $x^2 + y^2 = 10$ **11.** $(-5, -5)$ $x^2 + y^2 = 50$ **12.** $(-2, 4)$ $x^2 + y^2 = 20$

Graph the equation. Give the radius of the circle.   **13–18. See margin for graphs.**

13. $x^2 + y^2 = 36$ **6**   **14.** $x^2 + y^2 = 81$ **9**   **15.** $x^2 + y^2 = 32$ $4\sqrt{2}$

16. $x^2 + y^2 = 12$ $2\sqrt{3}$ **17.** $36x^2 + 36y^2 = 144$ **2**   **18.** $9x^2 + 9y^2 = 162$ $3\sqrt{2}$

19. 🌐 **SHOT PUT** A person throws a shot put from a circle that has a diameter of 7 feet. Write the standard form of the equation of the shot put circle if the center is the origin. $x^2 + y^2 = 12.25$

PRACTICE AND APPLICATIONS

STUDENT HELP
▶ **Extra Practice**
to help you master
skills is on p. 954.

MATCHING GRAPHS **Match the equation with its graph.**

20. $x^2 + y^2 = 16$ **C**   **21.** $x^2 + y^2 = 5$ **F**   **22.** $x^2 + y^2 = 4$ **D**

23. $x^2 + y^2 = 25$ **B**   **24.** $x^2 + y^2 = 100$ **E**   **25.** $x^2 + y^2 = 10$ **A**

A. 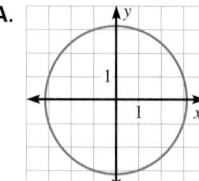   **B.** 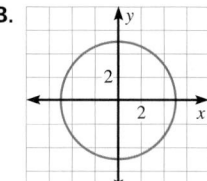   **C.**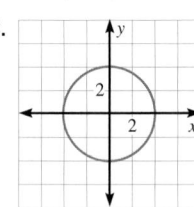

STUDENT HELP
▶ HOMEWORK HELP
Example 1: Exs. 20–46
Example 2: Exs. 47–70
Example 3: Exs. 71–79
Examples 4, 5: Exs. 81–87

D. 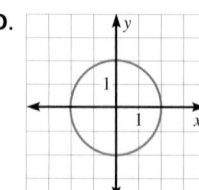   **E.** 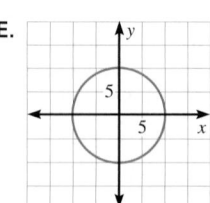   **F.**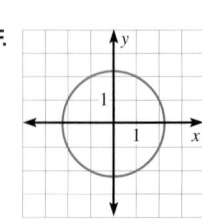

GRAPHING Graph the equation. Give the radius of the circle. 26–37. See margin for graphs.

26. $x^2 + y^2 = 1$ 1
27. $x^2 + y^2 = 49$ 7
28. $x^2 + y^2 = 64$ 8

29. $x^2 + y^2 = 20$ $2\sqrt{5} \approx 4.47$
30. $x^2 + y^2 = 8$ $2\sqrt{2} \approx 2.82$
31. $x^2 + y^2 = 10$ $\sqrt{10} \approx 3.16$

32. $x^2 + y^2 = 3$ $\sqrt{3} \approx 1.73$
33. $5x^2 + 5y^2 = 80$ 4
34. $24x^2 + 24y^2 = 96$ 2

35. $8x^2 + 8y^2 = 192$
$2\sqrt{6} \approx 4.90$
36. $9x^2 + 9y^2 = 135$
$\sqrt{15} \approx 3.87$
37. $4x^2 + 4y^2 = 52$
$\sqrt{13} \approx 3.61$

GRAPHING In Exercises 38–46, the equations of both circles and parabolas are given. Graph the equation.   38–46. See margin.

38. $x^2 + y^2 = 11$
39. $x^2 + y^2 = 1$
40. $x^2 + y = 0$

41. $\frac{1}{4}x^2 + \frac{1}{4}y^2 = 16$
42. $4x^2 + y = 0$
43. $9x^2 + 9y^2 = 441$

44. $-2x + 9y^2 = 0$
45. $\frac{3}{8}x^2 + \frac{3}{8}y^2 = 6$
46. $x^2 + 12y = 0$

WRITING EQUATIONS Write the standard form of the equation of the circle with the given radius and whose center is the origin.

47. 3 $x^2 + y^2 = 9$
48. 9 $x^2 + y^2 = 81$
49. 6 $x^2 + y^2 = 36$
50. 11 $x^2 + y^2 = 121$

51. $\sqrt{7}$ $x^2 + y^2 = 7$
52. $\sqrt{30}$
$x^2 + y^2 = 30$
53. $\sqrt{11}$
$x^2 + y^2 = 11$
54. $\sqrt{21}$ $x^2 + y^2 = 21$

55. $5\sqrt{6}$
$x^2 + y^2 = 150$
56. $4\sqrt{5}$
$x^2 + y^2 = 80$
57. $2\sqrt{7}$
$x^2 + y^2 = 28$
58. $3\sqrt{3}$ $x^2 + y^2 = 27$

WRITING EQUATIONS Write the standard form of the equation of the circle that passes through the given point and whose center is the origin.

59. $(0, -10)$
$x^2 + y^2 = 100$
60. $(8, 0)$
$x^2 + y^2 = 64$
61. $(-3, -4)$
$x^2 + y^2 = 25$
62. $(-4, -1)$
$x^2 + y^2 = 17$

63. $(5, -3)$
$x^2 + y^2 = 34$
64. $(-6, 4)$
$x^2 + y^2 = 52$
65. $(-6, 1)$
$x^2 + y^2 = 37$
66. $(-1, -9)$
$x^2 + y^2 = 82$

67. $(7, -4)$
$x^2 + y^2 = 65$
68. $(10, 2)$
$x^2 + y^2 = 104$
69. $(5, 8)$
$x^2 + y^2 = 89$
70. $(2, -12)$
$x^2 + y^2 = 148$

FINDING TANGENT LINES The equation of a circle and a point on the circle is given. Write an equation of the line that is tangent to the circle at that point.

71. $x^2 + y^2 = 10; (1, 3)$ $y = -\frac{1}{3}x + \frac{10}{3}$
72. $x^2 + y^2 = 5; (2, 1)$ $y = -2x + 5$

73. $x^2 + y^2 = 41; (-4, -5)$ $y = -\frac{4}{5}x - \frac{41}{5}$
74. $x^2 + y^2 = 145; (12, 1)$ $y = -12x + 145$

75. $x^2 + y^2 = 65; (-8, 1)$ $y = 8x + 65$
76. $x^2 + y^2 = 40; (-2, 6)$ $y = \frac{1}{3}x + \frac{20}{3}$

77. $x^2 + y^2 = 244; (-10, -12)$ $y = -\frac{5}{6}x - \frac{61}{3}$
78. $x^2 + y^2 = \frac{257}{4}; \left(\frac{1}{2}, -8\right)$ $y = \frac{1}{16}x - \frac{257}{32}$

79. CRITICAL THINKING Look back at Example 3. Find an equation of the line that is tangent to the circle at the point $(2, -3)$. Describe how the line is geometrically related to the line found in Example 3.   See margin.

80. *Writing* Describe how the equation of a circle is related to the Pythagorean theorem. Include a diagram to illustrate the relationship.   See margin.

81. 🌐 **EARTHQUAKES** Suppose an earthquake can be felt up to 80 miles from its epicenter. You are located at a point 60 miles west and 45 miles south of the epicenter. Do you feel the earthquake? If so, how many miles south would you have to travel to be out of the range of the earthquake?   Yes; about 7.92 mi

82. 🌐 **DESERT IRRIGATION** A circular field has an area of about 2,400,000 square yards. Write an equation that represents the boundary of the field. Let $(0, 0)$ represent the center of the field. $x^2 + y^2 = 764,000$ where x and y are in yards

10.3 *Circles*   **605**

79. $y = \frac{2}{3}x - \frac{13}{3}$; they have opposite slopes and intercepts.

80. A radius of the circle from $(0, 0)$ to a point on the circle (x, y) is the hypotenuse of a right triangle, so $x^2 + y^2 = r^2$. See margin for graph.

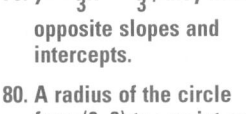
FOCUS ON
APPLICATIONS

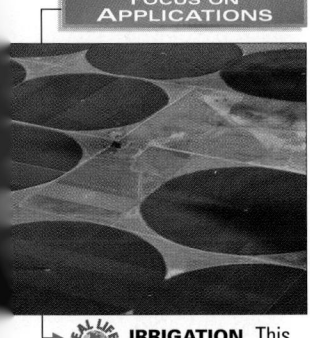

IRRIGATION This irrigation project in Colorado enables farmers to raise crops in the desert. Water from deep wells is pumped to sprinklers that rotate, forming circular patterns.

30–46, 80. See Additional Answers beginning on page AA1.

ENGLISH LEARNERS
EXERCISE 19 Students unfamiliar with track and field are unlikely to recognize the term *shot put*. Explain that the *shot put* is an event in which an athlete standing inside a *circle* throws, or *puts*, a heavy iron ball, or *shot*, as far as he or she can.

❗ **COMMON ERROR**
EXERCISES 26–37 When trying to find the radius of a circle from an equation that is in standard form, students may forget to take the square root of the value to find r, the radius. Remind students that the standard equation contains the radius squared and not the radius itself.

TEACHING TIPS
EXERCISES 26–46 When students are graphing a circle, the 4 easiest points to find and graph are those that are directly above, below, to the right, and to the left of the center of the circle. Once these 4 points are found and plotted, the rest of the circle can be drawn.

26.
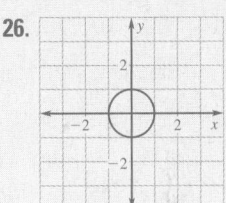

27.
$x^2 + y^2 = 49$

28.
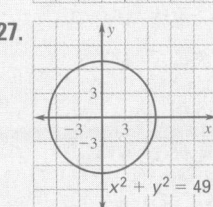

29.
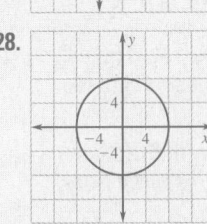
$x^2 + y^2 = 20$

CAREER NOTE
EXERCISES 86 AND 87

Additional information about a career as an Air Traffic Controller is available at **www.mcdougallittell.com**.

101.

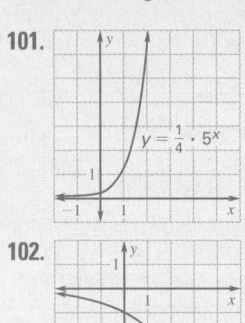

$y = \frac{1}{4} \cdot 5^x$

102.

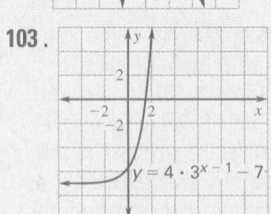

103.

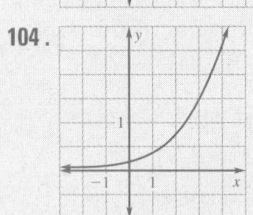

$y = 4 \cdot 3^{x-1} - 7$

104.

$105–106.$ See Additional Answers beginning on page AA1.

ADDITIONAL PRACTICE AND RETEACHING

For Lesson 10.3:

• Practice Levels A, B, and C (*Chapter 10 Resource Book*, p. 41)

• Reteaching with Practice (*Chapter 10 Resource Book*, p. 44)

• See Lesson 10.3 of the *Personal Student Tutor*

For more Mixed Review:

• Search the *Test and Practice Generator* for key words or specific lessons.

606

AIR TRAFFIC CONTROLLER

Air traffic controllers are responsible for making sure airplanes fly a safe distance away from one another. They also help keep flights on schedule.

▶ **CAREER LINK**
www.mcdougallittell.com

---

83. **RESIZING A RING** One way to resize a ring is to fit a bar into the ring, as shown. Suppose a ring that is 20 millimeters in diameter has to be resized to fit a finger 16 millimeters in diameter. What is the length of the bar that should be inserted in order to make the ring fit the finger? (*Hint:* Write an equation of the ring, assuming it is centered at the origin. Determine what the *y*-coordinate of the bar must be and then substitute this coordinate into the equation to find *x*.) **16 mm**

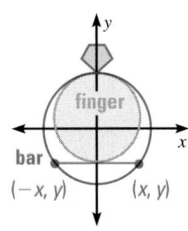

84. **LIFEGUARD** You are a lifeguard at a pond. The pond is a circle with a diameter of 360 feet. You want to rope off a section of the pond for swimming. If you want the rope to form a chord of the circle and have a maximum distance of 45 feet from shore, approximately how long will you need the rope to be? **238 ft**

85. **PHYSICAL THERAPY** A tilt-board is a physical therapy device that a person rocks back and forth on. Suppose the ends of a tilt-board are part of a circle with a radius of 30 inches. If the tilt-board has a depth of 6 inches, how wide is it?

▶ Source: *Steps to Follow* **36 in.**

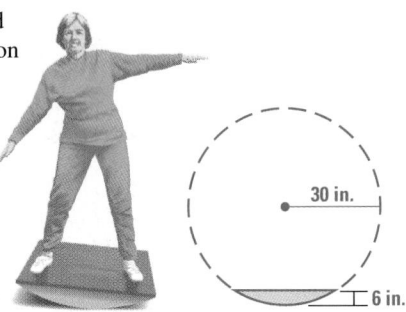

AIR TRAFFIC CONTROL In Exercises 86 and 87, use the following information.
An air traffic control tower can detect airplanes up to 50 miles away. You are in an airplane 42 miles east and 43 miles south of the control tower.

86. Write an inequality that describes the region in which planes can be detected by the control tower. Can the control tower detect your plane on its radar?
$x^2 + y^2 \le 2500$; no

87. Suppose a jet is 35 miles west and 66 miles north of the control tower and is traveling due south at a speed of 500 miles per hour. After how many minutes will the jet appear on the control tower's radar? **about 3.6 min**

Test
Preparation

88. **MULTIPLE CHOICE** What is the equation of the line that is tangent to the circle $x^2 + y^2 = 53$ at the point (7, 2)? **B**

  Ⓐ $y = -\frac{7}{2}x + \frac{45}{2}$ **Ⓑ** $y = -\frac{7}{2}x + \frac{53}{2}$

  Ⓒ $y = \frac{7}{2}x - \frac{45}{2}$ **Ⓓ** $y = -\frac{7}{2}x - \frac{45}{2}$

89. **MULTIPLE CHOICE** Suppose a signal from a television transmitter tower can be received up to 150 miles away. The following points represent the locations of houses near the transmitter tower with the origin representing the tower. Which point is *not* within the range of the tower? **C**

  Ⓐ (120, 20)   **Ⓑ** (40, 140)   **Ⓒ** (105, 120)   **Ⓓ** (10, 148)

★ **Challenge**

90. **ESTIMATING AREA** The *segment* of a circle is the region bounded by a chord and an arc. Estimate the area of the shaded segment by finding the area of △*ABC* and the area of △*ABD*, given that $\overline{AD}$ and $\overline{BD}$ are tangent to the circle. $8 \text{ units}^2 < \text{area} < 21\frac{1}{3} \text{ units}^2$

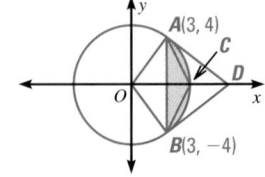

MIXED REVIEW

SOLVING SYSTEMS Solve the system using any algebraic method.
(Review 3.2)

91. $x - 9y = 25$ $(-2, -3)$ **92.** $9x - y = 8$ $\left(\frac{1}{3}, -5\right)$ **93.** $2x - 3y = 2$ $(-2, -2)$
     $6x - 5y = 3$ $3x + 10y = -49$ $-7x + 4y = 6$

94. $8x - 5y = 4$ $\left(\frac{1}{2}, 0\right)$ **95.** $-x + 5y = 3$ $(7, 2)$ **96.** $-9x + 4y = 15$ $\left(-\frac{1}{3}, 3\right)$
     $2x + y = 1$ $4x - 9y = 10$ $3x + 2y = 5$

COMPOSITION OF FUNCTIONS Find $f(g(x))$ and $g(f(x))$. (Review 7.3)

$f(g(x)) = 2x + 1;\ g(f(x)) = 2x + 2$ $f(g(x)) = 4x - 19;\ g(f(x)) = 4x - 4$
97. $f(x) = x + 1$ and $g(x) = 2x$ **98.** $f(x) = 4x + 1$ and $g(x) = x - 5$

99. $f(g(x)) = -x^2 - 10x -$ **99.** $f(x) = -x^2 - 1$ and $g(x) = x + 5$ **100.** $f(x) = x^2 - 7$ and $g(x) = 3x + 1$
$26;\ g(f(x)) = -x^2 + 4$ $f(g(x)) = 9x^2 + 6x - 6;\ g(f(x)) = 3x^2 - 20$

GRAPHING FUNCTIONS Graph the function. (Review 8.1)
101–106. See margin.

101. $y = \frac{1}{4} \cdot 5^x$ **102.** $y = -\left(\frac{5}{3}\right)^x$ **103.** $y = 4 \cdot 3^{x-1} - 7$

104. $y = 3 \cdot 2^{x-4}$ **105.** $y = 2^{x+3} - 1$ **106.** $y = \frac{1}{4} \cdot 8^{x+1}$

🌐 **BABYSITTING** In Exercises 107 and 108, use the following information.
In June you babysit 35 hours for the Johnsons and 52 hours for the Martins. In July
you babysit 112 hours for the Johnsons and 40 hours for the Martins. In August you
babysit 95 hours for the Johnsons and 63 hours for the Martins. (Review 4.1)

107. $\begin{bmatrix} 35 & 52 \\ 112 & 40 \\ 95 & 63 \end{bmatrix}$

107. Use a matrix to organize the information.

108. You charge $6 per hour for babysitting. Using your matrix from Exercise 107,
write a matrix that shows how much you earned over the summer vacation.

$6 \begin{bmatrix} 35 & 52 \\ 112 & 40 \\ 95 & 63 \end{bmatrix} = \begin{bmatrix} 210 & 312 \\ 672 & 240 \\ 570 & 378 \end{bmatrix}$; $2382

QUIZ 1

Self-Test for Lessons 10.1–10.3

**Find the distance between the two points. Then find the midpoint of the line
segment joining the two points.** (Lesson 10.1)
$5\sqrt{13} \approx 18.028;\ \left(1, -\frac{3}{2}\right)$

1. $(0, 0), (8, 6)$ $10;\ (4, 3)$ **2.** $(3, 3), (-3, -3)$ **3.** $(-2, 7), (4, -10)$
                                  $6\sqrt{2} \approx 8.485;\ (0, 0)$

4. $(3, -7), (-5, -9)$ **5.** $(8, 6), (-4, 4)$ **6.** $(-1, -13), (11, 15)$
   $2\sqrt{17} \approx 8.246;\ (-1, -8)$ $2\sqrt{37} \approx 12.166;\ (2, 5)$ $4\sqrt{58} \approx 30.463;\ (5, 1)$

Draw the parabola. Identify the focus and directrix. (Lesson 10.2)

7. $\left(\frac{3}{2}, 0\right); x = -\frac{3}{2}$ **7.** $y^2 = 6x$ **8.** $3y = x^2$ **9.** $-x^2 = 5y$ **10.** $-4y^2 = 6x$

8. $\left(0, \frac{3}{4}\right); y = -\frac{3}{4}$ **11.** $3x^2 = 7y$ **12.** $\frac{1}{2}x = 2y^2$ **13.** $x + \frac{1}{8}y^2 = 0$ **14.** $-x^2 - 12y = 0$

9. $\left(0, -\frac{5}{4}\right); y = \frac{5}{4}$ $(-2, 0); x = 2$ $(0, -3); y = 3$

10. $\left(-\frac{3}{8}, 0\right); x = \frac{3}{8}$ **Write the standard form of the equation of the circle that passes through the
given point and whose center is the origin.** (Lesson 10.3)

                    $x^2 + y^2 = 9$ $x^2 + y^2 = 25$ $x^2 + y^2 = 65$ $x^2 + y^2 = 29$

11. $\left(0, \frac{7}{12}\right); y = -\frac{7}{12}$ **15.** $(0, 3)$ **16.** $(-5, 0)$ **17.** $(4, 7)$ **18.** $(-2, -5)$

                  $x^2 + y^2 = 82$

12. $\left(\frac{1}{16}, 0\right); x = -\frac{1}{16}$ **19.** $(-1, 9)$ **20.** $(6, -3)$ **21.** $(6, -6)$ **22.** $(-7, 8)$
                             $x^2 + y^2 = 45$ $x^2 + y^2 = 72$ $x^2 + y^2 = 113$

23. No; $\sqrt{35^2 + 56^2} \approx 66$ mi   **23.** 🌐 **RADIO SIGNALS** The signals of a radio station can be received up to
65 miles away. Your house is 35 miles east and 56 miles south of the radio
station. Can you receive the radio station's signals? Explain. (Lesson 10.3)

10.3 *Circles*   **607**

Additional Test Preparation *Sample answer:*
1. Because the equation is in standard form, the center   2. It is centered at the origin with a radius of r.
   is the origin and the radius is $\sqrt{169} = 13$.

4 ASSESS

DAILY HOMEWORK QUIZ

🖥 *Transparency Available*

Graph the circle and give its radius.

1. $x^2 + y^2 = 3$ $\sqrt{3}$

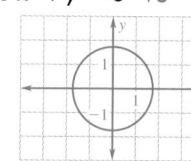

2. $6x^2 = 24 - 6y^2$ 2

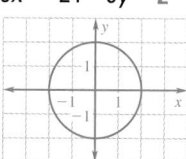

3. Write the standard form of the
equation for the circle that
passes through $(-8, 6)$ and
whose center is at the origin.
$x^2 + y^2 = 100$

4. Write an equation of the line
that is tangent to the circle
$x^2 + y^2 = 34$ at $(3, 5)$.
$y = -\frac{3}{5}x + \frac{34}{5}$

EXTRA CHALLENGE NOTE

→ Challenge problems for
Lesson 10.3 are available in
blackline format in the *Chapter 10
Resource Book,* p. 48 and at
www.mcdougallittell.com.

**ADDITIONAL TEST
PREPARATION**

1. WRITING Explain how to find
the center and the radius of the
circle represented by the equation $x^2 + y^2 = 169$. See margin.

2. WRITING Describe the circle
whose equation is $x^2 + y^2 = r^2$.
See margin.

ADDITIONAL RESOURCES
An alternative quiz for Lessons
10.1–10.3 is available in the
Chapter 10 Resource Book, p. 49.

7–14. See Additional Answers beginning on page AA1.

1 Planning the Activity

PURPOSE
To graph circles on a graphing calculator.

MATERIALS
- graphing calculator
- Keystroke blackline (*Chapter 10 Resource Book*, p. 40)

PACING
- Activity — 20 min

▶ LINK TO LESSON
Students can graph Example 1 using the techniques of this activity.

2 Managing the Activity

COOPERATIVE LEARNING
Students can work with a partner to solve the equations in the activity and find a square window. Also, if errors occur in entering the equations into the calculator or in graphing the equations, students can work with their partner to solve the problems.

3 Closing the Activity

★ KEY DISCOVERY
Circles can be graphed on a graphing calculator by writing them as two separate function equations.

ACTIVITY ASSESSMENT
Describe the steps in graphing a circle on a graphing calculator.
Sample answer: Solve the equation for *y*, then enter the function equations separately into the Y = list on the calculator. Then find a square window that contains a complete graph of the circle and graph it.

▶ ACTIVITY 10.3

Using Technology

Graphing Circles

When you use a graphing calculator to draw a circle, you need to remember two things. First, most graphing calculators cannot directly graph equations such as $x^2 + y^2 = 36$ because they are not functions. Second, to obtain a graph with true perspective (in which a circle looks like a circle) you must use a "square setting."

▶ EXAMPLE
Use a graphing calculator to draw the graph of $x^2 + y^2 = 36$.

▶ SOLUTION

1 Begin by solving the equation for *y*.

$$x^2 + y^2 = 36$$
$$y^2 = 36 - x^2$$
$$y = \pm\sqrt{36 - x^2}$$

Enter the two equations into the graphing calculator.

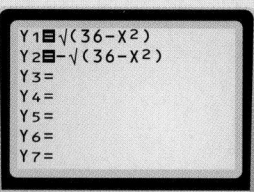

```
Y1=√(36-X²)
Y2=-√(36-X²)
Y3=
Y4=
Y5=
Y6=
Y7=
```

2 Next set the viewing window so that it has a "square setting." On some graphing calculators you can select a square setting, such as "ZSquare." On a graphing calculator whose viewing window's height is two thirds its width, you can obtain a "square setting" by choosing maximum and minimum values that satisfy this equation:

$$\frac{Ymax - Ymin}{Xmax - Xmin} = \frac{2}{3}$$

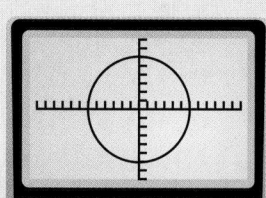

```
RANGE
Xmin=-12
Xmax=12
Xscl=1
Ymin=-8
Ymax=8
Yscl=1
```

3 The graph is shown at the right. (Some calculators may not connect the ends of the two graphs.)

▶ STUDENT HELP

INTERNET KEYSTROKE HELP

See keystrokes for several models of calculators at www.mcdougallittell.com

---

▶ EXERCISES

Use a graphing calculator to graph the equation. Write the setting of the viewing window that you used and verify that it is a square setting. **1–6. See margin.**

1. $x^2 + y^2 = 121$ **2.** $x^2 + y^2 = 50$ **3.** $x^2 + y^2 = 484$

4. $5x^2 + 5y^2 = 120$ **5.** $x^2 + y^2 = \dfrac{16}{9}$ **6.** $\dfrac{1}{2}x^2 + \dfrac{1}{2}y^2 = 72$

7. $\dfrac{4}{5}x^2 + \dfrac{4}{5}y^2 = 20$
$-9 \le x \le 9; -6 \le y \le 6$

8. $9x^2 + 9y^2 = 4$
$-1.5 \le x \le 1.5; -1 \le y \le 1$

9. $125x^2 + 125y^2 = 1000$
$-6 \le x \le 6; -4 \le y \le 4$

Sample answers are given for exercises 1–9.

1. $-18 \le x \le 18$;
   $-12 \le y \le 12$

2. $-12 \le x \le 12$;
   $-8 \le y \le 8$

3. $-36 \le x \le 36$;
   $-24 \le y \le 24$

4. $-9 \le x \le 9$;
   $-6 \le y \le 6$

5. $-3 \le x \le 3$;
   $-2 \le y \le 2$

6. $-24 \le x \le 24$;
   $-16 \le y \le 16$

10.4
Ellipses

What you should learn

GOAL 1 Graph and write equations of ellipses.

GOAL 2 Use ellipses in **real-life** situations, such as modeling the orbit of Mars in **Example 4**.

Why you should learn it

▼ To solve **real-life** problems, such as finding the area of an elliptical Australian football field in **Exs. 73–75.**

GOAL 1 GRAPHING AND WRITING EQUATIONS OF ELLIPSES

An **ellipse** is the set of all points P such that the sum of the distances between P and two distinct fixed points, called the **foci,** is a constant.

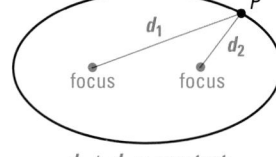

$$d_1 + d_2 = \text{constant}$$

The line through the foci intersects the ellipse at two points, the **vertices.** The line segment joining the vertices is the **major axis,** and its midpoint is the **center** of the ellipse. The line perpendicular to the major axis at the center intersects the ellipse at two points called the **co-vertices.** The line segment that joins these points is the **minor axis** of the ellipse. The two types of ellipses we will discuss are those with a horizontal major axis and those with a vertical major axis.

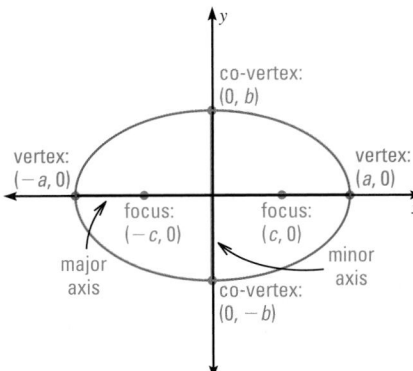

Ellipse with horizontal major axis

$$\frac{x^2}{a^2} + \frac{y^2}{b^2} = 1$$

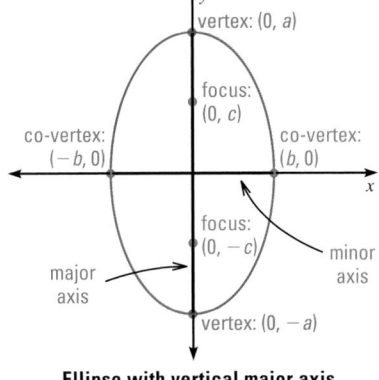

Ellipse with vertical major axis

$$\frac{x^2}{b^2} + \frac{y^2}{a^2} = 1$$

CHARACTERISTICS OF AN ELLIPSE (CENTER AT ORIGIN)

The **standard form of the equation of an ellipse** with center at $(0, 0)$ and major and minor axes of lengths $2a$ and $2b$, where $a > b > 0$, is as follows.

EQUATION	MAJOR AXIS	VERTICES	CO-VERTICES
$\frac{x^2}{a^2} + \frac{y^2}{b^2} = 1$	Horizontal	$(\pm a, 0)$	$(0, \pm b)$
$\frac{x^2}{b^2} + \frac{y^2}{a^2} = 1$	Vertical	$(0, \pm a)$	$(\pm b, 0)$

The foci of the ellipse lie on the major axis, c units from the center where $c^2 = a^2 - b^2$.

10.4 *Ellipses* **609**

---

1 PLAN

PACING
Basic: 1 day
Average: 1 day
Advanced: 1 day
Block Schedule: 0.5 block with 10.3

LESSON OPENER
ACTIVITY
An alternative way to approach Lesson 10.4 is to use the Activity Lesson Opener:
- Blackline Master (*Chapter 10 Resource Book,* p. 53)
-  Transparency (p. 68)

MEETING INDIVIDUAL NEEDS
- *Chapter 10 Resource Book*
  Prerequisite Skills Review (p. 5)
  Practice Level A (p. 54)
  Practice Level B (p. 55)
  Practice Level C (p. 56)
  Reteaching with Practice (p. 57)
  Absent Student Catch-Up (p. 59)
  Challenge (p. 61)
- *Resources in Spanish*
- *Personal Student Tutor*

NEW-TEACHER SUPPORT
See the Tips for New Teachers on pp. 1–2 of the *Chapter 10 Resource Book* for additional notes about Lesson 10.4.

WARM-UP EXERCISES

Transparency Available

Find c if $a = 5$ and $b = 3$.
1. $c^2 = a^2 + b^2$ $\sqrt{-34}, \sqrt{34}$
2. $c^2 = a^2 - b^2$ $-4, 4$
Find a if $c = 6$ and $b = 2$.
3. $c^2 = a^2 + b^2$ $-4\sqrt{2}, 4\sqrt{2}$
4. $c^2 = a^2 - b^2$ $-2\sqrt{10}, 2\sqrt{10}$

Ask students to name as many geometric shapes as they can. If and when someone mentions oval, use that to begin the discussion of an ellipse.

EXTRA EXAMPLE 1
Draw the ellipse given by $4x^2 + 25y^2 = 100$. Identify the foci.

foci: $\left(\sqrt{21}, 0\right), \left(-\sqrt{21}, 0\right)$

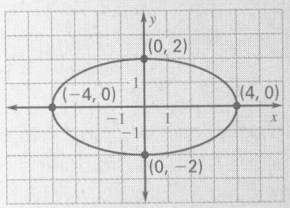

EXTRA EXAMPLE 2
Write an equation of the ellipse with the given characteristics and center at $(0, 0)$.
a. Vertex: $(0, 5)$
   Co-vertex: $(4, 0)$
   $\dfrac{x^2}{16} + \dfrac{y^2}{25} = 1$
b. Vertex: $(-6, 0)$
   Focus: $(3, 0)$
   $\dfrac{x^2}{36} + \dfrac{y^2}{27} = 1$

CHECKPOINT EXERCISES

For use after Example 1:
1. Draw the ellipse given by the equation $16x^2 + 4y^2 = 16$. Identify the foci.

foci: $\left(0, -\sqrt{3}\right), \left(0, \sqrt{3}\right)$

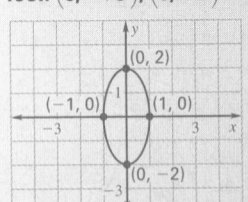

For use after Example 2:
2. Write an equation of the ellipse with center at $(0, 0)$, a vertex at $(3, 0)$, and a co-vertex at $(0, -1)$. $\dfrac{x^2}{9} + y^2 = 1$

STUDENT HELP

HOMEWORK HELP
Visit our Web site
www.mcdougallittell.com
for extra examples.

EXAMPLE 1 *Graphing an Equation of an Ellipse*

Draw the ellipse given by $9x^2 + 16y^2 = 144$. Identify the foci.

SOLUTION

First rewrite the equation in standard form.

$$\dfrac{9x^2}{144} + \dfrac{16y^2}{144} = \dfrac{144}{144} \qquad \text{Divide each side by 144.}$$

$$\dfrac{x^2}{16} + \dfrac{y^2}{9} = 1 \qquad \text{Simplify.}$$

Because the denominator of the x^2-term is greater than that of the y^2-term, the major axis is horizontal. So, $a = 4$ and $b = 3$. Plot the vertices and co-vertices. Then draw the ellipse that passes through these four points.

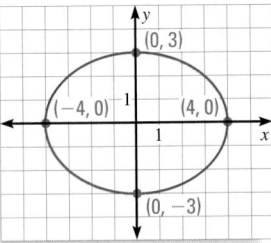

The foci are at $(c, 0)$ and $(-c, 0)$. To find the value of c, use the equation $c^2 = a^2 - b^2$.

$$c^2 = 4^2 - 3^2 = 16 - 9 = 7$$

$$c = \sqrt{7}$$

▶ The foci are at $\left(\sqrt{7}, 0\right)$ and $\left(-\sqrt{7}, 0\right)$.

EXAMPLE 2 *Writing Equations of Ellipses*

Write an equation of the ellipse with the given characteristics and center at $(0, 0)$.

a. Vertex: $(0, 7)$
   Co-vertex: $(-6, 0)$

b. Vertex: $(-4, 0)$
   Focus: $(2, 0)$

SOLUTION

In each case, you may wish to draw the ellipse so that you have something to check your final equation against.

a. Because the vertex is on the y-axis and the co-vertex is on the x-axis, the major axis is vertical with $a = 7$ and $b = 6$.

▶ An equation is $\dfrac{x^2}{36} + \dfrac{y^2}{49} = 1$.

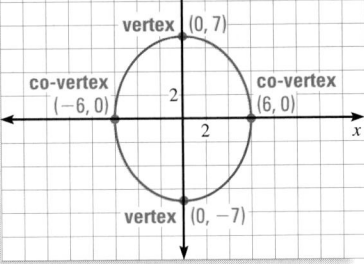

b. Because the vertex and focus are points on a horizontal line, the major axis is horizontal with $a = 4$ and $c = 2$. To find b, use the equation $c^2 = a^2 - b^2$.

$$2^2 = 4^2 - b^2$$

$$b^2 = 16 - 4 = 12$$

$$b = 2\sqrt{3}$$

▶ An equation is $\dfrac{x^2}{16} + \dfrac{y^2}{12} = 1$.

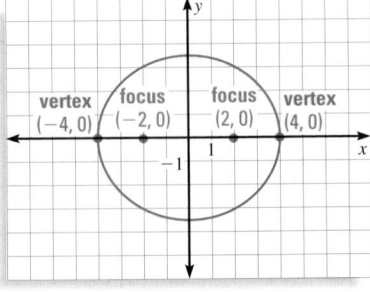

GOAL 2 USING ELLIPSES IN REAL LIFE

Both man-made objects, such as The Ellipse at the White House, and natural phenomena, such as the orbits of planets, involve ellipses.

Landscaping

EXAMPLE 3 Finding the Area of an Ellipse

A portion of the White House lawn is called The Ellipse. It is 1060 feet long and 890 feet wide.

a. Write an equation of The Ellipse.

b. The area of an ellipse is $A = \pi ab$. What is the area of The Ellipse at the White House?

SOLUTION

a. The major axis is horizontal with
$a = \dfrac{1060}{2} = 530$ and $b = \dfrac{890}{2} = 445.$

▶ An equation is $\dfrac{x^2}{530^2} + \dfrac{y^2}{445^2} = 1.$

b. The area is $A = \pi(530)(445) \approx 741{,}000$ square feet.

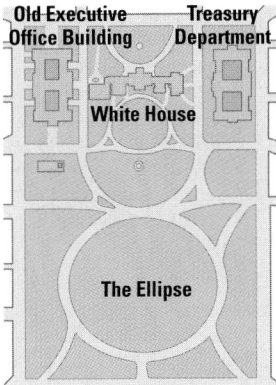

Astronomy

EXAMPLE 4 Modeling with an Ellipse

In its elliptical orbit, Mercury ranges from 46.04 million kilometers to 69.86 million kilometers from the center of the sun. The center of the sun is a focus of the orbit. Write an equation of the orbit.

SOLUTION

Using the diagram shown, you can write a system of linear equations involving a and c.

$a - c = 46.04$

$a + c = 69.86$

Adding the two equations gives $2a = 115.9$, so $a = \mathbf{57.95}$. Substituting this a-value into the second equation gives $57.95 + c = 69.86$, so $c = \mathbf{11.91}$.

From the relationship $c^2 = a^2 - b^2$, you can conclude the following:

$b = \sqrt{a^2 - c^2}$

$\quad = \sqrt{(57.95)^2 - (11.91)^2}$

$\quad \approx 56.71$

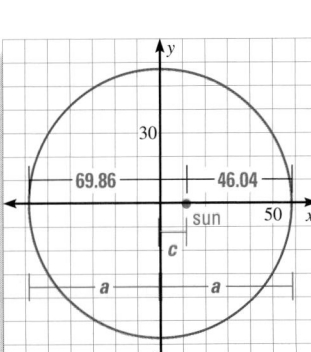

▶ An equation of the elliptical orbit is $\dfrac{x^2}{(57.95)^2} + \dfrac{y^2}{(56.71)^2} = 1$ where x and y are in millions of kilometers.

EXTRA EXAMPLE 3
An amusement park has an elliptical garden at its entrance. The garden is 32 ft long and 14 ft wide.
a. Write an equation of the ellipse. $\dfrac{x^2}{256} + \dfrac{y^2}{49} = 1$
b. The area of an ellipse is $A = \pi ab$. What is the area of the garden? $A = 112\pi \approx 351.86 \text{ ft}^2$

EXTRA EXAMPLE 4
In its elliptical orbit, the moon ranges from 252,000 mi to 221,500 mi from the center of Earth. The center of Earth is a focus of the orbit. Write an equation of the orbit.
$\dfrac{x^2}{(236{,}750)^2} + \dfrac{y^2}{(236{,}260)^2} = 1$

✔ **CHECKPOINT EXERCISES**
For use after Example 3:
1. The garden in Extra Example 3 has a smaller ellipse inside that is 2 ft long and 6 ft wide.
a. Write an equation for the smaller ellipse.
$x^2 + \dfrac{y^2}{9} = 1$
b. What is the area of the smaller ellipse?
$A = 3\pi \approx 9.42 \text{ ft}^2$

For use after Example 4:
2. The planet Jupiter ranges from 460.2 million miles away to 507.0 million miles away from the sun. If Jupiter's orbit is elliptical, write an equation for its orbit in millions of miles.
$\dfrac{x^2}{(483.6)^2} + \dfrac{y^2}{(483.0)^2} = 1$

CLOSURE QUESTION
In a standard equation for an ellipse, how can you tell which axis is the major axis and which axis is the minor axis? *Sample answer:* The variable whose denominator contains the larger value is the major axis. The other variable will be the minor axis.

ASSIGNMENT GUIDE

BASIC
Day 1: pp. 612–614 Exs. 18–24
even, 30–36 even, 42–46
even, 52–62 even, 69–71,
76, 79–93 odd

AVERAGE
Day 1: pp. 612–614 Exs. 18–38
even, 42–48 even, 52–64
even, 69–71, 76, 79–97 odd

ADVANCED
Day 1: pp. 612–614 Exs. 18–48
even, 52–64 even, 69–71,
73–77, 79–97 odd

BLOCK SCHEDULE
pp. 612–614 Exs. 18–38 even,
42–48 even, 52–64 even, 69–71,
76, 79–97 odd (with 10.3)

EXERCISE LEVELS
Level A: *Easier*
18–23, 30–40, 51–59
Level B: *More Difficult*
24–29, 40–50, 60–74, 76
Level C: *Most Difficult*
75, 77

✔ **HOMEWORK CHECK**
To quickly check student understanding of key concepts, go
over the following exercises:
Exs. 18, 24, 32, 46, 52. See also
the Daily Homework Quiz:

• Blackline Master (*Chapter 10
Resource Book,* p. 64)
• Transparency (p. 79)

4. The student incorrectly identified
a as 2 and *b* as 3. Since the major
axis is horizontal, $a = 3$ and $b = 2$.
The correct equation is $\frac{x^2}{9} + \frac{y^2}{4} = 1$.

11–17, 25–29.
See Additional Answers beginning on page AA1.

GUIDED PRACTICE

Vocabulary Check ✔

2. Write the equation in
standard form; if the
larger denominator is
under x^2, it is horizontal,
under y^2, it is vertical.

1. Complete each statement using the ellipse shown.

  a. The points $(-5, 0)$ and $(5, 0)$ are called the __?__.
         vertices
  b. The points $(0, -4)$ and $(0, 4)$ are called the __?__.
         co-vertices
  c. The points $(-3, 0)$ and $(3, 0)$ are called the __?__.
         foci
  d. The segment with endpoints $(-5, 0)$ and $(5, 0)$ is
     called the __?__. major axis

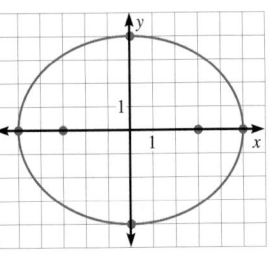

Ex. 1

Concept Check ✔

3. Find *a* and *b*, then use
$c = \sqrt{a^2 - b^2}$. The foci
are at $(\pm c, 0)$ for vertices
at $(\pm a, 0)$, and at $(0, \pm c)$
for vertices at $(0, \pm a)$.

7. $\frac{x^2}{49} + \frac{y^2}{9} = 1$

18. vertices: $(\pm 5, 0)$
co-vertices: $(0, \pm 4)$
foci: $(\pm 3, 0)$

2. How can you tell from the equation of an ellipse
whether the major axis is horizontal or vertical?

3. Explain how to find the foci of an ellipse given the
coordinates of its vertices and co-vertices.

4. **ERROR ANALYSIS** A student was asked to write
an equation of the ellipse shown at the right.

The student wrote the equation $\frac{x^2}{4} + \frac{y^2}{9} = 1$.

What did the student do wrong? What is the
correct equation? **See margin.**

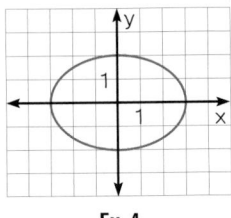

Ex. 4

Skill Check ✔

19. vertices: $(\pm 11, 0)$
co-vertices: $(0, \pm 10)$
foci: $(\pm\sqrt{21}, 0)$

20. vertices: $(0, \pm 3)$
co-vertices: $(\pm 2, 0)$
foci: $(0, \pm\sqrt{5})$

21. vertices: $(0, \pm 5)$
co-vertices: $(\pm 3, 0)$
foci: $(0, \pm 4)$

22. vertices: $(0, \pm 6)$
co-vertices: $(\pm 2\sqrt{3}, 0)$
foci: $(0, \pm 2\sqrt{6})$

23. vertices: $(\pm 2\sqrt{7}, 0)$
co-vertices: $(0, \pm 2\sqrt{5})$
foci: $(\pm 2\sqrt{2}, 0)$

Write an equation of the ellipse with the given characteristics and center at (0, 0).
See margin.

5. Vertex: $(0, 5)$ $\frac{x^2}{16} + \frac{y^2}{25} = 1$ **6.** Vertex: $(9, 0)$ $\frac{x^2}{81} + \frac{y^2}{4} = 1$ **7.** Vertex: $(-7, 0)$
   Co-vertex: $(-4, 0)$ Co-vertex: $(0, 2)$ Focus: $(-2\sqrt{10}, 0)$

8. Vertex: $(0, 13)$ **9.** Co-vertex: $(\sqrt{91}, 0)$ **10.** Co-vertex: $(0, \sqrt{33})$
   Focus: $(0, -5)$ $\frac{x^2}{144} + \frac{y^2}{169} = 1$ Focus: $(0, 3)$ $\frac{x^2}{91} + \frac{y^2}{100} = 1$ Focus: $(4, 0)$ $\frac{x^2}{49} + \frac{y^2}{33} = 1$

Draw the ellipse.
11–16. See margin.

11. $\frac{x^2}{49} + \frac{y^2}{25} = 1$ **12.** $\frac{x^2}{9} + \frac{y^2}{16} = 1$ **13.** $\frac{x^2}{30} + \frac{y^2}{4} = 1$

14. $\frac{x^2}{64} + \frac{y^2}{45} = 1$ **15.** $75x^2 + 36y^2 = 2700$ **16.** $81x^2 + 63y^2 = 5103$

17. 🌱 **GARDEN** An elliptical garden is 10 feet long and 6 feet wide. Write
an equation for the garden. Then graph the equation. Label the vertices,
co-vertices, and foci. Assume that the major axis of the garden is horizontal.

$$\frac{x^2}{25} + \frac{y^2}{9} = 1; \text{ See margin for graph.}$$

PRACTICE AND APPLICATIONS

STUDENT HELP

▸ **Extra Practice**
to help you master
skills is on p. 954.

24. $x^2 + \frac{y^2}{16} = 1$; vertices:
  $(0, \pm 4)$ co-vertices:
  $(\pm 1, 0)$; foci: $(0, \pm\sqrt{15})$

IDENTIFYING PARTS Write the equation in standard form (if not already). Then
identify the vertices, co-vertices, and foci of the ellipse. 18–29. See margin.

18. $\frac{x^2}{25} + \frac{y^2}{16} = 1$ **19.** $\frac{x^2}{121} + \frac{y^2}{100} = 1$ **20.** $\frac{x^2}{4} + \frac{y^2}{9} = 1$

21. $\frac{x^2}{9} + \frac{y^2}{25} = 1$ **22.** $\frac{x^2}{12} + \frac{y^2}{36} = 1$ **23.** $\frac{x^2}{28} + \frac{y^2}{20} = 1$

24. $16x^2 + y^2 = 16$ **25.** $49x^2 + 4y^2 = 196$ **26.** $9x^2 + 100y^2 = 900$

27. $x^2 + 10y^2 = 10$ **28.** $10x^2 + 25y^2 = 250$ **29.** $25x^2 + 15y^2 = 375$

GRAPHING Graph the equation. Then identify the vertices, co-vertices, and foci of the ellipse. 30–41. See margin.

30. $\dfrac{x^2}{16} + \dfrac{y^2}{36} = 1$ **31.** $\dfrac{x^2}{4} + \dfrac{y^2}{49} = 1$ **32.** $\dfrac{x^2}{36} + \dfrac{y^2}{64} = 1$

33. $\dfrac{x^2}{49} + \dfrac{y^2}{144} = 1$ **34.** $\dfrac{x^2}{196} + \dfrac{y^2}{100} = 1$ **35.** $\dfrac{x^2}{256} + \dfrac{y^2}{36} = 1$

36. $\dfrac{x^2}{225} + \dfrac{y^2}{81} = 1$ **37.** $\dfrac{x^2}{121} + \dfrac{y^2}{169} = 1$ **38.** $\dfrac{x^2}{144} + \dfrac{y^2}{400} = 1$

39. $\dfrac{x^2}{49} + \dfrac{y^2}{64} = 1$ **40.** $\dfrac{x^2}{4} + y^2 = 100$ **41.** $\dfrac{x^2}{4} + \dfrac{y^2}{25} = \dfrac{1}{4}$

GRAPHING In Exercises 42–50, the equations of parabolas, circles, and ellipses are given. Graph the equation. 42–50. See margin.

42. $x^2 + y^2 = 33^2$ **43.** $64x^2 + 25y^2 = 1600$ **44.** $24y + x^2 = 0$

45. $72x^2 = 144y$ **46.** $24x^2 + 24y^2 = 96$ **47.** $\dfrac{x^2}{81} + \dfrac{4y}{9} = 1$

48. $\dfrac{3x^2}{12} + \dfrac{5y^2}{500} = 1$ **49.** $\dfrac{x^2}{36} + \dfrac{y^2}{36} = 4$ **50.** $5x^2 + 9y^2 = 45$

WRITING EQUATIONS Write an equation of the ellipse with the given characteristics and center at (0, 0). 51–68. See margin.

51. Vertex: (0, 6)
Co-vertex: (5, 0)

52. Vertex: (0, 6)
Co-vertex: (−2, 0)

53. Vertex: (−4, 0)
Co-vertex: (0, 3)

54. Vertex: (0, −7)
Co-vertex: (−1, 0)

55. Vertex: (9, 0)
Co-vertex: (0, −8)

56. Vertex: (10, 0)
Co-vertex: (0, 4)

57. Vertex: (0, 7)
Focus: (0, 3)

58. Vertex: (−5, 0)
Focus: $(2\sqrt{6}, 0)$

59. Vertex: (0, 8)
Focus: $(0, -4\sqrt{3})$

60. Vertex: (15, 0)
Focus: (12, 0)

61. Vertex: (5, 0)
Focus: (−3, 0)

62. Vertex: (0, −30)
Focus: (0, 20)

63. Co-vertex: $(\sqrt{55}, 0)$
Focus: (0, −3)

64. Co-vertex: $(0, -\sqrt{3})$
Focus: (−1, 0)

65. Co-vertex: $(-2\sqrt{10}, 0)$
Focus: (0, 9)

66. Co-vertex: $(0, -3\sqrt{3})$
Focus: (3, 0)

67. Co-vertex: $(5\sqrt{11}, 0)$
Focus: (0, −7)

68. Co-vertex: $(0, -\sqrt{77})$
Focus: (−2, 0)

🌐 **WHISPERING GALLERY** In Exercises 69–71, use the following information.

Statuary Hall is an elliptical room in the United States Capitol in Washington, D.C. The room is also called the Whispering Gallery because a person standing at one focus of the room can hear even a whisper spoken by a person standing at the other focus. This occurs because any sound that is emitted from one focus of an ellipse will reflect off the side of the ellipse to the other focus. Statuary Hall is 46 feet wide and 97 feet long.

69. Find an equation that models the shape of the room. $\dfrac{x^2}{2352.25} + \dfrac{y^2}{529} = 1$

70. How far apart are the two foci?   **about 85.4 ft**

71. What is the area of the floor of the room?   **about 3500 ft²**

10.4 *Ellipses*   **613**

---

51. $\dfrac{x^2}{25} + \dfrac{y^2}{36} = 1$

52. $\dfrac{x^2}{4} + \dfrac{y^2}{36} = 1$

53. $\dfrac{x^2}{16} + \dfrac{y^2}{9} = 1$

54. $x^2 + \dfrac{y^2}{49} = 1$

55. $\dfrac{x^2}{81} + \dfrac{y^2}{64} = 1$

56. $\dfrac{x^2}{100} + \dfrac{y^2}{16} = 1$

57. $\dfrac{x^2}{40} + \dfrac{y^2}{49} = 1$

58. $\dfrac{x^2}{25} + y^2 = 1$

59. $\dfrac{x^2}{16} + \dfrac{y^2}{64} = 1$

60. $\dfrac{x^2}{225} + \dfrac{y^2}{81} = 1$

61. $\dfrac{x^2}{25} + \dfrac{y^2}{16} = 1$

62. $\dfrac{x^2}{500} + \dfrac{y^2}{900} = 1$

63. $\dfrac{x^2}{55} + \dfrac{y^2}{64} = 1$

64. $\dfrac{x^2}{4} + \dfrac{y^2}{3} = 1$

65. $\dfrac{x^2}{40} + \dfrac{y^2}{121} = 1$

66. $\dfrac{x^2}{36} + \dfrac{y^2}{27} = 1$

67. $\dfrac{x^2}{275} + \dfrac{y^2}{324} = 1$

68. $\dfrac{x^2}{81} + \dfrac{y^2}{77} = 1$

! **COMMON ERROR**

EXERCISES 18–29 Students may confuse the equation $c^2 = a^2 - b^2$, which is used to find the foci of an ellipse, with the Pythagorean theorem $c^2 = a^2 + b^2$. Remind students that the foci of the ellipse must be in the interior, and so c must be smaller than a.

TEACHING TIPS
EXERCISES 30–50 Drawing an ellipse is easiest when the vertices and co-vertices are plotted first. When students are graphing an ellipse, you may want to suggest that they plot these four points and then draw the ellipse.

30.

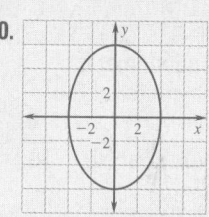

vertices: (0, ±6);
co-vertices: (±4, 0); foci: $(0, \pm 2\sqrt{5})$
31–50. See Additional Answers beginning on page AA1.

ADDITIONAL PRACTICE AND RETEACHING

For Lesson 10.4:
• Practice Levels A, B, and C (*Chapter 10 Resource Book,* p. 54)

• Reteaching with Practice (*Chapter 10 Resource Book,* p. 57)

• ⌨ See Lesson 10.4 of the *Personal Student Tutor*

For more Mixed Review:
• ⌨ Search the *Test and Practice Generator* for key words or specific lessons.

4 ASSESS

DAILY HOMEWORK QUIZ

📖 **Transparency Available**

1. Graph the equation and name the vertices, co-vertices, and foci of the ellipse $\frac{x^2}{36} + \frac{y^2}{16} = 1$.

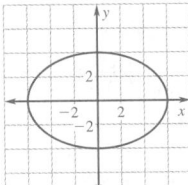

vertices: $(\pm 6, 0)$;
co-vertices: $(0, \pm 4)$;
foci: $\left(\pm 2\sqrt{5}, 0\right)$

2. Write an equation of the ellipse with center at the origin with a vertex at $(15, 0)$ and a co-vertex at $\left(0, -\sqrt{10}\right)$.
$\frac{x^2}{225} + \frac{y^2}{10} = 1$

EXTRA CHALLENGE NOTE

↳ Challenge problems for Lesson 10.4 are available in **blackline** format in the *Chapter 10 Resource Book,* p. 61 and at **www.mcdougallittell.com.**

ADDITIONAL TEST PREPARATION

1. WRITING Using the standard equation of an ellipse with center at $(0, 0)$, $\frac{x^2}{a^2} + \frac{y^2}{b^2} = 1$, explain what a and b represent and how they can be used to graph the ellipse. *Sample answer: a represents half of the length of the major axis and b represents half of the length of the minor axis; the endpoints of the ellipse will be $(a, 0)$, $(-a, 0)$, $(0, b)$, and $(0, -b)$.*

92–97. See Additional Answers beginning on page AA1.

614

---

72. $\frac{x^2}{4357.5^2} + \frac{y^2}{4351.7^2} = 1$

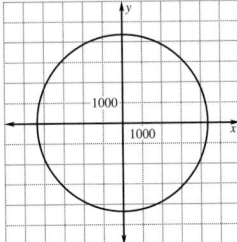

Test Preparation

76a. The greatest distance the boat can travel between islands is 20 mi, going straight out to a point and then straight to the other island. This is the definition of an ellipse.

76d. $\frac{x^2}{100} + \frac{y^2}{64} = 1$

★ Challenge

---

72. 🌐 **SPACE EXPLORATION** The first artificial satellite to orbit Earth was Sputnik I, launched by the Soviet Union in 1957. The orbit was an ellipse with Earth's center as one focus. The orbit's highest point above Earth's surface was 583 miles, and its lowest point was 132 miles. Find an equation of the orbit. (Use 4000 miles as the radius of Earth.) Graph your equation.

🏉 **AUSTRALIAN FOOTBALL** In Exercises 73–75, use the information below.
Australian football is played on an elliptical field. The official rules state that the field must be between 135 and 185 meters long and between 110 and 155 meters wide. ▶ Source: The Australian News Network

73. Write an equation for the largest allowable playing field. $\frac{x^2}{92.5^2} + \frac{y^2}{77.5^2} = 1$

74. Write an equation for the smallest allowable playing field. $\frac{x^2}{67.5^2} + \frac{y^2}{55^2} = 1$

75. Write an inequality that describes the possible areas of an Australian football field. $3710\pi \le A \le 7170\pi$

76. MULTI-STEP PROBLEM A tour boat travels between two islands that are 12 miles apart. For a trip between the islands, there is enough fuel for a 20-mile tour.

a. *Writing* The region in which the boat can travel is bounded by an ellipse. Explain why this is so.

b. Let $(0, 0)$ represent the center of the ellipse. Find the coordinates of each island. $(\pm 6, 0)$

c. Suppose the boat travels from one island, straight past the other island to the vertex of the ellipse, and back to the second island. How many miles does the boat travel? Use your answer to find the coordinates of the vertex. 20 mi; (10, 0)

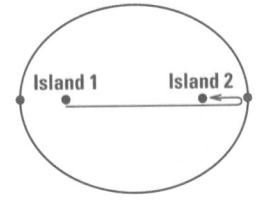

Island 1   Island 2

d. Use your answers to parts (b) and (c) to write an equation for the ellipse that bounds the region the boat can travel in.

77. LOGICAL REASONING Show that $c^2 = a^2 - b^2$ for any ellipse given by the equation $\frac{x^2}{a^2} + \frac{y^2}{b^2} = 1$ with foci at $(c, 0)$ and $(-c, 0)$. See margin.

MIXED REVIEW

77. The distance from one focus to a vertex is $a - c$. The distance from the vertex to the other focus is $a + c$. So the distance from focus to vertex to other focus is $(a + c) + (a - c)$, or $2a$. The distance from each focus to a co-vertex is $\sqrt{b^2 + c^2}$, so from focus to co-vertex to other focus is $2\sqrt{b^2 + c^2}$. By the definition of an ellipse, these distances must be equal, so $2a = 2\sqrt{b^2 + c^2}$, $a^2 = b^2 + c^2$, so $c^2 = a^2 - b^2$.

RATIONAL EXPONENTS Evaluate the expression without using a calculator. (Review 7.1)

78. $125^{2/3}$ 25   **79.** $-8^{5/3}$ -32 **80.** $4^{5/2}$ 32   **81.** $27^{-2/6}$ $\frac{1}{3}$

82. $4^{7/2}$ 128   **83.** $81^{3/4}$ 27   **84.** $64^{-2/3}$ $\frac{1}{16}$ **85.** $32^{4/5}$ 16

INVERSE VARIATION The variables x and y vary inversely. Use the given values to write an equation relating x and y. (Review 9.1)

86. $x = 3, y = -2$ $y = -\frac{6}{x}$ **87.** $x = 4, y = 6$ $y = \frac{24}{x}$ **88.** $x = 5, y = 1$ $y = \frac{5}{x}$

89. $x = 8, y = 9$ $y = \frac{72}{x}$ **90.** $x = 9, y = 2$ $y = \frac{18}{x}$ **91.** $x = 0.5, y = 24$ $y = \frac{12}{x}$

GRAPHING Graph the function. State the domain and range. (Review 9.2 for 10.5)
92–97. See margin.

92. $f(x) = \frac{9}{x}$ **93.** $f(x) = -\frac{9}{x}$ **94.** $f(x) = \frac{12}{x}$

95. $f(x) = \frac{24}{x}$ **96.** $f(x) = \frac{10}{x - 2}$ **97.** $f(x) = \frac{4}{x + 3}$

614   **Chapter 10**   *Quadratic Relations and Conic Sections*

10.5

Hyperbolas

What you should learn

GOAL 1 Graph and write equations of hyperbolas.

GOAL 2 Use hyperbolas to solve **real-life** problems, such as modeling a sundial in **Exs. 64–66**.

Why you should learn it

▼ To model **real-life** objects, such as a sculpture in **Example 4**.

GOAL 1 **GRAPHING AND WRITING EQUATIONS OF HYPERBOLAS**

The definition of a hyperbola is similar to that of an ellipse. For an ellipse, recall that the *sum* of the distances between a point on the ellipse and the two foci is constant. For a hyperbola, the *difference* is constant.

A **hyperbola** is the set of all points P such that the difference of the distances from P to two fixed points, called the **foci**, is constant. The line through the foci intersects the hyperbola at two points, the **vertices**. The line segment joining the vertices is the **transverse axis**, and its midpoint is the **center** of the hyperbola. A hyperbola has two branches and two asymptotes. The asymptotes contain the diagonals of a rectangle centered at the hyperbola's center, as shown below.

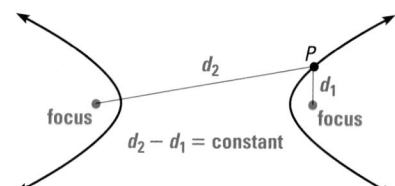

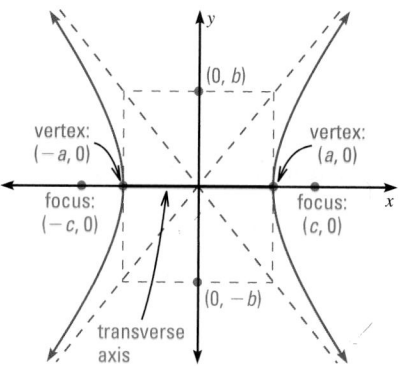

Hyperbola with horizontal transverse axis
$$\frac{x^2}{a^2} - \frac{y^2}{b^2} = 1$$

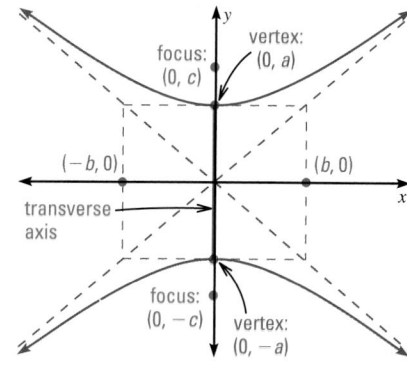

Hyperbola with vertical transverse axis
$$\frac{y^2}{a^2} - \frac{x^2}{b^2} = 1$$

CHARACTERISTICS OF A HYPERBOLA (CENTER AT ORIGIN)

The **standard form of the equation of a hyperbola** with center at (0, 0) is as follows.

EQUATION	TRANSVERSE AXIS	ASYMPTOTES	VERTICES
$\frac{x^2}{a^2} - \frac{y^2}{b^2} = 1$	Horizontal	$y = \pm\frac{b}{a}x$	$(\pm a, 0)$
$\frac{y^2}{a^2} - \frac{x^2}{b^2} = 1$	Vertical	$y = \pm\frac{a}{b}x$	$(0, \pm a)$

The foci of the hyperbola lie on the transverse axis, c units from the center where $c^2 = a^2 + b^2$.

10.5 *Hyperbolas* **615**

1 PLAN

PACING
Basic: 2 days
Average: 2 days
Advanced: 2 days
Block Schedule: 1 block

LESSON OPENER
VISUAL APPROACH
An alternative way to approach Lesson 10.5 is to use the Visual Approach Lesson Opener:

• Blackline Master (*Chapter 10 Resource Book*, p. 65)
• Transparency (p. 69)

MEETING INDIVIDUAL NEEDS
• *Chapter 10 Resource Book*
Prerequisite Skills Review (p. 5)
Practice Level A (p. 67)
Practice Level B (p. 68)
Practice Level C (p. 69)
Reteaching with Practice (p. 70)
Absent Student Catch-Up (p. 72)
Challenge (p. 74)

• *Resources in Spanish*
• *Personal Student Tutor*

NEW-TEACHER SUPPORT
See the Tips for New Teachers on pp. 1–2 of the *Chapter 10 Resource Book* for additional notes about Lesson 10.5.

WARM-UP EXERCISES

Transparency Available

Write the equation for the ellipse in standard form.

1. $64x^2 + 9y^2 = 576$ $\frac{x^2}{9} + \frac{y^2}{64} = 1$

2. $x^2 + 4y^2 = 4$ $\frac{x^2}{4} + y^2 = 1$

3. $2x^2 + 3y^2 = 1$ $\frac{x^2}{\left(\frac{1}{2}\right)} + \frac{y^2}{\left(\frac{1}{3}\right)} = 1$

State whether the major axis of the ellipse is vertical or horizontal.

4. $\frac{x^2}{12} + \frac{y^2}{16} = 1$ vertical

5. $\frac{x^2}{9} + \frac{y^2}{3} = 1$ horizontal

MOTIVATING THE LESSON
Write the standard equation for an ellipse on the board. Then ask students what they think will happen if you subtract the two fractions instead of adding them. Have students make a conjecture and explain their thinking. Then introduce the definition of hyperbola.

EXTRA EXAMPLE 1
Draw the hyperbola given by $9x^2 - 16y^2 = 144$.

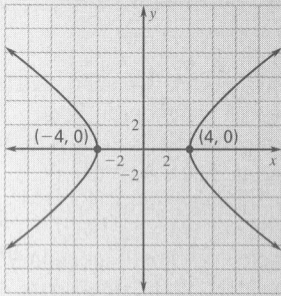

EXTRA EXAMPLE 2
Write an equation of the hyperbola with foci at (0, –5) and (0, 5) and vertices at (0, –3) and (0, 3).
$$\frac{y^2}{9} - \frac{x^2}{16} = 1$$

CHECKPOINT EXERCISES
For use after Example 1:
1. Draw the hyperbola given by $9y^2 - 16x^2 = 144$.

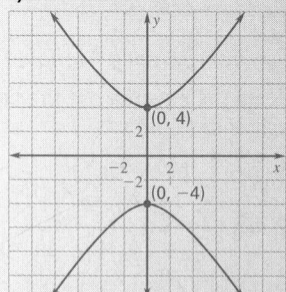

For use after Example 2:
2. Write an equation of the hyperbola with foci at (–2, 0) and (2, 0) and vertices at (–1, 0) and (1, 0). $x^2 - \frac{y^2}{3} = 1$

EXAMPLE 1 *Graphing an Equation of a Hyperbola*

Draw the hyperbola given by $4x^2 - 9y^2 = 36$.

SOLUTION

First rewrite the equation in standard form.

$$4x^2 - 9y^2 = 36 \qquad \text{Write original equation.}$$

$$\frac{4x^2}{36} - \frac{9y^2}{36} = \frac{36}{36} \qquad \text{Divide each side by 36.}$$

$$\frac{x^2}{9} - \frac{y^2}{4} = 1 \qquad \text{Simplify.}$$

Note from the equation that $a^2 = 9$ and $b^2 = 4$, so $a = 3$ and $b = 2$. Because the x^2-term is positive, the transverse axis is horizontal and the vertices are at $(-3, 0)$ and $(3, 0)$. To draw the hyperbola, first draw a rectangle that is centered at the origin, $2a = 6$ units wide and $2b = 4$ units high. Then show the asymptotes by drawing the lines that pass through opposite corners of the rectangle. Finally, draw the hyperbola.

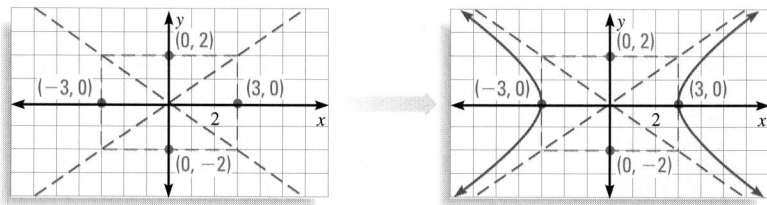

EXAMPLE 2 *Writing an Equation of a Hyperbola*

Write an equation of the hyperbola with foci at $(0, -3)$ and $(0, 3)$ and vertices at $(0, -2)$ and $(0, 2)$.

SOLUTION

The transverse axis is vertical because the foci and vertices lie on the y-axis. The center is the origin because the foci and the vertices are equidistant from the origin. Since the foci are each 3 units from the center, $c = 3$. Similarly, because the vertices are each 2 units from the center, $a = 2$.

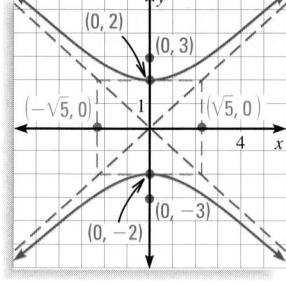

You can use these values of a and c to find b.

$$b^2 = c^2 - a^2$$

$$b^2 = 3^2 - 2^2 = 9 - 4 = 5$$

$$b = \sqrt{5}$$

Because the transverse axis is vertical, the standard form of the equation is as follows.

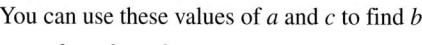

$$\frac{y^2}{2^2} - \frac{x^2}{(\sqrt{5})^2} = 1 \qquad \text{Substitute 2 for } a \text{ and } \sqrt{5} \text{ for } b.$$

$$\frac{y^2}{4} - \frac{x^2}{5} = 1 \qquad \text{Simplify.}$$

**PANORAMIC
CAMERAS**

A panoramic photograph
taken with the camera
shown above gives a 360°
view of a scene.

APPLICATION LINK
www.mcdougallittell.com

GOAL 2 USING HYPERBOLAS IN REAL LIFE

EXAMPLE 3 *Using a Real-Life Hyperbola*

PHOTOGRAPHY A hyperbolic mirror can be used to take panoramic photographs. A camera is pointed toward the vertex of the mirror and is positioned so that the lens is at one focus of the mirror. An equation for the cross section of the mirror is $\frac{y^2}{16} - \frac{x^2}{9} = 1$ where x and y are measured in inches. How far from the mirror is the lens?

SOLUTION

Notice from the equation that $a^2 = 16$ and $b^2 = 9$, so $a = 4$ and $b = 3$. Use these values and the equation $c^2 = a^2 + b^2$ to find the value of c.

$c^2 = a^2 + b^2$ **Equation relating *a*, *b*, and *c***

$c^2 = 16 + 9 = 25$ **Substitute for *a* and *b* and simplify.**

$c = 5$ **Solve for *c*.**

Since $a = 4$ and $c = 5$, the vertices are at $(0, -4)$ and $(0, 4)$ and the foci are at $(0, -5)$ and $(0, 5)$. The camera is below the mirror, so the lens is at $(0, -5)$ and the vertex of the mirror is at $(0, 4)$. The distance between these points is $4 - (-5) = 9$.

▶ The lens is 9 inches from the mirror.

Sculpture

EXAMPLE 4 *Modeling with a Hyperbola*

The diagram at the right shows the hyperbolic cross section of a sculpture located at the Fermi National Accelerator Laboratory in Batavia, Illinois.

a. Write an equation that models the curved sides of the sculpture.

b. At a height of 5 feet, how wide is the sculpture? (Each unit in the coordinate plane represents 1 foot.)

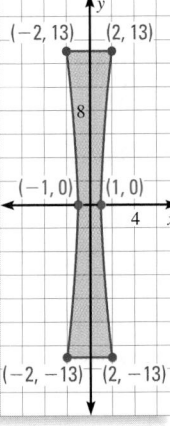

SOLUTION

a. From the diagram you can see that the transverse axis is horizontal and $a = 1$. So the equation has this form:

$$\frac{x^2}{1^2} - \frac{y^2}{b^2} = 1$$

Because the hyperbola passes through the point $(2, 13)$, you can substitute $x = 2$ and $y = 13$ into the equation and solve for b. When you do this, you obtain $b \approx 7.5$.

▶ An equation of the hyperbola is $\frac{x^2}{1^2} - \frac{y^2}{(7.5)^2} = 1$.

b. At a height of 5 feet above the ground, $y = -8$. To find the width of the sculpture, substitute this value into the equation and solve for x. You get $x \approx 1.46$.

▶ At a height of 5 feet, the width is $2x \approx 2.92$ feet.

EXTRA EXAMPLE 3
A hyperbolic mirror has a cross section represented by $\frac{y^2}{9} - \frac{x^2}{25} = 1$. How far is the mirror from the lens? $3 + \sqrt{34} \approx 8.83$ in.

EXTRA EXAMPLE 4
The diagram shows the hyperbolic cross section of a large hourglass.

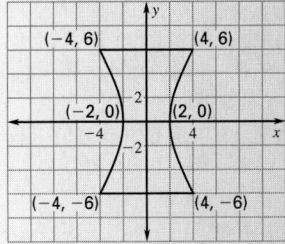

a. Write an equation that models the curved sides. $\frac{x^2}{4} - \frac{y^2}{12} = 1$

b. At a height of 7 in., how wide is the hourglass? about 4.16 in.

CHECKPOINT EXERCISES
For use after Examples 3 and 4:

1. The pendulum for a clock is shaped like a hyperbola. A cross section is shown below.

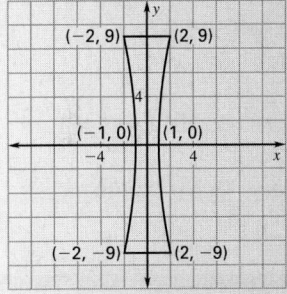

a. Write an equation that models the pendulum.
$x^2 - \frac{y^2}{27} = 1$

b. How wide is the pendulum at a height of 1 ft? (Each unit in the coordinate grid represents 2 in.)
about 4.62 in.

CLOSURE QUESTION
What are the asymptotes and vertices of a hyperbola modeled by $\frac{y^2}{a^2} - \frac{x^2}{b^2} = 1$? asymptotes: $y = \pm\frac{a}{b}x$; vertices: $(0, \pm a)$

ASSIGNMENT GUIDE

BASIC
Day 1: pp. 618–619 Exs. 15–18, 20–36 even, 43, 44, 49–52, 56–58
Day 2: pp. 619–621 Exs. 60–62, 64–65, 68, 71–85 odd, Quiz 2 Exs. 1–17

AVERAGE
Day 1: pp. 618–619 Exs. 15–18, 20–40 even, 43–45, 49–53, 56–60
Day 2: pp. 619–621 Exs. 61–65, 67–68, 71–91 odd, Quiz 2 Exs. 1–17

ADVANCED
Day 1: pp. 618–619 Exs. 15–18, 20–42 even, 43–45, 49–53, 56–60
Day 2: pp. 619–621 Exs. 61–66, 68–70, 71–91 odd, Quiz 2 Exs. 1–17

BLOCK SCHEDULE
pp. 618–621 Exs. 15–18, 20–36 even, 43, 44, 50–52, 56–58, 64–66, 68, 71–91 odd, Quiz 2 Exs. 1–17

EXERCISE LEVELS
Level A: *Easier*
15–27, 31–39
Level B: *More Difficult*
28–30, 40–68
Level C: *Most Difficult*
69, 70

✔ **HOMEWORK CHECK**
To quickly check student understanding of key concepts, go over the following exercises: Exs. 16, 20, 26, 32, 44, 50, 56. See also the Daily Homework Quiz:
• Blackline Master (*Chapter 10 Resource Book,* p. 78)
• 🖎 Transparency (p. 80)

4–9. See Additional Answers beginning on page AA1.

GUIDED PRACTICE

Vocabulary Check ✔

1. Complete these statements: The points $(0, -2)$ and $(0, 2)$ in the graph at the right are the _?_ of the hyperbola. The segment joining these two points is the _?_ . **vertices; transverse axis**

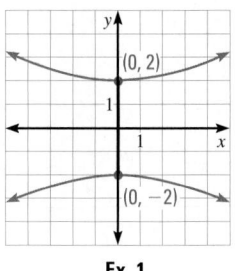

Ex. 1

Concept Check ✔

2. How are the definitions of ellipse and hyperbola alike? How are they different? **See margin.**

3. How do the asymptotes of a hyperbola help you draw the hyperbola? **Start at the vertices and draw curves that approach but don't meet the asymptotes.**

Skill Check ✔

2. Both involve all points a certain distance from 2 foci; For an ellipse, the sum of the distances is constant; for a hyperbola, the difference is constant.

4–9. See margin for graphs.

4. foci: $(\pm\sqrt{130}, 0)$ asymptotes:
$y = \pm\frac{9}{7}x$

5. foci: $(0, \pm5\sqrt{7})$
asymptotes:
$y = \pm\frac{2\sqrt{3}}{3}x$

6. foci: $(\pm\sqrt{65}, 0)$
asymptotes: $y = \pm\frac{1}{8}x$

7. foci: $(\pm2\sqrt{10}, 0)$
asymptotes: $y = \pm3x$

Graph the equation. Identify the foci and asymptotes. 4–9. See margin.

4. $\frac{x^2}{49} - \frac{y^2}{81} = 1$
5. $\frac{y^2}{100} - \frac{x^2}{75} = 1$
6. $\frac{x^2}{64} - y^2 = 1$
7. $36x^2 - 4y^2 = 144$
8. $12y^2 - 25x^2 = 300$
9. $y^2 - 9x^2 = 9$

Write an equation of the hyperbola with the given foci and vertices.

10. Foci: $(0, -5), (0, 5)$ Vertices: $(0, -3), (0, 3)$ $\frac{y^2}{9} - \frac{x^2}{16} = 1$

11. Foci: $(-8, 0), (8, 0)$ Vertices: $(-7, 0), (7, 0)$ $\frac{x^2}{49} - \frac{y^2}{15} = 1$

12. Foci: $(-\sqrt{34}, 0), (\sqrt{34}, 0)$ Vertices: $(-5, 0), (5, 0)$ $\frac{x^2}{25} - \frac{y^2}{9} = 1$

13. Foci: $(0, -9), (0, 9)$ Vertices: $(0, -3\sqrt{5}), (0, 3\sqrt{5})$ $\frac{y^2}{45} - \frac{x^2}{36} = 1$

14. 🌐 **PHOTOGRAPHY** Look back at Example 3. Suppose a mirror has a cross section modeled by the equation $\frac{x^2}{25} - \frac{y^2}{9} = 1$ where x and y are measured in inches. If you place a camera with its lens at the focus, how far is the lens from the vertex of the mirror? $5 + \sqrt{34} \approx 10.8$ in.

PRACTICE AND APPLICATIONS

STUDENT HELP

↳ **Extra Practice** to help you master skills is on p. 954.

8. foci: $(0, \pm\sqrt{37})$
asymptotes:
$= \pm\frac{5\sqrt{3}}{6}x$

9. foci: $(0, \pm\sqrt{10})$
asymptotes: $y = \pm3x$

STUDENT HELP

↳ **HOMEWORK HELP**
Example 1: Exs. 15–55
Example 2: Exs. 56–63
Example 3: Ex. 67
Example 4: Exs. 64–66

MATCHING Match the equation with its graph.

15. $\frac{x^2}{16} - \frac{y^2}{4} = 1$ **C** **16.** $\frac{y^2}{4} - \frac{x^2}{2} = 1$ **A** **17.** $\frac{y^2}{16} - \frac{x^2}{4} = 1$ **D** **18.** $\frac{x^2}{4} - \frac{y^2}{2} = 1$ **B**

A.

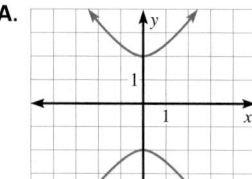

B.

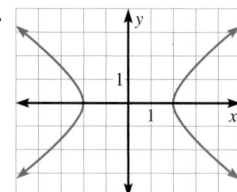

C.

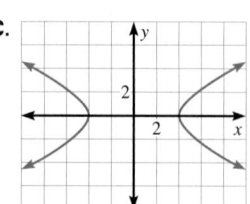

D.

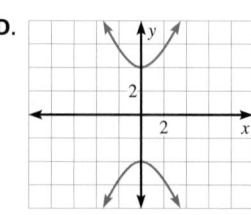

19. $\dfrac{x^2}{9} - \dfrac{y^2}{36} = 1$

20. $\dfrac{y^2}{81} - x^2 = 1$

21. $\dfrac{y^2}{\left(\frac{1}{4}\right)} - \dfrac{x^2}{\left(\frac{9}{4}\right)} = 1$

22. $\dfrac{x^2}{\left(\frac{1}{4}\right)} - \dfrac{y^2}{\left(\frac{9}{16}\right)} = 1$

23. $\dfrac{y^2}{4} - \dfrac{x^2}{144} = 1$

24. $\dfrac{x^2}{81} - \dfrac{y^2}{\left(\frac{81}{4}\right)} = 1$

25. vertices: $(\pm 3, 0)$
foci: $(\pm \sqrt{73}, 0)$

26. vertices: $(0, \pm 7)$
foci: $(0, \pm 5\sqrt{2})$

27. vertices: $(\pm 11, 0)$
foci: $(\pm 5\sqrt{5}, 0)$

28. vertices: $(0, \pm 9)$
foci: $(0, \pm \sqrt{85})$

29. vertices: $(0, \pm 2)$
foci: $(0, \pm \sqrt{29})$

STUDENT HELP

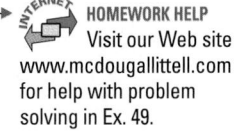

HOMEWORK HELP
Visit our Web site
www.mcdougallittell.com
for help with problem
solving in Ex. 49.

30. vertices: $\left(\pm\sqrt{10}, 0\right)$
foci: $(\pm\sqrt{46}, 0)$

43. $y = \pm\dfrac{6\sqrt{x^2 + 100}}{5}$

44. $y = \pm\dfrac{5\sqrt{x^2 - 16}}{4}$

45. $y = \pm\dfrac{8.5\sqrt{x^2 - 42.25}}{6.5}$

46. $y = \pm\sqrt{\dfrac{2.73(x^2 + 3.58)}{3.58}}$

47. $y = \pm\sqrt{\dfrac{22.3(x^2 - 10.1)}{10.1}}$

48. $y = \pm\sqrt{\dfrac{1.2x^2 - 4.6}{8.5}}$

STANDARD FORM Write the equation of the hyperbola in standard form.

19. $36x^2 - 9y^2 = 324$ **20.** $y^2 - 81x^2 = 81$ **21.** $36y^2 - 4x^2 = 9$

22. $16y^2 - 36x^2 + 9 = 0$ **23.** $y^2 - \dfrac{x^2}{36} = 4$ **24.** $\dfrac{x^2}{9} - \dfrac{4y^2}{9} = 9$

IDENTIFYING PARTS Identify the vertices and foci of the hyperbola. **25–30. See margin.**

25. $\dfrac{x^2}{9} - \dfrac{y^2}{64} = 1$ **26.** $\dfrac{y^2}{49} - x^2 = 1$ **27.** $\dfrac{x^2}{121} - \dfrac{y^2}{4} = 1$

28. $4y^2 - 81x^2 = 324$ **29.** $25y^2 - 4x^2 = 100$ **30.** $36x^2 - 10y^2 = 360$

GRAPHING Graph the equation. Identify the foci and asymptotes. **31–42. See margin.**

31. $\dfrac{x^2}{25} - \dfrac{y^2}{121} = 1$ **32.** $\dfrac{x^2}{36} - y^2 = 1$ **33.** $\dfrac{y^2}{25} - \dfrac{x^2}{49} = 1$

34. $\dfrac{y^2}{9} - \dfrac{x^2}{100} = 1$ **35.** $\dfrac{x^2}{169} - \dfrac{y^2}{16} = 1$ **36.** $\dfrac{y^2}{64} - x^2 = 1$

37. $\dfrac{16x^2}{25} - \dfrac{y^2}{81} = 1$ **38.** $\dfrac{x^2}{144} - \dfrac{y^2}{121} = 1$ **39.** $\dfrac{x^2}{64} - \dfrac{9y^2}{4} = 1$

40. $\dfrac{y^2}{25} - \dfrac{x^2}{16} = 16$ **41.** $100x^2 - 81y^2 = 8100$ **42.** $x^2 - 9y^2 = 25$

GRAPHING HYPERBOLAS Use a graphing calculator to graph the equation. Tell what two equations you entered into the calculator. **43–48. See margin.**

43. $\dfrac{y^2}{144} - \dfrac{x^2}{100} = 1$ **44.** $\dfrac{x^2}{16} - \dfrac{y^2}{25} = 1$ **45.** $\dfrac{x^2}{42.25} - \dfrac{y^2}{72.25} = 1$

46. $\dfrac{y^2}{2.73} - \dfrac{x^2}{3.58} = 1$ **47.** $\dfrac{x^2}{10.1} - \dfrac{y^2}{22.3} = 1$ **48.** $1.2x^2 - 8.5y^2 = 4.6$

49. CRITICAL THINKING Suppose you tried to graph an equation of a hyperbola on a graphing calculator. You enter one function correctly, but you forget to enter the other function. Sketch what your graph might look like if the transverse axis is horizontal. Then sketch what your graph might look like if the transverse axis is vertical. **See margin.**

GRAPHING CONIC SECTIONS In Exercises 50–55, the equations of parabolas, circles, ellipses, and hyperbolas are given. Graph the equation. **50–55. See margin.**

50. $\dfrac{x^2}{169} - \dfrac{y^2}{25} = 1$ **51.** $x^2 + y^2 = 30$ **52.** $\dfrac{y^2}{9} - \dfrac{x^2}{64} = 1$

53. $x^2 = 15y$ **54.** $\dfrac{x^2}{196} + \dfrac{y^2}{256} = 1$ **55.** $14x^2 + 14y^2 = 126$

WRITING EQUATIONS Write an equation of the hyperbola with the given foci and vertices.

56. Foci: $(0, -13), (0, 13)$ $\dfrac{y^2}{25} - \dfrac{x^2}{144} = 1$
Vertices: $(0, -5), (0, 5)$

57. Foci: $(-8, 0), (8, 0)$ $\dfrac{x^2}{36} - \dfrac{y^2}{28} = 1$
Vertices: $(-6, 0), (6, 0)$

58. Foci: $(-4, 0), (4, 0)$ $x^2 - \dfrac{y^2}{15} = 1$
Vertices: $(-1, 0), (1, 0)$

59. Foci: $(-6, 0), (6, 0)$ $\dfrac{x^2}{25} - \dfrac{y^2}{11} = 1$
Vertices: $(-5, 0), (5, 0)$

60. Foci: $(0, -7), (0, 7)$ $\dfrac{y^2}{9} - \dfrac{x^2}{40} = 1$
Vertices: $(0, -3), (0, 3)$

61. Foci: $(0, -9), (0, 9)$ $\dfrac{y^2}{64} - \dfrac{x^2}{17} = 1$
Vertices: $(0, -8), (0, 8)$

62. Foci: $(-8, 0), (8, 0)$
Vertices: $\left(-4\sqrt{3}, 0\right), \left(4\sqrt{3}, 0\right)$ $\dfrac{x^2}{48} - \dfrac{y^2}{16} = 1$

63. Foci: $\left(0, -5\sqrt{6}\right), \left(0, 5\sqrt{6}\right)$
Vertices: $(0, -4), (0, 4)$ $\dfrac{y^2}{16} - \dfrac{x^2}{134} = 1$

TEACHING TIPS
EXERCISES 31–42 When graphing hyperbolas, students need to remember that an asymptote is a line that a graph approaches but does not cross. For hyperbolas, the asymptotes can be seen as boundary lines for the graph.

GRAPHING CALCULATOR NOTE
EXERCISES 43–48 Students can review the method of graphing equations of circles using a graphing calculator by looking back to Activity 10.3 on page 608. Solving and entering an equation for a hyperbola is similar.

STUDENT HELP NOTES
→ **Homework Help** Students can find help for Ex. 49 at **www.mcdougallittell.com**. The information can be printed out for students who don't have access to the Internet.

! COMMON ERROR
EXERCISES 50–55 The equation for a hyperbola is very similar to that for an ellipse — the only difference is an operation symbol. Students may confuse the two equations and may also write an incorrect equation with the change of a positive or negative sign. Express the importance of being very careful when solving to graph the equations.

! COMMON ERROR
EXERCISES 56–63 When trying to find the transverse axis, students need to remember that the variable that is contained in the positive fraction is the transverse axis, and the hyperbola intersects that axis.

31–42, 49–55.
See Additional Answers beginning on page AA1.

APPLICATION NOTE
EXERCISES 64–66 Additional information about sundials is available at **www.mcdougallittell.com**.

66. lower branch summer solstice hyperbola:

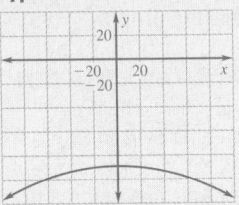

upper branch winter solstice hyperbola:

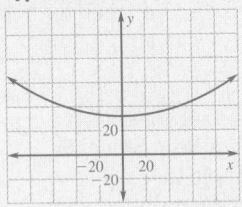

70. **No.** You could create a branch of a hyperbola such that Microphone A is a constant 2200 ft farther than Microphone B. The sound from any point on that branch would reach Microphone A 2 sec later. However, you would need more information to decide which point on the branch was the source of the sound.

ADDITIONAL PRACTICE AND RETEACHING

For Lesson 10.5:
• Practice Levels A, B, and C (*Chapter 10 Resource Book*, p. 67)
• Reteaching with Practice (*Chapter 10 Resource Book*, p. 70)
• ⊞ See Lesson 10.5 of the *Personal Student Tutor*

For more Mixed Review:
• ⊞ Search the *Test and Practice Generator* for key words or specific lessons.

620

FOCUS ON APPLICATIONS

SUNDIAL The Richard D. Swensen sundial, located at the University of Wisconsin – River Falls, gives the correct time to the minute.

APPLICATION LINK
www.mcdougallittell.com

68c. It decreases; the larger the outer circle of the ring, the more narrow the ring must become to keep the same area. If the inner circle is pushed out, so is the outer circle.

Test Preparation

69. $d_2 - d_1$ is a constant by the definition of a hyperbola. Find $d_2 - d_1$ at $(a, 0)$: $d_2 - d_1 = (c + a) - (c - a) = 2a$.

★ Challenge

EXTRA CHALLENGE
www.mcdougallittell.com

SUNDIAL In Exercises 64–66, use the following information.
The sundial at the left was designed by Professor John Shepherd. The shadow of the *gnomon* traces a hyperbola throughout the day. Aluminum rods form the hyperbolas traced on the summer solstice, June 21, and the winter solstice, December 21.

64. One focus of the summer solstice hyperbola is 207 inches above the ground. The vertex of the aluminum branch is 266 inches above the ground. If the x-axis is 355 inches above the ground and the center of the hyperbola is at the origin, write an equation for the summer solstice hyperbola. $\frac{y^2}{7921} - \frac{x^2}{13,983} = 1$

65. One focus of the winter solstice hyperbola is 419 inches above the ground. The vertex of the aluminum branch is 387 inches above the ground and the center of the hyperbola is at the origin. If the x-axis is 355 inches above the ground, write an equation for the winter solstice hyperbola. $\frac{y^2}{1024} - \frac{x^2}{3072} = 1$

66. Use your equations from Exercises 64 and 65 to draw the lower branch of the summer solstice hyperbola and the upper branch of the winter solstice hyperbola.
See margin for graph.

67. **AERONAUTICS** When an airplane travels faster than the speed of sound, the sound waves form a cone behind the airplane. If the airplane is flying parallel to the ground, the sound waves intersect the ground in a hyperbola with the airplane directly above its center. A sonic boom is heard along the hyperbola. If you hear a sonic boom that is audible along a hyperbola with the equation $\frac{x^2}{100} - \frac{y^2}{4} = 1$ where x and y are measured in miles, what is the shortest horizontal distance you could be to the airplane? **10 mi**

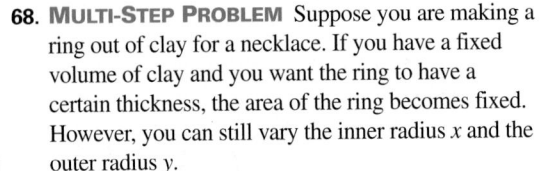
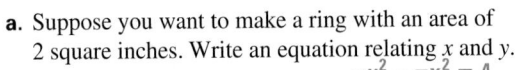
shock wave

ground

68. **MULTI-STEP PROBLEM** Suppose you are making a ring out of clay for a necklace. If you have a fixed volume of clay and you want the ring to have a certain thickness, the area of the ring becomes fixed. However, you can still vary the inner radius x and the outer radius y.

a. Suppose you want to make a ring with an area of 2 square inches. Write an equation relating x and y. $\pi y^2 - \pi x^2 = 4$

b. Find three coordinate pairs (x, y) that satisfy the relationship from part (a). Then find the width of the ring, y − x, for each coordinate pair.
Sample answer: (1.65, 2), 0.35 in.; (2.23, 2.5), 0.27 in.; (1.34, 1.75), 0.41 in.

c. *Writing* How does the width of the ring, y − x, change as x and y both increase? Explain why this makes sense. **See margin.**

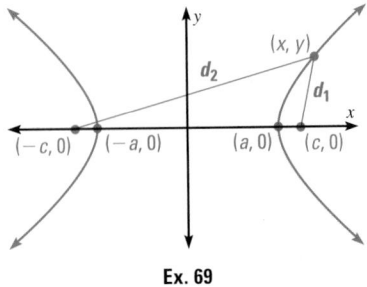
☐ fixed

69. Use the diagram at the right to show that $|d_2 - d_1| = 2a$. **See margin.**

70. **LOCATING AN EXPLOSION** Two microphones, 1 mile apart, record an explosion. Microphone A receives the sound 2 seconds after Microphone B. Is this enough information to decide where the sound came from? Use the fact that sound travels at 1100 feet per second. **See margin.**

Ex. 69

MIXED REVIEW

GRAPHING FUNCTIONS Graph the function. (Review 2.8, 5.1 for 10.6)
71–76. See margin.

77. $f(x) = x^3 - 6x^2 + 11x - 6$

71. $y = 2|x + 4| + 1$ **72.** $y = |x - 4| + 5$ **73.** $y = -|x - 6| - 8$

78. $f(x) = x^3 + 5x^2 - 17x - 21$

74. $y = 3(x - 1)^2 + 7$ **75.** $y = -2(x - 3)^2 - 6$ **76.** $y = \frac{1}{2}(x + 4)^2 + 5$

79. $f(x) = x^3 - 6x^2 - 4x + 24$

WRITING FUNCTIONS Write a polynomial function of least degree that has real coefficients, the given zeros, and a leading coefficient of 1. (Review 6.7)

1. $\frac{x^2}{9} + \frac{y^2}{49} = 1$

77. 3, 1, 2 See margin.   **78.** $-7, -1, 3$ See margin.   **79.** $6, -2, 2$ See margin.

2. $\frac{x^2}{36} + y^2 = 1$

80. $-6, 4, 2$
$f(x) = x^3 - 28x + 48$

81. $5, i, -i$
$f(x) = x^3 - 5x^2 + x - 5$

82. $3, -3, 2i$
$f(x) = x^4 - 5x^2 - 36$

EVALUATING LOGARITHMIC EXPRESSIONS Evaluate the expression without using a calculator. (Review 8.4)

3. $\frac{x^2}{100} + \frac{y^2}{64} = 1$

83. $\log 10,000$ 4   **84.** $\log_3 27$ 3   **85.** $\log_5 625$ 4   **86.** $\log_2 128$ 7

4. $\frac{x^2}{8} + \frac{y^2}{25} = 1$

87. $\log_4 64$ 3   **88.** $\log_3 243$ 5   **89.** $\log_6 216$ 3   **90.** $\log_{100} 100,000,000$ 4

5. $\frac{x^2}{15} + \frac{y^2}{12} = 1$

91. 🌎 **TEST SCORES** Find the mean, median, mode(s), and range of the following set of test scores. (Review 7.7)

6. $\frac{x^2}{81} + \frac{y^2}{97} = 1$

63, 67, 72, 75, 77, 78, 81, 81, 85, 86, 89, 89, 91, 92, 99
mean: 81.67; median: 81; modes: 81, 89; range: 36

QUIZ 2

Self-Test for Lessons 10.4 and 10.5

7. vertices: $(0, \pm 7)$
co-vertices: $(\pm 2, 0)$
foci: $(0, \pm 3\sqrt{5})$

Write an equation of the ellipse with the given characteristics and center at **(0, 0)**. (Lesson 10.4) 1–6. See margin.

8. vertices: $(\pm\sqrt{6}, 0)$
co-vertices: $(0, \pm 1)$
foci: $(\pm\sqrt{5}, 0)$

1. Vertex: $(0, 7)$
Co-vertex: $(-3, 0)$

2. Vertex: $(-6, 0)$
Co-vertex: $(0, -1)$

3. Vertex: $(-10, 0)$
Focus: $(6, 0)$

9. vertices: $(\pm 6, 0)$
co-vertices: $(0, \pm 2)$
foci: $(\pm 4\sqrt{2}, 0)$

4. Vertex: $(0, 5)$
Focus: $(0, \sqrt{17})$

5. Co-vertex: $(0, 2\sqrt{3})$
Focus: $(-\sqrt{3}, 0)$

6. Co-vertex: $(-9, 0)$
Focus: $(0, 4)$

14. vertices: $(0, \pm 5)$
foci: $(0, \pm\sqrt{61})$
asymptotes:
$y = \pm\frac{5}{6}x$

Graph the equation. Identify the vertices, co-vertices, and foci. (Lesson 10.4)
7–9. See margin for graphs.

7. $\frac{x^2}{4} + \frac{y^2}{49} = 1$ **8.** $\frac{x^2}{6} + y^2 = 1$ **9.** $x^2 + 9y^2 = 36$

15. vertices: $(0, \pm 2\sqrt{5})$
foci: $(0, \pm 2\sqrt{7})$
asymptotes:
$y = \pm\frac{\sqrt{10}}{2}x$

Write an equation of the hyperbola with the given characteristics. (Lesson 10.5)

10. Foci: $(0, -8), (0, 8)$
Vertices: $(0, -5), (0, 5)$
$\frac{y^2}{25} - \frac{x^2}{39} = 1$

11. Foci: $(-3, 0), (3, 0)$
Vertices: $(-1, 0), (1, 0)$
$x^2 - \frac{y^2}{8} = 1$

16. vertices: $(\pm\sqrt{2}, 0)$
foci: $(\pm\sqrt{11}, 0)$
asymptotes:
$y = \pm\frac{3\sqrt{2}}{2}x$

12. Foci: $(-6, 0), (6, 0)$
Vertices: $(-4, 0), (4, 0)$
$\frac{x^2}{16} - \frac{y^2}{20} = 1$

13. Foci: $(0, -2\sqrt{5}), (0, 2\sqrt{5})$
Vertices: $(0, -4), (0, 4)$
$\frac{y^2}{16} - \frac{x^2}{4} = 1$

17. $\frac{x^2}{4375^2} + \frac{y^2}{4369^2} = 1$

Graph the equation. Identify the vertices, foci, and asymptotes. (Lesson 10.5)
14–16. See margin for graphs.

14. $\frac{y^2}{25} - \frac{x^2}{36} = 1$ **15.** $8y^2 - 20x^2 = 160$ **16.** $18x^2 - 4y^2 = 36$

17. 🌎 **SPACE EXPLORATION** Suppose a satellite's orbit is an ellipse with Earth's center at one focus. If the satellite's least distance from Earth's surface is 150 miles and its greatest distance from Earth's surface is 600 miles, write an equation for the ellipse. (Use 4000 miles as Earth's radius.) (Lesson 10.4)

10.5 Hyperbolas   621

71–76, 7–9, 14–16. See Additional Answers beginning on page AA1.

DAILY HOMEWORK QUIZ

🖥 *Transparency Available*

1. Graph the equation and name the vertices, foci, and asymptotes of the hyperbola $16x^2 - 9y^2 = 576$.

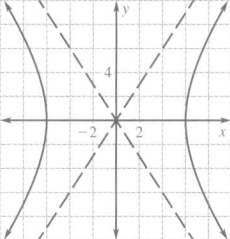

vertices: $(\pm 6, 0)$; foci: $(\pm 10, 0)$;
asymptotes: $y = \pm\frac{4}{3}x$

2. Write an equation of the hyperbola with center at the origin with foci at $(\pm 5, 0)$ and vertices $(\pm 4, 0)$. $\frac{x^2}{16} - \frac{y^2}{9} = 1$

EXTRA CHALLENGE NOTE

→ Challenge problems for Lesson 10.5 are available in **blackline** format in the *Chapter 10 Resource Book*, p. 74 and at **www.mcdougallittell.com**.

ADDITIONAL TEST PREPARATION

1. WRITING Explain how you can tell from the standard equation for a hyperbola if it opens to the left and right or if it opens up and down? *Sample answer:* If the standard equation is $\frac{x^2}{a^2} - \frac{y^2}{b^2} = 1$, the hyperbola opens left and right. If the standard equation is $\frac{y^2}{a^2} - \frac{x^2}{b^2} = 1$, the hyperbola opens up and down.

ADDITIONAL RESOURCES
An alternative quiz for Lessons 10.4–10.5 is available in the *Chapter 10 Resource Book*, p. 75.

1 Planning the Activity

PURPOSE
To explore the four conic sections.

MATERIALS
• flashlight
• graph paper
• Activity Support Master
  (*Chapter 10 Resource Book,*
  p. 79)

PACING
• Exploring the Concept — 15 min
• Drawing Conclusions — 10 min

 LINK TO LESSON
Students can relate the graphs
and equations of Examples 1–4
to the geometric concepts of the
conics presented here.

2 Managing the Activity

COOPERATIVE LEARNING
Students should take turns holding
the flashlight and drawing the
conics. They should work together
to find the equations.

ALTERNATIVE APPROACH
This activity can be done with
larger groups to produce fewer
circles and ellipses. If this is done,
the groups should get together to
compare all the circles and ellipses.

3 Closing the Activity

★ **KEY DISCOVERY**
All circles have the same general
equation, and all ellipses have the
same general equation.

ACTIVITY ASSESSMENT
Give the general forms of the
equations for circles and ellipses
centered at the origin.
circle: $x^2 + y^2 = r^2$;
ellipse: $\dfrac{x^2}{a^2} + \dfrac{y^2}{b^2} = 1$ or $\dfrac{x^2}{b^2} + \dfrac{y^2}{a^2} = 1$

▶ **ACTIVITY 10.6**

Developing Concepts

GROUP ACTIVITY
Work with a partner.

MATERIALS
• flashlight
• graph paper
• pencil

Exploring Conic Sections

▶ **QUESTION** How do a plane and a double-napped cone intersect to form
different conic sections?

▶ **EXPLORING THE CONCEPT** The reason that parabolas, circles, ellipses,
and hyperbolas are called *conics* or *conic sections* is that each can be formed
by the intersection of a plane and a double-napped cone, as shown below.

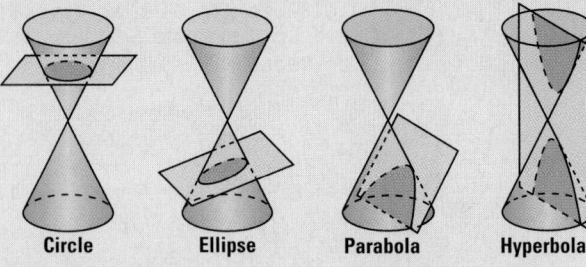

Circle Ellipse Parabola Hyperbola

The beam of light from a flashlight is a cone. When the light hits a flat surface such
as a wall, the edge of the beam of light forms a conic section.

Work in a group to find an equation of a conic formed by a flashlight beam.

❶ On a piece of graph paper, draw *x*- and *y*-axes to make a coordinate plane.

❷ Tape the paper to a wall.

❸ Aim a flashlight perpendicular to the
paper so that the light forms a circle.
Move the flashlight so that the circle is
centered on the origin of the coordinate
plane.

❹ Holding the flashlight very still, trace
the circle on the graph paper. Find the
radius of the circle and use it to write
the standard form of the equation of
the circle.

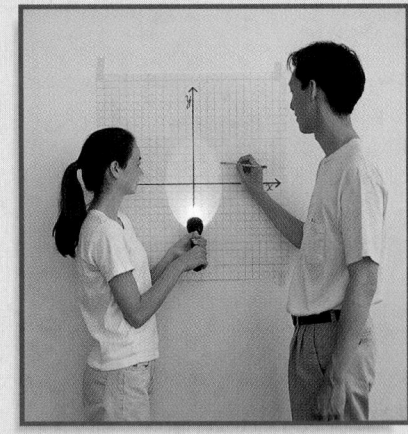

❺ Aim the flashlight at the paper to form
an ellipse with a vertical major axis and
center at the origin. Trace the ellipse and
find the standard form of its equation.

▶ **DRAWING CONCLUSIONS**

1. Compare the equation of your circle with the equations found by other groups.
   Are your equations all the same? Why or why not?
   Answers will vary; expect different equations.
2. Compare the equation of your ellipse with the equations found by other groups.
   Are your equations all the same? Why or why not?
   Answers will vary; expect different equations.

10.6
Graphing and Classifying Conics

What you should learn

GOAL 1 Write and graph an equation of a parabola with its vertex at (h, k) and an equation of a circle, ellipse, or hyperbola with its center at (h, k).

GOAL 2 Classify a conic using its equation, as applied in **Example 8**.

Why you should learn it

▼ To model **real-life** situations involving more than one conic, such as the circles that an ice skater uses to practice figure eights in **Ex. 64**.

GOAL 1 WRITING AND GRAPHING EQUATIONS OF CONICS

Parabolas, circles, ellipses, and hyperbolas are all curves that are formed by the intersection of a plane and a double-napped cone. Therefore, these shapes are called **conic sections** or simply **conics**.

In previous lessons you studied equations of parabolas with vertices at the origin and equations of circles, ellipses, and hyperbolas with centers at the origin. In this lesson you will study equations of conics that have been translated in the coordinate plane.

STANDARD FORM OF EQUATIONS OF TRANSLATED CONICS

In the following equations the point (h, k) is the *vertex* of the parabola and the *center* of the other conics.

	Horizontal axis	Vertical axis
CIRCLE	$(x - h)^2 + (y - k)^2 = r^2$	

	Horizontal axis	Vertical axis
PARABOLA	$(y - k)^2 = 4p(x - h)$	$(x - h)^2 = 4p(y - k)$
ELLIPSE	$\dfrac{(x - h)^2}{a^2} + \dfrac{(y - k)^2}{b^2} = 1$	$\dfrac{(x - h)^2}{b^2} + \dfrac{(y - k)^2}{a^2} = 1$
HYPERBOLA	$\dfrac{(x - h)^2}{a^2} - \dfrac{(y - k)^2}{b^2} = 1$	$\dfrac{(y - k)^2}{a^2} - \dfrac{(x - h)^2}{b^2} = 1$

EXAMPLE 1 *Writing an Equation of a Translated Parabola*

Write an equation of the parabola whose vertex is at $(-2, 1)$ and whose focus is at $(-3, 1)$.

SOLUTION

Choose form: Begin by sketching the parabola, as shown. Because the parabola opens to the left, it has the form

$$(y - k)^2 = 4p(x - h)$$

where $p < 0$.

Find h and k: The vertex is at $(-2, 1)$, so $h = -2$ and $k = 1$.

Find p: The distance between the vertex $(-2, 1)$ and the focus $(-3, 1)$ is

$$|p| = \sqrt{(-3 - (-2))^2 + (1 - 1)^2} = 1$$

so $p = 1$ or $p = -1$. Since $p < 0$, $p = -1$.

▶ The standard form of the equation is $(y - 1)^2 = -4(x + 2)$.

1 PLAN

PACING
Basic: 2 days
Average: 2 days
Advanced: 2 days
Block Schedule: 1 block

LESSON OPENER
GRAPHING CALCULATOR
An alternative way to approach Lesson 10.6 is to use the Graphing Calculator Lesson Opener:

- Blackline Master (*Chapter 10 Resource Book,* p. 80)
- Transparency (p. 70)

MEETING INDIVIDUAL NEEDS
- ***Chapter 10 Resource Book***
  Prerequisite Skills Review (p. 5)
  Practice Level A (p. 81)
  Practice Level B (p. 82)
  Practice Level C (p. 83)
  Reteaching with Practice (p. 84)
  Absent Student Catch-Up (p. 86)
  Challenge (p. 89)
- ***Resources in Spanish***
- *Personal Student Tutor*

NEW-TEACHER SUPPORT
See the Tips for New Teachers on pp. 1–2 of the *Chapter 10 Resource Book* for additional notes about Lesson 10.6.

WARM-UP EXERCISES
Transparency Available

State whether the graph is a *circle, parabola, ellipse,* or *hyperbola.*

1. $x = 2y^2$ parabola
2. $x^2 = 15 - y^2$ circle
3. $3x^2 + 12y^2 = 36$ ellipse
4. $3x^2 - 12y^2 = 36$ hyperbola

2 TEACH

EXTRA EXAMPLE 1
Write an equation of the parabola whose vertex is at (3, 2) and whose focus is at (4, 2).
$(y-2)^2 = 4(x-3)$

EXTRA EXAMPLE 2
Graph $(x-1)^2 + (y-4)^2 = 9$.

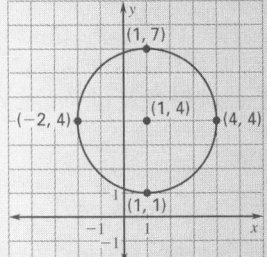

EXTRA EXAMPLE 3
Write an equation of the ellipse with foci at (4, 2) and (4, −6) and vertices at (4, 4) and (4, −8).
$\dfrac{(x-4)^2}{20} + \dfrac{(y+2)^2}{36} = 1$

 ## CHECKPOINT EXERCISES

For use after Example 1:

1. Write an equation of the parabola whose vertex is at (1, 5) and whose focus is at (1, 7). $(x-1)^2 = 8(y-5)$

For use after Example 2:

2. Graph $(x+4)^2 + (y+5)^2 = 25$.

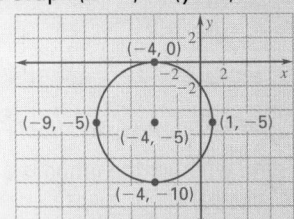

For use after Example 3:

3. Write an equation of the ellipse with foci at (2, 5) and (−8, 5) and vertices at (5, 5) and (−11, 5).
$\dfrac{(x+3)^2}{64} + \dfrac{(y-5)^2}{39} = 1$

STUDENT HELP

HOMEWORK HELP
Visit our Web site
www.mcdougallittell.com
for extra examples.

EXAMPLE 2 *Graphing the Equation of a Translated Circle*

Graph $(x-3)^2 + (y+2)^2 = 16$.

SOLUTION

Compare the given equation to the standard form of the equation of a circle:

$$(x-h)^2 + (y-k)^2 = r^2$$

You can see that the graph is a circle with center at $(h, k) = (3, -2)$ and radius $r = 4$.

Plot the center.

Plot several points that are each **4** units from the center:

$$(3+4, -2) = (7, -2)$$
$$(3-4, -2) = (-1, -2)$$
$$(3, -2+4) = (3, 2)$$
$$(3, -2-4) = (3, -6)$$

Draw a circle through the points.

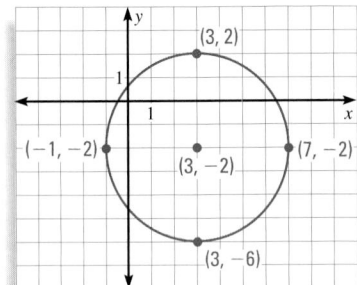

EXAMPLE 3 *Writing an Equation of a Translated Ellipse*

Write an equation of the ellipse with foci at (3, 5) and (3, −1) and vertices at (3, 6) and (3, −2).

SOLUTION

Plot the given points and make a rough sketch. The ellipse has a vertical major axis, so its equation is of this form:

$$\frac{(x-h)^2}{b^2} + \frac{(y-k)^2}{a^2} = 1$$

Find the center: The center is halfway between the vertices.

$$(h, k) = \left(\frac{3+3}{2}, \frac{6+(-2)}{2}\right) = (3, 2)$$

Find a: The value of a is the distance between the vertex and the center.

$$a = \sqrt{(3-3)^2 + (6-2)^2} = \sqrt{0 + 4^2} = 4$$

Find c: The value of c is the distance between the focus and the center.

$$c = \sqrt{(3-3)^2 + (5-2)^2} = \sqrt{0 + 3^2} = 3$$

Find b: Substitute the values of a and c into the equation $b^2 = a^2 - c^2$.

$$b^2 = 4^2 - 3^2$$
$$b^2 = 7$$
$$b = \sqrt{7}$$

▶ The standard form of the equation is $\dfrac{(x-3)^2}{7} + \dfrac{(y-2)^2}{16} = 1$.

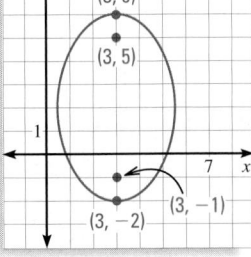

EXAMPLE 4 *Graphing the Equation of a Translated Hyperbola*

Graph $(y + 1)^2 - \dfrac{(x + 1)^2}{4} = 1$.

SOLUTION

The y^2-term is positive, so the transverse axis is vertical. Since $a^2 = 1$ and $b^2 = 4$, you know that $a = 1$ and $b = 2$.

Plot the center at $(h, k) = (-1, -1)$. Plot the vertices 1 unit above and below the center at $(-1, 0)$ and $(-1, -2)$.

Draw a rectangle that is centered at $(-1, -1)$ and is $2a = 2$ units high and $2b = 4$ units wide.

Draw the asymptotes through the corners of the rectangle.

Draw the hyperbola so that it passes through the vertices and approaches the asymptotes.

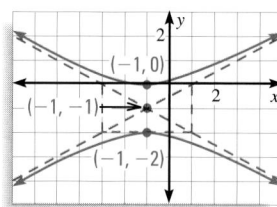

EXAMPLE 5 *Using Circular Models*

COMMUNICATIONS A cellular phone transmission tower located 10 miles west and 5 miles north of your house has a range of 20 miles. A second tower, 5 miles east and 10 miles south of your house, has a range of 15 miles.

  a. Write an inequality that describes each tower's range.

  b. Do the two regions covered by the towers overlap?

SOLUTION

  a. Let the origin represent your house. The first tower is at $(-10, 5)$ and the boundary of its range is a circle with radius 20. Substitute -10 for h, 5 for k, and 20 for r into the standard form of the equation of a circle.

$$(x - h)^2 + (y - k)^2 = r^2 \qquad \text{Standard form of a circle}$$

$$(x + 10)^2 + (y - 5)^2 < 400 \qquad \text{Region inside the circle}$$

The second tower is at $(5, -10)$. The boundary of its range is a circle with radius 15.

$$(x - h)^2 + (y - k)^2 = r^2 \qquad \text{Standard form of a circle}$$

$$(x - 5)^2 + (y + 10)^2 < 225 \qquad \text{Region inside the circle}$$

  b. One way to tell if the regions overlap is to graph the inequalities. You can see that the regions do overlap.

You can also check whether the distance between the two towers is less than the sum of the ranges.

$$\sqrt{(-10 - 5)^2 + (5 - (-10))^2} \overset{?}{<} 20 + 15$$

$$15\sqrt{2} \overset{?}{<} 35$$

$$21.2 < 35 \checkmark$$

▶ The regions do overlap.

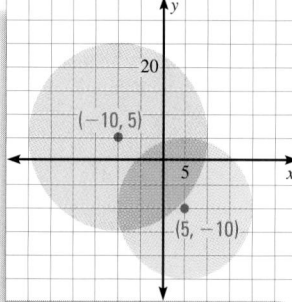

FOCUS ON APPLICATIONS

CELLULAR PHONES work only when there is a transmission tower nearby to retrieve the signal. Because of the need for many towers, they are often designed to blend in with the environment.

10.6 *Graphing and Classifying Conics* **625**

STUDENT HELP NOTES

↳ **Homework Help** Students can find extra examples at **www.mcdougallittell.com** that parallel the examples in the student edition.

EXTRA EXAMPLE 4

Graph $\dfrac{(y - 2)^2}{4} - \dfrac{(x + 3)^2}{9} = 1$.

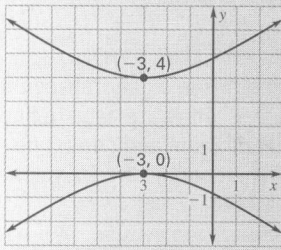

EXTRA EXAMPLE 5

A tower for a radio station is located 30 mi west and 15 mi south of your house and has a range of 60 mi. A second tower, 25 mi east and 40 mi north of your house, has a range of 50 mi.

  a. Write an inequality that describes each tower's range. Tower 1: $(x + 30)^2 + (y + 15)^2 \leq 3600$; Tower 2: $(x - 25)^2 + (y - 40)^2 \leq 2500$

  b. Do the two regions covered by the towers overlap? **Yes**

✓ CHECKPOINT EXERCISES

For use after Example 4:

1. Graph $(x - 5)^2 - \dfrac{(y + 2)^2}{16} = 1$.

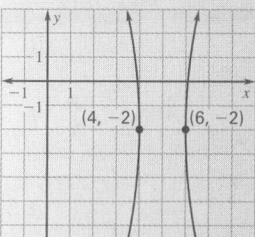

For use after Example 5:

2. Use the equations for the towers in Extra Example 5. If you are 20 mi east and 20 mi north of your house, which radio stations, if any, can you hear? **Tower 2**

TEACHING TIPS

The discriminant of a general second-degree equation is similar to the discriminant used in the quadratic formula. When $y = ax^2 + bx + c$, the discriminant in the formula

$$x = \frac{-b \pm \sqrt{b^2 - 4ac}}{2a}$$ is $b^2 - 4ac$.

You can relate the idea of both discriminants to help students understand the concept of classifying conics.

CONCEPT QUESTION

Which terms in a general second-degree equation are used to find the discriminant and determine what type of conic the equation represents? **the x^2-term, the y^2-term, and the xy-term**

STUDENT HELP NOTES

Look Back As students look back to page 282 for help on completing the square, remind them to add the correct value to both sides of the equation. Many students do not do this when they are completing the square because they forget to multiply by the factor outside the parentheses.

EXTRA EXAMPLE 6
a. Classify the conic given by $2x^2 + y^2 + 8y + 6 = 0$. **ellipse**
b. Graph the equation in part (**a**).

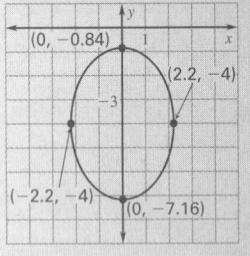

Checkpoint Exercises for Extra Example 6 on next page.

626

GOAL 2 **CLASSIFYING A CONIC FROM ITS EQUATION**

The equation of any conic can be written in the form

$$Ax^2 + Bxy + Cy^2 + Dx + Ey + F = 0$$

which is called a **general second-degree equation** in x and y. The expression $B^2 - 4AC$ is called the **discriminant** of the equation and can be used to determine which type of conic the equation represents.

CONCEPT SUMMARY **CLASSIFYING CONICS**

If the graph of $Ax^2 + Bxy + Cy^2 + Dx + Ey + F = 0$ is a conic, then the type of conic can be determined as follows.

DISCRIMINANT	TYPE OF CONIC
$B^2 - 4AC < 0$, $B = 0$, and $A = C$	Circle
$B^2 - 4AC < 0$ and either $B \neq 0$ or $A \neq C$	Ellipse
$B^2 - 4AC = 0$	Parabola
$B^2 - 4AC > 0$	Hyperbola

If $B = 0$, each axis of the conic is horizontal or vertical. If $B \neq 0$, the axes are neither horizontal nor vertical.

EXAMPLE 6 *Classifying a Conic*

a. Classify the conic given by $2x^2 + y^2 - 4x - 4 = 0$.

b. Graph the equation in part (a).

SOLUTION

a. Since $A = 2$, $B = 0$, and $C = 1$, the value of the discriminant is as follows:

$$B^2 - 4AC = 0^2 - 4(2)(1) = -8$$

▶ Because $B^2 - 4AC < 0$ and $A \neq C$, the graph is an ellipse.

b. To graph the ellipse, first complete the square as follows.

STUDENT HELP

▶ **Look Back**
For help with completing the square, see p. 282.

$$2x^2 + y^2 - 4x - 4 = 0$$
$$(2x^2 - 4x) + y^2 = 4$$
$$2(x^2 - 2x) + y^2 = 4$$
$$2(x^2 - 2x + \underline{\ ?\ }) + y^2 = 4 + 2(\underline{\ ?\ })$$
$$2(x^2 - 2x + 1) + y^2 = 4 + 2(1)$$
$$2(x - 1)^2 + y^2 = 6$$
$$\frac{(x-1)^2}{3} + \frac{y^2}{6} = 1$$

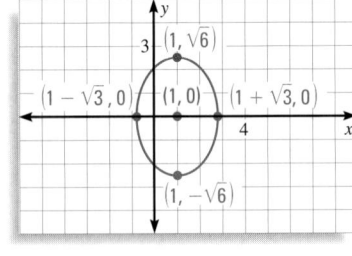

By comparing this equation to $\dfrac{(x - h)^2}{b^2} + \dfrac{(y - k)^2}{a^2} = 1$, you can see that $h = 1$, $k = 0$, $a = \sqrt{6}$, and $b = \sqrt{3}$. Use these facts to draw the ellipse.

EXAMPLE 7 *Classifying a Conic*

a. Classify the conic given by $4x^2 - 9y^2 + 32x - 144y - 548 = 0$.

b. Graph the equation in part (a).

SOLUTION

a. Since $A = 4$, $B = 0$, and $C = -9$, the value of the discriminant is as follows:

$$B^2 - 4AC = 0^2 - 4(4)(-9) = 144$$

▶ Because $B^2 - 4AC > 0$, the graph is a hyperbola.

b. To graph the hyperbola, first complete the square as follows.

$$4x^2 - 9y^2 + 32x - 144y - 548 = 0$$

$$(4x^2 + 32x) - (9y^2 + 144y) = 548$$

$$4(x^2 + 8x + \underline{?}) - 9(y^2 + 16y + \underline{?}) = 548 + 4(\underline{?}) - 9(\underline{?})$$

$$4(x^2 + 8x + 16) - 9(y^2 + 16y + 64) = 548 + 4(16) - 9(64)$$

$$4(x + 4)^2 - 9(y + 8)^2 = 36$$

$$\frac{(x + 4)^2}{3^2} - \frac{(y + 8)^2}{2^2} = 1$$

By comparing this equation to $\dfrac{(x - h)^2}{a^2} - \dfrac{(y - k)^2}{b^2} = 1$, you can see that $h = -4$, $k = -8$, $a = 3$, and $b = 2$.

To draw the hyperbola, plot the center at $(h, k) = (-4, -8)$ and the vertices at $(-7, -8)$ and $(-1, -8)$. Draw a rectangle $2a = 6$ units wide and $2b = 4$ units high and centered at $(-4, -8)$. Draw the asymptotes through the corners of the rectangle. Then draw the hyperbola so that it passes through the vertices and approaches the asymptotes.

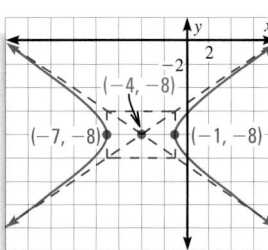

EXAMPLE 8 *Classifying Conics in Real Life*

Astronomy

The diagram at the right shows the mirrors in a Cassegrain telescope. The equations of the two mirrors are given below. Classify each mirror as parabolic, elliptical, or hyperbolic.

a. Mirror A: $y^2 - 72x - 450 = 0$

b. Mirror B: $88.4x^2 - 49.7y^2 - 4390 = 0$

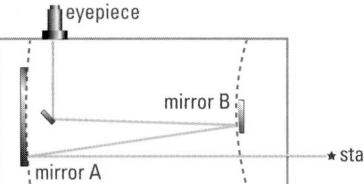

eyepiece
mirror B
★ star
mirror A

SOLUTION

EQUATION	$B^2 - 4AC$	TYPE OF MIRROR
a. $y^2 - 72x - 450 = 0$	$0^2 - 4(0)(1) = 0$	Parabolic
b. $88.4x^2 - 49.7y^2 - 4390 = 0$	$0^2 - 4(88.4)(-49.7) > 0$	Hyperbolic

10.6 *Graphing and Classifying Conics* **627**

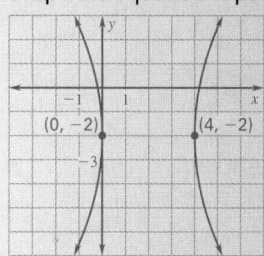

EXTRA EXAMPLE 7
a. Classify the conic given by $4x^2 - y^2 - 16x - 4y - 4 = 0$.
hyperbola
b. Graph the equation in part (**a**).

(0, −2) (4, −2)

EXTRA EXAMPLE 8
Classify each equation as parabolic, elliptical, or hyperbolic.
a. $604x^2 + 216y^2 - 826 = 0$
elliptical
b. $-12x^2 - 85y + 285 = 0$
parabolic

CHECKPOINT EXERCISES
For use after Examples 6–8:
1. a. Classify the conic given by $2x^2 + 2y^2 - 12x + 4y + 2 = 0$.
circle
b. Graph the equation in part (**a**).

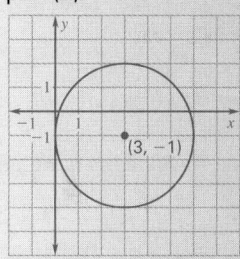

(3, −1)

FOCUS ON VOCABULARY
You may wish to have students create flash cards for the material in the property boxes on pages 623 and 626.

CLOSURE QUESTION
Using the equation for the circle $(x - 2)^2 + (y + 8)^2 = 18$, how many units from the origin and in what directions has the center of the circle been shifted? **2 units right, 8 units down**

DAILY PUZZLER
Write 1,000,000 as the product of two numbers that contain no zeros.
$5^6 \cdot 2^6 = 15,625 \cdot 64$

627

ASSIGNMENT GUIDE

BASIC
Day 1: pp. 628–631 Exs. 13–25, 63, 64, 68, 71–81 odd
Day 2: pp. 629–631 Exs. 30–40 even, 45–50, 52–60 even, 69, 83–87 odd

AVERAGE
Day 1: pp. 628–631 Exs. 13–28, 63–66, 68, 71–81 odd
Day 2: pp. 629–631 Exs. 30–40 even, 45–50, 52–62 even, 69, 83–87 odd

ADVANCED
Day 1: pp. 628–631 Exs. 13–28, 63–66, 71–87 odd
Day 2: pp. 629–631 Exs. 30–44 even, 45–50, 52–62 even, 67–70

BLOCK SCHEDULE
pp. 628–631 Exs. 13–28, 30–44 even, 45–50, 52–66 even, 68, 69, 73–87 odd

EXERCISE LEVELS

Level A: *Easier*
13–20, 29–50
Level B: *More Difficult*
21–28, 51–69
Level C: *Most Difficult*
70

✓ **HOMEWORK CHECK**
To quickly check student understanding of key concepts, go over the following exercises: Exs. 15, 18, 22, 26, 30, 46, 52. See also the Daily Homework Quiz:
• Blackline Master (*Chapter 10 Resource Book*, p. 92)
• Transparency (p. 80)

12, 21–28. See Additional Answers beginning on page AA1.

GUIDED PRACTICE

Vocabulary Check ✓
Concept Check ✓

1. **All are formed by intersecting a plane and a double-napped cone.**

2. **They are each circles of radius 5; the first is centered at (0, 0) and the second at (1, −2).**

3. **If the discriminant is greater than 0, it is a hyperbola. If it is 0, it is a parabola. If it is less than 0, it is a circle or an ellipse: a circle if $A = C$, an ellipse if $A \neq C$.**

1. Explain why circles, ellipses, parabolas, and hyperbolas are called conic sections.

2. How are the graphs of $x^2 + y^2 = 25$ and $(x - 1)^2 + (y + 2)^2 = 25$ alike? How are they different?   **See margin.**

3. How can the discriminant $B^2 - 4AC$ be used to classify the graph of $Ax^2 + Bxy + Cy^2 + Dx + Ey + F = 0$?   **See margin.**

Skill Check ✓

Write an equation for the conic section.

4. Circle with center at $(4, -1)$ and radius 7 $(x - 4)^2 + (y + 1)^2 = 49$

5. Ellipse with foci at $(2, -4)$ and $(5, -4)$ and vertices at $(-1, -4)$ and $(8, -4)$
   $\dfrac{(x - 3.5)^2}{20.25} + \dfrac{(y + 4)^2}{18} = 1$

6. Parabola with vertex at $(3, -2)$ and focus at $(3, -4)$ $(x - 3)^2 = -8(y + 2)$

7. Hyperbola with foci at $(5, 2)$ and $(5, -6)$ and vertices at $(5, 0)$ and $(5, -4)$
   $\dfrac{(y + 2)^2}{4} - \dfrac{(x - 5)^2}{12} = 1$

Classify the conic section.

8. $x^2 + 2x - 4y + 4 = 0$ **parabola**

9. $3x^2 - 5y^2 - 6x + y - 2 = 0$ **hyperbola**

10. $x^2 + y^2 + 7x - 4y - 8 = 0$ **circle**

11. $-5x^2 - 2y^2 + x - 3y + 1 = 0$ **ellipse**

12. 🌐 **COMMUNICATIONS** Look back at Example 5. Suppose there is a tower 25 miles east and 30 miles north of your house with a range of 25 miles. Does the region covered by this tower overlap the regions covered by the two towers in Example 5? Illustrate your answer with a graph.   **It overlaps one of the regions. See margin for graph.**

PRACTICE AND APPLICATIONS

STUDENT HELP

▶ **Extra Practice** to help you master skills is on p. 954.

17. $\dfrac{(x - 2)^2}{18} + \dfrac{(y - 1.5)^2}{20.25} = 1$

18. $\dfrac{(x - 1)^2}{9} + (y - 2)^2 = 1$

19. $\dfrac{y^2}{16} - \dfrac{(x - 5)^2}{20} = 1$

20. $\dfrac{(x + 1.5)^2}{6.25} - \dfrac{(y - 2)^2}{24} = 1$

STUDENT HELP

▶ **HOMEWORK HELP**
Examples 1, 3: Exs. 13–20
Examples 2, 4: Exs. 21–28
Example 5: Exs. 63, 64
Examples 6, 7: Exs. 29–62
Example 8: Exs. 65–67

WRITING EQUATIONS Write an equation for the conic section.

13. Circle with center at $(9, 3)$ and radius 4 $(x - 9)^2 + (y - 3)^2 = 16$

14. Circle with center at $(-4, 2)$ and radius 3 $(x + 4)^2 + (y - 2)^2 = 9$

15. Parabola with vertex at $(1, -2)$ and focus at $(1, 1)$ $(x - 1)^2 = 12(y + 2)$

16. Parabola with vertex at $(-3, 1)$ and directrix $x = -8$ $(y - 1)^2 = 20(x + 3)$

17. Ellipse with vertices at $(2, -3)$ and $(2, 6)$ and foci at $(2, 0)$ and $(2, 3)$

18. Ellipse with vertices at $(-2, 2)$ and $(4, 2)$ and co-vertices at $(1, 1)$ and $(1, 3)$

19. Hyperbola with vertices at $(5, -4)$ and $(5, 4)$ and foci at $(5, -6)$ and $(5, 6)$

20. Hyperbola with vertices at $(-4, 2)$ and $(1, 2)$ and foci at $(-7, 2)$ and $(4, 2)$

GRAPHING Graph the equation. Identify the important characteristics of the graph, such as the center, vertices, and foci.   **21–28. See margin.**

21. $(x - 6)^2 + (y - 2)^2 = 4$

22. $(x + 7)^2 = 12(y - 3)$

23. $\dfrac{(y - 8)^2}{16} - \dfrac{(x + 3)^2}{4} = 1$

24. $\dfrac{(x - 3)^2}{25} + \dfrac{(y + 6)^2}{49} = 1$

25. $\dfrac{(x + 1)^2}{16} + \dfrac{y^2}{9} = 1$

26. $\dfrac{x^2}{16} - (y + 4)^2 = 1$

27. $(x + 7)^2 + (y - 1)^2 = 1$

28. $(y - 4)^2 = 3(x + 2)$

51. $(y - 6)^2 = -4(x - 8)$

52. $(x - 3)^2 + (y - 4)^2 = 1$

53. $(x - 4)^2 - \dfrac{(y - 4)^2}{9} = 1$

54. $\dfrac{(x - 6)^2}{25} + \dfrac{(y - 2)^2}{100} = 1$

55. $\dfrac{(x - 1)^2}{4} + (y - 1)^2 = 1$

56. $(x - 6)^2 + (y - 12)^2 = 144$

57. $\dfrac{x^2}{4} - \dfrac{(y - 8)^2}{64} = 1$

58. $\dfrac{(x + 4)^2}{\left(\frac{85}{9}\right)} + \dfrac{\left(y + \frac{2}{9}\right)^2}{\left(\frac{85}{81}\right)} = 1$

59. $(x - 6)^2 + (y - 6)^2 = 36$

60. $(y - 10)^2 = 2(x + 3)$

61. $(x + 2)^2 = 8(y - 1)$

62. $\dfrac{(y - 2)^2}{36} - \dfrac{(x + 2)^2}{16} = 1$

CLASSIFYING Classify the conic section.

29. $9x^2 + 4y^2 + 36x - 24y + 36 = 0$ ellipse

30. $x^2 - 4y^2 + 3x - 26y - 30 = 0$ hyperbola

31. $4x^2 - 9y^2 + 18y + 3x = 0$ hyperbola

32. $x^2 + y^2 - 10x - 2y + 10 = 0$ circle

33. $36x^2 + 16y^2 - 25x + 22y + 2 = 0$ ellipse

34. $4x^2 + 4y^2 - 16x + 4y - 60 = 0$ circle

35. $9y^2 - x^2 + 2x + 54y + 62 = 0$ hyperbola

36. $16x^2 + 25y^2 - 18x - 20y + 8 = 0$ ellipse

37. $x^2 - 2x + 8y + 9 = 0$ parabola

38. $2y^2 - 8y - 4x + 10 = 0$ parabola

39. $12x^2 + 20y^2 - 12x + 40y - 37 = 0$ ellipse

40. $9x^2 - y^2 + 54x + 10y + 55 = 0$ hyperbola

41. $x^2 + y^2 - 4x - 2y - 4 = 0$ circle

42. $16x^2 + 9y^2 + 24x - 36y + 23 = 0$ ellipse

43. $16y^2 - x^2 + 2x + 64y + 63 = 0$ hyperbola

44. $x^2 - 4x + 16y + 17 = 0$ parabola

MATCHING Match the equation with its graph.

45. $9x^2 - 4y^2 + 36x - 24y - 36 = 0$ **E**

46. $y^2 - 2y - 4x + 9 = 0$ **A**

47. $9x^2 + 4y^2 + 36x + 24y + 36 = 0$ **D**

48. $y^2 - x^2 + 6y + 4x + 4 = 0$ **F**

49. $4x^2 + 9y^2 - 16x + 54y + 61 = 0$ **B**

50. $x^2 + y^2 - 4x + 6y + 4 = 0$ **C**

A.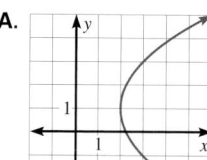

B.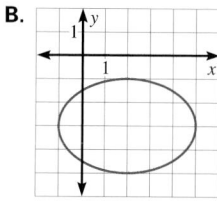

C.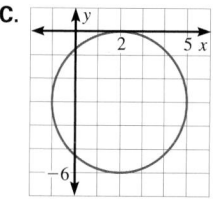

D.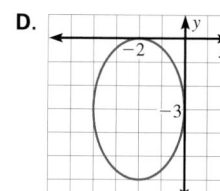

E.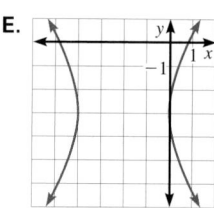

F.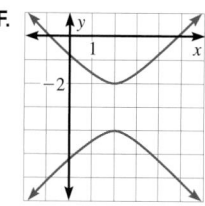

CLASSIFYING AND GRAPHING Classify the conic section and write its equation in standard form. Then graph the equation. 51–62. See margins for equations and graphs.

51. $y^2 - 12y + 4x + 4 = 0$ parabola

52. $x^2 + y^2 - 6x - 8y + 24 = 0$ circle

53. $9x^2 - y^2 - 72x + 8y + 119 = 0$ hyperbola

54. $4x^2 + y^2 - 48x - 4y + 48 = 0$ ellipse

55. $x^2 + 4y^2 - 2x - 8y + 1 = 0$ ellipse

56. $x^2 + y^2 - 12x - 24y + 36 = 0$ circle

57. $16x^2 - y^2 + 16y - 128 = 0$ hyperbola

58. $x^2 + 9y^2 + 8x + 4y + 7 = 0$ ellipse

59. $x^2 + y^2 - 12x - 12y + 36 = 0$ circle

60. $y^2 - 2x - 20y + 94 = 0$ parabola

61. $x^2 + 4x - 8y + 12 = 0$ parabola

62. $-9x^2 + 4y^2 - 36x - 16y - 164 = 0$ hyperbola

63.  **WHISPER DISHES** The whisper dish shown at the left can be seen at the Thronateeska Discovery Center in Albany, Georgia. Two dishes are positioned so that their vertices are 50 feet apart. The focus of each dish is 3 feet from its vertex. Write equations for the cross sections of the dishes so that the vertex of one dish is at the origin and the vertex of the other dish is on the positive x-axis. $y^2 = 12x$; $y^2 = -12(x - 50)$ (x, y in ft)

FOCUS ON APPLICATIONS

Whisper Dish

WHISPER DISHES are two parabolic dishes set up facing directly toward each other. A person listening at the focus of one dish is able to hear even the softest sound made at the focus of the other dish.

! COMMON ERROR

EXERCISES 21–28 Students may confuse the direction of translation when graphing conic equations. Remind them that a shift in the positive direction involves subtracting a positive *h* and *k*. A shift in the negative direction involves subtracting a negative *h* and *k*, which in essence means adding a positive *h* and *k*.

51. $(y - 6)^2 = -4(x - 8)$;

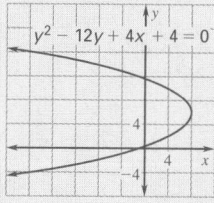

52. $(x - 3)^2 + (y - 4)^2 = 1$;

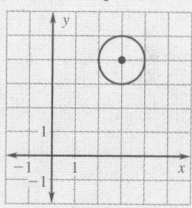

53. $\dfrac{(x - 4)^2}{1} - \dfrac{(y - 4)^2}{9} = 1$;

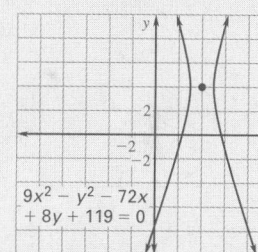

54. $\dfrac{(x - 6)^2}{25} + \dfrac{(y - 2)^2}{100} = 1$;

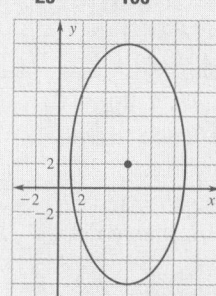

55–57. See page 630.

58–62. See Additional Answers beginning on page AA1.

55. $\dfrac{(x-1)^2}{4} + \dfrac{(y-1)^2}{1} = 1;$

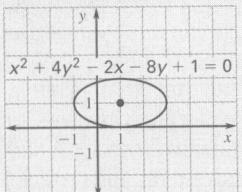

$x^2 + 4y^2 - 2x - 8y + 1 = 0$

56. $(x-6)^2 + (y-12)^2 = 144;$

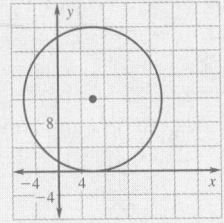

57. $\dfrac{x^2}{4} - \dfrac{(y-8)^2}{64} = 1;$

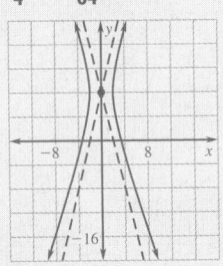

**ADDITIONAL PRACTICE
AND RETEACHING**

For Lesson 10.6:

• Practice Levels A, B, and C
(*Chapter 10 Resource Book,*
p. 81)

• Reteaching with Practice
(*Chapter 10 Resource Book,*
p. 84)

• See Lesson 10.6 of the
Personal Student Tutor

For more Mixed Review:

• Search the *Test and Practice
Generator* for key words or
specific lessons.

630

64. **FIGURE SKATING** To practice making a figure eight, a figure skater will skate along two circles etched in the ice. Write equations for two externally tangent circles that are each 6 feet in diameter so that the center of one circle is at the origin and the center of the other circle is on the positive y-axis. $x^2 + y^2 = 9;\ x^2 + (y-6)^2 = 9\ (x,\ y\ \text{in ft})$

65. **VISUAL THINKING** A new crayon has a cone-shaped tip. When it is used for the first time, a flat spot is worn on the tip. The edge of the flat spot is a conic section, as shown. What type(s) of conic could it be? **ellipse**

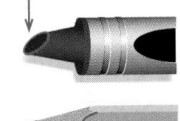

conic section

66. **VISUAL THINKING** When a pencil is sharpened the tip becomes a cone. On a pencil with flat sides, the intersection of the cone with each flat side is a conic section. What type of conic is it? **hyperbola branch**

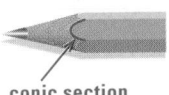

conic section

67. **ASTRONOMY** A Gregorian telescope contains two mirrors whose cross sections can be modeled by the equations $405x^2 + 729y^2 - 295{,}245 = 0$ and $-120y^2 - 1440x = 0$. What types of mirrors are they? **The first is elliptical, the second parabolic.**

68. **MULTIPLE CHOICE** Which of the following is an equation of the hyperbola with vertices at $(3, 5)$ and $(3, -1)$ and foci at $(3, 7)$ and $(3, -3)$? **E**

(A) $\dfrac{(x-3)^2}{25} - \dfrac{(y-2)^2}{9} = 1$　　(B) $\dfrac{(y-2)^2}{9} - \dfrac{(x-3)^2}{25} = 1$

(C) $\dfrac{(y-2)^2}{9} - \dfrac{(x-3)^2}{7} = 1$　　(D) $\dfrac{(x-3)^2}{9} - \dfrac{(y-2)^2}{16} = 1$

(E) $\dfrac{(y-2)^2}{9} - \dfrac{(x-3)^2}{16} = 1$

69. **MULTIPLE CHOICE** What conic does $25x^2 + y^2 - 100x - 2y + 76 = 0$ represent? **C**

(A) Parabola　　　(B) Circle　　　(C) Ellipse

(D) Hyperbola　　　(E) Not enough information

70. **DEGENERATE CONICS** A *degenerate* conic occurs when the intersection of a plane with a double-napped cone is something other than a parabola, circle, ellipse, or hyperbola.

a. Imagine a plane perpendicular to the axis of a double-napped cone. As the plane passes through the cone, the intersection is a circle whose radius decreases and then increases. At what point is the intersection something other than a circle? What is the intersection?
Where the cones meet, the intersection is a point.

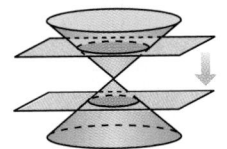

b. Imagine a plane parallel to the axis of a double-napped cone. As the plane passes through the cone, the intersection is a hyperbola whose vertices get closer together and then farther apart. At what point is the intersection something other than a hyperbola? What is the intersection? **When it passes through the point where the cones meet, the intersection is an X (2 intersecting lines).**

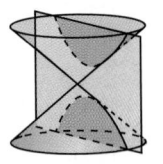

c. Imagine a plane parallel to the nappe passing through a double-napped cone. As the plane passes through the cone, the intersection is a parabola that gets narrower and then flips and gets wider. At what point is the intersection something other than a parabola? What is the intersection? **When it passes through the point where the cones meet, the intersection is a line.**

nappe

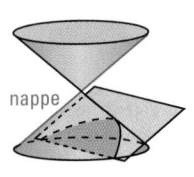

Test
Preparation

★ Challenge

EXTRA CHALLENGE
www.mcdougallittell.com

MIXED REVIEW

SYSTEMS Solve the system using any algebraic method. (Review 3.2 for 10.7)

71. $x - y = 10$ (5, −5)
$3x - 2y = 25$

72. $4x + 3y = 1$ (1, −1)
$-3x - 6y = 3$

73. $4x + y = 2$ (1, −2)
$6x + 3y = 0$

74. $2x - 3y = 0$ $\left(\frac{14}{5}, \frac{28}{15}\right)$
$x + 6y = 14$

75. $23x = 68$ $\left(\frac{68}{23}, \frac{123}{23}\right)$
$x + 3y = 19$

76. $x = y$ $\left(\frac{17}{105}, \frac{17}{105}\right)$
$123x - 18y = 17$

EVALUATING LOGARITHMIC EXPRESSIONS Evaluate the expression. (Review 8.4)

77. $\log_7 7^5$ 5

78. $\log_4 64$ 3

79. $\log_5 1$ 0

80. $\log_{1/3} 9$ −2

81. $\log_{25} 625$ 2

82. $\log 0.0001$ −4

SOLVING EQUATIONS Solve the equation. (Review 8.8)

83. $\frac{40}{1 + 6e^{-4x}} = 20$ about 0.45

84. $\frac{10}{1 + 9e^{-2x}} = 1$ 0

85. $\frac{8}{1 + 8e^{-x}} = 7$ about 4.03

86. $\frac{15}{1 + 3e^{-6x}} = 3$ about −0.048

87. $\frac{24}{1 + 5e^{-4x}} = 9$ about 0.27

88. $\frac{9}{1 + 2e^{-3x}} = 7$ about 0.65

MATH & History

History of Conic Sections

APPLICATION LINK
www.mcdougallittell.com

THEN

IN 200 B.C. conic sections were studied thoroughly for the first time by a Greek mathematician named Apollonius. Six hundred years later, the Egyptian mathematician Hypatia simplified the works of Apollonius, making it accessible to many more people. For centuries, conics were studied and appreciated only for their mathematical beauty rather than for their occurrence in nature or practical use.

NOW

TODAY astronomers know that the paths of celestial objects, such as planets and comets, are conic sections. For example, a comet's path can be parabolic, hyperbolic, or elliptical.

Tell what type of path each comet follows. Which comet(s) will pass by the sun more than once?

1. $3550x^2 + 14{,}200x + 7100y - 13{,}050 = 0$ parabolic

2. $2200x^2 + 4600y^2 - 13{,}200x - 18{,}400y + 12{,}900 = 0$ elliptical; will pass by the sun more than once.

3. $5000x^2 - 6500y^2 + 20{,}000x - 52{,}000y - 695{,}000 = 0$ hyperbolic

Apollonius studies conic sections.

200 B.C.

Hypatia simplifies Apollonius' *Conics.*

A.D. 400

Johannes Kepler discovers that the planets' orbits are elliptical.

1609

Debra Fischer discovers two planets.

1999

10.6 *Graphing and Classifying Conics*   **631**

DAILY HOMEWORK QUIZ

Transparency Available

Classify the conic section, write its equation in standard form, and graph it.

1. $4x^2 - 16x + 9y^2 - 54y + 61 = 0$

ellipse; $\dfrac{(x-2)^2}{9} + \dfrac{(y-3)^2}{4} = 1$

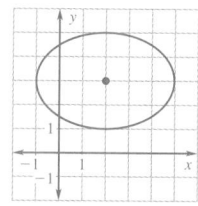

2. $x^2 - 2x - 8y - 23 = 0$

parabola; $(x-1)^2 = 8(y+3)$

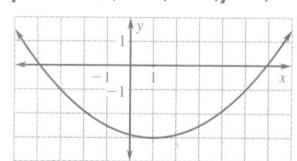

EXTRA CHALLENGE NOTE

→ Challenge problems for Lesson 10.6 are available in **blackline** format in the *Chapter 10 Resource Book,* p. 89 and at **www.mcdougallittell.com**.

ADDITIONAL TEST PREPARATION

1. WRITING In the general second-degree equation $Ax^2 + Bxy + Cy^2 + Dx + Ey + F = 0$, what is the discriminant and what does it tell you about the graph of the equation?
$B^2 - 4AC$; The discriminant tells you what type of conic equation you have.

ADDITIONAL RESOURCES

A **blackline** master with additional Math & History exercises is available in the *Chapter 10 Resource Book,* p. 88.

LESSON OPENER
VISUAL APPROACH
An alternative way to approach Lesson 10.7 is to use the Visual Approach Lesson Opener:
• Blackline Master (*Chapter 10 Resource Book,* p. 93)
• Transparency (p. 71)

MEETING INDIVIDUAL NEEDS
• *Chapter 3 Resource Book*
  Prerequisite Skills Review (p. 5)
  Practice Level A (p. 96)
  Practice Level B (p. 97)
  Practice Level C (p. 98)
  Reteaching with Practice (p. 99)
  Absent Student Catch-Up (p. 101)
  Challenge (p. 103)
• *Resources in Spanish*
• Personal Student Tutor

NEW-TEACHER SUPPORT
See the Tips for New Teachers on pp. 1–2 of the *Chapter 10 Resource Book* for additional notes about Lesson 10.7.

WARM-UP EXERCISES

Transparency Available

Solve the system of equations.
1. $3x + 2y = 11$
    $x - y = -3$ (1, 4)
2. $y = 4x - 9$
    $y = 6x - 13$ (2, –1)
3. $7x - 12y = -3$
    $5x + 3y = 21$ (3, 2)
4. $y = 12x + 7$
    $10x + 3y = -25$ (–1, –5)

a–d. See Additional Answers beginning on page AA1.

632

10.7 Solving Quadratic Systems

What you should learn

GOAL 1 Solve systems of quadratic equations.

GOAL 2 Use quadratic systems to solve **real-life** problems, such as determining when one car will catch up to another in **Ex. 58**.

Why you should learn it

▼ To model **real-life** situations with quadratic systems, such as finding the epicenter of an earthquake in **Example 4**.

GOAL 1 SOLVING A SYSTEM OF EQUATIONS

In Lesson 3.2 you studied two algebraic techniques for solving a system of linear equations. You can use the same techniques (substitution and linear combination) to solve quadratic systems.

EXAMPLE 1 *Finding Points of Intersection*

Find the points of intersection of the graphs of $x^2 + y^2 = 13$ and $y = x + 1$.

SOLUTION

To find the points of intersection, substitute $x + 1$ for y in the equation of the circle.

$x^2 + y^2 = 13$	Equation of circle
$x^2 + (x + 1)^2 = 13$	Substitute $x + 1$ for y.
$x^2 + x^2 + 2x + 1 = 13$	Expand the power.
$2x^2 + 2x - 12 = 0$	Combine like terms.
$2(x - 2)(x + 3) = 0$	Factor.
$x = 2$ or $x = -3$	Zero product property

You now know the x-coordinates of the points of intersection. To find the y-coordinates, substitute $x = 2$ and $x = -3$ into the linear equation and solve for y.

▶ The points of intersection are $(2, 3)$ and $(-3, -2)$.

✓ **CHECK** You can check your answer algebraically by substituting the coordinates of the points into each equation. Another way to check your answer is to graph the two equations. You can see from the graph shown that the line and the circle intersect in two points, at $(2, 3)$ and at $(-3, -2)$.

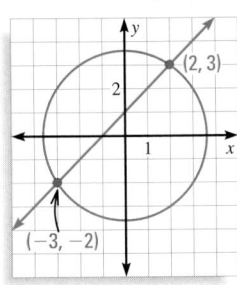

STUDENT HELP

Look Back
For help with solving systems, see p. 148.

▶ **ACTIVITY**
Developing Concepts

Investigating Points of Intersection

The circle and line in Example 1 intersect in two points. A circle and a line can also intersect in one point or no points. Sketch examples to illustrate the different numbers of points of intersection that the following graphs can have. a–d. See margin.
  a. Circle and parabola   **b.** Ellipse and hyperbola
  c. Circle and ellipse   **d.** Hyperbola and line

EXAMPLE 2 *Solving a System by Substitution*

Find the points of intersection of the graphs in the system.

$$x^2 + 4y^2 - 4 = 0 \qquad \text{Equation 1}$$
$$-2y^2 + x + 2 = 0 \qquad \text{Equation 2}$$

SOLUTION

Because Equation 2 has no x^2-term, solve that equation for x.

$$-2y^2 + x + 2 = 0$$
$$x = 2y^2 - 2$$

Next, substitute $2y^2 - 2$ for x in Equation 1 and solve for y.

$x^2 + 4y^2 - 4 = 0$	Equation 1
$(2y^2 - 2)^2 + 4y^2 - 4 = 0$	Substitute for x.
$4y^4 - 8y^2 + 4 + 4y^2 - 4 = 0$	Expand the power.
$4y^4 - 4y^2 = 0$	Combine like terms.
$4y^2(y^2 - 1) = 0$	Factor common monomial.
$4y^2(y - 1)(y + 1) = 0$	Difference of squares.
$y = 0, y = 1, \text{ or } y = -1$	Zero product property

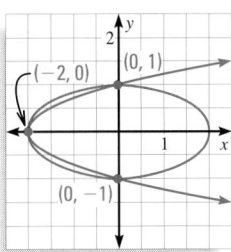

STUDENT HELP

▶ **Look Back**
For help with factoring, see p. 256.

The corresponding x-values are $x = -2$, $x = 0$, and $x = 0$.

▶ The graphs intersect at $(-2, 0)$, $(0, 1)$, and $(0, -1)$, as shown.

EXAMPLE 3 *Solving a System by Linear Combination*

Find the points of intersection of the graphs in the system.

$$x^2 + y^2 - 16x + 39 = 0 \qquad \text{Equation 1}$$
$$x^2 - y^2 - 9 = 0 \qquad \text{Equation 2}$$

SOLUTION

You can eliminate the y^2-term by adding the two equations. The resulting equation can be solved for x because it contains no other variables.

$$x^2 + y^2 - 16x + 39 = 0$$
$$\underline{x^2 - y^2 \qquad - 9 = 0}$$

$2x^2 \qquad - 16x + 30 = 0$	Add.
$2(x - 3)(x - 5) = 0$	Factor.
$x = 3 \text{ or } x = 5$	Zero product property

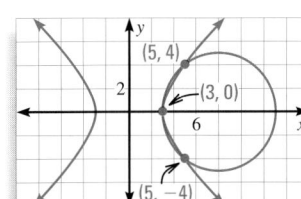

The corresponding y-values are $y = 0$ and $y = \pm 4$.

▶ The graphs intersect at $(3, 0)$, $(5, 4)$, and $(5, -4)$, as shown.

2 TEACH

MOTIVATING THE LESSON
Ask students to recall the methods they used to solve linear systems of equations. Tell them that similar methods can be used to solve quadratic systems.

ACTIVITY NOTE
Encourage students to draw several graphs for each part of the activity. It may help for them to think of and draw graphs with 0, 1, 2, 3, and 4 intersection points, and then make their conclusions from their drawings.

EXTRA EXAMPLE 1
Find the points of intersection of the graphs of $x^2 + y^2 = 5$ and $y = -x + 3$. **(1, 2) and (2, 1)**

EXTRA EXAMPLE 2
Find the points of intersection of the graphs in the system.
$x^2 + y^2 - 64 = 0$
$2x^2 + y - 8 = 0$
(0, 8), (−2.78, −7.5), (2.78, −7.5)

EXTRA EXAMPLE 3
Find the points of intersection of the graphs in the system.
$-x^2 + y^2 - 36 = 0$
$x^2 + y^2 - 32y + 156 = 0$
(0, 6), (8, 10), (−8, 10)

✔ **CHECKPOINT EXERCISES**
For use after Example 1:
1. Find the points of intersection of the graphs of $x^2 + y^2 = 17$ and $y = x + 3$. **(−4, −1) and (1, 4)**

For use after Examples 2 and 3:
2. Find the points of intersection of the graphs in the system.
$x^2 + y^2 - 25 = 0$
$x^2 + y - 13 = 0$
(3, 4), (−3, 4), (−4, −3), (4, −3)

EXTRA EXAMPLE 4
Use the following information given by three seismographs to find the epicenter of the earthquake.
Location 1: 1000 miles from the epicenter
Location 2: 200 miles east and 600 miles north of Location 1 and 600 miles from the epicenter
Location 3: 800 miles east and 400 miles north of Location 1 and 200 miles from the epicenter
The epicenter is 800 miles east and 600 miles north of Location 1.

CHECKPOINT EXERCISES
For use after Example 4:
1. Use the following information given by three seismographs to find the epicenter of the earthquake.
Location 1: 250 miles from the epicenter
Location 2: 50 miles east and 200 miles north of Location 1 and 200 miles from the epicenter
Location 3: 150 miles west and 50 miles south of Location 1 and 250 miles from the epicenter
The epicenter is 150 miles west and 200 miles north of Location 1.

CLOSURE QUESTION
How many points of intersection are possible when solving a quadratic system? 0, 1, 2, 3, or 4

DAILY PUZZLER
Insert only "+" and "−" between each number on the left side of the equation to make the equation true.
1 2 3 4 5 6 7 8 9 10 = 11
Sample answer:
$1 - 2 + 3 + 4 - 5 + 6 - 7 - 8 + 9 + 10$

FOCUS ON CAREERS

SEISMOLOGIST
A seismologist determines the location and intensity of an earthquake using an instrument which measures energy waves resulting from movements in the Earth's crust.

CAREER LINK
www.mcdougallittell.com

GOAL 2 SOLVING QUADRATIC SYSTEMS IN REAL LIFE

EXAMPLE 4 *Solving a System of Quadratic Models*

SEISMOLOGY A seismograph measures the intensity of an earthquake. Although a seismograph can determine the distance to the earthquake's epicenter, it cannot determine in what direction the epicenter is located. Use the following information from three seismographs to find an earthquake's epicenter.

Location 1: 500 miles from the epicenter

Location 2: 100 miles west and 400 miles south of Location 1
400 miles from the epicenter

Location 3: 300 miles east and 600 miles south of Location 1
200 miles from the epicenter

SOLUTION

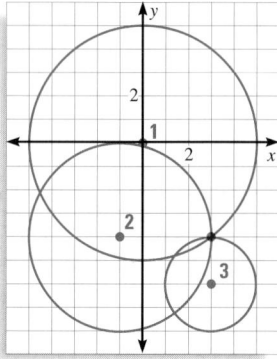

Let each unit represent 100 miles. If Location 1 is at $(0, 0)$, then Location 2 is at $(-1, -4)$ and Location 3 is at $(3, -6)$. Write the equation of each circle.

Location 1: $x^2 + y^2 = 25$

Location 2: $(x + 1)^2 + (y + 4)^2 = 16$, or
$x^2 + 2x + 1 + y^2 + 8y + 16 = 16$

Location 3: $(x - 3)^2 + (y + 6)^2 = 4$, or
$x^2 - 6x + 9 + y^2 + 12y + 36 = 4$

Subtract the equation for Location 1 from the equation for Location 2.

$$x^2 + 2x + 1 + y^2 + 8y + 16 = 16$$
$$\underline{- (x^2 + y^2 = 25)}$$
$$2x + 8y + 17 = -9$$
$$2x + 8y = -26, \text{ or } x + 4y = -13$$

Then subtract the equation for Location 1 from the equation for Location 3.

$$x^2 - 6x + 9 + y^2 + 12y + 36 = 4$$
$$\underline{- (x^2 + y^2 = 25)}$$
$$-6x + 12y + 45 = -21$$
$$-6x + 12y = -66, \text{ or } -x + 2y = -11$$

You are left with two linear equations. Solve this linear system to find the epicenter.

$$x + 4y = -13$$
$$\underline{-x + 2y = -11}$$
$$6y = -24$$
$$y = -4$$
$$x = 3$$

▶ The epicenter of the earthquake is 300 miles east and 400 miles south of Location 1.

GUIDED PRACTICE

Vocabulary Check ✓

1. Complete this statement: The equations $x^2 + 3y^2 - 2y = 4$ and $x^2 + y^2 = 5$ are an example of a(n) _?_ system.   quadratic

Concept Check ✓

2. Sketch an example of a circle and a line intersecting in a single point.   See margin.

3. Explain what method you would use to find the points of intersection of the graphs in the following system. Do not solve the system.   *Sample answer:* Linear combination since the y^2 terms can be eliminated.

$$4x^2 + y^2 - 16x = 0 \qquad \text{Equation 1}$$

$$x^2 - y^2 + 7 = 0 \qquad \text{Equation 2}$$

Skill Check ✓ **Find the points of intersection, if any, of the graphs in the system.**

4. $x^2 + y^2 = 17$ (−4, −1), (1, 4)
$y = x + 3$

5. $x^2 + y^2 + 8x - 20y + 7 = 0$ (−1, 0), (−7, 0)
$x^2 + 9y^2 + 8x + 4y + 7 = 0$

6. $x^2 + y^2 - 3x = 8$ (3, ±2√2)
$2x^2 - y^2 = 10$

7. $x^2 - 2x + 2y + 2 = 0$ (−2, −5), (4, −5)
$-x^2 + 2x - y + 3 = 0$

8. ⊕ **SEISMOLOGY** Look back at Example 4. Why are three (not just two) seismographs needed to determine the location of the epicenter?
Because 2 circles can intersect in 2 points.

PRACTICE AND APPLICATIONS

STUDENT HELP

▸ **Extra Practice**
to help you master
skills is on p. 955.

23. $\left(\dfrac{6 - \sqrt{6}}{5}, \dfrac{24 + \sqrt{6}}{10}\right),$

$\left(\dfrac{6 + \sqrt{6}}{5}, \dfrac{24 - \sqrt{6}}{10}\right)$

25. $(2 + \sqrt{6}, \sqrt{6} - 2),$
$(2 - \sqrt{6}, -\sqrt{6} - 2)$

26. $(−2, −1), (−1, 2)$

CHECKING POINTS OF INTERSECTION **Determine whether the given point is a point of intersection of the graphs in the system.**

9. $x^2 + y^2 = 25$
$y = -3$
Point: $(-3, 4)$ no

10. $x^2 + y^2 = 41$
$y = -x - 1$
Point: $(4, -5)$ yes

11. $x^2 + 4x - 4y - 16 = 0$
$-2x + y + 1 = 0$
Point: $(6, 11)$ yes

12. $3x^2 - 5y^2 + 2y = 45$
$y = 2x + 10$
Point: $(-3, 4)$ no

13. $2x^2 - 4y = 22$
$y = -2x + 3$
Point: $(-5, 7)$ no

14. $6x^2 - 5x + 8y^2 + y = 23$
$y = x - 1$
Point: $(2, 1)$ yes

SOLVING SYSTEMS **Find the points of intersection, if any, of the graphs in the system.**

15. $x^2 - y = 5$ (1, −4), (2, −1)
$-3x + y = -7$

16. $x^2 + y^2 = 18$ (3, 3), (−3, −3)
$x - y = 0$

17. $-3x^2 + y^2 = 9$ (3, 6), (−3, −6)
$-2x + y = 0$

18. $9x^2 + 4y^2 = 36$ none
$-x + y = -4$

19. $x^2 + y^2 = 5$
$y = -2x$ (1, −2), (−1, 2)

20. $x + 2y^2 = -6$ (−24, 3), (−8, 1)
$x + 8y = 0$

21. $5x^2 + 3y^2 = 17$
$-x + y = -1$ (−1, −2), $\left(\frac{7}{4}, \frac{3}{4}\right)$

22. $4x^2 - 5y^2 = 16$
$3x + y = 6$ (2, 0), $\left(\frac{98}{41}, -\frac{48}{41}\right)$

23. $2x^2 + 2y^2 = 15$
$x + 2y = 6$ See margin.

24. $x^2 + y^2 = 1$
$x + y = -1$ (0, −1), (−1, 0)

25. $x^2 + y^2 = 20$
$y = x - 4$ See margin.

26. $x^2 + y^2 = 5$
$y = 3x + 5$ See margin.

27. $x^2 = 6y$ (0, 0), (−6, 6)
$y = -x$

28. $x^2 + y^2 = 9$
$x - 3y = 3$ (3, 0), $\left(-\frac{12}{5}, -\frac{9}{5}\right)$

29. $x^2 + y^2 = 7$ none
$y = x - 7$

30. $y^2 - 2x^2 = 6$ (√3, −2√3),
$y = -2x$ (−√3, 2√3)

31. $6x^2 + 3y^2 = 12$
$y = -x + 2$ (0, 2), $\left(\frac{4}{3}, \frac{2}{3}\right)$

32. $3x^2 - y^2 = -6$
$y = 2x + 1$ (1, 3), (−5, −9)

STUDENT HELP

▸ **HOMEWORK HELP**
Example 1: Exs. 9–32
Examples 2, 3: Exs. 33–51
Example 4: Exs. 52–55,
      58–63

10.7 *Solving Quadratic Systems*   **635**

3 APPLY

○ **ASSIGNMENT GUIDE**

BASIC
Day 1: pp. 635–638 Exs. 9–20,
  33–37, 54, 57, 64, 65, 67–85
  odd, Quiz 3 Exs. 1–13

AVERAGE
Day 1: pp. 635–638 Exs. 9–23,
  33–40, 54, 56–58, 64, 65,
  67–85 odd, Quiz 3 Exs. 1–13

ADVANCED
Day 1: pp. 635–638 Exs. 9–23,
  33–40, 54–58, 62–66,
  67–85 odd, Quiz 3 Exs. 1–13

BLOCK SCHEDULE
pp. 635–638 Exs. 9–23, 33–40,
54–55, 56–58, 64, 65, 67–85 odd,
Quiz 3 Exs. 1–13

EXERCISE LEVELS
Level A: *Easier*
9–32
Level B: *More Difficult*
33–62, 64, 65
Level C: *Most Difficult*
63, 66

✓ **HOMEWORK CHECK**
To quickly check student under-
standing of key concepts, go
over the following exercises:
Exs. 12, 16, 34, 54, 57. See also
the Daily Homework Quiz:
 • Blackline Master (*Chapter 11
  Resource Book*, p. 11)
 • ◢ Transparency (p. 83)

2. *Sample answer:*

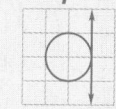

TEACHING TIPS
EXERCISES 15–55 Students may find it beneficial to graph the conic sections and then find the points of intersection. This can help them in understanding how many points of intersection there are and around where the points of intersection occur.

! COMMON ERROR
EXERCISES 15–55 Students may forget that any solution to a system of equations is a point, and therefore has an x-coordinate and a y-coordinate. Students may forget to find the other coordinate for the point once they have found either x-values or y-values that solve the equation.

CAREER NOTE
EXERCISE 58 Additional information about a career as a Police Officer is available at **www.mcdougallittell.com**.

57.
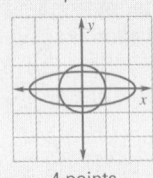

0 points   2 points

4 points

33. $\left(\pm\sqrt{\dfrac{5\sqrt{69}-15}{2}},\ \dfrac{-5+\sqrt{69}}{2}\right)$

34. $(3, \pm 6),\ \left(-\dfrac{9}{2}, \pm\dfrac{3\sqrt{21}}{2}\right)$

35. $\left(\dfrac{1+\sqrt{373}}{6},\ \pm\sqrt{\dfrac{7+\sqrt{373}}{18}}\right)$

41. $\left(\dfrac{9\sqrt{2}}{2}, -\dfrac{9\sqrt{2}}{2}\right),$
$\left(-\dfrac{9\sqrt{2}}{2}, \dfrac{9\sqrt{2}}{2}\right)$

48. $\left(\dfrac{3+\sqrt{129}}{4},\ \pm\dfrac{1}{2}\sqrt{\dfrac{-9+\sqrt{129}}{2}}\right)$

SOLVING SYSTEMS Find the points of intersection, if any, of the graphs in the system.

33. $x^2 + y^2 = 16$
$x^2 - 5y = 5$

34. $-3x^2 + y^2 - 3x = 0$
$x^2 - y^2 + 27 = 0$

35. $-x^2 + y^2 + 10 = 0$
$-3y^2 + x + 1 = 0$

36. $x^2 + 2y^2 - 10 = 0$ **none**
$4y^2 + x + 4 = 0$

37. $y^2 = 16x$ **none**
$4x - y = -24$

38. $10y = x^2$ $\left(\pm\sqrt{5}, \dfrac{1}{2}\right)$
$x^2 - 6 = -2y$

39. $y^2 + x = 2$ **none**
$3x + y = 8$

40. $x^2 - 16y^2 = 16$ **none**
$x^2 + y^2 = 9$

41. $x^2 + y^2 = 81$
$x + y = 0$

42. $16x^2 - y^2 + 16y - 128 = 0$
$y^2 - 48x - 16y - 32 = 0$
$(-2, 8), (5, 8 \pm 4\sqrt{21})$

43. $x^2 - y^2 - 8x + 8y - 24 = 0$ **none**
$x^2 + y^2 - 8x - 8y + 24 = 0$

44. $x^2 + 4y^2 - 4x - 8y + 4 = 0$
$x^2 + 4y - 4 = 0$ **(0, 1), (2, 0)**

45. $4x^2 - 56x + 9y^2 + 160 = 0$ **(4, 0)**
$4x^2 + y^2 - 64 = 0$

46. $x^2 + y^2 - 16x + 39 = 0$ **(3, 0), (5, ±4)**
$x^2 - y^2 - 9 = 0$

47. $x^2 - 4y^2 - 20x - 64y - 172 = 0$
$4x^2 + y^2 - 80x + 16y + 400 = 0$
$(6, -8), (14, -8)$

48. $x^2 - 2x + 4 + y^2 - 10 = 0$
$2y^2 - x + 3 = 0$

49. $4x^2 - y^2 - 8x + 6y - 9 = 0$ **(2, 3)**
$2x^2 - 3y^2 + 4x + 18y - 43 = 0$

50. $10x^2 - 25y^2 - 100x = -160$
$y^2 - 2x + 16 = 0$ **(8, 0)**

51. $x^2 - y - 4 = 0$ $(\pm\sqrt{6}, 2), (\pm\sqrt{3}, -1)$
$x^2 + 3y^2 - 4y - 10 = 0$

SYSTEMS OF THREE EQUATIONS Find the points, if any, that the graphs of all three equations have in common.

52. $x^2 + y^2 + 8x + 7 = 0$ **no intersection**
$x^2 + y^2 + 4x + 4y - 5 = 0$
$x^2 + y^2 = 1$

53. $x^2 + y^2 - 8 = 0$ **no intersection**
$x^2 + y^2 - 3x + y = 0$
$2x^2 + 2y^2 - 5x - 10 = 0$

54. $x^2 + 3y^2 = 16$ **(2, −2), (−2, 2)**
$3x^2 + y^2 = 16$
$y = -x$

55. $x^2 + y^2 - 4x - 4y = 26$ **(5, 7)**
$x^2 + y^2 - 4x = 54$
$y = 3x - 8$

56. **CRITICAL THINKING** Suppose a line intersects a circle whose center is at the origin, and the line passes through the origin. If you know one of the points of intersection, how do you know what the other point of intersection is without solving the system algebraically? **Use rotation by 180°: If one is at (a, b), the other is at $(-a, -b)$.**

57. **LOGICAL REASONING** Sketch examples to illustrate the different numbers of points of intersection that a circle and an ellipse can have if both are centered at the origin. **See margin.**

58. 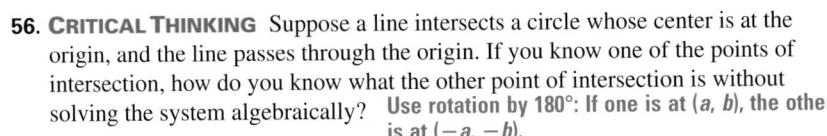 **LAW ENFORCEMENT** Suppose a car is traveling down the highway at a constant rate of 60 miles per hour. It passes a police car parked at the side of the road. To catch up to the car, the police officer accelerates at a constant rate. The distance d (in miles) the police car has traveled as a function of time t (in hours) since the other car has passed it is given by $d = 3600t^2$. Write and solve a system of equations to calculate how long it takes the police car to catch up to the other car. **6 min**

59. 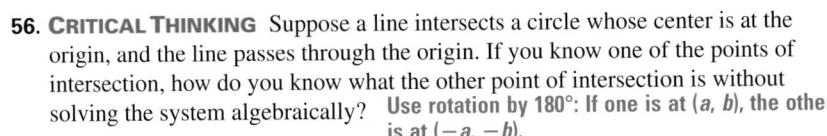 **COMMUNICATIONS** The range of a radio station is bounded by a circle given by the following equation:

$$x^2 + y^2 - 1620 = 0$$

A straight highway can be modeled by the following equation:

$$y = -\frac{1}{3}x + 30$$

Find the length of the highway that lies within the range of the radio station. **about 56.9 mi**

FOCUS ON
CAREERS

POLICE OFFICER
The duties of a police officer vary. An officer in a large city is often assigned to a specific type of duty, while an officer in a small community usually performs a variety of tasks.

CAREER LINK
www.mcdougallittell.com

60. BUS BOUNDARY To be eligible to ride the school bus to East High School, a student must live at least 1 mile from the school. How long is the portion of Clark Street for which the residents are not eligible to ride the school bus? (Use a coordinate plane in which the school is at (0, 0) and each unit represents one mile.)   **about 1.41 mi**

61. NAVIGATION LORAN (Long-Distance Radio Navigation) uses synchronized pulses sent out by pairs of transmitting stations. By calculating the difference in the times of arrival of the pulses from two stations, the LORAN equipment on a ship locates the ship on a hyperbola. By doing the same thing with a second pair of stations, LORAN locates the ship at the intersection of two hyperbolas. Suppose LORAN equipment indicates that a ship's location is the point of intersection of the graphs in the following system: $\left(4\sqrt{5}, \dfrac{6\sqrt{5}}{5}\right) \approx$ **(8.9, 2.7)**

$$xy - 24 = 0$$
$$x^2 - 25y^2 + 100 = 0$$

Find the ship's location given that it is north and east of the origin.

62. HYPERBOLIC MIRROR In a hyperbolic mirror, light rays directed to one focus will be reflected to the other focus. The mirror shown at the right has the following equation: **(6.38, 2.90)**

$$\frac{x^2}{36} - \frac{y^2}{64} = 1$$

At which point on the mirror will light from the point (0, 8) be reflected to the focus at (−10, 0)?

63. EARTHQUAKES An earthquake occurred in Peru on April 18, 1993. Use the following information to approximate the location of the epicenter.

▶ Source: U.S. Department of the Interior Geological Survey

Location 1: (Cayambe, Ecuador) The epicenter was 1300 kilometers away.

Location 2: (Cocohabamba, Bolivia, 1200 kilometers east and 1900 kilometers south of Cayambe) The epicenter was 1300 kilometers away.

Location 3: (Cerro El Oso, Venezuela, 1100 kilometers east and 1000 kilometers north of Cayambe) The epicenter was 2500 kilometers away.

64. MULTIPLE CHOICE How many points of intersection do the equations $x^2 + y^2 = 6$ and $2x^2 + 4y^2 = 7$ have?   **A**

Ⓐ 0   Ⓑ 1   Ⓒ 2   Ⓓ 3   Ⓔ 4

65. MULTIPLE CHOICE Which of the following is a point of intersection of the graphs of $25x^2 + 36y^2 - 900 = 0$ and $-2x^2 + y + 5 = 0$?   **E**

Ⓐ (−5, 0)   Ⓑ (0, 5)   Ⓒ (2, 5)   Ⓓ (1, 5)   Ⓔ (0, −5)

66. CRITICAL THINKING Write equations for three different conics that all intersect at the point (−4, 6).

STUDENT HELP NOTES

→ **Homework Help** Students can find help for Exs. 60–62 at **www.mcdougallittell.com**. The information can be printed out for students who don't have access to the Internet.

STUDENT HELP

INTERNET
HOMEWORK HELP
Visit our Web site www.mcdougallittell.com for help with problem solving in Exs. 60–62.

63. *Sample answer:* The epicenter of the earthquake was about 100 kilometers east and about 1300 kilometers south of Location 1.

Test
Preparation

66. *Sample answer.* $\dfrac{x^2}{16} +$ $\dfrac{(y-6)^2}{4} = 1,\ y = \dfrac{3}{8}x^2,$ $(x+4)^2 + (y-3)^2 = 9$

★ **Challenge**

ADDITIONAL PRACTICE AND RETEACHING

For Lesson 10.7:

• Practice Levels A, B, and C (*Chapter 10 Resource Book*, p. 96)

• Reteaching with Practice (*Chapter 10 Resource Book*, p. 99)

• See Lesson 10.7 of the *Personal Student Tutor*

For more Mixed Review:

• Search the *Test and Practice Generator* for key words or specific lessons.

Find the points of intersection of the graphs of the system.

1. $x^2 + y^2 = 25$

$y = -\dfrac{3}{4}x$ $(-4, 3), (4, -3)$

2. $y = -x^2 + 3$

$y = x^2 + 2$ $\left(-\dfrac{\sqrt{2}}{2}, \dfrac{5}{2}\right), \left(\dfrac{\sqrt{2}}{2}, \dfrac{5}{2}\right)$

3. $2x^2 - y^2 = 7$

$3x^2 + 2y^2 = 14$ $(2, \pm 1), (-2, \pm 1)$

EXTRA CHALLENGE NOTE

→ Challenge problems for Lesson 10.7 are available in **blackline** format in the *Chapter 10 Resource Book*, p. 103 and at **www.mcdougallittell.com**.

ADDITIONAL TEST PREPARATION

1. WRITING Explain how to solve

$y - x^2 + 9 = 0$
$x^2 + (y + 6)^2 - 9 = 0.$

Then solve.

Sample answer: Use substitution by writing $y = x^2 - 9$ for the first equation and then substituting $x^2 - 9$ for y in the second; $(0, -9)$, $\left(\sqrt{5}, -4\right), \left(-\sqrt{5}, -4\right)$

77.

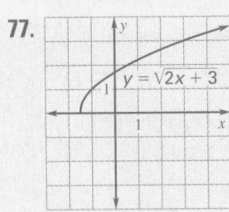

$y = \sqrt{2x + 3}$

78.

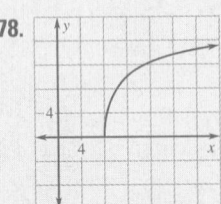

79–82. See Additional Answers beginning on page AA1.

MIXED REVIEW

71. $f(x) = x^3 - x^2 - 9x + 9$

72. $f(x) = x^4 - 8x^3 + 20x^2 - 16x$

75. $f(x) = x^5 - 2x^3 - 2x^2 - 3x - 2$

76. $f(x) = x^6 + 5x^5 + 8x^4 + 10x^3 + 13x^2 + 5x + 6$

77. domain: $x \ge -\dfrac{3}{2}$
range: $y \ge 0$

78. domain: $x \ge 8$
range: $y \ge 0$

79. domain: $x \ge -4$
range: $y \le 2$

80. domain: all reals
range: all reals

81. domain: all reals
range: all reals

82. domain: all reals
range: all reals

EVALUATING EXPRESSIONS Evaluate the expression for the given value of *x*. (Review 1.2 for 11.1)

67. $2x + 5$ when $x = 4$ 13

68. $\dfrac{1}{x^3} - 1$ when $x = 2$ $-\dfrac{7}{8}$

69. $(-2)^{x-1}$ when $x = 5$ 16

70. $\dfrac{3}{(-3)^{x-2}}$ when $x = 4$ $\dfrac{1}{3}$

WRITING FUNCTIONS Write a polynomial function of least degree that has real coefficients, the given zeros, and a leading coefficient of 1. (Review 6.7)
71, 72, 75, 76. See margin.

71. $3, -3, 1$

72. $0, 2, 2, 4$

73. $2i, -2i$ $f(x) = x^2 + 4$

74. $3 + i, 3 - i$ $f(x) = x^2 - 6x + 10$

75. $2, -1, -1 - i$

76. $-2, -3, i, i$

GRAPHING Graph the function. Then state the domain and range. (Review 7.5)
77–82. See margin.

77. $f(x) = \sqrt{2x + 3}$

78. $f(x) = 5\sqrt{x - 8}$

79. $f(x) = -(x + 4)^{1/2} + 2$

80. $f(x) = -3\sqrt[3]{x + 1}$

81. $f(x) = \sqrt[3]{4x + 1} + 2$

82. $f(x) = 5(x - 1)^{1/3}$

CLASSIFYING CONICS Classify the conic section. (Review 10.6)

83. $3x^2 + y^2 + 2x + 2y = 0$ ellipse

84. $4x^2 - y^2 - 8x + 4y - 9 = 0$ hyperbola

85. $x^2 + 6x - 2y + 13 = 0$ parabola

86. $x^2 + y^2 - 2x + 6y + 9 = 0$ circle

QUIZ 3   *Self-Test for Lessons 10.6 and 10.7*

2. $\dfrac{(x + 0.5)^2}{42.25} + \dfrac{(y - 2)^2}{22} = 1$

Write an equation for the conic section. (Lesson 10.6)

1. Circle with center at $(-3, -5)$ and radius 8 $(x + 3)^2 + (y + 5)^2 = 64$

2. Ellipse with vertices at $(-7, 2)$ and $(6, 2)$ and foci at $(4, 2)$ and $(-5, 2)$

3. Parabola with vertex at $(4, -1)$ and focus at $(7, -1)$ $(y + 1)^2 = 12(x - 4)$

4. Hyperbola with foci at $(2, -1)$ and $(2, 8)$ and vertices at $(2, 3)$ and $(2, 4)$
$\dfrac{(y - 3.5)^2}{0.25} - \dfrac{(x - 2)^2}{20} = 1$

Classify the conic section. (Lesson 10.6)

5. $x^2 + 4y^2 - 8x + 3y + 12 = 0$ ellipse

6. $-3x^2 - 3y^2 + 6x + 4y + 1 = 0$ circle

7. $-2y^2 + x + 5y + 26 = 0$ parabola

8. $-6x^2 + 4y^2 + 2x + 9 = 0$ hyperbola

Find the points of intersection, if any, of the graphs in the system. (Lesson 10.7)

9. $3x^2 - 4x - y + 2 = 0$ $\left(\dfrac{2}{3}, \dfrac{2}{3}\right), (-1, 9)$
$y = -5x + 4$

10. $-x^2 + y^2 + 4x - 6y + 4 = 0$ $(2, 2), (2, 4)$
$x^2 + y^2 - 4x - 6y + 12 = 0$

11. $x^2 + y^2 + 4y - 12 = 0$
$x^2 - 16y^2 - 64y - 80 = 0$
$(4, -2), (-4, -2)$

12. $y^2 - 6x - 2y - 3 = 0$ none
$2y^2 - 4y + x + 6 = 0$

13. 🌐 **SEISMOLOGY** A seismograph records the epicenter of an earthquake 50 miles away. A second seismograph, 50 miles west and 35 miles north of the first, records the epicenter as being 35 miles away. A third seismograph, 80 miles due west of the first, records the epicenter 30 miles away. Where was the earthquake's epicenter in relation to the first seismograph? (Lesson 10.7)

The epicenter of the earthquake is 50 miles due west of the first seismograph.

Extension

What you should learn

GOAL Find the eccentricity of a conic section.

Why you should learn it

▼ To write equations for **real-life** conics, such as the moon's orbit in **Example 3**.

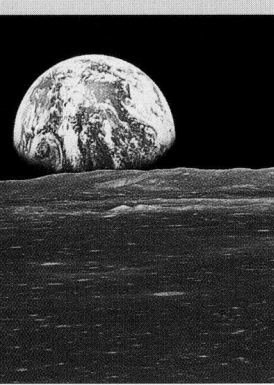

Earth, as seen from the moon

Eccentricity of Conic Sections

Some ellipses are more oval than others. In an ellipse that is nearly circular, the ratio $c{:}a$ is close to 0. In a more oval ellipse, $c{:}a$ is close to 1. This ratio is called the **eccentricity** of the ellipse. Every conic has an eccentricity e associated with it.

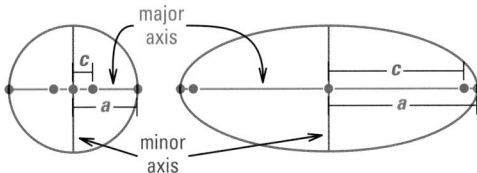

CONCEPT SUMMARY   **ECCENTRICITY OF CONIC SECTIONS**

Let c be the distance from each focus to the center of the conic section, and let a be the distance from each vertex to the center.

- The eccentricity of an ellipse is $e = \frac{c}{a}$, and $0 < e < 1$.

- The eccentricity of a hyperbola is $e = \frac{c}{a}$, and $e > 1$.

- The eccentricity of a parabola is $e = 1$.

- The eccentricity of a circle is $e = 0$.

EXAMPLE 1   *Finding Eccentricity*

Find the eccentricity of the conic section described by the equation.

a. $(x + 2)^2 = 4(y - 1)$ 　　　　　　　　　**b.** $25(x + 2)^2 - 36(y - 1)^2 = 900$

SOLUTION

a. This equation describes a parabola. By definition, the eccentricity is $e = 1$.

b. This equation describes a hyperbola with $a = \sqrt{36} = 6$, $b = \sqrt{25} = 5$, and $c = \sqrt{a^2 + b^2} = \sqrt{61}$. The eccentricity is $e = \frac{c}{a} = \frac{\sqrt{61}}{6} \approx 1.302$.

EXAMPLE 2   *Using Eccentricity to Write an Equation*

Find an equation of the hyperbola with center $(3, -5)$, vertex $(9, -5)$, and $e = 2$.

SOLUTION

Use the form $\frac{(x - h)^2}{a^2} - \frac{(y - k)^2}{b^2} = 1$. The vertex lies $9 - 3 = 6$ units from the center, so $a = 6$. Because $e = \frac{c}{a} = 2$, you know that $\frac{c}{6} = 2$, or $c = 12$. Therefore, $b^2 = c^2 - a^2 = 144 - 36 = 108$. The equation is $\frac{(x - 3)^2}{36} - \frac{(y + 5)^2}{108} = 1$.

Chapter 10 *Extension*   **639**

EXTRA EXAMPLE 1
Find the eccentricity of the conic section described by the equation.
a. $(x - 3)^2 + y^2 = 24$ 0
b. $4x^2 + 16y^2 = 16$ 0.866

EXTRA EXAMPLE 2
Find an equation of the ellipse with center at $(3, 1)$, vertex at $(3, -2)$, and $e = \frac{1}{3}$.
$\frac{(x - 3)^2}{8} + \frac{(y - 1)^2}{9} = 1$

CHECKPOINT EXERCISES
For use after Examples 1 and 2:
1. Find an equation of the hyperbola with a focus at $(5, 0)$, asymptote $y = x$, and $e = \sqrt{2}$.
$x^2 - y^2 = 25$

EXTRA EXAMPLE 3

Earth orbits the sun in an elliptical path, with the sun at one focus. The eccentricity of the orbit is 0.0167, and the length of the major axis is 186 million miles. Find an equation of Earth's orbit. $\dfrac{x^2}{93^2} + \dfrac{y^2}{92.987^2} = 1$

 ## CHECKPOINT EXERCISES

For use after Example 3:

1. An elliptical whispering gallery is 52 ft long and 20 ft wide. What is an equation for the ellipse? What is the eccentricity of the ellipse? $\dfrac{x^2}{26^2} + \dfrac{y^2}{10^2} = 1$; $e = 0.923$

MATHEMATICAL REASONING

If $e = 0$, explain why the equation of an "ellipse," $\dfrac{x^2}{a^2} + \dfrac{y^2}{b^2} = 1$, is the equation of a circle. If $\dfrac{c}{a} = 0$, then $c = 0$. Since $c^2 = a^2 - b^2$, $a^2 - b^2 = 0$, or $a^2 = b^2$. Therefore, $\dfrac{x^2}{a^2} + \dfrac{y^2}{b^2} = 1$ is equivalent to $x^2 + y^2 = a^2$, which is the equation of a circle with radius a.

9. $\dfrac{x^2}{25} + \dfrac{(y+1)^2}{16} = 1$

10. $\dfrac{(x-2)^2}{48} + \dfrac{y^2}{64} = 1$

11. $\dfrac{(x-2)^2}{60} + \dfrac{y^2}{64} = 1$

12. $\dfrac{x^2}{9} + \dfrac{(y-6)^2}{8.91} = 1$

13. $\dfrac{(y-1)^2}{\left(\frac{64}{9}\right)} - \dfrac{(x-3)^2}{\left(\frac{512}{9}\right)} = 1$

14. $\dfrac{(x+6)^2}{16} - \dfrac{(y-4)^2}{76.16} = 1$

15. $\dfrac{(y-2)^2}{9} - \dfrac{(x-3)^2}{23.49} = 1$

16. $(x+1)^2 - \dfrac{(y-2)^2}{24} = 1$

18. $\dfrac{x^2}{20,049.66} + \dfrac{y^2}{19,875.50} = 1$

(x, y in millions of miles)

19. In an ellipse the foci are always within the major axis, so $c < a$ and $\dfrac{c}{a} < 1$. In a hyperbola the foci are always outside the major axis, so $c > a$ and $\dfrac{c}{a} > 1$.

EXAMPLE 3   *Using Eccentricity to Write a Model*

The moon orbits Earth in an elliptical path with the center of Earth at one focus. The eccentricity of the orbit is $e = 0.055$ and the length of the major axis is about 768,800 kilometers. Find an equation of the moon's orbit.

SOLUTION

Let the major axis of the ellipse be horizontal. The equation of the orbit has the form $\dfrac{x^2}{a^2} + \dfrac{y^2}{b^2} = 1$. Using the length of the major axis, you know that $2a = 768,800$, or $a \approx 384,400$. Because $e = \dfrac{c}{a}$, you know that $0.055 = \dfrac{c}{384,400}$, or $c \approx 21,142$ and $b = \sqrt{a^2 - c^2} = \sqrt{384,400^2 - 21,142^2} = \sqrt{1.47 \times 10^{11}} \approx 383,800$. The equation of the moon's orbit is $\dfrac{x^2}{384,400^2} + \dfrac{y^2}{383,800^2} = 1$ where x and y are measured in kilometers.

EXERCISES

Find the eccentricity of the conic section.

1. $3x^2 - 5x + y + 20 = 0$ 1

2. $25(x - 3)^2 + 9(y + 6)^2 = 225$ 0.8

3. $x^2 + 16(y - 4)^2 = 16$ $\dfrac{\sqrt{15}}{4} \approx 0.968$

4. $\dfrac{(x-3)^2}{8} + \dfrac{(y-5)^2}{8} = 8$ 0

5. $\dfrac{(x+6)^2}{25} - \dfrac{(y-6)^2}{100} = 1$ $\sqrt{5} \approx 2.236$

6. $\dfrac{(x+2)^2}{49} + \dfrac{(y+2)^2}{16} = 1$ $\dfrac{\sqrt{33}}{7} \approx 0.821$

7. $4(x + 1)^2 - 8(y - 2)^2 = 16$ $\dfrac{\sqrt{6}}{2} \approx 1.225$

8. $(x - 4)^2 - (y - 3)^2 = 1$ $\sqrt{2} \approx 1.414$

Write an equation of the conic section.   9–16. See margin.

9. Ellipse with vertices at $(-5, -1)$ and $(5, -1)$, and $e = 0.6$

10. Ellipse with foci at $(2, -4)$ and $(2, 4)$, and $e = 0.5$

11. Ellipse with center at $(2, 0)$, focus at $(2, 2)$, and $e = 0.25$

12. Ellipse with center at $(0, 6)$, vertex at $(3, 6)$, and $e = 0.1$

13. Hyperbola with foci at $(3, -7)$ and $(3, 9)$, and $e = 3$

14. Hyperbola with vertices at $(-10, 4)$ and $(-2, 4)$, and $e = 2.4$

15. Hyperbola with center at $(3, 2)$, vertex at $(3, 5)$, and $e = 1.9$

16. Hyperbola with center at $(-1, 2)$, focus at $(4, 2)$, and $e = 5$

17. 🌐 **ASTRONOMY** Mercury orbits the sun in an elliptical path with the center of the sun at one focus. The eccentricity of Mercury's orbit is $e = 0.2056$. The length of the major axis of the orbit is 72 million miles. Find an equation of Mercury's orbit. $\dfrac{x^2}{1296} + \dfrac{y^2}{1241} = 1$ (x, y in millions of miles)

18. 🌐 **ASTRONOMY** Mars orbits the sun in an elliptical path with the center of the sun at one focus. The eccentricity of Mars' orbit is $e = 0.0932$. The *perihelion* of Mars' orbit is the point where the planet is closest to the sun. At the perihelion, Mars' distance from the sun is 128.4 million miles. Find an equation of Mars' orbit.

19. *Writing* Explain why the definition of eccentricity for ellipses and hyperbolas implies that $0 < e < 1$ for an ellipse and $e > 1$ for a hyperbola.

Chapter Summary

WHAT did you learn?

Find the distance between two points. **(10.1)**

Find the midpoint of the line segment connecting two points. **(10.1)**

Use distance and midpoint formulas in real-life situations. **(10.1)**

Graph and write equations of conics.
- parabolas **(10.2, 10.6)**
- circles **(10.3, 10.6)**
- ellipses **(10.4, 10.6)**

- hyperbolas **(10.5, 10.6)**

Classify a conic using its equation. **(10.6)**

Solve systems of quadratic equations. **(10.7)**

Use conics to solve real-life problems. **(10.2–10.7)**

WHY did you learn it?

Find the distance a medical helicopter must travel. **(p. 593)**

Find the diameter of a broken dish. **(p. 591)**

Design a city park. **(p. 593)**

Model a solar energy collector. **(p. 597)**
Model the region lit by a lighthouse. **(p. 603)**
Model the shape of an Australian football field. **(p. 614)**
Model the curved sides of a sculpture. **(p. 617)**

Classify mirrors in a Cassegrain telescope. **(p. 627)**

Find the epicenter of an earthquake. **(p. 634)**

Find the area of The Ellipse at the White House. **(p. 611)**

How does Chapter 10 fit into the BIGGER PICTURE of algebra?

In Chapter 5 you studied parabolas as graphs of quadratic functions, and in Chapter 9 you studied hyperbolas as graphs of rational functions. In a previous course you studied circles, and possibly ellipses, in the context of geometry. In Chapter 10 you studied all four conic sections (parabolas, hyperbolas, circles, and ellipses) as graphs of equations of the form $Ax^2 + Bxy + Cy^2 + Dx + Ey + F = 0$.

The conic sections are an important part of your study of algebra and geometry because they have many different real-life applications.

STUDY STRATEGY

How did you make and use a dictionary of graphs?

Here is an example of one entry for your dictionary of graphs, following the **Study Strategy** on page 588.

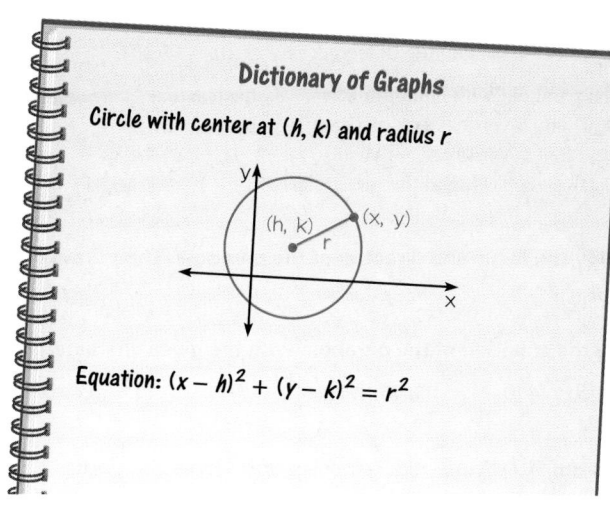

Dictionary of Graphs

Circle with center at (h, k) and radius r

(h, k) (x, y) r

Equation: $(x - h)^2 + (y - k)^2 = r^2$

641

Chapter Review

ADDITIONAL RESOURCES
The following resources are available to help review the material in this chapter.

• Chapter Review Games and Activities (*Chapter 10 Resource Book*, p. 104)

• *Instant Replay: Video Review Games*

• 🖥 *Personal Student Tutor*

• Cumulative Review, Chs. 1–10 (*Chapter 10 Resource Book*, p. 116)

5. focus: $(0, 1)$; directrix: $y = -1$

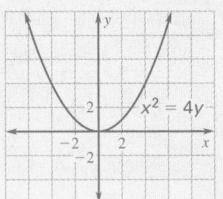

6. focus; $\left(0, -\dfrac{1}{2}\right)$; directrix: $y = \dfrac{1}{2}$

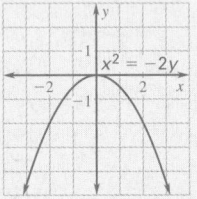

7. focus: $\left(-\dfrac{3}{2}, 0\right)$; directrix: $x = \dfrac{3}{2}$

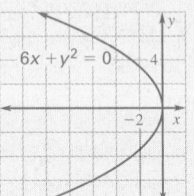

8. See Additional Answers beginning on page AA1.

642

• distance formula, p. 589
• midpoint formula, p. 590
• focus, p. 595, 609, 615
• directrix, p. 595
• circle, p. 601
• center, p. 601, 609, 615

• radius, p. 601
• equation of a circle, p. 601
• ellipse, p. 609
• vertex, p. 609, 615
• major axis, p. 609

• co-vertex, p. 609
• minor axis, p. 609
• equation of an ellipse, p. 609
• hyperbola, p. 615
• transverse axis, p. 615

• equation of a hyperbola, p. 615
• conic sections, p. 623
• general second-degree equation, p. 626
• discriminant, p. 626

10.1 THE DISTANCE AND MIDPOINT FORMULAS

Examples on pp. 589–591

EXAMPLES Let $A = (-2, 4)$ and $B = (2, -3)$.

Distance between A and $B = \sqrt{(x_2 - x_1)^2 + (y_2 - y_1)^2}$

$\qquad\qquad\qquad\qquad = \sqrt{(2 - (-2))^2 + (-3 - 4)^2}$

$\qquad\qquad\qquad\qquad = \sqrt{16 + 49} = \sqrt{65} \approx 8.06$

Midpoint of $\overline{AB} = M\left(\dfrac{x_1 + x_2}{2}, \dfrac{y_1 + y_2}{2}\right) = \left(\dfrac{(-2) + 2}{2}, \dfrac{4 + (-3)}{2}\right) = \left(0, \dfrac{1}{2}\right)$

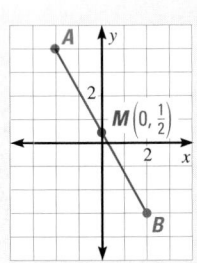

Find the distance between the two points. Then find the midpoint of the line segment connecting the two points.

1. $(-2, -3), (4, 2)$
$\sqrt{61} \approx 7.81; \left(1, -\dfrac{1}{2}\right)$

2. $(-5, 4), (10, -3)$
$\sqrt{274} \approx 16.6; \left(\dfrac{5}{2}, \dfrac{1}{2}\right)$

3. $(0, 0), (-4, 4)$
$4\sqrt{2} \approx 5.66; (-2, 2)$

4. $(-2, 0), (0, -8)$
$2\sqrt{17} \approx 8.25; (-1, -4)$

10.2 PARABOLAS

Examples on pp. 595–597

EXAMPLES The parabola with equation $y^2 = 8x$ has vertex $(0, 0)$ and a horizontal axis of symmetry. It opens to the right. Note that $y^2 = 4px = 8x$, so $p = 2$. The **focus** is $(p, 0) = (2, 0)$, and the **directrix** is $x = -p = -2$.

The parabola with equation $x^2 = -8y$ has **vertex** $(0, 0)$ and a vertical axis of symmetry. It opens down. Note that $x^2 = 4py = -8y$, so $p = -2$. The **focus** is $(0, p) = (0, -2)$, and the **directrix** is $y = -p = 2$.

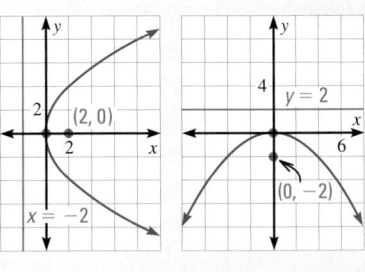

Identify the focus and directrix of the parabola. Then draw the parabola. 5–8. See margin.

5. $x^2 = 4y$

6. $x^2 = -2y$

7. $6x + y^2 = 0$

8. $y^2 - 12x = 0$

Write the equation of the parabola with the given characteristic and vertex (0, 0).

9. focus: $(4, 0)$ $y^2 = 16x$

10. focus: $(0, -3)$ $x^2 = -12y$

11. directrix: $y = -2$ $x^2 = 8y$

12. directrix: $x = 1$ $y^2 = -4x$

10.3 CIRCLES

Examples on pp. 601–603

EXAMPLE The circle with equation $x^2 + y^2 = 9$ has center at $(0, 0)$ and radius $r = \sqrt{9} = 3$.

Four points on the circle are $(3, 0)$, $(0, 3)$, $(-3, 0)$, and $(0, -3)$.

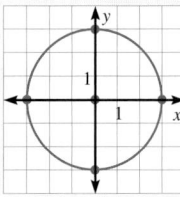

Graph the equation. 13–16. See margin.

13. $x^2 + y^2 = 16$ **14.** $x^2 + y^2 = 64$ **15.** $x^2 + y^2 = 6$ **16.** $3x^2 + 3y^2 = 363$

Write the standard form of the equation of the circle that has the given radius or passes through the given point and whose center is the origin.

17. radius: 5
$x^2 + y^2 = 25$

18. radius: $\sqrt{10}$
$x^2 + y^2 = 10$

19. point: $(-2, 3)$
$x^2 + y^2 = 13$

20. point: $(1, 8)$ $x^2 + y^2 = 65$

10.4 ELLIPSES

Examples on pp. 609–611

EXAMPLE The ellipse with equation $\dfrac{x^2}{9} + \dfrac{y^2}{4} = 1$ has a horizontal major axis because $9 > 4$.

Since $\sqrt{9} = 3$, the vertices are at $(-3, 0)$ and $(3, 0)$.

Since $\sqrt{4} = 2$, the co-vertices are at $(0, -2)$ and $(0, 2)$.

Since $9 - 4 = 5$, the foci are at $(-\sqrt{5}, 0)$ and $(\sqrt{5}, 0)$.

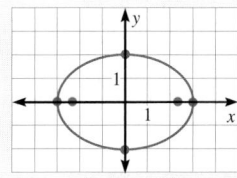

Graph the equation. 21–23. See margin.

21. $4x^2 + 81y^2 = 324$ **22.** $-9x^2 - 4y^2 = -36$ **23.** $49x^2 + 36y^2 = 1764$

Write an equation of the ellipse with the given characteristics and center at (0, 0).

24. Vertex: $(0, 5)$, Co-vertex: $(1, 0)$ $x^2 + \dfrac{y^2}{25} = 1$ **25.** Vertex: $(4, 0)$, Focus: $(-3, 0)$ $\dfrac{x^2}{16} + \dfrac{y^2}{7} = 1$

10.5 HYPERBOLAS

Examples on pp. 615–617

EXAMPLE The hyperbola with equation $\dfrac{y^2}{4} - \dfrac{x^2}{9} = 1$ has a vertical transverse axis because the y^2-term is positive.
Since $\sqrt{4} = 2$, vertices are $(0, -2)$ and $(0, 2)$.
Since $4 + 9 = 13$, foci are $(0, -\sqrt{13})$ and $(0, \sqrt{13})$.
Asymptotes are $y = \dfrac{2}{3}x$ and $y = -\dfrac{2}{3}x$.

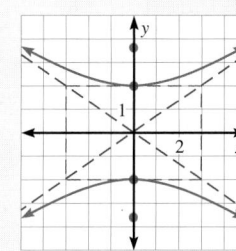

13.
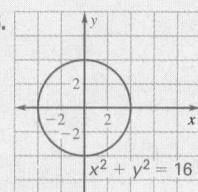
$x^2 + y^2 = 16$

14.
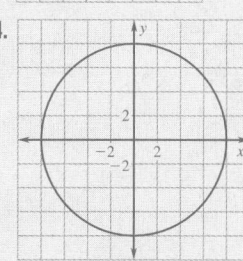

15.
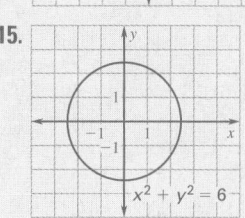
$x^2 + y^2 = 6$

16.
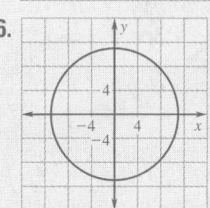

21.
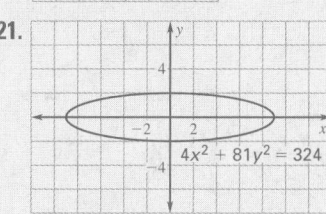
$4x^2 + 81y^2 = 324$

22, 23. See Additional Answers beginning on page AA1.

26.

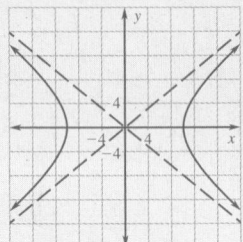

27.

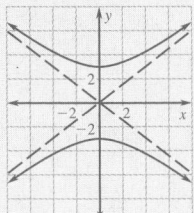

28.

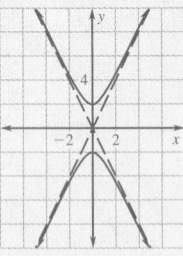

32. parabola: $(x+4)^2 = 8y$

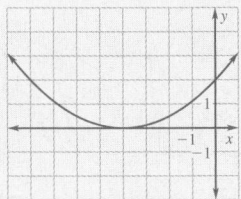

33. circle: $(x-5)^2 + (y+1)^2 = 100$

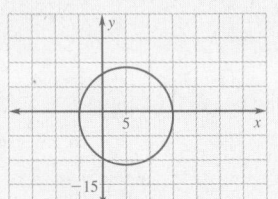

34, 35, 37. See Additional Answers beginning on page AA1.

10.5 continued

Graph the hyperbola. 26–28. See margin.

26. $\dfrac{x^2}{100} - \dfrac{y^2}{64} = 1$

27. $16y^2 - 9x^2 = 144$

28. $y^2 - 4x^2 = 4$

Write an equation of the hyperbola with the given foci and vertices.

29. Foci: $(0, -3), (0, 3)$
Vertices: $(0, -1), (0, 1)$
$y^2 - \dfrac{x^2}{8} = 1$

30. Foci: $(0, -4), (0, 4)$
Vertices: $(0, -2), (0, 2)$
$\dfrac{y^2}{4} - \dfrac{x^2}{12} = 1$

31. Foci: $(-5, 0), (5, 0)$
Vertices: $(-3, 0), (3, 0)$
$\dfrac{x^2}{9} - \dfrac{y^2}{16} = 1$

10.6 GRAPHING AND CLASSIFYING CONICS

Examples on pp. 623–627

EXAMPLE You can use the discriminant $B^2 - 4AC$ to classify a conic.

For the equation $x^2 + y^2 - 6x + 2y + 6 = 0$, the discriminant is $B^2 - 4AC = 0^2 - 4(1)(1) = -4$. Because $B^2 - 4AC < 0$, $B = 0$, and $A = C$, the equation represents a circle.

To graph the circle, complete the square as follows.

$$x^2 + y^2 - 6x + 2y + 6 = 0$$
$$(x^2 - 6x + 9) + (y^2 + 2y + 1) = -6 + 9 + 1$$
$$(x - 3)^2 + (y + 1)^2 = 4$$

The center of the circle is at $(h, k) = (3, -1)$ and $r = \sqrt{4} = 2$.

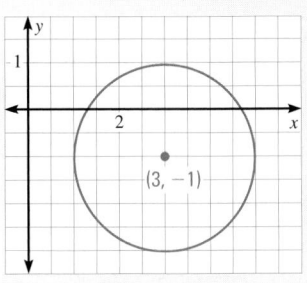

Classify the conic section and write its equation in standard form. Then graph the equation. 32–35. See margin.

32. $x^2 + 8x - 8y + 16 = 0$

33. $x^2 + y^2 - 10x + 2y - 74 = 0$

34. $9x^2 + y^2 + 72x - 2y + 136 = 0$

35. $y^2 - 4x^2 - 18y - 8x + 76 = 0$

10.7 SOLVING QUADRATIC SYSTEMS

Examples on pp. 632–634

EXAMPLE You can solve systems of quadratic equations algebraically.

$y^2 - 2x - 10y + 31 = 0$
$x - y + 2 = 0$ **Solve the second equation for y: $y = x + 2$.**

$(x + 2)^2 - 2x - 10(x + 2) + 31 = 0$ **Substitute into the first equation.**

$x^2 - 8x + 15 = 0$, so $x = 3$ or $x = 5$.   **Simplify and solve.**

The points of intersection of the graphs of the system are $(3, 5)$ and $(5, 7)$.

Find the points of intersection, if any, of the graphs in the system.

36. $x^2 + y^2 - 18x + 24y + 200 = 0$
$4x + 3y = 0$ $(6, -8), (12, -16)$

37. $5x^2 + 3x - 8y + 2 = 0$ See margin.
$3x + y - 6 = 0$

38. $4x^2 + y^2 - 48x - 2y + 129 = 0$ $(4, 1)$
$x^2 + y^2 - 2x - 2y - 7 = 0$

39. $9x^2 - 16y^2 + 18x + 153 = 0$ $(-1, -3)$ and $(-1, 3)$
$9x^2 + 16y^2 + 18x - 135 = 0$

ADDITIONAL RESOURCES
- **Chapter 10 Resource Book**
  Chapter Test (3 levels) (p. 105)
  SAT/ACT Chapter Test (p. 111)
  Alternative Assessment (p. 112)
- 🖥 *Test and Practice Generator*

Find the distance between the two points. Then find the midpoint of the line segment connecting the two points.

1. $(1, 9), (5, 3)$ $2\sqrt{13} \approx 7.21; (3, 6)$
2. $(-8, 3), (4, 7)$ $4\sqrt{10} \approx 12.6; (-2, 5)$
3. $(-4, -2), (3, 10)$ $\sqrt{193} \approx 13.9; \left(-\frac{1}{2}, 4\right)$

4. $(-11, -5), (-3, 7)$ $4\sqrt{13} \approx 14.4; (-7, 1)$
5. $(-1, 6), (2, 8)$ $\sqrt{13} \approx 3.61; \left(\frac{1}{2}, 7\right)$
6. $(3, -2), (4, 9)$ $\sqrt{122} \approx 11.0; \left(\frac{7}{2}, \frac{7}{2}\right)$

Graph the equation.   7–15. See margin.

7. $x^2 + y^2 = 36$
8. $y^2 = 16x$
9. $9y^2 - 81x^2 = 729$

10. $25x^2 + 9y^2 = 225$
11. $(x - 4)^2 = y + 7$
12. $(x - 3)^2 + (y + 2)^2 = 1$

13. $\dfrac{(x + 6)^2}{4} + \dfrac{(y - 7)^2}{1} = 1$
14. $\dfrac{(x - 4)^2}{16} - \dfrac{(y + 4)^2}{16} = 1$
15. $\dfrac{(y + 2)^2}{4} - \dfrac{(x + 1)^2}{16} = 1$

Write an equation for the conic section.

16. Parabola with vertex at $(0, 0)$ and directrix $x = 5$ $y^2 = -20x$

17. Parabola with vertex at $(3, -6)$ and focus at $(3, -4)$ $(x - 3)^2 = 8(y + 6)$

18. Circle with center at $(0, 0)$ and passing through $(4, 6)$ $x^2 + y^2 = 52$

19. Circle with center at $(-8, 3)$ and radius 5 $(x + 8)^2 + (y - 3)^2 = 25$

20. Ellipse with center at $(0, 0)$, vertex at $(4, 0)$, and co-vertex at $(0, 2)$ $\dfrac{x^2}{16} + \dfrac{y^2}{4} = 1$

21. Ellipse with vertices at $(3, -5)$ and $(3, -1)$ and foci at $(3, -4)$ and $(3, -2)$ $\dfrac{(x - 3)^2}{3} + \dfrac{(y + 3)^2}{4} = 1$

22. Hyperbola with vertices at $(-7, 0)$ and $(7, 0)$ and foci at $(-9, 0)$ and $(9, 0)$ $\dfrac{x^2}{49} - \dfrac{y^2}{32} = 1$

23. Hyperbola with vertex at $(4, 2)$, focus at $(4, 4)$, and center at $(4, -1)$ $\dfrac{(y + 1)^2}{9} - \dfrac{(x - 4)^2}{16} = 1$

Classify the conic section and write its equation in standard form.   24–35. See margin.

24. $x^2 + 4y^2 - 2x - 3 = 0$
25. $2x^2 + 20x - y + 41 = 0$
26. $5x^2 - 3y^2 - 30 = 0$

27. $x^2 + y^2 - 12x + 4y + 31 = 0$
28. $y^2 - 8x - 4y + 4 = 0$
29. $-x^2 + y^2 - 6x - 6y - 4 = 0$

30. $x^2 - 8x + 4y + 16 = 0$
31. $3x^2 + 3y^2 - 30x + 59 = 0$
32. $x^2 + 2y^2 - 8x + 7 = 0$

33. $4x^2 - y^2 + 16x + 6y - 3 = 0$
34. $3x^2 + y^2 - 4y + 3 = 0$
35. $x^2 + y^2 - 2x + 10y + 1 = 0$

Find the points of intersection, if any, of the graphs in the system.

36. $x^2 + y^2 = 64$ See margin.
$x - 2y = 17$

37. $x^2 + y^2 = 20$ no solution
$x^2 + 4y^2 - 2x - 2 = 0$

38. $x^2 = 8y$ $(-4, 2)$ and $(4, 2)$
$x^2 = 2y + 12$

39. 🌐 **ARCHITECTURE** The Royal Albert Hall in London is nearly elliptical in shape, about 230 feet long and 200 feet wide. Write an equation for the shape of the hall, assuming its center is at $(0, 0)$. Then graph the equation. $\dfrac{x^2}{13,225} + \dfrac{y^2}{10,000} = 1$; See margin for graph.

40. 🌐 **SEARCH TEAM** A search team of three members splits to search an area in the woods. Each member carries a family service radio with a circular range of 3 miles. They agree to communicate from their bases every hour. One member sets up base 2 miles north of the first member. Where should the other member set up base to be as far east as possible but within range of communication? $2\sqrt{2}$ or about 2.83 mi east of the midpoint between the other two members.

7.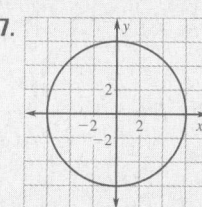

8.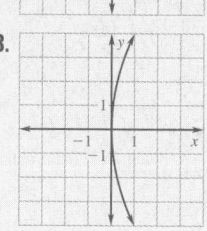

9.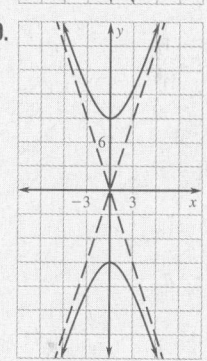

10.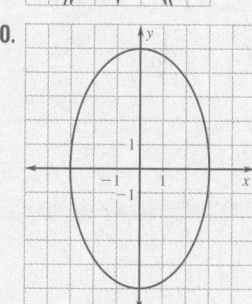

11.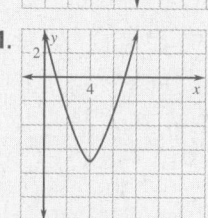

12–15, 24–35, 36, 39.
See Additional Answers beginning on page AA1.

Chapter Standardized Test

🔵 **TEST-TAKING STRATEGY** During the test, do not worry excessively about how much time you have left. Concentrate on the question in front of you.

ADDITIONAL RESOURCES
- *Chapter 10 Resource Book*
  Chapter Test (3 levels) (p. 105)
  SAT/ACT Chapter Test (p. 111)
  Alternative Assessment (p. 112)

- 🖥 *Test and Practice Generator*

1. **MULTIPLE CHOICE** What is the midpoint of the line segment connecting points $(0, 0)$ and $(-8, 2)$? **A**

   Ⓐ $(-4, 1)$ Ⓑ $(4, 1)$ Ⓒ $(4, -1)$

   Ⓓ $(1, 4)$ Ⓔ $(1, -4)$

2. **MULTIPLE CHOICE** Which equation represents the perpendicular bisector of the line segment connecting points $(-7, 1)$ and $(9, 13)$? **A**

   Ⓐ $y = -\frac{4}{3}x + \frac{25}{3}$ Ⓑ $y = \frac{3}{4}x + \frac{25}{4}$

   Ⓒ $y = \frac{4}{3}x + \frac{25}{3}$ Ⓓ $y = \frac{4}{3}x + \frac{17}{3}$

   Ⓔ $y = -\frac{4}{3}x + \frac{17}{3}$

3. **MULTIPLE CHOICE** Which equation is graphed? **D**

   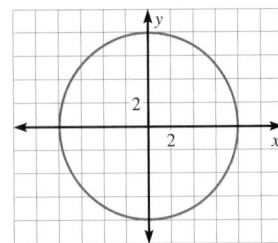

   Ⓐ $x^2 + y^2 = 8$ Ⓑ $x^2 - y^2 = 8$

   Ⓒ $x^2 + y^2 = 16$ Ⓓ $9x^2 + 9y^2 = 576$

   Ⓔ $9x^2 - 9y^2 = 576$

4. **MULTIPLE CHOICE** What is the standard form of the ellipse with center at $(0, 0)$, vertex at $(0, 9)$, and co-vertex at $(4, 0)$? **D**

   Ⓐ $\frac{x^2}{9} + \frac{y^2}{4} = 1$ Ⓑ $\frac{x^2}{4} + \frac{y^2}{9} = 1$

   Ⓒ $\frac{x^2}{81} + \frac{y^2}{16} = 1$ Ⓓ $\frac{x^2}{16} + \frac{y^2}{81} = 1$

   Ⓔ $\frac{x^2}{2} + \frac{y^2}{3} = 1$

5. **MULTIPLE CHOICE** What is the focus of the parabola with equation $2x^2 = -120y$? **E**

   Ⓐ $(0, 60)$ Ⓑ $(0, 15)$ Ⓒ $(0, -60)$

   Ⓓ $(0, 12)$ Ⓔ $(0, -15)$

6. **MULTIPLE CHOICE** What is the directrix of the parabola with equation $y^2 = 24x$? **B**

   Ⓐ $x = 6$ Ⓑ $x = -6$ Ⓒ $x = 24$

   Ⓓ $y = 6$ Ⓔ $y = -6$

7. **MULTIPLE CHOICE** Which graph represents the equation $\frac{y^2}{25} - \frac{x^2}{9} = 1$? **C**

   Ⓐ 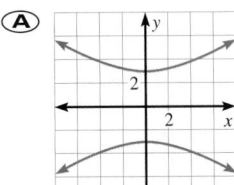   Ⓑ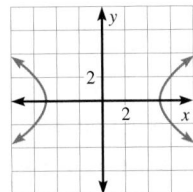

   Ⓒ 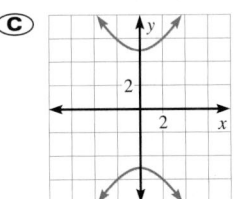   Ⓓ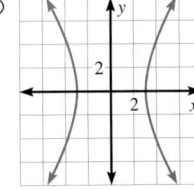

   Ⓔ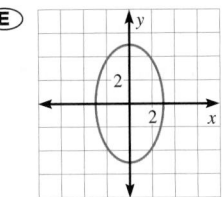

8. **MULTIPLE CHOICE** What conic does the equation $x^2 - 5x + 10y + 11 = 0$ represent? **D**

   Ⓐ circle   Ⓑ ellipse

   Ⓒ hyperbola   Ⓓ parabola

   Ⓔ none of the above

9. **MULTIPLE CHOICE** What point is the intersection of the graphs of $x^2 + y^2 = 41$ and $y = 3x - 7$? **B**

   Ⓐ $(-4, -4)$ Ⓑ $(4, 5)$ Ⓒ $(5, 8)$

   Ⓓ $(3, 2)$ Ⓔ $(-4, -19)$

In Exercises 10 and 11, choose the statement that is true about the given quantities.

(A) The quantity in column A is greater.

(B) The quantity in column B is greater.

(C) The two quantities are equal.

(D) The relationship cannot be determined from the given information.

	Column A	Column B	
10.	Distance between $(3, -2)$ and $(-5, 7)$	Distance between $(-8, -1)$ and $(0, 8)$	C
11.	Discriminant of $x^2 + y^2 - 6x + 1 = 0$	Discriminant of $3x^2 + y^2 - 2y + 5 = 0$	A

12. MULTI-STEP PROBLEM Let $(0, 0)$ represent a water fountain located in a city park. Each day Jane runs through the park along a path given by the equation $x^2 + y^2 - 200x - 52{,}500 = 0$ where x and y are measured in meters.

a. *Writing* What type of conic is Jane's path? How do you know?   a circle; $B^2 - 4AC < 0$, $B = 0$ and $A = C$

b. Write the equation of the conic in standard form. Then graph the equation.

$(x - 100)^2 + y^2 = 62{,}500$; See margin for graph.

c. After her run, Jane walks to the water fountain. If Jane stops running at $(-100, 150)$, how far must she walk for a drink of water?   about 180.28 m

13. MULTI-STEP PROBLEM The Mars Global Surveyor spacecraft followed an elliptical path with the center of Mars at one focus. The spacecraft's initial orbit had a low point of 262 kilometers above the northern hemisphere and a high point of 54,026 kilometers above the southern hemisphere. ▶ Source: NASA

a. *Writing* The radius of Mars is approximately 5400 kilometers. If $(0, 0)$ represents the center of Mars and the positive y-axis represents north, what are the coordinates of the other focus of the orbit? How do you know?

$(0, -53{,}764)$; The foci of an ellipse are equidistant from the vertices of the ellipse.

b. Write an equation for the spacecraft's initial orbit around Mars. $\dfrac{x^2}{336{,}470{,}012} + \dfrac{(y + 26{,}882)^2}{1{,}059{,}111{,}936} = 1$

c. In February, 1999, the spacecraft reached a nearly circular orbit, 410 kilometers above the surface of Mars. Write and graph an equation of the orbit. $x^2 + y^2 = 33{,}756{,}100$; See margin for graph.

14. MULTI-STEP PROBLEM Sara Peters is a mail carrier for a post office that receives mail for everyone living within a radius of 5 miles. Her route covers the portions of Anderson Road and Murphy Road that pass through this region.

a. Assume that the post office is located at the point $(0, 0)$. Write an equation for the circle that bounds the region where the mail is delivered. $x^2 + y^2 = 25$

b. Assuming Anderson Road follows one branch of a hyperbolic path given by $x^2 - y^2 - 4x - 23 = 0$, graph Anderson Road and the circular region where Sara delivers mail.   See margin.

c. *Writing* If Sara begins delivery on Anderson Road at the point $(-4, -3)$, where on Anderson Road does she end delivery? How do you know?

$(-4, 3)$; This point represents the second intersection of the circle and left branch of the hyperbola.

d. Sara finishes delivering on Anderson Road at the point where it intersects both the circular boundary and Murphy Road. At the intersection, she begins delivering on Murphy Road which is a straight road that cuts through the center of the circular region past the post office. Find the equation that represents Murphy Road. Where does Sara Peters end delivery on Murphy Road? $y = -\dfrac{3}{4}x$; $(4, -3)$

12b.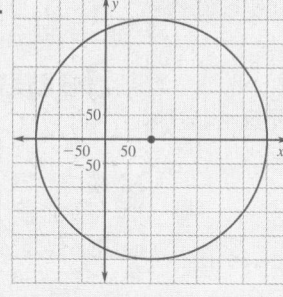

13c.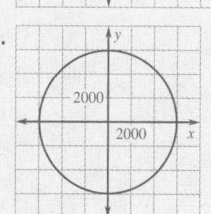

14b.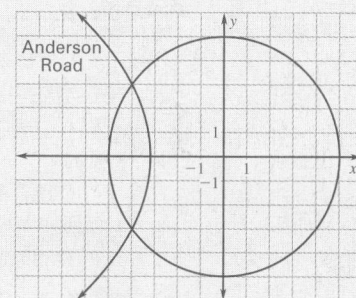

Chapter Standardized Test **647**

PLANNING THE CHAPTER

Sequences and Series

GOALS		NCTM	ITED	SAT9	Terra-Nova	Local
LESSON **11.1** *pp. 651–658*	**GOAL 1** Use and write sequences. **GOAL 2** Use summation notation to write series and find sums of series. **TECHNOLOGY ACTIVITY: 11.1** *Find the terms of a sequence, graph a sequence, and find the sum of a series on a graphing calculator.*	1, 2, 3, 8, 9, 10	RQGE	39	11	
11.2 *pp. 659–665*	**GOAL 1** Write rules for arithmetic sequences and find sums of arithmetic series. **GOAL 2** Use arithmetic series in real-life problems.	1, 2, 6, 8, 9, 10	MCWN, RQWN	39	11, 17	
11.3 *pp. 666–673*	**GOAL 1** Write rules for geometric sequences and find sums of geometric series. **GOAL 2** Use geometric sequences and series to model real-life quantities.	1, 2, 3, 8, 9, 10	RQGE	39	11, 14, 18	
11.4 *pp. 674–680*	**CONCEPT ACTIVITY: 11.4** *Investigate infinite geometric series.* **GOAL 1** Find sums of infinite series. **GOAL 2** Use infinite geometric series as models of real-life situations.	1, 2, 3, 8, 9, 10	RQGE	42	11, 14, 18	
11.5 *pp. 681–688*	**GOAL 1** Evaluate and write recursive rules for sequences. **GOAL 2** Use recursive rules to solve real-life problems. **TECHNOLOGY ACTIVITY: 11.5** *Evaluate a recursive rule on a spreadsheet.*	1, 2, 5, 6, 8, 9, 10	RQGE	1, 39	11	
Extension *pp. 689–690*	**GOAL** Use mathematical induction to prove statements about all positive integers.	1, 2, 7, 8, 9, 10	RQGE		14	

RESOURCES

CHAPTER RESOURCE BOOKLETS

CHAPTER SUPPORT

Tips for New Teachers	p. 1	Prerequisite Skills Review	p. 5
Parent Guide for Student Success	p. 3	Strategies for Reading Mathematics	p. 7

LESSON SUPPORT

	11.1	11.2	11.3	11.4	11.5
Lesson Plans (regular and block)	p. 9	p. 22	p. 35	p. 50	p. 65
Warm-Up Exercises and Daily Quiz	p. 11	p. 24	p. 37	p. 52	p. 67
Activity Support Masters				p. 53	
Lesson Openers	p. 12	p. 25	p. 38	p. 54	p. 68
Graphing Calculator Activities & Keystrokes	p. 13		p. 39	p. 55	p. 69
Practice (3 levels)	p. 14	p. 26	p. 41	p. 57	p. 70
Reteaching with Practice	p. 17	p. 29	p. 44	p. 60	p. 73
Quick Catch-Up for Absent Students	p. 19	p. 30	p. 46	p. 62	p. 75
Cooperative Learning Activities		p. 32			
Interdisciplinary Applications	p. 20		p. 47		p. 76
Real-Life Applications		p. 33		p. 63	
Math & History Applications					p. 77
Challenge: Skills and Applications	p. 21	p. 34	p. 48	p. 64	p. 78

REVIEW AND ASSESSMENT

Quizzes	pp. 49, 79	Alternative Assessment with Math Journal	p. 88
Chapter Review Games and Activities	p. 80	Project with Rubric	p. 90
Chapter Test (3 levels)	pp. 81–86	Cumulative Review	p. 92
SAT/ACT Chapter Test	p. 87	Resource Book Answers	p. A1

TRANSPARENCIES

	11.1	11.2	11.3	11.4	11.5
Warm-Up Exercises and Daily Quiz	p. 83	p. 84	p. 85	p. 86	p. 87
Alternative Lesson Opener Transparencies	p. 72	p. 73	p. 74	p. 75	p. 76
Examples/Standardized Test Practice	✓	✓	✓	✓	✓
Answer Transparencies	✓	✓	✓	✓	✓

TECHNOLOGY

- Electronic Teaching Tools
- Online Lesson Planner
- Internet Support
- Personal Student Tutor

- Test and Practice Generator
- Instant Replay: Video Review Games
- Electronic Lesson Presentations (Lesson 11.1)

ADDITIONAL RESOURCES

- Basic Skills Workbook: Diagnosis and Remediation
- Worked-Out Solution Key

- Resources in Spanish
- Standardized Test Practice Workbook
- Practice Workbook with Examples

PACING THE CHAPTER

REGULAR SCHEDULE

Day 1

11.1

STARTING OPTIONS
- Prereq. Skills Review
- Strategies for Reading
- Homework Check
- Warm-Up or Daily Quiz

TEACHING 11.1 OPTIONS
- Motivating the Lesson
- Les. Opener (Appl.)
- Graphing Calc. Activity
- Examples 1–6
- Technology Activity
- Closure Question
- Guided Practice Exs.

APPLY/HOMEWORK
- See Assignment Guide.
- See the CRB: Practice, Reteach, Apply, Extend

ASSESSMENT OPTIONS
- Checkpoint Exercises
- Daily Quiz (11.1)
- Stand. Test Practice

Day 2

11.2

STARTING OPTIONS
- Homework Check
- Warm-Up or Daily Quiz

TEACHING OPTIONS
- Motivating the Lesson
- Les. Opener (Visual)
- Examples 1–4
- Guided Practice Exs. 4–9

APPLY/HOMEWORK
- See Assignment Guide.
- See the CRB: Practice, Reteach, Apply, Extend

ASSESSMENT OPTIONS
- Checkpoint Exercises, pp. 660–661

Day 3

11.2 (cont.)

STARTING OPTIONS
- Homework Check

TEACHING OPTIONS
- Examples 5–7
- Closure Question
- Guided Practice Exs. 1–3, 10–14

APPLY/HOMEWORK
- See Assignment Guide.
- See the CRB: Practice, Reteach, Apply, Extend

ASSESSMENT OPTIONS
- Checkpoint Exercises, pp. 661–662
- Daily Quiz (11.2)
- Stand. Test Practice

Day 4

11.3

STARTING OPTIONS
- Homework Check
- Warm-Up or Daily Quiz

TEACHING OPTIONS
- Motivating the Lesson
- Les. Opener (Application)
- Graphing Calc. Activity
- Examples 1–3
- Guided Practice Exs. 1–2, 4–21

APPLY/HOMEWORK
- See Assignment Guide.
- See the CRB: Practice, Reteach, Apply, Extend

ASSESSMENT OPTIONS
- Checkpoint Exercises, p. 667

Day 5

11.3 (cont.)

STARTING OPTIONS
- Homework Check

TEACHING OPTIONS
- Examples 4–7
- Closure Question
- Guided Practice Exs. 3, 22–23

APPLY/HOMEWORK
- See Assignment Guide.
- See the CRB: Practice, Reteach, Apply, Extend

ASSESSMENT OPTIONS
- Checkpoint Exercises, pp. 667–669
- Daily Quiz (11.3)
- Stand. Test Practice
- Quiz (11.1–11.3)

Day 6

11.4

STARTING OPTIONS
- Homework Check
- Warm-Up or Daily Quiz

TEACHING OPTIONS
- Motivating the Lesson
- Concept Act. & Wksht.
- Les. Opener (Calculator)
- Graphing Calc. Activity
- Examples 1–4
- Closure Question
- Guided Practice Exs.

APPLY/HOMEWORK
- See Assignment Guide.
- See the CRB: Practice, Reteach, Apply, Extend

ASSESSMENT OPTIONS
- Checkpoint Exercises
- Daily Quiz (11.4)
- Stand. Test Practice

Day 9

Assess

DAY 9 START OPTIONS
- Homework Check

ASSESSMENT OPTIONS
- Chapter 11 Test
- SAT/ACT Ch. 11 Test
- Alternative Assessment

APPLY/HOMEWORK
- Skill Review, p. 700

Day 7

11.5

STARTING OPTIONS
- Homework Check
- Warm-Up or Daily Quiz

TEACHING OPTIONS
- Les. Opener (Activity)
- Graphing Calc. Activity
- Examples 1–5
- Technology Activity
- Closure Question
- Guided Practice Exs.

APPLY/HOMEWORK
- See Assignment Guide.
- See the CRB: Practice, Reteach, Apply, Extend

ASSESSMENT OPTIONS
- Checkpoint Exercises
- Daily Quiz (11.5)
- Stand. Test Practice
- Quiz (11.4–11.5)

Day 8

Review

DAY 8 START OPTIONS
- Homework Check

REVIEWING OPTIONS
- Chapter 11 Summary
- Chapter 11 Review
- Chapter Review Games and Activities

APPLY/HOMEWORK
- Chapter 11 Test (practice)
- Ch. Standardized Test (practice)

Day 1

Assess & 11.1
(Day 1 = Ch. 10 Day 6)

ASSESSMENT OPTIONS
- Chapter 10 Test
- SAT/ACT Ch. 10 Test
- Alternative Assessment

CH. 11 START OPTIONS
- Skill Review, p. 650
- Prereq. Skills Review
- Strategies for Reading

TEACHING 11.1 OPTIONS
- Warm-Up (Les. 11.1)
- Motivating the Lesson
- Les. Opener (Appl.)
- Graphing Calc. Activity
- Examples 1–6
- Technology Activity
- Closure Question
- Guided Practice Exs.

APPLY/HOMEWORK
- See Assignment Guide.
- See the CRB: Practice, Reteach, Apply, Extend

ASSESSMENT OPTIONS
- Checkpoint Exercises
- Daily Quiz (Les. 11.1)
- Stand. Test Practice

Day 2

11.2

DAY 2 START OPTIONS
- Homework Check
- Warm-Up or Daily Quiz

TEACHING 11.2 OPTIONS
- Motivating the Lesson
- Les. Opener (Visual)
- Examples 1–7
- Closure Question
- Guided Practice Exs.

APPLY/HOMEWORK
- See Assignment Guide.
- See the CRB: Practice, Reteach, Apply, Extend

ASSESSMENT OPTIONS
- Checkpoint Exercises
- Daily Quiz (Les. 11.2)
- Stand. Test Practice

Day 3

11.3

DAY 3 START OPTIONS
- Homework Check
- Warm-Up or Daily Quiz

TEACHING 11.3 OPTIONS
- Motivating the Lesson
- Les. Opener (Appl.)
- Graphing Calc. Activity
- Examples 1–7
- Closure Question
- Guided Practice Exs.

APPLY/HOMEWORK
- See Assignment Guide.
- See the CRB: Practice, Reteach, Apply, Extend

ASSESSMENT OPTIONS
- Checkpoint Exercises
- Daily Quiz (Les. 11.3)
- Stand. Test Practice
- Quiz (11.1–11.3)

Day 4

11.4 & 11.5

DAY 4 START OPTIONS
- Homework Check
- W-Up 11.4 or D. Quiz 11.3

TEACHING 11.4 OPTIONS
- Motivating the Lesson
- Concept Act. & Wksht.
- Les. Opener (Calc.)
- Graphing Calc. Activity
- Examples 1–4
- Closure Question
- Guided Practice Exs.

TEACHING 11.5 OPTIONS
- Warm-Up (Les. 11.5)
- Les. Opener (Activity)
- Technology Activities
- Examples 1–5
- Closure Question
- Guided Practice Exs.

APPLY/HOMEWORK
- See Assignment Guide.
- See the CRB: Practice, Reteach, Apply, Extend

ASSESSMENT OPTIONS
- Checkpoint Exercises
- Daily Quiz (11.4, 11.5)
- Stand. Test Practice
- Quiz (11.4–11.5)

Day 5

Review/Assess

DAY 5 START OPTIONS
- Homework Check

REVIEWING OPTIONS
- Chapter 11 Summary
- Chapter 11 Review
- Chapter Review Games and Activities
- Chapter 11 Test (practice)
- Ch. Standardized Test (practice)

ASSESSMENT OPTIONS
- Chapter 11 Test
- SAT/ACT Ch. 11 Test
- Alternative Assessment

APPLY/HOMEWORK
- Skill Review, p. 700

MEETING INDIVIDUAL NEEDS

BEFORE THE CHAPTER

The *Chapter 11 Resource Book* has the following materials to distribute and use before the chapter:

- **Parent Guide for Student Success**
- **Prerequisite Skills Review (pictured below)**
- **Strategies for Reading Mathematics**

PREREQUISITE SKILLS *Pages 5–6*

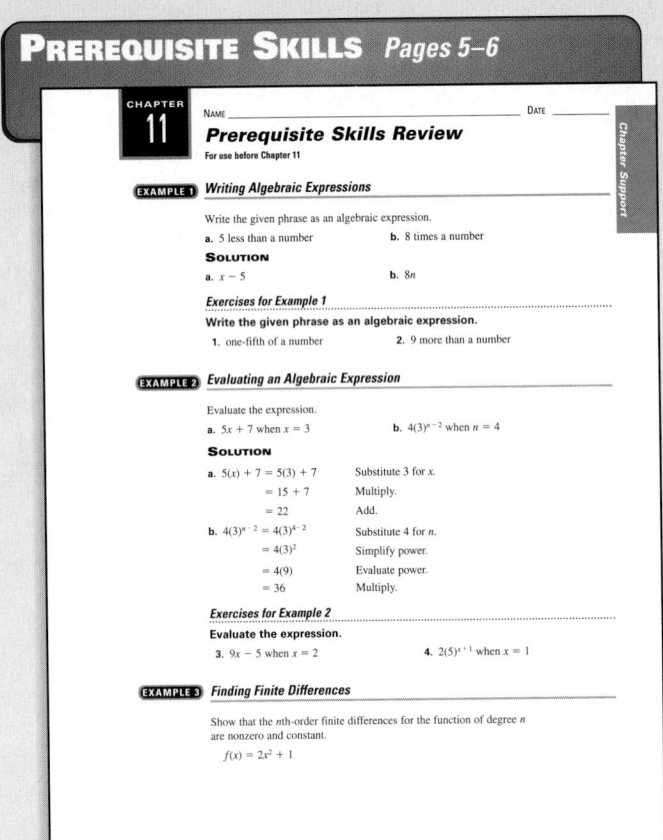

CHAPTER
11
NAME _____ DATE _____

Prerequisite Skills Review
For use before Chapter 11

Chapter Support

EXAMPLE 1 *Writing Algebraic Expressions*

Write the given phrase as an algebraic expression.

a. 5 less than a number   **b.** 8 times a number

SOLUTION

a. $x - 5$ **b.** $8n$

Exercises for Example 1

Write the given phrase as an algebraic expression.

1. one-fifth of a number   2. 9 more than a number

EXAMPLE 2 *Evaluating an Algebraic Expression*

Evaluate the expression.

a. $5x + 7$ when $x = 3$ **b.** $4(3)^{n-2}$ when $n = 4$

SOLUTION

a. $5(x) + 7 = 5(3) + 7$ Substitute 3 for x.
$= 15 + 7$ Multiply.
$= 22$ Add.

b. $4(3)^{n-2} = 4(3)^{4-2}$ Substitute 4 for n.
$= 4(3)^2$ Simplify power.
$= 4(9)$ Evaluate power.
$= 36$ Multiply.

Exercises for Example 2

Evaluate the expression.

3. $9x - 5$ when $x = 2$ 4. $2(5)^{x+1}$ when $x = 1$

EXAMPLE 3 *Finding Finite Differences*

Show that the nth-order finite differences for the function of degree n are nonzero and constant.

$f(x) = 2x^2 + 1$

PREREQUISITE SKILLS REVIEW These two pages support the Study Guide on page 650. They help students prepare for Chapter 11 by providing worked-out examples and practice for the following skills needed in the chapter:

- **Write algebraic expressions.**
- **Evaluate algebraic expressions.**
- **Find finite differences.**
- **Solve equations.**

TECHNOLOGY RESOURCE

Teachers can use the Real-Life Motivator clip for Chapter 11 from the video *Instant Replay: Video Review Games* as an alternative way to introduce the chapter.

DURING EACH LESSON

The *Chapter 11 Resource Book* has the following alternatives for introducing the lesson:

- **Lesson Openers (pictured below)**
- **Graphing Calculator Activities with Keystrokes**

LESSON OPENER *Page 38*

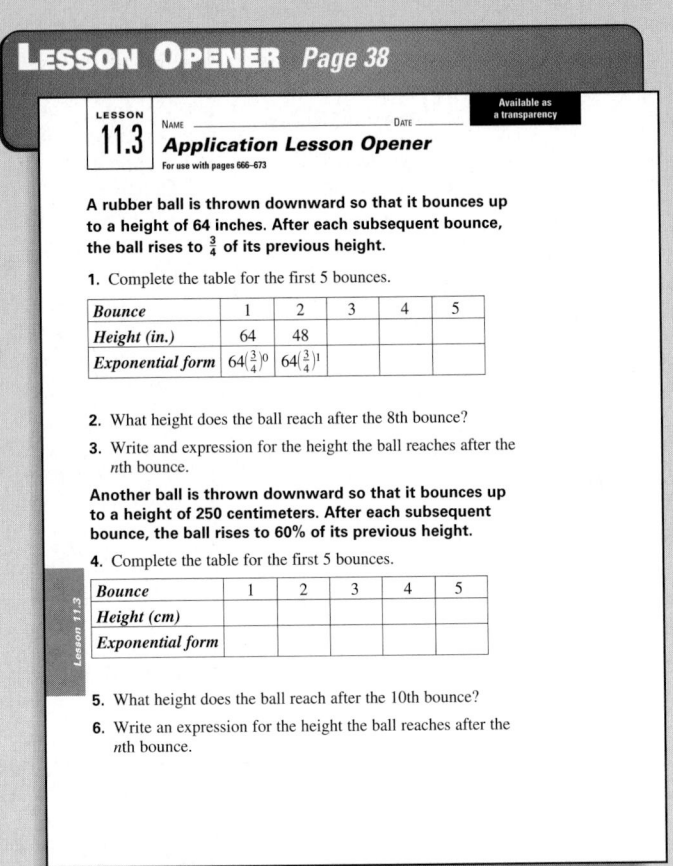

LESSON
11.3
NAME _____ DATE _____

Available as a transparency

Application Lesson Opener
For use with pages 666–673

A rubber ball is thrown downward so that it bounces up to a height of 64 inches. After each subsequent bounce, the ball rises to $\frac{3}{4}$ of its previous height.

1. Complete the table for the first 5 bounces.

Bounce	1	2	3	4	5
Height (in.)	64	48			
Exponential form	$64\left(\frac{3}{4}\right)^0$	$64\left(\frac{3}{4}\right)^1$			

2. What height does the ball reach after the 8th bounce?

3. Write and expression for the height the ball reaches after the nth bounce.

Another ball is thrown downward so that it bounces up to a height of 250 centimeters. After each subsequent bounce, the ball rises to 60% of its previous height.

4. Complete the table for the first 5 bounces.

Bounce	1	2	3	4	5
Height (cm)					
Exponential form					

5. What height does the ball reach after the 10th bounce?

6. Write an expression for the height the ball reaches after the nth bounce.

Lesson 11.3

APPLICATION LESSON OPENER This Lesson Opener provides an alternative way to start Lesson 11.3 in the form of a real-life application. Students are introduced to geometric sequences and series by determining the height of a bouncing ball.

The *Chapter 11 Resource Book* has a variety of materials to follow-up each lesson. They include the following:

- **Practice (3 levels)**
- **Reteaching with Practice**
- **Quick Catch-Up for Absent Students (pictured below)**
- **Challenge: Skills and Applications**
- **Interdisciplinary Applications**
- **Real-Life Applications**

The *Chapter 11 Resource Book* has the following review and assessment materials:

- **Quizzes**
- **Chapter Review Games and Activities**
- **Chapter Test (3 levels)**
- **SAT/ACT Chapter Test**
- **Alternative Assessment with Rubric and Math Journal**
- **Project with Rubric (pictured below)**
- **Cumulative Review**

QUICK CATCH-UP Page 19

LESSON 11.1

NAME _____ DATE _____

Quick Catch-Up for Absent Students
For use with pages 651–658

The items checked below were covered in class on (date missed) _____

Lesson 11.1: An Introduction to Sequences and Series

_____ **Goal 1:** Use and write sequences. (pp. 651–652)

Material Covered:
_____ Student Help: Look Back
_____ Example 1: Writing Terms of Sequences
_____ Student Help: Study Tip
_____ Example 2: Writing Rules for Sequences
_____ Student Help: Study Tip
_____ Example 3: Graphing a Sequence

Vocabulary:
terms of a sequence, p. 651   sequence, p. 651
finite sequence, p. 651   infinite sequence, p. 651

_____ **Goal 2:** Use summation notation to write series and find sums of series. (pp. 653–654)

Material Covered:
_____ Example 4: Writing Series with Summation Notation
_____ Example 5: Using Summation Notation
_____ Example 6: Using a Formula for a Sum.

Vocabulary:
series, p. 653   summation notation, p. 653
sigma notation, p. 653

Activity 11.1: Working with Sequences (p. 658)

_____ **Goal:** Use a graphing calculator to find the terms of a sequence, graph a sequence, and find the sum of a series.
_____ Student Help: Keystroke Help

_____ Other (specify) _____

Homework and Additional Learning Support
_____ Textbook (specify) pp. 655–657 _____

_____ *Reteaching with Practice* worksheet (specify exercises)_____
_____ *Personal Student Tutor* for Lesson 11.1

QUICK CATCH-UP FOR ABSENT STUDENTS You can use this form to let students know what they have missed when they've been absent from class. It allows you to quickly check off which Examples and other elements of Lesson 11.1 were covered on a given day and provides space for filling in the homework assignment.

 TECHNOLOGY RESOURCE

Students who have missed class can find keystroke and software help for the technology activities in Lessons 11.1 and 11.5 on the Internet at www.mcdougallittell.com.

PROJECT WITH RUBRIC Pages 90–91

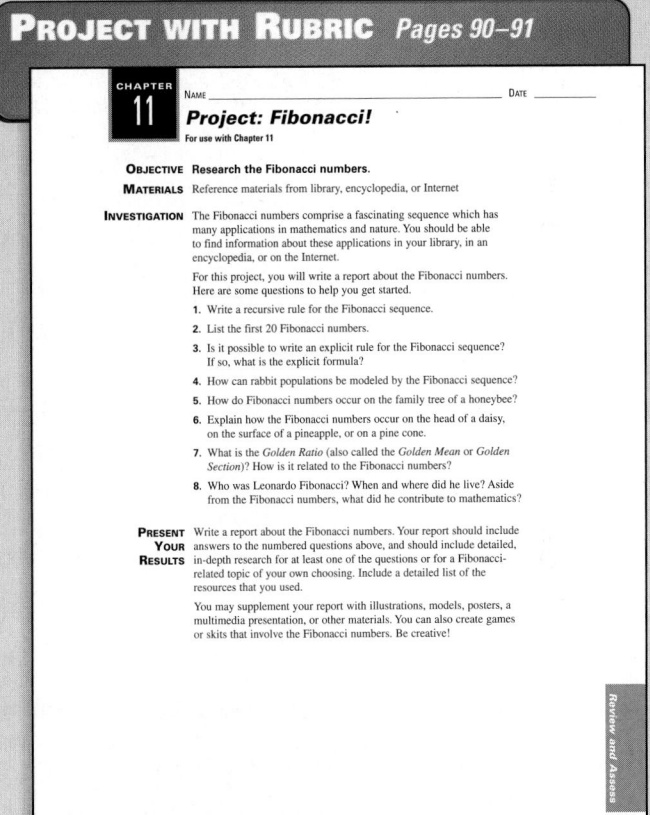

CHAPTER 11

NAME _____ DATE _____

Project: Fibonacci!
For use with Chapter 11

OBJECTIVE Research the Fibonacci numbers.

MATERIALS Reference materials from library, encyclopedia, or Internet

INVESTIGATION The Fibonacci numbers comprise a fascinating sequence which has many applications in mathematics and nature. You should be able to find information about these applications in your library, in an encyclopedia, or on the Internet.

For this project, you will write a report about the Fibonacci numbers. Here are some questions to help you get started.

1. Write a recursive rule for the Fibonacci sequence.
2. List the first 20 Fibonacci numbers.
3. Is it possible to write an explicit rule for the Fibonacci sequence? If so, what is the explicit formula?
4. How can rabbit populations be modeled by the Fibonacci sequence?
5. How do Fibonacci numbers occur on the family tree of a honeybee?
6. Explain how the Fibonacci numbers occur on the head of a daisy, on the surface of a pineapple, or on a pine cone.
7. What is the *Golden Ratio* (also called the *Golden Mean* or *Golden Section*)? How is it related to the Fibonacci numbers?
8. Who was Leonardo Fibonacci? When and where did he live? Aside from the Fibonacci numbers, what did he contribute to mathematics?

PRESENT YOUR RESULTS Write a report about the Fibonacci numbers. Your report should include answers to the numbered questions above, and should include detailed, in-depth research for at least one of the questions or for a Fibonacci-related topic of your own choosing. Include a detailed list of the resources that you used.

You may supplement your report with illustrations, models, posters, a multimedia presentation, or other materials. You can also create games or skits that involve the Fibonacci numbers. Be creative!

PROJECT WITH RUBRIC The Project for Chapter 11 provides students with the opportunity to apply the concepts they have learned in the chapter in a new way. In this project, students use what they have learned about series and sequences to research the Fibonacci numbers. Teacher's notes and a scoring rubric are provided on a separate sheet.

Students will use and write sequences, including arithmetic and geometric sequences and their rules. They will also graph sequences. Students will use summation notation to write a series and to find the sum of a series. They will write a rule for the nth term of arithmetic and geometric sequences and will find the nth term given either a term and the common difference or common ratio or two terms. They will find the sum of arithmetic and geometric sequences and series. Students will also find common ratios and will write repeating decimals as fractions. They will use infinite geometric series to model real-life situations. Students will complete the chapter by evaluating and writing recursive rules for arithmetic and geometric sequences and will use the rules to solve real-life problems. An extension will introduce students to the use of mathematical induction for proving statements about the set of positive integers.

APPLICATION NOTE
Benoit B. Mandelbrot is credited with the discovery of fractal geometry. He suggested that some curves, which are created by continuously applying the same rule so that the end of one step becomes the beginning of the next step, cannot be considered one-, two-, or three-dimensional. In fact their curves are so complex that they are considered to have a fractional dimension between 1 and 2. Consider a very irregular coastline for example. The curve of the coastline is so complex, that it is hard to imagine it as existing in just one dimension. Yet it does not exist in two dimensions. A figure produced by graphing in the complex plane is named the Mandelbrot set after the discoverer of fractal geometry.

Additional information about fractal geometry is available at **www.mcdougallittell.com.**

SEQUENCES AND SERIES

▶ *How is a fractal formed?*

648

APPLICATION: *Fractals*

Have you ever noticed how a fern leaf looks like a miniature version of the fern? When the parts of an object are similar to the whole object, the object is called *self-similar*. A *fractal* is a complex shape that basically looks the same at different levels of magnification and is generally self-similar.

Think & Discuss

The diagram shows the first three stages in the growth of a fractal plant.

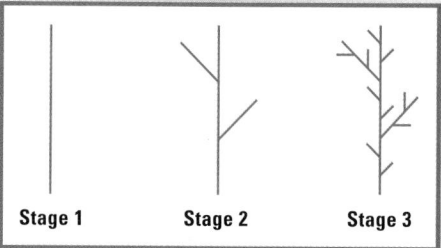

Stage 1   Stage 2   Stage 3

1. Draw the next stage in the growth process. **See margin.**
2. Describe in words how the fractal plant changes from one stage to the next. **See margin.**
3. What could you do to make the fractal plant look even more realistic? *Sample answer:* **Add new segments to the tips of existing segments at each stage.**

Learn More About It

You will find the number of branches in a fractal tree in Exercise 45 on p. 685.

 APPLICATION LINK Visit www.mcdougallittell.com for more information about fractals.

---

2. *Sample answer:* In each stage the lines that are there from the previous stage are divided into thirds by two lines, one drawn at a 60° angle on the left and another on the right.

ADDITIONAL RESOURCES
Another way to begin the chapter is to show the video clip of a real-life motivator for Chapter 11 on the *Instant Replay: Video Review Games.*

PROJECTS
A project covering Chapters 10–12 appears on pages 764–765 of the Student Edition. An additional project for Chapter 11 is available in the *Chapter 11 Resource Book,* p. 90.

TECHNOLOGY
 **Software**
- *Electronic Teaching Tools*
- *Online Lesson Planner*
- *Personal Student Tutor*
- *Test and Practice Generator*
- *Electronic Lesson Presentations* (Lesson 11.1)

Video
- *Instant Replay: Video Review Games*

Internet Connections
www.mcdougallittell.com
- **Application Links**
  649, 669, 687
- **Data Updates**
  669, 672
- **Student Help**
  656, 658, 664, 668, 676, 685, 688
- **Career Links**
  664, 671, 679, 683, 685
- **Extra Challenge**
  657, 665, 680, 686

1.
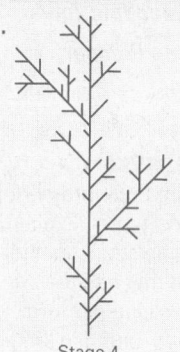
Stage 4

PREPARE

DIAGNOSTIC TOOLS

The **Skill Review** exercises can help you diagnose whether students have the following skills needed in Chapter 11:

• Evaluate algebraic expressions and write verbal phrases as algebraic expressions.

• Show that the *n*th-order finite difference for a function of degree *n* is nonzero and constant.

• Solve proportions and exponential equations.

The following resources are available for students who need additional help with these skills:

• Prerequisite Skills Review (*Chapter 11 Resource Book,* p. 5; *Warm-Up Transparencies,* p. 82)

• Reteaching with Practice (Chapter Resource Books for Lessons 1.2, 6.9, 8.6, 9.6)

• ⊞ *Personal Student Tutor*

ADDITIONAL RESOURCES

The following resources are provided to help you prepare for the upcoming chapter and customize review materials:

• *Chapter 11 Resource Book*
Tips for New Teachers (p. 1)
Parent Guide (p. 3)
Lesson Plans (every lesson)
Lesson Plans for Block Scheduling (every lesson)

• ⊞ *Electronic Teaching Tools*

• ⊞ *Online Lesson Planner*

• ⊞ *Test and Practice Generator*

ENGLISH LEARNERS

Inform students that the word *series* in the chapter title is plural. Explain that *series* is one of a small number of nouns in English that has the same plural form as singular form. (*Deer, moose, sheep,* and *fish* are some others.) So, instead of saying *two serieses,* you say *two series.*

7–9. See Additional Answers beginning on page AA1.

650

What's the chapter about?

Chapter 11 is about **sequences and series**. In Chapter 11 you'll learn

• how to find terms of sequences and write algebraic rules to define sequences.

• how to use summation notation and find sums of arithmetic and geometric series.

KEY VOCABULARY		
▶ **Review**	• finite sequence, p. 651	• arithmetic series, p. 661
• integers, p. 3	• infinite sequence, p. 651	• geometric sequence, p. 666
• finite differences, p. 380	• series, p. 653	• common ratio, p. 666
▶ **New**	• summation notation, p. 653	• geometric series, p. 668
• terms of a sequence, p. 651	• arithmetic sequence, p. 659	• explicit rule, p. 681
• sequence, p. 651	• common difference, p. 659	• recursive rule, p. 681

Are you ready for the chapter?

SKILL REVIEW Do these exercises to review key skills that you'll apply in this chapter. See the given **reference page** if there is something you don't understand.

STUDENT HELP

▶ **Study Tip**
"Student Help" boxes throughout the chapter give you study tips and tell you where to look for extra help in this book and on the Internet.

Write the given phrase as an algebraic expression. (Skills Review, p. 929)

1. 4 more than a number $n + 4$

2. 3 times a number $3n$

3. half of a number $\frac{n}{2}$

Evaluate the expression. (Review Example 3, p. 12)

4. $3x - 7$ when $x = 3$ 2

5. $\frac{x}{x + 1}$ when $x = 8$ $\frac{8}{9}$

6. $3(2)^{n-1}$ when $n = 4$ 24

Show that the *n*th-order finite differences for the function of degree *n* are nonzero and constant. (Review Example 2, p. 380)   7–9. See margin.

7. $f(x) = -3x^2 + 3$

8. $f(x) = x^3 + 2x^2$

9. $f(x) = x^4 - 5x + 1$

Solve the equation. (Review Example 3, p. 502; Example 4, p. 569)

10. $4^x = 16,384$ 7

11. $2^{x-1} = 32$ 6

12. $10 = \frac{5}{1-x}$ $\frac{1}{2}$

13. $24 = \frac{2}{1+x}$ $-\frac{11}{12}$

Here's a study strategy!

Learn by Teaching

Explain to a teacher, friend, or family member how to do an important skill in this chapter. Show an example and use words to describe your steps.

You can use a variation of this strategy when you are alone, too. Talk to yourself and explain your reasoning as you work toward an answer.

11.1

An Introduction to Sequences and Series

What you should learn

GOAL 1 Use and write sequences.

GOAL 2 Use summation notation to write series and find sums of series, as applied in **Example 6**.

Why you should learn it

▼ To model **real-life** situations, such as building a roof frame in **Exs. 65 and 66**.

GOAL 1 USING AND WRITING SEQUENCES

Saying that a collection of objects is listed "in sequence" means that the collection is ordered so that it has a first member, a second member, a third member, and so on. Below are two examples of sequences of numbers. The numbers in the sequences are called **terms**.

SEQUENCE 1:

3, 6, 9, 12, 15

SEQUENCE 2:

3, 6, 9, 12, 15, . . .

You can think of a **sequence** as a function whose domain is a set of consecutive integers. If a domain is not specified, it is understood that the domain starts with 1.

DOMAIN: 1 2 3 4 5   The domain gives the relative position of each term: 1st, 2nd, 3rd, and so on.

RANGE: 3 6 9 12 15   The range gives the terms of the sequence.

Sequence 1 above is a **finite sequence** because it has a last term. Sequence 2 is an **infinite sequence** because it continues without stopping. Both sequences have the general rule $a_n = 3n$ where a_n represents the nth term of the sequence. The general rule can also be written using function notation: $f(n) = 3n$.

EXAMPLE 1 *Writing Terms of Sequences*

Write the first six terms of the sequence.

a. $a_n = 2n + 3$ **b.** $f(n) = (-2)^{n-1}$

SOLUTION

a. $a_1 = 2(1) + 3 = 5$ 1st term

$a_2 = 2(2) + 3 = 7$ 2nd term

$a_3 = 2(3) + 3 = 9$ 3rd term

$a_4 = 2(4) + 3 = 11$ 4th term

$a_5 = 2(5) + 3 = 13$ 5th term

$a_6 = 2(6) + 3 = 15$ 6th term

b. $f(1) = (-2)^{1-1} = 1$ 1st term

$f(2) = (-2)^{2-1} = -2$ 2nd term

$f(3) = (-2)^{3-1} = 4$ 3rd term

$f(4) = (-2)^{4-1} = -8$ 4th term

$f(5) = (-2)^{5-1} = 16$ 5th term

$f(6) = (-2)^{6-1} = -32$ 6th term

STUDENT HELP

↪ **Look Back**
For help with evaluating expressions, see p. 12.

LESSON OPENER
APPLICATION
An alternative way to approach Lesson 11.1 is to use the Application Lesson Opener:

• Blackline Master (*Chapter 11 Resource Book,* p. 12)
• Transparency (p. 72)

MEETING INDIVIDUAL NEEDS
• *Chapter 11 Resource Book*
Prerequisite Skills Review (p. 5)
Practice Level A (p. 14)
Practice Level B (p. 15)
Practice Level C (p. 16)
Reteaching with Practice (p. 17)
Absent Student Catch-Up (p. 19)
Challenge (p. 21)
• *Resources in Spanish*
• *Personal Student Tutor*

NEW-TEACHER SUPPORT
See the Tips for New Teachers on pp. 1–2 of the *Chapter 11 Resource Book* for additional notes about Lesson 11.1.

WARM-UP EXERCISES

 *Transparency Available*

Find $f(3)$ for each function.

1. $f(x) = 3x^3$ 81
2. $f(x) = (-3)^{x-1}$ 9
3. $f(x) = 4x - 5$ 7

State the range of each function for the given domain.

4. $f(x) = 2x$; domain: whole numbers less than 5 0, 2, 4, 6, 8
5. $f(x) = 3^x$; domain: whole numbers 1, 3, 9, 27, 81, 243, ...

MOTIVATING THE LESSON
Have you ever heard the expression, "You are nothing but a number?" Some of the numbers that identify us, such as a social security number or our driver's license number, have no particular sequence of digits. Other identifiers, such as our DNA, contain particular sequences of information. Number sequences are the topic of this lesson.

EXTRA EXAMPLE 1
Write the first six terms of the sequence.
a. $a_n = 5 - n$ 4, 3, 2, 1, 0, −1
b. $f(n) = 5^n$ 5, 25, 125, 625, 3125, 15,625

EXTRA EXAMPLE 2
For each sequence, describe the pattern, write the next term, and write a rule for the nth term.
a. $\frac{2}{5}, \frac{2}{25}, \frac{2}{125}, \frac{2}{625}, \cdots$ $\frac{2}{5}, \frac{2}{5^2}, \frac{2}{5^3}, \frac{2}{5^4}, \ldots$; $a_5 = \frac{2}{3125}$; $a_n = \frac{2}{5^n}$
b. 3, 5, 7, 9, ... $2(1) + 1, 2(2) + 1, 2(3) + 1, 2(4) + 1, \ldots$; 11; $a_n = 2n + 1$

CHECKPOINT EXERCISES
For use after Example 1:
1. Write the first six terms of the sequence $a_n = \left(-\frac{1}{3}\right)^{n-1}$.

$1, -\frac{1}{3}, \frac{1}{9}, -\frac{1}{27}, \frac{1}{81}, -\frac{1}{243}$

For use after Example 2:
2. Describe the pattern, write the next term, and write a rule for the nth term of the sequence 3, 10, 29, 66, ... $1^3 + 2, 2^3 + 2, 3^3, + 2, 4^3 + 2, \ldots$; $a_5 = 5^3 + 2 = 127$; $a_n = n^3 + 2$

If the terms of a sequence have a recognizable pattern, then you may be able to write a rule for the nth term of the sequence.

STUDENT HELP
Study Tip
If you are given only the first several terms of a sequence, there is no *single* rule for the nth term. For instance, the sequence 2, 4, 8, ... can be given by $a_n = 2^n$ or $a_n = n^2 - n + 2$.

EXAMPLE 2   *Writing Rules for Sequences*

For each sequence, describe the pattern, write the next term, and write a rule for the nth term.

a. $-\frac{1}{3}, \frac{1}{9}, -\frac{1}{27}, \frac{1}{81}, \ldots$ b. 2, 6, 12, 20, ...

SOLUTION

a. You can write the terms as $\left(-\frac{1}{3}\right)^1, \left(-\frac{1}{3}\right)^2, \left(-\frac{1}{3}\right)^3, \left(-\frac{1}{3}\right)^4, \ldots$.

The next term is $a_5 = \left(-\frac{1}{3}\right)^5 = -\frac{1}{243}$. A rule for the nth term is $a_n = \left(-\frac{1}{3}\right)^n$.

b. You can write the terms as 1(2), 2(3), 3(4), 4(5), ...

The next term is $f(5) = 5(6) = 30$. A rule for the nth term is $f(n) = n(n + 1)$.

. . . .

You can graph a sequence by letting the horizontal axis represent the position numbers (the domain) and the vertical axis represent the terms (the range).

Retail Displays

EXAMPLE 3   *Graphing a Sequence*

You work in the produce department of a grocery store and are stacking oranges in the shape of a square pyramid with 10 layers.

a. Write a rule for the number of oranges in each layer.

b. Graph the sequence.

SOLUTION

a. The diagram below shows the first three layers of the stack. Let a_n represent the number of oranges in layer n.

STUDENT HELP
Study Tip
Although the plotted points in part (b) of Example 3 follow a curve, do *not* draw the curve because the sequence is defined only for integer values of n.

n	1	2	3
	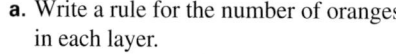		
a_n	$1 = 1^2$	$4 = 2^2$	$9 = 3^2$

From the diagram, you can see that $a_n = n^2$.

b. Plot the points (1, 1), (2, 4), (3, 9), ..., (10, 100). The graph is shown at the right.

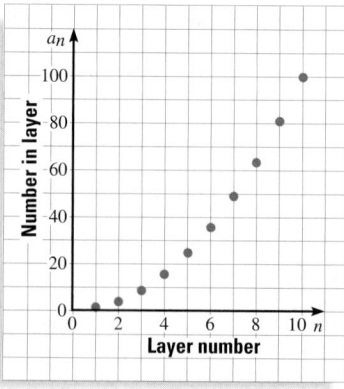

GOAL 2 USING SERIES

When the terms of a sequence are added, the resulting expression is a **series**. A series can be infinite or finite.

FINITE SEQUENCE	**INFINITE SEQUENCE**
3, 6, 9, 12, 15	3, 6, 9, 12, 15, . . .
FINITE SERIES	**INFINITE SERIES**
3 + 6 + 9 + 12 + 15	3 + 6 + 9 + 12 + 15 + · · ·

You can use **summation notation** to write a series. For example, for the finite series shown above, you can write

$$3 + 6 + 9 + 12 + 15 = \sum_{i=1}^{5} 3i$$

where i is the *index of summation*, 1 is the *lower limit of summation*, and 5 is the *upper limit of summation*. In this case the summation notation is read as "the sum from i equals 1 to 5 of $3i$." Summation notation is also called **sigma notation** because it uses the uppercase Greek letter *sigma,* written Σ.

Summation notation for an infinite series is similar to that for a finite series. For example, for the infinite series shown above, you can write:

$$3 + 6 + 9 + 12 + 15 + \cdots = \sum_{i=1}^{\infty} 3i$$

The infinity symbol, ∞, indicates that the series continues without end.

EXAMPLE 4 *Writing Series with Summation Notation*

Write each series with summation notation.

a. $5 + 10 + 15 + \cdots + 100$ **b.** $\frac{1}{2} + \frac{2}{3} + \frac{3}{4} + \frac{4}{5} + \cdots$

SOLUTION

a. Notice that the first term is 5(1), the second is 5(2), the third is 5(3), and the last is 5(20). So, the terms of the series can be written as:

$$a_i = 5i \text{ where } i = 1, 2, 3, \ldots, 20$$

▶ The summation notation for the series is $\sum_{i=1}^{20} 5i$.

b. Notice that for each term the denominator of the fraction is 1 more than the numerator. So, the terms of the series can be written as:

$$a_i = \frac{i}{i+1} \text{ where } i = 1, 2, 3, 4, \ldots$$

▶ The summation notation for the series is $\sum_{i=1}^{\infty} \frac{i}{i+1}$.

· · · · · · · · ·

The index of summation does not have to be i — any letter can be used. Also, the index does not have to begin at 1. For instance, in part (b) of Example 5 on the next page, the index begins at 3.

EXTRA EXAMPLE 3
Each swing of a pendulum is 3 in. shorter than the preceding swing. The first swing is 6 ft.
a. Write a rule for the length of each swing in inches.
$a_n = -3n + 75$
b. Graph the sequence.

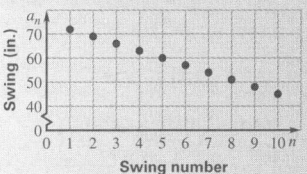

EXTRA EXAMPLE 4
Write each series with summation notation.

a. $4 + 8 + 12 + \ldots + 100$ $\sum_{i=1}^{25} 4i$

b. $2 + \frac{3}{4} + \frac{4}{9} + \frac{5}{16} + \ldots$ $\sum_{i=1}^{\infty} \frac{i+1}{i^2}$

✓ CHECKPOINT EXERCISES
For use after Example 3:
1. One year before a trip you begin making deposits to an account. The first month you deposit $40. For the next 11 months you deposit $3 more than the previous month.
  a. Write a rule for the amount of each monthly deposit.
  $a_n = 3n + 37$
  b. Graph the sequence.

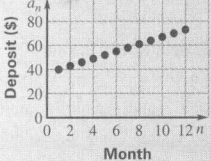

For use after Example 4:
2. Write the series $-6 + -12 + -18 + \ldots + -72$ with summation notation. $\sum_{i=1}^{12} -6i$

MATHEMATICAL REASONING
The sum of the series $1 + 2 + 3 + 4 + 5 + 6$ can be found using the formula $\frac{n(n+1)}{2} = \frac{6(7)}{2}$, or 3(7). How does this product represent the sum of the series? **There are three sums of seven in the series.**

EXTRA EXAMPLE 5
Find the sum of the series
$$\sum_{i=5}^{10} k^2 + 1. \quad 361$$

EXTRA EXAMPLE 6
How far does the pendulum in Extra Example 3 swing in 5 swings? $27\frac{1}{2}$ ft

CHECKPOINT EXERCISES
For use after Example 5:
1. Find the sum of the series
$$\sum_{i=1}^{5} 2i^2 - 1. \quad 105$$

For use after Example 6:
2. How much money do you have in the account described in the Checkpoint Exercise 1 after 12 months? $678

APPLICATION NOTE
EXAMPLE 6 By beginning with a square base for the oranges, the number of oranges in the base is maximized for a given perimeter of oranges.

FOCUS ON VOCABULARY
How is a series related to a sequence? Adding the terms of a sequence results in an expression called a series.

CLOSURE QUESTION
Describe the parts of sigma notation. The Greek letter Σ has an index of summation, i, a lower and an upper limit of summation, and a rule for the terms of the sequence to be added.

DAILY PUZZLER
You are writing a numerical code for the letters of the alphabet. The codes for the first seven letters of the alphabet are defined by this sequence: 2, 4, 8, 11, 33, 37, 148.... What are the codes for H, I, and J?
153, 765, 771

EXAMPLE 5 *Using Summation Notation*

Find the sum of the series.

a. $\displaystyle\sum_{i=1}^{6} 2i = 2(1) + 2(2) + 2(3) + 2(4) + 2(5) + 2(6)$

$\qquad\qquad = 2(1 + 2 + 3 + 4 + 5 + 6)$

$\qquad\qquad = 2(21)$

$\qquad\qquad = 42$

b. $\displaystyle\sum_{k=3}^{6} (2 + k^2) = (2 + 3^2) + (2 + 4^2) + (2 + 5^2) + (2 + 6^2)$

$\qquad\qquad\qquad = 11 + 18 + 27 + 38$

$\qquad\qquad\qquad = 94$

- - - - - - - - - -

STUDENT HELP

Study Tip
Notice that the first term in Example 5b occurs when $k = 3$ (not when $k = 1$) and that there are only 4 terms (not 6) in the series.

The sum of the terms of a finite sequence can be found by simply adding the terms. For sequences with many terms, however, adding the terms can be tedious. Formulas for finding the sum of the terms of three special types of sequences are given below.

CONCEPT SUMMARY   **FORMULAS FOR SPECIAL SERIES**

1. $\displaystyle\sum_{i=1}^{n} 1 = n$ **2.** $\displaystyle\sum_{i=1}^{n} i = \frac{n(n+1)}{2}$ **3.** $\displaystyle\sum_{i=1}^{n} i^2 = \frac{n(n+1)(2n+1)}{6}$

In words, the first formula gives the sum of n 1's. The second formula gives the sum of the positive integers from 1 to n. The third formula gives the sum of the squares of the positive integers from 1 to n.

EXAMPLE 6 *Using a Formula for a Sum*

FOCUS ON PEOPLE

THOMAS HALES, a mathematician at the University of Michigan, proved in 1998 that the arrangement of identical spheres illustrated in Examples 3 and 6 (using oranges) wastes less space than any other arrangement.

RETAIL DISPLAYS How many oranges are in the stack in Example 3?

SOLUTION

From Example 3 you know that the ith term of the series is given by $a_i = i^2$, where $i = 1, 2, 3, \ldots, 10$. Using summation notation and the third formula listed above, you can find the total number of oranges as follows.

$$\sum_{i=1}^{10} i^2 = 1^2 + 2^2 + \cdots + 10^2$$

$$= \frac{10(10+1)(2 \cdot 10 + 1)}{6}$$

$$= \frac{10(11)(21)}{6}$$

$$= 385$$

▶ There are 385 oranges in the stack. Check this by actually adding the number of oranges in each of the ten layers.

GUIDED PRACTICE

Vocabulary Check ✓

1. Explain the difference between a sequence and a series.

Concept Check ✓

2. Answer the following questions about the series $\sum\limits_{k=3}^{10} (k+2)$.

1. A sequence is a function whose domain is a set of consecutive integers. A series is the sum of a sequence.

    a. In words, how do you read the summation notation?   the sum from k equals 3 to 10 of $k+2$

    b. What is the index of summation? k

    c. What is the lower limit of summation?   3

    d. What is the upper limit of summation?   10

Skill Check ✓   **Write the first six terms of the sequence.**

2, 4, 6, 8, 10, 12   5, 4, 3, 2, 1, 0   4, 7, 10, 13, 16, 19   16, 32, 64, 128, 256, 512

3. $a_n = 2n$ **4.** $a_n = 6 - n$ **5.** $a_n = 3n + 1$ **6.** $f(n) = 2^{n+3}$

7. Find the sum of the series in Exercise 2.   68

8. 🌎 **STACKING** Find the total number of oranges in the stack in Example 3 if there are 12 layers.   650 oranges

PRACTICE AND APPLICATIONS

STUDENT HELP

▶ **Extra Practice**
to help you master
skills is on p. 955.

14. $-1, -8, -27, -64,$
$-125, -216$

15. 4, 7, 12, 19, 28, 39

16. 0, 1, 4, 9, 16, 25

17. $\frac{1}{2}, \frac{2}{3}, \frac{3}{4}, \frac{4}{5}, \frac{5}{6}, \frac{6}{7}$

18. $\frac{1}{2}, 1, \frac{3}{2}, 2, \frac{5}{2}, 3$

19. $\frac{3}{2}, 1, \frac{5}{6}, \frac{3}{4}, \frac{7}{10}, \frac{2}{3}$

20. $-3, -\frac{3}{2}, -1, -\frac{3}{4},$
$-\frac{3}{5}, -\frac{1}{2}$

WRITING TERMS Write the first six terms of the sequence.

2, 3, 4, 5, 6, 7   1, 4, 9, 16, 25, 36   2, 1, 0, -1, -2, -3 0, 7, 26, 63, 124, 215

9. $a_n = n + 1$ **10.** $a_n = n^2$ **11.** $a_n = 3 - n$ **12.** $a_n = n^3 - 1$

13. $a_n = (n+1)^2$ **14.** $a_n = (-n)^3$ **15.** $a_n = n^2 + 3$ **16.** $a_n = (n-1)^2$

4, 9, 16, 25, 36, 49

17. $f(n) = \dfrac{n}{n+1}$ **18.** $f(n) = \dfrac{n^2}{2n}$ **19.** $f(n) = \dfrac{n+2}{2n}$ **20.** $f(n) = \dfrac{3}{-n}$

WRITING RULES Write the next term in the sequence. Then write a rule for the nth term.

10,000; 10^{n-1} See margin.

21. 1, 3, 5, 7, . . .   9; $2n - 1$ **22.** 1, 10, 100, 1000, . . .   **23.** 2, -4, 8, -10, 14, . . .

24. -5, 10, -15, 20, . . .   **25.** $-\dfrac{1}{2}, -\dfrac{1}{4}, -\dfrac{1}{6}, -\dfrac{1}{8}, \ldots$ **26.** $\dfrac{1}{4}, \dfrac{2}{5}, \dfrac{3}{6}, \dfrac{4}{7}, \dfrac{5}{8}, \ldots$

-25; $(-1)^n(5n)$ 25, 26. See margin.

27. $\dfrac{1}{3}, \dfrac{2}{3}, \dfrac{3}{3}, \dfrac{4}{3}, \dfrac{5}{3}, \ldots$ 2; $\dfrac{n}{3}$ **28.** $\dfrac{1}{20}, \dfrac{2}{30}, \dfrac{3}{40}, \dfrac{4}{50}, \ldots$ **29.** 1.9, 2.7, 3.5, 4.3, 5.1, . . .

See margin.   5.9; $1.1 + 0.8n$

GRAPHING SEQUENCES Graph the sequence.   30–35. See margin.

30. 1, 4, 7, 10, . . . , 28   **31.** 3, 6, 12, 21, 33, 48   **32.** -1, -6, -11, -16, -21

33. 1, 4, 9, 16, 25, 36   **34.** $\dfrac{1}{9}, \dfrac{2}{8}, \dfrac{3}{7}, \dfrac{4}{6}, \ldots, \dfrac{9}{1}$ **35.** 3, -6, 9, -12, . . . , -36

WRITING SUMMATION NOTATION Write the series with summation notation.

36. $1 + 5 + 9 + 13 + 17$ $\sum\limits_{i=1}^{5} (4i - 3)$ **37.** $4 + 8 + 12 + 16 + 20$ $\sum\limits_{i=1}^{5} 4i$

38. $-3 + 3 + 9 + 15 + 21 + \cdots$ **39.** $1 - 2 + 3 - 4 + 5 - \cdots$ See margin.

See margin.

40. $-7 - 8 - 9 - 10 - 11$ $\sum\limits_{n=1}^{5} -(n + 6)$ **41.** $\dfrac{5}{6} + \dfrac{6}{7} + \dfrac{7}{8} + \dfrac{8}{9} + \cdots$ $\sum\limits_{n=5}^{\infty} \dfrac{n}{n+1}$

42. $1 + 0.1 + 0.01 + 0.001$ $\sum\limits_{k=1}^{4} \dfrac{1}{10^{k-1}}$ **43.** $1 + 4 + 9 + 16 + 25 + 36$ $\sum\limits_{i=1}^{6} i^2$

STUDENT HELP

▶ **HOMEWORK HELP**

Example 1: Exs. 9–20
Example 2: Exs. 21–29
Example 3: Exs. 30–35, 64–68
Example 4: Exs. 36–43
Example 5: Exs. 44–55
Example 6: Exs. 56–68

11.1 *An Introduction to Sequences and Series*   655

3 APPLY

○ **ASSIGNMENT GUIDE**

BASIC
Day 1: pp. 655–657 Exs. 10–32
even, 36–39, 44–48, 56–60,
64, 69–70, 73–83 odd

AVERAGE
Day 1: pp. 655–657 Exs. 10–34
even, 36–39, 44–54 even,
56–60, 64, 68–70,
73–83 odd

ADVANCED
Day 1: pp. 655–657 Exs. 10–34
even, 36–41, 44–56 even,
56–64, 67–72, 73–83 odd

BLOCK SCHEDULE
pp. 655–657 Exs. 10–34 even,
36–60, 64, 68–70, 73–83 odd
(with Ch. 10 Assess.)

EXERCISE LEVELS
Level A: *Easier*
9–35

Level B: *More Difficult*
36–70

Level C: *Most Difficult*
71–72

✔ **HOMEWORK CHECK**
To quickly check student understanding of key concepts, go
over the following exercises:
Exs. 14, 22, 30, 36, 44, 58. See
also the Daily Homework Quiz:

• Blackline Master (*Chapter 11
Resource Book*, p. 24)

• 📄 Transparency (p. 84)

23. -16; $a_n = 3n - 1$ if n is odd or $2 - 3n$ if n is even

25. $\dfrac{-1}{10}; \dfrac{-1}{2n}$

26. $\dfrac{2}{3}; \dfrac{n}{n+3}$

28. $\dfrac{5}{60}; \dfrac{n}{10(n+1)}$

38. $\sum\limits_{i=1}^{\infty} (6i - 9)$

39. $\sum\limits_{k=1}^{\infty} (-1)^{k-1}k$

30–35. See Additional Answers beginning on page AA1.

! **COMMON ERROR**
EXERCISE 40 If students describe the series with $\sum_{i=-7}^{-11} i$, remind them that the lower and upper bound are not the first and last terms but rather the number of the terms.

STUDENT HELP NOTES

→ **Homework Help** Students can find help for Exs. 56–63 at **www.mcdougallittell.com.** The information can be printed out for students who don't have access to the Internet.

APPLICATION NOTE
EXERCISE 67 If students repeat the series of moves for 1, 2, and 3 rings, they may discover that the moves required for n rings is the sum of the previous terms plus n.

71a. true; $\sum_{i=1}^{n} ka_i = ka_1 + ka_2 + \ldots +$
$ka_n = k(a_1 + a_2 + \ldots + a_n) =$
$k\sum_{i=1}^{n} a_i$

71b. true; $\sum_{i=1}^{n} (a_i + b_i) = (a_1 + b_1) +$
$(a_2 + b_2) + \ldots + (a_n + b_n) =$
$(a_1 + a_2 + \ldots + a_n) +$
$(b_1 + b_2 + \ldots + b_n) = \sum_{i=1}^{n} a_i + \sum_{i=1}^{n} b_i$

ADDITIONAL PRACTICE AND RETEACHING

For Lesson 11.1:
• Practice Levels A, B, and C (*Chapter 11 Resource Book*, p. 14)
• Reteaching with Practice (*Chapter 11 Resource Book*, p. 17)
• ▦ See Lesson 11.1 of the *Personal Student Tutor*

For more Mixed Review:
• ▦ Search the *Test and Practice Generator* for key words or specific lessons.

656

STUDENT HELP

▸ **HOMEWORK HELP**
Visit our Web site www.mcdougallittell.com for help with Exs. 56–63.

USING SUMMATION NOTATION Find the sum of the series.

44. $\sum_{i=1}^{6} 3i$ 63

45. $\sum_{i=0}^{5} 12i$ 180

46. $\sum_{n=0}^{4} n^2$ 30

47. $\sum_{n=1}^{3} 4n^3$ 144

48. $\sum_{k=1}^{5} (k^2 - 1)$ 50

49. $\sum_{n=0}^{4} (2n^2 + 1)$ 65

50. $\sum_{k=1}^{4} k(k + 2)$ 50

51. $\sum_{n=2}^{10} \frac{2}{n}$ $\frac{4861}{1260} \approx 3.858$

52. $\sum_{n=2}^{12} \frac{1}{n-1}$ $\frac{83,711}{27,720} \approx 3.0199$

53. $\sum_{n=1}^{5} \frac{n}{n+1}$ 3.55

54. $\sum_{i=2}^{6} \frac{i}{i-1}$ $\frac{437}{60} \approx 7.283$

55. $\sum_{n=1}^{\infty} \left(\frac{n}{n^2} - \frac{1}{n}\right)$ 0

USING FORMULAS Use one of the formulas for special series to find the sum of the series.

56. $\sum_{i=1}^{42} 1$ 42

57. $\sum_{n=1}^{5} n$ 15

58. $\sum_{i=1}^{18} i$ 171

59. $\sum_{k=1}^{20} k$ 210

60. $\sum_{n=1}^{6} n^2$ 91

61. $\sum_{i=1}^{10} i^2$ 385

62. $\sum_{i=1}^{12} i^2$ 650

63. $\sum_{k=1}^{35} k^2$ 14,910

64. **GEOMETRY** ▸ **CONNECTION** The degree measurement d_n in each angle at the tips of the six n-pointed stars shown at the right is given by:

$$d_n = \frac{180(n - 4)}{n}, n \geq 5$$

Write the first six terms of the sequence.
36°, 60°, 77.1°, 90°, 100°, 108°

 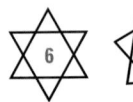

🌐 **CARPENTRY** In Exercises 65 and 66, use the following information.
The diagram shows part of a roof frame. The length (in feet) of each vertical support is given below the support. These lengths form an arithmetic sequence from each end to the middle.

65. Find the total length of the vertical supports from one end to the middle.   15 ft

66. Use your result from Exercise 65 to find the total length of the vertical supports from end to end.   25 ft

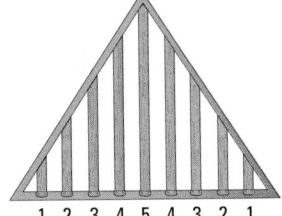
1 2 3 4 5 4 3 2 1

67. 🌐 **TOWER OF HANOI** In the puzzle called the Tower of Hanoi, the object is to use a series of moves to take the rings from one peg and stack them in order on another peg. A move consists of moving exactly one ring, and no ring may be placed on top of a smaller ring. The minimum number of moves required to move n rings is 1 for 1 ring, 3 for 2 rings, 7 for 3 rings, 15 for 4 rings, and 31 for 5 rings. Find a formula for the sequence. What is the minimum number of moves required to move 6 rings? $2^n - 1$; 63

68. 🌐 **PYRAMID STACK** Suppose you are stacking tennis balls in a pyramid as a display at a sports store. If the base is an equilateral triangle, then the number a_n of balls per layer would be $a_n = \frac{1}{2}n^2 + \frac{1}{2}n$ where $n = 1$ represents the top layer. How many balls are in the fifth layer? How many balls are in a stack with 5 layers?   15; 35

FOCUS ON APPLICATIONS

▸ **REAL LIFE** **TOWER OF HANOI** The puzzle was first described in print by the French mathematician Edouard Lucas in 1883 in his four volume book on recreational mathematics.

71c–d, 80. See Additional Answers beginning on page AA1.

QUANTITATIVE COMPARISON In Exercises 69 and 70, choose the statement that is true about the given quantities.

(A) The quantity in column A is greater.

(B) The quantity in column B is greater.

(C) The two quantities are equal.

(D) The relationship cannot be determined from the given information.

	Column A	Column B	
69.	The fifth term of the sequence $a_n = n^2 + 1$	$\sum_{n=1}^{5} (n^2 + 1)$	B
70.	The first term of the sequence $a_n = 5 - n$	$\sum_{n=4}^{8} (5 - n)$	A

★ **Challenge**

71. LOGICAL REASONING Tell whether the statement about summation notation is *true* or *false*. If the statement is true, prove it. If the statement is false, give a counterexample. **See margin.**

a. $\sum_{i=1}^{n} ka_i = k \sum_{i=1}^{n} a_i$

b. $\sum_{i=1}^{n} (a_i + b_i) = \sum_{i=1}^{n} a_i + \sum_{i=1}^{n} b_i$

c. $\sum_{i=1}^{n} a_i b_i = \left(\sum_{i=1}^{n} a_i \right)\left(\sum_{i=1}^{n} b_i \right)$

d. $\sum_{i=1}^{n} (a_i)^k = \left(\sum_{i=1}^{n} a_i \right)^k$

EXTRA CHALLENGE
www.mcdougallittell.com

72. Using the true statements from Exercise 71 and the special formulas from page 654, find a formula for the number of balls in n layers of the pyramid in Exercise 68. $\dfrac{n(n + 1)(n + 2)}{6}$

MIXED REVIEW

SOLVING EQUATIONS Solve the equation. Check your solution. (Review 1.3 for 11.2)

73. $17 = 3x + 5$ 4

74. $18 = -7 + x$ 25

75. $15 = -1 + 8x$ 2

76. $9 = 4 - 5x$ -1

77. $5 = 6 - 2x$ $\dfrac{1}{2}$

78. $24 = 10 + 7x$ 2

FINDING EXPONENTIAL MODELS Use the table of values to draw a scatter plot of ln y versus x. Then find an exponential model for the data. (Review 8.7)

79.

x	1	2	3	4	5	6	7	8	9
y	5	10	20	40	80	160	320	640	1280

$y = (2.5)2^x$; **See margin for graph.**

80.

x	1	2	3	4	5	6	7	8
y	3.2	9.6	28.8	86.4	259.2	777.6	2332.8	6998.4

$y = \left(\dfrac{16}{15} \right)3^x$; **See margin for graph.**

FINDING THE DISTANCE Find the distance between the points. (Review 10.1)

81. $(0, 0), (-4, -6)$

82. $(1, 4), (-3, -9)$

83. $(5, 2), (-1, 8)$

84. $(9, -1), (2, 9)$

85. $(3, -3), (11, -4)$

86. $(10, 30), (40, -20)$

81. $2\sqrt{13} \approx 7.211$
82. $\sqrt{185} \approx 13.601$
83. $6\sqrt{2} \approx 8.485$
84. $\sqrt{149} \approx 12.207$
85. $\sqrt{65} \approx 8.062$
86. $10\sqrt{34} \approx 58.310$

Additional Test Preparation *Sample answer:*

1. Each term is 3 more than the previous term in each sequence. Each term in the first sequence is one less than three times the term number. Each term in the second sequence is three times the term number.

2. The sequence 3, 6, 9, 12,18, 24, ... describes the numbers divisible by 3. Since $333 \times 3 = 999$, there are $333 - 5$ or 328 numbers between 20 and 1000 that are divisible by 3.

DAILY HOMEWORK QUIZ

Transparency Available

1. Write the first six terms of the sequence.

a. $a_n = 3n - 1$ 2, 5, 8, 11, 14, 17

b. $f(n) = (-2)^n$
   $-2, 4, -8, 16, -32, 64$

2. Write the next term in the sequence 3, 5, 7, 9, ... Then write a rule for the nth term.
   11, $a_n = 2n + 1$

3. Write the series $3 + 6 + 9 + 12 + ... + 75$ with summation notation.
   $\sum_{i=1}^{25} 3i$

4. Find the sum of the series.

a. $\sum_{i=2}^{6} \dfrac{i}{2}$ 10   **b.** $\sum_{k=1}^{5} k^2$ 55

EXTRA CHALLENGE NOTE

Challenge problems for Lesson 11.1 are available in **blackline** format in the *Chapter 11 Resource Book,* p. 21 and at **www.mcdougallittell.com.**

ADDITIONAL TEST PREPARATION

1. WRITING How are the sequences 2, 5, 8, 11, ... and 3, 6, 9, 12, ... alike? How are they different? **See margin.**

2. OPEN ENDED How could you use a sequence to determine how many numbers between 20 and 1000 are divisible by 3? **See margin.**

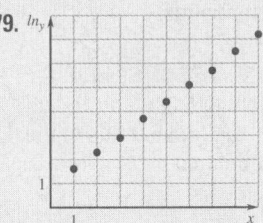
79.

80. **See Additional Answers beginning on page AA1.**

1 Planning the Activity

PURPOSE
To use a graphing calculator to graph a sequence and to find the terms and the sum of a sequence.

MATERIALS
- 1 graphing calculator per group
- Keystroke blackline (*Chapter 11 Resource Book*, p. 13)

PACING
- Activity — 25 min

▶ LINK TO LESSON
The ranges of x and y that are set in the viewing window correspond to the ranges on the n and a_n axes as seen in Example 3.

2 Managing the Activity

COOPERATIVE LEARNING
Have students work in groups of 4 using one calculator. Each group member should complete the keystrokes for one of the steps in the example. Members should take turns performing the steps as they complete the exercises.

CLASSROOM MANAGEMENT
As students complete Step 4 in each exercise, suggest that they check their results by calculating each sum.

3 Closing the Activity

★ KEY DISCOVERY
The graph of a sequence is a set of discrete points.

ACTIVITY ASSESSMENT
Use a graphing calculator to find the sum of the first 6 terms of the sequence defined by $a_n = 4n - 3$. 66

1–9. See Additional Answers beginning on page AA1.

● ACTIVITY 11.1

Using Technology

Working with Sequences

You can use a graphing calculator to find the terms of a sequence, graph a sequence, and find the sum of a series.

▶ EXAMPLE
Use a graphing calculator to perform the following.
- Find the first eight terms of the sequence $a_n = 3n - 1$.
- Graph the sequence.
- Find the sum of the first eight terms of the sequence.

▶ SOLUTION

STUDENT HELP

INTERNET
KEYSTROKE HELP
See keystrokes for several models of calculators at www.mcdougallittell.com

1. 5, 7, 9, 11, 13, 15, 17, 19, 21, 23; 140

2. 8, 12, 16, 20, 24, 28, 32, 36, 40, 44; 260

3. 48, 46, 44, 42, 40, 38, 36, 34, 32, 30; 390

4. 1, 2, 4, 8, 16, 32, 64, 128, 256, 512; 1023

5. $\frac{1}{2}, \frac{1}{4}, \frac{1}{8}, \frac{1}{16}, \frac{1}{32}, \frac{1}{64}, \frac{1}{128}, \frac{1}{256}, \frac{1}{512}, \frac{1}{1024}; \frac{1023}{1024}$

6. 3, 6, 11, 18, 27, 38, 51, 66, 83, 102; 405

7. $\frac{3}{4}$, 3, 12, 48, 192, 768, 3072, 12,288, 49,152, 196,608; 262,143.75

8. 12, 102, 1002, 10,002, 100,002, 1,000,002, 10,000,002, 100,000,002, 1,000,000,002, 10,000,000,002; 11,111,111,130

9. $\frac{4}{3}, 8\frac{1}{3}, 27\frac{1}{3}, 64\frac{1}{3}, 125\frac{1}{3}, 216\frac{1}{3}, 343\frac{1}{3}, 512\frac{1}{3}, 729\frac{1}{3}, 1000\frac{1}{3}; 3028\frac{1}{3}$

① Put the graphing calculator in *Sequence* mode and *Dot* mode. Enter the sequence. Note that the calculator uses $u(n)$ rather than a_n.

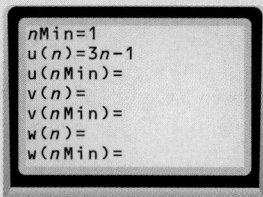

② Use the *Table* feature to view the terms of the sequence. The first eight terms are 2, 5, 8, 11, 14, 17, 20, and 23.

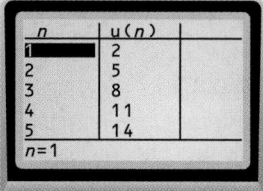

③ Set the viewing window so that $1 \leq n \leq 8$, $1 \leq x \leq 8$, and $0 \leq y \leq 25$. Graph the sequence. Use the *Trace* feature to view the terms of the sequence.

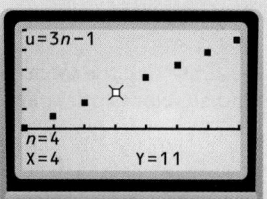

④ Use the *Summation* feature to find the sum of the first eight terms of the sequence. The screen shows that the sum is 100.

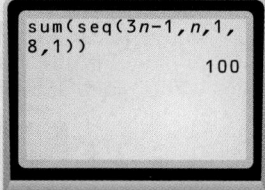

▶ EXERCISES

Use a graphing calculator to (a) find the first ten terms of the sequence, (b) graph the sequence, and (c) find the sum of the first ten terms of the sequence.

1–9. See margin for graphs.

1. $a_n = 2n + 3$

2. $a_n = 4(n + 1)$

3. $a_n = 50 - 2n$

4. $a_n = 2^{n-1}$

5. $a_n = \left(\frac{1}{2}\right)^n$

6. $a_n = 2 + n^2$

7. $a_n = 3 \cdot 4^{n-2}$

8. $a_n = 10^n + 2$

9. $a_n = \frac{1}{3} + n^3$

11.2
Arithmetic Sequences and Series

What you should learn

GOAL 1 Write rules for arithmetic sequences and find sums of arithmetic series.

GOAL 2 Use arithmetic sequences and series in **real-life** problems, such as finding the number of cells in a honeycomb in **Ex. 57**.

Why you should learn it

▼ To solve **real-life** problems, such as finding the number of seats in a concert hall in **Example 7**.

GOAL 1   USING ARITHMETIC SEQUENCES AND SERIES

In an **arithmetic sequence**, the difference between consecutive terms is constant. The constant difference is called the **common difference** and is denoted by d.

EXAMPLE 1   *Identifying Arithmetic Sequences*

Decide whether each sequence is arithmetic.

a. $-3, 1, 5, 9, 13, \ldots$ **b.** $2, 5, 10, 17, 26, \ldots$

SOLUTION

To decide whether a sequence is arithmetic, find the differences of consecutive terms.

a. $a_2 - a_1 = 1 - (-3) = 4$ **b.** $a_2 - a_1 = 5 - 2 = 3$

$a_3 - a_2 = 5 - 1 = 4$ $a_3 - a_2 = 10 - 5 = 5$

$a_4 - a_3 = 9 - 5 = 4$ $a_4 - a_3 = 17 - 10 = 7$

$a_5 - a_4 = 13 - 9 = 4$ $a_5 - a_4 = 26 - 17 = 9$

Each difference is 4, so the sequence is arithmetic.   The differences are not constant, so the sequence is not arithmetic.

RULE FOR AN ARITHMETIC SEQUENCE

The nth term of an arithmetic sequence with first term a_1 and common difference d is given by:

$$a_n = a_1 + (n - 1)d$$

EXAMPLE 2   *Writing a Rule for the nth Term*

Write a rule for the nth term of the sequence 50, 44, 38, 32, ... Then find a_{20}.

SOLUTION

The sequence is arithmetic with first term $a_1 = 50$ and common difference $d = 44 - 50 = -6$. So, a rule for the nth term is:

$a_n = a_1 + (n - 1)d$ **Write general rule.**

$ = 50 + (n - 1)(-6)$ **Substitute for a_1 and d.**

$ = 56 - 6n$ **Simplify.**

The 20th term is $a_{20} = 56 - 6(20) = -64$.

11.2 *Arithmetic Sequences and Series*   **659**

---

1 PLAN

PACING
Basic: 2 days
Average: 2 days
Advanced: 2 days
Block Schedule: 1 block

LESSON OPENER
VISUAL APPROACH
An alternative way to approach Lesson 11.2 is to use the Visual Approach Lesson Opener:

• Blackline Master (*Chapter 11 Resource Book*, p. 25)
• Transparency (p. 73)

MEETING INDIVIDUAL NEEDS
• *Chapter 11 Resource Book*
Prerequisite Skills Review (p. 5)
Practice Level A (p. 26)
Practice Level B (p. 27)
Practice Level C (p. 28)
Reteaching with Practice (p. 29)
Absent Student Catch-Up (p. 31)
Challenge (p. 34)

• *Resources in Spanish*

• Personal Student Tutor

NEW-TEACHER SUPPORT
See the Tips for New Teachers on pp. 1–2 of the *Chapter 11 Resource Book* for additional notes about Lesson 11.2.

WARM-UP EXERCISES

Transparency Available

Write the next term of the sequence. Then write a rule for the nth term.

1. 2, 3, 4, 5, …   6; $a_n = n + 1$
2. 0, 1, 2, 3   4; $a_n = n - 1$
3. $\dfrac{3}{2}, \dfrac{3}{4}, \dfrac{3}{8}, \dfrac{3}{16}, \ldots$ $\dfrac{3}{32}; a_n = \dfrac{3}{2^n}$

Find the sum of the series.

4. $1 + 2 + 3 + 4 + 5 + 6$ 21
5. $\displaystyle\sum_{i=1}^{4} 3i^2$ 90

A team is selling hats to raise money for summer camp. The hats will sell for $12 each. $8 from each sale covers the team's cost of a hat. For each hat the team sells, its camp fund grows by $4. The possible values of the fund can be calculated using an arithmetic series, which you will learn about in this lesson.

EXTRA EXAMPLE 1
Decide whether each sequence is arithmetic.
a. $-10, -6, -2, 0, 2, 6, 10, \ldots$ No
b. $5, 11, 17, 23, 29, \ldots$ Yes

EXTRA EXAMPLE 2
Write a rule for the nth term of the sequence 32, 47, 62, 77, ... Then find a_{12}. $a_n = 17 + 15n$; 197

EXTRA EXAMPLE 3
One term of an arithmetic sequence is $a_8 = 50$. The common difference is 0.25.
a. Write a rule for the nth term.
  $a_n = 48 + 0.25n$
b. Graph the sequence.

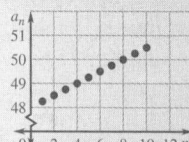

CHECKPOINT EXERCISES
For use after Examples 1 and 2:
1. Explain why the sequence 17, 11, 5, -1, -7, ... is arithmetic. Then write a rule for the nth term.   The difference between consecutive terms is constant; $a_n = 23 - 6n$

For use after Example 3:
2. In an arithmetic sequence $a_{20} = -111$. The common difference is $d = -6$. Write a rule for the nth term. $a_n = 9 - 6n$

EXAMPLE 3   *Finding the nth Term Given a Term and the Common Difference*

One term of an arithmetic sequence is $a_{13} = 30$. The common difference is $d = \frac{3}{2}$.

a. Write a rule for the nth term.   **b.** Graph the sequence.

SOLUTION

a. Begin by finding the first term as follows.

$$a_n = a_1 + (n-1)d \qquad \text{Write rule for } n\text{th term.}$$
$$a_{13} = a_1 + (13-1)d \qquad \text{Substitute 13 for } n.$$
$$30 = a_1 + 12\left(\frac{3}{2}\right) \qquad \text{Substitute for } a_{13} \text{ and } d.$$
$$12 = a_1 \qquad \text{Solve for } a_1.$$

So, a rule for the nth term is:

$$a_n = a_1 + (n-1)d \qquad \text{Write general rule.}$$
$$= 12 + (n-1)\frac{3}{2} \qquad \text{Substitute for } a_1 \text{ and } d.$$
$$= \frac{21}{2} + \frac{3}{2}n \qquad \text{Simplify.}$$

b. The graph is shown at the right. Notice that the points lie on a line. This is true for *any* arithmetic sequence.

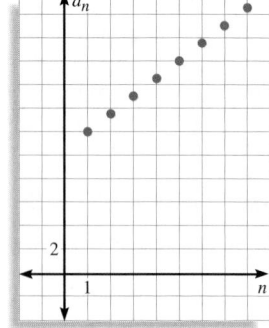

EXAMPLE 4   *Finding the nth Term Given Two Terms*

Two terms of an arithmetic sequence are $a_6 = 10$ and $a_{21} = 55$.

a. Find a rule for the nth term.   **b.** Find the value of n for which $a_n = 40$.

SOLUTION

a. *Write* a system of equations using $a_n = a_1 + (n-1)d$ and substituting 21 for n (Equation 1) and then 6 for n (Equation 2).

$$a_{21} = a_1 + (21-1)d \qquad \Longrightarrow \qquad 55 = a_1 + 20d \qquad \text{Equation 1}$$
$$a_6 = a_1 + (6-1)d \qquad \Longrightarrow \qquad 10 = a_1 + 5d \qquad \text{Equation 2}$$

Solve the system.
$$45 = 15d \qquad \text{Subtract equations.}$$
$$3 = d \qquad \text{Solve for } d.$$
$$55 = a_1 + 20(3) \qquad \text{Substitute for } d.$$
$$-5 = a_1 \qquad \text{Solve for } a_1.$$

Find a rule for a_n.
$$a_n = a_1 + (n-1)d \qquad \text{Write general rule.}$$
$$a_n = -5 + (n-1)3 \qquad \text{Substitute for } a_1 \text{ and } d.$$
$$a_n = -8 + 3n \qquad \text{Simplify.}$$

b. $a_n = -8 + 3n$ Use the rule for a_n from part (a).
$$40 = -8 + 3n \qquad \text{Substitute 40 for } a_n.$$
$$16 = n \qquad \text{Solve for } n.$$

FOCUS ON PEOPLE

KARL FRIEDRICH GAUSS, a famous nineteenth century mathematician, was a child prodigy. It is said that when Gauss was ten his teacher asked his class to add the numbers from 1 to 100. Almost immediately Gauss found the answer by mentally figuring the summation.

The expression formed by adding the terms of an arithmetic sequence is called an **arithmetic series**. The sum of the first n terms of an arithmetic series is denoted by S_n. To find a rule for S_n, you can write S_n in two different ways and add the results.

$$
\begin{array}{l}
S_n = a_1 \qquad\quad + (a_1 + d) + (a_1 + 2d) + \cdots + a_n \\
S_n = a_n \qquad\quad + (a_n - d) + (a_n - 2d) + \cdots + a_1 \\
\hline
2S_n = (a_1 + a_n) + (a_1 + a_n) + (a_1 + a_n) + \cdots + (a_1 + a_n)
\end{array}
$$

You can conclude that $2S_n = n(a_1 + a_n)$, which leads to the following result.

THE SUM OF A FINITE ARITHMETIC SERIES

The sum of the first n terms of an arithmetic series is:

$$S_n = n\left(\frac{a_1 + a_n}{2}\right)$$

In words, S_n is the mean of the first and nth terms, multiplied by the number of terms.

EXAMPLE 5 *Finding a Sum*

Consider the arithmetic series $4 + 7 + 10 + 13 + 16 + 19 + \cdots$.

  a. Find the sum of the first 30 terms.   **b.** Find n such that $S_n = 175$.

SOLUTION

  a. To begin, notice that $a_1 = 4$ and $d = 3$. So, a formula for the nth term is:

$a_n = a_1 + (n - 1)d$ Write rule for the nth term.

    $= 4 + (n - 1)3$ Substitute for a_1 and d.

    $= 1 + 3n$ Simplify.

The 30th term is $a_{30} = 1 + 3(30) = 91$. So, the sum of the first 30 terms is:

$S_{30} = 30\left(\dfrac{a_1 + a_{30}}{2}\right)$ Write rule for S_{30}.

    $= 30\left(\dfrac{4 + 91}{2}\right)$ Substitute for a_1 and a_{30}.

    $= 1425$ Simplify.

▶ The sum of the first 30 terms is 1425.

b. $n\left(\dfrac{4 + (1 + 3n)}{2}\right) = 175$ Use rule for S_n.

    $5n + 3n^2 = 350$ Multiply each side by 2.

    $3n^2 + 5n - 350 = 0$ Write in standard form.

    $(3n + 35)(n - 10) = 0$ Factor.

    $n = 10$ Choose positive solution.

▶ So, $S_n = 175$ when $n = 10$.

STUDENT HELP

▶ **Look Back**
For help with quadratic equations, see p. 256.

11.2 *Arithmetic Sequences and Series* **661**

 EXTRA EXAMPLE 4

Two terms of an arithmetic sequence are $a_5 = 10$ and $a_{30} = 110$.
a. Write a rule for the nth term. $a_n = -10 + 4n$
b. Write the value of n for which $a_n = -2$. $n = 2$

EXTRA EXAMPLE 5
Consider the arithmetic series $20 + 18 + 16 + 14 + \ldots$.
a. Find the sum of the first 25 terms. -100
b. Find n such that $S_n = -760$. $n = 40$

✓ **CHECKPOINT EXERCISES**

For use after Example 4:
1. Two terms of an arithmetic sequence are $a_8 = \dfrac{25}{2}$ and $a_{15} = 30$. Find the value of n for which $a_n = 50$. $n = 23$

For use after Example 5:
2. Find the sum of the first 18 terms of the arithmetic series $100 + 110 + 120 + 130 + \ldots$. 3330
3. Find n for the series $100 + 110 + 120 + 130 + \ldots$, such that $S_n = 600$. 5

MATHEMATICAL REASONING
One mathematical formula cannot contradict another mathematical formula. How is the formula for the sum of the first n integers $\sum_{i=1}^{n} i = \dfrac{n(n+1)}{2}$, which you learned in Lesson 11.1, related to the formula $S_n = \dfrac{n(a_1 + a_n)}{2}$? A series of the first n positive integers is an arithmetic series with $a_1 = 1$ and $d = 1$. Thus, in the first formula, $n + 1 = 1 + n = a_1 + a_n$.

EXTRA EXAMPLE 6

A construction company is laying a natural gas pipeline. Several sections of pipe have been laid in a pile at the construction site. There are 12 sections of pipe in the bottom row of the pile. Each row has one less pipe than the row below it. There are 8 rows of pipe.

a. Write a rule for the number of pipe sections in the nth row.

$a_n = 13 - n$

b. Which row has 6 pipe sections? **7th row**

EXTRA EXAMPLE 7

Use the information about the pipe sections in Extra Example 6.

a. What is the total number of pipe sections in the pile? **68**

b. Suppose three more rows of pipe are added to the pile. How many additional pipe sections will the pile have? **9**

 CHECKPOINT EXERCISES

For use after Examples 6 and 7:

Nicole takes a job with a starting salary of \$32,000. Her employer offers her a \$1500 raise for each of the next 6 years.

1. Write a rule for her salary in the nth year.

$a_n = 30,500 + 1500n$

2. What will her mean salary be for those 6 years? **\$37,250**

FOCUS ON VOCABULARY

How is a common difference related to an arithmetic sequence? In an arithmetic sequence, the difference between a term and the previous term is constant. This constant difference is called the common difference.

CLOSURE QUESTION

How can knowing the first term and the 10th term of an arithmetic series help you find the sum of the first 10 terms? The sum of a series is the product of the number of terms and half the sum of the first and last term.

PHILHARMONIE HALL in Berlin has 2335 seats. This hall was the first to use an architectural concept called "vineyard" design to reflect sound to the audience using several audience tiers at different heights.

GOAL 2 ARITHMETIC SEQUENCES AND SERIES IN REAL LIFE

EXAMPLE 6 *Writing an Arithmetic Sequence*

SEATING CAPACITY The first row of a concert hall has 25 seats, and each row after the first has one more seat than the row before it. There are 32 rows of seats.

a. Write a rule for the number of seats in the nth row.

b. Thirty-five students from a class want to sit in the same row. How close to the front can they sit?

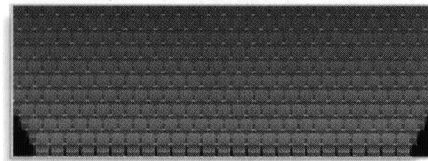

SOLUTION

a. Use $a_1 = 25$ and $d = 1$ to write a rule for a_n.

$$a_n = a_1 + (n - 1)d = 25 + (n - 1)(1) = 24 + n$$

b. Using the rule $a_n = 24 + n$, let $a_n = 35$ and solve for n.

$35 = 24 + n$ Substitute for a_n.

$11 = n$ Solve for n.

▶ The class can sit in the 11th row.

Seating Capacity

EXAMPLE 7 *Finding the Sum of an Arithmetic Series*

Use the information about the concert hall in Example 6.

a. What is the total number of seats in the concert hall?

b. Suppose 12 more rows of seats are built (where each row has one more seat than the row before it). How many additional seats will the concert hall have?

SOLUTION

a. Find the sum of an arithmetic series with $a_1 = 25$ and $a_{32} = 24 + 32 = 56$.

$$S_{32} = 32\left(\frac{a_1 + a_{32}}{2}\right)$$ Write rule for S_{32}.

$$= 32\left(\frac{25 + 56}{2}\right) = 1296$$ Substitute for a_1 and a_{32}.

▶ There are 1296 seats in the concert hall.

b. The expanded concert hall has $32 + 12 = 44$ rows of seats. Because $a_{44} = 24 + 44 = 68$, the *total* number of seats in the expanded hall is:

$$S_{44} = 44\left(\frac{a_1 + a_{44}}{2}\right)$$ Write rule for S_{44}.

$$= 44\left(\frac{25 + 68}{2}\right) = 2046$$ Substitute for a_1 and a_{44}.

▶ The number of *additional* seats is $S_{44} - S_{32} = 2046 - 1296 = 750$.

GUIDED PRACTICE

Vocabulary Check ✓

1. Complete this statement: The expression formed by adding the terms of an arithmetic sequence is called a(n) _?_.   arithmetic series

Concept Check ✓

2. What is the difference between an arithmetic sequence and an arithmetic series?
See margin.

3. Explain how to find the sum of the first n terms of an arithmetic series. Use the formula $S_n = n\left(\dfrac{a_1 + a_n}{2}\right)$.

Skill Check ✓

Write a rule for the nth term of the arithmetic sequence.

4. $d = 2, a_1 = 5$
$a_n = 3 + 2n$

5. $d = -3, a_2 = 18$
$a_n = 24 - 3n$

6. $d = \dfrac{1}{2}, a_5 = 20$
$a_n = \dfrac{35}{2} + \dfrac{n}{2}$

7. $a_8 = 12, a_{15} = 61$
$a_n = -44 + 7n$

8. $a_5 = 10, a_{12} = 24$
$a_n = 2n$

9. $a_{10} = 8, a_{16} = 32$
$a_n = -32 + 4n$

Find the sum of the first 10 terms of the arithmetic series.

10. $2 + 6 + 10 + 14 + 18 + \cdots$ 200

11. $3 + \dfrac{7}{2} + 4 + \dfrac{9}{2} + 5 + \cdots$ $\dfrac{105}{2}$

12. $6 + 3 + 0 + (-3) + (-6) + \cdots$ -75

13. $0.7 + 1.9 + 3.1 + 4.3 + 5.5 + \cdots$ 61

14. 🌐 **MOVIE THEATER** Suppose a movie theater has 42 rows of seats and there are 29 seats in the first row. Each row after the first has two more seats than the row before it. How many seats are in the theater?   2940

PRACTICE AND APPLICATIONS

STUDENT HELP

➤ **Extra Practice**
to help you master skills is on p. 955.

15. yes; constant difference is -3

16. no; difference is not constant

18. yes; constant difference is $\dfrac{1}{2}$

27. $a_n = \dfrac{41}{6} - \dfrac{4}{3}n; -\dfrac{53}{2}$

28. $a_n = \dfrac{19}{6} - \dfrac{2}{3}n; -\dfrac{27}{2}$

29. $a_n = -0.8 + 2.4n; 59.2$

IDENTIFYING ARITHMETIC SEQUENCES Decide whether the sequence is arithmetic. Explain why or why not.

15. $14, 11, 8, 5, 2, \ldots$

16. $1, 3, 9, 27, 81, \ldots$

17. $-5, -7, -11, -13, -15, \ldots$
no; difference is not constant

18. $0.5, 1, 1.5, 2, 2.5, \ldots$

19. $\dfrac{1}{5}, \dfrac{2}{5}, \dfrac{4}{5}, \dfrac{8}{5}, \dfrac{16}{5}, \ldots$
no; difference is not constant

20. $-\dfrac{5}{3}, -1, -\dfrac{1}{3}, \dfrac{1}{3}, 1, \ldots$
yes; constant difference is $\dfrac{2}{3}$

WRITING TERMS Write a rule for the nth term of the arithmetic sequence. Then find a_{25}.

21. $1, 3, 5, 7, 9, \ldots$
$a_n = -1 + 2n; 49$

22. $6, 14, 22, 30, 38, \ldots$
$a_n = -2 + 8n; 198$

23. $9, 23, 37, 51, 65, \ldots$
$a_n = -5 + 14n; 345$

24. $-1, 0, 1, 2, 3, \ldots$
$a_n = -2 + n; 23$

25. $4, 1, -2, -5, -8, \ldots$
$a_n = 7 - 3n; -68$

26. $\dfrac{1}{2}, 3, \dfrac{11}{2}, 8, \dfrac{21}{2}, \ldots$
$a_n = -2 + \dfrac{5}{2}n; \dfrac{121}{2}$

27. $\dfrac{11}{2}, \dfrac{25}{6}, \dfrac{17}{6}, \dfrac{3}{2}, \dfrac{1}{6}, \ldots$

28. $\dfrac{5}{2}, \dfrac{11}{6}, \dfrac{7}{6}, \dfrac{1}{2}, -\dfrac{1}{6}, \ldots$

29. $1.6, 4, 6.4, 8.8, 11.2, \ldots$

STUDENT HELP

➤ **HOMEWORK HELP**
Example 1: Exs. 15–20
Example 2: Exs. 21–29
Example 3: Exs. 30–32, 34, 38–44
Example 4: Exs. 33, 35–37
Example 5: Exs. 45–56
Examples 6, 7: Exs. 57–60

WRITING RULES Write a rule for the nth term of the arithmetic sequence.

30. $d = 4, a_{14} = 46$
$a_n = -10 + 4n$

31. $d = -12, a_1 = 80$
$a_n = 92 - 12n$

32. $d = \dfrac{5}{3}, a_8 = 24$
$a_n = \dfrac{32}{3} + \dfrac{5}{3}n$

33. $a_5 = 17, a_{15} = 77$
$a_n = -13 + 6n$

34. $d = -6, a_{12} = -4$
$a_n = 68 - 6n$

35. $a_2 = -28, a_{20} = 52$
$a_n = -\dfrac{332}{9} + \dfrac{40}{9}n$

36. $a_1 = -2, a_9 = -\dfrac{1}{6}$
$a_n = -\dfrac{107}{48} + \dfrac{11}{48}n$

37. $a_7 = 34, a_{18} = 122$
$a_n = -22 + 8n$

38. $d = -4.1, a_{16} = 48.2$
$a_n = 113.8 - 4.1n$

GRAPHING SEQUENCES Graph the arithmetic sequence.   39–44. See margin.

39. $a_n = 7 + 2n$

40. $a_n = -3 + 5n$

41. $a_n = 5 - 2n$

42. $a_n = 2 - \dfrac{1}{3}n$

43. $a_n = 4 - \dfrac{1}{2}n$

44. $a_n = -0.25 + 0.45n$

---

3 APPLY

ASSIGNMENT GUIDE

BASIC
Day 1: pp. 663–665 Exs. 15–20, 22–44 even, 65–81 odd
Day 2: pp. 663–665 Exs. 46–56 even, 57–58, 61–62

AVERAGE
Day 1: pp. 663–665 Exs. 15–29, 30–44 even, 65–81 odd
Day 2: pp. 663–665 Exs. 46–56 even, 57–59, 61–62

ADVANCED
Day 1: pp. 663–665 Exs. 15–29, 30–44, 65–81 odd
Day 2: pp. 663–665 Exs. 46–60 even, 61–63

BLOCK SCHEDULE
pp. 663–665 Exs. 15–29, 30–44 even, 46–56 even, 57–59, 61–62, 65–81 odd

EXERCISE LEVELS
Level A: *Easier*
15–26, 30–32, 34, 39–41, 57–58
Level B: *More Difficult*
27–29, 33, 35, 37–38, 42–56, 59–62
Level C: *Most Difficult*
63

✓ **HOMEWORK CHECK**
To quickly check student understanding of key concepts, go over the following exercises:
Exs. 16, 22, 30, 40, 46, 52. See also the Daily Homework Quiz:

• Blackline Master (*Chapter 11 Resource Book*, p. 37)
• 🖨 Transparency (p. 85)

39–44. See Additional Answers beginning on page AA1.

→ **Homework Help** Students
can find help for Exs. 45–50 at
www.mcdougallittell.com.
The information can be printed
out for students who don't have
access to the Internet.

APPLICATION NOTE

EXERCISE 57 The position of the
rings is important to the bee colony.
The queen bee remains central to
the hive by staying in the middle of
the honeycomb with her brood.
Pollen, which is a food source for
the immature bees, is kept in the
rings surrounding the queen. The
bees seal their honey in the outer-
most cells of the honeycomb.
Though the number of cells in each
ring can be modeled by a linear
function, the total number of cells
after each ring has been built can
be modeled by the quadratic function, $y = 3x^2 + 3x + 1$.

EXERCISE 59 By sewing a
square above the inner square,
and then sewing the rectangles
in a counterclockwise direction,
the quilting creates a spiral. Both
colors and shades of fabrics can be
used to create visual effects that
can direct the eye to see crosses,
stairs, triangles, diamonds, wheel
axles, barbells, or stars.

61 and 63. See Additional Answers
beginning on page AA1.

**ADDITIONAL PRACTICE
AND RETEACHING**

For Lesson 11.2:

• Practice Levels A, B, and C
(*Chapter 11 Resource Book*,
p. 26)

• Reteaching with Practice
(*Chapter 11 Resource Book*,
p. 29)

• ▣ See Lesson 11.2 of the
Personal Student Tutor

For more Mixed Review:

• ▣ Search the *Test and Practice
Generator* for key words or
specific lessons.

⟲ HOMEWORK HELP
Visit our Web site
www.mcdougallittell.com
for help with Exs. 45–50.

**FOCUS ON
CAREERS**

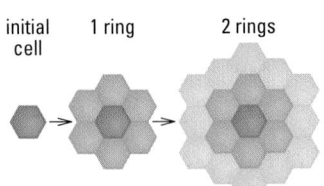

► **ENTOMOLOGIST**
Entomology is the
scientific study of insects.
Some entomologists study
bee population dynamics
and pollination of plants.

⟲ CAREER LINK
www.mcdougallittell.com

59. $1 + 4\sum_{i=1}^{4} 2i$; 81 ft²

FINDING SUMS For part (a), find the sum of the first *n* terms of the arithmetic
series. For part (b), find *n* for the given sum S_n.

45. $3 + 8 + 13 + 18 + 23 + \cdots$

   a. $n = 20$ **1010**   **b.** $S_n = 366$ **12**

46. $50 + 42 + 34 + 26 + 18 + \cdots$

   a. $n = 40$ **−4240**   **b.** $S_n = 182$ **7**

47. $-10 + (-5) + 0 + 5 + 10 + \cdots$

   a. $n = 19$ **665**   **b.** $S_n = 375$ **15**

48. $34 + 31 + 28 + 25 + 22 + \cdots$

   a. $n = 32$ **−400**   **b.** $S_n = -12$ **24**

49. $2 + 9 + 16 + 23 + 30 + \cdots$

   a. $n = 68$ **b.** $S_n = 1661$ **22**
   16,082

50. $2 + 16 + 30 + 44 + 58 + \cdots$

   a. $n = 24$ **3912**   **b.** $S_n = 2178$ **18**

USING SUMMATION NOTATION Find the sum of the series.

51. $\sum_{i=1}^{20} (3 + 5i)$ **1110**

52. $\sum_{i=1}^{34} (1 + 8i)$ **4794**

53. $\sum_{i=1}^{15} (-10 - 3i)$ **−510**

54. $\sum_{i=1}^{22} \left(6 - \frac{3}{4}i\right)$ **−57.75**

55. $\sum_{i=1}^{45} (11 + 4i)$ **4635**

56. $\sum_{i=1}^{18} (8.1 + 4.4i)$ **898.2**

57. 🌐 **HONEYCOMBS** Domestic bees make
their honeycomb by starting with a single
hexagonal cell, then forming ring after ring of
hexagonal cells around the initial cell, as
shown. The numbers of cells in successive
rings form an arithmetic sequence.

   ▶ Source: USDA's Carl Hayden Bee Research Lab

   initial cell   1 ring   2 rings
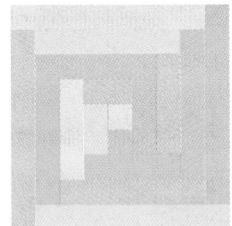

   a. Write a rule for the number of cells in the *n*th ring. $a_n = 6n$

   b. What is the total number of cells in the honeycomb after the 9th ring
is formed? (*Hint:* Do not forget to count the initial cell.) **271**

58. 🌐 **STACKING LOGS** Logs are stacked in a
pile, as shown at the right. The bottom row
has 21 logs and the top row has 15 logs. Each
row has one less log than the row below it.
How many logs are in the pile? **126**

59. 🌐 **QUILTING** A quilt is made up of strips of
cloth, starting with an inner square surrounded
by rectangles to form successively larger
squares. The inner square and all rectangles
have a width of 1 foot. Write an expression
using summation notation that gives the sum of
the areas of all the strips of cloth used to make
the quilt shown. Then evaluate the expression.

60. 🌐 **SEATING REVENUE** Suppose each seat in rows 1 through 11 of the concert
hall in Example 6 costs $24, each seat in rows 12 through 22 costs $18, and each
seat in rows 23 through 32 costs $12. How much money does the concert hall
take in for a sold-out event? **$22,218**

61. *Writing* Compare the graphs of $a_n = 2n + 1$ where *n* is a positive integer and
$f(x) = 2x + 1$ where *x* is a real number. Discuss how the graph of an arithmetic
sequence is similar to and different from the graph of a linear function.
See margin.

664   **Chapter 11** *Sequences and Series*

Test Preparation

62a. n: 1, 2, 3, 4; d_n (in.):
3, 3.008, 3.016, 3.024;
l_n (in.); 3π, 3.008π,
3.016π, 3.024π

62d. *Sample answer:* the
cost should be related
to the length of paper
on the roll. To create a
diameter of 11 in., you
would have to wrap
the paper around the
dowel 1000 times.
$S_{1000} \approx 1831.5$ ft, which
is about 2.8 times as
much as on a 7 in. roll.
So, you would expect
a cost of about $42.

63. The first man gets $\frac{7}{16}$ of
a hekat of barley, the
second $\frac{9}{16}$, the third $\frac{11}{16}$,
and so on with the tenth
man getting $\frac{25}{16}$ hekats.

★ Challenge

EXTRA CHALLENGE
→ www.mcdougallittell.com

62. MULTI-STEP PROBLEM A paper manufacturer sells paper rolled onto
cardboard dowels. The thickness of the paper is 0.004 inch. The diameter
of a dowel is 3 inches, and the total diameter of a roll is 7 inches.

n	d_n (in.)	l_n (in.)
1	3	3π
2	?	?
3	?	?
4	?	?

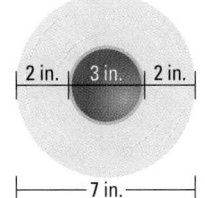

a. Let n be the number of times the paper is wrapped around the dowel, let d_n be
the diameter of the roll just before the nth wrap, and let l_n be the length of
paper added in the nth wrap. Copy and complete the table.

b. What can you say about the sequence $l_1, l_2, l_3, l_4, \ldots$? Write a formula for the
nth term of the sequence.   it is an arithmetic sequence; $a_n = 2.992\pi + 0.008n\pi$

c. Find the number of times the paper must be wrapped around the dowel to
create a roll with a 7 inch diameter. Use your answer and the formula from
part (b) to find the length of paper in a roll with a 7 inch diameter.
500; about 7847.70 in. or 654 ft

d. **LOGICAL REASONING** Suppose a roll with a 7 inch diameter costs $15.
How much would you expect to pay for a roll with an 11 inch diameter whose
dowel also has a diameter of 3 inches? Explain your reasoning and any
assumptions you make.

63. 🖋 **AHMES PAPYRUS** One of the major sources of our knowledge of Egyptian
mathematics is the Ahmes papyrus (also known as the Rhind papyrus), which is a
scroll copied in 1650 B.C. by an Egyptian scribe. The following problem is from
the Ahmes papyrus.

> *Divide 10 hekats of barley among 10 men so that the common difference
> is $\frac{1}{8}$ of a hekat of barley.*

Use what you know about arithmetic sequences and series to solve the problem.
See margin.

MIXED REVIEW

SOLVING RATIONAL EXPONENT EQUATIONS Solve the equation.
(Review 7.6 for 11.3)

64. $x^{1/2} = 5$ 25

65. $2x^{3/4} = 54$ 81

66. $x^{2/3} + 10 = 19$ 27

67. $(8x)^{1/2} + 6 = 0$
no solution

68. $x^{1/3} - 11 = 0$ 1331

69. $(2x)^{1/2} = x - 4$ 8

SOLVING EXPONENTIAL EQUATIONS Solve the exponential equation.
(Review 8.6 for 11.3)

$\log 2 \approx 0.3010$

70. $2^x = 4.5$ $\log_2 4.5 \approx 2.170$

71. $4^x - 3 = 5$ $\frac{3}{2}$

72. $10^{3x} + 7 = 15$

73. $6^x - 5 = 1$ 1

74. $25^x - 28 = 97$ $\frac{3}{2}$

75. $5(2)^{2x} - 4 = 13$
$\log_4 3.4 \approx 0.8828$

GRAPHING EQUATIONS Graph the equation. (Review 10.3)
76–81. See margin for graphs.

76. $x^2 + y^2 = 9$

77. $x^2 + y^2 = 24$

78. $x^2 + y^2 = 64$

79. $6x^2 + 6y^2 = 150$

80. $\frac{1}{2}x^2 + \frac{1}{2}y^2 = 4$

81. $20x^2 + 20y^2 = 400$

11.2 *Arithmetic Sequences and Series*   **665**

Additional Test Preparation *Sample answer:*

2. The first formula begins with the first term and adds
the common difference to each of the consecutive
terms until the last term is reached. The second
formula begins with the last term in the series and

subtracts the common difference from each of the
previous terms until the first term is reached.

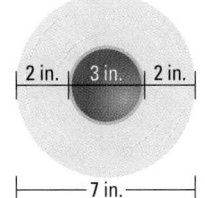

4 ASSESS

DAILY HOMEWORK QUIZ

📄 **Transparency Available**

1. Is the sequence 11, 22, 33, 44,
55, . . . arithmetic? **yes**

2. Write a rule for the nth term of
the sequence 30, 24, 18, 12, ...
Then find a_{20}. $36 - 6n$; -84

3. One term of an arithmetic
sequence is $a_{15} = 20$. The
common difference is $d = \frac{1}{2}$.
Write a rule for the nth term.
$a_n = \frac{25}{2} + \left(\frac{1}{2}\right)n$

4. Two terms of an arithmetic
sequence are $a_4 = 7$ and
$a_{15} = 40$. Find the rule for the
nth term. Then find the value
for n for which $a_n = 25$.
$a_n = -5 + 3n$; 10

5. Find the sum of the first 30
terms of the arithmetic series
$3 + 7 + 11 + 15 + 19 + \ldots$. Then
find n such that $S_n = 2485$.
1830; 35

EXTRA CHALLENGE NOTE
→ Challenge problems for
Lesson 11.2 are available in
blackline format in the *Chapter 11
Resource Book*, p. 34 and at
www.mcdougallittell.com.

**ADDITIONAL TEST
PREPARATION**

1. **OPEN ENDED** Use the formulas
$S_n = a_1 + (a_1 + d) + (a_1 + 2d) + \ldots + a_n$ and $S_n = a_n + (a_1 - d) + (a_1 - 2d) + \ldots + a_1$ to find the
sum of the first 10 integers.
$S_{10} = (1 + 10) + (2 + 9) + (3 + 8) + (4 + 7) + (5 + 6) = 5(11) = 55$

2. **WRITING** How do the formulas
$S_n = a_1 + (a_1 + d) + (a_1 + 2d) + \ldots + a_n$ and $S_n = a_n + (a_1 - d) + (a_1 - 2d) + \ldots + a_1$ represent
the terms in a finite arithmetic
series? See margin.

76–81. See Additional Answers
beginning on page AA1.

> **LESSON OPENER**
> **APPLICATION**
> An alternative way to approach Lesson 11.3 is to use the Application Lesson Opener:
> • Blackline Master (*Chapter 11 Resource Book,* p. 38)
> •  Transparency (p. 74)

MEETING INDIVIDUAL NEEDS
• *Chapter 11 Resource Book*
  Prerequisite Skills Review (p. 5)
  Practice Level A (p. 41)
  Practice Level B (p. 42)
  Practice Level C (p. 43)
  Reteaching with Practice (p. 44)
  Absent Student Catch-Up (p. 46)
  Challenge (p. 48)
• *Resources in Spanish*
• 🖥 *Personal Student Tutor*

NEW-TEACHER SUPPORT
See the Tips for New Teachers on pp. 1–2 of the *Chapter 11 Resource Book* for additional notes about Lesson 11.3.

WARM-UP EXERCISES

🖎 *Transparency Available*

Write a rule for the *n*th term.

1. $\frac{1}{2}, \frac{1}{4}, \frac{1}{8}, \frac{1}{16}, \ldots$ $a_n = \frac{1}{2^n}$

2. 1, 5, 25, 125, ... 5^{n-1}

3. Find the sum of the first 20 terms of the arithmetic sequence $4 + 5 + 6 + 7 + 8 + \ldots$
270

4. In an experiment a substance is modeled by $y = 300(0.8)^x$, where *y* is the amount in grams of the substance after *x* hours. How much remains after 8 hours?   **about 50 grams**

11.3 Geometric Sequences and Series

What you should learn

GOAL 1 Write rules for geometric sequences and find sums of geometric series.

GOAL 2 Use geometric sequences and series to model **real-life** quantities, such as monthly bills for cellular telephone service in **Example 6**.

Why you should learn it

▼ To solve **real-life** problems, such as finding the number of tennis matches played in **Exs. 70 and 71**.

GOAL 1 USING GEOMETRIC SEQUENCES AND SERIES

In a **geometric sequence,** the ratio of any term to the previous term is constant. This constant ratio is called the **common ratio** and is denoted by *r*.

EXAMPLE 1 *Identifying Geometric Sequences*

Decide whether each sequence is geometric.

a. 1, 2, 6, 24, 120, . . .   **b.** 81, 27, 9, 3, 1, . . .

SOLUTION

To decide whether a sequence is geometric, find the ratios of consecutive terms.

a. $\frac{a_2}{a_1} = \frac{2}{1} = 2$ $\frac{a_3}{a_2} = \frac{6}{2} = 3$ $\frac{a_4}{a_3} = \frac{24}{6} = 4$ $\frac{a_5}{a_4} = \frac{120}{24} = 5$

▶ The ratios are different, so the sequence is not geometric.

b. $\frac{a_2}{a_1} = \frac{27}{81} = \frac{1}{3}$ $\frac{a_3}{a_2} = \frac{9}{27} = \frac{1}{3}$ $\frac{a_4}{a_3} = \frac{3}{9} = \frac{1}{3}$ $\frac{a_5}{a_4} = \frac{1}{3}$

▶ The ratios are the same, so the sequence is geometric.

RULE FOR A GEOMETRIC SEQUENCE

The *n*th term of a geometric sequence with first term a_1 and common ratio *r* is given by:

$$a_n = a_1 r^{n-1}$$

EXAMPLE 2 *Writing a Rule for the nth Term*

Write a rule for the *n*th term of the sequence $-8, -12, -18, -27, \ldots$. Then find a_8.

SOLUTION

The sequence is geometric with first term $a_1 = -8$ and common ratio $r = \frac{-12}{-8} = \frac{3}{2}$. So, a rule for the *n*th term is:

$a_n = a_1 r^{n-1}$ Write general rule.

$\quad = -8\left(\frac{3}{2}\right)^{n-1}$ Substitute for a_1 and *r*.

The 8th term is $a_8 = -8\left(\frac{3}{2}\right)^{8-1} = -\frac{2187}{16}$.

EXAMPLE 3 *Finding the nth Term Given a Term and the Common Ratio*

One term of a geometric sequence is $a_3 = 5$. The common ratio is $r = 2$.

a. Write a rule for the nth term.　　　　**b.** Graph the sequence.

SOLUTION

a. Begin by finding the first term as follows.

$a_n = a_1 r^{n-1}$	**Write general rule.**
$a_3 = a_1 r^{3-1}$	**Substitute 3 for n.**
$5 = a_1 (2)^2$	**Substitute for a_3 and r.**
$1.25 = a_1$	**Solve for a_1.**

So, a rule for the nth term is:

$a_n = a_1 r^{n-1}$	**Write general rule.**
$= 1.25(2)^{n-1}$	**Substitute for a_1 and r.**

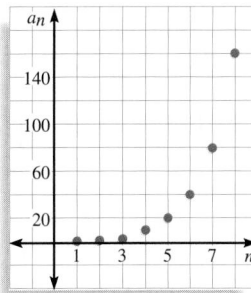

b. The graph is shown at the right. Notice that the points lie on an exponential curve. This is true for *any* geometric sequence with $r > 0$.

EXAMPLE 4 *Finding the nth Term Given Two Terms*

Two terms of a geometric sequence are $a_2 = 45$ and $a_5 = -1215$. Find a rule for the nth term.

SOLUTION

Write a system of equations using $a_n = a_1 r^{n-1}$ and substituting 2 for n (Equation 1) and then 5 for n (Equation 2).

$a_2 = a_1 r^{2-1}$	$45 = a_1 r$	**Equation 1**
$a_5 = a_1 r^{5-1}$	$-1215 = a_1 r^4$	**Equation 2**

Solve the system.

$\dfrac{45}{r} = a_1$	**Solve Equation 1 for a_1.**
$-1215 = \dfrac{45}{r}(r^4)$	**Substitute for a_1 in Equation 2.**
$-1215 = 45r^3$	**Simplify.**
$-27 = r^3$	**Divide each side by 45.**
$-3 = r$	**Take the cube root of each side.**
$45 = a_1(-3)$	**Substitute for r in Equation 1.**
$-15 = a_1$	**Solve for a_1.**

Find a rule for a_n.

$a_n = a_1 r^{n-1}$	**Write general rule.**
$a_n = -15(-3)^{n-1}$	**Substitute for a_1 and r.**

▶ A rule for the nth term is $a_n = -15(-3)^{n-1}$.

MOTIVATING THE LESSON
The amount and frequency of dosage for a medicine depends on how long the medicine remains in the bloodstream. The length of time that medicine remains in the bloodstream can be modeled using geometric series, a topic of today's lesson.

EXTRA EXAMPLE 1
Is the sequence geometric?
a. 4, –8, 16, –32, … Yes
b. 3, 9, –27, –81, 243, … No

EXTRA EXAMPLE 2
Write a rule for the nth term of the sequence 5, 2, 0.8, 0.32. Then find a_8. $a_n = 5(0.4)^{n-1}$; 0.008192

EXTRA EXAMPLE 3
One term of a geometric sequence is $a_4 = 3$. The common ratio is $r = 3$.
a. Write a rule for the nth term.
$a_n = \frac{1}{9}(3)^{n-1}$
b. Graph the sequence.

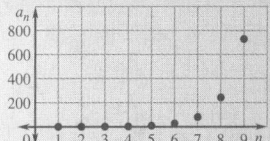

EXTRA EXAMPLE 4
Two terms of a geometric sequence are $a_2 = -4$ and $a_6 = -1024$. Write a rule for the nth term. $a_n = (-4)^{n-1}$

CHECKPOINT EXERCISES
For use after Examples 1–3:
1. One term of a geometric sequence is $a_3 = -2$. The common ratio is 3. Write a rule for the nth term. Then find a_8.
$a_n = \left(-\frac{2}{9}\right)(3)^{n-1}$; $a_8 = -486$

For use after Example 4:
2. Two terms of a geometric sequence are $a_2 = -18$ and $a_5 = \frac{2}{3}$. Write a rule for the nth term. $a_n = (54)\left(-\frac{1}{3}\right)^{n-1}$

EXTRA EXAMPLE 5
Consider the geometric series
$4 + 2 + 1 + \frac{1}{2} + \cdots$.
a. Find the sum of the first
   10 terms. $\frac{1023}{128}$
b. Find n such that $S_n = \frac{31}{4}$. 5

 CHECKPOINT EXERCISES
For use after Example 5:
Consider the geometric series
$-60 + 20 - \frac{20}{3} + \frac{20}{9} - \cdots$.
1. Find the sum of the first
   5 terms. $-\frac{1220}{27}$
2. Find n such that $S_n = -\frac{32800}{729}$.
   8

───────────────

STUDENT HELP NOTES

→**Homework Help** Students
can find extra examples at
www.mcdougallittell.com
that parallel the examples in
the student edition.

→**Look Back** As students look back
to p. 501 for a review of solving
logarithmic equations, remind
them that $5^n = 15{,}625$ can also be
solved by finding $\frac{\ln 15{,}625}{\ln 5}$.

CONCEPT QUESTION
EXAMPLE 5 How can the
geometric series be written using
summation notation? $\sum\limits_{i=1}^{\infty} 5^{i-1}$

MATHEMATICAL REASONING
Any term in a geometric sequence
is the geometric mean of the preceding term and the following term.
For example, in the geometric
sequence $-2, -6, -18, -54, \ldots, -6$
is the geometric mean between
-2 and -18 since $\frac{-2}{-6} = \frac{-6}{-18}$, or
$(-6)^2 = (-2)(-18)$. If $a_1 = x$ and
$a_3 = x^3$, what are the possible
values of a_2? x^2 and $-x^2$

The expression formed by adding the terms of a geometric sequence is called a
geometric series. As with an arithmetic series, the sum of the first n terms of a
geometric series is denoted by S_n. You can develop a rule for S_n as follows.

$$S_n = a_1 + a_1r + a_1r^2 + a_1r^3 + \cdots + a_1r^{n-1}$$
$$-rS_n = \qquad\quad -a_1r - a_1r^2 - a_1r^3 - \cdots - a_1r^{n-1} - a_1r^n$$
$$\overline{\quad S_n(1-r) = a_1 \qquad\qquad\qquad\qquad\qquad\qquad\quad - a_1r^n}$$

Therefore, $S_n(1-r) = a_1(1-r^n)$. If $r \ne 1$, you can divide both sides of this
equation by $1 - r$ to obtain the following rule for S_n.

───────────────

THE SUM OF A FINITE GEOMETRIC SERIES

The sum S_n of the first n terms of a geometric series with common ratio $r \ne 1$ is:

$$S_n = a_1\left(\frac{1 - r^n}{1 - r}\right)$$

───────────────

EXAMPLE 5 *Finding a Sum*

Consider the geometric series $1 + 5 + 25 + 125 + 625 + \cdots$.

STUDENT HELP
HOMEWORK HELP
Visit our Web site
www.mcdougallittell.com
for extra examples.

a. Find the sum of the first 10 terms.   **b.** Find n such that $S_n = 3906$.

SOLUTION

a. To begin, notice that $a_1 = 1$ and $r = 5$. Therefore:

$$S_{10} = a_1\left(\frac{1 - r^{10}}{1 - r}\right) \qquad \text{Write rule for } S_{10}.$$

$$= 1\left(\frac{1 - 5^{10}}{1 - 5}\right) \qquad \text{Substitute for } a_1 \text{ and } r.$$

$$= 2{,}441{,}406 \qquad \text{Simplify.}$$

▶ The sum of the first 10 terms is 2,441,406.

b. $a_1\left(\frac{1 - r^n}{1 - r}\right) = S_n$ Write general rule.

$$1\left(\frac{1 - 5^n}{1 - 5}\right) = 3906 \qquad \text{Substitute for } a_1, r, \text{ and } S_n.$$

$$\frac{1 - 5^n}{-4} = 3906 \qquad \text{Simplify.}$$

$$1 - 5^n = -15{,}624 \qquad \text{Multiply each side by } -4.$$

$$-5^n = -15{,}625 \qquad \text{Subtract 1 from each side.}$$

$$5^n = 15{,}625 \qquad \text{Divide each side by } -1.$$

$$n = \frac{\log 15{,}625}{\log 5} = 6 \qquad \text{Solve for } n.$$

STUDENT HELP
→ **Look Back**
For help with logarithmic
equations, see p. 501.

▶ So, $S_n = 3906$ when $n = 6$.

GOAL 2 **GEOMETRIC SEQUENCES AND SERIES IN REAL LIFE**

EXAMPLE 6 *Writing a Geometric Sequence*

CELLULAR TELEPHONES In 1990 the average monthly bill for cellular telephone service in the United States was $80.90. From 1990 through 1997, the average monthly bill decreased by about 8.6% per year. ▶ *Source: Statistical Abstract of the United States*

a. Write a rule for the average monthly cellular telephone bill a_n (in dollars) in terms of the year. Let $n = 1$ represent 1990.

b. What was the average monthly cellular telephone bill in 1993?

c. When did the average monthly cellular telephone bill fall to $50?

SOLUTION

a. Because the average monthly bill decreased by the same percent each year, the average monthly bills from year to year form a geometric sequence. Use $a_1 = 80.9$ and $r = 1 - 0.086 = 0.914$. A rule for the average monthly bill is:

$$a_n = 80.9(0.914)^{n-1}$$

b. In 1993, $n = 4$. So, the average monthly bill was $a_4 = 80.9(0.914)^3 \approx \$61.77$.

c. You want to find n such that $a_n = 50$.

$80.9(0.914)^{n-1} = 50$ Write equation using rule for a_n.

$(0.914)^{n-1} \approx 0.618$ Divide each side by 80.9.

$n - 1 \approx \dfrac{\log 0.618}{\log 0.914} \approx 5.35$ Solve for $n - 1$.

$n \approx 6$ Solve for n.

The average monthly cellular telephone bill reached $50 in 1995 (when $n = 6$).

CELLULAR TELEPHONES
In 1990 there were about 5 million cellular phone subscribers. By 1997 the number had grown to over 55 million.

APPLICATION LINK
www.mcdougallittell.com

Cellular Phones

EXAMPLE 7 *Finding the Sum of a Geometric Series*

Use the model for the average monthly cellular telephone bill in Example 6. On average, what did a person pay for cellular telephone service during 1990–1997?

SOLUTION

Because the model $a_n = 80.9(0.914)^{n-1}$ gives the average *monthly* bill, the model $b_n = 12(80.9)(0.914)^{n-1} = 970.8(0.914)^{n-1}$ gives the average *annual* bill. Using $a_1 = 970.8$ and $r = 0.914$, you can estimate a person's total cost for cellular telephone service during the 8 year period 1990–1997 to be:

$S_8 = a_1\left(\dfrac{1 - r^8}{1 - r}\right)$ Write rule for S_8.

$= 970.8\left(\dfrac{1 - (0.914)^8}{1 - 0.914}\right)$ Substitute for a_1 and r.

≈ 5790 Simplify.

▶ A person paid about $5790 for cellular telephone service during 1990–1997.

STUDENT HELP

DATA UPDATE
Visit our Web site
www.mcdougallittell.com

 11.3 *Geometric Sequences and Series*   **669**

 EXTRA EXAMPLE 6
You buy a new car for $25,000. The value of the car decreases by 16% each year.
a. Write a rule for the average yearly value of the car a_n (in dollars) in terms of the year. Let $n = $ the current year.
$a_n = 25{,}000(0.84)^{n-1}$
b. In about how many years will the value of the car fall to $10,455? **6 years**

EXTRA EXAMPLE 7
Suppose your computer system loses one-fifth of its value at the end of the first year. At the end of the second year, it loses one-fifth of the remaining four-fifths of its value, and so on. If you paid $2700 for the computer, what is its average value at the end of each year during the 5 years you own it? **about $1452**

 CHECKPOINT EXERCISES
For use after Examples 6 and 7:
1. Your father buys a car for $25,000. One year later your mother buys a car for $25,000. The next year you buy a car for $25,000. Four years after your father buys his car, all of you decide to trade in your cars. What is the total value of all three cars at trade-in time? **$53,457.60**

FOCUS ON VOCABULARY
How can you determine the common ratio of a geometric sequence from 2 consecutive terms? **Divide any term except the first term by its preceding term.**

CLOSURE QUESTION
What do the variables represent in the general rule for a geometric sequence, $a_n = a_1 r^{n-1}$? a_n is the *n*th term of a sequence, a_1 is the first term of the sequence, r is the common ratio, and n is the number of the term.

ASSIGNMENT GUIDE

BASIC
Day 1: pp. 670–673 Exs. 24–29, 33–37, 39–42, 45–48, 83–97 odd
Day 2: pp. 671–673 Exs. 54–56, 60–62, 64–66, 70–71, 81, Quiz 1 Exs. 1–14

AVERAGE
Day 1: pp. 670–673 Exs. 24–29, 33–37, 39–42, 45–48, 54–56, 83–97 odd
Day 2: pp. 671–673 Exs. 57–59, 60–68 even, 70–71, 80–81, Quiz 1 Exs. 1–14

ADVANCED
Day 1: pp. 670–673 Exs. 24–29, 33–37, 39–42, 45–48, 54–56, 83–97 odd
Day 2: pp. 671–673 Exs. 57–59, 60–68 even, 70–71, 76–82, Quiz 1 Exs. 1–14

BLOCK SCHEDULE
pp. 670–673 Exs. 24–52 even, 54–56, 60–68 even, 70–73, 78–81, 83–93 odd, Quiz 1 Exs. 1–14

EXERCISE LEVELS

Level A: *Easier*
24–28, 33–36, 39–42, 54–59,
Level B: *More Difficult*
29–32, 37–38, 43–53, 60–81
Level C: *Most Difficult*
82

✔ HOMEWORK CHECK

To quickly check student under-standing of key concepts, go over the following exercises: Exs. 24, 26, 34, 40, 46, 54, 60, 64. See also the Daily Homework Quiz:

• Blackline Master (*Chapter 11 Resource Book*, p. 52)
• 🖉 Transparency (p. 86)

20. $25\left(\frac{1}{5}\right)^{n-1}$ or $-25\left(-\frac{1}{5}\right)^{n-1}$

670

GUIDED PRACTICE

Vocabulary Check ✔

1. Complete this statement: The constant ratio in a geometric sequence is called the ? ratio and is denoted by ? . **common; r**

Concept Check ✔

2. What makes a sequence geometric? **See margin.**

3. State the rule for the sum of the first n terms of a geometric series. $S_n = a_1\left(\frac{1-r^n}{1-r}\right)$

Skill Check ✔

Find the common ratio of the geometric sequence.

2. If there is some quantity r such that the kth term is r times the $(k-1)$th term for every value of k.

24. geometric; common ratio of 4

25. neither; no common ratio or difference

26. arithmetic; common difference of 9

27. arithmetic; common difference of -4

28. arithmetic; common difference of 4

29. geometric; common ratio of 3

30. arithmetic; common difference of $\frac{1}{3}$

4. 4, 12, 36, 108, 324, . . . **3**   **5.** 1, 6, 36, 216, 1296, . . . **6**   **6.** 2, -6, 18, -54, 162, . . . **-3**

7. 7, 14, 28, 56, 128, . . . **2**   **8.** 64, -32, 16, -8, 4, . . . **$-\frac{1}{2}$**   **9.** 10, 5, $\frac{5}{2}$, $\frac{5}{4}$, $\frac{5}{8}$, . . . **$\frac{1}{2}$**

Write the next term and find a rule for the nth term of the geometric sequence.

10. 1, 3, 9, 27, . . . **81; 3^{n-1}**   **11.** 2, 8, 32, 128, **512; $2(4)^{n-1}$**   **12.** 1, -6, 36, -216, . . . **1296; $(-6)^{n-1}$**

13. 375, -75, 15, -3, **0.6; $375\left(-\frac{1}{5}\right)^{n-1}$**   **14.** $\frac{1}{2}$, $\frac{1}{4}$, $\frac{1}{8}$, $\frac{1}{16}$, . . . **$\frac{1}{32}$; $\frac{1}{2^n}$**   **15.** -28, 14, -7, $\frac{7}{2}$, $-\frac{7}{4}$, . . . **$\frac{7}{8}$; $-\frac{28}{(-2)^{n-1}}$**

Write a rule for the nth term of the geometric sequence.

16. $r = 3$, $a_1 = 2$ **$2(3)^{n-1}$**   **17.** $r = -2$, $a_1 = 6$ **$6(-2)^{n-1}$**   **18.** $r = -3$, $a_1 = 12$ **$12(-3)^{n-1}$**

19. $a_1 = \frac{1}{4}$, $a_3 = 6$ **$\frac{1}{4}(2\sqrt{6})^{n-1}$**   **20.** $a_2 = 5$, $a_4 = \frac{1}{5}$ **See margin.**   **21.** $a_2 = 28$, $a_5 = -1792$ **$-7(-4)^{n-1}$**

22. Find the sum of the first 8 terms of the geometric series $1 + 8 + 64 + 512 + \cdots$. **2,396,745**

23. 🌐 **CELLULAR PHONES** Use the model from Example 6 to find the average monthly bill for cellular telephone service in 1997. **$43.11**

PRACTICE AND APPLICATIONS

STUDENT HELP
↳ **Extra Practice**
to help you master skills is on p. 955.

43. $5\left(-\frac{1}{3}\right)^{n-1}$; $-\frac{5}{243}$

44. $2\left(\frac{2}{3}\right)^{n-1}$; $\frac{64}{243}$

STUDENT HELP
↳ **HOMEWORK HELP**
Example 1: Exs. 24–32
Example 2: Exs. 33–44
Example 3: Exs. 45–49, 54–59
Example 4: Exs. 50–53
Example 5: Exs. 60–69
Examples 6, 7: Exs. 70–79

CLASSIFYING SEQUENCES Decide whether the sequence is *arithmetic*, *geometric*, or *neither*. Explain your answer. **24–30. See margin.**

24. 6, 24, 96, 384, . . .   **25.** 1, 3, 7, 13, . . .   **26.** 4, 13, 22, 31, . . .

27. 3, -1, -5, -9, . . .   **28.** -11, -7, -3, 1, . . .   **29.** $\frac{1}{2}$, $\frac{3}{2}$, $\frac{9}{2}$, $\frac{27}{2}$, . . .

30. $\frac{1}{3}$, $\frac{2}{3}$, 1, $\frac{4}{3}$, . . .   **31.** $-\frac{3}{4}$, $\frac{1}{8}$, $-\frac{1}{16}$, $\frac{3}{32}$, . . .   **32.** $-\frac{3}{5}$, $\frac{4}{25}$, $\frac{5}{125}$, $\frac{6}{625}$, . . .

31, 32. neither; no common ratio or difference

FINDING COMMON RATIOS Find the common ratio of the geometric sequence.

33. 1, 4, 16, 64, . . . **4**   **34.** 3, 6, 12, 24, . . . **2**   **35.** -3, 6, -12, 24, . . . **-2**

36. 5, 40, 320, 2560, . . . **8**   **37.** 136, 68, 34, 17, . . . **$\frac{1}{2}$**   **38.** $-\frac{1}{4}$, $\frac{1}{8}$, $-\frac{1}{16}$, $\frac{1}{32}$, . . . **$-\frac{1}{2}$**

WRITING TERMS Write a rule for the nth term of the geometric sequence. Then find a_6.

39. 1, -4, 16, -64, . . . **$(-4)^{n-1}$; -1024**   **40.** 5, 10, 20, 40, . . . **$5(2)^{n-1}$; 160**   **41.** 2, 14, 98, 686, . . . **$2(7)^{n-1}$; 33,614**

42. 6, -30, 150, -750, . . . **$6(-5)^{n-1}$; $-18,750$**   **43.** 5, $-\frac{5}{3}$, $\frac{5}{9}$, $-\frac{5}{27}$, . . . **See margin.**   **44.** 2, $\frac{4}{3}$, $\frac{8}{9}$, $\frac{16}{27}$, . . . **See margin.**

WRITING RULES Write a rule for the *n*th term of the geometric sequence.

45. $r = 3, a_1 = 4$ $\quad 4(3)^{n-1}$ **46.** $r = \frac{1}{3}, a_1 = 45$ $45\left(\frac{1}{3}\right)^{n-1}$ **47.** $r = 6, a_3 = 72$ $\quad 2(6)^{n-1}$

$50.\ \left(-\frac{1}{2}\right)(\sqrt[3]{32})^{n-1}$

48. $r = \frac{1}{8}, a_1 = 4$ $\quad 4\left(\frac{1}{8}\right)^{n-1}$ **49.** $r = 8, a_1 = -2$ $\quad -2(8)^{n-1}$ **50.** $a_1 = -\frac{1}{2}, a_4 = -16$

$51.\ \left(\frac{10}{\sqrt[3]{900}}\right)(\sqrt[3]{30})^{n-1}$

51. $a_3 = 10, a_6 = 300$ **52.** $a_2 = -20, a_4 = -5$ **53.** $a_2 = -30, a_5 = 3750$ $\quad 6(-5)^{n-1}$

$52.\ -40\left(\frac{1}{2}\right)^{n-1}$ or $40\left(-\frac{1}{2}\right)^{n-1}$

GRAPHING SEQUENCES Graph the geometric sequence. 54–59. See margin.

54. $a_n = 4(2)^{n-1}$ **55.** $a_n = 3(5)^{n-1}$ **56.** $a_n = 2(3)^{n-1}$

57. $a_n = 8(3)^{n-1}$ **58.** $a_n = 5\left(\frac{1}{2}\right)^{n-1}$ **59.** $a_n = 4\left(\frac{3}{2}\right)^{n-1}$

FINDING SUMS For part (a), find the sum of the first *n* terms of the geometric series. For part (b), find *n* for the given sum S_n.

60. $1 + 4 + 16 + 64 + \cdots$
 a. $n = 14$ **b.** $S_n = 341$ $\quad 5$
 89,478, 485

61. $1 + 9 + 81 + 729 + \cdots$
 a. $n = 10$ **b.** $S_n = 820$ $\quad 4$
 435,848,050

62. $7 + (-21) + 63 + (-189) + \cdots$
 a. $n = 18$ **b.** $S_n = 3829$ $\quad 7$
 $-677{,}985{,}854$

63. $-90 + 30 + (-10) + \frac{10}{3} + \cdots$
 a. $n = 16$ **b.** $S_n = -66.67$ $\quad 4$
 -67.5

USING SUMMATION NOTATION Find the sum of the series.

64. $\sum_{i=1}^{10} 6(2)^{i-1}$ $\quad 6138$ **65.** $\sum_{i=1}^{8} 5(4)^{i-1}$ $\quad 109{,}225$ **66.** $\sum_{i=0}^{9} 12\left(-\frac{1}{2}\right)^i$ $\quad 7.9922$

67. $\sum_{i=1}^{10} 8\left(\frac{3}{4}\right)^{i-1}$ $\quad 30.198$ **68.** $\sum_{i=0}^{6} 4\left(\frac{3}{2}\right)^i$ $\quad 128.6875$ **69.** $\sum_{i=1}^{12} (-2)^{i-1}$ $\quad -1365$

🌐 **TENNIS** In Exercises 70 and 71, use the following information.
The men's U.S. Open tennis tournament is held annually in Flushing Meadow in New York City. In the first round of the tournament, 64 matches are played. In each successive round, the number of matches played decreases by one half.
▶ Source: United States Tennis Association

70. Find a rule for the number of matches played in the *n*th round. For what values of *n* does your rule make sense? $64\left(\frac{1}{2}\right)^{n-1}$; for $1 \le n \le 7$

71. Find the total number of matches played in the men's U.S. Open tennis tournament. **127**

🌐 **COMPUTER SCIENCE** In Exercises 72 and 73, use the following information.
When a computer must find an item in an ordered list of data (such as an alphabetical list of names), it may be programmed to perform a *binary search*. This search technique involves jumping to the middle of the list and deciding whether the item is there. If not, the computer decides whether the item comes before or after the middle. Half of the list is then ignored on the next pass through the list, and the computer jumps to the middle of the remaining list. This is repeated until the item is found.

72. An ordered list contains 1024 items. Find a rule for the number of items remaining after the *n*th pass through the list. $1024\left(\frac{1}{2}\right)^n$

73. In the worst case, the item to be found is the only one left in the list after *n* passes through the list. What is the worst-case value of *n* for a binary search of a list with 1024 items? **10**

FOCUS ON CAREERS

COMPUTER PROGRAMMER
Programmers write, test, and maintain computer programs. Programs are detailed lists of instructions that a computer must follow to perform its functions.

CAREER LINK
www.mcdougallittell.com

11.3 *Geometric Sequences and Series* **671**

! COMMON ERROR
EXERCISES 4–9 Students may divide incorrectly to determine the common ratio. Remind them the common ratio is the factor that is multiplied by a_n to get a_{n+1}, so the common ratio is the quotient $\frac{a_{n+1}}{a_n}$.

APPLICATION NOTE
EXERCISES 72 AND 73
A binary search is a search algorithm that avoids a slow sequential search by starting in the middle of an ordered database. Elegant algorithms, which are the principal challenge of a programmer, use the fewest steps possible. In the case of a binary search, since the list being searched is ordered, a logical operation such as determining whether the search item is less than the middle item allows the list being searched to immediately be halved.

CAREER NOTE
EXERCISES 72 AND 73
Additional information about computer programmers is available at **www.mcdougallittell.com**.

54.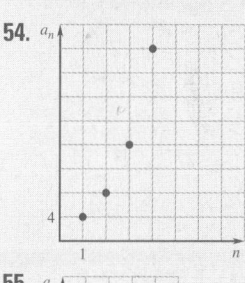

55.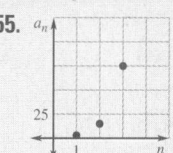

56.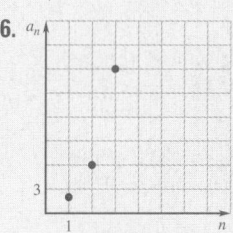

57–59. See Additional Answers beginning on page AA1.

APPLICATION NOTE

EXERCISES 74–77 Students should recognize that the pager sales *y* can also be modeled by the exponential growth function $y = 118(1.2)^n$. In this case, *n* represents the years since 1990.

GRAPHING CALCULATOR NOTE

EXERCISE 81 A graphing calculator will quickly allow students to compare the growth in revenue for the two companies and to see the difference in growth rate between an arithmetic sequence and a geometric sequence. The graphs, however, allow students to make a judgement about ownership based solely on revenue and will not show many other factors that could influence a person's preference for owning one company rather than the other. For instance, the size of the company or one's intent to own for only a short period are factors that might affect the preference. How well the rapid growth modeled by the geometric series is managed might also be a factor in choosing ownership.

81b. See Additional Answers beginning on page AA1.

ADDITIONAL PRACTICE AND RETEACHING

For Lesson 11.3:

• Practice Levels A, B, and C (*Chapter 11 Resource Book,* p. 41)

• Reteaching with Practice (*Chapter 11 Resource Book,* p. 44)

• 🖳 See Lesson 11.3 of the *Personal Student Tutor*

For more Mixed Review:

• 🖳 Search the *Test and Practice Generator* for key words or specific lessons.

672

79. $\left(\frac{\sqrt{3}}{4}\right)\left(\frac{3}{4}\right)^n$; 0.006

80. *Sample answer:* The graph of a_n is a set of discrete points in the first quadrant that all lie on the exponential curve $f(x) = 4(2)^{x-1}$. The graph of *f* itself is a smooth curve. In both cases, *y* increases continuously, and $y \to \infty$ as $x \to \infty$. In the same way, the graph of any geometric sequence is a set of discrete points that follow a related exponential curve.

81c. Company A: $4959.5 million; Company B: $2097.15 million

Test Preparation ✏️🖱️

81d. In the year 2002; *Sample answer:* Company A makes slow but steady progress throughout the time period, while Company B starts off much more slowly, but then increases more rapidly towards the end of the given time. In the near future, Company B yields a much higher revenue than A. Company B would be the better investment.

★ Challenge

🌐 PAGER SALES In Exercises 74–77, use the following information.

In 1990 factory sales of pagers in the United States totaled $118 million. From 1990 through 1996, the sales increased by about 20% per year.

🔗 DATA UPDATE of *Statistical Abstract of the United States* data at www.mcdougallittel.com.

74. Write a rule for pager sales a_n (in millions of dollars) in terms of the year. Let $n = 1$ represent 1990. $a_n = 118(1.2)^{n-1}$

75. What did factory sales of pagers total in 1992?   **$169.92 million**

76. When did factory sales of pagers reach $300 million?   **in 1995**

77. What was the total of factory sales of pagers for the period 1990–1996?
   about $1.524 billion

🌐 SIERPINSKI TRIANGLE In Exercises 78 and 79, use the following information.

The *Sierpinski triangle* is a design using equilateral triangles. The process involves removing smaller triangles from larger triangles by joining the midpoints of the sides of the larger triangles as shown below. Assume that the initial triangle is equilateral with sides 1 unit long.

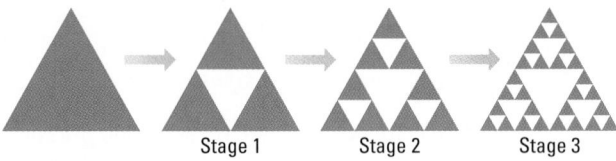

Stage 1   Stage 2   Stage 3

78. Let a_n be the number of triangles removed at the *n*th stage. Find a rule for a_n. Then find the total number of triangles removed through the 10th stage. 3^{n-1}; 19,683

79. Let b_n be the remaining area of the original triangle at the *n*th stage. Find a rule for b_n. Then find the remaining area of the original triangle at the 15th stage.
   See margin.

80. *Writing* Compare the graphs of $a_n = 4(2)^{n-1}$ where *n* is a positive integer and $f(x) = 4(2)^{x-1}$ where *x* is a real number. Discuss how the graph of a geometric sequence with $r > 0$ is similar to and different from the graph of an exponential function.   **See margin.**

81. 🖩 MULTI-STEP PROBLEM Suppose two computer companies, Company A and Company B, opened in 1991. The revenues of Company A increased arithmetically through 2000, while the revenues of Company B increased geometrically through 2000. In 1996 the revenue of Company A was $523.7 million. In 1996 the revenue of Company B was $65.6 million.

   a. The revenues of Company A have a common difference of 55.5. The revenues of Company B have a common ratio of 2. Find a rule for the revenues in the *n*th year of each company. Let a_1 represent 1991.   **Company A: $a_n = 190.7 + 55.5n$;**
   Company B: $a_n = 2.05(2)^{n-1}$

   b. Graph each sequence from part (a).   **See margin for graphs.**

   c. Find the sum of the revenues from 1991 through 2000 for each company.

   d. *Writing* Use a graphing calculator or spreadsheet to find when the revenue of Company B is greater than the revenue of Company A. Write a brief paragraph explaining which company you would rather own. Be sure to refer to your graphs from part (b).

82. WORKING WITH FRACTIONS Using the rule for the sum of the first *n* terms of a geometric series, write the polynomial as a rational expression.

   a. $1 + x + x^2 + x^3 + x^4$ $\frac{x^5 - 1}{x - 1}$ b. $3x + 6x^3 + 12x^5 + 24x^7$ $\frac{3x(16x^8 - 1)}{2x^2 - 1}$

MIXED REVIEW

ORDERING NUMBERS Plot the numbers on a number line. Write the numbers in increasing order. (Review 1.1 for 11.4)

83. $\frac{2}{5}, \frac{6}{7}, 1, \frac{7}{6}, \frac{3}{2}$

84. $-\frac{1}{5}, 0, 2, \sqrt{5}, 3$

85. $-3.2, -\frac{5}{2}, -2,$
    $-1, 1.5$

83. $\frac{3}{2}, \frac{2}{5}, \frac{7}{6}, 1, \frac{6}{7}$

84. $\sqrt{5}, 2, -\frac{1}{5}, 0, 3$

85. $-\frac{5}{2}, -1, 1.5, -3.2, -2$

SOLVING ALGEBRAICALLY Solve the inequality algebraically. (Review 5.7)

86. $x^2 + x - 2 \geq 0$
    $x \leq -2 \text{ or } x \geq 1$

87. $x^2 - 6x - 7 \leq 0$
    $-1 \leq x \leq 7$

88. $x^2 < 36$ $-6 < x < 6$

89. $-x^2 - 8x < 20$
    all real numbers

90. $3x^2 - 9x + 6 > 0$
    $x < 1 \text{ or } x > 2$

91. $\frac{1}{2}x^2 + 5x \leq -12$
    $-6 \leq x \leq -4$

SOLVING EQUATIONS Solve using any method. Check each solution.
(Review 9.6 for 11.4)

92. $\frac{3}{1+x} = 8$ $-\frac{5}{8}$

93. $\frac{4}{1-x} = 10$ $\frac{3}{5}$

94. $\frac{-12}{x+4} = -x$ $-6, 2$

95. $-\frac{24}{x} - x = 11$ $-8, -3$

96. $\frac{x}{x-8} = \frac{x}{24}$ $0, 32$

97. $x + 10 = \frac{x^2}{x-5}$ 10

QUIZ 1

Self-Test for Lessons 11.1–11.3

Write the next term in the sequence. Then write a rule for the nth term.
(Lesson 11.1)

1. $0, 2, 4, 6, \ldots$
   $8; 2(n-1)$

2. $3, 9, 27, 81, \ldots$ $243; 3^n$

3. $\frac{1}{5}, -\frac{1}{10}, \frac{1}{20}, -\frac{1}{40}, \ldots$
   $\frac{1}{80}; \left(\frac{1}{5}\right)\left(-\frac{1}{2}\right)^{n-1}$

Find the sum of the series. (Lesson 11.1)

4. $\sum_{k=0}^{4} k^4$ 354

5. $\sum_{m=1}^{6} (m^2 + 5)$ 121

6. $\sum_{n=1}^{5} (n^3 - 1)$ 220

Write a rule for the nth term of the arithmetic sequence. Then find a_{12}.
(Lesson 11.2)

7. $1, 5, 9, 13, \ldots$
   $-3 + 4n; 45$

8. $34, 25, 16, 7, -2, \ldots$
   $43 - 9n; -65$

9. $\frac{1}{2}, 1, \frac{3}{2}, 2, \ldots$ $\frac{n}{2}; 6$

10. Find the sum of the first 30 terms of the arithmetic series 694.5
    $1.4 + 2.9 + 4.4 + 5.9 + 7.4 + \cdots$. (Lesson 11.2)

Write a rule for the nth term of the geometric sequence. Then find a_{15}.
(Lesson 11.3)

11. $2, 10, 50, 250, \ldots$
    $2(5)^{n-1}; 12,207,031,250$

12. $-3, 12, -48, 192, \ldots$
    $-3(-4)^{n-1}; -805,306,368$

13. $12, 4, \frac{4}{3}, \frac{4}{9}, \ldots$
    $12\left(\frac{1}{3}\right)^{n-1}; 2.509 \times 10^{-6}$

14. **FAMILY TREE** A portion of John's parental family tree is shown at the right. Find a rule for the number of people in the nth generation. If 10 generations of his family have lived in this country, how many people is this? (Lesson 11.3) $2^{n-1}; 1023$

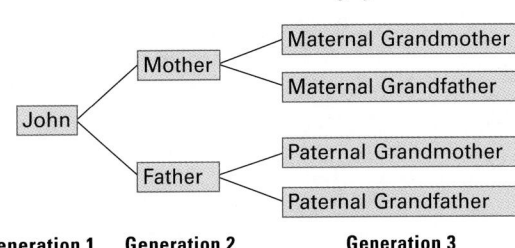

Generation 1   Generation 2   Generation 3

Additional Test Preparation *Sample answer:*

1. Using the formula $a_n = a_1(r)^{n-1}$, $1,048,576 = 1(2)^{n-1}$.
   Then $2^{n-1} = 1,048,576$, and $n - 1 = 20$.
   Thus, $2^{20} = 1,048,576$.

2. The common ratio for both sequences is the same.
   Using $a_n = a_1 r^{n-1}$, $648 = 24r^{4-1}$, and $r = 3$.

---

4 ASSESS

DAILY HOMEWORK QUIZ

🔲 *Transparency Available*

1. Is the sequence *arithmetic*, *geometric*, or *neither*?
   a. $2, 3, 5, 6, 8, 9, \ldots$ neither
   b. $\frac{1}{4}, \frac{1}{2}, \frac{3}{4}, 1, \ldots$ arithmetic
   c. $6, 2, \frac{2}{3}, \frac{2}{9}, \ldots$ geometric

2. Write a rule for the nth term of the geometric sequence $2, -8, 32, -128, \ldots$. Then find a_6.
   $a_n = 2(-4)^{n-1}, -2048$

3. Write a rule for the nth term of the geometric sequence $r = -2$, $a_3 = 16$. $a_n = 4(-2)^{n-1}$

4. Two terms of a geometric sequence are $a_2 = 28$ and $a_5 = -224$. Write a rule for the nth term. $a_n = -14(-2)^{n-1}$

5. Find the sum of the first 10 terms of the geometric series $1 - 4 + 16 - 64 + \ldots$. Then find n such that $S_n = 838,861$.
   $-209,715; 11$

EXTRA CHALLENGE NOTE

↪ Challenge problems for Lesson 11.3 are available in **blackline** format in the *Chapter 11 Resource Book,* p. 48 and at **www.mcdougallittell.com**.

ADDITIONAL TEST PREPARATION

1. **OPEN ENDED** Given the geometric sequence $1, 2, 4, 8, \ldots$ $1,048,576$, how can you determine which power of 2 is equal to $1,048,576$? See margin.

2. **WRITING** Suppose you know that in a geometric sequence $a_2 = 24$ and $a_5 = 648$. Explain why the sequence $24, \_\_\_, \_\_\_, 648$ can be used to find the common ratio. See margin.

ADDITIONAL RESOURCES
An alternative Quiz for Lessons 11.1–11.3 is available in the *Chapter 11 Resource Book,* p. 49.

83–85. See Additional Answers beginning on page AA1.

673

1 Planning the Activity

PURPOSE
To determine the rule and the sum for an infinite geometric series.

MATERIALS
- scissors
- piece of paper
- Activity Support Master (*Chapter 11 Resource Book*, p. 53)

PACING
- Exploring the Concept — 10 min
- Drawing Conclusions — 10 min

▶ LINK TO LESSON
The infinite geometric series presented in the activity is the same one that begins Lesson 11.4. Students can use their paper model to verify the mathematics presented on page 675.

2 Managing the Activity

ALTERNATIVE APPROACH
To save time you can demonstrate the activity at the front of the classroom and ask a student to record the entries in a table on the chalkboard.

CLASSROOM MANAGEMENT
To ensure that students begin the activity with a square, use graph paper.

3 Closing the Activity

★ KEY DISCOVERY
The sum of a geometric series in which each term is smaller than the previous term has a finite sum.

ACTIVITY ASSESSMENT
JOURNAL Explain how you could use the pieces of paper you cut to show that the sum of their areas gets closer and closer to 1 as you complete more and more folds.
See sample answer at right.

674

⊙ ACTIVITY 11.4
Developing Concepts

Investigating an Infinite Geometric Series

GROUP ACTIVITY
Work in a small group.

MATERIALS
- scissors
- piece of paper

Step 4. $\frac{1}{8}, \frac{1}{16}, \frac{1}{32}$; each area is $\frac{1}{2}$ the area before it, so the areas form a geometric sequence with a common ratio of $\frac{1}{2}$.

▶ **QUESTION** What is the sum of an infinite geometric series?

▶ **EXPLORING THE CONCEPT**

You can illustrate an infinite geometric series by cutting a piece of paper into smaller and smaller pieces. Start with a square piece of paper. Define the area of the paper to be 1 square unit.

① Fold the paper in half and cut along the fold. Place one half on a desktop and hold the remaining half.

② Fold the piece of paper you are holding in half and cut along the fold. Place one half on the desktop and hold the remaining half.

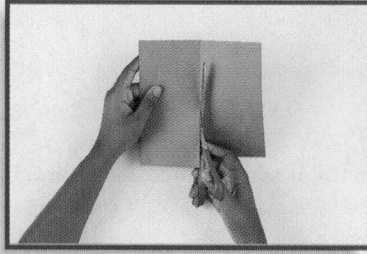

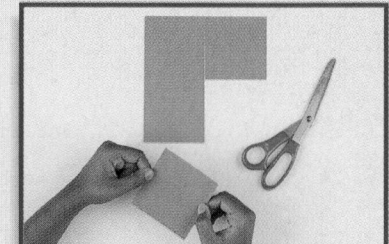

③ Repeat **Steps 1 and 2** until you find it too difficult to fold and cut the piece of paper you are holding.

④ The first piece of paper you placed on the desktop has an area of $\frac{1}{2}$ square unit. The second piece of paper has an area of $\frac{1}{4}$ square unit. Write the areas of the next three pieces of paper. Explain why these areas form a geometric sequence.

⑤ Copy and complete the table by recording the number of pieces of paper on the desktop and the combined area of the pieces at each step. $\frac{3}{4}, \frac{7}{8}, \frac{15}{16}, \frac{31}{32}$

Number of pieces on desktop	1	2	3	4	5	...
Combined area of pieces	$\frac{1}{2}$	$\frac{1}{2} + \frac{1}{4} = ?$	?	?	?	...

▶ **DRAWING CONCLUSIONS**

1. What number does the combined area of the pieces of paper appear to be approaching? **1**

2. The formula for the combined area after n cuts is $A_n = \frac{1}{2}\left(\dfrac{1 - \left(\frac{1}{2}\right)^n}{1 - \frac{1}{2}}\right)$.

   What happens to this formula as $n \to \infty$? (*Hint:* The only term with n in it, $\left(\frac{1}{2}\right)^n$, approaches 0 as $n \to \infty$.) **A_n approaches 1.**

674   **Chapter 11** *Sequences and Series*

Activity Assessment *Sample answer:*
Place the cut pieces of paper on a paper that is congruent to the 1-unit square. As you place more and more pieces on it, you get closer and closer to completely covering the square.

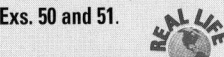

11.4

Infinite Geometric Series

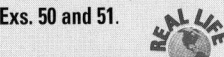

What you should learn

GOAL 1 Find sums of infinite geometric series.

GOAL 2 Use infinite geometric series as models of **real-life** situations, such as the distance traveled by a bouncing ball in **Example 4**.

Why you should learn it

▼ To solve **real-life** problems, such as finding the spending generated by tourists in Malaysia in **Exs. 50 and 51**.

GOAL 1 **USING INFINITE GEOMETRIC SERIES**

Consider the following infinite geometric series:

$$\frac{1}{2} + \frac{1}{4} + \frac{1}{8} + \frac{1}{16} + \frac{1}{32} + \cdots$$

Even though this series has infinitely many terms, it has a finite sum! To see this, compute and graph the sum of the first n terms for several values of n.

$S_1 = \frac{1}{2} = 0.5$

$S_2 = \frac{1}{2} + \frac{1}{4} = 0.75$

$S_3 = \frac{1}{2} + \frac{1}{4} + \frac{1}{8} \approx 0.88$

$S_4 = \frac{1}{2} + \frac{1}{4} + \frac{1}{8} + \frac{1}{16} \approx 0.94$

$S_5 = \frac{1}{2} + \frac{1}{4} + \frac{1}{8} + \frac{1}{16} + \frac{1}{32} \approx 0.97$

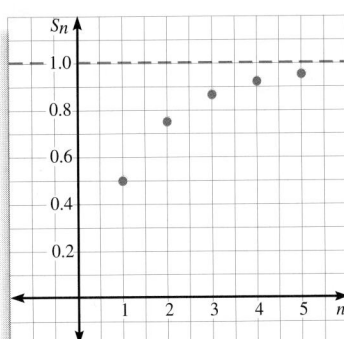

Notice that S_n appears to be approaching 1 as n increases. To see why this makes sense, consider the rule for S_n:

$$S_n = a_1\left(\frac{1 - r^n}{1 - r}\right) = \frac{1}{2}\left(\frac{1 - \left(\frac{1}{2}\right)^n}{1 - \frac{1}{2}}\right) = 1 - \left(\frac{1}{2}\right)^n$$

As n increases, $\left(\frac{1}{2}\right)^n$ gets closer and closer to 0, which means that S_n gets closer and closer to 1. The same is true for r^n provided r is between -1 and 1. Therefore, the formula for the sum of a *finite* geometric series, $S_n = a_1\left(\frac{1 - r^n}{1 - r}\right)$, approaches the formula below as n increases.

THE SUM OF AN INFINITE GEOMETRIC SERIES

The sum of an infinite geometric series with first term a_1 and common ratio r is given by

$$S = \frac{a_1}{1 - r}$$

provided $|r| < 1$. If $|r| \geq 1$, the series has no sum.

For the series described above, the sum is $S = \dfrac{\frac{1}{2}}{1 - \frac{1}{2}} = 1$, as expected.

11.4 *Infinite Geometric Series* **675**

1 PLAN

PACING
Basic: 1 day
Average: 1 day
Advanced: 1 day
Block Schedule: 0.5 block with 11.5

LESSON OPENER
GRAPHING CALCULATOR
An alternative way to approach Lesson 11.4 is to use the Graphing Calculator Lesson Opener:
• Blackline Master (*Chapter 11 Resource Book*, p. 54)
• Transparency (p. 75)

MEETING INDIVIDUAL NEEDS
• *Chapter 11 Resource Book*
  Prerequisite Skills Review (p. 5)
  Practice Level A (p. 57)
  Practice Level B (p. 58)
  Practice Level C (p. 59)
  Reteaching with Practice (p. 60)
  Absent Student Catch-Up (p. 62)
  Challenge (p. 64)
• *Resources in Spanish*
• *Personal Student Tutor*

NEW-TEACHER SUPPORT
See the Tips for New Teachers on pp. 1–2 of the *Chapter 11 Resource Book* for additional notes about Lesson 11.4.

WARM-UP EXERCISES
Transparency Available
1. Find the sum of the geometric series $8 + 4 + 2 + 1 + \ldots + \frac{1}{16}$. $\frac{255}{16}$
2. If $S_n = 93$, $a_1 = 3$, and $r = 2$, find n. **5**
3. A certain substance decomposes and loses 20% of its weight each hour. If the original quantity of the substance is 300 grams, how much remains after 7 hours? **about 63 grams**
4. Write $\frac{9}{11}$ as a decimal. **$0.\overline{81}$**

MOTIVATING THE LESSON
Suppose you form a group of 4 students who are willing to commit to recycling. Your goal is to double your group membership each month. How many members will you have after one month if you meet your goal? after 2 months? a year? You can use an infinite geometric series to predict how many members you can expect to have each month.

EXTRA EXAMPLE 1
Find the sum of the infinite geometric series.

a. $\sum_{i=1}^{\infty} 2(0.1)^{i-1}$ $\frac{20}{9}$

b. $12 + 4 + \frac{4}{3} + \frac{4}{9} + \ldots$ 18

EXTRA EXAMPLE 2
An infinite geometric series with first term $a_1 = 5$ has a sum of $\frac{27}{5}$. What is the common ratio of the series? $\frac{2}{27}$

EXTRA EXAMPLE 3
Write $0.416666\ldots$ as a fraction. $\frac{5}{12}$

CHECKPOINT EXERCISES

For use after Example 1:

1. Find the sum of the infinite geometric series $-30 + 15 - \frac{15}{2} + \frac{15}{4} - \ldots$. -20

For use after Example 2:

2. An infinite geometric series with first term $a_1 = -\frac{4}{3}$ has a sum of -2. What is the common ratio of the series? $\frac{1}{3}$

For use after Example 3:

3. Write $0.0272727\ldots$ as a fraction. $\frac{3}{110}$

STUDENT HELP

INTERNET
▸ HOMEWORK HELP
Visit our Web site
www.mcdougallittell.com
for extra examples.

EXAMPLE 1 *Finding Sums of Infinite Geometric Series*

Find the sum of the infinite geometric series.

a. $\sum_{i=1}^{\infty} 3(0.7)^{i-1}$

b. $1 - \frac{1}{4} + \frac{1}{16} - \frac{1}{64} + \cdots$

SOLUTION

a. For this series, $a_1 = 3$ and $r = 0.7$.

$$S = \frac{a_1}{1-r} = \frac{3}{1-0.7} = 10$$

b. For this series, $a_1 = 1$ and $r = -\frac{1}{4}$.

$$S = \frac{a_1}{1-r} = \frac{1}{1-\left(-\frac{1}{4}\right)} = \frac{4}{5}$$

EXAMPLE 2 *Finding the Common Ratio*

An infinite geometric series with first term $a_1 = 4$ has a sum of 10. What is the common ratio of the series?

SOLUTION

$$S = \frac{a_1}{1-r}$$ Write rule for sum.

$$10 = \frac{4}{1-r}$$ Substitute for S and a_1.

$$10(1-r) = 4$$ Multiply each side by $1-r$.

$$1 - r = \frac{2}{5}$$ Divide each side by 10.

$$r = \frac{3}{5}$$ Solve for r.

▸ The common ratio is $r = \frac{3}{5}$.

EXAMPLE 3 *Writing a Repeating Decimal as a Fraction*

Write $0.181818\ldots$ as a fraction.

SOLUTION

$$0.181818\ldots = 18(0.01) + 18(0.01)^2 + 18(0.01)^3 + \cdots$$

$$= \frac{a_1}{1-r}$$ Write rule for sum.

$$= \frac{18(0.01)}{1-0.01}$$ Substitute for a_1 and r.

$$= \frac{18}{99}$$ Write as a quotient of integers.

$$= \frac{2}{11}$$ Simplify.

▸ The repeating decimal $0.181818\ldots$ is $\frac{2}{11}$ as a fraction.

STUDENT HELP

↳ **Study Tip**
You can check the result in Example 3 by dividing 2 by 11 on a calculator.

GOAL 2 **INFINITE GEOMETRIC SERIES IN REAL LIFE**

EXAMPLE 4 *Using an Infinite Series as a Model*

BALL BOUNCE A ball is dropped from a height of 10 feet. Each time it hits the ground, it bounces to 80% of its previous height.

a. Find the total distance traveled by the ball.

b. On which bounce will the ball have traveled 85% of its total distance?

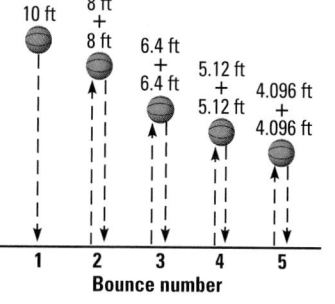

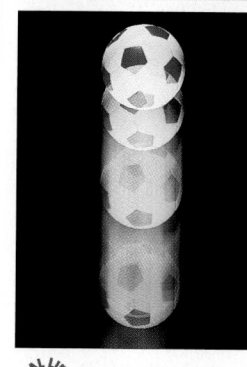

BALL BOUNCE
This photo of a ball bouncing was taken with time-lapse photography. The images of the ball get closer together as you move up, which means the ball's speed is decreasing.

SOLUTION

a. The total distance traveled by the ball is:

$$d = \underbrace{10}_{\text{down}} + \underbrace{10(0.8)}_{\text{up}} + \underbrace{10(0.8)}_{\text{down}} + \underbrace{10(0.8)^2}_{\text{up}} + \underbrace{10(0.8)^2}_{\text{down}} + \underbrace{10(0.8)^3}_{\text{up}} + \cdots$$

$$= 10 + 2[10(0.8)] + 2[10(0.8)^2] + 2[10(0.8)^3] + \cdots$$

$$= 10 + 20(0.8) + 20(0.8)^2 + 20(0.8)^3 + \cdots$$

$$= 10 + \frac{20(0.8)}{1 - 0.8} \qquad \text{Excluding first term, find sum of series.}$$

$$= 10 + 80 \qquad \text{Simplify fraction.}$$

$$= 90 \qquad \text{Simplify.}$$

▶ The ball travels a total distance of 90 feet.

b. Let n be the number of up-and-down bounces. The distance d_n the ball travels is:

$$d_n = \underbrace{10}_{\substack{\text{down-only} \\ \text{distance}}} + \underbrace{20(0.8)\left(\frac{1 - (0.8)^n}{1 - 0.8}\right)}_{\substack{\text{sum of } n \\ \text{up-and-down} \\ \text{bounces}}} \qquad \text{Write rule for } d_n.$$

$$0.85(90) = 10 + 20(0.8)\left(\frac{1 - (0.8)^n}{1 - 0.8}\right) \qquad \text{Substitute for } d_n.$$

$$76.5 = 10 + 16\left(\frac{1 - (0.8)^n}{1 - 0.8}\right) \qquad \text{Simplify.}$$

$$4.156 \approx \frac{1 - (0.8)^n}{0.2} \qquad \text{Isolate fraction.}$$

$$0.831 \approx 1 - (0.8)^n \qquad \text{Multiply each side by 0.2.}$$

$$(0.8)^n \approx 0.169 \qquad \text{Isolate exponential expression.}$$

$$n \approx \frac{\log 0.169}{\log 0.8} \approx 7.97 \qquad \text{Solve for } n.$$

▶ The ball travels 85% of its total distance after about 8 up-and-down bounces, or after 9 bounces including the first down-only bounce.

11.4 *Infinite Geometric Series* **677**

Checkpoint Exercise *Sample answer:*
190 ft; after about 21 up and down bounces or after 22 bounces including the first down-only bounce

Closure Question *Sample answer:*
Write the decimal as an infinite geometric series. Then find the sum of the series. The sum is the equivalent fraction.

 EXTRA EXAMPLE 4
A pendulum swings 10 feet going left to right. On its swing back, it swings 90% as far as the first swing. Each successive swing is 90% of its previous swing.
a. Find the total distance traveled by the pendulum when it finally stops. **100 ft**
b. When will the pendulum have traveled 50% of its total distance? **After its 7th swing, it will have traveled about 52 ft.**

✓ **CHECKPOINT EXERCISES**
For use after Example 4:
1. A ball is dropped from a height of 10 feet. Each time it hits the ground, it bounces to 90% of its previous height. Find the total distance traveled by the ball. On which bounce will the ball have traveled 90% of its total distance? **See margin.**

! **COMMON ERROR**
EXAMPLE 4 Students may forget to account for the first bounce or may count that distance twice. Suggest that they draw a diagram of the bounces and label each down and up distance with the correct expression.

FOCUS ON VOCABULARY
A finite geometric series has a sum. An infinite geometric series has a sum only if the constant ratio is between −1 and 1.

CLOSURE QUESTION
How can you use an infinite geometric series to convert a repeating decimal to a fraction? **See margin.**

DAILY PUZZLER
A farmer bought a chicken, a goat, a sheep, and a mule. He bought the goat for twice the amount of the chicken, the sheep for twice the amount of the goat, and the mule for twice the amount of the sheep. Lunch cost him $10 and he spent a total of $100 for the day. How much did each animal cost? **chicken: $6, goat: $12, sheep: $24, mule: $48**

ASSIGNMENT GUIDE

BASIC
Day 1: pp. 678–680 Exs. 13–22,
29–33, 38–40, 47, 53–54,
57–69 odd

AVERAGE
Day 1: pp. 678–680 Exs. 13–25,
29–34, 38–46 even, 47,
50–51, 53–54, 57–69 odd

ADVANCED
Day 1: pp. 678–680 Exs. 13–26,
29–37, 38–48 even, 50–56,
57–69 odd

BLOCK SCHEDULE
pp. 678–680 Exs. 13–25, 29–34,
38–46 even, 47, 50–51, 53–54,
57–69 odd (with 11.5)

EXERCISE LEVELS
Level A: *Easier*
13–24, 29–34

Level B: *More Difficult*
25–28, 35–54

Level C: *Most Difficult*
55

✓ **HOMEWORK CHECK**
To quickly check student under-
standing of key concepts, go
over the following exercises:
Exs. 16, 18, 30, 38. See also the
Daily Homework Quiz:
• Blackline Master (*Chapter 11
Resource Book*, p. 67)
• Transparency (p. 87)

GUIDED PRACTICE

Vocabulary Check ✓   **1.** Complete this statement: A(n) _?_ geometric series has infinitely many terms.
infinite

Concept Check ✓   **2.** Under what conditions will $\sum_{i=1}^{\infty} a_1 r^{i-1}$ have a sum?
if $-1 < r < 1$

3. What two things do you need to know to find the sum of an infinite geometric series? the first term and the common ratio

Skill Check ✓   **Find the sum of the infinite geometric series.**

4. $\sum_{n=1}^{\infty} 5\left(\frac{1}{4}\right)^{n-1}$ $\frac{20}{3}$ **5.** $-2 + \frac{1}{2} - \frac{1}{8} + \frac{1}{32} - \cdots$ $-\frac{8}{5}$

Find the common ratio of the infinite geometric series with the given sum and first term.

6. $S = 6, a_1 = 1$ $\frac{5}{6}$ **7.** $S = 12, a_1 = 2$ $\frac{5}{6}$ **8.** $S = 10\frac{1}{2}, a_1 = \frac{1}{2}$ $\frac{20}{21}$

Write the repeating decimal as a fraction.

9. $0.555\ldots$ $\frac{5}{9}$ **10.** $0.1212\ldots$ $\frac{4}{33}$ **11.** $245.245245\ldots$ $\frac{245,000}{999}$

12. 🏀 **BALL BOUNCE** A ball is dropped from a height of 5 feet. Each time it hits the ground, it bounces one half of its previous height.

   a. Find the total distance traveled by the ball.   15 ft

   b. On which bounce will the ball have traveled 75% of its total distance?
about halfway through the second double bounce

PRACTICE AND APPLICATIONS

STUDENT HELP
► **Extra Practice**
to help you master
skills is on p. 956.

IDENTIFYING A SUM Decide whether the infinite geometric series has a sum. Explain why or why not.   yes; $|r| = \frac{1}{3}, \frac{1}{3} < 1$

13. $\sum_{n=0}^{\infty} 3\left(\frac{3}{2}\right)^{n-1}$ no; $|r| = \frac{3}{2}, \frac{3}{2} > 1$ **14.** $\sum_{n=0}^{\infty} -5\left(\frac{1}{5}\right)^{n}$ yes; $|r| = \frac{1}{5}, \frac{1}{5} < 1$ **15.** $\sum_{n=1}^{\infty} \frac{3}{2}\left(\frac{1}{3}\right)^{n-1}$ **16.** $\sum_{n=0}^{\infty} \frac{1}{4}\left(\frac{4}{3}\right)^{n}$ no; $|r| = \frac{4}{3}, \frac{4}{3} > 1$

FINDING SUMS Find the sum of the infinite geometric series if it has one.

17. $\sum_{n=0}^{\infty} \left(\frac{1}{2}\right)^{n}$ 2   **18.** $\sum_{n=0}^{\infty} 3\left(\frac{2}{3}\right)^{n}$ 9   **19.** $\sum_{n=1}^{\infty} \left(-\frac{1}{2}\right)^{n-1}$ $\frac{2}{3}$ **20.** $\sum_{n=0}^{\infty} \frac{2}{7}(2)^{n}$ no sum

21. $\sum_{n=0}^{\infty} 4\left(\frac{1}{4}\right)^{n}$ $\frac{16}{3}$ **22.** $\sum_{n=1}^{\infty} \left(\frac{1}{10}\right)^{n-1}$ $\frac{10}{9}$ **23.** $\sum_{n=0}^{\infty} 2\left(\frac{6}{5}\right)^{n}$ no sum   **24.** $\sum_{n=0}^{\infty} 4\left(\frac{3}{7}\right)^{n}$ 7

25. $\sum_{n=0}^{\infty} -\frac{1}{8}\left(-\frac{1}{2}\right)^{n}$ $-\frac{1}{12}$ **26.** $\sum_{n=1}^{\infty} \frac{1}{2}\left(-\frac{2}{5}\right)^{n-1}$ $\frac{5}{14}$ **27.** $\sum_{n=0}^{\infty} \frac{1}{12}\left(-\frac{3}{25}\right)^{n}$ $\frac{25}{336}$ **28.** $\sum_{n=1}^{\infty} -\left(-\frac{2}{11}\right)^{n-1}$ $-\frac{11}{13}$

STUDENT HELP
► **HOMEWORK HELP**
Example 1: Exs. 13–28
Example 2: Exs. 29–37
Example 3: Exs. 38–46
Example 4: Exs. 47–51

FINDING COMMON RATIOS Find the common ratio of the infinite geometric series with the given sum and first term.

29. $S = 4, a_1 = 1$ $\frac{3}{4}$ **30.** $S = 10, a_1 = 1$ $\frac{9}{10}$ **31.** $S = 12, a_1 = 3$ $\frac{3}{4}$

32. $S = 8, a_1 = 2$ $\frac{3}{4}$ **33.** $S = 6, a_1 = 2$ $\frac{2}{3}$ **34.** $S = 50, a_1 = 4$ $\frac{23}{25}$

35. $S = -\frac{1}{9}, a_1 = -\frac{1}{6}$ $-\frac{1}{2}$ **36.** $S = -\frac{11}{13}, a_1 = -1$ $-\frac{2}{11}$ **37.** $S = 2\frac{2}{9}, a_1 = 4$ $-\frac{4}{5}$

WRITING REPEATING DECIMALS Write the repeating decimal as a fraction.

38. $0.444\ldots$ $\dfrac{4}{9}$

39. $0.777\ldots$ $\dfrac{7}{9}$

40. $0.999\ldots$ 1

41. $0.5151\ldots$ $\dfrac{17}{33}$

42. $0.2323\ldots$ $\dfrac{23}{99}$

43. $0.1616\ldots$ $\dfrac{16}{99}$

44. $63.6363\ldots$ $\dfrac{700}{11}$

45. $120.120120\ldots$ $\dfrac{40{,}000}{333}$

46. $297.297297\ldots$ $\dfrac{33{,}000}{111}$

47. 🌐 **PENDULUM** A pendulum is released to swing freely. On the first swing, the pendulum travels a distance of 18 inches. On each successive swing, the pendulum travels 90% of the distance of the previous swing. What is the total distance the pendulum swings? After how many swings has the pendulum traveled 80% of its total distance?   **180 in. = 15 ft; after 16 swings**

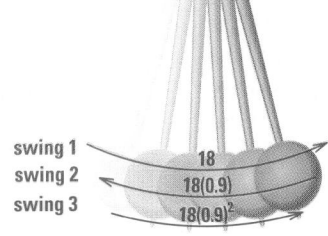

swing 1   18
swing 2   18(0.9)
swing 3   18(0.9)²

48. 🌐 **WINDOWS** Some types of windows are constructed with two parallel panes of glass, each of which reflects half of the sunlight that hits it from either side. The other half of the sunlight passes through the pane. How much of the sunlight will pass through *both* panes? $\dfrac{1}{3}$

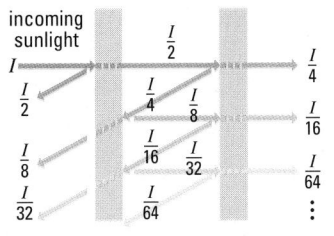

49. total distance $= \dfrac{20}{\left(1 - \frac{1}{2}\right)}$
$= 40$ ft; **total time**
$= \dfrac{1}{\left(1 - \frac{1}{2}\right)} = 2$ sec

49. 🌐 **ZENO'S PARADOX** Can the Greek hero Achilles, running at 20 feet per second, ever catch a tortoise, starting 20 feet away and running at 10 feet per second? The Greek mathematician Zeno said no. He reasoned as follows:

- When Achilles runs 20 feet the tortoise will be in a new spot, 10 feet away.

- Then, when Achilles gets to that spot, the tortoise will be 5 feet away.

- Achilles will keep cutting the distance in half but will never catch the tortoise.

In actuality, looking at the race as Zeno did, you can see that both the distances and the times required to achieve them form infinite geometric series. Using the table, show that both series have finite sums. What do these sums represent?

Distance (ft)	20	10	5	2.5	1.25	0.625	. . .
Time (sec)	1	0.5	0.25	0.125	0.0625	0.03125	. . .

FOCUS ON CAREERS

🌐 **ECONOMIST**
Economists study the ways a society distributes resources to produce goods and services. They conduct research, monitor economic trends, and make projections.

🖱 **CAREER LINK**
www.mcdougallittell.com

🌐 **TOURISM** In Exercises 50 and 51, use the following information.

In 1974 the Malaysian Tourist Development Corporation studied the economic impact of distributing tourist brochures. It was estimated that M\$4.72 ("M\$" means Malaysian dollars) in additional money was spent by tourists for every brochure distributed in the capital city of Kuala Lumpur. It was also estimated that for each M\$1 spent on goods or services, 80.5% of that would be re-spent, creating a "multiplier" effect. (That is, each Malaysian dollar spent would lead to total spending of M\$1 + (0.805)M\$1 + (0.805)(0.805)M\$1 + (0.805)(0.805)(0.805)M\$1 + · · · .)

50. What total spending was generated by a tourist spending M\$1 in 1974?   **about M\$5.13**

51. How much total spending would be generated by the average tourist who received a brochure?   **about M\$24.21**

52. LOGICAL REASONING Find two different infinite geometric series whose sum is 3.

Sample answer: $1 + \dfrac{2}{3} + \left(\dfrac{2}{3}\right)^2 + \left(\dfrac{2}{3}\right)^3 + \ldots;\ 4 - \dfrac{4}{3} + \dfrac{4}{9} - \dfrac{4}{27} + \dfrac{4}{81} - \cdots$

11.4 *Infinite Geometric Series*   **679**

!COMMON ERROR
EXERCISES 17–28 Students may confuse the first term with the common ratio. Before they begin evaluating, ask students to identify the first term and the common ratio of each series.

APPLICATION NOTE
EXERCISE 49 This is Zeno's paradox of motion, by which he intended to demonstrate the logical impossibility of motion. He argued that reason leads us to know that before a given distance can be traversed, half of the distance must be traversed. And before half the distance can be traversed, one-fourth of the distance must be traversed. And so on. Thus, he reasoned that the existence of motion is a deception of the senses and a true understanding of reality can be gained only from reason. The development of calculus with the notion of limits of sums was applied to problems in logic including logical paradoxes.

CAREER NOTE
EXERCISES 50 AND 51
Additional information about economists is available at **www.mcdougallittell.com**.

ADDITIONAL PRACTICE AND RETEACHING

For Lesson 11.4:
- Practice Levels A, B, and C (*Chapter 11 Resource Book*, p. 57)
- Reteaching with Practice (*Chapter 11 Resource Book*, p. 60)
- 🖥 See Lesson 11.4 of the *Personal Student Tutor*

For more Mixed Review:
- 🖥 Search the *Test and Practice Generator* for key words or specific lessons.

DAILY HOMEWORK QUIZ

📖 *Transparency Available*

1. Find the sum of the infinite geometric series $\sum_{i=1}^{\infty} 3(0.8)^{i-1}$.

   15

2. An infinite geometric series with a first term $a_1 = -6$ has a sum of -8. What is the common ratio of the series?

   $\frac{1}{4}$

3. Write $0.454545\ldots$ as a fraction.

   $\frac{5}{11}$

EXTRA CHALLENGE NOTE

↳ Challenge problems for Lesson 11.4 are available in **blackline** format in the *Chapter 11 Resource Book*, p. 64 and at **www.mcdougallittell.com**.

ADDITIONAL TEST PREPARATION

1. **OPEN ENDED** Why does the sum of an infinite geometric sequence not exist if the constant ratio is greater than or equal to one? **The sums get larger and larger as the number of terms gets larger and larger. Thus, they approach infinity.**

2. **WRITING** What do we mean when we say that the sum of the infinite geometric series $1 + \frac{1}{2} + \frac{1}{4} + \frac{1}{8} + \frac{1}{16} + \ldots$ is 2? **The sum of n terms as n gets larger and larger gets closer and closer to 2.**

62.
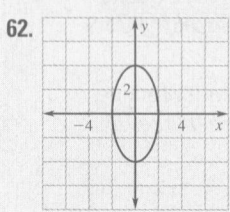

63 and 64. See Additional Answers beginning on page AA1.

53. **MULTIPLE CHOICE** An infinite geometric series with first term $a_1 = 24$ has a sum of 48. What is the common ratio of the series? **B**

Ⓐ $r = 1$ Ⓑ $r = \frac{1}{2}$ Ⓒ $r = 2$ Ⓓ $r = \frac{1}{24}$ Ⓔ $r = \frac{3}{8}$

54. **MULTIPLE CHOICE** The repeating decimal $18.181818\ldots$ is equivalent to what fraction? **D**

Ⓐ $\frac{891}{50}$ Ⓑ $\frac{2}{11}$ Ⓒ $\frac{181}{9}$ Ⓓ $\frac{200}{11}$ Ⓔ $\frac{1783}{99}$

★ **Challenge**

55. **GEOMETRY CONNECTION**

A *Koch snowflake* is created by starting with an equilateral triangle with sides 1 unit long. Then, on the middle third of each side of the triangle, a new equilateral triangle is constructed. This process is repeated as shown.

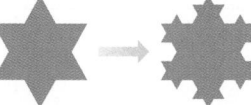

Figure 1   Figure 2   Figure 3

55a. Figure 3: 12, $\frac{\sqrt{3}}{324}$, $\frac{\sqrt{3}}{27}$;

Figure 4: 48, $\frac{\sqrt{3}}{2916}$, $\frac{\sqrt{3}}{60.75}$;

Figure n: $3(4)^{n-2}$, $\frac{\sqrt{3}}{4 \cdot 9^{n-1}}$,

$\left(\frac{3\sqrt{3}}{16}\right)\left(\frac{4}{9}\right)^{n-1}$ for $n \geq 2$

a. Recall that the area of an equilateral triangle with side length s can be found using the formula $A = \frac{s^2 \sqrt{3}}{4}$. Use this formula to complete the table below.

	Figure 1	Figure 2	Figure 3	Figure 4	...	Figure n
Number of new triangles	1	3	?	?	...	?
Area of each new triangle	$\frac{\sqrt{3}}{4}$	$\frac{\sqrt{3}}{36}$	?	?	...	?
Total new area	$\frac{\sqrt{3}}{4}$	$\frac{\sqrt{3}}{12}$	?	?	...	?

EXTRA CHALLENGE

↳ www.mcdougallittell.com

b. What is the total area of the Koch snowflake? (*Hint:* Add up the entries in the last row of the table.) **0.69282**

MIXED REVIEW

56. *x*-axis; *y*-axis; domain: $x \neq 0$; range: $y \neq 0$

57. *x*-axis; *y*-axis; domain: $x \neq 0$; range: $y \neq 0$

58. $y = -3$; *y*-axis; domain: $x \neq 0$; range: $y \neq -3$

59. $y = 1$; $x = -7$; domain: $x \neq -7$; range: $y \neq 1$

60. $y = -2$; $x = \frac{17}{2}$; domain: $x \neq \frac{17}{2}$; range: $y \neq -2$

61. $y = 2.2$; $x = 0.7$; domain: $x \neq 0.7$; range: $y \neq 2.2$

70. $-10\left(\frac{7}{8}\right)^{n-1}$

IDENTIFYING ASYMPTOTES Identify the horizontal and vertical asymptotes. State the domain and range of the function. (Review 9.2) 56–61. See margin.

56. $f(x) = \frac{5}{x}$ 57. $f(x) = -\frac{5}{x}$ 58. $f(x) = \frac{7}{x} - 3$

59. $f(x) = \frac{x-5}{x+7}$ 60. $f(x) = \frac{4x+3}{-2x+17}$ 61. $f(x) = \frac{2.2x+3.1}{x-0.7}$

GRAPHING Graph the equation. (Review 10.4) 62–64. See margin.

62. $\frac{x^2}{4} + \frac{y^2}{16} = 1$ 63. $\frac{25x^2}{4} + \frac{25y^2}{9} = 1$ 64. $\frac{x^2}{16} + y^2 = 1$

WRITING RULES Write a rule for the nth term of the sequence. Recall that d is the common difference of an arithmetic sequence and r is the common ratio of a geometric sequence. (Review 11.2, 11.3 for 11.5)

65. $d = 4$, $a_1 = 1$ $-3 + 4n$ 66. $d = 8$, $a_1 = -2$ $-10 + 8n$ 67. $d = -2$, $a_2 = 9$ $13 - 2n$

68. $r = 2$, $a_3 = -20$ $-5(2)^{n-1}$ 69. $r = 0.5$, $a_1 = 44\left(\frac{1}{2}\right)^{n-1}$ 70. $r = 0.875$, $a_1 = -10$

What you should learn

GOAL 1 Evaluate and write recursive rules for sequences.

GOAL 2 Use recursive rules to solve **real-life** problems, such as finding the number of fish in a lake in **Example 5**.

Why you should learn it

▼ To model **real-life** quantities, such as the number of trees on a tree farm in **Exs. 49 and 50**.

Recursive Rules for Sequences

GOAL 1 USING RECURSIVE RULES FOR SEQUENCES

So far in this chapter you have worked with *explicit rules* for the nth term of a sequence, such as $a_n = 3n - 2$ and $a_n = 3(2)^n$. An **explicit rule** gives a_n as a function of the term's position number n in the sequence.

In this lesson you will learn another way to define a sequence—by a *recursive rule*. A **recursive rule** gives the beginning term or terms of a sequence and then a *recursive equation* that tells how a_n is related to one or more preceding terms.

EXAMPLE 1 *Evaluating Recursive Rules*

Write the first five terms of the sequence.

a. *Factorial numbers:* $a_0 = 1$, $a_n = n \cdot a_{n-1}$

b. *Fibonacci sequence:* $a_1 = 1$, $a_2 = 1$, $a_n = a_{n-2} + a_{n-1}$

SOLUTION

a. $a_0 = 1$

$a_1 = 1 \cdot a_0 = 1 \cdot 1 = 1$

$a_2 = 2 \cdot a_1 = 2 \cdot 1 = 2$

$a_3 = 3 \cdot a_2 = 3 \cdot 2 = 6$

$a_4 = 4 \cdot a_3 = 4 \cdot 6 = 24$

b. $a_1 = 1$

$a_2 = 1$

$a_3 = a_1 + a_2 = 1 + 1 = 2$

$a_4 = a_2 + a_3 = 1 + 2 = 3$

$a_5 = a_3 + a_4 = 2 + 3 = 5$

· · · · · · · · · ·

The factorial numbers in part (a) of Example 1 are denoted by a special symbol, !, called a **factorial** symbol. The expression $n!$ is read "n factorial" and represents the product of all integers from 1 to n. Here are several factorial values.

$0! = 1$ (by definition) $1! = 1$ $2! = 2 \cdot 1 = 2$

$3! = 3 \cdot 2 \cdot 1 = 6$ $4! = 4 \cdot 3 \cdot 2 \cdot 1 = 24$ $5! = 5 \cdot 4 \cdot 3 \cdot 2 \cdot 1 = 120$

● ACTIVITY

Developing Concepts

Investigating Recursive Rules

❶ Find the first five terms of each sequence.

a. $a_1 = 3$ 3, 8, 13, 18, 23

$a_n = a_{n-1} + 5$

b. $a_1 = 3$ 3, 6, 12, 24, 48

$a_n = 2a_{n-1}$

❷ Based on the lists of terms you found in **Step 1**, what type of sequence is the sequence in part (a)? in part (b)?   **arithmetic; geometric**

11.5 *Recursive Rules for Sequences*   **681**

1 PLAN

PACING
Basic: 1 day
Average: 1 day
Advanced: 1 day
Block Schedule: 0.5 block with 11.4

LESSON OPENER
ACTIVITY
An alternative way to approach Lesson 11.5 is to use the Activity Lesson Opener:

• Blackline Master (*Chapter 11 Resource Book*, p. 68)

• Transparency (p. 76)

MEETING INDIVIDUAL NEEDS
• *Chapter 11 Resource Book*
  Prerequisite Skills Review (p. 5)
  Practice Level A (p. 70)
  Practice Level B (p. 71)
  Practice Level C (p. 72)
  Reteaching with Practice (p. 73)
  Absent Student Catch-Up (p. 75)
  Challenge (p. 78)

• *Resources in Spanish*

• Personal Student Tutor

NEW-TEACHER SUPPORT
See the Tips for New Teachers on pp. 1–2 of the *Chapter 11 Resource Book* for additional notes about Lesson 11.5.

WARM-UP EXERCISES

 Transparency Available

Write a rule for the nth term of the arithmetic sequence.

1. 4, 7, 11, 14, … $a_n = 1 + 3n$

2. $a_1 = -10$ and $d = -2$
$a_n = -8 - 2n$

Write a rule for the nth term of the geometric sequence.

3. −3, 6, −12, 24, −48, …
$a_n = -3(-2)^{n-1}$

4. $\dfrac{2}{3}, \dfrac{2}{9}, \dfrac{2}{27}, \dfrac{2}{81}$ $a_n = \dfrac{2}{3}\left(\dfrac{2}{3}\right)^{n-1}$

EXTRA EXAMPLE 1
Write the first five terms of the sequence.
a. $a_1 = 1$, $a_n = (a_{n-1})^2 + 1$ $1, 2, 5, 26, 677$
b. $a_1 = 2$, $a_2 = 2$, $a_n = a_{n-2} - a_{n-1}$ $2, 2, 0, 2, -2$

EXTRA EXAMPLE 2
Write the rule for the arithmetic sequence where $a_1 = 15$ and $d = 5$.
a. an explicit rule $a_n = 10 + 5n$
b. a recursive rule $a_1 = 15$, $a_n = a_{n-1} + 5$

EXTRA EXAMPLE 3
Write the rule for the geometric sequence where $a_1 = 4$ and $r = 0.2$.
a. an explicit rule $a_n = 4(0.2)^{n-1}$
b. a recursive rule $a_1 = 4$, $a_n = (0.2)a_{n-1}$

EXTRA EXAMPLE 4
Write a recursive rule for the sequence 1, 1, 4, 10, 28, 76.
$a_1 = 1$, $a_2 = 1$, $a_n = 2(a_{n-2} + a_{n-1})$

 CHECKPOINT EXERCISES

For use after Example 1:
1. Find the first 5 terms of the sequence where $a_1 = 3$, $a_n = \dfrac{a_{n-1}}{n}$. $3, \dfrac{3}{2}, \dfrac{1}{2}, \dfrac{1}{8}, \dfrac{1}{40}$

For use after Example 2:
2. Write the recursive rule for the arithmetic sequence where $a_1 = -2$ and $d = -10$.
$a_1 = -2$, $a_n = a_{n-1} - 10$

For use after Example 3:
3. Write the recursive rule for the geometric sequence where $a_1 = -2$ and $r = -3$.
$a_1 = -2$, $a_n = (-3)a_{n-1}$

For use after Example 4:
4. Write the recursive rule for the sequence 2, 3, 8, 63, 3968, 15,745,023, ... $a_1 = 2$, $a_n = (a_{n-1})^2 - 1$

EXAMPLE 2   *Writing a Recursive Rule for an Arithmetic Sequence*

Write the indicated rule for the arithmetic sequence with $a_1 = 4$ and $d = 3$.

a. an explicit rule   **b.** a recursive rule

SOLUTION

a. From Lesson 11.2 you know that an explicit rule for the nth term of the arithmetic sequence is:

$$a_n = a_1 + (n-1)d \qquad \text{General explicit rule for } a_n$$
$$= 4 + (n-1)3 \qquad \text{Substitute for } a_1 \text{ and } d.$$
$$= 1 + 3n \qquad \text{Simplify.}$$

b. To find the recursive equation, use the fact that you can obtain a_n by adding the common difference d to the previous term.

$$a_n = a_{n-1} + d \qquad \text{General recursive rule for } a_n$$
$$= a_{n-1} + 3 \qquad \text{Substitute for } d.$$

A recursive rule for the sequence is $a_1 = 4$, $a_n = a_{n-1} + 3$.

EXAMPLE 3   *Writing a Recursive Rule for a Geometric Sequence*

Write the indicated rule for the geometric sequence with $a_1 = 3$ and $r = 0.1$.

a. an explicit rule   **b.** a recursive rule

SOLUTION

a. From Lesson 11.3 you know that an explicit rule for the nth term of the geometric sequence is:

$$a_n = a_1 r^{n-1} \qquad \text{General explicit rule for } a_n$$
$$= 3(0.1)^{n-1} \qquad \text{Substitute for } a_1 \text{ and } r.$$

b. To write a recursive rule, use the fact that you can obtain a_n by multiplying the previous term by r.

$$a_n = r \cdot a_{n-1} \qquad \text{General recursive rule for } a_n$$
$$= (0.1)a_{n-1} \qquad \text{Substitute for } r.$$

A recursive rule for the sequence is $a_1 = 3$, $a_n = (0.1)a_{n-1}$.

EXAMPLE 4   *Writing a Recursive Rule*

Write a recursive rule for the sequence 1, 2, 2, 4, 8, 32, ...

SOLUTION

Beginning with the third term in the sequence, each term is the product of the two previous terms. Therefore, a recursive rule is given by:

$$a_1 = 1, \; a_2 = 2, \; a_n = a_{n-2} \cdot a_{n-1}$$

FOCUS ON CAREERS

FISHERY BIOLOGIST
Some fishery biologists manage fish hatcheries, conduct fish disease control programs, and work with organizations to restore and enhance fish habitats.

CAREER LINK
www.mcdougallittell.com

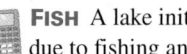

GOAL 2 USING RECURSIVE RULES IN REAL LIFE

EXAMPLE 5 Using a Recursive Rule

FISH A lake initially contains 5200 fish. Each year the population declines 30% due to fishing and other causes, and the lake is restocked with 400 fish.

a. Write a recursive rule for the number a_n of fish at the beginning of the nth year. How many fish are in the lake at the beginning of the fifth year?

b. What happens to the population of fish in the lake over time?

SOLUTION

a. Because the population declines 30% each year, 70% of the fish remain in the lake from one year to the next, and new fish are added.

VERBAL MODEL

$$\boxed{\text{Fish at start of }n\text{th year}} = 0.7 \boxed{\text{Fish at start of }(n-1)\text{st year}} + \boxed{\text{New fish added}}$$

LABELS

Fish at start of nth year = a_n

Fish at start of $(n-1)$st year = a_{n-1}

New fish added = **400**

ALGEBRAIC MODEL

$$a_n = (0.7)\,a_{n-1} + 400$$

A recursive rule is:

$$a_1 = 5200,\ a_n = (0.7)a_{n-1} + 400$$

You can use a graphing calculator to find a_5, the number of fish in the lake at the beginning of the fifth year. Enter the number of fish at the beginning of the first year, which is $a_1 = 5200$. Then enter the rule $0.7 \times$ Ans $+ 400$ to find a_2. Press **ENTER** three more times to find $a_5 \approx 2262$.

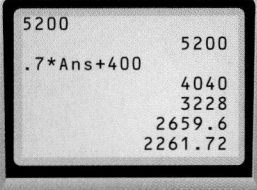

```
5200
                5200
.7*Ans+400
                4040
                3228
             2659.6
             2261.72
```

▶ There are about 2262 fish in the lake at the beginning of the fifth year.

b. To determine what happens to the lake's fish population over time, continue pressing **ENTER** on the calculator. The calculator screen at the right shows the fish populations for years 44–50. Observe that the numbers approach about 1333.

```
         1333.334178
         1333.333924
         1333.333747
         1333.333623
         1333.333536
         1333.333475
         1333.333433
```

▶ Over time, the population of fish in the lake stabilizes at about 1333 fish.

11.5 Recursive Rules for Sequences   **683**

Focus on Vocabulary *Sample answer:*
An explicit formula gives a_n as a function of n.
A recursive rule gives the beginning term or terms of a sequence. Then it relates the other terms to one or more preceding terms.

EXTRA EXAMPLE 5
A nursery initially had 500 trees. Each year it sells 70% of its stock and adds 500 new trees.
a. Write a recursive rule for the number a_n of trees at the beginning of the nth year. How many trees does it have at the beginning of the fifth year?
$a_1 = 500$, $a_n = (0.3)a_{n-1} + 500$; about 713
b. What happens to the number of trees at the nursery over time? It stabilizes at about 714.

CHECKPOINT EXERCISES
For use after Example 5:
1. A tree farm initially has 1500 fir trees for sale. Each year about 2% of the fir trees die and another 35% of the trees are cut and sold. Each year 300 new fir trees are planted. Write a recursive rule for the number a_n of fir trees at the beginning of the nth year.
$a_1 = 1500$, $a_n = (0.63)a_{n-1} + 300$
2. How many fir trees are on the farm at the beginning of the sixth year? about 879
3. What happens to the population of fir trees over time? It stabilizes at about 811.

FOCUS ON VOCABULARY
How is a recursive formula for a sequence different from an explicit formula? See margin.

CLOSURE QUESTION
Write a general rule that shows how each term of an arithmetic sequence except the first term is related to its preceding term.
$a_n = a_{n-1} + d$

DAILY PUZZLER
In a certain year I was half as old as my sister. She was half as old as our mother who was half as old as our grandmother. The sum of our ages was 165. My age that year was the same as the year in the 20th century that my grandmother was born. What year was my grandmother born? 1911

683

GUIDED PRACTICE

ASSIGNMENT GUIDE

BASIC
Day 1: pp. 684–687 Exs. 14–22,
26–32, 35–40, 44–46,
55–57, 59–73 odd,
Quiz 2 Exs. 1–17

AVERAGE
Day 1: pp. 684–687 Exs. 14–22,
26–32, 35–46, 55–57,
59–83 odd,
Quiz 2 Exs. 1–17

ADVANCED
Day 1: pp. 684–687 Exs. 14–34
even, 35–46, 51–58, 59–83
odd, Quiz 2 Exs. 1–17

BLOCK SCHEDULE
pp. 684–687 Exs. 14–22, 26–32,
35–46, 55–57, 59–83 odd,
Quiz 2 Exs. 1–17 (with 11.4)

EXERCISE LEVELS
Level A: *Easier*
14–25, 44
Level B: *More Difficult*
26–43, 45–57
Level C: *Most Difficult*
58

✔ **HOMEWORK CHECK**
To quickly check student understanding of key concepts, go
over the following exercises:
Exs. 14, 18, 26, 30, 36, 44. See
also the Daily Homework Quiz:

• Blackline Master (*Chapter 12 Resource Book*, p. 11)
• Transparency (p. 89)

2. *Sample answer:* An explicit rule
for a sequence gives an expression
for the nth term as a function of n.
A recursive rule gives the value of
one or more beginning terms and
an expression for the nth term as a
function of one or more preceding
terms.

Vocabulary Check ✔

1. Complete this statement: The expression __?__ represents the product of all integers from 1 to n. **$n!$**

Concept Check ✔

26. $2(10)^{n-1}$; $a_1 = 2$;
$a_n = (10)a_{n-1}$

2. Explain the difference between an explicit rule for a sequence and a recursive rule for a sequence. **See margin.**

3. Give an example of an explicit rule for a sequence and a recursive rule for a sequence. *Sample answer:* $a_n = n^2 - 4n$; $a_1 = 2$; $a_n = a_{n-1} + n$

Skill Check ✔

27. $-7 + 10n$; $a_1 = 3$;
$a_n = a_{n-1} + 10$

28. $10(2)^{n-1}$; $a_1 = 10$;
$a_n = (2)a_{n-1}$

29. $2 + 3n$; $a_1 = 5$;
$a_n = a_{n-1} + 3$

30. $1 - n$; $a_1 = 0$;
$a_n = a_{n-1} - 1$

31. $5(2.5)^{n-1}$; $a_1 = 5$;
$a_n = (2.5)a_{n-1}$

Write the first five terms of the sequence.

4. $a_1 = 1$ **1, 2, 3, 4, 5**
$a_n = a_{n-1} + 1$

5. $a_1 = 2$ **2, 8, 32, 128, 512**
$a_n = 4a_{n-1}$

6. $a_0 = 1$ **1, −1, −3, −5, −7**
$a_n = a_{n-1} - 2$

7. $a_1 = -1$
$a_n = -3a_{n-1}$
−1, 3, −9, 27, −81

8. $a_1 = 2$ **2, 1, −1, −5, −13**
$a_n = 2a_{n-1} - 3$

9. $a_0 = 3$
$a_n = (a_{n-1})^2 + 1$
3; 10; 101; 10,202; 104,080,805

Write a recursive rule for the sequence.

$a_1 = 21$; $a_n = a_{n-1} - 4$

$a_1 = 2$; $a_n = (3)a_{n-1}$

$a_1 = \frac{1}{2}$; $a_n = \left(\frac{1}{2}\right)a_{n-1}$

10. 21, 17, 13, 9, 5, . . .

11. 2, 6, 18, 54, 162, . . .

12. $\frac{1}{2}, \frac{1}{4}, \frac{1}{8}, \frac{1}{16}, \frac{1}{32}, \ldots$

13. 🌐 **FISH** Suppose each year the lake in Example 5 is restocked with 750 fish. How many fish are in the lake at the beginning of the fifth year? **about 3148**

PRACTICE AND APPLICATIONS

STUDENT HELP

► **Extra Practice**
to help you master
skills is on p. 956.

32. $13.5 + \left(\frac{1}{2}\right)n$; $a_1 = 14$;
$a_n = a_{n-1} + \frac{1}{2}$

33. $\left(\frac{1}{2}\right)4^{n-1}$; $a_1 = \frac{1}{2}$;
$a_n = (4)a_{n-1}$

34. $\frac{1}{2} - \left(\frac{3}{2}\right)n$; $a_1 = -1$;
$a_n = a_{n-1} - \frac{3}{2}$

STUDENT HELP

► **HOMEWORK HELP**
Example 1: Exs. 14–25
Examples 2, 3: Exs. 26–34
Example 4: Exs. 35–43
Example 5: Exs. 44–54

WRITING TERMS **Write the first five terms of the sequence.**

14. $a_0 = 1$ **1, 5, 9, 13, 17**
$a_n = a_{n-1} + 4$

15. $a_1 = 4$ **4, 12, 21, 31, 42**
$a_n = n + a_{n-1} + 6$

16. $a_0 = 0$ **0, −1, −5, −14, −30**
$a_n = a_{n-1} - n^2$

17. $a_0 = -4$
$a_n = a_{n-1} - 8$
−4, −12, −20, −28, −36

18. $a_1 = 2$
$a_n = (a_{n-1})^2 + 2$
2; 6; 38; 1446; 2,090,918

19. $a_0 = 5$ **5, −4, 8, 1, 15**
$a_n = n^2 - a_{n-1}$

20. $a_1 = 10$
$a_n = 3a_{n-1}$
10, 30, 90, 270, 810

21. $a_0 = 2$ **2, 1, 7, 8, 16**
$a_n = n^2 + 2n - a_{n-1}$

22. $a_0 = 3$
$a_n = (a_{n-1})^2 - 2$
3; 7; 47; 2207; 4,870,847

23. $a_0 = 48$
$a_n = \frac{1}{2}a_{n-1} + 2$
48, 26, 15, 9.5, 6.75

24. $a_0 = 4$, $a_1 = 2$
$a_n = a_{n-1} - a_{n-2}$
4, 2, −2, −4, −2

25. $a_1 = 1$, $a_2 = 3$
$a_n = a_{n-1} \cdot a_{n-2}$
1, 3, 3, 9, 27

WRITING RULES **Write an explicit rule and a recursive rule for the sequence. (Recall that d is the common difference of an arithmetic sequence and r is the common ratio of a geometric sequence.)** **26–34. See margin.**

26. $a_1 = 2$
$r = 10$

27. $a_1 = 3$
$d = 10$

28. $a_1 = 10$
$r = 2$

29. $a_1 = 5$
$d = 3$

30. $a_1 = 0$
$d = -1$

31. $a_1 = 5$
$r = 2.5$

32. $a_1 = 14$
$d = \frac{1}{2}$

33. $a_1 = \frac{1}{2}$
$r = 4$

34. $a_1 = -1$
$d = -\frac{3}{2}$

STUDENT HELP

HOMEWORK HELP
Visit our Web site
www.mcdougallittell.com
for help with Exs. 35–43.

WRITING RULES Write a recursive rule for the sequence. The sequence may be arithmetic, geometric, or neither. 35–43. See margin.

35. 1, 7, 13, 19, . . .

36. 66, 33, 16.5, 8.25, . . .

37. 41, 32, 23, 14, . . .

38. 3, 8, 63, 3968, . . .

39. 33, 11, $\frac{11}{3}$, $\frac{11}{9}$, . . .

40. 7.2, 3.2, −0.8, −4.8, . . .

41. 2, 5, 10, 50, 500, . . .

42. 6, 6$\sqrt{2}$, 12, 12$\sqrt{2}$, . . .

43. 48, 4.8, 0.48, 0.048, . . .

35. $a_1 = 1;\ a_n = a_{n-1} + 6$

36. $a_1 = 66;\ a_n = \frac{a_{n-1}}{2}$

37. $a_1 = 41;\ a_n = a_{n-1} - 9$

38. $a_1 = 3;$
$a_n = (a_{n-1})^2 - 1$

39. $a_1 = 33;\ a_n = \frac{a_{n-1}}{3}$

40. $a_1 = 7.2;\ a_n = a_{n-1} - 4$

41. $a_1 = 2;\ a_2 = 5;$
$a_n = a_{n-1} \cdot a_{n-2}$

42. $a_1 = 6;\ a_n = (\sqrt{2})a_{n-1}$

43. $a_1 = 48;\ a_n = \frac{a_{n-1}}{10}$

44. 🌐 **ON LAYAWAY** Suppose you buy a $500 camcorder on layaway by making a down payment of $150 and then paying $25 per month. Write a recursive rule for the total amount of money paid on the camcorder at the beginning of the *n*th month. How much will you have left to pay on the camcorder at the beginning of the twelfth month? $a_1 = 150;\ a_n = a_{n-1} + 25;\ \$75$

🌐 **FRACTAL TREE** In Exercises 45 and 46, use the following information.
A fractal tree starts with a single branch (the trunk). At each stage the new branches from the previous stage each grow two more branches, as shown.

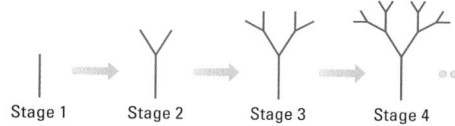

Stage 1   Stage 2   Stage 3   Stage 4

. . . mber of new branches in each of the first seven stages. What type of . . . mbers form? **1, 2, 4, 8, 16, 32, 64; geometric**

. . . rule for the sequence in Exercise 45.

. . . following information.
. . . d need to add chlorine to the
. . . k and 14 ounces every week
. . . pool evaporates.

. . . rine in the pool each week. How
. . . g of the sixth week?

. . . fter an extended period of time? **35 oz**

. . . se the following information.
. . . rees. Each year 10% of the trees are

. . . of trees on the tree farm at the beginning of
. . . at the beginning of the fourth year?
. . . 9
. . . es after an extended period of time? **8000 trees**

. . . se the following information.
. . . ms of a prescribed drug every four hours.
. . . oved from the bloodstream every four hours.

. . . mount of the drug in the bloodstream after *n* doses.

. . . el in the person's body approach after an extended
. . . called the *maintenance level*. **about 66.7 mg**

. . . doubled (to 40 milligrams), but the normal dosage is
. . . maintenance level from Exercise 52 change? **no**

. . . doubled. Does the maintenance level double as well? **yes**

FOCUS
CARE

illn
tre
th
l

CAREER
www.mcdougallittell.co

! **COMMON ERROR**
EXERCISES 35–43 Students often forget to define a_1 when they are writing recursive rules. Remind them that for any sequence, the first term must be determined.

CALCULATOR NOTE
EXERCISES 47–50 By using graphing calculators students can see how the terms of the sequence change without having to perform repetitive calculations, a task more efficiently performed by a calculator. Students with access to spreadsheet software can use the software to perform the calculations. An advantage to a spreadsheet is that more terms of the sequence can be viewed on one screen.

APPLICATION NOTE
EXERCISES 53 AND 54 As the results show, the purpose of increasing the initial dose is to reach a maintenance level for a given dose more quickly. A different dosage will introduce a different amount to the bloodstream, which will consequently have a different maintenance level.

ENGLISH LEARNERS
EXERCISES 44, 47–50 Some English learners may not be familiar with the terms *layaway*, *pool care*, and *tree farm*. You may want to have a knowledgeable student fluent in English explain the terms.

ADDITIONAL PRACTICE AND RETEACHING

For Lesson 11.5:
• Practice Levels A, B, and C (*Chapter 11 Resource Book*, p. 70)

• Reteaching with Practice (*Chapter 11 Resource Book*, p. 73)

• ▣ See Lesson 11.5 of the *Personal Student Tutor*

For more Mixed Review:
• ▣ Search the *Test and Practice Generator* for key words or specific lessons.

1. Write the first five terms of the sequence with $a_0 = 3$ and $a_1 = a_{n-1} + 4$. **3, 7, 11, 15, 19**

2. Write an explicit rule and a recursive rule for the sequence

  a. $a_1 = 3$, $r = 7$ $a_n = 3(7)^{n-1}$; $a_1 = 3$, $a_n = (7)a_{n-1}$

  b. $a_1 = 3$, $d = 7$ $a_n = -4 + 7n$; $a_1 = 3$, $a_n = a_{n-1} + 7$

3. Write a recursive rule for the sequence $55, 11, \frac{11}{5}, \frac{11}{25}, \ldots$

  $a_1 = 55$, $a_n = \left(\frac{1}{5}\right)a_{n-1}$

4. A lake initially contains 4,400 fish. Each year the population declines 30% and the lake is restocked with 300 fish. Write a recursive rule for the number of fish at the beginning of the nth year. How many fish are in the lake at the beginning of the fifth year?

  $a_1 = 4,400$, $a_n = (0.7)a_{n-1} + 300$, **about 1816 fish**

EXTRA CHALLENGE NOTE

↳ Challenge problems for Lesson 11.5 are available in **blackline** format in the *Chapter 11 Resource Book*, p. 78 and at **www.mcdougallittell.com**.

ADDITIONAL TEST PREPARATION

1. WRITING Explain why a logistic function is a good model for the sequence $a_1 = 5200$, $a_n = (0.7)a_{n-1} + 400$. As n increases to ∞, a_n gets closer and closer to 1333.

2. OPEN ENDED Find the lengths of a right triangle that are an arithmetic sequence. Any triangle with side lengths $3x$, $4x$, and $5x$, where x is a positive integer, is a right triangle.

---

55. CRITICAL THINKING Give an example of a sequence in which each term after the third term is a function of the three terms preceding it. Write a recursive rule for the sequence and find the first 8 terms. *Sample answer: $a_1 = 1$; $a_2 = 2$; $a_3 = 3$; $a_n = a_{n-1} + a_{n-3}$; 1, 2, 3, 4, 6, 9, 13, 19*

Test  **Preparation**

56. MULTIPLE CHOICE What is the fifth term of the sequence whose first term is $a_1 = 10$ and whose nth term is $a_n = 2a_{n-1} + 9$? **D**

  (A) 67   (B) 143   (C) 286   (D) 295   (E) 599

57. MULTIPLE CHOICE What is a recursive equation for the sequence $4, -6.6, 10.89, -17.9685, \ldots$? **B**

  (A) $a_n = (-2.6)a_{n-1}$ (B) $a_n = (-1.65)a_{n-1}$

  (C) $a_n = (2.6)a_{n-1}$ (D) $a_n = (1.65)a_{n-1}$

★ **Challenge**

58. PIECEWISE-DEFINED SEQUENCE You can define a sequence using a piecewise rule. The following is an example of a piecewise-defined sequence.

$$a_1 = 7, \; a_n = \begin{cases} \dfrac{a_{n-1}}{2}, & \text{if } a_{n-1} \text{ is even} \\ 3a_{n-1} + 1, & \text{if } a_{n-1} \text{ is odd} \end{cases}$$

a. Write the first ten terms of the sequence. **7, 22, 11, 34, 17, 52, 26, 13, 40, 20**

b. LOGICAL REASONING Choose three different values for a_1 (other than $a_1 = 7$). For each value of a_1, find the first ten terms of the sequence. What conclusions can you make about the behavior of this sequence?

Sample answer: Let $a_1 = 1$. The first 10 terms will be 1, 4, 2, 1, 4, 2, 1, 4, 2, 1. Let $a_1 = 2$. The first 10 terms will be 2, 1, 4, 2, 1, 4, 2, 1, 4, 2. Let $a_1 = 3$. The first 10 terms will be 3, 10, 5, 16, 8, 4, 2, 1, 4, 2. It appears that no matter what value you start with, eventually the series repeat the values 1, 4, 2.

EXTRA CHALLENGE
↳ www.mcdougallittell.com

MIXED REVIEW

EVALUATING POWERS Evaluate the power. **(Review 1.2 for 12.1)**

59. 2^5 **32**   **60.** 6^4 **1296**   **61.** 8^4 **4096**   **62.** 12^3 **1728**

63. 26^3 **17,576**   **64.** 10^5 **100,000**   **65.** 18^3 **5832**   **66.** 3^7 **2187**

OPERATIONS WITH RATIONAL EXPRESSIONS Perform the indicated operation and simplify. **(Review 9.5)**

  70. $\dfrac{2x^3 + 14x^2 - 42x - 70}{3x^2 + 26x + 35}$

  72. $\dfrac{x^3 + x^2 - 4x - 7}{x^2 + 3x + 2}$

67. $\dfrac{3}{5x} + \dfrac{3}{7x}$ $\dfrac{36}{35x}$ **68.** $\dfrac{-2}{7x} - \dfrac{5}{3x}$ $\dfrac{-41}{21x}$ **69.** $\dfrac{x+1}{x^2-9} - \dfrac{5}{x-3}$ $\dfrac{-(4x+14)}{x^2-9}$

70. $\dfrac{2x^2}{3x+5} - \dfrac{14}{x+7}$ **71.** $\dfrac{4x+1}{x^2-4} - \dfrac{3}{x-2}$ $\dfrac{x-5}{x^2-4}$ **72.** $\dfrac{x^2-1}{x+2} - \dfrac{3}{x+1}$

FINDING POINTS OF INTERSECTION Find the points of intersection, if any, of the graphs in the system. **(Review 10.7)**

  73. $(-1.272, 1.544)$; $(0.472, -1.944)$

  75. $(-0.980, -1.939)$; $(0.331, 1.993)$

  78. $(-0.877, -4.438)$; $(1.123, -3.439)$

  83. $\dfrac{1}{5}, \dfrac{1}{3}, \dfrac{3}{7}, \dfrac{1}{2}, \dfrac{5}{9}, \dfrac{3}{5}$

  84. $2, \dfrac{5}{3}, \dfrac{3}{2}, \dfrac{7}{5}, \dfrac{4}{3}, \dfrac{9}{7}$

73. $x^2 + y^2 = 4$
  $2x + y = -1$

74. $x^2 + y^2 = 25$ (4, 3); (-3, -4)
  $y = x - 1$

75. $x^2 + 4y^2 = 16$
  $y = 3x + 1$

76. $x^2 + y^2 = 10$
  $4x + y = 6$
  (0.731, 3.077); (2.093, -2.371)

77. $x^2 + y^2 = 30$
  $y = x + 2$
  (-4.742, -2.742); (2.742, 4.742)

78. $16x^2 + y^2 = 32$
  $\dfrac{1}{4}x - \dfrac{1}{2}y = 2$

WRITING TERMS Write the first six terms of the sequence. **(Review 11.1)**

79. $a_n = 8 - n$
  7, 6, 5, 4, 3, 2

80. $a_n = n^4$
  1, 16, 81, 256, 625, 1296

81. $a_n = n^2 + 9$
  10, 13, 18, 25, 34, 45

82. $a_n = (n + 3)^2$
  16, 25, 36, 49, 64, 81

83. $a_n = \dfrac{n}{n+4}$
  See margin.

84. $a_n = \dfrac{n+3}{n+1}$
  See margin.

Find the sum of the infinite geometric series if it has one. (Lesson 11.4)

1. $\sum_{n=0}^{\infty} 4\left(\frac{1}{9}\right)^n$ $\frac{9}{2}$ **2.** $\sum_{n=1}^{\infty} 5\left(-\frac{6}{7}\right)^{n-1}$ $\frac{35}{13}$ **3.** $\sum_{n=0}^{\infty} -\frac{3}{8}\left(\frac{4}{7}\right)^n$ $-\frac{7}{8}$ **4.** $\sum_{n=0}^{\infty} \frac{4}{5}\left(\frac{5}{4}\right)^n$ no sum

Find the common ratio of the infinite geometric series with the given sum and first term. (Lesson 11.4)

5. $S = 5, a_1 = 1$ $\frac{4}{5}$ **6.** $S = 12, a_1 = 1$ $\frac{11}{12}$ **7.** $S = 24, a_1 = 3$ $\frac{7}{8}$

Write the repeating decimal as a fraction. (Lesson 11.4)

8. $0.888\ldots$ $\frac{8}{9}$ **9.** $0.1515\ldots$ $\frac{5}{33}$ **10.** $126.126126\ldots$ $\frac{14{,}000}{111}$

Write the first five terms of the sequence. (Lesson 11.5)

11. $a_1 = 5$ 5, 8, 11, 14, 17   **12.** $a_0 = 1$ 1, 4, 16, 64, 256   **13.** $a_1 = 17$ 17, 19, 22, 26, 31
$a_n = a_{n-1} + 3$ $\qquad\qquad\quad$ $a_n = 4a_{n-1}$ $\qquad\qquad\qquad$ $a_n = a_{n-1} + n$

14. $a_1 = 1, a_2 = 2$ $\qquad$ **15.** $a_1 = 2, a_2 = 4$ $\qquad$ **16.** $a_1 = 10, a_2 = 10$
$a_n = a_{n-1} - a_{n-2}$ $\qquad\quad$ $a_n = a_{n-1} \cdot a_{n-2}$ $\qquad\quad$ $a_n = a_{n-2} + a_{n-1}$
$\quad$ 1, 2, 1, $-$1, $-$2 $\qquad\qquad\quad$ 2, 4, 8, 32, 256 $\qquad\qquad$ 10, 10, 20, 30, 50

17. **BALL BOUNCE** You drop a ball from a height of 8 feet. Each time it hits the ground, it bounces 40% of its previous height. Find the total distance traveled by the ball. (Lesson 11.4) $18\frac{2}{3}$ ft

---

MATH & History

The Fibonacci Sequence

 **APPLICATION LINK** www.mcdougallittell.com

THEN

IN 1202 the mathematician Leonardo Fibonacci wrote *Liber Abaci* in which he proposed the following rabbit problem.

Begin with a pair of newborn rabbits that never die. When a pair of rabbits is two months old, it begins producing a new pair of rabbits each month.

Month	1	2	3	4	5	6	. . .
Pairs at start of month	1	1	2	3	5	8	. . .

This problem can be represented by a sequence, known as the Fibonacci sequence. The numbers that make up the sequence are called Fibonacci numbers. The ratio of two Fibonacci numbers approximates the same number, denoted by Φ. The Greeks called this number the golden ratio.

2. 377 pairs of rabbits for a total of 754 rabbits; increases without bound

1. Draw a tree diagram to illustrate the sequence. See margin.

2. If the initial pair of rabbits produces their first pair of rabbits in January, how many pairs of rabbits will there be in December of that year? What happens to the rabbit population over time?

NOW

TODAY we know that Fibonacci numbers occur in nature, such as in the spiral patterns on the head of a sunflower or the surface of a pineapple.

 **2500 B.C.**
Golden Section used in Great Pyramid.

 **Fibonacci develops sequence.**
A.D. 1202

 **1999**
Fibonacci numbers recognized in nature.

11.5 *Recursive Rules for Sequences* **687**

ADDITIONAL RESOURCES
An alternative Quiz for Lessons 11.4 and 11.5 is available in the *Chapter 11 Resource Book*, p. 79.

MATH & HISTORY NOTE
A **blackline** master with additional Math & History exercises is available in the *Chapter 11 Resource Book*, p. 77.

APPLICATION NOTE
Additional information about the Fibonacci sequence is available at **www.mcdougallittell.com.**

1.

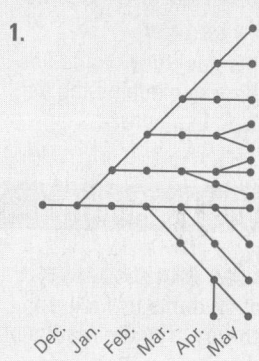

Dec. Jan. Feb. Mar. Apr. May

1 Planning the Activity

PURPOSE
To use a spreadsheet to evaluate a recursive rule.

MATERIALS
- spreadsheet software for each group
- Keystroke blackline (*Chapter 11 Resource Book,* p. 69)

PACING
- Activity — 25 min

▶ LINK TO LESSON
Ask students how they would use a spreadsheet to examine the fish populations in Example 5.

2 Managing the Activity

CLASSROOM MANAGEMENT
Suggest that students use a graphing calculator to verify the results of their spreadsheet formulas.

3 Closing the Activity

★ KEY DISCOVERY
The Fill Down feature on a spreadsheet can be used to evaluate a recursive rule.

ACTIVITY ASSESSMENT
In each cell after the first cell, how does the spreadsheet calculate the value for the cell? **The spreadsheet multiplies the value in the previous cell by the coefficient of a_{n-1} and then adds the constant term in the general rule.**

STUDENT HELP

INTERNET **SOFTWARE HELP**
Visit our Web site www.mcdougallittell.com to see instructions for several software packages.

1. 5100, 4465, 3893.5, 3379.15, 2916.24, 2499.61, 2124.65, 1787.19, 1483.47, 1210.12, 964.11, 742.70

2. 6150, 4843, 3771.26, 2892.43, 2171.80, 1580.87, 1096.32, 698.98, 373.16, 105.99, −113.09, −292.73

3. 3500, 2925, 2436.25, 2020.81, 1667.69, 1367.54, 1112.41, 895.55, 711.21, 554.53, 421.35, 308.15

4. 7500, 5637.50, 4194.06, 3075.40, 2208.43, 1536.54, 1015.82, 612.26, 299.50, 57.11, −130.74, −276.32

● ACTIVITY 11.5

Using Technology

Evaluating Recursive Rules

You can use a spreadsheet to evaluate a recursive rule.

▶ EXAMPLE

You owe $5000 to a credit card company that charges interest at a rate of 1.5% per month. Each month you make a payment of $100. Write a recursive rule for a_n, the balance of the account at the beginning of the nth month. How much will you still owe at the beginning of the 6th month? How long will it take to pay off the account?

▶ SOLUTION

A recursive rule for the balance of the account each month is:

$$a_1 = 5000, \quad a_n = (1.015)a_{n-1} - 100$$

You can use a spreadsheet to find how much you owe at the start of each month.

❶ Enter a_1 into cell A1.

A1	X ✓	5000	
	A	**B**	**C**
1	5000		
2			
3			

❷ In cell A2, enter the recursive equation.

A2		=1.015*A1−100	
	A	**B**	**C**
1	5000		
2	4975		
3			

❸ Use the *Fill Down* feature to copy the recursive equation into the rest of the column.

A2		=1.015*A1−100	
	A	**B**	**C**
1	5000		
2	4975		
3	4949.625		
4	4923.869375		
5	4897.727416		
6	4871.193327		
7	4844.261227		

A95		=1.015*A94−100	
	A	**B**	**C**
89	488.4879425		
90	395.8152616		
91	301.7524905		
92	206.2787779		
93	109.3729595		
94	11.01355393		
95	−88.8212428		

▶ At the beginning of the 6th month you still owe $4871.19, and it will take about 94 months, or 7 years and 10 months, to pay off the account.

▶ EXERCISES

Use a spreadsheet to find the first twelve terms of the sequence.

1. $a_1 = 5100, \ a_n = (0.9)a_{n-1} - 125$
2. $a_1 = 6150, \ a_n = (0.82)a_{n-1} - 200$

3. $a_1 = 3500, \ a_n = (0.85)a_{n-1} - 50$
4. $a_1 = 7500, \ a_n = (0.775)a_{n-1} - 175$

5. 🪙 **FINANCE** You owe $8000 on a credit card with 1% monthly interest. If you pay $125 each month, how long will it take to pay off the account?
103 months or 8 years, 7 months

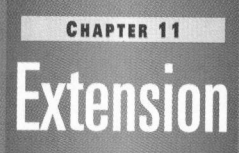

Extension

What you should learn

GOAL Use mathematical induction to prove statements about all positive integers.

Mathematical Induction

In Lesson 11.1 you saw that the rule for the sum of the first n positive integers is:

$$\sum_{i=1}^{n} i = 1 + 2 + \cdots + n = \frac{n(n+1)}{2}$$

Statements like the one above about all positive integers can be proven using a method called *mathematical induction*.

MATHEMATICAL INDUCTION

To show that a statement is true for all positive integers n, perform these steps.

BASIS STEP: Show that the statement is true for $n = 1$.

INDUCTIVE STEP: Assume that the statement is true for $n = k$ where k is any positive integer. Show that this implies the statement is true for $n = k + 1$.

Mathematical induction works as follows. If you know from the basis step that a statement is true for $n = 1$, then the inductive step implies that it is true for $n = 2$, and therefore, for $n = 3$, and so on for all positive integers n.

EXAMPLE 1 *Using Mathematical Induction*

Use mathematical induction to prove that $1 + 2 + \cdots + n = \frac{n(n+1)}{2}$.

SOLUTION

Basis step: Check to see if the formula works for $n = 1$.

$$1 \stackrel{?}{=} \frac{(1)(1+1)}{2} \longrightarrow 1 = 1 \checkmark$$

Inductive step: Assume that $1 + 2 + \cdots + k = \frac{k(k+1)}{2}$.

Show that $1 + 2 + \cdots + k + (k+1) = \frac{(k+1)[(k+1)+1]}{2}$.

$$1 + 2 + \cdots + k = \frac{k(k+1)}{2} \qquad \text{Assume true for } k.$$

$$1 + 2 + \cdots + k + (k+1) = \frac{k(k+1)}{2} + (k+1) \qquad \text{Add } k + 1 \text{ to each side.}$$

$$= \frac{k(k+1) + 2(k+1)}{2} \qquad \text{Add.}$$

$$= \frac{(k+1)(k+2)}{2} \qquad \text{Factor out } k + 1.$$

$$= \frac{(k+1)[(k+1)+1]}{2} \qquad \text{Rewrite } k + 2 \text{ as } (k+1) + 1.$$

Therefore, $1 + 2 + \cdots + n = \frac{n(n+1)}{2}$ for all positive integers n.

 EXTRA EXAMPLE 1

Use mathematical induction to prove that $\sum_{i=1}^{n} 2i = n^2 + n$.

Basis step: For $n = 1$, $2(1) = 2 = 1^2 + 1$. **Inductive step:** Assume that $\sum_{i=1}^{k} 2i = k^2 + k$. Show that $\sum_{i=1}^{k+1} 2i = (k+1)^2 + (k+1)$. $\sum_{i=1}^{k+1} 2i = k^2 + k + 2(k+1) = (k^2 + 2k + 1) + (k+1) = (k+1)^2 + (k+1)$. Therefore $\sum_{i=1}^{n} 2i = n^2 + n$ for all positive integers n.

CHECKPOINT EXERCISES

For use after Example 1:

1. Use mathematical induction to prove that $1 + 3 + 5 + 7 + \ldots + (2n - 1) = n^2$. **Basis step:** For $n = 1$, $2(1) - 1 = 1 = 1^2$. **Inductive step:** Assume that $1 + 3 + \ldots + (2k - 1) = k^2$. Show that $[1 + 3 + \ldots + (2k - 1)] + [2(k+1) - 1] = (k+1)^2$. $[1 + 3 + \ldots + (2k - 1)] + [2(k+1) - 1] = k^2 + [2(k+1) - 1] = k^2 + 2k + 1 = (k+1)^2$. Therefore $1 + 3 + \ldots + (2n - 1) = n^2$ for all positive integers n.

EXTRA EXAMPLE 2

Let $a_n = a_{n-1} + 3n(n-1)$ with $a_1 = 0$. Use mathematical induction to prove that an explicit rule for the nth term is $a_n = n^3 - n$.

Basis step: For $n = 1$, $1^3 - 1 = 0$.
Inductive step: Assume that $a_k = k^3 - k$. Show that $a_{k+1} = (k+1)^3 - (k+1)$. $a_{k+1} = a_k + 3(k+1)(k) = k^3 - k + 3(k+1)(k) = k^3 - k + 3k^2 + 3k = k^3 + 3k^2 + 3k - k + 1 - 1 = (k^3 + 3k^2 + 3k + 1) - (k+1) = (k+1)^3 - (k+1)$. Therefore an explicit rule for the nth term is $a_n = n^3 - n$ for all positive integers n.

CHECKPOINT EXERCISES

For use after Example 2:

1. Let $a_n = a_{n-1} + 3$ with $a_1 = 1$. Use mathematical induction to prove that an explicit rule for the nth term is $a_n = -2 + 3n$.
Basis step: For $n = 1$, $a_1 = -2 + 3(1) = 1$. Inductive step: Assume that $a_k = -2 + 3k$. Show that $a_{k+1} = -2 + 3(k+1)$. $a_{k+1} = a_k + 3 = -2 + 3k + 3 = -2 + 3(k+1)$. Therefore, an explicit rule for the nth term is $a_n = -2 + 3n$.

MATHEMATICAL REASONING

EXAMPLE 2 In the inductive step, why does $a_{k+1} = 3a_k + 1$?
If $a_n = 3a_{n-1} + 1$, then $a_k = 3a_{k-1} + 1$.
Then $a_{k+1} = 3a_{k+1-1} + 1 = 3a_k + 1$.

1. $\dfrac{1(1+1)(2 \cdot 1 + 1)}{6} = \dfrac{6}{6} = 1 = 1^2$, so the formula is true for $n = 1$. Suppose
$1^2 + 2^2 + \ldots k^2 = \dfrac{k(k+1)(2k+1)}{6}$.
Then $1^2 + 2^2 + \ldots k^2 + (k+1)^2 = \dfrac{k(k+1)(2k+1)}{6} + (k+1)^2 = $
$\dfrac{k(k+1)(2k+1) + 6(k+1)^2}{6} = $
$\dfrac{(k+1)(2k^2 + 7k + 6)}{6} = $
$\dfrac{(k+1)(k+2)(2k+3)}{6} = $
$\dfrac{(k+1)[(k+1)+1][2(k+1)+1]}{6}$,
and the formula is true for $n = k + 1$. Therefore, the formula is true for all positive integers.

2. $1(1+2) = 3 = 2 \cdot 1 + 1$, so the formula is true for $n = 1$.
Suppose $\sum_{i=1}^{k} (2i + 1) = k(k+2)$. Then $\sum_{i=1}^{k+1} (2i+1) = k(k+2) + [2(k+1) + 1] = k^2 + 4k + 3 = (k+1)(k+3) = (k+1)[(k+1)+2]$, and the formula is true for $n = k + 1$. Therefore, the formula is true for all positive integers.

3. $\dfrac{a_1(1-r^1)}{1-r} = a_1 \cdot r^{1-1}$, so the statement is true for $n = 1$. Assume it is true for $n = k$.

Then $\sum_{i=1}^{k} a_1 r^{i-1} = \dfrac{a_1(1-r^k)}{1-r}$, so
$\sum_{i=1}^{k+1} a_1 r^{i-1} = \dfrac{a_1(1-r^k)}{1-r} + a_1 r^{k+1-1} =$
$\dfrac{a_1(1-r^k) + a_1(r^k)(1-r)}{1-r} =$
$\dfrac{a_1[(1-r^k) + r^k(1-r)]}{1-r} =$
$\dfrac{a_1(1 - r^{k+1})}{1-r}$,
and the formula is true for $n = k + 1$. Therefore, the formula is true for all positive integers.

5. $\dfrac{5^{1+1} - 5}{4} = \dfrac{25 - 5}{4} = \dfrac{20}{4} = $
$5 = 5^1$, so the formula is true for $n = 1$. Suppose the formula is true for $n = k$. Then
$\sum_{i=1}^{k} 5^i = \dfrac{5^{k+1} - 5}{4}$. So $\sum_{i=1}^{k+1} 5^i = $
$\dfrac{5^{k+1} - 5}{4} + 5^{k+1} = $
$\dfrac{[(5^{k+1} - 5) + 4(5^{k+1})]}{4} = $
$\dfrac{5(5^{k+1}) - 5}{4} = \dfrac{5^{(k+1)+1} - 5}{4}$,
and the formula is true for $n = k + 1$. Therefore, the formula is true for all positive integers.

EXAMPLE 2 Using Mathematical Induction

Let $a_n = 3a_{n-1} + 1$ with $a_1 = 1$. Use mathematical induction to prove that an explicit rule for the nth term is $a_n = \dfrac{3^n - 1}{2}$.

SOLUTION

Basis step: Check to see that the rule works for $n = 1$.

$a_1 \overset{?}{=} \dfrac{3^1 - 1}{2} \longrightarrow 1 = 1 \checkmark$

Inductive step: Assume that $a_k = \dfrac{3^k - 1}{2}$. Show that $a_{k+1} = \dfrac{3^{k+1} - 1}{2}$.

$a_{k+1} = 3a_k + 1$ **Definition of a_n for $n = k + 1$**

$= 3\left(\dfrac{3^k - 1}{2}\right) + 1$ **Substitute for a_k.**

$= \dfrac{3^{k+1} - 3}{2} + 1$ **Multiply.**

$= \dfrac{3^{k+1} - 3 + 2}{2}$ **Add.**

$= \dfrac{3^{k+1} - 1}{2}$ **Simplify.**

Therefore, an explicit rule for the nth term is $a_n = \dfrac{3^n - 1}{2}$ for all positive integers n.

EXERCISES

Use mathematical induction to prove the statement.   1–6. See margin.

1. $\displaystyle\sum_{i=1}^{n} i^2 = \dfrac{n(n+1)(2n+1)}{6}$

2. $\displaystyle\sum_{i=1}^{n} (2i+1) = n(n+2)$

3. $\displaystyle\sum_{i=1}^{n} a_1 r^{i-1} = a_1\left(\dfrac{1 - r^n}{1-r}\right)$

4. $\displaystyle\sum_{i=1}^{n} (2i)^2 = \dfrac{2n(n+1)(2n+1)}{3}$

5. $\displaystyle\sum_{i=1}^{n} 5^i = \dfrac{5^{n+1} - 5}{4}$

6. $\displaystyle\sum_{i=1}^{n} \left(\dfrac{1}{2}\right)^i = 1 - \left(\dfrac{1}{2}\right)^n$

7. **GEOMETRY CONNECTION** The numbers 1, 5, 12, 22, 35, 51, . . . are called *pentagonal numbers* because they represent the numbers of dots used to make pentagons, as shown at the right. Prove that the nth pentagonal number P_n is given by:

$$P_n = \dfrac{n(3n-1)}{2}$$ See margin.

8. **LOGICAL REASONING** Let $f_1, f_2, \ldots, f_n, \ldots$ be the Fibonacci sequence. Prove that $f_1 + f_2 + \cdots + f_n = f_{n+2} - 1$. See margin.

Chapter Summary

WHAT did you learn?

Use summation notation to write a series. **(11.1)**

Find terms of sequences.
- defined by explicit rules **(11.1)**
- defined by recursive rules **(11.5)**

Graph and classify sequences. **(11.1–11.3)**

Write rules for *n*th terms of sequences.
- given some terms **(11.1–11.3)**

- arithmetic sequences **(11.2)**
- geometric sequences **(11.3)**

Find sums of series.
- by adding terms or using formulas **(11.1)**
- finite arithmetic series **(11.2)**
- finite geometric series **(11.3)**
- infinite geometric series **(11.4)**

Write recursive rules for sequences. **(11.5)**

Use sequences and series to solve real-life problems. **(11.1–11.5)**

WHY did you learn it?

Express the number of oranges in a stack. **(p. 654)**

Find angle measures at the tips of a star. **(p. 656)**
Find the number of fish in a stocked lake. **(p. 683)**

Compare the revenues of two companies. **(p. 672)**

Model the minimum number of moves in the Tower of Hanoi puzzle. **(p. 656)**
Model the number of seats in a concert hall. **(p. 662)**
Model the number of matches in a tennis tournament. **(p. 671)**

Find the number of tennis balls in a stack. **(p. 656)**
Find the number of cells in a honeycomb. **(p. 664)**
Find the cost of cellular telephone service. **(p. 669)**
Find the amount of money spent by tourists who receive a tourist brochure. **(p. 679)**

Model the number of trees on a tree farm. **(p. 685)**

Find the total length of the vertical supports used to build a roof. **(p. 656)**

How does Chapter 11 fit into the BIGGER PICTURE of algebra?

Since elementary school you have studied number patterns (sequences). Now you can use algebra to write and use rules for sequences and series. An arithmetic sequence has a common difference, so it is similar to a linear function. A geometric sequence has a common ratio, so it is similar to an exponential function. Recursive rules are used in computer programs and in spreadsheet formulas.

STUDY STRATEGY

How did you learn by teaching?

Here is an example of an explanation given by one student to another, following the **Study Strategy** on page 650.

Learn by Teaching

Write a rule for the *n*th term of this sequence:
2, 7, 12, 17, . . .

"First look for a common difference or a common ratio. $7 - 2 = 5$, $12 - 7 = 5$, and $17 - 12 = 5$, so the common difference is 5. So 5*n* will be part of the rule. If $n = 1$, $5n = 5$. But the first term is 2, which is 3 less than 5, so we need to subtract 3. Let's try $5n - 3$ and see if it works. $5(1) - 3 = 2$, $5(2) - 3 = 7$, $5(3) - 3 = 12$, and $5(4) - 3 = 17$. Yes, it works. So the rule is $a_n = 5n - 3$."

691

4. $\frac{2(1)(1+1)(2 \cdot 1 + 1)}{3} = \frac{12}{3} = 4 = (2 \cdot 1)^2$, so the formula is true for $n = 1$. Suppose the formula is true for $n = k$. Then $\sum_{i=1}^{k} (2i)^2 = \frac{2k(k+1)(2k+1)}{3}$. Then $\sum_{i=1}^{k+1} (2i)^2 = \frac{2k(k+1)(2k+1)}{3} + (2k+2)^2 = \frac{2k(k+1)(2k+1) + 3(2k+2)^2}{3} = \frac{(2k+2)[k(2k+1) + 3(2k+2)]}{3} = \frac{2(k+1)(2k^2 + 7k + 6)}{3} = \frac{2(k+1)(2k+3)(k+2)}{3} = \frac{2(k+1)[(k+1)+1][2(k+1)+1]}{3}$, and the formula is true for $n = k + 1$. Therefore, the formula is true for all positive integers.

6. $\left(\frac{1}{2}\right)^1 = 1 - \left(\frac{1}{2}\right)^1$, so the formula is true for $n = 1$. Suppose the formula is true for $n = k$. Then $\sum_{i=1}^{k} \left(\frac{1}{2}\right)^i = 1 - \left(\frac{1}{2}\right)^k$. Then $\sum_{i=1}^{k+1} \left(\frac{1}{2}\right)^i = 1 - \left(\frac{1}{2}\right)^k + \left(\frac{1}{2}\right)^{k+1} = 1 - \left(\frac{1}{2}\right)^k + \left(\frac{1}{2}\right)\left(\frac{1}{2}\right)^k = 1 - \left(\frac{1}{2}\right)\left(\frac{1}{2}\right)^k = 1 - \left(\frac{1}{2}\right)^{k+1}$, and the formula is true for $n = k + 1$. Therefore, the formula is true for all positive integers.

7. A recursive formula for the *n*th pentagonal number is $P_n = P_{n-1} + 3n - 2$. $\frac{1(3 \cdot 1 - 1)}{2} = 1 = P_1$, so the formula is true for $n = 1$. Suppose the formula is true for $n = k$. Then $P_k = \frac{k(3k-1)}{2}$, so $P_{k+1} = \frac{k(3k-1)}{2} + 3(k+1) - 2 = \frac{3k^2 - k + 6k + 6 - 4}{2} = \frac{3k^2 + 5k + 2}{2} = \frac{(k+1)(3k+2)}{2} = \frac{(k+1)[3(k+1) - 1]}{2}$, and the formula is true for $n = k + 1$. Therefore, the formula is true for all positive integers.

8. See Additional Answers beginning on page AA1.

Chapter Review

ADDITIONAL RESOURCES

The following resources are available to help review the material in this chapter.

- Chapter Review Games and Activities (*Chapter 11 Resource Book,* p. 80)
- *Instant Replay: Video Review Games*
- ⊞ *Personal Student Tutor*
- Cumulative Review, Chs. 1–11 (*Chapter 11 Resource Book,* p. 92)

VOCABULARY

- terms of a sequence, p. 651
- sequence, p. 651
- finite sequence, p. 651
- infinite sequence, p. 651
- series, p. 653
- summation notation, p. 653

- sigma notation, p. 653
- arithmetic sequence, p. 659
- common difference, p. 659
- arithmetic series, p. 661
- geometric sequence, p. 666

- common ratio, p. 666
- geometric series, p. 668
- explicit rule, p. 681
- recursive rule, p. 681
- factorial, p. 681

11.1   **AN INTRODUCTION TO SEQUENCES AND SERIES**

Examples on pp. 651–654

> **EXAMPLES**   You can find the first four terms of the sequence $a_n = 3n - 7$.
>
> $a_1 = 3(1) - 7 = -4$ ⟵ first term
>
> $a_2 = 3(2) - 7 = -1$ ⟵ second term
>
> $a_3 = 3(3) - 7 = 2$ ⟵ third term
>
> $a_4 = 3(4) - 7 = 5$ ⟵ fourth term
>
> The *sequence* defined by $a_n = 3n - 7$ is $-4, -1, 2, 5, \ldots$.
>
> The associated *series* is the sum of the terms of the sequence: $(-4) + (-1) + 2 + 5 + \cdots$.
>
> You can use summation notation to write the series $2 + 4 + 6 + 8 + 10$ as $\displaystyle\sum_{i=1}^{5} 2i$.
>
> You can find the sum of a series by adding the terms or by using formulas for special series.
>
> The sum of the series $\displaystyle\sum_{i=1}^{22} i^2$ is $\dfrac{n(n+1)(2n+1)}{6} = \dfrac{22(22+1)(2(22)+1)}{6} = 3795$.

Write the first six terms of the sequence.

6, 9, 14, 21, 30, 41   8, 27, 64, 125, 216, 343   4, 2, 0, −2, −4, −6 $\frac{1}{4}, \frac{2}{5}, \frac{1}{2}, \frac{4}{7}, \frac{5}{8}, \frac{2}{3}$

1. $a_n = n^2 + 5$ **2.** $a_n = (n+1)^3$ **3.** $a_n = 6 - 2n$ **4.** $a_n = \dfrac{n}{n+3}$

Write the next term in the sequence. Then write a formula for the nth term.

                             $-48; -3(-2)^{n-1}$

5. 2, 4, 6, 8, . . .   10; $2n$ **6.** $-3, 6, -12, 24, \ldots$ **7.** $\dfrac{1}{3}, \dfrac{1}{9}, \dfrac{1}{27}, \dfrac{1}{81}, \ldots$ $\dfrac{1}{243}; \left(\dfrac{1}{3}\right)^n$

Write the series with summation notation.

8. $4 + 8 + 12 + 16$ $\displaystyle\sum_{i=1}^{4} 4i$ **9.** $1 + 2 + 3 + 4 + \cdots$ $\displaystyle\sum_{i=1}^{\infty} i$ **10.** $0 + 3 + 6 + 9 + 12$ $\displaystyle\sum_{i=0}^{4} 3i$

Find the sum of the series.

11. $\displaystyle\sum_{n=1}^{25} n^2$ 5525   **12.** $\displaystyle\sum_{n=4}^{10} n(2n-1)$ 693   **13.** $\displaystyle\sum_{i=1}^{12} i$ 78   **14.** $\displaystyle\sum_{k=1}^{30} 4$ 120

ARITHMETIC SEQUENCES AND SERIES

Examples on pp. 659–662

EXAMPLES The sequence 4, 7, 10, 13, 16, ... is an arithmetic sequence because the difference between consecutive terms is constant:

$$7 - 4 = 3 \qquad 10 - 7 = 3 \qquad 13 - 10 = 3 \qquad 16 - 13 = 3$$

The common difference is 3, so $d = 3$.

A rule for the nth term of this arithmetic sequence is:

$$a_n = a_1 + (n - 1)d = 4 + (n - 1)3 = 3n + 1$$

The sum of the first 20 terms of this arithmetic series is:

$$S_{20} = 20\left(\frac{a_1 + a_{20}}{2}\right) = 20\left(\frac{4 + 61}{2}\right) = 650$$

Write a rule for the nth term of the arithmetic sequence.

15. 1, 7, 13, 19, 25, ... $-5 + 6n$ **16.** 4, 6, 8, 10, 12, ... $2 + 2n$ **17.** 3.5, 3, 2.5, 2, 1.5, ... $4 - \left(\frac{1}{2}\right)n$

18. $d = 5, a_1 = 13$ $8 + 5n$ **19.** $d = -2, a_9 = 3$ $21 - 2n$ **20.** $a_4 = 20, a_{13} = 65$ $5n$

Find the sum of the first n terms of the arithmetic series.

21. $8 + 20 + 32 + 44 + \cdots$; $n = 14$ 1204 **22.** $(-6) + (-2) + 2 + 6 + \cdots$; $n = 20$ 640

23. $0.5 + 0.9 + 1.3 + 1.7 + \cdots$; $n = 54$ 599.4 **24.** $(-12) + (-8) + (-4) + 0 + \cdots$; $n = 40$ 2640

GEOMETRIC SEQUENCES AND SERIES

Examples on pp. 666–669

EXAMPLES The sequence 5, 15, 45, 135, 405, ... is a geometric sequence because the ratio of any term to the previous term is constant:

$$\frac{15}{5} = 3 \qquad \frac{45}{15} = 3 \qquad \frac{135}{45} = 3 \qquad \frac{405}{135} = 3$$

The common ratio is 3, so $r = 3$.

A rule for the nth term of this geometric sequence is:

$$a_n = a_1 r^{n-1} = 5(3)^{n-1}$$

The sum of the first 8 terms of this geometric series is:

$$S_8 = a_1\left(\frac{1 - r^8}{1 - r}\right) = 5\left(\frac{1 - 3^8}{1 - 3}\right) = 16,400$$

Write a rule for the nth term of the geometric sequence.

25. 64, 32, 16, 8, 4, ... $64\left(\frac{1}{2}\right)^{n-1}$ **26.** 6, 12, 24, 48, ... $6 \cdot 2^{n-1}$ **27.** 200, 20, 2, 0.2, 0.02, ... $200\left(\frac{1}{10}\right)^{n-1}$

28. $r = 3, a_1 = 6$ $6(3)^{n-1}$ **29.** $r = -\frac{1}{4}, a_4 = 1$ $-64\left(-\frac{1}{4}\right)^{n-1}$ **30.** $a_2 = 50, a_6 = 0.005$ $500\left(\frac{1}{10}\right)^{n-1}$

Find the sum of the series.

31. $\displaystyle\sum_{i=1}^{5} 16(2)^{i-1}$ 496 **32.** $\displaystyle\sum_{i=1}^{10} 20(0.2)^{i-1}$ about 25 **33.** $\displaystyle\sum_{i=0}^{6} 10\left(\frac{1}{2}\right)^i$ 19.844 **34.** $\displaystyle\sum_{i=1}^{8} 2\left(\frac{3}{5}\right)^{i-1}$ 4.916

Chapter Review **693**

11.4 INFINITE GEOMETRIC SERIES

EXAMPLES You can find the sum of the infinite geometric series

$$\sum_{n=1}^{\infty} 4\left(\frac{3}{5}\right)^{n-1} \text{ because } |r| = \left|\frac{3}{5}\right| < 1: S = \frac{a_1}{1-r} = \frac{4}{1-\frac{3}{5}} = 10.$$

The infinite geometric series $\sum_{n=1}^{\infty} \frac{1}{2}(5)^{n-1}$ has no sum because $|r| = |5| \geq 1$.

Find the sum of the infinite geometric series.

35. $\sum_{n=1}^{\infty} 15\left(\frac{2}{9}\right)^{n-1}$ $\frac{135}{7}$ **36.** $\sum_{n=1}^{\infty} 3\left(\frac{3}{4}\right)^{n-1}$ 12 **37.** $\sum_{n=1}^{\infty} 5(0.8)^{n-1}$ 25 **38.** $\sum_{n=1}^{\infty} 4(-0.2)^{n-1}$ $\frac{10}{3}$

Find the common ratio of the infinite geometric series with the given sum and first term.

39. $S = 18, a_1 = 12$ $\frac{1}{3}$ **40.** $S = 2, a_1 = 0.5$ $\frac{3}{4}$ **41.** $S = 20, a_1 = 4$ $\frac{4}{5}$ **42.** $S = -5, a_1 = -2$ $\frac{3}{5}$

43. $S = -10, a_1 = -3$ $\frac{-3}{\frac{7}{10}}$ **44.** $S = 6, a_1 = \frac{1}{3}$ $\frac{17}{18}$ **45.** $S = \frac{1}{4}, a_1 = \frac{1}{16}$ $\frac{3}{4}$ **46.** $S = 3\frac{1}{3}, a_1 = 6$ $-\frac{4}{5}$

Write the repeating decimal as a fraction.

47. $0.222\ldots$ $\frac{2}{9}$ **48.** $0.4545\ldots$ $\frac{5}{11}$ **49.** $39.3939\ldots$ $\frac{1300}{33}$ **50.** $0.001001\ldots$ $\frac{1}{999}$

11.5 RECURSIVE RULES FOR SEQUENCES

EXAMPLES You can find the first five terms of the sequence defined by the recursive rule $a_1 = 3, a_n = a_{n-1} + n + 6$.

$a_1 = 3$ ←first term

$a_2 = a_{n-1} + n + 6 = a_1 + 2 + 6 = 3 + 2 + 6 = 11$ ←second term

$a_3 = a_{n-1} + n + 6 = a_2 + 3 + 6 = 11 + 3 + 6 = 20$ ←third term

$a_4 = a_{n-1} + n + 6 = a_3 + 4 + 6 = 20 + 4 + 6 = 30$ ←fourth term

$a_5 = a_{n-1} + n + 6 = a_4 + 5 + 6 = 30 + 5 + 6 = 41$ ←fifth term

The sequence is $3, 11, 20, 30, 41, \ldots$.

A recursive formula for the sequence $1, 5, 14, 30, \ldots$ is $a_1 = 1, a_n = a_{n-1} + n^2$.

Write the first six terms of the sequence.

10; 40; 160; 640; 2560; 10,240   −1; 4; 19; 364; 132,499; 17,555,985,004

51. $a_1 = 10$ **52.** $a_1 = 1$ **53.** $a_1 = 2$ **54.** $a_1 = -1$

$a_n = 4a_{n-1}$ $a_n = n \cdot a_{n-1}$ $a_n = a_{n-1} - n$ $a_n = (a_{n-1})^2 + 3$

     1, 2, 6, 24, 120, 720   2, 0, −3, −7, −12, −18

Write a recursive rule for the sequence. The sequence may be arithmetic, geometric, or neither.

$a_1 = 7; a_n = 2 \cdot a_{n-1}$ $a_1 = 4; a_n = a_{n-1} + n + 2$ $a_1 = 1; a_n = a_{n-1} + 5$

55. $7, 14, 28, 56, 112, \ldots$ **56.** $4, 8, 13, 19, 26, \ldots$ **57.** $1, 6, 11, 16, 21, \ldots$

58. $200, 100, 50, 25, \ldots$ **59.** $1, 2, 5, 26, 677, \ldots$ **60.** $-2, -6, -12, -20, \ldots$

$a_1 = 200; a_n = \left(\frac{1}{2}\right)a_{n-1}$ $a_1 = 1; a_n = (a_{n-1})^2 + 1$ $a_1 = -2; a_n = a_{n-1} - 2n$

694 Chapter 11 *Sequences and Series*

ADDITIONAL RESOURCES
- *Chapter 11 Resource Book*
  Chapter Test (3 levels) (p. 81)
  SAT/ACT Chapter Test (p. 87)
  Alternative Assessment (p. 88)
- 🔲 *Test and Practice Generator*

**Tell whether the sequence is *arithmetic, geometric,* or *neither.*
Explain your answer.** 1–4. See margin.

1. $-5, -3, -1, 1, \ldots$ **2.** $-4, -2, 2, 4, \ldots$ **3.** $12, 6, 3, \dfrac{3}{2}, \ldots$ **4.** $\dfrac{1}{3}, 1, 3, 9, \ldots$

Write the first six terms of the sequence.

5. $a_n = n^2 + 1$
2, 5, 10, 17, 26, 37

6. $a_n = 3n - 5$
-2, 1, 4, 7, 10, 13

7. $a_1 = 4$
$a_n = n + a_{n-1}$
4, 6, 9, 13, 18, 24

8. $a_1 = 1$
$a_n = 2a_{n-1}$
1, 2, 4, 8, 16, 32

Write the next term of the sequence, and then write a rule for the *n*th term.

9. 2, 4, 8, 16, ...   32; 2^n **10.** 4, 9, 14, 19, ... 24; $5n - 1$ **11.** 2, 10, 50, 250, ... 1250; $(2)5^{n-1}$ **12.** $-9, -10, -11, -12, \ldots$ -13; $-8 - n$

13. $5, -\dfrac{5}{2}, \dfrac{5}{4}, -\dfrac{5}{8}, \ldots$
See margin.

14. $\dfrac{2}{3}, \dfrac{3}{4}, \dfrac{4}{5}, \dfrac{5}{6}, \ldots$ $\dfrac{6}{7}$; $\dfrac{n+1}{n+2}$ **15.** $\dfrac{3}{2}, \dfrac{4}{4}, \dfrac{5}{6}, \dfrac{6}{8}, \ldots$ $\dfrac{7}{10}$; $\dfrac{(n+2)}{2n}$ **16.** 1.1, 2.2, 3.3, 4.4, ...
5.5; $(1.1)n$

**Write a recursive rule for the sequence. (Recall that *d* is the common difference
of an arithmetic sequence and *r* is the common ratio of a geometric sequence.)**

17. $r = 0.3$, $a_1 = 4$
$a_1 = 4$; $a_n = 0.3a_{n-1}$

18. $d = 4$, $a_1 = 1$
$a_1 = 1$; $a_n = a_{n-1} + 4$

19. 40, 20, 10, 5, ...
$a_1 = 40$; $a_n = \left(\dfrac{1}{2}\right)a_{n-1}$

20. 2, 8, 18, 32, 50, ...
$a_1 = 2$; $a_n = a_{n-1} + 4n - 2$

Find the sum of the series.

21. $\displaystyle\sum_{i=1}^{100} i$ 5050   **22.** $\displaystyle\sum_{i=2}^{5} \dfrac{1}{2}i^2$ 27   **23.** $\displaystyle\sum_{i=1}^{6} (i - 10)$ -39 **24.** $\displaystyle\sum_{i=1}^{20} (3i + 2)$ 670

25. $\displaystyle\sum_{i=1}^{5} 7(-2)^{i-1}$ 77   **26.** $\displaystyle\sum_{i=0}^{9} 5\left(\dfrac{1}{4}\right)^i$ 6.667   **27.** $\displaystyle\sum_{i=1}^{\infty} 64\left(-\dfrac{1}{2}\right)^{i-1}$ $42\dfrac{2}{3}$ **28.** $\displaystyle\sum_{i=1}^{\infty} 100\left(\dfrac{7}{10}\right)^{i-1}$ $333\dfrac{1}{3}$

29. Find the sum of the first 30 terms of the arithmetic sequence 3, 7, 11, 15, ... 1830

30. Find the sum of the infinite geometric series $2 + 1 + 0.5 + 0.25 + \cdots$. 4

31. Write the series $1 + 3 + 5 + 7 + 9 + 11$ with summation notation. $\displaystyle\sum_{i=1}^{6} (2i - 1)$

32. Write the repeating decimal 0.7575 ... as a fraction. $\dfrac{25}{33}$

33. 🌀 **FALLING OBJECT** An object is dropped from an airplane. During the
first second, the object falls 4.9 meters. During the second second, it falls
14.7 meters. During the third second, it falls 24.5 meters. During the fourth second,
it falls 34.3 meters. If this pattern continues, how far will the object fall during the
tenth second? Find the total distance the object will fall after 10 seconds.
93.1 m; 490 m

34. 🌀 **CELL DIVISION** In early growth of an embryo, a human cell divides into
two cells, each of which divides into two cells, and so on. The number a_n of new
cells formed after the *n*th division is $a_n = 2^{n-1}$. Find the sum of the first 9 terms
of the series to find the total number of new cells after the 8th division. 511

35. 🌀 **SPRING** The length of the first loop of a spring is 20 inches. The length of
the second loop is $\dfrac{9}{10}$ of the length of the first loop. The length of the third loop
is $\dfrac{9}{10}$ of the length of the second loop, and so on. If the spring could have infinitely
many loops, would its length be finite? If so, find the length. yes; 200 in.

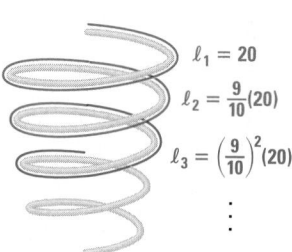

$\ell_1 = 20$
$\ell_2 = \dfrac{9}{10}(20)$
$\ell_3 = \left(\dfrac{9}{10}\right)^2(20)$
⋮

1. arithmetic; there is a common
   difference of 2
2. neither; there is neither a common ratio nor difference
3. geometric; there is a common
   ratio of $\dfrac{1}{2}$
4. geometric; there is a common
   ratio of 3
13. $\dfrac{5}{16}$; $5\left(-\dfrac{1}{2}\right)^{n-1}$

Chapter Standardized Test

▶ **TEST-TAKING STRATEGY** If the answers to a question are formulas, substitute the given numbers into the formulas to test the possible answers.

ADDITIONAL RESOURCES
- *Chapter 11 Resource Book*
  Chapter Test (3 levels) (p. 81)
  SAT/ACT Chapter Test (p. 87)
  Alternative Assessment (p. 88)
- ▦ *Test and Practice Generator*

1. MULTIPLE CHOICE What is the next term in the sequence 1, 4, 9, 16, 25, . . .? **C**

(A) 34 (B) 35 (C) 36

(D) 38 (E) 39

2. MULTIPLE CHOICE Which series is represented by

$$\sum_{i=1}^{4} (4i - 2)?\quad \mathbf{A}$$

(A) 2 + 6 + 10 + 14

(B) −2 + 2 + 6 + 10

(C) 4 + 8 + 12 + 16

(D) 6 + 10 + 14 + 18

(E) 2 + 6 + 10 + 14 + · · ·

3. MULTIPLE CHOICE What type of series is 32 + 16 + 8 + 4 + 2 + 1? **B**

(A) Finite arithmetic series

(B) Finite geometric series

(C) Infinite arithmetic series

(D) Infinite geometric series

(E) None of these

4. MULTIPLE CHOICE What is the sum of the series

$$\sum_{n=0}^{5} (n^3 + 3)?\quad \mathbf{E}$$

(A) 128 (B) 131 (C) 240

(D) 242 (E) 243

5. MULTIPLE CHOICE What is a rule for the nth term of the arithmetic sequence with $a_{14} = 9$ and common difference $d = 2$? **E**

(A) $a_n = 2n + 7$ (B) $a_n = 2n + 11$

(C) $a_n = 2n - 9$ (D) $a_n = 2n - 15$

(E) $a_n = 2n - 19$

6. MULTIPLE CHOICE What is the sum of the first 50 terms of the series 2 + 17 + 32 + 47 + · · ·? **C**

(A) 1600 (B) 18,235 (C) 18,475

(D) 18,800 (E) 19,125

7. MULTIPLE CHOICE What is a rule for the nth term of the geometric sequence with $a_3 = -12$ and common ratio $r = 3$? **A**

(A) $a_n = -\dfrac{4}{3}(3)^{n-1}$ (B) $a_n = -4(3)^{n-1}$

(C) $a_n = -\dfrac{3}{4}(3)^{n-1}$ (D) $a_n = -\dfrac{1}{3}(3)^{n-1}$

(E) $a_n = 4(3)^{n-1}$

8. MULTIPLE CHOICE What is the sum of the series

$$\sum_{i=0}^{9} 20\left(\dfrac{1}{2}\right)^i?\quad \mathbf{D}$$

(A) ≈ 11.74 (B) ≈ 13.30 (C) ≈ 13.32

(D) ≈ 39.96 (E) ≈ 29.97

9. MULTIPLE CHOICE What is the sum of the series

$$\sum_{i=1}^{\infty} 5(1.2)^{i-1}?\quad \mathbf{E}$$

(A) −30 (B) −25 (C) 25

(D) 30 (E) The series has no sum.

10. MULTIPLE CHOICE Which fraction is equivalent to the repeating decimal 0.3838 . . .? **E**

(A) $\dfrac{3}{10}$ (B) $\dfrac{3}{8}$ (C) $\dfrac{38}{100}$

(D) $\dfrac{383}{1000}$ (E) $\dfrac{38}{99}$

11. MULTIPLE CHOICE What is a recursive rule for the sequence 2, 6, 18, 54, . . .? **D**

(A) $a_n = 2(3)^{n-1}$

(B) $a_n = 3(2)^{n-1}$

(C) $a_1 = 2, a_n = a_{n-1} + 4$

(D) $a_1 = 2, a_n = 3a_{n-1}$

(E) $a_1 = 3, a_n = 2a_{n-1}$

12. MULTIPLE CHOICE What is the fourth term of the sequence defined by the recursive rule $a_1 = 3, a_n = n + a_{n-1} - 7$? **C**

(A) −1 (B) −6 (C) −9

(D) −10 (E) −11

QUANTITATIVE COMPARISON In Exercises 13 and 14, choose the statement that is true about the given quantities.

 (**A**) The quantity in column A is greater.

 (**B**) The quantity in column B is greater.

 (**C**) The two quantities are equal.

 (**D**) The relationship cannot be determined from the given information.

	Column A	Column B	
13.	The tenth term of the sequence defined by $a_n = 7 - 2n$	$\displaystyle\sum_{n=1}^{10} (7 - 2n)$	**A**
14.	$n!$ when n is an integer greater than 1	n^n when n is an integer greater than 1	**B**

15. 🌐 **MULTI-STEP PROBLEM** Use the pattern of checkerboard quilts at the right.

 a. What does n represent in each quilt? **The number of squares on a side.**

 b. What does a_n represent in each quilt? **The number of blue squares.**

 c. Draw the next four quilts in the pattern. **See margin.**

 d. Complete a table that gives n and a_n for $n = 1, 2, 3, 4, 5, 6, 7, 8$. **See margin.**

 e. Use the rule $a_n = \dfrac{n^2}{2} + \dfrac{1}{4}[1 - (-1)^n]$ to find a_n for $n = 1, 2, 3, 4, 5, 6, 7, 8$.
  $a_n = 1, 2, 5, 8, 13, 18, 25, 32$
  Compare with the results in your table. What can you conclude about the sequence defined by this rule? **they are the same; neither arithmetic or geometric**

$n = 1$
$a_1 = 1$

$n = 2$
$a_2 = 2$

16. **MULTI-STEP PROBLEM** Use the series $4 + 7 + 10 + 13 + 16 + 19 + 22 + 25$.

 a. Use a formula to find the sum of the series. Show your work. $\dfrac{8(4 + 25)}{2} = 116$

 b. Find the sum of the series without using a formula. Explain your method.
  Multiply the sum of the first and last terms by 4; 116

 c. Write the series with summation notation. Use 1 as the lower limit of summation. **See margin.**

 d. Write the series with summation notation. Use 0 as the lower limit of summation. **See margin.**

$n = 3$
$a_3 = 5$

 e. Write the series with summation notation. Use 4 as the lower limit of summation. **See margin.**

 f. *Writing* Compare your answers to parts (c), (d), and (e). Describe any similarities and differences. Which of these ways do you prefer to write the series? Explain your answer. **See margin.**

$n = 4$
$a_4 = 8$

17. **MULTI-STEP PROBLEM** Use the sequence $100, 50, 25, 12.5, \ldots$.

 a. Is this sequence arithmetic, geometric, or neither? Is it finite or infinite? **geometric; infinite**

 b. Write the next three terms of the sequence. **6.25, 3.125, 1.5625**

 c. Graph the sequence. Describe the curve on which the points lie. **See margin.**

 d. Write an explicit rule for the nth term of the sequence. $100\left(\dfrac{1}{2}\right)^{n-1}$

 e. Write a recursive rule for the sequence. $a_1 = 100; \ a_n = \left(\dfrac{1}{2}\right)a_{n-1}$

 f. Find the twelfth term of the sequence. Which rule from parts (d) and (e) did you use? Explain your choice. $\dfrac{25}{512}$; *Sample answer:* **I used the explicit rule from part (d) since it is the fastest way to find the answer.**

15c.

15d.

n	1	2	3	4	5	6	7	8
a_n	1	2	5	8	13	18	25	32

16c. $\displaystyle\sum_{i=1}^{8} (3i + 1)$

16d. $\displaystyle\sum_{i=0}^{7} (3i + 4)$

16e. $\displaystyle\sum_{i=4}^{11} (3i - 8)$

16f. *Sample answer:* **All answers have a sum of 8 terms, so the indices of summation are 7 apart. Each has a $3i$ term. The actual indices vary, as does the constant term.**

17c. *Sample answer:* **the points follow an exponential curve with a y-intercept of 200, y decreasing as x increases, becoming asymptotic to the x-axis as $x \to \infty$.**

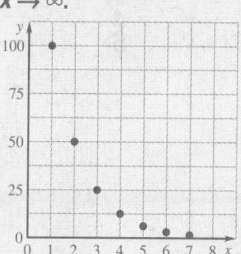

PLANNING THE CHAPTER

Probability and Statistics

LESSON	GOALS	NCTM	ITED	SAT9	Terra-Nova	Local
12.1 *pp. 701–707*	GOAL 1 Use the fundamental counting principle to count the number of ways an event can happen. GOAL 2 Use permutations to count the number of ways an event can happen.	1, 2, 5, 8	SAPP, MIP		12, 15, 18	
12.2 *pp. 708–715*	GOAL 1 Use combinations to count the number of ways an event can happen. GOAL 2 Use the binomial theorem to expand a binomial that is raised to a power.	1, 2, 5, 8	SAPP, MIP, SAPE		11, 12, 15, 18	
12.3 *pp. 716–723*	GOAL 1 Find theoretical and experimental probabilities. GOAL 2 Find geometric probabilities. TECHNOLOGY ACTIVITY: 12.3 *Use the random number generator of a graphing calculator to perform probability experiments.*	1, 2, 3, 5	SAPP, MIP, RQGE, RQPE	15, 16, 17, 29	11, 15, 50	
12.4 *pp. 724–729*	GOAL 1 Find probabilities of unions and intersections of two events. GOAL 2 Use complements to find the probability of an event.	1, 2, 5, 8	SAPP, MIP, RQPE	14	11, 15, 18, 50	
12.5 *pp. 730–737*	GOAL 1 Find the probability of independent events. GOAL 2 Find the probability of dependent events.	1, 2, 5	SAPP, MIP, RQPE	16	11, 15, 50	
12.6 *pp. 738–745*	CONCEPT ACTIVITY: 12.6 *Investigate binomial distributions.* GOAL 1 Find binomial probabilities and analyze binomial distributions. GOAL 2 Test a hypothesis. TECHNOLOGY ACTIVITY: 12.6 *Calculate binomial properties on a graphing calculator.*	1, 2, 5, 7, 8, 9	SAPP, MIP, RQPE, MCS	11, 16, 38	11, 15, 18, 50	
12.7 *pp. 746–752*	GOAL 1 Calculate probabilities using normal distributions. GOAL 2 Use normal distributions to approximate binomial distributions.	1, 2, 5, 8, 9	SAPP, MIP, RQPE, MCS		11, 15, 18, 50	
Extension *pp. 753–754*	GOAL 1 Find expected values of collections of outcomes.	1, 2, 5	SAPP, MIP, MCS		15	

RESOURCES

CHAPTER RESOURCE BOOKLETS

CHAPTER SUPPORT

Tips for New Teachers	p. 1	Prerequisite Skills Review	p. 5
Parent Guide for Student Success	p. 3	Strategies for Reading Mathematics	p. 7

LESSON SUPPORT

	12.1	12.2	12.3	12.4	12.5	12.6	12.7
Lesson Plans (regular and block)	p. 9	p. 24	p. 40	p. 53	p. 66	p. 80	p. 94
Warm-Up Exercises and Daily Quiz	p. 11	p. 26	p. 42	p. 55	p. 68	p. 82	p. 96
Activity Support Masters						p. 83	
Lesson Openers	p. 12	p. 27	p. 43	p. 56	p. 69	p. 84	p. 97
Graphing Calculator Activities & Keystrokes	p. 13	p. 28	p. 44	p. 57		p. 85	
Practice (3 levels)	p. 16	p. 31	p. 45	p. 58	p. 70	p. 86	p. 98
Reteaching with Practice	p. 19	p. 34	p. 48	p. 61	p. 73	p. 89	p. 101
Quick Catch-Up for Absent Students	p. 21	p. 36	p. 50	p. 63	p. 75	p. 91	p. 103
Cooperative Learning Activities							p. 104
Interdisciplinary Applications		p. 37		p. 64		p. 92	
Real-Life Applications	p. 22		p. 51		p. 76		p. 105
Math & History Applications					p. 77		
Challenge: Skills and Applications	p. 23	p. 38	p. 52	p. 65	p. 78	p. 93	p. 106

REVIEW AND ASSESSMENT

Quizzes	pp. 39, 79	Alternative Assessment with Math Journal	p. 115
Chapter Review Games and Activities	p. 107	Project with Rubric	p. 117
Chapter Test (3 levels)	pp. 108–113	Cumulative Review	p. 119
SAT/ACT Chapter Test	p. 114	Resource Book Answers	p. A1

TRANSPARENCIES

	12.1	12.2	12.3	12.4	12.5	12.6	12.7
Warm-Up Exercises and Daily Quiz	p. 89	p. 90	p. 91	p. 92	p. 93	p. 94	p. 95
Alternative Lesson Opener Transparencies	p. 77	p. 78	p. 79	p. 80	p. 81	p. 82	p. 83
Examples/Standardized Test Practice	✓	✓	✓	✓	✓	✓	✓
Answer Transparencies	✓	✓	✓	✓	✓	✓	✓

TECHNOLOGY

- Electronic Teaching Tools
- Online Lesson Planner
- Internet Support
- Personal Student Tutor
- Test and Practice Generator
- Instant Replay: Video Review Games
- Electronic Lesson Presentations (Lesson 12.3)

ADDITIONAL RESOURCES

- Basic Skills Workbook: Diagnosis and Remediation
- Worked-Out Solution Key
- Resources in Spanish
- Standardized Test Practice Workbook
- Practice Workbook with Examples

REGULAR SCHEDULE

Day 1

12.1

STARTING OPTIONS
- Prereq. Skills Review
- Strategies for Reading
- Homework Check
- Warm-Up or Daily Quiz

TEACHING OPTIONS
- Motivating the Lesson
- Les. Opener (Visual)
- Graphing Calc. Activity
- Examples 1–3
- Guided Practice
  Exs. 1–2, 5–8

APPLY/HOMEWORK
- See Assignment Guide.
- See the CRB: Practice, Reteach, Apply, Extend

ASSESSMENT OPTIONS
- Checkpoint Exercises, pp. 702–703

Day 2

12.1 (cont.)

STARTING OPTIONS
- Homework Check

TEACHING OPTIONS
- Examples 4–5
- Closure Question
- Guided Practice
  Exs. 3–4, 9–14

APPLY/HOMEWORK
- See Assignment Guide.
- See the CRB: Practice, Reteach, Apply, Extend

ASSESSMENT OPTIONS
- Checkpoint Exercises, p. 704
- Daily Quiz (12.1)
- Stand. Test Practice

Day 3

12.2

STARTING OPTIONS
- Homework Check
- Warm-Up or Daily Quiz

TEACHING OPTIONS
- Les. Opener (Visual)
- Graphing Calc. Activity
- Examples 1–5
- Guided Practice
  Exs. 1–2, 4–12

APPLY/HOMEWORK
- See Assignment Guide.
- See the CRB: Practice, Reteach, Apply, Extend

ASSESSMENT OPTIONS
- Checkpoint Exercises, pp. 709–711

Day 4

12.2 (cont.)

STARTING OPTIONS
- Homework Check

TEACHING OPTIONS
- Examples 6–8
- Closure Question
- Guided Practice
  Exs. 3, 13–17

APPLY/HOMEWORK
- See Assignment Guide.
- See the CRB: Practice, Reteach, Apply, Extend

ASSESSMENT OPTIONS
- Checkpoint Exercises, p. 711
- Daily Quiz (12.2)
- Stand. Test Practice
- Quiz (12.1–12.2)

Day 5

12.3

STARTING OPTIONS
- Homework Check
- Warm-Up or Daily Quiz

TEACHING OPTIONS
- Les. Opener (Application)
- Graphing Calc. Activity
- Examples 1–5
- Technology Activity
- Closure Question
- Guided Practice Exs.

APPLY/HOMEWORK
- See Assignment Guide.
- See the CRB: Practice, Reteach, Apply, Extend

ASSESSMENT OPTIONS
- Checkpoint Exercises
- Daily Quiz (12.3)
- Stand. Test Practice

Day 6

12.4

STARTING OPTIONS
- Homework Check
- Warm-Up or Daily Quiz

TEACHING OPTIONS
- Les. Opener (Activity)
- Graphing Calc. Activity
- Examples 1–5
- Closure Question
- Guided Practice Exs.

APPLY/HOMEWORK
- See Assignment Guide.
- See the CRB: Practice, Reteach, Apply, Extend

ASSESSMENT OPTIONS
- Checkpoint Exercises
- Daily Quiz (12.4)
- Stand. Test Practice

Day 9

12.7

STARTING OPTIONS
- Homework Check
- Warm-Up or Daily Quiz

TEACHING OPTIONS
- Motivating the Lesson
- Les. Opener (Calculator)
- Examples 1–3
- Closure Question
- Guided Practice Exs.

APPLY/HOMEWORK
- See Assignment Guide.
- See the CRB: Practice, Reteach, Apply, Extend

ASSESSMENT OPTIONS
- Checkpoint Exercises
- Daily Quiz (12.7)
- Stand. Test Practice
- Quiz (12.6–12.7)

Day 10

Review

DAY 10 START OPTIONS
- Homework Check

REVIEWING OPTIONS
- Chapter 12 Summary
- Chapter 12 Review
- Chapter Review Games and Activities

APPLY/HOMEWORK
- Chapter 12 Test (practice)
- Ch. Standardized Test (practice)

Day 11

Assess

DAY 11 START OPTIONS
- Homework Check

ASSESSMENT OPTIONS
- Chapter 12 Test
- SAT/ACT Ch. 12 Test
- Alternative Assessment

APPLY/HOMEWORK
- Skill Review, p. 768

Day 1

12.1

DAY 1 START OPTIONS
- Prereq. Skills Review
- Strategies for Reading
- Homework Check
- Warm-up or Daily Quiz

TEACHING 12.1 OPTIONS
- Motivating the Lesson
- Les. Opener (Visual)
- Graphing Calc. Activity
- Examples 1–5
- Closure Question
- Guided Practice Exs.

APPLY/HOMEWORK
- See Assignment Guide.
- See the CRB: Practice, Reteach, Apply, Extend

ASSESSMENT OPTIONS
- Checkpoint Exercises
- Daily Quiz (Les. 12.1)
- Stand. Test Practice

Day 2

12.2

DAY 2 START OPTIONS
- Homework Check
- Warm-up or Daily Quiz

TEACHING 12.2 OPTIONS
- Les. Opener (Visual)
- Graphing Calc. Activity
- Examples 1–8
- Closure Question
- Guided Practice Exs.

APPLY/HOMEWORK
- See Assignment Guide.
- See the CRB: Practice, Reteach, Apply, Extend

ASSESSMENT OPTIONS
- Checkpoint Exercises
- Daily Quiz (Les. 12.2)
- Stand. Test Practice
- Quiz (12.1–12.2)

Day 3

12.3 & 12.4

DAY 3 START OPTIONS
- Homework Check
- W-Up 12.3 or D. Quiz 12.2

TEACHING 12.3 OPTIONS
- Les. Opener (Appl.)
- Graphing Calc. Activity
- Examples 1–5
- Technology Activity
- Closure Question
- Guided Practice Exs.

TEACHING 12.4 OPTIONS
- Warm-Up (Les. 12.4)
- Les. Opener (Activity)
- Graphing Calc. Activity
- Examples 1–5
- Closure Question
- Guided Practice Exs.

APPLY/HOMEWORK
- See Assignment Guide.
- See the CRB: Practice, Reteach, Apply, Extend

ASSESSMENT OPTIONS
- Checkpoint Exercises
- Daily Quiz (12.3, 12.4)
- Stand. Test Practice

Day 4

12.5 & 12.6

DAY 4 START OPTIONS
- Homework Check
- W-Up 12.5 or D. Quiz 12.4

TEACHING 12.5 OPTIONS
- Les. Opener (Appl.)
- Examples 1–8
- Closure Question
- Guided Practice Exs.

TEACHING 12.6 OPTIONS
- Warm-Up (Les. 12.6)
- Concept Act. & Wksht.
- Les. Opener (Activity)
- Technology Activities
- Examples 1–4
- Closure Question
- Guided Practice Exs.

APPLY/HOMEWORK
- See Assignment Guide.
- See the CRB: Practice, Reteach, Apply, Extend

ASSESSMENT OPTIONS
- Checkpoint Exercises
- Daily Quiz (12.5, 12.6)
- Stand. Test Practice
- Quiz (12.3–12.5)

Day 5

12.7 & Review

DAY 5 START OPTIONS
- Homework Check
- Warm-Up or Daily Quiz

TEACHING 12.7 OPTIONS
- Motivating the Lesson
- Les. Opener (Calc.)
- Examples 1–3
- Closure Question
- Guided Practice Exs.

REVIEWING OPTIONS
- Chapter 12 Summary
- Chapter 12 Review
- Chapter Review Games and Activities

APPLY/HOMEWORK
- See Assignment Guide.
- See the CRB: Practice, Reteach, Apply, Extend
- Chapter 12 Test (prac.)
- Ch. Standardized Test (practice)

ASSESSMENT OPTIONS
- Checkpoint Exercises
- Daily Quiz (Les. 12.7)
- Stand. Test Practice
- Quiz (12.6–12.7)

Day 6

Assess & 13.1

(Day 6 = Ch. 13 Day 1)

ASSESSMENT OPTIONS
- Chapter 12 Test
- SAT/ACT Ch. 12 Test
- Alternative Assessment

CH. 13 START OPTIONS
- Skill Review, p. 768
- Prereq. Skills Review
- Strategies for Reading

TEACHING 13.1 OPTIONS
- Warm-Up (Les. 13.1)
- Motivating the Lesson
- Les. Opener (Visual)
- Graphing Calc. Activity
- Examples 1–5
- Closure Question
- Guided Practice Exs.

APPLY/HOMEWORK
- See Assignment Guide.
- See the CRB: Practice, Reteach, Apply, Extend

ASSESSMENT OPTIONS
- Checkpoint Exercises
- Daily Quiz (Les. 13.1)
- Stand. Test Practice

Day 7

12.5

STARTING OPTIONS
- Homework Check
- Warm-Up or Daily Quiz

TEACHING OPTIONS
- Les. Opener (Application)
- Examples 1–8
- Closure Question
- Guided Practice Exs.

APPLY/HOMEWORK
- See Assignment Guide.
- See the CRB: Practice, Reteach, Apply, Extend

ASSESSMENT OPTIONS
- Checkpoint Exercises
- Daily Quiz (12.5)
- Stand. Test Practice
- Quiz (12.3–12.5)

Day 8

12.6

STARTING OPTIONS
- Homework Check
- Warm-Up or Daily Quiz

TEACHING OPTIONS
- Concept Act. & Wksht.
- Les. Opener (Activity)
- Graphing Calc. Activity
- Examples 1–4
- Technology Activity
- Closure Question
- Guided Practice Exs.

APPLY/HOMEWORK
- See Assignment Guide.
- See the CRB: Practice, Reteach, Apply, Extend

ASSESSMENT OPTIONS
- Checkpoint Exercises
- Daily Quiz (12.6)
- Stand. Test Practice

MEETING INDIVIDUAL NEEDS

BEFORE THE CHAPTER

The *Chapter 12 Resource Book* has the following materials to distribute and use before the chapter:

- **Parent Guide for Student Success (pictured below)**
- **Prerequisite Skills Review**
- **Strategies for Reading Mathematics**

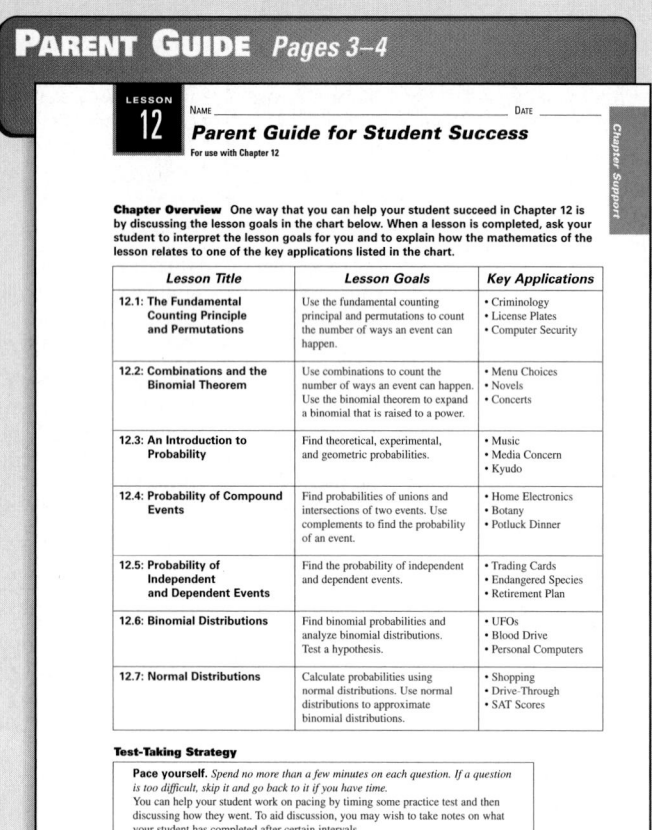

PARENT GUIDE *Pages 3–4*

LESSON 12 NAME _____ DATE _____

Parent Guide for Student Success
For use with Chapter 12

Chapter Overview One way that you can help your student succeed in Chapter 12 is by discussing the lesson goals in the chart below. When a lesson is completed, ask your student to interpret the lesson goals for you and to explain how the mathematics of the lesson relates to one of the key applications listed in the chart.

Lesson Title	Lesson Goals	Key Applications
12.1: The Fundamental Counting Principle and Permutations	Use the fundamental counting principal and permutations to count the number of ways an event can happen.	• Criminology • License Plates • Computer Security
12.2: Combinations and the Binomial Theorem	Use combinations to count the number of ways an event can happen. Use the binomial theorem to expand a binomial that is raised to a power.	• Menu Choices • Novels • Concerts
12.3: An Introduction to Probability	Find theoretical, experimental, and geometric probabilities.	• Music • Media Concern • Kyudo
12.4: Probability of Compound Events	Find probabilities of unions and intersections of two events. Use complements to find the probability of an event.	• Home Electronics • Botany • Potluck Dinner
12.5: Probability of Independent and Dependent Events	Find the probability of independent and dependent events.	• Trading Cards • Endangered Species • Retirement Plan
12.6: Binomial Distributions	Find binomial probabilities and analyze binomial distributions. Test a hypothesis.	• UFOs • Blood Drive • Personal Computers
12.7: Normal Distributions	Calculate probabilities using normal distributions. Use normal distributions to approximate binomial distributions.	• Shopping • Drive-Through • SAT Scores

Test-Taking Strategy

Pace yourself. *Spend no more than a few minutes on each question. If a question is too difficult, skip it and go back to it if you have time.*
You can help your student work on pacing by timing some practice test and then discussing how they went. To aid discussion, you may wish to take notes on what your student has completed after certain intervals.

PARENT GUIDE FOR STUDENT SUCCESS The first page summarizes the content of Chapter 12. Parents are encouraged to have their students explain how the material relates to key applications in the chapter, such as SAT scores. The second page (not shown) provides exercises and an activity that parents can do with their students. In the activity, parents and students use permutations to set up a dish-washing schedule for the family.

DURING EACH LESSON

The *Chapter 12 Resource Book* has the following alternatives for introducing the lesson:

- **Lesson Openers (pictured below)**
- **Graphing Calculator Activities with Keystrokes**

LESSON OPENER *Page 43*

LESSON 12.3 NAME _____ DATE _____   Available as a transparency

Application Lesson Opener
For use with pages 716–723

The *probability* of an event is a number between 0 and 1 that indicates the likelihood that the event will occur. Some sample probabilities and their meanings are given below.

$P = 0$ Event will not occur.

$P = 0.25$ Event probably will not occur.

$P = 0.5$ Event is equally likely to occur or not occur.

$P = 0.75$ Event probably will occur.

$P = 1$ Event is certain to occur.

State whether the probability that the event will occur is closer to 0, 0.25, 0.5, 0.75, or 1.

1. The next time you flip a coin, it will come up heads.
2. The first card you draw from a standard 52-card deck will be a heart.
3. The next baby born in your town will be a girl.
4. There will be a snowstorm in Ohio in July.
5. The next time you roll a number cube, you will get an odd number.
6. You will be assigned homework in the next week.
7. The first card you draw from a standard 52-card deck will be a 7, 8, or 9.
8. You will attend school on July 4.
9. The first card you draw from a standard 52-card deck will be a spade, club, or diamond.

APPLICATION LESSON OPENER This Lesson Opener provides an alternative way to start Lesson 12.3 in the form of a real-life application. Students see how probability relates to the likelihood of real-life events.

 TECHNOLOGY RESOURCE

Teachers can use the Electronic Lesson Presentations CD-ROM to present Lesson 12.3 using computer animation to step through the Examples.

The *Chapter 12 Resource Book* has a variety of materials to follow-up each lesson. They include the following:

- **Practice (3 levels)**
- **Reteaching with Practice**
- **Quick Catch-Up for Absent Students**
- **Challenge: Skills and Applications**
- **Interdisciplinary Applications (pictured below)**
- **Real-Life Applications**

APPLICATION *Page 37*

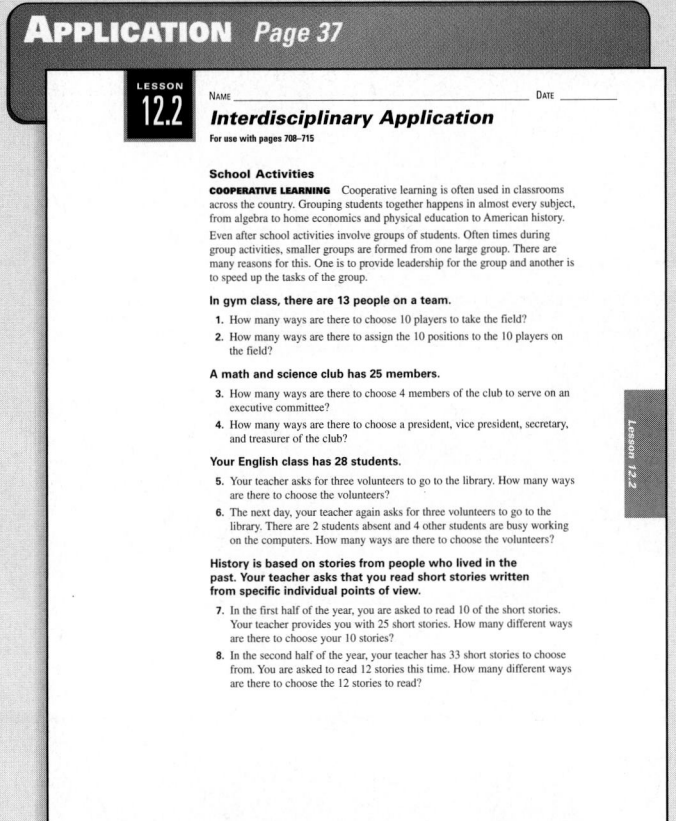

LESSON 12.2
NAME _____ DATE _____

Interdisciplinary Application
For use with pages 708–715

School Activities

COOPERATIVE LEARNING Cooperative learning is often used in classrooms across the country. Grouping students together happens in almost every subject, from algebra to home economics and physical education to American history.

Even after school activities involve groups of students. Often times during group activities, smaller groups are formed from one large group. There are many reasons for this. One is to provide leadership for the group and another is to speed up the tasks of the group.

In gym class, there are 13 people on a team.

1. How many ways are there to choose 10 players to take the field?
2. How many ways are there to assign the 10 positions to the 10 players on the field?

A math and science club has 25 members.

3. How many ways are there to choose 4 members of the club to serve on an executive committee?
4. How many ways are there to choose a president, vice president, secretary, and treasurer of the club?

Your English class has 28 students.

5. Your teacher asks for three volunteers to go to the library. How many ways are there to choose the volunteers?
6. The next day, your teacher again asks for three volunteers to go to the library. There are 2 students absent and 4 other students are busy working on the computers. How many ways are there to choose the volunteers?

History is based on stories from people who lived in the past. Your teacher asks that you read short stories written from specific individual points of view.

7. In the first half of the year, you are asked to read 10 of the short stories. Your teacher provides you with 25 short stories. How many different ways are there to choose your 10 stories?
8. In the second half of the year, your teacher has 33 short stories to choose from. You are asked to read 12 stories this time. How many different ways are there to choose the 12 stories to read?

Lesson 12.2

INTERDISCIPLINARY APPLICATION This application makes a connection between school activities and combinations, the topic of Lesson 12.2.

The *Chapter 12 Resource Book* has the following review and assessment materials:

- **Quizzes**
- **Chapter Review Games and Activities**
- **Chapter Test (3 levels) (pictured below)**
- **SAT/ACT Chapter Test**
- **Alternative Assessment with Rubrics and Math Journal**
- **Project with Rubric**
- **Cumulative Review**

CHAPTER TEST *Pages 110–111*

CHAPTER 12
NAME _____ DATE _____

Chapter Test B
For use after Chapter 12

Find the number of permutations or combinations.

1. $_4P_2$ 2. $_9P_7$ 3. $_4C_2$ 4. $_9C_7$

5. Find the number of distinguishable permutations of the letters in DALLAS.

Expand the power of the binomial.

6. $(x + y)^5$ 7. $(x + 2)^6$ 8. $(x^2 - 1)^4$ 9. $(2x - y)^3$

A card is drawn randomly from a standard 52-card deck. Find the probability of drawing the given card.

10. a red card   11. a red ace   12. a spade

13. the queen of diamonds   14. a jack

Find the indicated probability.

15. $P(A) = \frac{1}{4}$ 16. $P(A) = 30\%$ 17. $P(A) = 0.5$
    $P(A') = \underline{?}$ $P(B) = 70\%$ $P(B) = 0.3$
                              $P(A \text{ or } B) = 100\%$ $P(A \text{ or } B) = \underline{?}$
                              $P(A \text{ and } B) = \underline{?}$ $P(A \text{ and } B) = 0.2$

Answers

1. _____
2. _____
3. _____
4. _____
5. _____
6. _____
7. _____
8. _____
9. _____
10. _____
11. _____
12. _____
13. _____
14. _____
15. _____
16. _____
17. _____

Review and Assess

CHAPTER TEST There are three versions of this two-page Chapter Test, one each for basic (A), average (B), and advanced (C) students. Level B involves more advanced work with expanding the power of binomials than Level A, but has less advanced work with finding probabilities than Level C.

TECHNOLOGY RESOURCE

Teachers can use the Time-Saving Test and Practice Generator to create customized review and assessment materials for Chapter 12.

CHAPTER GOALS

This chapter begins with a study of counting techniques that are later used to calculate probabilities. The fundamental counting principle, permutations, and combinations are used to count the number of ways an event can happen. Students use what they have learned about combinations and the binomial theorem to expand a binomial that is raised to a power. The second part of the chapter is devoted to probability. Students find theoretical and experimental probabilities of events including those events involving the unions and intersections of events. They find probabilities of independent and dependent events. Complements are used to find the probability of an event and geometric probabilities are explored. Next, binomial and normal distributions are introduced. Students construct and interpret binomial distributions and use them to test a hypothesis. Normal distributions are used to calculate probabilities and approximate binomial distributions. Finally, in the Chapter Extension the concept of expected value is investigated.

APPLICATION NOTE

The theater shown on this page, known as The Hatch Shell, is in Boston, Massachusetts on the banks of the Charles River. This theater is home to various musical events throughout the summer. Every July 4th thousands gather on the grounds of the Hatch Shell and in boats on the river to listen to a concert by the Boston Pops orchestra. This event always culminates in a performance of the 1812 Overture accompanied by a fireworks display.

Additional information about concerts is available at **www.mcdougallittell.com.**

PROBABILITY AND STATISTICS

▶ *In how many ways can you attend part of a summer concert series?*

APPLICATION: Concerns

*M*any cities offer summer concert series. Some have municipal bands that tour the city and play at different parks throughout the summer. Other cities have outdoor theaters where a variety of professional musicians can come to perform concerts.

Think & Discuss

♪ Summer Concert Series ♪

June 5th	June 19th
Music by Beethoven	Music by Mozart
July 3rd	July 17th
Music by Dvorak	Music by Brahms
August 7th	August 21st
Music by Strauss	Music by Tchaikovsky

1. Suppose you plan to attend exactly two of the concerts. Make a list of the possible choices of two concerts you can attend. How many choices do you have? **See margin for list; 15**

2. Suppose you plan to attend either one or two of the concerts. How many choices do you have? **21**

Learn More About It

You will find the number of different combinations of concerts you can attend in Exercise 54 on p. 713.

 APPLICATION LINK Visit www.mcdougallittell.com for more information about concerts.

699

ADDITIONAL RESOURCES
Another way to begin the chapter is to show the video clip of a real-life motivator for Chapter 12 on the *Instant Replay: Video Review Games.*

PROJECTS
A project covering Chapters 10–12 appears on pages 764–765 of the Student Edition. An additional project for Chapter 12 is available in the *Chapter 12 Resource Book*, p. 117.

TECHNOLOGY

 **Software**
• *Electronic Teaching Tools*
• *Online Lesson Planner*
• *Personal Student Tutor*
• *Test and Practice Generator*
• *Electronic Lesson Presentations* (Lesson 12.3)

Video
• *Instant Replay: Video Review Games*

 Internet Connections
www.mcdougallittell.com
• **Application Links**
  699, 713, 721, 737, 741
• **Data Updates**
  719, 735, 751
• **Student Help**
  703, 704, 706, 709, 711, 718, 723, 726, 729, 735, 743, 745, 751, 754
• **Career Links**
  702, 728, 735, 748
• **Extra Challenge**
  707, 714, 722, 729, 736, 744, 751

1. (Jun 5, Jun 19), (Jun 5, Jul 3), (Jun 5, Jul 17), (Jun 5, Aug 7), (Jun 5, Aug 21), (Jun 19, Jul 3), (Jun 19, Jul 17), (Jun 19, Aug 7), (Jun 19, Aug 21), (Jul 3, Jul 17), (Jul 3, Aug 7), (Jul 3, Aug 21), (Jul 17, Aug 7), (Jul 17, Aug 21), (Aug 7, Aug 21)

Study Guide

DIAGNOSTIC TOOLS

The **Skill Review** exercises can help you diagnose whether students have the following skills needed in Chapter 12:

- Write fractions as decimals and percents.
- Find areas of figures.
- Solve exponential equations.

The following resources are available for students who need additional help with these skills:

- **Prerequisite Skills Review** (*Chapter 12 Resource Book,* p. 5; *Warm-Up Transparencies,* p. 88)
- ▣ *Personal Student Tutor*

ADDITIONAL RESOURCES

The following resources are provided to help you prepare for the upcoming chapter and customize review materials:

- *Chapter 12 Resource Book*
  Tips for New Teachers (p. 1)
  Parent Guide (p. 3)
  Lesson Plans (every lesson)
  Lesson Plans for Block Scheduling (every lesson)
- ▣ *Electronic Teaching Tools*
- ▣ *Online Lesson Planner*
- ▣ *Test and Practice Generator*

ENGLISH LEARNERS

Understanding relationships between mathematical terms and words used in everyday conversation can often help English learners remember the meaning of those mathematical terms. Point out that *probability* is a related form of the words *probably* and *probable.* Ask students what *probably* means ("likely to happen"). Then explain that probability has to do with *how* likely particular things are to happen.

PREVIEW

What's the chapter about?

Chapter 12 is about **probability and statistics**. In Chapter 12 you'll learn

- how to count the number of ways an event can happen.
- how to calculate and use probabilities.
- how to use binomial and normal distributions.

KEY VOCABULARY

▶ Review	▶ New	
• binomial, p. 256	• permutation, p. 703	• complement, p. 726
• mean, p. 445	• combination, p. 708	• independent events, p. 730
• standard deviation, p. 446	• binomial theorem, p. 710	• dependent events, p. 732
• factorial, p. 681	• probability, p. 716	• binomial distribution, p. 739
	• compound event, p. 724	• hypothesis testing, p. 741
		• normal distribution, p. 746

PREPARE

Are you ready for the chapter?

SKILL REVIEW Do these exercises to review key skills that you'll apply in this chapter. See the given **reference page** if there is something you don't understand.

STUDENT HELP

→ **Study Tip**
"Student Help" boxes throughout the chapter give you study tips and tell you where to look for extra help in this book and on the Internet.

Write the following as a decimal and as a percent. (Skills Review, p. 906)

1. $\frac{1}{2}$ 0.5, 50% **2.** $\frac{1}{5}$ 0.2, 20% **3.** $\frac{3}{20}$ 0.15, 15% **4.** $\frac{12}{25}$ 0.48, 48% **5.** $\frac{7}{36}$ 0.194, 19.4% **6.** $\frac{15}{32}$ 0.469, 46.9%

Find the area of the figure. (Skills Review, p. 914)

7. 50.27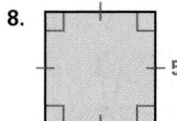

8. 25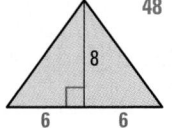

9. 48

Solve the exponential equation. (Review Examples 2 and 3, pp. 501 and 502)

10. $10^x = 0.5$ -0.301 **11.** $(0.5)^x + 3 = 3.75$ 0.415 **12.** $1 - 9^x = 0.25$ -0.131

STUDY STRATEGY

Here's a study strategy!

Connect to Your Life
Note real examples from your life that use the mathematics in this chapter. As you read the examples given in the textbook, listen to a teacher's explanation, and do the homework problems, think of problems in your own life that can be solved the same way. This will help you to understand and use the concepts you are learning.

12.1
The Fundamental Counting Principle and Permutations

What you should learn

GOAL 1 Use the fundamental counting principle to count the number of ways an event can happen.

GOAL 2 Use permutations to count the number of ways an event can happen, as applied in **Ex. 62.**

Why you should learn it

▼ To find the number of ways a **real-life** event can happen, such as the number of ways skiers can finish in an aerial competition in **Example 3**.

GOAL 1 THE FUNDAMENTAL COUNTING PRINCIPLE

In many real-life problems you want to count the number of possibilities. For instance, suppose you own a small deli. You offer 4 types of meat (ham, turkey, roast beef, and pastrami) and 3 types of bread (white, wheat, and rye). How many choices do your customers have for a meat sandwich?

One way to answer this question is to use a *tree diagram*, as shown below. From the list on the right you can see that there are 12 choices.

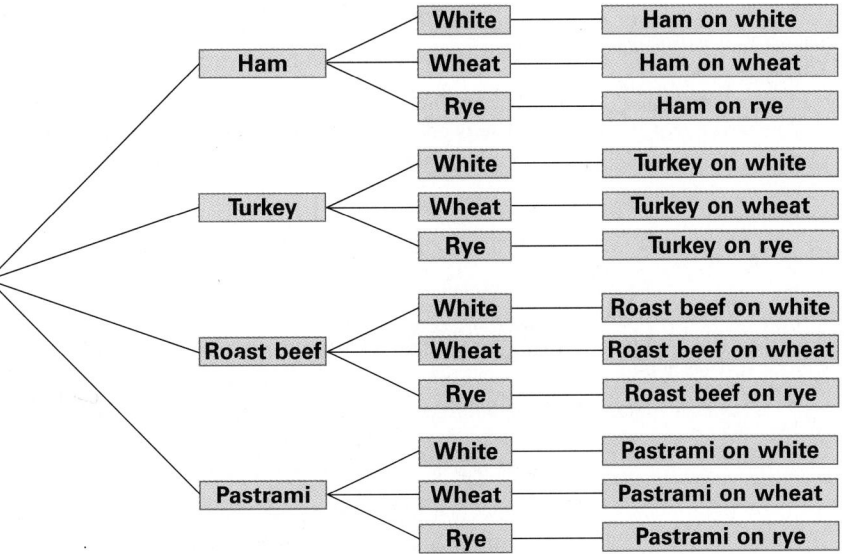

	White	Ham on white
Ham	Wheat	Ham on wheat
	Rye	Ham on rye
	White	Turkey on white
Turkey	Wheat	Turkey on wheat
	Rye	Turkey on rye
	White	Roast beef on white
Roast beef	Wheat	Roast beef on wheat
	Rye	Roast beef on rye
	White	Pastrami on white
Pastrami	Wheat	Pastrami on wheat
	Rye	Pastrami on rye

Another way to count the number of possible sandwiches is to use the *fundamental counting principle*. Because you have 4 choices for meat and 3 choices for bread, the total number of choices is $4 \cdot 3 = 12$.

FUNDAMENTAL COUNTING PRINCIPLE

TWO EVENTS If one event can occur in m ways and another event can occur in n ways, then the number of ways that *both* events can occur is $m \cdot n$.

For instance, if one event can occur in 2 ways and another event can occur in 5 ways, then both events can occur in $2 \cdot 5 = 10$ ways.

THREE OR MORE EVENTS The fundamental counting principle can be extended to three or more events. For example, if three events can occur in m, n, and p ways, then the number of ways that *all* three events can occur is $m \cdot n \cdot p$.

For instance, if three events can occur in 2, 5, and 7 ways, then all three events can occur in $2 \cdot 5 \cdot 7 = 70$ ways.

12.1 *The Fundamental Counting Principle and Permutations* **701**

1 PLAN

PACING
Basic: 2 days
Average: 2 days
Advanced: 2 days
Block Schedule: 1 block

LESSON OPENER
VISUAL APPROACH
An alternative way to approach Lesson 12.1 is to use the Visual Approach Lesson Opener:
- Blackline Master (*Chapter 12 Resource Book*, p. 12)
- Transparency (p. 77)

MEETING INDIVIDUAL NEEDS
- *Chapter 12 Resource Book*
  Prerequisite Skills Review (p. 5)
  Practice Level A (p. 16)
  Practice Level B (p. 17)
  Practice Level C (p. 18)
  Reteaching with Practice (p. 19)
  Absent Student Catch-Up (p. 21)
  Challenge (p. 23)
- *Resources in Spanish*
- *Personal Student Tutor*

NEW-TEACHER SUPPORT
See the Tips for New Teachers on pp. 1–2 of the *Chapter 12 Resource Book* for additional notes about Lesson 12.1.

WARM-UP EXERCISES
Transparency Available
Evaluate.
1. $3!$ 6   **2.** $7!$ 5040
3. $\dfrac{4!}{6!}$ $\dfrac{1}{30}$ **4.** $\dfrac{10!}{(10-4)!}$ 5040

Explain to students that new telephone area codes are added when areas run out of available telephone numbers. Telephone companies need to know how many phone numbers are possible for each area code so that they can plan for when a new area code will be needed. Listing every possible phone number would be time consuming. In today's lesson, students will learn how they could quickly determine the numbers of possible phone numbers.

EXTRA EXAMPLE 1
At a restaurant, you have a choice of 8 different entrees, 2 different salads, 12 different drinks, and 6 different desserts. How many different dinners consisting of 1 salad, 1 entree, 1 drink and 1 dessert can you choose? **1152**

EXTRA EXAMPLE 2
How many different 7 digit phone numbers are possible if the first digit cannot be 0 or 1? **8,000,000**

CHECKPOINT EXERCISES
For use after Example 1:
1. In a high school, there are 273 freshmen, 291 sophomores, 252 juniors and 237 seniors. In how many different ways can a committee of 1 freshman, 1 sophomore, 1 junior and 1 senior be chosen? **4,744,653,732**

For use after Example 2:
2. A multiple choice test has 10 questions with 4 answer choices for each question. In how many different ways could you complete the test? **1,048,576**

CAREER NOTE
EXAMPLE 1 Additional information about a career as a police detective is available at **www.mcdougallittell.com.**

POLICE DETECTIVE
A police detective is an officer who collects facts and evidence for criminal cases. Part of a detective's duties may include helping witnesses identify suspects.

CAREER LINK
www.mcdougallittell.com

License Plates

EXAMPLE 1   *Using the Fundamental Counting Principle*

CRIMINOLOGY Police use photographs of various facial features to help witnesses identify suspects. One basic identification kit contains 195 hairlines, 99 eyes and eyebrows, 89 noses, 105 mouths, and 74 chins and cheeks.
▶ Source: *Readers' Digest: How In The World?*

a. The developer of the identification kit claims that it can produce billions of different faces. Is this claim correct?

b. A witness can clearly remember the hairline and the eyes and eyebrows of a suspect. How many different faces can be produced with this information?

SOLUTION

a. You can use the fundamental counting principle to find the total number of different faces.

$$\text{Number of faces} = 195 \cdot 99 \cdot 89 \cdot 105 \cdot 74 = 13{,}349{,}986{,}650$$

▶ The developer's claim is correct since the kit can produce over 13 billion faces.

b. Because the witness clearly remembers the hairline and the eyes and eyebrows, there is only 1 choice for each of these features. You can use the fundamental counting principle to find the number of different faces.

$$\text{Number of faces} = 1 \cdot 1 \cdot 89 \cdot 105 \cdot 74 = 691{,}530$$

▶ The number of faces that can be produced has been reduced to 691,530.

EXAMPLE 2   *Using the Fundamental Counting Principle with Repetition*

The standard configuration for a New York license plate is 3 digits followed by 3 letters.
▶ Source: New York State Department of Motor Vehicles

NEW YORK
234 ABC

a. How many different license plates are possible if digits and letters can be repeated?

b. How many different license plates are possible if digits and letters cannot be repeated?

SOLUTION

a. There are 10 choices for each digit and 26 choices for each letter. You can use the fundamental counting principle to find the number of different plates.

$$\text{Number of plates} = 10 \cdot 10 \cdot 10 \cdot 26 \cdot 26 \cdot 26 = 17{,}576{,}000$$

▶ The number of different license plates is 17,576,000.

b. If you cannot repeat digits there are still 10 choices for the first digit, but then only 9 remaining choices for the second digit and only 8 remaining choices for the third digit. Similarly, there are 26 choices for the first letter, 25 choices for the second letter, and 24 choices for the third letter. You can use the fundamental counting principle to find the number of different plates.

$$\text{Number of plates} = 10 \cdot 9 \cdot 8 \cdot 26 \cdot 25 \cdot 24 = 11{,}232{,}000$$

▶ The number of different license plates is 11,232,000.

GOAL 2 USING PERMUTATIONS

An ordering of n objects is a **permutation** of the objects. For instance, there are six permutations of the letters A, B, and C: ABC, ACB, BAC, BCA, CAB, CBA.

The fundamental counting principle can be used to determine the number of permutations of n objects. For instance, you can find the number of ways you can arrange the letters A, B, and C by multiplying. There are 3 choices for the first letter, 2 choices for the second letter, and 1 choice for the third letter, so there are $3 \cdot 2 \cdot 1 = 6$ ways to arrange the letters.

In general, the number of permutations of n distinct objects is:

$$n! = n \cdot (n-1) \cdot (n-2) \cdot \ldots \cdot 3 \cdot 2 \cdot 1$$

Sports

EXAMPLE 3 *Finding the Number of Permutations*

Twelve skiers are competing in the final round of the Olympic freestyle skiing aerial competition.

a. In how many different ways can the skiers finish the competition? (Assume there are no ties.)

b. In how many different ways can 3 of the skiers finish first, second, and third to win the gold, silver, and bronze medals?

SOLUTION

a. There are 12! different ways that the skiers can finish the competition.

$$12! = 12 \cdot 11 \cdot 10 \cdot 9 \cdot 8 \cdot 7 \cdot 6 \cdot 5 \cdot 4 \cdot 3 \cdot 2 \cdot 1 = 479,001,600$$

b. Any of the 12 skiers can finish first, then any of the remaining 11 skiers can finish second, and finally any of the remaining 10 skiers can finish third. So, the number of ways that the skiers can win the medals is:

$$12 \cdot 11 \cdot 10 = 1320$$

· · · · · · · · · ·

STUDENT HELP

KEYSTROKE HELP
Visit our Web site www.mcdougallittell.com to see keystrokes for several models of calculators.

Some calculators have special keys to evaluate factorials. The solution to Example 3 is shown.

The number in part (b) of Example 3 is called the number of permutations of 12 objects taken 3 at a time, is denoted by $_{12}P_3$, and is given by $\frac{12!}{(12-3)!}$.

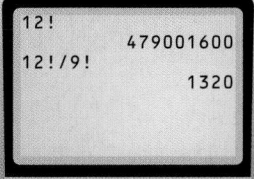

```
12!
                479001600
12!/9!
                     1320
```

STUDENT HELP

Derivations
For a derivation of the formula for the permutation of n objects taken r at a time, see p. 899.

PERMUTATIONS OF n OBJECTS TAKEN r AT A TIME

The number of permutations of r objects taken from a group of n distinct objects is denoted by $_nP_r$ and is given by:

$$_nP_r = \frac{n!}{(n-r)!}$$

STUDENT HELP

Study Tip
Recall from Lesson 11.5 that $n!$ is read as "n factorial." Also note that $0! = 1$ and $1! = 1$.

EXTRA EXAMPLE 3
You have homework assignments from 5 different classes to complete this weekend.
a. In how many different ways can you complete the assignments? **120**
b. In how many different ways can you choose 2 of the assignments to complete first and last? **20**

✓ CHECKPOINT EXERCISES

For use after Example 3:

1. There are 8 movies you would like to see currently showing in theaters.
  a. In how many different ways can you see all 8 of the movies? **40,320**
  b. In how many ways can you choose a movie to see this Saturday and one to see this Sunday? **56**

STUDENT HELP NOTES

Keystroke Help Keystrokes for several models of calculators are available in **blackline** format in the *Chapter 12 Resource Book*, p. 15 and at **www.mcdougallittell.com**.

ENGLISH LEARNERS
Some students may be confused by the uses of *order* and *ordering* in this chapter. Explain that *order* is used as a noun meaning "a certain sequence," and as a verb meaning "put in a certain sequence"; an *ordering* is a sequence someone has created. Next, preteach the terms *reordering* and *unordered*, which are introduced on page 708. You might then use a set of small objects to demonstrate the concepts of *ordering* and *permutations*.

EXTRA EXAMPLE 4
There are 12 books on the summer reading list. You want to read some or all of them. In how many orders can you read (a) 4 of the books or (b) all 12 of the books? **11,880; 479,001,600**

EXTRA EXAMPLE 5
Find the number of distinguishable permutations of the letters in (a) SUMMER and (b) WATERFALL. **360; 90,720**

 CHECKPOINT EXERCISES
For use after Example 4:
1. There are 9 players on a baseball team. (a) In how many ways can you choose the batting order for all 9 of the players? (b) In how many ways can you choose a pitcher, catcher, and shortstop from the 9 players? **362,880; 504**
2. Your dog has 8 puppies, 3 male and 5 female. One possible birth order is MMMFFFFF. How many different birth orders are possible? **56**

FOCUS ON VOCABULARY
Write a definition for *permutation* in your own words. *Sample answer:* a permutation is an ordering of objects.

CLOSURE QUESTION
Explain the difference between a situation in which you would use $n!$ and one in which you would use $\frac{n!}{(n-r)!}$. You would use $n!$ when you are ordering all n of a group of n objects. You would use $\frac{n!}{(n-r)!}$ when you are ordering only r of the n objects.

DAILY PUZZLER
A 12-letter word has one of its letters repeated at least twice and no other letters repeated. If there are 3,991,680 distinguishable permutations of the letters in the word, how many times is the one letter repeated? **5**

704

College Visits

EXAMPLE 4 *Finding Permutations of n Objects Taken r at a Time*

You are considering 10 different colleges. Before you decide to apply to the colleges, you want to visit some or all of them. In how many orders can you visit (a) 6 of the colleges and (b) all 10 colleges?

SOLUTION

a. The number of permutations of 10 objects taken 6 at a time is:

$$_{10}P_6 = \frac{10!}{(10-6)!} = \frac{10!}{4!} = \frac{3,628,800}{24} = 151,200$$

b. The number of permutations of 10 objects taken 10 at a time is:

$$_{10}P_{10} = \frac{10!}{(10-10)!} = \frac{10!}{0!} = 10! = 3,628,800$$

. . . .

STUDENT HELP
KEYSTROKE HELP
Visit our Web site www.mcdougallittell.com to see keystrokes for several models of calculators.

Some calculators have special keys that are programmed to evaluate $_nP_r$. The solution to Example 4 is shown.

So far you have been finding permutations of *distinct* objects. If some of the objects are repeated, then some of the permutations are not distinguishable. For instance, of the six ways to order the letters M, O, and M—

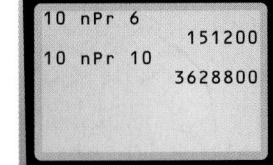

MOM   OMM   MMO
MOM   OMM   MMO

—only three are distinguishable without color: MOM, OMM, and MMO. In this case, the number of permutations is $\frac{3!}{2!} = \frac{6}{2} = 3$, not $3! = 6$.

PERMUTATIONS WITH REPETITION

The number of distinguishable permutations of n objects where one object is repeated q_1 times, another is repeated q_2 times, and so on is:

$$\frac{n!}{q_1! \cdot q_2! \cdot \ldots \cdot q_k!}$$

EXAMPLE 5 *Finding Permutations with Repetition*

Find the number of distinguishable permutations of the letters in (a) OHIO and (b) MISSISSIPPI.

SOLUTION

a. OHIO has 4 letters of which O is repeated 2 times. So, the number of distinguishable permutations is $\frac{4!}{2!} = \frac{24}{2} = 12$.

b. MISSISSIPPI has 11 letters of which I is repeated 4 times, S is repeated 4 times, and P is repeated 2 times. So, the number of distinguishable permutations is
$$\frac{11!}{4! \cdot 4! \cdot 2!} = \frac{39,916,800}{24 \cdot 24 \cdot 2} = 34,650.$$

GUIDED PRACTICE

Vocabulary Check ✓

Concept Check ✓

3. yes; $1! = 1$,

so $\frac{4!}{2!} = \frac{4!}{(2!)(1!)(1!)}$

Skill Check ✓

4. Because we are interested in all numbers with three even digits, those digits can be repeated. Therefore, the correct answer is $5 \times 5 \times 5 = 125$.

1. What is a permutation of n objects? **A permutation of n objects is an ordering of those objects.**

2. Explain how the fundamental counting principle can be used to justify the formula for the number of permutations of n distinct objects. **See margin.**

3. Rita found the number of distinguishable permutations of the letters in OHIO by evaluating the expression $\frac{4!}{2! \cdot 1! \cdot 1!}$. Does this method give the same answer as in part (a) of Example 5? Explain.

4. **ERROR ANALYSIS** Explain the error in calculating how many three-digit numbers from 000 to 999 have only even digits.

Number of 3-digit numbers
with only even digits
$= 5 \cdot 4 \cdot 3$
$= 60$

Find the number of permutations of n distinct objects.

5. $n = 2$ 2
6. $n = 6$ 720
7. $n = 1$ 1
8. $n = 4$ 24

Find the number of permutations of n objects taken r at a time.

9. $n = 6, r = 3$ 120
10. $n = 5, r = 1$ 5
11. $n = 3, r = 3$ 6
12. $n = 10, r = 2$ 90

Find the number of permutations of n objects where one or more objects are repeated the given number of times.

13. 7 objects with one object repeated 4 times   210

14. 5 objects with one object repeated 3 times and a second object repeated 2 times   10

PRACTICE AND APPLICATIONS

STUDENT HELP

▶ **Extra Practice**
to help you master skills is on p. 956.

FUNDAMENTAL COUNTING PRINCIPLE Each event can occur in the given number of ways. Find the number of ways all of the events can occur.

15. Event 1: 1 way, Event 2: 3 ways   3
16. Event 1: 3 ways, Event 2: 5 ways   15
17. Event 1: 2 ways, Event 2: 4 ways, Event 3: 5 ways   40
18. Event 1: 4 ways, Event 2: 6 ways, Event 3: 9 ways, Event 4: 7 ways   1512

LICENSE PLATES For the given configuration, determine how many different license plates are possible if (a) digits and letters can be repeated, and (b) digits and letters cannot be repeated.

19. 3 letters followed by 3 digits
a. 17,576,000   b. 11,232,000
20. 2 digits followed by 4 letters
a. 45,697,600   b. 32,292,000
21. 4 digits followed by 2 letters
a. 6,760,000   b. 3,276,000
22. 5 letters followed by 1 digit
a. 118,813,760   b. 78,936,000

STUDENT HELP

▶ **HOMEWORK HELP**
Example 1: Exs. 15–18, 55, 56
Example 2: Exs. 19–22, 57
Example 3: Exs. 23–30, 39–46, 59, 60
Example 4: Exs. 31–38, 61
Example 5: Exs. 47–54, 62, 63

FACTORIALS Evaluate the factorial.

23. 8!   40,320
24. 5!   120
25. 10!   3,628,800
26. 9!   362,880
27. 0!   1
28. 7!   5040
29. 3!   6
30. 12!   479,001,600

PERMUTATIONS Find the number of permutations.

31. $_3P_3$ 6
32. $_5P_2$ 20
33. $_2P_1$ 2
34. $_7P_6$ 5040
35. $_8P_5$ 6720
36. $_9P_4$ 3024
37. $_{12}P_3$ 1320
38. $_{16}P_0$ 1

12.1 *The Fundamental Counting Principle and Permutations* | 705

ASSIGNMENT GUIDE

BASIC
Day 1: pp. 705–706 Exs. 15–22, 24–30 even, 40–46 even, 55–57
Day 2: pp. 705–707 Exs. 31–38, 47–54, 64–66, 69–81 odd

AVERAGE
Day 1: pp. 705–706 Exs. 15–30, 39–44, 55–57, 59–60
Day 2: pp. 705–707 Exs. 31–38, 47–54, 58, 64–66, 69–85 odd, 86

ADVANCED
Day 1: pp. 705–706 Exs. 15–30, 39–46, 55–57, 59–60
Day 2: pp. 705–707 Exs. 31–38, 47–54, 58, 61–67, 69–85 odd, 86

BLOCK SCHEDULE
pp. 705–707 Exs. 15–22, 24–54 even, 55–59, 65–66, 69–85 odd, 86

EXERCISE LEVELS
Level A: *Easier*
15–57
Level B: *More Difficult*
58–65
Level C: *Most Difficult*
66–67

✓ **HOMEWORK CHECK**
To quickly check student understanding of key concepts, go over the following exercises: Exs. 18, 22, 26, 32, 42, 48. See also the Daily Homework Quiz:

• Blackline Master (*Chapter 12 Resource Book,* p. 26)
• Transparency (p. 90)

2. See Additional Answers beginning on page AA1.

STUDENT HELP NOTES

→ **Homework Help** Students can find help for Ex. 61 at **www.mcdougallittell.com**. The information can be printed out for students who don't have access to the Internet.

APPLICATION NOTE
EXERCISE 57 Have volunteers contribute other tips they have learned about creating passwords, such as not using the name of a pet or family member.

! **COMMON ERROR**
EXERCISE 63 Some students may be unable to decide whether or not there is repetition. They may be confused by their being several of each breed of dog as well as different breeds of dogs. If students represent each dog with a letter (CCCLLLLLPPPPBBB), they will recognize that this exercise is similar to problems about repeating letters.

ADDITIONAL PRACTICE AND RETEACHING

For Lesson 12.1:
• Practice Levels A, B, and C (*Chapter 12 Resource Book*, p. 16)
• Reteaching with Practice (*Chapter 12 Resource Book*, p. 19)
• See Lesson 12.1 of the *Personal Student Tutor*

For more Mixed Review:
• Search the *Test and Practice Generator* for key words or specific lessons.

706

FOCUS ON APPLICATIONS

PERMUTATIONS WITHOUT REPETITION Find the number of distinguishable permutations of the letters in the word.

39. HI 2 **40.** JET 6 **41.** IOWA 24 **42.** TEXAS 120

43. PENCIL 720 **44.** FLORIDA 5040 **45.** MAGNETIC 40,320 **46.** GOLDFINCH 362,880

PERMUTATIONS WITH REPETITION Find the number of distinguishable permutations of the letters in the word.

47. DAD 3 **48.** PUPPY 20 **49.** OREGON 360 **50.** LETTER 180

51. ALGEBRA 2520 **52.** ALABAMA 210 **53.** MISSOURI 10,080 **54.** CONNECTICUT 1,663,200

55. STEREO You are going to set up a stereo system by purchasing separate components. In your price range you find 5 different receivers, 8 different compact disc players, and 12 different speaker systems. If you want one of each of these components, how many different stereo systems are possible? 480

56. PIZZA A pizza shop runs a special where you can buy a large pizza with one cheese, one vegetable, and one meat for $9.00. You have a choice of 7 cheeses, 11 vegetables, and 6 meats. Additionally, you have a choice of 3 crusts and 2 sauces. How many different variations of the pizza special are possible? 2772

57. COMPUTER SECURITY To keep computer files secure, many programs require the user to enter a password. The shortest allowable passwords are typically six characters long and can contain both numbers and letters. How many six-character passwords are possible if (**a**) characters can be repeated and (**b**) characters cannot be repeated? a 2,176,782,336 b 1,402,410,240

58. CRITICAL THINKING Simplify the formula for $_nP_r$ when $r = 0$. Explain why this result makes sense.

59. CLASS SEATING A particular classroom has 24 seats and 24 students. Assuming the seats are not moved, how many different seating arrangements are possible? Write your answer in scientific notation. 6.20×10^{23}

60. RINGING BELLS "Ringing the changes" is a process where the bells in a tower are rung in all possible permutations. Westminster Abbey has 10 bells in its tower. In how many ways can its bells be rung? 3,628,800

61. PLAY AUDITIONS Auditions are being held for the play shown. How many ways can the roles be assigned if (**a**) 6 people audition and (**b**) 9 people audition? a. 720 b. 60,480

62. WINDOW DISPLAY A music store wants to display 3 identical keyboards, 2 identical trumpets, and 2 identical guitars in its store window. How many distinguishable displays are possible? 210

63. DOG SHOW In a dog show how many ways can 3 Chihuahuas, 5 Labradors, 4 poodles, and 3 beagles line up in front of the judges if the dogs of the same breed are considered identical? 12,612,600

64. CRITICAL THINKING Find the number of permutations of n objects taken $n - 1$ at a time for any positive integer n. Compare this answer with the number of permutations of all n objects. Does this make sense? Explain.

the **Drama Club**

is holding
Open Auditions
for parts in a one-act play
Parts available:

Student A	Teacher
Student B	Librarian
Principal	Coach

REAL LIFE **COMPUTER SECURITY** On the Internet there are three main ways to secure a site: restrict which addresses can access the site, use public key cryptography, or require a user name and password.

58. $_nP_0 = \dfrac{n!}{(n-0)!} = \dfrac{n!}{n!} = 1$.
This makes sense because there is only one way to choose zero objects from any set: take none.

STUDENT HELP

INTERNET **HOMEWORK HELP** Visit our Web site www.mcdougallittell.com for help with problem solving in Ex. 61.

64. $_nP_{n-1} =$
$\dfrac{n!}{[n-(n-1)]!} = \dfrac{n!}{1!} = n!$ is the same as the number of permutations of all n objects. This makes sense because it can be thought of as choosing one object to leave out (n ways) then doing permutations of the $n-1$ objects remaining [$(n-1)!$ ways]. Now we have $n \times (n-1)! = n!$.

QUANTITATIVE COMPARISON In Exercises 65 and 66, choose the statement that is true about the given quantities.

Ⓐ The quantity in column A is greater.

Ⓑ The quantity in column B is greater.

Ⓒ The two quantities are equal.

Ⓓ The relationship cannot be determined from the given information.

	Column A	Column B	
65.	$_5P_1$	$5!$	B
66.	The number of permutations of 12 objects taken 7 at a time	The number of permutations of 12 objects, one of which is repeated 5 times	C

★ **Challenge**

67. CIRCULAR PERMUTATIONS You have learned that $n!$ represents the number of ways that n objects can be placed in a *linear* order, where it matters which object is placed first. Now consider *circular* permutations, where objects are placed in a circle so it does *not* matter which object is placed first. Find a formula for the number of permutations of n objects placed in clockwise order around a circle when only the relative order of the objects matters. Explain how you derived your formula.

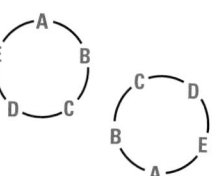

This is the same permutation.

EXTRA CHALLENGE
www.mcdougallittell.com

$\frac{n!}{n} = (n-1)!$; You can arrange n objects in a circle n different ways in the same order so divide $n!$ by n to find the number of unique circular permutations.

MIXED REVIEW

SPECIAL PRODUCTS Find the product. (Review 6.3 for 12.2)

68. $(x+9)(x-9)$ $x^2 - 81$ **69.** $(x^2+2)^2$ $x^4 + 4x^2 + 4$ **70.** $(2x-1)^3$ $8x^3 - 12x^2 + 6x - 1$

71. $(4x+5)(4x-5)$ **72.** $(2y+3x)^2$ **73.** $(8y-x)^2$
$16x^2 - 25$ $4y^2 + 12xy + 9x^2$ $64y^2 - 16xy + x^2$

GRAPHING Graph the equation of the parabola. (Review 10.2)
74–79. See margin.

74. $y^2 = 8x$ **75.** $x^2 = -10y$ **76.** $y^2 = -4x$

77. $x^2 = 26y$ **78.** $x + 14y^2 = 0$ **79.** $y - 2x^2 = 0$

FINDING SUMS Find the sum of the infinite geometric series if there is one. (Review 11.4)

80. $\sum_{n=0}^{\infty} 3\left(\frac{1}{2}\right)^n$ 6   **81.** $\sum_{n=0}^{\infty} -4\left(\frac{1}{4}\right)^n$ $-\frac{16}{3}$ **82.** $\sum_{n=0}^{\infty} 2\left(\frac{2}{3}\right)^n$ 6

83. $\sum_{n=1}^{\infty} 2\left(\frac{7}{5}\right)^{n-1}$ no sum   **84.** $\sum_{n=0}^{\infty} -5\left(\frac{1}{8}\right)^n$ $-\frac{40}{7}$ **85.** $\sum_{n=1}^{\infty} \frac{1}{2}(0.3)^{n-1}$ 0.714

86. SCIENCE CONNECTION Ohm's law states that the resistance R (in ohms) of a conductor varies directly with the potential difference V (in volts) between two points and inversely with the current I (in amperes). The constant of variation is 1. What is the resistance of a light bulb if there is a current of 0.80 ampere when the potential difference across the bulb is 120 volts? (Review 9.1) 150 ohms

12.1 *The Fundamental Counting Principle and Permutations*   **707**

---

4 ASSESS

DAILY HOMEWORK QUIZ

🗐 *Transparency Available*

How many different license plates are possible with 1 digit followed by 5 letters if

1. digits and letters can be repeated? 118,813,760

2. digits and letters cannot be repeated? 78,936,000

Find the number of distinguishable permutations of the letters in the word.

3. QUIZ 24

4. PLYWOOD 2520

EXTRA CHALLENGE NOTE
→ Challenge problems for Lesson 12.1 are available in **blackline** format in the *Chapter 12 Resource Book,* p. 23 and at **www.mcdougallittell.com.**

ADDITIONAL TEST PREPARATION

1. WRITING Use examples to explain how a problem involving permutations with repetition is different from one involving permutations without repetition. Show how to calculate the number of permutations in each case. Check explanations.

74.

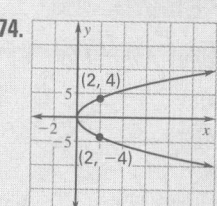

75.

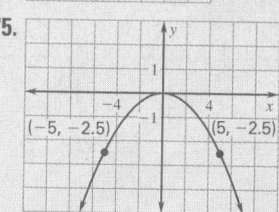

76–79. See Additional Answers beginning on page AA1.

LESSON OPENER
VISUAL APPROACH

An alternative way to approach Lesson 12.2 is to use the Visual Approach Lesson Opener:

- Blackline Master (*Chapter 12 Resource Book*, p. 27)
- Transparency (p. 78)

MEETING INDIVIDUAL NEEDS

- *Chapter 12 Resource Book*
  Prerequisite Skills Review (p. 5)
  Practice Level A (p. 31)
  Practice Level B (p. 32)
  Practice Level C (p. 33)
  Reteaching with Practice (p. 34)
  Absent Student Catch-Up (p. 36)
  Challenge (p. 38)
- *Resources in Spanish*
- Personal Student Tutor

NEW-TEACHER SUPPORT

See the Tips for New Teachers on pp. 1–2 of the *Chapter 12 Resource Book* for additional notes about Lesson 12.2.

WARM-UP EXERCISES

 **Transparency Available**

Find the product.

1. $(2x - 3)^2$ $4x^2 - 12x + 9$
2. $(x - 3y)^2$ $x^2 - 6xy + 9y^2$
3. $(x + y)^3$ $x^3 + 3x^2y + 3xy^2 + y^3$
4. $(1 - x)^3$ $-x^3 + 3x^2 - 3x + 1$

12.2 Combinations and the Binomial Theorem

What you should learn

GOAL 1 Use combinations to count the number of ways an event can happen, as applied in **Ex. 55**.

GOAL 2 Use the binomial theorem to expand a binomial that is raised to a power.

Why you should learn it

▼ To solve **real-life** problems, such as finding the number of different combinations of plays you can attend in **Example 3**.

GOAL 1 USING COMBINATIONS

In Lesson 12.1 you learned that order is important for some counting problems. For other counting problems, order is not important. For instance, in most card games the order in which your cards are dealt is not important. After your cards are dealt, reordering them does not change your card hand. These unordered groupings are called *combinations*. A **combination** is a selection of r objects from a group of n objects where the order is not important.

COMBINATIONS OF n OBJECTS TAKEN r AT A TIME

The number of combinations of r objects taken from a group of n distinct objects is denoted by $_nC_r$ and is given by:

$$_nC_r = \frac{n!}{(n - r)! \cdot r!}$$

For instance, the number of combinations of 2 objects taken from a group of 5 objects is $_5C_2 = \frac{5!}{3! \cdot 2!} = 10$.

EXAMPLE 1 *Finding Combinations*

A standard deck of 52 playing cards has 4 suits with 13 different cards in each suit as shown.

a. If the order in which the cards are dealt is not important, how many different 5-card hands are possible?

b. In how many of these hands are all five cards of the same suit?

SOLUTION

a. The number of ways to choose 5 cards from a deck of 52 cards is:

$$_{52}C_5 = \frac{52!}{47! \cdot 5!}$$

$$= \frac{52 \cdot 51 \cdot 50 \cdot 49 \cdot 48 \cdot 47!}{47! \cdot 5!}$$

$$= 2,598,960$$

b. For all five cards to be the same suit, you need to choose 1 of the 4 suits and then 5 of the 13 cards in the suit. So, the number of possible hands is:

$$_4C_1 \cdot {}_{13}C_5 = \frac{4!}{3! \cdot 1!} \cdot \frac{13!}{8! \cdot 5!} = \frac{4 \cdot 3!}{3! \cdot 1!} \cdot \frac{13 \cdot 12 \cdot 11 \cdot 10 \cdot 9 \cdot 8!}{8! \cdot 5!} = 5148$$

Standard 52-Card Deck

K ♠	K ♣	K ♦	K ♥
Q ♠	Q ♣	Q ♦	Q ♥
J ♠	J ♣	J ♦	J ♥
10 ♠	10 ♣	10 ♦	10 ♥
9 ♠	9 ♣	9 ♦	9 ♥
8 ♠	8 ♣	8 ♦	8 ♥
7 ♠	7 ♣	7 ♦	7 ♥
6 ♠	6 ♣	6 ♦	6 ♥
5 ♠	5 ♣	5 ♦	5 ♥
4 ♠	4 ♣	4 ♦	4 ♥
3 ♠	3 ♣	3 ♦	3 ♥
2 ♠	2 ♣	2 ♦	2 ♥
A ♠	A ♣	A ♦	A ♥

When finding the number of ways both an event *A and* an event *B* can occur, you need to multiply (as you did in part (b) of Example 1). When finding the number of ways that an event *A or* an event *B* can occur, you add instead.

Menu Choices

EXAMPLE 2 *Deciding to Multiply or Add*

A restaurant serves omelets that can be ordered with any of the ingredients shown.

a. Suppose you want *exactly* 2 vegetarian ingredients and 1 meat ingredient in your omelet. How many different types of omelets can you order?

b. Suppose you can afford *at most* 3 ingredients in your omelet. How many different types of omelets can you order?

Omelets $3.00	
(plus $.50 for each ingredient)	
Vegetarian	**Meat**
green pepper	ham
red pepper	bacon
onion	sausage
mushroom	steak
tomato	
cheese	

SOLUTION

a. You can choose 2 of 6 vegetarian ingredients and 1 of 4 meat ingredients. So, the number of possible omelets is:

$$_6C_2 \cdot {}_4C_1 = \frac{6!}{4! \cdot 2!} \cdot \frac{4!}{3! \cdot 1!} = 15 \cdot 4 = 60$$

b. You can order an omelet with 0, 1, 2, or 3 ingredients. Because there are 10 items to choose from, the number of possible omelets is:

$$_{10}C_0 + {}_{10}C_1 + {}_{10}C_2 + {}_{10}C_3 = 1 + 10 + 45 + 120 = 176$$

· · · · · · · · ·

STUDENT HELP

KEYSTROKE HELP
Visit our Web site
www.mcdougallittell.com
to see keystrokes for
several models of
calculators.

Some calculators have special keys to evaluate combinations. The solution to Example 2 is shown.

Counting problems that involve phrases like "at least" or "at most" are sometimes easier to solve by subtracting possibilities you do not want from the total number of possibilities.

```
(6 nCr 2)(4 nCr 1)
                 60
10 nCr 0+10 nCr 1
+10 nCr 2+10 nCr 3
                176
```

Theater

EXAMPLE 3 *Subtracting Instead of Adding*

A theater is staging a series of 12 different plays. You want to attend *at least* 3 of the plays. How many different combinations of plays can you attend?

SOLUTION

You want to attend 3 plays, or 4 plays, or 5 plays, and so on. So, the number of combinations of plays you can attend is $_{12}C_3 + {}_{12}C_4 + {}_{12}C_5 + \cdots + {}_{12}C_{12}$.

Instead of adding these combinations, it is easier to use the following reasoning. For each of the 12 plays, you can choose to attend or not attend the play, so there are 2^{12} total combinations. If you attend at least 3 plays you do not attend only 0, 1, or 2 plays. So, the number of ways you can attend at least 3 plays is:

$$2^{12} - \left({}_{12}C_0 + {}_{12}C_1 + {}_{12}C_2\right) = 4096 - (1 + 12 + 66) = 4017$$

12.2 *Combinations and the Binomial Theorem*

 3. $2^7 - {}_7C_0 = 127$

EXTRA EXAMPLE 1
Use a standard deck of 52 cards.
a. If the order is not important, how many different 7-card hands are possible?
$_{52}C_7 = 133{,}784{,}560$
b. How many of these hands have all 7 cards of the same suit? $_4C_1 \cdot {}_{13}C_7 = 6864$

EXTRA EXAMPLE 2
You are taking a vacation. You can visit as many as 5 different cities and 7 different attractions.
a. Suppose you want to visit exactly 3 different cities and 4 different attractions. How many different trips are possible? $_5C_3 \cdot {}_7C_4 = 350$
b. Suppose you want to visit at least 8 locations (cities or attractions). How many different types of trips are possible?
$_{12}C_8 + {}_{12}C_9 + {}_{12}C_{10} + {}_{12}C_{11} + {}_{12}C_{12} = 794$

EXTRA EXAMPLE 3
A restaurant offers 6 salad toppings. On a deluxe salad, you can have up to 4 toppings. How many different combinations of toppings can you have?
$2^6 - ({}_6C_5 + {}_6C_6) = 57$

CHECKPOINT EXERCISES
For use after Example 1:
1. How many possible 5-card hands contain exactly 3 kings?
$_4C_3 \cdot {}_{48}C_2 = 4512$

For use after Example 2:
2. From a group of 20 volunteers, you are choosing at least 18 to be peer counselors. In how many different ways can this be done?
$_{20}C_{18} + {}_{20}C_{19} + {}_{20}C_{20} = 211$

For use after Example 3:
3. Every committee of the school council must contain at least 1 senior. If there are 7 seniors on the school council, how many different combinations of seniors can be assigned to a committee?

ACTIVITY NOTE

This activity guides students to discover the relationship between Pascal's triangle and the coefficients of a binomial expansion.

EXTRA EXAMPLE 4

Expand $(a + 3)^5$. $a^5 + 15a^4 + 90a^3 + 270a^2 + 405a + 243$

CHECKPOINT EXERCISES

For use after Example 4:

1. Expand $(x + y)^6$.
$x^6 + 6x^5y + 15x^4y^2 + 20x^3y^3 + 15x^2y^4 + 6xy^5 + y^6$

CONCEPT QUESTION

EXAMPLE 4 After you expand an expression, what do you notice about the sum of the exponents? **The sum of the exponents for each term equals the value of the exponent in the expression before it was expanded.**

FOCUS ON
PEOPLE

BLAISE PASCAL developed his arithmetic triangle in 1653. The following year he and fellow mathematician Pierre Fermat outlined the foundations of probability theory.

Step 1a. $a^2 + 2ab + b^2$

Step 1b. $a^3 + 3a^2b + 3ab^2 + b^3$

Step 1c. $a^4 + 4a^3b + 6a^2b^2 + 4ab^3 + b^4$

Step 2. The coefficients for $(a + b)^n$ are the numbers in Pascal's triangle for $_nC_r$

Step 3. The exponents of a and b sum to n (where n is the exponent in $(a + b)^n$). The exponents of a decrease from n to 0 and the exponents of b increase from 0 to n.

GOAL 2 USING THE BINOMIAL THEOREM

If you arrange the values of $_nC_r$ in a triangular pattern in which each row corresponds to a value of n, you get what is called **Pascal's triangle**. It is named after the famous French mathematician Blaise Pascal (1623–1662).

$$
\begin{array}{ccccccc}
& & & _0C_0 & & & \\
& & _1C_0 & & _1C_1 & & \\
& _2C_0 & & _2C_1 & & _2C_2 & \\
_3C_0 & & _3C_1 & & _3C_2 & & _3C_3 \\
_4C_0 & _4C_1 & _4C_2 & _4C_3 & _4C_4 & & \\
_5C_0 & _5C_1 & _5C_2 & _5C_3 & _5C_4 & _5C_5 &
\end{array}
\qquad
\begin{array}{cccccc}
& & 1 & & & \\
& 1 & & 1 & & \\
1 & & 2 & & 1 & \\
1 & 3 & & 3 & & 1 \\
1 & 4 & 6 & 4 & 1 & \\
1 & 5 & 10 & 10 & 5 & 1
\end{array}
$$

Pascal's triangle has many interesting patterns and properties. For instance, each number other than 1 is the sum of the two numbers directly above it.

● ACTIVITY

Developing
Concepts

Investigating Pascal's Triangle

❶ Expand each expression. Write the terms of each expanded expression so that the powers of a decrease.

     a. $(a + b)^2$ **b.** $(a + b)^3$ **c.** $(a + b)^4$

❷ Describe the relationship between the coefficients in parts (a), (b), and (c) of **Step 1** and the rows of Pascal's triangle.

❸ Describe any patterns in the exponents of a and the exponents of b.

In the activity you may have discovered the following result, which is called the **binomial theorem**. This theorem describes the coefficients in the expansion of the binomial $a + b$ raised to the nth power.

THE BINOMIAL THEOREM

The binomial expansion of $(a + b)^n$ for any positive integer n is:

$$(a + b)^n = {_nC_0}a^nb^0 + {_nC_1}a^{n-1}b^1 + {_nC_2}a^{n-2}b^2 + \cdots + {_nC_n}a^0b^n$$

$$= \sum_{r=0}^{n} {_nC_r}a^{n-r}b^r$$

EXAMPLE 4 *Expanding a Power of a Simple Binomial Sum*

STUDENT HELP

➤ **Study Tip**
You can calculate combinations using either Pascal's triangle or the formula on p. 708.

Expand $(x + 2)^4$.

SOLUTION

$$(x + 2)^4 = {_4C_0}x^42^0 + {_4C_1}x^32^1 + {_4C_2}x^22^2 + {_4C_3}x^12^3 + {_4C_4}x^02^4$$

$$= (1)(x^4)(1) + (4)(x^3)(2) + (6)(x^2)(4) + (4)(x)(8) + (1)(1)(16)$$

$$= x^4 + 8x^3 + 24x^2 + 32x + 16$$

EXAMPLE 5 · Expanding a Power of a Binomial Sum

Expand $(u + v^2)^3$.

SOLUTION

$$(u + v^2)^3 = {}_3C_0 u^3(v^2)^0 + {}_3C_1 u^2(v^2)^1 + {}_3C_2 u^1(v^2)^2 + {}_3C_3 u^0(v^2)^3$$
$$= u^3 + 3u^2v^2 + 3uv^4 + v^6$$

· · · · · · · · · ·

To expand a power of a binomial difference, you can rewrite the binomial as a sum. The resulting expansion will have terms whose signs alternate between $+$ and $-$.

EXAMPLE 6 · Expanding a Power of a Simple Binomial Difference

Expand $(x - y)^5$.

SOLUTION

$$(x - y)^5 = [x + (-y)]^5$$
$$= {}_5C_0 x^5(-y)^0 + {}_5C_1 x^4(-y)^1 + {}_5C_2 x^3(-y)^2 + {}_5C_3 x^2(-y)^3 +$$
$$\quad {}_5C_4 x^1(-y)^4 + {}_5C_5 x^0(-y)^5$$
$$= x^5 - 5x^4y + 10x^3y^2 - 10x^2y^3 + 5xy^4 - y^5$$

EXAMPLE 7 · Expanding a Power of a Binomial Difference

Expand $(5 - 2a)^4$.

SOLUTION

$$(5 - 2a)^4 = [5 + (-2a)]^4$$
$$= {}_4C_0 5^4(-2a)^0 + {}_4C_1 5^3(-2a)^1 + {}_4C_2 5^2(-2a)^2 + {}_4C_3 5^1(-2a)^3 +$$
$$\quad {}_4C_4 5^0(-2a)^4$$
$$= (1)(625)(1) + (4)(125)(-2a) + (6)(25)(4a^2) + (4)(5)(-8a^3) +$$
$$\quad (1)(1)(16a^4)$$
$$= 625 - 1000a + 600a^2 - 160a^3 + 16a^4$$

EXAMPLE 8 · Finding a Coefficient in an Expansion

STUDENT HELP

HOMEWORK HELP
Visit our Web site
www.mcdougallittell.com
for extra examples.

Find the coefficient of x^4 in the expansion of $(2x - 3)^{12}$.

SOLUTION From the binomial theorem you know the following:

$$(2x - 3)^{12} = \sum_{r=0}^{12} {}_{12}C_r (2x)^{12-r}(-3)^r$$

The term that has x^4 is ${}_{12}C_8 (2x)^4(-3)^8 = (495)(16x^4)(6561) = 51,963,120x^4$.

▶ The coefficient is 51,963,120.

EXTRA EXAMPLE 5
Expand $(a + 2b^3)^4$. $a^4 + 8a^3b^3 + 24a^2b^6 + 32ab^9 + 16b^{12}$

EXTRA EXAMPLE 6
Expand $(x - 5)^4$. $x^4 - 20x^3 + 150x^2 - 500x + 625$

EXTRA EXAMPLE 7
Expand $(2x - y^2)^3$.
$8x^3 - 12x^2y^2 + 6xy^4 - y^6$

EXTRA EXAMPLE 8
Find the coefficient of x^7 in $(2 - 3x)^{10}$. $-2{,}099{,}520$

☑ CHECKPOINT EXERCISES

For use after Examples 5–8:
1. Expand $(3x + y)^4$. $81x^4 + 108x^3y + 54x^2y^2 + 12xy^3 + y^4$
2. Expand $(a - 2b^3)^3$.
   $a^3 - 6a^2b^3 + 12ab^6 - 8b^9$

STUDENT HELP NOTES

→ **Homework Help** Students can find extra examples at **www.mcdougallittell.com** that parallel the examples in the student edition.

FOCUS ON VOCABULARY
What is the relationship between Pascal's triangle and the binomial theorem? **Pascal's triangle contains the coefficients of the terms when an expression is expanded using the binomial theorem.**

CLOSURE QUESTION
Explain two ways that ${}_nC_r$ is used in this lesson. **${}_nC_r$ is used to find the number of combinations of n objects r at a time. It is also used to find the coefficients of the terms in a binomial expansion.**

DAILY PUZZLER
One of the terms in the expansion of $(x + 3y)^n$ is $945x^4y^3$. What is the value of n? **7**

ASSIGNMENT GUIDE

BASIC
Day 1: pp. 712–713 Exs. 18–30 even, 31, 32–40 even, 47–50
Day 2: pp. 712–713 Exs. 33–43 odd, 44–46, 55–58, 67, 75–87 odd, Quiz 1 Exs. 1–20

AVERAGE
Day 1: pp. 712–713 Exs. 18–30 even, 31, 32–42 even, 47–52
Day 2: pp. 712–713 Exs. 33–43 odd, 44–46, 55–63, 67, 75–89 odd, Quiz 1 Exs. 1–20

ADVANCED
Day 1: pp. 712–713 Exs. 18–30 even, 31, 32–42 even, 47–54
Day 2: pp. 712–713 Exs. 33–43 odd, 44–46, 57–73, 75–89 odd, 90, Quiz 1 Exs. 1–20

BLOCK SCHEDULE
pp. 712–715 Exs. 18–31, 32–50 even, 51, 55–58, 67, 75–87 odd, Quiz 1 Exs. 1–20

EXERCISE LEVELS

Level A: *Easier*
18–25, 31–43, 44–48
Level B: *More Difficult*
26–30, 49–58, 64–67
Level C: *Most Difficult*
59–63, 68–73

✓ HOMEWORK CHECK

To quickly check student understanding of key concepts, go over the following exercises:
Exs. 22, 28, 32, 36, 44. See also the Daily Homework Quiz:
- Blackline Master (*Chapter 12 Resource Book*, p. 42)
- 📷 Transparency (p. 91)

3–4. See Additional Answers beginning on page AA1.

GUIDED PRACTICE

Vocabulary Check ✓
Concept Check ✓

1. In a permutation, the order of the objects is important. In a combination, order does not matter.

Skill Check ✓

2a. Add combinations when finding the number of ways that event A or event B can happen.

2b. Multiply combinations when finding the number of ways that event A and event B can happen.

1. Explain the difference between a permutation and a combination.

2. Describe a situation in which to find the total number of possibilities you would **(a)** add two combinations and **(b)** multiply two combinations.

3. Write the expansions for $(x + y)^4$ and $(x - y)^4$. How are they similar? How are they different? See margin.

4. **ERROR ANALYSIS** What error was made in the calculation of $_{10}C_6$? Explain. See margin.

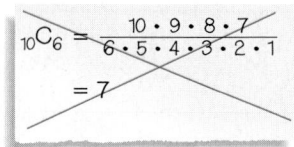

Find the number of combinations of _n_ objects taken _r_ at a time.

5. $n = 8, r = 2$ 28   6. $n = 6, r = 5$ 6   7. $n = 5, r = 1$ 5   8. $n = 9, r = 9$ 1

Expand the power of the binomial.
See margin.
9. $(x + y)^3$ $x^4 + 4x^3 + 6x^2 + 4x + 1$ See margin.
10. $(x + 1)^4$ 11. $(2x + 4)^3$

$32x^5 + 240x^4y + 720x^3y^2 + 1080x^2y^3 + 810xy^4 + 243y^5$
12. $(2x + 3y)^5$

13. $(x - y)^5$ 14. $(x - 2)^3$ 15. $(3x - 1)^4$ 16. $(4x - 4y)^3$
See margin. $x^3 - 6x^2 + 12x - 8$ 15, 16. See margin.

17. Complete this equation:

$$(x + 3y)^5 = x^5 + 15x^4y + 90x^3y^2 + \underline{} x^2y^3 + 405xy^4 + \underline{} y^5 \quad 270; 243$$

PRACTICE AND APPLICATIONS

STUDENT HELP

▸ **Extra Practice** to help you master skills is on p. 956.

9. $x^3 + 3x^2y + 3xy^2 + y^3$

11. $8x^3 + 48x^2 + 96x + 64$

13. $x^5 - 5x^4y + 10x^3y^2 - 10x^2y^3 + 5xy^4 - y^5$

15. $81x^4 - 108x^3 + 54x^2 - 12x + 1$

16. $64x^3 - 192x^2y + 192xy^2 - 64y^3$

STUDENT HELP

▸ **HOMEWORK HELP**
Example 1: Exs. 18–30, 47, 48
Example 2: Exs. 49–52
Example 3: Exs. 53, 54
Examples 4–7: Exs. 31–43
Example 8: Exs. 44–46

COMBINATIONS Find the number of combinations.

18. $_{10}C_2$ 45   19. $_8C_5$ 56   20. $_5C_2$ 10   21. $_8C_6$ 28

22. $_{12}C_4$ 495   23. $_{12}C_{12}$ 1   24. $_{14}C_6$ 3003   25. $_{11}C_3$ 165

CARD HANDS In Exercises 26–30, find the number of possible 5-card hands that contain the cards specified.

26. 5 face cards (either kings, queens, or jacks)   792

27. 4 aces and 1 other card   48

28. 1 ace and 4 other cards (none of which are aces)   778,320

29. 2 aces and 3 kings   24

30. 4 of one kind (kings, queens, and so on) and 1 of a different kind   624

31. **PASCAL'S TRIANGLE** Copy Pascal's triangle on page 710 and add the rows for $n = 6$ and $n = 7$ to it.   See margin.

PASCAL'S TRIANGLE Use the rows of Pascal's triangle from Exercise 31 to write the binomial expansion.   32–35. See margin.

32. $(x + 4)^6$ 33. $(x - 3y)^6$ 34. $(x^2 + y)^7$ 35. $(2x - y^3)^7$

BINOMIAL THEOREM Use the binomial theorem to write the binomial expansion.
36–43. See margin.
36. $(x - 2)^3$ 37. $(x + 4)^5$ 38. $(x + 3y)^4$ 39. $(2x - y)^6$

40. $(x^3 + 3)^5$ 41. $(3x^2 - 3)^4$ 42. $(2x - y^2)^7$ 43. $(x^3 + y^2)^3$

44. Find the coefficient of x^5 in the expansion of $(x - 3)^7$. **189**

45. Find the coefficient of x^4 in the expansion of $(x + 2)^8$. **1120**

46. Find the coefficient of x^6 in the expansion of $(x^2 + 4)^{10}$. **1,966,080**

47. 🌐 **NOVELS** Your English teacher has asked you to select 3 novels from a list of 10 to read as an independent project. In how many ways can you choose which books to read? **120**

48. 🌐 **GAMES** Your friend is having a party and has 15 games to choose from. There is enough time to play 4 games. In how many ways can you choose which games to play? **1365**

49. 🌐 **CARS** You are buying a new car. There are 7 different colors to choose from and 10 different types of optional equipment you can buy. You can choose only 1 color for your car and can afford only 2 of the options. How many combinations are there for your car? **315**

50. 🌐 **ART CONTEST** There are 6 artists each presenting 5 works of art in an art contest. The 4 works judged best will be displayed in a local gallery. In how many ways can these 4 works all be chosen from the same artist's collection? **30**

51. **LOGICAL REASONING** Look back at Example 2. Suppose you can afford at most 7 ingredients. How many different types of omelets can you order? **968**

52. 🌐 **AMUSEMENT PARKS** An amusement park has 20 different rides. You want to ride at least 15 of them. How many different combinations of rides can you go on? **21,700**

53. 🌐 **FISH** From the list of different species of fish shown, an aquarium enthusiast is interested in knowing how compatible any group of 3 or more different species are. How many different combinations are there to consider? **968**

> *On Sale This Month*
>
> **Freshwater Tropical Fish**
>
> | | |
> |---|---|
> | Neon Tetras | Black Mollies |
> | Tiger Barbs | Zebra fish |
> | Red Platys | Bala Sharks |
> | Angelfish | Lyretails |
> | Blue Gouramis | Catfish |

54. 🌐 **CONCERTS** A summer concert series has 12 different performing artists. You decide to attend at least 4 of the concerts. How many different combinations of concerts can you attend? **3797**

CRITICAL THINKING **Decide whether the problem requires combinations or permutations to find the answer. Then solve the problem.**

55. 🌐 **MARCHING BAND** Eight members of a school marching band are auditioning for 3 drum major positions. In how many ways can students be chosen to be drum majors? **combinations, 56**

56. 🌐 **YEARBOOK** Your school yearbook has an editor-in-chief and an assistant editor-in-chief. The staff of the yearbook has 15 students. In how many ways can students be chosen for these 2 positions? **permutations, 210**

57. 🌐 **RELAY RACES** A relay race has 4 runners who run different parts of the race. There are 16 students on your track team. In how many ways can your coach select students to compete in the race? **permutations, 43,680**

58. 🌐 **COLLEGE COURSES** You must take 6 elective classes to meet your graduation requirements for college. There are 12 classes that you are interested in. In how many ways can you select your elective classes? **combinations, 924**

38. $x^4 + 12x^3y + 54x^2y^2 + 108xy^3 + 81y^4$

39. $64x^6 - 192x^5y + 240x^4y^2 - 160x^3y^3 + 60x^2y^4 - 12xy^5 + y^6$

40. $x^{15} + 15x^{12} + 90x^9 + 270x^6 + 405x^3 + 243$

41. $81x^8 - 324x^6 + 486x^4 - 324x^2 + 81$

42. $128x^7 - 448x^6y^2 + 672x^5y^4 - 560x^4y^6 + 280x^3y^8 - 84x^2y^{10} + 14xy^{12} - y^{14}$

43. $x^9 + 3x^6y^2 + 3x^3y^4 + y^6$

CARS A 1998 survey showed that of 7 basic car colors, white is the most popular color for full-size cars with 18.8% of the vote. Green came in second with 16.4% of the vote.

🌐 **APPLICATION LINK**
www.mcdougallittell.com

! COMMON ERROR

EXERCISES 26–30 In problems such as these, students may not consider the possibilities for each card that is drawn. For example, in Exercise 27, students choose 4 aces from 4 aces which is $_4C_4 = 1$. They are also choosing 1 card from the remaining 48 cards. The answer is thus $_4C_4 \cdot {_{48}C_1} = 48$.

APPLICATION NOTE
EXERCISE 49 Additional information about cars is available at **www.mcdougallittell.com.**

31.
$$
\begin{array}{ccccccccc}
 & & & & 1 & & & & \\
 & & & 1 & & 1 & & & \\
 & & 1 & & 2 & & 1 & & \\
 & 1 & & 3 & & 3 & & 1 & \\
1 & & 4 & & 6 & & 4 & & 1 \\
\end{array}
$$
1 5 10 10 5 1
1 6 15 20 15 6 1
1 7 21 35 35 21 7 1

32. $x^6 + 24x^5 + 240x^4 + 1280x^3 + 3840x^2 + 6144x + 4096$

33. $x^6 - 18x^5y + 135x^4y^2 - 540x^3y^3 + 1215x^2y^4 - 1458xy^5 + 729y^6$

34. $x^{14} + 7x^{12}y + 21x^{10}y^2 + 35x^8y^3 + 35x^6y^4 + 21x^4y^5 + 7x^2y^6 + y^7$

35. $128x^7 - 448x^6y^3 + 672x^5y^6 - 560x^4y^9 + 280x^3y^{12} - 84x^2y^{15} + 14xy^{18} - y^{21}$

36. $x^3 - 6x^2 + 12x - 8$

37. $x^5 + 20x^4 + 160x^3 + 640x^2 + 1280x + 1024$

38–43. See margin below.

ADDITIONAL PRACTICE AND RETEACHING

For Lesson 12.2:

- Practice Levels A, B, and C (*Chapter 12 Resource Book,* p. 31)

- Reteaching with Practice (*Chapter 12 Resource Book,* p. 34)

- 📷 See Lesson 12.2 of the *Personal Student Tutor*

For more Mixed Review:

- 📷 Search the *Test and Practice Generator* for key words or specific lessons.

DAILY HOMEWORK QUIZ

📖 *Transparency Available*

Find the number of combinations.

1. $_{10}C_3$ 120   **2.** $_7C_4$ 35

Write the binomial expansion.

3. $(x+3)^4$
   $x^4 + 12x^3 + 54x^2 + 108x + 81$

4. $(3-y)^3$ $27 - 27y + 9y^2 - y^3$

EXTRA CHALLENGE NOTE

↳ Challenge problems for Lesson 12.2 are available in **blackline** format in the *Chapter 12 Resource Book*, p. 38 and at **www.mcdougallittell.com.**

ADDITIONAL TEST PREPARATION

1. WRITING Explain how Pascal's Triangle can be used to expand $(x+2y)^6$. *Sample answer:* Use the row of the triangle that starts 1, 6 to find the combinations used in the binomial theorem for this expansion. The value of the first combination in the expansion $_6C_0$ is 1, the value of the next combination $_6C_1$ is 6, and so on. Multiply this combination by the appropriate powers of x and $2y$ to get each term.

2. OPEN ENDED Make up a real-life problem for which the answer is a sum of combinations. Show how to find the answer. **Check problems and answers.**

60–61, 70–73.
   See Additional Answers beginning on page AA1.

62. The numbers in row $(n-1)$ of Pascal's triangle correspond to the number of ways to stack n blocks in r columns. The total number of ways is the sum of the numbers in row $(n-1)$ of Pascal's triangle.

64. The sum of the numbers in row n of Pascal's triangle is 2^n. Each number is the sum of the two numbers above it, so each internal number is added twice to the next row. For the ones on the ends, the additional ones on each end of the new row represents the second use of those values (this can be thought of as adding the one to an imaginary zero on the outside of the previous row).

Test Preparation

66. $S_n = S_{n-1} + S_{n-2} = 1, 1, 2, 3, 5, 8, \ldots,$ which are the Fibonacci numbers.

67b. $\dfrac{20!}{8! \times 12!} \times \dfrac{12!}{5! \times 7!} \times \dfrac{5!}{5! \times 0!}$
   $= 99,768,240$

67c. They are the same. They are two different ways to count the same thing. The additional factorials in the numerators and denominators of part b simplify to become the expression in part a.

★ **Challenge**

---

59. **CRITICAL THINKING** Write an equation that relates $_nP_r$ and $_nC_r$. $_nP_r = r! \times {_nC_r}$

STACKING CUBES In Exercises 60–63, use the diagram shown which illustrates the different ways to stack four cubes.

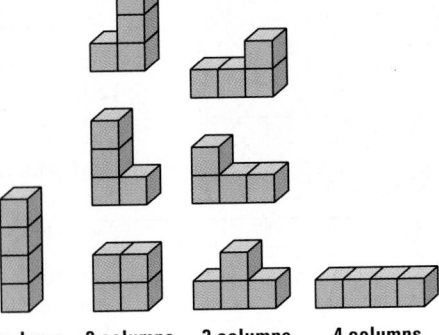

60. Sketch the different ways to stack three cubes.   **See margin.**

61. Sketch the different ways to stack five cubes.   **See margin.**

62. How does the number of ways to stack three, four, and five cubes relate to Pascal's triangle?

63. In how many different ways can you stack ten cubes?   **512**

1 column   2 columns   3 columns   4 columns

PASCAL'S TRIANGLE In Exercises 64–66, use the diagram of Pascal's triangle shown.

64. What is the sum of the numbers in row n of Pascal's triangle? Explain.

65. What is the sum of the numbers in rows 0 through 20 of Pascal's triangle?   **2,097,151**

66. **LOGICAL REASONING** Describe the pattern formed by the sums of the numbers along the diagonal segments of Pascal's triangle.

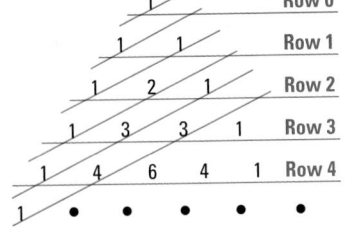

| | | | | | | | | | Row 0 |
1 | Row 0
1   1 | Row 1
1   2   1 | Row 2
1   3   3   1 | Row 3
1   4   6   4   1 | Row 4

67. **MULTI-STEP PROBLEM** A group of 20 high school students is volunteering to help elderly members of their community. Each student will be assigned a job based on requests received for help. There are 8 requests for raking leaves, 7 requests for running errands, and 5 requests for washing windows.

   a. One way to count the number of possible job assignments is to find the number of permutations of 8 L's (for "leaves"), 7 E's (for "errands"), and 5 W's (for "windows"). Use this method to write the number of possible job assignments first as an expression involving factorials and then as a simple number. $\dfrac{20!}{8! \times 7! \times 5!} = 99,768,240$

   b. Another way to count the number of possible job assignments is to first choose the 8 students who will rake leaves, then choose the 7 students who will run errands from the students who remain, and then choose the 5 students who will wash windows from the students who still remain. Use this method to write the number of possible job assignments first as an expression involving factorials and then as a simple number.

   c. *Writing* How do the answers to parts (a) and (b) compare to each other? Explain why this makes sense.

COMBINATORIAL IDENTITIES Verify the identity.   **70–73. See margin.**

68. $_nC_0 = 1$ $_nC_0 = \dfrac{n!}{n! \times 0!} = 1$

69. $_nC_n = 1$ $_nC_n = \dfrac{n!}{0! \times n!} = 1$

70. $_nC_1 = {_nP_1}$

71. $_nC_r = {_nC_{n-r}}$

72. $_nC_r \cdot {_rC_m} = {_nC_m} \cdot {_{n-m}C_{r-m}}$

73. $_{n+1}C_r = {_nC_r} + {_nC_{r-1}}$

MIXED REVIEW

FINDING AREA Find the area of the figure. (Skills Review, p. 914)

74. Circle with radius 18 centimeters   1017.88 cm²

75. Rectangle with sides 9.5 inches and 11.3 inches   107.35 in.²

76. Triangle with base 13 feet and height 9 feet   58.5 ft²

77. Trapezoid with bases 10 meters and 13 meters, and height 27 meters   310.5 m²

GRAPHING Graph the equation of the hyperbola. (Review 10.5)   78–83. See margin.

78. $\dfrac{x^2}{25} - \dfrac{y^2}{144} = 1$ **79.** $\dfrac{y^2}{100} - \dfrac{x^2}{36} = 1$ **80.** $x^2 - \dfrac{49y^2}{16} = 1$

81. $\dfrac{y^2}{4} - \dfrac{x^2}{9} = 9$ **82.** $64y^2 - x^2 = 64$ **83.** $9x^2 - 4y^2 = 144$

WRITING RULES Decide whether the sequence is arithmetic or geometric. Then write a rule for the *n*th term. (Review 11.2, 11.3)

84. 3, 9, 27, 81, 243, . . . **85.** 3, 10, 17, 24, 31, . . . **86.** 2, 10, 50, 250, 1250, . . .

87. 1, −2, 4, −8, 16, . . . **88.** 8, 6, 4, 2, 0, . . . **89.** −10, −5, 0, 5, 10, . . .

84. geometric, 3^n

85. arithmetic, $-4 + 7n$

86. geometric, $2(5)^{n-1}$

87. geometric, $(-2)^{n-1}$

88. arithmetic, $10 - 2n$

89. arithmetic, $-15 + 5n$

90. **POTTERY** A potter has 70 pounds of clay and 40 hours to make soup bowls and dinner plates to sell at a craft fair. A soup bowl uses 3 pounds of clay and a dinner plate uses 4 pounds of clay. It takes 3 hours to make a soup bowl and 1 hour to make a dinner plate. If the profit on a soup bowl is $25 and the profit on a dinner plate is $20, how many bowls and plates should the potter make in order to maximize profit? **(Review 3.4)**   10 bowls, 10 plates

QUIZ 1

Self-Test for Lessons 12.1 and 12.2

Find the number of distinguishable permutations of the letters in the word. (Lesson 12.1)

1. POP 3 **2.** JUNE 24 **3.** IDAHO 120 **4.** KANSAS 180

5. WYOMING 5040 **6.** THURSDAY 40,320 **7.** SEPTEMBER 60,480 **8.** CALIFORNIA 907,200

Write the binomial expansion. (Lesson 12.2)   9–16. See margin.

9. $(x + y)^6$ **10.** $(x + 2)^4$ **11.** $(x - 2y)^5$ **12.** $(3x - 4y)^3$

13. $(x^2 + 3y)^4$ **14.** $(4x^2 - 2)^6$ **15.** $(x^3 - y^3)^3$ **16.** $(2x^4 + 5y^2)^5$

17. Find the coefficient of x^3 in the expansion of $(x + 3)^5$. **(Lesson 12.2)**   90

18. Find the coefficient of y^4 in the expansion of $(5 - y^2)^3$. **(Lesson 12.2)**   15

19. **RESTAURANTS** You are eating dinner at a restaurant. The restaurant offers 6 appetizers, 12 main dishes, 6 side orders, and 8 desserts. If you order one of each of these, how many different dinners can you order? **(Lesson 12.1)**   3456

20. **FLOWERS** You are buying a flower arrangement. The florist has 12 types of flowers and 6 types of vases. If you can afford exactly 3 types of flowers and need only 1 vase, how many different arrangements can you buy? **(Lesson 12.2)**   1320

ADDITIONAL RESOURCES
An alternative Quiz for Lessons 12.1–12.2 is available in the *Chapter 12 Resource Book*, p. 39.

78.

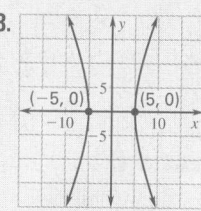

79.

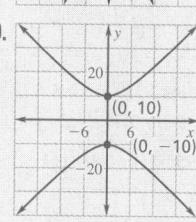

80.

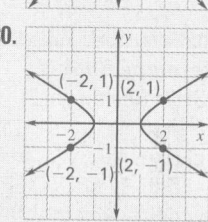

81.

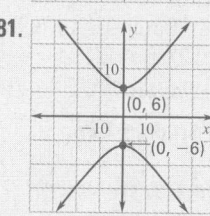

82.

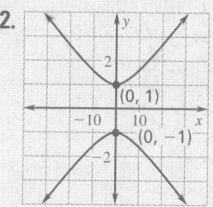

83.

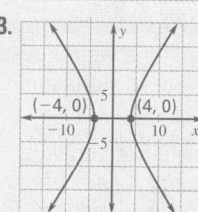

9–16. See Additional Answers beginning on page AA1.

1 PLAN

PACING
Basic: 1 day
Average: 1 day
Advanced: 1 day
Block Schedule: 0.5 block with 12.4

LESSON OPENER
APPLICATION
An alternative way to approach
Lesson 12.3 is to use the
Application Lesson Opener:
- Blackline Master (*Chapter 12
  Resource Book*, p. 43)
- Transparency (p. 79)

MEETING INDIVIDUAL NEEDS
- *Chapter 12 Resource Book*
  Prerequisite Skills Review (p. 5)
  Practice Level A (p. 45)
  Practice Level B (p. 46)
  Practice Level C (p. 47)
  Reteaching with Practice (p. 48)
  Absent Student Catch-Up (p. 50)
  Challenge (p. 52)
- *Resources in Spanish*
- *Personal Student Tutor*

NEW-TEACHER SUPPORT
See the Tips for New Teachers on
pp. 1–2 of the *Chapter 12 Resource
Book* for additional notes about
Lesson 12.3.

WARM-UP EXERCISES
Transparency Available

You are packing for a vacation.
At home you have 10 shirts and
7 pairs of shorts.

1. In how many different ways
   can you choose 4 pairs of
   shorts to take on vacation?
   $_7C_4 = 35$

2. In how many different ways
   can you choose two shirts to
   wear on the first and second
   days of vacation? $_{10}P_2 = 90$

3. If you bring 4 pairs of shorts
   and 6 shirts, how many different outfits can you make? **24**

What you should learn

GOAL 1 Find theoretical
and experimental
probabilities.

GOAL 2 Find geometric
probabilities, as applied in
Example 5.

Why you should learn it

▼ To solve **real-life**
problems, such as finding the
probability that an archer hits
the center of a target
in **Ex. 46**.

An Introduction to Probability

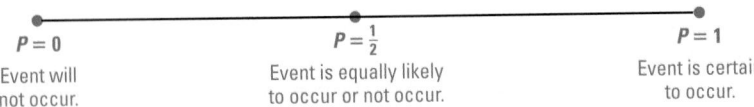

GOAL 1 **THEORETICAL AND EXPERIMENTAL PROBABILITY**

The **probability** of an event is a number between 0 and 1 that indicates the likelihood the event will occur. An event that is certain to occur has a probability of 1. An event that *cannot* occur has a probability of 0. An event that is equally likely to occur or not occur has a probability of $\frac{1}{2}$.

$P = 0$	$P = \frac{1}{2}$	$P = 1$
Event will not occur.	Event is equally likely to occur or not occur.	Event is certain to occur.

There are two types of probability: *theoretical* and *experimental*. Theoretical probability is defined below and experimental probability is defined on page 717.

THE THEORETICAL PROBABILITY OF AN EVENT

When all outcomes are equally likely, the **theoretical probability** that an event A will occur is:

$$P(A) = \frac{\text{number of outcomes in } A}{\text{total number of outcomes}}$$

The theoretical probability of an event is often simply called the probability of the event.

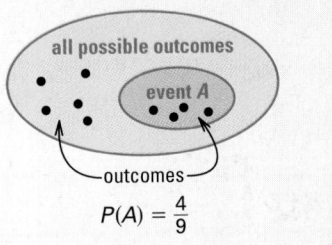

$$P(A) = \frac{4}{9}$$

EXAMPLE 1 *Finding Probabilities of Events*

You roll a six-sided die whose sides are numbered from 1 through 6. Find the probability of (**a**) rolling a 4, (**b**) rolling an odd number, and (**c**) rolling a number less than 7.

SOLUTION

a. Only one outcome corresponds to rolling a 4.

$$P(\text{rolling a 4}) = \frac{\text{number of ways to roll a 4}}{\text{number of ways to roll the die}} = \frac{1}{6}$$

b. Three outcomes correspond to rolling an odd number: rolling a 1, 3, or 5.

$$P(\text{rolling an odd number}) = \frac{\text{number of ways to roll an odd number}}{\text{number of ways to roll the die}} = \frac{3}{6} = \frac{1}{2}$$

c. All six outcomes correspond to rolling a number less than 7.

$$P(\text{rolling less than 7}) = \frac{\text{number of ways to roll less than 7}}{\text{number of ways to roll the die}} = \frac{6}{6} = 1$$

You can express a probability as a fraction, a decimal, or a percent. For instance, in part (b) of Example 1 the probability of rolling an odd number can be written as $\frac{1}{2}$, 0.5, or 50%.

Music

EXAMPLE 2 *Probabilities Involving Permutations or Combinations*

You put a CD that has 8 songs in your CD player. You set the player to play the songs at random. The player plays all 8 songs without repeating any song.

   a. What is the probability that the songs are played in the same order they are listed on the CD?

   b. You have 4 favorite songs on the CD. What is the probability that 2 of your favorite songs are played first, in any order?

SOLUTION

STUDENT HELP

▶ **Skills Review**
For help with converting decimals, fractions, and percents, see p. 906.

   a. There are 8! different *permutations* of the 8 songs. Of these, only 1 is the order in which the songs are listed on the CD. So, the probability is:

$$P(\text{playing 8 in order}) = \frac{1}{8!} = \frac{1}{40,320} \approx 0.0000248$$

   b. There are $_8C_2$ different *combinations* of 2 songs. Of these, $_4C_2$ contain 2 of your favorite songs. So, the probability is:

$$P(\text{playing 2 favorites first}) = \frac{_4C_2}{_8C_2} = \frac{6}{28} = \frac{3}{14} \approx 0.214$$

· · · · · · · · · ·

Sometimes it is not possible or convenient to find the theoretical probability of an event. In such cases you may be able to calculate an **experimental probability** by performing an experiment, conducting a survey, or looking at the history of the event.

Internet

EXAMPLE 3 *Finding Experimental Probabilities*

In 1998 a survey asked Internet users for their ages. The results are shown in the bar graph. Find the experimental probability that a randomly selected Internet user is **(a)** at most 20 years old, and **(b)** at least 41 years old.

▶ Source: GVU's WWW User Surveys™

Internet Users

Age (years)	Number of users
Under 21	1636
21–40	6617
41–60	3693
61–80	491
Over 80	6

SOLUTION The number of people surveyed was $1636 + 6617 + 3693 + 491 + 6 = 12{,}443$.

   a. Of the people surveyed, 1636 are at most 20 years old. So, the probability is:

$$P(\text{user is at most 20}) = \frac{1636}{12{,}443} \approx 0.131$$

   b. Of the people surveyed, $3693 + 491 + 6 = 4190$ are at least 41 years old. So, the probability is:

$$P(\text{user is at least 41}) = \frac{4190}{12{,}443} \approx 0.337$$

12.3 *An Introduction to Probability* 　717

---

2 TEACH

📖 **EXTRA EXAMPLE 1**
A spinner has 8 equal-size sectors numbered from 1 to 8. Find the probability of **(a)** spinning a 6 and **(b)** spinning a number greater than 5. $\frac{1}{8}; \frac{3}{8}$

EXTRA EXAMPLE 2
There are 9 students on the math team. You draw their names one by one to determine the order in which they answer questions at a math meet. What is the probability that 3 of the 5 seniors on the team will be chosen last, in any order? $\frac{_5C_3}{_9C_3} \approx 0.119$

EXTRA EXAMPLE 3
Ninth graders must enroll in one math class. The enrollments of ninth grade students during the previous year are shown in the bar graph. Find the probability that a randomly chosen student from this year's ninth grade class is enrolled in **(a)** Consumer Math **(b)** Algebra 1 or Introduction to Algebra. $\frac{36}{243} \approx 0.148; \frac{138}{243} \approx 0.568$

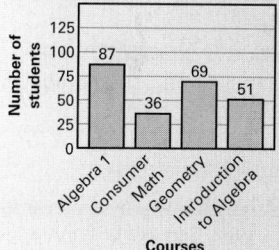

Ninth Grade Math Enrollment

✔ **CHECKPOINT EXERCISES**
For use after Examples 1 and 2:
1. Five cards are drawn from a standard 52-card deck. What is the probability that the first 2 cards are red? $\frac{_{26}C_2}{_{52}C_2} \approx 0.245$

For use after Example 3:
2. You made 15 of 21 free throw attempts. Find the probability that you will make your next free throw. $\frac{15}{21} \approx 0.714$

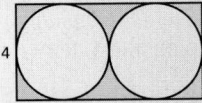

EXTRA EXAMPLE 4
Find the probability that a randomly thrown dart would hit the shaded portion of the target shown below. $\frac{32-8\pi}{32} \approx 0.215$

EXTRA EXAMPLE 5
A store is open from 8 A.M. to 8 P.M. The manager works from 9 A.M. to 4 P.M. What is the probability that the manager is there during the same time as a customer who arrives randomly during the store's hours of operation and stays for 15 minutes?
$\frac{6.75}{11.75} \approx 0.574$

✔ **CHECKPOINT EXERCISES**
For use after Exercises 4 and 5:
1. The target for a bean bag toss game is a rectangle 3 feet wide and 4 feet tall with 3 circular holes each with a radius of 1 foot. What is the probability that a bean bag thrown randomly at the target will pass through one of the holes? $\frac{\pi}{4} \approx 0.785$

FOCUS ON VOCABULARY
What is an experimental probability?
See margin.

CLOSURE QUESTION
How is a geometric probability different from other probabilities?
In a geometric probability, you use a ratio of lengths, areas, or volumes rather than a ratio of the number of outcomes.

DAILY PUZZLER
You have a sock drawer containing only black and white socks. There are 15 socks in the drawer. If the probability of choosing a pair of white socks is $\frac{1}{7}$, how many black socks are in the drawer?
9 black socks

718

Some probabilities are found by calculating a ratio of two lengths, areas, or volumes. Such probabilities are called **geometric probabilities**.

EXAMPLE 4 *Using Area to Find Probability*

You throw a dart at the board shown. Your dart is equally likely to hit any point inside the square board. Are you more likely to get 10 points or 0 points?

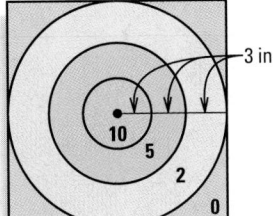

SOLUTION

The two probabilities are as follows.

$$P(10 \text{ points}) = \frac{\text{area of smallest circle}}{\text{area of entire board}}$$

$$= \frac{\pi \cdot 3^2}{18^2} = \frac{9\pi}{324} = \frac{\pi}{36} \approx 0.0873$$

$$P(0 \text{ points}) = \frac{\text{area outside largest circle}}{\text{area of entire board}}$$

$$= \frac{18^2 - (\pi \cdot 9^2)}{18^2} = \frac{324 - 81\pi}{324} = \frac{4 - \pi}{4} \approx 0.215$$

▶ You are more likely to get 0 points.

STUDENT HELP
↳ **Skills Review**
For help with area, see p. 914.

Entertainment

EXAMPLE 5 *Using Length to Find Probability*

You have recorded a 2 hour movie at the beginning of a videocassette that has 6 hours of recording time. Starting at a random location on the videocassette, your brother records a 30 minute television show. What is the probability that your brother's television show accidentally records over part of your movie?

SOLUTION

You can think of the videocassette as a number line from 0 to 6. The movie can be represented as a line segment 2 units long and the television show as a line segment 0.5 unit long. Because you know the movie starts at the beginning of the videocassette, the number line is as shown.

STUDENT HELP
HOMEWORK HELP
Visit our Web site www.mcdougallittell.com for extra examples.

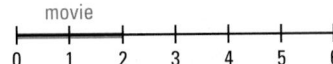

If the 30 minute, or half hour, television show is to fit on the tape, it must start somewhere between 0 and 5.5. If it records over part of the movie, it must start somewhere between 0 and 2. So, the probability of recording over part of the movie is:

$$P(\text{recording over movie}) = \frac{\text{length where show will record over movie}}{\text{length where show will fit on tape}}$$

$$= \frac{2 - 0}{5.5 - 0} = \frac{2}{5.5} = \frac{4}{11} \approx 0.364$$

Focus on Vocabulary *Sample answer:*
A probability that is calculated by performing an experiment, conducting a survey, or looking at the history of an event.

GUIDED PRACTICE

Vocabulary Check ✓

Concept Check ✓

2. *B*; The event with the higher probability is more likely to occur.

Skill Check ✓

3. A theoretical probability is based on the number of outcomes of the event and the total number of possible outcomes. An experimental probability is determined through an experiment, survey or historical data about an event.

Sample answer: The theoretical probability of rolling a 5 using a 6-sided die is $\frac{1}{6}$. If you actually rolled a 6-sided die 100 times, the experimental probability of rolling a 5 would be the number of fives you rolled divided by 100.

1. Complete this statement: A probability that involves length, area, or volume is called a(n) _?_ probability. **geometric**

2. $P(A) = 0.2$ and $P(B) = 0.6$. Which event is more likely to occur? Explain.

3. Explain the difference between theoretical probability and experimental probability. Give an example of each.

A jar contains 2 red marbles, 3 blue marbles, and 1 green marble. Find the probability of randomly drawing the given type of marble.

4. a red marble $\frac{1}{3}$

5. a green marble $\frac{1}{6}$

6. a blue or a green marble $\frac{2}{3}$

7. a red or a blue marble $\frac{5}{6}$

Find the probability that a dart thrown at the given target will hit the shaded region. Assume the dart is equally likely to hit any point inside the target. The targets and regions within are either squares, circles, or triangles.

8. 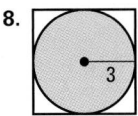 0.785

9. 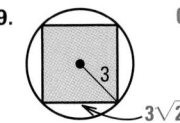 0.637

10. 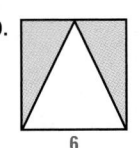 $\frac{1}{2}$

11. 🌐 **POPULATION** The bar graph shown gives the resident population (in thousands) of the United States in 1997. For a randomly selected person in the United States, find the probability of the given event.

DATA UPDATE of *Statistical Abstract of the United States* data at www.mcdougallittell.com

United States Population

Age (years)	Population (thousands)
24 and under	94,507
25 to 44	83,608
45 to 64	55,446
65 and older	34,076

a. The person is 24 years old or under. 0.353

b. The person is at least 45 years old. 0.334

PRACTICE AND APPLICATIONS

STUDENT HELP

► **Extra Practice**
to help you master skills is on p. 956.

CHOOSING NUMBERS You have an equally likely chance of choosing any integer from 1 through 20. Find the probability of the given event.

12. An odd number is chosen. 0.5

13. A number less than 7 is chosen. 0.3

14. A perfect square is chosen. 0.2

15. A prime number is chosen. 0.4

16. A multiple of 3 is chosen. 0.3

17. A factor of 240 is chosen. 0.6

CHOOSING CARDS A card is drawn randomly from a standard 52-card deck. Find the probability of drawing the given card.

STUDENT HELP

► **Look Back**
For help with a standard 52-card deck, see p. 708.

18. the ace of hearts 0.0192

19. any ace 0.0769

20. a diamond 0.25

21. a red card 0.5

22. a card other than 10 0.923

23. a face card (a king, queen, or jack) 0.231

12.3 *An Introduction to Probability* **719**

⬤ **ASSIGNMENT GUIDE**

BASIC
Day 1: pp. 719–722 Exs. 12–23, 24–27, 30–32, 35–38, 44, 48–49, 51–63 odd

AVERAGE
Day 1: pp. 719–722 Exs. 12–23, 24–36 even, 37–40, 44, 47–49, 51–67 odd

ADVANCED
Day 1: pp. 719–722 Exs. 12–23, 24–36 even, 37–44, 47–50, 51–67 odd

BLOCK SCHEDULE
pp. 719–722 Exs. 12–23, 24–36 even, 37–40, 44, 47–49, 51–67 odd (with 12.4)

EXERCISE LEVELS
Level A: *Easier*
12–23, 35–38, 44
Level B: *More Difficult*
24–34, 39–43, 45–49
Level C: *Most Difficult*
50

✓ **HOMEWORK CHECK**
To quickly check student understanding of key concepts, go over the following exercises: Exs. 14, 20, 24, 30, 36. See also the Daily Homework Quiz:
• Blackline Master (*Chapter 12 Resource Book,* p. 55)
• ✂ Transparency (p. 92)

MULTIPLE REPRESENTATIONS
As students begin this and other exercise sets, remind them that probability can be expressed as a fraction, as a decimal, or as a percent. In real life problems the form of the answer may depend on the form given in the exercise.

! **COMMON ERROR**

EXERCISES 35–40 Students often become so focused on the problem scenario that they forget to consider whether the situation is described by a combination or a permutation. Caution students to examine each situation to determine whether the order is important.

	Theoretical	Experimental
24.	0.167	0.217
25.	0.333	0.308
26.	0.5	0.492
27.	0.5	0.508
28.	0.667	0.725
29.	0.833	0.875

The two probabilities are not exactly the same, but they are very similar in every case.

STUDENT HELP

↳ **HOMEWORK HELP**
Example 1: Exs. 12–23
Example 2: Exs. 35–40
Example 3: Exs. 24–29, 41–43
Example 4: Exs. 30–34, 46, 47
Example 5: Exs. 44, 45

ROLLING A DIE The results of rolling a six-sided die 120 times are shown. Use the table to find the experimental probability of each event. Also find the theoretical probability. How do the probabilities compare? **24–29. See margin.**

Results from Rolling a Die 120 Times						
Roll	1	2	3	4	5	6
Number of occurrences	15	18	20	17	24	26

24. rolling a 6

25. rolling a 3 or 4

26. rolling an odd number

27. rolling an even number

28. rolling a number greater than 2

29. rolling anything but a 1

GEOMETRY CONNECTION Find the probability that a dart thrown at the square target shown will hit the given region. Assume the dart is equally likely to hit any point inside the target.

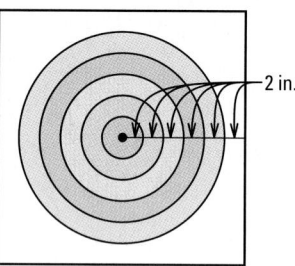

2 in.

24 in.

30. the red center   0.0218

31. the white border   0.455

32. the red center or the white border   0.477

33. the four rings or the red center   0.545

34. the yellow or green ring   0.262

🌐 **SPEECHES** **In Exercises 35 and 36, use the following information.**
Your English teacher is drawing names to see who will give the first speech. There are 26 students in the class and 4 speeches will be given each day.

35. What is the probability that you will give your speech first?   0.0385

36. What is the probability that you will give your speech on the first day?   0.154

🌐 **WORD GAMES** **In Exercises 37 and 38, use the following information.**
You and a friend are playing a word game that involves lettered tiles. The distribution of letters is shown at the right. At the start of the game you choose 7 letters.

Distribution of Letters			
A: 9	H: 2	O: 8	V: 2
B: 2	I: 9	P: 2	W: 2
C: 2	J: 1	Q: 1	X: 1
D: 4	K: 1	R: 6	Y: 2
E: 12	L: 4	S: 4	Z: 1
F: 2	M: 2	T: 6	Blank: 2
G: 3	N: 6	U: 4	

37. What is the probability that you will choose three vowels and four consonants? (Count "Y" as a vowel.)   0.262

38. What is the probability that you will choose the letters A, B, C, D, E, F, and G in order? 1.285×10^{-10}

🌐 **LOTTERIES** **In Exercises 39 and 40, find the probability of winning the lottery according to the given rules. Assume numbers are selected at random.**

39. You must correctly select 6 out of 51 numbers. The order of the numbers is not important. 5.6×10^{-8}

40. You must correctly select 3 numbers, each from 0 to 9. The order of the numbers is important.   0.001

FOCUS ON
APPLICATIONS

The Philadelphia Inquirer

**Dow Dives 508.32 Points
In Panic on Wall Street**

Billions lost in trading

STOCK MARKET
 The largest one-day stock-market loss in Wall Street history occurred on October 19, 1987. That day the Dow Jones fell 508.32 points, or 22.6%. The net change for the year, however, was a growth of 20.9%.

APPLICATION LINK
www.mcdougallittell.com

41. 🌐 **MEDIA CONCERN** In a 1998 survey, parents were asked what media influence on their children most concerned them. The results are shown in the bar graph. Find the experimental probability that a randomly selected parent is most concerned about the given topic. ▶ Source: Annenberg Public Policy Center

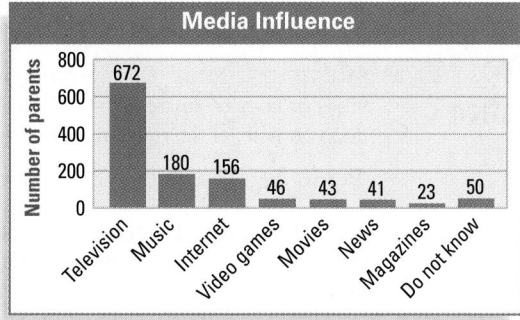

Media Influence

Number of parents: Television 672, Music 180, Internet 156, Video games 46, Movies 43, News 41, Magazines 23, Do not know 50

a. Television 0.555

b. Video games 0.0380

42. HISTORY CONNECTION The table shows how many years the stock market gained or lost a given percent over a recent 17 year period based on the Dow Jones Industrial Average. Find the experimental probability of the given event.

Range	Lost more than 0%	Gained 0% to 9%	Gained 10% to 19%	Gained 20% to 29%	Gained more than 29%
Years	4	2	4	4	3

a. The stock market has a loss. 0.235

b. The stock market gains at least 10%. 0.647

43. STATISTICS CONNECTION The table shows how people in the United States got to work in 1990. For a randomly selected person in the United States, find the probability that the person chose the given type of transportation.
▶ Source: *The World Almanac*

Means of transportation	Number
Automobile	99,592,932
Public transportation	6,069,589
Motorcycle	237,404
Bicycle	466,856
Other	5,297,468
None (work at home)	3,406,025

a. Used public transportation 0.0527

b. Drove to work (either in an automobile or on a motorcycle) 0.868

44. 🌐 **VIDEOCASSETTES** Look back at Example 5. Suppose you recorded your movie starting 1 hour into the videocassette. What is the probability that your brother's television show accidentally records over part of your movie? 0.364

45. 🌐 **CABLE INSTALLATION** You set up an appointment to have cable television installed between 12:00 P.M. and 4:00 P.M. The installer will wait 15 minutes if no one is home. Your cousin asks for a favor that would take you away from your home from 1:30 P.M. to 2:00 P.M. If you do the favor, what is the probability that you will miss the cable installer? 0.0625

46. 🌐 **KYUDO** *Kyudo* is a form of Japanese archery. The most common target is shown. Find the probability that an arrow shot at the target will hit the center circle. Assume the arrow is equally likely to hit any point inside the target. 0.04

3.6 cm
1.5 cm
3.3 cm
3.0 cm
3.0 cm
3.6 cm

12.3 *An Introduction to Probability* **721**

APPLICATION NOTE
EXERCISE 42 Explain to students that the Dow Jones is the average value of a select group of stocks. While the Dow Jones can be used to represent the overall trend of the stock market, it does not give the performance of any one stock.
 Additional information about the stock market is available at **www.mcdougallittell.com.**

ADDITIONAL PRACTICE AND RETEACHING

For Lesson 12.3:
• Practice Levels A, B, and C (*Chapter 12 Resource Book*, p. 45)
• Reteaching with Practice (*Chapter 12 Resource Book*, p. 48)
• 📺 See Lesson 12.3 of the *Personal Student Tutor*

For more Mixed Review:
• 📺 Search the *Test and Practice Generator* for key words or specific lessons.

DAILY HOMEWORK QUIZ

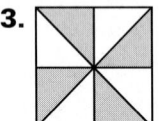 *Transparency Available*

Each of 20 students is randomly assigned to a group. Use the table of results to find the experimental probability of each event. Also find the theoretical probability.

No. of students in group					
Group	A	B	C	D	E
No. of students	4	6	2	5	3

1. a student is in group B 0.3; 0.2

2. a student is in group C 0.1; 0.2

Find the probability that a dart thrown at the square target will hit the shaded region. Assume the dart is equally likely to hit any point inside the target.

3. 0.5

EXTRA CHALLENGE NOTE

→ Challenge problems for Lesson 12.3 are available in **blackline** format in the *Chapter 12 Resource Book,* p. 52 and at **www.mcdougallittell.com.**

ADDITIONAL TEST PREPARATION

1. OPEN ENDED Design a dart board target. Calculate the probability that a randomly thrown dart lands in at least 2 different areas of the target. **Check designs and probabilities.**

2. WRITING You have 3 red balls, 5 green balls, and 2 white balls in a box. You draw 1 ball out at random. Show how to find the probability that the ball is red. Is this a theoretical or experimental probability? Explain. **See margin.**

---

47.  **CONTACT LENSES** You have just stepped into the tub to take a shower when one of your contact lenses falls out. (You have not yet turned on the shower.) Assuming that the lens is equally likely to land anywhere on or inside of the tub, what is the probability that it landed in the drain? 0.00242

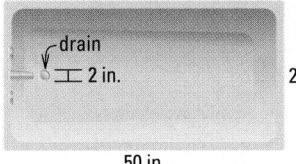

48. MULTIPLE CHOICE On a multiple choice question, you know that the answer is not B or D, but you are not sure about answers A, C, or E. What is the probability that you will get the right answer if you guess? **C**

  Ⓐ $\frac{1}{5}$ Ⓑ $\frac{4}{5}$ Ⓒ $\frac{1}{3}$ Ⓓ $\frac{2}{3}$ Ⓔ $\frac{3}{5}$

49. MULTIPLE CHOICE A dart thrown at the circular target shown is equally likely to hit any point inside the target. What is the probability that it hits the region outside the triangle? **E**

  Ⓐ 0.5   Ⓑ 0.75   Ⓒ 0.32

  Ⓓ 0.47   Ⓔ 0.68

★ Challenge

50. PROBABILITY Find the probability that the graph of $y = x^2 - 4x + c$ intersects the x-axis if c is a randomly chosen integer from 1 to 6. 0.667

MIXED REVIEW

DETERMINANTS Evaluate the determinant of the matrix. (Review 4.3)

51. $\begin{bmatrix} 2 & 7 \\ 5 & 9 \end{bmatrix}$ -17 **52.** $\begin{bmatrix} 6 & 0 \\ 1 & -3 \end{bmatrix}$ -18 **53.** $\begin{bmatrix} 3 & 8 \\ -2 & 1 \end{bmatrix}$ 19

54. $\begin{bmatrix} 1 & 2 & 3 \\ 2 & 3 & 1 \\ 3 & 1 & 2 \end{bmatrix}$ -18 **55.** $\begin{bmatrix} 0 & 1 & 5 \\ -3 & 4 & 7 \\ 2 & 1 & -4 \end{bmatrix}$ -53 **56.** $\begin{bmatrix} -1 & -2 & 2 \\ 4 & 3 & -4 \\ 2 & 4 & 6 \end{bmatrix}$ 50

MULTIPLYING Multiply the rational expressions. Simplify the result. (Review 9.4)

57. $\frac{6xy^2}{5x^3y} \cdot \frac{10y^4}{9xy}$ $\frac{4y^4}{3x^3}$ **58.** $\frac{x^2 + 3x + 2}{x^2 - x - 6} \cdot \frac{x^2 - 3x}{x^2 - x - 2}$ $\frac{x}{x-2}$

59. $\frac{25x^2 - 16}{5x - 4} \cdot \frac{x^2 - 4x - 21}{5x^3 - 31x^2 - 28x}$ $\frac{x+3}{x}$ **60.** $\frac{4x^2 - 12x}{27 - x^3} \cdot (x^2 + 3x + 9)$ $-4x$

WRITING TERMS Write the first five terms of the sequence. (Review 11.5)

61. 3, 10, 17, 24, 31

62. −1, −3, −9, −27, −81

63. 2; 8; 512; 134,217,728; 2.418 × 10²⁴

64. 1, 1, 2, 3, 5

65. −2, 0, 2, 2, 0

66. 1, −2, −2, 4, −8

61. $a_0 = 3$
$a_n = a_{n-1} + 7$

62. $a_0 = -1$
$a_n = 3 \cdot a_{n-1}$

63. $a_0 = 2$
$a_n = (a_{n-1})^3$

64. $a_0 = 1$
$a_1 = 1$
$a_n = a_{n-1} + a_{n-2}$

65. $a_0 = -2$
$a_1 = 0$
$a_n = a_{n-1} - a_{n-2}$

66. $a_0 = 1$
$a_1 = -2$
$a_n = a_{n-1} \cdot a_{n-2}$

67. 🌎 **TRADE SHOWS** You are attending a trade show that has booths from 20 different vendors. You hope to visit at least 5 of the booths. How many combinations of booths can you visit? (Review 12.2 for 12.4) 1,042,380

Additional Test Preparation *Sample answer:*

2. There are 3 red balls and a total of 10 balls, so the probability of drawing a red is $\frac{3}{10}$, or 0.3. This is a theoretical probability because it is not based on an experiment or the history of the event.

ACTIVITY 12.3

Using Technology

Generating Random Numbers

Most graphing calculators have a random number generator that you can use to perform probability experiments.

▶ EXAMPLE

Use the random number generator of a graphing calculator to simulate rolling a 6-sided die 120 times. Record the number of times you obtain 1, 2, 3, 4, 5, and 6.

▶ SOLUTION

STUDENT HELP

INTERNET

KEYSTROKE HELP

See keystrokes for several models of calculators at www.mcdougallittell.com

① In a list, enter randInt(1,6,120) to generate 120 random integers from 1 to 6.

② Put the list in ascending order using the *Sort* feature. Scroll through the list to count and record the frequency of each number.

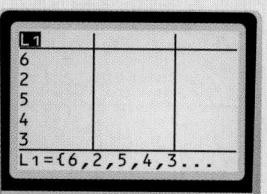

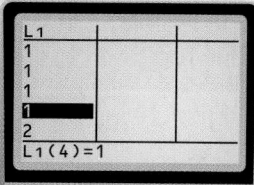

▶ EXERCISES 1–3. See margin.

1. Copy and complete the table by performing the experiment above. What is the theoretical probability of rolling each number? What is the experimental probability? Do your experimental results agree with the theoretical results?

Number	1	2	3	4	5	6
Frequency	?	?	?	?	?	?

2. Copy and complete the table by simulating drawing 52 cards with replacement from a standard deck. Let 1 represent an ace, 2–10 represent the cards 2–10, 11 represent a jack, 12 represent a queen, and 13 represent a king. What is the theoretical probability of drawing each card? What is the experimental probability? Do your experimental results agree with the theoretical results?

Card	1	2	3	4	5	6	7	8	9	10	11	12	13
Frequency	?	?	?	?	?	?	?	?	?	?	?	?	?

3. Copy and complete the table by simulating tossing a coin 10, 20, 50, 100, and 200 times. Let 0 represent heads and 1 represent tails. For each number of trials record the number of heads and tails. As the number of trials increases, how do the experimental results compare with the theoretical results?

Number of trials	10	20	50	100	200
Number of heads	?	?	?	?	?
Number of tails	?	?	?	?	?

---

1 **Planning the Activity**

PURPOSE
To simulate experimental probabilities using a random number generator.

MATERIALS
• graphing calculator
• Keystroke Help (*Chapter 12 Resource Book*, p. 44)

PACING
• Activity — 30 min

▶ *LINK TO LESSON*
Refer students back to Example 1 on page 716 when completing Exercise 1. Ask students why the results they get using the random number generator are experimental probabilities.

2 **Managing the Activity**

ALTERNATIVE APPROACH
This activity can be done with a computer that has random number generating software. You can also do this activity as a class using an overhead display with a graphing calculator or computer.

3 **Closing the Activity**

★ *KEY DISCOVERY*
As the number of trials increases, the experimental probability becomes closer to the theoretical probability.

ACTIVITY ASSESSMENT
A spinner has three sectors of equal size labeled *A*, *B*, and *C*. How could you use a random number generator to find the experimental probability of landing on each section with 50 spins? See sample answer at left.

1–3. See Additional Answers beginning on page AA1.

Activity Assessment *Sample answer:*
let 1 represent spinning an *A*, 2 represent spinning a *B*, and 3 represent spinning a *C*. Generate 50 random integers from 1 to 3 and record the frequency for each number. Then find the ratio of each frequency to 50 to find the experimental probabilities.

LESSON OPENER
ACTIVITY
An alternative way to approach Lesson 12.4 is to use the Activity Lesson Opener:

- Blackline Master (*Chapter 12 Resource Book*, p. 56)
- Transparency (p. 80)

MEETING INDIVIDUAL NEEDS
- ***Chapter 12 Resource Book***
  Prerequisite Skills Review (p. 5)
  Practice Level A (p. 58)
  Practice Level B (p. 59)
  Practice Level C (p. 60)
  Reteaching with Practice (p. 61)
  Absent Student Catch-Up (p. 63)
  Challenge (p. 65)
- ***Resources in Spanish***
- ***Personal Student Tutor***

NEW-TEACHER SUPPORT
See the Tips for New Teachers on pp. 1–2 of the *Chapter 12 Resource Book* for additional notes about Lesson 12.4.

WARM-UP EXERCISES

Transparency Available

The school council comprises 3 male students, 5 female students, and 2 teachers.

1. What is the probability that a person chosen at random from the school council is a female student? **0.5**
2. What is the probability that both members of a committee chosen at random from the school council are male students? $\frac{1}{15} \approx 0.067$

12.4

Probability of Compound Events

What you should learn

GOAL 1 Find probabilities of unions and intersections of two events.

GOAL 2 Use complements to find the probability of an event, as applied in **Example 5.**

Why you should learn it

▼ To solve **real-life** problems, such as finding the probability that friends will be in the same college dormitory in **Ex. 49.**

GOAL 1 PROBABILITIES OF UNIONS AND INTERSECTIONS

When you consider all the outcomes for either of two events A and B, you form the *union* of A and B. When you consider only the outcomes shared by both A and B, you form the *intersection* of A and B. The union or intersection of two events is called a **compound event**.

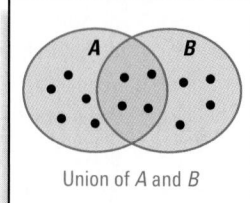
Union of A and B

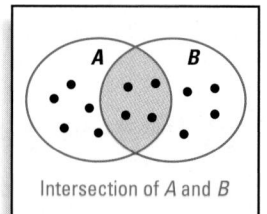
Intersection of A and B

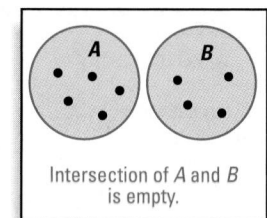
Intersection of A and B is empty.

To find $P(A \text{ or } B)$ you must consider what outcomes, if any, are in the intersection of A and B. If there are none, then A and B are **mutually exclusive events** and $P(A \text{ or } B) = P(A) + P(B)$. If A and B are not mutually exclusive, then the outcomes in the intersection of A and B are counted *twice* when $P(A)$ and $P(B)$ are added. So, $P(A \text{ and } B)$ must be subtracted *once* from the sum.

PROBABILITY OF COMPOUND EVENTS

If A and B are two events, then the probability of A or B is:
$$P(A \text{ or } B) = P(A) + P(B) - P(A \text{ and } B)$$
If A and B are mutually exclusive, then the probability of A or B is:
$$P(A \text{ or } B) = P(A) + P(B)$$

EXAMPLE 1 *Probability of Mutually Exclusive Events*

A card is randomly selected from a standard deck of 52 cards. What is the probability that it is an ace *or* a face card?

SOLUTION

Let event A be selecting an ace, and let event B be selecting a face card. Event A has 4 outcomes and event B has 12 outcomes. Because A and B are mutually exclusive, the probability is:

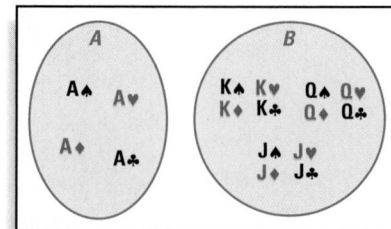

$$P(A \text{ or } B) = P(A) + P(B) = \frac{4}{52} + \frac{12}{52} = \frac{16}{52} = \frac{4}{13} \approx 0.308$$

EXAMPLE 2 Probability of a Compound Event

A card is randomly selected from a standard deck of 52 cards. What is the probability that the card is a heart *or* a face card?

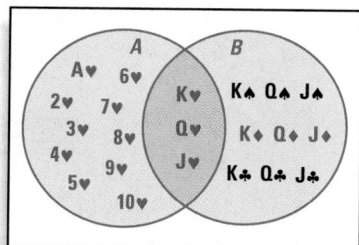

SOLUTION

Let event A be selecting a heart, and let event B be selecting a face card. Event A has 13 outcomes and event B has 12 outcomes. Of these, three outcomes are common to A and B. So, the probability of selecting a heart *or* a face card is:

$P(A \text{ or } B) = P(A) + P(B) - P(A \text{ and } B)$	Write general formula.
$= \dfrac{13}{52} + \dfrac{12}{52} - \dfrac{3}{52}$	Substitute known probabilities.
$= \dfrac{22}{52}$	Combine terms.
$= \dfrac{11}{26}$	Simplify.
≈ 0.423	Use a calculator.

Business

EXAMPLE 3 Using Intersection to Find Probability

Last year a company paid overtime wages *or* hired temporary help during 9 months. Overtime wages were paid during 7 months and temporary help was hired during 4 months. At the end of the year, an auditor examines the accounting records and randomly selects one month to check the company's payroll. What is the probability that the auditor will select a month in which the company paid overtime wages *and* hired temporary help?

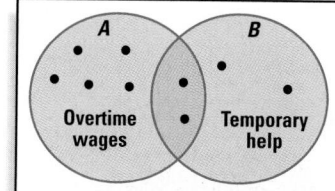

SOLUTION

Let event A represent paying overtime wages during a month, and let event B represent hiring temporary help during a month. From the given information you know that:

$$P(A) = \frac{7}{12}, P(B) = \frac{4}{12}, \text{ and } P(A \text{ or } B) = \frac{9}{12}$$

The probability that the auditor will select a month in which the company paid overtime wages *and* hired temporary help is $P(A \text{ and } B)$.

$P(A \text{ or } B) = P(A) + P(B) - P(A \text{ and } B)$	Write general formula.
$\dfrac{9}{12} = \dfrac{7}{12} + \dfrac{4}{12} - P(A \text{ and } B)$	Substitute known probabilities.
$P(A \text{ and } B) = \dfrac{7}{12} + \dfrac{4}{12} - \dfrac{9}{12}$	Solve for $P(A \text{ and } B)$.
$P(A \text{ and } B) = \dfrac{2}{12} = \dfrac{1}{6} \approx 0.167$	Simplify.

12.4 *Probability of Compound Events*   **725**

2 TEACH

EXTRA EXAMPLE 1
One six-sided die is rolled. What is the probability of rolling a multiple of 3 or 5? 0.5

EXTRA EXAMPLE 2
One six-sided die is rolled. What is the probability of rolling a multiple of 3 or a multiple of 2?
$\frac{2}{3} \approx 0.67$

EXTRA EXAMPLE 3
In a poll of high school juniors, 6 out of 15 took a French class and 11 out of 15 took a math class. Fourteen out of 15 students took French or math. What is the probability that a student took both French and math? $\frac{1}{5} = 0.20$

✓ **CHECKPOINT EXERCISES**
For use after Examples 1–3:
In a survey of 200 pet owners, 103 owned dogs, 88 owned cats, 25 owned birds, and 18 owned reptiles.

1. None of the respondents owned both a cat and a bird. What is the probability that they owned a cat or a bird?
$\frac{113}{200} = 0.565$

2. Of the respondents, 52 owned both a cat and a dog. What is the probability that a respondent owned a cat or a dog?
$\frac{139}{200} = 0.695$

3. Of the respondents, 119 owned a dog or a reptile. What is the probability that they owned a dog and a reptile? $\frac{1}{100} = 0.010$

MATHEMATICAL REASONING
Point out to students that the "or" used in this context and most mathematical contexts is called the inclusive or. That is, A or B means A or B or both A and B. The other form of "or," called the exclusive form, means A or B but not both.

EXTRA EXAMPLE 4
A card is randomly selected from a standard deck of 52 cards. Find the probability of the given event.
a. The card is not a king.
$\frac{48}{52} \approx 0.923$

b. The card is not an ace or a jack. $\frac{44}{52} \approx 0.846$

EXTRA EXAMPLE 5
One high school requires students to complete 30 hours of community service to graduate. There are 156 different community service options to choose from. What is the probability that in a group of 5 students, at least 2 of them will be doing the same community service? **0.063**

 CHECKPOINT EXERCISES
For use after Examples 4 and 5:
1. Seven prizes are being given in a raffle contest. 157 tickets are sold. After each prize is called, the winning ticket is returned to the drawing box and is eligible to be picked for another prize. What is the probability that at least one of the tickets is drawn twice? **0.127**

FOCUS ON VOCABULARY
Describe what it means for two events to be complementary.
Sample answer: The two events contain all the possible outcomes and no outcome is in both events.

CLOSURE QUESTION
What is $P(A$ and $B)$ if A and B are mutually exclusive? **0**

DAILY PUZZLER
In one community, it is never sunny and rainy at the same time. The probability of rain is 3 times the probability of sun and the probability that it rains or is sunny is 0.88. What is the probability that it will rain in this community? **0.66**

726

The event A', called the **complement** of event A, consists of all outcomes that are not in A. The notation A' is read as "A prime."

PROBABILITY OF THE COMPLEMENT OF AN EVENT

The probability of the complement of A is $P(A') = 1 - P(A)$.

EXAMPLE 4 *Probabilities of Complements*

STUDENT HELP

HOMEWORK HELP
Visit our Web site
www.mcdougallittell.com
for extra examples.

When two six-sided dice are tossed, there are 36 possible outcomes as shown. Find the probability of the given event.

a. The sum is not 8.

b. The sum is greater than or equal to 4.

SOLUTION

a. $P(\text{sum is not } 8) = 1 - P(\text{sum is } 8)$

$= 1 - \frac{5}{36}$

$= \frac{31}{36}$

≈ 0.861

b. $P(\text{sum} \geq 4) = 1 - P(\text{sum} < 4)$

$= 1 - \frac{3}{36}$

$= \frac{33}{36}$

$= \frac{11}{12}$

≈ 0.917

FOCUS ON APPLICATIONS

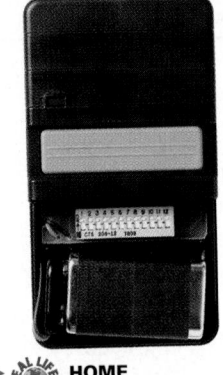

HOME ELECTRONICS
One type of garage door opener has 12 switches that can be set in one of two positions (off or on) to create a code. So for this type of garage door opener, there are 2^{12}, or 4096, possible transmitter codes.

EXAMPLE 5 *Using a Complement in Real Life*

HOME ELECTRONICS Four houses in a neighborhood have the same model of garage door opener. Each opener has 4096 possible transmitter codes. What is the probability that at least two of the four houses have the same code?

SOLUTION

The total number of ways to assign codes to the four openers is 4096^4. The number of ways to assign *different* codes to the four openers is $4096 \cdot 4095 \cdot 4094 \cdot 4093$. So, the probability that at least two of the four openers have the same code is:

$P(\text{at least 2 are the same}) = 1 - P(\text{none are the same})$

$= 1 - \dfrac{4096 \cdot 4095 \cdot 4094 \cdot 4093}{4096^4}$

$\approx 1 - 0.99854$

$= 0.00146$

GUIDED PRACTICE

Vocabulary Check ✓
Concept Check ✓

1. Describe what it means for two events to be mutually exclusive.

2. Write a formula for computing $P(A \text{ or } B)$ that applies to *any* events A and B. How can you simplify this formula when A and B are mutually exclusive?

3. Are the events A and A' mutually exclusive? Explain.

Skill Check ✓

Events A and B are mutually exclusive. Find $P(A \text{ or } B)$.

1. Two events are mutually exclusive if there are no outcomes shared by both of them.

2. $P(A \text{ or } B) =$ $P(A) + P(B) - P(A \text{ and } B)$. When A and B are mutually exclusive, $P(A \text{ and } B) = 0$, so the formula becomes $P(A \text{ or } B) = P(A) + P(B)$.

3. Yes. A' is defined as all outcomes not in A, so there is no intersection between events in A and events not in A.

4. $P(A) = 0.2$, $P(B) = 0.3$ **0.5**

5. $P(A) = 0.5$, $P(B) = 0.5$ **1**

6. $P(A) = \frac{3}{8}$, $P(B) = \frac{1}{8}$ $\frac{1}{2}$

7. $P(A) = \frac{1}{3}$, $P(B) = \frac{1}{4}$ $\frac{7}{12}$

Find $P(A \text{ or } B)$.

8. $P(A) = 0.5$, $P(B) = 0.4$, **0.6**
$P(A \text{ and } B) = 0.3$

9. $P(A) = \frac{2}{5}$, $P(B) = \frac{3}{5}$, $\frac{4}{5}$
$P(A \text{ and } B) = \frac{1}{5}$

Find $P(A \text{ and } B)$.

10. $P(A) = 0.7$, $P(B) = 0.2$, **0.1**
$P(A \text{ or } B) = 0.8$

11. $P(A) = \frac{5}{16}$, $P(B) = \frac{7}{16}$, $\frac{3}{16}$
$P(A \text{ or } B) = \frac{9}{16}$

Find $P(A')$.

12. $P(A) = 0.5$ **0.5**   **13.** $P(A) = 0.75$ **0.25**   **14.** $P(A) = \frac{1}{3}$ $\frac{2}{3}$ **15.** $P(A) = \frac{4}{7}$ $\frac{3}{7}$

PRACTICE AND APPLICATIONS

STUDENT HELP

▸ **Extra Practice**
to help you master
skills is on p. 957.

FINDING PROBABILITIES Find the indicated probability. State whether A and B are mutually exclusive.

16. $P(A) = 0.4$ **0.25, no**
$P(B) = 0.35$
$P(A \text{ or } B) = 0.5$
$P(A \text{ and } B) = \underline{?}$

17. $P(A) = 0.6$ **0.7, no**
$P(B) = 0.2$
$P(A \text{ or } B) = \underline{?}$
$P(A \text{ and } B) = 0.1$

18. $P(A) = 0.25$ **0.45, yes**
$P(B) = \underline{?}$
$P(A \text{ or } B) = 0.70$
$P(A \text{ and } B) = 0$

19. $P(A) = \frac{13}{17}$ $\frac{7}{17}$, **no**
$P(B) = \underline{?}$
$P(A \text{ or } B) = \frac{14}{17}$
$P(A \text{ and } B) = \frac{6}{17}$

20. $P(A) = \frac{1}{3}$ **0, yes**
$P(B) = \frac{1}{4}$
$P(A \text{ or } B) = \frac{7}{12}$
$P(A \text{ and } B) = \underline{?}$

21. $P(A) = \frac{3}{4}$ $\frac{5}{6}$, **no**
$P(B) = \frac{1}{3}$
$P(A \text{ or } B) = \underline{?}$
$P(A \text{ and } B) = \frac{1}{4}$

STUDENT HELP

▸ HOMEWORK HELP
Example 1: Exs. 16–24,
29–34, 42, 43
Example 2: Exs. 16–24,
29–34, 44, 45
Example 3: Exs. 16–24,
29–34, 46, 47
Example 4: Exs. 25–28,
35–40
Example 5: Exs. 48, 49

22. $P(A) = 5\%$ **34%, yes**
$P(B) = 29\%$
$P(A \text{ or } B) = \underline{?}$
$P(A \text{ and } B) = 0\%$

23. $P(A) = 30\%$ **30%, no**
$P(B) = \underline{?}$
$P(A \text{ or } B) = 50\%$
$P(A \text{ and } B) = 10\%$

24. $P(A) = 16\%$ **8%, no**
$P(B) = 24\%$
$P(A \text{ or } B) = 32\%$
$P(A \text{ and } B) = \underline{?}$

FINDING PROBABILITIES OF COMPLEMENTS Find $P(A')$.

25. $P(A) = 0.34$ **0.66**   **26.** $P(A) = 0$ **1**   **27.** $P(A) = \frac{3}{4}$ $\frac{1}{4}$ **28.** $P(A) = 1$ **0**

---

3 APPLY

⬤ **ASSIGNMENT GUIDE**

BASIC
Day 1: pp. 727–729 Exs. 16–21,
25–31, 35–38, 42, 45,
50–51, 57–71 odd

AVERAGE
Day 1: pp. 727–729 Exs. 16–40
even, 41–45, 50–51,
57–71 odd

ADVANCED
Day 1: pp. 727–729 Exs. 16–40
even, 41–45, 48–55,
57–71 odd

BLOCK SCHEDULE
pp. 727–729 Exs. 16–40 even,
41–45, 50–51, 57–71 odd
(with 12.3)

EXERCISE LEVELS
Level A: *Easier*
19–39

Level B: *More Difficult*
40–51

Level C: *Most Difficult*
52–55

✔ **HOMEWORK CHECK**
To quickly check student under-
standing of key concepts, go
over the following exercises:
Exs. 16, 25, 30, 38. See also the
Daily Homework Quiz:

• Blackline Master (*Chapter 12
Resource Book*, p. 68)

• 📖 Transparency (p. 93)

STUDENT HELP NOTES

→ **Look Back** As students look back to p. 708, remind them that a standard deck of cards has 4 suits — spades, hearts, diamonds, and clubs — each containing 13 cards.

→ **Look Back** As students look back to page 723, remind them that each possible outcome should be represented by a random number.

CAREER NOTE
EXERCISE 46 Additional information about a career as a botanist is available at **www.mcdougallittell.com.**

GRAPHING CALCULATOR NOTE
EXERCISE 51 Students create a table of probability values using a graphing calculator's *table* function. Students could create a similar table on a computer using spreadsheet software.

ENGLISH LEARNERS
EXERCISES 29–55 In order to translate word problems into equations successfully, students learning English must recognize key words—many of which are relatively unimportant when used in everyday speech. When they answer these problems, make sure students are identifying and understanding signal words and phrases such as *and, or, either . . . or, both . . . and, at least, odds in favor of,* and *odds against.*

ADDITIONAL PRACTICE AND RETEACHING

For Lesson 12.4:
• Practice Levels A, B, and C (*Chapter 12 Resource Book,* p. 58)
• Reteaching with Practice (*Chapter 12 Resource Book,* p. 61)
• ▨ See Lesson 12.4 of the *Personal Student Tutor*

For more Mixed Review:
• ▨ Search the *Test and Practice Generator* for key words or specific lessons.

STUDENT HELP

→ **Look Back**
For help with a standard 52-card deck in Exs. 29–34, see p. 708.

→ **Look Back**
For help with simulations on a graphing calculator in Ex. 41, see p. 723.

35. $\frac{17}{18}$, or about 0.944

36. $\frac{5}{6}$, or about 0.833

37. $\frac{7}{9}$, or about 0.778

38. $\frac{11}{12}$, or about 0.917

39. $\frac{35}{36}$, or about 0.972

40. $\frac{11}{18}$, or about 0.611

41. *Sample answer:*
Not 3: 0.933; ≥ 5: 0.825; not 3 or 7: 0.783; ≤ 10: 0.942; > 2: 0.983; < 8 or > 11: 0.600; The experimental results are very similar to the theoretical results.

CHOOSING CARDS A card is randomly drawn from a standard 52-card deck. Find the probability of the given event. (A face card is a king, queen, or jack.)

29. a queen and a heart $\frac{1}{52}$ 30. a queen or a heart $\frac{4}{13}$ 31. a heart or a diamond $\frac{1}{2}$

32. a five or a six $\frac{2}{13}$ 33. a five and a six 0 34. a three or a face card $\frac{4}{13}$

USING COMPLEMENTS Two six-sided dice are rolled. Find the probability of the given event. (Refer to Example 4 for a diagram of all possible outcomes.)

35. The sum is not 3.   36. The sum is greater than or equal to 5.

37. The sum is neither 3 nor 7.   38. The sum is less than or equal to 10.

39. The sum is greater than 2.   40. The sum is less than 8 or greater than 11.

41.  **EXPERIMENTAL PROBABILITIES** Simulate rolling two dice 120 times. Use separate lists for the results of each die, and a third list for the sum. Record the frequency of each sum in a table. Find the experimental probabilities of the events in Exercises 35–40. How do your experimental results compare with the theoretical results?

42. 🌎 **COMPANY MOVE** An employee of a large national company is promoted to management and will be moved within six months. The employee is told that there is a 33% probability of being moved to Denver, Colorado, and a 50% probability of being moved to Dallas, Texas. What is the probability that the employee will be moved to Dallas or Denver? **0.83**

43. 🌎 **CLASS ELECTIONS** You and your best friend are among several candidates running for class president. You estimate that there is a 40% chance you will win the election and a 35% chance your best friend will win. What is the probability that either you or your best friend wins the election? **0.75**

44. 🌎 **PARAKEETS** A pet store contains 35 light green parakeets (14 females and 21 males) and 44 sky blue parakeets (28 females and 16 males). You randomly choose one of the parakeets. What is the probability that it is a female or a sky blue parakeet? **0.734**

45. 🌎 **HONORS BANQUET** Of 162 students honored at an academic awards banquet, 48 won awards for mathematics and 78 won awards for English. Fourteen of these students won awards for both mathematics and English. One of the 162 students is chosen at random to be interviewed for a newspaper article. What is the probability that the student interviewed won an award for English or mathematics? **0.691**

46. [SCIENCE ▸ CONNECTION] A tree in a forest is not growing properly. A botanist determines that there is an 85% probability the tree has a disease or is being damaged by insects, a 45% probability it has a disease, and a 50% probability it is being damaged by insects. What is the probability that the tree both has a disease and is being damaged by insects? **0.1**

47. 🌎 **RAIN** A weather forecaster says that the probability it will rain on Saturday or Sunday is 50%, the probability it will rain on Saturday is 20%, and the probability it will rain on Sunday is 40%. What is the probability that it will rain on both Saturday and Sunday? **0.1**

48. 🌎 **POTLUCK DINNER** The organizer of a potluck dinner sends 5 people a list of 8 different recipes and asks each person to bring one of the items on the list. If all 5 people randomly choose a recipe from the list, what is the probability that at least 2 will bring the same thing? **0.795**

FOCUS ON CAREERS

→ **BOTANIST**
A botanist studies plants and their environment. Some botanists specialize in the causes and cures of plant illnesses, as discussed in Ex. 46.

CAREER LINK
www.mcdougallittell.com

50. *A*; *A'* is all outcomes not in *A*, so *A* has all outcomes not in *A'*.

49.  **CAMPUS HOUSING** Four high school friends will all be attending the same university next year. There are 14 dormitories on campus. Find the probability that at least 2 of the friends will be in the same dormitory.   **0.375**

50. CRITICAL THINKING What is the complement of *A'*? Explain.

51. 📱 **MULTI-STEP PROBLEM** Follow the steps below to explore a famous probability problem called the *birthday problem*.

   a. Suppose that 5 people are chosen at random. Find the probability that at least two share the same birthday (Assume that there are 365 possible birthdays). **0.0271**

   b. Suppose that 10 people are chosen at random. Find the probability that at least two of the people share the same birthday.   **0.117**

   c. Generalize the results from parts (a) and (b) by writing a formula that gives the probability $P(n)$ that in a group of n people at least two people share the same birthday. (*Hint:* Use $_nP_r$ notation.) $1 - \,_{365}P_n\left(\dfrac{1}{365^n}\right)$

   d. LOGICAL REASONING Enter the formula for $P(n)$ from part (c) into a graphing calculator put in sequence mode. Use the *Table* feature to make a table of values for $P(n)$. How large must a group be if the probability that at least two of the people share the same birthday exceeds 50%?
   23 (probability ≈ 0.507)

★ Challenge

ODDS **The odds in favor of an event occurring are the ratio of the probability that the event *will* occur to the probability that the event *will not* occur. The reciprocal of this ratio represents the odds *against* the event occurring.**

52. Five marbles in a jar are green. The odds against choosing a green marble are "4 to 1." How many marbles are in the jar?   **25**

53. If a jar contains 4 red marbles and 7 blue marbles, what are the odds in favor of choosing a red marble? What are the odds against choosing a red marble?
4 to 7; 7 to 4

54. $P(E) = \dfrac{\text{Odds in favor}}{1 + \text{Odds in favor}}$

54. Write a formula that converts the odds in favor of an event to the probability of the event.   **See margin.**

EXTRA CHALLENGE
www.mcdougallittell.com

55. Write a formula that converts the probability of an event to the odds in favor of the event.   Odds in favor of Event $E = \dfrac{P(E)}{P(E')} = \dfrac{P(E)}{1 - P(E)}$

STUDENT HELP

INTERNET **KEYSTROKE HELP** Visit our Web site www.mcdougallittell.com to see keystrokes for several models of calculators.

Test  **Preparation**

MIXED REVIEW

62. $x^2 + y^2 = 49$
63. $x^2 + y^2 = 25$
64. $x^2 + y^2 = 37$
65. $x^2 + y^2 = 68$
66. $x^2 + y^2 = 32$
67. $x^2 + y^2 = 109$
68. $x^2 + y^2 = 20$
69. $x^2 + y^2 = 256$

SOLVING EXPONENTIAL EQUATIONS **Solve the equation. (Review 8.6 for 12.5)**

56. $7^{2x} = 49^{16}$ **16**
57. $9^x = 3^{x+1}$ **1**
58. $2^{4x+8} = 32^{19}$ **21.75**

59. $5^x = 21$ **1.892**
60. $10^{3x-1} - 13 = 8$ **0.774**
61. $72 = 91e^{-0.023x} + 50$ **61.73**

WRITING EQUATIONS **Write the standard form of the equation of the circle that passes through the given point and whose center is the origin. (Review 10.3)**

62. $(0, 7)$
63. $(3, 4)$
64. $(-1, 6)$
65. $(8, -2)$

66. $(4, 4)$
67. $(3, 10)$
68. $(-4, -2)$
69. $(16, 0)$

LICENSE PLATES **For the given configuration, determine how many different license plates are possible if (a) digits and letters can be repeated, and (b) digits and letters cannot be repeated. (Review 12.1 for 12.5)**

70. 3 letters followed by 4 digits
   a. 175,760,000 **b.** 78,624,000

71. 4 letters followed by 3 digits
   a. 456,976,000 **b.** 258,336,000

DAILY HOMEWORK QUIZ

📖 *Transparency Available*

Find the indicated probability if $P(A) = 0.2$, $P(B) = 0.4$, and $P(A \text{ and } B) = 0.1$.

1. $P(A \text{ or } B)$ **0.5**
2. $P(A')$ **0.8**

Two 6-sided dice are rolled. Find the probability of the given event.

3. The sum is greater than 8.
$\dfrac{10}{36}$, or about 0.278

4. The sum is an even number less than 6.
$\dfrac{4}{36}$, or about 0.111

EXTRA CHALLENGE NOTE

→ Challenge problems for Lesson 12.4 are available in **blackline** format in the *Chapter 12 Resource Book*, p. 65 and at **www.mcdougallittell.com.**

ADDITIONAL TEST PREPARATION

1. WRITING For two events *A* and *B*, $P(A) = \dfrac{5}{27}$, $P(B) = \dfrac{1}{18}$, and $P(A \text{ or } B) = \dfrac{13}{54}$. Are *A* and *B* mutually exclusive? Explain.
Yes; since $P(A \text{ or } B) = P(A) + P(B)$, $P(A \text{ and } B) = 0$ so there must be no outcomes that are in both *A* and *B*.

2. OPEN ENDED Write a real-life problem you would solve using the probability of a complement. Find the probability of the event. **Check problems and answers.**

LESSON OPENER
APPLICATION
An alternative way to approach
Lesson 12.5 is to use the
Application Lesson Opener:
- Blackline Master (*Chapter 12 Resource Book*, p. 69)
- Transparency (p. 81)

MEETING INDIVIDUAL NEEDS
- *Chapter 12 Resource Book*
  Prerequisite Skills Review (p. 5)
  Practice Level A (p. 70)
  Practice Level B (p. 71)
  Practice Level C (p. 72)
  Reteaching with Practice (p. 73)
  Absent Student Catch-Up (p. 75)
  Challenge (p. 78)
- *Resources in Spanish*
- *Personal Student Tutor*

NEW-TEACHER SUPPORT
See the Tips for New Teachers on
pp. 1–2 of the *Chapter 12 Resource Book* for additional notes about
Lesson 12.5.

WARM-UP EXERCISES

 Transparency Available

One busy day, a drug store
serves 200 customers. Of the
200, 105 buy an over-the-counter
medication and 42 buy a prescription drug. Find the
probability that a randomly chosen customer buys each item.

1. an over the counter
medication $\frac{105}{200} = 0.525$

2. a prescription drug $\frac{42}{200} = 0.210$

12.5 Probability of Independent and Dependent Events

What you should learn

GOAL 1 Find the probability of independent events.

GOAL 2 Find the probability of dependent events, as applied in **Ex. 33**.

Why you should learn it

▼ To solve **real-life** problems, such as finding the probability that the Florida Marlins win three games in a row in **Example 2**.

GOAL 1 **PROBABILITIES OF INDEPENDENT EVENTS**

Two events are **independent** if the occurrence of one has no effect on the occurrence of the other. For instance, if a coin is tossed twice, the outcome of the first toss (heads or tails) has no effect on the outcome of the second toss.

PROBABILITY OF INDEPENDENT EVENTS

If A and B are independent events, then the probability that both A and B occur is $P(A \text{ and } B) = P(A) \cdot P(B)$.

EXAMPLE 1 *Probability of Two Independent Events*

You are playing a game that involves spinning the money wheel shown. During your turn you get to spin the wheel twice. What is the probability that you get more than $500 on your first spin and then go bankrupt on your second spin?

SOLUTION Let event A be getting more than $500 on the first spin, and let event B be going bankrupt on the second spin. The two events are independent. So, the probability is:

$$P(A \text{ and } B) = P(A) \cdot P(B) = \frac{8}{24} \cdot \frac{2}{24} = \frac{1}{36} \approx 0.028$$

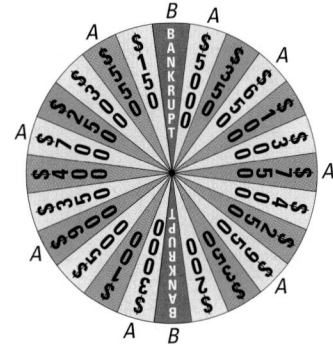

· · · · · · · · · ·

The formula given above for the probability of two independent events can be extended to the probability of three or more independent events.

EXAMPLE 2 *Probability of Three Independent Events*

BASEBALL During the 1997 baseball season, the Florida Marlins won 5 out of 7 home games and 3 out of 7 away games against the San Francisco Giants. During the 1997 National League Division Series with the Giants, the Marlins played the first two games at home and the third game away. The Marlins won all three games. Estimate the probability of this happening. ▶ Source: The Florida Marlins

SOLUTION Let events A, B, and C be winning the first, second, and third games. The three events are independent and have experimental probabilities based on the regular season games. So, the probability of winning the first three games is:

$$P(A \text{ and } B \text{ and } C) = P(A) \cdot P(B) \cdot P(C) = \frac{5}{7} \cdot \frac{5}{7} \cdot \frac{3}{7} = \frac{75}{343} \approx 0.219$$

Trading Cards

EXAMPLE 3   *Using a Complement to Find a Probability*

You collect hockey trading cards. For one team there are 25 different cards in the set, and you have all of them except for the starting goalie card. To try and get this card, you buy 8 packs of 5 cards each. All cards in a pack are different and each of the cards is equally likely to be in a given pack. Find the probability that you will get at least one starting goalie card.

SOLUTION

In one pack the probability of *not* getting the starting goalie card is:

$$P(\text{no starting goalie}) = \frac{_{24}C_5}{_{25}C_5}$$

Buying packs of cards are independent events, so the probability of getting at least one starting goalie card in the 8 packs is:

$$P(\text{at least one starting goalie}) = 1 - P(\text{no starting goalie in any pack})$$

$$= 1 - \left(\frac{_{24}C_5}{_{25}C_5}\right)^8$$

$$\approx 0.832$$

Manufacturing

EXAMPLE 4   *Solving a Probability Equation*

A computer chip manufacturer has found that only 1 out of 1000 of its chips is defective. You are ordering a shipment of chips for the computer store where you work. How many chips can you order before the probability that at least one chip is defective reaches 50%?

SOLUTION

Let n be the number of chips you order. From the given information you know that $P(\text{chip is } \textit{not} \text{ defective}) = \frac{999}{1000} = 0.999$. Use this probability and the fact that each chip ordered represents an independent event to find the value of n.

$P(\text{at least one chip is defective}) = 0.5$	**Write given assumption.**
$1 - P(\text{no chips are defective}) = 0.5$	**Use complement.**
$1 - (0.999)^n = 0.5$	**Substitute known probability.**
$-(0.999)^n = -0.5$	**Subtract 1 from each side.**
$(0.999)^n = 0.5$	**Divide each side by –1.**
$n = \dfrac{\log 0.5}{\log 0.999}$	**Solve for n.**
$n \approx 693$	**Use a calculator.**

▶ If you order 693 chips, you have a 50% chance of getting a defective chip. Therefore, you can order 692 chips before the probability that at least one chip is defective reaches 50%.

STUDENT HELP

Look Back
For help with solving exponential equations, see p. 501.

---

2 TEACH

 EXTRA EXAMPLE 1
A game machine claims that 1 in every 15 people win. What is the probability that you win twice in a row? $\frac{1}{225} \approx 0.004$

EXTRA EXAMPLE 2
In a survey 9 out of 11 men and 4 out of 7 women said they were satisfied with a product. If the next 3 customers are 2 women and a man, what is the probability that they will all be satisfied?
$\frac{144}{539} \approx 0.267$

EXTRA EXAMPLE 3
Refer to the survey above. If 4 men are the next customers, what is the probability that at least one of them is not satisfied with the product?
$1 - \left(\frac{9}{11}\right)^4 \approx 0.552$

EXTRA EXAMPLE 4
An auto repair company finds that 1 in 100 cars has had to return for the same problem. How many times can you bring your car to this company before the probability that you have to have your car repaired for the same problem at least once reaches 75%?   about 138 times

 CHECKPOINT EXERCISES
For use after Examples 1 and 2:
1. A bag contains 3 red marbles, 7 white marbles, and 5 blue marbles. You draw 3 marbles, replacing each one before drawing the next. What is the probability of drawing a red, then a blue, and then a white marble? $\frac{7}{225} \approx 0.031$

For use after Examples 3 and 4:
2. 1 in 10,000 cars has a defect. How many of these cars can a car dealer sell before the probability of selling at least one with a defect is 20%?
2231 cars

731

EXTRA EXAMPLE 5

The table shows the camp attendance for three age groups of students in one town. Find (a) the probability that a listed student attended camp and (b) the probability that a child in the 8–10 age group from the town did not attend camp.

Age	Attended Camp	No Camp
5–7	45	117
8–10	94	62
11–13	81	79

$\frac{220}{478} \approx 0.460$; $\frac{62}{156} \approx 0.397$

EXTRA EXAMPLE 6

You randomly select two cards from a standard 52-card deck. Find the probability that the first card is a diamond and the second card is red if (a) you replace the first card before selecting the second, and (b) you do not replace the first card.

$\frac{1}{8} = 0.125$; $\frac{25}{204} \approx 0.123$

✓ CHECKPOINT EXERCISES

Refer to Extra Example 5.

For use after Example 5:

1. What is the probability that a listed student was in the 11–13 age bracket?

$\frac{160}{478} \approx 0.335$

For use after Example 6:

2. What is the probability that a listed student did not attend camp? What is the probability that a student in the 11–13 age bracket did not attend camp?

$\frac{258}{478} \approx 0.540$; $\frac{79}{160} \approx 0.494$

STUDENT HELP NOTES

Look Back As students look back to p. 708, remind them that a standard deck of cards has 4 suits — spades, hearts, diamonds, and clubs — each containing 13 cards.

732

GOAL 2 PROBABILITIES OF DEPENDENT EVENTS

Two events A and B are **dependent events** if the occurrence of one affects the occurrence of the other. The probability that B will occur given that A has occurred is called the **conditional probability** of B given A and is written $P(B \mid A)$.

PROBABILITY OF DEPENDENT EVENTS

If A and B are dependent events, then the probability that both A and B occur is
$P(A \text{ and } B) = P(A) \cdot P(B \mid A)$.

Endangered Species

EXAMPLE 5 *Finding Conditional Probabilities*

The table shows the number of endangered and threatened animal species in the United States as of November 30, 1998. Find (a) the probability that a listed animal is a reptile and (b) the probability that an endangered animal is a reptile.

▶ Source: United States Fish and Wildlife Service

	Mammals	Birds	Reptiles	Amphibians	Other
Endangered	59	75	14	9	198
Threatened	8	15	21	7	69

SOLUTION

a. $P(\text{reptile}) = \dfrac{\text{number of reptiles}}{\text{total number of animals}} = \dfrac{35}{475} \approx 0.0737$

b. $P(\text{reptile} \mid \text{endangered}) = \dfrac{\text{number of endangered reptiles}}{\text{total number of endangered animals}}$

$= \dfrac{14}{355} \approx 0.0394$

EXAMPLE 6 *Comparing Dependent and Independent Events*

┌─ STUDENT HELP
└▶ **Look Back**
For help with a standard 52-card deck, see p. 708.

You randomly select two cards from a standard 52-card deck. What is the probability that the first card is not a face card (a king, queen, or jack) and the second card is a face card if (a) you replace the first card before selecting the second, and (b) you do *not* replace the first card?

SOLUTION

a. If you replace the first card before selecting the second card, then A and B are independent events. So, the probability is:

$$P(A \text{ and } B) = P(A) \cdot P(B) = \frac{40}{52} \cdot \frac{12}{52} = \frac{30}{169} \approx 0.178$$

b. If you do *not* replace the first card before selecting the second card, then A and B are dependent events. So, the probability is:

$$P(A \text{ and } B) = P(A) \cdot P(B \mid A) = \frac{40}{52} \cdot \frac{12}{51} = \frac{40}{221} \approx 0.181$$

The formula for finding probabilities of dependent events can be extended to three or more events, as shown in Example 7.

Dining Out

EXAMPLE 7 *Probability of Three Dependent Events*

You and two friends go to a restaurant and order a sandwich. The menu has 10 types of sandwiches and each of you is equally likely to order any type. What is the probability that each of you orders a different type?

SOLUTION

Let event A be that you order a sandwich, event B be that one friend orders a different type, and event C be that your other friend orders a third type. These events are dependent. So, the probability that each of you orders a different type is:

$$P(A \text{ and } B \text{ and } C) = P(A) \cdot P(B \mid A) \cdot P(C \mid A \text{ and } B)$$

$$= \frac{10}{10} \cdot \frac{9}{10} \cdot \frac{8}{10}$$

$$= \frac{18}{25} = 0.72$$

STUDENT HELP

↪ **Study Tip**
You can also use the fundamental counting principle to find the probability in Example 7.

$P(\text{all different})$

$= \dfrac{\text{no. of different orders}}{\text{no. of possible orders}}$

$= \dfrac{10 \cdot 9 \cdot 8}{10 \cdot 10 \cdot 10} = 0.72$

EXAMPLE 8 *Using a Tree Diagram to Find Conditional Probabilities*

HEALTH The American Diabetes Association estimates that 5.9% of Americans have diabetes. Suppose that a medical lab has developed a simple diagnostic test for diabetes that is 98% accurate for people who have the disease and 95% accurate for people who do not have it. If the medical lab gives the test to a randomly selected person, what is the probability that the diagnosis is correct?

SOLUTION

A probability tree diagram, where the probabilities are given along the branches, can help you see the different ways to obtain a correct diagnosis. Notice that the probabilities for all branches from the same point must sum to 1.

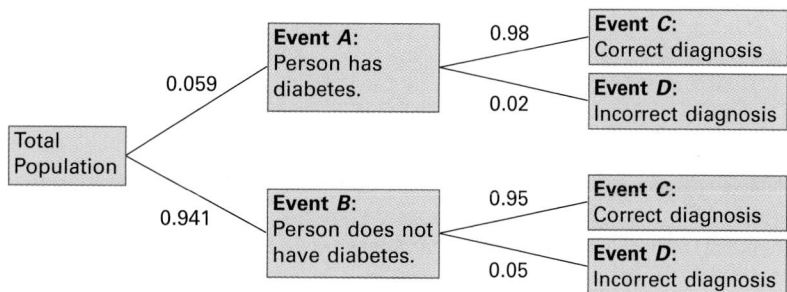

So, the probability that the diagnosis is correct is:

$P(C) = P(A \text{ and } C) + P(B \text{ and } C)$ Follow branches leading to C.

$= P(A) \cdot P(C \mid A) + P(B) \cdot P(C \mid B)$ Use formula for dependent events.

$= (0.059)(0.98) + (0.941)(0.95)$ Substitute.

≈ 0.952 Use a calculator.

EXTRA EXAMPLE 7
Three children have a choice of 12 summer camps that they can attend. If they each randomly choose which camp to attend, what is the probability that they attended all different camps? $\frac{110}{144} \approx 0.764$

EXTRA EXAMPLE 8
In one town 95% of the students graduate from high school. Suppose a study showed that at age 25, 81% of the high school graduates held full-time jobs while only 63% of those who did not graduate held full-time jobs. What is the probability that a randomly selected student from the town will have a full-time job at age 25? 0.801

✓ CHECKPOINT EXERCISES
For use after Example 7:
1. A family of 4 is each choosing 1 of 8 possible vacations. What is the probability that each family member picks a different vacation for his or her first choice? $\frac{210}{512} \approx 0.410$

For use after Example 8:
2. Suppose a survey of high school students showed that 47% of them worked during the summer. Of those who worked, 62% said they watched 2 hours or more of television per day during the summer. Of those who did not work, 79% watched 2 hours or more. What is the probability that a randomly chosen high school student watched fewer than 2 hours of television during the summer? 0.290

CLOSURE QUESTION
A basketball player is shooting two free throws. Let event A be making the first free throw and event B be making the second free throw. Are events A and B independent or dependent events? *Sample answer:* These are independent events.

ASSIGNMENT GUIDE

BASIC
Day 1: pp. 734–737 Exs. 12–16,
18–22, 24–25, 30–32,
37–38, 41–51 odd,
Quiz 2 Exs. 1–10

AVERAGE
Day 1: pp. 734–737 Exs. 12–16,
18–22, 24–28, 30–32,
35–38, 41–51 odd,
Quiz 2 Exs. 1–10

ADVANCED
Day 1: pp. 734–737 Exs. 12–22,
24–28, 30–39, 41–51 odd,
Quiz 2 Exs. 1–10

BLOCK SCHEDULE
pp. 734–737 Exs. 12–16, 18–22,
24–28, 30–32, 35–38, 41–51 odd,
Quiz 2 Exs. 1–10 (with 12.6)

EXERCISE LEVELS

Level A: *Easier*
12–17, 24–29

Level B: *More Difficult*
18–21, 30–34, 37–38

Level C: *Most Difficult*
35–36, 39

✓ HOMEWORK CHECK

To quickly check student understanding of key concepts, go
over the following exercises:
Exs. 14, 16, 18, 22, 24. See also
the Daily Homework Quiz:

• Blackline Master (*Chapter 12 Resource Book*, p. 82)

• 📖 Transparency (p. 94)

GUIDED PRACTICE

Vocabulary Check ✓
1. Explain the difference between dependent events and independent events, and give an example of each.

Concept Check ✓
2. If event A is drawing a queen from a deck of cards and event B is drawing a king from the remaining cards, are events A and B dependent or independent? **dependent**

3. If event A is rolling a two on a six-sided die and event B is rolling a four on a different six-sided die, are events A and B dependent or independent? **independent**

Skill Check ✓

1. Events are independent if the occurrence of one has no effect on the occurrence of the other. Events are dependent if the occurrence of one affects the occurrence of the other. *Sample answer:* Rolls of a die are independent events because the number you get on the first roll does not affect the number on the second roll. The sum of the numbers on two rolls of a die is dependent. For example, if the first roll is not a six, there is a 0 probability that the sum will be twelve.

Events A and B are independent. Find the indicated probability.

4. $P(A) = 0.3$ **0.27**
   $P(B) = 0.9$
   $P(A \text{ and } B) = \underline{?}$

5. $P(A) = \underline{?}$ **0.2**
   $P(B) = 0.3$
   $P(A \text{ and } B) = 0.06$

6. $P(A) = 0.75$ **0.2**
   $P(B) = \underline{?}$
   $P(A \text{ and } B) = 0.15$

Events A and B are dependent. Find the indicated probability.

7. $P(A) = 0.1$ **0.08**
   $P(B \mid A) = 0.8$
   $P(A \text{ and } B) = \underline{?}$

8. $P(A) = \underline{?}$ **0.5**
   $P(B \mid A) = 0.5$
   $P(A \text{ and } B) = 0.25$

9. $P(A) = 0.9$ **0.6**
   $P(B \mid A) = \underline{?}$
   $P(A \text{ and } B) = 0.54$

🌐 **READING LIST** In Exercises 10 and 11, use the following information.
Three friends are taking an English class that has a summer reading list. Each student is required to read one book from the list, which contains 3 biographies, 10 classics, and 5 historical novels.

10. Find the probability that the first friend chooses a biography, the second friend chooses a classic, and the third friend chooses a historical novel. **0.026**

11. Find the probability that the three friends each choose a different classic. **0.123**

PRACTICE AND APPLICATIONS

STUDENT HELP
↳ **Extra Practice**
to help you master
skills is on p. 957.

SPINNING A WHEEL You are playing a game that involves spinning the wheel shown. Find the probability of spinning the given colors.

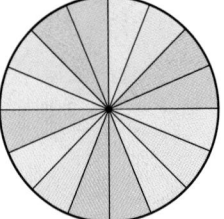

12. red, then blue **0.047**
13. red, then green **0.047**
14. yellow, then red **0.059**
15. green, then yellow **0.078**
16. blue, then yellow, then green **0.020**
17. green, then red, then blue **0.012**

STUDENT HELP
↳ **HOMEWORK HELP**
Example 1: Exs. 12–15
Example 2: Exs. 16, 17, 24, 25
Example 3: Exs. 26, 27
Example 4: Exs. 28, 29
Example 5: Exs. 30–32
Example 6: Exs. 18–21
Example 7: Exs. 22, 23, 33, 34
Example 8: Exs. 35, 36

DRAWING CARDS Find the probability of drawing the given cards from a standard 52-card deck (a) with replacement and (b) without replacement.

18. a heart, then a diamond
    a. 0.0625 b. 0.0637
19. a jack, then a king a. 0.0059 b. 0.0060
20. a 2, then a face card (K, Q, or J)
    a. 0.0178 b. 0.0181
21. a face card (K, Q, or J), then a 2
    a. 0.0178 b. 0.0181
22. an ace, then a 2, then a 3
    a. 0.00046 b. 0.00048
23. a heart, then a diamond, then another heart
    a. 0.0156 b. 0.0153
24. **GAMES** You are playing a game that involves drawing three numbers from a hat. There are 25 pieces of paper numbered 1 to 25 in the hat. Each number is replaced after it is drawn. What is the probability that each number is greater than 20 or less than 4? **0.0328**

25.  **LAWN CARE** The owner of a one-man lawn mowing business owns three old and unreliable riding mowers. As long as one of the three is working he can stay productive. From past experience, one of the mowers is unusable 12 percent of the time, one 6 percent of the time, and one 20 percent of the time. Find the probability that all three mowers are unusable on a given day. **0.00144**

26. **TRADING CARDS** You collect movie trading cards, which have different scenes from a movie. For one movie there are 90 different cards in the set, and you have all of them except the final scene. To try and get this card, you buy 10 packs of 8 cards each. All cards in a pack are different and each of the cards is equally likely to be in a given pack. Find the probability that you will get the final scene. **0.606**

27. **FREE THROWS** Chris Mullin of the Indiana Pacers led the National Basketball Association in free-throw percentage during the 1997–1998 season. He made 93.9% of his free-throw attempts. If he attempted 10 free throws in a game, what is the probability that he missed at least one? ▶ Source: NBA **0.467**

28. **MANUFACTURING** Look back at Example 4. Suppose the computer chip manufacturer has improved quality control so that only 1 out of 10,000 of its chips is defective. Now how many chips can you order before the probability that at least one chip is defective reaches 50%? **6931**

29. **LOTTERY** To win a state lottery, a player must correctly match six different numbers from 1 to 42. If a computer randomly assigns six numbers per ticket, how many tickets would a person have to buy to have a 1% chance of winning? **at least 52,722 tickets**

STATISTICS ▶ CONNECTION In Exercises 30 and 31, use the following information. The table, based on a Gallup Poll, shows the number of voters (in 1000's) by party affiliation who were expected to vote for Bill Clinton and Bob Dole in the 1996 Presidential election. ▶ Source: The Gallup Organization

	Democrat	Republican	Independent
Clinton	31,378	3,340	12,685
Dole	2,092	28,386	8,721

30. Find the probability that a randomly selected person voted for Clinton. **0.547**

31. Find the probability that a randomly selected Democrat voted for Clinton. **0.937**

32. **TEACHERS** In the United States during the 1993–1994 school year, 39.6% of all male teachers and 26.1% of all female teachers had twenty years or more of full-time teaching experience. That year 694,000 males and 1,867,000 females were teachers. What is the probability that a randomly chosen teacher in the United States that year was a female with twenty years or more of full-time teaching experience? **0.19**

DATA UPDATE of *Statistical Abstract of the United States* data at www.mcdougallittell.com

33. **COSTUMES** You and four of your friends go to the same store at different times to buy costumes for a costume party. There are 20 different costumes at the store, and the store has at least five duplicates of each costume. Find the probability that all five of you choose different costumes. **0.581**

34. **AIRPLANE MEALS** On a long flight an airline usually serves a meal. If there are 2 choices for the meal, what is the probability that all 6 people in the first row choose the same meal assuming choices are made independently? **0.0156**

STUDENT HELP

HOMEWORK HELP
Visit our Web site
www.mcdougallittell.com
for help with problem
solving in Ex. 29.

FOCUS ON CAREERS

TEACHER
In addition to teaching, teachers plan daily lessons and activities, assign and correct homework, and prepare and grade exams. They also monitor homerooms, study halls, and cafeterias, meet with parents, and supervise extracurricular activities.

CAREER LINK
www.mcdougallittell.com

! COMMON ERROR

EXERCISE 24 Some students need extra help differentiating between independent and dependent events. You may want to have them complete this exercise with replacing the slips of paper and without replacing the slips of paper to help them see the difference.

STUDENT HELP NOTES

Homework Help Students can find help for Ex. 29 at **www.mcdougallittell.com**. The information can be printed out for students who don't have access to the Internet.

DATA UPDATE
Updated data for Ex. 32 is available at **www.mcdougallittell.com**.

CAREER NOTE
Additional information about a career as a teacher is available at **www.mcdougallittell.com**.

ADDITIONAL PRACTICE AND RETEACHING

For Lesson 12.5:
• Practice Levels A, B, and C (*Chapter 12 Resource Book*, p. 70)
• Reteaching with Practice (*Chapter 12 Resource Book*, p. 73)
• See Lesson 12.5 of the *Personal Student Tutor*

For more Mixed Review:
• Search the *Test and Practice Generator* for key words or specific lessons.

DAILY HOMEWORK QUIZ

Transparency Available

Find the probability of drawing the given cards from a standard 52-card deck (**a**) with replacement and (**b**) without replacement.

1. a diamond, then a black card
0.125, 0.127

2. a 6, then a jack   0.0059, 0.006

EXTRA CHALLENGE NOTE

Challenge problems for Lesson 12.5 are available in **blackline** format in the *Chapter 12 Resource Book,* p. 78 and at **www.mcdougallittell.com.**

ADDITIONAL TEST PREPARATION

1. WRITING Suppose $P(A) = P(A \mid B)$. Are events A and B independent or dependent? Explain.   **independent; if A were dependent on B, then the occurrence of B would affect the occurrence of A so $P(A \mid B)$ would not equal $P(A)$.**

2. OPEN ENDED Write a real-life problem that you could solve using a tree diagram. Draw the diagram and show how to solve the problem.   **Check tree diagrams, problems, and solutions.**

40.
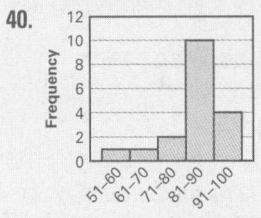

41.
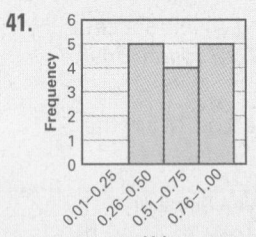

---

35. **RETIREMENT PLAN** At a particular company 64% of the employees are forty years old or over. Of those employees, 83% are enrolled in the company's retirement plan. Only 61% of the employees under forty years old are enrolled in the plan. Make a probability tree diagram and use it to find the probability that a randomly selected employee is enrolled in the company's retirement plan.   **0.751**

36. **FOCUS TESTING** A company is focus testing a new type of fruit drink. The focus group is 47% male. Of the males in the group, 40% said they would buy the fruit drink, and of the females, 54% said they would buy the fruit drink. Make a probability tree diagram and use it to find the probability that a randomly selected person would buy the fruit drink.   **0.474**

Test Preparation

QUANTITATIVE COMPARISON In Exercises 37 and 38, choose the statement that is true about the given quantities. Assume a standard 52-card deck is used.

(**A**) The quantity in column A is greater.

(**B**) The quantity in column B is greater.

(**C**) The two quantities are equal.

(**D**) The relationship cannot be determined from the given information.

	Column A	Column B
37.	The probability of drawing at least one heart when drawing 4 times with replacement	The probability of drawing at least one heart when drawing 4 times without replacement **B**
38.	The probability of not drawing any card twice when drawing 9 times with replacement	The probability of drawing any card 2 or more times when drawing 9 times with replacement **B**

★ Challenge

39. CONDITIONAL PROBABILITY Using the data from Example 8, find the conditional probability that a randomly selected person has diabetes given that the person is diagnosed incorrectly.   **0.0245**

MIXED REVIEW

40.	51–60	1
	61–70	1
	71–80	2
	81–90	10
	91–100	4
41.	0.01–0.25	0
	0.26–0.50	5
	0.51–0.75	4
	0.76–1.0	5

HISTOGRAMS Using the given intervals, make a frequency distribution of the data set. Then draw a histogram of the data set. (Review 7.7 for 12.6)
40–41. See margin for graphs.
40. Use five intervals beginning with 51–60.
56, 68, 73, 79, 82, 82, 83, 85, 85, 87, 88, 89, 90, 90, 91, 93, 95, 100

41. Use four intervals beginning with 0.01–0.25.
0.3, 0.3, 0.4, 0.4, 0.5, 0.6, 0.6, 0.6, 0.7, 0.9, 0.9, 0.9, 0.9, 0.9

SOLVING EQUATIONS Solve the equation. (Review 8.8)

42. $\dfrac{5}{1 + 2e^{-x}} = 4$ **2.08**   **43.** $\dfrac{10}{1 + 6e^{-x}} = 8$ **3.178**   **44.** $\dfrac{9}{1 + 3e^{-3x}} = 6$ **0.597**

45. $\dfrac{12}{1 + e^{-2x}} = 4$ **−0.347**   **46.** $\dfrac{1}{1 + 3e^{-5x}} = \dfrac{1}{2}$ **0.220**   **47.** $\dfrac{70}{1 + 12e^{-10x}} = \dfrac{2}{9}$ **−0.326**

BINOMIAL THEOREM Expand the power of the binomial. (Review 12.2 for 12.6)
48–51. See margin.
48. $(x + 1)^5$ **49.** $(2x - 1)^7$ **50.** $(x - 3y)^4$ **51.** $(x - 1)^6$

1. $\frac{12}{25}$, or 0.48

2. $\frac{6}{25}$, or 0.24

3. $\frac{18}{25}$, or 0.72

A jar contains 6 blue marbles, 12 green marbles, and 7 yellow marbles. Find the probability of randomly drawing the given marble. (Lesson 12.3)

1. a green marble 　　　　 2. a blue marble 　　　　 3. a green or a blue marble

Find the probability that a dart thrown at the circular target shown will hit the given region. Assume the dart is equally likely to hit any point inside the target. (Lesson 12.3)

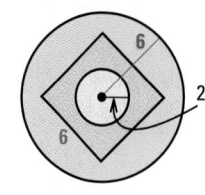

4. the center circle　0.111 　　　 5. outside the square　0.682

6. the area inside the square but outside the center circle　0.207

Find the indicated probability. (Lesson 12.4)

7. $P(A) = 0.7$　0.8
   $P(B) = 0.2$
   $P(A \text{ or } B) = \underline{?}$
   $P(A \text{ and } B) = 0.1$

8. $P(A) = 0.5$　0
   $P(B) = 0.4$
   $P(A \text{ or } B) = 0.9$
   $P(A \text{ and } B) = \underline{?}$

9. $P(A) = 0.25$　0.75
   $P(A') = \underline{?}$

10.  **FRUIT** You and four friends are in line at lunch and are each selecting a piece of fruit to eat. If there are 5 types of fruit available, what is the probability that you each select a different type? (Lesson 12.5)　0.0384

MATH & History ▶ Probability Theory

APPLICATION LINK
www.mcdougallittell.com

THEN

IN 1654 Blaise Pascal and Pierre Fermat solved the first probability problem, which asked how to divide the stakes of an interrupted game of chance between two players of equal ability. Suppose Player A has 2 points, Player B has 1 point, and the game is won by the first player to score 4 out of a possible 7 points.

1. If the stakes are divided based on which player is closest to winning, what fraction of the stakes should each player receive?

Since A is closer to winning, A should receive all of the stakes and B should receive none of the stakes.

2. If the stakes are divided based on the number of points each player has so far, what fraction of the stakes should each player receive? **See margin.**

3. What is the probability Player A will win the game? What is the probability Player B will win the game? Based on these probabilities, what fraction of the stakes should each player receive? **See margin.**

NOW

TODAY probability theory is used to decide more than just games of chance. Actuaries, for example, use probability theory to design financial plans, to calculate insurance rates, and to price corporate securities offerings.

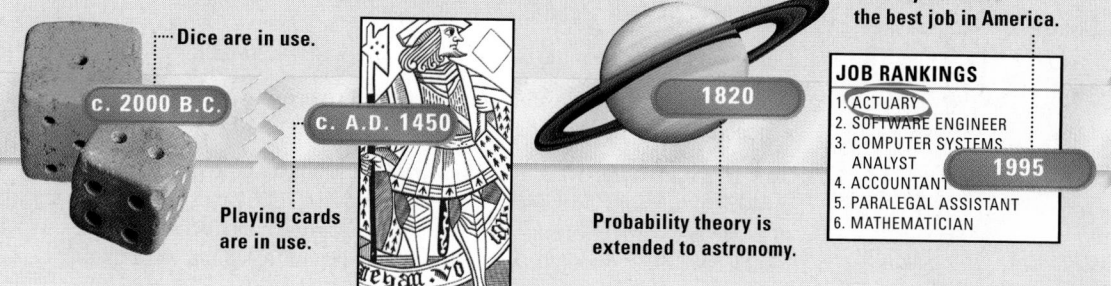

Dice are in use. — c. 2000 B.C.

Playing cards are in use. — c. A.D. 1450

Probability theory is extended to astronomy. — 1820

Actuary is named the best job in America. — 1995

JOB RANKINGS
1. ACTUARY
2. SOFTWARE ENGINEER
3. COMPUTER SYSTEMS ANALYST
4. ACCOUNTANT
5. PARALEGAL ASSISTANT
6. MATHEMATICIAN

12.5 *Probability of Independent and Dependent Events* 　**737**

ADDITIONAL RESOURCES
An alternative Quiz for Lessons 12.3–12.5 is available in the *Chapter 12 Resource Book*, p. 79.

A **blackline** master with additional Math & History exercises is available in the *Chapter 12 Resource Book*, p. 77.

48. $x^5 + 5x^4 + 10x^3 + 10x^2 + 5x + 1$

49. $128x^7 - 448x^6 + 672x^5 - 560x^4 + 280x^3 - 84x^2 + 14x - 1$

50. $x^4 - 12x^3y + 54x^2y^2 - 108xy^3 + 81y^4$

51. $x^6 - 6x^5 + 15x^4 - 20x^3 + 15x^2 - 6x + 1$

2. Since A has double the number of points, A should receive $\frac{2}{3}$ and B should receive $\frac{1}{3}$.

3. There are 10 ways in which the game can end: AA, ABA, BAA, ABBA, BABA, BBAA, BBB, BBAB, BABB, ABBB. However, the outcomes are not equally likely. For the outcome AA (with two more points scored), the probability is $\frac{1}{2} \cdot \frac{1}{2} = \frac{1}{4}$. For each outcome with three more points scored, the probability is $\frac{1}{2} \cdot \frac{1}{2} \cdot \frac{1}{2} = \frac{1}{8}$. For each outcome with four more points scored, the probability is $\left(\frac{1}{2}\right)^4 = \frac{1}{16}$. $P(\text{B wins}) = \frac{1}{8} + \frac{1}{16} + \frac{1}{16} + \frac{1}{16} = \frac{5}{16}$, so $P(\text{A wins}) = 1 - \frac{5}{16} = \frac{11}{16}$. A should get $\frac{11}{16}$ of the winnings and B should get $\frac{5}{16}$.

1 Planning the Activity

PURPOSE

To perform a binomial experiment and record the results in a histogram.

MATERIALS

- coins or a graphing calculator (per group)
- Activity Support Master (*Chapter 12 Resource Book,* p. 83)

PACING

- Exploring the Concept — 20 min
- Drawing Conclusions — 10 min

▶ **LINK TO LESSON**
Refer back to this activity when discussing the definition of binomial experiment on page 739. Ask students how the coin tossing activity meets each of the criteria for a binomial experiment.

2 Managing the Activity

CLASSROOM MANAGEMENT

You may want to pass out a sheet with the axes to be used to make the histograms. Students' histograms will thus all have the same scale and can be compared by different groups.

ALTERNATIVE APPROACH

Students can toss the coins individually and record their results in a table on the chalkboard. The histogram can then be made as a class using an overhead projector.

3 Closing the Activity

★ **KEY DISCOVERY**
The graph of a binomial distribution has a particular shape.

ACTIVITY ASSESSMENT

Describe the general shape of the histograms in this activity.
See sample answer at right.

Step 3–6. See Additional Answers
  beginning on page AA1.

▶ **ACTIVITY 12.6**

Developing Concepts

GROUP ACTIVITY
Work in a small group.

MATERIALS
- coins or a graphing
- calculator

1. **Sample Answer:** The shapes of the histograms of the experimental results and the theoretical probabilities look very similar. The theoretical probability histogram is perfectly symmetric and the experimental results are nearly, but not quite symmetric.

2. With more trials, the experimental results histogram would tend to look even more like the theoretical probability histogram. This is because with larger values of *n* you are likely to get results that are less skewed and closer to the theoretical distribution.

3. A histogram of results using weighted coins would differ from those using regular coins because the probabilities would be different. Since heads (60%) are more likely than tails (40%), you would expect to get more trials with one, two and three heads and fewer with 0 heads. This histogram would not be symmetric, or nearly symmetric, like the two histograms in steps 3 and 6.

Investigating Binomial Distributions

▶ **QUESTION** What are the characteristics of a histogram that displays the results of a *binomial experiment*?

▶ **EXPLORING THE CONCEPT** Performing an Experiment

1 Toss three coins and record the number of heads. Repeat this 25 times.

You can also use a random number generator on a graphing calculator to simulate the coin toss. Let 0 represent heads and let 1 represent tails.

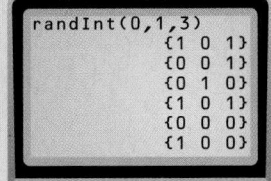

2 Record the results for the entire class in a table like the one shown below.

Number of Heads			
0	1	2	3
𝍇𝍇𝍇 𝍇𝍇	𝍇𝍇𝍇𝍇𝍇 𝍇𝍇𝍇𝍇	𝍇𝍇𝍇𝍇𝍇 𝍇𝍇𝍇 𝍇	𝍇𝍇𝍇

3 Construct a histogram for the class results for the experiment. Step 3–6. See margin.

▶ **EXPLORING THE CONCEPT** Calculating Theoretical Probabilities

4 Make a tree diagram of the results of tossing three coins simultaneously.

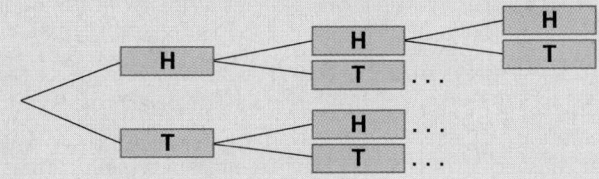

5 Calculate the theoretical probabilities of getting 0, 1, 2, or 3 heads.

6 Construct a histogram of the theoretical probabilities.

▶ **DRAWING CONCLUSIONS**

1. Compare the shapes of the two histograms from **Steps 3 and 6**.

2. Would the shape of the histogram from **Step 3** change if your class had conducted twice as many trials as it did? Explain.

3. Your friend has three weighted coins that land heads up 60% of the time and tails up 40% of the time. Suppose these coins are tossed 1000 times and the number of heads is recorded for each trial. Would a histogram constructed from the results have the same shape as the histograms from **Steps 3 and 6**? Explain.

738 Chapter 12 *Probability and Statistics*

Activity Assessment *Sample answer:*
The histogram is tallest in the middle and progressively shorter toward the left and right edges.

12.6

Binomial Distributions

What you should learn

GOAL 1 Find binomial probabilities and analyze binomial distributions.

GOAL 2 Test a hypothesis, as applied in **Example 4**.

Why you should learn it

▼ To solve **real-life** problems, such as determining whether a computer manufacturer's claim is correct in **Ex. 46**.

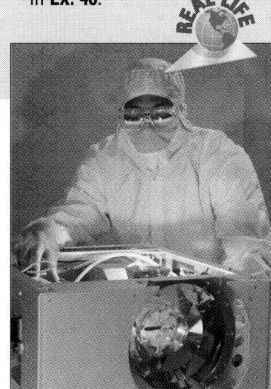

GOAL 1 FINDING BINOMIAL PROBABILITIES

There are many probability experiments in which the results of each trial can be reduced to two outcomes. If such an experiment satisfies the following conditions, then it is called a **binomial experiment**.

- There are n independent trials.

- Each trial has only two possible outcomes: success and failure.

- The probability of success is the same for each trial. This probability is denoted by p. The probability of failure is given by $1 - p$.

FINDING A BINOMIAL PROBABILITY

For a binomial experiment consisting of n trials, the probability of exactly k successes is

$$P(k \text{ successes}) = {}_nC_k p^k (1 - p)^{n - k}$$

where the probability of success on each trial is p.

EXAMPLE 1 *Finding a Binomial Probability*

UFOs According to a survey taken by *USA Today*, about 37% of adults believe that Unidentified Flying Objects (UFOs) really exist. Suppose you randomly survey 6 adults. What is the probability that exactly 2 of them believe that UFOs really exist?

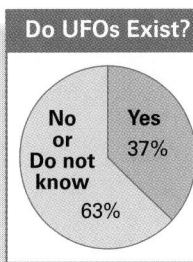

Do UFOs Exist?

No or Do not know 63%

Yes 37%

SOLUTION

Let $p = 0.37$ be the probability that a randomly selected adult believes that UFOs really exist. By surveying 6 adults, you are conducting $n = 6$ independent trials. The probability of getting exactly $k = 2$ successes is:

$$P(k = 2) = {}_6C_2 (0.37)^2 (1 - 0.37)^{6 - 2}$$

$$= \frac{6!}{4! \cdot 2!} (0.37)^2 (0.63)^4$$

$$\approx 0.323$$

▶ The probability that exactly 2 of the people surveyed believe that UFOs really exist is about 32%.

· · · · · · · · · ·

A **binomial distribution** shows the probabilities of all possible numbers of successes in a binomial experiment, as illustrated in Example 2 on the next page.

1 PLAN

PACING
Basic: 1 day
Average: 1 day
Advanced: 1 day
Block Schedule: 0.5 block with 12.5

LESSON OPENER
ACTIVITY
An alternative way to approach Lesson 12.6 is to use the Activity Lesson Opener:

- Blackline Master (*Chapter 12 Resource Book*, p. 84)
- Transparency (p. 82)

MEETING INDIVIDUAL NEEDS
- *Chapter 12 Resource Book*
  Prerequisite Skills Review (p. 5)
  Practice Level A (p. 86)
  Practice Level B (p. 87)
  Practice Level C (p. 88)
  Reteaching with Practice (p. 89)
  Absent Student Catch-Up (p.91)
  Challenge (p. 93)
- *Resources in Spanish*
- *Personal Student Tutor*

NEW-TEACHER SUPPORT
See the Tips for New Teachers on pp. 1–2 of the *Chapter 12 Resource Book* for additional notes about Lesson 12.6.

WARM-UP EXERCISES
Transparency Available
Decide whether the events are independent or dependent.

1. spinning a wheel 2 times
   independent
2. flipping a coin 2 times
   independent
3. drawing 2 cards from a standard 52-card deck without replacing them dependent
4. drawing 2 marbles out of a jar of different colored marbles if the first is replaced before the second is drawn independent

EXTRA EXAMPLE 1
At a college, 53% of students receive financial aid. In a random group of 9 students, what is the probability that exactly 5 of them receive financial aid?   **about 26%**

EXTRA EXAMPLE 2
Draw a histogram of the binomial distribution for the class of students in Extra Example 1 and find the probability that fewer than 3 students in the class receive financial aid.   **0.064**

EXTRA EXAMPLE 3
In one community, the probability that a child has a pet is 0.41. Draw a histogram of the binomial distribution based on the probability that exactly k children in a class of 10 have a pet and then find the most likely number of children in the class that have a pet.   **4 children**

☑ **CHECKPOINT EXERCISES**
For use after Examples 1–3:
1. In one community, 73% of women work full-time outside of the home. 14 women live in an apartment complex. Draw a histogram of the binomial distribution based on the probability that exactly k of the women work full-time outside of the home. Then find the probability that 10 or more of the women work outside the home, and the most likely number of women in the group that work outside the home.   **Check graphs; 0.681; 10 women**

 **UFOs**

EXAMPLE 2 | *Constructing a Binomial Distribution*

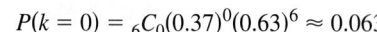

Draw a histogram of the binomial distribution for the survey in Example 1. Then find the probability that at most 2 of the people surveyed believe that UFOs really exist.

SOLUTION

$$P(k = 0) = {}_6C_0(0.37)^0(0.63)^6 \approx 0.063$$
$$P(k = 1) = {}_6C_1(0.37)^1(0.63)^5 \approx 0.220$$
$$P(k = 2) = {}_6C_2(0.37)^2(0.63)^4 \approx 0.323$$
$$P(k = 3) = {}_6C_3(0.37)^3(0.63)^3 \approx 0.253$$
$$P(k = 4) = {}_6C_4(0.37)^4(0.63)^2 \approx 0.112$$
$$P(k = 5) = {}_6C_5(0.37)^5(0.63)^1 \approx 0.026$$
$$P(k = 6) = {}_6C_6(0.37)^6(0.63)^0 \approx 0.003$$

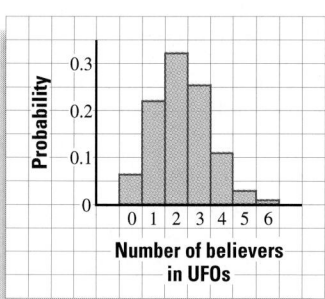

Number of believers in UFOs

The probability of getting at most $k = 2$ successes is:

$$P(k \leq 2) = P(2) + P(1) + P(0) \approx 0.323 + 0.220 + 0.063 = 0.606$$

▶ The probability that at most 2 of the people surveyed believe that UFOs really exist is about 61%.

STUDENT HELP
↳ **Study Tip**
You can check your calculations for a binomial distribution by adding all the probabilities. The sum should always be 1.

 **Egg Incubation**

EXAMPLE 3 | *Interpreting a Binomial Distribution*

For a science project you are incubating 12 chicken eggs. The probability that a chick is female is 0.5. Draw a histogram of the binomial distribution based on the probability that exactly k of the chicks are female. Then find the most likely number of female chicks.

SOLUTION Begin by calculating each binomial probability using the formula $P(k) = {}_nC_k(0.5)^k(0.5)^{n-k} = {}_nC_k(0.5)^n$ as shown in the table. Then draw the histogram.

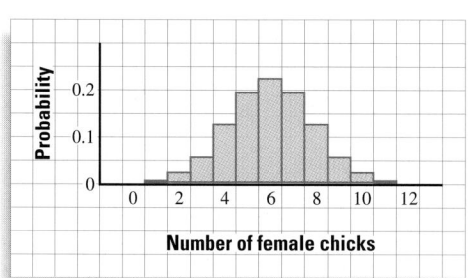

Number of female chicks

k	$P(k)$
0	0.000
1	0.003
2	0.016
3	0.054
4	0.121
5	0.193
6	0.226
7	0.193
8	0.121
9	0.054
10	0.016
11	0.003
12	0.000

The most likely number of female chicks is the value of k for which $P(k)$ is greatest. This probability is greatest for $k = 6$. So, the most likely number of female chicks is 6.

. . . .

The distribution in Example 3 is **symmetric**, which means that the left half of the histogram is a mirror image of the right half. A distribution that is not symmetric is called **skewed**. The distribution in Example 2 is an example of a skewed distribution.

GOAL 2 TESTING HYPOTHESES

Example 1 presented the claim that 37% of adults believe that UFOs really exist. If you wanted to test this claim, you could use a procedure from statistics called **hypothesis testing**. Here are the basic steps.

STUDENT HELP

Study Tip
Different standards are used by different statisticians, but "small" in Step 3 typically means that the probability is less than 0.1, 0.05, or 0.01.

HYPOTHESIS TESTING

STEP ❶ State the hypothesis you are testing. The hypothesis should make a statement about some statistical measure (mean, standard deviation, or proportion) of a population.

STEP ❷ Collect data from a random sample of the population and compute the statistical measure of the sample.

STEP ❸ Assume that the hypothesis is true and calculate the resulting probability of obtaining the sample statistical measure *or a more extreme* sample statistical measure. If this probability is small, you should reject the hypothesis.

EXAMPLE 4 *Testing a Hypothesis*

UFOS To test the claim that 37% of adults believe UFOs really exist, you conduct a survey of 10 randomly selected adults. In your survey only 2 of the adults believe that UFOs really exist. Should you reject the claim? Explain.

SOLUTION

❶ Assume the claim *37% of adults believe that UFOs really exist* is true.

❷ Form the binomial distribution for the probability of selecting exactly k people out of 10 who believe that UFOs really exist.

FOCUS ON APPLICATIONS

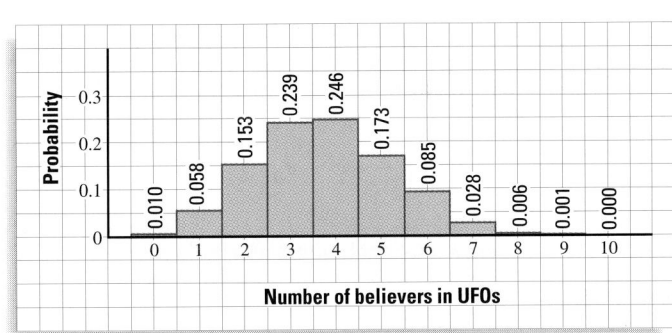

Number of believers in UFOs

UFOs In 1938 a radio adaptation of *The War of the Worlds* caused thousands of people to believe that UFOs had landed. *The War of the Worlds* was made into a feature film in 1953.

APPLICATION LINK
www.mcdougallittell.com

❸ Calculate the probability that you could randomly select 2 *or fewer* adults out of 10 who believe that UFOs really exist.

$$P(k \le 2) = 0.010 + 0.058 + 0.153 = 0.221$$

▶ If it is true that *37% of adults believe that UFOs really exist*, then there is a 22% probability of finding only 2 or fewer believers in a random sample of 10. With a probability this large, you should not reject the claim.

EXTRA EXAMPLE 4
A diet company claims that 78% of people who use their product lose weight. To test this hypothesis, you randomly choose 9 people and ask them to follow the diet for 3 months. Of the 9 people, only 4 people lose weight. Should you reject the claim? Explain your reasoning. $P(k \le 4) \approx 0.029$ which is fairly small, so you should reject the claim.

CHECKPOINT EXERCISES
For use after Example 4:

1. A fruit juice manufacturer makes the claim that 84% of adults prefer their brand of fruit juice over the other leading brand. To test this hypothesis, you conduct a test taste using 8 randomly selected adults. Six of the adults said they did prefer this fruit juice over the other leading brand. Should you reject the claim? Explain your reasoning. No, $P(k \le 6) \approx 0.374$ which is fairly large so you should accept the claim.

MATHEMATICAL REASONING
You may need to review the concepts of hypothesis and conclusion before discussing hypothesis testing. Students should understand that if the hypothesis is true, the conclusion must follow. Thus, if the conclusion is false, the hypothesis must also be false.

CLOSURE QUESTION
Explain how to use the formula $P(k) = {}_nC_k(p)^k(1-p)^{n-k}$ to calculate the probability in a binomial experiment. To find the probability of k successes in n trials of the binomial experiment, substitute the number of trials for n, the number of successes for k, and the probability of success for each trial for p.

ASSIGNMENT GUIDE

BASIC
Day 1: pp. 742–744 Exs. 10–13,
18–21, 26–27, 30–32,
36–37, 40–41, 49–50,
53–67 odd

AVERAGE
Day 1: pp. 742–744 Exs. 10–13,
18–21, 26–28, 30–33,
36–41, 44–46, 49–50,
53–67 odd

ADVANCED
Day 1: pp. 742–744 Exs. 10–13,
18–21, 26–28, 30–33,
36–46, 49–51, 53–67 odd

BLOCK SCHEDULE
pp. 742–744 Exs. 10–13, 18–21,
26–28, 30–33, 36–41, 44–46,
49–50, 53–67 odd (with 12.5)

EXERCISE LEVELS
Level A: *Easier*
10–25, 36–37
Level B: *More Difficult*
26–35, 38–45, 49–50
Level C: *Most Difficult*
46–48, 51

✓ HOMEWORK CHECK
To quickly check student understanding of key concepts, go
over the following exercises:
Exs. 12, 20, 26, 30, 36. See also
the Daily Homework Quiz:

- Blackline Master (*Chapter 12
  Resource Book*, p. 96)
- 📖 Transparency (p. 95)

2, 6–8, 30–35.
See Additional Answers beginning on page AA1.

GUIDED PRACTICE

Vocabulary Check ✓

1. Explain the difference between a binomial experiment and a binomial distribution. **See margin.**

Concept Check ✓

2. You draw ten cards from a standard 52-card deck without replacement. Is this a binomial experiment? Explain. **See margin.**

3. Consider the binomial distribution shown. Is the distribution skewed or symmetric? Explain. **See margin.**

k	P(k)
0	0.0625
1	0.25
2	0.375
3	0.25
4	0.0625

Ex. 3

Skill Check ✓

1. A binomial experiment is one in which each trial is independent and has only two outcomes – success or failure – with the probability of success constant.

3. Symmetric. The values for 0 and 1 are the same as for 4 and 3, so the left and right halves will be mirror images.

Calculate the probability of k successes for a binomial experiment consisting of n trials with probability p of success on each trial.

4. $k = 7, n = 12, p = 0.7$ **0.158**   **5.** $k \le 3, n = 14, p = 0.45$ **0.063**

A binomial experiment consists of n trials with probability p of success on each trial. Draw a histogram of the binomial distribution that shows the probability of exactly k successes. Then find the most likely number of successes.
6–8. See margin.

6. $n = 6, p = 0.5$ **7.** $n = 8, p = 0.33$ **8.** $n = 10, p = 0.25$

9. 🌐 **CLASS RINGS** You read an article that claims only 30% of graduating seniors will buy a class ring. To test this claim you survey 15 randomly selected seniors in your school and find that 4 are planning to buy class rings. Should you reject the claim? Explain. ▶ Source: *America by the Numbers*

No; The probability of 4 or fewer students buying rings is much greater than 0.1 if the claim is true. Therefore, you should not reject the claim.

PRACTICE AND APPLICATIONS

STUDENT HELP
▶ **Extra Practice**
to help you master
skills is on p. 957.

CALCULATING PROBABILITIES Calculate the probability of tossing a coin 20 times and getting the given number of heads.

10. 1 **0.0000191**   **11.** 3 **0.00109**   **12.** 5 **0.0148**   **13.** 9 **0.160**

14. 10 **0.176**   **15.** 11 **0.160**   **16.** 15 **0.0148**   **17.** 17 **0.00109**

BINOMIAL PROBABILITIES Calculate the probability of randomly guessing the given number of correct answers on a 30-question multiple-choice exam that has choices A, B, C, and D for each question.

18. 0 **0.000179**   **19.** 2 **0.00863**   **20.** 5 **0.105**   **21.** 10 **0.0909**

22. 15 **0.00193**   **23.** 20 **0.00000154**   **24.** 25 3.00×10^{-11} **25.** 30 8.67×10^{-19}

BINOMIAL DISTRIBUTIONS Calculate the probability of k successes for a binomial experiment consisting of n trials with probability p of success on each trial.

26. $k \ge 3, n = 5, p = 0.2$ **0.0579**   **27.** $k \le 2, n = 6, p = 0.5$ **0.344**

28. $k \le 1, n = 9, p = 0.15$ **0.599**   **29.** $k \le 5, n = 12, p = 0.64$ **0.097**

STUDENT HELP
▶ **HOMEWORK HELP**
Example 1: Exs. 10–25,
36, 37
Example 2: Exs. 26–29,
38–41
Example 3: Exs. 30–35,
42–45
Example 4: Exs. 46–48

HISTOGRAMS A binomial experiment consists of n trials with probability p of success on each trial. Draw a histogram of the binomial distribution that shows the probability of exactly k successes. Then find the most likely number of successes.
30–35. See margin.

30. $n = 2, p = 0.4$ **31.** $n = 4, p = 0.7$ **32.** $n = 5, p = 0.17$

33. $n = 8, p = 0.92$ **34.** $n = 9, p = 0.125$ **35.** $n = 12, p = 0.033$

FOCUS ON
PEOPLE

DR. CHARLES DREW was a leading expert in blood collection and storage, discussed in Exs. 42 and 43. Although he died in 1950 at the age of 45, his contributions to medicine continue to save lives.

36. 🌐 **COURT CASE** In a particular court case, legal analysts determine from the evidence presented that there is about a 90% probability a juror will vote that the defendant is guilty. Find the probability that all 12 jurors vote that the defendant is guilty. **0.282**

37. 🌐 **VIDEOCASSETTES** A product-quality researcher runs a study on a particular brand of videocassette tape and discovers that on a random basis one out of 400 tapes is defective. Find the probability that you buy 20 of these tapes and exactly 2 are defective. **0.00114**

🌐 **PUPPIES** **In Exercises 38 and 39, use the following information.**
A dog gives birth to a litter of 8 puppies. Assume that the probability of a puppy being female is 0.5.

38. Draw a histogram of the binomial distribution for the number of female puppies in the litter. **See margin.**

39. What is the probability of having at least 5 female puppies? **0.363**

🌐 **EGGS** **In Exercises 40 and 41, use the following information.**
A study states 34% of people prefer their eggs scrambled. You randomly select 7 people who had eggs for breakfast ▶ Source: *America by the Numbers*

40. Draw a histogram of the binomial distribution for the number of people who prefer their eggs scrambled. **See margin.**

41. What is the probability that at most 3 of these people prefer scrambled eggs?
0.816

🌐 **BLOOD DRIVE** **In Exercises 42 and 43, use the following information.**
A hospital is having a blood drive. The hospital is desperately in need of type O− blood, which 7% of the people in the United States have. Each hour at the hospital, an average of 12 people give blood. ▶ Source: American Red Cross

42. Draw a histogram of the binomial distribution for the number of people each hour who give type O− blood. **See margin.**

43. What is the most likely number of people each hour who give type O− blood?
0

🌐 **WORK HOURS** **In Exercises 44 and 45, use the following information.**
A recent survey found that 25% of women wish they had more flexible work hours. A small company has 24 female employees. ▶ Source: America by the Numbers

44. Draw a histogram of the binomial distribution for the number of female employees who want more flexible work hours. **See margin.**

45. At the small company, what is the most likely number of female employees who want more flexible work hours? **6**

46. 🌐 **PERSONAL COMPUTERS** A manufacturer of personal computers claims that under normal work use only 5% of its computers will fail to operate at some point in a month. A small business firm uses 30 of the manufacturer's computers under normal work use and has 2 failures in a month. Would you reject the manufacturer's claim? Explain. **See margin.**

47. 🌐 **STUDENT ATHLETES** In an article about athletes at a particular college, a journalist claims that 75% of the athletes who were given scholarships would still be attending the college if they had not been given scholarships. Curious, a professor polls a random sample of 10 students with sports scholarships. Three of the students said they had plans to attend the college with or without a scholarship. Would you reject the journalist's claim? Explain. **See margin.**

46. Do not reject the claim because the probability that two or more computers will fail is 0.446 which is much larger than 0.1.

47. Reject the claim because the probability that 3 or fewer students would have attended college anyway is 0.00351, which is much smaller than 0.01.

STUDENT HELP

🌐 INTERNET
HOMEWORK HELP
Visit our Web site www.mcdougallittell.com for help with problem solving in Ex. 47.

12.6 *Binomial Distributions* **743**

STUDENT HELP NOTES

➜ **Homework Help** Students can find help for Ex. 47 at **www.mcdougallittell.com.** The information can be printed out for students who don't have access to the Internet.

38.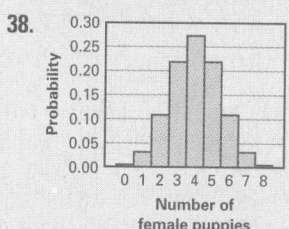
Number of female puppies

40.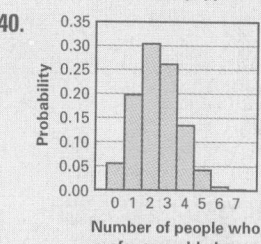
Number of people who prefer scrambled eggs

42, 44. See Additional Answers beginning on page AA1.

ADDITIONAL PRACTICE AND RETEACHING

For Lesson 12.6:
• Practice Levels A, B, and C (*Chapter 12 Resource Book,* p. 86)
• Reteaching with Practice (*Chapter 12 Resource Book,* p. 89)
• 🖥 See Lesson 12.6 of the *Personal Student Tutor*

For more Mixed Review:
• 🖥 Search the *Test and Practice Generator* for key words or specific lessons.

DAILY HOMEWORK QUIZ

 Transparency Available

Calculate the probability of tossing a coin 25 times and getting the given number of heads.

1. 10   0.097

2. 12   0.155

On any day, there is a 15% probability that a student will be late for school. What is the probability that the given number of students in a class of 20 are late for school today?

3. 6 students   0.045

4. 0 students   0.039

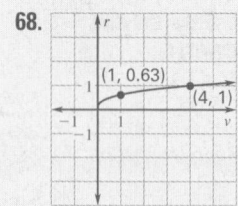

EXTRA CHALLENGE NOTE

→ Challenge problems for Lesson 12.6 are available in **blackline** format in the *Chapter 12 Resource Book,* p. 93 and at **www.mcdougallittell.com.**

ADDITIONAL TEST PREPARATION

1. WRITING A survey claims that 57% of adults buy a newspaper at least once a week. You take a poll of 15 randomly selected adults and find that only 7 of them have bought a newspaper in the last week. Use this information to test the survey claim. **See margin.**

68.

(1, 0.63)
(4, 1)

---

48. Reject the claim because the probability is 0.032 that 12 or fewer people would prefer the new juice and 0.032 < 0.1.

Test Preparation

★ Challenge

EXTRA CHALLENGE

→ www.mcdougallittell.com

---

48. 🌐 **BOTTLED JUICES** A company that makes bottled juices has created a new brand of apple juice. The company claims 80% of people prefer the new apple juice over a competitor's apple juice. A taste test is conducted to test this claim. Of 20 people sampled, 12 preferred the new apple juice. Would you reject the company's claim? Explain.

49. MULTIPLE CHOICE A baseball player's batting average is .310. What is the probability that the player will get 3 or more hits in a game in which the player has 5 official at-bats?   **D**

Ⓐ 0.500   Ⓑ 0.142   Ⓒ 0.035   Ⓓ 0.177   Ⓔ 0.600

50. MULTIPLE CHOICE Based on sales figures it is assumed that 60% of the students in a particular high school prefer Beverage A to Beverage B. In a survey of 10 randomly selected students, 3 preferred Beverage A. What is the probability of obtaining this sample result *or a more extreme* sample result?   **C**

Ⓐ 0.947   Ⓑ 0.043   Ⓒ 0.055   Ⓓ 0.300   Ⓔ 0.200

51. FINDING A FORMULA The mean of a binomial distribution is the sum of the products of the numbers of successes and the corresponding probabilities. Mathematically, the mean can be represented by:

$$\sum_{k=0}^{n} k \cdot P(k)$$

Find the mean of the binomial distribution in Example 2. Then divide the mean by n. What significance does this number have relative to the probability of success on any trial? Use this result to find a simpler formula for the mean of a binomial distribution.

mean = 2.221, $\frac{\text{mean}}{n}$ = 0.370 = probability of success on any trial.

So, mean = $n \times$ (probability of success) = np.

MIXED REVIEW

MEASURES OF DISPERSION Find the range and standard deviation of the data set. (Review 7.7 for 12.7)

52. 8, 9, 9, 9, 10   2, 0.632   **53.** 16, 18, 19, 21, 25, 27, 27   11, 4.155

54. 1.2, 1.3, 1.4, 1.7, 1.8   0.6, 0.232   **55.** 81, 87, 88, 91, 99, 100   19, 6.708

SOLVING SYSTEMS Find the points of intersection, if any, of the graphs in the system. (Review 10.7)

56. $\left(\frac{\sqrt{2}}{2}, \frac{\sqrt{2}}{2}\right),$ $\left(-\frac{\sqrt{2}}{2}, -\frac{\sqrt{2}}{2}\right)$

59. $\left(\pm\frac{1}{2}, \pm\frac{\sqrt{35}}{2}\right)$

	$(-7, -5), (5, 7)$	$\left(-\frac{3}{5}, \frac{2}{5}\right), (-1, 0)$

56. $x^2 + y^2 = 1$
$y = x$

57. $x^2 + y^2 = 74$
$x - y = -2$

58. $x^2 + 4y^2 = 1$
$y = x + 1$

59. $5x^2 + y^2 = 10$
$x^2 + y^2 = 9$

60. $x^2 - y^2 = 49$ ($\pm 9.899, 7$)
$y = 7$

61. $2x^2 - 3y^2 = 6$ none
$y = 3x + 1$

WRITING RULES Write a recursive rule for the sequence. The sequence may be arithmetic, geometric, or neither. (Review 11.5)

62. $a_1 = 3, a_n = a_{n-1} + 2(n-1)$

63. $a_1 = 4, a_n = a_{n-1} \times 10$

64. $a_1 = 80, a_n = a_{n-1} - 20$

65. $a_1 = 1, a_2 = 3, a_n = a_{n-1} \times a_{n-2}$

66. $a_1 = 160, a_n = \frac{1}{4} \times a_{n-1}$

67. $a_1 = 1, a_2 = 2, a_n = a_{n-1} + a_{n-2}$

62. 3, 5, 9, 15, ...   **63.** 4, 40, 400, 4000, ...   **64.** 80, 60, 40, 20, ...

65. 1, 3, 3, 9, ...   **66.** 160, 40, 10, 2.5, ...   **67.** 1, 2, 3, 5, ...

68. GEOMETRY ▶ CONNECTION   The radius r of a sphere can be approximated by $r = 0.62\sqrt[3]{V}$ where V is the volume of the sphere. Graph the model. Then determine the volume of a sphere with a radius of 3 inches. (Lesson 7.5) about 113 in.³; See margin for graph.

Additional Test Preparation *Sample answer:*

Assume that the claim is true and form the binomial distribution for the probability of selecting exactly k people out of 15 that bought a newspaper. Then calculate the probability that 7 or fewer of the 15 adults bought a newspaper.

If it is true that 57% of adults buy a newspaper at least once per week, there is a 29% probability of finding 7 or fewer people who bought a paper this week. With a probability this large, you should accept the claim.

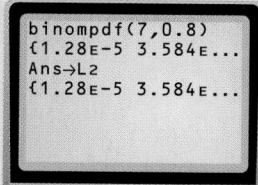

▶ ACTIVITY 12.6

Using Technology

Constructing a Binomial Distribution

Some calculators have a binomial probability distribution function that you can use to calculate binomial probabilities.

▶ **EXAMPLE**

According to a survey, oatmeal is found in 80% of all United States households. Suppose you survey 7 households at random. Draw a histogram of the binomial distribution showing the probability that oatmeal is found in exactly *k* households.

▶ **SOLUTION**

Let $p = 0.8$ be the probability that a household has oatmeal.

STUDENT HELP

INTERNET **KEYSTROKE HELP**

See keystrokes for several models of calculators at www.mcdougallittell.com

❶ Enter the *k*-values 0 through 7 into a list on your graphing calculator.

❷ Enter the binomial probability command to generate $P(k)$ for all eight *k*-values. Store the results in a second list.

```
L1      L2      L3
0
1
2
3
4
5
L1(1)=0
```

```
binompdf(7,0.8)
{1.28E-5 3.584E...
Ans→L2
{1.28E-5 3.584E...
```

❸ To set up the histogram, let the *x*-values be the *k*-values stored in List 1, and let the frequencies be the values of $P(k)$ stored in List 2.

❹ To draw the histogram shown, set the viewing window so that $-0.5 \le x \le 7.5$ with a scale of 1 and $0 \le y \le 0.5$ with a scale of 0.1.

```
Plot1 Plot2 Plot3
On Off
Type: ⊾ ⊿ ⊞
      ⊞ ⊞ ⊿
XList:L1
Freq:L2
```

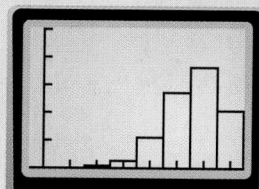

▶ **EXERCISES** 1, 2. See margin.

1. Suppose you conduct a survey similar to the one described in the example, except that you survey 14 households at random. Draw a histogram of the binomial distribution showing the probability that oatmeal is found in exactly *k* households.

2. According to a survey, 21% of teachers in kindergarten through grade 6 use the Internet with their students. You survey 10 teachers of kindergarten through grade 6 at random. Draw a histogram of the binomial distribution showing the probability that exactly *k* teachers use the Internet with their students.

12.6 *Technology Activity*   **745**

---

❶ Planning the Activity

PURPOSE
To construct a binomial probability distribution using a graphing calculator.

MATERIALS
• graphing calculator
• Keystroke Help (*Chapter 12 Resource Book*, p. 85)

PACING
• Example — 20 min
• Exercises — 10 min

▶ **LINK TO LESSON**
Ask students to generate the binomial distribution from Example 2 on page 740 using a graphing calculator.

❷ Managing the Activity

COOPERATIVE LEARNING
This activity can be done in pairs with a single graphing calculator. Have students take turns making the graphs and checking each other's work.

CLASSROOM MANAGEMENT
You may need to help students set the scales for the axes in Exercises 1 and 2. Start the *x*-axis at −0.5 with an increment of 1 and go to 0.5 greater than *n*. For the *y*-axis, students could start with $0 \le y \le 1$ and decrease the upper limit as needed.

❸ Closing the Activity

★ **KEY DISCOVERY**
A graphing calculator can quickly graph a binomial probability distribution.

ACTIVITY ASSESSMENT
Write the steps you use to make a binomial probability distribution on your calculator. **Check work.**

1–2. See Additional Answers beginning on page AA1.

PACING
Basic: 1 day
Average: 1 day
Advanced: 1 day
Block Schedule: 0.5 block with
Ch. Rev.

LESSON OPENER
GRAPHING CALCULATOR
An alternative way to approach Lesson 12.7 is to use the Graphing Calculator Lesson Opener:
• Blackline Master (*Chapter 12 Resource Book,* p. 97)
• Transparency (p. 83)

MEETING INDIVIDUAL NEEDS
• *Chapter 12 Resource Book*
Prerequisite Skills Review (p. 5)
Practice Level A (p. 98)
Practice Level B (p. 99)
Practice Level C (p. 100)
Reteaching with Practice (p. 101)
Absent Student Catch-Up (p. 103)
Challenge (p. 106)
• *Resources in Spanish*
• Personal Student Tutor

NEW-TEACHER SUPPORT
See the Tips for New Teachers on pp. 1–2 of the *Chapter 12 Resource Book* for additional notes about Lesson 12.7.

WARM-UP EXERCISES

Transparency Available

Find the mean and standard deviation of each data set.

1. 3, 3, 5, 8, 11, 11, 12 7.6, 3.6
2. 2.3, 2.3, 2.3, 2.7, 2.7, 2.9 2.5, 0.2
3. 42, 45, 45, 48, 50, 56 47.7, 4.5
4. 0, 0, 1, 1, 1, 2 0.8, 0.7

EXPLORING DATA AND STATISTICS

12.7

What you should learn

GOAL 1 Calculate probabilities using normal distributions.

GOAL 2 Use normal distributions to approximate binomial distributions, as applied in **Example 3**.

Why you should learn it

▼ To solve **real-life** problems, such as finding the probability that certain numbers of patients are nearsighted in **Exs. 53–55**.

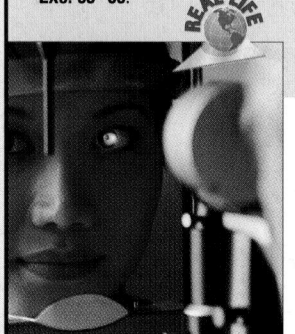

Normal Distributions

GOAL 1 **USING NORMAL DISTRIBUTIONS**

As statisticians began to study binomial distributions consisting of n trials with probability P of success on each trial, they discovered that when np and $n(1 - p)$ are both greater than or equal to five, the distributions all resemble one another. Here are two examples.

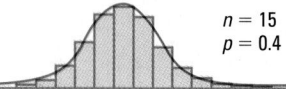

$n = 15$
$p = 0.4$

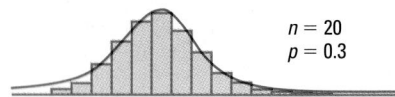

$n = 20$
$p = 0.3$

In both cases the binomial distribution can be approximated by a smooth, symmetrical, bell-shaped curve called a **normal curve**. Areas under this curve represent probabilities from **normal distributions**.

CONCEPT SUMMARY

AREAS UNDER A NORMAL CURVE

The mean $\overline{x}$ and standard deviation σ of a normal distribution determine the following areas.

• The total area under the curve is 1.
• 68% of the area lies within 1 standard deviation of the mean.
• 95% of the area lies within 2 standard deviations of the mean.
• 99.7% of the area lies within 3 standard deviations of the mean.

From the second bulleted statement above and the symmetry of a normal curve, you can deduce that 34% of the area lies within 1 standard deviation to the left of the mean, and 34% of the area lies within 1 standard deviation to the right of the mean. The diagram shows other partial areas (expressed as decimals rather than percents) based on the properties of a normal curve.

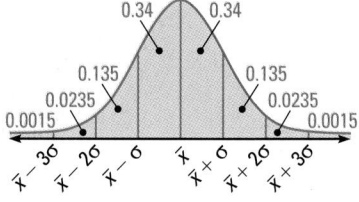

You can interpret these areas as probabilities. In a normal distribution, the probability that a randomly chosen x-value is between a and b is given by the area under the normal curve between a and b. For instance, the probability that a randomly selected x-value is between 1 and 2 standard deviations to the right of the mean is:

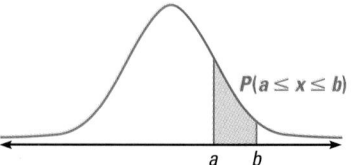

$$P(\overline{x} + \sigma \leq x \leq \overline{x} + 2\sigma) = 0.135$$

Many real-life distributions are normal or approximately normal.

Shopping

EXAMPLE 1 *Using a Normal Distribution*

A survey shows that the time spent by shoppers in supermarkets is normally distributed with a mean of 45 minutes and a standard deviation of 12 minutes.

a. What percent of the shoppers at a supermarket will spend between 33 and 57 minutes in the supermarket?

b. What is the probability that a randomly chosen shopper will spend between 45 and 69 minutes in the supermarket?

SOLUTION

STUDENT HELP

↳ **Study Tip**
When calculating probabilities using a normal distribution, refer to the diagram below the property box on p. 746.

a. The given times of 33 minutes and 57 minutes represent one standard deviation on either side of the mean, as shown below. So, 68% of the shoppers will spend between 33 and 57 minutes in the supermarket.

b. The time of 45 minutes is the mean and the time of 69 minutes is two standard deviations to the right of the mean, as shown below. So, the probability that a randomly chosen shopper will spend between 45 and 69 minutes in the supermarket is 0.475.

$P(45 \leq x \leq 69) = 0.34 + 0.135 = 0.475$

Health Statistics

EXAMPLE 2 *Using a Normal Distribution (Compound Event)*

According to a survey by the National Center for Health Statistics, the heights of adult men in the United States are normally distributed with a mean of 69 inches and a standard deviation of 2.75 inches. If you randomly choose 3 adult men, what is the probability that all three are 71.75 inches or taller?

SOLUTION

A height of 71.75 inches is one standard deviation to the right of the mean, as shown. The probability of randomly selecting a man who is this height or taller is:

$P(x \geq 71.75) = 0.135 + 0.0235 + 0.0015$

$= 0.16$

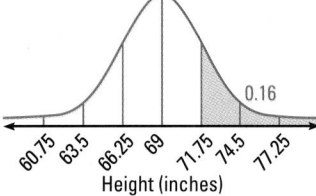

Randomly choosing men are independent events, so the probability that all three randomly chosen men are 71.75 inches or taller is:

$P(\text{all are 71.75 inches or taller}) = (0.16)^3$

≈ 0.00410

12.7 *Normal Distributions* **747**

2 TEACH

MOTIVATING THE LESSON
A class of 1000 students takes an exam with an average of 70%. Most students would probably score between 60% and 80%. How likely would it be for a student to score 100%? In today's lesson, students will learn how a normal distribution would be used to find the probability of students getting specific scores.

 EXTRA EXAMPLE 1
The price of sandals is normally distributed with a mean of $36 and a standard deviation of $9.
a. What percent were priced between $18 and $72? **97.5%**
b. What is the probability that a randomly chosen pair is priced between $9 and $18? **0.024**

EXTRA EXAMPLE 2
The weight of strawberries packed in a medium-sized container was normally distributed with a mean of 16.18 ounces and a standard deviation of 0.34 ounce. If you randomly choose two medium-sized containers of strawberries, what is the probability that both of them weigh less than 15.5 ounces? **about 0.000625**

 CHECKPOINT EXERCISES
For use after Example 1:
1. The number of people seeing a movie on a weeknight was normally distributed with a mean of 224 and a standard deviation of 17. What is the probability that between 241 and 275 people attend the movies on a randomly chosen weeknight? **0.159**

For use after Example 2:
2. Refer to Checkpoint Exercise 1. If you randomly choose 4 weeknights, what is the probability that the attendance is below 207 for all 4 nights? **about 0.00066**

747

EXTRA EXAMPLE 3
Suppose that in a nationwide poll, 72% of households owned at least one car. You are conducting a random survey of 129 households. What is the probability that at least 98 of them own at least one car?   0.16

 CHECKPOINT EXERCISES
For use after Example 3:
1. Based on research, you have concluded that 14% of the student population arrives late to school on any given day. You are conducting a survey of 75 randomly chosen students. What is the probability that fewer than 8 of them were late today?   0.16

CAREER NOTE
EXAMPLE 3 Additional information about a career as a market researcher is available at **www.mcdougallittell.com.**

FOCUS ON VOCABULARY
What is a normal curve?   It is a smooth symmetrical bell-shaped curve that can be used to approximate a binomial distribution.

CLOSURE QUESTION
Sketch a normal curve. Divide the area under the curve into the sections for 1, 2, and 3 standard deviations of the mean. Label each section.   Check curves and labels.

DAILY PUZZLER
Suppose that the age of people at a beach is normally distributed. If the probability is 95% that a randomly selected beachgoer is between the ages of 1 and 49 and the mean age of beachgoers is 25 years old, what is the standard deviation of the ages?   12

FOCUS ON CAREERS

MARKET RESEARCHER
A market researcher gathers data about the market potential of a product or service. The data collected are used to identify opportunities to improve a company's success in the marketplace.

CAREER LINK
www.mcdougallittell.com

When n is large it can be tedious to compute binomial probabilities using the formula $P(k) = {}_nC_k p^k (1 - p)^{n-k}$. In such cases you may be able to use a normal distribution to approximate a binomial distribution.

NORMAL APPROXIMATION OF A BINOMIAL DISTRIBUTION

Consider the binomial distribution consisting of n trials with probability p of success on each trial. If $np \geq 5$ and $n(1 - p) \geq 5$, then the binomial distribution can be approximated by a normal distribution with a mean of

$$\overline{x} = np$$

and a standard deviation of

$$\sigma = \sqrt{np(1 - p)}.$$

EXAMPLE 3   *Finding a Binomial Probability*

SURVEYS According to a survey conducted by the Harris Poll, 29% of adults in the United States say that they or someone in their family plays soccer regularly. You are conducting a random survey of 238 adults. What is the probability that you will find at most 55 adults who come from a family in which someone plays soccer regularly?

SOLUTION

To answer the question using the binomial probability formula, you would have to calculate the following:

$$P(x \leq 55) = P(0) + P(1) + P(2) + \cdots + P(55)$$

This would be tedious. Instead you can approximate the answer with a normal distribution having a mean of

$$\overline{x} = np = 238(0.29) \approx 69$$

and a standard deviation of

$$\sigma = \sqrt{np(1 - p)} = \sqrt{238(0.29)(0.71)} \approx 7.$$

For this normal distribution, 55 is two standard deviations to the left of the mean. So, the probability that you will find at most 55 people from families in which someone plays soccer regularly is:

$$P(x \leq 55) \approx 0.0015 + 0.0235$$

$$= 0.025$$

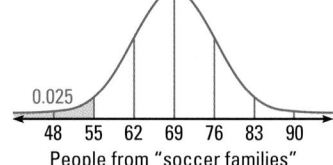

People from "soccer families"

· · · · · · · · · ·

If you had instead used the binomial probability formula in Example 3, you would have found the actual binomial probability to be approximately 0.0249. So, you can see that the normal approximation is a very good approximation.

GUIDED PRACTICE

Vocabulary Check ✔

1. Complete this statement: A(n) __?__ can be used to approximate a binomial distribution when np and $n(1 - p)$ are both greater than or equal to 5. **normal distribution**

Concept Check ✔

2. A normal curve is symmetric about what x-value? **the mean**

3. What percent of the area under a normal curve lies within 1 standard deviation of the mean? within 2 standard deviations of the mean? within 3 standard deviations of the mean? **68%, 95%, 99.7%**

Skill Check ✔

A normal distribution has a mean of 10 and a standard deviation of 1. Find the probability that a randomly selected x-value is in the given interval.

4. between 8 and 12 **0.95**   5. between 7 and 13 **0.997**   6. between 8 and 11 **0.815**

7. at most 10 **0.5**   8. at least 12 **0.025**   9. at most 9 **0.16**

Find the mean and standard deviation of a normal distribution that approximates a binomial distribution consisting of n trials with probability p of success on each trial.

10. $n = 10, p = 0.5$ **5, 1.58**   11. $n = 17, p = 0.3$ **5.1, 1.89**   12. $n = 28, p = 0.2$ **5.6, 2.12**

13. $n = 20, p = 0.25$ **5, 1.94**   14. $n = 12, p = 0.42$ **5.04, 1.71**   15. $n = 30, p = 0.17$ **5.1, 2.06**

🌐 **COLORBLINDNESS In Exercises 16–18, use the fact that approximately 2% of people are colorblind, and consider a class of 460 students.**

16. What is the probability that 15 or fewer students are colorblind? **0.975**

17. What is the probability that 12 or more students are colorblind? **0.16**

18. What is the probability that between 6 and 18 students are colorblind? **0.839**

PRACTICE AND APPLICATIONS

STUDENT HELP

▸ **Extra Practice**
to help you master
skills is on p. 957.

USING A NORMAL CURVE Give the percent of the area under a normal curve represented by the shaded region.

19. 50%

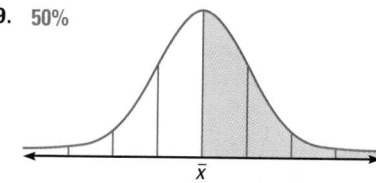

20. 15.85%

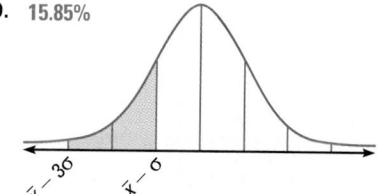

STUDENT HELP

▸ **HOMEWORK HELP**
Example 1: Exs. 19–28,
38–43
Example 2: Exs. 29–31,
44–49
Example 3: Exs. 32–37,
50–55

21. 2.5%

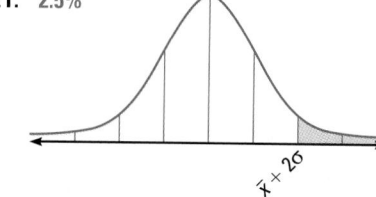

22. 27%

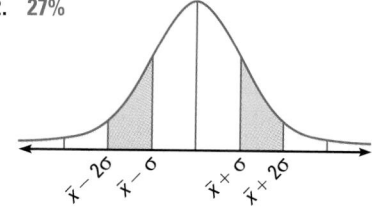

12.7 *Normal Distributions* **749**

---

3 APPLY

◯ **ASSIGNMENT GUIDE**

BASIC
Day 1: pp. 749–752 Exs. 19–28, 30, 32–33, 38–40, 53–56, 59–73 odd, Quiz 3 Exs. 1–22

AVERAGE
Day 1: pp. 749–752 Exs. 19–37 odd, 32–35, 38–40, 44–49, 53–56, 59–73 odd, Quiz 3 Exs. 1–22

ADVANCED
Day 1: pp. 749–752 Exs. 19–28, 30, 32–35, 38–40, 44–49, 53–57, 59–73 odd, Quiz 3 Exs. 1–22

BLOCK SCHEDULE
pp. 749–752 Exs. 19–28, 30, 32–35, 38–40, 44–49, 53–56, 59–73 odd, Quiz 3 Exs. 1–22 (with Ch. 12 Review)

EXERCISE LEVELS
Level A: *Easier*
19–25, 38–43

Level B: *More Difficult*
26–37, 44–56

Level C: *Most Difficult*
57

✔ **HOMEWORK CHECK**
To quickly check student understanding of key concepts, go over the following exercises: Exs. 20, 24, 28, 30, 32, 38. See also the Daily Homework Quiz:

• Blackline Master (*Chapter 13 Resource Book*, p. 11)

• 🖨 Transparency (p. 97)

APPLICATION NOTE

EXERCISES 41–43 Clarify that the life expectancy of lightbulbs depends on several factors such as how often the bulb is turned on and off.

STUDENT HELP NOTES

Homework Help Students can find help for Exs. 47–49 at **www.mcdougallittell.com.** The information can be printed out for students who don't have access to the Internet.

GRAPHING CALCULATOR NOTE

EXERCISE 57 A graphing calculator can save time when graphing the normal function. A computer with graphing software can also be used. Seeing these graphs will help students understand how the binomial distribution is approximated by the normal curve.

ADDITIONAL PRACTICE AND RETEACHING

For Lesson 12.7:

- Practice Levels A, B, and C (*Chapter 12 Resource Book*, p. 98)

- Reteaching with Practice (*Chapter 12 Resource Book*, p. 101)

- See Lesson 12.7 of the *Personal Student Tutor*

For more Mixed Review:

- Search the *Test and Practice Generator* for key words or specific lessons.

750

NORMAL DISTRIBUTIONS A normal distribution has a mean of 22 and a standard deviation of 3. Find the probability that a randomly selected x-value is in the given interval.

23. between 19 and 25   0.68   **24.** between 13 and 22 0.4985   **25.** between 16 and 31 0.9735

26. at most 25   0.84   **27.** at least 19   0.84   **28.** at most 28   0.975

FINDING PROBABILITIES A normal distribution has a mean of 64 and a standard deviation of 7. Find the given probability.

29. three randomly selected x-values are all 71 or greater   0.004096

30. four randomly selected x-values are all 50 or less   0.00000039

31. two randomly selected x-values are both between 57 and 78   0.664

APPROXIMATING BINOMIAL DISTRIBUTIONS Find the mean and standard deviation of a normal distribution that approximates a binomial distribution consisting of n trials with probability p of success on each trial.

32. $n = 18, p = 0.7$ 12.6, 1.94   **33.** $n = 50, p = 0.1$ 5, 2.12   **34.** $n = 32, p = 0.8$ 25.6, 2.263

35. $n = 49, p = 0.12$ 5.88, 2.27   **36.** $n = 24, p = 0.67$ 16.08, 2.304   **37.** $n = 140, p = 0.06$ 8.4, 2.810

DRIVE-THROUGH In Exercises 38–40, use the following information.
A certain bank is busiest during the Friday evening rush hours from 3:00 P.M. until 6:00 P.M. During these hours the waiting time for drive-through customers is normally distributed with a mean of 8 minutes and a standard deviation of 2 minutes.

38. What percent of drive-through customers will wait for 10 minutes or longer during the Friday evening rush hours?   16%

39. What is the probability that a customer will wait between 4 and 12 minutes during the Friday evening rush hours?   0.95

40. What is the probability that a customer will wait 2 minutes or less during the Friday evening rush hours?   0.0015

LIGHT BULBS In Exercises 41–43, use the following information.
A company produces light bulbs having a life expectancy that is normally distributed with a mean of 1000 hours and a standard deviation of 50 hours.

41. What percent of the bulbs will last for 1000 hours or more?   50%

42. What is the probability that a randomly chosen bulb will burn out in 900 hours or less?   0.025

43. What is the probability that a randomly chosen bulb will last between 850 and 1050 hours?   0.839

BIOLOGY **CONNECTION**   In Exercises 44–46, use the following information.
According to a survey by the National Center for Health Statistics, the heights of adult women in the United States are normally distributed with a mean of 64 inches and a standard deviation of 2.7 inches.

44. What is the probability that three randomly selected women are all 58.6 inches or shorter?   0.0000156

45. What is the probability that five randomly selected women are all between the heights of 61.3 and 66.7 inches?   0.145

46. What is the probability that four randomly selected women are all 72.1 inches or taller? 5.06×10^{-12}

REAL LIFE **LIGHT BULBS** Four compact fluorescent light bulbs use the same energy as one incandescent light bulb. Compact fluorescent light bulbs also last longer, with an average life expectancy of 10,000 hours.

STUDENT HELP

HOMEWORK HELP
Visit our Web site
www.mcdougallittell.com
for help with problem
solving in Exs. 47–49.

SAT SCORES In Exercises 47–49, use the following information.
In 1998 scores on the mathematics section of the SAT (Scholastic Aptitude Test) were normally distributed with a mean of 512 and a standard deviation of 112. Scores on the English section of the SAT were normally distributed with a mean of 505 and a standard deviation of 111. **DATA UPDATE** of SAT data at www.mcdougallittell.com

47. What is the probability that a randomly chosen student who took the SAT in 1998 scored at least 736 on the mathematics section and at least 727 on the English section? Assume the scores are independent. **0.000625**

48. What is the probability that five randomly chosen students who took the SAT in 1998 all scored at most 394 on the English section? **0.000105**

49. What is the probability that two randomly chosen students who took the SAT in 1998 both scored between 400 and 624 on the mathematics section? **0.462**

LEFT-HANDEDNESS In Exercises 50–52, use the fact that approximately 9% of people are left-handed, and consider a high school with 1221 students.

50. What is the probability that at least 140 students are left-handed? **0.0015**

51. What is the probability that at most 100 students are left-handed? **0.16**

52. What is the probability that between 80 and 130 students are left-handed? **0.974**

MYOPIA In Exercises 53–55, use the fact that myopia, or nearsightedness, is a condition that affects approximately 25% of the adult population in the United States, and consider a random sample of 192 people.

53. What is the probability that 42 or more people are nearsighted? **0.84**

54. What is the probability that between 36 and 60 people are nearsighted? **0.95**

55. What is the probability that 66 or fewer people are nearsighted? **0.999**

Test Preparation

56. MULTI-STEP PROBLEM In 1998 Ben took both the SAT (Scholastic Aptitude Test) and the ACT (American College Test). On the mathematics section of the SAT, he earned a score of 624. On the mathematics section of the ACT, he earned a score of 31. For the SAT the mean was 512 and the standard deviation was 112. For the ACT the mean was 21 and the standard deviation was 5.

a. What percent of students did Ben outscore on the math section of the SAT? **84%**

b. What percent of students did Ben outscore on the math section of the ACT? **97.5%**

c. On which exam did Ben score better? **ACT**

d. *Writing* Explain how you could translate ACT scores such as 15, 20, 25, and 30 into equivalent SAT scores if you know the mean and standard deviation of each exam.

56d. You could translate ACT scores into percentiles of the normal distribution. Those percentiles could then be transformed into SAT scores of the equivalent percentile.

★ **Challenge**

57. The normal distribution is a good approximation unless the binomial distribution is very skewed, such as when $p < 0.25$ or $p > 0.75$. The normal distribution approximates the binomial distribution best when $p = 0.5$; See margin.

EXTRA CHALLENGE
www.mcdougallittell.com

57. NORMAL CURVE A normal curve is defined by an equation whose general form is as follows:

$$y = \frac{1}{\sigma\sqrt{2\pi}} e^{-\frac{1}{2}\left(\frac{x-\bar{x}}{\sigma}\right)^2}$$

Use a graphing calculator to draw a histogram of a binomial distribution consisting of $n = 20$ trials with probability $p = 0.5$ of success. Also graph the normal curve that approximates this binomial distribution. Then graph other binomial distributions and normal curves in which you change p but leave n constant. When is the normal curve a good approximation of a binomial distribution and when is it a poor approximation? Why?

4 ASSESS

DAILY HOMEWORK QUIZ

Transparency Available

A class's test scores have a normal distribution with a mean of 80 and a standard deviation of 5. Find the given probability.

1. a student scored below 70 **0.025**

2. a student scored above 95 **0.0015**

3. two students selected at random both scored from 75 to 85. **0.4624**

EXTRA CHALLENGE NOTE
Challenge problems for Lesson 12.7 are available in **blackline** format in the *Chapter 12 Resource Book,* p. 106 and at **www.mcdougallittell.com.**

ADDITIONAL TEST PREPARATION

1. WRITING Suppose a survey showed that 21% of adults see a movie in a theater once a week. You take a random survey of 149 adults. Show how you can use the normal curve to approximate this binomial distribution and find the probability that between 36 and 41 of those surveyed saw a movie within the last week. Explain each step of your work. $n = 149, p = 0.21, \bar{x} = np \approx 31, \sigma \approx 5$. Between 36 and 41 people is equivalent to between 1 and 2 standard deviations above the mean. According to the normal probability distribution, the probability of this event is 0.135.

57. Check graphs.

9. 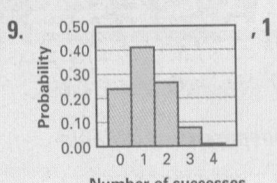 , 1

10. 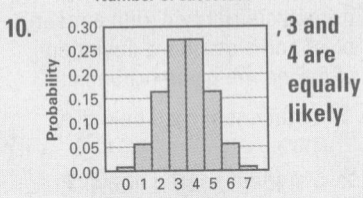 , 3 and 4 are equally likely

11. 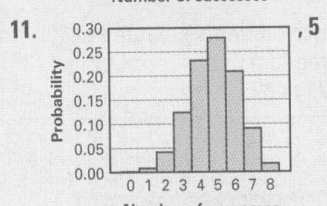 , 5

12. 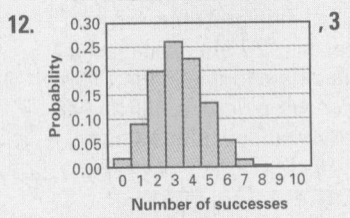 , 3

13. 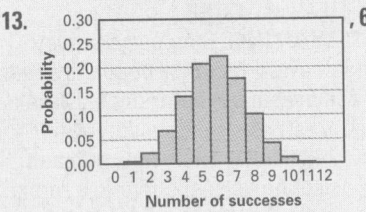 , 6

14.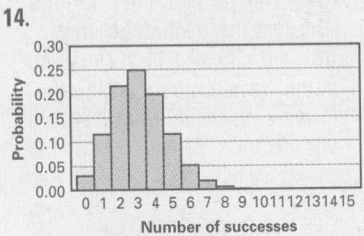

3

64. $(0, \pm 4); (\pm 2, 0); (0, \pm 2\sqrt{3})$
65. $(0, \pm 13); (\pm 12, 0); (0, \pm 5)$
66. $(0, \pm \sqrt{10}); (\pm \sqrt{5}, 0); (0, \pm \sqrt{5})$
67. $(0, \pm \sqrt{21}); (\pm \sqrt{6}, 0); (0, \pm \sqrt{15})$
68. $\frac{x^2}{9} + \frac{y^2}{4} = 1; (\pm 3, 0); (0, \pm 2); (\pm \sqrt{5}, 0)$
69. $\frac{x^2}{7} + \frac{y^2}{10} = 1; (0, \pm \sqrt{10}); (\pm \sqrt{7}, 0); (0, \pm \sqrt{3})$

21. Yes; There is a 0.083 chance of getting 19 or fewer out of 26 and 0.083 < 0.1 so reject the survey's findings.

MIXED REVIEW

EVALUATING EXPRESSIONS Evaluate the expression without using a calculator. (Review 7.1 for 13.1)

58. $9^{3/2}$ 27
59. $256^{3/4}$ 64
60. $49^{-1/2}$ $\frac{1}{7}$
61. $\sqrt[3]{125}$ 5
62. $\sqrt{81}$ 9
63. $\left(\sqrt[4]{625}\right)^2$ 25

IDENTIFYING PARTS Write the equation of the ellipse in standard form (if not already). Then identify the vertices, co-vertices, and foci of the ellipse. (Review 10.4)

64. $\frac{x^2}{4} + \frac{y^2}{16} = 1$
65. $\frac{x^2}{144} + \frac{y^2}{169} = 1$
66. $\frac{x^2}{5} + \frac{y^2}{10} = 1$
67. $\frac{x^2}{6} + \frac{y^2}{21} = 1$
68. $4x^2 + 9y^2 = 36$
69. $10x^2 + 7y^2 = 70$

USING COMPLEMENTS Two six-sided dice are rolled. Find the probability of the given event. (Review 12.4)

70. The sum is not 4. $\frac{11}{12}$
71. The sum is less than or equal to 10. $\frac{11}{12}$
72. The sum is not 3 or 12. $\frac{11}{12}$
73. The sum is greater than 3. $\frac{11}{12}$

QUIZ 3
Self-Test for Lessons 12.6 and 12.7

Calculate the probability of rolling a six-sided die 50 times and getting the given number of ones. (Lesson 12.6)

1. 0 0.000110
2. 1 0.00110
3. 8 0.151
4. 17 0.00142
5. 25 4.66×10^{-8}
6. 33 9.29×10^{-15}
7. 42 2.59×10^{-25}
8. 50 1.24×10^{-39}

A binomial experiment consists of *n* trials with probability *p* of success on each trial. Draw a histogram of the binomial distribution that shows the probability of exactly *k* successes. Then find the most likely number of successes. (Lesson 12.6) 9–14. See margin.

9. $n = 4, p = 0.3$
10. $n = 7, p = 0.5$
11. $n = 8, p = 0.6$
12. $n = 10, p = 0.33$
13. $n = 12, p = 0.48$
14. $n = 15, p = 0.21$

A normal distribution has a mean of 62 and a standard deviation of 4. Find the probability that a randomly selected *x*-value is in the given interval. (Lesson 12.7)

15. between 58 and 66 0.68
16. between 62 and 74 0.4985
17. between 50 and 70 0.9735
18. 62 or greater 0.50
19. 58 or less 0.16
20. 50 or less 0.0015

21. 🌐 **WELL-BEING** A survey that asked people in the United States about their feelings of personal well-being found that 85% are generally happy. To test this finding you survey 26 people at random and find that 19 consider themselves generally happy. Would you reject the survey's findings? Explain. (Lesson 12.6)

22. 🌐 **DRINKING WATER** Approximately 64% of people in the United States think that the nation's water supply is safe to drink. A town has 625 people. What is the probability that 400 or more people think that the nation's water supply is safe to drink? (Lesson 12.7) 0.50

Extension

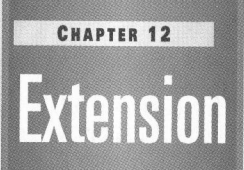

What you should learn

GOAL Find expected values of collections of outcomes.

Why you should learn it

▼ To solve **real-life** problems, such as finding the expected value of insurance coverage in **Example 2**.

Expected Value

GOAL **FINDING AN EXPECTED VALUE**

Suppose you and a friend are playing a game. You flip a coin. If the coin lands heads up, then your friend scores 1 point and you lose 1 point. If the coin lands tails up, then you score 1 point and your friend loses 1 point. After playing the game many times, would you expect to have more, fewer, or the same number of points as when you started? The answer is that you should expect to end up with about the same number of points. You can expect to lose a point about half the time and win a point about half the time. Therefore, the *expected value* for this game is 0.

EXPECTED VALUE

A collection of outcomes is partitioned into *n* events, no two of which have any outcomes in common. The probabilities of the *n* events occurring are $p_1, p_2, p_3, \ldots, p_n$ where $p_1 + p_2 + p_3 + \cdots + p_n = 1$. The values of the *n* events are $x_1, x_2, x_3, \ldots, x_n$. The **expected value** *V* of the collection of outcomes is the sum of the products of the events' probabilities and their values.

$$V = p_1x_1 + p_2x_2 + p_3x_3 + \cdots + p_nx_n$$

EXAMPLE 1 *Finding the Expected Value of a Game*

You and a friend each flip a coin. If both coins land heads up, then your friend scores 3 points and you lose 3 points. If one or both of the coins land tails up, then you score 1 point and your friend loses 1 point. What is the expected value of the game from your point of view?

SOLUTION

When two coins are tossed, four outcomes are possible: HH, HT, TH, and TT. Let event *A* be HH and event *B* be HT, TH, and TT. Note that all possible outcomes are listed, but no outcome is listed twice. The probabilities of the events are:

$$P(A) = \frac{1}{4} \qquad\qquad P(B) = \frac{3}{4}$$

From your point of view the values of the events are:

value of event $A = -3$ \qquad value of event $B = 1$

Therefore, the expected value of the game is:

$$V = \frac{1}{4}(-3) + \frac{3}{4}(1) = 0$$

· · · · · · · · · ·

If the expected value of the game is 0, as in Example 1, then the game is called a **fair game** .

Chapter 12 *Extension* **753**

 EXTRA EXAMPLE 1
You are playing a game with a spinner that has 6 equal size sectors. Two sectors are blue; the rest are red. If you spin blue, you get 2 points and if you spin red you lose a point. What is the expected value of the game? **0**

CHECKPOINT EXERCISES
For use after Example 1:
1. You are taking a test where every correct answer scores 4 points and every incorrect answer scores −1 point. There are 4 possible choices for each question. What is the expected value for one question if you guess the answer? $\frac{1}{4}$

EXTRA EXAMPLE 2

Suppose that in one year there were 4200 house fires in approximately 500,000 homes and that the average yearly premium for fire insurance for homeowners was $225. What was the expected value of insurance coverage for a person with homeowners fire insurance if the average claim paid by the insurance company was $10,000? **−$141**

CHECKPOINT EXERCISES

For use after Example 2:

1. Suppose that in one year, 113 of 10,000 insured diamond rings were lost and that the average claim on that loss was $4,000. The premium paid for loss of jewelry is $48 per year. What is the expected value of the insurance coverage? **−$2.80**

MATHEMATICAL REASONING

Example 2 gives students a good example of how mathematical reasoning can be used to make decisions in real life. Ask students why the expected value of the policy is negative. Also, how could they use the expected value to compare insurance coverage? *Sample answer:* The expected value is negative because the insurance company has to pay overhead charges and earn some profit. The coverage with the highest expected value is the best deal, although there may be other reasons for preferring one policy over another.

STUDENT HELP NOTES

→ **Homework Help** Students can find extra examples at **www.mcdougallittell.com** that parallel the examples in the student edition.

STUDENT HELP

HOMEWORK HELP
Visit our Web site www.mcdougallittell.com for extra examples.

1. Player A expected value:
$0 \cdot \frac{1}{3} + 1 \cdot \frac{1}{3} - 1 \cdot \frac{1}{3} = 0$;
Player B expected value:
$0 \cdot \frac{1}{3} - 1 \cdot \frac{1}{3} + 1 \cdot \frac{1}{3} = 0$;
yes

2. Player A:
expected value $= \frac{4}{9}$;

Player B:
expected value $= -\frac{4}{9}$.

The game is not fair because the expected value is not 0.

EXAMPLE 2 *Finding an Expected Value in Real Life*

In 1996 there were 124,600,000 cars in use in the United States. That year there were 13,300,000 automobile accidents. The average premium paid in 1996 for automobile collision insurance was $685 per car, and the average automobile collision claim paid by insurance companies was $2100 per car. What was the expected value of insurance coverage for a car with collision insurance in 1996?

SOLUTION

Let event A be having an automobile accident and event B be not having an automobile accident. Note that events A and B are mutually exclusive and that $B = A'$, so all outcomes are accounted for but no outcome is counted twice. The probabilities of the events are:

$$P(A) = \frac{13,300,000}{124,600,000} \approx 0.107 \qquad P(B) = 1 - P(A) \approx 1 - 0.107 = 0.893$$

You can calculate the values of the events as follows. If an insured car had an accident, the owner paid an average of $685 and received an average of $2100. If a car did not have an accident, the owner only paid an average of $685. So, the values of the events for an insured person are:

$$\text{value of event } A = -685 + 2100 = 1415 \qquad \text{value of event } B = -685$$

Therefore, the expected value of insurance coverage was:

$$V \approx 0.107(1415) + 0.893(-685) \approx -\$460$$

EXERCISES

EXPECTED VALUE In Exercises 1 and 2, consider a game in which two people each choose an integer from 1 to 3. Find the expected value of the game for each player. Is the game fair? (*Hint:* There may be more than two events to consider.)

1. If the two numbers are equal, then no points are received. If the numbers differ by one, then the player with the higher number wins 1 point and the other player loses 1 point. If the numbers differ by two, then the player with the lower number wins 1 point and the other player loses 1 point.

2. If the sum of the two numbers is odd, then player A loses that sum of points and player B wins that sum. If the sum of the two numbers is even, then player B loses 4 points and player A wins 4 points.

3. **LOTTERY** To win a certain state's weekly lottery, you must match 5 different numbers chosen from the integers 1 to 49 plus an additional number chosen from the integers 1 to 42. You purchase a ticket for $1. If the jackpot for that week is $45,000,000, what is the expected value of your ticket? **−$.44**

4. **CONTESTS** A fast-food restaurant chain is having a contest with five prizes. No purchase is necessary to enter. What is the expected value of a contest ticket? **$.0049**

Prize	Value	Probability of winning
Gift certificate	$5	0.0002
Home theater system	$3,000	0.0000004
Hawaiian vacation	$7,000	0.00000008
Car	$50,000	0.000000003
Cash	$1,000,000	0.000000002

Chapter Summary

WHAT did you learn?

Count the number of ways an event can happen.
- using the fundamental counting principle (12.1)
- using permutations (12.1)

- using combinations (12.2)

Expand a binomial that is raised to a power. (12.2)

Find theoretical, experimental, and geometric probabilities. (12.3)

Find probabilities of unions and intersections of two events. (12.4)

Use complements to find probabilities. (12.4)

Find probabilities of independent and dependent events. (12.5)

Find binomial probabilities and analyze binomial distributions. (12.6)

Test a hypothesis. (12.6)

Use normal distributions to calculate probabilities and to approximate binomial distributions. (12.7)

Use probability and statistics to solve real-life problems. (12.1–12.7)

WHY did you learn it?

Find the number of possible license plates. (p. 702)
Find the number of ways skiers can finish in an Olympic event. (p. 703)
Find the number of combinations of plays you can attend. (p. 709)

Apply Pascal's triangle to algebra. (p. 710)

Find the probability that an archer hits the center of a target. (p. 721)

Find the probability that it will rain on both Saturday and Sunday. (p. 728)

Find the probability that friends will be in the same college dormitory. (p. 729)

Find the probability that a baseball team wins three games in a row. (p. 730)

Find the most likely number of people who will give type O− blood. (p. 743)

Test the claim that only 5% of computers will fail in a month. (p. 743)

Find the probability that certain numbers of patients are nearsighted. (p. 751)

Find the probability of winning a lottery. (p. 720)

How does Chapter 12 fit into the BIGGER PICTURE of algebra?

In this chapter you saw how algebra is used in probability and statistics. In fact, every branch of mathematics uses algebra. You can use what you have learned in this and other chapters to make everyday decisions.

STUDY STRATEGY

How did you connect to your life?

Here is an example of a connection, following the **Study Strategy** on page 700.

Connect to Your Life

I just got a bank card and chose my 4-digit personal identification number (PIN).

There are $10 \cdot 10 \cdot 10 \cdot 10 = 10,000$ different PINs possible. (fundamental counting principle)

The probability that someone who finds my card will guess my PIN on the first try is $\frac{1}{10,000}$. (theoretical probability)

755

756

ADDITIONAL RESOURCES

The following resources are available to help review the material in this chapter.

- Chapter Review Games and Activities (*Chapter 12 Resource Book*, p. 107)
- *Instant Replay: Video Review Games*
- 🖥 *Personal Student Tutor*
- Cumulative Review, Chs. 1–12 (*Chapter 12 Resource Book*, p. 119)

16. $x^5 - 50x^4 + 1000x^3 - 10{,}000x^2 + 50{,}000x - 100{,}000$

17. $x^7 - 21x^6y + 189x^5y^2 - 945x^4y^3 + 2835x^3y^4 - 5103x^2y^5 + 5103xy^6 - 2187y^7$

18. $16x^4 + 32x^3y^2 + 24x^2y^4 + 8xy^6 + y^8$

21. experimental probability = 0.45; theoretical probability = 0.50. So, you got slightly fewer heads than expected.

Chapter Review

CHAPTER 12

VOCABULARY

- permutation, p. 703
- combination, p. 708
- Pascal's triangle, p. 710
- binomial theorem, p. 710
- probability, p. 716
- theoretical probability, p. 716

- experimental probability, p. 717
- geometric probability, p. 718
- compound event, p. 724
- mutually exclusive events, p. 724
- complement, p. 726

- independent events, p. 730
- dependent events, p. 732
- conditional probability, p. 732
- binomial experiment, p. 739
- binomial distribution, p. 739
- symmetric distribution, p. 740

- skewed distribution, p. 740
- hypothesis testing, p. 741
- normal curve, p. 746
- normal distribution, p. 746
- expected value, p. 753
- fair game, p. 753

12.1 THE FUNDAMENTAL COUNTING PRINCIPLE AND PERMUTATIONS

Examples on pp. 701–704

EXAMPLES You can use the fundamental counting principle and permutations to count the number of ways an event can happen.

The number of possible outfits you can make with 2 pairs of jeans and 5 shirts is:

$$2 \cdot 5 = 10 \text{ outfits}$$

The number of ways 4 members from a family of 5 can line up for a photo is:

$$_5P_4 = \frac{5!}{(5-4)!} = \frac{5!}{1!} = \frac{120}{1} = 120$$

1. How many different 5-digit zip codes are there if any of the digits 0–9 can be used? **100,000**

2. How many different ways can 4 friends stand in a cafeteria line? **24**

Find the number of permutations.

3. $_6P_6$ **720** 4. $_8P_4$ **1680** 5. $_5P_1$ **5** 6. $_9P_3$ **504** 7. $_{10}P_6$ **151,200** 8. $_4P_4$ **24**

12.2 COMBINATIONS AND THE BINOMIAL THEOREM

Examples on pp. 708–711

EXAMPLES You can use combinations to find the number of ways an event can happen when order is not important.

You must write reports on 3 of the 12 most recent Presidents of the United States for history class. The number of possible combinations of reports is:

$$_{12}C_3 = \frac{12!}{9! \cdot 3!} = \frac{12 \cdot 11 \cdot 10 \cdot 9!}{9! \cdot 3!} = \frac{1320}{6} = 220$$

You can use the binomial theorem to expand a binomial raised to a power.

$$(x + 6)^4 = {}_4C_0 x^4 6^0 + {}_4C_1 x^3 6^1 + {}_4C_2 x^2 6^2 + {}_4C_3 x^1 6^3 + {}_4C_4 x^0 6^4$$
$$= (1)(x^4)(1) + (4)(x^3)(6) + (6)(x^2)(36) + (4)(x)(216) + (1)(1)(1296)$$
$$= x^4 + 24x^3 + 216x^2 + 864x + 1296$$

Find the number of combinations.

9. $_9C_2$ 36 **10.** $_7C_1$ 7 **11.** $_5C_3$ 10 **12.** $_8C_7$ 8 **13.** $_{10}C_{10}$ 1 **14.** $_{13}C_5$ 1287

Use the binomial theorem to write the binomial expansion. 16–18. See margin.

15. $(x + 4)^3$ **16.** $(x - 10)^5$ **17.** $(x - 3y)^7$ **18.** $(2x + y^2)^4$
$x^3 + 12x^2 + 48x + 64$

12.3 AN INTRODUCTION TO PROBABILITY

Examples on pp. 716–718

EXAMPLES You can find the probability that an event will occur.

You toss two six-sided dice. The *theoretical* probability that the sum of the dice is 4 is

$$\frac{\text{number of ways sum can be 4}}{\text{number of possible outcomes}} = \frac{3}{36} = \frac{1}{12}.$$

You toss two 6-sided dice 100 times and record 8 times that the sum is 4. The *experimental* probability that the sum of the dice is 4 is

$$\frac{\text{number of times sum is 4}}{\text{number of times dice are tossed}} = \frac{8}{100} = \frac{2}{25}.$$

A dart thrown at the square target shown is equally likely to hit any point inside the target. The *geometric* probability that the dart hits the shaded square is $\dfrac{\text{area of shaded square}}{\text{area of entire target}} = \dfrac{4}{16} = \dfrac{1}{4}.$

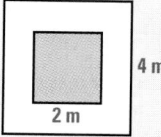
4 m
2 m

You toss a coin 3 times. Find the probability of the given event.

19. You toss exactly 1 tail. $\frac{3}{8}$ **20.** You toss at least 1 tail. $\frac{7}{8}$

21. You toss a coin 200 times and get heads 90 times. Find the experimental probability of getting heads. Compare this with the theoretical probability. See margin.

22. What is the probability that a dart hits the unshaded region of the target above? 0.75

12.4 PROBABILITY OF COMPOUND EVENTS

Examples on pp. 724–726

EXAMPLES You can find the probability that compound events will occur and the probability that the complement of an event will occur.

If A and B are two events and $P(A) = \frac{3}{4}$, $P(B) = \frac{2}{5}$, and $P(A \text{ and } B) = \frac{1}{4}$,

then $P(A \text{ or } B) = P(A) + P(B) - P(A \text{ and } B) = \frac{3}{4} + \frac{2}{5} - \frac{1}{4} = \frac{18}{20} = \frac{9}{10}.$

The probability of the complement of A is $P(A') = 1 - P(A) = 1 - \frac{3}{4} = \frac{1}{4}.$

Find the indicated probability.

23. $P(A) = 0.25$, $P(B) = 0.2$, $P(A \text{ and } B) = 0.15$, $P(A \text{ or } B) = \underline{\ ?\ }$ 0.3

24. $P(A) = \frac{2}{5}$, $P(B) = \frac{1}{10}$, $P(A \text{ and } B) = \underline{\ ?\ }$, $P(A \text{ or } B) = \frac{1}{2}$ 0

25. $P(A) = 99\%$, $P(A') = \underline{\ ?\ }$ 1%

Chapter Review **757**

757

> **EXAMPLES** You can find the probability that independent events will occur and the probability that dependent events will occur.
>
> Nine slips of paper numbered 1–9 are placed in a hat. You randomly draw two slips. What is the probability that the first number is odd (A) and the second is even (B)?
>
> If you replace the first slip of paper before selecting the second, A and B are *independent* events, and $P(A \text{ and } B) = P(A) \cdot P(B) = \frac{5}{9} \cdot \frac{4}{9} = \frac{20}{81} \approx 0.247$.
>
> If you do not replace the first slip of paper before selecting the second, A and B are *dependent* events, and $P(A \text{ and } B) = P(A) \cdot P(B \mid A) = \frac{5}{9} \cdot \frac{4}{8} = \frac{20}{72} = \frac{5}{18} \approx 0.278$.

Find the probability of randomly drawing the given marbles from a bag of 4 red, 6 green, and 2 blue marbles (a) with replacement and (b) without replacement.

26. a red, then a green
   a. 0.166   b. 0.182

27. a blue, then a red
   a. 0.056   b. 0.0606

28. a red, then a red
   a. 0.111   b. 0.0909

> **EXAMPLE** You can find the probability of getting exactly k successes for a binomial experiment.
>
> The probability of tossing a coin 10 times and getting exactly 7 heads is:
>
> $$P(k = 7) = {}_{10}C_7(0.5)^7(1 - 0.5)^3 = \frac{10!}{3! \cdot 7!}(0.5)^7(0.5)^3 \approx 0.117$$

Calculate the probability of tossing a coin 10 times and getting the given number of tails.

29. 3   0.117   **30.** 5   0.246   **31.** 9   0.00977   **32.** 6   0.205   **33.** 1   0.00977   **34.** 10   0.000977

> **EXAMPLE** You can use normal distributions to approximate binomial distributions.
>
> In 1990 about 1 in 43 births resulted in twins. If a town had 2157 births that year, what is the probability that between 29 and 50 of them were twins?
>
> $$\bar{x} = np = 2157\left(\frac{1}{43}\right) \approx 50 \text{ and } \sigma = \sqrt{np(1 - p)} = \sqrt{(2157)\left(\frac{1}{43}\right)\left(\frac{42}{43}\right)} \approx 7$$
>
> So, $P(29 \leq x \leq 50) = P(\bar{x} - 3\sigma \leq x \leq \bar{x}) = 0.0235 + 0.135 + 0.34 = 0.4985$, referring to the diagram on page 746.

A binomial distribution consists of 100 trials with probability 0.9 of success. Approximate the probability of getting the given numbers of successes.

35. between 87 and 93
   0.68

36. greater than 90
   0.5

37. less than 84
   0.025

38. between 81 and 84
   0.0235

Chapter Test

ADDITIONAL RESOURCES
• *Chapter 12 Resource Book*
Chapter Test (3 levels) (p. 108)
SAT/ACT Chapter Test (p. 114)
Alternative Assessment (p. 115)

• 🖥 *Test and Practice Generator*

Find the number of permutations or combinations.

1. $_4P_3$ 24   **2.** $_{11}P_5$ 55,440   **3.** $_{14}P_2$ 182   **4.** $_9C_6$ 84   **5.** $_{17}C_3$ 680   **6.** $_5C_4$ 5

7. Find the number of distinguishable permutations of the letters in MONTANA.   1260

Expand the power of the binomial.   **8–13. See margin.**

8. $(x + 4)^6$ **9.** $(2x - 2)^5$ **10.** $(x + 8)^3$ **11.** $(x^2 + 1)^4$ **12.** $(x + y^2)^5$ **13.** $(3x - y)^3$

8. $x^6 + 24x^5 + 240x^4 + 1280x^3 + 3840x^2 + 6144x + 4096$
9. $32x^5 - 160x^4 + 320x^3 - 320x^2 + 160x - 32$
10. $x^3 + 24x^2 + 192x + 512$
11. $x^8 + 4x^6 + 6x^4 + 4x^2 + 1$
12. $x^5 + 5x^4y^2 + 10x^3y^4 + 10x^2y^6 + 5xy^8 + y^{10}$
13. $27x^3 - 27x^2y + 9xy^2 - y^3$

A card is drawn randomly from a standard 52-card deck. Find the probability of drawing the given card. (For a listing of the deck, see page. 708.)

14. a black card   0.5   **15.** an ace   0.0769   **16.** a black ace   0.0385   **17.** a king   0.0769   **18.** a heart   0.25   **19.** the king of hearts   0.0192

Find the indicated probability.

20. $P(A) = 80\%$ 0%
$P(B) = 20\%$
$P(A \text{ or } B) = 100\%$
$P(A \text{ and } B) = \underline{?}$

21. $P(A) = \underline{?}$ 0.17
$P(B) = 0.7$
$P(A \text{ or } B) = 0.82$
$P(A \text{ and } B) = 0.05$

22. $P(A) = \dfrac{1}{4}$ $\dfrac{3}{4}$
$P(A') = \underline{?}$

23. *A* and *B* are independent events.
$P(A) = 0.25$
$P(B) = 0.75$
$P(A \text{ and } B) = \underline{?}$
0.1875

24. *A* and *B* are dependent events.
$P(A) = 30\%$
$P(B \mid A) = 40\%$
$P(A \text{ and } B) = \underline{?}$
12%

25. *A* and *B* are dependent events.
$P(A) = \underline{?}$
$P(B \mid A) = 0.8$
$P(A \text{ and } B) = 0.32$
0.4

26. Calculate the probability of randomly guessing at least 7 correct answers on a 10-question true-or-false quiz to get a passing grade.   0.172

27. What percent of the area under a normal curve lies within 1 standard deviation of the mean? What percent lies within 2 standard deviations of the mean?   68%, 95%

28. 🌐 SCHOOL SHIRTS A school shirt is available either long-sleeved or short-sleeved, in sizes small, medium, large, or extra large, and in one of two colors. How many different choices for a school shirt are there?   16

29. 🌐 SUPREME COURT The Supreme Court of the United States has 9 justices. On a certain case the justices voted 5 to 4 in favor of the defendant. In how many ways could this have happened?   126

30. 🌐 ASTRONOMY The surface area of Earth is about 197 million square miles. The land area is about 57 million square miles and the rest is water. What is the probability that a meteorite falling to Earth will hit land? What is the probability that it will hit water?   The probability is 0.289 that it will hit land and 0.711 that it will hit water.

31. 🌐 EMPLOYMENT AGENCY A temporary employment agency claims that it has a "no-show" rate of 1 out of 1000 workers. If fewer employees show up for a job than are requested, the difference is the number of "no-shows." A company hires the employment agency to supply 200 workers and only 198 show up. Would you reject the agency's claim about its "no-show" rate? Explain.   Reject the claim because the probability of this many no-shows is less than 0.05.

32. 🌐 HEALTH Health officials who have studied a particular virus say that 50% of all Americans have had the virus. If a random sample of 144 people is taken, what is the probability that fewer than 60 have had the virus?   0.025

Chapter Standardized Test

▶ **TEST-TAKING STRATEGY** Do not panic if you run out of time before answering all of the questions. You can still receive a high score on the SAT without answering every question.

ADDITIONAL RESOURCES
• *Chapter 12 Resource Book*
  Chapter Test (3 levels) (p. 108)
  SAT/ACT Chapter Test (p. 114)
  Alternative Assessment (p. 115)
• 🖥 *Test and Practice Generator*

1. MULTIPLE CHOICE In how many ways can a president and vice president be selected from a club of 20 students? **D**

ⓐ 20   ⓑ 39   ⓒ 40

ⓓ 380   ⓔ 400

2. MULTIPLE CHOICE In how many ways can 2 co-chairs be selected from a club of 20 students? **D**

ⓐ 10   ⓑ 39   ⓒ 40

ⓓ 190   ⓔ 380

3. MULTIPLE CHOICE What is the coefficient of x^5 in the expansion of $(2x + 5)^8$? **E**

ⓐ 6   ⓑ 56   ⓒ 240

ⓓ 1792   ⓔ 224,000

4. MULTIPLE CHOICE You have an equally likely chance of choosing any number from 1 to 10. What is the probability that you choose a number greater than 6? **B**

ⓐ $\frac{1}{10}$ ⓑ $\frac{2}{5}$ ⓒ $\frac{1}{2}$

ⓓ $\frac{3}{5}$ ⓔ $\frac{2}{3}$

5. MULTIPLE CHOICE A dart thrown at the square target shown is equally likely to hit anywhere inside the target. What is the probability that the dart hits the shaded semicircle? **C**

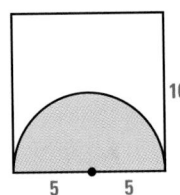

10

5   5

ⓐ $\frac{\pi}{20}$ ⓑ $\frac{\pi}{10}$ ⓒ $\frac{\pi}{8}$

ⓓ $\frac{\pi}{4}$ ⓔ $\frac{\pi}{2}$

6. MULTIPLE CHOICE $P(A) = 0.7$, $P(B) = 0.23$, and $P(A \text{ and } B) = 0.1$. What is $P(A \text{ or } B)$? **D**

ⓐ 0.13   ⓑ 0.47   ⓒ 0.6

ⓓ 0.83   ⓔ 0.93

7. MULTIPLE CHOICE A mother makes a list of 5 gift ideas for Mother's Day and gives the list to each of her 5 children. If each child buys one gift on the list at random, what is the probability that at least 2 of the gifts are the same? **D**

ⓐ 25%   ⓑ 50%   ⓒ 75%

ⓓ 96%   ⓔ 100%

8. MULTIPLE CHOICE Events A and B are independent, $P(A) = 0.8$, and $P(B) = 0.7$. What is $P(A \text{ and } B)$? **C**

ⓐ 1.5   ⓑ 0.75   ⓒ 0.56

ⓓ 0.28   ⓔ 0.1

9. MULTIPLE CHOICE Events A and B are dependent, $P(A) = 50\%$, and $P(B \mid A) = 50\%$. What is $P(A \text{ and } B)$? **A**

ⓐ 25%   ⓑ 30%   ⓒ 50%

ⓓ 75%   ⓔ 100%

10. MULTIPLE CHOICE What is the probability that in a family with 5 children exactly 2 are girls? Assume a boy and a girl are equally likely. **D**

ⓐ $\frac{1}{16}$ ⓑ $\frac{5}{32}$ ⓒ $\frac{1}{5}$

ⓓ $\frac{5}{16}$ ⓔ $\frac{2}{5}$

11. MULTIPLE CHOICE The time that it takes for a fire department to arrive at a particular address on an emergency call is normally distributed with a mean of 6 minutes and a standard deviation of 1 minute. What is the probability that the fire department takes longer than 8 minutes to arrive at a particular address on an emergency call? **B**

ⓐ 0.015   ⓑ 0.025   ⓒ 0.05

ⓓ 0.235   ⓔ 0.5

12. MULTIPLE CHOICE What is the standard deviation of the normal distribution that approximates a binomial distribution consisting of 119 trials with probability 0.7 of success? **B**

ⓐ ≈ 0.8   ⓑ ≈ 5   ⓒ ≈ 6

ⓓ ≈ 7   ⓔ ≈ 11

QUANTITATIVE COMPARISON In Exercises 13 and 14, choose the statement that is true about the given quantities.

 (A) The quantity in column A is greater.

 (B) The quantity in column B is greater.

 (C) The two quantities are equal.

 (D) The relationship cannot be determined from the given information.

	Column A	Column B	
13.	The number of ways 4 books can be arranged on a bookshelf	The number of ways 4 books can be chosen from a set of 6 books	A
14.	$_{12}C_7$	$_{12}C_5$	C

15. MULTI-STEP PROBLEM You and a friend are taking part in a fundraiser walk. You both agree to arrive for registration between 8:00 A.M. and 8:30 A.M. You will wait for each other at the registration table for up to 10 minutes.

 a. Let x be the number of minutes after 8:00 A.M. that you arrive, and let y be the number of minutes after 8:00 A.M. that your friend arrives. Let $x = 0$ and $y = 0$ represent 8:00 A.M. Write inequalities representing the time intervals in which you and your friend will arrive. $0 \le x \le 30; 0 \le y \le 30$

 b. If you and your friend are to meet, the difference between your arrival times must not exceed 10 minutes. Write two inequalities that show this fact. $x - y \le 10; y - x \le 10$

 c. Graph your inequalities from part (a) showing the times in which you and your friend will arrive. In the same coordinate plane, graph your inequalities from part (b) showing the times you and your friend could meet.   **See margin.**

 d. Using your graph from part (c), find the probability that you and your friend will meet at the registration table.   **0.556**

16. MULTI-STEP PROBLEM The *standard normal distribution* has a mean of 0 and a standard deviation of 1. To convert an x-value from a normal distribution with a mean of $\bar{x}$ and a standard deviation of σ to a z-value from a standard normal distribution, use this formula:

$$z = \frac{x - \bar{x}}{\sigma}$$

A z-value gives the number of standard deviations an x-value is from the mean $\bar{x}$. You can use z-values to compare x-values from different normal distributions.

 a. The scores of your class on a test have a mean of 70 and a standard deviation of 8. If your score on the test is x_{old}, what is your corresponding z-value? $z = \dfrac{x_{old} - 70}{8}$

 b. Your teacher wants to recenter the scores so that they have a mean of 85 and a standard deviation of 5, but does not want to change the z-value of any score. If your new score is to be x_{new}, what is your corresponding z-value? $z = \dfrac{x_{new} - 85}{5}$

 c. Set the formulas for the z-values from parts (a) and (b) equal to each other to get an equation in terms of x_{old} and x_{new}. Solve this equation for x_{new}. $x_{new} = \dfrac{5 \times (x_{old} - 70)}{8} + 85$

 d. If your score was 70, what is your new score? Does this make sense? Explain.   **16d. 85; This makes sense because your score was the mean originally (70), so it should also be the mean (85) in the new distribution.**

 e. What percent of your class would originally have had scores between 54 and 78? What will be the new range of scores for this part of the class? How can you answer this question without using the formula from part (c)?   **See margin.**

15c.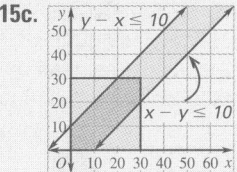

16e. 0.815; 75–90; The range is from two standard deviations below the mean to one standard deviation above the mean in the original distribution. Therefore, it should be two standard deviations below to one standard deviation above the mean in the new distribution. So, you could just subtract 5 twice from 85 and add 5 to 85 once to get the minimum and maximum values of the range.

Chapter Standardized Test **761**

9. $\frac{1}{8}\left(1 \pm i\sqrt{15}\right)$

17.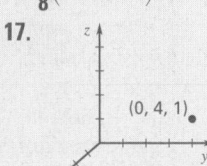

18.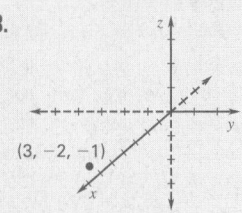

19.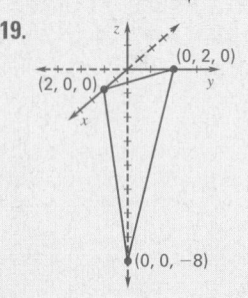

20.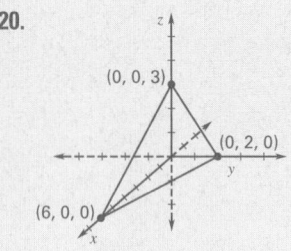

36.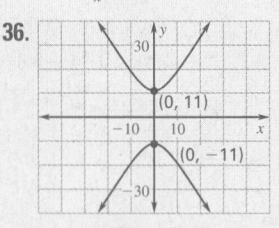

37.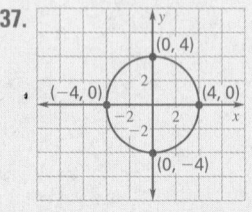

Solve the equation. (1.3, 1.7, 5.2–5.6, 6.4, 7.6, 8.6, 9.6)

1. $-4x + 5 = 33$ -7

2. $\frac{1}{4}(x - 7) = 2$ 15

3. $|x - 3| = 11$ $-8, 14$

4. $|8 - 3x| = 1$ $\frac{7}{3}, 3$

5. $x^2 + 7x + 10 = 0$ $-5, -2$

6. $5x^2 - 13 = 32$ ± 3

7. $-x^2 = 16$ $\pm 4i$

8. $x^2 + 6x - 5 = 0$
$-3 \pm \sqrt{14}$

9. $4x^2 - x + 1 = 0$
See margin.

10. $x^3 - 27 = 0$ 3

11. $x^3 + x^2 - 4x = 4$
$-2, -1, 2$

12. $\sqrt{x + 5} = 7$ 44

13. $8(x - 3)^{3/2} = 1$ 3.25

14. $4^{x+1} = 64$ 2

15. $\log 4x = 2$ 25

16. $\frac{1}{x - 4} = \frac{6}{x + 6}$ 6

Graph in a three-dimensional coordinate system. (3.5) 17–20. See margin.

17. $(0, 4, 1)$

18. $(3, -2, -1)$

19. $4x + 4y - z = 8$

20. $2x + 6y + 4z = 12$

Evaluate the determinant of the matrix. (4.3)

21. $\begin{bmatrix} 4 & -2 \\ -1 & 5 \end{bmatrix}$ 18

22. $\begin{bmatrix} -3 & 10 \\ -6 & 3 \end{bmatrix}$ 51

23. $\begin{bmatrix} 3 & -2 & 4 \\ -1 & 5 & 0 \\ 0 & 2 & 1 \end{bmatrix}$ 5

24. $\begin{bmatrix} -5 & -1 & 4 \\ 1 & 0 & 6 \\ 1 & 3 & 0 \end{bmatrix}$ 96

The variables *x* and *y* vary inversely. Use the given values to write an equation relating *x* and *y*. Then find *y* when *x* = 2. (9.1)

25. $x = -2, y = 20$
$y = -\frac{40}{x}, -20$

26. $x = \frac{1}{3}, y = 9$ $y = \frac{3}{x}, \frac{3}{2}$

$y = -\frac{16}{x}, -8$
27. $x = 20, y = -\frac{4}{5}$

28. $x = 1, y = 4$ $y = \frac{4}{x}, 2$

The variable *z* varies jointly with *x* and *y*. Use the given values to write an equation relating *x*, *y*, and *z*. Then find *z* when *x* = −1 and *y* = 5. (9.1)

$z = -2xy, 10$
29. $x = 2, y = 3, z = -4$
$z = -\frac{2}{3}xy, \frac{10}{3}$

30. $x = -2, y = 6, z = 24$

31. $x = \frac{1}{2}, y = \frac{1}{4}, z = \frac{3}{8}$ $z = 3xy, -15$

Find the distance between the two points. Then find the midpoint of the line segment connecting the two points. (10.1)

32. $(0, 0), (-9, 2)$
$\sqrt{85} \approx 9.22, (-4.5, 1)$

33. $(0, 8), (5, 0)$
$\sqrt{89} \approx 9.43, (2.5, 4)$

34. $(-5, 14), (3, -8)$
$2\sqrt{137} \approx 23.41, (-1, 3)$

35. $(-2, -3), (5, 1)$
$\sqrt{65} \approx 8.06, (1.5, -1)$

Graph the conic section. (10.2–10.6)
36–39. See margin.

36. $\frac{y^2}{121} - \frac{x^2}{49} = 1$

37. $x^2 + y^2 = 16$

38. $\frac{x^2}{81} + \frac{y^2}{36} = 1$

39. $(y + 4)^2 = x - 1$

Write an equation of the conic section. (10.2–10.6)

40. Parabola with vertex at $(0, 0)$ and directrix $y = -2$ $x^2 = 8y$

41. Circle with center at $(2, -2)$ and radius 3 $(x - 2)^2 + (y + 2)^2 = 9$

42. Ellipse with center at $(0, 0)$, vertex at $(8, 0)$, and co-vertex at $(0, 5)$ $\frac{x^2}{64} + \frac{y^2}{25} = 1$

43. Hyperbola with vertices at $(0, 2)$ and $(0, -2)$ and foci at $(0, 3)$ and $(0, -3)$ $\frac{y^2}{4} - \frac{x^2}{5} = 1$

Find the point(s) of intersection, if any, of the graphs in the system. (10.7)

44. $16x^2 + y^2 - 24y + 80 = 0$ $(0, 4)$
$16x^2 + 25y^2 - 400 = 0$

45. $x^2 + y^2 + 36x - 10y + 324 = 0$ $(-18, 0)$
$x^2 + y^2 + 36x - 20y + 324 = 0$

46. $x^2 + y^2 - 4x + 2y = 20$ none
$y^2 - 5x + 34 = 0$

47. $x^2 - y - 2 = 0$ $(-\sqrt{3}, 1), (\sqrt{3}, 1), \left(-\frac{\sqrt{6}}{2}, -\frac{1}{2}\right), \left(\frac{\sqrt{6}}{2}, -\frac{1}{2}\right)$
$x^2 + 4y^2 - 3y - 4 = 0$

Tell whether the sequence is *arithmetic, geometric,* or *neither*. Explain your answer. (11.2, 11.3) 48–51. See margin.

48. $-7, -1, 5, 11, \ldots$　　**49.** $7, 21, 63, 189, \ldots$　　**50.** $2, 3, 6, 11, \ldots$　　**51.** $1, 0.1, 0.01, 0.001, \ldots$

Write the first five terms of the sequence. (11.1, 11.5)

52. $a_n = 5n - 2$
$3, 8, 13, 18, 23$

53. $a_n = 10 - n^2$
$9, 6, 1, -6, -15$

54. $a_1 = 5$　$5, 11, 17, 23, 29$
$a_n = a_{n-1} + 6$

55. $a_1 = 1$　$1, 5, 14, 30, 55$
$a_n = a_{n-1} + n^2$

Write an explicit rule and a recursive rule for the sequence. (Recall that *d* is the common difference of an arithmetic sequence and *r* is the common ratio of a geometric sequence.) (11.2, 11.3, 11.5)

56. $r = 2, a_1 = 5$
$a_n = 5(2)^{n-1}; a_1 = 5, a_n = 2a^{n-1}$

57. $d = -6, a_1 = 1$
$a_n = 7 - 6n; a_1 = 1, a_n = a_{n-1} - 6$

58. $3, 5, 7, 9, \ldots$
$a_n = 2n + 1; a_1 = 3, a_n = a_{n-1} + 2$

59. $243, 81, 27, 9, \ldots$
$a_n = 243\left(\frac{1}{3}\right)^{n-1}; a_1 = 243, a_n = \frac{a_{n-1}}{3}$

Find the sum of the series. (11.1–11.4)

60. $\sum_{i=1}^{40} i$　820

61. $\sum_{i=1}^{5} (7 + i)$　50

62. $\sum_{i=1}^{6} \left(\frac{3}{4}\right)^{i-1}$　$\frac{3367}{1024}$

63. $\sum_{i=1}^{\infty} 8\left(\frac{1}{2}\right)^{i-1}$　16

Find the given number of permutations or combinations. (12.1, 12.2)

64. $_8P_5$　6720　**65.** $_6P_6$　720　**66.** $_{20}P_2$　380　**67.** $_8C_4$　70　**68.** $_5C_5$　1　**69.** $_7C_2$　21

Use the binomial theorem to write the binomial expansion. (12.2) 71–75. See margin.

70. $(x + 4)^5$　**71.** $(2x + 5)^3$　**72.** $(x + y)^6$　**73.** $(3x - 1)^4$　**74.** $(x + 2)^4$　**75.** $(x^2 - 4)^3$

Find the indicated probability. (12.4, 12.5)

76. $P(A) = 0.3$　0.05
$P(B) = 0.5$
$P(A \text{ or } B) = 0.75$
$P(A \text{ and } B) = \underline{\ ?\ }$

77. A and B are dependent events. 0.2
$P(B \mid A) = 0.5$
$P(A) = 0.4$
$P(A \text{ and } B) = \underline{\ ?\ }$

78. A and B are independent events. 9%
$P(A) = 90\%$
$P(B) = 10\%$
$P(A \text{ and } B) = \underline{\ ?\ }$

Calculate the probability of tossing a coin 5 times and getting exactly the given number of tails. (12.6)

79. 0　$\frac{1}{32}$　**80.** 1　$\frac{5}{32}$　**81.** 2　$\frac{5}{16}$　**82.** 3　$\frac{5}{16}$　**83.** 4　$\frac{5}{32}$　**84.** 5　$\frac{1}{32}$

85. ◉ **OVERNIGHT DELIVERY** The table at the right gives a company's overnight delivery charges for packages up to 10 pounds. Write and graph a piecewise function for this situation. (2.7) See margin.

86. ◉ **EARNING INTEREST** You deposit $1000 in an account that pays 4% annual interest compounded monthly. What is the balance after 5 years? (8.1) $1221.00

87. ◉ **FISH POPULATION** A lake initially contains 7000 fish. Each year the population declines 20% and the lake is restocked with 1000 new fish. Write a recursive rule for the number of fish in the lake after *n* years. What happens to the population of fish in the lake over time? (11.5)
$a_n = 0.8a_{n-1} + 1000$; it approaches 5000.

88. ◉ **PASSWORD** You need to select a four-character password for a computer account. Any digit 0–9 and any letter A–Z can be used for a character, and digits and letters can be repeated. How many possible passwords are there? (12.1)
1,679,616 passwords

89. ◉ **CLASS CLUB** A high school club has 10 members. The faculty advisor selects members at random to fill leadership positions for president, vice president, treasurer, and secretary. Find the probability that Mark, one of the club members, is selected for a leadership position. (12.3)　$\frac{2}{5}$

Package weight (lb)	Delivery charge ($)
0.5	11.75
1	14.00
2	15.75
3	18.50
4	21.25
5	24.00
6	26.25
7	28.00
8	30.25
9	31.00
10	32.75

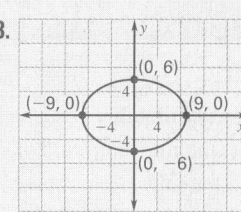

38.

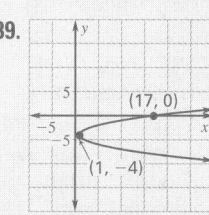

39.

48. arithmetic; each term is 6 more than the previous term.

49. geometric; each term is 3 times the previous term.

50. neither; no common ratio or difference.

51. geometric; each term is $\frac{1}{10}$ the previous term.

70. $x^5 + 20x^4 + 160x^3 + 640x^2 + 1280x + 1024$

71. $8x^3 + 60x^2 + 150x + 125$

72. $x^6 + 6x^5y + 15x^4y^2 + 20x^3y^3 + 15x^2y^4 + 6xy^5 + y^6$

73. $81x^4 - 108x^3 + 54x^2 - 12x + 1$

74. $x^4 + 8x^3 + 24x^2 + 32x + 16$

75. $x^6 - 12x^4 + 48x^2 - 64$

85.

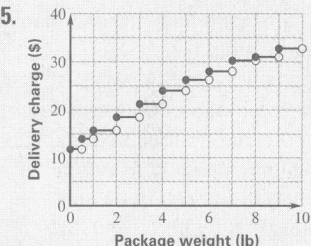

$$C(x) = \begin{cases} 11.75, & \text{if } 0 < x \le 0.5 \\ 14.00, & \text{if } 0.5 < x \le 1 \\ 15.75, & \text{if } 1 < x \le 2 \\ 18.50, & \text{if } 2 < x \le 3 \\ 21.25, & \text{if } 3 < x \le 4 \\ 24.00, & \text{if } 4 < x \le 5 \\ 26.25, & \text{if } 5 < x \le 6 \\ 28.00, & \text{if } 6 < x \le 7 \\ 30.25, & \text{if } 7 < x \le 8 \\ 31.00, & \text{if } 8 < x \le 9 \\ 32.75, & \text{if } 9 < x \le 10 \end{cases}$$

Monte Carlo Methods

OBJECTIVE Use a Monte Carlo method to estimate the area of an ellipse.

Materials: graph paper, calculator or computer

Some situations are too complex to be conveniently modeled by equations, graphs, or other standard mathematical tools. Even many simple-sounding problems, such as finding the average wait in line at an ATM machine, can be difficult to solve using traditional mathematics.

To solve such problems, you can often use *Monte Carlo* methods. A Monte Carlo method uses simulations based on random processes to model a real-life situation. The outcomes of these simulations are used to predict the outcomes of the real-life situation and to solve problems requiring knowledge of these outcomes. In this project you will estimate the area of an ellipse with a Monte Carlo method.

PROJECT GOALS

- Calculate experimental probability.
- Calculate geometric probability.
- Use a random number generator to simulate an experiment.

MANAGING THE PROJECT
CLASSROOM MANAGEMENT

The Chapter 12 Project may be completed by individual students or by students working in small groups. If students work in groups, each student can be responsible for doing at least one step of the Investigation while the other partners check the work. Each member of the group should test some of the points in Step 4 of the investigation and the group should discuss and make the conjecture in Step 7. The group can work on different parts of the report, but they should discuss each part beforehand and edit each others' work before a final draft is made.

ALTERNATIVE APPROACH

The Investigation portion of this project can be done as a class demonstration. A computer or calculator with an overhead display can be used to graph the ellipse. You can also plot the points on the display so that students can see if the point is contained in the ellipse. After the investigation, you can use the parts of the report as class discussion questions. The last four bulleted questions can be completed individually.

CONCLUDING THE PROJECT

Discuss the project as a class. Ask different individuals or groups to share parts of their reports. You can use some of these questions for class discussion.

- How does the number of points generated during the simulation affect the accuracy of the estimate of the ellipse's area?
- What is a good name for the sequence a_n?
- What are some of the advantages of the Monte Carlo method over the rectangle method?

INVESTIGATION

1. An ellipse with a horizontal major axis and center (0, 0) has an equation of the form $\frac{x^2}{a^2} + \frac{y^2}{b^2} = 1$. Write an inequality that must be satisfied if a point (x, y) is on or inside the ellipse (that is, (x, y) is part of the region bounded by the ellipse). $\frac{x^2}{a^2} + \frac{y^2}{b^2} \leq 1$

2. Write an equation for a particular ellipse with a horizontal major axis and center (0, 0). Choose integer values of a and b so that the equation will be easy to work with. Graph your equation.

3. On the graph, draw a rectangle extending horizontally from $-a$ to a and vertically from $-b$ to b that will contain your ellipse. An example is shown below.
Any equation of an ellipse where $a > b$.

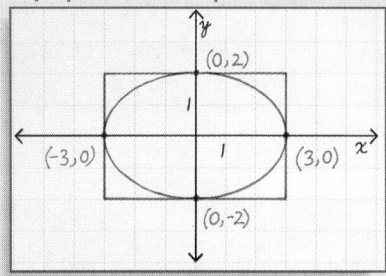

4. Use a random number generator on a calculator or computer to generate 20 points (x, y) where $-a \leq x \leq a$ and $-b \leq y \leq b$. Check if each point satisfies your inequality from **Step 1**. Make a table to keep track of your results.

Point (x, y)	On or inside ellipse?
$(-1.2, 0)$	Yes
$(2.6, 1.9)$	No

5. Use your table to find the ratio of the number of points on or inside the ellipse to the number of points on or inside the rectangle. To estimate the area of the ellipse, multiply the area of the rectangle by this ratio.

6. Use the formula for the area of an ellipse to find the actual area. Compare this with the area you found in **Step 5**. (*Hint:* See page 611 for help with the formula for the area of an ellipse.)

7. Repeat **Steps 4 and 5** using simulations with 10, 50, and 100 randomly generated points. Make a conjecture about the relationship between the number of points generated and the accuracy of your estimate for the ellipse's area.

Sample answer: the greater the number of points, the more accurate the estimated area is.

PRESENT YOUR RESULTS

Write a report to present your results.

- Explain how you used a Monte Carlo method to estimate the area of an ellipse.

- Is using the Monte Carlo method the same as calculating the area of the ellipse directly? Explain any differences.

- Explain why you estimated the area of the ellipse by multiplying the area of the rectangle by the ratio of the points on or inside the ellipse to those on or inside the rectangle.

- Include your graph.

- Include your table.

- Let n be the number of points used to estimate the area of the ellipse, and let a_n be a sequence defined as follows:

$$a_n = \left|\, \text{estimated area} - \text{actual area}\,\right|$$

Use your results from the investigation to graph the sequence defined by a_n for $n = 10, 25, 50,$ and 100. Describe what your graph shows.

- Modify the Monte Carlo method so that you generate random points only in the first quadrant. Explain how you can use these points and the ellipse's symmetry to estimate the entire area of the ellipse.

Now estimate the area of your ellipse using this method:

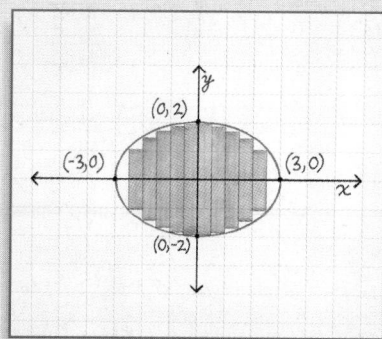

- Copy the ellipse you drew. Sketch small rectangles to fill it, as shown.

- Find the area of each rectangle. Add these areas together to estimate the area of the ellipse.

- Compare the areas given by this method and the Monte Carlo method. How would you make this method more accurate?

EXTENSION

Modify your Monte Carlo method to estimate the area of a region bounded (at least in part) by a different conic section. For example, you could estimate the area of a region bounded by one branch of a hyperbola and a line that intersects the branch in two points.

A Monte Carlo Method
for Estimating
Area

by
Kaylie Leite and
Amy Hill

Mrs. Kinney
Algebra 2
4th period

GRADING THE PROJECT
RUBRIC FOR CHAPTER PROJECT

4 Students' reports contain well-written explanations of the Monte Carlo method used. The report includes a complete and well-labeled graph of the ellipse and rectangle. The table is organized and is used to find the correct estimate of the area of the ellipse. The sequence a_n is graphed and the area of the ellipse is calculated using points only in the first quadrant as well. An estimate of the area using rectangles is given and compared to the estimate from the Monte Carlo method.

3 Students' graphs and table are included but may be lacking in some details. Explanations in the report may show misunderstandings of the Monte Carlo method. A description of the graph of the sequence a_n and a comparison of the rectangle and Monte Carlo method may be missing, unclear, or incomplete.

2 Some of the required graphs or table contain serious errors or are incomplete. Not all of the questions are answered thoroughly or correctly in the report.

1 The graphs and table are missing or contain serious errors. The Monte Carlo method is not used correctly to estimate the area of the ellipse. The explanations show that the student does not understand the mathematics involved in the project. The project should be returned with a new deadline for completion. The student should speak with the teacher as soon as possible so that the student understands the purpose and format of the project.

PLANNING THE CHAPTER

Trigonometric Ratios and Functions

GOALS

LESSON		NCTM	ITED	SAT9	Terra-Nova	Local
13.1 pp. 769–775	**GOAL 1** Use trigonometric relationships to evaluate trigonometric functions of acute angles. **GOAL 2** Use trigonometric functions to solve real-life problems.	1, 2, 3, 6, 8, 9, 10	RQGE, MIRA, RQRA	35, 36	11, 13, 14, 16, 17, 18, 48, 49, 51, 52	
13.2 pp. 776–783	**GOAL 1** Measure angles in standard position using degree measure and radian measure. **GOAL 2** Calculate arc lengths and areas of sectors.	1, 2, 3, 4	RQM	29, 40	11, 13, 14	
13.3 pp. 784–790	**GOAL 1** Evaluate trigonometric functions of any angle. **GOAL 2** Use trigonometric functions to solve real-life problems.	1, 2, 3, 4, 6, 8, 9, 10	RQGE, MIRA, RQRA	2, 38	11, 13, 14, 16, 17, 18, 48, 49, 51, 52	
13.4 pp. 791–798	**CONCEPT ACTIVITY: 13.4** *Investigate inverse trigonometric functions.* **GOAL 1** Evaluate inverse trigonometric functions. **GOAL 2** Use inverse trigonometric functions to solve real-life problems.	1, 2, 3, 6, 8, 9, 10	MIRA, RQRA	2	11, 13, 14, 16, 17, 18, 48, 49, 51, 52	
13.5 pp. 799–806	**GOAL 1** Use the law of sines to find the sides and angles of a triangle. **GOAL 2** Find the area of any triangle.	1, 2, 3, 4, 9	RQM, RQGC, RQRA	35, 36	11, 13, 14, 17, 48, 49, 51, 52	
13.6 pp. 807–812	**GOAL 1** Use the law of cosines to find the sides and angles of a triangle. **GOAL 2** Use Heron's formula to find the area of a triangle.	1, 2, 3, 4, 9	RQM, RQGE, RQRA	35, 36	11, 13, 14, 17, 48	
13.7 pp. 813–820	**GOAL 1** Use parametric equations to represent motion in a plane. **GOAL 2** Use parametric equations to represent projectile motion. **TECHNOLOGY ACTIVITY: 13.7** *Graph a set of parametric equations on a graphing calculator.*	1, 2, 3, 10	MIG, IIG, RQGE	2	14, 16, 49, 51, 52	

RESOURCES

CHAPTER RESOURCE BOOKLETS

CHAPTER SUPPORT

Tips for New Teachers	p. 1	Prerequisite Skills Review	p. 5
Parent Guide for Student Success	p. 3	Strategies for Reading Mathematics	p. 7

LESSON SUPPORT

	13.1	13.2	13.3	13.4	13.5	13.6	13.7
Lesson Plans (regular and block)	p. 9	p. 25	p. 38	p. 50	p. 64	p. 78	p. 90
Warm-Up Exercises and Daily Quiz	p. 11	p. 27	p. 40	p. 52	p. 66	p. 80	p. 92
Activity Support Masters				p. 53			
Lesson Openers	p. 12	p. 28	p. 41	p. 54	p. 67	p. 81	p. 93
Graphing Calculator Activities & Keystrokes	p. 13			p. 68			p. 94
Practice (3 levels)	p. 15	p. 29	p. 42	p. 55	p. 70	p. 82	p. 95
Reteaching with Practice	p. 18	p. 32	p. 45	p. 58	p. 73	p. 85	p. 98
Quick Catch-Up for Absent Students	p. 20	p. 34	p. 47	p. 60	p. 75	p. 87	p. 100
Cooperative Learning Activities	p. 21						
Interdisciplinary Applications	p. 22		p. 48		p. 76		p. 101
Real-Life Applications		p. 35		p. 61		p. 88	
Math & History Applications	p. 23						
Challenge: Skills and Applications	p. 24	p. 36	p. 49	p. 62	p. 77	p. 89	p. 102

REVIEW AND ASSESSMENT

Quizzes	pp. 37, 63	Alternative Assessment with Math Journal	p. 111
Chapter Review Games and Activities	p. 103	Project with Rubric	p. 113
Chapter Test (3 levels)	pp. 104–109	Cumulative Review	p. 115
SAT/ACT Chapter Test	p. 110	Resource Book Answers	p. A1

TRANSPARENCIES

	13.1	13.2	13.3	13.4	13.5	13.6	13.7
Warm-Up Exercises and Daily Quiz	p. 97	p. 98	p. 99	p. 100	p. 101	p. 102	p. 103
Alternative Lesson Opener Transparencies	p. 84	p. 85	p. 86	p. 87	p. 88	p. 89	p. 90
Examples/Standardized Test Practice	✓	✓	✓	✓	✓	✓	✓
Answer Transparencies	✓	✓	✓	✓	✓	✓	✓

TECHNOLOGY

- Electronic Teaching Tools
- Online Lesson Planner
- Internet Support
- Personal Student Tutor
- Test and Practice Generator
- Instant Replay: Video Review Games
- Electronic Lesson Presentations (Lesson 13.3)

ADDITIONAL RESOURCES

- Basic Skills Workbook: Diagnosis and Remediation
- Worked-Out Solution Key
- Resources in Spanish
- Standardized Test Practice Workbook
- Practice Workbook with Examples

CHAPTER
13

PACING THE CHAPTER

Resource Key
● STUDENT EDITION
● CHAPTER 13 RESOURCE BOOK
● TEACHER'S EDITION

REGULAR SCHEDULE

Day 1

13.1

STARTING OPTIONS
- Prereq. Skills Review
- Strategies for Reading
- Homework Check
- Warm-Up or Daily Quiz

TEACHING OPTIONS
- Motivating the Lesson
- Les. Opener (Visual)
- Graphing Calc. Activity
- Examples 1–5
- Closure Question
- Guided Practice Exs.

APPLY/HOMEWORK
- See Assignment Guide.
- See the CRB: Practice, Reteach, Apply, Extend

ASSESSMENT OPTIONS
- Checkpoint Exercises
- Daily Quiz (13.1)
- Stand. Test Practice

Day 2

13.2

STARTING OPTIONS
- Homework Check
- Warm-Up or Daily Quiz

TEACHING OPTIONS
- Motivating the Lesson
- Les. Opener (Visual)
- Examples 1–4
- Guided Practice Exs. 1–3, 5–20

APPLY/HOMEWORK
- See Assignment Guide.
- See the CRB: Practice, Reteach, Apply, Extend

ASSESSMENT OPTIONS
- Checkpoint Exercises, pp. 777–778

Day 3

13.2 (cont.)

STARTING OPTIONS
- Homework Check

TEACHING OPTIONS
- Examples 5–6
- Closure Question
- Guided Practice Exs. 4, 21–24

APPLY/HOMEWORK
- See Assignment Guide.
- See the CRB: Practice, Reteach, Apply, Extend

ASSESSMENT OPTIONS
- Checkpoint Exercises, p. 779
- Daily Quiz (13.2)
- Stand. Test Practice
- Quiz (13.1–13.2)

Day 4

13.3

STARTING OPTIONS
- Homework Check
- Warm-Up or Daily Quiz

TEACHING OPTIONS
- Motivating the Lesson
- Les. Opener (Calculator)
- Examples 1–6
- Closure Question
- Guided Practice Exs.

APPLY/HOMEWORK
- See Assignment Guide.
- See the CRB: Practice, Reteach, Apply, Extend

ASSESSMENT OPTIONS
- Checkpoint Exercises
- Daily Quiz (13.3)
- Stand. Test Practice

Day 5

13.4

STARTING OPTIONS
- Homework Check
- Warm-Up or Daily Quiz

TEACHING OPTIONS
- Motivating the Lesson
- Concept Act. & Wksht.
- Les. Opener (Application)
- Examples 1–3
- Guided Practice Exs. 1–16

APPLY/HOMEWORK
- See Assignment Guide.
- See the CRB: Practice, Reteach, Apply, Extend

ASSESSMENT OPTIONS
- Checkpoint Exercises, p. 793

Day 6

13.4 (cont.)

STARTING OPTIONS
- Homework Check

TEACHING OPTIONS
- Examples 4–5
- Closure Question
- Guided Practice Ex. 17

APPLY/HOMEWORK
- See Assignment Guide.
- See the CRB: Practice, Reteach, Apply, Extend

ASSESSMENT OPTIONS
- Checkpoint Exercises, p. 794
- Daily Quiz (13.4)
- Stand. Test Practice
- Quiz (13.3–13.4)

Day 9

13.6

STARTING OPTIONS
- Homework Check
- Warm-Up or Daily Quiz

TEACHING OPTIONS
- Motivating the Lesson
- Les. Opener (Application)
- Examples 1–5
- Closure Question
- Guided Practice Exs.

APPLY/HOMEWORK
- See Assignment Guide.
- See the CRB: Practice, Reteach, Apply, Extend

ASSESSMENT OPTIONS
- Checkpoint Exercises
- Daily Quiz (13.6)
- Stand. Test Practice

Day 10

13.7

STARTING OPTIONS
- Homework Check
- Warm-Up or Daily Quiz

TEACHING OPTIONS
- Les. Opener (Appl.)
- Graphing Calc. Activity
- Examples 1–4
- Technology Activity
- Closure Question
- Guided Practice Exs.

APPLY/HOMEWORK
- See Assignment Guide.
- See the CRB: Practice, Reteach, Apply, Extend

ASSESSMENT OPTIONS
- Checkpoint Exercises
- Daily Quiz (13.7)
- Stand. Test Practice
- Quiz (13.5–13.7)

Day 11

Review

DAY 11 START OPTIONS
- Homework Check

REVIEWING OPTIONS
- Chapter 13 Summary
- Chapter 13 Review
- Chapter Review Games and Activities

APPLY/HOMEWORK
- Chapter 13 Test (practice)
- Ch. Standardized Test (practice)

Day 12

Assess

DAY 12 START OPTIONS
- Homework Check

ASSESSMENT OPTIONS
- Chapter 13 Test
- SAT/ACT Ch. 13 Test
- Alternative Assessment

APPLY/HOMEWORK
- Skill Review, p. 830

Day 7

13.5

STARTING OPTIONS
- Homework Check
- Warm-Up or Daily Quiz

TEACHING OPTIONS
- Motivating the Lesson
- Les. Opener (Activity)
- Graphing Calc. Activity
- Examples 1–4
- Guided Practice
  Exs. 1–10

APPLY/HOMEWORK
- See Assignment Guide.
- See the CRB: Practice, Reteach, Apply, Extend

ASSESSMENT OPTIONS
- Checkpoint Exercises, pp. 800–801

Day 8

13.5 (cont.)

STARTING OPTIONS
- Homework Check

TEACHING OPTIONS
- Examples 5–6
- Closure Question
- Guided Practice
  Exs. 11–15

APPLY/HOMEWORK
- See Assignment Guide.
- See the CRB: Practice, Reteach, Apply, Extend

ASSESSMENT OPTIONS
- Checkpoint Exercises, p. 802
- Daily Quiz (13.5)
- Stand. Test Practice

Day 1

Assess & 13.1
(Day 1 = Ch. 12 Day 6)

ASSESSMENT OPTIONS
- Chapter 12 Test
- SAT/ACT Ch. 12 Test
- Alternative Assessment

CH. 13 START OPTIONS
- Skill Review, p. 768
- Prereq. Skills Review
- Strategies for Reading

TEACHING 13.1 OPTIONS
- Warm-Up (Les. 13.1)
- Motivating the Lesson
- Les. Opener (Visual)
- Graphing Calc. Activity
- Examples 1–5
- Closure Question
- Guided Practice Exs.

APPLY/HOMEWORK
- See Assignment Guide.
- See the CRB: Practice, Reteach, Apply, Extend

ASSESSMENT OPTIONS
- Checkpoint Exercises
- Daily Quiz (Les. 13.1)
- Stand. Test Practice

Day 2

13.2 & 13.3

DAY 2 START OPTIONS
- Homework Check
- W-Up 13.2 or D. Quiz 13.1

TEACHING 13.2 OPTIONS
- Motivating the Lesson
- Les. Opener (Visual)
- Examples 1–6
- Closure Question
- Guided Practice Exs.

TEACHING 13.3 OPTIONS
- Warm-Up (Les. 13.3)
- Motivating the Lesson
- Les. Opener (Calc.)
- Examples 1–6
- Closure Question
- Guided Practice Exs.

APPLY/HOMEWORK
- See Assignment Guide.
- See the CRB: Practice, Reteach, Apply, Extend

ASSESSMENT OPTIONS
- Checkpoint Exercises
- Daily Quiz (13.2, 13.3)
- Quiz (13.1–13.2)

Day 3

13.4

DAY 3 START OPTIONS
- Homework Check
- Warm-Up or Daily Quiz

TEACHING 13.4 OPTIONS
- Motivating the Lesson
- Concept Act. & Wksht.
- Les. Opener (Appl.)
- Examples 1–5
- Closure Question
- Guided Practice Exs.

APPLY/HOMEWORK
- See Assignment Guide.
- See the CRB: Practice, Reteach, Apply, Extend

ASSESSMENT OPTIONS
- Checkpoint Exercises
- Daily Quiz (Les. 13.4)
- Stand. Test Practice
- Quiz (13.3–13.4)

Day 4

13.5

DAY 4 START OPTIONS
- Homework Check
- Warm-Up or Daily Quiz

TEACHING 13.5 OPTIONS
- Motivating the Lesson
- Les. Opener (Activity)
- Graphing Calc. Activity
- Examples 1–6
- Closure Question
- Guided Practice Exs.

APPLY/HOMEWORK
- See Assignment Guide.
- See the CRB: Practice, Reteach, Apply, Extend

ASSESSMENT OPTIONS
- Checkpoint Exercises,
- Daily Quiz (Les. 13.5)
- Stand. Test Practice

Day 5

13.6 & 13.7

STARTING OPTIONS
- Homework Check
- W-Up 13.6 or D. Quiz 13.5

TEACHING 13.6 OPTIONS
- Motivating the Lesson
- Les. Opener (Appl.)
- Examples 1–5
- Closure Question
- Guided Practice Exs.

TEACHING 13.7 OPTIONS
- Warm-Up (Les. 13.7)
- Les. Opener (Appl.)
- Technology Activities
- Examples 1–4
- Closure Question
- Guided Practice Exs.

APPLY/HOMEWORK
- See Assignment Guide.
- See the CRB: Practice, Reteach, Apply, Extend

ASSESSMENT OPTIONS
- Checkpoint Exercises
- Daily Quiz (13.6, 13.7)
- Stand. Test Practice
- Quiz (13.5–13.7)

Day 6

Review/Assess

DAY 6 START OPTIONS
- Homework Check

REVIEWING OPTIONS
- Chapter 13 Summary
- Chapter 13 Review
- Chapter Review Games and Activities
- Chapter 13 Test (practice)
- Ch. Standardized Test (practice)

ASSESSMENT OPTIONS
- Chapter 13 Test
- SAT/ACT Ch. 13 Test
- Alternative Assessment

APPLY/HOMEWORK
- Skill Review, p. 830

MEETING INDIVIDUAL NEEDS

BEFORE THE CHAPTER

The *Chapter 13 Resource Book* has the following materials to distribute and use before the chapter:

- **Parent Guide for Student Success**
- **Prerequisite Skills Review (pictured below)**
- **Strategies for Reading Mathematics**

PREREQUISITE SKILLS *Pages 5–6*

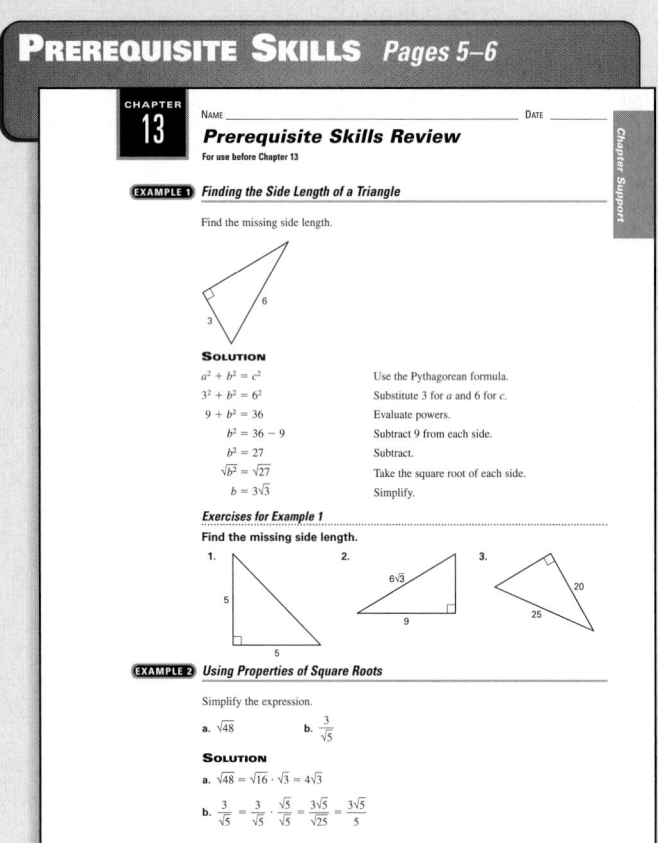

PREREQUISITE SKILLS REVIEW These two pages support the Study Guide on page 768. They help students prepare for Chapter 13 by providing worked-out examples and practice for the following skills needed in the chapter:

- **Find lengths of sides of triangles.**
- **Use properties of square roots to simplify expressions.**
- **Solve equations by cross multiplying.**

TECHNOLOGY RESOURCE
Students can use the Personal Student Tutor to find additional reteaching and practice for skills from earlier chapters that are used in Chapter 13.

DURING EACH LESSON

The *Chapter 13 Resource Book* has the following alternatives for introducing the lesson:

- **Lesson Openers (pictured below)**
- **Graphing Calculator Activities with Keystrokes**

LESSON OPENER *Page 28*

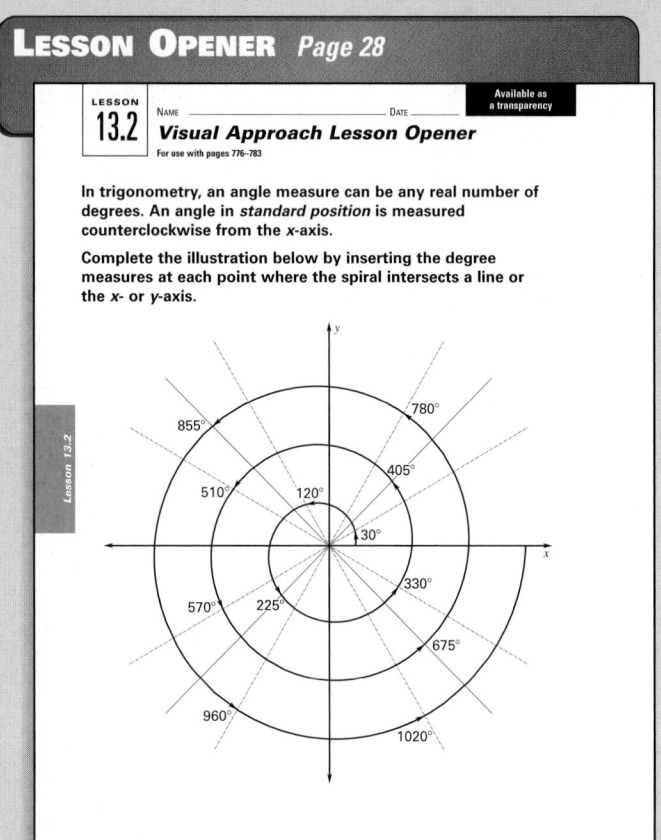

VISUAL APPROACH LESSON OPENER This Lesson Opener uses visuals as an alternative way to start Lesson 13.2. Students determine degree measures of a spiral as an introduction to angles.

TECHNOLOGY RESOURCE
Teachers can use the Electronic Lesson Presentations CD-ROM to present Lesson 13.3 using computer animation to step through the Examples.

FOLLOWING EACH LESSON

The *Chapter 13 Resource Book* has a variety of materials to follow-up each lesson. They include the following:

- **Practice (3 levels))**
- **Reteaching with Practice**
- **Quick Catch-Up for Absent Students**
- **Challenge: Skills and Applications (pictured below)**
- **Interdisciplinary Applications**
- **Real-Life Applications**

CHALLENGE *Page 62*

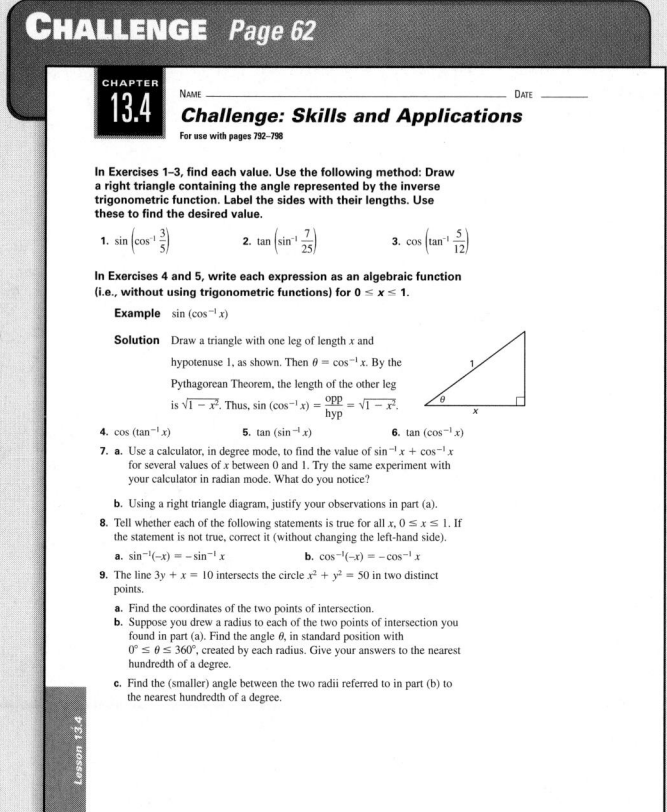

CHALLENGE: SKILLS AND APPLICATIONS Each lesson has a page of challenge exercises that extend the material in the lesson. The challenge page for Lesson 13.4 includes exercises where students rewrite trigonometric expressions in *x* in a form that does not involve trigonometric functions.

ASSESSING THE CHAPTER

The *Chapter 13 Resource Book* has the following review and assessment materials:

- **Quizzes**
- **Chapter Review Games and Activities**
- **Chapter Test (3 levels)**
- **SAT/ACT Chapter Test**
- **Alternative Assessment and Math Journal (below)**
- **Project with Rubric**
- **Cumulative Review**

ALTERNATIVE ASSESSMENT *Pages 111–112*

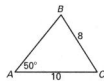

ALTERNATIVE ASSESSMENT WITH RUBRIC AND MATH JOURNAL The journal exercise asks students to demonstrate their understanding of trigonometric ratios in writing. The Multi-Step Problem has students pull together a variety of concepts from Chapter 13 to solve problems about triangles using the Law of Sines and the Law of Cosines. Answers and a scoring rubric are provided on a separate sheet.

CHAPTER GOALS

Students begin by evaluating trigonometric ratios of acute angles in right triangles. They consider general angles in standard position using both degree and radian measure, and find arc lengths and areas of sectors. Students use their knowledge of general angles to evaluate trigonometric functions of any angle. They also calculate projectile distance. Next, they evaluate and apply inverse trigonometric functions. All along, students apply their skill in solving right triangles to numerous real-life situations. They then use the law of sines and the law of cosines to solve general triangles. Finally, students explore parametric equations, and use them to model straight-line and projectile motion.

APPLICATION NOTE

A bridge that will withstand the heavy load and wear of all the vehicles that will use it requires careful design and engineering.

When the foundation of a bridge is being built under water, below the groundwater level, or in unstable soil, a shell called a *caisson* is sunk into the ground. The outer shell of the caisson protects workers inside against water pressure and soil collapse. Then the material inside the caisson is excavated and the interior filled with concrete. Large caissons may have a sealed-off working chamber filled with compressed air to keep soil and water from entering. Workers use air locks to enter the chamber.

Building the San Francisco-Oakland Bay Bridge required one of the largest caissons ever used. The caisson was sunk 279 feet below the water level.

Additional information about drawbridges is available at **www.mcdougallittell.com.**

TRIGONOMETRIC RATIOS AND FUNCTIONS

▶ *How can you find the width of the opening between two halves of a bridge?*

CHAPTER
13

APPLICATION: Drawbridges

The Chicago River System has 52 movable bridges, more than any other city in the world. One type of moveable bridge is a double-leaf drawbridge, a series of which are shown open in the photo at the left and closed in the photo at the right.

Think & Discuss

The diagram below gives the dimensions for a double-leaf drawbridge. Each leaf of the bridge is 65 feet long and has a maximum opening angle of 60°. For a 30°-60°-90° triangle, the ratio of the lengths of the sides, from shortest to longest, is $1 : \sqrt{3} : 2$.

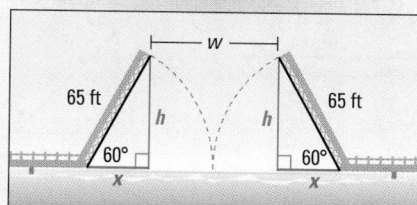

1. Find the height h to which the end of each leaf is lifted. **about 56.3 ft**

2. What is the maximum width w that a boat could have at the height of the opening and still fit through the opening? (*Hint*: Use the fact that $2x + w = 130$.) **65 ft**

Learn More About It

You will find the angle at which a drawbridge must open to allow a ship to pass in Ex. 56 on p. 796.

 APPLICATION LINK Visit www.mcdougallittell.com for more information on drawbridges.

ADDITIONAL RESOURCES
Another way to begin the chapter is to show the video clip of a real-life motivator for Chapter 13 on the *Instant Replay: Video Review Games.*

PROJECTS
A project covering Chapters 13 and 14 appears on pages 892 and 893 of the Student Edition. An additional project for Chapter 13 is available in the *Chapter 13 Resource Book,* p. 113.

TECHNOLOGY

 **Software**
- *Electronic Teaching Tools*
- *Online Lesson Planner*
- *Personal Student Tutor*
- *Test and Practice Generator*
- *Electronic Lesson Presentations* (Lesson 13.3)

Video
- *Instant Replay: Video Review Games*

Internet Connections
www.mcdougallittell.com
- **Application Links**
  767, 775, 778, 817
- **Data Updates**
  787, 815
- **Student Help**
  770, 777, 786, 796, 804, 811, 817, 820
- **Career Links**
  789, 801, 805, 811
- **Extra Challenge**
  774, 782, 790, 797, 806, 812, 818

767

PREPARE

DIAGNOSTIC TOOLS

The **Skill Review** exercises can help you diagnose whether students have the following skills needed in Chapter 13:

- Use the Pythagorean theorem to find a side of a triangle.
- Simplify radical expressions.
- Solve proportions.

The following resources are available for students who need additional help with these skills:

- Prerequisite Skills Review (*Chapter 13 Resource Book,* p. 5; *Warm-Up Transparencies,* p. 96)
- Reteaching with Practice (Chapter Resource Books for Lessons 5.3 and 9.6.)
- ▣ *Personal Student Tutor*

ADDITIONAL RESOURCES

The following resources are provided to help you prepare for the upcoming chapter and customize review materials:

- *Chapter 13 Resource Book* Tips for New Teachers (p. 1) Parent Guide (p. 3) Lesson Plans (every lesson) Lesson Plans for Block Scheduling (every lesson)
- ▣ *Electronic Teaching Tools*
- ▣ *Online Lesson Planner*
- ▣ *Test and Practice Generator*

ENGLISH LEARNERS

If students are not already keeping a glossary, encourage them to do so. In it they should describe in their own words the terms, properties, and rules of trigonometry they encounter in this chapter. Remind them to use this glossary as a reference guide.

PREVIEW

What's the chapter about?

Chapter 13 is about **trigonometry**. In Chapter 13 you'll learn

- how to evaluate trigonometric functions and inverse trigonometric functions.
- how to find side lengths, angle measures, and areas of triangles.
- how to use parametric equations.

KEY VOCABULARY

▶ Review
- reciprocal, p. 5
- inverse functions, p. 422

▶ New
- sine, p. 769
- cosine, p. 769

- tangent, p. 769
- cosecant, p. 769
- secant, p. 769
- cotangent, p. 769
- radian, p. 777
- sector, p. 779

- inverse sine, p. 792
- inverse cosine, p. 792
- inverse tangent, p. 792
- law of sines, p. 799
- law of cosines, p. 807
- parametric equations, p. 813

PREPARE

Are you ready for the chapter?

SKILL REVIEW Do these exercises to review key skills that you'll apply in this chapter. See the given **reference page** if there is something you don't understand.

STUDENT HELP

↳ **Study Tip**
"Student Help" boxes throughout the chapter give you study tips and tell you where to look for extra help in this book and on the Internet.

Find the missing side length. (Skills Review, p. 917)

1.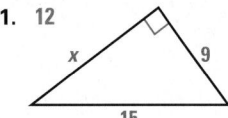

2.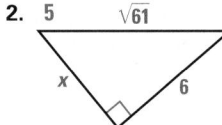

3.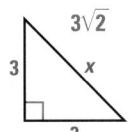

Simplify the expression. (Review Example 1, p. 264)

4. $\sqrt{96}$ $4\sqrt{6}$ 5. $\sqrt{18}$ $3\sqrt{2}$ 6. $\sqrt{200}$ $10\sqrt{2}$ 7. $\dfrac{2}{\sqrt{2}}$ $\sqrt{2}$ 8. $\dfrac{2}{\sqrt{3}}$ $\dfrac{2\sqrt{3}}{3}$ 9. $\sqrt{\dfrac{3}{4}}$ $\dfrac{\sqrt{3}}{2}$

Solve the equation. (Review Example 4, p. 569)

10. $\dfrac{3}{x} = \dfrac{6}{x-1}$ -1 11. $\dfrac{4x}{5} = \dfrac{5}{x}$ $\pm\dfrac{5}{2}$ 12. $\dfrac{7}{4} = \dfrac{x}{8}$ 14   13. $\dfrac{x+3}{x} = \dfrac{7}{10}$ -10

STUDY STRATEGY

Here's a study strategy!

Draw Diagrams

Drawing a diagram can help you figure out how to solve a problem from the given information. A diagram can help you with almost every problem (not just word problems) in this chapter. Whenever you are given lengths, angle measures, points, or ratios, try drawing and labeling a diagram.

13.1 Right Triangle Trigonometry

What you should learn

GOAL 1 Use trigonometric relationships to evaluate trigonometric functions of acute angles.

GOAL 2 Use trigonometric functions to solve **real-life** problems, such as finding the altitude of a kite in **Example 4**.

Why you should learn it

▼ To solve **real-life** problems, such as finding the length of a zip-line at a ropes course in **Ex. 50**.

GOAL 1 EVALUATING TRIGONOMETRIC FUNCTIONS

Consider a right triangle, one of whose acute angles is θ (the Greek letter *theta*). The three sides of the triangle are the *hypotenuse,* the side *opposite* θ, and the side *adjacent* to θ.

Ratios of a right triangle's three sides are used to define the six trigonometric functions: **sine, cosine, tangent, cosecant, secant,** and **cotangent**. These six functions are abbreviated sin, cos, tan, csc, sec, and cot, respectively.

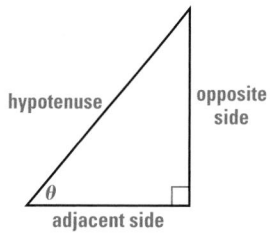

RIGHT TRIANGLE DEFINITION OF TRIGONOMETRIC FUNCTIONS

Let θ be an acute angle of a right triangle. The six trigonometric functions of θ are defined as follows.

$$\sin \theta = \frac{\text{opp}}{\text{hyp}} \qquad \cos \theta = \frac{\text{adj}}{\text{hyp}} \qquad \tan \theta = \frac{\text{opp}}{\text{adj}}$$

$$\csc \theta = \frac{\text{hyp}}{\text{opp}} \qquad \sec \theta = \frac{\text{hyp}}{\text{adj}} \qquad \cot \theta = \frac{\text{adj}}{\text{opp}}$$

The abbreviations *opp, adj,* and *hyp* represent the lengths of the three sides of the right triangle. Note that the ratios in the second row are the reciprocals of the ratios in the first row. That is:

$$\csc \theta = \frac{1}{\sin \theta} \qquad \sec \theta = \frac{1}{\cos \theta} \qquad \cot \theta = \frac{1}{\tan \theta}$$

EXAMPLE 1 Evaluating Trigonometric Functions

Evaluate the six trigonometric functions of the angle θ shown in the right triangle.

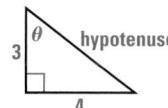

SOLUTION

From the Pythagorean theorem, the length of the hypotenuse is:

$$\sqrt{3^2 + 4^2} = \sqrt{25} = 5$$

Using adj = 3, opp = 4, and hyp = 5, you can write the following.

$$\sin \theta = \frac{\text{opp}}{\text{hyp}} = \frac{4}{5} \qquad \cos \theta = \frac{\text{adj}}{\text{hyp}} = \frac{3}{5} \qquad \tan \theta = \frac{\text{opp}}{\text{adj}} = \frac{4}{3}$$

$$\csc \theta = \frac{\text{hyp}}{\text{opp}} = \frac{5}{4} \qquad \sec \theta = \frac{\text{hyp}}{\text{adj}} = \frac{5}{3} \qquad \cot \theta = \frac{\text{adj}}{\text{opp}} = \frac{3}{4}$$

STUDENT HELP

→ **Skills Review**
For help with the Pythagorean theorem, see p. 917.

1 PLAN

PACING
Basic: 1 day
Average: 1 day
Advanced: 1 day
Block Schedule: 0.5 block with 13.2

LESSON OPENER
VISUAL APPROACH
An alternative way to approach Lesson 13.1 is to use the Visual Approach Lesson Opener:
• Blackline Master (*Chapter 13 Resource Book,* p. 12)
• Transparency (p. 84)

MEETING INDIVIDUAL NEEDS
• *Chapter 13 Resource Book*
Prerequisite Skills Review (p. 5)
Practice Level A (p. 15)
Practice Level B (p. 16)
Practice Level C (p. 17)
Reteaching with Practice (p. 18)
Absent Student Catch-Up (p. 20)
Challenge (p. 24)
• *Resources in Spanish*
• Personal Student Tutor

NEW-TEACHER SUPPORT
See the Tips for New Teachers on pp. 1–2 of the *Chapter 13 Resource Book* for additional notes about Lesson 13.1.

WARM-UP EXERCISES
Transparency Available
Simplify the expression.
1. $\sqrt{6^2 + 8^2}$ 10
2. $\sqrt{8^2 - 6^2}$ $2\sqrt{7}$
Solve each proportion.
3. $\frac{x}{6} = \frac{8}{15}$ $\frac{16}{5}$
4. $\frac{12}{x} = \frac{9}{5}$ $\frac{20}{3}$
5. $\frac{12}{70} = \frac{x}{28}$ $\frac{24}{5}$

Ask students how they could measure the height of a tall tree without climbing it. Tell them that the special properties of right triangles allow us to measure many quantities indirectly. For the tree, they need only know the length of the shadow and the angle from the shadow's end to the top of the tree.

EXTRA EXAMPLE 1
Evaluate the six trigonometric functions of the angle θ shown in the right triangle.

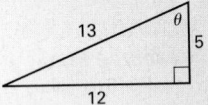

$\sin\theta = \dfrac{12}{13}$; $\cos\theta = \dfrac{5}{13}$; $\tan\theta = \dfrac{12}{5}$;

$\csc\theta = \dfrac{13}{12}$; $\sec\theta = \dfrac{13}{5}$; $\cot\theta = \dfrac{5}{12}$

EXTRA EXAMPLE 2
Find the value of x for the right triangle shown. **7.5**

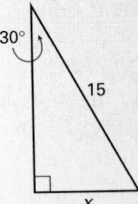

EXTRA EXAMPLE 3
Solve $\triangle ABC$.

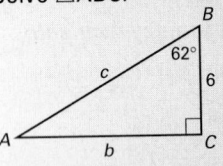

$A = 28°$; $b \approx 11.3$; $c \approx 12.8$

 CHECKPOINT EXERCISES

For use after Examples 1–3:
1. Solve $\triangle ABC$ without using a calculator.

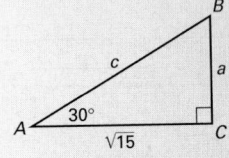

$B = 60°$; $a = \sqrt{5}$; $c = 2\sqrt{5}$

The angles 30°, 45°, and 60° occur frequently in trigonometry. The table below gives the values of the six trigonometric functions for these angles. To remember these values, you may find it easier to draw the triangles shown, rather than memorize the table.

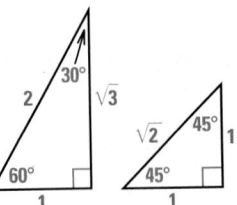

θ	$\sin\theta$	$\cos\theta$	$\tan\theta$	$\csc\theta$	$\sec\theta$	$\cot\theta$
30°	$\dfrac{1}{2}$	$\dfrac{\sqrt{3}}{2}$	$\dfrac{\sqrt{3}}{3}$	2	$\dfrac{2\sqrt{3}}{3}$	$\sqrt{3}$
45°	$\dfrac{\sqrt{2}}{2}$	$\dfrac{\sqrt{2}}{2}$	1	$\sqrt{2}$	$\sqrt{2}$	1
60°	$\dfrac{\sqrt{3}}{2}$	$\dfrac{1}{2}$	$\sqrt{3}$	$\dfrac{2\sqrt{3}}{3}$	2	$\dfrac{\sqrt{3}}{3}$

Trigonometric functions can be used to find a missing side length or angle measure of a right triangle. Finding *all* missing side lengths and angle measures is called **solving a right triangle**.

EXAMPLE 2 *Finding a Missing Side Length of a Right Triangle*

Find the value of x for the right triangle shown.

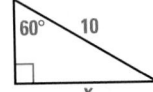

SOLUTION

Write an equation using a trigonometric function that involves the ratio of x and 10. Solve the equation for x.

$\sin 60° = \dfrac{\text{opp}}{\text{hyp}}$ **Write trigonometric equation.**

$\dfrac{\sqrt{3}}{2} = \dfrac{x}{10}$ **Substitute.**

$5\sqrt{3} = x$ **Multiply each side by 10.**

▶ The length of the side is $x = 5\sqrt{3} \approx 8.66$.

· · · · · · · · · ·

You can use a calculator to evaluate trigonometric functions of *any* angle, not just 30°, 45°, and 60°. Use the keys **SIN**, **COS**, and **TAN** for sine, cosine, and tangent. Use these keys and the reciprocal key for cosecant, secant, and cotangent. Before using the calculator be sure it is set in degree mode.

EXAMPLE 3 *Using a Calculator to Solve a Right Triangle*

Solve $\triangle ABC$.

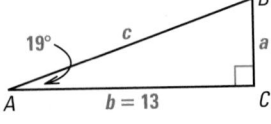

SOLUTION

Because the triangle is a right triangle, A and B are complementary angles, so $B = 90° - 19° = 71°$.

$\dfrac{a}{13} = \tan 19° \approx 0.3443$ $\dfrac{c}{13} = \sec 19° = \dfrac{1}{\cos 19°} \approx 1.058$

$a \approx 4.48$ $c \approx 13.8$

STUDENT HELP

HOMEWORK HELP
Visit our Web site
www.mcdougallittell.com
for extra examples.

STUDENT HELP

Study Tip
In Example 3, B is used to represent both the angle and its measure. Throughout this chapter, a capital letter is used to denote a vertex of a triangle and the same letter in lowercase is used to denote the side opposite that angle.

KITE FLYING
In the late 1800s and early 1900s, kites were used to lift weather instruments. In 1919 the German Weather Bureau set a kite-flying record. Eight kites on a single line, like those pictured above, were flown at an altitude of 9740 meters.

GOAL 2 **USING TRIGONOMETRY IN REAL LIFE**

EXAMPLE 4   *Finding the Altitude of a Kite*

KITE FLYING Wind speed affects the angle at which a kite flies. The table at the right shows the angle the kite line makes with a line parallel to the ground for several different wind speeds. You are flying a kite 4 feet above the ground and are using 500 feet of line. At what altitude is the kite flying if the wind speed is 35 miles per hour?

Wind speed (miles per hour)	Angle of kite line (degrees)
25	70
30	60
35	48
40	29
45	0

SOLUTION

At a wind speed of 35 miles per hour, the angle the kite line makes with a line parallel to the ground is 48°. Write an equation using a trigonometric function that involves the ratio of the distance d and 500.

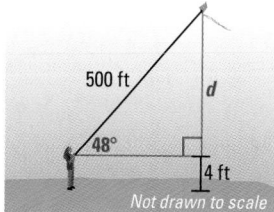

500 ft d

48°

4 ft

Not drawn to scale

$\sin 48° = \dfrac{d}{500}$ **Write trigonometric equation.**

$0.7431 \approx \dfrac{d}{500}$ **Simplify.**

$372 \approx d$ **Solve for d.**

▶ When you add 4 feet for the height at which you are holding the kite line, the kite's altitude is about 376 feet.

· · · · · · · · · ·

In Example 4 the angle the kite line makes with a line parallel to the ground is the **angle of elevation**. At the height of the kite, the angle from a line parallel to the ground to the kite line is the **angle of depression**. These two angles have the same measure.

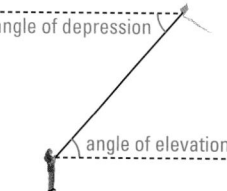

angle of depression

angle of elevation

Aviation

EXAMPLE 5   *Finding the Distance to an Airport*

An airplane flying at an altitude of 30,000 feet is headed toward an airport. To guide the airplane to a safe landing, the airport's landing system sends radar signals from the runway to the airplane at a 10° angle of elevation. How far is the airplane (measured along the ground) from the airport runway?

SOLUTION

Begin by drawing a diagram.

$\dfrac{x}{30{,}000} = \cot 10° = \dfrac{1}{\tan 10°} \approx 5.671$

$x \approx 170{,}100$

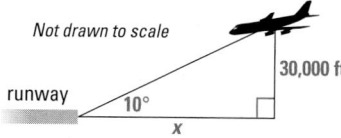

Not drawn to scale

30,000 ft

runway   10°

x

▶ The plane is about 170,100 feet (or 32.2 miles) from the airport.

EXTRA EXAMPLE 4
You are flying a kite 4 feet above the ground using 300 feet of line. With a wind speed of 40 miles per hour, the angle the kite line makes with the ground is 29°. How high is the kite flying?
about 149 ft

EXTRA EXAMPLE 5
An airplane flying at 20,000 feet is headed toward an airport. The airport's landing system sends radar signals from the runway to the airplane at a 5° angle of elevation. How many miles (measured along the ground) is the airplane from the runway?
about 43.3 mi

CHECKPOINT EXERCISES
For use after Example 4:
1. A support cable from a radio tower makes an angle of 56° with the ground. If the cable is 250 feet long, how far above the ground does it meet the tower?   about 207 ft

For use after Example 5:
2. You are standing at the end of the shadow of a giant sequoia, 150 feet from its base. The angle of elevation to the sun is 63°. How tall is the tree?   about 294 ft

FOCUS ON VOCABULARY
To help students understand *angle of elevation* and *angle of depression,* have them imagine holding a yardstick straight out parallel to the ground. If they rotate their arm and the yardstick upward, the angle through which they rotate it is the angle of elevation. If they rotate the yardstick downward, the angle through which they rotate it is the angle of depression.

CLOSURE QUESTION
What are the relationships of the sine, cosine, and tangent ratios to the cosecant, secant, and cotangent ratios?   Sine and cosecant are reciprocals; cosine and secant are reciprocals; and tangent and cotangent are reciprocals.

ASSIGNMENT GUIDE

BASIC
Day 1: pp. 772–775 Exs. 15–21,
22–32 even, 33–36, 42–45,
51, 55–63 odd

AVERAGE
Day 1: pp. 772–775 Exs. 15–21,
22–42 even, 43–46, 50, 51,
55–63 odd

ADVANCED
Day 1: pp. 772–775 Exs. 15–21,
22–42 even, 45–48, 51–54,
55–63 odd

BLOCK SCHEDULE WITH 13.2
pp. 772–775 Exs. 15–21, 22–42
even, 43–46, 50, 51, 55–63 odd

EXERCISE LEVELS
Level A: *Easier*
15–32, 55–58

Level B: *More Difficult*
33–51, 59–63

Level C: *Most Difficult*
52–54

✔ **HOMEWORK CHECK**
To quickly check student understanding of key concepts, go
over the following exercises:
Exs. 16, 22, 26, 34, 42. See also
the Daily Homework Quiz:

• Blackline Master (*Chapter 13
Resource Book*, p. 27)

• Transparency (p. 98)

TEACHING TIPS
EXERCISES 3–7, 15–21
Students should understand that
the symbol θ is used just like any
other variable, except that it is used
almost exclusively to represent an
angle measure.

15–21. See next page.

GUIDED PRACTICE

Vocabulary Check ✔

Concept Check ✔

2. No; all $30°-60°-90°$
triangles are similar to
one another, and you
need to know the length
of at least one side to
find the others.

Skill Check ✔

5. $\sin \theta = \frac{3}{5}$; $\cos \theta = \frac{4}{5}$;
$\tan \theta = \frac{3}{4}$; $\csc \theta = \frac{5}{3}$;
$\sec \theta = \frac{5}{4}$; $\cot \theta = \frac{4}{3}$

6. $\sin \theta = \frac{\sqrt{2}}{2}$; $\cos \theta = \frac{\sqrt{2}}{2}$;
$\tan \theta = 1$; $\csc \theta = \sqrt{2}$;
$\sec \theta = \sqrt{2}$; $\cot \theta = 1$

7. $\sin \theta = \frac{\sqrt{5}}{3}$; $\cos \theta = \frac{2}{3}$;
$\tan \theta = \frac{\sqrt{5}}{2}$; $\csc \theta = \frac{3\sqrt{5}}{5}$;
$\sec \theta = \frac{3}{2}$; $\cot \theta = \frac{2\sqrt{5}}{5}$

1. Explain what it means to solve a right triangle. **to find the measures of all unknown angles and sides**

2. Given a $30°$-$60°$-$90°$ triangle with only the measures of the angles labeled, can you find the lengths of any of the sides? Explain. **See margin.**

3. If you are given a right triangle with an acute angle θ, what two trigonometric functions of θ can you calculate using the lengths of the hypotenuse and the side opposite θ? **sine and cosecant**

4. For which acute angle θ is $\cos \theta = \frac{\sqrt{3}}{2}$? **30°**

Evaluate the six trigonometric functions of the angle θ. **5–7. See margin.**

5.

6.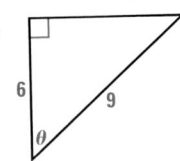

7.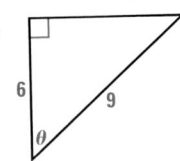

Solve $\triangle ABC$ using the diagram at the right and the given measurements.

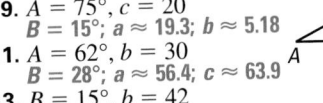

8. $A = 20°$, $a = 12$
$B = 70°$; $b \approx 33.0$; $c \approx 35.1$

9. $A = 75°$, $c = 20$
$B = 15°$; $a \approx 19.3$; $b \approx 5.18$

10. $B = 40°$, $c = 5$
$A = 50°$; $a \approx 3.83$; $b \approx 3.21$

11. $A = 62°$, $b = 30$
$B = 28°$; $a \approx 56.4$; $c \approx 63.9$

12. $B = 63°$, $a = 15$
$A = 27°$; $b \approx 29.4$; $c \approx 33.0$

13. $B = 15°$, $b = 42$
$A = 75°$; $a \approx 157$; $c \approx 162$

14. ⬡ **KITE FLYING** Look back at Example 4 on page 771. Suppose you are flying a kite 4 feet above the ground on a line that is 300 feet long. If the wind speed is 30 miles per hour, what is the altitude of the kite? **about 264 ft**

PRACTICE AND APPLICATIONS

STUDENT HELP

➤ **Extra Practice**
to help you master
skills is on p. 957.

EVALUATING FUNCTIONS Evaluate the six trigonometric functions of the angle θ. **15–21. See margin.**

15.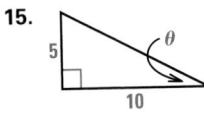

16.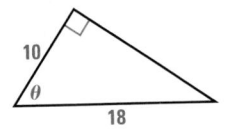

17.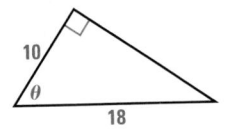

18.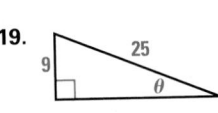

19.

20.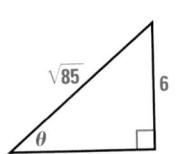

STUDENT HELP

➤ **HOMEWORK HELP**
Example 1: Exs. 15–21
Example 2: Exs. 22–24
Example 3: Exs. 25–40
Examples 4, 5: Exs. 43–50

21. **VISUAL THINKING** The lengths of the sides of a right triangle are 5 centimeters, 12 centimeters, and 13 centimeters. Sketch the triangle. Let θ represent the angle that is opposite the side whose length is 5 centimeters. Evaluate the six trigonometric functions of θ.

FINDING SIDE LENGTHS Find the missing side lengths *x* and *y*.

22.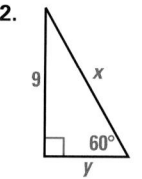
$6\sqrt{3}$; $3\sqrt{3}$

23.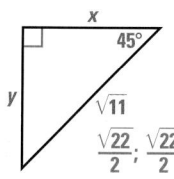
$\frac{\sqrt{22}}{2}$; $\frac{\sqrt{22}}{2}$

24.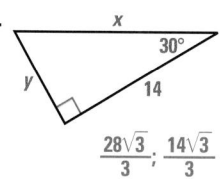
$\frac{28\sqrt{3}}{3}$; $\frac{14\sqrt{3}}{3}$

EVALUATING FUNCTIONS Use a calculator to evaluate the trigonometric function. Round the result to four decimal places.

25. sin 14°   0.2419   26. cos 31°   0.8572   27. tan 59°   1.6643   28. sec 23°   1.0864

29. csc 80°   1.0154   30. cot 36°   1.3764   31. csc 6°   9.5668   32. cot 11°   5.1446

SOLVING TRIANGLES Solve △*ABC* using the diagram and the given measurements. 33–40. See margin.

33. $B = 24°$, $a = 8$ 34. $A = 37°$, $c = 22$

35. $A = 19°$, $b = 4$ 36. $B = 41°$, $c = 18$

37. $A = 29°$, $b = 21$ 38. $B = 56°$, $a = 6.8$

39. $B = 65°$, $c = 12$ 40. $A = 70°$, $c = 30$

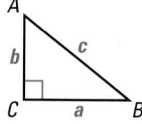

GEOMETRY CONNECTION Find the area of the regular polygon with point *P* at its center.

41.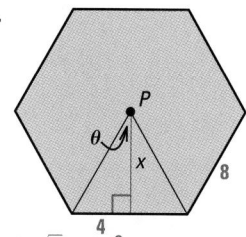
$96\sqrt{3}$ units², or about 166 units²

42.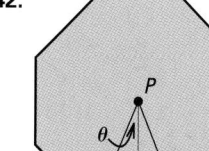
about 483 units²

🚠 **DUQUESNE INCLINE** In Exercises 43 and 44, use the following information.
The track of the Duquesne Incline is about 800 feet long and the angle of elevation is 30°. The average speed of the cable cars is about 320 feet per minute.

43. How high does the Duquesne Incline rise?   **400 ft**

44. What is the vertical speed of the cable cars (in feet per minute)?   **160 ft/min**

45. 🏔 **SKI SLOPE** A ski slope at a mountain has an angle of elevation of 25.2°. The vertical height of the slope is 1808 feet. How long is the ski slope?   **about 4250 ft**

46. 🚢 **BOARDING A SHIP** A gangplank is a narrow ramp used for boarding or leaving a ship. The maximum safe angle of elevation for a gangplank is 20°. Suppose a gangplank is 10 feet long. What is the closest a ship can come to the dock for the gangplank to be used?   **about 9.40 ft**

47. 🏙 **JIN MAO BUILDING** You are standing 75 meters from the base of the Jin Mao Building in Shanghai, China. You estimate that the angle of elevation to the top of the building is 80°. What is the approximate height of the building? Suppose one of your friends is at the top of the building. What is the distance between you and your friend? **about 425 m; about 432 m**

(Left margin answers)

33. $A = 66°$; $b \approx 3.56$; $c \approx 8.76$

34. $B = 53°$; $a \approx 13.2$; $b \approx 17.6$

35. $B = 71°$; $a \approx 1.38$; $c \approx 4.23$

36. $A = 49°$; $a \approx 13.6$; $b \approx 11.8$

37. $B = 61°$; $a \approx 11.6$; $c \approx 24.0$

38. $A = 34°$; $b \approx 10.1$; $c \approx 12.2$

39. $A = 25°$; $a \approx 5.07$; $b \approx 10.9$

40. $B = 20°$; $a \approx 28.2$; $b \approx 10.3$

FOCUS ON APPLICATIONS

DUQUESNE INCLINE Built in Pittsburgh in 1877, the Duquesne Incline transports people up and down the side of a mountain in cable cars. In 1877 the cost of a one-way trip was $.05. Today the cost is $1.

(Right margin)

❗ **COMMON ERROR**
EXERCISES 25–40 Remind students that their calculators must be in degree mode, not radian mode, which is often the default. Encourage them to inspect their answers to make sure they are reasonable.

15. $\sin \theta = \frac{\sqrt{5}}{5}$; $\cos \theta = \frac{2\sqrt{5}}{5}$; $\tan \theta = \frac{1}{2}$; $\csc \theta = \sqrt{5}$; $\sec \theta = \frac{\sqrt{5}}{2}$; $\cot \theta = 2$

16. $\sin \theta = \frac{4}{5}$; $\cos \theta = \frac{3}{5}$; $\tan \theta = \frac{4}{3}$; $\csc \theta = \frac{5}{4}$; $\sec \theta = \frac{5}{3}$; $\cot \theta = \frac{3}{4}$

17. $\sin \theta = \frac{2\sqrt{14}}{9}$; $\cos \theta = \frac{5}{9}$; $\tan \theta = \frac{2\sqrt{14}}{5}$; $\csc \theta = \frac{9\sqrt{14}}{28}$; $\sec \theta = \frac{9}{5}$; $\cot \theta = \frac{5\sqrt{14}}{28}$

18. $\sin \theta = \frac{4}{13}$; $\cos \theta = \frac{3\sqrt{17}}{13}$; $\tan \theta = \frac{4\sqrt{17}}{51}$; $\csc \theta = \frac{13}{4}$; $\sec \theta = \frac{13\sqrt{17}}{51}$; $\cot \theta = \frac{3\sqrt{17}}{4}$

19. $\sin \theta = \frac{9}{25}$; $\cos \theta = \frac{4\sqrt{34}}{25}$; $\tan \theta = \frac{9\sqrt{34}}{136}$; $\csc \theta = \frac{25}{9}$; $\sec \theta = \frac{25\sqrt{34}}{136}$; $\cot \theta = \frac{4\sqrt{34}}{9}$

20. $\sin \theta = \frac{6\sqrt{85}}{85}$; $\cos \theta = \frac{7\sqrt{85}}{85}$; $\tan \theta = \frac{6}{7}$; $\csc \theta = \frac{\sqrt{85}}{6}$; $\sec \theta = \frac{\sqrt{85}}{7}$; $\cot \theta = \frac{7}{6}$

21.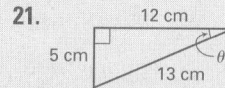
$\sin \theta = \frac{5}{13}$; $\cos \theta = \frac{12}{13}$; $\tan \theta = \frac{5}{12}$; $\csc \theta = \frac{13}{5}$; $\sec \theta = \frac{13}{12}$; $\cot \theta = \frac{12}{5}$

48. 🌎 **MEASURING RIVER WIDTH** To measure the width of a river you plant a stake on one side of the river, directly across from a boulder. You then walk 100 meters to the right of the stake and measure a 79° angle between the stake and the boulder. What is the width w of the river?   **about 514 m**

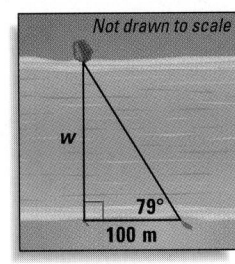

Not drawn to scale

w
79°
100 m

49. 🌎 **MOUNT COOK** You are climbing Mount Cook in New Zealand. You are below the mountain's peak at an altitude of 8580 feet. Using surveying instruments, you measure the angle of elevation to the peak to be 30.5°. The distance (along the face of the mountain) between you and the peak is 7426 feet. What is the altitude of the peak?   **about 12,350 ft**

51c. about 2180 ft; *Sample answer:* I found the distance from the helicopter to the point directly above the far side of the island by solving $\tan 27° = \dfrac{3000}{x + w}$ for $x + w$, and then subtracted the value for x from Part (b).

50. 🌎 **ROPES COURSE** You are designing a zip-line for a ropes course at a summer camp. A zip-line is a cable to which people can attach their safety harnesses and slide down to the ground. You want to attach one end of the cable to a pole 50 feet high and the other end to a pole 5 feet high. The maximum safe angle of elevation for the zip-line is 25°. Calculate the minimum length x of cable needed.   **about 106.5 ft**

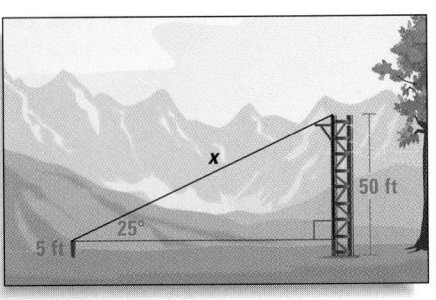

x
50 ft
5 ft
25°

Test Preparation

52. *Sample answer:* All are right triangles with angle A in common, so they are similar by the AA similarity postulate.

54. *Sample answer:* The ratios that define the other trigonometric functions will also be equal since the ratios of corresponding sides of similar triangles are equal.

51. MULTI-STEP PROBLEM You are a surveyor in a helicopter and are trying to determine the width of an island, as illustrated at the right.

a. What is the shortest distance d the helicopter would have to travel to land on the island?
   about 4770 ft, or about 0.90 mi

b. What is the horizontal distance x that the helicopter has to travel before it is directly over the nearer end of the island?   **about 3700 ft**

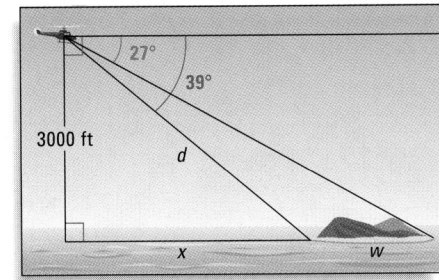

27°
39°
3000 ft
d
x
w

c. ✍️ *Writing* Find the width w of the island. Explain the process you used to find your answer.   **See margin.**

★ **Challenge**

ANALYZING SIMILAR TRIANGLES In Exercises 52–54, use the diagram below.

52. Explain why $\triangle ABC$, $\triangle ADE$, and $\triangle AFG$ are similar triangles.   **See margin.**

53. What does similarity imply about the ratios $\dfrac{BC}{AB}$, $\dfrac{DE}{AD}$, and $\dfrac{FG}{AF}$?

Does the value of sin A depend on which triangle from Exercise 52 is used to calculate it? Would the value of sin A change if it were found using a different right triangle that is similar to the three given triangles?
They are all equal; no; no.

54. Do your observations about sin A also apply to the other five trigonometric functions? Explain.   **Yes. See margin.**

STUDENT HELP

↳ **Skills Review**
For help with similar triangles, see p. 923.

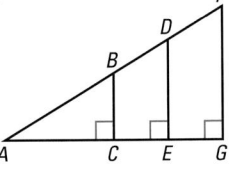

F
D
B
A   C   E   G

ADDITIONAL PRACTICE AND RETEACHING

For Lesson 13.1:
• Practice Levels A, B, and C (*Chapter 13 Resource Book*, p. 15)
• Reteaching with Practice (*Chapter 13 Resource Book*, p. 18)
• 🖥 See Lesson 13.1 of the *Personal Student Tutor*

For more Mixed Review:
• 🖥 Search the *Test and Practice Generator* for key words or specific lessons.

774

UNIT ANALYSIS Find the product. Give the answer with the appropriate unit of measure. (Review 1.1 for 13.2)

55. $(3.5 \text{ hours}) \cdot \dfrac{45 \text{ miles}}{1 \text{ hour}}$ **157.5 mi**

56. $(500 \text{ dollars}) \cdot \dfrac{12.2 \text{ schillings}}{1 \text{ dollar}}$ **6100 schillings**

57. $\dfrac{3 \text{ dollars}}{1 \text{ square foot}} \cdot (1222 \text{ square feet})$ **$3666**

58. $(12 \text{ seconds}) \cdot \dfrac{254 \text{ feet}}{1 \text{ second}}$ **3048 ft**

CLASSIFYING Classify the conic section. (Review 10.6)

59. $y^2 - 16x - 14y + 17 = 0$ **parabola**

60. $25x^2 + y^2 - 100x - 2y + 76 = 0$ **ellipse**

61. $x^2 + y^2 = 25$ **circle**

62. $x^2 - y^2 = 100$ **hyperbola**

63.  **ESSAY TOPICS** For a homework assignment you have to choose from 15 possible topics on which to write an essay. If all of the topics are equally interesting, what is the probability that you and your five friends will all choose different topics? (Review 12.5) $\dfrac{16,016}{50,625}$, or about 0.316

3. *Sample answer:* about 12,600 mi; Columbus's estimation of the radius of Earth at the equator was about 1100 mi short and the distance west nearly 10,000 mi short. The latitude he used for the Canary Islands was about 5° off, and for Japan was several more degrees off.

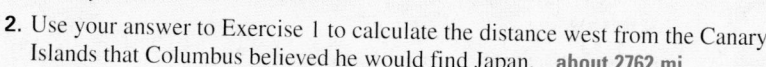

MATH & History

Columbus's Voyage

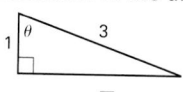 **APPLICATION LINK** www.mcdougallittell.com

THEN

IN 1492 Christopher Columbus set sail west from the Canary Islands intending to reach Japan. Due to miscalculations of Earth's circumference and the relative location of Japan, he instead sailed to the New World.

Columbus believed the distance west from the Canary Islands to Japan to be $\frac{1}{6}$ the circumference of Earth at that latitude. He supposed Earth's radius at the equator to be about 2865 miles.

circle of latitude at the Canary Islands

$67°$ r 2865 equator

1. Use the diagram at the right to calculate what Columbus believed to be the radius r of Earth at the latitude of the Canary Islands. **about 2637 mi**

2. Use your answer to Exercise 1 to calculate the distance west from the Canary Islands that Columbus believed he would find Japan. **about 2762 mi**

3. Use reference materials to find the true distance west from the Canary Islands to Japan. How far off were Columbus's calculations? **See margin.**

NOW

TODAY aerial photography and computers are used to make maps. Accurate maps in combination with satellite-based navigation make travel a more exact science.

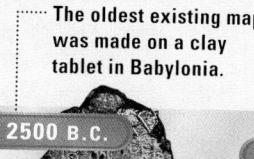

The oldest existing map was made on a clay tablet in Babylonia.

2500 B.C.

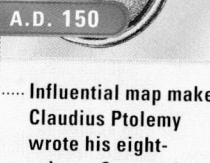

A.D. 150

Influential map maker Claudius Ptolemy wrote his eight-volume *Geography*.

1492

Columbus sails to the Bahama Islands and Cuba, intending to reach Japan.

1999

The Landsat 7 satellite was launched.

13.1 *Right Triangle Trigonometry* **775**

Additional Test Preparation *Sample answer:*
Since you know θ, you can find angle B because the sum of it and θ must be 90°. Then the ratio $\dfrac{b}{c}$ equals the sine of angle B. Writing this as a proportion, you can then solve for c.

4 ASSESS

DAILY HOMEWORK QUIZ

📝 *Transparency Available*

1. Evaluate the six trigonometric functions of the angle θ.

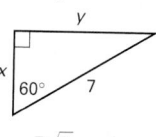

$\sin \theta = \dfrac{2\sqrt{2}}{3}$; $\cos \theta = \dfrac{1}{3}$;

$\tan \theta = 2\sqrt{2}$; $\csc \theta = \dfrac{3\sqrt{2}}{4}$;

$\sec \theta = 3$; $\cot \theta = \dfrac{\sqrt{2}}{4}$

2. Find the missing side lengths x and y.

y x $60°$ 7

3.5; $\dfrac{7\sqrt{3}}{2}$

3. Triangle ABC has a right angle at C. Given that $B = 41°$ and $c = 10$, solve the triangle.
$A = 49°$; $a \approx 7.55$; $b \approx 6.56$

EXTRA CHALLENGE NOTE

→ Challenge problems for Lesson 13.1 are available in **blackline** format in the *Chapter 13 Resource Book*, p. 24 and at **www.mcdougallittell.com**.

ADDITIONAL TEST PREPARATION

1. WRITING Explain how you can use the sine function to find the length c of the hypotenuse in the triangle shown if you know b and θ.

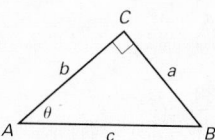

See sample answer at left.

ADDITIONAL RESOURCES
A **blackline** master with additional Math & History exercises is available in the *Chapter 13 Resource Book*, p. 23.

1 PLAN

PACING
Basic: 2 days
Average: 2 days
Advanced: 2 days
Block Schedule: 0.5 block with 13.1

LESSON OPENER
VISUAL APPROACH
An alternative way to approach Lesson 13.2 is to use the Visual Approach Lesson Opener:
• Blackline Master (*Chapter 13 Resource Book,* p. 28)
• Transparency (p. 85)

MEETING INDIVIDUAL NEEDS
• *Chapter 13 Resource Book*
   Prerequisite Skills Review (p. 5)
   Practice Level A (p. 29)
   Practice Level B (p. 30)
   Practice Level C (p. 31)
   Reteaching with Practice (p. 32)
   Absent Student Catch-Up (p. 34)
   Challenge (p. 36)
• *Resources in Spanish*
• *Personal Student Tutor*

NEW-TEACHER SUPPORT
See the Tips for New Teachers on pp. 1–2 of the *Chapter 13 Resource Book* for additional notes about Lesson 13.2.

WARM-UP EXERCISES

Transparency Available

Simplify.
1. $84° - 360°$ $-276°$
2. $125° - 360°$ $-235°$
3. $150 \cdot \frac{\pi}{180}$ $\frac{5\pi}{6}$
4. $345 \cdot \frac{\pi}{180}$ $\frac{23\pi}{12}$
5. $\frac{3\pi}{4} \cdot \frac{180}{\pi}$ 135

What you should learn

GOAL 1 Measure angles in standard position using degree measure and radian measure.

GOAL 2 Calculate arc lengths and areas of sectors, as applied in **Example 6**.

Why you should learn it

▼ To solve **real-life** problems, such as finding the angle generated by a rotating figure skater in **Exs. 77–79**.

GOAL 1 ANGLES IN STANDARD POSITION

In Lesson 13.1 you worked only with acute angles (angles measuring between 0° and 90°). In this lesson you will study angles whose measures can be any real numbers.

Recall that an angle is formed by two rays that have a common endpoint, called the vertex. You can generate any angle by fixing one ray, called the **initial side,** and rotating the other ray, called the **terminal side,** about the vertex. In a coordinate plane, an angle whose vertex is at the origin and whose initial side is the positive *x*-axis is in **standard position**.

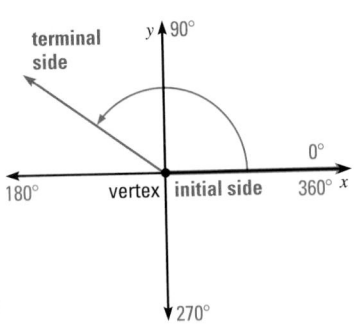

The measure of an angle is determined by the amount and direction of rotation from the initial side to the terminal side. The angle measure is positive if the rotation is counterclockwise, and negative if the rotation is clockwise. The terminal side of an angle can make more than one complete rotation.

EXAMPLE 1 *Drawing Angles in Standard Position*

Draw an angle with the given measure in standard position. Then tell in which quadrant the terminal side lies.

a. 210°   **b.** −45°   **c.** 510°

SOLUTION

a. Use the fact that $210° = 180° + 30°$. So, the terminal side is 30° counterclockwise past the negative *x*-axis.

b. Because −45° is negative, the terminal side is 45° clockwise from the positive *x*-axis.

c. Use the fact that $510° = 360° + 150°$. So, the terminal side makes one complete revolution counterclockwise and continues another 150°.

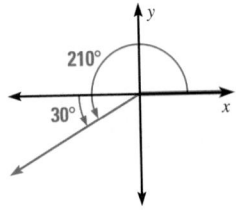

Terminal side in Quadrant III

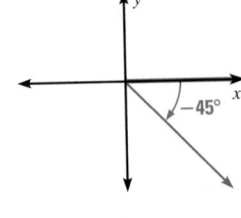

Terminal side in Quadrant IV

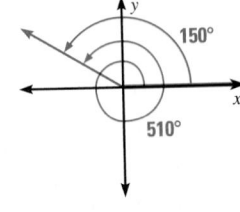

Terminal side in Quadrant II

In Example 1 the angles 510° and 150° are *coterminal*. Two angles in standard position are **coterminal** if their terminal sides coincide. An angle coterminal with a given angle can be found by adding or subtracting multiples of 360°.

EXAMPLE 2 *Finding Coterminal Angles*

STUDENT HELP

HOMEWORK HELP
Visit our Web site
www.mcdougallittell.com
for extra examples.

Find one positive angle and one negative angle that are coterminal with (**a**) −60° and (**b**) 495°.

SOLUTION

There are many such angles, depending on what multiple of 360° is added or subtracted.

a. Positive coterminal angle: −60° + 360° = 300°
Negative coterminal angle: −60° − 360° = −420°

b. Positive coterminal angle: 495° − 360° = 135°
Negative coterminal angle: 495° − 2(360°) = −225°

· · · · · · · · · ·

So far, all the angles you have worked with have been measured in degrees. You can also measure angles in *radians*. To define a radian, consider a circle with radius r centered at the origin. One **radian** is the measure of an angle in standard position whose terminal side intercepts an arc of length r.

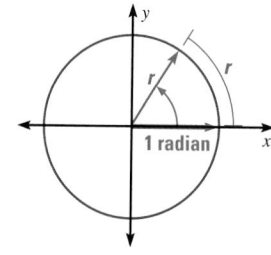

Because the circumference of a circle is $2\pi r$, there are 2π radians in a full circle. Degree measure and radian measure are therefore related by the equation 360° = 2π radians, or 180° = π radians.

The diagram shows equivalent radian and degree measures for special angles from 0° to 360° (0 radians to 2π radians).

You may find it helpful to memorize the equivalent degree and radian measures of special angles in the first quadrant and for $90° = \frac{\pi}{2}$ radians. All other special angles are just multiples of these angles.

You can use the following rules to convert degrees to radians and radians to degrees.

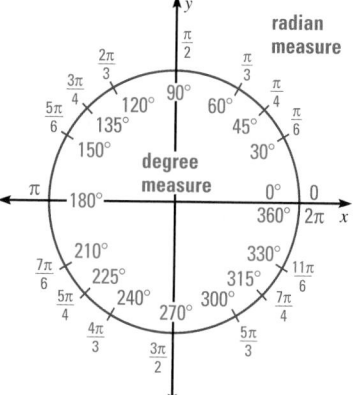

STUDENT HELP

Study Tip
When no units of angle measure are specified, radian measure is implied. For instance, $\theta = 2$ means that $\theta = 2$ radians.

CONVERSIONS BETWEEN DEGREES AND RADIANS

- To rewrite a degree measure in radians, multiply by $\frac{\pi \text{ radians}}{180°}$.

- To rewrite a radian measure in degrees, multiply by $\frac{180°}{\pi \text{ radians}}$.

---

2 TEACH

MOTIVATING THE LESSON
Ask students how many degrees are in a circle. Then ask how far it is around a circle with a radius of 1. This relationship will give a new way to measure angles that will make it easy to find the area of a pie-shaped portion of a circle.

EXTRA EXAMPLE 1
Draw an angle with the given measure in standard position. Then tell in which quadrant the terminal side lies.
a. −120°   Quadrant III

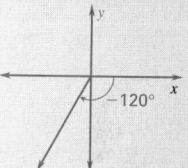

b. 400°   Quadrant I

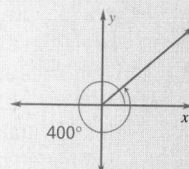

EXTRA EXAMPLE 2
Find one positive and one negative angle that are coterminal with (**a**) −100° and (**b**) 575°.
Sample answers: (a) 260° and −460°; (b) 215° and −145°

CHECKPOINT EXERCISES
For use after Example 1:
1. Draw and label an angle greater than 360° whose terminal side lies in Quadrant II.
Sample answer:

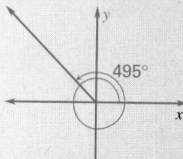

For use after Example 2:
2. Find three angles that are coterminal with −372°.
Sample answer: −12°, 348°, 708°

EXTRA EXAMPLE 3

a. Convert 320° to radians. $\frac{16\pi}{9}$

b. Convert $-\frac{5\pi}{12}$ radians to degrees. **−75°**

EXTRA EXAMPLE 4

In a touring bicycle's first gear, the chain passes over 32 teeth on the freewheel and 24 teeth on the chainwheel. If the chainwheel completes 3 rotations, through what angle does the freewheel turn? Give your answer in both degrees and radians.

810° or $\frac{9\pi}{2}$ radians

CHECKPOINT EXERCISES

For use after Example 3:

1. a. Convert −220° to radians.

$-\frac{11\pi}{9}$ radians

b. Convert $\frac{28\pi}{3}$ radians to degrees. **1680°**

For use after Example 4:

2. In second gear, a bicycle's chain passes over 26 teeth on the freewheel and 24 teeth on the chainwheel. If the chainwheel completes 13 rotations, through what angle does the freewheel turn? Give your answer in both degrees and radians. **4320° or 24π radians**

CONCEPT QUESTION

When finding the arc length of a sector of a circle, why must the angle θ be measured in radians? *Sample answer:* **Because radian measure is defined in terms of the length of the arc of a circle that the angle intercepts. Degree measure is not defined directly in terms of arc length, but in relation to 360° being assigned as the measure of a full rotation.**

APPLICATION NOTE

EXAMPLE 4 Additional information about bicycle gears is available at **www.mcdougallittell.com**.

EXAMPLE 3 · *Converting Between Degrees and Radians*

a. Convert 110° to radians.

b. Convert $-\frac{\pi}{9}$ radians to degrees.

SOLUTION

a. $110° = 110°\left(\frac{\pi \text{ radians}}{180°}\right)$

$= \frac{11\pi}{18}$ radians

✓ **CHECK** Check that your answer is reasonable:

The angle 110° is between the special angles 90° and 120°. The angle $\frac{11\pi}{18}$ is between the same special angles: $\frac{\pi}{2} = \frac{9\pi}{18}$ and $\frac{2\pi}{3} = \frac{12\pi}{18}$.

b. $-\frac{\pi}{9} = \left(-\frac{\pi}{9} \text{ radians}\right)\left(\frac{180°}{\pi \text{ radians}}\right)$

$= -20°$

✓ **CHECK** Check that your answer is reasonable:

The angle $-\frac{\pi}{9}$ is between the special angles 0 and $-\frac{\pi}{6}$. The angle −20° is between the same special angles: 0° and −30°.

Bicycles

EXAMPLE 4 · *Measuring an Angle for a Bicycle*

A bicycle's *gear ratio* is the number of times the freewheel turns for every one turn of the chainwheel. The table shows the number of teeth in the freewheel and chainwheel for the first 5 gears on an 18-speed touring bicycle. In fourth gear, if the chainwheel completes 3 rotations, through what angle does the freewheel turn? Give your answer in both degrees and radians. ▶ Source: *The All New Complete Book of Bicycling*

STUDENT HELP

APPLICATION LINK Visit our Web site www.mcdougallittell.com for more information about bicycle gears in Example 4.

Gear number	Number of teeth in freewheel	Number of teeth in chainwheel
1	32	24
2	26	24
3	22	24
4	32	40
5	19	24

freewheel

chainwheel

SOLUTION

In fourth gear, the gear ratio is $\frac{40}{32}$. For every one turn of the chainwheel in this gear, the freewheel makes 1.25 rotations. The measure of the angle θ through which the freewheel turns when the chainwheel completes 3 rotations is:

$$\theta = (3.75 \text{ rotations})\left(\frac{360°}{1 \text{ rotation}}\right) = 1350°$$

To find the angle measure in radians, multiply by $\frac{\pi \text{ radians}}{180°}$:

$$\theta = 1350°\left(\frac{\pi \text{ radians}}{180°}\right) = \frac{15\pi}{2} \text{ radians}$$

GOAL 2 ARC LENGTHS AND AREAS OF SECTORS

A **sector** is a region of a circle that is bounded by two radii and an arc of the circle. The **central angle** θ of a sector is the angle formed by the two radii. There are simple formulas for the arc length and area of a sector when the central angle is measured in radians.

STUDENT HELP

Derivations
For a derivation of the formula for arc length, see p. 900.

ARC LENGTH AND AREA OF A SECTOR

The arc length s and area A of a sector with radius r and central angle θ (measured in radians) are as follows.

Arc length: $s = r\theta$

Area: $A = \frac{1}{2}r^2\theta$

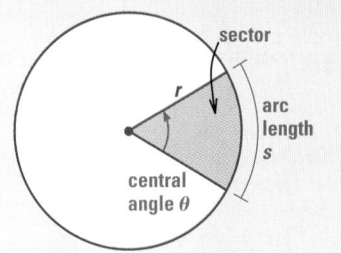

EXAMPLE 5   *Finding Arc Length and Area*

Find the arc length and area of a sector with a radius of 9 cm and a central angle of 60°.

SOLUTION

First convert the angle measure to radians.

$$\theta = 60°\left(\frac{\pi \text{ radians}}{180°}\right) = \frac{\pi}{3} \text{ radians}$$

Then use the formulas for arc length and area.

$$\text{Arc length: } s = r\theta = 9\left(\frac{\pi}{3}\right) = 3\pi \text{ centimeters}$$

$$\text{Area: } A = \frac{1}{2}r^2\theta = \frac{1}{2}(9^2)\left(\frac{\pi}{3}\right) = \frac{27\pi}{2} \text{ square centimeters}$$

EXAMPLE 6   *Finding an Angle and Arc Length*

SPACE NEEDLE Read the photo caption at the left. You go to dinner at the Space Needle and sit at a window table at 6:42 P.M. Your dinner ends at 8:18 P.M. Through what angle do you rotate during your stay? How many feet do you revolve?

SOLUTION

You spend 96 minutes at dinner. Because the Space Needle makes one complete revolution every 60 minutes, your angle of rotation is:

$$\theta = \frac{96}{60}(2\pi) = \frac{16\pi}{5} \text{ radians}$$

Because the radius is 47.25 feet, you move through an arc length of:

$$s = r\theta = 47.25\left(\frac{16\pi}{5}\right) \approx 475 \text{ feet}$$

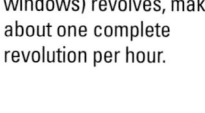

FOCUS ON APPLICATIONS

SPACE NEEDLE
The restaurant at the top of the Space Needle in Seattle, Washington, is circular and has a radius of 47.25 feet. The dining part of the restaurant (by the windows) revolves, making about one complete revolution per hour.

13.2 *General Angles and Radian Measure*   **779**

EXTRA EXAMPLE 5
Find the arc length and area of a sector with a radius of 5 centimeters and a central angle of 45°.
$\frac{5\pi}{4}$ cm; area: $\frac{25\pi}{8}$ cm²

EXTRA EXAMPLE 6
A carousel with a diameter of 25 ft takes 16 sec to make one rotation. If you ride the carousel for 2 min, through what angle do you rotate? How many feet does a point on the outer edge of the carousel revolve? 15π radians; about 589 ft

CHECKPOINT EXERCISES

For use after Example 5:
1. Find the arc length and area of a sector with a radius of 12 inches and a central angle of 120°. 8π in.; 48π in.²

For use after Example 6:
2. A ferris wheel with a diameter of 50 ft takes 30 sec to make a rotation. If you ride the ferris wheel for 2 min 45 sec, through what angle do you rotate? If your seat is at the outer edge, how many feet do you revolve? 11π radians; about 864 ft

FOCUS ON VOCABULARY
To reinforce students' understanding of coterminal angles, point out that the prefix *co-* means *together,* and *terminal* means *ending,* so *coterminal* angles are angles of rotation that end at the same place.

CLOSURE QUESTION
How can you convert an angle measured in degrees to radian measure? Multiply the degree measure by the fraction $\frac{\pi \text{ radians}}{180°}$.

DAILY PUZZLER
A dog is on an 8 foot leash attached to the corner of a building that is in the shape of a regular pentagon with 25 foot sides. How much area does the dog have in which to play? $\frac{224\pi}{5}$ ft², or about 141 ft²

ASSIGNMENT GUIDE

BASIC
Day 1: pp. 780–783 Exs. 25–35, 36–56 even, 77–81, 93–105 odd
Day 2: pp. 781–783 Exs. 60–76 even, 82, 83, 89, 90, 107–111 odd, Quiz 1 Exs. 1–18

AVERAGE
Day 1: pp. 780–783 Exs. 25–35, 36–56 even, 77–81, 93–105 odd
Day 2: pp. 781–783 Exs. 60–76 even, 82–86, 89, 90, 107–111 odd, Quiz 1 Exs. 1–18

ADVANCED
Day 1: pp. 780–783 Exs. 25–35, 36–56 even, 77–81, 93–105 odd
Day 2: pp. 781–783 Exs. 60–76 even, 82–92, 107–111 odd, Quiz 1 Exs. 1–18

BLOCK SCHEDULE WITH 13.1
pp. 780–783 Exs. 25–35, 36–56 even, 60–76 even, 77–83, 89, 90, 93–111 odd, Quiz 1 Exs. 1–18

EXERCISE LEVELS
Level A: *Easier*
25–35, 44–59, 69–79, 93–106
Level B: *More Difficult*
36–43, 60–68, 80–86, 89–90, 107–112
Level C: *Most Difficult*
87–88, 91–92

✔ HOMEWORK CHECK
To quickly check student understanding of key concepts, go over the following exercises: Exs. 26, 28, 34, 36, 40, 44, 52, 60. See also the Daily Homework Quiz:
• Blackline Master (*Chapter 13 Resource Book*, p. 40)
• Transparency (p. 99)

28–33. See next page.
5–12, 34, 35. See Additional Answers beginning on page AA1.

780

GUIDED PRACTICE

Vocabulary Check ✔
Concept Check ✔

1. In your own words, describe what a radian is.   See margin.
2. **ERROR ANALYSIS** An error has been made in finding the area of a sector with a radius of 5 inches and a central angle of 25°. Find and correct the error.   See margin.

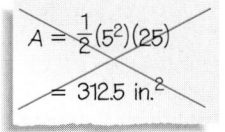
$$A = \frac{1}{2}(5^2)(25)$$
$$= 312.5 \text{ in.}^2$$

3. How does the sign of an angle's measure determine its direction of rotation?   See margin.
4. A circle has radius r. What is the length of the arc corresponding to a central angle of π radians? $r\pi$

Skill Check ✔

1. *Sample answer:* A radian is a unit of angle measure equal to the angle that intercepts an arc of length r units in a circle of radius r.

2. To find the area of a sector with this formula, you must use radian measure for the angle. The correct value is about 5.45 in.^{2}.

3. If the sign is positive, the terminal side is rotated counterclockwise. If the sign is negative, the terminal side is rotated clockwise.

Draw an angle with the given measure in standard position. Then find one positive and one negative coterminal angle.
5–12. Sample angles are given. See margin.
5. 60°   420°, −300°
6. −45°   315°, −405°
7. $\frac{7\pi}{4}$ $\frac{15\pi}{4}, -\frac{\pi}{4}$
8. 300°   660°, −60°
9. $-\frac{3\pi}{2}$ $\frac{\pi}{2}, -\frac{7\pi}{2}$
10. $\frac{7\pi}{8}$ $\frac{23\pi}{8}, -\frac{9\pi}{8}$
11. 150°   510°, −210°
12. $-\frac{5\pi}{4}$ $\frac{3\pi}{4}, -\frac{13\pi}{4}$

Rewrite each degree measure in radians and each radian measure in degrees.
13. 30° $\frac{\pi}{6}$
14. 100° $\frac{5\pi}{9}$
15. 260° $\frac{13\pi}{9}$
16. −320° $-\frac{16\pi}{9}$
17. $\frac{7\pi}{4}$ 315°
18. $\frac{18\pi}{4}$ 810°
19. $\frac{\pi}{12}$ 15°
20. $-\frac{5\pi}{2}$ −450°

Find the arc length and area of a sector with the given radius r and central angle θ.   21–23. See margin.
21. $r = 4$ in., $\theta = 55°$
22. $r = 5$ m, $\theta = 135°$
23. $r = 2$ cm, $\theta = 85°$

24. 🌐 **SPACE NEEDLE** Recall from Example 6 on page 779 that the circular restaurant at the Space Needle has a radius of 47.25 feet and rotates about once per hour. If you are seated at a window table from 6:00 P.M. to 8:10 P.M., through what angle do you rotate? How many feet do you revolve? $\frac{13\pi}{3}$; about 643 ft

PRACTICE AND APPLICATIONS

STUDENT HELP
↳ **Extra Practice**
to help you master skills is on p. 958.

21. $\frac{11\pi}{9}$ in.; $\frac{22\pi}{9}$ in.^{2}
22. $\frac{15\pi}{4}$ m; $\frac{75\pi}{8}$ m^2
23. $\frac{17\pi}{18}$ cm; $\frac{17\pi}{18}$ cm^2

VISUAL THINKING Match the angle measure with the angle.
25. −210°   C
26. 420°   B
27. $-\frac{13\pi}{3}$ A

A.
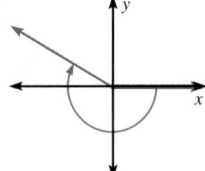

B.
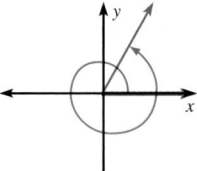

C.
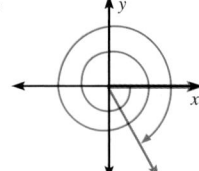

DRAWING ANGLES Draw an angle with the given measure in standard position.   28–35. See margin.
28. 144°
29. $\frac{2\pi}{9}$
30. −15°
31. $-\frac{7\pi}{6}$
32. $\frac{19\pi}{12}$
33. 1620°
34. −5π
35. $-\frac{13\pi}{4}$

STUDENT HELP

→ **HOMEWORK HELP**
Example 1: Exs. 25–35
Example 2: Exs. 36–43
Example 3: Exs. 44–59
Example 4: Exs. 77–81
Example 5: Exs. 60–68
Example 6: Exs. 82–88

FINDING COTERMINAL ANGLES Find one positive angle and one negative angle coterminal with the given angle. 36–43. Sample angles are given.

36. $55°$ $415°; -305°$ **37.** $210°$ $570°; -150°$ **38.** $420°$ $60°; -300°$ **39.** $780°$ $60°; -300°$

40. $\frac{13\pi}{2}$ $\frac{9\pi}{2}; -\frac{3\pi}{2}$ **41.** $\frac{17\pi}{4}$ $\frac{\pi}{4}; -\frac{7\pi}{4}$ **42.** $\frac{24\pi}{7}$ $\frac{10\pi}{7}; -\frac{4\pi}{7}$ **43.** $\frac{16\pi}{3}$ $\frac{4\pi}{3}; -\frac{2\pi}{3}$

CONVERTING MEASURES Rewrite each degree measure in radians and each radian measure in degrees.

44. $25°$ $\frac{5\pi}{36}$ **45.** $225°$ $\frac{5\pi}{4}$ **46.** $160°$ $\frac{8\pi}{9}$ **47.** $45°$ $\frac{\pi}{4}$

48. $-110°$ $-\frac{11\pi}{18}$ **49.** $325°$ $\frac{65\pi}{36}$ **50.** $400°$ $\frac{20\pi}{9}$ **51.** $-290°$ $-\frac{29\pi}{18}$

52. $\frac{7\pi}{3}$ $420°$ **53.** $-\frac{9\pi}{2}$ $-810°$ **54.** $\frac{\pi}{10}$ $18°$ **55.** $-\frac{5\pi}{12}$ $-75°$

56. $\frac{7\pi}{15}$ $84°$ **57.** $-\frac{15\pi}{4}$ $-675°$ **58.** $-\frac{5\pi}{6}$ $-150°$ **59.** $\frac{8\pi}{5}$ $288°$

FINDING ARC LENGTH AND AREA Find the arc length and area of a sector with the given radius *r* and central angle *θ*. 60–68. See margin.

60. $r = 3$ in., $\theta = \frac{\pi}{4}$ **61.** $r = 3$ ft, $\theta = \frac{\pi}{18}$ **62.** $r = 2$ cm, $\theta = \frac{9\pi}{20}$

63. $r = 12$ in., $\theta = 90°$ **64.** $r = 5$ m, $\theta = 120°$ **65.** $r = 15$ mm, $\theta = 175°$

66. $r = 4$ ft, $\theta = 200°$ **67.** $r = 16$ cm, $\theta = 50°$ **68.** $r = 20$ ft, $\theta = 270°$

EVALUATING FUNCTIONS Evaluate the trigonometric function using a calculator if necessary. If possible, give an exact answer.

69. $\sin \frac{\pi}{6}$ $\frac{1}{2}$ **70.** $\cos \frac{\pi}{4}$ $\frac{\sqrt{2}}{2}$ **71.** $\tan \frac{\pi}{3}$ $\sqrt{3}$ **72.** $\cos \frac{4\pi}{11}$ 0.4154

73. $\cot \frac{\pi}{5}$ 1.3764 **74.** $\sec \frac{\pi}{8}$ 1.0824 **75.** $\sin \frac{2\pi}{9}$ 0.6428 **76.** $\csc \frac{3\pi}{10}$ 1.2361

🌐 **FIGURE SKATING** In Exercises 77–79, use the following information.
The number of revolutions made by a figure skater for each type of Axel jump is given. Determine the measure of the angle generated as the skater performs the jump. Give the answer in both degrees and radians.

77. Single Axel: $1\frac{1}{2}$ $540°; 3\pi$ **78.** Double Axel: $2\frac{1}{2}$ $900°; 5\pi$ **79.** Triple Axel: $3\frac{1}{2}$ $1260°; 7\pi$

80. 🌐 **TIME IN SCHOOL** You are in school from 8:00 A.M. to 3:00 P.M. Draw a diagram that shows the number of rotations completed by the minute hand of a clock during this time. Find the measure of the angle generated by the minute hand. Give the answer in both degrees and radians. $2520°; 14\pi$

81. 🌐 **BICYCLE GEARS** Look back at Example 4 on page 778. In fifth gear, if the bicycle's chainwheel completes 4 rotations, through what angle does the freewheel turn? Give your answer in both degrees and radians. about $1820°$ or $\frac{91\pi}{9}$ radians

82. 🌐 **FARMING TECHNOLOGY** A sprinkler system on a farm rotates $140°$ and sprays water up to 35 meters. Draw a diagram that shows the region that can be irrigated with the sprinkler. Then find the area of the region. about 1497 m²

83. 🌐 **WINDSHIELD WIPERS** A car's rear windshield wiper rotates $120°$ as shown. Find the area covered by the wiper. about 528 in.²

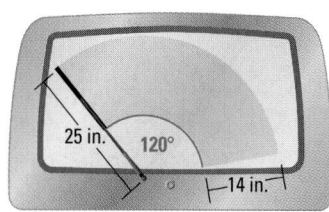

25 in. 120° 14 in.

60. $\frac{3\pi}{4}$ in.; $\frac{9\pi}{8}$ in.²

61. $\frac{\pi}{6}$ ft; $\frac{\pi}{4}$ ft²

62. $\frac{9\pi}{10}$ cm; $\frac{9\pi}{10}$ cm²

63. 6π in.; 36π in.²

64. $\frac{10\pi}{3}$ m; $\frac{25\pi}{3}$ m²

65. $\frac{175\pi}{12}$ mm; $\frac{875\pi}{8}$ mm²

66. $\frac{40\pi}{9}$ ft; $\frac{80\pi}{9}$ ft²

67. $\frac{40\pi}{9}$ cm; $\frac{320\pi}{9}$ cm²

68. 30π ft; 300π ft²

FOCUS ON PEOPLE

AXEL PAULSEN, a Norwegian speed skating champion, invented the Axel jump in 1882. It is the only jump in figure skating that requires taking off from a forward position.

! COMMON ERROR
EXERCISES 25–40 When converting between radians and degrees, students may choose the incorrect conversion factor. Remind them that one of the units must cancel when converting. For example, in the conversion $\left(\frac{2\pi}{3} \text{ radians}\right)\left(\frac{180°}{\pi \text{ radians}}\right)$, the radians cancel and you are left with degrees.

28.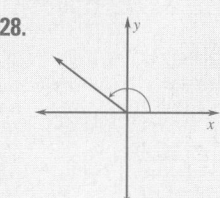

29.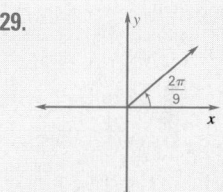

30.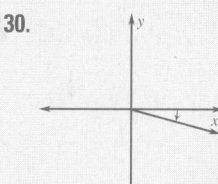

31.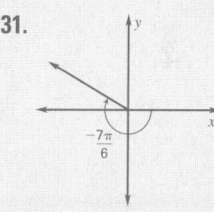

32.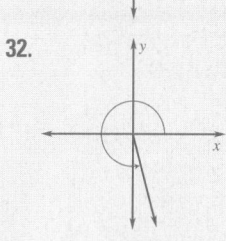

33.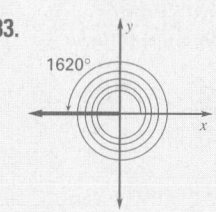

13.2 *General Angles and Radian Measure* **781**

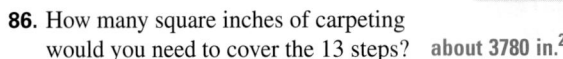 **SPIRAL STAIRS** In Exercises 84–86, use the following information.
A spiral staircase has 13 steps. Each step is a
sector with a radius of 36 inches and a central

angle of $\frac{\pi}{7}$.

84. What is the length of the arc formed by
the outer edge of each step? $\frac{36\pi}{7}$ in., or
about 16.16 in.

85. Through what angle would you rotate
by climbing the stairs? Include a
fourteenth turn for stepping up on the
landing. 2π

86. How many square inches of carpeting
would you need to cover the 13 steps? about 3780 in.^{2}

SNOW CONES In Exercises 87 and 88, use the following information.
You are starting a business selling homemade snow cones in paper cups. You cut out
a paper cup in the shape of a sector.

87. The sector has a central angle of 60° and a radius of
5 inches. When you shape the sector into a cone without
overlapping edges, what will the cone's diameter be? $\frac{5}{3}$ in.

88. Suppose you want to make a cone that has a diameter of
4 inches and a slant height of 6 inches. What should the
radius and central angle of the sector be? radius = 6 in.; central angle = $\frac{2\pi}{3}$ or 120°

Test Preparation

QUANTITATIVE COMPARISON In Exercises 89 and 90, choose the statement
that is true about the given quantities.

(A) The quantity in column A is greater.

(B) The quantity in column B is greater.

(C) The two quantities are equal.

(D) The relationship cannot be determined from the given information.

	Column A	Column B	
89.	Arc length of a sector with $r = 2$ inches and $\theta = 45°$	Arc length of a sector with $r = 2.5$ inches and $\theta = \frac{\pi}{5}$	C
90.	Area of a sector with $r = 2$ inches and $\theta = 45°$	Area of a sector with $r = 2.5$ inches and $\theta = \frac{\pi}{5}$	B

★ **Challenge**

DARTS In Exercises 91 and 92, use the following information.
A dart board is divided into 20 sectors. Each sector is worth a point
value from 1 to 20 and has shaded regions that double or triple this
value. The 20 point sector is shown at the right.

91. Find the area of the sector. Then find the areas of the double
region and the triple region in the sector.
6.89 in.^{2}; 0.758 in.^{2}; 0.464 in.^{2}

EXTRA CHALLENGE
➔ www.mcdougallittell.com

92. If you throw a dart and it randomly lands somewhere inside the
sector, what is the probability that it lands within the double
region? within the triple region? 0.110; 0.067

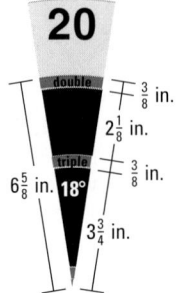

**ADDITIONAL PRACTICE
AND RETEACHING**

For Lesson 13.2:

• Practice Levels A, B, and C
(*Chapter 13 Resource Book,*
p. 29)

• Reteaching with Practice
(*Chapter 13 Resource Book,*
p. 32)

• See Lesson 13.2 of the
Personal Student Tutor

For more Mixed Review:

• Search the *Test and Practice
Generator* for key words or
specific lessons.

782

MIXED REVIEW

PROPERTIES OF SQUARE ROOTS Simplify the expression. (Review 5.3 for 13.3)

93. $\sqrt{275}$ $5\sqrt{11}$ **94.** $\sqrt{1216}$ $8\sqrt{19}$ **95.** $\sqrt{8}\cdot\sqrt{32}$ 16 **96.** $\sqrt{18}\cdot\sqrt{24}$ $12\sqrt{3}$

97. $\sqrt{\dfrac{7}{16}}$ $\dfrac{\sqrt{7}}{4}$ **98.** $\sqrt{\dfrac{11}{36}}$ $\dfrac{\sqrt{11}}{6}$ **99.** $\dfrac{\sqrt{8}}{\sqrt{7}}$ $\dfrac{2\sqrt{14}}{7}$ **100.** $\dfrac{\sqrt{12}}{\sqrt{5}}$ $\dfrac{2\sqrt{15}}{5}$

EVALUATING EXPRESSIONS Evaluate the expression $\dfrac{x^2}{2y+5}$ for the given values of x and y. (Review 1.2 for 13.3)

101. $x=6, y=11$ $11\tfrac{4}{3}$ **102.** $x=3, y=-3$ -9 **103.** $x=12, y=15$ $\dfrac{144}{35}$

104. $x=-1, y=-5$ $-\dfrac{1}{5}$ **105.** $x=-10, y=16$ $\dfrac{100}{37}$ **106.** $x=-20, y=-25$ $-\dfrac{80}{9}$

WRITING EQUATIONS Write the standard form of the equation of the parabola with the given focus and vertex at (0, 0). (Review 10.2)

107. $(5,0)$ $y^2=20x$ **108.** $(-3,0)$ $y^2=-12x$ **109.** $(6,0)$ $y^2=24x$

110. $(0,-12)$ $x^2=-48y$ **111.** $(0,-4.4)$ $x^2=-17.6y$ **112.** $(0,15)$ $x^2=60y$

QUIZ 1

Self-Test for Lessons 13.1 and 13.2

1. $\sin\theta=\dfrac{8}{17}$; $\cos\theta=\dfrac{15}{17}$;

$\tan\theta=\dfrac{8}{15}$; $\csc\theta=\dfrac{17}{8}$;

$\sec\theta=\dfrac{17}{15}$; $\cot\theta=\dfrac{15}{8}$

2. $\sin\theta=\dfrac{3\sqrt{58}}{58}$;

$\cos\theta=\dfrac{7\sqrt{58}}{58}$; $\tan\theta=\dfrac{3}{7}$;

$\csc\theta=\dfrac{\sqrt{58}}{3}$; $\sec\theta=\dfrac{\sqrt{58}}{7}$;

$\cot\theta=\dfrac{7}{3}$

3. $\sin\theta=\dfrac{6\sqrt{61}}{61}$;

$\cos\theta=\dfrac{5\sqrt{61}}{61}$;

$\tan\theta=\dfrac{6}{5}$; $\csc\theta=\dfrac{\sqrt{61}}{6}$;

$\sec\theta=\dfrac{\sqrt{61}}{5}$; $\cot\theta=\dfrac{5}{6}$

15. $\dfrac{242\pi}{9}$ in.; $\dfrac{2662\pi}{9}$ in.^{2}

18. The 6 in. slice has an area of 18.85 in.^{2} and costs about \$.08/in.^{2}, while the 7 in. slice has an area of about 19.24 in.^{2} and costs about \$.09/in.^{2}. The 6 in. slice has a lower unit price, so it is a better deal.

Evaluate the six trigonometric functions of θ. (Lesson 13.1) 1–3. See margin.

1. 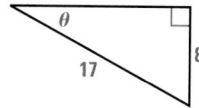   **2.** 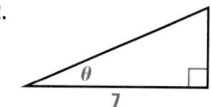   **3.**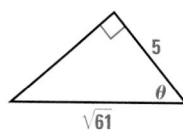

Solve $\triangle ABC$ using the diagram at the right and the given measurements. (Lesson 13.1)

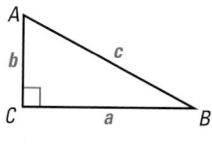

4. $B=50°, a=18$ **5.** $A=33°, c=12$
$A=40°; b\approx21.5; c\approx28.0$ $B=57°; a\approx6.54; b\approx10.1$
6. $A=10°, a=3$ **7.** $B=71°, c=2.3$
$B=80°; b\approx17.0; c\approx17.3$ $A=19°; a\approx0.749; b\approx2.17$

Find one positive angle and one negative angle coterminal with the given angle. (Lesson 13.2) 8–11. Sample answers are given.

8. $25°$ $385°; -335°$ **9.** $-\dfrac{14\pi}{3}$ $\dfrac{4\pi}{3}, -\dfrac{8\pi}{3}$ **10.** $\dfrac{33\pi}{4}$ $\dfrac{\pi}{4}, -\dfrac{7\pi}{4}$ **11.** $-6200°$ $280°; -80°$

Find the arc length and area of a sector with the given radius r and central angle θ. (Lesson 13.2)

12. $r=6$ m, $\theta=\dfrac{\pi}{3}$ **13.** $r=2$ ft, $\theta=\dfrac{5\pi}{6}$ **14.** $r=8$ cm, $\theta=20°$
2π m; 6π m^2 $\dfrac{5\pi}{3}$ ft; $\dfrac{5\pi}{3}$ ft^2 $\dfrac{8\pi}{9}$ cm; $\dfrac{32\pi}{9}$ cm^2

15. $r=22$ in., $\theta=220°$ **16.** $r=5$ ft, $\theta=75°$ **17.** $r=12$ mm, $\theta=160°$
See margin. $\dfrac{25\pi}{12}$ ft; $\dfrac{125\pi}{24}$ ft^2 $\dfrac{32\pi}{3}$ mm; 64π mm^2

18. 🌐 **THE BEST DEAL** Decide which of the two pizza slices shown is the best deal. Explain your reasoning. (Lesson 13.2)
See margin.

13.2 *General Angles and Radian Measure* **783**

---

4 ASSESS

DAILY HOMEWORK QUIZ

📄 *Transparency Available*

1. Draw an angle with a measure of $-\dfrac{11\pi}{3}$ in standard position.

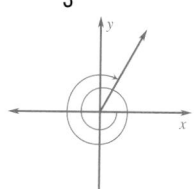

Find one positive angle and one negative angle coterminal with the given angle. Sample angles are given.

2. $-510°$ $-150°; 210°$

3. $\dfrac{15\pi}{4}$ $\dfrac{7\pi}{4}; -\dfrac{\pi}{4}$

4. Rewrite $\dfrac{7\pi}{6}$ radians in degrees. $210°$

5. Rewrite $-420°$ in radians. $-\dfrac{7\pi}{3}$

6. Find the arc length and area of a sector with radius 8 m and central angle 75°.
$\dfrac{10\pi}{3}$ m; $\dfrac{40\pi}{3}$ m^2

EXTRA CHALLENGE NOTE

→ Challenge problems for Lesson 13.2 are available in **blackline** format in the *Chapter 13 Resource Book*, p. 36 and at **www.mcdougallittell.com.**

ADDITIONAL TEST PREPARATION

1. WRITING Explain why the fraction $\dfrac{180°}{\pi\text{ radians}}$ is equivalent to 1. *Sample answer:* Since there are 360° or 2π radians in a circle, both 180° and π radians describe the angle of rotation that intercepts half of a circle.

ADDITIONAL RESOURCES

An alternative Quiz for Lessons 13.1 and 13.2 is available in the *Chapter 13 Resource Book*, p. 37.

Trigonometric Functions of Any Angle

PACING
Basic: 1 day
Average: 1 day
Advanced: 1 day
Block Schedule: 0.5 block with 13.4

LESSON OPENER
CALCULATOR ACTIVITY
An alternative way to approach Lesson 13.3 is to use the Calculator Activity Lesson Opener:
- Blackline Master (*Chapter 13 Resource Book,* p. 41)
- Transparency (p. 86)

MEETING INDIVIDUAL NEEDS
- **Chapter 13 Resource Book**
  Prerequisite Skills Review (p. 5)
  Practice Level A (p. 42)
  Practice Level B (p. 43)
  Practice Level C (p. 44)
  Reteaching with Practice (p. 45)
  Absent Student Catch-Up (p. 47)
  Challenge (p. 49)
- **Resources in Spanish**
- **Personal Student Tutor**

NEW-TEACHER SUPPORT
See the Tips for New Teachers on pp. 1–2 of the *Chapter 13 Resource Book* for additional notes about Lesson 13.3.

WARM-UP EXERCISES

✎ *Transparency Available*

Find the hypotenuse *c* in a right triangle with legs *a* and *b* indicated.
1. $a = 8$, $b = 15$ **17**
2. $a = 4$, $b = 6$ $2\sqrt{13}$

Give the radian measure for each angle.
3. 60° $\dfrac{\pi}{3}$
4. 90° $\dfrac{\pi}{2}$
5. 210° $\dfrac{7\pi}{6}$

What you should learn

GOAL 1 Evaluate trigonometric functions of any angle.

GOAL 2 Use trigonometric functions to solve **real-life** problems, such as finding the distance a soccer ball is kicked in **Ex. 71.**

Why you should learn it

▼ To solve **real-life** problems, such as finding distances for a marching band on a football field in **Example 6.**

GOAL 1 **EVALUATING TRIGONOMETRIC FUNCTIONS**

In Lesson 13.1 you learned how to evaluate trigonometric functions of an acute angle. In this lesson you will learn to evaluate trigonometric functions of *any* angle.

GENERAL DEFINITION OF TRIGONOMETRIC FUNCTIONS

Let θ be an angle in standard position and (x, y) be any point (except the origin) on the terminal side of θ. The six trigonometric functions of θ are defined as follows.

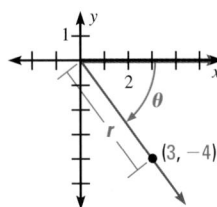

$$\sin \theta = \frac{y}{r} \qquad\qquad \csc \theta = \frac{r}{y},\, y \neq 0$$

$$\cos \theta = \frac{x}{r} \qquad\qquad \sec \theta = \frac{r}{x},\, x \neq 0$$

$$\tan \theta = \frac{y}{x},\, x \neq 0 \qquad \cot \theta = \frac{x}{y},\, y \neq 0$$

Pythagorean theorem gives
$$r = \sqrt{x^2 + y^2}.$$

For acute angles, these definitions give the same values as those given by the definitions in Lesson 13.1.

EXAMPLE 1 *Evaluating Trigonometric Functions Given a Point*

Let $(3, -4)$ be a point on the terminal side of an angle θ in standard position. Evaluate the six trigonometric functions of θ.

SOLUTION
Use the Pythagorean theorem to find the value of r.

$$r = \sqrt{x^2 + y^2}$$
$$= \sqrt{3^2 + (-4)^2}$$
$$= \sqrt{25}$$
$$= 5$$

Using $x = 3$, $y = -4$, and $r = 5$, you can write the following.

$$\sin \theta = \frac{y}{r} = -\frac{4}{5} \qquad\qquad \csc \theta = \frac{r}{y} = -\frac{5}{4}$$

$$\cos \theta = \frac{x}{r} = \frac{3}{5} \qquad\qquad \sec \theta = \frac{r}{x} = \frac{5}{3}$$

$$\tan \theta = \frac{y}{x} = -\frac{4}{3} \qquad\qquad \cot \theta = \frac{x}{y} = -\frac{3}{4}$$

If the terminal side of θ lies on an axis, then θ is a **quadrantal angle**. The diagrams below show the values of x and y for the quadrantal angles 0°, 90°, 180°, and 270°.

0° or 0 radians

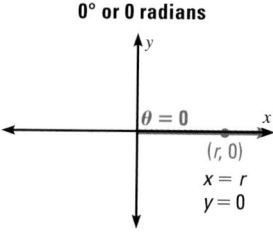

90° or $\frac{\pi}{2}$ radians

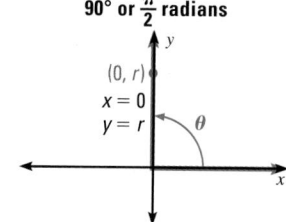

180° or π radians

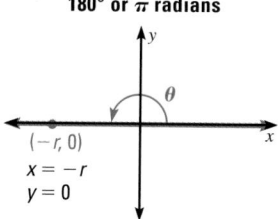

270° or $\frac{3\pi}{2}$ radians

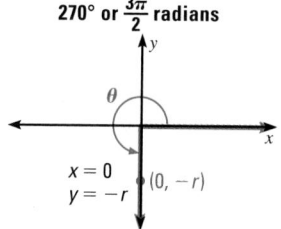

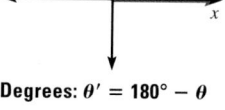 **EXAMPLE 2** *Trigonometric Functions of a Quadrantal Angle*

Evaluate the six trigonometric functions of $\theta = 180°$.

SOLUTION

When $\theta = 180°$, $x = -r$ and $y = 0$. The six trigonometric functions of θ are as follows.

$$\sin \theta = \frac{y}{r} = \frac{0}{r} = 0 \qquad\qquad \csc \theta = \frac{r}{y} = \frac{r}{0} = \text{undefined}$$

$$\cos \theta = \frac{x}{r} = \frac{-r}{r} = -1 \qquad\qquad \sec \theta = \frac{r}{x} = \frac{r}{-r} = -1$$

$$\tan \theta = \frac{y}{x} = \frac{0}{-r} = 0 \qquad\qquad \cot \theta = \frac{x}{y} = \frac{-r}{0} = \text{undefined}$$

. . . .

The values of trigonometric functions of angles greater than 90° (or less than 0°) can be found using corresponding acute angles called *reference angles*. Let θ be an angle in standard position. Its **reference angle** is the acute angle θ' (read *theta prime*) formed by the terminal side of θ and the x-axis. The relationship between θ and θ' is given below for nonquadrantal angles θ such that $90° < \theta < 360°$ $\left(\frac{\pi}{2} < \theta < 2\pi\right)$.

90° < θ < 180°;
$\frac{\pi}{2} < \theta < \pi$

Degrees: $\theta' = 180° - \theta$
Radians: $\theta' = \pi - \theta$

180° < θ < 270°;
$\pi < \theta < \frac{3\pi}{2}$

Degrees: $\theta' = \theta - 180°$
Radians: $\theta' = \theta - \pi$

270° < θ < 360°;
$\frac{3\pi}{2} < \theta < 2\pi$

Degrees: $\theta' = 360° - \theta$
Radians: $\theta' = 2\pi - \theta$

13.3 *Trigonometric Functions of Any Angle* **785**

MOTIVATING THE LESSON
Many students know that Earth is about 25,000 miles around at the equator. Ask if any of them know how far it is around Earth at the latitude where they live. Tell them that using the cosine ratio, they can easily find this distance.

EXTRA EXAMPLE 1
Let (–4, –3) be a point on the terminal side of θ. Evaluate the six trigonometric functions of θ.
$\sin \theta = -\frac{3}{5}$; $\cos \theta = -\frac{4}{5}$; $\tan \theta = \frac{3}{4}$;
$\csc \theta = -\frac{5}{3}$; $\sec \theta = -\frac{5}{4}$; $\cot \theta = \frac{4}{3}$

EXTRA EXAMPLE 2
Evaluate the six trigonometric functions of $\theta = 90°$. $\sin \theta = 1$; $\cos \theta = 0$; $\tan \theta$ is undefined; $\csc \theta = 1$; $\sec \theta$ is undefined; $\cot \theta = 0$

✔ CHECKPOINT EXERCISES
For use after Example 1:
1. Let (–5, 12) be a point on the terminal side of θ. Evaluate the six trigonometric functions of θ. $\sin \theta = \frac{12}{13}$; $\cos \theta = -\frac{5}{13}$; $\tan \theta = -\frac{12}{5}$; $\csc \theta = \frac{13}{12}$; $\sec \theta = -\frac{13}{5}$; $\cot \theta = -\frac{5}{12}$

For use after Example 2:
2. Evaluate the six trigonometric functions of $\theta = 270°$. $\sin \theta = -1$; $\cos \theta = 0$; $\tan \theta$ is undefined; $\csc \theta = -1$; $\sec \theta$ is undefined; $\cot \theta = 0$

ENGLISH LEARNERS
Point out that the prefix *quad-* in the word *quadrantal* indicates that it is associated with the number 4. Then challenge students to explain what this term has to do with the number 4. (The term describes angles with terminal sides on the axes. There are four possible positions for these angles; positive x-axis, positive y-axis, negative x-axis, negative y-axis.)

EXTRA EXAMPLE 3
Find the reference angle θ' for each angle θ.
a. $\theta = 140°$ $\theta' = 40°$
b. $\theta = -\dfrac{3\pi}{4}$ $\theta' = \dfrac{\pi}{4}$

EXTRA EXAMPLE 4
Evaluate **(a)** cos $(-150°)$ and **(b)** cot $\dfrac{16\pi}{3}$.
(a) $-\dfrac{\sqrt{3}}{2}$ **(b)** $\dfrac{\sqrt{3}}{3}$

 CHECKPOINT EXERCISES
For use after Example 3:
1. Find the reference angle θ' for each angle θ.
   a. $\theta = -250°$ $\theta' = 70°$
   b. $\theta = \dfrac{8\pi}{3}$ $\theta' = \dfrac{\pi}{3}$

For use after Example 4:
2. Evaluate **(a)** sin $225°$ and
   (b) sec $\left(\dfrac{7\pi}{6}\right)$.
   (a) $-\dfrac{\sqrt{2}}{2}$ **(b)** $-\dfrac{2\sqrt{3}}{3}$

MATHEMATICAL REASONING
In Lesson 13.1, students found trigonometric ratios for the acute angles of a right triangle. The extension to any angle may confuse them. Help students see that all nonquadrantal angles have a right triangle associated with them that contains the acute reference angle. For quadrantal angles, have students draw angles terminating on the unit circle for angles getting closer and closer to 90°. Ask them what is happening to the ratio of the side opposite the reference angle to the hypotenuse (it is getting closer to 1) and to the ratio of the adjacent side to the hypotenuse (it is getting closer to 0). Point out that having sin 90° = 1 and cos 90° = 0 are logical extensions of this. The same type of argument also holds for the other quadrantal angles.

STUDENT HELP NOTES
→ **Homework Help** Students can find extra examples at **www.mcdougallittell.com** that parallel the examples in the student edition.

EXAMPLE 3 *Finding Reference Angles*

Find the reference angle θ' for each angle θ.

a. $\theta = 320°$ **b.** $\theta = -\dfrac{5\pi}{6}$

SOLUTION

a. Because $270° < \theta < 360°$, the reference angle is $\theta' = 360° - 320° = 40°$.

b. Because θ is coterminal with $\dfrac{7\pi}{6}$ and $\pi < \dfrac{7\pi}{6} < \dfrac{3\pi}{2}$, the reference angle is $\theta' = \dfrac{7\pi}{6} - \pi = \dfrac{\pi}{6}$.

· · · · · · · · · ·

The signs of the trigonometric function values in the four quadrants can be determined from the function definitions. For instance, because $\cos\theta = \dfrac{x}{r}$ and r is always positive, it follows that $\cos\theta$ is positive wherever $x > 0$, which is in Quadrants I and IV.

> **CONCEPT SUMMARY**   **EVALUATING TRIGONOMETRIC FUNCTIONS**
>
> Use these steps to evaluate a trigonometric function of any angle θ.
>
> ❶ Find the reference angle θ'.
>
> ❷ Evaluate the trigonometric function for the angle θ'.
>
> ❸ Use the quadrant in which θ lies to determine the sign of the trigonometric function value of θ. (See the diagram at the right.)
>
> **Signs of Function Values**
>
>
> Quadrant II
> sin θ, csc θ: +
> cos θ, sec θ: −
> tan θ, cot θ: −
>
> Quadrant I
> sin θ, csc θ: +
> cos θ, sec θ: +
> tan θ, cot θ: +
>
> Quadrant III
> sin θ, csc θ: −
> cos θ, sec θ: −
> tan θ, cot θ: +
>
> Quadrant IV
> sin θ, csc θ: −
> cos θ, sec θ: +
> tan θ, cot θ: −

EXAMPLE 4 *Using Reference Angles to Evaluate Trigonometric Functions*

STUDENT HELP
HOMEWORK HELP
Visit our Web site www.mcdougallittell.com for extra examples.

Evaluate **(a)** tan $(-210°)$ and **(b)** csc $\dfrac{11\pi}{4}$.

SOLUTION

a. The angle $-210°$ is coterminal with $150°$. The reference angle is $\theta' = 180° - 150° = 30°$. The tangent function is negative in Quadrant II, so you can write:

$$\tan(-210°) = -\tan 30° = -\dfrac{\sqrt{3}}{3}$$

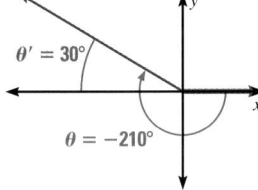

b. The angle $\dfrac{11\pi}{4}$ is coterminal with $\dfrac{3\pi}{4}$. The reference angle is $\theta' = \pi - \dfrac{3\pi}{4} = \dfrac{\pi}{4}$. The cosecant function is positive in Quadrant II, so you can write:

$$\csc\dfrac{11\pi}{4} = \csc\dfrac{\pi}{4} = \sqrt{2}$$

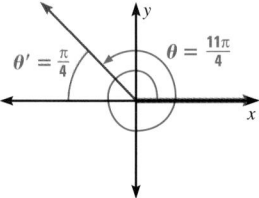

GOAL 2 **USING TRIGONOMETRIC FUNCTIONS IN REAL LIFE**

EXAMPLE 5 *Calculating Projectile Distance*

GOLF The horizontal distance d (in feet) traveled by a projectile with an initial speed v (in feet per second) is given by

$$d = \frac{v^2}{32} \sin 2\theta$$

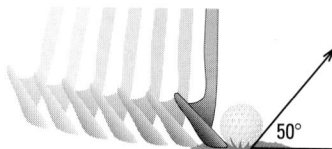

where θ is the angle at which the projectile is launched. Estimate the horizontal distance traveled by a golf ball that is hit at an angle of 50° with an initial speed of 105 feet per second. (This model neglects air resistance and wind conditions. It also assumes that the projectile's starting and ending heights are the same.)

SOLUTION

The horizontal distance given by the model is:

$$d = \frac{v^2}{32} \sin 2\theta \qquad \text{Write distance model.}$$

$$= \frac{105^2}{32} \sin (2 \cdot 50°) \approx 339 \text{ feet} \qquad \text{Substitute and use a calculator.}$$

▶ The golf ball travels a horizontal distance of about 339 feet.

Marching Band

EXAMPLE 6 *Modeling with Trigonometric Functions*

Your school's marching band is performing at halftime during a football game. In the last formation, the band members form a circle 100 feet wide in the center of the field. Your starting position is 100 feet from the goal line, where you will exit the field. How far from the goal line will you be after you have marched 300° around the circle?

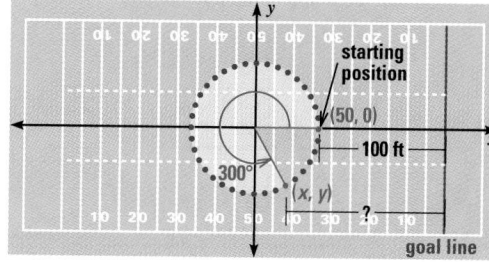

SOLUTION

The radius of the circle is $r = 50$. So, you can write:

$$\cos 300° = \frac{x}{r} \qquad \text{Use definition of cosine.}$$

$$\frac{1}{2} = \frac{x}{50} \qquad \text{Substitute.}$$

$$25 = x \qquad \text{Solve for } x.$$

▶ You will be $100 + (50 - 25) = 125$ feet from the goal line.

13.3 *Trigonometric Functions of Any Angle* **787**

EXTRA EXAMPLE 5
Using the formula $d = \frac{v^2}{32} \sin 2\theta$, estimate the horizontal distance traveled by a golf ball hit at an angle of 40° with an initial speed of 125 feet per second.
about 481 ft

EXTRA EXAMPLE 6
Your marching band's flag corps makes a circular formation. The circle is 20 feet wide in the center of the football field. Your starting position is 140 feet from the nearer goal line. How far from this goal line will you be after you have marched 120° counterclockwise around the circle?
155 ft

CHECKPOINT EXERCISES
For use after Example 5:
1. A golf club called a wedge is made to lift a ball high in the air. If a wedge has a 65° loft, how far does a ball hit with an initial speed of 100 feet per second travel? **about 239 ft**

For use after Example 6:
2. A circular clock gear is 2 inches wide. If the tooth at the farthest right edge of the gear starts 10 inches above the base of the clock, how far above the base is the tooth after the gear rotates 240° counterclockwise?
about 9.13 in.

FOCUS ON VOCABULARY
What is the *reference angle* for a nonquadrantal angle?
the angle formed by the terminal side of the angle and the x-axis

CLOSURE QUESTION
Which quadrantal angles coincide with the x-axis? with the y-axis?
0° or 0 radians and 180° or π radians; 90° or $\frac{\pi}{2}$ radians and 270° or $\frac{3\pi}{2}$ radians

GOLF BALLS
The dimples on a golf ball create pockets of air turbulence that keep the ball in the air for a longer period of time than if the ball were smooth. The longest drive of a golf ball on record is 473 yards, 2 feet, 6 inches.

DATA UPDATE
www.mcdougallittell.com

ASSIGNMENT GUIDE

BASIC
Day 1: pp. 788–790 Exs. 23–29,
34–56 even, 61–69, 71, 77,
78, 81–95 odd

AVERAGE
Day 1: pp. 788–790 Exs. 23–29,
34–56 even, 61–72, 77, 78,
81–95 odd

ADVANCED
Day 1: pp. 788–790 Exs. 23–29,
34–56 even, 61–80,
81–95 odd

BLOCK SCHEDULE WITH 13.4
pp. 788–790 Exs. 23–29, 34–56
even, 61–72, 77, 78, 81–95 odd

EXERCISE LEVELS
Level A: *Easier*
23–36, 61–68, 81–86
Level B: *More Difficult*
37–60, 69–74, 77, 78, 87–95
Level C: *Most Difficult*
75, 76, 79, 80

✔ **HOMEWORK CHECK**
To quickly check student understanding of key concepts, go
over the following exercises:
Exs. 24, 34, 40, 48, 54, 62. See
also the Daily Homework Quiz:

• Blackline Master (*Chapter 13
  Resource Book,* p. 52)
• 👆 Transparency (p. 100)

1. A quadrantal angle is one whose
   terminal side lies on one of the
   axes. A reference angle is the
   acute angle formed by the terminal
   ray of an angle and the x-axis.
2. *Sample answer:* In Quadrant III, the
   sine is negative. The reference
   angle for θ is the angle formed by
   θ's terminal side and the negative
   x-axis. The sine of θ will be the
   opposite of the sine of this reference angle.
3, 6–13, 25–44.
   See Additional Answers beginning on page AA1.

GUIDED PRACTICE

Vocabulary Check ✔

1. Define the terms quadrantal angle and reference angle.   **See margin.**

Concept Check ✔

2. Given an angle θ in Quadrant III, explain how you can use a reference angle to find sin θ.   **See margin.**

3. Explain why tan 270° is undefined.   **See margin.**

4. In which quadrant(s) must θ lie for cos θ to be positive?   **I or IV**

Skill Check ✔

5. Let $(-4, -5)$ be a point on the terminal side of an angle θ in standard position. Evaluate the six trigonometric functions of θ.

5. $\sin \theta = -\frac{5\sqrt{41}}{41}$;
   $\cos \theta = -\frac{4\sqrt{41}}{41}$;
   $\tan \theta = \frac{5}{4}$;
   $\csc \theta = -\frac{\sqrt{41}}{5}$;
   $\sec \theta = -\frac{\sqrt{41}}{4}$; $\cot \theta = \frac{4}{5}$

Sketch the angle. Then find its reference angle.   **6–13. See margin for graphs.**

6. $\frac{7\pi}{4}$ $\frac{\pi}{4}$ 7. $-120°$ $60°$ 8. $\frac{7\pi}{8}$ $\frac{\pi}{8}$ 9. $390°$ $30°$

10. $-\frac{2\pi}{3}$ $\frac{\pi}{3}$ 11. $-370°$ $10°$ 12. $\frac{2\pi}{3}$ $\frac{\pi}{3}$ 13. $230°$ $50°$

23. $\sin \theta = \frac{5}{13}$; $\cos \theta = -\frac{12}{13}$;
    $\tan \theta = -\frac{5}{12}$; $\csc \theta = \frac{13}{5}$;
    $\sec \theta = -\frac{13}{12}$;
    $\cot \theta = -\frac{12}{5}$

Evaluate the function without using a calculator.

14. $\cos\left(-\frac{4\pi}{3}\right)$ $-\frac{1}{2}$ 15. $\tan 240°$ $\sqrt{3}$ 16. $\sin \frac{7\pi}{4}$ $-\frac{\sqrt{2}}{2}$ 17. $\csc (-225°)$ $\sqrt{2}$

18. $\cot\left(-\frac{3\pi}{4}\right)$ 1 19. $\cos 240°$ $-\frac{1}{2}$ 20. $\sec \frac{11\pi}{6}$ $\frac{2\sqrt{3}}{3}$ 21. $\tan \frac{5\pi}{6}$ $-\frac{\sqrt{3}}{3}$

22. 🌐 **MARCHING BAND** Look back at Example 6 on page 787. Suppose you marched 135° around the circle from the same starting position. How far from the goal line would you be?   **about 185 ft**

PRACTICE AND APPLICATIONS

STUDENT HELP

↳ **Extra Practice**
to help you master
skills is on p. 958.

24. $\sin \theta = -\frac{4}{5}$; $\cos \theta = -\frac{3}{5}$;
    $\tan \theta = \frac{4}{3}$; $\csc \theta = -\frac{5}{4}$;
    $\sec \theta = -\frac{5}{3}$; $\cot \theta = \frac{3}{4}$

STUDENT HELP

↳ **HOMEWORK HELP**
Example 1: Exs. 23–33
Example 2: Exs. 34–36
Example 3: Exs. 37–44
Example 4: Exs. 45–60
Example 5: Exs. 69–71
Example 6: Exs. 72–76

USING A POINT Use the given point on the terminal side of an angle θ in standard position. Evaluate the six trigonometric functions of θ.   **23–33. See margin.**

23.   24.   25.

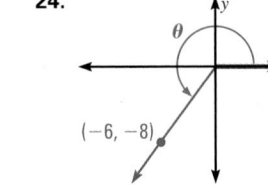

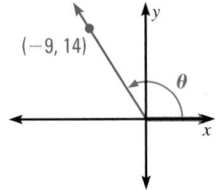

26. $(-12, -15)$ 27. $(-1, 1)$ 28. $(15, -8)$ 29. $(6, -9)$

30. $(7, 10)$ 31. $\left(1, -\sqrt{3}\right)$ 32. $(-3, -4)$ 33. $\left(-15, 5\sqrt{7}\right)$

QUADRANTAL ANGLES Evaluate the six trigonometric functions of θ.   **34–36.**
See margin.

34. $\theta = 90°$ 35. $\theta = 270°$ 36. $\theta = 0°$

FINDING REFERENCE ANGLES Sketch the angle. Then find its reference angle.
37–44. See margin for graphs.

37. $240°$ $60°$ 38. $-515°$ $25°$ 39. $-170°$ $10°$ 40. $315°$ $45°$

41. $-440°$ $80°$ 42. $-\frac{3\pi}{4}$ $\frac{\pi}{4}$ 43. $\frac{25\pi}{4}$ $\frac{\pi}{4}$ 44. $-\frac{11\pi}{3}$ $\frac{\pi}{3}$

EVALUATING FUNCTIONS Evaluate the function without using a calculator.

45. $\cos 315°$ $\frac{\sqrt{2}}{2}$ **46.** $\cos(-210°)$ $-\frac{\sqrt{3}}{2}$ **47.** $\csc(-240°)$ $\frac{2\sqrt{3}}{3}$ **48.** $\tan 210°$ $\frac{\sqrt{3}}{3}$

49. $\sec 780°$ 2 **50.** $\sin 225°$ $-\frac{\sqrt{2}}{2}$ **51.** $\cos(-225°)$ $-\frac{\sqrt{2}}{2}$ **52.** $\tan(-120°)$ $\sqrt{3}$

53. $\cot \frac{11\pi}{6}$ $-\sqrt{3}$ **54.** $\sec \frac{9\pi}{4}$ $\sqrt{2}$ **55.** $\sin\left(-\frac{5\pi}{6}\right)$ $-\frac{1}{2}$ **56.** $\cos \frac{5\pi}{3}$ $\frac{1}{2}$

57. $\sin\left(-\frac{17\pi}{6}\right)$ $-\frac{1}{2}$ **58.** $\sec \frac{23\pi}{6}$ $\frac{2\sqrt{3}}{3}$ **59.** $\csc \frac{17\pi}{3}$ $-\frac{2\sqrt{3}}{3}$ **60.** $\cot\left(-\frac{13\pi}{4}\right)$ -1

USING A CALCULATOR Use a calculator to evaluate the function. Round the result to four decimal places.

61. $\sec 137°$ -1.3673 **62.** $\cot 400°$ 1.1918 **63.** $\sin(-10°)$ -0.1736 **64.** $\csc 540°$ undefined

65. $\cot\left(-\frac{4\pi}{5}\right)$ 1.3764 **66.** $\sec \frac{11\pi}{2}$ undefined **67.** $\cos \frac{6\pi}{5}$ -0.8090 **68.** $\csc \frac{23\pi}{8}$ 2.6131

69. **SKATEBOARDING** A skate-boarder is setting up two ramps for a jump as shown. He wants to jump off one ramp and land on the other. If the ramps are placed 5 feet apart, at what speed must the skateboarder launch off the first ramp to land on the second ramp? **about 16.5 ft/sec**

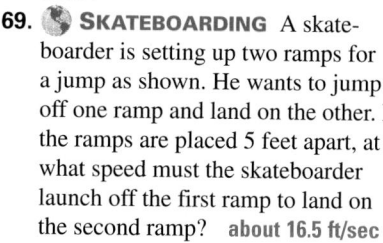

70. **VOLLEYBALL** While playing a game of volleyball, you set the ball to your teammate. You hit the ball with an initial speed of 24 feet per second at an angle of 70°. About how far away should your teammate be to receive your set? **about 11.6 ft**

71. **SOCCER** You and a friend are playing soccer. Both of you kick the ball with an initial speed of 42 feet per second. Your kick was projected at an angle of 45° and your friend's kick was projected at an angle of 60°. About how much farther will your soccer ball go than your friend's soccer ball? **about 7.4 ft**

72. **FERRIS WHEEL** The largest Ferris wheel in operation is the Cosmolock 21 at Yokohama City, Japan. It has a diameter of 328 feet. Passengers board the cars at the bottom of the wheel, about 16.5 feet above the ground. Imagine that you have boarded the Cosmolock 21. The wheel rotates 312° and then stops. How high above the ground are you? **about 70.8 ft**

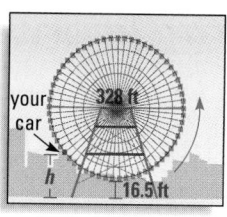

SOCIAL STUDIES **CONNECTION** In Exercises 73 and 74, use the information below.
The Tropic of Cancer is the circle of latitude farthest north of the equator where the sun can appear directly overhead. It lies 23.5° north of the equator, as shown below.

73. Find the circumference of the Tropic of Cancer using 3960 miles as Earth's approximate radius. **about 22,800 mi**

74. What is the distance between two points that lie directly across from each other (through the axis) on the Tropic of Cancer? **about 7260 mi**

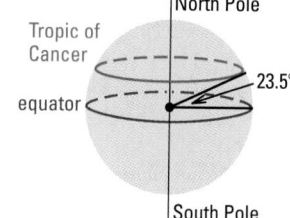

STUDENT HELP

Study Tip
Make sure your calculator is in radian mode when finding trigonometric functions of angles measured in radians.

FOCUS ON CAREERS

CARTOGRAPHER
Cartographers compile information from aerial photographs and satellite data to map Earth's surface. A map's circles of latitude and longitude, as discussed in Exs. 73 and 74, are used to describe location.

CAREER LINK
www.mcdougallittell.com

! **COMMON ERROR**
EXERCISES 45–60 Students may have trouble with the signs of the trigonometric functions in different quadrants. The mnemonic device "**A**ll **S**tudents **T**ake **C**alculus" may help. In Quadrant I, **A**ll functions are positive; in Quadrant II only the **S**ine function and its reciprocal are positive; in Quadrant III only the **T**angent function and its reciprocal are positive; and in Quadrant IV only the **C**osine function and its reciprocal are positive.

CAREER NOTE
EXERCISES 73 AND 74
Additional information about a career as a cartographer is available at **www.mcdougallittell.com**.

STUDENT HELP NOTES
Look Back As students look back to p. 589, remind them that to find the distance between two points you square the horizontal distance between the points, square the vertical distance between the points, add these two squares together, and then take the square root of the sum. Describing the process to the students or having them describe it to you will reinforce the meaning behind the formula.

ADDITIONAL PRACTICE AND RETEACHING

For Lesson 13.3:
• Practice Levels A, B, and C (*Chapter 13 Resource Book,* p. 42)

• Reteaching with Practice (*Chapter 13 Resource Book,* p. 45)

• See Lesson 13.3 of the *Personal Student Tutor*

For more Mixed Review:
• Search the *Test and Practice Generator* for key words or specific lessons.

DAILY HOMEWORK QUIZ

📝 *Transparency Available*

1. The point $(-5, -12)$ is on the terminal side of an angle θ in standard position. Evaluate the six trigonometric functions of θ. $\sin \theta = -\frac{12}{13}$; $\cos \theta = -\frac{5}{13}$; $\tan \theta = \frac{12}{5}$; $\csc \theta = -\frac{13}{12}$; $\sec \theta = -\frac{13}{5}$; $\cot \theta = \frac{5}{12}$

2. Evaluate the six trigonometric functions of $\theta = 180°$. $\sin \theta = 0$; $\cos \theta = -1$; $\tan \theta = 0$; $\csc \theta$ is undefined; $\sec \theta = -1$; $\cot \theta$ is undefined

3. Use a sketch to find the reference angle for $-\frac{17\pi}{6}$. $\frac{\pi}{6}$

Evaluate the function without using a calculator.

4. $\cot\left(-\frac{7\pi}{6}\right)$ $-\sqrt{3}$

5. $\csc 495°$ $\sqrt{2}$

EXTRA CHALLENGE NOTE

Challenge problems for Lesson 13.3 are available in **blackline** format in the *Chapter 13 Resource Book*, p. 49 and at **www.mcdougallittell.com.**

ADDITIONAL TEST PREPARATION

1. OPEN ENDED Name an angle in degrees for which the secant is negative and the cotangent is positive. **any angle θ for which $180° < \theta < 270°$**

81–86. See Additional Answers beginning on page AA1.

STUDENT HELP

▶ **Look Back**
For help with the distance formula, see p. 589.

SCIENCE ▶ **CONNECTION** In Exercises 75 and 76, use the following information. When two atoms in a molecule are bonded to a common atom, chemists are interested in both the bond angle and the bond length. A water molecule (H_2O) is made up of two hydrogen atoms bonded to an oxygen atom. The diagram below shows a coordinate plane superimposed on a cross section of a water molecule.

75. In the diagram, coordinates are given in picometers (pm). (Note: 1 pm $= 10^{-12}$ m). If the center of one hydrogen atom has coordinates $(96, 0)$, find the coordinates (x, y) of the center of the other hydrogen atom. **about $(-24, 93)$**

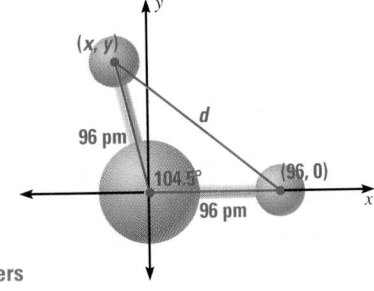

76. Use your answer to Exercise 75 and the distance formula to find the distance d (in picometers) between the centers of the two hydrogen atoms. **about 152 picometers**

Test Preparation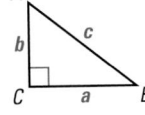

77. MULTIPLE CHOICE What is the value of $\sec\left(\frac{40\pi}{3}\right)$? **A**

(A) -2 (B) $-\sqrt{2}$ (C) $-\frac{\sqrt{2}}{2}$ (D) $-\frac{1}{2}$ (E) $\sqrt{2}$

78. MULTIPLE CHOICE What is the approximate horizontal distance traveled by a football that is kicked at an angle of $40°$ with an initial speed of 70 feet per second? **C**

(A) 98 feet (B) 142 feet (C) 151 feet (D) 157 feet (E) 280 feet

★ **Challenge**

79. $\sin \theta = \frac{2\sqrt{5}}{5}$; $\cos \theta = -\frac{\sqrt{5}}{5}$; $\csc \theta = \frac{\sqrt{5}}{2}$; $\sec \theta = -\sqrt{5}$; $\cot \theta = -\frac{1}{2}$

79. CRITICAL THINKING If θ is an angle in Quadrant II and $\tan \theta = -2$, find the values of the other five trigonometric functions of θ.

80. CRITICAL THINKING If θ is an angle in Quadrant III and $\cos \theta = -0.64$, find the values of the other five trigonometric functions of θ. $\sin \theta \approx -0.7684$; $\tan \theta \approx 1.2006$; $\csc \theta \approx -1.3014$; $\sec \theta \approx -1.5625$; $\cot \theta \approx 0.8329$

MIXED REVIEW

90. $B = 28°$; $a \approx 9.40$; $c \approx 10.7$

91. $A = 70°$; $a \approx 20.7$; $b \approx 7.52$

92. $A = 59°$; $b \approx 10.2$; $c \approx 19.8$

93. $B = 40°$; $a \approx 2.30$; $b \approx 1.93$

94. $A = 15°$; $a \approx 9.11$; $c \approx 35.2$

95. $B = 7°$; $b \approx 6.14$; $c \approx 50.4$

HORIZONTAL LINE TEST Graph the function. Then use the graph to determine whether the inverse of f is a function. *(Review 7.4 for 13.4)*

yes; See margin. yes; See margin. no; See margin.

81. $f(x) = x - 3$ **82.** $f(x) = 4x + 5$ **83.** $f(x) = 5x^2$

84. $f(x) = 5x^3$ **85.** $f(x) = 3x^2 - 7$ **86.** $f(x) = -|x + 2|$

yes; See margin.   no; See margin.   no; See margin.

CHOOSING CARDS A card is randomly drawn from a standard 52-card deck. Find the probability of the given event. (A face card is a king, queen, or jack.) *(Review 12.4)*

87. a king and a diamond $\frac{1}{52}$ **88.** a jack or a club $\frac{4}{13}$ **89.** a ten or a face card $\frac{4}{13}$

SOLVING TRIANGLES Solve $\triangle ABC$ using the diagram and the given measurements. *(Review 13.1)* 90–95. See margin.

90. $A = 62°$, $b = 5$ **91.** $B = 20°$, $c = 22$

92. $B = 31°$, $a = 17$ **93.** $A = 50°$, $c = 3$

94. $B = 75°$, $b = 34$ **95.** $A = 83°$, $a = 50$

Investigating Inverse Trigonometric Functions

GROUP ACTIVITY
Work in a small group.

MATERIALS
- paper
- pencil

▶ **QUESTION** Do the trigonometric functions sine, cosine, and tangent have inverses that are also functions?

▶ **EXPLORING THE CONCEPT**

1 Copy and complete the table to find the value of $f(x) = x^2$ for each of the given x-values.   16; 9; 4; 1; 0; 1; 4; 9; 16

x	−4	−3	−2	−1	0	1	2	3	4
$f(x) = x^2$	?	?	?	?	?	?	?	?	?

2. The function does not have an inverse over the domain $-4 \le x \le 4$ because most values of y correspond to two different x-values. For example, both 3 and −3 are mapped to 9.

2 For a function to have an inverse, it must be true that no two values of x are paired with the same value of y. Use your completed table to explain why the function $f(x) = x^2$ does not have an inverse on the domain $-4 \le x \le 4$.

3. *Sample answer:* Let $x \ge 0$. I chose this domain because it will give the usual square root function as the inverse.

3 Restrict the domain of $f(x) = x^2$ so that it does have an inverse. Explain why you chose the domain you did.

4. second row: 0; $-\frac{\sqrt{2}}{2}$; −1; $-\frac{\sqrt{2}}{2}$; 0; $\frac{\sqrt{2}}{2}$; 1; $\frac{\sqrt{2}}{2}$; 0

third row: −1; $-\frac{\sqrt{2}}{2}$; 0; $\frac{\sqrt{2}}{2}$; 1; $\frac{\sqrt{2}}{2}$; 0; $-\frac{\sqrt{2}}{2}$; −1

fourth row: 0; 1; undefined; −1; 0; 1; undefined; −1; 0

4 Copy and complete the table to find the values of $f(\theta) = \sin \theta$, $g(\theta) = \cos \theta$, and $h(\theta) = \tan \theta$ for each of the given values of θ.

θ	$-\pi$	$-\frac{3\pi}{4}$	$-\frac{\pi}{2}$	$-\frac{\pi}{4}$	0	$\frac{\pi}{4}$	$\frac{\pi}{2}$	$\frac{3\pi}{4}$	π
$f(\theta) = \sin \theta$	?	?	?	?	?	?	?	?	?
$g(\theta) = \cos \theta$	?	?	?	?	?	?	?	?	?
$h(\theta) = \tan \theta$	?	?	?	?	?	?	?	?	?

5. $f(\theta) = \sin \theta$ does not have an inverse over this domain since more than one value of θ maps to the same y-value. For example, both 0 and π map to 0.

5 Use the table to explain why $f(\theta) = \sin \theta$ does not have an inverse on the domain $-\pi \le \theta \le \pi$.

6 Does $g(\theta) = \cos \theta$ have an inverse on the domain $-\pi \le \theta \le \pi$? Explain why or why not. No; more than one value of θ maps to the same y-value. For example, both $-\pi$ and π map to −1.

7 Does $h(\theta) = \tan \theta$ have an inverse on the domain $-\pi \le \theta \le \pi$? Explain why or why not. No; more than one value of θ maps to the same y-value. For example, both $-\pi$ and π map to 0.

1. *Sample answer:*

Let $-\frac{\pi}{2} \le \theta \le \frac{\pi}{2}$.

This domain gives no repeated values for sin θ, and the domain cannot be extended any further and still allow an inverse.

2. *Sample answer:*

Let $0 \le \theta \le \pi$. This domain gives no repeated values for cos θ, and a longer interval cannot be used or the function will not be invertible.

▶ **DRAWING CONCLUSIONS**

1. Use the table you completed in **Step 4** to choose a restricted domain for which $f(\theta) = \sin \theta$ does have an inverse. Explain how you made your choice.

2. Write a restricted domain for which $g(\theta) = \cos \theta$ has an inverse. Explain how you chose the domain.

3. Write a restricted domain for which $h(\theta) = \tan \theta$ has an inverse. Explain how you chose the domain. See margin.

4. Are the domains that you wrote in Exercises 1–3 the *only* domains for which the trigonometric functions have inverses? Explain. See margin.

1 Planning the Activity

PURPOSE
To find restricted domains for which the sine, cosine, and tangent functions have inverse functions.

MATERIALS
- Activity Support Master (*Chapter 13 Resource Book*, p. 53)

PACING
- Exploring the Concept — 25 min
- Drawing Conclusions — 10 min

▶ **LINK TO LESSON**
The chart on page 792 of Lesson 13.4 summarizes the results of this Activity.

2 Managing the Activity

ALTERNATIVE APPROACH
Have students sketch a graph of $y = x^2$, and then rotate the paper 90° to the right so that the x- and y-axes are switched. Ask if this parabola is a function. Then ask if each branch is a function. Use the graphs to introduce the idea of restricting the domain of the original parabola so that the inverse is a function.

3 Closing the Activity

★ **KEY DISCOVERY**
If their domains are restricted properly, the inverses of the sine, cosine, and tangent functions are also functions.

ACTIVITY ASSESSMENT
Give a domain besides $-\frac{\pi}{2} \le \theta \le \frac{\pi}{2}$ for which the sine function has an inverse. $\frac{k\pi}{2} \le \theta \le \frac{(k+2)\pi}{2}$ for any odd integer k

3, 4. See Additional Answers beginning on page AA1.

LESSON OPENER
APPLICATION

An alternative way to approach
Lesson 13.4 is to use the
Application Lesson Opener:

• Blackline Master (*Chapter 13
  Resource Book,* p. 54)
• Transparency (p. 87)

MEETING INDIVIDUAL NEEDS

• *Chapter 13 Resource Book*
  Prerequisite Skills Review (p. 5)
  Practice Level A (p. 55)
  Practice Level B (p. 56)
  Practice Level C (p. 57)
  Reteaching with Practice (p. 58)
  Absent Student Catch-Up (p. 60)
  Challenge (p. 62)
• *Resources in Spanish*
• 🖳 *Personal Student Tutor*

NEW-TEACHER SUPPORT

See the Tips for New Teachers on
pp. 1–2 of the *Chapter 13 Resource
Book* for additional notes about
Lesson 13.4.

WARM-UP EXERCISES

🖳 **Transparency Available**

Name two angles such that
$0 \le \theta \le 2\pi$ with the given sine
value.

1. $\frac{1}{2}$ $\quad \frac{\pi}{6}, \frac{5\pi}{6}$

2. $-\frac{\sqrt{3}}{2}$ $\quad \frac{4\pi}{3}, \frac{5\pi}{3}$

Write the reciprocal function of
each trigonometric function.

3. $\sin \theta$ $\quad \csc \theta$

4. $\cos \theta$ $\quad \sec \theta$

5. $\tan \theta$ $\quad \cot \theta$

What you should learn

GOAL 1 Evaluate inverse
trigonometric functions.

GOAL 2 Use inverse
trigonometric functions to
solve **real-life** problems,
such as finding an angle of
repose in **Example 4**.

Why you should learn it

▼ To solve **real-life**
problems, such as finding
the angle at which to set
the arm of a crane in
Example 5.

13.4 Inverse Trigonometric Functions

GOAL 1 **EVALUATING AN INVERSE TRIGONOMETRIC FUNCTION**

In the first three lessons of this chapter, you learned to evaluate trigonometric
functions of a given angle. In this lesson you will study the reverse problem—
finding angles that correspond to a given value of a trigonometric function.

Suppose you were asked to find an angle θ whose sine is 0.5. After thinking about the
problem for a while, you would probably realize that there are *many* such angles. For
instance, the angles

$$\frac{\pi}{6}, \frac{5\pi}{6}, \frac{13\pi}{6}, \frac{17\pi}{6}, \text{ and } -\frac{7\pi}{6}$$

all have a sine value of 0.5. (Try checking this with a calculator.) Of these, the value
of the *inverse sine function* at 0.5 is defined to be $\frac{\pi}{6}$. General definitions of
inverse sine, inverse cosine, and inverse tangent are given below.

INVERSE TRIGONOMETRIC FUNCTIONS

• If $-1 \le a \le 1$, then the **inverse sine**
  of a is $\sin^{-1} a = \theta$ where $\sin \theta = a$
  and $-\frac{\pi}{2} \le \theta \le \frac{\pi}{2}$ (or $-90° \le \theta \le 90°$).

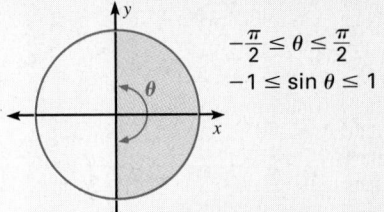

• If $-1 \le a \le 1$, then the **inverse cosine**
  of a is $\cos^{-1} a = \theta$ where $\cos \theta = a$
  and $0 \le \theta \le \pi$ (or $0° \le \theta \le 180°$).

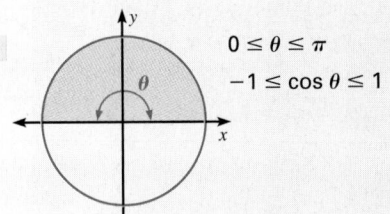

• If a is any real number, then the
  inverse tangent of a is $\tan^{-1} a = \theta$
  where $\tan \theta = a$ and $-\frac{\pi}{2} < \theta < \frac{\pi}{2}$
  (or $-90° < \theta < 90°$).

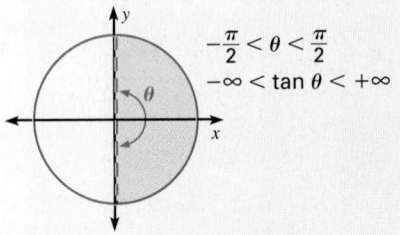

EXAMPLE 1   *Evaluating Inverse Trigonometric Functions*

Evaluate the expression in both radians and degrees.

a. $\sin^{-1} \dfrac{\sqrt{3}}{2}$ **b.** $\cos^{-1} 2$ **c.** $\tan^{-1}(-1)$

SOLUTION

a. When $-\dfrac{\pi}{2} \le \theta \le \dfrac{\pi}{2}$, or $-90° \le \theta \le 90°$, the angle whose sine is $\dfrac{\sqrt{3}}{2}$ is:

$$\theta = \sin^{-1} \frac{\sqrt{3}}{2} = \frac{\pi}{3} \qquad \text{or} \qquad \theta = \sin^{-1} \frac{\sqrt{3}}{2} = 60°$$

b. There is no angle whose cosine is 2. So, $\cos^{-1} 2$ is undefined.

c. When $-\dfrac{\pi}{2} < \theta < \dfrac{\pi}{2}$, or $-90° < \theta < 90°$, the angle whose tangent is -1 is:

$$\theta = \tan^{-1}(-1) = -\frac{\pi}{4} \qquad \text{or} \qquad \theta = \tan^{-1}(-1) = -45°$$

EXAMPLE 2   *Finding an Angle Measure*

Find the measure of the angle θ for the triangle shown.

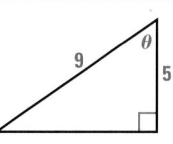

SOLUTION

In the right triangle, you are given the adjacent side and the hypotenuse. You can write:

$$\cos \theta = \frac{\text{adj}}{\text{hyp}} = \frac{5}{9}$$

This equation is asking you to find the acute angle whose cosine is $\dfrac{5}{9}$. Use a calculator to find the measure of θ.

$$\theta = \cos^{-1} \frac{5}{9} \approx 0.982 \text{ radians} \qquad \text{or} \qquad \theta = \cos^{-1} \frac{5}{9} \approx 56.3°$$

STUDENT HELP

▶ **Study Tip**
When approximating the value of an angle, make sure your calculator is set to radian mode if you want your answer in radians, or to degree mode if you want your answer in degrees.

EXAMPLE 3   *Solving a Trigonometric Equation*

Solve the equation $\sin \theta = -\dfrac{1}{4}$ where $180° < \theta < 270°$.

SOLUTION

In the interval $-90° < \theta < 90°$, the angle whose sine is $-\dfrac{1}{4}$ is $\sin^{-1}\left(-\dfrac{1}{4}\right) \approx -14.5°$. This angle is in Quadrant IV as shown. In Quadrant III (where $180° < \theta < 270°$), the angle that has the same sine value is:

$$\theta \approx 180° + 14.5° = 194.5°$$

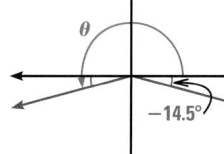

✓ **CHECK**   Use a calculator to check the answer.

$$\sin 194.5° \approx -0.25 \ \checkmark$$

---

2 TEACH

MOTIVATING THE LESSON
Students have used trigonometric functions to find a side when they know another side and an acute angle in a right triangle. If they know two sides and seek an angle, as in finding the angle to raise a 20 ft boom so that the camera at its end is 15 ft high, they must use an *inverse trigonometric function.*

EXTRA EXAMPLE 1
Evaluate the expression in both radians and degrees.
a. $\sin^{-1} \dfrac{\sqrt{2}}{2}$; $\dfrac{\pi}{4}$; 45°
b. $\cos^{-1} 3$ undefined
c. $\tan^{-1} 1$ $\dfrac{\pi}{4}$; 45°

EXTRA EXAMPLE 2
Find the measure of the angle θ for the triangle shown.

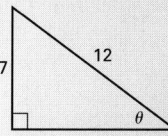

about 35.7° or 0.623 radians

EXTRA EXAMPLE 3
Solve the equation $\cos \theta = -\dfrac{2}{3}$ where $180° < \theta < 270°$.
about 228.2°

CHECKPOINT EXERCISES
For use after Example 1:
1. Evaluate $\cos^{-1}\left(-\dfrac{1}{2}\right)$ in both radians and degrees. $\dfrac{2\pi}{3}$; 120°

For use after Example 2:
2. Find the measure of the angle θ for the triangle shown.

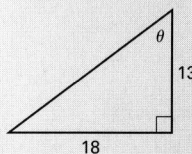

about 54.2° or 0.945 radians

For use after Example 3:
3. Solve $\tan \theta = -5$ where $90° < \theta < 180°$.   about 101.3°

793

EXTRA EXAMPLE 4

A sand pile in a yard is 4 feet high. The diameter of its base is 10 feet.

a. Find the angle of repose for this pile of sand. about 38.7°

b. How tall will a pile of this sand be if the base has a diameter of 40 feet? 16 ft

EXTRA EXAMPLE 5

A crane whose lower end is 4 feet off the ground has a 100 foot arm. The arm has to reach the top of a building 80 feet high. At what angle should the arm be set? about 49.5°

 CHECKPOINT EXERCISES

For use after Examples 4 and 5:

1. A 10 inch high pile of sand in a sandbox has a diameter of 35 inches.

  a. What is the angle of repose for this sand?
     about 29.7°

  b. How wide will a pile 18 inches high be? 63 in.

CONCEPT QUESTION

What is the difference between $\sin^{-1} \theta$ and $(\sin \theta)^{-1}$? The first is the angle whose sine is θ, and the second is the reciprocal of the sine of θ.

FOCUS ON VOCABULARY

What are you finding when you take the inverse of the sine, cosine, or tangent functions?
an angle measure

CLOSURE QUESTION

How do you solve the equation $\sin \theta = 0.5$ for θ? *Sample answer:*
Find the angle between $-\frac{\pi}{2}$ and $\frac{\pi}{2}$ whose sine is 0.5. This is the inverse sine of 0.5.

DAILY PUZZLER

What is $\sin (\sin^{-1} 0.8)$? Explain.
0.8; *Sample answer:* The expression describes the sine of the angle whose sine is 0.8, which is 0.8.

► **ROCK SALT**
Each year about 9 million tons of rock salt are poured on highways in North America to melt ice. Although rock salt is the best deicing material, it also eats away at cars and road surfaces.

GOAL 2 **USING INVERSE TRIGONOMETRIC FUNCTIONS IN REAL LIFE**

EXAMPLE 4 *Writing and Solving a Trigonometric Equation*

ROCK SALT Different types of granular substances naturally settle at different angles when stored in cone-shaped piles. This angle θ is called the *angle of repose*. When rock salt is stored in a cone-shaped pile 11 feet high, the diameter of the pile's base is about 34 feet. ▶ Source: Bulk-Store Structures, Inc.

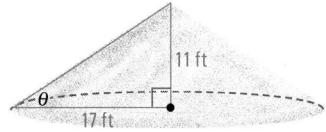

a. Find the angle of repose for rock salt.

b. How tall is a pile of rock salt that has a base diameter of 50 feet?

SOLUTION

a. In the right triangle shown inside the cone, you are given the opposite side and the adjacent side. You can write:

$$\tan \theta = \frac{\text{opp}}{\text{adj}} = \frac{11}{17}$$

This equation is asking you to find the acute angle whose tangent is $\frac{11}{17}$.

$$\theta = \tan^{-1} \frac{11}{17} \approx 32.9°$$

▶ The angle of repose for rock salt is about 32.9°.

b. The pile of rock salt has a base radius of 25 feet. From part (a) you know that the angle of repose for rock salt is about 32.9°. To find the height h (in feet) of the pile you can write:

$$\tan 32.9° = \frac{h}{25}$$

$$h = 25 \tan 32.9°$$

$$\approx 16.2$$

▶ The pile of rock salt is about 16.2 feet tall.

Construction

EXAMPLE 5 *Writing and Solving a Trigonometric Equation*

A crane has a 200 foot arm whose lower end is 5 feet off the ground. The arm has to reach the top of a building 130 feet high. At what angle θ should the arm be set?

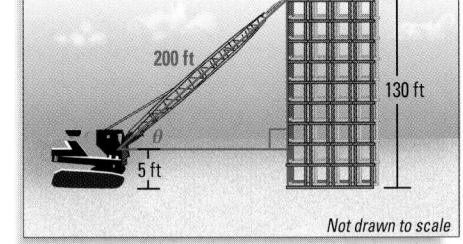

SOLUTION

In the right triangle in the diagram, you know the opposite side and the hypotenuse. You can write:

$$\sin \theta = \frac{\text{opp}}{\text{hyp}} = \frac{130 - 5}{200} = \frac{5}{8}$$

This equation is asking you to find the acute angle whose sine is $\frac{5}{8}$.

$$\theta = \sin^{-1} \frac{5}{8} \approx 38.7°$$

▶ The crane's arm should be set at an acute angle of about 38.7°.

GUIDED PRACTICE

Vocabulary Check ✓

Concept Check ✓

2. Some of the domain values map to the same *y*-value. For example, $-\frac{\pi}{4}$ and $\frac{\pi}{4}$ both are mapped to $\frac{\sqrt{2}}{2}$.

Skill Check ✓

3. The number 3 is in the range of tan θ, but not in the range of cos θ. The domain of $\tan^{-1}$ is $-\infty < x < \infty$, but the domain of $\cos^{-1}$ is $-1 \le x \le 1$.

4. The angle she obtained is in Quadrant II, not Quadrant III. She should have added the opposite of the angle to 180°, obtaining 198.8°.

1. Complete this statement: The __?__ sine of 1 equals $\frac{\pi}{2}$, or 90°. **inverse**

2. Explain why the domain of $y = \cos \theta$ cannot be restricted to $-\frac{\pi}{2} \le \theta \le \frac{\pi}{2}$ if the inverse is to be a function.

3. Explain why $\tan^{-1} 3$ is defined, but $\cos^{-1} 3$ is undefined. **See margin.**

4. **ERROR ANALYSIS** A student needed to find an angle θ in Quadrant III such that $\sin \theta = -0.3221$. She used a calculator to find that $\sin^{-1}(-0.3221) \approx -18.8°$. Then she added this result to 180° to get an answer of $\theta = 161.2°$. What did she do wrong? **See margin.**

Evaluate the expression without using a calculator.

5. $\tan^{-1}\sqrt{3}$ $\frac{\pi}{3}$, or 60° 6. $\cos^{-1}\frac{\sqrt{2}}{2}$ $\frac{\pi}{4}$, or 45° 7. $\sin^{-1}\frac{1}{2}$ $\frac{\pi}{6}$, or 30° 8. $\cos^{-1}\left(-\frac{1}{2}\right)$ $\frac{2\pi}{3}$, or 120°

Use a calculator to evaluate the expression in both radians and degrees. Round to three significant digits.

9. $\tan^{-1} 3.9$ 10. $\cos^{-1}(-0.94)$ 11. $\cos^{-1} 0.34$ 12. $\sin^{-1}(-0.4)$
   1.32; 75.6°   2.79; 160°   1.22; 70.1°   −0.412; −23.6°

Solve the equation for θ. Round to three significant digits.

13. $\sin \theta = -0.35$; $180° < \theta < 270°$ **200°** 14. $\tan \theta = 2.4$; $180° < \theta < 270°$ **247°**

15. $\cos \theta = 0.43$; $270° < \theta < 360°$ **295°** 16. $\sin \theta = 0.8$; $90° < \theta < 180°$ **127°**

17. 🌐 **CONSTRUCTION** A crane has a 150 foot arm whose lower end is 4 feet off the ground. The arm has to reach the top of a building 105 feet high. At what angle should the crane's arm be set? **about 42.3°**

PRACTICE AND APPLICATIONS

STUDENT HELP

→ **Extra Practice**
to help you master skills is on p. 958.

EVALUATING EXPRESSIONS Evaluate the expression without using a calculator. Give your answer in both radians and degrees.

18. $\sin^{-1}\frac{\sqrt{2}}{2}$ $\frac{\pi}{4}$; 45° 19. $\cos^{-1}\frac{1}{2}$ $\frac{\pi}{3}$; 60° 20. $\tan^{-1} 1$ $\frac{\pi}{4}$; 45° 21. $\sin^{-1} 0$ 0; 0°

22. $\cos^{-1}(-1)$ π; 180° 23. $\sin^{-1}(-1)$ $-\frac{\pi}{2}$; −90° 24. $\tan^{-1}\left(-\frac{\sqrt{3}}{3}\right)$ $-\frac{\pi}{6}$; −30° 25. $\cos^{-1}\left(-\frac{\sqrt{3}}{2}\right)$ $\frac{5\pi}{6}$; 150°

FINDING ANGLES Find the measure of the angle θ. Round to three significant digits.

STUDENT HELP

→ **HOMEWORK HELP**
Example 1: Exs. 18–25, 32–43
Example 2: Exs. 26–43
Example 3: Exs. 44–51
Examples 4, 5: Exs. 52–57

26. 26.6°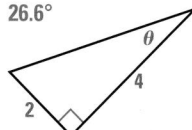

27. 48.2°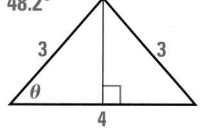

28. 34.8°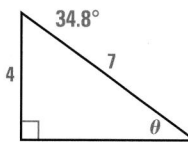

29. 120°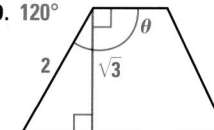

30. 36.9°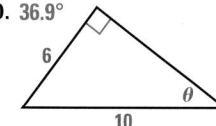

31. 18.4°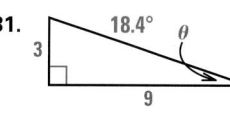

ASSIGNMENT GUIDE

BASIC
Day 1: pp. 795–798 Exs. 18–22, 26–29, 32–38, 44–48, 65–75 odd
Day 2: pp. 796–798 Exs. 49–53, 57, 63, 77–81 odd, Quiz 2 Exs. 1–31

AVERAGE
Day 1: pp. 795–798 Exs. 18–22, 26–38, 44–48, 65–75 odd
Day 2: pp. 796–798 Exs. 49–53, 57–60, 63, 77–81 odd, Quiz 2 Exs. 1–31

ADVANCED
Day 1: pp. 795–798 Exs. 18–22, 26–38, 44–51, 65–75 odd
Day 2: pp. 796–798 Exs. 52–56, 58–64, 77–81 odd, Quiz 2 Exs. 1–31

BLOCK SCHEDULE WITH 13.3
pp. 795–798 Exs. 18–22, 26–38, 44–48, 52–56, 63, 65–81 odd, Quiz 2 Exs. 1–31

EXERCISE LEVELS
Level A: *Easier*
18–42, 71–82
Level B: *More Difficult*
43–57, 63, 65–70
Level C: *Most Difficult*
58–62, 64

✔ HOMEWORK CHECK
To quickly check student understanding of key concepts, go over the following exercises: Exs. 18, 26, 34, 36, 44. See also the Daily Homework Quiz:
• Blackline Master (*Chapter 13 Resource Book*, p. 66)
• 💾 Transparency (p. 101)

! **COMMON ERROR**

EXERCISES 44–51 It is often difficult for students to find inverse trigonometric function values outside of their principal values. Remind them that the reference angle with the *x*-axis in the desired quadrant will have the same magnitude as does the original angle with the *x*-axis.

STUDENT HELP NOTES

→ **Homework Help** Students can find help for Exs. 44–51 at **www.mcdougallittell.com**. The information can be printed out for students who don't have access to the Internet.

EVALUATING EXPRESSIONS Use a calculator to evaluate the expression in both radians and degrees. Round to three significant digits.

32. $\tan^{-1} 3.9$
    1.32; 75.6°

33. $\cos^{-1} 0.24$
    1.33; 76.1°

34. $\cos^{-1} 0.34$
    1.22; 70.1°

35. $\sin^{-1} 0.75$
    0.848; 48.6°

36. $\sin^{-1} (-0.4)$
    −0.412; −23.6°

37. $\cos^{-1} (-0.6)$
    2.21; 127°

38. $\tan^{-1} (-0.2)$
    −0.197; −11.3°

39. $\tan^{-1} 2.25$
    1.15; 66.0°

40. $\cos^{-1} (-0.8)$
    2.50; 143°

41. $\sin^{-1} 0.99$
    1.43; 81.9°

42. $\tan^{-1} 12$
    1.49; 85.2°

43. $\cos^{-1} 0.55$
    0.988; 56.6°

SOLVING EQUATIONS Solve the equation for θ. Round to three significant digits.

STUDENT HELP

INTERNET **HOMEWORK HELP**
Visit our Web site
www.mcdougallittell.com
for help with Exs. 44–51.

44. $\sin \theta = -0.35$; $180° < \theta < 270°$ 200°

45. $\tan \theta = 2.4$; $180° < \theta < 270°$ 247°

46. $\cos \theta = 0.43$; $270° < \theta < 360°$ 295°

47. $\sin \theta = 0.8$; $90° < \theta < 180°$ 127°

48. $\tan \theta = -2.1$; $90° < \theta < 180°$ 115°

49. $\cos \theta = -0.72$; $180° < \theta < 270°$ 224°

50. $\sin \theta = 0.2$; $90° < \theta < 180°$ 168°

51. $\tan \theta = 0.9$; $180° < \theta < 270°$ 222°

52. 🌐 **SWIMMING POOL** The swimming pool shown in cross section at the right ranges in depth from 3 feet at the shallow end to 8 feet at the deep end. Find the angle of depression θ between the shallow end and the deep end. **about 26.6°**

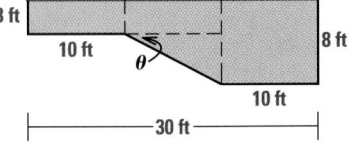

53. 🌐 **DUMP TRUCK** The dump truck shown has a 10 foot bed. When tilted at its maximum angle, the bed reaches a height of 7 feet above its original position. What is the maximum angle θ that the truck bed can tilt? **about 44.4°**

54. 🌐 **GRANULAR ANGLE OF REPOSE** Look back at Example 4 on page 794. When whole corn is stored in a cone-shaped pile 20 feet high, the diameter of the pile's base is about 82 feet. Find the angle of repose for whole corn. **about 26.0°**

55. 🌐 **ROAD DESIGN** Curves that connect two straight sections of a road are often constructed as arcs of circles. In the diagram, θ is the central angle of a circular arc that has a radius of 225 feet. Each radius line shown is perpendicular to one of the straight sections. The straight sections are therefore tangent to the arc. The extension of each straight section to their point of intersection is 158 feet in length. Find the degree measure of θ. **about 70.2°**

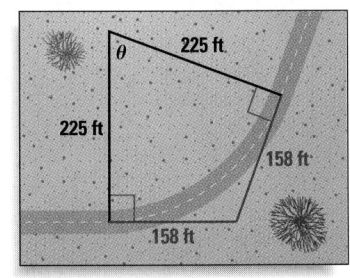

56. 🌐 **DRAWBRIDGE** The Park Street Bridge in Alameda County, California, is a double-leaf drawbridge. Each leaf of the bridge is 120 feet long. A ship that is 100 feet wide needs to pass through the bridge. What is the minimum angle θ that each leaf of the bridge should be opened to in order to ensure that the ship will fit?
▶ Source: Alameda County Drawbridges **about 54.3°**

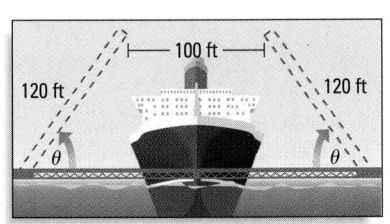

FOCUS ON
PEOPLE

JEFF GORDON
began racing in the top division of the National Association for Stock Car Auto Racing (NASCAR) in 1993. He has won three NASCAR division championships in the past four years.

57. 🌐 **RACEWAY** Suppose you are at a raceway and are sitting on the straightaway, 100 feet from the center of the track. If a car traveling 145 miles per hour passes directly in front of you, at what angle do you have to turn your head to see the car t seconds later? Assume that the car is still on the straightaway and is traveling at a constant speed. (*Hint:* First convert 145 miles per hour to a speed v in feet per second. The expression vt represents the distance in feet traveled by the car.)

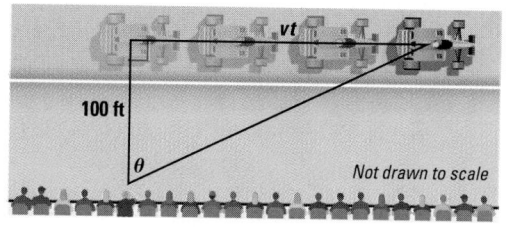

$$\theta = \tan^{-1}(2.127t)$$

100 ft

Not drawn to scale

GEOMETRY **CONNECTION** **In Exercises 58–62, use the following information.**
Consider a line with positive slope m that makes an angle θ with the x-axis (measuring counterclockwise from the x-axis).

58. Find the slope m of the line $y = 3x - 2$. **3**

59. Find θ for the line $y = 3x - 2$. **about 71.6°**

60. **CRITICAL THINKING** How could you have found θ for the line $y = 3x - 2$ by using the slope m of the line? Write an equation relating θ and m. $\theta = \tan^{-1} m$

61. Find an equation of the line that makes an angle of 58° with the x-axis and whose y-intercept is 3. $y = 1.6x + 3$

62. Find an equation of the line that makes an angle of 35° with the x-axis and whose x-intercept is 4. $y = 0.7x - 2.8$

Test ✏️ **Preparation**

63d. *Sample answer:* **As you move closer to the shell d must get smaller and smaller. As you walk closer, the angles θ_1 and θ_2 will decrease, along with the distance y, until you are standing directly over the shell, at which time y (and so also d) will be zero.**

63. **MULTI-STEP PROBLEM** If you stand in shallow water and look at an object below the surface of the water, the object will look farther away from you than it really is. This is because when light rays pass between air and water, the water *refracts*, or bends, the light rays. The *index of refraction* for seawater is 1.341. This is the ratio of the sine of θ_1 to the sine of θ_2 for angles θ_1 and θ_2 below.

a. You are standing in seawater that is 2 feet deep and are looking at a shell at angle $\theta_1 = 60°$ (measured from a line perpendicular to the surface of the water). Find θ_2.
about 40.2°

b. Find the distances x and y.
$x \approx 1.69$ ft; $y \approx 3.46$ ft

c. Find the distance d between where the shell is and where it appears to be. about 1.77 ft

d. *Writing* What happens to d as you move closer to the shell? Explain your reasoning. **See margin.**

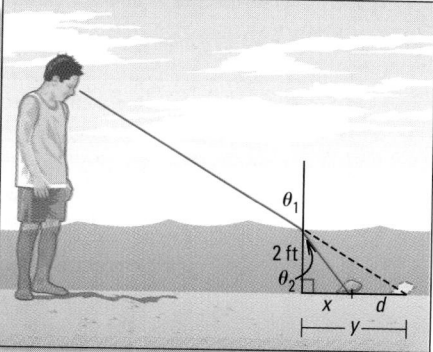

★ **Challenge**

64. **LENGTH OF A PULLEY BELT** Find the length of the pulley belt shown at the right. (*Hint:* Partition the belt into four parts: the two straight segments, the arc around the small wheel, and the arc around the large wheel.)
about 62.5 in.

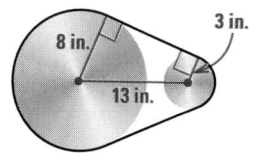

3 in.
8 in.
13 in.

ADDITIONAL PRACTICE AND RETEACHING

For Lesson 13.4:
• Practice Levels A, B, and C (*Chapter 13 Resource Book*, p. 55)
• Reteaching with Practice (*Chapter 13 Resource Book*, p. 58)
• 🖥 See Lesson 13.4 of the *Personal Student Tutor*

For more Mixed Review:
• 🖥 Search the *Test and Practice Generator* for key words or specific lessons.

4 ASSESS

DAILY HOMEWORK QUIZ

Transparency Available

1. Evaluate $\cos^{-1}\left(-\frac{\sqrt{2}}{2}\right)$ without a calculator in both radians and degrees. $\frac{3\pi}{4}$; 135°
2. Find the measure of θ. Round to three significant digits.

46.7°

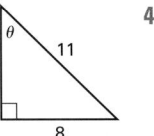

Use a calculator to evaluate the expression in both radians and degrees. Round to three significant digits.

3. $\cos^{-1}(-0.8)$ 2.50; 143°
4. $\tan^{-1} 4.8$ 1.37; 78.2°
5. Solve the equation for θ. Round to three significant digits. $\cos\theta = -0.11$; $180° < \theta < 270°$ 264°

EXTRA CHALLENGE NOTE

Challenge problems for Lesson 13.4 are available in **blackline** format in the *Chapter 13 Resource Book*, p. 62 and at **www.mcdougallittell.com**.

ADDITIONAL TEST PREPARATION

1. **WRITING** Explain why the sine function is restricted to the domain $-\frac{\pi}{2} \le \theta \le \frac{\pi}{2}$ to define the inverse sine function. See sample answer below.
2. **OPEN ENDED** If $\sin\theta = c$, and c is negative, give a possible value of θ where $360° < \theta \le 720°$. See sample answer below.

ADDITIONAL RESOURCES

An alternative Quiz for Lessons 13.3 and 13.4 is available in the *Chapter 13 Resource Book*, p. 63.

5–8. See Additional Answers beginning on page AA1.

1. $\sin\theta = -\frac{16\sqrt{337}}{337}$; $\cos\theta = -\frac{9\sqrt{337}}{337}$; $\tan\theta = \frac{16}{9}$; $\csc\theta = -\frac{\sqrt{337}}{16}$; $\sec\theta = -\frac{\sqrt{337}}{9}$; $\cot\theta = \frac{9}{16}$

2. $\sin\theta = -\frac{2\sqrt{53}}{53}$; $\cos\theta = \frac{7\sqrt{53}}{53}$; $\tan\theta = -\frac{2}{7}$; $\csc\theta = -\frac{\sqrt{53}}{2}$; $\sec\theta = \frac{\sqrt{53}}{7}$; $\cot\theta = -\frac{7}{2}$

3. $\sin\theta = \frac{5\sqrt{26}}{26}$; $\cos\theta = -\frac{\sqrt{26}}{26}$; $\tan\theta = -5$; $\csc\theta = \frac{\sqrt{26}}{5}$; $\sec\theta = -\sqrt{26}$; $\cot\theta = -\frac{1}{5}$

4. $\sin\theta = -\frac{11\sqrt{157}}{157}$; $\cos\theta = \frac{6\sqrt{157}}{157}$; $\tan\theta = -\frac{11}{6}$; $\csc\theta = -\frac{\sqrt{157}}{11}$; $\sec\theta = \frac{\sqrt{157}}{6}$; $\cot\theta = -\frac{6}{11}$

MIXED REVIEW

SOLVING EQUATIONS **Solve the rational equation. Check for extraneous solutions.** (Review 9.6 for 13.5)

65. $\frac{6}{x} = \frac{7}{x+3}$ 18
66. $\frac{3}{x-3} = \frac{7}{x}$ $\frac{21}{4}$
67. $\frac{-1}{4+x} = \frac{6}{2x}$ −3
68. $\frac{3}{x+3} + 7 = \frac{-4}{x+3}$ −4
69. $\frac{1}{x+2} = \frac{x}{2x+9}$ −3, 3
70. $\frac{3x}{x-2} = 2 + \frac{6}{x-2}$ no solution

CHOOSING NUMBERS **You have an equally likely chance of choosing any number 1 through 30. Find the probability of the given event.** (Review 12.3)

71. A multiple of 5 is chosen. $\frac{1}{5}$
72. A prime number is chosen. $\frac{1}{3}$
73. An even number is chosen. $\frac{1}{2}$
74. A factor of 90 is chosen. $\frac{1}{3}$
75. A number less than 12 is chosen. $\frac{11}{30}$
76. A number greater than 23 is chosen. $\frac{7}{30}$

EVALUATING FUNCTIONS **Use a calculator to evaluate the function. Round to four decimal places.** (Review 13.3 for 13.5)

77. $\sin 27°$ 0.4540
78. $\sin\frac{23\pi}{8}$ 0.3827
79. $\cos 67°$ 0.3907
80. $\sec\frac{53\pi}{9}$ 1.0642
81. $\tan 192°$ 0.2126
82. $\csc 219°$ −1.5890

QUIZ 2

Self-Test for Lessons 13.3 and 13.4

Use the given point on the terminal side of an angle θ in standard position. Evaluate the six trigonometric functions of θ. (Lesson 13.3) 1–8. See margin.

1. $(-9, -16)$
2. $(7, -2)$
3. $(-1, 5)$
4. $(6, -11)$
5. $(3, 6)$
6. $(-12, 3)$
7. $(9, -5)$
8. $(-7, -8)$

Evaluate the function without using a calculator. (Lesson 13.3)

9. $\sin(-135°)$ $-\frac{\sqrt{2}}{2}$
10. $\tan\frac{8\pi}{3}$ $-\sqrt{3}$
11. $\cos(-420°)$ $\frac{1}{2}$
12. $\tan\left(-\frac{2\pi}{3}\right)$ $\sqrt{3}$
13. $\sin\frac{5\pi}{3}$ $-\frac{\sqrt{3}}{2}$
14. $\cos 870°$ $-\frac{\sqrt{3}}{2}$
15. $\tan(-30°)$ $-\frac{\sqrt{3}}{3}$
16. $\sin\frac{23\pi}{6}$ $-\frac{1}{2}$

Use a calculator to evaluate the expression in both radians and degrees. Round to three significant digits. (Lesson 13.4)

17. $\tan^{-1} 2.3$ 1.16; 66.5°
18. $\sin^{-1}(-0.6)$ −0.644; −36.9°
19. $\cos^{-1} 0.95$ 0.318; 18.2°
20. $\sin^{-1} 0.23$ 0.232; 13.3°
21. $\tan^{-1}(-4)$ −1.33; −76.0°
22. $\cos^{-1}(-0.8)$ 2.50; 143°
23. $\sin^{-1} 0.1$ 0.100; 5.74°
24. $\tan^{-1} 10$ 1.47; 84.3°

Solve the equation for θ. Round to three significant digits. (Lesson 13.4)

25. $\sin\theta = 0.25$; $90° < \theta < 180°$ 166°
26. $\cos\theta = 0.21$; $270° < \theta < 360°$ 282°
27. $\tan\theta = 7$; $180° < \theta < 270°$ 262°
28. $\sin\theta = -0.44$; $180° < \theta < 270°$ 206°
29. $\cos\theta = -0.3$; $180° < \theta < 270°$ 253°
30. $\tan\theta = -4.5$; $90° < \theta < 180°$ 103°
31. **LACROSSE** A lacrosse player throws a ball at an angle of 55° and at an initial speed of 40 feet per second. How far away should her teammate be to catch the ball at the same height from which it was thrown? (Lesson 13.3) about 47 ft

Additional Test Preparation *Sample answers:*

1. Over this domain, the sine function takes on all of its possible values (from −1 to 1) without any of these values repeating. (If any values repeated, the inverse would not be a function.)
2. any value θ for which $540° < \theta < 720°$

13.5

What you should learn

GOAL 1 Use the law of sines to find the sides and angles of a triangle.

GOAL 2 Find the area of any triangle, as applied in **Example 6**.

Why you should learn it

▼ To solve **real-life** problems, such as finding the distance between the Empire State Building and the Statue of Liberty in **Ex. 60**.

The Law of Sines

GOAL 1 **USING THE LAW OF SINES**

In Lesson 13.1 you learned how to solve right triangles. To solve a triangle with no right angle, you need to know the measure of at least one side and any two other parts of the triangle. This breaks down into four possible cases.

1. Two angles and any side (AAS or ASA)

2. Two sides and an angle opposite one of them (SSA)

3. Three sides (SSS)

4. Two sides and their included angle (SAS)

The first two cases can be solved using the **law of sines**. The last two cases require the law of cosines, which you will study in Lesson 13.6.

LAW OF SINES

If $\triangle ABC$ has sides of length a, b, and c as shown, then:

$$\frac{\sin A}{a} = \frac{\sin B}{b} = \frac{\sin C}{c}$$

An equivalent form is $\frac{a}{\sin A} = \frac{b}{\sin B} = \frac{c}{\sin C}$.

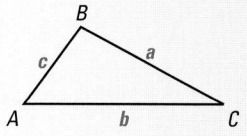

EXAMPLE 1 *The AAS or ASA Case*

Solve $\triangle ABC$ with $C = 103°$, $B = 28°$, and $b = 26$ feet.

SOLUTION

You can find the third angle of $\triangle ABC$ as follows.

$$A = 180° - 103° - 28° = 49°$$

By the law of sines you can write:

$$\frac{a}{\sin 49°} = \frac{26}{\sin 28°} = \frac{c}{\sin 103°}$$

You can then solve for a and c as follows.

$\dfrac{a}{\sin 49°} = \dfrac{26}{\sin 28°}$	**Write two equations, each with one variable.**	$\dfrac{c}{\sin 103°} = \dfrac{26}{\sin 28°}$
$a = \dfrac{26 \sin 49°}{\sin 28°}$	**Solve for the variable.**	$c = \dfrac{26 \sin 103°}{\sin 28°}$
$a \approx 41.8$ feet	**Use a calculator.**	$c \approx 54.0$ feet

STUDENT HELP

▶ **Derivations**
For a derivation of the law of sines, see p. 900.

---

1 PLAN

PACING
Basic: 2 days
Average: 2 days
Advanced: 2 days
Block Schedule: 1 block

LESSON OPENER
ACTIVITY
An alternative way to approach Lesson 13.5 is to use the Activity Lesson Opener:

• Blackline Master (*Chapter 13 Resource Book,* p. 67)
•  Transparency (p. 88)

MEETING INDIVIDUAL NEEDS
• *Chapter 13 Resource Book*
Prerequisite Skills Review (p. 5)
Practice Level A (p. 70)
Practice Level B (p. 71)
Practice Level C (p. 72)
Reteaching with Practice (p. 73)
Absent Student Catch-Up (p. 75)
Challenge (p. 77)

• *Resources in Spanish*
• *Personal Student Tutor*

NEW-TEACHER SUPPORT
See the Tips for New Teachers on pp. 1–2 of the *Chapter 13 Resource Book* for additional notes about Lesson 13.5.

WARM-UP EXERCISES

Transparency Available

Evaluate the expression in degrees.

1. $\sin^{-1}\left(-\dfrac{\sqrt{3}}{2}\right)$ −60°

2. $\sin^{-1}\dfrac{\sqrt{2}}{2}$ 45°

Solve for *x*.

3. $\dfrac{x}{\sin 50°} = 12$ 9.19

4. $\dfrac{\sin 23°}{x} = \dfrac{5}{2}$ 0.156

Have students imagine they are bordering a triangular garden with timbers. A 10 foot timber forms the "base," and the base angles are 80° and 60°. Ask them if any know a way to find how long the other timbers must be. Some may see that they can make a scale model and measure. Tell students, though, that this is one of the many cases where trigonometric ratios allow us to solve a nonright triangle.

EXTRA EXAMPLE 1
Solve $\triangle ABC$ with $B = 118°$, $C = 36°$, and $c = 14$ inches.
$A = 26°$, $a \approx 10.4$ in., $b \approx 21.0$ in.

EXTRA EXAMPLE 2
Solve $\triangle ABC$ with $C = 96°$, $b = 17$ centimeters, and $c = 19$ centimeters.
$A \approx 21.1°$, $B \approx 62.9°$, $a \approx 6.89$ cm

CHECKPOINT EXERCISES
For use after Example 1:
1. Solve $\triangle ABC$ with $A = 86°$, $B = 31°$, and $a = 10$ meters.
$C = 63°$, $b \approx 5.16$ m, $c \approx 8.93$ m

For use after Example 2:
2. Solve $\triangle ABC$ with $A = 79°$, $a = 16$ feet, and $b = 9$ feet.
$B \approx 33.5°$, $C \approx 67.5°$, $c \approx 15.1$ ft

! COMMON ERROR
EXAMPLE 2 When labeling a triangle, make sure that students understand that side a is opposite angle A, side b is opposite angle B, and side c is opposite angle C. This ensures that the ratios in the law of sines are written correctly.

STUDENT HELP NOTES
→ **Look Back** As students look back to p. 569, remind them that if a rational equation is a proportion, then cross multiplication is one method they can use to solve it.

Two angles and one side (AAS or ASA) determine exactly one triangle. Two sides and an angle opposite one of those sides (SSA) may determine no triangle, one triangle, or two triangles. The SSA case is called the *ambiguous case*.

POSSIBLE TRIANGLES IN THE SSA CASE

Consider a triangle in which you are given a, b, and A.

A IS OBTUSE.

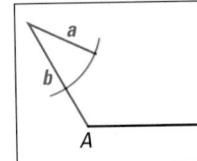

$a \leq b$
No triangle

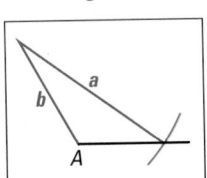

$a > b$
One triangle

A IS ACUTE.

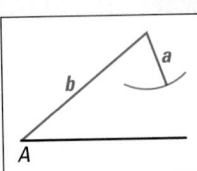

$b \sin A > a$
No triangle

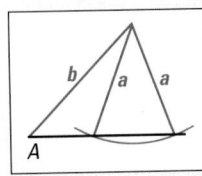

$b \sin A < a < b$
Two triangles

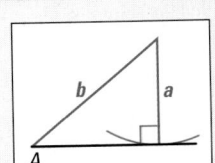

$b \sin A = a$
One triangle

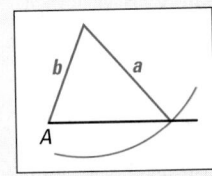

$a > b$
One triangle

EXAMPLE 2   *The SSA Case—One Triangle*

Solve $\triangle ABC$ with $C = 122°$, $a = 12$ cm, and $c = 18$ cm.

SOLUTION

First make a sketch. Because C is obtuse and the side opposite C is longer than the given adjacent side, you know that only one triangle can be formed. Use the law of sines to find A.

$$\frac{\sin A}{12} = \frac{\sin 122°}{18} \qquad \text{Law of sines}$$

$$\sin A = \frac{12 \sin 122°}{18} \qquad \text{Multiply each side by 12.}$$

$$\sin A \approx 0.5654 \qquad \text{Use a calculator.}$$

$$A \approx 34.4° \qquad \text{Use inverse sine function.}$$

You then know that $B \approx 180° - 122° - 34.4° = 23.6°$. Use the law of sines again to find the remaining side length b of the triangle.

$$\frac{b}{\sin 23.6°} = \frac{18}{\sin 122°}$$

$$b = \frac{18 \sin 23.6°}{\sin 122°} \approx 8.5 \text{ centimeters}$$

STUDENT HELP

→ **Look Back**
For help with solving rational equations, see p. 569.

EXAMPLE 3 *The SSA Case—No Triangle*

Solve $\triangle ABC$ with $a = 4$ inches, $b = 2.5$ inches, and $B = 58°$.

SOLUTION

Begin by drawing a horizontal line. On one end form a 58° angle (B) and draw a segment $(\overline{BC})$ 4 inches long. At vertex C, use a compass to draw an arc of radius 2.5 inches. This arc does not intersect the horizontal line, so it is not possible to draw the indicated triangle.

You can see that b needs to be at least 4 sin 58° ≈ 3.39 inches long to reach the horizontal side and form a triangle.

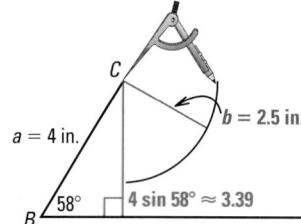

EXAMPLE 4 *The SSA Case—Two Triangles*

ASTRONOMY At certain times during the year, you can see Venus in the morning sky. The distance between Venus and the sun is approximately 67 million miles. The distance between Earth and the sun is approximately 93 million miles. Estimate the distance between Venus and Earth if the observed angle between the sun and Venus is 34°.

SOLUTION

Venus's distance from the sun, $e = 67$, is greater than $v \sin E = 93 \sin 34° \approx 52$ and less than Earth's distance from the sun, $v = 93$. Therefore, two possible triangles can be formed. Draw diagrams as shown. Use the law of sines to find the possible measures of V.

$$\frac{\sin 34°}{67} = \frac{\sin V}{93}$$

$$\sin V = \frac{93 \sin 34°}{67}$$

$$\sin V \approx 0.7762$$

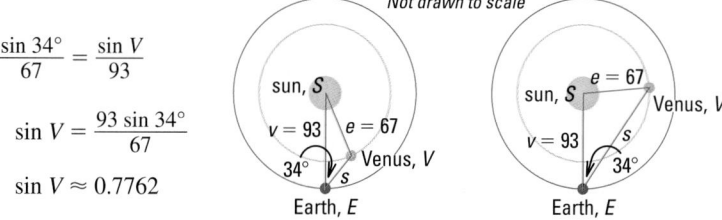

Not drawn to scale

There are two angles between 0° and 180° for which sin $V \approx 0.7762$. Use your calculator to find the angle between 0° and 90°: $\sin^{-1} 0.7762 \approx 50.9°$. To find the second angle, subtract the angle given by your calculator from 180°: $180° - 50.9° = 129.1°$. So, $V \approx 50.9°$ or $V \approx 129.1°$.

Because the sum of the angle measures in a triangle equals 180°, you know that $S \approx 95.1°$ when $V \approx 50.9°$ or $S \approx 16.9°$ when $V \approx 129.1°$. Finally, use the law of sines again to find the side length s.

$$\frac{s}{\sin 95.1°} = \frac{67}{\sin 34°} \qquad\qquad \frac{s}{\sin 16.9°} = \frac{67}{\sin 34°}$$

$$s = \frac{67 \sin 95.1°}{\sin 34°} \qquad\qquad s = \frac{67 \sin 16.9°}{\sin 34°}$$

$$\approx 119 \qquad\qquad\qquad \approx 34.8$$

▶ The approximate distance between Venus and Earth is either 119 million miles or 34.8 million miles.

FOCUS ON CAREERS

 **ASTRONOMER**
Astronomers study energy, matter, and natural processes throughout the universe. A doctoral degree and an aptitude for physics and mathematics are needed to become an astronomer.

CAREER LINK
www.mcdougallittell.com

EXTRA EXAMPLE 3
Solve $\triangle ABC$ with $C = 63°$, $b = 11$ inches, and $c = 8$ inches. Since 11 sin 63° = 9.80 > 8, no triangle is possible.

EXTRA EXAMPLE 4
The distance from Mercury to the sun is about 36 million miles. The distance from Earth to the sun is about 93 million miles. Estimate the distance from Earth to Mercury if the observed angle between the sun and Mercury is 19°. about 107 million mi or about 68.5 million mi

✔ CHECKPOINT EXERCISES

For use after Example 3:

1. Solve $\triangle ABC$ with $A = 104°$, $b = 22$ millimeters, and $a = 17$ millimeters. Since A is obtuse and $17 \le 22$, no triangle is possible.

For use after Example 4:

2. Solve $\triangle ABC$ with $a = 23$, $b = 14$, and $B = 33°$. $A \approx 63.5°$, $C \approx 83.5°$, $c \approx 25.5$ or $A \approx 116.5°$, $C \approx 30.5°$, $c \approx 13.0$

MATHEMATICAL REASONING
Using a triangle as shown below, have students write expressions for angles A and B using the sine function and the altitude x. Solving each of these expressions for x gives $x = b \sin A$ and $x = a \sin B$. By equating these expressions and dividing both sides by ab, students will obtain one of the proportions of the law of sines. Point out that similar reasoning involving angle C and its opposite side and with obtuse triangles can be used to verify the law of sines.

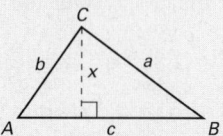

CAREER NOTE
EXAMPLE 4 Additional information about a career as an astronomer is available at **www.mcdougallittell.com**.

801

EXTRA EXAMPLE 5
Find the area of △ABC.

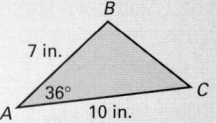

about 20.6 in.²

EXTRA EXAMPLE 6
You are buying the triangular piece of land shown at a price of $1500 per acre (1 acre = 4840 square yards). What is the cost of the land?

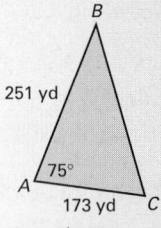

about $6500

 CHECKPOINT EXERCISES

For use after Examples 5 and 6:

1. You are covering the triangular piece of land shown with mulch. Each bag costs $3, and covers a square yard. What is the total cost for the mulch?

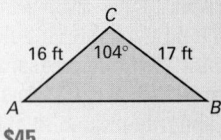

$45

FOCUS ON VOCABULARY
State the law of sines in your own words. **See sample answer below.**

CLOSURE QUESTION
What is the minimum amount of information that needs to be given in order to use the law of sines to solve a triangle? **the measure of two angles and any side or two sides and the angle opposite one of them**

DAILY PUZZLER
Find the area of a regular octagon with a side length of 10 cm. (Hint: Make a square by connecting every other vertex. Then find the areas of the square and the four triangles formed.) **about 483 cm²**

802

You can find the area of any triangle if you know the lengths of two sides and the measure of the included angle.

AREA OF A TRIANGLE

The area of any triangle is given by one half the product of the lengths of two sides times the sine of their included angle. For △ABC shown, there are three ways to calculate the area:

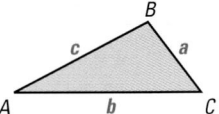

$$\text{Area} = \frac{1}{2}bc \sin A \qquad \text{Area} = \frac{1}{2}ac \sin B \qquad \text{Area} = \frac{1}{2}ab \sin C$$

EXAMPLE 5 *Finding a Triangle's Area*

Find the area of △ABC.

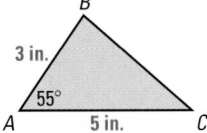

SOLUTION

Use the appropriate formula for the area of a triangle.

$$\text{Area} = \frac{1}{2}bc \sin A$$

$$= \frac{1}{2}(5)(3) \sin 55°$$

$$\approx 6.14 \text{ square inches}$$

Real Estate

EXAMPLE 6 *Calculating the Price of Land*

You are buying the triangular piece of land shown. The price of the land is $2000 per acre (1 acre = 4840 square yards). How much does the land cost?

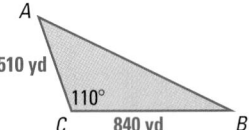

SOLUTION

The area of the land is:

$$\text{Area} = \frac{1}{2}ab \sin C$$

$$= \frac{1}{2}(840)(510) \sin 110°$$

$$\approx 201,000 \text{ square yards}$$

▶ The property contains 201,000 ÷ 4840 ≈ 41.5 acres. At $2000 per acre, the price of the land is about (2000)(41.5) = $83,000.

Focus on Vocabulary *Sample answer:*
For any triangle, the ratios of the sine of any angle to the side opposite that angle are equal.

GUIDED PRACTICE

Vocabulary Check ✓

1. What is the SSA case called? Why is it called this?
   The ambiguous case; because there may be no triangle, one triangle, or two triangles.

Concept Check ✓

2. Which two of the following cases can be solved using the law of sines?   **B and C**

   A. SSS   **B.** SSA   **C.** AAS or ASA   **D.** SAS

3. Suppose a, b, and A are given for $\triangle ABC$ where $A < 90°$. Under what conditions would you have no triangle? one triangle? two triangles?
   if $a < b \sin A$; if $a > b$ or $a = b \sin A$; if $b \sin A < a < b$

Skill Check ✓

Decide whether the given measurements can form exactly *one triangle*, exactly *two triangles*, or *no triangle*. (You do not need to solve the triangle.)

4. $C = 65°$, $c = 44$, $b = 32$ **one triangle**   5. $A = 140°$, $a = 5$, $c = 7$ **no triangle**

6. $A = 18°$, $a = 16$, $c = 10$ **one triangle**   7. $A = 70°$, $a = 155$, $c = 160$ **two triangles**

Solve △ABC.

8.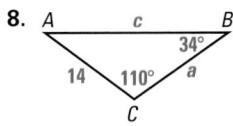

   $A \approx 35.8°$; $B \approx 49.2°$; $a = 14.7$

9.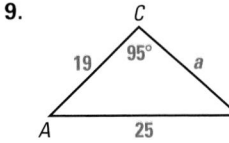

   $A = 36°$; $a \approx 14.7$; $c \approx 23.5$

10.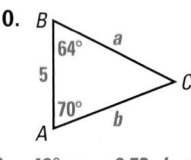

    $C = 46°$; $a \approx 6.53$; $b \approx 6.25$

Find the area of the triangle with the given side lengths and included angle.

11. $b = 2$, $c = 3$, $A = 47°$ **2.19 units^2**   12. $a = 23$, $b = 15$, $C = 51°$ **134 units^2**

13. $a = 13$, $c = 24$, $B = 127°$ **125 units^2**   14. $b = 12$, $c = 17$, $A = 103°$ **99.4 units^2**

15. 🌎 **REAL ESTATE** Suppose you are buying the triangular piece of land shown. The price of the land is $2200 per acre (1 acre = 4840 square yards). How much does the land cost?   **about $62,400**

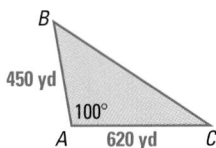

PRACTICE AND APPLICATIONS

STUDENT HELP

▶ **Extra Practice**
to help you master skills is on p. 958.

NUMBER OF SOLUTIONS Decide whether the given measurements can form exactly *one triangle*, exactly *two triangles*, or *no triangle*.

16. $C = 65°$, $c = 44$, $b = 32$ **one triangle**   17. $A = 140°$, $a = 5$, $c = 7$ **no triangle**

18. $A = 18°$, $a = 16$, $c = 10$ **one triangle**   19. $A = 70°$, $a = 155$, $c = 160$ **two triangles**

20. $C = 160°$, $c = 12$, $b = 15$ **no triangle**   21. $B = 105°$, $b = 11$, $a = 5$ **one triangle**

22. $B = 56°$, $b = 13$, $a = 14$ **two triangles**   23. $C = 25°$, $c = 6$, $b = 20$ **no triangle**

STUDENT HELP

▶ **HOMEWORK HELP**
Examples 1–3: Exs. 16–36
Example 4: Exs. 16–36, 56–62
Example 5: Exs. 37–52
Example 6: Exs. 63–67

SOLVING TRIANGLES Solve △ABC.

24.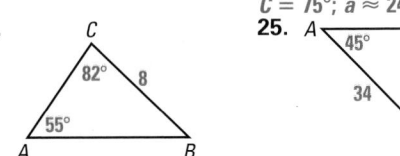

    $B = 43°$; $b \approx 6.66$; $c \approx 9.67$

25. $C = 75°$; $a \approx 24.9$; $b \approx 30.5$

26.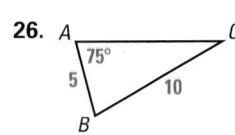

    $B \approx 76.1°$; $C \approx 28.9°$; $b \approx 10.1$

13.5 *The Law of Sines*   **803**

3 APPLY

◯ **ASSIGNMENT GUIDE**

BASIC
Day 1: pp. 803–805 Exs. 16–19, 20–26 even, 27–32, 56, 60
Day 2: pp. 804–806 Exs. 37–41, 47–50, 53–55, 63, 68, 71–81 odd

AVERAGE
Day 1: pp. 803–805 Exs. 16–19, 20–26 even, 27–32, 56–60
Day 2: pp. 804–806 Exs. 37–42, 47–55, 63–65, 68, 71–81 odd, 82

ADVANCED
Day 1: pp. 803–805 Exs. 16–26 even, 27–36, 56–62
Day 2: pp. 804–806 Exs. 37–42, 47–55, 63–69, 71–81 odd, 82

BLOCK SCHEDULE
pp. 803–806 Exs. 16–19, 20–26 even, 27–32, 37–42, 47–60, 63–65, 68, 71–81 odd, 82

EXERCISE LEVELS
Level A: *Easier*
16–26, 37–52, 76–84
Level B: *More Difficult*
27–36, 53–60, 63–68, 70–75
Level C: *Most Difficult*
61–62, 69

✔ **HOMEWORK CHECK**
To quickly check student understanding of key concepts, go over the following exercises:
Exs. 16, 22, 24, 28, 38, 48. See also the Daily Homework Quiz:

• Blackline Master (*Chapter 13 Resource Book*, p. 80)
• 📄 Transparency (p. 102)

803

➤**Homework Help** Students can find help for Exs. 27–36 at **www.mcdougallittell.com**. The information can be printed out for students who don't have access to the Internet.

GRAPHING CALCULATOR NOTE

EXERCISES 53–55 Using the table feature of a graphing calculator gives students an efficient way to look for patterns involving the areas of triangles with two common side lengths but different measures for the included angle.

STUDENT HELP

HOMEWORK HELP
Visit our Web site www.mcdougallittell.com for help with Exs. 27–36.

27. $A \approx 84.7°$; $C \approx 35.3°$; $a \approx 34.5$

28. $A = 40°$; $b \approx 21.9$; $c \approx 11.7$

29. no triangle

30. $B \approx 137.9°$; $C \approx 22.1°$; $b \approx 19.6$ or $B \approx 2.10°$; $C \approx 157.9°$; $b \approx 1.07$

31. $A \approx 62.3°$; $B \approx 22.7°$; $b \approx 3.48$

32. $C = 50°$; $a \approx 30.7$; $b \approx 28.3$

33. $A \approx 111.6°$; $B \approx 52.4°$; $a \approx 108$ or $A \approx 36.4°$; $B \approx 127.6°$; $a \approx 68.9$

34. $B = 40°$; $a \approx 1.35$; $c \approx 5.96$

35. $A \approx 15.4°$; $C \approx 129.6°$; $c \approx 34.9$

36. $A \approx 16.4°$; $B \approx 18.6°$; $a \approx 4.44$

SOLVING TRIANGLES Solve △*ABC*. (*Hint:* Some of the "triangles" have no solution and some have two solutions.) 27–36. See margin.

27. $B = 60°$, $b = 30$, $c = 20$

28. $B = 110°$, $C = 30°$, $a = 15$

29. $B = 130°$, $a = 10$, $b = 8$

30. $A = 20°$, $a = 10$, $c = 11$

31. $C = 95°$, $a = 8$, $c = 9$

32. $A = 70°$, $B = 60°$, $c = 25$

33. $C = 16°$, $b = 92$, $c = 32$

34. $A = 10°$, $C = 130°$, $b = 5$

35. $B = 35°$, $a = 12$, $b = 26$

36. $C = 145°$, $b = 5$, $c = 9$

FINDING AREA Find the area of the triangle with the given side lengths and included angle.

37. $B = 25°$, $a = 17$, $c = 33$ 119 units²

38. $C = 130°$, $a = 21$, $b = 17$ 137 units²

39. $C = 120°$, $a = 8$, $b = 5$ 17.3 units²

40. $A = 85°$, $b = 11$, $c = 18$ 98.6 units²

41. $A = 75°$, $b = 16$, $c = 21$ 162 units²

42. $B = 110°$, $a = 11$, $c = 24$ 124 units²

43. $C = 125°$, $a = 3$, $b = 8$ 9.83 units²

44. $B = 29°$, $a = 13$, $c = 13$ 41.0 units²

45. $B = 96°$, $a = 15$, $c = 9$ 67.1 units²

46. $A = 32°$, $b = 10$, $c = 12$ 31.8 units²

FINDING AREA Find the area of △*ABC*.

47.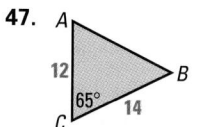
76.1 units²

48.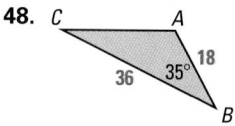
186 units²

49.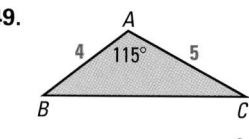
9.06 units²

50.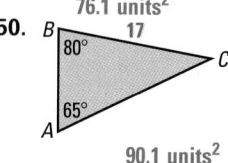
90.1 units²

51.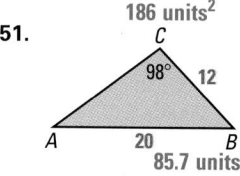
85.7 units²

52.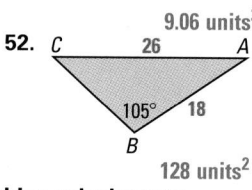
128 units²

🖩 **FINDING A PATTERN** In Exercises 53–55, use a graphing calculator to explore how the angle measure between two sides of a triangle affects the area of the triangle.

53. Choose a fixed length for each of two sides of a triangle. Let *x* represent the measure of the included angle. Enter an equation for the area of this triangle into the calculator.   *Sample answer:* Let the side lengths be 10 and 15. The equation is $A = 75 \sin x$.

54. Use the *Table* feature to look at the *y*-values for $0° < x < 180°$. Does area always increase for increasing values of *x*? Explain.   No; *A* reaches a maximum of 75 for $x = 90°$, and then decreases.

55. What value of *x* maximizes area?   90°

56. 🌐 **AQUEDUCT** A reservoir supplies water through an aqueduct to Springfield, which is 15 miles from the reservoir at 25° south of east. A pumping station at Springfield pumps water 7.5 miles to Centerville, which is due east from the reservoir. Plans have been made to build an aqueduct directly from the reservoir to Centerville. How long will the aqueduct be?   about 17.6 mi

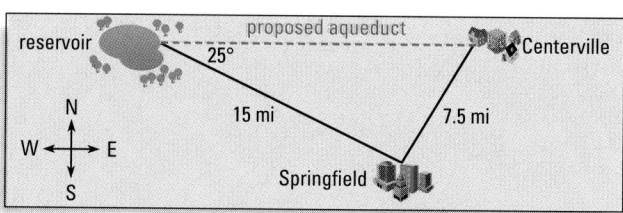

CAREER NOTE
EXERCISES 61 AND 62
Additional information about a
career as a sculptor is available
at **www.mcdougallittell.com.**

HISTORY **CONNECTION** **In Exercises 57–59, use the following information.**

In 1802 Captain William Lambton began what is known as the Great Trigonometrical
Survey of India. Lambton and his company systematically divided India into
triangles. They used trigonometry to find unknown distances from a known distance
they measured, called a *baseline*. The map below shows a section of the Great
Trigonometrical Survey of India.

57. Use the given measurements
to find the distance between
Júin and Amsot.
about 22.5 mi

58. Find the distance between
Júin and Rámpúr.
about 21.7 mi

59. *Writing* How could you
find the distance from Shí to
Dádú? Explain.

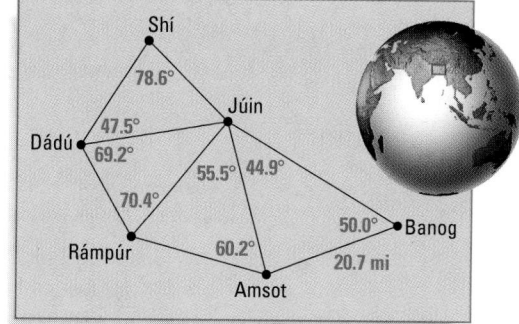

59. *Sample answer:* Use the
law of sines to find the
distance from Júin to
Amsot. Subtract 55.5° and
60.2° from 180° to find the
third angle of the next
triangle and then use the
law of sines to find the
distance from Júin to
Rámpúr. Use the law of
sines to find the distance
from Júin to Dádú.
Subtract 78.6° and 47.5°
from 180° to find the third
angle of the next triangle.
Use the law of sines to
find the distance from Shí
to Dádú.

60. 🌐 **NEW YORK CITY** You are on the observation deck of the Empire State
Building looking at the Chrysler Building. When you turn about 145° you see the
Statue of Liberty. You know that the Chrysler Building and the Empire State
Building are about 0.6 mile apart and that the Chrysler Building is about
5.7 miles from the Statue of Liberty. Find the approximate distance between the
Empire State Building and the Statue of Liberty. **about 5.2 mi**

ART **CONNECTION** **In Exercises 61 and 62, use the following information.**
You are creating a sculpture for an art show at your school. One 50 inch wooden
beam makes an angle of 70° with the base of your sculpture. You have another
wooden beam 48 inches long that you would like to attach to the top of the
50 inch beam and to the base of the sculpture, as shown below.

61. Find all possible angles θ that the
48 inch beam can make with the
50 inch beam. **31.8° or 8.2°**

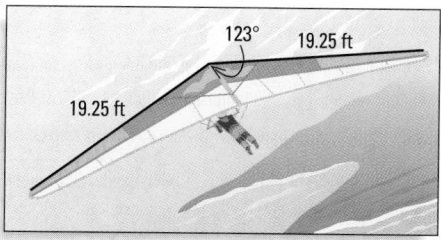

base of sculpture

62. Find all possible distances d that the
bottom of the 48 inch beam can be from
the left end of the base. **26.9 in. or 7.28 in.**

63. 🪂 **HANG GLIDER** A hang
glider is shown at the right. Use
the given nose angle and wing
measurements to approximate the
area of the sail. **about 155.4 ft**

123°   19.25 ft

19.25 ft

FOCUS ON
CAREERS

➤ **SCULPTOR**
Many sculptors
create large geometric
pieces that require precise
calculation of angle
measures and side lengths.
Shown above is *Center
Peace* by sculptor
Linda Howard.

📧 **CAREER LINK**
www.mcdougallittell.com

🌐 **COURTYARD** **In Exercises 64 and 65, use the following information.**
You are seeding a triangular courtyard. One side of the courtyard is 52 feet long and
another side is 46 feet long. The angle opposite the 52 foot side is 65°.

64. How long is the third side of the courtyard? **about 50.5 ft**

65. One bag of grass seed covers an area of 50 square feet. How many bags of grass
seed will you need to cover the courtyard? **21 bags**

13.5 *The Law of Sines* **805**

ADDITIONAL PRACTICE
AND RETEACHING

For Lesson 13.5:

• Practice Levels A, B, and C
(*Chapter 13 Resource Book*,
p. 70)

• Reteaching with Practice
(*Chapter 13 Resource Book*,
p. 73)

• 🖥 See Lesson 13.5 of the
Personal Student Tutor

For more Mixed Review:

• 🖥 Search the *Test and Practice
Generator* for key words or
specific lessons.

4 ASSESS

DAILY HOMEWORK QUIZ

🔲 **Transparency Available**

Decide whether the given measurements can form exactly *one triangle*, exactly *two triangles*, or *no triangle*.

1. $B = 77°$, $b = 26$, $c = 18$
   one triangle

2. $A = 61°$, $a = 31$, $b = 35$
   two triangles

3. Solve $\triangle ABC$.

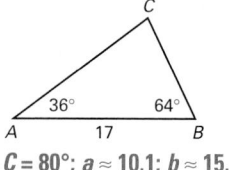

$C = 80°$; $a \approx 10.1$; $b \approx 15.5$

Solve $\triangle ABC$.

4. $B = 111°$, $a = 11$, $b = 14$
   $A \approx 47.2°$; $C \approx 21.8°$; $c \approx 5.57$

5. $C = 48°$, $a = 40$, $c = 26$
   no solution

6. Find the area of the triangle with the given side lengths and included angle. $C = 7°$, $a = 40$, $b = 50$ **122 units²**

EXTRA CHALLENGE NOTE
→ Challenge problems for Lesson 13.5 are available in **blackline** format in the *Chapter 13 Resource Book*, p. 77 and at **www.mcdougallittell.com**.

ADDITIONAL TEST PREPARATION

1. OPEN ENDED A triangle has a 30° angle. Give two pairs of sides adjacent to the angle so that the triangle will have an area of 30 square units.
any two of the following pairs: (1, 120), (2, 60), (3, 40), (4, 30), (5, 24), (6, 20), (8, 15), (10, 12)

69. See Additional Answers beginning on page AA1.

806

🌎 **BUYING PAINT** In Exercises 66 and 67, use the following information. You plan to paint the side of the house shown below. One gallon of paint will cover an area of 400 square feet.

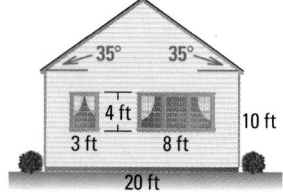

66. Find the area to be painted. Do not include the window area.   **about 226 ft²**

67. How many gallons of paint do you need?   **about 0.57 gal, so buy a single gallon can**

Test Preparation

68. MULTI-STEP PROBLEM You are at an unknown distance d from a mountain, as shown below. The angle of elevation to the top of the mountain is 65°. You step back 100 feet and measure the angle of elevation to be 60°.

   a. Find the height h of the mountain using the law of sines and right triangle trigonometry. (*Hint*: First find θ.)   **about 901 ft**

   b. Find the height h of the mountain using a system of equations. Set up one tangent equation involving the ratio of d and h, and another tangent equation involving the ratio of $100 + d$ and h, and then solve the system.
   about 901 ft

   c. *Writing* Which method was easier for you to use? Explain.   *Sample answer:* The method from Part (b); it was the first one that occurred to me and seemed more straightforward.

★ **Challenge**

EXTRA CHALLENGE
↳ www.mcdougallittell.com

69. DERIVING FORMULAS Using the triangle shown at the right as a reference, derive the formulas for the area of a triangle given in the property box on page 802. Then show how to derive the law of sines using the area formulas.
   See margin.

MIXED REVIEW

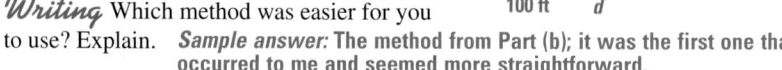

COMBINING EXPRESSIONS Perform the indicated operations. (Review 7.2 for 13.6)

70. $5\sqrt{11} + \sqrt{11} - 9\sqrt{11}$ $-3\sqrt{11}$

71. $2\sqrt{12} + 5\sqrt{12} + 3\sqrt{27}$ $23\sqrt{3}$

72. $\sqrt{125} - 7\sqrt{45} + 10\sqrt{40}$ $20\sqrt{10} - 16\sqrt{5}$

73. $\sqrt{7} + 5\sqrt{63} - 2\sqrt{112}$ $8\sqrt{7}$

74. $2\sqrt{486} - 5\sqrt{54} - 2\sqrt{150}$ $-7\sqrt{6}$

75. $\sqrt{72} + 6\sqrt{98} - 10\sqrt{8}$ $28\sqrt{2}$

FINDING COSINE VALUES Use a calculator to evaluate the trigonometric function. Round the result to four decimal places. (Review 13.1, 13.3 for 13.6)

76. $\cos 52°$ 0.6157
77. $\cos \dfrac{12\pi}{5}$ 0.3090
78. $\cos \dfrac{9\pi}{5}$ 0.8090
79. $\cos \dfrac{10\pi}{7}$ -0.2225

80. $\cos 20°$ 0.9397
81. $\cos 305°$ 0.5736
82. $\cos (-200°)$ -0.9397
83. $\cos 5°$ 0.9962

84. 🌎 **CAR HEADLIGHTS** In Massachusetts the low-beam headlights of cars are set to focus down 4 inches at a distance of 10 feet. At what angle θ are the beams directed? (Review 13.4)   **1.91°**

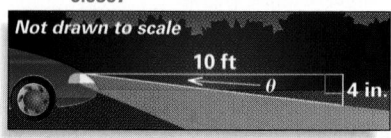

Not drawn to scale

13.6

The Law of Cosines

What you should learn

GOAL ① Use the law of cosines to find the sides and angles of a triangle.

GOAL ② Use Heron's formula to find the area of a triangle, as applied in **Example 5**.

Why you should learn it

▼ To solve **real-life** problems, such as finding the angle at which two swinging trapeze artists meet in **Ex. 50**.

GOAL ① USING THE LAW OF COSINES

You have not yet solved triangles for which two sides and the included angle (SAS) or three sides (SSS) are given. You can solve both of these cases using the **law of cosines**.

LAW OF COSINES

If $\triangle ABC$ has sides of length a, b, and c as shown, then:

$$a^2 = b^2 + c^2 - 2bc \cos A$$
$$b^2 = a^2 + c^2 - 2ac \cos B$$
$$c^2 = a^2 + b^2 - 2ab \cos C$$

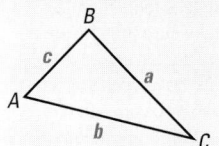

EXAMPLE 1 The SAS Case

Solve $\triangle ABC$ with $a = 12$, $c = 16$, and $B = 38°$.

SOLUTION

Begin by using the law of cosines to find the length b of the third side.

$b^2 = a^2 + c^2 - 2ac \cos B$	Write law of cosines.
$b^2 = 12^2 + 16^2 - 2(12)(16) \cos 38°$	Substitute for a, c, and B.
$b^2 \approx 97.4$	Simplify.
$b \approx \sqrt{97.4} \approx 9.87$	Take square root.

Now that you know all three sides and one angle, you can use the law of cosines *or* the law of sines to find a second angle.

$\dfrac{\sin A}{a} = \dfrac{\sin B}{b}$	Write law of sines.
$\dfrac{\sin A}{12} = \dfrac{\sin 38°}{9.87}$	Substitute for a, b, and B.
$\sin A = \dfrac{12 \sin 38°}{9.87}$	Multiply each side by 12.
$\sin A \approx 0.7485$	Simplify.
$A \approx \sin^{-1} 0.7485 \approx 48.5°$	Use inverse sine.

You can find the third angle as follows.

$$C \approx 180° - 38° - 48.5° = 93.5°$$

STUDENT HELP

▶ **Derivations**
For a derivation of the law of cosines, see p. 901.

---

1 PLAN

PACING
Basic: 1 day
Average: 1 day
Advanced: 1 day
Block Schedule: 0.5 block with 13.7

LESSON OPENER
APPLICATION
An alternative way to approach Lesson 13.6 is to use the Application Lesson Opener:

• Blackline Master (*Chapter 13 Resource Book*, p. 81)
• Transparency (p. 89)

MEETING INDIVIDUAL NEEDS
• *Chapter 13 Resource Book*
  Prerequisite Skills Review (p. 5)
  Practice Level A (p. 82)
  Practice Level B (p. 83)
  Practice Level C (p. 84)
  Reteaching with Practice (p. 85)
  Absent Student Catch-Up (p. 87)
  Challenge (p. 89)
• *Resources in Spanish*
• Personal Student Tutor

NEW-TEACHER SUPPORT
See the Tips for New Teachers on pp. 1–2 of the *Chapter 13 Resource Book* for additional notes about Lesson 13.6.

WARM-UP EXERCISES

Transparency Available
Use the law of sines to find B.
1. $A = 72°$, $a = 19$, $b = 12$ 36.9°
2. $C = 12°$, $c = 24$, $b = 35$ 17.7°
Solve for x, where x is positive.
3. $x^2 = y^2 + z^2 - 21$, where $y = 6$ and $z = 7$ 8
4. $x^2 = y^2 + z^2 - 19$, where $y = 10$ and $z = 12$ 15

Have students imagine that a triangular lot, measuring 40 yards by 50 yards by 60 yards, is donated for a recreation area. Have students sketch the triangle, and ask if any of them can find its area. Many will draw an altitude, but then realize that they need to know an angle of the triangle. Tell them that the *law of cosines* will help them find an angle, and will also lead to a formula for the area of a triangle given its side lengths.

 EXTRA EXAMPLE 1
Solve $\triangle ABC$ with $B = 81°$, $a = 22$, and $c = 19$. $A \approx 54.4°$, $C \approx 44.6°$, $b \approx 26.7$

EXTRA EXAMPLE 2
Solve $\triangle ABC$ with $a = 8$, $b = 12$, and $c = 10$. $A \approx 41.4°$, $B \approx 82.8°$, $C \approx 55.8°$

EXTRA EXAMPLE 3
In a junior baseball league, the pitcher's mound is 40 feet from home plate, and the bases are 55 feet apart on the square field. How far is the mound from first base?   about 38.9 ft

 CHECKPOINT EXERCISES

For use after Example 1:
1. Solve $\triangle ABC$ with $A = 132°$, $b = 8$, and $c = 9$. $B \approx 22.5°$, $C \approx 25.5°$, $a \approx 15.5$

For use after Example 2:
2. Solve $\triangle ABC$ with $a = 23$, $b = 14$, and $c = 19$. $A \approx 87.0°$, $B \approx 37.4°$, $C \approx 55.6°$

For use after Example 3:
3. A 15 foot length of pipe and a 20 foot length of pipe for a sprinkler system are connected in the ground at a 50° angle. How far apart are the pipes at their ends?   about 15.5 ft

---

EXAMPLE 2   *The SSS Case*

Solve $\triangle ABC$ with $a = 8$ feet, $b = 18$ feet, and $c = 13$ feet.

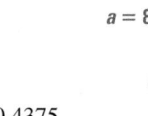

SOLUTION

First find the angle opposite the longest side, $\overline{AC}$.
Using the law of cosines, you can write:

$$\cos B = \frac{a^2 + c^2 - b^2}{2ac} = \frac{8^2 + 13^2 - 18^2}{2(8)(13)} = -0.4375$$

Using the inverse cosine function, you can find the measure of obtuse angle B:

$$B = \cos^{-1}(-0.4375) \approx 115.9°$$

Now use the law of sines to find A.

STUDENT HELP

Study Tip
In Example 2 the largest angle is found first to make sure that the other two angles are acute. This way, when you use the law of sines to find another angle measure, you will know that it is between 0° and 90°.

$\dfrac{\sin A}{a} = \dfrac{\sin B}{b}$	Write law of sines.
$\dfrac{\sin A}{8} = \dfrac{\sin 115.9°}{18}$	Substitute.
$\sin A = \dfrac{8 \sin 115.9°}{18}$	Multiply each side by 8.
$\sin A \approx 0.3998$	Simplify.
$A \approx \sin^{-1} 0.3998 \approx 23.6°$	Use inverse sine.

Finally, you can find the measure of angle C:

$$C \approx 180° - 23.6° - 115.9° = 40.5°$$

REAL LIFE

Softball

EXAMPLE 3   *The SAS Case*

The pitcher's mound on a softball field is 46 feet from home plate. The distance between the bases is 60 feet. How far is the pitcher's mound from first base?

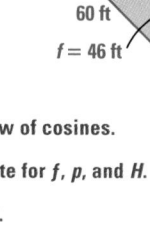

SOLUTION

Begin by forming $\triangle HPF$. In this triangle you know that $H = 45°$ because the line HP bisects the right angle at home plate. From the given information you know that $f = 46$ and $p = 60$. Using the law of cosines, you can solve for h.

$h^2 = f^2 + p^2 - 2fp \cos H$	Write law of cosines.
$h^2 = 46^2 + 60^2 - 2(46)(60) \cos 45°$	Substitute for f, p, and H.
$h^2 \approx 1812.8$	Simplify.
$h \approx \sqrt{1812.8}$	Take square root.
≈ 42.6 feet	Simplify.

▶ The distance between the pitcher's mound and first base is about 42.6 feet.

GOAL 2 USING HERON'S FORMULA

The law of cosines can be used to establish the following formula for the area of a triangle. This formula is credited to the Greek mathematician Heron (circa A.D. 100).

HERON'S AREA FORMULA

The area of the triangle with sides of length *a*, *b*, and *c* is

$$\text{Area} = \sqrt{s(s-a)(s-b)(s-c)}$$

where $s = \frac{1}{2}(a + b + c)$. The variable *s* is called the *semiperimeter*, or half-perimeter, of the triangle.

EXAMPLE 4 *Finding the Area of a Triangle*

Find the area of △*ABC*.

STUDENT HELP

Look Back
For help with simplifying radical expressions, see p. 264.

SOLUTION

Begin by finding the semiperimeter.

$$s = \frac{1}{2}(a + b + c) = \frac{1}{2}(22 + 40 + 50) = 56$$

Now use Heron's formula to find the area of △*ABC*:

$$\text{Area} = \sqrt{s(s-a)(s-b)(s-c)}$$
$$= \sqrt{56(56 - 22)(56 - 40)(56 - 50)}$$
$$= \sqrt{182{,}784} \approx 428 \text{ square units}$$

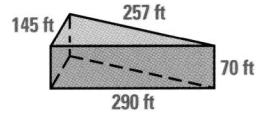

EXAMPLE 5 *Finding the Volume of a Building*

LANDAU BUILDING The dimensions of the Landau Building are given at the right. Find the volume of the building.

SOLUTION

Begin by finding the area of the base. The semiperimeter of the base is:

$$s = \frac{1}{2}(a + b + c) = \frac{1}{2}(145 + 257 + 290) = 346$$

So, the area of the base is:

$$\text{Area} = \sqrt{s(s-a)(s-b)(s-c)}$$
$$= \sqrt{346(346 - 145)(346 - 257)(346 - 290)}$$
$$\approx 18{,}600 \text{ square feet}$$

To find the volume, multiply this area by the building's height:

$$\text{Volume} = (\text{Area of base})(\text{Height}) \approx (18{,}600)(70) = 1{,}302{,}000 \text{ cubic feet}$$

FOCUS ON APPLICATIONS

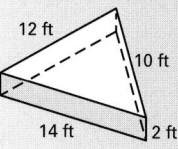

LANDAU BUILDING
The Landau Building, located in Cambridge, Massachusetts, was designed by architect I.M. Pei. Pei's work is known to have a sharp, geometric look.

Closure Question *Sample answer:*
the measures of all sides or the measures of two sides and the included angle

---

 EXTRA EXAMPLE 4
Find the area of △*ABC*.

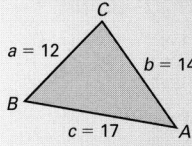

about 83.0 units²

EXTRA EXAMPLE 5
The triangular region shown is to be filled with concrete. Find the volume of concrete that is needed.

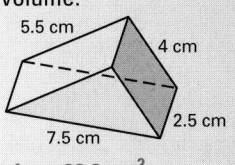

about 118 ft³

✓ CHECKPOINT EXERCISES
For use after Examples 4 and 5:
1. The dimensions of a child's block are shown. Find the volume.

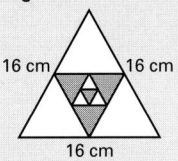

about 26.8 cm³

FOCUS ON VOCABULARY
What is the *semiperimeter* of a triangle? It is one half of its perimeter.

CLOSURE QUESTION
What given information indicates using the law of cosines to solve a triangle? See sample answer below.

DAILY PUZZLER
Each smaller central triangle is formed by connecting the midpoints of the sides of the next larger triangle. Find the area of the shaded region.

$13\sqrt{3}$, or about 22.5 cm²

ASSIGNMENT GUIDE

BASIC
Day 1: pp. 810–812 Exs. 15–23,
   30–33, 37–44, 49–51,
   55–56, 59–67 odd

AVERAGE
Day 1: pp. 810–812 Exs. 15–17,
   18–36 even, 37–44, 49–52,
   54–56, 59–67 odd, 68

ADVANCED
Day 1: pp. 810–812 Exs. 15–25,
   30–34, 37–44, 49–57,
   59–67 odd, 68

BLOCK SCHEDULE WITH 13.7
pp. 810–812 Exs. 15–17, 18–36
even, 37–44, 49–52, 54–56,
59–67 odd, 68

EXERCISE LEVELS
Level A: *Easier*
15–29, 38–48

Level B: *More Difficult*
30–37, 49–56, 58–68

Level C: *Most Difficult*
57

✔ **HOMEWORK CHECK**
To quickly check student understanding of key concepts, go
over the following exercises:
Exs. 16, 18, 30, 38, 44. See also
the Daily Homework Quiz:

- Blackline Master (*Chapter 13
  Resource Book,* p. 92)
- Transparency (p. 103)

! **COMMON ERROR**
EXERCISE 16 When using the
law of cosines in the SSS case to
solve for the cosine of an angle, it is
easy to make errors, especially
involving the signs of terms. If the
cosine of an angle does not come
out to be between −1 and 1, students
should immediately look for errors.

810

GUIDED PRACTICE

Vocabulary Check ✓

1. Complete this statement: In a triangle with sides of length a, b, and c,
$\frac{1}{2}(a + b + c)$ is called the _?_.   **semiperimeter**

Concept Check ✓

2. For each case, tell whether you would use the *law of sines* or the *law of cosines* to solve the triangle.

2a. law of cosines
2b. law of sines
2c. law of cosines
2d. law of sines
2e. law of sines

    a. SSS   **b.** SSA   **c.** SAS   **d.** ASA   **e.** AAS

3. If when using the law of cosines to find angle A in $\triangle ABC$, you get $\cos A < 0$, what type of angle is A?   **an obtuse angle**

4. Express Heron's formula in words.

Skill Check ✓

4. The area of a triangle with sides a, b, and c and semiperimeter s (one half the perimeter) is the square root of the product s times $(s - a)$ times $(s - b)$ times $(s - c)$.

Solve $\triangle ABC$.

5. $B = 20°$, $a = 120$, $c = 100$
$b \approx 43.0$; $A \approx 107.4°$; $C \approx 52.7°$

6. $C = 95°$, $a = 10$, $b = 12$
$c \approx 16.3$; $A \approx 37.7°$; $B \approx 47.3°$

7. $a = 25$, $b = 11$, $c = 24$
$A \approx 82.2°$; $B \approx 25.8°$; $C \approx 72.0°$

8. $a = 2$, $b = 4$, $c = 5$
$A \approx 22.3°$; $B \approx 49.5°$; $C \approx 108.2°$

Find the area of $\triangle ABC$ having the given side lengths.

9. $a = 25$, $b = 60$, $c = 45$ **510 units²**

10. $a = 9$, $b = 4$, $c = 11$ **17.0 units²**

11. $a = 100$, $b = 55$, $c = 61$ **1470 units²**

12. $a = 5$, $b = 27$, $c = 29$ **63.9 units²**

🌐 **BASEBALL** **In Exercises 13 and 14, use the following information.**
The pitcher's mound on a baseball field is 60.5 feet from home plate. The distance between the bases is 90 feet.

13. How far is the pitcher's mound from first base?
about 63.7 ft

14. Using Heron's formula, find the area of the triangle formed by the pitcher's mound, home plate, and first base. **about 1925 ft²**

PRACTICE AND APPLICATIONS

┌─ **STUDENT HELP**
└→ **Extra Practice**
to help you master
skills is on p. 959.

SOLVING TRIANGLES **Solve $\triangle ABC$.**

15.
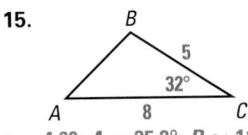
$c \approx 4.60$; $A \approx 35.2°$; $B \approx 112.8°$

16. $A \approx 52.9°$; $B \approx 103.6°$; $C \approx 23.5°$
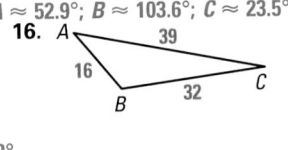

17.
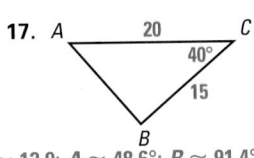
$c \approx 12.9$; $A \approx 48.6°$; $B \approx 91.4°$

SOLVING TRIANGLES **Solve $\triangle ABC$.**

┌─ **STUDENT HELP**
└→ **HOMEWORK HELP**
Examples 1, 2: Exs. 15–37
Example 3: Exs. 50, 51
Example 4: Exs. 38–48
Example 5: Exs. 52–54

18. $B = 20°$, $a = 120$, $c = 100$
$b \approx 43.0$; $A \approx 107.4°$; $C \approx 52.7°$

20. $a = 25$, $b = 11$, $c = 24$
$A \approx 82.2°$; $B \approx 25.8°$; $C \approx 72.0°$

22. $A = 78°$, $b = 2$, $c = 4$
$a \approx 4.08$; $B \approx 28.6°$; $C \approx 73.4°$

24. $B = 45°$, $a = 11$, $c = 22$
$b \approx 16.2$; $A \approx 28.7°$; $C \approx 106.3°$

26. $a = 9$, $b = 3$, $c = 11$
$A \approx 42.1°$; $B \approx 12.9°$; $C \approx 125.0°$

28. $a = 25$, $b = 26$, $c = 5$
$A \approx 73.0°$; $B \approx 96.0°$; $C \approx 11.0°$

19. $C = 95°$, $a = 10$, $b = 12$
$c \approx 16.3$; $A \approx 37.7°$; $B \approx 47.3°$

21. $a = 2$, $b = 4$, $c = 5$
$A \approx 22.3°$; $B \approx 49.5°$; $C \approx 108.2°$

23. $A = 60°$, $b = 30$, $c = 28$
$a \approx 29.1$; $B \approx 63.4°$; $C \approx 56.6°$

25. $C = 30°$, $a = 20$, $b = 20$
$c \approx 10.4$; $A \approx 75°$; $B \approx 75°$

27. $B = 15°$, $a = 12$, $c = 6$
$b \approx 6.40$; $A \approx 150.9°$; $C \approx 14.1°$

29. $a = 47$, $b = 30$, $c = 62$
$A \approx 47.0°$; $B \approx 27.8°$; $C \approx 105.1°$

HOMEWORK HELP
Visit our Web site
www.mcdougallittell.com
for help with Exs. 30–37.

CHOOSING A METHOD Use the law of sines, the law of cosines, or the Pythagorean theorem to solve △ABC.

30. $A = 96°$, $B = 39°$, $b = 13$
$C = 45°$; $a \approx 20.5$; $c \approx 14.6$

31. $B = 80°$, $C = 30°$, $b = 34$
$A = 70°$; $a \approx 32.4$; $c \approx 17.3$

32. $A = 34°$, $b = 17$, $c = 48$
$a \approx 35.2$; $B \approx 15.7°$; $C \approx 130.3°$

33. $C = 104°$, $b = 11$, $c = 32$
$a \approx 27.5$; $A \approx 56.5°$; $B \approx 19.5°$

34. $A = 48°$, $B = 51°$, $c = 36$
$C = 81°$; $a \approx 27.1$; $b \approx 28.3$

35. $a = 48$, $b = 51$, $c = 36$
$A \approx 64.3°$; $B \approx 73.2°$; $C \approx 42.5°$

36. $B = 10°$, $b = 5$, $c = 25$

37. $C = 90°$, $a = 4$, $b = 11$
$c \approx 11.7$; $A \approx 20.0°$; $B \approx 70.0°$

36. $a \approx 27.1$; $A \approx 109.7°$;
$C \approx 60.3°$ or $a \approx 22.1$;
$A \approx 50.3°$; $C \approx 119.7°$

FINDING AREA Find the area of △ABC.

38.
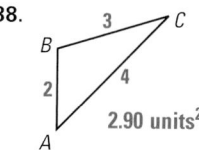
2.90 units²

39.
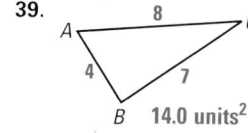
14.0 units²

40.
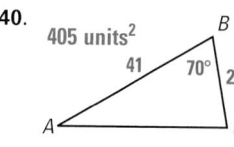
405 units²

49. Suppose △ABC is a right triangle with right angle at C. Then cos C = 0. By the law of cosines, $c^2 = a^2 + b^2 - 2ab \cos C$, so $c^2 = a^2 + b^2 - 2ab \cdot 0$. This gives $c^2 = a^2 + b^2$, which is the Pythagorean theorem.

FINDING AREA Find the area of △ABC having the given side lengths.

41. $a = 15$, $b = 20$, $c = 25$ 150 units²

42. $a = 13$, $b = 10$, $c = 4$ 15.0 units²

43. $a = 75$, $b = 68$, $c = 72$ 2210 units²

44. $a = 3$, $b = 19$, $c = 21$ 22.3 units²

45. $a = 4$, $b = 2$, $c = 4$ 3.87 units²

46. $a = 20$, $b = 21$, $c = 37$ 163 units²

47. $a = 8$, $b = 8$, $c = 8$ 27.7 units²

48. $a = 18$, $b = 15$, $c = 10$ 75.0 units²

49. CRITICAL THINKING Explain why the Pythagorean theorem is a special case of the law of cosines.   **See margin.**

50. 🌐 **TRAPEZE ARTISTS** The diagram shows the path of two trapeze artists who are both 5 feet long when hanging by their knees. The "flyer" on the left bar is preparing to make hand-to-hand contact with the "catcher" on the right bar. At what angle θ will the two meet? ▶ Source: Trapeze Arts, Inc.   **about 120°**

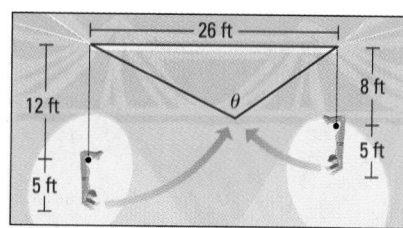

FOCUS ON
CAREERS

51. 🌐 **SURVEYING** You are a surveyor measuring the width of a pond from point A to point B, as shown. You set up your transit at point C and measure an angle of 73°. You also measure the distance from point C to points A and B, getting 56 feet and 68 feet, respectively. What is the width of the pond?   **about 74.4 ft**

SURVEYOR
A surveyor takes precise measurements to establish official land airspace, and water boundaries. A surveyor often uses an instrument called a *transit*, as pictured above, to measure angles.

CAREER LINK
www.mcdougallittell.com

52. 🌐 **GIBSON BLOCK** Built in 1913, the Gibson Block in Alberta, Canada, is shaped like a flat clothing iron of that time period. The approximately triangular base of the building has sides of length 18.3 meters, 37.1 meters, and 41.0 meters. The height of the Gibson Block is about 13.2 meters. Find the volume of the Gibson Block. ▶ Source: Stantec Architecture Ltd.
about 4480 m³

13.6 *The Law of Cosines*   **811**

STUDENT HELP NOTES

→ **Homework Help** Students can find help for Exs. 30–37 at **www.mcdougallittell.com**. The information can be printed out for students who don't have access to the Internet.

CAREER NOTE
EXERCISE 51 Additional information about a career as a surveyor is available at **www.mcdougallittell.com**.

ADDITIONAL PRACTICE AND RETEACHING

For Lesson 13.6:
• Practice Levels A, B, and C (*Chapter 13 Resource Book*, p. 82)
• Reteaching with Practice (*Chapter 13 Resource Book*, p. 85)
• See Lesson 13.6 of the *Personal Student Tutor*

For more Mixed Review:
• Search the *Test and Practice Generator* for key words or specific lessons.

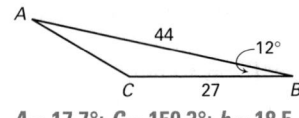
DAILY HOMEWORK QUIZ

📝 *Transparency Available*

1. Solve △*ABC*.

A ≈ 17.7°; *C* ≈ 150.3°; *b* ≈ 18.5

Solve △*ABC*.

2. *a* = 20, *b* = 25, *c* = 40
A ≈ 24.1°; *B* ≈ 30.8°; *C* ≈ 125.1°

3. *B* = 39°, *a* = 16, *c* = 22
A ≈ 46.5°; *C* ≈ 94.5°; *b* ≈ 13.9

Find the area of △*ABC*.

4. *a* = 35, *b* = 15, *c* = 30   224 units^2

5. *A* = 50°, *b* = 20, *c* = 24
184 units^2

EXTRA CHALLENGE NOTE

→ Challenge problems for Lesson 13.6 are available in **blackline** format in the *Chapter 13 Resource Book,* p. 89 and at **www.mcdougallittell.com**.

ADDITIONAL TEST PREPARATION

1. WRITING State the law of cosines in words.
Sample answer: The square of one side of a triangle equals the sum of the squares of the other sides minus twice the product of these two sides and the cosine of their included angle.

2. WRITING If you know the sides of a triangle, how can you find its area? *Sample answer:* Find half the perimeter, or semiperimeter. Then take the square root of the product of the semiperimeter and its differences with each of the side lengths.

---

53. 🦕 **DINOSAUR DIAMOND** In Utah and Colorado, an area called the Dinosaur Diamond is known for containing many dinosaur fossils. The map at the right shows the towns at the four vertices of the diamond. Use the given distances to find the area of the Dinosaur Diamond.
▶ Source: Dinomation   **about 7800 mi^2**

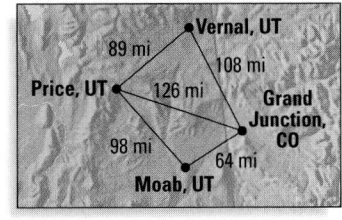

54. 🌍 **FERTILIZER** A farmer has a triangular field with sides that are 240 feet, 300 feet, and 360 feet long. He wants to apply fall fertilizer to the field. If it takes one 40 pound bag of fertilizer to cover 6000 square feet, how many bags does he need to cover the field?   **6 bags of fertilizer**

Test Preparation

55. MULTIPLE CHOICE Two airplanes leave an airport at the same time, the first headed due north and the second headed 37° east of north. At 2:00 P.M. the first airplane is 250 miles from the airport and the second airplane is 316 miles from the airport. How far apart are the two airplanes?   **A**

  Ⓐ about 190 miles   Ⓑ about 210 miles   Ⓒ about 200 miles

  Ⓓ about 310 miles   Ⓔ about 165 miles

56. MULTIPLE CHOICE Find the area of a triangle with sides of length 37 feet, 23 feet, and 42 feet.   **D**

  Ⓐ about 189 ft^2 Ⓑ about 134 ft^2 Ⓒ about 477 ft^2

  Ⓓ about 424 ft^2 Ⓔ about 777 ft^2

★ **Challenge**

57. 🪞 **MIRRORS** In the diagram, a beam of light is directed at the blue mirror, reflected to the red mirror, and then reflected back to the blue mirror. Find the distance *PT* that the light travels from the red mirror back to the blue mirror given that *OQ* = 6 feet and *OP* = 4.7 feet. (*Hint:* You will need to find θ. To do this, find ∠*OPQ* and use the fact that 2θ + *m*∠*TPQ* = 180°.)
2.01 ft

EXTRA CHALLENGE
→ www.mcdougallittell.com

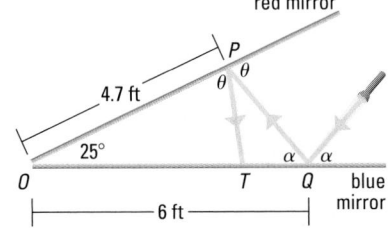

MIXED REVIEW

WRITING EQUATIONS Write an equation of the hyperbola with the given foci and vertices. (Review 10.5)

58. Foci: (−7, 0), (7, 0)
Vertices: (−2, 0), (2, 0) $\frac{x^2}{4} - \frac{y^2}{45} = 1$

59. Foci: (0, −11), (0, 11)
Vertices: (0, −3), (0, 3) $\frac{y^2}{9} - \frac{x^2}{112} = 1$

60. Foci: (−9, 0), (9, 0)
Vertices: (−5, 0), (5, 0) $\frac{x^2}{25} - \frac{y^2}{56} = 1$

61. Foci: $\left(0, -2\sqrt{5}\right), \left(0, 2\sqrt{5}\right)$
Vertices: (0, −1), (0, 1) $y^2 - \frac{x^2}{19} = 1$

CALCULATING PROBABILITIES Calculate the probability of rolling a die 30 times and getting the given number of 4's. (Review 12.6)

62. 1 0.0253   **63.** 3 0.137   **64.** 5 0.192   **65.** 6 0.160   **66.** 8 0.0631   **67.** 10 0.0130

68. 🏀 **VERTICAL MOTION** From a height of 120 feet, how long does it take a ball thrown downward at 20 feet per second to hit the ground? (Review 5.6 for 13.7)
about 2.18 sec

Parametric Equations and Projectile Motion

GOAL 1 USING PARAMETRIC EQUATIONS

What you should learn

GOAL 1 Use parametric equations to represent motion in a plane.

GOAL 2 Use parametric equations to represent projectile motion, as applied in **Example 4**.

Why you should learn it

▼ To solve **real-life** problems, such as modeling the path of a leaping dolphin in **Exs. 36–38**.

1. $x = 3t$; $y = 4t$

2. $x = 15$ and $y = 20$; the ant reaches the top edge first.

3. $5\frac{1}{4}$ sec

● ACTIVITY
Developing Concepts
Investigating Linear Motion

Suppose an ant starts at one corner of a picnic tablecloth and moves in a straight line, as shown. The ant's position (x, y) relative to the edges of the tablecloth is given for different times t (in seconds). **Steps 1–3. See margin.**

❶ Write two equations: one that gives the ant's horizontal position x as a function of t, and one that gives the ant's vertical position y as a function of t.

❷ What is the ant's position after 5 seconds?

❸ How long will it take the ant to reach an edge of the tablecloth?

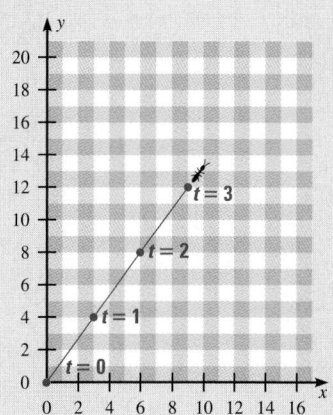

In the investigation you wrote a pair of equations that expressed x and y in terms of a third variable t. These equations, $x = f(t)$ and $y = g(t)$, are called **parametric equations**, and t is called the **parameter**.

EXAMPLE 1 *Graphing a Set of Parametric Equations*

Graph $x = 3t - 12$ and $y = -2t + 3$ for $0 \le t \le 5$.

SOLUTION

Begin by making a table of values.

t	0	1	2	3	4	5
x	−12	−9	−6	−3	0	3
y	3	1	−1	−3	−5	−7

Plot the points (x, y) given in the table:

$(-12, 3)$, $(-9, 1)$, $(-6, -1)$,
$(-3, -3)$, $(0, -5)$, $(3, -7)$

Then connect the points with a line segment as shown.

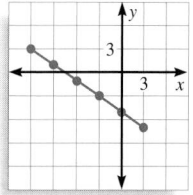

---

1 PLAN

PACING
Basic: 1 day
Average: 1 day
Advanced: 1 day
Block Schedule: 0.5 block with 13.6

LESSON OPENER
APPLICATION
An alternative way to approach Lesson 13.7 is to use the Application Lesson Opener:

• Blackline Master (*Chapter 13 Resource Book,* p. 93)
• Transparency (p. 90)

MEETING INDIVIDUAL NEEDS
• *Chapter 13 Resource Book*
  Prerequisite Skills Review (p. 5)
  Practice Level A (p. 95)
  Practice Level B (p. 96)
  Practice Level C (p. 97)
  Reteaching with Practice (p. 98)
  Absent Student Catch-Up (p. 100)
  Challenge (p. 102)
• *Resources in Spanish*
• *Personal Student Tutor*

NEW-TEACHER SUPPORT
See the Tips for New Teachers on pp. 1–2 of the *Chapter 13 Resource Book* for additional notes about Lesson 13.7.

WARM-UP EXERCISES

Transparency Available

Solve the equation for *t.*

1. $x = 3t + 4$ $t = \frac{1}{3}x - \frac{4}{3}$

2. $x = -2t - 6$ $t = -\frac{1}{2}x - 3$

3. Substitute $t = x + 7$ into $y = 8t - 3$ and simplify.
$y = 8x + 53$

4. Find the positive root of $y = -16x^2 + 22x + 52.5$.   2.625

A home-run ball is slammed over the outfield fence. Tell students that using *parametric equations,* they can find both the ball's height and horizontal position at any time.

 EXTRA EXAMPLE 1
Graph $x = 2t - 8$ and $y = -t + 4$ for $0 \le t \le 5$.

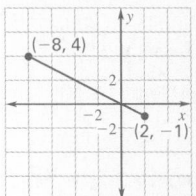

EXTRA EXAMPLE 2
Write an *xy*-equation for the parametric equations in Extra Example 1. State the domain for the equation. $y = -\frac{x}{2}; -8 \le x \le 2$

EXTRA EXAMPLE 3
A model airplane takes off and is 100 feet high and 400 feet away horizontally. Assume the plane is flying in a straight line at 51.5 ft/sec. Write a set of parametric equations for the airplane with *x* and *y* in feet and *t* in seconds.

$x \approx (51.5 \cos 14.0°)t$, or $x \approx 50t$;
$y \approx (51.5 \sin 14.0°)t$, or $y \approx 12.5t$

 CHECKPOINT EXERCISES
For use after Examples 1–3:
1. Graph $x = -2t + 1$ and $y = -4t + 6$ for $0 \le t \le 5$. Then rewrite the parametric equations as an *xy*-equation. State the domain.

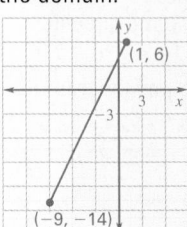

$y = 2x + 4; -9 \le x \le 1$

EXAMPLE 2 *Eliminating the Parameter*

Write an *xy*-equation for the parametric equations in Example 1: $x = 3t - 12$ and $y = -2t + 3$ for $0 \le t \le 5$. State the domain for the equation.

SOLUTION

First solve one of the parametric equations for *t*.

$$x = 3t - 12 \qquad \text{Write original equation.}$$

$$x + 12 = 3t \qquad \text{Add 12 to each side.}$$

$$\tfrac{1}{3}x + 4 = t \qquad \text{Multiply each side by } \tfrac{1}{3}.$$

Then substitute for *t* in the other parametric equation.

$$y = -2t + 3 \qquad \text{Write original equation.}$$

$$y = -2\left(\tfrac{1}{3}x + 4\right) + 3 \qquad \text{Substitute for } t.$$

$$y = -\tfrac{2}{3}x - 5 \qquad \text{Simplify.}$$

This process is called *eliminating the parameter* because the parameter *t* is not in the final equation. When $t = 0$, $x = -12$ and when $t = 5$, $x = 3$. So, the domain of the *xy*-equation is $-12 \le x \le 3$.

. . . .

Consider an object that is moving with constant speed *v* along a straight line that makes an angle θ measured counterclockwise from a line parallel to the *x*-axis. The position of the object at any time *t* can be represented by the parametric equations

$$x = (v \cos \theta)t + x_0$$

$$y = (v \sin \theta)t + y_0$$

where (x_0, y_0) is the object's location when $t = 0$.

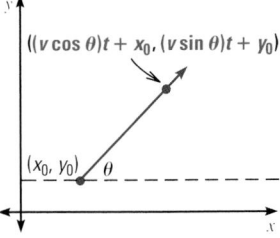

 **EXAMPLE 3** *Modeling Linear Motion*

Write a set of parametric equations for the airplane shown, given that its speed is 306 feet per second.

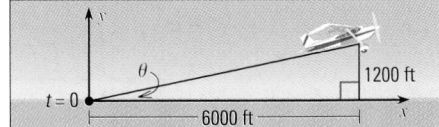

SOLUTION

The angle of elevation is $\theta = \tan^{-1}\left(\dfrac{1200}{6000}\right) \approx 11.3°$.

Using $v = 306$, $\theta = 11.3°$, and $(x_0, y_0) = (0, 0)$, you can write the following.

$$x = (v \cos \theta)t + x_0 \qquad \text{and} \qquad y = (v \sin \theta)t + y_0$$

$$x \approx (306 \cos 11.3°)t + 0 \qquad\qquad y \approx (306 \sin 11.3°)t + 0$$

$$\approx 300t \qquad\qquad\qquad\qquad \approx 60t$$

GOAL 2 **MODELING PROJECTILE MOTION**

Parametric equations can also be used to model nonlinear motion in a plane. For instance, consider an object that is projected into the air at an angle θ with an initial speed v. The object's parabolic path can be modeled with the parametric equations

$$x = (v \cos \theta)t + x_0$$

$$y = -\frac{1}{2}gt^2 + (v \sin \theta)t + y_0$$

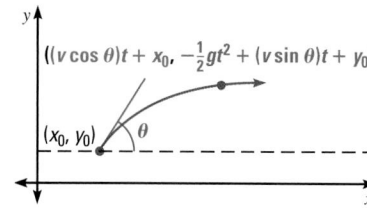

where (x_0, y_0) is the object's location when $t = 0$. The constant g is the acceleration due to gravity. Its value is 32 ft/sec^2 or 9.8 m/sec^2. (Note that this model neglects air resistance.)

PUMPKIN TOSSING

In the annual Morton, Illinois, pumpkin tossing contest, contestants use machines they built to throw pumpkins. In 1998 an air cannon entrant set a record by throwing a pumpkin 4491 feet.

DATA UPDATE
www.mcdougallittell.com

EXAMPLE 4 *Modeling Projectile Motion*

PUMPKIN TOSSING In a pumpkin tossing contest in Morton, Illinois, a contestant won the catapult competition by using two telephone poles, huge rubber bands, and a power winch. Suppose the pumpkin was launched with an initial speed of 125 feet per second, at an angle of 45°, and from an initial height of 25 feet.

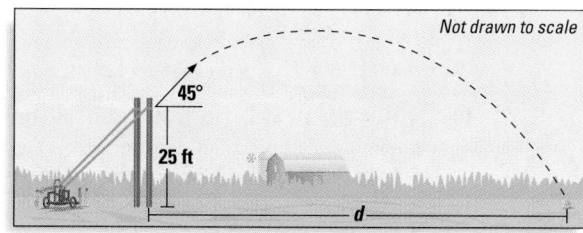

Not drawn to scale
45°
25 ft
d

a. Write a set of parametric equations for the motion of the pumpkin.

b. Use the equations to find how far the pumpkin traveled.

SOLUTION

a. Using $v = 125$ ft/sec, $\theta = 45°$, and $(x_0, y_0) = (0, 25)$, you can write the following.

$$x = (v \cos \theta)t + x_0 \qquad \text{and} \qquad y = -\frac{1}{2}gt^2 + (v \sin \theta)t + y_0$$

$$\approx 88.4t \qquad\qquad\qquad \approx -16t^2 + 88.4t + 25$$

b. The pumpkin hits the ground when $y = 0$.

$$-16t^2 + 88.4t + 25 = y \qquad \text{Write parametric equation for } y.$$

$$-16t^2 + 88.4t + 25 = 0 \qquad \text{Substitute 0 for } y.$$

$$t = \frac{-88.4 \pm \sqrt{(88.4)^2 - 4(-16)(25)}}{2(-16)} \qquad \text{Use the quadratic formula to find } t.$$

$$t \approx 5.8 \text{ seconds} \qquad \text{Simplify and choose positive } t\text{-value.}$$

When $t = 5.8$ seconds, the pumpkin's location will have an x-value of $x = (88.4)(5.8) \approx 513$ feet. So, the pumpkin traveled about 513 feet.

STUDENT HELP

Look Back
For help with the quadratic formula, see p. 291.

13.7 *Parametric Equations and Projectile Motion* **815**

Closure Question *Sample answer:*
For horizontal distance, the equation is a linear function of time, and the coefficient of *t* depends on the cosine of the initial angle. For vertical distance, the equation is a quadratic function of time, and the coefficient of *t* depends on the sine of the initial angle.

EXTRA EXAMPLE 4
At a watermelon seed-spitting contest, a seed is launched with an initial velocity of 35 feet per second, at an angle of 35°, and from an initial height of 5 feet.
a. Write a set of parametric equations for this motion.
$x \approx 28.7t$; $y \approx -16t^2 + 20.1t + 5$
b. Use the equations to find how far the seed travels.
about 42.1 ft

CHECKPOINT EXERCISES

For use after Example 4:
1. Assume the seed in Extra Example 4 is launched at the same velocity and from the same height, but at a 45° angle. How much farther does it travel? about 0.7 ft

DATA UPDATE
Updated data for Example 4 are available at
www.mcdougallittell.com.

STUDENT HELP NOTES
Look Back As students look back to p. 291, remind them that the values for *a, b,* and *c* are the coefficients of the t^2-term, the *t*-term, and the constant term, respectively.

FOCUS ON VOCABULARY
What is the *parameter* in a set of parametric equations for *x* and *y*?
Sample answer: It is a third variable in terms of which both *x* and *y* are expressed.

CLOSURE QUESTION
In the parametric equations for projectile motion, how do the equations for horizontal distance and vertical distance differ?
See sample answer at left.

DAILY PUZZLER
If $x = 6t^2$ and $y = 8t^2$ for the parameter *t,* express *z* in terms of *t* if $z = 0.5w + 5$, where $w = \sqrt{x^2 + y^2}$.
$z = 5t^2 + 5$

ASSIGNMENT GUIDE

BASIC
Day 1: pp. 816–819 Exs. 12, 15, 16, 20–24 even, 25–27, 36–38, 41, 42, 45–57 odd, Quiz 3 Exs. 1–18

AVERAGE
Day 1: pp. 816–819 Exs. 11–15, 16–24 even, 25–32, 36–38, 41, 42, 45–57 odd, Quiz 3 Exs. 1–18

ADVANCED
Day 1: pp. 816–819 Exs. 11–15, 16–24 even, 26–43, 45–57 odd, Quiz 3 Exs. 1–18

BLOCK SCHEDULE WITH 13.6
pp. 816–819 Exs. 11–15, 16–24 even, 25–32, 36–38, 41, 42, 45–57 odd, Quiz 3 Exs. 1–18

EXERCISE LEVELS

Level A: *Easier*
11–20, 41

Level B: *More Difficult*
21–40, 42, 44–57

Level C: *Most Difficult*
43

✔ **HOMEWORK CHECK**
To quickly check student understanding of key concepts, go over the following exercises: Exs. 12, 16, 18, 24, 26, 38. See also the Daily Homework Quiz:

• Blackline Master (*Chapter 14 Resource Book,* p. 11)
• Transparency (p. 105)

! COMMON ERROR
EXERCISES 16–20 Students may forget that when they eliminate the parameter and write an equation in *xy*-form, they must use the parameter to find the new domain. Though the equation no longer contains the parameter, that does not mean that any values for *x* and *y* can be used in the equation.

4–6, 11–13. See next page.
14, 15. See Additional Answers beginning on page AA1.

816

GUIDED PRACTICE

Vocabulary Check ✔

1. Complete this statement: Parametric equations express variables like x and y in terms of another variable such as t. In this case, t is called the __?__. **parameter**

Concept Check ✔

2. For an object moving in a straight line at a constant speed v, what do you need to know in order to write parametric equations describing the object's motion?
the object's starting position and the angle of its motion

3. *Sample answer:* The first set is for linear motion at a constant speed, and the second is for the motion of a projectile, where the motion is nonlinear, and varies in velocity because of the effect of gravity.

3. In this lesson you studied two parametric models for describing motion:

$$x = (v \cos \theta)t + x_0 \qquad \text{and} \qquad x = (v \cos \theta)t + x_0$$
$$y = (v \sin \theta)t + y_0 \qquad\qquad\qquad y = -\frac{1}{2}gt^2 + (v \sin \theta)t + y_0$$

Under what circumstances would you use each model?

Skill Check ✔

Graph the parametric equations. 4–6. See margin.

4. $x = 2t$ and $y = t$ for $0 \le t \le 4$

5. $x = 3t + 4$ and $y = t - 3$ for $0 \le t \le 5$

6. $x = (20 \cos 60°)t$ and $y = (20 \sin 60°)t$ for $2 \le t \le 6$

Write an *xy*-equation for the parametric equations. State the domain.

7. $x = 7t$ and $y = 3t - 2$ for $0 \le t \le 5$ $y = \frac{3}{7}x - 2; 0 \le x \le 35$

8. $x = -4t + 2$ and $y = 5t - 4$ for $0 \le t \le 6$ $y = -\frac{5}{4}x - \frac{3}{2}; -22 \le x \le 2$

9. $x = (11.5 \cos 72.1°)t$ and $y = (11.5 \sin 72.1°)t + 3$ for $0 \le t \le 10$
$y = (\tan 72.1°)x + 3$, or $y = 3.10x + 3; 0 \le x \le 35.3$

10. 🔵 **SOFTBALL CONTEST** At a softball throwing contest, you throw a softball with an initial speed of 60 feet per second, at an angle of 50°, and from an initial height of 5.5 feet. Write parametric equations for the softball's motion.
$x = (60 \cos 50°)t$, or $x = 38.6t$; $y = -16t^2 + (60 \sin 50°)t + 5.5$, or $y = -16t^2 + 46.0t + 5.5$

PRACTICE AND APPLICATIONS

STUDENT HELP

▸ **Extra Practice**
to help you master skills is on p. 959.

GRAPHING Graph the parametric equations. 11–15. See margin.

11. $x = 2t - 2$ and $y = -t + 3$ for $0 \le t \le 5$

12. $x = 5 - 5t$ and $y = 3t - 2$ for $0 \le t \le 5$

13. $x = 2t - 6$ and $y = t - 3$ for $3 \le t \le 8$

14. $x = 30t + 10$ and $y = 60t - 20$ for $0 \le t \le 4$

15. $x = (80.6 \cos 7.1°)t$ and $y = (80.6 \sin 7.1°)t$ for $0 \le t \le 5$

ELIMINATING THE PARAMETER Write an *xy*-equation for the parametric equations. State the domain.

STUDENT HELP

▸ **HOMEWORK HELP**
Example 1: Exs. 11–15
Example 2: Exs. 16–20
Example 3: Exs. 24–32
Example 4: Exs. 33–40

16. $x = 2t$ and $y = -4t$ for $0 \le t \le 5$ $y = -2x; 0 \le x \le 10$

17. $x = t + 1$ and $y = 2t - 3$ for $0 \le t \le 5$ $y = 2x - 5; 1 \le x \le 6$

18. $x = 3t + 6$ and $y = 5t - 1$ for $0 \le t \le 20$ $y = \frac{5}{3}x - 11; 6 \le x \le 66$

19. $x = (14.14 \cos 45°)t$ and $y = (14.14 \sin 45°)t$ for $0 \le t \le 10$ $y = x; 0 \le x \le 100$

20. $x = (111.8 \cos 63.43°)t$ and $y = (111.8 \sin 63.43°)t$ for $0 \le t \le 10$
$y = 2.00x; 0 \le x \le 500$

STUDENT HELP

INTERNET
HOMEWORK HELP
Visit our Web site
www.mcdougallittell.com
for help with Exs. 21–23.

21. $x = (20.0 \cos 71.6°)t$, or
$x = 6.31t$; $y =$
$(20.0 \sin 71.6°)t$, or
$y = 19.0t$; $0 \le t \le 3$

22. $x = (5.06 \cos 25.8°)t +$
18, or $x = 4.56t + 18$;
$y = (5.06 \sin 25.8°)t + 8$,
or $y = 2.20t + 8$;
$0 \le t \le 5$

23. $x = (13.0 \cos 80.0°)t +$
3, or $x = 2.26t + 3$;
$y = (13.0 \sin 80.0°)t + 2$,
or $y = 12.8t + 2$;
$0 \le t \le 5$

26. (x, y in ft, t in sec):
$x = (0.7 \cos 87.83°)t$, or
$x = 0.0265t$;
$y = (0.7 \sin 87.83°)t$, or
$y = 0.699t$

FOCUS ON
APPLICATIONS

REAL LIFE
INSTRUMENT
PANEL In addition
to gauges that give speed
and altitude readings, the
instrument panel of an
airplane also contains an
attitude indicator. This
instrument tells how the
airplane is tilted in relation
to Earth's horizon.

INTERNET
APPLICATION LINK
www.mcdougallittell.com

DESCRIBING LINEAR MOTION Use the given information to write parametric equations describing the linear motion. 21–23. See margin.

21. An object is at $(0, 0)$ at time $t = 0$ and then at $(19, 57)$ at time $t = 3$.

22. An object is at $(18, 8)$ at time $t = 4$ and then at $(40.8, 19.0)$ at time $t = 9$.

23. An object is at $(3, 2)$ at time $t = 0$ and then at $(14.3, 66.1)$ at time $t = 5$.

24. **ROWBOAT** You are trying to row a boat due east across a river that is 0.75 mile wide and flows due south. You reach the other side in 15 minutes, but the current has pulled you 1 mile downstream. Write a set of parametric equations to describe the path you traveled. Then write an xy-equation for the parametric equations. State the domain of the xy-equation. (x, y in mi, t in h): $x = [5 \cos (-53.1°)]t$, or $x = 3t$; $y = [5 \sin (-53.1°)]t$, or $y = -4t$; $y = -\frac{4}{3}x$; $0 \le x \le \frac{3}{4}$

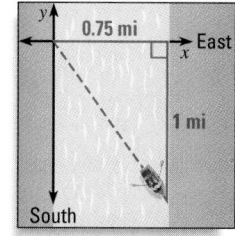

25. **BIKE PATH** A bike trail connects Park Street and Main Street as shown. You enter the trail 2 miles from the intersection of the streets and bike at a speed of 10 miles per hour. You reach Main Street 1.5 miles from the intersection. Write a set of parametric equations to describe your path.
$x = (10 \cos 143.13°)t + 2.0$; $y = (10 \sin 143.13°)t$

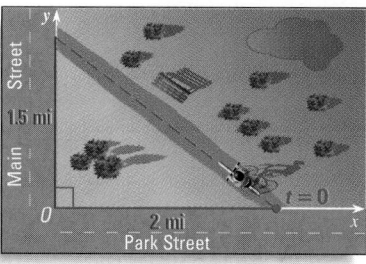

SWIMMING In Exercises 26–28, use the following information.
You are swimming in a race across a lake and back. Swimmers must swim to, and then back from, a buoy placed 2640 feet from the center of the start/finish line. You start the race 100 feet from the center of the start/finish line as shown.

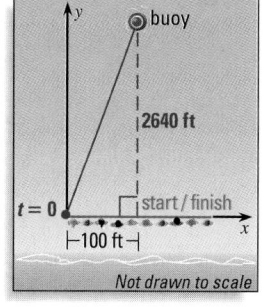

26. You swim to the buoy at a steady rate of 0.7 foot per second. Write a set of parametric equations for your path. See margin.

27. Use the equations to determine how long it takes you to reach the buoy. about 3774 sec, or about 63 min

28. If you continue to swim at a steady rate of 0.7 foot per second straight back to the center of the start/finish line, how long will it take you to complete the race? about 7545 sec, or about 126 min

LANDING A PLANE In Exercises 29–32, use the following information.
You are flying in a small airplane at an altitude of 10,000 feet. When you descend to land the plane, your horizontal air speed will be 260 feet per second (177 miles per hour) and your rate of descent will be 30 feet per second.

29. Write a set of parametric equations for the plane's descent. $x = 260t$; $y = 10,000 - 30t$

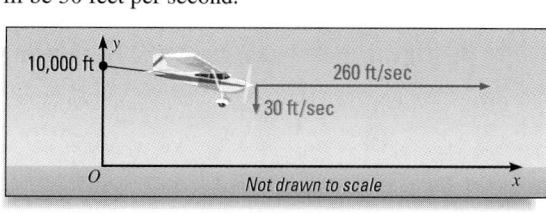

30. What is the angle of descent? about 6.58°

31. How long will it take for the plane to land? about 333 sec, or 5 min 33 sec

32. How far from the airport should you begin the descent? about 86,700 ft, or 16.4 mi

STUDENT HELP NOTES

→ **Homework Help** Students can find help for Exs. 21–23 at **www.mcdougallittell.com**. The information can be printed out for students who don't have access to the Internet.

**APPLICATION NOTE
EXERCISES 29–32**
Additional information about instrument panels is available at **www.mcdougallittell.com**.

4.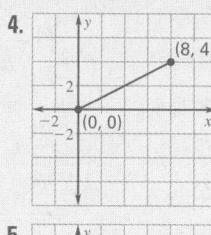

5.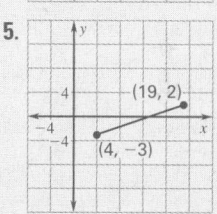

6.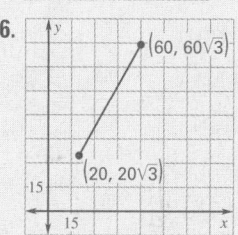

11.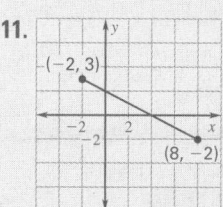

12.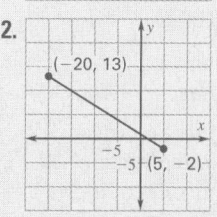

13.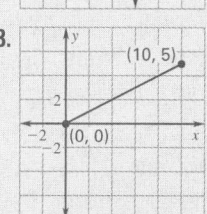

A water skier jumps off a ramp at a speed of 17.9 meters per second. The ramp's angle of elevation is 14.3°, and the height of the end of the ramp above the surface of the water is 1.71 meters. ▶ Source: American Water Ski Association

33. $x = (17.9 \cos 14.3°)t$, or $x = 17.3t$; $y = -4.9t^2 + (17.9 \sin 14.3°)t + 1.71$, or $y = -4.9t^2 + 4.42t + 1.71$

36. $x = (32 \cos 48°)t$, or $x = 21.4t$; $y = -16t^2 + (32 \sin 48°)t$, or $y = -16t^2 + 23.8t$

39. $x = (v \cos 43°)t$, or $x = 0.731vt$; $y = -16t^2 + (v \sin 43°)t + 6$, or $y = -16t^2 + 0.682vt + 6$

43. In the first case, $y = (\tan \theta)(x - x_0) + y_0$, while in the second case $y = -\frac{1}{2}g\left(\frac{x - x_0}{v \cos\theta}\right)^2 + (\tan \theta)(x - x_0) + y_0$. *Sample answer:* The first equation represents straight−line motion in a plane, and v is not in the equation. The object will follow the same straight path no matter what its speed.

Test Preparation

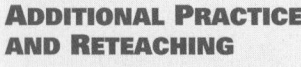

★ Challenge

EXTRA CHALLENGE
↪ www.mcdougallittell.com

33. Write a set of parametric equations for the water skier's jump.

34. For how many seconds is the water skier in the air?   **about 1.19 sec**

35. How far from the ramp does the water skier land?   **about 20.7 m**

14.3°   1.71 m

LEAPING DOLPHIN In Exercises 36–38, use the following information.
A dolphin is performing in a show at an oceanic park and makes a leap out of the water. The dolphin leaves the water traveling at a speed of 32 feet per second and at an angle of 48° with the surface of the water.

36. Write a set of parametric equations for the dolphin's motion.   **See margin.**

37. For how many seconds is the dolphin in the air?   **about 1.49 sec**

38. How far across the water does the dolphin travel in the air?   **about 31.8 ft**

SHOT PUT In Exercises 39 and 40, use the following information.
A shot put is thrown a distance of 54.5 feet at a high school track and field meet. The shot put was released from a height of 6 feet and at an angle of 43°.

39. Write a set of parametric equations for the path of the shot put.   **See margin.**

40. Use the equations to determine the speed of the shot put at the time of release.   **about 39.5 ft/sec**

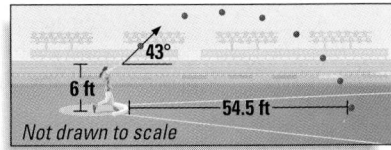

43°
6 ft
54.5 ft
Not drawn to scale

41. MULTIPLE CHOICE Which equation is an *xy*-equation for the parametric equations $x = 3t + 12$ and $y = 12t - 8$ where $0 \le t \le 20$?   **C**

(A) $y = 9x + 4$; $0 \le x \le 20$ **(B)** $y = 36x + 132$; $-8 \le x \le 132$

(C) $y = 4x - 56$; $12 \le x \le 72$ **(D)** $y = 15x + 4$; $4 \le x \le 304$

42. MULTIPLE CHOICE An airplane takes off at an angle of 10.6° with the ground and travels at a constant speed of 324 miles per hour. Which set of parametric equations describes the airplane's ascent?   **B**

(A) $x = 4670t$, $y = 874t$ **(B)** $x = 318t$, $y = 60t$

(C) $x = 5000t$, $y = 80t$ **(D)** $x = 800t$, $y = 150t$

43. CRITICAL THINKING Write the following pairs of equations in the form $y = f(x)$.

$$x = (v \cos \theta)t + x_0 \qquad\qquad x = (v \cos \theta)t + x_0$$
$$y = (v \sin \theta)t + y_0 \qquad\qquad y = -\frac{1}{2}gt^2 + (v \sin \theta)t + y_0$$

In which case is the path of the moving object *not* affected by changing the speed v? Explain why this makes sense.   **See margin.**

ADDITIONAL PRACTICE AND RETEACHING

For Lesson 13.7:
• Practice Levels A, B, and C (*Chapter 13 Resource Book,* p. 95)
• Reteaching with Practice (*Chapter 13 Resource Book,* p. 98)
• ▦ See Lesson 13.7 of the *Personal Student Tutor*

For more Mixed Review:
• ▦ Search the *Test and Practice Generator* for key words or specific lessons.

818

MIXED REVIEW

GRAPHING Graph the function. (Review 5.1, 7.5, 8.1 for 14.1) 44–49. See margin.

44. $y = 9x^2$

45. $y = -10x^2$

46. $y = 7\sqrt{x}$

47. $y = -6\sqrt{x}$

48. $y = 7 \cdot 2^x$

49. $y = -\dfrac{3}{4} \cdot 3^x$

FINDING SUMS Find the sum of the series. (Review 11.1, 11.4)

50. $\displaystyle\sum_{i=1}^{10} -3i$ -165

51. $\displaystyle\sum_{i=1}^{27} i^2$ 6930

52. $\displaystyle\sum_{n=1}^{\infty} 20\left(\dfrac{4}{5}\right)^{n-1}$ 100

53. $\displaystyle\sum_{n=1}^{\infty} -\dfrac{1}{6}\left(-\dfrac{1}{2}\right)^{n-1}$ $-\dfrac{1}{9}$

NORMAL DISTRIBUTIONS Find the probability that a randomly selected x-value is in the given interval. (Review 12.7)

54. to the left of the mean $\dfrac{1}{2}$

55. between the mean and 1 standard deviation to the left of the mean 0.3413

56. between 2 and 3 standard deviations from the mean 0.0430

57. more than 3 standard deviations to the right of the mean 0.0013

QUIZ 3

Self-Test for Lessons 13.5–13.7

Solve $\triangle ABC$. (Lessons 13.5 and 13.6)

1. $B = 70°$, $b = 30$, $c = 25$
$A \approx 58.5°$; $C \approx 51.5°$; $a \approx 27.2$

2. $B = 10°$, $C = 100°$, $a = 15$
$A = 70°$; $b \approx 2.77$; $c \approx 15.7$

3. $A = 40°$, $B = 110°$, $b = 30$
$C = 30°$; $a \approx 20.5$; $c \approx 16.0$

4. $A = 122°$, $a = 9$, $c = 13$ no triangle

5. $a = 45$, $b = 32$, $c = 24$
$A \approx 106.1°$; $B \approx 43.1°$; $C \approx 30.8°$

6. $A = 107°$, $b = 15$, $c = 28$
$a = 35.4$; $B = 23.9°$; $C = 49.1°$

Find the area of $\triangle ABC$. (Lessons 13.5 and 13.6)

7. $B = 95°$, $a = 12$, $c = 30$ 179 units²

8. $C = 103°$, $a = 41$, $b = 25$ 499 units²

9. $A = 117°$, $b = 16$, $c = 8$ 57.0 units²

10. $a = 7$, $b = 7$, $c = 5$ 16.3 units²

11. $a = 89$, $b = 55$, $c = 71$ 1950 units²

12. $a = 40$, $b = 21$, $c = 32$ 334 units²

Graph the parametric equations. (Lesson 13.7) 13–15. See margin.

13. $x = 4 - 2t$ and $y = 3t + 1$ for $0 \le t \le 5$

14. $x = 2t - 5$ and $y = 4t - 3$ for $3 \le t \le 7$

15. $x = (10.5 \cos 45°)t$ and $y = (10.5 \sin 45°)t + 4$ for $0 \le t \le 5$

Write an xy-equation for the parametric equations. State the domain. (Lesson 13.7)

16. $x = -5t + 3$ and $y = t - 6$ for $0 \le t \le 5$ $y = -\dfrac{1}{5}x - \dfrac{27}{5}$; $-22 \le x \le 3$

17. $x = (10 \cos 35°)t$ and $y = (10 \sin 35°)t$ for $0 \le t \le 30$ $y = 0.700x$; $0 \le x \le 246$

18. 🌐 **SOCCER** You are a goalie in a soccer game. You save the ball and then drop kick it as far as you can down the field. Your kick has an initial speed of 26 feet per second and starts at a height of 2 feet. If you kick the ball at an angle of 45°, how far down the field does the ball hit the ground? (Lesson 13.7) about 23.0 ft

13.7 *Parametric Equations and Projectile Motion* 819

Additional Test Preparation *Sample answer:*
The first case gives you the vertical distance traveled for a given horizontal distance traveled (or vice versa), but you don't know how these distances relate to time.

The second case gives you all the information of the first, but you also know *when* the ball is at a given position.

4 ASSESS

DAILY HOMEWORK QUIZ

📄 *Transparency Available*

1. Graph $x = -2t + 3$ and $y = t - 2$ for $0 \le t \le 5$.

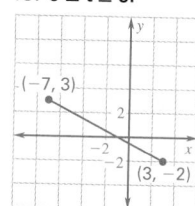

$(-7, 3)$

$(3, -2)$

Write an xy-equation for the parametric equations. State the domain.

2. $x = 3t + 3$ and $y = 6t - 2$ for $0 \le t \le 5$ $y = 2x - 8$; $3 \le x \le 18$

3. $x = -2t - 5$ and $y = 3t - 4$ for $0 \le t \le 10$
$y = -\dfrac{3}{2}x - \dfrac{23}{2}$; $-25 \le x \le -5$

4. The town of Jackson lies 40 mi east and 30 mi south of your home. It takes you 50 min to drive to Jackson on a highway that runs straight from your house to Jackson. Write a set of parametric equations for the path you travel.
(x, y in mi, t in h):
$x = (60 \cos (-36.9°))t$ or $x = 48t$;
$y = (60 \sin (-36.9°))t$ or $y = -36t$

EXTRA CHALLENGE NOTE
→ Challenge problems for Lesson 13.7 are available in **blackline** format in the *Chapter 13 Resource Book*, p. 102 and at **www.mcdougallittell.com.**

ADDITIONAL TEST PREPARATION

1. WRITING A ball has horizontal position x and vertical position y. What is the difference between a description of its path with y written as a function of x versus with x and y written in terms of the time parameter t?
See sample answer at left.

44–49, 13–15.
See Additional Answers beginning on page AA1.

1 Planning the Activity

PURPOSE
To use a graphing calculator to graph parametric equations of the path of a ball.

MATERIALS
- graphing calculator for each student
- Keystroke blackline (*Chapter 13 Resource Book*, p. 94)

PACING
- Example — 10 min
- Exercises — 15 min

▶ LINK TO LESSON
In Example 4 of Lesson 13.7, parametric equations are used to model projectile motion. Encourage students to return to this Example so they can graph the path of the pumpkin.

2 Managing the Activity

CLASSROOM MANAGEMENT
A graphing calculator for an overhead display is helpful for this activity. Also useful is an enlargement of the calculator face so keys and screens are easily shown to students who are having difficulty.

COOPERATIVE LEARNING
Students can work with a partner. For Exercise 1, students can graph a few or all of the equations at once, which will help them compare the paths more directly.

3 Closing the Activity

★ KEY DISCOVERY
A graphing calculator in parametric mode is useful for graphing and comparing parametric equations.

ACTIVITY ASSESSMENT
JOURNAL Explain how to graph parametric equations on a graphing calculator.
See sample answer at right.

820

▶ ACTIVITY 13.7

Using Technology

Graphing Parametric Equations

You can use a graphing calculator to graph a set of parametric equations.

▶ EXAMPLE

At a driving range, you hit a golf ball at ground level with an initial speed of 120 feet per second and at an angle of 45°. Use a graphing calculator to graph a set of parametric equations that describe the path of the ball. Then use the graphing calculator to estimate how far the ball travels.

▶ SOLUTION

Assume that $x_0 = 0$ and $y_0 = 0$. The parametric equations that describe the path of the golf ball are as follows.

$$x = (v \cos \theta)t + x_0$$
$$\approx 84.9t$$

$$y = -\frac{1}{2}gt^2 + (v \sin \theta)t + y_0$$
$$\approx -16t^2 + 84.9t$$

STUDENT HELP

INTERNET **KEYSTROKE HELP**
See keystrokes for several models of calculators at www.mcdougallittell.com

❶ Put your calculator in *Parametric* mode.

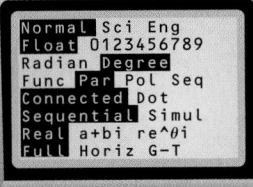

❷ Enter the parametric equations.

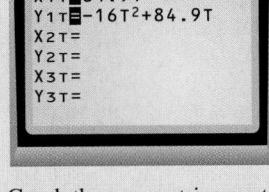

❸ Set the viewing window so that $0 \leq t \leq 10$, $0 \leq x \leq 500$, and $-50 \leq y \leq 150$.

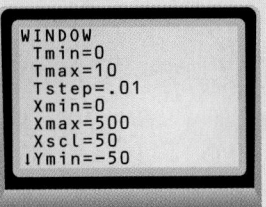

❹ Graph the parametric equations. Use the graphing calculator's *Trace* feature to find the value of x when $y = 0$.

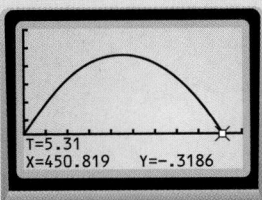

▶ An estimate of the distance the ball travels is about 450 feet, or 150 yards.

▶ EXERCISES

1. Graph a set of parametric equations that describe the path of the golf ball in the example when it is hit at different angles. Copy and complete the table.

Angle, θ	30°	35°	40°	50°	55°	60°
Horizontal distance (ft), x	?	?	?	?	?	?

390; 423; 443; 443; 423; 390

2. At what angle should you hit the ball so that it travels the maximum horizontal distance? Explain. 45°; the results look to be symmetric around the value $\theta = 45°$, with a maximum at that angle.

Chapter 13 *Trigonometric Ratios and Functions*

Activity Assessment *Sample answer:*
Set the calculator to parametric mode. Enter the *x*- and *y*-equations. Set an appropriate window for *x*, *y*, and the domain of the parameter, and graph. Tracing gives *x*- and *y*-values for a given *t*-value.

Chapter Summary

WHAT did you learn?

Evaluate trigonometric functions.
- of acute angles (13.1)
- of any angle (13.3)

Find the sides and angles of a triangle.
- solve right triangles (13.1)
- use the law of sines (13.5)
- use the law of cosines (13.6)

Measure angles using degree measure and radian measure. (13.2)

Find arc lengths and areas of sectors. (13.2)

Evaluate inverse trigonometric functions. (13.4)

Find the area of a triangle.
- using two sides and the included angle (13.5)

- using Heron's formula (13.6)

Use parametric equations to model linear or projectile motion. (13.7)

Use trigonometric and inverse trigonometric functions to solve real-life problems. (13.1, 13.3–13.7)

WHY did you learn it?

Find the altitude of a kite. (p. 771)
Find the horizontal distance traveled by a golf ball. (p. 787)

Find the length of a zip-line at a ropes course. (p. 774)
Find the distance between two buildings. (p. 805)
Find the angle at which two trapeze artists meet. (p. 811)

Find the angle generated by a figure skater performing a jump. (p. 781)

Find the area irrigated by a rotating sprinkler. (p. 781)

Find the angle at which to set the arm of a crane. (p. 794)

Find the amount of paint needed for the side of a house. (p. 806)
Find the area of the Dinosaur Diamond. (p. 812)

Model the path of a leaping dolphin. (p. 818)

Find distances for a marching band on a football field. (p. 787)

How does Chapter 13 fit into the BIGGER PICTURE of algebra?

Trigonometry is closely tied to both algebra and geometry. In this chapter you studied trigonometric functions of *angles*, defined by ratios of side lengths of right triangles.

In the next chapter you will study trigonometric functions of *real numbers*, used to model periodic behavior. You will see even more connections between trigonometry and algebra as you graph trigonometric functions in a coordinate plane.

STUDY STRATEGY

How did you draw diagrams?

Here is an example of a diagram drawn for Exercise 22 on page 810, following the **Study Strategy** on page 768.

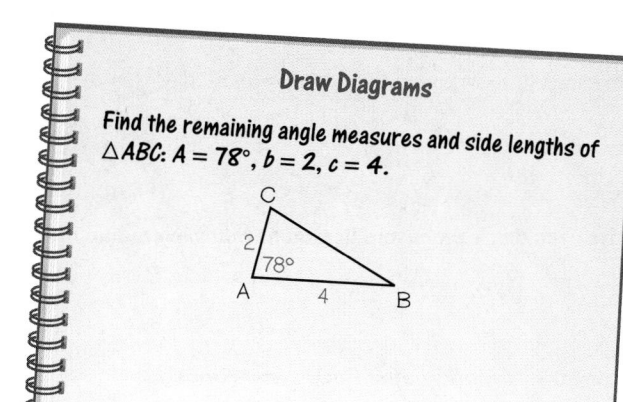

Draw Diagrams

Find the remaining angle measures and side lengths of △ABC: A = 78°, b = 2, c = 4.

821

ADDITIONAL RESOURCES

The following resources are available to help review the material in this chapter.

- Chapter Review Games and Activities (*Chapter 13 Resource Book*, p. 103)
- *Instant Replay: Video Review Games*
- 📺 *Personal Student Tutor*
- Cumulative Review, Chs. 1–13 (*Chapter 13 Resource Book*, p. 115)

1. $\sin \theta = \frac{3}{5}$, $\cos \theta = \frac{4}{5}$, $\tan \theta = \frac{3}{4}$, $\csc \theta = \frac{5}{3}$, $\sec \theta = \frac{5}{4}$, $\cot \theta = \frac{4}{3}$

2. $\sin \theta = \frac{3}{5}$, $\cos \theta = \frac{4}{5}$, $\tan \theta = \frac{3}{4}$, $\csc \theta = \frac{5}{3}$, $\sec \theta = \frac{5}{4}$, $\cot \theta = \frac{4}{3}$

3. $\sin \theta = \frac{\sqrt{2}}{2}$, $\cos \theta = \frac{\sqrt{2}}{2}$, $\tan \theta = 1$, $\csc \theta = \sqrt{2}$, $\sec \theta = \sqrt{2}$, $\cot \theta = 1$

4. $\sin \theta = \frac{12}{13}$, $\cos \theta = \frac{5}{13}$, $\tan \theta = \frac{12}{5}$, $\csc \theta = \frac{13}{12}$, $\sec \theta = \frac{13}{5}$, $\cot \theta = \frac{5}{12}$

Chapter Review

VOCABULARY

- sine, p. 769
- cosine, p. 769
- tangent, p. 769
- cosecant, p. 769
- secant, p. 769
- cotangent, p. 769
- solving a right triangle, p. 770

- angle of elevation, p. 771
- angle of depression, p. 771
- initial side of an angle, p. 776
- terminal side of an angle, p. 776
- standard position, p. 776
- coterminal angles, p. 777

- radian, p. 777
- sector, p. 779
- central angle, p. 779
- quadrantal angle, p. 785
- reference angle, p. 785
- inverse sine, p. 792

- inverse cosine, p. 792
- inverse tangent, p. 792
- law of sines, p. 799
- law of cosines, p. 807
- parametric equations, p. 813
- parameter, p. 813

13.1

RIGHT TRIANGLE TRIGONOMETRY

Examples on pp. 769–771

> **EXAMPLE** You can evaluate the six trigonometric functions of θ for the triangle shown. First find the hypotenuse length: $\sqrt{5^2 + 12^2} = \sqrt{169} = 13$.
>
> $$\sin \theta = \frac{\text{opp}}{\text{hyp}} = \frac{12}{13} \qquad \cos \theta = \frac{\text{adj}}{\text{hyp}} = \frac{5}{13} \qquad \tan \theta = \frac{\text{opp}}{\text{adj}} = \frac{12}{5}$$
>
> $$\csc \theta = \frac{\text{hyp}}{\text{opp}} = \frac{13}{12} \qquad \sec \theta = \frac{\text{hyp}}{\text{adj}} = \frac{13}{5} \qquad \cot \theta = \frac{\text{adj}}{\text{opp}} = \frac{5}{12}$$

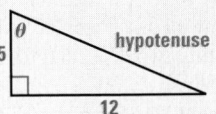

Evaluate the six trigonometric functions of θ. 1–4. See margin.

1.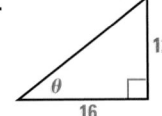

2.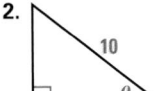

3.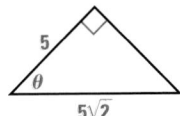

4.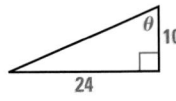

13.2

GENERAL ANGLES AND RADIAN MEASURE

Examples on pp. 776–779

> **EXAMPLES** You can measure angles using degree measure or radian measure.
>
> $$20° = 20°\left(\frac{\pi \text{ radians}}{180°}\right) = \frac{\pi}{9} \text{ radians} \qquad \frac{7\pi}{6} \text{ radians} = \left(\frac{7\pi}{6} \text{ radians}\right)\left(\frac{180°}{\pi \text{ radians}}\right) = 210°$$
>
> Arc length of the sector at the right: $s = r\theta = 8\left(\frac{2\pi}{3}\right) = \frac{16\pi}{3}$ inches
>
> Area of the sector at the right: $A = \frac{1}{2}r^2\theta = \frac{1}{2}(8^2)\left(\frac{2\pi}{3}\right) = \frac{64\pi}{3}$ square inches

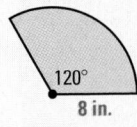

Rewrite each degree measure in radians and each radian measure in degrees.

5. $30°$ $\frac{\pi}{6}$ **6.** $225°$ $\frac{5\pi}{4}$ **7.** $-15°$ $-\frac{\pi}{12}$ **8.** $\frac{3\pi}{4}$ $135°$ **9.** $\frac{5\pi}{3}$ $300°$ **10.** $\frac{\pi}{3}$ $60°$

Find the arc length and area of a sector with the given radius *r* and central angle θ.

11. $r = 5$ ft, $\theta = \frac{\pi}{2}$ $\frac{5\pi}{2}$ ft, $\frac{25\pi}{4}$ ft²   **12.** $r = 12$ in., $\theta = 25°$ **13.** $r = 16$ cm, $\theta = 210°$

                                   $\frac{5\pi}{3}$ in., 10π in.² $\frac{56\pi}{3}$ cm, $\frac{448\pi}{3}$ cm²

13.3   TRIGONOMETRIC FUNCTIONS OF ANY ANGLE

Examples on pp. 784–787

> **EXAMPLE**   You can evaluate the six trigonometric functions of $\theta = 240°$ using a reference angle: $\theta' = \theta - 180° = 240° - 180° = 60°$.
>
> $\sin 240° = -\sin 60° = -\dfrac{\sqrt{3}}{2}$ $\csc 240° = -\csc 60° = -\dfrac{2\sqrt{3}}{3}$
>
> $\cos 240° = -\cos 60° = -\dfrac{1}{2}$ $\sec 240° = -\sec 60° = -2$
>
> $\tan 240° = +\tan 60° = \sqrt{3}$ $\cot 240° = +\cot 60° = \dfrac{\sqrt{3}}{3}$
>
>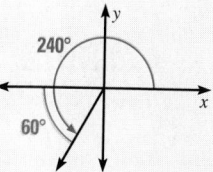

Evaluate the function without using a calculator.

14. $\tan \dfrac{11\pi}{4}$ -1 **15.** $\cos \dfrac{11\pi}{6}$ $\dfrac{\sqrt{3}}{2}$ **16.** $\sec 225°$ $-\sqrt{2}$ **17.** $\sin 390°$ $\dfrac{1}{2}$ **18.** $\csc(-120°)$ $-\dfrac{2\sqrt{3}}{3}$

13.4   INVERSE TRIGONOMETRIC FUNCTIONS

Examples on pp. 792–794

> **EXAMPLE**   You can find an angle within a certain range that corresponds to a given value of a trigonometric function.
>
> To find $\cos^{-1}\left(-\dfrac{\sqrt{2}}{2}\right)$, find θ so that $\cos\theta = -\dfrac{\sqrt{2}}{2}$ and $0° \le \theta \le 180°$.
>
> So, $\theta = \cos^{-1}\left(-\dfrac{\sqrt{2}}{2}\right) = 135°$ $\left(\text{or } \dfrac{3\pi}{4} \text{ radians}\right)$.
>
>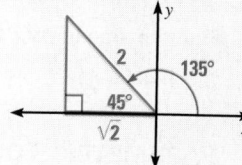

Evaluate the expression without using a calculator. Give your answer in both radians and degrees.

19. $\sin^{-1}\dfrac{\sqrt{2}}{2}$ $\dfrac{\pi}{4}, 45°$ **20.** $\tan^{-1}\dfrac{\sqrt{3}}{3}$ $\dfrac{\pi}{6}, 30°$ **21.** $\cos^{-1}0$ $\dfrac{\pi}{2}, 90°$ **22.** $\tan^{-1}(-1)$ $-\dfrac{\pi}{4}, -45°$ **23.** $\cos^{-1}\left(-\dfrac{1}{2}\right)$ $\dfrac{2\pi}{3}, 120°$

13.5   THE LAW OF SINES

Examples on pp. 799–802

> **EXAMPLE**   You can solve the triangle shown using the law of sines.
>
> The measure of the third angle is: $B = 180° - 105° - 48° = 27°$.
>
>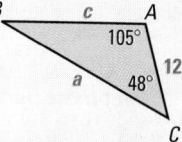
>
> $\dfrac{a}{\sin 105°} = \dfrac{12}{\sin 27°}$ $\dfrac{c}{\sin 48°} = \dfrac{12}{\sin 27°}$
>
> $a = \dfrac{12\sin 105°}{\sin 27°} \approx 25.5$ $c = \dfrac{12\sin 48°}{\sin 27°} \approx 19.6$
>
> Area of this triangle $= \dfrac{1}{2}bc\sin A = \dfrac{1}{2}(12)(19.6)\sin 105° \approx 114$ square units

25. $A \approx 29.3°$, $C \approx 132.7°$, $c \approx 28.5$ or
$A \approx 150.7°$, $C \approx 11.3°$, $c \approx 7.60$

36.

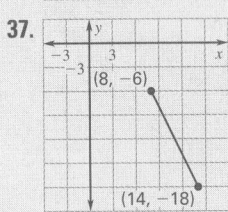

37.

Solve $\triangle ABC$. (*Hint:* Some of the "triangles" may have no solution and some may have two.)

24. $A = 45°$, $B = 60°$, $c = 44$ **25.** $B = 18°$, $b = 12$, $a = 19$ **26.** $C = 140°$, $c = 40$, $b = 20$
    $C = 75°$, $a \approx 32.2$, $b \approx 39.4$ **See margin.** $B \approx 18.7°$, $A \approx 21.3°$, $a \approx 22.6$

Find the area of the triangle with the given side lengths and included angle.

27. $C = 35°$, $b = 10$, $a = 22$ **28.** $A = 110°$, $b = 8$, $c = 7$ **29.** $B = 25°$, $a = 15$, $c = 31$
      63.1 units^2   **26.3 units^2**   **98.3 units^2**

13.6   THE LAW OF COSINES

Examples on pp. 807–809

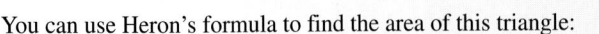

> **EXAMPLE** You can solve the triangle below using the law of cosines.
>
> Law of cosines: $b^2 = 35^2 + 37^2 - 2(35)(37) \cos 25° \approx 247$
>
> $b \approx 15.7$
>
> Law of sines: $\dfrac{\sin A}{35} \approx \dfrac{\sin 25°}{15.7}$, $\sin A \approx \dfrac{35 \sin 25°}{15.7}$, $A \approx 70.4°$
>
> $C \approx 180° - 25° - 70.4° = 84.6°$
>
> You can use Heron's formula to find the area of this triangle:
>
> $s \approx \dfrac{1}{2}(35 + 15.7 + 37) \approx 44$, so area $\approx \sqrt{44(44 - 35)(44 - 15.7)(44 - 37)} \approx 280$ square units

Solve $\triangle ABC$.

30. $a = 25$, $b = 18$, $c = 28$ **31.** $a = 6$, $b = 11$, $c = 14$ **32.** $B = 30°$, $a = 80$, $c = 70$
$A \approx 61.4°$, $B \approx 39.2°$, $C \approx 79.4°$ $A \approx 24.2°$, $B \approx 48.6°$, $C \approx 107.2°$ $b \approx 40.0°$, $A \approx 89.0°$, $C \approx 61.0°$

Find the area of $\triangle ABC$ having the given side lengths.

33. $a = 11$, $b = 2$, $c = 12$ **34.** $a = 4$, $b = 24$, $c = 26$ **35.** $a = 15$, $b = 8$, $c = 21$
      9.9 units^2   **43.2 units^2**   **46.4 units^2**

13.7   PARAMETRIC EQUATIONS AND PROJECTILE MOTION

Examples on pp. 813–815

> **EXAMPLE** You can graph the parametric equations $x = -3t$ and $y = -t$ for
> $0 \le t \le 3$. Make a table of values, plot the points (x, y), and connect the points.
>
> | t | 0 | 1 | 2 | 3 |
> |---|---|---|---|---|
> | x | 0 | -3 | -6 | -9 |
> | y | 0 | -1 | -2 | -3 |
>
>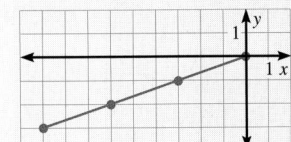
>
> To write an xy-equation for these parametric equations, solve the first equation for t:
> $t = -\dfrac{1}{3}x$. Substitute into the second equation: $y = \dfrac{1}{3}x$. The domain is $-9 \le x \le 0$.

Graph the parametric equations. 36–37. See margin for graphs.

36. $x = 3t + 1$ and $y = 3t + 6$ for $0 \le t \le 5$ **37.** $x = 2t + 4$ and $y = -4t + 2$ for $2 \le t \le 5$

Write an xy-equation for the parametric equations. State the domain.

38. $x = 5t$ and $y = t + 7$ for $0 \le t \le 20$ **39.** $x = 2t - 3$ and $y = -4t + 5$ for $0 \le t \le 8$
    $y = \dfrac{1}{5}x + 7$, $0 \le x \le 100$ $y = -2x - 1$, $-3 \le x \le 13$

Chapter Test

ADDITIONAL RESOURCES
• *Chapter 13 Resource Book*
 Chapter Test (3 levels) (p. 104)
 SAT/ACT Chapter Test (p. 110)
 Alternative Assessment (p. 111)

• 🖳 *Test and Practice Generator*

Evaluate the six trigonometric functions of θ. 1–4. See margin.

1.
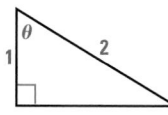

2.
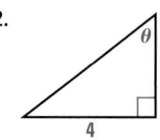

3.
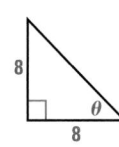

4.
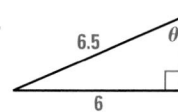

1. $\sin \theta = \dfrac{\sqrt{3}}{2}$, $\cos \theta = \dfrac{1}{2}$, $\tan \theta = \sqrt{3}$,

$\csc \theta = \dfrac{2\sqrt{3}}{3}$, $\sec \theta = 2$, $\cot \theta = \dfrac{\sqrt{3}}{3}$

2. $\sin \theta = \dfrac{4}{5}$, $\cos \theta = \dfrac{3}{5}$, $\tan \theta = \dfrac{4}{3}$,

$\csc \theta = \dfrac{5}{4}$, $\sec \theta = \dfrac{5}{3}$, $\cot \theta = \dfrac{3}{4}$

3. $\sin \theta = \dfrac{\sqrt{2}}{2}$, $\cos \theta = \dfrac{\sqrt{2}}{2}$, $\tan \theta = 1$,

$\csc \theta = \sqrt{2}$, $\sec \theta = \sqrt{2}$, $\cot \theta = 1$

4. $\sin \theta = \dfrac{12}{13}$, $\cos \theta = \dfrac{5}{13}$, $\tan \theta = \dfrac{12}{5}$,

$\csc \theta = \dfrac{13}{12}$, $\sec \theta = \dfrac{13}{5}$, $\cot \theta = \dfrac{5}{12}$

Rewrite each degree measure in radians and each radian measure in degrees.

5. $120°$ $\dfrac{2\pi}{3}$ **6.** $360°$ 2π **7.** $-60°$ $-\dfrac{\pi}{3}$ **8.** $\dfrac{\pi}{9}$ $20°$ **9.** 5π $900°$ **10.** $-\dfrac{5\pi}{4}$ $-225°$

Find the arc length and area of a sector with the given radius r and central angle θ.

11. $r = 4$ ft, $\theta = 240°$ $\dfrac{16\pi}{3}$ ft, $\dfrac{32\pi}{3}$ ft²   **12.** $r = 20$ cm, $\theta = 45°$ **13.** $r = 12$ in., $\theta = 150°$ 10π in., 60π in.²
                                             5π cm, 50π cm²

Evaluate the function without using a calculator.

14. $\cos 180°$ -1 **15.** $\sec(-30°)$ $\dfrac{2\sqrt{3}}{3}$ **16.** $\cot 495°$ -1 **17.** $\sin \dfrac{7\pi}{6}$ $-\dfrac{1}{2}$ **18.** $\tan\left(-\dfrac{\pi}{4}\right)$ -1 **19.** $\csc\left(-\dfrac{7\pi}{4}\right)$ $\sqrt{2}$

Evaluate the expression without using a calculator. Give your answer in both radians and degrees.

20. $\sin^{-1} 1$ $\dfrac{\pi}{2}$, $90°$ **21.** $\tan^{-1}\sqrt{3}$ **22.** $\cos^{-1}\dfrac{\sqrt{3}}{2}$ **23.** $\tan^{-1} 0$ 0, $0°$ **24.** $\cos^{-1} 1$ 0, $0°$ **25.** $\sin^{-1}\left(-\dfrac{\sqrt{2}}{2}\right)$
                                     $\dfrac{\pi}{3}$, $60°$ $\dfrac{\pi}{6}$, $30°$ $-\dfrac{\pi}{4}$, $-45°$

Solve $\triangle ABC$.

                                                   $A = 57°$, $b \approx 9.71$, $a \approx 8.17$

26.
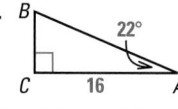

$B = 68°$, $a \approx 6.46$, $c \approx 17.3$

27.
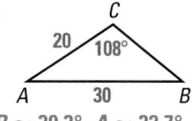

$B \approx 39.3°$, $A \approx 32.7°$, $a \approx 17.0$

28.
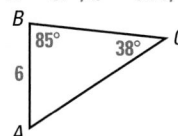

29.
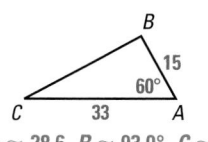

$a \approx 28.6$, $B \approx 93.0°$, $C \approx 27.0°$

30. $A = 120°$, $a = 14$, $b = 10$
$B \approx 38.2°$, $C \approx 21.8°$, $c \approx 6.00$

31. $B = 40°$, $a = 7$, $c = 10$
$b \approx 6.46$, $A \approx 44.1°$, $C \approx 95.9°$

32. $C = 105°$, $a = 4$, $b = 3$
$c \approx 5.59$, $A \approx 43.8°$, $B \approx 31.2°$

Find the area of $\triangle ABC$.

33. 28.3
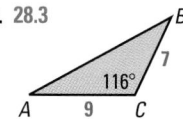

34. 76.1
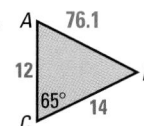

35. 195
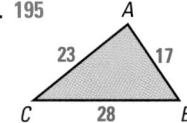

36. 42.8
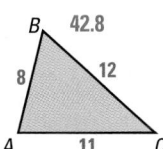

37.
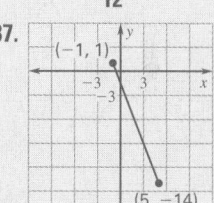

;

$y = -\dfrac{5}{2}x - \dfrac{3}{2}$, $-1 \le x \le 5$

38.
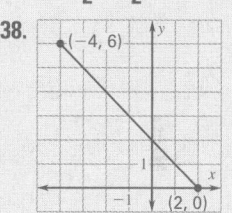

Graph the parametric equations. Then write an xy-equation and state the domain.

37. $x = 2t - 3$ and $y = -5t + 6$ for $1 \le t \le 4$ See margin. **38.** $x = t - 4$ and $y = -t + 6$ for $0 \le t \le 6$
                                                          $y = -x + 2$, $-4 \le x \le 2$; See margin for graph.

39. 🌐 **BOAT RIDE** A boat travels 50 miles due west before adjusting its course
$25°$ north of west and traveling an additional 35 miles. How far is the boat from
its point of departure? about 83 mi

40. 🌐 **PROJECTILE MOTION** You throw a ball at an angle of $50°$, from a height of
6 feet, and with an initial speed of 25 feet per second. Write a set of parametric
equations for the path of the ball. How far from you does the ball land? $x = 16.1t$, $y = -16t^2 + 19.2t + 6$, 23.4 ft

ADDITIONAL RESOURCES

• *Chapter 13 Resource Book*
Chapter Test (3 levels) (p. 104)
SAT/ACT Chapter Test (p. 110)
Alternative Assessment (p. 111)

• ⊞ *Test and Practice Generator*

CHAPTER 13

Chapter Standardized Test

▶ **TEST-TAKING STRATEGY** When taking a test, first tackle the questions that you know are easy for you to answer. Then go back and answer questions that you suspect will take you extra time and effort.

1. MULTIPLE CHOICE Given the diagram, which equation is correct? **B**

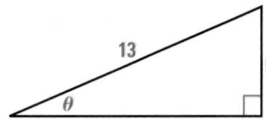

(A) $\sec \theta = \frac{12}{13}$ (B) $\cot \theta = \frac{12}{5}$

(C) $\cos \theta = \frac{5}{13}$ (D) $\csc \theta = \frac{12}{5}$

(E) $\sin \theta = \frac{13}{5}$

2. MULTIPLE CHOICE Suppose $(8, -15)$ is a point on the terminal side of an angle θ in standard position. Which equation is *not* true? **A**

(A) $\csc \theta = -\frac{15}{17}$ (B) $\cos \theta = \frac{8}{17}$

(C) $\tan \theta = -\frac{15}{8}$ (D) $\cot \theta = -\frac{8}{15}$

(E) $\sec \theta = \frac{17}{8}$

3. MULTIPLE CHOICE What is the value of $\cos 765°$? **A**

(A) $\frac{\sqrt{2}}{2}$ (B) $\frac{2\sqrt{3}}{3}$ (C) $\sqrt{2}$

(D) 2 (E) undefined

4. MULTIPLE CHOICE What is the value of $\sin\left(-\frac{13\pi}{6}\right)$? **C**

(A) $-\frac{\sqrt{3}}{2}$ (B) $-\frac{\sqrt{2}}{2}$ (C) $-\frac{1}{2}$

(D) $\frac{1}{2}$ (E) 1

5. MULTIPLE CHOICE What is the solution of the equation $\sin \theta = -\frac{3}{8}$, where $180° < \theta < 270°$? **D**

(A) $-202.02°$ (B) $-22.02°$ (C) $22.02°$

(D) $202.02°$ (E) $222.02°$

6. MULTIPLE CHOICE What is the approximate value of θ in the triangle shown? **C**

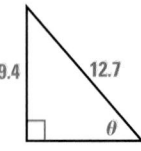

(A) $36.5°$ (B) $42.3°$ (C) $47.7°$

(D) $48.9°$ (E) $52.6°$

7. MULTIPLE CHOICE What is the area of the triangle shown? **B**

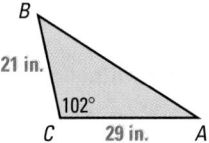

(A) about 182.4 in.^{2} (B) about 297.8 in.^{2}

(C) about 300.8 in.^{2} (D) about 304.5 in.^{2}

(E) about 595.7 in.^{2}

8. MULTIPLE CHOICE What is the value of a in the triangle shown? **B**

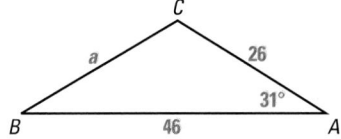

(A) about 24.6 (B) about 27.2

(C) about 28.9 (D) about 29.2

(E) about 30.5

9. MULTIPLE CHOICE Which equation is an *xy*-equation for the parametric equations $x = 5t - 9$ and $y = -3t + 11$? **E**

(A) $y = \frac{3}{5}x + \frac{28}{5}$ (B) $y = -\frac{3}{5}x + \frac{82}{5}$

(C) $y = -\frac{3}{5}x - \frac{82}{5}$ (D) $y = -\frac{3}{5}x - \frac{28}{5}$

(E) $y = -\frac{3}{5}x + \frac{28}{5}$

13d. *Sample answer:*

$$s = 3960 \tan\left(\cos^{-1}\left(\frac{3960}{3960 + \frac{h}{5280}}\right)\right)$$

QUANTITATIVE COMPARISON In Exercises 10–12, choose the statement that is true about the given quantities.

Ⓐ The quantity in column A is greater.

Ⓑ The quantity in column B is greater.

Ⓒ The two quantities are equal.

Ⓓ The relationship cannot be determined from the given information.

	Column A	Column B	
10.	Solution of $\tan\theta = -\sqrt{3}$, where $270° \le \theta \le 360°$	Solution of $\sin\theta = -\dfrac{\sqrt{2}}{2}$, where $270° \le \theta \le 360°$	B
11.	Area of a sector with $r = 5$ in. and $\theta = 60°$	Area of a sector with $r = 6$ in. and $\theta = 45°$	B
12.	Area of a triangle with side lengths 5 cm, 8 cm, and 11 cm	Area of a triangle with side lengths 7 cm, 7 cm, and 12 cm	B

13. **MULTI-STEP PROBLEM** You are enjoying the view at the top of a 200 foot tall building on a clear day. To find the distance you can see to the horizon, you draw the diagram at the right.

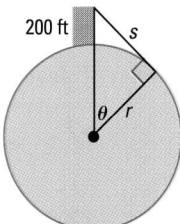

   a. Find the value of θ in the diagram. Use 3960 miles for the value of r. **about 0.251°**

   b. Use your answer to part (a) to find the distance s you can see to the horizon. **about 17.3 mi**

   c. How much farther could you see to the horizon if the building were 400 feet tall? **about 7.2 mi farther**

Not drawn to scale

   d. **CRITICAL THINKING** Write a general formula for finding the distance s you can see to the horizon from the top of a building that is h feet tall. **See margin.**

14. **MULTI-STEP PROBLEM** A trough can be made by folding a rectangular piece of metal in half and then enclosing the ends. The volume of water the trough can hold depends on how far you bend the metal.

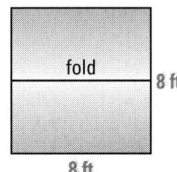

   a. Predict the value of θ that will maximize the volume of the trough shown. **Answers will vary.**

   b. Find the volume of the trough as a function of θ. (*Hint:* You will need to find the area of one of the triangular faces.) **$V = 64\sin\theta$**

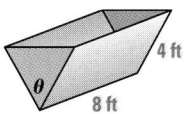

   c. **CRITICAL THINKING** Find the value of θ that maximizes the volume. What is the maximum volume? How close was your prediction? **90°; 64 ft³; Answers will vary.**

15. **MULTI-STEP PROBLEM** Two motorized toy boats are released in a pool at time $t = 0$. Boat 1 travels 65° north of east at a rate of 0.75 meter per second. Boat 2 travels due east at a rate of 0.5 meter per second.

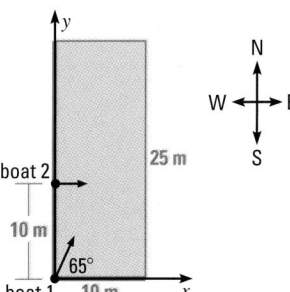

   a. Write a set of parametric equations to describe the path of each boat. **boat 1: $x = 0.317t$, $y = 0.680t$; boat 2: $x = 0.5t$, $y = 10$**

   b. At what point will each boat hit the east edge of the pool? **boat 1: (10, 21.24); boat 2: (10, 10)**

   c. At what point do the paths of the two boats cross? **(4.66, 10)**

   d. *Writing* If the two boats are released at the same time, will they collide? If so, how many seconds after the boats are released? If not, explain why not. **No, boat 1 passes the point of intersection after about 14.7 s, boat 2 after 9.3 s.**

PLANNING THE CHAPTER

Trigonometric Graphs, Identities, and Equations

GOALS		NCTM	ITED	SAT9	Terra-Nova	Local
LESSON						
14.1 *pp. 831–838*	**GOAL 1** Graph sine and cosine functions. **GOAL 2** Graph tangent functions. **TECHNOLOGY ACTIVITY: 14.1** *Graph trigonometric functions on a graphing calculator.*	2, 3	MIG, IIG, MIRA	22, 37, 43	14, 16, 17, 48	
14.2 *pp. 839–847*	**CONCEPT ACTIVITY: 14.2** *Investigate translating and reflecting trigonometric graphs.* **GOAL 1** Graph translations and reflections of sine and cosine graphs. **GOAL 2** Graph translations and reflections of tangent graphs.	2, 3	MIG, IIG, MIRA	22, 34, 37, 43	14, 16, 17, 48	
14.3 *pp. 848–854*	**GOAL 1** Use trigonometric identities to simplify trigonometric expressions and to verify other identities. **GOAL 2** Use trigonometric identities to solve real-life problems.	1, 2, 3, 6, 8, 9, 10	MIRA, RQRA		13, 14, 16, 17, 18, 48, 52	
14.4 *pp. 855–861*	**GOAL 1** Solve a trigonometric equation. **GOAL 2** Solve real-life trigonometric equations.	1, 2, 3, 8, 9, 10	MIRA, RQRA		11, 14, 16, 17, 18, 48, 49, 51, 52	
14.5 *pp. 862–868*	**GOAL 1** Model data with a sine or cosine function. **GOAL 2** Use technology to write a trigonometric model.	2, 3, 5, 8, 9, 10	MIRA, RQRA	21, 23, 27, 38	14, 16, 17, 18, 48	
14.6 *pp. 869–874*	**GOAL 1** Evaluate trigonometric functions of the sum or difference of two angles. **GOAL 2** Use sum and difference formulas to solve real-life problems.	1, 2, 3, 4, 6, 8, 9, 10	MIRA, RQRA, RQM		11, 13, 14, 16, 17, 18, 48, 49, 51, 52	
14.7 *pp. 875–882*	**GOAL 1** Evaluate expressions using double- and half-angle formulas. **GOAL 2** Use double- and half-angle formulas to solve real-life problems.	1, 2, 3, 4, 6, 8, 9, 10	MIRA, RQRA, RQM		11, 13, 14, 16, 18, 52	

RESOURCES

CHAPTER RESOURCE BOOKLETS

CHAPTER SUPPORT

Tips for New Teachers	p. 1	Prerequisite Skills Review	p. 5
Parent Guide for Student Success	p. 3	Strategies for Reading Mathematics	p. 7

LESSON SUPPORT

	14.1	14.2	14.3	14.4	14.5	14.6	14.7
Lesson Plans (regular and block)	p. 9	p. 22	p. 35	p. 49	p. 64	p. 78	p. 90
Warm-Up Exercises and Daily Quiz	p. 11	p. 24	p. 37	p. 51	p. 66	p. 80	p. 92
Activity Support Masters							
Lesson Openers	p. 12	p. 25	p. 38	p. 52	p. 67	p. 81	p. 93
Graphing Calculator Activities & Keystrokes	p. 13		p. 39	p. 53			
Practice (3 levels)	p. 14	p. 26	p. 41	p. 55	p. 68	p. 82	p. 94
Reteaching with Practice	p. 17	p. 29	p. 44	p. 58	p. 71	p. 85	p. 97
Quick Catch-Up for Absent Students	p. 19	p. 31	p. 46	p. 60	p. 73	p. 87	p. 99
Cooperative Learning Activities				p. 61			
Interdisciplinary Applications		p. 32		p. 62		p. 88	
Real-Life Applications	p. 20		p. 47		p. 74		p. 100
Math & History Applications					p. 75		
Challenge: Skills and Applications	p. 21	p. 33	p. 48	p. 63	p. 76	p. 89	p. 101

REVIEW AND ASSESSMENT

Quizzes	pp. 34, 77	Alternative Assessment with Math Journal	p. 110
Chapter Review Games and Activities	p. 102	Project with Rubric	p. 112
Chapter Test (3 levels)	pp. 103–108	Cumulative Review	p. 114
SAT/ACT Chapter Test	p. 109	Resource Book Answers	p. A1

TRANSPARENCIES

	14.1	14.2	14.3	14.4	14.5	14.6	14.7
Warm-Up Exercises and Daily Quiz	p. 105	p. 106	p. 107	p. 108	p. 109	p. 110	p. 111
Alternative Lesson Opener Transparencies	p. 91	p. 92	p. 93	p. 94	p. 95	p. 96	p. 97
Examples/Standardized Test Practice	✓	✓	✓	✓	✓	✓	✓
Answer Transparencies	✓	✓	✓	✓	✓	✓	✓

TECHNOLOGY

- Electronic Teaching Tools
- Online Lesson Planner
- Internet Support
- Personal Student Tutor
- Test and Practice Generator
- Instant Replay: Video Review Games
- Electronic Lesson Presentations (Lesson 14.2)

ADDITIONAL RESOURCES

- Basic Skills Workbook: Diagnosis and Remediation
- Worked-Out Solution Key
- Resources in Spanish
- Standardized Test Practice Workbook
- Practice Workbook with Examples

Resource Key
- ● STUDENT EDITION
- ● CHAPTER 14 RESOURCE BOOK
- ● TEACHER'S EDITION

REGULAR SCHEDULE

Day 1

14.1

STARTING OPTIONS
- Prereq. Skills Review
- Strategies for Reading
- Homework Check
- Warm-Up or Daily Quiz

TEACHING OPTIONS
- Motivating the Lesson
- Les. Opener (Visual)
- Graphing Calc. Activity
- Examples 1–2
- Guided Practice
  Exs. 1, 3–11,13–14

APPLY/HOMEWORK
- See Assignment Guide.
- See the CRB: Practice, Reteach, Apply, Extend

ASSESSMENT OPTIONS
- Checkpoint Exercises, pp. 832–833

Day 2

14.1 (cont.)

STARTING OPTIONS
- Homework Check

TEACHING OPTIONS
- Examples 3–4
- Technology Activity
- Closure Question
- Guided Practice
  Exs. 2, 12, 15–16

APPLY/HOMEWORK
- See Assignment Guide.
- See the CRB: Practice, Reteach, Apply, Extend

ASSESSMENT OPTIONS
- Checkpoint Exercises, pp. 833–834
- Daily Quiz (14.1)
- Stand. Test Practice

Day 3

14.2

STARTING OPTIONS
- Homework Check
- Warm-Up or Daily Quiz

TEACHING OPTIONS
- Concept Activity
- Les. Opener (Application)
- Examples 1–4
- Guided Practice
  Exs. 1–2, 4–11, 13–14

APPLY/HOMEWORK
- See Assignment Guide.
- See the CRB: Practice, Reteach, Apply, Extend

ASSESSMENT OPTIONS
- Checkpoint Exercises, pp. 841–842

Day 4

14.2 (cont.)

STARTING OPTIONS
- Homework Check

TEACHING OPTIONS
- Examples 5–7
- Closure Question
- Guided Practice
  Exs. 3, 12, 15–16

APPLY/HOMEWORK
- See Assignment Guide.
- See the CRB: Practice, Reteach, Apply, Extend

ASSESSMENT OPTIONS
- Checkpoint Exercises, pp. 842–843
- Daily Quiz (14.2)
- Stand. Test Practice
- Quiz (14.1–14.2)

Day 5

14.3

STARTING OPTIONS
- Homework Check
- Warm-Up or Daily Quiz

TEACHING OPTIONS
- Les. Opener (Visual)
- Graphing Calc. Activity
- Examples 1–4
- Guided Practice
  Exs. 1–2, 5–11

APPLY/HOMEWORK
- See Assignment Guide.
- See the CRB: Practice, Reteach, Apply, Extend

ASSESSMENT OPTIONS
- Checkpoint Exercises, pp. 849–850

Day 6

14.3 (cont.)

STARTING OPTIONS
- Homework Check

TEACHING OPTIONS
- Examples 5–7
- Closure Question
- Guided Practice
  Exs. 3–4, 12–15

APPLY/HOMEWORK
- See Assignment Guide.
- See the CRB: Practice, Reteach, Apply, Extend

ASSESSMENT OPTIONS
- Checkpoint Exercises, pp. 850–851
- Daily Quiz (14.3)
- Stand. Test Practice

Day 9

14.6

STARTING OPTIONS
- Homework Check
- Warm-Up or Daily Quiz

TEACHING OPTIONS
- Motivating the Lesson
- Les. Opener (Activity)
- Examples 1–6
- Closure Question
- Guided Practice Exs.

APPLY/HOMEWORK
- See Assignment Guide.
- See the CRB: Practice, Reteach, Apply, Extend

ASSESSMENT OPTIONS
- Checkpoint Exercises
- Daily Quiz (14.6)
- Stand. Test Practice

Day 10

14.7

STARTING OPTIONS
- Homework Check
- Warm-Up or Daily Quiz

TEACHING OPTIONS
- Motivating the Lesson
- Les. Opener (Application)
- Examples 1–4
- Guided Practice
  Exs. 1–15

APPLY/HOMEWORK
- See Assignment Guide.
- See the CRB: Practice, Reteach, Apply, Extend

ASSESSMENT OPTIONS
- Checkpoint Exercises, p. 877

Day 11

14.7 (cont.)

STARTING OPTIONS
- Homework Check

TEACHING OPTIONS
- Examples 5–8
- Closure Question
- Guided Practice
  Ex. 16

APPLY/HOMEWORK
- See Assignment Guide.
- See the CRB: Practice, Reteach, Apply, Extend

ASSESSMENT OPTIONS
- Checkpoint Exercises, pp. 877–878
- Daily Quiz (14.7)
- Stand. Test Practice
- Quiz (14.6–14.7)

Day 12

Review

DAY 12 START OPTIONS
- Homework Check

REVIEWING OPTIONS
- Chapter 14 Summary
- Chapter 14 Review
- Chapter Review Games and Activities

APPLY/HOMEWORK
- Chapter 14 Test (practice)
- Ch. Standardized Test (practice)

Day 13

Assess

DAY 13 START OPTIONS
- Homework Check

ASSESSMENT OPTIONS
- Chapter 14 Test
- SAT/ACT Ch. 14 Test
- Alternative Assessment

BLOCK SCHEDULE

Day 1

14.1

DAY 1 START OPTIONS
- Prereq. Skills Review
- Strategies for Reading
- Homework Check
- Warm-Up or Daily Quiz

TEACHING 14.1 OPTIONS
- Motivating the Lesson
- Les. Opener (Visual)
- Graphing Calc. Activity
- Examples 1–4
- Technology Activity
- Closure Question
- Guided Practice Exs.

APPLY/HOMEWORK
- See Assignment Guide.
- See the CRB: Practice, Reteach, Apply, Extend

ASSESSMENT OPTIONS
- Checkpoint Exercises
- Daily Quiz (Les. 14.1)
- Stand. Test Practice

Day 2

14.2

DAY 2 START OPTIONS
- Homework Check
- Warm-Up or Daily Quiz

TEACHING 14.2 OPTIONS
- Concept Activity
- Les. Opener (Appl.)
- Examples 1–7
- Closure Question
- Guided Practice Exs.

APPLY/HOMEWORK
- See Assignment Guide.
- See the CRB: Practice, Reteach, Apply, Extend

ASSESSMENT OPTIONS
- Checkpoint Exercises
- Daily Quiz (Les. 14.2)
- Stand. Test Practice
- Quiz (14.1–14.2)

Day 3

14.3

DAY 3 START OPTIONS
- Homework Check
- Warm-Up or Daily Quiz

TEACHING 14.3 OPTIONS
- Les. Opener (Visual)
- Graphing Calc. Activity
- Examples 1–7
- Closure Question
- Guided Practice Exs.

APPLY/HOMEWORK
- See Assignment Guide.
- See the CRB: Practice, Reteach, Apply, Extend

ASSESSMENT OPTIONS
- Checkpoint Exercises
- Daily Quiz (Les. 14.3)
- Stand. Test Practice

Day 4

14.4 & 14.5

DAY 4 START OPTIONS
- Homework Check
- W-Up 14.4 or D. Quiz 14.3

TEACHING 14.4 OPTIONS
- Motivating the Lesson
- Les. Opener (Calc.)
- Graphing Calc. Activity
- Examples 1–6
- Closure Question
- Guided Practice Exs.

TEACHING 14.5 OPTIONS
- Warm-Up (Les. 14.5)
- Les. Opener (Visual)
- Examples 1–4
- Closure Question
- Guided Practice Exs.

APPLY/HOMEWORK
- See Assignment Guide.
- See the CRB: Practice, Reteach, Apply, Extend

ASSESSMENT OPTIONS
- Checkpoint Exercises
- Daily Quiz (14.4, 14.5)
- Stand. Test Practice
- Quiz (14.3–14.5)

Day 5

14.6 & 14.7

DAY 5 START OPTIONS
- Homework Check
- W-Up 14.6 or D. Quiz 14.5

TEACHING 14.6 OPTIONS
- Motivating the Lesson
- Les. Opener (Activity)
- Examples 1–6
- Closure Question
- Guided Practice Exs.

TEACHING 14.7 OPTIONS
- Warm-Up (Les. 14.7)
- Motivating the Lesson
- Les. Opener (Appl.)
- Examples 1–8
- Closure Question
- Guided Practice Exs.

APPLY/HOMEWORK
- See Assignment Guide.
- See the CRB: Practice, Reteach, Apply, Extend

ASSESSMENT OPTIONS
- Checkpoint Exercises
- Daily Quiz (14.6, 14.7)
- Stand. Test Practice
- Quiz (14.6–14.7)

Day 6

Review/Assess

DAY 6 START OPTIONS
- Homework Check

REVIEWING OPTIONS
- Chapter 14 Summary
- Chapter 14 Review
- Chapter Review Games and Activities
- Chapter 14 Test (practice)
- Ch. Standardized Test (practice)

ASSESSMENT OPTIONS
- Chapter 14 Test
- SAT/ACT Ch. 14 Test
- Alternative Assessment

Day 7

14.4

STARTING OPTIONS
- Homework Check
- Warm-Up or Daily Quiz

TEACHING OPTIONS
- Motivating the Lesson
- Les. Opener (Calculator)
- Graphing Calc. Activity
- Examples 1–6
- Closure Question
- Guided Practice Exs.

APPLY/HOMEWORK
- See Assignment Guide.
- See the CRB: Practice, Reteach, Apply, Extend

ASSESSMENT OPTIONS
- Checkpoint Exercises
- Daily Quiz (14.4)
- Stand. Test Practice

Day 8

14.5

STARTING OPTIONS
- Homework Check
- Warm-Up or Daily Quiz

TEACHING OPTIONS
- Les. Opener (Visual)
- Examples 1–4
- Closure Question
- Guided Practice Exs.

APPLY/HOMEWORK
- See Assignment Guide.
- See the CRB: Practice, Reteach, Apply, Extend

ASSESSMENT OPTIONS
- Checkpoint Exercises
- Daily Quiz (14.5)
- Stand. Test Practice
- Quiz (14.3–14.5)

BEFORE THE CHAPTER

The *Chapter 14 Resource Book* has the following materials to distribute and use before the chapter:

- **Parent Guide for Student Success (pictured below)**
- **Prerequisite Skills Review**
- **Strategies for Reading Mathematics**

PARENT GUIDE *Pages 3–4*

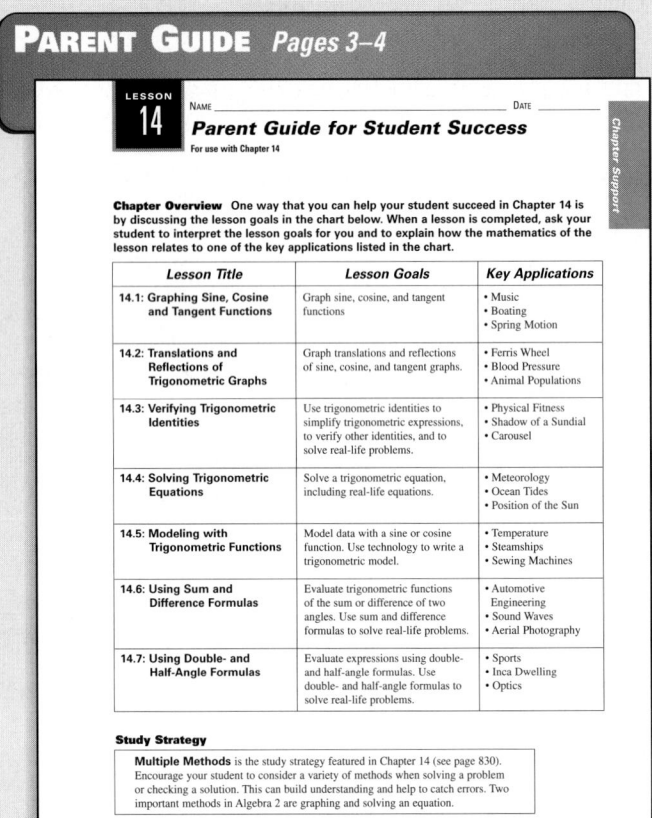

LESSON 14 *Chapter Support*

NAME _____ DATE _____

Parent Guide for Student Success
For use with Chapter 14

Chapter Overview One way that you can help your student succeed in Chapter 14 is by discussing the lesson goals in the chart below. When a lesson is completed, ask your student to interpret the lesson goals for you and to explain how the mathematics of the lesson relates to one of the key applications listed in the chart.

Lesson Title	Lesson Goals	Key Applications
14.1: Graphing Sine, Cosine and Tangent Functions	Graph sine, cosine, and tangent functions	• Music • Boating • Spring Motion
14.2: Translations and Reflections of Trigonometric Graphs	Graph translations and reflections of sine, cosine, and tangent graphs.	• Ferris Wheel • Blood Pressure • Animal Populations
14.3: Verifying Trigonometric Identities	Use trigonometric identities to simplify trigonometric expressions, to verify other identities, and to solve real-life problems.	• Physical Fitness • Shadow of a Sundial • Carousel
14.4: Solving Trigonometric Equations	Solve a trigonometric equation, including real-life equations.	• Meteorology • Ocean Tides • Position of the Sun
14.5: Modeling with Trigonometric Functions	Model data with a sine or cosine function. Use technology to write a trigonometric model.	• Temperature • Steamships • Sewing Machines
14.6: Using Sum and Difference Formulas	Evaluate trigonometric functions of the sum or difference of two angles. Use sum and difference formulas to solve real-life problems.	• Automotive Engineering • Sound Waves • Aerial Photography
14.7: Using Double- and Half-Angle Formulas	Evaluate expressions using double- and half-angle formulas. Use double- and half-angle formulas to solve real-life problems.	• Sports • Inca Dwelling • Optics

Study Strategy

Multiple Methods is the study strategy featured in Chapter 14 (see page 830). Encourage your student to consider a variety of methods when solving a problem or checking a solution. This can build understanding and help to catch errors. Two important methods in Algebra 2 are graphing and solving an equation.

PARENT GUIDE FOR STUDENT SUCCESS The first page summarizes the content of Chapter 14. Parents are encouraged to have their students explain how the material relates to key applications in the chapter, such as ocean tides. The second page (not shown) provides exercises and an activity that parents can do with their students. In the activity, parents and students develop a function for the height of a bicycle pedal over time.

DURING EACH LESSON

The *Chapter 14 Resource Book* has the following alternatives for introducing the lesson:

- **Lesson Openers (pictured below)**
- **Graphing Calculator Activities with Keystrokes**

LESSON OPENER *Page 52*

LESSON 14.4 *Lesson 14.4*

NAME _____ DATE _____ **Available as a transparency**

Graphing Calculator Lesson Opener
For use with pages 855–861

In Lesson 14.4, you will learn algebraic methods for solving an equation that contains trigonometric functions. You can use a graphing calculator to obtain the approximate solutions within a specified interval.

For example, to solve
$2 + 2 \cos x = \sin 2x$ in the interval
$0 \leq x < 2\pi$, graph
$y_1 = 2 + 2 \cos x$ and $y_2 = \sin 2x$.

Use the *Intersect* feature. The solutions in the given interval are
$x \approx 3.14$ (or $x = \pi$), and $x \approx 4.14$.

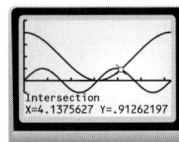

Intersection
X=4.1375627 Y=.91262197

Note that we found *only* the solutions within the interval we graphed, 0 to 2π. Because the functions involved are periodic there are actually infinitely many solutions.

Use a graphing calculator to solve the equation in the interval $0 \leq x < 2\pi$. If necessary, round to the nearest hundredth.

1. $3 \sin x = \cos (x - 1)$ 2. $\cos x = 2 \cos 3x$

3. $\tan x = 3 - 2x$ 4. $\sec x = x - 2$

5. $2 \cos x + \sin x = 4$ 6. $3 \sin x - 2 \cos x = 1$

7. $\tan x = \cos x$ 8. $\cot x - \sin 2x = 3$

9. $\cos (2x + 5) = \sin 3x$ 10. $3 + 2 \sin x = 3 \cos x$

GRAPHING CALCULATOR LESSON OPENER This Lesson Opener provides an alternative way to start Lesson 14.4 through the use of a graphing calculator. Students use the Intersect feature to estimate the solutions of equations to develop an understanding of solving trigonometric equations.

 TECHNOLOGY RESOURCE

Teachers can use the Electronic Lesson Presentations CD-ROM to present Lesson 14.2 using computer animation to step through the Examples.

The *Chapter 14 Resource Book* has a variety of materials to follow-up each lesson. They include the following:

- **Practice (3 levels)**
- **Reteaching with Practice**
- **Quick Catch-Up for Absent Students**
- **Challenge: Skills and Applications**
- **Interdisciplinary Applications**
- **Real-Life Applications (pictured below)**

APPLICATION *Page 47*

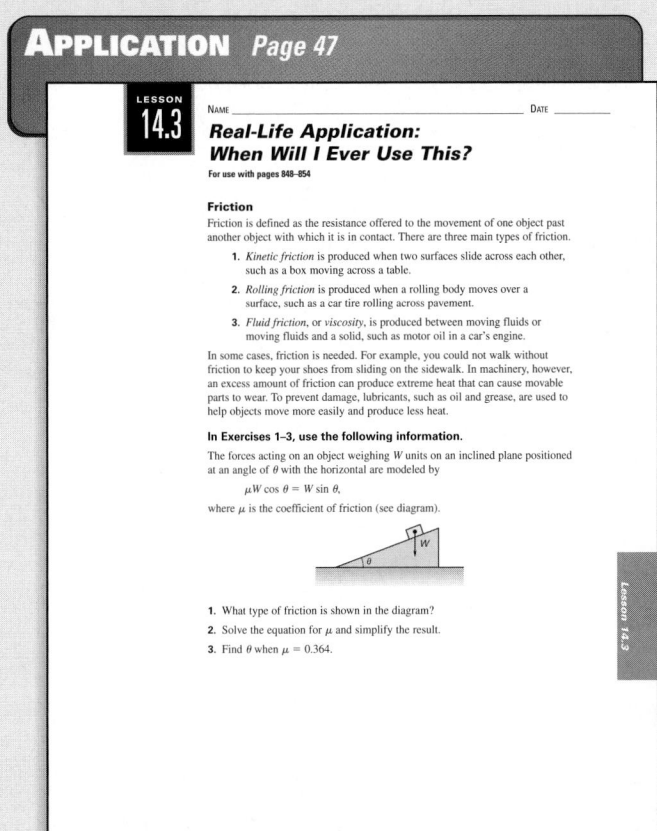

REAL-LIFE APPLICATION When students ask "When will I ever use trigonometric identities?," you can have them work through this Real-Life Application in which they use an identity to simplify a formula related to friction.

The *Chapter 14 Resource Book* has the following review and assessment materials:

- **Quizzes**
- **Chapter Review Games and Activities**
- **Chapter Test (3 levels)**
- **SAT/ACT Chapter Test**
- **Alternative Assessment with Rubrics and Math Journal**
- **Project with Rubric**
- **Cumulative Review (pictured below)**

CUMULATIVE REVIEW *Pages 114–115*

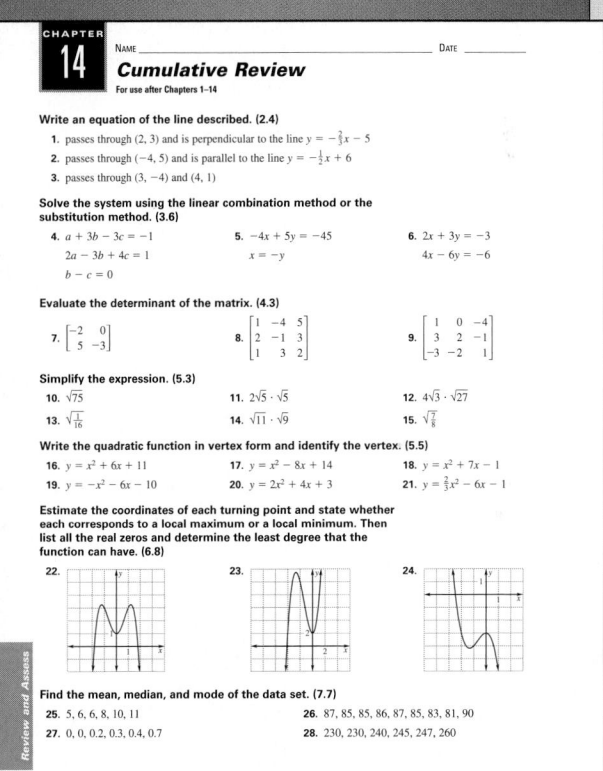

CUMULATIVE REVIEW The content of Chapters 1–14 is reviewed in this two-page cumulative review. Lesson references with each cluster of exercises guide students to the places in the textbook where they can go to review each concept.

TECHNOLOGY RESOURCE

Teachers can use the Time-Saving Test and Practice Generator to create customized review materials covering Chapters 1–14 of Algebra 2.

CHAPTER GOALS

Chapter 14 extends the previous work with trigonometry to include trigonometric functions, trigonometric identities, and trigonometric equations. First, sine, cosine, and tangent functions are graphed by identifying the amplitude and period. Translations and reflections of sine, cosine, and tangent graphs are then explored. Trigonometric functions are used to model real-life situations in which the frequency and amplitude are known. In Lesson 14.5, trigonometric models are derived from real-life data by hand and then with the use of technology.

Trigonometric identities are introduced and used to simplify trigonometric expressions and to verify other identities. Techniques for solving trigonometric equations such as solving a linear equation, using graphing, using factoring, and using the quadratic formula are examined so that students can solve real-life problems involving trigonometric equations. Finally, students evaluate trigonometric functions using sum or difference of two angles formulas or double- and half-angle formulas.

APPLICATION NOTE

The Ferris wheel is named after its inventor George Washington Gale Ferris. The motion of a person riding a Ferris wheel is a good example of circular motion. Other examples of circular motion include the motion of turning a carousel, a Compact Disc, and the wheels found on a steamship or at a mill. In this chapter, students will learn to model circular motion such as that of the Ferris wheel with a sine or cosine function.

Additional information about Ferris wheels is available at **www.mcdougallittell.com**.

TRIGONOMETRIC GRAPHS, IDENTITIES, AND EQUATIONS

▶ *How long does it take to go around a Ferris wheel?*

APPLICATION: Ferris Wheels

The Texas State Fair is home to America's tallest Ferris wheel. The Ferris wheel takes riders to a height of 212 feet and has a diameter of about 203 feet. The maximum speed of the Ferris wheel is 1.5 rotations per minute.

Think & Discuss

The graph below shows a person's height (in feet) above the ground while riding the Ferris wheel at maximum speed. Use the graph to answer the questions below.

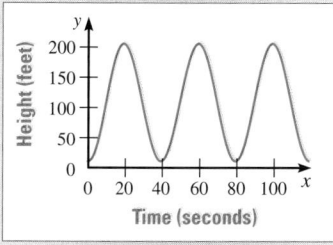

1. How many seconds does it take to reach the maximum height from the minimum height? **20 sec**

2. How long does it take to complete one revolution? **40 sec**

3. How would the graph change if the Ferris wheel rotated faster? if the Ferris wheel had a smaller diameter? **See margin.**

Learn More About It

In Example 5 on p. 842 you will use trigonometric functions to model a person's height above the ground while riding a Ferris wheel.

APPLICATION LINK Visit www.mcdougallittell.com for more information about Ferris wheels.

829

ADDITIONAL RESOURCES
Another way to begin the chapter is to show the video clip of a real-life motivator for Chapter 14 on the *Instant Replay: Video Review Games.*

PROJECTS
A project covering Chapters 13 and 14 appears on pages 892–893 of the Student Edition. An additional project for Chapter 14 is available in the *Chapter 14 Resource Book,* p. 112.

TECHNOLOGY
Software
• *Electronic Teaching Tools*
• *Online Lesson Planner*
• *Personal Student Tutor*
• *Test and Practice Generator*
• *Electronic Lesson Presentations* (Lesson 14.2)

Video
• *Instant Replay: Video Review Games*

Internet Connections
www.mcdougallittell.com
• **Application Links**
  829, 842, 860, 868, 878
• **Student Help**
  836, 838, 845, 851, 857, 862, 873, 877, 892
• **Career Links**
  836, 858, 871
• **Extra Challenge**
  861, 874, 881

3. *Sample answer:* If the Ferris wheel rotated faster, the maximum height would stay the same, but it would take less time to reach that height on each revolution. If the Ferris wheel had a smaller diameter, the maximum height would be smaller, but it would be reached in the same amount of time on each revolution.

Study Guide

PREPARE

PREPARE

DIAGNOSTIC TOOLS

The **Skill Review** exercises can help you diagnose whether students have the following skills needed in Chapter 14:

- Graph quadratic and absolute value functions and circle.

- Solve quadratic equations.

- Evaluate trigonometric functions and inverse trigonometric functions using both radians and degrees.

The following resources are available for students who need additional help with these skills:

- Prerequisite Skills Review (*Chapter 14 Resource Book*, p. 5; *Warm-Up Transparencies*, p. 104)

- Reteaching with Practice (Chapter Resource Books for Lessons 2.8, 5.2, 5.6, 10.6, 13.3 and 13.4)

- ▣ *Personal Student Tutor*

ADDITIONAL RESOURCES

The following resources are provided to help you prepare for the upcoming chapter and customize review materials:

- ***Chapter 14 Resource Book***
  Tips for New Teachers (p. 1)
  Parent Guide (p. 3)
  Lesson Plans (every lesson)
  Lesson Plans for Block Scheduling (every lesson)

- ▣ *Electronic Teaching Tools*

- ▣ *Online Lesson Planner*

- ▣ *Test and Practice Generator*

1.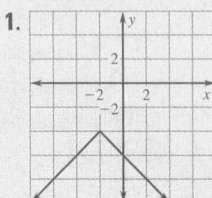

2, 3. See Additional Answers beginning on page AA1.

PREVIEW

What's the chapter about?

Chapter 14 is about **trigonometry**. In Chapter 14 you'll learn

- how to graph trigonometric functions and transformations of trigonometric graphs.
- how to use trigonometric identities and solve trigonometric equations.
- how to write and use trigonometric models.

KEY VOCABULARY

▶ **Review**
- identity, p. 13
- domain, p. 67
- range, p. 67
- *x*-intercept, p. 84
- quadratic form, p. 346

- local maximum, p. 374
- local minimum, p. 374
- asymptote, p. 465
- sine, p. 769
- cosine, p. 769
- tangent, p. 769

▶ **New**
- periodic function, p. 831
- cycle, p. 831
- period, p. 831
- amplitude, p. 831
- trigonometric identities, p. 848

PREPARE

Are you ready for the chapter?

SKILL REVIEW Do these exercises to review key skills that you'll apply in this chapter. See the given **reference page** if there is something you don't understand.

STUDENT HELP

▶ **Study Tip**
"Student Help" boxes throughout the chapter give you study tips and tell you where to look for extra help in this book and on the Internet.

Graph the function. (Review Example 1, p. 123; Example 2, p. 250; Example 2, p. 624)
1–3. See margin.

1. $y = -|x + 2| - 4$ **2.** $y = 2(x + 2)^2 + 3$ **3.** $(x + 4)^2 + (y - 1)^2 = 9$

Solve the equation. (Review Example 5, p. 258; Example 1, p. 291) $\frac{1}{6} \pm \frac{\sqrt{61}}{6}$

4. $x^2 + 7x - 8 = 0$ $-8, 1$ **5.** $9x^2 - 25 = 0$ $\pm\frac{5}{3}$ **6.** $3x^2 - x - 5 = 0$

Evaluate the function without using a calculator. (Review Example 4, p. 786)

7. $\sin 60°$ $\frac{\sqrt{3}}{2}$ **8.** $\tan 30°$ $\frac{\sqrt{3}}{3}$ **9.** $\cos \frac{\pi}{4}$ $\frac{\sqrt{2}}{2}$ **10.** $\sin \pi$ 0

Evaluate the expression without using a calculator. Give your answer in both radians and degrees. (Review Example 1, p. 793)

11. $\sin^{-1} \frac{\sqrt{2}}{2}$ **12.** $\cos^{-1} \frac{\sqrt{3}}{2}$ **13.** $\cos^{-1} 0$ **14.** $\tan^{-1}\left(-\sqrt{3}\right)$
$\frac{\pi}{4}$, 45° $\frac{\pi}{6}$, 30° $\frac{\pi}{2}$, 90° $-\frac{\pi}{3}$, $-60°$

STUDY STRATEGY

Here's a study strategy!

Multiple Methods

There is often more than one way to do an exercise. Multiple methods can help in three situations.
(1) If you get stuck using one method, try another.
(2) Do an exercise more than one way to reinforce your understanding. (3) Check your work by using a different method.

14.1

Graphing Sine, Cosine, and Tangent Functions

What you should learn

GOAL 1 Graph sine and cosine functions, as applied in **Example 3**.

GOAL 2 Graph tangent functions.

Why you should learn it

▼ To model repeating **real-life** patterns, such as the vibrations of a tuning fork in **Ex. 52**.

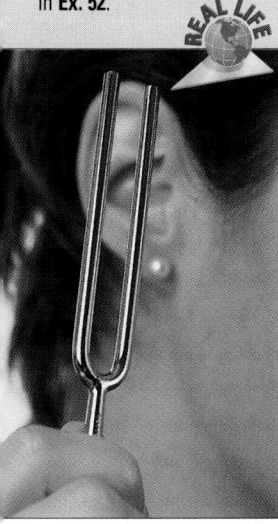

GOAL 1 **GRAPHING SINE AND COSINE FUNCTIONS**

In this lesson you will learn to graph functions of the form $y = a \sin bx$ and $y = a \cos bx$ where a and b are positive constants and x is in radian measure. The graphs of all sine and cosine functions are related to the graphs of

$$y = \sin x \qquad \text{and} \qquad y = \cos x$$

which are shown below.

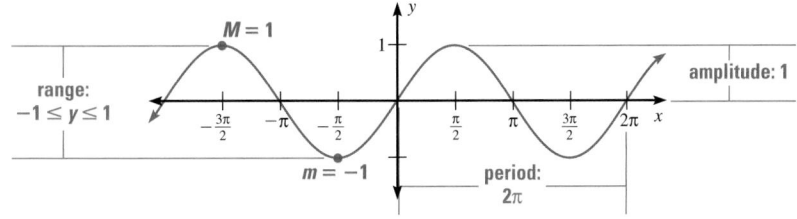

Graph of y = sin x

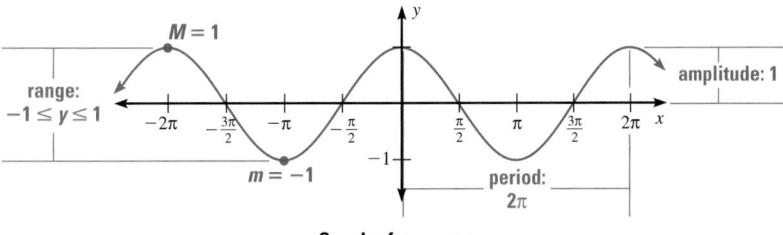

Graph of y = cos x

The functions $y = \sin x$ and $y = \cos x$ have the following characteristics.

1. The *domain* of each function is all real numbers.

2. The *range* of each function is $-1 \le y \le 1$.

3. Each function is **periodic,** which means that its graph has a repeating pattern that continues indefinitely. The shortest repeating portion is called a **cycle.** The horizontal length of each cycle is called the **period.** Each graph shown above has a period of 2π.

4. The maximum value of $y = \sin x$ is $M = 1$ and occurs when $x = \frac{\pi}{2} + 2n\pi$ where n is any integer. The maximum value of $y = \cos x$ is also $M = 1$ and occurs when $x = 2n\pi$ where n is any integer.

5. The minimum value of $y = \sin x$ is $m = -1$ and occurs when $x = \frac{3\pi}{2} + 2n\pi$ where n is any integer. The minimum value of $y = \cos x$ is also $m = -1$ and occurs when $x = (2n + 1)\pi$ where n is any integer.

6. The **amplitude** of each function's graph is $\frac{1}{2}(M - m) = 1$.

1 PLAN

PACING
Basic: 2 days
Average: 2 days
Advanced: 2 days
Block Schedule: 1 block

LESSON OPENER
VISUAL APPROACH
An alternative way to approach Lesson 14.1 is to use the Visual Approach Lesson Opener:
• Blackline Master (*Chapter 14 Resource Book,* p. 12)
•  Transparency (p. 91)

MEETING INDIVIDUAL NEEDS
• *Chapter 14 Resource Book*
  Prerequisite Skills Review (p. 5)
  Practice Level A (p. 14)
  Practice Level B (p. 15)
  Practice Level C (p. 16)
  Reteaching with Practice (p. 17)
  Absent Student Catch-Up (p. 19)
  Challenge (p. 21)
• *Resources in Spanish*
• 🖥 *Personal Student Tutor*

NEW-TEACHER SUPPORT
See the Tips for New Teachers on pp. 1–2 of the *Chapter 14 Resource Book* for additional notes about Lesson 14.1.

WARM-UP EXERCISES

🗒 *Transparency Available*

Evaluate each expression.
1. $\sin \frac{\pi}{2}$ **1**
2. $\cos \pi$ **−1**
3. $\cos \left(-\frac{3\pi}{2}\right)$ **0**
4. $\tan 2\pi$ **0**

The amplitude and period of the graphs of $y = a \sin bx$ and $y = a \cos bx$, where a and b are nonzero real numbers, are as follows:

$$\text{amplitude} = |a| \quad \text{and} \quad \text{period} = \frac{2\pi}{|b|}$$

Examples The graph of $y = 2 \sin 4x$ has amplitude 2 and period $\frac{2\pi}{4} = \frac{\pi}{2}$.

The graph of $y = \frac{1}{3} \cos 2\pi x$ has amplitude $\frac{1}{3}$ and period $\frac{2\pi}{2\pi} = 1$.

MOTIVATING THE LESSON
Ask students what happens to the temperature during a typical day starting at midnight. Work as a class to make a rough sketch of temperature vs. time on the chalkboard with time on the x-axis. Then extend the graph for another day. Explain that in today's lesson, they will learn about functions that can be used to model real-life behavior such as temperature.

For $a > 0$ and $b > 0$, the graphs of $y = a \sin bx$ and $y = a \cos bx$ each have five key x-values on the interval $0 \le x \le \frac{2\pi}{b}$: the x-values at which the **maximum** and **minimum** values occur and the **x-intercepts**.

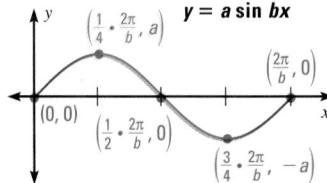
$y = a \sin bx$

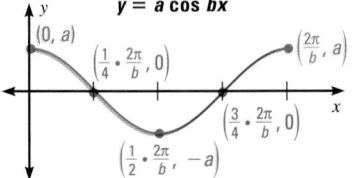
$y = a \cos bx$

EXTRA EXAMPLE 1
Graph the function.

a. $y = \frac{1}{2} \cos x$

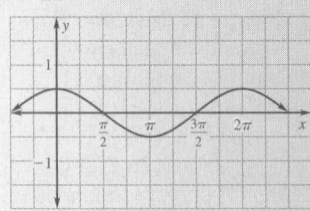

b. $y = \sin \frac{1}{2} x$

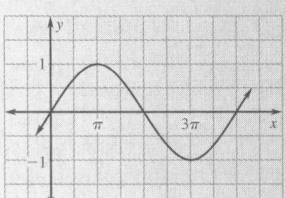

EXAMPLE 1 Graphing Sine and Cosine Functions

Graph the function.

a. $y = 2 \sin x$ **b.** $y = \cos 2x$

SOLUTION

a. The amplitude is $a = 2$ and the period is $\frac{2\pi}{b} = \frac{2\pi}{1} = 2\pi$. The five key points are:

Intercepts: $(0, 0)$; $(2\pi, 0)$;

$$\left(\frac{1}{2} \cdot 2\pi, 0\right) = (\pi, 0)$$

Maximum: $\left(\frac{1}{4} \cdot 2\pi, 2\right) = \left(\frac{\pi}{2}, 2\right)$

Minimum: $\left(\frac{3}{4} \cdot 2\pi, -2\right) = \left(\frac{3\pi}{2}, -2\right)$

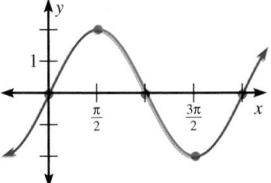

CHECKPOINT EXERCISES
For use after Example 1:
Give the amplitude, period, and five key points of the graph of each function.

1. $y = \sin \pi x$ 1; 2; $(0, 0)$, $\left(\frac{1}{2}, 1\right)$, $(1, 0)$, $\left(\frac{3}{2}, -1\right)$, $(2, 0)$

2. $y = 3 \cos x$ 3; 2π; $(0, 3)$, $\left(\frac{\pi}{2}, 0\right)$, $(\pi, -3)$, $\left(\frac{3\pi}{2}, 0\right)$, $(2\pi, 3)$

> **STUDENT HELP**
>
> **Study Tip**
> In Example 1 notice how changes in a and b affect the graphs of $y = a \sin bx$ and $y = a \cos bx$. When the value of a increases, the amplitude is greater. When the value of b increases, the period is shorter.

b. The amplitude is $a = 1$ and the period is $\frac{2\pi}{b} = \frac{2\pi}{2} = \pi$. The five key points are:

Intercepts: $\left(\frac{1}{4} \cdot \pi, 0\right) = \left(\frac{\pi}{4}, 0\right)$;

$$\left(\frac{3}{4} \cdot \pi, 0\right) = \left(\frac{3\pi}{4}, 0\right)$$

Maximums: $(0, 1)$; $(\pi, 1)$

Minimum: $\left(\frac{1}{2} \cdot \pi, -1\right) = \left(\frac{\pi}{2}, -1\right)$

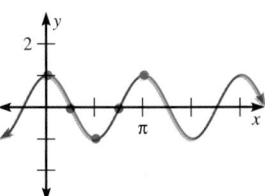

EXAMPLE 2 *Graphing a Cosine Function*

Graph $y = \frac{1}{3} \cos \pi x$.

SOLUTION

The amplitude is $a = \frac{1}{3}$ and the period is $\frac{2\pi}{b} = \frac{2\pi}{\pi} = 2$. The five key points are:

Intercepts: $\left(\frac{1}{4} \cdot 2, 0\right) = \left(\frac{1}{2}, 0\right)$;

$\left(\frac{3}{4} \cdot 2, 0\right) = \left(\frac{3}{2}, 0\right)$

Maximums: $\left(0, \frac{1}{3}\right)$; $\left(2, \frac{1}{3}\right)$

Minimum: $\left(\frac{1}{2} \cdot 2, -\frac{1}{3}\right) = \left(1, -\frac{1}{3}\right)$

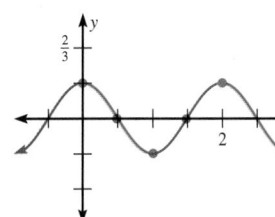

· · · · · · · · · ·

The periodic nature of trigonometric functions is useful for modeling *oscillating* motions or repeating patterns that occur in real life. Some examples are sound waves, the motion of a pendulum or a spring, and seasons of the year. In such applications, the reciprocal of the period is called the **frequency**. The frequency gives the number of cycles per unit of time.

EXAMPLE 3 *Modeling with a Sine Function*

MUSIC When you strike a tuning fork, the vibrations cause changes in the pressure of the surrounding air. A middle-A tuning fork vibrates with frequency $f = 440$ hertz (cycles per second). You strike a middle-A tuning fork with a force that produces a maximum pressure of 5 pascals.

  a. Write a sine model that gives the pressure P as a function of time t (in seconds).

  b. Graph the model.

FOCUS ON APPLICATIONS

OSCILLOSCOPE
The oscilloscope is a laboratory device invented by Karl Braun in 1897. This electrical instrument measures waveforms and is the forerunner of today's television.

SOLUTION

  a. In the model $P = a \sin bt$, the maximum pressure P is 5, so $a = 5$. You can use the frequency to find the value of b.

$$\text{frequency} = \frac{1}{\text{period}} \quad\Longrightarrow\quad 440 = \frac{b}{2\pi}$$

$$880\pi = b$$

  ▶ The pressure as a function of time is given by $P = 5 \sin 880\pi t$.

  b. The amplitude is $a = 5$ and the period is $\frac{1}{f} = \frac{1}{440}$. The five key points are:

Intercepts: $(0, 0)$; $\left(\frac{1}{440}, 0\right)$;

$\left(\frac{1}{2} \cdot \frac{1}{440}, 0\right) = \left(\frac{1}{880}, 0\right)$

Maximum: $\left(\frac{1}{4} \cdot \frac{1}{440}, 5\right) = \left(\frac{1}{1760}, 5\right)$

Minimum: $\left(\frac{3}{4} \cdot \frac{1}{440}, -5\right) = \left(\frac{3}{1760}, -5\right)$

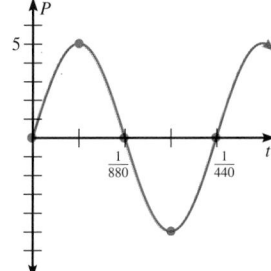

14.1 *Graphing Sine, Cosine, and Tangent Functions*

EXTRA EXAMPLE 2

Graph $y = 2 \sin \frac{x}{4}$.

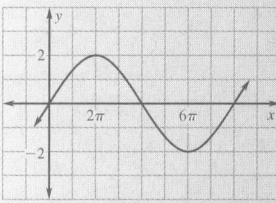

EXTRA EXAMPLE 3

A tuning fork vibrates with frequency $f = 880$ hertz (cycles per second.) You strike the tuning fork with a force that produces a maximum pressure of 4 pascals.
a. Write a sine model that gives the pressure P as a function of time t (in seconds).
  $P = 4 \sin 1760\pi t$
b. Graph the model.

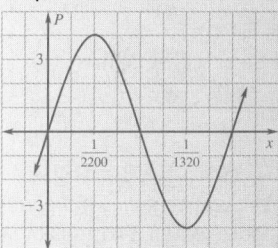

✓ CHECKPOINT EXERCISES

For use after Example 2:

1. Give the amplitude, period, and five key points of the graph of $y = 4 \cos \frac{\pi}{2}x$. 4; 4;
  (0, 4), (1, 0), (2, −4), (3, 0), (4, 4)

For use after Example 3:

2. You pluck the string on a violin so that it vibrates with frequency $f = 660$ hertz (cycles per second.) The force of the pluck produces a maximum pressure of 2 pascals. Write a sine model that gives the pressure P as a function of time t (in seconds). Then give the amplitude and period of the function's graph.
  $P = 2 \sin 1320\pi t$; 2; $\frac{1}{660}$

EXTRA EXAMPLE 4

Graph $y = 3 \tan \frac{\pi}{2}x$.

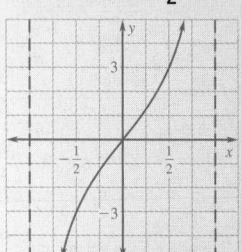

 CHECKPOINT EXERCISES

For use after Example 4:

1. Graph $y = \frac{1}{2} \tan \frac{1}{3}x$.

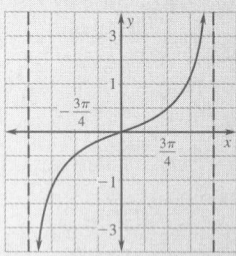

ENGLISH LEARNERS

English learners may not always recognize when a word is being used in its mathematical sense rather than in its ordinary sense. You might inform students that in the term *odd multiples,* the word *odd* is used in its mathematical sense: the numbers 1, 3, 5, and so on.

FOCUS ON VOCABULARY

What is the relationship between period and frequency? How is frequency related to cycle? Frequency is the reciprocal of the period of a periodic function. Frequency gives the number of cycles of the function per unit of time.

CLOSURE QUESTION

How do you find the amplitude and period of a sine, cosine, or tangent function from its equation? See margin.

DAILY PUZZLER

What is the equation of a function with twice the frequency as $y = 3 \cos 4x$? $y = 3 \cos 8x$

834

**GOAL 2** **GRAPHING TANGENT FUNCTIONS**

The graph of $y = \tan x$ has the following characteristics.

1. The domain is all real numbers except odd multiples of $\frac{\pi}{2}$. At odd multiples of $\frac{\pi}{2}$, the graph has vertical asymptotes.

2. The range is all real numbers.

3. The graph has a period of π.

CHARACTERISTICS OF Y = A TAN BX

If *a* and *b* are nonzero real numbers, the graph of $y = a \tan bx$ has these characteristics:

- The period is $\frac{\pi}{|b|}$.

- There are vertical asymptotes at odd multiples of $\frac{\pi}{2|b|}$.

Example The graph of $y = 5 \tan 3x$ has period $\frac{\pi}{3}$ and asymptotes at

$$x = (2n + 1)\frac{\pi}{2(3)} = \frac{\pi}{6} + \frac{n\pi}{3}$$ where *n* is any integer.

The graph at the right shows five key *x*-values that can help you sketch the graph of $y = a \tan bx$ for $a > 0$ and $b > 0$. These are the **x-intercept**, the *x*-values where the **asymptotes** occur, and the *x*-values **halfway between** the *x*-intercept and the asymptotes. At each halfway point, the function's value is either a or $-a$.

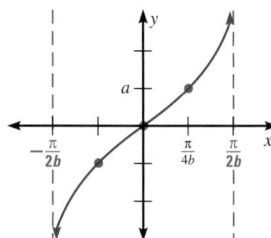

EXAMPLE 4 *Graphing a Tangent Function*

Graph $y = \frac{3}{2} \tan 4x$.

SOLUTION

The period is $\frac{\pi}{b} = \frac{\pi}{4}$.

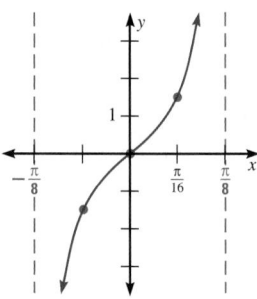

Intercept: $(0, 0)$

Asymptotes: $x = \frac{1}{2} \cdot \frac{\pi}{4}$, or $x = \frac{\pi}{8}$;

$x = -\frac{1}{2} \cdot \frac{\pi}{4}$, or $x = -\frac{\pi}{8}$

Halfway points: $\left(\frac{1}{4} \cdot \frac{\pi}{4}, \frac{3}{2}\right) = \left(\frac{\pi}{16}, \frac{3}{2}\right)$;

$\left(-\frac{1}{4} \cdot \frac{\pi}{4}, -\frac{3}{2}\right) = \left(-\frac{\pi}{16}, -\frac{3}{2}\right)$

834 **Chapter 14** *Trigonometric Graphs, Identities, and Equations*

Closure Question *Sample answer:*
Use the form $y = a \sin bx$, $y = a \cos bx$, or $y = a \tan bx$.
In all three cases, the amplitude is $|a|$. The period for the sine or cosine function is $\frac{2\pi}{|b|}$ and the period for the tangent function is $\frac{\pi}{|b|}$.

GUIDED PRACTICE

Vocabulary Check ✓

1. Define the terms cycle and period. cycle: the shortest repeating portion of the graph of a periodic function; period: the horizontal length of a cycle

Concept Check ✓

2. What are the domain and range of $y = a \sin bx$, $y = a \cos bx$, and $y = a \tan bx$? See margin.

3. Consider the two functions $y = 4 \sin \frac{x}{3}$ and $y = \frac{1}{3} \sin 4x$. Which function has the greater amplitude? Which function has the longer period? $y = 4 \sin \frac{x}{3}$, $y = 4 \sin \frac{x}{3}$

Skill Check ✓

Find the amplitude and period of the function.

amplitude: 6, period: 2π
4. $y = 6 \sin x$

amplitude: 3, period: 2
5. $y = 3 \cos \pi x$

6. $y = \frac{1}{4} \cos 3x$ See margin.

7. $y = \frac{2}{3} \sin \frac{\pi}{3}x$
amplitude: $\frac{2}{3}$, period: 6

8. $y = 5 \sin 3\pi x$
amplitude: 5, period: $\frac{2}{3}$

9. $y = \cos \frac{x}{2}$
amplitude: 1, period: 4π

Graph the function. 10–15. See margin.

10. $y = 3 \sin x$

11. $y = \cos 4x$

12. $y = \tan 3x$

13. $y = \frac{1}{4} \sin \pi x$

14. $y = 5 \cos \frac{2}{3}x$

15. $y = 2 \tan 4x$

16. 🌐 **PENDULUMS** The motion of a certain pendulum can be modeled by the function

$$d = 4 \cos 8\pi t$$

where d is the pendulum's horizontal displacement (in inches) relative to its position at rest and t is the time (in seconds). Graph the function. How far horizontally does the pendulum travel from its original position? **4 in.**
See margin for graph.

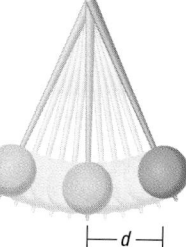

Sidebar (left margin):

2. $y = a \sin bx$, Domain: reals, Range: $-a \le y \le a$

$y = a \cos bx$, Domain: reals, Range: $-a \le y \le a$

$y = a \tan bx$, Domain: all reals except odd multiples of $\frac{\pi}{2b}$, Range: reals

6. amplitude: $\frac{1}{4}$, period: $\frac{2\pi}{3}$

PRACTICE AND APPLICATIONS

STUDENT HELP

▶ **Extra Practice**
to help you master skills is on p. 959.

MATCHING GRAPHS Match the function with its graph.

17. $y = 2 \sin \frac{1}{2}x$ **B**

18. $y = 2 \cos \frac{1}{2}x$ **E**

19. $y = 2 \sin 2x$ **D**

20. $y = 2 \tan \frac{1}{2}x$ **F**

21. $y = 2 \cos 2x$ **A**

22. $y = 2 \tan 2x$ **C**

A.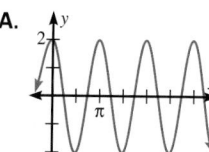

B.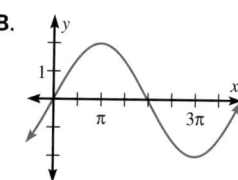

C.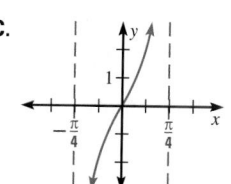

D.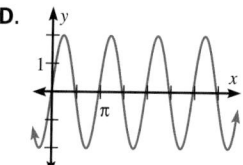

E.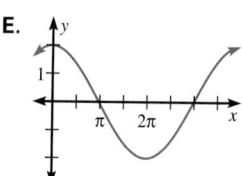

F.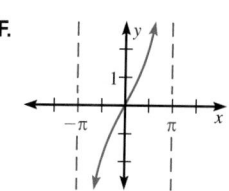

STUDENT HELP

▶ **HOMEWORK HELP**
Examples 1, 2: Exs. 17–49
Example 3: Exs. 51–55
Example 4: Exs. 17–22, 32–43

14.1 *Graphing Sine, Cosine, and Tangent Functions*   835

---

3 APPLY

⬤ ASSIGNMENT GUIDE

BASIC
Day 1: pp. 835–837 Exs. 17–19, 23–29, 44–46, 51, 61–69 odd
Day 2: pp. 835–837 Exs. 20–22, 32–39, 52–53, 56–57, 71–79 odd

AVERAGE
Day 1: pp. 835–837 Exs. 17–19, 23–29, 44–48, 51, 61–69 odd
Day 2: pp. 835–837 Exs. 20–22, 32–41, 52–57, 71–79 odd

ADVANCED
Day 1: pp. 835–837 Exs. 17–19, 23–31, 44–51, 61–69 odd
Day 2: pp. 835–837 Exs. 20–22, 32–43, 52–60, 71–79 odd

BLOCK SCHEDULE
pp. 835–837 Exs. 17–22, 23–29, 32–41, 44–48, 51–57, 61–79 odd

EXERCISE LEVELS
Level A: *Easier*
17–25, 51

Level B: *More Difficult*
26–50, 52–57

Level C: *Most Difficult*
58–60

✓ HOMEWORK CHECK
To quickly check student understanding of key concepts, go over the following exercises:
Exs. 18, 22, 24, 26, 32, 44. See also the Daily Homework Quiz:

• Blackline Master (*Chapter 14 Resource Book*, p. 24)

• 🖱 Transparency (p. 106)

10–16. See Additional Answers beginning on page AA1.

MATHEMATICAL REASONING
EXERCISES 32–43 The following reasoning may help students avoid errors when graphing periodic functions. If the coefficient of x is greater than 1, the graph changes more rapidly then the graph with coefficient of 1, thus it completes a cycle sooner and the period is *smaller*. If the coefficient of x is less than 1, the graph changes more slowly than the graph with a coefficient of 1, thus it completes a cycle later and the period is *greater*. Suggest students test points in the function to check their graphs.

STUDENT HELP NOTES

Homework Help Students can find help for Exs. 44–49 at **www.mcdougallittell.com**. The information can be printed out for students who don't have access to the Internet.

ENGLISH LEARNERS
EXERCISE 53 The ability to translate word problems into equations is an essential skill in algebra, but it is often a difficult task for English learners, particularly when problems include specialized vocabulary and difficult concepts. It may be helpful to students to have you talk them through Exercise 53 before they attempt to solve it independently. If possible, use a real spring to demonstrate the concepts of vertical displacement and elasticity.

ADDITIONAL PRACTICE AND RETEACHING

For Lesson 14.1:
- Practice Levels A, B, and C (*Chapter 14 Resource Book*, p. 14)
- Reteaching with Practice (*Chapter 14 Resource Book*, p. 17)
- See Lesson 14.1 of the *Personal Student Tutor*

For more Mixed Review:
- Search the *Test and Practice Generator* for key words or specific lessons.

836

STUDENT HELP

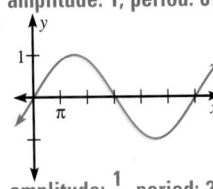

HOMEWORK HELP
Visit our Web site www.mcdougallittell.com for help with problem solving in Exs. 44–49.

50. Periodic motion is modeled by

$$y = a \sin \frac{2\pi t}{P} \text{ or}$$

$$y = a \cos \frac{2\pi t}{P}.$$

Since $f = \frac{1}{P}$, the motion is modeled by

$$y = a \sin 2\pi f t \text{ or}$$
$$y = a \cos 2\pi f t.$$

FOCUS ON CAREERS

MUSICIAN
Musicians may specialize in classical, rock, jazz, or many other types of music. This profession is very competitive and demands a high degree of discipline and talent in order to succeed.

CAREER LINK
www.mcdougallittell.com

ANALYZING FUNCTIONS In Exercises 23–31, find the amplitude and period of the graph of the function.

23. amplitude: 1, period: 6π

24. amplitude: 0.5, period: 2

25. amplitude: 4, period: π

amplitude: $\frac{1}{2}$, period: 2

26. $y = \frac{1}{2} \cos \pi x$

amplitude: 1, period: π

27. $y = \sin 2x$

amplitude: 3, period: 8π

28. $y = 3 \cos \frac{1}{4}x$

29. $y = 5 \cos \frac{1}{2}x$
amplitude: 5, period: 4π

30. $y = 2 \sin \frac{1}{2}\pi x$
amplitude: 2, period: 4

31. $y = \frac{1}{3} \sin 4\pi x$
amplitude: $\frac{1}{3}$, period: $\frac{1}{2}$

GRAPHING Draw one cycle of the function's graph.
32–43. See margin.

32. $y = \sin \frac{1}{4}x$

33. $y = \cos \frac{1}{5}x$

34. $y = \frac{1}{4} \tan \pi x$

35. $y = \frac{1}{4} \sin x$

36. $y = 4 \cos x$

37. $y = 4 \tan 2x$

38. $y = 3 \cos 2x$

39. $y = 8 \sin x$

40. $y = 2 \tan \frac{1}{3}x$

41. $y = \frac{1}{2} \sin \frac{1}{4}\pi x$

42. $y = \tan 4\pi x$

43. $y = 2 \cos 6\pi x$

WRITING EQUATIONS Write an equation of the form $y = a \sin bx$, where $a > 0$ and $b > 0$, so that the graph has the given amplitude and period.

44. Amplitude: 1
Period: 5 $y = \sin \frac{2\pi x}{5}$

45. Amplitude: 10
Period: 4 $y = 10 \sin \frac{\pi x}{2}$

46. Amplitude: 2
Period: 2π $y = 2 \sin x$

47. Amplitude: $\frac{1}{2}$
Period: 3π $y = \frac{1}{2} \sin \frac{2x}{3}$

48. Amplitude: 4
Period: $\frac{\pi}{6}$ $y = 4 \sin 12x$

49. Amplitude: 3
Period: $\frac{1}{2}$ $y = 3 \sin 4\pi x$

50. **LOGICAL REASONING** Use the fact that the frequency of a periodic function's graph is the reciprocal of the period to show that an oscillating motion with maximum displacement a and frequency f can be modeled by $y = a \sin 2\pi f t$ or $y = a \cos 2\pi f t$. **See margin.**

51. **BOATING** The displacement d (in feet) of a boat's water line above sea level as it moves over waves can be modeled by the function

$$d = 2 \sin 2\pi t$$

where t is the time (in seconds). Graph the height of the boat over a three second time interval. **See margin.**

52. **MUSIC** A tuning fork vibrates with a frequency of 220 hertz (cycles per second). You strike the tuning fork with a force that produces a maximum pressure of 3 pascals. Write a sine model that gives the pressure P as a function of the time t (in seconds). What is the period of the sound wave?
$P = 3 \sin 440\pi t, \frac{1}{220}$ sec

53. **SPRING MOTION** The motion of a simple spring can be modeled by $y = A \cos kt$ where y is the spring's vertical displacement (in feet) relative to its position at rest, A is the initial displacement (in feet), k is a constant that measures the elasticity of the spring, and t is the time (in seconds). Find the amplitude and period of a spring for which $A = 0.5$ foot and $k = 6$. amplitude: $\frac{1}{2}$ ft, period: $\frac{\pi}{3}$ sec

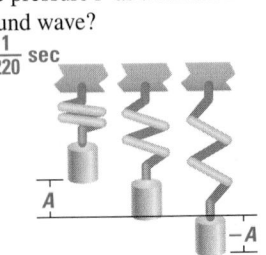

32–43, 51. See Additional Answers beginning on page AA1.

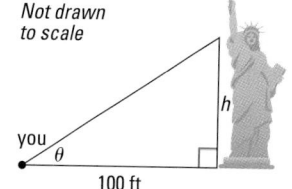

SIGHTSEEING In Exercises 54 and 55, use the following information.
Suppose you are standing 100 feet away from the base of the Statue of Liberty with a video camera. As you videotape the statue, you pan up the side of the statue at 5° per second.

Not drawn to scale

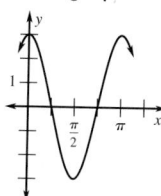

54. Write and graph an equation that gives the height h of the part of the statue seen through the video camera as a function of the time t. $h = 100 \tan \frac{\pi t}{36}$; See margin for graph.

55. Find the change in height from $t = 0$ to $t = 1$, from $t = 1$ to $t = 2$, and from $t = 2$ to $t = 3$. Briefly explain what happens to h as t increases.

55. 8.7 ft, 8.9 ft, 9.2 ft; $h_t - h_{t-1}$ increases as t increases.

56. MULTIPLE CHOICE Which function represents the graph shown? **B**

 Test Preparation

Ⓐ $y = \frac{1}{2} \tan 5x$ Ⓑ $y = 5 \tan \frac{1}{2}x$

Ⓒ $y = \tan 5x$ Ⓓ $y = 5 \tan 2x$

Ⓔ $y = 5 \tan \frac{1}{8}x$

58. Domain: all reals except multiples of π, Range: $y \geq 1$ or $y \leq -1$, period: 2π

59. Domain: all reals except odd multiples of $\frac{\pi}{2}$, Range: $y \geq 1$ or $y \leq -1$, period: 2π

57. MULTIPLE CHOICE Which of the following is an x-intercept of the graph of $y = \frac{1}{3} \sin \frac{\pi}{4}x$? **A**

Ⓐ 4   Ⓑ 2   Ⓒ -6 Ⓓ 1   Ⓔ 4π

★ **Challenge**

60. Domain: all reals except multiples of π, Range: all reals, period: π

SKETCHING GRAPHS Sketch the graph of the function by plotting points. Then state the function's domain, range, and period. 58–60. See margin for graphs.

58. $y = \csc x$ **59.** $y = \sec x$ **60.** $y = \cot x$

MIXED REVIEW

GRAPHING Graph the quadratic function. Label the vertex and axis of symmetry. (Review 5.1 for 14.2)   61–66. See margin.

61. $y = 2(x - 5)^2 + 4$ **62.** $y = -(x - 3)^2 - 7$ **63.** $y = 4(x + 2)^2 - 1$

64. $y = -3(x + 1)^2 + 6$ **65.** $y = \frac{3}{4}(x - 1)^2 - 2$ **66.** $y = 10(x + 4)^2 + 3$

CALCULATING PROBABILITY Find the probability of drawing the given numbers if the integers 1 through 30 are placed in a hat and drawn randomly without replacement. (Review 12.5) $\frac{15}{58}$

67. an even number, then an odd number   **68.** the number 30, then an odd number $\frac{1}{58}$

69. a multiple of 4, then an odd number $\frac{7}{58}$ **70.** the number 19, then the number 20 $\frac{1}{870}$

FINDING REFERENCE ANGLES Sketch the angle. Then find its reference angle. (Review 13.3)   71–78. See margin for graphs.

71. $220°$ $40°$ **72.** $-155°$ $25°$ **73.** $280°$ $80°$ **74.** $-510°$ $30°$

75. $\frac{35\pi}{3}$ $\frac{\pi}{3}$ **76.** $\frac{21\pi}{4}$ $\frac{\pi}{4}$ **77.** $-\frac{17\pi}{6}$ $\frac{\pi}{6}$ **78.** $-\frac{5\pi}{8}$ $\frac{3\pi}{8}$

79. 🌐 **PERSONAL FINANCE** You deposit $1000 in an account that pays 1.5% annual interest compounded continuously. How long will it take for the balance to double? (Review 8.6) **46.2 years**

14.1 *Graphing Sine, Cosine, and Tangent Functions*   **837**

Additional Test Preparation *Sample answer:*
The amplitude of the graph of $y = a \cos bx$ is larger than that of $y = \cos x$, which means that the graph has higher maximum points and lower minimum points. The period of the graph of $y = a \cos bx$ is shorter than that of $y = \cos x$, thus the graph completes a cycle before the graph of $y = \cos x$ does.

4 ASSESS

DAILY HOMEWORK QUIZ

📖 *Transparency Available*

1. Find the amplitude and period of the graph of the function.

amplitude: 3, period: π

2. Find the amplitude and period of the graph of the function $y = 5 \cos \frac{1}{4}x$.

amplitude: 5, period 8π

3. Draw one cycle of the graph of the function $y = \sin \frac{1}{2}x$.

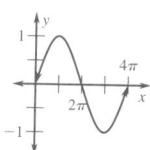

4. Write an equation of the form $y = a \sin bx$, where $a > 0$ and $b > 0$, so that the graph has an amplitude of 2 and a period of $\frac{1}{2}$. $y = 2 \sin 4\pi x$

EXTRA CHALLENGE NOTE
→ Challenge problems for Lesson 14.1 are available in **blackline** format in the *Chapter 14 Resource Book*, p. 21 and at **www.mcdougallittell.com**.

ADDITIONAL TEST PREPARATION

1. WRITING A function of the form $y = a \cos bx$ has $a > 1$ and $b > 1$. How does the graph of this function compare to the graph of $y = \cos x$? Use the terms amplitude and period in your discussion. See margin.

54, 58–66, 71–78.
See Additional Answers beginning on page AA1.

1 Planning the Activity

PURPOSE
To see the effects of changing a and b on the graphs of $y = a \sin bx$ and $y = a \cos bx$.

MATERIALS
- graphing calculator
- Keystroke blackline
  (*Chapter 14 Resource Book*, p. 13)

PACING
- Activity — 25 min

▶ LINK TO LESSON
Have students use the graphing calculator to check the graphs for Example 1 on p. 832.

2 Managing the Activity

COOPERATIVE LEARNING
Students can work on this activity with a partner and a single graphing calculator. Have them take turns entering the functions and adjusting the viewing window.

ALTERNATIVE APPROACH
This activity can be done as a class using a graphing calculator or computer with graphing software and an overhead display.

3 Closing the Activity

★ KEY DISCOVERY
As $|a|$ increases, the height of the graph of $y = a \sin x$ or $y = a \cos x$ increases. As $|b|$ increases, the number of cycles of the graph of $y = a \sin x$ or $y = a \cos x$ decreases.

ACTIVITY ASSESSMENT
Without graphing, describe the differences in the graphs of $y = \sin bx$ for $b = \frac{1}{4}, \frac{1}{3}, \frac{1}{2}$.
See sample answer at right.

▶ ACTIVITY 14.1
Using Technology

Graphing Trigonometric Functions

Using the *List* feature and the trigonometric viewing window of a graphing calculator, you can graph the trigonometric functions $y = a \sin bx$, $y = a \cos bx$, and $y = a \tan bx$ and observe the effects of changing the values of a and b.

▶ EXAMPLE

Graph $y = a \sin x$ for $a = 1, 2,$ and 4.

▶ SOLUTION

1 Use the graphing calculator's *List* feature to enter the function as $y_1 = \{1, 2, 4\} \sin x$. This represents the three functions $y = \sin x$, $y = 2 \sin x$, and $y = 4 \sin x$.

```
Y1={1,2,4}sin(X)
Y2=
Y3=
Y4=
Y5=
Y6=
Y7=
```

2 Select the trigonometric viewing window. In this window, each tick mark on the x-axis represents $\frac{\pi}{2}$ and each tick mark on the y-axis represents 1.

```
ZOOM MEMORY
1:ZBox
2:Zoom In
3:Zoom Out
4:ZDecimal
5:ZSquare
6:ZStandard
7:ZTrig
```

3 If the functions you graph have amplitudes much greater than 1 or periods longer than 4π, you may need to change the parameters of the viewing window. Then you can use the *Maximum*, *Minimum*, and *Zero* features to find the amplitude and the period of each function.

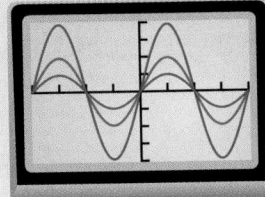

▶ EXERCISES

Use a graphing calculator's *List* feature to graph the functions for the given values of *a* in the same trigonometric viewing window. Find the amplitude and the period of each function's graph.

1. $y = a \sin x$ for $a = \frac{1}{3}, 3, 9$

2. $y = a \sin x$ for $a = 10, 20, 40$

3. $y = a \cos x$ for $a = \frac{1}{2}, 1, 2$

4. $y = a \cos x$ for $a = 1, 2, 4$

Use a graphing calculator's *List* feature to graph the functions for the given values of *b* in the same trigonometric viewing window. Find the amplitude and the period of each function's graph.

5. $y = \sin bx$ for $b = 1, 2, 4$

6. $y = \sin bx$ for $b = \frac{1}{3}, 1, 3$

7. $y = \cos bx$ for $b = \frac{1}{2}, 1, 2$

8. $y = \cos bx$ for $b = \frac{1}{3}, 1, 3$

STUDENT HELP

INTERNET **KEYSTROKE HELP**

See keystrokes for several models of calculators at www.mcdougallittell.com

1. amplitude: $\frac{1}{3}$, 3, 9; period: 2π

2. amplitude: 10, 20, 40; period: 2π

3. amplitude: $\frac{1}{2}$, 1, 2; period: 2π

4. amplitude: 1, 2, 4; period: 2π

5. amplitude: 1; period: $2\pi, \pi, \frac{\pi}{2}$

6. amplitude: 1; period: $6\pi, 2\pi, \frac{2\pi}{3}$

7. amplitude: 1; period: $4\pi, 2\pi, \pi$

8. amplitude: 1; period: $6\pi, 2\pi, \frac{2\pi}{3}$

Activity Assessment *Sample answer:*
The graphs will have different periods. For any portion of the x-axis, the graph of $y = \sin \frac{1}{2}x$ will complete the most cycles followed by the graphs with $a = \frac{1}{3}$ and then $a = \frac{1}{4}$. The maximum and minimum values of the functions are the same though they will occur at different x-values.

Translating and Reflecting Trigonometric Graphs

GROUP ACTIVITY
Work with a partner.

MATERIALS
graphing calculator

Step 1a. Second is first reflected in the x-axis.

Step 1b. Second is first reflected in the x-axis.

Step 1c. Second is first reflected in the x-axis.

Step 2a. Second is first shifted right $\frac{\pi}{2}$.

Step 2b. Second is first shifted left $\frac{\pi}{2}$.

Step 2c. Second is first shifted left π.

Step 3a. Second is first shifted up 1.

Step 3b. Second is first shifted down 1.

Step 3c. Second is first shifted down 2.

3a. shifted right 1 unit , shifted up 1 unit

3b. reflected in x-axis, shifted right 1 unit, shifted down 1 unit

3c. reflected in x-axis, shifted left 1 unit, shifted up 1 unit

3d. shifted left 1 unit, shifted down 1 unit

▶ **QUESTION** How can you graph reflections and horizontal and vertical translations of graphs of trigonometric functions?

▶ **EXPLORING THE CONCEPT**

① Use a graphing calculator to graph each pair of functions in the same viewing window. How are the graphs geometrically related?

a. $y = \cos x$

$y = -\cos x$

b. $y = 2 \sin x$

$y = -2 \sin x$

c. $y = \frac{3}{5} \sin 3x$

$y = -\frac{3}{5} \sin 3x$

② Use a graphing calculator to graph each pair of functions in the same viewing window. How are the graphs geometrically related?

a. $y = \cos x$

$y = \cos \left(x - \frac{\pi}{2}\right)$

b. $y = 2 \sin x$

$y = 2 \sin \left(x + \frac{\pi}{2}\right)$

c. $y = \sin 3x$

$y = \sin 3(x + \pi)$

③ Use a graphing calculator to graph each pair of functions in the same viewing window. How are the graphs geometrically related?

a. $y = \cos x$

$y = \cos x + 1$

b. $y = 2 \sin x$

$y = 2 \sin x - 1$

c. $y = \frac{1}{4} \sin 3x$

$y = \frac{1}{4} \sin 3x - 2$

▶ **DRAWING CONCLUSIONS**

1. Predict what the graph of each function looks like by making a sketch. Check your prediction by graphing the function on a graphing calculator.
1a–c. See margin.

a. $y = -3 \cos x$

b. $y = \sin \left(x + \frac{3\pi}{4}\right)$

c. $y = \sin x - 4$

2. Predict what the graph of each function looks like by making a sketch. Check your prediction by graphing the function on a graphing calculator.
2a–d. See margin.

a. $y = -\frac{1}{2} \cos x - 2$

b. $y = -3 \cos (x + \pi)$

c. $y = 2 \sin (x - \pi) + 3$

d. $y = -4 \sin \left(x - \frac{\pi}{4}\right) - 6$

3. **CRITICAL THINKING** Use the following phrases to describe how the graph of each function in parts (a)–(d) is related to the graph of $y = \sin x$.

- shifted up 1 unit
- shifted down 1 unit
- shifted left 1 unit
- shifted right 1 unit
- reflected in a horizontal line

a. $y = \sin (x - 1) + 1$

b. $y = -\sin (x - 1) - 1$

c. $y = -\sin (x + 1) + 1$

d. $y = \sin (x + 1) - 1$

Activity Assessment *Sample answer:*
The graph is reflected over the x-axis and shifted up two units.

---

① Planning the Activity

PURPOSE
To see the connection between a translation or reflection of a trigonometric graph and its equation.

MATERIALS
- graphing calculator per student

PACING
- Exploring the Concept — 10 min
- Drawing Conclusions — 15 min

▶ **LINK TO LESSON**
When discussing the transformations in the box on page 840, ask students for examples from the activity that incorporated a vertical shift, a horizontal shift, and a reflection.

② Managing the Activity

COOPERATIVE LEARNING
Students take turns using the graphing calculator. The students should work together to predict the graphs in the exercises, but each should draw the graphs.

CLASSROOM MANAGEMENT
Initially, have students select the trigonometric viewing window for each of these graphs. The window will then have to be adjusted for some of the graphs.

③ Closing the Activity

★ **KEY DISCOVERY**
Transformations of the graphs of trigonometric functions are seen in their equations in the same way that they are seen in the equations of other basic functions.

ACTIVITY ASSESSMENT
Describe how the graph of $y = -3 \cos x + 2$ is related to the graph of $y = 3 \cos x$.
See sample answer at left.

1a–d and 2 a–d.
See Additional Answers beginning on page AA1.

LESSON OPENER
APPLICATION

An alternative way to approach Lesson 14.2 is to use the Application Lesson Opener:

• Blackline Master (*Chapter 14 Resource Book,* p. 25)
• Transparency (p. 92)

MEETING INDIVIDUAL NEEDS

• *Chapter 14 Resource Book*
  Prerequisite Skills Review (p. 5)
  Practice Level A (p. 26)
  Practice Level B (p. 27)
  Practice Level C (p. 28)
  Reteaching with Practice (p. 29)
  Absent Student Catch-Up (p. 31)
  Challenge (p. 33)
• *Resources in Spanish*
• *Personal Student Tutor*

NEW-TEACHER SUPPORT

See the Tips for New Teachers on pp. 1–2 of the *Chapter 14 Resource Book* for additional notes about Lesson 14.2.

WARM-UP EXERCISES

Transparency Available

Describe how the graph of each function is related to the graph of $y = x^2$.

1. $y = -x^2$
  reflected over the *x*-axis
2. $y = (x + 2)^2$ shifted left 2 units
3. $y = x^2 - 4$ shifted down 4 units
4. $y = -(x - 3)^2 - 1$ reflected over the *x*-axis, shifted right 3 units and down 1 unit.

14.2 Translations and Reflections of Trigonometric Graphs

What you should learn

GOAL 1 Graph translations and reflections of sine and cosine graphs.

GOAL 2 Graph translations and reflections of tangent graphs, as applied in **Ex. 61**.

Why you should learn it

▼ To model **real-life** quantities, such as the height above the ground of a person rappelling down a cliff in **Example 7**.

GOAL 1 GRAPHING SINE AND COSINE FUNCTIONS

In previous chapters you learned that the graph of $y = a \cdot f(x - h) + k$ is related to the graph of $y = |a| \cdot f(x)$ by horizontal and vertical translations and by a reflection when a is negative. This also applies to sine, cosine, and tangent functions.

TRANSFORMATIONS OF SINE AND COSINE GRAPHS

To obtain the graph of

$$y = a \sin b(x - h) + k \qquad \text{or} \qquad y = a \cos b(x - h) + k,$$

transform the graph of $y = |a| \sin bx$ or $y = |a| \cos bx$ as follows.

VERTICAL SHIFT Shift the graph k units vertically.

HORIZONTAL SHIFT Shift the graph h units horizontally.

REFLECTION If $a < 0$, reflect the graph in the line $y = k$ after any vertical and horizontal shifts have been performed.

EXAMPLE 1 *Graphing a Vertical Translation*

Graph $y = -2 + 3 \sin 4x$.

SOLUTION

Because the graph is a transformation of the graph of $y = 3 \sin 4x$, the amplitude is **3** and the period is $\frac{2\pi}{4} = \frac{\pi}{2}$. By comparing the given equation to the general equation $y = a \sin b(x - h) + k$, you can see that $h = 0$, $k = -2$, and $a > 0$. Therefore, translate the graph of $y = 3 \sin 4x$ **down 2 units**.

The graph oscillates **3 units** up and down from its center line $y = -2$. Therefore, the maximum value of the function is $-2 + 3 = 1$ and the minimum value of the function is $-2 - 3 = -5$.

The five key points are:

On $y = k$: $(0, -2); \left(\frac{\pi}{4}, -2\right); \left(\frac{\pi}{2}, -2\right)$

Maximum: $\left(\frac{\pi}{8}, 1\right)$

Minimum: $\left(\frac{3\pi}{8}, -5\right)$

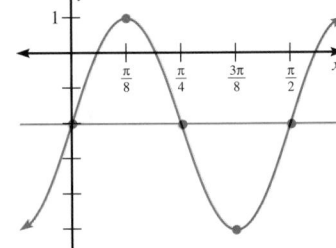

✓ **CHECK** You can check your graph with a graphing calculator. Use the *Maximum, Minimum,* and *Intersect* features to check the key points.

EXAMPLE 2 *Graphing a Horizontal Translation*

Graph $y = 2 \cos \frac{2}{3}\left(x - \frac{\pi}{4}\right)$.

SOLUTION

Because the graph is a transformation of the graph of $y = 2 \cos \frac{2}{3}x$, the amplitude is 2 and the period is $\frac{2\pi}{\frac{2}{3}} = 3\pi$. By comparing the given equation to the general equation $y = a \cos b(x - h) + k$, you can see that $h = \frac{\pi}{4}$, $k = 0$, and $a > 0$. Therefore, translate the graph of $y = 2 \cos \frac{2}{3}x$ **right $\frac{\pi}{4}$ unit**. (Notice that the maximum occurs $\frac{\pi}{4}$ unit to the right of the y-axis.)

STUDENT HELP

→ **Study Tip**
When graphing translations of functions, you may find it helpful to graph the basic function first and then translate the graph.

The five key points are:

On $y = k$: $\left(\left(\frac{1}{4} \cdot 3\pi\right) + \frac{\pi}{4}, 0\right) = (\pi, 0);$

$\left(\left(\frac{3}{4} \cdot 3\pi\right) + \frac{\pi}{4}, 0\right) = \left(\frac{5\pi}{2}, 0\right)$

Maximums: $\left(0 + \frac{\pi}{4}, 2\right) = \left(\frac{\pi}{4}, 2\right);$

$\left(3\pi + \frac{\pi}{4}, 2\right) = \left(\frac{13\pi}{4}, 2\right)$

Minimum: $\left(\left(\frac{1}{2} \cdot 3\pi\right) + \frac{\pi}{4}, -2\right) = \left(\frac{7\pi}{4}, -2\right)$

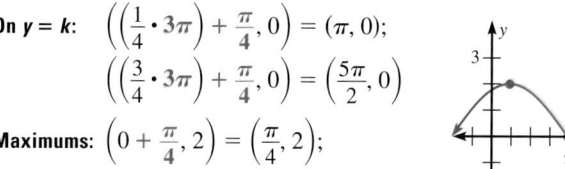

EXAMPLE 3 *Graphing a Reflection*

Graph $y = -3 \sin x$.

SOLUTION

Because the graph is a reflection of the graph of $y = 3 \sin x$, the amplitude is 3 and the period is 2π. When you plot the five key points on the graph, note that the intercepts are the same as they are for the graph of $y = 3 \sin x$. However, when the graph is reflected in the x-axis, the **maximum** becomes a **minimum** and the **minimum** becomes a **maximum**.

The five key points are:

On $y = k$: $(0, 0); (2\pi, 0);$

$\left(\frac{1}{2} \cdot 2\pi, 0\right) = (\pi, 0)$

Minimum: $\left(\frac{1}{4} \cdot 2\pi, -3\right) = \left(\frac{\pi}{2}, -3\right)$

Maximum: $\left(\frac{3}{4} \cdot 2\pi, 3\right) = \left(\frac{3\pi}{2}, 3\right)$

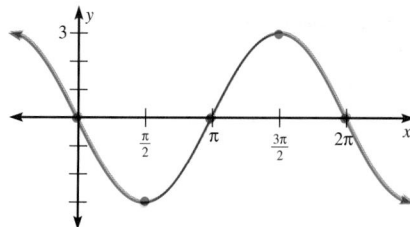

· · · · · · · · · ·

The next example shows how to graph a function when multiple transformations are involved.

14.2 *Translations and Reflections of Trigonometric Graphs* **841**

2 TEACH

📋 **EXTRA EXAMPLE 1**
Graph $y = 3 + \frac{1}{2} \cos \pi x$.

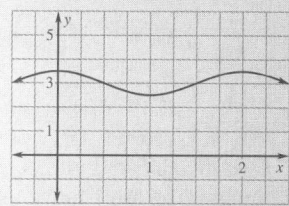

EXTRA EXAMPLE 2
Graph $y = 4 \sin 3\left(x + \frac{\pi}{3}\right)$.

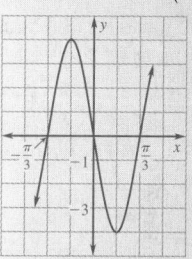

EXTRA EXAMPLE 3
Graph $y = -\cos 4x$.

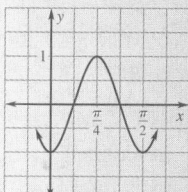

☑ **CHECKPOINT EXERCISES**
For use after Examples 1–3:
Write an equation for the graph described.

1. The graph of $y = \frac{1}{3} \cos 2x$ shifted up 4 units.
$y = 4 + \frac{1}{3} \cos 2x$

2. The graph of $y = 8 \sin \frac{1}{2}x$ shifted left 4 units.
$y = 8 \sin \frac{1}{2}(x + 4)$

3. The graph of $y = \frac{2}{3} \cos 3x$ reflected over the x-axis.
$y = -\frac{2}{3} \cos 3x$

EXTRA EXAMPLE 4
Graph $y = -3 \sin (4x - \pi) - 1$.

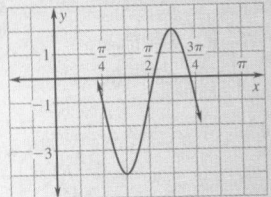

EXTRA EXAMPLE 5
You are riding a Ferris wheel. Your height h (in feet) above the ground at any time t (in seconds) can be modeled by the equation:
$h = 40 \cos \left(\frac{\pi}{20}t + \frac{\pi}{2} \right) + 50$. The
Ferris wheel turns for 160 sec before it stops to let the first passengers off.
a. Sketch the graph of your height with respect to t.

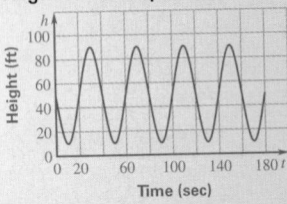

b. What are your minimum and maximum heights? **90 ft; 10 ft**

✔ **CHECKPOINT EXERCISES**
For use after Example 4:
1. Give the amplitude, period, and five key points of the graph of
$y = -\frac{1}{4} \sin \left(\frac{1}{2}x - \pi \right) + \frac{1}{4}$.

$\frac{1}{4}$; 4π; $\left(2\pi, \frac{1}{4}\right)$, $(3\pi, 0)$,
$\left(4\pi, \frac{1}{4}\right)$, $\left(5\pi, \frac{1}{2}\right)$, $\left(6\pi, \frac{1}{4}\right)$

For use after Example 5:
2. On another Ferris wheel, your height in feet is given by the equation
$h = 20 \sin \left(\frac{\pi}{30}t + \pi \right) + 23$.
a. How many cycles will this Ferris wheel make in 150 sec? **2.5 cycles**
b. What are your maximum and minimum heights? **43 ft, 3 ft**
c. What is your height after 150 sec? **23 ft**

842

EXAMPLE 4 Combining a Translation and a Reflection

Graph $y = -\frac{1}{2} \cos (2x + 3\pi) + 1$.

SOLUTION

Begin by rewriting the function in the form $y = a \cos b(x - h) + k$:

$$y = -\frac{1}{2} \cos (2x + 3\pi) + 1 = -\frac{1}{2} \cos 2\left[x - \left(-\frac{3\pi}{2}\right)\right] + 1$$

The amplitude is $\frac{1}{2}$ and the period is $\frac{2\pi}{2} = \pi$. Since $h = -\frac{3\pi}{2}$, $k = 1$, and $a < 0$, the

graph of $y = \frac{1}{2} \cos 2x$ is shifted **left** $\frac{3\pi}{2}$ **units** and **up 1 unit**, and then reflected in the

line $y = 1$. The five key points are:

On $y = k$: $\left(\left(\frac{1}{4} \cdot \pi\right) - \frac{3\pi}{2}, 1\right) = \left(-\frac{5\pi}{4}, 1\right)$;

$\left(\left(\frac{3}{4} \cdot \pi\right) - \frac{3\pi}{2}, 1\right) = \left(-\frac{3\pi}{4}, 1\right)$

Minimums: $\left(0 - \frac{3\pi}{2}, 1 - \frac{1}{2}\right) = \left(-\frac{3\pi}{2}, \frac{1}{2}\right)$;

$\left(\pi - \frac{3\pi}{2}, 1 - \frac{1}{2}\right) = \left(-\frac{\pi}{2}, \frac{1}{2}\right)$

Maximum: $\left(\left(\frac{1}{2} \cdot \pi\right) - \frac{3\pi}{2}, 1 + \frac{1}{2}\right) = \left(-\pi, \frac{3}{2}\right)$

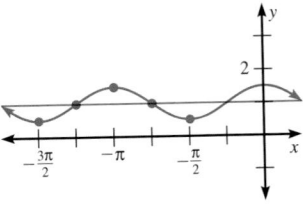

EXAMPLE 5 Modeling Circular Motion

FERRIS WHEEL You are riding a Ferris wheel. Your height h (in feet) above the ground at any time t (in seconds) can be modeled by the following equation:

$$h = 25 \sin \frac{\pi}{15}\left(t - 7.5\right) + 30$$

The Ferris wheel turns for 135 seconds before it stops to let the first passengers off.

a. Graph your height above the ground as a function of time.

b. What are your minimum and maximum heights above the ground?

SOLUTION

a. The amplitude is 25 and the period is $\frac{2\pi}{\frac{\pi}{15}} = 30$. The wheel turns $\frac{135}{30} = 4.5$ times

in 135 seconds, so the graph shows 4.5 cycles.

The five key points are $(7.5, 30)$, $(15, 55)$, $(22.5, 30)$, $(30, 5)$, and $(37.5, 30)$.

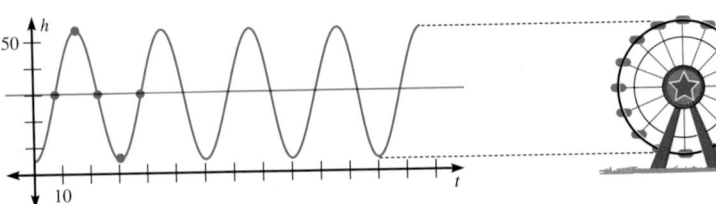

b. Since the amplitude is **25** and the graph is shifted **up 30 units**, the maximum height is $30 + 25 = 55$ feet and the minimum height is $30 - 25 = 5$ feet.

FOCUS ON APPLICATIONS

THE FERRIS WHEEL was invented in 1893 by George Washington Gale Ferris for the World's Columbian Exposition in Chicago. Each car held 60 people.

APPLICATION LINK
www.mcdougallittell.com

GOAL 2 GRAPHING TANGENT FUNCTIONS

Graphing tangent functions using translations and reflections is similar to graphing sine and cosine functions.

TRANSFORMATIONS OF TANGENT GRAPHS

To obtain the graph of $y = a \tan b(x - h) + k$, transform the graph of $y = |a| \tan bx$ as follows.
- Shift the graph k units vertically and h units horizontally.
- Then, if $a < 0$, reflect the graph in the line $y = k$.

EXAMPLE 6 *Combining a Translation and a Reflection*

Graph $y = -2 \tan \left(x + \dfrac{\pi}{4} \right)$.

SOLUTION

The graph is a transformation of the graph of $y = 2 \tan x$, so the period is π. By comparing the given equation to $y = a \tan b(x - h) + k$, you can see that $h = -\dfrac{\pi}{4}$, $k = 0$, and $a < 0$. Therefore, translate the graph of $y = 2 \tan x$ left $\dfrac{\pi}{4}$ unit and then reflect it in the x-axis.

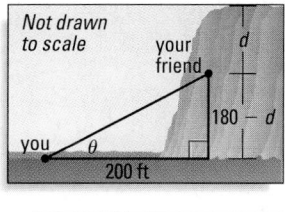

Asymptotes: $x = -\dfrac{\pi}{2 \cdot 1} - \dfrac{\pi}{4} = -\dfrac{3\pi}{4}$; $x = \dfrac{\pi}{2 \cdot 1} - \dfrac{\pi}{4} = \dfrac{\pi}{4}$

On $y = k$: $(h, k) = \left(-\dfrac{\pi}{4}, 0 \right)$

Halfway points: $\left(-\dfrac{\pi}{4 \cdot 1} - \dfrac{\pi}{4}, 2 \right) = \left(-\dfrac{\pi}{2}, 2 \right)$; $\left(\dfrac{\pi}{4 \cdot 1} - \dfrac{\pi}{4}, -2 \right) = (0, -2)$

Rappelling

EXAMPLE 7 *Modeling with a Tangent Function*

You are standing 200 feet from the base of a 180 foot cliff. Your friend is rappelling down the cliff. Write and graph a model for your friend's distance d from the top as a function of her angle of elevation θ.

SOLUTION

Use the tangent function to write an equation relating d and θ.

$\tan \theta = \dfrac{\text{opp}}{\text{adj}} = \dfrac{180 - d}{200}$ **Definition of tangent**

$200 \tan \theta = 180 - d$ **Multiply each side by 200.**

$d = -200 \tan \theta + 180$ **Solve for d.**

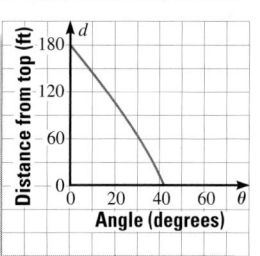

14.2 *Translations and Reflections of Trigonometric Graphs* **843**

Focus on Vocabulary *Sample answer:*
A translation is a type of transformation in which the graph is shifted horizontally or vertically. A transformation is a way that a graph is altered. Translations and reflections are two types of transformations.

Closure Question *Sample answer:*
The amplitude of the graph is *a*, and the period is $\dfrac{2\pi}{b}$. Sketch the cosine graph with this amplitude and period and shift it right *h* units and up *k* units.

843

EXTRA EXAMPLE 6
Graph $y = -2 + \tan \left(x - \dfrac{\pi}{2} \right)$.

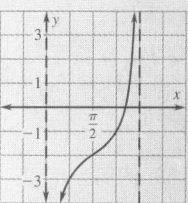

EXTRA EXAMPLE 7
You are standing 90 ft from where a balloon was launched. The balloon travels straight up to a maximum height of 120 ft. What is the angle of elevation of the balloon when it is 50 ft from the maximum height? about 38°

✔ CHECKPOINT EXERCISES
For use after Example 6:
1. Give the asymptotes, the halfway points, and center point of the graph of $y = 2 - 3 \tan 2x$. $x = -\dfrac{\pi}{4}$, $x = \dfrac{\pi}{4}$; $\left(-\dfrac{\pi}{8}, 5 \right)$, $\left(\dfrac{\pi}{8}, -1 \right)$; $(0, 2)$

For use after Example 7:
2. A balloon is launched 150 ft away from you. It can reach a maximum height of 200 ft. What is the angle of elevation of this balloon when it is 80 ft from the maximum height? about 39°

FOCUS ON VOCABULARY
What is the difference between a translation and a transformation?
See margin.

CLOSURE QUESTION
Describe how to graph $y = a \cos [b(x - h)] + k$ when $a > 0$.
See margin.

DAILY PUZZLER
What function do you get if you slide $y = \sin x$ to the right $\dfrac{\pi}{2}$ units? To the left $\dfrac{\pi}{2}$ units? $y = -\cos x$; $y = \cos x$

ASSIGNMENT GUIDE

BASIC
Day 1: pp. 844–847 Exs. 17–21, 26–37, 50–51, 56–57, 65–75 odd
Day 2: pp. 844–847 Exs. 44–49, 52, 55, 58–59, 63, 77–83 odd, Quiz 1 Exs. 1–19

AVERAGE
Day 1: pp. 844–847 Exs. 17–23, 26–40, 50–51, 53–54, 56–57, 65–75 odd
Day 2: pp. 844–847 Exs. 44–49, 52, 55, 58–60, 63, 77–83 odd, Quiz 1 Exs. 1–19

ADVANCED
Day 1: pp. 844–847 Exs. 17–43, 50–51, 53–54, 56–57, 65–75 odd
Day 2: pp. 844–847 Exs. 44–49, 52, 55, 58–64, 77–83 odd, Quiz 1 Exs. 1–19

BLOCK SCHEDULE
pp. 844–847 Exs. 17–23, 26–47, 50–57, 58–60, 63, 65–83 odd, Quiz 1 Exs. 1–19

EXERCISE LEVELS
Level A: *Easier*
17–31, 50–51
Level B: *More Difficult*
32–49, 52–63
Level C: *Most Difficult*
64

✔ **HOMEWORK CHECK**
To quickly check student understanding of key concepts, go over the following exercises: Exs. 18, 26, 32, 44, 50. See also the Daily Homework Quiz:
• Blackline Master (*Chapter 14 Resource Book,* p. 37)
• Transparency (p. 107)

10–16. See Additional Answers beginning on page AA1.

GUIDED PRACTICE

Vocabulary Check ✔

1. Complete this statement: A(n) _?_ shifts a graph horizontally or vertically.   **translation**

Concept Check ✔

2. How is the graph of $y = -2 \cos 3x$ related to the graph of $y = 2 \cos 3x$?
   reflection in x-axis

3. How is the graph of $y = \tan 2(x - \pi)$ related to the graph of $y = \tan 2x$?
   shifted right π units

Skill Check ✔

State whether the graph of the function is a *vertical shift*, a *horizontal shift*, and/or a *reflection* of the graph of $y = 4 \cos 2x$.

4. $y = 3 + 4 \cos 2x$
   vertical shift
5. $y = 4 \cos (2x + 1)$
   horizontal shift
6. $y = -4 \cos 2x$ **reflection**

7. $y = 4 \cos 2(x + 1)$
   horizontal shift
8. $y = 4 \cos (2x - 1) + 3$
   horizontal shift and vertical shift
9. $y = -3 - 4 \cos 2x$
   reflection, vertical shift

Graph the function.
10–15. See margin.

10. $y = 3 \sin (x + \pi)$
11. $y = -2 \cos x + 1$
12. $y = 2 \tan (x - \pi)$

13. $y = -\sin \pi(x - 2) + 3$
14. $y = 4 \cos 2(x - \pi) + 1$
15. $y = 5 - \tan 2(x - \pi)$

16. 🌐 **RAPPELLING** Look back at Example 7. Suppose the cliff is 250 feet high and you are 150 feet from the base. Write and graph an equation that gives your friend's distance from the top as a function of her angle of elevation.
    $d = -150 \tan \theta + 250$; **See margin for graph.**

PRACTICE AND APPLICATIONS

STUDENT HELP
➤ **Extra Practice**
to help you master skills is on p. 959.

TRANSFORMING GRAPHS Describe how the graph of $y = \sin x$ or $y = \cos x$ can be transformed to produce the graph of the given function.

17. $y = 2 + \sin x$
    shift up 2
18. $y = 5 - \cos x$
    reflect in x-axis and shift up 5
19. $y = -2 + \cos x$ **shift down 2**

20. $y = \cos \left(x + \dfrac{\pi}{2}\right)$
    shift left $\dfrac{\pi}{2}$
21. $y = -\sin (x + \pi)$
    reflect in x-axis and shift left π
22. $y = \sin \left(x - \dfrac{\pi}{2}\right)$
    shift right $\dfrac{\pi}{2}$

23. **reflect in x-axis, shift right $\dfrac{\pi}{4}$, shift up 5**
24. **reflect in x-axis, shift right π, shift down 2**
25. **shift left $\dfrac{3\pi}{4}$, shift up 3**

23. $y = 5 - \cos \left(x - \dfrac{\pi}{4}\right)$
24. $y = -2 - \sin (x - \pi)$
25. $y = 3 + \cos \left(x + \dfrac{3\pi}{4}\right)$

MATCHING Match the function with its graph.

26. $y = -2 + \sin (2x + \pi)$ **F**
27. $y = -\sin (x + \pi)$ **B**
28. $y = -3 + \cos x$ **E**

29. $y = \cos \left(x + \dfrac{\pi}{2}\right)$ **A**
30. $y = 1 + \sin \dfrac{1}{2}x$ **C**
31. $y = 1 + 2 \cos \left(\dfrac{1}{2}x + \dfrac{\pi}{2}\right)$ **D**

STUDENT HELP
➤ **HOMEWORK HELP**
Examples 1–4: Exs. 17–55
Example 5: Exs. 57–60
Example 6: Exs. 32–55
Example 7: Exs. 61, 62

A.
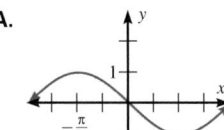

B.
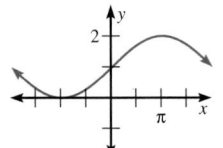

C.

D.
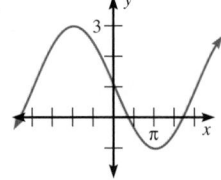

E.
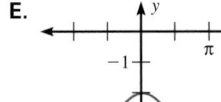

F.
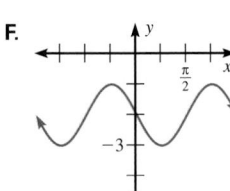

Graph the function. 32–49. See margin.

32. $y = 2 + \sin \frac{1}{2}x$ **33.** $y = \cos \left(x - \frac{\pi}{2}\right)$ **34.** $y = -4 \sin \frac{1}{4}x$

35. $y = 1 + \cos (x + \pi)$ **36.** $y = -1 + \cos (x - \pi)$ **37.** $y = 2 + 3 \sin (4x + \pi)$

38. $y = -4 + \sin 2x$ **39.** $y = 1 + 5 \cos (x - \pi)$ **40.** $y = 2 - \sin x$

41. $y = \sin \left(x - \frac{3\pi}{2}\right) - 2$ **42.** $y = \cos \left(2x + \frac{\pi}{2}\right) - 2$ **43.** $y = 6 + 4 \cos \left(x - \frac{\pi}{2}\right)$

44. $y = \frac{1}{2} \tan x$ **45.** $y = 2 - \tan \left(x + \frac{\pi}{2}\right)$ **46.** $y = 2 \tan \left(x - \frac{\pi}{2}\right)$

47. $y = -1 + \tan 2x$ **48.** $y = 3 + \tan (x - \pi)$ **49.** $y = 1 + \frac{1}{2} \tan \left(2x - \frac{\pi}{4}\right)$

WRITING EQUATIONS In Exercises 50–55, write an equation of the graph described.

50. The graph of $y = \sin 2\pi x$ translated down 5 units and right 2 units

$y = \sin 2\pi(x - 2) - 5$

51. The graph of $y = 3 \cos x$ translated up 3 units and left π units $y = 3 \cos (x + \pi) + 3$

52. The graph of $y = 5 \tan 4x$ translated left $\frac{\pi}{4}$ unit and then reflected in the x-axis

53. The graph of $y = \frac{1}{3} \sin 6x$ translated down 1 unit and then reflected in the line $y = -1$ $y = -\frac{1}{3} \sin 6x - 1$

54. The graph of $y = \frac{1}{2} \cos \pi x$ translated down $\frac{3}{2}$ units and left 1 unit, and then reflected in the line $y = -\frac{3}{2}$ $y = -\frac{1}{2} \cos \pi(x + 1) - \frac{3}{2}$

55. The graph of $y = 4 \tan \frac{\pi}{2}x$ translated up 6 units and right $\frac{1}{2}$ unit, and then reflected in the line $y = 6$ $y = -4 \tan \frac{\pi}{2}\left(x - \frac{1}{2}\right) + 6$

56. CRITICAL THINKING Explain how the graph of $y = \sin x$ can be translated to become the graph of $y = \cos x$.   Translate $\frac{\pi}{2}$ units left.

57. 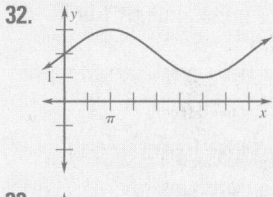 **HEIGHT OF A SWING** A swing's height h (in feet) above the ground is

$$h = -8 \cos \theta + 10$$

where the pivot is 10 feet above the ground, the rope is 8 feet long, and θ is the angle that the rope makes with the vertical. Graph the function. What is the height of the swing when θ is 45°?   4.3 ft; See margin for graph.

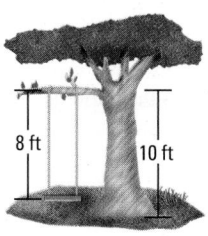

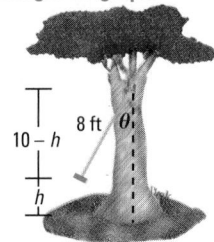

58. 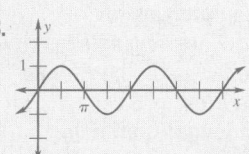 **BLOOD PRESSURE** The pressure P (in millimeters of mercury) against the walls of the blood vessels of a certain person is given by

$$P = 100 - 20 \cos \frac{8\pi}{3}t$$

where t is the time (in seconds). Graph the function. If one cycle is equivalent to one heartbeat, what is the person's pulse rate in heartbeats per minute?   80 beats/min; See margin for graph.

14.2 *Translations and Reflections of Trigonometric Graphs*   **845**

! COMMON ERROR

EXERCISES 32–49 Students may forget that when the coefficient of x is not 1, the function must be rewritten by factoring the coefficient outside of the parenthesis. If this step is not completed, the horizontal shift of the graph will be incorrect. Encourage students to test a few points from their graphs in the equations to check their work.

STUDENT HELP NOTES

Homework Help Students can find help for Exs. 50–55 at **www.mcdougallittell.com**. The information can be printed out for students who don't have access to the Internet.

32.

33.

34–49, 57, 58. See Additional Answers beginning on page AA1.

ADDITIONAL PRACTICE AND RETEACHING

For Lesson 14.2:
- Practice Levels A, B, and C (*Chapter 14 Resource Book*, p. 26)
- Reteaching with Practice (*Chapter 14 Resource Book*, p. 29)
- See Lesson 14.2 of the *Personal Student Tutor*

For more Mixed Review:
- Search the *Test and Practice Generator* for key words or specific lessons.

STUDENT HELP

HOMEWORK HELP Visit our Web site www.mcdougallittell.com for help with Exs. 50–55.

52. $y = -5 \tan 4\left(x + \frac{\pi}{4}\right)$

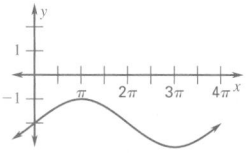
1. Describe how the graph of $y = \cos x$ can be transformed to produce the graph of $y = -3 - \cos x$. **reflect in x-axis and shift down 3**

2. Graph the function $y = -2 + \sin \frac{1}{2}x$.

3. Write an equation of the graph of $y = 4 \sin x$ translated down 3 units and left 1 unit. $y = 4 \sin(x + 1) - 3$

EXTRA CHALLENGE NOTE

Challenge problems for Lesson 14.2 are available in **blackline** format in the *Chapter 14 Resource Book,* p. 33 and at **www.mcdougallittell.com.**

ADDITIONAL TEST PREPARATION

1. WRITING Tell how to graph
$$y = -1 - 2\tan\left(x + \frac{\pi}{4}\right).$$
See margin.

2. OPEN ENDED Make up a real-life Ferris wheel problem. Find the amount of time needed for one cycle of the wheel, and the maximum and minimum heights of a passenger. **Check to see that answers are similar to Example 5.**

59.

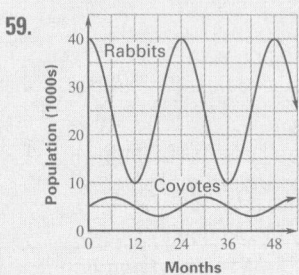

🌎 **ANIMAL POPULATIONS** In Exercises 59 and 60, use the following information.
Biologists use sine and cosine functions to model oscillations in predator and prey populations. The population R of rabbits and the population C of coyotes in a particular region can be modeled by

$$R = 25{,}000 + 15{,}000 \cos \frac{\pi}{12}t$$

$$C = 5000 + 2000 \sin \frac{\pi}{12}t$$

where t is the time in months.

59. Over the course of the first year, R falls while C rises and then falls. Over the second year R rises while C continues to fall and then rise. Both populations have the same period, 2 years, with the peak in R occurring 6 months before the peak in C, and the minimum in R 6 months before the minimum in C.

60. As the coyote population increases, they consume more rabbits. As the supply of food dwindles, the coyote population decreases which allows the rabbit population to replenish itself and the cycle begins anew.

 **Test Preparation**

59. Graph both functions in the same coordinate plane and describe how the characteristics of the graphs relate to the diagram shown. **See margin for graph.**

60. LOGICAL REASONING Look at the diagram above. Explain why each step leads to the next. **See margin.**

61. 🎢 **AMUSEMENT PARK** At an amusement park you watch your friend on a ride that simulates a free fall. You are standing 250 feet from the base of the ride and the ride is 100 feet tall. Write an equation that gives the distance d (in feet) that your friend has fallen as a function of the angle of elevation θ. State the domain of the function. Then graph the function. $d = -250 \tan \theta + 100$, where $0 \le \theta \le 21.8°$; **See margin for graph.**

62. 🌎 **WINDOW WASHERS** You are standing 80 feet from a 300 foot building, watching as a window washer lowers himself to the ground. Write an equation that gives the window washer's distance d (in feet) from the top of the building as a function of the angle of elevation θ. State the domain of the function. Then graph the function. $d = -80 \tan \theta + 300$, where $0 \le \theta \le 75°$; **See margin for graph.**

63. MULTI-STEP PROBLEM You are at the top of a 120 foot building that straddles a road. You are looking down at a car traveling straight toward the building.

  a. Write an equation of the car's distance from the base of the building as a function of the angle of depression from you to the car. $d = \dfrac{120}{\tan \theta}$

  b. Suppose the car is between you and a large road sign that you know is one mile (5280 feet) from the building. Write an equation for the distance between the road sign and the car as a function of the angle of depression from you to the car. $d = 5280 - \dfrac{120}{\tan \theta}$

  c. Graph the functions you wrote in parts (a) and (b) in the same coordinate plane. **See margin.**

  d. *Writing* Describe how the graphs you drew in part (c) are geometrically related. **Take the first, reflect it in the x-axis, shift up 5280, and you have the second.**

★ **Challenge**

64. 🌎 **FERRIS WHEEL** Suppose a Ferris wheel has a radius of 20 feet and operates at a speed of 3 revolutions per minute. The bottom car is 4 feet above the ground. Write a model for the height of a person above the ground whose height when $t = 0$ is $h = 44$. $h = 20 \cos\left(\dfrac{\pi t}{10}\right) + 24$

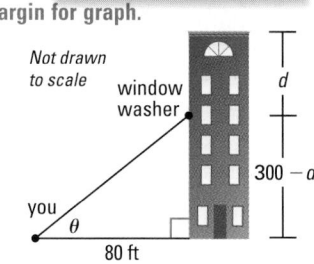

Not drawn to scale

Additional Test Preparation *Sample answer:*
1. Compare the function to $y = a \tan[b(x - h)] + k$. Since a is negative, the graph is a reflection of the basic tangent graph. The amplitude a is 2, and since $\quad h = -\frac{\pi}{4}$ and $k = -1$, the graph is shifted left $\frac{\pi}{4}$ and down 1.

MIXED REVIEW

CLASSIFYING CONICS Classify the conic section and write its equation in standard form. (Review 10.6 for 14.3)

65. $36x^2 + 25y^2 - 900 = 0$

66. $9x^2 - 16y^2 - 144 = 0$

67. $10x^2 + 10y^2 - 250 = 0$

68. $100x^2 + 81y^2 - 100 = 0$

COMBINATIONS Find the number of combinations. (Review 12.2)

69. $_8C_7$ 8

70. $_{142}C_1$ 142

71. $_{10}C_3$ 120

72. $_{14}C_5$ 2002

73. $_5C_3$ 10

74. $_6C_2$ 15

75. $_7C_6$ 7

76. $_{100}C_2$ 4950

EVALUATING FUNCTIONS Evaluate the six trigonometric functions of the angle θ. (Review 13.1 for 14.3)   77–82. See margin.

77.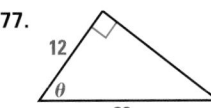

78.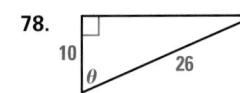

79.

80.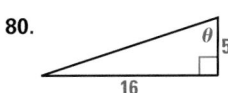

81.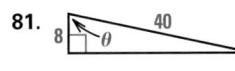

82.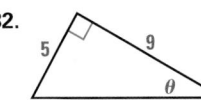

83. 🌐 **BOX LUNCHES** You have made eight different lunches for eight people. How many different ways can you distribute the lunches? (Review 12.1)   40,320

QUIZ 1

Self-Test for Lessons 14.1 and 14.2

Find the amplitude and period of the function. (Lesson 14.1)

1. $y = \frac{5}{2}\sin 7x$
amplitude: $\frac{5}{2}$, period: $\frac{2\pi}{7}$

2. $y = \cos 2x$
amplitude: 1, period: π

3. $y = \sin\frac{\pi}{2}x$
amplitude: 1, period: 4

4. $y = \frac{1}{4}\sin 2\pi x$
amplitude: $\frac{1}{4}$, period: 1

5. $y = 3\cos \pi x$
amplitude: 3, period: 2

6. $y = 4\cos\frac{3\pi}{2}x$
amplitude: 4, period: $\frac{4}{3}$

7. $y = \frac{7}{3}\cos 4x$
amplitude: $\frac{7}{3}$, period: $\frac{\pi}{2}$

8. $y = \frac{1}{3}\sin x$
amplitude: $\frac{1}{3}$, period: 2π

9. $y = 6\sin\frac{1}{8}x$
amplitude: 6, period: 16π

Graph the function. (Lessons 14.1, 14.2)
10–18. See margin.

10. $y = 2\sin \pi x$

11. $y = \frac{3}{2}\cos\frac{1}{2}\pi x$

12. $y = -\sin 2x$

13. $y = 4\tan\frac{1}{2}x$

14. $y = -4 + 2\sin 3x$

15. $y = 3\tan 2(x + \pi)$

16. $y = -3\cos(x + \pi)$

17. $y = -2 + \cos\frac{1}{2}(x - \pi)$

18. $y = 2 - 5\tan\left(x + \frac{\pi}{3}\right)$

19. 🌐 **GLASS ELEVATOR** You are standing 120 feet from the base of a 260 foot building. You are looking at your friend who is going up the side of the building in a glass elevator. Write and graph a function that gives your friend's distance d (in feet) above the ground as a function of her angle of elevation θ. What is her angle of elevation when she is 70 feet above the ground? (Lesson 14.2)
$d = 120\tan\theta,\ 0 \le \theta \le 65.2°$; 30.3°; See margin for graph.

14.2 *Translations and Reflections of Trigonometric Graphs*   **847**

---

65. ellipse, $\frac{x^2}{25} + \frac{y^2}{36} = 1$

66. hyperbola, $\frac{x^2}{16} - \frac{y^2}{9} = 1$

68. ellipse, $\frac{x^2}{1} + \frac{y^2}{\left(\frac{10}{9}\right)^2} = 1$

67. circle, $x^2 + y^2 = 25$

77. $\sin\theta = \frac{4}{5}$, $\cos\theta = \frac{3}{5}$,
$\tan\theta = \frac{4}{3}$, $\sec\theta = \frac{5}{3}$,
$\csc\theta = \frac{5}{4}$, $\cot\theta = \frac{3}{4}$

78. $\sin\theta = \frac{12}{13}$, $\cos\theta = \frac{5}{13}$,
$\tan\theta = \frac{12}{5}$, $\sec\theta = \frac{13}{5}$,
$\csc\theta = \frac{13}{12}$, $\cot\theta = \frac{5}{12}$

79. $\sin\theta = \frac{3}{10}$, $\cos\theta = \frac{\sqrt{91}}{10}$,
$\tan\theta = \frac{3\sqrt{91}}{91}$, $\sec\theta = \frac{10\sqrt{91}}{91}$, $\csc\theta = \frac{10}{3}$,
$\cot\theta = \frac{\sqrt{91}}{3}$

80. $\sin\theta = \frac{16\sqrt{281}}{281}$, $\cos\theta = \frac{5\sqrt{281}}{281}$, $\tan\theta = \frac{16}{5}$,
$\sec\theta = \frac{\sqrt{281}}{5}$, $\csc\theta = \frac{\sqrt{281}}{16}$, $\cot\theta = \frac{5}{16}$

81. $\sin\theta = \frac{2\sqrt{6}}{5}$, $\cos\theta = \frac{1}{5}$, $\tan\theta = 2\sqrt{6}$,
$\sec\theta = 5$, $\csc\theta = \frac{5\sqrt{6}}{12}$, $\cot\theta = \frac{\sqrt{6}}{12}$

82. $\sin\theta = \frac{5\sqrt{106}}{106}$, $\cos\theta = \frac{9\sqrt{106}}{106}$, $\tan\theta = \frac{5}{9}$,
$\sec\theta = \frac{\sqrt{106}}{9}$, $\csc\theta = \frac{\sqrt{106}}{5}$, $\cot\theta = \frac{9}{5}$

ENGLISH LEARNERS
EXERCISES 59–63 Some terms that are easily understood by most native speakers may be difficult for learners of English to comprehend. You may want to discuss the meanings of *predator, prey, simulates, free fall,* and *straddles* before students begin solving Exercises 59–63.

ADDITIONAL RESOURCES
An alternative Quiz for Lessons 14.1 and 14.2 is available in the *Chapter 14 Resource Book,* p. 34.

61.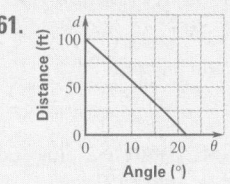

62.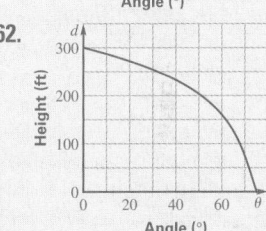

63c.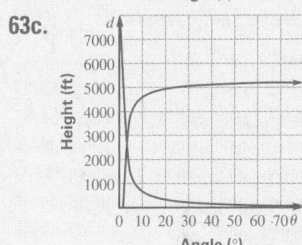

10.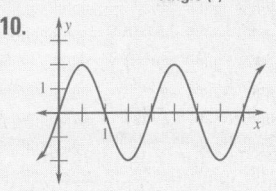

11.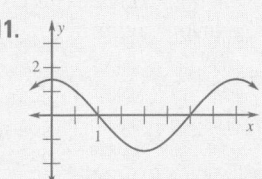

12–19. See Additional Answers beginning on page AA1.

LESSON OPENER
VISUAL APPROACH
An alternative way to approach Lesson 14.3 is to use the Visual Approach Lesson Opener:
- Blackline Master (*Chapter 14 Resource Book*, p. 38)
- Transparency (p. 93)

MEETING INDIVIDUAL NEEDS
- *Chapter 14 Resource Book*
  Prerequisite Skills Review (p. 5)
  Practice Level A (p. 41)
  Practice Level B (p. 42)
  Practice Level C (p. 43)
  Reteaching with Practice (p. 44)
  Absent Student Catch-Up (p. 46)
  Challenge (p. 48)
- *Resources in Spanish*
- *Personal Student Tutor*

NEW-TEACHER SUPPORT
See the Tips for New Teachers on pp. 1–2 of the *Chapter 14 Resource Book* for additional notes about Lesson 14.3.

WARM-UP EXERCISES
Transparency Available

Tell whether the trigonometric function will be positive or negative for the given values of θ.

1. $\cos \theta, 0 < \theta < \frac{\pi}{2}$ positive
2. $\csc \theta, \frac{3\pi}{2} < \theta < 2\pi$ negative
3. $\tan \theta, \frac{\pi}{2} < \theta < \pi$ negative
4. $\sin \theta, \frac{3\pi}{2} < \theta < 2\pi$ negative
5. $\cot \theta, \pi < \theta < \frac{3\pi}{2}$ positive

14.3 Verifying Trigonometric Identities

What you should learn

GOAL 1 Use trigonometric identities to simplify trigonometric expressions and to verify other identities.

GOAL 2 Use trigonometric identities to solve **real-life** problems, such as comparing the speeds at which people pedal exercise machines in **Example 7.**

Why you should learn it

▼ To simplify **real-life** trigonometric expressions, such as the parametric equations that describe a carousel's motion in **Ex. 65.**

GOAL 1 USING TRIGONOMETRIC IDENTITIES

In this lesson you will use *trigonometric identities* to evaluate trigonometric functions, simplify trigonometric expressions, and verify other identities.

● ACTIVITY
Developing Concepts **Investigating Trigonometric Identities**

Use a graphing calculator to graph each side of the equation in the same viewing window. What do you notice about the graphs? Is the equation true for (**a**) no x-values, (**b**) some x-values, or (**c**) all x-values? (Set your calculator in radian mode and use $-2\pi \le x \le 2\pi$ and $-2 \le y \le 2$.)

1. $\sin^2 x + \cos^2 x = 1$ all x-values   2. $\sin(-x) = -\sin x$ all x-values

3. $\sin x = -\cos x$ some x-values   4. $\cos x = 1.5$ no x-values

In the activity you may have discovered that some trigonometric equations are true for all values of x (in their domain). Such equations are called **trigonometric identities**. In Lesson 13.1 you used reciprocal identities to find the values of the cosecant, secant, and cotangent functions. These and other fundamental identities are listed below.

FUNDAMENTAL TRIGONOMETRIC IDENTITIES

RECIPROCAL IDENTITIES

$$\csc \theta = \frac{1}{\sin \theta} \qquad \sec \theta = \frac{1}{\cos \theta} \qquad \cot \theta = \frac{1}{\tan \theta}$$

TANGENT AND COTANGENT IDENTITIES

$$\tan \theta = \frac{\sin \theta}{\cos \theta} \qquad \cot \theta = \frac{\cos \theta}{\sin \theta}$$

PYTHAGOREAN IDENTITIES

$$\sin^2 \theta + \cos^2 \theta = 1 \qquad 1 + \tan^2 \theta = \sec^2 \theta \qquad 1 + \cot^2 \theta = \csc^2 \theta$$

COFUNCTION IDENTITIES

$$\sin\left(\frac{\pi}{2} - \theta\right) = \cos \theta \qquad \cos\left(\frac{\pi}{2} - \theta\right) = \sin \theta \qquad \tan\left(\frac{\pi}{2} - \theta\right) = \cot \theta$$

NEGATIVE ANGLE IDENTITIES

$$\sin(-\theta) = -\sin \theta \qquad \cos(-\theta) = \cos \theta \qquad \tan(-\theta) = -\tan \theta$$

EXAMPLE 1 *Finding Trigonometric Values*

Given that $\sin \theta = \frac{3}{5}$ and $\frac{\pi}{2} < \theta < \pi$, find the values of the other five trigonometric functions of θ.

SOLUTION

Begin by finding $\cos \theta$.

$$\sin^2 \theta + \cos^2 \theta = 1 \qquad \text{Write Pythagorean identity.}$$

$$\left(\frac{3}{5}\right)^2 + \cos^2 \theta = 1 \qquad \text{Substitute } \frac{3}{5} \text{ for } \sin \theta.$$

$$\cos^2 \theta = 1 - \left(\frac{3}{5}\right)^2 \qquad \text{Subtract } \left(\frac{3}{5}\right)^2 \text{ from each side.}$$

$$\cos^2 \theta = \frac{16}{25} \qquad \text{Simplify.}$$

$$\cos \theta = \pm \frac{4}{5} \qquad \text{Take square roots of each side.}$$

$$\cos \theta = -\frac{4}{5} \qquad \text{Because } \theta \text{ is in Quadrant II, } \cos \theta \text{ is negative.}$$

Now, knowing $\sin \theta$ and $\cos \theta$, you can find the values of the other four trigonometric functions.

$$\tan \theta = \frac{\sin \theta}{\cos \theta} = \frac{\frac{3}{5}}{-\frac{4}{5}} = -\frac{3}{4} \qquad\qquad \cot \theta = \frac{\cos \theta}{\sin \theta} = \frac{-\frac{4}{5}}{\frac{3}{5}} = -\frac{4}{3}$$

$$\csc \theta = \frac{1}{\sin \theta} = \frac{1}{\frac{3}{5}} = \frac{5}{3} \qquad\qquad \sec \theta = \frac{1}{\cos \theta} = \frac{1}{-\frac{4}{5}} = -\frac{5}{4}$$

EXAMPLE 2 *Simplifying a Trigonometric Expression*

Simplify the expression $\sec \theta \tan^2 \theta + \sec \theta$.

SOLUTION

$$\sec \theta \tan^2 \theta + \sec \theta = \sec \theta (\sec^2 \theta - 1) + \sec \theta \qquad \text{Pythagorean identity}$$

$$= \sec^3 \theta - \sec \theta + \sec \theta \qquad \text{Distributive property}$$

$$= \sec^3 \theta \qquad \text{Simplify.}$$

EXAMPLE 3 *Simplifying a Trigonometric Expression*

Simplify the expression $\cos\left(\frac{\pi}{2} - x\right) \cot x$.

SOLUTION

$$\cos\left(\frac{\pi}{2} - x\right) \cot x = \sin x \cot x \qquad \text{Cofunction identity}$$

$$= \sin x \left(\frac{\cos x}{\sin x}\right) \qquad \text{Cotangent identity}$$

$$= \cos x \qquad \text{Simplify.}$$

14.3 *Verifying Trigonometric Identities*   **849**

ACTIVITY NOTE
Students should recall that equations are true for no values if the graphs do not intersect, true for some values if the graphs intersect in only certain points, and true for all values if the graphs overlap completely.

EXTRA EXAMPLE 1
Given that $\cos \theta = -\frac{5}{13}$ and $\pi < \theta < \frac{3\pi}{2}$, find the values of the other five trigonometric functions of θ.
$\sin \theta = -\frac{12}{13}$, $\tan \theta = \frac{12}{5}$, $\cot \theta = \frac{5}{12}$, $\csc \theta = -\frac{13}{12}$, $\sec \theta = -\frac{13}{5}$

EXTRA EXAMPLE 2
Simplify the expression $\cos \theta - \cos \theta \sin^2 \theta$. $\cos^3 \theta$

EXTRA EXAMPLE 3
Simplify the expression $1 + \dfrac{\sin\left(\frac{\pi}{2} - \theta\right)}{\sec(-\theta)}$. $1 + \cos^2 \theta$

CHECKPOINT EXERCISES
For use after Example 1:

1. Given that $\tan \theta = -\frac{1}{2}$ and $\frac{3\pi}{2} < \theta < 2\pi$, find the values of the other five trigonometric functions of θ. $\sin \theta = -\frac{\sqrt{5}}{5}$, $\cos \theta = \frac{2\sqrt{5}}{5}$, $\sec \theta = \frac{\sqrt{5}}{2}$, $\csc \theta = -\sqrt{5}$, $\cot \theta = -2$

For use after Examples 2 and 3:
2. Simplify $\cot^2 \theta - \cot^2 \theta \cos^2 \theta$. $\cos^2 \theta$

3. Simplify $\cot \theta \cos\left(\frac{\pi}{2} - \theta\right)$. $\cos \theta$

EXTRA EXAMPLE 4
Verify sec $(-\theta)$ = sec (θ).
$$\sec(-\theta) = \frac{1}{\cos(-\theta)} = \frac{1}{\cos\theta} = \sec\theta$$

EXTRA EXAMPLE 5
Verify $\dfrac{\cos x \sec x}{1 + \tan^2 x} = \cos^2 x$.

$$\frac{\cos x \sec x}{1 + \tan^2 x} = \frac{\cos x \sec x}{\sec^2 x} = \frac{\cos x}{\sec x} =$$

$$\frac{\cos x}{\frac{1}{\cos x}} = \cos x \cos x = \cos^2 x$$

EXTRA EXAMPLE 6
Verify $\dfrac{\sec x}{1 + \sin x} = \dfrac{\sec x - \tan x}{\cos^2 x}$.

$$\frac{\sec x}{1 + \sin x} = \frac{\sec x(1 - \sin x)}{(1 + \sin x)(1 - \sin x)} =$$

$$\frac{\sec x - \sec x \sin x}{1 - \sin^2 x} =$$

$$\frac{\sec x - \frac{1}{\cos x}\sin x}{\cos^2 x} = \frac{\sec x - \tan x}{\cos^2 x}.$$

CHECKPOINT EXERCISES

For use after Examples 4–6:

1. Verify the identity $\dfrac{\sin^2 x}{\cot x} =$
tan x – sin x cos x.

$$\frac{\sin^2 x}{\cot x} = \frac{1 - \cos^2 x}{\cot x} = \frac{1}{\cot x} -$$

$$\frac{\cos^2 x}{\cot x} = \tan x - \cos^2 x \tan x =$$

$$\tan x - \cos^2 x \frac{\sin x}{\cos x} = \tan x -$$

$$\sin x \cos x.$$

2. Verify the identity

$$\frac{1 + \tan^2\left(\frac{\pi}{2} - x\right)}{\sin(-x)} = -\csc^3 x.$$

$$\frac{1 + \tan^2\left(\frac{\pi}{2} - x\right)}{\sin(-x)} = \frac{1 + \cot^2 x}{-\sin x} =$$

$$\frac{\csc^2 x}{-\sin x} = \csc^2 x\left(-\frac{1}{\sin x}\right) =$$

$$-\csc^3 x$$

MATHEMATICAL REASONING
After reading the first Study Tip on p. 850, students may ask why they cannot use properties of equality. To use properties of equality, you must know that the equation is true. When verifying an identity, you are trying to show that the equation is true.

850

You can use the fundamental identities on page 848 to *verify* new trigonometric identities. A *verification* of an identity is a chain of equivalent expressions showing that one side of the identity is equal to the other side. When verifying an identity, begin with the expression from one side and manipulate it algebraically until it is identical to the other side.

EXAMPLE 4 *Verifying a Trigonometric Identity*

Verify the identity cot $(-\theta) = -\cot\theta$.

SOLUTION

$$\cot(-\theta) = \frac{\cos(-\theta)}{\sin(-\theta)} \qquad \text{Cotangent identity}$$

$$= \frac{\cos\theta}{-\sin\theta} \qquad \text{Negative angle identities}$$

$$= -\cot\theta \qquad \text{Cotangent identity}$$

STUDENT HELP

▸ **Study Tip**
Verifying an identity is *not* the same as solving an equation. When verifying an identity you should *not* use any properties of equality, such as adding the same number or expression to both sides.

EXAMPLE 5 *Verifying a Trigonometric Identity*

Verify the identity $\dfrac{\cot^2 x}{\csc x} = \csc x - \sin x$.

SOLUTION

$$\frac{\cot^2 x}{\csc x} = \frac{\csc^2 x - 1}{\csc x} \qquad \text{Pythagorean identity}$$

$$= \frac{\csc^2 x}{\csc x} - \frac{1}{\csc x} \qquad \text{Write as separate fractions.}$$

$$= \csc x - \frac{1}{\csc x} \qquad \text{Simplify.}$$

$$= \csc x - \sin x \qquad \text{Reciprocal identity}$$

EXAMPLE 6 *Verifying a Trigonometric Identity*

Verify the identity $\dfrac{\sin x}{1 - \cos x} = \dfrac{1 + \cos x}{\sin x}$.

SOLUTION

$$\frac{\sin x}{1 - \cos x} = \frac{\sin x(1 + \cos x)}{(1 - \cos x)(1 + \cos x)} \qquad \text{Multiply by } \frac{1 + \cos x}{1 + \cos x}.$$

$$= \frac{\sin x(1 + \cos x)}{1 - \cos^2 x} \qquad \text{Simplify denominator.}$$

$$= \frac{\sin x(1 + \cos x)}{\sin^2 x} \qquad \text{Pythagorean identity}$$

$$= \frac{1 + \cos x}{\sin x} \qquad \text{Simplify.}$$

STUDENT HELP

▸ **Study Tip**
In Example 6, notice how multiplying by an expression equal to 1 allows you to write an expression in an equivalent form.

GOAL 2 USING TRIGONOMETRIC IDENTITIES IN REAL LIFE

In Lesson 13.7 you learned that parametric equations can be used to describe linear and parabolic motion. They can be used to describe other types of motion as well.

EXAMPLE 7 — *Using Parametric Equations in Real Life*

STUDENT HELP

HOMEWORK HELP
Visit our Web site
www.mcdougallittell.com
for extra examples.

PHYSICAL FITNESS You and Sara are riding exercise machines that involve pedaling. The following parametric equations describe the motion of your feet and Sara's feet:

YOU:	SARA:
$x = 8 \cos 4\pi t$	$x = 10 \cos 2\pi t$
$y = 8 \sin 4\pi t$	$y = 6 \sin 2\pi t$

In each case, x and y are measured in inches and t is measured in seconds.

a. Describe the paths followed by your feet and Sara's feet.

b. Who is pedaling faster (in revolutions per second)?

SOLUTION

a. Use the Pythagorean identity $\sin^2 \theta + \cos^2 \theta = 1$ to eliminate the parameter t.

YOU:	SARA:	
$\dfrac{x}{8} = \cos 4\pi t$	$\dfrac{x}{10} = \cos 2\pi t$	Isolate the cosine.
$\dfrac{y}{8} = \sin 4\pi t$	$\dfrac{y}{6} = \sin 2\pi t$	Isolate the sine.
$\cos^2 4\pi t + \sin^2 4\pi t = 1$	$\cos^2 2\pi t + \sin^2 2\pi t = 1$	Pythagorean identity
$\left(\dfrac{x}{8}\right)^2 + \left(\dfrac{y}{8}\right)^2 = 1$	$\left(\dfrac{x}{10}\right)^2 + \left(\dfrac{y}{6}\right)^2 = 1$	Substitute.
$x^2 + y^2 = 64$	$\dfrac{x^2}{100} + \dfrac{y^2}{36} = 1$	Simplify.

▶ Your feet follow a circle with a radius of 8 inches. Sara's feet follow an ellipse whose major axis is 20 inches long and whose minor axis is 12 inches long.

b. The number of revolutions per second for you and Sara is the reciprocal of the common period of the corresponding parametric functions.

YOU:	SARA:
$\dfrac{1}{\frac{2\pi}{4\pi}} = \dfrac{1}{\frac{1}{2}} = 2$	$\dfrac{1}{\frac{2\pi}{2\pi}} = \dfrac{1}{1} = 1$

▶ In one second, your feet travel around 2 times and Sara's feet travel around 1 time. So, you are pedaling faster.

✓ **CHECK** To check your results, set a graphing calculator to parametric, radian, and simultaneous modes. Enter both sets of parametric equations with $0 \le t \le 1$ and a t-step of 0.01. As the paths are graphed, you can see that your path is traced faster.

FOCUS ON APPLICATIONS

ELLIPTICAL TRAINERS provide excellent aerobic exercise by combining both lower and upper body movements. The smooth elliptical motion produces less impact than experienced on a treadmill.

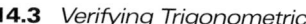 **14.3** *Verifying Trigonometric Identities* **851**

Closure Question *Sample answer:*
Start with one side of the equation and manipulate it algebraically until it is identical to the other side of the equation.

EXTRA EXAMPLE 7
Two children are playing with toy trains. The path of each train is given by the equations below where x and y are measured in feet and t is measured in minutes.
Train 1: $x = 3 \cos \pi t$, $y = 5 \sin \pi t$
Train 2: $x = 4 \cos \frac{\pi}{2}t$, $y = 4 \sin \frac{\pi}{2}t$

a. Describe the path followed by each train. Train 1 travels a path that is an ellipse with major axis 10 ft long and minor axis 6 ft long. Train 2 travels a circular path with radius 4 ft.

b. Which train is moving around its track faster (in revolutions per minute)? In one minute, Train 1 travels $\frac{1}{2}$ a revolution around its track while Train 2 travels $\frac{1}{4}$ of a revolution around its track, so Train 1 is going around its track faster.

CHECKPOINT EXERCISES

For use after Example 7:

1. Refer to Extra Example 7. Two other trains have the following parametric equations.
Train A: $x = 2 \cos 3\pi t$, $y = 2 \sin 3\pi t$
Train B: $x = 5 \cos \frac{\pi}{4}t$, $y = 8 \sin \frac{\pi}{4}t$

a. Describe the paths of these trains. Train A travels a circular path with a radius of 2 ft. Train B travels an elliptical path with major axis 16 ft and minor axis 10 ft.

b. Which of these two trains goes around its path fastest? Train A travels $1\frac{1}{2}$ times around its track in 1 minute while Train B ravels only $\frac{1}{8}$ of the way around its track in 1 minute so Train A is going around its track faster.

CLOSURE QUESTION
How do you verify a trigonometric identity? See margin.

ASSIGNMENT GUIDE

BASIC
Day 1: pp. 852–854 Exs. 16–19, 23–26, 28–36, 44–50 even, 54–56
Day 2: pp. 852–854 Exs. 40–43, 51–53, 57–61, 65, 69–85 odd

AVERAGE
Day 1: pp. 852–854 Exs. 16–24, 28–40, 44–50 even, 54–56
Day 2: pp. 852–854 Exs. 45–53 odd, 57–65, 69–85 odd

ADVANCED
Day 1: pp. 852–854 Exs. 16–25, 28–43, 44–50 even, 54–56
Day 2: pp. 852–854 Exs. 45–53 odd, 57–67, 69–85 odd

BLOCK SCHEDULE
pp. 852–854 Exs. 16–24, 28–40, 44–51, 54–57, 60–61, 65, 69–85 odd

EXERCISE LEVELS

Level A: *Easier*
16–39, 44–47, 54–59
Level B: *More Difficult*
40–43, 48–53, 60–65
Level C: *Most Difficult*
66–67

✔ HOMEWORK CHECK

To quickly check student understanding of key concepts, go over the following exercises: Exs. 16, 24, 28, 44, 54. See also the Daily Homework Quiz:

• Blackline Master (*Chapter 14 Resource Book*, p. 51)
• Transparency (p. 108)

13. $\cot x \tan (-x) = \dfrac{\cos x}{\sin x} \cdot \dfrac{\sin (-x)}{\cos (-x)} =$

$\dfrac{\cos x}{\sin x} \cdot \dfrac{-\sin x}{\cos x} = -1$

16–27. See additional Answers beginning on page AA1.

GUIDED PRACTICE

Vocabulary Check ✔

1. What is a trigonometric identity? A trigonometric equation that is true for all values of *x* in its domain.

Concept Check ✔

3. $1 - \sin^2 x \cot^2 x =$

$1 - \sin^2 x \dfrac{\cos^2 x}{\sin^2 x} =$

$1 - \cos^2 x = \sin^2 x$; yes;
Sample response: It's easiest to write everything in terms of sin and cos and then simplify.

4. $\sin^2 x = 1 - \cos^2 x$, not $1 + \cos^2 x$

2. Is sec $(-\theta)$ equal to sec θ or $-$sec θ? How do you know? sec θ, because $\sec \theta = \dfrac{1}{\cos \theta}$

3. Verify the identity $1 - \sin^2 x \cot^2 x = \sin^2 x$. Is there more than one way to verify the identity? If so, tell which way you think is easier and why.

4. ERROR ANALYSIS Describe what is wrong with the simplification shown.

$$\cos x - \cos x \sin^2 x = \cos x - \cos x (1 + \cos^2 x)$$
$$= \cos x - \cos x - \cos^3 x$$
$$= -\cos^3 x$$

Skill Check ✔

5. $\sin \theta = \dfrac{4}{5}$, $\tan \theta = -\dfrac{4}{3}$,
$\sec \theta = -\dfrac{5}{3}$, $\csc \theta = \dfrac{5}{4}$,
$\cot \theta = -\dfrac{3}{4}$

6. $\sin \theta = \dfrac{2\sqrt{13}}{13}$, $\cos \theta =$
$\dfrac{3\sqrt{13}}{13}$, $\sec \theta = \dfrac{\sqrt{13}}{3}$,
$\csc \theta = \dfrac{\sqrt{13}}{2}$, $\cot \theta = \dfrac{3}{2}$

7. $\sin \theta = -\dfrac{\sqrt{7}}{4}$, $\cos \theta = \dfrac{3}{4}$,
$\tan \theta = -\dfrac{\sqrt{7}}{3}$, $\csc \theta =$
$-\dfrac{4\sqrt{7}}{7}$, $\cot \theta = -\dfrac{3\sqrt{7}}{7}$

Find the values of the other five trigonometric functions of θ. 5–8. See margin.

5. $\cos \theta = -\dfrac{3}{5}, \dfrac{\pi}{2} < \theta < \pi$

6. $\tan \theta = \dfrac{2}{3}, 0 < \theta < \dfrac{\pi}{2}$

7. $\sec \theta = \dfrac{4}{3}, \dfrac{3\pi}{2} < \theta < 2\pi$

8. $\sin \theta = -\dfrac{1}{2}, \pi < \theta < \dfrac{3\pi}{2}$

Simplify the expression.

9. $\dfrac{(\sec x + 1)(\sec x - 1)}{\tan x}$ tan x

10. $\sin \left(\dfrac{\pi}{2} - x\right) \sec x$ 1

11. $\cos^2 \left(\dfrac{\pi}{2} - x\right) + \cos^2 (-x)$ 1

Verify the identity. 12–14. See margin.

12. $\dfrac{1}{\sin (-x)} = -\csc x$

13. $\cot x \tan (-x) = -1$

14. $\csc x \tan x = \sec x$

15. 🏃 **PHYSICAL FITNESS** Look back at Example 7 on page 851. Suppose your friend Pete starts riding another machine that involves pedaling. The motion of his feet is described by the equations $x = 10 \cos \dfrac{5\pi}{2}t$ and $y = 6 \sin \dfrac{5\pi}{2}t$. What type of path are his feet following? ellipse

8. $\cos \theta = -\dfrac{\sqrt{3}}{2}$, $\tan \theta =$
$\dfrac{\sqrt{3}}{3}$, $\sec \theta = -\dfrac{2\sqrt{3}}{3}$,
$\csc \theta = -2$, $\cot \theta = \sqrt{3}$

12. $\dfrac{1}{\sin (-x)} = \csc (-x) =$
$- \csc x$

14. $\csc x \tan x =$
$\dfrac{1}{\sin x} \cdot \dfrac{\sin x}{\cos x} = \dfrac{1}{\cos x} = \sec x$

PRACTICE AND APPLICATIONS

STUDENT HELP

↳ **Extra Practice**
to help you master skills is on p. 959.

FINDING VALUES **Find the values of the other five trigonometric functions of θ.** 16–27. See margin.

16. $\cos \theta = \dfrac{1}{\sqrt{5}}, 0 < \theta < \dfrac{\pi}{2}$

17. $\tan \theta = \dfrac{3}{8}, 0 < \theta < \dfrac{\pi}{2}$

18. $\sin \theta = \dfrac{5}{6}, 0 < \theta < \dfrac{\pi}{2}$

19. $\sin \theta = \dfrac{3}{5}, \dfrac{\pi}{2} < \theta < \pi$

20. $\cot \theta = -\dfrac{9}{4}, \dfrac{3\pi}{2} < \theta < 2\pi$

21. $\cos \theta = -\dfrac{11}{12}, \dfrac{\pi}{2} < \theta < \pi$

22. $\csc \theta = \dfrac{7}{5}, 0 < \theta < \dfrac{\pi}{2}$

23. $\sec \theta = -\dfrac{10}{3}, \pi < \theta < \dfrac{3\pi}{2}$

24. $\tan \theta = -\dfrac{1}{6}, \dfrac{\pi}{2} < \theta < \pi$

25. $\sec \theta = 2, \dfrac{3\pi}{2} < \theta < 2\pi$

26. $\csc \theta = -\dfrac{5}{3}, \pi < \theta < \dfrac{3\pi}{2}$

27. $\cot \theta = -\sqrt{3}, \dfrac{3\pi}{2} < \theta < 2\pi$

STUDENT HELP

→ HOMEWORK HELP

Example 1: Exs. 16–27
Examples 2, 3: Exs. 28–43
Examples 4–6: Exs. 44–53
Example 7: Exs. 61–64

SIMPLIFYING EXPRESSIONS Simplify the expression.

28. $\cot x \sec x$ $\csc x$

29. $\dfrac{\cos(-x)}{\sin(-x)}$ $-\cot x$

30. $\sec x \cos(-x) - \sin^2 x$ $\cos^2 x$

31. $\sin x\,(1 + \cot^2 x)$ $\csc x$

32. $1 - \sin^2\left(\dfrac{\pi}{2} - x\right)\sin^2 x$

33. $\dfrac{\tan\left(\dfrac{\pi}{2} - x\right)}{\csc x}$ $\cos x$

34. $\cos\left(\dfrac{\pi}{2} - x\right)\csc x$ 1

35. $\dfrac{\sin(-x)}{\csc x} + \cos^2(-x)$ $\cos^2 x - \sin^2 x$

36. $\dfrac{\cos^2 x \tan^2(-x) - 1}{\cos^2 x}$ -1

37. $\sec^2 x - \tan^2 x$ 1

38. $\dfrac{\tan\left(\dfrac{\pi}{2} - x\right)\sec x}{1 - \csc^2 x}$ $-\tan x \sec x$

39. $\dfrac{\cos\left(\dfrac{\pi}{2} - x\right) - 1}{1 + \sin(-x)}$ -1

40. $\dfrac{\cot x \cos x}{\tan(-x)\sin\left(\dfrac{\pi}{2} - x\right)}$ $-\cot^2 x$

41. $\dfrac{\sec x \sin x + \cos\left(\dfrac{\pi}{2} - x\right)}{1 + \sec x}$ $\sin x$

42. $\cot^2 x + \sin^2 x + \cos^2(-x)$ $\csc^2 x$

43. $\tan\left(\dfrac{\pi}{2} - x\right)\cot x - \csc^2 x$ -1

VERIFYING IDENTITIES Verify the identity.   44–53. See margin.

44. $\cos x \sec x = 1$

45. $\tan x \csc x \cos x = 1$

46. $\cos\left(\dfrac{\pi}{2} - x\right)\cot x = \cos x$

47. $2 - \sec^2 x = 1 - \tan^2 x$

48. $\sin x + \cos x \cot x = \csc x$

49. $\dfrac{\cos^2 x + \sin^2 x}{1 + \tan^2 x} = \cos^2 x$

50. $\dfrac{\sin^2(-x)}{\tan^2 x} = \cos^2 x$

51. $\dfrac{\sin\left(\dfrac{\pi}{2} - x\right) - 1}{1 - \cos(-x)} = -1$

52. $\dfrac{1 + \sin x}{\cos x} + \dfrac{\cos x}{1 + \sin x} = 2\sec x$

53. $\dfrac{\cos(-x)}{1 + \sin(-x)} = \sec x + \tan x$

IDENTIFYING CONICS Use a graphing calculator set in parametric mode to graph the parametric equations. Use a trigonometric identity to determine whether the graph is *a circle, an ellipse,* or *a hyperbola.* (Use a square viewing window.)   54–59. See margin.

54. $x = 6\cos t, \; y = 6\sin t$

55. $x = 5\sec t, \; y = \tan t$

56. $x = 2\cos t, \; y = 3\sin t$

57. $x = 8\cos \pi t, \; y = 8\sin \pi t$

58. $x = 2\cot 2t, \; y = 3\csc 2t$

59. $x = \cos\dfrac{t}{2}, \; y = 4\sin\dfrac{t}{2}$

60. CRITICAL THINKING A function f is *odd* if $f(-x) = -f(x)$. A function f is *even* if $f(-x) = f(x)$. Which of the six trigonometric functions are odd? Which of them are even?   odd: sin, csc, tan, cot; even: cos, sec

61. 🌎 **SHADOW OF A SUNDIAL** The length s of a shadow cast by a vertical *gnomon* (column or shaft on a sundial) of height h when the angle of the sun above the horizon is θ can be modeled by this equation:

$$s = \frac{h\sin(90° - \theta)}{\sin\theta}$$

This equation was developed by Abu Abdullah al-Battani (circa A.D. 920). Show that the equation is equivalent to $s = h\cot\theta$. ▶ Source: *Trigonometric Delights*

$$s = \frac{h\sin(90° - \theta)}{\sin\theta} = \frac{h\cos\theta}{\sin\theta} = h\cot\theta$$

14.3 *Verifying Trigonometric Identities*   **853**

(Left margin answers)

44. $\cos x \sec x =$
$\cos x \cdot \dfrac{1}{\cos x} = 1$

45. $\tan x \csc x \cos x =$
$\left(\dfrac{\sin x}{\cos x}\right)\left(\dfrac{1}{\sin x}\right)\cos x = 1$

46. $\cos\left(\dfrac{\pi}{2} - x\right)\cot x =$
$\sin x\,\dfrac{\cos x}{\sin x} = \cos x$

47. $2 - \sec^2 x = 1 +$
$(1 - \sec^2 x) = 1 - \tan^2 x$

48. $\sin x + \cos x \cot x =$
$\sin x + \cos x\,\dfrac{\cos x}{\sin x} =$
$\dfrac{\sin^2 x + \cos^2 x}{\sin x} = \dfrac{1}{\sin x} =$
$\csc x$

49. $\dfrac{\cos^2 x + \sin^2 x}{1 + \tan^2 x} =$
$\dfrac{1}{\sec^2 x} = \cos^2 x$

50. $\dfrac{\sin^2(-x)}{\tan^2 x} =$
$\dfrac{\sin^2 x \cos^2 x}{\sin^2 x} = \cos^2 x$

51. $\dfrac{\sin\left(\dfrac{\pi}{2} - x\right) - 1}{1 - \cos(-x)} =$
$\dfrac{\cos x - 1}{1 - \cos x} = -1$

54. $\sin^2 t + \cos^2 t =$
$\dfrac{x^2}{36} + \dfrac{y^2}{36} = 1$, circle

55. $1 = \sec^2 t - \tan^2 t =$
$\dfrac{x^2}{5} - \dfrac{y^2}{1}$, hyperbola

56. $1 = \sin^2 t + \cos^2 t =$
$\dfrac{y^2}{9} + \dfrac{x^2}{4}$, ellipse

57. $1 = \sin^2 \pi t + \cos^2 \pi t =$
$\dfrac{y^2}{64} + \dfrac{x^2}{64}$, circle

(Right sidebar)

❗ **COMMON ERROR**

EXERCISES 16–27 Sign errors are common in these problems. First have students determine the signs of each function in the given interval and then find the value.

❗ **COMMON ERROR**

EXERCISES 44–53 Some students may try to use the properties of equality to verify these identities. Refer them to the Study Tip on page 850 and remind them that we do not know these equations are true, that is what we are trying to prove.

52. $\dfrac{1 + \sin x}{\cos x} + \dfrac{\cos x}{1 + \sin x} =$
$\dfrac{(1 + \sin x)^2 + \cos^2 x}{\cos x\,(1 + \sin x)} =$
$\dfrac{1 + 2\sin x + \sin^2 x + \cos^2 x}{\cos x\,(1 + \sin x)} =$
$\dfrac{2(1 + \sin x)}{\cos x\,(1 + \sin x)} = 2\sec x$

53. $\dfrac{\cos(-x)}{1 + \sin(-x)} = \dfrac{\cos x}{1 - \sin x} =$
$\dfrac{\cos x\,(1 + \sin x)}{1 - \sin^2 x} = \dfrac{\cos x\,(1 + \sin x)}{\cos^2 x} =$
$\sec x + \tan x$

58. $1 = \csc^2 2t - \cot^2 2t = \dfrac{y^2}{9} - \dfrac{x^2}{4}$, hyperbola

59. $1 = \sin^2\dfrac{t}{2} + \cos^2\dfrac{t}{2} = \dfrac{y^2}{16} + \dfrac{x^2}{1}$, ellipse

62. See Additional Answers beginning on page AA1.

ADDITIONAL PRACTICE AND RETEACHING

For Lesson 14.3:
• Practice Levels A, B, and C (*Chapter 14 Resource Book,* p. 41)
• Reteaching with Practice (*Chapter 14 Resource Book,* p. 44)
• 🖥 See Lesson 14.3 of the *Personal Student Tutor*

For more Mixed Review:
• 🖥 Search the *Test and Practice Generator* for key words or specific lessons.

DAILY HOMEWORK QUIZ

📄 *Transparency Available*

1. Given that $\sin \theta = \frac{2}{5}$ and $\frac{\pi}{2} < \theta < \pi$, find the values of the other five trigonometric functions of θ. $\cos \theta = \frac{-\sqrt{21}}{5}$;

$\tan \theta = \frac{-2\sqrt{21}}{21}$; $\csc \theta = \frac{5}{2}$;

$\cot \theta = \frac{-\sqrt{21}}{2}$; $\sec \theta = \frac{-5\sqrt{21}}{21}$

2. Simplify the expression $\tan x \csc x$. $\sec x$

3. Verify the identify $\cot x \sec x \sin x = 1$

$\cot x \sec x \sin x =$
$\left(\frac{\cos x}{\sin x}\right)\left(\frac{1}{\cos x}\right) \sin x = 1$

4. Given the parametric equations $x = 6 \cos t$ and $y = 8 \sin t$, use a trigonometric identity to determine whether the graph is a *circle*, an *ellipse*, or a *hyperbola*. $1 = \sin^2 t + \cos^2 t = \frac{y^2}{64} + \frac{x^2}{36}$, ellipse

EXTRA CHALLENGE NOTE

↳ Challenge problems for Lesson 14.3 are available in **blackline** format in the *Chapter 14 Resource Book*, p. 48 and at **www.mcdougallittell.com.**

ADDITIONAL TEST PREPARATION

1. WRITING Verify the identity $\frac{\sec(-x)}{1 - \sin x} = \sec^3 x + \sec x \tan x$.
See margin.

80.
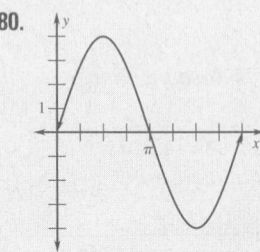

81–85. See Additional Answers beginning on page AA1.

854

---

63. Actual wheel is 18 ft wide, model is 1 ft wide. Actual wheel rotates once every 15 sec, model once every 8 sec.

65a. $x = 24 \cos \frac{\pi t}{6}$,

$y = 24 \sin \frac{\pi t}{6}$, x, y in ft, t in sec

Test Preparation ✏️

65c. *Sample response:* As speed increases, frequency increases and period decreases, so b in $y = a \sin bt$ and $y = a \cos bt$ increases. $a = r = \frac{diameter}{2}$, so as diameter increases, r increases.

★ **Challenge**

67. $\sin^2 \theta + \cos^2 \theta = 1$ original equation

$1 + \tan^2 \theta = \sec^2 \theta$ divide by $\cos^2 \theta$

$\sin^2 \theta + \cos^2 \theta = 1$ original equation

$1 + \cot^2 \theta = \csc^2 \theta$ divide by $\sin^2 \theta$

🖩 **GLEASTON WATER MILL** In Exercises 62–64, use the following information.
Suppose you have constructed a working scale model of the water wheel at the Gleaston Water Mill. The parametric equations that describe the motion of one of the paddles on each of the water wheels are as follows.

Actual waterwheel:	Scale model:
$x = 9 \cos 8\pi t$	$x = 0.5 \cos 15\pi t$
$y = 9 \sin 8\pi t$	$y = 0.5 \sin 15\pi t$

In each case, x and y are measured in feet and t is measured in minutes.

62. Use a graphing calculator to graph both sets of parametric equations. **See margin.**

63. How is your scale model different from the actual waterwheel? **See margin.**

64. How many revolutions does each wheel make in 5 minutes?
actual wheel: 20 revolutions; model wheel: 37.5 revolutions

65. MULTI-STEP PROBLEM The Dentzel Carousel in Glen Echo Park near Washington, D.C., is one of about 135 functioning antique carousels in the United States. The platform of the carousel is about 48 feet in diameter and makes about 5 revolutions per minute. ▶ Source: National Park Service

a. Find parametric equations that describe the ride's motion. **See margin.**

b. Suppose the platform of the carousel had a diameter of 38 feet and made about 4.5 revolutions per minute. Find parametric equations that would describe the ride's motion. $x = 19 \cos\left(\frac{3\pi}{20}t\right)$, $y = 19 \sin\left(\frac{3\pi}{20}t\right)$

c. *Writing* How are the parametric equations affected when the speed is changed? How are the equations affected when the diameter is changed? **See margin.**

66. Use the definitions of sine and cosine from Lesson 13.3 to derive the Pythagorean identity $\sin^2 \theta + \cos^2 \theta = 1$. path: $x^2 + y^2 = 1$. $\sin \theta = x$, $\cos \theta = y$, so $\sin^2 \theta + \cos^2 \theta = 1$.

67. Use the Pythagorean identity $\sin^2 \theta + \cos^2 \theta = 1$ to derive the other Pythagorean identities, $1 + \tan^2 \theta = \sec^2 \theta$ and $1 + \cot^2 \theta = \csc^2 \theta$. **See margin.**

MIXED REVIEW

QUADRATIC EQUATIONS Solve the equation by factoring. (Review 5.2 for 14.4)

68. $x^2 - 5x - 14 = 0$ $-2, 7$ **69.** $x^2 + 5x - 36 = 0$ $-9, 4$ **70.** $x^2 - 19x + 88 = 0$ $8, 11$

71. $2x^2 - 7x - 15 = 0$ $-\frac{3}{2}, 5$ **72.** $36x^2 - 16 = 0$ $-\frac{2}{3}, \frac{2}{3}$ **73.** $9x^2 - 1 = 0$ $-\frac{1}{3}, \frac{1}{3}$

EVALUATING EXPRESSIONS Evaluate the expression without using a calculator. Give your answer in both radians and degrees. (Review 13.4 for 14.4)

74. $\cos^{-1} \frac{\sqrt{2}}{2}$ $45°, \frac{\pi}{4}$ **75.** $\tan^{-1} \sqrt{3}$ $60°, \frac{\pi}{3}$ **76.** $\sin^{-1}\left(-\frac{\sqrt{3}}{2}\right)$ $-60°, -\frac{\pi}{3}$

77. $\sin^{-1} \frac{1}{2}$ $30°, \frac{\pi}{6}$ **78.** $\tan^{-1}(-1)$ $-45°, -\frac{\pi}{4}$ **79.** $\cos^{-1}\left(-\frac{1}{2}\right)$ $120°, \frac{2\pi}{3}$

GRAPHING Draw one cycle of the function's graph. (Review 14.1) 80–85. See margin.

80. $y = 4 \sin x$ **81.** $y = 2 \cos x$ **82.** $y = \tan 4\pi x$

83. $y = 3 \tan \pi x$ **84.** $y = 5 \cos 2x$ **85.** $y = 10 \sin 4x$

Additional Test Preparation *Sample answer:*

$$\frac{\sec(-x)}{1 - \sin x} = \frac{(\sec x)(1 + \sin x)}{(1 - \sin x)(1 + \sin x)} = \frac{\sec x + \sec x \sin x}{1 - \sin^2 x} = \frac{\sec x + \left(\frac{1}{\cos x}\right)(\sin x)}{\cos^2 x} =$$

$$(\sec x + \tan x)(\sec^2 x) = \sec^3 x + \sec^2 x \tan x$$

14.4

Solving Trigonometric Equations

What you should learn

GOAL 1 Solve a trigonometric equation.

GOAL 2 Solve **real-life** trigonometric equations, such as an equation for the number of hours of daylight in Prescott, Arizona, in **Example 6.**

Why you should learn it

▼ To solve many types of **real-life** problems, such as finding the position of the sun at sunrise in **Ex. 58.**

STUDENT HELP

⮕ **Look Back**
For help with inverse trigonometric functions, see p. 792.

GOAL 1 SOLVING A TRIGONOMETRIC EQUATION

In Lesson 14.3 you verified trigonometric identities. In this lesson you will solve trigonometric equations. To see the difference, consider the following equations:

$$\sin^2 x + \cos^2 x = 1 \qquad \textbf{Equation 1}$$

$$\sin x = 1 \qquad \textbf{Equation 2}$$

Equation 1 is an identity because it is true for all real values of x. Equation 2, however, is true only for some values of x. When you find these values, you are solving the equation.

EXAMPLE 1 *Solving a Trigonometric Equation*

Solve $2 \sin x - 1 = 0$.

SOLUTION

First isolate $\sin x$ on one side of the equation.

$2 \sin x - 1 = 0$	**Write original equation.**
$2 \sin x = 1$	**Add 1 to each side.**
$\sin x = \dfrac{1}{2}$	**Divide each side by 2.**

One solution of $\sin x = \dfrac{1}{2}$ in the interval $0 \le x < 2\pi$ is $x = \sin^{-1} \dfrac{1}{2} = \dfrac{\pi}{6}$. Another

such solution is $x = \pi - \dfrac{\pi}{6} = \dfrac{5\pi}{6}$.

Moreover, because $y = \sin x$ is a periodic function, there are infinitely many other solutions. You can write the general solution as

$$x = \dfrac{\pi}{6} + 2n\pi \qquad \text{or} \qquad x = \dfrac{5\pi}{6} + 2n\pi$$

where n is any integer.

✓ **CHECK** You can check your answer graphically. Graph $y = \sin x$ and $y = \dfrac{1}{2}$ in the same coordinate plane and find the points where the graphs intersect.

You can see that there are infinitely many such points.

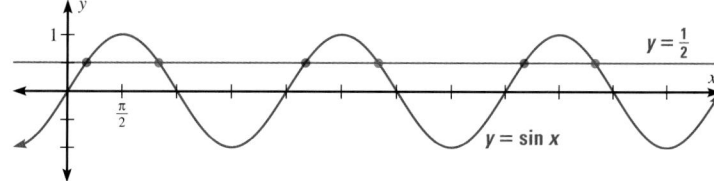

⟶ **LESSON OPENER**
GRAPHING CALCULATOR
An alternative way to approach Lesson 14.4 is to use the Graphing Calculator Lesson Opener:

• Blackline Master (*Chapter 14 Resource Book*, p. 52)
• Transparency (p. 94)

MEETING INDIVIDUAL NEEDS
• *Chapter 14 Resource Book*
Prerequisite Skills Review (p. 5)
Practice Level A (p. 55)
Practice Level B (p. 56)
Practice Level C (p. 57)
Reteaching with Practice (p. 58)
Absent Student Catch-Up (p. 60)
Challenge (p. 63)
• *Resources in Spanish*
• *Personal Student Tutor*

NEW-TEACHER SUPPORT
See the Tips for New Teachers on pp. 1–2 of the *Chapter 14 Resource Book* for additional notes about Lesson 14.4.

WARM-UP EXERCISES

Transparency Available
Evaluate each expression without using a calculator. Give your answer in radians.

1. $\sin^{-1} \dfrac{\sqrt{3}}{2}$ $\quad \dfrac{\pi}{3}$

2. $\tan^{-1} (-1)$ $\quad -\dfrac{\pi}{4}$

3. $\cos^{-1} \left(-\dfrac{\sqrt{2}}{2} \right)$ $\quad \dfrac{3\pi}{4}$

4. $\tan^{-1} \dfrac{\sqrt{3}}{3}$ $\quad \dfrac{\pi}{6}$

14.4 Solving Trigonometric Equations **855**

The number of hours of sunlight per day can be modeled using a trigonometric function. In today's lesson you will learn to solve trigonometric equations so that you can solve real-life problems using a trigonometric function such as finding on which day of the year there are 13 hours of sunlight in Prescott, Arizona.

EXTRA EXAMPLE 1
Solve $4 \cos x + 1 = -1$.
$x = \frac{2\pi}{3} + 2n\pi$ and $\frac{4\pi}{3} + 2n\pi$

EXTRA EXAMPLE 2
Solve $9 \sin^2 x + 2 = 3$ in the interval $0 \le x \le 2\pi$. $x \approx 0.340$, 2.80, 3.48, 5.94

EXTRA EXAMPLE 3
Solve $\sin x \tan^2 x = \sin x$.
$x = n\pi, -\frac{\pi}{4} + n\pi$, and $\frac{\pi}{4} + n\pi$

CHECKPOINT EXERCISES

For use after Example 1:

1. Solve $\sqrt{3} \tan x - 1 = 0$.
$x = \frac{\pi}{6} + n\pi$

For use after Example 2:

2. Solve $2 \cot x = \tan x$ in the interval $0 \le x \le 2\pi$.
$x \approx 0.955$, 2.19, 4.10, 5.33

For use after Example 3:

3. Solve $\cot^2 x = \cos x \csc x$.
$x = \frac{\pi}{4} + n\pi$

EXAMPLE 2 *Solving a Trigonometric Equation in an Interval*

Solve $4 \tan^2 x - 1 = 0$ in the interval $0 \le x < 2\pi$.

SOLUTION

$4 \tan^2 x - 1 = 0$	**Write original equation.**
$4 \tan^2 x = 1$	**Add 1 to each side.**
$\tan^2 x = \frac{1}{4}$	**Divide each side by 4.**
$\tan x = \pm \frac{1}{2}$	**Take square roots of each side.**

Use a calculator to find values of x for which $\tan x = \pm\frac{1}{2}$, as shown at the right.

The general solution of the equation is

$$x \approx 0.464 + n\pi$$

or

$$x \approx -0.464 + n\pi$$

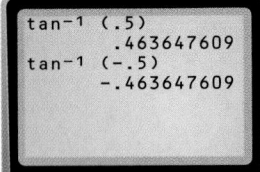

```
tan⁻¹ (.5)
            .463647609
tan⁻¹ (-.5)
           -.463647609
```

where n is any integer. The solutions that are in the interval $0 \le x < 2\pi$ are:

$x \approx 0.464$	$x \approx 0.464 + \pi \approx 3.61$
$x \approx -0.464 + \pi \approx 2.628$	$x \approx -0.464 + 2\pi \approx 5.82$

✓ **CHECK** Check these solutions by substituting them back into the original equation.

STUDENT HELP

↳ **Study Tip**
Note that to find the general solution of a trigonometric equation, you must add multiples of the period to the solutions in one cycle.

EXAMPLE 3 *Factoring to Solve a Trigonometric Equation*

Solve $\sin^2 x \cos x = 4 \cos x$.

SOLUTION

$\sin^2 x \cos x = 4 \cos x$	**Write original equation.**
$\sin^2 x \cos x - 4 \cos x = 0$	**Subtract 4 cos x from each side.**
$\cos x \,(\sin^2 x - 4) = 0$	**Factor out cos x.**
$\cos x \,(\sin x + 2)(\sin x - 2) = 0$	**Factor difference of squares.**

Set each factor equal to 0 and solve for x, if possible.

$\cos x = 0$	$\sin x + 2 = 0$	$\sin x - 2 = 0$
$x = \frac{\pi}{2}$ or $x = \frac{3\pi}{2}$	$\sin x = -2$	$\sin x = 2$

STUDENT HELP

↳ **Study Tip**
Remember not to divide both sides of an equation by a variable expression, such as cos x.

Because neither $\sin x = -2$ nor $\sin x = 2$ has a solution, the only solutions in the interval $0 \le x < 2\pi$ are $x = \frac{\pi}{2}$ and $x = \frac{3\pi}{2}$.

▶ The general solution is $x = \frac{\pi}{2} + 2n\pi$ or $x = \frac{3\pi}{2} + 2n\pi$ where n is any integer.

---

856   **Chapter 14** *Trigonometric Graphs, Identities, and Equations*

EXAMPLE 4 *Using the Quadratic Formula*

Solve $\cos^2 x - 4 \cos x + 1 = 0$ in the interval $0 \le x \le \pi$.

SOLUTION

STUDENT HELP

Look Back
For help with the quadratic formula, see p. 291.

Since the equation is in the form $au^2 + bu + c = 0$, you can use the quadratic formula to solve for $u = \cos x$.

$\cos^2 x - 4 \cos x + 1 = 0$	**Write original equation.**
$\cos x = \dfrac{4 \pm \sqrt{(-4)^2 - 4(1)(1)}}{2(1)}$	**Quadratic formula**
$= \dfrac{4 \pm \sqrt{12}}{2} = 2 \pm \sqrt{3}$	**Simplify.**
≈ 3.73 or 0.268	**Use a calculator.**
$x \approx \cos^{-1} 3.73$ or $x \approx \cos^{-1} 0.268$	**Use inverse cosine.**
No solution $\qquad \approx 1.30$	**Use a calculator if possible.**

▶ In the interval $0 \le x \le \pi$, the only solution is $x \approx 1.30$. Check this in the original equation.

· · · · · · · · · ·

When solving a trigonometric equation, it is possible to obtain extraneous solutions. Therefore, you should always check your solutions in the original equation.

EXAMPLE 5 *An Equation with Extraneous Solutions*

STUDENT HELP

HOMEWORK HELP
Visit our Web site
www.mcdougallittell.com
for extra examples.

Solve $1 - \cos x = \sqrt{3} \sin x$ in the interval $0 \le x < 2\pi$.

SOLUTION

$1 - \cos x = \sqrt{3} \sin x$	**Write original equation.**
$(1 - \cos x)^2 = (\sqrt{3} \sin x)^2$	**Square both sides.**
$1 - 2 \cos x + \cos^2 x = 3 \sin^2 x$	**Multiply.**
$1 - 2 \cos x + \cos^2 x = 3(1 - \cos^2 x)$	**Pythagorean identity**
$4 \cos^2 x - 2 \cos x - 2 = 0$	**Quadratic form**
$2 \cos^2 x - \cos x - 1 = 0$	**Divide each side by 2.**
$(2 \cos x + 1)(\cos x - 1) = 0$	**Factor.**
$2 \cos x + 1 = 0$ or $\cos x - 1 = 0$	**Zero product property**
$\cos x = -\dfrac{1}{2} \qquad \cos x = 1$	**Solve for cos x.**
$x = \dfrac{2\pi}{3}$ or $x = \dfrac{4\pi}{3} \qquad x = 0$	**Solve for x.**

▶ The apparent solution $x = \dfrac{4\pi}{3}$ does not check in the original equation. The only solutions in the interval $0 \le x < 2\pi$ are $x = 0$ and $x = \dfrac{2\pi}{3}$.

EXTRA EXAMPLE 4
Solve $\sin^2 x - 3 \sin x - 5 = 0$ in the interval $0 \le x < \pi$. **no solution**

EXTRA EXAMPLE 5
Solve $1 + \tan x = \sqrt{2} \sec x$ in the interval $0 \le x < 2\pi$. $x = \dfrac{\pi}{4}$

✓ CHECKPOINT EXERCISES
For use after Example 4:
1. Solve $2 \tan^2 x + 3 \tan x - 1 = 0$ in the interval $0 \le x < \pi$.
$x \approx 0.274$ and 2.08
For use after Example 5:
2. Solve $\cos x - 1 = 3 \sin x$ in the interval $0 \le x < 2\pi$.
$x = 0$ and about 3.78

STUDENT HELP NOTES
→ **Look Back** As students look back to p. 291, remind them to identify the values of *a*, *b*, and *c* before using the formula.

→ **Homework Help** Students can find extra examples at **www.mcdougallittell.com** that parallel the examples in the student edition.

CONCEPT QUESTION
EXAMPLE 5 Why was it necessary to square both sides of this equation? **You need to convert all the trigonometric functions to cosine so that the expression can be factored.**

EXTRA EXAMPLE 6
The number *T* of degrees (°F) for
the average daily temperature in
one city can be modeled by

$$T = 42 + 33.4 \sin\left(\frac{2\pi}{12}t + 5.2\right)$$

where *t* is measured in months,
with *t* = 0 representing January 1.
According to the model, on
which days of the year is the
average daily temperature 68°F?
about April 23 and about July 11

 CHECKPOINT EXERCISES
For use after Example 6:
1. Refer to Extra Example 6. On
which days of the year is the
average daily temperature
15°F? **about January 8 and
about October 27**

CAREER NOTE
EXAMPLE 6 Additional
information about a career as
a meteorologist is available at
www.mcdougallittell.com.

FOCUS ON VOCABULARY
What is an extraneous solution?
**An extraneous solution can arise
when one equation is algebraically
manipulated to be another equation.
An extraneous solution is a solution
to the last equation but not a solution
to the original equation.**

CLOSURE QUESTION
Name three techniques that you
might use to solve a trigonometric
equation. *Sample answer:* **Isolate a
single trigonometric function, use factoring, or use the quadratic formula.**

DAILY PUZZLER
For what value of *x* in the interval
$0 \le x < \pi$ is the value of cos *x* twice
the value of sin *x*? **x ≈ 0.464**

FOCUS ON CAREERS

METEOROLOGIST
Operational
meteorologists, the largest
group of specialists in the
field, use complex computer
models to forecast the
weather. Other types of
meteorologists include
physical meteorologists
and climatologists.

CAREER LINK
www.mcdougallittell.com

GOAL 2 **SOLVING TRIGONOMETRIC EQUATIONS IN REAL LIFE**

EXAMPLE 6 *Solving a Real-Life Trigonometric Equation*

METEOROLOGY The number *h* of hours of sunlight per day in Prescott, Arizona,
can be modeled by

$$h = 2.325 \sin\frac{\pi}{6}(t - 2.667) + 12.155$$

where *t* is measured in months and *t* = 0 represents January 1. On which days of the
year are there 13 hours of sunlight in Prescott? ▶ Source: Gale Research Company

SOLUTION

Method 1 Substitute 13 for *h* in the model and solve for *t*.

$$2.325 \sin\frac{\pi}{6}(t - 2.667) + 12.155 = 13$$

$$2.325 \sin\frac{\pi}{6}(t - 2.667) = 0.845$$

$$\sin\frac{\pi}{6}(t - 2.667) \approx 0.363$$

$$\frac{\pi}{6}(t - 2.667) \approx \sin^{-1}(0.363) \quad \text{or} \quad \frac{\pi}{6}(t - 2.667) \approx \pi - \sin^{-1}(0.363)$$

$$\frac{\pi}{6}(t - 2.667) \approx 0.371 \qquad\qquad \frac{\pi}{6}(t - 2.667) \approx \pi - 0.371 \approx 2.771$$

$$t - 2.667 \approx 0.709 \qquad\qquad t - 2.667 \approx 5.292$$

$$t \approx 3.38 \qquad\qquad\qquad t \approx 7.96$$

▶ The time *t* = 3.38 represents 3 full months plus (0.38)(30) ≈ 11 days, or April 11.
Likewise, the time *t* = 7.96 represents 7 full months plus (0.96)(31) ≈ 30 days, or
August 30. (Notice that these two days occur about 70 days before and after June
21, which is the date of the summer solstice, the longest day of the year.)

Method 2 Use a graphing calculator. Graph the equations

$$y = 2.325 \sin\frac{\pi}{6}(x - 2.667) + 12.155$$

$$y = 13$$

in the same viewing window. Then use the *Intersect* feature to find the points of
intersection.

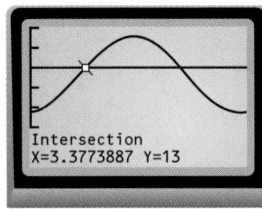

Intersection
X=3.3773887 Y=13

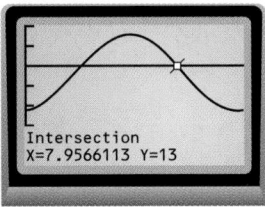

Intersection
X=7.9566113 Y=13

▶ From the screens above, you can see that *t* ≈ 3.38 or *t* ≈ 7.96. The time *t* = 3.38
is about April 11, and the time *t* = 7.96 is about August 30.

GUIDED PRACTICE

Vocabulary Check ✓

1. What is the difference between a trigonometric equation and a trigonometric identity? **See margin.**

Concept Check ✓

2. Name several techniques for solving trigonometric equations.

3. **ERROR ANALYSIS** Describe the error(s) in the calculations shown. $\cos x = 0$ is also a solution, and the general solutions should be $+ 2n\pi$, not $+ 2\pi$.

1. A trigonometric identity is true for all values of the variables for which the expressions are defined. A trigonometric equation is true for only some values of *x*.
2. substitute and solve, graphing calculator, factoring, quadratic formula

$$\cos^2 x = \frac{1}{2}\cos x$$

$$\cos x = \frac{1}{2}$$

$$x = \frac{\pi}{3} \text{ or } x = \frac{5\pi}{3} \qquad \text{Solution for } 0 \le x < 2\pi$$

$$x = \frac{\pi}{3} + 2\pi \text{ or } x = \frac{5\pi}{3} + 2\pi \qquad \text{General solution}$$

Skill Check ✓

10. $\frac{\pi}{4} + 2n\pi, \frac{3\pi}{4} + 2n\pi,$
$\frac{5\pi}{4} + 2n\pi, \frac{7\pi}{4} + 2n\pi$

21. $\frac{\pi}{6} + 2n\pi, \frac{5\pi}{6} + 2n\pi$

22. $\frac{\pi}{3} + 2n\pi, \frac{5\pi}{3} + 2n\pi$

26. $\frac{\pi}{4} + n\pi, \frac{\pi}{4} + 2n\pi,$
$\frac{3\pi}{4} + 2n\pi$

Solve the equation in the interval $0 \le x < 2\pi$.

4. $2 \cos x + 4 = 5 \quad \frac{\pi}{3}, \frac{5\pi}{3}$

5. $3 \sec^2 x - 4 = 0 \quad \frac{\pi}{6}, \frac{5\pi}{6}, \frac{7\pi}{6}, \frac{11\pi}{6}$

6. $\tan^2 x = \cos x \tan^2 x \quad 0, \pi$

7. $5 \cos x - \sqrt{3} = 3 \cos x \quad \frac{\pi}{6}, \frac{11\pi}{6}$

Find the general solution of the equation.

8. $3 \csc x + 5 = 0 \quad 5.64 + 2n\pi, 3.79 + 2n\pi$

9. $4 \sin x = \sqrt{3} \quad 0.45 + 2n\pi, 2.69 + 2n\pi$

10. $1 + \tan^2 x = 6 - 2 \sec^2 x$ **See margin.**

11. $2 - 2 \cos^2 x = 5 \sin x + 3 \quad \frac{7\pi}{6} + 2n\pi, \frac{11\pi}{6} + 2n\pi$

12. 🌐 **METEOROLOGY** Look back at the model in Example 6 on page 858. On which days of the year are there 10 hours of sunlight in Prescott, Arizona?
0.41, January 13, and 10.93, November 28

PRACTICE AND APPLICATIONS

STUDENT HELP

▶ **Extra Practice**
to help you master skills is on p. 960.

27. $\frac{\pi}{2} + 2n\pi, \frac{7\pi}{6} + 2n\pi,$
$\frac{11\pi}{6} + 2n\pi$

28. $1.36 + 2n\pi, 4.92 + 2n\pi$

29. $\frac{\pi}{2} + 2n\pi, \frac{11\pi}{6} + 2n\pi$

STUDENT HELP

▶ **HOMEWORK HELP**
Examples 1, 3: Exs. 13–32
Examples 2, 4, 5:
Exs. 13–18, 33–56
Example 6: Exs. 57–59

31. $\frac{\pi}{3} + 2n\pi, \frac{5\pi}{3} + 2n\pi$

CHECKING SOLUTIONS Verify that the given *x*-value is a solution of the equation.

13. $5 + 4 \cos x - 1 = 0, x = \pi$ **yes**

14. $\csc x - 2 = 0, x = \frac{5\pi}{6}$ **yes**

15. $4 \cos^2 x - 3 = 0, x = \frac{\pi}{6}$ **yes**

16. $3 \tan^3 x - 3 = 0, x = \frac{\pi}{4}$ **yes**

17. $2 \sin^4 x - \sin^2 x = 0, x = \frac{5\pi}{4}$ **yes**

18. $2 \cot^4 x - \cot^2 x - 15 = 0, x = \frac{13\pi}{6}$ **yes**

SOLVING Find the general solution of the equation.

19. $2 \cos x - 1 = 0 \quad \frac{\pi}{3} + 2n\pi, \frac{5\pi}{3} + 2n\pi$

20. $3 \tan x - \sqrt{3} = 0 \quad \frac{\pi}{6} + n\pi$

21. $\sin x = \sin(-x) + 1$ **See margin.**

22. $4 \cos x = 2 \cos x + 1$ **See margin.**

23. $4 \sin^2 x - 2 = 0 \quad \frac{\pi}{4} + \frac{\pi n}{2}$

24. $9 \tan^2 x - 3 = 0 \quad \frac{\pi}{6} + n\pi, \frac{5\pi}{6} + n\pi$

25. $\sin x \cos x - 2 \cos x = 0 \quad \frac{\pi}{2} + n\pi$

26. $\sqrt{2} \cos x \sin x - \cos x = 0$ **See margin.**

27. $2 \sin^2 x - \sin x = 1$ **See margin.**

28. $0 = \cos^2 x - 5 \cos x + 1$ **See margin.**

29. $1 - \sin x = \sqrt{3} \cos x$ **See margin.**

30. $\sqrt{\sin x} = 2 \sin x - 1 \quad \frac{\pi}{2} + 2n\pi$

31. $\cos x - 1 = -\cos x$ **See margin.**

32. $6 \sin x = \sin x + 3$
0.64 + 2n\pi, 2.50 + 2n\pi

14.4 *Solving Trigonometric Equations* **859**

○ **ASSIGNMENT GUIDE**

BASIC
Day 1: pp. 859–861 Exs. 14–24 even, 33–38, 44–54 even, 57–58, 60–61, 65–75 odd

AVERAGE
Day 1: pp. 859–861 Exs. 14–28 even, 33–40, 44–54 even, 57–61, 65–75 odd

ADVANCED
Day 1: pp. 859–861 Exs. 14–54 even, 57–63, 65–75 odd

BLOCK SCHEDULE
pp. 859–861 Exs. 14–28 even, 33–40, 44–54 even, 57–61, 65–75 odd (with 14.5)

EXERCISE LEVELS
Level A: *Easier*
13–16, 19–24, 33–36, 47–50
Level B: *More Difficult*
17–18, 25–32, 37–46, 51–59, 61
Level C: *Most Difficult*
62–63

✔ **HOMEWORK CHECK**
To quickly check student understanding of key concepts, go over the following exercises: Exs. 14, 18, 22, 36, 44, 48, 52. See also the Daily Homework Quiz:

• Blackline Master (*Chapter 14 Resource Book,* p. 66)
• 📖 Transparency (p. 109)

! **COMMON ERROR**
EXERCISES 33–46 Students may not include some of the solutions or include an extraneous solution. To be sure that the correct number of solutions is given, students can use a graphing calculator to graph both sides of the equation to see how many intersections there are in the desired interval.

APPLICATION NOTE
Additional information about the Bay of Fundy is available at **www.mcdougallittell.com**.

68.
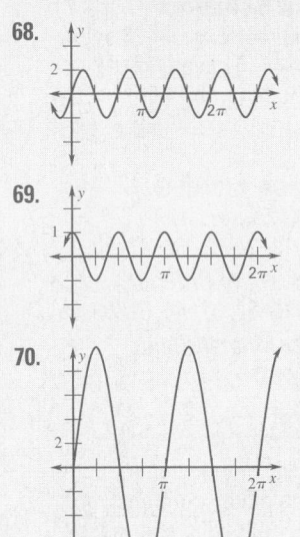

69.

70.

71–76. See Additional Answers beginning on page AA1.

ADDITIONAL PRACTICE AND RETEACHING

For Lesson 14.4:
• Practice Levels A, B, and C (*Chapter 14 Resource Book*, p. 55)
• Reteaching with Practice (*Chapter 14 Resource Book*, p. 58)
• See Lesson 14.4 of the *Personal Student Tutor*

For more Mixed Review:
• Search the *Test and Practice Generator* for key words or specific lessons.

860

SOLVING Solve the equation in the interval $0 \le x < 2\pi$. Check your solutions.

33. $5 \cos x - 3 = 0$ 0.93, 5.36

34. $3 \sin x = \sin x - 1$ $\frac{7\pi}{6}, \frac{11\pi}{6}$

35. $\tan^2 x - 3 = 0$ $\frac{\pi}{3}, \frac{2\pi}{3}, \frac{4\pi}{3}, \frac{5\pi}{3}$

36. $10 \tan x - 5 = 0$ 0.46, 3.61

37. $2 \cos^2 x - \sin x - 1 = 0$
37. $\frac{\pi}{6}, \frac{5\pi}{6}, \frac{3\pi}{2}$

38. $\cos^3 x = \cos x$ $0, \frac{\pi}{2}, \pi, \frac{3\pi}{2}$

39. $\sec^2 x - 2 = 0$ $\frac{\pi}{4}, \frac{7\pi}{4}$

40. $\tan^2 x = \sin x \sec x$ $0, \frac{\pi}{4}, \pi, \frac{5\pi}{4}$

41. $2 \cos x = \sec x$ $\frac{\pi}{4}, \frac{3\pi}{4}, \frac{5\pi}{4}, \frac{7\pi}{4}$

42. $\cos x \csc^2 x + 3 \cos x = 7 \cos x$
42. $\frac{\pi}{2}, \frac{3\pi}{2}, \frac{\pi}{6}, \frac{5\pi}{6}, \frac{7\pi}{6}, \frac{11\pi}{6}$

APPROXIMATING SOLUTIONS Use a graphing calculator to approximate the solutions of the equation in the interval $0 \le x < 2\pi$.

43. $3 \tan x + 1 = 13$ 1.33, 4.47

44. $8 \cos x + 3 = 4$ 1.45, 4.84

45. $4 \sin x = -2 \sin x - 5$ 4.13, 5.30

46. $3 \sin x + 5 \cos x = 4$ 1.36, 6.01

FINDING INTERCEPTS Find the *x*-intercepts of the graph of the given function in the interval $0 \le x < 2\pi$.

47. $y = 2 \sin x + 1$ $\frac{7\pi}{6}, \frac{11\pi}{6}$

48. $y = 2 \tan^2 x - 6$ $\frac{\pi}{3}, \frac{2\pi}{3}, \frac{4\pi}{3}, \frac{5\pi}{3}$

49. $y = \sec^2 x - 1$ 0

50. $y = -3 \cos x + \sin x$ 1.25, 4.39

FINDING INTERSECTION POINTS Find the points of intersection of the graphs of the given functions in the interval $0 \le x < 2\pi$.

51. $y = \sqrt{3} \tan^2 x$
  $y = \sqrt{3} - 2 \tan x$
51. $\frac{\pi}{6}, \frac{2\pi}{3}, \frac{7\pi}{6}, \frac{5\pi}{3}$

52. $y = 9 \cos^2 x$ $\frac{\pi}{3}, \frac{5\pi}{3}$
  $y = \cos^2 x + 8 \cos x - 2$

53. $y = \tan x \sin x$
  $y = \cos x$ $\frac{\pi}{4}, \frac{3\pi}{4}, \frac{5\pi}{4}, \frac{7\pi}{4}$

54. $y = \sin^2 x$
  $y = 2 \sin x - 1$ $\frac{\pi}{2}$

55. $y = 2 - \sin x \tan x$
  $y = \cos x$ $\frac{\pi}{3}, \frac{5\pi}{3}$

56. $y = 4 \cos^2 x$
  $y = 4 \cos x - 1$ $\frac{\pi}{3}, \frac{5\pi}{3}$

57. 🌊 **OCEAN TIDES** The *tide*, or depth of the ocean near the shore, changes throughout the day. The depth of the Bay of Fundy can be modeled by

$$d = 35 - 28 \cos \frac{\pi}{6.2} t$$

where d is the water depth in feet and t is the time in hours. Consider a day in which $t = 0$ represents 12:00 A.M. For that day, when do the high and low tides occur? At what time(s) is the water depth $3\frac{1}{2}$ feet?   **See margin.**

57. highs: 6:12 A.M. and 6:36 P.M., lows: 12:00 A.M. and 12:24 P.M. The water depth never goes below 7 ft.

58. 📱 **POSITION OF THE SUN** Cheyenne, Wyoming, has a latitude of 41°N. At this latitude, the position of the sun at sunrise can be modeled by

$$D = 31 \sin \left(\frac{2\pi}{365} t - 1.4 \right)$$

where t is the time in days and $t = 1$ represents January 1. In this model, D represents the number of degrees north or south of due east that the sun rises. Use a graphing calculator to determine the days that the sun is more than 20° north of due east at sunrise.   **days 122–223, or May 2 to August 11**

59. 🌐 **METEOROLOGY** A model for the average daily temperature T (in degrees Fahrenheit) in Kansas City, Missouri, is given by

$$T = 54 + 25.2 \sin \left(\frac{2\pi}{12} t + 4.3 \right)$$

where t is measured in months and $t = 0$ represents January 1. What months have average daily temperatures higher than 70°F? Do any months have average daily temperatures below 20°F? ▶ Source: National Climatic Data Center   **June, July, August; no**

FOCUS ON APPLICATIONS

🔺**THE BAY OF FUNDY** is the portion of the Atlantic Ocean that runs along the southern coast of New Brunswick in Canada. It is the site of the highest tides in the world.

APPLICATION LINK
www.mcdougallittell.com

60. MULTIPLE CHOICE What is the general solution of the equation $\cos x + \sqrt{2} = -\cos x$? Assume n is an integer. **D**

Ⓐ $x = \dfrac{\pi}{4} + 2n\pi$

Ⓑ $x = \dfrac{\pi}{4} + 2n\pi$ or $x = \dfrac{7\pi}{4} + 2n\pi$

Ⓒ $x = \dfrac{3\pi}{4} + 2n\pi$

Ⓓ $x = \dfrac{3\pi}{4} + 2n\pi$ or $x = \dfrac{5\pi}{4} + 2n\pi$

Ⓔ $x = \dfrac{3\pi}{4} + n\pi$ or $x = \dfrac{5\pi}{4} + n\pi$

61. MULTIPLE CHOICE Find the points of intersection of the graphs of $y = 2 + \sin x$ and $y = 3 - \sin x$ in the interval $0 \le x < 2\pi$. **A**

Ⓐ $\left(\dfrac{\pi}{6}, \dfrac{5}{2}\right), \left(\dfrac{5\pi}{6}, \dfrac{5}{2}\right)$

Ⓑ $\left(\dfrac{\pi}{6}, \dfrac{1}{2}\right), \left(\dfrac{5\pi}{6}, \dfrac{1}{2}\right)$

Ⓒ $\left(\dfrac{\pi}{3}, \dfrac{5}{2}\right), \left(\dfrac{4\pi}{3}, \dfrac{5}{2}\right)$

Ⓓ $\left(\dfrac{\pi}{3}, \dfrac{1}{2}\right), \left(\dfrac{4\pi}{3}, \dfrac{1}{2}\right)$

Ⓔ $\left(\dfrac{\pi}{6}, \dfrac{5}{2}\right), \left(\dfrac{11\pi}{6}, \dfrac{5}{2}\right)$

★ **Challenge**

MATRICES In Exercises 62 and 63, use the following information.
Matrix multiplication can be used to rotate a point (x, y) counter clockwise about the origin through an angle θ. The coordinates of the resulting point (x', y') are determined by the following matrix equation:

$$\begin{bmatrix} \cos\theta & -\sin\theta \\ \sin\theta & \cos\theta \end{bmatrix}\begin{bmatrix} x \\ y \end{bmatrix} = \begin{bmatrix} x' \\ y' \end{bmatrix}$$

62. The point $(4, 1)$ is rotated counter clockwise about the origin through an angle of $\dfrac{\pi}{3}$. What are the coordinates of the resulting point? $\left(\dfrac{4 - \sqrt{3}}{2}, \dfrac{4\sqrt{3} + 1}{2}\right)$

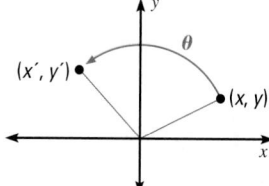

63. Through what angle θ must the point $(2, 4)$ be rotated to produce $(x', y') = \left(-2 + \sqrt{3}, 1 + 2\sqrt{3}\right)$? $\dfrac{\pi}{6}$

EXTRA CHALLENGE
▸ www.mcdougallittell.com

MIXED REVIEW

USING COMPLEMENTS Two six-sided dice are rolled. Find the probability of the given event. (Review 12.4)

64. The sum is less than or equal to 10. $\dfrac{33}{36}$ **65.** The sum is not 4. $\dfrac{33}{36}$

66. The sum is not 2 or 12. $\dfrac{17}{18}$ **67.** The sum is greater than 3. $\dfrac{33}{36}$

GRAPHING Graph the function. (Review 14.1, 14.2 for 14.5) 68–76. See margin.

68. $y = \sin 3x$

69. $y = 2\cos 4x$

70. $y = 10\sin 2x$

71. $y = 3\cos\dfrac{1}{2}x$

72. $y = \dfrac{1}{4}\tan x$

73. $y = 3\tan\dfrac{1}{2}x - 2$

74. $y = \cos\dfrac{1}{2}x + \pi$

75. $y = 3 + \tan\left(x - \dfrac{3\pi}{2}\right)$

76. $y = -\sin 3\pi(x + 4) + 1$

77. 🌐 **SURVEYING** Suppose you are trying to determine the width w of a small pond. You stand at a point 43 feet from one end of the pond and 50 feet from the other end. The angle formed by your lines of sight to each end of the pond measures $45°$. How wide is the pond? (Review 13.6) **36 ft**

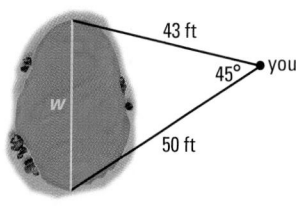

DAILY HOMEWORK QUIZ

📄 *Transparency Available*

1. Find the general solution of the equation $8\sin^2 x - 4 = 0$. $\dfrac{\pi}{4} + \dfrac{\pi n}{2}$

2. Is $\dfrac{\pi}{2}$ a solution to the equation $2\sin^2 \le 1 - \sin x = 1$? **yes**

3. Solve $\dfrac{1}{2}\sec x = \cos x$ in the interval $0 \le x < 2\pi$. $\dfrac{\pi}{4}, \dfrac{3\pi}{4}, \dfrac{5\pi}{4}, \dfrac{7\pi}{4}$

4. Use a graphing calculator to approximate the solutions of $6\tan x + 2 = 26$ in the interval $0 \le x \le 2\pi$. **1.33, 4.47**

5. Find the x-intercepts of the graph of $y = 3\cos\dfrac{1}{2}\theta$ in the interval $0 \le x < 2\pi$. π

6. Find the point(s) of intersection of the graphs of $y = 2\sin^2 x$ and $y = 4\sin x - 2$ in the interval $0 \le x < 2\pi$. $\dfrac{\pi}{2}$

EXTRA CHALLENGE NOTE

Challenge problems for Lesson 14.4 are available in **blackline** format in the *Chapter 14 Resource Book,* p. 63 and at **www.mcdougallittell.com.**

ADDITIONAL TEST PREPARATION

1. WRITING Describe how to use a graphing calculator to solve a trigonometric equation.
Sample answer: **Make two functions by setting *y* equal to the expression on either side of the equation. Enter these equations into the calculator. The solutions to the equation are the *x*-coordinates of the intersections of the two graphs. Use the Intersect feature to find these *x*-coordinates.**

LESSON OPENER
VISUAL APPROACH
An alternative way to approach Lesson 14.5 is to use the Visual Approach Lesson Opener:

• Blackline Master (*Chapter 14 Resource Book*, p. 67)
• Transparency (p. 95)

MEETING INDIVIDUAL NEEDS
• *Chapter 14 Resource Book*
  Prerequisite Skills Review (p. 5)
  Practice Level A (p. 68)
  Practice Level B (p. 69)
  Practice Level C (p. 70)
  Reteaching with Practice (p. 71)
  Absent Student Catch-Up (p. 73)
  Challenge (p. 76)
• *Resources in Spanish*
• 🖳 *Personal Student Tutor*

NEW-TEACHER SUPPORT
See the Tips for New Teachers on pp. 1–2 of the *Chapter 14 Resource Book* for additional notes about Lesson 14.5.

WARM-UP EXERCISES

🖳 **Transparency Available**

Give the amplitude, period, and horizontal or vertical shift of each function.

1. $y = 3 \cos 4x + 1$ $3; \frac{\pi}{2}$; up 1 unit

2. $y = \frac{1}{2} \sin (2x - \pi)$
   $\frac{1}{2}; \pi$; right $\frac{\pi}{2}$ units

3. $y = -1 + 2 \cos \left(x - \frac{2\pi}{3}\right)$ $2; 2\pi$;
   right $\frac{2\pi}{3}$ units; down 1 unit

4. $y = 2 + \frac{2}{3} \sin (\pi x + 3\pi)$
   $\frac{2}{3}; 2$; left 3 units; up 2 units

EXPLORING DATA AND STATISTICS

14.5

Modeling with Trigonometric Functions

GOAL 1 WRITING A TRIGONOMETRIC MODEL

What you should learn

GOAL 1 Model data with a sine or cosine function.

GOAL 2 Use technology to write a trigonometric model, as applied in **Example 4**.

Why you should learn it

▼ To model many types of **real-life** quantities, such as the temperature inside and outside an igloo in **Ex. 30**.

Graphs of sine and cosine functions are called *sinusoids*. When you write a sine or cosine function for a sinusoid, you need to find the values of $a, b > 0, h$, and k for

$$y = a \sin b(x - h) + k \qquad \text{or} \qquad y = a \cos b(x - h) + k$$

where $|a|$ is the amplitude, $\frac{2\pi}{b}$ is the period, h is the horizontal shift, and k is the vertical shift.

EXAMPLE 1 *Writing Trigonometric Functions*

Write a function for the sinusoid.

a.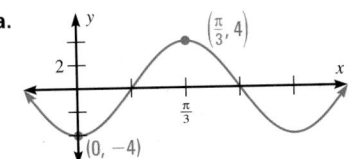

b.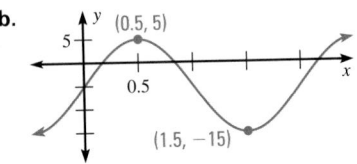

SOLUTION

a. Since the maximum and minimum values of the function occur at points equidistant from the *x*-axis, the curve has no vertical shift. And because the minimum occurs on the *y*-axis, the graph is a reflection of a cosine curve with no horizontal shift. Therefore, the function has the form $y = a \cos bx$.

The period is $\frac{2\pi}{b} = \frac{2\pi}{3}$, so $b = 3$.

The amplitude is $|a| = 4$. Since the graph is a reflection, $a = -4$.

▶ The function is $y = -4 \cos 3x$.

b. Since the maximum and minimum values of the function do *not* occur at points equidistant from the *x*-axis, the curve has a vertical shift. To find the value of k, add the maximum and minimum values and divide by 2:

$$k = \frac{M + m}{2} = \frac{5 + (-15)}{2} = \frac{-10}{2} = -5$$

Because the graph crosses the *y*-axis at $y = k$, the graph is a sine curve with no horizontal shift. Therefore, the function has the form $y = a \sin bx - 5$.

The period is $\frac{2\pi}{b} = 2$, so $b = \pi$.

The amplitude is $|a| = \frac{M - m}{2} = \frac{5 - (-15)}{2} = \frac{20}{2} = 10$.

Since the graph is not a reflection, $a = 10 > 0$.

▶ The function is $y = 10 \sin \pi x - 5$.

STUDENT HELP

 HOMEWORK HELP
Visit our Web site www.mcdougallittell.com for extra examples.

EXAMPLE 2 *Modeling a Sinusoid*

Meteorology

Write a trigonometric model for the average daily temperature in Birmingham, Alabama. ▶ Source: National Climatic Data Center

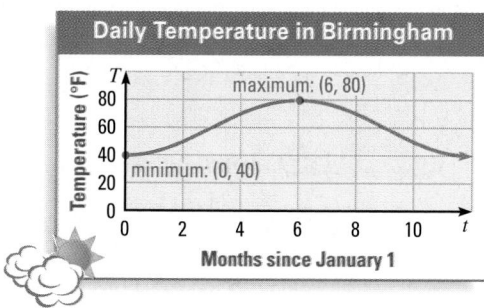

Daily Temperature in Birmingham

maximum: (6, 80)

minimum: (0, 40)

Months since January 1

SOLUTION

Notice that the graph crosses the *T*-axis at the minimum point. So if you model the temperature curve with a cosine function, there is a reflection but no horizontal shift. The mean of the maximum and minimum values is 60, so there is a vertical shift of $k = 60$.

The period is $\frac{2\pi}{b} = 12$, so $b = \frac{\pi}{6}$.

The amplitude is $|a| = 20$, and because the graph is a reflection it follows that $a = -20$.

▶ The model is $T = -20 \cos \frac{\pi}{6}t + 60$ where *t* is measured in months and $t = 0$ represents January 1.

EXAMPLE 3 *Modeling Circular Motion*

Ferris Wheel

A Ferris wheel with a radius of 25 feet is rotating at a rate of 3 revolutions per minute. When $t = 0$, a chair starts at the lowest point on the wheel, which is 5 feet above the ground. Write a model for the height *h* (in feet) of the chair as a function of the time *t* (in seconds).

SOLUTION

When the chair is at the bottom of the Ferris wheel, it is 5 feet above the ground, so $m = 5$. When it is at the top, it is $5 + 2(25) = 55$ feet above the ground, so $M = 55$.

The vertical shift for the model is $k = \frac{M + m}{2} = \frac{55 + 5}{2} = \frac{60}{2} = 30$.

When $t = 0$, the height is at its minimum, so the model is a cosine function with $a < 0$ and no horizontal shift.

The amplitude is $|a| = \frac{M - m}{2} = \frac{55 - 5}{2} = \frac{50}{2} = 25$. Because $a < 0$ it follows that $a = -25$.

Since the Ferris wheel is rotating at 3 revolutions per minute, it completes one revolution in 20 seconds. The period is $\frac{2\pi}{b} = 20$, so $b = \frac{\pi}{10}$.

▶ A model for the height of the chair as a function of time is $h = -25 \cos \frac{\pi}{10}t + 30$.

2 TEACH

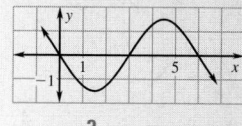

EXTRA EXAMPLE 1
Write trigonometric functions for the following graph.

$y = -\frac{3}{2} \sin \frac{\pi}{3}x$

EXTRA EXAMPLE 2
Write a trigonometric model for the average daily temperature.

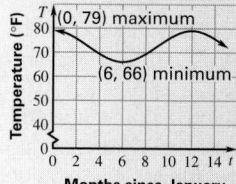

Daily Temperature

(0, 79) maximum

(6, 66) minimum

Months since January

$T = 6.5 \cos \frac{\pi}{6}t + 72.5$

EXTRA EXAMPLE 3
A Ferris wheel with a radius of 42 ft is rotating at a rate of 2 rev/min. When $t = 0$ (in seconds), a chair starts at the highest point on the wheel which is 87 ft above the ground. Write a model for the height *h* (in feet) of the chair as a function of time *t*.

$h = 42 \cos \frac{\pi}{15}t + 45$

CHECKPOINT EXERCISES
For use after Examples 1 and 2:

1. A sinusoidal curve has a maximum value of (0, 2), and a minimum value of (4, −4). Write an equation for this function. $y = 3 \cos \frac{\pi}{4}x - 1$

For use after Example 3:

2. A Ferris wheel has a radius of 15 ft and is rotating at a rate of 5 rev/min. When $t = 0$ (in seconds), a chair starts at the lowest point on the wheel, which is 4 ft above the ground. Write a model for the height *h* (in feet) of the chair as a function of time *t*.

$h = -15 \cos \frac{\pi}{6}x + 19$

EXTRA EXAMPLE 4

The average daily temperature T (in °F) is given in the table. Time t is measured in months, with $t = 0$ representing January 1. Write a trigonometric model that gives T as a function of t.

t	T	t	T
0.25	77	6.25	68
2.25	75	8.25	70
4.25	71	10.25	76

$T = 4.63 \sin (0.56x + 1.18) + 72.55$

CHECKPOINT EXERCISES

For use after Example 4:

1. At a bus stop a bus comes every 30 min. The number of people at the bus stop at any one time can be modeled by a sine or cosine function. You record the number of people at the bus stop at 5 min intervals beginning at noon for 30 min. Use the data to write a trigonometric model that gives the number of people N at the bus stop t min after noon.

t	T	t	T
0	14	20	4
5	18	25	8
10	15	30	11
15	7		

$T = 6.59 \sin (0.2x + 0.58) + 10.73$

FOCUS ON VOCABULARY

What is a sinusoid? **A sinusoid is the graph of a sine or cosine function.**

CLOSURE QUESTION

Describe how to find the vertical shift and amplitude of a sinusoid when you are given a maximum and minimum value. **See margin.**

DAILY PUZZLER

Jack and Jill each wrote an equation to model some sinusoidal data. Jack's function was a cosine function while Jill's was a sine function. Must one of them be wrong? Explain. **See margin.**

864

There are two ways you can model a set of data points whose scatter plot appears sinusoidal. One way is to estimate the minimum and maximum values and use the technique shown in Example 2. Another way is to use a graphing calculator that has a sinusoidal regression feature. The advantage of the second method is that it uses all of the data points to find the model.

EXAMPLE 4 *Using Sinusoidal Regression*

Meteorology

The average daily temperature T (in degrees Fahrenheit) in Fairbanks, Alaska, is given in the table. Time t is measured in months, with $t = 0$ representing January 1. Write a trigonometric model that gives T as a function of t.

t	0.5	1.5	2.5	3.5	4.5	5.5	6.5	7.5	8.5	9.5	10.5	11.5
T	-10.1	-3.6	11.0	30.7	48.6	59.8	62.5	56.8	45.5	25.1	2.7	-6.5

▶ Source: National Climatic Data Center

SOLUTION

Begin by entering the data in a graphing calculator and drawing a scatter plot.

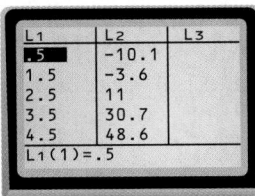 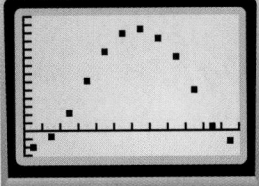

❶ Enter the data.　　　　❷ Draw a scatter plot.

Because the scatter plot appears sinusoidal, you can fit the data with a sine function to get the following model:

$$T = 37.4 \sin (0.518t - 1.72) + 26.5$$

 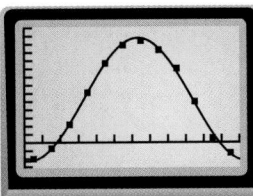

❸ Perform a sinusoidal regression.　　　　❹ Graph the model and the data in the same viewing window.

After obtaining the model, you should graph it in the same viewing window as the scatter plot to see how well the model fits the data. In this case, the model is a good fit.

✓ **CHECK** As a check on the reasonableness of the model, notice that the period is $\frac{2\pi}{0.518} \approx 12$, which is the number of months in a year.

Closure Question *Sample answer:*
The vertical shift is the average of the maximum and minimum values. The amplitude is the absolute value of the half of the difference of the maximum and minimum values.

Daily Puzzler *Sample answer:*
No, you can rewrite any cosine function as a sine function using a horizontal translation.

GUIDED PRACTICE

Vocabulary Check ✓

Concept Check ✓

1. Complete this statement: Graphs of sine and cosine functions are called _?_. **sinusoids**

2. Which two points are most useful when writing a sinusoidal model for a given graph or set of data? Explain. **See margin.**

3. Describe a characteristic of a sinusoidal graph that you would model with a cosine function rather than a sine function. **a graph that starts at a max or min at the y-intercept**

Skill Check ✓

2. Maximum and minimum points one-half period apart. With this you can find the amplitude, period, and horizontal and vertical shifts.

6. $y = 5 \cos x + 3$

7. $y = 3 \sin \frac{x}{2} + 7$

8. $y = -2 \sin \frac{\pi x}{4} + 3$

Write a function for the sinusoid.

4. $y = -\sin 3x$

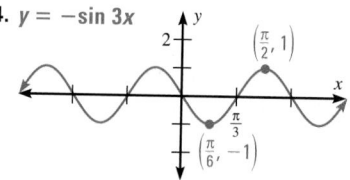

5. $y = \cos \pi x - 2$

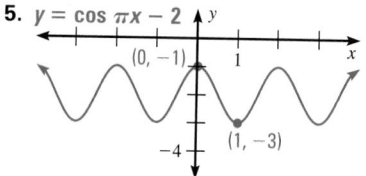

Write a function for the sinusoid with maximum at A and minimum at B.

6. $A(0, 8), B(\pi, -2)$ 7. $A(\pi, 10), B(3\pi, 4)$ 8. $A(6, 5), B(2, 1)$

9. ⊕ **FERRIS WHEEL** Look back at Example 3. Suppose the Ferris wheel rotates at a rate of 4 revolutions per minute and has a radius of 20 feet. Write a model for the height h (in feet) of the chair as a function of the time t (in seconds).

$h = -20 \cos\left(\frac{2\pi}{15}t\right) + 25$

PRACTICE AND APPLICATIONS

STUDENT HELP

▶ **Extra Practice**
to help you master skills is on p. 960.

WRITING FUNCTIONS Write a function for the sinusoid.

10. $y = 3 \cos 4x$

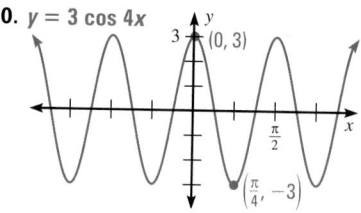

11. $y = 5 \sin 2x$

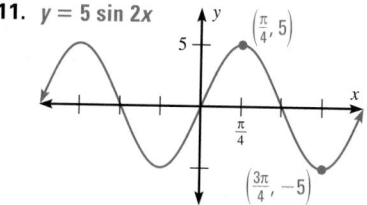

12. $y = -2 \sin 3x$

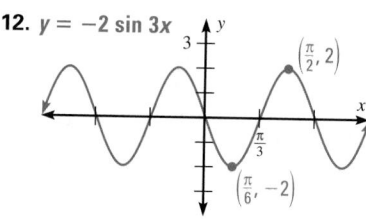

13. $y = -4 \cos \pi x$

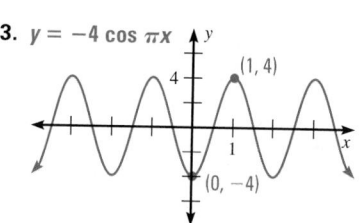

STUDENT HELP

▶ **HOMEWORK HELP**
Example 1: Exs. 10–27
Example 2: Exs. 30, 32
Example 3: Exs. 31, 33
Example 4: Exs. 34, 35

14. $y = \sin 2\pi x + 1$

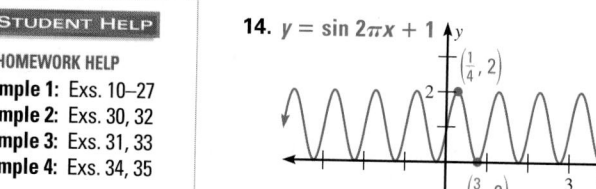

15. $y = 2 \cos \frac{\pi x}{2} - 4$

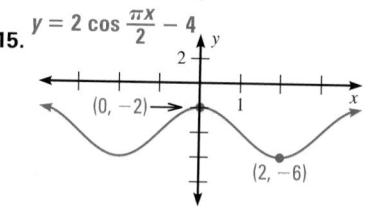

14.5 *Modeling with Trigonometric Functions* **865**

○ **ASSIGNMENT GUIDE**

BASIC
Day 1: pp. 865–867 Exs. 11–14, 17–22, 29, 31, 37–38, 41–55 odd, Quiz 2 Exs. 1–10

AVERAGE
Day 1: pp. 865–867 Exs. 11–16, 18–28 even, 29–31, 37–38, 41–55 odd, Quiz 2 Exs. 1–10

ADVANCED
Day 1: pp. 865–867 Exs. 11–16, 18–28 even, 29–30, 34–39, 41–55 odd, Quiz 2 Exs. 1–10

BLOCK SCHEDULE
pp. 865–867 Exs. 11–16, 18–28 even, 29–31, 37–38, 41–55 odd, Quiz 2 Exs. 1–10 (with 14.4)

EXERCISE LEVELS
Level A: *Easier*
11–14, 17–18, 29
Level B: *More Difficult*
15–16, 19–28, 30–38
Level C: *Most Difficult*
39

✓ **HOMEWORK CHECK**
To quickly check student understanding of key concepts, go over the following exercises: Exs. 12, 14, 18, 22. See also the Daily Homework Quiz:

• Blackline Master (*Chapter 14 Resource Book,* p. 80)

• 📖 Transparency (p. 110)

MATHEMATICAL REASONING
EXERCISES 28 AND 29 These exercises point out a fact usually overlooked by many students; the fact that sinusoids can be modeled by either a sine function or a cosine function. After completing these exercises students should realize the importance of choosing the more convenient function, but they should also realize that choosing the less convenient function does not necessarily lead to incorrect answers.

WRITING TRIGONOMETRIC FUNCTIONS Write a function for the sinusoid with maximum at A and minimum at B.

$y = -3 \sin \frac{x}{6} - 2$

16. $A(0, 3)$, $B(2\pi, -3)$
$y = 3 \cos \frac{x}{2}$

17. $A(\pi, 8)$, $B(3\pi, -8)$
$y = 8 \sin \frac{x}{2}$

18. $A(9\pi, 1)$, $B(3\pi, -5)$

19. $A\left(\frac{\pi}{3}, 4\right)$, $B(0, 2)$
$y = -\cos 3x + 3$

20. $A(2, 6)$, $B(6, 0)$
$y = 3 \sin \frac{\pi x}{4} + 3$

21. $A(3, 7)$, $B(1, -3)$
$y = -5 \sin \frac{\pi x}{2} + 2$

22. $A\left(\frac{2\pi}{3}, 9\right)$, $B(2\pi, 5)$
$y = 2 \sin \frac{3x}{4} + 7$

23. $A\left(\frac{\pi}{6}, 11\right)$, $B(0, -1)$
$y = -6 \cos 6x + 5$

24. $A(0, 5)$, $B(4, -13)$
$y = 9 \cos \frac{\pi x}{4} - 4$

25. $A(0, 0)$, $B(3\pi, -8)$
$y = 4 \cos \frac{x}{3} - 4$

26. $A(6, 2)$, $B(0, -4)$
$y = -3 \cos \frac{\pi x}{6} - 1$

27. $A\left(\frac{\pi}{4}, -2\right)$, $B\left(\frac{\pi}{12}, -6\right)$
$y = -2 \sin 6x - 4$

28. From graph: $|a| = 4$, a possible choice is $a = 4$

$b = \frac{2\pi}{P} = 3$, $h = \frac{\pi}{6}$,

$k = 0$; $4 \sin 3\left(x - \frac{\pi}{6}\right) =$

$4 \sin\left(3x - \frac{\pi}{2}\right) =$

$4 \sin\left(-\left(\frac{\pi}{2} - 3x\right)\right) =$

$-4 \sin\left(\frac{\pi}{2} - 3x\right) =$

$-4 \cos 3x$

29. A sine function, because the y-intercept of a sine function without a horizontal shift is midway between the maximum and minimum y-values.

30. $T = 10 \sin \frac{\pi t}{12} - 20$,

$T = 3 \sin \frac{\pi t}{12} + 23$,

$T = 2.5 \sin \frac{\pi t}{12} + 31.5$

32. $h = 4 \cos \frac{\pi}{8}(t - 5) + 6$

33. $h = -2.5 \cos \pi t + 6.5$

28. **CRITICAL THINKING** Any sinusoid can be modeled by either a sine function or a cosine function. Using the graph from part (a) of Example 1, find the values of a, b, h, and k for the model $y = a \sin b(x - h) + k$. Use identities to show that the model you found is equivalent to the model from part (a) of Example 1.

29. *Writing* Since any sinusoid can be modeled by either a sine function or a cosine function, you can choose the type of function that is more convenient. For a sinusoid whose y-intercept occurs halfway between the maximum and minimum values of the function, tell which type of function you would use to model the graph. Explain your answer.

30. 🌎 **CLIMATE CONTROL** Eskimos use igloos as temporary shelter from harsh winter weather. The graph shows the temperatures inside and outside an igloo throughout a typical winter day. Write a sinusoidal model for the outside temperature T (in degrees Fahrenheit) as a function of the time of day t (in hours since midnight). Then write sinusoidal models for the floor-level temperature and for the sleeping-platform temperature.

▶ Source: *Scientific American*

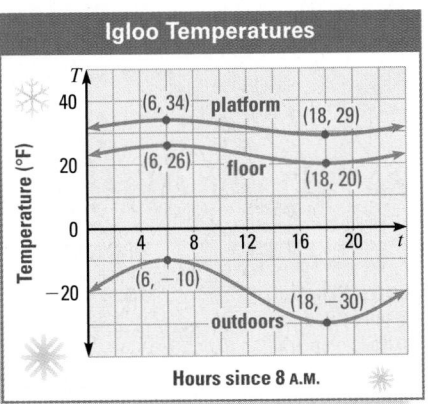

Igloo Temperatures

31. 🚢 **STEAMSHIPS** The paddle wheel of the S.S. Beaver was 13 feet in diameter and revolved 30 times per minute when moving at top speed. Using this speed and starting from a point at the very top of the wheel, write a model for the height h (in feet) of the end of a paddle relative to the water's surface as a function of the time t (in minutes). (Assume the paddle is 2 feet below the water's surface at its lowest point.) ▶ Source: *S.S. Beaver: The Ship That Saved the West*
$h = 6.5 \cos 60\pi t + 4.5$

32. 🌊 **OCEAN TIDES** The height of the water in a bay varies sinusoidally over time. On a certain day off the coast of Maine, a high tide of 10 feet occurred at 5:00 A.M. and a low tide of 2 feet occurred at 1:00 P.M. Write a model for the height h (in feet) of the water as a function of time t (in hours since midnight).
See margin.

33. 🧵 **SEWING MACHINES** In front of the Antique Sewing Machine Museum in Arlington, Texas, is the largest sewing machine in the world. The *flywheel*, which turns as the machine sews, is 5 feet in diameter. Write a model for the height h (in feet) of the handle on the flywheel as a function of the time t (in seconds), assuming that the wheel makes a complete revolution every 2 seconds and that the handle starts at its minimum height of 4 feet above the ground. See margin.

FOCUS ON APPLICATIONS

THE S.S. BEAVER was the first steamship on the Pacific Coast. It was built in the 1830s and sailed out of British Columbia, Canada, for over 50 years.

ADDITIONAL PRACTICE AND RETEACHING

For Lesson 14.5:
• Practice Levels A, B, and C (*Chapter 14 Resource Book*, p. 68)
• Reteaching with Practice (*Chapter 14 Resource Book*, p. 71)
• 🖥 See Lesson 14.5 of the *Personal Student Tutor*

For more Mixed Review:
• 🖥 Search the *Test and Practice Generator* for key words or specific lessons.

STATISTICS CONNECTION In Exercises 34 and 35, use a graphing calculator to find a sinusoidal model that fits the data.

34. $T = 28.6 \sin (0.47t - 1.53) + 46.6$

34.  **METEOROLOGY** The average daily temperature T (in degrees Fahrenheit) in Moline, Illinois, is given in the table. Time t is measured in months, with $t = 0$ representing January 1. Find a model for the data. ▶ Source: National Climatic Data Center

t	0.5	1.5	2.5	3.5	4.5	5.5	6.5	7.5	8.5	9.5	10.5	11.5
T	19.9	24.8	37.4	50.4	61.4	71.1	75.2	72.7	64.7	53.0	39.6	25.4

35. $T = 776.4 \sin (0.45t + 1.49) + 727.7$

35. **HEATING DEGREE-DAYS** For any given day, the number of degrees that the average temperature is below 65°F is called the *degree-days* for that day. This figure is used to calculate how much is spent on heating. The table below gives the total number T of degree-days for each month t in Dubuque, Iowa, with $t = 1$ representing January. Find a model for the data. ▶ Source: *Workshop Math*

t	1	2	3	4	5	6	7	8	9	10	11	12
T	1420	1204	1026	546	260	78	12	31	156	450	906	1287

Test Preparation

36. **MULTIPLE CHOICE** During one cycle, a sinusoid has a minimum at (18, 44) and a maximum at (30, 68). What is the amplitude of this sinusoid? **B**

(A) 6   (B) 12   (C) 22   (D) 24   (E) 48

37. **MULTIPLE CHOICE** During one cycle, a sinusoid has a maximum at (4, 12) and a minimum at (12, −2). What is the period of this sinusoid? **C**

(A) 8   (B) 8π (C) 16   (D) 16π (E) 32

★ Challenge

38. **FINDING A FUNCTION** Write a sinusoidal function whose graph has a minimum at $(\pi, 4)$ and a maximum at $\left(\frac{\pi}{2}, 7\right)$. $y = -1.5 \cos 2x + 5.5$

MIXED REVIEW

FINDING PROBABILITY Two six-sided dice are rolled. Find the probability of the given event. (Review 12.4)

39. The sum is 3. $\frac{1}{18}$ 40. The sum is 6. $\frac{5}{36}$ 41. The sum is 12. $\frac{1}{36}$

42. The sum is odd. $\frac{1}{2}$ 43. The sum is 5 or 6. $\frac{1}{4}$ 44. The sum is less than 7. $\frac{15}{36}$

EVALUATING FUNCTIONS Evaluate the function without using a calculator. (Review 13.3 for 14.6)

45. $\cos (-225°)$ $-\frac{\sqrt{2}}{2}$ 46. $\cos 240°$ $-\frac{1}{2}$ 47. $\sin (-30°)$ $-\frac{1}{2}$

48. $\tan 330°$ $-\frac{\sqrt{3}}{3}$ 49. $\tan \left(-\frac{7\pi}{6}\right)$ $-\frac{\sqrt{3}}{3}$ 50. $\sin \frac{5\pi}{4}$ $-\frac{\sqrt{2}}{2}$

FINDING AREA Find the area of $\triangle ABC$ having the given side lengths. (Review 13.6)

51. $a = 3, b = 8, c = 10$ 9.92   52. $a = 12, b = 20, c = 25$ 118.28

53. $a = 5, b = 9, c = 11$ 22.19   54. $a = 4, b = 6, c = 7$ 11.98

55. $a = 6, b = 11, c = 14$ 31.53   56. $a = 8, b = 13, c = 20$ 31.0

DAILY HOMEWORK QUIZ

📄 *Transparency Available*

1. Write a function for the sinusoid.

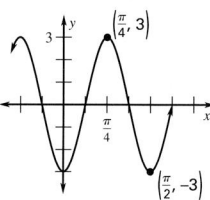

$y = -3 \cos 4x$

2. Write a function for the sinusoid with maximum at $A(0, 4)$ and minimum at $B(4\pi, -4)$.

$y = 4 \cos \frac{x}{4}$

3. Write a function for the sinusoid with maximum at $A(0, 0)$ and minimum at $B(2\pi, -6)$.

$y = 3 \cos \frac{x}{2} - 3$

EXTRA CHALLENGE NOTE

↳ Challenge problems for Lesson 14.5 are available in **blackline** format in the *Chapter 14 Resource Book,* p. 76 and at **www.mcdougallittell.com.**

ADDITIONAL TEST PREPARATION

1. **WRITING** Write a series of steps that you can use to find a trigonometric model.
   Sample answer: Step 1: Decide whether to use a sine or cosine function. Decide if the graph is a reflection of the chosen function. Step 2: Find the vertical shift and amplitude. Step 3: Find the period. Step 4: Find the horizontal shift. Step 5: Write the function in the form $y = a \sin b(x - h) + k$ or $y = a \cos b(x - h) + k$.

2. **OPEN ENDED** Write a Ferris wheel problem. Then write a model for this Ferris wheel. Check to see that answers are similar to Example 3.

867

APPLICATION NOTE
Additional information about music and math is available at **www.mcdougallittell.com**.

ADDITIONAL RESOURCES
An alternative Quiz for Lessons 14.3–14.5 is available in the *Chapter 14 Resource Book*, p. 77.
   A **blackline** master with additional Math & History exercises is available in the *Chapter 14 Resource Book*, p. 75.

Simplify the expression. (Lesson 14.3)

1. $\tan\left(\dfrac{\pi}{2} - x\right)\sec x \ \csc x$ 2. $1 - \sin^2 x + \dfrac{\cos x}{2\cos^2 x \ \sec x}$ 3. $\dfrac{\cos(-x)\cos\left(\dfrac{\pi}{2} - x\right)}{\sin x \cos x}$

Find the general solution of the equation. (Lesson 14.4)

4. $4\cos^2 x - 3 = 0$ $\dfrac{\pi}{6} + \pi n, \dfrac{5\pi}{6} + \pi n$
5. $3\sin^2 x - 8\sin x = 3$ $5.94 + 2\pi n, 3.48 + 2\pi n$
6. $\sqrt{3}\tan^2 x + 4\tan x = -\sqrt{3}$ $\dfrac{2\pi}{3} + \pi n, \dfrac{5\pi}{6} + \pi n$

Write a function for the sinusoid with maximum at A and minimum at B.
(Lesson 14.5)

7. $A\left(\dfrac{3\pi}{4}, 5\right), B\left(\dfrac{\pi}{4}, -5\right)$ 8. $A(0, 3), B(3\pi, 1)$ 9. $A(\pi, 6), B(0, 2)$
   $y = -5\sin 2x$ $y = \cos\dfrac{x}{3} + 2$ $y = -2\cos x + 4$

10. $T = 25.0\sin(0.50t - 1.76) + 47.3$

10. 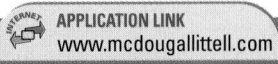 **METEOROLOGY** The average daily temperature T (in degrees Fahrenheit) in Detroit, Michigan, is given in the table. Time t is measured in months, with $t = 0$ representing January 1. Find a model for the data. (Lesson 14.5)

t	0.5	1.5	2.5	3.5	4.5	5.5	6.5	7.5	8.5	9.5	10.5	11.5
T	22.9	25.4	35.7	47.3	58.4	67.6	72.3	70.5	63.2	51.2	40.2	28.3

MATH & History   **Music and Math**   **APPLICATION LINK** www.mcdougallittell.com

THEN ▶ **MUSIC AND MATH** have been studied together for many centuries. For instance, 2500 years ago Pythagoras discovered that when the ratio of the lengths of two strings is a whole number, plucking the strings produces harmonious tones.

NOW ▶ **TODAY** we know that musical notes can be modeled by sine functions. The function $y = \sin 2\pi f x$ models a note with frequency f (in hertz) where x represents the length of time (in seconds) that the note is played. By adding the functions modeling two different notes, you can analyze the sound that occurs when the notes are played together.

1. Write functions that model notes with frequencies 10 hertz, 11 hertz, 12 hertz, 13 hertz, 14 hertz, and 15 hertz. $y = \sin 20\pi x, y = \sin 22\pi x, y = \sin 24\pi x, y = \sin 26\pi x,$ $y = \sin 28\pi x, y = \sin 30\pi x$

2. Use a graphing calculator to graph the sum of the models for 10 hertz and 15 hertz. How many cycles of the function occur in 1 second?   5

3. How many cycles per second are there in the function that describes what you hear when notes with frequencies of 10 and 14 hertz are played together? 10 and 13 hertz? 10 and 12 hertz?   2, 1, 2

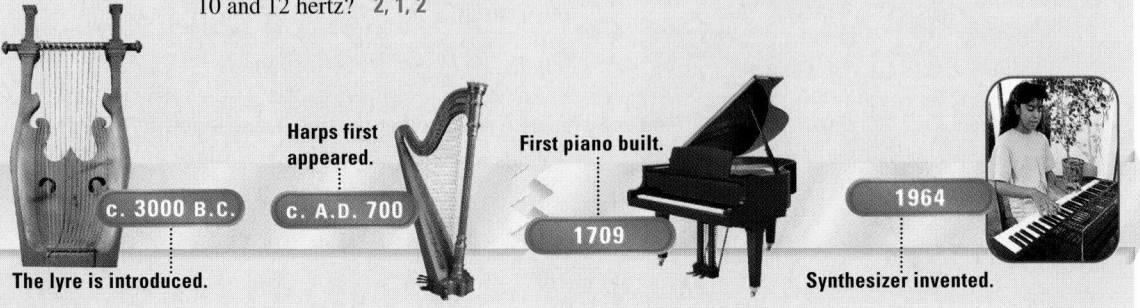

Harps first appeared.   First piano built.

c. 3000 B.C.   c. A.D. 700   1709   1964

The lyre is introduced.   Synthesizer invented.

868   **Chapter 14** *Trigonometric Graphs, Identities, and Equations*

14.6

Using Sum and Difference Formulas

What *you should learn*

GOAL 1 Evaluate trigonometric functions of the sum or difference of two angles.

GOAL 2 Use sum and difference formulas to solve **real-life** problems, such as determining when pistons in a car engine are at the same height in **Example 6**.

Why *you should learn it*

▼ To model **real-life** quantities, such as the size of an object in an aerial photograph in **Ex. 58**.

GOAL 1 SUM AND DIFFERENCE FORMULAS

In this lesson you will study formulas that allow you to evaluate trigonometric functions of the sum or difference of two angles.

SUM AND DIFFERENCE FORMULAS

SUM FORMULAS

$\sin (u + v) = \sin u \cos v + \cos u \sin v$

$\cos (u + v) = \cos u \cos v - \sin u \sin v$

$\tan (u + v) = \dfrac{\tan u + \tan v}{1 - \tan u \tan v}$

DIFFERENCE FORMULAS

$\sin (u - v) = \sin u \cos v - \cos u \sin v$

$\cos (u - v) = \cos u \cos v + \sin u \sin v$

$\tan (u - v) = \dfrac{\tan u - \tan v}{1 + \tan u \tan v}$

In general, $\sin (u + v) \neq \sin u + \sin v$. Similar statements can be made for the other trigonometric functions of sums and differences.

EXAMPLE 1 *Evaluating a Trigonometric Expression*

Find the exact value of **(a)** $\cos 75°$ and **(b)** $\tan \dfrac{\pi}{12}$.

SOLUTION

a. $\cos 75° = \cos (45° + 30°)$ Substitute $45° + 30°$ for $75°$.

$\qquad = \cos 45° \cos 30° - \sin 45° \sin 30°$ Sum formula for cosine

$\qquad = \dfrac{\sqrt{2}}{2}\left(\dfrac{\sqrt{3}}{2}\right) - \dfrac{\sqrt{2}}{2}\left(\dfrac{1}{2}\right)$ Evaluate.

$\qquad = \dfrac{\sqrt{6} - \sqrt{2}}{4}$ Simplify.

b. $\tan \dfrac{\pi}{12} = \tan \left(\dfrac{\pi}{3} - \dfrac{\pi}{4}\right)$ Substitute $\dfrac{\pi}{3} - \dfrac{\pi}{4}$ for $\dfrac{\pi}{12}$.

$\qquad = \dfrac{\tan \dfrac{\pi}{3} - \tan \dfrac{\pi}{4}}{1 + \tan \dfrac{\pi}{3} \tan \dfrac{\pi}{4}}$ Difference formula for tangent

$\qquad = \dfrac{\sqrt{3} - 1}{1 + (\sqrt{3})(1)}$ Evaluate.

$\qquad = 2 - \sqrt{3}$ Simplify.

✓ **CHECK** Try checking these results with a calculator. For instance, evaluate $\cos 75°$ and $\dfrac{\sqrt{6} - \sqrt{2}}{4}$ to see that both have the same value.

---

1 | PLAN

PACING
Basic: 1 day
Average: 1 day
Advanced: 1 day
Block Schedule: 0.5 block with 14.7

⟶ LESSON OPENER
ACTIVITY
An alternative way to approach Lesson 14.6 is to use the Activity Lesson Opener:
• Blackline Master (*Chapter 14 Resource Book,* p. 81)
• Transparency (p. 96)

MEETING INDIVIDUAL NEEDS
• *Chapter 14 Resource Book*
  Prerequisite Skills Review (p. 5)
  Practice Level A (p. 82)
  Practice Level B (p. 83)
  Practice Level C (p. 84)
  Reteaching with Practice (p. 85)
  Absent Student Catch-Up (p. 87)
  Challenge (p. 89)
• *Resources in Spanish*
• Personal Student Tutor

NEW-TEACHER SUPPORT
See the Tips for New Teachers on pp. 1–2 of the *Chapter 14 Resource Book* for additional notes about Lesson 14.6.

WARM-UP EXERCISES
Transparency Available
Evaluate the function without using a calculator.
1. $\tan 315°$ -1
2. $\cos 120°$ $-\dfrac{1}{2}$
3. $\sin (-135°)$ $-\dfrac{\sqrt{2}}{2}$
4. $\cos 210°$ $-\dfrac{\sqrt{3}}{2}$

Suppose you want to find tan 15° but you don't have a calculator. How can 15° be written as a sum or difference of two angles whose tangents you do know? The focus of today's lesson is formulas for the sine, cosine, or tangent of a sum or difference of two angles such as tan (45° − 30°).

 EXTRA EXAMPLE 1
Find the exact value of tan 15° and $\cos \frac{7\pi}{12}$. $2 - \sqrt{3}; \frac{\sqrt{2} - \sqrt{6}}{4}$

EXTRA EXAMPLE 2
Find $\cos (u - v)$, given that $\sin u = \frac{1}{5}$ with $0 \le u \le \frac{\pi}{2}$ and $\sin v = -\frac{5}{13}$ with $\pi < v \le \frac{3\pi}{2}$. $\frac{-24\sqrt{6} - 5}{65}$

EXTRA EXAMPLE 3
Simplify the expression $\sin \left(x + \frac{\pi}{2} \right)$. $\cos x$

EXTRA EXAMPLE 4
Solve $\cos \left(x + \frac{3\pi}{4} \right) + \cos \left(x - \frac{3\pi}{4} \right) = 1$ for $0 \le x < 2\pi$. $\frac{3\pi}{4}, \frac{5\pi}{4}$

 CHECKPOINT EXERCISES

For use after Example 1:
1. Find the exact value of sin 225°. $-\frac{\sqrt{2}}{2}$

For use after Example 2:
2. Find tan $(u + v)$, given that $\tan u = \frac{1}{4}$ with $\pi < u < \frac{3\pi}{2}$ and $\tan v = \frac{3}{7}$ with $0 < v < \frac{\pi}{2}$. $\frac{19}{25}$

For use after Examples 3–4:
3. Solve $2 \cos \left(x + \frac{\pi}{2} \right) = -1$ for $0 \le x \le 2\pi$. $\frac{\pi}{6}, \frac{5\pi}{6}$

EXAMPLE 2 *Using a Difference Formula*

Find $\sin (u - v)$ given that $\sin u = -\frac{3}{5}$ with $\pi < u < \frac{3\pi}{2}$ and $\cos v = \frac{12}{13}$ with $0 < v < \frac{\pi}{2}$.

SOLUTION
Using a Pythagorean identity and quadrant signs gives $\cos u = -\frac{4}{5}$ and $\sin v = \frac{5}{13}$.

$$\sin (u - v) = \sin u \cos v - \cos u \sin v \qquad \text{Difference formula for sine}$$

$$= -\frac{3}{5}\left(\frac{12}{13}\right) - \left(-\frac{4}{5}\right)\left(\frac{5}{13}\right) \qquad \text{Substitute.}$$

$$= -\frac{16}{65} \qquad \text{Simplify.}$$

EXAMPLE 3 *Simplifying an Expression*

Simplify the expression $\cos (x - \pi)$.

SOLUTION

$$\cos (x - \pi) = \cos x \cos \pi + \sin x \sin \pi \qquad \text{Difference formula for cosine}$$

$$= (\cos x)(-1) + (\sin x)(0) \qquad \text{Evaluate.}$$

$$= -\cos x \qquad \text{Simplify.}$$

EXAMPLE 4 *Solving a Trigonometric Equation*

Solve $\sin \left(x + \frac{\pi}{4} \right) + 1 = \sin \left(\frac{\pi}{4} - x \right)$ for $0 \le x < 2\pi$.

SOLUTION

$$\sin \left(x + \frac{\pi}{4} \right) + 1 = \sin \left(\frac{\pi}{4} - x \right) \qquad \text{Write original equation.}$$

$$\sin x \cos \frac{\pi}{4} + \cos x \sin \frac{\pi}{4} + 1 = \sin \frac{\pi}{4} \cos x - \cos \frac{\pi}{4} \sin x \qquad \text{Use formulas.}$$

$$\sin x \cos \frac{\pi}{4} + \cos x \sin \frac{\pi}{4} + 1 = \cos x \sin \frac{\pi}{4} - \sin x \cos \frac{\pi}{4} \qquad \text{Commutative property}$$

$$2 \sin x \cos \frac{\pi}{4} = -1 \qquad \text{Simplify.}$$

$$2(\sin x)\left(\frac{\sqrt{2}}{2}\right) = -1 \qquad \text{Evaluate.}$$

$$\sin x = -\frac{1}{\sqrt{2}} = -\frac{\sqrt{2}}{2} \qquad \text{Solve for sin } x.$$

▶ In the interval $0 \le x < 2\pi$, the solutions are $x = \frac{5\pi}{4}$ and $x = \frac{7\pi}{4}$.

✓**CHECK** You can check the solutions with a graphing calculator by graphing each side of the original equation and using the *Intersect* feature to determine the *x*-values for which the expressions are equal.

GOAL 2 SUM AND DIFFERENCE FORMULAS IN REAL LIFE

Biomechanics

EXAMPLE 5 Simplifying a Real-Life Formula

The force F (in pounds) on a person's back when he or she bends over at an angle θ is

$$F = \frac{0.6W \sin(\theta + 90°)}{\sin 12°}$$

where W is the person's weight (in pounds). Simplify this formula.

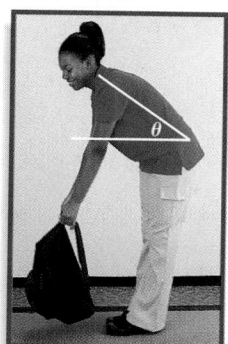

SOLUTION

Begin by expanding $\sin(\theta + 90°)$.

$$F = \frac{0.6W \sin(\theta + 90°)}{\sin 12°} \approx \frac{0.6W(\sin\theta \cos 90° + \cos\theta \sin 90°)}{0.208}$$

$$= \left(\frac{0.6}{0.208}\right)W\,[(\sin\theta)(0) + (\cos\theta)(1)] \approx 2.88W \cos\theta$$

EXAMPLE 6 Solving a Trigonometric Equation in Real Life

AUTOMOTIVE ENGINEERING The heights h (in inches) of pistons 1 and 2 in an automobile engine can be modeled by

$h_1 = 3.75 \sin 733t + 7.5$ and $h_2 = 3.75 \sin 733\left(t + \frac{4\pi}{3}\right) + 7.5$

where t is measured in seconds. How often are these two pistons at the same height?

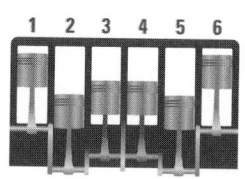

SOLUTION

Let $h_1 = h_2$ and solve for t.

$$3.75 \sin 733t + 7.5 = 3.75 \sin 733\left(t + \frac{4\pi}{3}\right) + 7.5$$

$$\sin 733t = \sin 733t \cos\frac{2932\pi}{3} + \cos 733t \sin\frac{2932\pi}{3}$$

$$\sin 733t = (\sin 733t)\left(-\frac{1}{2}\right) + (\cos 733t)\left(-\frac{\sqrt{3}}{2}\right)$$

$$\frac{3}{2}\sin 733t = -\frac{\sqrt{3}}{2}\cos 733t$$

$$\tan 733t = -\frac{\sqrt{3}}{3}$$

$$733t = \tan^{-1}\left(-\frac{\sqrt{3}}{3}\right) + n\pi$$

$$733t = -\frac{\pi}{6} + n\pi$$

$$t = -\frac{\pi}{4398} + \frac{n\pi}{733}$$

▶ The heights are equal once every $\frac{\pi}{733}$ second. So in one second, the heights are equal the following number of times: $1 \div \frac{\pi}{733} = \frac{733}{\pi} \approx 233.3$.

FOCUS ON CAREERS

AUTO MECHANIC
Auto mechanics inspect and repair mechanical and electrical systems of motor vehicles. They may specialize in areas such as transmission systems or diagnostic services.

CAREER LINK
www.mcdougallittell.com

 EXTRA EXAMPLE 5
Suppose the force F on your back when you bend over at an angle θ is given by
$F = \frac{0.3W \cos(\theta - 90°)}{\cos 75°}$ where W is your weight (in pounds). Simplify this formula. $F \approx 1.16W \sin\theta$.

EXTRA EXAMPLE 6
In an engine, the heights h (in inches) of two pistons can be modeled by: $h_1 = 2.5 \cos 26t$ and $h_2 = 2.5 \cos 26\left(t - \frac{\pi}{6}\right)$ where t is measured in seconds. How often are these two pistons at the same height? The heights are equal twice each cycle, which lasts $\frac{\pi}{13}$ sec. So, in 1 sec, the heights are equal $\frac{26}{\pi}$ times.

 CHECKPOINT EXERCISES
For use after Examples 5 and 6:
1. Simplify this formula.
$y = 3.4t \cos 3(x - 60°) + \sin 15°$
$y = -3.4t \cos 3x + 0.26$

CAREER NOTE
Additional information about a career as an auto mechanic is available at **www.mcdougallittell.com**.

FOCUS ON VOCABULARY
What is a trigonometric sum formula? It is a formula for the sine, cosine, or tangent of the sum of two angles.

CLOSURE QUESTION
What are some of the uses of the sum and difference formulas?
Sample answer: They can be used to simplify expressions and to find exact values of a trigonometric function without using a calculator.

DAILY PUZZLER
Simplify $\sin(x + 2\pi)$ using a sum formula. Explain your result.
$\sin(x + 2\pi) = \sin x$. Because the sine function is periodic with period 2π, the sine of any angle is the same as the sine of that angle plus 2π.

ASSIGNMENT GUIDE

BASIC
Day 1: pp. 872–874 Exs. 17–24,
29–31, 35–37, 43–46,
49–52, 55–56, 61,
65–75 odd

AVERAGE
Day 1: pp. 872–874 Exs. 17–31,
35–37, 43–46, 49–56, 61,
65–77 odd

ADVANCED
Day 1: pp. 872–874 Exs. 17–31,
35–37, 43–56, 58–64,
65–77 odd

BLOCK SCHEDULE
pp. 872–874 Exs. 17–31, 35–37,
43–46, 49–56, 61, 65–77 odd
(with 14.7)

EXERCISE LEVELS

Level A: *Easier*
17–28, 41–48

Level B: *More Difficult*
29–40, 49–61

Level C: *Most Difficult*
62–64

✔ **HOMEWORK CHECK**
To quickly check student understanding of key concepts, go over the following exercises:
Exs. 18, 24, 30, 36, 46, 50. See also the Daily Homework Quiz:

- Blackline Master (*Chapter 14 Resource Book*, p. 92)
- Transparency (p. 111)

GUIDED PRACTICE

Vocabulary Check ✓
Concept Check ✓

1. Give the sum and difference formulas for sine, cosine, and tangent.   **See margin.**

2. Fill in the blanks for each of the following equations.

  a. $\sin(45° − 30°) = \sin \underline{?} \cos 30° − \cos 45° \sin \underline{?}$ **45°, 30°**

  b. $\tan(90° + 60°) = \dfrac{\underline{?} + \tan 60°}{1 − \tan 90° \underline{?}}$ **tan 90°, tan 60°**

3. Explain how you can evaluate $\tan 105°$ using either the sum or difference formula for tangent.   **Use tan (60° + 45°) or tan (135° − 30°)**

Skill Check ✓

1. $\sin(u + v) = \sin u \cos v + \cos u \sin v$, $\sin(u − v) = \sin u \cos v − \cos u \sin v$

$\cos(u + v) = \cos u \cos v − \sin u \sin v$, $\cos(u − v) = \cos u \cos v + \sin u \sin v$

$\tan(u + v) = \dfrac{\tan u + \tan v}{1 − \tan u \tan v}$,

$\tan(u − v) = \dfrac{\tan u − \tan v}{1 + \tan u \tan v}$

Find the exact value of the expression.

4. $\cos 105°$ $\dfrac{\sqrt{2} − \sqrt{6}}{4}$

5. $\sin 15°$ $\dfrac{\sqrt{6} − \sqrt{2}}{4}$

6. $\tan 75°$ $2 + \sqrt{3}$

7. $\cos \dfrac{11\pi}{12}$ $−\dfrac{\sqrt{2} + \sqrt{6}}{4}$

8. $\sin \dfrac{23\pi}{12}$ $\dfrac{\sqrt{2} − \sqrt{6}}{4}$

9. $\tan \dfrac{7\pi}{12}$ $−2 − \sqrt{3}$

Solve the equation for $0 \le x < 2\pi$.

10. $2 \sin\left(x + \dfrac{\pi}{3}\right) = \tan \dfrac{\pi}{3}$ $0, \dfrac{\pi}{3}$

11. $\tan\left(x + \dfrac{\pi}{6}\right) = \tan\left(x + \dfrac{\pi}{4}\right)$ **none**

12. $\cos\left(x − \dfrac{\pi}{6}\right) = 1 + \cos\left(x + \dfrac{\pi}{6}\right)$ $\dfrac{\pi}{2}$

13. $\sin\left(x − \dfrac{4\pi}{3}\right) = 2 \sin\left(x − \dfrac{\pi}{3}\right)$ $\dfrac{\pi}{3}, \dfrac{4\pi}{3}$

14. $4 \sin(x + \pi) = 2 \cos\left(x + \dfrac{\pi}{2}\right) + 2$ $\dfrac{3\pi}{2}$

15. $−\cos x = 1 + 2 \cos(x − \pi)$ **0**

16. 🌐 **AUTOMOTIVE ENGINEERING** Look back at Example 6 on page 871. The height h_3 of piston 3 in the same engine can be modeled by

$$h_3 = 3.75 \sin 733\left(t + \dfrac{2\pi}{3}\right) + 7.5$$

where t is measured in seconds. How often is piston 3 the same height as piston 2? $\dfrac{\pi}{1466} + \dfrac{n\pi}{733}$

PRACTICE AND APPLICATIONS

STUDENT HELP

→ **Extra Practice**
to help you master skills is on p. 960.

FINDING VALUES Find the exact value of the expression.

17. $\cos 210°$ $−\dfrac{\sqrt{3}}{2}$

18. $\tan 195°$ $2 − \sqrt{3}$

19. $\tan 225°$ **1**

20. $\sin(−15°)$ $\dfrac{\sqrt{2} − \sqrt{6}}{4}$

21. $\cos(−225°)$ $−\dfrac{\sqrt{2}}{2}$

22. $\sin 165°$ $\dfrac{\sqrt{6} − \sqrt{2}}{4}$

23. $\tan \dfrac{11\pi}{12}$ $−2 + \sqrt{3}$

24. $\cos \dfrac{17\pi}{12}$ $\dfrac{\sqrt{2} − \sqrt{6}}{4}$

25. $\sin\left(−\dfrac{11\pi}{12}\right)$ $\dfrac{\sqrt{2} − \sqrt{6}}{4}$

26. $\cos \dfrac{\pi}{12}$ $\dfrac{\sqrt{2} + \sqrt{6}}{4}$

27. $\tan\left(−\dfrac{5\pi}{12}\right)$ $−2 − \sqrt{3}$

28. $\sin \dfrac{5\pi}{12}$ $\dfrac{\sqrt{2} + \sqrt{6}}{4}$

STUDENT HELP

→ **HOMEWORK HELP**
Example 1: Exs. 17–28
Example 2: Exs. 29–40
Example 3: Exs. 41–48
Example 4: Exs. 49–54
Examples 5, 6: Exs. 57–59

EVALUATING EXPRESSIONS Evaluate the expression given $\cos u = \dfrac{4}{7}$ with $0 < u < \dfrac{\pi}{2}$ and $\sin v = −\dfrac{9}{10}$ with $\pi < v < \dfrac{3\pi}{2}$.

29. $\sin(u + v)$ $−\dfrac{36 + \sqrt{627}}{70}$

30. $\cos(u + v)$ $\dfrac{9\sqrt{33} − 4\sqrt{19}}{70}$

31. $\tan(u + v)$ $\dfrac{\sqrt{627} + 36}{4\sqrt{19} − 9\sqrt{33}}$

32. $\sin(u − v)$ $\dfrac{36 − \sqrt{627}}{70}$

33. $\cos(u − v)$ $\dfrac{9\sqrt{33} + 4\sqrt{19}}{70}$

34. $\tan(u − v)$ $\dfrac{\sqrt{627} − 36}{4\sqrt{19} + 9\sqrt{33}}$

STUDENT HELP

INTERNET
HOMEWORK HELP
Visit our Web site
www.mcdougallittell.com
for help with problem
solving in Ex. 58.

EVALUATING EXPRESSIONS Evaluate the expression given that $\sin u = \frac{3}{5}$ with $\frac{\pi}{2} < u < \pi$ and $\cos v = -\frac{5}{6}$ with $\pi < v < \frac{3\pi}{2}$.

35. $\sin (u + v)$ $\frac{4\sqrt{11} - 15}{30}$ **36.** $\cos (u + v)$ $\frac{3\sqrt{11} + 20}{30}$ **37.** $\tan (u + v)$ $\frac{4\sqrt{11} - 15}{3\sqrt{11} + 20}$

38. $\sin (u - v)$ $-\frac{4\sqrt{11} + 15}{30}$ **39.** $\cos (u - v)$ $\frac{20 - 3\sqrt{11}}{30}$ **40.** $\tan (u - v)$ $\frac{4\sqrt{11} + 15}{3\sqrt{11} - 20}$

SIMPLIFYING EXPRESSIONS Simplify the expression.

41. $\tan (x - 2\pi)$ *tan x* **42.** $\tan (x + \pi) \tan x$ **43.** $\sin (x + \pi)$ *−sin x* **44.** $\cos (x + \pi)$ *−cos x*

45. $\sin \left(x - \frac{\pi}{2}\right)$ *−cos x* **46.** $\cos \left(x + \frac{3\pi}{2}\right)$ *sin x* **47.** $\cos \left(x + \frac{\pi}{2}\right)$ *−sin x* **48.** $\sin \left(x - \frac{3\pi}{2}\right)$ *cos x*

SOLVING TRIGONOMETRIC EQUATIONS Solve the equation for $0 \le x < 2\pi$.

49. $\cos \left(x + \frac{\pi}{6}\right) - 1 = \cos \left(x - \frac{\pi}{6}\right)$ $\frac{3\pi}{2}$ **50.** $\sin \left(x + \frac{3\pi}{4}\right) + \sin \left(x - \frac{3\pi}{4}\right) = 1$ $\frac{5\pi}{4}, \frac{7\pi}{4}$

51. $\sin \left(x + \frac{\pi}{6}\right) + \sin \left(x - \frac{\pi}{6}\right) = 0$ $0, \pi$ **52.** $\cos \left(x + \frac{\pi}{3}\right) + \cos \left(x - \frac{\pi}{3}\right) = 1$ 0

58. $\frac{WQ}{NA} = \frac{f(1 + \tan^2 t)}{h(1 + \tan t \tan \theta)}$;

$t = 0$: $\frac{WQ}{NA} =$

$\frac{f(1 + \tan^2 0)}{h(1 + \tan 0 \tan \theta)} =$

$\frac{f(1 + 0)}{h(1 + 0)} = \frac{f}{h}$

53. $\tan (x + \pi) + 2 \sin (x + \pi) = 0$ $0, \frac{\pi}{3}, \frac{5\pi}{3}$ **54.** $\tan (x + \pi) + \cos \left(x + \frac{\pi}{2}\right) = 0$ $0, \pi$

GEOMETRY **CONNECTION** In Exercises 55
and 56, use the following information.
In the figure shown, the acute angle of
intersection, $\theta_2 - \theta_1$, of two lines with
slopes m_1 and m_2 is given by:

$$\tan (\theta_2 - \theta_1) = \frac{m_2 - m_1}{1 + m_1 m_2}$$

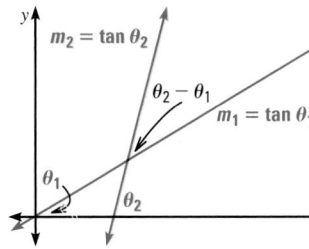

55. Find the acute angle of intersection of the
lines $y = \frac{1}{2}x + 3$ and $y = 2x - 3$. **36.9°**

56. Find the acute angle of intersection of the lines $y = x + 2$ and $y = 3x + 1$. **26.6°**

57. **SOUND WAVES** The pressure P of sound waves on a person's eardrum
can be modeled by

$$P = \frac{a}{r} \cos \left(\frac{2\pi r}{l} - 1100t\right)$$

where a is the maximum sound pressure (in pounds per square foot) at the
source, r is the distance (in feet) from the source, l is the length (in feet) of
the sound wave, and t is the time (in seconds). Simplify this formula when
$r = 16$ feet, $l = 4$ feet, and $a = 0.4$ pound per square foot. $P = \frac{1}{40} \cos 1100t$

FOCUS ON
PEOPLE

58. **AERIAL PHOTOGRAPHY** You are at a height h
taking aerial photographs. The ratio of the length WQ
of the image to the length NA of the actual object is

$$\frac{WQ}{NA} = \frac{f \tan (\theta - t) + f \tan t}{h \tan \theta}$$

where f is the focal length of the camera, θ is the angle
with the vertical made by the line from the camera to
point A and t is the tilt angle of the film. Use the
difference formula for tangent to simplify the
ratio. Then show that $\frac{WQ}{NA} = \frac{f}{h}$ when $t = 0$. **See margin.**

▶ Source: *Math Applied to Space Science*

BARRIE
ROKEACH has
more than 20 years of
experience as an aerial
photographer. He has had
more than one dozen
individual exhibitions in
museums and galleries
around the country.

14.6 *Using Sum and Difference Formulas* **873**

! **COMMON ERROR**
EXERCISES 29–40 Students
may use the wrong sign for the
angles. Suggest they identify the
quadrant of the angle before finding
its trigonometric value.

STUDENT HELP NOTES

→ **Homework Help** Students
can find help for Ex. 58 at
www.mcdougallittell.com.
The information can be printed
out for students who don't have
access to the Internet.

**ADDITIONAL PRACTICE
AND RETEACHING**

For Lesson 14.6:
• Practice Levels A, B, and C
(*Chapter 14 Resource Book*,
p. 82)

• Reteaching with Practice
(*Chapter 14 Resource Book*,
p. 85)

• ⌨ See Lesson 14.6 of the
Personal Student Tutor

For more Mixed Review:
• ⌨ Search the *Test and Practice
Generator* for key words or
specific lessons.

DAILY HOMEWORK QUIZ

📽 **Transparency Available**

1. Find the exact value of sin 15°.
$\frac{\sqrt{6} - \sqrt{2}}{4}$

2. Find the exact value of $\tan\left(\frac{5\pi}{12}\right)$. $2 + \sqrt{3}$

3. Given $\cos u = \frac{4}{5}$ with $0 < u < \frac{\pi}{2}$ and $\sin v = -\frac{3}{5}$ with $\pi < v < \frac{3\pi}{2}$, evaluate $\tan(u - v)$. **0**

4. Simplify $\cos(x - 2\pi)$. $\cos x$

5. Solve $\sin(x + \pi) + \sin(x - \pi) = 2$ for $0 \le x < 2\pi$. $\frac{3\pi}{2}$

EXTRA CHALLENGE NOTE
→ Challenge problems for Lesson 14.6 are available in **blackline** format in the *Chapter 14 Resource Book*, p. 89 and at **www.mcdougallittell.com**.

ADDITIONAL TEST PREPARATION

1. WRITING Show how to find a formula for sin 2x using a sum formula and the fact that sin 2x = sin (x + x).
sin 2x = sin (x + x) = sin x cos x + cos x sin x = 2 sin x cos x

2. OPEN ENDED Name an angle that can be expressed as a sum or difference of two basic angles. Then show how you can use a sum or difference formula to find the cosine of this angle. **Check that answers are similar to Example 1.**

---

60. The plus in the first goes with the minus in the second, and vice versa.
sin (u ± v) = sin u cos v ± cos u sin v; tan (u ± v) = $\frac{\tan u \pm \tan v}{1 \pm \tan u \tan v}$

Test Preparation

62. cos (u + v) =
cos (u −(−v)) =
cos u cos (−v) + sin u sin (−v) =
cos u cos v − sin u sin v.

★ Challenge

EXTRA CHALLENGE
→ www.mcdougallittell.com

63. tan (u − v) =
$\frac{\sin (u - v)}{\cos (u - v)} =$
$\frac{\sin u \cos v - \cos u \sin v}{\cos u \cos v + \sin u \sin v} \cdot$
$\frac{\frac{1}{\cos u \cos v}}{\frac{1}{\cos u \cos v}} = \frac{\tan u - \tan v}{1 + \tan u \tan v}$

68. $\begin{bmatrix} 0 & 10 \\ 0 & -8 \end{bmatrix}$

69. $\begin{bmatrix} 30 & 10 & -10 \\ -70 & -20 & 30 \end{bmatrix}$

70. $\begin{bmatrix} -6 & 0 & 6 \\ 15 & 3 & -9 \end{bmatrix}$

---

59. 🌐 **FERRIS WHEEL** The heights h (in feet) of two people in different seats on a Ferris wheel can be modeled by $0.26 + \frac{n\pi}{10}$

$$h_1 = 28 \cos 10t + 38 \quad \text{and} \quad h_2 = 28 \cos 10\left(t - \frac{\pi}{6}\right) + 38$$

where t is the time (in minutes). When are the two people at the same height?

60. CRITICAL THINKING You can write the sum and difference formulas for cosine as a single equation: $\cos(u \pm v) = \cos u \cos v \mp \sin u \sin v$. Explain why the symbol $\pm$ is used on the left side, but the symbol $\mp$ is used on the right side. Then use the symbols $\pm$ and $\mp$ to write the sum and difference formulas for sine and tangent as single equations. **See margin.**

61. 🖩 **MULTI-STEP PROBLEM** Suppose two middle-A tuning forks are struck at different times so that their vibrations are slightly out of phase. The combined pressure change P (in pascals) caused by the forks at time t (in seconds) is:

$$P = 3 \sin 880\pi t + 4 \cos 880\pi t$$

a. Graph the equation on a graphing calculator using a viewing window of $0 \le x \le 0.5$ and $-6 \le y \le 6$. What do you observe about the graph? **It is periodic.**

b. Write the given model in the form $y = a \cos b(x - h)$. $y = 5 \cos 880\pi\left(x - \frac{0.664}{880\pi}\right)$

c. Graph the model from part (b) to confirm that the graphs are the same. **yes**

VERIFYING FORMULAS **Use the difference formula for cosine to verify the following formulas.** **62, 63. See margin.**

62. The sum formula for cosine, by replacing v with $-v$

63. The difference formula for tangent, by using the identity $\tan \theta = \frac{\sin \theta}{\cos \theta}$

64. The difference formula for sine, by using a cofunction identity

$\sin(u - v) = \cos\left[\frac{\pi}{2} - (u - v)\right] = \cos\left[\left(\frac{\pi}{2} - u\right) + v\right] = \cos\left(\frac{\pi}{2} - u\right)\cos v -$
$\sin\left(\frac{\pi}{2} - u\right)\sin v = \left(\cos\frac{\pi}{2}\cos u + \sin\frac{\pi}{2}\sin u\right)\cos v - \cos u \sin v = \sin u \cos v -$
$\cos u \sin v$.

MIXED REVIEW

SIMPLIFYING EXPRESSIONS **Using the given matrices, simplify the expression. (Review 4.2)**

$A = \begin{bmatrix} 2 & 3 \\ -1 & -2 \end{bmatrix}, B = \begin{bmatrix} -1 \\ 3 \end{bmatrix}, C = \begin{bmatrix} 2 & 0 \\ -4 & 1 \end{bmatrix}, D = \begin{bmatrix} 3 & 1 & -1 \\ -2 & 0 & 2 \end{bmatrix}, E = \begin{bmatrix} 3 & 1 \\ -1 & 2 \\ 0 & -2 \end{bmatrix}$

65. $AD + D$ $\begin{bmatrix} 3 & 3 & 3 \\ -1 & -1 & -1 \end{bmatrix}$
66. $3(A + C)$ $\begin{bmatrix} 12 & 9 \\ -15 & -3 \end{bmatrix}$
67. $-2AB + B$ $\begin{bmatrix} -15 \\ 13 \end{bmatrix}$
68. $DE + AC$
69. $5CD$
70. $AD - CD$

SOLVING TRIANGLES **Solve** $\triangle ABC$. **(Review 13.5, 13.6)**

$C = 134°, a = 65.8, c = 153$ $B = 25°, a = 4.3, b = 2.1$
71. $A = 18°, B = 28°, b = 100$ **72.** $A = 60°, C = 95°, c = 5$

73. $a = 13, b = 4, c = 11$ **74.** $a = 2, b = 2.5, c = 3$
$A = 111°, B = 17°, C = 52°$ $C = 83°, B = 56°, A = 41°$

SOLVING TRIGONOMETRIC EQUATIONS **Solve the equation in the interval** $0 \le x < 2\pi$. **(Review 14.4 for 14.7)**

75. $\tan x + \sqrt{3} = 0$ $\frac{2\pi}{3}, \frac{5\pi}{3}$ **76.** $4 \cos^2 x - 3 = 0$ $\frac{\pi}{6}, \frac{5\pi}{6}, \frac{7\pi}{6}, \frac{11\pi}{6}$

77. $8 \tan x + 8 = 0$ $\frac{3\pi}{4}, \frac{7\pi}{4}$ **78.** $-5 + 8 \cos x = -1$ $\frac{\pi}{3}, \frac{5\pi}{3}$

14.7 Using Double- and Half-Angle Formulas

What you should learn

GOAL 1 Evaluate expressions using double- and half-angle formulas.

GOAL 2 Use double- and half-angle formulas to solve **real-life** problems, such as finding the mach number for an airplane in **Ex. 70**.

Why you should learn it

▼ To model **real-life** situations with double- and half-angle relationships, such as kicking a football in **Example 8**.

GOAL 1 DOUBLE- AND HALF-ANGLE FORMULAS

In this lesson you will use formulas for double angles (angles of measure $2u$) and half angles $\left(\text{angles of measure } \frac{u}{2}\right)$. The three formulas for $\cos 2u$ below are equivalent, as are the two formulas for $\tan \frac{u}{2}$. Use whichever formula is most convenient for solving a problem.

DOUBLE-ANGLE AND HALF-ANGLE FORMULAS

DOUBLE-ANGLE FORMULAS

$$\cos 2u = \cos^2 u - \sin^2 u \qquad\qquad \sin 2u = 2 \sin u \cos u$$

$$\cos 2u = 2\cos^2 u - 1 \qquad\qquad \tan 2u = \frac{2 \tan u}{1 - \tan^2 u}$$

$$\cos 2u = 1 - 2 \sin^2 u$$

HALF-ANGLE FORMULAS

$$\sin \frac{u}{2} = \pm\sqrt{\frac{1 - \cos u}{2}} \qquad\qquad \tan \frac{u}{2} = \frac{1 - \cos u}{\sin u}$$

$$\cos \frac{u}{2} = \pm\sqrt{\frac{1 + \cos u}{2}} \qquad\qquad \tan \frac{u}{2} = \frac{\sin u}{1 + \cos u}$$

The signs of $\sin \frac{u}{2}$ and $\cos \frac{u}{2}$ depend on the quadrant in which $\frac{u}{2}$ lies.

EXAMPLE 1 *Evaluating Trigonometric Expressions*

Find the exact value of **(a)** $\tan \frac{\pi}{8}$ and **(b)** $\cos 105°$.

SOLUTION

a. Use the fact that $\frac{\pi}{8}$ is half of $\frac{\pi}{4}$.

$$\tan \frac{\pi}{8} = \tan \frac{1}{2}\left(\frac{\pi}{4}\right) = \frac{1 - \cos \frac{\pi}{4}}{\sin \frac{\pi}{4}} = \frac{1 - \frac{\sqrt{2}}{2}}{\frac{\sqrt{2}}{2}} = \frac{2 - \sqrt{2}}{\sqrt{2}} = \sqrt{2} - 1$$

b. Use the fact that $105°$ is half of $210°$ and that cosine is negative in Quadrant II.

$$\cos 105° = \cos \frac{1}{2}(210°) = -\sqrt{\frac{1 + \cos 210°}{2}}$$

$$= -\sqrt{\frac{1 + \left(-\frac{\sqrt{3}}{2}\right)}{2}} = -\sqrt{\frac{2 - \sqrt{3}}{4}} = -\frac{\sqrt{2 - \sqrt{3}}}{2}$$

STUDENT HELP

▶ **Study Tip**
In Example 1 note that, in general, $\tan \frac{u}{2} \neq \frac{1}{2} \tan u$. Similar statements can be made for the other trigonometric functions of double and half angles.

---

1 PLAN

PACING
Basic: 2 days
Average: 2 days
Advanced: 2 days
Block Schedule: 0.5 block with 14.6

LESSON OPENER
APPLICATION
An alternative way to approach Lesson 14.7 is to use the Application Lesson Opener:

- Blackline Master (*Chapter 14 Resource Book*, p. 93)
- Transparency (p. 97)

MEETING INDIVIDUAL NEEDS
- **Chapter 14 Resource Book**
  Prerequisite Skills Review (p. 5)
  Practice Level A (p. 94)
  Practice Level B (p. 95)
  Practice Level C (p. 96)
  Reteaching with Practice (p. 97)
  Absent Student Catch-Up (p. 99)
  Challenge (p. 101)
- **Resources in Spanish**
- **Personal Student Tutor**

NEW-TEACHER SUPPORT
See the Tips for New Teachers on pp. 1–2 of the *Chapter 14 Resource Book* for additional notes about Lesson 14.7.

WARM-UP EXERCISES

Transparency Available
Solve each equation for the interval $0 \leq x < 2\pi$.

1. $\sin^2 x = \sin x$ $0, \frac{\pi}{2}, \pi$

2. $2 \tan^2 x = 2$ $\frac{\pi}{4}, \frac{3\pi}{4}, \frac{5\pi}{4}, \frac{7\pi}{4}$

3. $-2 + 3 \cos x = 5$ no solution

MOTIVATING THE LESSON
The angle at which you throw the ball affects the path of a ball. You can use a trigonometric function to model the path of a ball when you know the initial height, velocity, and angle of the ball. In today's lesson you will learn trigonometric formulas to help you simplify functions such as the one used to model the path of a ball.

 EXTRA EXAMPLE 1
Find the exact value of $\sin \frac{5\pi}{12}$ and $\cos 67.5°$. $\frac{\sqrt{2+\sqrt{3}}}{2}, \frac{\sqrt{2-\sqrt{2}}}{2}$

EXTRA EXAMPLE 2
Given that $\sin u = -\frac{1}{3}$ with $\frac{3\pi}{2} < u \le 2\pi$, find the following.
a. $\cos 2u$ $\frac{7}{9}$

b. $\cos \frac{u}{2}$ $-\sqrt{\frac{3 + 2\sqrt{2}}{6}}$

EXTRA EXAMPLE 3
Simplify $\frac{\sin 2x}{1 - \cos 2x}$. $\cot x$

EXTRA EXAMPLE 4
Verify the identity
$\left(\sin \frac{x}{2} + \cos \frac{x}{2}\right)^2 = 1 + \sin x$.

$\left(\sin \frac{x}{2} + \cos \frac{x}{2}\right)^2 = \sin^2 \frac{x}{2} +$

$2 \sin \frac{x}{2} \cos \frac{x}{2} + \cos^2 \frac{x}{2} =$

$\left(\sin^2 \frac{x}{2} + \cos^2 \frac{x}{2}\right) + 2 \sin \frac{x}{2} \cos \frac{x}{2} =$

$1 + 2 \sin \frac{x}{2} \cos \frac{x}{2} = 1 + \sin\left(2\left(\frac{x}{2}\right)\right) =$

$1 + \sin x$

Checkpoint Exercises on next page.

STUDENT HELP

▶ **Study Tip**
Because $\pi < u < \frac{3\pi}{2}$ in Example 2, you can multiply through the inequality by $\frac{1}{2}$ to get $\frac{\pi}{2} < \frac{u}{2} < \frac{3\pi}{4}$, so $\frac{u}{2}$ is in Quadrant II.

STUDENT HELP

▶ **Study Tip**
Because there are three formulas for cos 2u, you will want to choose the one that allows you to simplify the expression in which cos 2u appears, as illustrated in Example 3.

EXAMPLE 2 **Evaluating Trigonometric Expressions**

Given $\cos u = -\frac{3}{5}$ with $\pi < u < \frac{3\pi}{2}$, find the following.

a. $\sin 2u$ b. $\sin \frac{u}{2}$

SOLUTION

a. Use a Pythagorean identity to conclude that $\sin u = -\frac{4}{5}$.

$\sin 2u = 2 \sin u \cos u$

$= 2\left(-\frac{4}{5}\right)\left(-\frac{3}{5}\right) = \frac{24}{25}$

b. Because $\frac{u}{2}$ is in Quadrant II, $\sin \frac{u}{2}$ is positive.

$\sin \frac{u}{2} = \sqrt{\frac{1 - \cos u}{2}} = \sqrt{\frac{1 - \left(-\frac{3}{5}\right)}{2}} = \sqrt{\frac{4}{5}} = \frac{2\sqrt{5}}{5}$

EXAMPLE 3 **Simplifying a Trigonometric Expression**

Simplify $\frac{\cos 2\theta}{\sin \theta + \cos \theta}$.

SOLUTION

$\frac{\cos 2\theta}{\sin \theta + \cos \theta} = \frac{\cos^2 \theta - \sin^2 \theta}{\sin \theta + \cos \theta}$ Use a double-angle formula.

$= \frac{(\cos \theta - \sin \theta)(\cos \theta + \sin \theta)}{\sin \theta + \cos \theta}$ Factor difference of squares.

$= \cos \theta - \sin \theta$ Simplify.

EXAMPLE 4 **Verifying a Trigonometric Identity**

Verify the identity $\sin 3x = 3 \sin x - 4 \sin^3 x$.

SOLUTION

$\sin 3x = \sin (2x + x)$ Rewrite sin 3x as sin (2x + x).

$= \sin 2x \cos x + \cos 2x \sin x$ Use a sum formula.

$= (2 \sin x \cos x) \cos x + (1 - 2 \sin^2 x) \sin x$ Use double-angle formulas.

$= 2 \sin x \cos^2 x + \sin x - 2 \sin^3 x$ Multiply.

$= 2 \sin x (1 - \sin^2 x) + \sin x - 2 \sin^3 x$ Use a Pythagorean identity.

$= 2 \sin x - 2 \sin^3 x + \sin x - 2 \sin^3 x$ Distributive property

$= 3 \sin x - 4 \sin^3 x$ Combine like terms.

EXAMPLE 5 *Solving a Trigonometric Equation*

Solve $\tan 2x + \tan x = 0$ for $0 \le x < 2\pi$.

SOLUTION

$\tan 2x + \tan x = 0$	Write original equation.
$\dfrac{2\tan x}{1-\tan^2 x} + \tan x = 0$	Use a double-angle formula.
$2\tan x + \tan x\,(1 - \tan^2 x) = 0$	Multiply each side by $1 - \tan^2 x$.
$2\tan x + \tan x - \tan^3 x = 0$	Distributive property.
$3\tan x - \tan^3 x = 0$	Combine like terms.
$\tan x\,(3 - \tan^2 x) = 0$	Factor.

Set each factor equal to 0 and solve for x.

$\tan x = 0$ or $3 - \tan^2 x = 0$

$x = 0, \pi$ $3 = \tan^2 x$

$\pm\sqrt{3} = \tan x$

$x = \dfrac{\pi}{3}, \dfrac{2\pi}{3}, \dfrac{4\pi}{3}, \dfrac{5\pi}{3}$

✓ **CHECK** You can use a graphing calculator to check the solutions. Graph the following function:

$y = \tan 2x + \tan x$

Then use the *Zero* feature to find the x-values for which $y = 0$.

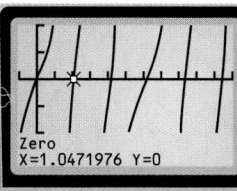

Zero
X=1.0471976 Y=0

. . . .

Some equations that involve double or half angles can be solved directly—without resorting to double- or half-angle formulas.

EXAMPLE 6 *Solving a Trigonometric Equation*

Solve $2\cos\dfrac{x}{2} + 1 = 0$.

SOLUTION

$2\cos\dfrac{x}{2} + 1 = 0$	Write original equation.
$2\cos\dfrac{x}{2} = -1$	Subtract 1 from each side.
$\cos\dfrac{x}{2} = -\dfrac{1}{2}$	Divide each side by 2.
$\dfrac{x}{2} = \dfrac{2\pi}{3} + 2n\pi \quad\text{or}\quad \dfrac{4\pi}{3} + 2n\pi$	General solution for $\dfrac{x}{2}$
$x = \dfrac{4\pi}{3} + 4n\pi \quad\text{or}\quad \dfrac{8\pi}{3} + 4n\pi$	General solution for x

14.7 *Using Double- and Half-Angle Formulas* **877**

Sidebar:

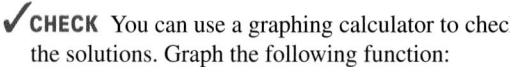

STUDENT HELP

HOMEWORK HELP
Visit our Web site
www.mcdougallittell.com
for extra examples.

✓ **CHECKPOINT EXERCISES**

For use after Example 1:

1. Find the exact value of $\tan 112.5°$. $-\sqrt{2}-1$

For use after Example 2:

2. Given $\sin u = -\dfrac{5}{13}$ with $\pi < u < \dfrac{3\pi}{2}$, find $\cos 2u$. $\dfrac{119}{169}$

For use after Examples 3 and 4:

3. Simplify $\dfrac{(1+\cot^2 x)(1-\cos 2x)}{2}$.   1

EXTRA EXAMPLE 5

Solve $\cos 2x = 5\sin^2 x - \cos^2 x$ for $0 \le x < 2\pi$. $\dfrac{\pi}{6}, \dfrac{5\pi}{6}, \dfrac{7\pi}{6}, \dfrac{11\pi}{6}$

EXTRA EXAMPLE 6

Solve $2\sin\dfrac{x}{2} - \sqrt{3} = 0$. $\dfrac{2\pi}{3} + 2\pi n$ or $\dfrac{4\pi}{3} + 2\pi n$

✓ **CHECKPOINT EXERCISES**

For use after Example 5:

1. Solve $\sin 2x \sec x + 2\cos x = 0$ for $0 \le x < 2\pi$. $\dfrac{3\pi}{4}$ or $\dfrac{7\pi}{4}$

For use after Example 6:

2. Solve $-4\sin\dfrac{x}{2} = 2$. $\dfrac{7\pi}{3} + 4\pi n$ or $\dfrac{11\pi}{3} + 4\pi n$

STUDENT HELP NOTES

→ **Homework Help** Students can find extra examples at **www.mcdougallittell.com** that parallel the examples in the student edition.

EXTRA EXAMPLE 7
A baseball is thrown from a height of 4 ft ($h_0 = 4$). Write a trigonometric model for this situation. Explain why you cannot use the zero-product property to simplify your model.
$y = -\dfrac{16}{v^2 \cos^2 \theta} x^2 + (\tan \theta)x + 4$; the factor x is not common to all three terms of the expression on the right side of the equation.

EXTRA EXAMPLE 8
You are kicking a football from ground level with an initial speed of 60 ft/sec. Can you make it travel 150 ft?
No, when you use the formula to find the angle, you get $\sin 2\theta = \dfrac{4}{3}$ which has no solution.

 CHECKPOINT EXERCISES
For use after Examples 7 and 8:
1. Suppose you kick a soccer ball from ground level with an initial speed of 50 ft/sec. Can you make it travel 65 ft?
Yes, if you kick it at an angle of about 28°.

FOCUS ON VOCABULARY
What is a double angle formula?
A double angle formula is a formula for the sine, cosine, or tangent of twice an angle.

CLOSURE QUESTION
What are some of the uses of double- and half-angle formulas?
Sample answer: They can be used to find exact values of trigonometric functions and to simplify formulas involving half- and double-angles.

DAILY PUZZLER
The sine of an acute angle is equal to the cosine of twice the angle. What is the angle? 30°

The path traveled by an object that is projected at an initial height of h_0 feet, an initial speed of v feet per second, and an initial angle θ is given by

$$y = -\frac{16}{v^2 \cos^2 \theta} x^2 + (\tan \theta)x + h_0$$

where x and y are measured in feet. (This model neglects air resistance.)

EXAMPLE 7 *Simplifying a Trigonometric Model*

SPORTS Find the horizontal distance traveled by a football kicked from ground level ($h_0 = 0$) at speed v and angle θ.

Not drawn to scale

SOLUTION

Using the model above with $h_0 = 0$, set y equal to 0 and solve for x.

$$-\frac{16}{v^2 \cos^2 \theta} x^2 + (\tan \theta)x = 0 \qquad \text{Let } y = 0.$$

$$(-x)\left(\frac{16}{v^2 \cos^2 \theta} x - \tan \theta\right) = 0 \qquad \text{Factor.}$$

$$\frac{16}{v^2 \cos^2 \theta} x - \tan \theta = 0 \qquad \begin{array}{l}\text{Zero product property}\\ \text{(Ignore } -x = 0.)\end{array}$$

$$\frac{16}{v^2 \cos^2 \theta} x = \tan \theta \qquad \text{Add } \tan \theta \text{ to each side.}$$

$$x = \frac{1}{16} v^2 \cos^2 \theta \tan \theta \qquad \text{Multiply each side by } \tfrac{1}{16} v^2 \cos^2 \theta.$$

$$x = \frac{1}{16} v^2 \cos \theta \sin \theta \qquad \text{Use } \cos \theta \tan \theta = \sin \theta.$$

$$x = \frac{1}{32} v^2 (2 \cos \theta \sin \theta) \qquad \text{Rewrite } \tfrac{1}{16} \text{ as } \tfrac{1}{32} \cdot 2.$$

$$x = \frac{1}{32} v^2 \sin 2\theta \qquad \text{Use a double-angle formula.}$$

EXAMPLE 8 *Using a Trigonometric Model*

SPORTS You are kicking a football from ground level with an initial speed of 80 feet per second. Can you make the ball travel 200 feet?

SOLUTION

$$200 = \frac{1}{32}(80)^2 \sin 2\theta \qquad \text{Substitute for } x \text{ and } v \text{ in the formula from Example 7.}$$

$$1 = \sin 2\theta \qquad \text{Divide each side by } \tfrac{1}{32}(80)^2 = 200.$$

$$90° = 2\theta \qquad \sin^{-1} 1 = 90°$$

$$45° = \theta \qquad \text{Solve for } \theta.$$

▶ You can make the football travel 200 feet if you kick it at an angle of 45°.

FOCUS ON APPLICATIONS

FIELD GOALS The longest professional field goal was 63 yards, made by Tom Dempsy in 1970. This record was tied by Jason Elam during the 1998–1999 season.

APPLICATION LINK www.mcdougallittell.com

GUIDED PRACTICE
2. $2\cos^2 u - 1 = \cos 2u$; Simplify and factor the numerator.

Vocabulary Check ✓

1. Complete this statement: $\sin 2u = 2 \sin u \cos u$ is called the $\underline{\ ?\ }$ formula for sine.

double angle

Concept Check ✓

2. Suppose you want to simplify $\dfrac{\cos 2\theta - \cos \theta}{\cos \theta - 1}$. Which double-angle formula for cosine would you use to rewrite $\cos 2\theta$? Explain.   **See margin.**

14. $\dfrac{\sin x}{2}$ for $0 \le x \le \pi$,

$-\dfrac{\sin x}{2}$ for $\pi < x < 2\pi$

26. $\sin \dfrac{u}{2} = \dfrac{\sqrt{5}}{5}$,

$\cos \dfrac{u}{2} = \dfrac{2\sqrt{5}}{5}$, $\tan \dfrac{u}{2} = \dfrac{1}{2}$

3. ERROR ANALYSIS Explain what is wrong in the calculations shown below.

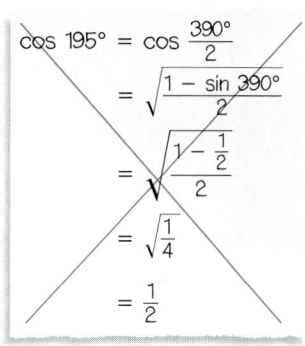

a.
$$\tan 15° = \tan \dfrac{30°}{2}$$
$$= \dfrac{1}{2} \tan 30°$$
$$= \dfrac{1}{2}\left(\dfrac{\sqrt{3}}{3}\right)$$
$$= \dfrac{\sqrt{3}}{6}$$

$\tan \dfrac{u}{2} \ne \dfrac{1}{2} \tan u$

b.
$$\cos 195° = \cos \dfrac{390°}{2}$$
$$= \sqrt{\dfrac{1 - \sin 390°}{2}}$$
$$= \sqrt{\dfrac{1 - \frac{1}{2}}{2}}$$
$$= \sqrt{\dfrac{1}{4}}$$
$$= \dfrac{1}{2}$$

$\cos 195° < 0$ and $1 - \sin 390°$ should be $1 + \cos 390°$

Skill Check ✓

27. $\sin \dfrac{u}{2} = \dfrac{\sqrt{6}}{6}$,

$\cos \dfrac{u}{2} = -\dfrac{\sqrt{30}}{6}$,

$\tan \dfrac{u}{2} = -\dfrac{\sqrt{5}}{5}$

28. $\sin \dfrac{u}{2} = \dfrac{\sqrt{50 + 5\sqrt{19}}}{10}$,

$\cos \dfrac{u}{2} = \dfrac{\sqrt{50 - 5\sqrt{19}}}{10}$,

$\tan \dfrac{u}{2} = \dfrac{\sqrt{50 + 5\sqrt{19}}}{\sqrt{50 - 5\sqrt{19}}}$

Given $\tan u = \dfrac{2}{5}$ with $\pi < u < \dfrac{3\pi}{2}$, find the exact value of the expression.

4. $\sin \dfrac{u}{2}$ $\sqrt{\dfrac{29 + 5\sqrt{29}}{58}}$

5. $\cos \dfrac{u}{2}$ $-\sqrt{\dfrac{29 - 5\sqrt{29}}{58}}$

6. $\tan \dfrac{u}{2}$ $-\dfrac{\sqrt{29} + 5}{2}$

7. $\sin 2u$ $\dfrac{20}{29}$

8. $\cos 2u$ $\dfrac{21}{29}$

9. $\tan 2u$ $\dfrac{20}{21}$

Simplify the expression.

10. $\cos^2 x - 2 \cos 2x + 1$ $3\sin^2 x$

11. $\dfrac{\sin 2x \tan \frac{x}{2}}{2 \cos x - 2 \cos^2 x}$

12. $\dfrac{\sin^2 x + \sin 2x + \cos^2 x}{1 + 2 \sin x \cos x}$

13. $\dfrac{\tan 2x}{\sec^2 x} \cdot \dfrac{2 \tan x}{1 - \tan^4 x}$

14. $\sin \dfrac{x}{2} \cos \dfrac{x}{2}$ **See margin.**

15. $\cos 2x \cot^2 x - \cot^2 x$ $-2\cos^2 x$

16. 🏈 **FOOTBALL** Look back at Example 8 on page 878. Through what range of angles can you kick the football to make it travel at least 150 feet?

0.42 to 1.15 radians, or 24° to 66°

PRACTICE AND APPLICATIONS

STUDENT HELP

▶ **Extra Practice**
to help you master skills is on p. 960.

29. $\sin \dfrac{u}{2} = \dfrac{\sqrt{5}}{5}$,

$\cos \dfrac{u}{2} = -\dfrac{2\sqrt{5}}{5}$,

$\tan \dfrac{u}{2} = -\dfrac{1}{2}$

EVALUATING TRIGONOMETRIC EXPRESSIONS Find the exact value of the expression.

17. $\tan 15°$ $2 - \sqrt{3}$

18. $\sin 22.5°$ $\dfrac{\sqrt{2 - \sqrt{2}}}{2}$

19. $\tan (-22.5°)$ $1 - \sqrt{2}$

20. $\cos 67.5°$ $\dfrac{\sqrt{2 - \sqrt{2}}}{2}$

21. $\tan (-75°)$ $-(2 + \sqrt{3})$

22. $\cos \dfrac{7\pi}{8}$ $-\dfrac{\sqrt{2 + \sqrt{2}}}{2}$

23. $\sin -\dfrac{7\pi}{12}$ **See margin.**

24. $\cos -\dfrac{5\pi}{12}$ $\dfrac{\sqrt{2 - \sqrt{3}}}{2}$

25. $\sin \dfrac{\pi}{12}$ **See margin.**

HALF-ANGLE FORMULAS Find the exact values of $\sin \dfrac{u}{2}$, $\cos \dfrac{u}{2}$, and $\tan \dfrac{u}{2}$.

26–29. See margin.

26. $\cos u = \dfrac{3}{5}, 0 < u < \dfrac{\pi}{2}$

27. $\cos u = \dfrac{2}{3}, \dfrac{3\pi}{2} < u < 2\pi$

28. $\sin u = \dfrac{9}{10}, \dfrac{\pi}{2} < u < \pi$

29. $\sin u = -\dfrac{4}{5}, \dfrac{3\pi}{2} < u < 2\pi$

14.7 *Using Double- and Half-Angle Formulas*   **879**

3 APPLY

ASSIGNMENT GUIDE

BASIC
Day 1: pp. 879–880 Exs. 17–23, 26–38 even, 43–46
Day 2: pp. 880–881 Exs. 51–56, 60–62, 70, 74–76, 79–95 odd, Quiz 3 Exs. 1–23

AVERAGE
Day 1: pp. 879–880 Exs. 17–23, 26–40 even, 43–48
Day 2: pp. 880–881 Exs. 51–56, 60–70, 74–76, 79–95 odd, Quiz 3 Exs. 1–23

ADVANCED
Day 1: pp. 879–880 Exs. 17–25, 26–42 even, 43–50
Day 2: pp. 880–881 Exs. 51–57, 60–68, 71–77, 79–95 odd, Quiz 3 Exs. 1–23

BLOCK SCHEDULE
pp. 879–881 Exs. 18–46 even, 52–68 even, 70, 74–76, 79–95 odd, Quiz 3 Exs. 1–23 (with 14.6)

EXERCISE LEVELS
Level A: *Easier*
17–33, 51–55, 60–62
Level B: *More Difficult*
32–50, 56–59, 63–72, 74–76
Level C: *Most Difficult*
73, 77

✔ **HOMEWORK CHECK**
To quickly check student understanding of key concepts, go over the following exercises:
Exs. 18, 26, 30, 36, 44, 52, 60. See also the Daily Homework Quiz:
• 📠 Transparency (p. 112)

23. $-\left(\dfrac{\sqrt{2 + \sqrt{3}}}{2}\right)$

25. $\dfrac{\sqrt{2 - \sqrt{3}}}{2}$

MATHEMATICAL REASONING
EXERCISES 66 AND 67

Students may question why it is necessary to have three equivalent double-angle formulas for cosine and two equivalent half-angle formulas for tangent. Students should realize that this allows them to choose the most convenient form when simplifying, verifying, or solving, thus eliminating unnecessary calculations.

53. $0, \frac{\pi}{4}, \frac{3\pi}{4}, \pi, \frac{5\pi}{4}, \frac{7\pi}{4}$

54. $\frac{\pi}{3}, \frac{2\pi}{3}, \frac{4\pi}{3}, \frac{5\pi}{3}$

56. $0, \frac{\pi}{3}, \frac{5\pi}{3}$

67. $\tan \frac{u}{2} = \frac{1 - \cos u}{\sin u} =$

$\frac{(1 - \cos u)(1 + \cos u)}{\sin u(1 + \cos u)} =$

$\frac{1 - \cos^2 u}{\sin u(1 + \cos u)} = \frac{\sin^2 u}{\sin u(1 + \cos u)} =$

$\frac{\sin u}{1 + \cos u}$

77. $\sin \frac{\theta}{2} = \frac{\text{opposite}}{\text{hypotenuse}}$. Since θ is acute, $\frac{\text{opposite}}{\text{hypotenuse}} =$

$\frac{\sin \theta}{\sqrt{\sin^2 \theta + (1 + \cos \theta)^2}} =$

$\sqrt{\frac{\sin^2 \theta}{2(\cos \theta + 1)}} = \sqrt{\frac{1 - \cos^2 \theta}{2(\cos \theta + 1)}} =$

$\sqrt{\frac{1 - \cos \theta}{2}}$.

ADDITIONAL PRACTICE AND RETEACHING

For Lesson 14.7:
- Practice Levels A, B, and C (*Chapter 14 Resource Book,* p. 94)
- Reteaching with Practice (*Chapter 14 Resource Book,* p. 97)
- ☐ See Lesson 14.7 of the *Personal Student Tutor*

For more Mixed Review:
- ☐ Search the *Test and Practice Generator* for key words or specific lessons.

880

STUDENT HELP

→ HOMEWORK HELP
Example 1: Exs. 17–25
Example 2: Exs. 26–33
Example 3: Exs. 34–42
Example 4: Exs. 43–50
Examples 5, 6: Exs. 51–65
Examples 7, 8: Exs. 69–73

32. $\sin 2x = \frac{4\sqrt{2}}{9}$,

$\cos 2x = -\frac{7}{9}$,

$\tan 2x = -\frac{4\sqrt{2}}{7}$

43. $(\sin x + \cos x)^2 =$
$\sin^2 x + \cos^2 x +$
$2 \sin x \cos x = 1 + \sin 2x$

45. $\cos \theta + 2 \sin^2 \frac{\theta}{2} =$

$\cos \theta + 2\left(\frac{1 - \cos \theta}{2}\right) =$

$\cos \theta + 1 - \cos \theta = 1$

46. $\sin \frac{\theta}{3} \cos \frac{\theta}{3} =$

$\frac{1}{2}\left(2 \sin \frac{\theta}{3} \cos \frac{\theta}{3}\right) =$

$\frac{1}{2} \sin \frac{2\theta}{3}$

47. $\cos 3x = \cos (2x + x) =$
$\cos 2x \cos x -$
$\sin 2x \sin x = \cos x \cdot$
$(\cos^2 x - \sin^2 x) -$
$\sin x (2 \sin x \cos x) =$
$\cos^3 x - 3 \sin^2 x \cos x$

48. $\sin 4\theta = 2 \sin 2\theta \cdot$
$\cos 2\theta = 2(2 \sin \theta \cos \theta) \cdot$
$(1 - 2\sin^2 \theta) =$
$4 \sin \theta \cos \theta (1 - 2 \sin^2 \theta)$

50. $\cot \theta + \tan \theta =$

$\frac{\cos \theta}{\sin \theta} + \frac{\sin \theta}{\cos \theta} =$

$\frac{\cos^2 \theta + \sin^2 \theta}{\sin \theta \cos \theta} =$

$\frac{2}{2 \sin \theta \cos \theta} =$

$\frac{2}{\sin 2\theta} = 2 \csc 2\theta$

63. $\pi + 2n\pi, \frac{\pi}{3} + 4n\pi,$
$\frac{5\pi}{3} + 4n\pi$

65. $\frac{3\pi}{2} + 2n\pi, \frac{7\pi}{6} + 2n\pi,$
$\frac{11\pi}{6} + 2n\pi$

DOUBLE-ANGLE FORMULAS Find the exact values of sin 2x, cos 2x, and tan 2x.

30. $\tan x = 2, 0 < x < \frac{\pi}{2}$
$\sin 2x = \frac{4}{5}, \cos 2x = -\frac{3}{5}, \tan 2x = -\frac{4}{3}$

31. $\tan x = -\frac{1}{2}, -\frac{\pi}{2} < x < 0$
$\sin 2x = -\frac{4}{5}, \cos 2x = \frac{3}{5}, \tan 2x = -\frac{4}{3}$

32. $\cos x = -\frac{1}{3}, \pi < x < \frac{3\pi}{2}$ See margin.

33. $\sin x = -\frac{3}{5}, \frac{3\pi}{2} < x < 2\pi$
$\sin 2x = -\frac{24}{25}, \cos 2x = \frac{7}{25}, \tan 2x = -\frac{24}{7}$

SIMPLIFYING TRIGONOMETRIC EXPRESSIONS Rewrite the expression without double angles or half angles, given that $0 < x < \frac{\pi}{2}$. Then simplify the expression.

34. $\sqrt{2 + 2 \cos x}\left(\cos \frac{x}{2}\right)$ $\frac{1 + \cos x}{}$

35. $\frac{\sin 2x}{\sin x}$ $2 \cos x$

36. $\tan 2x (1 + \tan x)$ $\frac{2 \tan x}{1 - \tan x}$

37. $\cos 2x - 3 \sin^2 x$ $1 - 5 \sin^2 x$

38. $\frac{\cos 2x}{\cos^2 x}$ $2 - \sec^2 x$

39. $\left(\frac{\sin x}{1 - \cos^2 x}\right) \tan \frac{x}{2}$ $\frac{1}{1 + \cos x}$

40. $\frac{(1 + \cos x)^2 \tan \frac{x}{2}}{\sin x + \sin x \cos x}$

41. $\frac{1 + \cos 2x}{\cot x}$ $2 \sin x \cos x$

42. $\frac{\sin \frac{x}{2} \tan \frac{x}{2}}{1 - \cos x}$ $\frac{\sqrt{2 - 2 \cos x}}{2 \sin x}$

VERIFYING IDENTITIES Verify the identity. 43, 45–48, 50. See margin.

43. $(\sin x + \cos x)^2 = 1 + \sin 2x$

44. $1 + \cos 10x = 2 \cos^2 5x$
$1 + \cos 10x = 1 + 2 \cos^2 5x - 1 = 2 \cos^2 5x$

45. $\cos \theta + 2 \sin^2 \frac{\theta}{2} = 1$

46. $\sin \frac{\theta}{3} \cos \frac{\theta}{3} = \frac{1}{2} \sin \frac{2\theta}{3}$

47. $\cos 3x = \cos^3 x - 3 \sin^2 x \cos x$

48. $\sin 4\theta = 4 \sin \theta \cos \theta (1 - 2 \sin^2 \theta)$

49. $\cos^2 2x - \sin^2 2x = \cos 4x$
$\cos^2 2x - \sin^2 2x = \cos 4x$

50. $\cot \theta + \tan \theta = 2 \csc 2\theta$

SOLVING TRIGONOMETRIC EQUATIONS Solve the equation for $0 \le x < 2\pi$. 53, 54, 56. See margin.

51. $\sin \frac{1}{2}x = -1$ no solution

52. $\cos x - \cos \frac{1}{2}x = 0$ $0, \frac{4\pi}{3}$

53. $\sin 2x \cos x = \sin x$

54. $\cos 2x = -2 \cos^2 x$

55. $\tan 2x - \tan x = 0$ $0, \pi$

56. $\sin \frac{x}{2} + \cos x = 1$

57. $\tan 2x = \frac{\cos 2x}{2}$
$0.21, 1.38, 3.36, 4.50$

58. $\tan \frac{x}{2} = \sin x$ $0, \frac{\pi}{2}, \frac{3\pi}{2}$

59. $\frac{\cos 2x}{\cos^2 x} = 1$ $0, \pi$

FINDING GENERAL SOLUTIONS Find the general solution of the equation.

60. $\cos 2x = -1$ $\frac{\pi}{2} + n\pi$

61. $\sin 2x + \sin x = 0$
$n\pi, \frac{2\pi}{3} + 2n\pi, \frac{4\pi}{3} + 2n\pi$

62. $\cos 2x - \cos x = 0$
$2n\pi, \frac{2\pi}{3} + 2n\pi, \frac{4\pi}{3} + 2n\pi$

63. $\cos \frac{x}{2} - \sin x = 0$ See margin.

64. $\sin \frac{x}{2} + \cos x = 0$
$\frac{7\pi}{3} + 4n\pi, \frac{11\pi}{3} + 4n\pi, \pi + 4n\pi$

65. $\cos 2x = 3 \sin x + 2$ See margin.

66. LOGICAL REASONING Show that the three double-angle formulas for cosine are equivalent. $\cos 2u = \cos^2 u - \sin^2 u = \cos^2 u - (1 - \cos^2 u) = 2 \cos^2 u - 1$

67. LOGICAL REASONING Show that the two half-angle formulas for tangent are equivalent. See margin.

68. *Writing* Use the formula at the top of page 878 to explain why the projection angle that maximizes the distance a projectile travels is $\theta = 45°$ when $h_0 = 0$. *x is a maximum when sin 2θ is a maximum, at sin 2θ = 1, or θ = 45°*

69. 🌐 **PROJECTILE HEIGHT** Find a formula for the maximum height of an object projected from ground level at speed v and angle θ. To do this, find half of the horizontal distance $\frac{1}{32}v^2 \sin 2\theta$ and then substitute it for x in the general model for the path of a projectile (where $h_0 = 0$) at the top of page 878. $y_{max} = \frac{1}{64} v^2 \sin^2 \theta$

880   **Chapter 14** *Trigonometric Graphs, Identities, and Equations*

FOCUS ON
APPLICATIONS

MACHU PICCHU is an ancient Incan city discovered by Hiram Bingham in 1911 and located in the Andes Mountains near Cuzco, Peru. The site consists of 5 square miles of terraced gardens linked by 3000 steps.

70. 🛩 **AERONAUTICS** An airplane's mach number M is the ratio of its speed to the speed of sound. When an airplane travels faster than the speed of sound, the sound waves form a cone behind the airplane (see Lesson 10.5, page 620).

The mach number is related to the apex angle θ of the cone by $\sin\frac{\theta}{2} = \frac{1}{M}$. Find the angle θ that corresponds to a mach number of 4.5.　$\theta = 26°$

🌎 **INCA DWELLING** **In Exercises 71 and 72, use the following information.**
Shown below is a drawing of an Inca dwelling found in Machu Picchu, about 50 miles northwest of Cuzco, Peru. All that remains of the ancient city today are stone ruins.

71. Express the area of the triangular portion of the side of the dwelling as a function of $\sin\frac{\theta}{2}$ and $\cos\frac{\theta}{2}$.　$A = 324 \sin\frac{\theta}{2} \cos\frac{\theta}{2}$

72. Express the area found in Exercise 71 as a function of $\sin\theta$. Then solve for θ assuming that the area is 132 square feet.
$A = 162 \sin\theta; \theta \approx 54.6°$

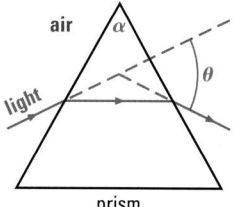

73. 🔬 **OPTICS** The index of refraction n of a transparent material is the ratio of the speed of light in a vacuum to the speed of light in the material. Some common materials and their indices are air (1.00), water (1.33), and glass (1.5). Triangular prisms are often used to measure the index of refraction based on this formula:

$$n = \frac{\sin\left(\frac{\theta}{2} + \frac{\alpha}{2}\right)}{\sin\frac{\theta}{2}}$$

For the prism shown, $\alpha = 60°$. Write the index of refraction as a function of $\cot\frac{\theta}{2}$. Then find θ if the prism is made of glass.　$n = \frac{\sqrt{3}}{2} + \frac{1}{2}\cot\frac{\theta}{2}, 77°$

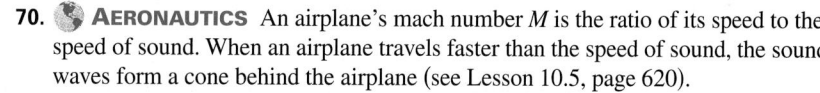
prism

Test Preparation

77. $\sin\frac{\theta}{2}$; See margin.

$\cos\frac{\theta}{2} = \frac{\text{adjacent}}{\text{hypotenuse}} =$

$\frac{\cos\theta + 1}{\sqrt{\sin^2\theta + (1 + \cos\theta)^2}} =$

$\sqrt{\frac{(\cos\theta + 1)^2}{2(\cos\theta + 1)}} = \sqrt{\frac{1 + \cos\theta}{2}};$

$\tan\frac{\theta}{2} = \frac{\text{opposite}}{\text{adjacent}} =$

$\frac{\sin\theta}{1 + \cos\theta}$

QUANTITATIVE COMPARISON **In Exercises 74–76, choose the statement that is true about the given quantities.**

  Ⓐ The quantity in column A is greater.

  Ⓑ The quantity in column B is greater.

  Ⓒ The two quantities are equal.

  Ⓓ The relationship cannot be determined from the given information.

	Column A	Column B	
74.	$\sin x$, with $45° < x < 90°$	$\sin 2x$	D
75.	$\cos x$, with $90° < x < 135°$	$\cos 2x$	D
76.	$\tan x$, with $45° < x < 90°$	$\tan 2x$	A

★ **Challenge**

77. DERIVING FORMULAS Use the diagram shown at the right to derive the formulas for $\sin\frac{\theta}{2}$, $\cos\frac{\theta}{2}$, and $\tan\frac{\theta}{2}$ when θ is an acute angle.
See margin.

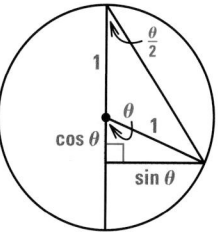

EXTRA CHALLENGE
www.mcdougallittell.com

DAILY HOMEWORK QUIZ

📄 **Transparency Available**

1. Find the exact value of $\tan 22.5°$.
$-1 + \sqrt{2}$

2. Given $\cos u = \frac{4}{5}$ with $0 < u < \frac{\pi}{2}$, find the exact values of $\sin\frac{u}{2}$, $\cos\frac{u}{2}$, and $\tan\frac{u}{2}$.　$\sin\frac{u}{2} = \frac{\sqrt{10}}{10}$, $\cos\frac{u}{2} = \frac{3\sqrt{10}}{10}$, and $\tan\frac{u}{2} = \frac{1}{3}$

3. Given $\tan x = -\frac{2}{3}$ with $-\frac{\pi}{2} < x < 0$, find the exact values of $\sin 2x$, $\cos 2x$, and $\tan 2x$.　$\sin 2x = \frac{-12}{13}$, $\cos 2x = \frac{5}{13}$, and $\tan 2x = \frac{-12}{5}$

4. Simplify $\frac{\sin 2x}{\sin^2 x}$.　$2\cot x$

5. Verify the identity $\cos 2x - \sin 2x + 2\sin^2 x = (\cos x - \sin x)^2$.
$\cos 2x - \sin 2x + 2\sin^2 x =$
$\cos^2 x - \sin^2 x - 2\sin x\cos x + 2\sin^2 x = \cos^2 x - 2\sin x\cos x + \sin^2 x = (\cos x - \sin x)(\cos x - \sin x) = (\cos x - \sin x)^2$

6. Find the general solution of $\frac{\sin 2x}{\cos x} = 1$.　$\frac{\pi}{6} + 2n\pi, \frac{5\pi}{6} + 2n\pi$

EXTRA CHALLENGE NOTE
→ Challenge problems for Lesson 14.7 are available in **blackline** format in the *Chapter 14 Resource Book*, p. 101 and at **www.mcdougallittell.com**.

ADDITIONAL TEST PREPARATION

1. OPEN ENDED Name an angle that can be expressed as a double or half angle. Show how you can use a double- or half-angle formula to find the sine of this angle. **Check to see that answers are similar to Example 1.**

14.7 *Using Double- and Half-Angle Formulas* **881**

MIXED REVIEW

FUNCTION OPERATIONS Let $f(x) = 4x + 1$ and $g(x) = 6x$. Perform the indicated operation and state the domain. (Review 7.3)

78. $f(x) + g(x)$ **79.** $f(x) - g(x)$ **80.** $f(x) \cdot g(x)$

81. $f(x) \div g(x)$ **82.** $f(g(x))$ **83.** $g(f(x))$

$f(g(x)) = 24x + 1$, reals $g(f(x)) = 24x + 6$, reals

CALCULATING PROBABILITIES Calculate the probability of rolling a die 30 times and getting the given number of 4's. (Review 12.6)

84. 1   0.025   **85.** 3   0.137   **86.** 5   0.192

87. 6   0.1601   **88.** 8   0.063   **89.** 10   0.013

SIMPLIFYING EXPRESSIONS Simplify the expression. (Review 14.6)

90. $\cos\left(x - \dfrac{3\pi}{2}\right)$ $-\sin x$ **91.** $\sin(x - \pi)$ $-\sin x$ **92.** $\tan\left(x - \dfrac{\pi}{3}\right)$

93. $\cos(x - \pi)$ $-\cos x$ **94.** $\sin\left(x + \dfrac{\pi}{2}\right)$ $\cos x$ **95.** $\tan\left(x + \dfrac{\pi}{4}\right)$

96. **MOVING** The truck you have rented for moving your furniture has a ramp. If the ramp is 20 feet long and the back of the truck is 3 feet above the ground, at what angle does the ramp meet the ground? (Review 13.4)   8.6°

Answers in left margin:

78. $f(x) + g(x) = 10x + 1$, reals

79. $f(x) - g(x) = -2x + 1$, reals

80. $f(x) \cdot g(x) = 24x^2 + 6x$, reals

81. $f(x) \div g(x) = \dfrac{4x + 1}{6x}$, reals except $x = 0$

92. $\dfrac{\sqrt{3}\tan^2 x - 4\tan x + \sqrt{3}}{3\tan^2 x - 1}$

95. $\dfrac{\tan x + 1}{1 - \tan x}$

QUIZ 3

Self-Test for Lessons 14.6 and 14.7

Find the exact value of the expression. (Lesson 14.6)

1. $\sin 105°$ $\dfrac{\sqrt{2 + \sqrt{3}}}{2}$ **2.** $\cos 285°$ $\dfrac{\sqrt{6} - \sqrt{2}}{4}$ **3.** $\tan 165°$ $-2 + \sqrt{3}$

4. $\sin\dfrac{17\pi}{12}$ $-\dfrac{\sqrt{6} + \sqrt{2}}{4}$ **5.** $\cos\dfrac{13\pi}{12}$ $-\dfrac{\sqrt{6} + \sqrt{2}}{4}$ **6.** $\tan\left(-\dfrac{\pi}{12}\right)$ $-2 + \sqrt{3}$

Given $\sin u = \dfrac{1}{3}$ **with** $\dfrac{\pi}{2} < u < \pi$, **find the exact value of the expression.** (Lesson 14.7)

7. $\sin\dfrac{u}{2}$ $\dfrac{\sqrt{18 + 12\sqrt{2}}}{6}$ **8.** $\cos\dfrac{u}{2}$ $\dfrac{\sqrt{18 - 12\sqrt{2}}}{6}$ **9.** $\tan\dfrac{u}{2}$ $\dfrac{\sqrt{3 + 2\sqrt{2}}}{\sqrt{3 - 2\sqrt{2}}}$

10. $\sin 2u$ $-\dfrac{4\sqrt{2}}{9}$ **11.** $\cos 2u$ $\dfrac{7}{9}$ **12.** $\tan 2u$ $-\dfrac{4\sqrt{2}}{7}$

Simplify the expression. (Lessons 14.6, 14.7)

13. $\sin(x + 3\pi)$ $-\sin x$ **14.** $\cos(\pi - x)$ $-\cos x$ **15.** $\tan\left(x + \dfrac{\pi}{4}\right)$ $\dfrac{\tan x + 1}{1 - \tan x}$

16. $\dfrac{\sin 2x}{2\cos x}$ $\sin x$ **17.** $2\cos^2\dfrac{x}{2} - \cos x$ 1 **18.** $\left(\dfrac{1 - \tan x}{2}\right)\tan 2x$ $\dfrac{\tan x}{1 + \tan x}$

Solve the equation. (Lessons 14.6, 14.7)   19, 21, 22. See margin.

19. $\sin 3x = 0.5$ **20.** $\cos 2x - \cos^2 x = 0$ $n\pi$

21. $\sin(2x - \pi) = \sin x$ **22.** $\tan(-2x) = 1$

23. **GOLF** Use the formula $x = \dfrac{1}{32}v^2 \sin 2\theta$ to find the horizontal distance x (in feet) that a golf ball will travel when it is hit at an initial speed of 50 feet per second and at an angle of 40°. (Lesson 14.7)   77 ft

Answers in left margin:

19. $\dfrac{\pi}{18} + \dfrac{2n\pi}{3}$, $\dfrac{5\pi}{18} + \dfrac{2n\pi}{3}$

21. $n\pi$; $\dfrac{2\pi}{3} + 2n\pi$; $\dfrac{4\pi}{3} + 2n\pi$

22. $\dfrac{3\pi}{8} + \dfrac{n\pi}{2}$

Chapter Summary

WHAT did you learn?

Graph sine, cosine, and tangent functions. **(14.1)**

Graph translations and reflections of sine, cosine, and tangent graphs. **(14.2)**

Use trigonometric identities to simplify expressions. **(14.3)**

Verify identities that involve trigonometric expressions. **(14.3)**

Solve trigonometric equations. **(14.4, 14.6, 14.7)**

Write sine and cosine models for graphs and data. **(14.5)**

Use sum and difference formulas. **(14.6)**

Use double- and half-angle formulas. **(14.7)**

Use trigonometric functions to solve real-life problems. **(14.1–14.7)**

WHY did you learn it?

Graph the height of a boat moving over waves. **(p. 836)**

Graph the height of a person rappelling down a cliff. **(p. 843)**

Simplify the parametric equations that describe a carousel's motion. **(p. 854)**

Show that two equations modeling the shadow of a sundial are equivalent. **(p. 853)**

Solve an equation that models the position of the sun at sunrise. **(p. 860)**

Write models for temperatures inside and outside an igloo. **(p. 866)**

Relate the length of an image to the length of an actual object when taking aerial photographs. **(p. 873)**

Find the angle at which you should kick a football to make it travel a certain distance. **(p. 878)**

Model real-life patterns, such as the vibrations of a tuning fork. **(p. 833)**

How does Chapter 14 fit into the BIGGER PICTURE of algebra?

In Chapter 14 you continued your study of trigonometry, focusing more on algebra connections than geometry connections. You graphed trigonometric functions and studied characteristics of the graphs, just as you have done with other types of functions during this course.

In this chapter you saw how some algebraic skills are used in trigonometry, such as in solving a trigonometric equation in quadratic form. If you go on to study higher-level algebra, you will see how trigonometry is used in algebra, such as in finding the complex nth roots of a real number.

STUDY STRATEGY

How did you use multiple methods?

Here is an example of two methods used for Example 2 on page 849 following the **Study Strategy** on page 830.

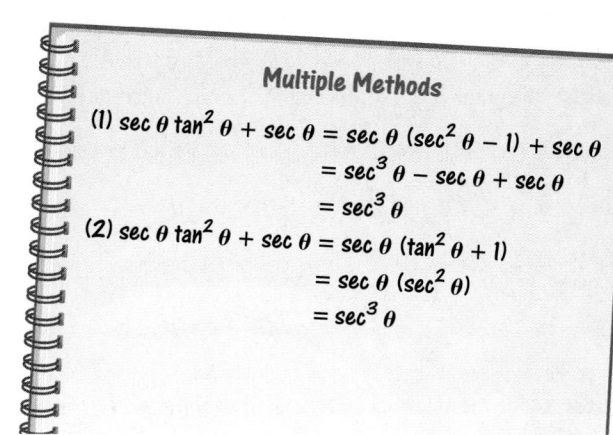

Multiple Methods

(1) $\sec \theta \tan^2 \theta + \sec \theta = \sec \theta (\sec^2 \theta - 1) + \sec \theta$
$= \sec^3 \theta - \sec \theta + \sec \theta$
$= \sec^3 \theta$

(2) $\sec \theta \tan^2 \theta + \sec \theta = \sec \theta (\tan^2 \theta + 1)$
$= \sec \theta (\sec^2 \theta)$
$= \sec^3 \theta$

883

ADDITIONAL RESOURCES

The following resources are available to help review the material in this chapter.

• Chapter Review Games and Activities (*Chapter 14 Resource Book,* p. 102)

• *Instant Replay: Video Review Games*

• ⊞ *Personal Student Tutor*

• Cumulative Review, Chs. 1–14 (*Chapter 14 Resource Book,* p. 114)

1.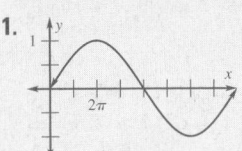

2.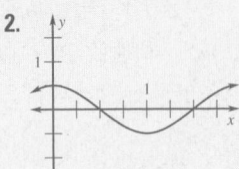

3.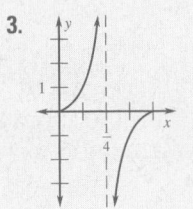

4.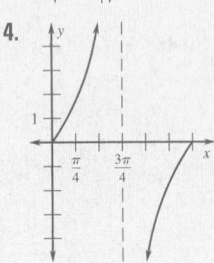

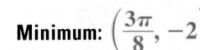

VOCABULARY

• periodic function, p. 831
• cycle, p. 831

• period, p. 831
• amplitude, p. 831

• frequency, p. 833
• trigonometric identities, p. 848

14.1   **GRAPHING SINE, COSINE, AND TANGENT FUNCTIONS**

Examples on pp. 831–834

EXAMPLES You can graph a trigonometric function by identifying the characteristics and key points of the graph.

$y = 2 \sin 4x$

Amplitude $= |2| = 2$ Period $= \dfrac{2\pi}{|4|} = \dfrac{\pi}{2}$

Intercepts: $(0, 0)$; $\left(\dfrac{\pi}{4}, 0\right)$; $\left(\dfrac{\pi}{2}, 0\right)$

Maximum: $\left(\dfrac{\pi}{8}, 2\right)$ **Minimum:** $\left(\dfrac{3\pi}{8}, -2\right)$

$y = \dfrac{1}{2} \tan 3x$

Period $= \dfrac{\pi}{|3|} = \dfrac{\pi}{3}$ **Intercept:** $(0, 0)$

Asymptotes: $x = \dfrac{\pi}{6}$, $x = -\dfrac{\pi}{6}$

Halfway points: $\left(\dfrac{\pi}{12}, \dfrac{1}{2}\right)$; $\left(-\dfrac{\pi}{12}, -\dfrac{1}{2}\right)$

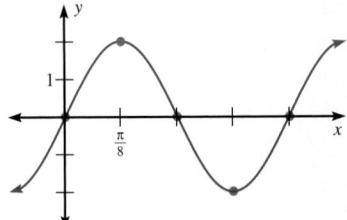

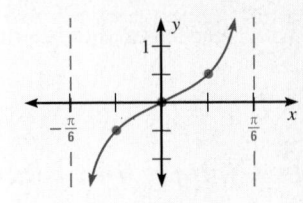

Draw one cycle of the function's graph.   1–4. See margin.

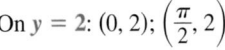

1. $y = \sin \dfrac{1}{4}x$ **2.** $y = \dfrac{1}{2} \cos \pi x$ **3.** $y = \tan 2\pi x$ **4.** $y = 3 \tan \dfrac{2}{3}x$

14.2   **TRANSLATIONS AND REFLECTIONS OF TRIGONOMETRIC GRAPHS**

Examples on pp. 840–843

EXAMPLE To graph $y = 2 - 4 \cos 2\left(x + \dfrac{\pi}{4}\right)$, start with the graph of $y = 4 \cos 2x$.

Translate the graph **left $\dfrac{\pi}{4}$ units** and **up 2 units**, and reflect it in the line $y = 2$.

Amplitude $= |-4| = 4$ Period $= \dfrac{2\pi}{|2|} = \pi$

On $y = 2$: $(0, 2)$; $\left(\dfrac{\pi}{2}, 2\right)$

Maximum: $\left(\dfrac{\pi}{4}, 6\right)$ **Minimums:** $\left(-\dfrac{\pi}{4}, -2\right)$; $\left(\dfrac{3\pi}{4}, -2\right)$

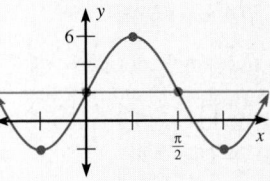

Graph the function. 5–7. See margin.

5. $y = 5 \sin (2x + \pi)$

6. $y = -4 \cos (x - \pi)$

7. $y = 2 + \tan \left(\frac{1}{2}x + \pi\right)$

14.3 VERIFYING TRIGONOMETRIC IDENTITIES

Examples on pp. 848–851

EXAMPLE You can verify identities such as $\sin x + \cot x \cos x = \csc x$.

$$\sin x + \cot x \cos x = \sin x + \left(\frac{\cos x}{\sin x}\right) \cos x \qquad \text{Reciprocal identity}$$

$$= \frac{\sin^2 x + \cos^2 x}{\sin x} \qquad \text{Write as one fraction.}$$

$$= \frac{1}{\sin x} \qquad \text{Pythagorean identity}$$

$$= \csc x \qquad \text{Reciprocal identity}$$

Simplify the expression.

8. $\tan (-x) \cos (-x)$ $-\sin x$

9. $\csc^2 (-x) \cos^2 \left(\frac{\pi}{2} - x\right)$ 1

10. $\sin^2 \left(\frac{\pi}{2} - x\right) - 2 \sin^2 x + 1$
$2 - 3 \sin^2 x$

Verify the identity.

11. $\sin^2 (-x) = \dfrac{\tan^2 x}{\tan^2 x + 1}$ See margin.

12. $1 - \cos^2 x = \tan^2 (-x) \cos^2 x$ See margin.

14.4 SOLVING TRIGONOMETRIC EQUATIONS

Examples on pp. 855–858

EXAMPLE You can find the general solution of a trigonometric equation or just the solution(s) in an interval.

$$3 \tan^2 x - 1 = 0 \qquad \text{Write original equation.}$$

$$3 \tan^2 x = 1 \qquad \text{Add 1 to each side.}$$

$$\tan^2 x = \frac{1}{3} \qquad \text{Divide each side by 3.}$$

$$\tan x = \pm \frac{\sqrt{3}}{3} \qquad \text{Take square roots of each side.}$$

There are two solutions in the interval $0 \le x < \pi$: $x = \dfrac{\pi}{6}$ and $x = \dfrac{5\pi}{6}$. The general

solution of the equation is: $x = \dfrac{\pi}{6} + n\pi$ or $x = \dfrac{5\pi}{6} + n\pi$ where n is any integer.

Find the general solution of the equation.

13. $2 \sin^2 x \tan x = \tan x$ $n\pi; \frac{\pi}{4} + \frac{n\pi}{2}$

14. $\sec^2 x - 2 = 0$ $\frac{\pi}{4} + \frac{n\pi}{2}$

15. $\cos 2x + 2 \sin^2 x - \sin x = 0$ $\frac{\pi}{2} + 2n\pi$

16. $\tan^2 3x = 3$ $\frac{\pi}{9} + \frac{n\pi}{3}, \frac{2\pi}{9} + \frac{n\pi}{3}$

17. $2 \sin x - 1 = 0$ $\frac{\pi}{6} + 2n\pi, \frac{5\pi}{6} + 2n\pi$

18. $\sin x (\sin x + 1) = 0$ $n\pi, \frac{3\pi}{2} + 2n\pi$

5.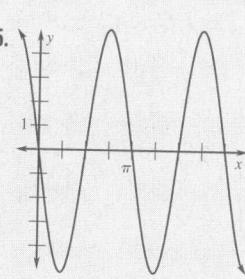

6.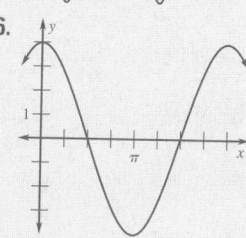

7.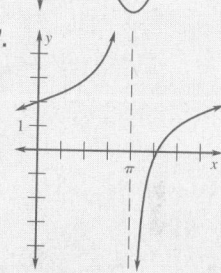

11. $\sin^2 (-x) =$
$\sin^2 x$ (negative angle identity) $=$
$\dfrac{\sin^2 x}{\cos^2 x} \cos^2 x \left(\text{multiply by } \dfrac{\cos^2 x}{\cos^2 x}\right) =$
$\dfrac{\tan^2 x}{\sec^2 x}$ (identities) $=$
$\dfrac{\tan^2 x}{1 + \tan^2 x}$ (Pythagorean identity)

12. $1 - \cos^2 x =$
$\sin^2 x$ (Pythagorean identity) $=$
$\dfrac{\sin^2 x}{\cos^2 x} \cos^2 x \left(\text{multiply by } \dfrac{\cos^2 x}{\cos^2 x}\right) =$
$\tan^2 x \cos^2 x$ (identity) $=$
$\tan^2 (-x) \cos^2 x$ (negative angle identity)

1.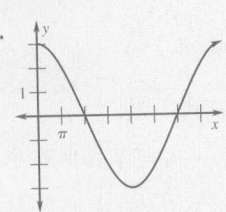

2.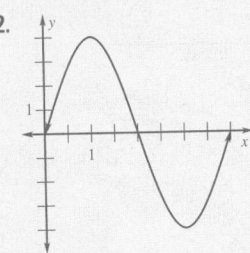

3.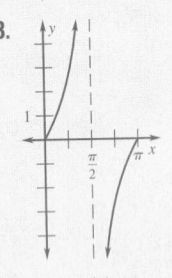

4.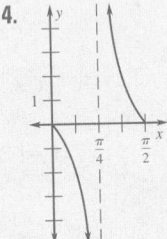

Examples on
pp. 862–864

14.5 **MODELING WITH TRIGONOMETRIC FUNCTIONS**

EXAMPLE You can write a model for the sinusoid at the right. Since the maximum and minimum values of the function do not occur at points equidistant from the x-axis, the curve has a vertical shift. To find the value of k, add the maximum and minimum values and divide by 2.

$$k = \frac{M + m}{2} = \frac{4 + (-2)}{2} = \frac{2}{2} = 1$$

The period is $\frac{2\pi}{b} = \pi$, so $b = 2$. Because the minimum occurs at $\frac{\pi}{4}$, the graph is a sine curve that involves a reflection but no horizontal shift. The amplitude is

$$|a| = \frac{M - m}{2} = \frac{4 - (-2)}{2} = 3.$$ Since $a < 0$, $a = -3$. The model is $y = 1 - 3 \sin 2x$.

Write a trigonometric function for the sinusoid with maximum at A and minimum at B.

19. $A\left(\frac{\pi}{2}, 2\right)$, $B\left(\frac{3\pi}{2}, -2\right)$
$y = 2 \sin x$

20. $A(0, 6)$, $B(2\pi, 0)$
$y = 3 \cos \frac{x}{2} + 3$

21. $A(0, 1)$, $B\left(\frac{\pi}{2}, -1\right)$
$y = \cos 2x$

Examples on
pp. 869–871

14.6 **USING SUM AND DIFFERENCE FORMULAS**

EXAMPLE You can use formulas to evaluate trigonometric functions of the sum or difference of two angles.

$$\sin 105° = \sin (45° + 60°) = \sin 45° \cos 60° + \cos 45° \sin 60°$$

$$= \frac{\sqrt{2}}{2} \cdot \frac{1}{2} + \frac{\sqrt{2}}{2} \cdot \frac{\sqrt{3}}{2} = \frac{\sqrt{2} + \sqrt{6}}{4}$$

Find the exact value of the expression.

22. $\sin 150°$ $\frac{1}{2}$

23. $\cos 195°$ $-\frac{\sqrt{6} + \sqrt{2}}{4}$

24. $\tan 15°$ $2 - \sqrt{3}$

25. $\tan \frac{7\pi}{12}$ $-2 - \sqrt{3}$

26. $\cos \frac{13\pi}{12}$ $-\frac{\sqrt{6} + \sqrt{2}}{4}$

Examples on
pp. 875–878

14.7 **USING DOUBLE- AND HALF-ANGLE FORMULAS**

EXAMPLE You can use formulas to evaluate some trigonometric functions.

$$\tan \frac{\pi}{12} = \tan \left(\frac{1}{2} \cdot \frac{\pi}{6}\right) = \frac{1 - \cos \frac{\pi}{6}}{\sin \frac{\pi}{6}} = \frac{1 - \frac{\sqrt{3}}{2}}{\frac{1}{2}} = 2 - \sqrt{3}$$

Find the exact value of the expression.

27. $\tan 165°$ $-2 + \sqrt{3}$

28. $\sin 67.5°$ $\frac{\sqrt{2 + \sqrt{2}}}{2}$

29. $\cos \frac{5\pi}{8}$ $-\frac{\sqrt{2 - \sqrt{2}}}{2}$

30. $\cos \frac{\pi}{12}$ $\frac{\sqrt{2 + \sqrt{3}}}{2}$

31. $\sin 6\pi$ 0

Chapter 14 *Trigonometric Graphs, Identities, and Equations*

Chapter Test

ADDITIONAL RESOURCES
• *Chapter 14 Resource Book*
Chapter Test (3 levels) (p. 103)
SAT/ACT Chapter Test (p. 109)
Alternative Assessment (p. 110)

• Test and Practice Generator

Draw one cycle of the function's graph. 1–8. See margin.

1. $y = 3 \cos \frac{1}{4}x$

2. $y = 4 \sin \frac{1}{2}\pi x$

3. $y = \frac{5}{2} \tan x$

4. $y = -2 \tan 2x$

5. $y = -3 + 2 \cos (x - \pi)$

6. $y = 1 - \cos x$

7. $y = 5 + \sin \frac{1}{2}x$

8. $y = 5 + 2 \tan (x + \pi)$

Simplify the expression.

9. $\cos \left(x - \frac{\pi}{2} \right) \; \sin x$

10. $\dfrac{\cos 2x + \sin^2 x}{\cos^2 x}$ 1

11. $\dfrac{\tan 2x}{2 \tan x} - \dfrac{\sec^2 x}{1 - \tan^2 x}$ $\dfrac{\tan^2 x}{\tan^2 x - 1}$

12. $\dfrac{4 \sin x \cos x - 2 \sin x \sec x}{2 \tan x}$ $2 \cos^2 x - 1$

Verify the identity. 13–15. See margin.

13. $-2 \cos^2 x \tan (-x) = \sin 2x$

14. $\tan \frac{x}{2} = \csc x - \cot x$

15. $\cos 3x = \cos^3 x - 3 \sin^2 x \cos x$

Solve the equation in the interval $0 \le x < 2\pi$. Check your solutions.

16. $-6 + 10 \cos x = -1$ $\frac{\pi}{3}, \frac{5\pi}{3}$

17. $\tan^2 x - 2 \tan x + 1 = 0$ $\frac{\pi}{4}, \frac{5\pi}{4}$

18. $\tan (x + \pi) + 2 \sin (x + \pi) = 0$ $\frac{\pi}{3}, \frac{5\pi}{3}, 0,$

Find the general solution of the equation.

19. $4 - 3 \sec^2 x = 0$ See margin.

20. $\cos x - \sin x \sin 2x = 0$ $\frac{\pi}{2} + n\pi, \frac{\pi}{4} + \frac{n\pi}{2}$

21. $\cos x \csc^2 x + 3 \cos x = 7 \cos x$ $\frac{\pi}{2} + n\pi, \frac{\pi}{6} + n\pi, \frac{5\pi}{6} + n\pi$

Find the exact value of the expression.

22. $\sin 345°$ See margin.

23. $\tan 112.5°$ $-\sqrt{2} - 1$

24. $\cos 375°$ $\frac{\sqrt{6} + \sqrt{2}}{4}$

25. $\tan \frac{13\pi}{12}$ $2 - \sqrt{3}$

26. $\sin \frac{\pi}{8}$ $\frac{\sqrt{2 - \sqrt{2}}}{2}$

27. $\cos \frac{41\pi}{12}$ $\frac{\sqrt{2} - \sqrt{6}}{4}$

Find the amplitude and period of the graph. Then write a trigonometric function for the graph.

28.

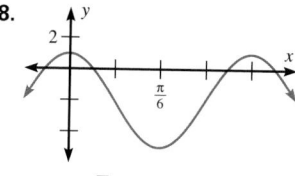

a: 3, P: $\frac{\pi}{3}$, $y = 3 \cos 6x - 2$

29.

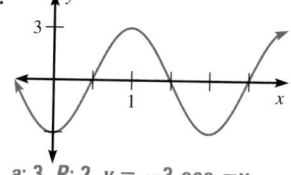

a: 3, P: 2, $y = -3 \cos \pi x$

30.

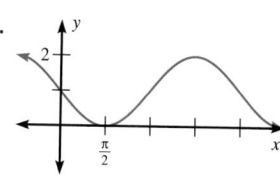

a: 1, P: 2π, $y = -\sin x + 1$

31. **TIDES** The depth of the ocean at a swim buoy reaches a maximum of 6 feet at 3 A.M. and a minimum of 2 feet at 9 A.M. Write a trigonometric function that models the water depth y (in feet) as a function of time t (in hours). Assume that $t = 0$ represents 12:00 A.M. $y = 2 \sin \frac{\pi t}{6} + 4$

32. **TEMPERATURES** The average daily temperature T (in degrees Fahrenheit) in Baltimore, Maryland, is given in the table. The variable t is measured in months, with $t = 0$ representing January 1. Use a graphing calculator to write a trigonometric model for T as a function of t. $T = 17 \sin (0.4t - 1) + 87.3$

▶ Source: U.S. National Oceanic and Atmospheric Administration

t	0.5	1.5	2.5	3.5	4.5	5.5	6.5	7.5	8.5	9.5	10.5	11.5
T	75	79	87	94	98	101	104	105	100	92	87	77

5.

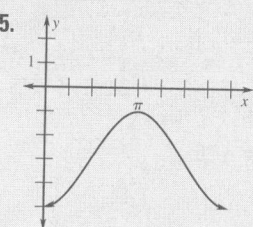

6.

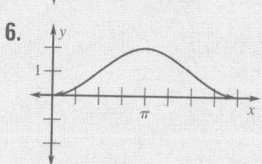

7.

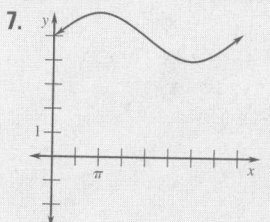

8.

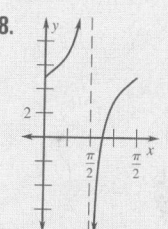

13. $-2 \cos^2 x \tan (-x) =$
$2 \cos^2 x \tan x = 2 \cos x \sin x =$
$\sin 2x$

14. $\tan \frac{x}{2} = \dfrac{1 - \cos x}{\sin x} = \dfrac{1}{\sin x} - \dfrac{\cos x}{\sin x} =$
$\csc x - \cot x$

15. $\cos 3x = \cos x \cos 2x -$
$\sin x \sin 2x = \cos x (\cos^2 x -$
$\sin^2 x) - \sin x (2 \sin x \cos x) =$
$\cos^3 x - 3 \sin^2 x \cos x$

19. $\frac{\pi}{6} + n\pi, \frac{5\pi}{6} + n\pi$

22. $\dfrac{\sqrt{2} - \sqrt{6}}{4}$

Chapter Test   **887**

ADDITIONAL RESOURCES
- *Chapter 14 Resource Book*
  Chapter Test (3 levels) (p. 103)
  SAT/ACT Chapter Test (p. 109)
  Alternative Assessment (p. 110)
- 🖥 *Test and Practice Generator*

CHAPTER 14

Chapter Standardized Test

▶ **TEST-TAKING STRATEGY** Long-term preparation for the SAT can be done throughout your high school career and can improve your overall abilities. If you keep up with your homework, both your problem-solving abilities and your vocabulary will improve. This type of long-term preparation will definitely affect not only your SAT scores, but your overall future academic performance as well.

1. MULTIPLE CHOICE Which function is graphed? **D**

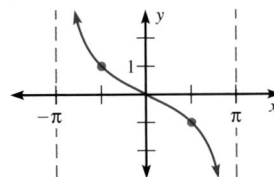

- (A) $y = \tan 2x$
- (B) $y = -\tan 2x$
- (C) $y = \tan \frac{1}{2}x$
- (D) $y = -\tan \frac{1}{2}x$
- (E) $y = -2 \tan x$

2. MULTIPLE CHOICE Which function is graphed? **E**

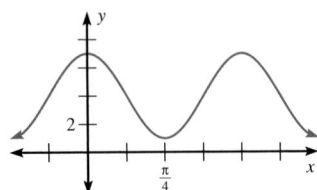

- (A) $y = 4 \cos \frac{1}{4}x$
- (B) $y = 3 + 4 \cos \frac{1}{4}x$
- (C) $y = 3 + 4 \cos 4x$
- (D) $y = 4 \cos 4x$
- (E) $y = 4 + 3 \cos 4x$

3. MULTIPLE CHOICE What is the simplified form of $\sin^2 \left(\frac{\pi}{2} - x \right) \tan^2 x + \cos^2 (-x) + \tan^2 x$? **A**
- (A) $\sec^2 x$
- (B) $2 \tan^2 x$
- (C) $\csc^2 x$
- (D) $1 + \sec^2 x$
- (E) $\sin^2 x + \tan^2 x$

4. MULTIPLE CHOICE Which of the following is a solution of the equation $\tan^2 x \cos x = \cos x$? **E**
- (A) $x = 0$
- (B) $x = \frac{\pi}{6}$
- (C) $x = \frac{\pi}{2}$
- (D) $x = \frac{5\pi}{6}$
- (E) $x = \frac{5\pi}{4}$

5. MULTIPLE CHOICE What is the exact value of $\tan \frac{5\pi}{12}$? **D**
- (A) $3 - 2\sqrt{3}$
- (B) $2 - \sqrt{3}$
- (C) $\sqrt{3} - 1$
- (D) $2 + \sqrt{3}$
- (E) $2\sqrt{3} + 3$

6. MULTIPLE CHOICE Given that $\cos \theta = -\frac{3}{5}$ and $\pi < \theta < \frac{3\pi}{2}$, which of the following is true? **E**
- (A) $\tan \theta = \frac{3}{4}$
- (B) $\csc \theta = \frac{5}{4}$
- (C) $\sec \theta = \frac{5}{3}$
- (D) $\tan \theta = -\frac{4}{3}$
- (E) $\cot \theta = \frac{3}{4}$

7. MULTIPLE CHOICE What trigonometric function has a graph with maximum $(\pi, 2)$ and minimum $(3\pi, -2)$? **C**
- (A) $y = 2 \sin 2x$
- (B) $y = 2 \cos 2x$
- (C) $y = 2 \sin \frac{1}{2}x$
- (D) $y = 2 \cos \frac{1}{2}x$
- (E) $y = 2 \sin \frac{\pi}{2}x$

8. MULTIPLE CHOICE Which of the following is a solution of the equation $\frac{\tan x \sin 2x + 2 \sin^2 x}{-1 + 4 \sin x} = 1$? **B**
- (A) $x = 0$
- (B) $x = \frac{5\pi}{6}$
- (C) $x = \frac{11\pi}{6}$
- (D) $x = \pi$
- (E) There is no solution.

9. MULTIPLE CHOICE What is the simplified form of the expression $\dfrac{2 \sin x \tan \frac{x}{2}}{\cos \left(\frac{\pi}{2} - x \right) \sin (-x) + 1}$? **C**
- (A) $\dfrac{1 - \cos x}{\cos^2 x}$
- (B) $2 \sec^2 x$
- (C) $\dfrac{2 - 2 \cos x}{\cos^2 x}$
- (D) $\dfrac{2}{\cos^2 x}$
- (E) $\dfrac{2 \sin x}{\cos^2 x}$

In Exercises 10 and 11, choose the statement that is true about the given quantities.

(A) The quantity in column A is greater.

(B) The quantity in column B is greater.

(C) The two quantities are equal.

(D) The relationship cannot be determined from the given information.

Column A	Column B	
10. Amplitude of the graph of $y = -5 + 4 \sin 3\pi x$	Amplitude of the graph of $y = 3 - 5 \sin 2\pi x$	B
11. Period of the graph of $y = \tan 4\pi x$	Period of the graph of $y = 3 \tan 4x$	B

12. ▦ **MULTI-STEP PROBLEM** The average daily time R of the sunrise and the average daily time S of the sunset for each month in Dallas, Texas, is given in the table. The variable t is measured in months, with $t = 0$ representing January 1.

t	0.5	1.5	2.5	3.5	4.5	5.5	6.5	7.5	8.5	9.5	10.5	11.5
R	7:29	7:10	6:37	5:58	5:29	5:20	5:31	5:51	6:11	6:32	6:58	7:21
S	17:45	18:12	18:36	18:58	19:20	19:36	19:35	19:11	18:33	17:54	17:27	17:24

a. Use a graphing calculator to find trigonometric models for R and S as functions of t. When entering the data into the calculator, you must convert the number of minutes into a fraction of an hour. For example, enter 7:27 as $7 + (27/60)$.
$$R = 1.1 \sin (.5t + 1.8) + 6.5; \; S = 1.1 \sin (0.5t - 1.1) + 18.5$$

b. Graph the functions you found in part (a). Use a viewing window of $0 \le x \le 48$ and $0 \le y \le 24$. Describe the periods, amplitudes, and locations of local maximums and minimums. How are the functions alike? How are they different?

c. Let $D = S - R$. What does D represent?   **hours of daylight**

d. Graph D in the same viewing window as R and S. How are the maximums and minimums of the three functions related? Explain the real-life significance of the relationships.

12b. periods: 12 months, amplitudes: just over 1 hour, local max: R in January, S in June, local min: R in June, S in Dec; Alike in that both have the same period and amplitude, different in that phase shift makes 1 increase as the other decreases and vice versa.

12d. D has local max and min at same times as S, and R has a min wherever D and S have a max, and vice versa. As sunrise gets earlier, the sun sets later and the hours of daylight are longer.

13. ▦ **MULTI-STEP PROBLEM** The average number of daylight hours D_M in Great Falls, Michigan, is given in the table. The variable t is measured in months, with $t = 0$ representing January 1.

t	0.5	1.5	2.5	3.5	4.5	5.5	6.5	7.5	8.5	9.5	10.5	11.5
D_M	8:57	10:18	11:55	13:38	15:07	15:54	15:31	14:14	12:34	10:51	9:21	8:32

a. Use a calculator to find a trigonometric model for D_M as a function of t. $D_m = 3.6 \sin (0.5t - 1.3) + 12.2$

b. Use the table given in Exercise 12. Subtract each R-value from its corresponding S-value to find the average number of hours of sunlight a day for each month in Dallas. Use a graphing calculator to find a trigonometric model for the data as a function of t. $D_T = 2.1 \sin (.5t - 1.4) + 12.1$

c. Graph the functions you found in parts (a) and (b). Use a viewing window of $0 \le x \le 48$ and $0 \le y \le 24$. Describe the periods, amplitudes, and locations of local maximums and minimums of the functions. How are the functions alike? How are they different? Do the graphs intersect? If so, where?

13c. periods: 12 months, amplitudes: D_T about 2, D_M about 4, local max in June, local min in December; Graphs have same periods, maxes, and mins, but D_M has greater amplitude. They intersect in March and September.

Chapter Standardized Test ▮ **889**

ADDITIONAL RESOURCES
A Cumulative Review covering
Chapters 1–14 is available in the
Chapter 14 Resource Book, p. 114.

7. $\begin{bmatrix} 11 & -8 \\ -14 & 9 \end{bmatrix}$

63.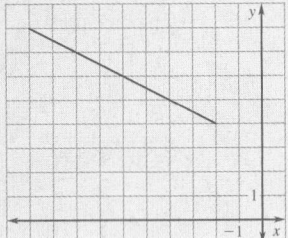

64.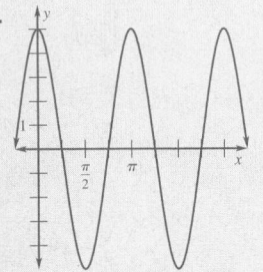

65.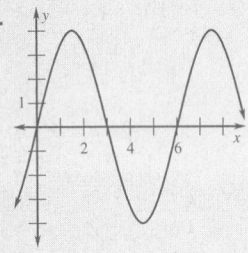

66.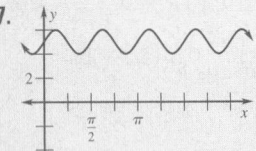

67.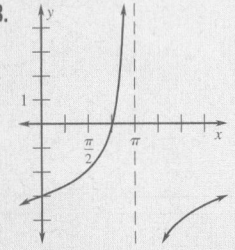

68.

Write an equation of the line with the given characteristics. (2.4)

1. slope: -2, y-intercept: 7
$y = -2x + 7$

2. points: $(5, 0)$, $(-3, 2)$
$y = -\frac{1}{4}x + \frac{5}{4}$

3. vertical line through $(4, 2)$ $x = 4$

Solve the system. (3.1, 3.2, 3.6, 4.3, 4.5, 10.7)

4. $x - 2y = 6$ $(2, -2)$
$3x + y = 4$

5. $x + y + z = 10$ $(10, 4, -4)$
$-x + 2y - z = 2$
$3x - y + 4z = 10$

6. $x^2 + y^2 = 16$
$x^2 + y^2 - 6x - 8y + 16 = 0$
$(0, 4), (3.84, 1.12)$

Solve the matrix equation. (4.4)

7. $\begin{bmatrix} 4 & 3 \\ -1 & -1 \end{bmatrix} X = \begin{bmatrix} 2 & -5 \\ 3 & -1 \end{bmatrix}$

8. $\begin{bmatrix} 5 & 3 \\ 7 & 4 \end{bmatrix} X = \begin{bmatrix} -1 & 6 \\ 2 & 0 \end{bmatrix} \begin{bmatrix} 10 & -24 \\ -17 & 42 \end{bmatrix}$

9. $\begin{bmatrix} 8 & -1 \\ -2 & 0 \end{bmatrix} X = \begin{bmatrix} 6 & 0 \\ 3 & -2 \end{bmatrix}$ $\begin{bmatrix} -\frac{3}{2} & 1 \\ -18 & 8 \end{bmatrix}$

See margin.

Perform the indicated operations. (6.3, 6.5, 9.4, 9.5)

10. $(-2x^2 - x + 4) - (3x + 10)$ $-2x^2 - 4x - 6$

11. $(x - 4)(2x^2 + 3x - 1)$ $2x^3 - 5x^2 - 13x + 4$

12. $(x^3 - 5x + 6) \div (x - 2)$ $x^2 + 2x - 1 + \dfrac{4}{x - 2}$

13. $\dfrac{x + 6}{8x + 10} \div \dfrac{x^2 - 36}{2x}$ $\dfrac{x}{4x^2 - 19x - 30}$

14. $\dfrac{6x}{x^2 + 3x - 10} + \dfrac{x - 4}{x - 2}$ $\dfrac{x^2 + 7x - 20}{x^2 + 3x - 10}$

15. $\dfrac{4x}{x - 7} - \dfrac{1}{x + 7}$ $\dfrac{4x^2 + 27x + 7}{x^2 - 49}$

Evaluate the expression without using a calculator. (7.1, 8.4)

16. $8^{2/3}$ 4

17. $125^{-1/3}$ $\frac{1}{5}$

18. $-9^{3/2}$ -27

19. $\sqrt[5]{-1}$ -1

20. $\sqrt[4]{10{,}000}$ 10

21. $\log_2 \frac{1}{16}$ -4

22. $\log_3 81$ 4

23. $\ln e^7$ 7

24. $\log 0.01$ -2

25. $\log_5 1$ 0

Find the distance between the two points. Then find the midpoint of the line segment connecting the two points. (10.1)

26. $(0, 0), (3, -8)$
8.54, $(1.5, -4)$

27. $(-5, 0), (0, 2)$
5.39, $(-2.5, 1)$

28. $(-1, -4), (2, 3)$
7.62, $(0.5, -0.5)$

29. $(7, 4), (0, -3)$
9.90, $(3.5, 0.5)$

Write the next term of the sequence. Then write a rule for the nth term. (11.1–11.3)

30. $1, 4, 9, 16, \ldots$ $25, n^2$

31. $8, 4, 2, 1, \ldots$ $\frac{1}{2}, 2^{4-n}$

32. $2, 6, 18, 54, \ldots$ $162, 2 \cdot 3^{n-1}$

33. $-6, -1, 4, 9, \ldots$ $14, 5n - 11$

Find the sum of the series. (11.1–11.4)

34. $\displaystyle\sum_{i=1}^{10} 16$ 160

35. $\displaystyle\sum_{i=1}^{5} (3i - 1)$ 40

36. $\displaystyle\sum_{i=0}^{4} 1000\left(\frac{1}{2}\right)^i$ 1937.5

37. $\displaystyle\sum_{n=1}^{\infty} 2\left(-\frac{1}{3}\right)^{n-1}$ $\frac{3}{2}$

Find the number of permutations or combinations. (12.1, 12.2)

38. $_6P_5$ 720

39. $_{10}P_2$ 90

40. $_3P_3$ 6

41. $_8C_1$ 8

42. $_4C_2$ 6

43. $_7C_4$ 35

Find the arc length and area of a sector with the given radius r and central angle θ. (13.2)

44. $r = 11$ cm, $\theta = 80°$
15.4 cm; 84.47 cm^2

45. $r = 6$ in., $\theta = 270°$
28.3 in.; 84.82 in.^{2}

46. $r = 3$ ft, $\theta = 120°$ 6.28 ft; 9.42 ft^2

Evaluate the function without using a calculator. (13.3)

47. $\tan 390°$ $\frac{\sqrt{3}}{3}$

48. $\sin(-45°)$ $-\frac{\sqrt{2}}{2}$

49. $\csc 90°$ 1

50. $\cot\left(-\frac{3\pi}{4}\right)$ 1

51. $\cos \frac{5\pi}{3}$ $\frac{1}{2}$

Evaluate the expression without using a calculator. Give your answer in both radians and degrees. (13.4)

52. $\cos^{-1} 0$ $\frac{\pi}{2}$, 90° 53. $\sin^{-1} \frac{1}{2}$ $\frac{\pi}{6}$, 30° 54. $\tan^{-1} 1$ $\frac{\pi}{4}$, 45° 55. $\cos^{-1}\left(-\frac{\sqrt{2}}{2}\right)$ $\frac{3\pi}{4}$, 135° 56. $\tan^{-1}\left(-\sqrt{3}\right)$ $-\frac{\pi}{3}$, −60°

Solve $\triangle ABC$. (13.5, 13.6)

57. $A = 65°, a = 7, b = 4$
 $B = 31°, C = 84°, c = 7.68$
 Find the area of $\triangle ABC$. (13.5, 13.6)

58. $B = 110°, a = 3, c = 8$
 $b = 9.46, A = 17°, C = 53°$

59. $a = 10, b = 9, c = 4$
 $A = 92°, B = 64°, C = 24°$

60. $A = 63°, c = 13, b = 20$ 115.8 61. $C = 98°, a = 34, b = 20$ 336.7 62. $a = 7, b = 4, c = 6$ 12.0

Graph the parametric equations. Then write an xy-equation and state the domain. (13.7) 63, 64. See margin for graphs.

63. $x = \frac{1}{4}t + 1, y = t - 3$ for $0 \le t \le 4$
 $y = 4x - 7, 1 \le x \le 2$

64. $x = -2t, y = t + 3$ for $1 \le t \le 5$
 $y = -\frac{1}{2}x + 3, -10 \le x \le -2$

Graph the function. (14.1, 14.2) 65–68. See margin.

65. $y = 5 \cos 2x$ 66. $y = 4 \sin \frac{1}{3}\pi x$ 67. $y = 5 + \sin 4x$ 68. $y = -3 + \tan \frac{1}{2}x$

Simplify the expression. (14.3)

69. $\tan(-x) + \tan x \sec^2 x$ $\tan^3 x$ 70. $\dfrac{\sin\left(\frac{\pi}{2} - x\right)}{\sin x}$ cot x 71. $\tan x \sec x - \csc x \sec^2 x$ −csc x

Find the general solution of the equation. (14.4)

72. $3 \sin x = \sqrt{3} + 5 \sin x$
 $\frac{4\pi}{3} + 2n\pi, \frac{5\pi}{3} + 2n\pi$

73. $2 \cos^2 \frac{x}{2} - 1 = 0$ $\frac{\pi}{2} + n\pi$

74. $\cos x \sin^2 x - \cos x = 0$ $\frac{\pi}{2} + n\pi$

Find the exact value of the expression. (14.6, 14.7)
75, 76. See margin.

75. $\sin 255°$ 76. $\sin 157.5°$ 77. $\tan 105°$ $-2 - \sqrt{3}$ 78. $\tan \frac{\pi}{12}$ $2 - \sqrt{3}$ 79. $\cos \frac{13\pi}{12}$ $-\frac{\sqrt{2} + \sqrt{6}}{4}$

80. **FRACTAL GEOMETRY** Tell whether $c = 1 + i$ is in the Mandelbrot set. Use absolute value to justify your answer. (5.4) No, because the absolute value of $f(z)$ increases.

81. 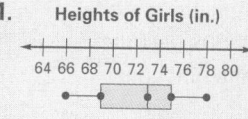 **GIRLS BASKETBALL** The heights (in inches) of the girls chosen for the first team on PARADE's 23rd annual All-America High School Girls Basketball Team are listed below. Find the mean, median, mode(s), range, and standard deviation of the heights. Draw a box-and-whisker plot for the heights.
 ▶ Source: Parade Magazine (7.7) mean: 72.4, median: 73, modes: 76, 74, range: 12, standard deviation: about 3.69

  76, 74, 76, 71, 72, 78, 66, 68, 74, 69 See margin for graph.

82. **EQUAL GENDERS** What is the probability that a family with four children has exactly two girls and two boys in any order? Assume that having a girl and having a boy are equally likely events. (12.6) $\frac{3}{8}$

83. **RIALTO TOWER** Suppose you are looking at the Rialto Tower in Melbourne, Australia, which reaches a height of 794 feet. Your angle of elevation to the top of the building is 39.8°. How far are you from the base of the building?
 ▶ Source: Council on Tall Buildings and Urban Habitat (13.1) 953 ft

84. **BICYCLING** As you pedal up a hill, the pedals on your mountain bike make one revolution every two seconds. The maximum height of the pedal is 19 inches above the ground and the minimum height is 5 inches above the ground. Write a trigonometric model for the height H of the pedal as a function of time t. (14.1, 14.2)

  $H = 7 \sin \pi t + 12$, where t is in seconds

Cumulative Practice 891

---

75. $-\dfrac{\sqrt{2} + \sqrt{6}}{4}$

76. $\dfrac{\sqrt{2 - \sqrt{2}}}{2}$

81. Heights of Girls (in.)

PROJECT GOALS

- Model data with a sine function.
- Use technology to write a trigonometric model.
- Graph sine functions.

MANAGING THE PROJECT
CLASSROOM MANAGEMENT
The Chapter 14 Project may be done in pairs or groups of 3 students. One student should work with the bottle and water to produce the note while another student uses the CBL and CBL microphone to collect the data. Group members should switch roles for the second sine function. The group can write the functions together. They should then discuss the parts of the report and either write the parts individually or together.

GUIDING STUDENTS' WORK
If this is the first time that students have used a CBL, you may want to demonstrate its use to the class as a whole prior to completing this experiment. It may also be helpful to set up a station for each group of students with all the necessary materials prior to the start of class. Each group will need a container of water that is easy to pour from such as a pitcher. A funnel may also be helpful. To avoid getting water on the desktop or on the equipment, provide a tray over which the students can pour the water. Also have paper towels available.

CONCLUDING THE PROJECT
Have each group choose two members to explain their sine waves to the class. As a class, compare sine waves of the same note. Ask students to describe how the sine waves are the same and how they are different. Pick sine waves of as many different notes as possible and have students hold them up in order. Ask how these waves change for notes from middle C to C. You might invite a music teacher or other musician to help the class complete the Music Connection part of this project.

PROJECT
Applying Chapters 13 and 14

The Mathematics of Music

OBJECTIVE Explore the relationship between music and trigonometric functions.

Materials: plastic or glass bottle, container of water, CBL, CBL microphone, TI-82 or TI-83 graphing calculator with cable to link to the CBL

STUDENT HELP

TECHNOLOGY HELP

For suggestions about doing this project see www.mcdougallittell.com

Sound is a variation in pressure transmitted through air, water, or other matter. Sound travels as a wave. The sound of a pure note can be represented using a sine wave (or a cosine wave; recall that a cosine wave is just a sine wave shifted horizontally). More complicated sounds can be modeled by the sum of several sine waves.

The pitch of a sound wave is determined by the wave's frequency. The greater the frequency, the higher the pitch.

Note	middle C	D	E	F	G	A	B	C
Frequency (cycles/second)	262	294	330	349	392	440	494	523

INVESTIGATION

1. **The frequency decreases as the water level decreases. You can tell because the pitch decreases as you remove water from the bottle, and we know that the smaller the frequency, the lower the pitch.**

1. Fill a 12–20 ounce plastic or glass bottle almost to the top with water. Blow across the top and listen to the note produced. Pour a small amount of water out and repeat. Continue to remove water and blow notes until the bottle is empty. What happens to the frequency of the notes as the water level decreases? How can you tell? **See margin.**

2. Fill the bottle partway with water and blow across the top to create a note with constant pitch and volume. If you have trouble producing a steady stream of air, use a straw to blow across the bottle top. Use the CBL and the CBL microphone to collect the sound data and store it in the graphing calculator. Use the graphing calculator to graph the pressure of the sound as a function of time. The graph should resemble a sine wave.

3. Use the graph of the sound data to calculate the frequency of the note—the number of complete cycles in one second.

4. Write a sine function to describe the note.

5. Choose a note from the table. Try producing the note as follows: Adjust the water level in the bottle, blow across the top, and use the CBL and graphing calculator to find the frequency of the resulting note. Repeat this process until the frequency of the note you produce is approximately equal to the frequency of the note you chose from the table.

6. Write a sine function to describe the note you chose from the table.
   Middle C: $y = \sin 524 \, \pi x$, D: $y = \sin 588 \, \pi x$
   E: $y = \sin 660 \, \pi x$, F: $y = \sin 698 \, \pi x$
   G: $y = \sin 784 \, \pi x$, A: $y = \sin 880 \, \pi x$
   B: $y = \sin 988 \, \pi x$, C: $y = \sin 1046 \, \pi x$

PRESENT YOUR RESULTS

Write a report to present your results.

- Explain how you used the CBL.

- Explain how you found the frequency of a note from the note's sine wave.

- Explain how you wrote the sine functions in Exercises 4 and 6.

- Include a sketch of the water level in your bottle for both notes you produced in Exercises 2 and 5.

- Include a graph of the sine wave for each of the two notes.

- Consider including a recording of the notes you produced. You might repeat the experiment to produce several different notes.

- Describe how you used your knowledge of trigonometric functions in this project.

The Mathematics of Music

Laurie Hernandez
Paul Green
Algebra 2
Mrs. Cheung

EXTENSION

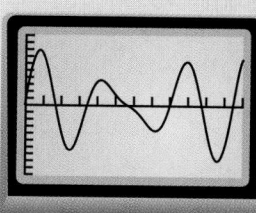

Choose a note to play and have a classmate also choose a note. Find sine functions $y = f(x)$ and $y = g(x)$ that model the two notes (as you did in Exercises 4 and 6 of the investigation). Then play the notes *simultaneously* and use the CBL and graphing calculator to graph the resulting sound wave. Compare this graph with the graph of $y = f(x) + g(x)$. What do you notice?

MUSIC CONNECTION

Bring in musical instruments and play a pure note on each. Use the CBL and graphing calculator to find the sine function that corresponds to each note. Observe what happens when you change the volume of the notes. Also, compare the sine waves for different instruments playing the same note.

Project **893**

GRADING THE PROJECT
RUBRIC FOR CHAPTER PROJECT

4 Students' sketch of the sine wave for each of the two notes is correct and complete. The equation of each graph is provided. Clear explanations of using the CBL, finding the frequency and writing the sine function are included. The written report demonstrates an understanding of how a pure musical note is modeled by a sine function.

3 Students' graphs and equations are included but may contain minor errors. Using the CBL, finding the frequency of the note, or writing the sine function may not be fully explained in the report. The written report explains how the project involved trigonometric functions.

2 The graphs and equations contain serious errors or are incomplete. Not all of the questions are answered in the report. It is not clear from the report how trigonometry was used in the project.

1 The graphs and equations are inaccurate and show that the student does not have an understanding of the way music notes can be modeled using sine functions. The report is incomplete or not understandable. The project should be returned with a new deadline for completion. The student should speak with the teacher as soon as possible to review his/her work and to make a new start on the project.

1. Completing the square:

$3x^2 + 5x + 2 = 0$

(Write original equation.)

$x^2 + \frac{5}{3}x + \frac{2}{3} = 0$

(Divide each side by 3.)

$x^2 + \frac{5}{3}x = -\frac{2}{3}$

$\left(\text{Subtract } \frac{2}{3} \text{ from each side.} \right)$

$x^2 + \frac{5}{3}x + \left(\frac{5}{6}\right)^2 = -\frac{2}{3} + \left(\frac{5}{6}\right)^2$

(Complete the square by adding the square of half the coefficient of x to each side.)

$\left(x + \frac{5}{6}\right)^2 = \frac{-24 + 25}{36}$

(Write the left side as the square of a binomial. Write the right side as a single fraction.)

$x + \frac{5}{6} = \pm\sqrt{\frac{1}{36}}$

(Take square roots of each side.)

$x = -\frac{5}{6} \pm \frac{1}{6}$

(Subtract $\frac{5}{6}$ from each side and simplify.)

$x = -1$ or $x = -\frac{2}{3}$ (Simplify.)

Using the quadratic formula:

$x = \frac{-b \pm \sqrt{b^2 - 4ac}}{2a}$

(Quadratic formula)

$x = \frac{-5 \pm \sqrt{5^2 - 4(3)(2)}}{2(3)}$

(Substitute values into formula.)

$x = \frac{-5 \pm 1}{6}$ (Simplify.)

$x = -1$ or $x = -\frac{2}{3}$

(Answers agree with those obtained from completing the square.)

Contents of Student Resources

DERIVATIONS OF KEY FORMULAS   pages 895–904

SKILLS REVIEW HANDBOOK   pages 905–939

- **Real Numbers** Operations with Signed Numbers, **905–906**; Converting Decimals, Fractions, and Percents, **906**; Calculating Percents, **907–908**; Factors and Multiples, **908–909**; Writing Ratios and Solving Proportions, **910–911**; Significant Digits, **911–912**; Scientific Notation, **913**   905–913

- **Geometry** Perimeter, Area, and Volume, **914–917**; Triangle Relationships, **917–918**; Symmetry, **919–920**; Transformations, **921–922**; Similar Figures, **923**   914–923

- **Logical Reasoning** Logical Argument, **924–925**; If-Then Statements, **926–927**; Counterexamples, **927–928**; Justify Reasoning, **928–929**   924–929

- **Problem Solving** Translating Phrases into Algebraic Expressions, **929–930**; Additional Problem Solving Strategies, **930–932**   929–932

- **Graphing** Points in the Coordinate Plane, **933**; Bar, Circle, and Line Graphs, **934–936**   933–936

- **Algebra** Opposites, **936–937**; Multiplying Binomials, **937**; Factoring, **938**; Least Common Denominator, **939**   936–939

EXTRA PRACTICE FOR CHAPTERS 1–14   pages 940–960

TABLES   pages 961–970

- **Symbols**   961
- **Measures**   962
- **Formulas**   963–968
- **Properties**   969–970

GLOSSARY   pages 971–980

INDEX   pages 981–998

SELECTED ANSWERS   pages SA1–SA64

Derivations of Key Formulas

The derivations of various key formulas presented in this book are given below. Follow-up exercises allow you to understand the derivations better by applying them to specific situations and/or repeating them under different conditions.

The Quadratic Formula .... *Lesson 5.6, page 291*

You can derive the quadratic formula by completing the square for the general quadratic equation $ax^2 + bx + c = 0$ (where $a \neq 0$).

$$ax^2 + bx + c = 0$$ Standard form of general equation

$$x^2 + \frac{b}{a}x + \frac{c}{a} = 0$$ Divide each side by a.

$$x^2 + \frac{b}{a}x = -\frac{c}{a}$$ Subtract $\frac{c}{a}$ from each side.

$$x^2 + \frac{b}{a}x + \left(\frac{b}{2a}\right)^2 = -\frac{c}{a} + \left(\frac{b}{2a}\right)^2$$ Complete the square by adding the square of half the coefficient of x to each side.

$$\left(x + \frac{b}{2a}\right)^2 = \frac{b^2 - 4ac}{4a^2}$$ Write the left side as the square of a binomial. Write the right side as a single fraction.

$$x + \frac{b}{2a} = \pm\sqrt{\frac{b^2 - 4ac}{4a^2}}$$ Take square roots of each side.

$$x = -\frac{b}{2a} \pm \sqrt{\frac{b^2 - 4ac}{4a^2}}$$ Subtract $\frac{b}{2a}$ from each side.

$$x = \frac{-b \pm \sqrt{b^2 - 4ac}}{2a}$$ Simplify.

Exercises 1–3. See margin.

1. Solve the quadratic equation $3x^2 + 5x + 2 = 0$ by completing the square and by using the quadratic formula. Check to see that you get the same solutions.

2. Derive a formula for the solution of a quadratic equation of the form $x^2 + mx + n = 0$. Check to see that your formula works for $x^2 - 4x + 3 = 0$.

3. Show that the function $f(x) = ax^2 + bx + c$ can be written in intercept form as

$$f(x) = a\left(x - \frac{-b + \sqrt{b^2 - 4ac}}{2a}\right)\left(x - \frac{-b - \sqrt{b^2 - 4ac}}{2a}\right)$$

by multiplying the factors in the intercept form and simplifying.

Equation of a Parabola .... *Lesson 10.2, page 596*

Using the geometric definition of a parabola and the distance formula (page 589), you can derive the equation of a parabola.

DEFINITION A parabola is the set of points (x, y) that are equidistant from a fixed line, called the *directrix*, and a fixed point, called the *focus*.

---

1. See the previous page.

2. $x^2 + mx + n = 0$
(Write the original equation.)
$x^2 + mx = -n$
(Subtract n from each side.)
$x^2 + mx + \left(\frac{m}{2}\right)^2 = -n + \left(\frac{m}{2}\right)^2$
(Complete the square by adding the square of half the coefficient of x to each side.)
$\left(x + \frac{m}{2}\right)^2 = \frac{m^2 - 4n}{4}$
(Write the left side as the square of a binomial. Write the right side as a single fraction.)
$x + \frac{m}{2} = \pm\sqrt{\frac{m^2 - 4n}{4}}$
(Take square roots of each side.)
$x = \frac{-m \pm \sqrt{m^2 - 4n}}{2}$
(Subtract $\frac{m}{2}$ from each side, and simplify.)
Now use the derived formula to solve the equation $x^2 - 4x + 3 = 0$.
$x = \frac{-m \pm \sqrt{m^2 - 4n}}{2}$
(Formula for solving, $x^2 + mx + n = 0$)
$x = \frac{-(-4) \pm \sqrt{(-4)^2 - 4(3)}}{2}$
(Substitute values for m and n.)
$x = 3$ or $x = 1$ (Simplify.)
Now substitute 1 and 3 back into the equation $x^2 - 4x + 3 = 0$.
$(1)^2 - 4(1) + 3 \overset{?}{=} 0$
(Substitute 1 in for x.)
$1 - 4 + 3 = 0$ (Simplify.)
$0 = 0$ (Solution checks.)
$(3)^2 - 4(3) + 3 \overset{?}{=} 0$
(Substitute 3 in for x.)
$9 - 12 + 3 = 0$ (Simplify.)
$0 = 0$ (Solution checks.)

3. See Additional Answers beginning on page AA1.

Derivations of Key Formulas **895**

896

1. distance between (x, y) and $(0, 5)$ = distance between (x, y) and $y = -5$
(Definition of a parabola)

$\sqrt{(x-0)^2 + (y-5)^2} = \sqrt{(x-x)^2 + (y-(-5))^2}$
(Distance formula)

$\sqrt{x^2 + (y-5)^2} = \sqrt{(y+5)^2}$
(Simplify.)

$x^2 + (y-5)^2 = (y+5)^2$
(Square each side.)

$x^2 + y^2 - 10y + 25 = y^2 + 10y + 25$
(Multiply.)

$x^2 = 20y$
(Add $10y - 25 - y^2$ to each side.)

2. For any point (x, y) on the parabola:

distance between (x, y) and $(p, 0)$ = distance between (x, y) and $x = -p$
(Definition of a parabola)

$\sqrt{(x-p)^2 + (y-0)^2} = \sqrt{(x-(-p))^2 + (y-y)^2}$
(Distance formula)

$\sqrt{(x-p)^2 + y^2} = \sqrt{(x+p)^2}$
(Simplify.)

$(x-p)^2 + y^2 = (x+p)^2$
(Square each side.)

$x^2 - 2px + p^2 + y^2 = x^2 + 2px + p^2$
(Multiply.)

$y^2 = 4px$
(Subtract $x^2 - 2px + p^2$ from each side.)

In the preceding derivation, p was assumed to be positive. If p were negative, the focus would be to the left of the y-axis and the directrix would be to the right of it. There would be no change in the derivation, however, so the equation $x^2 = 4py$ describes a parabola that opens to the right when $p > 0$ and a parabola that opens to the left when $p < 0$.

Equation of a Parabola (*continued*)

For $p > 0$, let the coordinates of the focus be $(0, p)$ and the equation of the directrix be $y = -p$ as in the diagram shown. Notice that $(0, 0)$ is p units from the focus and also p units from the directrix. Therefore, $(0, 0)$ is a point on the parabola.

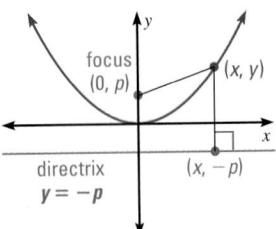

For any point (x, y) on the parabola:

distance between (x, y) and $(0, p)$ = distance between (x, y) and $y = -p$

Definition of a parabola

$\sqrt{(x-0)^2 + (y-p)^2} = \sqrt{(x-x)^2 + (y-(-p))^2}$

Distance formula

$\sqrt{x^2 + (y-p)^2} = \sqrt{(y+p)^2}$

Simplify.

$x^2 + (y-p)^2 = (y+p)^2$

Square each side.

$x^2 + y^2 - 2py + p^2 = y^2 + 2py + p^2$

Multiply.

$x^2 = 4py$

Subtract $y^2 - 2py + p^2$ from each side.

In the preceding derivation, p was assumed to be positive. If p were negative, the focus would be below the x-axis and the directrix would be above it, as shown. There would be no change in the derivation, however, so the equation $x^2 = 4py$ describes a parabola that opens up when $p > 0$ and a parabola that opens down when $p < 0$.

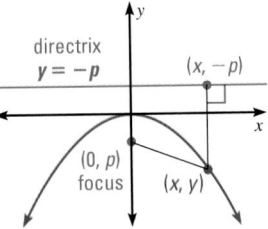

Exercises   1–2. See margin.

1. Derive an equation for the set of all points (x, y) that are the same distance from $(0, 5)$ as they are from $y = -5$. Compare your work with each step of the derivation shown above.

2. Show that $y^2 = 4px$ describes a parabola that opens to the right when $p > 0$ and a parabola that opens to the left when $p < 0$.

Equation of an Ellipse ······························· *Lesson 10.4, page 609*

Using the geometric definition of an ellipse and the distance formula (page 589), you can derive the equation of an ellipse.

DEFINITION An ellipse is the set of points (x, y) such that the sum of the distances between (x, y) and two distinct points, called the *foci*, is constant.

Let the foci have coordinates $(-c, 0)$ and $(c, 0)$, and let $(0, b)$ be the point where the ellipse intersects the positive y-axis. Let the distance between $(0, b)$ and each focus be a. Then the sum of the distances from any point (x, y) on the ellipse to the two foci must be $2a$.

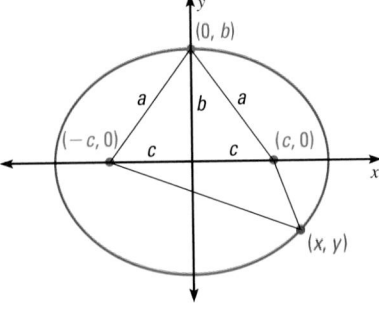

Student Resources

$$\underset{\text{(x, y) and (-c, 0)}}{\text{distance between}} + \underset{\text{(x, y) and (c, 0)}}{\text{distance between}} = 2a \qquad \textbf{Definition of an ellipse}$$

$$\sqrt{(x-(-c))^2 + (y-0)^2} + \sqrt{(x-c)^2 + (y-0)^2} = 2a \qquad \textbf{Distance formula}$$

$$\sqrt{(x+c)^2 + y^2} + \sqrt{(x-c)^2 + y^2} = 2a \qquad \textbf{Simplify.}$$

$$\sqrt{(x+c)^2 + y^2} = 2a - \sqrt{(x-c)^2 + y^2} \qquad \begin{array}{l}\textbf{Subtract } \sqrt{(x-c)^2 + y^2} \\ \textbf{from each side.}\end{array}$$

At this point, square each side of the equation twice to eliminate the two radicals.

$$(x+c)^2 + y^2 = 4a^2 - 4a\sqrt{(x-c)^2 + y^2} + (x-c)^2 + y^2 \qquad \textbf{Square each side.}$$

$$4cx = 4a^2 - 4a\sqrt{(x-c)^2 + y^2} \qquad \begin{array}{l}\textbf{Subtract } (x-c)^2 + y^2 \textbf{ from} \\ \textbf{each side, and simplify.}\end{array}$$

$$cx - a^2 = -a\sqrt{(x-c)^2 + y^2} \qquad \begin{array}{l}\textbf{Subtract } 4a^2 \textbf{ from each side,} \\ \textbf{and divide each side by 4.}\end{array}$$

$$c^2x^2 - 2a^2cx + a^4 = a^2\left[(x-c)^2 + y^2\right] \qquad \textbf{Square each side again.}$$

$$c^2x^2 - 2a^2cx + a^4 = a^2x^2 - 2a^2cx + a^2c^2 + a^2y^2 \qquad \textbf{Multiply.}$$

$$a^4 = a^2x^2 - c^2x^2 + a^2c^2 + a^2y^2 \qquad \textbf{Subtract } c^2x^2 - 2a^2cx \textbf{ from each side.}$$

$$a^2(a^2 - c^2) = (a^2 - c^2)x^2 + a^2y^2 \qquad \textbf{Subtract } a^2c^2 \textbf{ from each side, and factor.}$$

Since a, b, and c are the lengths of the sides of a right triangle (see diagram on previous page), you have $a^2 = b^2 + c^2$, or $a^2 - c^2 = b^2$, by the Pythagorean theorem. Therefore:

$$a^2b^2 = b^2x^2 + a^2y^2 \qquad \textbf{Substitute } b^2 \textbf{ for } a^2 - c^2.$$

$$1 = \frac{x^2}{a^2} + \frac{y^2}{b^2} \qquad \textbf{Divide each side by } a^2b^2.$$

You now have the equation of an ellipse whose center is $(0, 0)$. You know that the length of the vertical axis of the ellipse—the distance between $(0, b)$ and $(0, -b)$—is $2b$. But what is the length of the other axis?

Let $(d, 0)$ be the intersection of the ellipse with the positive x-axis as shown. Then the distance between $(d, 0)$ and $(-c, 0)$ plus the distance between $(d, 0)$ and $(c, 0)$ is $2a$. That is:

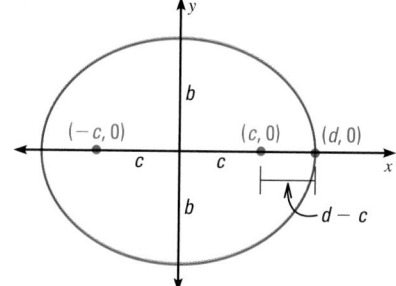

$$[c + c + (d - c)] + (d - c) = 2a$$

$$2d = 2a$$

$$d = a$$

Therefore, the ellipse intersects the x-axis at $(-a, 0)$ and $(a, 0)$, and the length of the horizontal axis of the ellipse is $2a$.

Exercises 1–2. See margin.

1. Derive an equation for the set of all points (x, y) such that the sum of the distances between (x, y) and the two points $(-3, 0)$ and $(3, 0)$ is 10. Compare your work with each step of the derivation shown above.

2. Show that the equation $\frac{x^2}{b^2} + \frac{y^2}{a^2} = 1$ describes an ellipse where the two foci have coordinates $(0, -c)$ and $(0, c)$ and where $(b, 0)$ is the point where the ellipse intersects the positive x-axis. (Let the distance between $(b, 0)$ and each focus be a.)

---

1. $$\underset{\text{(x, y) and (-3, 0)}}{\text{distance between}} + \underset{\text{(x, y) and (3, 0)}}{\text{distance between}} = 10 \text{ (Definition of an ellipse)}$$

$$\sqrt{(x-(-3))^2 + (y-0)^2} + \sqrt{(x-3)^2 + (y-0)^2} = 10$$
(Distance formula)

$$\sqrt{(x+3)^2 + y^2} + \sqrt{(x-3)^2 + y^2} = 10$$
(Simplify.)

$$\sqrt{(x+3)^2 + y^2} = 10 - \sqrt{(x-3)^2 + y^2}$$
(Subtract $\sqrt{(x-3)^2 + y^2}$ from each side.)

$$(x+3)^2 + y^2 = 100 - 20\sqrt{(x-3)^2 + y^2} + (x-3)^2 + y^2$$
(Square each side.)

$$12x - 100 = -20\sqrt{(x-3)^2 + y^2}$$
(Subtract $y^2 + (x-3)^2 - 100$ from each side, and simplify.)

$$3x - 25 = -5\sqrt{(x-3)^2 + y^2}$$
(Divide each side by 4.)

$$9x^2 - 150x + 625 = 25(x-3)^2 + 25y^2$$
(Square each side.)

$$9x^2 - 150x + 625 = 25x^2 - 150x^2 + 225 + 25y^2 \text{ (Multiply.)}$$

$$400 = 16x^2 + 25y^2$$
(Subtract $9x^2 - 150x + 255$ from each side.)

$$1 = \frac{x^2}{25} + \frac{y^2}{16}$$
(Divide each side by 400.)

2. See Additional Answers beginning on page AA1.

1. $\begin{matrix}\text{distance} \\ \text{between} \\ (x, y) \text{ and } (-5, 0)\end{matrix} - \begin{matrix}\text{distance} \\ \text{between} \\ (x, y) \text{ and } (5, 0)\end{matrix} =$

± 8 (Definition of a hyperbola)

$\sqrt{(x-(-5))^2+(y-0)^2} - \sqrt{(x-5)^2+(y-0)^2} = \pm 8$

(Distance formula)

$\sqrt{(x+5)^2+y^2} - \sqrt{(x-5)^2+y^2} = \pm 8$

(Simplify.)

$\sqrt{(x+5)^2+y^2} = \pm 8 + \sqrt{(x-5)^2+y^2}$

(Add $\sqrt{(x-5)^2+y^2}$ to each side.)

$(x+5)^2+y^2 = 64 \pm 16\sqrt{(x-5)^2+y^2} + (x-5)^2+y^2$

(Square each side.)

$20x = 64 \pm 16\sqrt{(x-5)^2+y^2}$

(Subtract $(x-5)^2+y^2$ from each side and simplify.)

$5x - 16 = \pm 4\sqrt{(x-5)^2+y^2}$

(Subtract 64 from each side and divide by 4.)

$25x^2 - 160x + 256 = 16[(x-5)^2+y^2]$

(Square each side again.)

$25x^2 - 160x + 256 = 16x^2 - 160x + 400 + 16y^2$

(Multiply.)

$9x^2 - 16y^2 = 144$

(Subtract $16x^2 + 16y^2 - 160x + 256$ from each side.)

$\frac{x^2}{16} - \frac{y^2}{9} = 1$

(Divide each side by 144.)

Equation of a Hyperbola .... *Lesson 10.5, page 615*

Using the geometric definition of a hyperbola and the distance formula (page 589), you can derive the equation of a hyperbola.

DEFINITION A hyperbola is the set of points (x, y) such that the difference of the distances between (x, y) and two distinct points, called the *foci*, is constant.

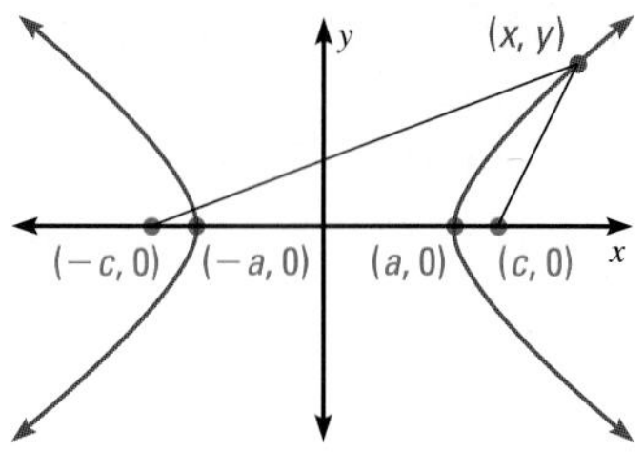

Let the two foci have coordinates $(-c, 0)$ and $(c, 0)$ where $c > 0$, and let the coordinates of the points, called the *vertices*, where the hyperbola intersects the x-axis have coordinates $(-a, 0)$ and $(a, 0)$ where $a > 0$.

The distance between either focus and the vertex farther from it is $c + a$, while the distance between either focus and the vertex closer to it is $c - a$. The difference in the distances is $c + a - (c - a) = 2a$ or $c - a - (c + a) = -2a$. Therefore, $\pm 2a$ is the common difference in the distances to the foci from any point on the hyperbola.

$\begin{matrix}\text{distance between} \\ (x, y) \text{ and } (-c, 0)\end{matrix} - \begin{matrix}\text{distance between} \\ (x, y) \text{ and } (c, 0)\end{matrix} = \pm 2a$	**Definition of a hyperbola**
$\sqrt{(x-(-c))^2+(y-0)^2} - \sqrt{(x-c)^2+(y-0)^2} = \pm 2a$	**Distance formula**
$\sqrt{(x+c)^2+y^2} - \sqrt{(x-c)^2+y^2} = \pm 2a$	**Simplify.**
$\sqrt{(x+c)^2+y^2} = \pm 2a + \sqrt{(x-c)^2+y^2}$	**Add $\sqrt{(x-c)^2+y^2}$ to each side.**

At this point, square each side of the equation twice to eliminate the two radicals.

$(x+c)^2+y^2 = 4a^2 \pm 4a\sqrt{(x-c)^2+y^2} + (x-c)^2+y^2$	**Square each side.**
$4cx = 4a^2 \pm 4a\sqrt{(x-c)^2+y^2}$	**Subtract $(x-c)^2+y^2$ from each side, and simplify.**
$cx - a^2 = \pm a\sqrt{(x-c)^2+y^2}$	**Subtract $4a^2$ from each side, and divide each side by 4.**
$c^2x^2 - 2a^2cx + a^4 = a^2[(x-c)^2+y^2]$	**Square each side again.**
$c^2x^2 - 2a^2cx + a^4 = a^2x^2 - 2a^2cx + a^2c^2 + a^2y^2$	**Multiply.**
$c^2x^2 = a^2x^2 + a^2c^2 + a^2y^2 - a^4$	**Subtract $-2a^2cx + a^4$ from each side.**
$(c^2 - a^2)x^2 - a^2y^2 = a^2(c^2 - a^2)$	**Subtract $a^2x^2 + a^2y^2$ from each side, and factor.**

By drawing perpendiculars to the x-axis at the vertices, you can form right triangles each having one leg (along the x-axis) of length a and a hypotenuse of length c as shown at the top of the next page. Let b be the length of the other leg. Then $a^2 + b^2 = c^2$, or $b^2 = c^2 - a^2$, by the Pythagorean theorem. Therefore:

$b^2x^2 - a^2y^2 = a^2b^2$	**Substitute b^2 for $c^2 - a^2$.**
$\frac{x^2}{a^2} - \frac{y^2}{b^2} = 1$	**Divide each side by a^2b^2.**

You now have the equation of a hyperbola whose center is $(0, 0)$. The lines that contain the hypotenuses of the right triangles have equations $y = \pm\frac{b}{a}x$. To see that these lines are asymptotes of the hyperbola, solve the equation of the hyperbola for y:

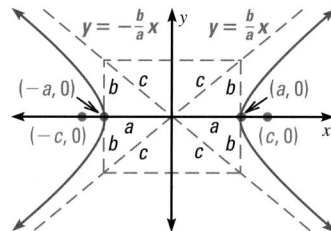

$$\frac{x^2}{a^2} - \frac{y^2}{b^2} = 1 \qquad \text{Equation of a hyperbola}$$

$$b^2x^2 - a^2y^2 = a^2b^2 \qquad \text{Multiply each side by } a^2b^2.$$

$$-a^2y^2 = a^2b^2 - b^2x^2 \qquad \text{Subtract } b^2x^2 \text{ from each side.}$$

$$y^2 = \frac{b^2}{a^2}x^2 - b^2 \qquad \text{Divide each side by } -a^2.$$

$$y^2 = \frac{b^2}{a^2}(x^2 - a^2) \qquad \text{Factor out } \frac{b^2}{a^2}.$$

$$y = \pm\frac{b}{a}\sqrt{x^2 - a^2} \qquad \text{Take square roots of each side.}$$

As $x \rightarrow +\infty$, the value of x^2 becomes much greater than the value of a^2 (a constant) so that $\sqrt{x^2 - a^2} \rightarrow \sqrt{x^2} = x$. Therefore, the graph of $y = \pm\frac{b}{a}\sqrt{x^2 - a^2}$ approaches the graph of $y = \pm\frac{b}{a}x$ as $x \rightarrow +\infty$. You can make a similar argument for $x \rightarrow -\infty$.

Exercises   1–2. See margin.

1. Derive an equation for the set of all points (x, y) such that the difference of the distances between (x, y) and the two points $(-5, 0)$ and $(5, 0)$ is ± 8. Compare your work with each step of the derivation shown on the previous page.

2. Show that the equation $\frac{y^2}{a^2} - \frac{x^2}{b^2} = 1$ describes a hyperbola where the two foci have coordinates $(0, -c)$ and $(0, c)$ and where the hyperbola intersects the y-axis at $(0, -a)$ and $(0, a)$.

Permutations of n Objects Taken r at a Time
$\qquad\qquad$ *Lesson 12.1, page 703*

Using the fundamental counting principle (page 701) and the definition of factorial (page 681), you can derive the formula for the number of permutations of n objects taken r at a time.

| Number of permutations | = | Number of ways to choose 1st object | · | Number of ways to choose 2nd object | · | Number of ways to choose 3rd object | · . . . · | Number of ways to choose rth object |

$$_nP_r = n \cdot (n-1) \cdot (n-2) \cdot \ldots \cdot (n-r+1)$$

$$= \frac{n \cdot (n-1) \cdot (n-2) \cdot \ldots \cdot (n-r+1) \cdot (n-r) \cdot (n-r-1) \cdot \ldots \cdot 2 \cdot 1}{(n-r) \cdot (n-r-1) \cdot \ldots \cdot 2 \cdot 1} \qquad \begin{array}{l}\text{Multiply and divide}\\ \text{by } (n-r)!.\end{array}$$

$$= \frac{n!}{(n-r)!} \qquad \text{Definition of factorial}$$

Derivations of Key Formulas **899**

1. See the previous page.
2. $\underset{(x,\,y)\text{ and }(0,\,-c)}{\underbrace{\text{distance between}}} - \underset{(x,\,y)\text{ and }(0,\,c)}{\underbrace{\text{distance between}}} = \pm 2a$ (Definition of a hyperbola)

$\sqrt{(x-0)^2 + (y-(-c))^2} - \sqrt{(x-0)^2 + (y-c)^2} = \pm 2a$
(Distance formula)

$\sqrt{x^2 + (y+c)^2} - \sqrt{x^2 + (y-c)^2} = \pm 2a$
(Simplify.)

$\sqrt{x^2 + (y+c)^2} = \pm 2a + \sqrt{x^2 + (y-c)^2}$
(Add $\sqrt{x^2 + (y-c)^2}$ to each side.)

$x^2 + (y+c)^2 = 4a^2 \pm 4a\sqrt{x^2 + (y-c)^2} + x^2 + (y-c)^2$
(Square each side.)

$4cy = 4a^2 \pm 4a\sqrt{x^2 + (y-c)^2}$
(Subtract $x^2 + (y-c)^2$ and simplify.)

$cy - a^2 = \pm a\sqrt{x^2 + (y-c)^2}$
(Subtract $4a^2$ from each side and divide each side by 4.)

$c^2y^2 - 2a^2cy + a^4 = a^2[x^2 + (y-c)^2]$
(Square each side again.)

$c^2y^2 - 2a^2cy + a^4 = a^2x^2 + a^2y^2 - 2a^2cy + a^2c^2$
(Multiply.)

$c^2y^2 = a^2y^2 + a^2c^2 + a^2x^2 - a^4$
(Subtract $-2a^2cy + a^4$ from each side.)

$(c^2 - a^2)y^2 - a^2x^2 = a^2(c^2 - a^2)$
(Subtract $a^2y^2 + a^2x^2$ from each side and factor.)

In a hyperbola, $a^2 = b^2 + c^2$.
Therefore: $b^2y^2 - a^2x^2 = a^2b^2$
(Substitute b^2 for $c^2 - a^2$.)

$\dfrac{y^2}{a^2} - \dfrac{x^2}{b^2} = 1$
(Divide each side by a^2b^2.)

899

Margin notes (left column):

1. The number of ways to choose the rth object is $n - r + 1$ because you started with n objects and $(r - 1)$ objects have already been chosen, so at the time you want to choose the rth object, there are $n - (r - 1)$, or $n - r + 1$, objects to choose from.

2a. For any group of r objects there exists $r!$ different ways of arranging the objects.

2b. For each combination of r objects, those objects can be arranged $r!$ ways. Thus, every possible combination, $_nC_r$ must be multiplied by $r!$ in order to yield the number of permutations of those objects.

2c. $_nP_r = r! \cdot {}_nC_r$ (Formula from part (b))

$\dfrac{n!}{(n-r)!} = r! \cdot {}_nC_r$

(Substitute formula for $_nP_r$.)

$\dfrac{n!}{(n-r)!\,r!} = {}_nC_r$

(Divide each side by $r!$.)

---

1. In a semicircle, $\theta = \pi$.

$s = r\theta$ (Formula for arc length)

$s = r\pi$ (Substitute π for θ.)

This is a true statement because the circumference of a circle is $2\pi r$ and a semicircle is half of that length, or πr, which is equal to the value the formula yields.

2. Let r be the radius of a circle, and let θ be the radian measure of a central angle that defines a sector of area A. Knowing that the area of a circle is πr^2 and that there are 2π radians in a full circle, you can write the following proportion:

$\dfrac{\text{area of sector}}{\text{area of circle}} =$

$\dfrac{\text{radian measure of central angle}}{\text{radian measure of full circle}}$

(Write a proportion.)

$\dfrac{A}{\pi r^2} = \dfrac{\theta}{2\pi}$ (Substitute.)

$A = \dfrac{\theta r^2}{2}$

(Multiply each side by πr^2.)

900

Main content (right column):

Permutations of n Objects Taken r at a Time (continued)

Exercises 1–2. See margin.

1. In the derivation shown on the previous page, explain why the number of ways to choose the rth object is $n - r + 1$.

2. You know that order is important when arranging a group of objects. You will learn in Lesson 12.2 that a *combination* is a grouping of objects where the order of the objects is not important.

   a. For any group of r objects, how many ways are there of arranging the objects?

   b. Let $_nC_r$ denote the number of combinations of n objects taken r at a time. Use your answer from part (a) to write $_nP_r$ in terms of $_nC_r$. (That is, complete this statement: $_nP_r = (\underline{?})({}_nC_r)$ because for each combination of r objects there are $\underline{?}$ permutations of those objects.)

   c. Using the equation from part (b) and the formula for $_nP_r$, derive a formula for $_nC_r$ in terms of n and r. (This is the formula given on page 708.)

Formula for Arc Length .... *Lesson 13.2, page 779*

By using the circumference and radian measure of a circle (page 777), you can write a proportion to derive the formula for the length of a circular arc.

Let r be the radius of a circle, and let θ be the radian measure of a central angle that intercepts an arc of length s. Knowing that the circumference of the circle is $2\pi r$ and that there are 2π radians in a full circle, you can write the following proportion:

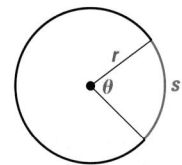

$\dfrac{\text{length of arc}}{\text{circumference of circle}} = \dfrac{\text{radian measure of central angle}}{\text{radian measure of full circle}}$ **Write a proportion.**

$\dfrac{s}{2\pi r} = \dfrac{\theta}{2\pi}$ **Substitute.**

$s = r\theta$ **Multiply each side by $2\pi r$.**

So, the length of the arc is just the product of the circle's radius and the radian measure of the central angle that intercepts the arc.

Exercises 1–2. See margin.

1. Show that the arc length formula gives a correct result for a semicircular arc.

2. Derive the formula for the area of a sector formed by a central angle θ (measured in radians) in a circle of radius r. Your derivation should involve setting up and solving a proportion, as above.

The Law of Sines .... *Lesson 13.5, page 799*

By using the right triangle definition of sine (page 769), you can derive the law of sines, which applies to any triangle.

Let a, b, and c be the lengths of the sides of $\triangle ABC$ as shown. Introduce altitude $\overline{CD}$ having length h.

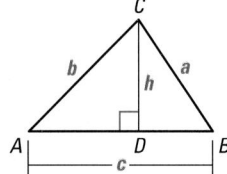

From right $\triangle ACD$ you have $\sin A = \dfrac{h}{b}$, or $h = b \sin A$, by the definition of sine.

Likewise, from right $\triangle BCD$ you have $\sin B = \dfrac{h}{a}$, or $h = a \sin B$, also by the definition of sine. Therefore:

$$b \sin A = a \sin B \qquad \text{Equate expressions for } h.$$

$$\dfrac{\sin A}{a} = \dfrac{\sin B}{b} \qquad \text{Divide each side by } ab.$$

This establishes one of the three equalities from the law of sines.

Exercises   1–2. See margin.

1. Introduce a different altitude in $\triangle ABC$ and derive another equality from the law of sines. How does this result, combined with the one above, imply the third equality from the law of sines?

2. Derive the three formulas for the area of a triangle given on page 802 using an argument similar to the one above for the law of sines.

The Law of Cosines   Lesson 13.6, page 807

By using the Pythagorean theorem and the right triangle definition of cosine (page 769), you can derive the law of cosines, which applies to any triangle.

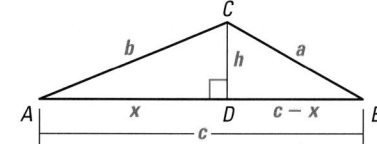

Let a, b, and c be the lengths of the sides of $\triangle ABC$ as shown. Introduce altitude $\overline{CD}$ having length h.

Let $AD = x$; it follows that $DB = c - x$. Use the Pythagorean theorem to find two different expressions for h^2:

RIGHT $\triangle BCD$	**RIGHT $\triangle ACD$**
$h^2 + (c - x)^2 = a^2$	$x^2 + h^2 = b^2$
$h^2 = a^2 - (c - x)^2$	$h^2 = b^2 - x^2$

Therefore:

$$a^2 - (c - x)^2 = b^2 - x^2 \qquad \text{Equate expressions for } h^2.$$

$$a^2 - c^2 + 2cx - x^2 = b^2 - x^2 \qquad \text{Multiply.}$$

$$a^2 = b^2 + c^2 - 2cx \qquad \text{Add } c^2 - 2cx + x^2 \text{ to each side.}$$

From right $\triangle ACD$ you have $\cos A = \dfrac{x}{b}$, or $x = b \cos A$, by the definition of cosine. Therefore:

$$a^2 = b^2 + c^2 - 2c(b \cos A) \qquad \text{Substitute } b \cos A \text{ for } x.$$

$$a^2 = b^2 + c^2 - 2bc \cos A \qquad \text{Commutative property of multiplication}$$

This establishes one of the three forms of the law of cosines.

Exercises   1–2. See margin.

1. Using a different altitude in $\triangle ABC$, derive another form of the law of cosines.

2. Solve $a^2 = b^2 + c^2 - 2bc \cos A$ for $\cos A$.

Derivations of Key Formulas   901

1–2. See Additional Answers
    beginning on page AA1.

1.

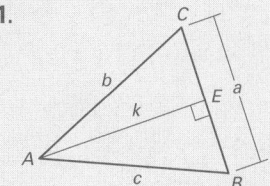

Let a, b, and c be the lengths of the sides of $\triangle ABC$ as shown. Introduce altitude AE having length k. From right $\triangle ACE$ you have

$\sin C = \dfrac{k}{b}$, or $k = b \sin C$, by the definition of sine. Likewise, from right $\triangle ABE$ you have $\sin B = \dfrac{k}{c}$, or $k = c \sin B$, also by the definition of sine. Therefore: $b \sin C = c \sin B$ (Equate expressions for k.)

$\dfrac{\sin C}{c} = \dfrac{\sin B}{b}$

(Divide each side by bc.) To derive the third equality, simply substitute $\dfrac{\sin C}{c}$ for $\dfrac{\sin B}{b}$ in the equality $\dfrac{\sin A}{a} = \dfrac{\sin B}{b}$. The third equality is $\dfrac{\sin A}{a} = \dfrac{\sin C}{c}$.

2. Each of the expressions $a \sin B$, $b \sin C$, and $c \sin A$ corresponds to one of the three altitudes in $\triangle ABC$. Multiplying one of those altitudes with its corresponding base gives the area of a rectangle with the height and width of the triangle. Thus, dividing by 2 yields the area of the triangle.

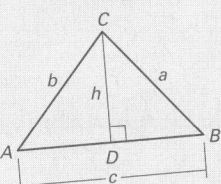

Area $= \dfrac{1}{2} \cdot$ base $\cdot$ height

Area $= \dfrac{1}{2} cb \sin A$

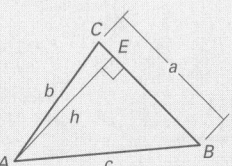

Area $= \dfrac{1}{2} \cdot$ base $\cdot$ height

Area $= \dfrac{1}{2} ac \sin B$

Solution continues on next page.

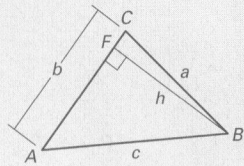

$$\text{Area} = \frac{1}{2} \cdot \text{base} \cdot \text{height}$$

$$\text{Area} = \frac{1}{2}ba \sin C$$

1. Refer to the diagram for sin $(-\theta)$ = $-\sin \theta$, where $P'(a, -b)$ is on the terminal side of the angle $-\theta$.

$$\cos(-\theta) = \frac{a}{r} = \cos \theta$$

(definition of cosine)

$$\tan(-\theta) = \frac{-b}{a} = -\frac{b}{a} = -\tan \theta$$

(definition of tangent)

2.

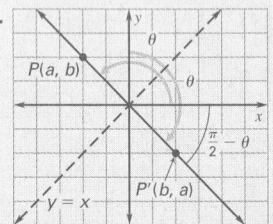

If $P(a, b)$ is a point on the terminal side of θ, then $P'(b, a)$ is a point on the terminal side of $\frac{\pi}{2} - \theta$.

If r is the distance $OP = OP'$, then by the definitions of sine and cosine you have:

$$\sin\left(\frac{\pi}{2} - \theta\right) = \frac{a}{r} = \cos \theta$$

Negative Angle Identities Lesson 14.3, page 848

By using a geometric argument, you can establish the negative angle identities.

Draw an angle θ in standard position. Let $P(a, b)$ be a point (other than the origin) on the terminal side of θ. The angle $-\theta$ has the same amount of rotation as θ but the direction of rotation is clockwise rather than counterclockwise from the positive x-axis. The terminal side of $-\theta$ is therefore a reflection of the terminal side of θ in the x-axis. This means that the point $P'(a, -b)$ is on the terminal side of $-\theta$. If r is the distance $OP = OP'$, then by the definition of sine (see page 784) you have:

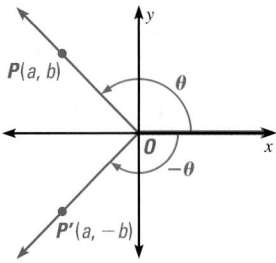

$$\sin(-\theta) = \frac{-b}{r} = -\frac{b}{r} = -\sin \theta$$

Exercises 1–2. See margin.

1. Continue the argument presented above to establish the identities $\cos(-\theta) = \cos \theta$ and $\tan(-\theta) = -\tan \theta$.

2. Use a geometric argument to establish the cofunction identity $\sin\left(\frac{\pi}{2} - \theta\right) = \cos \theta$.

(*Hint:* The terminal side of $\frac{\pi}{2} - \theta$ is the reflection of the terminal side of θ in the line $y = x$.)

The Difference Formula for Cosine Lesson 14.6, page 869

By using the right triangle definitions of sine and cosine (page 769), the distance formula (page 589), the law of cosines (page 807), and a Pythagorean identity (page 848), you can derive the difference formula for cosine.

Draw two angles in standard position. Let v be the measure of the smaller angle and u be the measure of the larger angle. Choose points P and Q on the terminal sides of the angles so that the points are each 1 unit from the origin.

Draw perpendiculars from P and Q to the x-axis, and let R and S be the points of intersection of the perpendiculars with the x-axis. Since $\triangle PRO$ and $\triangle QSO$ are right triangles, the lengths of their legs are $\sin u$ and $\cos u$ (for $\triangle PRO$) and $\sin v$ and $\cos v$ (for $\triangle QSO$) by the right triangle definitions of sine and cosine. Therefore, the coordinates of P are $(\cos u, \sin u)$, and the coordinates of Q are $(\cos v, \sin v)$.

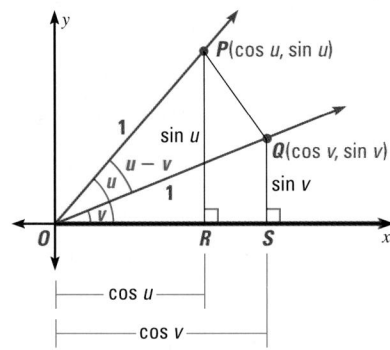

By the distance formula, you have:

$$PQ = \sqrt{(\cos u - \cos v)^2 + (\sin u - \sin v)^2},$$

or $PQ^2 = (\cos u - \cos v)^2 + (\sin u - \sin v)^2$.

By the law of cosines applied to $\triangle PQO$, you have:

$$PQ^2 = 1^2 + 1^2 - 2(1)(1)\cos(u - v)$$

Equating the two expressions for PQ^2, using a Pythagorean identity, and solving for $\cos(u - v)$, you have:

$$2 - 2\cos(u - v) = (\cos u - \cos v)^2 + (\sin u - \sin v)^2$$

$$2 - 2\cos(u - v) = \cos^2 u - 2\cos u \cos v + \cos^2 v + \sin^2 u - 2\sin u \sin v + \sin^2 v$$

$$2 - 2\cos(u - v) = (\cos^2 u + \sin^2 u) + (\cos^2 v + \sin^2 v) - 2(\cos u \cos v + \sin u \sin v)$$

$$2 - 2\cos(u - v) = 1 + 1 - 2(\cos u \cos v + \sin u \sin v)$$

$$-2\cos(u - v) = -2(\cos u \cos v + \sin u \sin v)$$

$$\cos(u - v) = \cos u \cos v + \sin u \sin v$$

The same argument can be used for angles in any quadrants, not just Quadrant I. You can verify the cofunction identities (page 848) using the difference formula for cosine. And you can in turn derive the other sum and difference formulas on page 869 using the difference formula for cosine and the cofunction and negative angle identities (for example, see Exercises 62–64 on page 874).

Exercises   1–3. See margin.

1. a. Verify the cofunction identity $\cos\left(\dfrac{\pi}{2} - \theta\right) = \sin\theta$ using the difference formula for cosine.

   b. Use the cofunction identity from part (a) to verify $\sin\left(\dfrac{\pi}{2} - \theta\right) = \cos\theta$.

2. Use the cofunction identities from Exercise 1 to derive the sum formula for sine. Begin by writing $\sin(u + v) = \cos\left(\dfrac{\pi}{2} - (u + v)\right) = \cos\left(\left(\dfrac{\pi}{2} - u\right) - v\right)$.

3. Use the sum formula for sine (see Exercise 2) and the negative angle identities to derive the difference formula for sine.

The Double- and Half-Angle Formulas .... *Lesson 14.7, page 875*

The double-angle formulas are obtained directly from the sum formulas, and the half-angle formulas are obtained directly from the double-angle formulas.

To establish one of the double-angle formulas for cosine, simply let $v = u$ in the sum formula for cosine:

$$\cos 2u = \cos(u + u) \qquad \text{Write } 2u \text{ as a sum.}$$

$$= \cos u \cos u - \sin u \sin u \qquad \text{Use sum formula for cosine.}$$

$$= \cos^2 u - \sin^2 u \qquad \text{Simplify.}$$

The other variations of the double-angle formula for cosine are obtained using the Pythagorean identity $\sin^2 u + \cos^2 u = 1$. For instance, replace $\cos^2 u$ with $1 - \sin^2 u$ in $\cos 2u = \cos^2 u - \sin^2 u$ to obtain:

$$\cos 2u = (1 - \sin^2 u) - \sin^2 u \qquad \text{Substitute.}$$

$$= 1 - 2\sin^2 u \qquad \text{Simplify.}$$

Likewise, replace $\sin^2 u$ with $1 - \cos^2 u$ in $\cos 2u = \cos^2 u - \sin^2 u$ to obtain:

$$\cos 2u = \cos^2 u - (1 - \cos^2 u) \qquad \text{Substitute.}$$

$$= \cos^2 u - 1 + \cos^2 u \qquad \text{Distribute.}$$

$$= 2\cos^2 u - 1 \qquad \text{Simplify.}$$

Derivations of Key Formulas **903**

1a. $\cos\left(\dfrac{\pi}{2} - \theta\right) = \cos\dfrac{\pi}{2}\cos\theta + \sin\dfrac{\pi}{2}\sin\theta$
(Difference formula for cosine)
$= 0 \cdot \cos\theta + 1 \cdot \sin\theta$
$\left(\text{Substitute 0 for } \cos\dfrac{\pi}{2} \text{ and } 1 \text{ for } \sin\dfrac{\pi}{2}.\right)$
$= \sin\theta$ (Simplify.)

1b. $\sin\theta = \cos\left(\dfrac{\pi}{2} - \theta\right)$
(Identity from part (a))
$\sin\left(\dfrac{\pi}{2} - \theta\right) = \cos\left(\dfrac{\pi}{2} - \left(\dfrac{\pi}{2} - \theta\right)\right)$
$\left(\text{Substitute } \dfrac{\pi}{2} - \theta \text{ for } \theta.\right)$
$= \cos\left(\dfrac{\pi}{2} - \dfrac{\pi}{2} + \theta\right)$ (Distribute.)
$= \cos\theta$ (Simplify.)

2. $\sin(u + v) = \cos\left(\dfrac{\pi}{2} - (u + v)\right)$
(Cofunction identity)
$\sin(u + v) = \cos\left(\left(\dfrac{\pi}{2} - u\right) - v\right)$
(Distributive and associative properties)
$\sin(u + v) = \cos\left(\dfrac{\pi}{2} - u\right)\cos v + \sin\left(\dfrac{\pi}{2} - u\right)\sin v$
(Difference formula for cosine)
$\sin(u + v) = \sin u \cos v + \cos u \sin v$
(Cofunction identities)

3. $\sin(u + v) = \sin u \cos v + \cos u \sin v$
(Sum formula for sine.)
$\sin(u + (-v)) = \sin u \cos(-v) + \cos u \sin(-v)$
(Substitute $-v$ for v.)
$\sin(u + (-v)) = \sin u \cos v + \cos u (-\sin v)$
(Negative angle identities)
$\sin(u - v) = \sin u \cos v - \cos u \sin v$
(Simplify.)

1. $\sin 2u = \sin(u + u)$
(Write $2u$ as $u + u$.)
$\sin 2u = \sin u \cos u + \cos u \sin u$
(Sum formula for sine)
$\sin 2u = 2 \sin u \cos u$ (Simplify.)

2. $\cos 2\theta = 1 - 2\sin^2\theta$
(Rewrite original equation)
$\cos u = 1 - 2\sin^2\frac{u}{2}$

$\left(\text{Substitute }\frac{u}{2}\text{ for }\theta\text{ and simplify.}\right)$

$-1 + \cos u = -2\sin^2\frac{u}{2}$
(Subtract 1 from each side.)
$\frac{1 - \cos u}{2} = \sin^2\frac{u}{2}$
(Divide each side by -2.)
$\pm\sqrt{\frac{1 - \cos u}{2}} = \sin\frac{u}{2}$
(Take square roots of each side.)

The Double- and Half-Angle Formulas (continued)

To establish the half-angle formula for cosine, use the double-angle formula $\cos 2\theta = 2\cos^2\theta - 1$:

$$\cos 2\theta = 2\cos^2\theta - 1 \qquad \text{Double-angle formula for cosine}$$

$$\cos 2\left(\frac{u}{2}\right) = 2\cos^2\frac{u}{2} - 1 \qquad \text{Substitute } \frac{u}{2} \text{ for } \theta.$$

$$\cos u = 2\cos^2\frac{u}{2} - 1 \qquad \text{Simplify.}$$

$$1 + \cos u = 2\cos^2\frac{u}{2} \qquad \text{Add 1 to each side.}$$

$$\frac{1 + \cos u}{2} = \cos^2\frac{u}{2} \qquad \text{Divide each side by 2.}$$

$$\pm\sqrt{\frac{1 + \cos u}{2}} = \cos\frac{u}{2} \qquad \text{Take square roots of each side.}$$

Exercises   1–2. See margin.

1. Use the sum formula for sine to establish the double-angle formula for sine.

2. Use the double-angle formula $\cos 2\theta = 1 - 2\sin^2\theta$ to establish the half-angle formula for sine.

Skills Review Handbook

▶ Real Numbers
OPERATIONS WITH SIGNED NUMBERS

When adding signed numbers, you may find using a number line helpful. When subtracting signed numbers, remember that you can add the opposite because $a - b = a + (-b)$.

EXAMPLE Simplify the expression.

a. $3 + (-5)$ **b.** $(-2) - (-1)$

SOLUTION

a. $3 + (-5) = -2$ **b.** $(-2) - (-1) = -2 + 1$

$$= -1$$

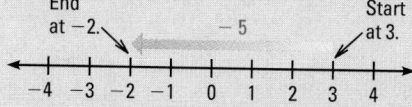   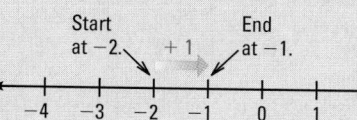

When multiplying and dividing signed numbers, use the following rules.

- Two positive numbers have a positive product or dividend.

- Two negative numbers have a positive product or dividend.

- A positive number and a negative number have a negative product or dividend.

EXAMPLE Perform the operation.

  a. $3 \cdot 5$ **b.** $(-4) \cdot (-3)$ **c.** $(-6) \div (-2)$ **d.** $(-4) \div 2$ **e.** $2 \cdot (-5)$

SOLUTION

  a. $3 \cdot 5 = 15$ **b.** $(-4) \cdot (-3) = 12$ **c.** $(-6) \div (-2) = 3$ **d.** $(-4) \div 2 = -2$ **e.** $2 \cdot (-5) = -10$

PRACTICE

Simplify the expression.

1. $1 + (-3)$ −2	**2.** $3 + 12$ 15	**3.** $(-4) + 4$ 0	**4.** $(-8) + (-3)$ −11
5. $7 + (-8)$ −1	**6.** $(-3) + (-9)$ −12	**7.** $(-4) + 10$ 6	**8.** $4 + (-12)$ −8
9. $6 + (-16)$ −10	**10.** $(-18) + 2$ −16	**11.** $(-13) + (-8)$ −21	**12.** $(-3) + (-22)$ −25
13. $8 - 2$ 6	**14.** $(-5) - 8$ −13	**15.** $1 - 5$ −4	**16.** $0 - (-3)$ 3
17. $7 - (-2)$ 9	**18.** $(-8) - (-8)$ 0	**19.** $6 - 11$ −5	**20.** $(-1) - 4$ −5
21. $(-2) - 1$ −3	**22.** $(-11) - (-3)$ −8	**23.** $12 - (-3)$ 15	**24.** $(-11) - (-5)$ −6

Simplify the expression.

25. $8 \cdot 3$ 24   **26.** $(-7) \cdot 2$ -14 **27.** $4 \cdot (-6)$ -24 **28.** $(-3) \cdot (-3)$ 9

29. $(-6) \cdot (-5)$ 30   **30.** $2 \cdot (-3)$ -6 **31.** $5 \cdot 5$ 25   **32.** $(-9) \cdot 2$ -18

33. $5 \div (-1)$ -5 **34.** $(-9) \div (-3)$ 3   **35.** $16 \div 2$ 8   **36.** $(-4) \div 4$ -1

37. $(-36) \div 9$ -4 **38.** $21 \div (-7)$ -3 **39.** $(-12) \div (-4)$ 3   **40.** $(-36) \div (-3)$ 12

Perform the indicated operation.

41. $(-5) \cdot 4$ -20 **42.** $3 - (-4)$ 7   **43.** $(-9) + 7$ -2 **44.** $27 \div 3$ 9

45. $(-30) \div 10$ -3 **46.** $45 + (-5)$ 40   **47.** $17 - (-12)$ 29   **48.** $(-8) \cdot (-6)$ 48

49. $14 \div (-2)$ -7 **50.** $(-5) + 4$ -1 **51.** $(-9) \cdot (-15)$ 135   **52.** $(-20) - 12$ -32

53. $(-42) - (-7)$ -35 **54.** $7 \cdot (-3)$ -21 **55.** $18 \div (-6)$ -3 **56.** $(-13) + (-6)$ -19

57. $(-11) + 18$ 7   **58.** $12 \cdot (-8)$ -96 **59.** $(-14) - (-7)$ -7 **60.** $(-24) \div (-3)$ 8

61. $63 \div (-7)$ -9 **62.** $(-7) - (-26)$ 19   **63.** $-12 \cdot (-11)$ 132   **64.** $(-27) + (-15)$ -42

CONVERTING DECIMALS, FRACTIONS, AND PERCENTS

Percent means "per hundred." It is a ratio (see page 910) that compares a number to 100.

EXAMPLE Write as a percent.

a. 0.3 **b.** $\dfrac{4}{5}$ **c.** 1.6

SOLUTION

a. $0.3 = \dfrac{3}{10} = \dfrac{30}{100} = 30\%$ **b.** $\dfrac{4}{5} = \dfrac{4 \cdot 20}{5 \cdot 20} = \dfrac{80}{100} = 80\%$ **c.** $1.6 = 1\dfrac{6}{10} = \dfrac{16}{10} = \dfrac{160}{100} = 160\%$

EXAMPLE Write as a decimal.

a. 66% **b.** $\dfrac{17}{25}$ **c.** 125%

SOLUTION

a. $66\% = \dfrac{66}{100} = 0.66$ **b.** $\dfrac{17}{25} = 17 \div 25 = 0.68$ **c.** $125\% = \dfrac{125}{100} = 1.25$

PRACTICE

Write as a percent.

1. 0.20 20%   **2.** 0.15 15%   **3.** 0.55 55%   **4.** 1.34 134%   **5.** 0.87 87%

6. $\dfrac{9}{10}$ 90%   **7.** $\dfrac{2}{5}$ 40%   **8.** $\dfrac{15}{50}$ 30%   **9.** $\dfrac{3}{5}$ 60%   **10.** $\dfrac{21}{20}$ 105%

Write as a decimal.

11. 50% 0.5   **12.** 120% 1.2   **13.** 2% 0.02   **14.** 85% 0.85   **15.** 40% 0.4

16. $\dfrac{3}{5}$ 0.6   **17.** $\dfrac{9}{25}$ 0.36   **18.** $\dfrac{11}{20}$ 0.55   **19.** $\dfrac{75}{50}$ 1.5   **20.** $\dfrac{9}{10}$ 0.9

CALCULATING PERCENTS

To calculate a percent of a number, write the percent as a fraction or decimal and multiply.

EXAMPLE   **a.** Find 14% of 150.   **b.** Find 80% of 200.

SOLUTION   **a.** 14% of 150 = $\frac{14}{100}(150) = 21$ **b.** 80% of 200 = $0.8 \cdot 200 = 160$

To find the percent one number is of another, divide.

EXAMPLE   **a.** What percent is 2 of 8?   **b.** What percent is 12 of 9?

SOLUTION   **a.** $\frac{2}{8} = 0.25 = 25\%$ **b.** $\frac{12}{9} = 1.\overline{3} = 133\frac{1}{3}\%$

To find a percent increase or decrease, find the difference between the two numbers and divide by the first number.

EXAMPLE   A television is marked down from $500 to $400. Find the percent increase or decrease.

SOLUTION   Percent change in price = $\dfrac{\text{New price} - \text{Old price}}{\text{Old price}} = \dfrac{400 - 500}{500} = \dfrac{-100}{500} = \dfrac{-20}{100} = -20\%$

▶ The negative sign indicates that the percent change is a decrease. Therefore, the price of a television decreased 20%.

PRACTICE

Find the number.

1. 15% of 20   3

2. 50% of $\frac{2}{3}$ $\frac{1}{3}$

3. 10% of 3   0.3

4. 20% of $\frac{1}{2}$ 0.1

5. 60% of 50   30

6. 12% of 18.5   2.22

7. 9% of 6   0.54

8. 2% of 100   2

9. 100% of 12   12

10. 25% of $\frac{3}{5}$ 0.15

11. 1% of $\frac{3}{8}$ 0.00375

12. 85% of $\frac{1}{10}$ 0.085

13. 5% of 0.5   0.025

14. 20% of 90   18

15. 10% of 0.84   0.084

16. 38% of 16   6.08

17. 200% of 7   14

18. 33% of 15   4.95

19. 0.5% of 1   0.005

20. 125% of 1.2   1.5

Find the answer.

21. What percent is 15 of 30?   50%   **22.** What percent is 3 of 12?   25%   **23.** What percent is 10 of 10?   100%

24. What percent is 1 of 20?   5%   **25.** What percent is 14 of 40?   35%   **26.** What percent is 300 of 200?   150%

27. What percent is 4 of 18?   about 22%   **28.** What percent is 6 of 16?   about 37.5%   **29.** What percent is 2 of 85?   about 2.4%

30. What percent is 4 of 100?   4%   **31.** What percent is 8 of 40?   20%   **32.** What percent is 90 of 50?   180%

33. What percent is 0.4 of 200?   0.2%   **34.** What percent is 80 of 5?   1600%   **35.** What percent is 0.22 of 50?   0.44%

Find the percent increase or decrease.

36. 50 votes increased to 200 votes   **300% increase**

37. $80 decreased to $56   **30% decrease**

38. 300 fish increased to 360 fish   **20% increase**

39. $400 increased to $600   **50% increase**

40. 15 feet decreased to 12 feet   **20% decrease**

41. 4500 units sold increased to 4800 units sold
about 6.67% increase

42. 100 students increased to 108 students **8% increase**

43. A 40 minute run decreased to a 35 minute run
12.5% decrease

FACTORS AND MULTIPLES

Factors are numbers or variable expressions that are multiplied together. A **prime number** is a whole number greater than 1 that has exactly two factors, itself and 1. To write the **prime factorization** of a number, write the number as a product of prime numbers.

Prime numbers less than 100
2, 3, 5, 7, 11, 13, 17, 19, 23, 29, 31, 37, 41, 43, 47, 53, 59, 61, 67, 71, 73, 79, 83, 89, 97

EXAMPLE   Write the prime factorization of 24.

SOLUTION   Use a tree diagram: 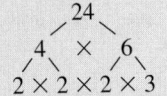   Write 24 as a product.

Write 4 and 6 as products.

A **common factor** of two whole numbers is a whole number that is a factor of each number. The **greatest common factor (GCF)** of two whole numbers is the greatest whole number that is a factor of each number. The **least common multiple (LCM)** of two whole numbers is the smallest whole number (other than zero) that is a multiple of each number.

EXAMPLE   What is the greatest common factor of 16 and 24?

SOLUTION

Method 1

Make a list of each number's factors.

16: 1, 2, 4, 8, 16
24: 1, 2, 3, 4, 6, 8, 12, 24

The common factors are 1, 2, 4, and 8.
The GCF of 16 and 24 is 8.

Method 2

The GCF of two whole numbers is equal to the product of all common prime factors of the numbers.

$16 = 2 \cdot 2 \cdot 2 \cdot 2$ **Prime factorization of 16**
$24 = 2 \cdot 2 \cdot 2 \cdot 3$ **Prime factorization of 24**

The common prime factors are 2, 2, and 2, so the GCF of 16 and 24 is $2 \cdot 2 \cdot 2$, or 8.

EXAMPLE   What is the least common multiple of 16 and 24?

SOLUTION

Method 1

Make a list of each number's multiples.

Multiples of 16: 16, 32, 48, 64, . . .
Multiples of 24: 24, 48, 72, . . .

The LCM of 16 and 24 is 48.

Method 2

The LCM of two whole numbers is the product of the highest power of each prime number that appears in the factorization of either number.

$16 = 2^4$ **Prime factorization of 16**
$24 = 2^3 \cdot 3$ **Prime factorization of 24**

The LCM of 16 and 24 is $2^4 \cdot 3 = 48$.

The **least common denominator (LCD)** of two fractions is the least common multiple of the denominators. To add or subtract two fractions with unlike denominators, first write equivalent fractions using the LCD, then add or subtract the numerators.

EXAMPLE Add: $\dfrac{3}{16} + \dfrac{7}{24}$

SOLUTION The LCD is 48. Rewrite fractions: $\dfrac{3}{16} = \dfrac{3 \cdot 3}{16 \cdot 3} = \dfrac{9}{48}$ and $\dfrac{7}{24} = \dfrac{7 \cdot 2}{24 \cdot 2} = \dfrac{14}{48}$

Add the rewritten fractions: $\dfrac{9}{48} + \dfrac{14}{48} = \dfrac{23}{48}$

PRACTICE

Write the prime factorization of the number. If the number is prime, write *prime*.

1. 8 $2 \times 2 \times 2$
2. 100 $2 \times 2 \times 5 \times 5$
3. 64 $2 \times 2 \times 2 \times 2 \times 2 \times 2$
4. 21 3×7
5. 17 Prime

6. 9 3×3
7. 12 $2 \times 2 \times 3$
8. 56 $2 \times 2 \times 2 \times 7$
9. 22 2×11
10. 50 $2 \times 5 \times 5$

11. 41 Prime
12. 30 $2 \times 3 \times 5$
13. 31 Prime
14. 46 2×23
15. 25 5×5

Give the greatest common factor (GCF) and least common multiple (LCM) of the pair of numbers.

16. 15, 25 5, 75
17. 4, 7 1, 28
18. 10, 15 5, 30
19. 20, 6 2, 60
20. 3, 13 1, 39

21. 20, 40 20, 40
22. 12, 9 3, 36
23. 18, 24 6, 72
24. 8, 10 2, 40
25. 48, 36 12, 144

26. 4, 6 2, 12
27. 2, 3 1, 6
28. 22, 11 11, 22
29. 6, 18 6, 18
30. 45, 25 5, 225

Find the least common denominator.

31. $\dfrac{7}{48}, \dfrac{5}{24}$ 48
32. $\dfrac{3}{7}, \dfrac{5}{6}$ 42
33. $\dfrac{3}{13}, \dfrac{1}{2}$ 26
34. $\dfrac{7}{2}, \dfrac{3}{4}$ 4
35. $\dfrac{6}{10}, \dfrac{5}{2}$ 10

36. $\dfrac{1}{3}, \dfrac{7}{8}$ 24
37. $\dfrac{5}{6}, \dfrac{11}{12}$ 12
38. $\dfrac{3}{10}, \dfrac{17}{30}$ 30
39. $\dfrac{7}{15}, \dfrac{7}{12}$ 60
40. $\dfrac{7}{4}, \dfrac{3}{5}$ 20

41. $\dfrac{5}{6}, \dfrac{3}{4}, \dfrac{1}{2}$ 12
42. $\dfrac{2}{3}, \dfrac{5}{12}, \dfrac{4}{9}$ 36
43. $\dfrac{7}{8}, \dfrac{5}{6}, \dfrac{4}{5}$ 120
44. $\dfrac{1}{2}, \dfrac{1}{4}, \dfrac{1}{5}$ 20
45. $\dfrac{8}{12}, \dfrac{6}{4}$ 12

46. $\dfrac{2}{11}, \dfrac{4}{9}, \dfrac{5}{3}$ 99
47. $\dfrac{5}{6}, \dfrac{11}{20}, \dfrac{14}{15}$ 60
48. $\dfrac{4}{7}, \dfrac{3}{4}, \dfrac{1}{28}$ 28
49. $\dfrac{1}{5}, \dfrac{5}{12}, \dfrac{1}{4}$ 60
50. $\dfrac{5}{3}, \dfrac{1}{2}$ 6

Perform the indicated operation(s). Simplify the result.

51. $\dfrac{3}{8} + \dfrac{5}{12}$ $\dfrac{19}{24}$
52. $\dfrac{7}{4} - \dfrac{1}{12}$ $\dfrac{5}{3}$
53. $-\dfrac{3}{4} - \dfrac{1}{2}$ $-\dfrac{5}{4}$
54. $\dfrac{7}{8} - \dfrac{1}{2}$ $\dfrac{3}{8}$

55. $\dfrac{2}{5} - \dfrac{1}{3} - \dfrac{1}{6}$ $-\dfrac{1}{10}$
56. $\dfrac{8}{9} + \dfrac{2}{3} + \dfrac{1}{2}$ $\dfrac{37}{18}$
57. $\dfrac{5}{16} + \dfrac{2}{5} - \dfrac{3}{10}$ $\dfrac{33}{80}$
58. $\dfrac{4}{9} + \dfrac{2}{3}$ $\dfrac{10}{9}$

59. $\dfrac{1}{2} + \dfrac{1}{3} + \dfrac{1}{4}$ $\dfrac{13}{12}$
60. $-\dfrac{5}{24} + \dfrac{2}{3} - \dfrac{1}{6}$ $\dfrac{7}{24}$
61. $\dfrac{9}{11} - \dfrac{5}{3} - \dfrac{5}{6}$ $-\dfrac{111}{66}$
62. $-\dfrac{6}{7} - \dfrac{2}{14}$ -1

63. $\dfrac{2}{6} - \dfrac{1}{3} + \dfrac{1}{2}$ $\dfrac{1}{2}$
64. $\dfrac{2}{8} - \dfrac{3}{4} + \dfrac{1}{2}$ 0
65. $-\dfrac{3}{12} + \dfrac{4}{10} - \dfrac{1}{5}$ $-\dfrac{1}{20}$
66. $\dfrac{1}{3} + \dfrac{5}{6} - \dfrac{2}{9}$ $\dfrac{17}{18}$

67. $\dfrac{3}{2} - \dfrac{4}{8} + \dfrac{1}{6}$ $\dfrac{7}{6}$
68. $\dfrac{1}{12} - \dfrac{5}{6} + \dfrac{4}{9}$ $-\dfrac{11}{36}$
69. $\dfrac{7}{15} - \dfrac{4}{5} + \dfrac{2}{3}$ $\dfrac{1}{3}$
70. $\dfrac{1}{2} - \dfrac{8}{10} + \dfrac{5}{4}$ $\dfrac{19}{20}$

Skills Review Handbook **909**

WRITING RATIOS AND SOLVING PROPORTIONS

A *ratio* compares two numbers using division. If a and b are two quantities measured in the same units, then the **ratio of a to b** can be written in three ways:

a to b $\qquad\qquad$ $a : b$ $\qquad\qquad$ $\dfrac{a}{b}$

EXAMPLE Write the ratio 4 to 3 in two other ways.

SOLUTION $4 : 3$ $\qquad$ $\dfrac{4}{3}$

To write a ratio in lowest terms, divide out any common factors.

EXAMPLE Write the ratio 12 to 18 in lowest terms.

SOLUTION 6 is the greatest common factor, so divide each number by 6.

▶ In lowest terms, the ratio 12 to 18 is 2 to 3.

A **proportion** is an equation stating that two ratios are equivalent. If a proportion contains a variable, you can cross multiply to solve for the variable.

EXAMPLE Solve the proportion $\dfrac{5}{6} = \dfrac{10}{x}$.

SOLUTION

$\dfrac{5}{6} = \dfrac{10}{x}$	**Rewrite proportion.**
$5 \cdot x = 6 \cdot 10$	**Cross multiply.**
$5x = 60$	**Simplify.**
$x = 12$	**Solve for x.**

PRACTICE

Write the ratio in two other ways.

1. 4 to 5 $\quad$ $4 : 5, \dfrac{4}{5}$

2. 1 : 1 $\quad$ 1 to 1, $\dfrac{1}{1}$

3. 2 to 6 $\quad$ $2 : 6, \dfrac{2}{6}$

4. 3 : 5 $\quad$ 3 to 5, $\dfrac{3}{5}$

5. $\dfrac{1}{5}$ $\quad$ 1 to 5, 1 : 5

6. 10 to 1 $\quad$ $10 : 1, \dfrac{10}{1}$

7. $\dfrac{8}{5}$ $\quad$ 8 to 5, 8 : 5

8. 5 : 4 $\quad$ 5 to 4, $\dfrac{5}{4}$

9. 3 : 1 $\quad$ 3 to 1, $\dfrac{3}{1}$

10. $\dfrac{2}{3}$ $\quad$ 2 to 3, 2 : 3

11. $\dfrac{3}{4}$ $\quad$ 3 to 4, 3 : 4

12. 6 to 3 $\quad$ $6 : 3, \dfrac{6}{3}$

Write the ratio in lowest terms.

13. 2 to 8 $\quad$ 1 to 4

14. 5 : 10 $\quad$ 1 : 2

15. 4 : 16 $\quad$ 1 : 4

16. 80 to 100 $\quad$ 4 to 5

17. $\dfrac{2}{10}$ $\quad$ $\dfrac{1}{5}$

18. $\dfrac{3}{27}$ $\quad$ $\dfrac{1}{9}$

19. 25 to 15 $\quad$ 5 to 3

20. 9 : 3 $\quad$ 3 : 1

21. $\dfrac{12}{20}$ $\quad$ $\dfrac{3}{5}$

22. 4 : 24 $\quad$ 1 : 6

23. $\dfrac{24}{18}$ $\quad$ $\dfrac{4}{3}$

24. 20 : 35 $\quad$ 4 : 7

Solve the proportion.

25. $\frac{x}{4} = \frac{2}{8}$ 1

26. $\frac{5}{7} = \frac{a}{28}$ 20

27. $\frac{6}{b} = \frac{3}{8}$ 16

28. $\frac{20}{4} = \frac{10}{y}$ 2

29. $\frac{2}{1} = \frac{6}{c}$ 3

30. $\frac{5}{4} = \frac{x}{10}$ 12.5

31. $\frac{3}{7} = \frac{9}{b}$ 21

32. $\frac{36}{r} = \frac{12}{3}$ 9

33. $\frac{p}{2} = \frac{5}{2}$ 5

34. $\frac{8}{5} = \frac{w}{25}$ 40

35. $\frac{9}{4} = \frac{k}{12}$ 27

36. $\frac{n}{3} = \frac{3}{1}$ 9

37. $\frac{2}{11} = \frac{x}{99}$ 18

38. $\frac{80}{48} = \frac{10}{s}$ 6

39. $\frac{z}{5} = \frac{8}{2}$ 20

40. $\frac{1}{8} = \frac{5}{j}$ 40

41. $\frac{16}{3} = \frac{2a}{6}$ 16

42. $\frac{c}{15} = \frac{4}{3}$ 20

43. $\frac{60}{40} = \frac{12}{m}$ 8

44. $\frac{3x}{4} = \frac{27}{12}$ 3

45. $\frac{2}{9} = \frac{y}{27}$ 6

46. $\frac{13}{w} = \frac{39}{9}$ 3

47. $\frac{x}{20} = \frac{3}{60}$ 1

48. $\frac{x}{3} = \frac{40}{6}$ 20

SIGNIFICANT DIGITS

Significant digits indicate how precisely a number is known. Use the following guidelines to determine the number of significant digits.

• All nonzero digits are significant.
• All zeros that appear between two nonzero digits are significant.
• For a decimal, all zeros that appear after the last nonzero digit are significant. For a whole number, you cannot tell whether any zeros after the last nonzero digit are significant, so you should assume that they are not significant (unless you know otherwise).

Sometimes calculations involve measurements that have various numbers of significant digits. In this case, a general rule is to carry all digits through the calculation and then round the result to the same number of significant digits as the measurement with the *fewest* number of significant digits.

EXAMPLE   Add: $76.33 + 22.0 + 1500$

SOLUTION   Of the three numbers, 1500 has the fewest number of significant digits. Add all three numbers, then round the sum to two significant digits.

$$76.33 + 22.0 + 1500 = 1598.33$$
$$\approx 1600$$

EXAMPLE   Perform the indicated operation.

a. $0.004 \cdot 3.22$

b. $374{,}039.8 \div 305$

SOLUTION   **a.** Since the zeros before the 4 in 0.004 are not significant, round the answer to one significant digit.

$$0.004 \cdot 3.22 = 0.01288$$
$$\approx 0.01$$

b. Since the zero between the 3 and the 5 in 305 is significant, round the answer to three significant digits.

$$374{,}039.8 \div 305 = 1226.36$$
$$\approx 1230$$

Note that some units, such as number of people, cannot be divided into fractional parts. In that case, use the significant digits of the other numbers to round the answer.

EXAMPLE A bill of $98.80 is divided among 8 people. How much does each person pay?

SOLUTION The number of people is exact, so the fact that it is a one-digit number is irrelevant. Use the significant digits for the money to round your answer.

$$\$98.80 \div 8 = \$12.35$$

▶ Each person should pay $12.35.

PRACTICE

Simplify the expression. Write your answer with the appropriate number of significant digits.

1. $8244 + 3.6$ **8200**

2. $-25 - 3$ **−28**

3. $2.50 \cdot 3.80$ **9.50**

4. $0.95 \div 4.25$ **0.22**

5. $30.82 - 2.6690$ **28.15**

6. $16 \div 7$ **2**

7. $700 + 20$ **700**

8. $60 \div 24$ **3**

9. $50 \div 4.5$ **10**

10. $2.64 + 3.0008$ **5.64**

11. $38.25 \div 52$ **0.74**

12. $6 - 3.4$ **3**

13. $5.0 - 1.8$ **3.2**

14. $0.74 \cdot 2.15$ **1.6**

15. $25.000 \div 25$ **1.0**

16. $13.36 + 40.58$ **53.94**

17. $200 - 3.5$ **200**

18. $40 \div 0.368$ **100**

19. $14.85 + 5.00 + 4.8$ **24.7**

20. $0.0036 + 0.017 + 0.0249$ **0.046**

21. $23.89 - 2.5 - 3.74$ **18**

22. $100 - 21 - 2.9 - 3.62$ **70**

23. $27.5 \cdot 9.8 \cdot 0.332$ **89**

24. $0.783 \cdot 2.11 \cdot 4.51$ **7.45**

25. $2.48 \cdot 16.4 \div 56.25$ **0.723**

26. $42.6 \cdot 2.05 \div 0.0068$ **12,800**

27. $60 \div (52.4 \cdot 20)$ **0.06**

28. $388 \cdot 16 \cdot 108 \cdot 27$ **18,000,000**

29. $13,720 + 2800 - 513$ **16,000**

30. $(200 \cdot 45) \div (36 \cdot 15)$ **20**

Perform the calculation. Write your answer with the appropriate number of significant digits.

31. $1.50 per card • 5 cards   **$7.50**

32. 324 pens ÷ 36 students   **9 pens/student**

33. $39.95 per sweater • 6 sweaters   **$239.70**

34. $.40 per orange • 10 oranges   **$4.00**

35. 89 miles ÷ 6.8 gallons   **13 mi/gal**

36. 282 books ÷ 47 students   **6 books/student**

37. 101 gallons of milk + 8.75 gallons of milk − 6.9 gallons of milk   **100 gal of milk**

38. 210 pounds of sand − 16.25 pounds of sand − 1.5 pounds of sand   **200 lbs of sand**

39. 20.3 milliliters of water + 1.08 milliliters of hydrochloric acid   **21.4 mL**

40. 8.0 liters of juice − 5 liters of juice   **3 liters of juice**

41. 38,050 computers ÷ 52 computer stores   **730 computers/store**

42. 3000 kilogram car + 65.50 kilogram passenger + 2.37 kilogram groceries   **3000 kg**

43. 13.2 milligrams of rice + 0.015 milligram of saffron + 1.25 milligrams of salt   **15 mg**

44. 325 milligrams of Vitamin C + 5.50 milligrams of Vitamin C − 24.3 milligrams of Vitamin C   **306 mg of Vitamin C**

45. 7.55 inches of rain in March + 12.25 inches of rain in April + 6.08 inches of rain in May   **25.9 in. of rain**

SCIENTIFIC NOTATION

Numbers written in scientific notation have the form $c \times 10^n$ where $1 \le c < 10$ and n is an integer. Recall that $10^0 = 1$.

EXAMPLE Write each number in scientific notation.

    a. 721,000,000   **b.** 0.001046

SOLUTION **a.** Move the decimal point 8 places to the left.

$$721{,}000{,}000 = 7.21 \times 10^8$$

                b. Move the decimal point three places to the right.

$$0.001046 = 1.046 \times 10^{-3}$$

EXAMPLE Write each number in standard form.

    a. 5.23×10^7 **b.** 2.600×10^{-4}

SOLUTION **a.** Move the decimal point 7 places to the right.

$$5.23 \times 10^7 = 52{,}300{,}000$$

                b. Move the decimal point 4 places to the left. Note that 4 significant digits are kept.

$$2.600 \times 10^{-4} = 0.0002600$$

PRACTICE

Write each number in scientific notation.

1. 0.4 4×10^{-1} **2.** 0.34 3.4×10^{-1} **3.** 0.09 9×10^{-2} **4.** 30.58 3.058×10^1

5. 4 4×10^0 **6.** 0.0000000025 2.5×10^{-9} **7.** 0.0000926 9.26×10^{-5} **8.** 4,983,200,000 4.9832×10^9

9. 211.111 2.11111×10^2 **10.** 4193 4.193×10^3 **11.** 0.005 5×10^{-3} **12.** 21,040 2.104×10^4

13. 98,400 9.84×10^4 **14.** 0.00002 2×10^{-5} **15.** 204.89 2.0489×10^2 **16.** 295 2.95×10^2

17. 0.00037 3.7×10^{-4} **18.** 0.2000 2.000×10^{-1} **19.** 59.8 5.98×10^1 **20.** 5,000,000 5×10^6

21. 23,085,600   **22.** 0.0000004 4×10^{-7} **23.** 0.000100 1.00×10^{-4} **24.** 0.101001 1.01001×10^{-1}
      2.30856×10^7

Write each number in standard form.

25. 9×10^2 900   **26.** 2.52×10^{-1} 0.252   **27.** 3.1×10^3 3100

28. 6×10^5 600,000   **29.** 2.90×10^{-1} 0.290   **30.** 9.1×10^0 9.1

31. 1.001×10^4 10,010   **32.** 5.273×10^{-3} 0.005273   **33.** 7.926×10^6 7,926,000

34. 8.13×10^{-1} 0.813   **35.** 3.84×10^{-4} 0.000384   **36.** 4.6000×10^8 460,000,000

37. 3.7×10^{-5} 0.000037   **38.** 1.11×10^{-2} 0.0111   **39.** 4.9831×10^{-3} 0.0049831

40. 7.05×10^{-7} 0.000000705   **41.** 3.9502×10^5 395,020   **42.** 1.0063×10^0 1.0063

43. 2.64095×10^3 2640.95   **44.** 3.03×10^{-7} 0.000000303   **45.** 4.55×10^{-4} 0.000455

46. 5.0×10^3 5000   **47.** 5.9438×10^{-2} 0.059438   **48.** 6.105×10^{-6} 0.000006105

▶ GEOMETRY
PERIMETER, AREA, AND VOLUME

The **perimeter** of a two-dimensional figure is the sum of the lengths of the edges, or the distance around the figure.

> **EXAMPLE** Find the perimeter of the figure.

a.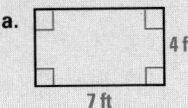

b.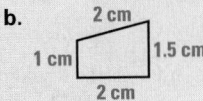

SOLUTION

a. $4 + 4 + 7 + 7 = 22$

▶ The perimeter is 22 feet.

b. $2 + 1.5 + 2 + 1 = 6.5$

▶ The perimeter is 6.5 cm.

The perimeter of a circle, called its **circumference**, is the distance around the circle. The formula for circumference is $C = 2\pi r$ where r is the **radius**. Because the **diameter** is twice the radius, the formula for circumference can also be written as $C = \pi d$ where d is the diameter.

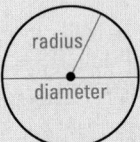

> **EXAMPLE** Find the circumference of the circle with radius 3 inches.

SOLUTION $C = 2\pi r = 2\pi(3) = 6\pi \approx 6(3.14) = 18.84$

▶ The circumference is 6π inches or about 18.84 inches.

> **EXAMPLE** Find the circumference of the circle with diameter 4 meters.

SOLUTION $C = \pi d = \pi(4) = 4\pi \approx 4(3.14) = 12.56$

▶ The circumference is 4π meters or about 12.56 meters.

The **area** of a two-dimensional figure is the number of square units enclosed within the boundary of the figure.

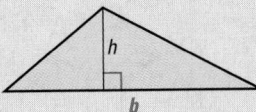

Area of a triangle: $A = \frac{1}{2}bh$

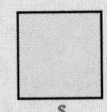

Area of a square: $A = s^2$

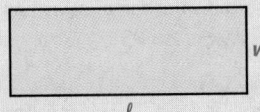

Area of a rectangle: $A = \ell w$

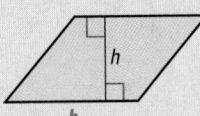

Area of a parallelogram: $A = bh$

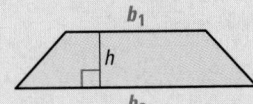

Area of a trapezoid: $A = \frac{1}{2}(b_1 + b_2)h$

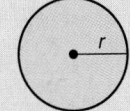

Area of a circle: $A = \pi r^2$

EXAMPLE Find the area of the figure.

a.

4 ft 5 ft
3 ft

b.

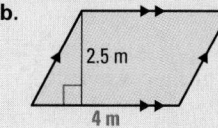

2.5 m
4 m

c.

6 cm

SOLUTION

a. $A = \frac{1}{2}bh$

$= \frac{1}{2}(3)(4)$

$= 6$ square feet

b. $A = bh$

$= (4)(2.5)$

$= 10$ square meters

c. $A = \pi r^2$

$= \pi(6)^2$

$= 36\pi$

≈ 113 square centimeters

A **prism** is a three-dimensional figure with two congruent faces, called *bases*, that lie in parallel planes. The **surface area** of a prism is the sum of the areas of all the faces of the prism. Surface area is measured in square units.

EXAMPLE Find the surface area of the prism or cylinder.

a.

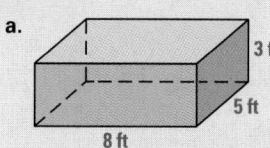

3 ft
5 ft
8 ft

b.

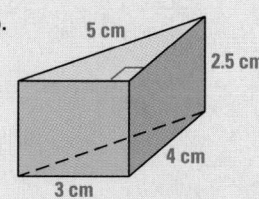

5 cm
2.5 cm
4 cm
3 cm

c.

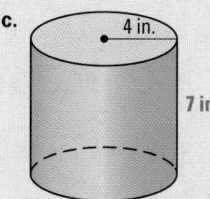

4 in.
7 in.

SOLUTION

a. A rectangular prism has three pairs of identical rectangular faces.

Surface area $= 2(8 \cdot 5) + 2(8 \cdot 3) + 2(5 \cdot 3)$

$= 80 + 48 + 30$

$= 158$

▶ The prism has a surface area of 158 square feet.

b. Surface area $=$ area of bases $+$ area of faces

$= 2\left[\frac{1}{2}(3)(4)\right] + (2.5)(3) + (2.5)(4) + (2.5)(5)$

$= 12 + 7.5 + 10 + 12.5$

$= 42$

▶ The prism has a surface area of 42 square centimeters.

c. Surface area $=$ area of bases $+$ (circumference)(height)

$= 2(\pi r^2) + (2\pi r)(h)$

$= 2[\pi(4)^2] + [2\pi(4)](7)$

$= 32\pi + 56\pi$

$= 88\pi$

≈ 276

▶ The cylinder has a surface area of about 276 square inches.

The **volume** of a solid is a measure of how much it will hold and is measured in cubic units. The volume of a prism is calculated by multiplying the area of the base by the height.

EXAMPLE Find the volume of the three solids in the previous example.

SOLUTION

a. Volume = $(8 \cdot 5) \cdot 3$
  = 120

b. Volume = $\left(\frac{1}{2} \cdot 3 \cdot 4\right) \cdot 2.5$
  = 15

c. Volume = $\left[\pi(4)^2\right] \cdot 7$
  = 112π
  ≈ 352

▶ The prism has a volume of 120 ft^3.

▶ The prism has a volume of 15 cm^3.

▶ The cylinder has a volume of about 352 in.^{3}

PRACTICE

Find the perimeter or circumference of the figure.

1. about 6.28 m
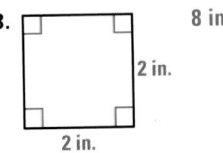
1 m

2.
19 cm
6 cm   6 cm
7 cm

3.
8 in.
2 in.
2 in.

4.
18 ft
4 ft
6 ft

5.
13 m
2 m
2 m   3 m
2 m
4 m

6.
36 in.
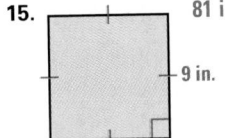
6 in.

7. A square 3 ft on each side   **12 ft**

8. A circle with diameter 10 in.   **about 31.4 in.**

9. A rectangle with sides of 4 cm and 7 cm   **22 cm**

10. A regular pentagon with side length 2.5 m   **12.5 m**

11. A triangle with sides of length 8 cm, 3 cm, and 7 cm   **18 cm**

12. A parallelogram with sides 3.5 m and 5.8 m   **18.6 m**

13. A circle with a diameter of 22 in.   **about 69 in.**

14. A rectangle with sides of 5 ft and 8 ft   **26 ft**

Find the area of the figure.

15.
81 in.^{2}
9 in.

16.
24 m^2
6 m
8 m

17.
21 cm^2
7 cm
3 cm

18.
12 in.^{2}
3 in.
3 in.
5 in.

19. A trapezoid with bases 4 in. and 8 in. and height 4 in.   **24 in.^{2}**

20. A square with side length 7 ft   **49 ft^2**

21. A circle with radius 0.5 in.   **about 0.79 in.^{2}**

22. A parallelogram with height 6 m and base 9 m   **54 m^2**

23. A 2 mi by 6 mi rectangle   **12 mi^2**

24. A circle with radius 9 mm   **about 254 mm^2**

25. A triangle with a base of 5 in. and height of 4 in.   **10 in.^{2}**

26. A circle with a radius of 10 ft   **about 314 ft^2**

Find the surface area of the prism or cylinder.

27.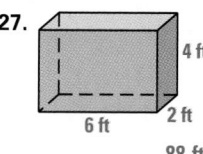
4 ft
6 ft
2 ft
88 ft²

28.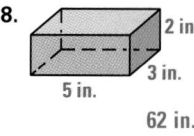
2 in.
3 in.
5 in.
62 in.²

29.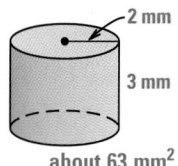
2 mm
3 mm
about 63 mm²

30.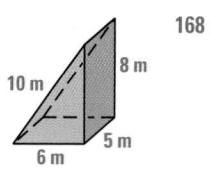
168 m²
8 m
10 m
5 m
6 m

31. A cylinder with radius 2 in. and height 14 in. **201 in.²** 32. A cube with side length 3 cm **54 cm²**

33. A rectangular prism 4 cm by 6 cm by 12 cm **288 cm²** 34. A cylinder with a radius of 50 ft and height of 200 ft
about 78,500 ft²

Find the volume of the prism or cylinder.

35.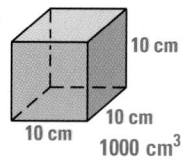
10 cm
10 cm
10 cm
1000 cm³

36.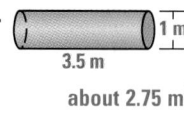
1 m
3.5 m
about 2.75 m³

37.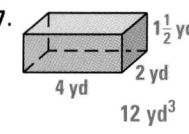
$1\frac{1}{2}$ yd
2 yd
4 yd
12 yd³

38.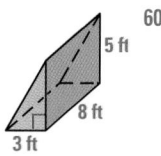
60 ft³
5 ft
8 ft
3 ft

39. A rectangular prism 1 ft by 1 ft by 5 ft **5 ft³**

40. A cube 8 in. on each side **512 in.³**

41. A cylinder with diameter 9 in. and height 2 in. **about 127 in.³**

42. A rectangular prism with base 12.8 m² and height 3 m **38.4 m³**

TRIANGLE RELATIONSHIPS

The sum of the angles of a triangle is 180°.

EXAMPLE Find the value of x.

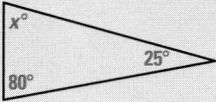
$x°$
$80°$
$25°$

SOLUTION $180° = 25° + 80° + x$

$x = 75°$

The **Pythagorean theorem** states that in a right triangle with legs of length a and b and hypotenuse of length c, $c^2 = a^2 + b^2$.

a
c
b

EXAMPLE Find the length of the unknown side.

a.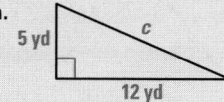
5 yd
c
12 yd

b.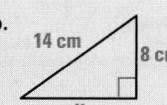
14 cm
8 cm
x

SOLUTION

a. $5^2 + 12^2 = c^2$
$25 + 144 = c^2$
$169 = c^2$
$c = 13$
▶ The hypotenuse is 13 yards long.

b. $8^2 + x^2 = 14^2$
$64 + x^2 = 196$
$x^2 = 132$
$x \approx 11.5$
▶ The length of the second leg is about 11.5 centimeters.

The sum of the lengths of the two shorter sides of a triangle must be greater than the length of the third side.

EXAMPLE Can you form a triangle with the given side lengths? Write *yes* or *no*.

a. 2, 3, 8

b. 9, 11, 19

c. 3, 18, 21

SOLUTION

a. No, because $2 + 3 < 8$.

b. Yes, because $9 + 11 > 19$.

c. No, because $18 + 3 = 21$.

PRACTICE

Find the value of x.

1. 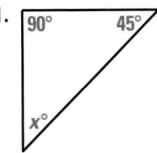 45

2. 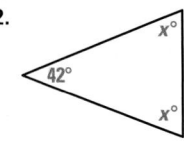 69

3. 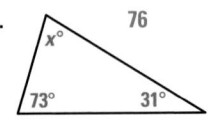 76

4. 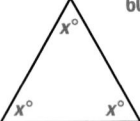 60

5. A triangle with angles $x°$, 5°, and 10°   165

6. A triangle with angles $x°$, 48°, and 22°   110

7. A triangle with angles $x°$, 38°, and 82°   60

8. A triangle with angles $x°$, 25°, and 63°   92

Can a triangle have the following angle measures? Write *yes* or *no*.

9. 60°, 60°, 60°   yes

10. 136°, 19°, 45°   no

11. 112°, 15°, 43°   no

12. 45°, 67°, 68°   yes

13. 47°, 90°, 23°   no

14. 59°, 60°, 61°   yes

15. 31°, 78°, 91°   no

16. 25°, 30°, 125°   yes

17. 160°, 5°, 5°   no

18. 55°, 75°, 60°   no

19. 40°, 50°, 90°   yes

20. 113°, 14°, 53°   yes

21. 17°, 52°, 111°   yes

22. 20°, 140°, 20°   yes

23. 70°, 60°, 50°   yes

24. 43°, 56°, 101°   no

Find the length of the unknown side.

25. about 5.7 mm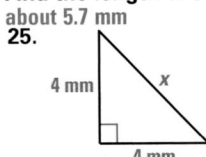

26. about 6.7 ft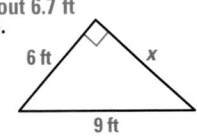

27. 2.5 m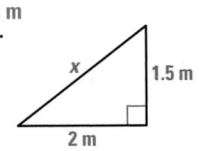

28. about 3.6 in.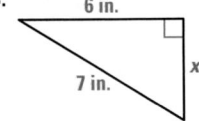

29. about 3.6 m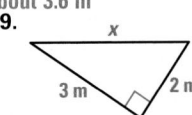

30. 15 yd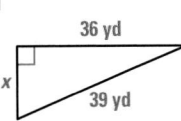

31. 8 cm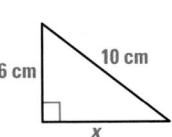

32. about 5.7 ft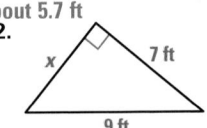

33. A right triangle with two sides 5 in. long   about 7.1 in.

34. A right triangle with one side 8 ft and hypotenuse 10 ft   6 ft

Can you form a triangle with the given side lengths? Write *yes* or *no*.

35. 8, 3, 7   yes

36. 10, 10, 10   yes

37. 16, 5, 11   no

38. 3, 6, 8   yes

39. 4, 4, 5   yes

40. 2, 5, 2   no

41. 6, 5, 4   yes

42. 17, 9, 8   no

43. 85, 19, 51   no

44. 12, 7, 4   no

45. 10, 24, 26   yes

46. 46, 22, 17   no

SYMMETRY

A figure has **line symmetry** if it can be divided by a line into two parts, each of which is the mirror image of the other. The line that divides the figure into two parts is called the **line of symmetry**.

EXAMPLE Identify the lines of symmetry in the figure.

a.

b.

c.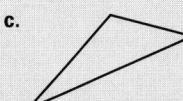

SOLUTION **a.** This figure has a vertical line of symmetry and a horizontal line of symmetry.

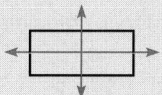

b. This figure has a horizontal line of symmetry.

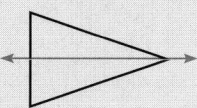

c. This figure is not symmetric. It has no line of symmetry.

A figure has **rotational symmetry** if it coincides with itself after rotating 180° or less, either clockwise or counterclockwise, about a point. The point of rotation is usually the center of the figure.

EXAMPLE Identify any rotational symmetry in the figure.

a.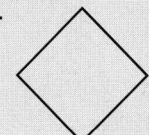

b.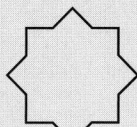

c.

SOLUTION **a.** This figure has rotational symmetry. It will coincide with itself after being rotated 90° or 180° in either direction. Notice that the point of rotation is located at the center of the figure.

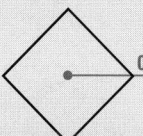 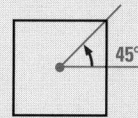 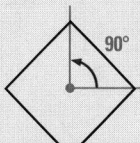

b. This figure has rotational symmetry. It will coincide with itself after being rotated 45°, 90°, 135°, or 180° in either direction. The point of rotation is located at the center of the figure.

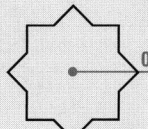 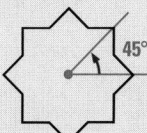

c. This figure has no rotational symmetry.

1. Line symmetry and rotational symmetry; 4 lines of symmetry, 90° or 180°

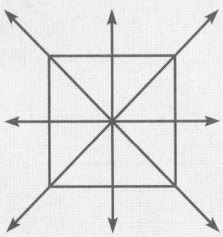

2. Line symmetry; 1 line of symmetry

3. Line symmetry and rotational symmetry; 6 lines of symmetry, 60°, 120°, or 180°

4. Line symmetry and rotational symmetry; 2 lines of symmetry, 180°

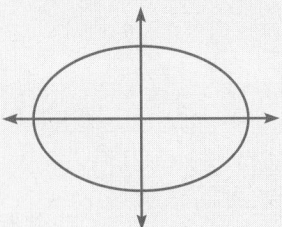

5. Line symmetry and rotational symmetry; 5 lines of symmetry, 72° or 144°

EXAMPLE The figure at the right has line symmetry. Find the coordinates of point *A*.

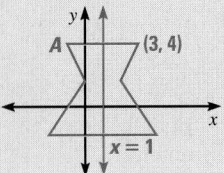

SOLUTION The line of symmetry is $x = 1$. Point $(3, 4)$ is 2 units to the right of the line of symmetry so *A* must be an equal distance to the left. Therefore, *A* is at $(-1, 4)$.

PRACTICE

State whether the figure has line symmetry or rotational symmetry. Then identify the line(s) of symmetry or the angle of rotation.   1–8. See margin.

1.

2.
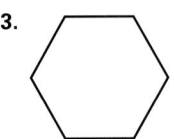

3.
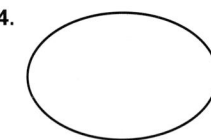

4.
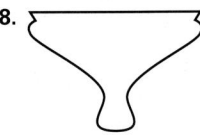

5.

6.
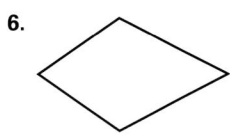

7.
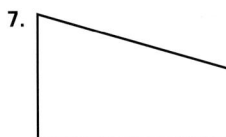

8.

The figure is symmetric. The line of symmetry is shown in red. Find the coordinates of point *A*.

9. $(-3, 2)$
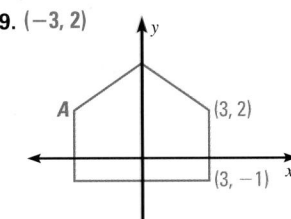

10. $(3, -2)$
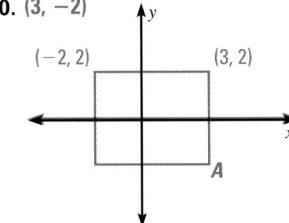

11. $(4, 1)$
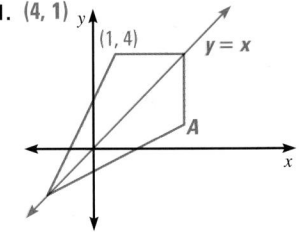

12. $(5, 0)$
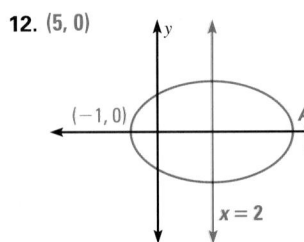

13. $(4, 6)$
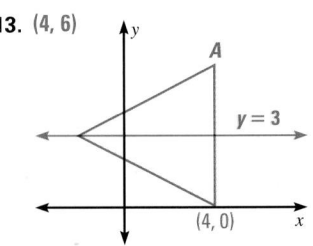

14. $(-5, -5)$
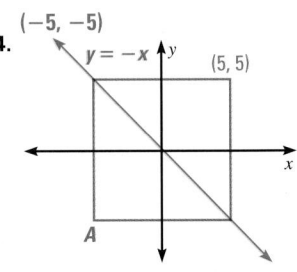

TRANSFORMATIONS

A **transformation** is a change made to the size or position of a figure. A **translation** is a transformation that slides every point of a figure the same distance in the same direction while preserving its size and orientation.

EXAMPLE Translate $\overline{AB}$ 2 units to the left and 5 units up.

SOLUTION To shift $\overline{AB}$ left 2 units, subtract 2 from each *x*-coordinate. To shift $\overline{AB}$ up 5 units, add 5 to each *y*-coordinate. You can describe this transformation as $\overline{AB}$ *is mapped onto* $\overline{A'B'}$, written symbolically as $\overline{AB} \rightarrow \overline{A'B'}$. In particular:

$$A(3, 0) \rightarrow A'(1, 5)$$
$$B(8, 2) \rightarrow B'(6, 7)$$

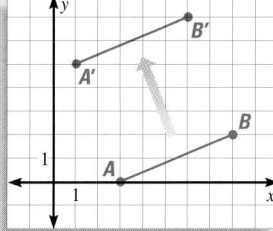

A **reflection** is a transformation in which each point of a figure has an image that is the same distance from the **line of reflection** as the original point but on the opposite side. A reflection preserves the size of a figure but not its orientation.

EXAMPLE **a.** Reflect the blue triangle over the *x*-axis. **b.** Reflect the blue kite over the *y*-axis.

SOLUTION **a.**

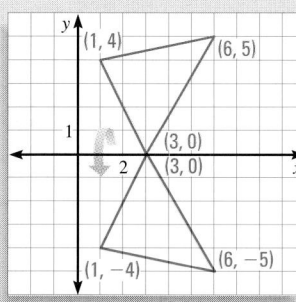

Change each *y*-coordinate to its opposite.

b.

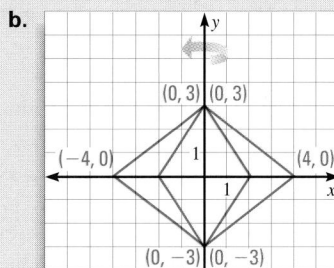

Change each *x*-coordinate to its opposite.

A **rotation** is a transformation in which every point moves along a circular path around a fixed point. Rotations preserve both size and orientation.

EXAMPLE Rotate the blue quadrilateral 270° about the origin.

SOLUTION When a point (x, y) is rotated counterclockwise around the origin use the following patterns:

90° rotation: $A(x, y) \rightarrow A'(-y, x)$
180° rotation: $A(x, y) \rightarrow A'(-x, -y)$
270° rotation: $A(x, y) \rightarrow A'(y, -x)$

In this case, use the pattern for a 270° rotation about the origin. So, $(-3, 5) \rightarrow (5, 3)$, $(-3, 2) \rightarrow (2, 3)$, $(0, 2) \rightarrow (2, 0)$, and $(1, 5) \rightarrow (5, -1)$.

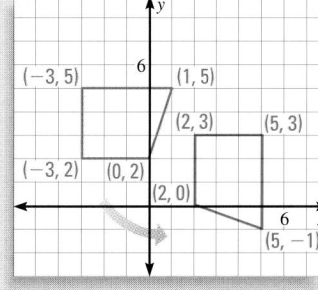

Skills Review Handbook 921

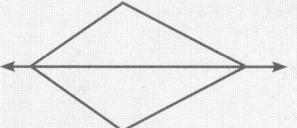
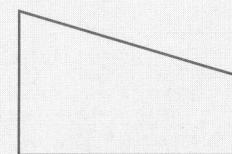
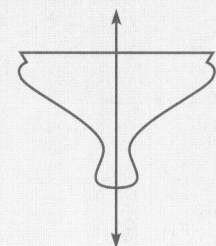
921

11.

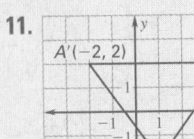

12.

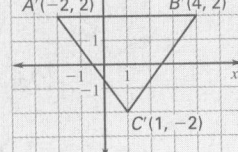

13.

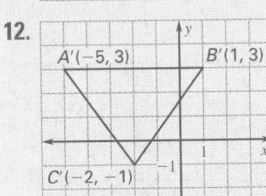

14.

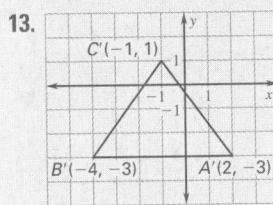

15.

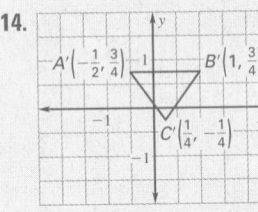

16.

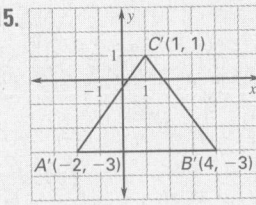

17.

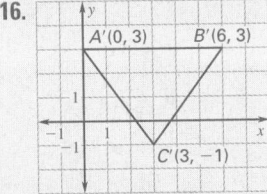

18.

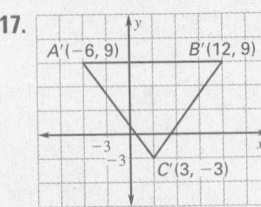

A **dilation** is a transformation in which every point of a figure is multiplied by a **scale factor** to create a similar image (see page 923). Dilations preserve the orientation of the original figure while enlarging or reducing the size.

EXAMPLE Dilate $\overline{AB}$ by a scale factor of $\frac{2}{3}$.

SOLUTION Multiply the coordinates of A and B by the scale factor $\frac{2}{3}$. Then graph points A' and B'.

$$A' = \left(\frac{2}{3} \cdot 6, \frac{2}{3} \cdot 0\right) = (4, 0)$$

$$B' = \left(\frac{2}{3} \cdot 18, \frac{2}{3} \cdot 12\right) = (12, 8)$$

$\overline{A'B'}$ is $\frac{2}{3}$ as long as $\overline{AB}$.

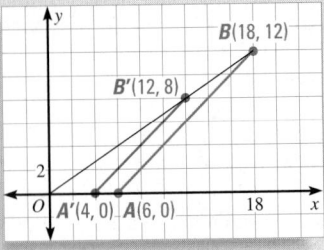

PRACTICE

Give the coordinates of $A(3, -6)$ after the following transformations. For rotations, rotate about the origin.

1. Reflect in y-axis. (−3, −6) **2.** Dilate by $\frac{2}{3}$. (2, −4) **3.** Dilate by $\frac{3}{2}$. $\left(\frac{9}{2}, -9\right)$ **4.** Translate left 6 units. (−3, −6)

5. Rotate 90°. (6, 3) **6.** Translate up 6 units. (3, 0) **7.** Reflect in x-axis. (3, 6) **8.** Rotate 180°. (−3, 6)

9. Translate left 3 units and up 5 units. (0, −1) **10.** Translate right 2 units and up 1 unit. (5, −5)

Transform $\triangle ABC$. Graph the result. For rotations, rotate about the origin. 11–18. See margin.

11. Translate down 1 unit. **12.** Translate left 3 units.

13. Rotate 180°. **14.** Dilate by $\frac{1}{4}$.

15. Reflect in the x-axis. **16.** Translate right 2 units.

17. Dilate by 3. **18.** Rotate 90°.

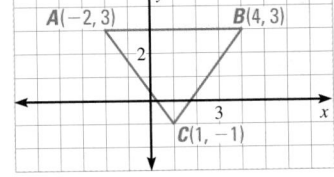

Transform $MNOP$. Graph the result. For rotations, rotate about the origin. 19–26. See margin.

19. Dilate by $\frac{1}{2}$. **20.** Translate left 1 unit and down 2 units.

21. Rotate 270°. **22.** Translate down 4 units.

23. Reflect in the x-axis. **24.** Dilate by $\frac{5}{2}$.

25. Reflect in the y-axis. **26.** Reflect in the line $y = x$.

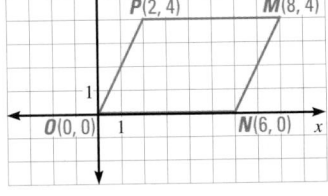

Transform $DEFG$. Graph the result. For rotations, rotate about the origin. 27–32. See margin.

27. Translate up 4 units. **28.** Reflect in the y-axis.

29. Rotate 90°. **30.** Translate right 9 units.

31. Translate left 5 units. **32.** Dilate by 4.

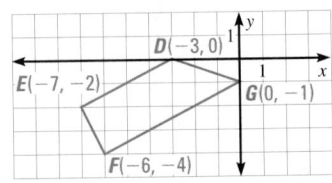

SIMILAR FIGURES

Two figures are **congruent** if they are exactly the same shape and the same size. Triangles *ABC* and *DEF* are congruent. Corresponding angles are marked with the same symbol.

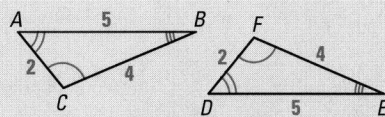

Two figures are **similar** if corresponding angles are congruent and the lengths of corresponding sides are in proportion.

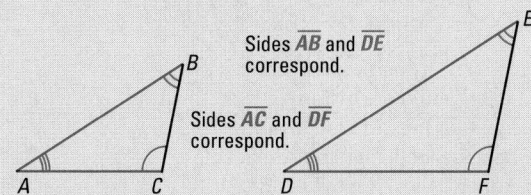

Sides $\overline{AB}$ and $\overline{DE}$ correspond.

Sides $\overline{AC}$ and $\overline{DF}$ correspond.

The ratios of the lengths of corresponding sides of similar figures are equal.

For example, in the similar triangles above, $\dfrac{AB}{DE} = \dfrac{AC}{DF} = \dfrac{BC}{EF}$.

EXAMPLE The two polygons are similar. Find the values of *x* and *y*.

SOLUTION Write and solve a proportion to find each unknown length.

$$\frac{AB}{EF} = \frac{CD}{GH} \quad \text{and} \quad \frac{AD}{EH} = \frac{BC}{FG}$$

$$\frac{6}{4} = \frac{6}{x} \qquad\qquad \frac{9}{6} = \frac{4}{y}$$

$$6x = 24 \qquad\qquad 9y = 24$$

$$x = 4 \qquad\qquad y = 2\frac{2}{3}$$

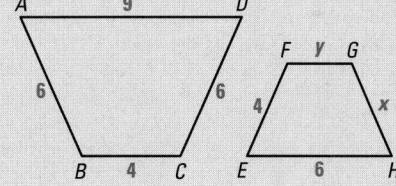

PRACTICE

The two polygons are similar. Find the value of *x*.

1.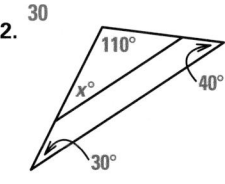

2.

3.

4.

5.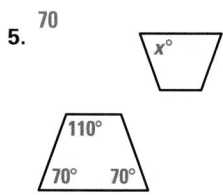

6.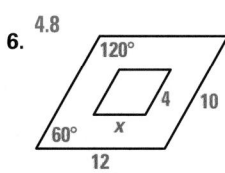

7.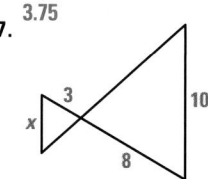

8.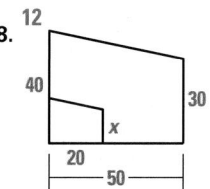

Skills Review Handbook **923**

19.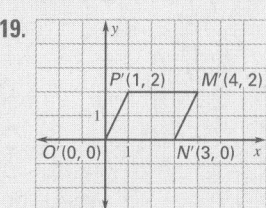

20.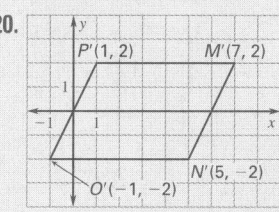

21.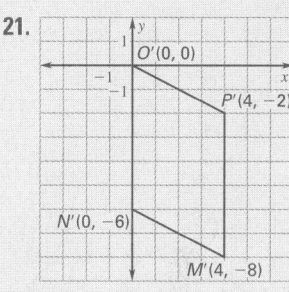

22.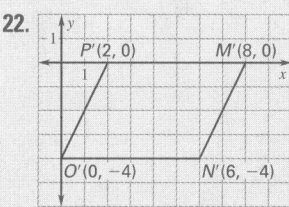

23.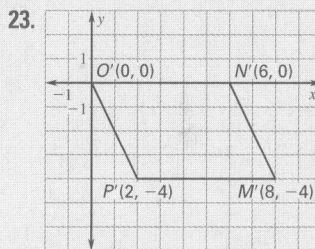

24.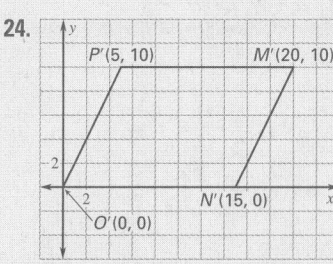

25.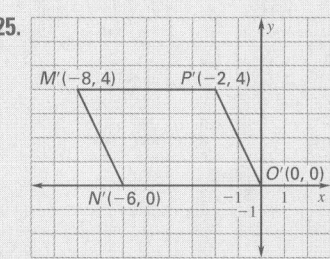

923

26.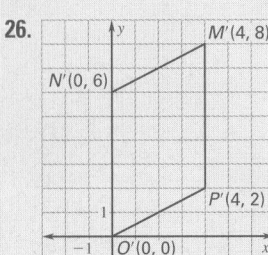

27.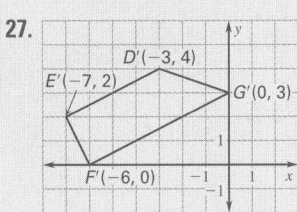

28.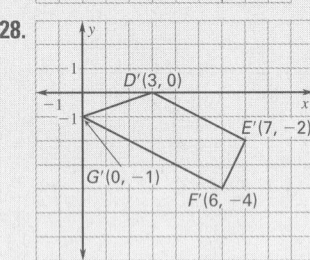

29.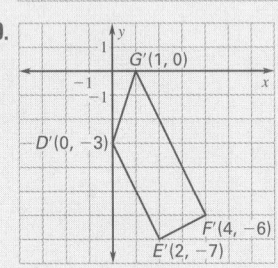

30.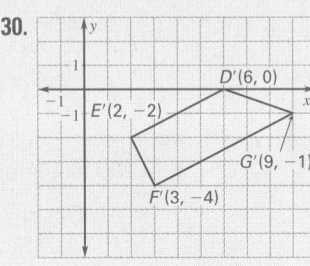

31.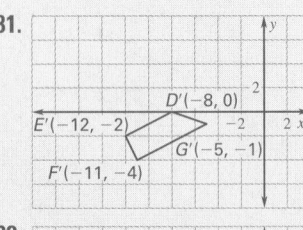

32.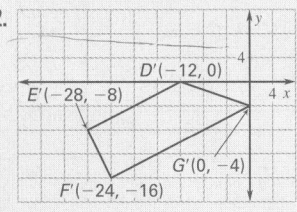

▶ **Logical Reasoning**
LOGICAL ARGUMENT

A logical argument has two given statements, called **premises**, and a statement, called a **conclusion**, that follows them.

If a figure is a rhombus, then it is a parallelogram.	**Premise**
JKLM is a rhombus.	**Premise**
Therefore, JKLM is a parallelogram.	**Conclusion**

There are five types of logical arguments that can be made using these statements. Arguments that use these patterns correctly will have a **valid conclusion**. Arguments that use these patterns incorrectly will have an **invalid conclusion**. The letters p and q are often used to write an argument symbolically. In the examples below, p and q are given the following meanings.

p: a figure is a rhombus
q: a figure is a parallelogram

Direct Argument
If p is true, then q is true.
p is true.
Therefore, q is true.

Example:
If *JKLM* is a rhombus, then it is a parallelogram.
JKLM is a rhombus.
Therefore, *JKLM* is a parallelogram.

Indirect Argument
If p is true, then q is true.
q is not true.
Therefore, p is not true.

Example:
If *JKLM* is a rhombus, then it is a parallelogram.
JKLM is not a parallelogram.
Therefore, *JKLM* is not a rhombus.

Chain Rule
If p is true, then q is true.
If q is true, then r is true.
Therefore, if p, then r.

Example:
If *JKLM* is a rhombus, then it is a parallelogram.
If *JKLM* is a parallelogram, then it is a quadrilateral.
Therefore, if *JKLM* is a rhombus, then it is a quadrilateral.

***Or* Rule**
p is true or q is true.
p is not true.
Therefore, q is true.

Example:
JKLM is a rhombus or a parallelogram.
JKLM is not a rhombus.
Therefore, *JKLM* is a parallelogram.

***And* Rule**
p and q are not both true.
But q is true.
Therefore, p is not true.

Example:
JKLM is not both a rhombus and a parallelogram.
JKLM is a parallelogram.
Therefore, *JKLM* is not a rhombus.

EXAMPLE   Use logical reasoning to decide whether the conclusion is *valid* or *invalid*. State the type of logical argument used to arrive at the conclusion.

    a. If $x = 2$, then $3x - 1 = 5$. $3x - 1 \neq 5$. Therefore, $x \neq 2$.

    b. If *ABCD* is a square, then it is a rectangle. *ABCD* is a rectangle.
       Therefore, *ABCD* is a square.

SOLUTION   **a.** The conclusion is valid. This is an example of indirect argument.

       b. The conclusion is invalid. This does not follow the pattern for a direct argument.

A compound statement has two or more parts joined by *or* or *and*. For an *and* statement to be true, each part must be true. For an *or* statement to be true, at least one part must be true.

EXAMPLE   Tell whether the compound statement is *true* or *false*.
     a. $2 < 3$ and $1 < 2$
     b. $8 > 7$ and $7 > 9$
     c. $-2 < 1$ or $-1 < 1$
     d. $2 > 3$ or $1 > 2$

SOLUTION   **a.** True; both parts are true.
     b. False; only one part is true, but both must be true for *and*.
     c. True; at least one part is true, as required for *or*.
     d. False; no part is true.

PRACTICE

Use logical reasoning to decide whether the conclusion is *valid* or *invalid*. State the type of logical argument used to arrive at the conclusion.   1–10. See margin.

1. If Harold can drive, he has a license.
If Harold has a license, he is at least 16.
Therefore, if Harold is at least 16, he can drive.

2. If triangle ABC is equilateral, it is also isosceles.
Triangle ABC is equilateral.
Therefore, triangle ABC is isosceles.

3. John is in his room or John is in the kitchen.
John is not in the kitchen.
Therefore, John is in his room.

4. If Grace is 19, her twin brother Michael is also 19.
Michael is 17.
Therefore, Grace is not 19.

5. If $x = 4$, then $y = 1$.
If $y = 1$, then $z = 8$.
Therefore, if $x = 4$, then $z = 8$.

6. If an animal has a backbone, it is a vertebrate.
A horse has a backbone.
Therefore, a horse is a vertebrate.

7. If $x = 2$, then $2x = 4$.
$2x = 6$.
Therefore, $x = 2$.

8. If an apple is a Granny Smith, it is green.
The apple is green.
Therefore, the apple is a Granny Smith.

9. It is impossible for my watch to be right and that we are late. We are late.
Therefore, my watch cannot be right.

10. Triangle ABC is equilateral or scalene.
Triangle ABC is not scalene.
Therefore, triangle ABC is not equilateral.

State whether each compound statement is *true* or *false*.

11. $1 < 2$ and $8 \geq 5$ True

12. $5 < 1$ or $3 < 4$ True

13. $-6 \geq -6$ or $3 < -3$ True

14. $2 \leq 5$ and $2 \leq 1$ False

15. $-8 < 5$ and $-5 < 8$ True

16. $3 < 1$ or $3 < -1$ False

17. $4 = 4$ or $4 = 5$ or $4 = -4$ True

18. $-1 < 1$ and $1 \geq 0$ and $-1 < 0$ True

19. $8 < 9$ and $9 < 14$ and $14 < 20$ True

20. $3 > 4$ or $3 > 7$ or $3 > 6$ False

21. $-10 < -8$ and $-7 < -4$ and $7 > 4$ True

22. $-15 < -35$ or $0 \geq 1$ or $26 \geq 26$ True

23. $159 \leq 100$ or $100 < 159$ True

24. $47 \leq 48$ and $48 < 49$ and $49 > 47$ True

25. $95 \neq 95$ or $95 > -96$ or $95 > 94$ True

26. $5 \cdot 6 = 30$ or $-6 \cdot 5 = 30$ True

1. The conclusion is invalid. This does not follow the chain rule.
2. The conclusion is valid. This is an example of a direct argument.
3. The conclusion is valid. This is not an example of the OR rule.
4. The conclusion is valid. This is an example of an indirect argument.
5. The conclusion is valid. This is an example of the chain rule.
6. The conclusion is valid. This is an example of a direct argument.
7. The conclusion is invalid. This is not an example of an indirect argument.
8. The conclusion is invalid. This is not an example of a direct argument.
9. The conclusion is valid. This is an example of the AND rule.
10. The conclusion is invalid. This does not follow the OR rule.

9. If a rectangle has four equal
   sides, then it is a square.
10. If he has sales over $10,000 in a
    month, then he earns a bonus.
11. If a curve is described by $y = x^2$,
    then it is a parabola.
12. If a figure is a circle, then the
    circumference is π times the
    diameter.

IF-THEN STATEMENTS

The **conditional** statement "if p, then q" has a **hypothesis** p and a **conclusion** q.

EXAMPLE   Identify the hypothesis and conclusion.

        a. $y = 6$ when $x = 5$.   **b.** Raspberries are a red fruit.

SOLUTION   Rewrite the statement as an if-then statement.

        a. If $x = 5$, then $y = 6$.   **b.** If a fruit is a raspberry, then the fruit is red.
           Hypothesis: $x = 5$ Hypothesis: a fruit is a raspberry
           Conclusion: $y = 6$ Conclusion: the fruit is red

The **converse** of the **conditional** statement "if p, then q" is "if q, then p."

EXAMPLE   Give the converse of each statement. State whether the converse is *true* or *false*.

        a. If $x = 8$, then $2x = 16$.   **b.** If a fruit is a raspberry, then the fruit is red.

SOLUTION   Reverse the hypothesis and conclusion of each statement.

        a. If $2x = 16$, then $x = 8$. True   **b.** If a fruit is red, then the fruit is a raspberry. False

---

When a conditional statement and its converse are combined by "if and only if," the resulting statement is called a **biconditional statement**. The biconditional "*p if and only if q*" is true only when the conditional "if p, then q" and its converse "if q, then p" are *both* true.

EXAMPLE   Tell whether the statement is *true* or *false*. If false, tell why.

        a. A parallelogram is a rectangle if and only if it has four right angles.

        b. A triangle is an equilateral triangle if and only if it has two equal sides.

SOLUTION   **a.** True

        b. False. It is true that an equilateral triangle has two equal sides
           (it has three), but not that a triangle with two equal sides must be
           equilateral (for example, a 5-5-2 triangle).

PRACTICE

Rewrite the statement as an if-then statement.   9–12. See margin.

1. The rain in Spain falls on the plain.
   If it rains in Spain, then it falls on the plain.
2. A rhombus is a parallelogram with four equal sides.
   If a parallelogram has four equal sides, then it is a rhombus.
3. $3x^2 = 48$ when $x = 4$.
   If $x = 4$, then $3x^2 = 48$.
4. The area of a square is given by the formula $A = s^2$.
   If a shape is a square, then the area is given by the formula $A = s^2$.
5. You can go out tonight if you finish cleaning.
   If you finish cleaning, then you can go out tonight.
6. Luis will earn $50 for baby-sitting 12 hours.
   If Luis baby-sits for 12 hours, then he will earn $50.
7. $y = 16$ when $x = 3$.
   If $x = 3$, then $y = 16$.
8. Corresponding angles of similar figures are congruent.
   If figures are similar, then corresponding angles are congruent.
9. A square is a rectangle with four equal sides.
10. He earns a bonus for sales over $10,000 each month.
11. The graph of $y = x^2$ is a parabola.
12. The circumference of a circle is π times the diameter.

Give the converse of each statement. State whether the converse is *true* or *false*.

13. If $x = 4$, then $x^2 = 16$.   If $x^2 = 16$, then $x = 4$. False.

14. If you live in Ohio, then you live in the United States. If you live in the United States, then you live in Ohio. False.

15. If a line is vertical, then its slope is undefined.   If a line's slope is undefined, then it is a vertical line. True.

16. If an animal is a pigeon, then it is a bird.   If an animal is a bird, then it is a pigeon. False.

17. If a figure has two pairs of opposite congruent sides, then it is a parallelogram.
   If a figure is a parallelogram, then it has two pairs of opposite congruent sides. True.
18. If you add two odd numbers, then the answer will be an even number.
   If the answer is an even number, then you added two odd numbers. False.
19. If you are in Minnesota in January, then you will be cold.
   If you are cold, then you are in Minnesota in January. False.
20. If an animal is a dog, then it has four legs.   If an animal has four legs, then it is a dog. False.

21. If Margot won the election, then she got more votes than her opponent.
   If Margot got more votes than her opponent, then she won the election. True.
22. If a triangle has three sides of different lengths, then it is a scalene triangle.
   If a triangle is scalene, then it has three sides of different lengths. True.
23. If a convex polygon has five equal sides, then it is a regular pentagon.
   If a convex polygon is a regular pentagon, then it has five equal sides. True.
24. If $x = 3$, then $x - 2 = 1$.   If $x - 2 = 1$, then $x = 3$. True.

Determine whether the statement is *true* or *false*. If false, tell why.

25. A figure is a square if and only if it has four equal sides.
   False. A square has four equal sides, and fair 90° angles.
26. Eric will win the election if and only if he receives 60% of the vote.
   False. Eric can win with the greatest percentage of the vote.
27. $x^2 = 25$ if and only if $x = -5$.   False. $x^2 = 25$ for $x = 5, -5$.

28. A quadrilateral is a trapezoid if and only if it has exactly one pair of parallel opposite sides.   True.

29. $2x + 6 = 6$ if and only if $x = 0$.   True.

30. An animal is an cat if and only if it is a mammal.   False. Cats are not the only animals that are mammals.

31. Corresponding angles of figures are congruent if and only if the figures are similar.   True.

32. A triangle is isosceles if and only if it has two equal angles.   True.

33. Your team wins at basketball if and only if your team scores more points than your opponents.   True.

34. $x^3 = 8$ if and only if $x = 2$.   True.

35. You are north of the equator if and only if you are in the northern hemisphere.   True.

36. A polygon is a decagon if and only if it has 10 sides.   True.

COUNTEREXAMPLES

A **counterexample** disproves a logical statement.

EXAMPLE   Is it true that when $|a| < |b|$, $a < b$?

SOLUTION   No, because when $a = 1$ and $b = -2$, $|a| < |b|$, but $a > b$.

PRACTICE

Determine whether each statement is *true* or *false*. If false, give a counterexample.

1. If a quadrilateral is a parallelogram, then it is a rectangle.   **False.** Any parallelogram with angles that are not right angles is not a rectangle.

2. If Joe has $5, then he earned it mowing lawns.   **False.** Joe may have earned $5 baby-sitting.

3. If the last digit of a number is 6, then it is divisible by 3.   **False.** The last digit of the number 16 is 6, but 16 is not divisible by 3.

4. If a triangle is equilateral, then it is equiangular.   **True.**

5. If a triangle contains a 90° angle, then it contains another 90° angle.   **False.** No triangle has two 90 degree angles because there must be a third angle and together the three must total 180 degrees.

6. If a parallelogram has four right angles, then it is a square.   **False.** Any rectangle that has adjacent sides of different lengths has four right angles, but is not a square.

7. If an animal is black, then it is a dog.   **False.** Cats can also be black.

8. If you live in California, then you live in the Pacific time zone.   **True.**

9. If two lines are perpendicular, then they form right angles.   **True.**

10. If two lines in the same plane are intersected by a transversal and alternate interior angles are equal in measure, then the lines are parallel.   **True.**

11. If $a > 0$, then $3a - 4 > 0$.
False. If $a = 1$, then $3a - 4 = -1 < 0$.

12. If $a < b$, then $2a < 2b$.   **True.**

13. If $c = d$, then $c - 2 = d - 2$.   **True.**

14. If $x > 0$, then $x^2 > x$.   **False.** If $x = 1$, then $x^2 = x$.

15. If $x \leq 0$, then $x^2 \geq x$.   **True.**

16. If $x \leq 0$, then $2x^2 \geq -4x$.
False. If $x = -1$, then $2x^2 = 2 < 4 = -4x$.

JUSTIFY REASONING

Algebraic reasoning can be justified using the postulates of algebra.

Postulates of Algebra	Statement of Postulate	Example
ADDITION/SUBTRACTION PROPERTY OF EQUALITY	If the same number is added to (or subtracted from) equal numbers, then the sums (differences) are equal.	$x - 2 = 4$ $x - 2 + 2 = 4 + 2$
MULTIPLICATION/DIVISION PROPERTY OF EQUALITY	If equal numbers are multiplied by (or divided by) the same number, then the products (quotients) are equal.	$3x = -9$ $\dfrac{3x}{3} = \dfrac{-9}{3}$
SUBSTITUTION PROPERTY	If values are equal, then one value may be substituted for the other.	$x = y - 1$ and $x = 2$ $2 = y - 1$
DISTRIBUTIVE PROPERTY	$a(b + c) = ab + ac$	$3(2x - 1) = 3(2x) + 3(-1)$

You may also use algebraic definitions, such as the definition of raising to a power, to justify algebraic reasoning.

EXAMPLE   Solve the equation $2x + 5 = 3$ and justify each step.

SOLUTION
$2x + 5 = 3$ Given
$2x = -2$ Subtraction property of equality
$x = -1$ Division property of equality

PRACTICE

Identify the property that justifies the statement.

1. If $2x = 8$, then $x = 4$.
Division property of equality

2. If $4x - 1 = 7$, then $4x = 8$.
Addition property of equality

3. If $\frac{4}{5}x = 8$, then $4x = 40$.
Multiplication property of equality

4. If $x^2 = 36$, then $x = 6$ or $x = -6$.
Definition of a square root

5. If $3x = 9$, then $3x + 2 = 11$.
Addition property of equality

6. If $3x^2 - 6 = 21$, then $x^2 - 2 = 7$.
Division property of equality

7. If $x = 4$, then $3x = 12$.
Multiplication property of equality

8. If $x(2x - 3) = 7$, then $2x^2 - 3x = 7$.
Distributive property

9. If $\sqrt{x} = 4$, then $x = 16$.
Definition of raising to a power (2)

10. If $-5x = 0$, then $x = 0$.
Division property of equality

11. If $4x + 7 = 9$, then $4x = 2$.
Subtraction property of equality

12. If $x^5 - 1 = 0$, then $x^5 = 1$.
Addition property of equality

13. If $\frac{1}{6}x = \frac{2}{3}$, then $x = 4$.
Multiplication property of equality

14. If $\sqrt{x} = \frac{3}{4}$, then $x = \frac{9}{16}$.
Definition of raising to a power (2)

15. If $6 = x - 1$, then $x = 7$.
Addition property of equality

16. If $4(2x + 2) = 12$, then $2x + 2 = 3$.
Division property of equality

17. If $\frac{2x}{9} = 3$, then $2x = 27$.
Multiplication property of equality

18. If $2 + x = 5$, then $1 + x = 4$.
Subtraction property of equality

19. If $2(3 - 5x) = 1$, then $6 - 10x = 1$.
Distributive property

20. If $6x = 1$, then $6x - 3 = -2$.
Subtraction property of equality

Solve each equation for x and justify each step of the solution. 21–28. See margin.

21. $9x = 27$

22. $8 + x = 8$

23. $\frac{x}{2} + 5 = 0$

24. $2(3 - x) = 1$

25. $\frac{5x}{2} - 2 = -5$

26. $3x - 8 = 13$

27. $\frac{3x}{4} = 6$

28. $6 = \frac{1}{2}(x - 2)$

▶ Problem Solving

TRANSLATING PHRASES INTO ALGEBRAIC EXPRESSIONS

To solve a problem algebraically, you often must translate a phrase into an algebraic expression.

EXAMPLE Write the given phrase as an algebraic expression.

a. 3 less than a number

b. twice a number

c. the quotient of x and y

SOLUTION

a. "Less than" indicates subtraction:

$$x - 3$$

b. "Twice" indicates multiplication by 2:

$$2x$$

c. "Quotient of" indicates division:

$$\frac{x}{y}$$

EXAMPLE Write an algebraic expression to answer the question.

a. You buy x pounds of apples at \$1.49 per pound. How much do you spend?

b. You have x dollars and spend \$17.38 on dinner. How much do you have left?

SOLUTION

a. The amount you spent is the product of the number of pounds of apples and the price per pound:

$$1.49x$$

b. The amount you have left is the difference between the amount you had before dinner and the cost of the dinner:

$$x - 17.38$$

Skills Review Handbook 929

---

21. $9x = 27$ Given
$x = 3$ Division property of equality

22. $8 + x = 8$ Given
$x = 0$ Subtraction property of equality

23. $\frac{x}{2} + 5 = 0$ Given
$\frac{x}{2} = -5$ Subtraction property of equality
$x = -10$ Multiplication property of equality

24. $2(3 - x) = 1$ Given
$6 - 2x = 1$ Distributive property
$-2x = -5$ Subtraction property of equality
$x = \frac{5}{2}$ Division property of equality

25. $\frac{5x}{2} - 2 = -5$ Given
$5x - 4 = -10$ Multiplication property of equality
$5x = -6$ Addition property of equality
$x = \frac{-6}{5}$ Division property of equality

26. $3x - 8 = 13$ Given
$3x = 21$ Addition property of equality
$x = 7$ Division property of equality

27. $\frac{3x}{4} = 6$ Given
$3x = 24$ Multiplication property of equality
$x = 8$ Division property of equality

28. $6 = \frac{1}{2}(x - 2)$ Given
$12 = x - 2$ Multiplication property of equality
$14 = x$ Addition property of equality

PRACTICE

Write the given phrase as an algebraic expression.

1. 8 more than a number $x + 8$
2. 3 times a number $3x$
3. a number minus 49 $x - 49$

4. a number divided by 100 $\frac{x}{100}$
5. a number multiplied by 7 $7x$
6. one sixth of a number $\frac{x}{6}$

7. $\frac{3}{4}$ of a number $\frac{3}{4}x$
8. $\frac{7}{8}$ more than a number $x + \frac{7}{8}$
9. 90% of a number $0.90x$

10. 5 times a number $5x$
11. 4 less than a number $x - 4$
12. the cube of a number x^3

13. 3 more than a number, all divided by 2 $\frac{x + 3}{2}$
14. the square root of the product of 8 and a number $\sqrt{8x}$

Write an expression to answer the question.

15. A triangle has base b and height 6. What is its area? $3b$

16. Violet pays $50 for a membership fee and $25 per dress rental. How much does she spend altogether for x rentals? $50 + 25x$

17. Serge has $67.39. He spends x dollars for a new book. How much does he have left? $67.39 - x$

18. How much is 4.5% sales tax on an item that costs x dollars? $0.045x$

19. You bicycle at x miles per hour for 2.5 hours. How far do you go? $2.5x$

20. You buy s sandwiches at $4.25 apiece and d drinks at $1 apiece. How much do you spend all together? $4.25s + d$

ADDITIONAL PROBLEM SOLVING STRATEGIES

When solving mathematical or real-life problems, you will find that different strategies are appropriate for different types of problems. Refer to Lesson 1.5 for examples of problem solving based on the strategies *use a verbal model*, *draw a diagram*, *look for a pattern*, and *guess, check, and revise*.

Make a List or Table

An organized list or table is helpful when enumerating possibilities.

EXAMPLE   Andrea, Brian, and Colleen are going to have their picture taken. In how many ways can the three friends line up for the picture?

SOLUTION   You may find a tree diagram useful in listing the possibilities.

PERSON ON LEFT	PERSON IN MIDDLE	PERSON ON RIGHT	POSSIBLE ARRANGEMENT
Andrea	Brian	Colleen	Andrea, Brian, Colleen
	Colleen	Brian	Andrea, Colleen, Brian
Brian	Andrea	Colleen	Brian, Andrea, Colleen
	Colleen	Andrea	Brian, Colleen, Andrea
Colleen	Andrea	Brian	Colleen, Andrea, Brian
	Brian	Andrea	Colleen, Brian, Andrea

▶ There are six ways that the three friends can line up for the picture.

Use a Formula

You may be given a formula or know one that applies to the situation.

EXAMPLE Carol biked 25 kilometers in 2 hours. What was her average speed?

SOLUTION The formula for speed is $s = \dfrac{d}{t}$ where s is speed, d is distance, and t is time.

$$s = \frac{d}{t} = \frac{25 \text{ km}}{2 \text{ h}} = 12.5$$

▶ Carol's average speed was 12.5 kilometers per hour.

Break into Simpler Parts

You may want to break a difficult problem into more easily managed parts or cases. Be sure the parts or cases are *mutually exclusive* (that is, they do not overlap) and *collectively exhaustive* (that is, they cover all the possibilities).

EXAMPLE Find the area of the pentagon shown at the right.

SOLUTION Break the figure into a square and two right triangles as shown at the bottom right. Using the Pythagorean theorem, you find that the length of the leg shared by the two triangles is 4 units.

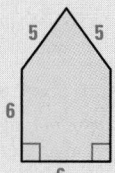

Area of pentagon = area of square + area of triangles

$$= 6^2 + 2\left[\frac{1}{2}(3)(4)\right]$$
$$= 36 + 12$$
$$= 48$$

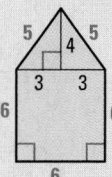

▶ The area of the pentagon is 48 square units.

Solve a Simpler Problem

You may try solving simpler problems and looking for a pattern in their solutions.

EXAMPLE When 15 diameters are drawn in a circle, into how many wedges is the circle divided?

SOLUTION Look for a pattern in the number of wedges when 1, 2, 3, and 4 diameters are drawn in a circle.

1 diameter **2 diameters** **3 diameters** **4 diameters**

2 wedges **4 wedges** **6 wedges** **8 wedges**

▶ The number of wedges is always twice the number of diameters. So, when 15 diameters are drawn in a circle, the circle is divided into 30 wedges.

PRACTICE

Use a list or table to solve the problem.

1. A frozen yogurt stand offers walnuts, peanuts, sprinkles, chocolate chips, and toffee bits as toppings. How many combinations of two different toppings are possible? **10**

2. Kyle packs 4 pairs of pants and 6 shirts for a trip. How many different outfits are possible? **24**

3. At a small theater, tickets for adults cost $12 and tickets for children cost $8. At one performance ticket sales were $480. How many people may have attended the performance? **From 40 to 60 people**

Use a formula to solve the problem.

4. What is the area of a trapezoid with base lengths 7 inches and 11 inches and height 3 inches? **27 square inches**

5. The formula $s = 32t$ gives the speed s (in feet per second) of an object falling without air resistance after t seconds have elapsed. How fast is a rock falling after 4 seconds? **128 feet per second**

6. You took two trips in your new car. The first trip covered 136 miles and used 6.4 gallons of gas. The second trip covered 285 miles and used 12.5 gallons of gas. On which trip did you get better gas mileage? **Second trip**

Solve the problem by breaking it into simpler parts or cases.

7. You throw three darts at the target shown. All three darts hit the target, and your score is the sum of the points that correspond to the regions where the darts land. What are the possibilities for your score? **3, 6, 9, 11, 12, 14, 17, 19, 22, 27**

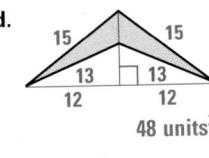

8. A painter is stenciling numbers, starting with 1, on the parking spaces in a parking lot. If there are 200 spaces, how many times does the painter stencil the digit 7? **40**

9. In a best-of-five series, the first team to win three games wins the series. How many ways are there for a team to win a best-of-five series? **10**

10. Find the area of the figure.

a.
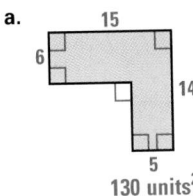
15
6
14
5
130 units²

b.
4
6
24 + 2π ≈ 30.28 units²

c.
18 units²
6
4
3
3
4
6

d.
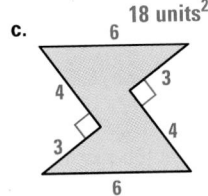
15   15
13   13
12   12
48 units²

Solve the problem by first solving simpler problems.

11. By moving only up and to the right, find the number of paths that lead from point A to point B on the grid shown. **35**

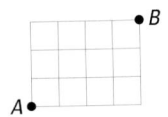

B

A

12. How many diagonals does a convex polygon with 12 sides have? **54**

13. How many squares of any integral size can you draw on an 8 × 8 grid? **204**

14. Without using a calculator, find the value of $(100,000,001)^2$. **10,000,000,200,000,001**

▶ Graphing

POINTS IN THE COORDINATE PLANE

A **coordinate plane** is divided into four regions by the x-axis and y-axis. Each region is called a **quadrant**. A point in a coordinate plane can be represented by an **ordered pair** of numbers. The **x-coordinate** gives the horizontal position of the point. The **y-coordinate** gives the vertical position of the point. You can tell which quadrant a point is in by looking at the signs of its coordinates.

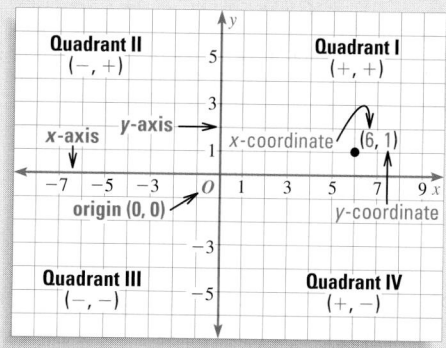

EXAMPLE   Graph the point $(4, -1)$. What quadrant is it in?

SOLUTION   To graph the point, start at the origin and move 4 units to the right and 1 unit down.

$(4, -1)$ is in Quadrant IV.

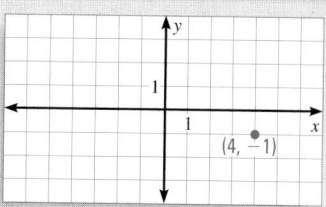

PRACTICE

Graph the point in a coordinate plane.   1–16. See margin.

1. $A(3, 4)$	**2.** $B(0, -4)$	**3.** $C(3, -1)$	**4.** $D(-4, 5)$
5. $E(-1, -1)$	**6.** $F(1, 1)$	**7.** $G(-6, -6)$	**8.** $H(0, 2)$
9. $J(1, 5)$	**10.** $K(-1, 0)$	**11.** $L(5, -2)$	**12.** $M(-2, -4)$
13. $N(0, 5)$	**14.** $P(-3, 2)$	**15.** $Q(3, 0)$	**16.** $R(-1, -3)$

Give the coordinates and quadrant of each of the following points.   17–37. See margin.

17. A	**18.** B	**19.** C
20. D	**21.** E	**22.** F
23. G	**24.** H	**25.** J
26. K	**27.** L	**28.** M
29. N	**30.** O	**31.** P
32. Q	**33.** R	**34.** S
35. T	**36.** U	**37.** V

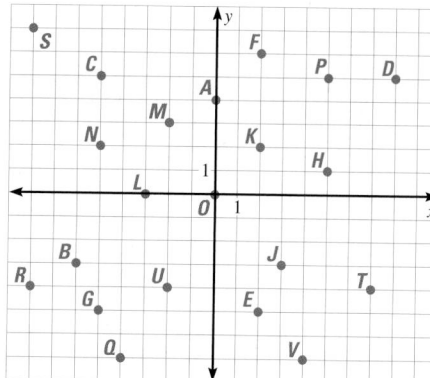

Skills Review Handbook   **933**

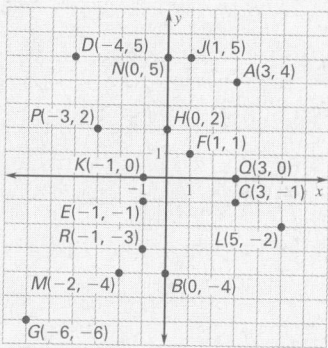

17. $(0, 4)$, y-axis
18. $(-6, -3)$, Quadrant III
19. $(-5, 5)$, Quadrant II
20. $(8, 5)$, Quadrant I
21. $(2, -5)$, Quadrant IV
22. $(2, 6)$, Quadrant I
23. $(-5, -5)$, Quadrant III
24. $(5, 1)$, Quadrant I
25. $(3, -3)$, Quadrant IV
26. $(2, 2)$, Quadrant I
27. $(-3, 0)$, x-axis
28. $(-2, 3)$, Quadrant II
29. $(-5, 2)$, Quadrant II
30. $(0, 0)$, origin
31. $(5, 5)$, Quadrant I
32. $(-4, -7)$, Quadrant III
33. $(-8, -4)$, Quadrant III
34. $(-8, 7)$, Quadrant II
35. $(7, -4)$, Quadrant IV
36. $(-2, -4)$, Quadrant III
37. $(4, -7)$, Quadrant IV

BAR, CIRCLE, AND LINE GRAPHS

A **bar graph** is used to represent data that fall into distinct categories.

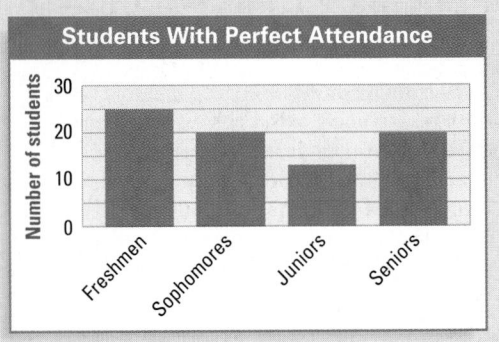

Students With Perfect Attendance

EXAMPLE According to the graph, how many juniors have perfect attendance?

SOLUTION The third bar represents juniors. Approximately 13 juniors have perfect attendance.

EXAMPLE Make a bar graph of the number of patients of a veterinary clinic. Represent each category with a bar.

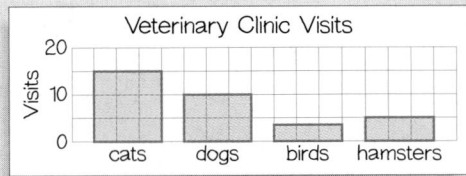

Patients	Cats	Dogs	Birds	Hamsters
Visits	15	10	3	5

SOLUTION

Veterinary Clinic Visits

A **circle graph** is used to show parts of a whole.

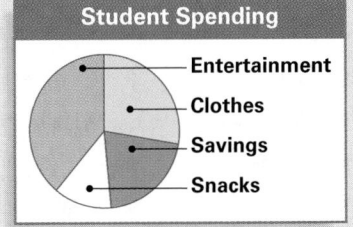

Student Spending

EXAMPLE According to the graph, for what category do students spend the most money? the least money?

SOLUTION Students spend the most money on entertainment and the least money on snacks.

EXAMPLE Make a circle graph of the favorite books that students read in English class this year.

Favorite Book	Students
Othello	27
The Odyssey	13
Jane Eyre	10

SOLUTION There are 50 students altogether. Find the percent who chose each book, and use this to find the measure of the central angle for each category. Then draw the circle graph.

OTHELLO $\frac{27}{50} = 54\%$ and $0.54 \cdot 360° \approx 194°$

THE ODYSSEY $\frac{13}{50} = 26\%$ and $0.26 \cdot 360° \approx 94°$

JANE EYRE $\frac{10}{50} = 20\%$ and $0.20 \cdot 360° = 72°$

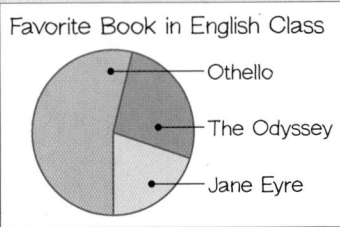

Favorite Book in English Class

A **line graph** is often used to show change over time.

EXAMPLE   Make a line graph of the number of scholarship awards given.

Year	1998	1999	2000	2001	2002
Awards	6	7	4	8	7

SOLUTION   Graph the data in the table. Connect the data points from year to year.

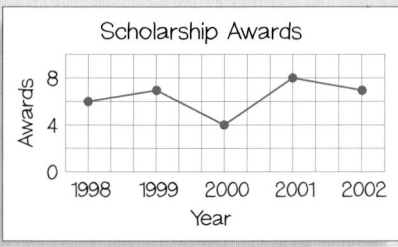

EXAMPLE   According to the graph, how many juniors took AP English in 1999?

SOLUTION   Find 1999 on the time axis, move up to the graph, and then move over to the student axis. Eleven juniors took AP English in 1999.

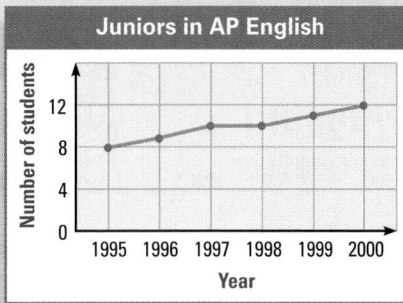

PRACTICE

Use the graphs to answer the questions.

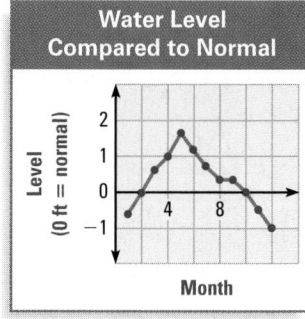

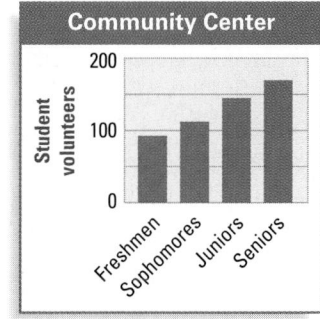

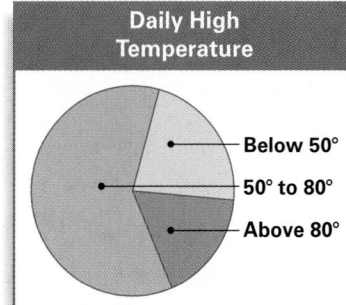

1. What was the water level in June?   **about 1.2 ft**

2. In what month was water level lowest?   **December**

3. How many sophomores volunteer at the community center?   **about 120**

4. How many more juniors than sophomores volunteer at the community center?   **about 70**

5. About what percent of the year is the average daily temperature above 80°?   **about 20%**

6. Does the temperature graph support the statement that during most of the year the average temperature is above 50°?   **Yes**

7. Students with Each Major

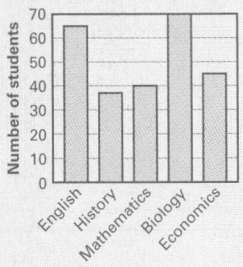

8. Students' Favorite Sports

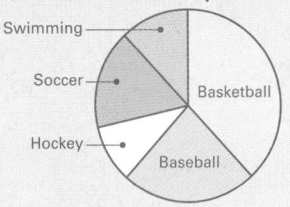

9. Visitors to the Zoo

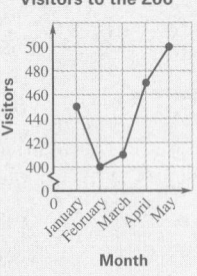

7. Make a bar graph to represent choices of college majors. **7–9. See margin.**

College majors	English	History	Mathematics	Biology	Economics
Number of students	65	37	40	70	45

8. Make a circle graph to represent the favorite sports of students in a class.

Favorite sports	Basketball	Baseball	Hockey	Soccer	Swimming
Number of students	16	10	4	7	5

9. Make a line graph to represent the number of visitors to a city zoo.

Month	January	February	March	April	May
Number of visitors	450	400	410	470	500

▶ Algebra
OPPOSITES

Two numbers that have the same absolute value but opposite signs are **opposites**.
Multiplication by -1 changes a number to its opposite.

EXAMPLE Find the opposite of the number.

  a. 13 **b.** -4

SOLUTION **a.** $13(-1) = -13$ **b.** $(-4)(-1) = 4$

To find the opposite of an expression, use the distributive property to multiply each
term in the expression by -1.

EXAMPLE Simplify the expression $-(3 - 2x)$.

SOLUTION $-(3 - 2x) = -1(3 - 2x) = (-1)(3) + (-1)(-2x)$
$$= -3 + 2x$$

PRACTICE

Find the opposite of the number.

1. 3 -3 **2.** -27 27 **3.** 150 -150 **4.** -13 13

5. 4.3 -4.3 **6.** -9.28 9.28 **7.** $\frac{3}{5}$ $-\frac{3}{5}$ **8.** $-\frac{1}{2}$ $\frac{1}{2}$

Simplify the expression.

9. $-(2a + b)$ $-2a - b$ **10.** $-(y - x)$ $x - y$ **11.** $-(a - b - c)$ $-a + b + c$ **12.** $-(2y - 3x)$ $3x - 2y$

13. $-(2 + x)$ $-2 - x$ **14.** $-(-3 - 11x)$ $3 + 11x$ **15.** $-(4x - 1)$ $1 - 4x$ **16.** $-(-x + 13)$ $x - 13$

Simplify the expression.

17. $-(x^2 + 2x - 4)$
$-x^2 - 2x + 4$

18. $-(x - (-2y))$
$-x - 2y$

19. $-x - (3y + x)$
$-2x - 3y$

20. $-(-(x - y + 3))$
$x - y + 3$

21. $-(3x - y + 11z)$
$-3x + y - 11z$

22. $x - (7y - 4x)$
$5x - 7y$

23. $-9(-4x + 6y)$
$36x - 54y$

24. $5x - (-(x - y))$
$6x - y$

25. $-(2x + 7y) + 3x$
$x - 7y$

26. $4x - (-x + y)$
$5x - y$

27. $a - (3b + 2a) + 6b$
$-a + 3b$

28. $-2(-x + 3y - 6)$
$2x - 6y + 12$

MULTIPLYING BINOMIALS

A **binomial** is an expression that has two terms. You may find using a geometric model helpful when multiplying two binomials.

EXAMPLE Simplify $(x + 1)(x + 2)$.

SOLUTION Draw a rectangle and divide the width and length into two parts: x and 1 for the width, and x and 2 for the length. Divide the rectangle into four regions and find the area of each region. The product of $x + 1$ and $x + 2$ is the sum of the areas of the four regions.

$$(x + 1)(x + 2) = x^2 + 2x + x + 2$$
$$= x^2 + 3x + 2$$

The **FOIL** method can also be used to multiply two binomials. **FOIL** stands for First, Outer, Inner, and Last, which is the order in which you multiply terms.

$$\underset{\text{F} \quad \text{O} \quad \text{I} \quad \text{L}}{(2x + 3)(x - 1)} = 2x(x) + 2x(-1) + 3(x) + 3(-1)$$
$$= 2x^2 - 2x + 3x - 3$$
$$= 2x^2 + x - 3$$

EXAMPLE Simplify $(3x + 2)(x - 1)$.

SOLUTION $(3x + 2)(x - 1) = 3x \cdot x + 3x \cdot (-1) + 2 \cdot x + 2 \cdot (-1)$ Multiply using FOIL.
$$= 3x^2 - 3x + 2x - 2$$ Simplify.
$$= 3x^2 - x - 2$$ Combine like terms.

PRACTICE

Simplify.

1. $(x + 1)(x + 1)$ $x^2 + 2x + 1$

2. $(2 - 4x)(1 + 2x)$ $2 - 8x^2$

3. $(4x + 1)(2 + x)$ $4x^2 + 9x + 2$

4. $(3x + 2)(x - 1)$ $3x^2 - x - 2$

5. $(1 - 2x)(x + 3)$ $-2x^2 - 5x + 3$

6. $(-2x + 1)(3x - 4)$ $-6x^2 + 11x - 4$

7. $(2x - 5)(2x + 5)$ $4x^2 - 25$

8. $(6x + 3)(1 - x)$ $-6x^2 + 3x + 3$

9. $(5x + 3)(x - 2)$ $5x^2 - 7x - 6$

10. $(a + 1)(b + 1)$ $ab + a + b + 1$

11. $(y + 3)(2y - 3)$ $2y^2 + 3y - 9$

12. $(-x + 2)(-3x - 2)$ $3x^2 - 4x - 4$

13. $(a + b)(c + d)$
$ac + ad + bc + bd$

14. $(x + 2)(x + 3)$ $x^2 + 5x + 6$

15. $(2x - 1)(-2x - 1)$ $-4x^2 + 1$

16. $(x + 2)(2x - y)$
$2x^2 - xy + 4x - 2y$

17. $(x - y)(x + y)$ $x^2 - y^2$

18. $(3y - a)(y + 3a)$ $3y^2 + 8ay - 3a^2$

19. $(3x + 5)(x - 2)$ $3x^2 - x - 10$

20. $(x + 0)(x - 1)$ $x^2 - x$

21. $(-4x + 12)(3x + 8)$ $-12x^2 + 4x + 96$

FACTORING

To factor a polynomial of the form $x^2 + bx + c$, you may find a geometric model helpful.

As when multiplying binomials, draw a rectangle and divide the width and length into two parts: x and m for the width, and x and n for the length. Divide the rectangle into four regions and find the area of each region. You can see that m and n must be factors of c and that the sum of m and n must be equal to b.

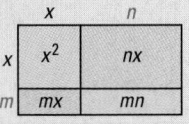

$$(x + m)(x + n) = x^2 + \underbrace{nx + mx}_{} + \underbrace{mn}_{}$$
$$= x^2 + \quad bx \quad + c$$

EXAMPLE Factor $x^2 + 5x + 4$.

SOLUTION

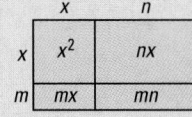

You want $mx + nx = 5x$ and $mn = 4$.

The integral factors of 4 are 1 and 4, -1 and -4, 2 and 2, and -2 and -2. Since $1x + 4x = 5x$, $m = 1$ and $n = 4$.

$$x^2 + 5x + 4 = (x + 1)(x + 4)$$

You can factor polynomials of the form $x^2 + bx + c$ without using a geometric model by listing the factors of c and finding the pair of factors that has a sum equal to b.

EXAMPLE Factor $x^2 - 2x - 15$.

SOLUTION You want $x^2 - 2x - 15 = (x + m)(x + n)$ where $mn = -15$ and $m + n = -2$.

Factors of -15	$1, -15$	$-1, 15$	$3, -5$	$-3, 5$
Sum of factors $(m + n)$	-14	14	-2	2

▶ The table shows that the values of m and n you want are $m = 3$ and $n = -5$. So, $x^2 - 2x - 15 = (x + 3)(x - 5)$.

PRACTICE

Factor.

1. $x^2 + 5x + 6$ $(x + 3)(x + 2)$
2. $x^2 + 7x + 10$ $(x + 5)(x + 2)$
3. $x^2 + 9x + 20$ $(x + 4)(x + 5)$
4. $x^2 - 7x + 10$ $(x - 5)(x - 2)$
5. $x^2 + 6x + 9$ $(x + 3)(x + 3)$
6. $x^2 - 10x + 21$ $(x - 7)(x - 3)$
7. $x^2 - 5x - 24$ $(x + 3)(x - 8)$
8. $x^2 + 3x - 28$ $(x + 7)(x - 4)$
9. $x^2 + 3x + 2$ $(x + 2)(x + 1)$
10. $x^2 - 4x - 12$ $(x - 6)(x + 2)$
11. $x^2 + x - 6$ $(x + 3)(x - 2)$
12. $x^2 + 6x - 16$ $(x + 8)(x - 2)$
13. $x^2 + 14x + 49$ $(x + 7)(x + 7)$
14. $x^2 + 5x - 6$ $(x + 6)(x - 1)$
15. $x^2 - 8x - 20$ $(x - 10)(x + 2)$
16. $x^2 + 9x - 36$ $(x + 12)(x - 3)$
17. $x^2 - 18x + 81$ $(x - 9)(x - 9)$
18. $x^2 + 5x + 4$ $(x + 4)(x + 1)$
19. $x^2 - 8x + 15$ $(x - 5)(x - 3)$
20. $x^2 + 10x + 9$ $(x + 9)(x + 1)$
21. $x^2 - 21x + 80$ $(x - 16)(x - 5)$
22. $x^2 - 4x - 5$ $(x - 5)(x + 1)$
23. $x^2 - 3x - 4$ $(x - 4)(x + 1)$
24. $x^2 + 8x + 12$ $(x + 6)(x + 2)$
25. $x^2 - 9x + 20$ $(x - 5)(x - 4)$
26. $x^2 + 8x + 16$ $(x + 4)(x + 4)$
27. $x^2 + 10x + 25$ $(x + 5)(x + 5)$
28. $x^2 + x - 30$ $(x + 6)(x - 5)$
29. $x^2 + 6x + 8$ $(x + 4)(x + 2)$
30. $x^2 + 4x - 21$ $(x + 7)(x - 3)$

LEAST COMMON DENOMINATOR

To add or subtract rational expressions with unlike denominators, first find the **least common denominator (LCD)** of the original rational expressions. To find the least common denominator of two rational expressions, follow these steps.

❶ Factor each denominator. If a constant factor is negative, multiply the numerator and denominator of the expression by -1.

❷ Find the *least common multiple (LCM)* of the constant factors in the factored denominators. (For help with LCM, see page 908.)

❸ For each different variable factor that appears in any denominator, write the factor as many times as it appears in the denominator having the greatest number of that factor.

❹ Write the LCD as the product of the results of Steps 2 and 3.

EXAMPLE Find the LCD of the pair of rational expressions.

a. $\dfrac{5}{6x^2}, \dfrac{x}{4x^2 - 12x}$

b. $\dfrac{x + 1}{x^2 + 4x + 4}, \dfrac{2}{x^2 - 4}$

SOLUTION

a. **❶** $6x^2 = 6 \cdot x \cdot x$

$4x^2 - 12x = 4 \cdot x \cdot (x - 3)$

❷ LCM of 6 and 4 is 12.

❸ LCM of variable factors is $x \cdot x \cdot (x - 3)$.

❹ LCD is $12x^2(x - 3)$.

▶ LCD of $\dfrac{5}{6x^2}$ and $\dfrac{x}{4x^2 - 12x}$ is $12x^2(x - 3)$.

b. **❶** $x^2 + 4x + 4 = (x + 2) \cdot (x + 2)$

$x^2 - 4 = (x + 2) \cdot (x - 2)$

❷ LCM of 1 and 1 is 1.

❸ LCM of variable factors is $(x + 2) \cdot (x + 2) \cdot (x - 2)$.

❹ LCD is $(x + 2)^2(x - 2)$.

▶ LCD of $\dfrac{x + 1}{x^2 + 4x + 4}$ and $\dfrac{2}{x^2 - 4}$ is $(x + 2)^2(x - 2)$.

PRACTICE

Find the least common denominator of the pair of rational expressions.

1. $\dfrac{1}{2x}, \dfrac{1}{2}$ $2x$

2. $\dfrac{1}{3y}, \dfrac{3}{-4y}$ $-12y$

3. $\dfrac{2}{45k}, \dfrac{-1}{30k^2}$ $90k^2$

4. $\dfrac{9}{z(z + 1)}, \dfrac{15}{z^3}$ $z^3(z + 1)$

5. $\dfrac{4}{6x}, \dfrac{6}{2y}$ $6xy$

6. $\dfrac{5}{12a}, \dfrac{a}{9}$ $36a$

7. $\dfrac{10}{3z}, \dfrac{1}{z^2}$ $3z^2$

8. $\dfrac{-4}{9k}, \dfrac{1}{3k^2}$ $9k^2$

9. $\dfrac{b}{b - 1}, \dfrac{3}{(b + 1)^2}$
$(b - 1)(b + 1)^2$

10. $\dfrac{7}{12d}, \dfrac{d + 4}{-12d^2}$ $12d^2$

11. $\dfrac{-3n + 4}{2n + 4}, \dfrac{5}{n + 2}$
$2(n + 2)$

12. $\dfrac{w}{w + 9}, \dfrac{w - 1}{w^2 + 18w + 81}$
$(w + 9)^2$

13. $\dfrac{x - 11}{6 + 9x}, \dfrac{-x + 11}{18x + 12}$
$6(2 + 3x)$

14. $\dfrac{5g}{3g^3 - 21}, \dfrac{g^3 + 3}{2g^3 - 14}$
$6(g^3 - 7)$

15. $\dfrac{h}{20 - 15h}, \dfrac{9h}{6h - 8}$
$10(3h - 4)$

16. $\dfrac{7 - q^2}{q - 8q^3}, \dfrac{5q^2}{24q^3 - 3q}$
$3q(8q^2 - 1)$

17. $\dfrac{7e}{6 - 10e}, \dfrac{e + 9}{12 - 20e}$
$12 - 20e$

18. $\dfrac{4x}{5x^3 - 20x}, \dfrac{9 - x}{4x - x^3}$
$5x^3 - 20x$

19. $\dfrac{12c + 1}{c^3 - 4c^2}, \dfrac{c^3 - 3}{c^2 - c^3}$
$c^2(c - 1)(c - 4)$

20. $\dfrac{5}{6 - 2p^5}, \dfrac{7}{-2p^5 + 6}$
$6 - 2p^5$

21. $\dfrac{7}{36x^2}, \dfrac{11}{-9x}$ $36x^2$

22. $\dfrac{16}{x - 5}, \dfrac{6}{7x - 35}$
$7(x - 5)$

23. $\dfrac{x + 4}{3x^2 + 2x}, \dfrac{6}{3x + 2}$
$3x^2 + 2x$

24. $\dfrac{6x + 9}{2x^3 + 5x^2}, \dfrac{15}{-4x^2}$
$4x^2(2x + 5)$

Extra Practice

CHAPTER 1

Graph the numbers on a number line. Then write the numbers in increasing order. (Lesson 1.1) 1–6. See margin.

1. $-3, \sqrt{8}, \frac{4}{7}, -2.8, \frac{9}{5}$

2. $4, \frac{2}{3}, -\sqrt{6}, -\frac{2}{3}, 0$

3. $0.4, \frac{3}{5}, -1.6, 0, \sqrt{3}$

4. $-5, 0, -\frac{17}{3}, -\sqrt{5}, 1$

5. $3.4, 0.3, -\frac{2}{5}, \sqrt{5}, \frac{5}{2}$

6. $\sqrt{2}, 1.4, 1.5, 0.5, 1$

Identify the property shown. (Lesson 1.1)

7. $(3 + 7)5 = 5(3 + 7)$
commutative property of multiplication

8. $(19 \cdot 4) \cdot 4 = 19 \cdot (4 \cdot 4)$
associative property of multiplication

9. $-6 + 6 = 0$
inverse property of addition

10. $2(13 + 11) = 2 \cdot 13 + 2 \cdot 11$
distributive property

11. $150 + 11 = 11 + 150$
commutative property of addition

12. $(4 + 8) + 9 = 4 + (8 + 9)$
associative property of addition

Evaluate the expression. (Lesson 1.2)

13. $4 + 7 - 8 \div 4$ 9

14. $3 \cdot 9 - (15 - 7)$ 19

15. $8 - (4 + 3)^2 + 5$ -36

16. $2x - 8$ when $x = 6$ 4

17. $x^3 - 5x$ when $x = 3$ 12

18. $3 + 12x - x^2$ when $x = -2$ -25

Simplify the expression. (Lesson 1.2)

19. $5x^2 - 3x + 7x^2 - 10x$
$12x^2 - 13x$

20. $7x - y + 9x - 2y$ $16x - 3y$

21. $-3x^2 + 2x - 6x^2$ $-9x^2 + 2x$

22. $4(x - 5) - 3(2x + 7)$ $-2x - 41$

23. $2(x - 1) + 3(x + 2)$ $5x + 4$

24. $6(x - y) + 3(y + 2x)$ $12x - 3y$

Solve the equation. Check your solution. (Lesson 1.3)

25. $3n - 4 = 17$ 7

26. $m + 14 = 8 - 2m$ -2

27. $5x + 17 = 2x - 10$ -9

28. $-5(2x - 1) = 3(x + 4)$ $-\frac{7}{13}$

29. $4.7a + 6.2 = -4.61$ -2.3

30. $\frac{1}{3}(x - 6) = -\frac{2}{5}x + \frac{14}{15}$ 4

Solve the equation for y. (Lesson 1.4)

31. $3x + 4y = 12$ $y = \frac{-3x + 12}{4}$

32. $3y - 5x = -13$ $y = \frac{5x - 13}{3}$

33. $-6y + 7x = -9$ $y = \frac{7x + 9}{6}$

34. $3xy + x = 15$ $y = \frac{-x + 15}{3x}$

35. $\frac{4}{5}x - 10y = -3$ $y = \frac{2}{25}x + \frac{3}{10}$

36. $\frac{1}{3}x - \frac{2}{5}y = -10$ $y = \frac{5}{6}x + 25$

POSTAGE In Exercises 37–39, the cost of sending an overnight package from Speedy Air is $15.00 for the first pound and $3.00 for each additional pound. How much will it cost to send a 7 pound package? Use the following verbal model. (Lesson 1.5)

| Total cost | = | Cost of first pound | + | Cost per pound of additional pounds | · | Number of additional pounds |

37. Assign labels to the parts of the verbal model.

37. C = Total cost (dollars), 15 = Cost of first pound (dollars), 3 = Cost per pound of each additional pound (dollars per pound), 6 = Number of additional pounds

38. Use the labels to translate the verbal model into an algebraic model. $C = 15 + 3 \cdot 6$

39. Solve the algebraic model. Answer the question. $C = 33$; it will cost $33 to send a 7 pound package.

40. SCHOOL BAND The school band is planning a carnival to raise money. They plan to sell 500 tickets. Adult tickets will be $4.50 and student tickets will be $2.50. They need to collect $1650 in ticket sales to meet their goal. How many adult and student tickets do they need to sell? (Lesson 1.5)   The band must sell 200 adult tickets and 300 student tickets.

Margin answers

1.
$-3, -2.8, \frac{4}{7}, \frac{9}{5}, \sqrt{8}$

2.
$-\sqrt{6}, -\frac{2}{3}, 0, \frac{2}{3}, 4$

3.
$-1.6, 0, 0.4, \frac{3}{5}, \sqrt{3}$

4.
$-\frac{17}{3}, -5, -\sqrt{5}, 0, 1$

5.
$-\frac{2}{5}, 0.3, \sqrt{5}, \frac{5}{2}, 3.4$

6.
$0.5, 1, 1.4, \sqrt{2}, 1.5$

41.

42.

43.

44.

45.

46.

47.

48.

49.

59.

60.

61.

62.

63.

64.

65.

66.

67.

Solve the inequality. Then graph your solution. (Lesson 1.6) 41–49. See margin for graphs.

41. $3x + 7 > 28$ $x > 7$ **42.** $-m - 3 < 3m + 5$ $m > -2$ **43.** $2.3x - 5.9 > -1.3$ $x > 2$

44. $-7(n + 3) \geq 0$ $n \leq -3$ **45.** $4 \leq x + 2 \leq 12$ $2 \leq x \leq 10$ **46.** $-6 \leq 3x + 2 \leq 11$ $-\frac{8}{3} \leq x \leq 3$

47. $6x + 4 < 22$ or $5x - 8 \geq 32$ **48.** $5n + 16 \leq 31$ or $8 + 4n > 48$ **49.** $3x - 7 \leq 16$ or $2x - 1 > 23$
     $x < 3$ or $x \geq 8$ $n \leq 3$ or $n > 10$ $x \leq \frac{23}{3}$ or $x > 12$

Solve the equation. (Lesson 1.7)

50. $|x + 3| = 6$ $3, -9$ **51.** $|2x - 6| = 50$ $28, -22$ **52.** $|x - 7| = 3$ $10, 4$

53. $|10x - 73| = 29$ $10.2, 4.4$ **54.** $|9 - 3x| = 15$ $8, -2$ **55.** $|20 - 7x| = 42$ $\frac{62}{7}, -\frac{22}{7}$

56. $\left|\frac{1}{4}x + 5\right| = 21$ $64, -104$ **57.** $\left|\frac{1}{2}x - 1\right| = 0$ 2 **58.** $\left|10 + \frac{1}{3}x\right| = 16$ $18, -78$

Solve the inequality. Then graph the solution. (Lesson 1.7) 59–67. See margin for graphs.

59. $|x + 3| > 4$ $x < -7$ or $x > 1$ **60.** $|4 - 8x| \geq 100$ $x \leq -12$ or $x \geq 13$ **61.** $|7x + 7| < 14$ $-3 < x < 1$

62. $|y + 8| \leq 15$ $-23 \leq y \leq 7$ **63.** $|2y - 5| < 1$ $2 < y < 3$ **64.** $|2a - 6| > 0$ $a < 3$ or $a > 3$

65. $|3x + 1| \geq 16$ **66.** $|4a + 7| \leq 13$ **67.** $|-2y + 3| > 5$
    $x \leq -\frac{17}{3}$ or $x \geq 5$ $-5 \leq a \leq \frac{3}{2}$ $y < -1$ or $y > 4$

CHAPTER 2

Use a mapping diagram to represent the relation. Then tell whether the relation is a function. (Lesson 2.1) 1–3. See margin for graphs.

1. 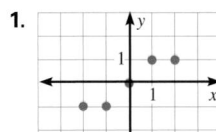 yes   **2.** 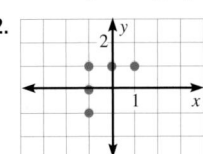 no   **3.** 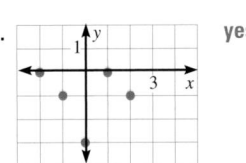 yes

Evaluate the function when $x = -2$. (Lesson 2.1)

4. $f(x) = x + 17$ 15 **5.** $f(x) = -x + 3$ 5 **6.** $f(x) = -5 + 8x$ -21

7. $f(x) = -x - 48$ -46 **8.** $f(x) = |x + 3| - 9$ -8 **9.** $f(x) = 2x^3 - 7x^2 + 8$ -36

Tell whether the lines are *parallel*, *perpendicular*, or *neither*. (Lesson 2.2)

10. Line 1: through $(3, 4)$ and $(1, 6)$ neither   **11.** Line 1: through $(1, 5)$ and $(-4, -5)$ parallel
Line 2: through $(-1, 0)$ and $(3, 5)$ Line 2: through $(-1, -9)$ and $(2, -3)$

12. Line 1: through $(-6, 7)$ and $(-3, 6)$ perpendicular **13.** Line 1: through $(0, 0)$ and $(5, 2)$ perpendicular
Line 2: through $(-1, -9)$ and $(1, -3)$ Line 2: through $(0, -4)$ and $(-2, 1)$

14. Line 1: through $(1, 8)$ and $(-3, -4)$ neither   **15.** Line 1: through $(0, -2)$ and $(2, -2.5)$ parallel
Line 2: through $(-2, -5)$ and $(3, 5)$ Line 2: through $(-4, 6)$ and $(0, 5)$

Draw the line with the given slope and y-intercept. (Lesson 2.3) 16–21. See margin.

16. $m = 2, b = -4$ **17.** $m = 0, b = 4$ **18.** $m = -3, b = -2$

19. $m = -1, b = 0$ **20.** $m = \frac{1}{2}, b = 2$ **21.** $m = -\frac{4}{5}, b = -1$

Find the slope and y-intercept of the line. (Lesson 2.3)

22. $y = 2x$ $2; 0$ **23.** $x = -1$ undefined; none   **24.** $y = 5$ $0; 5$ **25.** $y = 2x - 5$ $2; -5$

26. $y = 3x + 7$ $3; 7$ **27.** $-2x + y = 10$ $2; 10$ **28.** $5x - y = 12$ $5; -12$ **29.** $x - 3y = -8$ $\frac{1}{3}; \frac{8}{3}$

1.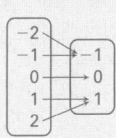

2.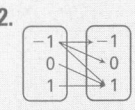

3.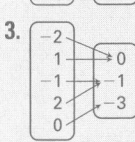

16.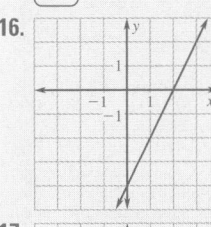

17.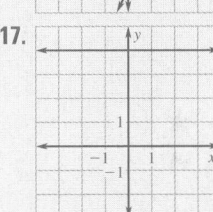

18.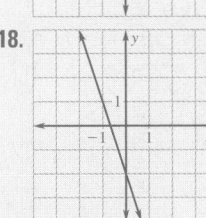

19.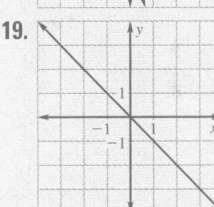

20.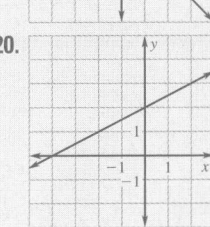

21.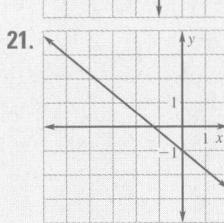

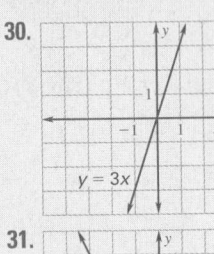
30.

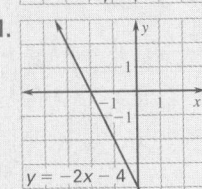
31.

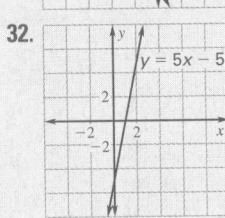
32.

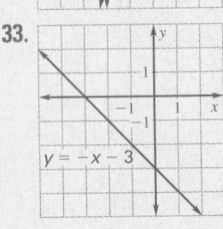
33.

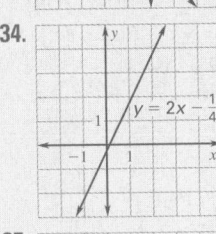
34.

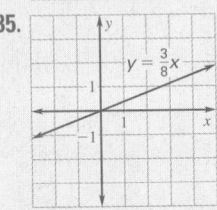
35.

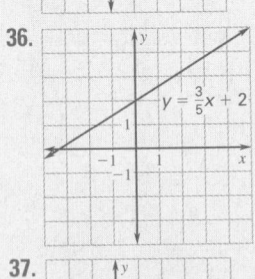
36.

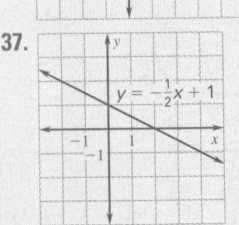
37.

Graph the equation. (Lesson 2.3)   30–37. See margin.

30. $y = 3x$

31. $y = -2x - 4$

32. $y = 5x - 5$

33. $y = -x - 3$

34. $y = 2x - \frac{1}{4}$

35. $y = \frac{3}{8}x$

36. $y = \frac{3}{5}x + 2$

37. $y = -\frac{1}{2}x + 1$

Write an equation of a line that has the given properties. (Lesson 2.4)

38. slope: 2, y-intercept: -4 $y = 2x - 4$

39. slope: 0, y-intercept: 2 $y = 2$

40. slope: $\frac{4}{5}$, y-intercept: 5 $y = \frac{4}{5}x + 5$

41. slope: 2, passes through $(1, -3)$ $y = 2x - 5$

42. slope: $\frac{1}{2}$, passes through $(-1, -1)$ $y = \frac{1}{2}x - \frac{1}{2}$

43. slope: $-\frac{5}{2}$, passes through $(3, -4)$ $y = -\frac{5}{2}x + \frac{7}{2}$

44. passes through $(1, -6)$ and $(4, -3)$ $y = x - 7$

45. passes through $(-3, 3)$ and $(2, -7)$ $y = -2x - 3$

46. passes through $(-5, 3)$ and $(5, -3)$ $y = -\frac{3}{5}x$

47. passes through $(2, 6)$ and $(-7, 6)$ $y = 6$

48. FLYING TIME The table below gives the distance (in miles) and flying time (in hours) to Atlanta, Georgia, from various U.S. cities. Draw a scatter plot of the data and approximate an equation of the best-fitting line. Then predict the flying time for a city that is 900 miles from Atlanta. (Lesson 2.5)

$y = 0.0031x + 0.94$; about 3.73 hours; See margin for graph.

	Mobile	Little Rock	Chicago	Dallas	Austin	Colorado Springs	Denver	Los Angeles
Distance	302	459	585	717	817	1185	1204	1944
Time	2.23	2.5	2.82	3.03	3.2	4.83	4.92	7.33

Graph the inequality in a coordinate plane. (Lesson 2.6)   49–60. See margin.

49. $y \geq 2$

50. $x < 3.6$

51. $x > -3$

52. $3x > 12$

53. $-2y \leq 8$

54. $y > 2.5$

55. $y < x - 2$

56. $y \geq 3x + 4$

57. $y > 4x - 7$

58. $y \leq 2x + 3$

59. $6x + 12y \leq -24$

60. $\frac{1}{2}x + \frac{3}{4}y > 0$

Evaluate the function for the given value of x. $f(x) = \begin{cases} 3x + 2, & \text{if } x \leq 1 \\ x + 4, & \text{if } x > 1 \end{cases}$
(Lesson 2.7)

61. $f(-2)$ -4 **62.** $f(1)$ 5   **63.** $f(5)$ 9   **64.** $f(-1)$ -1 **65.** $f(0)$ 2

Graph the function. (Lesson 2.7)   66–67. See margin.

66. $f(x) = \begin{cases} \frac{1}{2}x - 5, & \text{if } x < -2 \\ 5x + 4, & \text{if } x \geq -2 \end{cases}$

67. $f(x) = \begin{cases} -1, & \text{if } x < 0 \\ 1, & \text{if } 0 \leq x < 3 \\ 3, & \text{if } x \geq 3 \end{cases}$

Graph the function. Then identify the vertex, tell whether the graph opens up or down, and tell whether the graph is *wider, narrower,* or the *same width* as the graph of $y = |x|$. (Lesson 2.8)   68–76. See margin for graphs and width changes.

68. $y = |x| - 4$ $(0, -4)$; up

69. $y = 2|x| + 5$ $(0, 5)$; up

70. $y = -|x| + 1$ $(0, 1)$; down

71. $y = |x + 3|$ $(-3, 0)$; up

72. $y = -2|x|$ $(0, 0)$; down

73. $y = 3|x| - 4$ $(0, -4)$; up

74. $y = |1 - x| + 3$ $(1, 3)$; up

75. $y = \frac{1}{2}|x| + 2$ $(0, 2)$; up

76. $y = \frac{1}{3}|x|$ $(0, 0)$; up

48–60, 66–76 See Additional Answers beginning on page AA1.

CHAPTER 3

Check whether the ordered pair is a solution of the system. (Lesson 3.1)

1. $(2, 1)$ no
$3x - 2y = 4$
$-2x + 2y = 3$

2. $(0, 5)$ yes
$5x + y = 5$
$9x - 4y = -20$

3. $(-3, -2)$ no
$-7x + 12y = -22$
$-4x + y = 10$

4. $(-1, -8)$ yes
$10x + 5y = -50$
$3x - 7y = 53$

Graph the linear system and tell how many solutions it has. If there is exactly one solution, estimate the solution and check it algebraically. (Lesson 3.1) 5–12. Estimates may vary; See margin.

5. $y = 3$ 1; (4, 3)
$x + y = 7$

6. $3x + y = 10$ 1; (3, 1)
$y = 2x - 5$

7. $y = 2x - 5$ infinitely many
$6x - 3y = 15$

8. $2x - y = 4$ 1; (2.$\overline{7}$, 2.$\overline{5}$)
$5x + 2y = 17$

9. $5x - y = 7$ none
$y = 5x + 6$

10. $y = \frac{1}{3}x - 4$ none
$y = \frac{1}{3}x + 9$

11. $\frac{1}{2}x + 3y = 2$
$\frac{1}{5}x - 2y = -4$
1; (−5, 1.5)

12. $\frac{1}{3}x + y = 0$ 1; (6, −2)
$\frac{1}{6}x - 4y = 9$

Solve the system using any algebraic method. (Lesson 3.2)

13. $3x - 2y = 4$
$-2x + 2y = 3$ $\left(7, \frac{17}{2}\right)$

14. $5x + y = 5$ (0, 5)
$9x - 4y = -20$

15. $12x - 7y = -22$
$-4x + y = 10$ (−3, −2)

16. $8x - y = 1$ (1, 7)
$-x + 4y = 27$

17. $y = 2x - 4$ (2, 0)
$-2y = x - 2$

18. $x + 2y = 5$ (3, 1)
$-2x + 3y = -3$

19. $x - y = 1$ (−8, −9)
$9x - 8y = 0$

20. $-2x + 3y = 10$
$5x + 6y = -16$ $\left(-4, \frac{2}{3}\right)$

Graph the system of linear inequalities. (Lesson 3.3) 21–28. See margin.

21. $x > -6$
$y < x + 4$

22. $y \geq 2x + 3$
$y < -3x + 5$

23. $x + y \geq -2$
$-4x + y \leq -5$

24. $2x - y > 1$
$-5x + y \leq 4$

25. $x > 3$
$y < -10$
$x + y < 7$

26. $3x + y < -4$
$y \geq 2x + 1$
$-x + y < 4$

27. $x < -10$
$y < 3$
$y \geq x + 5$
$y \leq -x - 8$

28. $x > 1$
$y < 5$
$y \geq x + 2$
$y \leq -x + 5$

Find the minimum and maximum values of the objective function subject to the given constraints. (Lesson 3.4)

29. Objective function: −25; 48
$C = 4x + 5y$

Constraints:
$x \geq 0$
$y \leq 4$
$x \leq 7$
$y \geq -5$

30. Objective function: 5; 35
$C = x + 3y$

Constraints:
$x \leq 0$
$x \geq -4$
$y \geq 3$
$y \leq -2x + 5$

31. Objective Function: −27; 25
$C = 5x - 3y$

Constraints:
$x \geq 0$
$y \geq 0$
$y \leq -x + 9$
$x \leq 5$

Sketch the graph of the equation. Label the points where the graph crosses the x-, y-, and z-axes. (Lesson 3.5) 32–37. See margin.

32. $x + y + z = 10$

33. $3x + y + 2z = 12$

34. $4x + 5y + 2z = 20$

35. $5x + 5y + 3z = 15$

36. $6x - 4y + 3z = 16$

37. $3x + 5y + z = -9$

Write the linear equation as a function of x and y. Then evaluate the function for the given values. (Lesson 3.5)

$f(x, y) = -\frac{1}{3}x - \frac{1}{3}y + 3; 3$

$f(x, y) = \frac{1}{2}x + \frac{1}{3}y - \frac{2}{3}; -\frac{1}{6}$

38. $3x + 2y + 4z = 12, f(2, 3)$
$f(x, y) = -\frac{3}{4}x - \frac{1}{2}y + 3; 0$

39. $x + y + 3z = 9, f(-3, 3)$

40. $9x + 6y - 18z = 12, f\left(-\frac{1}{3}, 2\right)$

41. $-2x + 5y - 2z = 10, f(1, 4)$
$f(x, y) = -x + \frac{5}{2}y - 5; 4$

42. $8x - y - z = 16, f\left(-\frac{1}{4}, -8\right)$
$f(x, y) = 8x - y - 16; -10$

43. $-x + 4y + 7z = -31, f\left(5, \frac{1}{2}\right)$
$f(x, y) = \frac{1}{7}x - \frac{4}{7}y - \frac{31}{7}; -4$

Extra Practice 943

21–28, 32–37 See Additional Answers beginning on page AA1.

5.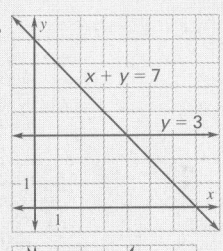

6.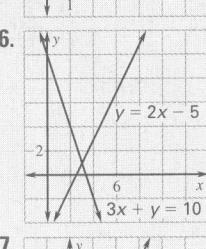

7.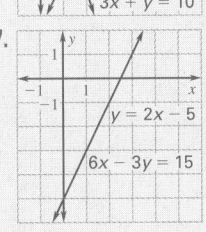

8.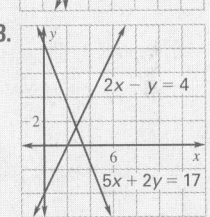

9.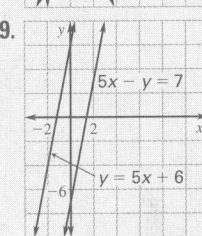

10.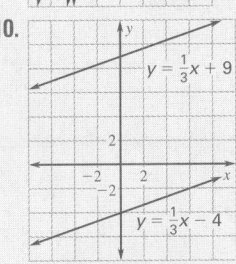

11.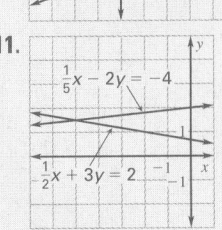

12.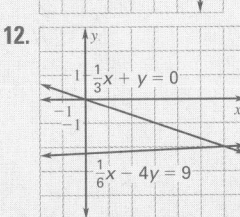

4. $\begin{bmatrix} -7 & 0 \\ 5 & -1 \\ 4 & 1 \end{bmatrix}$

7. $\begin{bmatrix} 5 & -6 \\ 17 & 27 \end{bmatrix}$

8. $\begin{bmatrix} -17 & -7 & -3 \\ -18 & 29 & 20 \end{bmatrix}$

12. Not defined; the number of columns in the first matrix does not equal the number of rows in the second matrix.

13. $\begin{bmatrix} 10.83 & -12.65 \\ 66.62 & 23.31 \end{bmatrix}$

14. $\begin{bmatrix} 5 & -15 \\ -2 & 7.2 \\ 0.15 & 1.46 \end{bmatrix}$

15. $\begin{bmatrix} -14 & 33 & 57 \\ -66 & -3 & -36 \\ -74 & 23 & -12 \end{bmatrix}$

31. $\begin{bmatrix} 2 & -1 \\ 7 & 4 \\ -\dfrac{3}{3} & 3 \end{bmatrix}$

1.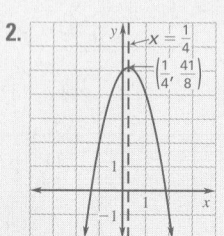

2.

Solve the system using any algebraic method. (Lesson 3.6)

44. $3x + 2y = 12$ (2, 3, 1)
    $2y - 5z = 1$
    $x + y + z = 6$

45. $2x + y = -2$ (0, -2, 5)
    $-x + 3y - 4z = -26$
    $5x - 6y + z = 17$

46. $x + y + z = 3$ no solution
    $3x + 3y + 3z = 10$
    $x - 3y + 4z = 6$

47. $-2x + 3y + z = 20$ (-3, 4, 2)
    $7x - 5y + 3z = -35$
    $4x + 4y + 4z = 12$

48. $x + y + z = 3$ infinitely many
    $x + y - z = 3$ solutions
    $2x + 2y + z = 6$

49. $2x - 4y - z = -18$ $\left(\dfrac{1}{2}, 3, 7\right)$
    $-6x - 3y + 2z = 2$
    $4x + y - 6z = -37$

50. **APPLES** You have $20.75 to spend on picking 15 pounds of three different types of apples in an orchard. The Red Rome apples cost $1.29 per pound, the Granny Smith apples cost $1.49 per pound, and the Empire apples cost $1.09 per pound. You want twice as many Granny Smith apples as the other two kinds combined. How many pounds of each type of apples should you buy? (Lesson 3.6)
    She should buy 2 pounds of Red Rome apples, 10 pounds of Granny Smith apples, and 3 pounds of Empire apples.

CHAPTER 4

Perform the indicated operation. (Lesson 4.1)

1. $\begin{bmatrix} 3 & 6 \\ -4 & -2 \end{bmatrix} + \begin{bmatrix} 1 & -4 \\ 0 & 6 \end{bmatrix}$ $\begin{bmatrix} 4 & 2 \\ -4 & 4 \end{bmatrix}$

2. $\begin{bmatrix} -6 & 5 \\ 7 & 9 \end{bmatrix} + \begin{bmatrix} 1 & 0 \\ 3 & -1 \end{bmatrix}$ $\begin{bmatrix} -5 & 5 \\ 10 & 8 \end{bmatrix}$

3. $\begin{bmatrix} -1 & 3 & 5 \\ -2 & 6 & -3 \end{bmatrix} + \begin{bmatrix} 0 & -4 & 2 \\ 3 & -6 & -1 \end{bmatrix}$ $\begin{bmatrix} -1 & -1 & 7 \\ 1 & 0 & -4 \end{bmatrix}$

4. $\begin{bmatrix} 2 & -3 \\ 3 & 4 \\ 4 & 5 \end{bmatrix} - \begin{bmatrix} 9 & -3 \\ -2 & 5 \\ 0 & 4 \end{bmatrix}$
See margin.

5. $\begin{bmatrix} 4 & \frac{2}{5} \\ 6 & \frac{1}{3} \end{bmatrix} - \begin{bmatrix} \frac{1}{3} & \frac{1}{5} \\ 1 & \frac{2}{3} \end{bmatrix}$ $\begin{bmatrix} 3\frac{2}{3} & \frac{1}{5} \\ 5 & -\frac{1}{3} \end{bmatrix}$

6. $-6 \begin{bmatrix} 2 & -1 & 0 \\ -3 & 4 & 7 \end{bmatrix}$ $\begin{bmatrix} -12 & 6 & 0 \\ 18 & -24 & -42 \end{bmatrix}$

7. $2 \begin{bmatrix} 4 & 0 \\ 1 & 3 \end{bmatrix} + 3 \begin{bmatrix} -1 & -2 \\ 5 & 7 \end{bmatrix}$
See margin.

8. $5 \begin{bmatrix} -2 & 0 & 1 \\ -3 & 7 & 4 \end{bmatrix} - \begin{bmatrix} 7 & 7 & 8 \\ 3 & 6 & 0 \end{bmatrix}$
See margin.

9. $\dfrac{1}{2} \begin{bmatrix} 5.2 & 7.4 & 9.8 \\ 4.6 & 6.8 & 8.4 \end{bmatrix}$ $\begin{bmatrix} 2.6 & 3.8 & 4.9 \\ 2.3 & 3.4 & 4.2 \end{bmatrix}$

Find the product. If it is not defined, state the reason. (Lesson 4.2)

10. $\begin{bmatrix} 2 & -6 \\ 3 & 1 \end{bmatrix} \begin{bmatrix} 0 & 4 \\ -1 & -5 \end{bmatrix}$ $\begin{bmatrix} 6 & 38 \\ -1 & 7 \end{bmatrix}$

11. $\begin{bmatrix} 1.6 & 3 & 9 \end{bmatrix} \begin{bmatrix} 2 \\ 6.4 \\ -2 \end{bmatrix}$ $\begin{bmatrix} 4.4 \end{bmatrix}$

12. $\begin{bmatrix} -8 & 0 & 5 \\ 3 & 6 & 7 \end{bmatrix} \begin{bmatrix} 4 & -3 & 9 \\ 8 & 6 & 9 \end{bmatrix}$ See margin.

13. $\begin{bmatrix} -4 & 4.5 & 3.8 \\ 2 & 1.7 & 7.5 \end{bmatrix} \begin{bmatrix} 9 & 8 \\ 6.1 & 4.3 \\ 5.1 & 0 \end{bmatrix}$
See margin.

14. $\begin{bmatrix} -3.0 & 10 \\ 0 & -4 \\ -2.0 & 0.3 \end{bmatrix} \begin{bmatrix} 0 & -1 \\ 0.5 & -1.8 \end{bmatrix}$
See margin.

15. $\begin{bmatrix} 5 & 1 & 5 \\ -3 & 0 & 6 \\ -3 & -8 & 11 \end{bmatrix} \begin{bmatrix} 6 & 5 & 12 \\ -4 & -2 & -3 \\ -8 & 2 & 0 \end{bmatrix}$
See margin.

BEVERAGE MACHINE In Exercises 16–18, you refill the beverage machines at work and you record the money received as income from each machine every day. There are three machines with four types of beverages. Juice is $.85, fruit punch is $.75, lemonade is $.65, and water is $.60. The matrix shows how many of each item were sold today. (Lesson 4.2)

	J	FP	L	W
MACHINE 1	12	13	18	14
MACHINE 2	15	15	21	22
MACHINE 3	8	16	9	33

	Price
J	$.85
FP	$.75
L	$.65
W	$.60

16. Write the matrix that gives the price of each item.

17. Use matrix multiplication to determine how much money is in each machine.

	Total
Machine 1	$40.05
Machine 2	$50.85
Machine 3	$44.45

18. Which machine has the most money?   Machine 2

Evaluate the determinant of the matrix. (Lesson 4.3)

19. $\begin{bmatrix} 4 & -3 \\ 7 & 2 \end{bmatrix}$ 29

20. $\begin{bmatrix} 0 & 6 \\ 1 & -4 \end{bmatrix}$ −6

21. $\begin{bmatrix} -3 & 5 \\ 9 & 2 \end{bmatrix}$ −51

22. $\begin{bmatrix} 6 & 3 & 1 \\ 0 & 0 & -1 \\ 13 & 9 & 12 \end{bmatrix}$ 15

23. $\begin{bmatrix} 21 & 7 & 2 \\ -6 & 10 & 9 \\ 1 & 0 & 3 \end{bmatrix}$ 799

24. $\begin{bmatrix} -9 & 5 & -6 \\ 0 & 3 & 10 \\ -10 & 17 & 4 \end{bmatrix}$ 742

Use Cramer's rule to solve the linear system. (Lesson 4.3)

25. $3x + y = 3$ (2, −3)
$4x + 5y = -7$

26. $4x + 5y = 30$ (−15, 18)
$-3x - 3y = -9$

27. $8x - 10y = -8$ (−6, −4)
$9x + 2y = -62$

28. $2x + z = 6$ (1, 3, 4)
$3x - 2y + 4z = 13$
$-y - 3z = -15$

29. $x + y + 2z = 0$ (5, −1, −2)
$2x - 6y + 5z = 6$
$-x + 3y - 7z = 6$

30. $3x + 4y + 2z = 12$ $\left(-7, 8, \frac{1}{2}\right)$
$-2x - 3y - 4z = -12$
$5x + 5y + 6z = 8$

Find the inverse of the matrix. (Lesson 4.4)

31. $\begin{bmatrix} 4 & 3 \\ 7 & 6 \end{bmatrix}$ See margin.

32. $\begin{bmatrix} 0 & 6 \\ 1 & -4 \end{bmatrix}$ $\begin{bmatrix} \frac{2}{3} & 1 \\ \frac{1}{6} & 0 \end{bmatrix}$

33. $\begin{bmatrix} -3 & 6 \\ 1 & 2 \end{bmatrix}$ $\begin{bmatrix} -\frac{1}{6} & \frac{1}{2} \\ \frac{1}{12} & \frac{1}{4} \end{bmatrix}$

34. $\begin{bmatrix} -1 & 7 \\ 2 & -5 \end{bmatrix}$ $\begin{bmatrix} \frac{5}{9} & \frac{7}{9} \\ \frac{2}{9} & \frac{1}{9} \end{bmatrix}$

35. $\begin{bmatrix} 1 & 2 \\ 4 & -8 \end{bmatrix}$ $\begin{bmatrix} \frac{1}{2} & \frac{1}{8} \\ \frac{1}{4} & -\frac{1}{16} \end{bmatrix}$

36. $\begin{bmatrix} 6 & 2 \\ -8 & 1 \end{bmatrix}$ $\begin{bmatrix} \frac{1}{22} & \frac{-1}{11} \\ \frac{4}{11} & \frac{3}{11} \end{bmatrix}$

37. $\begin{bmatrix} -9 & 7 \\ 4 & -3 \end{bmatrix}$ $\begin{bmatrix} 3 & 7 \\ 4 & 9 \end{bmatrix}$

38. $\begin{bmatrix} 3 & -1 \\ -2 & 9 \end{bmatrix}$ $\begin{bmatrix} \frac{9}{25} & \frac{1}{25} \\ \frac{2}{25} & \frac{3}{25} \end{bmatrix}$

Use an inverse matrix to solve the linear system. (Lesson 4.5)

39. $2x + 3y = 13$ (5, 1)
$x - 5y = 0$

40. $-4x - 3y = -2$
$2x + y = 2$ (2, −2)

41. $6x - 3y = -3$
$-4x + 7y = -3$
(−1, −1)

42. $5x + 2y = 8$ (4, −6)
$-2x - 9y = 46$

43. $3x - 8y = 16$ (0, −2)
$-2x + 5y = -10$

44. $-7x - 2y = -8$
$3x - 6y = 0$ $\left(1, \frac{1}{2}\right)$

45. $-5x - y = 2$
$10x + 3y = 1$ $\left(-\frac{7}{5}, 5\right)$

46. $-6x + 5y = -2$ (2, 2)
$4x - 3y = 2$

47. Use the given inverse of the coefficient matrix to solve the linear system. (Lesson 4.5)

$x + 2z = 5$
$-2x + 3y + 4z = -8$
$2x - y + 2z = 10$

$A^{-1} = \begin{bmatrix} 5 & -1 & -3 \\ 6 & -1 & -4 \\ -2 & 0.5 & 1.5 \end{bmatrix}$ (3, −2, 1)

CHAPTER 5

Graph the quadratic function. Label the vertex and axis of symmetry. (Lesson 5.1) 1–8. See margin.

1. $y = x^2 + 3x - 4$

2. $y = -2x^2 + x + 5$

3. $y = (x + 3)^2 - 4$

4. $y = (x + 1)(x - 4)$

5. $y = \frac{1}{2}(x - 4)^2 + 2$

6. $y = 3(x + 4)(x - 1)$

7. $y = (x + 8)(x - 3)$

8. $y = -\frac{1}{3}(x + 2)(x - 1)$

9. SWIMMING The drag force F (in pounds) of water on a swimmer can be modeled by $F = 1.35s^2$ where s is the swimmer's speed (in miles per hour). At what speed is the force minimized? (Lesson 5.1) 0 miles per hour

Factor the trinomial. If the trinomial cannot be factored, say so. (Lesson 5.2)

10. $x^2 + 8x + 15$
$(x + 5)(x + 3)$

11. $m^2 - 9m + 20$
$(m - 4)(m - 5)$

12. $3x^2 + 11x - 4$
$(3x - 1)(x + 4)$

13. $6x^2 + 5x - 6$
$(3x - 2)(2x + 3)$

14. $9a^2 - 56a + 12$
$(9a - 2)(a - 6)$

15. $4u^2 - 4u - 35$
$(2u + 5)(2u - 7)$

16. $n^2 - 49$
$(n + 7)(n - 7)$

17. $x^2 - 10x + 25$
$(x - 5)^2$

18. $16m^2 - 24m + 9$
$(4m - 3)^2$

19. $4x^2 - 2x - 20$
$2(2x - 5)(x + 2)$

20. $3p^2 + 15p - 42$
$3(p + 7)(p - 2)$

21. $6x^2 + 13x - 25$
cannot be factored

3.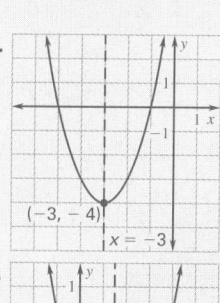
$(-3, -4)$ $x = -3$

4.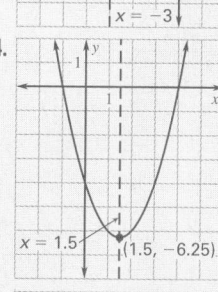
$x = 1.5$ $(1.5, -6.25)$

5.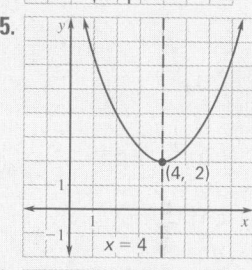
$(4, 2)$ $x = 4$

6.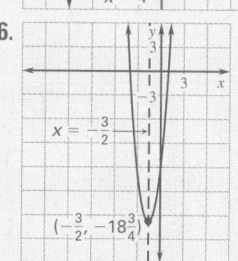
$x = -\frac{3}{2}$ $\left(-\frac{3}{2}, -18\frac{3}{4}\right)$

7.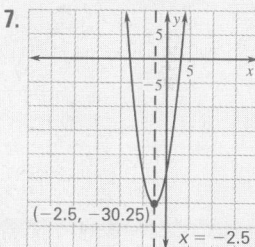
$(-2.5, -30.25)$ $x = -2.5$

8.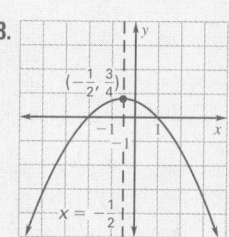
$\left(-\frac{1}{2}, \frac{3}{4}\right)$ $x = -\frac{1}{2}$

80. $-1+\frac{2\sqrt{6}}{3}, -1-\frac{2\sqrt{6}}{3}$

87. $\frac{3}{7}+\frac{\sqrt{61}}{7}i, \frac{3}{7}-\frac{\sqrt{61}}{7}i$

88. $-\frac{1}{4}+\frac{\sqrt{41}}{4}, -\frac{1}{4}-\frac{\sqrt{41}}{4}$

95.
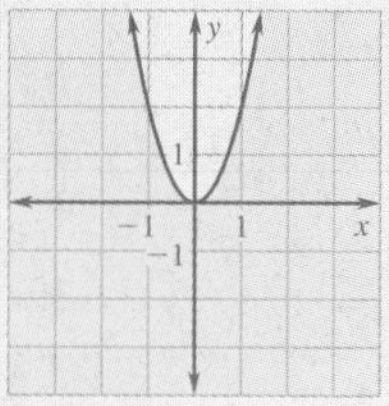

96.
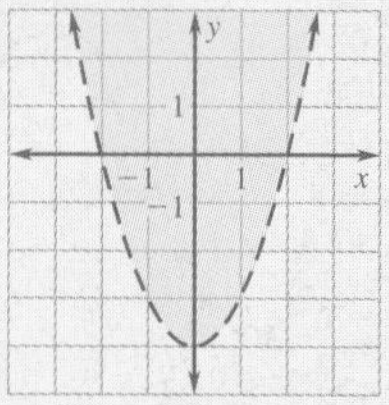

97.
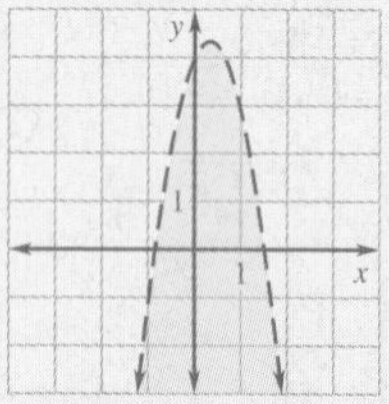

98.
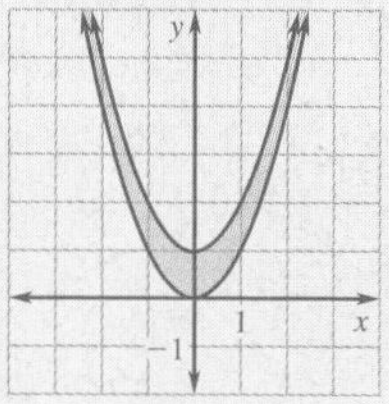

99.
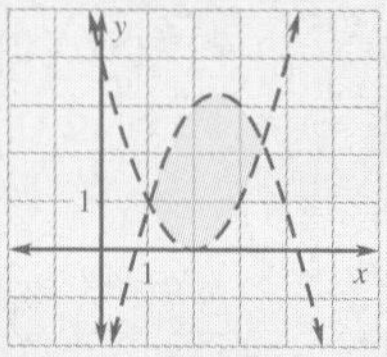

100.
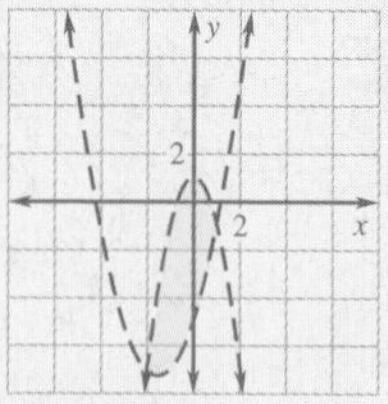

6. $\frac{49}{9}$; negative exponent, power of a quotient

7. $\frac{1}{512}$; *Sample answer:* zero exponent, negative exponent

9. $\frac{1}{46{,}656}$; *Sample answer:* product of powers, power of a quotient

10. 729; *Sample answer:* power of a power, negative exponent

Solve the equation. (Lesson 5.2)

22. $x^2 + 10x + 21 = 0$ $-7, -3$
23. $2x^2 - 13x - 7 = 0$ $-\frac{1}{2}, 7$
24. $3x^2 - 24x - 27 = 0$ $9, -1$
25. $25m^2 - 20m + 4 = 0$ $\frac{2}{5}$
26. $x^2 - 8x = -15$ $3, 5$
27. $8k^2 + 5k = 2k^2 + 4$ $-\frac{4}{3}, \frac{1}{2}$
28. $10x^2 - 3x = -2x^2 + 36$ $-1.61, 1.86$
29. $2(q^2 - 20) + 17q = -10q^2$ $-\frac{8}{3}, \frac{5}{4}$

Write the quadratic function in intercept form and give the function's zeros. (Lesson 5.2)

30. $y = x^2 + 10x + 9$ $y = (x + 1)(x + 9)$; $-1, -9$
31. $y = x^2 - 5x$ $y = x(x - 5)$; $0, 5$
32. $y = 2x^2 + 3x - 2$ $y = (2x - 1)(x + 2)$; $\frac{1}{2}, -2$
33. $y = 6x^2 - 24$ $y = 6(x + 2)(x - 2)$; $2, -2$
34. $y = 4x^2 - 12x + 8$ $y = 4(x - 1)(x - 2)$; $1, 2$
35. $y = 5x^2 - 13x + 6$ $y = (5x - 3)(x - 2)$; $\frac{3}{5}, 2$
36. $y = 4x^2 + 22x + 24$ $y = 2(2x + 3)(x + 4)$; $-\frac{3}{2}, -4$
37. $y = 7x^2 - 63$ $y = 7(x + 3)(x - 3)$; $3, -3$

Simplify the expression. (Lesson 5.3)

38. $\sqrt{32}$ $4\sqrt{2}$
39. $\sqrt{125}$ $5\sqrt{5}$
40. $3\sqrt{27} \cdot \sqrt{3}$ 27
41. $\sqrt{243}$ $9\sqrt{3}$
42. $\sqrt{15} \cdot \sqrt{3}$ $3\sqrt{5}$
43. $\sqrt{\frac{81}{125}}$ $\frac{9\sqrt{5}}{25}$
44. $6\sqrt{5} \cdot \sqrt{5}$ 30
45. $\sqrt{\frac{16}{25}}$ $\frac{4}{5}$

Solve the equation. (Lesson 5.3)

46. $x^2 = 144$ $12, -12$
47. $x^2 = 160$ $4\sqrt{10}, -4\sqrt{10}$
48. $2x^2 = 400$ $10\sqrt{2}, -10\sqrt{2}$
49. $-4(x + 2)^2 = -20$ $\sqrt{5} - 2, -\sqrt{5} - 2$
50. $\frac{x^2}{9} - 1 = 5$ $3\sqrt{6}, -3\sqrt{6}$
51. $7x^2 = 175$ $5, -5$
52. $x^2 - 100 = -82$ $3\sqrt{2}, -3\sqrt{2}$
53. $\frac{1}{3}(x - 4)^2 = 3$ $7, 1$

Solve the equation. (Lesson 5.4)

54. $x^2 = -16$ $4i, -4i$
55. $x^2 = -10$ $i\sqrt{10}, -i\sqrt{10}$
56. $3x^2 = -27$ $3i, -3i$
57. $5x^2 = -125$ $5i, -5i$
58. $(y - 3)^2 = -49$ $3 + 7i, 3 - 7i$
59. $6x^2 = -216$ $6i, -6i$
60. $4(x + 5)^2 = -8$ $-5 + i\sqrt{2}, -5 - i\sqrt{2}$
61. $-\frac{1}{4}(r + 1)^2 = 5$ $-1 + 2i\sqrt{5}, -1 - 2i\sqrt{5}$

Write the expression as a complex number in standard form. (Lesson 5.4)

62. $(3 + 5i) + (2 + i)$ $5 + 6i$
63. $(-6 + 4i) + (2 - 7i)$ $-4 - 3i$
64. $(4 + 3i)^2$ $7 + 24i$
65. $(15 - 7i) - (15 - 7i)$ 0
66. $i(5 + i)$ $-1 + 5i$
67. $-2i(3 - 2i)$ $-4 - 6i$
68. $(9 - 2i)(9 + 2i)$ 85
69. $(9 - 5i) - (-2 + 6i)$ $11 - 11i$
70. $(10 - 7i)^2$ $51 - 140i$
71. $\frac{3}{5 + i}$ $\frac{15}{26} - \frac{3}{26}i$
72. $\frac{2i}{4 - i}$ $-\frac{2}{17} + \frac{8}{17}i$
73. $\frac{1 - i}{1 + i}$ $-i$

Solve the equation by completing the square. (Lesson 5.5)

74. $x^2 - 6x = 7$ $7, -1$
75. $x^2 - 4x + 8 = 0$ $2 + 2i, 2 - 2i$
76. $x^2 - 10x = 1$ $5 + \sqrt{26}, 5 - \sqrt{26}$
77. $m^2 + 2.6m - 3 = 0$ $0.866, -3.47$
78. $2n^2 - 5n = 7$ $\frac{7}{2}, -1$
79. $3n^2 - 4n = 4$ $-\frac{2}{3}, 2$
80. $3y^2 + 2y = 5 - 4y$ See margin.
81. $5n^2 + 6n = 8$ $\frac{4}{5}, -2$

82. **VEGETABLE GARDEN** You are planning to create a vegetable garden behind your house. Your house will be one side of the rectangular garden, and the garden will have a fence on its other sides. You bought 40 feet of fencing and enough mulch to cover 140 square feet. If the back of the house is 30 feet wide, what should the garden's dimensions be? (Lesson 5.5) about 15.5 feet long and 9 feet wide

Use the quadratic formula to solve the equation. (Lesson 5.6)

83. $4x^2 + x = 3$ $\frac{3}{4}, -1$
84. $x^2 + 10x + 25 = 0$ -5
85. $x^2 + 3x - 8 = 0$ $-\frac{3}{2} + \frac{\sqrt{41}}{2}, -\frac{3}{2} - \frac{\sqrt{41}}{2}$
86. $x^2 - 4x + 5 = 0$ $2 + i, 2 - i$
87. $7m^2 - 6m + 10 = 0$ See margin.
88. $2(m + 1)^2 = 3m + 7$ See margin.

Find the discriminant of the quadratic equation and give the number and type of solutions of the equation. (Lesson 5.6)

89. $x^2 + 7x + 12 = 0$ 1; 2; real
90. $x^2 - 8x + 16 = 0$ 0; 1; real
91. $5m^2 + 3m + 10 = 0$ -191; 2; imaginary
92. $x^2 + 5x - 6 = 0$ 49; 2; real
93. $2x^2 - 4x + 7 = 0$ -40; 2; imaginary
94. $4x^2 + 3x - 15 = 0$ 249; 2; real

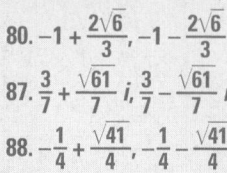

Graph the inequality or system of inequalities. (Lesson 5.7) 95–100. See margin.

95. $y \le 2x^2$

96. $y > x^2 - 4$

97. $y < -3x^2 + 2x + 4$

98. $y \le x^2 + 1$
$y \ge x^2$

99. $y > x^2 - 4x + 4$
$y < -x^2 + 5x - 3$

100. $y > x^2 + 3x - 5$
$y < -2x^2 + 1$

Write a quadratic function in vertex form for the parabola whose graph has the given vertex and passes through the given point. (Lesson 5.8)

101.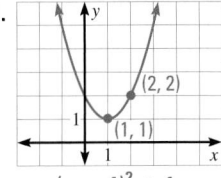
$y = (x - 1)^2 + 1$

102.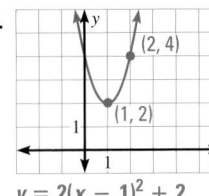
$y = 2(x - 1)^2 + 2$

103.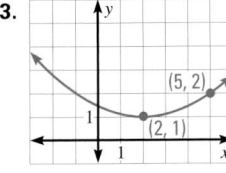
$y = \frac{1}{9}(x - 2)^2 + 1$

104. vertex: (3, 0)
point: (2, 1)
$y = (x - 3)^2$

105. vertex: (−3, −5)
point: (1, 27)
$y = 2(x + 3)^2 - 5$

106. vertex: (−1, −4)
point: (−2, −6)
$y = -2(x + 1)^2 - 4$

Write a quadratic function in intercept form whose graph has the given x-intercepts and passes through the given point. (Lesson 5.8)

107. x-intercepts: 2, 6
point: (5, −3)
$y = (x - 2)(x - 6)$

108. x-intercepts: −1, 3
point: (2, 3)
$y = -(x + 1)(x - 3)$

109. x-intercepts: 4, 0
point: (1, −6)
$y = 2x(x - 4)$

110. x-intercepts: −2, 3
point: (2, 1)
$y = -\frac{1}{4}(x + 2)(x - 3)$

111. x-intercepts: 1, 2
point: (5, 9)
$y = \frac{3}{4}(x - 1)(x - 2)$

112. x-intercepts: −1, 4
point: (0, −1)
$y = \frac{1}{4}(x + 1)(x - 4)$

113. x-intercepts: 5, −2
point: (2, 2)
$y = -\frac{1}{6}(x - 5)(x + 2)$

114. x-intercepts: −3, −3
point: (1, 48)
$y = 3(x + 3)(x + 3)$

CHAPTER 6

Evaluate the expression. Tell which properties of exponents you used. (Lesson 6.1)

1. $5^2 \cdot 5^2$ 625; product of powers

2. $(-4)^3(-4)$ 256; product of powers

3. $(2^3)^3$ 512; power of a power

4. 6^{-2} $\frac{1}{36}$; negative exponent

5. $\left(\frac{4}{5}\right)^2$ $\frac{16}{25}$; power of a quotient

6. $\left(\frac{3}{7}\right)^{-2}$ See margin.

7. $8^0 \cdot 8^{-3}$ See margin.

8. $\frac{3^{-2}}{3^{-4}}$ 9; *Sample answer:* quotient of powers

9. $\left(\frac{1}{6}\right)^3\left(\frac{1}{6}\right)^3$ See margin.

10. $\left(\left(\frac{1}{3}\right)^2\right)^{-3}$ See margin.

11. $\frac{6^3}{4^0 \cdot 6^2}$ 6; zero exponent, quotient of powers

12. $5^5 \cdot 5^0 \cdot 5^{-3}$ See margin.

Simplify the expression. Tell which properties of exponents you used. (Lesson 6.1) 13–24, See margin for properties.

13. $(32x^2)^4$ $1{,}048{,}576x^8$

14. $(x^2y^2)^{-3}$ $\frac{1}{x^6y^6}$

15. $\frac{x^8}{x^5}$ x^3

16. $\frac{4x^4y^7}{8x^5y^3}$ $\frac{y^4}{2x}$

17. $(6x^3y^4)^{-2}$ $\frac{1}{36x^6y^8}$

18. $-4(x^{-5}y^2)^2$ $\frac{-4y^4}{x^{10}}$

19. $(-3x^9y^3)^{-7}$ $\frac{-1}{2187x^{63}y^{21}}$

20. $(6x^{-3}y^{-1})^{-8}$ $\frac{x^{24}y^8}{1{,}679{,}616}$

21. $(8(x^3y^4)^2)^{-2}$ $\frac{1}{64x^{12}y^{16}}$

22. $\frac{2x^{-3}y^{-5}}{3x^{-6}y^{-3}}$ $\frac{2x^3}{3y^2}$

23. $\frac{x^{10}}{3y^4} \cdot \frac{9x^2y^2}{x^4y^3}$ $\frac{3x^8}{y^5}$

24. $\frac{15xy^4}{8x^3y^0} \cdot \frac{16x^5y^2}{5y^4}$ $6x^3y^2$

Use synthetic division to evaluate the polynomial function for the given value of x. (Lesson 6.2)

25. $f(x) = 2x^3 + 3x^2 - 5x + 1; x = 2$ 19

26. $f(x) = 10x^3 - 5x^2 + 4; x = -1$ −11

27. $f(x) = x^5 - 3x^3 - 2x; x = -2$ −4

28. $f(x) = -x^4 + 7x - 12; x = 3$ −72

Graph the polynomial function. (Lesson 6.2) 29–34. See margin.

29. $f(x) = x^3$

30. $f(x) = x^4 + 1$

31. $f(x) = 3 - x^3$

32. $f(x) = x^4 - 3x$

33. $f(x) = -x^5 - 2$

34. $f(x) = x^5 + 2x^3 + 3$

12. 25; *Sample answer:* zero exponent, product of powers
13. power of a product, power of a power
14. power of a product, power of a power, negative exponent
15. quotient of powers
16. quotient of powers
17. power of a product, power of a power, negative exponent
18. power of a product, power of a power, negative exponent
19. power of a product, power of a power, negative exponent
20. power of a product, power of a power, negative exponent
21. power of a product, power of a power, negative exponent
22. quotient of powers, negative exponent
23. product of powers, quotient of powers, negative exponent
24. *Sample answer:* product of powers, quotient of powers, zero exponent

29.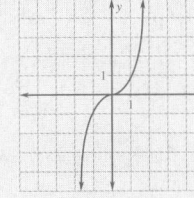

30.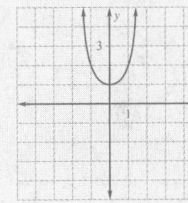

31.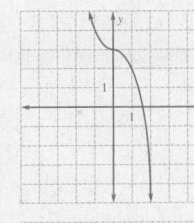

32.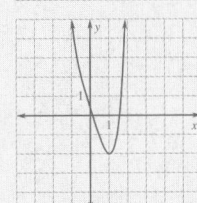

33–34. See Additional Answers beginning on page AA1.

62. $5x + 10 + \dfrac{14}{x-2}$

63. $3x^3 - 29x^2 + 129x - 540 + \dfrac{2176}{x+4}$

78. $x^3 - 9x^2 + 23x - 15$

79. $x^3 + 5x^2 + 8x + 4$

80. $x^3 - 3x^2 - 46x + 168$

81. $x^3 - 3x^2 + x - 3$

82. $x^4 + 17x^2 + 16$

83. $x^3 - 4x^2 + 6x - 4$

84. $x^4 + 72x^2 + 1296$

85. $x^4 + x^3 - 6x^2 - 14x - 12$

86. $x^5 - 20x^4 + 159x^3 - 624x^2 +$
    $1198x - 884$

87. $x^4 - 4x^3 + 41x^2 - 144x + 180$

88. $x^4 - 10x^3 + 38x^2 - 64x + 40$

89. $x^6 + 36x^4 - 625x^2 - 22{,}500$

90. $f(x) = (x+3)(x-4)(x+1)$
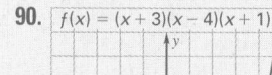
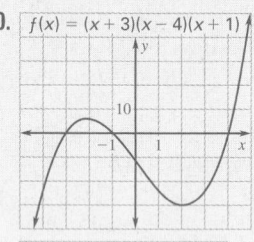

91.
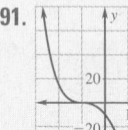
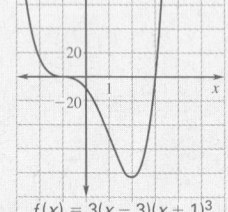
$f(x) = 3(x-3)(x+1)^3$

92. $f(x) = -(x-1)(x+1)(x-5)$
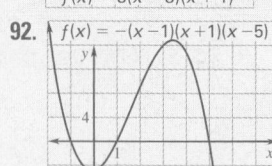

93.
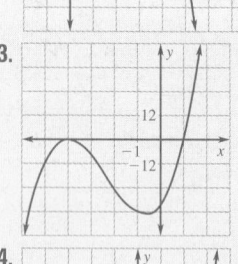

94.
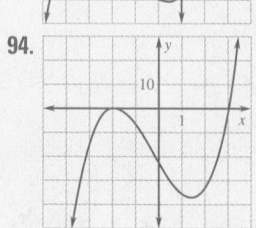

Find the sum or difference. (Lesson 6.3)

35. $(2x^2 + 6x + 3) + (3x^2 + 4x + 4)$ $5x^2 + 10x + 7$

36. $(4x - 3) + (3 - 8x)$ $-4x$

37. $(5x^3 - 2x^2 + 7) - (8x^2 - 11)$ $5x^3 - 10x^2 + 18$

38. $(9x^3 - 7x^2 + 8) + (-8x^3 + 5x^2 - 15)$ $x^3 - 2x^2 - 7$

39. $(29x - 8) + (15x^3 + 9x^2 - 8)$
    $15x^3 + 9x^2 + 29x - 16$

40. $(6x^3 - 7x^4 + 10x) - (4x^3 - 6x^2)$
    $-7x^4 + 2x^3 + 6x^2 + 10x$

Find the product of the polynomials. (Lesson 6.3)

41. $(x + 7)(x - 5)$ $x^2 + 2x - 35$

42. $(x - 3)^2$ $x^2 - 6x + 9$

43. $(5 - 3x)(x + 1)(x + 6)$
    $-3x^3 - 16x^2 + 17x + 30$

44. $(-x^3 - 3)(x^2 - 5x + 4)$
    $-x^5 + 5x^4 - 4x^3 - 3x^2 + 15x - 12$

45. $6x(2x^3 - 4x^2 + 7)$
    $12x^4 - 24x^3 + 42x$

46. $(x + 12)(2x^2 - 3x + 5)$
    $2x^3 + 21x^2 - 31x + 60$

47. $(2x + 8)^3$
    $8x^3 + 96x^2 + 384x + 512$

48. $(x + 1)(3x + 3)(2x + 3)$
    $6x^3 + 21x^2 + 24x + 9$

49. $(x + y)^3$
    $x^3 + 3x^2y + 3xy^2 + y^3$

Factor the polynomial. (Lesson 6.4)

50. $x^3 - 27$
    $(x - 3)(x^2 + 3x + 9)$

51. $2x^3 + 250$
    $2(x + 5)(x^2 - 5x + 25)$

52. $256x^5 - 81x^3$
    $x^3(16x + 9)(16x - 9)$

53. $x^3 + 7x^2 + 15x + 9$
    $(x + 1)(x + 3)^2$

54. $x^3 - x^2 - 14x + 24$
    $(x - 2)(x + 4)(x - 3)$

55. $3x^3 - 24$
    $3(x - 2)(x^2 + 2x + 4)$

56. $x^3 + 5x^2 + 8x + 40$
    $(x^2 + 8)(x + 5)$

57. $2x^3 + 18x^2 - 5x - 45$
    $(2x^2 - 5)(x + 9)$

58. $3x^5 + 6x^3 - 45x$
    $3x(x^2 + 5)(x^2 - 3)$

59. **PACKAGING** A factory needs a box that has a volume of 6 cubic inches. The width should be 1 inch less than the height and the length should be 3 inches greater than the height. What should the dimensions of the box be? (Lesson 6.4) $(\sqrt{3} + 3)$ inches by $(\sqrt{3} - 1)$ inches by $\sqrt{3}$ inches

Divide. Use synthetic division when possible. (Lesson 6.5)

60. $(x^3 - 2x^2 - 8x + 5) \div (x - 1)$
    $x^2 - x - 9 - \dfrac{4}{x-1}$

61. $(x^3 - 10x^2 + 27x - 12) \div (x - 4)$ $x^2 - 6x + 3$

62. $(5x^2 - 6) \div (x - 2)$
    See margin.

63. $(3x^4 - 17x^3 + 13x^2 - 24x + 16) \div (x + 4)$
    See margin.

64. $(x^4 + x^3 - 3x - 3) \div (x + 1)$ $x^3 - 3$

65. $(4x^4 - 5x^3 + 2x^2 - x + 5) \div (x - 2)$
    $4x^3 + 3x^2 + 8x + 15 + \dfrac{35}{x - 2}$

Find all the real zeros of the polynomial function. (Lesson 6.6)

66. $f(x) = x^3 - 2x^2 - 11x + 12$ $-3, 1, 4$

67. $f(x) = x^4 + 5x^3 + 10x^2 + 20x + 24$ $-3, -2$

68. $f(x) = 2x^3 - 3x^2 - 23x + 12$ $-3, \frac{1}{2}, 4$

69. $f(x) = x^5 + x^4 + 3x^3 - 8x^2 - 8x - 24$ 2

70. $f(x) = 3x^4 - 5x^3 - 5x^2 + 5x + 2$ $-1, -\frac{1}{3}, 1, 2$

71. $f(x) = 16x^3 + 80x^2 + x + 5$ -5

Find all the zeros of the polynomial function. (Lesson 6.7)

72. $f(x) = x^3 - x^2 + 4x - 4$ $1, 2i, -2i$

73. $f(x) = x^4 - 7x^3 + 17x^2 - 17x + 6$ $1, 2, 3$

74. $f(x) = x^3 + x^2 + 9x + 9$ $-1, 3i, -3i$

75. $f(x) = x^4 + 2x^3 - 12x^2 - 40x - 32$ $-2, 4$

76. $f(x) = x^3 - 7x^2 - x + 7$ $-1, 1, 7$

77. $f(x) = x^4 - 6x^2 + 5$ $-1, 1, \sqrt{5}, -\sqrt{5}$

Write a polynomial function of least degree that has real coefficients, the given zeros, and a leading coefficient of 1. (Lesson 6.7) 78–89. See margin.

78. $3, 1, 5$

79. $-1, -2, -2$

80. $4, 6, -7$

81. $i, -i, 3$

82. $i, -4i, 4i$

83. $2, 1 + i$

84. $6i, 6i$

85. $3, -2, -1 + i$

86. $4 - i, 5 - i, 2$

87. $2 + i, 6i$

88. $2, 2, 3 - i$

89. $5, -5, -6i, 5i$

Graph the function. (Lesson 6.8) 90–98. See margin.

90. $f(x) = (x + 3)(x - 4)(x + 1)$

91. $f(x) = 3(x - 3)(x + 1)^3$

92. $f(x) = -(x - 1)(x + 1)(x - 5)$

93. $f(x) = 2(x - 1)(x + 4)^2$

94. $f(x) = 2(x - 3)(x + 2)^2$

95. $f(x) = -3(x + 1)(x - 1)(x - 2)$

96. $f(x) = 2(x - 1)(x + 2)^3$

97. $f(x) = 5(x + 3)(x - 2)^2$

98. $f(x) = 2(x - 1)(x - 2)(x + 3)$

Find a polynomial that fits the data. (Lesson 6.9) 99, 100. See margin.

99.

x	1	2	3	4	5	6
f(x)	2	5	9	14	20	27

100.

x	1	2	3	4	5	6
f(x)	5	16	43	92	169	280

101.

x	1	2	3	4	5	6
f(x)	3	3	9	27	63	123

$y = x^3 - 3x^2 + 2x + 3$

102.

x	1	2	3	4	5	6
f(x)	3	15	55	141	291	523

$y = 3x^3 - 4x^2 + 3x + 1$

CHAPTER 7

Evaluate the expression without using a calculator. (Lesson 7.1)

1. $\sqrt[3]{27}$ 3
2. $\sqrt[3]{-125}$ −5
3. $16^{-1/2}$ $\frac{1}{4}$
4. $64^{2/3}$ 16

5. $-(25^{3/2})$ −125
6. $-(243^{3/5})$ −27
7. $(\sqrt[4]{81})^{-2}$ $\frac{1}{9}$
8. $\sqrt[5]{32}$ 2

9. $8^{1/3}$ 2
10. $(-216)^{-1/3}$ $-\frac{1}{6}$
11. $(\sqrt[3]{64})^{1/2}$ 2
12. $(\sqrt[3]{729})^{1/2}$ 3

Simplify the expression. (Lesson 7.2)

13. $5^{1/4} \cdot 5^{3/4}$ 5
14. $(3^{1/3})^{2/5}$ $3^{2/15}$
15. $2^{1/4} \cdot 8^{1/4}$ 2
16. $\frac{12^{3/5}}{12^{1/5}}$ $12^{2/5}$

17. $\frac{80^{1/2}}{16^{1/2}}$ $5^{1/2}$
18. $\sqrt{25} \cdot \sqrt[3]{25}$ $5\sqrt[3]{25}$
19. $(\sqrt[3]{7} \cdot \sqrt[4]{7})^2$ $7\sqrt[6]{7}$
20. $\frac{\sqrt{10}}{\sqrt[3]{10}}$ $\sqrt[4]{10}$

Simplify the expression. Assume all variables are positive. (Lesson 7.2)

21. $x^{1/2} \cdot x^{1/5}$ $x^{7/10}$
22. $(x^3)^{1/2}$ $x^{3/2}$
23. $\sqrt[4]{81x^6y^8}$ $3xy^2\sqrt{x}$
24. $\sqrt{\frac{16x^4y^5}{25z^4}}$ $\frac{4x^2y^2\sqrt{y}}{5z^2}$

25. $\frac{\sqrt[3]{x} \cdot \sqrt{x^3}}{\sqrt{16x^{12}}}$ $\frac{\sqrt[6]{x^5}}{4x^5}$
26. $\frac{\sqrt[5]{x^4}}{\sqrt[8]{x^3}}$ $\sqrt[40]{x^{17}}$
27. $\sqrt[3]{\frac{8x^6y^{12}}{27}}$ $\frac{2x^2y^4}{3}$
28. $\sqrt[6]{9xy^6} \cdot \sqrt[6]{6x^{12}}$ $x^2y\sqrt[6]{54x}$

Let $f(x) = x^2 - 4x + 5$ and $g(x) = x^2 - 9$. Perform the indicated operation and state the domain. (Lesson 7.3)

$2x^2 - 4x - 4$; all real numbers
29. $f(x) + g(x)$

30. $f(x) - g(x)$
−4x + 14; all real numbers

$2x^2 - 4x - 4$; all real numbers
31. $g(x) + f(x)$

32. $g(x) - f(x)$
4x − 14; all real numbers

33. $f(x) + f(x)$
$2x^2 - 8x + 10$; all real numbers

34. $f(x) - f(x)$
0; all real numbers

35. $g(x) + g(x)$
$2x^2 - 18$; all real numbers

36. $g(x) - g(x)$
0; all real numbers

Let $f(x) = 3x^{1/3}$ and $g(x) = x^{1/2}$. Perform the indicated operation and state the domain. (Lesson 7.3)

$3x^{5/6}$; nonnegative reals
37. $f(x) \cdot g(x)$

38. $\frac{f(x)}{g(x)}$ $\frac{3x^{5/6}}{x}$; positive reals

39. $g(x) \cdot f(x)$ $3x^{5/6}$; nonnegative reals

40. $\frac{g(x)}{f(x)}$ $\frac{x^{1/6}}{3}$; nonnegative reals

41. $f(g(x))$
$3x^{1/6}$; nonnegative reals

42. $g(f(x))$
$3^{1/2}x^{1/6}$; nonnegative reals

43. $f(f(x))$ $3^{4/3}x^{1/9}$; nonnegative reals

44. $g(g(x))$ $x^{1/4}$; nonnegative reals

Find the inverse function. (Lesson 7.4) 45–56. See margin.

45. $f(x) = 3x + 1$
46. $f(x) = -2x - 1$
47. $f(x) = -x - 4$
48. $f(x) = 5x - 7$

49. $f(x) = 2x + 3$
50. $f(x) = -4x - 5$
51. $f(x) = \frac{1}{2}x - 4$
52. $f(x) = 3x^3 + 2$

53. $f(x) = -\frac{1}{3}x + 5$
54. $f(x) = 2x^4$; $x \geq 0$
55. $f(x) = x^4 - \frac{1}{8}$; $x \geq 0$
56. $f(x) = \frac{1}{2}x^2 - 5$; $x \geq 0$

57. **AREA** The area A of a circular object is $A = \pi r^2$ where r is the radius of the object. Find r in terms of A. (Lesson 7.4) $r = \frac{\sqrt{\pi A}}{\pi}$

95.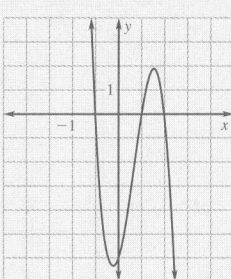

96.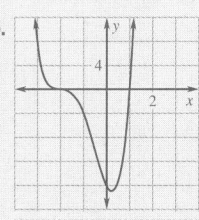

97.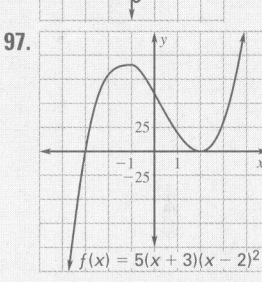
$f(x) = 5(x + 3)(x - 2)^2$

98.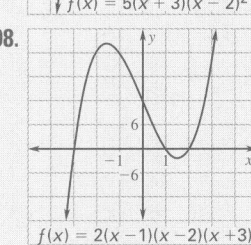
$f(x) = 2(x - 1)(x - 2)(x + 3)$

99. $y = \frac{1}{2}x^2 + \frac{3}{2}x$

100. $y = x^3 + 2x^2 - 2x + 4$

45. $f^{-1}(x) = \frac{x-1}{3}$

46. $f^{-1}(x) = -\frac{x+1}{2}$

47. $f^{-1}(x) = -x - 4$

48. $f^{-1}(x) = \frac{x+7}{5}$

49. $f^{-1}(x) = \frac{x-3}{2}$

50. $f^{-1}(x) = -\frac{x+5}{4}$

51. $f^{-1}(x) = 2x + 8$

52. $f^{-1}(x) = \frac{\sqrt[3]{9x - 18}}{3}$

53. $f^{-1}(x) = -3x + 15$

54. $f^{-1}(x) = \frac{\sqrt[4]{8x}}{2}$

55. $f^{-1}(x) = \sqrt[4]{x + \frac{1}{8}}$

56. $f^{-1}(x) = \sqrt{2x + 10}$

58.

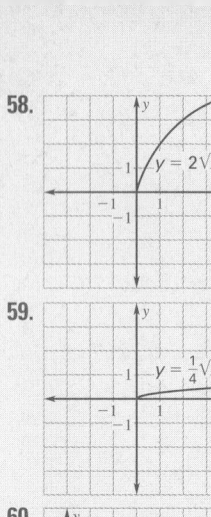

59.

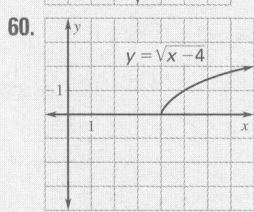

60.

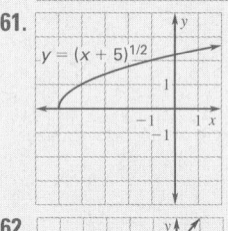

61.

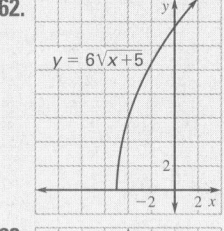

62.

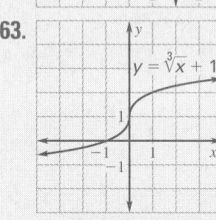

63.

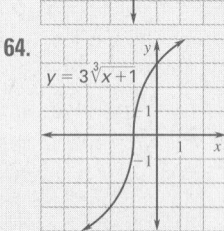

64.

65.

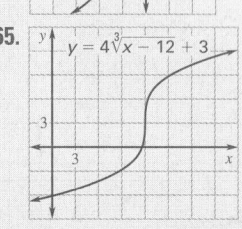

66–69, 76, 1–20, 33–40. See Additional
Answers beginning on page AA1.

950

Graph the function. Then state the domain and range. (Lesson 7.5)   58–69. See margin for graphs.

58. $y = 2\sqrt{x}$ $x \geq 0$; $y \geq 0$ **59.** $y = \frac{1}{4}\sqrt{x}$ $x \geq 0$; $y \geq 0$ **60.** $y = \sqrt{x - 4}$ $x \geq 4$; $y \geq 0$ **61.** $y = (x + 5)^{1/2}$ $x \geq -5$; $y \geq 0$

62. $y = 6\sqrt{x + 5}$ $x \geq -5$; $y \geq 0$ **63.** $y = \sqrt[3]{x} + 1$ all reals; all reals   **64.** $y = 3\sqrt[3]{x} + 1$ all reals; all reals   **65.** $y = 4\sqrt[3]{x - 12} + 3$ all reals; all reals

66. $y = \frac{1}{2}\sqrt{x + 2}$ $x \geq -2$; $y \geq 0$ **67.** $y = \sqrt[3]{x - 5} + 1$ all reals; all reals   **68.** $y = \frac{1}{3}\sqrt[3]{x - 8} - 5$ all reals; all reals   **69.** $y = -2(x + 2)^{1/3} - 4$ all reals; all reals

Solve the equation. Check for extraneous solutions. (Lesson 7.6)

70. $x^{1/4} = \frac{1}{256}$ $\frac{1}{4,294,967,296}$ **71.** $x^{1/4} - 81 = 0$ 43,046,721   **72.** $2(x + 1)^{2/3} = 6$ 4.20, −6.20   **73.** $\sqrt{x} + 1 = \frac{1}{16}$ no solution

74. $x^{2/3} = 16$ 64   **75.** $\sqrt[3]{x} + 4 = 2$ −8   **76.** $\sqrt{11x + 3} = 2x$ See margin.   **77.** $\sqrt{x - 13} = 2\sqrt{x + 7}$ no solution

78. $\sqrt{5x + 1} = x - 4$ 11.72   **79.** $\sqrt{x + 3} = \sqrt{2x - 7}$ 10   **80.** $2\sqrt{x - 2} = \sqrt{x}$ $\frac{8}{3}$ **81.** $4\sqrt{3x - 7} = 2\sqrt{-x + 73}$ $\frac{101}{13}$

Find the mean, median, mode, range, and standard deviation of the data set. (Lesson 7.7)

82. 8, 9, 9, 10, 11, 10, 12, 8, 9, 11   9.7; 9.5; 9; 4; 1.27   **83.** 52, 56, 57, 58, 58, 73, 55, 58, 57, 58   58.2; 57.5; 58; 21; 5.25

84. 2.3, 2.7, 2.8, 2.8, 2.8, 4.7, 4.9, 5.2   3.53; 2.8; 2.8; 2.9; 1.11   **85.** 21.4, 18.6, 15.3, 62, 21.9, 18.6, 21.3   25.6; 21.3; 18.6; 46.7; 15.0

CHAPTER 8

Graph the function. State the domain and range. (Lesson 8.1)   1–12. See margin.

1. $y = 3^x$ **2.** $y = 3 \cdot 4^x$ **3.** $y = 5(1.5)^x$ **4.** $y = 4(2)^x$

5. $y = 2 \cdot 7^{x - 1}$ **6.** $y = -\frac{1}{2}(2.5)^x$ **7.** $y = 3^{x - 1}$ **8.** $y = 3^{x - 2} + 1$

9. $y = 2^{x - 2} + 4$ **10.** $y = 4 \cdot 5^{x - 1} - 2$ **11.** $y = 3 \cdot 2^{x + 2}$ **12.** $y = 5 \cdot 2^{x - 3}$

Graph the function. State the domain and range. (Lesson 8.2)   13–20. See margin.

13. $y = \left(\frac{1}{2}\right)^x$ **14.** $y = 2\left(\frac{1}{3}\right)^x$ **15.** $y = -3\left(\frac{1}{4}\right)^x$ **16.** $y = -\left(\frac{1}{5}\right)^x$

17. $y = (0.25)^x$ **18.** $y = -2\left(\frac{1}{4}\right)^x - 1$ **19.** $y = \left(\frac{2}{3}\right)^x + 3$ **20.** $y = -5\left(\frac{1}{2}\right)^x$

Simplify the expression. (Lesson 8.3)

21. $e^4 \cdot e^3$ e^7 **22.** $e^{-6} \cdot e^7$ e **23.** $4e^{3x} \cdot 4e^{3x}$ $16e^{6x}$ **24.** $\left(7e^{-x}\right)^{-2}$ $\frac{e^{2x}}{49}$

25. $\frac{10e^x}{e^{3x}}$ $\frac{10}{e^{2x}}$ **26.** $\sqrt[3]{64e^{6x}}$ $4e^{2x}$ **27.** $e^{2x} \cdot e^{4x - 1}$ $e^{6x - 1}$ **28.** $\frac{e^x}{5e}$ $\frac{e^{x - 1}}{5}$

29. $\frac{20e^{4x}}{5e}$ $4e^{4x - 1}$ **30.** $\left(6e^{-2x}\right)^3$ $216e^{-8x^3}$ **31.** $\sqrt{16e^{8x}}$ $4e^{4x}$ **32.** $\left(\frac{1}{3}e^{-3}\right)^{-3}$ $27e^9$

Graph the function. State the domain and range. (Lesson 8.3)   33–40. See margin.

33. $y = e^{0.5x}$ **34.** $y = e^{-0.75x}$ **35.** $y = 2e^{-(x - 1)}$ **36.** $y = 0.5e^{-x}$

37. $y = \frac{1}{2}e^{x - 3} + 1$ **38.** $y = 3e^{x - 2} - 4$ **39.** $y = \frac{1}{3}e^{-2(x - 1)} - 2$ **40.** $y = 0.1e^{2x} - 3$

41. MOUNT FUJI The relationship between air pressure and altitude can be modeled by $P = 14.7e^{-0.00004h}$ where P is the air pressure (in pounds per square inch) and h is the altitude (in feet above sea level). Mount Fuji in Japan rises to a height of 12,388 feet above sea level. What is the air pressure at the peak of Mount Fuji? (Lesson 8.3) about 8.96 pounds per square inch

Evaluate the expression without using a calculator. (Lesson 8.4)

42. $\log_2 16$ 4
43. $\log_5 25$ 2
44. $\log_{11} 1$ 0
45. $\log_{1/4} 2$ $-\frac{1}{2}$

46. $\log_3 3^{-2.16}$ -2.16
47. $\log_7 343$ 3
48. $\log_{29} 29$ 1
49. $\log_9 9^3$ 3

Find the inverse of the function. (Lesson 8.4)

50. $y = \log_4 x$ $y = 4^x$
51. $y = \log_{1/3} x$ $y = \left(\frac{1}{3}\right)^x$
52. $y = \log_6 36^x$ $y = \frac{1}{2}x$

53. $y = \ln 3x$ $y = \frac{1}{3}e^x$
54. $y = \ln(x + 1)$ $y = e^x - 1$
55. $y = \ln(x - 3)$ $y = e^x + 3$

Graph the function. State the domain and range. (Lesson 8.4)   56–63. See margin.

56. $y = \log_3 x$
57. $y = \ln x - 2$
58. $y = \log x + 4$
59. $y = \ln(x - 3)$

60. $y = \log_5(x + 2)$
61. $y = \ln x + 7$
62. $y = \log_{1/2} x + 1$
63. $y = \log_5 x + 2$

Use a property of logarithms to evaluate the expression. (Lesson 8.5)

64. $\log_2(4 \cdot 8)$ 5
65. $\ln e^3$ 3
66. $\log_2 8^2$ 6
67. $\log_6 216$ 3

68. $\log \frac{1}{100}$ -2
69. $\ln \frac{1}{e^5}$ -5
70. $\log 0.001$ -3
71. $\log_3 27^2$ 6

Expand the expression. (Lesson 8.5)

72. $\log_3 9x$ $2 + \log_3 x$
73. $\log 3x^4$ $\log 3 + 4 \log x$
74. $\log_6 x^5$ $5 \log_6 x$
75. $\ln 15x$ $\ln 15 + \ln x$

76. $\log_7 49x^2$ $2 + 2 \log_7 x$
77. $\log \sqrt{9x}$ $\log 3 + \frac{1}{2}\log x$
78. $\ln x^{1/3}y^4$ $\frac{1}{3}\ln x + 4 \ln y$
79. $\log x^2y^3z^4$
  $2 \log x + 3 \log y + 4 \log z$

Condense the expression. (Lesson 8.5)

80. $\log_4 7 + \log_4 10 - \log_4 2$
   $\log_4 35$
81. $4 \ln x + 6 \ln y + 3 \ln z$
   $\ln x^4y^6z^3$
82. $5 \log_4 3 + 6 \log_4 x + 7 \log_4 y$
   $\log_4 243x^6y^7$

83. $\frac{1}{4}(\ln 9 - \ln x) + \frac{1}{4}\ln 3$
   $\ln \left(\frac{27}{x}\right)^{1/4}$
84. $6(\ln 3 + \ln x) + \frac{1}{4}\ln 3$
   $\ln 3^{25/4}x^6$
85. $3(\log_5 10 - \log_5 2) + \frac{1}{2}\log_5 \frac{1}{100}$
   $\log_5 5^3 10^{-1}$

Solve the equation. Check for extraneous solutions. (Lesson 8.6)

86. $3^x = 10$ 2.10
87. $4^x - 3 = 11$ 1.90
88. $3^{x+2} = 9^{x+1}$ 0
89. $10^x + 4 = 10$ 0.778

90. $\ln 8x = 4$ 6.82
91. $\ln(5 - x) = 12$
  $-163,000$
92. $\log_3 x = 4$ 81
93. $\log_5(2x + 10) = \log_5 4x$ 5

Write an exponential function of the form $y = ab^x$ whose graph passes through the given points. (Lesson 8.7)

94. (2, 18), (1, 6) $y = 2(3)^x$
95. (0, 0.5), (3, 4) $y = 0.5(2)^x$
96. (−1, 6), (1, 0.5)
   $y = 1.73(0.289)^x$
97. (−2, 0.01), (1, 1.25)
   $y = 0.25(5)^x$

98. $(3, 9), \left(8, \frac{25}{4}\right)$
   $y = 11.2(0.93)^x$
99. $\left(-1, \frac{1}{4}\right), \left(2, \frac{3}{8}\right)$
   $y = 0.286(1.14)^x$
100. $(2, 27), \left(-2, \frac{1}{3}\right)$
   $y = 3(3)^x$
101. (1, −8), (0, −2)
   $y = -2(4)^x$

102. EXPANDING BUSINESS The table below shows the number s of stores owned by a company from 1987 to 1998 where t represents the number of years since 1987. Find an exponential model for the data. Then use the model to predict how many more stores there will be in 2006. (Lesson 8.7) $s = 21.2(1.53)^t$; 68,466 stores

t	0	1	2	3	4	5	6	7	8	9	10	11
s	17	33	55	84	116	165	272	425	676	1015	1412	1900

Write a power function of the form $y = ax^b$ whose graph passes through the given points. (Lesson 8.7)

103. (−2, −8), (3, 27)
   $y = x^3$
104. (1, 5), (4, 10) $y = 5x^{1/2}$
105. (1, 2), (4, 4) $y = 2x^{1/2}$
106. (2, 4), (3, 37)
   $y = 0.0892x^{5.49}$

107. (−1, −3), (3, 81)
   $y = 3x^3$
108. (2, 1), (6, 9) $y = \frac{1}{4}x^2$
109. (−1, 0.5), (4, 8)
   $y = 0.5x^2$
110. (−5, −8), (−10, −32)
   $y = -\frac{8}{25}x^2$

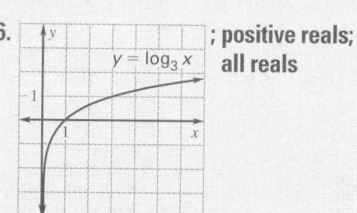
56. ; positive reals; all reals $y = \log_3 x$

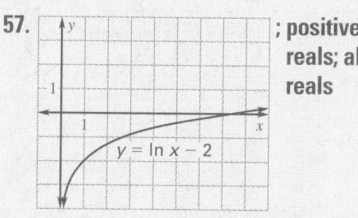
57. ; positive reals; all reals $y = \ln x - 2$

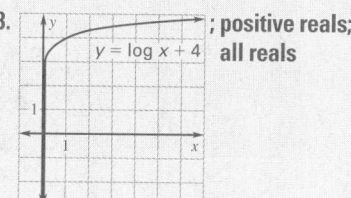
58. ; positive reals; all reals $y = \log x + 4$

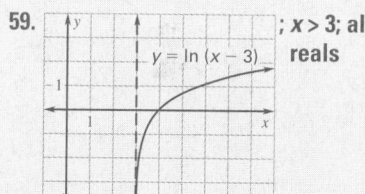
59. ; $x > 3$; all reals $y = \ln(x - 3)$

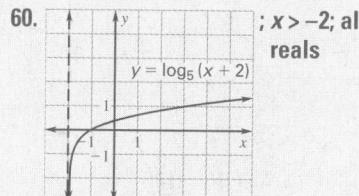
60. ; $x > -2$; all reals $y = \log_5(x + 2)$

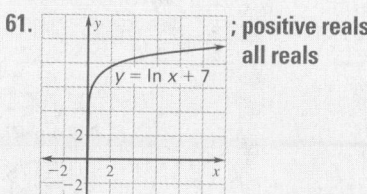
61. ; positive reals; all reals $y = \ln x + 7$

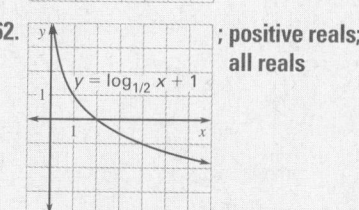
62. ; positive reals; all reals $y = \log_{1/2} x + 1$

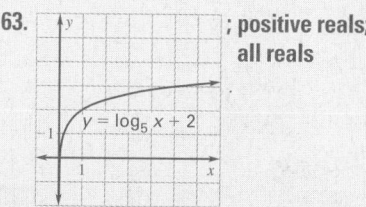
63. ; positive reals; all reals $y = \log_5 x + 2$

Extra Practice   951

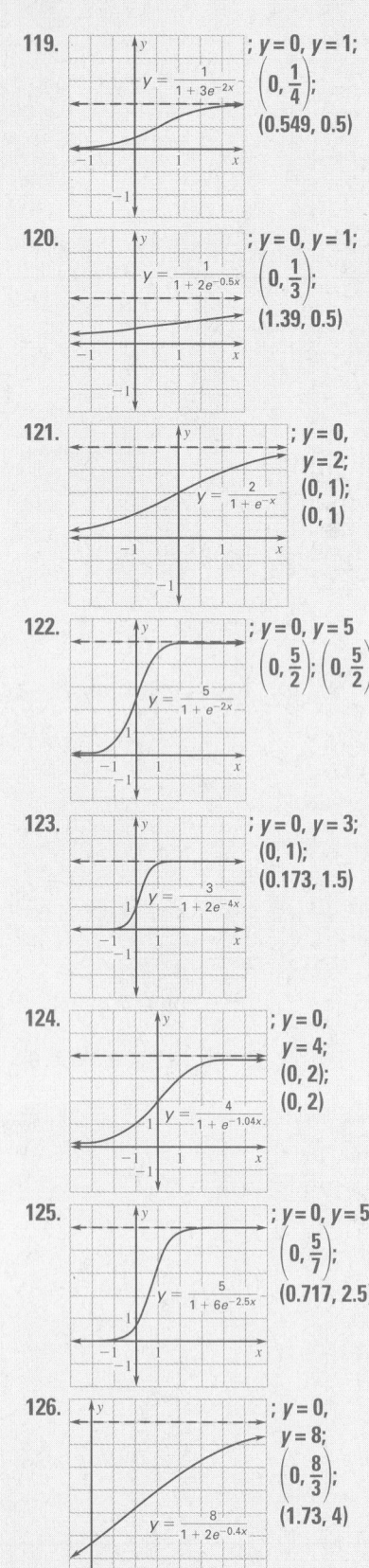

119. ; $y = 0$, $y = 1$; $\left(0, \frac{1}{4}\right)$; $y = \frac{1}{1 + 3e^{-2x}}$ (0.549, 0.5)

120. ; $y = 0$, $y = 1$; $\left(0, \frac{1}{3}\right)$; $y = \frac{1}{1 + 2e^{-0.5x}}$ (1.39, 0.5)

121. ; $y = 0$, $y = 2$; $(0, 1)$; $y = \frac{2}{1 + e^{-x}}$ (0, 1)

122. ; $y = 0$, $y = 5$; $\left(0, \frac{5}{2}\right)$; $\left(0, \frac{5}{2}\right)$; $y = \frac{5}{1 + e^{-2x}}$

123. ; $y = 0$, $y = 3$; $(0, 1)$; $y = \frac{3}{1 + 2e^{-4x}}$ (0.173, 1.5)

124. ; $y = 0$, $y = 4$; $(0, 2)$; $(0, 2)$; $y = \frac{4}{1 + e^{-1.04x}}$

125. ; $y = 0$, $y = 5$; $\left(0, \frac{5}{7}\right)$; $y = \frac{5}{1 + 6e^{-2.5x}}$ (0.717, 2.5)

126. ; $y = 0$, $y = 8$; $\left(0, \frac{8}{3}\right)$; (1.73, 4); $y = \frac{8}{1 + 2e^{-0.4x}}$

Evaluate the function $f(x) = \dfrac{6}{1 + 4e^{-x}}$ **for the given value of x.** (Lesson 8.8)

111. $f(1)$ 2.43
112. $f(2)$ 3.89
113. $f(-2)$ 0.196
114. $f(3)$ 5.00

115. $f(0)$ 1.2
116. $f(4.1)$ 5.63
117. $f(-0.6)$ 0.724
118. $f\left(\frac{1}{4}\right)$ 1.46

Graph the function. Identify the asymptotes, y-intercept, and the point of maximum growth. (Lesson 8.8) 119–126. See margin.

119. $y = \dfrac{1}{1 + 3e^{-2x}}$
120. $y = \dfrac{1}{1 + 2e^{-0.5x}}$
121. $y = \dfrac{2}{1 + e^{-x}}$
122. $y = \dfrac{5}{1 + e^{-2x}}$

123. $y = \dfrac{3}{1 + 2e^{-4x}}$
124. $y = \dfrac{4}{1 + e^{-1.04x}}$
125. $y = \dfrac{5}{1 + 6e^{-2.5x}}$
126. $y = \dfrac{8}{1 + 2e^{-0.4x}}$

CHAPTER 9

The variables x and y vary inversely. Use the given values to write an equation relating x and y. Then find y when x = 4. (Lesson 9.1)

1. $x = 3, y = 6$ $y = \frac{18}{x}$; 4.5
2. $x = 2, y = 8$ $y = \frac{16}{x}$; 4
3. $x = -1, y = 4$ $y = -\frac{4}{x}$; -1
4. $x = 2, y = 6$ $y = \frac{12}{x}$; 3

5. $x = -\frac{1}{3}, y = 9$ $y = -\frac{3}{x}$; $-\frac{3}{4}$
6. $x = \frac{1}{2}, y = 6$ $y = \frac{3}{x}$; $\frac{3}{4}$
7. $x = \frac{1}{2}, y = \frac{1}{8}$ $y = \frac{1}{16x}$; $\frac{1}{64}$
8. $x = \frac{2}{5}, y = \frac{1}{10}$ $y = \frac{1}{25x}$; $\frac{1}{100}$

The variable z varies jointly with x and y. Use the given values to write an equation relating x, y, and z. Then find z when x = 4 and y = 7. (Lesson 9.1)

9. $x = 2, y = 3, z = 6$ $z = xy$; 28
10. $x = -3, y = 6, z = 18$ $z = -xy$; -28
11. $x = 10, y = -15, z = 5$ $z = -\frac{xy}{30}$; $-\frac{14}{15}$

12. $x = -1, y = 2, z = 4$ $z = -2xy$; -56
13. $x = \frac{3}{4}, y = \frac{1}{2}, z = 8$ $z = \frac{64xy}{3}$; $\frac{1792}{3}$
14. $x = \frac{1}{5}, y = \frac{7}{8}, z = \frac{1}{2}$ $z = \frac{20xy}{7}$; 80

Graph the function. State the domain and range. (Lesson 9.2) 15–22. See margin.

15. $y = \dfrac{3}{x}$
16. $y = \dfrac{2}{x - 3} + 1$
17. $y = \dfrac{5}{2x + 1} - 4$
18. $y = \dfrac{x + 1}{x + 6}$

19. $y = \dfrac{x - 2}{x + 1}$
20. $y = \dfrac{x + 3}{x + 4}$
21. $y = \dfrac{x}{2x - 5}$
22. $y = \dfrac{4x}{-x - 3}$

FUNDRAISER In Exercises 23–25, your school is publishing a calendar to raise money for a local charity. The total cost of using the photos in the calendar is $710. In addition to this "one-time" charge, the unit cost of printing each calendar is $4.50. (Lesson 9.2)

23. Write a model that gives the average cost per calendar as a function of the number of calendars printed. $C = \dfrac{4.50n + 710}{n}$

24. Graph the model and use the graph to estimate the number of calendars you need to print before the average cost decreases to $6 per calendar. about 475 calendars; See margin for graph.

25. Describe what happens to the average cost as the number of calendars printed increases. The average cost decreases as the number of calendars printed increases.

Graph the function. (Lesson 9.3) 26–33. See margin.

26. $y = \dfrac{2x^2 + 1}{x + 3}$
27. $y = \dfrac{7}{x^2 + 5}$
28. $y = \dfrac{x^2 + 9}{x^2 + 2}$
29. $y = \dfrac{x^2 - 5x - 6}{x - 6}$

30. $y = \dfrac{5 - x}{3x^2 - 2x + 1}$
31. $y = \dfrac{x^2 + 8x + 15}{2x}$
32. $y = \dfrac{-3x^2}{x^2 - 9}$
33. $y = \dfrac{x^2 - 3x - 10}{2x}$

15–22, 24, 26–33 See Additional Answers beginning on page AA1.

Perform the indicated operation. Simplify the result. (Lesson 9.4)

34. $\dfrac{3xy^5}{x^2y^3} \cdot \dfrac{y^2}{6x} \cdot \dfrac{y^4}{2x^2}$

35. $\dfrac{20x^5}{y^2} \cdot \dfrac{x^2y^2}{10x^3}$ $2x^4$

36. $\dfrac{x^2-4}{x-3} \cdot \dfrac{x+2}{8x-16}$ $\dfrac{(x+2)^2}{8(x-3)}$

37. $\dfrac{x^3+3x^2}{2x} \cdot \dfrac{5x^3}{x^2+5x+6}$ $\dfrac{5x^4}{2x+4}$

38. $\dfrac{7x^2-14x}{x^3} \div \dfrac{5x-10}{x^5}$ $\dfrac{7x^3}{5}$

39. $\dfrac{x^2-x-20}{x+4} \cdot \dfrac{x-3}{x^2-2x-15}$ $\dfrac{x-3}{x+3}$

40. $(x^2+5x-36) \div \dfrac{5x^2+45x}{x-6}$ $\dfrac{(x-4)(x-6)}{5x}$

41. $(x^3+8) \cdot \dfrac{6x^3-9x^2}{3x^3-12x}$ $\dfrac{x(x^2-2x+4)(2x-3)}{x-2}$

42. $\dfrac{x^2+2x-35}{x^2-7x+12} \div \dfrac{x^2-13x+40}{3x^2-12x}$ $\dfrac{3x(x+7)}{(x-3)(x-8)}$

Perform the indicated operation and simplify. (Lesson 9.5)

43. $\dfrac{3}{5x} + \dfrac{9}{5x}$ $\dfrac{12}{5x}$

44. $\dfrac{15}{6x^2} - \dfrac{8}{6x^2}$ $\dfrac{7}{6x^2}$

45. $\dfrac{4}{3x} + \dfrac{2}{5x}$ $\dfrac{26}{15x}$

46. $\dfrac{3}{2(x-1)} + \dfrac{x+1}{4}$ $\dfrac{x^2+5}{4x-4}$

47. $\dfrac{2x+1}{x^2-4} + \dfrac{5}{x-2}$ $\dfrac{7x+11}{(x+2)(x-2)}$

48. $\dfrac{4-9x}{x+5} + \dfrac{1}{2x-1}$ $\dfrac{-18x^2+18x+1}{(x+5)(2x-1)}$

49. $\dfrac{7}{x^2+8x+15} - \dfrac{3}{x+5}$ $\dfrac{-3x-2}{(x+3)(x+5)}$

50. $\dfrac{8x-1}{x^2+x-6} - \dfrac{4}{x-2}$ $\dfrac{4x-13}{(x-2)(x+3)}$

Simplify the complex fraction. (Lesson 9.5)

51. $\dfrac{\dfrac{4}{x}-4}{2+\dfrac{1}{x}}$ $\dfrac{-4x+4}{2x+1}$

52. $\dfrac{\dfrac{9}{x+1}}{\dfrac{1}{3}-\dfrac{6}{x+1}}$ $\dfrac{27}{x-17}$

53. $\dfrac{\dfrac{7}{5x+2}-\dfrac{3}{2(5x+2)}}{\dfrac{x^2}{5x+2}}$ $\dfrac{11}{2x^2}$

54. $\dfrac{\dfrac{2}{3x^2-3}}{\dfrac{1}{x+1}+\dfrac{3x}{x^2-2x-3}}$ See margin.

55. $\dfrac{\dfrac{2}{3x-1}-\dfrac{5}{4(3x+1)}}{\dfrac{x}{9x^2-1}}$ $\dfrac{9x+13}{4x}$

56. $\dfrac{\dfrac{2}{x^2-4}+\dfrac{1}{x-2}}{\dfrac{5}{x-2}+\dfrac{3}{x+2}}$ $\dfrac{x+4}{8x+4}$

57. $\dfrac{\dfrac{8}{x^2-49}}{\dfrac{5}{3x^2-21x}-\dfrac{6}{x-7}}$ See margin.

58. $\dfrac{\dfrac{2}{3x^2+6x+12}+\dfrac{x}{x^3-8}}{\dfrac{3x}{2x^2+4}-\dfrac{x-2}{4x^2+8}}$ See margin.

Solve the equation using any method. Check each solution. (Lesson 9.6)

59. $\dfrac{7}{x} + \dfrac{1}{2} = 4$ 2

60. $\dfrac{x}{4} + \dfrac{1}{2} = 5$ 18

61. $\dfrac{4}{x} + \dfrac{1}{3} = 10$ $\dfrac{12}{29}$

62. $\dfrac{1}{2x} + \dfrac{x}{3} = 7$ $0.0717, 20.9$

63. $\dfrac{-2}{x+3} = \dfrac{1}{x+1}$ $-\dfrac{5}{3}$

64. $\dfrac{4}{x+2} = \dfrac{-3}{x-3}$ $\dfrac{6}{7}$

65. $\dfrac{-4}{x+1} = \dfrac{2}{x-1}$ $\dfrac{1}{3}$

66. $\dfrac{3}{x+4} = \dfrac{9}{x-2}$ -7

67. $\dfrac{4x}{x-1} = \dfrac{x}{x^2-1}$ $-\dfrac{3}{4}, 0$

68. $\dfrac{5x}{10-x} = \dfrac{x^2}{x-10}$ $-5, 0$

69. $\dfrac{3}{x^2-9} = \dfrac{6}{x+3}$ $\dfrac{7}{2}$

70. $\dfrac{3}{x^2-4} = \dfrac{2}{x+2} + \dfrac{x}{x-2}$ $-5.32, 1.32$

CHAPTER 10

Find the distance between the two points. Then find the midpoint of the line segment joining the two points. (Lesson 10.1)

1. $(0, 0), (6, 8)$ $10; (3, 4)$

2. $(0, 5), (-2, 0)$ $5.39; \left(-1, \dfrac{5}{2}\right)$

3. $(-4, -2), (1, -5)$ $5.83; \left(-\dfrac{3}{2}, -\dfrac{7}{2}\right)$

4. $(3, 3), (3, 6)$ $3; \left(3, \dfrac{9}{2}\right)$

5. $(4.5, 2), (1.5, 6)$ $5; (3, 4)$

6. $(-3, 5), \left(\dfrac{1}{2}, 6\right)$ See margin.

7. $(-5, 2.3), (-3, 4.7)$ See margin.

8. $\left(-\dfrac{1}{4}, 6\right), \left(8, -\dfrac{3}{4}\right)$ See margin.

Graph the equation. Identify the focus and directrix of the parabola. (Lesson 10.2) 9–16. See margin.

9. $y^2 = 10x$

10. $x^2 = -4y$

11. $y^2 = -6x$

12. $y^2 = 11x$

13. $x^2 - 16y = 0$

14. $6x^2 = 5y$

15. $x - \dfrac{1}{8}y^2 = 0$

16. $x + \dfrac{1}{10}y^2 = 0$

Write the standard form of the equation of the parabola with the given focus or directrix and vertex at (0, 0). (Lesson 10.2)

17. $(3, 0)$ $y^2 = 12x$

18. $(0, -4)$ $x^2 = -16y$

19. $\left(\dfrac{1}{2}, 0\right)$ $y^2 = 2x$

20. $\left(0, -\dfrac{1}{8}\right)$ $x^2 = -\dfrac{1}{2}y$

21. $y = 6$ $x^2 = -24y$

22. $x = -2$ $y^2 = 8x$

23. $y = \dfrac{3}{4}$ $x^2 = -3y$

24. $x = -\dfrac{7}{8}$ $y^2 = \dfrac{7}{2}x$

54. $\dfrac{2(x-3)}{3(x-1)(4x-3)}$

57. $\dfrac{24x}{(x+7)(5-18x)}$

58. $\dfrac{4(5x-4)(x^2+2)}{3(x^3-8)(5x+2)}$

6. $3.64; \left(-\dfrac{5}{4}, \dfrac{11}{2}\right)$

7. $3.12; \left(-4, \dfrac{7}{2}\right)$

8. $10.7; \left(\dfrac{31}{8}, \dfrac{21}{8}\right)$

9. 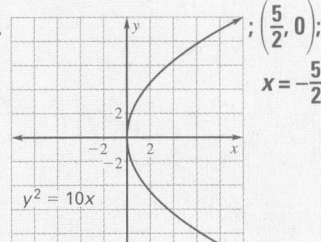 $; \left(\dfrac{5}{2}, 0\right); \quad x = -\dfrac{5}{2}$

$y^2 = 10x$

10. 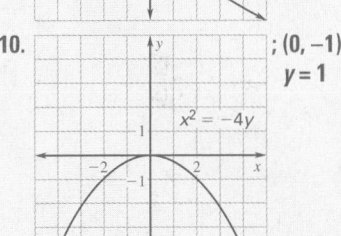 $; (0, -1); \quad y = 1$

$x^2 = -4y$

11. 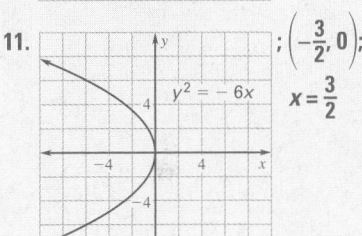 $; \left(-\dfrac{3}{2}, 0\right); \quad x = \dfrac{3}{2}$

$y^2 = -6x$

12. 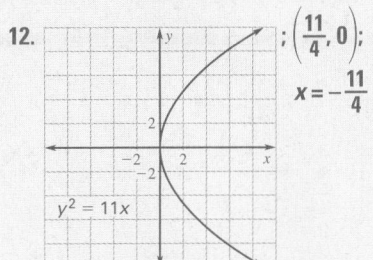 $; \left(\dfrac{11}{4}, 0\right); \quad x = -\dfrac{11}{4}$

$y^2 = 11x$

13. 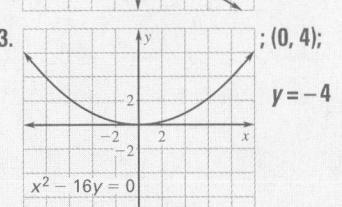 $; (0, 4); \quad y = -4$

$x^2 - 16y = 0$

14–16. See Additional Answers beginning on page AA1.

953

25.

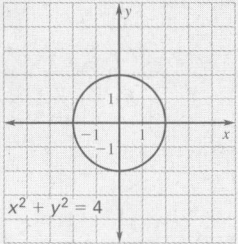

$x^2 + y^2 = 4$

26.

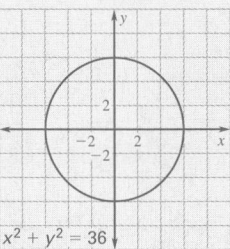

$x^2 + y^2 = 36$

27.

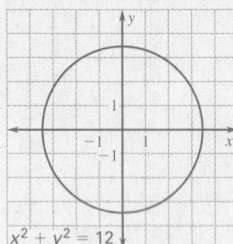

$x^2 + y^2 = 12$

28.

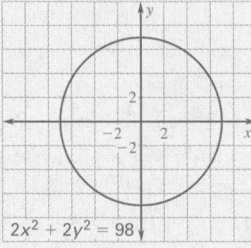

$2x^2 + 2y^2 = 98$

29.

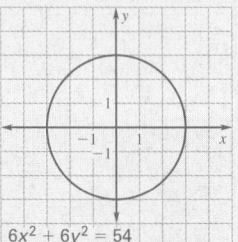

$6x^2 + 6y^2 = 54$

30.

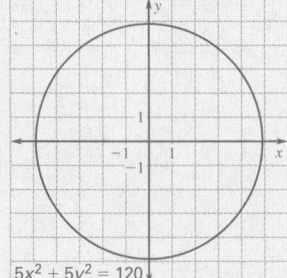

$5x^2 + 5y^2 = 120$

45–50, 55–57, 59–66. See Additional Answers beginning on page AA1.

954

Graph the equation. (Lesson 10.3) 25–32. See margin.

25. $x^2 + y^2 = 4$ **26.** $x^2 + y^2 = 36$ **27.** $x^2 + y^2 = 12$ **28.** $2x^2 + 2y^2 = 98$

29. $6x^2 + 6y^2 = 54$ **30.** $5x^2 + 5y^2 = 120$ **31.** $9x^2 + 9y^2 = 126$ **32.** $15x^2 + 15y^2 = 300$

Write the standard form of the equation of the circle that has the given radius or passes through the given point and whose center is the origin. (Lesson 10.3)

33. $r = 4$ $x^2 + y^2 = 16$ **34.** $r = 8$ $x^2 + y^2 = 64$ **35.** $r = \sqrt{13}$ $x^2 + y^2 = 13$ **36.** $r = 2\sqrt{5}$ $x^2 + y^2 = 20$

37. $(0, 6)$ $x^2 + y^2 = 36$ **38.** $(-4, 1)$ $x^2 + y^2 = 17$ **39.** $(2, 5)$ $x^2 + y^2 = 29$ **40.** $(-6, -2)$ $x^2 + y^2 = 40$

OCEAN NAVIGATION In Exercises 41 and 42, the beam of a lighthouse can be seen for up to 10 miles. You are on a ship that is 5 miles east and 7 miles north of the lighthouse. **(Lesson 10.3)**

41. Write an inequality to describe the region lit by the lighthouse beam. $x^2 + y^2 \le 100$

42. Can you see the lighthouse beam from the ship? yes

Graph the equation. Then identify the vertices, co-vertices, and foci of the ellipse. (Lesson 10.4) 43–50. See margin.

43. $\dfrac{x^2}{9} + \dfrac{y^2}{16} = 1$ **44.** $\dfrac{x^2}{81} + \dfrac{y^2}{36} = 1$ **45.** $\dfrac{x^2}{25} + \dfrac{y^2}{49} = 1$ **46.** $\dfrac{x^2}{25} + \dfrac{y^2}{144} = 1$

47. $\dfrac{x^2}{169} + \dfrac{y^2}{225} = 1$ **48.** $x^2 + \dfrac{y^2}{4} = 1$ **49.** $\dfrac{x^2}{16} + y^2 = 4$ **50.** $\dfrac{x^2}{4} + y^2 = 81$

Write an equation of the ellipse with the given characteristics and center at (0, 0). (Lesson 10.4)

51. Vertex: $(0, 8)$ $\dfrac{x^2}{25} + \dfrac{y^2}{4} = 1$ Co-vertex: $(4, 0)$ $\dfrac{x^2}{16} + \dfrac{y^2}{64} = 1$
52. Vertex: $(5, 0)$ Co-vertex: $(0, -3)$
53. Vertex: $(-7, 0)$ $\dfrac{x^2}{49} + \dfrac{y^2}{4} = 1$ Co-vertex: $(0, -2)$
54. Vertex: $(-2, 0)$ $\dfrac{x^2}{4} + y^2 = 1$ Focus: $(-\sqrt{3}, 0)$

55. Vertex: $(16, 0)$ Focus: $(2\sqrt{39}, 0)$ See margin.
56. Vertex: $(0, 13)$ Focus: $(0, 12)$ See margin.
57. Co-vertex: $(-2, 0)$ Focus: $(0, 2\sqrt{99})$ See margin.
58. Co-vertex: $(0, -3)$ Focus: $(-\sqrt{7}, 0)$ $\dfrac{x^2}{16} + \dfrac{y^2}{9} = 1$

Graph the equation. Identify the foci and asymptotes. (Lesson 10.5) 59–66. See margin.

59. $\dfrac{x^2}{49} - \dfrac{y^2}{64} = 1$ **60.** $\dfrac{x^2}{25} - y^2 = 1$ **61.** $\dfrac{x^2}{10} - \dfrac{y^2}{6} = 1$ **62.** $\dfrac{x^2}{81} - \dfrac{y^2}{25} = 1$

63. $\dfrac{y^2}{36} - x^2 = 1$ **64.** $y^2 - 25x^2 = 25$ **65.** $x^2 - 16y^2 = 144$ **66.** $100x^2 - 49y^2 = 4900$

Write an equation for the conic section. (Lesson 10.6)

67. Circle with center at $(3, 4)$ and radius 5 $(x - 3)^2 + (y - 4)^2 = 25$
68. Parabola with vertex at $(2, -1)$ and focus at $(2, 1)$ $(x - 2)^2 = 8(y + 1)$

69. Circle with center at $(2, -5)$ and radius 7 $(x - 2)^2 + (y + 5)^2 = 49$
70. Parabola with vertex at $(-2, 5)$ and focus at $(3, 5)$ $(y - 5)^2 = 20(x + 2)$
71. Ellipse with vertices at $(-2, 3)$ and $(8, 3)$ and foci at $(-1, 3)$ and $(7, 3)$ $\dfrac{(x - 3)^2}{25} + \dfrac{(y - 3)^2}{9} = 1$

72. Ellipse with vertices at $(-9, 1)$ and $(5, 1)$ and co-vertices at $(-2, 6)$ and $(-2, -4)$ $\dfrac{(x + 2)^2}{49} + \dfrac{(y - 1)^2}{25} = 1$

73. Hyperbola with vertices at $(5, 6)$ and $(-1, 6)$ and foci at $(-2, 6)$ and $(6, 6)$ $\dfrac{(x - 2)^2}{9} - \dfrac{(y - 6)^2}{7} = 1$

74. Hyperbola with vertices at $(4, -2)$ and $(4, -6)$ and foci at $(4, 1)$ and $(4, -9)$ $\dfrac{(y + 4)^2}{4} - \dfrac{(x - 4)^2}{21} = 1$

Classify the conic section. (Lesson 10.6)

75. $x^2 - 3y + 10 = 0$ parabola   **76.** $8x^2 + 8y^2 + 64x + 32y - 160 = 0$ circle

77. $16x^2 - 9y^2 + 96x + 18y - 135 = 0$ hyperbola   **78.** $25x^2 + 16y^2 + 100x - 128y - 44 = 0$ ellipse

Find the points of intersection, if any, of the graphs in the system. (Lesson 10.7)

79. $x^2 + y^2 = 20$
$x - y = -2$ (2, 4),
$(5, 22), (-3, 6)$

80. $x^2 - y^2 = 3$
$2x - y = -12$
$(-3, 6)$

81. $x^2 + y^2 = 6$ none
$x - y = 5$

82. $4x^2 + y^2 = 16$
$y = x - 2$
$\left(-\dfrac{6}{5}, -\dfrac{16}{5}\right), (2, 0)$

83. $x^2 - y^2 = 9$ $(-4, -2)$
$y = 2x - 6$ (3, 0), (5, 4)

84. $2x^2 + y^2 = 10$
$y = x - 3$
$(2.15, -0.845), (-0.155, -3.15)$

85. $x^2 + y^2 = 8$
$\dfrac{1}{4}x + \dfrac{1}{2}y = \dfrac{1}{2}$
$\left(\dfrac{14}{5}, -\dfrac{2}{5}\right), (-2, 2)$

86. $5x^2 + 4y^2 = 12$
$y = \dfrac{1}{3}x + 1$
$(0.992, 1.33), (-1.48, 0.506)$

CHAPTER 11

Write the next term in the sequence. Then write a rule for the nth term. (Lesson 11.1)

1. 2, 5, 8, 11, . . .   14; $a_n = 3n - 1$

2. 5, 10, 20, 40, . . .   80; $a_n = 5(2)^{n-1}$

3. 3, -1, -5, -9, . . .   -13; $a_n = -4n + 7$

4. $\dfrac{1}{3}, 1, \dfrac{5}{3}, \dfrac{7}{3}, \ldots$ 3; $a_n = \dfrac{2}{3}n - \dfrac{1}{3}$

5. -1, 4, -16, 64, . . .
-256; $a_n = -(-4)^{n-1}$

6. $\dfrac{2}{3}, \dfrac{3}{6}, \dfrac{4}{9}, \dfrac{5}{12}, \ldots$ $\dfrac{6}{15}$; $a_n = \dfrac{n+1}{3n}$

7. $\dfrac{1}{4}, \dfrac{1}{16}, \dfrac{1}{64}, \dfrac{1}{256}, \ldots$
$\dfrac{1}{1024}$; $a_n = \dfrac{1}{4^n}$

8. $\dfrac{1}{3}, 3, 27, 243, \ldots$
2187; $a_n = 3^{2n-3}$

Find the sum of the series. (Lesson 11.1)

9. $\displaystyle\sum_{i=1}^{8} 2i$ 72

10. $\displaystyle\sum_{i=1}^{4} (6i + 1)$ 64

11. $\displaystyle\sum_{i=0}^{5} i^2$ 55

12. $\displaystyle\sum_{k=1}^{7} 3k^2$ 420

13. $\displaystyle\sum_{n=2}^{6} n^3$ 440

14. $\displaystyle\sum_{n=1}^{3} \dfrac{n}{n+1}$ $\dfrac{23}{12}$

15. $\displaystyle\sum_{k=1}^{6} \dfrac{k+1}{k}$ $\dfrac{169}{20}$

16. $\displaystyle\sum_{k=4}^{8} k(k-1)$ 160

Write a rule for the nth term of the arithmetic sequence. Then find a_{10}. (Lesson 11.2)

17. 1, 4, 7, 10, 13, . . .
$a_n = -2 + 3n$; 28

18. -3, 2, 7, 12, 17, . . .
$a_n = -8 + 5n$; 42

19. 8, -2, -12, -22, -32, . . .
$a_n = 18 - 10n$; -82

20. 2.4, 3.5, 4.6, 5.7, . . .
$a_n = 1.3 + 1.1n$; 12.3

21. $\dfrac{9}{4}, \dfrac{7}{4}, \dfrac{5}{4}, \dfrac{3}{4}, \dfrac{1}{4}, \ldots$ $a_n = \dfrac{11}{4} - \dfrac{1}{2}n$; $-\dfrac{9}{4}$

22. $d = 3$, $a_1 = 3.5$
$a_n = 0.5 + 3n$; 30.5

23. $d = -2$, $a_4 = 0$
$a_n = 2 - 2n$; -18

24. $d = 1.75$, $a_6 = 10.75$
$a_n = 0.25 + 1.75n$; 17.75

25. $a_1 = -5$, $a_8 = 23$
$a_n = -9 + 4n$; 31

SEATING CAPACITY In Exercises 26 and 27, the first row of a concert hall has 20 seats, and each row after the first has one more seat than the row before it. There are 30 rows of seats. (Lesson 11.2)

26. Write a rule for the number of seats in the nth row. $a_n = 19 + n$

27. Forty students from a class want to sit in the same row. How close to the front can they sit?   the 21st row

Write a rule for the nth term of the geometric sequence. Then find a_8. (Lesson 11.3)

28. 3, 6, 12, 24, . . .
$a_n = 3(2)^{n-1}$; 384

29. $-1, -\dfrac{1}{2}, -\dfrac{1}{4}, -\dfrac{1}{8}, \ldots$
$a_n = -\left(\dfrac{1}{2}\right)^{n-1}$; $-\dfrac{1}{128}$

30. 6, 42, 294, 2058, . . .
$a_n = 6(7)^{n-1}$; 4,941,258

31. $-10, 1, -\dfrac{1}{10}, \dfrac{1}{100}, \ldots$
$a_n = \left(-\dfrac{1}{10}\right)^{n-2}$; $\dfrac{1}{1,000,000}$

32. $r = 3$, $a_1 = 3$ $a_n = 3^n$; 6561

33. $r = 6$, $a_2 = -18$ $a_n = -\dfrac{1}{2}(6)^n$; -839,808

34. $r = 9$, $a_1 = -27$
$a_n = -3(9)^n$; -129,140,163

35. $a_1 = 150$, $a_3 = 6$ $a_n = 750\left(\dfrac{1}{5}\right)^n$; $\dfrac{6}{3125}$

36. $a_2 = 20$, $a_6 = 5120$ $a_n = 5(4)^{n-1}$; 81,920

Find the sum of the first n terms of the geometric series. (Lesson 11.3)

37. $1 + 5 + 25 + 125 + \cdots$
$n = 12$ 61,035,156

38. $5 + 10 + 20 + 40 + \cdots$
$n = 10$ 5115

39. $4 + (-12) + 36 + (-108) + \cdots$
$n = 6$ -728

40. $100 + 50 + 25 + \dfrac{25}{2} + \cdots$
$n = 8$ $\dfrac{6375}{32}$

41. $60 + 10 + \dfrac{10}{6} + \dfrac{10}{36} + \cdots$
$n = 10$ $\dfrac{60,466,175}{839,808}$

42. $6 + (-12) + 24 + (-48) + \cdots$
$n = 7$ 258

31.
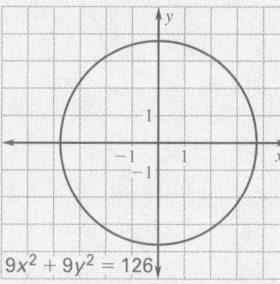
$9x^2 + 9y^2 = 126$

32.
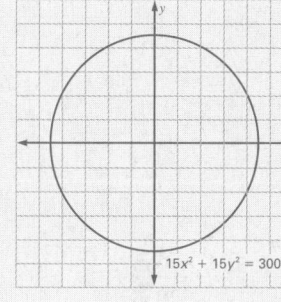
$15x^2 + 15y^2 = 300$

43.
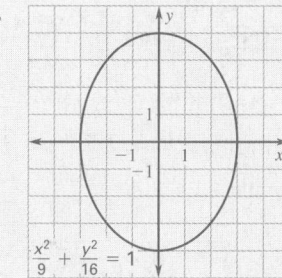
$\dfrac{x^2}{9} + \dfrac{y^2}{16} = 1$

(0, 4), (0, -4);
(-3, 0), (3, 0);
$(0, \sqrt{7}), (0, -\sqrt{7})$

44.
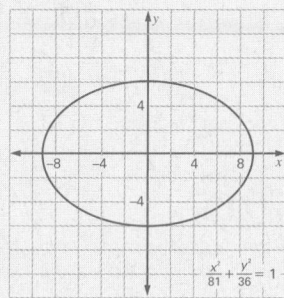
$\dfrac{x^2}{81} + \dfrac{y^2}{36} = 1$

(-9, 0), (9, 0);
(0, 6), (0, -6);
$(-3\sqrt{5}, 0), (3\sqrt{5}, 0)$

22. $x^4 + 12x^3 + 54x^2 + 108x + 81$
23. $x^7 + 7x^6y + 21x^5y^2 + 35x^4y^3 +$
    $35x^3y^4 + 21x^2y^5 + 7xy^6 + y^7$
24. $32x^5 - 80x^4y + 80x^3y^2 - 40x^2y^3 +$
    $10xy^4 - y^5$
25. $x^{12} + 12x^{10}y + 60x^8y^2 + 160x^6y^3 +$
    $240x^4y^4 + 192x^2y^5 + 64y^6$
26. $x^{12} + 12x^8 + 48x^4 + 64$
27. $243x^{10} - 2025x^8 + 6750x^6 -$
    $11{,}250x^4 + 9375x^2 - 3125$
28. $81x^4 - 108x^3y^2 + 54x^2y^4 - 12xy^6 +$
    y^8
29. $x^9 + 3x^6y^3 + 3x^3y^6 + y^9$

Find the sum of the infinite geometric series if it has one. (Lesson 11.4)

43. $\sum\limits_{n=0}^{\infty} \left(\dfrac{1}{3}\right)^n$ $\dfrac{3}{2}$

44. $\sum\limits_{n=1}^{\infty} 2\left(\dfrac{1}{2}\right)^n$ 2

45. $\sum\limits_{n=0}^{\infty} \dfrac{4}{5}(3)^n$ none

46. $\sum\limits_{n=1}^{\infty} 2\left(\dfrac{1}{5}\right)^n$ $\dfrac{1}{2}$

47. $\sum\limits_{n=0}^{\infty} 3\left(\dfrac{7}{6}\right)^n$ none

48. $\sum\limits_{n=0}^{\infty} \left(\dfrac{1}{8}\right)^{n-1}$ $\dfrac{64}{7}$

49. $\sum\limits_{n=1}^{\infty} \dfrac{1}{3}\left(-\dfrac{1}{5}\right)^n$ $-\dfrac{1}{18}$

50. $\sum\limits_{n=0}^{\infty} \left(-\dfrac{1}{4}\right)^{n-1}$ $-\dfrac{16}{5}$

Write a recursive rule for the sequence. The sequence may be arithmetic, geometric, or neither. (Lesson 11.5)

51. $2, 6, 10, 14, \ldots$
    $a_1 = 2, a_n = a_{n-1} + 4$

52. $77, 11, \dfrac{11}{7}, \dfrac{11}{49}, \ldots$
    $a_1 = 77, a_n = \dfrac{a_{n-1}}{7}$

53. $2, 7, 22, 67, 202, \ldots$
    $a_1 = 2, a_n = a_{n-1} + 5(3)^{n-2}$

54. $11.6, 10.1, 8.6, 7.1, \ldots$
    $a_1 = 11.6, a_n = a_{n-1} - 1.5$

55. $-6, -9, 54, -486, \ldots$
    $a_1 = -6, a_2 = -9, a_n = a_{n-1} \cdot a_{n-2}$

56. $8, 8\sqrt{3}, 24, 24\sqrt{3}, \ldots$
    $a_1 = 8, a_n = \sqrt{3} \cdot a_{n-1}$

CHAPTER 12

Each event can occur in the given number of ways. Find the number of ways all of the events can occur. (Lesson 12.1)

1. Event 1: 2 ways, Event 2: 1 way 2
2. Event 1: 2 ways, Event 2: 3 ways, Event 3: 4 ways 24
3. Event 1: 3 ways, Event 2: 4 ways 12
4. Event 1: 3 ways, Event 2: 3 ways, Event 3: 6 ways, Event 4: 5 ways 270
5. **FOOD** At your school cafeteria you can order a taco with one meat filling and one cheese filling. You have a choice of 3 meats and 4 cheeses. How many ways can you order a taco with meat and cheese? (Lesson 12.1) 12

Find the number of distinguishable permutations of the letters in the word. (Lesson 12.1)

6. MATH 24
7. DOG 6
8. HELLO 60
9. MAINE 120
10. SCHOOL 360
11. STATISTICS 50,400
12. GEOMETRY 20,160
13. SISTERS 840

Find the number of combinations. (Lesson 12.2)

14. $_{10}C_3$ 120
15. $_6C_2$ 15
16. $_{11}C_5$ 462
17. $_8C_8$ 1
18. $_{15}C_4$ 1365
19. $_1C_1$ 1
20. $_9C_8$ 9
21. $_{20}C_3$ 1140

Expand the binomial. (Lesson 12.2) 22–29. See margin.

22. $(x + 3)^4$
23. $(x + y)^7$
24. $(2x - y)^5$
25. $(x^2 + 2y)^6$
26. $(x^4 + 4)^3$
27. $(3x^2 - 5)^5$
28. $(3x - y^2)^4$
29. $(x^3 + y^3)^3$

A card is drawn randomly from a standard 52-card deck. Find the probability of drawing the given card. (Lesson 12.3)

30. ace of diamonds $\dfrac{1}{52}$
31. any king $\dfrac{1}{13}$
32. a club $\dfrac{1}{4}$
33. a red card $\dfrac{1}{2}$
34. a card other than 7 $\dfrac{12}{13}$
35. a face card (king, queen, jack) $\dfrac{3}{13}$

The results of rolling a six-sided die 200 times are shown in the table below. Use the table to find the experimental probability of each event. (Lesson 12.3)

Roll on die	1	2	3	4	5	6
Number of occurrences	20	40	33	37	34	36

36. rolling a 1 $\dfrac{1}{10}$
37. rolling an even number $\dfrac{113}{200}$
38. rolling a number greater than 1 $\dfrac{9}{10}$
39. rolling an odd number $\dfrac{87}{200}$
40. rolling a 5 or a 6 $\dfrac{7}{20}$
41. rolling a number other than 2 or 3 $\dfrac{127}{200}$

Find the probability that the spinner will stop on a given region. (Lesson 12.3)

42. $1\ \frac{1}{8}$ **43.** $3\ \frac{1}{8}$

44. even $\frac{1}{2}$ **45.** odd $\frac{1}{2}$

46. blue $\frac{1}{4}$ **47.** yellow $\frac{3}{8}$

Find the indicated probability. State whether A and B are mutually exclusive.
(Lesson 12.4)

48. $P(A) = 0.3$ 0; yes
$P(B) = 0.55$
$P(A \text{ or } B) = 0.85$
$P(A \text{ and } B) = \underline{?}$

49. $P(A) = 0.4$ 0.5; no
$P(B) = 0.2$
$P(A \text{ or } B) = \underline{?}$
$P(A \text{ and } B) = 0.1$

50. $P(A) = \frac{8}{15}$ $\frac{4}{15}$; yes
$P(B) = \underline{?}$
$P(A \text{ or } B) = \frac{12}{15}$
$P(A \text{ and } B) = 0$

51. $P(A) = 6\%$ 43%; yes
$P(B) = 37\%$
$P(A \text{ or } B) = \underline{?}$
$P(A \text{ and } B) = 0\%$

52. $P(A) = 40\%$ 32%; no
$P(B) = \underline{?}$
$P(A \text{ or } B) = 12\%$
$P(A \text{ and } B) = 60\%$

53. $P(A) = 38\%$ 4%; no
$P(B) = 6\%$
$P(A \text{ or } B) = 40\%$
$P(A \text{ and } B) = \underline{?}$

Find $P(A')$. (Lesson 12.4)

54. $P(A) = \frac{1}{2}$ $\frac{1}{2}$ **55.** $P(A) = 0$ 1 **56.** $P(A) = 1$ 0 **57.** $P(A) = \frac{3}{4}$ $\frac{1}{4}$

58. $P(A) = 0.6$ 0.4 **59.** $P(A) = 0.2$ 0.8 **60.** $P(A) = \frac{1}{12}$ $\frac{11}{12}$ **61.** $P(A) = \frac{15}{16}$ $\frac{1}{16}$

You are drawing marbles from a bag. There are 6 green, 4 yellow, and 5 blue marbles. Find the probability for the event if you replace the marble after each draw. (Lesson 12.5)

62. blue, then blue $\frac{1}{9}$ **63.** green, then yellow $\frac{8}{75}$ **64.** blue, then green $\frac{2}{15}$ **65.** yellow, then blue $\frac{4}{45}$

66. green, then green $\frac{4}{25}$ **67.** yellow, then green $\frac{8}{75}$ **68.** blue, then yellow $\frac{4}{45}$ **69.** green, then blue $\frac{2}{15}$

Calculate the probability of rolling a six-sided die 10 times and getting the given result. (Lesson 12.6)

70. exactly 3 sixes 0.155 **71.** exactly 4 ones 0.0543 **72.** all odd numbers 0.000977 **73.** exactly 3 evens 0.117

74. no odd numbers 0.000977 **75.** exactly 6 threes 0.00217 **76.** exactly 7 fives 0.000248 **77.** 8 rolls greater than four 0.00305

A normal distribution has a mean of 36 and a standard deviation of 5. Find the probability that a randomly selected x-value is in the given interval. (Lesson 12.7)

78. between 31 and 41 0.68 **79.** between 21 and 36 0.4985 **80.** between 31 and 46 0.815 **81.** less than 41 0.84

82. greater than 26 0.975 **83.** less than 51 0.9985 **84.** between 21 and 46 0.9735 **85.** greater than 36 0.5

CHAPTER 13

Evaluate the six trigonometric functions of the angle θ. (Lesson 13.1) 1–4. See margin.

1. 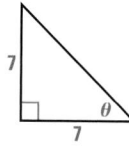 **2.** 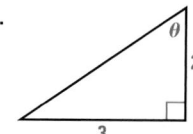 **3.** 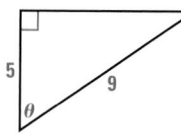 **4.**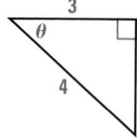

1. $\sin \theta = \frac{\sqrt{2}}{2}$; $\cos \theta = \frac{\sqrt{2}}{2}$, $\tan \theta = 1$;
$\sec \theta = \sqrt{2}$; $\csc \theta = \sqrt{2}$; $\cot \theta = 1$

2. $\sin \theta = \frac{3\sqrt{13}}{13}$; $\cos \theta = \frac{2\sqrt{13}}{13}$,
$\tan \theta = \frac{3}{2}$; $\sec \theta = \frac{\sqrt{13}}{2}$; $\csc \theta = \frac{\sqrt{13}}{3}$;
$\cot \theta = \frac{2}{3}$

3. $\sin \theta = \frac{2\sqrt{14}}{9}$; $\cos \theta = \frac{5}{9}$,
$\tan \theta = \frac{2\sqrt{14}}{5}$; $\sec \theta = \frac{9}{5}$;
$\csc \theta = \frac{9\sqrt{14}}{28}$; $\cot \theta = \frac{5\sqrt{14}}{28}$

4. $\sin \theta = \frac{\sqrt{7}}{4}$; $\cos \theta = \frac{3}{4}$, $\tan \theta = \frac{\sqrt{7}}{3}$;
$\sec \theta = \frac{4}{3}$; $\csc \theta = \frac{4\sqrt{7}}{7}$; $\cot \theta = \frac{3\sqrt{7}}{7}$

17. $\frac{4\pi}{3}$ in.; $\frac{8\pi}{3}$ in.^{2}

18. $\frac{259\pi}{180}$ ft; $\frac{1813\pi}{360}$ ft^2

19. $\frac{21\pi}{2}$ cm; $\frac{147\pi}{2}$ cm^2

20. $\frac{334\pi}{3}$ m; 6680π m^2

21. $\frac{\pi}{5}$ cm; $\frac{9\pi}{10}$ cm^2

22. $\frac{98\pi}{3}$ in.; $\frac{1372\pi}{3}$ in.^{2}

23. $\sin\theta = \frac{5\sqrt{41}}{41}$; $\cos\theta = \frac{4\sqrt{41}}{41}$;
$\tan\theta = \frac{5}{4}$; $\sec\theta = \frac{\sqrt{41}}{4}$
$\csc\theta = \frac{\sqrt{41}}{5}$; $\cot\theta = \frac{4}{5}$

24. $\sin\theta = -\frac{\sqrt{17}}{17}$; $\cos\theta = \frac{4\sqrt{17}}{17}$;
$\tan\theta = -\frac{1}{4}$; $\sec\theta = \frac{\sqrt{17}}{4}$;
$\csc\theta = -\sqrt{17}$; $\cot\theta = -4$

25. $\sin\theta = -\frac{3\sqrt{10}}{10}$; $\cos\theta = -\frac{\sqrt{10}}{10}$;
$\tan\theta = 3$; $\sec\theta = -\sqrt{10}$;
$\csc\theta = -\frac{\sqrt{10}}{3}$; $\cot\theta = \frac{1}{3}$

26. $\sin\theta = \frac{1}{7}$; $\cos\theta = -\frac{4\sqrt{3}}{7}$;
$\tan\theta = -\frac{\sqrt{3}}{12}$; $\sec\theta = -\frac{7\sqrt{3}}{12}$;
$\csc\theta = 7$; $\cot\theta = -4\sqrt{3}$

27. $\sin\theta = \frac{\sqrt{5}}{5}$; $\cos\theta = \frac{2\sqrt{5}}{5}$;
$\tan\theta = \frac{1}{2}$; $\sec\theta = \frac{\sqrt{5}}{2}$;
$\csc\theta = \sqrt{5}$; $\cot\theta = 2$

28. $\sin\theta = -\frac{12}{13}$; $\cos\theta = \frac{5}{13}$;
$\tan\theta = -\frac{12}{5}$; $\sec\theta = \frac{13}{5}$;
$\csc\theta = -\frac{13}{12}$; $\cot\theta = -\frac{5}{12}$

29. $\sin\theta = -\frac{2\sqrt{67}}{67}$; $\cos\theta = \frac{3\sqrt{469}}{67}$;
$\tan\theta = -\frac{2\sqrt{7}}{21}$; $\sec\theta = \frac{\sqrt{469}}{21}$;
$\csc\theta = -\frac{\sqrt{67}}{2}$; $\cot\theta = -\frac{3\sqrt{7}}{2}$

30. $\sin\theta = \frac{\sqrt{2}}{2}$; $\cos\theta = \frac{\sqrt{2}}{2}$; $\tan\theta = 1$;
$\sec\theta = \sqrt{2}$; $\csc\theta = \sqrt{2}$; $\cot\theta = 1$

31–34, 52–55. See Additional Answers beginning on page AA1.

Find one positive angle and one negative angle coterminal with the given angle. (Lesson 13.2) 5–16. *Sample answers* are given.

5. 35° 395°; −325°
6. −70° 290°; −430°
7. 125° 485°; −235°
8. 2° 362°; −358°
9. −45° 315°; −405°
10. 315° 675°; −45°
11. 585° 225°; −135°
12. 600° 240°; −120°
13. $\frac{2\pi}{3}$ $\frac{8\pi}{3}$; $-\frac{4\pi}{3}$
14. $\frac{11\pi}{2}$ $\frac{3\pi}{2}$; $-\frac{\pi}{2}$
15. $\frac{16\pi}{5}$ $\frac{6\pi}{5}$; $-\frac{4\pi}{5}$
16. $\frac{7\pi}{13}$ $\frac{33\pi}{13}$; $-\frac{19\pi}{13}$

Find the arc length and area of a sector with the given radius r and central angle θ. (Lesson 13.2) 17–22. See margin.

17. $r = 4$ in., $\theta = 60°$
18. $r = 7$ ft, $\theta = 37°$
19. $r = 14$ cm, $\theta = 135°$
20. $r = 120$ m, $\theta = 167°$
21. $r = 9$ cm, $\theta = 4°$
22. $r = 28$ in., $\theta = 210°$

Use the given point on the terminal side of an angle θ in standard position. Evaluate the six trigonometric functions of θ. (Lesson 13.3) 23–34. See margin.

23. (4, 5)
24. (4, −1)
25. (−2, −6)
26. $(-12, \sqrt{3})$
27. (6, 3)
28. (5, −12)
29. $(3\sqrt{7}, -2)$
30. $(\sqrt{2}, \sqrt{2})$
31. (3, −5)
32. (7, −8)
33. (−3, 6)
34. $(\sqrt{5}, 3)$

Evaluate the function without using a calculator. (Lesson 13.3)

35. $\sin(-390°)$ $-\frac{1}{2}$
36. $\sec 120°$ -2
37. $\cos 315°$ $\frac{\sqrt{2}}{2}$
38. $\tan(-150°)$ $\frac{\sqrt{3}}{3}$
39. $\cos\frac{7\pi}{4}$ $\frac{\sqrt{2}}{2}$
40. $\tan\frac{7\pi}{6}$ $\frac{\sqrt{3}}{3}$
41. $\sin\left(-\frac{2\pi}{3}\right)$ $-\frac{\sqrt{3}}{2}$
42. $\csc\frac{13\pi}{4}$ $-\sqrt{2}$

Evaluate the expression without using a calculator. Give your answer in both radians and degrees. (Lesson 13.4)

43. $\tan^{-1}(-1)$ −45°; $-\frac{\pi}{4}$
44. $\cos^{-1} 0$ 90°; $\frac{\pi}{2}$
45. $\sin^{-1}\left(-\frac{1}{2}\right)$ −30°; $-\frac{\pi}{6}$
46. $\cos^{-1}\left(-\frac{\sqrt{2}}{2}\right)$ 135°; $\frac{3\pi}{4}$
47. $\tan^{-1}\frac{\sqrt{3}}{3}$ 30°; $\frac{\pi}{6}$
48. $\sin^{-1}\frac{1}{2}$ 30°; $\frac{\pi}{6}$
49. $\tan^{-1}\sqrt{3}$ 60°; $\frac{\pi}{3}$
50. $\sin^{-1}\frac{\sqrt{3}}{2}$ 60°; $\frac{\pi}{3}$

51. **PLAYGROUND EQUIPMENT** Two slides 12 feet long will be installed at the local playground. At what angle to the ground should the first slide be set if the top of the slide is 8 feet off the ground? At what angle should the top of the second slide be set if the slide must fit into a space that is only 6 feet wide? (Lesson 13.4) 41.8°; 30°

Solve △*ABC*. (Lesson 13.5) 52–55. See margin.

52.
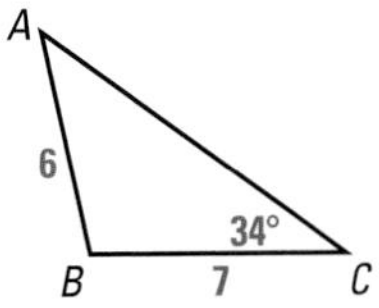

53.
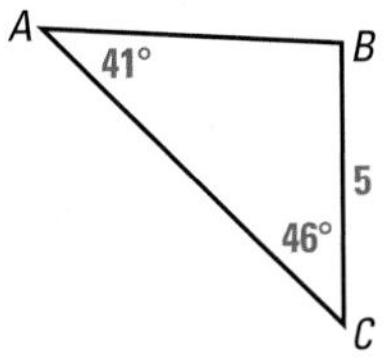

54.
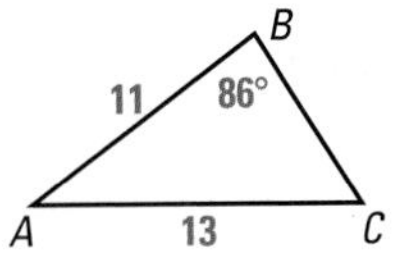

55.
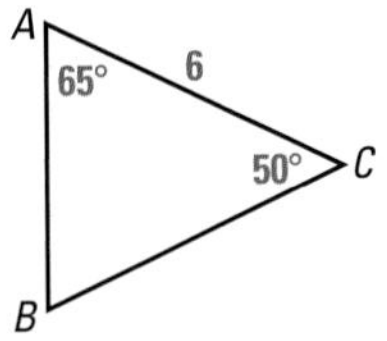

Find the area of the triangle with the given side lengths and included angle. (Lesson 13.5)

56. $C = 120°, a = 12, b = 20$ about 104
57. $A = 55°, b = 7, c = 12$ about 34.4
58. $B = 30°, a = 18, c = 13$ 58.5
59. $A = 80°, b = 120, c = 70$ about 4140
60. $C = 20°, a = 10, b = 16$ about 27.4
61. $B = 35°, a = 50, c = 120$ about 1720

Find the area of △ABC. (Lesson 13.6)

62.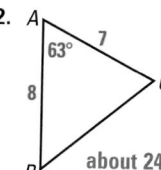
about 24.9

63.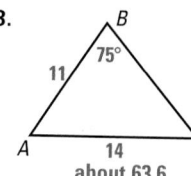
about 63.6

64.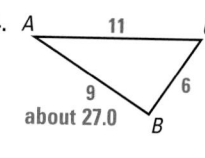
about 27.0

65. about 283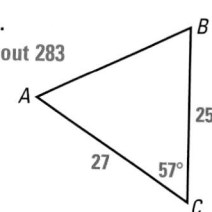

Write an *xy*-equation for the parametric equations. State the domain.
(Lesson 13.7)

66. $x = 12t - 5$ and $y = 10t - 3$ for $0 \le t \le 60$ $\quad y = \dfrac{5x + 7}{6}; -5 \le x \le 715$

67. $x = 42t - 7$ and $y = 21t - 7$ for $0 \le t \le 128$ $\quad y = \dfrac{x - 7}{2}; -7 \le x \le 5369$

68. $x = (17.54 \sin 20°)t$ and $y = 3t$ for $0 \le t \le 24$ $\quad y = \dfrac{x}{2}; 0 \le x \le 144$

69. $x = (10 \cos 1°)t$ and $y = (10 \cos 1°)t$ for $0 \le t \le 50$ $\quad y = x; 0 \le x \le 500$

CHAPTER 14

Find the amplitude and period of the graph of the function. (Lesson 14.1)

1. $y = 6 \sin \frac{1}{2}x$ $\;$ 6; 4π
2. $y = 2 \sin \frac{1}{8}x$ $\;$ 2; 16π
3. $y = \frac{1}{7} \cos \pi x$ $\;$ $\frac{1}{7}$; 2
4. $y = \frac{1}{2} \cos 2\pi x$ $\;$ $\frac{1}{2}$; 1

5. $y = \sin 2\pi x$ $\;$ 1, 1
6. $y = \frac{1}{3} \cos \frac{1}{12}x$ $\;$ $\frac{1}{3}$; 24π
7. $y = \frac{2}{5} \cos \frac{1}{4}x$ $\;$ $\frac{2}{5}$; 8π
8. $y = \frac{1}{3} \sin \frac{1}{2}\pi x$ $\;$ $\frac{1}{3}$; 4

Draw one cycle of the function's graph. (Lesson 14.1) $\quad$ 9–16. See margin.

9. $y = 3 \sin 2x$
10. $y = \frac{1}{2} \cos 5x$
11. $y = 2 \cos \pi x$
12. $y = 3 \sin 2\pi x$

13. $y = 5 \tan 2x$
14. $y = 5 \tan \frac{1}{6}x$
15. $y = 7 \cos 2\pi x$
16. $y = \frac{1}{3} \sin 3\pi x$

Describe how the graph of $y = \sin x$ or $y = \cos x$ can be transformed to produce the graph of the given function. (Lesson 14.2) $\quad$ 17–24. See margin.

17. $y = 3 + \cos x$
18. $y = 7 + \cos x$
19. $y = 4 - \cos x$
20. $y = \sin (x - \pi)$

21. $y = \cos (x - \pi)$
22. $y = \sin \left(x + \frac{\pi}{4}\right)$
23. $y = \cos \left(x - \frac{\pi}{2}\right)$
24. $y = -1 - \sin \left(x + \frac{3\pi}{4}\right)$

Graph the function. (Lesson 14.2) $\quad$ 25–32. See margin.

25. $y = 2 + \cos (x + \pi)$
26. $y = 1 - \tan \left(x + \frac{\pi}{2}\right)$
27. $y = 1 + \sin (x + \pi)$
28. $y = 1 + \sin \left(\frac{1}{2}x + \frac{3\pi}{4}\right)$

29. $y = -2 + \tan x$
30. $y = -2 \sin \frac{1}{2}x$
31. $y = \cos \left(x - \frac{3\pi}{2}\right)$
32. $y = 2 - \tan (x - \pi)$

Simplify the expression. (Lesson 14.3)

33. $\csc x \tan x$ $\quad$ sec *x*

34. $\sin \left(\frac{\pi}{2} - x\right) \tan (-x)$ $\quad$ $-\sin x$ $\quad$ **35.** $1 - \sin \left(\frac{\pi}{2} - x\right) \cos x$ $\quad$ $\sin^2 x$

36. $\dfrac{\sin^2 (-x)}{\sec^2 x - 1} - \dfrac{1}{\tan^2 x + 1}$ $\quad$ 0 $\quad$ **37.** $\dfrac{\sin (-x)}{-\tan x}$ $\quad$ cos *x* $\quad$ **38.** $\dfrac{\cos^4 x + \sin^2 x \cos^2 x + \sin^2 x}{\cos^2 x}$ $\quad$ $\sec^2 x$

39. $\dfrac{\sin x}{1 + \cos x} + \dfrac{1 + \cos x}{\sin x}$ $\quad$ $\dfrac{2}{\sin x}$ $\quad$ **40.** $\dfrac{\sec x - \cos x}{\tan^2 x}$ $\quad$ cos *x* $\quad$ **41.** $\dfrac{\cos x \sin^2 x - \cos x}{\cos x \cot x}$ $\quad$ $-\cos x \sin x$

9.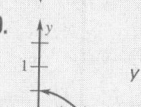
$y = 3 \sin 2x$

10.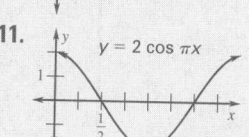
$y = \frac{1}{2} \cos 5x$

11.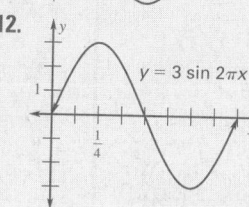
$y = 2 \cos \pi x$

12.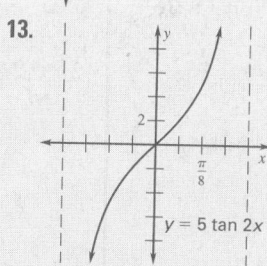
$y = 3 \sin 2\pi x$

13.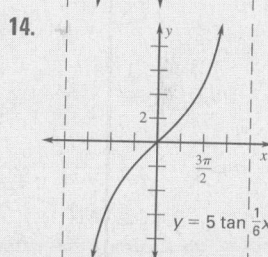
$y = 5 \tan 2x$

14.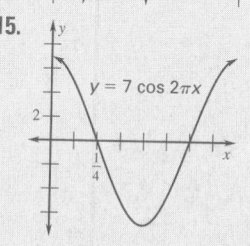
$y = 5 \tan \frac{1}{6}x$

15.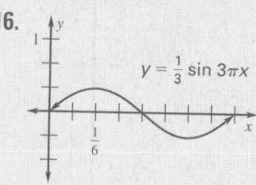
$y = 7 \cos 2\pi x$

16.
$y = \frac{1}{3} \sin 3\pi x$

17. Shift the graph of $y = \cos x$ up 3 units.

18–32. See Additional Answers beginning on page AA1.

45. $\frac{\pi}{3} + 2n\pi, \frac{2\pi}{3} + 2n\pi, \frac{4\pi}{3} + 2n\pi,$
$\frac{5\pi}{3} + 2n\pi$

47. $2n\pi, \frac{2\pi}{3} + 2n\pi, \pi + 2n\pi,$
$\frac{4\pi}{3} + 2n\pi$

49. $\frac{\pi}{4} + 2n\pi, \frac{7\pi}{4} + 2n\pi, \frac{3\pi}{4} + 2n\pi,$
$\frac{5\pi}{4} + 2n\pi$

60. *Sample answer:* $y = 4 \cos \frac{1}{2} x + 6$

61. *Sample answer:*
$y = \frac{1}{2} \cos \left(\frac{1}{2} (x + \pi) \right) + \frac{3}{2}$

62. *Sample answer:*
$y = 3 \cos \left(\frac{\pi}{2} (x + 2) \right) + 4$

63. *Sample answer:*
$y = 4 \cos \left(3 \left(x + \frac{\pi}{3} \right) \right) - 4$

64. *Sample answer:* $y = \frac{1}{2} \cos x - \frac{1}{2}$

65. *Sample answer:*
$y = \cos \left(\pi (x - 1) \right) + 22$

95. $\sin 2x = \frac{9\sqrt{19}}{50}; \cos 2x = \frac{31}{50};$
$\tan 2x = \frac{9\sqrt{19}}{31}$

96. $\sin 2x = -\frac{24}{25}; \cos 2x = -\frac{7}{25};$
$\tan 2x = \frac{24}{7}$

97. $\sin 2x = \frac{24}{25}; \cos 2x = \frac{7}{25};$
$\tan 2x = \frac{24}{7}$

98. $\sin 2x = -\frac{5\sqrt{39}}{32}; \cos 2x = \frac{7}{32};$
$\tan 2x = -\frac{5\sqrt{39}}{7}$

99. $\sin 2x = \frac{11\sqrt{23}}{72}; \cos 2x = \frac{49}{72};$
$\tan 2x = \frac{11\sqrt{23}}{49}$

100. $\sin 2x = -\frac{24}{25}; \cos 2x = \frac{7}{25};$
$\tan 2x = -\frac{24}{7}$

101. $\sin 2x = -\frac{24}{25}; \cos 2x = -\frac{7}{25};$
$\tan 2x = \frac{24}{7}$

102. $\sin 2x = \frac{3\sqrt{7}}{8}; \cos 2x = \frac{1}{8};$
$\tan 2x = 3\sqrt{7}$

103. $\sin 2x = -\frac{4\sqrt{5}}{9}; \cos 2x = \frac{1}{9};$
$\tan 2x = -4\sqrt{5}$

Find the general solution of the equation. (Lesson 14.4)

42. $2 \sin x - \sqrt{3} = 0$ $\frac{\pi}{3} + 2n\pi, \frac{2\pi}{3} + 2n\pi$

43. $-1 - 2 \sin x = 0$ $\frac{7\pi}{6} + 2n\pi, \frac{11\pi}{6} + 2n\pi$

44. $10 \cos x = 9 \cos x + 1$ $2n\pi$

45. $4 \cos^2 x - 1 = 0$ See margin.

46. $\cos x = 2 \cos x + 1$ $\pi + 2n\pi$

47. $2 \cos x \sin x + \sin x = 0$ See margin.

48. $\cos^2 x = \tan x = \sin^2 x$ $\frac{\pi}{4} + n\pi$

49. $6 \cos^2 x - 3 = 0$ See margin.

50. $\tan^2 x - 2 \tan x + 1 = 0$ $\frac{\pi}{4} + n\pi$

Solve the equation in the interval $0 \le x < 2\pi$**. Check your solutions.** (Lesson 14.4)

51. $2 \tan^2 x - 1 = 0$ 0.615, 3.76

52. $\sec^2 x - 4 = 0$ $\frac{\pi}{3}, \frac{2\pi}{3}, \frac{4\pi}{3}, \frac{5\pi}{3}$

53. $2 \sin x = \sin x - 1$ $\frac{3\pi}{2}$

54. $\sin^3 x = \sin x$ $0, \frac{\pi}{2}, \pi, \frac{3\pi}{2}$

55. $4 \cos x - 2 = 0$ $\frac{\pi}{3}, \frac{5\pi}{3}$

56. $6 \cos x = 3 \sec x$ $\frac{\pi}{4}, \frac{3\pi}{4}, \frac{5\pi}{4}, \frac{7\pi}{4}$

57. $2 \tan^2 x = \sin x \sec x$ 0, 0.464, π, 3.61

58. $1 - \sin x = \sqrt{2} \cos x$ $\frac{\pi}{2}$, 5.94

59. $\cos^2 x \sin x = 3 \sin x$ $0, \pi$

Write a trigonometric function for the sinusoid with maximum at *A* **and minimum at** *B.* (Lesson 14.5)   60–65. See margin.

60. $A(0, 10), B(2\pi, 2)$

61. $A(-\pi, 2), B(\pi, 1)$

62. $A(-2, 7), B(0, 1)$

63. $A\left(\frac{\pi}{3}, 0 \right), B(0, -8)$

64. $A(0, 0), B(\pi, -1)$

65. $A(1, 23), B(2, 21)$

66. CARNIVAL RIDE You and your friend are riding on a Ferris wheel with a diameter of 30 feet. When $t = 0$, your chair starts at the lowest point on the wheel, which is 6 feet above the ground. If the Ferris wheel is rotating at a rate of 4 revolutions per minute, write a model for the height h (in feet) of the chair as a function of the time t (in seconds). (Lesson 14.5)   *Sample answer:* $y = -15 \cos \frac{2\pi}{15} x + 21$

Find the exact value of the expression. (Lesson 14.6)

67. $\sin 225°$ $-\frac{\sqrt{2}}{2}$

68. $\cos (-15°)$ $\frac{\sqrt{6} + \sqrt{2}}{4}$

69. $\cos 195°$ $\frac{-\sqrt{6} - \sqrt{2}}{4}$

70. $\sin 555°$ $\frac{-\sqrt{6} + \sqrt{2}}{4}$

71. $\tan \left(-\frac{\pi}{12} \right)$ $-2 + \sqrt{3}$

72. $\sin \frac{19\pi}{2}$ -1

73. $\tan \frac{7\pi}{6}$ $\frac{\sqrt{3}}{3}$

74. $\sin \frac{\pi}{12}$ $\frac{\sqrt{6} - \sqrt{2}}{4}$

Evaluate the expression given $\sin u = \frac{2}{3}$ **with** $0 < u < \frac{\pi}{2}$ **and** $\cos v = -\frac{2}{7}$ **with** $\pi < v < \frac{3\pi}{2}$**.** (Lesson 14.6)

75. $\sin (u + v)$ $-\frac{19}{21}$

76. $\cos (u + v)$ $\frac{4\sqrt{5}}{21}$

77. $\tan (u + v)$ $\frac{-19\sqrt{5}}{20}$

78. $\sin (u - v)$ $\frac{11}{21}$

79. $\cos (u - v)$ $-\frac{8\sqrt{5}}{21}$

80. $\tan (u - v)$ $-\frac{11\sqrt{5}}{40}$

81. $\sin (v - u)$ $-\frac{11}{21}$

82. $\cos (v - u)$ $-\frac{8\sqrt{5}}{21}$

Find the exact value of the expression. (Lesson 14.7)

83. $\tan 105°$ $-2 - \sqrt{3}$

84. $\sin (-22.5°)$ $-\frac{\sqrt{2 - \sqrt{2}}}{2}$

85. $\cos (-112.5°)$ $-\frac{\sqrt{2 - \sqrt{2}}}{2}$

86. $\cos 165°$ $-\frac{\sqrt{2 + \sqrt{3}}}{2}$

87. $\sin \frac{7\pi}{8}$ $\frac{\sqrt{2 - \sqrt{2}}}{2}$

88. $\cos \left(-\frac{\pi}{8} \right)$ $\frac{\sqrt{2 + \sqrt{2}}}{2}$

89. $\tan (-75°)$ $-2 - \sqrt{3}$

90. $\cos \left(-\frac{7\pi}{8} \right)$ $-\frac{\sqrt{2 + \sqrt{2}}}{2}$

91. $\sin 67.5°$ $\frac{\sqrt{2 + \sqrt{2}}}{2}$

92. $\tan 22.5°$ $\sqrt{2} - 1$

93. $\cos 105°$ $-\frac{\sqrt{2 - \sqrt{3}}}{2}$

94. $\sin (-112.5°)$ $-\frac{\sqrt{2 + \sqrt{2}}}{2}$

Find the exact values of $\sin 2x$, $\cos 2x$, **and** $\tan 2x$**.** (Lesson 14.7)   95–103. See margin.

95. $\cos x = \frac{9}{10}, 0 < x < \frac{\pi}{2}$

96. $\sin x = \frac{4}{5}, \frac{\pi}{2} < x < \pi$

97. $\sin x = \frac{3}{5}, 0 < x < \frac{\pi}{2}$

98. $\sin x = \frac{5}{8}, \frac{\pi}{2} < x < \pi$

99. $\cos x = \frac{11}{12}, 0 < x < \frac{\pi}{2}$

100. $\cos x = \frac{4}{5}, \frac{3\pi}{2} < x < 2\pi$

101. $\sin x = -\frac{4}{5}, \frac{3\pi}{2} < x < 2\pi$

102. $\cos x = \frac{3}{4}, 0 < x < \frac{\pi}{2}$

103. $\sin x = \frac{2}{3}, \frac{3\pi}{2} < x < 2\pi$

Table of Symbols

Symbol		Page
$\ldots$	and so on	3
$\approx$	is approximately equal to	3
$<$	is less than	4
$>$	is greater than	4
$\leq$	is less than or equal to	4
$\geq$	is greater than or equal to	4
$\cdot$	multiplication, times	5
$-a$	opposite of a	5
$\dfrac{1}{a}$	reciprocal of a, $a \neq 0$	5
$\neq$	not equal to	5
π	pi; irrational number ≈ 3.14	28
$\lvert x \rvert$	absolute value of x	50
(x, y)	ordered pair	67
$f(x)$	f of x, or the value of f at x	69
m	slope	75
x_1	x sub 1	75
$\llbracket x \rrbracket$	greatest integer less than or equal to x	115
(x, y, z)	ordered triple	170
$f(x, y)$	function of two variables	171
$\begin{bmatrix} 1 & 0 \\ 0 & 1 \end{bmatrix}$	matrix	199
$\lvert A \rvert$	determinant of matrix A	214
A^{-1}	inverse of matrix A	223
$\sqrt{a}$	the nonnegative square root of a	264
i	imaginary unit equal to $\sqrt{-1}$	272
$\lvert z \rvert$	absolute value of complex number z	275
$+\infty$	positive infinity	331
$-\infty$	negative infinity	331
$x \to +\infty$	x approaches positive infinity	331

Symbol		Page
$f(x) \to +\infty$	f of x approaches positive infinity	331
$\sqrt[n]{a}$	nth root of a	401
f^{-1}	inverse of function f	423
$\overline{x}$	x-bar; the mean of a data set	445
σ	sigma; the standard deviation of a data set	446
e	irrational number ≈ 2.718	480
$\log_b y$	base-b logarithm of y	486
$\log x$	base-10 logarithm of x	487
$\ln x$	base-e logarithm of x	487
Σ	summation	653
S_n	sum of the first n terms of an arithmetic or geometric series	661, 668
$n!$	n factorial; number of permutations of n objects	703
$_nP_r$	number of permutations of r objects from n distinct objects	703
$_nC_r$	number of combinations of r objects from n distinct objects	708
$P(A)$	probability of event A	716
$P(B \mid A)$	probability of event B given that event A has occurred	732
θ	theta; name of an angle, or measure of an angle	769
$\sin$	sine	769
$\cos$	cosine	769
$\tan$	tangent	769
$\csc$	cosecant	769
$\sec$	secant	769
$\cot$	cotangent	769
$\sin^{-1}$	inverse sine	792
$\cos^{-1}$	inverse cosine	792
$\tan^{-1}$	inverse tangent	792

Table of Measures

Time

$$
\begin{array}{ll}
60 \text{ seconds (sec)} = 1 \text{ minute (min)} \\
60 \text{ minutes} = 1 \text{ hour (h)} \\
24 \text{ hours} = 1 \text{ day} \\
7 \text{ days} = 1 \text{ week} \\
4 \text{ weeks (approx.)} = 1 \text{ month}
\end{array}
$$

$$
\left.\begin{array}{l}
365 \text{ days} \\
52 \text{ weeks (approx.)} \\
12 \text{ months}
\end{array}\right\} = 1 \text{ year}
$$

10 years = 1 decade

100 years = 1 century

Metric

Length

10 millimeters (mm) = 1 centimeter (cm)

$\left.\begin{array}{l} 100 \text{ cm} \\ 1000 \text{ mm} \end{array}\right\} = 1 \text{ meter (m)}$

1000 m = 1 kilometer (km)

Area

100 square millimeters = 1 square centimeter

$\left(\text{mm}^2\right) \qquad \left(\text{cm}^2\right)$

$10{,}000 \text{ cm}^2 = 1$ square meter $\left(\text{m}^2\right)$

$10{,}000 \text{ m}^2 = 1$ hectare (ha)

Volume

1000 cubic millimeters = 1 cubic centimeter

$\left(\text{mm}^3\right) \qquad \left(\text{cm}^3\right)$

$1{,}000{,}000 \text{ cm}^3 = 1$ cubic meter $\left(\text{m}^3\right)$

Liquid Capacity

$\left.\begin{array}{l} 1000 \text{ milliliters (mL)} \\ 10 \text{ deciliters (dL)} \end{array}\right\} = 1 \text{ liter (L)}$

1000 L = 1 kiloliter (kL)

Mass

1000 milligrams (mg) = 1 gram (g)

1000 g = 1 kilogram (kg)

1000 kg = 1 metric ton (t)

Temperature — Degrees Celsius (°C)

0°C = freezing point of water

37°C = normal body temperature

100°C = boiling point of water

United States Customary

Length

12 inches (in.) = 1 foot (ft)

$\left.\begin{array}{l} 36 \text{ in.} \\ 3 \text{ ft} \end{array}\right\} = 1 \text{ yard (yd)}$

$\left.\begin{array}{l} 5280 \text{ ft} \\ 1760 \text{ yd} \end{array}\right\} = 1 \text{ mile (mi)}$

Area

144 square inches $\left(\text{in.}^2\right) = 1$ square foot $\left(\text{ft}^2\right)$

$9 \text{ ft}^2 = 1$ square yard $\left(\text{yd}^2\right)$

$\left.\begin{array}{l} 43{,}560 \text{ ft}^2 \\ 4840 \text{ yd}^2 \end{array}\right\} = 1 \text{ acre (A)}$

Volume

1728 cubic inches $\left(\text{in.}^3\right) = 1$ cubic foot $\left(\text{ft}^3\right)$

$27 \text{ ft}^3 = 1$ cubic yard $\left(\text{yd}^3\right)$

Liquid Capacity

8 fluid ounces (fl oz) = 1 cup (c)

2 c = 1 pint (pt)

2 pt = 1 quart (qt)

4 qt = 1 gallon (gal)

Weight

16 ounces (oz) = 1 pound (lb)

2000 lb = 1 ton (t)

Temperature — Degrees Fahrenheit (°F)

32°F = freezing point of water

98.6°F = normal body temperature

212°F = boiling point of water

Table of Formulas

Formulas from Coordinate Geometry

Slope of a line	$m = \dfrac{y_2 - y_1}{x_2 - x_1}$ where m is the slope of the nonvertical line through (x_1, y_1) and (x_2, y_2)
Parallel and perpendicular lines	If line l_1 has slope m_1 and line l_2 has slope m_2, then: $$l_1 \parallel l_2 \text{ if and only if } m_1 = m_2$$ $$l_1 \perp l_2 \text{ if and only if } m_1 = -\frac{1}{m_2} \text{ or } m_1 m_2 = -1$$
Distance formula	$d = \sqrt{(x_2 - x_1)^2 + (y_2 - y_1)^2}$ where d is the distance between points (x_1, y_1) and (x_2, y_2)
Midpoint formula	$M\left(\dfrac{x_1 + x_2}{2}, \dfrac{y_1 + y_2}{2}\right)$ is the midpoint of the line segment joining points (x_1, y_1) and (x_2, y_2)

Formulas from Matrix Algebra

Determinant of a 2×2 matrix	$\det \begin{bmatrix} a & b \\ c & d \end{bmatrix} = ad - cb$
Determinant of a 3×3 matrix	$\det \begin{bmatrix} a & b & c \\ d & e & f \\ g & h & i \end{bmatrix} = (aei + bfg + cdh) - (gec + hfa + idb)$
Area of a triangle	The area of a triangle with vertices (x_1, y_1), (x_2, y_2), and (x_3, y_3) is given by $$\text{Area} = \pm\frac{1}{2}\begin{vmatrix} x_1 & y_1 & 1 \\ x_2 & y_2 & 1 \\ x_3 & y_3 & 1 \end{vmatrix}$$ where the appropriate sign ($\pm$) should be chosen to yield a positive value.
Cramer's rule	Let $A = \begin{bmatrix} a & b \\ c & d \end{bmatrix}$ be the coefficient matrix of this linear system: $$ax + by = e$$ $$cx + dy = f$$ If $\det A \neq 0$, then the system has exactly one solution. The solution is: $$x = \frac{\begin{vmatrix} e & b \\ f & d \end{vmatrix}}{\det A} \quad \text{and} \quad y = \frac{\begin{vmatrix} a & e \\ c & f \end{vmatrix}}{\det A}$$ Cramer's rule can be extended to a linear system of 3 equations in 3 variables.
Inverse of a 2×2 matrix	The inverse of the matrix $A = \begin{bmatrix} a & b \\ c & d \end{bmatrix}$ is $$A^{-1} = \frac{1}{|A|}\begin{bmatrix} d & -b \\ -c & a \end{bmatrix} = \frac{1}{ad - cb}\begin{bmatrix} d & -b \\ -c & a \end{bmatrix}$$ provided $ad - cb \neq 0$.

Formulas and Theorems from Algebra

Special product patterns	**Sum and difference:** $\quad (a + b)(a - b) = a^2 - b^2$ **Square of a binomial:** $\quad (a + b)^2 = a^2 + 2ab + b^2$ $\qquad\qquad\qquad\qquad\quad (a - b)^2 = a^2 - 2ab + b^2$ **Cube of a binomial:** $\quad\; (a + b)^3 = a^3 + 3a^2b + 3ab^2 + b^3$ $\qquad\qquad\qquad\qquad\quad (a - b)^3 = a^3 - 3a^2b + 3ab^2 - b^3$
Special factoring patterns	Each of the patterns above can be read from right to left as a factoring pattern. In addition, there are two other special factoring patterns: **Sum of two cubes:** $\qquad\quad a^3 + b^3 = (a + b)(a^2 - ab + b^2)$ **Difference of two cubes:** $\quad a^3 - b^3 = (a - b)(a^2 + ab + b^2)$
Quadratic formula	$x = \dfrac{-b \pm \sqrt{b^2 - 4ac}}{2a}$ where x is a solution of $ax^2 + bx + c = 0$ and a, b, and c are real numbers such that $a \neq 0$
Discriminant of a quadratic equation	The expression $b^2 - 4ac$ is called the discriminant of the associated equation $ax^2 + bx + c = 0$. The value of the discriminant can be positive, zero, or negative, which corresponds to an equation having two real solutions, one real solution, or two imaginary solutions, respectively.
Remainder theorem	If a polynomial $f(x)$ is divided by $x - k$, then the remainder is $r = f(k)$.
Factor theorem	A polynomial $f(x)$ has a factor $x - k$ if and only if $f(k) = 0$.
Rational zero theorem	If $f(x) = a_n x^n + \cdots + a_1 x + a_0$ has *integer* coefficients, then every rational zero of f has this form: $\dfrac{p}{q} = \dfrac{\text{factor of constant term } a_0}{\text{factor of leading coefficient } a_n}$
Fundamental theorem of algebra	If $f(x)$ is a polynomial of degree n where $n > 0$, then the equation $f(x) = 0$ has at least one solution in the set of complex numbers.

Formulas from Statistics

Mean of a data set	$\bar{x} = \dfrac{x_1 + x_2 + \cdots + x_n}{n}$ where $\bar{x}$ (read "x-bar") is the mean of the data $x_1, x_2, \ldots, x_n$
Standard deviation of a data set	$\sigma = \sqrt{\dfrac{(x_1 - \bar{x})^2 + (x_2 - \bar{x})^2 + \cdots + (x_n - \bar{x})^2}{n}}$ where σ (read "sigma") is the standard deviation of the data $x_1, x_2, \ldots, x_n$
Areas under a normal curve	The mean $\bar{x}$ and standard deviation σ of a normal distribution determine the following areas under the corresponding normal curve. • The total area under the curve is 1. • 68% of the area lies within 1 standard deviation of the mean. • 95% of the area lies within 2 standard deviations of the mean. • 99.7% of the area lies within 3 standard deviations of the mean.
Normal approximation of a binomial distribution	Consider the binomial distribution consisting of n trials with a probability p of success on each trial. If $np \geq 5$ and $n(1 - p) \geq 5$, then the binomial distribution can be approximated by a normal distribution with a mean of $\bar{x} = np$ and a standard deviation of $\sigma = \sqrt{np(1 - p)}$.

Formulas for Sequences and Series

Explicit rule for an arithmetic sequence	The nth term of an arithmetic sequence with first term a_1 and common difference d is: $$a_n = a_1 + (n-1)d$$
Explicit rule for a geometric sequence	The nth term of a geometric sequence with first term a_1 and common ratio r is: $$a_n = a_1 r^{n-1}$$
Sum of a finite arithmetic series	The sum of the first n terms of an arithmetic series is: $$S_n = n\left(\frac{a_1 + a_n}{2}\right)$$
Sum of a finite geometric series	The sum of the first n terms of a geometric series with common ratio $r \neq 1$ is: $$S_n = a_1\left(\frac{1 - r^n}{1 - r}\right)$$
Sum of an infinite geometric series	The sum of an infinite geometric series with first term a_1 and common ratio r is $$S = \frac{a_1}{1 - r}$$ provided $\lvert r \rvert < 1$. If $\lvert r \rvert \geq 1$, the series has no sum.
Formulas for sums of special series	1. $\displaystyle\sum_{i=1}^{n} 1 = n$ 2. $\displaystyle\sum_{i=1}^{n} i = \frac{n(n+1)}{2}$ 3. $\displaystyle\sum_{i=1}^{n} i^2 = \frac{n(n+1)(2n+1)}{6}$

Formulas from Combinatorics

Fundamental counting principle	If one event can occur in m ways and another event can occur in n ways, then the number of ways that *both* events can occur is $m \cdot n$.
Permutations of n objects taken r at a time	The number of permutations of r objects taken from a group of n distinct objects is denoted by $_nP_r$ and is given by: $$_nP_r = \frac{n!}{(n-r)!}$$
Permutations with repetition	The number of distinguishable permutations of n objects where one object is repeated q_1 times, another is repeated q_2 times, and so on is: $$\frac{n!}{q_1! \cdot q_2! \cdot \ldots \cdot q_k!}$$
Combinations of n objects taken r at a time	The number of combinations of r objects taken from a group of n distinct objects is denoted by $_nC_r$ and is given by: $$_nC_r = \frac{n!}{(n-r)! \cdot r!}$$
Pascal's triangle	If you arrange the values of $_nC_r$ in a triangular pattern in which each row corresponds to a value of n, you get what is called Pascal's triangle. $$\begin{array}{ccccccccc} & & & & _0C_0 & & & & \\ & & & _1C_0 & & _1C_1 & & & \\ & & _2C_0 & & _2C_1 & & _2C_2 & & \\ & _3C_0 & & _3C_1 & & _3C_2 & & _3C_3 & \\ _4C_0 & & _4C_1 & & _4C_2 & & _4C_3 & & _4C_4 \end{array}$$ $$\begin{array}{ccccccccc} & & & & 1 & & & & \\ & & & 1 & & 1 & & & \\ & & 1 & & 2 & & 1 & & \\ & 1 & & 3 & & 3 & & 1 & \\ 1 & & 4 & & 6 & & 4 & & 1 \end{array}$$
Binomial theorem	The binomial expansion of $(a+b)^n$ for any positive integer n is: $$(a+b)^n = {}_nC_0 a^n b^0 + {}_nC_1 a^{n-1} b^1 + {}_nC_2 a^{n-2} b^2 + \cdots + {}_nC_n a^0 b^n$$ $$= \sum_{r=0}^{n} {}_nC_r a^{n-r} b^r$$

Formulas from Probability

Theoretical probability of an event	When all outcomes are equally likely, the theoretical probability that an event A will occur is: $$P(A) = \frac{\text{number of outcomes in } A}{\text{total number of outcomes}}$$
Probability of compound events	If A and B are two events, then the probability of A or B is: $$P(A \text{ or } B) = P(A) + P(B) - P(A \text{ and } B)$$ If A and B are mutually exclusive, then the probability of A or B is: $$P(A \text{ or } B) = P(A) + P(B)$$
Probability of the complement of an event	The probability of the complement of event A, denoted A', is: $$P(A') = 1 - P(A)$$
Probability of independent events	If A and B are independent, then the probability that both A and B occur is: $$P(A \text{ and } B) = P(A) \cdot P(B)$$
Probability of dependent events	If A and B are dependent, then the probability that both A and B occur is: $$P(A \text{ and } B) = P(A) \cdot P(B \mid A)$$
Binomial probabilities	For a binomial experiment consisting of n trials where the probability of success on each trial is p, the probability of exactly k successes is: $$P(k \text{ successes}) = {}_nC_k p^k (1 - p)^{n - k}$$

Formulas and Identities from Trigonometry

Conversion between degrees and radians	To rewrite a degree measure in radians, multiply by $\dfrac{\pi \text{ radians}}{180°}$. To rewrite a radian measure in degrees, multiply by $\dfrac{180°}{\pi \text{ radians}}$.
Definition of trigonometric functions	Let θ be an angle in standard position and (x, y) be any point (except the origin) on the terminal side of θ. Let $r = \sqrt{x^2 + y^2}$. Then the six trigonometric functions of θ are: $\sin \theta = \dfrac{y}{r}$ $\cos \theta = \dfrac{x}{r}$ $\tan \theta = \dfrac{y}{x}, x \neq 0$ $\csc \theta = \dfrac{r}{y}, y \neq 0$ $\sec \theta = \dfrac{r}{x}, x \neq 0$ $\cot \theta = \dfrac{x}{y}, y \neq 0$
Law of sines	If $\triangle ABC$ has sides of length a, b, and c, then: $$\frac{\sin A}{a} = \frac{\sin B}{b} = \frac{\sin C}{c}$$
Area of a triangle (given two sides and the included angle)	If $\triangle ABC$ has sides of length a, b, and c, then the area is: $$\text{Area} = \frac{1}{2}bc \sin A \qquad \text{Area} = \frac{1}{2}ac \sin B \qquad \text{Area} = \frac{1}{2}ab \sin C$$
Law of cosines	If $\triangle ABC$ has sides of length a, b, and c, then: $$a^2 = b^2 + c^2 - 2bc \cos A$$ $$b^2 = a^2 + c^2 - 2ac \cos B$$ $$c^2 = a^2 + b^2 - 2ab \cos C$$

Formulas and Identities from Trigonometry (continued)

Heron's area formula	The area of the triangle with sides of length a, b, and c is $$\text{Area} = \sqrt{s(s-a)(s-b)(s-c)}$$ where $s = \frac{1}{2}(a+b+c)$.
Reciprocal identities	$$\csc \theta = \frac{1}{\sin \theta} \qquad \sec \theta = \frac{1}{\cos \theta} \qquad \cot \theta = \frac{1}{\tan \theta}$$
Tangent and cotangent identities	$$\tan \theta = \frac{\sin \theta}{\cos \theta} \qquad \cot \theta = \frac{\cos \theta}{\sin \theta}$$
Pythagorean identities	$$\sin^2 \theta + \cos^2 \theta = 1 \qquad 1 + \tan^2 \theta = \sec^2 \theta \qquad 1 + \cot^2 \theta = \csc^2 \theta$$
Cofunction identities	$$\sin\left(\frac{\pi}{2} - \theta\right) = \cos \theta \qquad \cos\left(\frac{\pi}{2} - \theta\right) = \sin \theta \qquad \tan\left(\frac{\pi}{2} - \theta\right) = \cot \theta$$
Negative angle identities	$$\sin(-\theta) = -\sin \theta \qquad \cos(-\theta) = \cos \theta \qquad \tan(-\theta) = -\tan \theta$$
Sum formulas	$$\sin(u+v) = \sin u \cos v + \cos u \sin v$$ $$\cos(u+v) = \cos u \cos v - \sin u \sin v$$ $$\tan(u+v) = \frac{\tan u + \tan v}{1 - \tan u \tan v}$$
Difference formulas	$$\sin(u-v) = \sin u \cos v - \cos u \sin v$$ $$\cos(u-v) = \cos u \cos v + \sin u \sin v$$ $$\tan(u-v) = \frac{\tan u - \tan v}{1 + \tan u \tan v}$$
Double-angle formulas	$$\cos 2u = \cos^2 u - \sin^2 u \qquad\qquad \sin 2u = 2 \sin u \cos u$$ $$\cos 2u = 2 \cos^2 u - 1 \qquad\qquad \tan 2u = \frac{2 \tan u}{1 - \tan^2 u}$$ $$\cos 2u = 1 - 2 \sin^2 u$$
Half-angle formulas	$$\sin \frac{u}{2} = \pm\sqrt{\frac{1 - \cos u}{2}} \qquad\qquad \tan \frac{u}{2} = \frac{1 - \cos u}{\sin u}$$ $$\cos \frac{u}{2} = \pm\sqrt{\frac{1 + \cos u}{2}} \qquad\qquad \tan \frac{u}{2} = \frac{\sin u}{1 + \cos u}$$ The signs of $\sin \frac{u}{2}$ and $\cos \frac{u}{2}$ depend on the quadrant in which $\frac{u}{2}$ lies.

Formulas from Mathematical Modeling

Projectile motion	**Height as a function of time:** $h = -16t^2 + v_0 t + h_0$ where h is the height (in feet) of the object t seconds after launch, h_0 is the object's initial height, and v_0 is the object's initial vertical velocity (in feet per second) **Parametric equations for a projectile's path:** $x = (v \cos \theta)t + x_0$ and $y = -\frac{1}{2}gt^2 + (v \sin \theta)t + y_0$ where θ is the angle at which the projectile is launched, v is the initial speed, and (x_0, y_0) is the projectile's location at time $t = 0$. (The constant g is the acceleration due to gravity; its value is 32 ft/sec^2 or 9.8 m/sec^2.)
Compound interest	**Compounded n times per year:** $A = P\left(1 + \frac{r}{n}\right)^{nt}$ where A is the amount in the account after t years, P is the initial deposit (called the principal), and r is the annual interest rate (expressed as a decimal) **Compounded continuously:** $A = Pe^{rt}$ where A is the amount in the account after t years, P is the initial deposit (called the principal), and r is the annual interest rate (expressed as a decimal)

Formulas from Geometry

Basic geometric figures	See page 914 for area formulas for basic two-dimensional geometric figures.
Area of an equilateral triangle	Area $= \frac{\sqrt{3}}{4}s^2$ where s is the length of a side
Arc length and area of a sector	Arc length $= r\theta$ where r is the radius and θ is the radian measure of the central angle that intercepts the arc Area $= \frac{1}{2}r^2\theta$
Area of an ellipse	Area $= \pi ab$ where a and b are half the lengths of the major and minor axes of the ellipse
Volume and surface area of a right rectangular prism	Volume $= \ell wh$ where ℓ is the length, w is the width, and h is the height Surface area $= 2(\ell w + wh + \ell h)$
Volume and surface area of a right cylinder	Volume $= \pi r^2 h$ where r is the base radius and h is the height Lateral surface area $= 2\pi rh$ Surface area $= 2\pi r^2 + 2\pi rh$
Volume and surface area of a right regular pyramid	Volume $= \frac{1}{3}Bh$ where B is the area of the base and h is the height Lateral surface area $= \frac{1}{2}n\ell s$ where n is the number of sides on the base, l is the length of a side of the base, and s is the slant height Surface area $= B + \frac{1}{2}n\ell s$
Volume and surface area of a right circular cone	Volume $= \frac{1}{3}\pi r^2 h$ where r is the base radius and h is the height Lateral surface area $= \pi rs$ where s is the slant height Surface area $= \pi r^2 + \pi rs$
Volume and surface area of a sphere	Volume $= \frac{4}{3}\pi r^3$ where r is the radius Surface area $= 4\pi r^2$

Table of Properties

Properties of Real Numbers

Let a, b, and c be real numbers.

	Addition	**Multiplication**
Closure Property	$a + b$ is a real number.	ab is a real number.
Commutative Property	$a + b = b + a$	$ab = ba$
Associative Property	$(a + b) + c = a + (b + c)$	$(ab)c = a(bc)$
Identity Property	$a + 0 = a, 0 + a = a$	$a \cdot 1 = a, 1 \cdot a = a$
Inverse Property	$a + (-a) = 0$	$a \cdot \dfrac{1}{a} = 1, a \neq 0$

Distributive Property
The distributive property involves both addition and multiplication:
$$a(b + c) = ab + ac$$

Zero Product Property
Let A and B be real numbers or algebraic expressions. If $AB = 0$, then $A = 0$ or $B = 0$.

Properties of Matrices

Let A, B, and C be matrices, and let c be a scalar.

Associative Property of Addition	$(A + B) + C = A + (B + C)$
Commutative Property of Addition	$A + B = B + A$
Distributive Property of Addition	$c(A + B) = cA + cB$
Distributive Property of Subtraction	$c(A - B) = cA - cB$
Associative Property of Matrix Multiplication	$(AB)C = A(BC)$
Left Distributive Property of Matrix Multiplication	$A(B + C) = AB + AC$
Right Distributive Property of Matrix Multiplication	$(A + B)C = AC + BC$
Associative Property of Scalar Multiplication	$c(AB) = (cA)B = A(cB)$

Multiplicative Identity
An $n \times n$ matrix with 1's on the main diagonal and 0's elsewhere is an identity matrix, denoted I. For any $n \times n$ matrix A, $AI = IA = A$.

Inverse Matrices
If the determinant of an $n \times n$ matrix A is nonzero, then A has an inverse, denoted A^{-1}, such that $AA^{-1} = A^{-1}A = I$.

Properties of Exponents

Let a and b be real numbers, and let m and n be integers.

Product of Powers Property	$a^m \cdot a^n = a^{m + n}$
Power of a Power Property	$(a^m)^n = a^{mn}$
Power of a Product Property	$(ab)^m = a^m b^m$
Negative Exponent Property	$a^{-m} = \dfrac{1}{a^m}, a \neq 0$
Zero Exponent Property	$a^0 = 1, a \neq 0$
Quotient of Powers Property	$\dfrac{a^m}{a^n} = a^{m - n}, a \neq 0$
Power of a Quotient Property	$\left(\dfrac{a}{b}\right)^m = \dfrac{a^m}{b^m}, b \neq 0$

Properties of Radicals and Rational Exponents

Number of Real nth Roots	Let n be an integer greater than 1, and let a be a real number.
	  • If n is odd, then a has one real nth root: $\sqrt[n]{a} = a^{1/n}$
	  • If n is even and $a > 0$, then a has two real nth roots: $\pm\sqrt[n]{a} = \pm a^{1/n}$
	  • If n is even and $a = 0$, then a has one nth root: $\sqrt[n]{0} = 0^{1/n} = 0$
	  • If n is even and $a < 0$, then a has no real nth roots.

Radicals and Rational Exponents	Let $a^{1/n}$ be an nth root of a, and let m be a positive integer.
	  • $a^{m/n} = \left(a^{1/n}\right)^m = \left(\sqrt[n]{a}\right)^m$
	  • $a^{-m/n} = \dfrac{1}{a^{m/n}} = \dfrac{1}{\left(a^{1/n}\right)^m} = \dfrac{1}{\left(\sqrt[n]{a}\right)^m},\ a \neq 0$

Properties of Rational Exponents	All of the properties of exponents listed on the previous page apply to rational exponents as well as integer exponents.
Product and Quotient Properties of Radicals	Let n be an integer greater than 1, and let a and b be positive real numbers. Then: $$\sqrt[n]{a \cdot b} = \sqrt[n]{a} \cdot \sqrt[n]{b} \quad \text{and} \quad \sqrt[n]{\frac{a}{b}} = \frac{\sqrt[n]{a}}{\sqrt[n]{b}}$$

Properties of Logarithms

	Let b, c, x, y, u, and v be positive real numbers such that $b \neq 1$ and $c \neq 1$.
Logarithms and Exponents	$\log_b y = x$ if and only if $b^x = y$
Special Logarithm Values	$\log_b 1 = 0$ because $b^0 = 1$ and $\log_b b = 1$ because $b^1 = b$
Common and Natural Logarithms	$\log_{10} x = \log x$ and $\log_e x = \ln x$
Product Property of Logarithms	$\log_b uv = \log_b u + \log_b v$
Quotient Property of Logarithms	$\log_b \dfrac{u}{v} = \log_b u - \log_b v$
Power Property of Logarithms	$\log_b u^n = n \log_b u$
Change of Base	$\log_c u = \dfrac{\log_b u}{\log_b c}$

Properties of Functions

Operations on Functions	Let f and g be any two functions. A new function h can be defined using any of the following operations: **Addition:** $\quad h(x) = f(x) + g(x)$ **Subtraction:** $\quad h(x) = f(x) - g(x)$ **Multiplication:** $\quad h(x) = f(x) \cdot g(x)$ **Division:** $\quad h(x) = \dfrac{f(x)}{g(x)}$ **Composition:** $\quad h(x) = f(g(x))$ For addition, subtraction, multiplication, and division, the domain of h consists of the x-values that are in the domains of both f and g. Additionally, the domain of a quotient does not include x-values for which $g(x) = 0$. For composition, the domain of h is the set of all x-values such that x is in the domain of g and $g(x)$ is in the domain of f.
Inverse Functions	Functions f and g are inverses of each other provided: $$f(g(x)) = x \quad \text{and} \quad g(f(x)) = x$$

Glossary

A

absolute value of a complex number (p. 275) If $z = a + bi$, then the absolute value of z, denoted $|z|$, is a nonnegative real number defined as $|z| = \sqrt{a^2 + b^2}$. Geometrically, the absolute value of a complex number is the number's distance from the origin in the complex plane.

absolute value of a real number (p. 50) The distance the number is from 0 on a number line. The absolute value of a number x is written $|x|$.

algebraic expression (p. 12) An expression with variables.

algebraic model (p. 33) A mathematical statement that represents a real-life problem.

amplitude (p. 831) The amplitude of the graph of a sine or cosine function is $\frac{1}{2}(M - m)$ where M is the maximum value of the function and m is the minimum value of the function.

angle of depression (p. 771) The angle from a horizontal line through an object A to a line connecting object A and a lower object B.

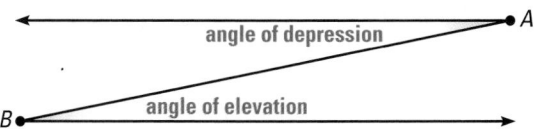

angle of elevation (p. 771) The angle from a horizontal line through an object B to a line connecting B and a higher object A.

arithmetic sequence (p. 659) A sequence in which the difference between consecutive terms is constant.

arithmetic series (p. 661) The expression formed by adding the terms of an arithmetic sequence.

asymptote (p. 465) A line that a graph approaches as you move away from the origin.

augmented matrix (p. 237) A matrix containing the coefficient matrix and the matrix of constants for a system of linear equations. The augmented matrix of the linear system $ax + by = e, cx + dy = f$ is $\begin{bmatrix} a & e \vdots d \\ b & c \vdots f \end{bmatrix}$.

axis of symmetry of a parabola (pp. 249, 595) The line perpendicular to the parabola's directrix and passing through its focus. In particular, the axis of symmetry is the vertical line through the vertex of the graph of a quadratic function.

B

base of an exponential function (p. 465) *See* exponential function.

base of a power (p. 11) The number in a power that is used as a factor. The base of the expression 2^5 is the number 2. *See also* exponent *and* power.

best-fitting quadratic model

best-fitting quadratic model (p. 308) The model given by performing quadratic regression on a graphing calculator, which uses all the data points entered.

binomial (p. 256) An expression with two terms, such as $x + 3$.

binomial distribution (p. 739) The set of probabilities of all possible numbers of successes in a binomial experiment.

binomial experiment (p. 739) An experiment that satisfies the following three conditions. (1) There are n independent trials. (2) Each trial has only two possible outcomes, success and failure. (3) The probability of success is the same for each trial. This probability is denoted by p. The probability of failure is given by $1 - p$.

binomial theorem (p. 710) The binomial expansion of $(a + b)^n$ for any positive integer n is $(a + b)^n = {}_nC_0 a^n b^0 + {}_nC_1 a^{n-1} b^1 + {}_nC_2 a^{n-2} b^2 + \cdots + {}_nC_n a^0 b^n = \sum_{r=0}^{n} {}_nC_r a^{n-r} b^r$.

box-and-whisker plot (p. 447) A type of statistical graph in which a "box" encloses the middle half of the data set and "whiskers" extend to the minimum and maximum data values. An example is shown.

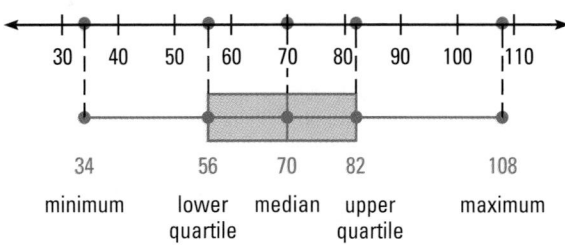

branches of a hyperbola (p. 540) Two symmetrical parts of a hyperbola. *See also* hyperbola.

C

center of a circle (p. 601) *See* circle.

center of a hyperbola (p. 615) The midpoint of the transverse axis of a hyperbola. *See also* hyperbola.

center of an ellipse (p. 609) The midpoint of the major axis of an ellipse. *See also* ellipse.

central angle of a sector (p. 779) An angle formed by two radii of a circle. *See also* sector.

circle (p. 601) The set of all points (x, y) that are equidistant from a fixed point, called the center of the circle. The distance r between the center of the circle and any point (x, y) on the circle is the radius.

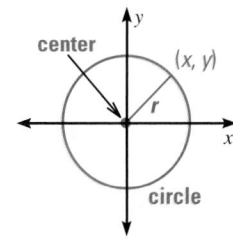

coefficient (p. 13) When a term is the product of a number and a power of a variable, such as $2x$ or $4x^3$, the number is the coefficient of the power. The coefficient of $2x$ is 2.

coefficient matrix (p. 216) The coefficient matrix of the linear system $ax + by = e$, $cx + dy = f$ is $\begin{bmatrix} a & b \\ c & d \end{bmatrix}$.

combination (p. 708) A selection of r objects from a group of n objects where the order is not important. The number of combinations of r objects taken from a group of n distinct objects is denoted $_nC_r$.

common difference (p. 659) The constant difference between consecutive terms of an arithmetic sequence.

common logarithm (p. 487) The logarithm with base 10. It is denoted by $\log_{10}$ or simply by log.

common ratio (p. 666) The constant ratio between consecutive terms of a geometric sequence.

complement (p. 726) The complement of event A, denoted A', consists of all outcomes that are not in A.

completing the square (p. 282) A process in which you write an expression of the form $x^2 + bx$ as the square of a binomial by adding the square of half the x-coefficient to the expression: $x^2 + bx + \left(\dfrac{b}{2}\right)^2 = \left(x + \dfrac{b}{2}\right)^2$. The process can be used to solve any quadratic equation.

complex conjugates (p. 274) Two complex numbers of the form $a + bi$ and $a - bi$. The product of complex conjugates is always a real number.

complex fraction (p. 564) A fraction that contains a fraction in its numerator or denominator.

complex number (p. 272) A number $a + bi$ where a and b are real numbers and i is the imaginary unit. The number a is the real part of the complex number, and the number bi is the imaginary part.

complex plane (p. 273) A coordinate plane where each point (a, b) represents a complex number $a + bi$. The complex plane has a horizontal real axis and a vertical imaginary axis.

composition (p. 416) The composition of the function f with the function g is $h(x) = f(g(x))$. The domain of h is the set of all x-values such that x is in the domain of g and $g(x)$ is in the domain of f.

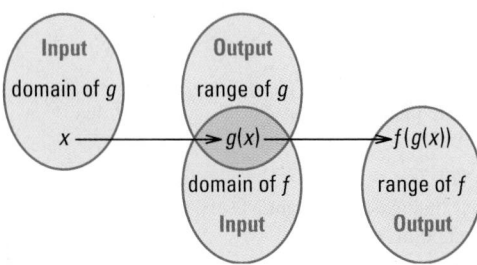

compound event (p. 724) The union or intersection of two events.

compound inequality (p. 43) Two simple inequalities joined by "and" or "or."

conditional probability (p. 732) The probability that event B will occur depending on whether event A has occurred. This is called the conditional probability of B given A and is written $P(B \mid A)$.

conic (p. 623) *See* conic section.

conic section (p. 623) A curve formed by the intersection of a plane and a double-napped cone. Examples include parabolas, circles, ellipses, and hyperbolas.

constant of variation (pp. 94, 534) The nonzero constant (usually denoted k) in a direct variation equation ($y = kx$), an inverse variation equation $\left(y = \dfrac{k}{x}\right)$, or a joint variation equation ($z = kxy$).

constant term (pp. 13, 329) A term that has no variable part, such as -4 or 2. *See also* polynomial function.

constraints (p. 163) In linear programming, the linear inequalities that form a system. *See also* linear programming.

coordinate (p. 3) The number that corresponds to a point on a number line.

coordinate plane (p. 67) A plane divided into four quadrants by the x-axis and the y-axis. It is used to plot ordered pairs of the form (x, y).

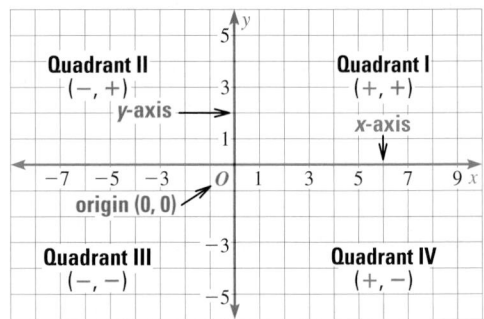

cosecant function (p. 769) If θ is an acute angle of a right triangle, the cosecant of θ is $\csc \theta = \dfrac{\text{hyp}}{\text{opp}}$ where *hyp* represents the length of the hypotenuse and *opp* represents the length of the side opposite θ.

cosine function (p. 769) If θ is an acute angle of a right triangle, the cosine of θ is $\cos \theta = \dfrac{\text{adj}}{\text{hyp}}$ where *adj* represents the length of the side adjacent to θ and *hyp* represents the length of the hypotenuse.

cotangent function (p. 769) If θ is an acute angle of a right triangle, the cotangent of θ is $\cot \theta = \dfrac{\text{adj}}{\text{opp}}$ where *adj* represents the length of the side adjacent to θ and *opp* represents the length of the side opposite θ.

coterminal angles (p. 777) Two angles in standard position with terminal sides that coincide.

co-vertices of an ellipse (p. 609) The points of intersection of an ellipse and the line perpendicular to the major axis at the center. *See also* ellipse.

Cramer's rule (p. 216) A method for solving a system of linear equations which uses determinants of matrices.

cross multiplying (p. 569) A method of solving a simple rational equation for which each side of the equation is a single rational expression. Equal products are formed by multiplying the numerator of each expression by the denominator of the other.

cubic function (p. 329) A polynomial function of degree 3.

cycle (p. 831) The shortest repeating portion of a periodic function.

D......................................

decay factor (p. 476) The quantity $1 - r$ in the exponential decay model $y = a(1 - r)^t$ where a is the initial amount and r is the percent decrease expressed as a decimal.

degree of a polynomial (p. 329) *See* polynomial function.

dependent events (p. 732) Two events such that the occurrence of one affects the occurrence of the other. *See also* conditional probability.

dependent variable (p. 69) The output variable in an equation, which depends on the value of the input variable. *See also* independent variable.

determinant (p. 214) A real number associated with any square matrix A, denoted by det A or by $|A|$. The determinant of a 2×2 matrix is the difference of the products of the entries on the diagonals.

dimensions of a matrix (p. 199) The number m of rows of a matrix by the number n of columns of the matrix, written $m \times n$.

directrix of a parabola (p. 595) *See* parabola.

direct variation (p. 94) Two variables x and y show direct variation provided $y = kx$ where k is a nonzero constant.

discriminant of a general second-degree equation (p. 626) The expression $B^2 - 4AC$ for the equation $Ax^2 + Bxy + Cy^2 + Dx + Ey + F = 0$. Used to determine which type of conic the equation represents.

discriminant of a quadratic equation (p. 293) The expression $b^2 - 4ac$ for the quadratic equation $ax^2 + bx + c = 0$; also the expression under the radical sign in the quadratic formula. Used to find the number and type of solutions of a quadratic equation.

distance formula (p. 589) The distance d between the points (x_1, y_1) and (x_2, y_2) is $d = \sqrt{(x_2 - x_1)^2 + (y_2 - y_1)^2}$.

domain of a relation (p. 67) The set of input values for a relation.

E......................................

eccentricity of a conic section (p. 639) The eccentricity of a hyperbola or an ellipse is $e = \dfrac{c}{a}$ where c is the distance from each focus to the center and a is the distance from each vertex to the center. The eccentricity of a parabola is $e = 1$. The eccentricity of a circle is $e = 0$.

ellipse (p. 609) The set of all points P such that the sum of the distances between P and two distinct fixed points, called foci, is a constant.

The ellipse shown below has a horizontal major axis.

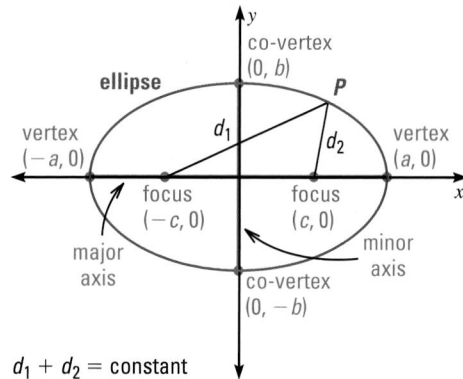

$d_1 + d_2 = $ constant

end behavior (p. 331) The behavior of the graph of a function as x approaches positive infinity or negative infinity.

entries of a matrix (p. 199) The numbers in a matrix.

equal matrices (p. 199) Matrices that have the same dimensions and equal entries in corresponding positions.

equation (p. 19) A statement in which two expressions are equal.

equation in two variables (p. 69) An equation such as $y = 2x - 7$.

equivalent algebraic expressions (p. 13) Expressions that have the same value for all values of their variable(s).

equivalent equations (p. 19) Equations that have the same solutions.

Euler number (p. 480) *See* natural base e.

expected value (p. 753) A collection of outcomes is partitioned into n events, no two of which have any outcomes in common. The probabilities of the n events occurring are $p_1, p_2, p_3, \ldots, p_n$ where $p_1 + p_2 + p_3 + \cdots + p_n = 1$. The values of the n events are $x_1, x_2, x_3, \ldots, x_n$. The expected value, V, of the collection of outcomes is the sum of the products of the events, probabilities and their values:
$$V = p_1 x_1 + p_2 x_2 + p_3 x_3 + \cdots + p_n x_n.$$

experimental probability (p. 717) A calculation of the probability of an event based on performing an experiment, conducting a survey, or looking at the history of an event.

explicit rule (p. 681) A rule for a sequence that gives a_n as a function of the term's position number n in the sequence.

exponent (p. 11) The number in a power that represents the number of times the base is used as a factor. The exponent of the expression 2^5 is the number 5. *See also* base of a power *and* power.

exponential decay function (p. 474) A function of the form $f(x) = ab^x$ where $a > 0$ and $0 < b < 1$.

exponential function (p. 465) A function that involves the expression b^x where the base b is a positive number other than 1.

exponential growth function (p. 466) A function of the form $f(x) = ab^x$ where $a > 0$ and $b > 1$.

extraneous solution (p. 439) A solution of a transformed equation that is not a valid solution of the original equation.

factor by grouping (p. 346) A method used to factor some polynomials with pairs of terms that have a common monomial factor: $ra + rb + sa + sb = r(a + b) + s(a + b) = (r + s)(a + b)$.

factorial (p. 681) The expression $n!$ is read "n factorial" and represents the product of all integers from 1 to n. Example: $4! = 4 \cdot 3 \cdot 2 \cdot 1 = 24$.

factoring (p. 256) A process used to write a polynomial as a product of other polynomials having equal or lesser degree. Example: $x^2 + 8x + 15 = (x + 3)(x + 5)$.

fair game (p. 753) A game for which the expected value is 0.

feasible region (p. 163) In linear programming, the graph of the system of constraints. *See also* linear programming.

finite differences (p. 380) The first-order differences of a polynomial function $f(x)$ are found by subtracting function values for equally spaced x-values. The second-order differences are found by subtracting consecutive first-order differences. The third-order differences are found by subtracting consecutive second-order differences, and so on.

finite sequence (p. 651) A sequence that has a last term.

foci of a hyperbola (p. 615) *See* hyperbola.

foci of an ellipse (p. 609) *See* ellipse.

focus of a parabola (p. 595) *See* parabola.

frequency distribution (p. 448) A table that shows the frequencies for the intervals into which data are grouped.

frequency of a periodic function (p. 833) The reciprocal of the period. Frequency is the number of cycles per unit of time.

frequency of data values (p. 448) The number of data values in an interval. *See also* frequency distribution.

function (p. 67) A relation with exactly one output for each input.

function notation (p. 69) Use of the symbol $f(x)$ for the dependent variable of a function. For example, the linear function $y = mx + b$ can be written $f(x) = mx + b$.

function of two variables (p. 171) A relationship in which one variable depends on two other variables. A linear equation in x, y, and z can be written as a function of two variables by solving for z and then replacing z with $f(x, y)$.

general second-degree equation in x and y (p. 626) The form $Ax^2 + Bxy + Cy^2 + Dx + Ey + F = 0$.

geometric probability (p. 718) A type of probability found by calculating a ratio of two lengths, areas, or volumes.

geometric sequence (p. 666) A sequence in which the ratio of any term to the previous term is constant.

geometric series (p. 668) The expression formed by adding the terms of a geometric sequence.

graph of an equation in two variables (p. 69) The collection of all points (x, y) whose coordinates are solutions of the equation.

graph of an inequality in one variable (p. 41) All points on a real number line that correspond to solutions of the inequality.

graph of an inequality in two variables (p. 108) The graph of all solutions of the inequality.

graph of a real number (p. 3) The point on a number line that corresponds to a real number.

graph of a system of linear inequalities (p. 156) The graph of all solutions of the system.

growth factor (p. 467) The quantity $1 + r$ in the exponential growth model $y = a(1 + r)^t$ where a is the initial amount and r is the percent increase expressed as a decimal.

half-planes (p. 108) The two regions of a coordinate plane that are separated by the boundary line of an inequality. One region contains the points that are solutions of the inequality, and the other region contains the points that are not.

histogram (p. 448) A special type of bar graph in which data are grouped into intervals of equal width.

hyperbola (pp. 540, 615) The set of all points P such that the difference of the distances from P to two fixed points, called the foci, is constant. The hyperbola below has a horizontal transverse axis.

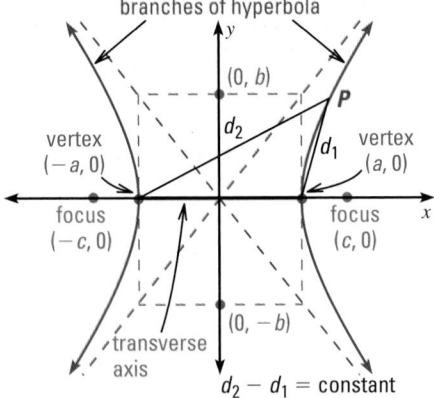

The graphs of rational functions of the form $y = \dfrac{a}{x - h} + k$ are hyperbolas.

hypothesis testing (p. 741) A three-step procedure from statistics for testing a claim. (1) State the hypothesis you are testing. The hypothesis should make a statement about some statistical measure (mean, standard deviation, or proportion) of a population. (2) Collect data from a random sample of the population and compute the statistical measure of the sample. (3) Assume that the hypothesis is true and calculate the resulting probability of obtaining the sample statistical measure or a more extreme sample statistical measure. If this probability is small, you should reject the hypothesis.

I

identity (p. 13) A statement such as $7x + 4x = 11x$ that equates two equivalent expressions.

identity matrix (p. 223) The $n \times n$ matrix that has 1's on the main diagonal and 0's elsewhere. The 2×2 identity matrix is $\begin{bmatrix} 1 & 0 \\ 0 & 1 \end{bmatrix}$.

imaginary number (p. 272) A complex number $a + bi$ where $b \neq 0$.

imaginary unit i (p. 272) The imaginary unit i is defined as $i = \sqrt{-1}$, so that $i^2 = -1$.

independent events (p. 730) Two events such that the occurrence of one has no effect on the occurrence of the other.

independent variable (p. 69) The input variable in an equation. *See also* dependent variable.

index of a radical (p. 401) The integer n (greater than 1) in the expression $\sqrt[n]{a}$.

infinite sequence (p. 651) A sequence that continues without stopping.

initial side of an angle (p. 776) You can generate any angle by fixing one ray, called the initial side, and rotating the other ray, called the terminal side, about the vertex.

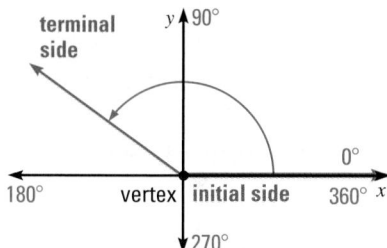

intercept form of a quadratic function (p. 250) The form $y = a(x - p)(x - q)$ where the x-intercepts of the graph are p and q and the axis of symmetry is halfway between $(p, 0)$ and $(q, 0)$.

inverse cosine function (p. 792) If $-1 \leq a \leq 1$, then the inverse cosine of a is $\cos^{-1} a = \theta$ where $\cos \theta = a$ and $0 \leq \theta \leq \pi$ (or $0° \leq \theta \leq 180°$).

inverse functions (p. 422) A relation and its inverse relation whenever both relations are functions. Functions f and g are inverses of each other provided $f(g(x)) = x$ and $g(f(x)) = x$. *See also* inverse relation.

inverse matrices (p. 223) Two $n \times n$ matrices are inverses of each other if their product (in both orders) is the $n \times n$ identity matrix. *See also* identity matrix.

inverse relation (p. 422) A relation that maps the output values of an original relation back to their original input values. The graph of an inverse relation is the reflection of the graph of the original relation, with $y = x$ as the line of reflection.

inverse sine function (p. 792) If $-1 \leq a \leq 1$, then the inverse sine of a is $\sin^{-1} a = \theta$ where $\sin \theta = a$ and $-\dfrac{\pi}{2} \leq \theta \leq \dfrac{\pi}{2}$ (or $-90° \leq \theta \leq 90°$).

inverse tangent function (p. 792) If a is any real number, then the inverse tangent of a is $\tan^{-1} a = \theta$ where $\tan \theta = a$ and $-\dfrac{\pi}{2} < \theta < \dfrac{\pi}{2}$ (or $-90° < \theta < 90°$).

inverse variation (p. 534) Two variables x and y show inverse variation provided $y = \dfrac{k}{x}$ where k is a nonzero constant.

J

joint variation (p. 536) A relationship that occurs when a quantity varies directly as the product of two or more other quantities. For instance, if $z = kxy$ where the constant $k \neq 0$, then z varies jointly with x and y.

L

law of cosines (p. 807) If $\triangle ABC$ has sides of length a, b, and c as shown below, then $a^2 = b^2 + c^2 - 2bc \cos A$, $b^2 = a^2 + c^2 - 2ac \cos B$, and $c^2 = a^2 + b^2 - 2ab \cos C$.

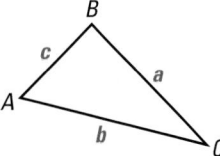

law of sines (p. 799) If $\triangle ABC$ has sides of length a, b, and c as shown above, then $\dfrac{\sin A}{a} = \dfrac{\sin B}{b} = \dfrac{\sin C}{c}$.

leading coefficient (p. 329) *See* polynomial function.

like radicals (p. 408) Two radical expressions that have the same index and the same radicand.

like terms (p. 13) Terms that have the same variable part, such as $3x^2$ and $-5x^2$.

linear equation in one variable (p. 19) An equation that can be written in the form $ax = b$ where a and b are constants and $a \neq 0$.

linear equation in three variables (p. 171) An equation of the form $ax + by + cz = d$ where x, y, and z are variables and a, b, and c are not all zero. The solution of a linear equation in three variables is an ordered triple (x, y, z), and the graph is a plane.

linear function (p. 69) A function of the form $y = mx + b$ where m and b are constants. The graph of a linear function is a line.

linear inequality in one variable (p. 41) An inequality such as $x \leq 1$ or $2n - 3 > 9$. Note that an inequality symbol is placed between two expressions.

linear inequality in two variables (p. 108) An inequality that can be written in one of the following forms: $Ax + By < C$, $Ax + By \leq C$, $Ax + By > C$, or $Ax + By \geq C$.

linear programming (p. 163) The process of optimizing a linear objective function subject to a system of linear inequalities called constraints. The graph of the system of constraints is called the feasible region.

local maximum (p. 374) The y-coordinate of a turning point of the graph of a function if the point is higher than all nearby points.

local minimum (p. 374) The y-coordinate of a turning point of the graph of a function if the point is lower than all nearby points.

logarithm of y with base b (p. 486) Let b and y be positive numbers with $b \neq 1$. The logarithm of y with base b is denoted by $\log_b y$ and is defined as $\log_b y = x$ if and only if $b^x = y$. The expression $\log_b y$ is read as "log base b of y."

logistic growth function (p. 517) A function of the form $y = \dfrac{c}{1 + ae^{-rx}}$ where a, c, and r are all positive constants. Used to model real-life quantities whose growth levels off because the rate of growth changes—from an increasing growth rate to a decreasing growth rate.

lower quartile (p. 447) The median of the lower half of a data set. *See also* box-and-whisker plot.

M.

major axis of an ellipse (p. 609) The line segment joining the vertices of an ellipse. *See also* ellipse.

mathematical model (p. 12) A mathematical representation of a real-life situation.

matrix (p. 199) A rectangular arrangement of numbers in rows and columns.

matrix of constants (p. 230) The matrix of constants of the linear system $ax + by = e$, $cx + dy = f$ is $\begin{bmatrix} e \\ f \end{bmatrix}$.

matrix of variables (p. 230) The matrix of variables of the linear system $ax + by = e$, $cx + dy = f$ is $\begin{bmatrix} x \\ y \end{bmatrix}$.

mean (p. 445) The sum of n numbers divided by n. Also called *average*.

measures of central tendency (p. 445) Three commonly used statistics: the mean, the median, and the mode of a set of numbers.

measures of dispersion (p. 446) Commonly used statistics that tell you how spread out the data are. They include the range and the standard deviation.

median (p. 445) The middle number when n numbers are written in order. (If n is even, the median is the mean of the two middle numbers.)

midpoint formula (p. 590) The midpoint of the line segment joining $A(x_1, y_1)$ and $B(x_2, y_2)$ is $M\left(\dfrac{x_1 + x_2}{2}, \dfrac{y_1 + y_2}{2}\right)$. Each coordinate of M is the mean of the corresponding coordinates of A and B.

minor axis of an ellipse (p. 609) The line segment joining the co-vertices of an ellipse. *See also* ellipse.

mode (p. 445) The number or numbers that occur most frequently in a set of n numbers. There may be one mode, no mode, or more than one mode.

monomial (p. 257) An expression with one term, such as $7x$.

mutually exclusive events (p. 724) Events A and B are mutually exclusive if the intersection of A and B is empty.

N.

natural base e (p. 480) An irrational number defined as follows: As n approaches $+\infty$, the value of $\left(1 + \dfrac{1}{n}\right)^n$ approaches $e \approx 2.718281828459$.

natural logarithm (p. 487) The logarithm with base e. It can be denoted by $\log_e$, but it is more often denoted by ln.

negative correlation (p. 100) The relationship between paired data when y tends to decrease as x increases, as shown by a scatter plot where the plotted points generally fall from left to right.

normal curve (p. 746) A smooth, symmetrical, bell-shaped curve that can model normal distributions and approximate some binomial distributions. *See also* normal distribution *and* binomial distribution.

normal distribution (p. 746) A distribution for which the mean and the standard deviation determine the following areas under a normal curve. (1) The total area under the curve is 1. (2) 68% of the area lies within 1 standard deviation of the mean. (3) 95% of the area lies within 2 standard deviations of the mean. (4) 99.7% of the area lies within 3 standard deviations of the mean. *See also* normal curve.

nth root of a (p. 401) For an integer n greater than 1, if $b^n = a$, then b is an nth root of a. Written as $\sqrt[n]{a}$.

numerical expression (p. 11) An expression that consists of numbers, operations, and grouping symbols.

O.

objective function (p. 163) In linear programming, the linear function that is optimized. *See also* linear programming.

octants (p. 170) *See* three-dimensional coordinate system.

opposite (p. 5) The opposite, or additive inverse, of any number a is $-a$.

optimization (p. 163) A process in which you find the maximum or minimum value of some variable quantity. One type of optimization process is linear programming.

ordered pair (p. 67) A pair of numbers of the form (x, y) that represents a point in the coordinate plane.

ordered triple (p. 170) A set of three numbers of the form (x, y, z) that represents a point in space. *See also* three-dimensional coordinate system.

order of operations (p. 11) A set of rules that gives the order in which operations should be performed when evaluating expressions.

origin of a coordinate plane (p. 67) The point $(0, 0)$ where the x-axis and y-axis intersect on a coordinate plane. *See also* coordinate plane.

origin of a real number line (p. 3) The point labeled O on a real number line.

P

parabola (pp. 249, 595) The set of all points equidistant from a point called the focus and a line called the directrix. The focus lies on the axis of symmetry, and the directrix is perpendicular to the axis of symmetry.

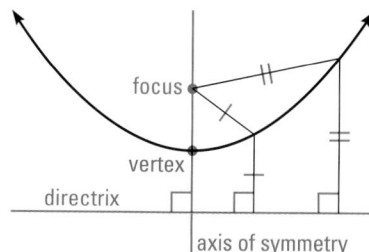

The graph of a quadratic function $y = ax^2 + bx + c$ is a parabola.

parallel lines (p. 77) Two lines in a plane that do not intersect.

parameter (p. 813) A variable, usually denoted t, upon which two other variables depend. *See also* parametric equations.

parametric equations (p. 813) Equations that express two variables in terms of a third variable, called the parameter.

Pascal's triangle (p. 710) An arrangement of the values of $_nC_r$ in a triangular pattern in which each row corresponds to a value of n. Each number other than 1 in Pascal's triangle is the sum of the two numbers directly above it.

$$_0C_0 \qquad\qquad 1$$
$$_1C_0 \ _1C_1 \qquad\qquad 1 \quad 1$$
$$_2C_0 \ _2C_1 \ _2C_2 \qquad\qquad 1 \quad 2 \quad 1$$
$$_3C_0 \ _3C_1 \ _3C_2 \ _3C_3 \qquad 1 \quad 3 \quad 3 \quad 1$$
$$_4C_0 \ _4C_1 \ _4C_2 \ _4C_3 \ _4C_4 \qquad 1 \quad 4 \quad 6 \quad 4 \quad 1$$
$$_5C_0 \ _5C_1 \ _5C_2 \ _5C_3 \ _5C_4 \ _5C_5 \quad 1 \quad 5 \quad 10 \quad 10 \quad 5 \quad 1$$

period (p. 831) The horizontal length of each cycle of a periodic function.

periodic function (p. 831) A function whose graph has a repeating pattern that continues indefinitely.

permutation (p. 703) An ordering of objects. The number of permutations of r objects taken from a group of n distinct objects is denoted $_nP_r$.

perpendicular lines (p. 77) Two lines in a plane that intersect to form a right angle.

piecewise function (p. 114) A function represented by a combination of equations, each corresponding to a part of the domain. Example: $f(x) = \begin{cases} 2x - 1, & \text{if } x \le 1 \\ 3x + 1, & \text{if } x > 1 \end{cases}$

polynomial function (p. 329) A function of the form $f(x) = a_nx^n + a_{n-1}x^{n-1} + \cdots + a_1x + a_0$ where $a_n \ne 0$, $a_0, a_1, a_2, \ldots a_n$ are real numbers, and the exponents are all whole numbers. For this polynomial function, a_n is the leading coefficient, a_0 is the constant term, and n is the degree.

polynomial long division (p. 352) A method used to divide polynomials similar to the way you divide numbers.

positive correlation (p. 100) The relationship between paired data when y tends to increase as x increases, as shown by a scatter plot where the plotted points generally rise from left to right.

power (p. 11) An expression such as 2^5, which represents $2 \cdot 2 \cdot 2 \cdot 2 \cdot 2 = 32$.

power function (p. 415) A function of the form $y = ax^b$ where a is a real number and b is a rational number.

probability (p. 716) A number between 0 and 1 that indicates the likelihood an event will occur.

pure imaginary number (p. 272) A complex number $a + bi$ where $a = 0$ and $b \ne 0$.

Q

quadrantal angle (p. 785) An angle in standard position with its terminal side on an axis. Examples: $0°, 90°, 180°,$ and $270°$.

quadrants (p. 67) The four regions that result when the x-axis and y-axis divide a coordinate plane. *See also* coordinate plane.

quadratic equation in one variable (p. 257) An equation that can be written in the form $ax^2 + bx + c = 0$ where $a \ne 0$.

quadratic form (p. 346) The form $au^2 + bu + c$ where u is any expression in x.

quadratic formula (p. 291) A formula that gives the solutions of any quadratic equation. If a, b, and c are real numbers with $a \ne 0$, the solutions of $ax^2 + bx + c = 0$ are $x = \dfrac{-b \pm \sqrt{b^2 - 4ac}}{2a}$.

quadratic function (p. 249) A function of the form $y = ax^2 + bx + c$ where $a \ne 0$.

quadratic inequality in one variable (p. 301) An inequality of the form $ax^2 + bx + c < 0$, $ax^2 + bx + c > 0$, $ax^2 + bx + c \le 0$, or $ax^2 + bx + c \ge 0$.

quadratic inequality in two variables (p. 299) An inequality of the form $y < ax^2 + bx + c$, $y > ax^2 + bx + c$, $y \le ax^2 + bx + c$, or $y \ge ax^2 + bx + c$.

quartic function (p. 329) A polynomial function of degree 4.

R

radian (p. 777) In a circle with radius r and center at the origin, one radian is the measure of an angle in standard position whose terminal side intercepts an arc of length r.

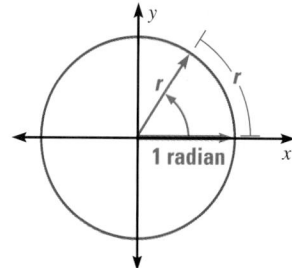

radical (p. 264) An expression of the form $\sqrt{s}$ or $\sqrt[n]{s}$ where s is a number or expression.

radical function (p. 431) A function that contains a radical, such as $y = \sqrt{x}$ or $y = \sqrt[3]{x}$.

radical symbol (p. 264) The symbol $\sqrt{}$ or $\sqrt[n]{}$, which denotes a square root or nth root, respectively.

radicand (p. 264) The number or expression beneath a radical sign. The radicand of $\sqrt{5}$ is 5 and the radicand of $\sqrt[3]{7x}$ is $7x$.

radius of a circle (p. 601) The distance from the center of a circle to a point on the circle, or the line segment that connects the center of a circle to a point on the circle. *See also* circle.

range of a relation (p. 67) The set of output values for a relation.

range of data values (p. 446) The difference between the greatest and least data values.

rational function (p. 540) A function of the form $f(x) = \dfrac{p(x)}{q(x)}$ where $p(x)$ and $q(x)$ are polynomials and $q(x) \neq 0$.

rationalizing the denominator (p. 265) The process of eliminating a radical in the denominator of a fraction by multiplying both the numerator and the denominator by an appropriate radical.

reciprocal (p. 5) The reciprocal, or multiplicative inverse, of any nonzero real number a is $\dfrac{1}{a}$.

recursive rule (p. 681) A rule for a sequence that gives the beginning term or terms of a sequence and then a recursive equation that tells how a_n is related to one or more preceding terms.

reference angle (p. 785) If θ is an angle in standard position, its reference angle is the acute angle θ' formed by the terminal side of θ and the x-axis. An example is shown.

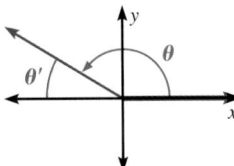

relation (p. 67) A mapping, or pairing, of input values with output values.

relatively no correlation (p. 100) The relationship between paired data when a scatter plot of the data shows no linear pattern.

repeated solution (p. 366) For the equation $f(x) = 0$, k is a repeated solution if and only if the factor $(x - k)$ has degree greater than 1 when f is factored completely.

scalar (p. 200) A real number by which you multiply a matrix.

scalar multiplication (p. 200) The process of multiplying each entry in a matrix by a scalar.

scatter plot (p. 100) A graph of ordered pairs used to determine whether there is a relationship between paired data.

scientific notation (p. 325) A number is expressed in scientific notation if it is in the form $c \times 10^n$ where $1 \le c < 10$ and n is an integer.

secant function (p. 769) If θ is an acute angle of a right triangle, the secant of θ is $\sec \theta = \dfrac{\text{hyp}}{\text{adj}}$ where *hyp* represents the length of the hypotenuse and *adj* represents the length of the side adjacent to θ.

sector (p. 779) A region of a circle that is bounded by two radii and an arc of the circle.

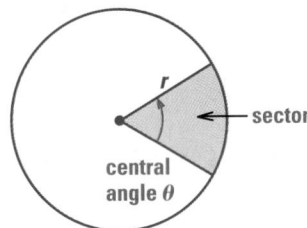

sequence (p. 651) A function whose domain is a set of consecutive integers. The domain gives the relative position of each term of the sequence: 1st, 2nd, 3rd, and so on. The range gives the terms of the sequence.

series (p. 653) The expression that results when the terms of a sequence are added.

sigma notation (p. 653) *See* summation notation.

simplest form of a radical (p. 408) A radical expression after you apply the properties of radicals, remove any perfect nth powers, and rationalize any denominators.

simplified form of a rational expression (p. 554) A rational expression in which the numerator and denominator have no common factors (other than ± 1).

sine function (p. 769) If θ is an acute angle of a right triangle, the sine of θ is $\sin \theta = \dfrac{\text{opp}}{\text{hyp}}$ where *opp* represents the length of the side opposite θ and *hyp* represents the length of the hypotenuse.

skewed distribution (p. 740) A distribution that is not symmetric. *See also* symmetric distribution.

slope (p. 75) The ratio of vertical change (the rise) to horizontal change (the run) for a nonvertical line. The slope of a nonvertical line passing through the points (x_1, y_1) and (x_2, y_2) is $m = \dfrac{y_2 - y_1}{x_2 - x_1} = \dfrac{\text{rise}}{\text{run}}$.

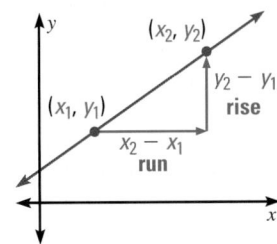

slope-intercept form (p. 82) A linear equation written in the form $y = mx + b$ where m is the slope and b is the y-intercept of the line.

solution of an equation in one variable (p. 19) A number that, when substituted for the variable, makes the equation a true statement.

solution of an equation in two variables (p. 69) An ordered pair (x, y) that makes the equation a true statement when the values of x and y are substituted in the equation.

solution of an inequality in one variable (p. 41) A value of the variable that makes the inequality true.

solution of an inequality in two variables (p. 108) An ordered pair (x, y) that, when x and y are substituted in the inequality, gives a true statement.

solution of a system of linear equations (p. 139) An ordered pair (x, y) that satisfies each equation of the system.

solution of a system of linear inequalities (p. 156) An ordered pair that is a solution of each inequality in the system.

solution of a system of three linear equations (p. 177) An ordered triple (x, y, z) that is a solution of all three equations of the system.

solving a right triangle (p. 770) Finding all missing side lengths and angle measures of a right triangle.

square root (p. 264) A number r is a square root of a number s if $r^2 = s$.

standard deviation (p. 446) The typical difference (or *deviation*) between the mean and a data value. The standard deviation σ of $x_1, x_2, \ldots, x_n$ is
$$\sigma = \sqrt{\frac{(x_1 - \bar{x})^2 + (x_2 - \bar{x})^2 + \cdots + (x_n - \bar{x})^2}{n}}.$$

standard form of a complex number (p. 272) The form $a + bi$ where a and b are real numbers and i is the imaginary unit.

standard form of a linear equation (p. 84) A linear equation written in the form $Ax + By = C$ where A and B are not both zero.

standard form of a polynomial function (p. 329) The form of a polynomial function when the terms are written in descending order of exponents from left to right.

standard form of a quadratic equation (p. 257) The form $ax^2 + bx + c = 0$ where $a \neq 0$.

standard form of a quadratic function (p. 250) The form $y = ax^2 + bx + c$ where $a \neq 0$.

standard form of the equation of a circle (pp. 601, 623) If a circle has center (h, k) and radius r, its equation is $(x - h)^2 + (y - k)^2 = r^2$. *See also* circle.

standard form of the equation of a hyperbola (p. 615) If a hyperbola has center (h, k), its equation is as follows:
$$\frac{(x - h)^2}{a^2} - \frac{(y - k)^2}{b^2} = 1 \text{ (horizontal transverse axis) or}$$
$$\frac{(y - k)^2}{a^2} - \frac{(x - h)^2}{b^2} = 1 \text{ (vertical transverse axis)}.$$
See also hyperbola.

standard form of the equation of an ellipse (p. 609) If an ellipse has center (h, k) and major and minor axes of lengths $2a$ and $2b$, where $a > b > 0$, its equation is as follows:
$$\frac{(x - h)^2}{a^2} + \frac{(y - k)^2}{b^2} = 1 \text{ (horizontal major axis) or}$$
$$\frac{(x - h)^2}{b^2} + \frac{(y - k)^2}{a^2} = 1 \text{ (vertical major axis)}. \text{ See also ellipse.}$$

standard form of the equation of a parabola (pp. 596, 623) If a parabola has vertex (h, k), its equation is as follows: $(y - k)^2 = 4p(x - h)$ (horizontal axis) or $(x - h)^2 = 4p(y - k)$ (vertical axis).

standard position of an angle (p. 776) In a coordinate plane, the position of an angle whose vertex is at the origin and whose initial side is the positive x-axis. *See also* initial side of an angle.

statistics (p. 445) Numerical values used to summarize and compare sets of data.

step function (p. 115) A piecewise function whose graph resembles a set of stair steps. *See also* piecewise function.

summation notation (p. 653) Notation for a series that uses the uppercase Greek letter sigma, Σ. For example, you can write $3 + 6 + 9 + 12 + 15 = \sum_{i=1}^{5} 3i$ where i is the index of summation, 1 is the lower limit of summation, and 5 is the upper limit of summation.

symmetric distribution (p. 740) A distribution in which the left half of the histogram representing the distribution is a mirror image of the right half.

synthetic division (p. 353) A method used to divide a polynomial by an expression of the form $x - k$.

synthetic substitution (p. 330) A method used to evaluate a polynomial function.

system of linear inequalities in two variables (p. 156) A system made up of two linear inequalities in two variables. *See also* linear inequality in two variables.

system of three linear equations (p. 177) A system made up of three linear equations in three variables. *See also* linear equation in three variables.

system of two linear equations (p. 139) Two equations of the form $Ax + By = C$ and $Dx + Ey = F$ where x and y are variables, A and B are not both zero, and D and E are not both zero.

 T

tangent function (p. 769) If θ is an acute angle of a right triangle, the tangent of θ is $\tan \theta = \dfrac{opp}{adj}$ where *opp* represents the length of the side opposite θ and *adj* represents the length of the side adjacent to θ.

terminal side of an angle (p. 776) *See* initial side of an angle.

terms of an expression (p. 13) The parts of an algebraic expression that are added together. The terms of $2x + 3$ are $2x$ and 3. The terms of $2x - 3 = 2x + (-3)$ are $2x$ and -3.

terms of a sequence (p. 651) For a sequence of numbers, the numbers in the sequence are called terms. *See also* sequence.

theoretical probability (p. 716) When all outcomes are equally likely, the theoretical probability that an event A will occur is $P(A) = \dfrac{\text{number of outcomes in } A}{\text{total number of outcomes}}$. The theoretical probability of an event is often simply called the probability of the event.

three-dimensional coordinate system (p. 170) A coordinate system determined by three mutually perpendicular axes. When taken pairwise, these axes form three coordinate planes that divide space into eight parts called octants. A point in space is represented by an ordered triple of the form (x, y, z). The ordered triples $(-5, 3, 4)$, $(0, 0, 0)$, and $(2, -2, -3)$ are plotted below.

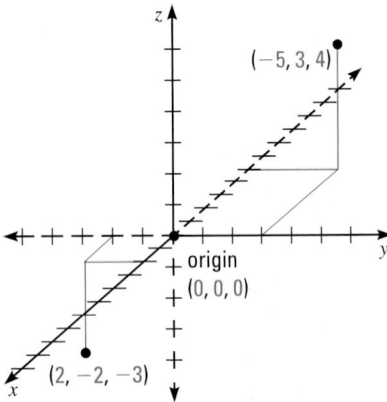

transverse axis of a hyperbola (p. 615) The line segment joining the vertices of a hyperbola. *See also* hyperbola.

trigonometric identity (p. 848) A trigonometric equation that is true for all domain values.

trinomial (p. 256) An expression with three terms, such as $x^2 + 8x + 15$.

U··

upper quartile (p. 447) The median of the upper half of a data set. *See also* box-and-whisker plot.

V··

value of an expression (p. 12) The result when the variables in an algebraic expression are replaced by numbers and the expression is simplified.

value of a variable (p. 12) Any number used to replace a variable.

variable (p. 12) A letter that is used to represent one or more numbers.

verbal model (p. 33) A word equation that represents a real-life problem.

vertex form of a quadratic function (p. 250) The form $y = a(x - h)^2 + k$ where the vertex of the graph is (h, k) and the axis of symmetry is $x = h$.

vertex of an absolute value graph (p. 122) The corner point of the graph of an absolute value function.

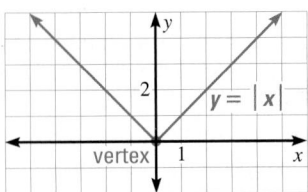

vertex of a parabola (pp. 249, 595) The point on a parabola that lies on the axis of symmetry. This point is the lowest or highest point on a parabola with a vertical axis of symmetry and the leftmost or rightmost point on a parabola with a horizontal axis of symmetry. *See also* parabola.

vertices of a hyperbola (p. 615) The points of intersection of a hyperbola and the line through the foci of the hyperbola. *See also* hyperbola.

vertices of an ellipse (p. 609) The points of intersection of an ellipse and the line through the foci of the ellipse. *See also* ellipse.

X··

x-axis (p. 67) The horizontal axis in the coordinate plane. *See also* coordinate plane.

x-coordinate (p. 67) The first number in an ordered pair.

x-intercept of a line (p. 84) The x-coordinate of the point where a line intersects the x-axis. Given an equation of the line, it is the value of x when $y = 0$.

Y··

y-axis (p. 67) The vertical axis in the coordinate plane. *See also* coordinate plane.

y-coordinate (p. 67) The second number in an ordered pair.

y-intercept (p. 82) If the graph of an equation intersects the y-axis at the point $(0, b)$, then the number b is the y-intercept of the graph. Given an equation of the graph, it is the value of y when $x = 0$.

Z··

z-axis (p. 170) A vertical line through the origin and perpendicular to the xy-coordinate plane in a three-dimensional coordinate system. *See also* three-dimensional coordinate system.

zero of a function (p. 354) A number k is a zero of a function f if $f(k) = 0$.

Answers and Index

The Teacher's Edition supplements the answers and index material in the Student Edition. The credits are also included.

▶ **Selected Answers** . . . . *SA1 – SA64*

The Selected Answers that appear at the back of the Student Edition are included in their entirety.

▶ **Teacher's Edition Index** . . . . *IN1 – IN18*

The Teacher's Edition Index includes the Student Edition Index along with references (in blue type) to material in the Teacher's Edition margins and to material on the interleaved pages at the beginning of each chapter.

▶ **Additional Answers** . . . . *AA1 – AA57*

The Additional Answers include all answers that do not appear at point of use.

Credits

Cover Credits

Earth Imaging/Tony Stone Images (background); Don Hamerman/New England Stock Photo (t); David Perdew/Stock South/PNI/PictureQuest (r); Steve Vidler/Leo deWys, Inc. (b); Stuart Westmorland/Tony Stone Images (t).

Photography

i, ii Earth Imaging/Tony Stone Images (background); Don Hamerman/Tony Stone Images (t); David Perdew/Stock South/PNI/PictureQuest (r); Steve Vidler/Leo deWys, Inc. (b); Stuart Westmorland/Tony Stone Images (t); **iv** Mark C. Burnett/Stock Boston (l); Dallas & John Heaton/Stock Boston (r); **v** CORBIS/Kevin Schafer (l); A. Witte/C. Mahaney/Tony Stone Images (r); **vi** CORBIS/Colin Garratt: Milepost 92 1/2; **vii** courtesy, City Year; **viii** CORBIS/Sergio Carmona; **ix** Bob Daemmrich/The Image Works & David Young-Wolff/PhotoEdit; **x** Paul Chesley/Tony Stone Images; **xi** courtesy, NASA; **xii** CORBIS/Michael S. Yamashita; **xiii** Paul Souders/Tony Stone Images; **xiv** Vandystadt/Photo Researchers, Inc.; **xv** CORBIS/Roger Ressmeyer; **xvi** Craig Tuttle/The Stock Market; **xvii** Bill Gallery/Stock Boston; **xviii** Duvall/Liaison Agency; **xix** Siegfried Lavda/Tony Stone Images; **xxi** Jon Riley/Tony Stone Images (cl); CORBIS/Jonathan Blair (c); Photodisc, Inc. (cr); Terry Vine/Tony Stone Images (bl); L.D. Gordon/Image Bank; **xxxiv** CORBIS/Colin Garratt: Milepost 92 1/2; **1** CORBIS/Tim Hawkins; Eye Ubiquitous; **3** Kathy Squires; **6** Kathy Squires; **8** Matthew Stockham/Allsport; **9** Patrick Aventurier/Liaison Agency; **11** Peter Timmermans/Tony Stone Images; **13** courtesy, Sony Corporation; **16** Terry Vine/Tony Stone Images; **21** L.D. Gordon/Image Bank; **19** Charles D. Winters/Photo Researchers, Inc.; **23** courtesy, Sony Corporation; **26** CORBIS; **27** Paul S. Howell/Liaison Agency; **30** Otto Greule/Allsport; **32** Andy Sacks/Tony Stone Images; **33** Dallas & John Heaton/Stock Boston; **35** CORBIS/Bettmann; **36** CORBIS/Ecoscene Graham Neden; **38** Bassignac/Liaison Agency; **40** Art Resource, NY (cr); bachmann/Stock Boston (bl); CORBIS/Paul Almasy (bcl); The Granger Collection (bcr); CERN/SPL/Photo Researchers, Inc. and PhotoDisc, Inc. (montage br); **41** Steve Vidler/Leo de Wys Inc.; **42** Aredea London Ltd.; **44** François Darmigny/Liaison Agency; **46** National Geographic Society Image Collection; **49** RMIP/Richard Haynes (all); **50** Tony Duffy/NBC/Allsport; **52** Mark Cooper/The Stock Market; **54** Giovanni Lunardi/International Stock Photo; **64, 65** courtesy, City Year; **67** Matin de Lausanne/Sygma; **68** Tom Stewart/The Stock Market; **70** Becker Jean/Sygma; **73** A. Witte/C. Mahaney/Tony Stone Images; **75** Superstock; **78** Michael Dwyer/Stock Boston; **80** UPI/CORBIS/Bettmann; **82** Frank Cezus/Tony Stone Images; **87** Sue Cunningham/Tony Stone Images; **89** Science Photo Library/Photo Researchers, Inc. (cr); CORBIS/Bettmann (bcl); John Coletti/Stock Boston (bc); Jim Bourg/Liaison Agency, Inc. (br); **91** Paul Conklin/PhotoEdit; **93** Terry Ashe/Liaison Agency; **97** National Center for Atmospheric Research/National Science Foundation; **99** RMIP/Richard Haynes (all); **100** Barbara Filet/Tony Stone Images; **102** Rob Lewine/The Stock Market; **104** CORBIS/Dewitt Jones; **108** Ronnie Kaufman/The Stock Market; **112** Ed Young/Science Photo Library/Photo Researchers, Inc.; **114** Spencer Grant/PhotoEdit; **119** Ann-Marie Louvet/Explorer/Photo Researchers, Inc.; **122** Bob Daemmrich/Stock Boston; **126** BPI Communications, Inc. Used with permission from BILLBOARD magazine; **127** Steve Elmore/The Stock Market; **136, 137** CORBIS/Sergio Carmona; **139** Tony Arruza/Tony Stone Images; **144** School Division, Houghton Mifflin Company; **148** Michael A. Keller/The Stock Market; **151** Richard Pasley/Stock Boston; **154** Mickey Pfleger/Photo 20-20; **156** Ron Sanford/The Stock Market;

158 Jon Feingersh/The Stock Market; **161** Joe Mann/Allsport; **163** Andy Sacks/Tony Stone Images; **165** Alain Le Garsmeur/Tony Stone Images; **167** Kim Butler; **169** Hulton Getty/Liaison Agency (cr); Photo by Edward W. Souza, courtesy of Stanford University (bl); CORBIS/Bettmann-UPI (bcl, both); Copyright ©1979 by the New York Times Co. Reprinted by permission.(bcr); courtesy of Lucent Technologies, Bell Labs Innovations (br); **170** Mark C. Burnett/Stock Boston; **172** Michael Rosenfeld/Tony Stone Images; **174** CORBIS/Kevin Fleming; **177** Bob Daemmrich/Stock Boston; **183** Bob Daemmrich/Stock Boston; **194** *The Avenue at Middelharnis*, 1689 (oil on canvas) by Meindert Hobbema (1638–1709), National Gallery, London, UK/Bridgeman Art Library (l); *Corner of George and Hunter Streets, Sydney*, 1849 (w/c) by A. Torning (19th Century), Dixson Galleries, State Library of New South Wales/Bridgeman Art Library (r); **196** Bob Daemmrich/The Image Works & David Young-Wolff/PhotoEdit; **197** Bob Daemmrich/The Image Works; **199** Fotex/Shooting Star; **202** James Prince/Photo Researchers, Inc.; **208** Bill Devine/David Madison; **210** Rick T. Wilking/Reuters/Archive Photos; **212** Mike Dowell/Allsport; **214** Phyllis Picardi/Stock Boston; **215** courtesy, Naval Historical Foundation; **217** Rick Brady/Uniphoto; **219** David Weintraub/Stock Boston; **223** Jim Sanborn; **225** Archive Photos; **226** King's College/Cambridge University Library; **228** The Granger Collection; **230** CORBIS/Ted Spiegel; **232** Nancy Sheehan/PhotoEdit; **234** Larry Gatz/Image Bank; **236** Lawrence Migdale/Stock Boston (bl); CORBIS/Bettmann (bc); UPI/CORBIS/Bettmann (br); **246** Joe Carini/Pacific Stock; **247** Paul Chesley/Tony Stone Images; **249** R. Berenholtz/The Stock Market; **252** Jon Riley/Tony Stone Images; **254** Lewis Portnoy/The Stock Market; **256** CORBIS/Morton Beebe, S.F.; **262** David Young-Wolff/PhotoEdit; **264** Hideo Kurihara/Tony Stone Images; **268** Courtesy, NASA; **270** CORBIS/Jim Sugar (bl); The Granger Collection (bcl, bcr); courtesy of Professor Borra, Dept. Physique, Université Laval (br); **275** Raphael Gaillarde/Liaison Agency, Inc.; **279** Jean Higgins/Unicorn Stock Photo; **281** RMIP/Richard Haynes (all); **282** Stuart Westmorland/Tony Stone Images; **288** Tony Duffy/Allsport; **291** Nick Gunderson/Tony Stone Images; **296** Photodisc, Inc.; **299** Charlie Westerman/International Stock Photo; **302** Stephen Frisch/Stock Boston; **303** Paul Steel/The Stock Market; **304** Anna E. Zuckerman/PhotoEdit; **306** David Madison; **308** Kevin Horan/Tony Stone Images; **311** Vincent Laforet/Allsport; **320** courtesy, NASA; **321** NASA/Rainbow/PNI/PictureQuest; **323** Fred Hirschmann/Tony Stone Images; **325** NASA/Science Photo Library/Photo Researchers, Inc., NASA and NASA/PNI (montage); **327** David Weintraub/Stock Boston; **329** Clive Brunskill/Allsport; **330** CORBIS/Sue Garcia; **335** Charles Gupton/Stock Boston; **336** Tim Davis/Photo Researchers, Inc.; **338** David Perdew/Stock South/PNI; **340** Andy Sacks/Tony Stone Images; **342** Frank Siteman/Stock Boston; **345** Avner Raban/Combined Caesarea Expeditions; **347** Garo Nalbandian/Photo Researchers, Inc.; **350** Sculpture at Goodwood, sculpture by David Nash, *Charred Sphere, Cube and Pyramid, 1997*; **351** The British Museum (cr); North Wind Picture Archives (bl); CORBIS/Burstein Collection (bcl); The Granger Collection (bcr); courtesy, NASA (br); **352** EDGE Productions; **355** Frank Siteman/Tony Stone Images; **357** Joshua and Kaia Tickell/Green Teach; **359** Stephen Johnson/Tony Stone Images; **361** Armen Kachaturian/Liaison Agency, Inc.; **363** Volker Steger/Science Photo Library/Photo Researchers, Inc.; **364** Chad Slattery; **366** Lawrence Migdale; **370** Robert Brenner/PhotoEdit; **373** Claus Meyer/Black Star/PNI; **377** CORBIS/Bettmann; **380** NASA/Rainbow/PNI; **382** P. Plisson/Photo Researchers, Inc.; **385** courtesy, NASA; **397** Foto Marburg/Art Resource; **398, 399** CORBIS/Michael S. Yamashita; **401** CORBIS/Kevin Schafer; **403** Alexandra Guest/Trireme Trust (bl); Mark Maxwell (tr); **405** Jim Wark/Peter Arnold, Inc.; **407**

TEACHER'S EDITION CREDITS
T4 Mark C. Burnett/Stock Boston (l); Dallas & John Heaton/Stock Boston (r); T5 CORBIS/Kevin Schafer (l); A. Witte/C. Mahaney/Tony Stone Images (r); T6 CORBIS/Colin Garratt: Milepost 92$\frac{1}{2}$; T7 courtesy, City Year; T8 CORBIS/Sergio Carmona; T9 Bob Daemmrich/The Image Works & David Young-Wolff/PhotoEdit; T10 Paul Chesley/Tony Stone Images; T11 courtesy, NASA; T12 CORBIS/Michael S. Yamashita; T13 Paul Souders/Tony Stone Images; T14 Vandystadt/Photo Researchers, Inc.; T15 CORBIS/Roger Ressmeyer; T16 Craig Tuttle/The Stock Market; T17 Bill Gallery/Stock Boston; T18 Duvall/Liaison Agency; T19 Siegfried Lavda/Tony Stone Images; T21 Jon Riley/Tony Stone Images (tl); CORBIS/Jonathan Blair (tc); Photodisc, Inc. (tr); Terry Vine/Tony Stone Images (bl); L.D. Gordon/The Image Bank (br); T24 RMIP/Richard Haynes; T25 RMIP/Richard Haynes (t); Paul Steel/The Stock Market (b); T37 Nancy Sheehan/PhotoEdit.

CORBIS/Catherine Karnow; **410** Stephen J. Krasemann/DRK Photo; **415** CORBIS/Tom Bean; **417** Joe McDonald/Animals Animals; **419** CORBIS/Jonathan Blair; **421** RMIP/Richard Haynes (all); **422** Michael Newman/PhotoEdit; **425** Frank Rossotto/The Stock Market; **427** CORBIS/Robert Maass; **428** Herb Segars/Animals Animals; **431** M. Colbeck/Animals Animals; **433** Courtesy of Chance Rides, Inc. (all); **435** CORBIS/Jim Sugar; **437** CORBIS/Nick Rains; Cordaiy Photo Library Ltd.; **440** CORBIS/James L. Amos; **442** Courtesy of Dr. Alexa Canady/The Detroit Medical Center (cl); Kathy Squires (br); **444** The Granger Collection (bl); Smithsonian Institution (bcl); Courtesy of NOAA/National Geophysical Data Center (bcr); PMEL/NOAA, Seattle, Washington (br); **445** Todd Warshaw/Allsport; **447** The Detroit News, photograph by Robin Buckson; **450** William Taufic/The Stock Market; **462** Galen Rowell/Mountain Light; **463** Paul Souders/Tony Stone Images; **465** Alan G. Nelson/Animals Animals/Earth Scenes; **467** Mark Richards/PhotoEdit; **468** Gary Conner/PhotoEdit; **473** RMIP/Richard Haynes; **474** James Wilson/Woodfin Camp and Associates; **480** Michio Hoshino/Minden Pictures; **481** The Granger Collection; **482** Douglas Faulkner/Photo Researchers, Inc.; **484** Archive Photos; **486** Herb Segars/Animals Animals/Earth Scenes; **489** E.H. Degginger/Photo Researchers, Inc. (l); Richard Kolar/Earth Scenes (r); **491** A.T. Willett/Image Bank; **493** Uniphoto; **495** Stephen Frisch/Stock Boston; **497** RMIP/Richard Haynes; **498** Courtesy of Jean Young; **499** School Division, Houghton Mifflin Company (cr, br); Science & Society Picture Library (bl); Science Photo Library/Photo Researchers, Inc. (bcl); The Granger Collection (bcr); **498** Courtesy of Jean Young; **501** Phil Degginger/Animals Animals; **504** UPI/CORBIS/Bettmann; **507** Jerry Schad/Photo Researchers, Inc.; **509** Reproduced with permission from Hamilton Projects, Inc. and the U.S. Postal Service. Photo taken by School Division, Houghton Mifflin Company; **512** Science Photo Library/Photo Researchers, Inc.; **515** Akira Uchiyama/Photo Researchers, Inc.; **517** Richard B. Levine; **521** Mitch Kezar/Tony Stone Images; **530** Vandystadt/Photo Researchers, Inc.; **531** Superstock; **533** RMIP/Richard Haynes (all); **534** Françoise Sauze/SPL/Photo Researchers, Inc.; **535** Ian Beames/Ardea London Limited; **538** Mike Yamashita/Woodfin Camp and Associates; **540** Mug Shots/The Stock Market; **542** Gregory Sams/Photo Researchers, Inc.; **544** Keith Kent/Photo Researchers, Inc.; **547** Gerard Lacz/Peter Arnold, Inc.; **551** Feingersh/The Stock Market; **554** Jeremy Walker/Tony Stone Images; **557** Courtesy of Gregory Robertson; **559** Zane Williams/Tony Stone Images; **562** Chromosohm/Sohm/Photo Researchers, Inc.; **566** Bruce Ayres/Tony Stone Images; **568** Bob Daemmrich/Stock Boston; **570** Bill Gallery/Stock Boston; **572** John Marshall/Tony Stone Images; **574** Macduff Everton (cr); The Granger Collection (bl); CORBIS/Bettmann-UPI (bcl); John (New York) Tee-Van/National Geographic Image Collection (bcr); Charles Nicklin/Al Giddings Images (br); **584** Bob Daemmrich/Stock Boston; **585** RMIP/Richard Haynes (all); **586, 587** CORBIS/Roger Ressmeyer; **589** Richard Pasley/Stock Boston; **591** CORBIS/Lowell Georgia; **593** Brian G. Miller/Illinois State Police; **595** Ken Biggs/Tony Stone Images; **597** Sonia Balcer (tr); Douglas Kirkland (bl); **599** CORBIS/Bettmann; **601** CORBIS/Ron Watts; **603** Stock Montage, Inc.; **605** CORBIS/James L. Amos; **606** Peter Menzel/Stock Boston; **609** Gatha Ashvin/Leo de Wys Inc.; **613** David Burnett/Contact Press/PNI; **615** Fermilab Visual Media Services; **617** Photo courtesy of Carnegie Mellon University; **620** John P. Shepherd; **622** RMIP/Richard Haynes; **623** Nathan Bilow/Allsport; **625** Richard B. Levine; **629** courtesy, Thronateeska Heritage Center, Albany, Georgia; **631** Rev. Ronald Royer/Science Photo Library/Photo Researchers, Inc. (cr); CORBIS/Ruggero Vanni (bl); CORBIS/Bettmann (bcl); PhotoDisc, Inc. (frame bcl); courtesy, Debra Fischer, San Francisco State University (br); **632** Kevin Schafer/Allstock/PNI; **634** Andrew Rafkind/Tony Stone Images; **636** Spencer Grant III/Stock Boston; **640** courtesy, NASA; **648** (clockwise from left) Bruce Forster/Tony Stone Images; Craig Tuttle/The Stock Market; Daniel W. Gotshall/Visuals Unlimited; Gerben Oppermans/Tony Stone Images; **649** TC Nature/Animals Animals; **651** Lance Nelson/The Stock Market; **654** Bill

Wood/University of Michigan, Photo Services Department; **656** RMIP/Richard Haynes; **659** Lincoln Russell/Stock Boston; **661** The Granger Collection; **662** CORBIS/Andrea Jemolo; **664** Stephen Frisch/Stock Boston; **666** Al Bello/Allsport; **669** Mike Yamashita/Woodfin Camp and Associates; **671** David Young-Wolff/PhotoEdit; **674** RMIP/Richard Haynes (all); **675** Jose Fuste Raga/The Stock Market; **677** Andy Washnik/The Stock Market; **679** Jon Feingersh/The Stock Market; **681** Mark Burnett/Stock Boston; **683** Bob Daemmrich/Tony Stone Images; **685** Peter Holden/Visuals Unlimited; **687** CORBIS/Larry Lee (bl); The Granger Collection (bc); Darrell Gulin/DRK Photo & CORBIS (montage, br); **698** Bill Gallery/Stock Boston; **699** CORBIS/Kevin Fleming; **701** Nathan Bilow/Allsport; **702** Richard Pasley/Stock Boston; **706** Bob Daemmrich/Stock Boston; **708** Bonnie Kamin/PhotoEdit; **710** The Granger Collection; **713** CORBIS/Kevin Fleming; **716** Paolo Negri/Tony Stone Images; **721** Carlin/Archive Photos; **724** Andy Levin/Photo Researchers, Inc.; **726** School Division, Houghton Mifflin Company; **728** Gary Retherford/Science Source/Photo Researchers, Inc.; **730** Andy Lyons/Allsport; **731** School Division, Houghton Mifflin Company; **735** Barros & Barros/The Image Bank; **737** National Museum, Karachi/E.T. Archive (bl); (Bibl. Imp., Paris.)/North Wind Picture Archives (bcl); World Perspectives/Tony Stone Images (bcr); **739** CORBIS/Richard T. Nowitz; **741** Photofest; **743** Betsy G. Reyneau/The Granger Collection; **746** Bruce Ayres/Tony Stone Images; **748** Bob Daemmrich/Stock Boston; **750** CORBIS/Roger Ressmeyer; **753** Tony Freeman/PhotoEdit; **766** Duvall/Liaison Agency; **767** Mark Segal/Tony Stone Images; **769** Amy C. Etra/PhotoEdit; **771** CORBIS/Richard Bickel; **773** Mark C. Burnett/Stock Boston; **775** CORBIS/Bettmann (bl, bcl); The Granger Collection (bcr); courtesy, NASA (br); **776** Matthew Stockman/Allsport; **779** Tom Dietrich/Tony Stone Images; **782** CORBIS/Kevin R. Morris; **784** Myrleen Ferguson/PhotoEdit; **787** Tom Young/The Stock Market; **789** David Sailors/The Stock Market; **792** Randy Masser/International Stock; **794** DiMaggio/The Stock Market; **797** Robert Laberge/Allsport; **799** Photodisc, Inc.; **801** Paul Souders/Tony Stone Images; **805** Linda Howard, *Centerpeace* 14'h x 24' w x 16' d, Aluminum, 1991, Bradley University, Peoria, Illinois; **807** Addison Geary/Stock Boston; **809** Landsman Photography/Courtesy of Pei Cobb Freed & Partners; **811** Bob Daemmrich/Stock Boston (l); Robert B. McGouey/Spectrum (r); **813** Stuart Westmorland/Tony Stone Images; **815** Courtesy of the Morton Punkin Chuckin' Contest; **817** Jeff Corwin/Tony Stone Images; **829, 830** Siegfried Lavda/Tony Stone Images; **831** Llewllyn/Uniphoto; **833** Leonard Lessin, FBPA/Photo Researchers, Inc.; **836** Mark Burnett/Stock Boston; **840** CORBIS/Galen Rowell; **842** The Granger Collection; **846** Hans Reinhard/Tony Stone Images (l); Alan D. Carey/Photo Researchers, Inc.(r); **848** CORBIS/The Purcell Team; **851** courtesy, Precor and EFX Elliptical Fitness Crosstrainer (all); **854** courtesy, Michael & Vicky Brereton/The Gleaston Water Mill (www.watermill.co.uk); **855** Jon Ortner/Tony Stone Images; **858** CORBIS/Tony Aruzza; **860** Jeff Greenberg/dMRp/Photo Researchers, Inc.; **862** CORBIS/Stuart Westmorland; **866** British Columbia Archives, A-00009; **868** PhotoDisc, Inc. (cr); CORBIS/Archivo Iconografico, S.A. (bl); Artville, LLC. (bcl, bcr); Bob Daemmrich/Stock Boston/PNI/PictureQuest (br); **869** Barrie Rokeach; **873** Jonathan Reichele/Barrie Rokeach; **871** Bob Daemmrich/Stock Boston; **875** Tim Davis/Photo Researchers, Inc.; **878** Michael Dwyer/Stock Boston; **881** Jeremy Horner/Tony Stone Images; **892** RMIP/Richard Haynes; **893** School Division, Houghton Mifflin Company (t); RMIP/Richard Haynes (b).

Illustration

Steve Cowden **606, 781**
Laurie O'Keefe **268**
School Division, Houghton Mifflin Company **597, 599, 600, 767, 774 (all), 778, 789, 796, 797** (t), **804, 805** (b), **811** (all), **812, 814, 815, 817, 818**
Doug Stevens **294** (all)**, 368, 797** (b)

Selected Answers

CHAPTER 1

SKILL REVIEW (p. 2) **1.** 11 **2.** −70 **3.** 8 **4.** 9 **5.** 24 **6.** −7 **7.** −10 **8.** −8 **9.** 60 units^2 **10.** 121 units^2 **11.** 165 units^2 **12.** 20.25π units^2, or about 63.6 units^2

1.1 PRACTICE (pp. 7–10)

5.

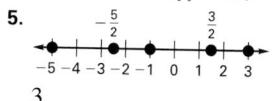

3

7.

3.2

9. inverse property of addition **11.** commutative property of multiplication **13.** inverse property of multiplication

15.

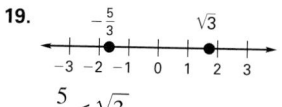

$\dfrac{1}{2}$; $\dfrac{1}{2} > -5$ **17.** $2.3 > -0.6$

19.

$-\dfrac{5}{3} < \sqrt{3}$

21.

$-\dfrac{9}{4} > -3$

23. $\sqrt{5} > 2$ **25.** $\sqrt{8} > 2.5$ **27.** $-6, -3, -\dfrac{1}{2}, 2, \dfrac{13}{4}$

29.

$; -\dfrac{5}{2}, -\sqrt{5}, -\dfrac{1}{3}, 0, 3$

31. $-\sqrt{12}, -\dfrac{12}{5}, -1.5, 0, 0.3$ **33.** inverse property of addition **35.** commutative property of multiplication **37.** identity property of multiplication **39.** Yes; the associative property of addition is true for all real numbers a, b, and c. **41.** Yes; the associative property of multiplication is true for all real numbers a, b, and c. **43.** $32 + (-7) = 25$ **45.** $-5 - 8 = -13$ **47.** $9 \cdot (-4) = -36$ **49.** $-5 \div \left(-\dfrac{1}{2}\right) = 10$ **51.** 13 ft **53.** $612.50 **55.** Honolulu, HI; New Orleans, LA; Jackson, MS; Seattle-Tacoma, WA; Norfolk, VA; Atlanta, GA; Detroit, MI; Milwaukee, WI; Albany, NY; Helena, MT; three **57.** Yes; the result of performing the given operations is 9, the check digit. **59.** Sky Central Plaza: 352 yd, 12,672 in., 0.2 mi; Petronas Tower I: about 494.3 yd, 17,796 in., about 0.2809 mi **61.** yes **63.** $214 **65.** −15°F

1.1 MIXED REVIEW (p. 10) **69.** 63 **71.** −30 **73.** 19 **75.** −34 **77.** $x - 3$ **79.** $\dfrac{1}{4}x$ **81.** 10.5 in.^{2} **83.** 750 in.^{2}

1.2 PRACTICE (pp. 14–16) **7.** 5 **9.** 27 **11.** $9x + 9y$ **13.** $8x^2 - 8x$ **15.** 8^3 **17.** 5^n **19.** 256 **21.** −32 **23.** 125 **25.** 256 **27.** 24 **29.** 19 **31.** 0 **33.** −5 **35.** 125 **37.** −8 **39.** 76 **41.** $\dfrac{9}{5}$ **43.** $-\dfrac{5}{13}$ **45.** 16 **47.** $6x^2 - 28x$ **49.** $16n - 88$ **51.** $-5x - y$ **53.** $\dfrac{1}{2}n(n + 10)$; 1000 **55.** $(x + y)^2$; 289

57. about 1,200,000; about 238,000 **59.** $149 + 3.85(12)n$, where n is the number of movies rented each month; $426.20 **61.** $[4n + 8(3 - n)]15$, or $360 - 60n$, where n is the number of hours spent walking; $240

1.2 MIXED REVIEW (p. 17) **69.** 20 **71.** 15 **73.** 105

75.

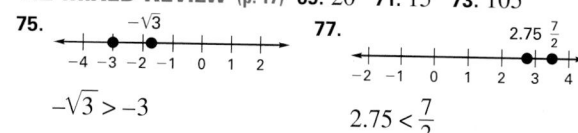

$-\sqrt{3} > -3$

77.

$2.75 < \dfrac{7}{2}$

79. inverse property of addition **81.** identity property of multiplication **83.** $\dfrac{8}{7}$ **85.** $-\dfrac{4}{5}$ **87.** −9 **89.** $-\dfrac{1}{14}$

QUIZ 1 (p. 17)

1.

$-2.5, -\dfrac{3}{4}, 0, 1, \dfrac{9}{2}$

2.

$-1.5, -0.25, 0.8, \dfrac{15}{8}, \dfrac{10}{3}$

3. distributive property **4.** associative property of addition **5.** 15 **6.** $-\dfrac{17}{3}$ **7.** −14 **8.** 76 **9.** −124 **10.** $8x - 11y + 4$ **11.** $2x - 10$ **12.** $-2x^2 + 5x - 6$ **13.** $-2x^2 + 14x$ **14.** $0.35n + 13.95(15 - n)$, or $209.25 - 13.60n$, where n is the number of regular floppy disks bought

TECHNOLOGY ACTIVITY 1.2 (p. 18) **1.** $(-4)^2 - 5$; 11 **3.** $(1 + 4)^6$; 15,625 **5.** 4.32 **7.** 160.989 **9.** 7.833 **11.** 5912.099 **13.** 0.81

1.3 PRACTICE (pp. 22–24) **7.** 5 **9.** 5 **11.** $\dfrac{5}{4}$ **13.** −3 **15.** 28 **17.** Subtract 5 from each side. **19.** Multiply each side by $-\dfrac{7}{4}$. **21.** Subtract 2 from each side; then multiply each side by 3. **23.** 5 **25.** $\dfrac{7}{2}$ **27.** $\dfrac{4}{5}$ **29.** −1 **31.** 0 **33.** 4 **35.** $\dfrac{85}{12}$ **37.** 3.2 **39.** 7.5 **41.** length: 36, width: 14 **43.** −78.5°C **45.** 5 h **47.** $635,000 **49.** 16.25 ft

1.3 MIXED REVIEW (p. 24) **57.** 25π in.^{2}, or about 78.5 in.^{2} **59.** 49π in.^{2}, or about 154 in.^{2} **61.** 8 **63.** 21 **65.** 11 **67.** −28 **69.** $21 - 5x$ **71.** $7x - 6$ **73.** $x + 35$ **75.** $3x^2 - x + 11$ **77.** $4x^2 + 16x$

TECHNOLOGY ACTIVITY 1.3 (p. 25) **1.** False; $y_1 = y_2$ when $x = -2$, not when $x = 2$. **3.** −2 **5.** 1 **7.** 1

1.4 PRACTICE (pp. 29–32) **5.** $y = \dfrac{5}{3}x - 3$ **7.** $y = -\dfrac{3}{20}x + 4$ **9.** $y = \dfrac{4}{3}x - 24$ **11.** 20 in. **13.** −1 **15.** $\dfrac{16}{9}$ **17.** $\dfrac{35}{3}$ **19.** 1 **21.** −4 **23.** $\dfrac{11}{2}$ **25.** $h = \dfrac{3V}{\pi r^2}$ **27.** $P = \dfrac{I}{rt}$ **29.** $b_2 = \dfrac{2A}{h} - b_1$ **31.** $h = \dfrac{S - 2\pi r^2}{2\pi r}$; $\dfrac{35 - 6\pi}{2\pi}$, or about 2.57 in. **33.** $L = \dfrac{T}{m} + 21$

SELECTED ANSWERS

35. $W \approx \dfrac{TR^2}{R^2 + A^2}$ **37.** $R = p_1 V + p_2 C$ **39.** *Sample answer:*
210 sun visors, 550 baseball caps; 490 sun visors,
430 baseball caps; 700 sun visors, 340 baseball caps
41. a. $A = \dfrac{\sqrt{3}}{4} b^2$ **b.** $A = \dfrac{\sqrt{3}}{3} h^2$

1.4 MIXED REVIEW (p. 32) **47.** $30 - x$ **49.** $250 + x$ **51.** $2x$
53. 8736 h **55.** $4\frac{3}{8}$ L **57.** \$165 **59.** -6 **61.** 4 **63.** -7
65. 40 **67.** 3

1.5 PRACTICE (pp. 37–39) **3.** The diagram helps you see
how to express the numbers of gallons used in town in
terms of x, the label given to the number of gallons used on
the highway. **5.** water pressure = 2184 (lb/ft^2); pressure
per ft of depth = 62.4 (lb/ft^2 per ft); depth = d (ft) **7.** 35 ft
9. $547 = 32t$ **11.** about 17 h **13.** $80t = (180)(3)$ **15.** total
calories = (calories/gram of fat)(number of grams of fat) +
(calories/gram of protein)(number of grams of protein) +
(calories/gram of carbohydrate)(number of grams of
carbohydrate) **17.** 4.1 g **19.** Great Britain: 22.4 km,
France: 15.5 km; Dec. 1, 1990 **21.** \$1.68 per page
23. length: 135 ft, width: 105 ft **25.** 4.5 m **27.** 4 bounces

1.5 MIXED REVIEW (p. 39) **31.** true **33.** false **35.** -55,
$-10, -5, -1, 4$ **37.** $-2.9, -2.1, -1.2, 2, 2.09$ **39.** 2 **41.** $\dfrac{4}{7}$

QUIZ 2 (p. 40) **1.** 4 **2.** -8 **3.** $\dfrac{17}{3}$ **4.** 160 **5.** $y = -\dfrac{3}{5}x + \dfrac{9}{5}; \dfrac{3}{5}$
6. $y = \dfrac{4}{3}x - \dfrac{14}{3}; -2$ **7.** $d_1 = \dfrac{2A}{d_2}$ **8.** 49 boxes

1.6 PRACTICE (pp. 45–47)
5. $x \geq 5$;   **7.** $x \leq 12$;
9. $x > 2$;   **11.**
13. C **15.** D **17.** F **19.** no **21.** no **23.** yes **25.** $x > 5$
27. $x \leq -11$;   **29.** $x < 6$;
31. $x > 3$;   **33.** $x < 6$ **35.** $x < 0$
37. $5 \leq x \leq 18$ **39.** $-6 \leq n \leq -1$;
41. $-1 < x < 1$;   **43.** $x \leq 3$ or $x \geq 6$;
45. $x < -5$ or $x > -0.52$;
47. $0.5 \leq x < 2.5$
49. Your sales must be greater than or equal to \$5000.
51. Her score must be between 93 and 100, inclusive.
53. $184 \leq K \leq 242$ **55.** $c > 2.83$

1.6 MIXED REVIEW (p. 47) **61.** associative property of
multiplication **63.** commutative property of addition
65. $-\dfrac{10}{7}$ **67.** -1 **69.** $1\frac{1}{5}$ h, or 1 h 12 min

TECHNOLOGY ACTIVITY 1.6 (p. 48) **1.** $x \leq 4$ **3.** $x > 3$
5. $x \leq -6$ **7.** $x < 2$ **9.** $x < 6$ **11.** $x \leq 9$ **13.** $x < -7$

1.7 PRACTICE (pp. 53–55) **5.** yes **7.** no **9.** no
11. $11 - 2x \leq -13$ or $11 - 2x \geq 13$ **13.** $-9 \leq x + 5 \leq 9$
15. $-18 < \dfrac{1}{4}x + 10 < 18$ **17.** $x - 8 = 11$ or $x - 8 = -11$
19. $6n + 1 = \dfrac{1}{2}$ or $6n + 1 = -\dfrac{1}{2}$ **21.** $2x + 1 = 5$ or $2x + 1 = -5$
23. $15 - 2x = 8$ or $15 - 2x = -8$ **25.** $\dfrac{2}{3}x - 9 = 18$ or
$\dfrac{2}{3}x - 9 = -18$ **27.** no **29.** no **31.** yes **33.** 2, 3 **35.** 6, -1
37. $\dfrac{26}{7}, \dfrac{34}{7}$ **39.** 12, -18 **41.** $-15 \leq 3 + 4x \leq 15$
43. $-7 < 3x + 2 < 7$ **45.** $-18 \leq 8 - 3n \leq 18$
47. $-9 < x < 7$;   **49.** $x \leq 6$ or $x \geq 26$
51. $3 \leq x \leq 13$;   **53.** $x < -\dfrac{4}{3}$ or $x > \dfrac{32}{3}$;
55. $x \leq -\dfrac{15}{2}$ or $x \geq \dfrac{1}{2}$;
57. $-4 < x < \dfrac{18}{7}$ **59.** $-4 < x < 2$
61. $x < -3$ or $x > 7$ **63.** $x < 1$ or $x > 4$
65. $|p - 3.49| \leq 0.26$;
67. $|x - p| \leq \dfrac{3}{16}$; between $8\frac{15}{16}$ in. and $9\frac{5}{16}$ in., inclusive.
69. $|t - 98.6| \leq 1$ **71.** 393.6 oz; 374.4 oz; $|c - 384| \leq 9.6$
73. volleyball: $|v - 270| > 10$, basketball: $|b - 625| > 25$,
water polo: $|w - 425| > 25$, lacrosse: $|l - 145.5| > 3.5$,
football: $|f - 14.5| > 0.5$ **75.** 2 L: $|c - 2000| > 9$,
1 L: $|c - 1000| > 5$, 500 mL: $|c - 500| > 2$

1.7 MIXED REVIEW (p. 56) **91.** False; if $x = -7$, then $2x = 2(-7) = -14$, not 14. **93.** 21 **95.** -27 **97.** -14 **99.** 10
101. $x > \dfrac{1}{3}$ **103.** $x \geq -5$ **105.** $-14 < x < -2$

QUIZ 3 (p. 56)
1. $x \leq 5$;   **2.** $x < -7$;
3. $-4 < x < 6$;   **4.** $x \leq -2$ or $x > 3$;
5. $-1, -9$ **6.** 5, 1 **7.** -3, 15 **8.** 5, $-\dfrac{3}{2}$ **9.** $\dfrac{16}{3}, -8$ **10.** 1, 9
11. $y \leq -5$ or $y \geq 1$;   **12.** $-10 < x < -2$;

SELECTED ANSWERS

13. $x < -4$ or $x > 10$;

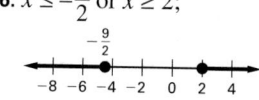

14. $1 \le y \le 4$;

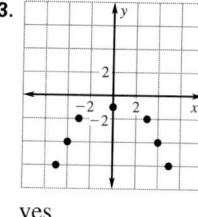

15. $x < 1$ or $x > 2$;

16. $x \le -\dfrac{9}{2}$ or $x \ge 2$;

17. $20 \le e \le 28$; between 320 mi and 448 mi, inclusive

18. $|d - 30| \le 0.045$; between 29.955 mm and 30.045 mm, inclusive

CHAPTER 1 REVIEW (pp. 58–60)

1. 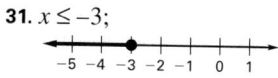 ; $-\pi, -\sqrt{6}, -2, 0.2, \dfrac{6}{5}$

3. distributive property **5.** -18 **7.** 4 **9.** $5x + 4y$

11. $11x^2 - x$ **13.** -3 **15.** -32 **17.** 4 **19.** $y = 5x - 10$

21. $y = -0.2x + 7$ **23.** $y = \dfrac{5}{6}x + 2$ **25.** $l = \dfrac{P - 2w}{2}$

27. about 5 h 55 min **29.** $x > 8$;

31. $x \le -3$;

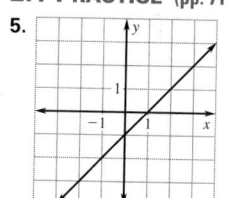

33. $-2 \le y \le 2$;

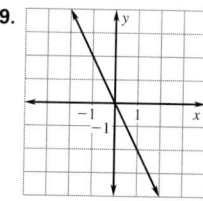

35. $-5, 3$ **37.** $-\dfrac{8}{3}, 6$ **39.** $-2 < x < 7$

CHAPTER 2

SKILL REVIEW (p. 66) **1.** 2 **2.** 2 **3.** 3 **4.** $y = -3x + 4$

5. $y = \dfrac{1}{2}x - 5$ **6.** $y = -\dfrac{5}{6}x - 10$ **7.** $x < \dfrac{9}{2}$ **8.** $y \ge -26$ **9.** $x < \dfrac{5}{2}$

2.1 PRACTICE (pp. 71–74)

5.

9.

11. 3 **13.** 9 **15.** 1

17. domain: $0 \le t \le 8$; range: $0 \le g \le 16$;

Gasoline Remaining

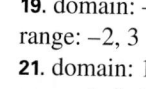

19. domain: $-1, 2, 5, 6$; range: $-2, 3$

21. domain: $1, 2, 3, 4$; range: $1, 2, 3, 4$

23. 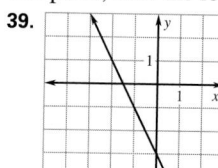 ; yes

25. Input Output; 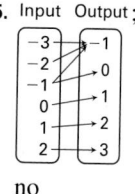 no

27. Input Output; 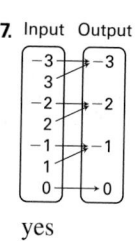 yes

29. If a relation is a function, then no vertical line intersects the graph of the relation at more than one point. If no vertical line intersects the graph of a relation at more than one point, then the relation is a function. **31.** yes

39. 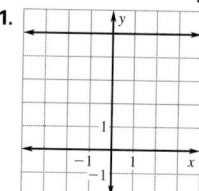 **41.**

43. linear; -7 **45.** not linear; 1 **47.** not linear; -25

49. 125; the volume of a cube with sides of length 5 units

51. No. *Sample answer:* Not every age corresponds to exactly one place. For example, there were 24-year-olds with finishes of first and third.

53. domain: 1, 5, 6, 10, 12, 25; range: 1, 2, 3, 4, 6, 9;

Jazz Shooting

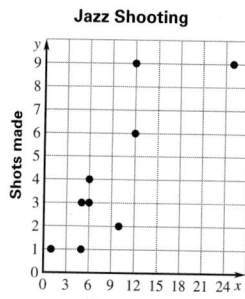

55. domain: $0 \le d \le 130$; range: $1 \le p \le 4\dfrac{31}{33}$;

Pressure Versus Depth

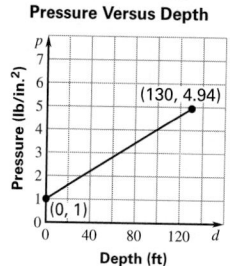

57. domain: $20\dfrac{7}{8} \le c \le 25$; range: $6\dfrac{5}{8} \le s \le 8$;

Cap Size

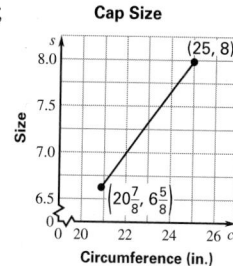

2.1 MIXED REVIEW (p. 74) **65.** 1 **67.** $\dfrac{1}{2}$ **69.** $\dfrac{1}{4}$ **71.** -7.5

73. $-4\dfrac{11}{16}$ **75.** $-\dfrac{12}{11}$ **77.** yes **79.** yes **81.** yes

2.2 PRACTICE (pp. 79–81) **5.** undefined; vertical **7.** -1; falls

9. 2; rises **11.** line 2 **13.** neither **15.** parallel **17.** 1

19. undefined **21.** 10; rises **23.** $\dfrac{1}{2}$; rises **25.** -1; falls

27. undefined; vertical **29.** $-\dfrac{1}{2}$; falls **31.** undefined; vertical **33.** C **35.** A **37.** line 1 **39.** line 2 **41.** parallel

43. perpendicular **45.** 6; dollars/h **47.** 3; in./year **49.** 10.75 **51.** 0.062 ft/year; this is the ratio of the number of vertical feet the volcano must grow to the length of time it will take to grow that high.

2.2 MIXED REVIEW (p. 81) **59.** additive inverse property **61.** distributive property **63.** $15 - 8x$ **65.** $8 - \frac{4}{3}x$ **67.** $-8, -1$ **69.** $-1, \frac{5}{3}$ **71.** about \$.45/oz

2.3 PRACTICE (pp. 86–88) **5.** $-2; -7$ **7.** x-intercept: 11; y-intercept: -11 **9.** x-intercept: 3; y-intercept: -15 **17.** A

19. 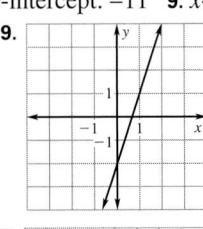 **23.**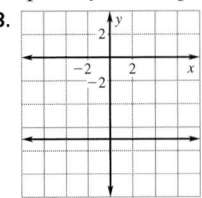

27. 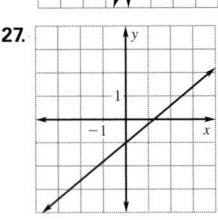 **29.**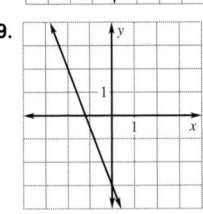

31. 6; 10 **33.** 0; 100 **35.** 4; -7 **37.** B **39.** A

41. 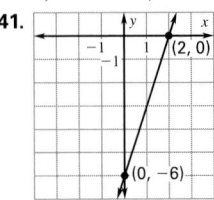 **45.**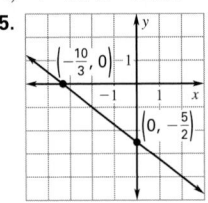

47. 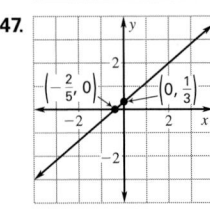 **51.**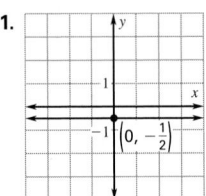

53.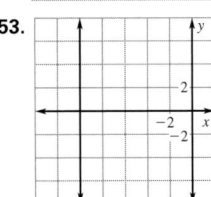

59.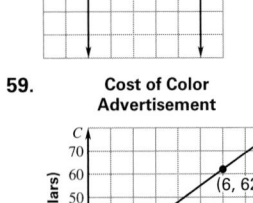

The slope, 7, represents the price of each line in the ad, while the intercept, 20, represents the initial cost of placing a colored ad.

61. $8w + 12x = 3464;$

Car Washes

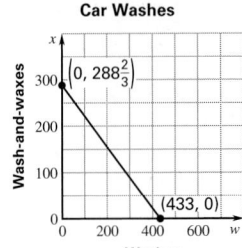

63. $2.5s + 6a = 7000;$

Ticket Sales

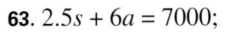
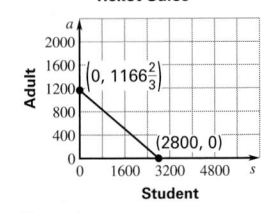

Sample answer:
1600 student tickets, 500 adult; 880 student, 800 adult; 400 student, 1000 adult

2.3 MIXED REVIEW (p. 88)

69. $x > -12;$ 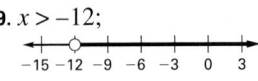 **71.** $x \le 45;$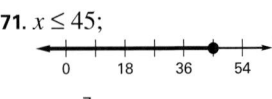

73. $x \le -7$ or $x \ge 7;$

75. 12 **77.** 8 **79.** -16 **81.** $-\frac{6}{7}$ **83.** undefined **85.** -2

QUIZ 1 (p. 89) **1.** domain: $-2, -1, 0, 1, 2$; range: $-2, 1$; function **2.** domain: 1, 2, 3, 4; range: 1, 2, 3, 4; not a function **3.** domain: $-3, -1, 0, 1, 2$; range: $-3, -2, 0, 1$; function **4.** -21 **5.** 139 **6.** perpendicular **7.** neither

8. 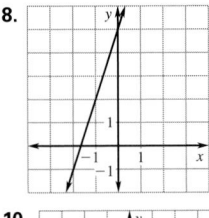 **9.**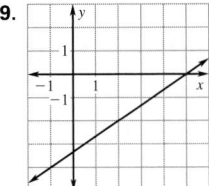

10. 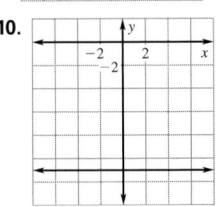 **11.** about 8.36 mi/h

TECHNOLOGY ACTIVITY 2.3 (p. 90)

1. 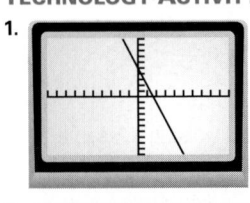 **3.**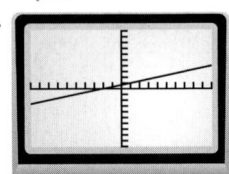

5. 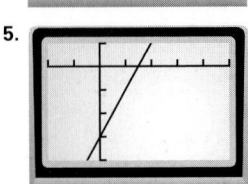 **7.**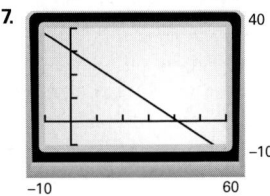

2.4 PRACTICE (pp. 95–98) **5.** $y = 2x - 4$ **7.** $y = -\frac{3}{4}x - \frac{21}{4}$

9. $y = \frac{2}{5}x + 2$ **11.** $y = 5x - 6$ **13.** $y = 5x - 3$ **15.** $y = -4x$

17. $y = \frac{3}{5}x + 6$ **19.** $y = 2x + 4$ **21.** $y = 5$ **23.** $y = -\frac{4}{3}x + 2$

25. $y = 2x - 3$ **27.** $x = 2$ **29.** $y = \frac{3}{2}x - \frac{1}{2}$ **31.** $y = -\frac{1}{2}x - \frac{15}{2}$

33. $y = -x + 8$ **35.** $y = 3x - 19$ **37.** $y = -\frac{7}{8}x + 1$

39. $y = x + 10$ **41.** $3 = -\frac{1}{2}(2) + b; 3 = -1 + b; b = 4.$ The

equation is $y = -\frac{1}{2}x + 4$, the same as in Example 2. The

slope-intercept equation of a line is unique. **43.** $y = \frac{7}{2}x; 28$

45. $y = -3x; -24$ **47.** $y = \frac{1}{2}x; 4$ **49.** $y = \frac{1}{2}x; -10$

51. $y = \frac{1}{5}x; -25$ **53.** $y = \frac{1}{2}x; -10$ **55.** yes; $y = \frac{1}{2}x$

57. yes; $y = -x$ **59.** $P = 60,300t + 2,842,200; 4,289,400$

61. $s = 0.629t + 7.4$; about \$21.2 billion **63.** $h = \frac{1}{7}l; 38.5$ ft

65. $r = \frac{1}{240}t; 11$ min **67.** no

2.4 MIXED REVIEW (p. 98) **71.** $-7, 27$ **73.** $-10, -8$

75. $-\frac{38}{55}, \frac{8}{55}$ **77.** 14 **79.** 2 **81.** 0 **83.** -2 **85.** 1

87. 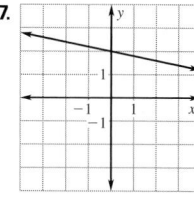 **91.**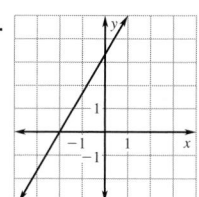

93.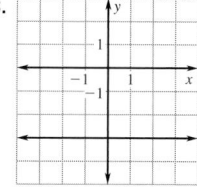

2.5 PRACTICE (pp. 103–105) **5.** about 1.4 m

7. *Sample answer:* about 8830 **9.** positive correlation

11. 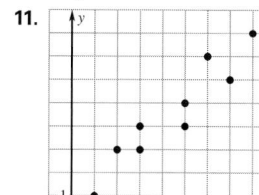 ; **13.** ;

positive correlation   negative correlation

15. *Sample answer:* List the data points so that the values
of x are in increasing order. If the y-values mostly increase
along with the x-values, there is a positive correlation. If
the y-values mostly decrease as the x-values increase, there
is a negative correlation. Otherwise, there is relatively no
correlation. **17.** *Sample answer:* $y = -0.86x - 0.05$

19. 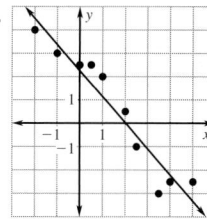 ; *Sample answer:*
$y = -1.11x + 2.27$
21. *Sample answer:*
$y = -0.73x + 2.47$

23. 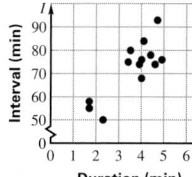 ; positive correlation
25. about 2290
27. about 84.3 years

Old Faithful Eruptions

2.5 MIXED REVIEW (p. 106)

31. $x < -\frac{11}{4}$; **33.** $x < 4$ or $x \geq 10$;

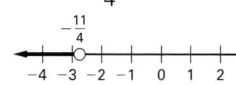

35. line 2 **37.** line 1 **39.**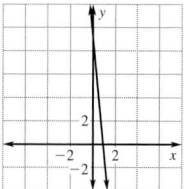

41. 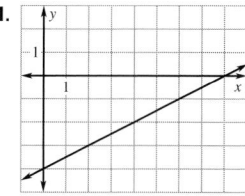 **43.**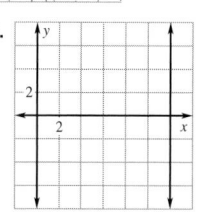

QUIZ 2 (p. 106) **1.** $y = \frac{2}{3}x + 6$ **2.** $y = 2x + 5$ **3.** $y = -\frac{1}{5}x - \frac{33}{5}$

4. $y = 2x - 4$ **5.** relatively no correlation **6.** negative
correlation **7.** positive correlation **8.** $d = 1.3h; 4$ ft

9. 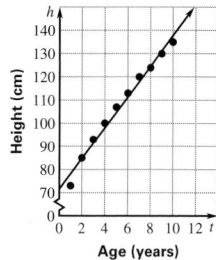 ; *Sample answer:* $h = 6.63t + 71.5$

Heights of Children

TECHNOLOGY ACTIVITY 2.5 (p. 107)
1. $y = 0.0028x + 0.32$;

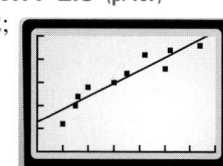

2.6 PRACTICE (pp. 111–113)

7. 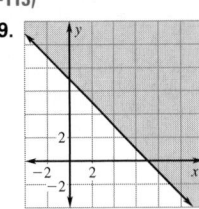 **9.**

11.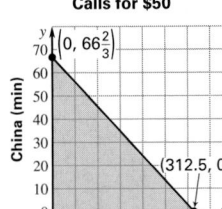

13. $0.16x + 0.75y \leq 50$;

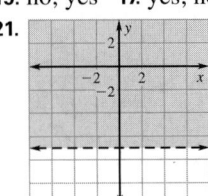

; One possible solution is to spend 50 min on calls to China and 78 min on calls in the United States, for a total cost of $49.98. Another solution would be to spend 50 min on calls within the United States and 56 min on calls to China; this uses exactly $50. A third solution is 100 min on calls within the United States and 45 min on calls to China. This solution uses a total of $49.75.

15. no; yes **17.** yes; no

21.   **23.** 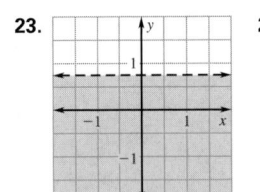 **25.** C

27. 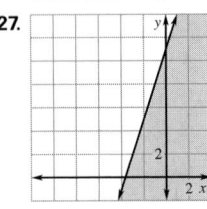 **31.** 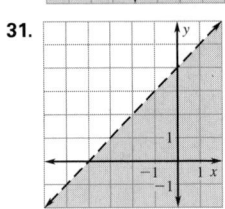 **33.** C
35. B

41. 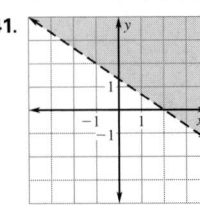 **45.** $y < 0.9x$;

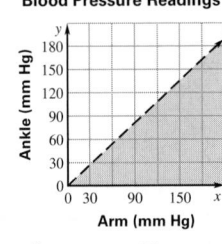

47. about 1.77 cups **49.** *Sample answer:* You can attend 5 matinees and no evening showings for a total of $22.50, 2 of each for a total cost of $24, or 3 evening showings at a cost of $22.50.

51. *Sample answer:* 9 touchdowns and no field goals for 63 points; 5 touchdowns and 1 field goal for 38 points; 2 touchdowns and 3 field goals for 23 points; 3 touchdowns and 3 field goals for 30 points; 4 touchdowns and 6 field goals for 46 points

2.6 MIXED REVIEW (p. 113)
57. 1.65×10^9 **59.** 6.7×10^{-4} **61.** 8.08×10^{-2}

63. 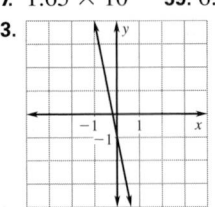 **67.**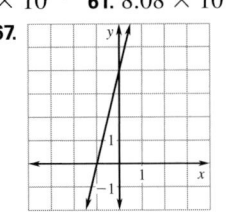

69. $y = -\dfrac{6}{5}x + 7$ **71.** $x = 3$ **73.** $y = -8$

2.7 PRACTICE (pp. 117–120) **5.** 27 **7.** 11

11. $f(x) = -\dfrac{4}{3}x + 6$, if $0 \leq x < 3$, $f(x) = -\dfrac{2}{5}x + \dfrac{16}{5}$, if $3 \leq x \leq 8$ **13.** -21 **15.** -9 **17.** -9.5 **19.** -7

23. 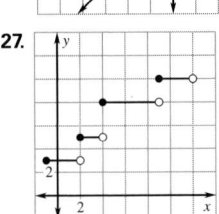 **25.**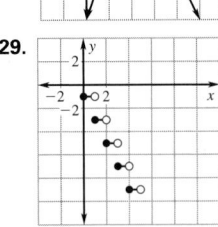

27. 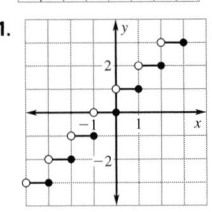 **29.**

31. 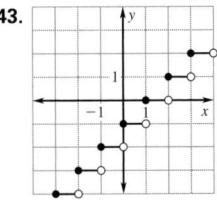 ; *Sample answer:* The function graphs each x-value to the smallest integer that is not less than it, giving a sort of upper limit to the x-values in each interval.

35. $f(x) = \begin{cases} x, & \text{if } x < 0 \text{ (or } x \leq 0) \\ 2x, & \text{if } x \geq 0 \text{ (or } x > 0) \end{cases}$

37. $f(x) = \begin{cases} \dfrac{3}{2}x + \dfrac{9}{2}, & \text{if } x < -1 \\ -1, & \text{if } x \geq -1 \end{cases}$

39. $f(x) = \begin{cases} x + 2, & \text{if } x \leq -1 \\ x + 3, & \text{if } -1 < x < 1 \\ x + 1, & \text{if } 1 \leq x \end{cases}$

43. 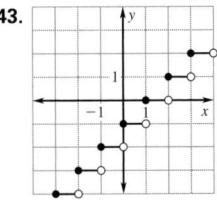 **45.**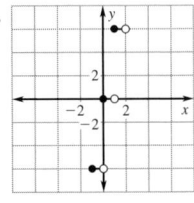

47. 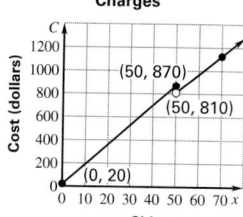 **49.**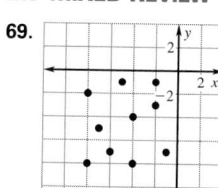

51. domain: $0 < x \le 80$; range: 11.75, 15.75, 18.50, 21.25, 24.00 **53.** 450 photocopies cost more than 501 would.

55.

Charges

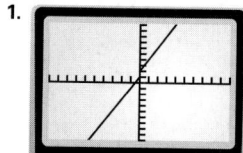

57. $1860 **59.** 15 in.

2.7 MIXED REVIEW (p. 120) **63.** $\frac{3}{2}, -6$ **65.** 6, 15 **67.** $-12, 32$

69. 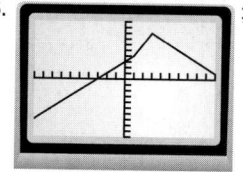 ; relatively no correlation

71. $n = -\frac{1}{40}T + 2.5$; 2.5 in.

TECHNOLOGY ACTIVITY 2.7 (p. 121)

1. 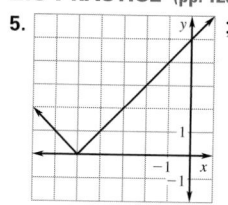 ; 6 **3.** 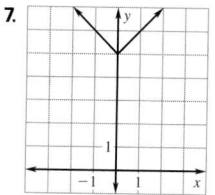 ; 2

5. 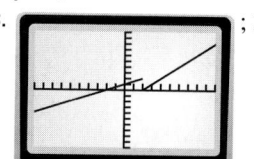 ; 6

2.8 PRACTICE (pp. 125–127)

5. 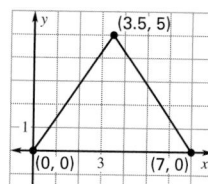 ; **7.** 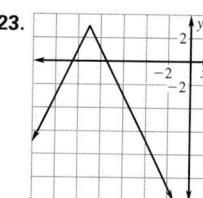 ;

$(-5, 0)$; opens up; same width

$(0, 5)$; opens up; same width

9. $\left(\frac{1}{2}, -14\right)$; opens down; same width

11. *Sample answer:* $y = -\frac{10}{7}|x - 3.5| + 5$;

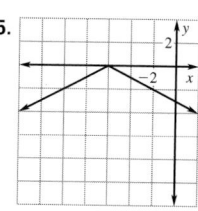

; domain: $0 \le x \le 7$; range: $0 \le y \le 5$

13. C **15.** C **17.** B

19. $(0, 9)$; opens up; same width

21. $(-2, 11)$; opens down; same width

23. 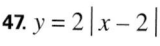 ; **25.** 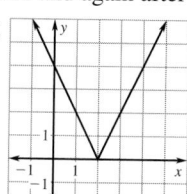 ;

$(-9, 3)$; opens down; narrower

$(-6, 0)$; opens down; wider

27. $-23, -5$ **29.** $-\frac{39}{7}, \frac{31}{7}$ **31.** $-2.8125, 2.8125$ **33.** 1.5, 4.5

35. $y = -|x - 3| + 1$ **37.** $y = 2|x + 1| - 1$

39. $y = -4|x| + 20$ **41.** 40,000

43. 2 h; 1 h after the rain started

45. after 2 measures and again after 6 measures

47. $y = 2|x - 2|$;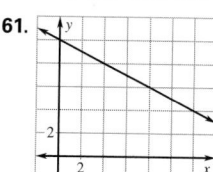

2.8 MIXED REVIEW (p. 128) **57.** $y = -3x - \frac{9}{2}$

61. 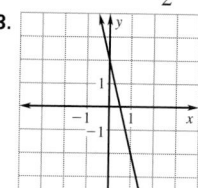 **63.**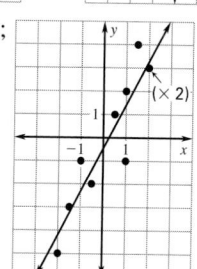

65. $y = 1.87x - 0.46$;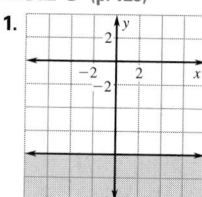

QUIZ 3 (p. 128)

1. 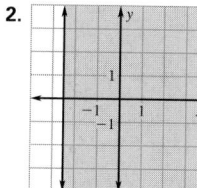 **2.**

SELECTED ANSWERS

3. 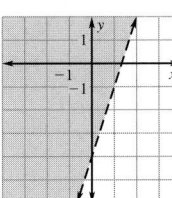 **4.** 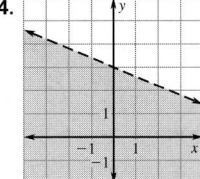   **5.** 7 **6.** 5

7. 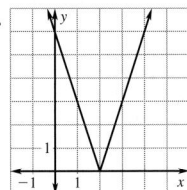 **8.**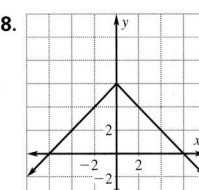

9.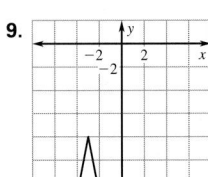

10. $y = \frac{3}{2}|x - 2|$

11. $y = -|x + 2| + 2$

12. $y = \frac{1}{3}|x + 1| + 2$

13. $2.5p + 1.25d \le 15$;

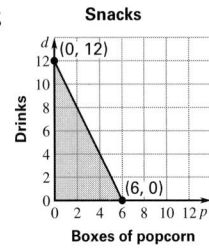

14. $f(x) = \begin{cases} 200, & \text{if } 0 < x \le 1000 \\ 0.2x, & \text{if } x > 1000 \end{cases}$; $240

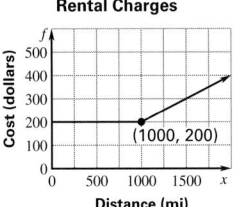

Chapter 2 Review (pp. 130–132)

1. 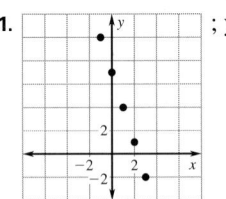 ; yes **3.** $\frac{2}{3}$ **5.** -1

7. 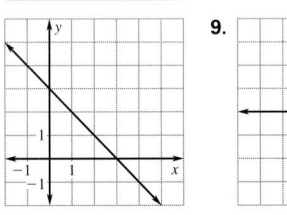 **9.**

11. $y = -x + 2$ **13.** $y = 2x - 14$

15. 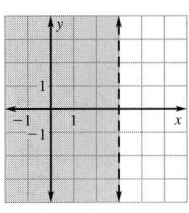 **17.**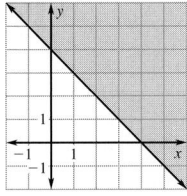

19. 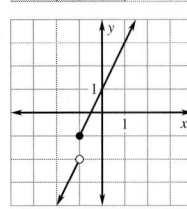 **21.**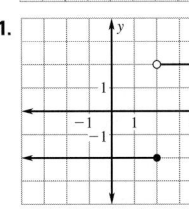

23. 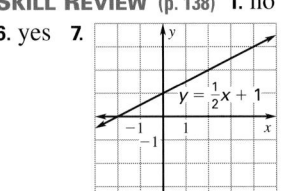 **25.**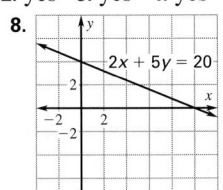

CHAPTER 3

Skill Review (p. 138) **1.** no **2.** yes **3.** yes **4.** yes **5.** no

6. yes **7.** 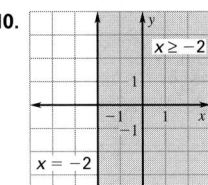 **8.**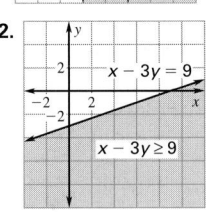

9. 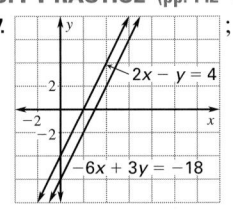 **10.**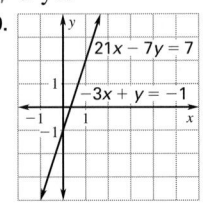

11. 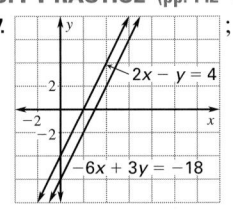 **12.**

3.1 Practice (pp. 142–145) **5.** yes

7. ; **9.** ;

0   infinitely many

11. yes **13.** no **15.** yes **17.** no **19.** no

21.
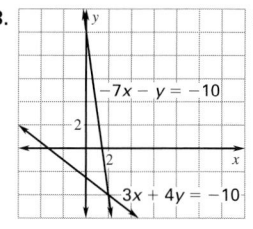
(7, 1)

; **23.**

(2, −4)

29.
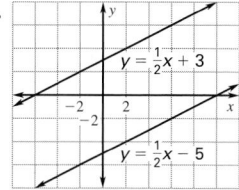

; (2, −2) **33.** no solution; the two lines are parallel and have no points in common **35.** E; 1 **37.** B; infinitely many **39.** A; 0

43.
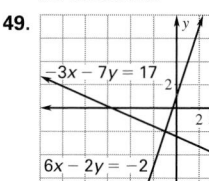
no solutions

; **45.**
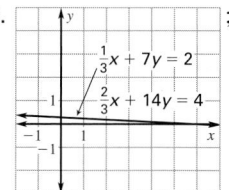
infinitely many solutions

;

49.
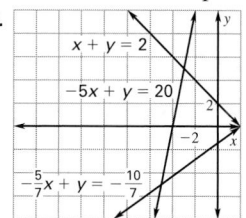

; one solution; (−1, −2)
55. $l + m = 125$; $0.1l + 0.5m = 32.5$; buy 75 latex balloons and 50 mylar balloons. **57.** $d + 1.25h = 6$; $720d + 1440h = 6480$; you can buy 4 high density disks and 1 double density disk.

59. Let f = the travel time in hours of the first bus; let s = the travel time in hours of the second bus; $f = s + \frac{1}{12}$; 10 miles from the airport; $30f = 40s$.

61. consistent and independent

63.
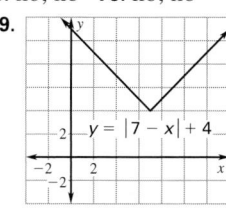

; a triangle; (−3, 5), (−5, −5), and (2, 0); *Sample answer:* I graphed the lines carefully and found the apparent points of their intersections from the graph. It was easy to see that two of the lines had the same x-intercept, so that was one point. The other points I checked algebraically in the equations to make sure they were solutions.

MIXED REVIEW (p. 145)
67. 36 **69.** −0.3 **71.** −2 **73.** no; no **75.** no; no

77.
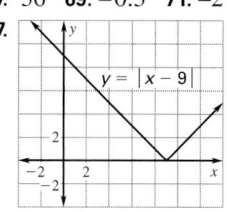

79.
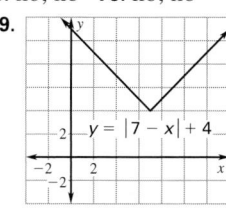

TECHNOLOGY ACTIVITY 3.1 (p. 146)

1. (−1, 3) **3.** $\left(\frac{141}{19}, \frac{119}{19}\right)$, or about (7.42, 6.26)

5. $\left(-\frac{116}{21}, \frac{47}{21}\right)$, or about (−5.52, 2.24)

3.2 PRACTICE (pp. 152–154) **5.** (4, −1) **7.** (6, 6) **9.** (3, 4)

11. (4, −1) **13.** (3, 3) **15.** $\left(0, \frac{5}{2}\right)$ **17.** (−2.4, 10.2)

19. (3, −10) **21.** (−2, 2) **23.** $\left(-\frac{11}{3}, -1\right)$ **25.** $\left(0, \frac{4}{5}\right)$

27. infinitely many solutions **29.** $\left(\frac{1}{3}, 1\right)$ **31.** $\left(\frac{18}{41}, \frac{605}{82}\right)$, or about (0.439, 7.378) **33.** no solution **35.** (−5, −2)

37. (5, 0) **39.** no solution **41.** $\left(-\frac{69}{11}, \frac{65}{11}\right)$ **43.** $\left(-\frac{25}{4}, 2.5\right)$

45. (20, 3) **47.** no solution **49.** (9, 6) **51.** (2, 3) **53.** (2, 2)
55. $12; *Sample answer:* let x = the cost per foot of the cable itself and y = the cost of one connector. Then $6x + 2y = 15.5$ and $3x + 2y = 10.25$. Subtracting the second equation from the first, find $x = 1.75$. Then a 4-foot cable with connectors will cost $10.25 + 1.75 = 12. **57.** inline skating: 25 min; swimming: 15 min

59.
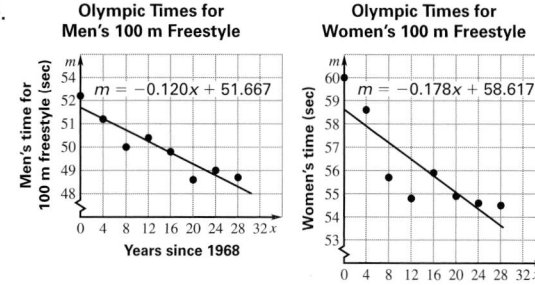

61. (119.83, 37.288); 120 years after 1968, in the year 2088 summer olympics, the men's and women's times in the 100 m freestyle will both be about 37.3 sec.

3.2 MIXED REVIEW (p. 155)
67. −8, −2 **69.** $-\frac{3}{2}$, 1 **71.** 24, −4 **73.** $y = 2x − 3$

75.
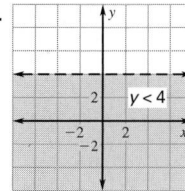

77.
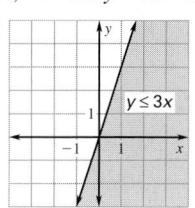

79.
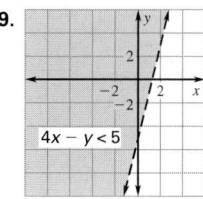

81. $12x + 25 \le 60$; $x \le \frac{35}{12}$

SELECTED ANSWERS

SELECTED ANSWERS

QUIZ 1 (p. 155) **1.** $(-2, 1)$ **2.** $(1, -3)$ **3.** no solutions
4. $\left(\dfrac{7}{3}, -\dfrac{8}{3}\right)$ **5.** $(1, 4)$ **6.** $(-1, -1)$ **7.** infinitely many
solutions **8.** 1 **9.** no solutions **10.** 1 **11.** 1 **12.** infinitely
many solutions **13.** $\left(-\dfrac{5}{4}, -\dfrac{15}{4}\right)$ **14.** $(6, 6)$ **15.** infinitely
many solutions **16.** $\left(-4, \dfrac{7}{2}\right)$ **17.** no solution **18.** $\left(-\dfrac{33}{29}, \dfrac{13}{29}\right)$
19. 371; 566

3.3 GUIDED PRACTICE (pp. 159–161) **5.** no **7.** yes
9.
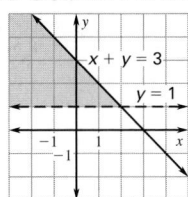
11. $18 \le x \le 55$; $60 \le y \le 74$;
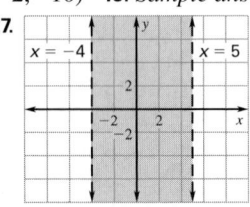

13. no **15.** *Sample answer:* (13, 10) **17.** *Sample answer:*
$(-2, -10)$ **19.** *Sample answer:* (4, 2) **21.** C **23.** F **25.** A
27.
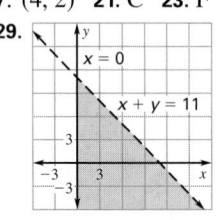
29.

33.
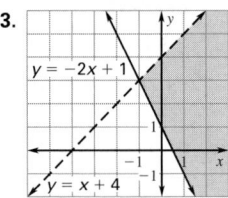
37.
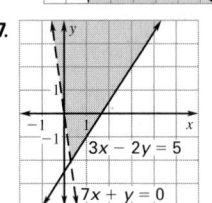

41.
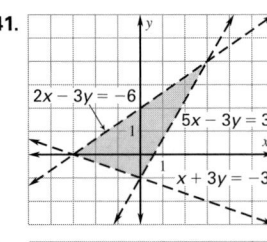
43.
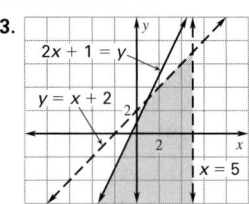

49.
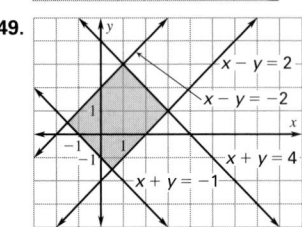

51. $7.4 \le p \le 7.6$, $1.0 \le c \le 1.5$;
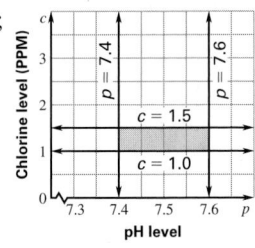

53.
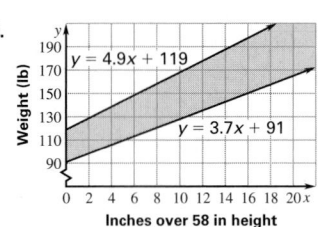
Inches over 58 in height
55. $0.75x \le y$; $y \le 0.9x$; $20 \le x \le 80$

57. $s > 205.5$; $j \le 262.5$; $s + j > 465.0$;
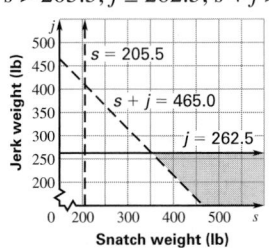
Snatch weight (lb)

3.3 MIXED REVIEW (p. 162) **67.** 27 **69.** -13 **71.** relatively
no correlation **73.** $\left(\dfrac{58}{57}, -\dfrac{128}{57}\right)$ **75.** no solution **77.** $(-8, 2)$

3.4 PRACTICE (pp. 166–167) **5.** Minimum is 0; maximum
is 38. **7.** max of 31 at (17, 3); min of -20 at (0, 20)
9. min of -40 at (0, 40); max of 40 at (40, 0) **11.** min of 10
at (2, 1); no max—feasible region is unbounded.
13. min of 6 at (2, 1); max of 29 at (5, 6) **15.** min of 0 at
(0, 0); max of 740 at (60, 20) **17.** no min, since feasible
region is unbounded; max of 132 at (15, 12) **19.** min of 6
at (0, 2); max of 29 at (5, 3) **21.** Make 37.5 gallons of
Orangeade and 31.25 gallons of Berry-fruity for a profit
of $31.25. **23.** Make 14 jars of tomato sauce and 4 jars of
salsa for a profit of $34.

3.4 MIXED REVIEW (p. 168)
29.
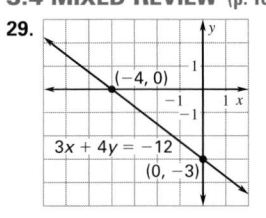
31.
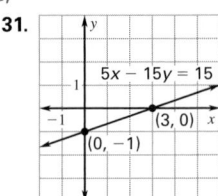

33.
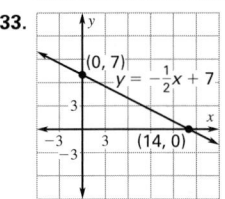
35. -7 **37.** -6 **39.** 35 **41.** 15

43.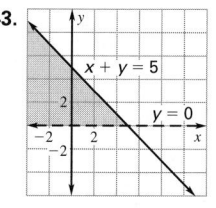

45.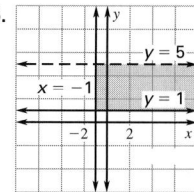

47.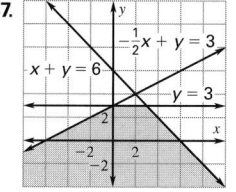

QUIZ 2 (p. 169)

1.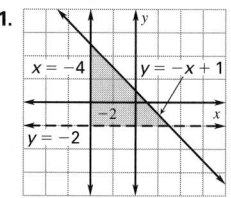

2.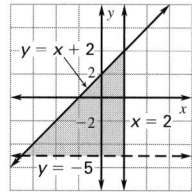

3.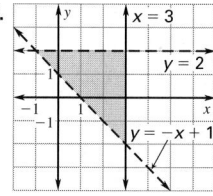

4. min of −18 at (−4, 1); max of 2 at (−2, 6)   **5.** min of 19 at (3, 2); max of 24 at (4, 2)   **6.** min of 0 at (0, 0); max of 70 at (14, 0)

7. 6 small boxes and 6 large boxes

3.5 PRACTICE (pp. 173–174)

5.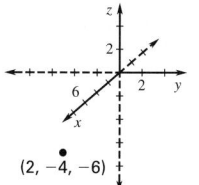

7.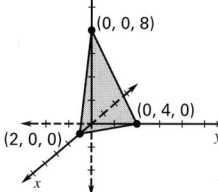

11.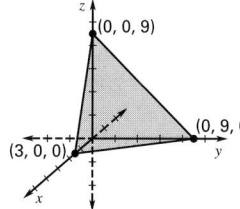

15. $f(x, y) = -2x - \frac{1}{2}y - 4$; −17

17. $C = 2.25r + 2.95p + 2.65$; $37.50

19.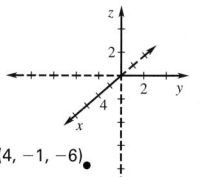

23.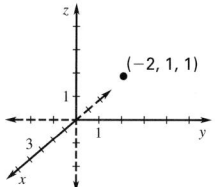

27.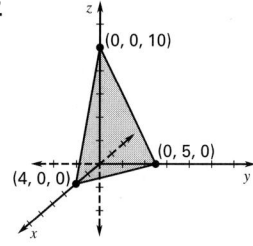

33.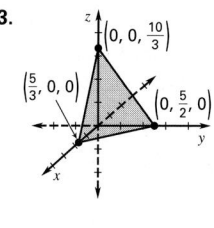

37.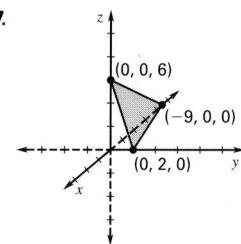

39. $f(x, y) = \frac{2}{5}x + y + 3$; $\frac{8}{5}$

41. $f(x, y) = -\frac{6}{5}x + \frac{3}{10}y + \frac{18}{5}$; 12

43. $f(x, y) = -\frac{1}{6}x - \frac{1}{4}y + \frac{1}{5}$; $\frac{1}{2}$

45. $f(x, y) = -\frac{1}{9}x + \frac{2}{3}y - \frac{4}{3}$; $\frac{121}{18}$ **47.** 60

49. $C = 1.5n + p + 16$; *Sample answer:*

	Number of Colors				
Price of Pottery	1	2	3	4	5
$8	$22.50	$27.00	$28.50	$30.00	$31.50
$18	$35.50	$37.00	$38.50	$40.00	$41.50
$28	$45.50	$47.00	$48.50	$50.00	$51.50
$38	$55.50	$57.00	$58.50	$60.00	$61.50
$48	$65.50	$67.00	$68.50	$70.00	$71.50

51. $C = 0.9e + 0.25s + 20$; $29.70; *Sample answer:*

	Number of Subway Trips				
Number of Express Bus Trips	2	4	6	8	10
2	$22.30	$22.80	$23.30	$23.80	$24.30
4	$24.10	$24.60	$25.10	$25.60	$26.10
6	$25.90	$26.40	$26.90	$27.40	$27.90
8	$27.70	$28.20	$28.70	$29.20	$29.70
10	$29.50	$30.00	$30.50	$31.00	$31.50

3.5 MIXED REVIEW (p. 175)

57. $x \le 14$;

59. $x > -2$;

61. $18 \le x \le 21$;   **63.** neither

65. parallel   **67.** $3.95r + 3.1p = 48.5$; $r + p = 14$; buy 6 red oak boards and 8 poplar boards.

TECHNOLOGY ACTIVITY 3.5 (p. 176)   **1.** −14   **3.** 0.4   **5.** 21.6

3.6 PRACTICE (pp. 181-183) **5.** no **7.** no **9.** $(5, -1, 1)$
11. She should invest $2000 in savings, $12,000 in CDs, and $6000 in bonds. **13.** $(2, 1, -1)$ **15.** $(6, 0, -3)$
17. $(1, -4, 2)$ **19.** $(4, 3, -3)$ **21.** $(-3, 2, 5)$ **23.** $(7, 3, 5)$
25. $\left(-\frac{2}{7}, 0, -\frac{29}{14}\right)$ **27.** $(2, 1, 2)$ **29.** $(-1, 1, -1)$ **31.** $(6, 6, -4)$
33. $\left(\frac{128}{13}, -\frac{113}{26}, 13.5\right)$ **35.** $f + s + t = 20; 5f + 3s + t = 68;$
$s = f + t$; there were 7 first-place finishers, 10 second-place finishers, and 3 third-place finishers. **37.** $s + l = 1300$;
$s + 2c = 1400; s + l + c = 1600$ **39.** Democrat: 50 million, Republican: 40 million, Other parties: 10 million
41. Sample answers are given.
a. $x + y + z = 3; 2x - 2y + 5z = 23; 4x + 3z = 1$
b. $x + y + z = 3; 2x - 2y + 5z = 23; 4x - 4y + 10z = 11$
c. $x + y + z = 3; 2x - 2y + 5z = 23; 3x - y + 6z = 26$

3.6 MIXED REVIEW (p. 184)
45. 11 **47.** 84 **49.** -16 **51.** $\frac{3}{10}$ **53.** $-\frac{9}{4}$
55. $x \le -14.5$ or $x \ge 11.5$;

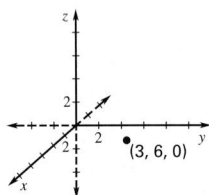

59. $-\frac{29}{3} \le x \le \frac{31}{3}$;

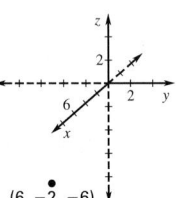

63.

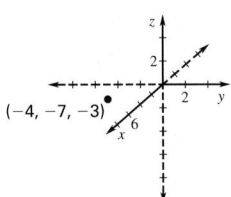

67.

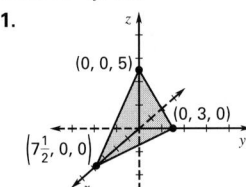

71.

QUIZ 3 (p. 184)
1.

2.

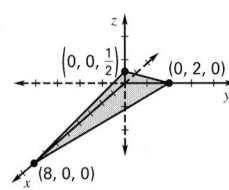

3.

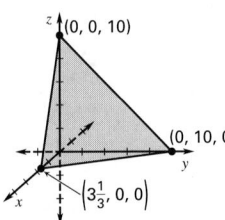

4.

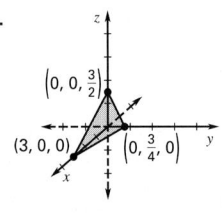

5.

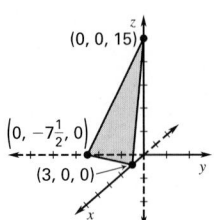

6.

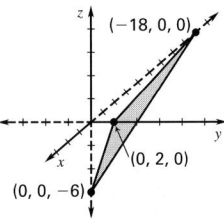

7. $f(x, y) = \frac{1}{3}x - \frac{1}{6}y + 6; \frac{20}{3}$ **8.** $f(x, y) = \frac{1}{2}x + y + 2; 4$
9. $f(x, y) = 20x - 3y - 15; 66$ **10.** $f(x, y) = \frac{1}{3}x - \frac{1}{6}y + 4; \frac{41}{6}$
11. $(5, 0, 0)$ **12.** $(2, -4, -1)$ **13.** no solutions
14. 3 string players, 10 woodwinds, and 2 percussionists were selected.

CHAPTER 3 REVIEW (pp. 186–188)
1.

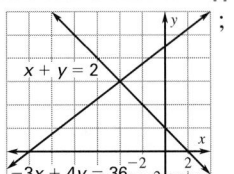

; **3.**

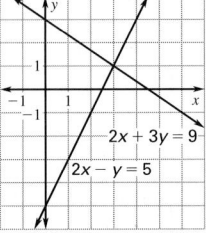

;

one solution; $(-4, 6)$

one solution; $(3, 1)$

5. $(0, 6)$ **7.** $(-2, -1)$
9.

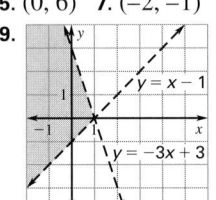

11.

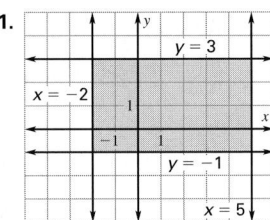

13. max of 50 at $(10, 0)$; min of 0 at $(0, 0)$
15. max of 38 at $(4, 9)$; min of 5 at $(1, 0)$
19.

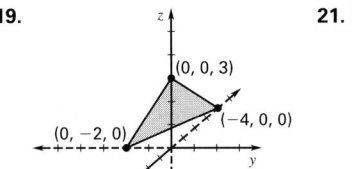

21. $\left(-\frac{1}{2}, 1, 2\right)$

CUMULATIVE PRACTICE (pp. 192–193)
1.

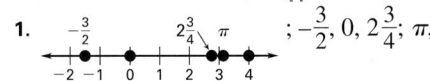

; $-\frac{3}{2}, 0, 2\frac{3}{4}; \pi, 4$
5. distributive property **7.** -22 **9.** 16 **11.** $16a + 11$
13. $n^2 + 2n$ **15.** -8 **17.** -4 **19.** $-10, 9.5$ **21.** 10 **23.** $h = \frac{V}{\pi r^2}$

25. $x < 4$;

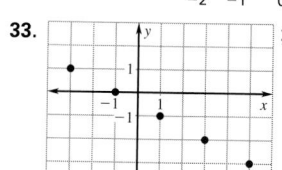

27. $x \le \frac{2}{3}$ or $x > 2$;

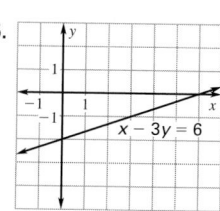

31. $-2 < x < 2$;

33.

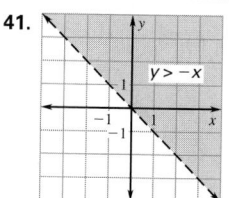

; yes **35.**

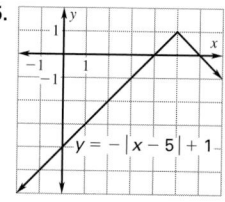

$x - 3y = 6$

41.

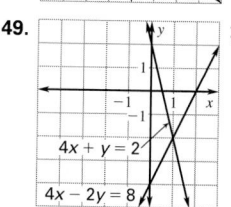

$y > -x$

45.

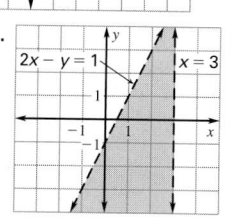

$y = -|x - 5| + 1$

49.

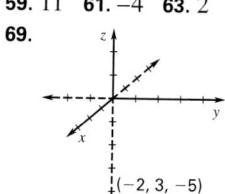

$4x + y = 2$
$4x - 2y = 8$

one solution at $(1, -2)$

51.

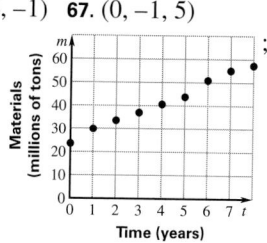

$2x - y = 1$ $x = 3$

Solution region is to the right of $2x - y = 1$ and to the left of $x = 3$.

53. perpendicular **55.** $y = -3x + 7$ **57.** $y = \frac{1}{2}x + 1$

59. 11 **61.** -4 **63.** 2 **65.** $(4, -1)$ **67.** $(0, -1, 5)$

69.

z axis, $(-2, 3, -5)$

75.

Sample answer: $y = 4.20t + 24.5$; about 83.3 million tons

Time (years), Materials (millions of tons)

77. Order 100 lb of vegetables and 50 lb of beef at a total cost of $228.50.

CHAPTER 4

SKILL REVIEW (p. 198) **1.** -1 **2.** -13 **3.** -14 **4.** 40
5. commutative property of multiplication **6.** commutative property of addition **7.** distributive property **8.** $(15, 3)$

9. $(-3, -10)$ **10.** $\left(\frac{112}{5}, -\frac{4}{5}\right)$ **11.** $(-2, -2)$

4.1 PRACTICE (pp. 203–205) **7.** $\begin{bmatrix} -7 & -12 & 12 \\ -5 & 12 & -10 \end{bmatrix}$

9. $\begin{bmatrix} -25 & -6 \\ -8 & 15 \end{bmatrix}$ **11.** not equal **13.** not equal **15.** $\begin{bmatrix} 4 & 1 \\ -12 & 4 \end{bmatrix}$

17. $\begin{bmatrix} -4 & -7 \\ 5 & 5 \end{bmatrix}$ **19.** $\begin{bmatrix} 5.3 & 12.2 \\ 2.8 & 10.4 \end{bmatrix}$ **21.** Not possible; the two matrices do not have the same dimensions.

23. $\begin{bmatrix} 4 & 12 & -28 \\ 16 & 0 & -24 \end{bmatrix}$ **25.** $\begin{bmatrix} 4 & 12 & 36 \\ -20 & 20 & 60 \\ -12 & -20 & -44 \end{bmatrix}$ **27.** $\begin{bmatrix} -1 & -1 & -2 \\ \frac{1}{8} & \frac{3}{11} & -5 \end{bmatrix}$

29. $\begin{bmatrix} 8 & -8 \\ 12 & -3 \\ -16 & 23 \end{bmatrix}$ **31.** $\begin{bmatrix} 22 & -30 \\ -22 & -18 \end{bmatrix}$ **33.** $x = -3, y = -8$

35. $x = -2, y = 44$ **37–41.** Matrices can also be written with the rows and columns switched.

37.

	Before Wins	Before Losses	After Wins	After Losses
Atlanta Braves	59	29	47	27
Seattle Mariners	37	51	39	34
Chicago Cubs	48	39	42	34

39.

1996	No. of units shipped (in mil)	$ Value (in mil)
CDs	20,779	$268,441
Cassettes	15,299	$122,329
Music Videos	45	$916

1997	No. of units shipped (in mil)	$ Value (in mil)
CDs	26,277	$344,697
Cassettes	17,799	$144,645
Music Videos	70	$1,260

41. $\begin{bmatrix} 5,498 & \$76,256 \\ 2,500 & \$22,316 \\ 25 & \$344 \end{bmatrix}$ **43.** $2V + M$; $\begin{bmatrix} 146.8 & 148.4 \\ 146.1 & 147.8 \\ 146.8 & 148.4 \\ 146.2 & 148.1 \end{bmatrix}$

45.

Percent of Population in 1991	0–17	18–65	over 65
Northeast	4.8	12.6	2.8
Midwest	6.3	14.5	3.1
South	8.9	21.2	4.3
Mountain	1.6	3.4	0.6
Pacific	4.2	9.9	1.7

Percent of Population in 2010	0–17	18–65	over 65
Northeast	4.2	11.4	2.5
Midwest	5.3	13.8	3.0
South	8.5	22.6	5.0
Mountain	1.7	4.2	0.9
Pacific	4.6	10.5	1.9

47. South: 18–65, over 65, Mountain: 0–17, 18–65, over 65, Pacific: 0–17, 18–65, over 65

4.1 MIXED REVIEW (p. 206)

51.

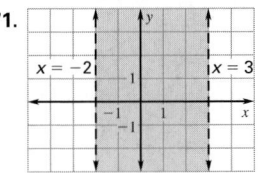

53. 20 **55.** 7 **57.** $\frac{5}{14}$

59. no, yes **61.** no, yes

63. *Sample answer:* (1, 2)

65. *Sample answer:* (5, 5)

TECHNOLOGY ACTIVITY 4.1 (p. 207)

1. $\begin{bmatrix} 6.6 & -6.1 \\ 15.33 & 1.72 \end{bmatrix}$ **3.** $\begin{bmatrix} 6.4666 & 1.6688 \\ 23.0503 & 7.301 \end{bmatrix}$

5. $\begin{bmatrix} -8 & -1 & 0 & -1 \\ -3 & -2 & -1 & 0 \end{bmatrix}$; none; Rock CDs, Country CDs,

Easy Listening CDs, Rock tapes, Country tapes, Jazz tapes

4.2 PRACTICE (pp. 211–212) **5.** defined; 3×3

7. $\begin{bmatrix} 2 & 0 \\ -5 & -3 \end{bmatrix}$ **9.** $\begin{bmatrix} -9 & -3 \\ 7 & 2 \\ 2 & 1 \end{bmatrix}$ **11.** defined; 1×2

13. not defined **15.** defined; 3×1 **17.** [2] **19.** $\begin{bmatrix} 4 & 11 \\ 12 & 3 \end{bmatrix}$

21. Not defined; the number of columns in the left matrix (3) does not equal the number of rows in the right matrix (2).

23. $\begin{bmatrix} -1.3 \\ 0.9 \end{bmatrix}$ **25.** $\begin{bmatrix} -32 & 0 & 32 \\ 12 & -26 & 1 \\ 20 & -30 & -5 \end{bmatrix}$ **27.** $\begin{bmatrix} 16 & -16 \\ 16 & -8 \end{bmatrix}$

29. $\begin{bmatrix} 8 & -5 & 8 \\ -1 & 1 & 1 \\ 7 & -30 & -35 \end{bmatrix}$ **31.** $\begin{bmatrix} 0 & -30 \\ 12 & -51 \end{bmatrix}$ **33.** $x = 2, y = 8$

35. $\begin{bmatrix} 0.201 & 0.348 & 0.180 \\ 0.220 & 0.215 & 0.017 \\ 0.073 & 0.001 & 0.005 \\ 0.113 & 0.014 & 0.405 \end{bmatrix}$ **37.** Matrix B $\begin{bmatrix} 6 \\ 5 \\ 4 \end{bmatrix}$

39. Team 3; 62 points

4.2 MIXED REVIEW (p. 213) **45.** 180 m^2 **47.** 9π ft^2, or
about 28.26 ft^2 **49.** $y = -\frac{1}{4}x + 4$ **51.** $y = 3x + 2$

53. $y = \frac{3}{2}x - 6$ **55.** (−7, 5) **57.** no solution **59.** (0, −5)

61. $\left(-\frac{49}{37}, -\frac{52}{37}\right)$

4.3 PRACTICE (pp. 218–220) **5.** −6 **7.** 28 **9.** (−5, 1)
11. 1750 in.^{2} **13.** 24 **15.** 63 **17.** −31 **19.** 24 **21.** −77
23. 360 **25.** 116 **27.** 81 **29.** −732 **31.** 6 **33.** 11 **35.** 6
37. (−2, −5) **39.** (4, −1) **41.** (6, 2) **43.** $\left(\frac{584}{11}, \frac{480}{11}\right)$
45. (0, 5, 4) **47.** $\left(-\frac{2}{3}, -34, -12\right)$ **49.** (4, 3, −2)

51. $\left(\frac{1}{11}, \frac{34}{11}, \frac{19}{11}\right)$ **53.** $\left(-\frac{1}{44}, -\frac{69}{22}, -\frac{481}{88}\right)$ **55.** 144 ft^2

57. 4 in.^{2} **59.** regular: $1.03 per gal, premium: $1.15 per gal

61. The determinant is multiplied by −1. Proof for

2×2 matrices: $-1\begin{vmatrix} a & b \\ c & d \end{vmatrix} = -1(ad - bc) = bc - ad = \begin{vmatrix} b & a \\ d & c \end{vmatrix}$

4.3 MIXED REVIEW (p. 221) **65.** −3 **67.** 4 **69.** $\frac{5}{4}$

71.

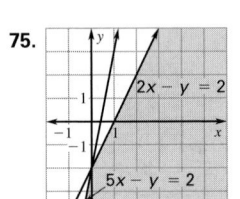

73.

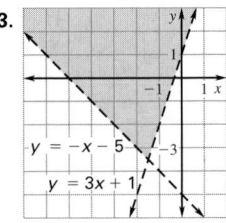

75.

77. $\begin{bmatrix} -24 & 14 \\ 33 & -8 \end{bmatrix}$ **79.** $\begin{bmatrix} -104 & 35 \\ 32 & -4 \end{bmatrix}$

81. $\begin{bmatrix} 12 & 2.7 \\ 4 & 0.92 \end{bmatrix}$

QUIZ 1 (p. 221) **1.** $\begin{bmatrix} -5 & 4 & 15 \\ 2 & -14 & 1 \end{bmatrix}$ **2.** $\begin{bmatrix} -5 & -7 \\ 0 & -1 \end{bmatrix}$

3. $\begin{bmatrix} -2 & -2 \\ -18 & -12 \end{bmatrix}$ **4.** $\begin{bmatrix} -4 & -2 & 22 \\ 3 & -18 & 20 \\ -17 & -4 & 1 \end{bmatrix}$ **5.** $\begin{bmatrix} 26 & 56 \\ 22 & 42 \end{bmatrix}$

6. $\begin{bmatrix} 5 & -15 \\ 38 & -12 \end{bmatrix}$ **7.** 10 **8.** 0 **9.** 70 **10.** −15 **11.** (1, 2)

12. $\left(\frac{4}{9}, -\frac{13}{3}\right)$ **13.** $\left(2, \frac{1}{2}\right)$ **14.** $\left(\frac{5}{2}, 1, -\frac{3}{2}\right)$ **15.** $\left(\frac{7}{3}, 10, -\frac{4}{3}\right)$

16. (0, −4, 3) **17.** 12 ft^2

4.4 PRACTICE (pp. 227–228) **7.** $\begin{bmatrix} -\frac{1}{3} & -\frac{2}{3} \\ 0 & -1 \end{bmatrix}$ **9.** $\begin{bmatrix} \frac{2}{65} & -\frac{32}{65} \\ \frac{16}{65} & \frac{4}{65} \end{bmatrix}$

11. $\begin{bmatrix} -0.0329 & 0.3289 \\ 0.5263 & -0.2632 \end{bmatrix}$ **13.** $\begin{bmatrix} 4 & 5 \\ 3 & 4 \end{bmatrix}$ **15.** $\begin{bmatrix} -7 & 8 \\ 1 & -1 \end{bmatrix}$

17. $\begin{bmatrix} 1 & -2 \\ -3 & 7 \end{bmatrix}$ **19.** $\begin{bmatrix} 1 & \frac{7}{2} \\ -1 & -3 \end{bmatrix}$ **21.** $\begin{bmatrix} \frac{1}{2} & \frac{1}{2} \\ \frac{3}{2} & \frac{11}{6} \end{bmatrix}$ **23.** $\begin{bmatrix} 5 & -1.25 \\ -4 & 1.1 \end{bmatrix}$

25. $\begin{bmatrix} \frac{37}{25} & -\frac{1}{5} \\ -\frac{4}{5} & 0 \end{bmatrix}$ **27.** $\begin{bmatrix} -4 & 2 & -7 \\ 3 & -1 & 5 \end{bmatrix}$ **29.** $\begin{bmatrix} \frac{17}{5} & \frac{136}{5} \\ -\frac{8}{5} & -\frac{64}{5} \end{bmatrix}$

31. $\begin{bmatrix} 11 & -2 \\ 8 & -1.5 \end{bmatrix}$ **33.** no **35.** yes

37. $\begin{bmatrix} -0.0654 & -0.0131 & 0.1634 \\ 0.0131 & 0.2026 & -0.0327 \\ 0.1503 & -0.1699 & 0.1242 \end{bmatrix}$ **39.** $\begin{bmatrix} 12 & -7 & 3 \\ -20 & 12 & -5 \\ 1.5 & -1 & 0.5 \end{bmatrix}$

41. 39, 98, 26, 77, 20, 60, 13, 31, 23, 51
43. 36, −14, 16, 0, 125, −50, −26, 14, 10, 4, 24, −8, −95, 48
45. KARNAK TEMPLE **47.** THE GREAT SPHINX

49. a. $\begin{bmatrix} -1 & -4 & -2 \\ 1 & 2 & 3 \end{bmatrix}$; $\begin{bmatrix} -1 & -2 & -3 \\ -1 & -4 & -2 \end{bmatrix}$;

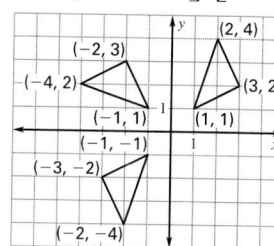

; 90° rotation

b. *Sample answer:* Find A^{-1} and then multiply AAT by A^{-1} on the left: $A^{-1}AAT = IAT = AT$. Now multiply AT by A^{-1} on the left: $A^{-1}AT = IT = T$.

4.4 MIXED REVIEW (p. 229)

55. all real numbers **57.** $(4, 0, -2)$ **59.** $\left(\dfrac{1}{2}, 4, \dfrac{1}{4}\right)$

61. Not possible; the matrices have different dimensions.

63. $\begin{bmatrix} 17 & -3 & -1 \\ 0 & 25 & 31 \end{bmatrix}$ **65.** $\begin{bmatrix} 2 & 5 & 1 \\ 3 & 4 & 8 \end{bmatrix}$

4.5 PRACTICE (pp. 233–235)

5. $\begin{bmatrix} 1 & 3 \\ 4 & -2 \end{bmatrix}\begin{bmatrix} x \\ y \end{bmatrix} = \begin{bmatrix} 9 \\ 7 \end{bmatrix}$ **7.** $(-5, 7)$ **9.** $\left(\dfrac{21}{13}, -\dfrac{2}{13}\right)$

11. $\begin{bmatrix} 1 & 1 \\ 3 & -4 \end{bmatrix}\begin{bmatrix} x \\ y \end{bmatrix} = \begin{bmatrix} 5 \\ 8 \end{bmatrix}$ **13.** $\begin{bmatrix} 5 & -3 \\ -4 & 2 \end{bmatrix}\begin{bmatrix} x \\ y \end{bmatrix} = \begin{bmatrix} 9 \\ 10 \end{bmatrix}$

15. $\begin{bmatrix} 1 & 8 \\ 4 & -5 \end{bmatrix}\begin{bmatrix} x \\ y \end{bmatrix} = \begin{bmatrix} 4 \\ -11 \end{bmatrix}$ **17.** $\begin{bmatrix} 1 & -4 & 5 \\ 2 & 1 & -7 \\ -4 & 5 & 2 \end{bmatrix}\begin{bmatrix} x \\ y \\ z \end{bmatrix} = \begin{bmatrix} -4 \\ -23 \\ 38 \end{bmatrix}$

19. $\begin{bmatrix} 0.5 & 3.1 & -0.2 \\ 1.2 & -2.5 & 0.7 \\ 0.3 & 4.8 & -4.3 \end{bmatrix}\begin{bmatrix} x \\ y \\ z \end{bmatrix} = \begin{bmatrix} 5.9 \\ 2.2 \\ 4.8 \end{bmatrix}$

21. $\begin{bmatrix} 0 & 8 & -10 \\ 0 & 6 & -12 \\ -9 & 0 & 5 \end{bmatrix}\begin{bmatrix} x \\ y \\ z \end{bmatrix} = \begin{bmatrix} -23 \\ 14 \\ 0 \end{bmatrix}$ **23.** $(5, -7)$ **25.** $(5, -9)$

27. $(1, -7)$ **29.** $(-1, -4)$ **31.** $(-3, -14)$ **33.** $(-61, 179, -83)$
35. $(4, 3, 1)$ **37.** $(2, 3, -2)$ **39.** $(3, -2, 6)$ **41.** 2239.8 g of A, 1313.6 g of B, 4067.6 g of C **43.** transformer: $10.00, wire: $.20 per ft, light: $1.00

4.5 MIXED REVIEW (p. 235) **47.** -2 **49.** $-\dfrac{19}{2}$ **51.** 5 **53.** -3

55. 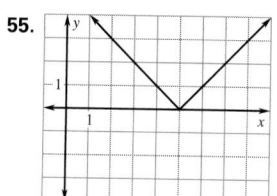 **57.**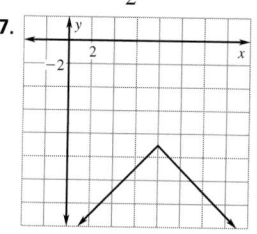

61. $\begin{bmatrix} 3 & 4 \\ 5 & 7 \end{bmatrix}$ **65.** $\begin{bmatrix} 1 & -2 \\ -\dfrac{3}{2} & \dfrac{7}{2} \end{bmatrix}$

QUIZ 2 (p. 236)

1. $\begin{bmatrix} 2 & -1 \\ -7 & 4 \end{bmatrix}$ **2.** $\begin{bmatrix} -3 & -5 \\ -4 & -7 \end{bmatrix}$ **3.** $\begin{bmatrix} -\dfrac{1}{3} & -\dfrac{1}{9} \\ -1 & -\dfrac{2}{3} \end{bmatrix}$ **4.** $\begin{bmatrix} \dfrac{7}{2} & -\dfrac{5}{2} \\ -4 & 3 \end{bmatrix}$

---

5. $(-1, 4)$ **6.** $(4, 3)$ **7.** $(3, -3)$ **8.** place setting: $35.50, serving set: $67.00

CHAPTER 4 EXTENSION (p. 238) **1.** $(-2, 5)$ **3.** $(-1, -4)$

5. $(4, -5)$ **7.** $(2, 1)$ **9.** $\left(0, \dfrac{1}{5}\right)$ **11.** $(16, -5, 2)$ **13.** $(-5, 2, 0)$

15. $(-16, 12, 10)$

CHAPTER 4 REVIEW (pp. 240–242)

1. $\begin{bmatrix} 15 & -5 \\ 1 & 5 \end{bmatrix}$ **3.** $\begin{bmatrix} 8 & 11 \\ 9 & 13 \\ 8 & 6 \end{bmatrix}$ **5.** $\begin{bmatrix} 8 & 12 & -2 \\ 20 & -10 & 4 \\ 0 & 22 & 2 \end{bmatrix}$ **7.** $x = -1, y = 10$

9. $x = -1, y = 5$ **11.** $\begin{bmatrix} -120 & -84 \\ 40 & 28 \end{bmatrix}$ **13.** $\begin{bmatrix} 17 & -29 & 64 \\ 18 & -36 & 72 \end{bmatrix}$

15. 12 **17.** 4 **19.** $(-1, -1)$ **21.** $(6, 0, -3)$

23. $\begin{bmatrix} \dfrac{3}{4} & -\dfrac{1}{2} \\ -\dfrac{1}{4} & \dfrac{1}{2} \end{bmatrix}$ **25.** $\begin{bmatrix} 1 & 1 \\ 5 & 6 \end{bmatrix}$ **27.** $\begin{bmatrix} -3 & -2 \\ 4 & 3 \end{bmatrix}$ **29.** $\left(\dfrac{5}{2}, \dfrac{3}{2}\right)$

31. $(4, 1, 0)$ **33.** $(-3, 2, 4)$

CHAPTER 5

SKILL REVIEW (p. 248) **1.** $\dfrac{5}{3}$ **2.** -3 **3.** 2

4. 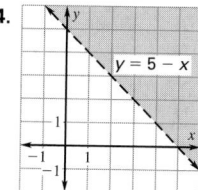 **5.**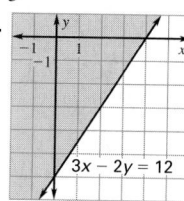

6. 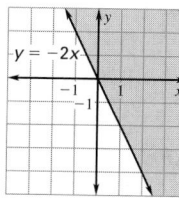 **7.**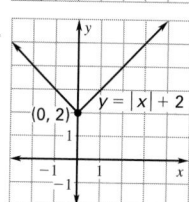

8. 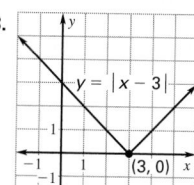 **9.**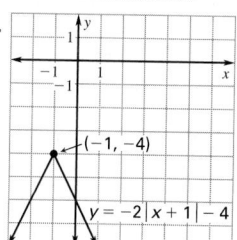

5.1 PRACTICE (pp. 253–254)

5. 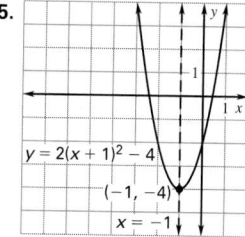 **7.**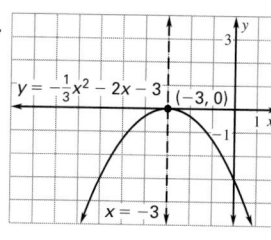

SELECTED ANSWERS

9.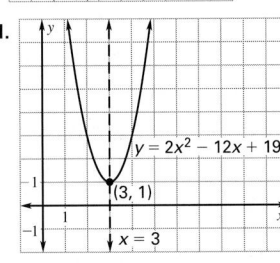

11. $y = -2x^2 - 2x + 24$
13. $y = -x^2 - 4x - 11$
15. $y = \frac{2}{3}x^2 - 12x + 50$
17. C   **19.** B

21.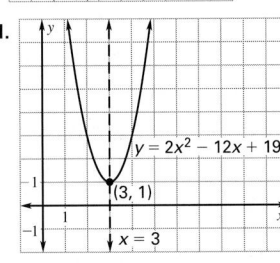

23.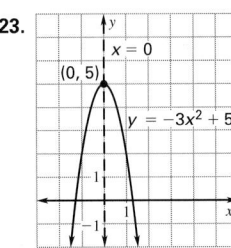

25.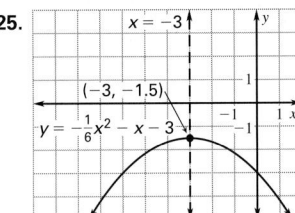

27.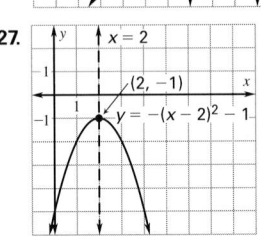

29.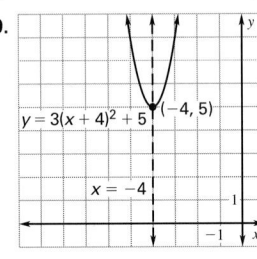

31.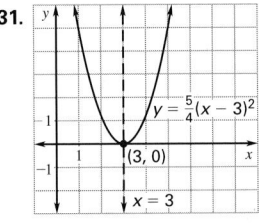

33.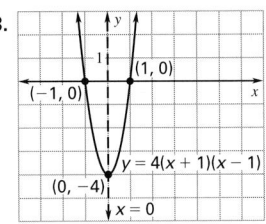

35.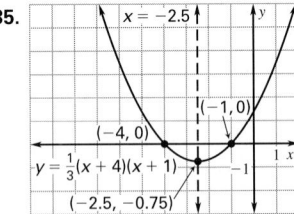

37.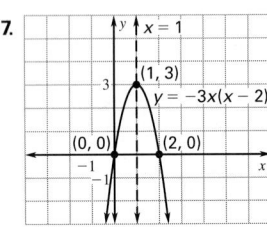

39. $y = -x^2 + x + 12$
41. $y = -3x^2 + 9x + 84$
43. $y = x^2 + 6x + 11$
45. $y = -6x^2 + 24x - 33$
47. $y = -81x^2 - 32x - 4$
49. $y = 32x^2 - 8x - 1$

51. about 3,090 revolutions per min; about 74.7 foot-pounds   **53.** *Sample answer:* The energy use decreases until about 90 meters per minute and then increases.

5.1 MIXED REVIEW (p. 255)   **57.** 2   **59.** -7 **61.** -5 **63.** 7
65. -3 **67.**

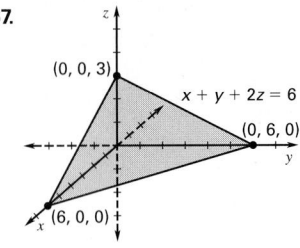

69.

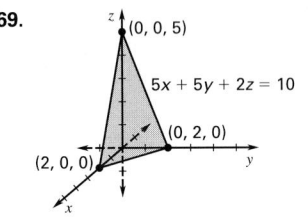

71.

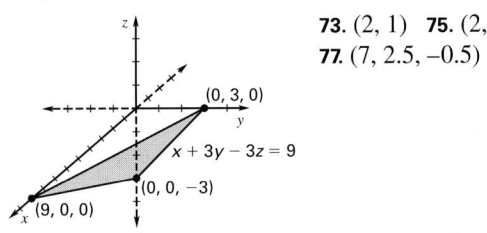

73. $(2, 1)$ **75.** $(2, -4, 1)$
77. $(7, 2.5, -0.5)$

5.2 PRACTICE (pp. 260–262)   **5.** $(2x + 3)(x - 1)$ **7.** $(y + 1)^2$
9. $q(q + 1)$ **11.** $-2, 4$ **13.** $-\frac{1}{2}, \frac{1}{2}$ **15.** $0, 6$
17. $y = (x + 4)(x + 2); -4, -2$ **19.** $y = (x + 5)^2; -5$
21. $y = (3x - 2)(x - 2); \frac{2}{3}, 2$ **23.** $(x + 4)(x + 1)$
25. $(x + 5)(x + 8)$ **27.** $(x - 6)(x - 2)$ **29.** $(a + 5)(a - 2)$
31. $(c + 10)(c - 8)$ **33.** cannot be factored
35. $(2x + 1)(x + 3)$ **37.** $(4x + 3)(2x + 3)$ **39.** cannot be
factored   **41.** $(3k - 1)(k + 11)$ **43.** $(3n - 2)(6n + 7)$
45. $(3v - 7)(4v + 1)$ **47.** $(x - 5)(x + 5)$ **49.** $(x - 3)^2$
51. $(3s + 2)^2$ **53.** $(7 - 10a)(7 + 10a)$ **55.** $(9c + 11)^2$
57. $2(3x - 1)(3x + 1)$ **59.** $4(2y + 3)(y - 5)$ **61.** $u(u + 7)$
63. $-(v - 1)^2$ **65.** $-1, 4$ **67.** $\frac{3}{5}, 2$ **69.** -12 **71.** $-\frac{4}{9}, \frac{4}{9}$
73. $-5, 6$ **75.** $\frac{1}{4}$ **77.** $-1, \frac{8}{3}$ **79.** $-\frac{9}{2}, 0$ **81.** $y = (x + 4)(x + 3)$;
$-4, -3$ **83.** $y = (x - 2)(x + 2); -2, 2$ **85.** $y = x(x - 3); 0, 3$
87. $y = -(x - 8)^2; 8$ **89. a.** $m + n = 0, mn = 9$ **b.** If $m + n = 0$,
then $m = -n$. Substituting in $mn = 9$, $(-n)(n) = 9$, $-n^2 = 9$,
and $n^2 = -9$. There is no number such that $n^2 = -9$.
Therefore, $x^2 + 9$ is not factorable.   **91.** 60 ft   **93.** 7   **95.** 6
97. 2.5 ft   **99.** $80; $12,800   **101.** about 70 mi; about 24 mi

5.2 MIXED REVIEW (p. 263)   **107.** $-4, 8$ **109.** $-2, 3.6$
111. $-4 < x < 2$ **113.** $x < -3$ or $x > 11$

SELECTED ANSWERS

115.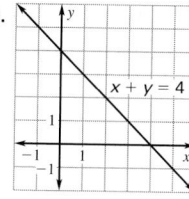

119.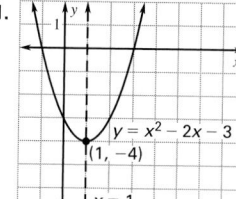

123.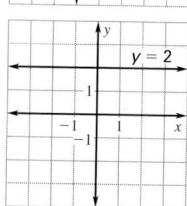

125.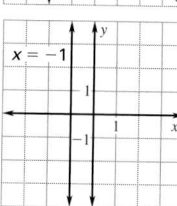

127.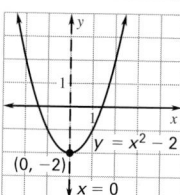

129.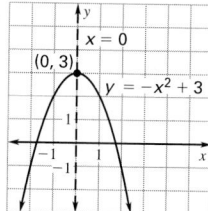

131.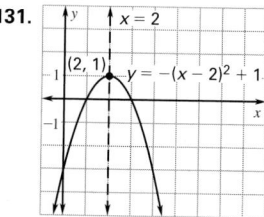

133.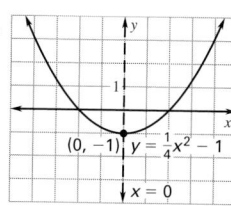

135.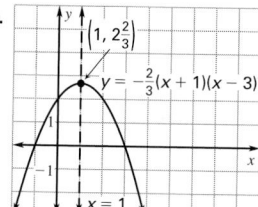

5.3 PRACTICE (pp. 267–268) **5.** $2\sqrt{3}$ **7.** 9 **9.** $\dfrac{\sqrt{7}}{3}$ **11.** $\dfrac{\sqrt{10}}{2}$
13. $-5, 5$ **15.** $-2\sqrt{3}, 2\sqrt{3}$ **17.** $-2\sqrt{7} - 8, 2\sqrt{7} - 8$
19. $3\sqrt{2}$ **21.** $3\sqrt{3}$ **23.** $6\sqrt{2}$ **25.** $7\sqrt{2}$ **27.** 14 **29.** 6
31. $2\sqrt{6}$ **33.** $12\sqrt{7}$ **35.** $\dfrac{1}{3}$ **37.** $\dfrac{6}{5}$ **39.** $\dfrac{\sqrt{3}}{4}$ **41.** $\dfrac{5\sqrt{3}}{6}$
43. $\dfrac{2\sqrt{3}}{3}$ **45.** $\dfrac{\sqrt{30}}{5}$ **47.** $\dfrac{\sqrt{14}}{4}$ **49.** $\dfrac{3\sqrt{10}}{8}$ **51.** $-11, 11$
53. $-6, 6$ **55.** $-5\sqrt{3}, 5\sqrt{3}$ **57.** $-10\sqrt{3}, 10\sqrt{3}$ **59.** $-12, 12$
61. $-6, 4$ **63.** $-3\sqrt{3} + 7, 3\sqrt{3} + 7$ **65.** $-1, 13$ **67.** $-2, 7$
69. about 3.3 sec **71.** Earth: 3.5 sec; Mars: 5.8 sec;
Jupiter: 2.2 sec; Neptune: 3.3 sec; Pluto: 13.8 sec
73. 16.2 in. by 21.6 in. **75. a.** about 60.6 sec **b.** 146 sec
c. *Sample answer:* The water drains more slowly as the
time increases.

5.3 MIXED REVIEW (p. 269) **77.** $(1, 2)$ **79.** $(-3, -5)$

81. $(6, -2)$ **83.** $\begin{bmatrix} 13 & -1 \\ -11 & 1 \end{bmatrix}$ **85.** $\begin{bmatrix} 81 & 57 \\ -40 & -31 \end{bmatrix}$

87. $y = x^2 - 9x + 8$ **89.** $y = 16x^2 - 81$
91. $y = 5x^2 + 60x + 168$

QUIZ 1 (p. 270)

1.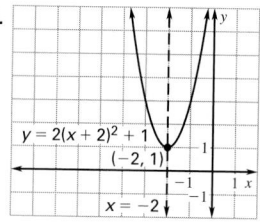

2.

3.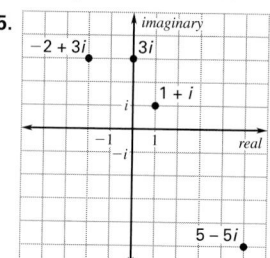

4. $-3, 9$ **5.** $-4, -\dfrac{5}{4}$ **6.** $\dfrac{1}{2}$
7. $3\sqrt{6}$ **8.** $14\sqrt{5}$
9. $\dfrac{6\sqrt{5}}{5}$ **10.** $\dfrac{2\sqrt{3}}{3}$
11. about 2.7 mi/h

TECHNOLOGY ACTIVITY 5.3 (p. 271) **1.** $-1.53, 1.53$
3. $-2.45, 2.45$ **5.** $-2.73, 0.73$ **7.** $-3.65, 1.65$
9. $48\pi = 6\pi r^2$; $r \approx 2.8$ in.

5.4 PRACTICE (pp. 277–279) **5.** $-2i\sqrt{2}, 2i\sqrt{2}$ **7.** $7 + 3i$
9. $9 - 5i$ **11.** $\sqrt{2}$ **13.** $\sqrt{13}$ **15.**
17. $-2i, 2i$ **19.** $-3i\sqrt{3}, 3i\sqrt{3}$
21. $-i\sqrt{3}, i\sqrt{3}$ **23.** $-i, i$
25. $2 + 4i, 2 - 4i$
27. $-3 - 2i\sqrt{14}, -3 + 2i\sqrt{14}$

29–35 odd: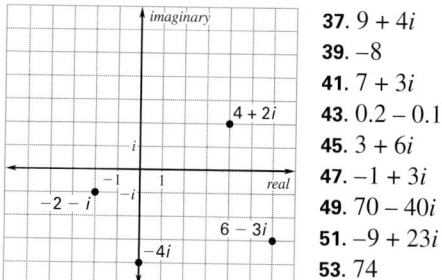

37. $9 + 4i$
39. -8
41. $7 + 3i$
43. $0.2 - 0.1i$
45. $3 + 6i$
47. $-1 + 3i$
49. $70 - 40i$
51. $-9 + 23i$
53. 74

55. $161 - 240i$ **57.** $-1 + i$ **59.** $\dfrac{4}{5} + \dfrac{3}{5}i$ **61.** $-\dfrac{87}{97} + \dfrac{26}{97}i$
63. $\dfrac{17}{19} - \dfrac{6\sqrt{2}}{19}i$ **65.** 13 **67.** $5\sqrt{2}$ **69.** $4\sqrt{5}$ **71.** 4

73. *Sample answer:* It does because the absolute values are
equal to or less than $N = 1$. **75.** *Sample answer:* It does not
because the absolute values become infinitely large.
77. *Sample answer:* It does not because the absolute values
become infinitely large. **79.** *Sample answer:* It does
because the absolute values are less than $N = 1$.

81. true
83. false; *Sample answer:* $(6 + 3i) + (-5 - 3i) = 1$, which is
not imaginary. **85.** true

87. true; true **89.** false; false **91.** false; false **95. a.** $2 - 2i$
b. $12 - 7i$ **c.** $8 - 4i$

5.4 MIXED REVIEW (p. 280) **101.** 11 **103.** 3 **105.** $(1, 2)$
107. $(4, -3)$ **109.** $-8, 4$ **111.** $5 + \sqrt{10}, 5 - \sqrt{10}$
113. $6 + \sqrt{7}, 6 - \sqrt{7}$

5.5 PRACTICE (pp. 286–289) **5.** $49; (x + 7)^2$ **7.** $25; (x - 5)^2$
9. $\frac{169}{4}; \left(x - \frac{13}{2}\right)^2$ **11.** $1 - \sqrt{5}, 1 + \sqrt{5}$ **13.** $-4 - \sqrt{7}, -4 + \sqrt{7}$
15. $2 - 3i\sqrt{3}, 2 + 3i\sqrt{3}$ **17.** $y = (x - 2)^2 + 3; (2, 3)$
19. $y = (x + 5)^2 - 8; (-5, -8)$ **21.** $y = 2(x + 1)^2 - 6; (-1, -6)$
23. $(x + 8)^2$ **25.** $(x - 12)^2$ **27.** $(x + 0.5)^2$ **29.** $\left(x - \frac{3}{2}\right)^2$
31. $\left(x - \frac{2}{9}\right)^2$ **33.** $81; (x + 9)^2$ **35.** $484; (x - 22)^2$ **37.** $\frac{121}{4}$;
$\left(x - \frac{11}{2}\right)^2$ **39.** $\frac{225}{4}; \left(x + \frac{15}{2}\right)^2$ **41.** $8.41; (x - 2.9)^2$
43. $22.09; (x + 4.7)^2$ **45.** $\frac{25}{9}; \left(x + \frac{5}{3}\right)^2$ **47.** $-1 + \sqrt{10}$,
$-1 - \sqrt{10}$ **49.** $-10 + 2i, -10 - 2i$ **51.** $3 - 2\sqrt{11}, 3 + 2\sqrt{11}$
53. $-0.9 - \sqrt{2.31}, -0.9 + \sqrt{2.31}$ **55.** $3 + \sqrt{2}, 3 - \sqrt{2}$
57. $-7 - i, -7 + i$ **59.** $\frac{5 - 2\sqrt{3}}{2}, \frac{5 + 2\sqrt{3}}{2}$ **61.** $\frac{-1 - i}{2}, \frac{-1 + i}{2}$
63. $-6, 2$ **65.** $-\frac{\sqrt{23}}{3}, \frac{\sqrt{23}}{3}$ **67.** $\frac{1 - i\sqrt{71}}{6}, \frac{1 + i\sqrt{71}}{6}$
69. $-1 - 4\sqrt{2}, -1 + 4\sqrt{2}$ **71.** $11 - 13i, 11 + 13i$
73. $y = (x - 3)^2 + 2; (3, 2)$ **75.** $y = (x + 8)^2 - 50; (-8, -50)$
77. $y = \left(x - \frac{3}{2}\right)^2 - \frac{17}{4}; \left(\frac{3}{2}, -\frac{17}{4}\right)$ **79.** $y = -(x - 10)^2 + 20$;
$(10, 20)$ **81.** $y = 3(x - 2)^2 - 11; (2, -11)$
83. $y = 1.4(x + 2)^2 - 2.6; (-2, -2.6)$ **85.** $-5 + 5\sqrt{5}$,
or ≈ 6.18 **87.** $\sqrt{39} - 2$, or ≈ 4.24 **89.** $d = 0.08(30)^2 +$
$1.1(30) = 105$ ft; about 25.5 mi/h **91.** 45.50 ft; 161.16 ft
93. about 1 cm **95.** 507.5°F; 3.91 Btu/ft³

5.5 MIXED REVIEW (p. 289) **101.** 17 **103.** 52 **105.** 0
107. $y = 2x - 5$ **109.** $y = -5x - 25$
111. $y = \frac{1}{3}x + 7$ **113.**

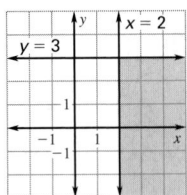

115. **117.**

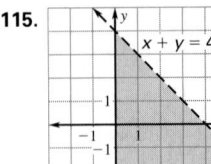 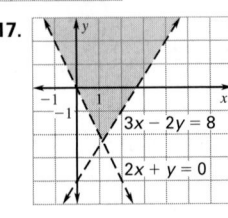

TECHNOLOGY ACTIVITY 5.5 (p. 290)
1–9 odd: Estimates may vary. **1.** minimum; -4.25; 2.5
3. minimum; 4; -3 **5.** maximum; 8.125; -0.75
7. minimum; 2.375; 3.75 **9.** maximum; 8.65; 2.29

5.6 PRACTICE (pp. 295–297)
5. $\frac{-1 + \sqrt{5}}{2}, \frac{-1 - \sqrt{5}}{2}$ **7.** $\frac{-1 + \sqrt{2}}{3}, \frac{-1 - \sqrt{2}}{3}$
9. $\frac{1}{2} + 3i, \frac{1}{2} - 3i$ **11.** -16; 2 imaginary **13.** -47; 2 imaginary
15. 261; 2 real **17.** $-2, 7$ **19.** $1 + \sqrt{5}, 1 - \sqrt{5}$ **21.** $-3 - 7i$,
$-3 + 7i$ **23.** $\frac{-3 + \sqrt{29}}{10}, \frac{-3 - \sqrt{29}}{10}$ **25.** $\frac{-1 + i\sqrt{7}}{4}, \frac{-1 - i\sqrt{7}}{4}$
27. $-1, \frac{9}{7}$ **29.** $\frac{-9 + \sqrt{33}}{8}, \frac{-9 - \sqrt{33}}{8}$ **31.** $-\frac{2}{5} + \frac{\sqrt{26}}{10}$,
$-\frac{2}{5} - \frac{\sqrt{26}}{10}$ **33.** $-9, 11$ **35.** $4 + i\sqrt{19}, 4 - i\sqrt{19}$
37. $-8 + 3\sqrt{2}, -8 - 3\sqrt{2}$ **39.** $\frac{1}{2} + \frac{\sqrt{6}}{4}, \frac{1}{2} - \frac{\sqrt{6}}{4}$ **41.** $5 + \frac{i}{2}$,
$5 - \frac{i}{2}$ **43.** $\frac{1}{3}, -\frac{5}{3}$ **45.** $\frac{-9.5 + \sqrt{218.17}}{7.8}, \frac{-9.5 - \sqrt{218.17}}{7.8}$
47. $\frac{3 + \sqrt{69}}{2}, \frac{3 - \sqrt{69}}{2}$ **49.** 2, 16 **51.** $-4 + 3i, -4 - 3i$
53. $\frac{\sqrt{3}}{2}, -\frac{\sqrt{3}}{2}$ **55.** $-\frac{3}{2}, \frac{1}{7}$ **57.** 33; 2 real **59.** 160; 2 real
61. -7; 2 imaginary **63.** -19; 2 imaginary **65.** zero
67. positive **69.** $c < 4; c = 4; c > 4$ **71.** $c < 16; c = 16$;
$c > 16$ **73.** $c < 36; c = 36; c > 36$ **75.** about 2.56 sec
77. about 0.17 sec **79.** 1993

5.6 MIXED REVIEW (p. 298)
85. $x > 2$ **87.** $x \geq -13$ **89.** $3 \leq x \leq 8$
91. **93.**

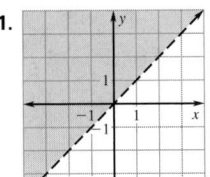 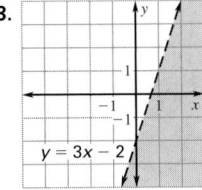

95. **97.**

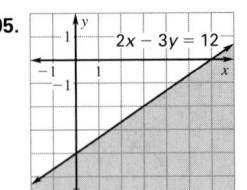 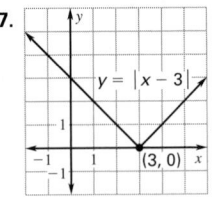

99. **101.**

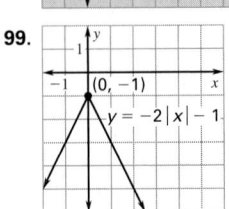 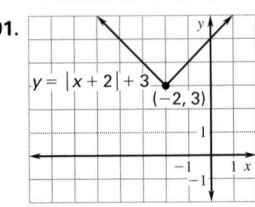

QUIZ 2 (p. 298) **1.** $5 + 16i$ **2.** $-4 + 10i$ **3.** $31 + 22i$
4. $\dfrac{1}{13} - \dfrac{8}{13}i$ **5–10.**

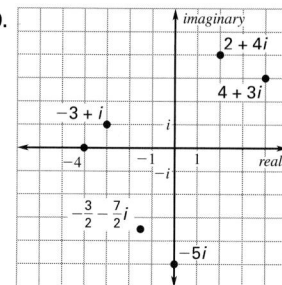

5. $2\sqrt{5}$ **6.** 5
7. $\sqrt{10}$ **8.** 5
9. 4 **10.** $\dfrac{\sqrt{58}}{2}$

11. $-4 + \sqrt{2},\ -4 - \sqrt{2}$ **12.** $1 + 4i,\ 1 - 4i$ **13.** $5 + 3\sqrt{3},$
$5 - 3\sqrt{3}$ **14.** $-2 + \dfrac{\sqrt{5}}{5},\ -2 - \dfrac{\sqrt{5}}{5}$ **15.** $y = (x + 3)^2 - 8$
16. $y = (x - 9)^2 - 31$ **17.** $y = -2(x - 2)^2 + 1$ **18.** $-1 + \sqrt{11},$
$-1 - \sqrt{11}$ **19.** $8 + 3i,\ 8 - 3i$ **20.** $\dfrac{3 + i\sqrt{7}}{2},\ \dfrac{3 - i\sqrt{7}}{2}$
21. $\dfrac{-4 + 2\sqrt{6}}{5},\ \dfrac{-4 - 2\sqrt{6}}{5}$ **22.** about 1 sec

5.7 PRACTICE (pp. 303–305)

5.

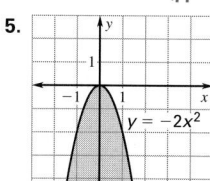

7.

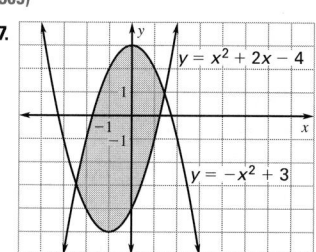

9.

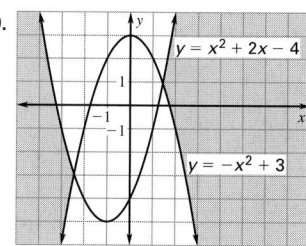

11. $x \le -2$ or $x \ge 2$
13. about 55.1 m and
447.3 m **15.** C

17.

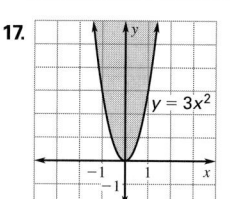

19.

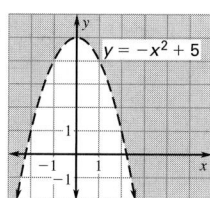

21.

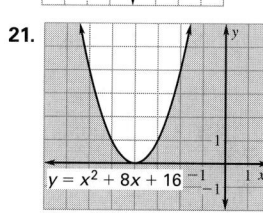

23.

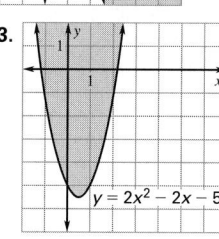

25.

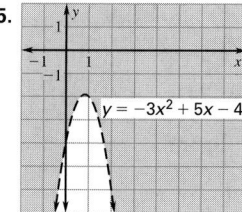

27.

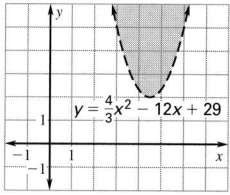

29.

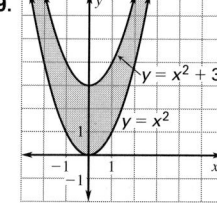

31.

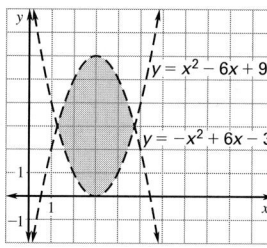

33.

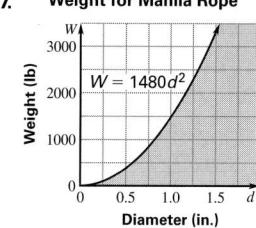

35. $-2 < x < 1$
37. $x \le -4$ or $x \ge 2$
39. $x \le -5.5$ or $x \ge -2.5$
41. $x \le -6$ or $x \ge 3$
43. $-\dfrac{5}{2} < x < \dfrac{5}{2}$
45. $x < -0.9$ or $x > 2.9$

47.

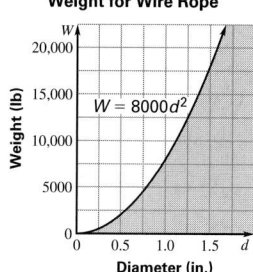

49.

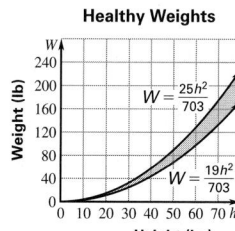

; $121 \le W \le 160$

51. about 39 to 61 years old

5.7 MIXED REVIEW (p. 305) **55.** $y = 4x - 5$ **57.** $y = -\dfrac{11}{4} - \dfrac{1}{2}x$
59. $y = -9x$ **61.** $(2, 3, -4)$ **63.** -6 **65.** $6 - 5i$ **67.** $29 - 29i$
69. $\dfrac{6}{17} - \dfrac{7}{17}i$

5.8 PRACTICE (pp. 309–311) **3.** $y = -1(x - 1)^2 + 3$
5. $y = x^2 + 3x - 2$ **7.** $y = (x - 2)^2 - 2$ **9.** $y = -\dfrac{3}{4}(x - 1)^2$
11. $y = \dfrac{1}{3}(x + 4)^2 + 6$ **13.** $y = -3x^2$ **15.** $y = -\dfrac{3}{2}(x + 6)^2 - 7$
17. $y = 3(x + 2)(x - 1)$ **19.** $y = -1(x - 1)(x - 4)$
21. $y = 2(x + 1)(x - 6)$ **23.** $y = \dfrac{7}{5}(x - 3)(x - 9)$

SELECTED ANSWERS

25. $y = -x^2 + x + 4$ **27.** $y = -\frac{3}{4}x^2 - \frac{11}{4}x + 1$

29. $y = -x^2 + 5x - 2$ **31.** $y = -2x^2 - 4x + 9$

33. $y = \frac{5}{2}x^2 + 6x - 8$ **35.** $y = -0.00168(x - 0)(x - 24)$

37. $s = -0.0807p^2 + 55.2p + 330;$
$k = -0.0000609p^2 + 0.626p + 125$

5.8 MIXED REVIEW (p. 312)
41. 5   **43.** -182 **45.** $(3, -1)$ **47.** $(-4, 5)$

QUIZ 3 (p. 312)

1.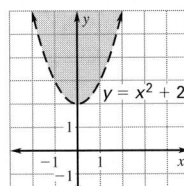

2.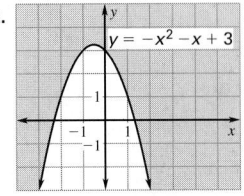

3.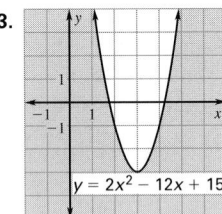

4.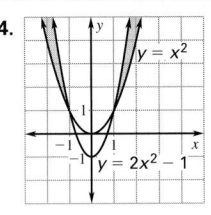

5.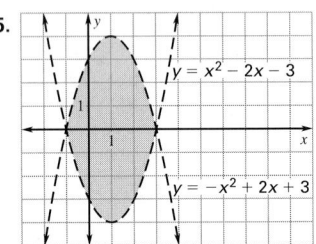

6.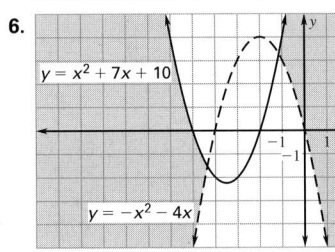

7. $y = 2(x - 5)^2 - 2$

8. $y = -1(x + 3)(x - 1)$

9. $y = \frac{3}{4}x^2 + x$

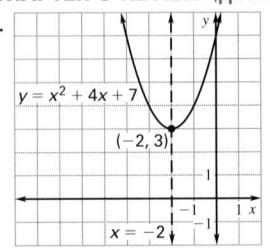

10. $0.00339N^2 + 0.00143N - 5.95 < 1000; \; 0 < N < 544$

CHAPTER 5 REVIEW (pp. 314-316)

1.

3.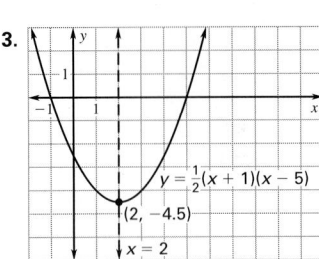

5. 4   **7.** $-3, \frac{5}{3}$

9. $-10, 10$

11. $-6 - 2\sqrt{10},$
$-6 + 2\sqrt{10}$

13. $5 + i$ **15.** $102 + 13i$

17. $3\sqrt{13}$ **19.** $5 + i, 5 - i$

21. $y = (x - 4)^2 + 1; \; (4, 1)$

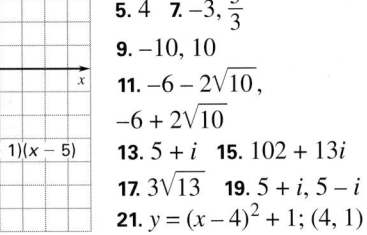

23. $y = 4(x + 2)^2 + 7; \; (-2, 7)$ **25.** $-\frac{7}{18} - \frac{\sqrt{85}}{18}, \; -\frac{7}{18} + \frac{\sqrt{85}}{18}$

27.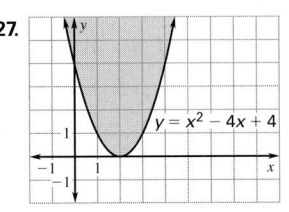

29.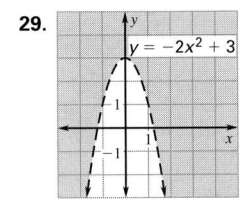

31. $x \le \frac{-7 - \sqrt{33}}{4}$ or $x \ge \frac{-7 + \sqrt{33}}{4}$ **33.** $y = (x - 6)^2 + 1$

35. $y = 0.5x^2 + 1.5x - 4$

CHAPTER 6

SKILL REVIEW (p. 322)   **1.** $3x^2 - x$ **2.** $-3x + 10$

3. $-5x^4 - 4x^3 + 7x^2$

4.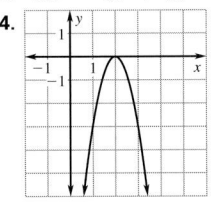

5.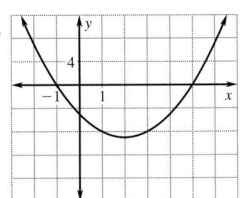

6.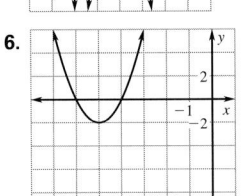

7. $y = x^2 - 2x - 6$

8. $y = 2x^2 + 16x + 32$

9. $y = -x^2 - 6x + 16$

10. $-9, 3$ **11.** -10 **12.** $-4, \frac{3}{2}$

6.1 PRACTICE (pp. 326–328)   **3.** 216   **5.** 64   **7.** $\frac{25}{9}$ **9.** 1

11. $\frac{1}{16x^6}$ **13.** $3y^3$ **15.** sun's volume: 1.41×10^{18} km³;

Earth's volume: 1.09×10^{12} km³; ratio is about 1,298,000;

the results match.   **17.** $\frac{1}{15,625}$ **19.** 262,144   **21.** $\frac{27}{343}$

23. $\frac{1}{121}$ **25.** 4096   **27.** 2048   **29.** $\frac{1}{6}$ **31.** $\frac{15,625}{64}$

33. $32,768x^{10}$ **35.** x^7 **37.** $\frac{1}{x^{12}y^{21}}$ **39.** $-\frac{3}{x^4}$ **41.** $\frac{y^3}{x^2}$

43. $\frac{1}{3}xy^2$ **45.** $-\frac{y^{12}}{9x^4}$ **47.** $3x^2y^2$ **49.** $A = 16\pi x^2$

51. $V = \frac{4}{81}\pi x^3$

SELECTED ANSWERS

53.

Country	Per capita GDP
France	$\$2.13 \times 10^4$
Germany	$\$2.24 \times 10^4$
Ireland	$\$1.95 \times 10^4$
Luxembourg	$\$3.24 \times 10^4$
The Netherlands	$\$2.14 \times 10^4$
Sweden	$\$2.00 \times 10^4$

55. about 7.48×10^3 days

6.1 MIXED REVIEW (p. 328)

61.
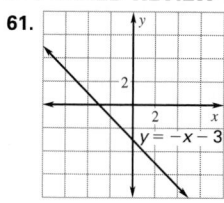

63.
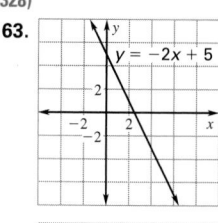

65.
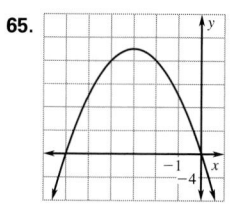

67.
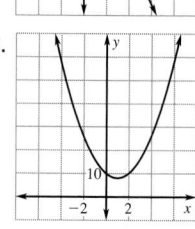

69. ± 4 **71.** $\pm \dfrac{4}{5}$ **73.** ± 1 **75.** $\pm\sqrt{5}$ **77.** $\pm\sqrt{3}$ **79.** $-3 + 4i$
81. $2 - 7i$ **83.** $26 + 12i$

6.2 PRACTICE (pp. 333–336)   **5.** no   **7.** yes; -2
9. $f(x) \to +\infty$ as $x \to -\infty$ and $f(x) \to -\infty$ as $x \to +\infty$
11. $f(x) \to -\infty$ as $x \to -\infty$ and $f(x) \to +\infty$ as $x \to +\infty$
13. $f(x) \to -\infty$ as $x \to -\infty$ and $f(x) \to -\infty$ as $x \to +\infty$
15. yes; $f(x) = -5x + 12$, 1, linear, -5 **17.** yes;
$f(x) = x + \pi$, 1, linear, 1   **19.** no   **21.** yes; $f(x) = x^2 - x + 1$,
2, quadratic, 1   **23.** yes; $f(x) = x^4 - x^3 + 36x^2$, 4, quartic, 1
25. yes; $f(x) = 3x^3$, 3, cubic, 3   **27.** 4   **29.** 36   **31.** 4   **33.** 2
35. 7930   **37.** 73   **39.** -91 **41.** -31 **43.** -7 **45.** 255

47.

Function	As $x \to -\infty$	As $x \to +\infty$
$f(x) = -5x^3$	$f(x) \to +\infty$	$f(x) \to -\infty$
$f(x) = -x^3 + 1$	$f(x) \to +\infty$	$f(x) \to -\infty$
$f(x) = 2x - 3x^3$	$f(x) \to +\infty$	$f(x) \to -\infty$
$f(x) = 2x^2 - x^3$	$f(x) \to +\infty$	$f(x) \to -\infty$

49. C   **51.** B
53. $f(x) \to -\infty$ as $x \to -\infty$ and $f(x) \to -\infty$ as $x \to +\infty$
55. $f(x) \to -\infty$ as $x \to -\infty$ and $f(x) \to +\infty$ as $x \to +\infty$
57. $f(x) \to -\infty$ as $x \to -\infty$ and $f(x) \to -\infty$ as $x \to +\infty$
59. $f(x) \to +\infty$ as $x \to -\infty$ and $f(x) \to -\infty$ as $x \to +\infty$
61. $f(x) \to +\infty$ as $x \to -\infty$ and $f(x) \to +\infty$ as $x \to +\infty$
63. $f(x) \to +\infty$ as $x \to -\infty$ and $f(x) \to -\infty$ as $x \to +\infty$

65.
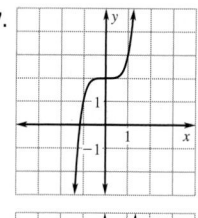

67.
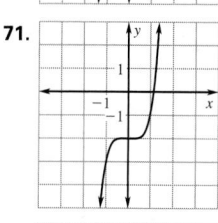

69.
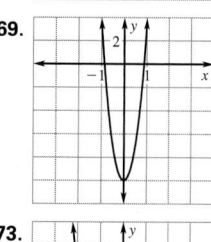

71.
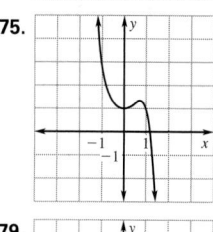

73.
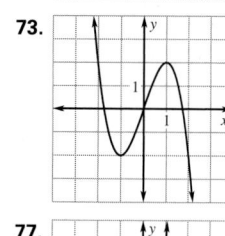

75.
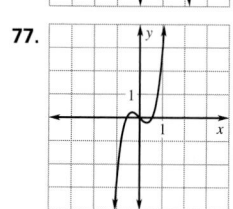

77.
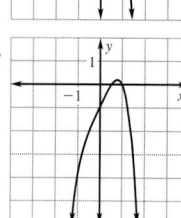

79.

81. about 4272.9 million ft^2 **83.** $f(x) \to -\infty$ as $x \to -\infty$ and
$f(x) \to -\infty$ as $x \to +\infty$; less; the graph will tend to go down
over time.   **85.** $f(x) \to +\infty$ as $x \to -\infty$ and $f(x) \to +\infty$ as
$x \to +\infty$; more; the graph will tend to go up over time.

6.2 MIXED REVIEW (p. 336)   **91.** $7x$ **93.** $x^2 + 4x - 11$
95. $-x^2 - x + 2$ **97.** $y = -2x^2 - 2x + 60$ **99.** $y = 4x^2 -$
$24x + 12$ **101.** $y = -3x^2 + 30x - 72$ **103.** $\pm\sqrt{5}\,i$ **105.** $\pm\sqrt{3}\,i$
107. $\pm\dfrac{\sqrt{6}}{6}i$ **109.** $\pm\dfrac{\sqrt{10}}{2}i$

TECHNOLOGY ACTIVITY 6.2 (p. 337) **1–7.** Ranges may vary.
1. $-10 \le x \le 10$, $-10 \le y \le 100$ **3.** $-5 \le x \le 5$, $-5 \le y \le 10$
5. $-5 \le x \le 5$, $0 \le y \le 20$ **7.** $0 \le x \le 16$, $0 \le y \le 300{,}000$

6.3 PRACTICE (pp. 341–343)   **5.** $2x^3 - 5x^2 - 3x + 6$
7. $-2x^2 + 4x - 2$ **9.** $4x^4 + 10x^3 + 27x^2 - 41x - 70$
11. $-27x^3 + 27x^2 - 9x + 1$ **13.** $11x^2 - 1$ **15.** $-7x + 7$
17. $-8x^3 - 4x^2 + x - 4$ **19.** $4x^2 - 6x - 21$ **21.** $-7x^3 - x^2 +$
$2x - 11$ **23.** $9x^3 - 3x^2 + 3x - 1$ **25.** $x^3 + 7x^2 + 8x + 14$
27. $x^3 + 6x^2 - 7x$ **29.** $-4x^3 + 32x^2 - 12x$ **31.** $x^2 - 11x + 28$
33. $x^3 - x^2 - 3x + 27$ **35.** $6x^4 + 13x^3 - 3x^2 + 5x$
37. $x^3 + 6x^2 - 46x + 99$ **39.** $x^4 + x^3 - 2x^2 + 2x - 2$
41. $3x^4 + 12x^3 + 7x^2 - 8x - 6$ **43.** $2x^4 + x^3 + 8x^2 - 3x + 4$
45. $x^3 - 67x + 126$ **47.** $-x^3 - 11x^2 - 23x + 35$
49. $3x^3 - 31x^2 + 32x + 36$ **51.** $6x^3 + 29x^2 + 21x + 4$
53. $x^2 - 49$ **55.** $64x^3 - 144x^2 + 108x - 27$ **57.** $x^4 - 12x^2 +$
36 **59.** $27x^3 + 189x^2 + 441x + 343$ **61.** $8x^3 + 36x^2y +$
$54xy^2 + 27y^3$ **63.** $V = 2x^3 + 5x^2 + 3x$

SELECTED ANSWERS

65. $y = -0.8246t^4 + 27.57t^3 - 268.42t^2 + 2797t + 219,260$; about 252 million people **67.** $W = -0.0004128t^5 - 0.03414t^4 + 1.3539t^3 - 12.8387t^2 + 51.9t + 833$; about 1,086,000 degrees **69.** $4000(1 + r)^3 + 5000(1 + r)^2 + 7000(1 + r)$; $10,000r^3 + 43,000r^2 + 72,000r + 39,000$

6.3 MIXED REVIEW (p. 344) **73.** ± 3 **75.** -8 **77.** $-\frac{3}{2}, 5$ **79.** $y = -\frac{6}{5}x^2 - \frac{12}{5}x + \frac{48}{5}$ **81.** $y = \frac{1}{3}x^2 - 12$ **83.** x^3 **85.** $-\frac{1}{25}$ **87.** $\frac{1}{2}x^4y^{11}$

QUIZ 1 (p. 344) **1.** $\frac{1}{125}$ **2.** $\frac{81}{16}$ **3.** $\frac{25}{81}$ **4.** $\frac{1}{16}$ **5.** 1 **6.** $\frac{1}{648}$ **7.** $\frac{1}{25}$ **8.** $\frac{1}{9x^6y^{12}}$ **9.** $\frac{x^7}{y^3}$ **10.** $\frac{x^3}{y}$ **11.** $\frac{y^6}{8x^3}$ **12.** $\frac{x^7}{y^7}$

13. 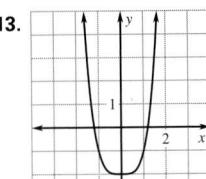 **14.**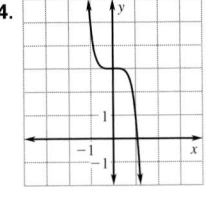

15. 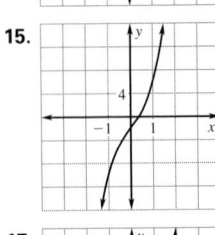 **16.**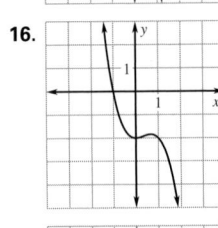

17. 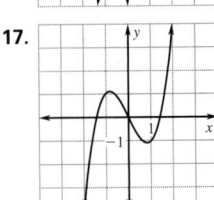 **18.**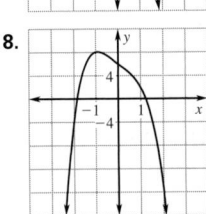

19. $7x^3 + 3x^2 + 7x - 3$ **20.** $3x^2 + 3x - 11$ **21.** $2x^2 + 18x - 2$ **22.** $x^3 + 3x^2 + 2x - 6$ **23.** $4x^3 + 19x^2 - 6x - 5$ **24.** $2x^3 + 3x^2 - 17x - 30$ **25.** $x^3 - 18x^2 + 108x - 216$ **26.** $4x^4 + 12x^2 + 9$ **27.** about 1.98×10^4 hours (about 825 days)

6.4 PRACTICE (pp. 348–350) **5.** $(x^2 + 5)(x^4 - 5x^2 + 25)$ **7.** $(x + 1)(x - 1)(x^2 + 1)$ **9.** $5(x - 4)(x^2 + 4x + 16)$ **11.** 3 **13.** $-2, \pm 3$ **15.** $\pm \frac{\sqrt{6}}{3}$ **17.** 1998 **19.** $3x^3$ **21.** $6x$ **23.** 1 **25.** $3x^3$ **27.** C **29.** F **31.** E **33.** $(x - 2)(x^2 + 2x + 4)$ **35.** $(6x + 1)(36x^2 - 6x + 1)$ **37.** $(10x + 3)(100x^2 - 30x + 9)$ **39.** $4(2x - 1)(4x^2 + 2x + 1)$ **41.** $(x + 1)(x^2 + 1)$ **43.** $(x + 3)(x^2 + 10)$ **45.** $(2x - 5)(x^2 + 9)$ **47.** $(x - 2)(3x^2 + 1)$ **49.** $(3x - 2)(x^2 - 3)$ **51.** $(x^2 + 1)(x^2 + 2)$ **53.** $(3x - 4)(3x + 4)(9x^2 + 16)$ **55.** $(x^2 + 2)(x^2 + 8)$ **57.** $2x^2(2x - 1)(2x + 1)(4x^2 + 1)$ **59.** $(2x^2 + 3)(9x - 1)$ **61.** $(2x + 1)(2x - 1)(x^2 + 10)$ **63.** $8(x - 2)(x^2 + 2x + 4)$ **65.** $3x(x - 2)(x^2 + 2x + 4)$ **67.** $x(3x^2 + 1)(x + 3)$ **69.** $0, 3$ **71.** -3 **73.** $-7, 2$ **75.** $0, \pm 3$ **77.** $\frac{1}{2}$ **79.** 5 **81.** ± 1 **83.** none **85.** $0, \pm 2, \pm \sqrt{2}$ **87.** about 3.16 in. by 1.16 in. by 8.16 in.

89. 6 ft by 3 ft by 1 ft **91.** base: 5 ft by 5 ft, height: 30 ft

6.4 MIXED REVIEW (p. 351) **99.** $\frac{y^{11}}{6}$ **101.** y^4 **103.** 481

6.5 PRACTICE (pp. 356–358) **5.** $x^2 + x - 4 + \frac{14}{x + 4}$ **7.** $-x + 2 + \frac{-3x + 5}{x^2 - 1}$ **9.** $x^3 - 4x^2 + 1$ **11.** $x + 9 + \frac{16}{x - 2}$ **13.** $-2, -1$ **15.** $x + 9 + \frac{13}{x - 2}$ **17.** $2x - 5 + \frac{19}{x + 4}$ **19.** $x + 15 + \frac{147}{x - 10}$ **21.** $2x^2 + 2 + \frac{9}{x^2 - 1}$ **23.** $3x - 4 + \frac{5}{2x + 3}$ **25.** $5x^2 - x + 3$ **27.** $x^2 + 2x - 3 - \frac{12}{x - 2}$ **29.** $4x + 1 - \frac{5}{x + 1}$ **31.** $2x + 11 + \frac{30}{x - 2}$ **33.** $x - 4 + \frac{26}{x + 4}$ **35.** $10x^3 - 5x^2 + 9x - 9$ **37.** $2x^3 + x - \frac{3}{x - 3}$ **39.** $(x + 2)(x - 3)(x - 4)$ **41.** $(x - 10)(x - 4)(x + 2)$ **43.** $(x + 5)(x - 3)^2$ **45.** $(x - 1)(2x + 3)(2x - 3)$ **47.** $-\frac{1}{9}, 1$ **49.** $-5, -\frac{1}{2}$ **51.** $\frac{5 \pm \sqrt{17}}{2}$ **53.** $1 \pm i\sqrt{7}$ **55.** $3x - 10$ **57.** $(-2, 6), (-1, 5), (1, -3)$ **59.** $5x^3 - 3x^2 + 21x - 8$; I multiplied $5x^2 - 13x + 47$ by $x + 2$ and added -102. **61.** Answers may vary depending on rounding. $C = 0.0031x^2 + 0.1578x + 11.155 + \frac{6398}{8.4x - 580}$; about 144 million cars

6.5 MIXED REVIEW (p. 358) **67.** Both are solutions. **69.** $(1, 4)$ is a solution, but $(2, 0)$ is not a solution. **71.** $4 \pm \sqrt{13}$ **73.** $\frac{7 \pm \sqrt{33}}{8}$ **75.** $\frac{-1 \pm \sqrt{41}}{10}$ **77.** $\frac{-1 \pm i\sqrt{159}}{10}$ **79.** $-4x + 9$ **81.** $-14x^3 - 2x^2 + x + 4$ **83.** 82 guests

6.6 PRACTICE (pp. 362–364) **5.** $\pm 1, \pm 2, \pm 3, \pm 4, \pm 6, \pm 8, \pm 9, \pm 12, \pm 18, \pm 24, \pm 36, \pm 72$ **7.** $\pm 1, \pm 2, \pm 5, \pm 10, \pm \frac{1}{5}, \pm \frac{2}{5}$ **9.** $-4, -1, 1$ **11.** $-3, \frac{3}{2}, 2$ **13.** $-5, -1, 1$ **15.** $\pm 1, \pm 2, \pm 3, \pm 4, \pm 6, \pm 8, \pm 12, \pm 24$ **17.** $\pm \frac{1}{2}, \pm 1, \pm 2, \pm 4, \pm 8, \pm 16$ **19.** $\pm 1, \pm 2, \pm 5, \pm 10, \pm \frac{1}{2}, \pm \frac{5}{2}, \pm \frac{1}{3}, \pm \frac{2}{3}, \pm \frac{5}{3}, \pm \frac{10}{3}, \pm \frac{1}{6}, \pm \frac{5}{6}$ **21.** $\pm 1, \pm 3, \pm \frac{1}{2}, \pm \frac{3}{2}, \pm \frac{1}{4}, \pm \frac{3}{4}, \pm \frac{1}{8}, \pm \frac{3}{8}$ **23.** $-2, 2$ **25.** $-2, -1$ **27.** $-1, 1$ **29.** none **31.** $-2, -1, 1, 2$ **33.** $-3, 1, 10$ **35.** $-2, 4, 5$ **37.** $-4, 3, 6$ **39.** $-1, 2$ **41.** $-3, -2, 1, 3$ **43.** $-3, -2, 3$ **45.** $-1, \frac{3}{2}, \frac{5}{2}$ **47.** $-2, -1, 1$ **49.** $-1, \frac{3}{2}, 2$ **51.** $-4, \frac{1}{2}, 4$ **53.** $-\frac{5}{2}, 1$ **55.** $-1, 1$ **57.** $-2, -\frac{1}{2}, 2$ **59.** 1993 **61.** 2 in. by 2 in. by 5 in. **63.** 5 ft deep, 10 ft wide, 40 ft long

6.6 MIXED REVIEW (p. 365) **71.** 3 **73.** 1 **75.** 10 **77.** $y = -\frac{5}{9}(x + 3)(x - 3)$ **79.** $y = -2(x + 1)(x - 5)$ **81.** $y = -\frac{1}{63}(x + 12)(x + 6)$ **83.** $y = -\frac{1}{3}(x - 4)(x - 10)$ **85.** $y = (x + 1)(x + 9)$

QUIZ 2 (p. 365) **1.** $5(x + 3)(x^2 - 3x + 9)$ **2.** $6(x + 2)(x^2 + 2)$ **3.** $4x(x^2 + 2)(x^2 - 2)$ **4.** $(x^2 - 5)(3x - 1)$ **5.** $0, \pm 6$ **6.** $0, \frac{3}{2}$ **7.** $0, 3$ **8.** $-\frac{5}{2}, -2, 2$ **9.** $x + 11$ **10.** $x - \frac{10}{3} + \frac{80}{3(3x + 2)}$

11. $4x - 7 + \dfrac{11x - 11}{x^2 - 3}$ **12.** $12x^3 - 7x^2 + 10x - 10 + \dfrac{5}{x + 1}$

13. $x + \dfrac{2x^2 + 6x + 6}{x^3 - 3}$ **14.** $5x^3 - 23x^2 + 115x - 576 + \dfrac{2875}{x + 5}$

15. $\pm\sqrt{7}, 4$ **16.** 2 **17.** $-5, -3, \dfrac{1}{2}$ **18.** $-6, \dfrac{1}{2}, 2$

19. 16 ft by 16 ft by 0.5 ft

6.7 PRACTICE (pp. 369–371) **5.** $\pm\sqrt{3}, \pm2i$ **7.** $-1, 2, \pm2i$
9. $f(x) = x^4 - 2x^3 + 2x^2 - 2x + 1$ **11.** $f(x) = x^5 - 3x^4 - 5x^3 + 15x^2 + 4x - 12$ **13.** $f(x) = x^4 + 32x^2 + 256$ **15.** yes
17. no **19.** yes **21.** $-3, -2, -1, 1$ **23.** $0, 1, 3$ **25.** $-5, -4, -1, 3$ **27.** $1, \pm7i$ **29.** $-5, -1, \pm3i$ **31.** $-2, 3, \pm i$ **33.** $-3, -1, 3, 4.5$ **35.** $f(x) = x^3 - 7x^2 + 14x - 8$ **37.** $f(x) = x^3 - 2x^2 - 33x + 90$ **39.** $f(x) = x^3 + 13x^2 + 50x + 56$ **41.** $f(x) = x^3 - 5x^2 + 9x - 45$ **43.** $f(x) = x^4 + 10x^2 + 9$ **45.** $f(x) = x^4 - 12x^3 + 53x^2 - 104x + 80$ **47.** $-2.09, 0.57, 2.51$ **49.** -0.47
51. $-1.27, 2.86$ **53.** $-0.75, 0.75$ **55.** 1988 **57.** Yes; there were 2 such years, 1988 and 1993, because the graph intersects the line $S = 2000$ when t is about 1.6 and when t is about 6.3. **59.** 1965

6.7 MIXED REVIEW (p. 371)
65. 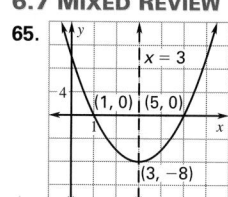 **67.**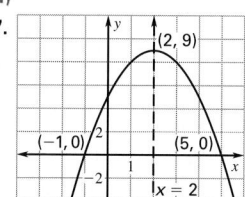

69. 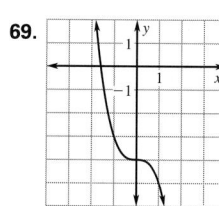 **71.**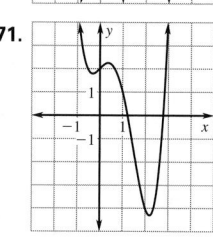

TECHNOLOGY ACTIVITY 6.7 (p. 372) **1.** $-0.640, 1.135, 5.505$ **3.** 5 **5.** $-2.334, -0.742, 0.742, 2.334$
7. $-1.088, -0.668, 1.191$ **9.** $-7.349, 16.429, 30.921$; yes

6.8 PRACTICE (pp. 376–378)
5. 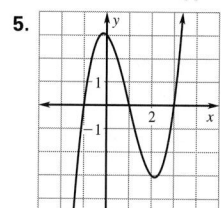 **7.**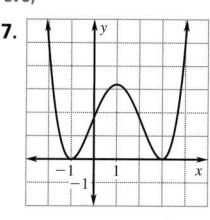

9. x-intercepts: $-0.41, 1, 2.41$; local maximum: $(0.18, 1.09)$; local minimum: $(1.82, -1.09)$ **11.** x-intercepts: $0, 1, 1.51$; local maximums: $(-1.59, -3.23), (0.49, 1.35)$; local minimums: $(-1, -4), (1.30, -0.79)$

13. 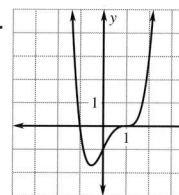 **15.**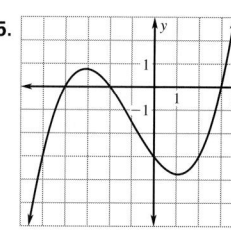

17. 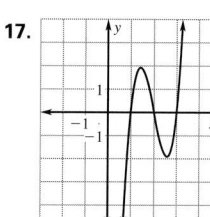 **19.**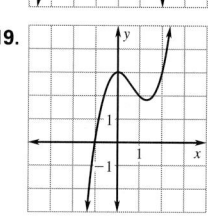

21. 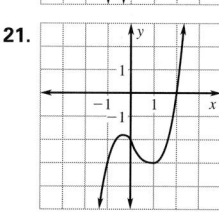 **23.** $(-0.5, 0.5)$ max, $(0.5, -0.3)$ min; $-0.9, 0, 0.6$; 3
25. $(-2, 1)$ min, $(0, 2)$ max; 1.4; 3
27. $(-2, -1)$ max, $(0, -2.2)$ min, $(1, -2)$ max; none; 4

29. x-intercepts: $-1.79, 0.11, 1.67$; local maximum: $(-1, 7)$; local minimum: $(1, -5)$ **31.** x-intercepts: $-2.83, 0, 2.83$; local maximums: $(-2, 4), (2, 4)$; local minimum: $(0, 0)$
33. x-intercepts: $-2, -1, 0, 1, 2$; local maximums: $(-1.64, 3.63), (0.54, 1.42)$; local minimums: $(-0.54, -1.42), (1.64, -3.63)$

35. 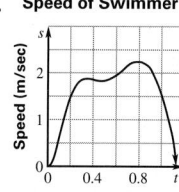 ; at about $t = 0.8$ sec into the stroke

37. $l = \dfrac{600 - \pi r^2}{\pi r}$

39. 1600 ft³; $r \approx 7.98$ ft, $l \approx 15.97$ ft, or about 16 ft long, 16 ft wide, and 8 ft high

6.8 MIXED REVIEW (p. 378) **45.** $y = 7x$ **47.** $y = \dfrac{1}{4}x$
49. $y = -\dfrac{3}{5}x$ **51.** yes; 4×1 **53.** no **55.** $y = -(x - 1)^2 + 4$
57. $y = \dfrac{5}{24}(x + 5)(x - 5)$ **59.** 10 in./day

6.9 PRACTICE (pp. 383–385)
5.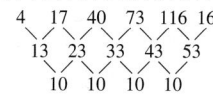

$f(1)$	$f(2)$	$f(3)$	$f(4)$	$f(5)$	$f(6)$	
4	17	40	73	116	169	Values
	13	23	33	43	53	First-order differences
		10	10	10	10	Second-order differences

7.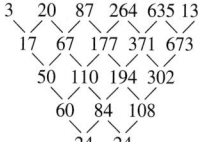

$f(1)$	$f(2)$	$f(3)$	$f(4)$	$f(5)$	$f(6)$	
3	20	87	264	635	1308	Values
	17	67	177	371	673	First-order differences
		50	110	194	302	Second-order differences
			60	84	108	Third-order differences
				24	24	Fourth-order differences

9. 3 **11.** $f(x) = -x^3 + 5x^2 + x + 1$ **13.** $d(n) = \dfrac{1}{2}n^2 - \dfrac{3}{2}n$

15. $f(x) = -\dfrac{1}{2}(x + 1)(x - 2)(x - 3)$ **17.** $f(x) = -\dfrac{1}{2}x(x + 1)(x + 2)$

19. $f(x) = -\frac{1}{4}(x-1)(x-3)(x+2)$

21. $f(x) = (x-3)(x-2)(x+1)$

23.

$f(1)$	$f(2)$	$f(3)$	$f(4)$	$f(5)$	$f(6)$	
5	5	7	11	17	25	Values
	0	2	4	6	8	First-order differences
		2	2	2	2	Second-order differences

25.

$f(1)$	$f(2)$	$f(3)$	$f(4)$	$f(5)$	$f(6)$	
-3	-3	-9	-27	-63	-123	Values
	0	-6	-18	-36	-60	First-order differences
		-6	-12	-18	-24	Second-order differences
			-6	-6	-6	Third-order differences

27.

$f(1)$	$f(2)$	$f(3)$	$f(4)$	$f(5)$	$f(6)$	
-18	-8	102	432	1150	2472	Values
	10	110	330	718	1322	First-order differences
		100	220	388	604	Second-order differences
			120	168	216	Third-order differences
				48	48	Fourth-order differences

29.

$f(1)$	$f(2)$	$f(3)$	$f(4)$	$f(5)$	$f(6)$	
4	4	-36	-176	-500	-1116	Values
	0	-40	-140	-324	-616	First-order differences
		-40	-100	-184	-292	Second-order differences
			-60	-84	-108	Third-order differences
				-24	-24	Fourth-order differences

31.

$f(1)$	$f(2)$	$f(3)$	$f(4)$	$f(5)$	$f(6)$	
3	-2	-13	-30	-53	-82	Values
	-5	-11	-17	-23	-29	First-order differences
		-6	-6	-6	-6	Second-order differences

33. $f(x) = -3x^2 + 20x$ **35.** $f(x) = x^3 - 4x^2 + x$
37. $f(x) = x^3 + 4x^2 - x - 2$ **39.** $y = 2x^3 - 16x^2 + 37x - 25$
41. $f(x) = -x^3 + 10x^2 - 30x + 23$ **43.** $f(x) = -x^4 + 13x^3 - 58x^2 + 104x - 58$ **47.** $f(t) = 0.641t^3 - 4.93t^2 + 25.8t + 232$
where t is the number of years since 1989; 772,000 Girl Scouts
49. $y = 0.007t^3 - 0.740t^2 + 49.0t - 236$; about 101 sec

6.9 MIXED REVIEW (p. 386) **53.** $\pm\frac{1}{2}$ **55.** $\pm\frac{\sqrt{78}}{6}$ **57.** $\pm\frac{\sqrt{2}}{2}$

59. $-3 \pm \sqrt{33}$ **61.** $-2 \pm \frac{i\sqrt{6}}{2}$ **63.** $3 \pm \frac{i\sqrt{15}}{3}$

65. $(3x+2)(9x^2 - 6x + 4)$ **67.** $(2x-5)(4x^2 + 10x + 25)$
69. $8(x+3)(x^2 - 3x + 9)$ **71.** $3(x+3)(x^2 - 3x + 9)$

QUIZ 3 (p. 386) **1.** $-2.61, -0.74, 3.86$ **2.** $-2, \frac{-1 \pm i\sqrt{3}}{2}$

3. $-1, 4, \pm i\sqrt{2}$ **4.** $-\frac{3}{2}, -1, 1, 2$ **5.** $y = x^3 + 2x^2 - 4x - 8$
6. $y = x^3 + 2x^2 - 3x$ **7.** $y = x^3 - 8x^2 + 21x - 20$
8. $y = x^4 - 7x^3 + 11x^2 - 7x + 10$ **9.** $y = x^3 - 8x^2 + 29x - 52$
10. $y = x^4 - 6x^3 + 18x^2 - 24x + 16$
11. local max $(0.79, 8.21)$, local min $(-2.12, -4.06)$
12. local max $(-0.50, 0.56)$, local min $(-1.62, -1)$, $(0.62, -1)$
13. local max $(2.42, 0.77)$, local min $(3.58, -0.77)$
14. local max $(-3, 0)$, local min $(-1.67, -1.19)$ **15.** $f(x) = -\frac{1}{3}(x+2)(x+4)(x-2)$ **16.** $f(x) = -\frac{1}{70}(x+1)(x-4)(x-2)$
17. $f(x) = x(x-3)(x-5)$ **18.** $f(x) = 2(x-1)(x+3)(x+5)$

19. $f(x) = x^3 - 3x^2 + x - 4$ **20.** $f(x) = x^3 - 4x^2 + 2x$
21. $N = -3.75x^3 + 50.9x^2 - 97.3x + 3210$ where x is the number of years since 1988

CHAPTER 6 REVIEW (pp. 388–390)
1. $\frac{96x^3}{y^3}$; negative exponent, power of a quotient, power of a product, and power of a power property

3. $-\frac{7}{2}x^3y^6$; quotient of powers property **5.** 25

7. 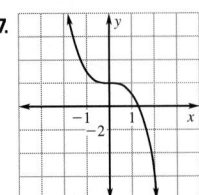 **9.**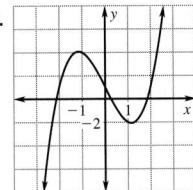

11. $x^3 - 2x^2 - 10x + 21$ **13.** -4 **15.** $-3, -1, 1$
17. $x^2 + \frac{5}{2} + \frac{33}{2(2x-5)}$ **19.** $-2, 1$

21. 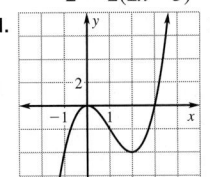 ; x-intercepts: 0, 3; local max: $(0, 0)$; local min: $(2, -4)$

23.

$f(1)$	$f(2)$	$f(3)$	$f(4)$	$f(5)$	$f(6)$	
2	9	28	65	126	217	Values
	7	19	37	61	91	First-order differences
		12	18	24	30	Second-order differences
			6	6	6	Third-order differences

CUMULATIVE PRACTICE (pp. 394–395) **1.** -5 **3.** $-4, 8$
5. $x < 3$; **7.** $-2 < x < 8$;

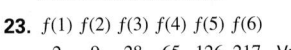
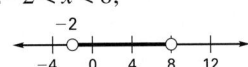

9. 0 **11.** 4

13. 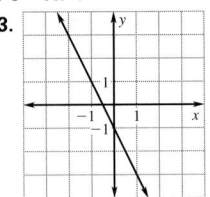 **15.**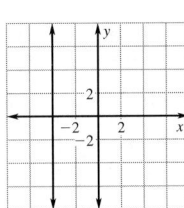

17. 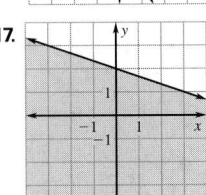 **19.**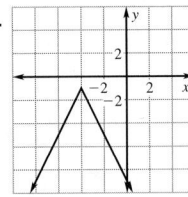

SA24

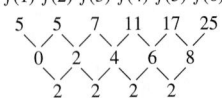

21.
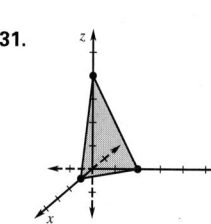

23. $y = -4x + 5$ **25.** $(3, 5)$
27. $(1, 0, 3)$

31.
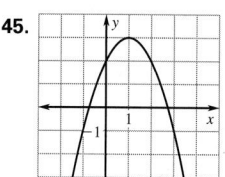

33. $\begin{bmatrix} -11 & 8 \\ 1 & -2 \end{bmatrix}$ **35.** $\begin{bmatrix} 17 & -7 & -27 \\ 3 & -9 & 69 \end{bmatrix}$

37. 3 **39.** -55 **41.** $\begin{bmatrix} 7 & 2 \\ -4 & -1 \end{bmatrix}$

43. no inverse

45.
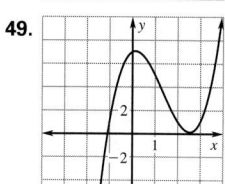

47.
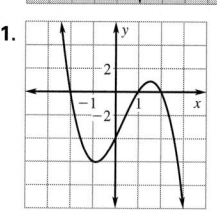

49.
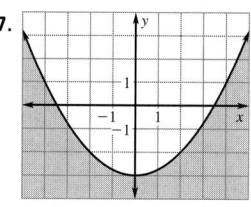

51.

53. $\pm\sqrt{13}$ **55.** $\pm 8i$ **57.** $-10 \le x \le 10$ **59.** $\pm 2, \pm 1$

61. $-2, \pm\dfrac{\sqrt{6}}{2}$ **63.** $32 + 20i$ **65.** $9 + 2i$ **67.** $y = -(x+3)(x+2)$

69. $36x^2y^6$ **71.** $\dfrac{16}{25}$ **73.** $x^4 - 5x^3 + 11x^2 - 27x + 36$

75. $x^3 - 5x^2 + 18x - 36 + \dfrac{70}{x+2}$ **77.** $\pm\sqrt{5}, \pm\sqrt{5}\,i$

79. $f(x) = (x+4)(x+1)(x-1)$ **81.** $r = \dfrac{I}{Pt}$, 5.5%

83. 8 min **85.** about 5.45 h

CHAPTER 7

SKILL REVIEW (p. 400) **1.** $y = \dfrac{3x - 12}{2}$ **2.** $y = 10 - 2x$

3. $y = \dfrac{x+1}{4}$ **4.** $(x+7)(x+3)$ **5.** $(x+9)(x-4)$

6. $2(x-3)(x-5)$ **7.** $a^4b^4c^8$ **8.** x^2 **9.** $\dfrac{x^4}{y^2}$ **10.** $\dfrac{3x^3}{4y^6}$

11. $5x^3 - 40x^2$ **12.** $9y^2 - 12y + 4$ **13.** $7x^2 - 5x + 4$

7.1 PRACTICE (pp. 404–406) **5.** -7 **7.** 25 **9.** -1 **11.** ± 10

13. $14^{1/4}$ **15.** $5^{2/7}$ **17.** $2^{11/8}$ **19.** $\sqrt[4]{7}$ **21.** $\left(\sqrt[5]{5}\right)^2$ **23.** ± 10

25. -2 **27.** none **29.** 4 **31.** -2 **33.** 1 **35.** 4 **37.** 0

39. 16 **41.** -7 **43.** 4 **45.** 0.56 **47.** 0.0019 **49.** 1.82

51. 0.087 **53.** 3 **55.** 0 **57.** -1.69 **59.** -9.24 **61.** ± 1.40

63. 1247.73 ft^3/sec **65.** 1.58 ft **67.** about 37 species

7.1 MIXED REVIEW (p. 406) **73.** $x = 3, y = -4$ **75.** $x = \dfrac{16}{5}$,

$y = \dfrac{3}{10}$ **77.** $x = \dfrac{13}{11}, y = -\dfrac{13}{11}$ **79.** $\dfrac{1}{x^{15}}$; power of a power and

negative exponent properties **81.** $\dfrac{5}{x^2}$; negative exponent
and zero exponent properties

83. $\dfrac{1}{x^4y^2}$; negative exponent and power of a quotient

properties **85.** $4x^2y$; product of powers and quotient of
powers properties **87.** $-1, 2, 3, -5$ **89.** $1, \pm 3i$

7.2 PRACTICE (pp. 411–413) **5.** 3 **7.** 4 **9.** $\dfrac{2}{3}$ **11.** $3\sqrt[7]{8}$

13. x^2 **15.** $2a^3$ **17.** $\dfrac{x^2}{y}$ **19.** $-4a^{1/5}$ **21.** 1333.78 cm^2

23. $5^{1/3}$ **25.** 6 **27.** $5^{1/3}$ **29.** $\dfrac{8}{5}$ **31.** $5^{3/4}$ **33.** $\dfrac{1}{64{,}000}$

35. 2 **37.** $6^7 = 279{,}936$ **39.** $\dfrac{1}{2}$ **41.** 3 **43.** $3\sqrt[5]{5}$ **45.** $30\sqrt[3]{3}$

47. $\dfrac{2\sqrt[3]{3}}{3}$ **49.** $\sqrt[15]{2}$ **51.** $-2\sqrt[7]{5}$ **53.** $3\sqrt{10}$ **55.** $9\sqrt[4]{11}$

57. $y^{1/2}$ **59.** $x^{5/4}$ **61.** $\dfrac{x^3}{y}$ **63.** $y^{5/3}$ **65.** $\dfrac{x^{1/2}y}{z}$ **67.** $\dfrac{1}{3y^2}$

69. $xy^2z^2\sqrt[4]{10xz^2}$ **71.** $y^2z^2\sqrt{2xz}$ **73.** $\dfrac{x\sqrt[3]{y}}{y}$ **75.** $x^{1/35}$

77. $7x^{1/5}$ **79.** $2x^3y^{1/3}$ **81.** $(2x-1)y\sqrt[3]{3x^2}$ **83.** y^2

85. $\dfrac{1}{4^{\sqrt{7}}}$ **87.** $\dfrac{x}{y^2}$ **89.** $-2xy^{\sqrt{11}}$ **91.** $\dfrac{\sqrt{3}}{2}$ **93.** 0.45 mm

95. Higher notes have frequencies twice as high as lower
notes of the same letter. **97.** $2^{2/3}$

7.2 MIXED REVIEW (p. 414) **101.** $\dfrac{441}{4}, \left(x - \dfrac{21}{2}\right)^2$

103. 24.5025, $(x + 4.95)^2$ **105.** $\dfrac{1}{64}, \left(x - \dfrac{1}{8}\right)^2$

107. $8x^3 + 9x^2 + 52x + 1$ **109.** $4x^2 + 28x + 49$

111. $(4x - 1) - \dfrac{2}{x+1}$ **113.** $x^3 + 3x^2 + 15x + 5 + \dfrac{45}{x-5}$

QUIZ 1 (p. 414) **1.** 4 **2.** $\dfrac{1}{8}$ **3.** -3 **4.** 16 **5.** 1.58 **6.** ± 1.12

7. ± 1.90 **8.** -4.47 **9.** $4^{1/4}$ or $2^{1/2}$ **10.** $\dfrac{2\sqrt[4]{27}}{3}$ **11.** 4

12. $3\sqrt{5}$ **13.** 7 **14.** $3\sqrt[5]{8}$ **15.** $x^{11/12}$ **16.** $x^{1/2}$ **17.** $x^{1/4}y^{5/2}$

18. $xy\sqrt[3]{5y^2}$ **19.** $\dfrac{6\sqrt{xy}}{y^2}$ **20.** $2xy^{1/2}$ **21.** about 30,000

horsepower **22.** No; The surface area of the Labrador
retriever is about 2.08 times the surface area of the
Scottish terrier.

7.3 PRACTICE (pp. 418–420) **5.** $5x - 1$; all real numbers
7. $4x^2 - 4x$; all real numbers **9.** $4x - 4$; all real numbers
11. $g(f(x))$; The bonus is 0.02 times the amount over
\$200,000 $(x - 200{,}000)$, so calculate amount first and then
take 2%. **13.** $2x^2 - 5x + 4$; all real numbers **15.** $2x^2 - 8$;
all real numbers **17.** $5x - 12$; all real numbers **19.** 0; all
real numbers **21.** $6x^{7/6}$; nonnegative real numbers

23. $9x$; nonnegative real numbers **25.** $\dfrac{3}{2x^{1/6}}$; positive real

numbers **27.** 1; positive real numbers **29.** $2^{3/2}x^{-15/4}$;
positive real numbers **31.** $x^{9/16}$; nonnegative real numbers

33. $9x - 4$; all real numbers **35.** $\dfrac{10x}{x+4}$; all real numbers

SELECTED ANSWERS

except -4 **37.** $10x + 4$; all real numbers   **39.** $x + 8$; all real numbers   **41.** $x^{1/2}$; nonnegative real numbers
43. $x^2 - x - 8$; all real numbers   **45.** $4x^3 - 16x^2$; all real numbers   **47.** $x - 5$; all real numbers except 0
49. $x^4 - 6x^2 + 10$; all real numbers
51. $81x - 20$; all real numbers   **53.** $r(w) = 220w^{-0.266}$; about 134 breaths per minute; about 18 breaths per minute; about 11 breaths per minute

7.3 MIXED REVIEW (p. 420)
69. $y = \dfrac{-2x - 8}{3}$ **71.** $y = \dfrac{5}{x}$ **73.** $y = \dfrac{c - ax}{b}$

75. 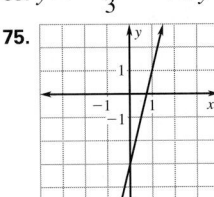   **77.**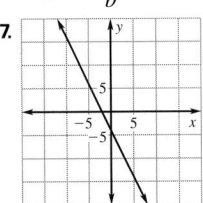

79. 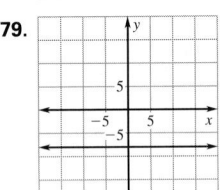   **81.** 3   **83.** $-6, -2$

7.4 PRACTICE (pp. 426–428)
5.

x	2	1	0	1	2
y	-4	-2	0	2	4

7. $y = \dfrac{x + 1}{2}$
9. Both compositions equal x.

11. $\dfrac{\sqrt[4]{27x}}{3}$ **13.** No; horizontal lines, such as $y = 0$, cross the graph more than once.   **15.**

x	0	3	-2	2	-1
y	1	-2	4	2	-2

17. $y = \dfrac{x + 3}{3}$

19. $y = -\dfrac{5}{4}(x - 11)$ **21.** $y = \dfrac{-x + 7}{12}$ **23.** $y = \dfrac{x + 13}{8}$
33. A   **35.** B   **37.** $f^{-1}(x) = \sqrt[6]{-x}$ **39.** $f^{-1}(x) = 2\sqrt[5]{x}$

41. $f^{-1}(x) = -\dfrac{2}{3}\sqrt{-x}$ **43.** $f^{-1}(x) = \sqrt[5]{-\dfrac{1}{2}x + \dfrac{1}{6}}$

45. $f^{-1}(x) = \sqrt[3]{\dfrac{5}{3}x + 15}$ **47.** $f^{-1}(x) = \sqrt[5]{6x - 4}$

49. 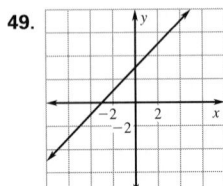 ;   **51.** 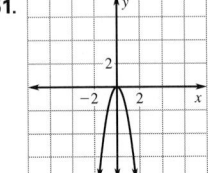 ;

Yes, inverse is a function.   No, inverse is not a function.

53. 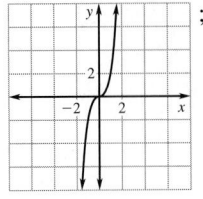 ;   **55.** 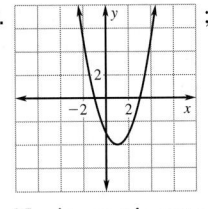 ;

Yes, inverse is a function.   No, inverse is not a function.
57. $D_{US} = 0.65677D_C$ **59.** $a = 200 - 1.11h$; 170
61. $l = \sqrt[3]{106723.59w}$; 41.69 cm

7.4 MIXED REVIEW (p. 429)
69. 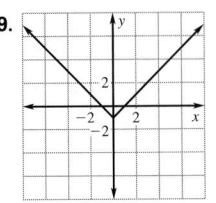   **71.**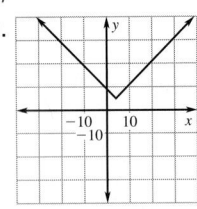

73. 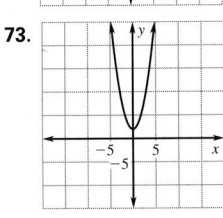   **75.**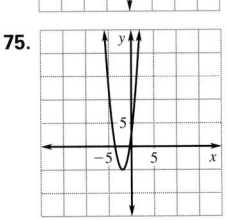

77. 2   **79.** $\dfrac{1}{5y}$ **81.** $5\sqrt[7]{5}$ **83.** $\$.65$

QUIZ 2 (p. 429)
1. $f(x) + g(x) = 6x^2 + x^{1/2}$; nonnegative real numbers
2. $f(x) - g(x) = 6x^2 - 3x^{1/2}$; nonnegative real numbers
3. $f(x) \cdot g(x) = 2x(6x^{3/2} - 1)$; nonnegative real numbers
4. $\dfrac{f(x)}{g(x)} = 3x^{3/2} - \dfrac{1}{2}$; positive real numbers
5. $f(g(x)) = \dfrac{3}{x - 8}$; real numbers except 8

6. $g(f(x)) = \dfrac{3}{x} - 8$; real numbers except 0   **7.** $f(f(x)) = x$; real numbers except 0   **8.** $g(g(x)) = x - 16$; all real numbers
9. Both compositions equal x.   **10.** Both compositions equal x.

11. $f^{-1}(x) = x - 8$ **12.** $f^{-1}(x) = \dfrac{-\sqrt[4]{8x}}{2}$ **13.** $f^{-1}(x) = \sqrt[5]{6 - x}$

14. 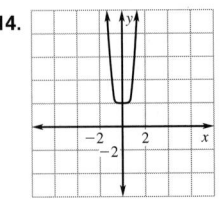 ;   **15.** 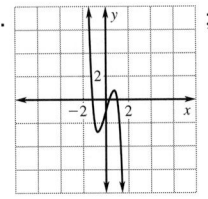 ;

No, inverse is not a function.   No, inverse is not a function.

SELECTED ANSWERS

16. 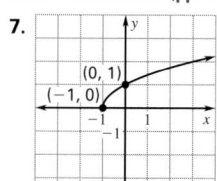 ; Yes, inverse is a function.

17. $A(t) = 0.36\pi t^2$; about 4.52 ft^2

35. 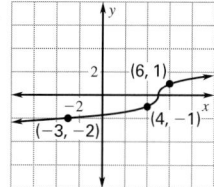 ;

37. 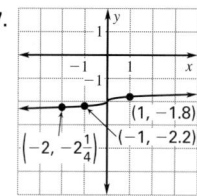 ;

x, y are all real numbers. x, y are all real numbers.

39. 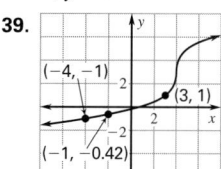 ; x, y are all real numbers.

47. 2.36 square units,   **49.** 80.15 nautical miles

TECHNOLOGY ACTIVITY 7.4 (p. 430) **1.** Yes; the inverse passes the vertical line test. **3.** Yes; the inverse passes the vertical line test. **5.** No; the inverse does not pass the vertical line test. **7.** Yes; the inverse passes the vertical line test. **9.** Yes; the inverse passes the vertical line test. **11.** No; the inverse does not pass the vertical line test.

7.5 PRACTICE (pp. 434–436) **5.** Shift the graph 10 units down.

7. 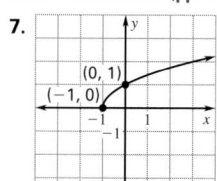 ; **9.** 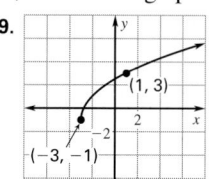 ;

$x \geq -1$, $y \geq 0$ $x \geq -3$, $y \geq -1$

11. 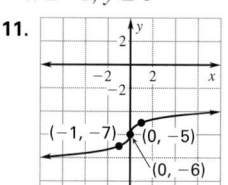 ; **13.** 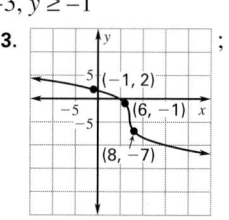 ;

x, y: all real numbers. x, y: all real numbers.

15. Shift graph 14 units left. **17.** Shift graph 10 units down.

19. B **21.** C

23. 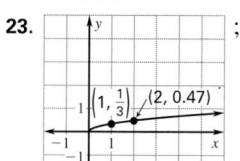 ; **25.** 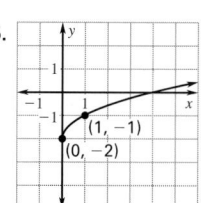 ;

$x \geq 0$, $y \geq 0$ $x \geq 0$, $y \geq -2$

27. 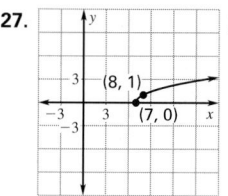 ; **29.** 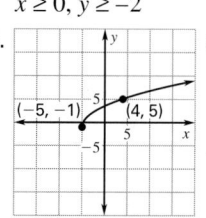 ;

$x \geq 7$, $y \geq 0$ $x \geq -5$, $y \geq -1$

31. 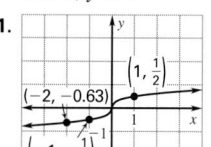 ; **33.** 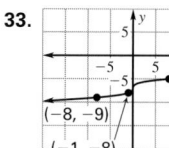 ;

x, y are all real numbers. x, y are all real numbers.

7.5 MIXED REVIEW (p. 436) **55.** $\pm\sqrt{10} - 7$ **57.** ± 6

59. $\dfrac{\pm\sqrt{33}}{2} + \dfrac{1}{4}$ **61.** $x^2 - 18xy + 81y^2$ **63.** $9x^2 - 24xy^4 +$ $16y^8$ **65.** $1 + 4x^2 + 4x^4$ **67.** $f(g(x)) = 2x - 5$; $g(f(x)) = 2x - 2$ **69.** $f(g(x)) = 9x^2 - 18x + 16$; $g(f(x)) = 3x^2 + 18$

7.6 PRACTICE (pp. 441–443) **5.** 1 **7.** 8 **9.** -5 **11.** $\dfrac{64}{3}$

13. 2, 3 **15.** no solution **17.** yes **19.** yes **21.** no **23.** 4

25. 27 **27.** 81 **29.** $\dfrac{11}{2}$ **31.** $\dfrac{406}{81}$ **33.** 216 **35.** 200

37. no solution **39.** $\dfrac{12}{7}$ **41.** 36 **43.** $-\dfrac{2}{3}$ **45.** 1, 3 **47.** 5

49. $-\dfrac{1}{6}$ **51.** no solution **53.** 5 **55.** -18.96296 **57.** 0.10345

59. 11.099 **61.** no solution **63.** 0.146 in. **65.** 1991

67. 34.078 mi/h **69.** 4.90

7.6 MIXED REVIEW (p. 444) **81.** 20 **83.** -78 **85.** 19

87. -0.95; no local maximums or minimums

89. 0, ±1.41; $(-0.914, 4.08)$; $(0.914, -4.08)$

7.7 PRACTICE (pp. 449–451) **5.** 31

7. 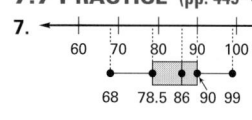 **9.**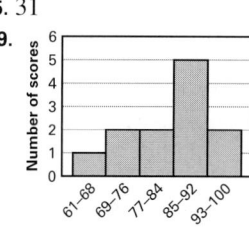

11. 49.57, 47, 47 **13.** about 249, 230, 230 **15.** 0.356; 0.3; 0, 0.5 (two modes) **17.** 8, 2.73 **19.** 417, 143

21. 12.1, 3.82 **23.**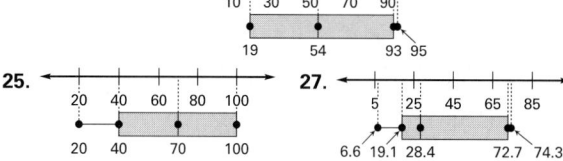

25. 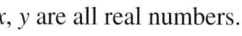 **27.**

29.

10–19	5
20–29	5
30–39	7
40–49	2
50–59	1

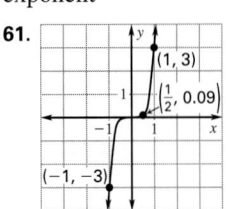

31.

0–0.4	0	4.0–4.4	1
0.5–0.9	0	4.5–4.9	1
1.0–1.4	0	5.0–5.4	2
1.5–1.9	0	5.5–5.9	2
2.0–2.4	6	6.0–6.4	1
2.5–2.9	0	6.5–6.9	1
3.0–3.4	0	7.0–7.4	1
3.5–3.9	0		

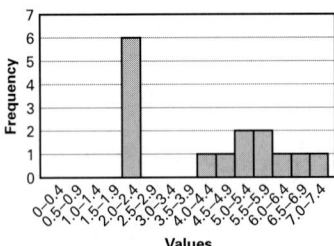

33. machine 1: 2.59, 2.59, none; machine 2: 2.59, 2.59, none
37. $645,000; $213,243.66 **39.** The mode is the most appropriate measure because it would indicate that most people have a positive opinion on the issue. Because the categories are not part of an ordered scale, means and medians are not meaningful.

41.

Age	Pres	VP
30–39	0	1
40–49	8	12
50–59	24	21
60–69	10	12
70–79	0	1

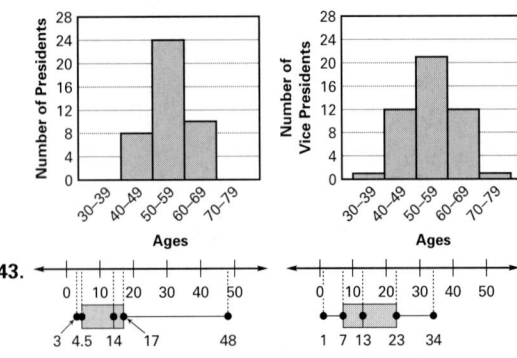

43.

45. *Sample answer:* You cannot conclude that one conference consistently has larger (or smaller) margins of victory than the other.

51. 24 **53.** –326
55. 2187; product of powers **57.** $\frac{1}{4}$; product of powers, negative exponent **59.** $\frac{1}{100}$; zero exponent; negative exponent

61.   **63.**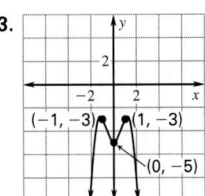

65.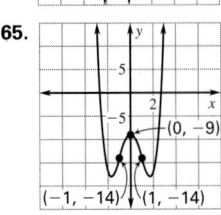

QUIZ 3 (p. 452)

1. 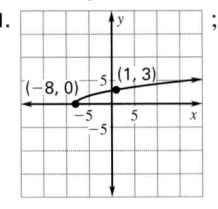 ; **2.** 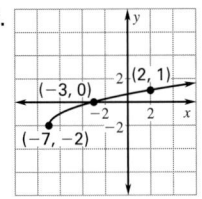 ;

$x \geq -8, y \geq 0$ $x \geq -7, y \geq -2$

3. 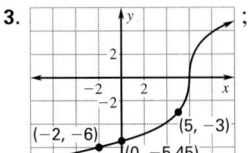 ; x and y are all real numbers.
4. 312.5
5. 6 (–1 is an extraneous solution.)
6. 0 **7.** 4.4, 5.5, 6, 9, 2.8
8. 23.9, 21, none, 31, 9.99

9. 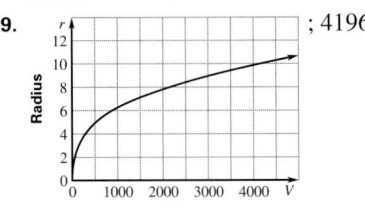 ; 4196 cubic units

10. 228.24 million km **11.**

12.

1750–1799	16
1800–1849	14
1850–1899	15
1900–1949	3
1950–1999	2

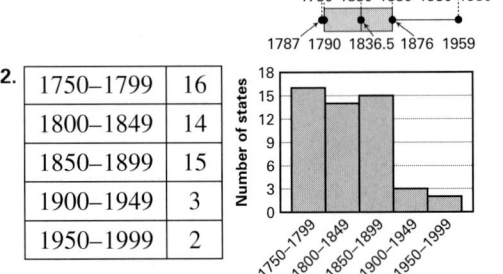

TECHNOLOGY ACTIVITY 7.7 (p. 454)
1. 17.3, 17.5, 22, 5.71 **5.** The second restaurant's

SELECTED ANSWERS

sandwiches have fewer calories than the sandwiches at the first restaurant. The histograms show that half of the sandwiches in the 1st restaurant contain over 500 calories while only 1 out of 10 sandwiches in the second restaurant contain over 500 calories.

CHAPTER 7 REVIEW (pp. 456–458) **1.** 2 **3.** $\frac{1}{243}$ **5.** -2

7. -1 **9.** $\frac{1}{25}$ **11.** $\frac{\sqrt[3]{2}}{5}$ **13.** $3x^{1/4}$ **15.** $xyz\sqrt[6]{6yz^4}$ **17.** $3x - 6$

17. $3x - 6$ **19.** $2x^2 - 8x + 8$ **21.** $2x - 8$ **23.** $f^{-1}(x) = (-x)^{1/4}$, $x \le 0$ **25.** Both compositions equal x.

27. 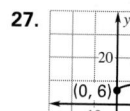 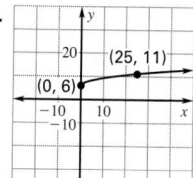 ; **29.** 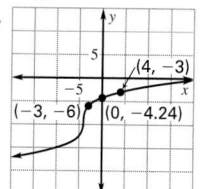 ;

$x \ge 0; y \ge 6$ x and y are all real numbers.
31. -3 **33.** 40.9, 42, 51, 42, 11.3

CHAPTER 8

SKILL REVIEW (p. 464) **1.** $\frac{1}{64}$ **2.** $\frac{1}{9}$ **3.** 1 **4.** -25 **5.** $\frac{2}{5}$

6. $f(x) \to -\infty$ as $x \to -\infty$; $f(x) \to +\infty$ as $x \to +\infty$
7. $f(x) \to -\infty$ as $x \to -\infty$; $f(x) \to -\infty$ as $x \to +\infty$
8. $f(x) \to +\infty$ as $x \to -\infty$; $f(x) \to +\infty$ as $x \to +\infty$
9. $f(x) \to +\infty$ as $x \to -\infty$; $f(x) \to -\infty$ as $x \to +\infty$
10. *Sample answer:* $y = 0.403x + 2.013$

8.1 PRACTICE (pp. 469–471)

5. 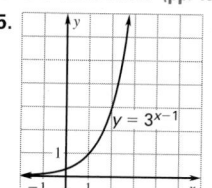 ; **7.** 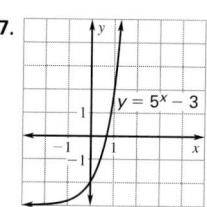 ;

domain: all real numbers;   domain: all real numbers;
range: all positive   range: $y > -3$
real numbers

9. 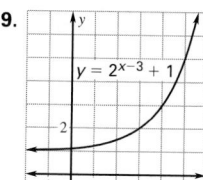 ; domain: all real numbers;
range: $y > 1$
11. 6191; 4% **13.** 1; the x-axis
15. 4; the x-axis **17.** $\frac{3}{2}$; the x-axis
19. C **21.** B **23.** F

25. 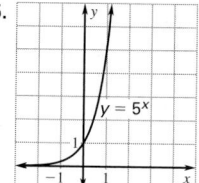 **27.**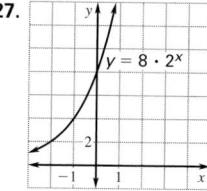

29. 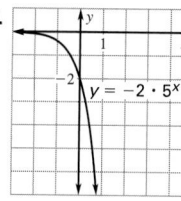 **31.**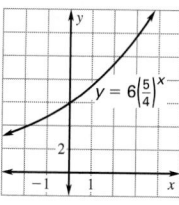

33. 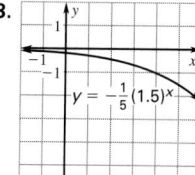 **35.** 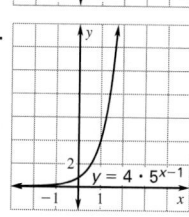 ;

domain: all real numbers;
range: $y > 0$

39. 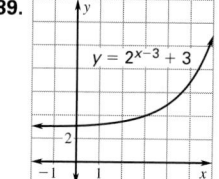 ; **41.** 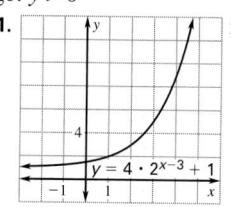 ;

domain: all real numbers;   domain: all real numbers;
range: $y > 3$ range: $y > 1$
43. 2.91 trillion ft³; 1.07; 7% **45.** 8.03 trillion ft³
47.   **49.** $E = 5(1.59)^t$;
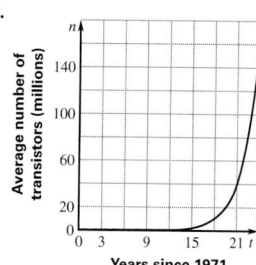   about 32 gigawatt-hours
                                 51. $t \approx 5.98$;
                                 near the end of 1985

53.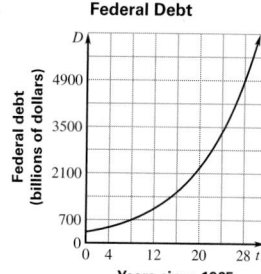

Federal Debt

55. a. $2600 **b.** $3041.63
c. ANS + ANS $\times$ 0.01; push "ENTER" four times. **d.** $3050.48; this is $8.85 more.
57. $A = 400(1.005)^{4t}$ where t is the number of years **59.** $1724.48
61. $1799.78 **63.** $2402.21

8.1 MIXED REVIEW (p. 472) **71.** $\frac{1}{8}$ **73.** $\frac{1}{32}$ **75.** $\frac{343}{1728}$

77. $\frac{16}{25}$ **79.** 2.18 **81.** -3 **83.** 3.16 **85.** 3 **87.** 3.04 **89.** 1.73

91. $4x^2 + 6x - 11$; all real numbers **93.** $24x^3 - 44x^2$; all real numbers **95.** $24x^2 - 11$; all real numbers

97. $\frac{6x - 11}{4x^2}$; all nonzero real numbers

99. $36x - 77$; all real numbers

SELECTED ANSWERS

8.2 PRACTICE (pp. 477–479)

5.

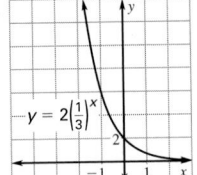

$y = 2\left(\frac{1}{3}\right)^x$

domain: all real numbers;
range: $y > 0$

7.

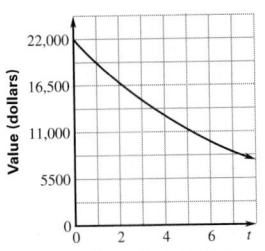

$y = -5\left(\frac{2}{3}\right)^{x-2}$

domain: all real numbers;
range: $y < 0$

9.

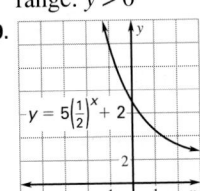

$y = 5\left(\frac{1}{2}\right)^x + 2$

; domain: all real numbers;
range: $y > 2$

11. exponential decay
13. exponential decay
15. exponential growth
17. exponential decay

19. F   **21.** D   **23.** C

25.

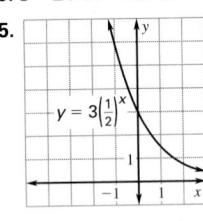

$y = 3\left(\frac{1}{2}\right)^x$

27.

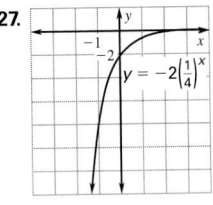

$y = -2\left(\frac{1}{4}\right)^x$

33.

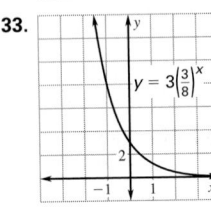

$y = 3\left(\frac{3}{8}\right)^x$

35.

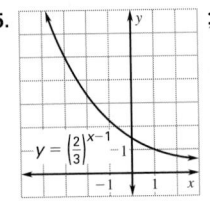

$y = \left(\frac{2}{3}\right)^{x-1} - 1$

;

domain: all real numbers;
range: $y > 0$

37.

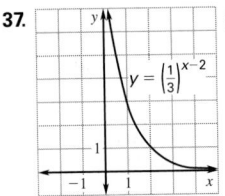

$y = \left(\frac{1}{3}\right)^{x-2}$

;

domain: all real numbers;
range: $y > 0$

39.

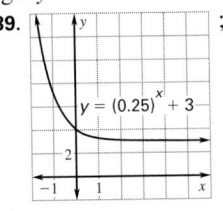

$y = (0.25)^x + 3$

;

domain: all real numbers;
range: $y > 3$

41.

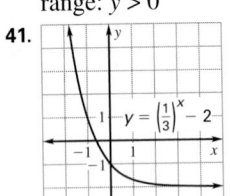

$y = \left(\frac{1}{3}\right)^x - 2$

; domain: all real numbers;
range: $y > -2$

43. $V = 780(0.95)^t$
45. $i = 400(0.71)^h$
47. 265; 0.39; 61%
49. about 1988

51.

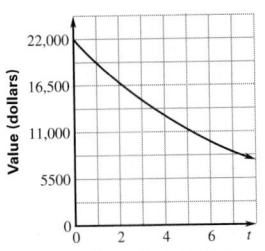

Value of Car

53. $V = 2100(0.5)^t$; \$525
55. after about 22 months

57. a. $V = 18{,}354(0.83)^t$

b. $A(n) = \left(18{,}354 - \dfrac{280}{\frac{0.085}{12}}\right)\left(1 + \dfrac{0.085}{12}\right)^n + \dfrac{280}{\frac{0.085}{12}}$

8.2 MIXED REVIEW (p. 479)

59.

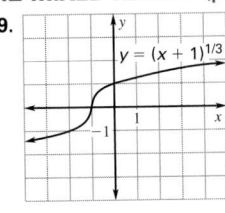

$y = (x + 1)^{1/3}$

61.

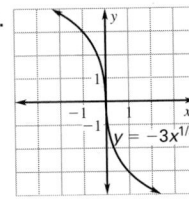

$y = -3x^{1/3}$

63.

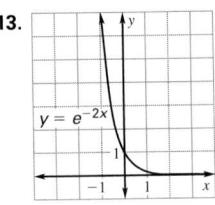

$y = -\sqrt{x} + 5$

65. 16; 15; 15; 12
67. a. \$2639.86   **b.** \$2441.79

8.3 PRACTICE (pp. 483–485)   **5.** $3e^5$ **7.** $\dfrac{64}{e^6}$ **9.** $6e^{2x}$ **11.** $\dfrac{e^6}{3}$

13.

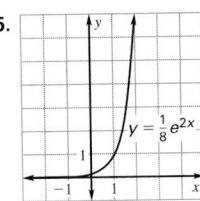

$y = e^{-2x}$

15.

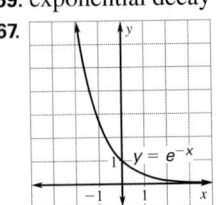

$y = \frac{1}{8}e^{2x}$

17. e^6 **19.** $\dfrac{e^{3x}}{3}$ **21.** $3e^4$ **23.** e^{-2x+5} **25.** $\dfrac{1}{10{,}000e^x}$

27. $\dfrac{e^{x-1}}{2}$ **29.** $3e^{2x}$ **31.** $\dfrac{3}{2}e^{3x-1}$ **33.** 20.086   **35.** 5.474

37. 0.779   **39.** 2980.958   **41.** 0.018   **43.** −0.199
45. −178.096   **47.** $4.34 \cdot 10^{-20}$ **49.** exponential decay
51. exponential decay   **53.** exponential growth
55. exponential growth   **57.** exponential decay
59. exponential decay   **61.** C   **63.** F   **65.** D

67.

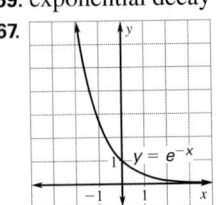

$y = e^{-x}$

;

domain: all real numbers;
range: $y > 0$

73.

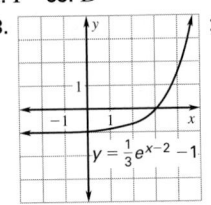

$y = \frac{1}{3}e^{x-2} - 1$

;

domain: all real numbers;
range: $y > -1$

SELECTED ANSWERS

75. 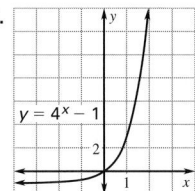 ; domain: all real numbers; range: $y > -2$

77. \$2650; \$2652.25; \$2653.41; \$2654.19; \$2654.59; *Sample answer:* The extra amount of interest earned with more and more compoundings decreases drastically, with the difference between compounding monthly and continuously being only 40¢, 0.016% of the amount initially invested. **79.** about 4.603 lb/in.^{2}

8.3 MIXED REVIEW (p. 485) **85.** $f^{-1}(x) = \dfrac{x-7}{6}$
87. $f^{-1}(x) = 2x + 20$ **89.** $f^{-1}(x) = -5x - 65$ **91.** 6.2 **93.** 2 **95.** no solution

QUIZ 1 (p. 485)

1. 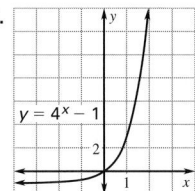 ;
domain: all real numbers; range: $y > -1$

2. 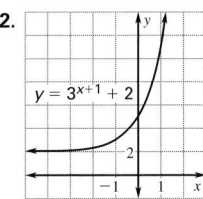 ;
domain: all real numbers; range: $y > 2$

3. 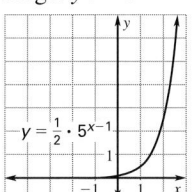 ;
domain: all real numbers; range: $y > 0$

4. 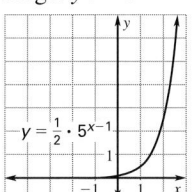 ;
domain: all real numbers; range: $y < 0$

5. 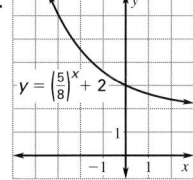 ;
domain: all real numbers; range: $y > 2$

6. 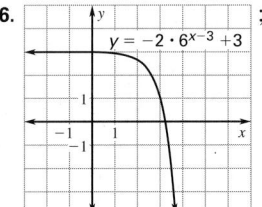 ;
domain: all real numbers; range: $y < 3$

7. $2e^7$ **8.** $4e^2$ **9.** $9e^{4x}$ **10.** $\dfrac{e^{12x}}{5^{4x}}$ **11.** $\dfrac{3}{4}e^{x-1}$ **12.** $\dfrac{6}{e^{4x}}$
13. $4e\sqrt{x}$ **14.** $5e^{2x}$ **15.**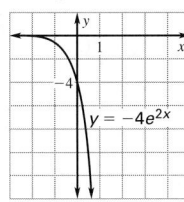

16. **Amount of Radium Left from a 100 g Sample** ; about 1.357 g

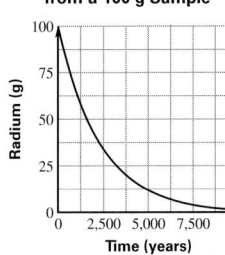

8.4 PRACTICE (pp. 490–492) **5.** $3^2 = 9$ **7.** $\left(\dfrac{1}{2}\right)^{-2} = 4$ **9.** 6
11. 0 **13.** 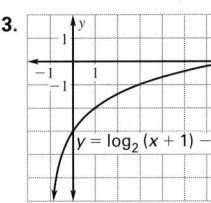 ; domain: $x > -1$; range: all real numbers
15. about 0.316 mm
17. $5^{-1} = \dfrac{1}{5}$
19. $8^3 = 512$

21. $14^2 = 196$ **23.** $105^2 = 11{,}025$ **25.** 3 **27.** 1 **29.** 2
31. 4 **33.** -0.38 **35.** -2 **37.** 2.303 **39.** 0.571 **41.** -0.523
43. 0.544 **45.** 5.011 **47.** 3.114 **49.** x **51.** x **53.** x **55.** $3x$
57. $y = \left(\dfrac{1}{4}\right)^x$ **59.** $y = \left(\dfrac{1}{2}\right)^x$ **61.** $y = \dfrac{e^x}{6}$ **63.** $y = -2 + e^x$

65. 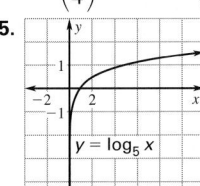 ;
domain: $x > 0$; range: all real numbers

67. 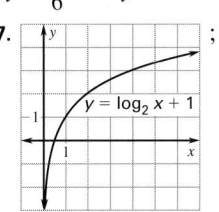 ;
domain: $x > 0$; range: all real numbers

71. 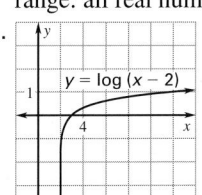 ;
domain: $x > 2$; range: all real numbers

73. 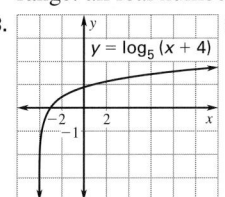 ;
domain: $x > -4$; range: all real numbers

75. 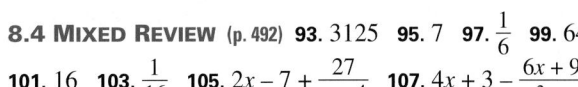 ; domain: $x > 0$; range: all real numbers
77. a. 2.4 **b.** 3 **c.** 3.5
79. about 8 (7.9982…)
81. about 205 mi

8.4 MIXED REVIEW (p. 492) **93.** 3125 **95.** 7 **97.** $\dfrac{1}{6}$ **99.** 64
101. 16 **103.** $\dfrac{1}{16}$ **105.** $2x - 7 + \dfrac{27}{x+4}$ **107.** $4x + 3 - \dfrac{6x+9}{x^2+2}$
109. $y = -\dfrac{1}{6}x(x-2)(x+3)$ **111.** $y = \dfrac{1}{75}(x-4)(x-6)(x+4)$

SELECTED ANSWERS

8.5 PRACTICE (pp. 496–498) **5.** 3 **7.** −1 **9.** 1.58 **11.** 7.2
13. about 26 decibels **15.** −2 **17.** 3 **19.** −1 **21.** −6
23. 1.398 **25.** 2.097 **27.** 2.352 **29.** −0.477 **31.** $\ln 22 + \ln x$
33. $6 \log_6 x$ **35.** $2 \log_3 5$ **37.** $\ln 3 + \ln x + 3 \ln y$
39. $2 + 2 \log_8 x$ **41.** $\frac{5}{6} \log_3 12 + 9 \log_3 x$
43. $\ln 3 + 4 \ln y - 3 \ln x$ **45.** $1 + \frac{1}{2} \log_2 x$ **47.** $\ln 4$
49. $\log_{16} 1296$ **51.** $\log_4 128 x^5 y^3$ **53.** $\log_3 2\sqrt{y}$ **55.** $\ln \frac{3}{x^2}$
57. $\log_5 \frac{1}{6}$ **59.** 1.277 **61.** 1.465 **63.** 1.226 **65.** 2.153
67. 1.774 **69.** 1.585 **71.** −0.529 **73.** 1.471

75.

f	s
1.414	1.000
2.000	2.000
2.828	3.000
4.000	4.000
5.657	5.000
8.000	6.000
11.314	7.000
16.000	8.000

; The first row of the table shows successive powers of $\sqrt{2}$, and the second row shows the integers, beginning with 1.
77. $E = 1.4 \log \frac{C_2}{C_1}$
79. about 1.089 kcal/g-molecule
81. about 95 decibels; between subway train and boiler shop
83. about 92 decibels

85. 10 log 0.5, or about 3 decibels less

8.5 MIXED REVIEW (p. 499) **93.** y^{12} **95.** $9x^4$
97. $\frac{x^2}{y^2}$ **99.** $\frac{xy^8}{2}$ **101.** 7 **103.** 500 **105.** 6.14×10^{-6}
107. 3.581×10^{-3} **109.** 0.238 **111.** 1.773

TECHNOLOGY ACTIVITY 8.5 (p. 500)
Points may vary. Points given are sample responses.
1. 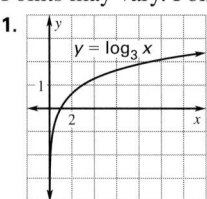 ; **7.** 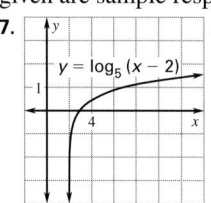 ;
(1, 0); $x = 0$ (3, 0); $x = 2$
9. 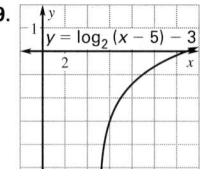 ; (6, −3); $x = 5$

13. *Sample answer:* The domain of $y = \log x$ is all real numbers greater than 0, while that of $y = \log |x|$ is all real numbers except 0. The graph of $y = \log |x|$ is the graph of $y = \log x$ and its reflection in the y-axis.

8.6 PRACTICE (pp. 505–507) **5.** 1.292 **7.** 1 **9.** $\frac{\log 28 + 1}{3} \approx$
0.816 **11.** 1000 **13.** 39.121 **15.** $-1 + \frac{\sqrt{39}}{3} \approx 1.082$

17. $e^{\log_2 5x} \neq 5x$, since e^x and $\log_2 x$ are not inverse functions. **19.** yes **21.** no **23.** yes **25.** 1 **27.** $-\frac{7}{5}$ **29.** $\frac{16}{3}$
31. 3.907 **33.** $\frac{3}{2}$ **35.** $\frac{\log 5}{2} \approx 0.3495$ **37.** 1
39. $-\frac{1}{12} \log 94 \approx -0.164$ **41.** 20 **43.** 2 **45.** 2187 **47.** 2916
49. $-e^{7/2}$ **51.** $1 + \sqrt{1 + e} \approx 2.928$ **53.** no solution
55. $\frac{1}{3} e^5$ **57.** 47.158 **59.** no solution
63. a little over 9 years **65.** about 27.7 years
67. Subantarctic: 8°; Antarctic intermediate: 4°; North Atlantic deep: 2°; Antarctic bottom: 0° **69.** 100 mm

8.6 MIXED REVIEW (p. 508) **77.** Lines may vary.;
$y = 0.305x + 1.780$ **79.** (4, 5) **81.** (0, −6) **83.** no solution
85. $(3x^2 + 4)(x − 2)$ **87.** $(x^2 + 5)(7x + 4)$

QUIZ 2 (p. 508) **1.** 3 **2.** 4 **3.** 3 **4.** $y = e^x − 3$
5. 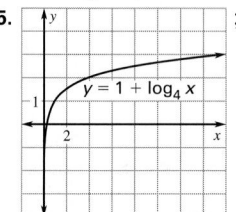 ; **6.** 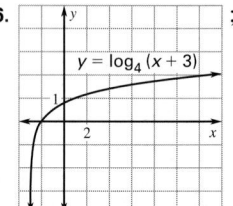 ;
domain: $x > 0$;   domain: $x > −3$;
range: all real numbers   range: all real numbers
7. 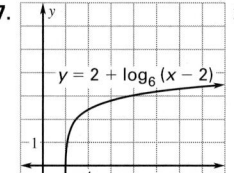 ; domain: $x > 2$;
range: all real numbers
8. 4 **9.** −1 **10.** 2
11. $\frac{1}{2} \log_4 x + 4 \log_4 y$
12. $\log_6 28x^3$ **13.** 2.230 **14.** $\ln 5$
15. $2^{28/3}$ **16.** no solution

17. about 87 billion ergs

8.7 PRACTICE (pp. 513–516) **5.** $y = \frac{2}{9} \cdot 3^x$
7. $y = \frac{2704}{350} \left(\frac{35}{52}\right)^x$ **9.** $y = \frac{1}{\sqrt{3}} \left(\frac{\sqrt{3}}{2}\right)^x$ **11.** $y = 2x^2$
13. $y = 4x^{0.631}$ **15.** $y = 0.417x^{0.263}$ **17.** $y = \left(\frac{4}{3}\right) 3^x$
19. $y = \left(\frac{1}{512}\right) 4^x$ **21.** $y = 2^x$ **23.** $y = 7\left(\frac{2}{3}\right)^x$ **25.** $y = \left(\frac{1}{4}\right) 5^x$

27. 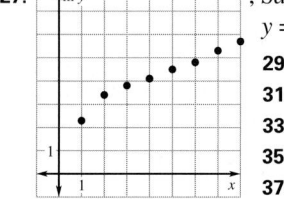 ; *Sample answer:*
$y = 9.715(1.550)^{-x}$
29. $y = 0.362x^{1.465}$
31. $y = 0.358x^{2.181}$
33. $y = 6.325x^{0.661}$
35. $y = 7.109x^{0.482}$
37. $y = 2.481x^{0.954}$

39. 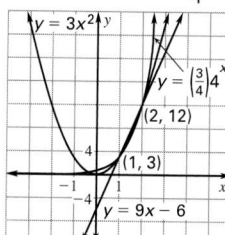 ; $y = 1.193x^{1.962}$

41. $y = 31,623(1.738)^x$
43. $y = 54.598e^x$
45. $y = 0.283x^{-0.48}$
47. $y = 2.664(0.0926)^x$
49. $y = 12.182(0.223)^x$
51. $y = x^{5/3}$

53. $y = 9x - 6$; $y = \frac{3}{4} \cdot 4^x$; $y = 3x^2$;

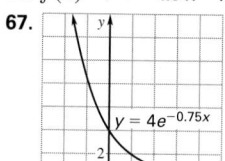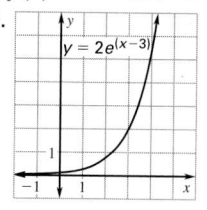 ; *Sample answer:* The linear function grows the slowest, the quadratic is in the middle, and the exponential function grows at the fastest rate.

55. a. yes   **b.** $C = 250.31(1.104)^t$; about 35,232

57. a. yes   **b.** $y = 2.022x^{-0.582}$; about 354,000

8.7 MIXED REVIEW (p. 516)
61. $f(x) \to +\infty$ as $x \to -\infty$; $f(x) \to -\infty$ as $x \to +\infty$
63. $f(x) \to -\infty$ as $x \to -\infty$; $f(x) \to -\infty$ as $x \to +\infty$
65. $f(x) \to +\infty$ as $x \to -\infty$; $f(x) \to +\infty$ as $x \to +\infty$

67. 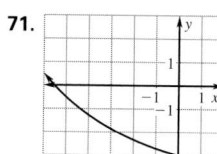   **69.**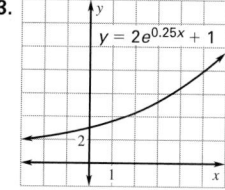

71. 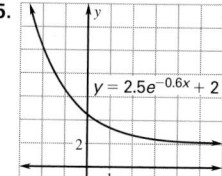   **73.**

75. 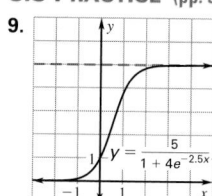   **77.** $\log 27$ **79.** $\ln \dfrac{x^2}{4}$
81. $\log_7 3840$

8.8 PRACTICE (pp. 520–522)   **5.** 0.0438   **7.** 0.822
9. 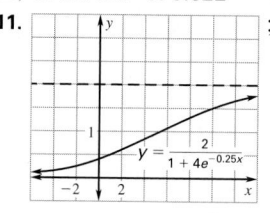 ;   **11.** ;

x-axis and $y = 5$; 1; (0.555, 2.5)

x-axis and $y = 2$; $\dfrac{2}{5}$; (5.545, 1)

13. 0.693   **15.** $h = \dfrac{117}{1 + 18e^{-0.73t}}$ **17.** 6.090   **19.** 0.00578

21. 2.896   **23.** 0.835   **25.** A

27. 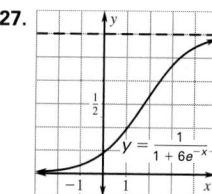 ; asymptotes: x-axis and $y = 1$; y-intercept: $\dfrac{1}{7}$; pt of max growth: $\left(1.792, \dfrac{1}{2}\right)$

29. 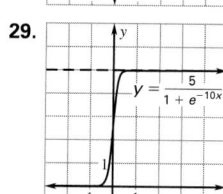 ; asymptotes: x-axis and $y = 5$; y-intercept: $\dfrac{5}{2}$; pt of max growth: $\left(0, \dfrac{5}{2}\right)$

31. 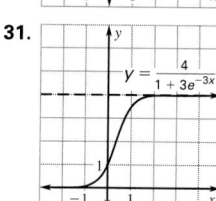 ; asymptotes: x-axis and $y = 4$; y-intercept: 1; pt of max growth: (0.366, 2)

33. 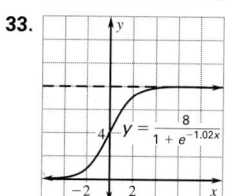 ; asymptotes: x-axis and $y = 8$; y-intercept: 4; pt of max growth: (0, 4)

37. $\dfrac{\ln 18}{4} \approx 0.723$ **39.** 1.741
41. 0.356   **43.** -3.942
45. during 1994

47. to approach 91.86 million households

49.  ;   **51.** $y = \dfrac{721}{1 + 72e^{-0.526t}}$

1987

8.8 MIXED REVIEW (p. 522)   **55.** $y = -2x$ **57.** $y = \dfrac{1}{8}x$
59. $y = 0.2x$ **61.** $y = 2.560(0.0872)^x$ **63.** $y = 0.0174x^{-0.75}$

QUIZ 3 (p. 522)   **1.** $y = 1.191(1.587)^x$ **2.** $y = 9.541(1.677)^x$
3. $y = 0.936(1.573)^x$ **4.** $y = 10.693x^{1.389}$ **5.** $y = \dfrac{1}{2}x^{2.547}$
6. $y = 1.429x^{2.070}$ **7.** ; when $t \approx 10.65$, or after about $10\frac{1}{2}$ days

SELECTED ANSWERS

CHAPTER 8 REVIEW (pp. 524–526)

1. 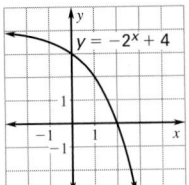 ; **3.** 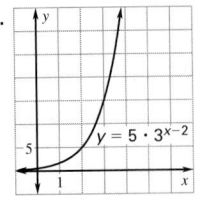 ;

domain: all real numbers;   domain: all real numbers;
range: $y < 4$ range: $y > 0$

5. exponential decay   **7.** exponential decay

9. 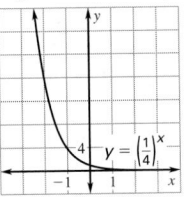 ; **11.** 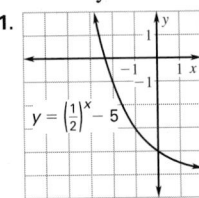 ;

domain: all real numbers;   domain: all real numbers;
range: $y > 0$ range: $y > -5$

13. 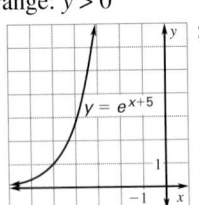 ; **15.** 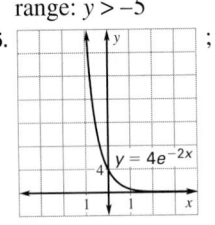 ;

domain: all real numbers;   domain: all real numbers;
range: $y > 0$ range: $y > 0$

17. 3   **19.** -2

21. 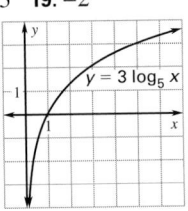 ; **23.** 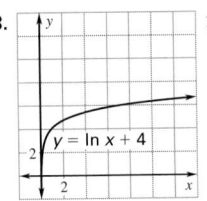 ;

domain: $x > 0$;   domain: $x > 0$;
range: all real numbers   range: all real numbers

25. $\log_3 6 + \log_3 x + \log_3 y$ **27.** $\log 5 + 3 \log x$

29. $\ln \frac{9}{5}$ **31.** $\log 18$ **33.** -1.466 **35.** 160.49

37. $y = 3.9605(1.499)^x$ **39.** $y = 2.099x^{0.696}$

41. $y = 3.188x^{1.673}$

43. 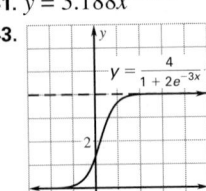 ; asymptotes: x-axis and $y = 4$;
y-intercept: $\frac{4}{3}$; pt of max growth:
$(0.231, 2)$

CHAPTER 9

SKILL REVIEW (p. 532)
1. $y = \frac{5}{2}x$ **2.** $y = \frac{1}{10}x$

3. $y = -\frac{1}{4}x$ **4.** $y = -4x$ **5.** $15x - 5$ **6.** $x^3 + 7x^2 + 8x - 16$

7. $-x^3 + 5x$ **8.** $x^3 + 7x^2 - 8x$ **9.** $(x - 3)^2$
10. $4(x - 1)(x^2 + x + 1)$ **11.** $2x(2x - 9)(2x + 9)$
12. $(2x - 1)(3x + 5)$ **13.** $0, -2$ **14.** $-5, 3$ **15.** $-1, 4$

9.1 PRACTICE (pp. 537–539)
5. direct variation   **7.** inverse
variation   **9.** inverse variation   **11.** neither   **13.** yes   **15.** yes
17. yes   **19.** yes   **21.** inverse variation   **23.** neither
25. inverse variation   **27.** direct variation   **29.** $y = -\frac{10}{x}$; -5
31. $y = \frac{7}{x}$; 3.5 **33.** $y = -\frac{4}{x}$; -2 **35.** inverse variation
37. neither   **39.** $z = \frac{1}{4}xy$; -7 **41.** $z = 15xy$; -420
43. $z = 32xy$; -896 **45.** $x = \frac{kz}{y}$ **47.** $w = \frac{kyz}{x}$
49. yes; $l = \frac{45\pi}{8A}$ **51.** $D = \frac{k\sqrt{L}}{T^2}$ **53.** 139,000,000 km
55. $W = \frac{49}{5}mh$; 1470 joules   **57.** 285 watts

9.1 MIXED REVIEW (p. 539)
61.  ; domain: all real numbers x
such that $x \geq -2$; range: all real
numbers y such that $y \geq 0$

63. ; domain: all real numbers x
such that $x \geq -1$; range: all real
numbers y such that $y \geq -3$
65. 128   **67.** 113   **69.** 7

9.2 PRACTICE (pp. 543–545)
5. $y = 2$; $x = -4$ **7.** $y = \frac{1}{2}$; $x = 2$
9. $y = -5$; $x = 6$ **11.** $y = 2$; $x = 0$; domain: all real numbers
except 0; range: all real numbers except 2   **13.** $y = -2$;
$x = -3$; domain: all real numbers except -3; range: all real
numbers except -2 **15.** $y = \frac{2}{3}$; $x = -\frac{1}{3}$; domain: all real
numbers except $-\frac{1}{3}$; range: all real numbers except $\frac{2}{3}$
17. $y = -17$; $x = -43$; domain: all real numbers except -43;
range: all real numbers except -17 **19.** $y = 19$; $x = 6$;
domain: all real numbers except 6; range: all real numbers
except 19   **21.** C

23. ; **25.** ;

domain: all real numbers   domain: all real numbers
except 0; range: all real   except -5; range: all real
numbers except 0   numbers except -8

SELECTED ANSWERS

27.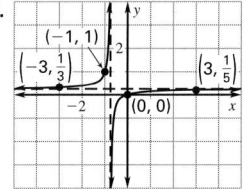

domain: all real numbers except −2; range: all real numbers except −6

33. ;

domain: all real numbers except −$\frac{3}{4}$; range: all real numbers except $\frac{1}{4}$

35.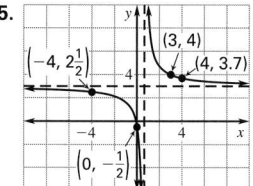

; domain: all real numbers except $\frac{2}{3}$; range: all real numbers except 3

41. *Sample answer:*

$y = \frac{1}{x+4} + 3$ **43.** 30

45.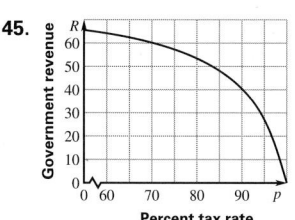

47. $f = \frac{1,480,000}{740 - r}$

9.2 MIXED REVIEW (p. 545)

53. 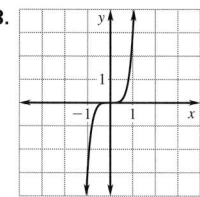   **55.**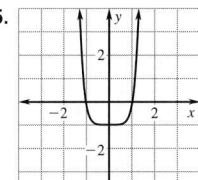

57.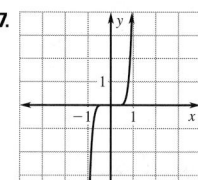

59. $(2x - 5)(4x^2 + 10x + 25)$

61. $(x + 3)(x^2 + 3)$

63. $(3x - 1)(3x + 1)(9x^2 + 1)$

65. $\frac{1}{5}e^{x-1}$ **67.** e^{5x+1}

69. e^{4-x}

TECHNOLOGY ACTIVITY 9.2 (p. 546)

7. $A = \frac{2 + 8n}{n}$;

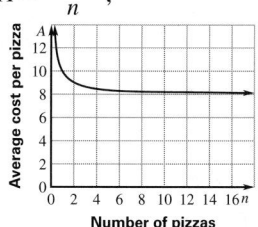

The average cost approaches $8.

9.3 PRACTICE (pp. 550–552)

5. 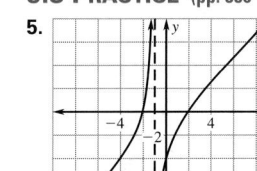   **7.**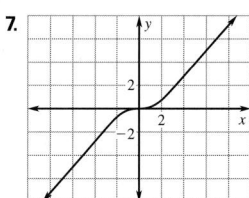

9.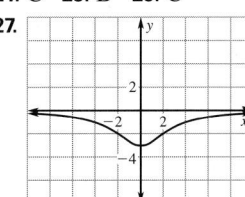

11. x-intercept: 0; vertical asymptotes: $x = -3$, $x = 3$

13. x-intercepts: $-\frac{1}{2}$, 5; vertical asymptotes: $x = -4$, $x = 4$

15. x-intercepts: −5, 1; vertical asymptote: $x = 6$

17. x-intercept: −4; vertical asymptotes: $x = -\sqrt{3}$, $x = \sqrt{3}$

19. x-intercept: 3; vertical asymptote: $x = 0$

21. C   **23.** B   **25.** C

27. 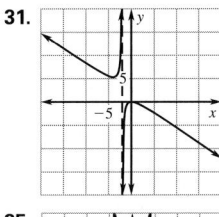   **29.**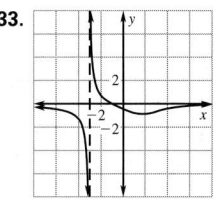

31. 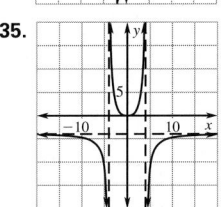   **33.**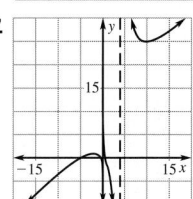

35. 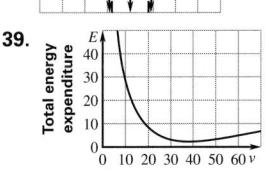   **37.**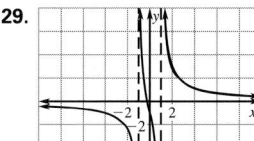

39.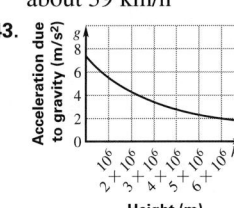

; **41.** No; this model predicts an average daily cost close to zero after 2005, and this is not realistic.

about 39 km/h

43. 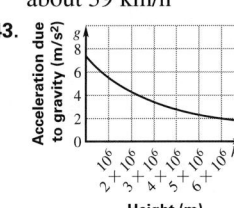   **45.** g' decreases as h increases.

9.3 MIXED REVIEW (p. 553)   **51.** x^5y^4 **53.** $\frac{4x^7}{y^7}$ **55.** $\frac{x^6}{125y^6}$

57. $z = -\frac{3xy}{40}$; $\frac{9}{20}$ **59.** $z = 8xy$; −48

SELECTED ANSWERS

61. $f(g(x)) = f\left(-\frac{1}{3}x + \frac{2}{3}\right) = -3\left(-\frac{1}{3}x + \frac{2}{3}\right) + 2 = x - 2 +$
$2 = x$; $g(f(x)) = g(-3x + 2) = -\frac{1}{3}(-3x + 2) + \frac{2}{3} = x - \frac{2}{3} + \frac{2}{3} = x$ **63.** $f(g(x)) = f\left(\frac{\sqrt[4]{x}}{2}\right) = 16\left(\frac{\sqrt[4]{x}}{2}\right)^4 = 16\left(\frac{x}{16}\right) = x$;
$g(f(x)) = g(16x^4) = \frac{\sqrt[4]{16x^4}}{2} = \frac{2x}{2} = x$

QUIZ 1 (p. 553) **1.** $y = -\frac{12}{x}$; 4 **2.** $y = \frac{66}{x}$; -22 **3.** $y = \frac{6}{x}$; -2
4. $x = -\frac{yz}{6}$; -24 **5.** $x = 4yz$; 1 **6.** $x = -\frac{5yz}{4}$; $-\frac{16}{5}$

7.
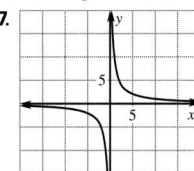

8.
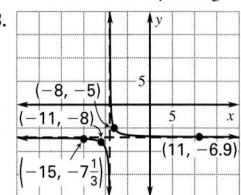

9.
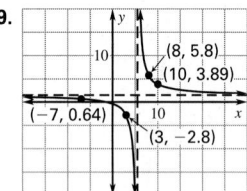

10.
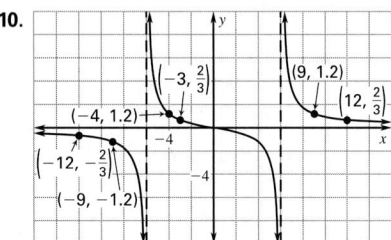

11.
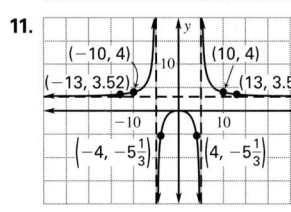

12.
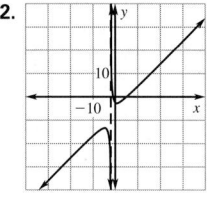

13.
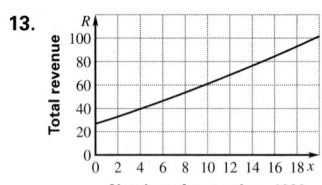
; 1992

9.4 PRACTICE (p. 558–560) **3.** $\frac{x}{x^2 + 3}$ **5.** not possible

7. not possible **9.** $\frac{x^5}{25y^2}$ **11.** $\frac{x - 2}{x^2}$ **13.** $\frac{6y^3}{x^5}$

15. with: 6.9; without: 9.3 **17.** not possible

19. not possible **21.** $\frac{3(x + 1)}{x + 2}$ **23.** $\frac{x - 3}{x}$ **25.** not possible

27. $\frac{x^2 - 2}{x^2 - 3x + 9}$ **29.** $\frac{16x^3}{y^2}$ **31.** $\frac{3(x + 4)}{x + 3}$ **33.** $2(x - 1)(x - 3)$

35. $\frac{-(x + 1)^2}{x^2}$ **37.** $\frac{1}{y^2}$ **39.** $\frac{1}{3x}$ **41.** $\frac{(x + 4)(x - 2)}{x + 2}$

43. $\frac{(x - 3)(x + 5)}{3x}$ **45.** $\frac{(x - 4)(x + 2)}{4x^2}$ **47.** $9(x + 3)$

49. $3(x + 2)$ **51.** $H = \frac{k_2}{k_1 V^2}$ or HV^2 is a constant. A shorter runner can run faster than a taller runner and still have the heat being generated equal the heat being released, so a shorter runner has an advantage. **53.** 468.5 acres
55. about $4,400 million

9.4 MIXED REVIEW (p. 560) **61.** 15; 1320 **63.** 12; 504
65. 120; 2400 **67.** $x^2 + 6x - 7$ **69.** $x^3 + 6x^2 + 11x + 6$
71. $-6x^6 + 24x^4 + 5x^3 - 20x$

73.
; in 6.5 years

TECHNOLOGY ACTIVITY 9.4 (p. 561)
1. $\frac{x}{x + 4}$ **3.** $\frac{x - 2}{x + 1}$ **5.** $\frac{2x}{3}$

9.5 PRACTICE (pp. 565–567)
5. $\frac{2x + 7}{x + 5}$ **7.** $\frac{x^2 - 3x + 24}{(x - 4)(x + 3)}$ **9.** $\frac{x(x - 23)}{20(2x + 1)}$

11. $\frac{Pi}{1 - \left(\frac{1}{1 + i}\right)^{12t}} = \frac{Pi(1 + i)^{12t}}{\left(1 - \left(\frac{1}{(1 + i)^{12t}}\right)\right)(1 + i)^{12t}} = \frac{Pi(1 + i)^{12t}}{(1 + i)^{12t} - 1}$

13. $\frac{23 - x}{10x^2}$ **15.** $\frac{5x(x + 1)}{x + 8}$ **17.** $\frac{1}{x}$ **19.** $21x^2(x - 5)$

21. $x(x + 3)(x - 6)$ **23.** $(x - 7)(x + 2)(x + 4)$ **25.** Always; each denominator must be a factor of the LCD, so the LCD must have degree greater than or equal to each of the separate denominators. **27.** $\frac{-47}{21x}$ **29.** $\frac{10x + 13}{(x - 3)(x + 3)}$

31. $\frac{11 - x}{(x - 2)(x + 4)}$ **33.** $\frac{-3(5x^2 + x + 2)}{(x - 10)(3x + 2)}$ **35.** $\frac{2(x^2 - 5x - 8)}{(x - 4)(x + 4)^2}$

37. $\frac{49x^2 + 24x - 5}{6x(x - 1)(x + 1)}$ **39.** $\frac{80}{x - 27}$ **41.** $\frac{-(x^3 - x - 1)}{3(x + 1)}$ **43.** $\frac{-2}{3x}$

45. $\frac{3x(x - 4)}{(13x + 8)(x^2 - 4x + 16)}$

47. $M = \frac{357t^3 + 5500t^2 - 37,100t + 485,000}{(0.00418t^2 + 1)(-0.0580t + 1)}$

49. $A = \frac{391(t - 1)^2 + 0.112}{0.218(t - 1)^4 + 0.991(t - 1)^2 + 1}$

51. about 1.2 hours after the second dose **53.** $\frac{24}{7}$ ohms

9.5 MIXED REVIEW (p. 567) **57.** 24 **59.** $\frac{16}{3}$ **61.** -66
63. $-\frac{102}{23}$ **65.** 72 **67.** $\pm 2\sqrt{5}$ **69.** 2, 8 **71.** $-7, \frac{1}{2}$

9.6 PRACTICE (pp. 571–573) **5.** $-\dfrac{8}{3}$ **7.** $\dfrac{3}{2}$ **9.** -5 **11.** $-15, 0$
13. 0 **15.** no **17.** no **19.** yes **21.** 2 **23.** $-1, \dfrac{1}{4}$ **25.** $-\dfrac{2}{3}, 2$
27. $-\dfrac{3}{2}, 2$ **29.** $\dfrac{5}{7}, 3$ **31.** -3 **33.** $\dfrac{6}{17}$ **35.** $-4, 4$ **37.** $2, 5$
39. 4 **41.** $-\dfrac{3}{2}, 5$ **43.** -5 **45.** no solution **47.** $-2, 0$ **49.** $2, 6$
51. Always; when you solve by cross multiplying, you get $x = 1$ or $x = a$ and $x = a$ makes both fractions undefined.
53. Always; when you multiply each side of the equation by $x^2 - a^2$, you get $x = a$, making the fractions undefined.
55. 87 **57.** about 2198 flies/m^3 **59.** $\$16.50$

9.6 MIXED REVIEW (p. 573) **63.** $1; -1$ **65.** $-\dfrac{2}{3}; \dfrac{3}{2}$ **67.** $\dfrac{1}{2}; -2$
69. $4\sqrt{3}$ **71.** $6\sqrt{3}$ **73.** $3\sqrt{30}$ **75.** 15 **77.** 6.796

QUIZ 2 (p. 574) **1.** $\dfrac{5x^5y}{3}$ **2.** $\dfrac{x-8}{5x}$ **3.** $\dfrac{2x^2}{(x-9)(x+4)}$
4. $\dfrac{16x^2 - 5x + 6}{2(5x-6)(5x+6)}$ **5.** $\dfrac{-6(11x+8)}{6x-1}$ **6.** -6
7. $\dfrac{-3(x-3)(2x-1)(2x+1)}{(x-1)(x+1)}$ **8.** $\dfrac{2x}{(x-5)(x+1)}$ **9.** 20 dozen

CHAPTER 9 REVIEW (pp. 576–578)
1. $y = \dfrac{5}{x}; 2.5$ **3.** $y = \dfrac{2}{x}; 1$ **5.** $z = \dfrac{1}{3}xy; -10$ **7.** $z = 3xy; -90$

9. 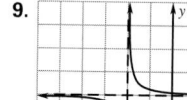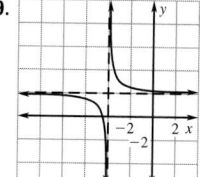 ;
11. 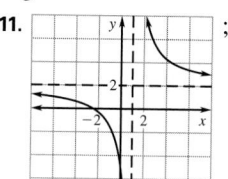 ;

domain: all real numbers; except -4; range: all real numbers except 2

domain: all real numbers; except 1; range: all real numbers except 2

13. 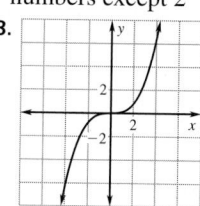 **15.**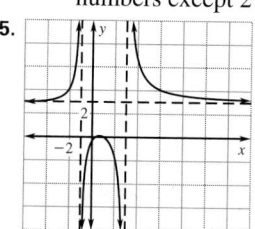

17. $5(x-6)(x+3)(x-3)$ **19.** $\dfrac{x^3 + 5}{x^2(x-2)}$
21. $\dfrac{-9x^2 + 18x - 10}{5x(x-1)(x+5)}$ **23.** $\dfrac{x(x-8)}{2(9x+2)}$ **25.** $\dfrac{12}{5}$
27. $\dfrac{3}{2}$ **29.** no solution **31.** $-4, 1$

CUMULATIVE PRACTICE (pp. 582–583)
1. $y = 3x - \dfrac{7}{2}$ **3.** $y = -\dfrac{5}{6}x + 25$ **5.** parallel

7. 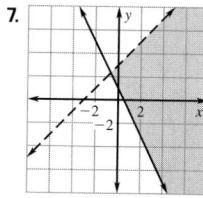 **9.**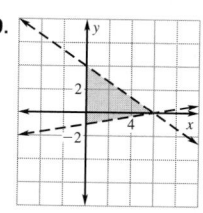

11. $\begin{bmatrix} 12 & -2 \\ -12 & 2 \end{bmatrix}$ **13.** $\begin{bmatrix} -2 & -8 \\ 17 & 30 \end{bmatrix}$

15–19. 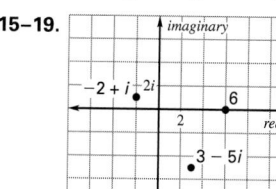 **15.** $\sqrt{5}$ **17.** 6 **19.** $\sqrt{34}$
21. 5 **23.** $\dfrac{3x^4}{10}$
25. $2ab\sqrt[4]{bc}$ **27.** $\dfrac{1}{8e^6}$
29. 2 **31.** $\dfrac{1}{5}$ **33.** $\dfrac{1}{2}$

35. $-x^2 + 2x + 13$; all real numbers
37. $-2x^2 - 15$; all real numbers **39.** $f^{-1}(x) = 2(x+6)$
41. $f^{-1}(x) = 5^x$ **43.** $\log(3x^2y^3)$ **45.** $\ln(x^2y^2)$

47. 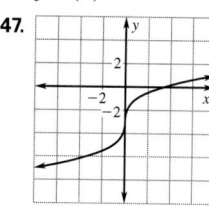 **49.**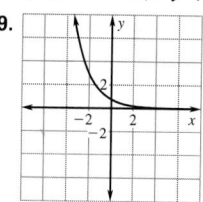

51. 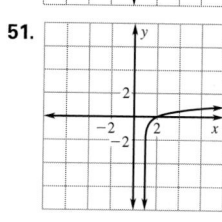 **53.**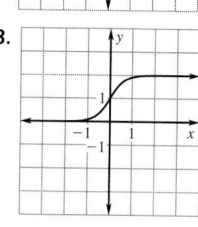

55. 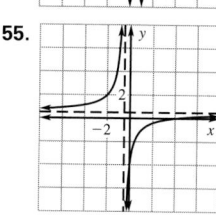 **57.**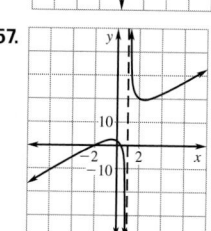

59. 10 **61.** $-\dfrac{9\sqrt{3}}{2}$ **63.** $\ln 8 \approx 2.079$ **65.** $-\dfrac{9}{5}$
67. $y = \dfrac{5}{32}(2)^x$ **69.** $y = 0.759(1.737)^x$ **71.** $y = 1.651x^{0.861}$
73. $y = 1.704x^{0.231}$ **75.** $\dfrac{6x^3 + 7x^2 - 20x - 9}{2x(x-1)(3x+1)}$
77. about 3.5 sec **81.** $f = \dfrac{kq_1q_2}{r^2}$

CHAPTER 10

SKILL REVIEW (p. 588) **1.** $y = 2x + 4$ **2.** $y = \dfrac{1}{3}x - \dfrac{8}{3}$
3. $y = -\dfrac{3}{4}x - 2$ **4.** $(2, 3)$ **5.** $(-1, 5)$ **6.** $(4, 9)$
7. 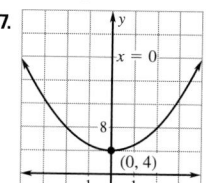 **8.**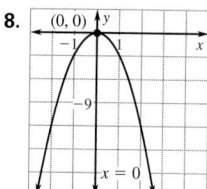

9.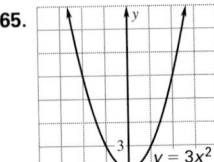
10. $-4 \pm \sqrt{2}$ **11.** $-\frac{3}{2} \pm \frac{\sqrt{11}}{2} i$
12. $-3 \pm \sqrt{23}$

9. 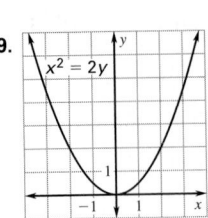 ; $\left(0, \frac{1}{2}\right)$; $y = -\frac{1}{2}$

11. $y^2 = 20x$ **13.** $y^2 = -16x$
15. $x^2 = -32y$ **17.** B **19.** E **21.** C
23. down **25.** right **27.** left **29.** up
31. $\left(\frac{1}{8}, 0\right)$; $x = -\frac{1}{8}$

10.1 PRACTICE (pp. 592–594) **5.** 5 **7.** $3\sqrt{5} \approx 6.71$
9. $3\sqrt{34} \approx 17.49$ **11.** (2, 6) **13.** (2, 7) **15.** $\left(-\frac{9}{2}, -\frac{1}{2}\right)$
17. 5; $\left(\frac{3}{2}, 2\right)$ **19.** $\sqrt{113} \approx 10.63$; $\left(4, \frac{1}{2}\right)$ **21.** $5\sqrt{5} \approx 11.18$;
$\left(2, \frac{3}{2}\right)$ **23.** $2\sqrt{58} \approx 15.23$; (−2, −1) **25.** $2\sqrt{13} \approx 7.21$;
(5, 1) **27.** $\sqrt{115.25} \approx 10.74$; (1.25, −1.3) **29.** 2.5; (−6.25, 3)
31. $\sqrt{\frac{377}{8}} \approx 6.86$; $\left(\frac{17}{8}, \frac{1}{8}\right)$ **33.** isosceles **35.** scalene
37. scalene **39.** scalene **41.** $y = -\frac{1}{3}x + \frac{28}{3}$ **43.** $y = \frac{4}{15}x + \frac{61}{30}$
45. $y = \frac{2}{15}x - 2.22$ **47.** −5; 5 **49.** −15; −1 **51.** $\left(\frac{25}{2}, \frac{35}{2}\right)$;
$\left(\frac{75}{2}, \frac{35}{2}\right)$ **53.** about 18.97 mi **55.** about 11.40 mi
57. r is about 58.56 m, v is about 20 m/sec.

33. $\left(-\frac{5}{2}, 0\right)$; $x = \frac{5}{2}$ **35.** (0, −9); $y = 9$ **37.** (0, 7); $y = -7$

39. 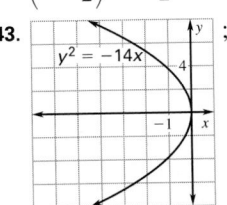 ; **41.** 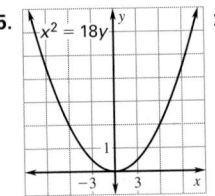 ;

$\left(0, -\frac{3}{2}\right)$; $y = \frac{3}{2}$ (6, 0); $x = -6$

10.1 MIXED REVIEW (p. 594)
65. 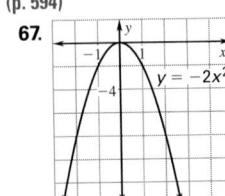 **67.**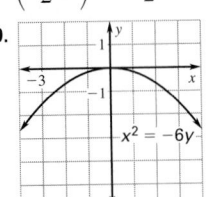

69. 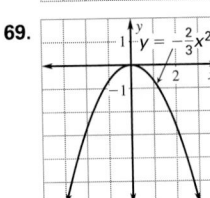 **71.**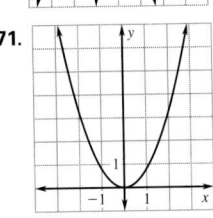

43. 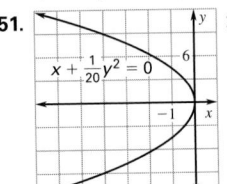 ; **45.** 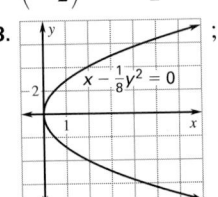 ;

$\left(-\frac{7}{2}, 0\right)$; $x = \frac{7}{2}$ $\left(0, \frac{9}{2}\right)$; $y = -\frac{9}{2}$

51. 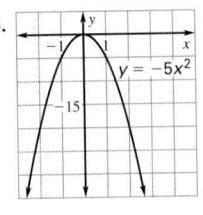 ; **53.** 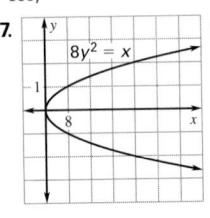 ;

(−5, 0); $x = 5$ (2, 0); $x = -2$
55. $y^2 = -8x$ **57.** $x^2 = 4y$ **59.** $x^2 = -12y$ **61.** $y^2 = -20x$
63. $x^2 = -\frac{3}{2}y$ **65.** $y^2 = \frac{5}{3}x$ **67.** $x^2 = 12y$ **69.** $y^2 = -24x$
71. $x^2 = 4y$ **73.** $x^2 = -16y$ **75.** $y^2 = -3x$ **77.** $x^2 = \frac{1}{3}y$
79. $y^2 = 6x$; 2.04 in. **81.** 2.25 in.

73. 525 **75.** 1 **77.** $4^{2/3} \approx 2.52$ **79.** $\frac{x+6}{3x^2}$ **81.** $\frac{-x^2+4x+9}{x^2+3x}$
83. $\frac{-6x^2+x-11}{(x-6)(2x+1)}$

10.2 MIXED REVIEW (p. 600) **85.** $\frac{4}{7}$ **87.** about 1.209
89. no solution **91.** $\frac{y^3}{2x^3}$ **93.** $x + 3$ **95.** $\frac{1}{6x^2}$
97. $3\sqrt{2} \approx 4.243$ **99.** $\sqrt{569} \approx 23.854$ **101.** $\sqrt{1733} \approx 41.629$
103. $A = 1.5p$

10.2 PRACTICE (pp. 598–600)

5. 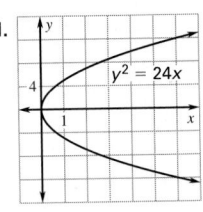 ; **7.** 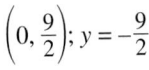 ;

$\left(0, -\frac{1}{20}\right)$; $y = \frac{1}{20}$ $\left(\frac{1}{32}, 0\right)$; $x = -\frac{1}{32}$

10.3 PRACTICE (pp. 604–606) **5.** $x^2 + y^2 = 16$
7. $x^2 + y^2 = 100$ **9.** $x^2 + y^2 = 117$ **11.** $x^2 + y^2 = 50$

SELECTED ANSWERS

13. 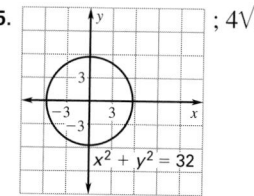 ; 6 **15.** 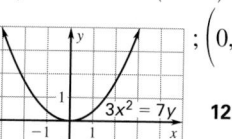 ; $4\sqrt{2}$

17. 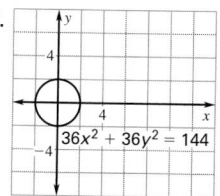 ; 2 **19.** $x^2 + y^2 = 12.25$
21. F **23.** B **25.** A

27. 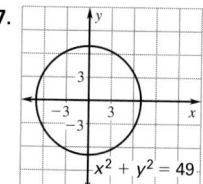 ; **29.** 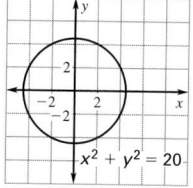 ;

7                    $2\sqrt{5} \approx 4.47$

35. 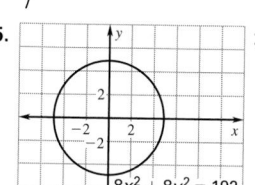 ; **41.**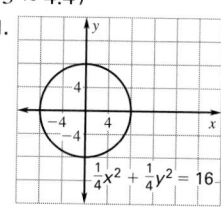

$2\sqrt{6} \approx 4.90$

47. $x^2 + y^2 = 9$ **49.** $x^2 + y^2 = 36$ **51.** $x^2 + y^2 = 7$
53. $x^2 + y^2 = 11$ **55.** $x^2 + y^2 = 150$ **57.** $x^2 + y^2 = 28$
59. $x^2 + y^2 = 100$ **61.** $x^2 + y^2 = 25$ **63.** $x^2 + y^2 = 34$
65. $x^2 + y^2 = 37$ **67.** $x^2 + y^2 = 65$ **69.** $x^2 + y^2 = 89$
71. $y = -\frac{1}{3}x + \frac{10}{3}$ **73.** $y = -\frac{4}{5}x - \frac{41}{5}$ **75.** $y = 8x + 65$
77. $y = -\frac{5}{6}x - \frac{61}{3}$ **79.** $y = \frac{2}{3}x - \frac{13}{3}$; they have opposite
slopes and intercepts. **81.** yes; about 7.92 mi **83.** 16 mm
85. 36 in. **87.** about 3.6 min

10.3 Mixed Review (p. 607)
91. $(-2, -3)$ **93.** $(-2, -2)$ **95.** $(7, 2)$
97. $f(g(x)) = 2x + 1$; $g(f(x)) = 2x + 2$
99. $f(g(x)) = -x^2 - 10x - 26$; $g(f(x)) = -x^2 + 4$
101. 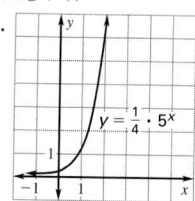 **103.**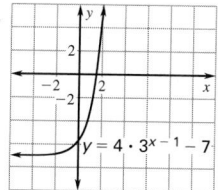

107. $\begin{bmatrix} 35 & 52 \\ 112 & 40 \\ 95 & 63 \end{bmatrix}$

Quiz 1 (p. 607) **1.** 10; (4, 3) **2.** $6\sqrt{2} \approx 8.485$; (0, 0) **3.** $5\sqrt{13} \approx$
18.028; $\left(1, -\frac{3}{2}\right)$ **4.** $2\sqrt{17} \approx 8.246$; (−1, −8) **5.** $2\sqrt{37} \approx$
12.166; (2, 5) **6.** $4\sqrt{58} \approx 30.463$; (5, 1) **7.** $\left(\frac{3}{2}, 0\right)$; $x = -\frac{3}{2}$
8. $\left(0, \frac{3}{4}\right)$; $y = -\frac{3}{4}$ **9.** $\left(0, -\frac{5}{4}\right)$; $y = \frac{5}{4}$ **10.** $\left(-\frac{3}{8}, 0\right)$; $x = \frac{3}{8}$
11. 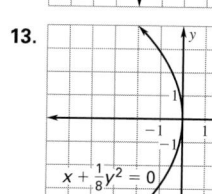 ; $\left(0, \frac{7}{12}\right)$; $y = -\frac{7}{12}$
12. $\left(\frac{1}{16}, 0\right)$; $x = -\frac{1}{16}$

13. 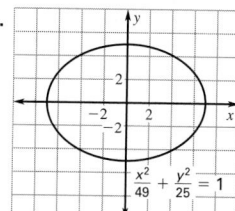 ; (−2, 0); $x = 2$ **14.** (0, −3); $y = 3$
15. $x^2 + y^2 = 9$ **16.** $x^2 + y^2 = 25$
17. $x^2 + y^2 = 65$ **18.** $x^2 + y^2 = 29$
19. $x^2 + y^2 = 82$ **20.** $x^2 + y^2 = 45$
21. $x^2 + y^2 = 72$
22. $x^2 + y^2 = 113$

23. no; $\sqrt{35^2 + 56^2} \approx 66$ mi

Technology Activity 10.3 (p. 608) **1–9:** Sample answers
are given. **1.** $-18 \le x \le 18$; $-12 \le y \le 12$ **3.** $-36 \le x \le 36$;
$-24 \le y \le 24$ **5.** $-3 \le x \le 3$; $-2 \le y \le 2$ **7.** $-9 \le x \le 9$;
$-6 \le y \le 6$ **9.** $-6 \le x \le 6$; $-4 \le y \le 4$

10.4 Practice (pp. 612–614)
5. $\frac{x^2}{16} + \frac{y^2}{25} = 1$ **7.** $\frac{x^2}{49} + \frac{y^2}{9} = 1$ **9.** $\frac{x^2}{91} + \frac{y^2}{100} = 1$
11. 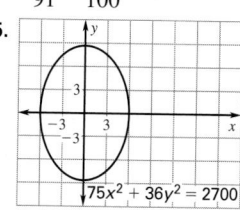 **15.**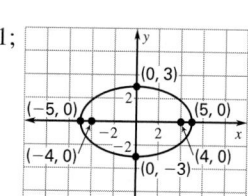

17. $\frac{x^2}{25} + \frac{y^2}{9} = 1$;

19. vertices: (±11, 0); co-vertices: (0, ±10); foci: $\left(\pm\sqrt{21}, 0\right)$
21. vertices: (0, ±5); co-vertices: (±3, 0); foci: (0, ±4)
23. vertices: $\left(\pm 2\sqrt{7}, 0\right)$; co-vertices: $\left(0, \pm 2\sqrt{5}\right)$;
foci: $\left(\pm 2\sqrt{2}, 0\right)$ **25.** $\frac{x^2}{4} + \frac{y^2}{49} = 1$; vertices: (0, ±7);
co-vertices: (±2, 0); foci: $\left(0, \pm 3\sqrt{5}\right)$ **27.** $\frac{x^2}{10} + y^2 = 1$;
vertices: $\left(\pm\sqrt{10}, 0\right)$; co-vertices: (0, ±1); foci: (±3, 0)

SELECTED ANSWERS

29. $\dfrac{x^2}{15} + \dfrac{y^2}{25} = 1$; vertices: $(0, \pm 5)$; co-vertices: $\left(\pm\sqrt{15}, 0\right)$; foci: $\left(0, \pm\sqrt{10}\right)$

31.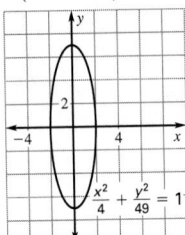
vertices: $(0, \pm 7)$;
co-vertices: $(\pm 2, 0)$;
foci: $\left(0, \pm 3\sqrt{5}\right)$

35.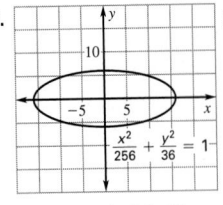
vertices: $(\pm 16, 0)$;
co-vertices: $(0, \pm 6)$;
foci: $\left(\pm 2\sqrt{55}, 0\right)$

37.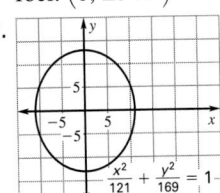
vertices: $(0, \pm 13)$;
co-vertices: $(\pm 11, 0)$;
foci: $\left(0, \pm 4\sqrt{3}\right)$

41.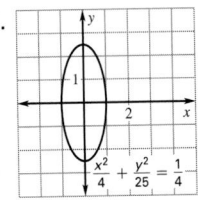
vertices: $(0, \pm 2.5)$;
co-vertices: $(\pm 1, 0)$;
foci: $\left(0, \pm\dfrac{\sqrt{21}}{2}\right)$

43.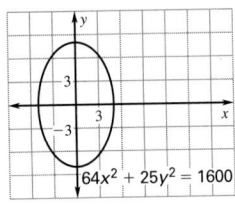

45.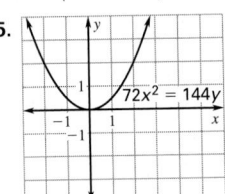

49.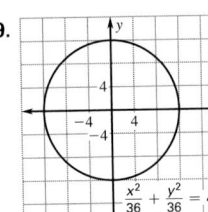

51. $\dfrac{x^2}{25} + \dfrac{y^2}{36} = 1$

53. $\dfrac{x^2}{16} + \dfrac{y^2}{9} = 1$

55. $\dfrac{x^2}{81} + \dfrac{y^2}{64} = 1$

56. $\dfrac{x^2}{100} + \dfrac{y^2}{16} = 1$

57. $\dfrac{x^2}{9} + \dfrac{y^2}{49} = 1$

59. $\dfrac{x^2}{16} + \dfrac{y^2}{64} = 1$

61. $\dfrac{x^2}{25} + \dfrac{y^2}{16} = 1$

63. $\dfrac{x^2}{55} + \dfrac{y^2}{64} = 1$

65. $\dfrac{x^2}{40} + \dfrac{y^2}{121} = 1$

67. $\dfrac{x^2}{275} + \dfrac{y^2}{324} = 1$

69. $\dfrac{x^2}{2352.25} + \dfrac{y^2}{529} = 1$

71. about 3500 ft^2

73. $\dfrac{x^2}{92.5^2} + \dfrac{y^2}{77.5^2} = 1$

75. $3710\pi \le A \le 7170\pi$

10.4 MIXED REVIEW (p. 614) **79.** -32 **81.** $\dfrac{1}{3}$ **83.** 27

85. 16 **87.** $y = \dfrac{24}{x}$ **89.** $y = \dfrac{72}{x}$ **91.** $y = \dfrac{12}{x}$

93.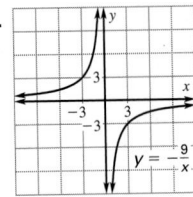
domain: all real numbers
except 0; range: all real
numbers except 0

97.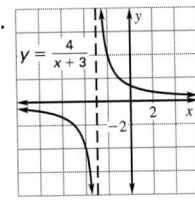
domain: all real numbers
except -3; range: all real
numbers except 0

10.5 PRACTICE (pp. 618–620)

5.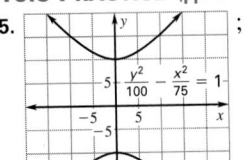
foci: $\left(0, \pm 5\sqrt{7}\right)$;
asymptotes: $y = \pm\dfrac{2\sqrt{3}}{3}x$

7.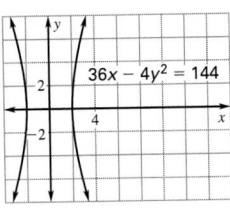
foci: $\left(\pm 2\sqrt{10}, 0\right)$;
asymptotes: $y = \pm 3x$

9.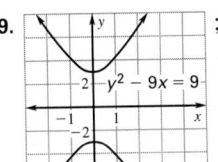
foci: $\left(0, \pm\sqrt{10}\right)$;
asymptotes: $y = \pm 3x$

11. $\dfrac{x^2}{49} - \dfrac{y^2}{15} = 1$ **13.** $\dfrac{y^2}{45} - \dfrac{x^2}{36} = 1$

15. C **17.** D **19.** $\dfrac{x^2}{9} - \dfrac{y^2}{36} = 1$

21. $\dfrac{y^2}{\left(\frac{1}{4}\right)} - \dfrac{x^2}{\left(\frac{9}{4}\right)} = 1$ **23.** $\dfrac{y^2}{4} - \dfrac{x^2}{144} = 1$

25. vertices: $(\pm 3, 0)$; foci: $\left(\pm\sqrt{73}, 0\right)$ **27.** vertices: $(\pm 11, 0)$; foci: $\left(\pm 5\sqrt{5}, 0\right)$ **29.** vertices: $(0, \pm 2)$; foci: $\left(0, \pm\sqrt{29}\right)$

31.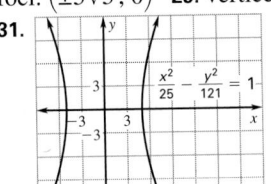
foci: $\left(\pm\sqrt{146}, 0\right)$;
asymptotes: $y = \pm\dfrac{11}{5}x$

33.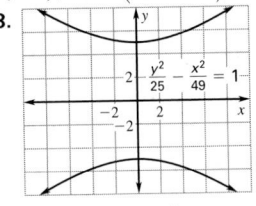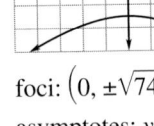
foci: $\left(0, \pm\sqrt{74}\right)$;
asymptotes: $y = \pm\dfrac{5}{7}x$

35.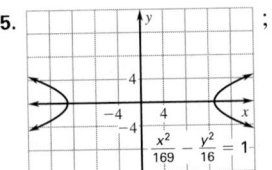
foci: $\left(\pm\sqrt{185}, 0\right)$;
asymptotes: $y = \pm\dfrac{4}{13}x$

41.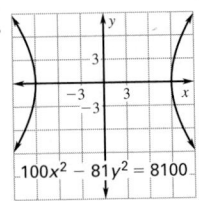
foci: $\left(\pm\sqrt{181}, 0\right)$;
asymptotes: $y = \pm\dfrac{10}{9}x$

SELECTED ANSWERS

43. $y = \pm \dfrac{6\sqrt{x^2 + 100}}{5}$ **45.** $y = \pm \dfrac{8.5\sqrt{x^2 - 42.25}}{6.5}$

47. $y = \pm \sqrt{\dfrac{22.3(x^2 - 10.1)}{10.1}}$

49. *Sample answer:*

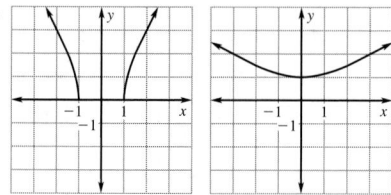

53.

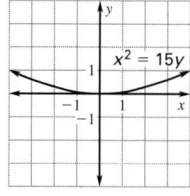

55.

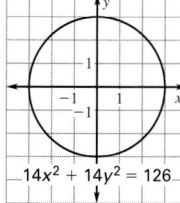

57. $\dfrac{x^2}{36} - \dfrac{y^2}{28} = 1$ **59.** $\dfrac{x^2}{25} - \dfrac{y^2}{11} = 1$ **61.** $\dfrac{y^2}{64} - \dfrac{x^2}{17} = 1$

63. $\dfrac{y^2}{16} - \dfrac{x^2}{134} = 1$ **65.** $\dfrac{y^2}{1024} - \dfrac{x^2}{3070} = 1$ **67.** 10 mi

10.5 MIXED REVIEW (p. 621)

71.

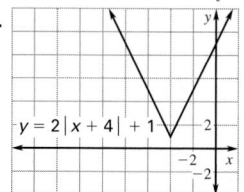

73.

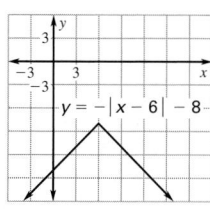

75.

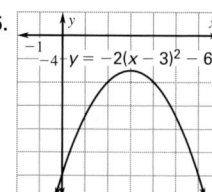

77. $f(x) = x^3 - 6x^2 + 11x - 6$
79. $f(x) = x^3 - 6x^2 - 4x + 24$
81. $f(x) = x^3 - 5x^2 + x - 5$
83. 4   **85.** 4   **87.** 3   **89.** 3
91. mean: 81.67; median: 81; modes: 81, 89; range: 36

QUIZ 2 (p. 621)

1. $\dfrac{x^2}{9} + \dfrac{y^2}{49} = 1$ **2.** $\dfrac{x^2}{36} + y^2 = 1$ **3.** $\dfrac{x^2}{100} + \dfrac{y^2}{64} = 1$

4. $\dfrac{x^2}{8} + \dfrac{y^2}{25} = 1$ **5.** $\dfrac{x^2}{15} + \dfrac{y^2}{12} = 1$ **6.** $\dfrac{x^2}{81} + \dfrac{y^2}{97} = 1$

7.

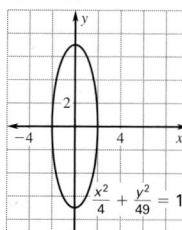

;   **8.**

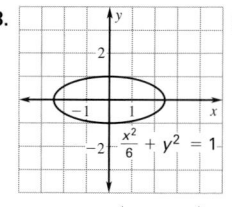

;

vertices: $(0, \pm 7)$;
co-vertices: $(\pm 2, 0)$;
foci: $\left(0, \pm 3\sqrt{5}\right)$

vertices: $\left(\pm\sqrt{6}, 0\right)$;
co-vertices: $(0, \pm 1)$;
foci: $\left(\pm\sqrt{5}, 0\right)$

9.

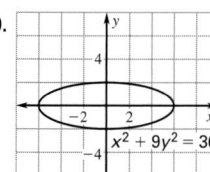

; vertices: $(\pm 6, 0)$;
co-vertices: $(0, \pm 2)$; foci: $\left(\pm 4\sqrt{2}, 0\right)$

10. $\dfrac{y^2}{25} - \dfrac{x^2}{39} = 1$ **11.** $x^2 - \dfrac{y^2}{8} = 1$

12. $\dfrac{x^2}{16} - \dfrac{y^2}{20} = 1$ **13.** $\dfrac{y^2}{16} - \dfrac{x^2}{4} = 1$

14.

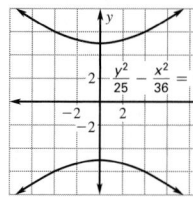

;   **15.**

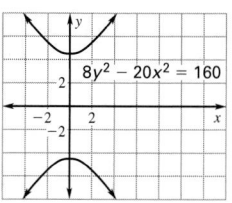

;

vertices: $(0, \pm 5)$;
foci: $\left(0, \pm\sqrt{61}\right)$;
asymptotes: $y = \pm\dfrac{5}{6}x$

vertices: $\left(0, \pm 2\sqrt{5}\right)$;
foci: $\left(0, \pm 2\sqrt{7}\right)$;
asymptotes: $y = \pm\dfrac{\sqrt{10}}{2}x$

16.

; vertices: $\left(\pm\sqrt{2}, 0\right)$;
foci: $\left(\pm\sqrt{11}, 0\right)$;
asymptotes: $y = \pm\dfrac{3\sqrt{2}}{2}x$

17. $\dfrac{x^2}{4375^2} + \dfrac{y^2}{4369^2} = 1$

10.6 PRACTICE (pp. 628–630)   **5.** $\dfrac{(x - 3.5)^2}{20.25} + \dfrac{(y + 4)^2}{18} = 1$

7. $\dfrac{(y + 2)^2}{4} - \dfrac{(x - 5)^2}{12} = 1$ **9.** hyperbola   **11.** ellipse

13. $(x - 9)^2 + (y - 3)^2 = 16$ **15.** $(x - 1)^2 = 12(y + 2)$

17. $\dfrac{(x - 2)^2}{18} + \dfrac{(y - 1.5)^2}{20.25} = 1$ **19.** $\dfrac{y^2}{16} - \dfrac{(x - 5)^2}{20} = 1$

21.

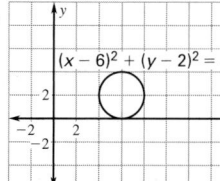

center: $(6, 2)$;
points: $(6, 4), (6, 0), (4, 2), (8, 2)$

23.

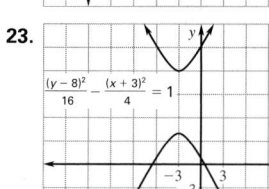

center $(-3, 8)$;
vertices: $(-3, 4), (-3, 12)$;
foci: $\left(-3, 8 \pm 2\sqrt{5}\right)$

25.

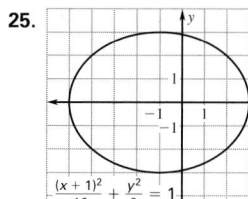

center $(-1, 0)$;
vertices: $(-5, 0), (3, 0)$;
co-vertices: $(-1, -3), (-1, 3)$;
foci: $\left(-1 \pm \sqrt{7}, 0\right)$
29. ellipse   **31.** hyperbola
33. ellipse   **35.** hyperbola   **37.** parabola   **39.** ellipse

SELECTED ANSWERS

41. circle **43.** hyperbola **45.** E **47.** D **49.** B
51. parabola;
$(y-6)^2 = -4(x-8)$;

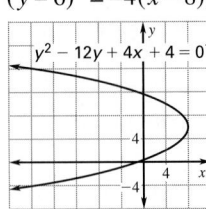

53. hyperbola;
$(x-4)^2 - \dfrac{(y-4)^2}{9} = 1$;

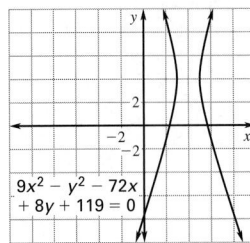

55. ellipse;
$\dfrac{(x-1)^2}{4} + (y-1)^2 = 1$;

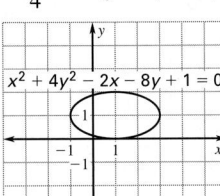

59. circle;
$(x-6)^2 + (y-6)^2 = 36$;

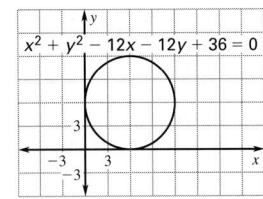

61. parabola;
$(x+2)^2 = 8(y-1)$;

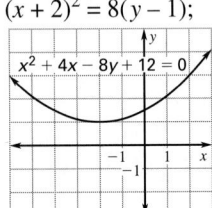

63. $y^2 = 12x$; $y^2 = -12(x-50)$ $(x, y$ in ft)
65. ellipse
67. The first is elliptical, the second is parabolic.

10.6 MIXED REVIEW (p. 631) **71.** $(5, -5)$ **73.** $(1, -2)$
75. $\left(\dfrac{68}{23}, \dfrac{123}{23}\right)$ **77.** 5 **79.** 0 **81.** 2 **83.** about 0.45
85. about 4.03 **87.** about 0.27

10.7 PRACTICE (pp. 635–637) **5.** $(-1, 0)$, $(-7, 0)$ **7.** $(-2, -5)$, $(4, -5)$ **9.** no **11.** yes **13.** no **15.** $(1, -4)$, $(2, -1)$
17. $(3, 6)$, $(-3, -6)$ **19.** $(1, -2)$, $(-1, 2)$ **21.** $(-1, -2)$, $\left(\dfrac{7}{4}, \dfrac{3}{4}\right)$
23. $\left(\dfrac{6-\sqrt6}{5}, \dfrac{24+\sqrt6}{10}\right)$, $\left(\dfrac{6+\sqrt6}{5}, \dfrac{24-\sqrt6}{10}\right)$
25. $\left(2+\sqrt6, \sqrt6 - 2\right)$, $\left(2-\sqrt6, -\sqrt6 - 2\right)$
27. $(0, 0)$, $(-6, 6)$ **29.** none **31.** $(0, 2)$, $\left(\dfrac{4}{3}, \dfrac{2}{3}\right)$
33. $\left(\pm\sqrt{\dfrac{5\sqrt{69}-15}{2}}, \dfrac{-5+\sqrt{69}}{2}\right)$
35. $\left(\dfrac{1+\sqrt{373}}{6}, \pm\sqrt{\dfrac{7+\sqrt{373}}{18}}\right)$ **37.** none **39.** none
41. $\left(\dfrac{9\sqrt2}{2}, -\dfrac{9\sqrt2}{2}\right)$, $\left(-\dfrac{9\sqrt2}{2}, \dfrac{9\sqrt2}{2}\right)$ **43.** none **45.** $(4, 0)$

47. $(6, -8)$, $(14, -8)$ **49.** $(2, 3)$ **51.** $\left(\pm\sqrt6, 2\right)$, $\left(\pm\sqrt3, -1\right)$
53. no intersection **55.** $(5, 7)$ **59.** about 56.9 mi
61. $\left(4\sqrt5, \dfrac{6\sqrt5}{5}\right) \approx (8.9, 2.7)$ **63.** *Sample answer:* The epicenter of the earthquake is about 100 kilometers east and about 1300 kilometers south of Location 1.

10.7 MIXED REVIEW (p. 638) **67.** 13 **69.** 16
71. $f(x) = x^3 - x^2 - 9x + 9$ **73.** $f(x) = x^2 + 4$
75. $f(x) = x^5 - 2x^3 - 2x^2 - 3x - 2$

77. 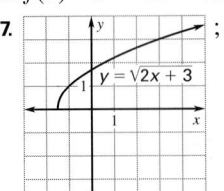 ; **79.** 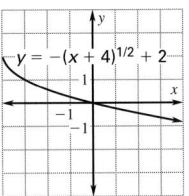 ;

domain: $x \geq -\dfrac{3}{2}$;
range: $y \geq 0$

domain: $x \geq -4$;
range: $y \leq 2$

81. 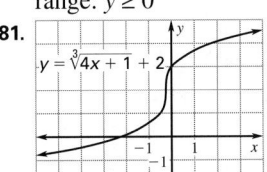 ; domain: all reals; range: all reals **83.** ellipse **85.** parabola

QUIZ 3 (p. 638) **1.** $(x+3)^2 + (y+5)^2 = 64$
2. $\dfrac{(x+0.5)^2}{42.25} + \dfrac{(y-2)^2}{22} = 1$ **3.** $(y+1)^2 = 12(x-4)$
4. $\dfrac{(y-3.5)^2}{0.25} - \dfrac{(x-2)^2}{20} = 1$ **5.** ellipse **6.** circle **7.** parabola
8. hyperbola **9.** $\left(\dfrac{2}{3}, \dfrac{2}{3}\right)$, $(-1, 9)$ **10.** $(2, 2)$, $(2, 4)$
11. $(4, -2)$, $(-4, -2)$ **12.** none **13.** The epicenter of the earthquake is 50 mi due west of the first seismograph.

CHAPTER 10 EXTENSION (p. 640) **1.** 1 **3.** $\dfrac{\sqrt{15}}{4} \approx 0.968$
5. $\sqrt5 \approx 2.236$ **7.** $\dfrac{\sqrt6}{2} \approx 1.225$ **9.** $\dfrac{x^2}{25} + \dfrac{(y+1)^2}{16} = 1$
11. $\dfrac{(x-2)^2}{60} + \dfrac{y^2}{64} = 1$ **13.** $\dfrac{(y-1)^2}{\left(\frac{64}{9}\right)} - \dfrac{(x-3)^2}{\left(\frac{512}{9}\right)} = 1$
15. $\dfrac{(y-2)^2}{9} - \dfrac{(x-3)^2}{23.49} = 1$ **17.** $\dfrac{x^2}{1296} + \dfrac{y^2}{1241} = 1$
$(x, y$ in millions of miles) **19.** In an ellipse, the foci are always within the major axis, so $c < a$ and $\dfrac{c}{a} < 1$. In a hyperbola, the foci are always outside the major axis, so $c > a$ and $\dfrac{c}{a} > 1$.

CHAPTER 10 REVIEW (pp. 642–644)

1. $\sqrt{61} \approx 7.81$; $\left(1, -\frac{1}{2}\right)$ **3.** $4\sqrt{2} \approx 5.66$; $(-2, 2)$

5. focus: $(0, 1)$; directrix: $y = -1$;

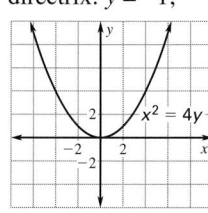

7. focus: $\left(-\frac{3}{2}, 0\right)$; directrix: $x = \frac{3}{2}$;

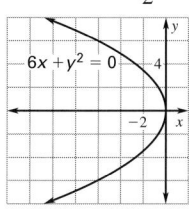

9. $y^2 = 16x$ **11.** $x^2 = 8y$

13.

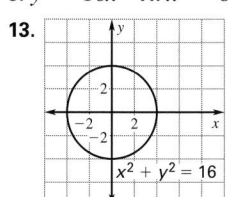

15.

17. $x^2 + y^2 = 25$ **19.** $x^2 + y^2 = 13$

21.

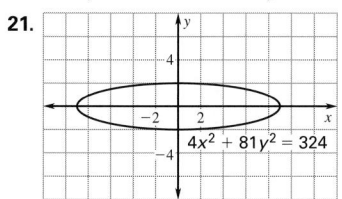

23.

25. $\frac{x^2}{16} + \frac{y^2}{7} = 1$ **27.**

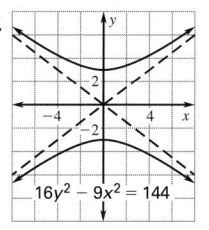

29. $y^2 - \frac{x^2}{8} = 1$

31. $\frac{x^2}{9} - \frac{y^2}{16} = 1$

33. circle; $(x - 5)^2 + (y + 1)^2 = 100$;

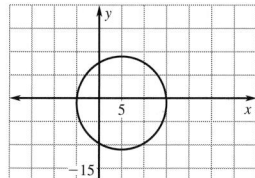

35. hyperbola;

$(y - 9)^2 - \dfrac{(x + 1)^2}{\left(\frac{1}{4}\right)} = 1$;

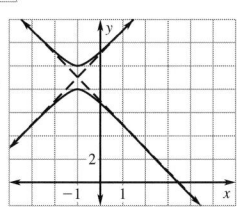

37. $\left(\dfrac{-27 - \sqrt{1649}}{10}, \dfrac{141 + 3\sqrt{1649}}{10}\right)$;

$\left(\dfrac{-27 + \sqrt{1649}}{10}, \dfrac{141 - 3\sqrt{1649}}{10}\right) \approx (-6.761, 26.28)$;

$(1.361, 1.918)$ **39.** $(-1, -3)$ and $(-1, 3)$

CHAPTER 11

SKILL REVIEW (p. 650)   **1.** $n + 4$ **2.** $3n$ **3.** $\frac{n}{2}$ **4.** 2   **5.** $\frac{8}{9}$ **6.** 24

7.

	$f(0)$	$f(1)$	$f(2)$	$f(3)$	$f(4)$	$f(5)$	$f(6)$
function values	3	0	−9	−24	−45	−72	−105
1st order differences	−3	−9	−15	−21	−27	−33	
2nd order differences	−6	−6	−6	−6	−6		

8.

	$f(1)$	$f(2)$	$f(3)$	$f(4)$	$f(5)$	$f(6)$
function values	3	16	45	96	175	288
1st order differences	13	29	51	79	113	
2nd order differences	16	22	28	34		
3rd order differences	6	6	6			

9.

	$f(1)$	$f(2)$	$f(3)$	$f(4)$	$f(5)$	$f(6)$
function values	−3	7	67	237	601	1267
1st order differences	10	60	170	364	666	
2nd order differences	50	110	194	302		
3rd order differences	60	84	108			
4th order differences	24	24				

10. 7   **11.** 6   **12.** $\frac{1}{2}$ **13.** $-\frac{11}{12}$

11.1 PRACTICE (pp. 655–657)   **3.** 2, 4, 6, 8, 10, 12   **5.** 4, 7, 10, 13, 16, 19   **7.** 68   **9.** 2, 3, 4, 5, 6, 7   **11.** 2, 1, 0, −1, −2, −3   **13.** 4, 9, 16, 25, 36, 49   **15.** 4, 7, 12, 19, 28, 39   **17.** $\frac{1}{2}, \frac{2}{3}, \frac{3}{4}, \frac{4}{5}, \frac{5}{6}, \frac{6}{7}$ **19.** $\frac{3}{2}, 1, \frac{5}{6}, \frac{3}{4}, \frac{7}{10}, \frac{2}{3}$ **21.** 9; $2n - 1$ **23.** −16; $a_n = 3n - 1$ if n is odd or $2 - 3n$ if n is even.   **25.** $-\frac{1}{10}$; $-\frac{1}{2n}$ **27.** 2; $\frac{n}{3}$ **29.** 5.9; $1.1 + 0.8n$

31.

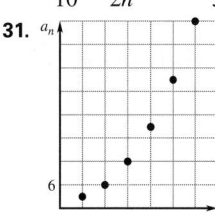

33.

35.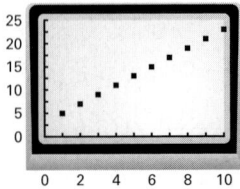

37. $\displaystyle\sum_{i=1}^{5} 4i$

39. $\displaystyle\sum_{k=1}^{\infty} (-1)^{k-1}k$, or $\displaystyle\sum_{k=1}^{\infty} (-1)^{k+1}k$

41. $\displaystyle\sum_{n=5}^{\infty} \frac{n}{n+1}$ **43.** $\displaystyle\sum_{i=1}^{6} i^2$ **45.** 180

47. 144 **49.** 65 **51.** $\dfrac{4861}{1260} \approx 3.858$

53. 3.55 **55.** 0 **57.** 15 **59.** 210 **61.** 385 **63.** 14,910
65. 15 ft **67.** $2^n - 1$; 63 **69.** B

71. a. true; $\displaystyle\sum_{i=1}^{n} ka_i = ka_1 + ka_2 + \ldots + ka_n =$

$k(a_1 + a_2 + \ldots + a_n) = k\displaystyle\sum_{i=1}^{n} a_i$

b. true; $\displaystyle\sum_{i=1}^{n} (a_i + b_i) = (a_1 + b_1) + (a_2 + b_2) + \ldots +$

$(a_n + b_n) = (a_1 + a_2 + \ldots + a_n) + (b_1 + b_2 + \ldots + b_n) =$

$\displaystyle\sum_{i=1}^{n} a_i + \sum_{i=1}^{n} b_i$

c. false; $\displaystyle\sum_{i=1}^{4} i(i+1) = 1(2) + 2(3) + 3(4) + 4(5) =$

$2 + 6 + 12 + 20 = 40$, but $\displaystyle\sum_{i=1}^{4} i = 10$ and $\displaystyle\sum_{i=1}^{4} (i+1) = 14$

and $10 \times 14 = 140 \neq 40$.

d. false; $\displaystyle\sum_{i=1}^{5} (i)^2 = 1 + 4 + 9 + 16 + 25 = 55$,

but $\left(\displaystyle\sum_{i=1}^{5} i\right)^2 = 15^2 = 225$.

11.1 MIXED REVIEW (p. 657) **73.** 4 **75.** 2 **77.** $\dfrac{1}{2}$

79. $y = (2.5)2^x$;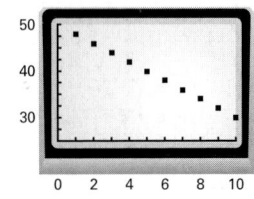

81. $2\sqrt{13} \approx 7.211$
83. $6\sqrt{2} \approx 8.485$
85. $\sqrt{65} \approx 8.062$

TECHNOLOGY ACTIVITY 11.1 (p. 658)

1. 5, 7, 9, 11, 13, 15, 17, 19, 21, 23; 140 **3.** 48, 46, 44, 42, 40, 38, 36, 34, 32, 30; 390

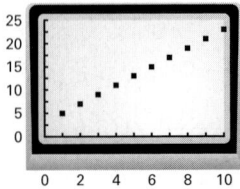

5. $\dfrac{1}{2}, \dfrac{1}{4}, \dfrac{1}{8}, \dfrac{1}{16}, \dfrac{1}{32}, \dfrac{1}{64}, \dfrac{1}{128}$, **7.** $\dfrac{3}{4}$, 3, 12, 48, 192, 768,

$\dfrac{1}{256}, \dfrac{1}{512}, \dfrac{1}{1024}, \dfrac{1023}{1024}$ 3072, 12,288, 49,152, 196,608; 262,143.75

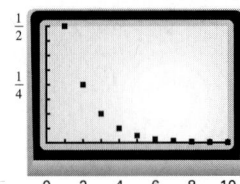 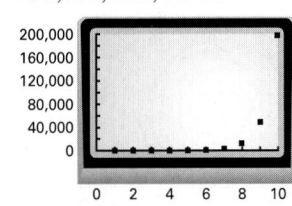

9. $\dfrac{4}{3}, 8\dfrac{1}{3}, 27\dfrac{1}{3}, 64\dfrac{1}{3}, 125\dfrac{1}{3}$,

$216\dfrac{1}{3}, 343\dfrac{1}{3}, 512\dfrac{1}{3}, 729\dfrac{1}{3}$,

$1000\dfrac{1}{3}; 3028\dfrac{1}{3}$

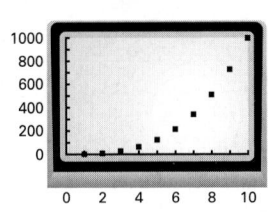

11.2 PRACTICE (pp. 663–664) **5.** $a_n = 24 - 3n$

7. $a_n = -44 + 7n$ **9.** $a_n = -32 + 4n$ **11.** $\dfrac{105}{2}$ **13.** 61 **15.** Yes;

constant difference is -3. **17.** No; difference is not constant.
19. No; difference is not constant. **21.** $a_n = -1 + 2n$; 49
23. $a_n = -5 + 14n$; 345 **25.** $a_n = 7 - 3n$; -68
27. $a_n = \dfrac{41}{6} - \dfrac{4}{3}n$; $-\dfrac{53}{2}$ **29.** $a_n = -0.8 + 2.4n$; 59.2
31. $a_n = 92 - 12n$ **33.** $a_n = -13 + 6n$ **35.** $a_n = -\dfrac{332}{9} + \dfrac{40}{9}n$
37. $a_n = -22 + 8n$

39. 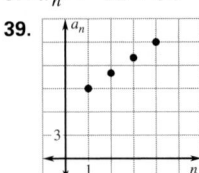   **41.**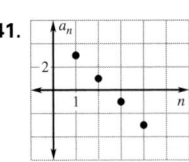

43.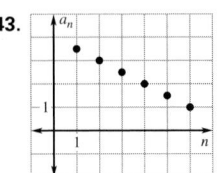

45. a. 1010 **b.** 12 **47. a.** 665 **b.** 15
49. a. 16,082 **b.** 22 **51.** 1110
53. -510 **55.** 4635 **57. a.** $a_n = 6n$
b. 271 **59.** $1 + 4\displaystyle\sum_{i=1}^{4} 2i$; 81 ft^2

11.2 MIXED REVIEW (p. 665) **65.** 81 **67.** no solution **69.** 8
71. $\dfrac{3}{2}$ **73.** 1 **75.** $\log_4 3.4 \approx 0.8828$

77. 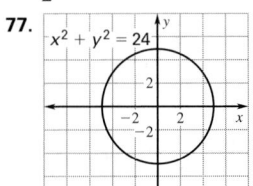   **79.**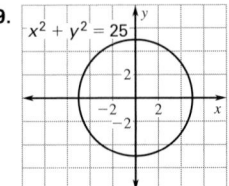

81.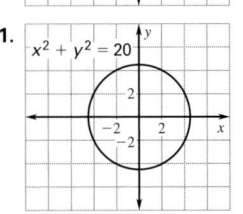

SELECTED ANSWERS

11.3 PRACTICE (pp. 670–672) **5.** 6 **7.** 2 **9.** $\frac{1}{2}$

11. 512; $2(4)^{n-1}$ **13.** 0.6; $375\left(-\frac{1}{5}\right)^{n-1}$ **15.** $\frac{7}{8}$; $\frac{-28}{(-2)^{n-1}}$

17. $6(-2)^{n-1}$ **19.** $\left(\frac{1}{4}\right)\left(2\sqrt{6}\right)^{n-1}$ **21.** $-7(-4)^{n-1}$

23. $43.11 **25.** neither; no common ratio or difference
27. arithmetic; common difference of -4 **29.** geometric; common ratio of 3 **31.** neither; no common ratio or difference **33.** 4 **35.** -2 **37.** $\frac{1}{2}$ **39.** $(-4)^{n-1}$; -1024

41. $2(7)^{n-1}$; 33,614 **43.** $5\left(-\frac{1}{3}\right)^{n-1}$; $-\frac{5}{243}$ **45.** $4(3)^{n-1}$

47. $2(6)^{n-1}$ **49.** $-2(8)^{n-1}$ **51.** $\left(\frac{10}{\sqrt[3]{900}}\right)\left(\sqrt[3]{30}\right)^{n-1}$

53. $6(-5)^{n-1}$ **55.** 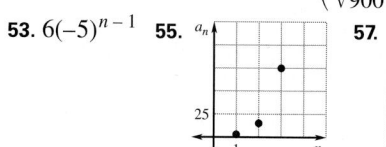 **57.**

59. 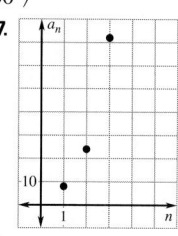 **61. a.** 435,848,050 **b.** 4
63. a. -67.5 **b.** 4 **65.** 109,225
67. 30.198 **69.** -1365 **71.** 127
73. 10 **75.** $169.92 million

77. about $1.524 billion **79.** $\left(\frac{\sqrt{3}}{4}\right)\left(\frac{3}{4}\right)^{n}$; 0.006

11.3 MIXED REVIEW (p. 673)
83.

85.

$\frac{2}{5}, \frac{6}{7}, 1, \frac{7}{6}, \frac{3}{2}$ $-3.2, -\frac{5}{2}, -2, -1, 1.5$
87. $-1 \le x \le 7$ **89.** all real numbers **91.** $-6 \le x \le -4$
93. $\frac{3}{5}$ **95.** $-8, -3$ **97.** 10

QUIZ 1 (p. 673) **1.** 8; $2(n-1)$ **2.** 243; 3^n
3. $\frac{1}{80}$; $\left(\frac{1}{5}\right)\left(-\frac{1}{2}\right)^{n-1}$ **4.** 354 **5.** 121 **6.** 220 **7.** $-3 + 4n$; 45
8. $43 - 9n$; -65 **9.** $\frac{n}{2}$; 6 **10.** 694.5 **11.** $2(5)^{n-1}$;
12,207,031,250 **12.** $-3(-4)^{n-1}$; $-805,306,368$
13. $12\left(\frac{1}{3}\right)^{n-1}$; 2.509×10^{-6} **14.** 2^{n-1}; 1023

11.4 PRACTICE (pp. 678–679) **5.** $-\frac{8}{5}$ **7.** $\frac{5}{6}$ **9.** $\frac{5}{9}$

11. $\frac{245,000}{999}$ **13.** no; $|r| = \frac{3}{2}, \frac{3}{2} > 1$ **15.** yes; $|r| = \frac{1}{3}$,

$\frac{1}{3} < 1$ **17.** 2 **19.** $\frac{2}{3}$ **21.** $\frac{16}{3}$ **23.** no sum **25.** $-\frac{1}{12}$

27. $\frac{25}{336}$ **29.** $\frac{3}{4}$ **31.** $\frac{3}{4}$ **33.** $\frac{2}{3}$ **35.** $-\frac{1}{2}$ **37.** $-\frac{4}{5}$ **39.** $\frac{7}{9}$ **41.** $\frac{17}{33}$

43. $\frac{16}{99}$ **45.** $\frac{40,000}{333}$ **47.** 180 in. = 15 ft; after 16 swings

49. total distance $= \dfrac{20}{1 - \frac{1}{2}} = 40$ ft; total time $= \dfrac{1}{1 - \frac{1}{2}} = 2$ sec

51. about M$24.21

11.4 MIXED REVIEW (p. 680)
57. x-axis; y-axis; domain: $x \ne 0$; range: $y \ne 0$
59. $y = 1$; $x = -7$; domain: $x \ne -7$; range: $y \ne 1$
61. $y = 2.2$; $x = 0.7$; domain: $x \ne 0.7$; range: $y \ne 2.2$
63. 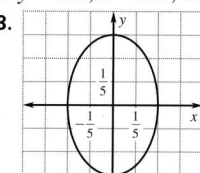 **65.** $-3 + 4n$ **67.** $13 - 2n$
69. $4\left(\frac{1}{2}\right)^{n-1}$

11.5 PRACTICE (pp. 684–685) **5.** 2, 8, 32, 128, 512
7. $-1, 3, -9, 27, -81$ **9.** 3; 10; 101; 10,202; 104,080,805
11. $a_1 = 2$; $a_n = (3)a_{n-1}$ **13.** about 3148 **15.** 4, 12, 21, 31, 42 **17.** $-4, -12, -20, -28, -36$ **19.** 5, -4, 8, 1, 15
21. 2, 1, 7, 8, 16 **23.** 48, 26, 15, 9.5, 6.75 **25.** 1, 3, 3, 9, 27
27. $-7 + 10n$; $a_1 = 3$; $a_n = a_{n-1} + 10$ **29.** $2 + 3n$; $a_1 = 5$;
$a_n = a_{n-1} + 3$ **31.** $5(2.5)^{n-1}$; $a_1 = 5$; $a_n = (2.5)a_{n-1}$
33. $\left(\frac{1}{2}\right)4^{n-1}$; $a_1 = \frac{1}{2}$; $a_n = (4)a_{n-1}$ **35.** $a_1 = 1$;
$a_n = a_{n-1} + 6$ **37.** $a_1 = 41$; $a_n = a_{n-1} - 9$ **39.** $a_1 = 33$;
$a_n = \dfrac{a_{n-1}}{3}$ **41.** $a_1 = 2$; $a_2 = 5$; $a_n = a_{n-1} \cdot a_{n-2}$

43. $a_1 = 48$; $a_n = \dfrac{a_{n-1}}{10}$ **45.** 1, 2, 4, 8, 16, 32, 64;
geometric **47.** $a_1 = 32$; $a_n = 0.6a_{n-1} + 14$; about 34.77 oz
49. $a_1 = 9000$; $a_n = (0.9)a_{n-1} + 800$; 8729,
51. $a_1 = 20$; $a_n = (0.7)a_{n-1} + 20$ **53.** no

11.5 MIXED REVIEW (p. 686) **59.** 32 **61.** 4096 **63.** 17,576
65. 5832 **67.** $\frac{36}{35x}$ **69.** $\frac{-(4x+14)}{x^2-9}$ **71.** $\frac{x-5}{x^2-4}$
73. $(-1.272, 1.544)$; $(0.472, -1.944)$ **75.** $(-0.980, -1.939)$;
$(0.331, 1.993)$ **77.** $(-4.742, -2.742)$; $(2.742, 4.742)$
79. 7, 6, 5, 4, 3, 2 **81.** 10, 13, 18, 25, 34, 45
83. $\frac{1}{5}, \frac{1}{3}, \frac{3}{7}, \frac{1}{2}, \frac{5}{9}, \frac{3}{5}$

QUIZ 2 (p. 687) **1.** $\frac{9}{2}$ **2.** $\frac{35}{13}$ **3.** $-\frac{7}{8}$ **4.** no sum **5.** $\frac{4}{5}$ **6.** $\frac{11}{12}$
7. $\frac{7}{8}$ **8.** $\frac{8}{9}$ **9.** $\frac{5}{33}$ **10.** $\frac{14,000}{111}$ **11.** 5, 8, 11, 14, 17 **12.** 1, 4, 16, 64, 256 **13.** 17, 19, 22, 26, 31 **14.** 1, 2, 1, -1, -2
15. 2, 4, 8, 32, 256 **16.** 10, 10, 20, 30, 50 **17.** $18\frac{2}{3}$ ft

TECHNOLOGY ACTIVITY 11.5 (p. 688) **1.** 5100, 4465,
3893.5, 3379.15, 2916.24, 2499.61, 2124.65, 1787.19,
1483.47, 1210.12, 964.11, 742.70
3. 3500, 2925, 2436.25, 2020.81, 1667.69, 1367.54,
1112.41, 895.55, 711.21, 554.53, 421.35, 308.15
5. 103 months or 8 years, 7 months

SELECTED ANSWERS

CHAPTER 11 EXTENSION (p. 689–690)

1. $\dfrac{1(1+1)(2\cdot 1+1)}{6} = \dfrac{6}{6} = 1 = 1^2$, so the formula is true for $n = 1$. Suppose $1^2 + 2^2 + \ldots k^2 = \dfrac{k(k+1)(2k+1)}{6}$. Then

$1^2 + 2^2 + \ldots k^2 + (k+1)^2 = \dfrac{k(k+1)(2k+1)}{6} + (k+1)^2 =$

$\dfrac{k(k+1)(2k+1)+6(k+1)^2}{6} = \dfrac{(k+1)(2k^2+k+6k+6)}{6} =$

$\dfrac{(k+1)(2k^2+7k+6)}{6} = \dfrac{(k+1)(k+2)(2k+3)}{6} =$

$\dfrac{(k+1)[(k+1)+1][2(k+1)+1]}{6}$, and the formula is true for $n = k + 1$. Therefore, the formula is true for all positive integers.

3. $\dfrac{a_1(1-r^1)}{1-r} = a_1 \cdot r^{1-1}$, so the statement is true for $n = 1$.

Assume it is true for $n = k$. Then $\displaystyle\sum_{i=1}^{k} a_1 r^{i-1} = \dfrac{a_1(1-r^k)}{1-r}$,

so $\displaystyle\sum_{i=1}^{k+1} a_1 r^{i-1} = \dfrac{a_1(1-r^k)}{1-r} + a_1 r^{k+1-1} =$

$\dfrac{a_1(1-r^k)+a_1(r^k)(1-r)}{1-r} = \dfrac{a_1[(1-r^k)+r^k(1-r)]}{1-r} =$

$\dfrac{a_1(1-r^{k+1})}{1-r}$, and the formula is true for $n = k + 1$.
Therefore, the formula is true for all positive integers.

5. $\dfrac{5^{1+1}-5}{4} = \dfrac{25-5}{4} = \dfrac{20}{4} = 5 = 5^1$, so the formula is true for $n = 1$. Suppose the formula is true for $n = k$.

Then $\displaystyle\sum_{i=1}^{k} 5^i = \dfrac{5^{k+1}-5}{4}$. So $\displaystyle\sum_{i=1}^{k+1} 5^i = \dfrac{5^{k+1}-5}{4} + 5^{k+1} =$

$\dfrac{[(5^{k+1}-5)+4(5^{k+1})]}{4} = \dfrac{5(5^{k+1})-5}{4} = \dfrac{5^{(k+1)+1}-5}{4}$, and

the formula is true for $n = k + 1$. Therefore, the formula is true for all positive integers.

7. A recursive formula for the nth pentagonal number is $P_n = P_{n-1} + 3n - 2$. $\dfrac{1(3\cdot 1-1)}{2} = 1 = P_1$, so the formula is true for $n = 1$. Suppose the formula is true for $n = k$. Then

$P_k = \dfrac{k(3k-1)}{2}$, so $P_{k+1} = \dfrac{k(3k-1)}{2} + 3(k+1) - 2 =$

$\dfrac{3k^2-k+6k+6-4}{2} = \dfrac{3k^2+5k+2}{2} = \dfrac{(k+1)(3k+2)}{2} =$

$\dfrac{(k+1)[3(k+1)-1]}{2}$, and the formula is true for $n = k + 1$.
Therefore, the formula is true for all positive integers.

CHAPTER 11 REVIEW (pp. 692–694) **1.** 6, 9, 14, 21, 30, 41

3. 4, 2, 0, −2, −4, −6 **5.** 10; $2n$ **7.** $\dfrac{1}{243}$; $\left(\dfrac{1}{3}\right)^n$ **9.** $\displaystyle\sum_{i=1}^{\infty} i$

11. 5525 **13.** 78 **15.** $-5 + 6n$ **17.** $4 - \dfrac{1}{2}n$ **19.** $21 - 2n$

21. 1204 **23.** 599.4 **25.** $64\left(\dfrac{1}{2}\right)^{n-1}$ **27.** $200\left(\dfrac{1}{10}\right)^{n-1}$

29. $-64\left(-\dfrac{1}{4}\right)^{n-1}$ **31.** 496 **33.** 19.844 **35.** $\dfrac{135}{7}$ **37.** 25

39. $\dfrac{1}{3}$ **41.** $\dfrac{4}{5}$ **43.** $\dfrac{7}{10}$ **45.** $\dfrac{3}{4}$ **47.** $\dfrac{2}{9}$ **49.** $\dfrac{1300}{33}$ **51.** 10; 40; 160; 640; 2560; 10,240 **53.** 2, 0, −3, −7, −12, −18

55. $a_1 = 7$; $a_n = 2 \cdot a_{n-1}$ **57.** $a_1 = 1$; $a_n = a_{n-1} + 5$

59. $a_1 = 1$; $a_n = (a_{n-1})^2 + 1$

CHAPTER 12

SKILL REVIEW (p. 700) **1.** 0.5, 50% **2.** 0.2, 20% **3.** 0.15, 15% **4.** 0.48, 48% **5.** 0.194, 19.4% **6.** 0.469, 46.9% **7.** 50.27 **8.** 25 **9.** 48 **10.** −0.301 **11.** 0.415 **12.** −0.131

12.1 PRACTICE (pp. 705–707) **5.** 2 **7.** 1 **9.** 120 **11.** 6 **13.** 210 **15.** 3 **17.** 40 **19. a.** 17,576,000 **b.** 11,232,000 **21. a.** 6,760,000 **b.** 3,276,000 **23.** 40,320 **25.** 3,628,800 **27.** 1 **29.** 6 **31.** 6 **33.** 2 **35.** 6720 **37.** 1320 **39.** 2 **41.** 24 **43.** 720 **45.** 40,320 **47.** 3 **49.** 360 **51.** 2520 **53.** 10,080 **55.** 480 **57. a.** 2,176,782,336 **b.** 1,402,410,240 **59.** 6.20×10^{23} **61. a.** 720 **b.** 60,480 **63.** 12,612,600

12.1 MIXED REVIEW (p. 707) **69.** $x^4 + 4x^2 + 4$ **71.** $16x^2 - 25$ **73.** $64y^2 - 16xy + x^2$

75.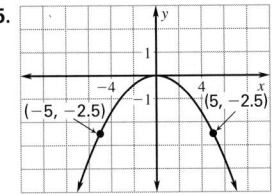

77. 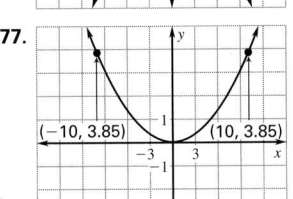 **79.**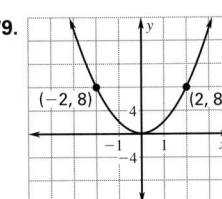

81. $-\dfrac{16}{3}$ **83.** no sum **85.** 0.714

12.2 PRACTICE (pp. 712–714) **5.** 28 **7.** 5 **9.** $x^3 + 3x^2y + 3xy^2 + y^3$ **11.** $8x^3 + 48x^2 + 96x + 64$ **13.** $x^5 - 5x^4y + 10x^3y^2 - 10x^2y^3 + 5xy^4 - y^5$ **15.** $81x^4 - 108x^3 + 54x^2 - 12x + 1$ **17.** 270; 243 **19.** 56 **21.** 28 **23.** 1 **25.** 165 **27.** 48 **29.** 24

31.
```
                1
              1   1
            1   2   1
          1   3   3   1
        1   4   6   4   1
      1   5  10  10   5   1
    1   6  15  20  15   6   1
  1   7  21  35  35  21   7   1
```
33. $x^6 - 18x^5y + 135x^4y^2 - 540x^3y^3 + 1215x^2y^4 - 1458xy^5 + 729y^6$

35. $128x^7 - 448x^6y^3 + 672x^5y^6 - 560x^4y^9 + 280x^3y^{12} - 84x^2y^{15} + 14xy^{18} - y^{21}$ **37.** $x^5 + 20x^4 + 160x^3 + 640x^2 + 1280x + 1024$ **39.** $64x^6 - 192x^5y + 240x^4y^2 - 160x^3y^3 + 60x^2y^4 - 12xy^5 + y^6$ **41.** $81x^8 - 324x^6 + 486x^4 - 324x^2 + 81$ **43.** $x^9 + 3x^6y^2 + 3x^3y^4 + y^6$ **45.** 1120 **47.** 120 **49.** 315 **51.** 968 **53.** 968

61.

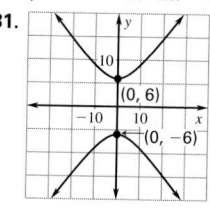

1 col. 2 col. 3 col. 4 col. 5 col.

63. 512 **65.** 2,097,151

12.2 MIXED REVIEW (p. 715) **75.** 107.35 in.^{2} **77.** 310.5 m^2

79.

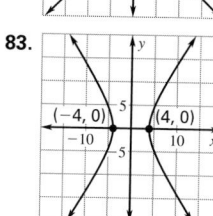

(0, 10)
(0, −10)

81.

(0, 6)
(0, −6)

83.

(−4, 0) (4, 0)

85. arithmetic; $-4 + 7n$
87. geometric; $(-2)^{n-1}$
89. arithmetic; $-15 + 5n$

QUIZ 1 (p. 715) **1.** 3 **2.** 24 **3.** 120 **4.** 180 **5.** 5040
6. 40,320 **7.** 60,480 **8.** 907,200 **9.** $x^6 + 6x^5y + 15x^4y^2 + 20x^3y^3 + 15x^2y^4 + 6xy^5 + y^6$ **10.** $x^4 + 8x^3 + 24x^2 + 32x + 16$ **11.** $x^5 - 10x^4y + 40x^3y^2 - 80x^2y^3 + 80xy^4 - 32y^5$ **12.** $27x^3 - 108x^2y + 144xy^2 - 64y^3$
13. $x^8 + 12x^6y + 54x^4y^2 + 108x^2y^3 + 81y^4$ **14.** $4096x^{12} - 12,288x^{10} + 15,360x^8 - 10,240x^6 + 3840x^4 - 768x^2 + 64$
15. $x^9 - 3x^6y^3 + 3x^3y^6 - y^9$ **16.** $32x^{20} + 400x^{16}y^2 + 2000x^{12}y^4 + 5000x^8y^6 + 6250x^4y^8 + 3125y^{10}$ **17.** 90
18. 15 **19.** 3456 **20.** 1320

12.3 PRACTICE (pp. 719–722) **5.** $\frac{1}{6}$ **7.** $\frac{5}{6}$ **9.** 0.637
11. a. 0.353 **b.** 0.334 **13.** 0.3 **15.** 0.4 **17.** 0.6
19. 0.0769 **21.** 0.5 **23.** 0.231

	theor prob	exp prob	The two probabilities are
25.	0.333	0.308	not exactly the same, but
27.	0.5	0.508	they are very similar in
29.	0.833	0.875	every case.

31. 0.455 **33.** 0.545 **35.** 0.0385 **37.** 0.262 **39.** 5.6×10^{-8}
41. a. 0.555 **b.** 0.0380 **43. a.** 0.0527 **b.** 0.868 **45.** 0.0625
47. 0.00242

12.3 MIXED REVIEW (p. 722) **51.** −17 **53.** 19 **55.** −53
57. $\frac{4y^4}{3x^3}$ **59.** $\frac{x+3}{x}$ **61.** 3, 10, 17, 24, 31 **63.** 2; 8; 512;
134,217,728; 2.418×10^{24} **65.** −2, 0, 2, 2, 0 **67.** 1,042,380

TECHNOLOGY ACTIVITY 12.3 (p. 723)
1.

number	1	2	3	4	5	6
freq	19	24	20	16	20	21
theor prob	0.167	0.167	0.167	0.167	0.167	0.167
exp prob	0.158	0.200	0.167	0.133	0.167	0.175

The experimental and theoretical probabilities are close but not the same.

3.

trials	10	20	50	100	200
heads	4	13	32	52	100
tails	6	7	18	48	100

As the number of trials increases, the experimental results get closer to the theoretical results.

12.4 PRACTICE (pp. 727–729) **5.** 1 **7.** $\frac{7}{12}$ **9.** $\frac{4}{5}$ **11.** $\frac{3}{16}$
13. 0.25 **15.** $\frac{3}{7}$ **17.** 0.7; no **19.** $\frac{7}{17}$; no **21.** $\frac{5}{6}$; no
23. 30%; no **25.** 0.66 **27.** $\frac{1}{4}$ **29.** $\frac{1}{52}$ **31.** $\frac{1}{2}$ **33.** 0 **35.** $\frac{17}{18}$,
or about 0.944 **37.** $\frac{7}{9}$, or about 0.778 **39.** $\frac{35}{36}$, or about 0.972
41. *Sample answers:* not 3: 0.933; ≥ 5: 0.825; not 3 or 7: 0.783; ≤ 10: 0.942; > 2: 0.983; < 8 or > 11: 0.600; The experimental results are very similar to the theoretical results. **43.** 0.75 **45.** 0.691 **47.** 0.1 **49.** 0.375

12.4 MIXED REVIEW (p. 729) **57.** 1 **59.** 1.892 **61.** 61.73
63. $x^2 + y^2 = 25$ **65.** $x^2 + y^2 = 68$ **67.** $x^2 + y^2 = 109$
69. $x^2 + y^2 = 256$ **71. a.** 456,976,000 **b.** 258,336,000

12.5 PRACTICE (pp. 734–736) **5.** 0.2 **7.** 0.08 **9.** 0.6
11. 0.123 **13.** 0.047 **15.** 0.078 **17.** 0.012 **19. a.** 0.0059
b. 0.0060 **21. a.** 0.0178 **b.** 0.0181 **23. a.** 0.0156
b. 0.0153 **25.** 0.00144 **27.** 0.467 **29.** at least 52,722 tickets
31. 0.937 **33.** 0.581 **35.** 0.751

12.5 MIXED REVIEW (p. 736)
41.

0.01–0.25	0
0.26–0.50	5
0.51–0.75	4
0.76–1.00	5

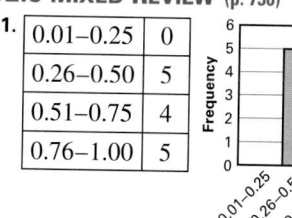

Value

43. 3.178
45. −0.347
47. −0.326

49. $128x^7 - 448x^6 + 672x^5 - 560x^4 + 280x^3 - 84x^2 + 14x - 1$ **51.** $x^6 - 6x^5 + 15x^4 - 20x^3 + 15x^2 - 6x + 1$

QUIZ 2 (p. 737) **1.** $\frac{12}{25}$, or 0.48 **2.** $\frac{6}{25}$, or 0.24
3. $\frac{18}{25}$, or 0.72 **4.** 0.111 **5.** 0.682 **6.** 0.207 **7.** 0.8 **8.** 0
9. 0.75 **10.** 0.0384

SELECTED ANSWERS

12.6 PRACTICE (pp. 742–744) **5.** 0.063

7. 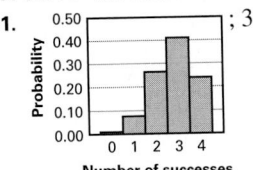 ; 2   **9.** No; the probability of 4 or fewer students buying rings is much greater than 0.1 if the claim is true. Therefore, you should not reject the claim.

11. 0.00109 **13.** 0.160 **15.** 0.160 **17.** 0.00109
19. 0.00863 **21.** 0.0909 **23.** 0.00000154 **25.** 8.67×10^{-19}
27. 0.344 **29.** 0.097

31. 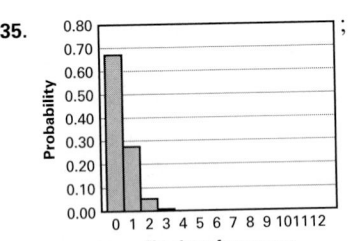 ; 3   **33.** ; 8

35. 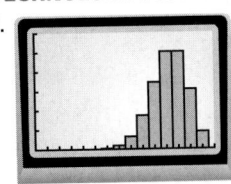 ; 0   **37.** 0.00114
39. 0.363
41. 0.816
43. 0 **45.** 6

47. Reject the claim because the probability that 3 or fewer students would have attended college anyway is 0.00351, which is much smaller than 0.01.

12.6 MIXED REVIEW (p. 744) **53.** $11, 4.155$ **55.** $19, 6.708$
57. $(-7, -5), (5, 7)$ **59.** $\left(\pm\dfrac{1}{2}, \pm\dfrac{\sqrt{35}}{2}\right)$ **61.** none
63. $a_1 = 4, a_n = a_{n-1} \times 10$ **65.** $a_1 = 1, a_2 = 3, a_n = a_{n-1} \times a_{n-2}$ **67.** $a_1 = 1, a_2 = 2, a_n = a_{n-1} + a_{n-2}$

TECHNOLOGY ACTIVITY 12.6 (p. 745)

1.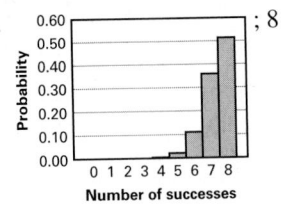

12.7 PRACTICE (pp. 749–751) **5.** 0.997 **7.** 0.5 **9.** 0.16
11. $5.1, 1.89$ **13.** $5, 1.94$ **15.** $5.1, 2.06$ **17.** 0.16 **19.** 50%
21. 2.5% **23.** 0.68 **25.** 0.9735 **27.** 0.84 **29.** 0.004096
31. 0.664 **33.** $5, 2.12$ **35.** $5.88, 2.27$ **37.** $8.4, 2.810$
39. 0.95 **41.** 50% **43.** 0.839 **45.** 0.145 **47.** 0.000625
49. 0.462 **51.** 0.16 **53.** 0.84 **55.** 0.999

12.7 MIXED REVIEW (p. 752) **59.** 64 **61.** 5 **63.** 25
65. $(0, \pm13); (\pm12, 0); (0, \pm5)$ **67.** $\left(0, \pm\sqrt{21}\right); \left(\pm\sqrt{6}, 0\right);$ $\left(0, \pm\sqrt{15}\right)$ **69.** $\dfrac{x^2}{7} + \dfrac{y^2}{10} = 1; \left(0, \pm\sqrt{10}\right); \left(\pm\sqrt{7}, 0\right); (0, \pm\sqrt{3})$
71. $\dfrac{11}{12}$ **73.** $\dfrac{11}{12}$

QUIZ 3 (p. 752) **1.** 0.000110 **2.** 0.00110 **3.** 0.151

4. 0.0014 **5.** 4.66×10^{-8} **6.** 9.29×10^{-15}
7. 2.59×10^{-25} **8.** 1.24×10^{-39}

9. 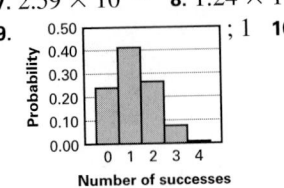 ; 1   **10.** 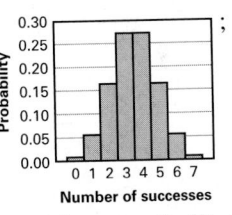 ;

3 and 4 are equally likely.

11. 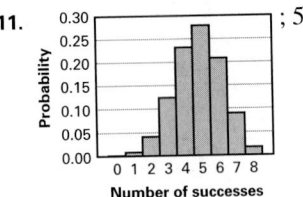 ; 5

12. 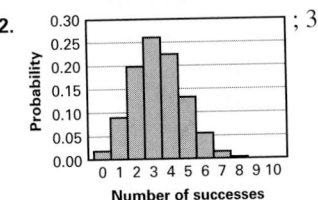 ; 3

13. 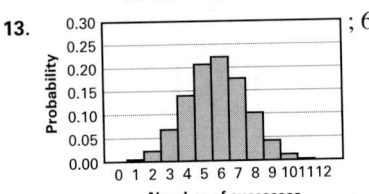 ; 6

14. 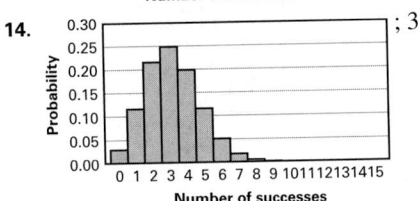 ; 3

15. 0.68 **16.** 0.4985 **17.** 0.9735 **18.** 0.50 **19.** 0.16
20. 0.0015 **21.** Yes; there is a 0.083 chance of getting 19 or fewer out of 26 and $0.083 < 0.1$, so reject the survey's findings.   **22.** 0.50

CHAPTER 12 EXTENSION (p. 754) **1.** Player A expected value: $0 \cdot \dfrac{1}{3} + 1 \cdot \dfrac{1}{3} - 1 \cdot \dfrac{1}{3} = 0$; Player B expected value: $0 \cdot \dfrac{1}{3} - 1 \cdot \dfrac{1}{3} + 1 \cdot \dfrac{1}{3} = 0$; Yes, the game is fair.   **3.** $-\$.44$

CHAPTER 12 REVIEW (pp. 756–758) **1.** $100,000$ **3.** 720 **5.** 5
7. $151,200$ **9.** 36 **11.** 10 **13.** 1 **15.** $x^3 + 12x^2 + 48x + 64$
17. $x^7 - 21x^6y + 189x^5y^2 - 945x^4y^3 + 2835x^3y^4 - 5103x^2y^5 + 5103xy^6 - 2187y^7$ **19.** $\dfrac{3}{8}$ **21.** experimental probability = 0.45; theoretical probability = 0.50; you got slightly fewer heads than expected.   **23.** 0.3
25. 1% **27. a.** 0.056 **b.** 0.0606 **29.** 0.117 **31.** 0.00977
33. 0.00977 **35.** 0.68 **37.** 0.025

CUMULATIVE PRACTICE (pp. 762–763) **1.** -7 **3.** $14, -8$

SELECTED ANSWERS

5. $-2, -5$ **7.** $4i, -4i$ **9.** $\frac{1}{8} \pm \frac{\sqrt{15}}{8} i$ **11.** $-1, 2, -2$ **13.** 3.25

15. 25 **17.**

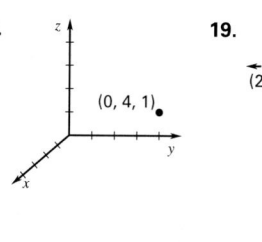

19.

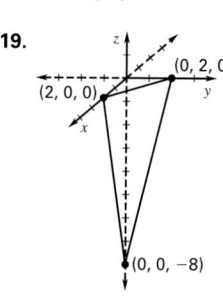

21. 18 **23.** 5 **25.** $xy = -40; y = -20$ **27.** $xy = -16;$
$y = -8$ **29.** $z = -\frac{2}{3}xy; z = \frac{10}{3}$ **31.** $z = 3xy; z = -15$
33. $\sqrt{89} \approx 9.43; (2.5, 4)$ **35.** $\sqrt{65} \approx 8.06; (1.5, -1)$

37.

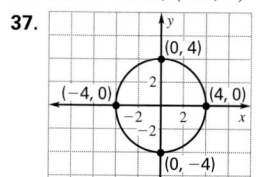

39.

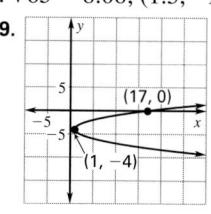

41. $(x - 2)^2 + (y + 2)^2 = 9$ **43.** $\frac{y^2}{4} - \frac{x^2}{5} = 1$ **45.** $(-18, 0)$
47. $\left(\pm\frac{\sqrt{6}}{2}, -\frac{1}{2}\right), \left(\pm\sqrt{3}, 1\right)$ **49.** geometric; Each term is
3 times the previous term. **51.** geometric; Each term is $\frac{1}{10}$
the previous term. **53.** $9, 6, 1, -6, -15$ **55.** $1, 5, 14, 30, 55$
57. $a_n = 7 - 6n; a_1 = 1; a_n = a_{n-1} - 6$ **59.** $a_n = 243\left(\frac{1}{3}\right)^{n-1};$
$a_1 = 243; a_n = \frac{1}{3}a_{n-1}$ **61.** 50 **63.** 16 **65.** 720 **67.** 70
69. 21 **71.** $8x^3 + 60x^2 + 150x + 125$ **73.** $81x^4 - 108x^3 +$
$54x^2 - 12x + 1$ **75.** $x^6 - 12x^4 + 48x^2 - 64$ **77.** 0.2
79. $\frac{1}{32}$ **81.** $\frac{5}{16}$ **83.** $\frac{5}{32}$

85.
$$C(x) = \begin{cases} 11.75, & \text{if } 0 < x \le 0.5 \\ 14.00, & \text{if } 0.5 < x \le 1 \\ 15.75, & \text{if } 1 < x \le 2 \\ 18.50, & \text{if } 2 < x \le 3 \\ 21.25, & \text{if } 3 < x \le 4 \\ 24.00, & \text{if } 4 < x \le 5 \\ 26.25, & \text{if } 5 < x \le 6 \\ 28.00, & \text{if } 6 < x \le 7 \\ 30.25, & \text{if } 7 < x \le 8 \\ 31.00, & \text{if } 8 < x \le 9 \\ 32.75, & \text{if } 9 < x \le 10 \end{cases}$$

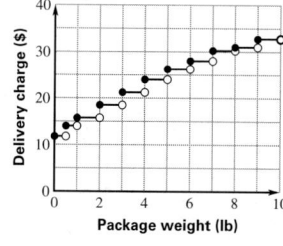

87. $a_n = 0.8a_{n-1} + 1000;$
It approaches a limit of 5000 fish. **89.** $\frac{2}{5}$

CHAPTER 13

SKILL REVIEW (p. 768) **1.** 12 **2.** 5 **3.** $3\sqrt{2}$ **4.** $4\sqrt{6}$
5. $3\sqrt{2}$ **6.** $10\sqrt{2}$ **7.** $\sqrt{2}$ **8.** $\frac{2\sqrt{3}}{3}$ **9.** $\frac{\sqrt{3}}{2}$ **10.** -1 **11.** $-\frac{5}{2}, \frac{5}{2}$
12. 14 **13.** -10

13.1 PRACTICE (pp. 772–774) **5.** $\sin\theta = \frac{3}{5}; \cos\theta = \frac{4}{5};$
$\tan\theta = \frac{3}{4}; \csc\theta = \frac{5}{3}; \sec\theta = \frac{5}{4}; \cot\theta = \frac{4}{3}$ **7.** $\sin\theta = \frac{\sqrt{5}}{3};$
$\cos\theta = \frac{2}{3}; \tan\theta = \frac{\sqrt{5}}{2}; \csc\theta = \frac{3\sqrt{5}}{5}; \sec\theta = \frac{3}{2}; \cot\theta = \frac{2\sqrt{5}}{5}$
9. $B = 15°; a \approx 19.3; b \approx 5.18$ **11.** $B = 28°; a \approx 56.4; c \approx 63.9$
13. $A = 75°; a \approx 157; c \approx 162$ **15.** $\sin\theta = \frac{\sqrt{5}}{5}; \cos\theta = \frac{2\sqrt{5}}{5};$
$\tan\theta = \frac{1}{2}; \csc\theta = \sqrt{5}; \sec\theta = \frac{\sqrt{5}}{2}; \cot\theta = 2$
17. $\sin\theta = \frac{2\sqrt{14}}{9}; \cos\theta = \frac{5}{9}; \tan\theta = \frac{2\sqrt{14}}{5}; \csc\theta = \frac{9\sqrt{14}}{28};$
$\sec\theta = \frac{9}{5}; \cot\theta = \frac{5\sqrt{14}}{28}$ **19.** $\sin\theta = \frac{9}{25}; \cos\theta = \frac{4\sqrt{34}}{25};$
$\tan\theta = \frac{9\sqrt{34}}{136}; \csc\theta = \frac{25}{9}; \sec\theta = \frac{25\sqrt{34}}{136}; \cot\theta = \frac{4\sqrt{34}}{9}$

21.

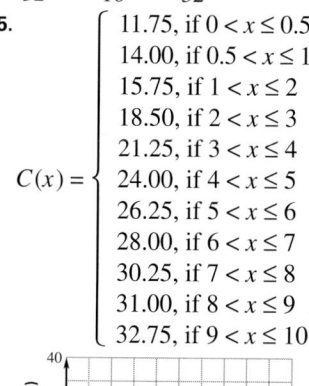

$; \sin\theta = \frac{5}{13}; \cos\theta = \frac{12}{13}; \tan\theta = \frac{5}{12};$
$\csc\theta = \frac{13}{5}; \sec\theta = \frac{13}{12}; \cot\theta = \frac{12}{5}$
23. $\frac{\sqrt{22}}{2}; \frac{\sqrt{22}}{2}$ **25.** 0.2419 **27.** 1.6643 **29.** 1.0154
31. 9.5668 **33.** $A = 66°; b \approx 3.56; c \approx 8.76$ **35.** $B = 71°;$
$a \approx 1.38; c \approx 4.23$ **37.** $B = 61°; a \approx 11.6; c \approx 24.0$
39. $A = 25°; a \approx 5.07; b \approx 10.9$ **41.** $96\sqrt{3}$ units^2,
or about 166 units^2 **43.** about 400 ft **45.** about 4250 ft
47. about 425 m; about 432 m **49.** about 12,350 ft

13.1 MIXED REVIEW (p. 775) **55.** 157.5 mi **57.** $\$3666$
59. parabola **61.** circle **63.** $\frac{16,016}{50,625}$, or about 0.316

13.2 PRACTICE (pp. 780–782) **5–11.** Sample angles are given.
5.

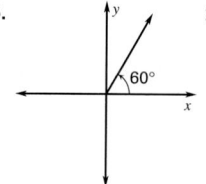

; **7.**

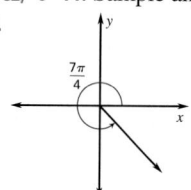

;

$420°, -300°$ $\qquad\qquad \frac{15\pi}{4}, -\frac{\pi}{4}$

9.

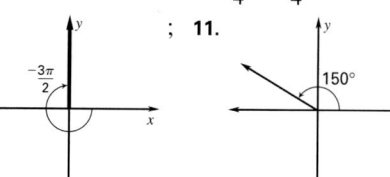

; **11.** **;**

$\frac{\pi}{2}, -\frac{7\pi}{2}$ $\qquad\qquad 510°, -210°$
13. $\frac{\pi}{6}$ **15.** $\frac{13\pi}{9}$ **17.** $315°$ **19.** $15°$ **21.** $\frac{11\pi}{9}$ in.; $\frac{22\pi}{9}$ in.^{2}

SELECTED ANSWERS

23. $\frac{17\pi}{18}$ cm; $\frac{17\pi}{18}$ cm^2 25. C 27. A

29. 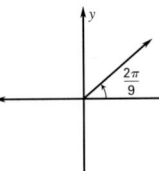 35.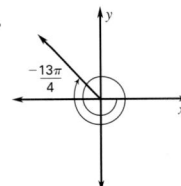

37–43. Sample angles are given. 37. 570°; −150°
39. 60°; −300° 41. $\frac{\pi}{4}$; $-\frac{7\pi}{4}$ 43. $\frac{4\pi}{3}$; $-\frac{2\pi}{3}$ 45. $\frac{5\pi}{4}$ 47. $\frac{\pi}{4}$
49. $\frac{65\pi}{36}$ 51. $-\frac{29\pi}{18}$ 53. −810° 55. −75° 57. −675°
59. 288° 61. $\frac{\pi}{6}$ ft; $\frac{\pi}{4}$ ft^2 63. 6π in.; 36π in.^{2} 65. $\frac{175\pi}{12}$ mm; $\frac{875\pi}{8}$ mm^2 67. $\frac{40\pi}{9}$ cm; $\frac{320\pi}{9}$ cm^2 69. $\frac{1}{2}$ 71. $\sqrt{3}$
73. 1.3764 75. 0.6428 77. 540°; 3π 79. 1260°; 7π
81. about 1820° or $\frac{91\pi}{9}$ radians 83. about 528 in.^{2}
85. 2π 87. $\frac{5}{3}$ in.

13.2 MIXED REVIEW (p. 783) 93. $5\sqrt{11}$ 95. 16 97. $\frac{\sqrt{7}}{4}$
99. $\frac{2\sqrt{14}}{7}$ 101. $\frac{4}{3}$ 103. $\frac{144}{35}$ 105. $\frac{100}{37}$ 107. $y^2 = 20x$
109. $y^2 = 24x$ 111. $x^2 = -17.6y$

QUIZ 1 (p. 783) 1. $\sin\theta = \frac{8}{17}$; $\cos\theta = \frac{15}{17}$; $\tan\theta = \frac{8}{15}$; $\csc\theta = \frac{17}{8}$; $\sec\theta = \frac{17}{15}$; $\cot\theta = \frac{15}{8}$ 2. $\sin\theta = \frac{3\sqrt{58}}{58}$; $\cos\theta = \frac{7\sqrt{58}}{58}$; $\tan\theta = \frac{3}{7}$; $\csc\theta = \frac{\sqrt{58}}{3}$; $\sec\theta = \frac{\sqrt{58}}{7}$; $\cot\theta = \frac{7}{3}$ 3. $\sin\theta = \frac{6\sqrt{61}}{61}$; $\cos\theta = \frac{5\sqrt{61}}{61}$; $\tan\theta = \frac{6}{5}$; $\csc\theta = \frac{\sqrt{61}}{6}$; $\sec\theta = \frac{\sqrt{61}}{5}$; $\cot\theta = \frac{5}{6}$ 4. $A = 40°$; $b \approx 21.5$; $c \approx 28.0$ 5. $B = 57°$; $a \approx 6.54$; $b \approx 10.1$ 6. $B = 80°$; $b \approx 17.0$; $c \approx 17.3$ 7. $A = 19°$; $a \approx 0.749$; $b \approx 2.17$
8–11. Sample answers are given. 8. 385°; −335°
9. $\frac{4\pi}{3}$; $-\frac{8\pi}{3}$ 10. $\frac{\pi}{4}$; $-\frac{7\pi}{4}$ 11. 280°; −80° 12. 2π m; 6π m^2
13. $\frac{5\pi}{3}$ ft; $\frac{5\pi}{3}$ ft^2 14. $\frac{8\pi}{9}$ cm; $\frac{32\pi}{9}$ cm^2 15. $\frac{242\pi}{9}$ in.; $\frac{2662\pi}{9}$ in.^{2} 16. $\frac{25\pi}{12}$ ft; $\frac{125\pi}{24}$ ft^2 17. $\frac{32\pi}{3}$ mm; 64π mm^2
18. The 6 in. slice has an area of 18.85 in.^{2} and costs about $.80/in.^{2}, while the 7 in. slice has an area of about 19.24 in.^{2} and costs about $.09/in.^{2}. The 6 in. slice has a lower unit price, so it is a better deal.

13.3 PRACTICE (pp. 788–790)
5. $\sin\theta = -\frac{5\sqrt{41}}{41}$; $\cos\theta = -\frac{4\sqrt{41}}{41}$; $\tan\theta = \frac{5}{4}$; $\csc\theta = -\frac{\sqrt{41}}{5}$; $\sec\theta = -\frac{\sqrt{41}}{4}$; $\cot\theta = \frac{4}{5}$

9. 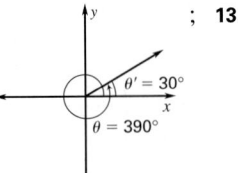 ; 13. 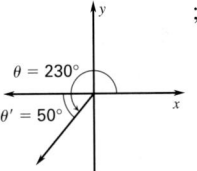 ;

30° 50°

15. $\sqrt{3}$ 17. $\sqrt{2}$ 19. $-\frac{1}{2}$ 21. $-\frac{\sqrt{3}}{3}$ 23. $\sin\theta = \frac{5}{13}$; $\cos\theta = -\frac{12}{13}$; $\tan\theta = -\frac{5}{12}$; $\csc\theta = \frac{13}{5}$; $\sec\theta = -\frac{13}{12}$; $\cot\theta = -\frac{12}{5}$ 25. $\sin\theta = \frac{14\sqrt{277}}{277}$; $\cos\theta = \frac{-9\sqrt{277}}{277}$; $\tan\theta = -\frac{14}{9}$; $\csc\theta = \frac{\sqrt{277}}{14}$; $\sec\theta = -\frac{\sqrt{277}}{9}$; $\cot\theta = -\frac{9}{14}$ 27. $\sin\theta = \frac{\sqrt{2}}{2}$; $\cos\theta = -\frac{\sqrt{2}}{2}$; $\tan\theta = -1$; $\csc\theta = \sqrt{2}$; $\sec\theta = -\sqrt{2}$; $\cot\theta = -1$ 29. $\sin\theta = -\frac{3\sqrt{13}}{13}$; $\cos\theta = \frac{2\sqrt{13}}{13}$; $\tan\theta = -\frac{3}{2}$; $\csc\theta = -\frac{\sqrt{13}}{3}$; $\sec\theta = \frac{\sqrt{13}}{2}$; $\cot\theta = -\frac{2}{3}$ 31. $\sin\theta = -\frac{\sqrt{3}}{2}$; $\cos\theta = \frac{1}{2}$; $\tan\theta = -\sqrt{3}$; $\csc\theta = -\frac{2\sqrt{3}}{3}$; $\sec\theta = 2$; $\cot\theta = -\frac{\sqrt{3}}{3}$ 33. $\sin\theta = \frac{\sqrt{7}}{4}$; $\cos\theta = -\frac{3}{4}$; $\tan\theta = -\frac{\sqrt{7}}{3}$; $\csc\theta = \frac{4\sqrt{7}}{7}$; $\sec\theta = -\frac{4}{3}$; $\cot\theta = -\frac{3\sqrt{7}}{7}$ 35. $\sin 270° = -1$; $\cos 270° = 0$; $\tan 270°$ is undefined; $\csc 270° = -1$; $\sec 270°$ is undefined; $\cot 270° = 0$.

39. 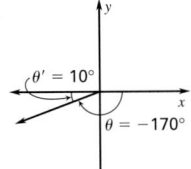 ; 41. 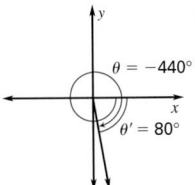 ;

10° 80°

43. 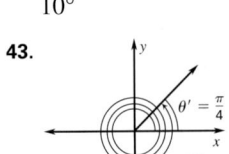 ; $\frac{\pi}{4}$ 45. $\frac{\sqrt{2}}{2}$ 47. $\frac{2\sqrt{3}}{3}$ 49. 2 51. $-\frac{\sqrt{2}}{2}$ 53. $-\sqrt{3}$ 55. $-\frac{1}{2}$ 57. $-\frac{1}{2}$ 59. $-\frac{2\sqrt{3}}{3}$ 61. −1.3673 63. −0.1736 65. 1.3764
67. −0.8090 69. about 16.5 ft/sec 71. about 7.4 ft
73. about 22,800 mi 75. about (−24, 93)

13.3 MIXED REVIEW (p. 790)
81. 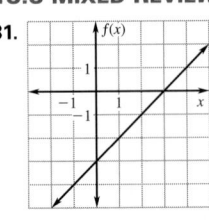 ; 83. 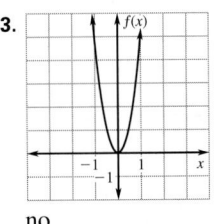 ;
yes no

85. 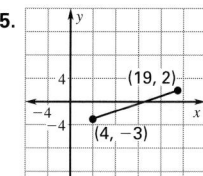 ; no **87.** $\frac{1}{52}$ **89.** $\frac{4}{13}$

91. $A = 70°$; $a \approx 20.7$; $b \approx 7.52$
93. $B = 40°$; $a \approx 2.30$; $b \approx 1.93$
95. $B = 7°$; $b \approx 6.14$; $c \approx 50.4$

13.4 PRACTICE (pp. 795–797)
5. $\frac{\pi}{3}$, or 60° **7.** $\frac{\pi}{6}$, or 30°
9. 1.32; 75.6° **11.** 1.22; 70.1° **13.** 200° **15.** 295°
17. about 42.3° **19.** $\frac{\pi}{3}$; 60° **21.** 0; 0° **23.** $-\frac{\pi}{2}$; −90°
25. $\frac{5\pi}{6}$; 150° **27** 48.2° **29.** 120° **31.** 18.4° **33.** 1.33; 76.1°
35. 0.848; 48.6° **37.** 2.21; 127° **39.** 1.15; 66.0°
41. 1.43; 81.9° **43.** 0.988; 56.6° **45.** 247° **47.** 127°
49. 224° **51.** 222° **53.** about 44.4° **55.** about 70.2°
57. $\theta = \tan^{-1}(2.127t)$ **59.** about 71.6° **61.** $y = 1.6x + 3$

13.4 MIXED REVIEW (p. 798)
65. 18 **66.** $\frac{21}{4}$ **67.** −3
68. −4 **69.** −3, 3 **70.** no solution **71.** $\frac{1}{5}$ **72.** $\frac{1}{3}$ **73.** $\frac{1}{2}$
74. $\frac{1}{3}$ **75.** $\frac{11}{30}$ **76.** $\frac{7}{30}$ **77.** 0.4540 **78.** 0.3827 **79.** 0.3907
80. 1.0642 **81.** 0.2126 **82.** −1.5890

QUIZ 2 (p. 798)
1. $\sin\theta = -\frac{16\sqrt{337}}{337}$; $\cos\theta = -\frac{9\sqrt{337}}{337}$;
$\tan\theta = \frac{16}{9}$; $\csc\theta = -\frac{\sqrt{337}}{16}$; $\sec\theta = -\frac{\sqrt{337}}{9}$; $\cot\theta = \frac{9}{16}$
2. $\sin\theta = -\frac{2\sqrt{53}}{53}$; $\cos\theta = \frac{7\sqrt{53}}{53}$; $\tan\theta = -\frac{2}{7}$;
$\csc\theta = -\frac{\sqrt{53}}{2}$; $\sec\theta = \frac{\sqrt{53}}{7}$; $\cot\theta = -\frac{7}{2}$ **3.** $\sin\theta = \frac{5\sqrt{26}}{26}$;
$\cos\theta = -\frac{\sqrt{26}}{26}$; $\tan\theta = -5$; $\csc\theta = \frac{\sqrt{26}}{5}$; $\sec\theta = -\sqrt{26}$;
$\cot\theta = -\frac{1}{5}$ **4.** $\sin\theta = -\frac{11\sqrt{157}}{157}$; $\cos\theta = \frac{6\sqrt{157}}{157}$;
$\tan\theta = -\frac{11}{6}$; $\csc\theta = -\frac{\sqrt{157}}{11}$; $\sec\theta = \frac{\sqrt{157}}{6}$; $\cot\theta = -\frac{6}{11}$
5. $\sin\theta = \frac{2\sqrt{5}}{5}$; $\cos\theta = \frac{\sqrt{5}}{5}$; $\tan\theta = 2$; $\csc\theta = \frac{\sqrt{5}}{2}$;
$\sec\theta = \sqrt{5}$; $\cot\theta = \frac{1}{2}$ **6.** $\sin\theta = \frac{\sqrt{17}}{17}$; $\cos\theta = -\frac{4\sqrt{17}}{17}$;
$\tan\theta = -\frac{1}{4}$; $\csc\theta = \sqrt{17}$; $\sec\theta = -\frac{\sqrt{17}}{4}$; $\cot\theta = -4$
7. $\sin\theta = -\frac{5\sqrt{106}}{106}$; $\cos\theta = \frac{9\sqrt{106}}{106}$; $\tan\theta = -\frac{5}{9}$;
$\csc\theta = -\frac{\sqrt{106}}{5}$; $\sec\theta = \frac{\sqrt{106}}{9}$; $\cot\theta = -\frac{9}{5}$
8. $\sin\theta = -\frac{8\sqrt{113}}{113}$; $\cos\theta = -\frac{7\sqrt{113}}{113}$; $\tan\theta = \frac{8}{7}$;
$\csc\theta = -\frac{\sqrt{113}}{8}$; $\sec\theta = -\frac{\sqrt{113}}{7}$; $\cot\theta = \frac{7}{8}$ **9.** $-\frac{\sqrt{2}}{2}$
10. $-\sqrt{3}$ **11.** $\frac{1}{2}$ **12.** $\sqrt{3}$ **13.** $-\frac{\sqrt{3}}{2}$ **14.** $-\frac{\sqrt{3}}{2}$ **15.** $-\frac{\sqrt{3}}{3}$
16. $-\frac{1}{2}$ **17.** 1.16; 66.5° **18.** −0.644; −36.9° **19.** 0.318; 18.2°
20. 0.232; 13.3° **21.** −1.33; −76.0° **22.** 2.50; 143°
23. 0.100; 5.74° **24.** 1.47; 84.3° **25.** 166° **36.** 282°
27. 262° **28.** 206° **29.** 253° **30.** 103° **31.** about 47 ft

13.5 PRACTICE (pp. 803–806)
5. no triangle **7.** two triangles
9. $A \approx 35.8°$; $B \approx 49.2°$; $a = 14.7$ **11.** 2.19 units^2
13. 125 units^2 **15.** about $62,400 **17.** no triangle
19. two triangles **21.** one triangle **23.** no triangle
25. $C = 75°$; $a \approx 24.9$; $b \approx 30.5$ **27.** $A \approx 84.7°$; $C \approx 35.3°$;
$a \approx 34.5$ **29.** no triangle **31.** $A \approx 62.3°$; $B \approx 22.7°$; $b \approx 3.48$
33. $A \approx 111.6°$; $B \approx 52.4°$; $a \approx 108$, or $A \approx 36.4°$; $B \approx 127.6°$;
$a \approx 68.9$ **35.** $A \approx 15.4°$; $C \approx 129.6°$ $c \approx 34.9$ **37.** 119 units^2
39. 17.3 units^2 **41.** 162 units^2 **43.** 9.83 units^2
45. 67.1 units^2 **47.** 76.1 units^2 **49.** 9.06 units^2
51. 85.7 units^2 **53.** *Sample answer:* Let the side lengths
be 10 and 15. The equation is $A = 75 \sin x$. **55.** 90°
57. about 22.5 mi **61.** 31.8°, or 8.2° **63.** about 155.4 ft
65. 21 bags **67.** about 0.57 gal, so buy a single gallon can

13.5 MIXED REVIEW (p. 806)
71. $23\sqrt{3}$ **73.** $8\sqrt{7}$
75. $28\sqrt{2}$ **77.** 0.3090 **79.** −0.2225 **81.** 0.5736 **83.** 0.9962

13.6 PRACTICE (pp. 810–812)
5. $b \approx 43.0$; $A \approx 107.4°$;
$C \approx 52.7°$ **7.** $A \approx 82.2°$; $B \approx 25.8°$; $C \approx 72.0°$
9. 510 units^2 **11.** 1470 units^2 **13.** about 63.7 ft
15. $c \approx 4.60$; $A \approx 35.2°$; $B \approx 112.8°$ **17.** $c \approx 12.9$; $A \approx 48.6°$;
$B \approx 91.4°$ **19.** $c \approx 16.3$; $A \approx 37.7°$; $B \approx 47.3°$ **21.** $A \approx 22.3°$;
$B \approx 49.5°$; $C \approx 108.2°$ **23.** $a \approx 29.1$; $B \approx 63.4°$; $C \approx 56.6°$
25. $c \approx 10.4$; $A \approx 75°$; $B \approx 75°$ **27.** $b \approx 6.40$; $A \approx 150.9°$;
$C \approx 14.1°$ **29.** $A \approx 47.0°$; $B \approx 27.8°$; $C \approx 105.1°$
31. $A = 70°$; $a \approx 32.4$; $c \approx 17.3$ **33.** $a \approx 27.5$; $A \approx 56.5°$;
$B \approx 19.5°$ **35.** $A \approx 64.3°$; $B \approx 73.2°$; $C \approx 42.5°$
37. $c \approx 11.7$; $A \approx 20.0°$; $B \approx 70.0°$ **39.** 14.0 units^2
41. 150 units^2 **43.** 2210 units^2 **45.** 3.87 units^2
47. 27.7 units^2 **51.** about 74.4 ft **53.** about 7800 mi^2

13.6 MIXED REVIEW (p. 812)
59. $\frac{y^2}{9} - \frac{x^2}{112} = 1$
61. $y^2 - \frac{x^2}{19} = 1$ **63.** 0.137 **65.** 0.160 **67.** 0.0130

13.7 PRACTICE (pp. 816–818)
5. 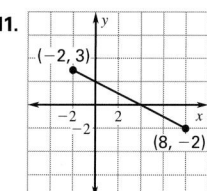 **7.** $y = \frac{3}{7}x - 2$; $0 \le x \le 35$
9. $y = (\tan 72.1°)x + 3$, or
$y = 3.10x + 3$; $0 \le x \le 35.3$

11. 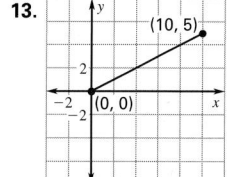 **13.**

15. 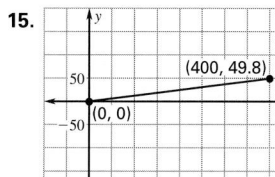 **17.** $y = 2x - 5$; $1 \le x \le 6$
19. $y = x$; $0 \le x \le 100$
21. $x = (20.0 \cos 71.6°)t$, or
$x = 6.31t$;
$y = (20.0 \sin 71.6°)t$, or
$y = 19.0t$; $0 \le t \le 3$

SELECTED ANSWERS

23. $x = (13.0 \cos 80.0°)t + 3$, or $x = 2.26t + 3$; $y = (13.0 \sin 80.0°)t + 2$, or $y = 12.8t + 2$; $0 \le t \le 5$
25. $x = (10 \cos 143.13°)t + 2.0$; $y = (10 \sin 143.13°)t$
27. about 3774 sec, or about 63 min **29.** $x = 260t$; $y = 10,000 - 30t$ **31.** about 333 sec, or 5 min 33 sec
33. $x = (17.9 \cos 14.3°)t$, or $x = 17.3t$; $y = -4.9t^2 + (17.9 \sin 14.3°)t + 1.71$, or $y = -4.9t^2 + 4.42t + 1.71$
35. about 20.7 m **37.** about 1.49 sec **39.** $x = (v \cos 43°)t$, or $x = 0.731vt$; $y = -16t^2 + (v \sin 43°)t + 6$, or $y = -16t^2 + 0.682vt + 6$

13.7 MIXED REVIEW (p. 819)

45.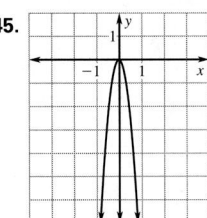

47.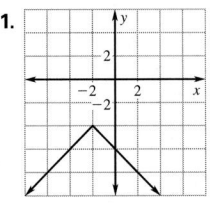

49.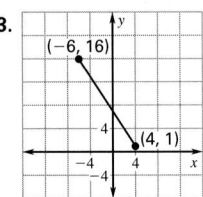

51. 6930 **53.** $-\dfrac{1}{9}$
55. 0.3413 **57.** 0.0013

QUIZ 3 (p. 819) **1.** $A \approx 58.5°$; $C \approx 51.5°$; $a \approx 27.2$
2. $A = 70°$; $b \approx 2.77$; $c \approx 15.7$ **3.** $C = 30°$; $a \approx 20.5$; $c \approx 16.0$
4. no triangle **5.** $A \approx 106.1°$; $B \approx 43.1°$; $C \approx 30.8°$
6. $a = 35.4$; $B = 23.9°$; $C = 49.1°$ **7.** 179 units^2
8. 499 units^2 **9.** 57.0 units^2 **10.** 16.3 units^2
11. 1950 units^2 **12.** 334 units^2

13.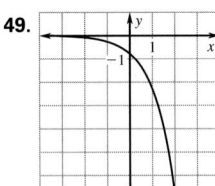

14.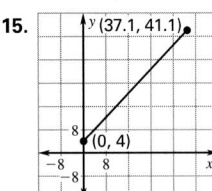

15.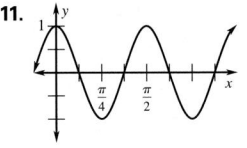

16. $y = -\dfrac{1}{5}x - \dfrac{27}{5}$; $-22 \le x \le 3$
17. $y = 0.700x$; $0 \le x \le 246$
18. about 23.0 ft

TECHNOLOGY ACTIVITY 13.7 (p. 820) **1.** 390; 423; 443; 443; 423; 390 **2.** 45°; the results look to be symmetric around the value $\theta = 45°$, with a maximum at that angle.

CHAPTER REVIEW (pp. 822–824) **1.** $\sin \theta = \dfrac{3}{5}$; $\cos \theta = \dfrac{4}{5}$; $\tan \theta = \dfrac{3}{4}$; $\csc \theta = \dfrac{5}{3}$; $\sec \theta = \dfrac{5}{4}$; $\cot \theta = \dfrac{4}{3}$

3. $\sin \theta = \dfrac{\sqrt{2}}{2}$; $\cos \theta = \dfrac{\sqrt{2}}{2}$; $\tan \theta = 1$; $\csc \theta = \sqrt{2}$; $\sec \theta = \sqrt{2}$; $\cot \theta = 1$ **5.** $\dfrac{\pi}{6}$ **7.** $-\dfrac{\pi}{12}$ **9.** 300° **11.** $\dfrac{5\pi}{2}$ ft, $\dfrac{25\pi}{4}$ ft^2 **13.** $\dfrac{56\pi}{3}$ cm, $\dfrac{448\pi}{3}$ cm^2 **15.** $\dfrac{\sqrt{3}}{2}$ **17.** $\dfrac{1}{2}$ **19.** $\dfrac{\pi}{4}$, 45°
21. $\dfrac{\pi}{2}$, 90° **23.** $\dfrac{2\pi}{3}$, 120° **25.** $A = 29.3°$; $C = 132.7°$; $c = 28.5$ or $A = 150.7°$; $C = 11.3°$; $c = 7.60$ **27.** 63.1 units^2
29. 98.3 units^2 **31.** $C = 107°$, $A = 24°$, $B = 49°$ **33.** 9.9 units^2
35. 46.4 units^2 **39.** $y = -2x - 1$; $-3 \le x \le 13$

CHAPTER 14

SKILL REVIEW (p. 830)

1.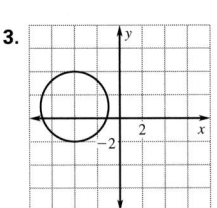

2.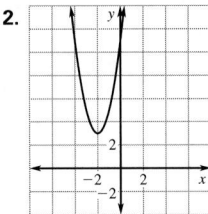

3.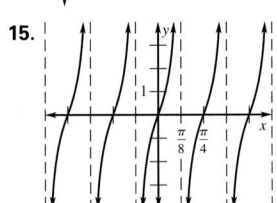

4. -8, 1 **5.** $\pm\dfrac{5}{3}$ **6.** $\dfrac{1}{6} \pm \dfrac{\sqrt{61}}{6}$
7. $\dfrac{\sqrt{3}}{2}$ **8.** $\dfrac{\sqrt{3}}{3}$ **9.** $\dfrac{\sqrt{2}}{2}$ **10.** 0

11. $\dfrac{\pi}{4}$, 45° **12.** $\dfrac{\pi}{6}$, 30° **13.** $\dfrac{\pi}{2}$, 90° **14.** $-\dfrac{\pi}{3}$, $-60°$

14.1 PRACTICE (pp. 835–837) **5.** amplitude: 3, period: 2
7. amplitude: $\dfrac{2}{3}$, period: 6 **9.** amplitude: 1, period: 4π

11.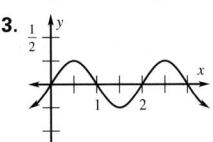

13.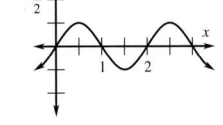

15.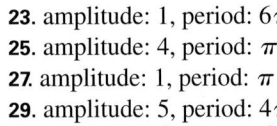

17. B **19.** D **21.** A
23. amplitude: 1, period: 6π
25. amplitude: 4, period: π
27. amplitude: 1, period: π
29. amplitude: 5, period: 4π
31. amplitude: $\dfrac{1}{3}$, period: $\dfrac{1}{2}$

33.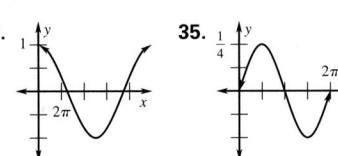

35.

37.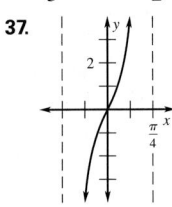

39. 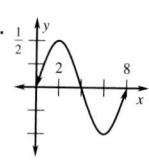 **41.** 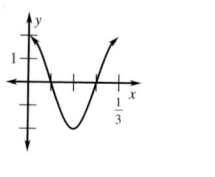 **43.**

45. $y = 10 \sin \dfrac{\pi x}{2}$ **47.** $y = \dfrac{1}{2} \sin \dfrac{2}{3}x$ **49.** $y = 3 \sin 4\pi x$

53. amplitude: $\dfrac{1}{2}$ ft, period: $\dfrac{\pi}{3}$ sec **55.** 8.7 ft, 8.9 ft, 9.2 ft; $h_t - h_{t-1}$ increases as t increases.

14.1 MIXED REVIEW (p. 837)

61. 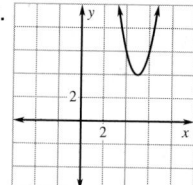 **63.**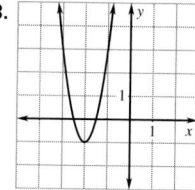

67. $\dfrac{15}{58}$ **69.** $\dfrac{7}{58}$

71. 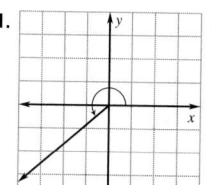 ; 40° **73.** 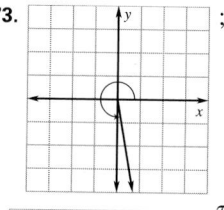 ; 80°

75. 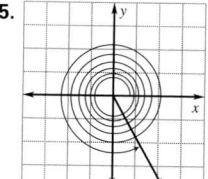 ; $\dfrac{\pi}{3}$ **77.** 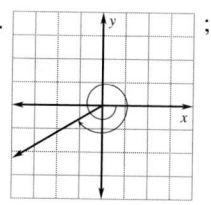 ; $\dfrac{\pi}{6}$

79. 46.2 years

TECHNOLOGY ACTIVITY 14.1 (p. 838)
1. amplitude: $\dfrac{1}{3}$, 3, 9; period: 2π **3.** amplitude: $\dfrac{1}{2}$, 1, 2; period: 2π

5. amplitude: 1; period: 2π, π, $\dfrac{\pi}{2}$

7. amplitude: 1; period: 4π, 2π, π

14.2 PRACTICE (pp. 844–846)
1. translation **3.** shifted right π units **5.** horizontal shift **7.** horizontal shift **9.** reflection, vertical shift

11. 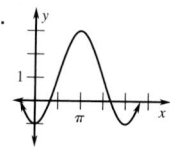 **13.**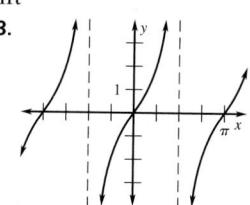

15.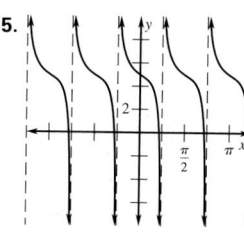

17. shift up 2 **19.** shift down 2 **21.** reflect in x-axis and shift left π **23.** reflect in x-axis, shift right $\dfrac{\pi}{4}$, shift up 5 **25.** shift left $\dfrac{3\pi}{4}$, shift up 3

27. B **29.** A **31.** D

33. 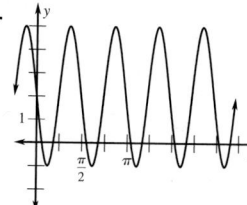 **35.**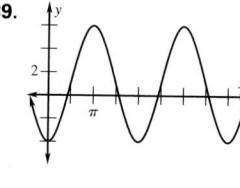

37. 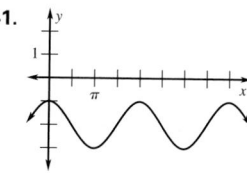 **39.**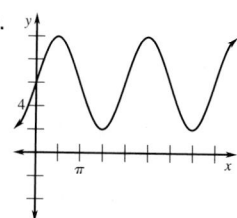

41. 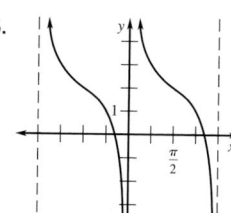 **43.**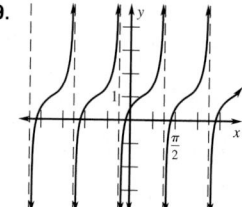

45. 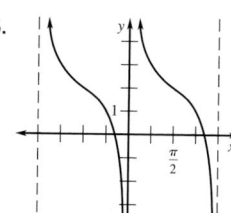 **49.**

51. $y = 3 \cos(x + \pi) + 3$ **53.** $y = -\dfrac{1}{3} \sin 6x - 1$

55. $y = -4 \tan \dfrac{\pi}{2}\left(x - \dfrac{1}{2}\right) + 6$

57. 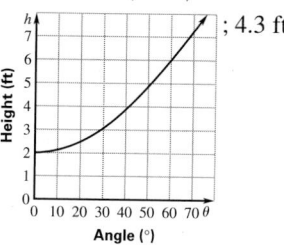 ; 4.3 ft

59. 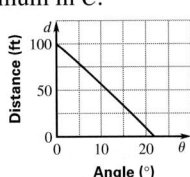 ; Over the course of the first year, R falls while C rises and then falls. Over the second year, R rises while C continues to fall and then rise. Both populations have the same period, 2 years, with the peak in R occurring 6 months before the peak in C, and the minimum in R 6 months before the minimum in C.

61. $d = -250 \tan \theta + 100$, where $0 \le \theta \le 21.8°$;

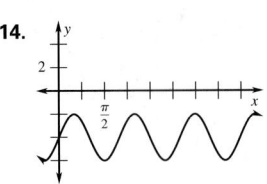 (distance vs angle graph)

14.2 MIXED REVIEW (p. 847) **65.** ellipse, $\dfrac{x^2}{25} + \dfrac{y^2}{36} = 1$
67. circle, $x^2 + y^2 = 25$ **69.** 8 **71.** 120 **73.** 10 **75.** 7
77. $\sin \theta = \dfrac{4}{5}$, $\cos \theta = \dfrac{3}{5}$, $\tan \theta = \dfrac{4}{3}$, $\sec \theta = \dfrac{5}{3}$,

$\csc \theta = \dfrac{5}{4}$, $\cot \theta = \dfrac{3}{4}$ **79.** $\sin \theta = \dfrac{3}{10}$, $\cos \theta = \dfrac{\sqrt{91}}{10}$,

$\tan \theta = \dfrac{3\sqrt{91}}{91}$, $\sec \theta = \dfrac{10\sqrt{91}}{91}$, $\csc \theta = \dfrac{10}{3}$, $\cot \theta = \dfrac{\sqrt{91}}{3}$

81. $\sin \theta = \dfrac{2\sqrt{6}}{5}$, $\cos \theta = \dfrac{1}{5}$, $\tan \theta = 2\sqrt{6}$, $\sec \theta = 5$,

$\csc \theta = \dfrac{5\sqrt{6}}{12}$, $\cot \theta = \dfrac{\sqrt{6}}{12}$ **83.** 40,320

QUIZ 1 (p. 847) **1.** amplitude: $\dfrac{5}{2}$, period: $\dfrac{2\pi}{7}$ **2.** amplitude: 1,
period: π **3.** amplitude: 1, period: 4 **4.** amplitude: $\dfrac{1}{4}$,
period: 1 **5.** amplitude: 3, period: 2 **6.** amplitude: 4,
period: $\dfrac{4}{3}$ **7.** amplitude: $\dfrac{7}{3}$, period: $\dfrac{\pi}{2}$ **8.** amplitude: $\dfrac{1}{3}$,
period: 2π **9.** amplitude: 6, period: 16π

10. 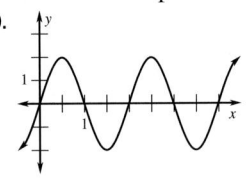 **11.**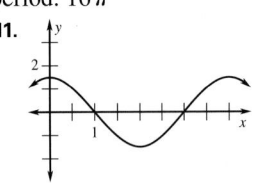

12. 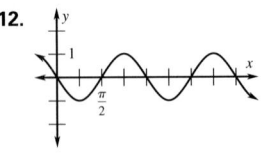 **13.**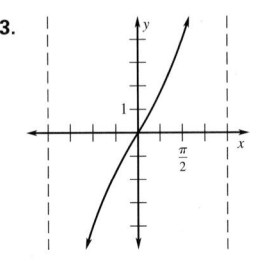

14. 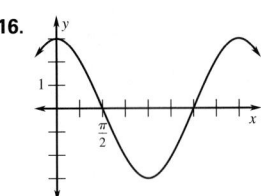 **15.**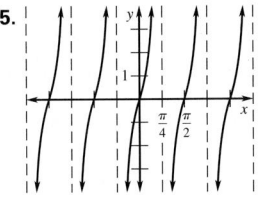

16. 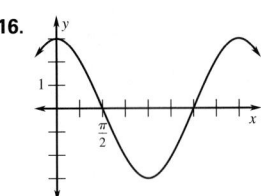 **17.** (graph)

18. 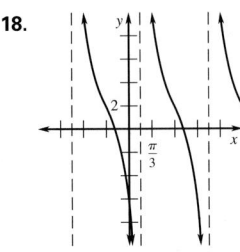 **19.** $d = 120 \tan \theta$
$0° \le \theta \le 65.2°$; $30.3°$;

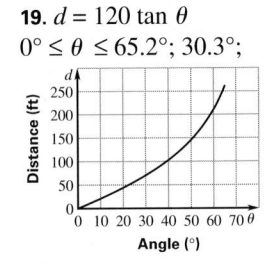 (distance vs angle graph)

14.3 PRACTICE (pp. 852–854) **5.** $\sin \theta = \dfrac{4}{5}$, $\tan \theta = -\dfrac{4}{3}$,
$\sec \theta = -\dfrac{5}{3}$, $\csc \theta = \dfrac{5}{4}$, $\cot \theta = -\dfrac{3}{4}$ **7.** $\sin \theta = -\dfrac{\sqrt{7}}{4}$, $\cos \theta = \dfrac{3}{4}$,
$\tan \theta = -\dfrac{\sqrt{7}}{3}$, $\csc \theta = -\dfrac{4\sqrt{7}}{7}$, $\cot \theta = -\dfrac{3\sqrt{7}}{7}$ **9.** $\tan x$ **11.** 1

13. $\cot x \tan(-x) = \dfrac{\cos x}{\sin x} \cdot \dfrac{\sin(-x)}{\cos(-x)} = \dfrac{\cos x}{\sin x} \cdot \dfrac{-\sin x}{\cos x} = -1$

15. ellipse **17.** $\sin \theta = \dfrac{3\sqrt{73}}{73}$, $\cos \theta = \dfrac{8\sqrt{73}}{73}$, $\sec \theta = \dfrac{\sqrt{73}}{8}$,
$\csc \theta = \dfrac{\sqrt{73}}{3}$, $\cot \theta = \dfrac{8}{3}$ **19.** $\cos \theta = -\dfrac{4}{5}$, $\tan \theta = -\dfrac{3}{4}$,
$\sec \theta = -\dfrac{5}{4}$, $\csc \theta = \dfrac{5}{3}$, $\cot \theta = -\dfrac{4}{3}$ **21.** $\sin \theta = \dfrac{\sqrt{23}}{12}$,
$\tan \theta = -\dfrac{\sqrt{23}}{11}$, $\sec \theta = -\dfrac{12}{11}$, $\csc \theta = \dfrac{12\sqrt{23}}{23}$,
$\cot \theta = -\dfrac{11\sqrt{23}}{23}$ **23.** $\sin \theta = -\dfrac{\sqrt{91}}{10}$, $\cos \theta = -\dfrac{3}{10}$,
$\tan \theta = \dfrac{\sqrt{91}}{3}$, $\csc \theta = -\dfrac{10\sqrt{91}}{91}$, $\cot \theta = \dfrac{3\sqrt{91}}{91}$

25. $\sin \theta = -\dfrac{\sqrt{3}}{2}$, $\cos \theta = \dfrac{1}{2}$, $\tan \theta = -\sqrt{3}$, $\csc \theta = -\dfrac{2\sqrt{3}}{3}$,
$\cot \theta = -\dfrac{\sqrt{3}}{3}$ **27.** $\sin \theta = -\dfrac{1}{2}$, $\cos \theta = \dfrac{\sqrt{3}}{2}$, $\tan \theta = -\dfrac{\sqrt{3}}{3}$,
$\sec \theta = \dfrac{2\sqrt{3}}{3}$, $\csc \theta = -2$ **29.** $-\cot x$ **31.** $\csc x$

33. $\cos x$ **35.** $\cos^2 x - \sin^2 x$ **37.** 1 **39.** -1 **41.** $\sin x$

43. -1 **45.** $\tan x \csc x \cos x = \left(\dfrac{\sin x}{\cos x}\right)\left(\dfrac{1}{\sin x}\right) \cos x = 1$

47. $2 - \sec^2 x = 1 + (1 - \sec^2 x) = 1 - \tan^2 x$

49. $\dfrac{\cos^2 x + \sin^2 x}{1 + \tan^2 x} = \dfrac{1}{\sec^2 x} = \cos^2 x$

51. $\dfrac{\sin\left(\dfrac{\pi}{2} - x\right) - 1}{1 - \cos(-x)} = \dfrac{\cos x - 1}{1 - \cos x} = -1$

SELECTED ANSWERS

53. $\dfrac{\cos(-x)}{1+\sin(-x)} = \dfrac{\cos x}{1-\sin x} = \dfrac{\cos x\,(1+\sin x)}{1-\sin^2 x} =$

$\dfrac{\cos x\,(1+\sin x)}{\cos^2 x} = \sec x + \tan x$ **55.** $1 = \sec^2 t - \tan^2 t =$

$\dfrac{x^2}{5} - \dfrac{y^2}{1}$, hyperbola **57.** $1 = \sin^2 \pi t + \cos^2 \pi t = \dfrac{y^2}{64} + \dfrac{x^2}{64}$,

circle **59.** $1 = \sin^2 \dfrac{t}{2} + \cos^2 \dfrac{t}{2} = \dfrac{y^2}{16} + \dfrac{x^2}{1}$, ellipse

61. $s = \dfrac{h \sin(90° - \theta)}{\sin \theta} = \dfrac{h \cos \theta}{\sin \theta} = h \cot \theta$

63. Actual wheel is 18 ft wide, model is 1 ft wide. Actual wheel rotates once every 15 sec, model once every 8 sec.

14.3 MIXED REVIEW (p. 854) **69.** $-9, 4$ **71.** $-\dfrac{3}{2}, 5$

73. $-\dfrac{1}{3}, \dfrac{1}{3}$ **75.** $60°, \dfrac{\pi}{3}$ **77.** $30°, \dfrac{\pi}{6}$ **79.** $120°, \dfrac{2\pi}{3}$

81.  **83.** 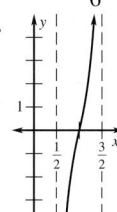 **85.**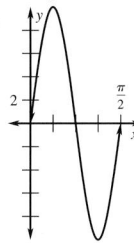

14.4 PRACTICE (pp. 859–860) **5.** $\dfrac{\pi}{6}, \dfrac{5\pi}{6}, \dfrac{7\pi}{6}, \dfrac{11\pi}{6}$ **7.** $\dfrac{\pi}{6}, \dfrac{11\pi}{6}$

9. $0.45 + 2n\pi, 2.69 + 2n\pi$ **11.** $\dfrac{7\pi}{6} + 2n\pi, \dfrac{11\pi}{6} + 2n\pi$

13. yes **15.** yes **17.** yes **19.** $\dfrac{\pi}{3} + 2n\pi, \dfrac{5\pi}{3} + 2n\pi$

21. $\dfrac{\pi}{6} + 2n\pi, \dfrac{5\pi}{6} + 2n\pi$ **23.** $\dfrac{\pi}{4} + \dfrac{n\pi}{2}$ **25.** $\dfrac{\pi}{2} + n\pi$

27. $\dfrac{\pi}{2} + 2n\pi, \dfrac{7\pi}{6} + 2n\pi, \dfrac{11\pi}{6} + 2n\pi$

29. $\dfrac{\pi}{2} + 2n\pi, \dfrac{11\pi}{6} + 2n\pi$ **31.** $\dfrac{\pi}{3} + 2n\pi, \dfrac{5\pi}{3} + 2n\pi$

33. $0.93, 5.36$ **35.** $\dfrac{\pi}{3}, \dfrac{2\pi}{3}, \dfrac{4\pi}{3}, \dfrac{5\pi}{3}$ **37.** $\dfrac{\pi}{6}, \dfrac{5\pi}{6}, \dfrac{3\pi}{2}$

39. $\dfrac{\pi}{4}, \dfrac{7\pi}{4}$ **41.** $\dfrac{\pi}{4}, \dfrac{3\pi}{4}, \dfrac{5\pi}{4}, \dfrac{7\pi}{4}$ **43.** $1.33, 4.47$ **45.** $4.13, 5.30$

47. $\dfrac{7\pi}{6}, \dfrac{11\pi}{6}$ **49.** $0, \pi$ **51.** $\dfrac{\pi}{6}, \dfrac{2\pi}{3}, \dfrac{7\pi}{6}, \dfrac{5\pi}{3}$

53. $\dfrac{\pi}{4}, \dfrac{3\pi}{4}, \dfrac{5\pi}{4}, \dfrac{7\pi}{4}$ **55.** $\dfrac{\pi}{3}, \dfrac{5\pi}{3}$ **57.** highs: 6:12 A.M. and 6:36 P.M., lows: 12:00 A.M. and 12:24 P.M., the water depth never goes below 7 ft. **59.** June, July, August; no

14.4 MIXED REVIEW (p. 861) **65.** $\dfrac{33}{36}$ **67.** $\dfrac{33}{36}$

69. 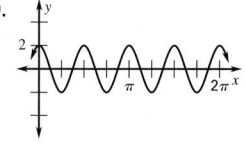 **71.**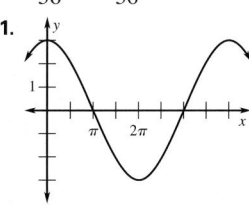

73. 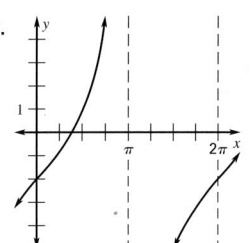 **77.** 36 ft

14.5 PRACTICE (pp. 865–867) **5.** $y = \cos \pi x - 2$

7. $y = 3 \sin \dfrac{x}{2} + 7$ **9.** $h = -20 \cos\left(\dfrac{2\pi}{15}t\right) + 25$

11. $y = 5 \sin 2x$ **13.** $y = -4 \cos \pi x$ **15.** $y = 2 \cos \dfrac{\pi x}{2} - 4$

17. $y = 8 \sin \dfrac{x}{2}$ **19.** $y = -\cos 3x + 3$ **21.** $y = -5 \sin \dfrac{\pi x}{2} + 2$

23. $y = -6 \cos 6x + 5$ **25.** $y = 4 \cos \dfrac{x}{3} - 4$

27. $y = -2 \sin 6x - 4$ **31.** $h = 6.5 \cos 60\pi t + 4.5$

33. $h = -2.5 \cos \pi t + 6.5$

35. $T = 776.4 \cdot \sin\left(0.45t + 1.49\right) + 727.7$

14.5 MIXED REVIEW (p. 867) **41.** $\dfrac{1}{36}$ **43.** $\dfrac{1}{4}$ **45.** $-\dfrac{\sqrt{2}}{2}$

47. $-\dfrac{1}{2}$ **49.** $-\dfrac{\sqrt{3}}{3}$ **51.** 9.92 **53.** 22.19 **55.** 31.53

QUIZ 2 (p. 868) **1.** $\csc x$ **2.** $2\cos^2 x$ **3.** $\sin x \cos x$

4. $\dfrac{\pi}{6} + \pi n, \dfrac{5\pi}{6} + \pi n$ **5.** $5.94 + 2\pi n, 3.48 + 2\pi n$

6. $\dfrac{2\pi}{3} + \pi n, \dfrac{5\pi}{6} + \pi n$ **7.** $y = -5 \sin 2x$ **8.** $y = \cos \dfrac{x}{3} + 2$

9. $y = -2 \cos x + 4$ **10.** $T = 25.0 \sin(0.50t - 1.76) + 47.3$

14.6 PRACTICE (pp. 872–874) **5.** $\dfrac{\sqrt{6} - \sqrt{2}}{4}$ **7.** $-\dfrac{\sqrt{2} + \sqrt{6}}{4}$

9. $-2 - \sqrt{3}$ **11.** none **13.** $\dfrac{\pi}{3}, \dfrac{4\pi}{3}$ **15.** 0 **17.** $-\dfrac{\sqrt{3}}{2}$

19. 1 **21.** $-\dfrac{\sqrt{2}}{2}$ **23.** $-2 + \sqrt{3}$ **25.** $\dfrac{\sqrt{2} - \sqrt{6}}{4}$ **27.** $-2 - \sqrt{3}$

29. $-\dfrac{36 + \sqrt{627}}{70}$ **31.** $\dfrac{\sqrt{627} + 36}{4\sqrt{19} - 9\sqrt{33}}$ **33.** $-\dfrac{9\sqrt{33} + 4\sqrt{19}}{70}$

35. $\dfrac{4\sqrt{11} - 15}{30}$ **37.** $\dfrac{4\sqrt{11} - 15}{3\sqrt{11} + 20}$ **39.** $\dfrac{20 - 3\sqrt{11}}{30}$ **41.** $\tan x$

43. $-\sin x$ **45.** $-\cos x$ **47.** $-\sin x$ **49.** $\dfrac{3\pi}{2}$ **51.** $0, \pi$

53. $0, \dfrac{\pi}{3}, \dfrac{5\pi}{3}$ **55.** $36.9°$ **57.** $P = \dfrac{1}{40} \cos 1100t$

59. $0.26 + \dfrac{n\pi}{10}$

14.6 MIXED REVIEW (p. 874) **65.** $\begin{bmatrix} 3 & 3 & 3 \\ -1 & -1 & -1 \end{bmatrix}$ **67.** $\begin{bmatrix} -15 \\ 13 \end{bmatrix}$

69. $\begin{bmatrix} 30 & 10 & -10 \\ -70 & -20 & 30 \end{bmatrix}$ **71.** $C = 134°, a = 65.8, c = 153$

73. $A = 111°, B = 17°, C = 52°$ **75.** $\dfrac{2\pi}{3}, \dfrac{5\pi}{3}$ **77.** $\dfrac{3\pi}{4}, \dfrac{7\pi}{4}$

14.7 PRACTICE (pp. 879–881) **5.** $-\sqrt{\dfrac{29 - 5\sqrt{29}}{58}}$ **7.** $\dfrac{20}{29}$ **9.** $\dfrac{20}{21}$

11. $2 \cos x - 2 \cos^2 x$ **13.** $\dfrac{2 \tan x}{1 - \tan^4 x}$ **15.** $-2 \cos^2 x$

17. $2 - \sqrt{3}$ **19.** $1 - \sqrt{2}$ **21.** $-\left(2 + \sqrt{3}\right)$

23. $-\left(\dfrac{\sqrt{2 + \sqrt{3}}}{2}\right)$ **25.** $\dfrac{\sqrt{2 - \sqrt{3}}}{2}$

27. $\sin \dfrac{u}{2} = \dfrac{\sqrt{6}}{6}$, $\cos \dfrac{u}{2} = -\dfrac{\sqrt{30}}{6}$, $\tan \dfrac{u}{2} = -\dfrac{\sqrt{5}}{5}$

29. $\sin \dfrac{u}{2} = \dfrac{\sqrt{5}}{5}$, $\cos \dfrac{u}{2} = -\dfrac{2\sqrt{5}}{5}$, $\tan \dfrac{u}{2} = -\dfrac{1}{2}$

31. $\sin 2x = -\dfrac{4}{5}$, $\cos 2x = \dfrac{3}{5}$, $\tan 2x = -\dfrac{4}{3}$

33. $\sin 2x = -\dfrac{24}{25}$, $\cos 2x = \dfrac{7}{25}$, $\tan 2x = -\dfrac{24}{7}$ **35.** $2 \cos x$

37. $1 - 5 \sin^2 x$ **39.** $\dfrac{1}{1 + \cos x}$ **41.** $2 \sin x \cos x$

43. $(\sin x + \cos x)^2 = \sin^2 x + \cos^2 x + 2 \sin x \cos x =$
$1 + \sin 2x$ **45.** $\cos \theta + 2 \sin^2 \dfrac{\theta}{2} = \cos \theta + 2\left(\dfrac{1 - \cos \theta}{2}\right) =$
$\cos \theta + 1 - \cos \theta = 1$ **47.** $\cos 3x = \cos (2x + x) =$
$\cos 2x \cos x - \sin 2x \sin x = \cos x (\cos^2 x - \sin^2 x) -$
$\sin x (2 \sin x \cos x) = \cos^3 x - 3 \sin^2 x \cos x$
49. $\cos^2 2x - \sin^2 2x = \cos 4x$ **51.** no solution

53. $0, \dfrac{\pi}{4}, \dfrac{3\pi}{4}, \pi, \dfrac{5\pi}{4}, \dfrac{7\pi}{4}$ **55.** $0, \pi$ **57.** $0.21, 1.38, 3.36, 4.50$

59. $0, \pi$ **61.** $n\pi, \dfrac{2\pi}{3} + 2n\pi, \dfrac{4\pi}{3} + 2n\pi$ **63.** $\pi + 2n\pi$,
$\dfrac{\pi}{3} + 4n\pi, \dfrac{5\pi}{3} + 4n\pi$ **65.** $\dfrac{3\pi}{2} + 2n\pi, \dfrac{7\pi}{6} + 2n\pi, \dfrac{11\pi}{6} + 2n\pi$

67. $\tan \dfrac{u}{2} = \dfrac{1 - \cos u}{\sin u} = \dfrac{(1 - \cos u)(1 + \cos u)}{\sin u (1 + \cos u)} =$
$\dfrac{1 - \cos^2 u}{\sin u (1 + \cos u)} = \dfrac{\sin^2 u}{\sin u (1 + \cos u)} = \dfrac{\sin u}{1 + \cos u}$

69. $y_{max} = \dfrac{1}{64} v^2 \sin^2 \theta$ **71.** $A = 324 \sin \dfrac{\theta}{2} \cos \dfrac{\theta}{2}$

73. $n = \dfrac{\sqrt{3}}{2} + \dfrac{1}{2} \cot \dfrac{\theta}{2}$; $77°$

14.7 MIXED REVIEW (p. 882) **79.** $f(x) - g(x) = -2x + 1$,
all real numbers **81.** $f(x) \div g(x) = \dfrac{4x + 1}{6x}$, all real numbers
except $x = 0$ **83.** $g(f(x)) = 24x + 6$, all real numbers
85. 0.137 **87.** 0.1601 **89.** 0.013 **91.** $-\sin x$ **93.** $-\cos x$
95. $\dfrac{\tan x + 1}{1 - \tan x}$

QUIZ 3 (p. 882) **1.** $\dfrac{\sqrt{2 + \sqrt{3}}}{2}$ **2.** $\dfrac{\sqrt{6} - \sqrt{2}}{4}$ **3.** $-2 + \sqrt{3}$

4. $-\dfrac{\sqrt{6} + \sqrt{2}}{4}$ **5.** $-\dfrac{\sqrt{6} + \sqrt{2}}{4}$ **6.** $-2 + \sqrt{3}$ **7.** $\dfrac{\sqrt{18 + 12\sqrt{2}}}{6}$

8. $\dfrac{\sqrt{18 + 12\sqrt{2}}}{6}$ **9.** $\dfrac{\sqrt{3 + 2\sqrt{2}}}{\sqrt{3 - 2\sqrt{2}}}$ **10.** $-\dfrac{4\sqrt{2}}{9}$ **11.** $\dfrac{7}{9}$ **12.** $-\dfrac{4\sqrt{2}}{7}$

13. $-\sin x$ **14.** $-\cos x$ **15.** $\dfrac{\tan x + 1}{1 - \tan x}$ **16.** $\sin x$ **17.** 1

18. $\dfrac{\tan x}{1 + \tan x}$ **19.** $\dfrac{\pi}{18} + \dfrac{2n\pi}{3}, \dfrac{5\pi}{18} + \dfrac{2n\pi}{3}$ **20.** $n\pi$

21. $n\pi; \dfrac{2\pi}{3} + 2n\pi; \dfrac{4\pi}{3} + 2n\pi$ **22.** $\dfrac{3\pi}{8} + \dfrac{n\pi}{2}$

23. 77 ft

CHAPTER 14 REVIEW (pp. 884–886)

1. 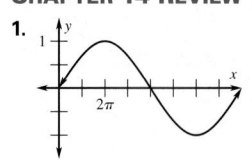 **3.**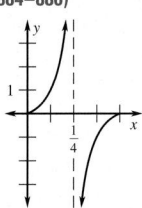

5. 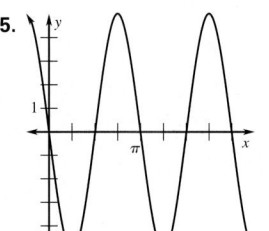 **7.** 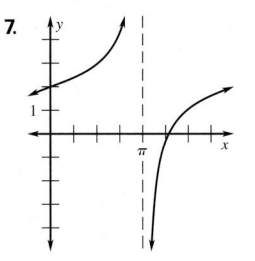 **9.** 1

11. $\sin^2 (-x) = \sin^2 x$ Negative angle identity
$= \dfrac{\sin^2 x}{\cos^2 x} \cos^2 x$ Multiply by $\dfrac{\cos^2 x}{\cos^2 x}$.
$= \dfrac{\tan^2 x}{\sec^2 x}$ Identities
$= \dfrac{\tan^2 x}{1 + \tan^2 x}$ Pythagorean identity

13. $\dfrac{\pi}{4} + \dfrac{n\pi}{2}, n\pi$ **15.** $\dfrac{\pi}{2} + 2n\pi$ **17.** $\dfrac{\pi}{6} + 2n\pi, \dfrac{5\pi}{6} + 2n\pi$

19. $y = 2 \sin x$ **21.** $y = \cos 2x$ **23.** $-\dfrac{\sqrt{6} + \sqrt{2}}{4}$ **25.** $-2 - \sqrt{3}$

27. $-2 + \sqrt{3}$ **29.** $-\dfrac{\sqrt{2 - \sqrt{2}}}{2}$ **31.** 0

CUMULATIVE PRACTICE (pp. 890–891) **1.** $y = -2x + 7$

3. $x = 4$ **5.** $(10, 4, -4)$ **7.** $\begin{bmatrix} 11 & -8 \\ -14 & 9 \end{bmatrix}$ **9.** $\begin{bmatrix} -\dfrac{3}{2} & 1 \\ -18 & 8 \end{bmatrix}$

11. $2x^3 - 5x^2 - 13x + 4$ **13.** $\dfrac{x}{4x^2 - 19x - 30}$

15. $\dfrac{4x^2 + 27x + 7}{x^2 - 49}$ **17.** $\dfrac{1}{5}$ **19.** -1 **21.** -4 **23.** 7 **25.** 0

27. $5.39, (-2.5, 1)$ **29.** $9.90, (3.5, 0.5)$ **31.** $\dfrac{1}{2}, 2^{4 - n}$

33. $14, 5n - 11$ **35.** 40 **37.** $\dfrac{3}{2}$ **39.** 90 **41.** 8 **43.** 35

45. 28.3 in.; 84.82 in.^{2} **47.** $\dfrac{\sqrt{3}}{3}$ **49.** 1 **51.** $\dfrac{1}{2}$ **53.** $\dfrac{\pi}{6}, 30°$

55. $\dfrac{3\pi}{4}, 135°$ **57.** $B = 31°, C = 84°, c = 7.68$

59. $A = 92°, B = 64°, C = 24°$ **61.** 336.7

63. 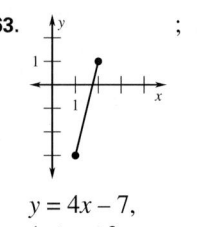 **; 65.**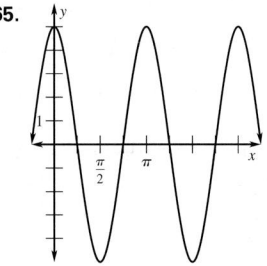

$y = 4x - 7,$
$1 \le x \le 2$

SELECTED ANSWERS

67.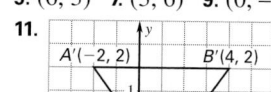

69. $\tan^3 x$ **71.** $-\csc x$

73. $\dfrac{\pi}{2} + n\pi$ **75.** $-\dfrac{\sqrt{2}+\sqrt{6}}{4}$

77. $-2-\sqrt{3}$ **79.** $-\dfrac{\sqrt{2}+\sqrt{6}}{4}$

81. mean: 72.4; median: 73; modes: 76, 74; range: 12; standard deviation: 3.7; **83.** 953 ft

Heights of Girls (in.)

64 66 68 70 72 74 76 78 80

SKILLS REVIEW HANDBOOK

OPERATIONS WITH SIGNED NUMBERS (p. 905) **1.** -2 **3.** 0
5. -1 **7.** 6 **9.** -10 **11.** -21 **13.** 6 **15.** -4 **17.** 9 **19.** -5
21. -3 **23.** 15 **25.** 24 **27.** -24 **29.** 30 **31.** 25 **33.** -5
35. 8 **37.** -4 **39.** 3 **41.** -20 **43.** -2 **45.** -3 **47.** 29 **49.** -7
51. 135 **53.** -35 **55.** -3 **57.** 7 **59.** -7 **61.** -9 **63.** 132

CONVERTING DECIMALS, FRACTIONS, AND PERCENTS
(p. 906) **1.** 20% **3.** 55% **5.** 87% **7.** 40% **9.** 60% **11.** 0.5
13. 0.02 **15.** 0.4 **17.** 0.36 **19.** 1.5

CALCULATING PERCENTS (p. 907) **1.** 3 **3.** 0.3 **5.** 30
7. 0.54 **9.** 12 **11.** 0.00375 **13.** 0.025 **15.** 0.084 **17.** 14
19. 0.005 **21.** 50% **23.** 100% **25.** 35% **27.** about 22%
29. about 2.4% **31.** 20% **33.** 0.2% **35.** 0.44%

LEAST COMMON DENOMINATOR (p. 909) **1.** $2 \times 2 \times 2$
3. $2 \times 2 \times 2 \times 2 \times 2 \times 2$ **5.** prime **7.** $2 \times 2 \times 3$
9. 2×11 **11.** prime **13.** prime **15.** 5×5 **17.** 1, 28
19. 2, 60 **21.** 20, 40 **23.** 6, 72 **25.** 12, 144 **27.** 1, 6
29. 6, 18 **31.** 48 **33.** 26 **35.** 10 **37.** 12 **39.** 60 **41.** 12
43. 120 **45.** 12 **47.** 60 **49.** 60 **51.** $\dfrac{19}{24}$ **53.** $-\dfrac{5}{4}$ **55.** $-\dfrac{1}{10}$
57. $\dfrac{33}{80}$ **59.** $\dfrac{13}{12}$ **61.** $-\dfrac{111}{66}$ **63.** $\dfrac{1}{2}$ **65.** $-\dfrac{1}{20}$ **67.** $\dfrac{7}{6}$ **69.** $\dfrac{1}{3}$

WRITING RATIOS AND SOLVING PROPORTIONS (p. 910)
1. $4:5, \dfrac{4}{5}$ **3.** $2:6, \dfrac{2}{6}$ **5.** 1 to 5, 1:5 **7.** 8 to 5, 8:5
9. 3 to 1, $\dfrac{3}{1}$ **11.** 3 to 4, 3:4 **13.** 1 to 4 **15.** 1:4 **17.** $\dfrac{1}{5}$
19. 5 to 3 **21.** $\dfrac{3}{5}$ **23.** $\dfrac{4}{3}$ **25.** 1 **27.** 16 **29.** 3 **31.** 21
33. 5 **35.** 27 **37.** 18 **39.** 20 **41.** 16 **43.** 8 **45.** 6 **47.** 1

SIGNIFICANT DIGITS (p. 912) **1.** 8200 **3.** 9.50 **5.** 28.15
7. 700 **9.** 10 **11.** 0.74 **13.** 3.2 **15.** 1.0 **17.** 200 **19.** 24.7
21. 17.7 **23.** 89 **25.** 0.723 **27.** 0.06 **29.** 16,000
31. \$7.50 **33.** \$239.70 **35.** 13 mi/gal
37. 100 gal of milk **39.** 230 mL **41.** 730 computers/store
43. 15 mg **45.** 25.9 in. of rain

SCIENTIFIC NOTATION (p. 913) **1.** 4×10^{-1} **3.** 9×10^{-2}
5. 4×10^{0} **7.** 9.26×10^{-5} **9.** 2.11111×10^{2} **11.** 5×10^{-3}
13. 9.84×10^{4} **15.** 2.0489×10^{2} **17.** 3.7×10^{-4}
19. 5.98×10^{1} **21.** 2.30856×10^{7} **23.** 1.00×10^{-4}
25. 900 **27.** 3100 **29.** 0.290 **31.** 10,010 **33.** 7,926,000
35. 0.000384 **37.** 0.000037 **39.** 0.0049831 **41.** 395,020
43. 2640.95 **45.** 0.000455 **47.** 0.059438

PERIMETER, AREA, AND VOLUME (p. 916) **1.** about 6.28 m
3. 8 in. **5.** 13 m **7.** 12 ft **9.** 22 cm **11.** 18 cm **13.** about
69 in. **15.** 81 in.^{2} **17.** 21 cm^2 **19.** 24 in.^{2} **21.** about
0.79 in.^{2} **23.** 12 mi^2 **25.** 10 in.^{2} **27.** 88 ft^2 **29.** about
63 mm^2 **31.** 201 in.^{2} **33.** 288 cm^2 **35.** 1000 cm^3
37. 12 yd^3 **39.** 5 ft^3 **41.** about 127 in.^{3} **43.** 38.4 m^3

TRIANGLE RELATIONSHIPS (p. 918) **1.** 45 **3.** 76 **5.** 165 **7.**
60 **9.** yes **11.** no **13.** no **15.** no **17.** no **19.** yes **21.** yes
23. yes **25.** about 5.7 mm **27.** 2.5 m **29.** about 3.6 m
31. 8 cm **33.** about 7.1 in. **35.** yes **37.** no **39.** yes **41.** yes
43. no **45.** yes

SYMMETRY (p. 920) **1.** line symmetry: 4 lines of symmetry;
rotational symmetry: 90° or 180° in either direction
3. line symmetry: 6 lines of symmetry; rotational
symmetry: 60°, 120°, or 180° in either direction
5. line symmetry: 5 lines of symmetry; rotational
symmetry: 72° or 144° in either direction **7.** no line or
rotational symmetry **9.** $(-3, 2)$ **11.** $(4, 1)$ **13.** $(4, 6)$

TRANSFORMATIONS (p. 922) **1.** $(-3, -6)$ **3.** $\left(\dfrac{9}{2}, -9\right)$
5. $(6, 3)$ **7.** $(3, 6)$ **9.** $(0, -1)$

11.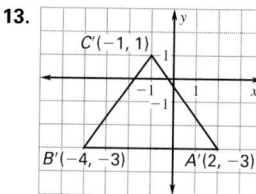

13.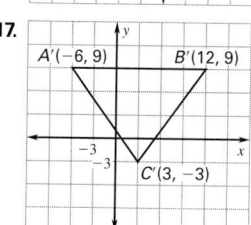

15.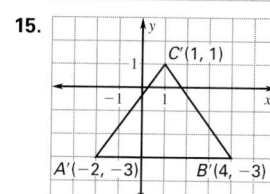

17.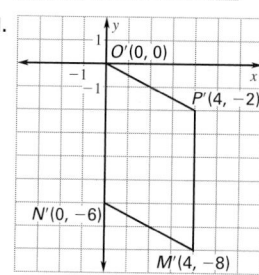

19.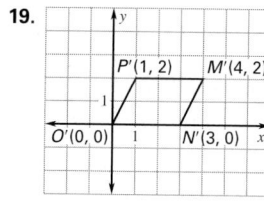

21.

23.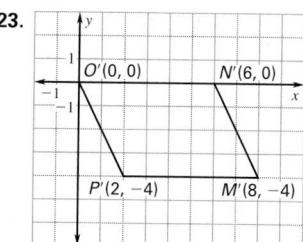

SELECTED ANSWERS

27.

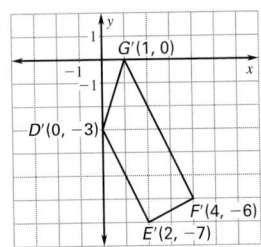

29.

(graph with G'(1, 0), D'(0, −3), F'(4, −6), E'(2, −7))

SIMILAR FIGURES (p. 923) **1.** 2.5 **3.** 1 **5.** 70 **7.** 3.75

LOGICAL ARGUMENT (p. 925) **1.** The conclusion is invalid. This does not follow the chain rule. **3.** The conclusion is valid. This is not an example of the *Or* rule. **5.** The conclusion is valid. This is an example of the chain rule. **7.** The conclusion is invalid. This is not an example of an indirect argument. **9.** The conclusion is valid. This is an example of the AND rule. **11.** true **13.** true **17.** true **19.** true **21.** true **23.** true **25.** true

IF-THEN STATEMENTS (p. 926) **1.** If it rains in Spain, then it falls on the plain. **3.** If $x = 4$, then $3x^2 = 48$. **5.** If you finish cleaning, then you can go out tonight. **7.** If $x = 3$, then $y = 16$. **9.** If a rectangle has four equal sides, then it is a square. **11.** If a curve is described by $y = x^2$, then it is a parabola. **13.** If $x^2 = 16$, then $x = 4$; false. **15.** If a line's slope is undefined, then it is a vertical line; true. **17.** If a figure is a parallelogram, then it has two pairs of opposite congruent sides; true. **19.** If you are cold, then you are in Minnesota in January; false. **21.** If Margot got more votes than her opponent, then she won the election; true. **23.** If a convex polygon is a regular pentagon, then it has five equal sides; true. **25.** False; a square has four equal sides and four 90° angles. **27.** False; $x^2 = 25$ for $x = 5, -5$. **29.** true **31.** true **33.** true **35.** true

COUNTEREXAMPLES (p. 928) **1.** False; any parallelogram with angles that are not right angles is not a rectangle. **3.** False; the last digit of the number 16 is 6, but 16 is not divisible by 3. **5.** False; no triangle has two 90° angles because there must be a third angle and together the three must total 180°. **7.** False; cats can also be black. **9.** true **11.** False; if $a = 1$, then $3a - 4 = -1 < 0$. **13.** true **15.** true

JUSTIFY REASONING (p. 929) **1.** Division property of equality **3.** Multiplication property of equality **5.** Addition property of equality **7.** Multiplication property of equality **9.** Definition of raising to a power (2) **11.** Subtraction property of equality

13. Multiplication property of equality **15.** Addition property of equality **17.** Multiplication property of equality **19.** Distributive property

21. $9x = 27$ Given
$x = 3$ Division property of equality

23. $\frac{x}{2} + 5 = 0$ Given
$\frac{x}{2} = -5$ Subtraction property of equality
$x = -10$ Multiplication property of equality

25. $\frac{5x}{2} - 2 = -5$ Given
$5x - 4 = -10$ Multiplication property of equality
$5x = -6$ Addition property of equality
$x = -\frac{6}{5}$ Division property of equality

27. $\frac{3x}{4} = 6$ Given
$3x = 24$ Multiplication property of equality
$x = 8$ Division property of equality

TRANSLATING PHRASES INTO ALGEBRAIC EXPRESSIONS (p. 930) **1.** $x + 8$ **3.** $x - 49$ **5.** $7x$ **7.** $\frac{3}{4}x$ **9.** $0.90x$ **11.** $x - 4$ **13.** $\frac{x + 3}{2}$ **15.** $3b$ **17.** $67.39 - x$ **19.** $2.5x$

ADDITIONAL PROBLEM SOLVING STRATEGIES (p. 932) **1.** 10 **3.** from 40 to 60 people **5.** 128 ft/sec **7.** 3, 6, 9, 11, 12, 14, 17, 19, 22, 27 **9.** 10 **11.** 35 **13.** 204

POINTS IN THE COORDINATE PLANE (p. 933)
1–15 odd:

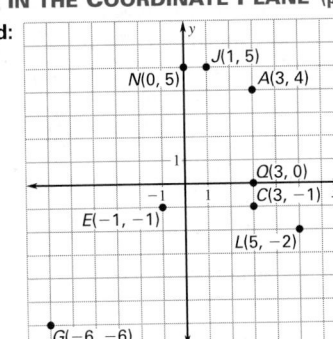

17. (0, 4), y-axis
19. (−5, 5), Quadrant II
21. (2, −5), Quadrant IV
23. (−5, −5), Quadrant III
25. (3, −3), Quadrant IV
27. (−3, 0), x-axis **29.** (−5, 2), Quadrant II
31. (5, 5), Quadrant I **33.** (−8, −4), Quadrant III
35. (7, −4), Quadrant IV **37.** (4, −7), Quadrant IV

BAR, CIRCLE, AND LINE GRAPHS (p. 935)
1. about 1.2 ft **3.** about 120 **5.** about 20%
7. Students with Each Major **9.** Visitors to the Zoo

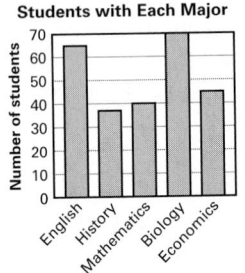

OPPOSITES (p. 936) **1.** −3 **3.** −150 **5.** −4.3 **7.** $-\frac{3}{5}$ **9.** $-2a - b$ **11.** $-a + b + c$ **13.** $-2 - x$ **15.** $1 - 4x$

17. $-x^2 - 2x + 4$ **19.** $-2x - 3y$ **21.** $-3x + y - 11z$
23. $36x - 54y$ **25.** $x - 7y$ **27.** $-a + 3b$

MULTIPLYING BINOMIALS (p. 937) **1.** $x^2 + 2x + 1$ **3.** $4x^2 + 9x + 2$ **5.** $-2x^2 - 5x + 3$ **7.** $4x^2 - 25$ **9.** $5x^2 - 7x - 6$
11. $2y^2 + 3y - 9$ **13.** $ac + ad + bc + bd$ **15.** $-4x^2 + 1$
17. $x^2 - y^2$ **19.** $3x^2 - x - 10$ **21.** $-12x^2 + 4x + 96$

FACTORING (p. 938) **1.** $(x + 3)(x + 2)$ **3.** $(x + 4)(x + 5)$
5. $(x + 3)(x + 3)$ **7.** $(x + 3)(x - 8)$ **9.** $(x + 2)(x + 1)$
11. $(x + 3)(x - 2)$ **13.** $(x + 7)(x + 7)$ **15.** $(x - 10)(x + 2)$
17. $(x - 9)(x - 9)$ **19.** $(x - 5)(x - 3)$ **21.** $(x - 16)(x - 5)$
23. $(x - 4)(x + 1)$ **25.** $(x - 5)(x - 4)$ **27.** $(x + 5)(x + 5)$
29. $(x + 4)(x + 2)$

LEAST COMMON DENOMINATOR (p. 939)
1. $2x$ **3.** $90k^2$ **5.** $6xy$ **7.** $3z^2$ **9.** $(b - 1)(b + 1)^2$
11. $2(n + 2)$ **13.** $6(2 + 3x)$ **15.** $10(3h - 4)$ **17.** $12 - 20e$
19. $c^2(c - 1)(c - 4)$ **21.** $36x^2$ **23.** $3x^2 + 2x$

EXTRA PRACTICE

CHAPTER 1 (p. 940)

1.
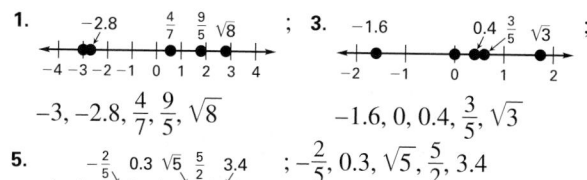

$-3, -2.8, \dfrac{4}{7}, \dfrac{9}{5}, \sqrt{8}$ $-1.6, 0, 0.4, \dfrac{3}{5}, \sqrt{3}$

5.
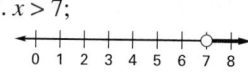
$; -\dfrac{2}{5}, 0.3, \sqrt{5}, \dfrac{5}{2}, 3.4$

7. commutative property of multiplication **9.** inverse property of addition **11.** commutative property of addition
13. 9 **15.** -36 **17.** 12 **19.** $12x^2 - 13x$ **21.** $-9x^2 + 2x$
23. $5x + 4$ **25.** 7 **27.** -9 **29.** -2.3 **31.** $y = \dfrac{-3x + 12}{4}$
33. $y = \dfrac{7x + 9}{6}$ **35.** $y = \dfrac{2}{25}x + \dfrac{3}{10}$ **37.** C = Total cost (dollars),
15 = Cost of first pound (dollars), 3 = Cost per pound of each additional pound (dollars per pound), 6 = Number of additional pounds (pounds) **39.** $C = 33$; it will cost \$33 to send a 7 pound package.
41. $x > 7$;

43. $x > 2$;

45. $2 \le x \le 10$;

47. $x < 3$ or $x \ge 8$;

49. $x \le \dfrac{23}{3}$ or $x > 12$;

51. $28, -22$ **53.** $10.2, 4.4$ **55.** $\dfrac{62}{7}, -\dfrac{22}{7}$ **57.** 2
59. $x < -7$ or $x > 1$;   **61.** $-3 < x < 1$;

63. $2 < y < 3$;   **65.** $x \le -\dfrac{17}{3}$ or $x \ge 5$;

67. $y < -1$ or $y > 4$;

CHAPTER 2 (p. 941)

1. 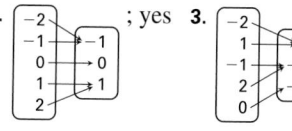 ; yes **3.** 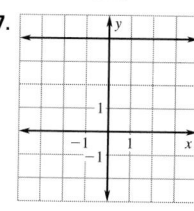 ; yes **5.** 5 **7.** -46
9. -36
11. parallel
13. perpendicular

15. parallel **17.** **19.**

21.   **23.** undefined; none **25.** $2; -5$
27. $2; 10$ **29.** $\dfrac{1}{3}; \dfrac{8}{3}$

31. 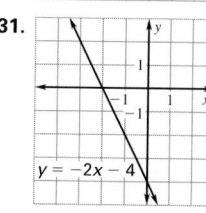 **33.**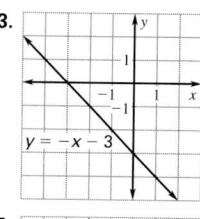
$y = -2x - 4$ $y = -x - 3$

35. 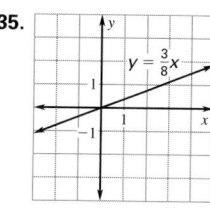 **37.**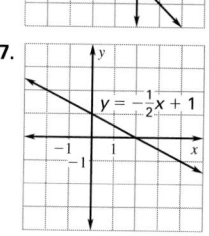
$y = \dfrac{3}{8}x$ $y = -\dfrac{1}{2}x + 1$

39. $y = 2$ **41.** $y = 2x - 5$ **43.** $y = -\dfrac{5}{2}x + \dfrac{7}{2}$
45. $y = -2x - 3$ **47.** $y = 6$
49. 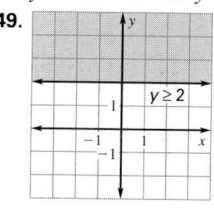 **51.**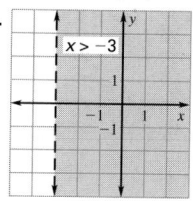
$y \ge 2$ $x > -3$

53. 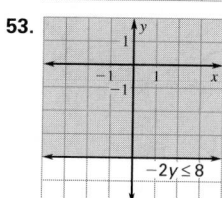 **55.**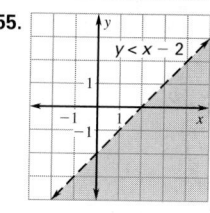
$-2y \le 8$ $y < x - 2$

61. -4 **63.** 9 **65.** 2 **67.**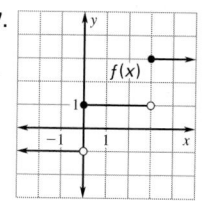
$f(x)$

69.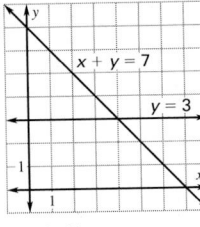

$y = 2|x| + 5$

(0, 5); up; narrower

71.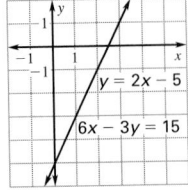

$y = |x + 3|$

(−3,0); up; same

CHAPTER 3 (p. 943) **1.** no **3.** no **5–11.** Estimates may vary.

5.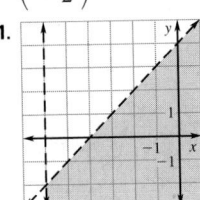

$x + y = 7$
$y = 3$

1; (4, 3)

7.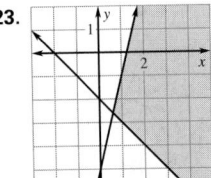

$y = 2x − 5$
$6x − 3y = 15$

infinitely many

9.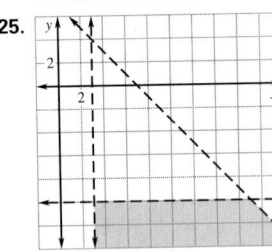

$5x − y = 7$
$y = 5x + 6$

11.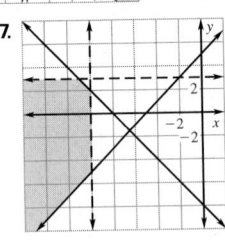

$\frac{1}{5}x − 2y = −4$
$\frac{1}{2}x + 3y = 2$

1; (−5, 1.5)

13. $\left(7, \frac{17}{2}\right)$ **15.** (−3, −2) **17.** (2, 0) **19.** (−8, −9)

21.
23.

25.
27.

29. −25; 48 **31.** −27; 25

33.

$(0, 0, 6)$
$3x + y + 2z = 12$
$(4, 0, 0)$
$(0, 12, 0)$

39. $f(x, y) = -\frac{1}{3}x - \frac{1}{3}y + 3; 3$

41. $f(x, y) = -x + \frac{5}{2}y - 5; 4$

43. $f(x, y) = \frac{1}{7}x - \frac{4}{7}y - \frac{31}{7}; -4$

45. (0, −2, 5) **47.** (−3, 4, 2)

49. $\left(\frac{1}{2}, 3, 7\right)$

CHAPTER 4 (p. 944)

1. $\begin{bmatrix} 4 & 2 \\ -4 & 4 \end{bmatrix}$ **3.** $\begin{bmatrix} -1 & -1 & 7 \\ 1 & 0 & -4 \end{bmatrix}$ **5.** $\begin{bmatrix} 3\frac{2}{3} & \frac{1}{5} \\ 5 & -\frac{1}{3} \end{bmatrix}$ **7.** $\begin{bmatrix} 5 & -6 \\ 17 & 27 \end{bmatrix}$

9. $\begin{bmatrix} 2.6 & 3.8 & 4.9 \\ 2.3 & 3.4 & 4.2 \end{bmatrix}$ **11.** [4.4] **13.** $\begin{bmatrix} 10.83 & -23.2 \\ 66.62 & 23.31 \end{bmatrix}$

15. $\begin{bmatrix} -14 & 33 & 57 \\ -66 & -3 & -36 \\ -74 & 23 & -12 \end{bmatrix}$ **17.**

	Total
Machine 1	$40.05
Machine 2	$50.85
Machine 3	$44.45

19. 29

21. −51 **23.** 799 **25.** (2, −3) **27.** (−6, −4) **29.** (5, −1, −2)

31. $\begin{bmatrix} 2 & -1 \\ -\frac{7}{3} & \frac{4}{3} \end{bmatrix}$ **33.** $\begin{bmatrix} -\frac{1}{6} & \frac{1}{2} \\ \frac{1}{12} & \frac{1}{4} \end{bmatrix}$ **35.** $\begin{bmatrix} \frac{1}{2} & \frac{1}{8} \\ \frac{1}{4} & -\frac{1}{16} \end{bmatrix}$ **37.** $\begin{bmatrix} 3 & 7 \\ 4 & 9 \end{bmatrix}$

39. (5, 1) **41.** (−1, −1) **43.** (0, −2) **45.** $\left(-\frac{7}{5}, 5\right)$

47. (3, −2, 1)

CHAPTER 5 (p. 945)

1.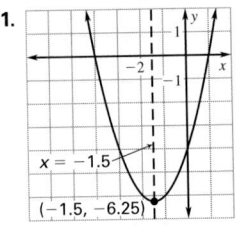

$x = -1.5$
$(-1.5, -6.25)$

3.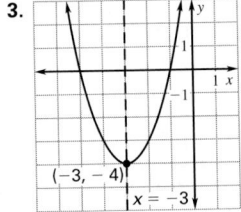

$(-3, -4)$
$x = -3$

5.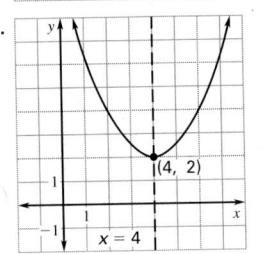

$(4, 2)$
$x = 4$

7.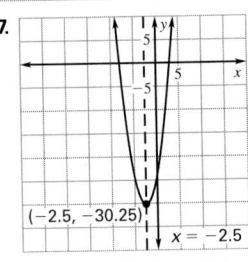

$(-2.5, -30.25)$
$x = -2.5$

9. 0 mi/h **11.** $(m − 4)(m − 5)$ **13.** $(3x − 2)(2x + 3)$
15. $(2u + 5)(2u − 7)$ **17.** $(x − 5)^2$ **19.** $2(2x − 5)(x + 2)$
21. cannot be factored **23.** $-\frac{1}{2}, 7$ **25.** $\frac{2}{5}$ **27.** $-\frac{4}{3}, \frac{1}{2}$
29. $-\frac{8}{3}, \frac{5}{4}$ **31.** $y = x(x − 5); 0, 5$ **33.** $y = 6(x + 2)(x − 2); 2, −2$
35. $y = (5x − 3)(x − 2); \frac{3}{5}, 2$ **37.** $y = 7(x + 3)(x − 3); 3, −3$
39. $5\sqrt{5}$ **41.** $9\sqrt{3}$ **43.** $\frac{9\sqrt{5}}{25}$ **45.** $\frac{4}{5}$ **47.** $4\sqrt{10}, -4\sqrt{10}$
49. $\sqrt{5} − 2, -\sqrt{5} − 2$ **51.** 5, −5 **53.** 7, 1 **55.** $i\sqrt{10}, -i\sqrt{10}$
57. $5i, -5i$ **59.** $6i, -6i$ **61.** $-1 + 2i\sqrt{5}, -1 - 2i\sqrt{5}$
63. $-4 − 3i$ **65.** 0 **67.** $-4 − 6i$ **69.** $11 − 11i$ **71.** $\frac{15}{26} - \frac{3}{26}i$
73. $-i$ **75.** $2 + 2i, 2 − 2i$ **77.** 0.866, −3.47 **79.** $-\frac{2}{3}, 2$
81. $\frac{4}{5}, -2$ **83.** $\frac{3}{4}, -1$ **85.** $-\frac{3}{2} + \frac{\sqrt{41}}{2}, -\frac{3}{2} - \frac{\sqrt{41}}{2}$
87. $\frac{3}{7} + \frac{\sqrt{61}}{7}i, \frac{3}{7} - \frac{\sqrt{61}}{7}i$ **89.** 1; 2; real
91. −191; 2; imaginary **93.** −40; 2; imaginary

SELECTED ANSWERS

99.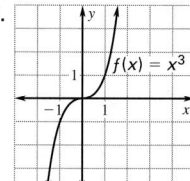

101. $y = (x - 1)^2 + 1$

103. $y = \frac{1}{9}(x - 2)^2 + 1$

105. $y = 2(x + 3)^2 - 5$

107. $y = (x - 2)(x - 6)$

109. $y = 2x(x - 4)$

111. $y = \frac{3}{4}(x - 1)(x - 2)$

113. $y = -\frac{1}{6}(x - 5)(x + 2)$

CHAPTER 6 (p. 947) **1.** 625; product of powers **3.** 512; power of a power **5.** $\frac{16}{25}$; power of a quotient **7.** $\frac{1}{512}$; *Sample answer:* zero exponent, negative exponent **9.** $\frac{1}{46,656}$; *Sample answer:* product of powers, power of a quotient **11.** 6; zero exponent, quotient of powers **13.** $1,048,576x^8$; power of a product, power of a power **15.** x^3; quotient of powers **17.** $\frac{1}{36x^6y^8}$; power of a product, power of a power, negative exponent **19.** $-\frac{1}{2187x^{63}y^{21}}$; power of a product, power of a power, negative exponent **21.** $\frac{1}{64x^{12}y^{16}}$; power of a product, power of a power, negative exponent **23.** $\frac{3x^8}{y^5}$; product of powers, quotient of powers, negative exponent **25.** 19 **27.** −4

29.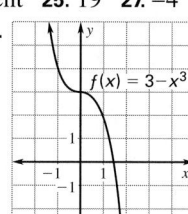

31.

33.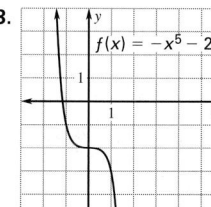

35. $5x^2 + 10x + 7$

37. $5x^3 - 10x^2 + 18$

39. $15x^3 + 9x^2 + 29x - 16$

41. $x^2 + 2x - 35$

43. $-3x^3 - 16x^2 + 17x + 30$

45. $12x^4 - 24x^3 + 42x$

47. $8x^3 + 96x^2 + 384x + 512$

49. $x^3 + 3x^2y + 3xy^2 + y^3$

51. $2(x + 5)(x^2 - 5x + 25)$ **53.** $(x + 1)(x + 3)^2$
55. $3(x - 2)(x^2 + 2x + 4)$ **57.** $(2x^2 - 5)(x + 9)$ **59.** $(\sqrt{3} + 3)$
inches by $(\sqrt{3} - 1)$ inches by $\sqrt{3}$ inches **61.** $x^2 - 6x + 3$

63. $3x^3 - 29x^2 + 129x - 540 + \frac{2176}{x + 4}$

65. $4x^3 + 3x^2 + 8x + 15 + \frac{35}{x - 2}$

67. −3, −2 **69.** 2 **71.** −5 **73.** 1, 2, 3 **75.** −2, 4 **77.** −1, 1, $\sqrt{5}, -\sqrt{5}$ **79.** $x^3 + 5x^2 + 8x + 4$ **81.** $x^3 - 3x^2 + x - 3$
83. $x^3 - 4x^2 + 6x - 4$ **85.** $x^4 + x^3 - 6x^2 - 14x - 12$
87. $x^4 - 4x^3 + 41x^2 - 144x + 180$
89. $x^6 + 36x^4 - 625x^2 - 22,500$

91.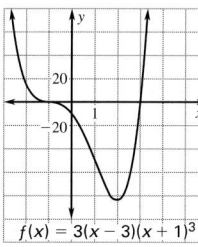
$f(x) = 3(x - 3)(x + 1)^3$

93.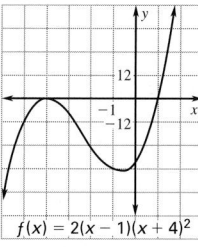
$f(x) = 2(x - 1)(x + 4)^2$

99. $y = \frac{1}{2x^2} + \frac{3}{2x}$ **101.** $y = x^3 - 3x^2 + 2x + 3$

CHAPTER 7 (p. 949) **1.** 3 **3.** $\frac{1}{4}$ **5.** −125 **7.** $\frac{1}{9}$ **9.** 2 **11.** 2
13. 5 **15.** 2 **17.** $5^{1/2}$ **19.** $7\sqrt[6]{7}$ **21.** $x^{7/10}$ **23.** $3xy^2\sqrt{x}$
25. $\frac{\sqrt[6]{x^5}}{4x^5}$ **27.** $\frac{2x^2y^4}{3}$ **29.** $2x^2 - 4x - 4$; all real numbers
31. $2x^2 - 4x - 4$; all real numbers **33.** $2x^2 - 8x + 10$;
all real numbers **35.** $2x^2 - 18$; all real numbers
37. $3x^{5/6}$; nonnegative reals **39.** $3x^{5/6}$; nonnegative reals
41. $3x^{1/6}$; nonnegative reals **43.** $3^{4/3}x^{1/9}$; nonnegative reals
45. $f^{-1}(x) = \frac{x - 1}{3}$ **47.** $f^{-1}(x) = -x - 4$ **49.** $f^{-1}(x) = \frac{x - 3}{2}$
51. $f^{-1}(x) = 2x + 8$ **53.** $f^{-1}(x) = -3x + 15$
55. $f^{-1}(x) = \sqrt[4]{x + \frac{1}{8}}$ **57.** $r = \frac{\sqrt{\pi A}}{\pi}$

59. 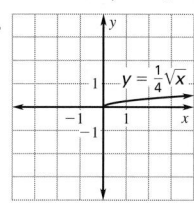 ; **61.** 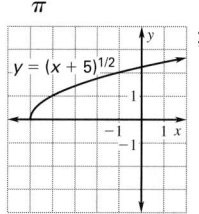 ;

$x \geq 0$; $y \geq 0$ $x \geq -5$; $y \geq 0$

63. 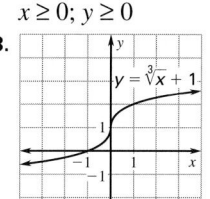 ; **65.** 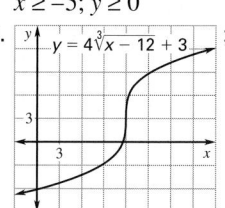 ;

all reals; all reals   all reals; all reals

67. ; **69.** ;

all reals; all reals

all reals; all reals

71. 43,046,721 **73.** no solution **75.** −8 **77.** no solution
79. 10 **81.** $\frac{101}{13}$ **83.** 58.2; 57.5; 58; 21; 5.25
85. 25.6; 21.3; 18.6; 46.7; 15.0

CHAPTER 8 (p. 950)

1.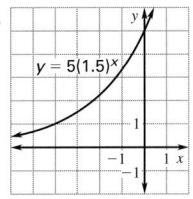

all reals; positive reals

3.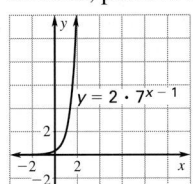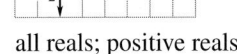

all reals; positive reals

5.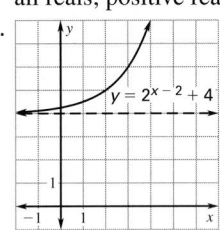

all reals; positive reals

7.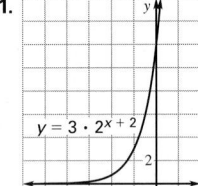

all reals; positive reals

9.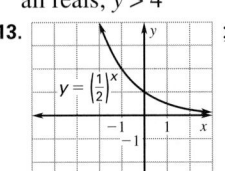

all reals; $y > 4$

11.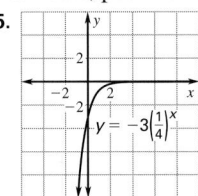

all reals; positive reals

13.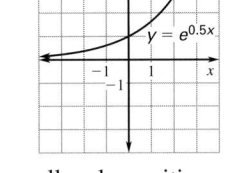

all reals; positive reals

15.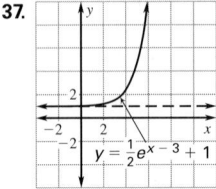

all reals; negative reals

21. e^7 **23.** $16e^{6x}$ **25.** $\dfrac{10}{e^{2x}}$ **27.** e^{6x-1} **29.** $4e^{4x-1}$ **31.** $4e^{4x}$

33.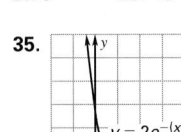

all reals; positive reals

35.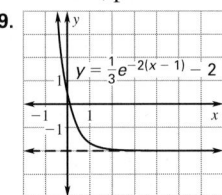

all reals; positive reals

37.

all reals; $y > 1$

39.

all reals; $y > -2$

41. about 8.96 lb per square inch **43.** 2 **45.** $-\dfrac{1}{2}$ **47.** 3

49. 3 **51.** $y = \left(\dfrac{1}{3}\right)^x$ **53.** $y = \dfrac{1}{3}e^x$ **55.** $y = e^x + 3$

57.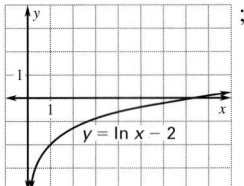

positive reals; all reals

63.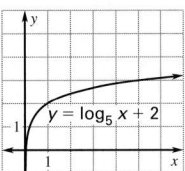

positive reals; all reals

65. 3 **67.** 3 **69.** -5 **71.** 6 **73.** $\log 3 + 4\log x$

75. $\ln 15 + \ln x$ **77.** $\log 3 + \dfrac{1}{2}\log x$

79. $2\log x + 3\log y + 4\log z$ **81.** $\ln x^4 y^6 z^3$ **83.** $\ln\left(\dfrac{27}{x}\right)^{1/4}$

85. $\log_5 5^3 10^{-1}$ **87.** 1.90 **89.** 0.778 **91.** $-163,000$ **93.** 5
95. $y = 0.5(2)^x$ **97.** $y = 0.25(5)^x$ **99.** $y = 0.286(1.14)^x$
101. $y = -2(4)^x$ **103.** $y = x^3$ **105.** $y = 2x^{1/2}$ **107.** $y = 3x^3$
109. $y = 0.5x^2$ **111.** 2.43 **113.** 0.196 **115.** 1.2 **117.** 0.724

119.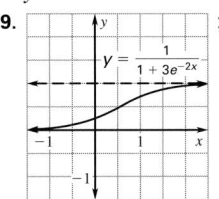

$y = 0, y = 1; \left(0, \dfrac{1}{4}\right);$
$(0.549, 0.5)$

121.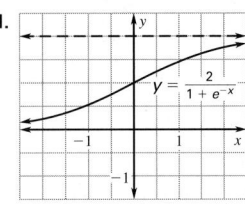

$y = 0, y = 2; (0, 1); (0, 1)$

CHAPTER 9 (p. 952)

1. $y = \dfrac{18}{x}; 4.5$ **3.** $y = -\dfrac{4}{x}; -1$

5. $y = -\dfrac{3}{x}; -\dfrac{3}{4}$ **7.** $y = \dfrac{1}{16x}; \dfrac{1}{64}$ **9.** $z = xy; 28$

11. $z = -\dfrac{xy}{30}; -\dfrac{14}{15}$ **13.** $z = \dfrac{64xy}{3}; \dfrac{1792}{3}$

15.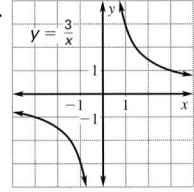

all reals except 0;
all reals except 0

17.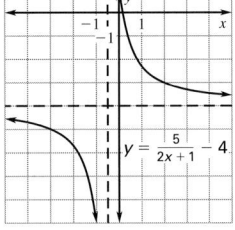

all reals except $-\dfrac{1}{2}$;
all reals except -4

19.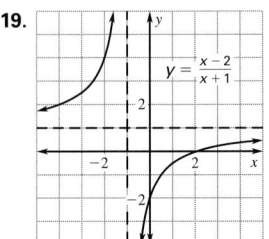

all reals except -1;
all reals except 1

21.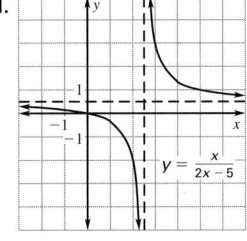

all reals except $\dfrac{5}{2}$;
all reals except $\dfrac{1}{2}$

23. $C = \dfrac{4.50n + 710}{n}$ **25.** The average cost decreases as the number of calendars printed increases.

27. 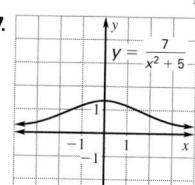 **29.**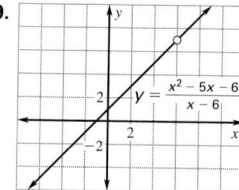

31. 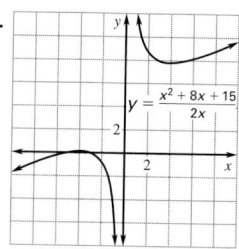 **33.**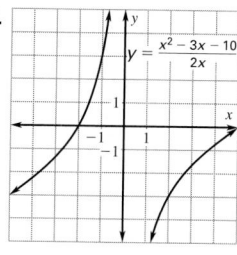

35. $2x^4$ **37.** $\dfrac{5x^4}{2x+4}$ **39.** $\dfrac{x-3}{x+3}$ **41.** $\dfrac{x(x^2-2x+4)(2x-3)}{x-2}$

43. $\dfrac{12}{5x}$ **45.** $\dfrac{26}{15x}$ **47.** $\dfrac{7x+11}{(x+2)(x-2)}$ **49.** $\dfrac{-3x-2}{(x+3)(x+5)}$

51. $\dfrac{-4x+4}{2x+1}$ **53.** $\dfrac{11}{2x^2}$ **55.** $\dfrac{9x+13}{4x}$ **57.** $\dfrac{24x}{(x+7)(5-18x)}$

59. 2 **61.** $\dfrac{12}{29}$ **63.** $-\dfrac{5}{3}$ **65.** $\dfrac{1}{3}$ **67.** $-\dfrac{3}{4}, 0$ **69.** $\dfrac{7}{2}$

CHAPTER 10 (p. 953)

1. 10; (3, 4) **3.** 5.83; $\left(-\dfrac{3}{2}, -\dfrac{7}{2}\right)$ **5.** 5; (3, 4) **7.** 3.12; $\left(-4, \dfrac{7}{2}\right)$

9. 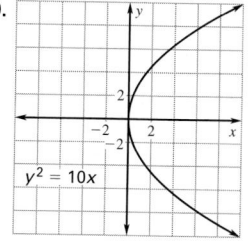 ; **11.** 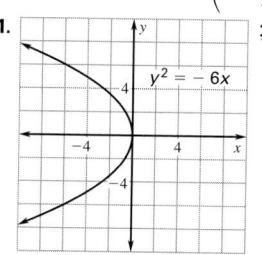 ;

$\left(\dfrac{5}{2}, 0\right); x = -\dfrac{5}{2}$ $\left(-\dfrac{3}{2}, 0\right); x = \dfrac{3}{2}$

13. 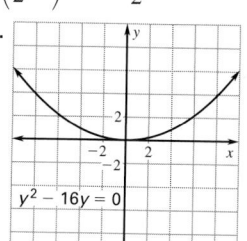 ; **15.** 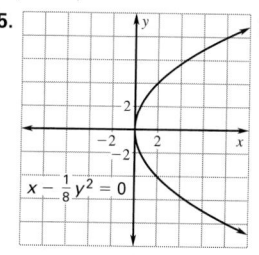 ;

$(0, 4); y = -4$ $(2, 0); y = -2$

17. $y^2 = 12x$ **19.** $y^2 = 2x$ **21.** $x^2 = -24y$ **23.** $x^2 = -3y$

25. 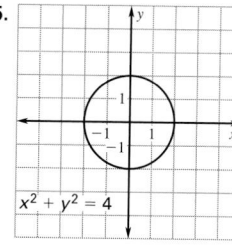 **31.**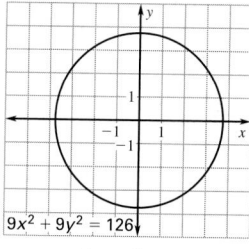

33. $x^2 + y^2 = 16$ **35.** $x^2 + y^2 = 13$ **37.** $x^2 + y^2 = 36$
39. $x^2 + y^2 = 29$ **41.** $x^2 + y^2 \le 100$

43. 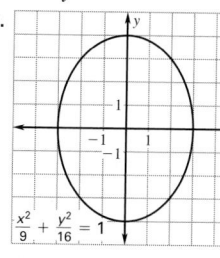 **49.**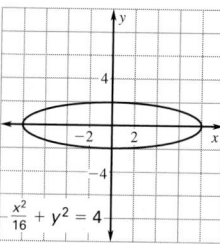

(0, 4), (0, −4); (−8, 0), (8, 0);
(−3, 0), (3, 0); (0, 2), (0, −2);
$\left(0, \sqrt{7}\right), \left(0, -\sqrt{7}\right)$ $\left(-2\sqrt{15}, 0\right), \left(2\sqrt{15}, 0\right)$

51. $\dfrac{x^2}{16} + \dfrac{y^2}{64} = 1$ **53.** $\dfrac{x^2}{49} + \dfrac{y^2}{4} = 1$

55. $\dfrac{x^2}{256} + \dfrac{y^2}{100} = 1$ **57.** $\dfrac{x^2}{4} + \dfrac{y^2}{400} = 1$

59. 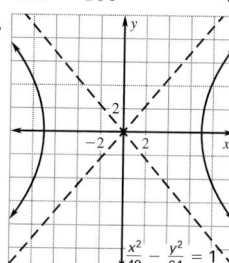 **63.**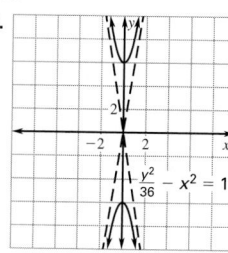

$\left(-\sqrt{113}, 0\right), \left(\sqrt{113}, 0\right);$ $\left(0, \sqrt{37}\right), \left(0, -\sqrt{37}\right);$
$y = \dfrac{8}{7}x, y = -\dfrac{8}{7}x$ $y = 6x, y = -6x$

67. $(x-3)^2 + (y-4)^2 = 25$ **69.** $(x-2)^2 + (y+5)^2 = 49$

71. $\dfrac{(x-3)^2}{25} + \dfrac{(y-3)^2}{9} = 1$ **73.** $\dfrac{(x-2)^2}{9} - \dfrac{(y-6)^2}{7} = 1$

75. parabola **77.** hyperbola **79.** (2, 4), (−4, −2) **81.** none

83. (3, 0), (5, 4) **85.** $\left(\dfrac{14}{5}, -\dfrac{2}{5}\right)$, (−2, 2)

CHAPTER 11 (p. 955) **1.** 14; $a_n = 3n - 1$ **3.** −13;

$a_n = -4n + 7$ **5.** −256; $a_n = -(-4)^{n-1}$ **7.** $\dfrac{1}{1024}$; $a_n = \dfrac{1}{4^n}$

9. 72 **11.** 55 **13.** 440 **15.** $\dfrac{169}{20}$ **17.** $a_n = -2 + 3n$; 28

19. $a_n = 18 - 10n$; −82 **21.** $a_n = \dfrac{11}{4} - \dfrac{1}{2}n$; $-\dfrac{9}{4}$

23. $a_n = 2 - 2n$; −18 **25.** $a_n = -9 + 4n$; 31 **27.** the 21st row

29. $a_n = -\left(\dfrac{1}{2}\right)^{n-1}$; $-\dfrac{1}{128}$ **31.** $a_n = \left(-\dfrac{1}{10}\right)^{n-2}$; $\dfrac{1}{1,000,000}$

33. $a_n = -\dfrac{1}{2}(6)^n$; −839,808 **35.** $a_n = 750\left(\dfrac{1}{5}\right)^n$; $\dfrac{6}{3125}$

37. $61{,}035{,}156$ **39.** -728 **41.** $\dfrac{60{,}466{,}175}{839{,}808}$ **43.** $\dfrac{3}{2}$ **45.** none

47. none **49.** $-\dfrac{1}{18}$ **51.** $a_1 = 2,\ a_n = a_{n-1} + 4$

53. $a_1 = 2,\ a_n = a_{n-1} + 5(3)^{n-2}$ **55.** $a_1 = -6,\ a_2 = -9,$
$a_n = a_{n-1} \cdot a_{n-2}$

CHAPTER 12 (p. 956) **1.** 2 **3.** 12 **5.** 12 **7.** 6 **9.** 120
11. 50,400 **13.** 840 **15.** 15 **17.** 1 **19.** 1 **21.** 1140
23. $x^7 + 7x^6 y + 21x^5 y^2 + 35x^4 y^3 + 35x^3 y^4 + 21x^2 y^5 +$
$7xy^6 + y^7$ **25.** $x^{12} + 12x^{10}y + 60x^8 y^2 + 160x^6 y^3 +$
$240x^4 y^4 + 192x^2 y^5 + 64y^6$ **27.** $243x^{10} - 2025x^8 +$
$6750x^6 - 11{,}250x^4 + 9375x^2 - 3125$ **29.** $x^9 + 3x^6 y^3 +$
$3x^3 y^6 + y^9$ **31.** $\dfrac{1}{13}$ **33.** $\dfrac{1}{2}$ **35.** $\dfrac{3}{13}$ **37.** $\dfrac{113}{200}$ **39.** $\dfrac{87}{200}$

41. $\dfrac{127}{200}$ **43.** $\dfrac{1}{8}$ **45.** $\dfrac{1}{2}$ **47.** $\dfrac{3}{8}$ **49.** 0.5; no **51.** 43%; yes

53. 4%; no **55.** 1 **57.** $\dfrac{1}{4}$ **59.** 0.8 **61.** $\dfrac{1}{16}$ **63.** $\dfrac{8}{75}$ **65.** $\dfrac{4}{45}$

67. $\dfrac{8}{75}$ **69.** $\dfrac{2}{15}$ **71.** 0.0543 **73.** 0.117 **75.** 0.00217

77. 0.00305 **79.** 0.4985 **81.** 0.84 **83.** 0.9985 **85.** 0.5

CHAPTER 13 (p. 957) **1.** $\sin\theta = \dfrac{\sqrt{2}}{2}$; $\cos\theta = \dfrac{\sqrt{2}}{2}$, $\tan\theta = 1$;

$\sec\theta = \sqrt{2}$; $\csc\theta = \sqrt{2}$; $\cot\theta = 1$ **3.** $\sin\theta = \dfrac{2\sqrt{14}}{9}$;

$\cos\theta = \dfrac{5}{9}$, $\tan\theta = \dfrac{2\sqrt{14}}{5}$; $\sec\theta = \dfrac{9}{5}$; $\csc\theta = \dfrac{9\sqrt{14}}{28}$;

$\cot\theta = \dfrac{5\sqrt{14}}{28}$ **5–15.** Sample answers are given.
5. $395°$; $-325°$ **7.** $485°$; $-235°$ **9.** $315°$; $-405°$
11. $225°$; $-135°$ **13.** $\dfrac{8\pi}{3}$; $-\dfrac{4\pi}{3}$ **15.** $\dfrac{6\pi}{5}$; $-\dfrac{4\pi}{5}$

17. $\dfrac{4\pi}{3}$ in.; $\dfrac{8\pi}{3}$ in.^{2} **19.** $\dfrac{21\pi}{2}$ cm; $\dfrac{147\pi}{2}$ cm^2

21. $\dfrac{\pi}{5}$ cm; $\dfrac{9\pi}{10}$ cm^2 **23.** $\sin\theta = \dfrac{5\sqrt{41}}{41}$; $\cos\theta = \dfrac{4\sqrt{41}}{41}$;

$\tan\theta = \dfrac{5}{4}$; $\sec\theta = \dfrac{\sqrt{41}}{4}$; $\csc\theta = \dfrac{\sqrt{41}}{5}$; $\cot\theta = \dfrac{4}{5}$

25. $\sin\theta = -\dfrac{3\sqrt{10}}{10}$; $\cos\theta = -\dfrac{\sqrt{10}}{10}$; $\tan\theta = 3$;

$\sec\theta = -\sqrt{10}$; $\csc\theta = -\dfrac{\sqrt{10}}{3}$; $\cot\theta = \dfrac{1}{3}$

27. $\sin\theta = \dfrac{\sqrt{5}}{5}$; $\cos\theta = \dfrac{2\sqrt{5}}{5}$; $\tan\theta = \dfrac{1}{2}$; $\sec\theta = \dfrac{\sqrt{5}}{2}$;

$\csc\theta = \sqrt{5}$; $\cot\theta = 2$ **29.** $\sin\theta = -\dfrac{2\sqrt{67}}{67}$; $\cos\theta = \dfrac{3\sqrt{469}}{67}$;

$\tan\theta = -\dfrac{2\sqrt{7}}{21}$; $\sec\theta = \dfrac{\sqrt{469}}{21}$; $\csc\theta = -\dfrac{\sqrt{67}}{2}$;

$\cot\theta = -\dfrac{3\sqrt{7}}{2}$ **31.** $\sin\theta = -\dfrac{5\sqrt{34}}{34}$; $\cos\theta = \dfrac{3\sqrt{34}}{34}$;

$\tan\theta = -\dfrac{5}{3}$; $\sec\theta = \dfrac{\sqrt{34}}{3}$; $\csc\theta = -\dfrac{\sqrt{34}}{5}$; $\cot\theta = -\dfrac{3}{5}$

33. $\sin\theta = \dfrac{2\sqrt{5}}{5}$; $\cos\theta = -\dfrac{\sqrt{5}}{5}$; $\tan\theta = -2$; $\sec\theta = -\sqrt{5}$;

$\csc\theta = \dfrac{\sqrt{5}}{2}$; $\cot\theta = -\dfrac{1}{2}$ **35.** $-\dfrac{1}{2}$ **37.** $\dfrac{\sqrt{2}}{2}$ **39.** $\dfrac{\sqrt{2}}{2}$

41. $-\dfrac{\sqrt{3}}{2}$ **43.** $-45°$; $-\dfrac{\pi}{4}$ **45.** $-30°$; $-\dfrac{\pi}{6}$ **47.** $30°$; $\dfrac{\pi}{6}$

49. $60°$; $\dfrac{\pi}{3}$ **51.** $41.8°$; $30°$ **53.** $B \approx 93°$; $b \approx 7.61$; $c \approx 5.48$

55. $B \approx 65°$; $a \approx 6$; $c \approx 5.07$ **57.** about 34.4 **59.** about 4140

61. about 1720 **63.** about 63.6 **65.** about 283

67. $y = \dfrac{x-7}{2}$; $-7 \le x \le 5369$ **69.** $y = x$; $0 \le x \le 500$

CHAPTER 14 (p. 959) **1.** 6; 4π **3.** $\dfrac{1}{7}$; 2 **5.** 1, 1 **7.** $\dfrac{2}{5}$; 8π

9.
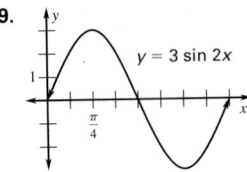

11.
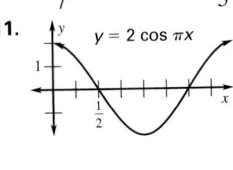

13.
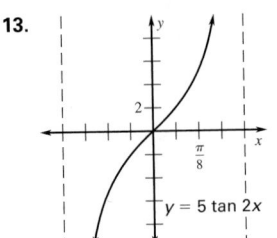

15.
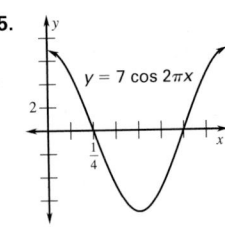

17. Shift the graph of $y = \cos x$ up 3 units. **19.** Shift the
graph of $y = \cos x$ up 4 units and reflect the graph in the
line $y = 4$. **21.** Shift the graph of $y = \cos x$ right π units.
23. Shift the graph of $y = \cos x$ right $\dfrac{\pi}{2}$ units.

27.
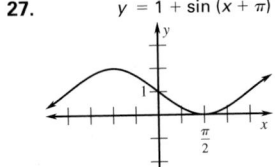

29.
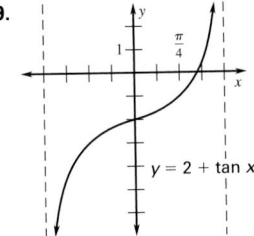

33. $\sec x$ **35.** $\sin^2 x$ **37.** $\cos x$ **39.** $\dfrac{2}{\sin x}$ **41.** $-\cos x \sin x$

43. $\dfrac{7\pi}{6} + 2n\pi$, $\dfrac{11\pi}{6} + 2n\pi$ **45.** $\dfrac{\pi}{3} + 2n\pi$, $\dfrac{2\pi}{3} + 2n\pi$, $\dfrac{4\pi}{3} +$

$2n\pi$, $\dfrac{5\pi}{3} + 2n\pi$ **47.** $2n\pi$, $\dfrac{2\pi}{3} + 2n\pi$, $\pi + 2n\pi$, $\dfrac{4\pi}{3} + 2n\pi$

49. $\dfrac{\pi}{4} + 2n\pi$, $\dfrac{7\pi}{4} + 2n\pi$, $\dfrac{3\pi}{4} + 2n\pi$, $\dfrac{5\pi}{4} + 2n\pi$

51. 0.615, 3.76 **53.** $\dfrac{3\pi}{2}$ **55.** $\dfrac{\pi}{3}$, $\dfrac{5\pi}{3}$ **57.** 0, 0.464, π, 3.61

59. 0, π **61–65.** Sample answers are given.
61. $y = \dfrac{1}{2}\cos\left(\dfrac{1}{2}\left(x + \pi\right)\right) + \dfrac{3}{2}$ **63.** $y = 4\cos\left(3\left(x + \dfrac{\pi}{3}\right)\right) - 4$

65. $y = \cos\left(\pi(x - 1)\right) + 22$ **67.** $-\dfrac{\sqrt{2}}{2}$ **69.** $\dfrac{-\sqrt{6} - \sqrt{2}}{4}$

71. $-2 + \sqrt{3}$ **73.** $\dfrac{\sqrt{3}}{3}$ **75.** $\dfrac{11}{21}$ **77.** $\dfrac{11\sqrt{5}}{40}$ **79.** $\dfrac{4\sqrt{5}}{21}$

81. $\dfrac{19}{21}$ **83.** $-2 - \sqrt{3}$ **85.** $-\dfrac{\sqrt{2 - \sqrt{2}}}{2}$ **87.** $\dfrac{\sqrt{2 - \sqrt{2}}}{2}$

89. $-2 - \sqrt{3}$ **91.** $\dfrac{\sqrt{2 + \sqrt{2}}}{2}$ **93.** $-\dfrac{\sqrt{2 - \sqrt{3}}}{2}$

95. $\sin 2x = \dfrac{9\sqrt{19}}{50}$; $\cos 2x = \dfrac{31}{50}$; $\tan 2x = \dfrac{9\sqrt{19}}{31}$

97. $\sin 2x = \dfrac{24}{25}$; $\cos 2x = \dfrac{7}{25}$; $\tan 2x = \dfrac{24}{7}$

TEACHER'S EDITION INDEX

A

Absolute value
of a complex number, 275
definition of, 50
equations, 49–55, 60
functions, 122–127, 132
graphing, 49–51
inequalities, 49–55, 60
Activity Notes, 27, 42, 157, 164, 265, 324, 330, 346, 367, 466, 518, 633, 710, 849
Activity Assessment, 18, 25, 48, 49, 90, 99, 107, 121, 146, 147, 176, 207, 222, 271, 281, 290, 337, 372, 379, 421, 430, 453, 473, 500, 533, 546, 561, 608, 622, 658, 674, 688, 723, 738, 745, 791, 820, 838, 839
Alternative Approach, 48, 49, 90, 146, 147, 176, 207, 222, 271, 281, 372, 379, 421, 430, 500, 533, 546, 622, 674, 723, 738, 791, 838
Classroom Management, 25, 48, 99, 107, 121, 146, 271, 290, 337, 421, 430, 453, 473, 500, 533, 561, 658, 674, 688, 738, 745, 820, 839
Cooperative Learning, 18, 90, 99, 107, 147, 176, 207, 222, 281, 337, 453, 473, 546, 561, 608, 622, 658, 745, 820, 838, 839
Key Discovery, 18, 25, 48, 49, 90, 99, 107, 121, 146, 147, 176, 207, 222, 271, 281, 290, 337, 372, 379, 421, 430, 453, 473, 500, 533, 546, 561, 608, 622, 658, 674, 688, 723, 738, 745, 791, 820, 838, 839
Link to Lesson, 18, 25, 48, 49, 90, 99, 107, 121, 146, 147, 176, 207, 222, 271, 281, 290, 337, 372, 379, 421, 430, 453, 473, 500, 533, 546, 561, 608, 622, 658, 674, 688, 723, 738, 745, 791, 820, 838, 839

Addition
combinations and, 709
of complex numbers, 273, 315
of functions, 415
inequalities and, 41
of matrices, 200, 240
of polynomials, 338, 341–343
properties of, 5, 19, 201, 928
of radicals, 408–409
of rational expressions, 562–563, 565–567, 578
of roots, 408–409

Additional Test Preparation, 10, 17, 24, 32, 39, 47, 56, 74, 81, 88, 98, 106, 113, 120, 128, 145, 155, 162, 168, 175, 184, 206, 213, 220, 229, 235, 255, 263, 269, 280, 289, 297, 305, 311, 328, 336, 343, 350, 358, 364, 371, 378, 385, 406, 414, 420, 429, 436, 443, 452, 472, 479, 485, 492, 499, 508, 516, 522, 539, 545, 553, 560, 567, 573, 594, 600, 607, 614, 621, 631, 638, 657, 665, 673, 680, 686, 707, 714, 722, 729, 736, 744, 751, 775, 783, 790, 798, 806, 812, 819, 837, 846, 854, 861, 867, 874, 881

Additive inverse, 5
of a complex number, 279
Ahmes papyrus, 40, 665
Algebra tiles, modeling completing the square with, 281
Algorithm
for approximating a best-fitting line, 101
for converting between degrees and radians, 777
for drawing an object in two-point perspective, 194
for evaluating an algebraic expression, 12
for evaluating trigonometric functions of any angle, 786
for graphing equations in slope-intercept form, 82
for graphing equations in standard form, 84
for graphing equations in two variables, 69
for graphing a linear inequality, 109
for graphing a quadratic inequality in two variables, 299
for graphing a radical function, 431
for hypothesis testing, 741
for order of operations, 11
for solving linear systems using the linear combination method, 149
for solving linear systems using the substitution method, 148
for solving a problem algebraically, 33
for transforming sine and cosine graphs, 840
for transforming a tangent graph, 843
for using the linear combination method with three variables, 178
Amplitude, of a function, 831
***And* rule,** 924–925

Angles, *See also* **Trigonometric functions**
central, 779
coterminal, 777
of depression, 771
of elevation, 771
finding the measure of, 793–798
initial side of, 776
quadrantal, 785
radian measure of, 777–778, 822–823
reference, 785–786
of repose, 794
in standard position, 776–778
terminal side of, 776
Application Note, xxiv, 35, 38, 46, 64, 73, 80, 97, 104, 105, 112, 119, 126, 136, 141, 144, 169, 174, 196, 219, 232, 246, 254, 268, 302, 310, 320, 327, 335, 342, 351, 357, 361, 370, 375, 377, 398, 403, 405, 419, 425, 427, 428, 440, 442, 450, 451, 462, 470, 478, 484, 491, 497, 498, 506, 507, 511, 530, 538, 544, 559, 572, 586, 620, 648, 654, 656, 664, 671, 672, 679, 685, 687, 698, 706, 713, 721, 750, 766, 778, 817, 828, 860, 868
Applications, *See also* **Connections; Multicultural connections**
acoustics, 495, 496, 498
advertising, 87, 744
agriculture, 212, 285, 310, 311, 340, 605, 685, 781, 812
amusement park rides, 133, 297, 829, 842, 846, 854, 863, 865, 874
anatomy, 54, 55, 73, 106, 751
animals, 30, 42, 174, 219, 254, 268, 288, 405, 410, 419, 428, 515, 516, 535, 683, 684, 706, 728, 732, 740, 743, 818
archeology, 347, 591
architecture, 9, 36, 80, 127, 303, 377, 613, 645, 662, 773, 779, 780, 782, 804, 809, 811, 847, 881, 891
astronomy, 268, 325, 326, 344, 425, 452, 507, 512, 538, 587, 611, 627, 630, 640, 759, 801
automobiles, 16, 23, 32, 44, 45, 56, 128, 254, 302, 307, 342, 357, 435, 476, 573, 702, 705, 713, 729, 781, 806, 871, 872
aviation, 70, 297, 606, 620, 771, 772, 814, 817, 881
baseball, 30, 204, 311, 583, 730, 808, 810, 816

basketball, 54, 61, 73, 88, 295, 445–448, 571, 735, 891

bicycling, 89, 342, 778, 817, 891

billiards, 124, 135

biology, 101, 254, 288, 332, 391, 399, 410, 417, 419, 433, 497, 519, 521, 522, 527, 535, 551, 559, 563, 695, 740, 846

boats and boating, 87, 219, 382, 403, 414, 443, 572, 817, 825, 836, 866

bowling, 52, 428

business, 34, 39, 87, 97, 113, 119, 152, 153, 167, 189, 193, 259, 262, 317, 335, 348, 355, 356, 357, 369, 370, 417, 450, 542, 543, 552, 553, 559, 579, 664, 722, 725, 735, 754, 759, 763

camping, 124, 125, 599

chemistry, 161, 234, 477, 478, 485, 506, 507, 570, 685

collectibles, 471

communications, 110, 111, 510, 513, 606, 625, 628, 636, 669, 670, 672

computers and internet, 23, 83, 144, 296, 312, 450, 467, 470, 479, 514, 671, 706, 717, 743

construction, 23, 35, 38, 54, 78, 80, 81, 175, 234, 235, 261, 300, 364, 442, 472, 537, 538, 656, 679, 794, 795, 796

consumer economics, 12, 13, 14, 16, 17, 40, 61, 81, 83, 97, 116, 117, 119, 128, 142, 144, 145, 153, 154, 155, 161, 174, 182, 189, 193, 220, 229, 234, 236, 243, 307, 309, 333, 335, 351, 358, 370, 378, 395, 419, 429, 444, 478, 546, 572, 574, 600, 685, 783, 802, 803, 806

cooking, 167, 173, 502, 506

crafts, 218, 243, 258, 288, 350, 361, 362

criminology, 702

cryptography, 225–229, 243, 395, 420

design, 284, 286, 350, 363, 364, 365, 375, 550, 664, 706, 796

earth science, 104, 106, 247, 297, 317, 487, 489, 491, 504, 505, 573, 605, 634, 635, 637, 638, 860, 866, 887

economics, 325, 327, 405, 471, 521, 544, 551, 679

education, 38, 46, 205, 337, 343, 370, 539, 566, 704, 713, 728, 734, 743, 751, 775

electronics, 262, 279, 280, 538, 567, 726

employment, 15, 16, 21, 22, 23, 39, 46, 116, 135, 169, 175, 263, 288, 305, 418, 606, 607, 652, 728, 735, 743

endangered species, 482, 483, 732

engineering, 252, 266, 284, 288, 319, 405, 599, 767, 774, 796, 871

entertainment, 16, 46, 112, 155, 168, 234, 266, 268, 294, 298, 304, 358, 370, 459, 495, 570, 663, 706, 709, 713, 718, 721, 811, 815

figure skating, 630, 781

finance, 6, 9, 38, 119, 181, 190, 232, 233, 343, 395, 427, 469, 476, 478, 479, 482, 484, 506, 527, 560, 565, 688, 721, 736, 763

fine arts, 97, 350, 617, 713, 715

food, 478, 706, 709, 715, 735, 736, 737, 743, 847

football, 112, 254, 572, 583, 614, 878, 879

fractal geometry, 275, 277, 278, 436, 649, 672, 680, 685, 891

fundraising, 16, 27, 30, 40, 85, 166

games and puzzles, 124, 428, 656, 712, 713, 720, 730, 734, 735, 753, 754, 782

gardening, 28, 38, 113, 135, 172, 221, 262, 284, 286, 310, 378, 520, 551, 611, 612, 805

geography, 4, 104, 215, 812, 860, 881

geometry, 21, 24, 28, 78, 413

golf, 8, 127, 787, 820, 882

government, 87, 119, 470, 728, 759

health and fitness, 102, 112, 137, 144, 154, 158, 161, 191, 193, 202, 203, 212, 254, 304, 312, 363, 454, 733, 747, 750, 759, 851, 852, 871

hiking, 592

history, 35, 133, 225, 262, 263, 268, 403

jewelry, 94, 606

law, 636, 743

manufacturing, 52, 53, 56, 165, 271, 349, 355, 363, 375, 376, 442, 549, 731, 735, 743, 750

measurement, 34, 54, 55, 133, 144, 167, 182, 258, 260, 261, 284, 286, 288, 349, 363, 365, 377, 412, 413, 428, 496, 497, 620, 774, 781, 794, 797, 805, 812

medicine, 54, 335, 368, 442, 478, 484, 566, 606, 685, 743, 749, 751, 845

mountain climbing, 296, 463, 484, 774, 843, 844

music, 13, 37, 126, 127, 184, 197, 204, 207, 413, 478, 699, 713, 717, 787, 788, 833, 836, 892

natural resources, 34, 304, 357, 559, 685, 763

navigation, 603, 637

nutrition, 38, 167, 454

oceanography, 80, 268, 435, 506, 535, 551

photography, 330, 564, 873

physics, 37, 39, 73, 97, 127, 269, 423, 677, 678, 679, 695, 825, 835, 836, 845, 873, 880, 881

politics, 73, 93, 183, 450

population, 16, 97, 205, 342, 370, 469, 515, 719

publishing, 340, 343

racquetball, 544

retail display, 38, 652, 654, 655, 656

skateboarding, 789

skiing, 703, 818

skydiving, 531, 557, 558

soccer, 819

social studies, 95, 328, 385, 386, 395, 470, 521, 673, 747, 752

softball, 210, 211

space exploration, 46, 321, 325, 327, 385, 403, 614, 621

surveys, 739, 740, 741, 747, 748, 811, 861

swimming and diving, 120, 154, 270, 296, 377, 817

technology, 9, 44, 74, 507, 597, 607, 617, 618, 620, 629, 637, 833, 866

telecommunications, 103, 110, 111, 297, 599, 645

temperature, 8, 23, 46, 61, 78, 81, 104, 120, 252, 253, 427, 539, 544, 864, 866, 867, 868, 887

tennis, 319, 335, 671

track and field, 24, 72, 182, 288, 311, 404, 604, 713, 818

transportation, 1, 33, 35, 174, 263, 290, 307, 308, 310, 593, 773

travel, 9, 10, 37, 45, 47, 71, 79, 141, 144, 193, 228

volleyball, 297, 789

weather, 36, 46, 97, 119, 126, 255, 440, 441, 443, 491, 728, 771, 859, 860, 867

weightlifting, 161

youth services, 65, 105, 385

Approximation See **Estimation; Prediction**

Arc length, 779

derivation of the formula for, 900

Area, See also **Formulas**

of a circle, 28, 914

of an ellipse, 29, 764–765

Heron's formula for, 809

of a parallelogram, 914–915

of a rectangle, 28, 914

of a sector, 779

of a square, 914

of a trapezoid, 28, 914

of a triangle, 28, 802–806, 809, 914

Area model, for completing the square, 282
Arithmetic sequence, 659–665, 693
  recursive rule for, 682
Arithmetic series, 661–665, 693
Assessment, *See also* **Projects; Reviews**
  Chapter Standardized Test, 62–63, 134–135,
    190–191, 244–245, 318–319, 392–393,
    460–461, 528–529, 580–581, 646–647,
    696–697, 760–761, 826–827, 888–889
  Chapter Test, 61, 133, 189, 243, 317, 391,
    459, 527, 579, 645, 695, 759, 825, 887
  Quiz, 17, 40, 56, 89, 106, 128, 155, 169,
    184, 221, 236, 270, 298, 312, 344, 365,
    386, 414, 429, 452, 485, 508, 522, 553,
    574, 607, 621, 638, 673, 687, 715, 737,
    752, 783, 798, 819, 847, 868, 882
Assignment Guide *Appears at the beginning
of each Exercise set*
Associative property, 5, 201
  of matrix multiplication, 209
  of scalar multiplication, 209
Asymptotes, 465
  horizontal, 540
  of logarithmic functions, 489
  of logistic growth functions, 518
  of rational functions, 540–541, 547–548
  of tangent functions, 834
  vertical, 540
Augmented matrix, 237–238
Average, rate of change and, 78
Axis (Axes)
  coordinate plane, 67
  imaginary, 273
  real, 273
  of symmetry, of a parabola, 249, 595–596
  three-dimensional coordinate, 170

B

Bar graph, 934
Base, of an exponential expression, 11, 465
al-Battani, Abu Abdullah, 853
Best-fitting line, 99–105, 131–132
  linear regression and, 107
Best-fitting quadratic model, 308
Binomial distribution, 738–744,
  758
  constructing, 745
  mean of, 744
  normal approximation of, 748
Binomial experiment, 739
Binomial probability, 739–744

Binomial theorem, 710–714, 756
Binomials, 256, *See also* **Polynomials**
  dividing by, 556
  expansion, 710–714, 756
  multiplying, 251, 339, 937
Bounded region, 163
Box-and-whisker plot, 447, 453,
  458
Branches, of graphs of rational functions,
  540

C

Calculator, *See also* **Graphing calculator;
  Student Help; Technology Activities**
  evaluating expressions, 18
  evaluating factorials, 703, 704,
    709
  exercises, 404, 471, 789
  exponent key, 402
  root key, 402
  trigonometric functions, 770
Careers
  accident reconstructionist, 593
  accountant, 355
  air traffic controller, 606
  amusement ride designer, 433
  archeologist, 347, 591
  astronomer, 801
  auto mechanic, 871
  automotive designer, 308
  biotechnician, 172
  botanist, 728
  cartographer, 789
  caterer, 151
  chemical engineer, 450
  chemist, 217
  civil engineer, 252
  coast guard, 435
  computer programmer, 671
  dentist, 234
  economist, 679
  electrician, 279
  entomologist, 664
  farmer, 559
  financial planner, 468
  fishery biologist, 683
  forester, 68
  gerontologist, 342
  health services manager, 202
  hospital administrator, 551
  investment banker, 427
  marine biologist, 482

  market researcher, 748
  meteorologist, 858
  musician, 836
  nurse, 335
  nutritionist, 112
  ornithologist, 327
  paleontologist, 419
  pediatrician, 102
  personal trainer, 158
  pharmacist, 566
  photographer, 330
  physical therapist, 16
  physician, 685
  police detective, 702
  police officer, 636
  real estate broker, 21
  sculptor, 805
  seismologist, 634
  set designer, 304
  sound technician, 495
  sports statistician, 30
  stockbroker, 23
  surveyor, 811
  teacher, 735
  web developer, 296
Ceiling function, 118
Center, of an ellipse, 609
Central angle, 779
Central tendency, measures of, 445, 453, 458,
  See also **Mean; Median; Mode**
Chain rule, 924
Challenge exercises, 10, 16, 24, 32, 39, 47, 55,
  74, 81, 88, 98, 105, 113, 120, 127, 145,
  154, 162, 168, 175, 183, 206, 213, 220,
  229, 235, 255, 263, 269, 280, 289, 297,
  305, 312, 328, 336, 343, 350, 358, 364,
  371, 378, 385, 406, 413, 420, 428, 436,
  443, 451, 472, 479, 485, 492, 498, 507,
  516, 522, 539, 545, 552, 560, 567, 573,
  594, 600, 606, 614, 620, 630, 637, 657,
  665, 672, 680, 686, 707, 714, 722, 729,
  744, 751, 774, 782, 790, 797, 806, 812,
  818, 837, 846, 854, 861, 867, 874, 881
Change-of-base formula, 494
Chapter Review, *See* **Reviews**
Chapter Summary, *See* **Reviews**
Chapter Test, *See* **Assessment**
Checking solutions
  algebraically, 149, 292, 440
  using a calculator, 793, 869
  by graphing, 92, 149, 259, 265,
    283, 291, 292, 382, 424, 440, 596, 632,
    840, 851, 855, 870, 877

using inverse operations, 352
using logical reasoning, 123, 778, 864
using substitution, 19, 42, 48, 108, 139, 148, 216, 223, 224, 265, 283, 292, 367, 438, 439, 501, 502, 503, 632, 856

Choosing a method
for factoring a polynomial, 349
for solving linear systems, 149–150
for solving a quadratic function, 297
for solving a rational equation, 572

Circle graph, 934
Circles
area of, 28, 914, 915
central angle of, 779
circumference of, 28, 915
degree measure and, 777–778
eccentricity of, 639
equation of, 601–608, 643
translated, 624–626, 628–630
graph of an equation of, 601–608, 643
radian measure and, 777–778
radius of, 601, 915
sector of, 779
standard form of the equation of, 601

Circular motion, modeling, 842, 863
Circular permutations, 707
Circumference, 914
Classification
of conics, 626–627
of lines by slope, 76
of parallel and perpendicular lines, 77
of triangles, 589

Closure property, 5
Coefficient, 13
leading, 329
Cofunction identities, 848
Column matrix, 199
Combinations, 708–709, 756
probability and, 717
Common difference, 659
Common Error, 8, 14, 22, 29, 35, 46, 54, 55, 72, 80, 87, 95, 111, 118, 127, 143, 153, 160, 167, 204, 219, 261, 287, 304, 324, 334, 349, 357, 363, 370, 384, 405, 419, 427, 435, 442, 449, 469, 470, 483, 490, 496, 503, 514, 520, 551, 552, 559, 572, 593, 599, 605, 613, 619, 629, 636, 656, 671, 677, 679, 685, 706, 713, 720, 735, 773, 781, 789, 796, 800, 810, 816, 845, 853, 860, 873
Common factor, 908
Common logarithm, 487
Common ratio, 666

Communication, *See also* **Critical Thinking; Logical Reasoning**
discussion, 1, 49, 65, 82, 108, 137, 147, 195, 197, 222, 281, 321, 345, 379, 397, 399, 421, 463, 465, 473, 517, 531, 533, 587, 622, 649, 674, 699, 738, 765, 791, 829, 839, 893
presenting, 195, 397, 585, 745, 765, 893
writing, 47, 63, 72, 81, 88, 98, 145, 153, 175, 183, 191, 195, 206, 213, 229, 235, 245, 255, 269, 279, 280, 305, 328, 336, 349, 371, 385, 393, 397, 406, 413, 420, 436, 459, 461, 479, 484, 498, 506, 507, 516, 529, 539, 545, 560, 581, 585, 600, 605, 614, 620, 640, 664, 672, 697, 714, 751, 765, 774, 797, 805, 806, 827, 846, 854, 866, 880, 893
Commutative property, 5, 201
Comparing
matrices, 199
methods for dividing polynomials, 358
real numbers, 4
Complement of an event, 726
probability and, 731
Completing the square, 282–289, 315
area model for, 282
using algebra tiles, 281
Complex conjugates, 274, 367
Complex fractions, 564, 566–567
Complex numbers, 272–280, 315
absolute value of, 275, 315
adding and subtracting, 273, 315
additive inverse of, 279
dividing, 274, 315
multiplicative inverse of, 279
multiplying, 274, 315
plotting, 273
standard form of, 272
Complex plane, 273
Composition of functions, 416–417
Compound event, 724–729, 757
normal distribution and, 747
Compound interest, 468
Computer, *See also* **Calculator; Graphing Calculator; Technology Activities**
evaluating recursive rules, 688
graphing linear equations in three variables, 176
Concept Activities
Absolute Value Equations and Inequalities, 49
Combining Equations in a Linear System, 147

The Difference of Two Cubes, 345
Equations with More than One Variable, 26
Exploring Conic Sections, 622
Exploring Finite Differences, 379
Exploring Inverse Functions, 421
Exponential Growth and Decay, 473
Fitting a Line to a Set of Data, 99
Graphs of Absolute Value Functions, 122
Graphs of Logistic Growth Functions, 517
Investigating Binomial Distributions, 738
Investigating End Behavior, 331
Investigating the Graph of an Inequality, 108
Investigating Graphs of Exponential Functions, 465
Investigating Graphs of Radical Functions, 431
Investigating Graphs of Rational Functions, 540
Investigating Graphs of Systems of Inequalities, 156
Investigating Identity and Inverse Matrices, 222
Investigating an Infinite Geometric Series, 674
Investigating Inverse Trigonometric Functions, 791
Investigating Inverse Variation, 533
Investigating Linear Motion, 813
Investigating Linear Programming, 163
Investigating Matrix Equations, 230
Investigating the Natural Base e, 480
Investigating the Number of Solutions, 366
Investigating Parabolas, 249
Investigating Pascal's Triangle, 710
Investigating Points of Intersection, 632
Investigating Polynomial Division, 353
Investigating Properties of Inequalities, 41
Investigating Properties of Square Roots, 264
Investigating a Property of Logarithms, 493
Investigating Recursive Rules, 681
Investigating Slope and y-intercept, 82
Investigating Trigonometric Identities, 848
Products and Quotients of Powers, 323
Translating and Reflecting Trigonometric Graphs, 839
Using Algebra Tiles to Complete the Square, 281
Writing a Quadratic in Standard Form, 307
Concept Summary
Areas Under a Normal Curve, 746
Classifying Conics, 626
Eccentricity of Conic Sections, 639

End Behavior for Polynomial Functions, 331
Evaluating Trigonometric Functions, 786
Graph of a Quadratic Function, 249
Number of Solutions of a Linear System, 140
Operations on Functions, 415
Properties of Exponents, 323
Properties of Rational Exponents, 407
Writing an Equation of a Line, 91
Zeros, Factors, Solutions, and Intercepts, 373
Condensing a logarithmic expression, 494, 525
Conditional probability, 732–733, 736
Conics, 622–630, 644
  degenerate, 630
  eccentricity of, 639–640
Connections, *See also* **Applications; Math & History; Multicultural connections**
  art, 262, 805
  biology, 105, 161, 254, 288, 327, 405, 410, 411, 412, 414, 459, 572, 750
  geometry, 10, 15, 23, 24, 30, 31, 61, 72, 97, 145, 153, 174, 206, 228, 262, 287, 305, 312, 327, 341, 342, 350, 357, 383, 405, 412, 435, 443, 452, 491, 539, 560, 579, 581, 592, 600, 656, 680, 690, 720, 744, 773, 797, 873
  history, 9, 80, 262, 266, 268, 451, 452, 506, 599, 721, 805
  science, 37, 39, 46, 55, 97, 153, 217, 220, 253, 269, 289, 395, 442, 459, 479, 487, 489, 491, 497, 515, 536, 538, 552, 573, 579, 583, 707, 728, 790
  social studies, 9, 16, 183, 220, 327, 342, 789
  statistics, 8, 73, 98, 280, 451, 594, 721, 735, 867
Consistent linear system, 145
Constant term, 13, 329
Constant of variation, 94, 534
Constraints, 163
Continuously compounded interest, 482
Coordinate
  plane, 67
  of a point, 3
  three dimensional system, 170
Correlation, determining, 100–105, 131
Corresponding sides of similar figures, 923
Cosecant, 769, *See also* **Trigonometric functions**
Cosine, 769
  difference formula for, 869–874
    derivation of, 902–903

double-angle formula for, 875–881
  derivation of, 903
half-angle formula for, 875–881
  derivation of, 904
inverse, 792
law of, 807–812, 824
  derivation of, 901
sum formula for, 869–874
Cosine functions, *See also* **Trigonometric functions**
  graphing, 831–833, 835–838, 884
  transformations of graphs of, 839–842, 844–846
Cotangent, 769, *See also* **Trigonometric functions**
Cotangent identity, 848
Coterminal angles, 777
Counterexample, 927–928
Co-vertices, of an ellipse, 609
Cramer's rule, 216–220, 241
Critical Thinking, 24, 47, 72, 81, 87, 105, 113, 118, 144, 154, 162, 183, 269, 279, 289, 296, 319, 335, 364, 378, 413, 419, 435, 443, 450, 471, 479, 485, 492, 498, 544, 552, 567, 605, 619, 636, 637, 686, 706, 713, 714, 729, 790, 797, 811, 818, 827, 839, 845, 853, 866, 874, *See also* **Logical Reasoning**
Critical *x*-values, 302
Cross multiplying, 569–573
Cross-curriculum connections, *See* **Connections**
Cube root function, 431–436, 457
Cubes
  of a binomial, 339
  difference of two, 345–346
  sum of, 345
Cubic regression, 382
Cumulative Practice, *See* **Reviews**

Daily Homework Quiz, 10, 17, 24, 32, 39, 47, 56, 74, 81, 88, 98, 106, 113, 120, 128, 145, 155, 162, 168, 175, 184, 206, 213, 220, 229, 235, 255, 263, 269, 280, 289, 297, 305, 311, 328, 336, 343, 350, 358, 364, 371, 378, 385, 406, 414, 420, 429, 436, 443, 452, 472, 479, 485, 492, 499, 508, 516, 522, 539, 545, 553, 560, 567, 573, 594, 600, 607, 614, 621, 631, 638, 657, 665, 673, 680, 686, 707, 714, 722, 729, 736, 744, 751, 775, 783, 790, 798,

806, 812, 819, 837, 846, 854, 861, 867, 874, 881
Dantzig, George, 169
Data, *See also* **Graphs; Modeling**
  fitting a curve to, 307–308
  fitting a line to, 99–105
  fitting a model to, 509–516, 584–585, 862–867
  interpreting
    using inverse variation, 535
    using measures of central tendency, 445
    using measures of dispersion, 446, 458
    using range, 446, 458
    using standard deviation, 446, 458
  organizing
    using a box-and-whisker plot, 447
    using a frequency distribution, 448
    using a histogram, 448, 738
    using matrices, 202
    using a table, 738
    using a tree diagram, 701, 733
  using a polynomial to model, 380–385
  using a quadratic function to model, 307–308
  using a trigonometric function to model, 862–867
Decay functions, exponential, 473–479, 524
Decimals, converting, 906
Decision making, *See* **Choosing a method; Critical Thinking; Logical Reasoning**
Deductive reasoning, *See* **Logical Reasoning**
Degree
  measure of a circle, 777–778
  of a polynomial function, 329
Dependent events, 732–733, 758
Dependent linear system, 145
Dependent variable, 69
Derivation
  of the difference formula for cosine, 902–903
  of the double-angle formula for cosine, 903
  of the equation of an ellipse, 896–897
  of the equation of a hyperbola, 898–899
  of the equation of a parabola, 895–896
  of the formula for arc length, 900
  of the half-angle formula for cosine, 904
  of the law of cosines, 901
  of the law of sines, 900–901
  of negative angle identities, 902
  of the quadratic formula, 895
Determinant, 214–220, 241
Difference
  formulas, 869–874, 886
    derivation of cosine, 902–903

of two cubes, 345–346
of two squares, 257
Dilation, 922
Dimensions, of a matrix, 199
Direct argument, 924–925
Direct variation, 94, 534
Directrix, of a parabola, 595, 642
Discrete mathematics, *See also* **Algorithm;**
     Logical Reasoning
  counting methods, 701–707, 756
  greatest common factor, 908
  inductive reasoning, 689–690
  linear programming, 163–167, 187–188
  matrices, 199–242
  recursion, 681–686, 694
  scatter plot, 100–105
  sequences, 651–652, 659–672, 692
  tree diagram, 701, 733, 738
Discriminant, 293
  classification of conics and, 626
Dispersion, measures of, 446, 458
Distance, between two points, 589, 591–594,
     642
Distribution
  binomial, 738–744, 758
  normal, 746–751, 758
Distributive property, 5, 13, 928
  matrix operations and, 201, 209
  for solving linear equations, 20, 59
  Division
  of complex numbers, 274, 315
  definition of, 5
  of functions, 415–416
  inequalities and, 41
  by a polynomial, 556
  polynomial, 352–358
  property of equality, 19, 928
  with rational expressions, 556–561, 577
  synthetic, 353
  Domain
  of a function, 67
     restricted, 424
Double-angle formulas, 875–881, 886
  derivation for cosine, 903

E

Eccentricity, of conic sections, 639–640
Eliminating the parameter, 814
Ellipse
  area of, 764–765
  eccentricity of, 639–640
  equation of, 609–614, 643

derivation of, 896–897
  graphing, 609–614, 643
  standard form, 609
  translated, 624, 626–630
End behavior, for polynomial functions, 331
English Learners, 2, 8, 15, 75, 82, 137, 144,
     149, 199, 208, 234, 249, 256, 323, 353,
     401, 437, 464, 503, 532, 538, 588, 599,
     605, 650, 685, 700, 703, 728, 768, 785,
     834, 836, 847
Enrichment, *See* **Challenge exercises;**
     Extensions
Equations, *See also* **Formulas; Functions;**
     Inequalities; Linear equations;
     Quadratic equations; Polynomials;
     Rational equations
  absolute value, 49–55, 60
  of circles, 601–608, 624–626, 628–630,
     639–640
  of conic sections, 623–630, 639–640
  definition of, 19
  of ellipses, 609–614, 624, 626–630,
     639–640
  equivalent, 19
  exponential, 501–507
  general second-degree, 626
  of hyperbolas, 615–620, 625–630, 639–640
  involving nth roots, 402
  logarithmic, 486, 501–507, 526
  matrix, 201, 230
  of parabolas, 249–255, 595–600, 623,
     626–627, 639–640
  parametric, 813–818, 820, 824
  quadratic, 256–271, 282–284, 291–298
  radical, 437–443, 458
  rational, 402–406, 566–573, 578
  recursive, 681–688
  trigonometric, 793–798, 855–861, 870–874,
     877–881, 885
  in two variables, 69
Equivalent equations, 19
Equivalent expressions, 13
Error analysis, 14, 22, 103, 125, 181, 218,
     260, 277, 286, 326, 341, 348, 411, 418,
     434, 490, 505, 543, 604, 612, 705, 712,
     780, 795, 852, 859
Estimation, *See also* **Prediction**
  approximating best-fitting line, 99–106,
     131–133
  approximating the coordinates of turning
     points, 374
  approximating maximum volume, 375
  of area, 606

of nth roots, 402, 404, 406
of real zeros, 368
of square root, 264, 266
using the distance formula, 591
using an exponential growth model, 467,
     470, 471, 476, 478, 479
using a formula, 452
using function operations, 417
using a geometric series, 669
using graphs of functions, 247, 433, 434
using an inverse function, 425, 428
using joint variation, 536
using the law of sines, 799–801, 803–805
using a logarithmic function, 487, 489, 491,
     495, 507
using matrices, 207
using the Monte Carlo method, 764–765
using a normal curve, 746–751
using parametric equations, 814–815, 820
using a polynomial model, 332
using properties of rational exponents, 410,
     412, 413
using a quadratic model, 284, 285
using a rational model, 542, 543, 544, 570
using regression, 382
using a scatter plot, 321, 399
using a square root function, 433, 434
using a trigonometric equation, 856–858,
     860
using a trigonometric function, 787, 789,
     790
Euler, Leonhard, 480, 481
Euler number, 480
Exercises, types of
  Checkpoint Exercises, 4–6, 12, 13, 20, 21,
     27, 28, 34–36, 42–44, 51, 52, 68–70,
     76–78, 83–85, 92–94, 101, 102, 109,
     110, 115, 116, 123, 124, 140, 141, 149,
     150, 157, 158, 164, 165, 171, 172,
     178–180, 200–202, 209, 210, 215–217,
     224–226, 231, 232, 237, 250–252,
     257–259, 265, 266, 273–276, 283–285,
     292–294, 301, 302, 307, 308, 324, 325,
     330–332, 339, 340, 346, 347, 353–355,
     360, 361, 367, 368, 374, 375, 381, 382,
     402, 403, 408–410, 416, 417, 423–425,
     432, 433, 438–440, 446–448, 466–468,
     475, 476, 481, 482, 487–489, 494, 495,
     502–504, 510–512, 518, 519, 535, 536,
     541, 542, 548, 549, 555–557, 563, 564,
     569, 570, 590, 591, 596, 597, 602, 603,
     610, 611, 616, 617, 624, 625, 627, 633,
     634, 639, 640, 652–654, 660–662,

667–669, 676, 677, 682, 683, 689, 690, 702–704, 709–711, 717, 718, 725, 726, 731–733, 740, 741, 747, 748, 753, 754, 770, 771, 777–779, 785–787, 793, 794, 800–802, 808, 809, 814, 815, 832–834, 841–843, 849–851, 856–858, 863, 864, 870, 871, 877, 878

Closure Question, 6, 13, 21, 28, 36, 44, 52, 70, 78, 85, 94, 102, 110, 116, 124, 141, 151, 158, 165, 172, 180, 202, 210, 217, 226, 232, 252, 259, 266, 276, 285, 294, 302, 308, 325, 332, 340, 347, 355, 361, 368, 375, 382, 403, 410, 417, 425, 433, 440, 448, 468, 476, 482, 489, 495, 504, 512, 519, 536, 542, 549, 557, 564, 570, 591, 597, 603, 611, 617, 627, 634, 654, 662, 669, 677, 683, 704, 711, 718, 726, 733, 741, 748, 771, 779, 787, 794, 802, 809, 815, 834, 843, 851, 858, 864, 871, 878

Concept Question, 76, 274, 293, 360, 409, 488, 494, 502, 626, 668, 710, 778, 794, 857

Daily Puzzler, 6, 13, 21, 28, 52, 94, 102, 116, 141, 151, 165, 172, 180, 210, 217, 226, 232, 259, 266, 276, 285, 294, 302, 325, 347, 355, 361, 368, 375, 382, 403, 410, 417, 425, 440, 448, 468, 495, 504, 557, 603, 627, 634, 654, 677, 683, 704, 711, 718, 726, 748, 779, 794, 802, 809, 815, 834, 843, 858, 864, 871, 878

Exercise Levels, 7, 14, 22, 29, 37, 45, 53, 71, 79, 86, 95, 103, 111, 117, 125, 142, 152, 159, 166, 173, 181, 203, 211, 218, 227, 233, 253, 260, 267, 277, 286, 295, 303, 309, 326, 333, 341, 348, 356, 362, 369, 376, 383, 404, 411, 418, 426, 434, 441, 449, 469, 477, 483, 490, 496, 505, 513, 520, 537, 543, 550, 558, 565, 571, 592, 598, 604, 612, 618, 628, 635, 655, 663, 670, 678, 684, 705, 712, 719, 727, 734, 742, 749, 772, 780, 788, 795, 803, 810, 816, 835, 844, 852, 859, 865, 872, 879

Vocabulary, Focus on, 6, 13, 21, 28, 36, 44, 70, 78, 85, 94, 102, 110, 141, 151, 165, 172, 180, 217, 226, 232, 259, 266, 276, 294, 302, 308, 347, 361, 368, 375, 382, 403, 410, 417, 425, 433, 440, 448, 468, 489, 495, 504, 512, 557, 570, 597, 627, 654, 662, 669, 677, 683, 704, 711, 718, 726, 748, 771, 779, 787, 794, 802, 809, 815, 834, 843, 858, 864, 871, 878

Warm-Up Exercises, 3, 11, 19, 26, 33, 41, 50, 67, 75, 82, 91, 100, 108, 114, 122, 139, 148, 156, 163, 170, 177, 199, 208, 214, 223, 230, 249, 256, 264, 272, 282, 291, 299, 306, 323, 329, 338, 345, 352, 359, 366, 373, 380, 401, 407, 415, 422, 431, 437, 445, 465, 474, 480, 486, 493, 501, 509, 517, 534, 540, 547, 554, 562, 568, 589, 595, 601, 609, 615, 623, 632, 651, 659, 666, 675, 681, 701, 708, 716, 724, 730, 739, 746, 769, 776, 784, 792, 799, 807, 813, 831, 840, 848, 855, 862, 869, 875

Expanding a logarithmic expression, 494, 525

Expected value, 753–754

Experimental probability, 717, 757

Explicit rule, 681

Exponential equations, solving, 501–507, 526

Exponential expressions, evaluating, 324

Exponential functions
  decay, 473–479, 524
  graphing, 466, 475
  growth, 465–472, 524
  modeling with, 509–516, 526
  natural base, 481–485, 525
  writing, 509

Exponents
  definition of, 11
  evaluating, 11–16
  irrational, 412, 472
  properties of, 323
    using, 323–328, 388
  rational, 401–413

Expressions
  equivalent, 13
  evaluating, 11–16, 58
  exponential, 324
    with natural base e, 480–481
  logarithmic, 487, 494
  numerical, 11
  polynomial, factoring, 345–350
  quadratic, factoring, 256–263
  radical, simplifying, 407–413
  rational, 402, 407–413, 554–567, 577–578
  square root, simplifying, 264–265
  terms of, 13
  trigonometric
    evaluating, 869–870, 872–873, 875–876, 879, 880
    simplifying, 849, 870, 873
  value of, 12

Extensions
  Additive and Multiplicative Inverses of Complex Numbers, 279
  Eccentricity of Conic Sections, 639
  Expected Value, 753–754
  Irrational Exponents, 412
  Mathematical Induction, 689–690
  Solving Systems Using Augmented Matrices, 237–238

Extra Challenge Note, 10, 17, 24, 32, 39, 47, 56, 74, 81, 88, 98, 106, 113, 120, 128, 145, 155, 162, 168, 175, 184, 206, 213, 220, 229, 235, 255, 263, 269, 280, 289, 297, 305, 311, 328, 336, 343, 350, 358, 364, 371, 378, 385, 406, 414, 420, 429, 436, 443, 452, 472, 479, 485, 492, 499, 508, 516, 522, 539, 545, 553, 560, 567, 573, 594, 600, 607, 614, 621, 631, 638, 657, 665, 673, 680, 686, 707, 714, 722, 729, 736, 744, 751, 775, 783, 790, 798, 806, 812, 819, 837, 846, 854, 861, 867, 874, 881

Extra Examples, *Occur throughout*

Extra Practice, *See* Reviews

Extraneous solutions
  for a logarithmic equation, 504
  for a radical equation, 439
  for a trigonometric equation, 857

Factor, 908
  common, 908

Factor theorem, 354

Factorial values, 681, *See also* Combinations; Permutations

Factoring
  the difference of two squares, 257
  by grouping, 346
  polynomial expressions, 345–350, 354, 938
  quadratic expressions, 256–263, 314
  the sum or difference of cubes, 345–346
  to solve a trigonometric equation, 856
  zero product property and, 257

Factorization, prime, 908

Fair game, 753

False position, rule of, 40

Feasible region, 163

Fermat, Pierre, 737

Fibonacci, Leonardo, 687

Fibonacci sequence, 687
  recursive rule for, 681

TEACHER'S EDITION INDEX

Finite differences, 379–385, 390
 properties of, 381
Finite sequence, 651
First-order differences, 379
Focus (Foci)
 of an ellipse, 609
 of a hyperbola, 615
 of a parabola, 595, 642
Focus on
 Applications, 6, 9, 27, 35, 36, 38, 44, 46, 52,
  54, 73, 78, 87, 97, 104, 119, 126, 144,
  154, 165, 174, 183, 205, 212, 215, 219,
  225, 232, 254, 262, 268, 302, 325, 340,
  357, 363, 364, 368, 370, 377, 382, 403,
  410, 417, 425, 440, 467, 470, 487, 491,
  497, 498, 507, 515, 521, 535, 536, 544,
  564, 570, 591, 599, 603, 605, 617, 620,
  625, 629, 656, 662, 669, 677, 706, 713,
  721, 726, 741, 750, 771, 773, 779, 787,
  794, 809, 815, 817, 833, 842, 851, 860,
  866, 878, 881
 Careers, 16, 21, 23, 30, 68, 102, 112, 151,
  158, 172, 202, 217, 234, 252, 279, 296,
  304, 308, 327, 330, 335, 342, 347, 355,
  419, 427, 433, 435, 450, 468, 482, 495,
  551, 559, 566, 593, 606, 634, 636, 664,
  671, 679, 683, 685, 702, 728, 735, 748,
  789, 801, 805, 811, 836, 858, 871
 People, 8, 13, 70, 80, 93, 161, 167, 210, 226,
  275, 288, 311, 361, 385, 442, 447, 481,
  484, 504, 512, 538, 557, 597, 654, 661,
  710, 743, 781, 789, 797, 873
FOIL method, 251
Formulas
 area
  of a circle, 28, 914–915
  of an ellipse, 29
  of a parallelogram, 914, 915
  of a rectangle, 28, 914
  of a square, 914
  of a trapezoid, 28, 914
  of a triangle, 28, 914, 915
 Beaufort number, 440
 Beaufort wind scale, 440
 change-of-base, 494
 circumference of a circle, 24, 28
 combining, 28
 distance, 28, 589, 642
 double-angle, 875
 Doyle log rule, 406
 Fahrenheit/Celsius, 23, 28, 427
 half-angle, 875
 height of falling objects, 266, 268

height of launched/thrown objects, 294
Heron's area, 809
hexagonal number, 384
interest
 compound, 468
 continuously compounded, 482
 simple, 28
length of a standard nail, 442
mean of a binomial distribution, 744
midpoint, 590, 642
normal curve, 751
nth term of an arithmetic sequence, 659
nth term of a geometric sequence, 666
number of combinations, 708
number of permutations, 703
pentagonal number, 384
perimeter, of a rectangle, 28
pH of soil, 573
probability
 binomial, 739
 compound events, 724
 dependent events, 732
 independent events, 730
 mutually exclusive events, 724
quadratic, 291
radius
 of a cone, 435
 of a sphere, 452
rewriting, 28–32, 59
slant height of a truncated pyramid, 443
for special series, 654
standard deviation, 446, 451
sum of an infinite geometric series, 675
sum of n terms of an arithmetic series, 661
sum of n terms of a geometric series, 668
surface area
 of a cylinder, 413
 of a sphere, 413
Table of, 963–968
theoretical probability, 716–717
triangular number, 380
trigonometric difference, 869
trigonometric sum, 869
volume
 of a cube, 72
 of a cylinder, 271, 413
 of a dodecahedron, 405
 of an icosahedron, 405
 of a pyramid, 361
 of a sphere, 72
water pressure at depth, 73
Fractal geometry, 275–276
 Koch snowflake, 680

 Mandelbrot set, 275–276
 Sierpinski triangle, 672
Fractions
 complex, 564, 566–567
 converting, 906
 linear equations with, 20
 writing a repeating decimal as, 676
Frequency, of a periodic function, 833
Frequency distribution, 448
Function notation, 69
Functions
 absolute value, 122–127, 132
 composition of, 416–417
 cosine, 831–833, 835–842, 844–846
 evaluating, 69–70
 exponential
  decay, 473–479, 524
  growth, 465–472, 524
  modeling with, 509–516, 526
  natural base, 481–485, 525
 graphing, 69–70
  absolute value, 122–127, 132
  exponential, 466, 475, 524
  general rational, 547–552, 577
  logarithmic, 488–489, 500, 525
  logistic growth, 517–518, 526
  natural base, 481–482, 525,
  polynomial, 331–332, 373–378, 388, 390
  quadratic, 249–255, 314
  radical, 431–436
  simple rational, 540–546
  trigonometric, 831–837, 884
 horizontal line test for, 424
 identifying, 67
 inverse, 421–429, 457
 logarithmic, 486–492, 525
 logistic growth, 517–522, 526
 natural base exponential, 481–485, 525
 objective, 163
 operations on, 415–420, 457
 periodic, 831
 piecewise, 114–121, 132
 power, 415–420, 457, 511–516, 526
 problem solving, 70, 73
 quadratic, 249–255, 285, 306–312, 314–316
  intercept form of, 250, 251, 259, 306
  standard form of, 250–252, 307
  vertex form of, 250–252, 285, 306
 radical, 431–436, 457
 rational, 540–552, 576–577
 representing, 67, 130
 sine, 831–833, 835–842, 844–846
 step, 115, 116

ceiling function, 118
  greatest integer function, 115
  rounding function, 118
tangent, 834–837
trigonometric
  of any angle, 784–790, 823
  inverse, 791–798, 823
  of a quadrantal angle, 785
  right triangle, 769–774, 822
of two variables, 171–172, 174–175
vertical line test for, 68
Fundamental counting principle, 701–702, 756
  permutation and, 703–707, 756
Fundamental theorem of algebra, 366–371

Gauss, Karl Friedrich, 661
General rational functions, 547–552
General second-degree equation, 626
Geometric probability, 718, 757
Geometric sequence, 666–672, 693
  recursive rule for, 682
Geometric series, 668–672, 693
  infinite, 674–680, 694
Geometry, *See also* **Angles; Applications;**
      Circles; Connections; Formulas;
      Triangles; Trigonometric functions
  conics, 622–630, 639–640, 644
  parallel lines, 77, 92, 140
  perimeter, area, and volume, 914–917
  perpendicular bisector, 590–591
  Pythagorean theorem, 917–918
  reflection, 839–846, 921
  similar figures, 923
  symmetry, 919–920
  transformations, 921–922
  triangle relationships, 917–919
Glossary, 971–980
Graphing calculator, *See also* **Calculator;**
      Student Help; Technology Activities
  absolute value feature, 126
  binomial probability distribution function, 745
  common logarithm function, 487, 500
  draw inverse feature, 430
  evaluate feature, 176
  evaluating expressions, 18
  evaluating recursive rules, 683
  e^x function, 481, 482
  exercises, 54, 118, 126, 154, 161, 228, 234, 254, 268, 304, 305, 309, 311, 336, 370,

376, 434, 435, 442, 483, 490, 514, 520, 529, 551, 566, 567, 619, 685, 728, 751, 804, 853, 854, 860, 867, 868, 887
  exponential regression feature, 510
  graphing feature, 90, 121, 146, 249, 271, 290, 308, 331, 372, 374, 382, 430, 433, 500, 517, 608, 658, 839, 848, 858, 864, 870
  graphing modes, 546, 561, 848
  greatest integer function, 118
  intersect feature, 146, 372, 382, 433, 435, 442, 482, 489, 570, 858, 870
  linear regression feature, 107
  list feature, 308, 723, 745, 838, 864
  logistic regression feature, 519
  matrix feature, 207, 224, 231, 232
  maximum feature, 290, 308, 374, 375, 838
  minimum feature, 290, 374, 549, 838
  natural logarithm function, 487, 500
  parametric mode, 820
  power regression feature, 512
  quadratic regression feature, 308
  random number generator, 723, 738
  root feature, 271, 368, 838
  sequence mode, 658
  sinusoidal regression feature, 864
  sort feature, 723
  stat calc feature, 107, 382, 453
  stat edit feature, 107, 453, 454
  stat plot feature, 107, 308, 382, 453, 454, 745
  summation feature, 658
  table feature, 25, 268, 510, 561, 658
  test feature, 48
  trace feature, 121, 126, 453, 454, 517, 658, 820
  window, 90, 337, 372, 453, 546, 608, 658, 820, 838
  zero feature, 271, 368, 374
Graphing Calculator Note, 54, 154, 161, 225, 228, 231, 234, 268, 304, 384, 442, 514, 521, 551, 619, 672, 685, 728, 750, 804
Graphs
  of absolute value equations and inequalities, 49–51
  of absolute value functions, 122–127, 132
  bar, 934
  box-and-whisker plot, 447, 453, 458
  circle, 934
  of a complex number, 273
  of a cosine function, 831–833, 835–838, 884
    transformations of, 839–842, 844–846, 884–885

of an equation of a circle, 601–608, 643
of an equation of an ellipse, 609–614, 643
of an equation of a hyperbola, 615–620, 643–644
of an equation of a parabola, 595–596, 598–600
of exponential functions, 466, 475, 524
of a geometric sequence, 667
histogram, 448, 458, 738, 740–741
of a horizontal line, 85
line, 935
of a linear equation, 82–88, 131
  system, 139–146, 186
  in three variables, 170–176, 188
of a linear function, 69–70, 130
of a linear inequality
  in one variable, 41–43
  system, 156–162
  in two variables, 108–113, 132
of a logarithmic function, 488–489, 500, 525
of logistic growth functions, 517–518, 526
of a natural base function, 481–482, 525
of parametric equations, 813, 820
of a piecewise function, 114–121, 132
of points in the coordinate plane, 933
of polynomial functions, 331–332, 373–378, 390
of quadratic functions, 249–255, 306–312, 314, 316
of quadratic inequalities, 299–305, 316
of radical functions, 431–436
of rational functions
  general, 547–552, 577
  simple, 540–546, 576–577
of a real number, 3
of a relation, 68
scatter plot, 100–105
of a sequence, 652
of a sine function, 831–833, 835–838, 884
  transformations of, 839–842, 844–846, 884–885
of a system of quadratic inequalities, 300
of a tangent function, 834–837, 884
  transformations of, 843–846, 884–885
three dimensional, 170–176, 188
of a vertical line, 85
Greatest common factor (GCF), 908
Greatest integer function, 115
Growth factor, 467
Growth functions
  exponential, 465–472, 524
  logistic, 517–522

H

Hales, Thomas, 654
Half-angle formulas, 875–881, 886
  derivation for cosine, 904
Half-plane, 108
Heron's area formula, 809
Hexagonal numbers, 384
Histogram, 448, 458, 738, 740–741
History, *See* **Applications; Connections;
  Math & History; Mathematics history**
Homework Check, 7, 14, 22, 29, 37, 45, 53,
  71, 79, 86, 95, 103, 111, 117, 125, 142,
  152, 159, 166, 173, 181, 203, 211, 218,
  227, 233, 253, 260, 267, 277, 286, 295,
  303, 309, 326, 333, 341, 348, 356, 362,
  369, 376, 383, 404, 411, 418, 426, 434,
  441, 449, 469, 477, 483, 490, 496, 505,
  513, 520, 537, 543, 550, 558, 565, 571,
  592, 598, 604, 612, 618, 628, 635, 655,
  663, 670, 678, 684, 705, 712, 719, 727,
  734, 742, 749, 772, 780, 788, 795, 803,
  810, 816, 835, 844, 852, 859, 865, 872,
  879
Horizontal asymptote, 540
Horizontal line
  graph of, 85
  slope of, 76
Horizontal line test, 424
Horizontal translation, of a trigonometric
  graph, 839–846
Hyperbola
  eccentricity of, 639–640
  equation of, 615–620, 643–644
    derivation of, 898–899
    translated, 625–630
  as graph of a function, 540
Hypothesis testing, 741–744

I

Identity, 13
  trigonometric, 848–854
Identity matrix, 222–229, 242
Identity property
  of addition, 5
  of multiplication, 5
If-then statements, 926–927
Imaginary axis, 273
Imaginary number, 272
Imaginary unit *i*, 272
  powers of, 280
Inconsistent linear system, 145

Independent events, 730–731, 758
Independent variable, 69
Index of a radical, 401
Index of summation, 653
Indirect argument, 924–925
Indirect variation, 94
Induction, mathematical, 689–690
Inductive reasoning, 689–690
Inequalities, *See also* **Linear inequalities**
  absolute value, 49–55, 60
  compound, 43–44
  quadratic, 299–305, 316
  real number, 4
  systems of, 156–162, 187, 300
Infinite geometric series, 674–680, 694
Infinite sequence, 651
Infinity, positive and negative, 331
Initial side, of an angle, 776
Integers, 3
Intercept form
  of a quadratic function, 250, 251
    writing, 259, 306
Internet Connections
  Application Links, 1, 35, 38, 40, 46, 65, 78,
    89, 112, 119, 126, 137, 169, 174, 188,
    197, 210, 225, 232, 236, 247, 254, 268,
    270, 311, 321, 325, 351, 357, 377, 399,
    403, 425, 440, 444, 463, 467, 482, 491,
    497, 499,

    507, 531, 536, 574, 587, 597, 617, 620,
    631, 649, 669, 687, 699, 713, 721, 737,
    741, 767, 775, 778, 817, 829, 842, 860,
    868, 878
  Career Links, 16, 21, 23, 30, 68, 97, 102,
    112, 151, 158, 172, 202, 217, 234, 252,
    279, 296, 304, 308, 327, 330, 335, 342,
    347, 355, 419, 427, 433, 435, 450, 468,
    482, 495, 551, 559, 566, 593, 606, 634,
    636, 664, 671, 679, 683, 685, 702, 728,
    735, 748, 789, 801, 805, 811, 836, 858,
    871
  Data Updates, 6, 16, 93, 119, 154, 161, 205,
    212, 280, 297, 305, 309, 325, 327, 342,
    370, 385, 395, 427, 478, 522, 552, 572,
    669, 672, 719, 735, 751, 787, 815
  Extra Challenge, 10, 16, 24, 32, 39, 55, 74,
    113, 120, 127, 145, 162, 168, 175, 183,
    213, 220, 229, 235, 255, 263, 269, 280,
    289, 297, 305, 312, 328, 336, 343, 350,
    358, 364, 371, 378, 385, 406, 420, 428,
    436, 443, 451, 472, 492, 498, 507, 552,
    560, 567, 573, 600, 620, 630, 657, 665,

    680, 686, 707, 729, 744, 751, 782, 797,
    806, 812, 818, 861, 874, 881
  Student Help, 6, 15, 18, 20, 25, 31, 38, 43,
    48, 51, 54, 70, 88, 90, 92, 110, 118, 119,
    126, 127, 140, 153, 160, 167, 171, 172,
    178, 204, 207, 209, 216, 224, 234, 254,
    259, 265, 271, 279, 285, 290, 292, 304,
    308, 310, 324, 332, 337, 343, 349, 354,
    372, 405, 408, 409, 419, 428, 430, 435,
    438, 446, 453, 467, 471, 476, 484, 487,
    497, 500, 502, 515, 518, 519, 538, 541,
    546, 551, 556, 561, 564, 568, 593, 596,
    602, 608, 610, 619, 624, 637, 656, 658,
    664, 668, 676, 685, 688, 703, 704, 706,
    711, 718, 723, 726, 729, 735, 743, 745,
    751, 754, 770, 777, 786, 796, 804, 807,
    811, 817, 820, 836, 838, 845, 851, 857,
    862, 873, 877, 892
Intersection
  of graphs of quadratic systems, 632–637
  of sets, probability and, 724–725
Inverse functions, 421–429, 457
  trigonometric, 791–798, 823
Inverse matrix, 222–229, 242
Inverse property
  additive, 5
    of a complex number, 279
  for a logarithmic function, 488
  multiplicative, 5
    of a complex number, 279
Inverse relation, 422
Inverse variation, 533–535, 576
Investigations, *See* **Concept Activities;
  Technology Activities**
Irrational exponent, 472
Irrational number, definition of, 3

J

Joint variation, 536–539, 576

K

Karmarkar, Narendra, 169
Kepler, Johannes, 512, 631
Koch snowflake, 680

L

Latus rectum, of a parabola, 600
Law of cosines, 807–812, 824
  derivation of, 901
Law of sines, 799–806, 823–824

derivation of, 900–901
Leading coefficient, 329
Least common denominator (LCD),
  908–909, 939
Least common multiple (LCM), 908
Left distributive property, 209
Lesson Goals, 1A, 64A, 136A, 196A, 246A,
  320A, 398A, 462A, 530A, 586A, 648A,
  698A, 766A, 828A
Like terms, 13
Line graph, 935
Line symmetry, 919
Linear combination method
  for solving linear systems, 149–150, 187
    in three variables, 178
  for solving quadratic systems, 633–637
Linear equations
  absolute value, 49–55, 60
  for a best-fitting line, 99–105, 131–132
  definition of, 19
  direct variation, 94
  of the form $y = f(x)$, 90
  given two points, 91, 93
  graphing, 82–88, 131
  point-slope form of, 91
  rewriting, 26–32, 59
  slope-intercept form of, 82–83, 91
  solving, 19–24, 59
  standard form of, 84–85
  subscripts and, 27
  systems of
    solving algebraically, 148–154, 177–183,
      186–187, 188
    solving graphically, 139–146, 170–176,
      186
    in three variables
      graphing, 170–176
      solving algebraically, 177–183
  writing, 21, 91–98, 131
Linear functions
  definition of, 69
  evaluating, 69–70
  graphing, 69–70
  inverse of, 422–423
Linear inequalities
  absolute value, 49–55, 60
  compound, 43–44
  constraints and, 163
  graphing
    in one variable, 41–43
    in two variables, 108–113, 132
  solving, 41–48
  systems of, 156–162, 187

Linear motion, modeling, 814
Linear programming, 163–167, 187–188
Linear regression, 107
  best-fitting line and, 107
Linear systems
  consistent, 145
  definition of, 139
  dependent, 145
  inconsistent, 145
  with many solutions, 140, 150, 179
  with no solutions, 140, 150, 179
  number of solutions of, 140
  solving
    algebraically, 148–154, 177–183,
      186–187
    by graphing, 139–146, 170–176, 186
    using augmented matrices, 237–238
    using Cramer's rule, 216–220, 241
    using elementary row operations,
      237–238
    using the linear combination method,
      149–150, 178, 187
    using matrices, 230–235, 242
    using the substitution method, 148,
      186
  in three variables
    graphing, 170–176
    solving algebraically, 177–183
Lines
  classifying by slope, 76
  horizontal, slope and, 76
  parallel, slope and, 77
  perpendicular, slope and, 77
  of reflection, 921
  slope of, 75–81
  of symmetry, 919
  vertical, slope of, 76
Local maximum, 374
Local minimum, 374
Logarithmic equations, solving,
  501–507, 526
Logarithmic expressions
  condensing, 494, 525
  evaluating, 487, 525
  expanding, 494, 525
Logarithmic functions, 486–492, 525
  graphing, 488–489, 500, 525
Logarithms
  change-of-base formula, 494
  common, 487
  definition of, 486
  natural, 487
  properties of, 493–498, 525

Logical Reasoning, *See also* **Critical**
  Thinking
  and rule, 924–925
  chain rule, 924
  counterexample, 927–928
  direct argument, 924–925
  exercises, 8, 10, 56, 80, 96, 104, 212, 220,
    255, 261, 278, 304, 357, 413, 418, 471,
    498, 539, 565, 572, 614, 636, 657, 665,
    679, 686, 713, 714, 729, 836, 846, 880
  false solutions, 439
  hypothesis testing, 741–744
  if-then statements, 926–927
  indirect argument, 924–925
  justify reasoning, 928–929
  logical argument, 924–925
  mathematical induction, 689–690
  or rule, 924–925
Logistic growth functions, 517–522
  graphing, 517–518, 526
Lucas, Edouard, 656, 687

M

Major axis, of an ellipse, 609
Mandelbrot, Benoit, 275
Mandelbrot set, 275–276
Manheim, Amendee, 499
Manipulatives
  algebra tiles, 281
  coins, 738
  flashlight, 622
  index cards, 49
  ruler, 99, 147, 194, 533
  straightedge, 421
Mapping, 67
Math & History
  Columbus's Voyage, 775
  Deep Water Diving, 574
  The Fibonacci Sequence, 687
  History of Conic Sections, 631
  Linear Programming in World
    War II, 169
  Logarithms, 499
  Music and Math, 868
  Probability Theory, 737
  Problem Solving, 40
  Solving Polynomial Equations, 351
  Systems of Equations, 236
  Telescopes, 270
  Transatlantic Voyages, 89
  Tsunamis, 444
Mathematical induction, 689–690

Mathematical model, *See also* **Modeling**
  definition of, 12
Mathematical Reasoning, 5, 15, 54, 77, 96,
    151, 164, 216, 237, 251, 293, 353, 360,
    403, 410, 413, 427, 450, 467, 470, 478,
    488, 491, 497, 502, 503, 506, 510, 514,
    521, 569, 640, 653, 661, 668, 690, 725,
    741, 754, 786, 801, 836, 850, 866, 880
Mathematics history
  Ahmes papyrus, 665
  Archimedes' burning mirror, 599
  Chiu chang suan shu, 263
  magic squares, 396
  natural base e, 480–481
  sum of an arithmetic series, 661
  Tower of Hanoi, 656
  Great trigonometric survey of
      India, 805
  Zeno's paradox, 679
Matrix (Matrices)
  adding and subtracting, 200, 240
  augmented, 237–238
  comparing, 199
  of constants, 230
  Cramer's rule and, 216–220, 241
  definition of, 199
  determinants and, 214–220, 241
  dimensions of, 199
  equations, 201, 240
    writing, 230
  identity, 222–229, 242
  inverse, 222–229, 242
  multiplying, 208–213, 241
    by a scalar, 200
  properties
    of multiplication, 209
    of operations, 201
  rotational, 213
  row operations, 237–238
  for solving linear systems, 216–220,
      230–235, 237–238
  square, 199, 214–220
  of variables, 230
  zero, 199
Maximum value
  of a quadratic function, 290
  of sine and cosine functions, 832
Mean, 445, 453, 458
  of a binomial distribution, 744
  standard deviation and, 746–751
Measures of dispersion, 446, 458
Measures, Table of, 962
Median, 445, 453, 458

Midpoint, of a line segment, 590–594, 642
Minimum value
  of a quadratic function, 290
  of sine and cosine functions, 832
Minor axis, of an ellipse, 609
Mode, 445, 458
Modeling, *See also* **Concept Activities;**
      Manipulatives; Technology Activities
  best-fitting quadratic model, 308
  circular motion, 842, 863
  with cubic regression, 382
  with an ellipse, 611, 613–614
  with finite differences, 381
  fitting a model to data, 509–516, 584–585,
      862–867
  with a hyperbola, 617, 620
  with parametric equations, 814–815,
      817–818, 851
  with a power function, 511–516, 526
  a sinusoid, 863
  with a tangent function, 843
  using absolute value functions, 124–127
  using an algebraic model, 12, 13, 21, 27,
      33–39, 52, 60, 85, 93, 110, 141, 151,
      172, 180, 232, 258, 347, 355, 361, 375,
      417, 542, 570, 683
  using an area model, 262, 282, 284, 718, 937
  using a circular model, 603, 605–606, 625
  using combinations, 708–709
  using a direct variation equation, 94
  using an equation of a parabola, 597
  using an exponential model, 325, 467–468,
      476, 478, 479, 482, 509–516
  using a function of two variables, 172,
      174–175
  using a geometric sequence, 669
  using a geometric series, 669
  using graphs of linear equations, 83, 85,
      87–88
  using a histogram, 448, 458, 738, 740–741
  using an infinite series, 677
  using an inverse model, 423, 425
  using a linear equation, 21, 27, 93
  using a linear expression, 13
  using a linear function, 70, 73
  using a linear inequality
    in one variable, 44, 46–47
    in two variables, 110, 112–113
  using linear programming, 165, 167
  using a linear system, 141, 144–145, 151,
      153–154, 217, 232, 234–235
    in three variables, 180, 182–183

  using logarithms, 489, 495, 497, 498, 504,
      509–516
  using a logistic growth model, 519
  using mapping, 67
  using matrices, 202, 204–205, 210, 212,
      215, 232, 234–235
  using the Monte Carlo method, 764–765
  using a normal distribution, 747–748,
      750–751
  using nth roots, 403, 405–406
  using a number line, 3–4, 716, 718
  using permutations, 702–704
  using a piecewise function, 116
  using a polynomial equation, 347, 349–350
  using a polynomial function, 330, 332, 368
  using polynomial models, 340, 342–343,
      355, 357, 361, 375, 380–385
  using probability, 717–718, 726, 731–733
  using a quadratic inequality, 300, 302,
      304–305
  using a quadratic model, 252–255, 258–262,
      266–269, 284–286, 288–289, 294–298,
      300, 302–305, 307–312, 316, 634
  using a radical function, 433, 435
  using rational exponents, 410
  using a rational model, 542, 549, 557, 563,
      570
  using a recursive rule, 683
  using scientific notation, 325
  using a sequence, 662
  using a series, 662
  using a system of linear inequalities, 158,
      161
  using a tree diagram, 701, 733
  using a trigonometric model, 787, 789, 790,
      858, 860, 862–867, 878, 886
  using unit analysis, 6, 12, 33
  using a variation model, 535, 536
  using a Venn diagram, 716, 724–725
  using a verbal model, 12, 13, 15, 21, 27,
      33–39, 52, 60, 85, 93, 110, 141, 151,
      172, 180, 232, 258, 347, 355, 361, 375,
      417, 542, 570, 683
Monomials, 257
  multiplying rational expressions involving,
      555
Motivating the Lesson, 4, 12, 20, 27, 34, 42,
      51, 68, 76, 83, 92, 101, 109, 123, 140,
      149, 157, 164, 178, 224, 250, 257, 265,
      273, 283, 292, 300, 324, 330, 346, 353,
      360, 367, 374, 402, 438, 446, 487, 494,
      502, 518, 555, 569, 602, 610, 616, 633,
      652, 660, 667, 676, 702, 747, 770, 777,

785, 793, 800, 808, 814, 832, 856, 870, 876

Multicultural connections, 1, 6, 8, 9, 13, 33, 37, 38, 40, 61, 70, 80, 87, 110, 111, 161, 165, 169, 204, 220, 225, 236, 263, 325, 327, 351, 396, 427, 444, 499, 502, 515, 535, 614, 679, 773, 781, 789, 811, 866, 881, 891

Multiple representations, 12, 13, 21, 26, 27, 33–36, 41–43, 49–52, 68, 69, 84, 92, 93, 101, 102, 107, 110–112, 114–116, 122–125, 139–141, 156–158, 202, 215, 249–252, 258, 259, 265, 266, 273, 281–283, 291–293, 299–302, 306–308, 314, 316, 347, 360, 367–368, 373–375, 379–382, 388, 424–425, 431–433, 439, 440, 445–448, 453–454, 465, 466, 467, 469, 481–482, 484, 486, 488, 489, 500, 509–512, 524, 525, 526, 540–542, 546, 547–549, 563, 590, 596, 601–603, 632–634, 652, 667, 701, 703, 710–711, 738, 740, 746–748, 831–835, 838–843, 858, 862–864, *See also* **Manipulatives; Modeling**

Multiple Representations, 153, 369, 504, 590, 719

Multiplication
combinations and, 709
of complex numbers, 274, 315
cross multiplying, 569–573
of functions, 415–416
inequalities and, 41
linear combination method and, 149–150
of a matrix by a scalar, 200
by a polynomial, 555, 937
of polynomials, 338–343
properties of, 5, 19, 209, 257, 928
with rational expressions, 554–561, 577
of two matrices, 208–213, 241

Multiplicative inverse, 5
of a complex number, 279

Mutually exclusive events, 724

N

Napier, John, 499
Natural base e, 480–485, 525
Natural base exponential functions, 481–485, 525
Natural logarithm, 487
Negative angle identities, 848
derivation of, 902
Negative correlation, 100

Negative exponent property, 323, 407
Normal curve, 746
general equation of, 751
Normal distribution, 746–751, 758
*n*th roots, 401–406, 456
Number line
graphing linear inequalities on, 41–43
graphing real numbers on, 3
ordering real numbers on, 4
Numerical expression, 11

O

Objective function, 163
Octants, 170
Odds of an event, 729
Opposite, of a number, 5, 936–937
Optimal solution, 163
Optimization, 163–167
Or rule, 924–925
Order of operations, 11
Ordered pair, 67
Ordered triple, 170
Ordering, real numbers, 4
Origin
of a coordinate graph, 67, 68
of a coordinate plane, 67
of a real number line, 3
of a three-dimensional coordinate system, 170
Oughtred, William, 499

P

Pacing, T36–T37, 1C–1D, 64C–64D, 136C–136D, 196C–196D, 246C–246D, 320C–320D, 398C–398D, 462C–462D, 530C–530D, 586C–586D, 648C–648D, 698C–698D, 766C–766D, 828C–828D

Parabola, 249
axis of symmetry, 249, 595–596
directrix of, 595, 642
eccentricity of, 639
equation of, 249–255, 595–600
derivation of, 895–896
translated, 623, 626–627
focus of, 595, 642
graph of, 249–255, 595–596
latus rectum of, 600
standard equation of, 596
vertex of, 595, 642
Parallel lines
linear systems and, 140

slope and, 77
writing equations of, 92
Parallelogram, area of, 914, 915
Parameter, 813
eliminating, 814
Parametric equations, 813–818, 824, 851, 854
eliminating the parameter, 814
graphing, 813, 820
Pascal, Blaise, 710, 737
Pascal's triangle, 710
Patterns
angle measure and triangle area, 804
binomial products, 339
difference of two cubes, 345
end behavior, 334
factoring, 257
Pascal's triangle, 712
powers of i, 280
sum of two cubes, 345
Pentagonal numbers, 384, 690
Percents
calculating, 907–907
converting, 906
Perfect square trinomial, 257
Perimeter
of a polygon, 914
of a rectangle, 28
Periodic function, 831
Permutation, 703–707, 756
circular, 707
probability and, 717
Perpendicular bisector, 590–591
Perpendicular lines
slope and, 77
writing equations of, 92
Perspective, drawing, 194–195
Piecewise function, 114–121, 132
Piecewise-defined sequence, 686
Point-slope form, 91
Polynomial equations
fundamental theorem of algebra and, 366–371
number of solutions of, 366
solving
by factoring, 347–350, 389
by graphing, 368, 372
Polynomial functions
analyzing graphs of, 373–378, 390
definition of, 329
degree of, 329, 379–385
end behavior for, 331

evaluating, 329–330, 388
  using synthetic substitution, 330
finding rational zeros of, 359–364, 390
finite differences and, 379–385
graphing, 331–332, 388
power function and, 415
standard form of, 329
turning points of, 374–378
using zeros to write, 367
zeros of, 354
Polynomial long division, 352
Polynomials
adding, 338, 341–343
dividing, 352–358
factoring, 345–350, 354, 938
multiplying, 338–343
multiplying by, 555
multiplying rational expressions involving, 555
quadratic form of, 346
subtracting, 338, 340–343, 389
Positive correlation, 100
Postulates of algebra, 928–929
Power functions, 415–420, 457, 511–516, 526
inverses of, 424–425
Power of a power property, 323, 407
Power of a product property, 323, 407
Power property of logarithms, 493
Power of a quotient property, 323, 407
Powers, 11, *See also* **Exponential functions; Exponents**
Powers property of equality, 437
Prediction, *See also* **Estimation**
using a best-fitting line, 99, 104
using a graph, 65
using inverse variation, 533
using a linear equation, 93
using the Monte Carlo method, 764–765
using a polynomial model, 385
using transformations of trigonometric graphs, 839
Prime factorization, 908
Prime number, 908
Probability
binomial, 739–744, 748
binomial distribution and, 738–744
combination, 708–709, 756
complement of an event, 726
compound events, 724–729, 757
conditional, 732–733, 736
definition of, 716
dependent events, 732–733, 758
expected value, 753–754

experimental, 717, 757
fundamental counting principle, 701–702, 756
geometric, 718, 757
independent events, 730–731, 758
intersection and, 724–725
mutually exclusive events, 724
normal distribution and, 746–751
number of possible events, 701–702
odds of an event, 729
permutation, 703–707, 756
random number generation, 723
standard deviation and, 746–751
theoretical, 716–717, 757
union and, 724–725
using permutations and combinations, 717
Probability Theory, 737
Problem-Solving Strategies, *See also* **Choosing a method; Modeling**
break a problem into simpler parts, 931
draw a diagram, 21, 35, 768, 821
guess, check, and revise, 36
look for a pattern, 36
make a list or table, 930
solve a simpler problem, 931
translating phrases into algebraic expressions, 929–930
use an algebraic model, 12, 13, 21, 27, 33–39, 52, 60, 85, 93, 110, 141, 151, 258, 355, 361, 375, 542
use a formula, 931
use a verbal model, 12, 13, 21, 27, 33–39, 52, 60, 85, 93, 110, 141, 151, 258, 355, 361, 375, 542
write an equation, 21, 27, 33–39, 93
write an expression, 12, 13
Product of powers property, 323, 407
Product property
of logarithms, 493
of square roots, 264
Programming, linear, 163–167
Project, notes and rubric, 1, 65, 137, 194–195, 197, 247, 321, 396–397, 399, 463, 531, 584–585, 587, 649, 699, 764–765, 767, 829, 892–893
Projects
Drawing with Linear Perspective, 194–195
Magic Squares, 396–397
Mathematical Models of Learning, 584–585
The Mathematics of Music, 892–893
Monte Carlo Methods, 764–765
Properties
of addition, 5, 19, 201, 928

of division, 19, 928
of equality, 19, 928
of finite differences, 381
of integer exponents, 323
of linear inequalities, 41
of logarithms, 493–498, 525
of matrix multiplication, 209
of matrix operations, 201
of multiplication, 5, 19, 209, 257, 928
powers property of equality, 437
of rational exponents, 407
reflexive property of equality, 928
of square roots, 264
of a negative number, 272
of subtraction, 19, 201, 928
Table of, 969–970
zero product, 257
Proportions, *See also* **Variation**
solving, 910–911
Pure imaginary number, 272
Pythagorean identities, 848
Pythagorean theorem, 917–918

Quadrantal angle, 785
Quadrants, of a coordinate plane, 67
Quadratic equations
with complex solutions, 272, 283, 292–293
number and type of solutions of, 293
solving
  by completing the square, 281–289, 315
  by factoring, 256–263, 314
  by finding square roots, 264–270, 314
  by graphing, 271
  with the quadratic formula, 291–298
  using the discriminant, 293
standard form of, 257
systems of, 632–637, 644
Quadratic expressions, factoring, 256–263
Quadratic form, 346
Quadratic formula, 291–298, 316
derivation of, 895
discriminant and, 293
trigonometric equations and, 857
Quadratic functions
graphing, 249–255, 314
intercept form of, 250, 251, 259, 306
maximum and minimum values of, 290
modeling with, 306–312, 316
standard form of, 250–252, 307
vertex form of, 250–252, 285, 306

writing, 306–312
zeros of, 259
Quadratic inequalities
in one variable, 301–302, 316
system of, 300
in two variables, 299–300, 316
Quadratic regression, 308
Quotient of powers property, 323, 407
Quotient property
of logarithms, 493
of square roots, 264

R

Radian
conversion to degrees, 777–778, 822–823
measure of an angle, 777–778, 822–823
Radical equations, solving, 437–443, 458
Radical functions, graphing, 431–436, 457
Radicals, 264
adding, 408–409
index of, 401
notation, 264, 401
simplest form of, 264, 408
simplifying, 264–265, 407–413
subtracting, 408–409
Radicand, 264
Radius, of a circle, 601, 914
Random number generation, 723, 738
Range
of a data set, 446, 453
of a function, 67
Rate, of change, 78
Rational equations, solving, 402–406, 568–573
Rational exponents, 401–406, 456
notation, 401–402
properties of, 407–413, 456
solving an equation with, 437
Rational expressions
adding, 562–563, 565–567, 578
dividing, 556–561, 577
evaluating, 402, 939
multiplying, 554–561, 577
simplifying, 407–413, 554
subtracting, 562–563, 565–567, 578
Rational functions
graphing
general, 547–552, 577
simple, 540–546, 576–577
Rational numbers, definition of, 3
Rational zero theorem, 359
Rationalizing the denominator, 265

Ratios
common, 666
trigonometric, 769
writing, 910–911
Real axis, 273
Real numbers
calculating percents, 907–908
converting decimals, fractions, and percents, 906
graphing, 3
least common denominator, 908–909
nth roots, 401
operations with, 5–6, 58
signed numbers, 905–906
ordering, 4
properties of, 5
scalar, 200
scientific notation, 913
significant digits, 911–912
writing ratios and solving proportions, 910–911
Reasoning, *See* **Critical Thinking; Logical Reasoning**
Reciprocal, 5
Reciprocal identities, 848
Rectangle
area of, 28, 914
perimeter of, 28
Recursive rules, for sequences, 681–686, 694
Reference angle, 785–786
Reflections, 921
of trigonometric graphs, 839–846
Reflexive property of equality, 928
Relation, 67
inverse, 422
representing, 67
Remainder theorem, 353
Repeated solution, 366
Repeating decimal, 676
Resource Material, 1B, 64B, 136B, 196B, 246B, 320B, 398B, 462B, 530B, 586B, 648B, 698B, 766B, 828B
Reviews, *See also* **Assessment; Projects**
Chapter Review, 58–60, 130–132, 186–188, 240–242, 314–316, 388–390, 456–458, 524–526, 576–578, 642–644, 692–694, 756–758, 822–824, 884–886
Chapter Summary, 57, 129, 185, 239, 313, 387, 455, 523, 575, 641, 691, 755, 821, 883
Cumulative Practice, 192–193, 394–395, 582–583, 762–763, 890–891
Extra Practice, 940–960

Mixed Review, 10, 17, 24, 32, 39, 47, 56, 74, 81, 88, 98, 106, 113, 120, 128, 145, 155, 162, 168, 175, 184, 206, 213, 221, 229, 235, 255, 263, 269, 280, 289, 298, 305, 312, 328, 336, 344, 351, 358, 365, 371, 378, 386, 406, 414, 420, 429, 436, 444, 452, 472, 479, 485, 492, 499, 508, 516, 522, 539, 545, 553, 560, 567, 573, 594, 600, 607, 614, 621, 631, 638, 657, 665, 673, 680, 686, 707, 715, 722, 729, 736, 744, 752, 775, 783, 790, 798, 806, 812, 819, 837, 847, 854, 861, 867, 874, 882
Skill Review, 2, 66, 138, 198, 248, 322, 400, 464, 532, 588, 650, 700, 768, 830
Skills Review Handbook
Algebra, 936–939
Geometry, 914–923
Graphing, 933–936
Logical Reasoning, 924–929
Problem Solving, 929–932
Real Numbers, 905–913
Study Guide, 2, 66, 138, 198, 248, 322, 400, 464, 532, 588, 650, 700, 768, 830
Richter, Charles, 504
Right distributive property, 209
Right triangle trigonometry, 769–774, 882
Roots, *See also* **Radicals; Solutions**
adding, 408–409
nth, 401–406
properties of, 264, 272
square, 264–270, 272
subtracting, 408–409
Rotation, 921
Rotational symmetry, 919
Rounding function, 118
Row matrix, 199
Row operations, 237–238
Rule of false position, 40

S

Scalar, definition of, 200
Scalar multiplication, 200
Scale factor, 922
Scatter plot, 100–105
Science, *See* **Applications; Connections**
Scientific notation, 325, 913
Secant, 769, *See also* **Trigonometric functions**
Second-order differences, 379
Sector, of a circle, 779
Semiperimeter, of a triangle, 809
Sequences
arithmetic, 659–665, 693

definition of, 651

finite, 651

geometric, 666–672, 693

graphing, 652

infinite, 651

piecewise-defined, 686

recursive rules for, 681–686, 694

terms of, 651, 692

writing rules for, 652

writing terms of, 651

Series

arithmetic, 661–665, 693

definition of, 653

geometric, 668–672, 693

infinite geometric, 674–680, 694

summation notation and, 653–657, 692

Sierpinski triangle, 672

Sigma notation, 653

Significant digits, 911–912

Similar figures, 923

Simple interest, formula, 28

Simplest form, of a radical, 408

Sine

definition of, 769

difference formula for, 869–874

double-angle formula for, 875–881

half-angle formula for, 875–881

inverse, 792

law of, 799–806, 823–824

derivation of, 900–901

sum formula for, 869–874

Sine functions, *See also* **Trigonometric functions**

graphing, 831–833, 835–838, 884

transformations of graphs of, 839–842, 844–846

Sinusoidal regression, 864

Sinusoids, 862–866

Skewed distribution, 740

Skill Review, *See* **Reviews**

Skills Review Handbook, *See* **Reviews**

Slope

classification of lines by, 76

definition of, 75

finding, 75

linear equations and, 92

Slope-intercept form, linear equations and, 82–83, 91

Solution

of a linear equation, 19

in two variables, 69

of a linear inequality, 41

in two variables, 108

of a linear system, 139

in three variables, 177

of a polynomial equation, 373

of a quadratic equation, 293

repeated, 366

of a system of linear inequalities, 156

Spreadsheet, 172, 336

for evaluating recursive rules, 688

exercises, 672

Square

of a binomial, 339

difference of two, 257

Square matrix, 199

determinants and, 214–220

Square pyramidal numbers, 384

Square root, 264

of a negative number, 272

properties of, 264

solving quadratic equations by finding, 264–270, 314

Square root function, 431–436

Standard deviation, 446, 453, 458

normal curve and, 746–751

Standard form

of a complex number, 272

of the equation of a circle, 601

of the equation of an ellipse, 609

of the equation of a hyperbola, 615

of the equation of a parabola, 596

of equations of translated conics, 623

of a linear equation, 84–85

of a polynomial function, 329

of a quadratic equation, 257

of a quadratic function, 250–252

Standard position, for an angle, 776–778

Statistics, *See also* **Data; Graphs; Probability**

best-fitting line, 99–105, 131–132

best-fitting quadratic model, 308

binomial distribution, 738–744

definition of, 445

direct and indirect variation, 94

frequency distribution, 448

measures of central tendency, 445, 453, 458

measures of dispersion, 446, 458

normal distribution, 746–751

range, 446, 458

standard deviation, 446, 453

statistical graphs, 447–448, 453–454

Step function, 115, 116

ceiling function, 118

greatest integer function, 115

rounding function, 118

Student Help, *See also* **Internet Connections;**

Reviews; Study Strategies; Test-Taking Strategies

Keystroke Help, 18, 25, 48, 54, 90, 107, 118, 121, 126, 146, 161, 176, 207, 224, 271, 290, 308, 337, 372, 430, 453, 471, 500, 519, 546, 561, 608, 658, 703, 704, 709, 723, 729, 745, 820, 838, 892

Look Back, 75, 84, 109, 121, 157, 179, 201, 223, 231, 250, 258, 300, 301, 307, 331, 338, 339, 407, 416, 422, 424, 439, 466, 488, 504, 510, 536, 541, 548, 549, 555, 563, 566, 590, 591, 596, 626, 632, 633, 651, 661, 668, 719, 728, 731, 732, 764, 790, 800, 809, 815, 855, 857

Software Help, 688

Study Tips, 2, 5, 13, 25, 34, 42, 43, 67, 69, 76, 85, 109, 138, 149, 157, 164, 171, 198, 200, 231, 248, 257, 259, 283, 290, 322, 324, 330, 47, 353, 402, 416, 423, 437, 453, 454, 494, 544, 570, 588, 602, 650, 652, 676, 700, 703, 710, 733, 740, 741, 747, 768, 770, 777, 789, 793, 808, 830, 832, 841, 850, 856, 875, 876

Study Strategies

Building on Previous Skills, 138, 185

Connect to Your Life, 700, 755

Dictionary of Functions, 532, 575

Dictionary of Graphs, 588, 641

Draw Diagrams, 768, 821

Learn by Teaching, 650, 691

Making a Flow Chart, 322, 387

Multiple Methods, 830, 883

Quiz Yourself, 400, 455

Skills File, 66, 129

Study Group, 464, 523

Troubleshoot, 248, 313

Vocabulary File, 2, 57

Writing Out the Steps, 198, 239

Substitution method

for checking solutions, 19, 42, 48, 108, 139, 148, 216, 223, 224, 265, 283, 292, 367, 438, 439, 501, 502, 503, 632, 856

for solving linear systems, 148, 186

for solving quadratic systems, 633–637

Substitution property, 928

Subtraction

of complex numbers, 273

definition of, 5

of functions, 415

inequalities and, 41

of matrices, 200, 240

of polynomials, 338, 340–343

properties of, 19, 201, 928

of radicals, 408–409

with rational expressions, 562–563, 565–567, 578

of roots, 408–409

Sum formulas, 869–874, 886

Sum of two cubes, 345

Summation notation, 653–654

Surface area, 915

Symbols

approximately equal to, 3

common ratio, 666

factorial, 681

function notation, 69

imaginary unit, 272

inequality, 4

infinity, 653

negative infinity, 331

plus or minus, 265

positive infinity, 331

radical sign, 264

sigma, 653

Table of, 961

theta, 769

Symmetric distribution, 740

Symmetry, 919–920

Synthetic division, 353

Synthetic substitution, 330

T

Table

frequency distribution, 448

organizing data with, 738

solving linear equations with, 25

Tables

Formulas, 963–968

Measures, 962

Properties, 969–970

Symbols, 961

Tangent

definition of, 769

difference formula for, 869–874

double-angle formula for, 875–881

half-angle formulas for, 875–881

identity, 848

inverse, 792

sum formula for, 869–874

Tangent functions, *See also* **Trigonometric functions**

graphing, 834–837, 884

transformations of graphs of, 843–846

Teaching Tips, 15, 178, 182, 261, 278, 296, 335, 349, 363, 412, 435, 488, 511, 551,

572, 599, 605, 613, 619, 626, 636, 772

Technology, *See* **Calculator; Graphing calculator; Technology Activities**

Technology Activities

Constructing a Binomial Distribution, 745

Evaluating Expressions, 18

Evaluating Recursive Rules, 688

Finding Maximums and Minimums, 290

Generating Random Numbers, 723

Graphing Circles, 608

Graphing Equations, 90

Graphing Inverse Functions, 430

Graphing Linear Equations in Three Variables, 176

Graphing Logarithmic Functions, 500

Graphing Parametric Equations, 820

Graphing Piecewise Functions, 121

Graphing Rational Functions, 546

Graphing Systems of Equations, 146

Graphing Trigonometric Functions, 838

Operations with Rational Expressions, 561

Setting a Good Viewing Window, 337

Solving an Inequality, 48

Solving Polynomial Equations, 372

Solving Quadratic Equations, 271

Statistics and Statistical Graphs, 453–454

Using Linear Regression, 107

Using Matrix Operations, 207

Using Tables to Solve Equations, 25

Working with Sequences, 658

Terminal side of an angle, 776

Terms

constant, 13, 329

of an expression, 13

like, 13

of a sequence, 651, 692

Test Preparation

multi-step problem, 10, 24, 47, 81, 98, 105, 113, 145, 175, 183, 206, 235, 255, 269, 305, 311, 328, 336, 371, 385, 406, 413, 436, 479, 507, 516, 522, 539, 560, 567, 600, 614, 620, 665, 672, 714, 729, 751, 774, 797, 806, 846, 854, 874

multiple choice, 39, 55, 88, 127, 162, 168, 213, 229, 263, 289, 343, 350, 358, 378, 443, 451, 472, 485, 498, 545, 552, 606, 630, 637, 680, 686, 722, 744, 790, 812, 818, 837, 861, 867

quantitative comparison, 16, 31, 74, 120, 154, 220, 280, 297, 364, 420, 428, 492, 573, 594, 657, 707, 736, 782, 881

Test-Taking Strategies, 62, 134, 190, 244, 318, 392, 460, 528, 580, 646, 696, 760,

826, 888

Theorems

binomial theorem, 710–714, 756

factor theorem, 354

fundamental theorem of algebra, 366–371

Pythagorean theorem, 917–918

rational zero theorem, 359

remainder theorem, 353

Theoretical probability, 716–717, 757

Third-order differences, 379

Three-dimensional coordinate system, 170

Transformations, 921–922

of absolute value inequalities, 51

that produce equivalent equations, 19

that produce equivalent inequalities, 41

of trigonometric graphs, 839–846

Translations, 921

of trigonometric graphs, 839–846

Transverse axis, of a hyperbola, 615

Trapezoid, area of, 28, 914

Tree diagram

for finding conditional probability, 733

for finding theoretical probability, 738

fundamental counting principle and, 701

Triangles

AAS, 799

area of, 28, 215, 802–806, 809–812, 914, 915

Heron's area formula and, 809

law of cosines and, 807–812

law of sines and, 799–806

relationships, 917–919

right-triangle trigonometric, 769–774

SAS, 807, 808

semiperimeter of, 809

SSA, 800–801

SSS, 808

Triangular numbers, 380

Triangular pyramidal numbers, 381

Trigonometric equations, solving, 793–798, 855–861, 870–874, 877–881, 885

Trigonometric expressions

evaluating, 869–870, 872–873, 875–876, 879, 880

simplifying, 849, 854, 870, 873, 876, 879, 880

Trigonometric functions

of any angle, 784–790, 823

graphing, 831–838

inverse, 791–798, 823

modeling with, 862–867

of a quadrantal angle, 785

right triangle, 769–774, 822
writing, 862
Trigonometric identities, 848–854, 876, 885
verification of, 850, 853, 876, 885
Trinomials, 256, *See also* **Polynomials**
irreducible, 257
perfect square, 257
Turing, Alan, 226
Turning points, of a polynomial function, 374–378

Unbounded region, 163, 164
Union, of sets, probability and, 724–725
Unit analysis, 6, 12, 33
exercises, 8, 32

Value
of an expression, 12
of a variable, 12
Variable
on both sides
of an equation, 20
of an inequality, 42
definition of, 12
dependent, 69
equations with more than one, 26–27, 69
independent, 69
on one side
of an equation, 19
of an inequality, 42
subscripts and, 27
value of, 12

Variation
constant of, 94, 534
direct, 94, 534
inverse, 533–535, 537–539, 576
joint, 536–539, 576
Venn diagram, compound events and, 724–725
Verbal model, 12, 13, 15, 21, 27, 33–39, 52, 60, 85, 93, 110, 141, 151, 172, 180, 232, 258, 347, 355, 361, 375, 417, 542, 570, 683
Verification, of trigonometric identities, 850, 853, 876
Vertex
of an ellipse, 609
of the graph of an absolute value function, 122
of the graph of a quadratic function, 249, 250
of a hyperbola, 615
of a parabola, 595, 642
Vertex form
of a quadratic function, 250–252, 285
writing, 306
Vertical asymptote, 540
Vertical line
graph of, 85
slope of, 76
Vertical line test, 68
Vertical translation, of trigonometric graphs, 839–840, 842–846
Visual Thinking, 96, 113, 119, 162, 254, 262, 278, 296, 406, 427, 428, 451, 514, 630, 772, 780, *See also* **Graphs; Manipulatives; Modeling; Multiple representations**
Volume, 916, *See also* **Formulas**

Whole numbers, 3

x-coordinate, 67
x-intercept, 84
of the graph of a polynomial function, 373
of sine and cosine functions, 832
of a tangent function, 834
x-values, critical, 302

y-coordinate, 67
of a turning point, 374
y-intercept, 82
linear equations and, 91

z-axis, 170
Zeno's paradox, 679
Zero exponent property, 323, 407
Zero matrix, 199
Zero product property, 257
Zeros
approximating, 368
complex, 367
of a polynomial function, 354, 373, 390
finding, 367
finding the number of, 366
of a quadratic function, 259

ADDITIONAL ANSWERS

CHAPTER 1

1.1 PRACTICE AND APPLICATIONS (pp. 7–10)

15.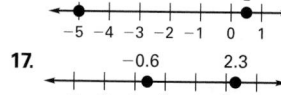

16.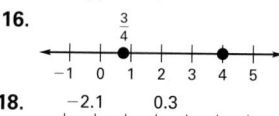

17.
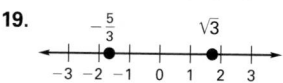

18.

19.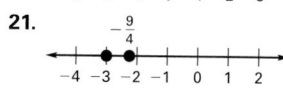

20.

21.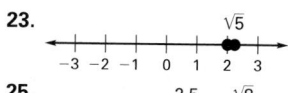

22.

23.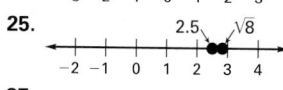

24.

25.

26.

27.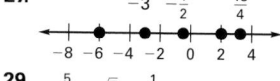

28.

29.

30.

31.

32.

QUIZ 1 (p. 17)

1.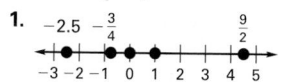

2.

1.3 GUIDED PRACTICE (p. 22) **5.** In going from the second to the third line, $3x$ was added on the left and subtracted on the right; it should have been subtracted from both sides to give $-x + 6 = -3$. The other steps are then "$-x = -9$" and "$x = 9$."

1.3 PRACTICE AND APPLICATIONS (pp. 22–24) **20.** Add 9 to each side; then divide each side by 2. **21.** Subtract 2 from each side; then multiply each side by 3. **22.** Add 5 to each side; then divide each side by −1.

1.6 GUIDED PRACTICE (p. 45)

9.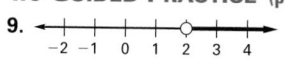

10.

11.

1.6 PRACTICE AND APPLICATIONS (pp. 45–47)

25.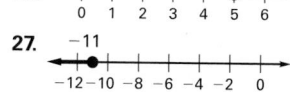

26.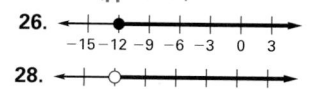

27.

28.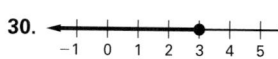

29.

30.

31.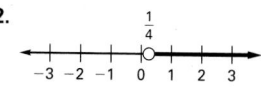

32.

33.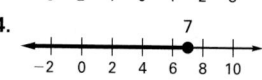

34.

35.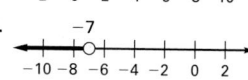

36.

46.

47.

48.

58. a. $10 < d < 160$, where d is the distance from Sonora to Lake Tahoe.
b. *Sample answer:* By the triangle inequality theorem, the distance from Sonora to Lake Tahoe must be less than 85 mi + 75 mi = 160 mi.
c. C; *Sample answer:* One of the distances (78) is greater than the sum of the other two distances (49 + 28 = 77). This violates the triangle inequality theorem.

1.7 CONCEPT ACTIVITY (p. 49)

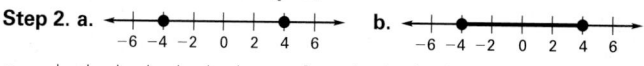

Step 2. a.
b.
c.
d.
e.
f.
g.
h.
i.

1.7 PRACTICE AND APPLICATIONS (pp. 53–55)

57. $-4 < x < \dfrac{18}{7}$;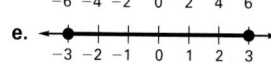

58. $-\dfrac{29}{3} \le x \le 6$;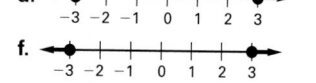

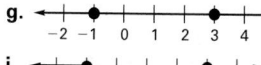

65.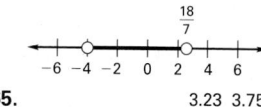

66.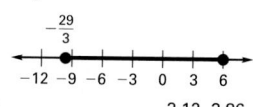

CHAPTER 2

2.1 GUIDED PRACTICE (p. 71)

7.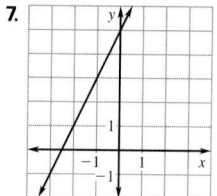

8.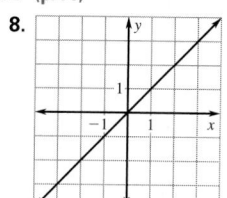

9.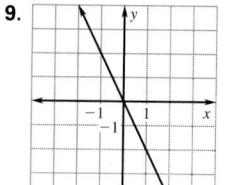

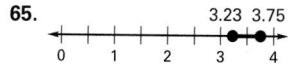

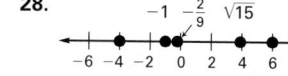

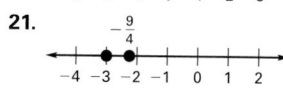

10.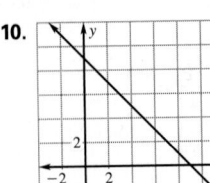

17.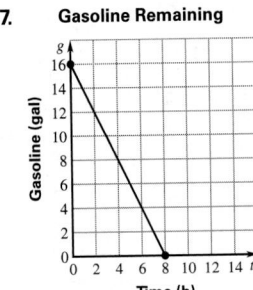

Gasoline Remaining

2.1 PRACTICE AND APPLICATIONS (pp. 71–74)

22.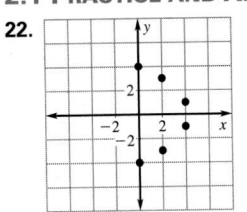

23.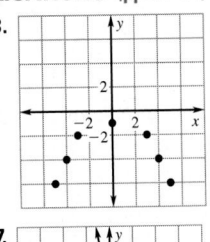

24.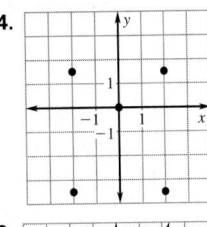

36.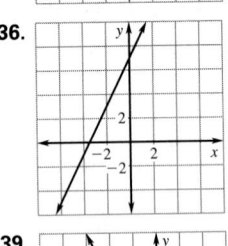

37.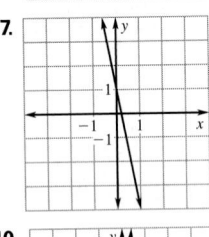

38.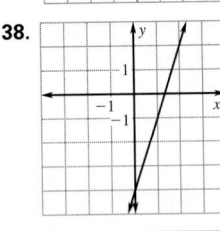

39.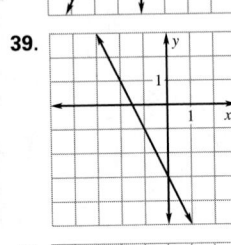

40.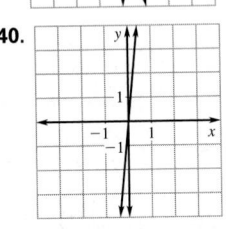

41.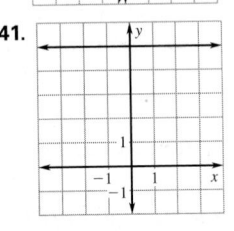

42.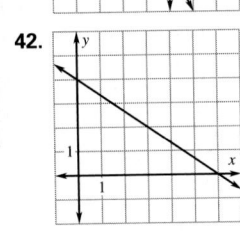

63.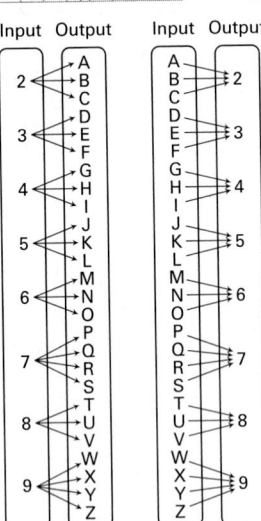

2.3 ACTIVITY (p. 82)

1. Row 1: (0, 3), (1, 5); 2; 3
Row 2: (0, 2), (1, 1); −1; 2
Row 3: (0, −4), (1, −3.5); $\frac{1}{2}$; −4
Row 4: (0, 0), (1, −2); −2; 0
Row 5: (0, 7), (1, 7); 0; 7

2.3 GUIDED PRACTICE (p. 86)

10.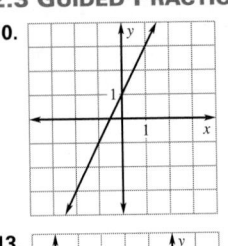

11.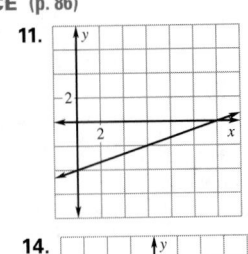

12.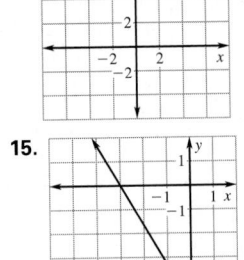

13.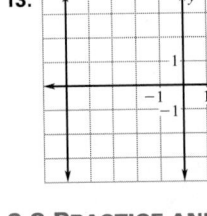

14.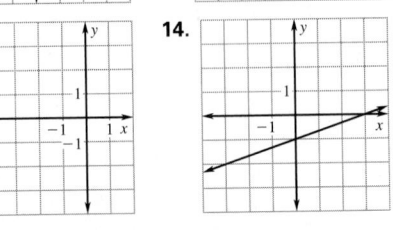

15.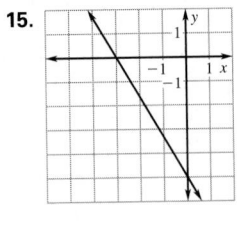

2.3 PRACTICE AND APPLICATIONS (pp. 86–88)

19.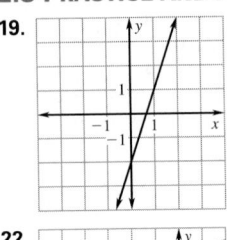

20.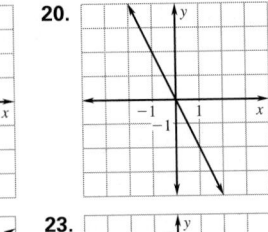

21.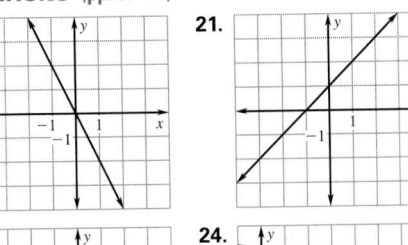

22.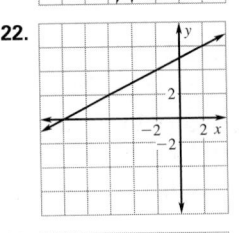

23.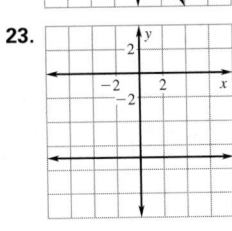

24.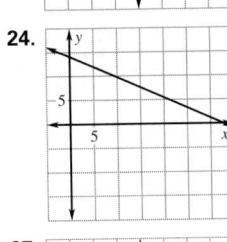

25.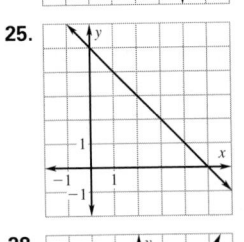

26.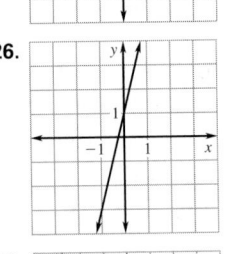

27.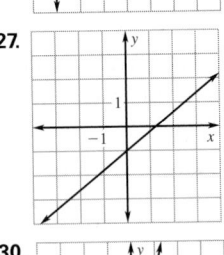

28.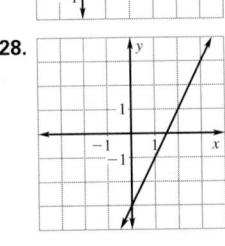

29.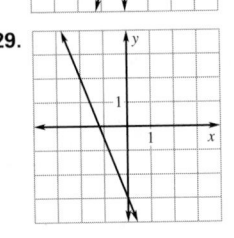

30.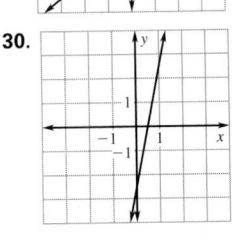

ADDITIONAL ANSWERS

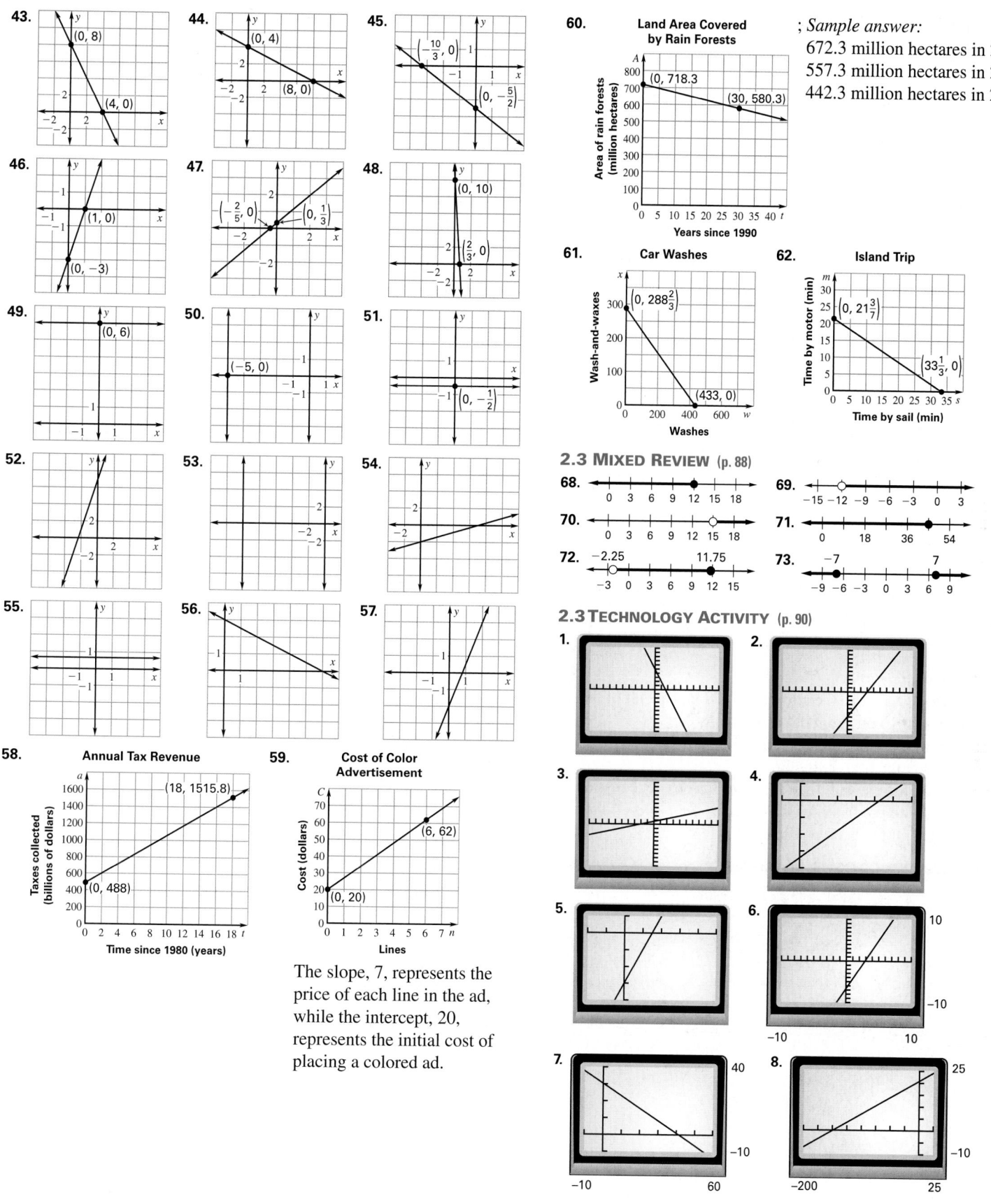

43. (0, 8), (4, 0)

44. (0, 4), (8, 0)

45. $\left(-\frac{10}{3}, 0\right)$, $\left(0, -\frac{5}{2}\right)$

46. (1, 0), (0, −3)

47. $\left(-\frac{2}{5}, 0\right)$, $\left(0, \frac{1}{3}\right)$

48. (0, 10), $\left(\frac{2}{3}, 0\right)$

49. (0, 6)

50. (−5, 0)

51. $\left(0, -\frac{1}{2}\right)$

52.

53.

54.

55.

56.

57.

58. Annual Tax Revenue
(0, 488), (18, 1515.8)

59. Cost of Color Advertisement
(0, 20), (6, 62)

The slope, 7, represents the price of each line in the ad, while the intercept, 20, represents the initial cost of placing a colored ad.

60. Land Area Covered by Rain Forests
(0, 718.3), (30, 580.3)

; *Sample answer:*
672.3 million hectares in 2000,
557.3 million hectares in 2025,
442.3 million hectares in 2050

61. Car Washes
$\left(0, 288\frac{2}{3}\right)$, (433, 0)

62. Island Trip
$\left(0, 21\frac{3}{7}\right)$, $\left(33\frac{1}{3}, 0\right)$

2.3 MIXED REVIEW (p. 88)

68.

69.

70.

71.

72. −2.25, 11.75

73. −7, 7

2.3 TECHNOLOGY ACTIVITY (p. 90)

1.

2.

3.

4.

5.

6.

7.

8.

2.4 PRACTICE AND APPLICATIONS (pp. 95–98)

69. a. 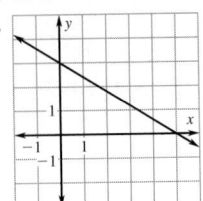 **b.**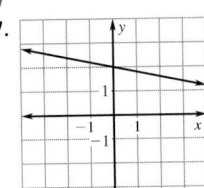

2.4 MIXED REVIEW (p. 98)

86.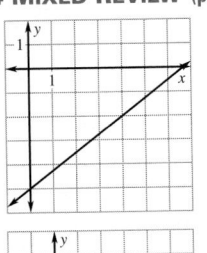

87.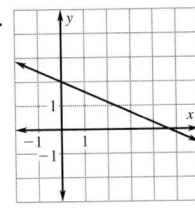

88.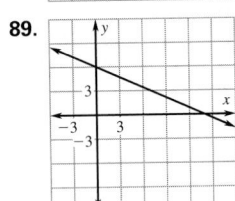

89.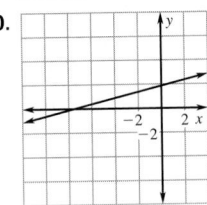

90.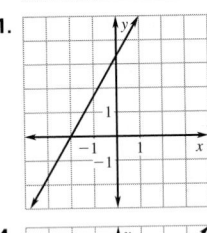

91.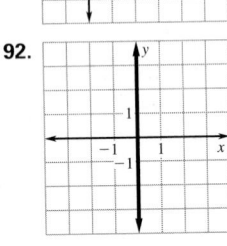

92.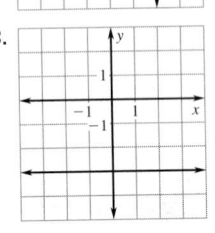

93.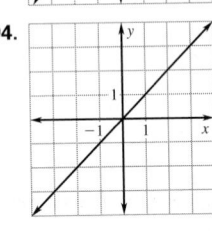

94.

2.5 CONCEPT ACTIVITY (p. 99)

Good responses to the 4 steps and 6 exercises should include all of these:
- a complete table with 10 different data points
- an accurate scatter plot of the data
- a reasonable guess at the best-fitting line
- correct calculation of slope and y-intercept, with a correct equation
- correct use of model to predict y for $x = 300$ cm
- an actual measurement to check prediction

2.5 PRACTICE AND APPLICATIONS (pp. 103–105)

11.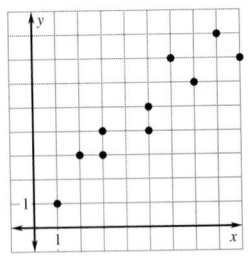

12.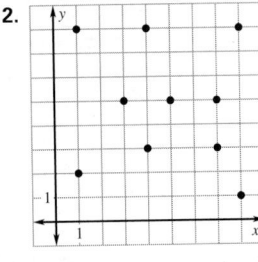

22. High Altitude Temperatures

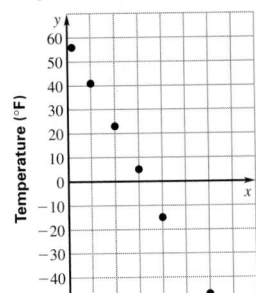

23. Old Faithful Eruptions

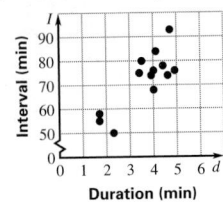

2.5 MIXED REVIEW (p. 106)

30.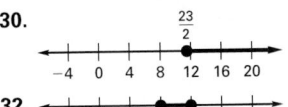

31.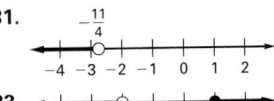

32.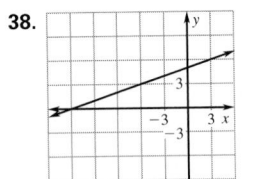

33.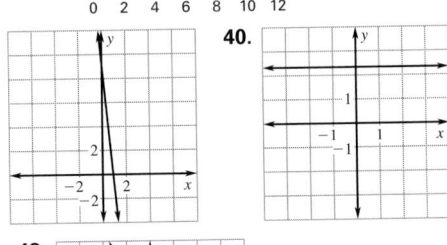

38.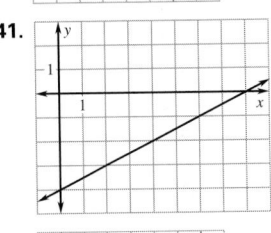

39.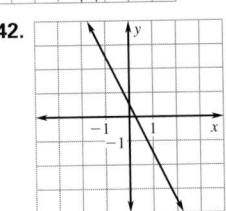

40.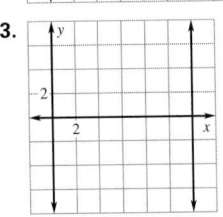

41.

42.

43.

QUIZ 2 (p. 106)

9. Heights of Children

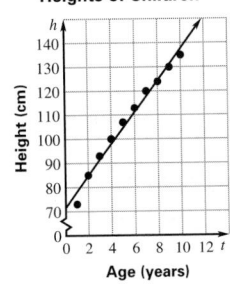

2.5 TECHNOLOGY ACTIVITY (p. 107)

1. 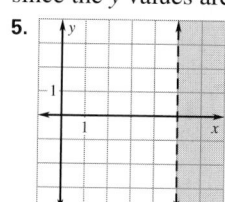   **2.**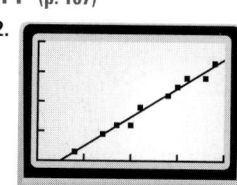

2.6 GUIDED PRACTICE (p. 111)

4. True; for points (x, y) on the line, $y = 3x + 5$. For points (x, y) below the line, the inequality is satisfied, since the y values are smaller.

5. 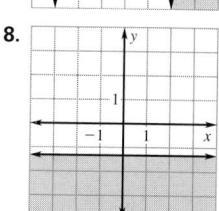   **6.** 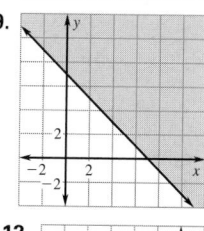   **7.**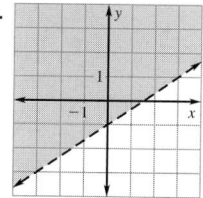

8. 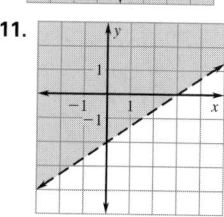   **9.** 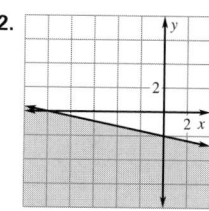   **10.**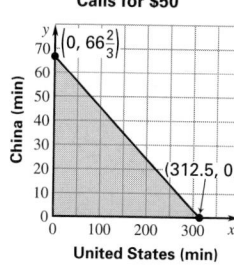

11. 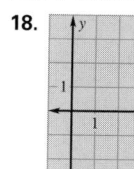   **12.**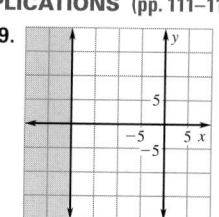

13. $0.16x + 0.75y \le 50$;

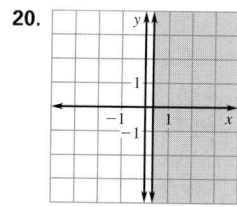

Calls for $50

; One possible solution is to spend 50 min on calls to China and 78 min on calls in the United States, for a total cost of $49.98. Another solution would be to spend 50 min on calls within the United States and 56 min on calls to China; this uses exactly $50. A third solution is 100 min on calls within the United States and 45 min on calls to China. This solution uses a total of $49.75.

2.6 PRACTICE AND APPLICATIONS (pp. 111–113)

18. 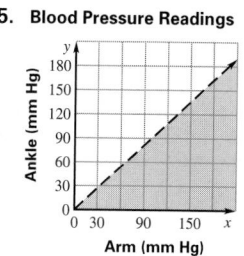   **19.** 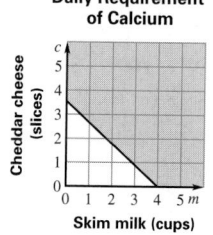   **20.**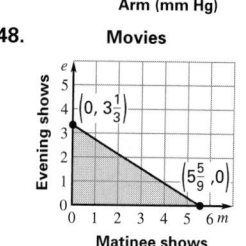

21. 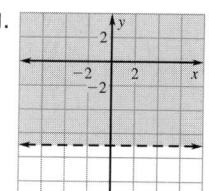   **22.** 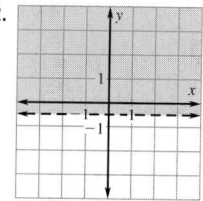   **23.**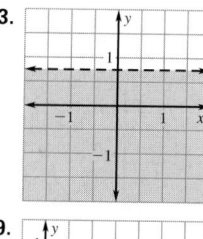

27. 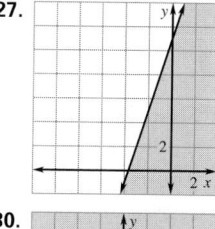   **28.** 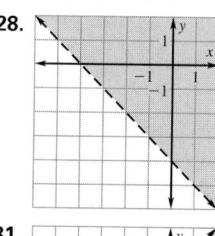   **29.**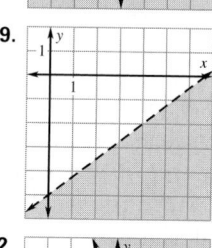

30. 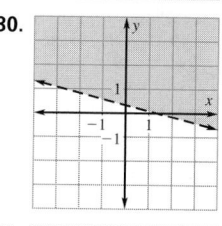   **31.** 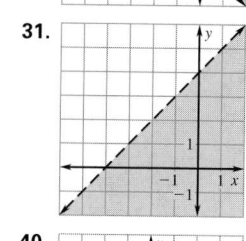   **32.**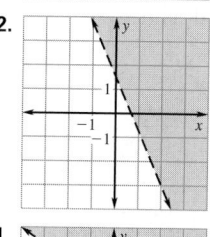

39. 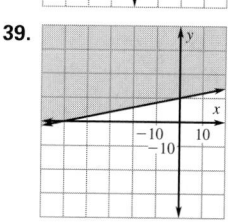   **40.** 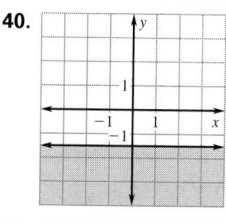   **41.**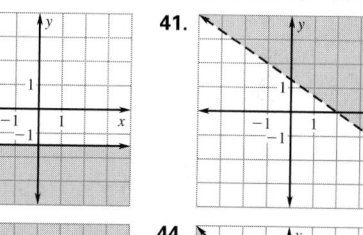

42. 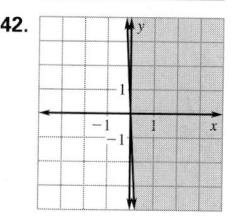   **43.** 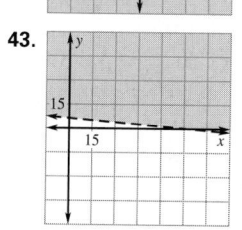   **44.**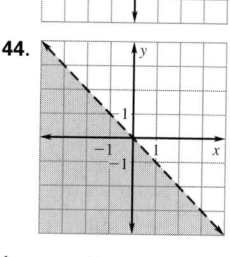

45. Blood Pressure Readings   **46.** Daily Requirement of Calcium

48. Movies   **50.** Football Scoring

52. a.

Truck Rental

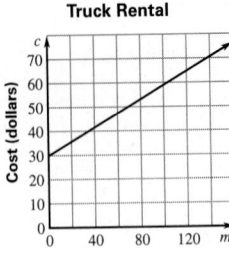

b.

Truck Rental

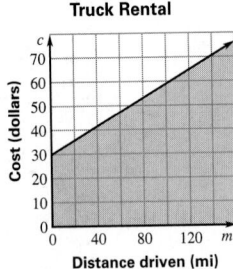

c. *Sample answer:* I would charge a lower flat fee, say $27.77, and not change the per mile charge, because customers see that number up front and will know that the rental cost will be lower whether or not they plan to drive many miles with the truck. Also, maintenance costs on the trucks will be better covered by using the standard per mile charge that the other businesses have found to work for them.

d.

Truck Rental

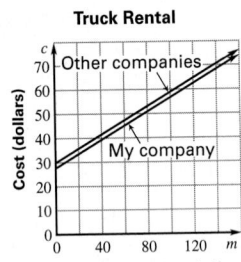

e. *Sample answer:* If a customer is only going to drive a few miles, the lower per mile rate will not offset the higher flat rate.
54. *Sample answer:* Find the equation of the line that is the edge of the half-plane, and note that the solution set consists of those points on the line and below it.

2.6 MIXED REVIEW (p. 113)

62.

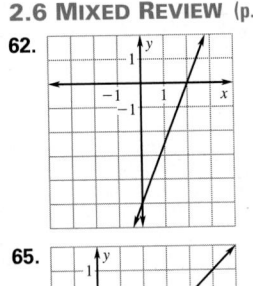

63.

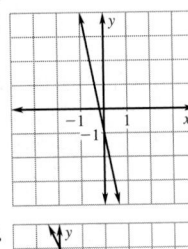

64.

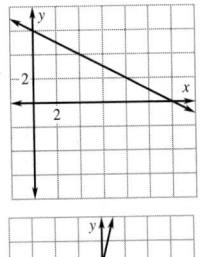

65.

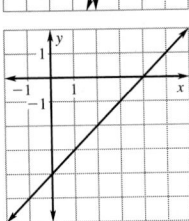

66.

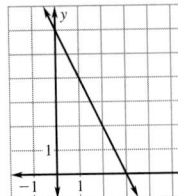

67.

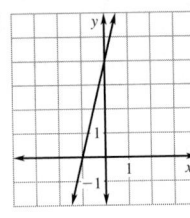

74.

Bloom Times

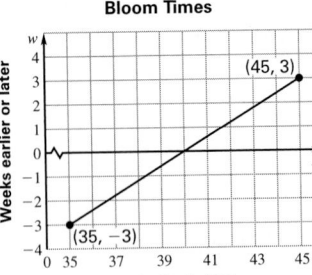

2.7 GUIDED PRACTICE (p. 117)

2. *Sample answer:* A solid dot indicates that an endpoint is part of the graph of that step. An open dot indicates that an endpoint is not part of the graph of that step.

3. False; *Sample answer:* The separate pieces are graphs of different functions. The graphs don't have to be connected. For example, a step function is a piecewise function, but the "steps" of its graph aren't connected.   **4.** True; *Sample answer:* For the first piece, $[x] = 1$; for the second piece, $[x] = 2$; and for the third piece, $[x] = 3$. The domain and range of the new function therefore match those of the piecewise-defined function.

9.

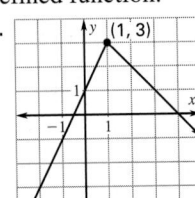

10.

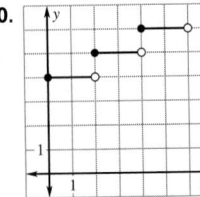

11. $f(x) = \begin{cases} -\dfrac{4}{3}x + 6, & \text{if } 0 \le x < 3 \\ -\dfrac{2}{5}x + \dfrac{16}{5}, & \text{if } 3 \le x \le 8 \end{cases}$

12.

Parking Rates

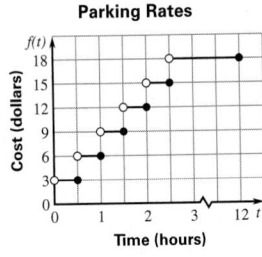

2.7 PRACTICE AND APPLICATIONS (pp. 117–120)

21.

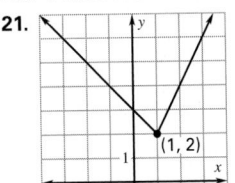

22.

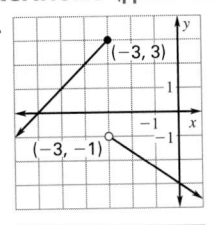

23.

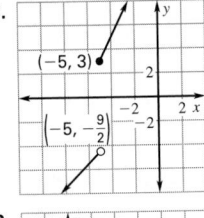

24.

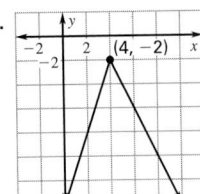

25.

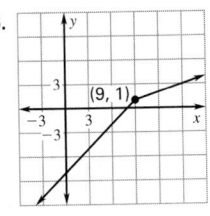

26.

32.

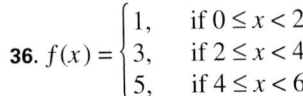

35. $f(x) = \begin{cases} x, & \text{if } x < 0 \ (\text{or } x \le 0) \\ 2x, & \text{if } x \ge 0 \ (\text{or } x > 0) \end{cases}$

36. $f(x) = \begin{cases} 1, & \text{if } 0 \le x < 2 \\ 3, & \text{if } 2 \le x < 4 \\ 5, & \text{if } 4 \le x < 6 \end{cases}$

37. $f(x) = \begin{cases} \dfrac{3}{2}x + \dfrac{9}{2}, & \text{if } x < -1 \\ -1, & \text{if } x \ge -1 \end{cases}$ **38.** $f(x) = \begin{cases} 3x + 10, & \text{if } x \le -2 \\ 4, & \text{if } -2 < x \le 2 \\ -3x + 10, & \text{if } 2 < x \end{cases}$

ADDITIONAL ANSWERS

39. $f(x) = \begin{cases} x + 2, & \text{if } x \le -1 \\ x + 3, & \text{if } -1 < x < 1 \\ x + 1, & \text{if } 1 \le x \end{cases}$ **40.** $f(x) = \begin{cases} 5, & \text{if } -5 < x \le -4 \\ 4, & \text{if } -4 < x \le -3 \\ 3, & \text{if } -3 < x \le -2 \\ 2, & \text{if } -2 < x \le -1 \\ 1, & \text{if } -1 < x \le 0 \end{cases}$

62. $t(d) = \begin{cases} 600 - 10d, & \text{if } 40 \le d \le 53\frac{1}{3} \\ 120 - d, & \text{if } 53\frac{1}{3} < d < 90 \\ 75 - \frac{1}{2}d, & \text{if } 90 \le d \le 130 \end{cases}$

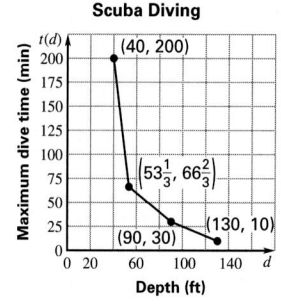

41. 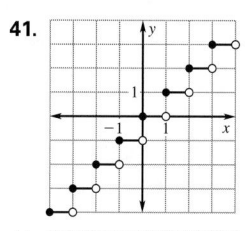 **42.** 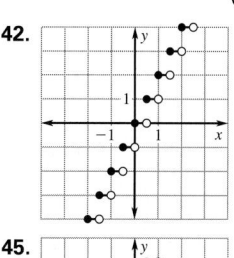 **43.**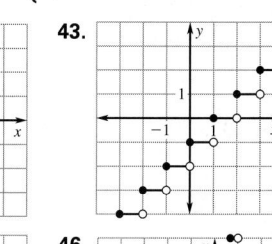

44. 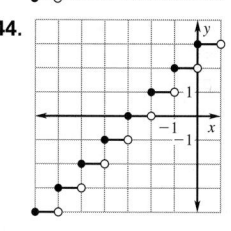 **45.** 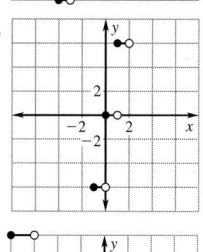 **46.**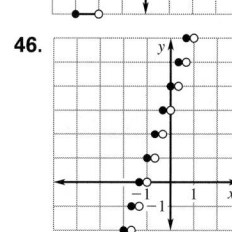

47. 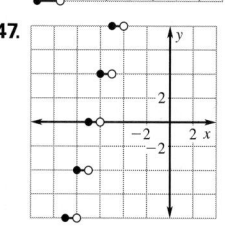 **48.** 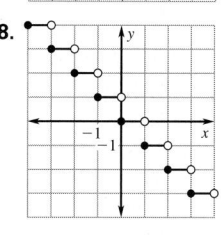 **49.**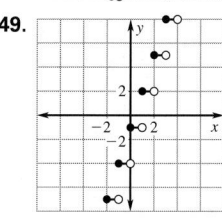

55.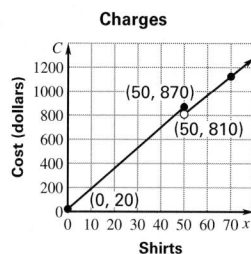

56. $f(x) = \begin{cases} 0.062x, & \text{if } 0 < x < 72,600 \\ 4501.20, & \text{if } x \ge 72,600 \end{cases}$

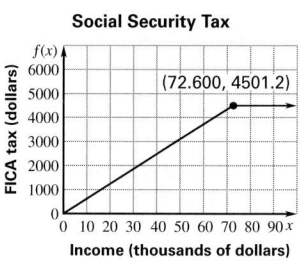

58. $d(t) = \begin{cases} t, & \text{if } 0 \le t \le 2 \\ 2t - 2, & \text{if } 2 < t \le 8 \\ t + 6, & \text{if } 8 < t \le 9 \end{cases}$

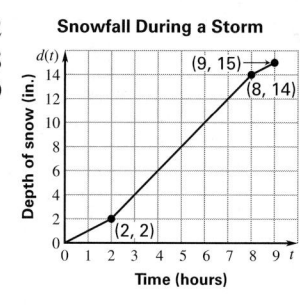

2.7 MIXED REVIEW (p. 120)

69. 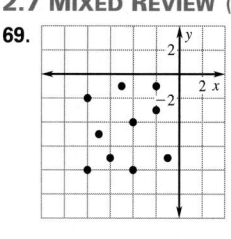 **70.**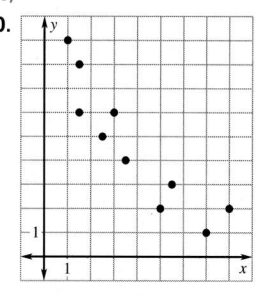

2.7 TECHNOLOGY ACTIVITY (p. 121)

1. 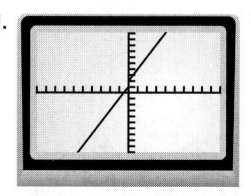 **2.**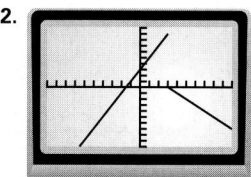

3. 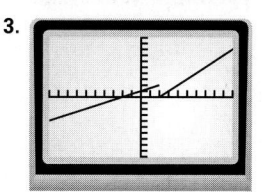 **4.**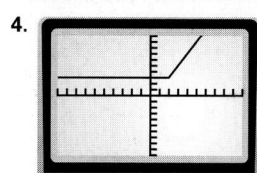

5. 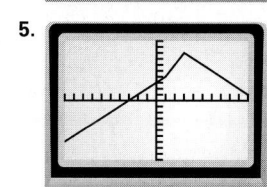 **6.**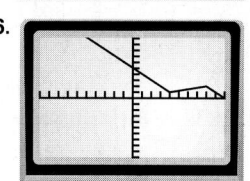

2.8 ACTIVITY (p. 122)

1. 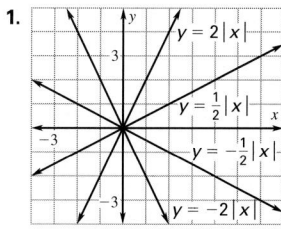 **2.**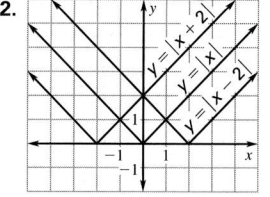

ADDITIONAL ANSWERS

3.

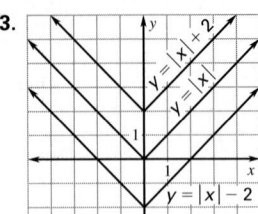

$y = |x| + 2$
$y = |x|$
$y = |x| - 2$

2.8 GUIDED PRACTICE (p. 125)

5.

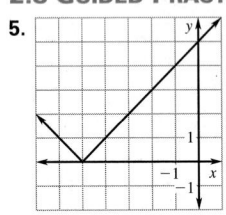

6.

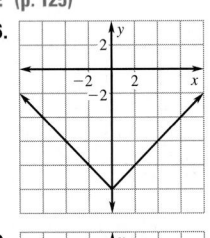

7.

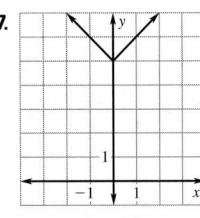

8.

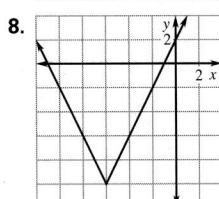

9.

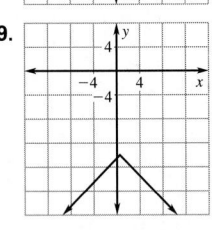

11.

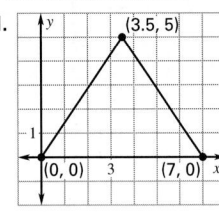

(3.5, 5)
(0, 0) 3 (7, 0)

2.8 PRACTICE AND APPLICATIONS (pp. 125–127)

23.

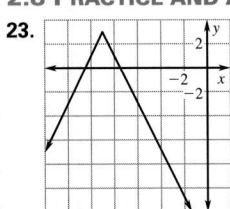

24.

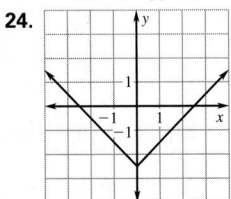

25.

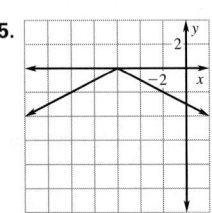

40. Music Single Sales

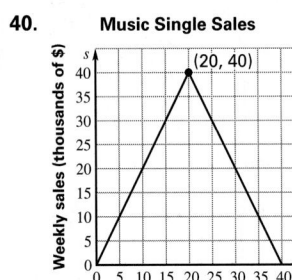

(20, 40)

Weekly sales (thousands of $)
Time (weeks)

42. Rainstorm Log

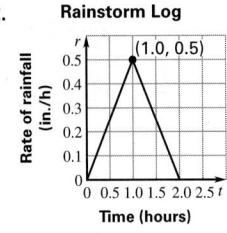

(1.0, 0.5)

Rate of rainfall (in./h)
Time (hours)

51.

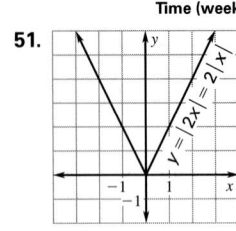

$y = |2x| = 2|x|$

52.

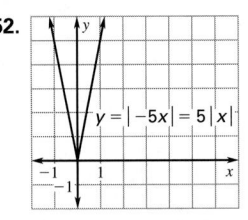

$y = |-5x| = 5|x|$

53.

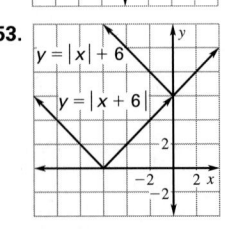

$y = |x| + 6$
$y = |x + 6|$

54.

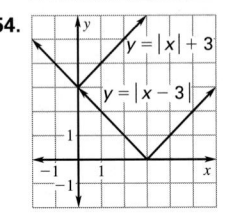

$y = |x| + 3$
$y = |x - 3|$

2.8 MIXED REVIEW (p. 128)

66.

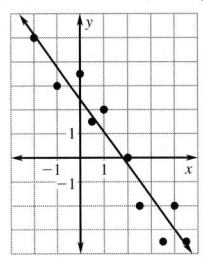

QUIZ 3 (p. 128)

1.

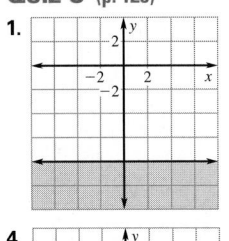

2.

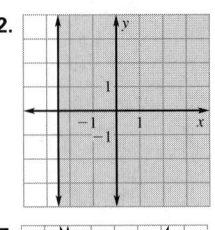

3.

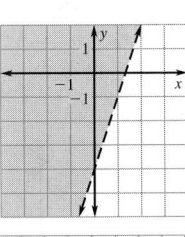

4.

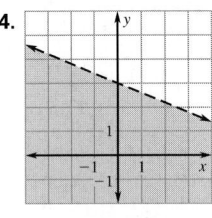

7.

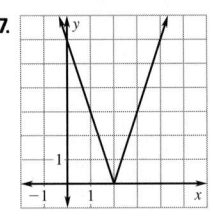

8.

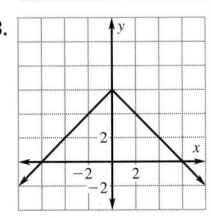

9.

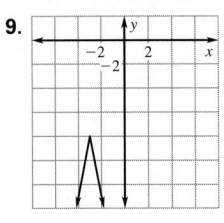

13. Snacks

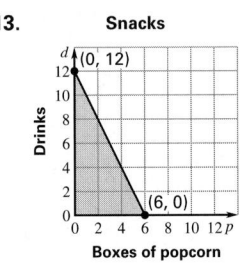

(0, 12)
(6, 0)

Drinks
Boxes of popcorn

14. Rental Charges

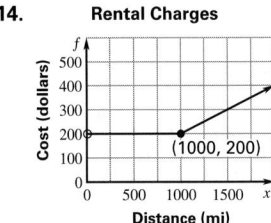

(1000, 200)

Cost (dollars)
Distance (mi)

CHAPTER 2 REVIEW (pp. 130–132)

22.

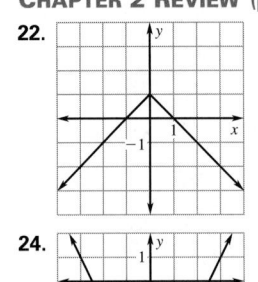

23.

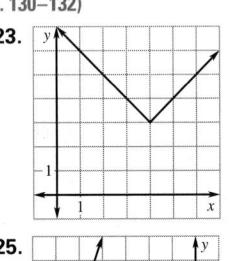

24.

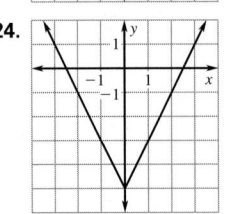

25.

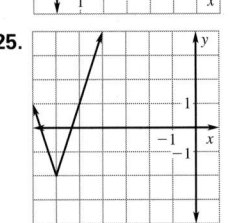

ADDITIONAL ANSWERS

21.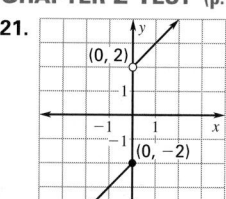

22.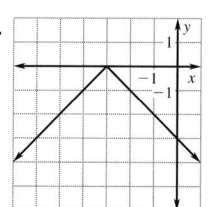

23.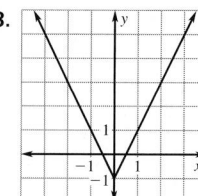

24.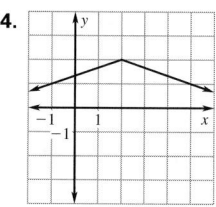

27.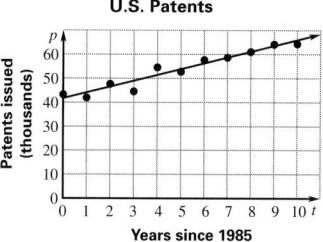
U.S. Patents

CHAPTER 3

SKILL REVIEW (p. 138)

7.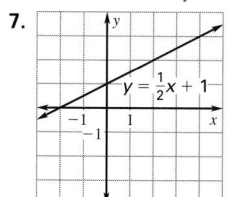

8.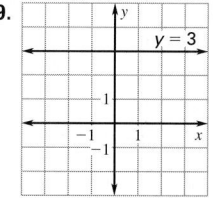

9.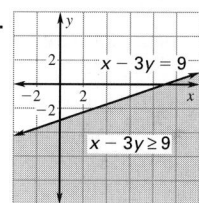

10.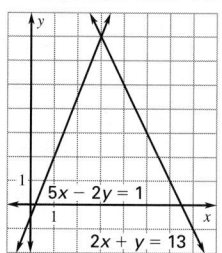

11.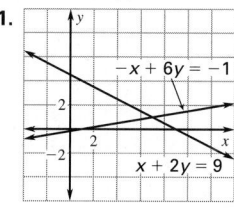

12.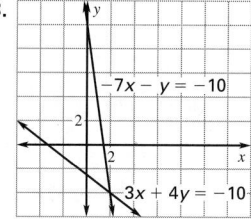

3.1 PRACTICE AND APPLICATIONS (pp. 142–145)

20.

21.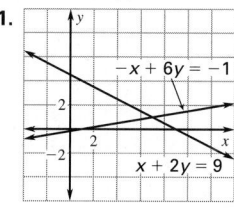

22.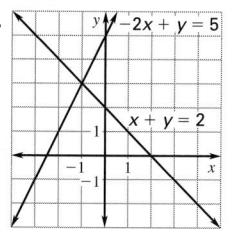

23.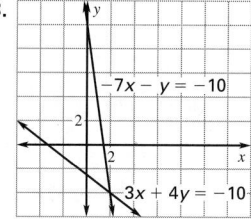

24.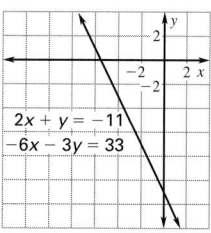
$2x + y = -11$
$-6x - 3y = 33$

25.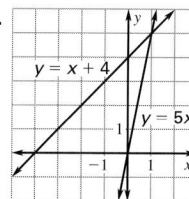
$y = x + 4$
$y = 5x$

26.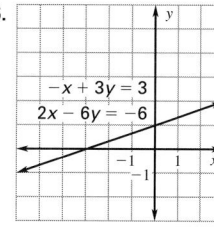
$-x + 3y = 3$
$2x - 6y = -6$

27.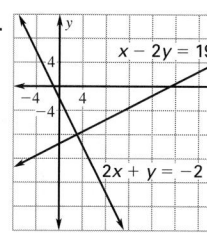
$x - 2y = 19$
$2x + y = -2$

28.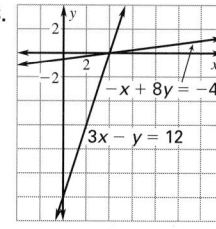
$-x + 8y = -4$
$3x - y = 12$

29.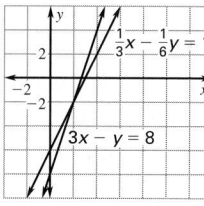
$\frac{1}{3}x - \frac{1}{6}y = 1$
$3x - y = 8$

30.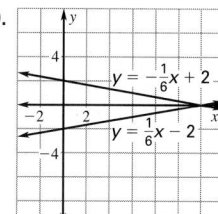
$y = -\frac{1}{6}x + 2$
$y = \frac{1}{6}x - 2$

31.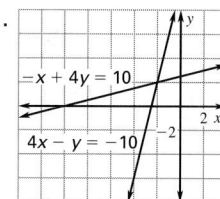
$-x + 4y = 10$
$4x - y = -10$

41.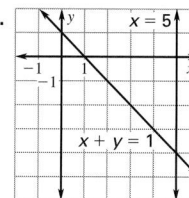
$x = 5$
$x + y = 1$

42.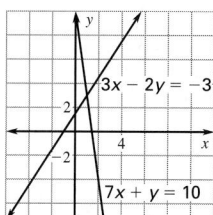
$3x - 2y = -3$
$7x + y = 10$

43.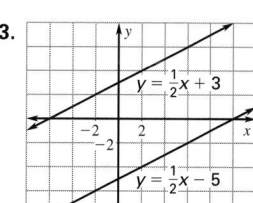
$y = \frac{1}{2}x + 3$
$y = \frac{1}{2}x - 5$

44.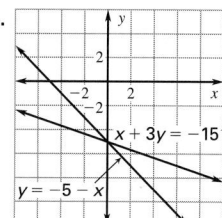
$x + 3y = -15$
$y = -5 - x$

45.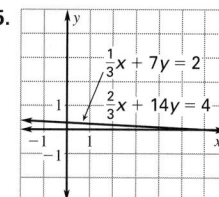
$\frac{1}{3}x + 7y = 2$
$\frac{2}{3}x + 14y = 4$

46.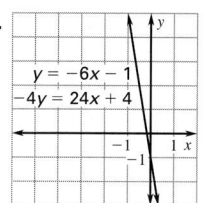
$y = -6x - 1$
$-4y = 24x + 4$

ADDITIONAL ANSWERS

47.

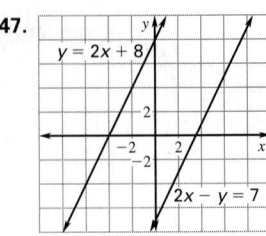

$y = 2x + 8$

$2x - y = 7$

48.

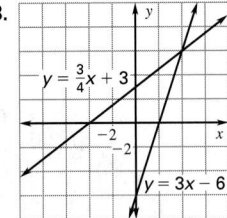

$y = \frac{3}{4}x + 3$

$y = 3x - 6$

49.

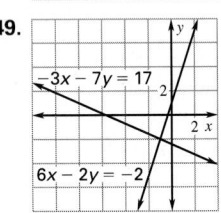

$-3x - 7y = 17$

$6x - 2y = -2$

50.

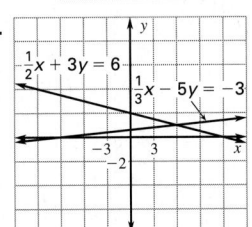

$\frac{1}{2}x + 3y = 6$

$\frac{1}{3}x - 5y = -3$

51.

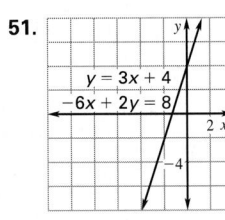

$y = 3x + 4$

$-6x + 2y = 8$

52.

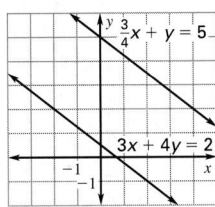

$\frac{3}{4}x + y = 5$

$3x + 4y = 2$

63.

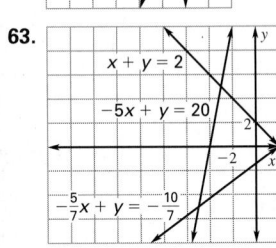

$x + y = 2$

$-5x + y = 20$

$-\frac{5}{7}x + y = -\frac{10}{7}$

64. a. **Long-Distance Phone Service**

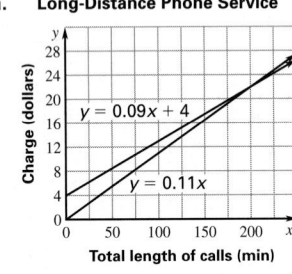

$y = 0.09x + 4$

$y = 0.11x$

Charge (dollars)

Total length of calls (min)

65. **Cost of a Digital Camera**

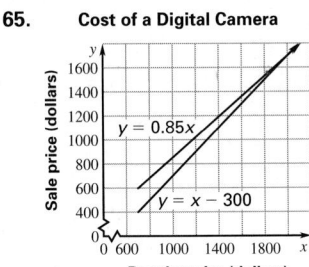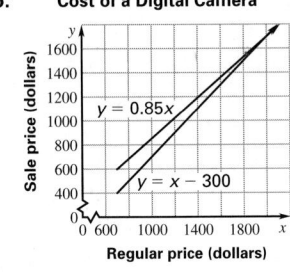

$y = 0.85x$

$y = x - 300$

Sale price (dollars)

Regular price (dollars)

3.1 MIXED REVIEW (p. 145)

76.

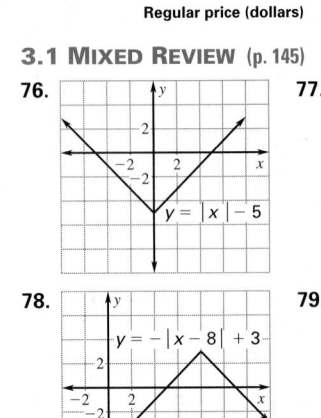

$y = |x| - 5$

77.

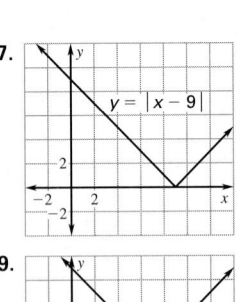

$y = |x - 9|$

78.

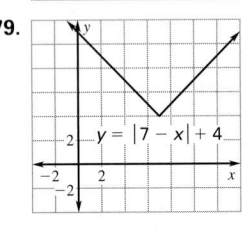

$y = -|x - 8| + 3$

79.

$y = |7 - x| + 4$

3.2 CONCEPT ACTIVITY (p. 147)

1. a.

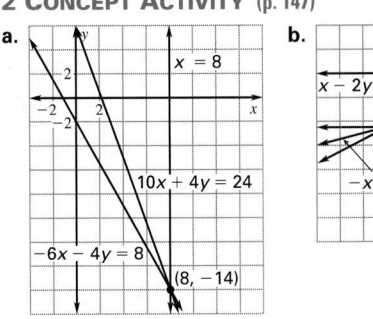

$x = 8$

$10x + 4y = 24$

$-6x - 4y = 8$

$(8, -14)$

b.

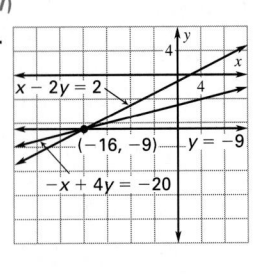

$x - 2y = 2$

$(-16, -9)$ $y = -9$

$-x + 4y = -20$

c.

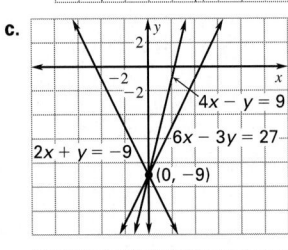

$4x - y = 9$

$6x - 3y = 27$

$2x + y = -9$ $(0, -9)$

d.

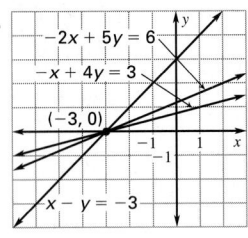

$-2x + 5y = 6$

$-x + 4y = 3$

$(-3, 0)$

$x - y = -3$

e.

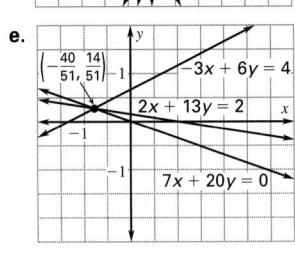

$\left(-\frac{40}{51}, \frac{14}{51}\right)$ $-3x + 6y = 4$

$2x + 13y = 2$

$7x + 20y = 0$

f.

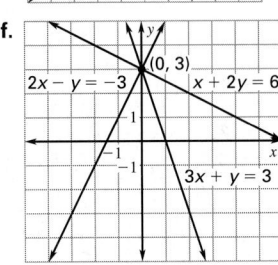

$2x - y = -3$ $(0, 3)$ $x + 2y = 6$

$3x + y = 3$

3.2 MIXED REVIEW (p. 155)

75.

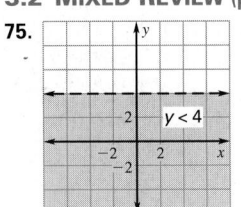

$y < 4$

76.

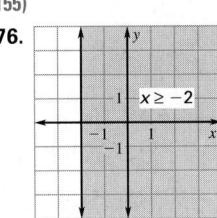

$x \geq -2$

77.

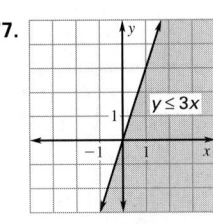

$y \leq 3x$

78.

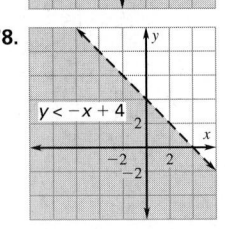

$y < -x + 4$

79.

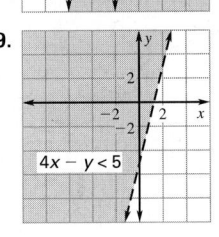

$4x - y < 5$

80.

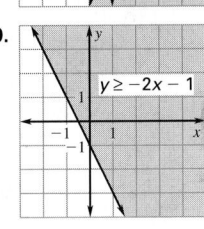

$y \geq -2x - 1$

3.3 GUIDED PRACTICE (p. 159)

9.

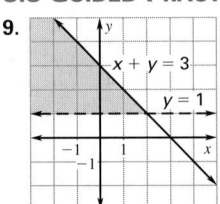

$x + y = 3$

$y = 1$

10.

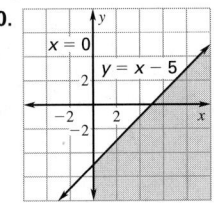

$x = 0$

$y = x - 5$

11. $18 \le x \le 55$; $60 \le y \le 74$;

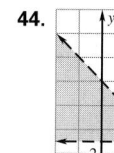

3.3 PRACTICE AND APPLICATIONS (pp. 159–162)

29.
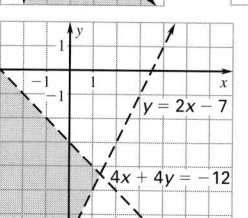

30.
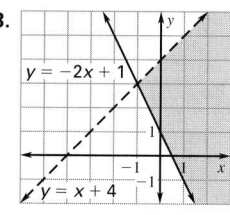

31.
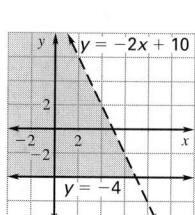

32.
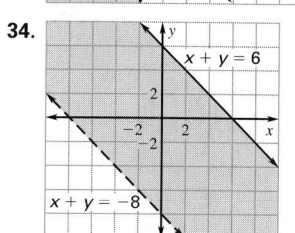

33.
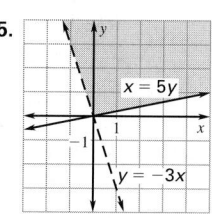

34.
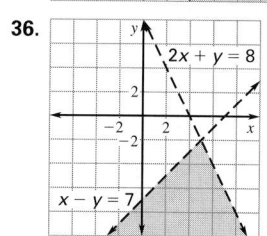

35.
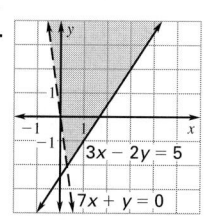

36.
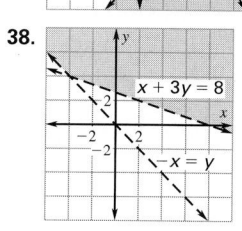

37.
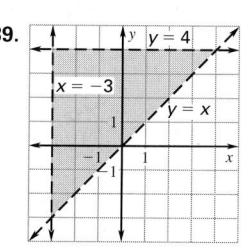

38.
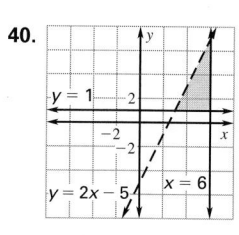

39.
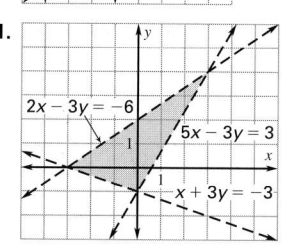

40.

41.

42.
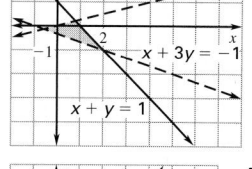

43.
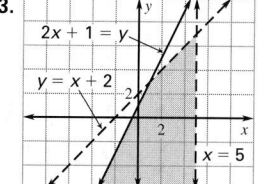

44.

45.
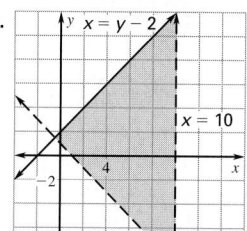

46.
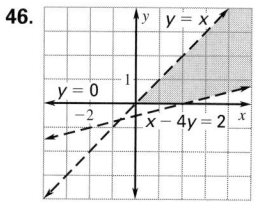

47.
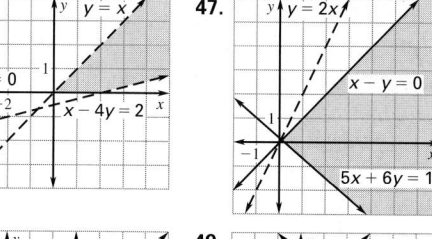

48.
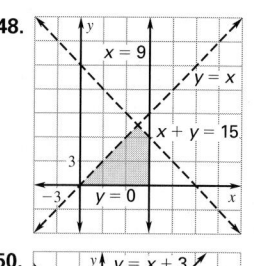

49.
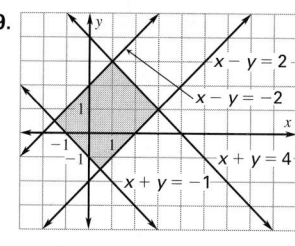

50.
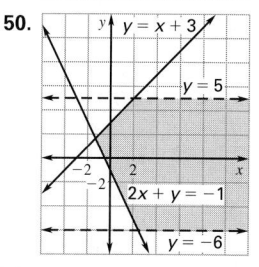

53.
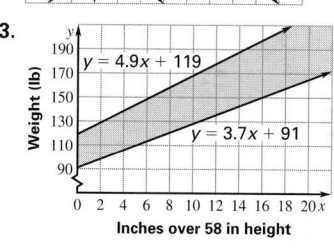

56.
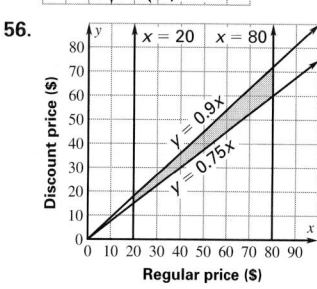

57.
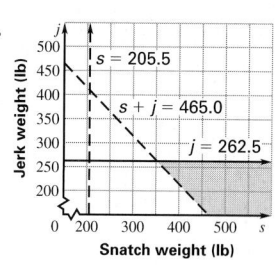

ADDITIONAL ANSWERS

58.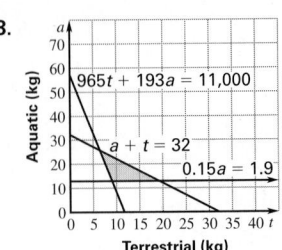

59.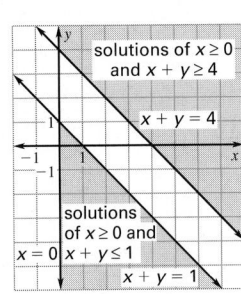

solutions of $x \geq 0$ and $x + y \geq 4$

$x + y = 4$

solutions of $x \geq 0$ and $x = 0$ $x + y \leq 1$

$x + y = 1$

47.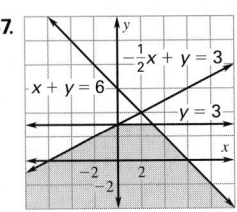

$-\frac{1}{2}x + y = 3$

$x + y = 6$

$y = 3$

3.4 PRACTICE AND APPLICATIONS (pp. 166–168)

27. a. at $(0, 4)$, $C = 8$; at $(1, 4)$, $C = 10$; at $(1.5, 4)$, $C = 11$; at $(2, 4)$, $C = 12$; at $(2, 4)$, $C = 12$; at $(3, 3)$, $C = 12$; at $(4, 2)$, $C = 12$; at $(5, 1)$, $C = 12$; at $(5, 1)$, $C = 12$; at $(5, 0.5)$, $C = 11$; at $(5, 0.25)$, $C = 10.5$; at $(5, 0)$, $C = 10$ **b.** at $(-1, -1)$, $C = -4$; at $(0, 4)$, $C = -4$; at $(1, 9)$, $C = -4$; at $(3, 19)$, $C = -4$; at $(3, 19)$, $C = -4$; at $(3, 10)$, $C = 5$; at $(3, 0)$, $C = 15$; at $(3, -1)$, $C = 16$; at $(3, -1)$, $C = 16$; at $(2, -1)$, $C = 11$; at $(0, -1)$, $C = -1$; at $(-1, -1)$, $C = -4$; Possible conclusions: If the objective function has a certain value at two vertices, it has that same value at each point of the edge connecting them. If an edge is parallel to the objective function, the value of C is constant all along that edge. The value of the objective function at points along an edge is between the values at the vertices at its endpoints.

3.4 MIXED REVIEW (p. 168)

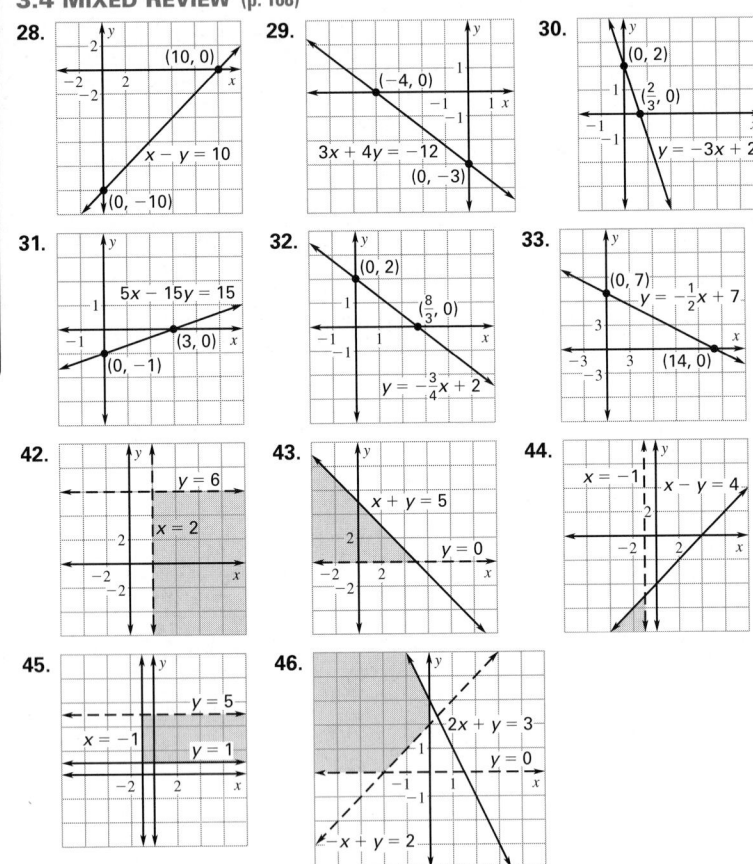

28. $x - y = 10$; $(10, 0)$; $(0, -10)$

29. $3x + 4y = -12$; $(-4, 0)$; $(0, -3)$

30. $(0, 2)$; $\left(\frac{2}{3}, 0\right)$; $y = -3x + 2$

31. $5x - 15y = 15$; $(3, 0)$; $(0, -1)$

32. $(0, 2)$; $\left(\frac{8}{3}, 0\right)$; $y = -\frac{3}{4}x + 2$

33. $(0, 7)$; $y = -\frac{1}{2}x + 7$; $(14, 0)$

42. $y = 6$; $x = 2$

43. $x + y = 5$; $y = 0$

44. $x = -1$; $x - y = 4$

45. $y = 5$; $x = -1$; $y = 1$

46. $2x + y = 3$; $y = 0$; $-x + y = 2$

3.5 GUIDED PRACTICE (p. 173)

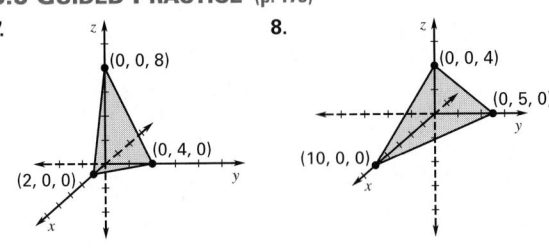

7. $(0, 0, 8)$; $(0, 4, 0)$; $(2, 0, 0)$

8. $(0, 0, 4)$; $(0, 5, 0)$; $(10, 0, 0)$

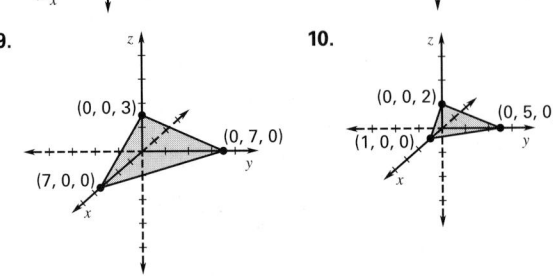

9. $(0, 0, 3)$; $(0, 7, 0)$; $(7, 0, 0)$

10. $(0, 0, 2)$; $(0, 5, 0)$; $(1, 0, 0)$

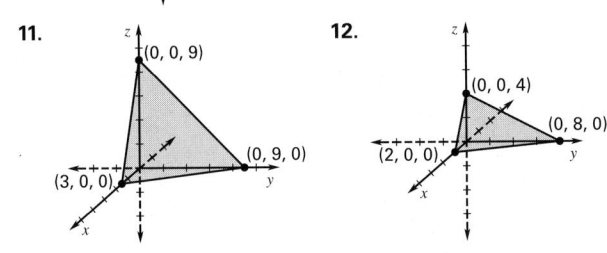

11. $(0, 0, 9)$; $(0, 9, 0)$; $(3, 0, 0)$

12. $(0, 0, 4)$; $(0, 8, 0)$; $(2, 0, 0)$

3.5 PRACTICE AND APPLICATIONS (pp. 173–175)

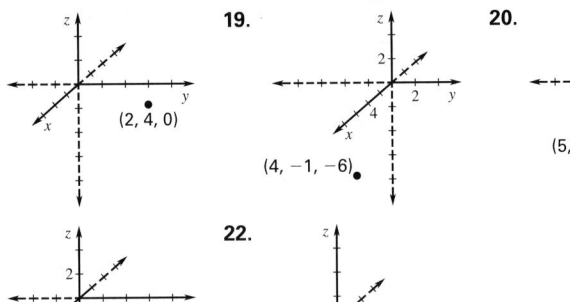

18. $(2, 4, 0)$

19. $(4, -1, -6)$

20. $(5, -2, -2)$

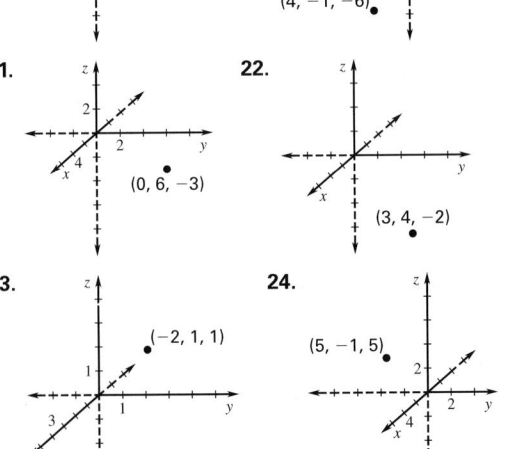

21. $(0, 6, -3)$

22. $(3, 4, -2)$

23. $(-2, 1, 1)$

24. $(5, -1, 5)$

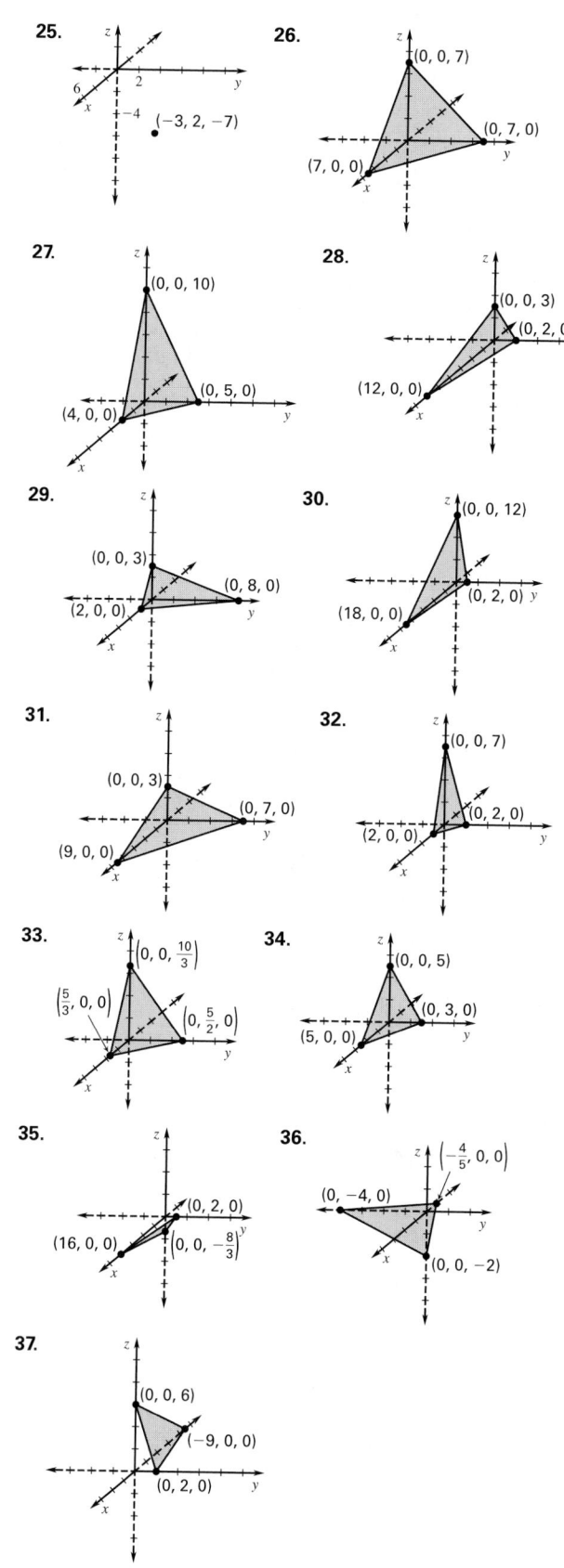

48. *Sample answer:*

Number of angel fish	Number of goldfish			
	2	4	6	8
2	$73.80	$74.60	$75.40	$76.20
4	$81.80	$82.60	$83.40	$84.20
6	$89.80	$90.60	$91.40	$92.20
8	$97.80	$98.60	$99.40	$100.20

49. *Sample answer:*

Price of pottery	Number of colors				
	1	2	3	4	5
$8	$25.50	$27.00	$28.50	$30.00	$31.50
$18	$35.50	$37.00	$38.50	$40.00	$41.50
$28	$45.50	$47.00	$48.50	$50.00	$51.50
$38	$55.50	$57.00	$58.50	$60.00	$61.50
$48	$65.50	$67.00	$68.50	$70.00	$71.50

50. *Sample answer:*

Number of carnations	Number of tulips				
	4	6	8	10	12
4	$16.00	$17.40	$18.80	$20.20	$21.60
6	$16.60	$18.00	$19.40	$20.80	$22.20
8	$17.20	$18.60	$20.00	$21.40	$22.80
10	$17.80	$19.20	$20.60	$22.00	$23.40
12	$18.40	$19.80	$21.20	$22.60	$24.00

51. *Sample answer:*

Number of express bus trips	Number of subway trips				
	2	4	6	8	10
2	$22.30	$22.80	$23.30	$23.80	$24.30
4	$24.10	$24.60	$25.10	$25.60	$26.10
6	$25.90	$26.40	$26.90	$27.40	$27.90
8	$27.70	$28.20	$28.70	$29.20	$29.70
10	$29.50	$30.00	$30.50	$31.00	$31.50

52.

Hours babysitting	Hours lifeguarding				
	2	4	6	8	10
2	$38	$54	$70	$86	$102
4	$50	$66	$82	$98	$114
6	$62	$78	$94	$110	$126
8	$74	$90	$106	$122	$138
10	$86	$102	$118	$134	$150

53. b.

Peak commercial spots	Off-peak commercial spots					
	0	2	4	6	8	10
0	–	$700	$900	$1100	$1300	$1500
2	$1200	$1400	$1600	$1800	$2000	$2200
4	$1900	$2100	$2300	$2500	$2700	$2900
6	$2600	$2800	$3000	$3200	$3400	$3600
8	$3300	$3500	$3700	$3900	$4100	$4300
10	$4000	$4200	$4400	$4600	$4800	$5000

3.6 MIXED REVIEW (p. 184)

54. -9 to 31

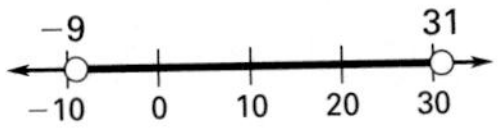

55. -14.5 to 11.5

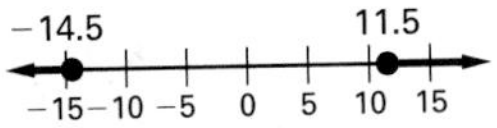

56. -16

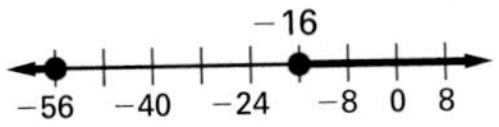

57. $-\frac{3}{2}$, $-\frac{1}{4}$

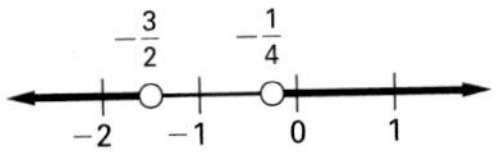

58.

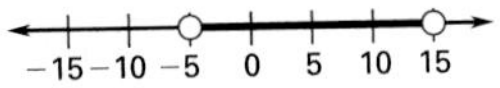

59. $-\frac{29}{3}$, $\frac{31}{3}$

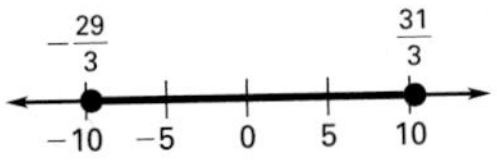

60.

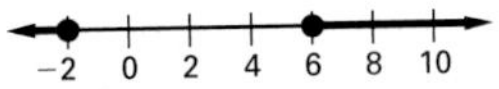

61. $-\frac{22}{3}$

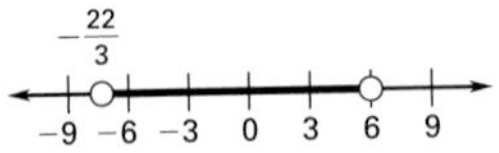

62.

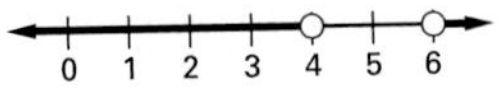

63. (3, 6, 0)

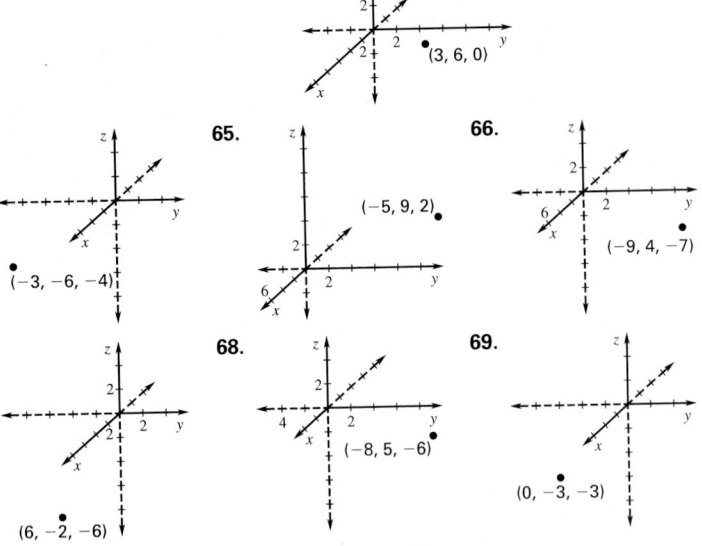

64. (−3, −6, −4)

65. (−5, 9, 2)

66. (−9, 4, −7)

67. (6, −2, −6)

68. (−8, 5, −6)

69. (0, −3, −3)

70. (2, 2, −2)

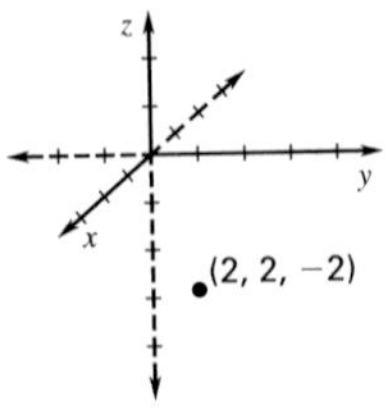

71. (−4, −7, −3)

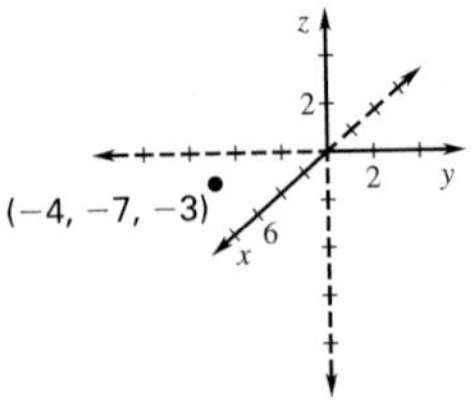

QUIZ 3 (p. 184)

1. (0, 0, 5); (0, 3, 0); $\left(7\frac{1}{2}, 0, 0\right)$

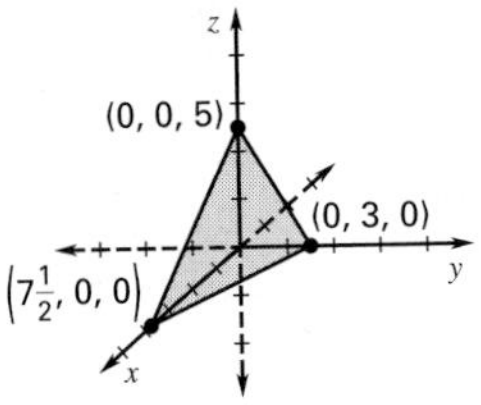

2. $\left(0, 0, \frac{1}{2}\right)$; (0, 2, 0); (8, 0, 0)

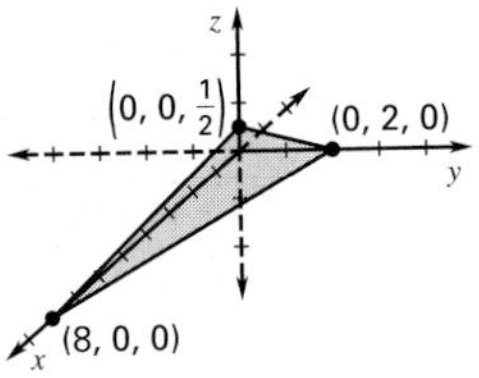

3. (0, 0, 10); (0, 10, 0); $\left(3\frac{1}{3}, 0, 0\right)$

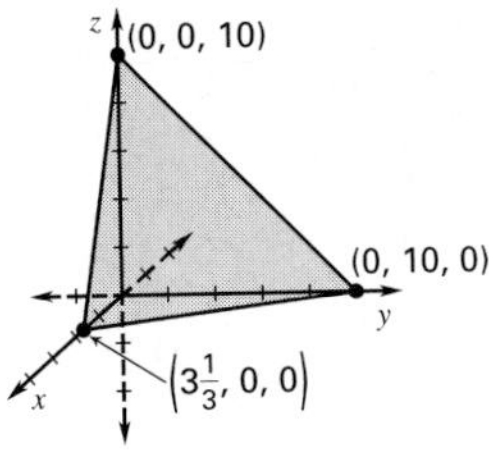

4. $\left(0, 0, \frac{3}{2}\right)$; (3, 0, 0); $\left(0, \frac{3}{4}, 0\right)$

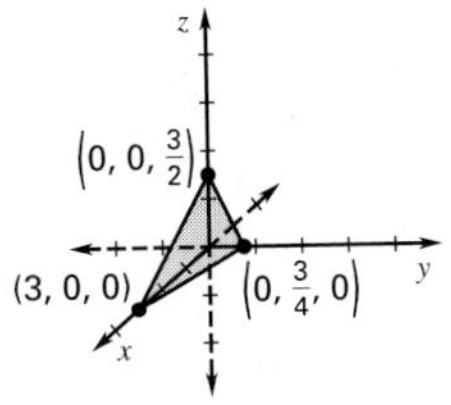

5. (0, 0, 15); $\left(0, -7\frac{1}{2}, 0\right)$; (3, 0, 0)

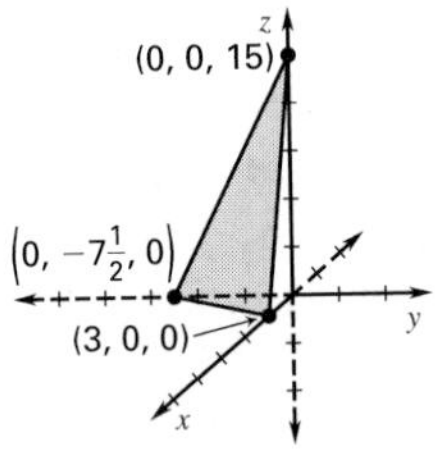

6. (−18, 0, 0); (0, 2, 0); (0, 0, −6)

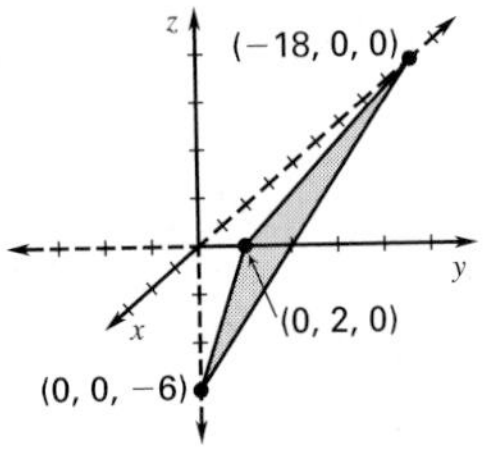

CHAPTER 3 TEST (p. 189)

9. $2x + y = 1$; $x = 3$

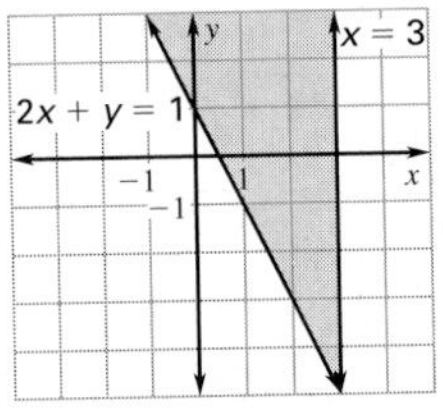

10. $x = 0$; $y = x$; $y = -x$

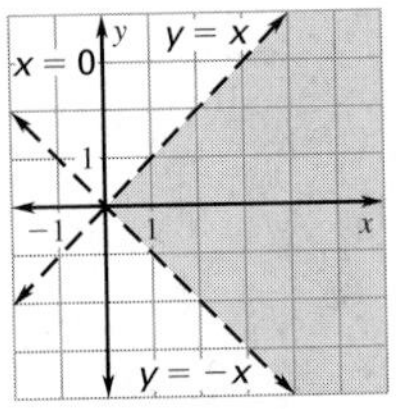

11. $x + 2y = 2$; $y = -1$; $x + 2y = -6$

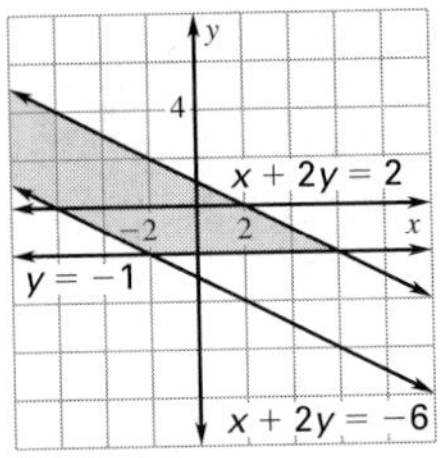

12. $x = -2$; $2x - y = 5$; $x + y = 7$

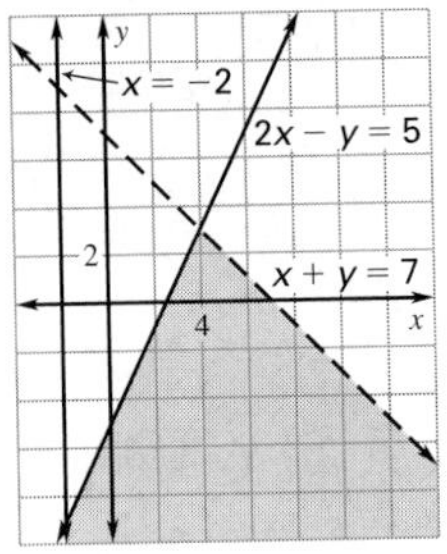

15. (−1, 3, 2)

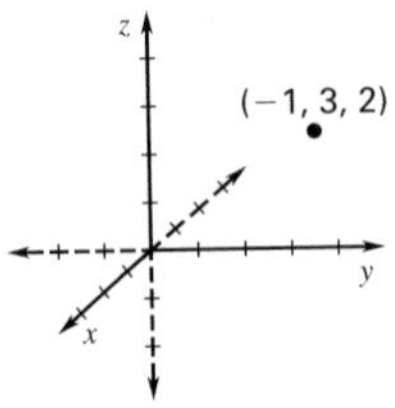

16. (0, 4, −2)

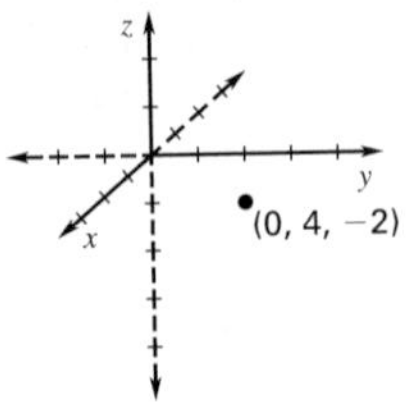

17. (−5, −1, 2)

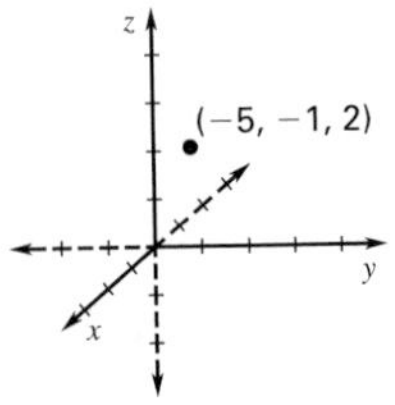

18. (6, −2, 1)

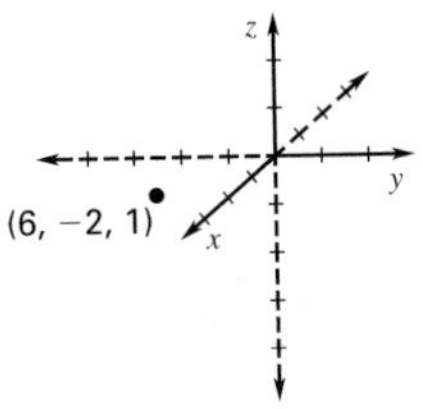

19. (0, 0, 6); (0, 10, 0); (15, 0, 0)

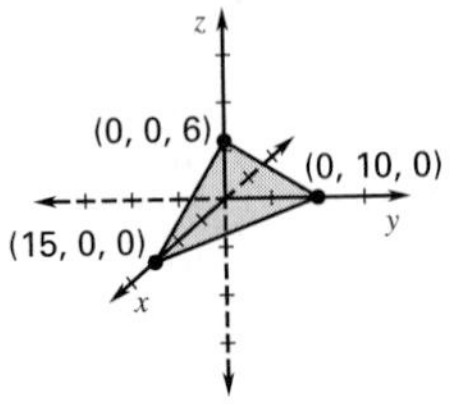

ADDITIONAL ANSWERS

20.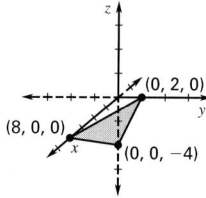

21.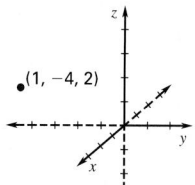

38.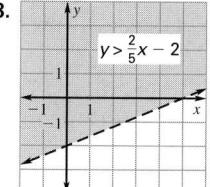

$y > \frac{2}{5}x - 2$

39.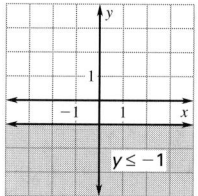

$y \le -1$

40.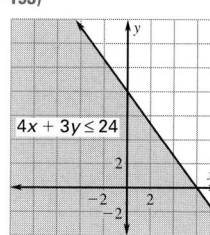

$4x + 3y \le 24$

41.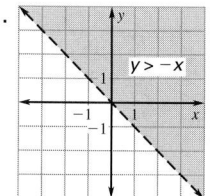

$y > -x$

42.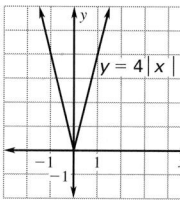

$y = 4|x|$

43.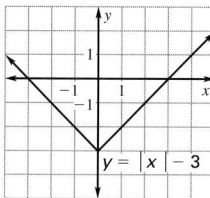

$y = |x| - 3$

44.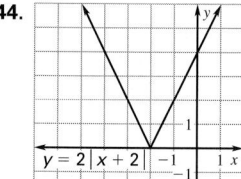

$y = 2|x + 2|$

45.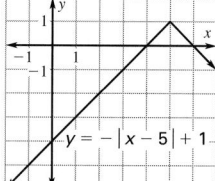

$y = -|x - 5| + 1$

46.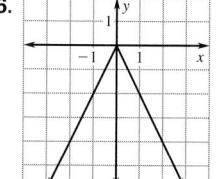

47.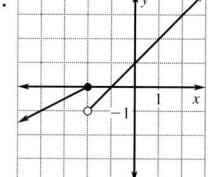

48.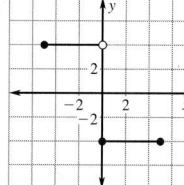

49.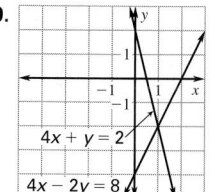

$4x + y = 2$

$4x - 2y = 8$

one solution at $(1, -2)$

50.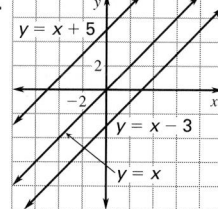

$y = x + 5$

$y = x - 3$

$y = x$

no solution

51.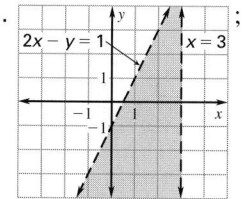

$2x - y = 1$ $x = 3$

solution region is to the
right of $2x - y = 1$ and
to the left of $x = 3$.

52.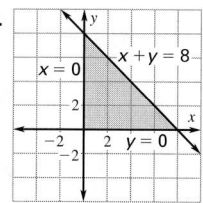

$x + y = 8$

$x = 0$

$y = 0$

solution region is the triangular
region bounded by the
x- and y-axes and $x + y = 8$.

68.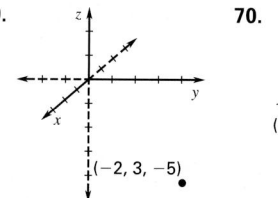

$(1, -4, 2)$

69.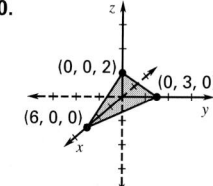

70.

$(0, 0, 2)$ $(0, 3, 0)$

$(6, 0, 0)$

$(-2, 3, -5)$

71.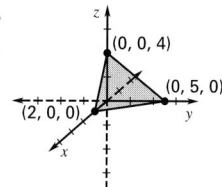

$(0, 0, 4)$

$(0, 5, 0)$

$(2, 0, 0)$

73.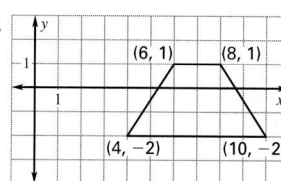

98.6

97 98 99 100

75.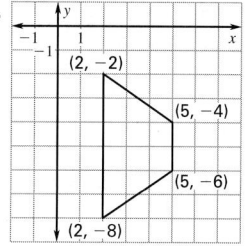

CHAPTER 4

4.1 PRACTICE AND APPLICATIONS (pp. 203–206)

48. c. A good answer should include these points:
  • The number of volumes sold increased for law and music books.
  • The number of volumes sold decreased for art and travel books.
  • The average price per volume increased for art and law books.
  • The average price per volume decreased for music and travel books.
  • There is no direct relationship between the number of volumes sold
    and the average price per volume: art book sales decreased, while the
    average price increased; law book sales increased, while the average
    price increased; music book sales increased, while the average price
    decreased; and travel book sales decreased, while the average price
    decreased.

4.1 MIXED REVIEW (p. 206)

50.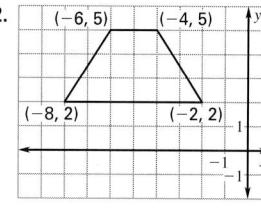

$(2, -2)$

$(5, -4)$

$(5, -6)$

$(2, -8)$

51.

$(6, 1)$ $(8, 1)$

$(4, -2)$ $(10, -2)$

52.

$(-6, 5)$ $(-4, 5)$

$(-8, 2)$ $(-2, 2)$

ADDITIONAL ANSWERS

4.2 PRACTICE AND APPLICATIONS (pp. 211–213)

24. $\begin{bmatrix} -8 & 7 & -10 \\ -29 & 20 & 38 \\ -50 & -11.1 & 32.3 \end{bmatrix}$ **25.** $\begin{bmatrix} -32 & 0 & 32 \\ 12 & -26 & 1 \\ 20 & -30 & -5 \end{bmatrix}$ **26.** $\begin{bmatrix} 3 & 12 & 6 \\ 25 & -73 & 18 \\ -7 & 59 & -14 \end{bmatrix}$

41. *Sample answer:* Check that the product is defined. To find the entry in the nth row, mth column, use the nth row of the left matrix and the mth column of the right matrix. Multiply each pair of corresponding entries and find the sum of these products.

44. a.

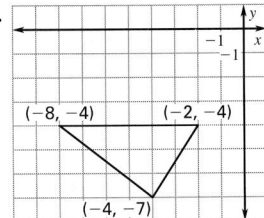

4.4 CONCEPT ACTIVITY (p. 222)

1. $AI = IA = A$; if $A = \begin{bmatrix} a & b \\ c & d \end{bmatrix}$, then $AI = \begin{bmatrix} a & b \\ c & d \end{bmatrix}\begin{bmatrix} 1 & 0 \\ 0 & 1 \end{bmatrix} = \begin{bmatrix} a+0 & 0+b \\ c+0 & 0+d \end{bmatrix} =$

$\begin{bmatrix} a & b \\ c & d \end{bmatrix} = A$, and $IA = \begin{bmatrix} 1 & 0 \\ 0 & 1 \end{bmatrix}\begin{bmatrix} a & b \\ c & d \end{bmatrix} = \begin{bmatrix} a+0 & b+0 \\ 0+c & 0+d \end{bmatrix} = \begin{bmatrix} a & b \\ c & d \end{bmatrix} = A.$

5. No; sample answer: Consider the matrix $\begin{bmatrix} 1 & 1 \\ 0 & 0 \end{bmatrix}$. If $\begin{bmatrix} 1 & 1 \\ 0 & 0 \end{bmatrix}$ has inverse

$\begin{bmatrix} a & b \\ c & d \end{bmatrix}$, then $\begin{bmatrix} 1 & 1 \\ 0 & 0 \end{bmatrix}\begin{bmatrix} a & b \\ c & d \end{bmatrix} = \begin{bmatrix} 1 & 0 \\ 0 & 1 \end{bmatrix}$. But multiplying $\begin{bmatrix} 1 & 1 \\ 0 & 0 \end{bmatrix}$ by

$\begin{bmatrix} a & b \\ c & d \end{bmatrix}$ will yield a matrix with a row of zero entries: $\begin{bmatrix} 1 & 1 \\ 0 & 0 \end{bmatrix}\begin{bmatrix} a & b \\ c & d \end{bmatrix} =$

$\begin{bmatrix} a+c & b+d \\ 0 & 0 \end{bmatrix}$; thus $\begin{bmatrix} 1 & 1 \\ 0 & 0 \end{bmatrix}$ does not have an inverse.

4.5 PRACTICE AND APPLICATIONS (pp. 233–235)

11. $\begin{bmatrix} 1 & 1 \\ 3 & -4 \end{bmatrix}\begin{bmatrix} x \\ y \end{bmatrix} = \begin{bmatrix} 5 \\ 8 \end{bmatrix}$ **12.** $\begin{bmatrix} 1 & 2 \\ 4 & -1 \end{bmatrix}\begin{bmatrix} x \\ y \end{bmatrix} = \begin{bmatrix} 6 \\ 5 \end{bmatrix}$ **13.** $\begin{bmatrix} 5 & -3 \\ -4 & 2 \end{bmatrix}\begin{bmatrix} x \\ y \end{bmatrix} = \begin{bmatrix} 9 \\ 10 \end{bmatrix}$

14. $\begin{bmatrix} 2 & -5 \\ -3 & 7 \end{bmatrix}\begin{bmatrix} x \\ y \end{bmatrix} = \begin{bmatrix} -11 \\ 5 \end{bmatrix}$ **15.** $\begin{bmatrix} 1 & 8 \\ 4 & -5 \end{bmatrix}\begin{bmatrix} x \\ y \end{bmatrix} = \begin{bmatrix} 4 \\ -11 \end{bmatrix}$ **16.** $\begin{bmatrix} 2 & -5 \\ 1 & -3 \end{bmatrix}\begin{bmatrix} x \\ y \end{bmatrix} = \begin{bmatrix} 4 \\ 1 \end{bmatrix}$

17. $\begin{bmatrix} 1 & -4 & 5 \\ 2 & 1 & -7 \\ -4 & 5 & 2 \end{bmatrix}\begin{bmatrix} x \\ y \\ z \end{bmatrix} = \begin{bmatrix} -4 \\ -23 \\ 38 \end{bmatrix}$ **18.** $\begin{bmatrix} 3 & -1 & 4 \\ 2 & 4 & -1 \\ 1 & -1 & 3 \end{bmatrix}\begin{bmatrix} x \\ y \\ z \end{bmatrix} = \begin{bmatrix} 16 \\ 10 \\ 31 \end{bmatrix}$

19. $\begin{bmatrix} 0.5 & 3.1 & -0.2 \\ 1.2 & -2.5 & 0.7 \\ 0.3 & 4.8 & -4.3 \end{bmatrix}\begin{bmatrix} x \\ y \\ z \end{bmatrix} = \begin{bmatrix} 5.9 \\ 2.2 \\ 4.8 \end{bmatrix}$ **20.** $\begin{bmatrix} 1 & 0 & 1 \\ -1 & -1 & 2 \\ 2 & 7 & -1 \end{bmatrix}\begin{bmatrix} x \\ y \\ z \end{bmatrix} = \begin{bmatrix} 9 \\ 6 \\ -4 \end{bmatrix}$

21. $\begin{bmatrix} 0 & 8 & -10 \\ 0 & 6 & -12 \\ -9 & 0 & 5 \end{bmatrix}\begin{bmatrix} x \\ y \\ z \end{bmatrix} = \begin{bmatrix} -23 \\ 14 \\ 0 \end{bmatrix}$ **22.** $\begin{bmatrix} 1 & 1 & -1 \\ 2 & 0 & -1 \\ 0 & 1 & 1 \end{bmatrix}\begin{bmatrix} x \\ y \\ z \end{bmatrix} = \begin{bmatrix} 0 \\ 1 \\ 2 \end{bmatrix}$

4.5 MIXED REVIEW (p. 235)

55. 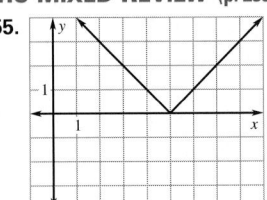 **56.**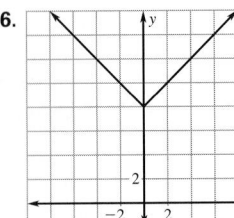

57. 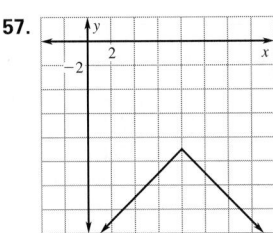 **58.**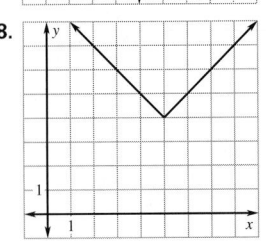

59. 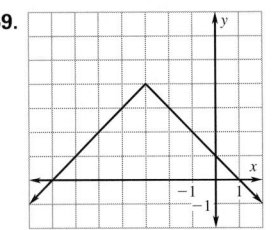 **60.**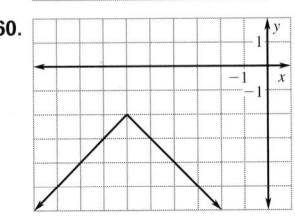

CHAPTER 5

SKILL REVIEW (p. 248)

6. 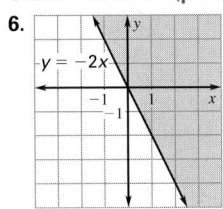 **7.** 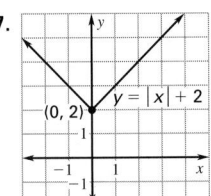 **8.**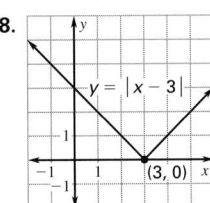

9.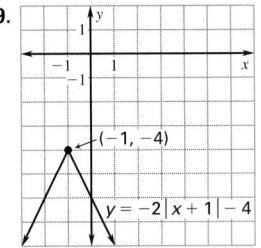

5.1 ACTIVITY (p. 249)

Step 1. 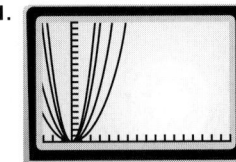 **Step 2.**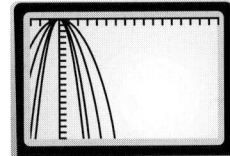

ADDITIONAL ANSWERS

5.1 GUIDED PRACTICE (p. 253)

4.
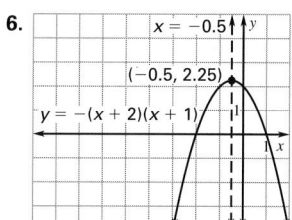
$y = x^2 - 4x + 7$
(2, 3)
$x = 2$

5.
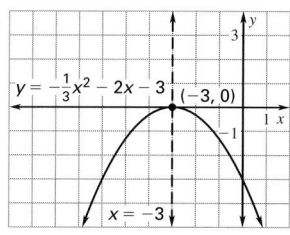
$y = 2(x + 1)^2 - 4$
(-1, -4)
$x = -1$

6.
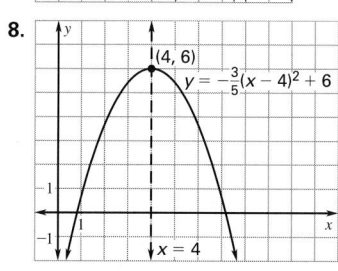
$x = -0.5$
(-0.5, 2.25)
$y = -(x + 2)(x + 1)$

7.
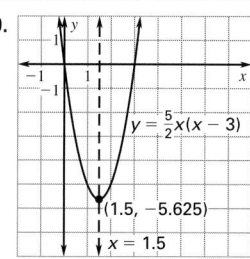
$y = -\frac{1}{3}x^2 - 2x - 3$ (-3, 0)
$x = -3$

8.
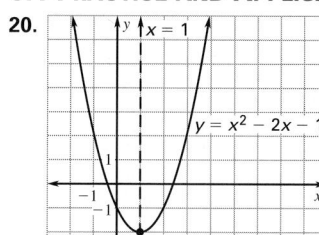
(4, 6)
$y = -\frac{3}{5}(x - 4)^2 + 6$
$x = 4$

9.
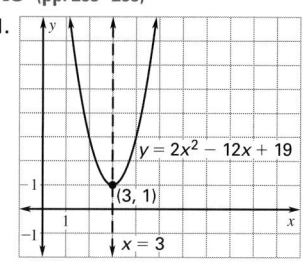
$y = \frac{5}{2}x(x - 3)$
(1.5, -5.625)
$x = 1.5$

5.1 PRACTICE AND APPLICATIONS (pp. 253–255)

20.
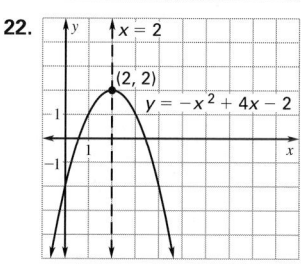
$x = 1$
$y = x^2 - 2x - 1$
(1, -2)

21.
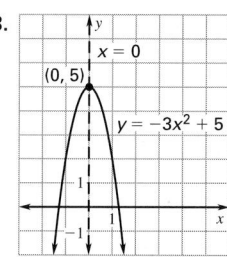
$y = 2x^2 - 12x + 19$
(3, 1)
$x = 3$

22.
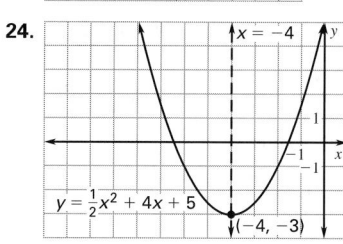
$x = 2$
(2, 2)
$y = -x^2 + 4x - 2$

23.
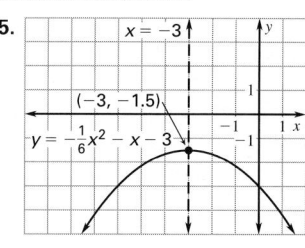
$x = 0$
(0, 5)
$y = -3x^2 + 5$

24.

$x = -4$
$y = \frac{1}{2}x^2 + 4x + 5$
(-4, -3)

25.
$x = -3$
(-3, -1.5)
$y = -\frac{1}{6}x^2 - x - 3$

26.
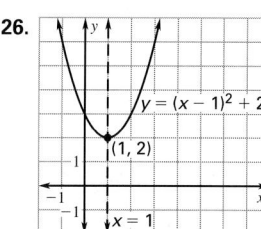
$y = (x - 1)^2 + 2$
(1, 2)
$x = 1$

27.
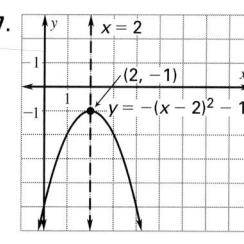
$x = 2$
(2, -1)
$y = -(x - 2)^2 - 1$

28.
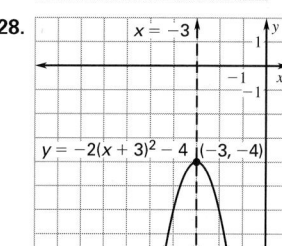
$x = -3$
$y = -2(x + 3)^2 - 4$ (-3, -4)

29.
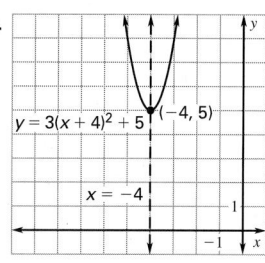
$y = 3(x + 4)^2 + 5$ (-4, 5)
$x = -4$

30.
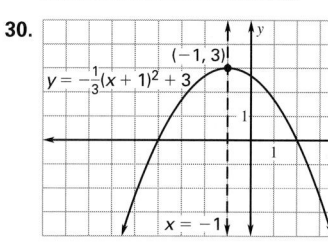
(-1, 3)
$y = -\frac{1}{3}(x + 1)^2 + 3$
$x = -1$

31.
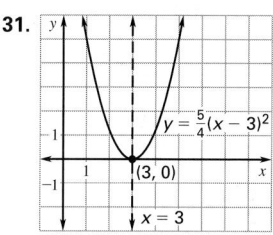
$y = \frac{5}{4}(x - 3)^2$
(3, 0)
$x = 3$

35.
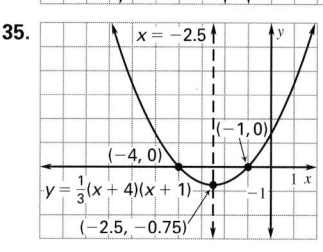
$x = -2.5$
(-4, 0)   (-1, 0)
$y = \frac{1}{3}(x + 4)(x + 1)$
(-2.5, -0.75)

36.
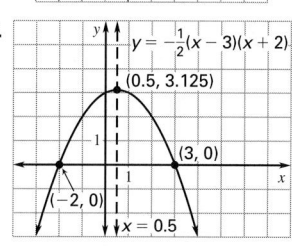
$y = -\frac{1}{2}(x - 3)(x + 2)$
(0.5, 3.125)
(3, 0)
(-2, 0)
$x = 0.5$

37.
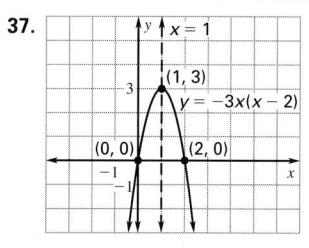
$x = 1$
(1, 3)
$y = -3x(x - 2)$
(0, 0)   (2, 0)

55. c.
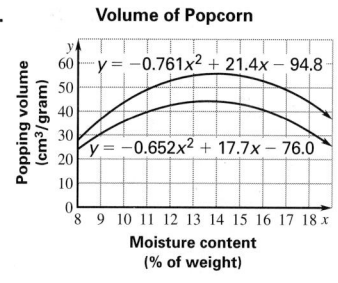
Volume of Popcorn
$y = -0.761x^2 + 21.4x - 94.8$
$y = -0.652x^2 + 17.7x - 76.0$
Popping volume (cm³/gram)
Moisture content (% of weight)

56. $y = ax^2 - 2ahx + ah^2 + k$ and $y = ax^2 - apx - aqx + apq$;

For $y = ax^2 - 2ahx + ah^2 + k$, $a = a$ and $b = -2ah$. Then $x = -\frac{b}{2a}$

(the x-coordinate of the vertex) $= -\frac{-2ah}{2a} = h$. For $y = ax^2 - apx - aqx + apq$,

$a = a$ and $b = -a(p + q)$. Then $x = -\frac{b}{2a}$ (the x-coordinate of the vertex) $=$

$-\frac{-a(p + q)}{2a} = \frac{p + q}{2}$.

ADDITIONAL ANSWERS

5.1 MIXED REVIEW (p. 255)

66.

$(0, 0, 4)$
$x + y + z = 4$
$(0, 4, 0)$
$(4, 0, 0)$

67.

$(0, 0, 3)$
$x + y + 2z = 6$
$(0, 6, 0)$
$(6, 0, 0)$

68.

$(0, 0, 12)$
$3x + 4y + z = 12$
$(0, 3, 0)$
$(4, 0, 0)$

69.

$(0, 0, 5)$
$5x + 5y + 2z = 10$
$(0, 2, 0)$
$(2, 0, 0)$

70.

$(0, 0, 14)$
$2x + 7y + 3z = 42$
$(0, 6, 0)$
$(21, 0, 0)$

71.

$(0, 3, 0)$
$x + 3y - 3z = 9$
$(0, 0, -3)$
$(9, 0, 0)$

5.2 MIXED REVIEW (p. 263)

119.

$x + y = 4$

120.

$2x - y = 6$

121.

$3x + 4y = -12$

122.

$-5x + 3y = 15$

123.

$y = 2$

124.

$y = -3$

125.

$x = -1$

126.

$x = 4$

127.

$y = x^2 - 2$
$(0, -2)$
$x = 0$

128.

$y = 2x^2 - 5$
$(0, -5)$
$x = 0$

129.

$x = 0$
$(0, 3)$
$y = -x^2 + 3$

130.

$y = (x + 1)^2 - 4$
$(-1, -4)$
$x = -1$

131.

$x = 2$
$(2, 1)$
$y = -(x - 2)^2 + 1$

132.

$x = -3$
$(-3, 7)$
$y = -3(x + 3)^2 + 7$

133.

$(0, -1)$ $y = \frac{1}{4}x^2 - 1$
$x = 0$

134.

$x = 4$
$(4, -6)$
$y = \frac{1}{2}(x - 4)^2 - 6$

135.

$\left(1, 2\frac{2}{3}\right)$
$y = -\frac{2}{3}(x + 1)(x - 3)$
$x = 1$

5.4 MIXED REVIEW (p. 280)

114.

Number of DVD players sold (in thousands)
$N = 35.33t - 14.58$
Time (in months)

5.5 MIXED REVIEW (p. 289)

115.

$x + y = 4$
$x = 0$

116.

$y = x - 2$
$x - 3y = 6$

117.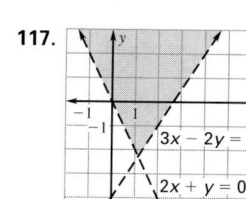

$3x - 2y = 8$
$2x + y = 0$

118.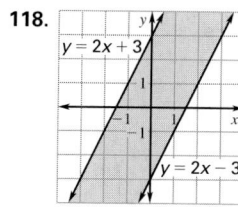

$y = 2x + 3$
$y = 2x - 3$

5.6 MIXED REVIEW (p. 298)

98.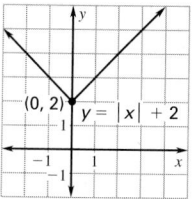

$(0, 2)$
$y = |x| + 2$

99.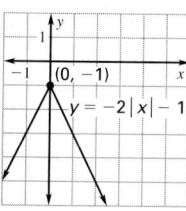

$(0, -1)$
$y = -2|x| - 1$

100.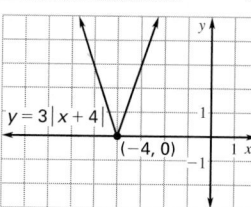

$y = 3|x + 4|$
$(-4, 0)$

101.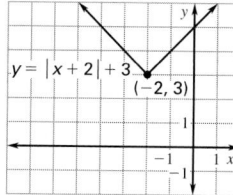

$y = |x + 2| + 3$
$(-2, 3)$

102.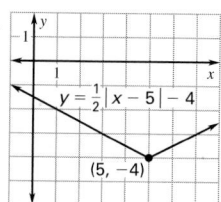

$y = \frac{1}{2}|x - 5| - 4$
$(5, -4)$

QUIZ 2 (p. 298)

5–10.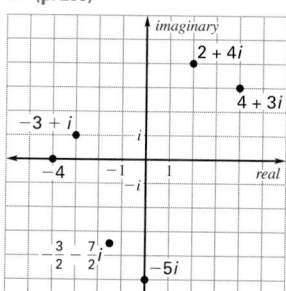

imaginary
$2 + 4i$
$4 + 3i$
$-3 + i$
-4
$-\frac{3}{2} - \frac{7}{2}i$
$-5i$
real

5.7 GUIDED PRACTICE (p. 303)

4.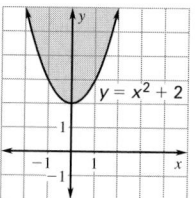

$y = x^2 + 2$

5.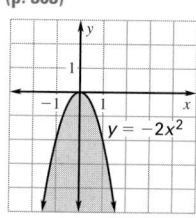

$y = -2x^2$

6.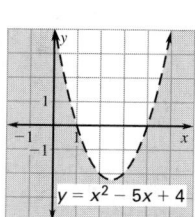

$y = x^2 - 5x + 4$

7.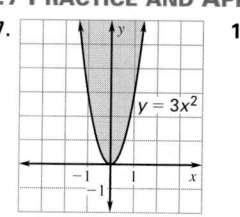

$y = x^2 + 2x - 4$
$y = -x^2 + 3$

8.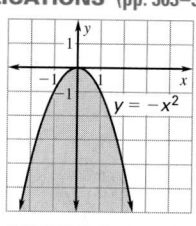

$y = x^2 + 2x - 4$
$y = -x^2 + 3$

9.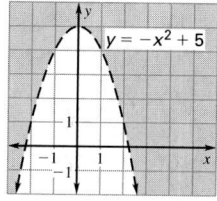

$y = x^2 + 2x - 4$
$y = -x^2 + 3$

5.7 PRACTICE AND APPLICATIONS (pp. 303–305)

17.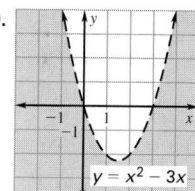

$y = 3x^2$

18.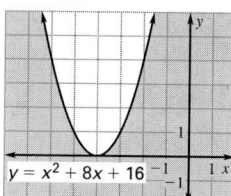

$y = -x^2$

19.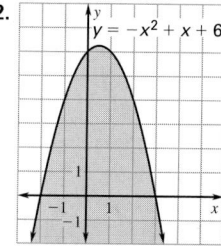

$y = -x^2 + 5$

20.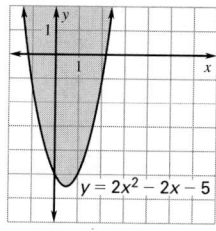

$y = x^2 - 3x$

21.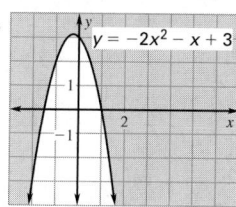

$y = x^2 + 8x + 16$

22.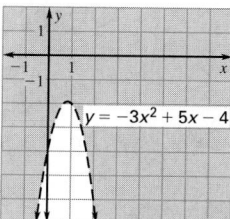

$y = -x^2 + x + 6$

23.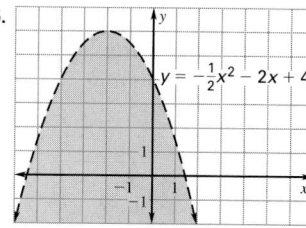

$y = 2x^2 - 2x - 5$

24.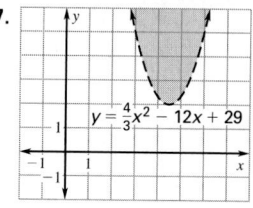

$y = -2x^2 - x + 3$

25.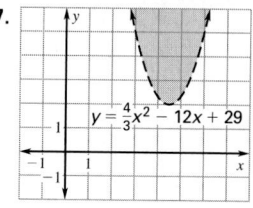

$y = -3x^2 + 5x - 4$

26.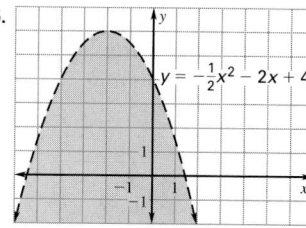

$y = -\frac{1}{2}x^2 - 2x + 4$

27.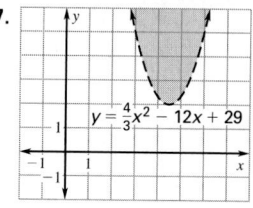

$y = \frac{4}{3}x^2 - 12x + 29$

ADDITIONAL ANSWERS

28.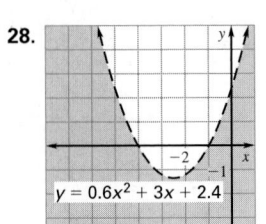

29.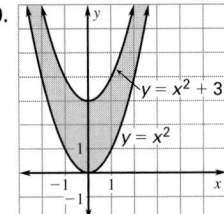

30.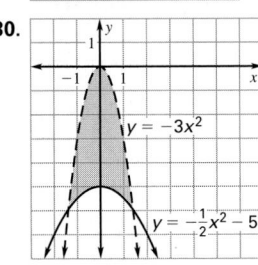

31.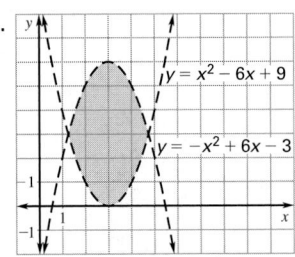

32.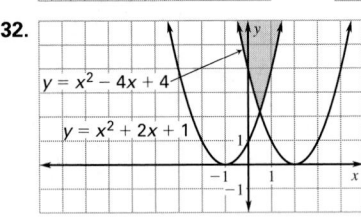

33.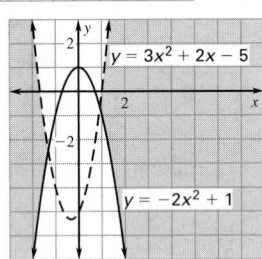

34.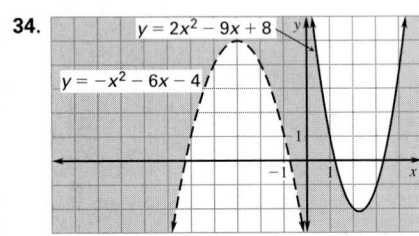

47.

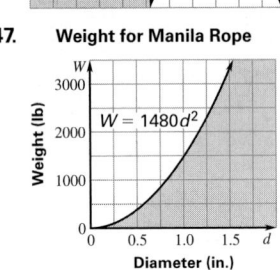
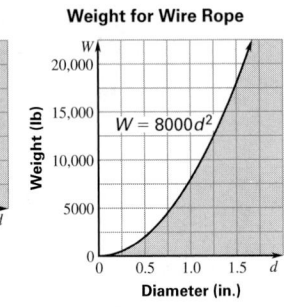

49.

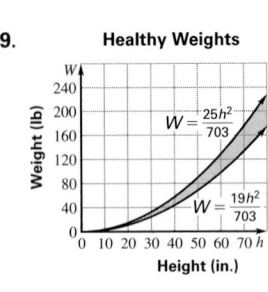

52. a.

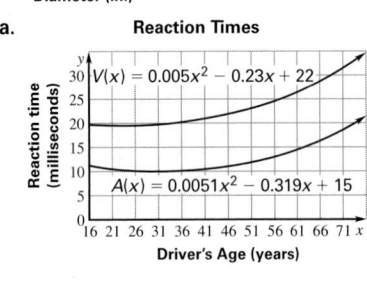

CHAPTER 6

SKILL REVIEW (p. 322)

4. 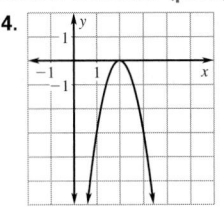 **5.**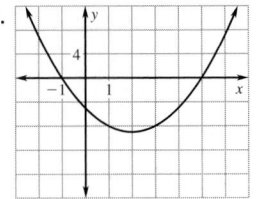

6.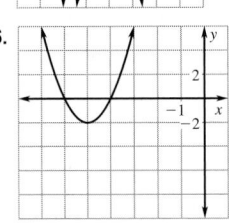

6.1 PRACTICE AND APPLICATIONS (pp. 326–328)

57. a–b.

State	Total area (acres)	Amount of park space (acres)	Park space / Total area
AK	3.937472×10^8	3.25×10^6	0.00825
CA	1.01676×10^8	1.345×10^6	0.01323
CT	3.548×10^6	1.76×10^5	0.04961
KS	5.266×10^7	2.9×10^4	0.00055
OH	2.869×10^7	2.04×10^5	0.00711
PA	2.9477×10^7	2.83×10^5	0.00960

c. A good answer should include the fact that the writer's state has 0.78% parkland, and compare that percentage to the percentages for other states.

58. $\dfrac{a^m}{a^n} = a^{m-n}$ Quotient of powers property

$\dfrac{a^0}{a^n} = a^{0-n}$ Substitute 0 for m.

$\dfrac{1}{a^n} = a^{-n}$ Zero exponent property

59. $\dfrac{a^m}{a^n} = a^m \cdot \dfrac{1}{a^n}$ Multiplication of fractions

$= a^m a^{-n}$ Negative exponent property

$= a^{m+(-n)}$ Product of powers property

$= a^{m-n}$ Simplify exponent.

6.1 MIXED REVIEW (p. 328)

60. 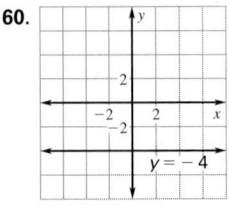 **61.** 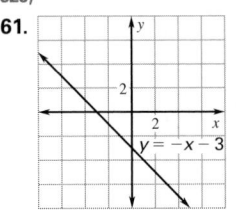 **62.**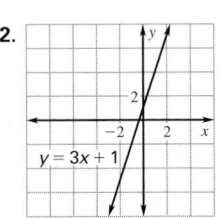

ADDITIONAL ANSWERS

63.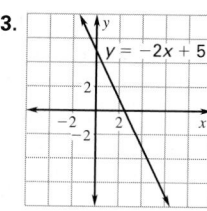
$y = -2x + 5$

64.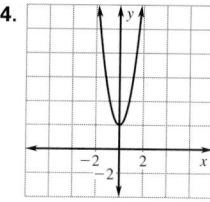

65.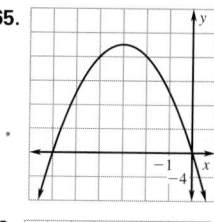

66.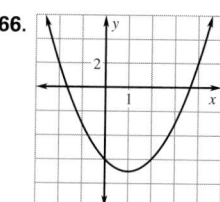

67.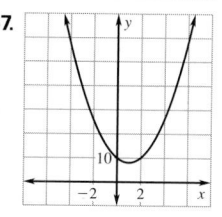

68.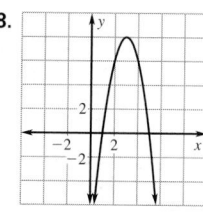

71.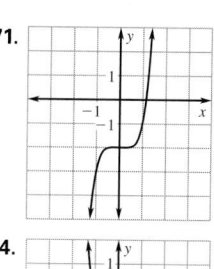

72.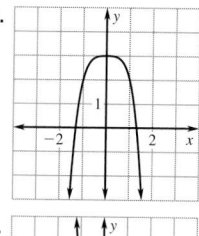

73.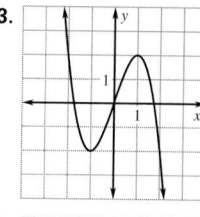

74.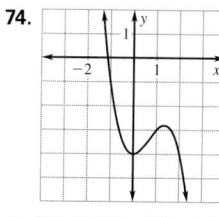

75.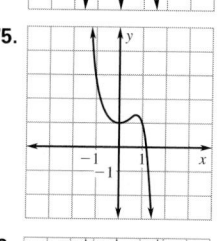

76.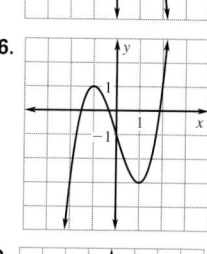

77.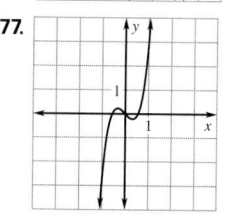

78.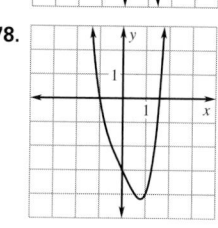

79.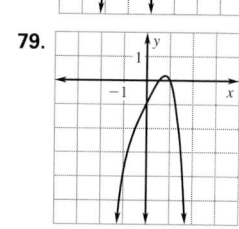

6.2 GUIDED PRACTICE (p. 333)

2.

```
-2 | 3   1   -9    2
   |    -6   10   -2
     3  -5    1    0
```

Bring down the 3; multiply $-2 \cdot 3$ and write -6 in the second column; add $1 + (-6)$ and write -5 at the bottom of the column; multiply $-2 \cdot (-5)$ and write 10 in the next column; add $-9 + 10$ and write 1 at the bottom of the column; multiply $-2 \cdot 1$ and write -2 in the next column; add $2 + (-2)$ to get 0.

6.2 PRACTICE AND APPLICATIONS (pp. 333–336)

47.

Function	As $x \to -\infty$	As $x \to +\infty$
$f(x) = -5x^3$	$f(x) \to +\infty$	$f(x) \to -\infty$
$f(x) = -x^3 + 1$	$f(x) \to +\infty$	$f(x) \to -\infty$
$f(x) = 2x - 3x^3$	$f(x) \to +\infty$	$f(x) \to -\infty$
$f(x) = 2x^2 - x^3$	$f(x) \to +\infty$	$f(x) \to -\infty$

48.

Function	As $x \to -\infty$	As $x \to +\infty$
$f(x) = x^4 + 3x^3$	$f(x) \to +\infty$	$f(x) \to +\infty$
$f(x) = x^4 + 2$	$f(x) \to +\infty$	$f(x) \to +\infty$
$f(x) = x^4 - 2x - 1$	$f(x) \to +\infty$	$f(x) \to +\infty$
$f(x) = 3x^4 - 5x^2$	$f(x) \to +\infty$	$f(x) \to +\infty$

65.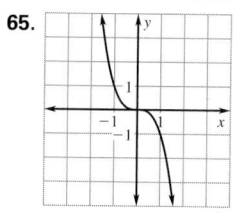

66.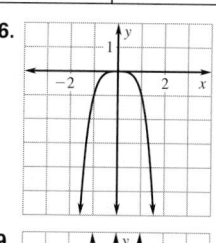

67.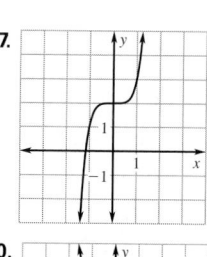

68.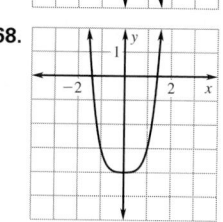

69.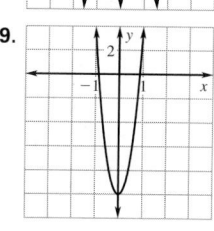

70.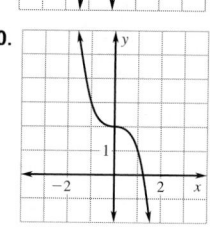

84.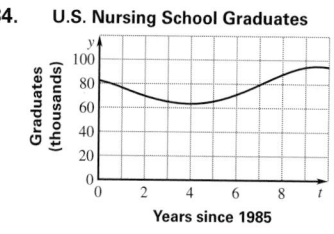
U.S. Nursing School Graduates

86.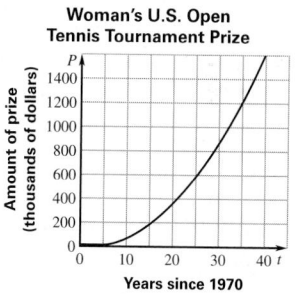
Woman's U.S. Open Tennis Tournament Prize

87. c.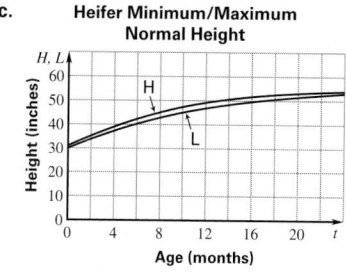
Heifer Minimum/Maximum Normal Height

d. Answers may vary. *Sample answer:* about 7.5 months; if a heifer's height is modeled by the maximum normal height function, then the heifer will be 43 in. high at about 6.6 months. If a heifer's height is modeled by the minimum normal height function, then the heifer will be 43 in. high at about 8.3 months. So I used the average of 6.6 and 8.3 months, which is about 7.5 months.

88.

x	$f(x)$	$g(x)$	$\dfrac{f(x)}{g(x)}$
50	125,000	120,205	1.03989
100	1,000,000	980,405	1.01999
500	1.25×10^8	124,502,005	1.004
1000	10^9	998,004,005	1.002
5000	1.25×10^{11}	1.2495×10^{11}	1.0004

QUIZ 1 (p. 344)

18.

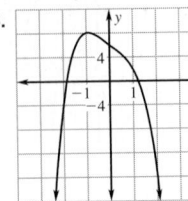

6.4 PRACTICE AND APPLICATIONS (pp. 348–350)

41. $(x + 1)(x^2 + 1)$ **42.** $(x + 2)(10x^2 + 1)$ **43.** $(x + 3)(x^2 + 10)$
44. $(x - 2)(x^2 + 4)$ **45.** $(2x - 5)(x^2 + 9)$ **46.** $(2x^2 + 3)(-x - 2)$
47. $(x - 2)(3x^2 + 1)$ **48.** $(2x - 1)(x^2 + 1)$ **49.** $(3x - 2)(x^2 - 3)$
50. $(2x - 1)(2x + 1)(4x^2 + 1)$ **51.** $(x^2 + 1)(x^2 + 2)$ **52.** $(x - 3)(x + 3)(x^2 + 9)$
53. $(3x - 4)(3x + 4)(9x^2 + 16)$ **54.** $(2x - 3)(2x + 3)(x^2 + 1)$
55. $(x^2 + 2)(x^2 + 8)$ **56.** $(3 - 2x)(3 + 2x)(9 + 4x^2)$
57. $2x^2(2x - 1)(2x + 1)(4x^2 + 1)$ **58.** $3x(x + 3)(x - 3)(2x^2 + 1)$
59. $(2x^2 + 3)(9x - 1)$ **60.** $3x(x + 1)(2x + 5)$ **61.** $(2x + 1)(2x - 1)(x^2 + 10)$
62. $(2x - 1)(2x + 1)(2x - 3)$ **63.** $8(x - 2)(x^2 + 2x + 4)$
64. $3x^2(x + 10)(x - 10)$ **65.** $3x(x - 2)(x^2 + 2x + 4)$ **66.** $(5x^2 + 1)(x^2 + 6)$
67. $x(3x^2 + 1)(x + 3)$

6.5 PRACTICE AND APPLICATIONS (pp. 356–358)   **38.** $4x^3 + x^2 + x - 1$

39. $(x + 2)(x - 3)(x - 4)$ **40.** $(x - 6)(x + 1)(x + 2)$ **41.** $(x - 10)(x - 4)(x + 2)$
42. $(x - 9)(x - 2)(x - 7)$ **43.** $(x + 5)(x - 3)^2$ **44.** $(x - 8)(x - 5)(x + 2)$
45. $(x - 1)(2x + 3)(2x - 3)$ **46.** $(x + 6)(2x + 1)(x - 3)$ **65.** $6x^2 - 7x + 6 -$
$\dfrac{4}{2x + 1}$, $4x^2 - 4x + 3 - \dfrac{1}{3x + 1}$, $3x^2 - \dfrac{11}{4}x + \dfrac{31}{16} + \dfrac{1}{16(4x + 1)}$;

$12x^2 - 14x + 12 - \dfrac{4}{x + \frac{1}{2}}$, $12x^2 - 12x + 9 - \dfrac{1}{x + \frac{1}{3}}$, $12x^2 - 11x + \dfrac{31}{4} +$

$\dfrac{1}{16\left(x + \frac{1}{4}\right)}$; Remainders of each pair are the same; the coefficients are 2, 3, and 4 times greater with the fractions.

6.5 MIXED REVIEW (p. 358)   **66.** Both are solutions.   **67.** Both are solutions.
68. $(2, 2)$ is a solution, but $(-1, -4)$ is not a solution.   **69.** $(1, 4)$ is a solution, but $(2, 0)$ is not a solution.

6.6 PRACTICE AND APPLICATIONS (pp. 362–364)   **21.** $\pm 1, \pm 3, \pm\frac{1}{2}, \pm\frac{3}{2}$,
$\pm\frac{1}{4}, \pm\frac{3}{4}, \pm\frac{1}{8}, \pm\frac{3}{8}$ **22.** $\pm 1, \pm 3, \pm 5, \pm 15, \pm\frac{1}{3}, \pm\frac{5}{3}$ **70.** No, no; if a cubic polynomial had 4 or more distinct real zeros, then there would be 4 or more binomials of the form $x - a$ that divide the polynomial to give a zero remainder. This would imply that the polynomial has degree 4 or greater. However, this is impossible since the polynomial is a cubic polynomial. So a cubic polynomial has at most 3 real zeros. As $x \to -\infty$ and $x \to +\infty$, the values of a cubic polynomial approach $-\infty$ and $+\infty$, respectively, or else $+\infty$ and $-\infty$. At some value of x, therefore, the graph is below the x-axis, and at some other value of x, the graph is above the x-axis. This means that the graph crosses the x-axis somewhere between these two values, and the x-coordinate of the point where the graph crosses the x-axis is a zero.

ADDITIONAL ANSWERS

6.7 GUIDED PRACTICE (p. 369)   **2.** Suppose there is a complex root $a + bi$ with $b \neq 0$. Since $a - bi$ must also be a zero, $f(x)$ has four zeros. But this is not possible, since the degree of $f(x)$ is 3. Therefore the remaining root is a complex number $a + bi$ for which $b = 0$, which means that the root is a real number.   **8.** $f(x) = x^3 - x^2 - 6x$ **9.** $f(x) = x^4 - 2x^3 + 2x^2 - 2x + 1$
10. $f(x) = x^3 - 9x^2 + 33x - 65$ **11.** $f(x) = x^5 - 3x^4 - 5x^3 + 15x^2 + 4x - 12$
12. $f(x) = x^4 + x^3 - 6x^2 - 14x - 12$ **13.** $f(x) = x^4 + 32x^2 + 256$

6.7 MIXED REVIEW (p. 371)

64.

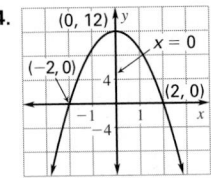

65.

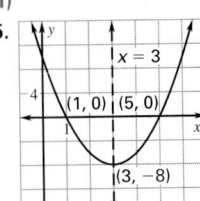

66.

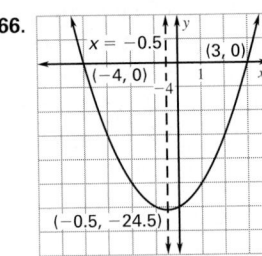

67.

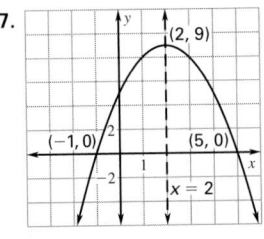

68.

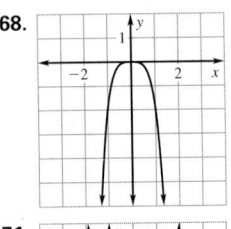

69.

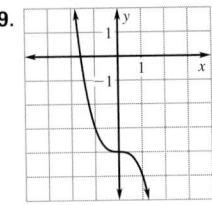

70.

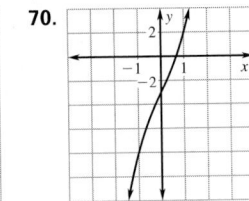

71.

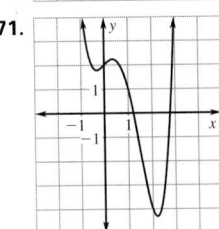

6.8 GUIDED PRACTICE (p. 376)

5.

6.

7.

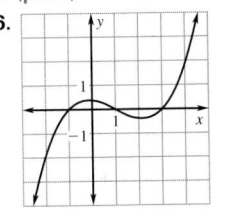

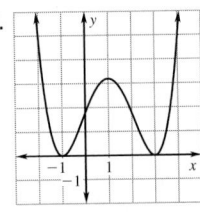

6.8 PRACTICE AND APPLICATIONS (pp. 376–378)

13.

14.

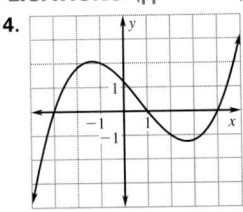

AA22 *Additional Answers*

15. 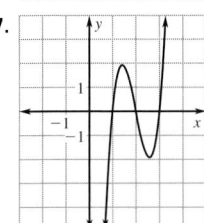 **16.**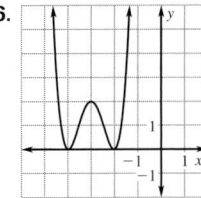

17. 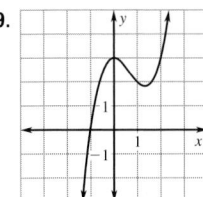 **18.**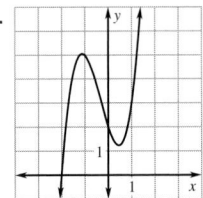

19. 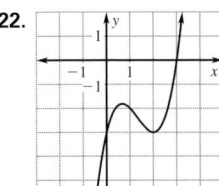 **20.** 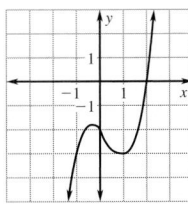 **21.**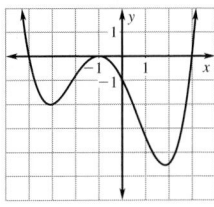

22.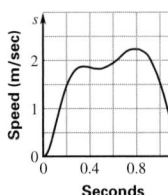

29. x-intercepts: $-1.79, 0.11, 1.67$; local maximum: $(-1, 7)$; local minimum: $(1, -5)$

30. x-intercepts: $-2, 1$; local maximum: $(1, 0)$; local minimum: $\left(-1, -\dfrac{4}{3}\right)$

31. x-intercepts: $-2.83, 0, 2.83$; local maximums: $(-2, 4)$, $(2, 4)$; local minimum: $(0, 0)$

32. x-intercepts: $-1.73, 0, 1.73$; local maximums: $(-1.73, 0)$, $(0.77, 4.46)$; local minimums: $(-0.77, -4.46)$, $(1.73, 0)$ **33.** x-intercepts: $-2, -1, 0, 1, 2$; local maximums: $(-1.64, 3.63)$, $(0.54, 1.42)$; local minimums: $(-0.54, -1.42)$, $(1.64, -3.63)$ **34.** x-intercepts: $-1.53, -0.35, 1.88, 2$; local maximum: $(0.61, 3.62)$; local minimums: $(-1.05, -3.03)$, $(1.94, -0.03)$

35. Speed of Swimmer

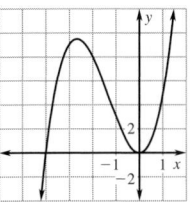

36. reaches a local maximum at $(1.85, 14.04)$ and a minimum at $(4.25, 11.98)$; from 1991 until about 1993, the number of pounds of oranges eaten per person increased to about 14 lb. This amount then declined to about 12 lb in 1995, when it began to increase again.

44. 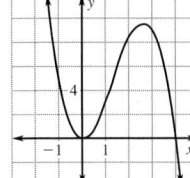 ; The graph of $f(x)$ reflected over the x-axis is the graph of $-f(x)$. Same x-intercepts; local max (x, y) becomes local min $(x, -y)$ and local min (x, y) becomes local max $(x, -y)$.

; The graph of $f(x)$ reflected over the y-axis is the graph of $f(-x)$. The x-intercepts of $f(-x)$ are the opposites of the x-intercepts of $f(x)$, local max (x, y) is now local max $(-x, y)$, and local min (x, y) is now local min $(-x, y)$.

6.9 CONCEPT ACTIVITY (p. 379)

Steps 1–4.
For the function $g(n)$: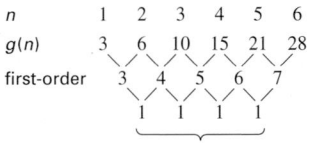

Second-order differences are constant.

For the function $h(n)$: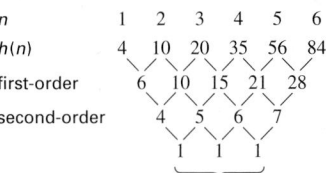

Third-order differences are constant.

Drawing Conclusions 1. a. $f(1)\ f(2)\ f(3)\ f(4)\ f(5)\ f(6)$

      4   7   10   13   16   19
        3   3   3   3   3

b. $f(1)\ f(2)\ f(3)\ f(4)\ f(5)\ f(6)$

   0   5   12 21 32 45
    5   7   9 11 13
     2   2   2   2

c. $f(1)\ f(2)\ f(3)\ f(4)\ f(5)\ f(6)$

   5 18 55 128 249 430
  13 37 73 121 181
   24 36 48 60
    12 12 12

3. It should fit the pattern of the previous examples; fourth-order differences have a constant value of 24.

$f(1)\ f(2)\ f(3)\ f(4)\ f(5)\ f(6)$

  2 18 84 260 630 1302
   16 66 176 370 672
    50 110 194 302
     60 84 108
      24 24

6.9 GUIDED PRACTICE (p. 383)

6. $f(1)\ f(2)\ f(3)\ f(4)\ f(5)\ f(6)$

  1 11 35 79 149 251
   10 24 44 70 102
    14 20 26 32
     6 6 6

7. $f(1)\ f(2)\ f(3)\ f(4)\ f(5)\ f(6)$

  3 20 87 264 635 1308
   17 67 177 371 673
    50 110 194 302
     60 84 108
      24 24

8. $f(1)\ f(2)\ f(3)\ f(4)\ f(5)\ f(6)$

 $-12\ -39\ -66\ -81\ -72\ -27$
  $-27\ -27\ -15\ \ 9\ \ 45$
   $0\ \ 12\ \ 24\ \ 36$
    $12\ \ 12\ \ 12$

6.9 PRACTICE AND APPLICATIONS (pp. 383–385)

17. $f(x) = -\dfrac{1}{2}x(x + 1)(x + 2)$ **18.** $f(x) = -\dfrac{1}{8}(x - 3)(x - 2)(x + 3)$

19. $f(x) = -\dfrac{1}{4}(x - 1)(x - 3)(x + 2)$ **20.** $f(x) = -\dfrac{3}{16}(x + 1)(x + 4)(x - 4)$

21. $f(x) = (x - 3)(x - 2)(x + 1)$ **22.** $f(x) = \dfrac{3}{14}x(x + 3)(x - 5)$

23. $f(1)\ f(2)\ f(3)\ f(4)\ f(5)\ f(6)$

  5 5 7 11 17 25
   0 2 4 6 8
    2 2 2 2

24. $f(1)\ f(2)\ f(3)\ f(4)\ f(5)\ f(6)$

 $-4\ \ -6\ \ 6\ \ 44\ \ 120\ \ 246$
  $-2\ \ 12\ \ 38\ \ 76\ \ 126$
   $14\ \ 26\ \ 38\ \ 50$
    $12\ \ 12\ \ 12$

ADDITIONAL ANSWERS

25. f(1) f(2) f(3) f(4) f(5) f(6)

−3 −3 −9 −27 −63 −123
  0 −6 −18 −36 −60
   −6 −12 −18 −24
    −6 −6 −6

26. f(1) f(2) f(3) f(4) f(5) f(6)

−2 −8 0 64 250 648
 −6 8 64 186 398
  14 56 122 212
   42 66 90
    24 24

27. f(1) f(2) f(3) f(4) f(5) f(6)

−18 −8 102 432 1150 2472
  10 110 330 718 1322
   100 220 388 604
    120 168 216
     48 48

28. f(1) f(2) f(3) f(4) f(5) f(6)

3 −8 −27 −54 −89 −132
 −11 −19 −27 −35 −43
  −8 −8 −8 −8

29. f(1) f(2) f(3) f(4) f(5) f(6)

4   4 −36 −176 −500 −1116
  0 −40 −140 −324 −616
  −40 −100 −184 −292
   −60 −84 −108
    −24 −24

30. f(1) f(2) f(3) f(4) f(5) f(6)

−4 2 34 110 248 466
  6 32 76 138 218
  26 44 62 80
   18 18 18

31. f(1) f(2) f(3) f(4) f(5) f(6)

3 −2 −13 −30 −53 −82
 −5 −11 −17 −23 −29
  −6 −6 −6 −6

CHAPTER 6 TEST (p. 391)

6. $f(x) \to +\infty$ as $x \to -\infty$, $f(x) \to +\infty$ as $x \to +\infty$

x	−4	−3	−2	−1	0	1	2	3	4
$f(x)$	227	65	9	−1	−1	−3	5	59	219

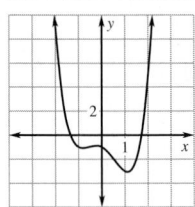

7. $f(x) \to +\infty$ as $x \to -\infty$, $f(x) \to -\infty$ as $x \to +\infty$

x	−4	−3	−2	−1	0	1	2	3	4
$f(x)$	96	27	0	−3	0	−9	−48	−135	−288

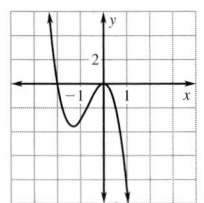

8. $f(x) \to -\infty$ as $x \to -\infty$, $f(x) \to +\infty$ as $x \to +\infty$

x	−4	−3	−2	−1	0	1	2	3	4
$f(x)$	−42	−12	0	0	−6	−12	−12	0	30

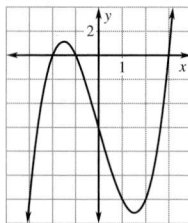

CUMULATIVE PRACTICE, CHAPTERS 1–6 (pp. 394–395)

29.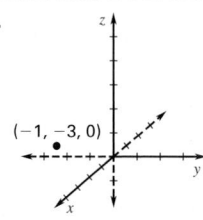
(−1, −3, 0)

30.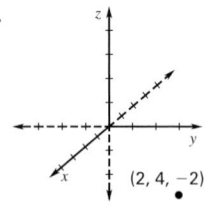
(2, 4, −2)

31.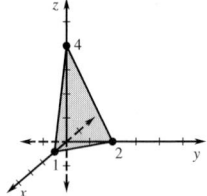

32.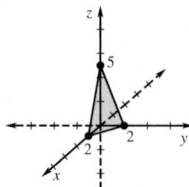

33. $\begin{bmatrix} -11 & 8 \\ 1 & -2 \end{bmatrix}$

35. $\begin{bmatrix} 17 & -7 & -27 \\ 3 & -9 & 69 \end{bmatrix}$

44.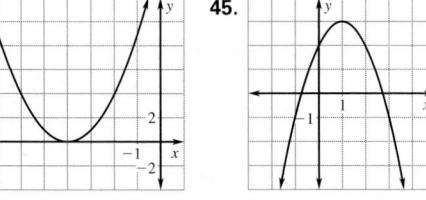

45.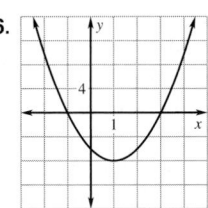

46.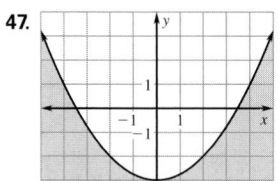

47.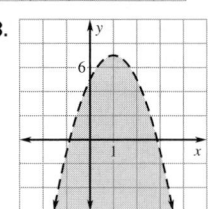

48.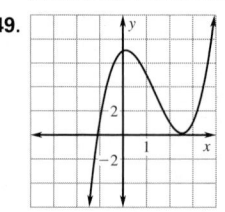

49.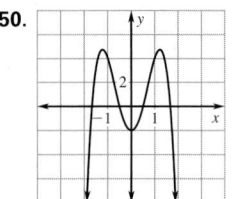

50.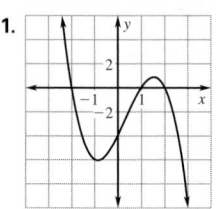

51.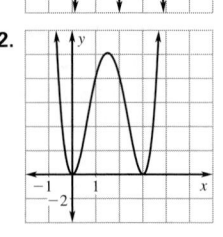

52.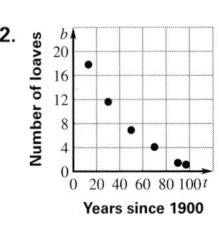

82.
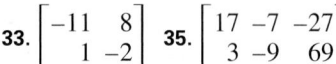
Number of loaves / Years since 1900

CHAPTER 7

THINK & DISCUSS (p. 399) **2.** *Sample answer:* I found an equation for the line containing (348, 1400) and (504, 3800). Then I found the second coordinate of the point on this line with *x*-coordinate 400.

7.3 PRACTICE AND APPLICATIONS (pp. 418–420) **12.** $2x^2 - 5x + 4$; all real numbers **13.** $2x^2 - 5x + 4$; all real numbers **14.** $2x^2 - 10x + 16$; all real numbers **15.** $2x^2 - 8$; all real numbers **16.** $-5x + 12$; all real numbers **17.** $5x - 12$; all real numbers **20.** $6x^{7/6}$; nonnegative real numbers **21.** $6x^{7/6}$; nonnegative real numbers **22.** $4x^{4/3}$; all real numbers **23.** $9x$; nonnegative real numbers **24.** $\dfrac{2x^{1/6}}{3}$; positive real numbers **25.** $\dfrac{3}{2x^{1/6}}$; positive real numbers **26.** 1; all real numbers except 0 **27.** 1; positive real numbers **28.** $4x^{-15/4}$; positive real numbers **29.** $2^{3/2}x^{-15/4}$; positive real numbers **30.** $\dfrac{x^{25}}{256}$; all real numbers except 0 **31.** $x^{9/16}$; nonnegative real numbers **32.** $11x + 4$; all real numbers **33.** $9x - 4$; all real numbers **34.** $10x^2 + 40x$; all real numbers **35.** $\dfrac{10x}{x + 4}$; all real numbers except -4 **36.** $10x + 40$; all real numbers **37.** $10x + 4$; all real numbers **38.** $100x$; all real numbers **39.** $x + 8$; all real numbers

7.3 MIXED REVIEW (p. 420)

74. 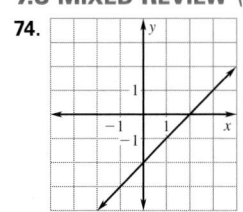 **75.** 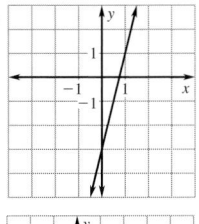 **76.**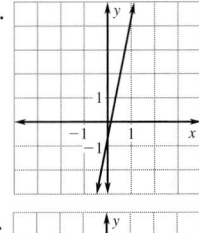

77. 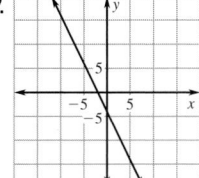 **78.** 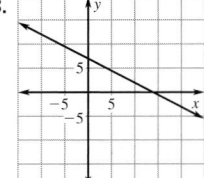 **79.**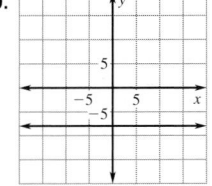

7.4 CONCEPT ACTIVITY (p. 421)

Step 1. 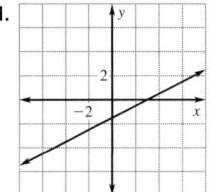 **Step 2.**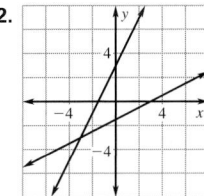

For $f(x) = 2x + 5$:

1. 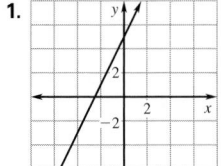 ; 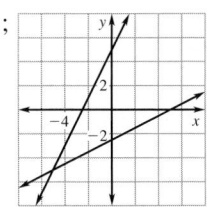 ; $g(x) = \dfrac{x - 5}{2}$

For $f(x) = \dfrac{x - 2}{4}$:

1. 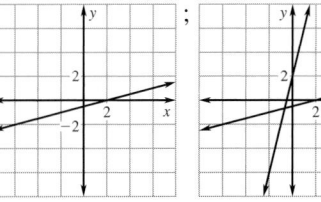 ; $g(x) = 4x + 2$

For $f(x) = 5 - \dfrac{5}{2}x$:

1. 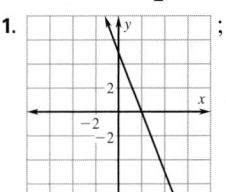 ; $g(x) = \dfrac{2}{5}(5 - x)$

7.4 GUIDED PRACTICE (p. 426)

4.

x	-1	-2	-3	-4	-5
y	1	2	3	4	5

5.

x	2	1	0	1	2
y	-4	-2	0	2	4

7.4 PRACTICE AND APPLICATIONS (pp. 426–428)

14.

x	3	-1	6	-3	9
y	1	4	1	0	1

15.

x	0	3	-2	2	-1
y	1	-2	4	2	-2

48. 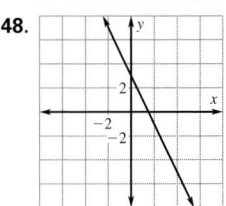 **49.** 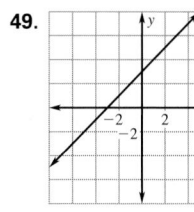 **50.**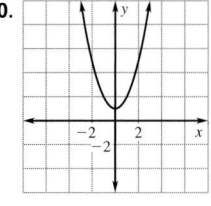

51. 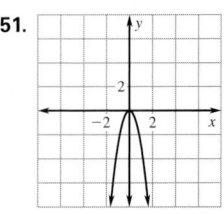 **52.** 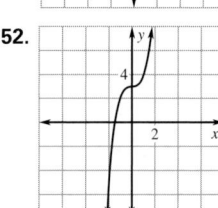 **53.**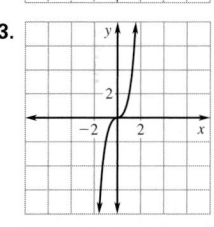

54. 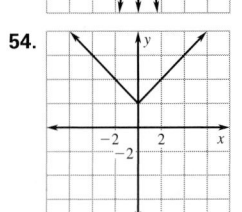 **55.** 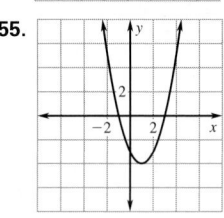 **56.**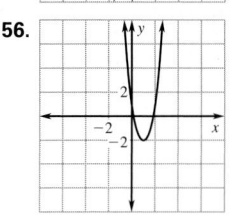

7.4 MIXED REVIEW (p. 429)

69. 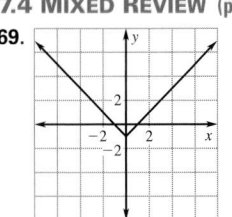 **70.** 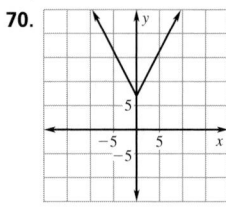 **71.**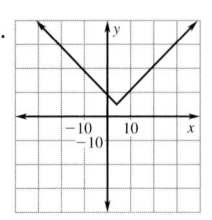

ADDITIONAL ANSWERS

72. 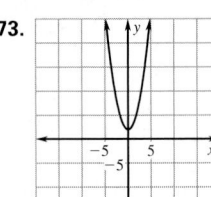 **73.** 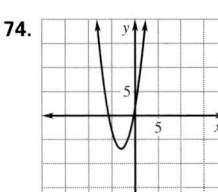 **74.**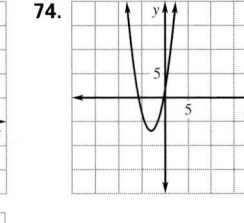

75. 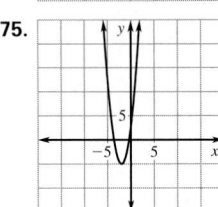 **76.**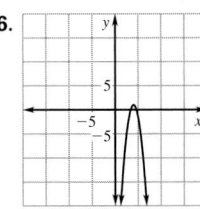

9.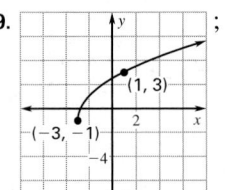

$x \geq -3,\ y \geq -1$

10.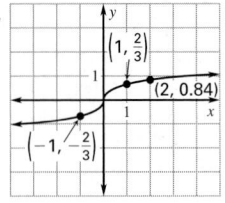

$x,\ y$: all real numbers.

11.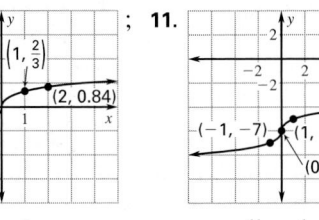

$x,\ y$: all real numbers.

12.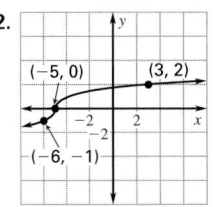

$x,\ y$: all real numbers.

13.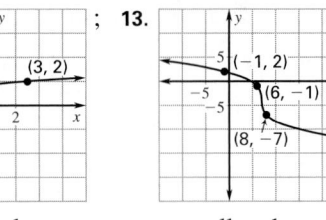

$x,\ y$: all real numbers.

QUIZ 2 (p. 429)

14. 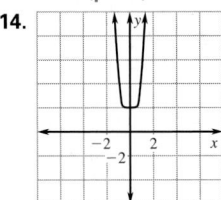 **15.** 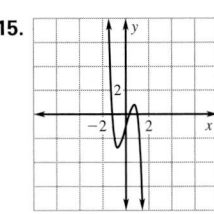 **16.**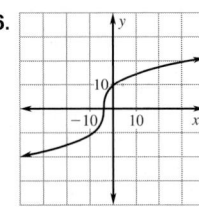

7.5 PRACTICE AND APPLICATIONS (pp. 434–436)

15. Shift graph 14 units left. **16.** Shift graph 10 units right and 3 units down. **17.** Shift graph 10 units down. **18.** Shift graph 6 units left and 5 units down.

22.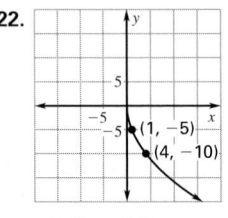

$x \geq 0,\ y \leq 0$

23.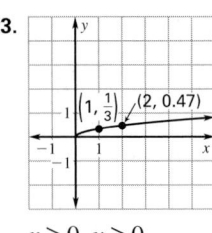

$x \geq 0,\ y \geq 0$

24.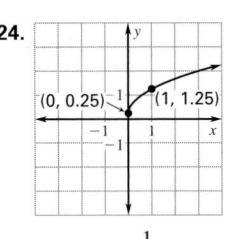

$x \geq 0,\ y \geq \dfrac{1}{4}$

7.5 ACTIVITY (p. 431)

1. 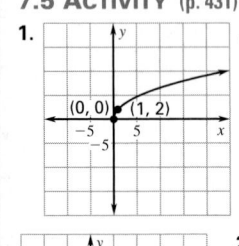 ,

2. 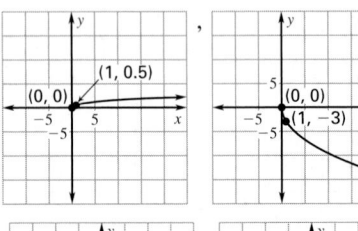 ,

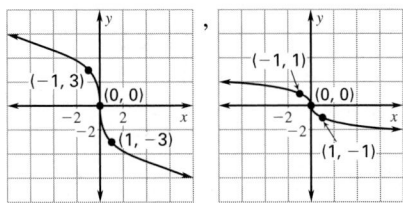 ,

25.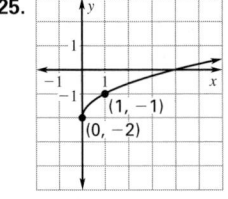

$x \geq 0,\ y \geq -2$

26.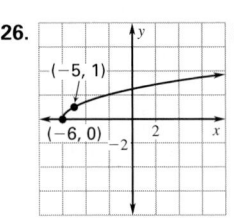

$x \geq -6,\ y \geq 0$

27.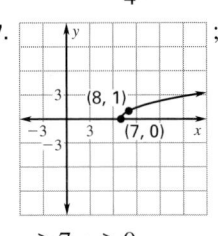

$x \geq 7,\ y \geq 0$

28.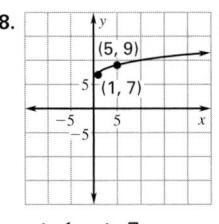

$x \geq 1,\ y \geq 7$

29.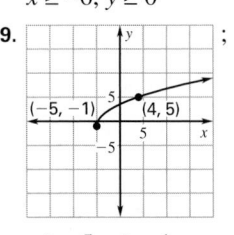

$x \geq -5,\ y \geq -1$

30.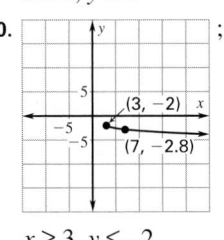

$x \geq 3,\ y \leq -2$

7.5 GUIDED PRACTICE (p. 434)
2. The graph should begin at (1, 2) and pass through (2, 3). **3.** The graph should pass through (–3, –4), (–2, –3), and (–1, –2). **4.** Shift the graph 5 units left. **5.** Shift the graph 10 units down.

6. 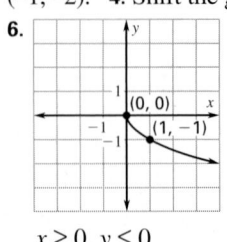 **7.** 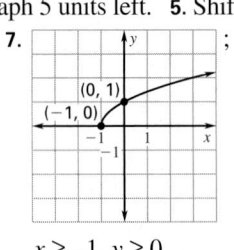 **8.**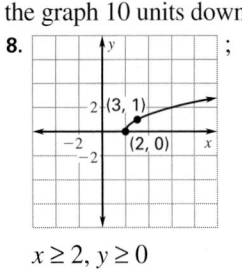

$x \geq 0,\ y \leq 0$ $x \geq -1,\ y \geq 0$ $x \geq 2,\ y \geq 0$

31. 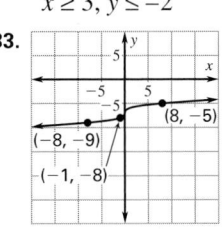 **32.**  **33.**

AA26 *Additional Answers*

34.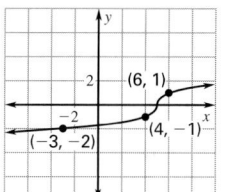

35.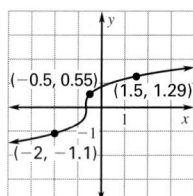

36.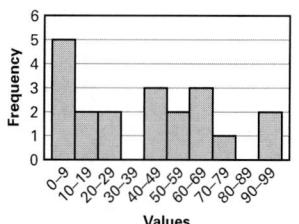

30.

0–9	5	50–59	2
10–19	2	60–69	3
20–29	2	70–79	1
30–39	0	80–89	0
40–49	3	90–99	2

37.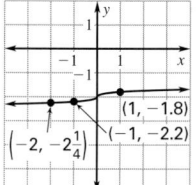

38.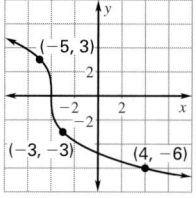

39.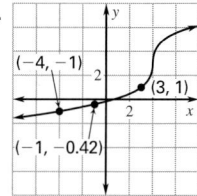

31.

0–0.4	0	4.0–4.4	1
0.5–0.9	0	4.5–4.9	1
1.0–1.4	0	5.0–5.4	2
1.5–1.9	0	5.5–5.9	2
2.0–2.4	6	6.0–6.4	1
2.5–2.9	0	6.5–6.9	1
3.0–3.4	0	7.0–7.4	1
3.5–3.9	0		

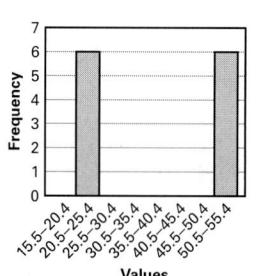

50. a.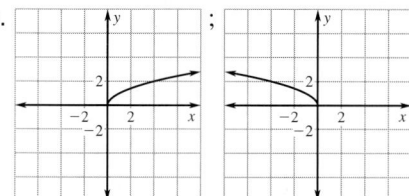

b.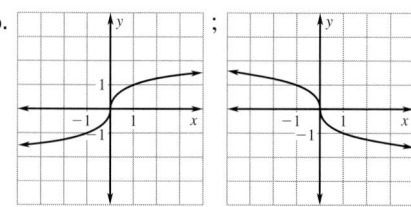

c.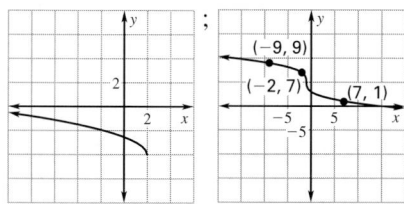

32.

15.5–20.4	0	35.5–40.4	0
20.5–25.4	6	40.5–45.4	0
25.5–30.4	0	45.5–50.4	0
30.5–35.4	0	50.5–55.4	6

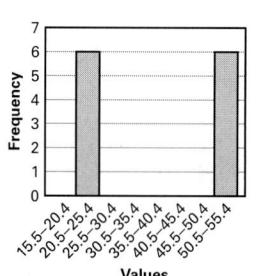

33. machine 1: 2.59, 2.59, none; machine 2: 2.59, 2.59, none

34. machine 1: 0.25, 0.09; machine 2: 3.01, 1.11

43.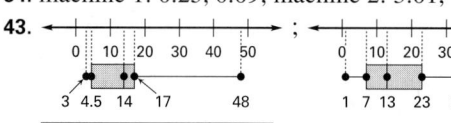

44.

Point Margin	AFC	NFC
1–10	13	12
11–20	15	12
21–30	3	7
31–40	1	2
41–50	1	0

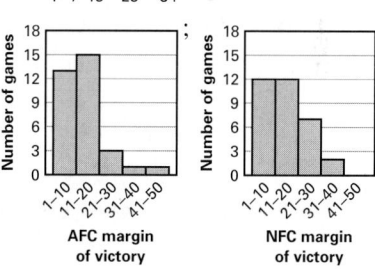

7.7 GUIDED PRACTICE (p. 449)

7.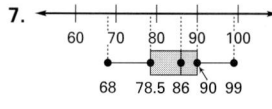

8.

61–68	1
69–76	2
77–84	2
85–92	5
93–100	2

9.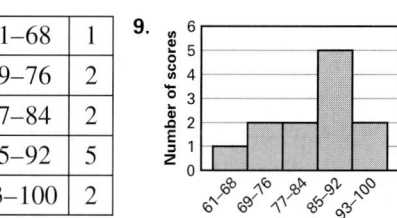

50.
$$\sigma = \sqrt{\frac{(x_1 - \overline{x})^2 + (x_2 - \overline{x})^2 + (x_3 - \overline{x})^2}{3}} =$$

$$\sqrt{\frac{x_1^2 - 2x_1\overline{x} + \overline{x}^2 + x_2^2 - 2x_2\overline{x} + \overline{x}^2 + x_3^2 - 2x_3\overline{x} + \overline{x}^2}{3}} =$$

$$\sqrt{\frac{x_1^2 + x_2^2 + x_3^2}{3} + \frac{3\overline{x}^2}{3} - \frac{2}{3}(x_1\overline{x} + x_2\overline{x} + x_3\overline{x})} =$$

$$\sqrt{\frac{x_1^2 + x_2^2 + x_3^2}{3} + \overline{x}^2 - \frac{2}{3}(x_1 + x_2 + x_3)\overline{x}}$$

We know that $(x_1 + x_2 + x_3) = 3\overline{x}$, so $\sqrt{\frac{x_1^2 + x_2^2 + x_3^2}{3} + \overline{x}^2 - \frac{2}{3}(3\overline{x}^2)} =$

$\sqrt{\frac{x_1^2 + x_2^2 + x_3^2}{3} - \overline{x}^2}$, which is the formula given in Exercise 50.

7.7 PRACTICE AND APPLICATIONS (pp. 449–451)

22.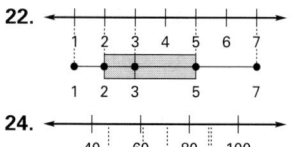

23.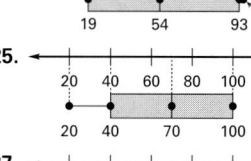

24.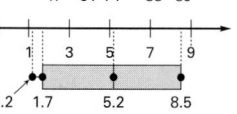

25.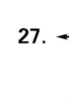

26.

27.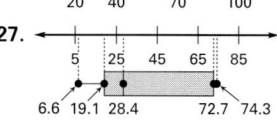

61.
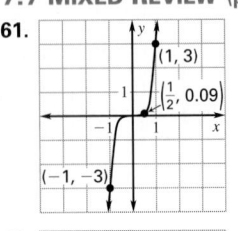

62.
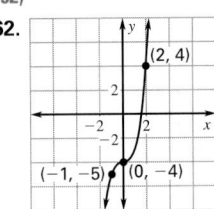

63.
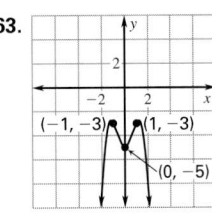

64.
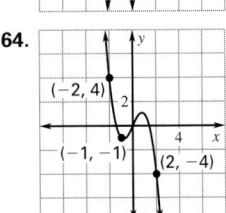

65.
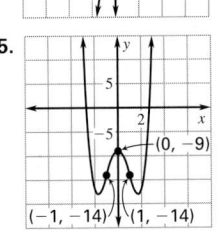

66.
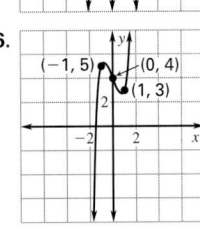

QUIZ 3 (p. 452)

1.
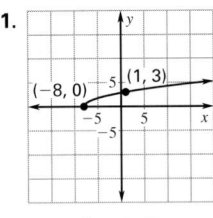

; **2.**
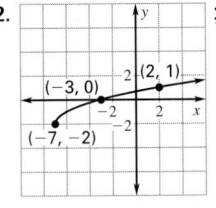

; **3.**
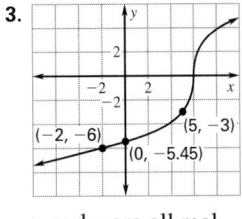
;

$x \geq -8, \ y \geq 0$ $\qquad$ $x \geq -7, \ y \geq -2$ $\qquad$ x and y are all real numbers.

9.
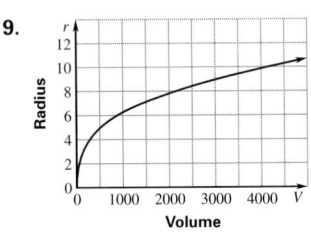

11.
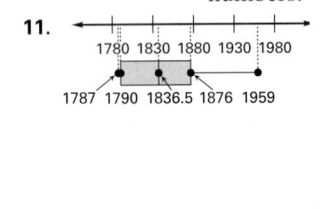

CHAPTER 7 TEST (p. 459)

32.

	Actresses	Actors
21–30	4	1
31–40	7	6
41–50	5	5
51–60	0	5
61–70	1	1
71–80	2	1

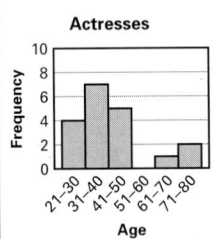

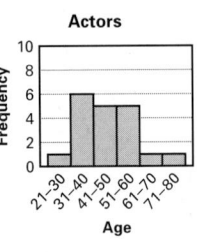

33. A good answer should include references to statistics and graphs.
Sample answers:
- On average (both mean and median), winning actors are older than actresses.
- Ages of actresses have more variability (both a larger range and standard deviation; see box-and-whisker plots).
- Both histograms show a cluster in the middle (30s and 40s), but more younger actresses (20s) and older actors (50s) win.
- Few people older than 60 win in either category.

CHAPTER 8

8.1 ACTIVITY (p. 465)

1.
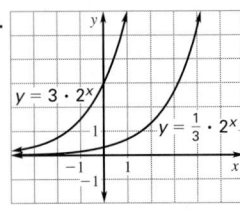

The graph of $y = \frac{1}{3} \cdot 2^x$ lies below that of $y = 2^x$ and has a y-intercept of $\frac{1}{3}$, while that of $y = 3 \cdot 2^x$ lies above $y = 2^x$ and has a y-intercept of 3. All three graphs exhibit the same end behavior and have the same general shape.

2.
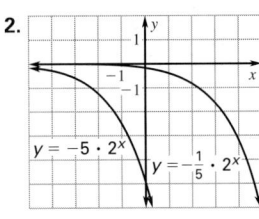

The graph of $y = -\frac{1}{5} \cdot 2^x$ lies closer to the x-axis than that of $y = 2^x$ and lies below the x-axis, so that y approaches the x-axis from below as $x \to -\infty$, and $y \to -\infty$ as $x \to +\infty$. The graph of $y = -5 \cdot 2^x$ also lies below the x-axis and exhibits the same end behavior as that of $y = -\frac{1}{5} \cdot 2^x$. If the graph of $y = 2^x$ were reflected in the x-axis, it would lie between the graphs of $y = -\frac{1}{5} \cdot 2^x$ and $y = -5 \cdot 2^x$.

3. If $0 < a < 1$, then the graph of $y = a \cdot 2^x$ lies below that of $y = 2^x$, while if $a > 1$, the graph of $y = a \cdot 2^x$ lies above that of $y = 2^x$. In either case, the graph has the same end behavior and general shape. If $-1 < a < 0$, then the graph of $y = a \cdot 2^x$ lies closer to the x-axis than that of $y = 2^x$, but below the axis instead of above it. If $a < -1$, the graph of $y = a \cdot 2^x$ lies below the x-axis, but grows away from the axis more quickly than that of $y = 2^x$. In both cases where a is negative, the graph approaches the x-axis asymptotically as $x \to -\infty$, and $y \to -\infty$ as $x \to +\infty$. In all cases, the y-intercept of the graph is a.

8.1 GUIDED PRACTICE (p. 469)

4.
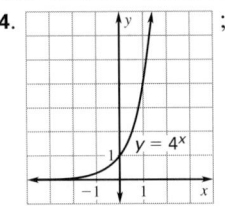
;

domain: all real numbers;
range: all positive real numbers

5.
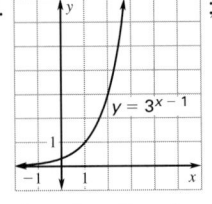
;

domain: all real numbers;
range: all positive real numbers

6.
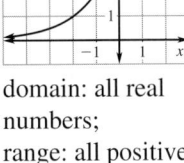
;

domain: all real numbers;
range: all positive real numbers

7.
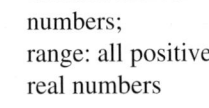
;

domain: all real numbers;
range: $y > -3$

8.
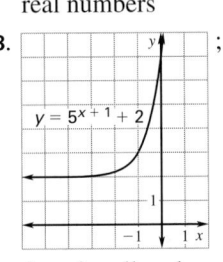
;

domain: all real numbers;
range: $y > 2$

9.
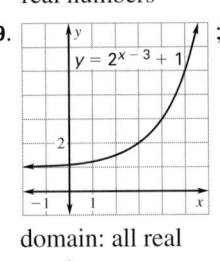
;

domain: all real numbers;
range: $y > 1$

ADDITIONAL ANSWERS

8.1 PRACTICE AND APPLICATIONS (pp. 469–472)

33.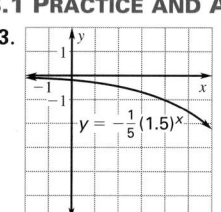
$y = -\frac{1}{5}(1.5)^x$

34.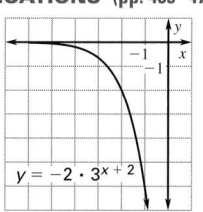
$y = -2 \cdot 3^{x+2}$

35.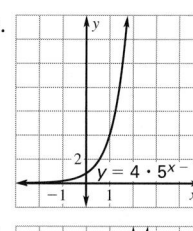
$y = 4 \cdot 5^{x-1}$

36.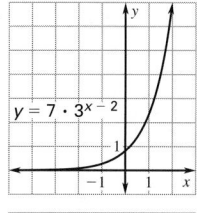
$y = 7 \cdot 3^{x-2}$

37.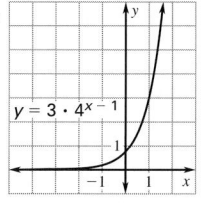
$y = 3 \cdot 4^{x-1}$

38.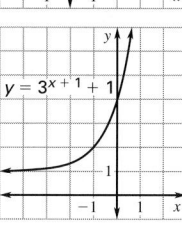
$y = 3^{x+1} + 1$

39.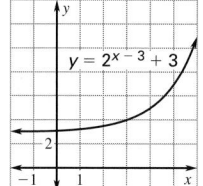
$y = 2^{x-3} + 3$

40.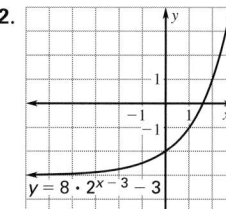
$y = -3 \cdot 6^{x+2} - 2$

41.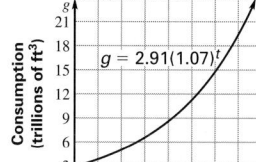
$y = 4 \cdot 2^{x-3} + 1$

42.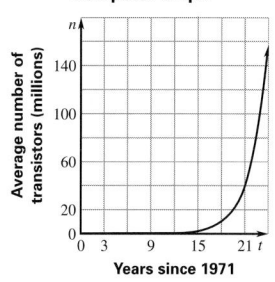
$y = 8 \cdot 2^{x-3} - 3$

44. Natural Gas Consumption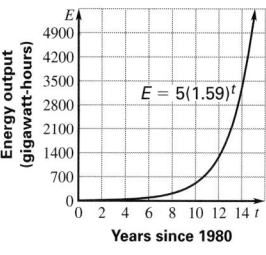
$g = 2.91(1.07)^t$

47. Computer Chips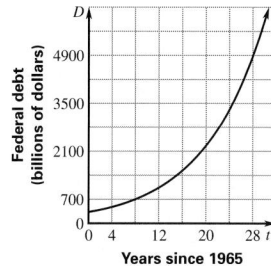

50. Wind Energy Generation
$E = 5(1.59)^t$

53. Federal Debt

8.2 CONCEPT ACTIVITY (p. 473)

Step 4.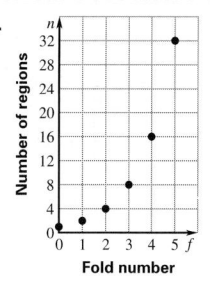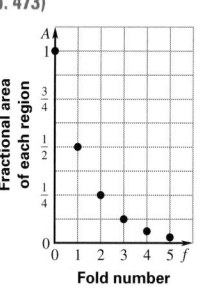

8.2 GUIDED PRACTICE (p. 477)

6.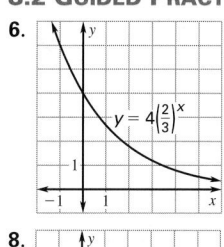
$y = 4\left(\frac{2}{3}\right)^x$

7.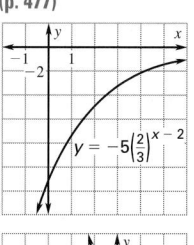
$y = -5\left(\frac{2}{3}\right)^{x-2}$

8.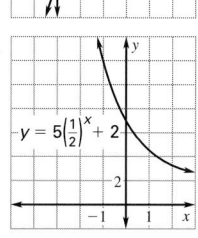
$y = -4(0.25)^{x+1}$

9.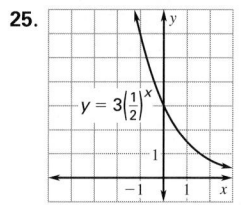
$y = 5\left(\frac{1}{2}\right)^x + 2$

8.2 PRACTICE AND APPLICATIONS (pp. 477–479)

25.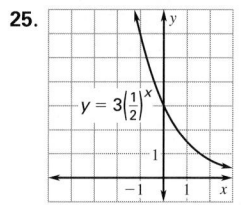
$y = 3\left(\frac{1}{2}\right)^x$

26.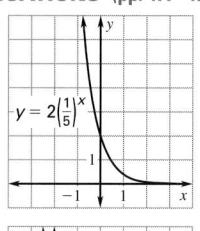
$y = 2\left(\frac{1}{5}\right)^x$

27.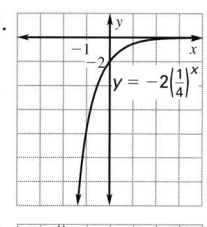
$y = -2\left(\frac{1}{4}\right)^x$

28.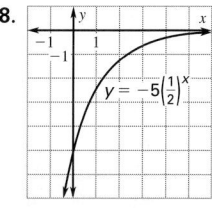
$y = -5\left(\frac{1}{2}\right)^x$

29.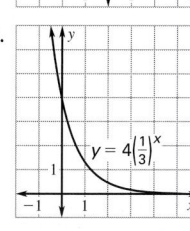
$y = 4\left(\frac{1}{3}\right)^x$

30.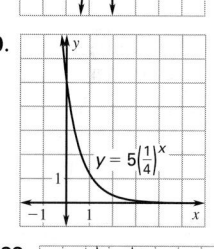
$y = 5\left(\frac{1}{4}\right)^x$

31.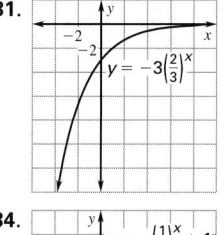
$y = -3\left(\frac{2}{3}\right)^x$

32.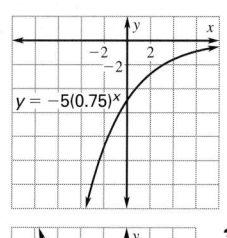
$y = -5(0.75)^x$

33.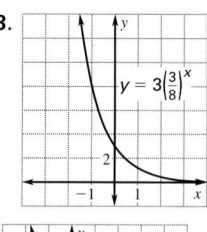
$y = 3\left(\frac{3}{8}\right)^x$

34.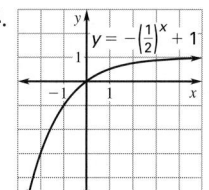
$y = -\left(\frac{1}{2}\right)^x + 1$

35.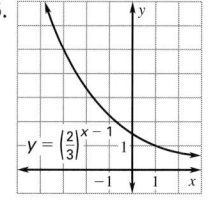
$y = \left(\frac{2}{3}\right)^{x-1}$

36.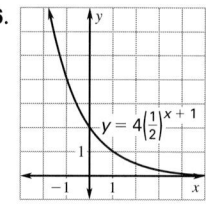
$y = 4\left(\frac{1}{2}\right)^{x+1}$

ADDITIONAL ANSWERS

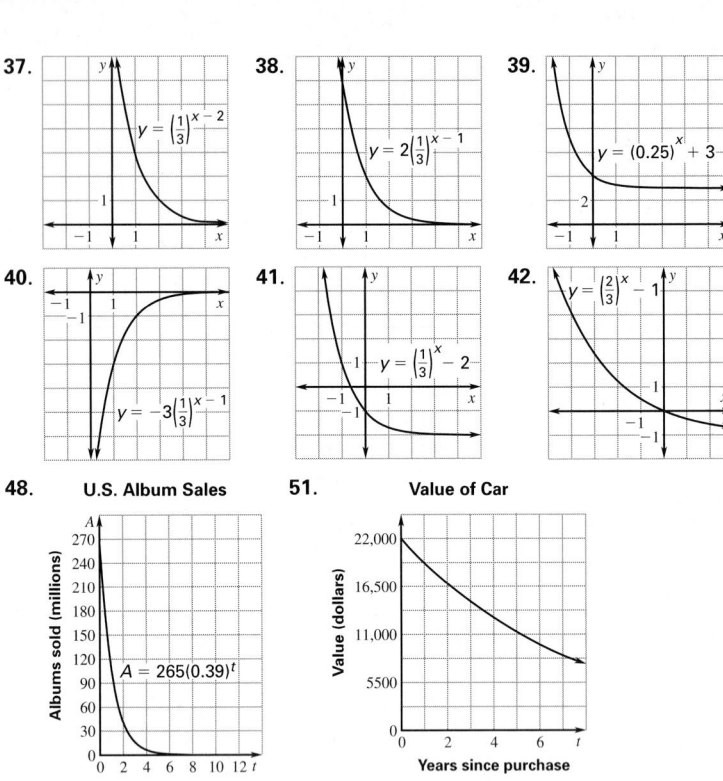

37. $y = \left(\frac{1}{3}\right)^{x-2}$

38. $y = 2\left(\frac{1}{3}\right)^{x-1}$

39. $y = (0.25)^x + 3$

40. $y = -3\left(\frac{1}{3}\right)^{x-1}$

41. $y = \left(\frac{1}{3}\right)^x - 2$

42. $y = \left(\frac{2}{3}\right)^x - 1$

48.

U.S. Album Sales

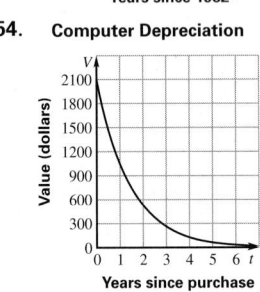

$A = 265(0.39)^t$

51.

Value of Car

54.

Computer Depreciation

57. c.

Year	V	$A(n)$
0	$18,354	$18,354
1	$15,233.82	$16,482.29
2	$12,644.07	$14,445.13
3	$10,494.58	$12,227.91
4	$8710.50	$9814.70
5	$7229.72	$7188.19

58. *Sample answer:* An exponential decay function has the form $y = ab^x$, where $a > 0$ and $0 < b < 1$. Given two exponential decay functions, say $y = ab^x$ and $y = cd^x$, their product is the function $y = (ac)(bd)^x$. Since a and c are both positive, $ac > 0$. Since b and d are both between 0 and 1, so is bd. So, the product function is an exponential decay function. The quotient of two exponential decay functions may not be an exponential decay function. For example, $\frac{4(0.75)^x}{2(0.25)^x} = 2(3)^x$, which is an exponential growth function.

8.2 MIXED REVIEW (p. 479)

59. $y = (x+1)^{1/3}$

60. $y = \sqrt[3]{x} + 1$

61. $y = -3x^{1/3}$

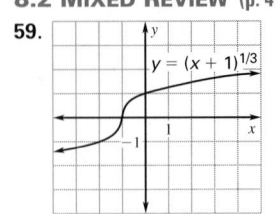 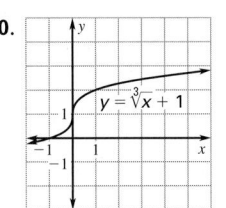 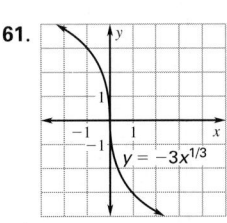

62. $y = \sqrt{x} + 4$

63. $y = -\sqrt{x} + 5$

64. $y = \sqrt[3]{x} + \frac{1}{4}$

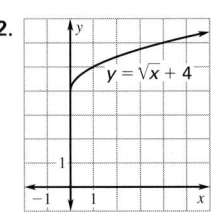 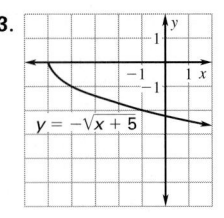 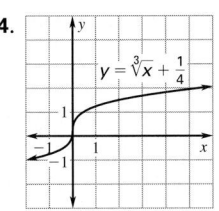

8.3 GUIDED PRACTICE (p. 483)

13. $y = e^{-2x}$

14. $y = \frac{1}{2}e^x$

15. $y = \frac{1}{8}e^{2x}$

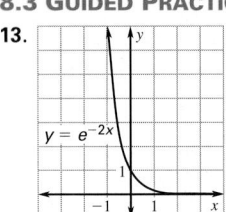 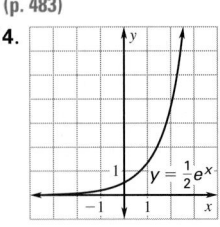 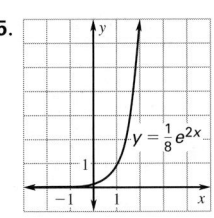

8.3 PRACTICE AND APPLICATIONS (pp. 483–485)

68. $y = 4e^x$;

69. $y = \frac{1}{3}e^x$;

70. $y = 3e^{2x} + 2$;

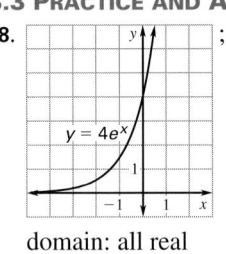 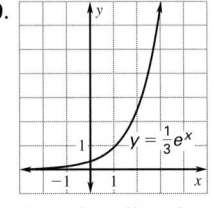 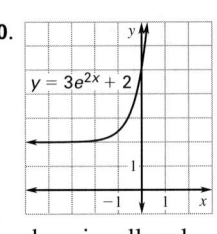

domain: all real numbers; range: $y > 0$

domain: all real numbers; range: $y > 0$

domain: all real numbers; range: $y > 2$

71. $y = 1.5e^{-0.5x}$;

72. $y = 0.1e^{2x} - 4$;

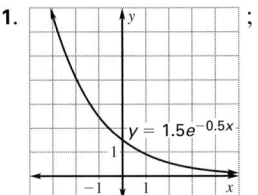 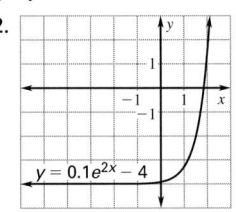

domain: all real numbers; range: $y > 0$

domain: all real numbers; range: $y > -4$

73. $y = \frac{1}{3}e^{x-2} - 1$;

74. $y = \frac{4}{3}e^{x-3} + 1$;

75. $y = 0.5e^{-2(x-1)} - 2$;

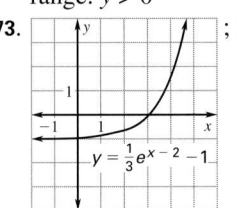 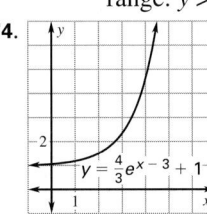 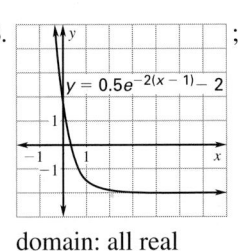

domain: all real numbers; range: $y > -1$

domain: all real numbers; range: $y > 1$

domain: all real numbers; range: $y > -2$

78. Answers may vary. A good answer will include:
- several comparisons of the amount of interest earned using different values of r and perhaps of P
- summary of results or conclusion stating that the difference between the amounts of interest earned by the two formulas tends to be insignificant

ADDITIONAL ANSWERS

83. $n = 10^{10}$; *Sample answer:* From the table made in the Activity on p. 480, I noticed that using a value of $n = 10^k$, the answer is accurate to $k - 1$ decimal places, with an error in the kth decimal place. I tried $n = 10^{10}$, which worked. I tried numbers close to 10^{10} to see if a smaller value would work, but my calculator couldn't give a more accurate result due to the rounding error.

QUIZ 1 (p. 485)

1.
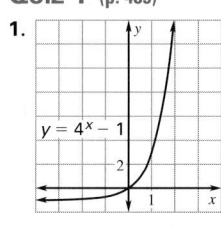
$y = 4^x - 1$

2.
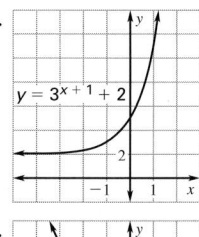
$y = 3^{x+1} + 2$

3.
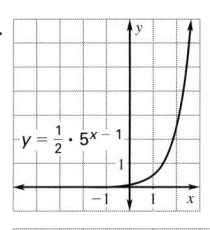
$y = \frac{1}{2} \cdot 5^{x-1}$

4.
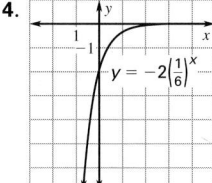
$y = -2\left(\frac{1}{6}\right)^x$

5.
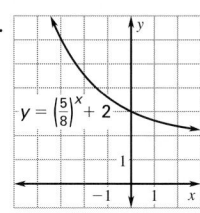
$y = \left(\frac{5}{8}\right)^x + 2$

6.
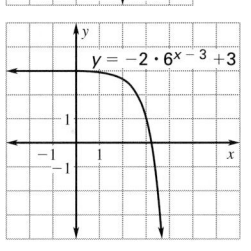
$y = -2 \cdot 6^{x-3} + 3$

15.
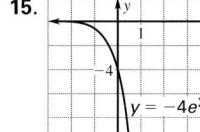
$y = -4e^{2x}$

16.
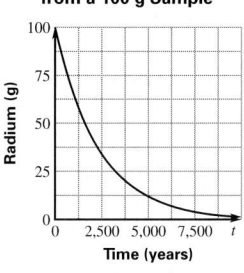
Amount of Radium Left from a 100 g Sample

8.4 GUIDED PRACTICE (p. 490)

2. Logarithms are not defined for negative numbers, since there is no real number x such that $3^x = -1$, for example. The number 1 cannot be used as the base for a logarithmic function, since 1 to any power is equal to 1. **3.** $\log_b y$ is the number x such that $b^x = y$. **4.** $\log_2 25$ is the number x such that $2^x = 25$. The base of the logarithmic function is 2, not 5.

13.
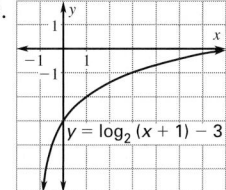
$y = \log_2 (x + 1) - 3$

14.
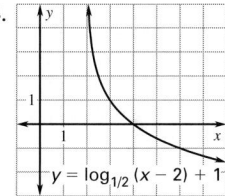
$y = \log_{1/2} (x - 2) + 1$

8.4 PRACTICE AND APPLICATIONS (pp. 490–492)

66.
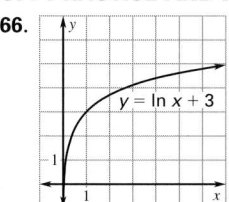
$y = \ln x + 3$

67.
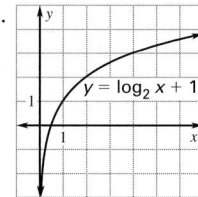
$y = \log_2 x + 1$

68.
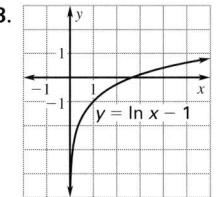
$y = \ln x - 1$

69.
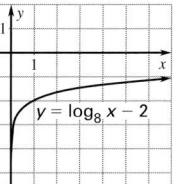
$y = \log_8 x - 2$

70.
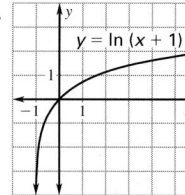
$y = \ln (x + 1)$

71.
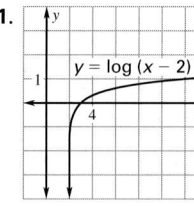
$y = \log (x - 2)$

72.
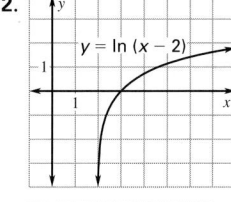
$y = \ln (x - 2)$

73.
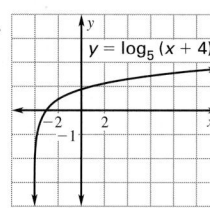
$y = \log_5 (x + 4)$

74.
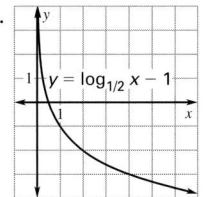
$y = \log_{1/2} x - 1$

75.
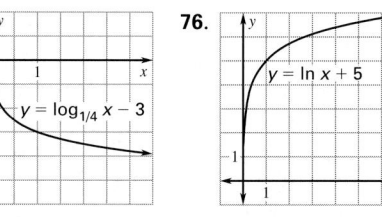
$y = \log_{1/4} x - 3$

76.
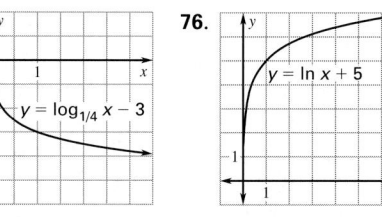
$y = \ln x + 5$

8.5 PRACTICE AND APPLICATIONS (pp. 496–498)

91. a. Let $x = \log_b u$ and $y = \log_b v$. Then $u = b^x$ and $v = b^y$, so that $\log_b uv = \log_b (b^x b^y) = \log_b (b^{x+y}) = x + y = \log_b u + \log_b v$. **b.** Let $x = \log_b u$ and $y = \log_b v$. Then $u = b^x$ and $v = b^y$, so that $\log_b \frac{u}{v} = \log_b \frac{b^x}{b^y} = \log_b (b^{x-y}) = x - y = \log_b u - \log_b v$. **c.** Let $x = \log_b u$. Then $u = b^x$ and $u^n = b^{nx}$, so that $\log_b u^n = \log_b b^{nx} = nx = n \log_b u$. **d.** Let $x = \log_b u$, $y = \log_b c$, and $z = \log_c u$. Then $u = b^x$, $c = b^y$, and $u = c^z$, so that $b^x = c^z$. Thus, $x = \log_b u = \log_b b^x = \log_b c^z = z \log_b c = zy$. Thus, $x = yz$, so $z = \frac{x}{y}$, or $\log_c u = \frac{\log_b u}{\log_b c}$.

8.5 TECHNOLOGY ACTIVITY (p. 500)

1.
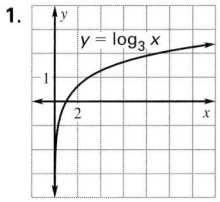
$y = \log_3 x$

2.
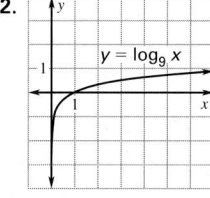
$y = \log_9 x$

3.
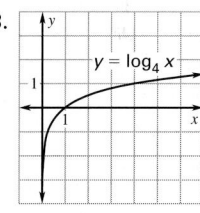
$y = \log_4 x$

4.
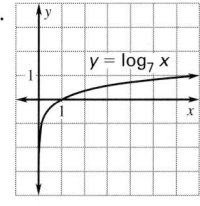
$y = \log_7 x$

5.
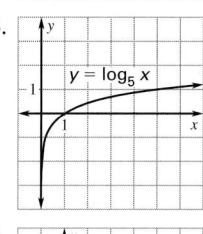
$y = \log_5 x$

6.
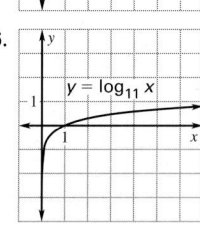
$y = \log_{11} x$

7.
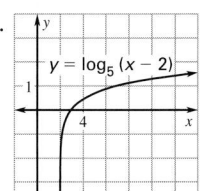
$y = \log_5 (x - 2)$

8.
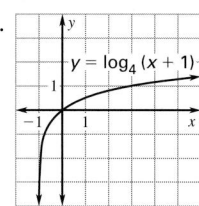
$y = \log_4 (x + 1)$

9.
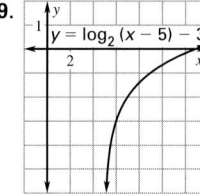
$y = \log_2 (x - 5) - 3$

ADDITIONAL ANSWERS

10.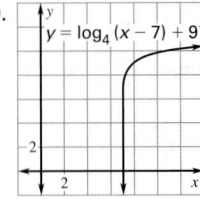
$y = \log_4(x - 7) + 9$

11.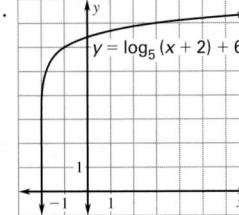
$y = \log_5(x + 2) + 6$

12.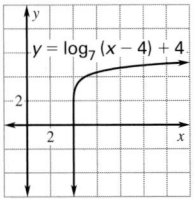
$y = \log_7(x - 4) + 4$

8.6 MIXED REVIEW (p. 508)

77.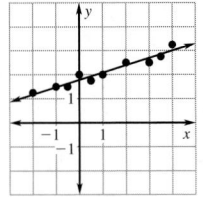

78.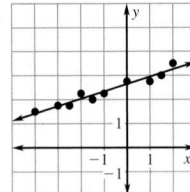

QUIZ 2 (p. 508)

5.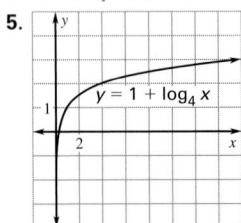
$y = 1 + \log_4 x$

6.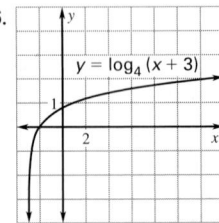
$y = \log_4(x + 3)$

7.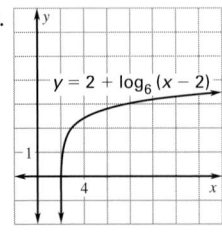
$y = 2 + \log_6(x - 2)$

8.7 PRACTICE AND APPLICATIONS (pp. 513–516)

28.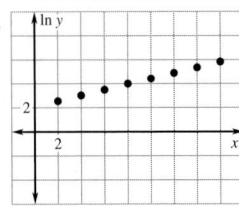

40.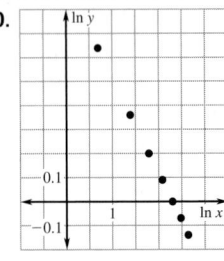

53.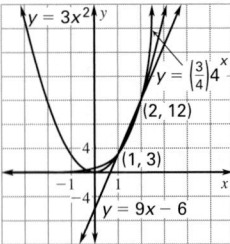
$y = 3x^2$
$y = \left(\frac{3}{4}\right)4^x$
$(2, 12)$
$(1, 3)$
$y = 9x - 6$

58. a.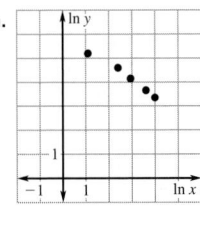

59. a.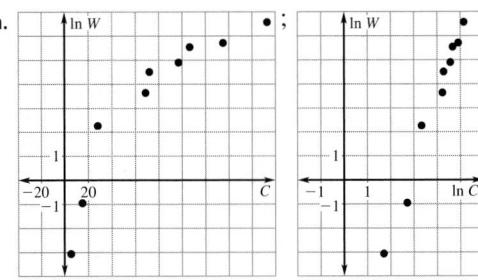

8.7 MIXED REVIEW (p. 516)

67.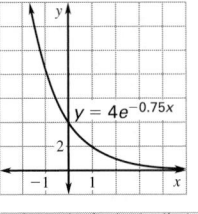
$y = 4e^{-0.75x}$

68.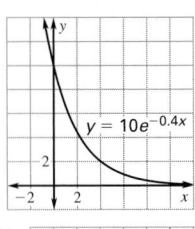
$y = 10e^{-0.4x}$

69.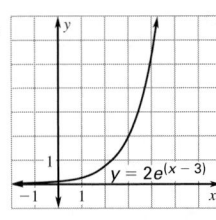
$y = 2e^{(x - 3)}$

70.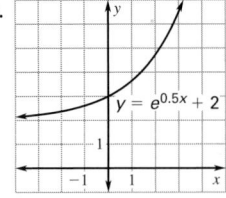
$y = e^{0.5x} + 2$

71.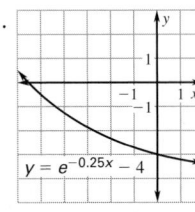
$y = e^{-0.25x} - 4$

72.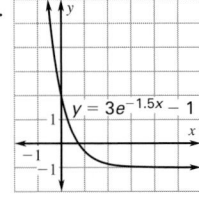
$y = 3e^{-1.5x} - 1$

73.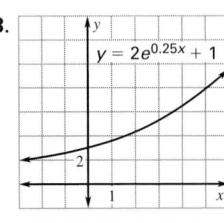
$y = 2e^{0.25x} + 1$

74.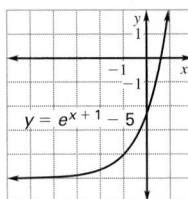
$y = e^{x + 1} - 5$

75.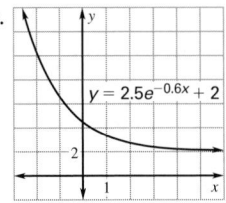
$y = 2.5e^{-0.6x} + 2$

8.8 ACTIVITY (p. 517)

1.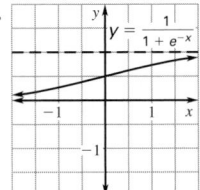
$f(x) = \dfrac{100}{1 + 9e^{-2x}}$

2. a.
$y = \dfrac{1}{1 + e^{-x}}$

b.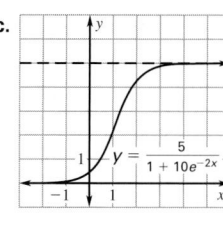
$y = \dfrac{10}{1 + 5e^{-2x}}$

c.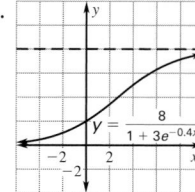
$y = \dfrac{5}{1 + 10e^{-2x}}$

8.8 GUIDED PRACTICE (p. 520)

9.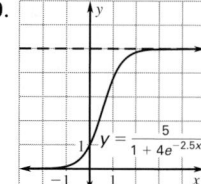
$y = \dfrac{5}{1 + 4e^{-2.5x}}$

10.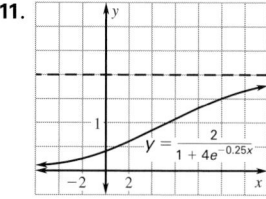
$y = \dfrac{8}{1 + 3e^{-0.4x}}$

11.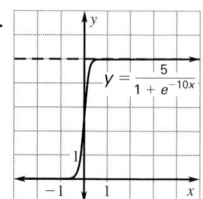
$y = \dfrac{2}{1 + 4e^{-0.25x}}$

8.8 PRACTICE AND APPLICATIONS (pp. 520–522)

29. 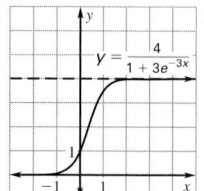 ;
$y = \dfrac{5}{1 + e^{-10x}}$

asymptotes: *x*-axis and *y* = 5;
y-intercept: $\dfrac{5}{2}$;
pt. of max. growth: $\left(0, \dfrac{5}{2}\right)$

30. 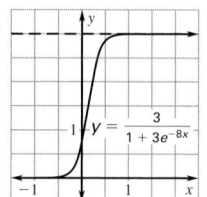 ;
$y = \dfrac{4}{1 + 0.08e^{-2.1x}}$

asymptotes: *x*-axis and *y* = 4;
y-intercept: 3.704;
pt. of max. growth: (−1.203, 2)

31. 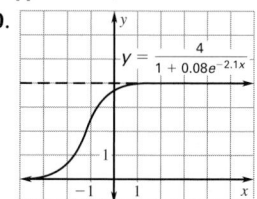 ;
$y = \dfrac{4}{1 + 3e^{-3x}}$

asymptotes: *x*-axis and *y* = 4;
y-intercept: 1;
pt. of max. growth: (0.366, 2)

32. ;
$y = \dfrac{3}{1 + 3e^{-8x}}$

asymptotes: *x*-axis and *y* = 3;
y-intercept: $\dfrac{3}{4}$;
pt. of max. growth: $\left(0.137, \dfrac{3}{2}\right)$

33. 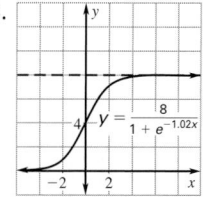 ;
$y = \dfrac{8}{1 + e^{-1.02x}}$

asymptotes: *x*-axis and *y* = 8;
y-intercept: 4;
pt. of max. growth: (0, 4)

34. 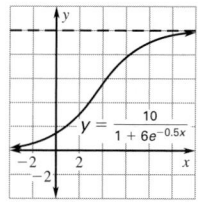 ;
$y = \dfrac{10}{1 + 6e^{-0.5x}}$

asymptotes: *x*-axis and *y* = 10;
y-intercept: $\dfrac{10}{7}$;
pt. of max. growth: (3.584, 5)

35. 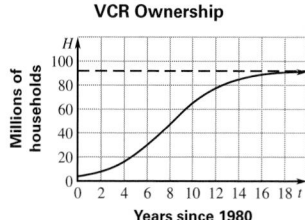 ; asymptotes: *x*-axis and *y* = 6;
$y = \dfrac{6}{1 + 0.8e^{-2x}}$
y-intercept: $\dfrac{10}{3}$;
pt. of max. growth: (−0.112, 3)

46. VCR Ownership
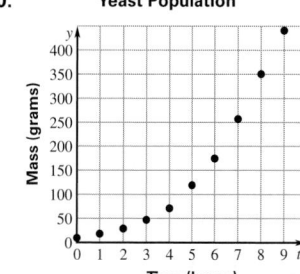

49. Gross Domestic Product of the U.S.

50. Yeast Population
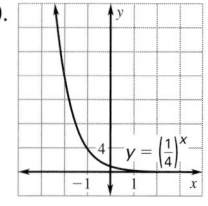

52. a. U.S. Population 1800–1870
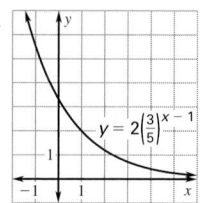
$P = 5.36(1.03)^t$
$P = \dfrac{186.45}{1 + 35.37e^{-0.03t}}$

CHAPTER 8 REVIEW (pp. 524–526)

9. 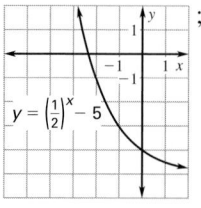 ;
$y = \left(\dfrac{1}{4}\right)^x$

domain: all real numbers;
range: *y* > 0

10. ;
$y = 2\left(\dfrac{3}{5}\right)^{x-1}$

domain: all real numbers;
range: *y* > 0

11. ;
$y = \left(\dfrac{1}{2}\right)^x - 5$

domain: all real numbers;
range: *y* > −5

12. 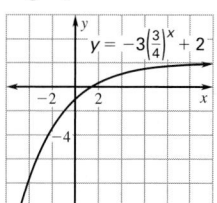 ; domain: all real numbers;
$y = -3\left(\dfrac{3}{4}\right)^x + 2$
range: *y* < 2

ADDITIONAL ANSWERS

ADDITIONAL ANSWERS

CHAPTER 8 TEST (p. 527)

6. 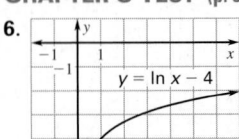 ;
7. 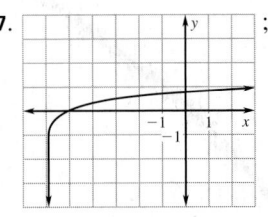 ;

domain: $x > 0$;
range: all real numbers

domain: $x > -6$;
range: all real numbers

8. 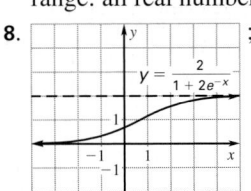 ;
33. a.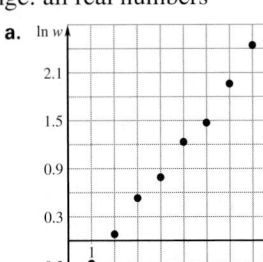

domain: all real numbers;
range: $0 < y < 2$

CHAPTER 9

9.1 MIXED REVIEW (p. 539)

63.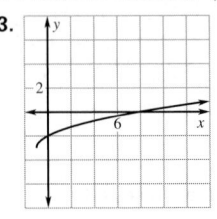
70. b.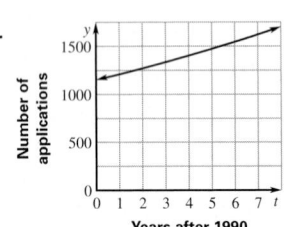

9.2 ACTIVITY (p. 540)

Step 1. a.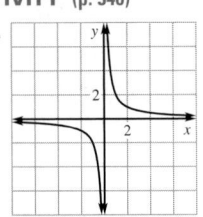
b.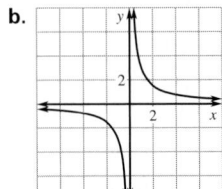

c.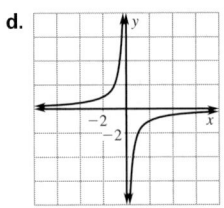
d.

9.2 GUIDED PRACTICE (p. 543)

10.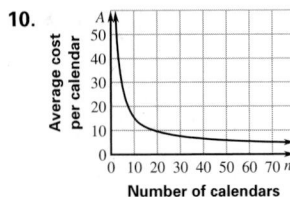

9.2 PRACTICE AND APPLICATIONS (pp. 543–545)
17. $y = -17$; $x = -43$; domain: all real numbers except -43; range: all real numbers except -17
18. $y = \frac{17}{8}$; $x = -\frac{1}{4}$; domain: all real numbers except $-\frac{1}{4}$; range: all real numbers except $\frac{17}{8}$
19. $y = 19$; $x = 6$; domain: all real numbers except 6; range: all real numbers except 19

26.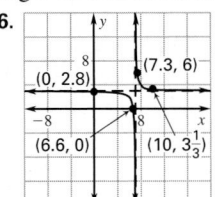
27.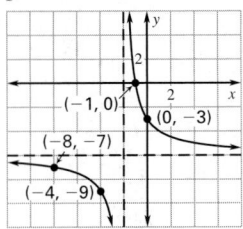

28.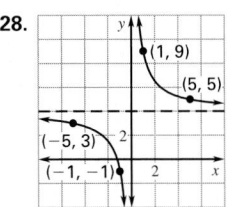
29.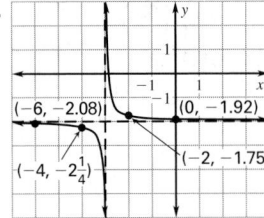

30. 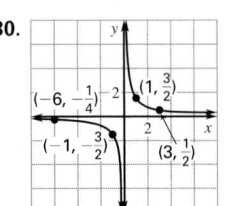 ;
31. 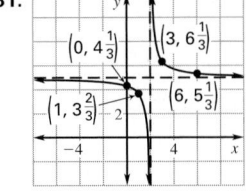 ;

domain: all real numbers
except 0; range: all real
numbers except 0

domain: all real numbers
except 2; range: all real
numbers except 5

32. 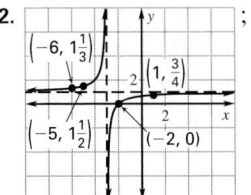 ;
33. 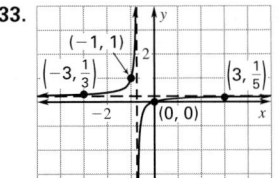 ;

domain: all real numbers
except -3; range: all real
numbers except 1

domain: all real numbers except $-\frac{3}{4}$;
range: all real numbers except $\frac{1}{4}$

34. 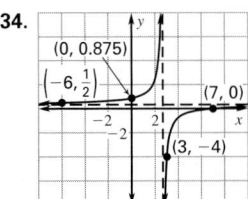 ;
35. 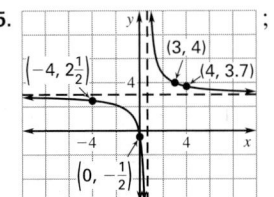 ;

domain: all real numbers
except $\frac{8}{3}$; range: all real
numbers except $\frac{1}{3}$

domain: all real numbers
except $\frac{2}{3}$; range: all real
numbers except 3

36.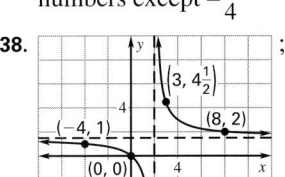

domain: all real numbers except 3; range: all real numbers except $-\frac{3}{4}$

37.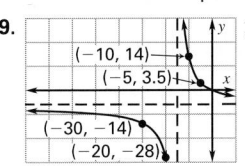

domain: all real numbers except 0; range: all real numbers except $\frac{5}{4}$

38.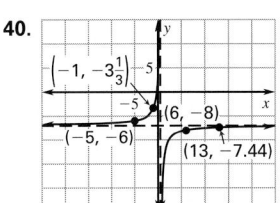

domain: all real numbers except 2; range: all real numbers except $\frac{3}{2}$

39.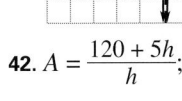

domain: all real numbers except -15; range: all real numbers except -7

40. 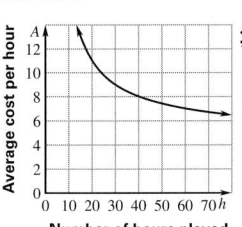 ; domain: all real numbers except $\frac{1}{2}$; range: all real numbers except -7

41. *Sample Answer:* $y = \dfrac{1}{x+4} + 3$

42. $A = \dfrac{120 + 5h}{h}$; 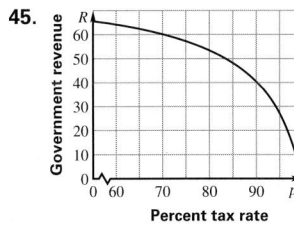 ; $A = 5$; the average cost approaches \$5 per hour as the number of hours played increases.

45.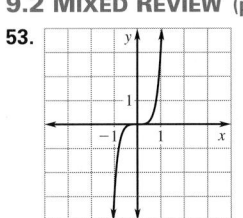

48.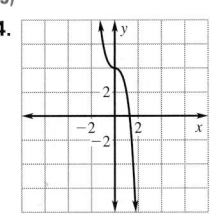

9.2 MIXED REVIEW (p. 545)

53.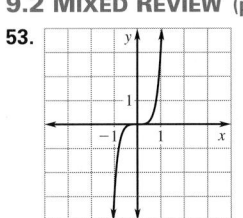

54.

55.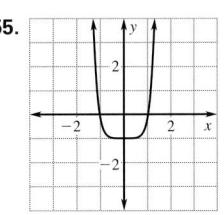

56.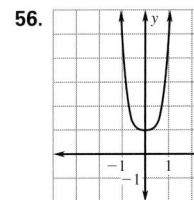

57.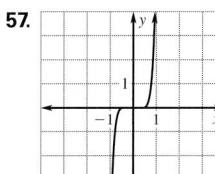

58.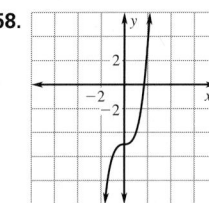

9.2 TECHNOLOGY ACTIVITY (p. 546)

7.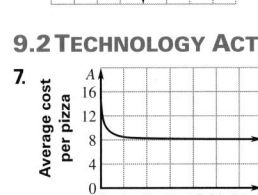

9.3 GUIDED PRACTICE (p. 550)

4.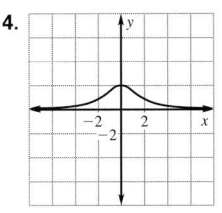

5.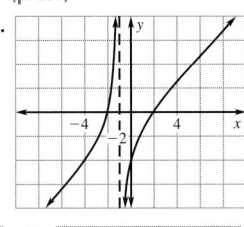

6.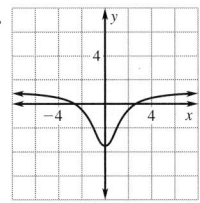

7.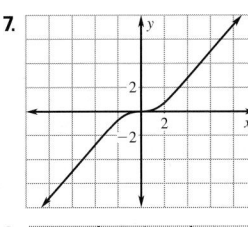

8.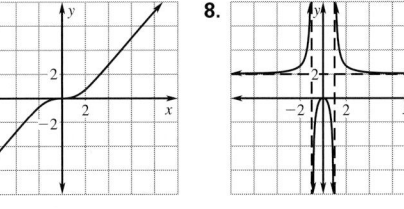

9.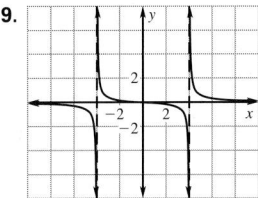

9.3 PRACTICE AND APPLICATIONS (pp. 550–552)

29.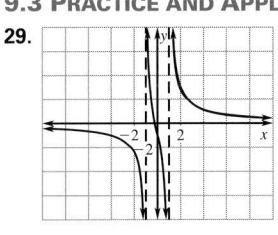

30.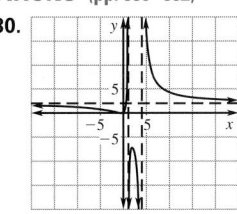

31.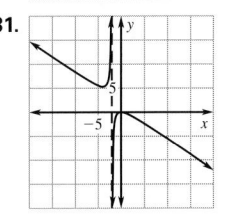

32.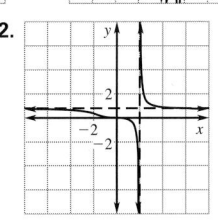

33.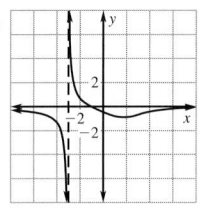

34.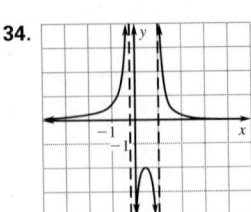

35.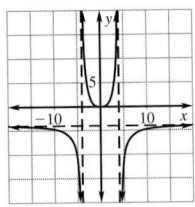

36.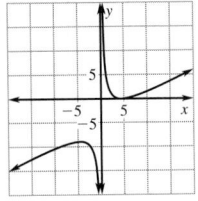

9.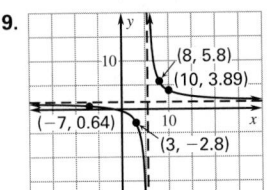

10.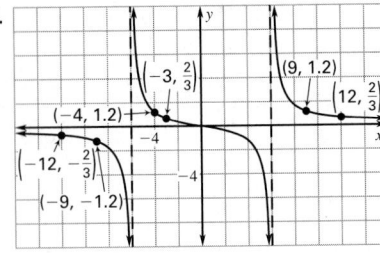

37.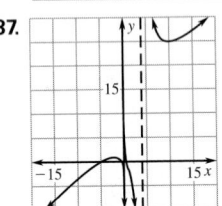

39.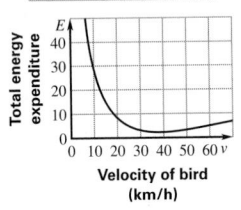

11.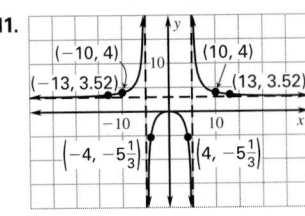

12.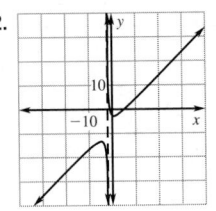

40.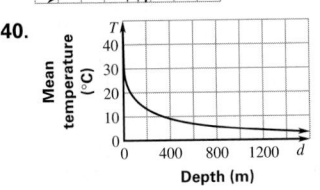

41.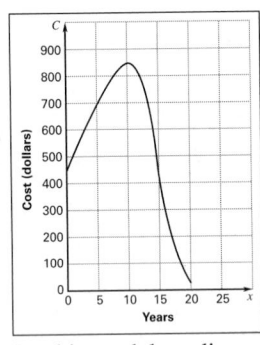

13.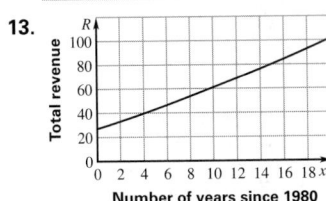

No; this model predicts an average daily cost close to zero after 2005, and this is not realistic.

9.5 PRACTICE AND APPLICATIONS (pp. 565–567)

50. $A = \dfrac{391t^2 + 0.112}{0.218t^4 + 0.991t^2 + 1} + \dfrac{391(t-1)^2 + 0.112}{0.218(t-1)^4 + 0.991(t-1)^2 + 1}$

Check graphs.

9.6 GUIDED PRACTICE (p. 571)
3. (1) Multiply each term on both sides of the equation by the LCD of the terms. Simplify and solve the resulting polynomial equation. (2) Cross multiply. The first method always works. It always gives a polynomial to be solved. Cross multiplying only works when each side of the equation is a single rational expression.

CUMULATIVE PRACTICE, CHAPTERS 1–9 (pp. 582–583)

56.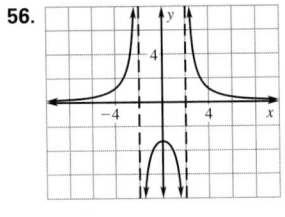

57.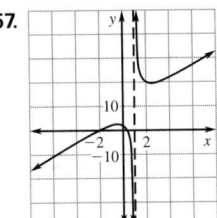

79.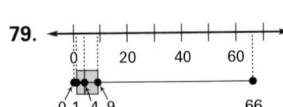

49. a. *Sample table:*

x	$f(x)$	$g(x)$
2.95	190.76	−2.415
2.96	240.71	−2.431
2.97	323.99	−2.448
2.98	490.60	−2.465
2.99	990.55	−2.483
3	undefined	undefined
3.01	−1010	−2.518
3.02	−509.6	−2.535
3.03	−343	−2.553
3.04	−259.7	−2.571
3.05	−209.8	−2.590

b. graph of $f(x)$:

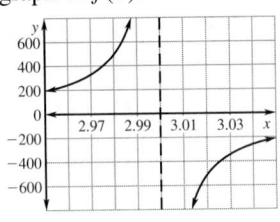

graph of $g(x)$:

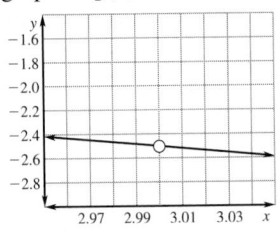

80.

Chicago Cubs		
Interval	Tally	Freq.
0–15	JHT JHT JHT IIII	19
16–30	II	2
31–45	I	1
46–60		0
61–75	I	1

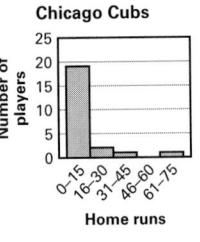

QUIZ 1 (p. 553)

7.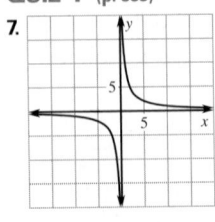

8.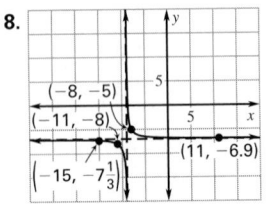

ADDITIONAL ANSWERS

CHAPTER 10

SKILL REVIEW (p. 588)

7.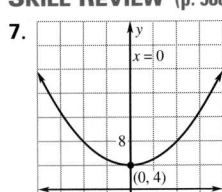

8.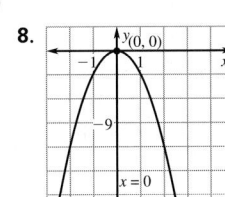

9.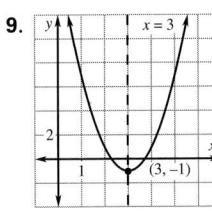

10.1 MIXED REVIEW (p. 594)

65.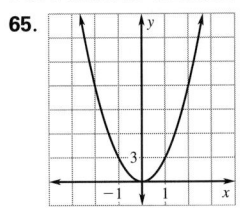

66.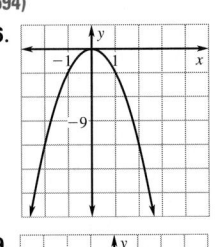

67.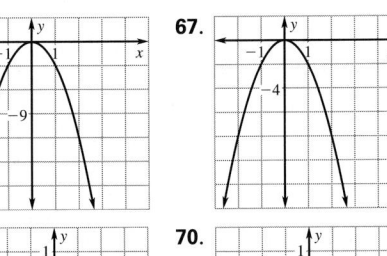

68.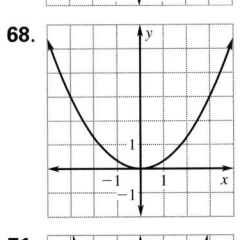

69.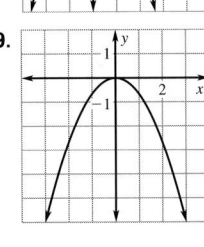

70.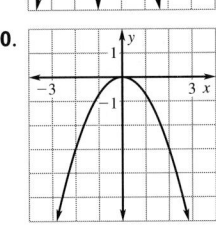

71.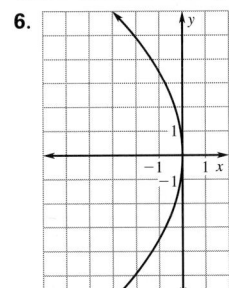

10.2 GUIDED PRACTICE (p. 598)

6.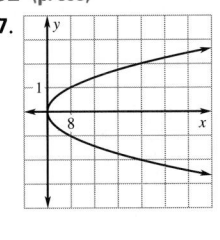

7.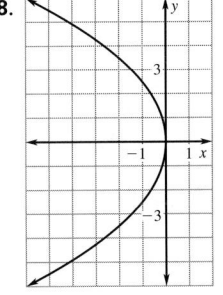

8.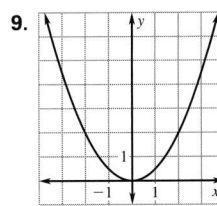

9.

10.2 PRACTICE AND APPLICATIONS (pp. 598–600)

39.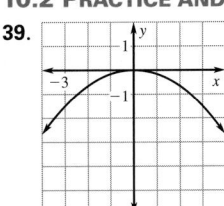

40.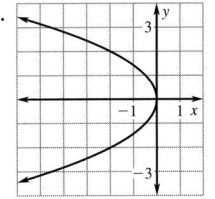

41.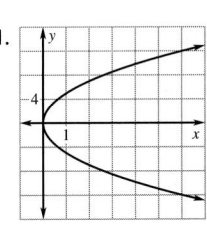

42.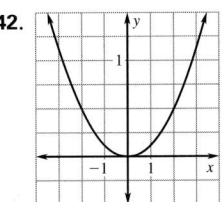

43.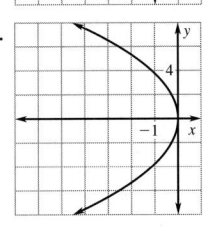

44.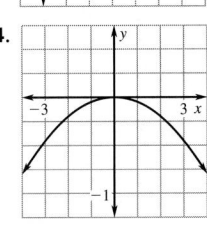

45.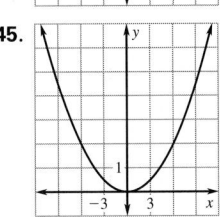

46.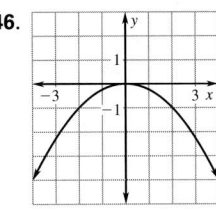

47.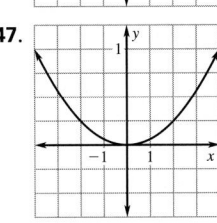

48.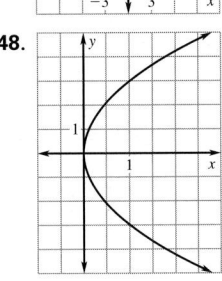

49.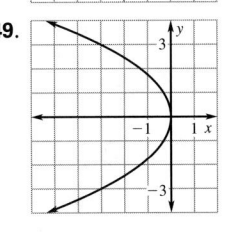

50.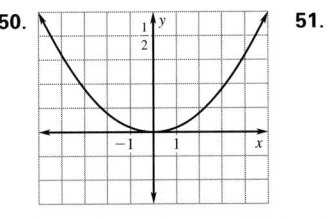

51.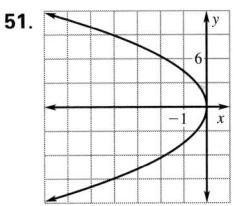

52.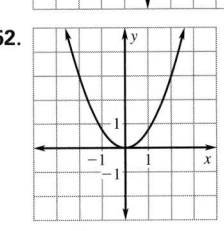

53.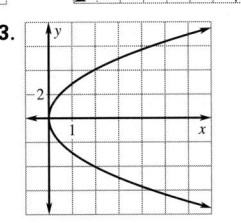

10.3 GUIDED PRACTICE (p. 604)

14.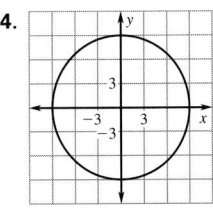

15.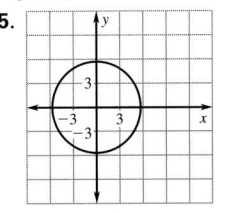

16.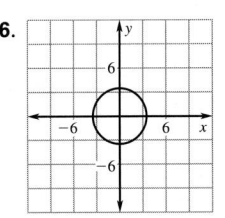

ADDITIONAL ANSWERS

17.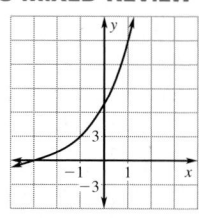

18.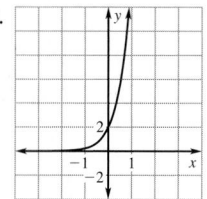

105.

106.

10.3 PRACTICE AND APPLICATIONS (pp. 604–606)

30.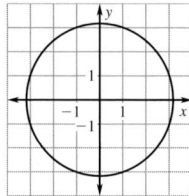

31.

32.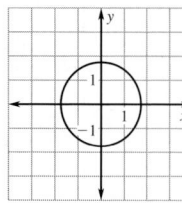

QUIZ 1 (p. 607)

7.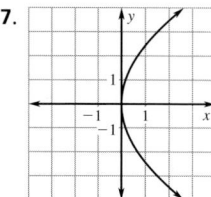

8.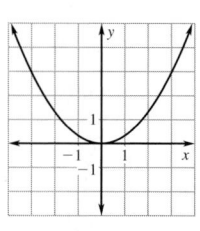

9.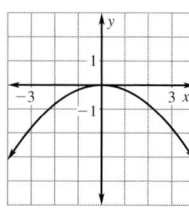

33.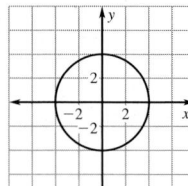

34.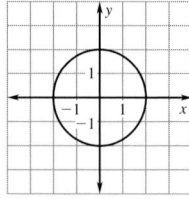

35.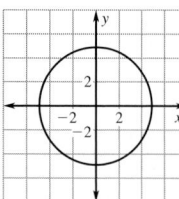

10.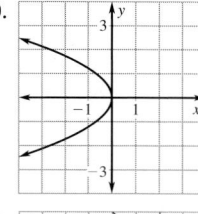

11.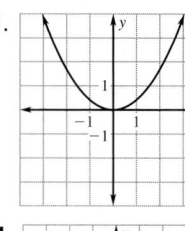

12.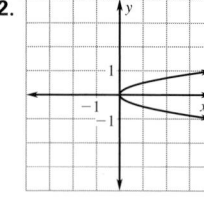

36.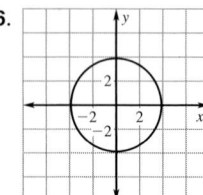

37.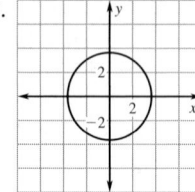

38.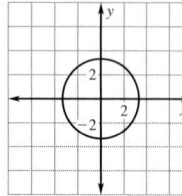

13.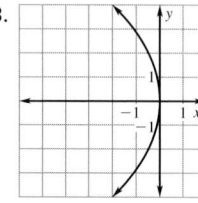

14.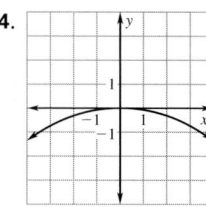

39.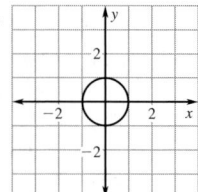

40.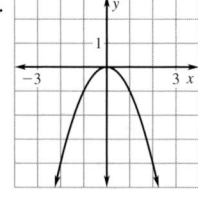

41.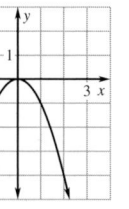

10.4 GUIDED PRACTICE (p. 612)

11.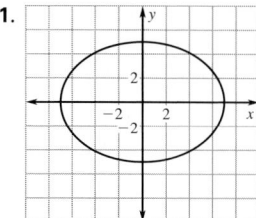

12.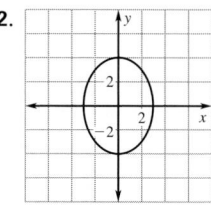

42.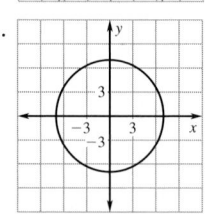

43.

44.

13.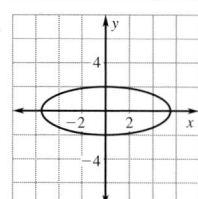

14.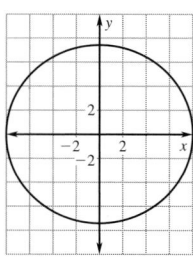

15.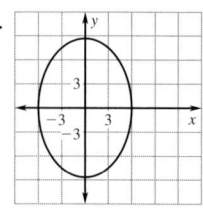

45.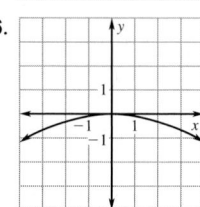

46.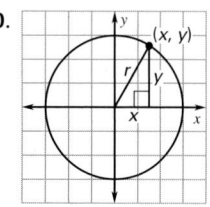

80.

16.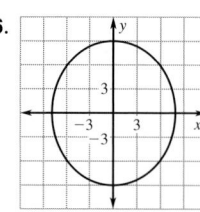

17.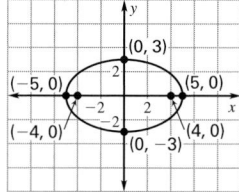

25. $\dfrac{x^2}{4} + \dfrac{y^2}{49} = 1$; vertices: $(0, \pm7)$; co-vertices: $(\pm2, 0)$; foci: $\left(0, \pm3\sqrt{5}\right)$

26. $\dfrac{x^2}{100} + \dfrac{y^2}{9} = 1$; vertices: $(\pm10, 0)$; co-vertices: $(0, \pm3)$; foci: $\left(\pm\sqrt{91}, 0\right)$

27. $\dfrac{x^2}{10} + y^2 = 1$; vertices: $\left(\pm\sqrt{10}, 0\right)$; co-vertices: $(0, \pm1)$; foci: $(\pm3, 0)$

28. $\dfrac{x^2}{25} + \dfrac{y^2}{10} = 1$; vertices: $(\pm5, 0)$; co-vertices: $\left(\pm\sqrt{10}, 0\right)$; foci: $\left(\pm\sqrt{15}, 0\right)$

29. $\dfrac{x^2}{15} + \dfrac{y^2}{25} = 1$; vertices: $(0, \pm5)$; co-vertices: $\left(\pm\sqrt{15}, 0\right)$; foci: $\left(0, \pm\sqrt{10}\right)$

31. 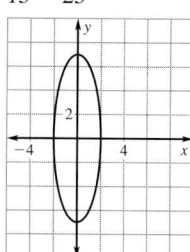 ;

vertices: $(0, \pm7)$; co-vertices: $(\pm2, 0)$; foci: $\left(0, \pm3\sqrt{5}\right)$

32. 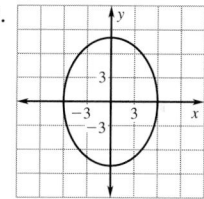 ;

vertices: $(0, \pm8)$; co-vertices: $(\pm6, 0)$; foci: $\left(0, \pm2\sqrt{7}\right)$

33. 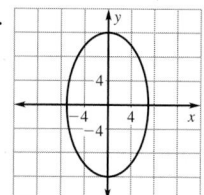 ;

vertices: $(0, \pm12)$; co-vertices: $(\pm7, 0)$; foci: $\left(0, \pm\sqrt{95}\right)$

34. 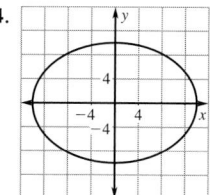 ;

vertices: $(\pm14, 0)$; co-vertices: $(0, \pm10)$; foci: $\left(\pm4\sqrt{6}, 0\right)$

35. 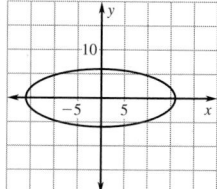 ;

vertices: $(\pm16, 0)$; co-vertices: $(0, \pm6)$; foci: $\left(\pm2\sqrt{55}, 0\right)$

36. 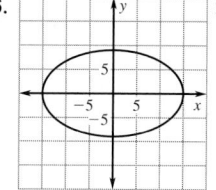 ;

vertices: $(\pm15, 0)$; co-vertices: $(0, \pm9)$; foci: $(\pm12, 0)$

37. 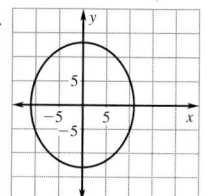 ;

vertices: $(0, \pm13)$; co-vertices: $(\pm11, 0)$; foci: $\left(0, \pm4\sqrt{3}\right)$

38. 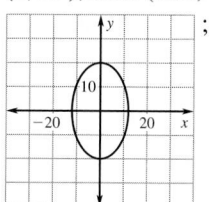 ;

vertices: $(0, \pm20)$; co-vertices: $(\pm12, 0)$; foci: $(0, \pm16)$

39. 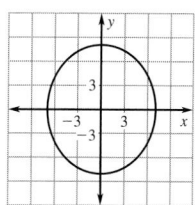 ;

vertices: $(0, \pm8)$; co-vertices: $(\pm7, 0)$; foci: $\left(0, \pm\sqrt{15}\right)$

40. 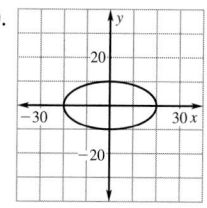 ;

vertices: $(\pm20, 0)$; co-vertices: $(0, \pm10)$; foci: $\left(\pm10\sqrt{3}, 0\right)$

41. 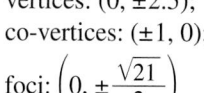 ; vertices: $(0, \pm2.5)$; co-vertices: $(\pm1, 0)$; foci: $\left(0, \pm\dfrac{\sqrt{21}}{2}\right)$

42.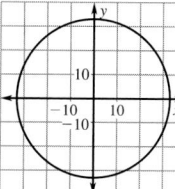

43.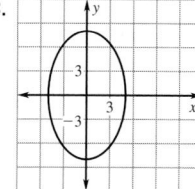

44.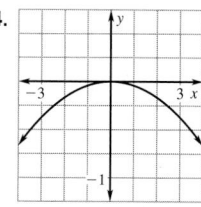

45.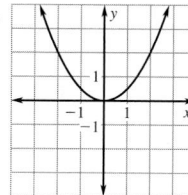

46.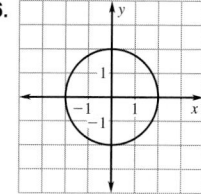

47.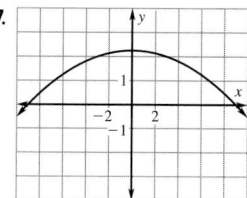

48.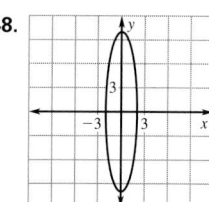

49.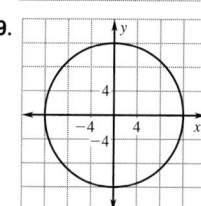

50.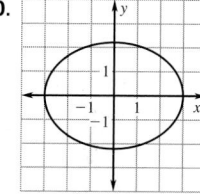

92. 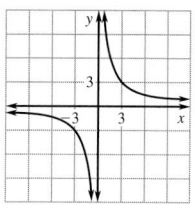 ;

domain: all reals except 0; range: all reals except 0

93. 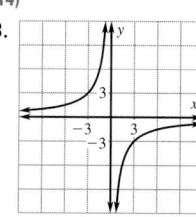 ;

domain: all reals except 0; range: all reals except 0

94. 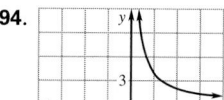 ;

domain: all reals except 0; range: all reals except 0

ADDITIONAL ANSWERS

95. 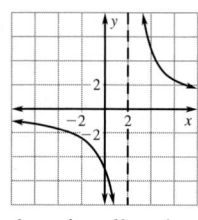 ;

domain: all reals except 0; range: all reals except 0

96. 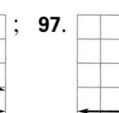 ;

domain: all reals except 2; range: all reals except 0

97. 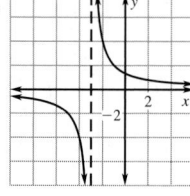 ;

domain: all reals except −3; range: all reals except 0

33. 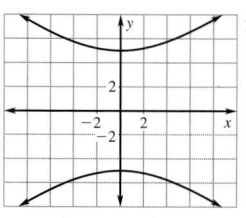 ;

foci: $\left(0, \pm\sqrt{74}\right)$; asymptotes: $y = \pm\frac{5}{7}x$

34. ;

foci: $\left(0, \pm\sqrt{109}\right)$; asymptotes: $y = \pm\frac{3}{10}x$

10.5 GUIDED PRACTICE (p. 618)

4. ;

foci: $\left(\pm\sqrt{130}, 0\right)$; asymptotes: $y = \pm\frac{9}{7}x$

5. ;

foci: $\left(0, \pm5\sqrt{7}\right)$; asymptotes: $y = \pm\frac{2\sqrt{3}}{3}x$

35. 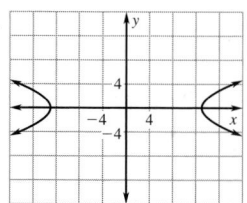 ;

foci: $\left(\pm\sqrt{185}, 0\right)$; asymptotes: $y = \pm\frac{4}{13}x$

36. 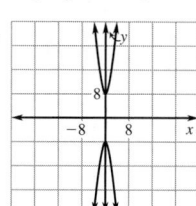 ;

foci: $\left(0, \pm\sqrt{65}\right)$; asymptotes: $y = \pm8x$

6. ;

foci: $\left(\pm\sqrt{65}, 0\right)$; asymptotes: $y = \pm\frac{1}{8}x$

7. ;

foci: $\left(\pm2\sqrt{10}, 0\right)$; asymptotes: $y = \pm3x$

37. 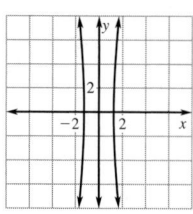 ;

foci: $\left(\pm\frac{\sqrt{1321}}{4}, 0\right)$; asymptotes: $y = \pm\frac{36}{5}x$

38. ;

foci: $\left(\pm\sqrt{265}, 0\right)$; asymptotes: $y = \pm\frac{11}{12}x$

8. ;

foci: $\left(0, \pm\sqrt{37}\right)$; asymptotes: $y = \pm\frac{5\sqrt{3}}{6}x$

9. ;

foci: $\left(0, \pm\sqrt{10}\right)$; asymptotes: $y = \pm3x$

39. ;

foci: $\left(\pm\frac{2\sqrt{145}}{3}, 0\right)$; asymptotes: $y = \pm\frac{1}{12}x$

40. ; foci: $\left(0, \pm4\sqrt{41}\right)$; asymptotes: $y = \pm\frac{5}{4}x$

10.5 PRACTICE AND APPLICATIONS (pp. 618–620)

31. 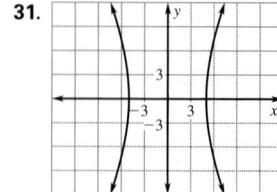 ;

foci: $\left(\pm\sqrt{146}, 0\right)$; asymptotes: $y = \pm\frac{11}{5}x$

32. 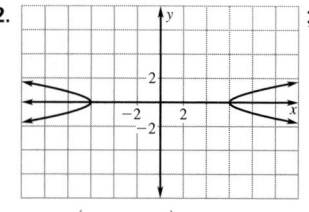 ;

foci: $\left(\pm\sqrt{37}, 0\right)$; asymptotes: $y = \pm\frac{1}{6}x$

41. ;

foci: $\left(\pm\sqrt{181}, 0\right)$; asymptotes: $y = \pm\frac{10}{9}x$

42. ;

foci: $\left(\pm\frac{5\sqrt{10}}{3}, 0\right)$; asymptotes: $y = \pm\frac{1}{3}x$

ADDITIONAL ANSWERS

49. *Sample answer:* 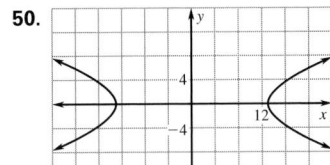 ;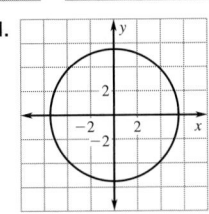

50.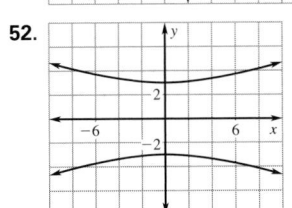

51.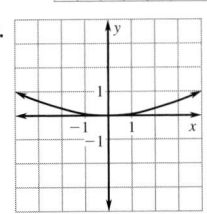

52.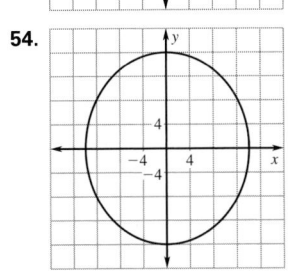

53.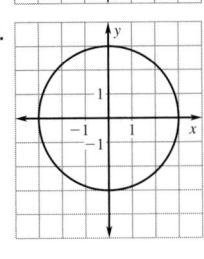

54.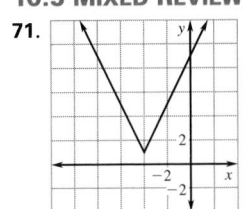

55.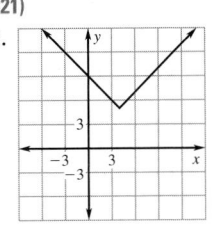

10.5 MIXED REVIEW (p. 621)

71.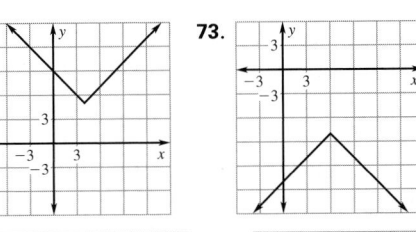

72.

73.

74.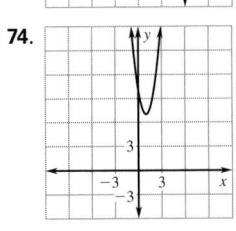

75.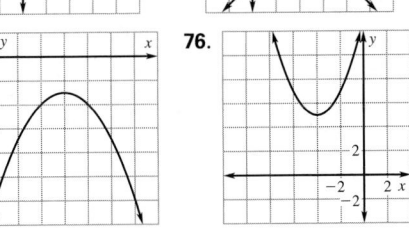

76.

QUIZ 2 (p. 621)

7.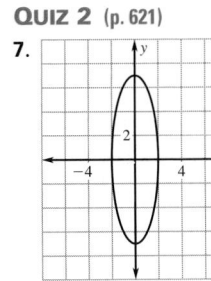

8.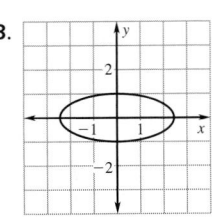

9.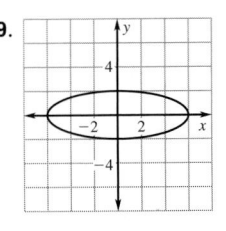

14.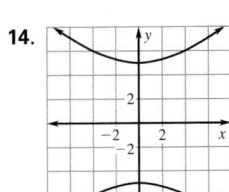

15.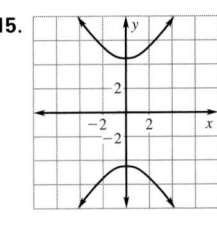

16.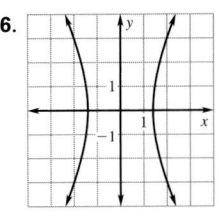

10.6 GUIDED PRACTICE (p. 628)

12.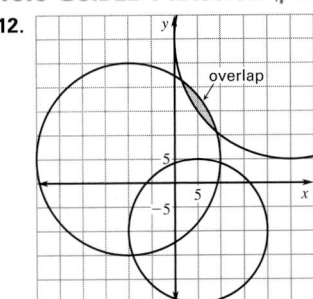

overlap

10.6 PRACTICE AND APPLICATIONS (pp. 628–630)

21. 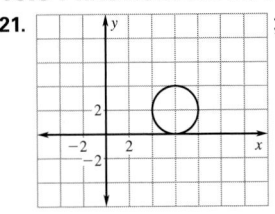 ;

center: (6, 2); points: (6, 4), (6, 0), (4, 2), (8, 2)

22. 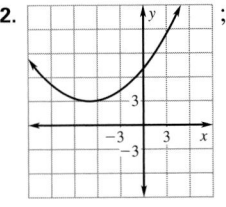 ;

vertex: (−7, 3); focus: (−7, 6); directrix: $y = 0$

23. 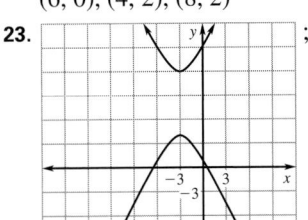 ;

center: (−3, 8);
vertices: (−3, 4), (−3, 12);
foci: $\left(-3, 8 \pm 2\sqrt{5}\right)$

24. 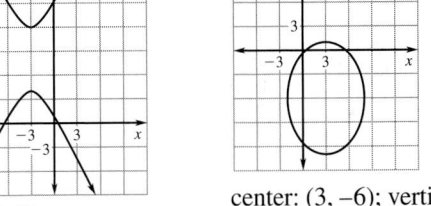 ;

center: (3, −6); vertices: (3, −13), (3, 1); co-vertices: (−2, −6), (8, −6);
foci: $\left(3, -6 \pm 2\sqrt{6}\right)$

25. 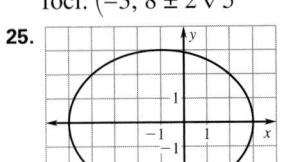 ;

center: (−1, 0); vertices: (−5, 0), (3, 0); co-vertices: (−1, −3), (−1, 3); foci: $\left(-1 \pm \sqrt{7}, 0\right)$

26. 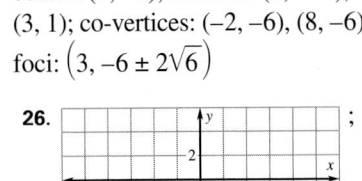 ;
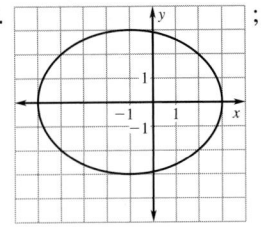

center: (0, −4);
vertices: (4, −4), (−4, −4);
foci: $\left(\pm\sqrt{17}, -4\right)$

ADDITIONAL ANSWERS

27.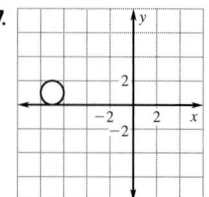

;

center: (−7, 1); points: (−8, 1), (−6, 1), (−7, 0), (−7, 2)

28.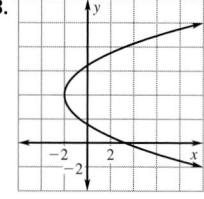

;

vertex: (−2, 4); focus: $\left(-\dfrac{5}{4}, 4\right)$; directrix: $x = -\dfrac{11}{4}$

34. ellipse; $(x + 4)^2 + \dfrac{(y - 1)^2}{9} = 1$;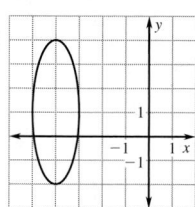

35. hyperbola; $(y - 9)^2 - \dfrac{(x + 1)^2}{\left(\frac{1}{4}\right)} = 1$;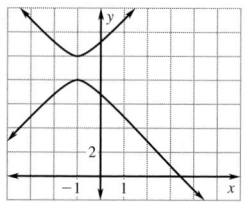

37. $\left(\dfrac{-27 - \sqrt{1649}}{10}, \dfrac{141 + 3\sqrt{1649}}{10}\right); \left(\dfrac{-27 + \sqrt{1649}}{10}, \dfrac{141 - 3\sqrt{1649}}{10}\right) \approx$ (−6.761, 26.28); (1.361, 1.918)

58.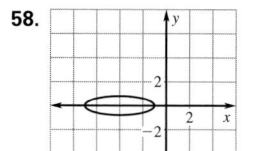

59.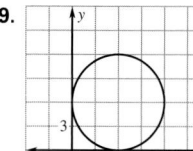

60.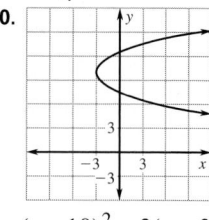

$\dfrac{9(x + 4)^2}{85} + \dfrac{81\left(y + \frac{2}{9}\right)^2}{85} = 1$; $(x - 6)^2 + (y - 6)^2 = 36$; $(y - 10)^2 = 2(x + 3)$

61.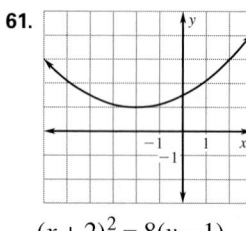

$(x + 2)^2 = 8(y - 1)$

62.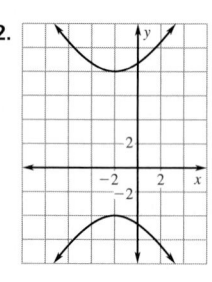

$\dfrac{(y - 2)^2}{36} - \dfrac{(x + 2)^2}{16} = 1$

CHAPTER 10 TEST (p. 645)

12.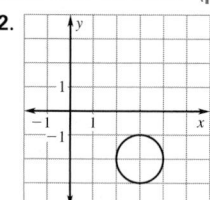

13.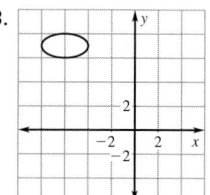

14.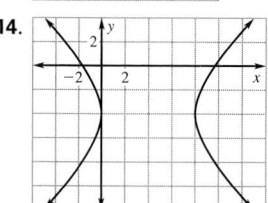

15.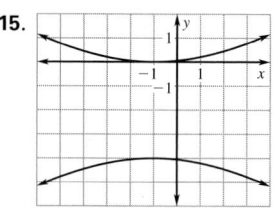

24. ellipse; $\dfrac{(x - 1)^2}{4} + y^2 = 1$ **25.** parabola; $(x + 5)^2 = \dfrac{1}{2}(y + 9)$

26. hyperbola; $\dfrac{x^2}{6} - \dfrac{y^2}{10} = 1$ **27.** circle; $(x - 6)^2 + (y + 2)^2 = 9$

28. parabola; $(y - 2)^2 = 8x$ **29.** hyperbola; $\dfrac{(y - 3)^2}{4} - \dfrac{(x + 3)^2}{4} = 1$

30. parabola; $(x - 4)^2 = -4y$ **31.** circle; $(x - 5)^2 + y^2 = \dfrac{16}{3}$

32. ellipse; $\dfrac{(x - 4)^2}{9} + \dfrac{y^2}{\left(\frac{9}{2}\right)} = 1$ **33.** hyperbola; $\dfrac{(x + 2)^2}{\left(\frac{5}{2}\right)} - \dfrac{(y - 3)^2}{10} = 1$

34. ellipse; $\dfrac{x^2}{\left(\frac{1}{3}\right)} + (y - 2)^2 = 1$ **35.** circle; $(x - 1)^2 + (y + 5)^2 = 25$

36. $\left(\dfrac{17 - 2\sqrt{31}}{5}, \dfrac{-34 - \sqrt{31}}{5}\right); \left(\dfrac{17 + 2\sqrt{31}}{5}, \dfrac{-34 + \sqrt{31}}{5}\right) \approx$ (1.173, −7.914); (5.627, −5.686) **39.**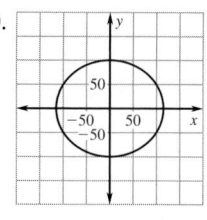

10.7 ACTIVITY (p. 632) *Sample answers:*

a.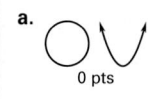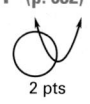

0 pts   2 pts   4 pts

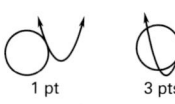

1 pt   3 pts

b.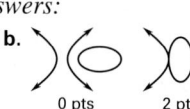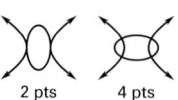

0 pts   2 pts   4 pts

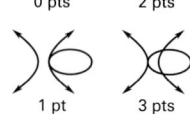

1 pt   3 pts

c.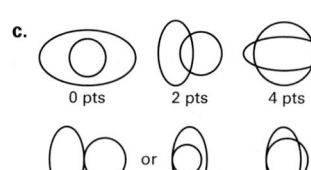

0 pts   2 pts   4 pts

or

1 pt   3 pts

d.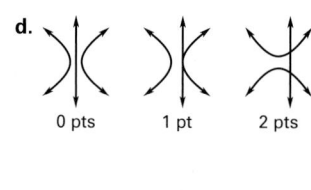

0 pts   1 pt   2 pts

CHAPTER 10 REVIEW (pp. 642–644)

8. focus: (3, 0); directrix: $x = -3$;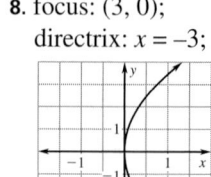

22.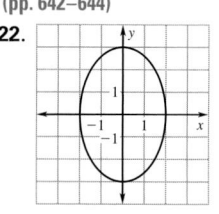

23.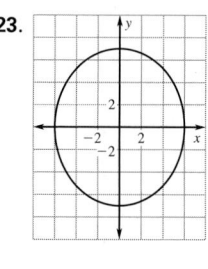

CHAPTER 11

7.

	$f(0)$	$f(1)$	$f(2)$	$f(3)$	$f(4)$	$f(5)$	$f(6)$
function values	3	0	–9	–24	–45	–72	–105
1st order differences	–3	–9	–15	–21	–27	–33	
2nd order differences	–6	–6	–6	–6	–6		

8.

	$f(1)$	$f(2)$	$f(3)$	$f(4)$	$f(5)$	$f(6)$
function values	3	16	45	96	175	288
1st order differences	13	29	51	79	113	
2nd order differences	16	22	28	34		
3rd order differences	6	6	6			

9.

	$f(1)$	$f(2)$	$f(3)$	$f(4)$	$f(5)$	$f(6)$
function values	–3	7	67	237	601	1267
1st order differences	10	60	170	364	666	
2nd order differences	50	110	194	302		
3rd order differences	60	84	108			
4th order differences	24	24				

11.1 PRACTICE AND APPLICATIONS (pp. 655–657)

30.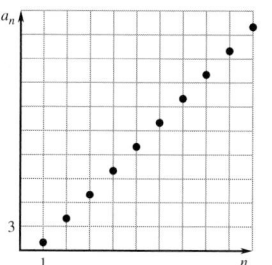

31.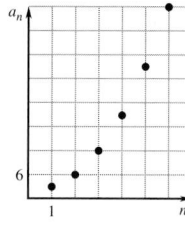

32.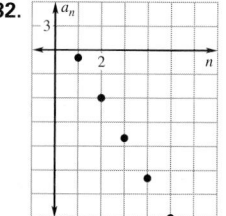

33.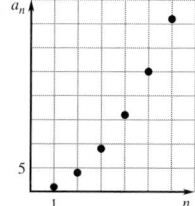

34.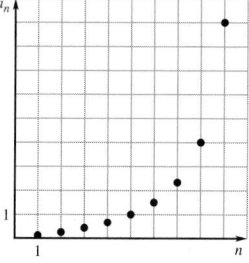

35.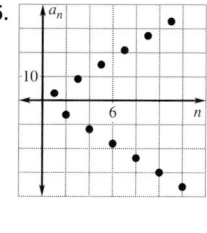

71. c. false; $\sum_{i=1}^{4} i(i+1) = 1(2) + 2(3) + 3(4) + 4(5) = 2 + 6 + 12 + 20 = 40$,

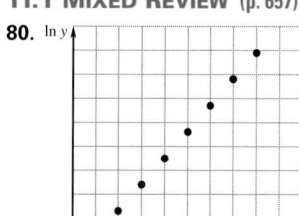

but $\sum_{i=1}^{4} i = 10$ and $\sum_{i=1}^{4} (i+1) = 14$ and $10 \times 14 = 140 \neq 40$.

d. false; $\sum_{i=1}^{5} (i)^2 = 1 + 4 + 9 + 16 + 25 = 55$, but $\left(\sum_{i=1}^{5} i \right)^2 = 15^2 = 225$.

11.1 MIXED REVIEW (p. 657)

80.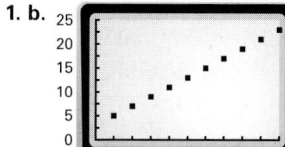

11.1 TECHNOLOGY ACTIVITY (p. 658)

1. b.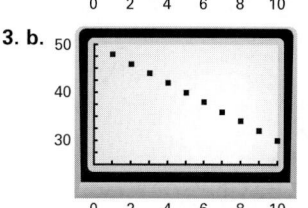

2. b.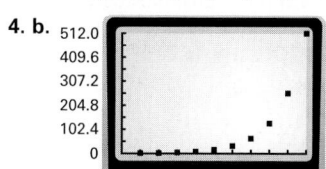

3. b.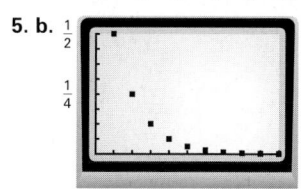

4. b.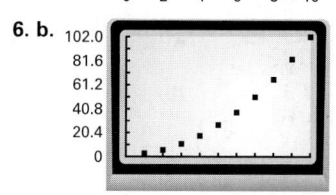

5. b.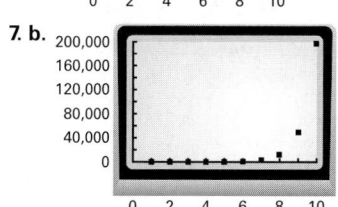

6. b.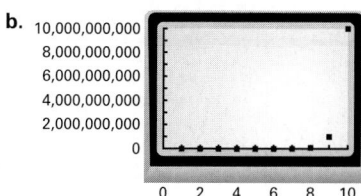

7. b.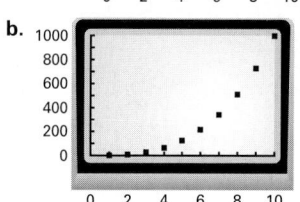

8. b.

9. b.

39. 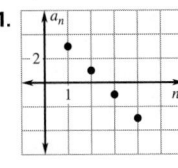 **40.** **41.**

42. 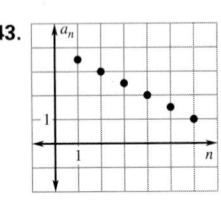 **43.** 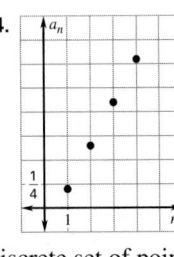 **44.**

61. *Sample answer:* The graph of $a_n = 2n + 1$ is a discrete set of points, while that of $f(x) = 2x + 1$ is a continuous straight line. Each point on the graph of a_n is also a point on the line f. The graph of a_n is located entirely in the 1st quadrant, while the graph of the line also contains points in the 2nd and 3rd quadrants. The graph of an arithmetic sequence is always a discrete set of collinear points, evenly spaced from one another. The graph of an infinite sequence will extend indefinitely in one direction, while that of a line extends in two directions.

63. The first man gets $\frac{7}{16}$ of a hekat of barley, the second $\frac{9}{16}$, the third $\frac{11}{16}$, and so on, with the tenth man getting $\frac{25}{16}$ hekats.

11.2 MIXED REVIEW (p. 665)

76. 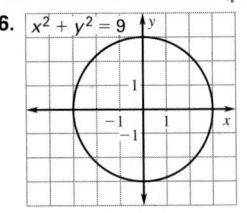 $x^2 + y^2 = 9$

77. 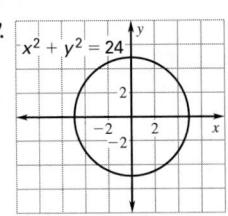 $x^2 + y^2 = 24$

78. $x^2 + y^2 = 64$

79. $x^2 + y^2 = 25$

80. 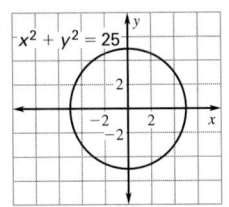 $x^2 + y^2 = 8$

81. 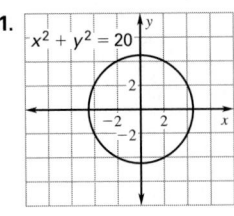 $x^2 + y^2 = 20$

57. 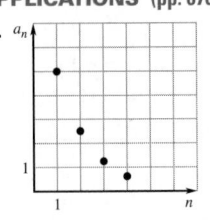 **58.** 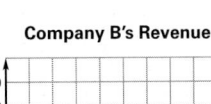 **59.**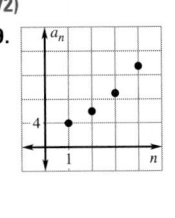

11.3 MIXED REVIEW (p. 673)

81. b.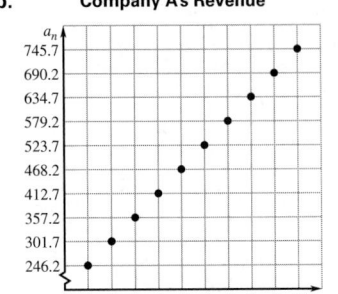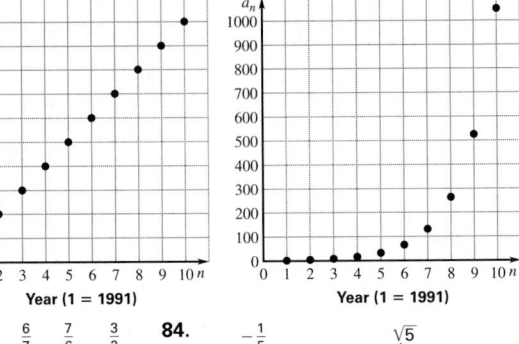

Company A's Revenue   Company B's Revenue

Year (1 = 1991)   Year (1 = 1991)

83. $\frac{2}{5} \quad \frac{6}{7} \quad \frac{7}{6} \quad \frac{3}{2}$ **84.** $-\frac{1}{5} \quad \sqrt{5}$

85. $-3.2 \quad -\frac{5}{2} \, -2 \, -1 \qquad 1.5$

11.4 MIXED REVIEW (p. 680)

63. 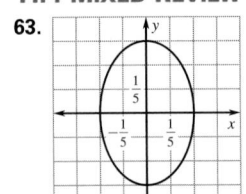 **64.**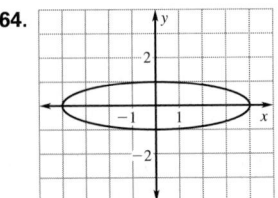

CHAPTER 11 EXTENSION (pp. 689–690)

8. The first several terms of the Fibonocci sequence are 1, 1, 2, 3, 5, 8, 13, . . . , and for each $n > 2$, $f_n = f_{n-1} + f_{n-2}$. $f_1 = 1 = 2 - 1 = f_3 - 1$, so the formula is true for $n = 1$. Suppose the formula is true for $n = k$.

Then $\sum_{i=1}^{k} f_i = f_{n+2} - 1$. So $\sum_{i=1}^{k+1} f_i = f_{k+2} - 1 + f_{k+1} = (f_{k+2} + f_{k+1}) - 1 = f_{k+3} - 1 = f_{(k+1)+2} - 1$, and the formula is true for $n = k + 1$. Therefore, the formula is true for all positive integers.

CHAPTER 12

12.1 GUIDED PRACTICE (p. 705) **2.** The number of permutations of n distinct objects is $n! = n \cdot (n-1) \cdot (n-2) \cdot \ldots \cdot 3 \cdot 2 \cdot 1$. Any of the n objects can be in the first position, any of the remaining $n - 1$ objects can be in the second position, and so on until there is only one object left for the last position. Thus, there are n ways to fill position 1, times $n - 1$ ways to fill position 2, times $n - 2$ ways to fill position 3, and so on. By the fundamental counting principle for three or more events, we know that the number of ways this can be done is $n(n-1)(n-2) \ldots 1$.

ADDITIONAL ANSWERS

12.1 MIXED REVIEW (p. 707)

76.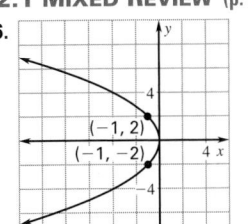

77.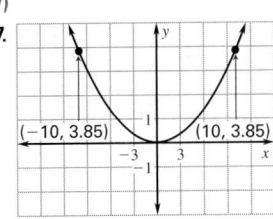

78.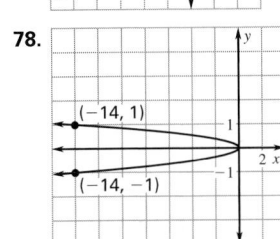

79.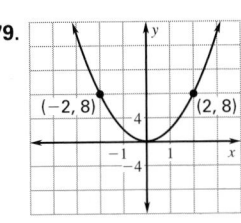

12.2 GUIDED PRACTICE (p. 712)

3. $x^4 + 4x^3y + 6x^2y^2 + 4xy^3 + y^4$; $x^4 - 4x^3y + 6x^2y^2 - 4xy^3 + y^4$; They are similar in the exponents of x and y and the magnitude of the coefficients. They are different in the signs of the coefficients for each term in which y is raised to an odd power (y and y^3).

4. $_{10}C_6 = \dfrac{10!}{4! \cdot 6!} = \dfrac{10 \cdot 9 \cdot 8 \cdot 7}{4 \cdot 3 \cdot 2} = 210$. In the incorrect version, the denominator is 6! when it should be 4!.

12.2 PRACTICE AND APPLICATIONS (pp. 712–714)

60.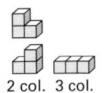

1 col. 2 col. 3 col.

61.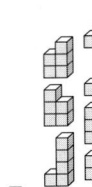

1 col. 2 col. 3 col. 4 col. 5 col.

70. $_nC_1 = \dfrac{n!}{(n-1)! \cdot 1!} = \dfrac{n!}{(n-1)!} = {}_nP_1$

71. $_nC_r = \dfrac{n!}{(n-(n-r))! \cdot (n-r)!} = \dfrac{n!}{r! \cdot (n-r)!} = {}_nC_{n-r}$

72. $_nC_r \cdot {}_rC_m = \dfrac{n!}{r! \cdot (n-r)!} \cdot \dfrac{r!}{m! \cdot (r-m)!} = \dfrac{n!}{m! \cdot (n-m)!} \cdot$

$\dfrac{(n-m)!}{(r-m)! \cdot (n-m-(r-m))!} = {}_nC_m \cdot {}_{n-m}C_{r-m}$

73. $_nC_r + {}_nC_{r-1} = \dfrac{n!}{(n-r)! \cdot r!} + \dfrac{n!}{(n-(r-1))! \cdot (r-1)!} =$

$\dfrac{n! \cdot (n+1-r)}{(n-r)! \cdot r! \cdot (n+1-r)} + \dfrac{n! \cdot r}{(n+1-r)! \cdot (r-1)! \cdot r} =$

$\dfrac{n! \cdot (n+1-r) + n! \cdot r}{(n+1-r)! \cdot r!} = \dfrac{n! \cdot (n+1-r+r)}{(n+1-r)! \cdot r!} =$

$\dfrac{n! \cdot (n+1)}{r! \cdot (n+1-r)!} = \dfrac{(n+1)!}{r! \cdot (n+1-r)!} = {}_{n+1}C_r$

QUIZ 1 (p. 715)

9. $x^6 + 6x^5y + 15x^4y^2 + 20x^3y^3 + 15x^2y^4 + 6xy^5 + y^6$

10. $x^4 + 8x^3 + 24x^2 + 32x + 16$ **11.** $x^5 - 10x^4y + 40x^3y^2 - 80x^2y^3 + 80xy^4 - 32y^5$ **12.** $27x^3 - 108x^2y + 144xy^2 - 64y^3$ **13.** $x^8 + 12x^6y + 54x^4y^2 + 108x^2y^3 + 81y^4$ **14.** $4096x^{12} - 12{,}288x^{10} + 15{,}360x^8 - 10{,}240x^6 + 3840x^4 - 768x^2 + 64$ **15.** $x^9 - 3x^6y^3 + 3x^3y^6 - y^9$

16. $32x^{20} + 400x^{16}y^2 + 2000x^{12}y^4 + 5000x^8y^6 + 6250x^4y^8 + 3125y^{10}$

12.3 TECHNOLOGY ACTIVITY (p. 723)

1.

number	1	2	3	4	5	6
freq	19	24	20	16	20	21
theor prob	0.167	0.167	0.167	0.167	0.167	0.167
exp prob	0.158	0.200	0.167	0.133	0.167	0.175

The experimental and theoretical probabilities are close but not the same.

2.

card	1	2	3	4	5	6	7
freq	11	3	3	3	1	7	4
theor prob	0.077	0.077	0.077	0.077	0.077	0.077	0.077
exp prob	0.212	0.058	0.058	0.058	0.019	0.135	0.077

card	8	9	10	11	12	13
freq	1	2	2	3	7	5
theor prob	0.077	0.077	0.077	0.077	0.077	0.077
exp prob	0.019	0.038	0.038	0.058	0.135	0.096

The experimental and theoretical probabilities are close but not the same.

3.

trials	10	20	50	100	200
heads	4	13	32	52	100
tails	6	7	18	48	100

As the number of trials increases, the experimental results get closer to the theoretical results.

12.6 CONCEPT ACTIVITY (p. 738)

Step 3. *Sample answer:*

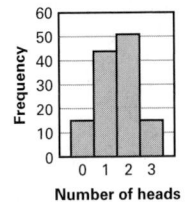

Step 4.

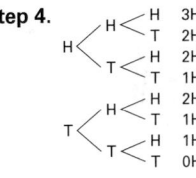

Step 5.

Number of heads	Theoretical Probability
0	$\dfrac{1}{8}$
1	$\dfrac{3}{8}$
2	$\dfrac{3}{8}$
3	$\dfrac{1}{8}$

Step 6.

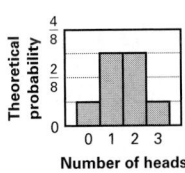

ADDITIONAL ANSWERS

12.6 GUIDED PRACTICE (p. 742)

2. No; it is not a binomial experiment because you are not replacing the cards. The probability of success differs for each trial.

6. 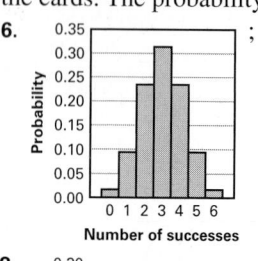 ; 3 **7.** 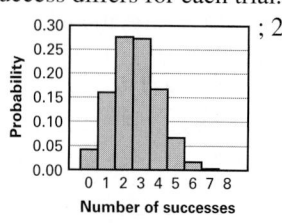 ; 2

8. 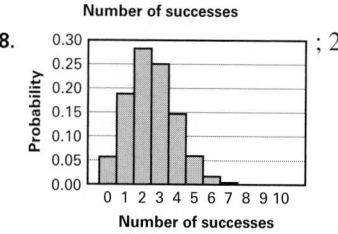 ; 2

12.6 PRACTICE AND APPLICATIONS (pp. 742–744)

30. 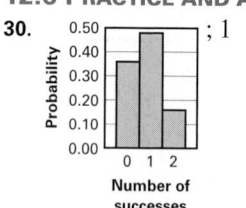 ; 1 **31.** 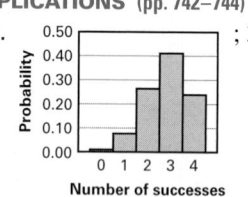 ; 3

32. 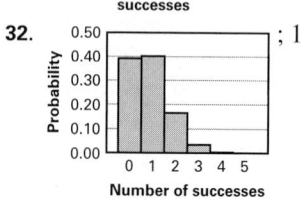 ; 1 **33.** 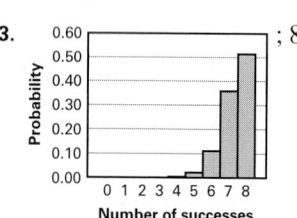 ; 8

34. 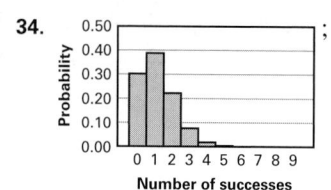 ; 1 **35.** 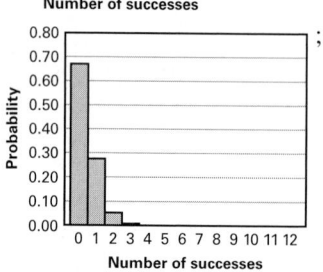 ; 0

42.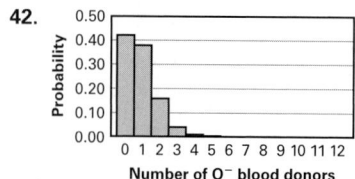

44.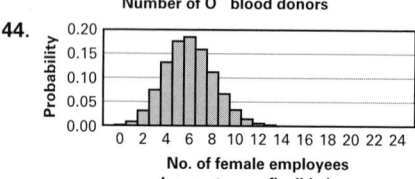

12.6 TECHNOLOGY ACTIVITY (p. 745)

1. 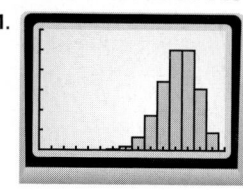 **2.**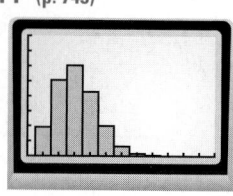

CHAPTER 13

13.2 GUIDED PRACTICE (p. 780)

5. 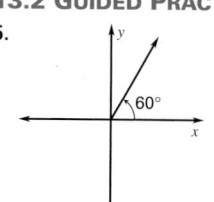 **6.** 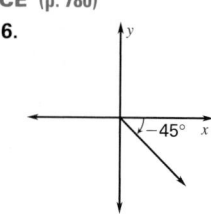 **7.**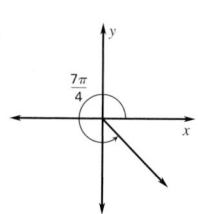

8. 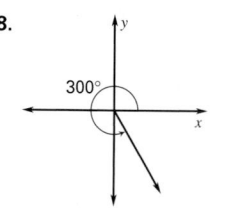 **9.** 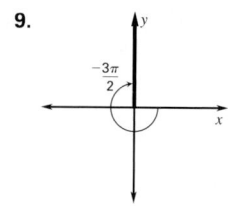 **10.**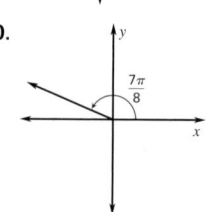

11. 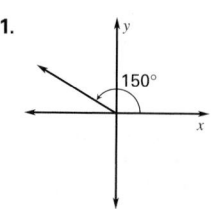 **12.**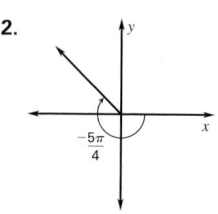

13.2 PRACTICE AND APPLICATIONS (pp. 780–782)

34. 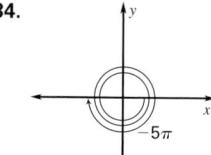 **35.**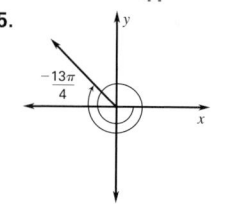

13.3 GUIDED PRACTICE (p. 788)

3. *Sample answer:* The tangent of an angle θ is defined as the ratio $\frac{y}{x}$ for some point (x, y) other than the origin on the angle's terminal side. If $\theta = 270°$, the terminal side is the negative y-axis, where all the x-coordinates are zero. Division by zero is undefined.

6. 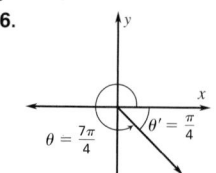 **7.** 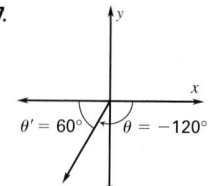 **8.**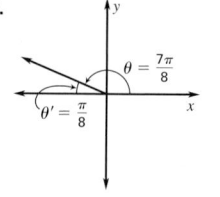

9.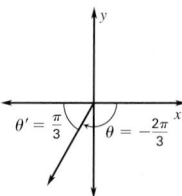

10.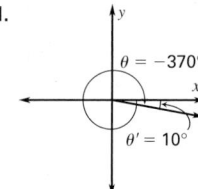

11.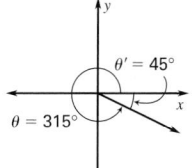

40.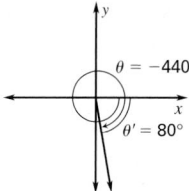

41.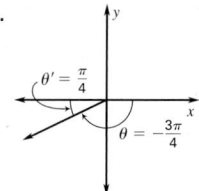

42.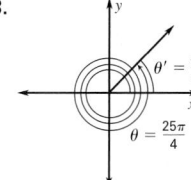

12.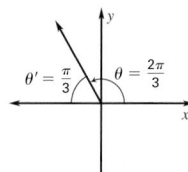

13.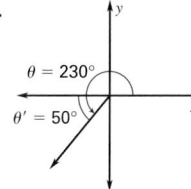

43.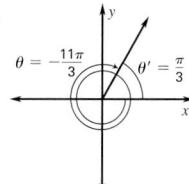

44.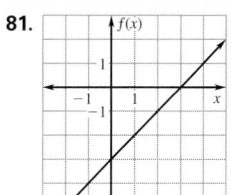

13.3 PRACTICE AND APPLICATIONS (pp. 788–790)

25. $\sin\theta = \dfrac{14\sqrt{277}}{277}$; $\cos\theta = -\dfrac{9\sqrt{277}}{277}$; $\tan\theta = -\dfrac{14}{9}$;

$\csc\theta = \dfrac{\sqrt{277}}{14}$; $\sec\theta = -\dfrac{\sqrt{277}}{9}$; $\cot\theta = -\dfrac{9}{14}$ **26.** $\sin\theta = -\dfrac{5\sqrt{41}}{41}$;

$\cos\theta = -\dfrac{4\sqrt{41}}{41}$; $\tan\theta = \dfrac{5}{4}$; $\csc\theta = -\dfrac{\sqrt{41}}{5}$; $\sec\theta = -\dfrac{\sqrt{41}}{4}$; $\cot\theta = \dfrac{4}{5}$

27. $\sin\theta = \dfrac{\sqrt{2}}{2}$; $\cos\theta = -\dfrac{\sqrt{2}}{2}$; $\tan\theta = -1$; $\csc\theta = \sqrt{2}$; $\sec\theta = -\sqrt{2}$;

$\cot\theta = -1$ **28.** $\sin\theta = -\dfrac{8}{17}$; $\cos\theta = \dfrac{15}{17}$; $\tan\theta = -\dfrac{8}{15}$; $\csc\theta = -\dfrac{17}{8}$;

$\sec\theta = \dfrac{17}{15}$; $\cot\theta = -\dfrac{15}{8}$ **29.** $\sin\theta = -\dfrac{3\sqrt{13}}{13}$; $\cos\theta = \dfrac{2\sqrt{13}}{13}$; $\tan\theta = -\dfrac{3}{2}$;

$\csc\theta = -\dfrac{\sqrt{13}}{3}$; $\sec\theta = \dfrac{\sqrt{13}}{2}$; $\cot\theta = -\dfrac{2}{3}$ **30.** $\sin\theta = \dfrac{10\sqrt{149}}{149}$;

$\cos\theta = \dfrac{7\sqrt{149}}{149}$; $\tan\theta = \dfrac{10}{7}$; $\csc\theta = \dfrac{\sqrt{149}}{10}$; $\sec\theta = \dfrac{\sqrt{149}}{7}$; $\cot\theta = \dfrac{7}{10}$

31. $\sin\theta = -\dfrac{\sqrt{3}}{2}$; $\cos\theta = \dfrac{1}{2}$; $\tan\theta = -\sqrt{3}$; $\csc\theta = -\dfrac{2\sqrt{3}}{3}$; $\sec\theta = 2$;

$\cot\theta = -\dfrac{\sqrt{3}}{3}$ **32.** $\sin\theta = -\dfrac{4}{5}$; $\cos\theta = -\dfrac{3}{5}$; $\tan\theta = \dfrac{4}{3}$; $\csc\theta = -\dfrac{5}{4}$;

$\sec\theta = -\dfrac{5}{3}$; $\cot\theta = \dfrac{3}{4}$ **33.** $\sin\theta = \dfrac{\sqrt{7}}{4}$; $\cos\theta = -\dfrac{3}{4}$; $\tan\theta = -\dfrac{\sqrt{7}}{3}$;

$\csc\theta = \dfrac{4\sqrt{7}}{7}$; $\sec\theta = -\dfrac{4}{3}$; $\cot\theta = -\dfrac{3\sqrt{7}}{7}$

34. $\sin 90° = 1$; $\cos 90° = 0$; $\tan 90°$ is undefined; $\csc 90° = 1$; $\sec 90°$ is undefined; $\cot 90° = 0$.   **35.** $\sin 270° = -1$; $\cos 270° = 0$; $\tan 270°$ is undefined; $\csc 270° = -1$; $\sec 270°$ is undefined; $\cot 270° = 0$.
36. $\sin 0° = 0$; $\cos 0° = 1$; $\tan 0° = 0$; $\csc 0°$ is undefined; $\sec 0° = 1$; $\cot 0°$ is undefined.

37.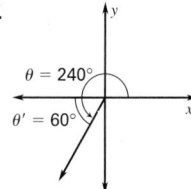

38.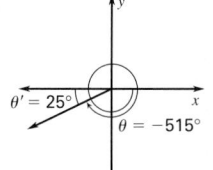

39.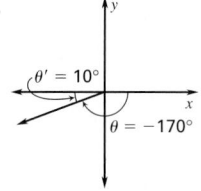

13.3 MIXED REVIEW (p. 790)

81.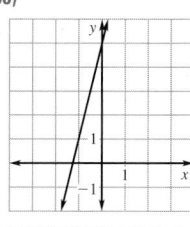

82.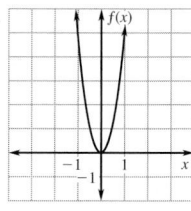

83.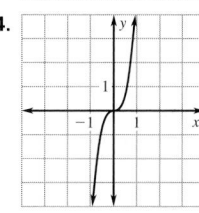

84.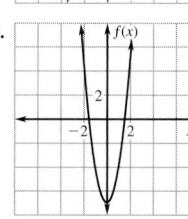

85.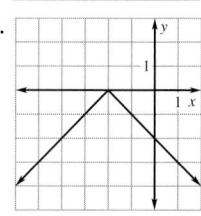

86.

13.4 CONCEPT ACTIVITY (p. 791)
3. *Sample answer:* Let $-\dfrac{\pi}{2} < \theta < \dfrac{\pi}{2}$. The tangent function is defined for each point in this interval, and there are no repeated values for $\tan\theta$. The domain cannot be extended without including values of θ for which $\tan\theta$ is undefined and/or repeated values of $\tan\theta$.   **4.** *Sample answer:* No; each of these functions is periodic and infinitely many different domains are possible.

QUIZ 2 (p. 798)   **5.** $\sin\theta = \dfrac{2\sqrt{5}}{5}$; $\cos\theta = \dfrac{\sqrt{5}}{5}$; $\tan\theta = 2$; $\csc\theta = \dfrac{\sqrt{5}}{2}$;

$\sec\theta = \sqrt{5}$; $\cot\theta = \dfrac{1}{2}$ **6.** $\sin\theta = \dfrac{\sqrt{17}}{17}$; $\cos\theta = -\dfrac{4\sqrt{17}}{17}$; $\tan\theta = -\dfrac{1}{4}$;

$\csc\theta = \sqrt{17}$; $\sec\theta = -\dfrac{\sqrt{17}}{4}$; $\cot\theta = -4$ **7.** $\sin\theta = -\dfrac{5\sqrt{106}}{106}$;

$\cos\theta = \dfrac{9\sqrt{106}}{106}$; $\tan\theta = -\dfrac{5}{9}$; $\csc\theta = -\dfrac{\sqrt{106}}{5}$; $\sec\theta = \dfrac{\sqrt{106}}{9}$; $\cot\theta = -\dfrac{9}{5}$

8. $\sin\theta = -\dfrac{8\sqrt{113}}{113}$; $\cos\theta = -\dfrac{7\sqrt{113}}{113}$; $\tan\theta = \dfrac{8}{7}$; $\csc\theta = -\dfrac{\sqrt{113}}{8}$;

$\sec\theta = -\dfrac{\sqrt{113}}{7}$; $\cot\theta = \dfrac{7}{8}$

ADDITIONAL ANSWERS

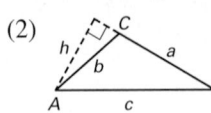

13.5 PRACTICE AND APPLICATIONS (pp. 803–806)

69. area formulas: (1) Area $= \frac{1}{2}ch$, where $\frac{h}{b} = \sin A$, or $h = b \sin A$.

So, area $= \frac{1}{2}c(b \sin A) = \frac{1}{2}bc \sin A$.

(2) $\qquad$ Area $= \frac{1}{2}ah$, where $\frac{h}{b} = \sin(180° - C)$.

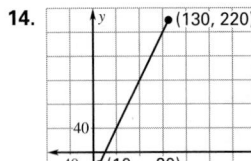

Because $\sin(180° - C) = \sin C$, $h = b \sin C$.

So, area $= \frac{1}{2}a(b \sin C) = \frac{1}{2}ab \sin C$.

(3) Area $= \frac{1}{2}ch$, where $\frac{h}{a} = \sin B$, or $h = a \sin B$.

So, area $= \frac{1}{2}c(a \sin B) = \frac{1}{2}ac \sin B$.

law of sines:

(1) $\frac{1}{2}bc \sin A = \frac{1}{2}ac \sin B$, so $b \sin A = a \sin B$, or $\frac{\sin A}{a} = \frac{\sin B}{b}$.

(2) $\frac{1}{2}bc \sin A = \frac{1}{2}ab \sin C$, so $c \sin A = a \sin C$, or $\frac{\sin A}{a} = \frac{\sin C}{c}$.

(3) $\frac{1}{2}ab \sin C = \frac{1}{2}ac \sin B$, so $b \sin C = c \sin B$, or $\frac{\sin C}{c} = \frac{\sin B}{b}$.

13.7 PRACTICE AND APPLICATIONS (pp. 816–818)

14. 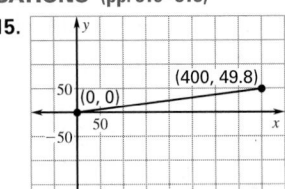 **15.**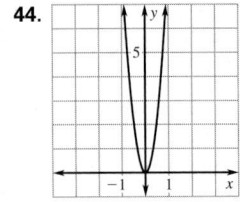

13.7 MIXED REVIEW (p. 819)

44. 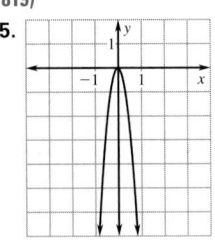 **45.** 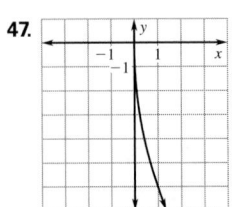 **46.**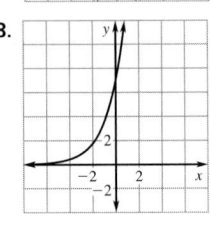

47. 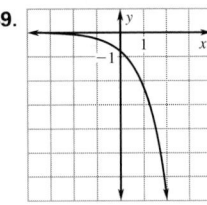 **48.** 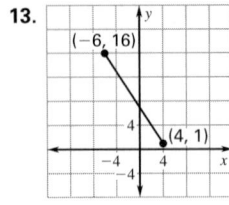 **49.**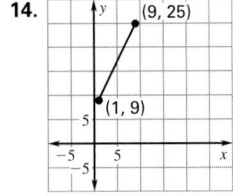

QUIZ 3 (p. 819)

13. 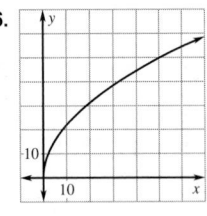 **14.** 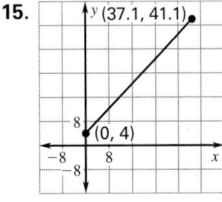 **15.**

CHAPTER 14

SKILL REVIEW (p. 830)

2. 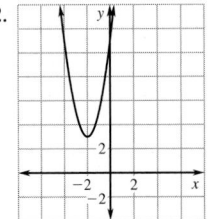 **3.**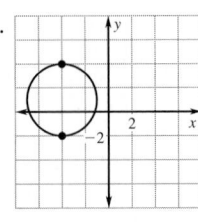

14.1 GUIDED PRACTICE (p. 835)

10. 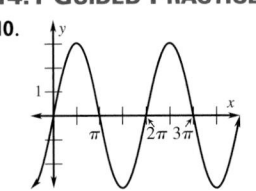 **11.**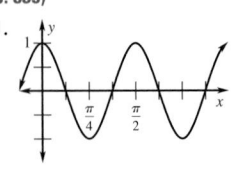

12. 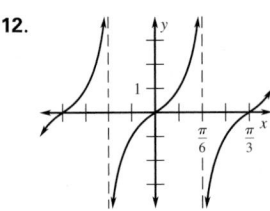 **13.**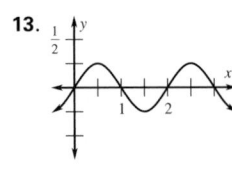

14. 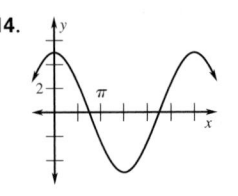 **15.**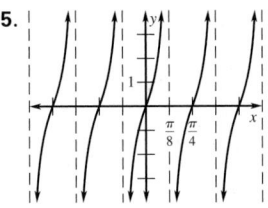

16.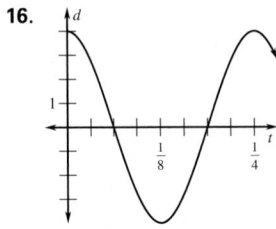

14.1 PRACTICE AND APPLICATIONS (pp. 835–837)

32. 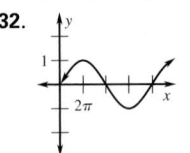 **33.** 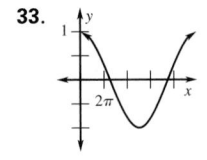 **34.**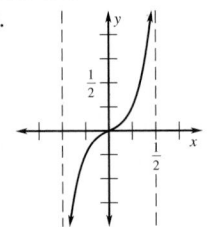

35. 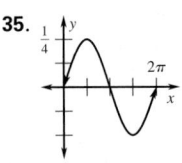 **36.** 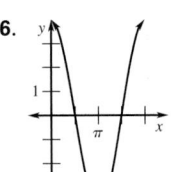 **37.**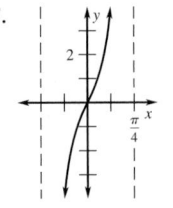

38.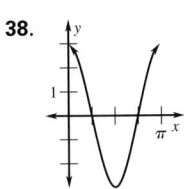

39.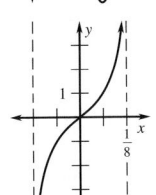

40.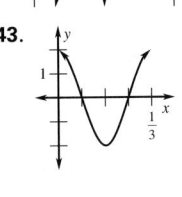

41.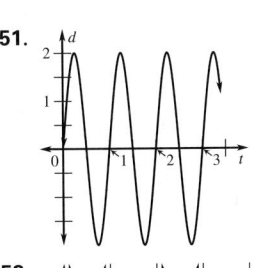

42.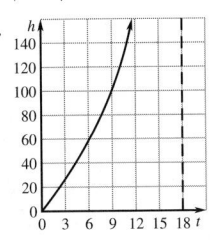

43.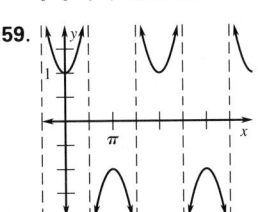

51.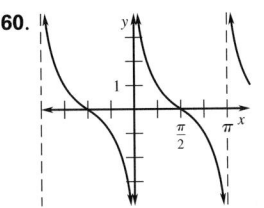

54.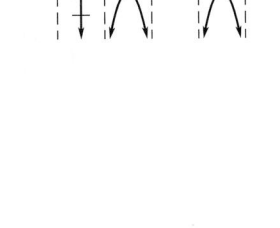

58.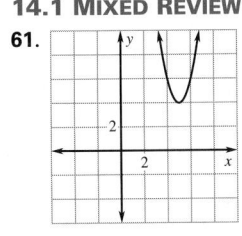

59.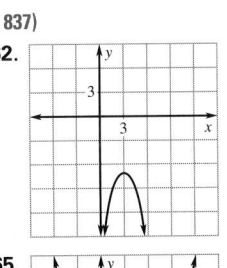

60.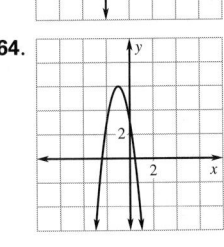

71.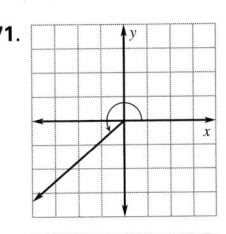

72.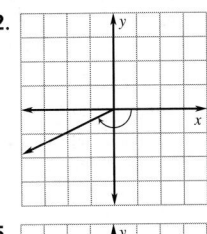

73.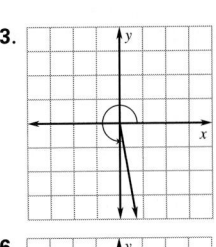

74.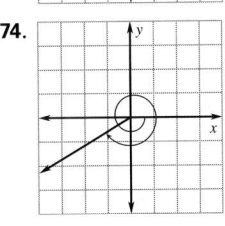

75.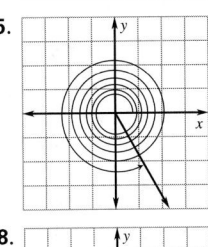

76.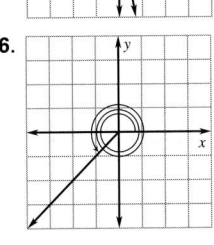

77.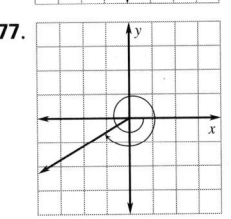

78.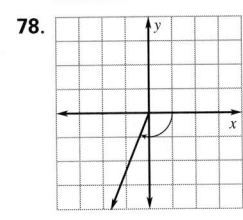

14.2 CONCEPT ACTIVITY (p. 839)

1. a.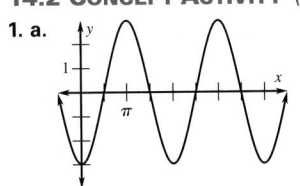

b.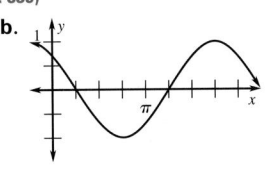

c.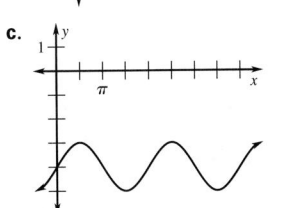

2. a.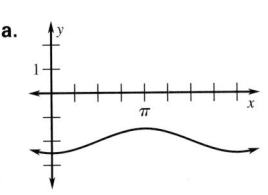

b.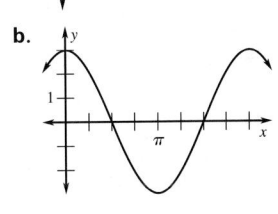

c.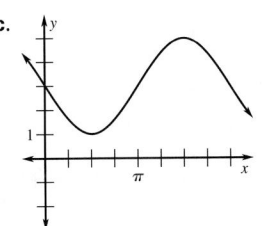

d.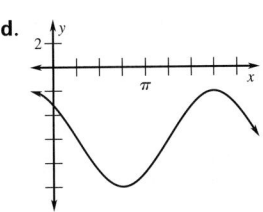

14.1 MIXED REVIEW (p. 837)

61.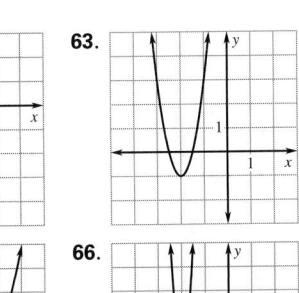

62.

63.

64.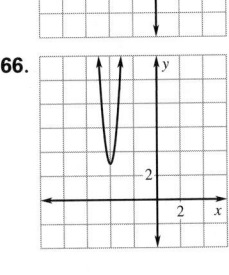

65.

66.

14.2 GUIDED PRACTICE (p. 844)

10.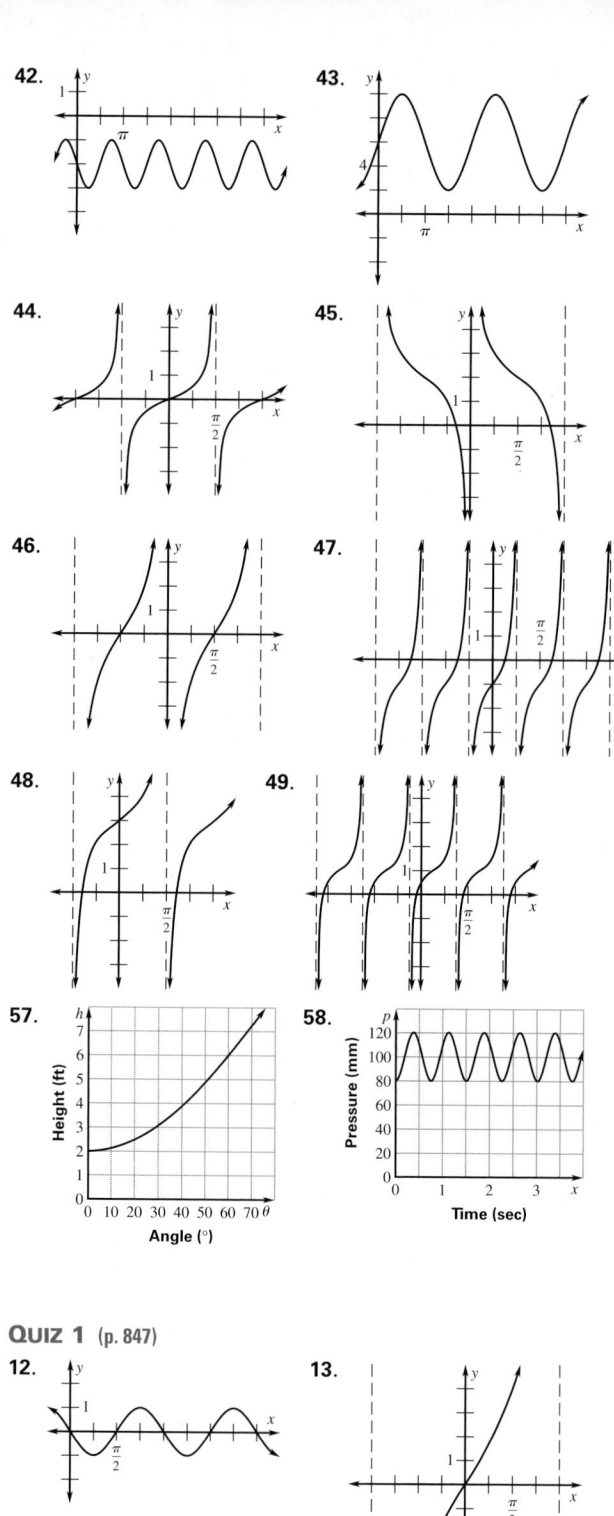

11.

12.

13.

14.

15.

16.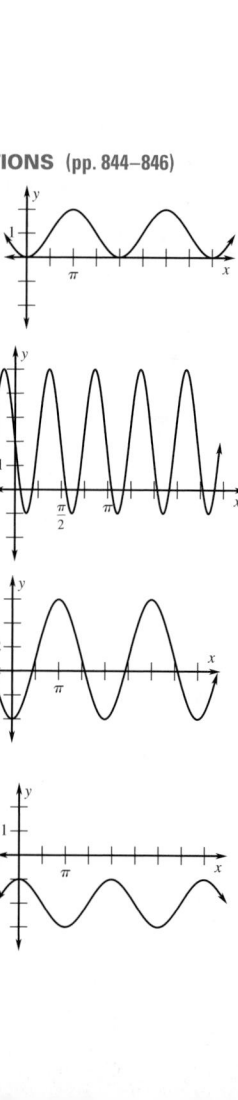

14.2 PRACTICE AND APPLICATIONS (pp. 844–846)

34.

35.

36.

37.

38.

39.

40.

41.

42.

43.

44.

45.

46.

47.

48.

49.

57.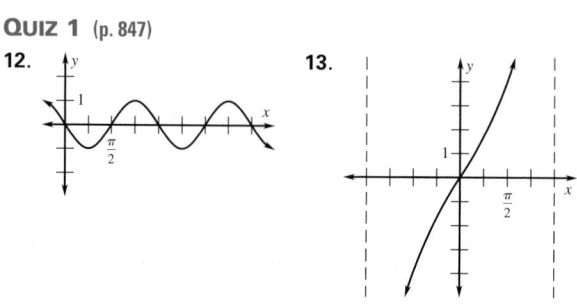

58.

QUIZ 1 (p. 847)

12.

13.

14.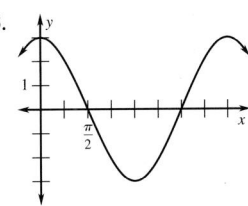

15.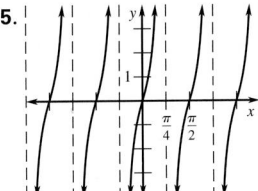

16.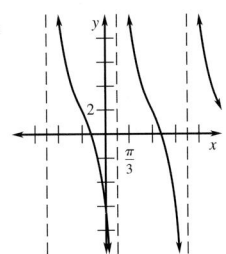

17.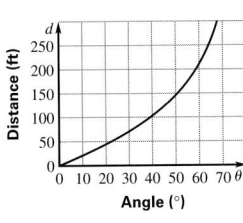

18.

19.

62.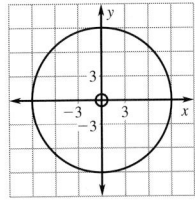

81.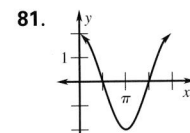

82.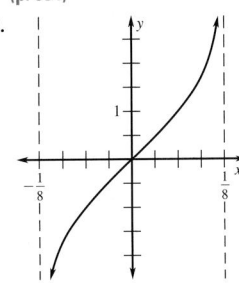

83.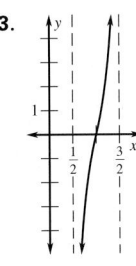

84.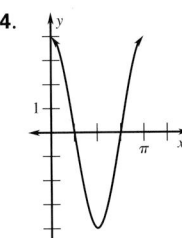

85.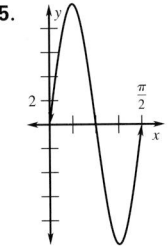

14.3 PRACTICE AND APPLICATIONS (pp. 852–854)

16. $\sin \theta = \dfrac{2\sqrt{5}}{5}$, $\tan \theta = 2$, $\sec \theta = \sqrt{5}$, $\csc \theta = \dfrac{\sqrt{5}}{2}$, $\cot \theta = \dfrac{1}{2}$

17. $\sin \theta = \dfrac{3\sqrt{73}}{73}$, $\cos \theta = \dfrac{8\sqrt{73}}{73}$, $\sec \theta = \dfrac{\sqrt{73}}{8}$, $\csc \theta = \dfrac{\sqrt{73}}{3}$, $\cot \theta = \dfrac{8}{3}$

18. $\cos \theta = \dfrac{\sqrt{11}}{6}$, $\tan \theta = \dfrac{5\sqrt{11}}{11}$, $\sec \theta = \dfrac{6\sqrt{11}}{11}$, $\csc \theta = \dfrac{6}{5}$, $\cot \theta = \dfrac{\sqrt{11}}{5}$

19. $\cos \theta = -\dfrac{4}{5}$, $\tan \theta = -\dfrac{3}{4}$, $\sec \theta = -\dfrac{5}{4}$, $\csc \theta = \dfrac{5}{3}$, $\cot \theta = -\dfrac{4}{3}$

20. $\sin \theta = -\dfrac{4\sqrt{97}}{97}$, $\cos \theta = \dfrac{9\sqrt{97}}{97}$, $\tan \theta = -\dfrac{4}{9}$, $\sec \theta = \dfrac{\sqrt{97}}{9}$, $\csc \theta = -\dfrac{\sqrt{97}}{4}$ **21.** $\sin \theta = \dfrac{\sqrt{23}}{12}$, $\tan \theta = -\dfrac{\sqrt{23}}{11}$, $\sec \theta = -\dfrac{12}{11}$, $\csc \theta = \dfrac{12\sqrt{23}}{23}$, $\cot \theta = -\dfrac{11\sqrt{23}}{23}$ **22.** $\sin \theta = \dfrac{5}{7}$, $\cos \theta = \dfrac{2\sqrt{6}}{7}$, $\tan \theta = \dfrac{5\sqrt{6}}{12}$, $\sec \theta = \dfrac{7\sqrt{6}}{12}$, $\cot \theta = \dfrac{2\sqrt{6}}{5}$ **23.** $\sin \theta = -\dfrac{\sqrt{91}}{10}$, $\cos \theta = -\dfrac{3}{10}$, $\tan \theta = \dfrac{\sqrt{91}}{3}$, $\csc \theta = -\dfrac{10\sqrt{91}}{91}$, $\cot \theta = \dfrac{3\sqrt{91}}{91}$ **24.** $\sin \theta = \dfrac{\sqrt{37}}{37}$, $\cos \theta = -\dfrac{6\sqrt{37}}{37}$, $\sec \theta = -\dfrac{\sqrt{37}}{6}$, $\csc \theta = \sqrt{37}$, $\cot \theta = -6$

25. $\sin \theta = -\dfrac{\sqrt{3}}{2}$, $\cos \theta = \dfrac{1}{2}$, $\tan \theta = -\sqrt{3}$, $\csc \theta = -\dfrac{2\sqrt{3}}{3}$, $\cot \theta = -\dfrac{\sqrt{3}}{3}$

26. $\sin \theta = -\dfrac{3}{5}$, $\cos \theta = -\dfrac{4}{5}$, $\tan \theta = \dfrac{3}{4}$, $\sec \theta = -\dfrac{5}{4}$, $\cot \theta = \dfrac{4}{3}$

27. $\sin \theta = -\dfrac{1}{2}$, $\cos \theta = \dfrac{\sqrt{3}}{2}$, $\tan \theta = -\dfrac{\sqrt{3}}{3}$, $\sec \theta = \dfrac{2\sqrt{3}}{3}$, $\csc \theta = -2$

71.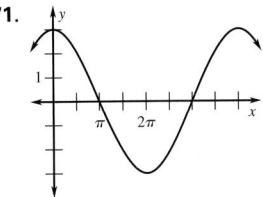

72.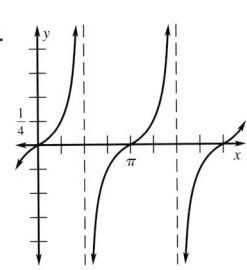

73.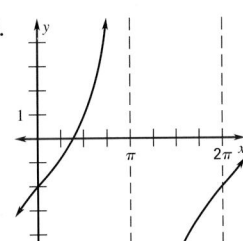

74.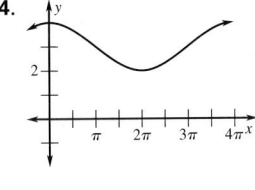

75.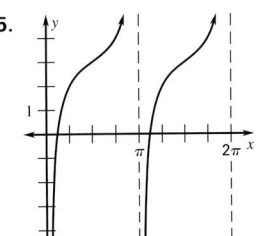

76.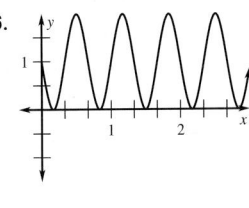

ADDITIONAL ANSWERS

DERIVATIONS OF KEY FORMULAS

THE QUADRATIC FORMULA (p. 895)

3. $f(x) = a\left(x - \dfrac{-b + \sqrt{b^2 - 4ac}}{2a}\right)\left(x - \dfrac{-b - \sqrt{b^2 - 4ac}}{2a}\right)$

$f(x) = a\left(x^2 - \left(\dfrac{-b - \sqrt{b^2 - 4ac}}{2a}\right)x - \left(\dfrac{-b + \sqrt{b^2 - 4ac}}{2a}\right)x + \right.$

$\left(\dfrac{-b + \sqrt{b^2 - 4ac}}{2a}\right)\left(\dfrac{-b - \sqrt{b^2 - 4ac}}{2a}\right)\Big)$

$f(x) = a\left(x^2 - \left(\dfrac{-2b}{2a}\right)x + \dfrac{1}{4a^2}\left[\left(-b + \sqrt{b^2 - 4ac}\right)\left(-b - \sqrt{b^2 - 4ac}\right)\right]\right)$

$f(x) = a\left(x^2 + \dfrac{b}{a}x + \dfrac{1}{4a^2}[b^2 - (b^2 - 4ac)]\right)$

$f(x) = a\left(x^2 + \dfrac{b}{a}x + \dfrac{4ac}{4a^2}\right)$

$f(x) = ax^2 + bx + c$

EQUATION OF AN ELLIPSE (pp. 896–897)

2. $\begin{array}{c}\text{distance between}\\(x, y) \text{ and } (0, -c)\end{array} + \begin{array}{c}\text{distance between}\\(x, y) \text{ and } (0, c)\end{array} = 2a$ (Definition of an ellipse)

$\sqrt{(x-0)^2 + (y-(-c))^2} + \sqrt{(x-0)^2 + (y-c)^2} = 2a$ (Distance formula)

$\sqrt{x^2 + (y+c)^2} + \sqrt{x^2 + (y-c)^2} = 2a$ (Simplify.)

$\sqrt{x^2 + (y+c)^2} = 2a - \sqrt{x^2 + (y-c)^2}$

$\left(\text{Subtract } \sqrt{x^2 + (y-c)^2} \text{ from each side.}\right)$

$x^2 + (y+c)^2 = 4a^2 - 4a\sqrt{x^2 + (y-c)^2} + x^2 + (y-c)^2$
    (Square each side.)

$4cy = 4a^2 - 4a\sqrt{x^2 + (y-c)^2}$
    (Subtract $x^2 + (y-c)^2$ from each side and simplify.)

$cy - a^2 = -a\sqrt{x^2 + (y-c)^2}$
    (Subtract $4a^2$ from each side and divide by 4.)

$c^2y^2 - 2a^2cy + a^4 = a^2(x^2 + (y-c)^2)$ (Square each side again.)

$c^2y^2 - 2a^2cy + a^4 = a^2x^2 + a^2y^2 - 2a^2cy + a^2c^2$ (Multiply.)

$a^4 = a^2x^2 + a^2y^2 - c^2y^2 + a^2c^2$
    (Subtract $c^2y^2 - 2a^2cy$ from each side.)

$a^2(a^2 - c^2) = a^2x^2 + (a^2 - c^2)y^2$
    (Subtract a^2c^2 from each side and factor.)

In an ellipse, $a^2 = b^2 + c^2$. Therefore:

$a^2b^2 = a^2x^2 + b^2y^2$ (Substitute b^2 for $a^2 - c^2$.)

$1 = \dfrac{x^2}{b^2} + \dfrac{y^2}{a^2}$ (Divide each side by a^2b^2.)

THE LAW OF COSINES (p. 901)

1. 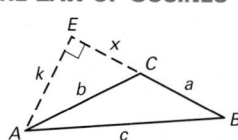 Let a, b, and c be the lengths of the sides of $\triangle ABC$ shown. Introduce altitude AE having length k. Let $CE = x$. Use the Pythagorean theorem to find two different expressions for k^2.

For right $\triangle ACE$, $k^2 + x^2 = b^2$ and so $k^2 = b^2 - x^2$. For right $\triangle ABE$, $k^2 + (a+x)^2 = c^2$ and so $k^2 = c^2 - (a+x)^2$. Therefore:

$b^2 - x^2 = c^2 - (a+x)^2$ (Equate expressions for k.)

$b^2 - x^2 = c^2 - a^2 - 2ax - x^2$ (Multiply.)

$c^2 = b^2 + a^2 + 2ax$ (Add $a^2 + 2ax + x^2$ to each side.)

$c^2 = a^2 + b^2 + 2ax$ (Commutative property of addition)

Use the definition of cosine for right $\triangle ACE$ and reference angles.

$\cos \angle ACE = \dfrac{x}{b}$ (Definition of cosine)

$\cos \angle ACE = -\cos \angle ACB$
    ($\angle ACB$ is the reference $\angle$ for $\angle ACE$ in Quadrant 2.)

$\dfrac{x}{b} = -\cos \angle ACB$ (Substitute $\dfrac{x}{b}$ for $\cos \angle ACE$.)

$x = -b \cos \angle ACB$ (Multiply each side by b.)

Therefore:

$c^2 = a^2 + b^2 + 2a(-b \cos \angle ACB)$
    (Substitute $-b \cos \angle ACB$ for x in the equation $c^2 = a^2 + b^2 + 2ax$.)

$c^2 = a^2 + b^2 + (2a \cdot (-b)) \cos \angle ACB$ (Associative property of mult.)

$c^2 = a^2 + b^2 - 2ab \cos \angle ACB$ (Simplify.)

$c^2 = a^2 + b^2 - 2ab \cos \angle C$ ($\angle ACB = \angle C$ of the original $\triangle$.)

2. $a^2 = b^2 + c^2 - 2bc \cos A$ (Rewrite original equation.)

$a^2 - b^2 - c^2 = -2bc \cos A$ (Subtract $b^2 + c^2$ from each side.)

$\dfrac{b^2 + c^2 - a^2}{2bc} = \cos A$ (Divide each side by $-2bc$.)

SKILLS REVIEW HANDBOOK

TRANSFORMATIONS (p. 922)

26.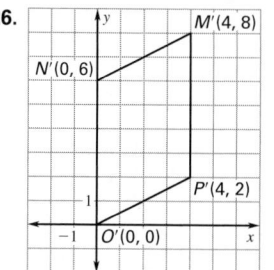

27.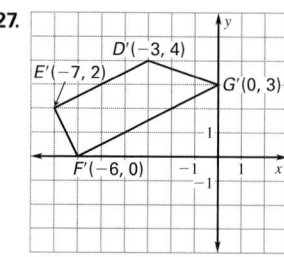

28.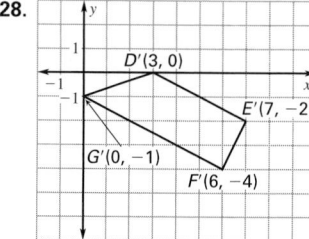

29.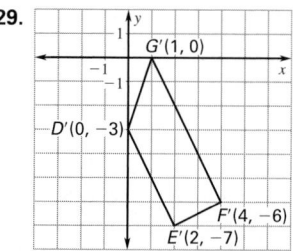

ADDITIONAL ANSWERS

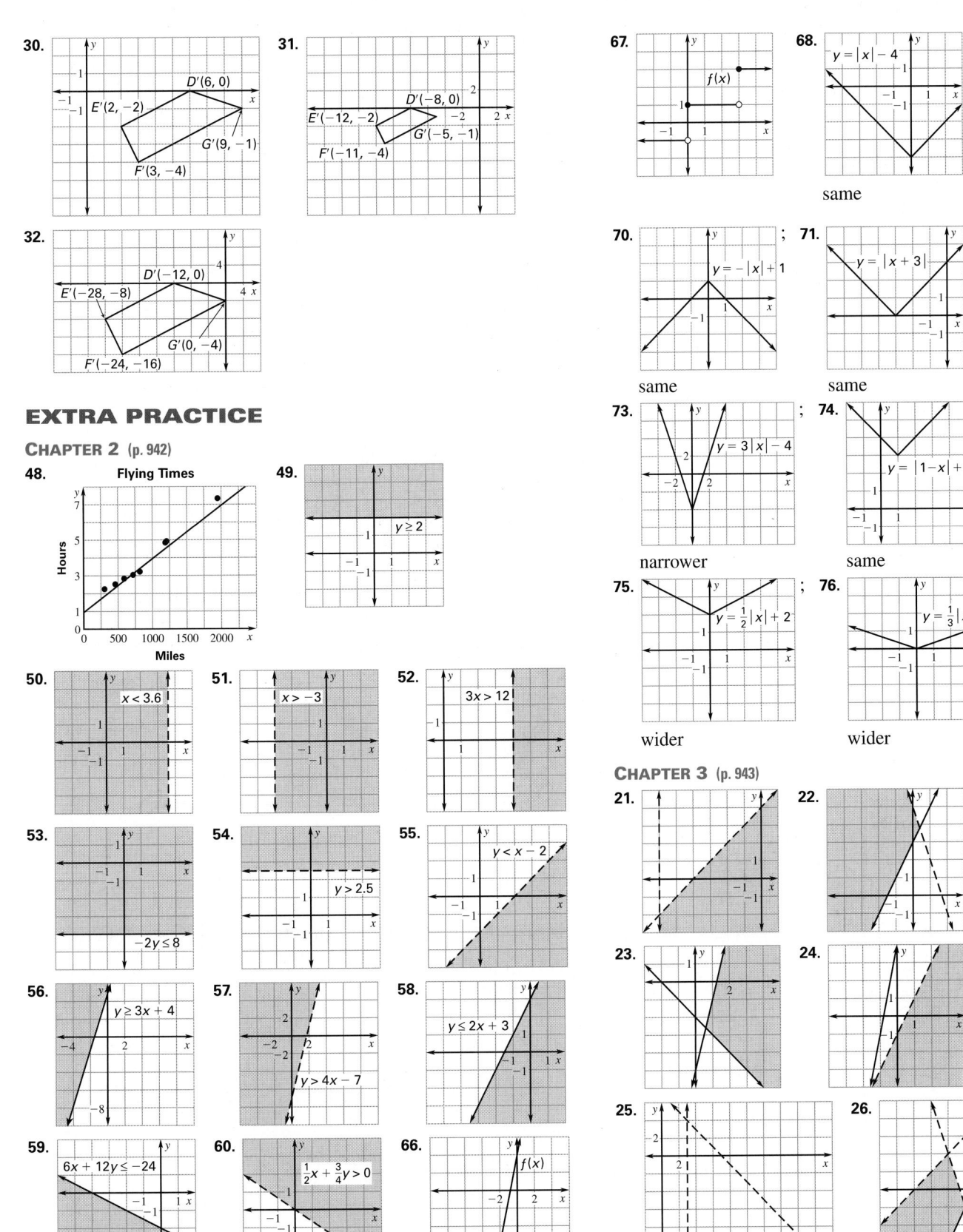

30. D'(6, 0) E'(2, −2) G'(9, −1) F'(3, −4)

31. D'(−8, 0) E'(−12, −2) G'(−5, −1) F'(−11, −4)

32. D'(−12, 0) E'(−28, −8) G'(0, −4) F'(−24, −16)

EXTRA PRACTICE

CHAPTER 2 (p. 942)

48. Flying Times

49. $y \geq 2$

50. $x < 3.6$

51. $x > -3$

52. $3x > 12$

53. $-2y \leq 8$

54. $y > 2.5$

55. $y < x - 2$

56. $y \geq 3x + 4$

57. $y > 4x - 7$

58. $y \leq 2x + 3$

59. $6x + 12y \leq -24$

60. $\frac{1}{2}x + \frac{3}{4}y > 0$

66. $f(x)$

67. $f(x)$

68. $y = |x| - 4$; same

69. $y = 2|x| + 5$; narrower

70. $y = -|x| + 1$; same

71. $y = |x + 3|$; same

72. $y = -2|x|$; narrower

73. $y = 3|x| - 4$; narrower

74. $y = |1 - x| + 3$; same

75. $y = \frac{1}{2}|x| + 2$; wider

76. $y = \frac{1}{3}|x|$; wider

CHAPTER 3 (p. 943)

21.

22.

23.

24.

25.

26.

ADDITIONAL ANSWERS

27.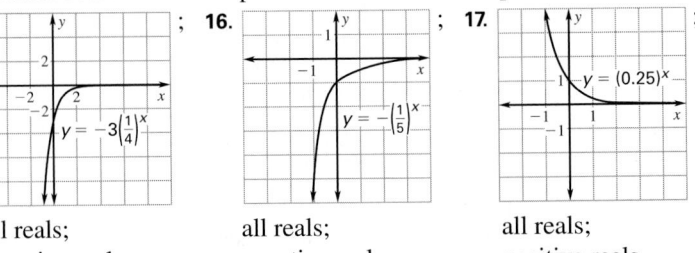

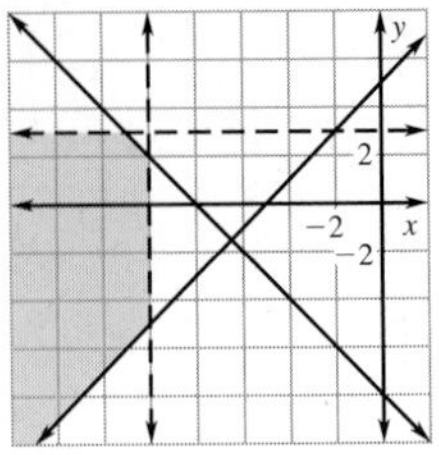

28.

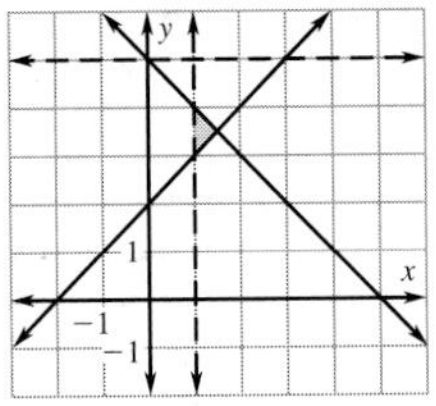

32.

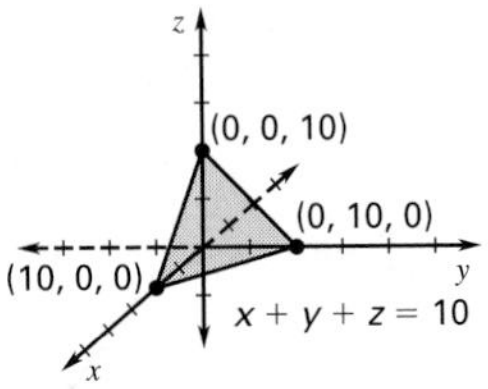

$(0, 0, 10)$
$(0, 10, 0)$
$(10, 0, 0)$
$x + y + z = 10$

33.

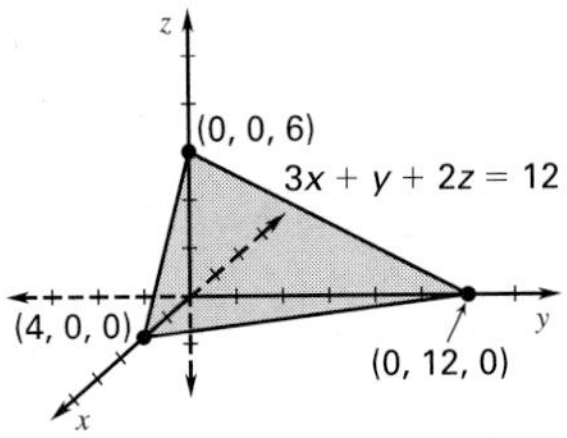

$(0, 0, 6)$
$3x + y + 2z = 12$
$(4, 0, 0)$
$(0, 12, 0)$

34.

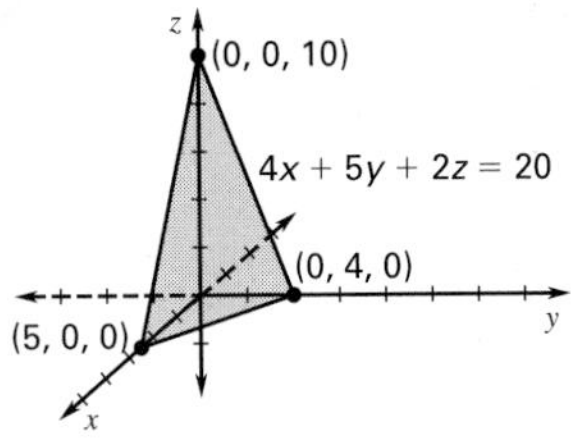

$(0, 0, 10)$
$4x + 5y + 2z = 20$
$(0, 4, 0)$
$(5, 0, 0)$

35.

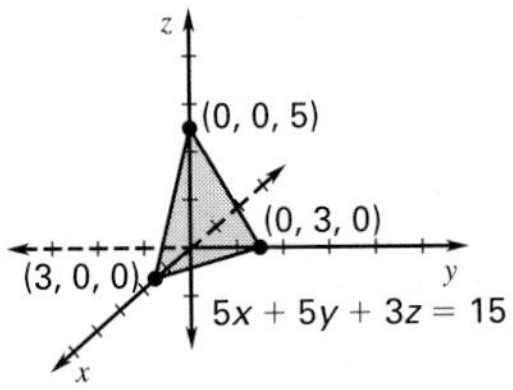

$(0, 0, 5)$
$(0, 3, 0)$
$(3, 0, 0)$
$5x + 5y + 3z = 15$

36.

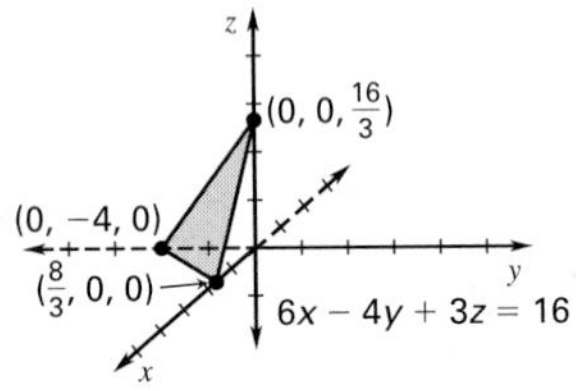

$(0, 0, \frac{16}{3})$
$(0, -4, 0)$
$(\frac{8}{3}, 0, 0)$
$6x - 4y + 3z = 16$

37.

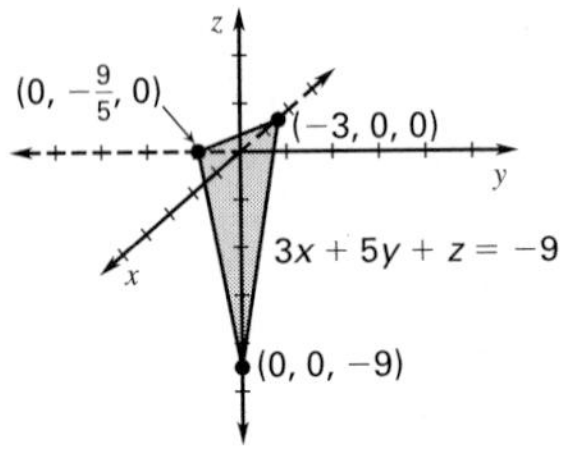

$(0, -\frac{9}{5}, 0)$
$(-3, 0, 0)$
$3x + 5y + z = -9$
$(0, 0, -9)$

CHAPTER 6 (p. 947)

33.

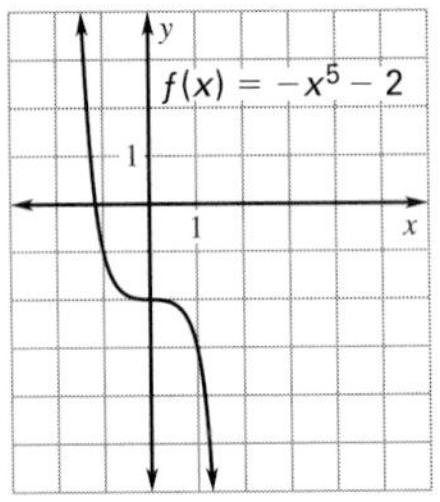

$f(x) = -x^5 - 2$

34.

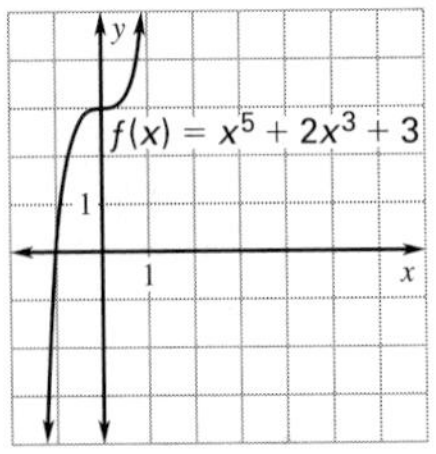

$f(x) = x^5 + 2x^3 + 3$

CHAPTER 7 (p. 950)

66.

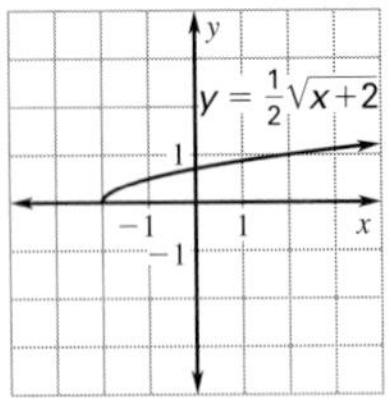

$y = \frac{1}{2}\sqrt{x+2}$

67.

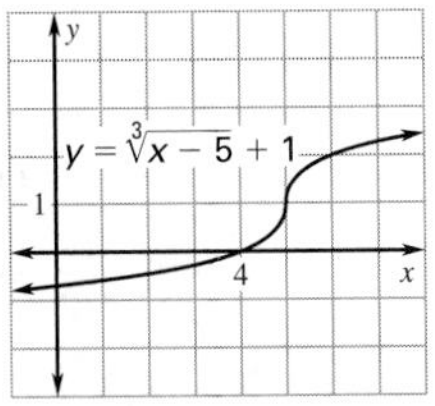

$y = \sqrt[3]{x - 5} + 1$

68.

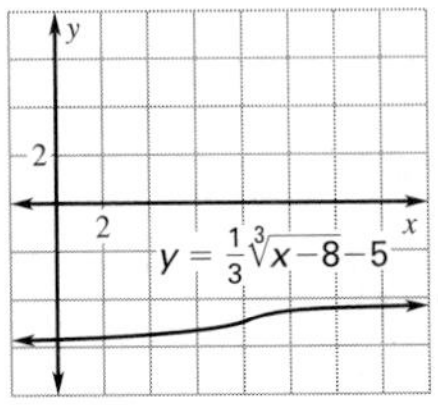

$y = \frac{1}{3}\sqrt[3]{x-8} - 5$

69.

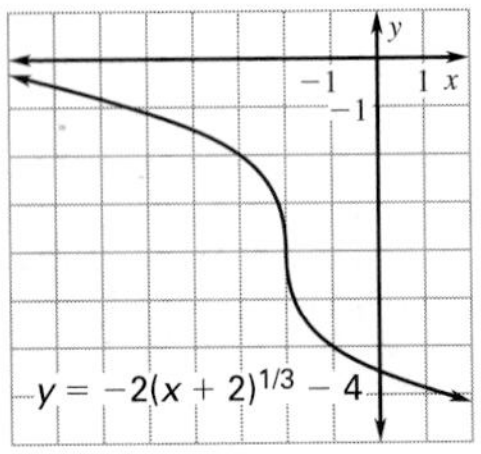

$y = -2(x + 2)^{1/3} - 4$

76. 3

CHAPTER 8 (p. 950)

1. 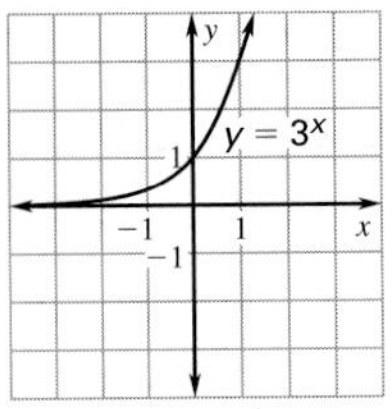 ;

$y = 3^x$

all reals;
positive reals

2. 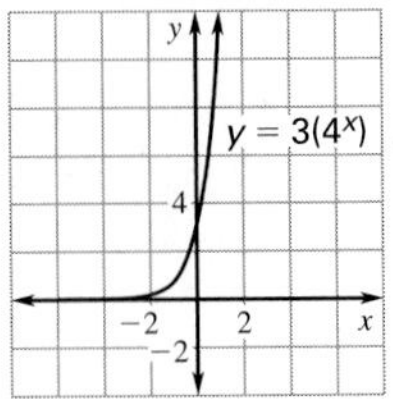 ;

$y = 3(4^x)$

all reals;
positive reals

3. 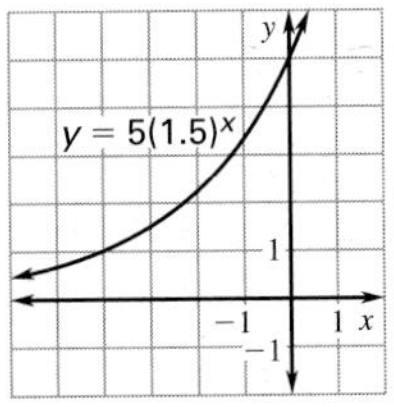 ;

$y = 5(1.5)^x$

all reals;
positive reals

4. 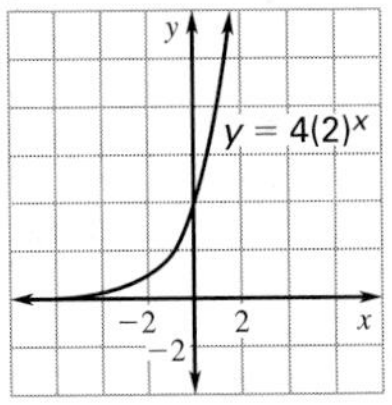 ;

$y = 4(2)^x$

all reals;
positive reals

5. 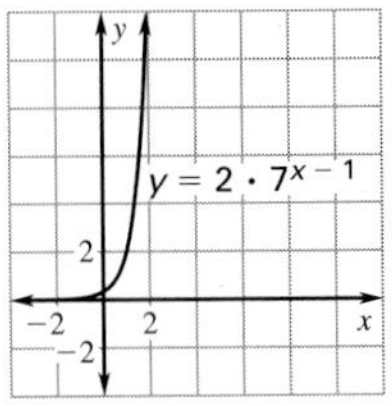 ;

$y = 2 \cdot 7^{x-1}$

all reals;
positive reals

6. 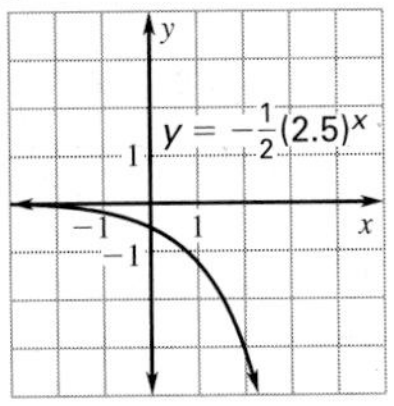 ;

$y = -\frac{1}{2}(2.5)^x$

all reals;
negative reals

7. 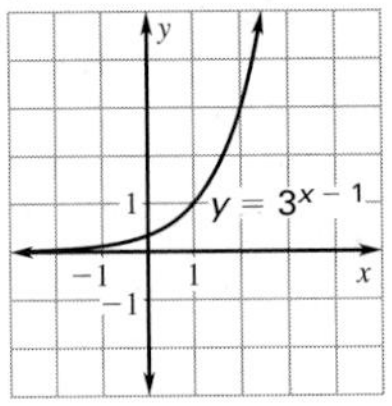 ;

$y = 3^{x-1}$

all reals;
positive reals

8. 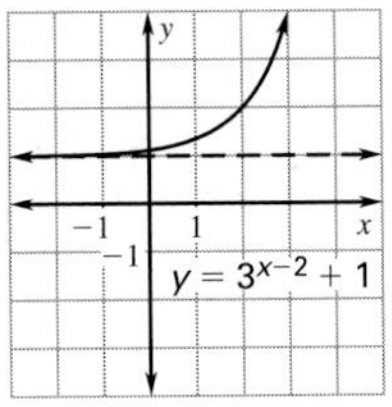 ;

$y = 3^{x-2} + 1$

all reals; $y > 1$

9. 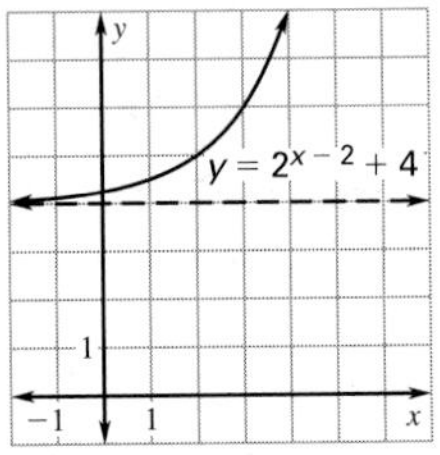 ;

$y = 2^{x-2} + 4$

all reals; $y > 4$

10. 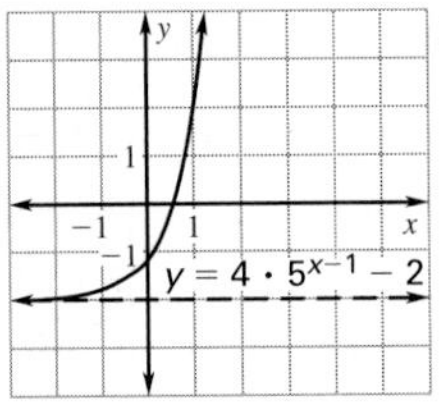 ;

$y = 4 \cdot 5^{x-1} - 2$

all reals;
$y > -2$

11. 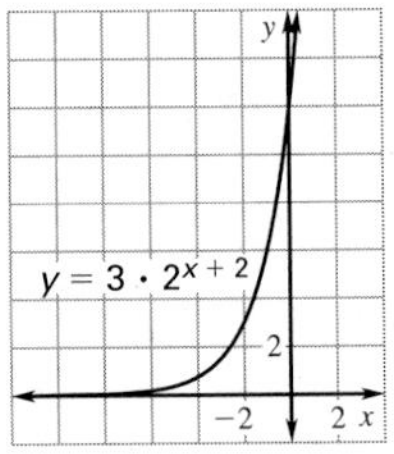 ;

$y = 3 \cdot 2^{x+2}$

all reals;
positive reals

12. 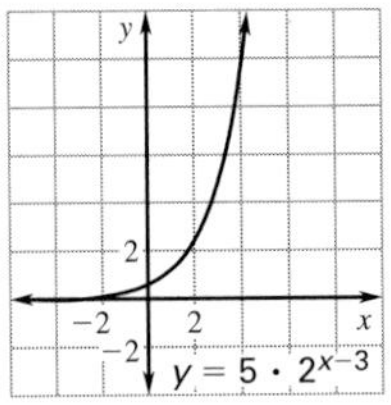 ;

$y = 5 \cdot 2^{x-3}$

all reals;
positive reals

13. 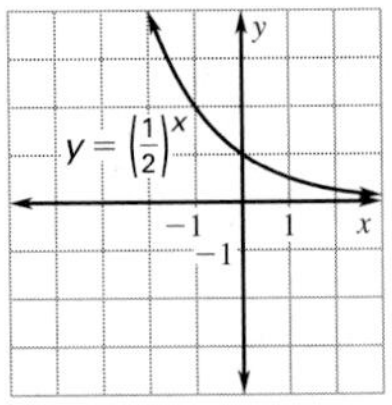 ;

$y = \left(\frac{1}{2}\right)^x$

all reals;
positive reals

14. 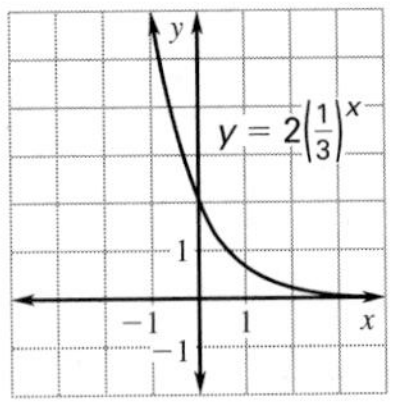 ;

$y = 2\left(\frac{1}{3}\right)^x$

all reals;
positive reals

15. ;

$y = -3\left(\frac{1}{4}\right)^x$

all reals;
negative reals

16. ;

$y = -\left(\frac{1}{5}\right)^x$

all reals;
negative reals

17. ;

$y = (0.25)^x$

all reals;
positive reals

ADDITIONAL ANSWERS

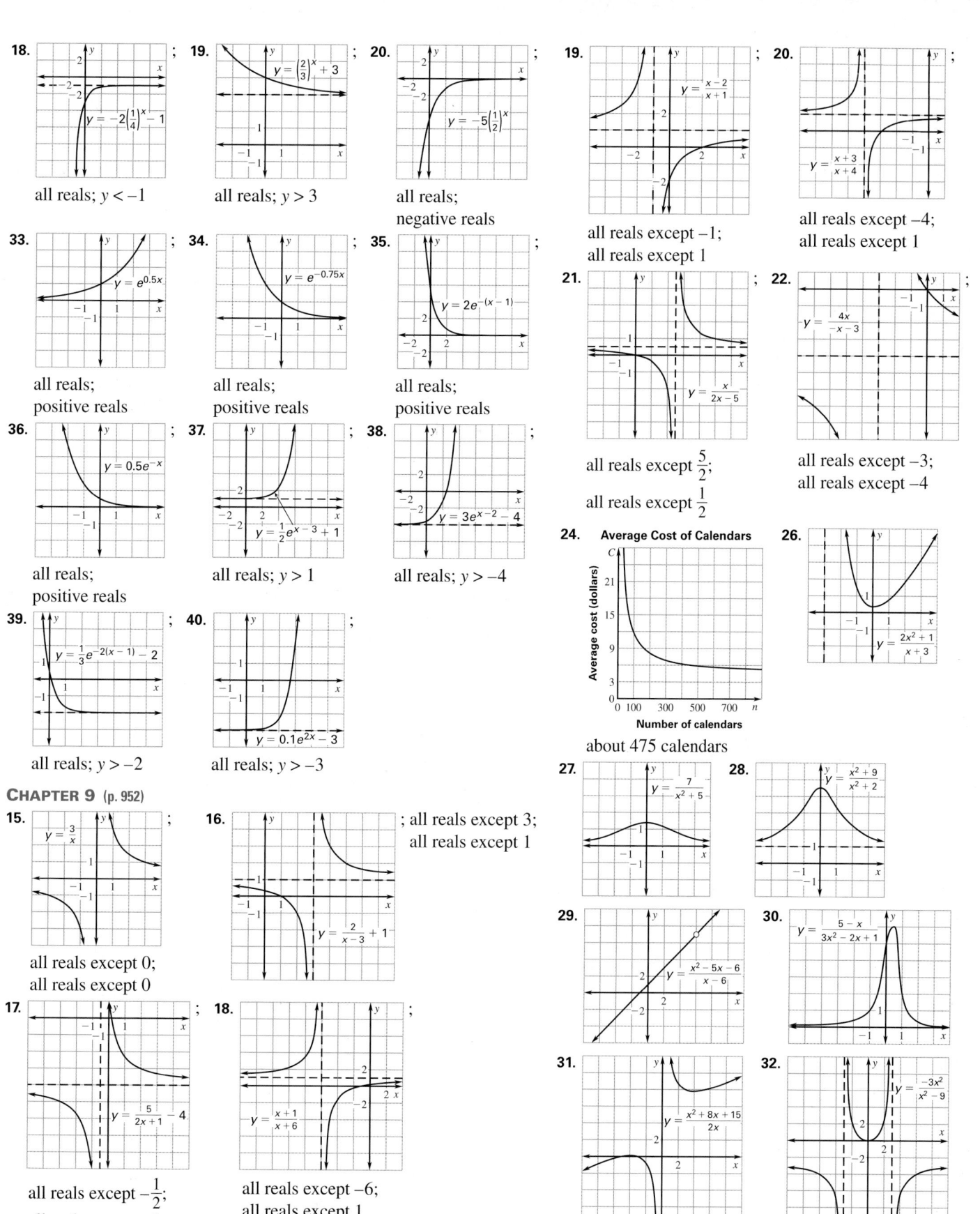

18. $y = -2\left(\frac{1}{4}\right)^x - 1$
all reals; $y < -1$

19. $y = \left(\frac{2}{3}\right)^x + 3$
all reals; $y > 3$

20. $y = -5\left(\frac{1}{2}\right)^x$
all reals;
negative reals

33. $y = e^{0.5x}$
all reals;
positive reals

34. $y = e^{-0.75x}$
all reals;
positive reals

35. $y = 2e^{-(x-1)}$
all reals;
positive reals

36. $y = 0.5e^{-x}$
all reals;
positive reals

37. $y = \frac{1}{2}e^{x-3} + 1$
all reals; $y > 1$

38. $y = 3e^{x-2} - 4$
all reals; $y > -4$

39. $y = \frac{1}{3}e^{-2(x-1)} - 2$
all reals; $y > -2$

40. $y = 0.1e^{2x} - 3$
all reals; $y > -3$

CHAPTER 9 (p. 952)

15. $y = \frac{3}{x}$
all reals except 0;
all reals except 0

16. $y = \frac{2}{x-3} + 1$
; all reals except 3;
all reals except 1

17. $y = \frac{5}{2x+1} - 4$
all reals except $-\frac{1}{2}$;
all reals except -4

18. $y = \frac{x+1}{x+6}$
all reals except -6;
all reals except 1

19. $y = \frac{x-2}{x+1}$
all reals except -1;
all reals except 1

20. $y = \frac{x+3}{x+4}$
all reals except -4;
all reals except 1

21. $y = \frac{x}{2x-5}$
all reals except $\frac{5}{2}$;
all reals except $\frac{1}{2}$

22. $y = \frac{4x}{-x-3}$
all reals except -3;
all reals except -4

24. Average Cost of Calendars
about 475 calendars

26. $y = \frac{2x^2 + 1}{x + 3}$

27. $y = \frac{7}{x^2 + 5}$

28. $y = \frac{x^2 + 9}{x^2 + 2}$

29. $y = \frac{x^2 - 5x - 6}{x - 6}$

30. $y = \frac{5 - x}{3x^2 - 2x + 1}$

31. $y = \frac{x^2 + 8x + 15}{2x}$

32. $y = \frac{-3x^2}{x^2 - 9}$

ADDITIONAL ANSWERS

33.

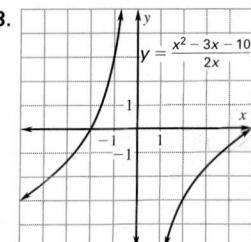

$$y = \frac{x^2 - 3x - 10}{2x}$$

13.

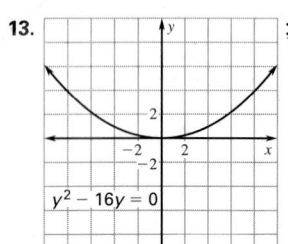

$; \left(0, 4\right);$
$y = -4$
$y^2 - 16y = 0$

14.

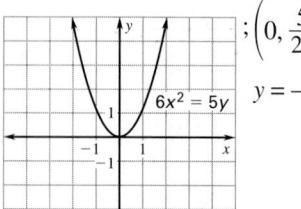

$; \left(0, \frac{5}{24}\right);$
$y = -\frac{5}{24}$
$6x^2 = 5y$

15.

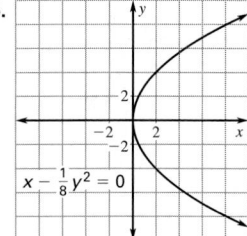

$; (2, 0);$
$x = -2$
$x - \frac{1}{8}y^2 = 0$

16.

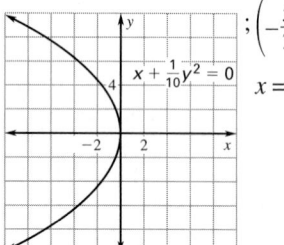

$; \left(-\frac{5}{2}, 0\right);$
$x = \frac{5}{2}$
$x + \frac{1}{10}y^2 = 0$

45.

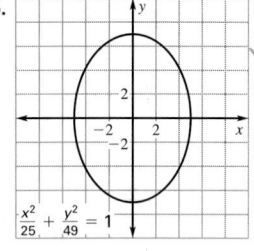

;
$$\frac{x^2}{25} + \frac{y^2}{49} = 1$$
$(0, 7), (0, -7); (-5, 0), (5, 0);$
$\left(0, 2\sqrt{6}\right), \left(0, -2\sqrt{6}\right)$

46.

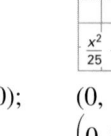

;
$$\frac{x^2}{25} + \frac{y^2}{144} = 1$$
$(0, 12), (0, -12); (-5, 0), (5, 0);$
$\left(0, \sqrt{119}\right), \left(0, -\sqrt{119}\right)$

47.

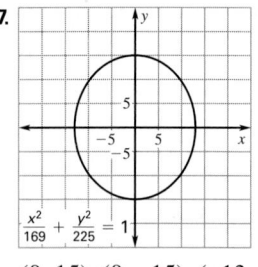

;
$$\frac{x^2}{169} + \frac{y^2}{225} = 1$$
$(0, 15), (0, -15); (-13, 0),$
$(13, 0); \left(0, 2\sqrt{14}\right), \left(0, -2\sqrt{14}\right)$

48.

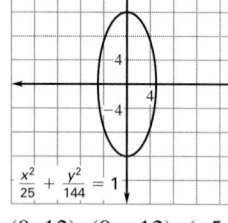

;
$$-x^2 + \frac{y^2}{4} = 1$$
$(0, 2), (0, -2); (-1, 0), (1, 0);$
$\left(0, \sqrt{3}\right), \left(0, -\sqrt{3}\right)$

49.

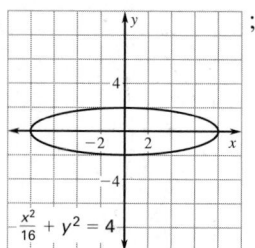

;
$$\frac{x^2}{16} + y^2 = 4$$
$(-8, 0), (8, 0); (0, 2), (0, -2);$
$\left(-2\sqrt{15}, 0\right), \left(2\sqrt{15}, 0\right)$

50.

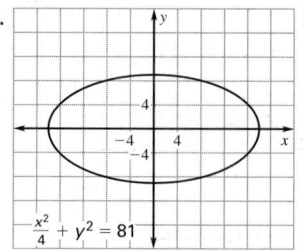

;
$$\frac{x^2}{4} + y^2 = 81$$
$(-18, 0), (18, 0); (0, 9), (0, -9);$
$\left(-9\sqrt{3}, 0\right), \left(9\sqrt{3}, 0\right)$

55. $\dfrac{x^2}{256} + \dfrac{y^2}{100} = 1$ **56.** $\dfrac{x^2}{25} + \dfrac{y^2}{169} = 1$ **57.** $\dfrac{x^2}{4} + \dfrac{y^2}{400} = 1$

59.

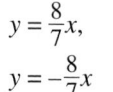

$; \left(-\sqrt{113}, 0\right),$
$\left(\sqrt{113}, 0\right);$
$y = \frac{8}{7}x,$
$y = -\frac{8}{7}x$
$$\frac{x^2}{49} - \frac{y^2}{64} = 1$$

60.

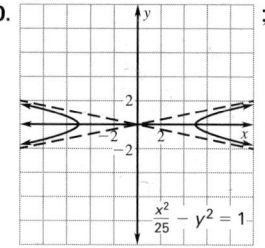

;
$$\frac{x^2}{25} - y^2 = 1$$
$\left(-\sqrt{26}, 0\right), \left(\sqrt{26}, 0\right);$
$y = \frac{1}{5}x, y = -\frac{1}{5}x$

61.

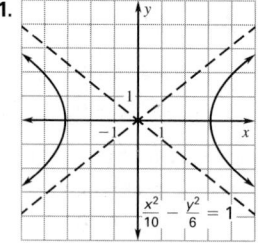

;
$$\frac{x^2}{10} - \frac{y^2}{6} = 1$$
$(-4, 0), (4, 0);$
$y = \frac{\sqrt{15}}{5}x, y = -\frac{\sqrt{15}}{5}x$

62.

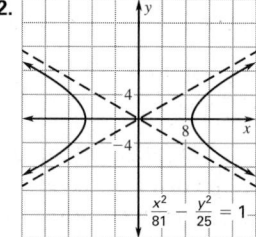

;
$$\frac{x^2}{81} - \frac{y^2}{25} = 1$$
$\left(-\sqrt{106}, 0\right), \left(\sqrt{106}, 0\right);$
$y = \frac{5}{9}x, y = -\frac{5}{9}x$

63.

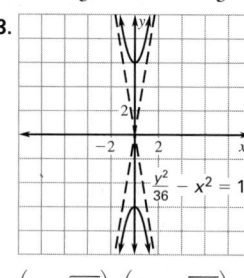

;
$$\frac{y^2}{36} - x^2 = 1$$
$\left(0, \sqrt{37}\right), \left(0, -\sqrt{37}\right);$
$y = 6x, y = -6x$

64.

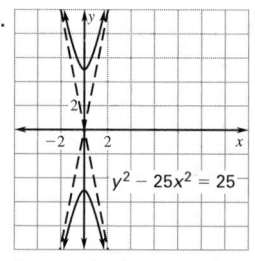

;
$$y^2 - 25x^2 = 25$$
$\left(0, \sqrt{26}\right), \left(0, -\sqrt{26}\right);$
$y = 5x, y = -5x$

ADDITIONAL ANSWERS

65.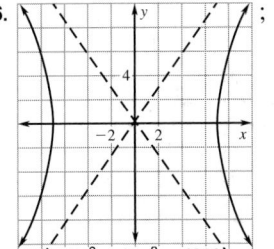
;

$x^2 - 16y^2 = 144$

$\left(-3\sqrt{17}, 0\right), \left(3\sqrt{17}, 0\right);$

$y = \dfrac{1}{4}x, \; y = -\dfrac{1}{4}x$

66.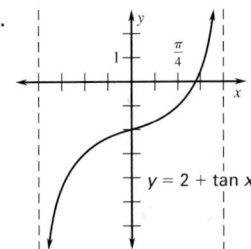
;

$100x^2 - 49y^2 = 4900$

$\left(-\sqrt{149}, 0\right), \left(\sqrt{149}, 0\right);$

$y = \dfrac{10}{7}x, \; y = -\dfrac{10}{7}x$

29.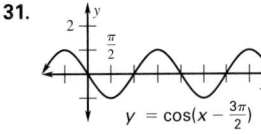

$y = 2 + \tan x$

30.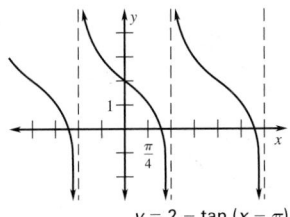

$y = -2 \sin \dfrac{1}{2}x$

31.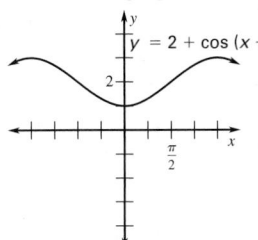

$y = \cos\left(x - \dfrac{3\pi}{2}\right)$

32.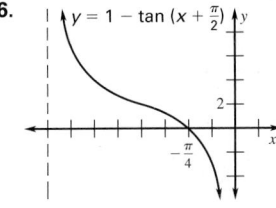

$y = 2 - \tan(x - \pi)$

CHAPTER 13 (p. 958)

31. $\sin \theta = -\dfrac{5\sqrt{34}}{34}$; $\cos \theta = \dfrac{3\sqrt{34}}{34}$; $\tan \theta = -\dfrac{5}{3}$; $\sec \theta = \dfrac{\sqrt{34}}{3}$;

$\csc \theta = -\dfrac{\sqrt{34}}{5}$; $\cot \theta = -\dfrac{3}{5}$ **32.** $\sin \theta = -\dfrac{8\sqrt{113}}{113}$; $\cos \theta = \dfrac{7\sqrt{113}}{113}$;

$\tan \theta = -\dfrac{8}{7}$; $\sec \theta = \dfrac{\sqrt{113}}{7}$; $\csc \theta = -\dfrac{\sqrt{113}}{8}$; $\cot \theta = -\dfrac{7}{8}$ **33.** $\sin \theta = \dfrac{2\sqrt{5}}{5}$;

$\cos \theta = -\dfrac{\sqrt{5}}{5}$; $\tan \theta = -2$; $\sec \theta = -\sqrt{5}$; $\csc \theta = \dfrac{\sqrt{5}}{2}$; $\cot \theta = -\dfrac{1}{2}$

34. $\sin \theta = \dfrac{3\sqrt{14}}{14}$; $\cos \theta = \dfrac{\sqrt{70}}{14}$; $\tan \theta = \dfrac{3\sqrt{5}}{5}$; $\sec \theta = \dfrac{\sqrt{70}}{5}$;

$\csc \theta = \dfrac{\sqrt{14}}{3}$; $\cot \theta = \dfrac{\sqrt{5}}{3}$ **52.** $m\angle A \approx 40.7°$; $m\angle B \approx 105.3°$; $b \approx 10.4$ or

$m\angle A \approx 139.3°$; $m\angle B \approx 6.7°$; $b \approx 1.25$ **53.** $m\angle B \approx 93°$; $b \approx 7.61$; $c \approx 5.48$

54. $m\angle C \approx 57.6°$; $m\angle A \approx 36.4°$; $a \approx 7.74$ **55.** $m\angle B \approx 65°$; $a \approx 6$; $c \approx 5.07$

CHAPTER 14 (p. 959) **17.** Shift the graph of $y = \cos x$ up 3 units.

18. Shift the graph of $y = \cos x$ up 7 units. **19.** Shift the graph of $y = \cos x$ up 4 units and reflect the graph in the line $y = 4$. **20.** Shift the graph of $y = \sin x$ right π units. **21.** Shift the graph of $y = \cos x$ right π units.

22. Shift the graph of $y = \sin x$ left $\dfrac{\pi}{4}$ units. **23.** Shift the graph of $y = \cos x$ right $\dfrac{\pi}{2}$ units. **24.** Shift the graph of $y = \sin x$ down 1 unit and left $\dfrac{3\pi}{4}$ units and reflect the graph in the line $y = -1$.

25.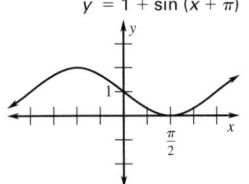

$y = 2 + \cos(x + \pi)$

26.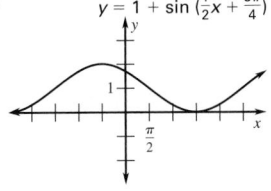

$y = 1 - \tan\left(x + \dfrac{\pi}{2}\right)$

27. $y = 1 + \sin(x + \pi)$

28. $y = 1 + \sin\left(\dfrac{1}{2}x + \dfrac{3\pi}{4}\right)$

ADDITIONAL ANSWERS